FOURTH EDITION

Matter & Interactions

VOLUME I
Modern Mechanics

VOLUME II
Electric and Magnetic Interactions

RUTH W. CHABAY
BRUCE A. SHERWOOD

North Carolina State University

WILEY

VICE PRESIDENT & PUBLISHER Petra Recter
EXECUTIVE EDITOR Jessica Fiorillo
ASSOCIATE EDITOR Aly Rentrop
EDITORIAL ASSISTANT Amanda Rillo
SENIOR MARKETING MANAGER Kristy Ruff
MARKETING ASSISTANT Claudine Scrivanich
SENIOR CONTENT MANAGER Kevin Holm
SENIOR PRODUCTION EDITOR Elizabeth Swain
DESIGN DIRECTOR Harry Nolan
TEXT/COVER DESIGN Maureen Eide
COVER IMAGE Ruth Chabay

Cover Description: The cover image is a snapshot from a VPython program that models the motion of a mass-spring system in 3D (see Computational Problem P70 in Chapter 4).

This book was set in 10/12 Times Ten Roman in LaTex by MPS and printed and bound by Quad Graphics. The cover was printed by Quad Graphics.

This book is printed on acid-free paper.

Complete text ISBN 978-1-118-87586-5
Complete text binder version ISBN 978-1-118-91451-9

Volume 1 ISBN 978-1-118-91449-6
Volume 1 Binder version ISBN 978-1-118-91452-6

Volume 2 ISBN 978-1-118-91450-2
Volume 2 Binder version ISBN 978-1-118-91453-3

Printed in the United States of America
SKY10028947_081021

Brief Contents

VOLUME I Modern Mechanics

VOLUME II Electric and Magnetic Interactions

The Supplements can be found at the web site, www.wiley.com/college/chabay

Contents

VOLUME I
Modern Mechanics

CHAPTER 1 *Interactions and Motion* 1

CHAPTER 2 *The Momentum Principle* 45

CHAPTER 3 *The Fundamental Interactions* 88

CHAPTER 4 *Contact Interactions* 130

The Supplements can be found at the web site, www.wiley.com/college/chabay

Preface

TO THE STUDENT

This textbook emphasizes a 20th-century perspective on introductory physics. Contemporary physicists build models of the natural world that are based on a small set of fundamental physics principles and on an understanding of the microscopic structure of matter, and they apply these models to explain and predict a very broad range of physical phenomena. In order to involve students of introductory physics in the contemporary physics enterprise, this textbook emphasizes:

- Reasoning directly from a small number of fundamental physics principles, rather than from a large set of special-case equations.
- Integrating contemporary insights, such as atomic models of matter, quantized energy, and relativistic dynamics, throughout the curriculum.
- Engaging in the full process of creating and refining physical models (idealizing, making approximations, explicitly stating assumptions, and estimating quantities).
- Reasoning iteratively about the time-evolution of system behavior, both on paper and through the construction and application of computational models.

Because the physical world is 3-dimensional, we work in 3D throughout the text. Many students find the approach to 3D vectors used in this book easier than standard treatments of 2D vectors.

Textbook and Supplemental Resources

Modern Mechanics (Volume 1, Chapters 1–12) focuses on the atomic structure of matter and interactions between material objects. It emphasizes the wide applicability and utility of a small number of fundamental principles: the Momentum Principle, the Energy Principle, and the Angular Momentum Principle, and the Fundamental Assumption of Statistical Mechanics. We study how to explain and predict the behavior of systems as different as elementary particles, molecules, solid metals, and galaxies.

Electric and Magnetic Interactions (Volume 2, Chapters 13–23) emphasizes the somewhat more abstract concepts of electric and magnetic fields and extends the study of the atomic structure of matter to include the role of electrons. The principles of electricity and magnetism are the foundation for much of today's technology, from cell phones to medical imaging.

Additional resources for students are freely available at this site:

www.wiley.com/college/chabay

The web resources include several supplements. A copy of Chapter 1 is provided for students who are currently using Volume 2 but whose previous physics course did not use Volume 1. This chapter introduces 3D vectors and vector algebra, and includes an introduction to computational modeling in VPython, which is used throughout the textbook.

Supplement S1 treats the kinetic theory of gases and heat engines, and can be used by students who have completed Chapter 12 on Entropy. Supplement S2 explains the basic principles of PN junctions in semiconductor devices, and can be used by students who have completed Chapter 21: Patterns of Field in Space. Supplement S3 includes a more mathematically sophisticated treatment of mechanical and electromagnetic waves and wave phenomena, and can be used by students who have completed Chapter 23 on Electromagnetic Radiation.

Answers to odd-numbered problems may be found at the end of the book.

The new Student Solutions Manual is available for purchase as a printed supplement and contains fully worked solutions for a subset of end of chapter problems.

Prerequisites

This book is intended for introductory calculus-based college physics courses taken by science and engineering students. It requires a basic knowledge of derivatives and integrals, which can be obtained by studying calculus concurrently.

Modeling

Matter & Interactions places a major emphasis on constructing and using physical models. A central aspect of science is the modeling of complex real-world phenomena. A physical model is based on what we believe to be fundamental principles; its intent is to predict or explain the most important aspects of an actual situation. Modeling necessarily involves making approximations and simplifying assumptions that make it possible to analyze a system in detail.

Computational Modeling

Computational modeling is now as important as theory and experiment in contemporary science and engineering. We introduce you to serious computer modeling right away to help you build a strong foundation in the use of this important tool.

In this course you will construct simple computational models based on fundamental physics principles. You do not need any prior programming experience–this course will teach you the small number of computational concepts you will need. Using VPython, a computational environment based on the Python programming language, you will find that after less than an hour you can write a simple computational model that produces a navigable 3D animation as a side effect of your physics code.

Computational modeling allows us to analyze complex systems that would otherwise require very sophisticated mathematics or that could not be analyzed at all without a computer. Numerical calculations based on the Momentum Principle give us the opportunity to watch the dynamical evolution of the behavior of a system. Simple models frequently need to be refined and extended. This can be done straightforwardly with a computer model but is often impossible with a purely analytical (non-numerical) model.

VPython is free, and runs on Windows, MacOS, and Linux. Instructions in Chapter 1 tell you how to install it on your own computer, and how to find a set of instructional videos that will help you learn to use VPython.

Questions

As you read the text, you will frequently come to a question that looks like this:

> QUESTION What should I do when I encounter a question in the text?

A question invites you to stop and think, to make a prediction, to carry out a step in a derivation or analysis, or to apply a principle. These questions are answered in the following paragraphs, but it is important that you make a serious effort to answer the questions on your own before reading further. Be honest in comparing your answers to those in the text. Paying attention to surprising or counterintuitive results can be a useful learning strategy.

Checkpoints

Checkpoints at the end of some sections ask you to apply new concepts or techniques. These may involve qualitative reasoning or simple calculations. You should complete these checkpoints when you come to them, before reading further. The goal of a checkpoint is to help you consolidate your understanding of the material you have just read, and to make sure you are ready to continue reading. Answers to checkpoints are found at the end of each chapter.

Conventions Used in Diagrams

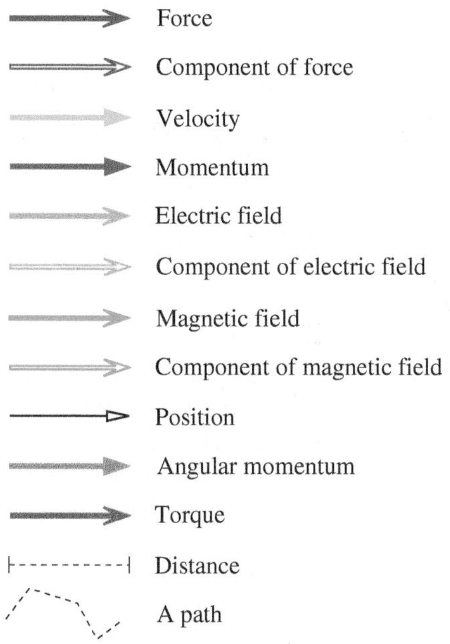

The conventions most commonly used to represent vectors and scalars in diagrams in this text are shown in the margin. In equations and text, a vector will be written with an arrow above it: $\vec{p}$.

TO THE INSTRUCTOR

The approach to introductory physics in this textbook differs significantly from that in most textbooks. Key emphases of the approach include:

- Starting from fundamental principles rather than secondary formulas
- Atomic-level description and analysis
- Modeling the real world through idealizations and approximations
- Computational modeling of physical systems
- Unification of mechanics and thermal physics
- Unification of electrostatics and circuits
- The use of 3D vectors throughout

Web Resources for Instructors

Instructor resources are available at this web site:

www.wiley.com/college/chabay

Resources on this site include lecture-demo software, textbook figures, clicker questions, test questions, lab activities including experiments and computational modeling, a computational modeling guide, and a full solutions manual. Contact your Wiley representative for information about this site.

Electronic versions of the homework problems are available in WebAssign:

www.webassign.net

Some instructor resources are available through WebAssign as well.

Other information may be found on the authors' *Matter & Interactions* web site:

matterandinteractions.org

Also on the authors' website are reprints of published articles about *Matter & Interactions*, including these:

- Chabay, R. & Sherwood, B. (1999). Bringing atoms into first-year physics. *American Journal of Physics* 67, 1045–1050.
- Chabay, R. W. & Sherwood, B. (2004). Modern mechanics. *American Journal of Physics* 72, 439–445.
- Chabay, R. W. & Sherwood, B. (2006). Restructuring Introductory E&M. *American Journal of Physics* 74, 329–336.
- Chabay, R. & Sherwood, B. (2008) Computational physics in the introductory calculus-based course. *American Journal of Physics* 76(4&5), 307–313.
- Beichner, R., Chabay, R., & Sherwood, B. (2010) Labs for the Matter & Interactions curriculum. *American Journal of Physics* 78(5), 456–460.

Computational Homework Problems

Some important homework problems require the student to write a simple computer program. The textbook and associated instructional videos teach VPython, which is based on the Python programming language, and which generates real-time 3D animations as a side effect of simple physics code written by students. Such animations provide powerfully motivating and instructive visualizations of fields and motions. VPython supports true vector computations, which encourages students to begin thinking about vectors as much more than mere components. VPython can be obtained at no cost for Windows, Macintosh, and Linux at vpython.org.

In the instructor resources section of matterandinteractions.org is "A Brief Guide to Computational Modeling in Matter & Interactions" which explains how to incorporate computation into the curriculum in a way that is easy for instructors to manage and which is entirely accessible to students with no prior programming experience. There you will also find a growing list of advanced computational physics textbooks that use VPython, which means that introducing students to Python and VPython in the introductory physics course can be of direct utility in later courses. Python itself is now widely used in technical fields.

Desktop Experiment Kit for Volume 2

On the authors' web site mentioned above is information about a desktop experiment kit for E&M that is distributed by PASCO. The simple equipment in this kit allows students to make key observations of electrostatic, circuit, and magnetic phenomena, tightly integrated with the theory (www.pasco.com, search for EM-8675). Several chapters contain optional experiments that can be done with this kit. This does not preclude having other, more complex laboratory experiences associated with the curriculum. For example, one such lab that we use deals with Faraday's law and requires signal generators, large coils, and oscilloscopes. You may have lab experiments already in place that will go well with this textbook.

What's New in the 4th Edition

The 4th edition of this text includes the following major new features:

- Increased support for computational modeling throughout, including sample code.
- Discussion throughout the text contrasting iterative and analytical problem solutions.
- Many new computational modeling problems (small and large).
- Improved discussion throughout the text of the contrast between models of a system as a point particle and as an extended system.
- An improved discussion of the Momentum Principle throughout Volume 1, emphasizing that the future momentum depends on two elements: the momentum now, and the impulse applied.
- Improved treatment of polarization surface charge in electrostatics (Chapter 14) and circuits (Chapter 18) based on the results of detailed 3D computational models.
- A more extensive set of problems at the end of each chapter, with improved indication of difficulty level.

In order to reduce cost and weight, some materials that have seen little use by instructors have been moved to the Wiley web site (www.wiley.com/college/chabay) where they are freely available. These materials include Supplement S1 (Chapter 13 in the 3rd Edition: kinetic theory of gases, thermal processes, and heat engines), Supplement S2 on PN junctions (formerly an optional section in Chapter 22 in the 3rd Edition), and Supplement S3 (a significantly extended version of Chapter 25 in the 3rd Edition: electromagnetic interference and diffraction, wave-particle duality, and a new section on mechanical waves and the wave equation).

Additional changes in the 4th Edition include:

- In Chapter 5, improved treatment of curving motion and an added section on the dynamics of multiobject systems.
- An improved sequence of topics in Chapter 6, with an explicit discussion of the role of energy in computational models, and an improved treatment of path independence, highlighting its limitation to point particles.
- A new section in Chapter 7 on the effect of the choice of reference frame on the form of the Energy Principle, and explicit instruction on how to model several kinds of friction in a computational model.
- In Chapter 8, discussion of the lifetime of excited states and on the probabilistic nature of energy transitions.
- In Chapter 9, now renamed "Translational, Rotational, and Vibrational Energy," improved treatment of the energetics of deformable systems.
- In Chapter 11, analysis of a physical pendulum.
- A detailed discussion in Chapter 16 of how to calculate potential difference by numerical path integration.
- An improved treatment of motional emf in the case of a bar dragged along rails (Chapter 20).

Suggestions for Condensed Courses

In a large course for engineering and science students with three 50-minute lectures and one 110-minute small-group studio lab per week, or in a studio format with five 50-minute sessions per week, it is possible to complete most but not all of the mechanics and E&M material in two 15-week semesters. In an honors course, or a course for physics majors, it is possible to do almost everything. You may be able to go further or deeper if your course has a weekly recitation session in addition to lecture and lab.

What can be omitted if there is not enough time to do everything? In mechanics, the one thing we feel should not be omitted is the introduction to entropy in terms of the statistical mechanics of the Einstein solid (Chapter 12). This is a climax of the integration of mechanics and thermal physics. One approach to deciding what mechanics topics can be omitted is to be guided by what foundation is required for Chapter 12. See other detailed suggestions below.

In E&M, one should not omit electromagnetic radiation and its effects on matter (Chapter 23). This is the climax of the whole E&M enterprise. One way to decide what E&M topics can be omitted is to be guided by what foundation is required for Chapter 23. See other detailed suggestions below.

Any starred section (*) can safely be omitted. Material in these sections is not referenced in later work. In addition, the following sections may be omitted:

Chapter 3 (The Fundamental Interactions): The section on determinism may be omitted.

Chapter 4 (Contact Interactions): Buoyancy and pressure may be omitted (one can return to these topics by using Supplement S1 on gases).

Chapter 7 (Internal Energy): If you are pressed for time, you might choose to omit the second half of the chapter on energy dissipation, beginning with Section 7.10.

Chapter 9 (Translational, Rotational, and Vibrational Energy): The formalism of finding the center of mass may be skipped, because the important applications have obvious locations of the center of mass. Although they are very instructive, it is possible to omit the sections contrasting point-particle with extended system models; you may also omit the analysis of sliding friction.

Chapter 10 (Collisions): A good candidate for omission is the analysis of collisions in the center-of-mass frame. Since there is a basic introduction to collisions in Chapter 3 (before energy is introduced), one could omit all of Chapter 10. On the other hand, the combined use of the Momentum Principle and the Energy Principle can illuminate both fundamental principles.

Chapter 11 (Angular Momentum): The main content of this chapter should not be omitted, as it introduces the third fundamental principle of mechanics, the Angular Momentum Principle. One might choose to omit most applications involving nonzero torque.

Chapter 12 (Entropy: Limits on the Possible): The second half of this chapter, on the Boltzmann distribution, may be omitted if necessary.

Chapter 15 (Electric Field of Distributed Charges): It is important that students acquire a good working knowledge of the patterns of electric field around some standard charged objects (rod, ring, disk, capacitor, sphere). If however they themselves are to acquire significant expertise in setting up physical integrals, they need extensive practice, and you might decide that the amount of time necessary for acquiring this expertise is not an appropriate use of the available course time.

Chapter 16 (Electric Potential): The section on dielectric constant can be omitted if necessary.

Chapter 17 (Magnetic Field): In the sections on the atomic structure of magnets, you might choose to discuss only the first part, in which one finds that the magnetic moment of a bar magnet is consistent with an atomic model. Omitting the remaining sections on spin and domains will not cause significant difficulties later.

Chapter 19 (Circuit Elements): The sections on series and parallel resistors and on internal resistance, meters, quantitative analysis of RC circuits, and multiloop circuits can be omitted. Physics and engineering students who need to analyze complex multiloop circuits will later take specialized courses on the topic; in the introductory physics course the emphasis should be on giving all students a good grounding in the fundamental mechanisms underlying circuit behavior.

Chapter 20 (Magnetic Force): We recommend discussing Alice and Bob and Einstein, but it is safe to omit the sections on relativistic field transformations. However, students often express high interest in the relationship between electric fields and magnetic fields, and here is an opportunity to satisfy some aspects of their curiosity. Motors and generators may be omitted or downplayed. The case study on sparks in air can be omitted, because nothing later depends critically on this topic, though it provides an introductory-level example of a phenomenon where an intuitively appealing model fails utterly, while a different model predicts several key features of the phenomenon. Another possibility is to discuss sparks near the end of the course, because it can be a useful review of many aspects of E&M.

Chapter 22 (Faraday's Law): Though it can safely be omitted, we recommend retaining the section on superconductors, because students are curious about this topic. The section on inductance may be omitted.

Chapter 23 (Electromagnetic Radiation): The treatment of geometrical optics may be omitted.

Acknowledgments

We owe much to the unusual working environment provided by the Department of Physics and the former Center for Innovation in Learning at Carnegie Mellon, which made it possible during the 1990s to carry out the research and development leading to the first edition of this textbook in 2002. We are grateful for the open-minded attitude of our colleagues in the Carnegie Mellon physics department toward curriculum innovations.

We are grateful to the support of our colleagues Robert Beichner and John Risley in the Physics Education Research and Development group at North Carolina State University, and to other colleagues in the NCSU physics department.

We thank Fred Reif for emphasizing the role of the three fundamental principles of mechanics, and for his view on the reciprocity of electric and gravitational forces. We thank Robert Bauman, Gregg Franklin, and Curtis Meyer for helping us think deeply about energy.

Much of Chapter 12 on quantum statistical mechanics is based on an article by Thomas A. Moore and Daniel V. Schroeder, "A different approach to introducing statistical mechanics," *American Journal of Physics*, vol. 65, pp. 26–36 (January 1997). We have benefited from many stimulating conversations with Thomas Moore, author of another introductory textbook that takes a contemporary view of physics, *Six Ideas that Shaped Physics*. Michael Weissman and Robert Swendsen provided particularly helpful critiques on some aspects of our implementation of Chapter 12.

We thank Hermann Haertel for opening our eyes to the fundamental mechanisms of electric circuits. Robert Morse, Priscilla Laws, and Mel Steinberg stimulated our thinking about desktop experiments. Bat-Sheva Eylon offered important guidance at an early stage. Ray Sorensen provided deep analytical critiques that influenced our thinking in several important areas. Randall Feenstra taught us about semiconductor junctions. Thomas Moore showed us a useful way to present the differential form of Maxwell's equations. Fred Reif helped us devise an assessment of student learning of basic E&M concepts. Uri Ganiel suggested the high-voltage circuit used to demonstrate the reality of surface charge. The unusual light bulb circuits at the end of Chapter 22 are based on an article by P. C. Peters, "The role of induced emf's in simple circuits," *American Journal of Physics* 52, 1984, 208–211. Thomas Ferguson gave us unusually detailed and useful feedback on the E&M chapters. Discussions with John Jewett about energy transfers were helpful. We thank Seth Chabay for help with Latin.

We thank David Andersen, David Scherer, and Jonathan Brandmeyer for the development of tools that enabled us and our students to write associated software.

The research of Matthew Kohlmyer, Sean Weatherford, and Brandon Lunk on student engagement with computational modeling has made major contributions to our instruction on computational modeling. Lin Ding developed an energy assessment instrument congruent with the goals of this curriculum.

We thank our colleagues David Brown, Krishna Chowdary, Laura Clarke, John Denker, Norman Derby, Ernst-Ludwig Florin, Thomas Foster, Jon D.H. Gaffney, Chris Gould, Mark Haugan, Joe Heafner, Robert Hilborn, Eric Hill, Andrew Hirsch, Leonardo Hsu, Barry Luokkala, Sara Majetich, Jonathan Mitschele, Arjendu Pattanayak, Jeff Polak, Prabha Ramakrishnan, Vidhya Ramachandran, Richard Roth, Michael Schatz, Robert Swendsen, Aaron Titus, Michael Weissman, and Hugh Young.

We thank a group of reviewers assembled by the publisher, who gave us useful critiques on the second edition of this textbook: Kelvin Chu, Michael Dubson, Tom Furtak, David Goldberg, Javed Iqbal, Shawn Jackson, Craig Ogilvie, Michael Politano, Norris Preyer, Rex Ramsier, Tycho Sleator, Robert Swendsen, Larry Weinstein, and Michael Weissman. We also thank the group who offered useful critiques on the third edition: Alex Small, Bereket Behane, Craig Wiegert, Galen Pickett, Ian Affleck, Jeffrey Bindel, Jeremy King, Paula Heron, and Surenda Singh.

We have benefited greatly from the support and advice of Stuart Johnson and Jessica Fiorillo of John Wiley & Sons. Elizabeth Swain of John Wiley & Sons was exceptionally skilled in managing the project. Helen Walden did a superb job of copyediting; any remaining errors are ours.

This project was supported, in part, by the National Science Foundation (grants MDR-8953367, USE-9156105, DUE-9554843, DUE-9972420, DUE-0320608, DUE-0237132, and DUE-0618504). We are grateful to the National Science Foundation and its reviewers for their long-term support of this challenging project. Opinions expressed are those of the authors, and not necessarily those of the Foundation.

How the Figures Were Made

Almost all of the figures in this book were produced by us (for the third edition the Aptara studio created the figures that show human figures, and the studio added full color to our two-color versions from the second edition). Our main tool was Adobe Illustrator. The many 3D computer-generated images were made using VPython, with optional processing in POV-Ray using a module written by Ruth Chabay to generate a POV-Ray scene description file corresponding to a VPython scene, followed by editing in Adobe Photoshop before exporting to Illustrator. We used TeXstudio for editing LaTeX, with a package due in part to the work of Aptara. All the computer work was done on Windows computers.

Ruth Chabay and Bruce Sherwood
Santa Fe, New Mexico, July 2014

Biographical Background

Ruth Chabay earned a Ph.D in physical chemistry from the University of Illinois at Urbana-Champaign; her undergraduate degree was in chemistry from the University of Chicago. She is Professor Emerita in the Department of Physics at North Carolina State University and was Weston Visiting Professor, Department of Science Teaching, at the Weizmann Institute of Science in Rehovot, Israel. She has also taught at the University of Illinois at Urbana-Champaign and Carnegie Mellon University. She is a Fellow of the American Physical Society.

Bruce Sherwood's Ph.D is in experimental particle physics from the University of Chicago; his undergraduate degree was in engineering science from Purdue University, after which he studied physics for one year at the University of Padua, Italy. He is Professor Emeritus in the Department of Physics at North Carolina State University. He has also taught at Caltech, the University of Illinois at Urbana-Champaign, and Carnegie Mellon University. He is a Fellow of the American Physical Society and of the American Association for the Advancement of Science.

Chabay and Sherwood have been joint recipients of several educational awards. At Carnegie Mellon University they received the Ashkin Award for Teaching in the Mellon College of Science in 1999 and the Teaching Award of the National Society of Collegiate Scholars in 2001. At North Carolina State University they received the Margaret Cox Award for excellence in teaching and learning with technology in 2005. In 2014 the American Association of Physics Teachers presented them with the David Halliday and Robert Resnick Award for Excellence in Undergraduate Physics Teaching.

CHAPTER

1

Interactions and Motion

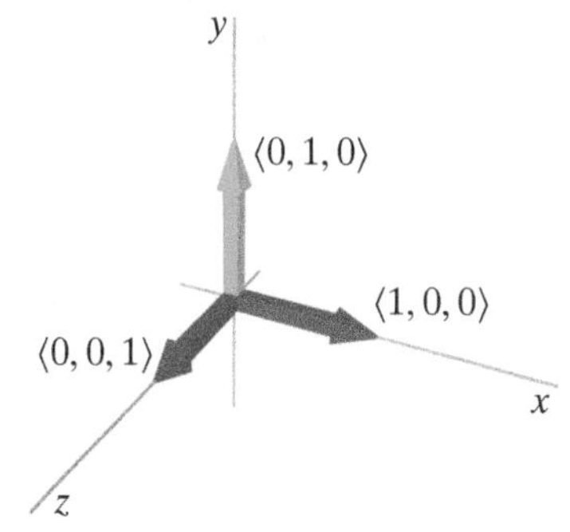

This textbook deals with the nature of matter and its interactions. The main goal of this textbook is to have you engage in a process central to science: constructing and applying physical models based on a small set of powerful fundamental physical principles and the atomic structure of matter. The variety of phenomena that we will be able to model, explain, and predict is very wide, including the orbit of stars around a black hole, nuclear fusion, the formation of sparks in air, and the speed of sound in a solid. This first chapter deals with the physical idea of interactions.

OBJECTIVES

After studying this chapter you should be able to

- Deduce from observations of an object's motion whether or not it has interacted with its surroundings.
- Mathematically describe position and motion in three dimensions.
- Mathematically describe momentum and change of momentum in three dimensions.
- Read and modify a simple computational model of motion at constant velocity.

1.1 KINDS OF MATTER

We will deal with material objects of many sizes, from subatomic particles to galaxies. All of these objects have certain things in common.

Atoms and Nuclei

Ordinary matter is made up of tiny atoms. An atom isn't the smallest type of matter, for it is composed of even smaller objects (electrons, protons, and neutrons), but many of the ordinary everyday properties of ordinary matter can be understood in terms of atomic properties and interactions. As you probably know from studying chemistry, atoms have a very small, very dense core, called the nucleus, around which is found a cloud of electrons. The nucleus contains protons and neutrons, collectively called nucleons. Electrons are kept close to the nucleus by electric attraction to the protons (the neutrons hardly interact with the electrons).

> QUESTION Recall your previous studies of chemistry. How many protons and electrons are there in a hydrogen atom? In a helium or carbon atom?

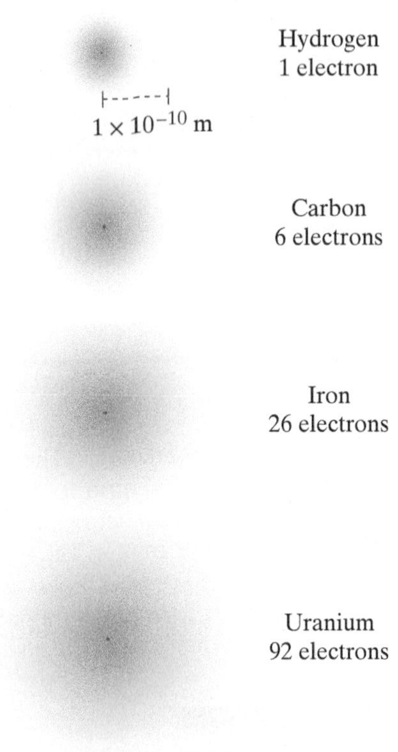

Figure 1.1 Atoms of hydrogen, carbon, iron, and uranium. The gray blur represents the electron cloud surrounding the nucleus. The black dot shows the location of the nucleus. On this scale, however, the nucleus would be much too small to see.

When you encounter a question in the text, you should think for a moment before reading on. Active reading contributes to significantly greater understanding. In the case of the questions posed above, if you don't remember the properties of these atoms, it may help to refer to the periodic table on the inside front cover of this textbook.

Hydrogen is the simplest atom, with just one proton and one electron. A helium atom has two protons and two electrons. A carbon atom has six protons and six electrons. Near the other end of the chemical periodic table, a uranium atom has 92 protons and 92 electrons. Figure 1.1 shows the relative sizes of the electron clouds in atoms of several elements but cannot show the nucleus to the same scale; the tiny dot marking the nucleus in the figure is much larger than the actual nucleus.

The radius of the electron cloud for a typical atom is about 1×10^{-10} meter. The reason for this size can be understood using the principles of quantum mechanics, a major development in physics in the early 20th century. The radius of a proton is about 1×10^{-15} meter, very much smaller than the radius of the electron cloud.

Nuclei contain neutrons as well as protons (Figure 1.2). The most common form or "isotope" of hydrogen has no neutrons in the nucleus. However, there exist isotopes of hydrogen with one or two neutrons in the nucleus (in addition to the proton). Hydrogen atoms containing one or two neutrons are called deuterium or tritium. The most common isotope of helium has two neutrons (and two protons) in its nucleus, but a rare isotope has only one neutron; this is called helium-3.

The most common isotope of carbon has six neutrons together with the six protons in the nucleus (carbon-12), whereas carbon-14 with eight neutrons is an isotope that plays an important role in dating archaeological objects.

Near the other end of the periodic table, uranium-235, which can undergo a fission chain reaction, has 92 protons and 143 neutrons, whereas uranium-238, which does not undergo a fission chain reaction, has 92 protons and 146 neutrons.

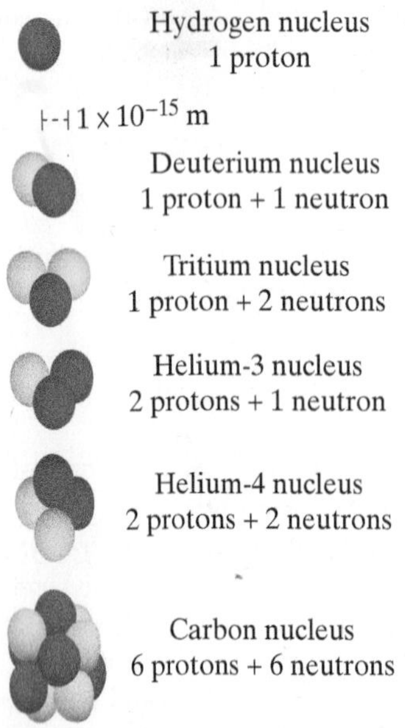

Figure 1.2 Nuclei of hydrogen, helium, and carbon. Note the *very* much smaller scale than in Figure 1.1!

Molecules and Solids

When atoms come in contact with each other, they may stick to each other ("bond" to each other). Several atoms bonded together can form a molecule—a substance whose physical and chemical properties differ from those of the constituent atoms. For example, water molecules (H_2O) have properties quite different from the properties of hydrogen atoms or oxygen atoms.

An ordinary-sized rigid object made of bound-together atoms and big enough to see and handle is called a solid, such as a bar of aluminum. A new kind of microscope, the scanning tunneling microscope (STM), is able to map the locations of atoms on the surface of a solid, which has provided new techniques for investigating matter at the atomic level. Two such images appear in Figure 1.3. You can see that atoms in a crystalline solid are arranged in a regular three-dimensional array. The arrangement of atoms on the surface depends on the direction along which the crystal is cut. The irregularities in the bottom image reflect "defects," such as missing atoms, in the crystal structure.

Liquids and Gases

When a solid is heated to a higher temperature, the atoms in the solid vibrate more vigorously about their normal positions. If the temperature is raised high enough, this thermal agitation may destroy the rigid structure of the solid. The atoms may become able to slide over each other, in which case the substance is a liquid.

At even higher temperatures the thermal motion of the atoms or molecules may be so large as to break the interatomic or intermolecular bonds completely, and the liquid turns into a gas. In a gas the atoms or molecules are quite free to move around, only occasionally colliding with each other or the walls of their container.

We will learn how to analyze many aspects of the behavior of solids and gases. We won't have much to say about liquids, because their properties are much harder to analyze. Solids are simpler to analyze than liquids because the atoms stay in one place (though with thermal vibration about their usual positions). Gases are simpler to analyze than liquids because between collisions the gas molecules are approximately unaffected by the other molecules. Liquids are the awkward intermediate state, where the atoms move around rather freely but are always in contact with other atoms. This makes the analysis of liquids very complex.

Figure 1.3 Two different surfaces of a crystal of pure silicon. The images were made with a scanning tunneling microscope. (Images courtesy of Randall Feenstra, IBM Corp.)

Planets, Stars, Solar Systems, and Galaxies

In our brief survey of the kinds of matter that we will study, we make a giant leap in scale from atoms all the way up to planets and stars, such as our Earth and Sun. We will see that many of the same principles that apply to atoms apply to planets and stars. By making this leap we bypass an important physical science, geology, whose domain of interest includes the formation of mountains and continents. We will study objects that are much bigger than mountains, and we will study objects that are much smaller than mountains, but we don't have time to apply the principles of physics to every important kind of matter.

Our Sun and its accompanying planets constitute our Solar System. It is located in the Milky Way galaxy, a giant rotating disk-shaped system of stars. On a clear dark night you can see a band of light (the Milky Way) coming from the huge number of stars lying in this disk, which you are looking at from a position in the disk, about two-thirds of the way out from the center of the disk. Our galaxy is a member of a cluster of galaxies that move around each other much as the planets of our Solar System move around the Sun (Figure 1.4). The Universe contains many such clusters of galaxies.

Figure 1.4 Our Solar System exists inside a galaxy, which itself is a member of a cluster of galaxies. (Photos courtesy NASA/JPL-Caltech)

Point Particles

It is common in physics to talk about the motion of a "point particle." What we mean by a particle is an object whose size, shape, and internal structure are not important to us in the current context, and which we can consider to be located at a single point in space. In modeling the motion of a real object (whether it is a galaxy or a proton), we often choose to make the simplifying assumption that it is a point particle, as if Superman or a giant space alien had come along and squeezed the object until it was compressed into a very tiny, structureless microscopic speck with the full mass of the original object.

Of course, there are many situations in which it would be absurd to use this approximation. The Earth, for example, is a large, complex object, with a core of turbulent molten rock, huge moving continents, and massive sloshing oceans. Radioactivity keeps its core hot; electromagnetic radiation from the Sun warms its surface; and thermal energy is also radiated away into space. If we are interested in energy flows or continental motion or earthquakes we need to consider the detailed structure and composition of the Earth. However, if what we want to do is model the motion of the Earth as it interacts with other objects in our Solar System, it works quite well to ignore this complexity, and to treat the Earth, the Sun, the Moon, and the other planets as if they were point particles.

Even most very tiny objects, such as atoms, protons, and neutrons, are not truly point particles—they do have finite size, and they have internal structure, which can influence their interactions with other objects. By contrast, electrons may really be point particles—they appear to have no internal structure, and attempts to measure the radius of an electron have not produced a definite number (recent experiments indicate only that the radius of an electron is less than 2×10^{-20} m, much smaller than a proton).

As we consider various aspects of matter and its interactions, it will be important for us to state explicitly whether or not we are modeling material objects as point particles, or as extended and perhaps deformable macroscopic chunks of matter. In Chapters 1–3 we will emphasize systems that can usefully be modeled as particles. In Chapter 4 we will begin to consider the detailed internal structure of material objects.

1.2 DETECTING INTERACTIONS

Objects made of different kinds of matter interact with each other in various ways: gravitationally, electrically, magnetically, and through nuclear interactions. How can we detect that an interaction has occurred? In this section we consider various kinds of observations that indicate the presence of interactions.

> **QUESTION** Before you read further, take a moment to think about your own ideas of interactions. How can you tell that two objects are interacting with each other?

Change of Direction of Motion

Figure 1.5 A proton moves through space, far from almost all other objects. The initial direction of the proton's motion is upward, as indicated by the arrow. The ×'s represent the position of the proton at equal time intervals.

Suppose that you observe a proton moving through a region of outer space, far from almost all other objects. The proton moves along a path like the one shown in Figure 1.5. The arrow indicates the initial direction of the proton's motion, and the ×'s in the diagram indicate the position of the proton at equal time intervals.

> **QUESTION** Do you see evidence in Figure 1.5 that the proton is interacting with another object?

Evidently a change in direction is a vivid indicator of interactions. If you observe a change in direction of the motion of a proton, you will find another object somewhere that has interacted with this proton.

> **QUESTION** Suppose that the only other object nearby was another proton. What was the approximate initial location of this second proton?

Since two protons repel each other electrically, the second proton must have been located to the right of the bend in the first proton's path.

Change of Speed

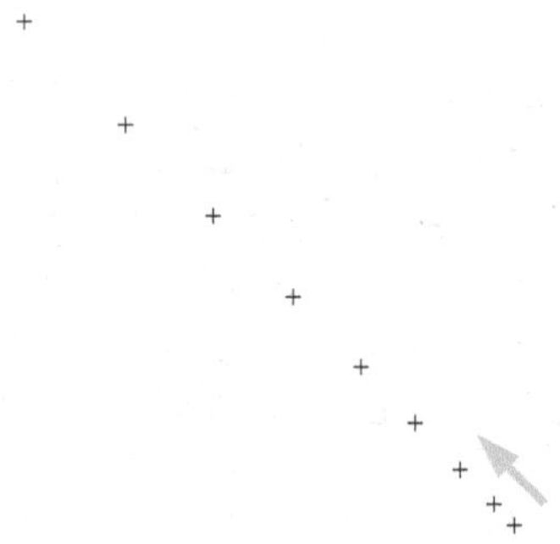

Figure 1.6 An electron moves through space, far from almost all other objects. The initial direction of the electron's motion is upward and to the left, as indicated by the arrow. The ×'s represent the position of the electron at equal time intervals.

Suppose that you observe an electron traveling in a straight line through outer space far from almost all other objects (Figure 1.6). The path of the electron is shown as though a camera had taken multiple exposures at equal time intervals.

> **QUESTION** Where is the electron's speed largest? Where is the electron's speed smallest?

The speed is largest at the upper left, where the ×'s are farther apart, which means that the electron has moved farthest during the time interval between exposures. The speed is smallest at the bottom right, where the ×'s are closer together, which means that the electron has moved the least distance during the time interval between exposures.

> QUESTION Suppose that the only other object nearby was another electron. What was the approximate initial location of this other electron?

The other electron must have been located directly just below and to the right of the starting location, since electrons repel each other electrically.

Evidently a change in speed is an indicator of interactions. If you observe a change in speed of an electron, you will find another object somewhere that has interacted with the electron.

Velocity Includes Both Speed and Direction

In physics, the word "velocity" has a special technical meaning that is different from its meaning in everyday speech. In physics, the quantity called "velocity" denotes a combination of speed and direction. Even if the speed or direction of motion is changing, the velocity has a precise value (speed and direction) at any instant. In contrast, in everyday speech, "speed" and "velocity" are often used as synonyms. In physics and other sciences, however, words have rather precise meanings and there are few synonyms.

For example, consider an airplane that at a particular moment is flying with a speed of 1000 kilometers/hour in a direction that is due east. We say the velocity is 1000 km/h, east, where we specify both speed and direction. An airplane flying west with a speed of 1000 km/h would have the same speed but a different velocity.

Figure 1.7 Two successive positions of a particle (indicated by a dot), with arrows indicating the velocity of the particle at each location. The shorter arrow indicates that the speed of the particle at location 2 is less than its speed at location 1.

We have seen that a change in an object's speed, or a change in the direction of its motion, indicates that the object has interacted with at least one other object. The two indicators of interaction, change of speed and change of direction, can be combined into one compact statement:

> A change of velocity (speed or direction or both) indicates the existence of an interaction.

In physics diagrams, the velocity of an object is represented by an arrow: a line with an arrowhead. The tail of the arrow is placed at the location of the object at a particular instant, and the arrow points in the direction of the motion of the object at that instant. The length of the arrow is proportional to the speed of the object. Figure 1.7 shows two successive positions of a particle at two different times, with velocity arrows indicating a change in speed of the particle (it's slowing down). Figure 1.8 shows three successive positions of a different particle at three different times, with velocity arrows indicating a change in direction but no change in speed. Note that the arrows themselves are straight; even if the path of the particle curves over time, at any instant the particle may be considered to be traveling in a specific direction.

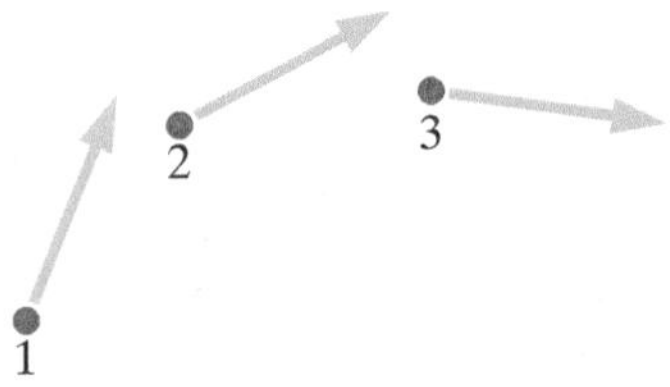

Figure 1.8 Three successive positions of a particle (indicated by a dot), with arrows indicating the velocity of the particle at each location. The arrows are the same length, indicating the same speed, but they point in different directions, indicating a change in direction and therefore a change in velocity.

We will see a little later that velocity is only one example of a physical quantity that has a "magnitude" (an amount or a size) and a direction. Other examples of such quantities are position relative to an origin in 3D space, changes in position or velocity, and force. In Section 1.4 we will see how to represent such quantities as vectors: single mathematical entities that combine information about magnitude and direction.

Uniform Motion

Suppose that you observe a rock moving along in outer space far from all other objects. We don't know what made it start moving in the first place; presumably a long time ago an interaction gave it some velocity and it has been coasting through the vacuum of space ever since.

Figure 1.9 "Uniform motion"—no change in speed or direction.

It is an observational fact that such an isolated object moves at constant, unchanging speed, in a straight line. Its velocity does not change (neither its direction nor its speed changes). We call motion with unchanging velocity "uniform motion" (Figure 1.9). Other terms for uniform motion include "uniform velocity" and "constant velocity," since velocity refers to both speed and direction.

QUESTION Is an object at rest in uniform motion?

If an object remains at rest, then neither the speed nor direction of the object's velocity changes. This is a special case of uniform motion: the object's speed is constant (zero is a valid value of speed) and the direction of motion, while undefined, is not changing.

QUESTION If we observe an object in uniform motion, can we conclude that it has no interactions with its surroundings?

When we observe an object in uniform motion, one possibility is that it has no interactions at all with its surroundings. However, there is another possibility: the object may be experiencing multiple interactions that cancel each other out. In either case, we can correctly deduce that the "net" (total) interaction of the object with its surroundings is zero.

Checkpoint 1 **(a)** Which of the following do you see moving with constant velocity? (1) A ship sailing northeast at a speed of 5 meters per second (2) The Moon orbiting the Earth (3) A tennis ball traveling across the court after having been hit by a tennis racket (4) A can of soda sitting on a table (5) A person riding on a Ferris wheel that is turning at a constant rate. **(b)** In which of the following situations is there observational evidence for significant interaction between two objects? How can you tell? (1) A ball bounces off a wall with no change in speed. (2) A baseball that was hit by a batter flies toward the outfield. (3) A communications satellite orbits the Earth. (4) A space probe travels at constant speed toward a distant star. (5) A charged particle leaves a curving track in a particle detector.

1.3 NEWTON'S FIRST LAW OF MOTION

The basic relationship between change of velocity and interaction is summarized qualitatively by what is known as Newton's "first law of motion," though it was originally discovered by Galileo. In his original Latin, Newton said, "Corpus omne perseverare in statu suo quiescendi vel movendi uniformiter in directum, nisi quatenus a viribus impressis cogitur statum illum mutare." A literal translation is "Every body persists in its state of resting or of moving uniformly in a direction, except to the extent that it is compelled to change that state by forces pressed upon it." Expressing this in more modern language, we have this:

NEWTON'S FIRST LAW OF MOTION

Every body persists in its state of rest or of moving with constant speed in a constant direction, except to the extent that it is compelled to change that state by forces acting on it.

"Force" is the way in which the amount of interaction is quantified, and we'll discuss force in detail in Chapter 2. The words "except to the extent" imply that the stronger the interaction, the more change there will be in direction and/or speed. The weaker the interaction, the less change. If there is no net (total) interaction at all, the object's motion will be uniform (constant speed and direction); this could happen either because there are no interactions or because there are interactions that cancel each other, such as equally strong pushes to the left and right. It is important to remember that if an object is not moving at all, its velocity is not changing, so it too may be considered to be in uniform motion.

Newton's first law of motion is only qualitative, because it doesn't give us a way to calculate quantitatively how much change in speed or direction will be produced by a certain amount of interaction, a subject we will take up in the next chapter. Nevertheless, Newton's first law of motion is important in providing a conceptual framework for thinking about the relationship between interaction and motion.

This law represented a major break with ancient tradition, which assumed that constant pushing was required to keep something moving. This law says something radically different: no interactions at all are needed to keep something moving!

> QUESTION To move a box across a table at constant speed in a straight line, you must keep pushing it. Does this contradict Newton's first law?

Since a constant interaction is required to keep the box moving, we might be tempted to conclude that Newton's first law of motion does not apply in many everyday situations. However, what matters is the *net* interaction of the box with its surroundings, which could be zero if there are multiple interactions that cancel each other out.

> QUESTION In addition to your hand, what other objects in the surroundings interact with the box?

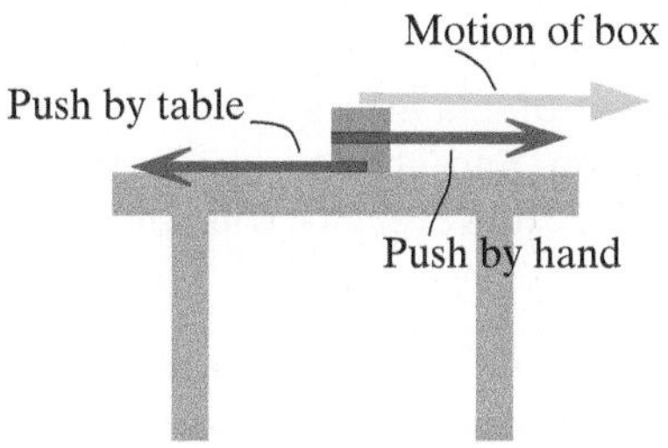

Figure 1.10 The red arrows represent the magnitude and direction of the pushes the box gets from your hand and from the friction with the table. If these pushes add up to zero, the box moves with constant speed in a straight line, indicated by the green arrow.

The table also interacts with the box, in a way that we call friction. If you push just hard enough to compensate exactly for the table friction, the sum of all the interactions is zero, and the box moves at constant speed as predicted by Newton's first law (Figure 1.10). (If you push harder than the table does, the box's speed steadily increases.)

It is difficult to observe motion without friction in everyday life, because objects almost always interact with many other objects, including air, flat surfaces, and so on. You may be able to think of situations in which you have seen an object keep moving at constant (or nearly constant) velocity, without being pushed or pulled. One example of a nearly friction-free situation is a hockey puck sliding on ice. The puck slides a long way at nearly constant speed in a straight line (constant velocity) because there is little friction with the ice. An even better example is the uniform motion of an object in outer space, far from all other objects.

> QUESTION Is a change of position an indicator of an interaction?

Not necessarily. If the change of position occurs simply because a particle is moving at constant speed and direction, then a mere change of position is not an indicator of an interaction, since uniform motion is an indicator of zero net interaction. We need to know the object's velocity at each observation to be able to make further deductions.

> QUESTION If you observe an object at rest in one location, and later you observe it again at rest but in a different location, can you conclude that an interaction took place?

Yes. You can infer that there must have been an interaction to give the object some velocity to move the object toward the new position, and another interaction to slow the object to a stop in its new position.

> QUESTION Is it possible to deduce the existence of an interaction even though you do not observe a change?

As we saw when we considered pushing a box across a table at constant speed, sometimes we may find indirect evidence for an additional interaction.

When something doesn't change although we would normally expect a change due to a known interaction, we can logically deduce that an additional interaction must be occurring. For example, consider a helium-filled balloon that hovers motionless in the air despite the downward gravitational pull of the Earth. Evidently there is some additional kind of interaction that opposes the gravitational interaction. In this case, interactions with air molecules have the net effect of pushing up on the balloon ("buoyancy"). The lack of change implies that the effect of the air molecules exactly compensates for the gravitational interaction with the Earth.

The stability of the nucleus of an atom is another example of indirect evidence for an additional interaction. The nucleus contains positively charged protons that repel each other electrically, yet the nucleus remains intact. We conclude that there must be some other kind of interaction present, a nonelectric attractive interaction that overcomes the electric repulsion. This is evidence for a nonelectric interaction called the "strong interaction," which as we will see acts among protons and neutrons to hold the nucleus together. We will discuss the strong interaction in Chapter 3.

Other Indicators of Interaction

Change of velocity is not the only indication that an object has interacted with its surroundings, but it is the only change possible for a single object that is modeled as a point particle, which has neither shape nor internal structure. In later chapters we will examine other kinds of changes, such as change of temperature, change of shape or configuration, and change of identity (for example, in nuclear reactions). In Chapters 1–3, however, we will concentrate on how interactions change motion.

Checkpoint 2 (a) Apply Newton's first law to each of the following situations. In which situations can you conclude that the object is undergoing a net interaction with one or more other objects? (1) A book slides across the table and comes to a stop. (2) A proton in a particle accelerator moves faster and faster. (3) A car travels at constant speed around a circular race track. (4) A spacecraft travels at a constant speed toward a distant star. (5) A hydrogen atom remains at rest in outer space. **(b)** A spaceship far from all other objects uses its rockets to attain a speed of 1×10^4 m/s. The crew then shuts off the power. According to Newton's first law, which of the following statements about the motion of the spaceship after the power is shut off are correct? (Choose all statements that are correct.) (1) The spaceship will move in a straight line. (2) The spaceship will travel on a curving path. (3) The spaceship will enter a circular orbit. (4) The speed of the spaceship will not change. (5) The spaceship will gradually slow down. (6) The spaceship will stop suddenly.

1.4 DESCRIBING THE 3D WORLD: VECTORS

Physical phenomena take place in the 3D world around us. In order to be able to make quantitative predictions and give detailed, quantitative explanations, we need tools for describing precisely the positions and velocities of objects in 3D, and the changes in position and velocity due to interactions. These tools are mathematical entities called 3D "vectors." A symbol denoting a vector is written with an arrow over it:

$$\vec{r} \text{ is a vector}$$

In three dimensions a vector is a triple of numbers $\langle x, y, z \rangle$. Quantities like the position or velocity of an object can be represented as vectors:

$$\vec{r}_1 = \langle 3.2, -9.2, 66.3 \rangle \text{ m} \quad \text{(a position vector)}$$
$$\vec{v}_1 = \langle -22.3, 0.4, -19.5 \rangle \text{ m/s} \quad \text{(a velocity vector)}$$

Many vectors have units associated with them, such as meters or meters per second. In this course, we will work with the following important physical quantities that are vectors: position, velocity, rate of change of velocity (acceleration), momentum, rate of change of momentum, force, angular momentum, torque, electric field, magnetic field, energy flow, and momentum flow. All of these vectors have associated physical units.

We use the notation $\langle x, y, z \rangle$ for vectors because it emphasizes the fact that a vector is a single entity, and because it is easy to work with. This notation appears in many calculus textbooks; you will probably encounter other ways of expressing vectors mathematically as well.

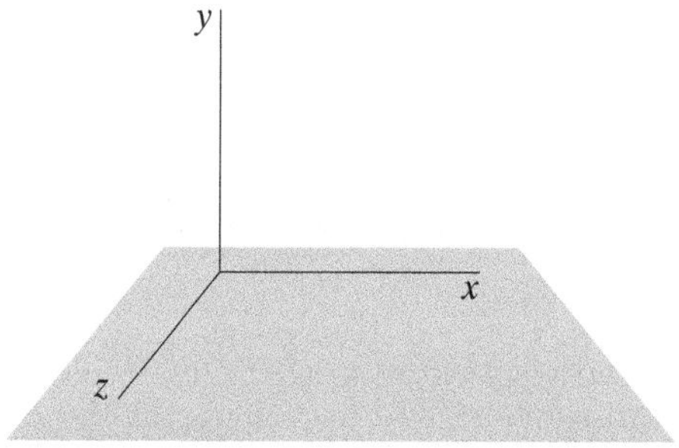

Figure 1.11 Right-handed 3D coordinate system. The xy plane is in the plane of the page, and the z axis projects out of the page, toward you.

Position Vectors

A position vector is a simple example of a physical vector quantity. We will use a 3D Cartesian coordinate system to specify positions in space and other vector quantities. Usually we will orient the axes of the coordinate system as shown in Figure 1.11: $+x$ axis to the right, $+y$ axis upward, and $+z$ axis coming out of the page, toward you. This is a "right-handed" coordinate system: if you hold the thumb, first, and second fingers of your right hand perpendicular to each other, and align your thumb with the x axis and your first finger with the y axis, your second finger points along the z axis. In some math textbook discussions of 3D coordinate systems, the x axis points out, the y axis points to the right, and the z axis points up. This is the same right-handed coordinate system, viewed from a different "camera position." Since we will sometimes consider motion in a single plane, it makes sense to orient the xy plane in the plane of a vertical page or computer display, so we will use the viewpoint in which the y axis points up.

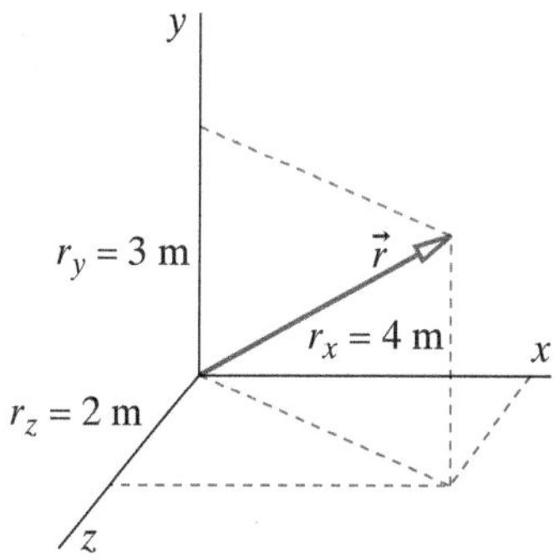

Figure 1.12 A position vector $\vec{r} = \langle 4,3,2 \rangle$ m and its x, y, and z components.

A position in 3D space can be considered to be a vector, called a *position vector*, pointing from an origin to that location. Figure 1.12 shows a position vector, represented by an arrow with its tail at the origin, that might represent your final position if you started at the origin and walked 4 meters along the x axis, then 2 meters parallel to the z axis, then climbed a ladder so you were 3 meters above the ground. Your new position relative to the origin is a vector that can be written like this:

$$\vec{r} = \langle 4,3,2 \rangle \text{ m}$$

Each of the numbers in the triple is called a "component" of the vector, and is associated with a particular axis. Usually the components of a vector are denoted symbolically by the subscripts x, y, and z:

$$\vec{v} = \langle v_x, v_y, v_z \rangle \quad \text{(a velocity vector)}$$
$$\vec{r} = \langle r_x, r_y, r_z \rangle \quad \text{(a position vector)}$$
$$\vec{r} = \langle x, y, z \rangle \quad \text{(alternative notation for a position vector)}$$

The components of the position vector $\vec{r} = \langle 4,3,2 \rangle$ m are:

$$r_x = 4\text{ m} \quad \text{(the } x \text{ component)}$$
$$r_y = 3\text{ m} \quad \text{(the } y \text{ component)}$$
$$r_z = 2\text{ m} \quad \text{(the } z \text{ component)}$$

The x component of the vector $\vec{v}$ is the number v_x. The z component of the vector $\vec{v}_1 = \langle -22.3, 0.4, -19.5 \rangle$ m/s is -19.5 m/s. A component such as v_x is not a vector, since it is only one number.

QUESTION Can a vector be zero?

The zero vector $\langle 0,0,0\rangle$ is a legal vector, which we will sometimes write as $\vec{0}$. A zero position vector describes the position of an object located at the origin. A zero velocity vector describes the velocity of an object that is at rest at a particular instant.

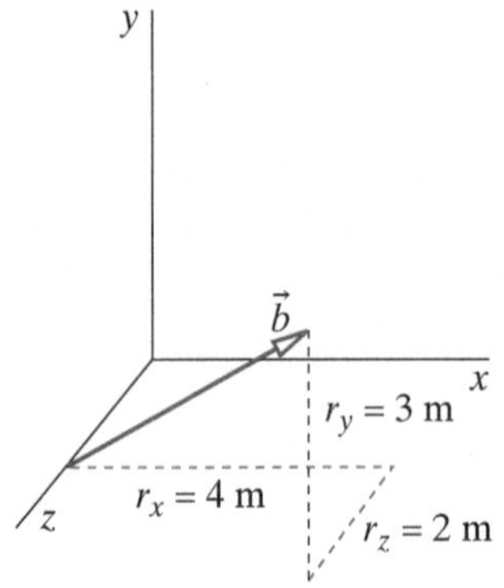

Figure 1.13 The arrow represents the vector $\vec{b} = \langle 4,3,2\rangle$ m, drawn with its tail at location $\langle 0,0,2\rangle$.

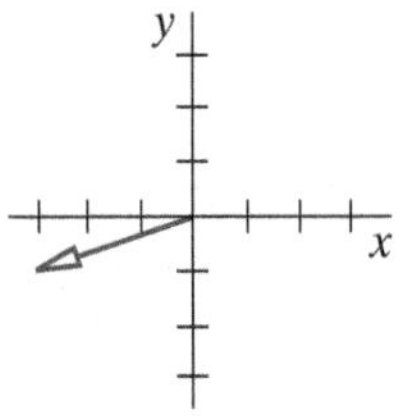

Figure 1.14 The position vector $\langle -3,-1,0\rangle$, drawn at the origin, in the *xy* plane. The components of the vector specify the displacement from the tail to the tip. The *z* axis, which is not shown, comes out of the page, toward you.

Drawing Vectors

A position vector is special in that its tail is always at the origin of a coordinate system, but this is not the case for other vectors. It is important to note that the x component of a vector specifies the difference between the x coordinate of the tail of the vector and the x coordinate of the tip of the vector. It does not give any information about the location of the tail of the vector (compare Figures 1.12 and 1.13). By convention, arrows representing vector quantities such as velocity are usually drawn with the tail of the arrow at the location of the object.

In Figure 1.12 we represented your position vector relative to the origin graphically by an arrow whose tail is at the origin and whose arrowhead is at your position. The length of the arrow represents the distance from the origin, and the direction of the arrow represents the direction of the vector, which is the direction of a direct path from the initial position to the final position (the "displacement"; by walking and climbing you "displaced" yourself from the origin to your final position).

Since it is difficult to draw a 3D diagram on paper, when working on paper you will usually be asked to draw vectors that all lie in a single plane. Figure 1.14 shows an arrow in the xy plane representing the vector $\langle -3,-1,0\rangle$.

Scalars

A quantity that is represented by a single number is called a *scalar*. A scalar quantity does not have a direction. Examples include the mass of an object, such as 50 kg, or the temperature, such as −20 °C. Vectors and scalars are very different entities; a vector can never be equal to a scalar, and a scalar cannot be added to a vector. Scalars can be positive, negative, or zero:

$$m = 50\,\text{kg}$$
$$T = -20\,°\text{C}$$

Vector Operations

Vectors are mathematical entities, and have their own mathematical operations. Some of these operations are the same as those you already know for scalars. Others, such as multiplication, are quite different, and division by a vector is not legal. Here are the vector operations that we will discuss and use in this textbook:

VECTOR OPERATIONS

Mathematical operations that are defined for vectors:

- Multiply or divide a vector by a scalar: $2\vec{a}$, $\vec{v}/5$
- Find the magnitude of a vector: $|\vec{a}|$
- Find a unit vector giving direction: $\hat{a}$
- Add one vector to another: $\vec{a}+\vec{b}$
- Subtract one vector from another: $\vec{a}-\vec{b}$
- Differentiate a vector: $d\vec{r}/dt$
- Dot product of two vectors (result is a scalar): $\vec{a}\bullet\vec{b}$
- Cross product of two vectors (result is a vector): $\vec{a}\times\vec{b}$

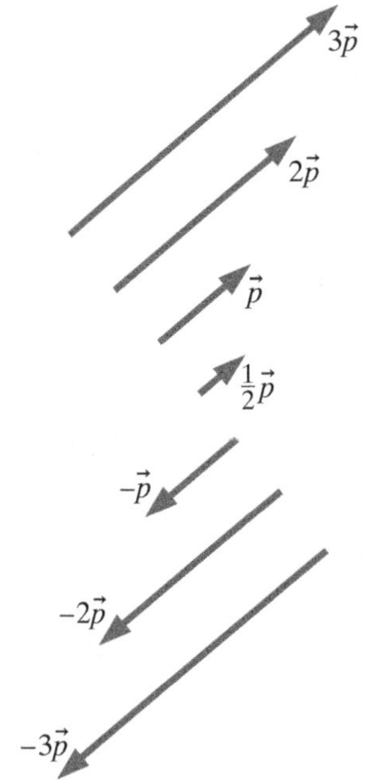

Figure 1.15 Multiplying a vector by a scalar changes the magnitude of the vector. Multiplying by a negative scalar reverses the direction of the vector.

The dot product will be introduced in Chapter 5, and the cross product in Chapter 11.

There are certain operations that are neither legal nor meaningful for vectors:

- A vector cannot be set equal to a scalar.
- A vector cannot be added to or subtracted from a scalar.
- A vector cannot occur in the denominator of an expression. (Although you can't divide by a vector, note that you can legally divide by the *magnitude* of a vector, which is a scalar.)
- As with scalars, you can't add or subtract vectors that have different units.

Multiplying a Vector by a Scalar

A vector can be multiplied (or divided) by a scalar. If a vector is multiplied by a scalar, each of the components of the vector is multiplied by the scalar:

$$\text{If } \vec{r} = \langle x, y, z \rangle, \text{ then } a\vec{r} = \langle ax, ay, az \rangle$$

$$\text{If } \vec{v} = \langle v_x, v_y, v_z \rangle, \text{ then } \frac{\vec{v}}{b} = \left\langle \frac{v_x}{b}, \frac{v_y}{b}, \frac{v_z}{b} \right\rangle$$

$$\frac{1}{2}\langle 6, -20, 9 \rangle = \langle 3, -10, 4.5 \rangle$$

Multiplication by a scalar "scales" a vector, keeping its direction the same but making its magnitude larger or smaller (Figure 1.15). Multiplying by a negative scalar reverses the direction of a vector.

$$(-1)\langle 0, 0, 4 \rangle = \langle 0, 0, -4 \rangle$$

> **Checkpoint 3** You stand at location $\vec{r} = \langle 2, -3, 5 \rangle$ m. Your friend stands at location $\vec{r}/2$. What is your friend's position vector?

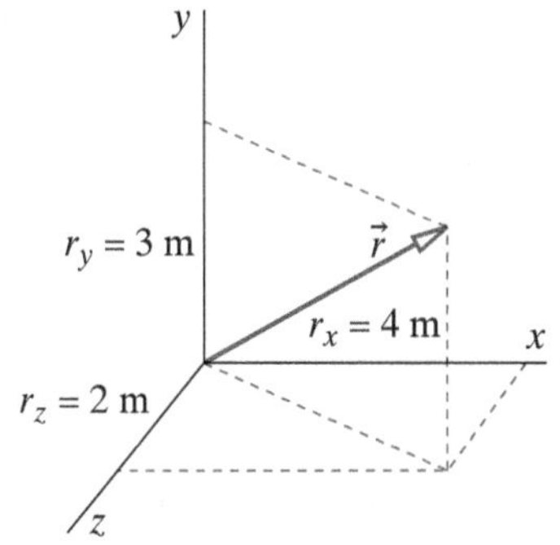

Figure 1.16 A vector representing a displacement from the origin.

Magnitude

Figure 1.16 shows a vector representing a displacement of $\langle 4, 3, 2 \rangle$ m from the origin. What is the distance from the tip of this vector to the origin? Using a 3D extension of the Pythagorean theorem for right triangles (Figure 1.17), we find that

$$\sqrt{(4\,\text{m})^2 + (3\,\text{m})^2 + (2\,\text{m})^2} = \sqrt{29}\,\text{m} = 5.39\,\text{m}$$

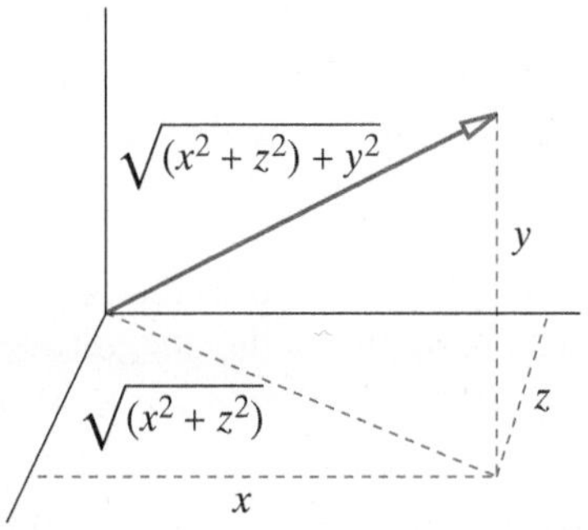

Figure 1.17 The magnitude of a vector is the square root of the sum of the squares of its components (3D version of the Pythagorean theorem).

We say that the *magnitude* $|\vec{r}|$ of the position vector $\vec{r}$ is

$$|\vec{r}| = 5.39\,\text{m}$$

The magnitude of a vector is written either with absolute-value bars around the vector as $|\vec{r}|$, or simply by writing the symbol for the vector without the little arrow above it, r.

MAGNITUDE OF A VECTOR

> If the vector $\vec{r} = \langle r_x, r_y, r_z \rangle$ then $|\vec{r}| = \sqrt{r_x^2 + r_y^2 + r_z^2}$ (a scalar).
>
> The magnitude of a vector is always a positive number. The magnitude of a vector is a single number, not a triple of numbers, and it is a scalar, not a vector.

You may wonder how to find the magnitude of a quantity like $-3\vec{r}$, which involves the product of a scalar and a vector. This expression can be factored:

$$|-3\vec{r}| = |-3| \cdot |\vec{r}|$$

The magnitude of a scalar is its absolute value, so:

$$|-3\vec{r}| = |-3| \cdot |\vec{r}| = 3\sqrt{r_x^2 + r_y^2 + r_z^2}$$

Checkpoint 4 If $\vec{v} = \langle 2, -3, 5 \rangle$ m/s, what is $\left|-\frac{1}{2}\vec{v}\right|$?

Unit Vectors

One way to describe the direction of a vector is by specifying a *unit vector*. A unit vector is a vector of magnitude 1, pointing in some direction. A unit vector is written with a "hat" (caret) over it instead of an arrow. The unit vector $\hat{a}$ is called "a-hat."

QUESTION Is the vector $\langle 1,1,1 \rangle$ a unit vector?

The magnitude of $\langle 1,1,1 \rangle$ is $\sqrt{1^2+1^2+1^2} = 1.73$, so this is not a unit vector.

The vector $\langle 1/\sqrt{3},\ 1/\sqrt{3},\ 1/\sqrt{3} \rangle$ is a unit vector, since its magnitude is 1:

$$\sqrt{\left(\frac{1}{\sqrt{3}}\right)^2 + \left(\frac{1}{\sqrt{3}}\right)^2 + \left(\frac{1}{\sqrt{3}}\right)^2} = 1$$

Note that every component of a unit vector must be less than or equal to 1.

In our 3D Cartesian coordinate system, there are three special unit vectors, oriented along the three axes. They are called i-hat, j-hat, and k-hat, and they point along the x, y, and z axes, respectively (Figure 1.18):

$$\hat{\imath} = \langle 1,0,0 \rangle$$
$$\hat{\jmath} = \langle 0,1,0 \rangle$$
$$\hat{k} = \langle 0,0,1 \rangle$$

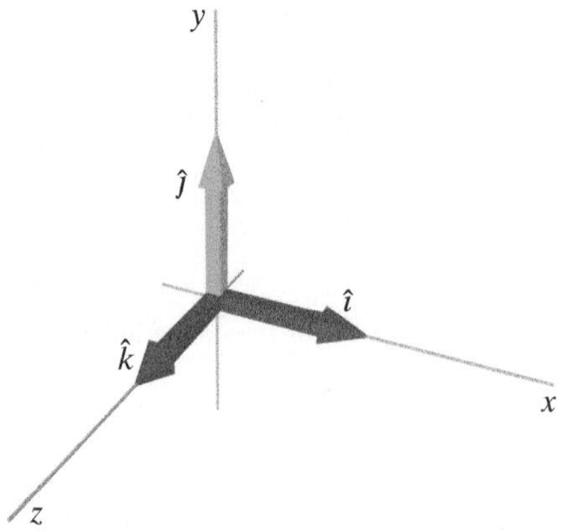

Figure 1.18 The unit vectors $\hat{\imath}$, $\hat{\jmath}$, $\hat{k}$.

One way to express a vector is in terms of these special unit vectors:

$$\langle 0.02, -1.7, 30.0 \rangle = 0.02\hat{\imath} + (-1.7)\hat{\jmath} + 30.0\hat{k}$$

Not all unit vectors point along an axis, as shown in Figure 1.19. For example, the vectors

$$\hat{g} = \langle 0.5774, 0.5774, 0.5774 \rangle \quad \text{and} \quad \hat{r} = \langle 0.424, 0.566, 0.707 \rangle$$

are both approximately unit vectors, since the magnitude of each is approximately equal to 1. Again, note that every component of a unit vector is less than or equal to 1.

Any vector may be factored into the product of a unit vector in the direction of the vector, multiplied by a scalar equal to the magnitude of the vector.

$$\vec{w} = |\vec{w}| \cdot \hat{w}$$

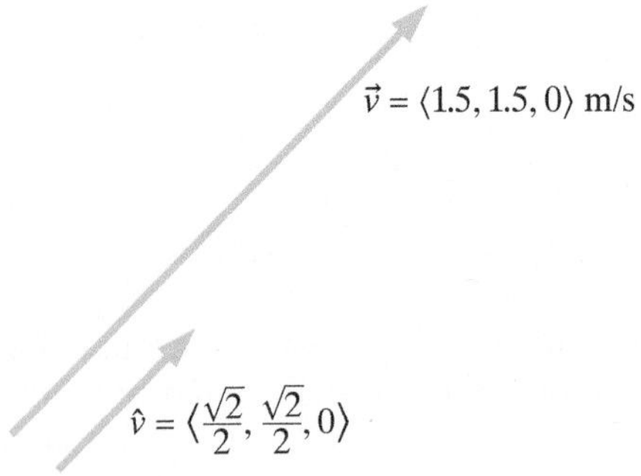

Figure 1.19 The unit vector $\hat{v}$ has the same direction as the vector $\vec{v}$, but its magnitude is 1, and it has no physical units.

For example, a vector of magnitude 5, aligned with the y axis, could be written as:

$$\langle 0,5,0 \rangle = 5\langle 0,1,0 \rangle$$

Therefore, to find a unit vector in the direction of a particular vector, we just divide the vector by its magnitude:

FINDING A UNIT VECTOR

$$\hat{r} = \frac{\vec{r}}{|\vec{r}|} = \frac{\langle x,y,z \rangle}{\sqrt{(x^2+y^2+z^2)}}$$

$$\hat{r} = \left\langle \frac{x}{\sqrt{(x^2+y^2+z^2)}}, \frac{y}{\sqrt{(x^2+y^2+z^2)}}, \frac{z}{\sqrt{(x^2+y^2+z^2)}} \right\rangle$$

EXAMPLE **Magnitude and Direction**

Factor the vector $\vec{v} = \langle -22.3, 0.4, -19.5 \rangle$ m/s into a magnitude times a unit vector.

Solution

$$|\vec{v}| = \sqrt{(-22.3)^2 + (0.4)^2 + (-19.5)^2}\ \text{m/s} = 29.6\ \text{m/s}$$

$$\hat{v} = \frac{\vec{v}}{|\vec{v}|} = \frac{\langle -22.3, 0.4, -19.5 \rangle\ \text{m/s}}{29.6\ \text{m/s}} = \langle -0.753, 0.0135, -0.658 \rangle$$

$$\vec{v} = (29.6\ \text{m/s})\langle -0.753, 0.0135, -0.658 \rangle$$

We can now explain algebraically why multiplying a vector by a scalar changes the magnitude but not the direction of a vector. If we write the original vector as the product of a magnitude and a unit vector, after multiplying by a scalar the unit vector is unchanged, but the magnitude is increased or decreased:

$$\begin{aligned} \vec{a} &= \langle 3, -2, 4 \rangle = (5.385)\langle 0.577, -0.371, 0.743 \rangle \\ 2 \cdot \vec{a} &= (2)(5.385)\langle 0.577, -0.371, 0.743 \rangle \\ &= 10.770\langle 0.577, -0.371, 0.743 \rangle \\ &= \langle 6, -4, 8 \rangle \end{aligned}$$

EQUALITY OF VECTORS

A vector is equal to another vector if and only if all the components of the vectors are equal.

$$\vec{w} = \vec{r} \quad \text{means that}$$

$$w_x = r_x \quad \text{and} \quad w_y = r_y \quad \text{and} \quad w_z = r_z$$

The magnitudes and directions of two equal vectors are the same:

$$|\vec{w}| = |\vec{r}| \quad \text{and} \quad \hat{w} = \hat{r}$$

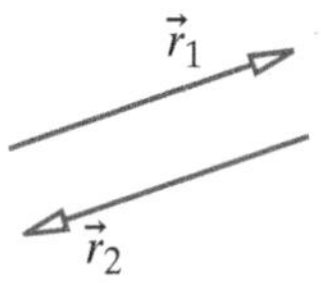

Figure 1.20 Are these two vectors equal? (Checkpoint 5)

Checkpoint 5 (a) Consider the vectors $\vec{r}_1$ and $\vec{r}_2$ represented by arrows in Figure 1.20. Are these two vectors equal? **(b)** If $\vec{a} = \langle 400, 200, -100 \rangle$ m/s^2, and $\vec{c} = \vec{a}$, what is the unit vector $\hat{c}$ in the direction of $\vec{c}$?

Vector Addition and Subtraction

Vectors may be added, and one vector may be subtracted from another vector. However, a scalar cannot be added to or subtracted from a vector.

ADDING AND SUBTRACTING VECTORS

The sum or difference of two vectors is another vector, obtained by adding or subtracting the components of the vectors. Given two vectors $\vec{A} = \langle A_x, A_y, A_z \rangle$ and $\vec{B} = \langle B_x, B_y, B_z \rangle$, then

$$\begin{aligned} \vec{A} + \vec{B} &= \langle (A_x + B_x), (A_y + B_y), (A_z + B_z) \rangle \\ \langle 1,2,3 \rangle + \langle -4,5,6 \rangle &= \langle -3,7,9 \rangle \\ \vec{A} - \vec{B} &= \langle (A_x - B_x), (A_y - B_y), (A_z - B_z) \rangle \\ \langle 1,2,3 \rangle - \langle -4,5,6 \rangle &= \langle 5,-3,-3 \rangle \end{aligned}$$

If $\vec{C} = \vec{A} + \vec{B}$, then $\vec{C} - \vec{B} = \vec{A}$ and so on, just as in scalar addition and subtraction. Note also that $\vec{A} - \vec{B} = \vec{A} + (-\vec{B})$, which is sometimes useful in the context of graphical subtraction (see below).

QUESTION Is adding the magnitudes of two vectors equivalent to adding two vectors, then taking the magnitude?

No. The magnitude of a vector is *not* in general equal to the sum of the magnitudes of the two original vectors! For example, the magnitude of the vector $\langle 3,0,0\rangle$ is 3, and the magnitude of the vector $\langle -2,0,0\rangle$ is 2, but the magnitude of the vector $(\langle 3,0,0\rangle + \langle -2,0,0\rangle)$ is 1, not 5!

Checkpoint 6 If $\vec{F}_1 = \langle 300, 0, -200\rangle$ and $\vec{F}_2 = \langle 150, -300, 0\rangle$, calculate the following quantities and make the requested comparisons: **(a)** $\vec{F}_1 + \vec{F}_2$ **(b)** $|\vec{F}_1 + \vec{F}_2|$ **(c)** $|\vec{F}_1| + |\vec{F}_2|$ **(d)** Is $|\vec{F}_1 + \vec{F}_2| = |\vec{F}_1| + |\vec{F}_2|$? **(e)** $\vec{F}_1 - \vec{F}_2$ **(f)** $|\vec{F}_1 - \vec{F}_2|$ **(g)** $|\vec{F}_1| - |\vec{F}_2|$ **(h)** Is $|\vec{F}_1 - \vec{F}_2| = |\vec{F}_1| - |\vec{F}_2|$?

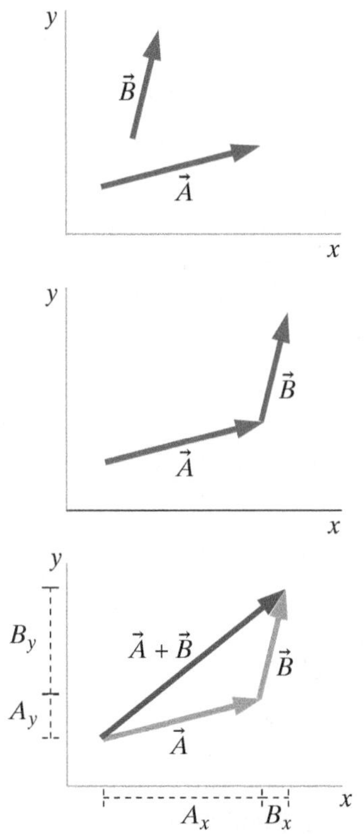

Figure 1.21 The procedure for adding two vectors graphically: To add $\vec{A} + \vec{B}$ graphically, move $\vec{B}$ so the tail of $\vec{B}$ is at the tip of $\vec{A}$, then draw a new arrow starting at the tail of $\vec{A}$ and ending at the tip of $\vec{B}$.

The sum of two vectors has a geometric interpretation. In Figure 1.21 you first walk along displacement vector $\vec{A}$, followed by walking along displacement vector $\vec{B}$. What is your net displacement vector $\vec{C} = \vec{A} + \vec{B}$? The x component C_x of your net displacement is the sum of A_x and B_x. Similarly, the y component C_y of your net displacement is the sum of A_y and B_y.

GRAPHICAL ADDITION OF VECTORS

To add two vectors $\vec{A}$ and $\vec{B}$ graphically (Figure 1.21):

- Draw the first vector $\vec{A}$.
- Move the second vector $\vec{B}$ (without rotating it) so its tail is located at the *tip* of the first vector.
- Draw a new vector from the tail of vector $\vec{A}$ to the tip of vector $\vec{B}$.

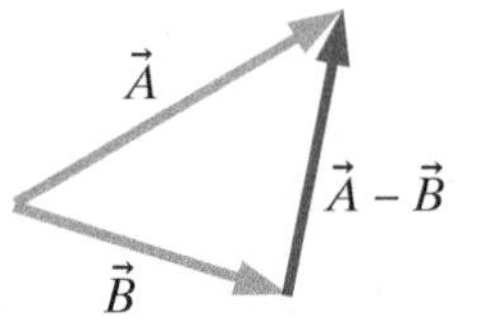

Figure 1.22 The procedure for subtracting vectors graphically: Draw vectors tail to tail; draw a new vector from the tip of the second vector to the tip of the first vector.

GRAPHICAL SUBTRACTION OF VECTORS

To subtract one vector $\vec{B}$ from another vector $\vec{A}$ graphically (Figure 1.22):

- Draw the first vector $\vec{A}$.
- Move the second vector $\vec{B}$ (without rotating it) so its tail is located at the *tail* of the first vector.
- Draw a new vector from the tip of vector $\vec{B}$ to the tip of vector $\vec{A}$.

Note that you can check this algebraically and graphically. As shown in Figure 1.22, since the tail of $\vec{A} - \vec{B}$ is located at the tip of $\vec{B}$, then the vector $\vec{A}$ should be the sum of $\vec{B}$ and $\vec{A} - \vec{B}$, as indeed it is:

$$\vec{B} + (\vec{A} - \vec{B}) = \vec{A}$$

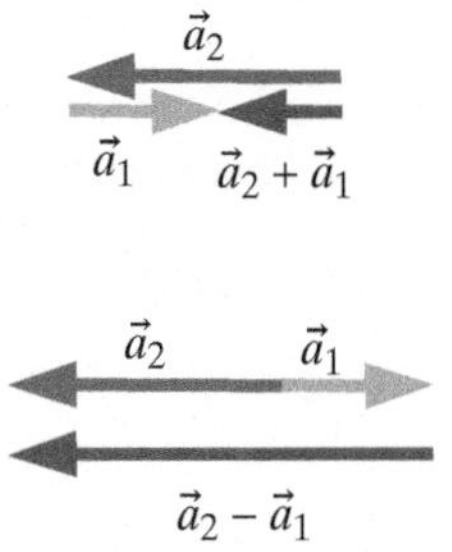

Figure 1.23 To add (top diagram) and subtract (bottom diagram) collinear vectors graphically, we offset the arrows slightly for clarity.

Graphical addition and subtraction of collinear vectors would be messy and difficult to interpret if we actually drew the arrows on top of each other. To make diagrams easier to interpret, we typically offset arrows slightly so we can see the results (Figure 1.23).

Checkpoint 7 Which of the following statements about the three vectors in Figure 1.24 are correct?
(a) $\vec{s} = \vec{t} - \vec{r}$ **(b)** $\vec{r} = \vec{t} - \vec{s}$ **(c)** $\vec{r} + \vec{t} = \vec{s}$ **(d)** $\vec{s} + \vec{t} = \vec{r}$ **(e)** $\vec{r} + \vec{s} = \vec{t}$

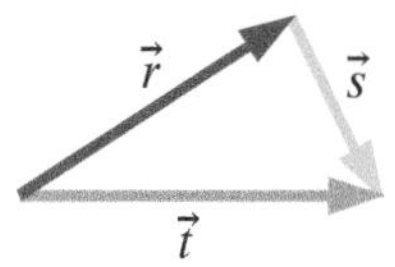

Figure 1.24 Checkpoint 7.

Commutativity and Associativity

Vector addition is commutative:

$$\vec{A}+\vec{B}=\vec{B}+\vec{A}$$

Vector subtraction is *not* commutative:

$$\vec{A}-\vec{B}\neq\vec{B}-\vec{A}$$

The associative property holds for vector addition and subtraction:

$$(\vec{A}+\vec{B})-\vec{C}=\vec{A}+(\vec{B}-\vec{C})$$

Applications of Vector Subtraction

Since we are interested in changes caused by interactions, we will frequently need to calculate the change in a vector quantity. For example, we may want to know the change in a moving object's position or the change in its velocity during some time interval. Finding such changes requires vector subtraction.

The Greek letter Δ (capital delta suggesting "D for Difference") is traditionally used to denote the change in a quantity (either a scalar or a vector). We use the subscript i to denote an *initial* value of a quantity, and the subscript f to denote the *final* value of a quantity.

Δ (DELTA) IS THE SYMBOL FOR A CHANGE

The symbol Δ (delta) means "final minus initial." If a vector $\vec{r}_i$ denotes the initial position of an object relative to the origin (its position at the beginning of a time interval), and $\vec{r}_f$ denotes the final position of the object, then

$$\Delta\vec{r}=\vec{r}_f-\vec{r}_i$$

$\Delta\vec{r}$ means "change of $\vec{r}$" or $\vec{r}_f-\vec{r}_i$ (displacement).
Δt means "change of t" or t_f-t_i (time interval).

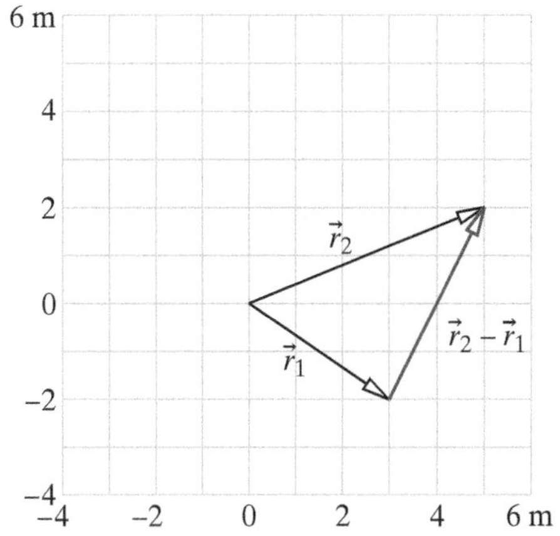

Figure 1.25 Relative position vector.

Since subtraction is not commutative, the order of the quantities matters: the symbol Δ (delta) always means "final minus initial," not "initial minus final." For example, when a child's height changes from 1.1 m to 1.2 m, the change is $\Delta y=+0.1$ m, a positive number. If your bank account dropped from \$150 to \$130, what was the change in your balance? Δ (bank account) $=-20$ dollars.

Another important application of vector subtraction is the calculation of relative position vectors, vectors that represent the position of one object relative to another object.

RELATIVE POSITION VECTOR

If object 1 is at location $\vec{r}_1$ and object 2 is at location $\vec{r}_2$ (Figure 1.25), the position of 2 relative to 1 is:

$$\vec{r}_{2\text{ relative to }1}=\vec{r}_2-\vec{r}_1$$

Checkpoint 8 At 10:00 AM you are at location $\langle -3,2,5\rangle$ m. By 10:02 AM you have walked to location $\langle 6,4,25\rangle$ m. **(a)** What is $\Delta\vec{r}$, the change in your position? **(b)** What is Δt, the time interval during which your position changed?

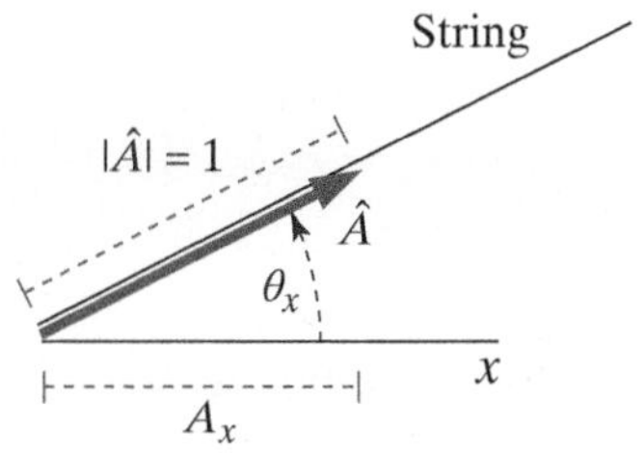

Figure 1.26 A unit vector whose direction is at a known angle from the $+x$ axis.

Unit Vectors and Angles

Suppose that a taut string is at an angle θ_x to the $+x$ axis, and we need a unit vector in the direction of the string. Figure 1.26 shows a unit vector $\hat{A}$ pointing along the string. What is the x component of this unit vector? Consider the triangle whose base is A_x and whose hypotenuse is $|\hat{A}| = 1$. From the definition of the cosine of an angle we have this:

$$\cos\theta = \frac{\text{adjacent}}{\text{hypotenuse}} = \frac{A_x}{1}, \text{ so } A_x = \cos\theta_x$$

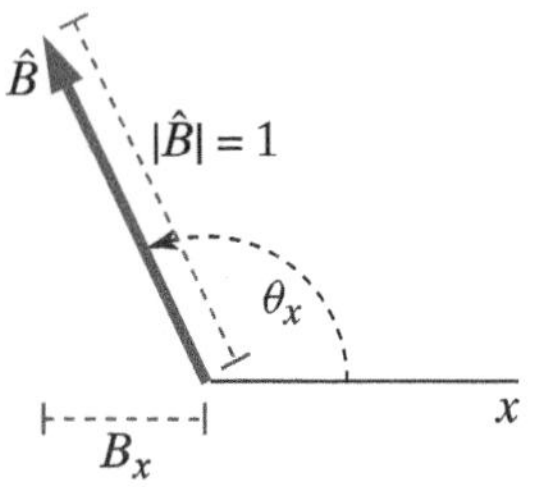

Figure 1.27 A unit vector in the second quadrant from the $+x$ axis.

In Figure 1.26 the angle θ_x is shown in the first quadrant (θ_x less than 90°), but this works for larger angles as well. For example, in Figure 1.27 the angle from the $+x$ axis to a unit vector $\hat{B}$ is in the second quadrant (θ_x greater than 90°) and $\cos\theta_x$ is negative, which corresponds to a negative value of B_x.

What is true for x is also true for y and z. Figure 1.28 shows a 3D unit vector $\hat{r}$ and indicates the angles between the unit vector and the x, y, and z axes. Evidently we can write

$$\hat{r} = \langle\cos\theta_x, \cos\theta_y, \cos\theta_z\rangle$$

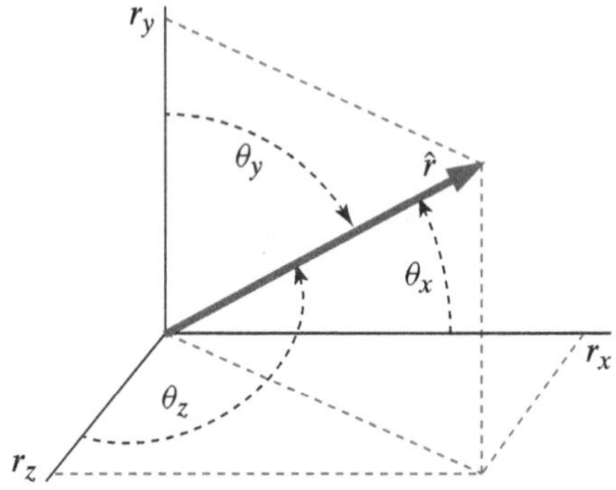

Figure 1.28 A 3D unit vector and its angles to the x, y, and z axes.

These three cosines of the angles between a vector (or unit vector) and the coordinate axes are called the "direction cosines" of the vector. The cosine function is never greater than 1, just as no component of a unit vector can be greater than 1.

A common special case is that of a unit vector lying in the xy plane, with zero z component (Figure 1.29). In this case $\theta_x + \theta_y = 90°$, so that $\cos\theta_y = \cos(90° - \theta_x) = \sin\theta_x$, therefore you can express the cosine of θ_y as the sine of θ_x, which is often convenient. However, in the general 3D case shown in Figure 1.28 there is no such simple relationship among the direction angles or among their cosines.

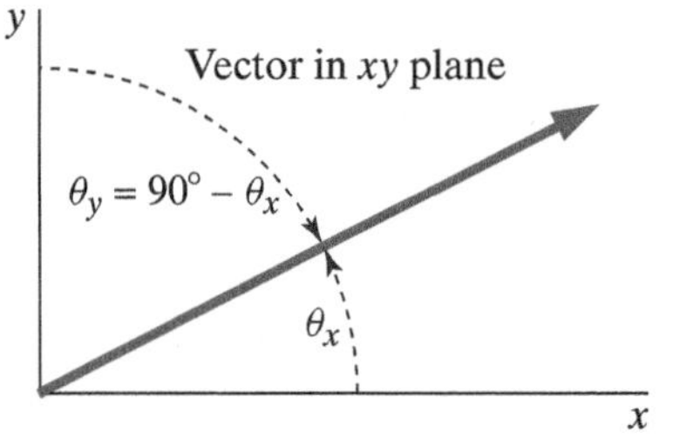

Figure 1.29 If a vector lies in the xy plane, $\cos\theta_y = \sin\theta_x$.

FINDING A UNIT VECTOR FROM ANGLES

To find a unit vector if angles are given:

- Redraw the vector of interest with its tail at the origin, and determine the angles between this vector and the axes.
- Imagine the vector $\langle 1,0,0\rangle$, which lies on the $+x$ axis. θ_x is the angle through which you would rotate the vector $\langle 1,0,0\rangle$ until its direction matched that of your vector. θ_x is positive, and $\theta_x \leq 180°$.
- θ_y is the angle through which you would rotate the vector $\langle 0,1,0\rangle$ until its direction matched that of your vector. θ_y is positive, and $\theta_y \leq 180°$.
- θ_z is the angle through which you would rotate the vector $\langle 0,0,1\rangle$ until its direction matched that of your vector. θ_z is positive, and $\theta_z \leq 180°$.

EXAMPLE **From Unit Vector to Angles**

A vector $\vec{r}$ points from the origin to the location $\langle -600, 0, 300\rangle$ m. What is the angle that this vector makes to the x axis? To the y axis? To the z axis?

Solution

$$\hat{r} = \frac{\langle -600, 0, 300\rangle \text{ m}}{\sqrt{(-600)^2 + (0)^2 + (300)^2} \text{ m}} = \langle -0.894, 0, 0.447\rangle$$

But we also know that $\hat{r} = \langle\cos\theta_x, \cos\theta_y, \cos\theta_z\rangle$, so $\cos\theta_x = -0.894$, and the arccosine gives $\theta_x = 153.4°$. Similarly,

$$\cos\theta_y = 0, \text{ so } \theta_y = 90° \quad \text{(which checks; no } y \text{ component)}$$

$$\cos\theta_z = 0.447, \text{ so } \theta_z = 63.4°$$

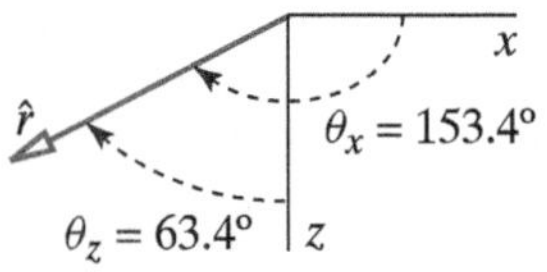

Figure 1.30 Look down on the xz plane. The difference in the two angles is 90°, as it should be.

Looking down on the xz plane in Figure 1.30, you can see that the difference between $\theta_x = 153.4°$ and $\theta_z = 63.4°$ is 90°, as it should be.

EXAMPLE **From Angle to Unit Vector**

A rope lying in the xy plane, pointing up and to the right, supports a climber at an angle of 20° to the vertical (Figure 1.31). What is the unit vector pointing up along the rope?

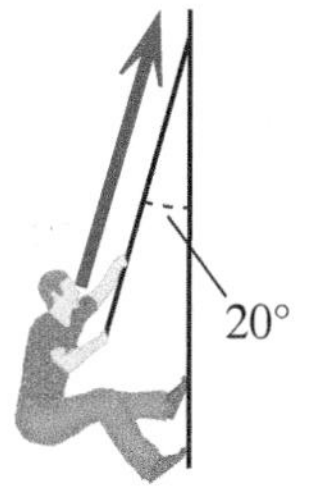

Figure 1.31 A climber supported by a rope.

Solution Follow the procedure given above for finding a unit vector from angles. In Figure 1.32 we redraw the vector with its tail at the origin, and we determine the angles between the vector and the axes.

If we rotate the unit vector $\langle 1,0,0\rangle$ from along the $+x$ axis to the vector of interest, we see that we have to rotate through an angle $\theta_x = 70°$. To rotate the unit vector $\langle 0,1,0\rangle$ from along the $+y$ axis to the vector of interest, we have to rotate through an angle of $\theta_y = 20°$. The angle from the $+z$ axis to our vector is $\theta_z = 90°$. Therefore the unit vector that points along the rope is:

$$\langle \cos 70°, \cos 20°, \cos 90°\rangle = \langle 0.342, 0.940, 0\rangle$$

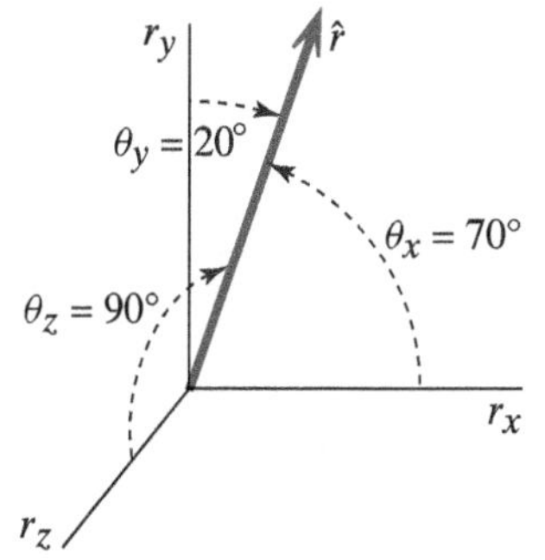

Figure 1.32 Redraw the vector with its tail at the origin. Identify the angles between the positive axes and the vector. In this example the vector lies in the xy plane.

You may have noticed that the y component of the unit vector can also be calculated as $\sin 70° = 0.940$, and it can be useful to recognize that a vector component can be obtained using sine instead of cosine. There is, however, an advantage in consistently calculating in terms of direction cosines. This is a method that always works, especially in 3D, and that helps avoid errors due to choosing the wrong trig function.

Checkpoint 9 **(a)** A unit vector lies in the xy plane, at an angle of 160° from the $+x$ axis, with a positive y component. What is the unit vector? (It helps to draw a diagram.) **(b)** A string runs up and to the left in the xy plane, making an angle of 40° to the vertical. Determine the unit vector that points along the string.

Reorienting the Coordinate Axes

In order to describe position and displacement we had to choose an origin and a set of axes. What if we had made different choices? Certain quantities related to vectors change when a different orientation is chosen for the coordinate axes, but others remain the same. Scalar quantities such as mass and temperature do not change. The magnitude of a vector remains the same when axes are oriented differently, even though the components of the vector do change—the x component of velocity will have a different value if the x axis is chosen to have a different orientation. (Because of this, a vector component is not considered mathematically to be a true scalar, even though it is a single number. However, this distinction is not going to be important in the context of the physics we will study in this course.)

Quantity	Unit	Symbol
mass	kilogram	kg
distance	meter	m
time	second	s
charge	coulomb	C

Figure 1.33 Basic SI units.

1.5 SI UNITS

In this book we use the SI (Système Internationale) unit system, as is customary in technical work. The SI unit of mass is the kilogram (kg), the unit of distance is the meter (m), and the unit of time is the second (s) (Figure 1.33). In later chapters we will encounter other SI units, such as the newton (N), which is a unit of force. Many quantities combine SI units (for example, velocity, which has units of m/s).

It is essential to use SI units in physics equations; this may require that you convert from some other unit system to SI units. Common metric prefixes are shown in Figure 1.34. If mass is known in grams, you need to divide by 1000 and use the mass in kilograms. If a distance is given in centimeters, you need

1×10^{3}	kilo	1×10^{-3}	milli
1×10^{6}	mega	1×10^{-6}	micro
1×10^{9}	giga	1×10^{-9}	nano
1×10^{12}	tera	1×10^{-12}	pico

Figure 1.34 Common metric prefixes.

to divide by 100 to convert the distance to meters. If the time is measured in minutes, you need to multiply by 60 to use a time in seconds. A convenient way to do such conversions is to multiply by factors that are equal to 1, such as (1 min)/(60 s) or (100 cm)/(1 m). As an example, consider converting 60 miles per hour to SI units, meters per second. Start with the 60 mi/h and multiply by factors of 1:

$$\left(60\frac{\cancel{\text{mi}}}{\cancel{\text{h}}}\right)\left(\frac{1\,\cancel{\text{h}}}{60\,\cancel{\text{min}}}\right)\left(\frac{1\,\cancel{\text{min}}}{60\,\text{s}}\right)\left(\frac{5280\,\cancel{\text{ft}}}{1\,\cancel{\text{mi}}}\right)\left(\frac{12\,\cancel{\text{in}}}{1\,\cancel{\text{ft}}}\right)\left(\frac{2.54\,\cancel{\text{cm}}}{1\,\cancel{\text{in}}}\right)\left(\frac{1\,\text{m}}{100\,\cancel{\text{cm}}}\right)=26.8\frac{\text{m}}{\text{s}}$$

Observe how most of the units cancel, leaving final units of m/s.

Checkpoint 10 A snail moved 80 cm (80 centimeters) in 5 min. What was its average speed in SI units? Write out the factors as was done above.

1.6 SPEED AND VELOCITY

Because it has a magnitude and a direction, velocity is a vector quantity. Velocity is unusual among vector quantities because its magnitude has a special name: "speed." In everyday conversation the words "velocity" and "speed" are often used interchangeably, but because the vector quantity velocity (represented by $\vec{v}$) and the scalar quantity speed ($|\vec{v}|$) have quite different meanings in physics, we will need to use these terms precisely.

Motion is important in everyday life, and it may be somewhat surprising to reflect on the fact that neither velocity nor speed can be measured directly in a single measurement. To determine the speed of a moving object it is necessary to measure two times and two positions, and to calculate the speed by dividing the distance traveled by the time elapsed, as reflected in the familiar expression $v = d/t$. Even the radar guns commonly used to measure the speed of cars or baseballs do not make instantaneous measurements. The wavelength of radar waves reflected from a moving object is different from that of the original waves; the original and reflected waves must be compared over a period of time to determine the speed of the moving object.

Since the measurement interval is finite, it is of course possible that the moving object is speeding up or slowing down during that interval. So what speed do we actually measure, if we are not measuring the exact speed at one instant? We refer to the results of our measurements as "average speed" or, if we include direction, "average velocity."

EXAMPLE **A Sprinter's Average Speed**

At the 2008 summer Olympics in Beijing, the Jamaican sprinter Usain Bolt won the gold medal in the 100 m race, finishing in a time of 9.69 s, and setting a new world record for the event. What was his average speed? From videos of the event one can determine that he reached the 60 meter mark at a time 5.73 s after the start of the race. Was his speed constant?

Solution Bolt's average speed over the entire 100 m was

$$|\vec{v}_{\text{avg}}| = \frac{100\,\text{m}}{9.69\,\text{s}} = 10.32\ \text{m/s}$$

However, his average speed over the first 60 m was

$$|\vec{v}_{\text{avg}}| = \frac{60\,\text{m}}{5.73\,\text{s}} = 10.47\ \text{m/s}$$

Bolt clearly slowed down at the end of the race, beginning to celebrate his win even before reaching the finish line.

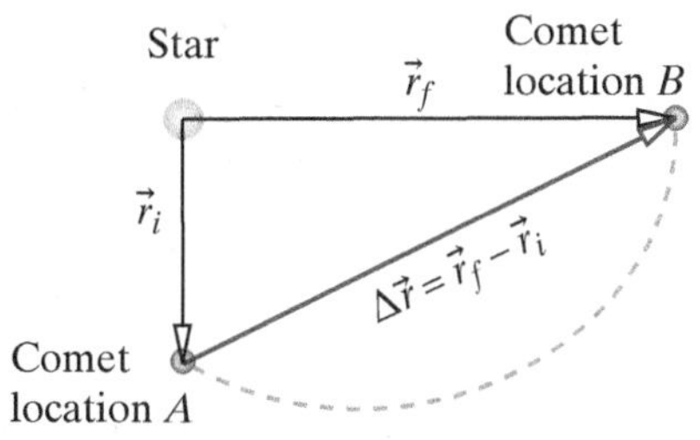

Figure 1.35 A comet orbiting a star moves from location A to location B in 5 years. The vector $\Delta\vec{r}$ is the displacement of the comet during this time period.

If we wanted to know Bolt's speed as accurately as possible at the instant he reached the 60 m mark, we would need to know his position at two times: just before he reached the mark, and just afterward. The smaller the time interval, the closer the average speed we calculate is to the actual instantaneous speed.

Vector Velocity

As with speed, to determine velocity we need two measurements separated in time. With this data we can find the "displacement," or change of position $\Delta\vec{r} = \vec{r}_f - \vec{r}_i$ of an object during a time interval, where $\vec{r}_i$ is the initial 3D position and $\vec{r}_f$ is the final 3D position (note that as with relative position vectors, we always calculate "final minus initial"). For example, Figure 1.35 shows the displacement of a comet over a 5 month period. Dividing the (vector) displacement by the (scalar) time interval $t_f - t_i$ (final time minus initial time) gives the average (vector) velocity of the object:

DEFINITION: AVERAGE VELOCITY

$$\vec{v}_{\text{avg}} = \frac{\vec{r}_f - \vec{r}_i}{t_f - t_i}$$

A more compact way of writing this expression, using the "Δ" symbol (capital Greek delta, defined in Section 1.4), is

$$\vec{v}_{\text{avg}} = \frac{\Delta\vec{r}}{\Delta t}$$

Note that we are using vector subtraction here to find the displacement of a single object from one time to another, while earlier we used vector subtraction to find the relative position of one object with respect to a second object at a single time. The mathematical operation is the same, but the physical meaning is different.

EXAMPLE

Average Velocity of a Bee

Consider a bee in flight. At time $t_i = 15.0$ s after 9:00 AM, the bee's position vector was $\vec{r}_i = \langle 2,4,0\rangle$ m. At time $t_f = 15.1$ s after 9:00 AM, the bee's position vector was $\vec{r}_f = \langle 3,3.5,0\rangle$ m. What was the average velocity of the bee during this interval? Express this vector as the product of the magnitude of the velocity (speed) and a unit vector in the direction of the velocity.

Solution

On the diagram shown in Figure 1.36, we draw and label three arrows representing the vectors $\vec{r}_i$, $\vec{r}_f$, and $\vec{r}_f - \vec{r}_i$. The tail of the latter arrow is placed at the bee's initial position. The vector $\vec{r}_f - \vec{r}_i$, which points in the direction of the bee's motion, is the displacement of the bee during this time interval.

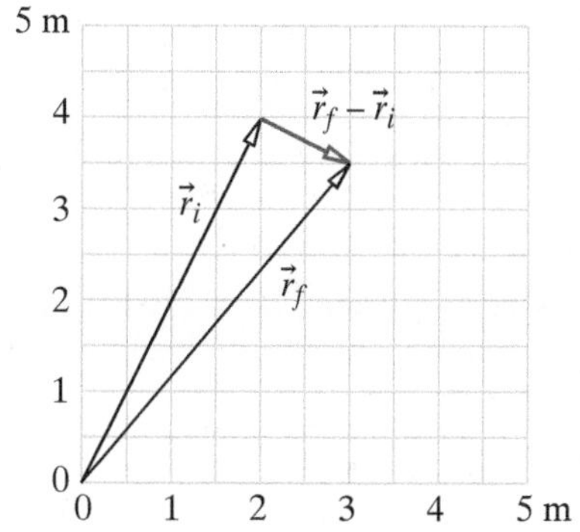

Figure 1.36 The displacement vector points from initial position to final position.

We calculate the bee's displacement vector numerically by taking the difference of the two vectors, final minus initial:

$$\vec{r}_f - \vec{r}_i = \langle 3,3.5,0\rangle\,\text{m} - \langle 2,4,0\rangle\,\text{m} = \langle 1,-0.5,0\rangle\,\text{m}$$

This numerical result should be consistent with our graphical construction. Look at the components of $\vec{r}_f - \vec{r}_i$ in Figure 1.36. Do you see that this vector has an x component of $+1$ and a y component of -0.5 m?

The average velocity of the bee, a vector quantity, is the (vector) displacement $\vec{r}_f - \vec{r}_i$ divided by the (scalar) time interval $t_f - t_i$.

$$\vec{v}_{\text{avg}} = \frac{\vec{r}_f - \vec{r}_i}{t_f - t_i} = \frac{\langle 1,-0.5,0\rangle\,\text{m}}{(15.1 - 15.0)\,\text{s}} = \frac{\langle 1,-0.5,0\rangle\,\text{m}}{0.1\,\text{s}} = \langle 10,-5,0\rangle\,\text{m/s}$$

Since we divided $\vec{r}_f - \vec{r}_i$ by a scalar $(t_f - t_i)$, the average velocity $\vec{v}_{avg}$ points in the direction of the bee's motion, assuming that the bee flew in a straight line.

The average speed of the bee is the magnitude of its velocity:

$$|\vec{v}_{avg}| = \sqrt{10^2 + (-5)^2 + 0^2}\ \text{m/s} = 11.18\ \text{m/s}$$

The direction of the bee's motion, expressed as a unit vector, is:

$$\hat{v}_{avg} = \frac{\vec{v}_{avg}}{|\vec{v}_{avg}|} = \frac{\langle 10, -5, 0\rangle\ \text{m/s}}{11.18\ \text{m/s}} = \langle 0.894, -0.447, 0\rangle$$

Note that the "m/s" units cancel; the result is dimensionless. We can check that this really is a unit vector:

$$\sqrt{0.894^2 + (-0.447)^2 + 0^2} = 0.9995$$

This is not quite 1.0 due to rounding the velocity coordinates and speed to three significant figures. To check, we can put the pieces back together and see what we get. The original vector factors into the product of the magnitude times the unit vector:

$$|\vec{v}|\hat{v} = (11.18\ \text{m/s})\langle 0.894, -0.447, 0\rangle = \langle 10, -5, 0\rangle\ \text{m/s}$$

This is the same as the original vector $\vec{v}$.

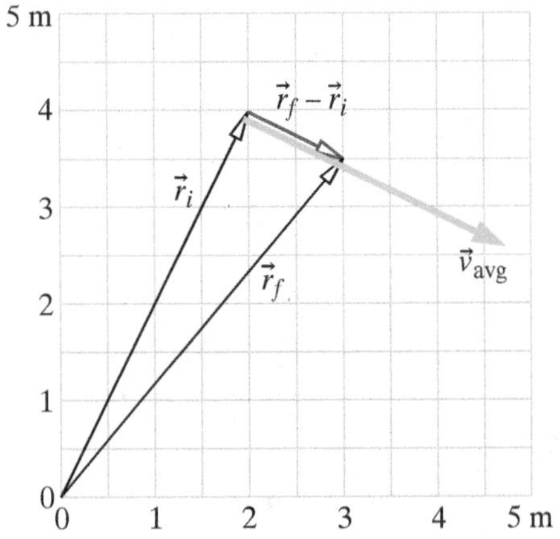

Figure 1.37 Average velocity vector: displacement divided by time interval.

Scaling a Vector to Fit on a Graph

We can plot the average velocity vector on the same graph that we use for showing the vector positions of the bee (Figure 1.37). However, note that velocity has units of meters per second whereas positions have units of meters, so we are mixing quantities on this diagram.

Moreover, the magnitude of the vector, 11.18 m/s, doesn't fit on a graph that is only 5 units wide (in meters). It is standard practice in such situations to scale down the arrow representing the vector to fit on the graph, preserving the correct direction. In Figure 1.37 we've scaled down the arrow representing the velocity vector by about a factor of 3 to make the arrow fit on the graph. Of course if there is more than one velocity vector we use the same scale factor for all the velocity vectors. The same kind of scaling is used with other physical quantities that are vectors, such as force and momentum, which we will encounter later.

> **Checkpoint 11** At a time 0.2 s after it has been hit by a tennis racket, a tennis ball is located at $\langle 5,7,2\rangle$ m, relative to an origin in one corner of a tennis court. At a time 0.7 s after being hit, the ball is located at $\langle 9,2,8\rangle$ m. **(a)** What is the average velocity of the tennis ball? **(b)** What is the average speed of the tennis ball? **(c)** What is the unit vector in the direction of the ball's average velocity?

1.7 PREDICTING A NEW POSITION

We can rewrite the velocity relationship in the form

$$(\vec{r}_f - \vec{r}_i) = \vec{v}_{avg}(t_f - t_i)$$

That is, the (vector) displacement of an object is its average (vector) velocity times the time interval. This is just the vector version of the simple notion that if you run at a speed of 7 m/s for 5 s you move a distance of (7 m/s)(5 s) = 35 m, or that a car going 50 mi/h for 2 h goes (50 mi/h)(2 h) = 100 mi.

Is $(\vec{r}_f - \vec{r}_i) = \vec{v}_{avg}(t_f - t_i)$ a valid vector relation? Yes, multiplying a vector $\vec{v}_{avg}$ times a scalar $(t_f - t_i)$ yields a vector. We make a further rearrangement to obtain a relation for updating the position when we know the velocity:

THE POSITION UPDATE EQUATION

$$\vec{r}_f = \vec{r}_i + \vec{v}_{avg}(t_f - t_i)$$

or

$$\vec{r}_f = \vec{r}_i + \vec{v}_{avg}\Delta t$$

This equation says that if we know the starting position, the average velocity, and the time interval, we can predict the final position. This equation will be important throughout our work.

The position update equation $\vec{r}_f = \vec{r}_i + \vec{v}_{avg}\,\Delta t$ is a vector equation, so we can write out its full component form:

$$\langle x_f, y_f, z_f\rangle = \langle x_i, y_i, z_i\rangle + \langle v_{avg,x}, v_{avg,y}, v_{avg,z}\rangle\,\Delta t$$

Because the x component on the left of the equation must equal the x component on the right (and similarly for the y and z components), this compact vector equation represents three separate component equations:

$$x_f = x_i + v_{avg,x}\,\Delta t$$
$$y_f = y_i + v_{avg,y}\,\Delta t$$
$$z_f = z_i + v_{avg,z}\,\Delta t$$

EXAMPLE **Predicting the Position of a Ball**

At time $t_i = 12.18$ s after 1:30 PM a ball's position vector is $\vec{r}_i = \langle 20, 8, -12\rangle$ m. The ball's velocity at that moment is $\vec{v} = \langle 9, -4, 6\rangle$ m/s. At time $t_f = 12.21$ s after 1:30 PM, where will the ball be, assuming that its velocity hardly changes during this short time interval?

Solution

$$\vec{r}_f = \langle 20, 8, -12\rangle \text{ m} + (\langle 9, -4, 6\rangle \text{ m/s})(12.21 - 12.18)\text{ s}$$
$$\vec{r}_f = \langle 20, 8, -12\rangle \text{ m} + \langle 0.27, -0.12, 0.18\rangle \text{ m}$$
$$\vec{r}_f = \langle 20.27, 7.88, -11.82\rangle \text{ m}$$

Checkpoint 12 A proton traveling with a velocity of $\langle 3 \times 10^5, 2 \times 10^5, -4 \times 10^5\rangle$ m/s passes the origin at a time 9.0 s after a proton detector is turned on. Assuming that the velocity of the proton does not change, what will be its position at time 9.7 s?

Instantaneous Velocity

Figure 1.39 shows the path of a ball, with positions marked at 1 s intervals, and the table in Figure 1.38 lists the position information. While the ball is in the air, its velocity is constantly changing, due to interactions with the Earth (gravity) and with the air (air resistance).

Loc.	t (s)	Position (m)
A	0.0	$\langle 0,0,0\rangle$
B	1.0	$\langle 22.3,26.1,0\rangle$
C	2.0	$\langle 40.1,38.1,0\rangle$
D	3.0	$\langle 55.5,39.2,0\rangle$
E	4.0	$\langle 69.1,31.0,0\rangle$
F	5.0	$\langle 80.8,14.8,0\rangle$

Figure 1.38 Table showing elapsed time and position of the ball at each location marked by a dot in Figure 1.39.

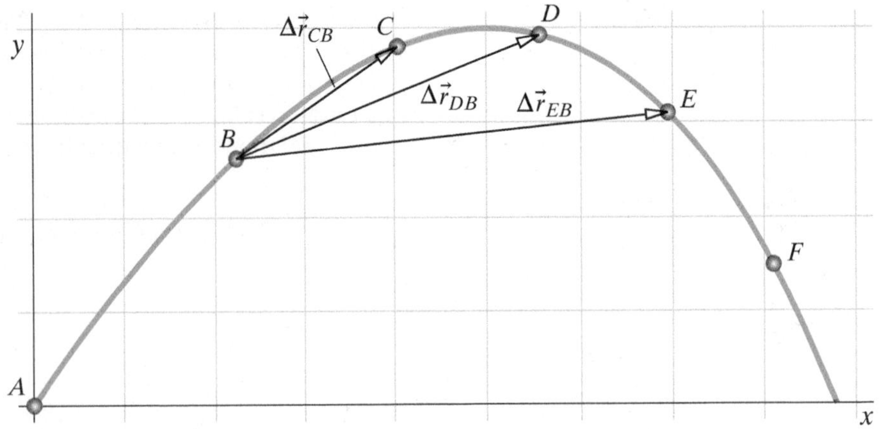

Figure 1.39 The trajectory of a ball through air. The axes represent the x and y distance from the ball's initial location; each square on the grid corresponds to 10 meters. The position of the ball at intervals of 1 s is represented by the colored dots. Three different displacements, corresponding to three different time intervals, are indicated by arrows on the diagram.

Suppose we ask: What is the velocity of the ball at the precise instant that it reaches location B? This quantity would be called the "instantaneous velocity" of the ball. We can start by approximating the instantaneous velocity of the ball by finding its average velocity over some larger time interval.

We can use the position and time data in Figure 1.38 to calculate the average velocity of the ball over three different intervals, by finding the ball's displacement during each interval, and dividing by the appropriate Δt for that interval:

$$\vec{v}_{EB} = \frac{\Delta\vec{r}_{EB}}{\Delta t} = \frac{\vec{r}_E - \vec{r}_B}{t_E - t_B} = \frac{(\langle 69.1,31.0,0\rangle - \langle 22.3,26.1,0\rangle)\,\text{m}}{(4.0-1.0)\,\text{s}}$$
$$= \langle 15.6,1.6,0\rangle \frac{\text{m}}{\text{s}}$$

$$\vec{v}_{DB} = \frac{\Delta\vec{r}_{DB}}{\Delta t} = \frac{\vec{r}_D - \vec{r}_B}{t_D - t_B} = \frac{(\langle 55.5,39.2,0\rangle - \langle 22.3,26.1,0\rangle)\,\text{m}}{(3.0-1.0)\,\text{s}}$$
$$= \langle 16.6,6.55,0\rangle \frac{\text{m}}{\text{s}}$$

$$\vec{v}_{CB} = \frac{\Delta\vec{r}_{CB}}{\Delta t} = \frac{\vec{r}_C - \vec{r}_B}{t_C - t_B} = \frac{(\langle 40.1,38.1,0\rangle - \langle 22.3,26.1,0\rangle)\,\text{m}}{(2.0-1.0)\,\text{s}}$$
$$= \langle 17.8,12.0,0\rangle \frac{\text{m}}{\text{s}}$$

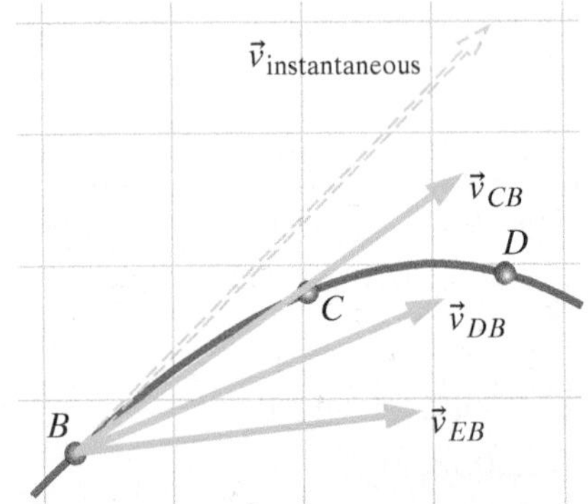

Figure 1.40 A segment of the trajectory shown in Figure 1.39. The three different average velocity vectors calculated above are shown by three arrows, each with its tail at location B. The three arrows representing average velocities are drawn with their tails at the location of interest. The dashed arrow represents the actual instantaneous velocity of the ball at location B. Note that since the units of velocity are m/s, these arrows use a different scale from the distance scale used for the path of the ball.

Not surprisingly, the average velocities over these different time intervals are not the same, because both the direction of the ball's motion and the speed of the ball were changing continuously during its flight. The three average velocity vectors that we calculated are shown in Figure 1.40.

QUESTION Which of the three average velocity vectors depicted in Figure 1.40 best approximates the instantaneous velocity of the ball at location B?

Simply by looking at the diagram, we can tell that $\vec{v}_{CB}$ is closest to the actual instantaneous velocity of the ball at location B, because its direction is closest to the direction in which the ball is actually traveling. Because the direction of the instantaneous velocity is the direction in which the ball is moving at a particular instant, the instantaneous velocity is tangent to the ball's path. Of the three average velocity vectors we calculated, $\vec{v}_{CB}$ best approximates a tangent to the path of the ball. Evidently $\vec{v}_{CB}$, the velocity calculated with the shortest time interval, $t_C - t_B$, is the best approximation to the instantaneous velocity at location B. If we used even smaller values of Δt in our calculation of average velocity, such as 0.1 s, or 0.01 s, or 0.001 s, we would presumably have better and better estimates of the actual instantaneous velocity of the object at the instant when it passes location B.

We can draw two important conclusions about instantaneous velocity:

- The direction of the instantaneous velocity of an object is tangent to the path of the object's motion.
- Smaller time intervals yield more accurate estimates of instantaneous velocity.

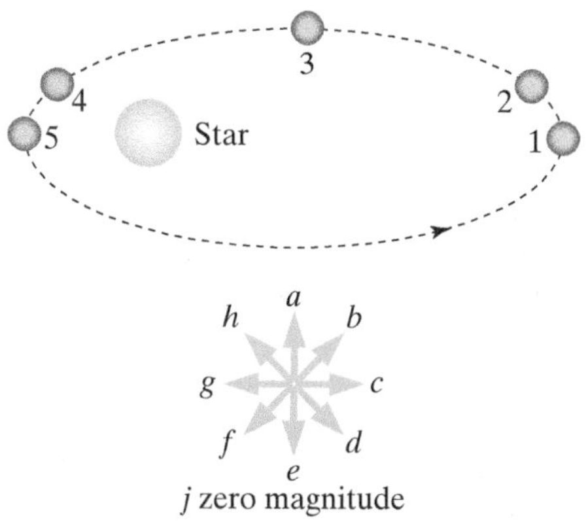

Figure 1.41 A comet goes around a star.

Checkpoint 13 A comet travels in an elliptical path around a star, in the direction shown in Figure 1.41. Which arrow best indicates the direction of the comet's instantaneous velocity vector at each of the numbered locations in the orbit?

Velocity Is a Rate of Change

You may already have learned about derivatives in calculus. Because instantaneous velocity is the rate of change of position, it is a derivative, the limit of $\Delta\vec{r}/\Delta t$ as the time interval Δt used in the calculation gets closer and closer to zero:

$$\vec{v} = \lim_{\Delta t \to 0} \frac{\Delta\vec{r}}{\Delta t}, \text{ which is written as } \vec{v} = \frac{d\vec{r}}{dt}$$

In Figure 1.40, the process of taking the limit is illustrated graphically. As smaller values of Δt are used in the calculation, the average velocity vectors approach the limiting value: the actual instantaneous velocity.

The rate of change (derivative) of a vector is also a vector. Differentiating a vector simply requires differentiating each component:

$$\vec{v} = \frac{d\vec{r}}{dt} = \frac{d}{dt}\langle x, y, z\rangle = \left\langle \frac{dx}{dt}, \frac{dy}{dt}, \frac{dz}{dt} \right\rangle = \langle v_x, v_y, v_z\rangle$$

The derivative of the position vector $\vec{r}$ gives components that are the components of the velocity, as we should expect.

Informally, you can think of $d\vec{r}$ as a very small ("infinitesimal") displacement, and dt as a very small ("infinitesimal") time interval. It is as though we had continued the process illustrated in Figure 1.40 to smaller and smaller time intervals, down to an extremely tiny time interval dt with a correspondingly tiny displacement $d\vec{r}$. The ratio of these tiny quantities is the instantaneous velocity.

The ratio of these two tiny quantities need not be small. For example, suppose that an object moves in the x direction a tiny distance of 1×10^{-15} m, the radius of a proton, in a very short time interval of 1×10^{-23} s:

$$\vec{v} = \frac{\langle 1 \times 10^{-15}, 0, 0\rangle \text{ m}}{1 \times 10^{-23} \text{ s}} = \langle 1 \times 10^{8}, 0, 0\rangle \text{ m/s}$$

which is one-third the speed of light (3×10^8 m/s)!

Acceleration

Velocity is the time rate of change of position: $\vec{v} = d\vec{r}/dt$. Similarly, we define "acceleration" as the time rate of change of velocity: $\vec{a} = d\vec{v}/dt$. Acceleration, which is itself a vector quantity, has units of meters per second per second, written as m/s/s or m/s^2.

DEFINITION: ACCELERATION

Instantaneous acceleration is the time rate of change of velocity:

$$\vec{a} = \frac{d\vec{v}}{dt}$$

Average acceleration can be calculated from a change in velocity:

$$\vec{a}_{\text{avg}} = \frac{\Delta \vec{v}}{\Delta t}$$

The units of acceleration are m/s^2.

EXAMPLE **Acceleration of a Car**

A car traveling in the $+x$ direction speeds up from 20 m/s to 26 m/s in 3 s. What is its average acceleration?

Solution

$$\vec{a}_{\text{avg}} = \frac{\Delta \vec{v}}{\Delta t} = \frac{(\langle 26,0,0\rangle - \langle 20,0,0\rangle)\ \text{m/s}}{3\ \text{s}} = \langle 2,0,0\rangle\ \text{m/s/s}$$

For another example, if you drop a rock near the surface of the Earth, its speed increases 9.8 m/s every second, so its acceleration is 9.8 m/s/s, as long as air resistance is negligible.

Because velocity is a vector, there are two parts to its time derivative, acceleration:

$$\vec{v} = |\vec{v}|\hat{v}$$

$$\vec{a} = \frac{d\vec{v}}{dt} = \frac{d|\vec{v}|}{dt}\hat{v} + |\vec{v}|\frac{d\hat{v}}{dt}$$

As we'll see in later chapters, these two parts of the acceleration are associated with pushing or pulling parallel to the motion (changing the speed) or perpendicular to the motion (changing the direction).

Checkpoint 14 **(a)** Powerful sports cars can go from zero to 25 m/s (about 60 mi/h) in 5 s. (1) What is the magnitude of the average acceleration? (2) How does this compare with the acceleration of a rock falling near the Earth's surface? **(b)** Suppose the position of an object at time t is $\langle 3+5t, 4t^2, 2t-6t^3\rangle$. (1) What is the instantaneous velocity at time t? (2) What is the instantaneous acceleration at time t? (3) What is the instantaneous velocity at time $t=0$? (4) What is the instantaneous acceleration at time $t=0$?

1.8 MOMENTUM

In trying to model the real world, physicists look for powerful ideas that are very general—that is, that apply to a very large range of systems and phenomena. Some of the most powerful and general principles involve "hidden" quantities—things we do not perceive directly. Momentum is such a quantity.

We have discussed velocity, a vector quantity that describes motion and can be determined from measurements of position and time; position, time, speed, and now 3D velocity are all familiar quantities. However, velocity is not the whole story. Consider the following thought experiment:

Suppose you gently toss a tennis ball to a friend, in such a way that just before the ball reaches her hands, its velocity is $\langle 0.3, -0.2, 0\rangle$ m/s. When your friend catches the ball, she must interact with the ball to stop its motion, changing its velocity from $\langle 0.3, -0.2, 0\rangle$ m/s to $\vec{0}$ m/s.

Figure 1.42 Catching a bowling ball requires a larger interaction than catching a tennis ball with the same velocity.

Now suppose that you again toss a ball to your friend with the same velocity; this time, however, the ball is a bowling ball! When the ball reaches her hands with velocity $\langle 0.3, -0.2, 0 \rangle$ m/s, your friend must interact much more strongly with the ball to change its velocity to $\vec{0}$ m/s (Figure 1.42). Even though the change in the velocities of the two balls is identical, the amount of interaction needed to cause this change is very different. Evidently the mass of the moving object must explicitly be taken into account.

The larger the mass of the object, the stronger the interaction required to change its motion. Since the same is true for the velocity of the object (your friend would have had to interact more strongly to stop a tennis ball with velocity $\langle 40, 0, 0 \rangle$ m/s), we will surmise that it is the product of mass and velocity that is important. This quantity is called "momentum"; since it is the product of a scalar and a vector, momentum is a vector. For reasons lost in the mists of history, the symbol used to represent momentum is $\vec{p}$. Momentum is of fundamental importance not only in classical (prequantum) mechanics but also in relativity and quantum mechanics. In Chapter 3 we will discuss the fact that momentum is a "conserved" quantity; the total momentum of the universe is constant.

APPROXIMATE DEFINITION OF MOMENTUM

$$\vec{p} \approx m\vec{v}$$

The units of momentum are kg · m/s. We will see in Section 1.10 that this expression is a good approximation for the momentum of objects traveling at speeds less than about one-tenth of the speed of light.

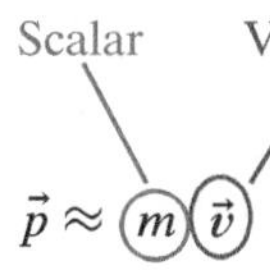

Figure 1.43 The approximate expression for momentum is the product of a scalar times a vector. The scalar factor, mass, must be positive, so the direction of an object's momentum is the same as the direction of its velocity.

Although we can see and compare velocities, momentum is a quantity that we can't see directly. We will encounter other important quantities that aren't directly visible, such as energy, angular momentum, and electric and magnetic fields. Like velocity, momentum is a vector quantity, so it has a magnitude and a direction. Since the mass m must be a positive number, this scalar factor cannot change the direction of the vector (Figure 1.43). Therefore the direction of an object's momentum is the same as the direction of its velocity.

Change of Momentum

Looking back at Newton's first law of motion, we can see that the idea that a body "persists in its state of rest or of moving with constant speed in a constant direction..." can be stated compactly as "the momentum of a body remains constant..." In Chapter 2 we will relate momentum change to interaction mathematically, using the concept of "force" to quantify interactions. This will allow us to predict quantitatively the motion of objects whose momentum is changed by interactions with their surroundings.

Change in momentum, therefore, is an important quantity. We have just noted that it was harder for the person to change the momentum of a bowling ball than to change the momentum of a tennis ball with the same velocity. Calculating a change of momentum requires vector subtraction.

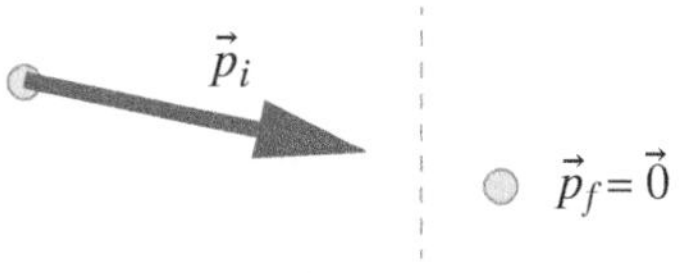

Figure 1.44 The system is the ball. The initial state is just before touching your friend's hands, and the final state is just after the ball has come to a stop in her hands.

EXAMPLE **Magnitude of Momentum Change**

Consider the tennis ball and bowling ball discussed above. What is the magnitude of the change in the momentum of each ball when your friend catches it?

Solution The mass of a regulation tennis ball is about 58 g, or 0.058 kg in S.I. units. The momentum of the tennis ball just before it reaches your friend's hands (Figure 1.44) is:

$$\vec{p}_i = (0.058\,\text{kg})\langle 0.3, -0.2, 0 \rangle\ \text{m/s} = \langle 0.0174, -0.0116, 0 \rangle\ \text{kg} \cdot \text{m/s}$$

The final momentum of the tennis ball is $\vec{0}$ kg·m/s, and the change in the tennis ball's momentum is:

$$\Delta\vec{p} = \vec{p}_f - \vec{p}_i = \langle 0,0,0\rangle \text{ kg}\cdot\text{m/s} - \langle 0.0174,-0.0116,0\rangle \text{ kg}\cdot\text{m/s}$$
$$= \langle -0.0174, 0.0116, 0\rangle \text{ kg}\cdot\text{m/s}$$

The magnitude of the tennis ball's change in momentum is:

$$|\Delta\vec{p}| = \sqrt{-0.0174^2 + 0.0116^2} \text{ kg}\cdot\text{m/s} = 0.0209 \text{ kg}\cdot\text{m/s}$$

For a bowling ball of mass 5.8 kg (about 13 lb):

$$\vec{p}_i = (5.8\text{ kg})\langle 0.3,-0.2,0\rangle \text{ m/s} = \langle 1.74,-1.16,0\rangle \text{ kg}\cdot\text{m/s}$$
$$\Delta\vec{p} = \langle 0,0,0\rangle \text{ kg}\cdot\text{m/s} - \langle 1.74,-1.16,0\rangle \text{ kg}\cdot\text{m/s}$$
$$= \langle -1.74, 1.16, 0\rangle \text{ kg}\cdot\text{m/s}$$
$$|\Delta\vec{p}| = \sqrt{-1.74^2 + 1.16^2} \text{ kg}\cdot\text{m/s} = 2.09 \text{ kg}\cdot\text{m/s}$$

The change in velocity of each ball was the same, but the magnitude of the change in momentum of the bowling ball was 100 times larger than the change of momentum of the tennis ball.

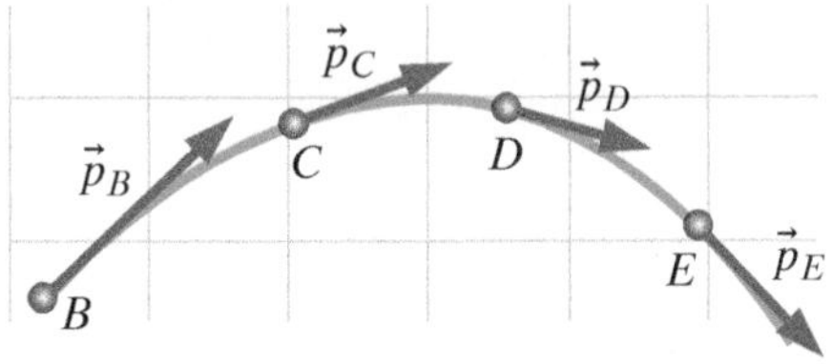

Figure 1.45 A portion of the trajectory of a ball moving through air, subject to gravity and air resistance. The arrows represent the momentum of the ball at the locations indicated by letters.

Momentum is a vector quantity and proportional to the velocity, so just as was the case with velocity, there are two aspects of momentum that can change: magnitude and direction. A mathematical description of change of momentum must include either a change in the magnitude of the momentum, or a change in the direction of the momentum, or both.

EXAMPLE **Change in Magnitude and Direction of Momentum**

Figure 1.45 shows a portion of the trajectory of a ball in air, subject to gravity and air resistance. At location B, the ball's momentum is $\vec{p}_B = \langle 3.03, 2.83, 0\rangle$ kg·m/s. At location C, the ball's momentum is $\vec{p}_C = \langle 2.55, 0.97, 0\rangle$ kg·m/s. Find the change in the ball's momentum between these locations, and show it on the diagram. What changed: the direction of the ball's momentum, the magnitude of the ball's momentum, or both?

Solution

$$\Delta\vec{p} = \vec{p}_C - \vec{p}_B = \langle 2.55, 0.97, 0\rangle \text{ kg}\cdot\text{m/s} - \langle 3.03, 2.83, 0\rangle \text{ kg}\cdot\text{m/s}$$
$$= \langle -0.48, -1.86, 0\rangle \text{ kg}\cdot\text{m/s}$$

Both the x and y components of the ball's momentum decreased, so $\Delta\vec{p}$ has negative x and y components. This is consistent with the graphical subtraction shown in Figure 1.46.

It is clear from the diagram that both the magnitude and direction of the ball's momentum changed. The arrow representing $\vec{p}_B$ is longer than the arrow representing $\vec{p}_C$, and the directions of the arrows are different.

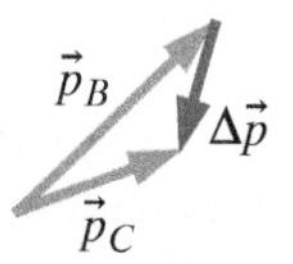

Figure 1.46 Graphical calculation of $\Delta\vec{p}$.

EXAMPLE **A Ball Bounces Off a Wall**

A tennis ball of mass 58 g travels with velocity $\langle 50,0,0\rangle$ m/s toward a wall. After bouncing off the wall, the tennis ball is observed to be moving at nearly the same speed, in the opposite direction. **(a)** Draw a diagram showing the initial and final momentum of the tennis ball. **(b)** What is the change in the momentum of the tennis ball? **(c)** Compare the change in the magnitude of the tennis ball's momentum to the magnitude of the change of the ball's momentum.

Solution **(a)** The initial and final momenta of the ball are shown in Figure 1.47.

(b)

$$\vec{p}_i = (0.058\,\text{kg})\langle 50,0,0\rangle\ \text{m/s} = \langle 2.9,0,0\rangle\ \text{kg}\cdot\text{m/s}$$
$$\vec{p}_f = (0.058\,\text{kg})\langle -50,0,0\rangle\ \text{m/s} = \langle -2.9,0,0\rangle\ \text{kg}\cdot\text{m/s}$$
$$\Delta\vec{p} = \langle -2.9,0,0\rangle\ \text{kg}\cdot\text{m/s} - \langle 2.9,0,0\rangle\ \text{kg}\cdot\text{m/s}$$
$$\Delta\vec{p} = \langle -5.80,0,0\rangle\ \text{kg}\cdot\text{m/s}$$

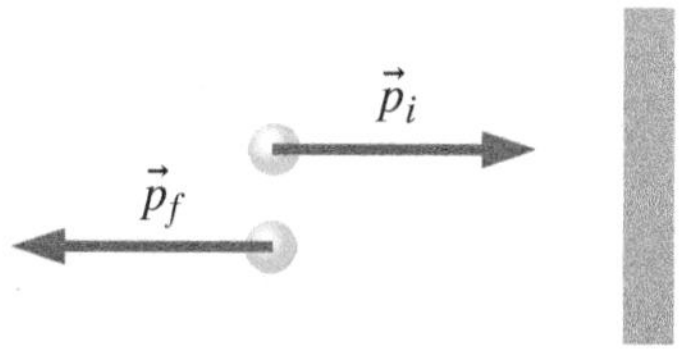

Figure 1.47 The initial and final momentum of the tennis ball.

(c) The change in the magnitude of the ball's momentum was:

$$\Delta|\vec{p}| = |\vec{p}_f| - |\vec{p}_i|$$
$$\Delta|\vec{p}| = \sqrt{-2.9^2+0^2+0^2}\ \text{kg}\cdot\text{m/s} - \sqrt{2.9^2+0^2+0^2}\ \text{kg}\cdot\text{m/s}$$
$$\Delta|\vec{p}| = 0\ \text{kg}\cdot\text{m/s}$$

The magnitude of the change in the ball's momentum was:

$$|\Delta\vec{p}| = \sqrt{-5.8^2+0^2+0^2} = 5.8\ \text{kg}\cdot\text{m/s}$$

How do we make sense of this difference? The interaction with the wall made a large change in the (vector) momentum of the ball; the magnitude of this change is twice as large as the magnitude of the ball's original momentum. However, because the change in the ball's speed was negligible, the change in the magnitude of its momentum was also negligible. We will see in Chapter 2 that this distinction is important, because it is the change in the vector momentum that is proportional to the strength of an interaction with the surroundings. In discussing momentum change we will almost always be interested in $\Delta\vec{p}$ and its magnitude ($|\Delta\vec{p}|$), rather than in the change in the magnitude ($\Delta|\vec{p}|$).

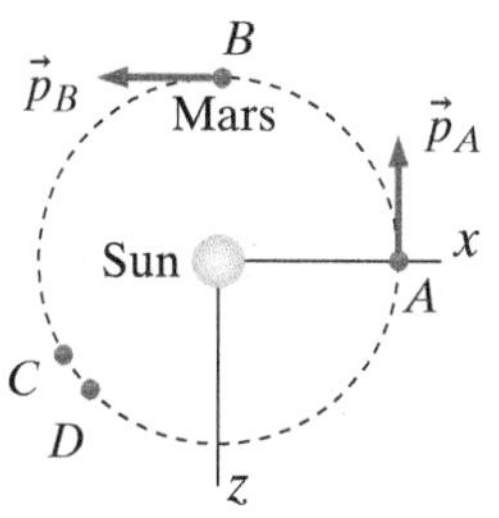

Figure 1.48 The nearly circular orbit of Mars around the Sun, viewed from above the orbital plane ($+x$ to the right, $+z$ down the page). Not to scale: The sizes of the Sun and Mars are exaggerated.

Checkpoint 15 The planet Mars has a mass of 6.4×10^{23} kg, and travels in a nearly circular orbit around the Sun, as shown in Figure 1.48. When it is at location A, the velocity of Mars is $\langle 0,0,-2.5\times10^4\rangle$ m/s. When it reaches location B, the planet's velocity is $\langle -2.5\times10^4,0,0\rangle$ m/s. We're looking down on the orbit from above the north poles of the Sun and Mars, with $+x$ to the right and $+z$ down the page. **(a)** What is $\Delta\vec{p}$, the change in the momentum of Mars between locations A and B? **(b)** On a copy of the diagram in Figure 1.48, draw two arrows representing the momentum of Mars at locations C and D, paying attention to both the length and direction of each arrow. **(c)** What is the direction of the change in the momentum of Mars between locations C and D? Draw the vector $\Delta\vec{p}$ on your diagram.

1.9 USING MOMENTUM TO UPDATE POSITION

If you know the momentum of an object, you can calculate the change in position of the object over a given time interval. This is straightforward if the object is traveling at a speed low enough that the approximate expression for momentum can be used, since in this case $\vec{v} \approx \vec{p}/m$.

EXAMPLE **Displacement of an Ice Skater**

An ice skater whose mass is 50 kg moves with constant momentum $\langle 400,0,300\rangle$ kg·m/s. At a particular instant in her skating program she passes location $\langle 0,0,3\rangle$ m. What was her location at a time 3 s earlier?

Solution

$$\vec{v} \approx \frac{\vec{p}}{m} = \frac{\langle 400,0,300\rangle\,\text{kg}\cdot\text{m/s}}{50\,\text{kg}} = \langle 8,0,6\rangle\,\text{m/s}$$

$$(\vec{r}_f - \vec{r}_i) = \vec{v}\,\Delta t$$

$$\begin{aligned}\vec{r}_i &= \vec{r}_f - \vec{v}\,\Delta t\\ &= \langle 0,0,3\rangle\,\text{m} - (\langle 8,0,6\rangle\,\text{m/s})(3\,\text{s})\\ &= \langle -24,0,-15\rangle\,\text{m}\end{aligned}$$

Checkpoint 16 At time $t_1 = 12$ s, a car with mass 1300 kg is located at $\langle 94,0,30\rangle$ m and has momentum $\langle 4500,0,-3000\rangle$ kg·m/s. The car's momentum is not changing. At time $t_2 = 17$ s, what is the position of the car?

1.10 MOMENTUM AT HIGH SPEEDS

■ The largest particle accelerator currently in operation is the Large Hadron Collider at CERN, in Geneva, Switzerland (http://cern.ch/), where the Higgs boson was recently found. There are many other accelerators in both the United States and other countries.

Although most of the objects we encounter in our daily lives move at speeds that are much less than the speed of light, motion at very high speeds is not unusual. Every day many particles enter the Earth's atmosphere traveling at speeds near the speed of light. Some of these particles are protons, which react with atomic nuclei in the atmosphere to produce showers of high-speed particles that rain down on the Earth. To study high-speed particles in a controlled way, scientists use machines called "particle accelerators."

Experiments on particles moving at very high speeds, close to the speed of light $c = 3 \times 10^8$ m/s, show that changes in $m\vec{v}$ (the approximate momentum) are not really proportional to the strength of the interactions. As we keep applying a force to a particle near the speed of light, the speed of the particle barely increases, and it is not possible to increase a particle's speed beyond the speed of light.

Because of these experiments we can define momentum in a precise (not approximate) way. We observe that changes in the following quantity are truly proportional to the amount of interaction:

DEFINITION OF MOMENTUM

For a particle of mass m, momentum is defined as the product of mass times velocity, multiplied by a proportionality factor gamma:

$$\vec{p} = \gamma m \vec{v}$$

The proportionality factor γ (lowercase Greek gamma) is defined as

$$\gamma = \frac{1}{\sqrt{1 - \left(\frac{|\vec{v}|}{c}\right)^2}}$$

In these equations $\vec{p}$ represents momentum, m is the mass of the object, $\vec{v}$ is the velocity of the object, and c is the speed of light (3×10^8 m/s). Momentum has units of kg·m/s. The factor γ is a positive number that is always greater than or equal to one, and it has no units.

This is the "relativistic" definition of momentum. Albert Einstein in 1905 in his Special Theory of Relativity predicted that this would be the appropriate definition for momentum at high speeds, a prediction that has been abundantly verified in a wide range of experiments. In Chapter 6 we will see that the factor γ is also important in expressions for the energy of objects in motion.

EXAMPLE **Momentum of a Fast-Moving Proton**

Suppose that a proton (mass 1.7×10^{-27} kg) in an accelerator at CERN (the large particle physics laboratory in Geneva, Switzerland) is traveling with a velocity of $\langle 2 \times 10^7, 1 \times 10^7, -3 \times 10^7 \rangle$ m/s. **(a)** What is the momentum of the proton? **(b)** What is the magnitude of the momentum of the proton?

Solution

(a)
$$|\vec{v}| = \sqrt{(2 \times 10^7)^2 + (1 \times 10^7)^2 + (-3 \times 10^7)^2}\ \text{m/s} = 3.7 \times 10^7\ \text{m/s}$$
$$\frac{|\vec{v}|}{c} = \frac{3.7 \times 10^7\ \text{m/s}}{3 \times 10^8\ \text{m/s}} = 0.12$$
$$\gamma = \frac{1}{\sqrt{1-(0.12)^2}} = 1.007$$
$$\vec{p} = (1.007)(1.7 \times 10^{-27}\ \text{kg})\langle 2 \times 10^7, 1 \times 10^7, -3 \times 10^7 \rangle\ \text{m/s}$$
$$\vec{p} = \langle 3.4 \times 10^{-20}, 1.7 \times 10^{-20}, -5.1 \times 10^{-20} \rangle\ \text{kg} \cdot \text{m/s}$$

(b)
$$|\vec{p}| = \sqrt{(3.4 \times 10^{-20})^2 + (1.7 \times 10^{-20})^2 + (-5.1 \times 10^{-20})^2}\ \text{kg} \cdot \text{m/s}$$
$$|\vec{p}| = 6.4 \times 10^{-20}\ \text{kg} \cdot \text{m/s}$$

$\lvert\vec{v}\rvert$ m/s	$\lvert\vec{v}\rvert/c$	γ
0	0	1.0000
3	1×10^{-8}	1.0000
300	1×10^{-6}	1.0000
3×10^6	0.01	1.0001
3×10^7	0.1	1.0050
1.5×10^8	0.5	1.1547
2.997×10^8	0.999	22.3663
2.9997×10^8	0.9999	70.7124
3×10^8	1	Infinite! Impossible!

Figure 1.49 Values of γ calculated for some speeds. γ is shown to four decimal places, which is more accuracy than we will usually need.

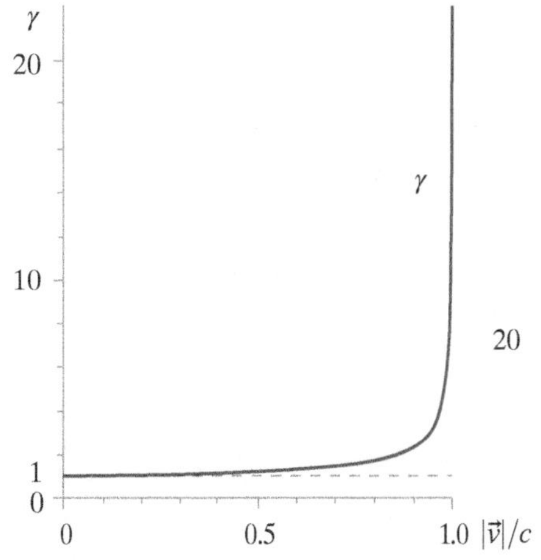

Figure 1.50 γ as a function of $|\vec{v}|/c$. At low speeds, γ is approximately equal to 1. At very high speeds, $|\vec{v}| \approx c$, γ increases rapidly.

Approximate Expression for Momentum

In the example above, we found that $\gamma = 1.007$. Since in that calculation we used only two significant figures, we could have used the approximation that $\gamma \approx 1.0$ without affecting our answer. Let's examine the expression for γ to see whether we can come up with a guideline for when it is reasonable to use the approximate expression

$$\vec{p} \approx 1 \cdot m\vec{v}$$

Looking at the expression for γ,

$$\gamma = \frac{1}{\sqrt{1-\left(\frac{|\vec{v}|}{c}\right)^2}}$$

we see that it depends only on the ratio of the speed of the object to the speed of light (the object's mass doesn't appear in this expression).

If $|\vec{v}|/c$ is a very small number, then $1-(|\vec{v}|/c)^2 \approx 1 - 0 \approx 1$, so $\gamma \approx 1$.

APPROXIMATION: MOMENTUM AT LOW SPEEDS

When $|\vec{v}| \ll c$ then $\gamma \approx 1$, and

$$\vec{p} \approx 1 \cdot m\vec{v} \approx m\vec{v}$$

Some values of $(|\vec{v}|/c)$ and γ are displayed in Figure 1.49. From this table you can see that even at the very high speed where $|\vec{v}|/c=0.1$, which means that $|\vec{v}|=3 \times 10^7$ m/s, the relativistic factor γ is only slightly different from 1.0. For large-scale objects such as a space rocket, whose speed is typically only about 1×10^4 m/s, we can ignore the factor γ, and momentum is to a good approximation $\vec{p} \approx m\vec{v}$. It is only for high-speed cosmic rays or particles produced in high-speed particle accelerators that we need to use the full relativistic definition for momentum, $\vec{p} = \gamma m\vec{v}$.

Figure 1.50 displays graphically the data shown in Figure 1.49. For speeds up to about half the speed of light, γ is very nearly equal to 1, but at very high speeds, approaching the speed of light, γ increases rapidly. From examining Figure 1.50, you can see why it is not possible to exceed the speed of light. As you make a particle go faster and faster, approaching the speed of light, additional increases in the speed become increasingly difficult, because a tiny increase in speed means a huge increase in γ and a corresponding huge increase in momentum, requiring huge amounts of interaction. In fact, for the speed to equal the speed of light, the momentum would have to increase to be infinite! There is a cosmic speed limit in the Universe, 3×10^8 m/s.

The Role of Approximations and Models in Physics

QUESTION Earlier in this chapter we used the approximate expression $\vec{p} \approx m\vec{v}$ for momentum. Is this legitimate? Shouldn't we always use the exact equation?

The table of values of γ in Figure 1.49 shows that for many real-world situations in which objects are traveling at speeds small compared to the speed of light, our calculators will give the same result whether we actually calculate γ or simply use the approximation that $\gamma \approx 1$. Our measurements and calculations simply aren't precise enough to make the distinction useful.

You may have heard physics described as an "exact science." This is not a good description of what physics is actually about. Physics and physicists try to describe and understand what happens in the real world, which is a messy, complicated place. For example, we'll see in the next chapter that it is reasonably easy to predict the motion of a thrown ball in the highly simplified case where there is no air resistance or deflection by wind. Even if we can really get rid of air resistance by throwing the ball in a big vacuum chamber, it makes sense to simplify the analysis by neglecting the tiny but real effects of gravitational attraction by nearby mountains, the Moon, and Mars, and also to ignore the fact that the gravitational attraction changes very slightly as the ball gets closer to or farther from the center of the Earth. Moreover, to predict *exactly* the future motion of the ball, we would need to know *exactly* its initial position and velocity. This is impossible, because meter sticks and clocks aren't exact.

QUESTION If physics isn't exact, what is the point?

We can often learn a great deal by carrying out an approximate analysis of a complex situation. In some cases the differences between an approximate analysis and a hypothetical exact analysis may be negligible. For example, it certainly makes sense to ignore the gravitational attraction of Mars when predicting the motion of a ball thrown on Earth. In other cases, a simplified model may allow us to identify the most important interactions and effects in a complex situation. A comparison of the predictions of such a model with data from real-world observations can suggest refinements to our model that would make its predictions more accurate.

A good way of describing what physics can (and can't) do is that with physics we construct, analyze, and refine simplified, idealized, approximate "models" of real-world phenomena, in the hope that such analyses will give us useful but necessarily approximate understanding of the real world.

Updating Position at High Speeds

For speeds near the speed of light, we cannot use the approximation $\vec{v} \approx \vec{p}/m$. Instead, we need to start with the definition of momentum $\vec{p} = \gamma m\vec{v}$, and solve for $\vec{v}$. The detailed derivation is given at the end of the chapter. The result is:

$$\vec{v} = \frac{\vec{p}/m}{\sqrt{1+\left(\frac{|\vec{p}|}{mc}\right)^2}}$$

EXAMPLE **Displacement of a Fast Proton**

A proton with constant momentum $\langle 0,0,2.72\times10^{-19}\rangle\,\text{kg}\cdot\text{m/s}$ leaves the origin 10.0 s after an accelerator experiment is started. What is the location of the proton 2 ns later? (ns = nanosecond $= 1\times10^{-9}$ s)

Solution

$$\vec{v} = \frac{\vec{p}/m}{\sqrt{1+\left(\dfrac{|\vec{p}|}{mc}\right)^2}} = \frac{\langle 0,0,2.72\times10^{-19}\rangle\,\text{kg}\cdot\text{m/s}}{(1.7\times10^{-27}\,\text{kg})\sqrt{1+\left(\dfrac{2.72\times10^{-19}\,\text{kg}\cdot\text{m/s}}{(1.7\times10^{-27}\,\text{kg})(3\times10^{8}\,\text{m/s})}\right)^2}}$$

$$= \langle 0,0,1.4\times10^{8}\rangle\,\text{m/s}$$

$$\vec{r}_f = \vec{r}_i + \vec{v}_{\text{avg}}\,\Delta t$$

$$= \langle 0,0,0\rangle\,\text{m} + \langle 0,0,1.4\times10^{8}\,\text{m/s}\rangle(2\times10^{-9}\,\text{s})$$

$$= \langle 0,0,0.28\rangle\,\text{m}$$

The proton traveled 28 cm in 2 ns.

Checkpoint 17 What is the momentum of an electron traveling at a velocity of $\langle 0,0,-2\times10^{8}\rangle$ m/s? (Masses of particles are given on the inside back cover of this textbook.) What is the magnitude of the momentum of the electron?

1.11 COMPUTATIONAL MODELING

Computational modeling plays an important role not only in physics theory and experiment, but in nearly all other scientific and engineering fields. Creating simple computational models based on fundamental physics principles can allow us to see more clearly how these principles govern the real-time behavior of all physical systems. Such models can allow us to apply physics principles to complicated systems. Appropriate computational tools can let us visualize the time evolution of the behavior of 3D physical systems, and can also help us visualize vector quantities such as velocity, momentum, and other quantities we will encounter in later chapters. For these reasons, computational modeling is included as an integral part of this textbook.

QUESTION But what if I don't know how to program?

Don't worry! No prior experience is required. You will be able to learn all you need to know to build simple but powerful computational models as we go along in the course.

VPython

We have chosen to use VPython to build our computational models. VPython is an extension of the widely used Python programming language. VPython supports 3D vector algebra, and allows you to create real-time dynamic 3D animations as a side effect of physics calculations. VPython is free and open source, and runs on Windows, MacOS, and Linux.

Here is an example of a small VPython program. The display generated by the program is shown in Figure 1.51.

Figure 1.51 This is the display generated when you run the VPython program shown at the right.

```
from visual import *
ball = sphere(pos = vector(-3,-3,0),
       color = color.orange,
       radius = 0.5)
velocity = vector(3,1.5,4)
arrow(pos = ball.pos,
      axis = velocity,
      color = color.green)
```

QUESTION In the code above, the sphere named "ball" is created at the position $\langle -3,-3,0\rangle$. Considering this, look at the display in Figure 1.51, and decide where you think the origin of the coordinate system is located.

■ To install VPython on your computer, go to http://vpython.org and follow the directions for your operating system.
See homework problems P68–P77 at the end of this chapter to find the instructional videos for VPython.

The origin of the 3D Cartesian coordinate system, $\langle 0,0,0\rangle$, is in the center of the display. The orientation of the axes is the same as the one we have been using, with $+x$ to the right, $+y$ up, and $+z$ out of the plane of the display, toward you. The location of the ball, below and to the left of the origin, is consistent with the position $\langle -3,-3,0\rangle$ assigned to the sphere in the program.

QUESTION What could we change in the code to make the ball larger?

The sphere named "ball" is created with a radius of 0.5. To make the ball larger, we could specify a larger radius.

QUESTION How could the program be extended to animate the motion of the ball?

In the example above a velocity has already been given. To use this velocity to update the position of the ball, we need to instruct the computer to do exactly the same calculation you have been doing by hand. A translation of these calculations into VPython syntax would look like this:

$\Delta t = 0.1$	`delta_t = 0.1`
$\vec{r}_f = \vec{r}_i + \vec{v}\Delta t$	`ball.pos = ball.pos + velocity * delta_t`

A statement such as `x = x + 1` is not wrong in a computer language. Such a statement instructs the computer to pick up the current value of the variable x, add 1 to it, and replace the value of x with this new value. Subscripts identifying "current" and "new" values are not needed, since anything to the right of the equal sign refers to the current value, and anything to the left of the equal sign is the new value. Therefore the statement

```
ball.pos = ball.pos + velocity * delta_t
```

tells the computer to pick up the current position of the ball (`ball.pos`), add the quantity `velocity * delta_t` to it, and update `ball.pos` to this new value. To animate the motion of the ball over a period of 4 seconds, we can instruct the computer to repeat this calculation over and over again, using a "while loop":

```
delta_t = 0.1
t = 0
while t < 4:
    rate(100)
    ball.pos = ball.pos + velocity * delta_t
    t = t + delta_t
```

To get an animation of the object's motion as a side effect of our calculations, we had to add the statement `rate(100)` inside the while loop. This statement tells VPython to allow only 100 repetitions of the while loop to be done each second of real time, and allows it to render a new image of the scene many times per second, producing a movie-like animation as a side effect of our physics calculations.

For clarity, we can refer to the time step Δt as "virtual time" and to the time required to do a calculation as "real time." If you were doing the calculation by hand, it might take you 60 real seconds to calculate the new position of an object at a time 0.1 virtual second in the future. In contrast, it takes a computer a very short time—about 5 microseconds (5×10^{-6} s). In neither case is the time taken to do the calculation equal to the "virtual" elapsed time of 0.1 second.

It is much easier to learn to write simple programs in VPython by watching videos than by reading text. There is a series of YouTube videos that were developed to help physics students learn VPython. Homework problems P68–P77 at the end of this chapter ask you to watch these videos and do the challenge tasks at the end of each video. Problem P78 includes a complete program to study and modify. Your instructor may also assign more extended computational activities both to help you get started and to accompany future topics. In future chapters we will assume that you have watched these videos and done these simple introductory exercises.

1.12 *THE PRINCIPLE OF RELATIVITY

Sections marked with an asterisk ("*") are optional. They provide additional information and context, but later sections of the textbook don't depend critically on them. This optional section deals with some deep issues about the "reference frame" from which you observe motion. Newton's first law of motion only applies in an "inertial reference frame," which we will discuss here in the context of the principle of relativity.

A great variety of experimental observations has led to the establishment of the following principle, first recognized by Galileo:

THE PRINCIPLE OF RELATIVITY

Physical laws work in the same way for observers in uniform motion as for observers at rest.

This principle is called "the principle of relativity." (Einstein's extensions of this principle are known as "special relativity" and "general relativity.") Phenomena observed in a room in uniform motion (for example, on a train moving with constant speed on a smooth straight track) obey the same physical laws in the same way as experiments done in a room that is not moving. According to this principle, Newton's first law of motion should be true both for an observer moving at constant velocity and for an observer at rest.

For example, suppose that you're riding in a car moving with constant velocity, and you're looking at a map lying on the dashboard. As far as you're concerned, the map isn't moving, and no interactions are required to hold it still on the dashboard. Someone standing at the side of the road sees the car go by, sees the map moving at a high speed in a straight line, and can see that no interactions are required to hold the map still on the dashboard. Both you and the bystander agree that Newton's first law of motion is obeyed: the bystander sees the map moving with constant velocity in the absence of interactions, and you see the map not moving at all (a zero constant velocity) in the absence of interactions.

On the other hand, if the car suddenly speeds up, it moves out from under the map, which ends up in your lap. To you it looks like "the map sped up in the backwards direction" without any interactions to cause this to happen, which looks like a violation of Newton's first law of motion. The problem is that you're strapped to the car, which is an accelerated reference frame, and Newton's first law of motion applies only to nonaccelerated reference frames, called "inertial" reference frames. Similarly, if the car suddenly turns to the right, moving out from under the map, the map tends to keep going in its original direction, and to you it looks like "the map moved to the left" without any interactions. So a change of speed or a change of direction of the car (your reference frame) leads you to see the map behave in a strange way.

The bystander, who is in an inertial (nonaccelerating) reference frame, doesn't see any violation of Newton's first law of motion. The bystander's reference frame is an inertial frame, and the map behaves in an understandable way, tending to keep moving with unchanged speed and direction when the car changes speed or direction.

The Cosmic Microwave Background

The principle of relativity, and Newton's first law of motion, apply only to observers who have a constant speed and direction (or zero speed) relative to the "cosmic microwave background," which provides the only backdrop and frame of reference with an absolute, universal character. It used to be that the basic reference frame was loosely called "the fixed stars," but stars and galaxies have their own individual motions within the Universe and do not constitute an adequate reference frame with respect to which to measure motion.

The cosmic microwave background is low-intensity electromagnetic radiation with wavelengths in the microwave region, which pervades the Universe, radiating in all directions. Measurements show that our galaxy is moving through this microwave radiation with a large, essentially constant velocity, toward a cluster of a large number of other galaxies. The way we detect our motion relative to the microwave background is through the "Doppler shift" of the frequencies of the microwave radiation, toward higher frequencies in front of us and lower frequencies behind. This is essentially the same phenomenon as that responsible for a fire engine siren sounding at a higher frequency when it is approaching us and a lower frequency when it is moving away from us.

The discovery of the cosmic microwave background provided major support for the "Big Bang" theory of the formation of the Universe. According to the Big Bang theory, the early Universe must have been an extremely hot mixture of charged particles and high-energy, short-wavelength electromagnetic radiation (visible light, x-rays, gamma rays, etc.). Electromagnetic radiation interacts strongly with charged particles, so light could not travel very far without interacting, making the Universe essentially opaque. Also, the Universe was so hot that electrically neutral atoms could not form without the electrons immediately being stripped away again by collisions with other fast-moving particles.

As the Universe expanded, the temperature dropped. Eventually the temperature was low enough for neutral atoms to form. The interaction of electromagnetic radiation with neutral atoms is much weaker than with individual charged particles, so the radiation was now essentially free, dissociated from the matter, and the Universe became transparent. As the Universe continued to expand (the actual space between clumps of matter got bigger), the wavelengths of the electromagnetic radiation got longer, until today this fossil radiation has wavelengths in the relatively low-energy, long-wavelength microwave portion of the electromagnetic spectrum.

Inertial Frames of Reference

It is an observational fact that in reference frames that are in uniform motion with respect to the cosmic microwave background, far from other objects (so that interactions are negligible), an object maintains uniform motion. Such frames are called "inertial frames" and are reference frames in which

Newton's first law of motion is valid. All of these reference frames are equally valid; the cosmic microwave background simply provides a concrete example of such a reference frame.

QUESTION Is the surface of the Earth an inertial frame?

No! The Earth is rotating on its axis, so the velocity of an object sitting on the surface of the Earth is constantly changing direction, as is a coordinate frame tied to the Earth (Figure 1.52). Moreover, the Earth is orbiting the Sun, and the Solar System itself is orbiting the center of our Milky Way galaxy, and our galaxy is moving toward other galaxies. So the motion of an object sitting on the Earth is actually quite complicated and definitely not uniform with respect to the cosmic microwave background.

However, for many purposes the surface of the Earth can be considered to be (approximately) an inertial frame. For example, it takes 6 hours for the rotation of the Earth on its axis to make a 90° change in the direction of the velocity of a "fixed" point. When a hockey puck slides across an ice rink with negligible friction, during these few seconds the puck moves in nearly a straight line at constant speed, and if the puck is struck its velocity change is much larger than the very small velocity change of the approximate inertial frame of the Earth's surface.

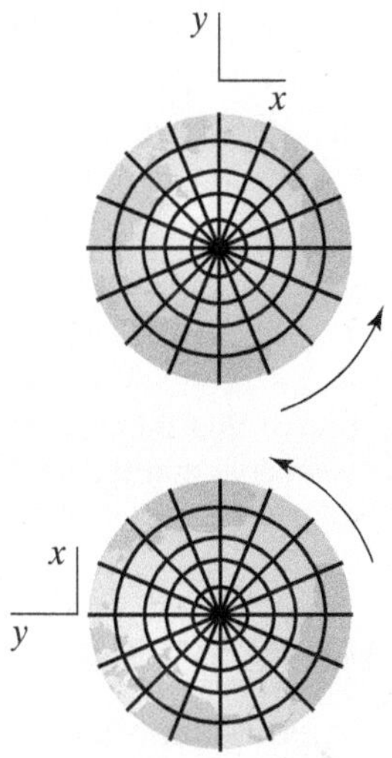

Figure 1.52 Axes tied to the Earth rotate through 90° in a quarter of a day (6 h).

Similarly, although the Earth is in orbit around the Sun, it takes 365 days to go around once, so for a period of a few days or even weeks the Earth's orbital motion is nearly in a straight line at constant speed. For our purposes we will consider the Earth's surface to represent an approximately inertial reference frame.

The Special Theory of Relativity

Einstein's Special Theory of Relativity (published in 1905) built on the basic principle of relativity introduced by Galileo but added the conjecture that the speed of a beam of light must be the same as measured by observers in different frames of reference in uniform motion with respect to each other. In Figure 1.53, observers on each spaceship measure the speed of the light c emitted by the ship at the top to be the same ($c = 3 \times 10^8$) m/s, despite the fact that they are moving at different velocities.

This additional condition seems peculiar and has far-reaching consequences. After all, the map on the dashboard of your car has different speeds relative to different observers, depending on the motion of the observer. Yet a wide range of experiments has confirmed Einstein's conjecture: all observers measure the same speed for the same beam of light, $c = 3 \times 10^8$ m/s. (The color of the light is different for the different observers, but the speed is the same.)

On the other hand, if someone on the ship at the top throws a ball or a proton or some other piece of matter, the speed of the object will be different for observers on the three ships; it is only light whose speed is independent of the observer.

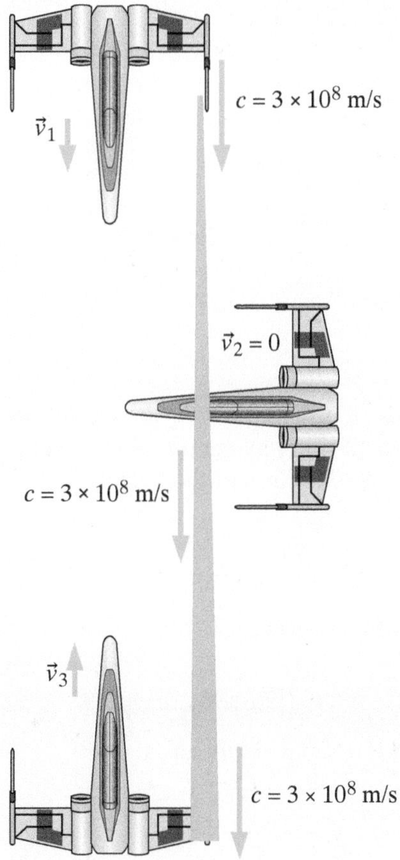

Figure 1.53 Light emitted by the top spaceship is measured to have the same speed by observers in all three ships.

Einstein's theory has interesting consequences. For example, it predicts that time will be measured to run at different rates in different frames of reference. These predictions have been confirmed by many experiments. These unusual effects are large only at very high speeds (a sizable fraction of the speed of light), which is why we don't normally observe these effects in everyday life, and why we can use nonrelativistic calculations for low-speed phenomena.

However, for the Global Positioning System (GPS) to give adequate accuracy it is necessary to take Einstein's Special Theory of Relativity into account. The atomic clocks on the satellites run slower than our clocks, due to the speed of the satellites, and the difference in clock rate depends on γ. Although γ for the GPS satellites orbiting the Earth is nearly 1, it differs just enough that making the approximation $\gamma \approx 1$ would make the GPS hopelessly inaccurate and useless. It is even necessary to apply corrections based on Einstein's General Theory of Relativity, which correctly predicts an additional change in clock rate due to the gravity of the Earth being weaker at the high altitude of the GPS satellites.

Checkpoint 18 A spaceship at rest with respect to the cosmic microwave background emits a beam of red light. A different spaceship, moving at a speed of 2.5×10^8 m/s toward the first ship, detects the light. Which of the following statements are true for observers on the second ship? (More than one statement may be correct.) **(a)** They observe that the light travels at 3×10^8 m/s. **(b)** The light is not red. **(c)** They observe that the light travels at 5.5×10^8 m/s. **(d)** They observe that the light travels at 2.5×10^8 m/s.

1.13 *UPDATING POSITION AT HIGH SPEED

If $v \ll c$, $\vec{p} \approx m\vec{v}$ and $\vec{v} \approx \vec{p}/m$. But at high speed it is more complicated to determine the velocity from the (relativistic) momentum. Here is a way to solve for $\vec{v}$ in terms of $\vec{p}$:

$$|\vec{p}| = \frac{1}{\sqrt{1-(|\vec{v}|/c)^2}} m|\vec{v}|$$

Divide by m and square: $\dfrac{|\vec{p}|^2}{m^2} = \dfrac{|\vec{v}|^2}{1-(|\vec{v}|/c)^2}$

Multiply by $(1-(|\vec{v}|/c)^2)$: $\dfrac{|\vec{p}|^2}{m^2} - \left(\dfrac{|\vec{p}|^2}{m^2c^2}\right)|\vec{v}|^2 = |\vec{v}|^2$

Collect terms: $\left(1+\dfrac{|\vec{p}|^2}{m^2c^2}\right)|\vec{v}|^2 = \dfrac{|\vec{p}|^2}{m^2}$

$$|\vec{v}| = \frac{|\vec{p}|/m}{\sqrt{1+\left(\dfrac{|\vec{p}|}{mc}\right)^2}}$$

The expression above gives the magnitude of $\vec{v}$, in terms of the magnitude of $\vec{p}$. To get an expression for the vector $\vec{v}$, recall that any vector can be factored into its magnitude times a unit vector in the direction of the vector, so

$$\vec{p} = |\vec{p}|\hat{p} \quad \text{and} \quad \vec{v} = |\vec{v}|\hat{v}$$

But since $\vec{p}$ and $\vec{v}$ are in the same direction, $\hat{v} = \hat{p}$, so

$$\vec{v} = |\vec{v}|\hat{p} = \frac{|\vec{p}|/m}{\sqrt{1+\left(\dfrac{|\vec{p}|}{mc}\right)^2}}\hat{p} = \frac{(|\vec{p}|\hat{p})/m}{\sqrt{1+\left(\dfrac{|\vec{p}|}{mc}\right)^2}}$$

$$\vec{v} = \frac{\vec{p}/m}{\sqrt{1+\left(\dfrac{|\vec{p}|}{mc}\right)^2}}$$

THE RELATIVISTIC POSITION UPDATE EQUATION

$$\vec{r}_f = \vec{r}_i + \frac{1}{\sqrt{1+\left(\frac{|\vec{p}|}{mc}\right)^2}}\left(\frac{\vec{p}}{m}\right)\Delta t \quad \text{(for small } \Delta t\text{)}$$

Note that at low speeds $|\vec{p}| \approx m|\vec{v}|$, and the denominator is

$$\sqrt{1+\left(\frac{|\vec{v}|}{c}\right)^2} \approx 1$$

so the equation becomes the familiar $\vec{r}_f = \vec{r}_i + (\vec{p}/m)\,\Delta t$.

SUMMARY

Interactions
Interactions are indicated by

- Change of velocity (change of direction and/or change of speed)
- Change of identity
- Change of shape of multiparticle system
- Change of temperature of multiparticle system
- Indirect evidence for interactions

Newton's first law of motion
Every body persists in its state of rest or of moving with constant speed in a constant direction, except to the extent that it is compelled to change that state by forces acting on it.

Vectors
A 3D *vector* is a quantity with magnitude and a direction, which can be expressed as a triple $\langle x, y, z\rangle$. A vector is indicated by an arrow: $\vec{r}$.
A *scalar* is a single number.
Legal mathematical operations involving vectors include:

- Adding one vector to another vector
- Subtracting one vector from another vector
- Multiplying or dividing a vector by a scalar
- Finding the magnitude of a vector
- Taking the derivative of a vector

Operations that are *not* legal with vectors include:

- A vector cannot be added to a scalar.
- A vector cannot be set equal to a scalar.
- A vector cannot appear in the denominator (you can't divide by a vector).

A unit vector $\hat{r} = \vec{r}/|\vec{r}|$ has magnitude 1.
A vector can be factored using a unit vector: $\vec{F} = |\vec{F}|\hat{F}$.
Direction cosines: $\hat{r} = \langle\cos\theta_x, \cos\theta_y, \cos\theta_z\rangle$

The symbol Δ
The symbol Δ (delta) means "change of": $\Delta t = t_f - t_i$, $\Delta\vec{r} = \vec{r}_f - \vec{r}_i$.
Δ always means "final minus initial."

Velocity and change of position
Definition of average velocity

$$\vec{v}_{\text{avg}} = \frac{\Delta\vec{r}}{\Delta t} = \frac{\vec{r}_f - \vec{r}_i}{t_f - t_i}$$

Velocity is a vector. $\vec{r}$ is the position of an object (a vector). t is the time. Average velocity is equal to the change in position divided by the time elapsed. SI units of velocity are meters per second (m/s).
The position update equation

$$\vec{r}_f = \vec{r}_i + \vec{v}_{\text{avg}}\,\Delta t$$

The final position (vector) is the vector sum of the initial position plus the product of the average velocity and the elapsed time.
Definition of instantaneous velocity

$$\vec{v} = \lim_{\Delta t \to 0}\frac{\Delta\vec{r}}{\Delta t} = \frac{d\vec{r}}{dt}$$

The instantaneous velocity is the limiting value of the average velocity as the time elapsed becomes very small.
Velocity in terms of momentum

$$\vec{v} = \frac{\vec{p}/m}{\sqrt{1+\left(\frac{|\vec{p}|}{mc}\right)^2}} \quad \text{or } \vec{v} \approx \vec{p}/m \text{ at low speeds}$$

Acceleration
Acceleration is the time rate of change of velocity: $\vec{a} = d\vec{v}/dt$

Momentum

Definition of momentum

$$\vec{p} = \gamma m \vec{v}$$

where $\gamma = \dfrac{1}{\sqrt{1-(|\vec{v}|/c)^2}}$ (lowercase Greek gamma)

Momentum (a vector) is the product of the relativistic factor "gamma" (a scalar), mass, and velocity.

Combined into one equation: $\vec{p} = \dfrac{1}{\sqrt{1-(|\vec{v}|/c)^2}} m\vec{v}$.

Approximation for momentum at low speeds

$$\vec{p} \approx m\vec{v} \text{ at speeds such that } |\vec{v}| \ll c$$

Useful numbers:

Radius of a typical atom $\approx 1 \times 10^{-10}$ meter
Radius of a proton or neutron $\approx 1 \times 10^{-15}$ meter
Speed of light: 3×10^8 m/s

These and other useful data and conversion factors are given on the inside back cover of the textbook.

QUESTIONS

Q1 Why do we use a spaceship in outer space, far from other objects, to illustrate Newton's first law? Why not a car or a train? (More than one of the following statements may be correct.) (1) A car or train touches other objects, and interacts with them. (2) A car or train can't travel fast enough. (3) The spaceship has negligible interactions with other objects. (4) A car or train interacts gravitationally with the Earth. (5) A spaceship can never experience a gravitational force.

Q2 In the periodic table on the inside front cover of this book (or one you find on the internet), for each element there is given the "atomic number," the number of protons or electrons in an atom, and the "atomic mass," which is essentially the number of nucleons, protons plus neutrons, in the nucleus, averaged over the various isotopes of the element, which differ in the number of neutrons. Make a graph of the number of neutrons vs. the number of protons in the elements. You needn't graph every element, just enough to see the trend. What do you observe about the data? (This reflects the need for more neutrons in proton-rich nuclei in order to prevent the electric repulsion of the protons of each other from destroying the nucleus.)

Q3 Which of the following observers might observe something that appears to violate Newton's first law of motion? Explain why. (1) A person standing still on a street corner (2) A person riding on a roller coaster (3) A passenger on a starship traveling at $0.75c$ toward the nearby star Alpha Centauri (4) An airplane pilot doing aerobatic loops (5) A hockey player coasting across the ice.

Q4 Place a ball on a book and walk with the book in uniform motion. Note that you don't really have to do anything to the ball to keep the ball moving with constant velocity (relative to the ground) or to keep the ball at rest (relative to you). Then stop suddenly, or abruptly change your direction or speed. What does Newton's first law of motion predict for the motion of the ball (assuming that the interaction between the ball and the book is small)? Does the ball behave as predicted? It may help to take the point of view of a friend who is standing still, watching you.

Q5 Which of the following statements about the velocity and momentum of an object are correct? (1) The momentum of an object is always in the same direction as its velocity. (2) The momentum of an object can be either in the same direction as its velocity or in the opposite direction. (3) The momentum of an object is perpendicular to its velocity. (4) The direction of an object's momentum is not related to the direction of its velocity. (5) The direction of an object's momentum is tangent to its path.

Q6 Answer the following questions about the factor γ (gamma) in the full relativistic equation for momentum: **(a)** Is γ a scalar or a vector quantity? **(b)** What is the minimum possible value of γ? **(c)** Does γ reach its minimum value when an object's speed is high or low? **(d)** Is there a maximum possible value for γ? **(e)** Does γ become large when an object's speed is high or low? **(f)** Does the approximation $\gamma \approx 1$ apply when an object's speed is low or when it is high?

Q7 In which of these situations is it reasonable to use the approximate equation for the momentum of an object, instead of the full relativistically correct equation? (1) A car traveling on an interstate highway (2) A commercial jet airliner flying between New York and Seattle (3) A neutron traveling at 2700 meters per second (4) A proton in outer space traveling at 2×10^8 m/s (5) An electron in a television tube traveling 3×10^6 m/s

Q8 Moving objects left the traces labeled *A–F* in Figure 1.54. The dots were deposited at equal time intervals (for example, one dot each second). In each case the object starts from the square. Which trajectories show evidence that the moving object was interacting with another object somewhere? If there is evidence of an interaction, what is the evidence?

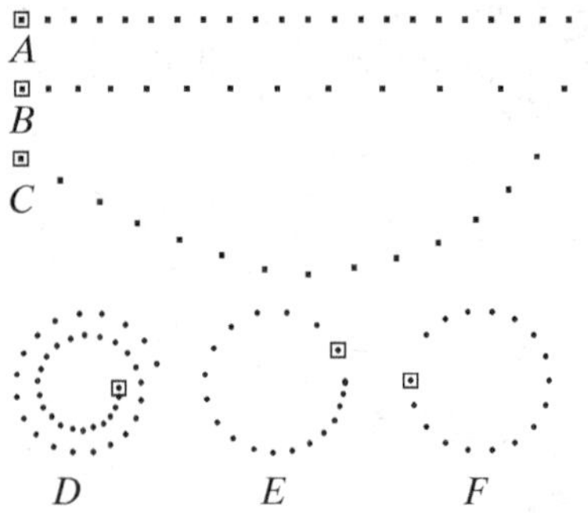

Figure 1.54

Q9 A car moves along a straight road. It moves at a speed of 50 km/h for 4 minutes, then during 4 minutes it gradually speeds up to 100 km/h, continues at this speed for 4 minutes, then during

4 minutes gradually slows to a stop. Make a sketch like the figures in Section 1.2, marking dots for the position along the road every minute.

Q10 A spaceship far from all other objects uses its thrusters to attain a speed of 1×10^4 m/s. The crew then shuts off the power. According to Newton's first law, what will happen to the motion of the spaceship from then on?

Q11 Which of the following are vectors? **(a)** $\vec{r}/2$ **(b)** $|\vec{r}|/2$ **(c)** $\langle r_x, r_y, r_z \rangle$ **(d)** $5 \cdot \vec{r}$

PROBLEMS

The difficulty of a problem is represented by the number of dots preceding the problem number.

Section 1.4

•P12 Figure 1.55 shows several arrows representing vectors in the *xy* plane. **(a)** Which vectors have magnitudes equal to the magnitude of $\vec{a}$? **(b)** Which vectors are equal to $\vec{a}$?

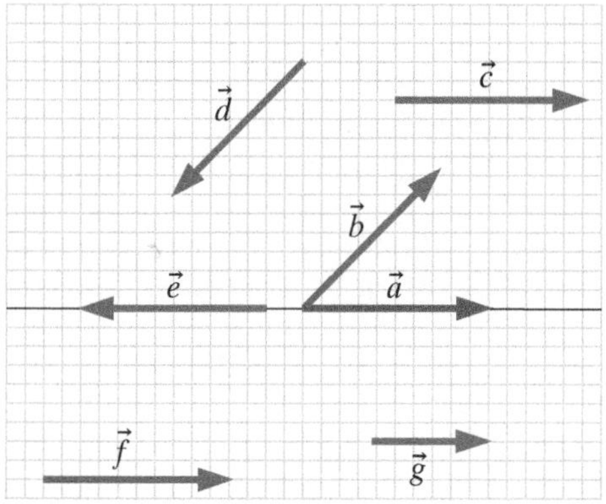

Figure 1.55

•P13 What is the magnitude of the vector $\vec{v}$, where $\vec{v} = \langle 8 \times 10^6, 0, -2 \times 10^7 \rangle$ m/s?

•P14 In Figure 1.56 three vectors are represented by arrows in the *xy* plane. Each square in the grid represents one meter. For each vector, write out the components of the vector, and calculate the magnitude of the vector.

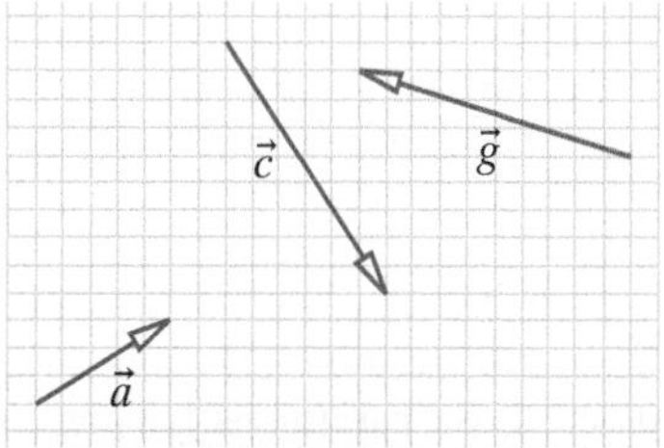

Figure 1.56

•P15 The following questions refer to the vectors depicted by arrows in Figure 1.57. **(a)** What are the components of the vector $\vec{a}$? (Note that since the vector lies in the *xy* plane, its *z* component is zero.) **(b)** What are the components of the vector $\vec{b}$? **(c)** Is this statement true or false? $\vec{a} = \vec{b}$ **(d)** What are the components of the vector $\vec{c}$? **(e)** Is this statement true or false? $\vec{c} = -\vec{a}$ **(f)** What are the components of the vector $\vec{d}$? **(g)** Is this statement true or false? $\vec{d} = -\vec{c}$

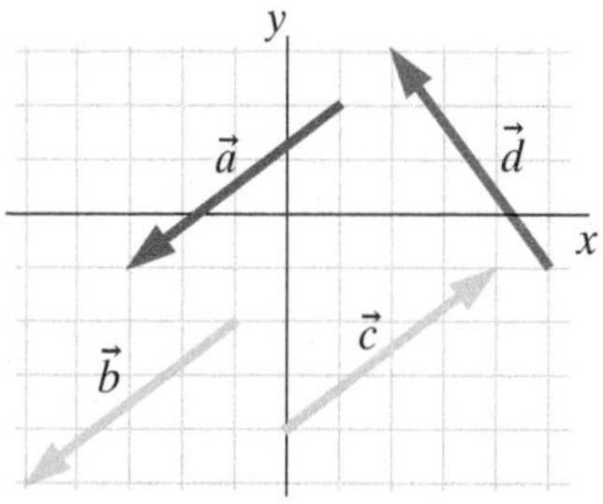

Figure 1.57

•P16 On a piece of graph paper, draw arrows representing the following vectors. Make sure the tip and tail of each arrow you draw are clearly distinguishable. **(a)** Placing the tail of the vector at $\langle 5,2,0 \rangle$, draw an arrow representing the vector $\vec{p} = \langle -7,3,0 \rangle$. Label it $\vec{p}$. **(b)** Placing the tail of the vector at $\langle -5,8,0 \rangle$, draw an arrow representing the vector $-\vec{p}$. Label it $-\vec{p}$.

•P17 What is the result of multiplying the vector $\vec{a}$ by the scalar f, where $\vec{a} = \langle 0.02, -1.7, 30.0 \rangle$ and $f = 2.0$?

•P18 **(a)** In Figure 1.58, what are the components of the vector $\vec{d}$? **(b)** If $\vec{e} = -\vec{d}$, what are the components of $\vec{e}$? **(c)** If the tail of vector $\vec{d}$ were moved to location $\langle -5,-2,4 \rangle$ m, where would the tip of the vector be located? **(d)** If the tail of vector $-\vec{d}$ were placed at location $\langle -1,-1,-1 \rangle$ m, where would the tip of the vector be located?

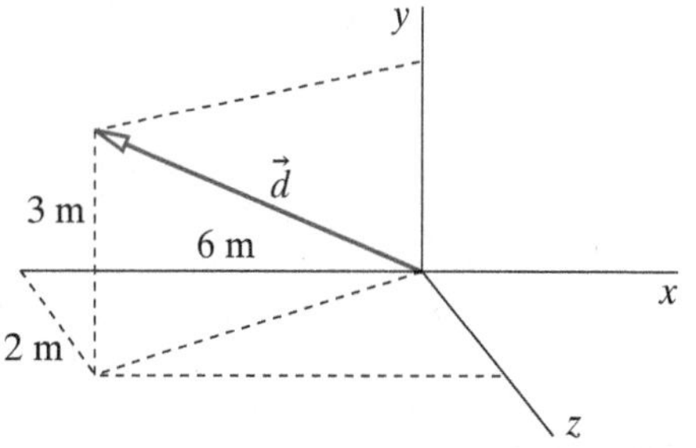

Figure 1.58

•P19 What is the unit vector in the direction of $\langle 2,2,2 \rangle$? What is the unit vector in the direction of $\langle 3,3,3 \rangle$?

•P20 **(a)** On a piece of graph paper, draw the vector $\vec{f} = \langle -2,4,0 \rangle$, putting the tail of the vector at $\langle -3,0,0 \rangle$. Label the vector $\vec{f}$. **(b)** Calculate the vector $2\vec{f}$, and draw this vector on the graph, putting its tail at $\langle -3,-3,0 \rangle$, so you can compare it to the original vector. Label the vector $2\vec{f}$. **(c)** How does the magnitude of $2\vec{f}$ compare to the magnitude of $\vec{f}$? **(d)** How does the direction of $2\vec{f}$ compare to the direction of $\vec{f}$? **(e)** Calculate the vector $\vec{f}/2$, and draw this vector on the graph, putting its tail at $\langle -3,-6,0\rangle$, so you can compare it to the other vectors. Label the vector $\vec{f}/2$. **(f)** How does the magnitude of $\vec{f}/2$ compare to the magnitude of $\vec{f}$? **(g)** How does the direction of $\vec{f}/2$ compare to the direction of $\vec{f}$? **(h)** Does multiplying a vector by a scalar change the magnitude of the vector? **(i)** The vector $a(\vec{f})$ has a magnitude three times as great as that of $\vec{f}$, and its direction is opposite to the direction of $\vec{f}$. What is the value of the scalar factor a?

•P21 Write the vector $\vec{a} = \langle 400,200,-100\rangle$ m/s^2 as the product $|\vec{a}| \cdot \hat{a}$.

•P22 (a) On a piece of graph paper, draw the vector $\vec{g} = \langle 4,7,0\rangle$ m. Put the tail of the vector at the origin. **(b)** Calculate the magnitude of $\vec{g}$. **(c)** Calculate $\hat{g}$, the unit vector pointing in the direction of $\vec{g}$. **(d)** On the graph, draw $\hat{g}$. Put the tail of the vector at $\langle 1,0,0\rangle$ m so you can compare $\hat{g}$ and $\vec{g}$. **(e)** Calculate the product of the magnitude $|\vec{g}|$ times the unit vector $\hat{g}$, $(|\vec{g}|)(\hat{g})$.

•P23 A proton is located at $\langle 3\times 10^{-10},-3\times 10^{-10},8\times 10^{-10}\rangle$ m. **(a)** What is $\vec{r}$, the vector from the origin to the location of the proton? **(b)** What is $|\vec{r}|$? **(c)** What is $\hat{r}$, the unit vector in the direction of $\vec{r}$?

•P24 In Figure 1.59, the vector $\vec{r}_1$ points to the location of object 1 and $\vec{r}_2$ points to the location of object 2. Both vectors lie in the xy plane. **(a)** Calculate the position of object 2 relative to object 1, as a relative position vector. **(b)** Calculate the position of object 1 relative to object 2, as a relative position vector.

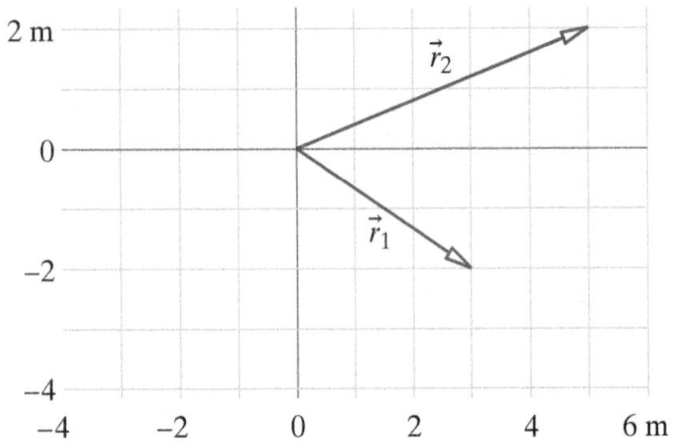

Figure 1.59

•P25 (a) What is the vector whose tail is at $\langle 9.5,7,0\rangle$ m and whose head is at $\langle 4,-13,0\rangle$ m? **(b)** What is the magnitude of this vector?

•P26 A man is standing on the roof of a building with his head at the position $\langle 12,30,13\rangle$ m. He sees the top of a tree, which is at the position $\langle -25,35,43\rangle$ m. **(a)** What is the relative position vector that points from the man's head to the top of the tree? **(b)** What is the distance from the man's head to the top of the tree?

•P27 A star is located at $\langle 6\times 10^{10},8\times 10^{10},6\times 10^{10}\rangle$ m. A planet is located at $\langle -4\times 10^{10},-9\times 10^{10},6\times 10^{10}\rangle$ m. **(a)** What is the vector pointing from the star to the planet? **(b)** What is the vector pointing from the planet to the star?

•P28 A planet is located at $\langle -1\times 10^{10},8\times 10^{10},-3\times 10^{10}\rangle$. A star is located at $\langle 6\times 10^{10},-5\times 10^{10},1\times 10^{10}\rangle$. **(a)** What is $\vec{r}$, the vector from the star to the planet? **(b)** What is the magnitude of $\vec{r}$? **(c)** What is $\hat{r}$, the unit vector (vector with magnitude 1) in the direction of $\vec{r}$?

•P29 A proton is located at $\langle x_p,y_p,z_p\rangle$. An electron is located at $\langle x_e,y_e,z_e\rangle$. What is the vector pointing from the electron to the proton? What is the vector pointing from the proton to the electron?

••P30 A cube is 3 cm on a side, with one corner at the origin. What is the unit vector pointing from the origin to the diagonally opposite corner at location $\langle 3,3,3\rangle$ cm? What is the angle from this diagonal to one of the adjacent edges of the cube?

Section 1.6

•P31 A "slow" neutron produced in a nuclear reactor travels from location $\langle 0.2,-0.05,0.1\rangle$ m to location $\langle -0.202,0.054,0.098\rangle$ m in 2 microseconds ($1\,\mu s = 1\times 10^{-6}$ s). **(a)** What is the average velocity of the neutron? **(b)** What is the average speed of the neutron?

•P32 The position of a baseball relative to home plate changes from $\langle 15,8,-3\rangle$ m to $\langle 20,6,-1\rangle$ m in 0.1 s. As a vector, write the average velocity of the baseball during this time interval.

•P33 You jog at a steady speed of 2 m/s. You start from the location $\langle 0,0,0\rangle$ and for the first 200 s your direction is given by the unit vector $\langle 1,0,0\rangle$. Next you jog for 300 s in the direction given by the unit vector $\langle \cos 45°,0,\cos 45°\rangle$. Finally you jog for 150 s in the direction given by the unit vector $\langle \cos 60°,0,\cos 30°\rangle$. **(a)** Now what is your position? **(b)** What was your average velocity?

•P34 The position of a golf ball relative to the tee changes from $\langle 50,20,30\rangle$ m to $\langle 53,18,31\rangle$ m in 0.1 second. As a vector, write the velocity of the golf ball during this short time interval.

•P35 The crew of a stationary spacecraft observe an asteroid whose mass is 4×10^{17} kg. Taking the location of the spacecraft as the origin, the asteroid is observed to be at location $\langle -3\times 10^3,-4\times 10^3,8\times 10^3\rangle$ m at a time 18.4 s after lunchtime. At a time 21.4 s after lunchtime, the asteroid is observed to be at location $\langle -1.4\times 10^3,-6.2\times 10^3,9.7\times 10^3\rangle$ m. Assuming that the velocity of the asteroid does not change during this time interval, calculate the vector velocity $\vec{v}$ of the asteroid.

••P36 A spacecraft traveling at a velocity of $\langle -20,-90,40\rangle$ m/s is observed to be at a location $\langle 200,300,-500\rangle$ m relative to an origin located on a nearby asteroid. At a later time the spacecraft is at location $\langle -380,-2310,660\rangle$ m. **(a)** How long did it take the spacecraft to travel between these locations? **(b)** How far did the spacecraft travel? **(c)** What is the speed of the spacecraft? **(d)** What is the unit vector in the direction of the spacecraft's velocity?

••P37 Here are the positions at three different times for a bee in flight (a bee's top speed is about 7 m/s).

Time	6.3 s	6.8 s	7.3 s
Position	$\langle -3.5,9.4,0\rangle$ m	$\langle -1.3,6.2,0\rangle$ m	$\langle 0.5,1.7,0\rangle$ m

(a) Between 6.3 s and 6.8 s, what was the bee's average velocity? Be careful with signs. **(b)** Between 6.3 s and 7.3 s, what was the bee's average velocity? Be careful with signs. **(c)** Of the two average velocities you calculated, which is the best estimate of the bee's instantaneous velocity at time 6.3 s? **(d)** Using the best information available, what was the displacement of the bee during the time interval from 6.3 s to 6.33 s?

Section 1.7

•P38 At time $t_1 = 12$ s, a car is located at $\langle 84,78,24\rangle$ m and has velocity $\langle 4,0,-3\rangle$ m/s. At time $t_2 = 18$ s, what is the position of the car? (The velocity is constant in magnitude and direction during this time interval.)

•**P39** An electron passes location $\langle 0.02, 0.04, -0.06\rangle$ m, and 2 μs later is detected at location $\langle 0.02, 1.84, -0.86\rangle$ m (1 microsecond is 1×10^{-6} s). **(a)** What is the average velocity of the electron? **(b)** If the electron continues to travel at this average velocity, where will it be in another 5 μs?

•**P40** After World War II the U.S. Air Force carried out experiments on the amount of acceleration a human can survive. These experiments, led by John Stapp, were the first to use crash dummies as well as human subjects, especially Stapp himself, who became an effective advocate for automobile safety belts. In one of the experiments Stapp rode a rocket sled that decelerated from 140 m/s (about 310 mi/h) to 70 m/s in just 0.6 s. **(a)** What was the absolute value of the (negative) average acceleration? **(b)** The acceleration of a falling object if air resistance is negligible is 9.8 m/s/s, called "one g." What was the absolute value of the average acceleration in g's? (Stapp eventually survived a test at 46 g's!)

••**P41** At a certain instant a ball passes location $\langle 7, 21, -17\rangle$ m. In the next 3 s, the ball's average velocity is $\langle -11, 42, 11\rangle$ m/s. At the end of this 3 s time interval, what is the height y of the ball?

••**P42** You throw a ball. Assume that the origin is on the ground, with the $+y$ axis pointing upward. Just after the ball leaves your hand its position is $\langle 0.06, 1.03, 0\rangle$ m. The average velocity of the ball over the next 0.7 s is $\langle 17, 4, 6\rangle$ m/s. At time 0.7 s after the ball leaves your hand, what is the height of the ball above the ground?

••**P43** Figure 1.60 shows the trajectory of a ball traveling through the air, affected by both gravity and air resistance. Here are the positions of the ball at several successive times:

Location	t(s)	Position (m)
A	0.0	$\langle 0,0,0\rangle$
B	1.0	$\langle 22.3, 26.1, 0\rangle$
C	2.0	$\langle 40.1, 38.1, 0\rangle$

(a) What is the average velocity of the ball as it travels between location A and location B? **(b)** If the ball continued to travel at the same average velocity during the next second, where would it be at the end of that second? (That is, where would it be at time t=2 s?) **(c)** How does your prediction from part (b) compare to the actual position of the ball at t=2 s (location C)? If the predicted and observed locations of the ball are different, explain why.

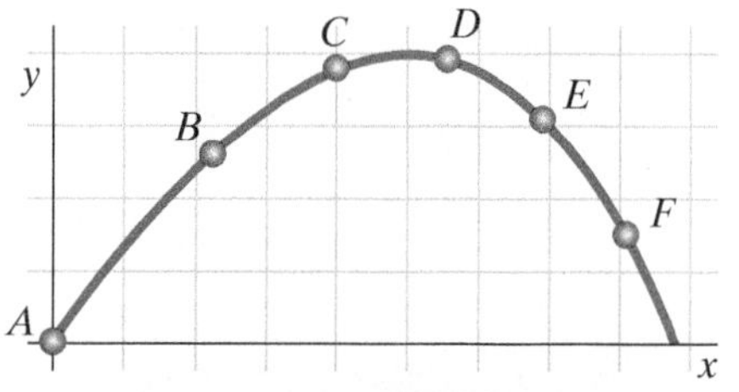

Figure 1.60

••**P44** At 6 s after 3:00, a butterfly is observed leaving a flower whose location is $\langle 6, -3, 10\rangle$ m relative to an origin on top of a nearby tree. The butterfly flies until 10 s after 3:00, when it alights on a different flower whose location is $\langle 6.8, -4.2, 11.2\rangle$ m relative to the same origin. What was the location of the butterfly at a time 8.5 s after 3:00? What assumption did you have to make in calculating this location?

Section 1.8

•**P45** A baseball has a mass of 0.155 kg. A professional pitcher throws a baseball 90 mi/h, which is 40 m/s. What is the magnitude of the momentum of the pitched baseball?

•**P46** A hockey puck with a mass of 0.4 kg has a velocity of $\langle 38, 0, -27\rangle$ m/s. What is the magnitude of its momentum, $|\vec{p}|$?

•**P47** What is the magnitude (in kg·m/s) of the momentum of a 1000 kg airplane traveling at a speed of 500 mi/h? (Note that you need to convert speed to meters per second.)

•**P48** A baseball has a mass of about 155 g. What is the magnitude of the momentum of a baseball thrown at a speed of 100 miles per hour? (Note that you need to convert mass to kilograms and speed to meters per second. See the inside back cover of the textbook for conversion factors.)

•**P49** If a particle has momentum $\vec{p} = \langle 4, -5, 2\rangle$ kg·m/s, what is the magnitude $|\vec{p}|$ of its momentum?

•**P50** An object with mass 1.6 kg has momentum $\langle 0, 0, 4\rangle$ kg·m/s. **(a)** What is the magnitude of the momentum? **(b)** What is the unit vector corresponding to the momentum? **(c)** What is the speed of the object?

•**P51** A tennis ball of mass m traveling with velocity $\langle v_x, 0, 0\rangle$ hits a wall and rebounds with velocity $\langle -v_x, 0, 0\rangle$. **(a)** What was the change in momentum of the tennis ball? **(b)** What was the change in the magnitude of the momentum of the tennis ball?

•**P52** A basketball has a mass of 570 g. Heading straight downward, in the $-y$ direction, it hits the floor with a speed of 5 m/s and rebounds straight up with nearly the same speed. What was the momentum change $\Delta\vec{p}$?

••**P53** A basketball has a mass of 570 g. Moving to the right and heading downward at an angle of 30° to the vertical, it hits the floor with a speed of 5 m/s and bounces up with nearly the same speed, again moving to the right at an angle of 30° to the vertical. What was the momentum change $\Delta\vec{p}$?

•**P54** The first stage of the giant Saturn V rocket reached a speed of 2300 m/s at 170 s after liftoff. **(a)** What was the average acceleration in m/s/s? **(b)** The acceleration of a falling object if air resistance is negligible is 9.8 m/s/s, called "one g." What was the average acceleration in g's?

••**P55** A 50 kg child is riding on a carousel (merry-go-round) at a constant speed of 5 m/s. What is the magnitude of the change in the child's momentum $|\Delta\vec{p}|$ in going all the way around (360°)? In going halfway around (180°)? It is very helpful to draw a diagram, and to do the vector subtraction graphically.

••**P56** Figure 1.61 shows a portion of the trajectory of a ball traveling through the air. Arrows indicate its momentum at several locations.

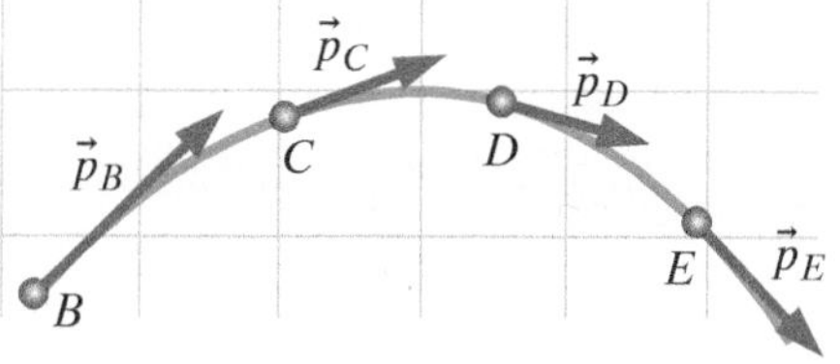

Figure 1.61

At various locations, the ball's momentum is:

$\vec{p}_B = \langle 3.03, 2.83, 0\rangle$ kg·m/s
$\vec{p}_C = \langle 2.55, 0.97, 0\rangle$ kg·m/s
$\vec{p}_D = \langle 2.24, -0.57, 0\rangle$ kg·m/s

$\vec{p}_E = \langle 1.97, -1.93, 0 \rangle$ kg · m/s
$\vec{p}_F = \langle 1.68, -3.04, 0 \rangle$ kg · m/s
(a) Calculate the change in the ball's momentum between each pair of adjacent locations. **(b)** On a copy of the diagram, draw arrows representing each $\Delta\vec{p}$ you calculated in part (a). **(c)** Between which two locations is the magnitude of the change in momentum greatest?

Section 1.9

•P57 What is the velocity of a 3 kg object when its momentum is $\langle 60, 150, -30 \rangle$ kg · m/s?

•P58 A 1500 kg car located at $\langle 300, 0, 0 \rangle$ m has a momentum of $\langle 45000, 0, 0 \rangle$ kg · m/s. What is its location 10 s later?

••P59 An ice hockey puck of mass 170 g enters the goal with a momentum of $\langle 0, 0, -6.3 \rangle$ kg · m/s, crossing the goal line at location $\langle 0, 0, -26 \rangle$ m relative to an origin in the center of the rink. The puck had been hit by a player 0.4 s before reaching the goal. What was the location of the puck when it was hit by the player, assuming negligible friction between the puck and the ice? (Note that the ice surface lies in the xz plane.)

••P60 A space probe of mass 400 kg drifts past location $\langle 0, 3 \times 10^4, -6 \times 10^4 \rangle$ m with momentum $\langle 6 \times 10^3, 0, -3.6 \times 10^3 \rangle$ kg · m/s. Assuming the momentum of the probe does not change, what will be its position 2 minutes later?

Section 1.10

•P61 A proton in an accelerator attains a speed of $0.88c$. What is the magnitude of the momentum of the proton?

•P62 An electron with a speed of $0.95c$ is emitted by a supernova, where c is the speed of light. What is the magnitude of the momentum of this electron?

•P63 A "cosmic-ray" proton hits the upper atmosphere with a speed $0.9999c$, where c is the speed of light. What is the magnitude of the momentum of this proton? Note that $|\vec{v}|/c = 0.9999$; you don't actually need to calculate the speed $|\vec{v}|$.

•P64 A proton in a particle accelerator is traveling at a speed of $0.99c$. (Masses of particles are given on the inside back cover of this textbook.) **(a)** If you use the approximate nonrelativistic equation for the magnitude of momentum of the proton, what answer do you get? **(b)** What is the magnitude of the correct relativistic momentum of the proton? **(c)** The approximate value (the answer to part a) is significantly too low. What is the ratio of magnitudes you calculated (correct/approximate)?

•P65 When the speed of a particle is close to the speed of light, the factor γ, the ratio of the correct relativistic momentum $\gamma m\vec{v}$ to the approximate nonrelativistic momentum $m\vec{v}$, is quite large. Such speeds are attained in particle accelerators, and at these speeds the approximate nonrelativistic equation for momentum is a very poor approximation. Calculate γ for the case where $|\vec{v}|/c = 0.9996$.

•P66 An electron travels at speed $|\vec{v}| = 0.996c$, where $c = 3 \times 10^8$ m/s is the speed of light. The electron travels in the direction given by the unit vector $\hat{v} = \langle 0.655, -0.492, -0.573 \rangle$. The mass of an electron is 9×10^{-31} kg. **(a)** What is the value of γ? You can simplify the calculation if you notice that $(|\vec{v}|/c)^2 = (0.996)^2$. **(b)** What is the speed of the electron? **(c)** What is the magnitude of the electron's momentum? **(d)** What is the vector momentum of the electron? Remember that any vector can be "factored" into its magnitude times its unit vector, so that $\vec{v} = |\vec{v}|\hat{v}$.

•P67 If $|\vec{p}|/m$ is $0.85c$, what is $|\vec{v}|$ in terms of c?

COMPUTATIONAL PROBLEMS

These problems are intended to introduce you to using a computer to model matter, interactions, and motion in 3D. You do not need to know how to program; you will learn what you need to know by doing these problems. In later chapters you will build on these small calculations to build models of physical systems.

To install the free 3D programming environment VPython, go to http://vpython.org and (carefully) follow the instructions for your operating system (Windows, MacOS, or Linux). Note the instructions given there on how to zoom and rotate the "camera" when viewing a 3D scene you have created.

More detailed and extended versions of some of these computational modeling problems may be found in the lab activities included in the *Matter & Interactions 4th Edition* resources for instructors.

•P68 Watch the first introductory VPython video, *VPython Instructional Videos 1: 3D Objects*, at vpython.org/video01.html and complete the challenge activity at the end of the video.

•P69 (a) Write a VPython program that creates eight spheres, each placed at one corner of a cube centered on the origin. The length of a side of the cube should be 6 units, and the radius of each sphere should be 0.5. Use at least two different colors for the spheres. **(b)** Add to the program an arrow whose tail is at one corner of the cube and whose tip is at the corner diagonally opposite. Figure 1.62 shows the display from one possible solution to this problem.

Figure 1.62

•P70 Write a VPython program that represents the x, y, and z axes by three cylinders of different colors. The display from one possible solution is shown in Figure 1.63.

Figure 1.63

•**P71** Write a VPython program that represents the x, y, and z axes by three boxes (rectangular solids) of different colors. The display from one possible solution is shown in Figure 1.64.

Figure 1.64

•**P72** Watch the second introductory VPython video, *VPython Instructional Videos 2: Variable Assignment*, at vpython.org/video02.html and complete the challenge activity at the end of the video.

•**P73** Based on Problem P72, write a program that shows an arrow pointing from one small box to another in such a way that when you change only the position of the first box, making no other changes, the arrow and the other box move too, so that the two boxes remain linked by the arrow.

•**P74** Watch the third introductory VPython video, *VPython Instructional Videos 3: Beginning Loops*, at vpython.org/video03.html and complete the challenge activity at the end of the video.

••**P75** **(a)** Write a VPython program that uses three while loops to create a display in which each of the axes (x, y, z) is represented by a linear array of boxes, with spaces between the boxes. Figure 1.65 shows a possible example. **(b)** (Optional) Rewrite your program to produce the same display using only one while loop.

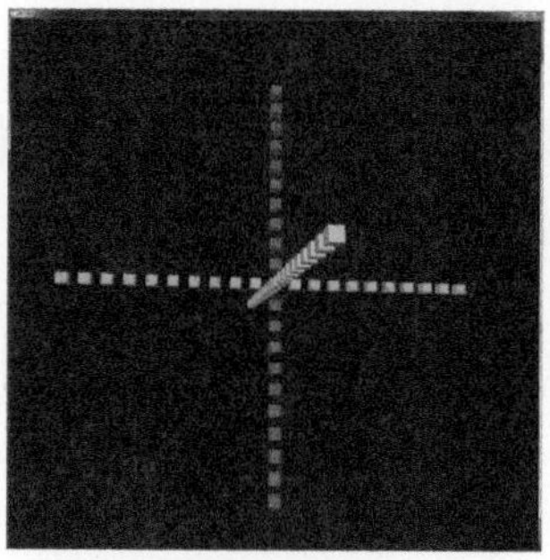

Figure 1.65

••**P76** Write a VPython program that creates three circles of spheres: one in the xy plane, one in the yz plane, and one in the xz plane. Each ring should be centered on the origin. Use either three while loops or one while loop. A possible solution is shown in Figure 1.66.

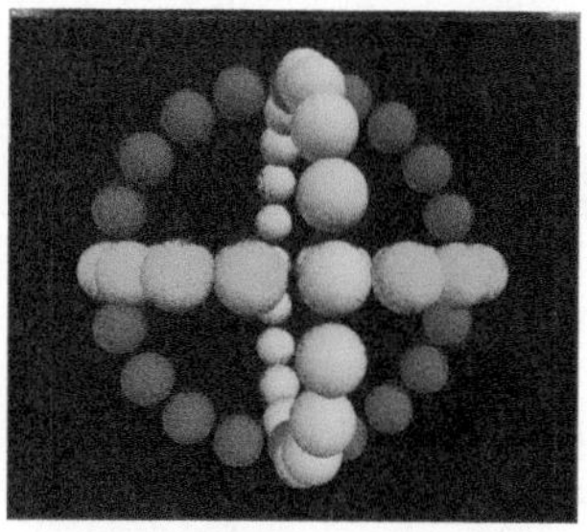

Figure 1.66

•**P77** Watch the fourth introductory VPython video, *VPython Instructional Videos 4: Loops and Animation*, at vpython.org/video04.html and complete the challenge activity at the end of the video.

••**P78** Consider the VPython program shown below. **(a)** An important skill is being able to read and understand an existing program, in order to be able to make useful modifications. *Before running the program*, study the program carefully line by line, then answer the following questions: (1) What is the initial velocity of the particle? (2) Is the particle initially located in front of the box or behind it? (3) In which line of code is the position of the particle updated? (4) What is the value of the time step Δt? (5) Will the particle bounce off of the red box, or travel through it? **(b)** Now run the program, and see if your answers were correct. **(c)** Modify the program to start the particle from an initial position on the $+x$ axis, to the right of and in front of the red box. Give the particle a velocity that will make it travel to the left, along the x axis, passing in front of the box.

```
from visual import *
box(pos=vector(0,0,-1),
    size=(5,5,0.5),
    color=color.red,
    opacity = 0.4)
particle = sphere(pos=vector(-5,0,-5),
                  radius=0.3,
                  color=color.cyan,
                  make_trail = True)
v = vector(0.5,0,0.5)
delta_t = 0.05
t = 0
while t < 20:
    rate(100)
    particle.pos = particle.pos + v * delta_t
    t = t + delta_t
```

•••**P79** Modify the program shown above to make the particle bounce off the red box instead of passing through it. On the Help menu available in IDLE (the VPython editor), choose "Python docs" or search the web for "Python if" to find out how to use an `if` statement. You may also find it helpful to look at the example program `bounce2.py`, included in the examples installed with VPython.

ANSWERS TO CHECKPOINTS

Numerical answers are given to three significant figures.

1 (a) 1, 4, **(b)** 1: change of direction, 2: change of speed and direction, 3: change of direction, 5: change of direction

2 (a) 1: change of speed, 2: change of speed, 3: change of direction, **(b)** 1, 4

3 $\langle 1, -1.5, 2.5 \rangle$ m

4 3.08 m/s

5 (a) no, **(b)** $\langle 0.873, 0.436, -0.218 \rangle$

6 (a) $\langle 450, -300, -300 \rangle$, **(b)** 577, **(c)** 696, **(d)** No, **(e)** $\langle 150, 300, -200 \rangle$, **(f)** 391, **(g)** 25.1, **(h)** No

7 a, b, e

8 (a) $\langle 9, 2, 20 \rangle$ m, **(b)** 120 s

9 (a) $\langle -0.949, 0.342, 0 \rangle$, **(b)** $\langle -0.643, 0.766, 0 \rangle$

10 2.67×10^{-3} m/s

11 (a) $\langle 8, -10, 12 \rangle$ m/s, **(b)** 17.5 m/s, **(c)** $\langle 0.456, -0.570, 0.684 \rangle$

12 $\langle 2.1 \times 10^5, 1.4 \times 10^5, -2.8 \times 10^5 \rangle$ m

13 (1) a, (2) h, (3) g, (4) f, (5) e

14 (a) (1) 5 m/s/s, (2) about half as big **(b)** (1) $\langle 5, 8t, 2 - 18t^2 \rangle$, (2) $\langle 0, 8, -36t \rangle$, (3) $\langle 5, 0, 2 \rangle$, (4) $\langle 0, 8, 0 \rangle$

15 (a) $\langle -1.6 \times 10^{28}, 0, -1.6 \times 10^{28} \rangle$ kg·m/s, **(b and c)** If you make a careful diagram, with the two arrows tail to tail, you'll see that the arrow representing $\Delta\vec{p}$ from C to D points up and to the right on the page (toward the Sun).

16 $\langle 111.31, 0, 18.46 \rangle$ m

17 $\langle 0, 0, -2.415 \times 10^{-22} \rangle$ kg·m/s, 2.415×10^{-22} kg·m/s

18 a, b

CHAPTER

2

The Momentum Principle

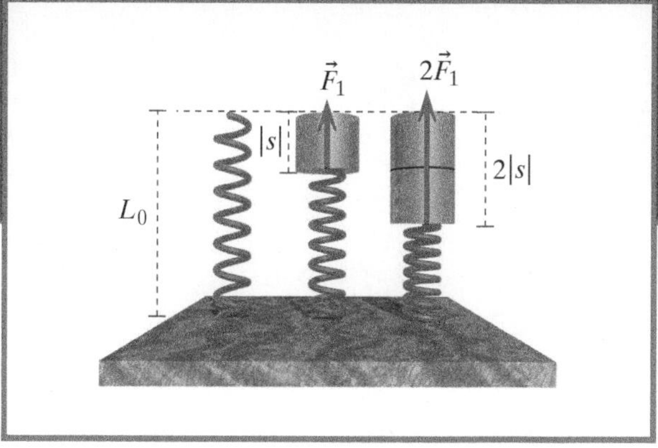

OBJECTIVES

After studying this chapter, you should be able to

- Use both iterative and analytical techniques to predict the future motion of a system that is subjected to a constant net force.
- Use an iterative approach to predict the future momentum of an object that is subjected to a varying net force.
- Draw and interpret graphs of the components of position or velocity *vs.* time.
- Calculate the force exerted by a spring on an object in contact with it.
- Calculate the approximate gravitational force on an object near the Earth's surface.

2.1 THE MOMENTUM PRINCIPLE

In many real situations, the momentum (and therefore the velocity) of a moving object are continually changing due to interactions of the object with its surroundings. In order to predict the motion of such a system, whether it is a meteor, a tennis ball, or an electron, we need to be able to express mathematically the relationship between interaction and momentum change. The Momentum Principle makes a quantitative connection between amount of interaction and change of momentum.

The Momentum Principle is the first of three fundamental principles of mechanics that together make it possible to predict and explain a very broad range of real-world phenomena. The other two fundamental principles we will encounter are the Energy Principle (Chapter 6) and the Angular Momentum Principle (Chapter 11). A fundamental principle is powerful because it applies in absolutely every situation. In its most general form, a fundamental principle expresses the idea that certain important quantities are *conserved*; that is, that the total value of the quantity in the Universe does not change. However, in order to predict the details of what will happen to a system we will often write the principle in a form relating the change in a quantity to the details of the interaction causing the change.

System and Surroundings

The Momentum Principle relates an interaction quantitatively to a change in momentum. To analyze a change in momentum, we need to specify clearly the objects whose momentum change we wish to know. One or more objects can be considered to be a "system." Everything that is not included in the system

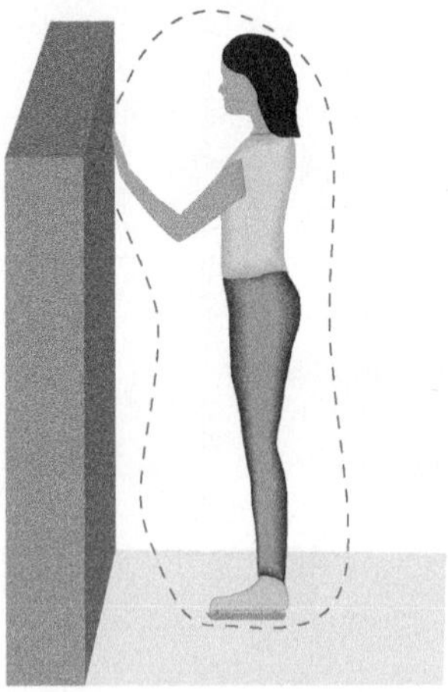

Figure 2.1 A skater pushes off from a wall. If we choose the skater as the system (indicated by the dashed line), the wall, the ice, the air, and the Earth are included in the surroundings.

is part of the "surroundings." For example, the red dashed line in Figure 2.1 indicates that we have chosen the skater as the system, and that therefore all other objects, including the wall, the ice, the air, and the Earth, are part of the surroundings. The Momentum Principle relates the change in momentum of a system to the amount of interaction with its surroundings. No matter what system we choose, the Momentum Principle will correctly predict the behavior of the system.

In this chapter we will always consider a single object to be the system, and everything else in the Universe to be the surroundings. However, in later chapters we will see that it is possible, and often useful, to choose a system comprising several objects. A fundamental principle such as the Momentum Principle applies to any system, no matter how complex.

Cause and Effect

The Momentum Principle is a fundamental principle that is also known as Newton's second law. It restates and extends Newton's first law of motion in a quantitative, causal form that can be used to predict the behavior of objects. The validity of the Momentum Principle has been verified through a very wide variety of observations and experiments, involving large and small objects, moving slowly or at speeds near the speed of light. It is a summary of the way interactions affect motion in the real world.

THE MOMENTUM PRINCIPLE

$$\Delta\vec{p} = \vec{F}_{\text{net}}\,\Delta t$$

The change of momentum of a system (the effect) is equal to the net force acting on the system times the duration of the interaction (the cause). The time interval Δt must be small enough that the net force is nearly constant during this time interval.

The Momentum Principle is a fundamental principle because: it applies to every possible system, no matter how large or small (from clusters of galaxies to subatomic particles), and no matter how fast it is moving; it is true for every kind of interaction (electric, gravitational, etc.); it relates an effect (change in momentum) to a cause (an interaction).

To begin to understand this equation, we will consider each quantity involved.

$$\Delta\vec{p} = \vec{F}_{\text{net}}\,\Delta t$$

Change of Momentum $\Delta\vec{p}$

As we saw in the previous chapter, the change in momentum $\Delta\vec{p} = \vec{p}_f - \vec{p}_i$ of a system can involve a change in the magnitude of momentum, the direction of momentum, or both magnitude and direction.

$$\Delta\vec{p} = \vec{F}_{\text{net}}\,\Delta t$$

Force $\vec{F}$

Scientists and engineers employ the concept of "force" to quantify interactions between two objects. Force is a vector quantity because a force has a magnitude and is exerted in a particular direction. Examples of forces include the following:

- The repulsive electric force a proton exerts on another proton
- The attractive gravitational force the Earth exerts on you
- The force that a compressed spring exerts on your hand
- The force on a spacecraft by expanding gases in a rocket engine
- The force of the surrounding air on the propeller of an airplane

Measuring a Force

A simple way to measure force is to use the stretch or compression of a spring. In Figure 2.2 we hang a block from a spring, and note that the spring is stretched a distance s. Then we hang two such blocks from the spring, and we see that the spring is stretched twice as much. By experimentation, we find that any spring made of the same material and produced to the same specifications behaves in the same way.

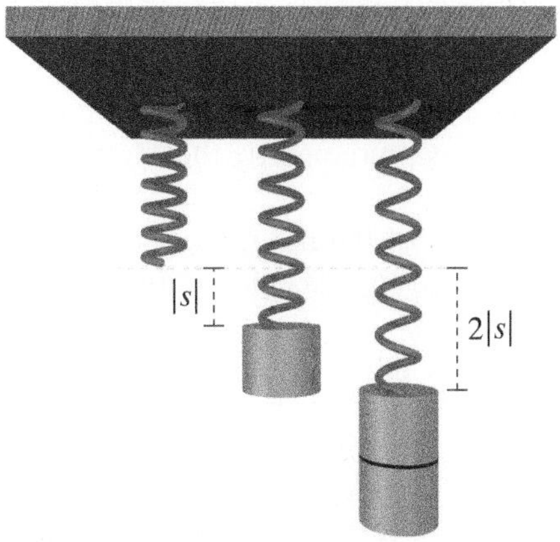

Figure 2.2 Stretching of a spring is a measure of force.

Similarly, we can observe how much the spring compresses when the same blocks are supported by it (Figure 2.3). We find that one block compresses the spring by the same distance $|s|$, and two blocks compress it by $2|s|$. (Compression can be considered negative stretch, because the length of the spring decreases.)

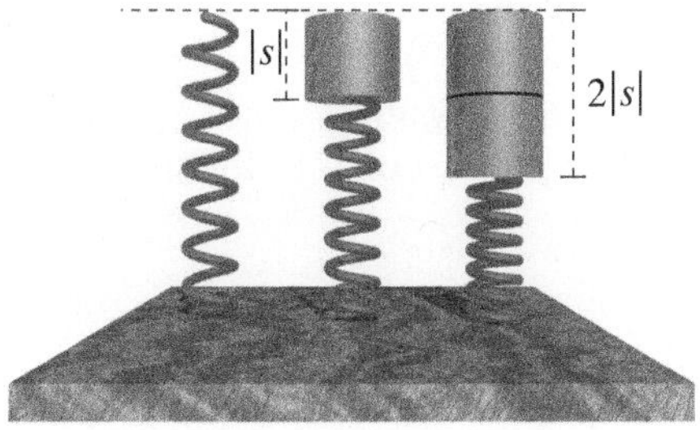

Figure 2.3 Compression of a spring is also a measure of force.

Unit of Force

We can use a spring to make a scale for measuring forces, calibrating it in terms of how much force is required to produce a given stretch. The SI unit of force is the newton, abbreviated as N. One newton is a rather small force. A newton is approximately the downward gravitational force of the Earth on a small apple, or about a quarter of a pound. If you hold a small apple at rest in your hand, you apply an upward force of about one newton, compensating for the downward pull of the Earth.

The Meaning of *net*

In physics the word *net* has a precise meaning: it is the sum of all contributions to a quantity, both pluses and minuses.

Because a force has a magnitude and a direction, it is a vector, so the *net* force on a system is the vector sum of all the forces exerted on the system by all the objects in the surroundings. It is the net force $\vec{F}_{\text{net}}$ on a system, acting for some time Δt, that causes a change of momentum.

$$\Delta\vec{p} = \vec{F}_{\text{net}}\Delta t$$

DEFINITION OF NET FORCE

$$\vec{F}_{\text{net}} \equiv \vec{F}_1 + \vec{F}_2 + \cdots$$

The "net" force acting on a system at an instant is the vector sum of *all* the forces exerted on the system by all the objects in the surroundings at that instant.

The net force acting on a system at an instant is the vector sum of all of the forces exerted on the system by all the objects in the surroundings, which are called "external" forces (Figure 2.4). There may be forces internal to the system, exerted by one object in the system on another object in the system, but such internal forces cannot change the momentum of the system. We will see in detail why this is so in Chapter 3, but the basic reason is that internal forces cancel each other's effects. A force that object 1 in a system exerts on object 2 in the system does change the momentum of object 2, but object 2 exerts a force in the opposite direction on object 1 that changes the momentum of object 1 in the opposite way, so that the changes in momenta of the two objects add up to zero.

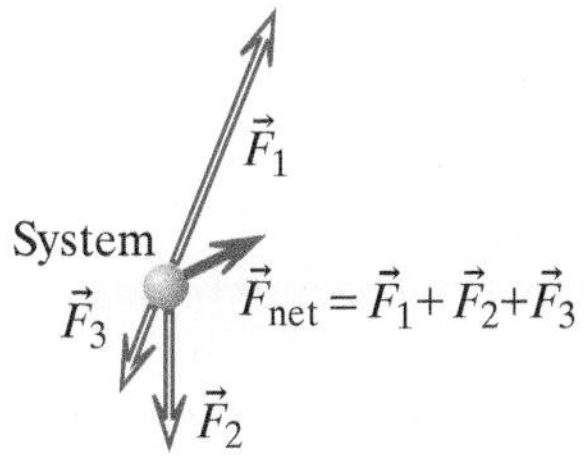

Figure 2.4 The net force on the system is the vector sum of all the individual forces acting on the system.

QUESTION A ball falling toward the Earth consists of many atoms. Each atom in the ball exerts forces on neighboring atoms in the ball, the Earth exerts forces on each atom in the ball, and the air exerts forces on atoms at the surface of the ball. Taking the ball as the system, and the Earth and air as the surroundings, which of these forces are external forces and which are internal forces?

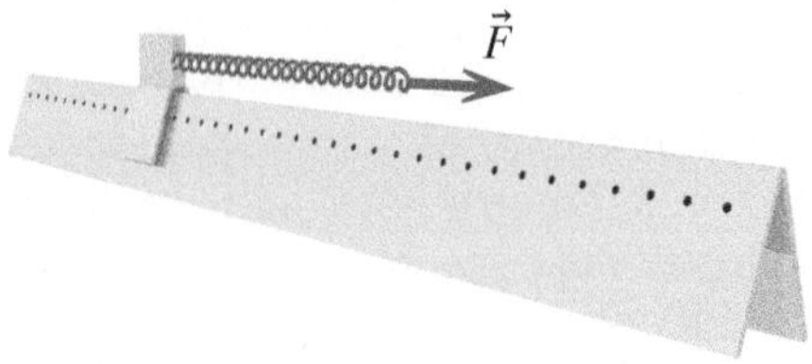

Figure 2.5 A block mounted on a nearly frictionless air track, pulled by a spring.

The Earth and the air are part of the surroundings, so the force exerted by the Earth and the force exerted by the air are both external forces. They are forces exerted on the chosen system (the ball) by objects in the surroundings (the Earth and the air). Forces that atoms in the ball exert on neighboring atoms in the ball are internal forces. The internal forces don't contribute to the net force $\vec{F}_{net}$ that changes the ball's momentum as predicted by the Momentum Principle, $\Delta\vec{p} = \vec{F}_{net}\,\Delta t$.

We find experimentally that the magnitude of the net force acting on an object affects the magnitude of the change in its momentum. Many introductory physics laboratories have air tracks like the one illustrated in Figure 2.5. The long triangular base has many small holes in it, and air under pressure is blown out through these holes. The air forms a cushion under the glider, allowing it to coast smoothly with very little friction.

Suppose that we place a block on a glider on a long air track, and attach a calibrated spring to it (Figure 2.5). We pull on the spring so that it is stretched a distance s, so that it exerts a force F on the block, and we pull for a short time. Choose the block and glider as the system, so that the net force on the system is just the spring, since friction with the track is very small due to the air cushion.

We observe that the momentum of the system of block and glider increases from zero to an amount mv. If instead we pull on the spring so that it stretches a distance $2s$, the spring exerts a force $2F$ on the system, and we observe that the system's momentum increases to $2mv$ in the same amount of time. Apparently the magnitude of the change in momentum is proportional to the magnitude of the net force applied to the object.

$$\Delta\vec{p} = \vec{F}_{net}\,\Delta t$$

Duration of Interaction Δt

Another experiment will show us that not only the magnitude of the force, but also the length of time during which it acts on an object, affects the change of momentum of the object. Suppose that we again place a block on an air track and attach a spring to it. We pull on the spring for a time Δt with a force F, and we observe that the momentum of the system of block plus glider increases from zero to an amount mv. However, if we repeat the experiment but pull for a time twice as long with the same force F, the system's momentum increases to $2mv$; the change in its momentum is twice as great. We observe that the magnitude of the change in momentum is directly proportional to the length of time $\Delta t = t_2 - t_1$ during which the force acts on the object.

$$\Delta\vec{p} = \vec{F}_{net}\,\Delta t$$

Impulse $\vec{F}_{net}\,\Delta t$

The amount of interaction affecting an object includes both the strength of the interaction (expressed as the net force $\vec{F}_{net}$) and the duration Δt of the interaction. Either a bigger force or applying the force for a longer time will cause more change of momentum. The product of a force and a time interval is called "impulse."

DEFINITION OF IMPULSE

$$\text{Impulse} = \vec{F}\Delta t$$

Impulse has units of N · s (newton-seconds)

The time interval Δt must be small enough that the net force is nearly constant during this time interval.

Now we can state the Momentum Principle this way:

The change of momentum of a system
is equal to the net impulse applied to the system.

QUESTION Are the units of impulse and momentum the same? If not, how are they related?

The units of impulse are N · s (a force of 1 N acting during a time of 1 second gives an impulse of 1 N · s). The units of momentum are kg · m/s, so evidently 1 N · s = 1 kg · m/s, and therefore 1 N = 1 kg · m/s^2.

EXAMPLE **Impulse Applied to a Baseball**

Just before it touches the bat, the momentum of a baseball is $\langle 0,0,3\rangle$ kg · m/s. After the batter has hit the ball, and the ball is no longer in contact with the bat, the momentum of the baseball is $\langle -2,2,-1\rangle$ kg · m/s. What was the impulse applied to the ball during its contact with the bat?

Solution

$$\Delta\vec{p} = \vec{F}_{\text{net}}\,\Delta t$$

$$\langle -2,2,-1\rangle \text{ kg}\cdot\text{m/s} - \langle 0,0,3\rangle \text{ kg}\cdot\text{m/s} = \vec{F}_{\text{net}}\,\Delta t$$

$$\langle -2,2,-4\rangle \text{ kg}\cdot\text{m/s} = \vec{F}_{\text{net}}\,\Delta t$$

Without more information we cannot find the values of $\vec{F}_{\text{net}}$ or Δt separately.

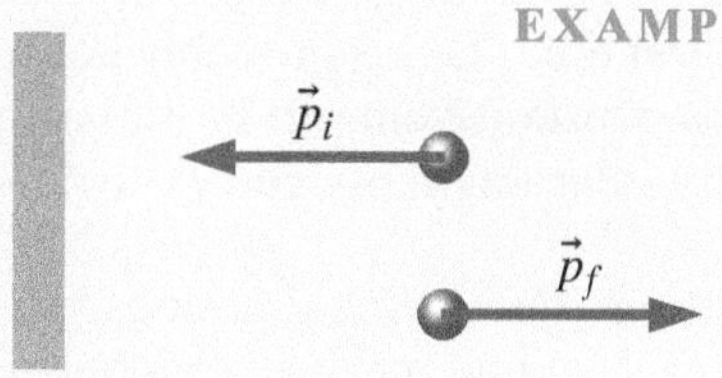

Figure 2.6 A ball hits a wall and rebounds.

EXAMPLE **Collision with a Wall**

A tennis ball traveling in the $-x$ direction with magnitude of momentum 2.8 kg · m/s hits a wall and rebounds (Figure 2.6). After the collision the magnitude of the ball's momentum is nearly the same, but the ball is traveling in the $+x$ direction. What was the impulse applied to the ball by the wall?

Solution Although the magnitude of the ball's momentum did not change significantly, its direction did change, so the impulse is not zero.

$$\Delta\vec{p} = \vec{F}_{\text{net}}\,\Delta t$$

$$(\langle 2.8,0,0\rangle \text{ kg}\cdot\text{m/s} - \langle -2.8,0,0\rangle \text{ kg}\cdot\text{m/s}) = \vec{F}_{\text{net}}\,\Delta t$$

$$\langle 5.6,0,0\rangle \text{ kg}\cdot\text{m/s} = \vec{F}_{\text{net}}\,\Delta t$$

Note that in this collision the magnitude of the change in the ball's momentum (which is equal to the magnitude of the impulse) was about twice as large as the ball's original momentum.

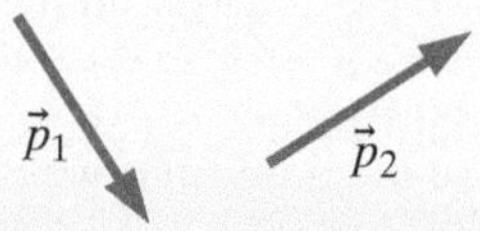

Figure 2.7 Initial and final momenta of a hockey puck. Both momenta lie in the xz plane; this is a top view, looking down on the ice from above.

EXAMPLE **Finding Direction of Impulse Graphically**

A hockey puck is sliding across the ice with momentum $\vec{p}_1$ (shown in Figure 2.7) when a hockey player hits it sharply with a hockey stick. Afterward, the puck has momentum $\vec{p}_2$. Both $\vec{p}_1$ and $\vec{p}_2$ lie in the xz plane. What was the direction of the impulse applied to the puck by the stick? What was the direction of the net force exerted on the puck during its short contact with the hockey stick?

Solution We can find the direction of $\Delta\vec{p}$ graphically, as shown in Figure 2.8. From the vector equation

$$\Delta\vec{p} = \vec{F}_{\text{net}}\,\Delta t$$

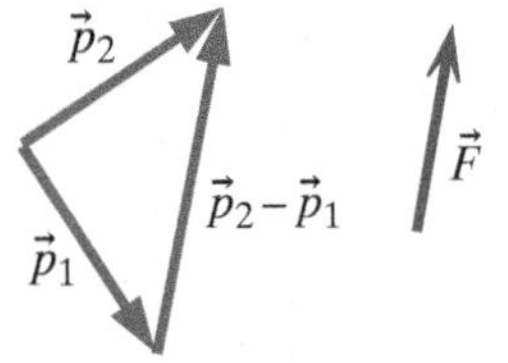

Figure 2.8 To find $\Delta\vec{p}$, the vectors $\vec{p}_1$ and $\vec{p}_2$ are placed tail to tail. The net force on the puck was in the direction of the observed change in momentum $\Delta\vec{p}$.

we see that $\Delta\vec{p}$ and $\vec{F}_{net}\,\Delta t$ must have the same directions. Since Δt is a positive scalar, the direction of the net force is the same as the direction of the impulse, because multiplying the vector $\vec{F}_{net}$ by the positive scalar Δt can't change the direction of the vector.

Checkpoint 1 (1) Two external forces, $\langle 40,-70,0\rangle$ N and $\langle 20,10,0\rangle$ N, act on a system. What is the net force acting on the system? (2) A hockey puck initially has momentum $\langle 0,2,0\rangle$ kg·m/s. It slides along the ice, gradually slowing down, until it comes to a stop. **(a)** What was the impulse applied by the ice and the air to the hockey puck? **(b)** It took 3 seconds for the puck to come to a stop. During this time interval, what was the net force on the puck by the ice and the air (assuming that this force was constant)?

2.2 LARGE FORCES AND SHORT TIMES

The Momentum Principle can be a powerful tool for analyzing real events. We'll consider two situations in which there is a sudden impact: a very large force, applied to a system for a very short time, that produces a dramatic change in momentum. In one case we will deduce the approximate magnitude of the force, and in the other case we will infer the duration of the interaction. In both cases we will need to make the simplifying assumption that the net force on the system is constant during the short interaction time. This is probably not really true in either case, but by considering this constant value to be the average force exerted on the system during the short interaction, we can extract useful information about the impact.

By making these simplifying assumptions and approximations, we are making a "model" of each physical system. Making and using models is an activity central to physics, and the criteria for a "good" physical model will depend on how we intend to use the model. In fact, one of the most important problems a scientist or engineer faces is deciding what interactions must be included in a model of a real physical, chemical, or biological system, and what interactions can reasonably be ignored.

EXAMPLE

Colliding Students

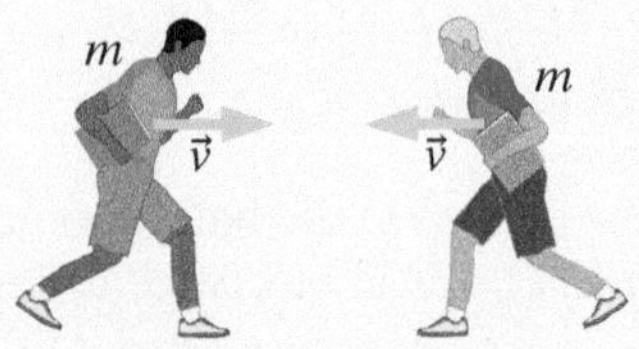

Figure 2.9 Two students running fast collide head-on.

Two students who are late for tests are running to classes in opposite directions as fast as they can. They turn a corner, run into each other head-on, and come to a complete stop (Figure 2.9). Using physics principles, estimate the force that one student exerts on the other during the collision. You will need to estimate some quantities; give reasons for your choices and provide checks showing that your estimates are physically reasonable.

This problem is rather ill-defined and doesn't seem much like a "textbook" problem. No numbers have been given, yet you're asked to estimate the force of the collision.

QUESTION How can our answer possibly be meaningful if we have to estimate quantities?

Since we have no idea how big such a force might be, even an approximate answer will be informative. This kind of problem is typical of the kinds of problems engineers and scientists encounter in their professional work.

To begin, we'll make a diagram of the forces due to objects in the environment acting on a system consisting of the student on the left. Such a diagram is called a "free-body diagram." If we are modeling the system as a point particle, we usually represent the system as a dot.

Solution

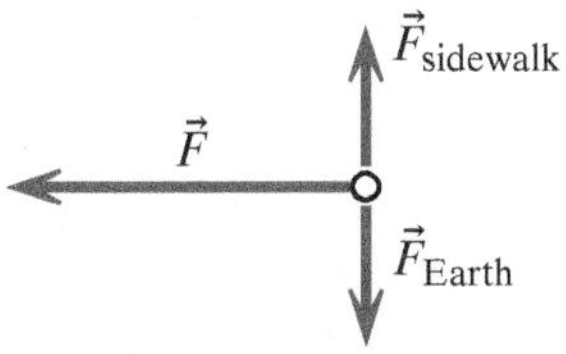

Figure 2.10 Free-body diagram showing forces on the left student. Forces that we assume to be negligible have been omitted.

System: The student on the left (we could choose either student)
Surroundings: Other student, Earth, sidewalk, air
Free-body diagram: Figure 2.10
Initial time: Just before impact
Final time: When speeds become zero
Simplifying assumptions:

- The students have the same mass m and same initial speed $|\vec{v}_i|$.
- Friction and air resistance are negligible during this short Δt.

> QUESTION We don't know the upward force exerted on the student by the sidewalk, so how can we solve this problem?

We know that the y component of the student's momentum does not change during the collision. From this we can deduce that the y component of the net force on the student is zero, and therefore the downward gravitational force exerted by the Earth and the upward force exerted by the sidewalk are equal in magnitude: $\vec{F}_{\text{sidewalk}} + \vec{F}_{\text{Earth}} = \vec{0}$. Therefore, the only force we need to consider is the unknown force $\vec{F}$ exerted by the other student.

Applying the Momentum Principle to the system of the student on the left, we get:

$$\Delta\vec{p} = \vec{F}_{\text{net}}\,\Delta t$$
$$(\langle 0,0,0\rangle - \langle mv_i,0,0\rangle) = \vec{F}_{\text{net}}\,\Delta t$$

To go further we must estimate some quantities. We need the initial momentum of the student on the left, $\langle mv_i,0,0\rangle$, and we need the duration of the collision, Δt. Let's choose these values:

$$m \approx 60\text{ kg (1 kg is about 2.2 lb)}$$
$$|\vec{v}_i| \approx 6\text{ m/s (elite sprinters run 100 m in about 10 s)}$$
$$\Delta t \approx ???$$

We know the collision occurs during a short time, but how short? One way to estimate a time is to estimate the distance traveled during the collision. In this case, this distance is the amount by which the student's body is compressed by the impact. Let's guess that the student's body is compressed by about 5 cm during the collision, which will be painful. During this collision, the student's speed drops from 6 m/s to 0, so we'll estimate the average speed as the "arithmetic average" $(6\,\text{m/s} + 0\,\text{m/s})/2 = 3\,\text{m/s}$ (more about this later in the chapter).

All motion is in the x direction, so:

$$\Delta x = v_{\text{avg},x}\,\Delta t$$
$$0.05\text{ m} \approx 3\text{ m/s}\,\Delta t$$
$$\Delta t \approx 0.017\text{ s}$$

The whole interaction lasts only about 0.02 s! The force exerted by one student on the other must be very large. Let's see about how large it is:

$$(\langle 0,0,0\rangle\,\text{kg}\cdot\text{m/s} - \langle mv_i,0,0\rangle\,\text{kg}\cdot\text{m/s}) = \vec{F}_{\text{net}}\,\Delta t$$
$$(\langle 0,0,0\rangle\,\text{kg}\cdot\text{m/s} - \langle 60\cdot 6,0,0\rangle\,\text{kg}\cdot\text{m/s}) \approx \vec{F}_{\text{net}} \times 0.017\,\text{s}$$
$$\vec{F}_{\text{net}} \approx \langle -21100,0,0\rangle\text{ N}$$

Check:

- The units check ($\text{N}\cdot\text{s} = \text{kg}\cdot\text{m/s}$, so $(\text{kg}\cdot\text{m/s})/\text{s} = \text{N}$).
- Is the result reasonable?

21,000 N is a very large force. As we will see in Section 2.4, the gravitational force on a 60 kg student (the "weight") is only about 600 N. The force of the impact is about 35 times the weight of the student! It's like having a stack of 35 students sitting on you. If the students hit heads instead of stomachs, the squeeze might be less than 1 cm, and the force would be over 5 times as large! This is why heads can break in such a collision. Our result of 21,000 N shows why collisions are so dangerous. Collisions involve very large forces acting for very short times, giving impulses of ordinary magnitude.

QUESTION Estimating masses or distances is usually not difficult, but most people find it more challenging to estimate time durations, especially if the times are very short. Why would a guess of 1 second for Δt have been a poor guess?

At an average speed of 3 m/s, each student would have traveled 3 meters during a one-second impact. This is clearly not physically possible—they would have had to pass through each other to do this!

QUESTION We estimated various quantities to get this result. Is the result still useful?

A useful model should omit extraneous detail but retain the most important features of the real-world situation. When we make many estimates, our goal is often to determine the order of magnitude of the answer: is the force on a colliding student closer to 0.0001 newton or 10,000 newtons? Knowing the order of magnitude to expect in an answer can be critically important in solving a real problem, designing an experiment, or designing a crash helmet. The designer of a crash helmet needs an approximate value for the maximum force it must withstand. Actual collisions will vary, so there is no "right" or simple answer.

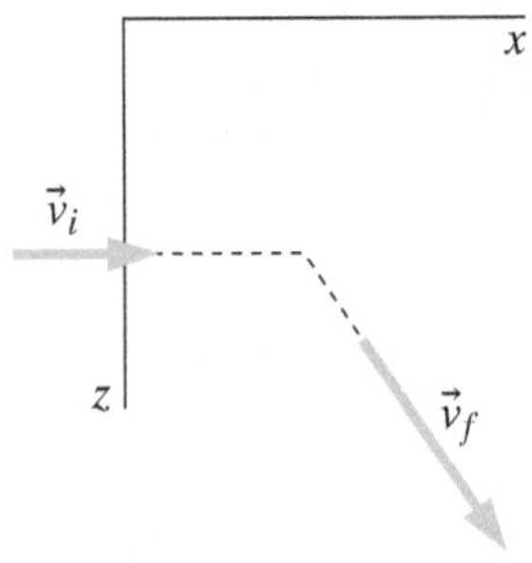

Figure 2.11 Top view of the path of a hockey puck that is hit by a hockey stick. (Assuming our usual coordinate system with y up, the ice surface lies in the xz plane.) The impact occurs at the location where the trajectory changes direction.

QUESTION What simplifications did we make in our model of the colliding students?

Our model of the colliding student situation is good enough for many purposes. However, we left out some aspects of the actual motion. For example, we mostly ignored the flexible structure of the students, how much their shoes slip on the ground during the collision, and so on. We have also quite sensibly neglected the gravitational force of Mars on the students, because it is so tiny compared to the force of one student on the other.

EXAMPLE **Striking a Hockey Puck**

In a video of a hockey game, we see a 0.16 kg hockey puck sliding across the ice, as shown in Figure 2.11. A player hits the puck sharply with a hockey stick; the stick breaks, and the direction of motion of the puck changes. From examining the video frame by frame, we find that before it was hit the velocity of the puck was $\langle 20,0,0\rangle$ m/s, and after the impact the velocity of the puck was $\langle 20,0,31.7\rangle$ m/s. When we pile weights on the side of a similar hockey stick we find that the stick breaks under a force of about 1000 N (this is roughly 225 lb). How long was the puck in contact with the hockey stick? Explain what assumptions were made in the solution.

Solution This problem can't be solved simply by plugging numbers into an equation. Instead, as will often be the case, we need to start from a fundamental principle, and work forward until the solution emerges. We will start by writing down the Momentum Principle, which relates change in momentum of a system to net impulse applied. If we can find both the change in momentum of the puck and the net force applied to it, we will be able to deduce the contact time Δt.

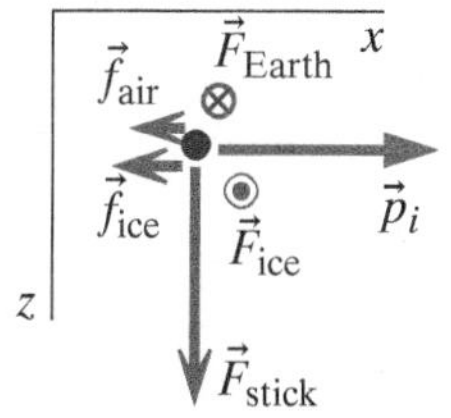

Figure 2.12 Free-body diagram for the hockey puck during the impact. The viewpoint is above the ice, looking down at the xz plane.

System: The hockey puck
Surroundings: Earth, ice, hockey stick, air (free-body diagram: Figure 2.12)
Initial time: When stick first makes contact with puck
Final time: When stick breaks

$$\vec{p}_f - \vec{p}_i = \vec{F}_{net}\,\Delta t$$

$$0.16\,\text{kg}\cdot\langle 20,0,31.7\rangle\ \text{m/s} - 0.16\,\text{kg}\cdot\langle 20,0,0\rangle\ \text{m/s} = \vec{F}_{net}\,\Delta t$$

$$\langle 0,0,5.072\rangle\ \text{m/s} = \vec{F}_{net}\,\Delta t$$

Only the z component of the puck's momentum changed, so the net impulse must have been in the z direction. (This implies that the downward gravitational force by the Earth and the upward force by the ice are equal in magnitude and opposite in direction. In Chapter 4 we will see how the solid ice can exert an upward force.) Assuming that the forces by the air and the ice are small compared to the force on the puck by the hockey stick, we can now simplify the free-body diagram, as shown in Figure 2.13.

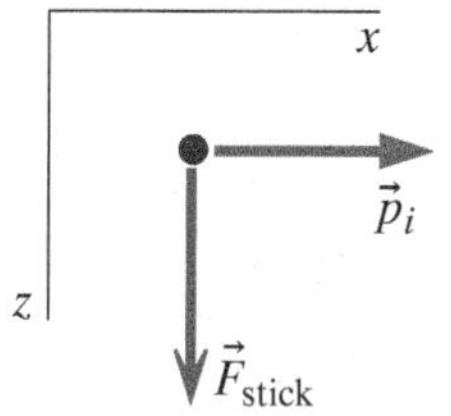

Figure 2.13 Simplified free-body diagram, taking into account assumptions and deductions. The viewpoint is above the ice, looking down at the xz plane.

The direction of the force is the unit vector $\hat{F}$:

$$\hat{F} = \langle 0,0,1\rangle$$

Because the stick broke, the magnitude of the force must have been about 1000 N. Therefore

$$\vec{F}_{net} = \left|\vec{F}\right|\hat{F} = \langle 0,0,1000\rangle\ \text{N}$$

From the Momentum Principle:

$$\langle 0,0,5.072\rangle\ \text{m/s} = \langle 0,0,1000\rangle\ \text{N}\ \Delta t$$

QUESTION Since we can't divide by a vector, how do we solve for Δt?

Recall that a vector equation can be written as three separate equations, one for each component. To solve for Δt we must consider the z component separately:

$$p_{fz} - p_{iz} = F_{net,z}\,\Delta t$$

$$\Delta t = \frac{p_{fz} - p_{iz}}{F_{net,z}} = \frac{5.072\ \text{kg}\cdot\text{m/s}}{1000\ \text{N}}$$

$$\Delta t = 0.005\ \text{s}$$

We assumed that:

- The force by the hockey stick on the puck was constant during the contact time.
- The friction force due to the ice surface was negligible during the contact time.
- Air resistance was negligible during the contact time.

QUESTION How good were our assumptions?

The neglect of sliding friction and air resistance is probably pretty good, since a hockey puck slides for long distances on ice with nearly constant speed.

We know the hockey stick exerts a maximum force of $F_{stick} = 1000$ N, because we observe that the stick breaks. We approximate the force as nearly constant during contact. Actually, this force grows quickly from zero at first contact to 1000 N, then abruptly drops to zero when the stick breaks.

Our analysis of the stick contact time (0.005 s) isn't exact. However, it is adequate to get a reasonably good determination of the short interaction time, something that we wouldn't know without using the Momentum Principle.

QUESTION How can we decide if our result is reasonable? Could the actual contact time have been much longer?

The very short duration of the impact is consistent with what we hear: a sharp, short crack. For further verification, let's consider how far the puck would move during contacts of differing times.

During 0.005 s the puck moves (20 m/s)(0.005 s) = 0.1 m in the x direction, sliding along the blade of the stick. Also during this time v_z increases from 0 to 31.7 m/s, with an average value of about 15.8 m/s, so the z displacement is about (15.8 m/s)(0.005 s) = 0.08 m during contact. A more accurate sketch of the path of the puck might show a bend as in Figure 2.14.

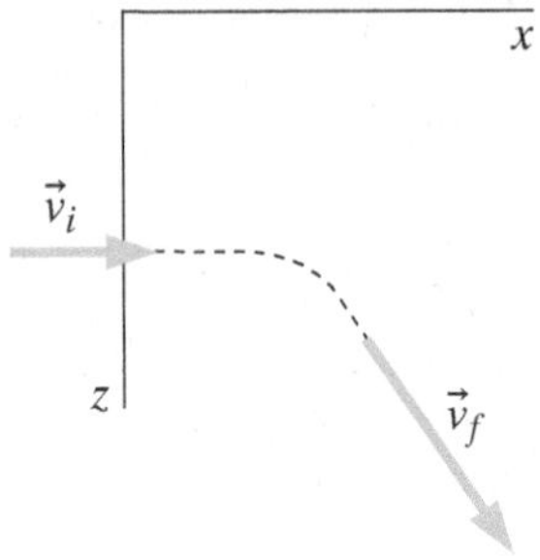

Figure 2.14 A more accurate overhead view of the path of the hockey puck, showing the bend during impact.

QUESTION Suppose we were to guess that the contact time between the stick and the puck was about one second. We know that the original speed of the puck was 20 m/s. At this speed, how far would the puck have traveled in 1 s?

$$(20 \text{ m/s})(1 \text{ s}) = 20 \text{ m}$$

It is clear that the puck did not move 20 m during the impact. We must conclude that the impact took much less than 1 s.

QUESTION If we guess that the puck may have slid 20 cm along the stick during the impact, what would the contact time be?

$$\Delta t \approx \frac{(0.20 \text{ m})}{(20 \text{ m/s})} = 0.01 \text{ s}$$

This estimate of the contact time differs from our result by only a factor of around 2, instead of a factor of 100, so it would be a much better estimate. Using an object's speed and estimating the distance traveled during an interval allows us to come up with a much better estimate of interaction times than we get simply by guessing.

Idealized Models

In the previous two examples, we made "idealized" models of real physical situations. By idealized models, we mean models that involve simple, clean, stripped-down situations, free of messy complexities (Figure 2.15). Ideally, a hockey puck will slide forever on a smooth ice surface, but a real puck sliding on real ice eventually comes to a stop. An ideal gas is a fictitious gas in which the molecules don't interact at all with each other. In contrast, the molecules of a real gas do interact, but only when they come close to each other. A model of a single Earth orbiting a Sun is an idealized model because it leaves out the effects of the other bodies in the Solar System.

The behavior of idealized models allows us to investigate simple patterns of motion, and learn what factors are important in determining these patterns. Once we understand these factors, we can revise and extend our models, including more interactions and complexities, to see what effects these have.

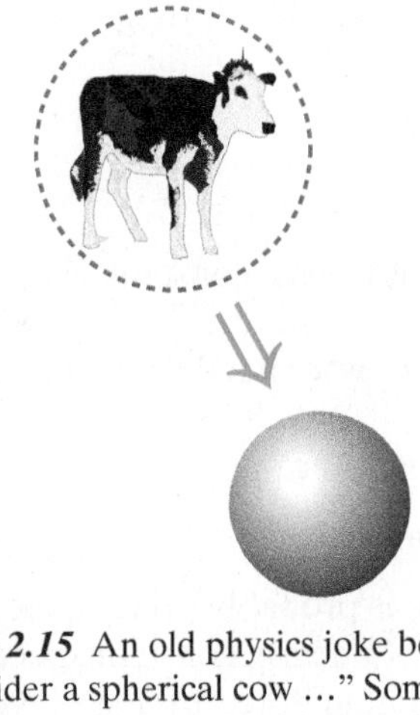

Figure 2.15 An old physics joke begins, "Consider a spherical cow ..." Sometimes this degree of idealization is actually appropriate.

An important aspect of physical modeling is that we make appropriate approximations and assumptions to simplify the messy, real-world situation enough to permit us to apply fundamental principles such as the Momentum Principle, the Energy Principle (see Chapter 6), and the Angular Momentum Principle (see Chapter 11).

Checkpoint 2 **(a)** In the colliding students example, how was Δt, the duration of the collision, estimated? **(b)** In the hockey puck example, why were only the z components of $\Delta\vec{p}$ and $\vec{F}_{\text{net}}$ used to get Δt?

2.3 PREDICTING THE FUTURE

Since $\Delta\vec{p} = \vec{p}_f - \vec{p}_i$ ("final minus initial"), we can rearrange the Momentum Principle algebraically to the update form:

THE MOMENTUM PRINCIPLE (UPDATE FORM)

$$\vec{p}_f = \vec{p}_i + \vec{F}_{\text{net}}\,\Delta t$$

or, choosing "now" as the initial time, and "a short time in the future" as the final time, we can write:

$$\vec{p}_{\text{future}} = \vec{p}_{\text{now}} + \vec{F}_{\text{net,now}}\,\Delta t$$

The time interval Δt must be small enough that the net force is nearly constant during this time interval.

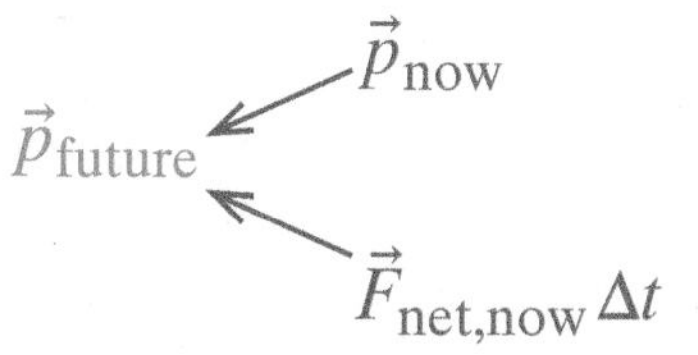

Figure 2.16 The future momentum of a system depends on two things: its momentum now and the net force acting on it now.

This way of writing the Momentum Principle emphasizes the important idea that the momentum that an object will have a short time in the future depends on *two* things: its momentum right now and the total force acting on the object right now. Figure 2.16 summarizes this graphically.

Even without numerical values, we can use the update form of the Momentum Principle to reason qualitatively, but rigorously, about physical situations.

EXAMPLE

A Skydiver

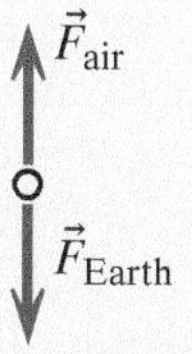

Figure 2.17 At a particular instant the upward force of the air becomes equal in magnitude to the downward force of the Earth on a skydiver. At this instant the net force on the skydiver becomes zero.

A skydiver jumps out of an airplane and her parachute opens. For a while she falls faster and faster. Then at a particular instant, which we will choose to be *now*, the upward air resistance force acting on her and her parachute (this force increases as her speed increases) gets large enough to be exactly equal in magnitude and opposite in direction to the downward gravitational force due to the Earth (Figure 2.17). Therefore, at this instant the net force on the skydiver becomes zero. Which of the following answers correctly predicts the skydiver's motion a short time in the future from now? **(a)** She will float without moving up or down. **(b)** She will fall faster and faster. **(c)** She will fall at a constant speed.

Solution

Even though this problem does not involve numbers, we must use the Momentum Principle to reason formally about the future momentum of the skydiver. This principle tells us that $\vec{p}_{\text{future}}$ depends on two things: her current momentum $\vec{p}_{\text{now}}$, and the net impulse $\vec{F}_{\text{net,now}}\Delta t$ acting on her during the next short time interval, as indicated in Figure 2.18. The net impulse will be zero because the net force on her is zero. Therefore, her momentum a short time in the future will be exactly equal to her momentum right now. Because her momentum does not change, her speed does not change, and she will fall at constant speed. Answer (c) is correct.

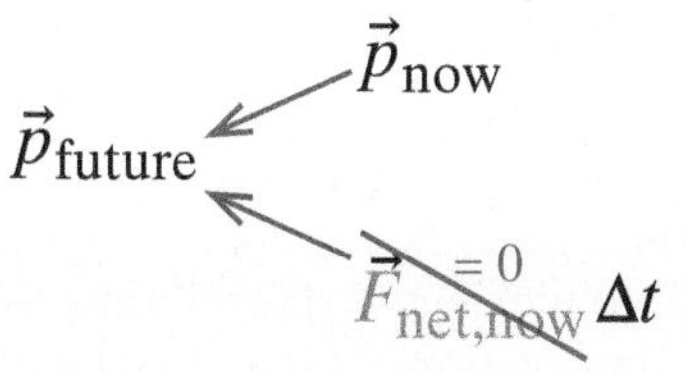

Figure 2.18 At the moment of interest, the momentum of the skydiver now is the only nonzero contribution to the future momentum of the skydiver.

Almost everyone who is learning physics occasionally makes mistakes when thinking about problems like this one. Many of these mistakes are due to forgetting about one of the two things that determine an object's future momentum. For example, if you picked answer (a) above, perhaps you considered only the net force acting on the skydiver, and forgot to take her current momentum into account. You can avoid many mistakes by starting from the Momentum Principle, and by making sure you have taken both current momentum and net impulse into account.

How Do We Take the Past into Account?

The momentum of an object at any particular instant depends on its history—the interactions it underwent in the past. However, the Momentum Principle refers only to the momentum of an object *now* and in the *future*—it does not mention the past.

For example, suppose you hit a ball with a tennis racket. At a time one-tenth of a second after the ball has left the racket, it is no longer interacting with the racket, so at that moment the net force on the ball includes forces by the Earth and by the air, but not by the racket. This may seem counterintuitive: the fact that the racket interacted with the ball in the past is important!

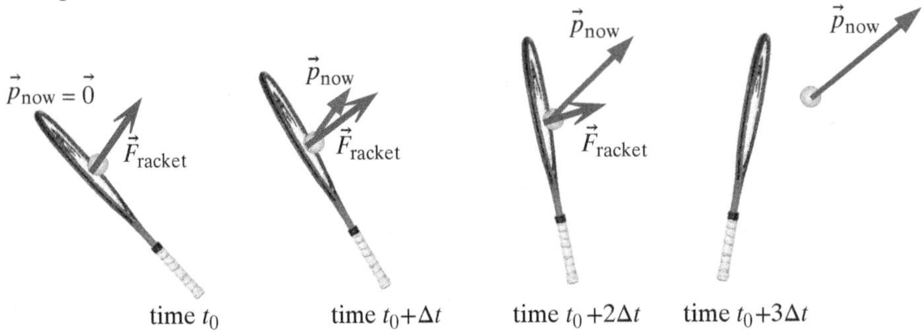

Figure 2.19 Several steps in the process of hitting a tennis ball. At each time, the momentum of the ball $\vec{p}_{now}$ includes the effects of all prior impulses delivered by the racket. Only the force by the racket is shown; forces on the ball by the Earth and the air are not shown.

If we reflect for a moment, though, we see that the effects of all past interactions have already been added up into the object's momentum at the current instant, $\vec{p}_{now}$. Imagine what happened a few seconds ago (Figure 2.19). Suppose that you tossed the ball straight up in the air, and that your racket first made contact with the ball at time t_0 when the ball had reached the top of its trajectory and was momentarily motionless. At that time (the first image in Figure 2.19) the ball's momentum was zero, but because at that instant the racket exerted a force on the ball, after a fraction of a second at time $t_0 + \Delta t$ (the second image in Figure 2.19) the ball's momentum was zero plus the net impulse—a nonzero quantity. Considering the next time step, we see that the ball's initial momentum was now not zero, and it increased even more because of the impulse delivered by your racket in the next small time step. This continued, step by step, until the ball lost contact with your racket. At that instant, the momentum of the ball was the sum of all the impulses delivered by the racket during the time it was in contact with the ball, and the racket no longer exerted a force on the ball.

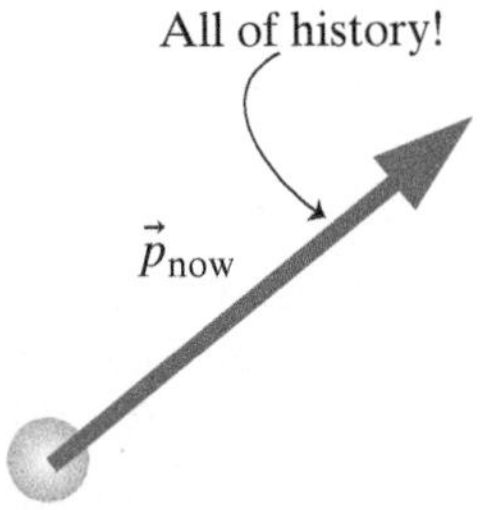

Figure 2.20 The entire history of an object—all impulses that have ever been applied to it—has been added up into the current momentum of the object.

The fact that all of the past history of an object and its interactions has been incorporated into its current momentum, $\vec{p}_{now}$, is extremely important (Figure 2.20). Because this is true, we are free to pick any instant of time as the initial time of interest, and consider only the interactions of the object with its surroundings at this instant, in order to predict its future momentum. As long as we know an object's current momentum, we do not need to know anything about how the object came to be moving this way. If we had to know all of the details of all of an object's past interactions, predicting the future motion of an object would almost never be possible!

In the following section we will work through a process like this quantitatively, dividing a time interval into very small steps and adding up all the impulses exerted on an object, to predict its motion over an extended time period.

Checkpoint 3 (1) You drop a piece of paper, and observe that it eventually falls at a constant speed. Which of the following statements about this situation is based on a fundamental physics principle? **(a)** Because the paper is moving downward, we know that it experiences a nonzero net downward force. **(b)** Since the momentum of the paper does not change from one instant to the next, $\vec{p}_{\text{future}} = \vec{p}_{\text{now}}$, and therefore the net force on the paper must be zero. (2) You give a push to a toy car, which rolls away smoothly on a wooden floor. Why does the car keep moving after your hand is no longer touching it? **(a)** The momentum of the car just after it leaves your hand reflects the total impulse given to the car by your hand. **(b)** Your hand continues to exert a force on the car even after the car and hand no longer touch.

2.4 ITERATIVE PREDICTION: CONSTANT NET FORCE

The Momentum Principle states that if the momentum of a system changes, a nonzero net force must have been applied to the system. The simplest case of a nonzero net force is that of a constant force: a force whose magnitude and direction do not change even though the position and momentum of the system change. In the real world, there are very few situations in which the net force on a system is truly constant. However, this can be a useful approximate model for some real phenomena. One example of an approximately constant force is the gravitational force exerted by the Earth on an object near the Earth's surface.

Approximate Gravitational Force

In general, the gravitational interaction between two objects depends on the distance between the objects, and its magnitude and direction vary as the positions of the objects vary. In Chapter 3 we will discuss the general form of the force law that describes the gravitational attraction that all objects in the Universe have for each other. However, anywhere near the Earth's surface it is approximately true that a 1 kg object is attracted to the Earth by a gravitational force of magnitude 9.8 N, and a 2 kg object is attracted by a force of magnitude $2 \times 9.8\,\text{N}$ (Figure 2.21). We define $g = +9.8\,\text{N/kg}$ to be the strength of this interaction in newtons per kilogram, so we can say that a mass m near the surface of the Earth is pulled toward the Earth by a force whose magnitude is mg. We can treat this force as approximately constant in magnitude and direction, even if the object moves.

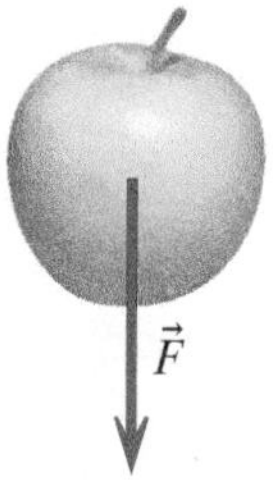

Figure 2.21 The magnitude of the gravitational force by the Earth on a small apple is about 1 newton.

APPROXIMATE GRAVITATIONAL FORCE

$$|\vec{F}_{\text{grav}}| \approx mg$$

where $g = +9.8\,\text{N/kg}$

for an object of mass m near the surface of the Earth.

Iterative Prediction of Motion

The most general method for predicting the motion of an object that interacts with its surroundings is to use an iterative approach. To "iterate" means "to repeat" or "to perform again"; the word "iteration" implies repeating the same process many times. The basic idea we will use in predicting the motion of interacting systems is to divide up the time interval of interest into many very small time intervals, and iteratively apply the Momentum Principle over successive intervals. The time intervals must be small enough that neither the velocity of the object nor the net force acting on it changes significantly in one time step. We will update momentum and position iteratively like this:

ITERATIVE PREDICTION OF MOTION

Repeat:

- Calculate the net force $\vec{F}_{\text{net,now}}$ acting on the system.
- Update momentum: $\vec{p}_{\text{future}} = \vec{p}_{\text{now}} + \vec{F}_{\text{net,now}}\Delta t$.
- Update position: $\vec{r}_{\text{future}} = \vec{r}_{\text{now}} + \vec{v}_{\text{avg}}\,\Delta t$.

In the final step we use the approximation $\vec{v}_{\text{avg}} \approx \vec{p}_{\text{future}}/m$.

This general procedure can be used to predict the entire path along which an object will travel. As we will see in Section 2.6, it can be used even if the net force is not constant. The order of the steps turns out to be important; to get good accuracy it is necessary to update the momentum of the system before updating its position.

EXAMPLE **A Ball Without Air Resistance**

Let's apply this procedure to calculate the trajectory of a ball moving without air resistance. It is necessary to neglect air resistance in order to assume a constant force, because the air resistance force depends on the velocity of the object. At low speeds, this can be a reasonable approximation.

Solution In outline, here is the calculation we will do. We need two initial values:

- The initial position of the ball
- The initial momentum of the ball

We will do the following calculation iteratively (repeatedly):

- Find the net force $\vec{F}_{\text{net,now}}$ on the ball.
- Apply the Momentum Principle to find $\vec{p}_{\text{future}}$.
- Use $(\vec{p}_{\text{future}}/m)$ to approximate the average velocity, and use this to find the new position of the ball.

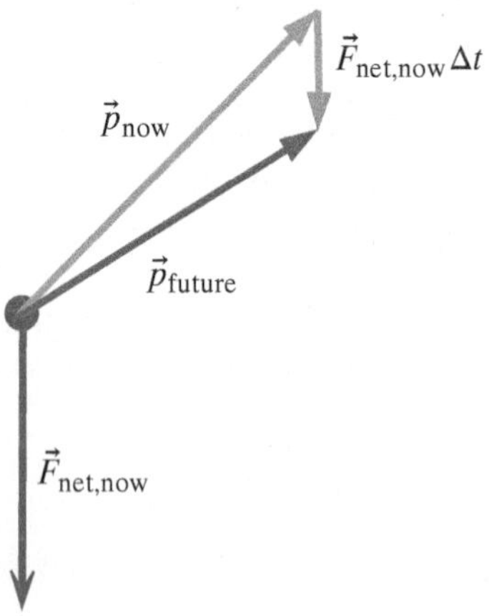

Figure 2.22 Time step 1 in the calculation of the motion of a ball without air resistance.

We'll consider a tennis ball of mass 0.06 kg that is hit by a racket so its initial velocity just after losing contact with the racket is $\langle 10,10,0\rangle$ m/s, so its initial momentum is $\langle 0.6,0.6,0\rangle$ kg · m/s. The initial position of the ball is $\langle 0,2,0\rangle$ m. We'll use a time step of 0.4 s. In the absence of air resistance, the net force on the ball is just the downward gravitational force exerted by the Earth. The result of the first step in our calculation is shown graphically in Figure 2.22, and the actual calculations are shown below.

Time Step 1:

$$\begin{aligned}
\vec{F}_{\text{net,now}} &= \langle 0,-9.8\times 0.06,0\rangle\ \text{N/kg} = \langle 0,-0.588,0\rangle\ \text{N}\\
\vec{p}_{\text{future}} &= \langle 0.6,0.6,0\rangle\ \text{kg}\cdot\text{m/s} + \langle 0,-0.588,0\rangle\ \text{N}\times 0.4\,\text{s}\\
&= \langle 0.6,0.365,0\rangle\ \text{kg}\cdot\text{m/s}\\
\vec{r}_{\text{future}} &= \langle 0,2,0\rangle\ \text{m} + \left(\frac{\langle 0.6,0.365,0\rangle\ \text{kg}\cdot\text{m/s}}{0.06\,\text{kg}}\right)\times 0.4\,\text{s}\\
&= \langle 4,4.32,0\rangle\ \text{m}
\end{aligned}$$

QUESTION The y component of the ball's momentum changed, but the x and z components did not. Why not?

The y component of the net force is nonzero, but the other components are zero. The y component of a force can only affect the y component of an object's momentum. Therefore the x and z components of the ball's momentum (and hence the x and z components of its velocity) are not changed by the gravitational force.

> **QUESTION** Why did we use $\vec{p}_{\text{future}}$ to obtain a value for the average velocity of the ball, instead of an average of $\vec{p}_{\text{future}}$ and $\vec{p}_{\text{now}}$?

We will see later that although in the special case of a constant force an average of $\vec{p}_{\text{future}}$ and $\vec{p}_{\text{now}}$ would provide a more accurate estimate of $\vec{v}_{\text{avg}}$, in the general case of a changing force this is often not true. As long as we take small enough time steps, so that the changes in momentum are small, $\vec{p}_{\text{future}}/m$ provides an adequate estimate of $\vec{v}_{\text{avg}}$.

> **QUESTION** What value should we use for $\vec{p}_{\text{now}}$ in the second time step?

To take the next time step, we advance the "clock." If our initial time was t_0, the time is now $t_0 + \Delta t$. The value of $\vec{p}_{\text{now}}$ at this time is the value we previously called $\vec{p}_{\text{future}}$: $\langle 0.6, 0.365, 0\rangle$ kg · m/s, and the value of $\vec{r}_{\text{now}}$ at this time is the value we previously called $\vec{r}_{\text{future}}$: $\langle 4, 4.32, 0\rangle$ m.

Time Step 2:

$$\begin{aligned}
\vec{F}_{\text{net,now}} &= \langle 0, -9.8 \times 0.06, 0\rangle \text{ N/kg} = \langle 0, -0.588, 0\rangle \text{ N} \\
\vec{p}_{\text{future}} &= \langle 0.6, 0.365, 0\rangle \text{ kg}\cdot\text{m/s} + \langle 0, -0.588, 0\rangle \text{ N} \times 0.4 \text{ s} \\
&= \langle 0.6, 0.130, 0\rangle \text{ kg}\cdot\text{m/s} \\
\vec{r}_{\text{future}} &= \langle 4, 4.32, 0\rangle \text{ m} + \left(\frac{\langle 0.6, 0.365, 0\rangle \text{ kg}\cdot\text{m/s}}{0.06 \text{ kg}}\right) \times 0.4 \text{ s} \\
&= \langle 8, 5.30, 0\rangle \text{ m}
\end{aligned}$$

Figure 2.23 displays the results of four steps in the calculation of the path of the ball. Even with only four steps the trajectory looks roughly like what is observed in the real world. If we used smaller time steps, the straight line segments representing displacement would be shorter, and the trajectory would look more and more like the actual curving path of the ball.

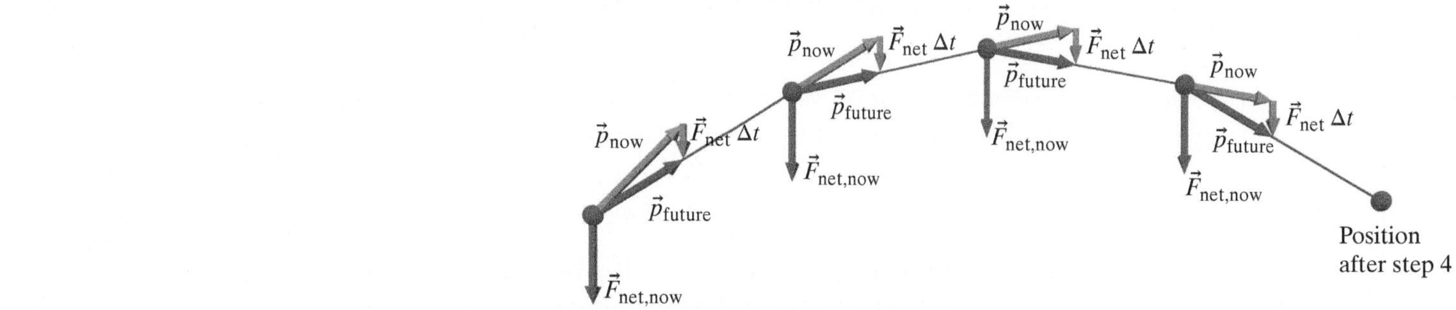

Figure 2.23 Four steps in the iterative prediction of the path of a ball in the absence of air resistance. In each step the future momentum of the ball is calculated, then this new momentum is used to update the position of the ball.

Checkpoint 4 (1) For the third time step in the iterative solution above, what value should you use for $\vec{p}_{\text{now}}$? **(a)** $\langle 0.6, 0.6, 0\rangle$ kg·m/s **(b)** $\langle 0.6, 0.365, 0\rangle$ kg · m/s **(c)** $\langle 0.6, 0.130, 0\rangle$ kg · m/s **(d)** A value that has not yet been calculated. (2) What value should you use for $\vec{r}_{\text{now}}$? **(a)** $\langle 0, 2, 0\rangle$ m **(b)** $\langle 4, 4.32, 0\rangle$ m **(c)** $\langle 8, 5.30, 0\rangle$ m **(d)** A value that has not yet been calculated.

2.5 ANALYTICAL PREDICTION: CONSTANT NET FORCE

In contrast to an iterative solution, an analytical solution to a problem yields a mathematical function that gives the value of a quantity of interest as a function of time. If such a function exists, one can use it to find the position or velocity of a system at a given time in one step, instead of taking many small steps to get this information. It is not always possible to get an analytical solution, so iterative methods are of more general applicability. One case in which an iterative solution is possible is the case of constant net force. We will derive this analytical solution to illustrate the approach and the nature of the results.

Approximations for Average Velocity

In the previous iterative solution for the motion of a tennis ball, we used the approximation

$$\frac{\vec{p}_{\text{future}}}{m} \approx \vec{v}_{\text{future}} \approx \vec{v}_{\text{avg}}$$

for the average velocity over each time interval Δt.

> **QUESTION** Why didn't we calculate the average velocity directly from the definition?

We can't calculate the average velocity directly from the definition

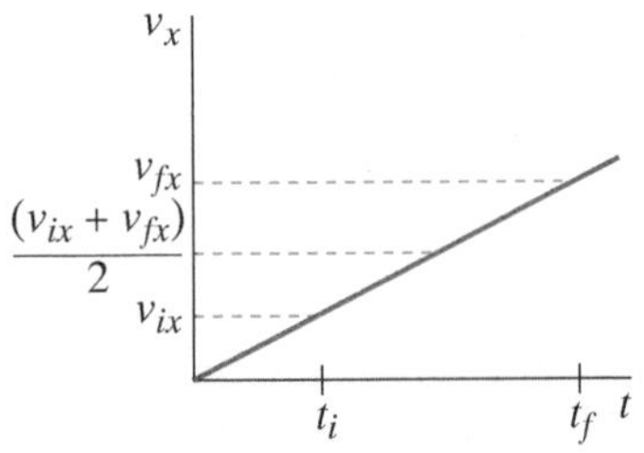

Figure 2.24 v_x is changing at a constant rate (linearly with time), so the arithmetic average is equal to $v_{\text{avg},x}$.

$$\vec{v}_{\text{avg}} \equiv \frac{\Delta\vec{r}}{\Delta t} = \frac{\vec{r}_f - \vec{r}_i}{\Delta t}$$

because the final location $\vec{r}_f$ of the moving ball is unknown. Since we do not have any other information that would allow us to determine the average velocity, we need to find an approximate value for the average velocity in this situation. There are two possible approaches:

(1) If the change in velocity is small over the time interval of interest, we can use the final velocity (or the initial velocity) as an approximation to the average velocity. (2) Alternatively, we can average two values of velocity to approximate the average velocity.

> **QUESTION** What might be a better approximation for $\vec{v}_{\text{avg}}$ if the net force on a system is constant?

Let's consider using the arithmetic average:

$$\vec{v}_{\text{avg}} \approx \frac{\vec{v}_{\text{future}} + \vec{v}_{\text{now}}}{2}$$

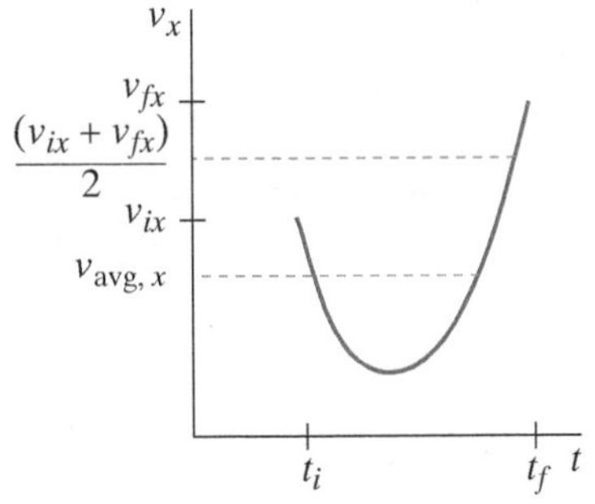

Figure 2.25 When v_x does not change linearly with time, the arithmetic average may be significantly different from the true average value of $v_{\text{avg},x}$.

The arithmetic average lies between the two values. For example, the arithmetic average of 6 and 8 is $(6+8)/2 = 14/2 = 7$, halfway between 6 and 8. The arithmetic average is often a good approximation, but it is not necessarily equal to the true average $\vec{v}_{\text{avg}} = \Delta\vec{r}/\Delta t$. The arithmetic average does not give the true average unless each component of $\vec{v}$ is changing at a constant rate, as in Figure 2.24, which is the case only if the net force is constant. The detailed proof that $\vec{v}_{\text{avg}} = (\vec{v}_i + \vec{v}_f)/2$ when $\vec{v}$ changes at a constant rate is more complicated than one might expect, and is given in optional Section 2.8.

> **QUESTION** When is the arithmetic average a poor approximation for $\vec{v}_{\text{avg}}$?

The arithmetic average is a poor approximation for $\vec{v}_{\text{avg}}$ if the rate of change of velocity (the acceleration $\vec{a} = d\vec{v}/dt$) is not constant. Figure 2.25 shows an example of such a situation. At low speeds ($v \ll c$), the acceleration is directly proportional to the rate of change of momentum because $\vec{F}_{net} = \Delta\vec{p}/\Delta t \approx m(d\vec{v}/dt)$, so we can conclude that $d\vec{v}/dt$ will be constant if the net force on the system is constant.

EXAMPLE **The Arithmetic Average Can Be a Poor Approximation**

If you drive 50 mi/h in the $+x$ direction for four hours, and then 20 mi/h in the $+x$ direction for an hour, how far have you traveled (Figure 2.26)?

Solution If we use v_x at the beginning and at the end of this 5 h time interval to determine the arithmetic average velocity, you would calculate that you have traveled 175 mi:

$$v_{avg,x} \approx \frac{50+20}{2} \text{ mi/h} = 35 \text{ mi/h}$$

$$x_f - x_i \approx (35 \text{ mi/h})(5 \text{ h}) = 175 \text{ mi}$$

However, you have actually traveled significantly farther:

$$x_f - x_i = (50 \text{ mi/h})(4 \text{ h}) + (20 \text{ mi/h})(1 \text{ h}) = 220 \text{ mi}$$

The x component of your actual average velocity was:

$$v_{avg,x} = \frac{220 \text{ mi}}{5 \text{ h}} = 44 \text{ mi/h}$$

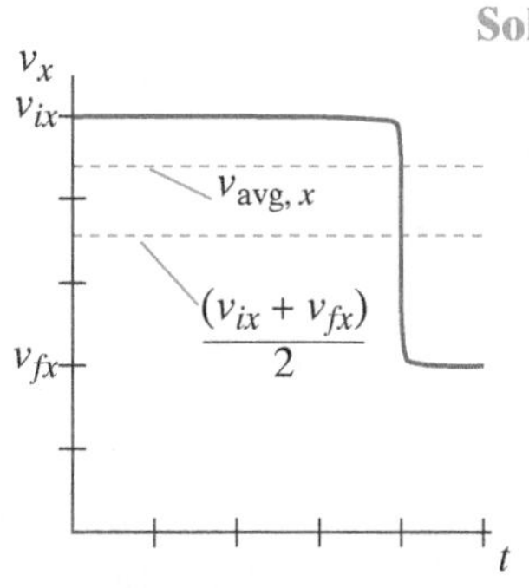

Figure 2.26 Because v_x of the car does not change linearly with time, $v_{avg,x}$ is significantly different from the arithmetic average of $v_{i,x} + v_{f,x}$.

APPROXIMATE AVERAGE VELOCITY

$$\vec{v}_{avg} \approx \frac{\vec{v}_{future} + \vec{v}_{now}}{2}$$

This is exactly correct if all components of $\vec{v}$ change linearly with time (acceleration is constant). However, it may be a poor approximation if the net force is not constant.

Analytical Solution for Constant Net Force

The term *analytical* describes a solution that can be written symbolically (algebraically). In many cases it is not possible to find an analytical solution for the motion of a system. However, if the net force on a system is constant, we can easily find an analytical solution of the form $\vec{r}(t) = \ldots$.

Consider a system on which there is a constant net force in the $+x$ direction during the time Δt. For ordinary speeds $\gamma \approx 1$, so $v_x \approx p_x/m$. Using (a) the Momentum Principle in its update form, (b) the position update equation, and (c) the special result for average velocity in the case of constant net force, we can easily derive equations for the position and velocity as a function of time.

For convenience we'll start measuring the time t from 0, and write t to represent Δt:

$$p_{xf} = p_{xi} + F_{net,x}t$$

$$v_{xf} = v_{xi} + \frac{F_{net,x}}{m}t$$

$$v_{avg,x} = \frac{\left(v_{xi} + \frac{F_{net,x}}{m}t\right) + v_{xi}}{2}$$

$$v_{avg,x} = v_{xi} + \frac{1}{2}\frac{F_{net,x}}{m}t$$

$$x_f = x_i + v_{\text{avg},x}\, t$$
$$x_f = x_i + v_{xi} t + \frac{1}{2}\frac{F_{\text{net},x}}{m} t^2$$

To emphasize that we now have an *analytical solution,* we will write equations that express position and velocity as a function of time, given that the net force is constant:

ANALYTICAL SOLUTION: CONSTANT NET FORCE

$$x(t) = x_i + v_{xi} t + \frac{1}{2}\frac{F_{\text{net},x}}{m} t^2$$
$$v_x(t) = v_{xi} + \frac{F_{\text{net},x}}{m} t$$

If the net force had acted in the y or z direction the solution would be the same, except for the subscripts indicating direction. The time t can be arbitrarily large as long as the net force on the system is constant during the time interval 0 to t.

The time derivative of v_x is the x component of acceleration, which is constant since $F_{\text{net},x}$ is constant:

$$\frac{dv_x}{dt} = \frac{F_{\text{net},x}}{m} = a_x$$

EXAMPLE **Constant Net Force on a Fan Cart**

Figure 2.27 A fan cart on a track. (Ruth Chabay/Bruce Sherwood)

An easy way to apply a nearly constant force is to mount a battery-powered fan on a cart (Figure 2.27). If the fan directs air backwards, the interaction with the air pushes the cart forward with a nearly constant force. (Airboats used in the very shallow Florida Everglades are built in a similar way, with large fans on top of the boats propelling them through the swamp.) Suppose that you have a fan cart whose mass is 0.5 kg. You turn on the fan, then give the cart a shove so that when it is at location $\langle 0.2, 0, 0\rangle$ m it is traveling with velocity $\langle 0.3, 0, 0\rangle$ m/s. At a time 0.9 seconds later, you observe that the cart is at location $\langle 0.146, 0, 0\rangle$ m. What was the force exerted by the air on the fan cart?

Solution Since both net force and motion are parallel to the track, we only need to consider the x components of position, velocity, and force. Using the equations derived above for the motion of a system that experiences a constant net force:

$$x_f = x_i + v_{xi} t + \frac{1}{2}\frac{F_{\text{net},x}}{m} t^2$$
$$0.0.146\,\text{m} = 0.2\,\text{m} + (0.3\ \text{m/s})(0.9\ \text{s}) + \frac{F_{\text{net},x}}{2(0.5\ \text{kg})}(0.9\ \text{s})^2$$
$$F_{\text{net},x} = -0.4\ \text{N}$$

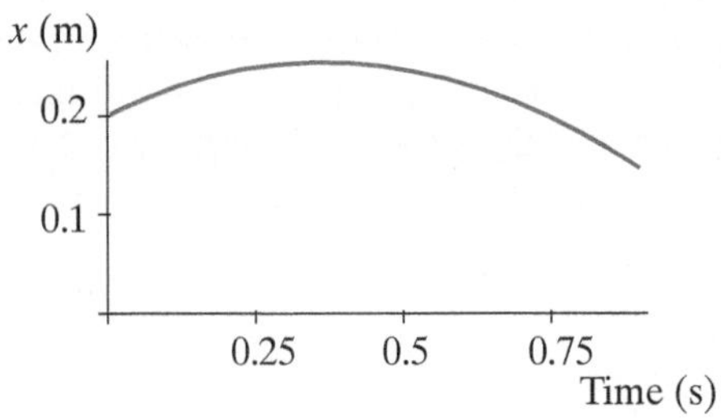

Figure 2.28 A plot of x vs. t for the fan cart. Note that $v_x(t)$ is the derivative of this function. Where the slope is zero, v_x is zero; this occurs at the time when v_x changes sign, from positive to negative.

We see that the net force was in the $-x$ direction, opposite to the initial direction of motion of the cart. We conclude that the cart initially moved in the $+x$ direction, slowing down, coming to a stop, then moving at increasing speed in the $-x$ direction. A graph of x vs. t for the cart is shown in Figure 2.28. (We used a computer to calculate all the (x,t) values needed for this graph.)

EXAMPLE **Kicking a Ball on the Moon**

A soccer ball, whose mass is 0.45 kg, is initially at rest on the surface of the Moon, at location $\langle 0,0,0\rangle$ m. On the surface of the Moon, the constant $g \approx 1.6$N/kg. An astronaut kicks the ball, giving it an initial velocity of $\langle 15,15,0\rangle$ m/s; this speed of $\sqrt{15^2+15^2} = 21$ m/s can be compared to the top speed of about 35 m/s achieved by professional soccer players. What are the coordinates of the highest position the ball reaches? How long is the ball in the air before hitting the ground? How far does the ball travel before hitting the ground?

Solution It is not immediately obvious how to find the maximum height of the ball, because this is not a quantity that appears in the equations we derived above. If we think about the ball's motion, we note that at the top of the ball's trajectory, v_y is instantaneously zero. This is a useful observation because we can solve for the time at which this occurs, and then we can calculate the x and y coordinates of the ball at that time. Because the Moon is airless, there is truly no air resistance, and the net force on the ball is the downward gravitational force $\langle 0,-mg,0\rangle$, which does not affect the x component of the ball's momentum. Again setting the initial time $t_0 = 0$, and substituting t for Δt, we get:

$$
\begin{aligned}
v_{yf} &= v_{yi} + \frac{(-mg)}{m}t \\
0 &= 15\text{ m/s} + (-1.6\text{N/kg})t \\
t &= 9.375\text{ s} \\
y_f &= y_i + v_{yi}t + \frac{1}{2}\frac{(-mg)}{m}t^2 \\
y_f &= 0 + (15\text{ m/s})(9.375\text{ s}) + (0.5)(-1.6\text{N/kg})(9.375\text{ s})^2 \\
y_f &= 70.3\text{ m} \\
x_f &= x_i + v_{xi}t \\
x_f &= 0 + (15\text{ m/s})(9.375\text{ s}) \\
x_f &= 140.6\text{ m}
\end{aligned}
$$

The coordinates of the highest position of the ball are $\langle 140.6, 70.3, 0\rangle$ m, assuming we choose our coordinate axes so that the motion is in the $z = 0$ plane. Note that the maximum height of 70 m is about the height of a 20-story building.

Next we find the time of flight and the distance the ball goes. One way to find the flight time is to note that on level ground $y_f = 0$ when the ball hits:

$$
\begin{aligned}
y_f &= y_i + v_{yi}t + \frac{1}{2}\frac{(-mg)}{m}t^2 \\
0 &= 0 + (15\text{ m/s})t + (0.5)(-1.6\text{N/kg})t^2 \\
t &= 18.75\text{ s} \\
x_f &= x_i + v_{xi}t \\
x_f &= 0 + (15\text{ m/s})(18.75\text{ s}) \\
x_f &= 281.3\text{ m}
\end{aligned}
$$

On Earth, where $g = 9.8$ N/kg, even if there were no atmosphere this soccer ball would go only about 46 m, less than half the length of a soccer field. The "hang time" of 18.75 s is very long; on Earth (in the absence of air resistance) it would be only 3 s.

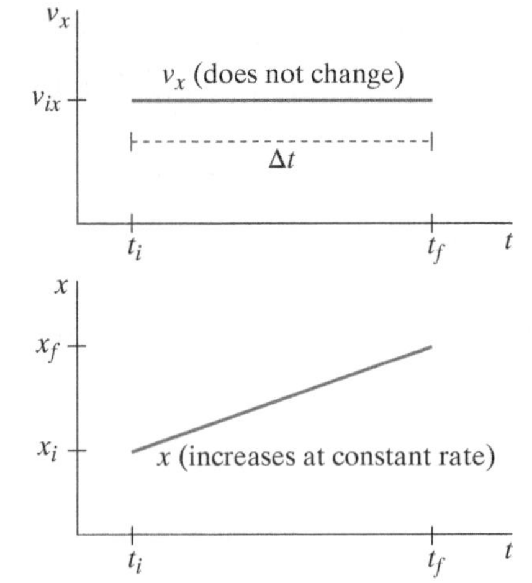

Figure 2.29 Motion graphs for the kicked ball. Top: v_x vs. t; bottom: x vs. t. v_x does not change because the net force acted only in the y direction.

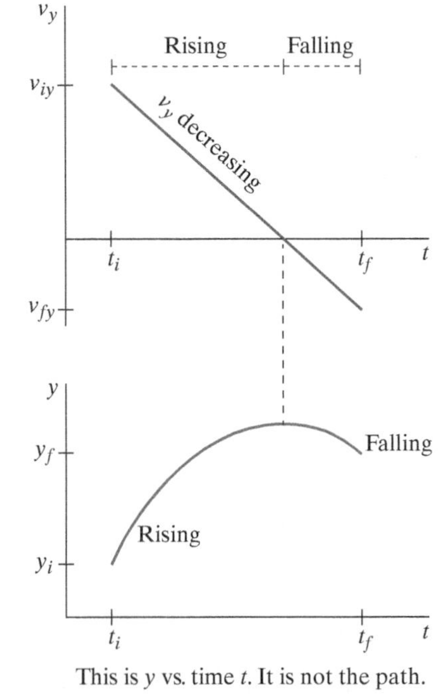

Figure 2.30 Motion graphs for the kicked ball. Top: v_y vs. t; bottom: y vs. t.

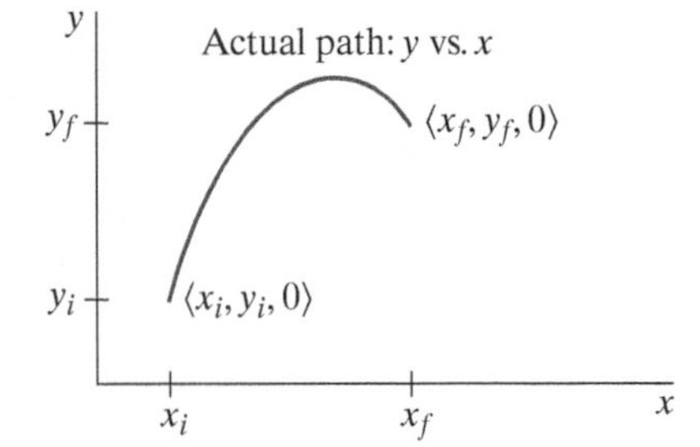

Figure 2.31 The actual trajectory of the thrown ball, with negligible air resistance (y vs. x).

The approach we took above is equivalent to a more formal calculus-based approach. We could differentiate the equation giving y as a function of t and find its maximum by setting $dy/dt = 0$. Either approach will yield a correct answer.

Graphs of Motion

Graphs can be a compact way of representing particular aspects of motion. It is difficult to graph all components of a system's position, velocity, or momentum at once, so typically we make one graph for each component. For example, Figures 2.29 and 2.30 show graphs of position and velocity components vs. time for the soccer ball in the preceding example.

QUESTION In Figure 2.29, why is the graph of v_x vs. t a horizontal line?

Since the x component of the net force on the ball is zero, v_x doesn't change, and the graph of v_x vs. t is simply a horizontal line with a slope of zero. (The slope of v_x vs. t is the x component of the acceleration: $a_x = dv_x/dt$. In this case $a_x = 0$.)

QUESTION How is the graph of v_x vs. time related to the graph of x vs. time?

The graph of x vs. time is a straight line, rising because the ball is moving in the $+x$ direction. The slope of the x vs. t graph is equal to v_x, since v_x is the rate of change of x ($v_x = dx/dt$). Similarly, the slope of the y vs. t graph at any time is equal to v_y at that time.

QUESTION What is the significance of the point at which $v_y = 0$ in Figure 2.30?

When v_y is equal to zero, the slope of y vs. t is momentarily zero, and y is a maximum. At this instant the ball is neither rising nor falling. Before that point the slope is positive, corresponding to $v_y > 0$, and after that point the slope is negative, corresponding to $v_y < 0$. In Figure 2.30 the graph of v_y has a constant negative slope (the top graph), because the y component of the force has the constant value $-mg$, which makes the y component of momentum decrease at a constant rate. In the next section we will see that the equation for y is a quadratic function in the time, which is consistent with the parabolic form of the graph of y vs. time t.

QUESTION How would a plot of p_y vs. time compare to the plot of v_y vs. time (Figure 2.30)?

Since at speeds that are low compared to the speed of light $\vec{p} \approx m\vec{v}$, $p_y \approx mv_y$. A graph of p_y vs. time would look very similar to the graph of v_y vs. time, but the units of the y axis would be different.

The actual path of the ball, the graph of y vs. x (Figure 2.31), is also an inverted parabola. Since x increases linearly with t, whether we plot y vs. t or y vs. x we'll see a similar curve. The units of the horizontal axis are different, of course (meters instead of seconds).

Checkpoint 5 A ball is kicked on Earth from a location $\langle 9, 0, -5\rangle$ m (on the ground) with initial velocity $\langle -10, 13, -5\rangle$ m/s. Neglecting air resistance: **(a)** What is the velocity of the ball 0.6 s after being kicked? **(b)** What is the location of the ball 0.6 s after being kicked? **(c)** What is the maximum height reached by the ball?

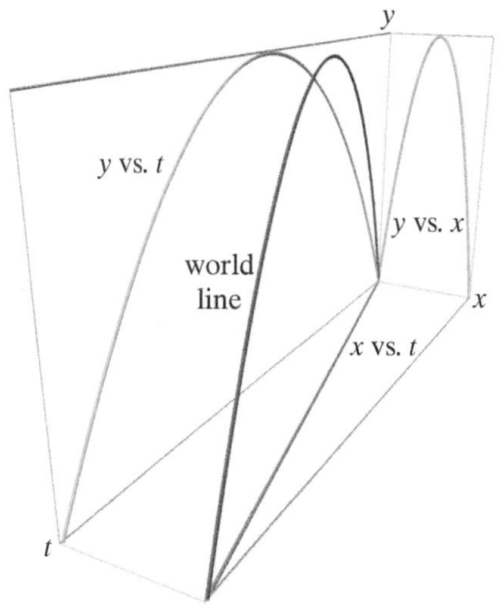

Figure 2.32 A "world line" and its projections.

Projections of a "World Line"

Figure 2.32 illustrates an interesting way to think of the graphs of y vs. x, y vs. t, and x vs. t. The red curve in Figure 2.32 is a graphical description of the motion in x, y, and t of a ball that moves in the xy plane. Imagine shining a light on the red curve to project a shadow onto the xy plane, or the yt plane, or the xt plane. These "projections" are shown in cyan in the figure. Compare these projections with the individual graphs shown in Figures 2.29, 2.30, and 2.31.

The red curve, a function of x, y, and t, is called a "world line." World lines play an important role in textbooks on special relativity. In reality, a world line should be a function of x, y, z, and t, but it's very difficult even with a computer to draw a four-dimensional curve!

2.6 ITERATIVE PREDICTION: VARYING NET FORCE

A familiar example of a force that is not constant is the force exerted by a spring that is stretched or compressed. The magnitude of this force depends on the amount of stretch or compression of the spring. The force is directed along the line of the spring.

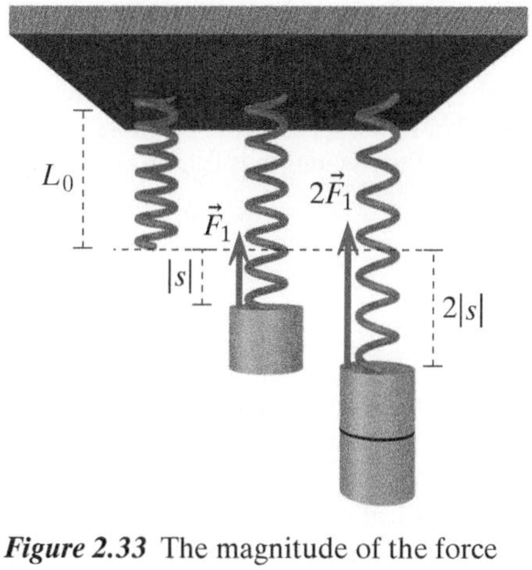

Figure 2.33 The magnitude of the force exerted by a spring is proportional to the absolute value of the stretch of the spring. For an elongated spring, stretch is positive.

The Spring Force

A "force law" describes mathematically how a force depends on the situation. For a spring, it is determined experimentally that the force exerted by a spring on an object attached to the spring is given by the following equation:

MAGNITUDE OF THE SPRING FORCE

$$|\vec{F}_{\text{spring}}| = k_s|s|$$

$|s|$ is the absolute value of the stretch: $s = L - L_0$.
L_0 is the length of the relaxed spring.
L is the length of the spring when stretched or compressed.
k_s is the "spring stiffness" (also called "spring constant").
The force acts in a direction to restore the spring to its relaxed length.

The constant k_s is a positive number, and is a property of the particular spring: the stiffer the spring, the larger the spring stiffness, and the larger the force needed to stretch the spring. Note that s is positive if the spring is stretched ($L > L_0$, Figure 2.33) and negative if the spring is compressed ($L < L_0$, Figure 2.34). This equation is sometimes called "Hooke's law." It is valid as long as the spring is not stretched or compressed too much.

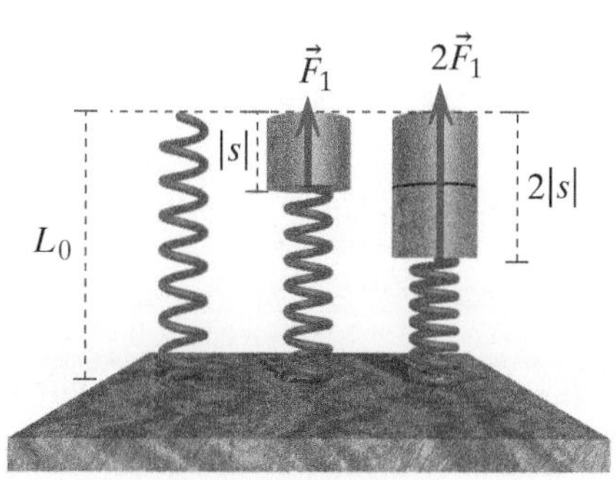

Figure 2.34 The magnitude of the force exerted by a spring is proportional to the absolute value of the stretch of the spring. For a compressed spring, stretch is negative.

QUESTION Suppose a certain spring has been calibrated so that we know that its spring stiffness k_s is 500 N/m. You pull on the spring and observe that it is 0.01 m (1 cm) longer than it was when relaxed. What is the magnitude of the force exerted by the spring on your hand?

The force law gives $|\vec{F}_{\text{spring}}| = (500\,\text{N/m})(|+0.01\,\text{m}|) = 5\,\text{N}$. Note that the total length of the spring doesn't matter; it's just the amount of stretch or compression that matters. Because of the "reciprocity" of the electric forces between the protons and electrons in the spring and those in your hand, as we'll study in Chapter 3, the force you exert on the spring is equal in magnitude and opposite in direction to the force the spring exerts on your hand.

QUESTION Suppose that instead of pulling on the spring, you push on it, so the spring becomes shorter than its relaxed length. If the relaxed length of the spring is 10 cm, and you compress the spring to a length of 9 cm, what is the magnitude of the force exerted by the spring on your hand?

The stretch of the spring in SI units is

$$s = L - L_0 = (0.09\,\text{m} - 0.10\,\text{m}) = -0.01\,\text{m}$$

The force law gives

$$|\vec{F}_{\text{spring}}| = (500\,\text{N/m})(|-0.01\,\text{m}|) = 5\,\text{N}$$

The magnitude of the force is the same as in the previous case. Of course the direction of the force exerted by the spring on your hand is now different, so we will need to write a full vector equation to incorporate this information.

Checkpoint 6 (1) You push on a spring whose stiffness is 11 N/m, compressing it until it is 2.5 cm shorter than its relaxed length. What is the magnitude of the force the spring now exerts on your hand? (2) A different spring is 0.17 m long when it is relaxed. **(a)** When a force of magnitude 250 N is applied, the spring becomes 0.24 m long. What is the stiffness of this spring? **(b)** This spring is compressed so that its length is 0.15 m. What magnitude of force is required to do this?

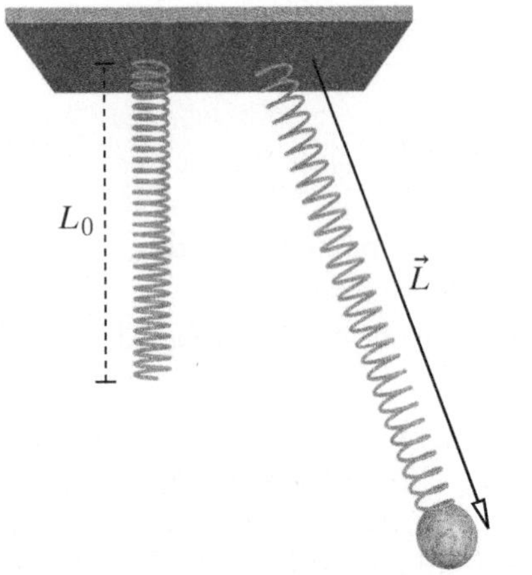

Figure 2.35 The vector $\vec{L}$ points from the location where the spring is attached to the ceiling to the location of the mass. The arrow is offset for clarity in the diagram. L_0 is the length of the relaxed spring.

The Spring Force as a Vector

To write an equation for the spring force as a vector, combining magnitude and direction into one expression, we need to consider the general case, in which the spring may be stretched both vertically and horizontally. Since the force exerted by the spring will be directed along the axis of the spring, we need a unit vector in that direction. We will get this unit vector from the vector $\vec{L}$, shown in Figure 2.35, which is a relative position vector: it specifies the position of the movable end of the spring relative to the fixed end of the spring. $\vec{L}$ extends from the point at which the spring is attached to a support to the mass at the other end of the spring. We can factor the relative position vector $\vec{L}$ into the product of a magnitude (the length of the spring) and a unit vector:

$$\vec{L} = |\vec{L}|\hat{L}$$

The stretch of the spring can now be written as

$$s = |\vec{L}| - L_0$$

- When the spring is longer than its relaxed length, the value of s will be positive, but the direction of the force will be given by $-\hat{L}$.
- When the spring is shorter than its relaxed length, the value of s will be negative, but the direction of the force will be given by $+\hat{L}$.

The general expression for the vector spring force is therefore:

THE VECTOR SPRING FORCE

$$\vec{F} = -k_s s \hat{L}$$

The stretch $s = |\vec{L}| - L_0$ and may be positive or negative.
The scalar L_0 is the length of the relaxed spring.
The vector $\vec{L}$ extends from the point of attachment of the spring to the movable end.
k_s is the "spring stiffness" (also called "spring constant").

QUESTION Consider the two stretched springs in Figure 2.33. What is the unit vector $\hat{L}$ for each stretched spring (assuming the usual coordinate system)? Does $\hat{L}$ depend on how much the spring is stretched?

In each case $\hat{L} = \langle 0,-1,0 \rangle$. Although the magnitude of $\vec{L}$ depends on the stretch, the unit vector $\hat{L}$, which has magnitude 1, is the same in both cases. Similarly, for both compressed springs in Figure 2.34, $\hat{L} = \langle 0,1,0 \rangle$.

EXAMPLE **Force Due to a Stretched Spring**

Suppose that the stiffness of the rightmost spring shown in Figure 2.35 is 9 N/m, and its relaxed length is 21 cm. At the instant shown, the location of the green mass is $\langle 0.07,-0.33,0 \rangle$ m relative to an origin at the point of attachment of the spring. What is the force exerted by the spring on the green mass at this instant?

Solution

$$\vec{L} = \langle 0.07,-0.33,0 \rangle \text{ m} - \langle 0,0,0 \rangle \text{ m} = \langle 0.07,-0.33,0 \rangle \text{ m}$$

$$|\vec{L}| = \sqrt{(0.07\text{ m})^2 + (-.33\text{ m})^2} = 0.337\text{ m}$$

$$\hat{L} = \frac{\langle 0.07,-0.33,0 \rangle \text{ m}}{0.337\text{ m}} = \langle 0.208,-.979,0 \rangle$$

$$s = 0.337\text{ m} - 0.21\text{ m} = 0.127\text{ m}$$

$$\vec{F} = -(9\text{ N/m})(0.127\text{ m})\langle 0.208,-.979,0 \rangle = \langle -0.238,1.12,0 \rangle \text{ N/m}$$

Check: The x component of the force is negative, and the y component is positive, as they should be.

Checkpoint 7 (1) A spring of stiffness 13 N/m, with relaxed length 20 cm, stands vertically on a table as shown in Figure 2.36. Use the usual coordinate system, with $+x$ to the right, $+y$ up, and $+z$ out of the page, toward you. **(a)** When the spring is compressed to a length of 13 cm, what is the unit vector $\hat{L}$? **(b)** When the spring is stretched to a length of 24 cm, what is the unit vector $\hat{L}$? (2) A different spring of stiffness 95 N/m, and with relaxed length 15 cm, stands vertically on a table, as shown in Figure 2.36. With your hand you push straight down on the spring until your hand is only 11 cm above the table. Find **(a)** the vector $\vec{L}$, **(b)** the magnitude of $\vec{L}$, **(c)** the unit vector $\hat{L}$, **(d)** the stretch s, **(e)** the force $\vec{F}$ exerted on your hand by the spring.

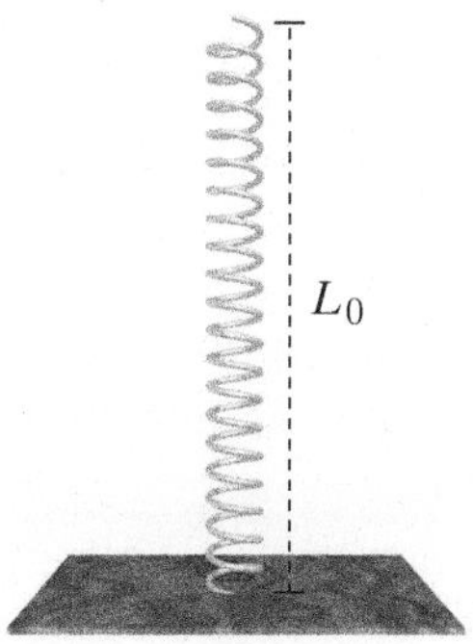

Figure 2.36 A relaxed vertical spring. The tabletop lies in the xz plane, and y is up, as usual.

Motion of a Block–Spring System

If we attach a block to the top of a spring, push down on the block, and then release it, the block will oscillate up and down. (This repetitive motion is described as "periodic.") As the spring stretches and compresses, the force exerted on the block by the spring changes in magnitude and direction. There is also a constant gravitational force on the block. Because the net force on the block is continually changing, we can't use a one-step calculation to predict its motion. We need to apply the Momentum Principle iteratively to predict the location and velocity of the block at any instant.

QUESTION Can we use the equations we derived for the constant force situation?

No, those equations would give us the wrong answer here, because the net force on the block is not constant. We need to solve the problem iteratively, using the same procedure we used to find the path of a ball.

We need two initial values:

- The initial position of the block
- The initial momentum of the block

We will do the following calculation iteratively (repeatedly):

- Find the net force $\vec{F}_{\text{net,now}}$ on the block.
- Apply the Momentum Principle to find $\vec{p}_{\text{future}}$.
- Use $(\vec{p}_{\text{future}}/m)$ to approximate the average velocity, and use this to find the new position of the block.

QUESTION Why is it important to calculate the net force for each iteration? Couldn't we just calculate it once and use this value in each iteration?

As the block moves up and down, the length of the spring changes, so the force exerted by the spring on the block is different each time we take a step.

EXAMPLE **Block on Spring: 1D, Nonconstant Net Force**

A spring has a relaxed length of 20 cm (0.2 m) and its spring stiffness is 8 N/m (Figure 2.36). You glue a 60 g block (0.06 kg) to the top of the spring, and push the block down, compressing the spring so its total length is 10 cm (Figure 2.37). You make sure the block is at rest, then you quickly move your hand away. The block begins to move upward, because the upward force on the block by the spring is greater than the downward force on the block by the Earth. Make a graph of y vs. time for the block during a 0.3 s interval after you release the block.

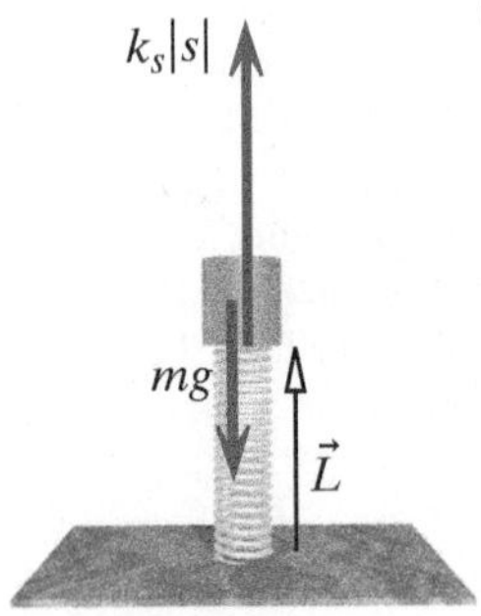

Figure 2.37 You compress the spring, make sure the block is at rest, then release the block. The red arrows show the forces on the block, due to the surroundings, at the instant just after you release the block.

Solution System: Block
Surroundings: Spring, Earth
Diagram: Figure 2.37

For convenience we place the origin at the base of the spring. We will use the shorter notation $\vec{p}_i$ and $\vec{p}_f$ rather than $\vec{p}_{\text{now}}$ and $\vec{p}_{\text{future}}$ to refer to the momentum at the beginning and end of each time step.

$$\vec{L} = \langle 0, 0.1, 0\rangle - \langle 0,0,0\rangle = \langle 0, 0.1, 0\rangle \text{ m}$$

$$|\vec{L}| = 0.1$$

$$\hat{L} = \langle 0,1,0\rangle$$

$$\vec{F}_{\text{spring}} = -k_s(|\vec{L}| - L_0)\langle 0,1,0\rangle = \langle 0, -k_s(|\vec{L}| - L_0), 0\rangle$$

$$\vec{F}_{\text{Earth}} = \langle 0, -mg, 0\rangle$$

The initial momentum of the block is zero, since it is at rest when you release it:

$$\vec{p}_i = \langle 0,0,0 \rangle$$

The net force on the block will be the sum of the forces on the block by the Earth and by the spring. Because both the force by the spring and the gravitational force by the Earth act in the $\pm y$ direction, and the initial x and z components of the block's momentum are zero, we could consider only the y components of force, momentum, and position in our solution. However, writing out all the vectors helps considerably in avoiding sign errors in forces and momenta.

To get an approximate answer, let's divide the 0.3 s time interval into three intervals each 0.1 s long. (It would be better to use even shorter intervals, but this would be unduly tedious if done by hand.)

First time step (Figure 2.38):

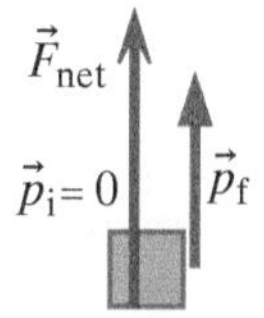

Figure 2.38 At the beginning of time step 1, the net force on the block is in the $+y$ direction, so the y component of the block's momentum will increase.

$$|\vec{L}| = 0.1\text{ m}$$
$$s = 0.1\text{ m} - 0.2\text{ m} = -0.1\text{ m}$$
$$\vec{F}_{spring} = -8\text{ N/m}(-0.1\text{ m})\langle 0,1,0 \rangle = \langle 0,+0.8,0 \rangle\text{ N}$$
$$\vec{F}_{Earth} = \langle 0,-0.06\text{ kg}\cdot 9.8\text{ N/kg},0 \rangle = \langle 0,-0.588,0 \rangle\text{ N}$$
$$\vec{F}_{net} = \langle 0,+0.212,0 \rangle\text{ N}$$
$$\vec{p}_f = \langle 0,0,0 \rangle + \langle 0,+0.212,0 \rangle\text{ N }(0.1\text{ s}) \quad \text{(Momentum Principle)}$$
$$\vec{p}_f = \langle 0,+0.0212,0 \rangle\text{ kg}\cdot\text{m/s}$$
$$\vec{v}_{avg} \approx \vec{v}_f$$
$$\vec{v}_f = \frac{\vec{p}_f}{m} = \frac{\langle 0,+0.0212,0 \rangle\text{ kg}\cdot\text{m/s}}{0.06\text{ kg}} = \langle 0,+0.353,0 \rangle\text{ m/s}$$
$$\vec{r}_f = \langle 0,0.1,0 \rangle\text{ m} + \langle 0,+0.353,0 \rangle\text{ m/s }(0.1\text{ s}) \quad \text{(position update)}$$
$$\vec{r}_f = \langle 0,0.135,0 \rangle\text{ m}$$

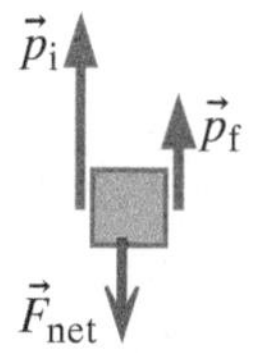

Figure 2.39 At the beginning of time step 2, the net force on the block is now in the $-y$ direction. The block will move upward, but its upward momentum will decrease.

Second time step (Figure 2.39):
Now we advance the clock. At the beginning of time step 2, we need to recalculate $F_{spring,y}$, because the length of the spring has changed. We find that both the magnitude and the direction of the net force are different from the values we found at the beginning of step 1. The momentum of the block at this time ("now") reflects the forces that the block experienced during step 1.

$$|\vec{L}| = 0.135\text{ m}$$
$$s = -0.0647\text{ m}$$
$$\vec{F}_{spring} = \langle 0,+0.520,0 \rangle\text{ N}$$
$$\vec{F}_{net} = \langle 0,-0.0707,0 \rangle\text{ N}$$
$$\vec{p}_f = \langle 0,0.0141,0 \rangle\text{ kg}\cdot\text{m/s}$$
$$\vec{r}_f = \langle 0,0.159,0 \rangle\text{ m}$$

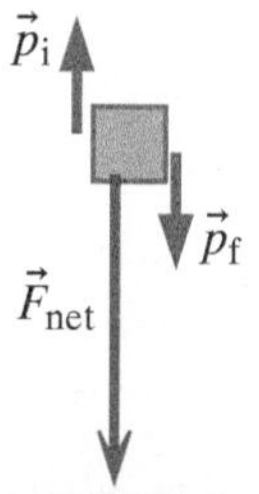

Figure 2.40 At the beginning of time step 3, the net force on the block is in the $-y$ direction, and the final momentum of the block will be downward.

QUESTION What value did we use for $\vec{p}_i$ in time step 2?

$\vec{p}_f$ from step 1 became $\vec{p}_i$ in step 2, as we advanced the clock.

Third time step (Figure 2.40):
At the beginning of the third time step we find that the net force on the block has changed again. The momentum of the block reflects the impulses it has experienced during the first two steps:

$$|\vec{L}| = 0.159\text{ m}$$
$$s = -0.0411\text{ m}$$
$$\vec{F}_{\text{net}} = \langle 0, -0.259, 0\rangle\text{ N}$$
$$\vec{p}_f = \langle 0, -0.0118, 0\rangle\text{ kg}\cdot\text{m/s}$$
$$\vec{r}_f = \langle 0, 0.139, 0\rangle\text{ m}$$

The graph of y vs. time for our three-step calculation is shown in Figure 2.41.

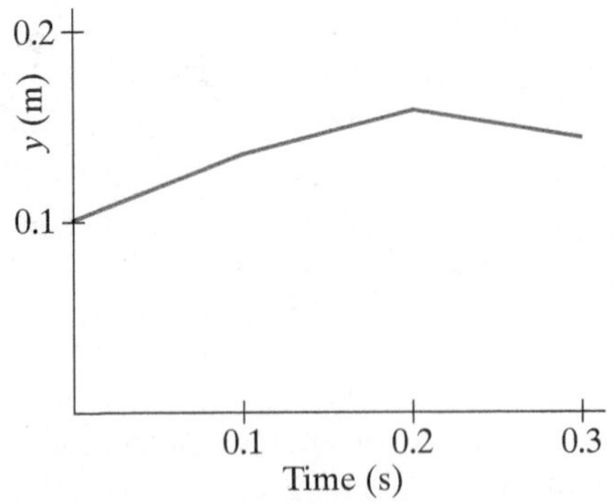

Figure 2.41 A graph of the y component of the block's position vs. time for the three-step iterative calculation.

QUESTION Is this result reasonable?

Yes, because we expect the block to oscillate up and down on the spring.

QUESTION What approximations were made in this calculation?

We made the approximation that the net force did not change significantly over each small time step. During each time step, we also used the final velocity as an approximation for the average velocity. (This is simpler than computing the arithmetic average, and not necessarily worse in a situation where the force is changing. In this case, it turns out that using the arithmetic average would actually have given a less accurate answer.)

Figure 2.42 shows the graph of y vs. t produced when the iterative calculation above is carried out for 10 time steps of 0.1 s.

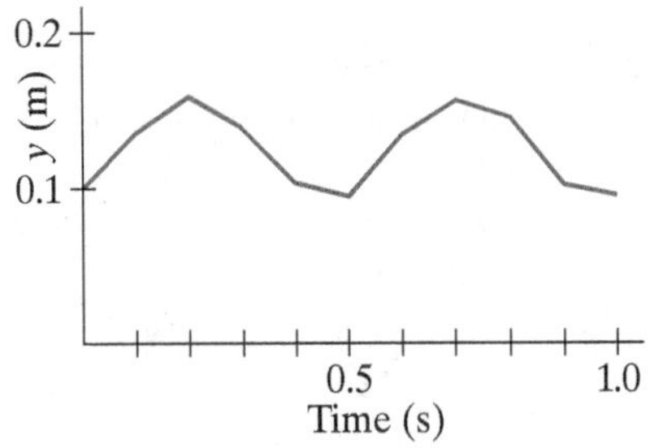

Figure 2.42 A graph of the y component of the block's position vs. time for an iterative calculation carried out for 10 steps of 0.1 s.

QUESTION What features of the graph in Figure 2.42 reflect inaccuracies in the calculations?

Although the graph does show the oscillatory motion of the block, its jagged lines reflect the fact that 0.1 s is too large a step size to produce an accurate result. There are two issues:

- The graph is jagged because we're connecting temporally distant computed points by long straight lines.
- The computed points themselves are inaccurate because we used a rather long time step of 0.1 s, which is clearly large compared to the time scale of the changes occurring in the motion.

Since the motion of the block–spring system repeats over and over, this motion is called *periodic* motion. The time interval between maxima in the plot of y vs. t is called the *period* of the motion, and is usually represented by the symbol T. From the graph in Figure 2.42 we can see that for this particular mass–spring system the period T appears to be around half a second. To get a better estimate of the period, we would need a better graph.

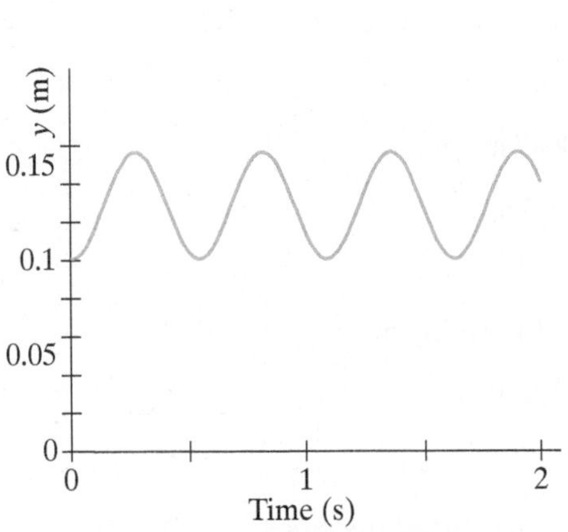

Figure 2.43 A graph of the y component of the block's position vs. time for an iterative calculation using a step size of 0.02 s.

Checkpoint 8 (a) In step 2 of the mass–spring example above, the net force on the block was downward, but the block moved upward. Explain why this was possible, and what the effect of the net force was during this time step. **(b)** In step 3 of the mass–spring example above, only the results of the calculations are given. Carry out the calculations for this step yourself, to be sure you understand in detail the procedure used. Compare your results to the values given above.

Improving Accuracy

QUESTION How could we improve the accuracy of the preceding calculations?

In general, decreasing the time step size will produce more accurate results, because the assumption that $\vec{F}_{net}$ is constant during the time Δt will be more valid. Using a step size of 0.02 s produces a smoothly oscillating graph of y vs. t, as shown by the curve in Figure 2.43. Now the straight lines connecting the computed points are short, making the graph look more smooth, and the computed points themselves are more accurate thanks to using a shorter time step, so that the force and velocity are more nearly constant during each time interval. This prediction is in agreement with experiments with masses oscillating on springs.

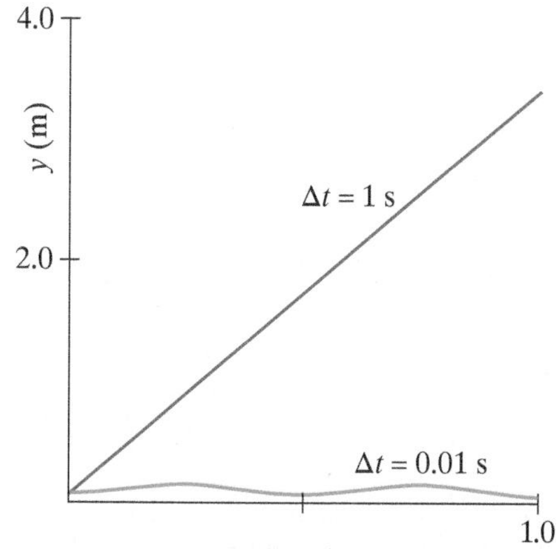

Figure 2.44 If we use a time step of 1 s to predict the motion of the mass–spring system, we predict that the mass will be more than 3 m above the floor! Note that the vertical scale of this graph is very different from the scale of the preceding graphs.

QUESTION Suppose that we had chosen to use a Δt of 1 s. What effect would this have had on our prediction of the block's motion?

The periodicity of the system's motion is due to the particular way in which the net force on the block is changing with time. Since 1 s is actually longer than the period of the motion, assuming that the net force is constant over this interval will lead to an extremely inaccurate result. The resulting graph of y vs. t, shown in Figure 2.44, shows that the predicted y position of the mass after 1 s would be more than 3 m above the floor! A 1 s time step is clearly too large for this system; if the motion we want to predict is periodic, we need a time step that is much smaller than the period of the motion.

There is no single rule for picking an appropriate time step, and sometimes we will need to use a guess-and-check strategy to refine our choice. One approach to deciding if we have chosen an appropriate value for Δt is to try reducing the step size significantly, and see if the predicted motion changes. However, it will be much less tedious to do calculations for many small time steps if we can instruct a computer to do the calculation for us.

QUESTION Would you expect the iterative calculation to be more accurate if we updated the position using $(\vec{v}_i + \vec{v}_f)/2$ instead of using the velocity at the end of the time interval?

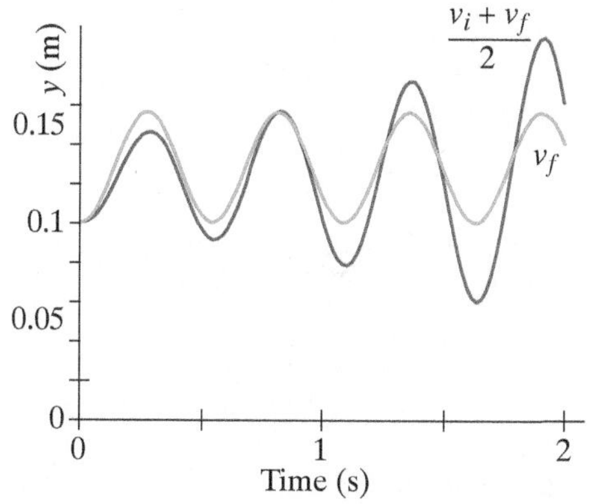

Figure 2.45 Two graphs of the y component of the block's position vs. time for an iterative calculation using a step size of 0.02 s. Blue curve: the velocity at the end of the time interval, $\vec{v}_f$, was used to update the position. Red curve: the arithmetic average, $(\vec{v}_i + \vec{v}_f)/2$, was used. In this calculation the arithmetic average gives significantly worse results.

For the blue curve in Figure 2.45 the velocity at the end of the time interval, $\vec{v}_f$, was used to update the position, whereas for the red curve the arithmetic average, $(\vec{v}_i + \vec{v}_f)/2$, was used. Using the arithmetic average (red curve) clearly produces a very inaccurate prediction: the oscillation will not grow larger and larger in the real world! Here we see an example of the fact that the arithmetic average isn't necessarily the best way to estimate the average velocity in an iterative calculation. In more advanced courses on computational modeling one learns special numerical techniques for optimizing speed and accuracy. For our purposes in this introductory course, we get good results by calculating the net force, then updating the momentum, then updating the position using the final velocity.

Why Not Just Use Calculus?

You might wonder why we don't simply use calculus to predict the motion of physical systems. There are two answers to this question:

First: In fact, we actually are using calculus, in its most fundamental form. Step by step, we add up a large number of small increments of the momentum of an object, and a large number of small displacements of the object, to calculate a large change in its momentum and position over a long time, and this corresponds to a numerical evaluation of an integral:

$$\Delta\vec{p}_1 + \Delta\vec{p}_2 + \cdots \approx \int_i^f d\vec{p} = \int_i^f \vec{F}_{net}\, dt$$

$$\Delta\vec{r}_1 + \Delta\vec{r}_2 + \cdots \approx \int_i^f d\vec{r} = \int_i^f \vec{v}_{avg}\, dt$$

When we calculate these summations with finite time steps, we call the process *numerical integration* (Figures 2.46 and 2.47). The integral sign used in calculus is a distorted "S" meaning "sum" of an infinite number of infinitesimal quantities, from the initial time i to the final time f. Evaluating a definite integral in calculus corresponds to taking an infinite number of infinitesimally small time steps in a summation.

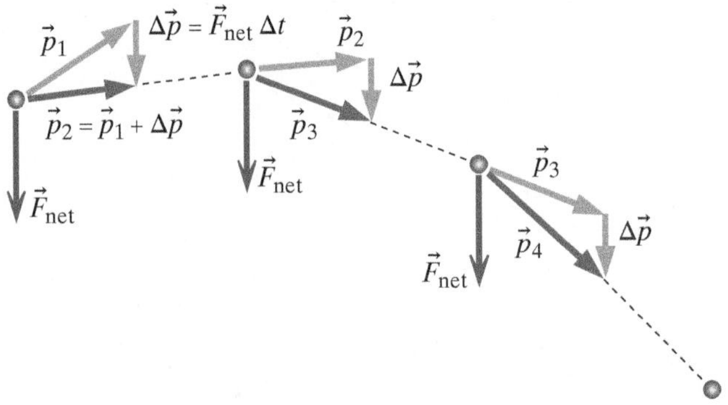

Figure 2.46 When we predict motion iteratively, we are numerically adding a large number of small increments ($\Delta\vec{p}_i$) to the starting momentum to get the momentum at some future time, as shown in Figure 2.47. This is called "numerical integration."

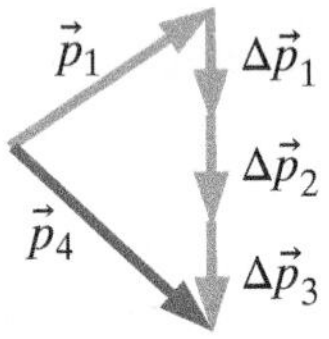

Figure 2.47 Numerical integration: adding all the $\Delta\vec{p}_i$s to $\vec{p}_1$ (Figure 2.46) gives the future momentum $\vec{p}_4$.

Second: A more interesting answer is that the motion of most physical systems actually *cannot* be predicted using calculus in any way other than by numerical integration. In a few special cases calculus does give a general result without carrying out a numerical integration. For example, we have seen that an object subjected to a constant force has a constant rate of change of momentum and velocity, and calculus can be used to obtain a prediction for the position as a function of time. An analytical solution can be derived for the motion in one dimension of a mass attached to a very low mass spring, in the absence of air resistance or friction, as we will see in Chapter 4. The elliptical orbits of two stars around each other can be predicted mathematically without an iterative approach, although the math is quite challenging.

However, in only slightly more complicated situations, an analytical solution may not be possible at all. For example, the general motion of three stars around each other has never been successfully analyzed in this way. The basic problem is that it is usually relatively easy to take the derivative of a known function, but it is often impossible to determine in algebraic form the integral of a known function, which is what would be involved in long-term prediction (adding up a large number of small momentum increments due to known forces).

In contrast, a step-by-step procedure of the kind we carried out for the mass–spring system can easily be extended to three or more bodies in three dimensions. It is also possible to include the effects of various kinds of friction and damping in an iterative calculation. This is why we learn to use the step-by-step prediction method: because it is a powerful technique of increasing importance in modern science and engineering, thanks to the availability of powerful computers to do the repetitive work for us.

2.7 ITERATIVE CALCULATIONS ON A COMPUTER

While the iterative scheme is very general, doing it by hand is tedious. It is not difficult to program a computer to do these calculations repetitively. Computers are now fast enough that it is possible to get high accuracy simply by taking very short time steps, so that during each step the net force and velocity do not change much.

A computer program is simply a sequence of instructions that specify how to perform a calculation. There are many programming languages that can be used to do this, but the basic organization of the program will be the same in almost all cases:

STRUCTURING ITERATIVE CALCULATIONS ON A COMPUTER

- Define the values of constants such as g or k_s to use in the program.
- Specify the masses, initial positions, and initial momenta of the interacting objects.
- Specify an appropriate value for Δt, small enough that the objects don't move very far during one update.
- Create a "loop" structure for repetitive calculations:

Repeat:
- Calculate the net (vector) force $\vec{F}_{\text{net}}$ acting on the system.
- Update the momentum of the system: $\vec{p}_f = \vec{p}_i + \vec{F}_{\text{net}}\,\Delta t$.
- Update the position: $\vec{r}_f = \vec{r}_i + \vec{v}_{\text{avg}}\,\Delta t$.

As before, we use the approximation $\vec{v}_{\text{avg}} \approx \vec{p}_f/m$.

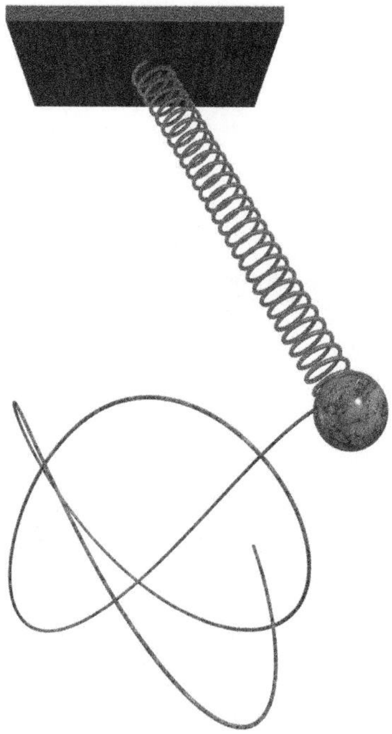

Figure 2.48 A snapshot from a computer program to calculate and animate the motion of a mass–spring system in 3D. This program was written in the language VPython (http://vpython.org), which generates real-time 3D visualizations of motion.

Figure 2.48 shows a display created by such a computer program.

QUESTION Do computer calculations give answers that are exactly correct?

Neither a computer nor your own calculator give "exact" answers, for several reasons. First, when we take a finite time step Δt, we are making the approximation that $\vec{F}_{\text{net}}$ is constant over this time interval. In a situation in which the net force is varying, this approximation may never be exactly true. However, if we take small enough time steps, this approximation can give very good results.

Second, real numbers (typically called "floating-point numbers" in a computational context) cannot be represented infinitely precisely inside a calculator or computer. When arithmetic operations are done with floating-point numbers there is always a small round-off error. For example, if you start with zero and repeatedly add to it the number 1×10^{-5}, doing this 1×10^{5} times, you may get the result 0.99999999998 instead of 1.0.

Although round-off error does accumulate as the number of steps taken increases, it turns out that taking more, smaller steps does make calculations more accurate. The increased accuracy of a single smaller step more than compensates for the accumulation of round-off error in many steps.

Time Step Size in a Computer Calculation

When doing a calculation by hand, there is a trade-off between accuracy and time required to do many iterations (the smaller the time step, the more calculations must be done). Since computers are quite fast, it is reasonable to use much smaller time steps in a computer calculation than one would use by hand. However, even a fast computer can take a very long time to do calculations that use an unnecessarily tiny time step—it would not be reasonable to use a time step of 1×10^{-20} s in a prediction of the Earth's motion around the Sun.

If the motion is periodic, as in the case of an oscillating mass–spring system or a planet orbiting a star, it is important to use a time step that is much shorter than the period of the motion (the repetition time).

A standard method of checking the accuracy of a computer calculation is to decrease the time step and repeat the calculation. If the results do not change significantly, the original time step was adequately small.

Checkpoint 9 Jupiter goes around the Sun in 4333 Earth days. Which of the following would be a reasonable value to try for Δt in a *computer* calculation of the orbit? **(a)** 1 day **(b)** 4333 days **(c)** 0.01 second **(d)** 800 days

The Euler–Cromer Method of Numerical Integration

If you have progressed far enough in your study of calculus, you may have recognized that the iterative procedures we have been using to predict motion involve the Euler method of numerical integration. Actually we have been using a variation on this method, called the Euler–Cromer method, which much improves the accuracy of these iterative predictions. The key feature of this method is the order of the steps:

- First calculate the net force using current positions
- Second, use this force to update the momentum of the system
- Third, use this new momentum to update the position

Doing these calculations in a different order gives much less accurate predictions. (The proof of this is beyond the scope of this textbook; if you are interested you can find more information online.)

Iterative Calculations in VPython

■ The computational homework problems at the end of Chapter 1, and the instructional videos referenced there, introduce you to the key concepts needed to understand VPython code.

The most important part of an iterative computer model is the calculation loop, which contains all the instructions that need to be repeated for each time step. Consider the case of a fan cart moving under the influence of the nearly constant force of the air on the fan. Our computational loop might look like this code, which is an excerpt from a complete program:

```
while True:
    rate(100)
    F_fan = vector(-0.4,0,0)
    F_net = F_fan
    p_cart = p_cart + F_net * deltat
    cart.pos = cart.pos + (p_cart/m_cart) * deltat
```

In contrast, the computational loop in a program modeling a block hanging from a vertical spring might look like this:

```
while True:
    rate(100)
    F_grav = vector(0, -g*m_block, 0)
    L = block.pos - spring.pos
    Lhat = L/mag(L)
    s = mag(L) - L0
    F_spring = -ks * s * Lhat
    F_net = F_grav + F_spring
    p_block = p_block + F_net * deltat
    block.pos = block.pos + (p_block/m_block) * deltat
```

QUESTION What code is essentially the same in both loops?

The last two lines of code are essentially the same. In the next-to-last line, although the objects involved have different names (one is a block, one is a cart), the Momentum Principle is used to update the momentum of each object. In the last line, the position update equation is used to update the position of each object by approximating the average velocity by $\vec{p}_f/m$.

QUESTION What is the major difference between these two computations?

The forces on the two objects are very different. The fan cart is subject only to a constant force, while the net force on the block is the sum of two forces: a constant gravitational force and a spring force that varies with the stretch of the spring.

The instructions that specify how to calculate the spring force mirror what you would do on paper, using your calculator. If $\vec{r}_0$ is the location of the fixed end of the spring, then here is a translation of the algebraic expressions to VPython expressions:

$\vec{L} = \vec{r} - \vec{r}_0$	`L = block.pos - spring.pos`
$\hat{L} = \vec{L}/\lvert\vec{L}\rvert$	`Lhat = L/mag(L)`
$s = \lvert\vec{L}\rvert - L0$	`s = mag(L) - L0`
$\vec{F} = -k_s \cdot s \cdot \hat{L}$	`F_spring = -ks * s * Lhat`

QUESTION Where is $|\vec{L}|$ calculated?

VPython provides a function for calculating the magnitude of a vector. In the code above, the `mag()` function is used to get the magnitude of $\vec{L}$. We could have calculated it separately instead, giving it a name such as Lmag, as is done below:

```
Lmag = sqrt(L.x**2 + L.y**2 + L.z**2)
Lhat = L/Lmag
```

Checkpoint 10 Some code would need to be added in front of each computational loop discussed above in order to make a runnable program. What things would you need to instruct the computer to do before beginning the loop? **(a)** Create objects **(b)** Define constants **(c)** Set initial positions **(d)** Set initial momentum **(e)** Calculate the net force

2.8 *DERIVATION: SPECIAL-CASE AVERAGE VELOCITY

Here are two proofs, one geometric and one algebraic (using calculus), for the following special-case result concerning average velocity:

$$v_{\text{avg},x} = \frac{(v_{ix} + v_{fx})}{2} \quad \text{if } v_x \text{ changes at a constant rate.}$$

The results are similar for $v_{\text{avg},y}$ and $v_{\text{avg},z}$.

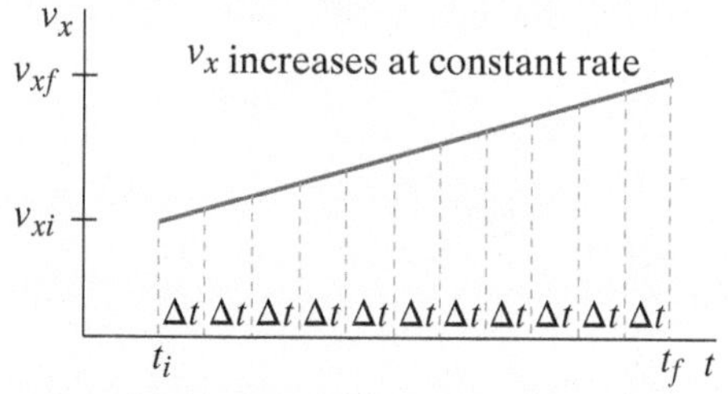

Figure 2.49 Graph of v_x vs. t (constant force), divided into narrow vertical slices each of height v_x and width Δt.

Geometric Proof

If $F_{\text{net},x}$ is constant, $p_{fx} = p_{ix} + F_{\text{net},x}\Delta t$ implies that p_x changes at a constant rate. At speeds small compared to the speed of light, $v_x \approx p_x/m$, so a graph of v_x vs. time is a straight line, as in Figure 2.49. Using this graph, we form narrow vertical slices, each of height v_x and narrow width Δt.

Within each narrow slice v_x changes very little, so the change in position during the brief time Δt is approximately $\Delta x = v_x\Delta t$. Therefore the change in x is approximately equal to the area of the slice of height v_x and width Δt (Figure 2.50).

If we add up the areas of all these slices, we get approximately the area under the line in Figure 2.49, and this is also equal to the total displacement $\Delta x_1 + \Delta x_2 + \Delta x_3 + \cdots = x_f - x_i$. If we go to the limit of an infinite number of slices, each with infinitesimal width, the sum of slices really *is* the area, and this area we have shown to be equal to the change in position. This kind of sum of an infinite number of infinitesimal pieces is called an "integral" in calculus.

The area under the line is a trapezoid, and from geometry we know that the area of a trapezoid is the average of the two bases times the altitude:

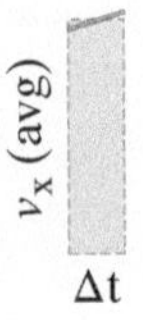

Figure 2.50 One narrow slice has an area given approximately by $v_x\Delta t$. This is equal to Δx, the displacement of the object.

$$\text{area} = \frac{(\text{top}+\text{bottom})}{2}(\text{altitude})$$

Turn Figure 2.50 on its side, as in Figure 2.51, and you see that the top and bottom have lengths v_{ix} and v_{fx}, while the altitude of the trapezoid is the total time $(t_f - t_i)$. Therefore we have the following result:

$$\text{trapezoid area } = x_f - x_i = \frac{(v_{ix}+v_{fx})}{2}(t_f - t_i)$$

Dividing by $(t_f - t_i)$, we have this:

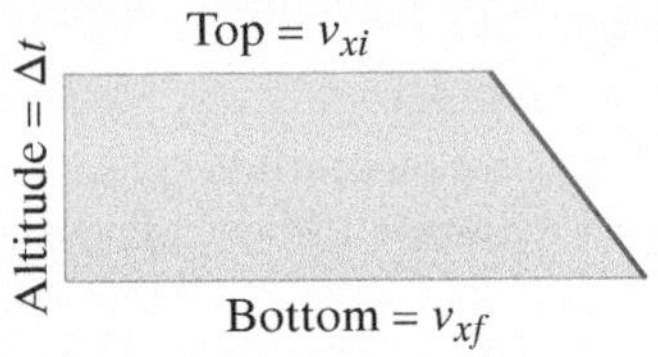

Figure 2.51 The area of the whole trapezoid is equal to the total displacement. The figure in Figure 2.50 has been rotated 90 degrees clockwise.

$$\frac{x_f - x_i}{t_f - t_i} = \frac{(v_{ix}+v_{fx})}{2}$$

By definition, however, the x component of average velocity is the change in x divided by the total time, so we have proved that

$$v_{\text{avg},x} = \frac{(v_{ix}+v_{fx})}{2} \quad \text{if } v_x \text{ changes linearly with time}$$

The proof depended critically on the straight-line ("linear") change in velocity, which occurs if $F_{\text{net},x}$ is constant (and $v \ll c$). Otherwise we wouldn't have a trapezoidal area. That's why the result isn't true in general; it's an important but special case.

Algebraic Proof Using Calculus

An algebraic proof using calculus can also be given. We will use the x component of the derivative version of the Momentum Principle (more about this in a later chapter):

$$\Delta\vec{p} = \vec{F}_{\text{net}}\,\Delta t \quad \text{implies that} \quad \frac{\Delta\vec{p}}{\Delta t} = \vec{F}_{\text{net}}$$

In the limit we have

$$\lim_{\Delta t\to 0}\frac{\Delta\vec{p}}{\Delta t} = \frac{d\vec{p}}{dt} = \vec{F}_{\text{net}} \quad \text{and} \quad \frac{dp_x}{dt} = F_{\text{net},x}$$

If $F_{\text{net},x}$ is a constant, the time derivative of p_x is a constant, so we have

$$p_x = F_{\text{net},x}t + p_{ix} \quad \text{since } p_x = p_{ix} \text{ when } t = 0$$

You can check this by taking the derivative with respect to time t, which gives the original equation $dp_x/dt = F_{\text{net},x}$. At speeds small compared to the speed of light, $v_x \approx p_x/m$, so we can write

$$v_x = \frac{F_{\text{net},x}}{m}t + v_{ix} \quad \text{since } v_x = v_{ix} \text{ when } t = 0$$

However, the x component of velocity is the rate at which x is changing:

$$v_x = \frac{dx}{dt} = \frac{F_{\text{net},x}}{m}t + v_{ix}$$

Now the question is, can you think of a function of x that has this time derivative? Since the time derivative of t^2 is $2t$, the following equation for x has the appropriate derivative:

$$x = \frac{1}{2}\frac{F_{\text{net},x}}{m}t^2 + v_{ix}t + x_i \quad \text{since } x = x_i \text{ when } t = 0$$

You can check this by taking the derivative with respect to t, which gives the equation for v_x, since $d(\frac{1}{2}t^2)/dt = t$ and $d(t)/dt = 1$.

The average velocity that we seek is the change in position divided by the total time:

$$v_{\text{avg},x} = \frac{x - x_i}{t} = \frac{1}{2}\frac{F_{\text{net},x}}{m}t + v_{ix} = \frac{1}{2}(v_{fx} - v_{ix}) + v_{ix}$$

where we have used the equation we previously derived for the velocity:

$$v_{fx} = v_x = \frac{F_{\text{net},x}}{m}t + v_{ix}$$

Simplifying the expression for $v_{\text{avg},x}$ we have the proof:

$$v_{\text{avg},x} = \frac{(v_{ix} + v_{fx})}{2} \quad \text{if } v_x \text{ changes at a constant rate}$$

2.9 *RELATIVISTIC MOTION

Inertial Frames of Reference

Newton's first law is valid only in an "inertial frame" of reference, one in uniform motion (or at rest) with respect to the pervasive "cosmic microwave background" (see the optional discussion at the end of Chapter 1). Since the Momentum Principle is a quantitative version of Newton's first law, we expect the Momentum Principle to be valid in an inertial reference frame, but not in a reference frame that is not in uniform motion. Let's check that this is true.

If you view some objects from a space ship that is moving uniformly with velocity $\vec{v}_s$ with respect to the cosmic microwave background, all of the velocities of those objects have the constant $\vec{v}_s$ subtracted from them, as far as you are concerned. For example, a rock moving at the same velocity as your spacecraft would have $\vec{v}_{\text{rock}} = (\vec{v} - \vec{v}_s) = \vec{0}$ in your reference frame: it would appear to be stationary as it coasted along beside your spacecraft.

With a constant space ship velocity, we have $\Delta\vec{v}_s = \vec{0}$, and the change of momentum of the moving object reduces to the following (for speeds small compared to c):

$$\Delta[m(\vec{v} - \vec{v}_s)] = \Delta(m\vec{v}) - \Delta(m\vec{v}_s) = \Delta(m\vec{v}) \quad \text{since } \Delta(m\vec{v}_s) = \vec{0}$$

Therefore,

$$\Delta[m(\vec{v} - \vec{v}_s)] = \Delta(m\vec{v}) = \vec{F}_{\text{net}}\,\Delta t$$

If the velocity $\vec{v}_s$ of the space ship doesn't change (it represents an inertial frame of reference), the form (and validity) of the Momentum Principle is unaffected by the motion of the space ship.

However, if your space ship increases its speed, or changes direction, $\Delta\vec{v}_s \neq \vec{0}$, an object's motion relative to you changes without any force acting on it. In that case the Momentum Principle is not valid for the object, because you are not in an inertial frame.

Fast-Moving Objects

Applying the Momentum Principle to an object moving at a speed near the speed of light is a bit more complex algebraically, but conceptually exactly the same as what we have done previously. The derivation of the equation for speed in terms of momentum for a fast-moving particle may be found in Chapter 1.

EXAMPLE **Fast Proton: 1D, Constant Net Force, Relativistic**

A proton in a particle accelerator is moving with velocity $\langle 0.960c, 0, 0\rangle$, so the speed is $0.960 \times 3 \times 10^8$ m/s $= 2.88 \times 10^8$ m/s. A constant electric force is applied to the proton to speed it up, $\vec{F}_{\text{net}} = \langle 5 \times 10^{-12}, 0, 0\rangle$ N. What is the proton's speed as a fraction of the speed of light after 20 nanoseconds (1 ns $= 1 \times 10^{-9}$ s)?

Solution System: The proton
Surroundings: Electric charges in the accelerator
Free-body diagram: Figure 2.52
Momentum Principle:

Figure 2.52 The free-body diagram for the proton.

$$\vec{p}_f = \vec{p}_i + \vec{F}_{\text{net}}\,\Delta t$$

$$\langle p_{fx}, 0, 0\rangle = \langle \gamma_i m v_{ix}, 0, 0\rangle + \langle F_{\text{net},x}, 0, 0\rangle\,\Delta t$$

Find the x component of the initial momentum:

$$\gamma = \frac{1}{\sqrt{1 - \left(\frac{0.960c}{c}\right)^2}} = 3.57$$

$$\begin{aligned} p_{ix} &= \gamma_i m v_{ix} = 3.57 \cdot (1.7 \times 10^{-27}\ \text{kg})(0.96 \cdot 3 \times 10^8\ \text{m/s}) \\ &= 1.75 \times 10^{-18}\ \text{kg}\cdot\text{m/s} \end{aligned}$$

Find the x component of the impulse:

$$F_{\text{net},x}\,\Delta t = 5 \times 10^{-12}\ \text{N} \cdot (20 \times 10^{-9}\ \text{s}) = 1 \times 10^{-19}\,\text{N}\cdot\text{s}$$

Apply the Momentum Principle:

$$\begin{aligned} p_{fx} &= (1.75 \times 10^{-18}\ \text{kg}\cdot\text{m/s}) + (1 \times 10^{-19}\,\text{N}\cdot\text{s}) \\ &= 1.85 \times 10^{-18}\ \text{kg}\cdot\text{m/s} \end{aligned}$$

Find the final speed:

$$\frac{v_{fx}}{c} = \frac{\dfrac{p_{fx}}{mc}}{\sqrt{1+\left(\dfrac{p_{fx}}{mc}\right)^2}}$$

$$\frac{p_{fx}}{mc} = \frac{1.85\times 10^{-18}\ \text{kg}\cdot\text{m/s}}{(1.7\times 10^{-27}\ \text{kg})(3\times 10^{8}\ \text{m/s})} = 3.62 \quad \text{(no units)}$$

$$\frac{v_{fx}}{c} = \frac{3.62}{\sqrt{1+3.62^2}} = 0.964$$

$$v_{fx} = 0.964c$$

Although the magnitude of the momentum increased from 1.75×10^{-18} kg·m/s to 1.85×10^{-18} kg·m/s, the speed didn't increase very much, because the proton's initial speed, $0.960c$, was already close to the cosmic speed limit, c. Because the speed hardly changed, the distance the proton moved during the 20 ns was approximately equal to

$$(0.964 \times 3 \times 10^8\ \text{m/s})(20 \times 10^{-9}\ \text{s}) = 5.8\ \text{m}$$

2.10 *MEASUREMENTS AND UNITS

Using the Momentum Principle requires a consistent way to measure length, time, mass, and force, and a consistent set of units. We state the definitions of the standard Système Internationale (SI) units, and we briefly discuss some subtle issues underlying this choice of units.

Units: Meters, Seconds, Kilograms, Coulombs, and Newtons

Originally the meter was defined as the distance between two scratches on a platinum bar in a vault in Paris, and a second was 1/86,400th of a "mean solar day." Now, however, the second is defined in terms of the frequency of light emitted by a cesium atom, and the meter is defined as the distance light travels in 1/299,792,458th of a second, or about 3.3×10^{-9} s (3.3 ns). The speed of light is defined to be exactly 299,792,458 m/s (very close to 3×10^8 m/s). As a result of these modern redefinitions, it is really speed (of light) and time that are the internationally agreed-upon basic units, not length and time.

By international agreement, one kilogram is the mass of a platinum block kept in that same vault in Paris. As a practical matter, other masses are compared to this standard kilogram by using a balance-beam or spring weighing scale (more about this in a moment). The newton, the unit of force, is defined as that force which acting for 1 s imparts to 1 kg a velocity change of 1 m/s. We could make a scale for force by calibrating the amount of stretch of a spring in terms of newtons.

The coulomb, the SI unit of electric charge, is defined in terms of electric currents. The charge of a proton is 1.6×10^{-19} C.

Some Subtle Issues

What we have just said about SI units is sufficient for practical purposes to predict the motion of objects, but here are some questions that might bother you. Is it legitimate to measure the mass that appears in the Momentum Principle by seeing how that mass is affected by gravity on a balance-beam scale? Is it legitimate to use the Momentum Principle to define the units of force, when the concept of force is itself associated with the same law? Is this all circular reasoning, and the Momentum Principle merely a definition with little content? Here is a chain of reasoning that addresses these issues.

Measuring Inertial Mass

When we use balance-beam or spring weighing scales to measure mass, what we're really measuring is the "gravitational mass"—that is, the mass that appears in the law of gravitation and is a measure of how much this object is affected by the gravity of the Earth. In principle, this "gravitational mass" could be different from the so-called "inertial mass"—the mass that appears in the definition of momentum. It is possible to compare the inertial masses of two objects, and we find experimentally that inertial and gravitational mass seem to be entirely equivalent.

Here is a way to compare two inertial masses directly, without involving gravity. Starting from rest, pull on the first object with a spring stretched by some amount s for an amount of time Δt, and measure the increase of speed Δv_1. Then, starting from rest, pull on the second object with the same spring stretched by the same amount s for the same amount of time Δt, and measure the increase of speed Δv_2. We *define* the ratio of the inertial masses as $m_1/m_2 = \Delta v_2/\Delta v_1$. Since one of these masses could be the standard kilogram kept in Paris, we now have a way of measuring inertial mass in kilograms. Having defined inertial mass this way, we find experimentally that the Momentum Principle is obeyed by both of these objects in all situations, not just in the one special experiment we used to compare the two masses.

Moreover, we find to extremely high precision that the inertial mass in kilograms measured by this comparison experiment is exactly the same as the gravitational mass in kilograms obtained by comparing with a standard kilogram on a balance-beam scale (or using a calibrated spring scale), and that it doesn't matter what the objects are made of (wood, copper, glass, etc.). This justifies the convenient use of ordinary weighing scales to determine inertial mass.

Is This Circular Reasoning?

The definitions of force and mass may sound like circular reasoning, and the Momentum Principle may sound like just a kind of definition, with no real content, but there is real power in the Momentum Principle. Forget for a moment the definition of force in newtons and mass in kilograms. The experimental fact remains that any object if subjected to a single force by a spring with constant stretch experiences a change of momentum (and velocity) proportional to the duration of the interaction. Note that it is not a change of *position* proportional to the time (that would be a constant speed), but a change of *velocity*. That's real content. Moreover, we find that the change of velocity is proportional to the amount of stretch of the spring. That too is real content.

Then we find that a different object undergoes a different rate of change of velocity with the same spring stretch, but after we've made one single comparison experiment to determine the mass relative to the standard kilogram, the Momentum Principle works in all situations. That's real content.

Finally come the details of setting standards for measuring force in newtons and mass in kilograms, and we use the Momentum Principle in helping set these standards. Logically, however, this comes after having established the law itself.

SUMMARY

System is a portion of the Universe acted on by the *surroundings*.

Force is a quantitative measure of interactions; units are newtons (N).

Net force $\vec{F}_{\text{net}}$ is the vector sum of all the forces acting on a system.

Impulse is the product of force times time $\vec{F}\Delta t$; momentum change equals net impulse (the impulse due to the net force).

Physical models are tractable approximations/idealizations of the real world.

The Momentum Principle

$$\Delta\vec{p} = \vec{F}_{\text{net}}\,\Delta t$$

$$\vec{p}_f = \vec{p}_i + \vec{F}_{\text{net}}\,\Delta t \quad \text{(update form)}$$

or, choosing "now" as the initial time, and "a short time in the future" as the final time, we can write:

$$\vec{p}_{\text{future}} = \vec{p}_{\text{now}} + \vec{F}_{\text{net,now}}\,\Delta t$$

The time interval Δt must be small enough that the net force is nearly constant during this time interval.

ITERATIVE PREDICTION OF MOTION

Repeat:
- Calculate the net (vector) force $\vec{F}_{\text{net}}$ acting on the system.
- Update the momentum of the system: $\vec{p}_f = \vec{p}_i + \vec{F}_{\text{net}}\,\Delta t$.
- Update the position: $\vec{r}_f = \vec{r}_i + \vec{v}_{\text{avg}}\,\Delta t$.

Use $\vec{p}_f/m$ to approximate the $\vec{v}_{\text{avg}}$ in each step.

Special-case result for average velocity if net force is constant

$$v_{\text{avg},x} = \frac{(v_{ix} + v_{fx})}{2} \quad \text{if } v_x \text{ changes linearly with time}$$

Results are similar for $v_{\text{avg},y}$ and $v_{\text{avg},z}$.

Analytical solution for motion with constant net force

$$x(t) = x_i + v_{xi}t + \frac{1}{2}\frac{F_{\text{net},x}}{m}t^2$$

$$v_x(t) = v_{xi} + \frac{F_{\text{net},x}}{m}t$$

Specific forces

Force by a spring

$$\vec{F} = -k_s s\hat{L}$$

The stretch $s = |\vec{L}| - L_0$ and may be positive or negative.
The scalar L_0 is the length of the relaxed spring.
The vector $\vec{L}$ extends from the point of attachment of the spring to the movable end.
k_s is the "spring stiffness" (also called "spring constant").

Approximate gravitational force near the Earth's surface

$$|\vec{F}_{\text{grav}}| \approx mg \quad \text{where } g = +9.8\,\text{N/kg}$$

for an object of mass m near the surface of the Earth.

QUESTIONS

Q1 Because the *change* of the momentum is equal to the net impulse, the relationship of momentum itself to the net force is somewhat indirect, as can be seen in this question. An object is initially moving in the $+x$ direction with magnitude of momentum p, with a net force of magnitude F acting on the object in either the $+x$ or $-x$ direction. After a very short time, say whether the magnitude of the momentum increases, decreases, or stays the same in each of the following situations: **(a)** The net force acts in the $+x$ direction and F is constant. **(b)** The net force acts in the $+x$ direction and F is increasing. **(c)** The net force acts in the $+x$ direction and F is decreasing. **(d)** The net force acts in the $-x$ direction and F is constant. **(e)** The net force acts in the $-x$ direction and F is increasing. **(f)** The net force acts in the $-x$ direction and F is decreasing.

Q2 An object is moving in the $+y$ direction. Which, if any, of the following statements might be true? Check all that apply. **(a)** The net force on the object is zero. **(b)** The net force on the object is in the $-y$ direction. **(c)** The net force on the object is in the $+y$ direction.

Q3 You observe three carts moving to the left. Cart A moves to the left at nearly constant speed. Cart B moves to the left, gradually speeding up. Cart C moves to the left, gradually slowing down. Which cart or carts, if any, experience a net force to the left?

Q4 In order to pull a sled across a level field at constant velocity you have to exert a constant force. Doesn't this violate Newton's first and second laws of motion, which imply that no force is required to maintain a constant velocity? Explain this seeming contradiction.

Q5 In a lab experiment you observe that a pendulum swings with a "period" (time for one round trip) of 2 s. In an iterative calculation of the motion, which of the following would NOT be a reasonable choice for Δt, for either hand or computer iterative calculations? **(a)** 1 s **(b)** 0.1 s **(c)** 0.05 s **(d)** 0.01 s

Q6 A comet passes near the Sun. When the comet is closest to the Sun, it is 9×10^{10} m from the Sun. You need to choose a time step to use in predicting the comet's motion. Which of the following would be a reasonable distance for the comet to move in one time step, doing an iterative calculation *by hand*? **(a)** 1×10^2 m **(b)** 1×10^{10} m **(c)** 1×10^{11} m **(d)** 1×10^9 m

Q7 A ball moves in the direction of the arrow labeled c in Figure 2.53. The ball is struck by a stick that briefly exerts a force on the ball in the direction of the arrow labeled e. Which arrow best describes the direction of $\Delta\vec{p}$, the change in the ball's momentum?

Figure 2.53

PROBLEMS

Section 2.1

•P8 A system is acted upon by two forces, $\langle 18, 47, -23 \rangle$ N, and $\langle -20, -13, 41 \rangle$ N. What is the net force acting on the system?

•P9 A truck driver slams on the brakes and the momentum of the truck changes from $\langle 65{,}000, 0, 0 \rangle$ kg·m/s to $\langle 26{,}000, 0, 0 \rangle$ kg·m/s in 4.1 s due to a constant force of the road on the wheels of the truck. As a vector, write the net force exerted on the truck by the surroundings.

•P10 At a certain instant a particle is moving in the $+x$ direction with momentum $+8$ kg·m/s. During the next 0.13 s a constant force acts on the particle, with $F_x = -7$ N and $F_y = +5$ N. What is the *magnitude* of the momentum of the particle at the end of this 0.13 s interval?

•P11 At $t = 16.0$ s an object with mass 4 kg was observed to have a velocity of $\langle 9, 29, -10 \rangle$ m/s. At $t = 16.2$ s its velocity was $\langle 18, 20, 25 \rangle$ m/s. What was the average net force acting on the object?

••P12 A proton (mass 1.7×10^{-27} kg) interacts electrically with a neutral HCl molecule located at the origin. At a certain time t, the proton's position is $\langle 1.6 \times 10^{-9}, 0, 0 \rangle$ m and the proton's velocity is $\langle 3600, 600, 0 \rangle$ m/s. The force exerted on the proton by the HCl molecule is $\langle -1.12 \times 10^{-11}, 0, 0 \rangle$ N. At a time $t + 3.4 \times 10^{-14}$ s, what is the approximate velocity of the proton? (You may assume that the force was approximately constant during this interval.)

••P13 A Ping-Pong ball is acted upon by the Earth, air resistance, and a strong wind. Here are the positions of the ball at several times.
Early time interval:

- At $t = 12.35$ s, the position was $\langle 3.17, 2.54, -9.38 \rangle$ m.
- At $t = 12.37$ s, the position was $\langle 3.25, 2.50, -9.40 \rangle$ m.

Late time interval:

- At $t = 14.35$ s, the position was $\langle 11.25, -1.50, -11.40 \rangle$ m.
- At $t = 14.37$ s, the position was $\langle 11.27, -1.86, -11.42 \rangle$ m.

(a) In the early time interval, from $t = 12.35$ s to $t = 12.37$ s, what was the average momentum of the ball? The mass of the Ping-Pong ball is 2.7 grams (2.7×10^{-3} kg). Express your result as a vector. **(b)** In the late time interval, from $t = 14.35$ s to $t = 14.37$ s, what was the average momentum of the ball? Express your result as a vector. **(c)** In the time interval from $t = 12.35$ s (the start of the early time interval) to $t = 14.35$ s (the start of the late time interval), what was the average net force acting on the ball? Express your result as a vector.

••P14 A 0.7 kg block of ice is sliding by you on a very slippery floor at 2.5 m/s. As it goes by, you give it a kick perpendicular to its path. Your foot is in contact with the ice block for 0.003 s. The block eventually slides at an angle of 22 degrees from its original direction. The overhead view shown in Figure 2.54 is approximately to scale. The arrow represents the average force your toe applies briefly to the block of ice.
(a) Which of the possible paths shown in the diagram corresponds to the correct overhead view of the block's path? **(b)** Which components of the block's momentum are changed by the impulse applied by your foot? (Check all that apply. The diagram shows a top view, looking down on the xz plane.) **(c)** What is the unit vector $\hat{p}$ in the direction of the block's momentum after the kick? **(d)** What is the x component of the block's momentum after the kick? **(e)** Remember that $\vec{p} = |\vec{p}|\hat{p}$. What is the magnitude of the block's momentum after the kick? **(f)** Use your answers to the preceding questions to find the z component of the block's momentum after the kick (drawing a diagram is helpful). **(g)** What was the magnitude of the average force you applied to the block?

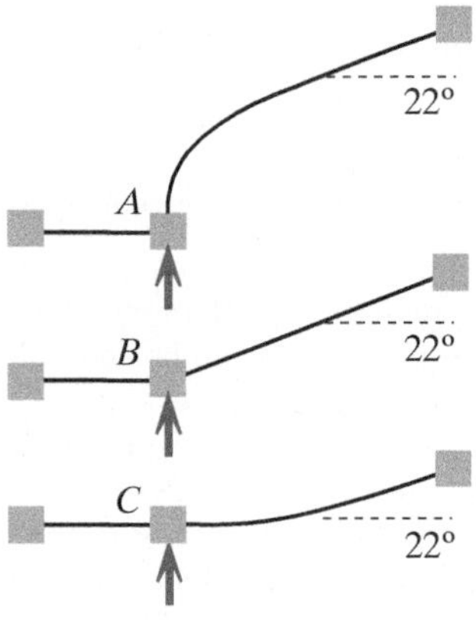

Figure 2.54

••P15 In outer space a rock of mass 5 kg is acted on by a constant net force $\langle 29, -15, 40 \rangle$ N during a 4 s time interval. At the end of this time interval the rock has a velocity of

$\langle 114, 94, 112 \rangle$ m/s. What was the rock's velocity at the beginning of the time interval?

Section 2.2

••P16 A steel safe with mass 2200 kg falls onto concrete. Just before hitting the concrete its speed is 40 m/s, and it smashes without rebounding and ends up being 0.06 m shorter than before. What is the approximate magnitude of the force exerted on the safe by the concrete? How does this compare with the gravitational force of the Earth on the safe? Explain your analysis carefully, and justify your estimates on physical grounds.

••P17 In a crash test, a truck with mass 2500 kg traveling at 24 m/s smashes head-on into a concrete wall without rebounding. The front end crumples so much that the truck is 0.72 m shorter than before. **(a)** What is the average speed of the truck during the collision (that is, during the interval between first contact with the wall and coming to a stop)? **(b)** About how long does the collision last? (That is, how long is the interval between first contact with the wall and coming to a stop?) **(c)** What is the magnitude of the average force exerted by the wall on the truck during the collision? **(d)** It is interesting to compare this force to the weight of the truck. Calculate the ratio of the force of the wall to the gravitational force mg on the truck. This large ratio shows why a collision is so damaging. **(e)** What approximations did you make in your analysis?

••P18 A tennis ball has a mass of 0.057 kg. A professional tennis player hits the ball hard enough to give it a speed of 50 m/s (about 120 mi/h). The ball hits a wall and bounces back with almost the same speed (50 m/s). As indicated in Figure 2.55, high-speed photography shows that the ball is crushed 2 cm (0.02 m) at the instant when its speed is momentarily zero, before rebounding.

Making the very rough approximation that the large force that the wall exerts on the ball is approximately constant during contact, determine the approximate magnitude of this force. *Hint:* Think about the approximate amount of time it takes for the ball to come momentarily to rest. (For comparison note that the gravitational force on the ball is quite small, only about (0.057 kg)(9.8 N/kg) $\approx$ 0.6 N. A force of 5 N is approximately the same as a force of one pound.)

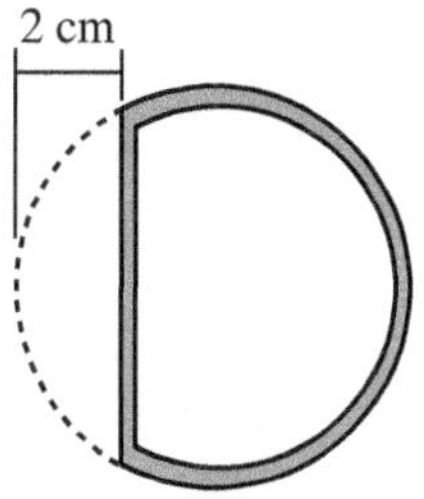

Figure 2.55

••P19 An object is on a collision course with the Earth and is predicted to hit in the center of the target, coming in vertically. The object is roughly spherical, with an approximate diameter of 100 m (the meteor that damaged Chelyabinsk, Russia, in February 2013 had a diameter of about 20 m; the object that killed off the dinosaurs 65 million years ago is thought to have had a diameter of about 10 km). If we start immediately, we can rendezvous with the object when it is 2.5×10^{11} m from the Sun, traveling toward the Sun at a speed of 30 km/s. There is a proposal to implant a rocket engine on the surface of the object to deflect the object enough to miss the Earth. Perform an approximate feasibility analysis of this critical mission. Give details of your design, including the estimates, assumptions, and idealizations you make. Note that the first stage of the most powerful rocket ever used, the Saturn V, had a thrust of 3.4×10^7 N and burned for 170 s. Can the Earth be saved?

Section 2.4

For all problems in this section, use the approximation that $\vec{v}_{avg} \approx \vec{p}_f/m$ for each time step.

••P20 Suppose that you are navigating a spacecraft far from other objects. The mass of the spacecraft is 1.5×10^5 kg (about 150 tons). The rocket engines are shut off, and you're coasting along with a constant velocity of $\langle 0, 20, 0 \rangle$ km/s. As you pass the location $\langle 12, 15, 0 \rangle$ km you briefly fire side-thruster rockets, so that your spacecraft experiences a net force of $\langle 6 \times 10^4, 0, 0 \rangle$ N for 20 s. After turning off the thrusters, you then continue coasting with the rocket engines turned off. (The ejected gases have a mass that is very small compared to the mass of the spacecraft.) **(a)** Using a step size of 20 seconds, predict where you will be 40 s after you began firing the thrusters. (Note that for the second step the thrusters are off.) **(b)** Where would you have been if you had not fired the thrusters?

••P21 You throw a metal block of mass 0.25 kg into the air, and it leaves your hand at time $t = 0$ at location $\langle 0, 2, 0 \rangle$ m with velocity $\langle 3, 4, 0 \rangle$ m/s. At this low velocity air resistance is negligible. Using the iterative method shown in Section 2.4 with a time step of 0.05 s, calculate step by step the position and velocity of the block at $t = 0.05$ s, $t = 0.10$ s, and $t = 0.15$ s.

••P22 A small space probe, of mass 240 kg, is launched from a spacecraft near Mars. It travels toward the surface of Mars, where it will land. At a time 20.7 s after it is launched, the probe is at the location $\langle 4.30 \times 10^3, 8.70 \times 10^2, 0 \rangle$ m, and at this same time its momentum is $\langle 4.40 \times 10^4, -7.60 \times 10^3, 0 \rangle$ kg·m/s. At this instant, the net force on the probe due to the gravitational pull of Mars plus the air resistance acting on the probe is $\langle -7 \times 10^3, -9.2 \times 10^2, 0 \rangle$ N. Assuming that the net force on the probe is approximately constant over this time interval, what are the momentum and position of the probe 20.9 s after it is launched? Divide the interval into two time steps, and use the approximation $\vec{v}_{avg} \approx \vec{p}_f/m$.

••P23 A soccer ball of mass 0.43 kg is rolling with velocity $\langle 0, 0, 2.2 \rangle$ m/s, when you kick it. Your kick delivers an impulse of magnitude 1.3 N·s in the $-x$ direction. The net force on the rolling ball, due to the air and the grass, is 0.25 N in the direction opposite to the direction of the ball's momentum. Using a time step of 0.5 s, find the position of the ball at a time 1.5 s after you kick it, assuming that the ball is at the origin at the moment it is kicked. Use the approximation $\vec{v}_{avg} \approx \vec{p}_f/m$.

••P24 As your spaceship coasts toward Mars, you need to move a heavy load of 1200 kg along a hallway of the spacecraft that has a 90° right turn, without touching the walls, floor, or ceiling, by working remotely, using devices attached to the load that can be programmed to fire blasts of compressed air for up to 1.0 s in any desired direction. During a blast the load is subjected to a force of 20 N. The center of the load must move 3 m along the first section of the hallway, starting from rest, then 4 m along the second section, ending at rest. Let the starting point be $\langle 0, 0, 0 \rangle$ m, with the first section ending at $\langle 0, 3, 0 \rangle$ m and the second section ending at $\langle 4, 3, 0 \rangle$ m. Using just three blasts of compressed air, choose the times when these blasts should be scheduled, their durations, and their directions. How long does it take to complete the entire move?

Section 2.5

•P25 A runner starts from rest and in 3 s reaches a speed of 8 m/s. Assume that her speed changed at a constant rate (constant net force). **(a)** What was her average speed during this 3 s interval? **(b)** How far did she go in this 3 s interval?

•P26 The driver of a car traveling at a speed of 18 m/s slams on the brakes and comes to a stop in 4 s. If we assume that the car's speed changed at a constant rate (constant net force): **(a)** What was the car's average speed during this 4 s interval? **(b)** How far did the car go in this 4 s interval?

•P27 On a straight road with the $+x$ axis chosen to point in the direction of motion, you drive for 3 h at a constant 30 mi/h, then in a few seconds you speed up to 60 mi/h and drive at this speed for 1 h. **(a)** What was the x component of average velocity for the 4 h period, using the fundamental definition of average velocity, which is the displacement divided by the time interval? **(b)** Suppose that instead you use the equation $v_{avg,x} = (v_{ix} + v_{fx})/2$. What do you calculate for the x component of average velocity? **(c)** Why does the equation used in part (b) give the wrong answer?

•P28 A ball of mass 0.4 kg flies through the air at low speed, so that air resistance is negligible. **(a)** What is the net force acting on the ball while it is in motion? **(b)** Which components of the ball's momentum will be changed by this force? **(c)** What happens to the x component of the ball's momentum during its flight? **(d)** What happens to the y component of the ball's momentum during its flight? **(e)** What happens to the z component of the ball's momentum during its flight? **(f)** In this situation, why is it legitimate to use the expression for average y component of velocity, $v_{avg,y} = (v_{iy} + v_{fy})/2$, to update the y component of position?

•P29 For each graph of v_x vs. t numbered 1–6 in Figure 2.56, choose the letter (a–i) corresponding to the appropriate description of motion of a fan cart moving along a track. Not all descriptions will be used. Assume the usual coordinate system ($+x$ to the right, $+y$ up, $+z$ out of the page).

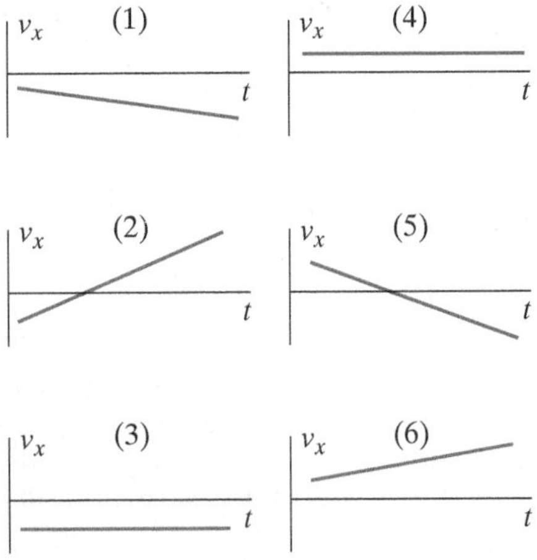

Figure 2.56

(a) A cart moves to the left, gradually slowing down. **(b)** A cart moves to the right, gradually speeding up. **(c)** A cart moves to the left at constant speed. **(d)** A cart moves to the left, gradually slowing down, stops, and moves to the right, speeding up. **(e)** A cart remains stationary and does not move. **(f)** A cart moves to the right, gradually slowing down. **(g)** A cart moves to the right, gradually slowing down, stops, and moves to the left, speeding up. **(h)** A cart moves to the left, gradually speeding up. **(i)** A cart moves to the right at constant speed.

•P30 A cart rolls with low friction on a track. A fan is mounted on the cart, and when the fan is turned on, there is a constant force acting on the cart. Three different experiments are performed: **(a)** Fan off: The cart is originally at rest. You give it a brief push, and it coasts a long distance along the track in the $+x$ direction, slowly coming to a stop. **(b)** Fan forward: The fan is turned on, and you hold the cart stationary. You then take your hand away, and the cart moves forward, in the $+x$ direction. After traveling a long distance along the track, you quickly stop and hold the cart. **(c)** Fan backward: The fan is turned on facing the "wrong" way, and you hold the cart stationary. You give it a brief push, and the cart moves forward, in the $+x$ direction, slowing down and then turning around, returning to the starting position, where you quickly stop and hold the cart. Figure 2.57 displays four graphs of p_x (numbered 1–4), the x component of momentum, vs. time. The graphs start when the cart is at rest, and end when the cart is again at rest. Match the experiment with the correct graph.

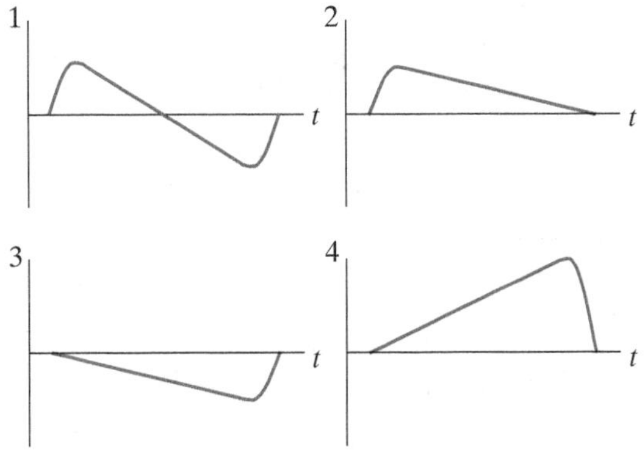

Figure 2.57

•P31 Consider the three experiments described in Problem 30. Figure 2.58 displays four graphs of $F_{net,x}$, the x component of the net force acting on the cart, vs. time. The graphs start when the cart is at rest, and end when the cart is again at rest. Match the experiment with the graph.

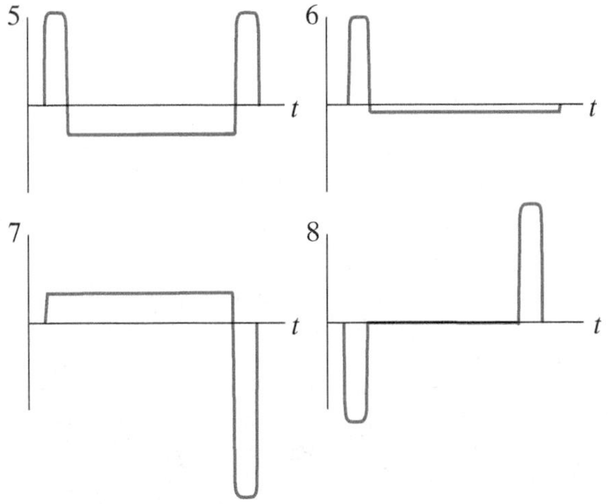

Figure 2.58

•P32 Consider the three experiments described in Problem 30. Figure 2.59 displays four graphs of x, positioned along the track, vs. time. The graphs start when the cart is at rest, and end when the cart is again at rest. Match the experiment with the graph.

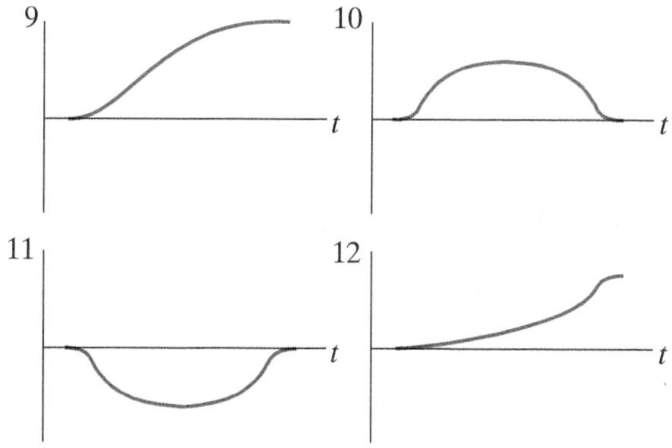

Figure 2.59

••**P33** You are a detective investigating why someone was hit on the head by a falling flowerpot. One piece of evidence is a home video taken in a 4th-floor apartment, which happens to show the flowerpot falling past a tall window. Inspection of individual frames of the video shows that in a span of 6 frames the flowerpot falls a distance that corresponds to 0.85 of the window height seen in the video. (*Note:* Standard video runs at a rate of 30 frames per second.) You visit the apartment and measure the window to be 2.2 m high. What can you conclude? Under what assumptions? Give as much detail as you can.

••**P34** A soccer ball is kicked at an angle of 59° to the horizontal with an initial speed of 20 m/s. Assume for the moment that we can neglect air resistance. **(a)** For how much time is the ball in the air? **(b)** How far does it go (horizontal distance along the field)? **(c)** How high does it go?

••**P35** A ball is kicked from a location $\langle 9,0,-6\rangle$ (on the ground) with initial velocity $\langle -11,16,-6\rangle$ m/s. The ball's speed is low enough that air resistance is negligible. **(a)** What is the velocity of the ball 0.5 s after being kicked? (Use the Momentum Principle!) **(b)** In this situation (constant force), which velocity will give the most accurate value for the location of the ball 0.5 s after it is kicked: the arithmetic average of the initial and final velocities, the final velocity of the ball, or the initial velocity of the ball? **(c)** What is the average velocity of the ball over this time interval (a vector)? **(d)** Use the average velocity to find the location of the ball 0.5 s after being kicked.

Now consider a different time interval: the interval between the initial kick and the moment when the ball reaches its highest point. We want to find how long it takes for the ball to reach this point, and how high the ball goes. **(e)** What is the y component of the ball's velocity at the instant when the ball reaches its highest point (the end of this time interval)? **(f)** Fill in the known quantities in the update form of the Momentum Principle, $mv_{yf} = mv_{yi} + F_{\text{net},y}\,\Delta t$, leaving as symbols anything that is unknown. **(g)** How long does it take for the ball to reach its highest point? **(h)** Knowing this time, first find the y component of the average velocity during *this* time interval, then use it to find the maximum height attained by the ball.

Now take a moment to reflect on the reasoning used to solve this problem. You should be able to do a similar problem on your own, without prompting. Note that the only equations needed were the Momentum Principle and the expression for the arithmetic average velocity.

••**P36** A small dense ball with mass 1.5 kg is thrown with initial velocity $\langle 5,8,0\rangle$ m/s at time $t=0$ at a location we choose to call the origin ($\langle 0,0,0\rangle$). Air resistance is negligible. **(a)** When the ball reaches its maximum height, what is its velocity (a vector)? It may help to make a simple diagram. **(b)** When the ball reaches its maximum height, what is t? You know how v_y depends on t, and you know the initial and final velocities. **(c)** Between the launch at $t=0$ and the time when the ball reaches its maximum height, what is the average velocity (a vector)? You know how to determine average velocity when velocity changes at a constant rate. **(d)** When the ball reaches its maximum height, what is its location (a vector)? You know how average velocity and displacement are related. **(e)** At a later time the ball's height y has returned to zero, which means that the average value of v_y from $t=0$ to this time is zero. At this instant, what is the time t? **(f)** At the time calculated in part (e), when the ball's height y returns to zero, what is x? (This is called the "range" of the trajectory.) **(g)** At the time calculated in part (e), when the ball's height y returns to zero, what is v_y? **(h)** What was the angle to the x axis of the initial velocity? **(i)** What was the angle to the x axis of the velocity at the time calculated in part (e), when the ball's height y returned to zero?

••**P37** Apply the general results obtained in the full analysis of motion under the influence of a constant force in Section 2.5 to answer the following questions. You hold a small metal ball of mass m a height h above the floor. You let go, and the ball falls to the floor. Choose the origin of the coordinate system to be on the floor where the ball hits, with y up as usual. Just after release, what are y_i and v_{iy}? Just before hitting the floor, what is y_f? How much time Δt does it take for the ball to fall? What is v_{fy} just before hitting the floor? Express all results in terms of m, g, and h. How would your results change if the ball had twice the mass?

••**P38** In a cathode ray tube (CRT) used in older television sets, a beam of electrons is steered to different places on a phosphor screen, which glows at locations hit by electrons. The CRT is evacuated, so there are few gas molecules present for the electrons to run into. Electric forces are used to accelerate electrons of mass m to a speed $v_0 \ll c$, after which they pass between positively and negatively charged metal plates that deflect the electron in the vertical direction (upward in Figure 2.60, or downward if the sign of the charges on the plates is reversed).

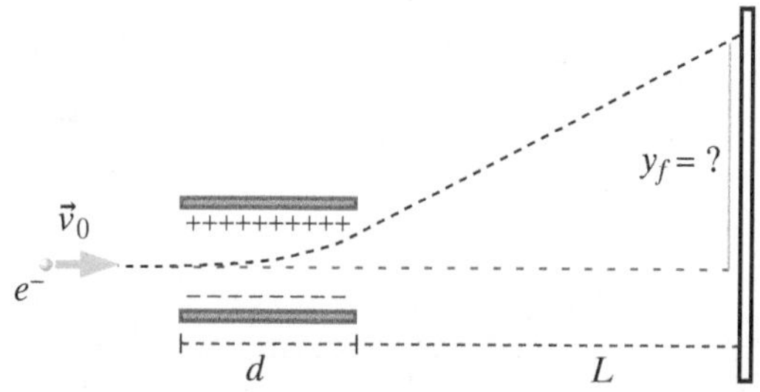

Figure 2.60

While an electron is between the plates, it experiences a uniform vertical force F, but when the electron is outside the plates there is negligible force on it. The gravitational force on the electron is negligibly small compared to the electric force in this situation. The length of the metal plates is d, and the phosphor screen is a distance L from the metal plates. Where does the electron hit the screen? (That is, what is y_f?)

••P39 The performance of two different cars, car 1 and car 2, was measured on a long horizontal test track. Car 1 started from rest and ran with constant acceleration until it was halfway down the track and then stopped accelerating, continuing to run at the attained speed to the end of the track. Car 2 started from rest and ran with a constant acceleration for the entire distance. It was observed that both cars covered the test distance in the same amount of time. **(a)** What was the ratio of the average speed of car 1 to that of car 2? **(b)** What was the ratio of the initial acceleration of car 1 to that of car 2? **(c)** What was the ratio of the final speed of car 1 to that of car 2?

••P40 A driver starts from rest on a straight test track that has markers every 0.1 km. The driver presses on the accelerator and for the entire period of the test holds the car at constant acceleration. The car passes the 0.1 km post at 8 s after starting the test. **(a)** What was the car's acceleration? **(b)** What was the car's speed as it passed the 0.1 km post? **(c)** What does the speedometer read at the 0.2 km post? **(d)** When does the car pass the 0.2 km post?

Section 2.6

For all problems in this section, use the approximation that $\vec{v}_{avg} \approx \vec{p}_f/m$ for each time step.

•P41 The stiffness of a particular spring is 40 N/m. One end of the spring is attached to a wall. When you pull on the other end of the spring with a steady force of 2 N, the spring elongates to a total length of 18 cm. What was the relaxed length of the spring? (Remember to convert to SI units.)

••P42 A spring with a relaxed length of 25 cm and a stiffness of 11 N/m stands vertically on a table. A block of mass 70 g is attached to the top of the spring. You pull the block upward, stretching the spring until its length is now 28 cm, hold the block at rest for a moment, and then release it. Using a time step of 0.1 s, predict the position and momentum of the block at a time 0.2 s after you release the block.

••P43 A block is attached to the top of a spring that stands vertically on a table. The spring stiffness is 55 N/m, its relaxed length is 23 cm, and the mass of the block is 350 g. The block is oscillating up and down as the spring stretches and compresses. At a particular time you observe that the velocity of the block is $\langle 0, 0.0877, 0 \rangle$ m/s, and the position of the block is $\langle 0, 0.0798, 0 \rangle$ m, relative to an origin at the base of the spring. Using a time step of 0.1 s, determine the position of the block 0.2 s later.

••P44 A paddle ball toy consists of a flat wooden paddle and a small rubber ball that are attached to each other by an elastic band (Figure 2.61). You have a paddle ball toy for which the mass of the ball is 0.015 kg, the stiffness of the elastic band is 0.9 N/m, and the relaxed length of the elastic band is 0.30 m. You are holding the paddle so the ball hangs suspended under it, when your cat comes along and bats the ball around, setting it in motion. At a particular instant the the momentum of the ball is $\langle -0.02, -0.01, -0.02 \rangle$ kg · m/s, and the moving ball is at location $\langle -0.2, -0.61, 0 \rangle$ m relative to an origin located at the point where the elastic band is attached to the paddle. **(a)** Determine the position of the ball 0.1 s later, using a Δt of 0.1 s. **(b)** Starting with the same initial position ($\langle -0.2, -0.61, 0 \rangle$ m) and momentum ($\langle -0.02, -0.01, -0.02 \rangle$ kg · m/s), determine the position of the ball 0.1 s later, using a Δt of 0.05 s. **(c)** If your answers are different, explain why.

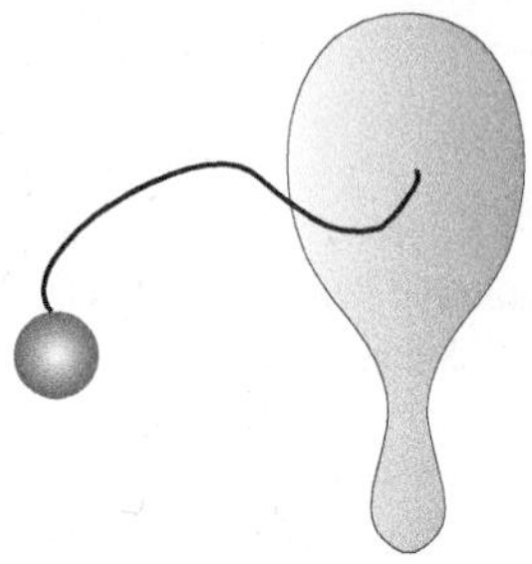

Figure 2.61

••P45 A block of mass 0.5 kg is placed at rest on a relaxed vertical spring-like device of length 0.1 m which exerts an upward force of $F_y = -2 \times 10^5 \times s^3$, where s is the stretch. With a time step of 0.04 s, calculate the length L of the device and the y component of the velocity of the block at $t = 0.04s$ and $t = 0.08s$.

Section 2.9

•P46 A proton has mass 1.7×10^{-27} kg. What is the magnitude of the impulse required to increase its speed from $0.990c$ to $0.994c$?

••P47 SLAC, the Stanford Linear Accelerator Center, located at Stanford University in Palo Alto, California, accelerates electrons through a vacuum tube 2 mi long (it can be seen from an overpass of the Junipero Serra freeway that goes right over the accelerator). Electrons that are initially at rest are subjected to a continuous force of 2×10^{-12} newton along the entire length of 2 mi (1 mi is 1.6 km) and reach speeds very near the speed of light. **(a)** Determine how much time is required to increase the electrons' speed from $0.93c$ to $0.99c$. (That is, the quantity $|\vec{v}|/c$ increases from 0.93 to 0.99.) **(b)** Approximately how far does the electron go in this time? What is approximate about your result?

COMPUTATIONAL PROBLEMS

You should do the introductory computational problems at the end of Chapter 1 before doing these problems.

More detailed and extended versions of some of these computational modeling problems may be found in the lab activities included in the *Matter & Interactions, 4th Edition*, resources for instructors.

A note on graphs: Graphs in VPython appear in a separate window. A graph is dynamic; it appears point by point as the program runs, and scales itself automatically. To make a graph in VPython requires three things, as shown in the program shown below:

(1) At the beginning of the program, type these two lines:
```
from visual import *
from visual.graph import *
```
(2) Create one or more `gcurve` objects, such as the one called `speed` in the program shown below.

(3) Inside the computational loop, add one or more `plot` operations specifying the pair of values to plot, such as the statement `speed.plot( pos=(t, v) )` in the program shown below.

Consult the VPython help for full details on graphing.

```
from visual import *
# Import graphing:
from visual.graph import *
...
# Create a graphing curve:
speed = gcurve(color=color.yellow)
...
while ship.pos.x < L:
    rate(20000)
    # update position and momentum:
    ....
    # Calculate speed v, and
    # add to the graphing curve:
    speed.plot( pos=(t, v) )
    # Update the time:
    t = t + dt
```

••P48 Write an iterative computational model that predicts and displays (as a real-time animation) the motion of a fan cart on a low-friction track. A typical track is 2 m long, and a typical fan cart (which can be represented by a box object) is about 10 cm long and has a mass of about 0.8 kg (including the fan). Use a time step $\Delta t = 0.01$ s. Run your program often while writing it. It is much easier to find and fix errors when creating the program incrementally. **(a)** Start the cart at the left edge of the track, and give it an initial velocity of $\langle 0.5, 0, 0 \rangle$ m/s. **(b)** With the fan off ($\vec{F}_{\text{net}} = \vec{0}$) determine how long it takes for the cart to reach the right end of the track. **(c)** To model the behavior of the cart with the fan on, add a constant force to your model. By experimenting with your program, find values for the force and the initial velocity that lead to this behavior: The cart starts at the right end of the track and moves to the left, gradually slowing down. It stops at the left end of the track, and moves back to the right, gradually speeding up. **(d)** What happens when the initial velocity of the cart has a nonzero y component? Make the cart leave a trail so you can see its path clearly. Explain why your model predicts this behavior.

••P49 Starting with the program you wrote in Problem P48, **(a)** Add commands to create a graph of the x component of the cart's position (`cart.pos.x`) vs. time as the cart moves. Explain the shape of the graph. **(b)** Change your graph to plot the x component of the cart's momentum vs. time. Explain the shape of the graph.

••P50 Write an iterative computational model that predicts and displays the 3D motion of a 55 g tennis ball that leaves a tennis racket with a speed of 55 m/s. Neglect air resistance (at this speed, this approximation is actually not realistic). Experiment by changing the initial velocity of the ball to see what kinds of different trajectories can be produced.

•••P51 The nearest stars are a group of three stars orbiting each other, Alpha Centauri A, Alpha Centauri B, and Proxima Centauri, located about 4.3 light years from Earth (one light year is the distance light travels in one Earth year). Suppose a space tug is able to pull a cargo ship with a constant force of $g = 9.8$ N/kg times the mass of the cargo ship, for many years. Starting from rest, speed up the cargo ship until you're halfway to the nearest stars, then pull back with the same force to slow the cargo ship back to rest when you reach the nearest stars. Determine the maximum speed attained (which occurs at the halfway point) and how many Earth years the trip takes. You must use the relativistic position update equation, because the speed approaches the speed of light. It is interesting to make a graph of the speed as a function of time, which does not consist of straight lines as it would for nonrelativistic motion. Time passes slowly on board the cargo ship, a relativistic effect that we'll discuss in Chapter 20.

ANSWERS TO CHECKPOINTS

1 (1) $\langle 60, -60, 0 \rangle$ N; **(2) (a)** $\langle 0, -2, 0 \rangle$ kg·m/s **(b)** $\langle 0, -0.667, 0 \rangle$ N
2 (a) By estimating the distance one student "traveled" during the collision–that is, how much the student's body was compressed. Using the approximation that $v_{\text{avg}} \approx (v_i + 0)/2$ we divided distance by average speed to estimate contact time. **(b)** Only the z component of $\vec{p}$ changed, so we were able to deduce that $\vec{F}_{\text{net}}$ was in the z direction.
3 (1) b **(2)** a
4 (1) c **(2)** c
5 (a) $\langle -10, 7.12, -5 \rangle$ m/s **(b)** $\langle 3, 6.04, -8 \rangle$ m **(c)** 8.62 m
6 (1) 0.275 N **(2) (a)** 3571 N/m **(b)** 71.4 N
7 (1) **(a)** $\langle 0, 1, 0 \rangle$, **(b)** $\langle 0, 1, 0 \rangle$; (2) **(a)** $\langle 0, 0.11, 0 \rangle$ m, **(b)** 0.11 m, **(c)** $\langle 0, 1, 0 \rangle$, **(d)** -0.04 m, **(e)** $\langle 0, 3.8, 0 \rangle$ N
8 (a) The momentum of the block a short time in the future depends on two things: its momentum now and the net force acting on it now. At the beginning of time step 2 the net force is downward but the momentum is upward. The sum of the momentum now plus the downward impulse over the next Δt gives a new momentum that is smaller in magnitude, but still upward, so the block moves upward.
9 a
10 a, b, c, d

CHAPTER

3

The Fundamental Interactions

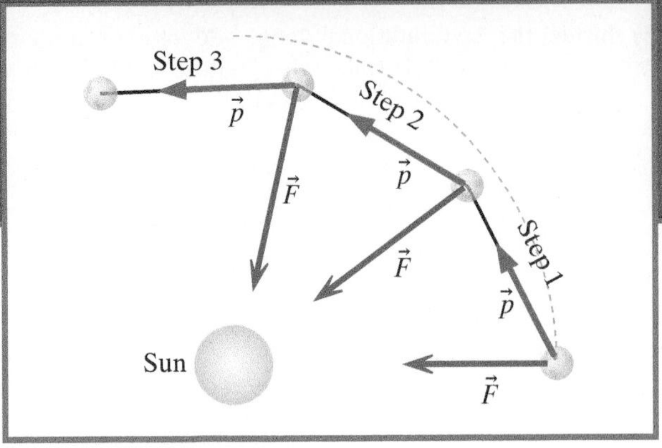

OBJECTIVES

After studying this chapter you should be able to

- Calculate the 3D gravitational or electric force exerted on a system by objects in its surroundings.
- Iteratively predict the motion of an object that interacts gravitationally or electrically with its surroundings, by hand or with a computer.
- Analyze simple collisions by applying the Momentum Principle to a system of more than one particle.

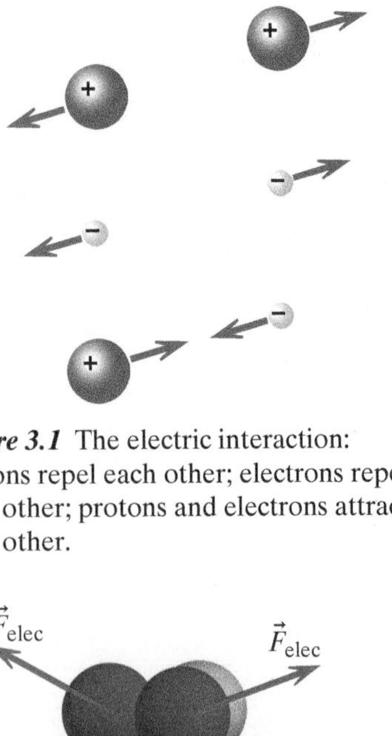

Figure 3.1 The electric interaction: protons repel each other; electrons repel each other; protons and electrons attract each other.

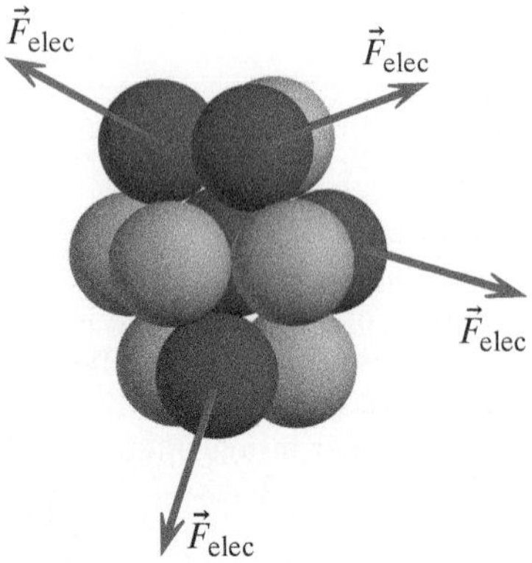

Figure 3.2 The strong interaction: the protons in the nucleus of an atom exert repulsive electric forces on each other, but the strong interaction (which involves neutrons as well as protons) holds the nucleus together despite this electric repulsion.

3.1 THE FUNDAMENTAL INTERACTIONS

In the world we see what seem to be many different kinds of interactions. Planets orbit stars. A falling leaf whirls in the wind. You bend a metal rod. Carbon and oxygen react to form carbon dioxide. In a nuclear power station, uranium nuclei split, making water boil, which drives electric generators. Despite the variety of effects we observe, it became clear in the 20th century that all the changes we see are due to just four different kinds of fundamental interactions: gravitational, electromagnetic, "strong" (also referred to as the nuclear interaction), and "weak."

- The *gravitational* interaction is responsible for an attraction between objects that have mass. For example, the Earth exerts a gravitational force on the Moon, and the Moon exerts a gravitational force on the Earth.
- The *electromagnetic* interaction is responsible for attraction or repulsion between objects that have electric charge. Electric forces are responsible for sparks, static cling, and the behavior of electronic circuits, and magnetic forces are responsible for the operation of motors driven by electric current. Protons repel each other electrically, as do electrons, whereas protons and electrons attract each other (Figure 3.1). Electric forces bind protons and electrons to each other in atoms, and are responsible for the chemical bonds between atoms in molecules. The force of a stretched or compressed spring is due to electric forces between the atoms that make up the spring.
- The *strong* or nuclear interaction occurs between objects made of quarks, such as protons and neutrons, which are held together in the nucleus of an atom despite the large mutual electric repulsion of the protons (Figure 3.2). (The neutrons are not electrically charged and don't exert electric forces.)
- The *weak* interaction affects all kinds of elementary particles but is much weaker than the strong and electromagnetic interactions. An example of its effects is seen in the instability of a neutron. If a neutron is removed from a nucleus, with an average lifetime of about 15 minutes the neutron decays

into a proton, an electron, and a ghostly particle called the antineutrino. This change is brought about by the weak interaction.

In this chapter we will be concerned primarily with gravitational and electric interactions. In later chapters dealing with energy we will encounter situations in which the strong interaction plays an important role. An example of the weak interaction appears in the section on conservation of momentum later in this chapter. The weak interaction is not important in most everyday interactions, so we will not discuss it extensively. We will defer a discussion of magnetic interaction, the other part of the electromagnetic interaction, until later chapters.

Although it continues to be fruitful to classify interactions into four types, it was found in the second half of the 20th century that the electromagnetic interaction and the weak interaction can be considered to be different manifestations of one type of interaction, now called the "electroweak" interaction. Soon after this discovery, it became possible to unify the strong interaction and the electroweak interaction within one powerful theory, the "Standard Model," which also explains the nature of subatomic particles such as the proton and neutron. At present there seem to be really only two fundamental categories of interactions: those explained by the Standard Model and those explained by the gravitational interaction. Physicists are aggressively searching for ways to unify the Standard Model with gravity; this is one of the major scientific quests of the present era.

3.2 THE GRAVITATIONAL FORCE

The motion of stars and planets is in important ways simpler than other mechanical phenomena, because there is no friction to worry about. These massive bodies interact through the gravitational force, which is always an attractive force. Studying how to predict the motion of stars and planets is one of the most direct ways to understand in general how the Momentum Principle determines the behavior of objects in the real world. The basic ideas used to predict the motion of stars and planets can be applied to a wide range of everyday and atomic phenomena.

In the 1600s Isaac Newton deduced that there must be an attractive force associated with a gravitational interaction between any pair of objects. The gravitational force acts along a line connecting the two objects (Figure 3.3), is proportional to the mass of one object and to the mass of the other object, and is inversely proportional to the square of the distance between the *centers* of the two objects (not the gap between their surfaces). The gravitational force exerted on object 2 by object 1 is expressed by this compact vector equation:

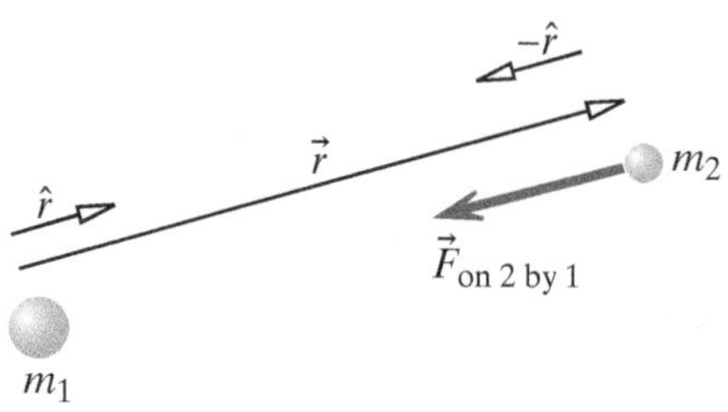

Figure 3.3 The gravitational force exerted on object 2 by object 1.

THE GRAVITATIONAL FORCE

$$\vec{F}_{\text{grav on 2 by 1}} = -G\frac{m_1 m_2}{|\vec{r}|^2}\hat{r}$$

m_1 and m_2 are the masses of the objects.
$\vec{r} = \vec{r}_2 - \vec{r}_1$ extends from the center of object 1 to the center of object 2.

The universal constant $G = 6.7 \times 10^{-11} \dfrac{\text{N}\cdot\text{m}^2}{\text{kg}^2}$.

This is often called Newton's "universal law of gravitation," reflecting its central importance in the development of human understanding of the Universe. The word "universal" is used because all massive objects in the Universe attract all other massive objects in the Universe.

The gravitational force involves a lot of different symbols and may look pretty intimidating at first. Let's take the expression apart and look at the individual pieces to try to make sense of the expression.

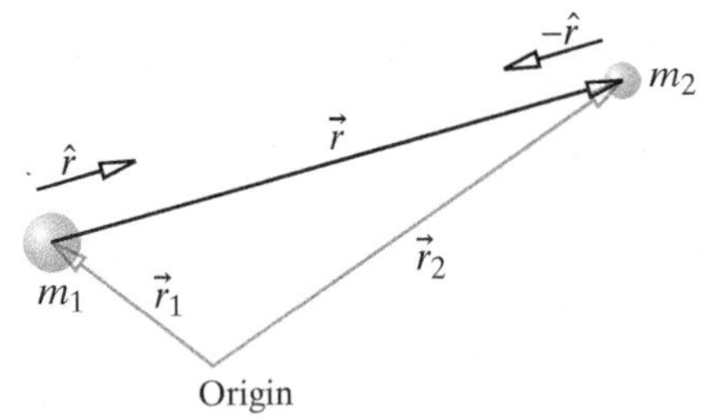

Figure 3.4 $\vec{r} = \vec{r}_2 - \vec{r}_1$, the location of object 2 relative to object 1: "final minus initial." The gravitational force on object 2 is in the direction of $-\hat{r}$.

The Relative Position Vector

The relative position vector $\vec{r}$ extends from the center of object 1 to the center of object 2, as shown in Figure 3.4. (This vector can also be written as $\vec{r}_{2-1}$; we will use the notation $\vec{r}$, which is more compact, but requires that you remember that object 1 is the initial location and object 2 is the final location.) The unit vector $\hat{r}$ points in the same direction as $\vec{r}$, but has magnitude 1. The magnitude of $\vec{r}$ is the distance between the centers of the two objects.

The Direction of the Gravitational Force

The direction of the gravitational force on object 2 by object 1 is specified by $-\hat{r}$ (Figure 3.4), which is the unit vector $\hat{r}$, in combination with the minus sign, as indicated in Figure 3.5.

$$\vec{F}_{\text{on 2 by 1}} = -G\frac{m_1 m_2}{|\vec{r}|^2}\hat{r}$$

Figure 3.5 The direction of the gravitational force is opposite to the direction of the unit vector $\hat{r}$, which points from object 1 to object 2.

QUESTION Why is the minus sign necessary?

The minus sign is necessary because the force on object 2 due to object 1 is in the direction opposite to $\hat{r}$ (Figure 3.4).

Mass and Magnitude of the Gravitational Force

As highlighted in Figure 3.6, the gravitational force is proportional to the product of the two masses, $m_1 m_2$. If you double either of these masses, keeping the other one the same, the force will be twice as big.

$$\vec{F}_{\text{on 2 by 1}} = -G\frac{m_1 m_2}{|\vec{r}|^2}\hat{r}$$

Figure 3.6 The magnitude of the gravitational force depends on the masses of both interacting objects.

QUESTION If both masses are doubled, how much larger will the force be?

If you double both of the masses, the force will be four times as big.

Note that since $m_2 m_1 = m_1 m_2$, the magnitude of the force exerted on object 1 by object 2 is exactly the same as the magnitude of the force exerted on object 2 by object 1, but the direction is opposite. This important property of the gravitational force is called "reciprocity" and is discussed in more detail in Section 3.4.

Distance and Magnitude of the Gravitational Force

$$\vec{F}_{\text{on 2 by 1}} = -G\frac{m_1 m_2}{|\vec{r}|^2}\hat{r}$$

Figure 3.7 The gravitational force is an "inverse square" force.

The gravitational force is an "inverse square" force. As highlighted in Figure 3.7, the square of the center-to-center distance appears in the denominator. This means that the gravitational force depends very strongly on the distance between the objects. For example, if you double the distance between them, the only thing that changes is the denominator, which gets four times bigger (2 squared is 4), so the force is only 1/4 as big as before:

$$|\vec{F}_d| = \frac{Gm_1 m_2}{|\vec{r}|^2}$$

$$|\vec{F}_{2d}| = \frac{Gm_1 m_2}{|2\vec{r}|^2} = \frac{1}{4}\frac{Gm_1 m_2}{|\vec{r}|^2} = \frac{1}{4}|\vec{F}_d|$$

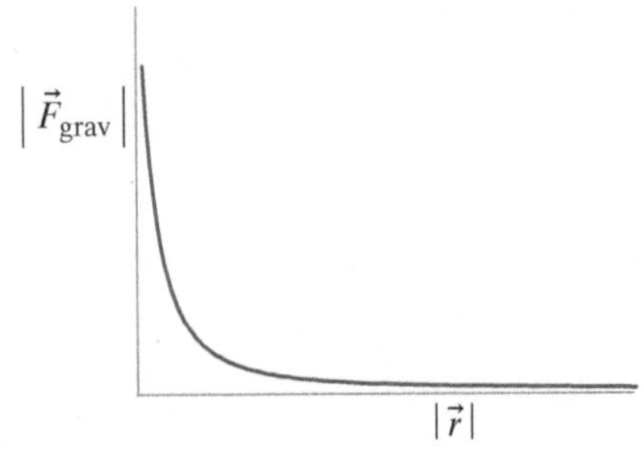

Figure 3.8 The magnitude of an inverse square force is large at small distances, and very small at large distances.

QUESTION If you move the masses 10 times farther apart than they were originally, how does the gravitational force change?

The force goes down by a factor of 100. Evidently when two objects are very far apart, the gravitational forces they exert on each other will be vanishingly small: big denominator, small force (Figure 3.8).

The Gravitational Constant *G*

$$G = 6.7 \times 10^{-11} \frac{\text{N} \cdot \text{m}^2}{\text{kg}^2}$$

We say that the gravitational constant G is "universal" because it is the same for any pair of interacting masses, no matter how big or small they are, or where they are located in the Universe. Because G is universal, it can be measured for a particular pair of objects and then used with other pairs of objects. As we describe in Section 3.17, Cavendish was the first person to make such a measurement.

Checkpoint 1 A star exerts a gravitational force of magnitude 4×10^{25} N on a planet. **(a)** What is the magnitude of the gravitational force that the planet exerts on the star? **(b)** If the mass of the planet were twice as large, what would be the magnitude of the gravitational force on the planet? **(c)** If the distance between the star and planet (with their original masses) were three times larger, what would be the magnitude of this force?

Calculating Gravitational Force

A useful way to think about the gravitational force is to factor it into a scalar part and a direction, like this:

A vector is a magnitude times a direction: $\vec{F}_{\text{grav}} = |\vec{F}_{\text{grav}}|\hat{F}_{\text{grav}}$

- Magnitude: $|\vec{F}_{\text{grav}}| = G\dfrac{m_1 m_2}{|\vec{r}|^2}$
- Direction (unit vector): $\hat{F}_{\text{grav}} = -\hat{r}$

It is often simplest to calculate the magnitude and direction separately, then combine them to get the vector force. That way you can focus on one thing at a time rather than getting confused (or intimidated!) by the full complexity of the vector force.

Calculating the gravitational force on an object due to another object requires several steps. Use the equation for gravitational force as a guide that tells you to carry out these steps:

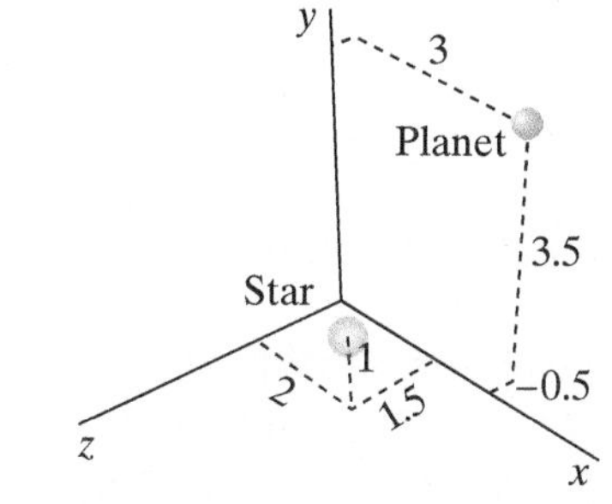

Figure 3.9 A star and a planet interact gravitationally. Distances shown should be multiplied by 1×10^{11} m.

CALCULATING GRAVITATIONAL FORCE

- Calculate $\vec{r} = \vec{r}_2 - \vec{r}_1$, the position of the center of object 2 relative to the center of object 1.
- Calculate $|\vec{r}|$, the center-to-center distance between the objects.
- Calculate $Gm_1m_2/|\vec{r}|^2$, the magnitude of the force.
- Calculate $-\hat{r} = -\vec{r}/|\vec{r}|$, the direction of the force.
- Multiply the magnitude times the direction to get the vector force.

EXAMPLE **The Force on a Planet by a Star**

Figure 3.9 shows a star of mass 4×10^{30} kg located at this moment at position $\langle 2 \times 10^{11}, 1 \times 10^{11}, 1.5 \times 10^{11} \rangle$ m and a planet of mass 3×10^{24} kg located at position $\langle 3 \times 10^{11}, 3.5 \times 10^{11}, -0.5 \times 10^{11} \rangle$ m. (These are typical values for stars and planets. Notice that the mass of the star is much greater than that of the planet.) In Figure 3.9 we show the x, y, and z components of the positions, to be multiplied by 1×10^{11} m. Make sure you understand how the numbers on the diagram correspond to the positions given as vectors. **(a)** Calculate the gravitational force exerted on the planet by the star. **(b)** Calculate the gravitational force exerted on the star by the planet.

Solution

QUESTION Which object is object 1, and which is object 2?

In this case **(a)** the planet is object 2 (we want the force on the planet), and the star is object 1.

Relative position vector:

$$\vec{r} = \vec{r}_2 - \vec{r}_1$$
$$\vec{r} = \langle 3\times10^{11}, 3.5\times10^{11}, -0.5\times10^{11}\rangle\ \text{m} - \langle 2\times10^{11}, 1\times10^{11}, 1.5\times10^{11}\rangle\ \text{m}$$
$$\vec{r} = \langle 1\times10^{11}, 2.5\times10^{11}, -2\times10^{11}\rangle\ \text{m}$$

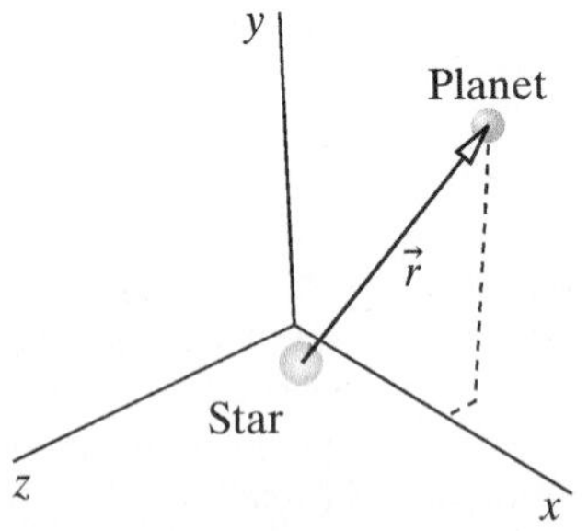

Figure 3.10 The position vector of the planet relative to the star. Initial location: center of star. Final location: center of planet.

Check: Do the signs of the components make sense? Yes, see Figure 3.10.

Distance:

$$|\vec{r}| = \sqrt{(1\times10^{11})^2 + (2.5\times10^{11})^2 + (-2\times10^{11})^2}\ \text{m}$$
$$|\vec{r}| = 3.35\times10^{11}\ \text{m}$$

Magnitude of force:

$$|\vec{F}_{\text{on}\,P\,\text{by}\,S}| = G\frac{m_1 m_2}{|\vec{r}|^2}$$
$$|\vec{F}_{\text{on}\,P\,\text{by}\,S}| = \left(6.7\times10^{-11}\frac{\text{N}\cdot\text{m}^2}{\text{kg}^2}\right)\frac{(3\times10^{24}\ \text{kg})(4\times10^{30}\ \text{kg})}{(3.35\times10^{11}\ \text{m})^2}$$
$$|\vec{F}_{\text{on}\,P\,\text{by}\,S}| = 7.15\times10^{21}\ \text{N}$$

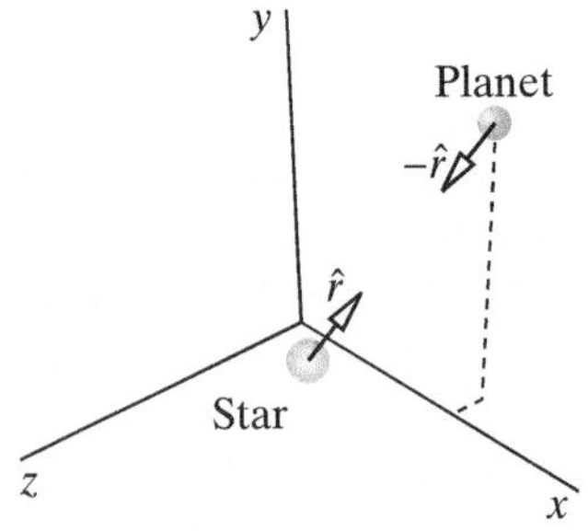

Figure 3.11 Unit vectors.

Direction of force on planet:

$$\hat{F}_{\text{on}\,P\,\text{by}\,S} = -\hat{r} = -\frac{\vec{r}}{|\vec{r}|}$$
$$\hat{F}_{\text{on}\,P\,\text{by}\,S} = -\frac{\langle 1\times10^{11}, 2.5\times10^{11}, -2\times10^{11}\rangle\ \text{m}}{3.35\times10^{11}\ \text{m}}$$
$$\hat{F}_{\text{on}\,P\,\text{by}\,S} = \langle -0.298, -0.745, 0.596\rangle$$

Check: Units? Okay, unit vector has no units. Direction? Figure 3.11 shows that $\hat{F}_{\text{on}\,P\,\text{by}\,S}$ points from the planet toward the star, which is correct for the direction of the force. Magnitude of the unit vector? $\sqrt{(-0.298)^2 + (-0.745)^2 + (0.596)^2} = 0.9995$, okay to three significant figures.

Calculate force as a vector:

$$\vec{F}_{\text{on}\,P\,\text{by}\,S} = |\vec{F}_{\text{on}\,P\,\text{by}\,S}|\hat{F}_{\text{on}\,P\,\text{by}\,S}$$
$$\vec{F}_{\text{on}\,P\,\text{by}\,S} = (7.15\times10^{21}\ \text{N})\langle -0.298, -0.745, 0.596\rangle$$
$$\vec{F}_{\text{on}\,P\,\text{by}\,S} = \langle -2.13\times10^{21}, -5.33\times10^{21}, 4.26\times10^{21}\rangle\ \text{N}$$

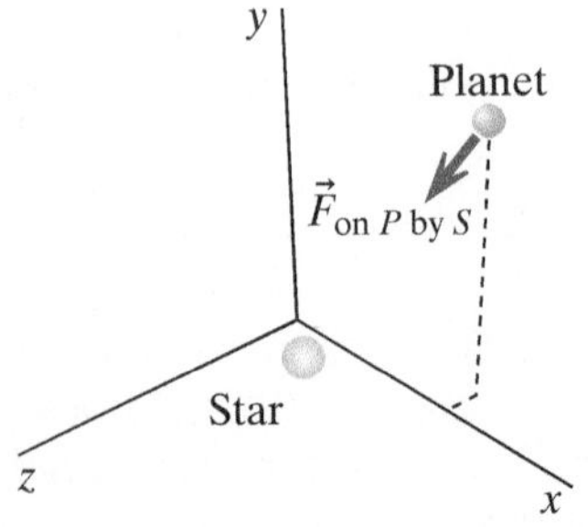

Figure 3.12 Gravitational force on the planet by the star.

Check: Direction? Figure 3.12 shows the force in the correct direction. Magnitude? Do a rough "order of magnitude" check: the distance between star and planet is *very* roughly 1×10^{11} m, the mass of the star is *very* roughly 1×10^{30} kg, the mass of the planet is *very* roughly 1×10^{24} kg, and the gravitational constant is *very* roughly $1\times10^{-11}\ \text{N}\cdot\text{m}^2/\text{kg}^2$. Therefore we expect the magnitude of the force to be *very* roughly this:

$$|\vec{F}_{\text{on}\,P\,\text{by}\,S}| \approx \left(1\times10^{-11}\frac{\text{N}\cdot\text{m}^2}{\text{kg}^2}\right)\frac{(1\times10^{24}\ \text{kg})(1\times10^{30}\ \text{kg})}{(1\times10^{11}\ \text{m})^2} = 1\times10^{21}\ \text{N}$$

The fact that we get within an order of magnitude of the result 7.15×10^{21} N is evidence that we haven't made any huge mistakes.

The magnitude of the force on the star by the planet (part **b**) will be the same. The only change is in the direction, because $\vec{r}_{S-P} = -\vec{r}_{P-S}$.

$$\vec{F}_{\text{on } S \text{ by } P} = \langle 2.13 \times 10^{21}, 5.33 \times 10^{21}, -4.26 \times 10^{21} \rangle \text{ N}$$

Note that the signs of the force components for the force on the star are consistent with Figure 3.13. You might be puzzled that the planet pulls just as hard on the star as the star pulls on the planet. We'll discuss this in more detail later in this chapter.

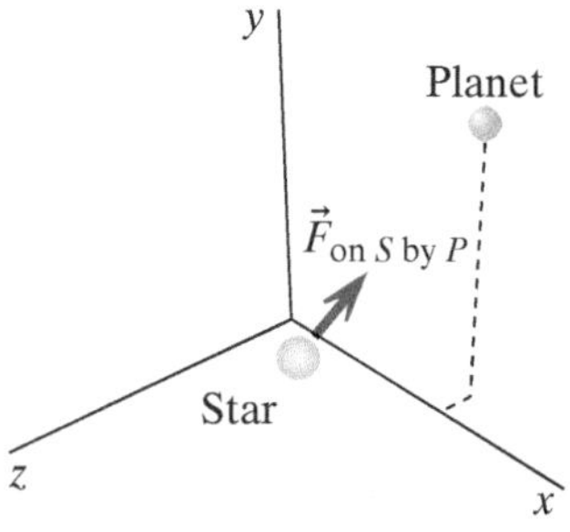

Figure 3.13 The gravitational force exerted on the star by the planet.

Checkpoint 2 At a particular instant Ganymede and Europa, two of the moons of Jupiter, are aligned as shown in Figure 3.14. Coordinate axes are shown in the diagram. **(a)** In calculating the gravitational force on Ganymede by Europa: (1) What is the direction of $\vec{r}$? (2) What is the direction of $-\hat{r}$? (3) What is the direction of the gravitational force? **(b)** In calculating the gravitational force on Europa by Ganymede: (1) What is the direction of $\vec{r}$? (2) What is the direction of $-\hat{r}$? (3) What is the direction of the gravitational force?

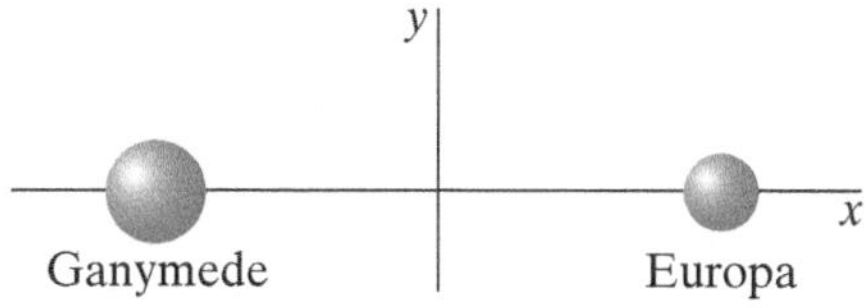

Figure 3.14 Two of Jupiter's moons, aligned along the x axis.

3.3 APPROXIMATE GRAVITATIONAL FORCE NEAR THE EARTH'S SURFACE

Earlier we used the expression mg to represent the magnitude of the gravitational force on an object near the Earth's surface. This is an approximation to the actual force, but it is a good one. The magnitude of the gravitational force that the Earth exerts on an object of mass m near the Earth's surface (Figure 3.15) is

$$F_g = G\frac{M_E m}{(R_E + y)^2}$$

where y is the distance of the object above the surface of the Earth, R_E is the radius of the Earth, and M_E is the mass of the Earth.

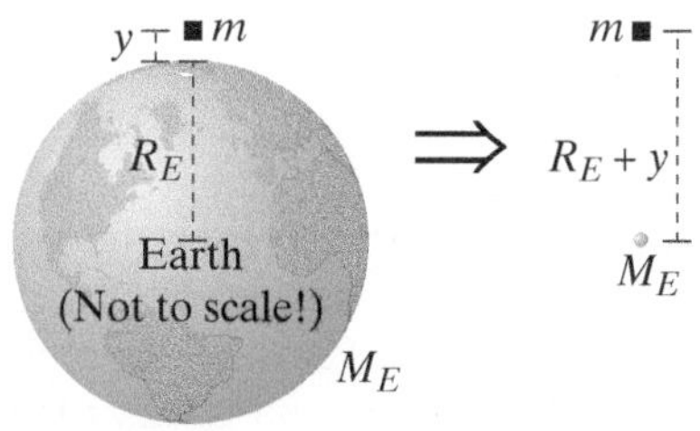

Figure 3.15 Determining the gravitational force by the Earth on an object a height y above the surface. The nearly spherical Earth acts as if all its mass were concentrated at its center.

A spherical object of uniform density can be treated as if all its mass were concentrated at its center (see Section 3.16 for a discussion of this). As a result, we can treat the effect of the Earth on the object as though the Earth were a tiny, very dense ball a distance $R_E + y$ away.

Consider the difference in the gravitational force exerted by the Earth on an object when it is one meter above the surface compared with when it is two meters above the surface. The radius of the Earth is about 6.4×10^6 m. Then

$$F_{1\text{ m}} = G\frac{M_E m}{(6.400001 \times 10^6 \text{ m})^2} \quad \text{whereas} \quad F_{2\text{ m}} = G\frac{M_E m}{(6.400002 \times 10^6 \text{ m})^2}$$

For most purposes this difference is not significant. In fact, for all interactions of objects near the surface of the Earth, it makes sense to use the same approximate value, $R_E = 6.4 \times 10^6$ m, for the distance from the center of the Earth to the object. This simplifies calculation of gravitational forces by allowing us to combine all the constants into a single lumped constant, g:

$$F_g \approx \left(G\frac{M_E}{R_E^2}\right) m = gm$$

$$g = G\frac{M_E}{R_E^2}$$

QUESTION The mass of the Earth is 6×10^{24} kg and the radius of the Earth is 6.4×10^6 m. What is the value of g?

$$g = \left(6.7\times10^{-11}\frac{\text{N}\cdot\text{m}^2}{\text{kg}^2}\right)\frac{(6\times10^{24}\,\text{kg})}{(6.4\times10^{6}\,\text{m})^2}$$
$$= 9.8\frac{\text{N}}{\text{kg}}$$

The constant g, called "the magnitude of the gravitational field," has the value $g = +9.8$ newtons/kilogram near the Earth's surface. Note that g is a positive number.

APPROXIMATE GRAVITATIONAL FORCE NEAR EARTH'S SURFACE

$$F_g \approx mg$$

where $g = G\frac{M_E}{R_E^2} = +9.8\frac{\text{N}}{\text{kg}}$.

M_E is the mass of the Earth, and R_E is the radius of the Earth. g is the magnitude of the gravitational field near the Earth's surface. The units of g are newtons/kg.

Figure 3.16 The "gravitational field" around the Earth. At a position where the field is $\vec{g}$, an object of mass m experiences a force $m\vec{g}$. The magnitude g of the field decreases with distance away from the Earth.

The "gravitational field" $\vec{g}$ at a location in space is defined to be the vector gravitational force per kg, so the force on a 1 kg mass placed at that location would be $(1\,\text{kg})\vec{g}$. A 2 kg mass would experience a force $(2\,\text{kg})\vec{g}$ at that location. In general, a mass m will experience a force $m\vec{g}$ of magnitude mg. The gravitational field around the Earth is shown in Figure 3.16.

The gravitational field is a vector quantity. The Earth's gravitational field at any location points toward the center of the Earth. Assuming the usual coordinate system, if you stand at the North Pole, which is located on the $+y$ axis, you would experience a gravitational field $\vec{g} = \langle 0, -g, 0\rangle$. Note that g is a *positive* number, the magnitude of the gravitational field $\vec{g}$.

If the Earth consisted of layers of uniform-density spherical shells, the Earth's gravitational field would be perfectly spherically symmetric, and g would be truly constant all over the surface of the Earth. However, the Earth due to its spin bulges out at the equator, so g is smaller at the equator than at the poles because you're farther from the center of the Earth. Also, regions in the Earth of higher or lower density make g above these regions vary slightly from the average value (and the direction may also be affected), and geologists use these small variations to find valuable minerals and oil. The fact that a reference frame tied to the spinning Earth is not an inertial frame further complicates measurements of $\vec{g}$, though the effect is negligibly small for the purposes of this course.

We will use the concept of field extensively when we study electric and magnetic fields in later chapters. Until then, we will refer to it only in the context of the gravitational field near the Earth's surface ($g = +9.8\,\text{N/kg}$).

EXAMPLE **A Falling Object**

Drop a rock with mass m near the surface of the Earth. What is its (vector) acceleration?

Solution Use the Momentum Principle: $\Delta\vec{p} = \vec{F}_{\text{net}}\,\Delta t$. Since the speed is small compared to the speed of light, and the mass isn't changing, we can write $\vec{p} \approx m\vec{v}$. If we neglect air resistance, the only force is the gravitational force, so $\Delta(m\vec{v}) = m\vec{g}\,\Delta t$. Divide through by m and by Δt and we get this:

$$\frac{\Delta\vec{v}}{\Delta t} = \vec{g}$$

However, the acceleration of the rock is the rate of change of the velocity, $\frac{\Delta\vec{v}}{\Delta t}$, so we have

$$\vec{a} = \vec{g} = \langle 0, -g, 0 \rangle$$

The rock accelerates toward the ground (the $-y$ direction) at a rate of 9.8 m/s/s. Evidently the units of N/kg are equivalent to the units of m/s/s.

The reason that the mass dropped out of the calculation is that both the momentum and the gravitational force are proportional to m, which is not true for other kinds of force. This has the interesting effect that if the gravitational force is the only force acting, a heavy rock and a light rock will fall in exactly the same way. However, the air resistance force depends on how wide the rock is, and how blunt it is, and how fast the rock is moving, and is not proportional to m, so if the air resistance force is significant, different masses will fall differently. An obvious example is that if you drop a metal ball and a partially crumpled sheet of paper, the metal ball will hit the ground before the paper does.

Checkpoint 3 A roughly spherical asteroid has a mass of 3.1×10^{20} kg and a radius of 270 km. **(a)** What is the value of the constant g at a location on the surface of the asteroid? **(b)** What would be the magnitude of the gravitational force exerted by the asteroid on a 70 kg astronaut standing on the asteroid's surface? **(c)** How does this compare to the gravitational force on the same astronaut when standing on the surface of the Earth?

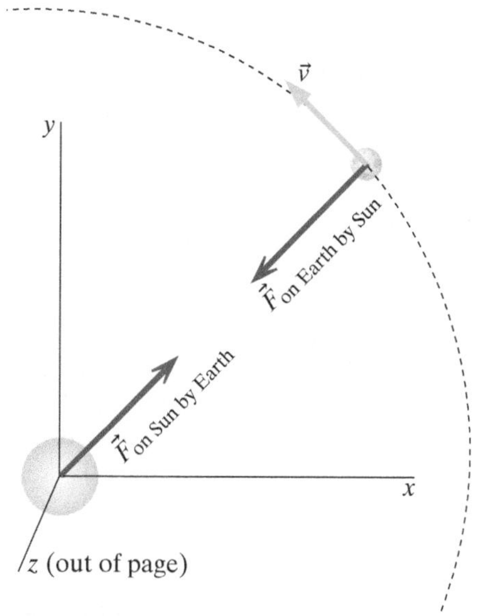

Figure 3.17 The Sun and the Earth exert equal and opposite forces on each other.

3.4 RECIPROCITY

An important aspect of the gravitational interaction is that the force that object 1 exerts on object 2 is equal and opposite to the force that object 2 exerts on object 1 (Figure 3.17). That the magnitudes must be equal is clear from the algebraic form of the gravitational force, because $m_1 m_2 = m_2 m_1$. The directions of the forces are along the line connecting the centers, and in opposite directions.

This property of gravitational interactions is called "reciprocity" or "Newton's third law of motion" and as we will see later in this chapter it is also a property of electric interactions, including the electric interactions between neighboring atoms.

RECIPROCITY

$$\vec{F}_{\text{on 1 by 2}} = -\vec{F}_{\text{on 2 by 1}}$$

(gravitational and electric forces)

The force that the Earth exerts on the massive Sun in Figure 3.17 is just as big as the force that the Sun exerts on the Earth, so in the same time interval the momentum changes are equal in magnitude and opposite in direction ("equal and opposite" for short):

$$\Delta\vec{p}_1 = \vec{F}_{\text{on 1 by 2}}\Delta t = -\Delta\vec{p}_2$$

However, the velocity change $\Delta\vec{v} = \Delta\vec{p}/m$ of the Sun is extremely small compared to the velocity change of the Earth, because the mass of the Sun is enormous in comparison with the mass of the Earth. The mass of our Sun, which is a rather ordinary star, is 2×10^{30} kg. This is enormous compared to the mass of the Earth (6×10^{24} kg) or even to the mass of the largest planet in our Solar System, Jupiter (2×10^{27} kg). Nevertheless, very accurate measurements of small velocity changes of distant stars have been used to infer the presence of unseen planets orbiting those stars.

Figure 3.18 Reciprocity: There are six forces acting on the 3 g object and six forces acting on the 2 g object.

Magnetic forces do not have the property of reciprocity. Two electrically charged particles that are both moving can interact magnetically as well as electrically, and the magnetic forces that these two particles exert on each other need not be equal in magnitude or opposite in direction. Reciprocity applies to gravitational and electric forces, but not in general to magnetic forces acting between individual moving charges.

As we will see in Chapter 4, the forces associated with contact between two objects are actually electric forces at the microscopic level. Thus contact forces also have the property of reciprocity.

Why Reciprocity?

The algebraic form of the expression for the gravitational force indicates that reciprocity should hold. A diagram may be helpful in explaining *why* the forces behave this way.

Consider two small objects, a 2 g object made up of two 1 g balls, and a 3 g object made up of three 1 g balls (Figure 3.18). The distance between centers of the two objects is large compared to the size of either object, so the distances between pairs of 1 g balls are about the same for all pairs. You can see that in the 2 g object at the top of Figure 3.18 each ball has three gravitational forces exerted on it by the distant balls in the 3 g object.

> QUESTION How many of these individual forces act on the 2 g object?

There is a net force of $2 \times 3 = 6$ times the force associated with one pair of balls.

> QUESTION Similarly, consider the forces acting on the 3 g object at the bottom of Figure 3.18. How many of these individual forces act on the 3 g object?

There is again a net force of $3 \times 2 = 6$ times the force associated with one pair of balls.

The effect is that the force exerted by the 3 g object on the 2 g object has the same magnitude as the force exerted by the 2 g object on the 3 g object. The same reciprocity holds, for similar reasons, for the electric forces between a lithium nucleus containing 3 protons and a helium nucleus containing 2 protons.

Checkpoint 4 A 60 kg person stands on the Earth's surface. **(a)** What is the approximate magnitude of the gravitational force on the person by the Earth? **(b)** What is the magnitude of the gravitational force on the Earth by the person?

3.5 PREDICTING MOTION OF GRAVITATIONALLY INTERACTING OBJECTS

In Chapter 2 we used an iterative approach to predict the motion of a system that is interacting with its surroundings:

Repeat:

- Calculate the net (vector) force $\vec{F}_{\text{net}}$ acting on the system.
- Update the momentum of the system: $\vec{p}_f = \vec{p}_i + \vec{F}_{\text{net}}\,\Delta t$.
- Update the position: $\vec{r}_f = \vec{r}_i + \vec{v}_{\text{avg}}\,\Delta t$.

The same iterative approach can be used to predict the motion of objects that interact gravitationally. The only difference is in the force we use: the full $1/r^2$ gravitational force. The gravitational force changes magnitude and direction as the objects move, so at the beginning of every time step we need to recalculate the gravitational force, both magnitude and direction.

EXAMPLE

Earth and Sun

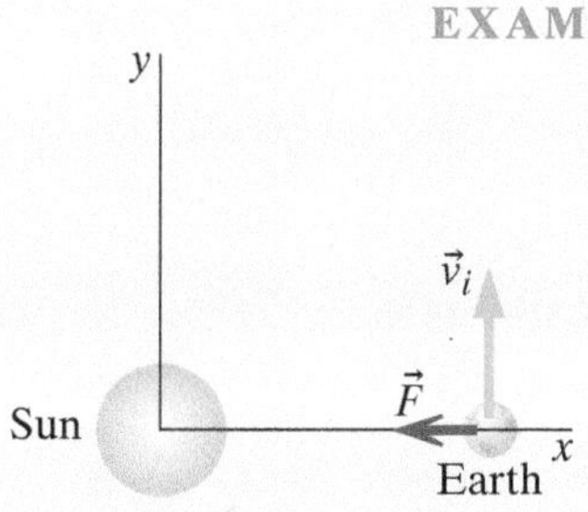

Figure 3.19 Initial location and momentum of the Earth. The origin of the coordinate system is at the center of the Sun. Sizes of the Sun and Earth are exaggerated.

Let's try to predict approximately the motion of the Earth around the Sun. Choose a coordinate system whose origin is at the center of the Sun, and in which the Earth orbits the Sun in the xy plane, as shown in Figure 3.19. The mass of the Sun is 2×10^{30} kg; the mass of the Earth is 6×10^{24} kg. Suppose that at a particular instant the Earth has these "initial conditions": position $\langle 1.5 \times 10^{11}, 0, 0 \rangle$ m and velocity $\langle 0, 3 \times 10^4, 0 \rangle$ m/s (Figure 3.19). (This is close to the actual speed of the Earth in its nearly circular orbit; to calculate this speed divide the circumference of the orbit by the orbital period.) What does the Momentum Principle predict will be the location of the Earth 3 months later? How well does the prediction match the actual orbit?

Solution System: Earth
Surroundings: Sun
We will make the approximation that the effect of the Earth on the Sun is negligible, because the Sun's mass is so much larger than the mass of the Earth. Therefore we can neglect changes in the position of the Sun. We will also neglect the interactions of the Earth with all other planets and stars.

To get an approximate answer, we'll divide the 3 mo time interval into three intervals each 1 mo long. This will give us only a rough approximation to the motion, because the gravitational force acting on the Earth changes quite a lot in one month, which is one-twelfth of the way around the circular orbit. We need to convert one month into SI units (seconds):

$$\Delta t = (1\text{ mo})\left(30\frac{\text{days}}{\text{mo}}\right)\left(24\frac{\text{hr}}{\text{day}}\right)\left(60\frac{\text{min}}{\text{hr}}\right)\left(60\frac{\text{sec}}{\text{min}}\right) = 2.6 \times 10^6\text{ s}$$

Initial conditions (position and momentum of the Earth):

$$\vec{r}_E = \langle 1.5 \times 10^{11}, 0, 0 \rangle\text{ m}$$
$$\vec{p}_i = (6 \times 10^{24}\text{ kg})\langle 0, 3 \times 10^4, 0 \rangle\text{ m/s} = \langle 0, 1.8 \times 10^{29}, 0 \rangle\text{ kg}\cdot\text{m/s}$$

A note on significant figures: To make the example easier to read, we show all numbers to only two significant figures. In the actual calculations, however, many significant figures were used, because this is how a computer would do the calculations. In a computer calculation, rounding is typically done only on the final results.

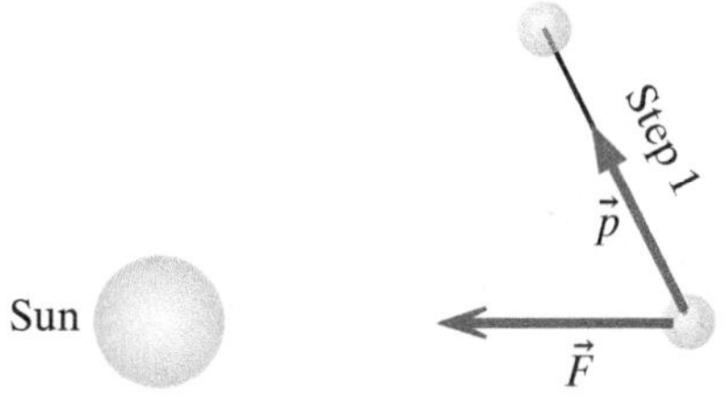

Figure 3.20 Step 1 of three-step iterative calculation. Gravitational force on Earth at beginning of step and final momentum of Earth are shown.

Time step 1 (Figure 3.20):

$$\vec{r} = \langle 1.5 \times 10^{11}, 0, 0 \rangle\text{ m} - \langle 0, 0, 0 \rangle\text{ m} = \langle 1.5 \times 10^{11}, 0, 0 \rangle\text{ m}$$
$$|\vec{r}| = 1.5 \times 10^{11}\text{ m}$$
$$\hat{r} = \langle 1, 0, 0 \rangle$$

$$\vec{F}_{\text{on }E\text{ by }S} = (6.7 \times 10^{-11}\text{ N/ kg})\frac{(6 \times 10^{24}\text{ kg})(2 \times 10^{30}\text{ kg})}{(1.5 \times 10^{11}\text{ m})^2}\langle -1, 0, 0 \rangle$$
$$= \langle -3.6 \times 10^{22}, 0, 0 \rangle\text{ N}$$
$$\vec{p}_f = \langle 0, 1.8 \times 10^{29}, 0 \rangle\text{ kg}\cdot\text{m/s} + \langle -3.6 \times 10^{22}, 0, 0 \rangle\text{ N}\cdot(2.6 \times 10^6\text{ s})$$
$$= \langle -9.3 \times 10^{28}, 1.8 \times 10^{29}, 0 \rangle\text{ kg}\cdot\text{m/s}$$

$$\vec{v}_{avg} \approx \vec{v}_f \approx \frac{\vec{p}_f}{m} = \frac{\langle -9.3\times 10^{28}, 1.8\times 10^{29}, 0\rangle \text{ kg}\cdot\text{m/s}}{6\times 10^{24}\text{ kg}}$$
$$= \langle -1.6\times 10^{4}, 3\times 10^{4}, 0\rangle \text{ m/s}$$
$$\vec{r}_{E,f} = \langle 1.5\times 10^{11}, 0, 0\rangle \text{ m} + \langle -1.6\times 10^{4}, 3\times 10^{4}, 0\rangle \text{ m/s}\cdot(2.6\times 10^{6}\text{ s})$$
$$= \langle 1.1\times 10^{11}, 7.8\times 10^{10}, 0\rangle \text{ m}$$

Time Step 2 (Figure 3.21):

$$\vec{r} = \langle 1.1\times 10^{11}, 7.8\times 10^{10}, 0\rangle \text{ m}$$
$$|\vec{r}| = 1.348\times 10^{11}\text{ m}$$
$$\hat{r} = \langle 0.816, 0.578, 0\rangle$$
$$\vec{F}_{\text{on } E \text{ by } S} = \langle -3.6\times 10^{22}, -2.6\times 10^{22}, 0\rangle \text{ N}$$
$$\vec{p}_f = \langle -1.9\times 10^{29}, 1.1\times 10^{29}, 0\rangle \text{ kg}\cdot\text{m/s}$$
$$\vec{r}_{E,f} = \langle 2.9\times 10^{10}, 1.3\times 10^{11}, 0\rangle \text{ m}$$

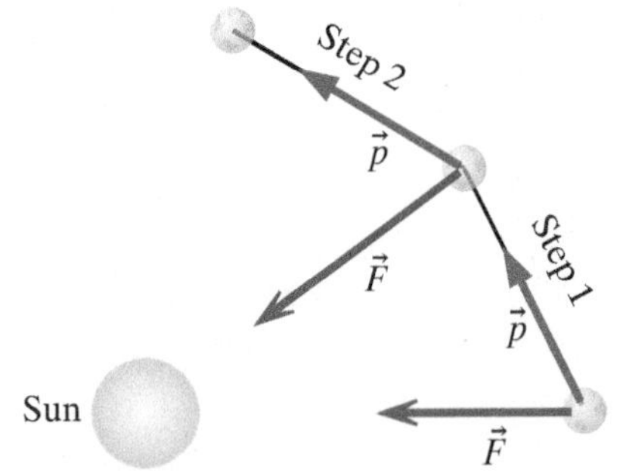

Figure 3.21 Results of first two steps of the iterative calculation. The gravitational force on the Earth is different in the two steps.

QUESTION Why was it necessary to recalculate the gravitational force in time step 2?

The position of the Earth changed, so the direction (and magnitude) of the gravitational force were not the same as in step 1.

Time Step 3 (Figure 3.22):

Following the same procedure, we must again calculate $\vec{r}$ and $\vec{F}_{\text{on } E \text{ by } S}$, then update the momentum and the position. The results of the third iteration are shown in Figure 3.22.

$$\vec{r}_{E,f} = \langle -6.3\times 10^{10}, 1.2\times 10^{11}, 0\rangle \text{ m}$$

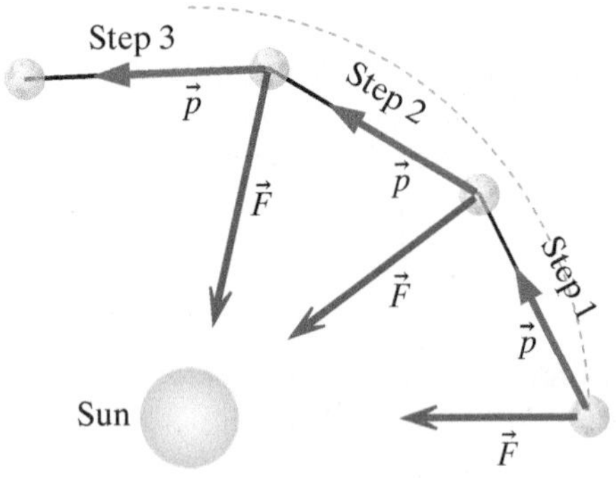

Figure 3.22 The results of all three steps in the iterative calculation of the motion of the Earth under the influence of the Sun. The dashed arc shows the actual orbit of the Earth in three months.

Check: Is the path reasonable? Yes. Although this was a rough calculation, it does predict that the Earth will orbit around the Sun in a roughly circular orbit. However, one indication of the sizable inaccuracy in our calculation is that in 3 mo the Earth should have gone only one-quarter of the way around the circle, but it has gone significantly farther than this in Figure 3.22.

QUESTION How could we improve the accuracy of this calculation?

When we use the Momentum Principle in the form $\vec{p}_f = \vec{p}_i + \vec{F}_{net}\Delta t$, we're assuming that $\vec{F}_{net}$ doesn't change much during the interval Δt. This isn't true for a gravitational force if the position changes a lot. If we use smaller time steps, the displacement of the Earth will be smaller in each step, and the gravitational force will change much less. Using smaller time steps should make our predicted path much closer to the actual orbit.

You might wonder about the effect of the approximation that $\vec{v}_{avg} \approx \vec{p}_f/m$. It turns out in this case, as in many others, that this approximation gives a significantly more accurate result than we would get using the arithmetic average velocity.

Clearly, the procedure used above is very tedious if done by hand. The natural way to carry out an iterative prediction of motion is to program a computer to do the repetitive calculations. In a computer program one would choose a step size much smaller than 1 mo, to ensure that the change in $\vec{F}_{net}$ is small over the interval Δt.

QUESTION How would you have to change the calculation described above to include in the prediction how the Sun will move?

You would need to update the momentum and position of the Sun each time you update the momentum and position of the Earth. Fortunately, the force acting on the Sun is just the opposite of the force acting on the Earth: $\vec{F}_{\text{on } S \text{ by } E} = -\vec{F}_{\text{on } E \text{ by } S}$ thanks to the reciprocity of gravitational forces (Section 3.4). So it is unnecessary to repeat all of the usual steps in calculating the gravitational force that acts on the Sun; just reverse the force vector.

Because its mass is so large, you would find that the Sun moves very little. Nevertheless, in recent years many planets have been discovered orbiting distant stars by observing the very small wobbles of the stars caused by the gravitational forces of the planets acting on the stars.

EXAMPLE **If the Step Size Is Too Big**

By taking a step size of 1 mo, we predicted that the Earth would orbit the Sun, though our prediction was rather imprecise, because in 1 mo the gravitational force and the momentum change a lot. It is instructive to see what happens if we take an even bigger step size, making the results even more imprecise. Taking a step size of 3 mo instead of 1 mo, what will we predict the location of the Earth to be 3 mo later?

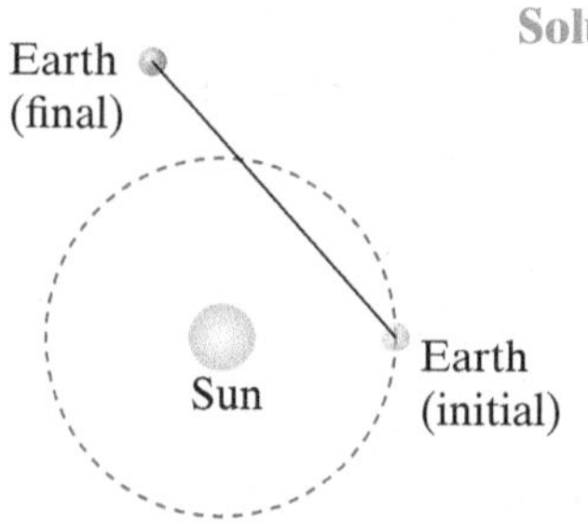

Figure 3.23 The displacement of the Earth predicted by a one-step calculation using the initial value of gravitational force, and the final velocity. The dashed line shows the actual path of the Earth around the Sun.

Solution System: Earth
Surroundings: Sun (ignore all other planets, stars)

$$\Delta t = (3\,\text{mo})\left(30\frac{\text{days}}{\text{mo}}\right)\left(24\frac{\text{hr}}{\text{day}}\right)\left(60\frac{\text{min}}{\text{hr}}\right)\left(60\frac{\text{sec}}{\text{min}}\right) = 7.8 \times 10^6\,\text{s}$$

$$\vec{p}_i = \langle 0, 1.8 \times 10^{29}, 0\rangle\,\text{kg}\cdot\text{m/s}$$

$$\vec{p}_f = \langle 0, 1.8 \times 10^{29}, 0\rangle\,\text{kg}\cdot\text{m/s} + \langle -3.6 \times 10^{22}, 0, 0\rangle\,\text{N}\cdot(7.8 \times 10^6\,\text{s})$$
$$= \langle -2.8 \times 10^{29}, 1.8 \times 10^{29}, 0\rangle\,\text{kg}\cdot\text{m/s}$$

$$\vec{v}_f \approx \frac{\vec{p}_f}{m} = \frac{\langle -2.8 \times 10^{29}, 1.8 \times 10^{29}, 0\rangle\,\text{kg}\cdot\text{m/s}}{6 \times 10^{24}\,\text{kg}}$$
$$= \langle -4.6 \times 10^4, 3 \times 10^4, 0\rangle\,\text{m/s}$$

$$\vec{r}_{E,f} = \langle 1.5 \times 10^{11}, 0, 0\rangle\,\text{m} + \langle -4.6 \times 10^4, 3 \times 10^4, 0\rangle\,\text{m/s}\cdot(7.8 \times 10^6\,\text{s})$$
$$= \langle -2e11, 2.3e11, 0\rangle\,\text{m}$$

Check: Units are ok (meters). Is the location reasonable? NO! The path we predicted, from initial location to final location, is shown in Figure 3.23. This disagrees very much with our more accurate calculation using a time step of 1 mo. Even our 1 mo calculation was imprecise; in a computer calculation we would choose a much shorter time step than 1 mo.

QUESTION What went wrong? In this calculation we used the same procedure we used before.

The major error was the tacit assumption that the force on the Earth by the Sun was constant in magnitude and direction over the 3 mo interval. The direction and magnitude of the gravitational force between two objects changes when the relative positions of the objects change. Also, because the time interval was so long, the final velocity was a poor approximation to the actual average velocity over the interval.

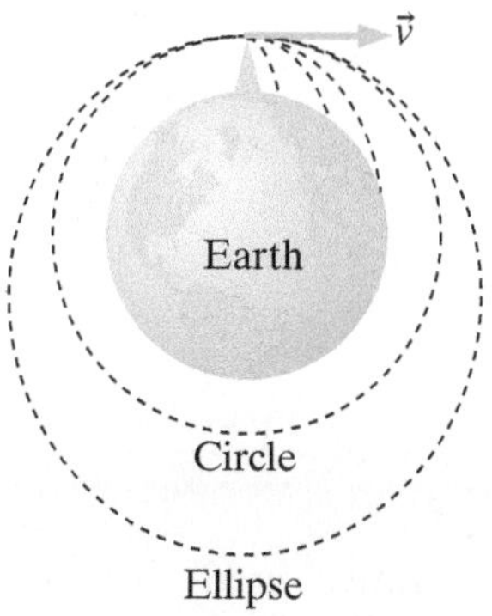

Figure 3.24 Newton imagined throwing a rock horizontally with different speeds from a high mountain resulting in different ellipses. Only one special speed makes a circular orbit.

Elliptical Orbits

The Earth's orbit around the Sun isn't really a circle; it's actually an ellipse that is very nearly circular. Newton himself drew an instructive diagram

(Figure 3.24) to show how the launch speed of an object affects the shape of the orbit, by imagining throwing a rock horizontally from the top of a mountain so tall as to stick up above the atmosphere (so there will be no air resistance). In this figure you can see that getting a circular orbit requires throwing the rock at just the right speed. Any other speed will give a noncircular orbit, which Newton was able to prove would be an ellipse (or portion of an ellipse if the rock hits the ground).

EXAMPLE **Analyzing an Elliptical Orbit**

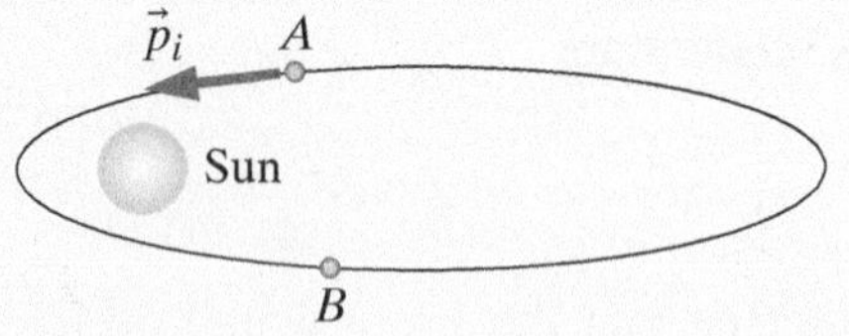

Figure 3.25 A comet approaches the Sun.

It is useful to "think iteratively" about a highly elliptical orbit like that of many comets that come near the Sun at one extreme of the orbit and go far beyond Pluto at the other extreme. At some instant the comet is at location A, headed toward the Sun (Figure 3.25). Graphically and qualitatively, carry out one step in the iterative updating of momentum and position, starting with the comet at position A with momentum $\vec{p}_i$ tangent to the orbit. Will the magnitude of the momentum (and therefore the speed) increase or decrease? (Think about the relative directions of $\vec{p}_{now}$ and $\vec{F}_{net,now}$.)

Solution

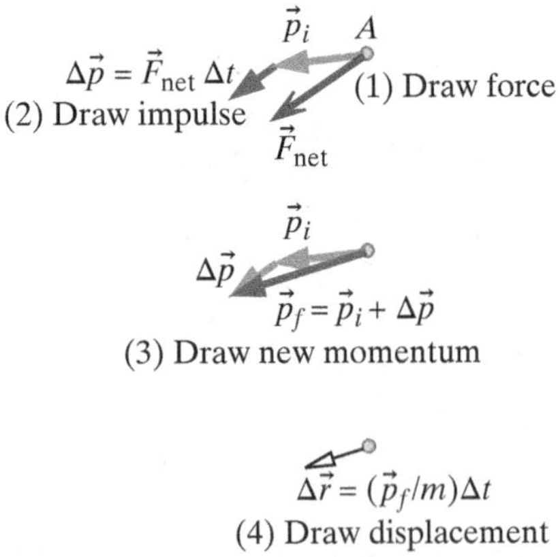

Figure 3.26 Steps in graphical updating of momentum and position. (1) Draw the net force, (2) draw the impulse, (3) draw the new momentum, (4) use the new momentum to draw the displacement.

Consider a short time interval Δt and apply the Momentum Principle to the system of the comet, which experiences a gravitational force exerted by the Sun. To update the momentum and position graphically, follow the procedure illustrated in Figure 3.26: (1) Draw the net force acting on the comet, due to the Sun. (2) Draw the impulse $\vec{F}_{net}\,\Delta t$ for a short time interval Δt. The change in momentum $\Delta\vec{p}$ is equal to the impulse. (3) Draw the new momentum $\vec{p}_f = \vec{p}_i + \Delta\vec{p}$. (4) Approximate the average velocity by the new velocity $\vec{v}_{avg} \approx \vec{p}_f/m$ and indicate the displacement $\Delta\vec{r} = \vec{r}_f - \vec{r}_i = (\vec{p}_f/m)\Delta t$.

Since part of the force acts in the same direction as the momentum of the comet, the magnitude of the comet's momentum increases.

QUESTION A student said, "The reason why the comet speeds up is because the force gets bigger as the comet moves toward the Sun. Bigger force, bigger speed." What's wrong with this analysis?

This statement ignores the fact that $\vec{p}_{future}$ depends on two things: $\vec{p}_{now}$ and $\vec{F}_{net,now}$, not just on $\vec{F}_{net,now}$. If, over the next Δt, $\vec{F}_{net}$ is at least partly in the same direction as the current momentum, then the new momentum will be larger. This would be true even if the magnitude of the net force on the comet were constant or decreasing.

Checkpoint 5 Now consider the situation when the comet is at position B in Figure 3.25, heading away from the Sun. Mentally take one step in the iterative update of momentum and position. What happens to the magnitude of the momentum, and the speed? What happens to the direction of the momentum? Explain briefly in terms of $\vec{p}_{future}$, $\vec{p}_{now}$, and $\vec{F}_{net,now}$.

3.6 GRAVITATIONAL FORCE IN COMPUTATIONAL MODELS

In Chapter 2 we saw how to structure an iterative prediction of motion on a computer, for a mass–spring system. To predict the motion of objects interacting gravitationally we will use the same basic structure, but we need to calculate the gravitational force instead of a spring force.

In writing a program to tell a computer how to calculate a gravitational force, we use exactly the same organization as we use when doing the calculation with a calculator. Let's assume that our program creates two objects named rock1 and rock2, each represented by a sphere at an appropriate location in 3D space. The variables m1 and p1 represent the mass and momentum, respectively, of rock1; assume that they have been given appropriate values, and the relevant constants have been defined. Recall that the pos attribute of an object in VPython is a vector. A translation of the mathematical expressions (shown in red) into code (shown in blue) for calculating the gravitational force on rock1 due to rock2 into VPython code might look like this:

$\vec{r} = \vec{r}_1 - \vec{r}_2$	`r = rock1.pos - rock2.pos`
$\hat{r} = \vec{r}/\lvert\vec{r}\rvert$	`rhat = r / mag(r)`
$\lvert\vec{F}\rvert = Gm_1m_2/\lvert\vec{r}\rvert^2$	`Fmag = G * m1 * m2 / mag(r)**2`
$\vec{F} = \lvert\vec{F}\rvert \cdot (-\hat{r})$	`F_on1 = Fmag * (-rhat)`

QUESTION Is the quantity r calculated in the first line of code above a vector or a scalar?

It is a vector, since it is the difference of two position vectors.

QUESTION The code for calculating the force on rock1 above was used in a program that generated the display shown in Figure 3.27. Suppose that you wanted to calculate and display the force on rock2 due to rock1. What code would you write to do this? If the mass of rock2 is less than the mass of rock1, how should the magnitudes and directions of the two forces compare?

Figure 3.27 Display generated by a VPython program to calculate the gravitational force on one object due to another.

Because of the reciprocity of gravitational forces, the direction of the force on rock2 will be opposite to the direction of the force on rock1, but their magnitudes will be the same, no matter what their masses may be. So we could say:

```
F_on2 = -F_on1
```

QUESTION If you write a program to model the motion of gravitationally interacting objects, should you put the calculation of the gravitational force before the iterative loop, or should it be done inside the loop?

The gravitational force calculation must be done inside the loop. Every time one of the objects moves, the relative positions of the objects change, and therefore the gravitational force changes. It is necessary to recalculate the gravitational force in every time step. The key lines of code inside the iterative loop might look like this:

```
while True:
    rate(100)
    r = rock1.pos - rock2.pos
    rhat = r/mag(r)
    Fmag = G * m1 * m2 / mag(r)**2
    F_on1 = Fmag * (-rhat)
    p1 = p1 + F_on1 * deltat
    rock1.pos = rock1.pos + (p1/m1) * deltat
    F_on2 = -F_on1
    p2 = p2 + F_on2 * deltat
    rock2.pos = rock2.pos + (p2/m2) * deltat
```

QUESTION In a program that incorporates the while loop shown above, will both objects move?

Yes, because the momentum and position of both rocks are updated.

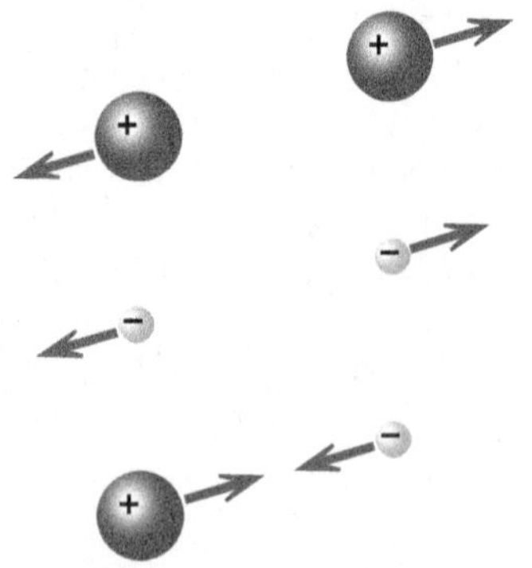

Figure 3.28 Protons repel each other; electrons repel each other; protons and electrons attract each other.

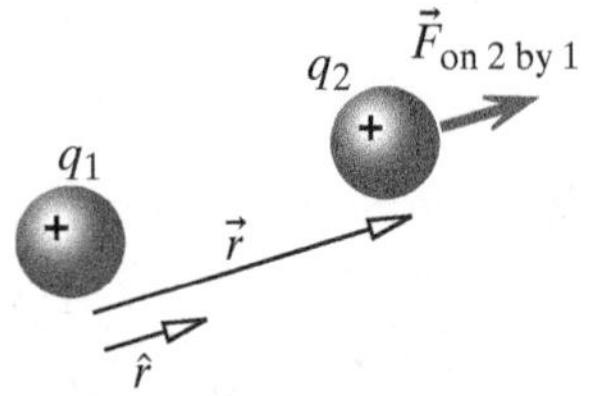

Figure 3.29 The electric force exerted on object 2 by object 1. (The force exerted on object 2 by object 1 has the same magnitude but opposite direction.)

3.7 THE ELECTRIC FORCE

Particles such as the protons and electrons found in atoms are electrically charged. It is observed that two protons repel each other, as do two electrons, while a proton and an electron attract each other (Figure 3.28). Protons are said to have "positive electric charge" and electrons have "negative electric charge."

The force corresponding to this electric interaction is known as "Coulomb's law" to honor the French scientist who established this behavior of electric forces in the late 1700s, during the same period when Cavendish measured the gravitational constant G. See Figure 3.29.

THE ELECTRIC FORCE (COULOMB'S LAW)

$$\vec{F}_{\text{elec on 2 by 1}} = \frac{1}{4\pi\varepsilon_0}\frac{q_1 q_2}{|\vec{r}|^2}\hat{r}$$

q_1 and q_2 are the electric charges of the objects.
$\vec{r} = \vec{r}_2 - \vec{r}_1$ is the position of 2 relative to 1.

The universal constant $\dfrac{1}{4\pi\varepsilon_0} = 9 \times 10^9 \dfrac{\text{N}\cdot\text{m}^2}{\text{C}^2}$.

The SI unit of electric charge is the coulomb, abbreviated as C. The proton has a charge of $+1.6 \times 10^{-19}$ C and the electron has a charge of -1.6×10^{-19} C. The symbol e is often used to represent the charge of a proton; $e = +1.6 \times 10^{-19}$ C. We can see from the equations describing the electric and gravitational forces that there are similarities between these forces. The most striking similarity is that the magnitude of both forces is inversely proportional to the square of the distance between the centers of the objects, and that the force acts along the line connecting the objects, as indicated by the presence of the unit vector $\hat{r}$:

$$\vec{F}_{\text{electric}} = \frac{1}{4\pi\varepsilon_0}\frac{q_1 q_2}{|\vec{r}|^2}\hat{r} \qquad \vec{F}_{\text{grav}} = -G\frac{m_1 m_2}{|\vec{r}|^2}\hat{r}$$

However, there are important differences between the electric and gravitational forces as well. Electric forces may be attractive or repulsive, whereas gravitational forces are always attractive. Electric charge may have positive or negative values, while mass has only positive values. As indicated in Figure 3.28, particles with the same sign of charge repel each other, and particles with opposite-signed charges attract each other.

QUESTION What quantities in the equation for electric force determine the direction of the force?

There are three quantities we need to consider: the unit vector $\hat{r}$, and the charges q_1 and q_2, which may be positive or negative. If the product of q_1 and q_2 is a positive number (that is, if q_1 and q_2 have the same signs), then the force will be in the direction of $\hat{r}$. If q_1 and q_2 have different signs, their product will be negative, and the force will be opposite to $\hat{r}$.

QUESTION Is the following quantity the magnitude of the electric force?

$$\frac{1}{4\pi\varepsilon_0}\frac{q_1 q_2}{|\vec{r}|^2}$$

No. It is a scalar quantity, but it may be negative, whereas a magnitude is always positive. When calculating an electric force, it is more useful to think about this scalar quantity than about the magnitude.

EXAMPLE **An Electron and an Alpha Particle**

An electron is located at $\langle -3\times 10^{-9}, 2\times 10^{-9}, 0\rangle$ m and an alpha particle (two protons and two neutrons) is located at $\langle 6\times 10^{-9}, -4\times 10^{-9}, 0\rangle$ m (Figure 3.30). Find the force exerted on the electron by the alpha particle.

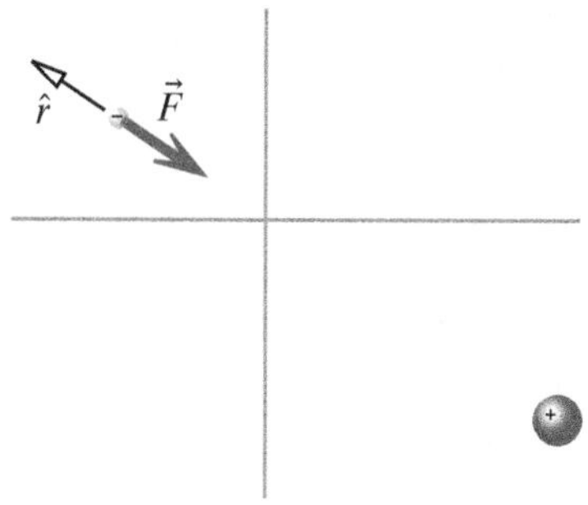

Figure 3.30 Force on an electron due to an alpha particle.

Solution

$$\vec{r} = \langle -9\times 10^{-9}, 6\times 10^{-9}, 0\rangle \text{ m}$$
$$|\vec{r}| = 1.08\times 10^{-8}\text{ m}$$
$$\hat{r} = \langle -0.832, 0.555, 0\rangle$$
$$\vec{F} = \left(9\times 10^{9}\frac{\text{N}\cdot\text{m}^2}{\text{C}^2}\right)\frac{(2\cdot 1.6\times 10^{-19}\text{ C})(-1.6\times 10^{-19}\text{ C})}{(1.08\times 10^{-8}\text{ m})^2}\langle -0.832, 0.555, 0\rangle$$
$$= \langle 3.28\times 10^{-12}, -2.19\times 10^{-12}, 0\rangle \text{ N}$$

QUESTION Does the direction of the force make sense?

Yes, because the electron and the alpha particle have opposite signs, so the electron should be attracted to the alpha particle.

QUESTION Isn't this a very small force?

To be able to answer that question, we need to rephrase it in terms of a comparison—small relative to what? In the next example we explore this issue.

EXAMPLE **Comparison of Electric and Gravitational Forces**

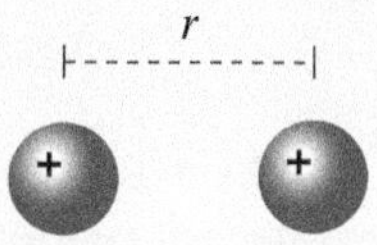

Figure 3.31 Two protons a distance *r* apart.

Two protons are a distance r from each other, center to center, as shown in Figure 3.31. Find the ratio of the magnitude of the electric force one proton exerts on the other to the magnitude of the gravitational force one proton exerts on the other.

Solution

$$F_{\text{elec}} = (9\times 10^{9}\text{ N}\cdot\text{m}^2/\text{C}^2)\frac{(1.6\times 10^{-19}\text{C})^2}{r^2}$$
$$F_{\text{grav}} = (6.7\times 10^{-11}\text{ N}\cdot\text{m}^2/\text{kg}^2)\frac{(1.7\times 10^{-27}\text{ kg})^2}{r^2}$$
$$\frac{F_{\text{elec}}}{F_{\text{grav}}} = 1.2\times 10^{36}$$

Notice that r canceled in this ratio, because both the electric and gravitational forces are proportional to the same factor $1/r^2$. The units of newtons also canceled.

This ratio is astoundingly huge. The electric repulsion of two protons is greater than their gravitational attraction by a factor of

1,200,000,000,000,000,000,000,000,000,000,000,000 !!

As a result, when considering the interactions of subatomic charges, it is reasonable to ignore their extremely tiny gravitational interactions with one another.

QUESTION If electric interactions are intrinsically so enormously much stronger than gravitational interactions, why is it that we are so much more aware of gravitational forces in our daily lives?

At the microscopic level, electric forces are extremely important. The electrons and protons that make up a material object are bound together by electric forces. The fact that you can stand up despite the gravitational pull of the entire Earth, instead of having all the electrons and protons in your body end up in a puddle on the Earth's surface, indicates how strong these electric interactions are. However, at the macroscopic level, it is rare to find large chunks of matter that have a nonzero electric charge, since under most circumstances matter is electrically "neutral"—that is, an object typically has equal numbers of electrons and protons.

It is occasionally possible to transfer some charge from one object to another by rubbing, leaving one object with a tiny excess of electrons and the other with a tiny excess of protons; when this occurs you may see electric effects such as "static cling." We'll explore such effects in depth in later chapters on electric interactions.

The forces that objects exert on each other when they touch are also due to electric interactions. Although matter is typically electrically neutral, atoms on the surfaces of two objects that come into contact deform each other in such a way that the electric forces do not entirely cancel out. In the next chapter we will examine interatomic forces in more detail.

Checkpoint 6 A moving electron passes near the nucleus of a gold atom, which contains 79 protons and 118 neutrons. At a particular moment the electron is a distance of 3×10^{-9} m from the gold nucleus. Recall that the electron charge is $-e = -1.6 \times 10^{-19}$ C, the proton charge is $+e = +1.6 \times 10^{-19}$ C, and the neutron charge is 0. **(a)** What is the magnitude of the force exerted by the gold nucleus on the electron? **(b)** What is the magnitude of the force exerted by the electron on the gold nucleus?

3.8 THE STRONG INTERACTION

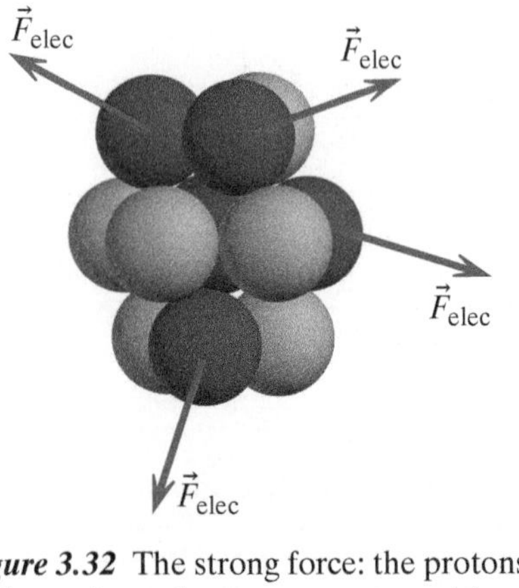

Figure 3.32 The strong force: the protons in the nucleus of an atom exert repulsive electric forces on each other as shown in the diagram. The strong interaction (which involves neutrons as well as protons) holds the nucleus together despite this electric repulsion.

The strong interaction involves particles made of quarks, so leptons such as electrons do not experience this force, because they are not made of quarks. The strong interaction holds the nucleus of an atom together. Inside an atomic nucleus the protons repel each other electrically, yet the protons and neutrons remain packed closely together due to the strong interaction (Figure 3.32). This is a very short-range interaction. The strong (or nuclear) interactions among protons and neutrons are essentially zero unless the particles touch each other, which means that the center-to-center distance must be about equal to the diameter of a proton or neutron (the radius of a proton or neutron is about 1×10^{-15} m, very much smaller than the radius of an atom, which is about 1×10^{-10} m). At these very short distances, the electric force is extremely large, since it is inversely proportional to the tiny distance squared, yet the strong force is even larger in stable nuclei.

There is a delicate balance between the electric repulsion and the strong attraction that determines whether a nucleus can exist at all. For example, a nucleus consisting of just two protons and no neutrons does not exist in nature because the attractive strong force between the two protons is smaller than the repulsive electric force. This unstable helium nucleus would be called He-2, meaning the total number of protons and neutrons is 2. He-3 (2 protons and 1 neutron) and He-4 (2 protons and 2 neutrons) are stable "isotopes" of helium (different isotopes of the same element have the same number of protons but different numbers of neutrons). In these stable nuclei the protons experience attractive strong interactions with the neutrons as well as with the other proton, and this is enough to overcome the electric repulsion between the two protons.

At the start of the periodic table you'll find about the same number of neutrons as protons in the nuclei, as in He-4. For example, the most common isotope of oxygen has 8 protons and 8 neutrons (O-16). As you go farther in the periodic table you'll see that more and more neutrons are needed to offset the proton repulsions. The most common isotope of uranium has

92 protons but many more neutrons, 146 in U-238. In uranium not only are there more neutrons for the protons to interact with but also the many neutrons dilute the concentration of protons, so that on average the protons are farther apart, thereby decreasing the electric repulsion.

While there are (relatively) simple equations for the gravitational force and the electric force, it is very difficult to write a "force law" equation for the strong interaction. Usually models of the strong interaction involve energy (which we will encounter in Chapter 6) rather than force, and require significant mathematical sophistication. We will not try to work quantitatively with the strong interaction in this text.

■ The large particle physics laboratory called CERN, near Geneva, Switzerland, is the home of many collaborative international physics projects that seek answers to questions about fundamental particles and interactions. You may have heard of the Large Hadron Collider (LHC) at CERN, which was featured in the movie *Particle Fever*, and is an important tool for these investigations. You can find out more at the CERN website.

QUESTION How strong is the "strong" interaction?

If we "rank" the fundamental interactions, we find that the gravitational force is intrinsically extremely weak. You certainly notice the gravitational force when you fall, but it is significant only because you're interacting with the entire Earth. The electric force, which we have seen to be very strong, and the strong force are comparable when two protons touch each other, but if the protons are not in contact the strong force is zero.

Antimatter

For every fundamental particle, there is an "antiparticle" whose mass is the same, but whose other properties (such as charge) are opposite. These particles are referred to as "antimatter." For example, an antiproton has the same mass as a proton, but it has a negative charge. When a particle meets its own antiparticle (for example, when a proton encounters an antiproton), the two particles annihilate each other, producing various other particles. A particle such as a proton is composed of quarks, whereas an antiproton is composed of antiquarks. However, there are interesting particles called "mesons" that are composed of a quark and a different kind of antiquark. An example of a meson is a pion, which is represented by the Greek letter π. Antiparticles and mesons, along with ordinary particles, may be produced in nuclear reactions.

One situation in which we can observe the strong interaction is the reaction in which a high-speed proton hits a stationary proton (which can be a proton in a nucleus) so hard that an additional particle, a pion, is produced. The pion has a mass about 15% that of a proton. Depending on which quark and antiquark are involved, a pion can have an electric charge of $-e$, 0, or e (where e is the electric charge of the proton: $e = +1.6 \times 10^{-19}$ C). Here are examples of such reactions, where p is a proton, n is a neutron, π is a pion, and a superscripted $+$ is shorthand for an amount of charge $+e$.

$$p^+ + p^+ \rightarrow p^+ + p^+ + \pi^0$$
$$p^+ + p^+ \rightarrow p^+ + n^0 + \pi^+$$

Conservation

Many fundamental ideas in physics can be expressed as "conservation" principles. In the nuclear reactions above, as in chemical reactions you may have studied in chemistry, charge is "conserved": the Universe has the same net (total) charge before and after the interaction. The total charge is calculated by adding up all the $+e$ and $-e$ charges. In a process involving any of the four fundamental interactions, not only is charge conserved, but the number of quarks is also conserved. A proton or a neutron consists of three quarks, and a pion consists of a quark and an antiquark. The Universe has the same "quark number" before and after an interaction, where this number is calculated by adding up all the quarks and subtracting all the antiquarks. In the reactions shown above, the newly created pion doesn't change the quark number, because it consists of a quark and an antiquark; one quark minus one antiquark is zero.

Later in this chapter we will see that momentum is also conserved in all processes involving any of the four fundamental interactions. The total momentum of the Universe does not change.

Checkpoint 7 Look at the periodic table on the inside front cover of this textbook. As is standard practice, each entry gives the "atomic number" (the number of protons in the nucleus; 8 for oxygen, 92 for uranium) and the "atomic mass" in grams per mole, which is equal to the number of protons plus neutrons in a nucleus, averaged over the various isotopes for that element, which contain different numbers of neutrons. **(a)** How many protons and neutrons are in a typical silicon nucleus (Si)? **(b)** In a tin nucleus (Sn)? **(c)** In a gold nucleus (Au)? **(d)** In a thorium nucleus (Th)? **(e)** Do you see a trend?

3.9 THE WEAK INTERACTION

The fourth fundamental interaction is called the "weak interaction" because it involves nuclear interactions that are much weaker than the strong interaction or the electromagnetic interaction. A distinctive feature of many weak interactions is the production of neutrinos, ghostly particles that interact so weakly with ordinary matter that almost all of the neutrinos emitted in fusion reactions in the Sun pass right through the Earth as though the Earth were completely transparent. Although about 100 trillion (1×10^{14}) solar neutrinos pass through your body every second, on average only one of the neutrinos will undergo a weak interaction with your body in your entire lifetime! A lowercase Greek letter nu (ν) is used to represent a neutrino.

The weak interaction is responsible for the instability of a "free neutron," a neutron that is isolated outside of a nucleus. With an average lifetime of about 15 minutes, a free neutron decays into a proton, an electron, and an antineutrino (the bar above the ν indicates an antiparticle):

$$n^0 \rightarrow p^+ + e^- + \overline{\nu}^0$$

This same weak interaction process occurs in some radioactive nuclei, in which a neutron turns into a proton with the emission of an electron and antineutrino, and the new proton remains bound into the nucleus. This means that the nucleus has changed into that of a different element, with one more proton (and one less neutron) than before. Such nuclear change is called "beta decay" because in early studies of this kind of radioactivity it wasn't at first clear that the negatively charged particles were electrons, and they were called "beta rays." Beta decay occurs in heavy elements that have excess neutrons.

In the Sun, fusion reactions lead to the emission of neutrinos. Solar neutrinos have been detected on Earth despite the extremely low probability of an interaction between a neutrino and an ordinary object. For this purpose, in the late 1960s the American physicist Raymond Davis built a huge vat containing the liquid dry-cleaning chemical perchlorethylene, which contains chlorine atoms, deep in the Homestake mine in Lead, South Dakota (to shield the detector from unwanted cosmic rays). Neutrinos from the Sun occasionally interacted with a chlorine nucleus to change a neutron into a proton, thereby changing the nucleus into an argon nucleus. Here is the reaction concerning the neutron, which you can see is related to the neutron decay reaction shown earlier:

$$\nu^0 + n^0 \rightarrow p^+ + e^-$$

The corresponding change in the nucleus from chlorine to argon was $\nu^0 + \text{Cl} \rightarrow \text{Ar} + e^-$. Davis collected the argon produced in the vat as an indication of how many neutrinos had interacted with chlorine nuclei in the vat.

■ Raymond Davis won the Nobel Prize in 2002 for his work on neutrinos. You can see a video of Davis's Nobel lecture here: http://www.nobelprize.org/

The Mystery of the Missing Neutrinos

Given what we know about the fusion reactions taking place in the Sun, it was possible to predict how many neutrino events Davis should have observed, but he detected only a third as many neutrinos as expected. This was a deep mystery for several decades until several theoretical physicists proposed that on their way from the Sun to the Earth the neutrinos repeatedly change identity, sometimes existing as "electron neutrinos," which Davis's apparatus could detect, and sometimes as "muon neutrinos," a different kind of neutrino that the apparatus could not detect.

At our particular distance from the Sun, these "neutrino oscillations" are such that an electron neutrino detector finds only one-third of the neutrinos that were headed our way. An electron neutrino detector closer to or farther from the Sun would find a different fraction of the total. You can read more about this scientific mystery and its resolution by searching the Internet for "Raymond Davis neutrinos." Davis shared the 2002 Nobel Prize for physics for his work.

Later experiments in many countries confirmed and extended Davis's work and provided solid support for the neutrino oscillation theory. For example, in the MINOS experiment begun in 2003, neutrinos produced by the Fermilab accelerator outside Chicago traveled through 735 km of rock to an underground detector in the Soudan mine in northern Minnesota, and the neutrino oscillation theory was again shown to be able to explain the observations.

As with the other three fundamental interactions, in the weak interaction charge is conserved, quark number is conserved, and lepton number is conserved. In the Universe the number of leptons (mainly electrons and neutrinos) minus the number of antileptons (mainly positrons and antineutrinos) does not change. However, a unique feature of the weak interaction is that it doesn't conserve "parity," a measure of right- or left-handedness. You can read about this by searching the Internet for "parity nonconservation."

3.10 CONSERVATION OF MOMENTUM

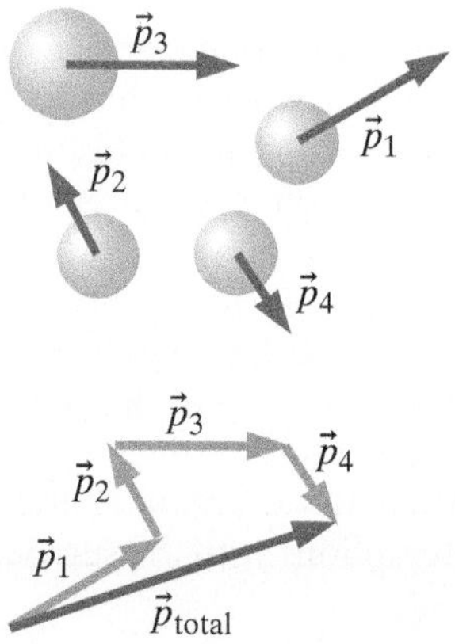

Figure 3.33 The total momentum of the system of four objects is the sum of the individual momenta of each object.

Now that we understand the reciprocity of gravitational and electric forces, we are in a position to understand the principle of Conservation of Momentum, which is a very general statement of the Momentum Principle.

The Momentum Principle, $\Delta\vec{p} = \vec{F}_{\text{net}}\,\Delta t$, predicts how the momentum of a *system* will change, due to forces exerted on the system by objects in the surroundings of that system. But what happens to the momentum of objects in the *surroundings*? For all four kinds of interactions, gravitational, electromagnetic, strong, and weak, it is observed that momentum gained by the system is lost by the surroundings, and we say that momentum is "conserved." That is, you can't create or destroy momentum, you can only move it from one object to another:

CONSERVATION OF MOMENTUM

$$\Delta\vec{p}_{\text{sys}} + \Delta\vec{p}_{\text{surr}} = \vec{0}$$

The momentum of the system is defined as the (vector) sum of the momenta of all objects in the system (Figure 3.33):

$$\vec{p}_{\text{sys}} = \vec{p}_1 + \vec{p}_2 + \vec{p}_3 + \cdots$$

Similarly, the momentum of the surroundings is the (vector) sum of the momenta of all the objects in the surroundings. This principle is true regardless of the speed of the objects involved; that is, whether or not $\gamma \approx 1$.

Momentum Can Flow Between System and Surroundings

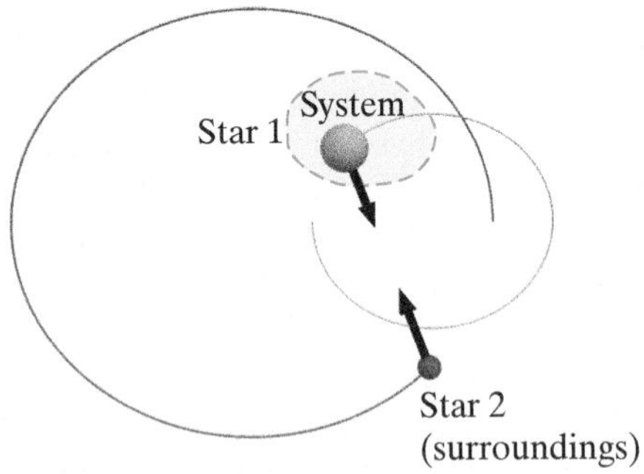

Figure 3.34 A binary star far from other objects. The gray lines show the trajectories of the individual stars. Choose star 1 as the system; star 2 is the surroundings.

Consider the case of two stars orbiting each other, a "binary star" far from all other objects (Figure 3.34). If you choose as the system of interest star 1, star 2 is the surroundings and exerts a force that changes the first star's momentum. Star 2, the surroundings, experiences a force of equal magnitude and opposite direction (reciprocity of gravitational forces). In an amount of time Δt, the momentum change of the system (star 1) is $\Delta\vec{p}_1 = \vec{F}_{\text{on 1 by 2}}\Delta t$, and the momentum change of the surroundings (star 2) is $\Delta\vec{p}_2 = -\vec{F}_{\text{on 1 by 2}}\Delta t = -\Delta\vec{p}_1$. Whatever momentum the system gains, the surroundings lose: $\Delta\vec{p}_{\text{sys}} + \Delta\vec{p}_{\text{surr}} = \vec{0}$. We say momentum is a "conserved" quantity.

The Momentum of an Isolated System Does Not Change

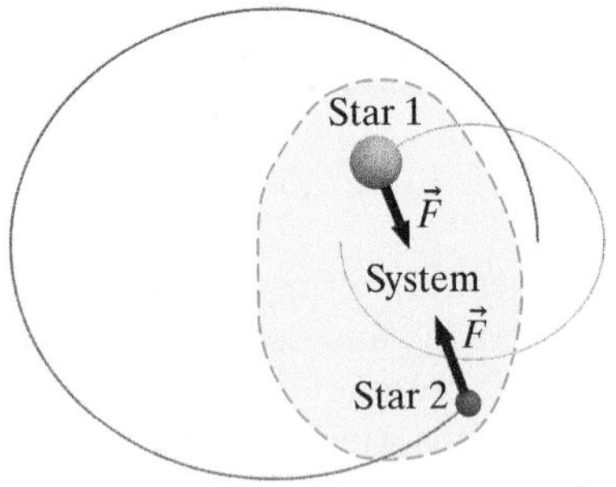

Figure 3.35 A binary star: choose both stars as the system. The momentum of the combined system doesn't change.

If on the other hand you choose both stars as your system (Figure 3.35), the surroundings consist of other stars, which may be so far away as to have negligible effects on the binary star. In that case the net force exerted by the surroundings on the system is nearly zero. The momentum of the combined system is $\vec{p}_{\text{sys}} = \vec{p}_1 + \vec{p}_2$. Let's see how the momentum of the system consisting of both stars changes:

$$\begin{aligned}\Delta\vec{p}_{\text{sys}} &= \Delta\vec{p}_1 + \Delta\vec{p}_2\\ &= \vec{F}_{\text{on 1 by 2}}\Delta t + \vec{F}_{\text{on 2 by 1}}\Delta t\\ &= \vec{F}_{\text{on 1 by 2}}\Delta t + (-\vec{F}_{\text{on 1 by 2}}\Delta t)\\ &= \vec{0}\end{aligned}$$

For the two-star system, the Momentum Principle $\vec{p}_f = \vec{p}_i + \vec{F}_{\text{net}}\Delta t$ reduces to $\vec{p}_{\text{sys},f} = \vec{p}_{\text{sys},i}$ and the momentum of the system $\vec{p}_{\text{sys}} = \vec{p}_1 + \vec{p}_2$ remains constant in magnitude and direction. This is an important special case: the momentum of an isolated system, a system with negligible interactions with the surroundings, doesn't change but stays constant. The momentum of each star changes, due to interactions with the other star, but momentum lost by one star is gained by the other star, and the total momentum of the system doesn't change. (In the case of the interactions of charged particles, some of the momentum may be associated with electric or magnetic "fields," in addition to the momentum of the particles.)

Stating the Momentum Principle in terms of the conservation of momentum highlights another aspect of the predictive power of this fundamental principle. It allows us to predict and explain aspects of the behavior of complex systems without knowing the details of the large number of interactions inside the systems. It can be particularly useful in analyzing the behavior of systems of many particles whose interactions with the surroundings are negligible during the time interval of interest.

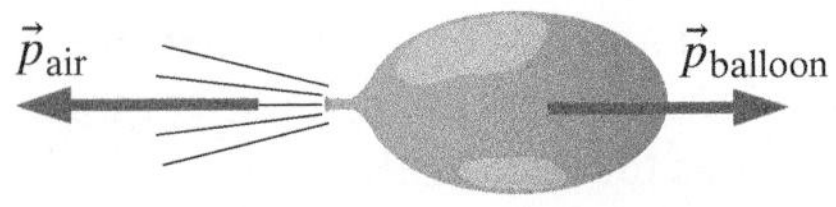

Figure 3.36 When the balloon is released, the total momentum of the system must still be zero.

EXAMPLE **A Balloon Rocket**

Suppose that you blow up a balloon, and hold it closed with your fingers. When you release your grip, air rushes out the opening in the balloon. If the balloon is oriented horizontally, as shown in Figure 3.36, so the outflowing air travels in the $-x$ direction, what must happen to the balloon?

Solution System: Balloon plus all the air initially inside it
Surroundings: Everything else
Initial time: Just before the balloon is released
Final time: Just after the balloon is released
Assumption: Negligible interactions with surroundings during this short Δt
Momentum Principle:

$$\Delta\vec{p}_{\text{sys}} + \Delta\vec{p}_{\text{surr}} = \vec{0}$$

Let $\vec{p}$ represent $\vec{p}_{\text{sys}}$, the total momentum of the system; that is, it is the sum of the momenta of the air inside the balloon plus the momentum of the rubber balloon itself.

$$\vec{p} = \vec{p}_{\text{air}} + \vec{p}_{\text{rubber}}$$

Initially the balloon and its contents are at rest, so

$$\vec{p}_i = \vec{0}$$

Because the interaction with the surroundings was negligible over this short interval, it follows that

$$\vec{p}_f - \vec{p}_i = \vec{0}$$
$$\vec{p}_f = \vec{0}$$

In the final state the total momentum of the system must still be zero. However, the momentum of the air coming out of the balloon is not zero; it is in the $-x$ direction. Therefore the momentum of the rubber balloon must be nonzero, in the $+x$ direction, so the balloon will move in the $+x$ direction. In this way the x components of the momenta of all of the objects in the system add up to zero.

EXAMPLE **Two Blocks Collide**

A system consists of a 1 kg block moving with velocity $\langle 5,2,0\rangle$ m/s and a 3 kg block moving with velocity $\langle -3,4,0\rangle$ m/s. The two-block system is nearly isolated from the surroundings. **(a)** What is the momentum of this two-block system? **(b)** The blocks collide and bounce off each other, changing both the magnitude and direction of the velocity of each block. What, if anything, can you conclude about the momentum of the two-block system after the collision?

Solution **(a)** The momentum of the system is simply the vector sum of the momenta of each piece of the system:

$$\vec{p}_{\text{system}} = \vec{p}_1 + \vec{p}_2 = (1\text{ kg})(\langle 5,2,0\rangle\text{ m/s}) + (3\text{ kg})(\langle -3,4,0\rangle\text{ m/s})$$
$$= \langle -4,14,0\rangle\text{ kg}\cdot\text{m/s}$$

(b) The total momentum of the system is still $\langle -4,14,0\rangle$ kg · m/s.

EXAMPLE **The Weak Interaction: Neutron Decay**

A free neutron, one that is outside a nucleus, is unstable and decays into a positively charged proton, a negatively charged electron, and an uncharged antineutrino:

$$n \rightarrow p^+ + e^- + \bar{\nu}$$

This decay of a free neutron is caused by the weak interaction. The precise time that the neutron will decay is governed by quantum mechanical laws of probability and the nature of the weak interaction and is not predictable for any particular neutron, but the average lifetime of a neutron is about 15 minutes. Also probabilistic are the magnitudes and directions of the momenta of the decay products (proton, electron, and antineutrino), and Figure 3.37 shows some decays of an isolated free neutron that was at rest. Which of the decays shown in Figure 3.37 is actually impossible, if the neutron was at rest?

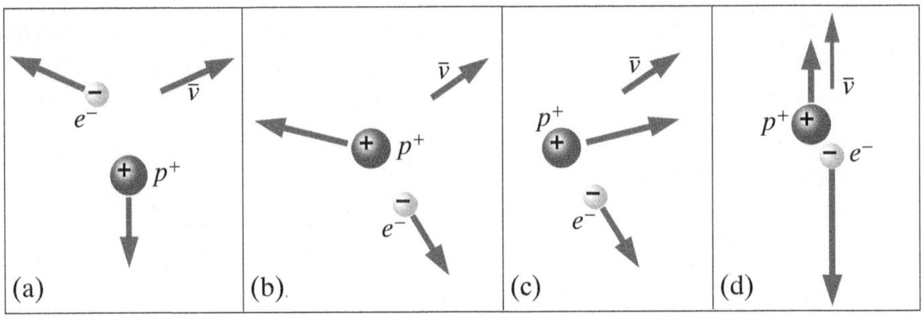

Figure 3.37 Which of these decays of a neutron at rest is impossible? The arrows represent momentum.

Solution The decay shown in Figure 3.37c is not actually possible. Conservation of momentum applies to all of the four fundamental interactions, including the weak interaction. The momentum of the isolated neutron was zero, so the momenta of the decay products must also add up to zero. In Figure 3.37c, however, the total momentum of the proton, electron, and antineutrino has a positive x component, which is not possible.

What's interesting about this is that we can rule out the possibility of the decay shown in Figure 3.37c without knowing anything about the details of the weak interaction. In fact, in quantum mechanical calculations at the atomic level concerned with electromagnetism, the strong interaction, or the weak interaction, force isn't even a particularly useful concept, yet conservation of momentum continues to play an important role even in the subatomic world.

Relativistic Momentum Conservation

Particle accelerators produce beams of particles such as electrons, protons, and pions at speeds very close to the speed of light. When these high-speed particles interact with other particles, experiments show that $\Delta\vec{p}_{\text{sys}} + \Delta\vec{p}_{\text{surr}} = \vec{0}$, but only if the momentum of each particle is defined in the way Einstein proposed, $\vec{p} \equiv \gamma m\vec{v}$. When the speed v approaches the speed of light c, the low-speed approximation $\gamma \approx 1$ ($\vec{p} \approx m\vec{v}$) is not valid. The quantity $m\vec{v}$ is not a conserved quantity when the speed v approaches the speed of light c, but the momentum $\vec{p} = \gamma m\vec{v}$ is conserved.

3.11 THE MULTIPARTICLE MOMENTUM PRINCIPLE

In our applications of the Momentum Principle we have often chosen a single object as the system of interest. However, we have seen that in some situations it can be very useful to choose a system that consists of two or more interacting objects. (Note that we can choose any object or set of objects as the system. Sometimes one choice is more useful than another, but any choice is valid.)

We will call forces "external" if they are exerted on objects in the system by objects in the surroundings. We call forces "internal" if they are forces exerted on objects in the system by other objects that are also in the system. What is remarkable is that the internal forces do not appear in the Momentum Principle for multiparticle systems:

MOMENTUM PRINCIPLE FOR MULTIPARTICLE SYSTEMS

$$\Delta\vec{p}_{\text{sys}} = (\vec{p}_{\text{sys},f} - \vec{p}_{\text{sys},i}) = \vec{F}_{\text{net}}\,\Delta t$$

$$\vec{p}_{\text{sys}} = \vec{p}_1 + \vec{p}_2 + \vec{p}_3 + \cdots$$

$$\vec{F}_{\text{net}} = \vec{F}_1 + \vec{F}_2 + \vec{F}_3 + \cdots \quad \text{(sum of all \textit{external} forces)}$$

This is an interesting statement of the Momentum Principle: it says that if we know the external forces acting on a multi-object system, forces exerted on objects in the chosen system by objects in the surroundings, we can draw conclusions about the change of momentum of the system over some time interval without worrying about any of the details of the interactions of the objects inside the system with each other. This can greatly simplify the analysis of the motion of some very complex systems. In a way, it should not be surprising that internal forces alone cannot change a system's total momentum. If they could, you could lift yourself off the ground by pulling up on your own feet!

We have implicitly been using this multiparticle version of the Momentum Principle when we have treated macroscopic objects like humans, spacecraft, and planets as if they were single point-like objects.

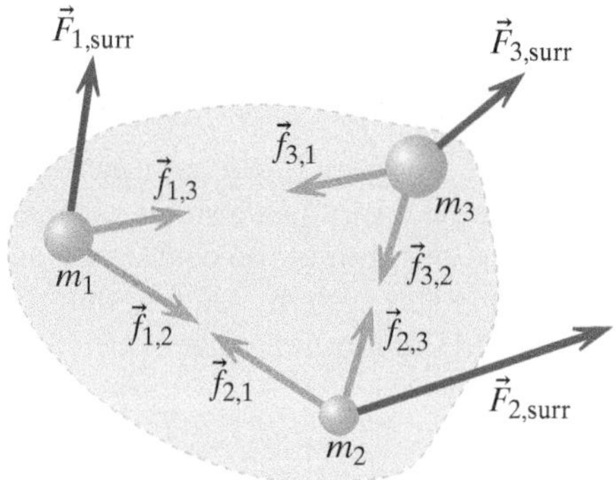

Figure 3.38 External and internal forces acting on a system of three particles.

Internal Forces Cancel

In a three-particle system (Figure 3.38) we show all of the forces acting on each particle, where the lowercase $\vec{f}$'s are forces the particles exert on each other (so-called "internal" forces), and the uppercase $\vec{F}$'s are forces exerted by objects in the surroundings that are not shown and are not part of our chosen system (these are so-called "external" forces, such as the gravitational attraction of the Earth, or a force that you exert by pulling on one of the particles).

We will use the following shorthand notation: $\vec{f}_{1,3}$ will denote the force exerted on particle 1 by particle 3; $\vec{F}_{2,\text{surr}}$ will denote the force on particle 2 exerted by objects in the surroundings, and so on. We start by applying the Momentum Principle to each of the three particles separately:

$$\Delta\vec{p}_1 = (\vec{F}_{1,\text{surr}} + \vec{f}_{1,2} + \vec{f}_{1,3})\Delta t$$
$$\Delta\vec{p}_2 = (\vec{F}_{2,\text{surr}} + \vec{f}_{2,1} + \vec{f}_{2,3})\Delta t$$
$$\Delta\vec{p}_3 = (\vec{F}_{3,\text{surr}} + \vec{f}_{3,1} + \vec{f}_{3,2})\Delta t$$

Nothing new so far—but now we add up these three equations. That is, we create a new equation by adding up all the terms on the left sides of the three equations, and adding up all the terms on the right sides, and setting them equal to each other:

$$\Delta\vec{p}_1 + \Delta\vec{p}_2 + \Delta\vec{p}_3 = (\vec{F}_{1,\text{surr}} + \vec{f}_{1,2} + \vec{f}_{1,3} + \vec{F}_{2,\text{surr}} + \vec{f}_{2,1} + \vec{f}_{2,3} + \vec{F}_{3,\text{surr}} + \vec{f}_{3,1} + \vec{f}_{3,2})\Delta t$$

Many of these terms cancel. By the principle of reciprocity (see Section 3.4), which is obeyed by gravitational and electric interactions, we have this:

$$\vec{f}_{1,2} = -\vec{f}_{2,1}$$
$$\vec{f}_{1,3} = -\vec{f}_{3,1}$$
$$\vec{f}_{2,3} = -\vec{f}_{3,2}$$

Thanks to reciprocity, all that remains after the cancellations is this:

$$\Delta\vec{p}_1 + \Delta\vec{p}_2 + \Delta\vec{p}_3 = (\vec{F}_{1,\text{surr}} + \vec{F}_{2,\text{surr}} + \vec{F}_{3,\text{surr}})\Delta t$$

The total momentum of the system is $\vec{p}_{\text{sys}} = \vec{p}_1 + \vec{p}_2 + \vec{p}_3$, so we have:

$$\Delta\vec{p}_{\text{sys}} = \vec{F}_{\text{net, surr}}\,\Delta t$$

The importance of this equation is that reciprocity has eliminated all of the internal forces (the forces that the particles in the system exert on each other); internal forces cannot affect the motion of the system as a whole. All that matters in determining the change of (total) momentum is the net external force. The equation has the same form as the Momentum Principle for a single particle.

The Center of Mass

The momentum of a multiparticle system such as a binary star can be expressed in terms of the velocity of the "center of mass" of the system. The position of the center of mass of a system of particles is defined as a weighted average of their positions:

$$\vec{r}_{\text{CM}} = \frac{m_1\vec{r}_1 + m_2\vec{r}_2 + m_3\vec{r}_3 + \cdots}{M_{\text{total}}}$$

$M_{\text{total}} = m_1 + m_2 + m_3 + \cdots$ is the total mass of the system. It can be shown that this definition is also valid if object 1 is a large object (not just a small particle) as long as m_1 is the mass of object 1 and $\vec{r}_1$ is the position of the center of mass of object 1. If you apply this definition to a binary star system, you find that the center of mass is between the two stars, closer to the more massive star.

EXAMPLE **Two Rocks**

A 10 kg rock is located at $\langle 3,0,0\rangle$ m, and a 2 kg rock is located at $\langle 8,2,0\rangle$ m (Figure 3.39). Find the center of mass of the two-rock system.

Solution

$$\vec{r}_{\text{CM}} = \frac{(10\,\text{kg})\langle 3,0,0\rangle\,\text{m} + (2\,\text{kg})\langle 8,2,0\rangle\,\text{m}}{(10\,\text{kg} + 2\,\text{kg})}$$
$$= \langle 3.83, 0.33, 0\rangle\,\text{m}$$

The center of mass of the system is closer to the more massive rock, as we would expect.

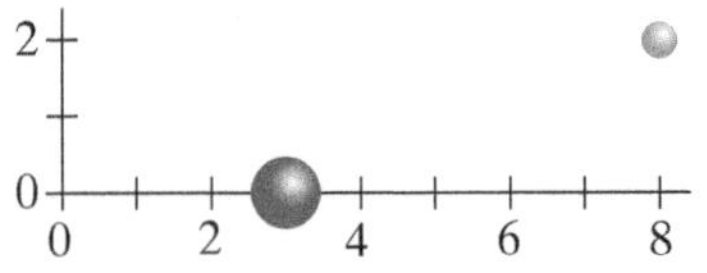

Figure 3.39 A 10 kg rock is located at $\langle 2,0,0\rangle$ m and a 2 kg rock is located at $\langle 8,2,0\rangle$ m. Where is the center of mass?

The Velocity of the Center of Mass

Taking the derivative with respect to time of the definition of center of mass, and assuming no changes in the masses, we find the following (since $\vec{v} = d\vec{r}/dt$):

$$\vec{v}_{\text{CM}} = \frac{m_1\vec{v}_1 + m_2\vec{v}_2 + m_3\vec{v}_3 + \cdots}{M_{\text{total}}}$$

Multiplying through by the total mass, we have this:

$$M_{\text{total}}\vec{v}_{\text{CM}} = m_1\vec{v}_1 + m_2\vec{v}_2 + m_3\vec{v}_3 + \cdots$$

However, if the speeds are small compared to the speed of light ($\gamma \approx 1$), the quantity $m_1\vec{v}_1 + m_2\vec{v}_2 + m_3\vec{v}_3 + \cdots$ is the total momentum of the system, so we can express the momentum of a system in terms of its center-of-mass velocity:

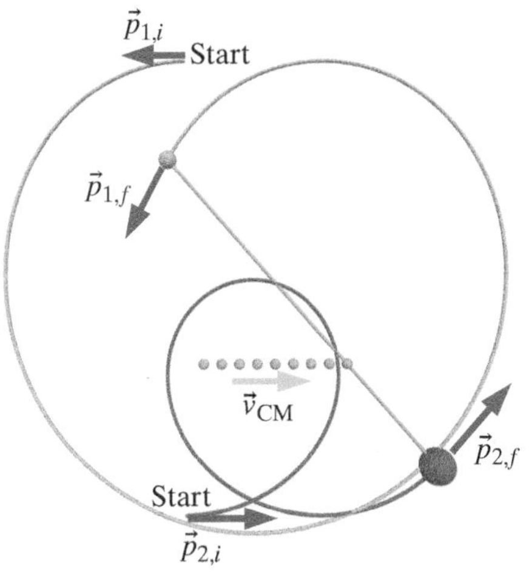

Figure 3.40 A binary star drifts through space. The center of mass moves with constant velocity. A line connecting the two stars passes through the center of mass, which is closer to the more massive star.

Figure 3.41 The total momentum of the two-star system does not change, as can be seen by the graphical sum of the vector momenta.

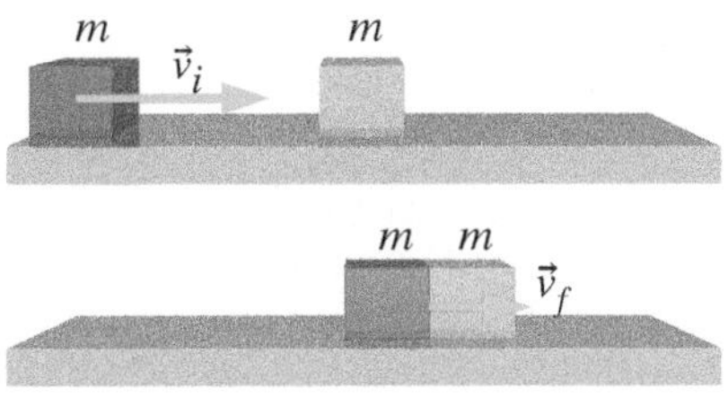

Figure 3.42 A sticking collision between blocks of equal mass. Top: Before the collision. Bottom: After the collision.

MOMENTUM AND CENTER OF MASS VELOCITY

$$\vec{p}_{sys} = M_{total}\vec{v}_{CM}$$

(If the speeds are small compared to the speed of light.)

If there are negligible interactions with the surroundings of a binary star, the velocity of the center of mass does not change but is constant as the binary star drifts through space (or is zero if the binary star's total momentum is zero). Figure 3.40 shows the motion of the center of mass for a binary star that is drifting through space. The two stars orbit around the (moving) center-of-mass point. A line connecting the two stars passes through the center of mass, which is closer to the more massive star.

The initial momentum of the two-star system is $\vec{p}_{1,i} + \vec{p}_{2,i}$, which is also equal to $(m_1 + m_2)\vec{v}_{CM}$. Later, both momenta of the stars have changed, but the vector sum $\vec{p}_{1,f} + \vec{p}_{2,f}$ is still the same and still equal to $(m_1 + m_2)\vec{v}_{CM}$. Figure 3.41 shows graphical addition of the momenta of the two stars to make the total system momentum, which does not change.

We can characterize the momentum of a multiparticle system simply as $\vec{p}_{sys} = M_{total}\vec{v}_{CM}$, as though the system were a point particle. Moreover, if an object is a sphere whose density is only a function of radius, the object exerts a gravitational force on other objects as though the sphere were a point particle. We can predict the motion of a system of star or a planet or an asteroid as though it were a single point particle of large mass.

Checkpoint 8 A system consists of a 2 kg block moving with velocity $\langle -3, 4, 0\rangle$ m/s and a 5 kg block moving with velocity $\langle 2, 6, 0\rangle$ m/s. Calculate the velocity of the center of mass of the two-block system, given that the momentum of the system is $M_{total}\vec{v}_{CM}$.

3.12 COLLISIONS: NEGLIGIBLE EXTERNAL FORCES

An event is called a "collision" if it involves an interaction that takes place in a relatively short time and has a large effect on the momenta of the objects compared to the effects of other interactions during that short time. A collision does not necessarily involve actual physical contact between objects, which may be interacting via long-range forces like the gravitational force or the electric force. A collision can involve attractive forces. For example, a spacecraft is deflected as it passes close to Jupiter on its way to Saturn, and Jupiter exerts a large gravitational force for a relatively short time. During that short time the effects of the other planets are negligible.

EXAMPLE **What Is *Not* Conserved in a Collision?**

In later chapters we will see that energy and angular momentum are also conserved quantities. However, most quantities, including velocity, are not conserved quantities. Consider a collision between two blocks on ice (so that there is low friction), each with mass m. One block is initially at rest and the other is moving with initial velocity $\langle v_i, 0, 0\rangle$. In the collision they stick together and move with final velocity $\langle v_f, 0, 0\rangle$ (Figure 3.42). What is the final velocity? Is velocity a conserved quantity?

Solution As the blocks slide on the ice there are no changes in momentum in the vertical direction (the y components of momentum are zero at all times). In the xz plane of the ice there are negligible interactions with the surroundings, so the momentum of the surroundings doesn't change, and $\Delta\vec{p}_{sys} = \vec{0}$. Therefore $\vec{p}_{sys} = \vec{p}_1 + \vec{p}_2$ doesn't change. Before the collision the total momentum is

$$\vec{p}_{sys} = m\langle v_i, 0, 0\rangle + m\vec{0} = \langle mv_i, 0, 0\rangle$$

After the collision the total momentum is

$$\vec{p}_{\text{sys}} = m\langle v_f, 0, 0\rangle + m\langle v_f, 0, 0\rangle = \langle 2mv_f, 0, 0\rangle$$

Since $\vec{p}_{\text{sys}}$ doesn't change, we have

$$\langle 2mv_f, 0, 0\rangle = \langle mv_i, 0, 0\rangle$$

$$v_f = \frac{1}{2}v_i$$

The final speed (and velocity) is half the initial value. Nothing in the surroundings changed, so $\Delta\vec{v}_{\text{system}} + \Delta\vec{v}_{\text{surr}} \neq \vec{0}$. We say that velocity is not a conserved quantity, but momentum is. This is another reason why momentum is such an important physical quantity.

EXAMPLE **Different Choices of System**

Consider the head-on collision of two identical bowling balls (Figure 3.43). Each ball has a mass of 6 kg. Ball A with velocity $\langle 5,0,0\rangle$ m/s strikes ball B, which was at rest. Then ball A stops and ball B moves with the same velocity ($\langle 5,0,0\rangle$ m/s) that ball A had initially. For the following three different choices of system, find the momentum change of the system and the momentum change of the surroundings. **(a)** Choose a system consisting only of ball A. **(b)** Choose a system consisting only of ball B. **(c)** Choose a system consisting of both balls.

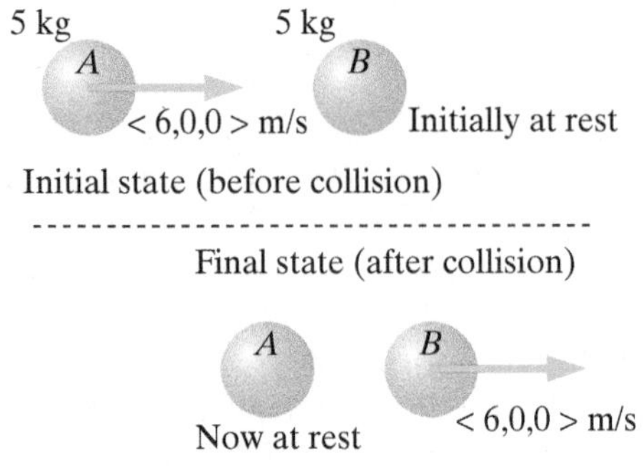

Figure 3.43 A head-on collision between two identical bowling balls.

Solution **(a)** The initial momentum of ball A is $\vec{p}_{A,i} = \langle 30,0,0\rangle$ kg·m/s, so if we choose ball A as the system:

$$\Delta\vec{p}_{\text{sys}} = \langle 0,0,0\rangle \text{ kg}\cdot\text{m/s} - \langle 30,0,0\rangle \text{ kg}\cdot\text{m/s} = \langle -30,0,0\rangle \text{ kg}\cdot\text{m/s}$$

Since ball A is the system, ball B is the surroundings:

$$\Delta\vec{p}_{\text{surr}} = \langle 30,0,0\rangle \text{ kg}\cdot\text{m/s} - \langle 0,0,0\rangle \text{ kg}\cdot\text{m/s} = \langle +30,0,0\rangle \text{ kg}\cdot\text{m/s}$$

(b) If we choose ball B as the system:

$$\Delta\vec{p}_{\text{sys}} = \langle +30,0,0\rangle \text{ kg}\cdot\text{m/s}$$
$$\Delta\vec{p}_{\text{surr}} = \langle -30,0,0\rangle \text{ kg}\cdot\text{m/s}$$

(c) If we choose both balls as the system, then:

$$\vec{p}_{\text{sys},i} = \langle 30,0,0\rangle \text{ kg}\cdot\text{m/s} + \langle 0,0,0\rangle \text{ kg}\cdot\text{m/s} = \langle 30,0,0\rangle \text{ kg}\cdot\text{m/s}$$
$$\vec{p}_{\text{sys},f} = \langle 0,0,0\rangle \text{ kg}\cdot\text{m/s} + \langle 30,0,0\rangle \text{ kg}\cdot\text{m/s} = \langle 30,0,0\rangle \text{ kg}\cdot\text{m/s}$$
$$\Delta\vec{p}_{\text{sys}} = \langle 0,0,0\rangle \text{ kg}\cdot\text{m/s}$$

Often it can be useful to analyze collisions by choosing a system that includes all of the colliding objects, as we will see in the following example.

EXAMPLE **Two Lumps of Clay Collide**

Two lumps of clay travel through the air toward each other, at speeds much less than the speed of light (Figure 3.44), rotating as they move. When the lumps collide they stick together. The mass of lump 1 is 0.2 kg and its initial velocity is $\langle 6,0,0\rangle$ m/s, and the mass of lump 2 is 0.5 kg and its initial velocity is $\langle -5,4,0\rangle$ m/s. What is the final velocity of the center of mass of the stuck-together lumps?

Solution The Momentum Principle applies to the motion of the center of mass, even though the objects are rotating. We can neglect air resistance and gravitational force (which are negligible compared to the large contact forces during the very short time of the collision).

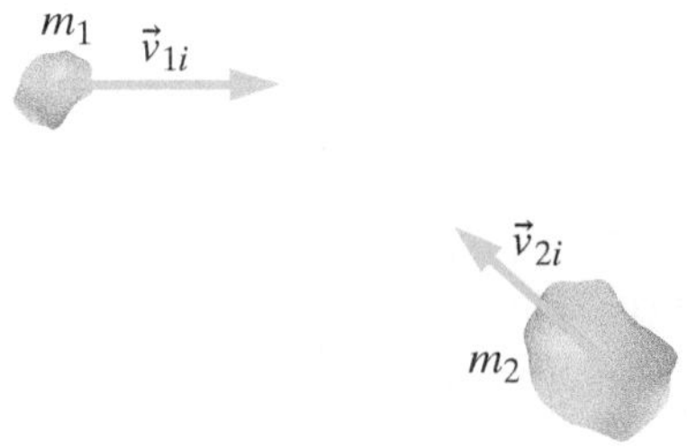

Figure 3.44 Two lumps of clay traveling through the air, just before colliding.

System: Both lumps
Surroundings: Air, Earth

$$\vec{p}_{\text{total},f} = \vec{p}_{\text{total},i} + \vec{F}_{\text{net}}\,\Delta t$$
$$\vec{p}_{\text{total},f} \approx m_1\vec{v}_{1i} + m_2\vec{v}_{2i} + \langle 0,0,0\rangle \Delta t$$
$$(m_1 + m_2)\vec{v}_f = m_1\vec{v}_{1i} + m_2\vec{v}_{2i}$$
$$\vec{v}_f = \frac{m_1\vec{v}_{1i} + m_2\vec{v}_{2i}}{m_1 + m_2}$$
$$\vec{v}_f = \frac{(0.2\text{ kg})\langle 6,0,0\rangle + (0.5\text{ kg})\langle -5,4,0\rangle}{(0.7\text{ kg})}\text{ m/s}$$
$$\vec{v}_f = \langle -1.86, 2.86, 0\rangle\text{ m/s}$$

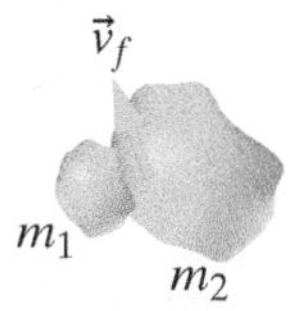

Figure 3.45 The momentum of the stuck-together lumps of clay just after the collision is equal to the sum of the initial momenta of the two lumps of clay. The object may rotate.

Check: The units are correct.

By choosing both lumps as the system, we were able to find the final momentum of the system without needing to know anything about the details of the complex forces that the lumps exerted on each other during the collision.

QUESTION Why doesn't it matter that the lumps may be rotating after the collision (Figure 3.45)?

The rotation of an object doesn't affect its momentum; the total momentum of the system is still the sum of the individual momenta of the objects within the system and equal to $M_{\text{total}}\vec{v}_{\text{CM}}$. However, the Momentum Principle does not tell us anything about how fast the stuck-together objects will be rotating; for this, we will have to apply the Angular Momentum Principle, which is discussed in a later chapter.

The initial velocities are the center-of-mass velocities of each lump, and the final velocity is the center-of-mass velocity of the combined system after the two lumps stick together. We can also speak of the velocity of the center of mass just before the collision, which is the velocity of a mathematical point between the two lumps, calculated as the weighted average of their positions.

QUESTION What was the center-of-mass velocity of the two lumps just before they collided?

It was $\langle -1.86, 2.86, 0\rangle$ m/s, the same as after the collision, since the total momentum can be calculated as $M_{\text{total}}\vec{v}_{\text{CM}}$ both before and after the collision.

Some Problems Require More than One Principle

Because the lumps in the previous example stuck together, we had enough information to solve for the final velocity of the stuck-together lumps—we had one equation with only one unknown (vector) quantity. In more complex situations, we will find that we sometimes do not have enough information to solve for all of the unknown quantities by applying only the Momentum Principle.

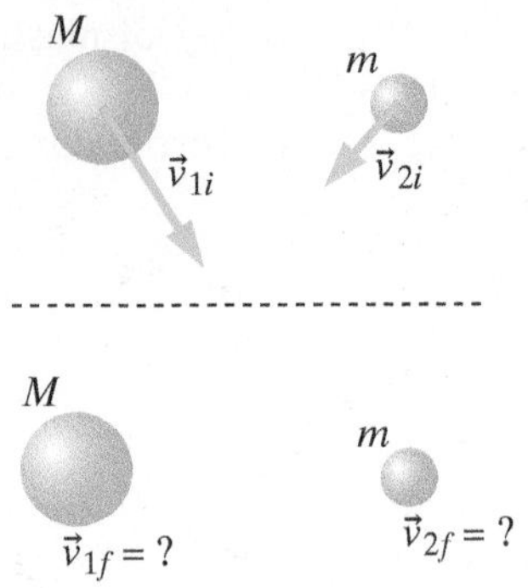

Figure 3.46 The initial momenta of the colliding objects are known, but the final momenta after the collision are unknown.

For example, consider a collision between two balls that bounce off each other, as shown in Figure 3.46. The Momentum Principle tells us that

$$\vec{p}_{\text{total},f} = \vec{p}_{\text{total},i}$$
$$\vec{p}_{1f} + \vec{p}_{2f} = \vec{p}_{1i} + \vec{p}_{2i}$$

but we are left with two unknown (vector) quantities, $\vec{p}_{1f}$ and $\vec{p}_{2f}$, and only one equation. We will not be able to analyze problems of this kind fully until we can invoke the Energy Principle (Chapter 6) and the Angular Momentum Principle (Chapter 11), along with the Momentum Principle.

Checkpoint 9 You and a friend each hold a lump of wet clay. Each lump has a mass of 20 g. You each toss your lump of clay into the air, where the lumps collide and stick together. Just before the impact, the velocity of one lump was $\langle 5, 2, -3 \rangle$ m/s, and the velocity of the other lump was $\langle -3, 0, -2 \rangle$ m/s. **(a)** What was the total momentum of the lumps just before impact? **(b)** What is the momentum of the stuck-together lump just after the impact? **(c)** What is its velocity?

3.13 NEWTON AND EINSTEIN

Newton devised a particular explanatory scheme in which the analysis of motion is divided into two distinct parts:

1. Quantify the interaction in terms of a concept called "force." Specific examples are Newton's gravitational force and Coulomb's electric force.
2. Quantify the change of motion in terms of the change in a quantity called "momentum." The change in the momentum is equal to the force times Δt (for sufficiently small values of Δt).

This scheme, called the "Newtonian synthesis," has turned out to be extraordinarily successful in explaining a huge variety of diverse physical phenomena, from the fall of an apple to the orbiting of the Moon. Yet we have no way of asking whether the Universe *really* works this way. It seems unlikely that the Universe actually uses the human concepts of force and momentum in the unfolding motion of an apple or the Moon.

We refer to the Newtonian synthesis both to identify and honor the particular, highly successful analysis scheme introduced by Newton, but also to remind ourselves that this is not the only possible way to view and analyze the Universe.

Einstein's Alternative View

Newton identified the gravitational force and represented it algebraically but could give no explanation for it. He was content with showing that it correctly predicted the motion of the planets, and this was a huge advance, the real beginning of modern science.

Einstein made a further huge advance by giving a deeper explanation for gravity, as a part of his Theory of General Relativity. He realized that it was possible to think of the massive Sun bending space and time (!) in such a way as to make the planets move the way they do. The equations in Einstein's General Theory of Relativity make it possible to calculate the curvature of space and time due to massive objects, and to predict how other objects will move in this altered space and time.

Moreover, Einstein's Theory of General Relativity accurately predicts some tiny effects that Newton's gravitational force does not, such as the slight bending of light as it passes near the Sun. General Relativity also explains some bizarre large-scale phenomena such as black holes and the observed expansion of the space between the galaxies.

Einstein's earlier Theory of Special Relativity established that nothing, not even information, can travel faster than light. Because Newton's gravitational force law depends only on the distance between objects, not on the time, something's wrong with Newton's law, since this implies that if an object were suddenly yanked away, its force on another object would vanish instantaneously, thus giving (in principle) a way to send information from one place to another instantaneously. Einstein's Theory of General Relativity doesn't have this problem.

Since the equations of General Relativity are very difficult to work with, and Newton's gravitational force works very well for most purposes, in this textbook we will use Newton's approach to gravity. However, you should be aware that for the most precise calculations one must use the Theory of General Relativity. For example, the highly accurate atomic clocks in the satellites that make up the Global Positioning System (GPS) have to be continually corrected using Einstein's Theory of General Relativity. Otherwise, GPS positions would be wrong by several kilometers after just one day of operation!

Figure 3.47 Our Solar System orbits around the center of our galaxy, which interacts with other galaxies. (Courtesy NASA/JPL-Caltech)

3.14 PREDICTING THE FUTURE OF COMPLEX SYSTEMS

We can use the Newtonian method and numerical integration to try to predict the future of a group of objects that interact with each other mainly gravitationally, such as the Solar System consisting of the Sun, planets, moons, asteroids, comets, and so on. There are other, nongravitational interactions present. Radiation pressure from sunlight makes a comet's tail sweep away from the Sun. Streams of charged particles (the "solar wind") from the Sun hit the Earth and contribute to the Northern and Southern Lights (auroras). However, the main interactions within the Solar System are gravitational.

The Solar System as a group is orbiting around the center of our Milky Way galaxy, and our galaxy is interacting gravitationally with other galaxies in a local cluster of galaxies (Figure 3.47), but changes in these speeds and directions take place very slowly compared to those within the Solar System. To a very good approximation we can neglect these interactions while studying the inner workings of the Solar System itself.

We have modeled the motion of planets and stars using the Momentum Principle and the gravitational force. We can use the same approach to model the gravitational interactions of more complex systems involving three, four, or any number of gravitationally interacting objects. All that is necessary is to include the forces associated with all pairs of objects.

In principle we could use these techniques to predict the future of our entire Solar System. Of course our prediction would be only approximate, because we take finite time steps that introduce numerical errors, and we are neglecting various small interactions. We would need to investigate whether the prediction of our model is a reasonable approximation to the real motion of our real Solar System.

Our Solar System contains a huge number of objects. Most of the planets have moons, and Jupiter and Saturn have many moons. There are thousands of asteroids and comets, not to mention an uncounted number of tiny specks of dust. This raises a practical limitation: the fastest computers would take a very long time to carry out even one time step for all the pieces of the Solar System.

The Three-Body Problem

As we have seen, in order to carry out a numerical integration we need to know the position and velocity of each object at some time t, from which we can

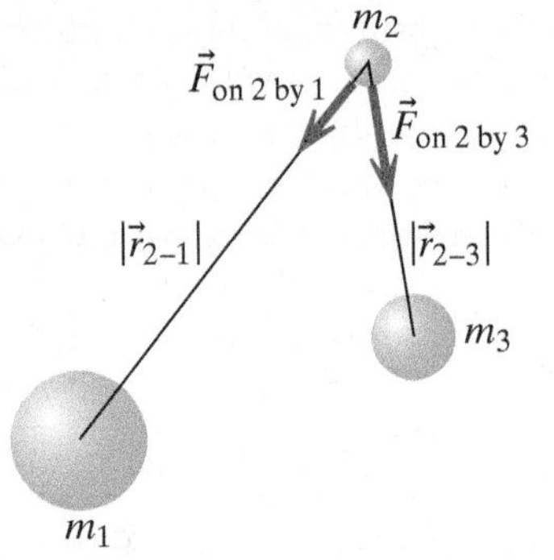

Figure 3.48 In the "three-body problem" each of the three objects interacts with two other objects.

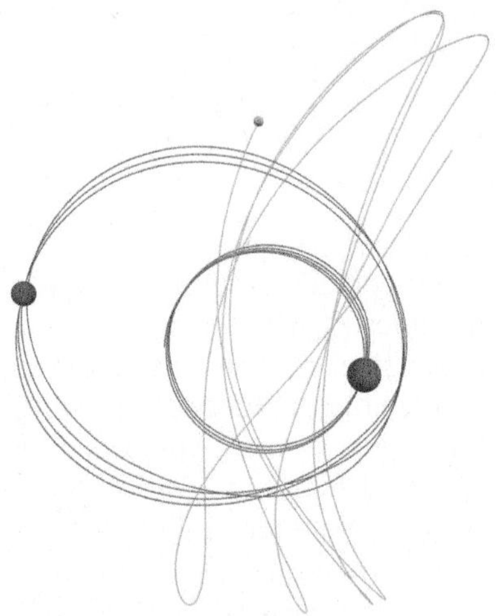

Figure 3.49 An example of three bodies interacting gravitationally.

calculate the forces on each object at that time. This enables us to calculate the position and momentum for each object at a slightly later time $t+\Delta t$. We need the net force on an object, which we get from vector addition of all the pairwise forces calculated from Newton's gravitational force.

Suppose that there are three gravitating objects, as shown in Figure 3.48. The force on object m_2, which is acted upon gravitationally by objects m_1 and m_3, is a natural extension of what we did earlier for the analysis of a star and a planet. We simply use the superposition principle to include the interaction with m_3 as well as the interaction with m_1:

$$\vec{F}_{\text{net on } m_2} = -G\frac{m_2 m_1}{|\vec{r}_{2-1}|^2}\hat{r}_{2-1} - G\frac{m_2 m_3}{|\vec{r}_{2-3}|^2}\hat{r}_{2-3}$$

We use this force to update the momentum of object m_2. Similar computations apply to the other masses, m_1 and m_3. These calculations are tedious, but let the computer do them! We can carry out a numerical integration for each mass: given the current locations of all the masses we can calculate the vector net force exerted on each mass at a time t, and then we can calculate the new values of momentum and position at a slightly later time $t+\Delta t$.

Reciprocity simplifies these calculations somewhat. For example, for the pair of interacting objects 2 and 3 you only need to calculate fully the force of object 2 on object 3. The force of object 3 on object 2 is just the opposite: $\vec{F}_{3 \text{ on } 2} = -\vec{F}_{2 \text{ on } 3}$.

While there exist analytical (nonnumerical) solutions for two-body gravitational motion, except for some very special cases the "three-body problem" has not been solved analytically. However, in a numerical integration, the computer doesn't mind that there are lots of quantities and a lot of repetitive calculations to be done. You yourself can solve the three-body problem numerically, by adding additional calculations to your two-body computations (see the homework problems P66 and P74).

For a single object orbiting a very massive stationary object, or for two objects orbiting each other (with zero total momentum), the possible trajectories are only a circle, ellipse, parabola, hyperbola, or straight line. The motion of a three-body system can be vastly more complex and diverse. Figure 3.49 shows the numerical integration of a complicated three-body trajectory for one particular set of initial conditions. Adding just one more object opens up a vast range of complex behaviors.

Imagine how complex the motion of a galaxy can be!

Checkpoint 10 Suppose you have four stars with given initial conditions (positions and momenta). The net force on each star is the vector sum of three forces exerted by the other three stars, for a total of 12 forces that must be evaluated in each iterative step. In each step, how many forces would you have to calculate fully, and how many forces could you calculate simply by reversing vectors already calculated?

Sensitivity to Initial Conditions

For a two-body system, slight changes in the initial conditions make only slight changes in the orbit, such as changing a circle into an ellipse, or an ellipse into a slightly different ellipse, but you never get anything other than a simple trajectory.

The situation is very different in systems with three or more interacting objects. Consider a low-mass object orbiting two massive objects, which we imagine somehow to be nailed down so that they cannot move. (This could also represent two positive charges that are fixed in position, with a low-mass negative charge orbiting around them.) In Figure 3.50 are two very different orbits (one shown in red, the other in blue), starting from rest with slightly

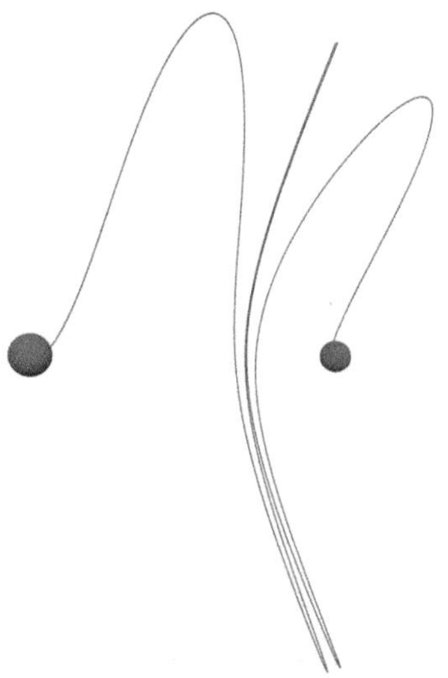

Figure 3.50 Two different orbits (one red, the other blue), starting from rest but from slightly different initial values of x for a low-mass object. The two orbits end with crashes on the two different massive objects. The mass of the red object is twice the mass of the blue object; both of these objects are somehow nailed down and cannot move.

different initial values of x. The two orbits end with crashes on the two different massive objects. The trajectories are wildly different!

This sensitivity to initial conditions generally becomes more extreme as you add more objects, and the Solar System contains lots of objects, big and small. Also, we can anticipate that if small errors in specifying the initial conditions can have large effects, so too it is likely that failing to take into account the tiny force exerted by a small asteroid might make a big difference after a long integration time.

The Solar System is actually fairly predictable, because there is one giant mass (the Sun), and the other, much smaller masses are very far apart. This is unlike the example shown here, which deliberately emphasizes the sensitivity observed when there are large masses near each other.

3.15 DETERMINISM

If you know the net force on an object as a function of its position and momentum (or velocity), you can predict the future motion of the object by simple step-by-step calculations based on the Momentum Principle. Newton's demonstration of the power of this approach induced many 17th and 18th century philosophers and scientists to adopt the view already proposed by Descartes, Boyle, and others, that the Universe is a giant clockwork device, whose future is completely determined by the present positions and momenta of all the macroscopic and microscopic objects in it. Just turn the crank, and predict the future! This point of view is called "determinism," and taken to its extreme it raises the question of whether humans actually have any free will, or whether all of our actions are predetermined. Scientific and technological advances in the 20th century, however, have led us to see that although the Newtonian approach can be used to predict the long-term future of a simple system or the short-term future of a complicated system, there are both practical and theoretical limitations on predictions in some systems.

Practical Limitations

One reason we may be unable to predict the long-term future of even a simple system is a practical limitation in our ability to measure its initial conditions with sufficient accuracy. Over time, even small inaccuracies in initial conditions can lead to large cumulative errors in calculations. This is a practical limitation rather than a theoretical one, because in principle if we could measure initial conditions more accurately, we could perform more accurate calculations.

Another practical limitation is our inability to account for all interactions in our model. Every object in the Universe interacts with every other object. In constructing simplified models, we ignore interactions whose magnitude is extremely small. However, over time, even very small interactions can lead to significant effects. Even the tiny effects of sunlight have been shown to affect noticeably the motion of an asteroid over a long time. With larger and faster computers, we can include more and more interactions in our models, but our models can never completely reflect the complexity of the real world.

A branch of current research that focuses on how the detailed interactions of a large number of atoms or molecules lead to the bulk properties of matter is called "molecular dynamics." For example, one way to test our understanding of the nature of the interactions of water molecules is to create a computational model in which the forces between molecules and the initial state of a large number of molecules—several million—are described, and then see whether running the model produces a virtual fluid that actually behaves like water. However, even with an accurate expression for the (electric) forces between atoms or molecules, it is not feasible to compute and track the motions of the 10^{25} molecules in a glass of water. Significant efforts are under way to build faster and faster computers, and to develop more efficient computer algorithms, to permit realistic numerical integration of more complex systems.

Chaos

A second kind of limitation occurs in systems that display an extreme sensitivity to initial conditions such as the situation seen in Figure 3.50, where small changes in initial conditions produce large changes in behavior. In recent years scientists have discovered systems in which *any* change in the initial conditions, no matter how small (infinitesimal), can lead to complete loss of predictability: the difference in the two possible future motions of the system diverges exponentially with time. Such systems are called "chaotic." It is thought that over a long time period the weather may be literally unpredictable in this sense.

An interesting popular science book about this new field of research is *Chaos: Making a New Science*, by James Gleick (Penguin, 1988). The issue of small changes in initial conditions is a perennial concern of time travelers in science fiction stories. Perhaps the most famous such story is Ray Bradbury's "The Sound of Thunder," in which a time traveler steps on a butterfly during the Jurassic era, and returns to an eerily changed present time.

Breakdown of Newton's Laws

There are other situations of high interest in which we cannot usefully apply Newton's laws of motion, because these classical laws do not adequately describe the behavior of physical systems. To model systems composed of very small particles such as protons and electrons, quarks (the constituents of protons and neutrons), and photons, it is necessary to use the laws of quantum mechanics. To model in detail the gravitational interactions between massive objects, it is necessary to apply the principles of General Relativity. Given the principles of quantum mechanics and General Relativity, one might assume that it would be possible to follow a procedure similar to the one we have followed in this chapter, and predict in detail the future of these systems. However, there appear to be more fundamental limitations on what we can know about the future.

Probability and Uncertainty

Our understanding of the atomic world of quantum mechanics suggests that there are fundamental limits to our ability to predict the future, because at the atomic and subatomic levels the Universe itself is nondeterministic. We cannot know exactly what will happen at a given time, but only the probability that certain events will occur within a given time frame.

A simple example may make this clear. A free neutron (one not bound into a nucleus) is unstable and eventually decays into a positively charged proton, a negatively charged electron, and an electrically neutral antineutrino:

$$n \to p^+ + e^- + \overline{\nu}$$

The average lifetime of the free neutron is about 15 minutes. Some neutrons survive longer than this, and some last a shorter amount of time. All of our experiments and all of our theory are consistent with the notion that *it is not possible to predict when a particular neutron will decay*. There is only a (known) probability that the neutron will decay in the next microsecond, or the next, or the next.

If this is really how the Universe works, as a great deal of evidence demonstrates, then there is an irreducible lack of predictability and determinism of the Universe itself. As far as our own predictions using the Momentum Principle are concerned, consider the following simple scenario. An electron is traveling with constant velocity through nearly empty space, but there is a free neutron in the vicinity (Figure 3.51).

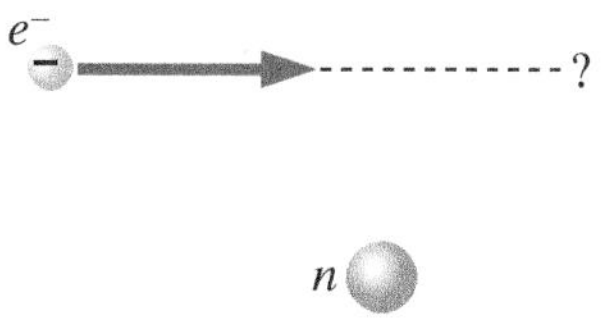

Figure 3.51 The electron may continue to travel in a straight line, but we cannot be certain that it will, because we do not know when the neutron will decay.

We might predict that the electron will move in a straight line, and some fraction of the time we will turn out to be right (the electrically uncharged neutron exerts no electric force on the electron, there is a magnetic interaction but it is quite small, and the gravitational interaction is tiny). However, if the neutron happens (probabilistically) to decay just as our electron passes nearby, suddenly our electron is subjected to large electric forces due to the decay proton and electron, and will deviate from a straight path. Moreover, the directions in which the decay products move are also only probabilistic, so we can't even predict whether the proton or the electron will come closest to our electron (Figures 3.52 and 3.53).

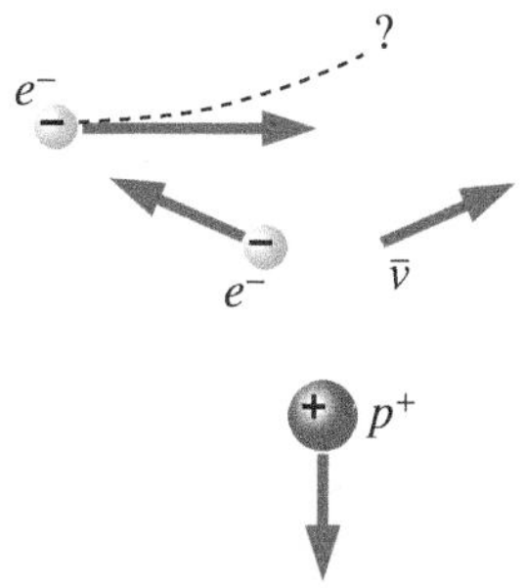

Figure 3.52 One possible path of the electron if the neutron does decay. The arrows represent momentum.

This is a simple but dramatic example of how quantum indeterminacy can lead to indeterminacy even in the context of the Momentum Principle.

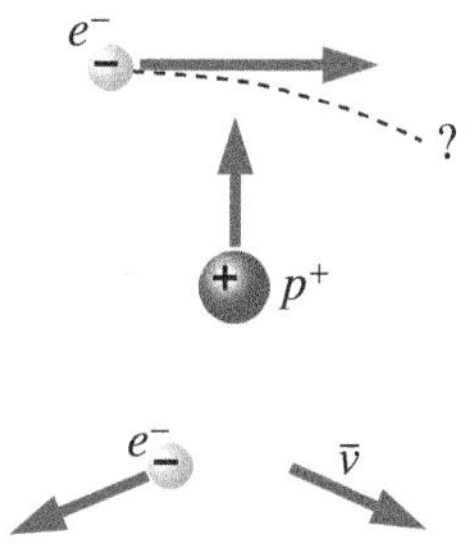

Figure 3.53 Another possible path of the electron if the neutron does decay. The trajectory of the electron depends on the directions of the momenta of the decay products. The arrows represent momentum.

The Heisenberg Uncertainty Principle

The Heisenberg uncertainty principle states that there are actual theoretical limits to our knowledge of the state of physical systems. One quantitative formulation states that the position and the momentum of a particle cannot both be simultaneously measured exactly (here Δ means variation, not difference):

$$\Delta x \Delta p_x \geq h$$

This relation says that the product of the uncertainty in position Δx and the uncertainty in momentum Δp_x is greater than or equal to a constant h, called Planck's constant. Planck's constant is tiny ($h = 6.6 \times 10^{-34}$ kg·m^2/s), so this limitation is not noticeable for macroscopic systems, but for objects small enough to require a quantum mechanical description, the uncertainty principle puts fundamental limits on how accurately we can know the initial conditions, and therefore how well we can predict the future.

3.16 POINTS AND SPHERES

The expression for the gravitational force applies to objects that are "point-like" (very small compared to the center-to-center distance between the objects). In later chapters we will be able to show that any hollow spherical shell with uniform density acts gravitationally on external objects as though all the mass of the shell were concentrated at its center. The density of the Earth is not uniform, because the central iron core has higher density than the outer layers. However, by considering the Earth as layers of hollow spherical shells, like an onion, with each shell of nearly uniform density, we get the result that the Earth can be modeled for most purposes as a point mass located at the center of the Earth (for very accurate calculations one must take into account small irregularities in the Earth's density from place to place). Similar statements can be made about other planets and stars. In Figure 3.54, the gravitational force is correctly calculated using the center-to-center distance $|\vec{r}|$.

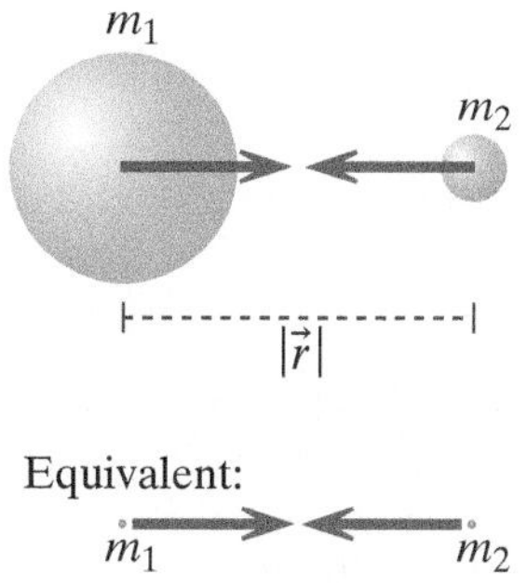

Figure 3.54 A sphere whose density depends only on distance from the center acts like a point particle with all the mass at the center.

This is not an obvious result. After all, in Figure 3.54 some of the atoms are closer and some farther apart than the center-to-center distance $|\vec{r}|$, but the net effect after adding up all the interactions of the individual atoms is as though the two objects had collapsed down to points at their centers. This is a very special property of $1/r^2$ forces, both gravitational and electric, and is not true for forces that have a different dependence on distance.

3.17 MEASURING THE GRAVITATIONAL CONSTANT G

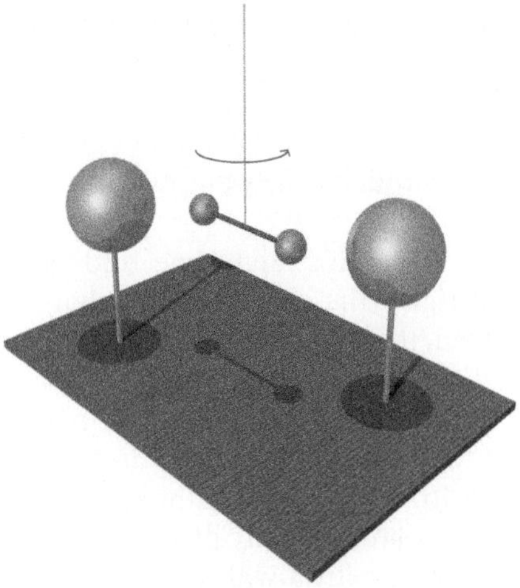

Figure 3.55 The Cavendish experiment.

In order to make quantitative predictions and analyses of physical phenomena involving gravitational interactions, it is necessary to know the universal gravitational constant G. In 1797–1798 Henry Cavendish performed the first experiment to determine a precise value for G (Figure 3.55). In this kind of experiment, a bar with metal spheres at each end is suspended from a thin quartz fiber that constitutes a "torsional" spring. From other measurements, it is known how large a tangential force measured in newtons is required to twist the fiber through a given angle. Large balls are brought near the suspended spheres, and one measures how much the fiber twists due to the gravitational interactions between the large balls and the small spheres.

If the masses are measured in kilograms, the distance in meters, and the force in newtons, the gravitational constant G has been measured in such experiments to be

$$G = 6.7 \times 10^{-11} \frac{\text{N} \cdot \text{m}^2}{\text{kg}^2}$$

This extremely small number reflects the fact that gravitational interactions are inherently very weak compared with electromagnetic interactions. The only reason that gravitational interactions are significant in our daily lives is that objects interact with the entire Earth, which has a huge mass. It takes sensitive measurements such as the Cavendish experiment to observe gravitational interactions between two ordinary-sized objects.

SUMMARY

Four fundamental interactions:

- Gravitational interactions (all objects attract each other gravitationally)
- Electromagnetic interactions (electric and magnetic interactions, closely related to each other); interatomic forces are electric in nature
- "Strong" interactions (inside the nucleus of an atom)
- "Weak" interactions (neutron decay, for example)

The gravitational force (Figure 3.56)

$$\vec{F}_{\text{grav on 2 by 1}} = -G\frac{m_1 m_2}{|\vec{r}|^2}\hat{r}$$

$\vec{r} = \vec{r}_2 - \vec{r}_1$ is the position of 2 relative to 1.

G is a universal constant: $G = 6.7 \times 10^{-11} \frac{\text{N} \cdot \text{m}^2}{\text{kg}^2}$.

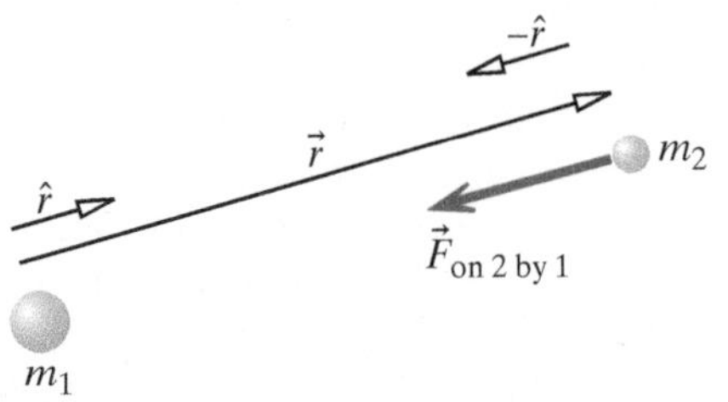

Figure 3.56 Near the Earth's surface, $|\vec{F}_{\text{grav}}| = mg$, where $g = +9.8\,\text{N/kg}$.

The electric force (Figure 3.57)

$$\vec{F}_{\text{elec on 2 by 1}} = \frac{1}{4\pi\varepsilon_0}\frac{q_1 q_2}{|\vec{r}|^2}\hat{r}$$

$\vec{r} = \vec{r}_2 - \vec{r}_1$ is the position of 2 relative to 1.

$\frac{1}{4\pi\varepsilon_0}$ is a universal constant: $\frac{1}{4\pi\varepsilon_0} = 9 \times 10^9 \frac{\text{N} \cdot \text{m}^2}{\text{C}^2}$.

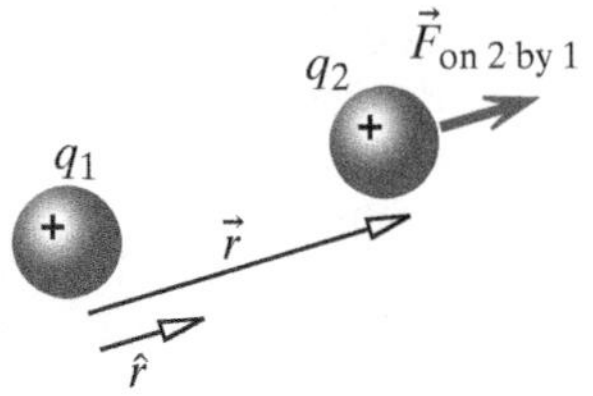

Figure 3.57

Reciprocity
$\vec{F}_{\text{on 1 by 2}} = -\vec{F}_{\text{on 2 by 1}}$ (gravitational and electric forces).
This is also called "Newton's third law of motion."

Conservation of momentum

$$\Delta\vec{p}_{\text{sys}} + \Delta\vec{p}_{\text{surr}} = \vec{0}$$

The multiparticle Momentum Principle

$$\Delta\vec{p}_{\text{sys}} = \vec{F}_{\text{net,surr}}\Delta t$$

Additional results
Uniform-density spheres act as though all the mass were at the center.

The center of mass

$$\vec{r}_{CM} = \frac{m_1\vec{r}_1 + m_2\vec{r}_2 + m_3\vec{r}_3 + \cdots}{M_{total}}$$

$$\vec{v}_{CM} = \frac{m_1\vec{v}_1 + m_2\vec{v}_2 + m_3\vec{v}_3 + \cdots}{M_{total}}$$

$$\vec{p}_{sys} = M_{total}\vec{v}_{cm}$$

Complex systems; determinism; chaos

QUESTIONS

Q1 Which fundamental interaction (gravitational, electromagnetic, strong, or weak) is responsible for each of these processes? How do you know? **(a)** A neutron outside a nucleus decays into a proton, electron, and antineutrino. **(b)** Protons and neutrons attract each other in a nucleus. **(c)** The Earth pulls on the Moon. **(d)** Protons in a nucleus repel each other.

Q2 Why is the value of the constant g different on Earth and on the Moon? Explain in detail.

Q3 You hold a tennis ball above your head, then open your hand and release the ball, which begins to fall. At this instant (ball starting to fall, no longer in contact with your hand), what can you conclude about the relative magnitudes of the force on the ball by the Earth and the force on the Earth by the ball? Explain.

Q4 Suppose that you are going to program a computer to carry out an iterative calculation of motion involving electric forces. Assume that as usual we use the final velocity in each time interval as the approximate average velocity during that interval. Which of the following calculations should be done *before* starting the repetitive calculation loop? Which of the calculations listed above should go *inside* the repetitive loop, and in what order?

(a) Define constants such as $\frac{1}{4\pi\varepsilon_0}$.
(b) Update the (vector) position of each object.
(c) Calculate the (vector) forces acting on the objects.
(d) Specify the initial (vector) momentum of each object.
(e) Specify an appropriate value for the time step.
(f) Specify the mass of each object.
(g) Update the (vector) momentum of each object.
(h) Specify the initial (vector) position of each object.

Q5 A bullet traveling horizontally at a very high speed embeds itself in a wooden block that is sitting at rest on a very slippery sheet of ice. You want to find the speed of the block just after the bullet embeds itself in the block. **(a)** What should you choose as the system to analyze? **(b)** Which of the following statements is true? (1) After the collision, the speed of the block with the bullet stuck in it is the same as the speed of the bullet before the collision. (2) The momentum of the block with the bullet stuck in it is the same as the momentum of the bullet before the collision. (3) The initial momentum of the bullet is greater than the momentum of the block with the bullet stuck in it.

Q6 You hang from a tree branch, then let go and fall toward the Earth. As you fall, the y component of your momentum, which was originally zero, becomes large and negative. **(a)** Choose yourself as the system. There must be an object in the surroundings whose y momentum must become equally large, and positive. What object is this? **(b)** Choose yourself and the Earth as the system. The y component of your momentum is changing. Does the total momentum of the system change? Why or why not?

Q7 One kind of radioactivity is called "alpha decay." For example, the nucleus of a radium-220 atom can spontaneously split into a radon-216 nucleus plus an alpha particle (a helium nucleus containing two protons and two neutrons). Consider a radium-220 nucleus that is initially at rest. It spontaneously decays, and the alpha particle travels off in the $+z$ direction. What can you conclude about the motion of the new radon-216 nucleus? Be as precise as you can, and explain your reasoning.

Q8 A bowling ball is initially at rest. A Ping-Pong ball moving in the $+z$ direction hits the bowling ball and bounces off it, traveling back in the $-z$ direction. Consider a time interval Δt extending from slightly before to slightly after the collision. **(a)** In this time interval, what is the sign of Δp_z for the system consisting of both balls? **(b)** In this time interval, what is the sign of Δp_z for the system consisting of the bowling ball alone?

Q9 The windshield of a speeding car hits a hovering insect. Consider the time interval from just before the car hits the insect to just after the impact. **(a)** For which choice of system is the change of momentum zero? **(b)** Is the magnitude of the change of momentum of the bug bigger than, the same as, or smaller than that of the car? **(c)** Is the magnitude of the change of velocity of the bug bigger than, the same as, or smaller than that of the car?

PROBLEMS

Section 3.2

•P10 At a particular instant the magnitude of the gravitational force exerted by a planet on one of its moons is 3×10^{23} N. If the mass of the moon were three times as large, what would be the magnitude of the force? If instead the distance between the moon and the planet were three times as large (no change in mass), what would be the magnitude of the force?

•P11 Masses M and m attract each other with a gravitational force of magnitude F. Mass m is replaced with a mass $3m$, and it is moved four times farther away. Now what is the magnitude of the force?

•**P12** A 3 kg ball and a 5 kg ball are 2 m apart, center to center. What is the magnitude of the gravitational force that the 3 kg ball exerts on the 5 kg ball? What is the magnitude of the gravitational force that the 5 kg ball exerts on the 3 kg ball?

•**P13** The mass of the Earth is 6×10^{24} kg, and the mass of the Moon is 7×10^{22} kg. At a particular instant the Moon is at location $\langle 2.8\times 10^8, 0, -2.8\times 10^8\rangle$ m, in a coordinate system whose origin is at the center of the Earth. **(a)** What is $\vec{r}_{M-E}$, the relative position vector from the Earth to the Moon? **(b)** What is $|\vec{r}_{M-E}|$? **(c)** What is the unit vector $\hat{r}_{M-E}$? **(d)** What is the gravitational force exerted by the Earth on the Moon? Your answer should be a vector.

•**P14** A star exerts a gravitational force of magnitude $|\vec{F}|$ on a planet. The distance between the star and the planet is r. If the planet were three times farther away (that is, if the distance between the bodies were $3r$), by what factor would the force on the planet due to the star change?

•**P15** A planet exerts a gravitational force of magnitude 7×10^{22} N on a star. If the planet were four times closer to the star (that is, if the distance between the star and the planet were 1/4 what it is now), what would be the magnitude of the force on the star due to the planet?

•**P16** A moon orbits a planet in the xy plane, as shown in Figure 3.58. You want to calculate the force on the moon by the planet at each location labeled by a letter (A, B, C, D). At each of these locations, what are: **1.** the unit vector $\hat{r}$, **2.** the unit vector $\hat{F}$ in the direction of the force?

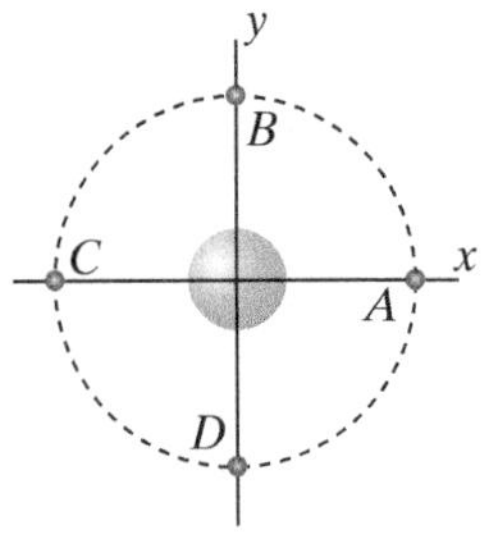

Figure 3.58

•**P17** The mass of the Sun is 2×10^{30} kg, and the mass of Mercury is 3.3×10^{23} kg. The distance from the Sun to Mercury is 4.8×10^{10} m. **(a)** Calculate the magnitude of the gravitational force exerted by the Sun on Mercury. **(b)** Calculate the magnitude of the gravitational force exerted by Mercury on the Sun.

•**P18** Measurements show that Jupiter's gravitational force on a mass of 1 kg near Jupiter's surface would be 24.9 N. The radius of Jupiter is 71500 km (71.5×10^6 m). From these data determine the mass of Jupiter.

•**P19** A planet of mass 4×10^{24} kg is at location $\langle 5e11, -2e11, 0\rangle$ m. A star of mass 5×10^{30} kg is at location $\langle -2e11, 3e11, 0\rangle$ m. It will be useful to draw a diagram of the situation, including the relevant vectors. **(a)** What is the relative position vector $\vec{r}$ pointing from the planet to the star? **(b)** What is the distance between the planet and the star? **(c)** What is the unit vector $\hat{r}$ in the direction of $\vec{r}$? **(d)** What is the magnitude of the force exerted on the planet by the star? **(e)** What is the magnitude of the force exerted on the star by the planet? **(f)** What is the (vector) force exerted on the planet by the star? **(g)** What is the (vector) force exerted on the star by the planet?

•**P20** A planet of mass 4×10^{24} kg is at location $\langle -6e11, 3e11, 0\rangle$ m. A star of mass 6×10^{30} kg is at location $\langle 5e11, -3e11, 0\rangle$ m. What is the force exerted on the planet by the star? (It will probably be helpful to draw a diagram, including the relevant vectors.)

••**P21** Two copies of this textbook are standing right next to each other on a bookshelf. Make a rough estimate of the magnitude of the gravitational force that the books exert on each other. Explicitly list all quantities that you had to estimate, and all simplifications and approximations you had to make to do this calculation. Compare your result to the gravitational force on the book by the Earth.

Section 3.3

•**P22** The mass of Mars is 6.4×10^{23} kg and its radius is 3.4×10^6 m. What is the value of the constant g on Mars?

•**P23** At what height above the surface of the Earth is there a 1% difference between the approximate magnitude of the gravitational field (9.8 N/kg) and the actual magnitude of the gravitational field at that location? That is, at what height y above the Earth's surface is $GM_E/(R_E+y)^2 = 0.99\,GM_E/R_E^2$?

•**P24** Calculate the approximate gravitational force exerted by the Earth on a human standing on the Earth's surface. Compare with the approximate gravitational force of a human on another human at a distance of 3 m. What approximations or simplifying assumptions must you make?

•••**P25** A steel ball of mass m falls from a height h onto a scale calibrated in newtons. The ball rebounds repeatedly to nearly the same height h. The scale is sluggish in its response to the intermittent hits and displays an *average* force F_{avg}, such that $F_{avg}T = F\Delta t$, where $F\Delta t$ is the brief impulse that the ball imparts to the scale on every hit, and T is the time between hits.

Calculate this average force in terms of m, h, and physical constants. Compare your result with the scale reading if the ball merely rests on the scale. Explain your analysis carefully (but briefly).

Section 3.5

•**P26** The mass of the Sun is 2×10^{30} kg, the mass of the Earth is 6×10^{24} kg, and their center-to-center distance is 1.5×10^{11} m. Suppose that at some instant the Sun's momentum is zero (it's at rest). Ignoring all effects but that of the Earth, what will the Sun's speed be after one day? (Very small changes in the velocity of a star can be detected using the "Doppler" effect, a change in the frequency of the starlight, which has made it possible to identify the presence of planets in orbit around a star.)

••**P27** At $t = 532.0$ s after midnight, a spacecraft of mass 1400 kg is located at position $\langle 3 \times 10^5, 7 \times 10^5, -4 \times 10^5\rangle$ m, and at that time an asteroid whose mass is 7×10^{15} kg is located at position $\langle 9 \times 10^5, -3 \times 10^5, -12 \times 10^5\rangle$ m. There are no other objects nearby. **(a)** Calculate the (vector) force acting on the spacecraft. **(b)** At $t = 532.0$ s the spacecraft's momentum was $\vec{p}_i$, and at the later time $t = 538.0$ s its momentum was $\vec{p}_f$. Calculate the (vector) change of momentum $\vec{p}_f - \vec{p}_i$.

••**P28** **(a)** In outer space, far from other objects, block 1 of mass 40 kg is at position $\langle 7, 11, 0\rangle$ m, and block 2 of mass 1000 kg is at position $\langle 18, 11, 0\rangle$ m. What is the (vector) gravitational force acting on block 2 due to block 1? It helps to make a sketch of the situation. **(b)** At 4.6 s after noon both blocks were at rest at the positions given above. At 4.7 s after noon, what is the (vector) momentum of block 2? **(c)** At 4.7 s after noon, what is the (vector) momentum of block 1? **(d)** At 4.7 s after noon, which one of the following statements is true? A. Block 1 and block 2 have the same speed. B. Block 2 is moving faster than block 1. C. Block 1 is moving faster than block 2.

••**P29** A star of mass 7×10^{30} kg is located at $\langle 5 \times 10^{12}, 2 \times 10^{12}, 0 \rangle$ m. A planet of mass 3×10^{24} kg is located at $\langle 3 \times 10^{12}, 4 \times 10^{12}, 0 \rangle$ m and is moving with a velocity of $\langle 0.3 \times 10^4, 1.5 \times 10^4, 0 \rangle$ m/s. **(a)** At a time 1×10^6 s later, what is the new velocity of the planet? **(b)** Where is the planet at this later time? **(c)** Explain briefly why the procedures you followed in parts (a) and (b) were able to produce usable results but wouldn't work if the later time had been 1×10^9 s instead of 1×10^6 s after the initial time. Explain briefly how you could use a computer to get around this difficulty.

••**P30** At $t = 0$ a star of mass 4×10^{30} kg has velocity $\langle 7 \times 10^4, 6 \times 10^4, -8 \times 10^4 \rangle$ m/s and is located at $\langle 2.00 \times 10^{12}, -5.00 \times 10^{12}, 4.00 \times 10^{12} \rangle$ m relative to the center of a cluster of stars. There is only one nearby star that exerts a significant force on the first star. The mass of the second star is 3×10^{30} kg, its velocity is $\langle 2 \times 10^4, -1 \times 10^4, 9 \times 10^4 \rangle$ m/s, and this second star is located at $\langle 2.03 \times 10^{12}, -4.94 \times 10^{12}, 3.95 \times 10^{12} \rangle$ m relative to the center of the cluster of stars. **(a)** At $t = 1 \times 10^5$ s, what is the approximate momentum of the first star? **(b)** Discuss briefly some ways in which your result for (a) is approximate, not exact. **(c)** At $t = 1 \times 10^5$ s, what is the approximate position of the first star? **(d)** Discuss briefly some ways in which your result for (b) is approximate, not exact.

•••**P31** In June 1997 the NEAR spacecraft ("Near Earth Asteroid Rendezvous"; see http://near.jhuapl.edu/), on its way to photograph the asteroid Eros, passed within 1200 km of asteroid Mathilde at a speed of 10 km/s relative to the asteroid. (See Figure 3.59.) From photos transmitted by the 805 kg spacecraft, Mathilde's size was known to be about 70 km by 50 km by 50 km. The asteroid is presumably made of rock. Rocks on Earth have a density of about 3000 kg/m^3 (3 grams/cm^3). **(a)** Make a rough diagram to show qualitatively the effect on the spacecraft of this encounter with Mathilde. Explain your reasoning. **(b)** Make a very rough estimate of the change in momentum of the spacecraft that would result from encountering Mathilde. Explain how you made your estimate. **(c)** Using your result from part (b), make a rough estimate of how far off course the spacecraft would be, one day after the encounter. **(d)** From actual observations of the location of the spacecraft one day after encountering Mathilde, scientists concluded that Mathilde is a loose arrangement of rocks, with lots of empty space inside. What was it about the observations that must have led them to this conclusion?

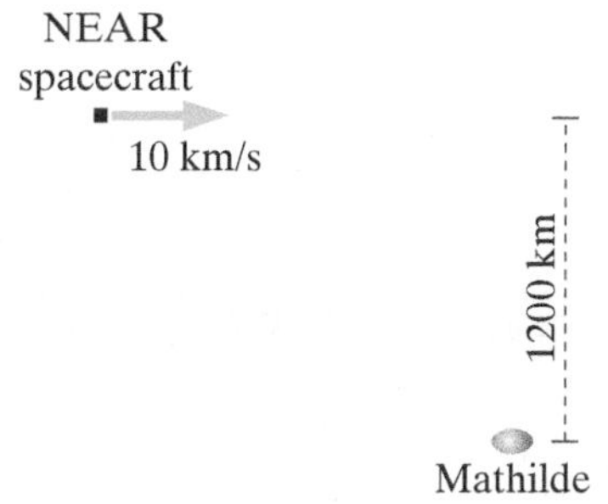

Figure 3.59

Experimental background: The position was tracked by very accurate measurements of the time that it takes for a radio signal to go from Earth to the spacecraft followed immediately by a radio response from the spacecraft being sent back to Earth. Radio signals, like light, travel at a speed of 3×10^8 m/s, so the time measurements had to be accurate to a few nanoseconds (1 ns $= 1 \times 10^{-9}$ s).

Section 3.7

•**P32** Figure 3.60 shows two positively charged objects (with the same charge) and one negatively charged object. What is the direction of the net electric force on the negatively charged object? If the net force is zero, state this explicitly.

Figure 3.60

•**P33** Figure 3.61 shows two negatively charged objects (with the same charge) and one positively charged object. What is the direction of the net electric force on the positively charged object? If the net force is zero, state this explicitly.

Figure 3.61

•**P34** The left side of Figure 3.62 shows a proton and an electron. **(a)** What is the direction of the electric force on the electron by the proton? **(b)** What is the direction of the electric force on the proton by the electron? **(c)** How do the magnitudes of these forces compare? The right side of Figure 3.62 shows two electrons. **(d)** What is the direction of the electric force on electron *A* due to electron *B*? **(d)** What is the direction of the electric force on electron *B* due to electron *A*? **(d)** How do the magnitudes of these forces compare?

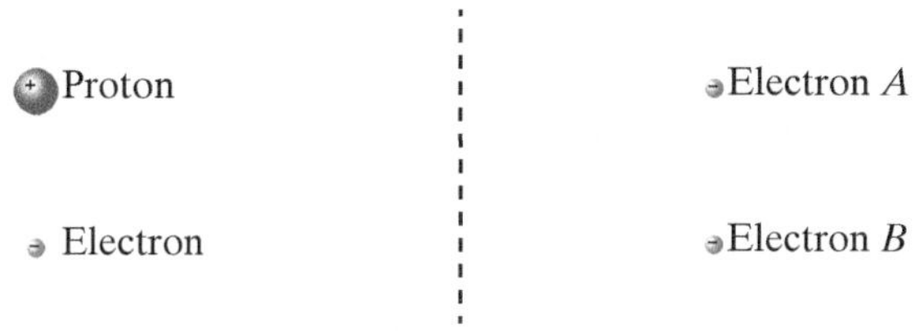

Figure 3.62

•**P35** An alpha particle contains two protons and two neutrons, and has a net charge of $+2e$. The alpha particle is 1 mm away from a single proton, which has a charge of $+e$. Which statement about the magnitudes of the electric forces between the particles is correct? **(a)** The force on the proton by the alpha particle is larger than the force on the alpha particle by the proton. **(b)** The force on the alpha particle by the proton is larger than the force on the proton by the alpha particle. **(c)** The forces are equal in magnitude. **(d)** Not enough information is given.

••**P36** A proton and an electron are separated by 1×10^{-10} m, the radius of a typical atom. Calculate the magnitude of the electric force that the proton exerts on the electron, and the magnitude of the electric force that the electron exerts on the proton.

••**P37** Two thin hollow plastic spheres, about the size of a Ping-Pong ball with masses ($m_1 = m_2 = 2 \times 10^{-3}$ kg), have been rubbed with wool. Sphere 1 has a charge $q_1 = -2 \times 10^{-9}$ C and is at location $\langle 0.50, -0.20, 0\rangle$ m. Sphere 2 has a charge $q_2 = -4 \times 10^{-9}$ C and is at location $\langle -0.40, 0.40, 0\rangle$ m. It will be useful to draw a diagram of the situation, including the relevant vectors. **(a)** What is the relative position vector $\vec{r}$ pointing from q_1 to q_2? **(b)** What is the distance between q_1 and q_2? **(c)** What is the unit vector $\hat{r}$ in the direction of $\vec{r}$? **(d)** What is the magnitude of the gravitational force exerted on q_2 by q_1? **(e)** What is the (vector) gravitational force exerted on q_2 by q_1? **(f)** What is the magnitude of the electric force exerted on q_2 by q_1? **(g)** What is the (vector) electric force exerted on q_2 by q_1? **(h)** What is the ratio of the magnitude of the electric force to the magnitude of the gravitational force? **(i)** If the two masses were four times farther away (that is, if the distance between the masses were $4\vec{r}$), what would be the ratio of the magnitude of the electric force to the magnitude of the gravitational force now?

••P38 A proton is located at $\langle 0, 0, -2 \times 10^{-9}\rangle$ m, and an alpha particle (consisting of two protons and two neutrons) is located at $\langle 1.5 \times 10^{-9}, 0, 2 \times 10^{-9}\rangle$ m. **(a)** Calculate the force the proton exerts on the alpha particle. **(b)** Calculate the force the alpha particle exerts on the proton.

••P39 Use data from the inside back cover to calculate the gravitational and electric forces two electrons exert on each other when they are 1×10^{-10} m apart (about one atomic radius). Which interaction between two electrons is stronger, the gravitational attraction or the electric repulsion? If the two electrons are at rest, will they begin to move toward each other or away from each other? Note that since both the gravitational and electric forces depend on the inverse square distance, this comparison holds true at all distances, not just at a distance of 1×10^{-10} m.

••P40 At a particular instant a proton exerts an electric force of $\langle 0, 5.76 \times 10^{-13}, 0\rangle$ N on an electron. How far apart are the proton and the electron?

Section 3.10

•P41 Two balls of mass 0.3 kg and 0.5 kg are connected by a low-mass spring (Figure 3.63). This device is thrown through the air with low speed, so air resistance is negligible. The motion is complicated: the balls whirl around each other, and at the same time the system vibrates, with continually changing stretch of the spring. At a particular instant, the 0.3 kg ball has a velocity $\langle 4, -3, 2\rangle$ m/s and the 0.5 kg ball has a velocity $\langle 2, 1, 4\rangle$ m/s. **(a)** At this instant, what is the total momentum of the device? **(b)** What is the net gravitational (vector) force exerted by the Earth on the device? **(c)** At a time 0.1 s later, what is the total momentum of the device?

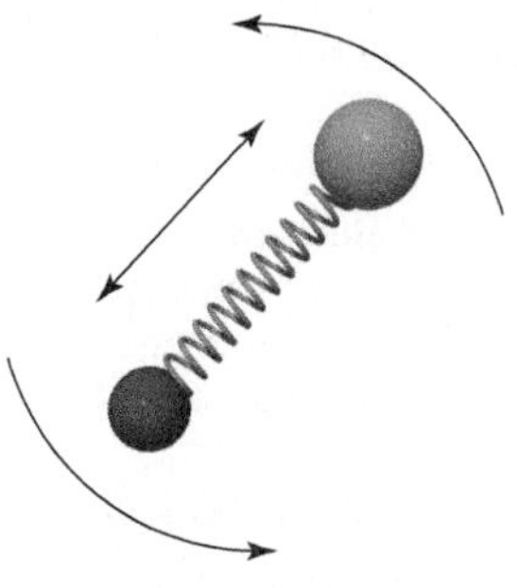

Figure 3.63

Section 3.11

•P42 At a certain instant object 1 is at location $\langle 10, -8, 6\rangle$ m, moving with velocity $\langle 4, 6, -2\rangle$ m/s. At the same instant object 2 is at location $\langle 3, 0, -2\rangle$ m, moving with velocity $\langle -8, 2, 7\rangle$ m/s. **(a)** What is the location of the center of mass of the two equal-mass objects? **(b)** What is the velocity of the center of mass?

•P43 The mass of the Earth is 6×10^{24} kg, the mass of the Moon is 7×10^{22} kg, and the center-to-center distance is 4×10^{8} m. How far from the center of the Earth is the center of mass of the Earth–Moon system? Note that the Earth's radius is 6.4×10^{6} m.

•P44 The mass of the Sun is 2×10^{30} kg, the mass of the Earth is 6×10^{24} kg, and the center-to-center distance is 1.5×10^{11} m. How far from the center of the Sun is the center of mass of the Sun–Earth system? Note that the Sun's radius is 7×10^{8} m.

••P45 Two rocks are tied together with a string of negligible mass and thrown into the air. At a particular instant, rock 1, which has a mass of 0.1 kg, is headed straight up with a speed of 5 m/s, and rock 2, which has a mass of 0.25 kg, is moving parallel to the ground, in the $+x$ direction, with a speed of 7.5 m/s. **(a)** What is the total momentum of the system consisting of both rocks and the string? **(b)** What is the velocity of the center of mass of the system?

••P46 A tennis ball of mass 0.06 kg traveling at a velocity of $\langle 9, -2, 13\rangle$ m/s is about to collide with an identical tennis ball whose velocity is $\langle 4, 5, -10\rangle$ m/s. **(a)** What is the total momentum of the system of the two tennis balls? **(b)** What is the velocity of the center of mass of the two tennis balls?

Section 3.12

•P47 In outer space, far from other objects, two rocks collide and stick together. Before the collision their momenta were $\langle -10, 20, -5\rangle$ kg·m/s and $\langle 8, -6, 12\rangle$ kg·m/s. What was their total momentum before the collision? What must be the momentum of the combined object after the collision?

•P48 When they are far apart, the momentum of a proton is $\langle 3.4 \times 10^{-21}, 0, 0\rangle$ kg·m/s as it approaches another proton that is initially at rest. The two protons repel each other electrically, without coming close enough to touch. When they are once again far apart, one of the protons now has momentum $\langle 2.4 \times 10^{-21}, 1.55 \times 10^{-21}, 0\rangle$ kg·m/s. At that instant, what is the momentum of the other proton?

••P49 You and a friend each hold a lump of wet clay. Each lump has a mass of 30 g. You each toss your lump of clay into the air, where the lumps collide and stick together. Just before the impact, the velocity of one lump was $\langle 3, 3, -3\rangle$ m/s, and the velocity of the other lump was $\langle -3, 0, -3\rangle$ m/s. **(a)** What was the total momentum of the lumps just before the impact? **(b)** What is the momentum of the stuck-together lump just after the collision? **(c)** What is the velocity of the stuck-together lump just after the collision?

••P50 A car of mass 2800 kg collides with a truck of mass 4700 kg, and just after the collision the car and truck slide along, stuck together. The car's velocity just before the collision was $\langle 40, 0, 0\rangle$ m/s, and the truck's velocity just before the collision was $\langle -14, 0, 29\rangle$ m/s. **(a)** What is the velocity of the stuck-together car and truck just after the collision? **(b)** In your analysis in part (a), why can you neglect the effect of the force of the road on the car and truck?

••P51 A car of mass 2045 kg moving in the x direction at a speed of 29 m/s strikes a hovering mosquito of mass

2.5 mg, and the mosquito is smashed against the windshield. The interaction between the mosquito and the windshield is an electric interaction between the electrons and protons in the mosquito and those in the windshield. **(a)** What is the approximate momentum change of the mosquito? Give magnitude and direction. Explain any approximations you make. **(b)** At a particular instant during the impact, when the force exerted on the mosquito by the car is F, what is the magnitude of the force exerted on the car by the mosquito? **(c)** What is the approximate momentum change of the car? Give magnitude and direction. Explain any approximations you make. **(d)** Qualitatively, why is the collision so much more damaging to the mosquito than to the car?

••**P52** A bullet of mass 0.105 kg traveling horizontally at a speed of 300 m/s embeds itself in a block of mass 2 kg that is sitting at rest on a nearly frictionless surface. What is the speed of the block after the bullet embeds itself in the block?

••**P53** Object A has mass $m_A = 8$ kg and initial momentum $\vec{p}_{A,i} = \langle 20, -5, 0\rangle$ kg·m/s, just before it strikes object B, which has mass $m_B = 11$ kg. Just before the collision object B has initial momentum $\vec{p}_{B,i} = \langle 5, 6, 0\rangle$ kg·m/s. **(a)** Consider a system consisting of both objects A and B. What is the total initial momentum of this system just before the collision? **(b)** The forces that A and B exert on each other are very large but last for a very short time. If we choose a time interval from just before to just after the collision, what is the approximate value of the impulse applied to the two-object system due to forces exerted on the system by objects outside the system? **(c)** Therefore, what does the Momentum Principle predict that the total final momentum of the system will be just after the collision? **(d)** Just after the collision, object A is observed to have momentum $\vec{p}_{A,f} = \langle 18, 5, 0\rangle$ kg·m/s. What is the momentum of object B just after the collision?

••**P54** In outer space a small rock with mass 5 kg traveling with velocity $\langle 0, 1800, 0\rangle$ m/s strikes a stationary large rock head-on and bounces straight back with velocity $\langle 0, -1500, 0\rangle$ m/s. After the collision, what is the vector momentum of the large rock?

••**P55** Two rocks collide in outer space. Before the collision, one rock had mass 9 kg and velocity $\langle 4100, -2600, 2800\rangle$ m/s. The other rock had mass 6 kg and velocity $\langle -450, 1800, 3500\rangle$ m/s. A 2 kg chunk of the first rock breaks off and sticks to the second rock. After the collision the 7 kg rock has velocity $\langle 1300, 200, 1800\rangle$ m/s. After the collision, what is the velocity of the other rock, whose mass is 8 kg?

••**P56** Two rocks collide with each other in outer space, far from all other objects. Rock 1 with mass 5 kg has velocity $\langle 30, 45, -20\rangle$ m/s before the collision and $\langle -10, 50, -5\rangle$ m/s after the collision. Rock 2 with mass 8 kg has velocity $\langle -9, 5, 4\rangle$ m/s before the collision. Calculate the final velocity of rock 2.

••**P57** In outer space two rocks collide and stick together. Here are the masses and initial velocities of the two rocks:

Rock 1: mass = 15 kg, initial velocity = $\langle 10, -30, 0\rangle$ m/s
Rock 2: mass = 32 kg, initial velocity = $\langle 15, 12, 0\rangle$ m/s

What is the velocity of the stuck-together rocks after colliding?

••**P58** A bullet of mass m traveling horizontally at a very high speed v embeds itself in a block of mass M that is sitting at rest on a nearly frictionless surface. What is the speed of the block after the bullet embeds itself in the block?

••**P59** A satellite that is spinning clockwise has four low-mass solar panels sticking out as shown. A tiny meteor traveling at high speed rips through one of the solar panels and continues in the same direction but at reduced speed. Afterward, calculate the v_x and v_y components of the center-of-mass velocity of the satellite. In Figure 3.64 $\vec{v}_1$, and $\vec{v}_2$ are the initial and final velocities of the meteor, and $\vec{v}$ is the initial velocity of the center of mass of the satellite, in the x direction.

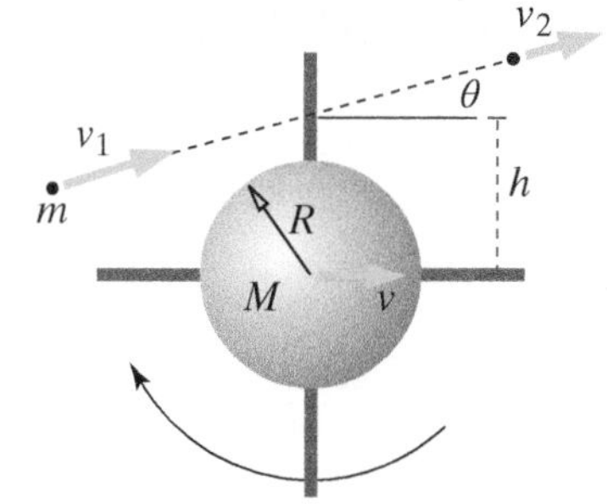

Figure 3.64

••**P60** A tiny piece of space junk of mass m strikes a glancing blow to a spinning satellite. Before the collision the satellite was moving and rotating as shown in Figure 3.65. After the collision the space junk is traveling in a new direction and moving more slowly. The velocities of the space junk before and after the collision are shown in the diagram. The satellite has mass M and radius R. Just after the collision, what are the components of the center-of-mass velocity of the satellite (v_x and v_y)?

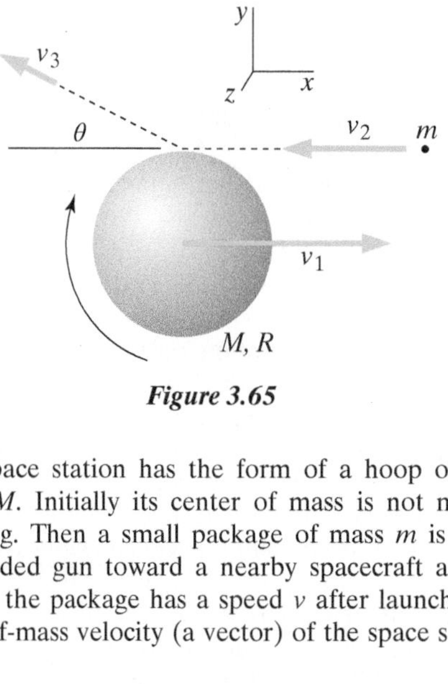

Figure 3.65

••**P61** A space station has the form of a hoop of radius R, with mass M. Initially its center of mass is not moving, but it is spinning. Then a small package of mass m is thrown by a spring-loaded gun toward a nearby spacecraft as shown in Figure 3.66; the package has a speed v after launch. Calculate the center-of-mass velocity (a vector) of the space station after the launch.

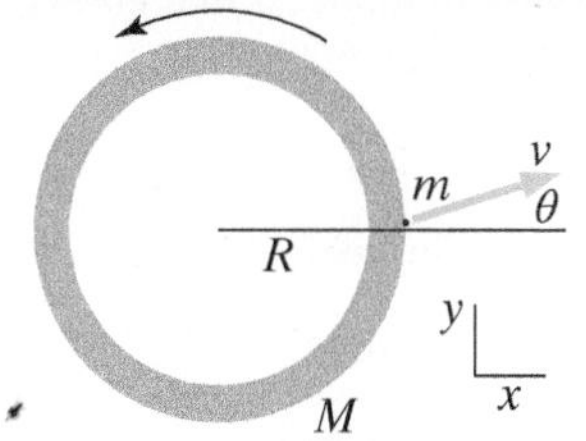

Figure 3.66

••**P62** A ball of mass 0.05 kg moves with a velocity $\langle 17,0,0\rangle$ m/s. It strikes a ball of mass 0.1 kg that is initially at rest. After the collision, the heavier ball moves with a velocity of $\langle 3,3,0\rangle$ m/s. **(a)** What is the velocity of the lighter ball after impact? **(b)** What is the impulse delivered to the 0.05 kg ball by the heavier ball? **(c)** If the time of contact between the balls is 0.03 s, what is the force exerted by the heavier ball on the lighter ball?

••**P63** Suppose that all the people of the Earth go to the North Pole and, on a signal, all jump straight up. Estimate the recoil speed of the Earth. The mass of the Earth is 6×10^{24} kg, and there are about 6 billion people (6×10^{9}).

COMPUTATIONAL PROBLEMS

More detailed and extended versions of some of these computational modeling problems may be found in the lab activities included in the *Matter & Interactions*, Fourth Edition, resources for instructors.

••**P64** To use an `arrow` object to visualize a force, it is usually necessary to scale the length of the arrow in order to make it fit on the screen with the objects exerting and experiencing the force. Watch VPython Instructional Video 5: Scalefactors, at vpython.org/video05.html to learn how to do this.

Now, write a VPython program to do the following: **(a)** Create a sphere representing a planet at the origin, with radius 6.4×10^{6} m. The mass of this planet is 6×10^{24} kg. **(b)** Create five spheres representing 5 spacecraft, at locations $\langle -13\times 10^{7}, 6.5\times 10^{7}, 0\rangle$ m, $\langle -6.5\times 10^{7}, 6.5\times 10^{7}, 0\rangle$ m, $\langle 0, 6.5\times 10^{7}, 0\rangle$ m, $\langle 6.5\times 10^{7}, 6.5\times 10^{7}, 0\rangle$ m, and $\langle 13\times 10^{7}, 6.5\times 10^{7}, 0\rangle$ m. You will have to exaggerate the radius of each spacecraft to make it visible; try 3×10^{6} m. The mass of each spacecraft is 15×10^{3} kg. **(c)** For each spacecraft, have your program calculate the gravitational force exerted on the spacecraft by the planet, and visualize it with an arrow whose tail is at the center of the spacecraft. Use the same scalefactor for all arrows. Check that the display your program produces makes physical sense. **(d)** For each spacecraft, have your program calculate the gravitational force exerted on the planet by the spacecraft, and visualize it with an arrow whose tail is at the center of the planet. Use the same scalefactor as you used in the previous part. Check that the display your program produces makes physical sense.

••**P65** Write a computer program to model the motion of a spacecraft of mass 15000 kg that is launched from a location 10 Earth radii from the center of the Earth (Figure 3.67). (Data for the Earth are given on the inside back cover of the textbook. Start with a Δt of 60 s and an initial speed of 2×10^{3} m/s in a direction perpendicular to the line between the spacecraft and the Earth.) **(a)** Vary the initial speed (but not the direction), and have the spacecraft leave a trail. What trajectories can you produce? **(b)** Find an initial speed that produces an elliptical orbit. **(c)** For the elliptical orbit, display arrows indicating the directions of the momentum of the spacecraft and the net force on the spacecraft as it moves. **(d)** Find an initial speed that produces a circular orbit. **(e)** Experiment by increasing and decreasing the time step Δt. What is the largest value of Δt that gives enough accuracy to produce a closed circular orbit?

Figure 3.67

••**P66** Extend the program you wrote in the previous problem by including the effect of the Moon, placing the Moon on the opposite side of the Earth from the spacecraft's initial location (Figure 3.68). (Relevant data are given on the inside back cover of the textbook.) To simplify the model, keep both the Earth and the Moon fixed. (This is called a "restricted three-body problem.") **(a)** Find an initial speed for the spacecraft that results in an orbit around both Earth and Moon. **(b)** By adjusting the initial speed of the spacecraft, can you produce a figure-eight trajectory? **(c)** What other interesting trajectories can you produce by varying the initial speed? (Small variations may produce large effects.)

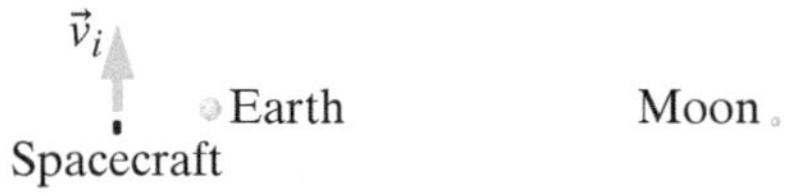

Figure 3.68

•••**P67** Once you have completed the previous problem, allow the Earth and Moon to move. Use what you know about the period of their orbits to determine appropriate initial velocities for the Earth and the Moon.

••**P68** Model the motion of Mars around the Sun. You can find data for masses and distances online. Use the known period of Mars's orbit to determine an approximate initial velocity. Display a trail, so you can see the shape of the orbit. Determine an appropriate value for Δt. Display arrows representing the momentum of Mars and the net force on Mars. Determine the period of the orbit of the planet in your model. How can you produce noncircular orbits?

••**P69** About a half of the visible "stars" are actually systems consisting of two stars orbiting each other, called "binary stars." In your computer model of a planet and Sun (Problem P68), replace the planet with a star whose mass is half the mass of our Sun, and take into account the gravitational effects that the second star has on the Sun. **(a)** Give the second star the speed of the actual Earth, and give the Sun zero initial momentum. What happens? Try a variety of other initial conditions. What kinds of orbits do you find? **(b)** Choose initial conditions so that the total momentum of the two-star system is zero, but the stars are not headed directly at each other. What is special about the motion you observe in this case?

•••**P70** Modify your orbit computation to use a different force, such as a force that is proportional to $1/r$ or $1/r^3$, or a constant force, or a force proportional to r (this represents the force of a spring whose relaxed length is nearly zero). How do orbits with these forces differ from the circles and ellipses that result from a $1/r^2$ force? If you want to keep the magnitude of the force roughly the same as before, you will need to adjust the force constant G.

••**P71** The first U.S. spacecraft to photograph the Moon close up was the unmanned *Ranger 7* photographic mission in 1964. The spacecraft, shown in Figure 3.69, contained television cameras that transmitted close-up pictures of the Moon back to Earth as the spacecraft approached the Moon. The spacecraft did not have retro-rockets to slow itself down, and it eventually simply crashed onto the Moon's surface, transmitting its last photos immediately before impact.

Figure 3.69 (Image courtesy of NASA)

Figure 3.70 is the first image of the Moon taken by a U.S. spacecraft, *Ranger 7*, on July 31, 1964, about 17 minutes before impact on the lunar surface. To find out more about the actual *Ranger* lunar missions, see http://nssdc.gsfc.nasa.gov/planetary/lunar/ranger.html.

Figure 3.70 (Image courtesy of NASA)

Create a computational model of the *Ranger 7* mission. Start your model when the spacecraft, whose mass is = 173 kg, has been brought to 50 km above the Earth's surface (5×10^4 m) by several stages of large rockets, and has a speed of around 1×10^4 m/s. All fuel has been used up, and the spacecraft now coasts toward the Moon. Data for the Earth and Moon may be found on the inside back cover of the textbook.

For this simple model, keep the Earth and Moon fixed in space during the mission, and ignore the effect of the Sun. **(a)** Compute and display the path of the spacecraft, having it leave a trail. **(b)** Determine experimentally the approximate *minimum* initial speed needed to reach the Moon, to three significant figures (this is the speed that the spacecraft obtained from the multistage rocket, at the time of release above the Earth's atmosphere). **(c)** Check your result by decreasing the time step size until your results do not change significantly. **(d)** Use a launch speed 10% larger than the approximate minimum value found in part (b). How long does it take to go to the Moon, in hours or days? **(e)** What is the "impact speed" of the spacecraft (its speed just before it hits the Moon's surface)? Make sure that your spacecraft crashes on the surface of the Moon, and not at its center!

•••P72 In the *Ranger 7* model, take into account the motion of the Moon around the Earth and the motion of the Earth around the Sun. In addition, the Sun and other planets exert gravitational forces on the spacecraft.

•••P73 In the *Ranger 7* analysis (the Moon voyage), you used a simplified model in which you neglected among other things the effect of Venus. An important aspect of physical modeling is making estimates of how large the neglected effects might be. Venus and the Earth have similar size and mass. At its closest approach to the Earth, Venus is about 40 million km away (4×10^{10} m). In the real world, Venus would attract the Earth and the Moon as well as the spacecraft, but to get an idea of the size of the effects, imagine that the Earth, the Moon, and Venus are all fixed in position. (See Figure 3.71.)

Figure 3.71

If we take Venus into account, make a rough estimate of whether the spacecraft will miss the Moon entirely. How large a sideways deflection of the crash site will there be? Explain your reasoning and approximations. If you expect a significant effect, modify your program to include the effects of Venus.

•••P74 Create a computational model of the motion of a three-body gravitational system, with all three objects free to move, and plot the trajectories, leaving trails behind the objects. Calculate all of the forces before using these forces to update the momenta and positions of the objects. Otherwise the calculations of gravitational forces would mix positions corresponding to different times.

Try different initial positions and initial momenta. Find at least one set of initial conditions that produces a long-lasting orbit, one set of initial conditions that results in a collision with a massive object, and one set of initial conditions that allows one of the objects to wander off without returning. Report the masses and initial conditions that you used.

ANSWERS TO CHECKPOINTS

1 (a) 4×10^{25} N, **(b)** 8×10^{25} N, **(c)** 4.4×10^{24} N
2 (a) (1) $\langle -1,0,0\rangle$, (2) $\langle 1,0,0\rangle$, (3) $\langle 1,0,0\rangle$, **(b)** (1) $\langle 1,0,0\rangle$, (2) $\langle -1,0,0\rangle$, (3) $\langle -1,0,0\rangle$
3 (a) 0.285 N/kg, **(b)** 19.9 N, **(c)** 0.029 as much (the astronaut's weight on the asteroid is about 3% of what it would be on Earth)
4 (a) 588 N, **(b)** 588 N
5 The magnitude of $\vec{p}_{\text{future}}$ will be less than the magnitude of $\vec{p}_{\text{now}}$ (so speed decreases) because part of $\vec{F}_{\text{net,now}}$ is opposite to $\vec{p}_{\text{now}}$. The direction of $\vec{p}_{\text{future}}$ will also be different (it will have a larger $+y$ component).
6 (a) 2.02×10^{-9} N, **(b)** 2.02×10^{-9} N
7 (a) Si: 14 p and 14 n, **(b)** Sn: 50 p and 69 n, **(c)** Au: 79 p and 118 n, **(d)** Th: 90 p and 142 n, **(e)** The farther you go in the periodic table, the more "excess" neutrons with their strong interactions are needed to offset the proton repulsions.
8 $\langle 0.57, 5.43, 0\rangle$ kg·m/s
9 (a) $\langle 0.04, 0.04, -0.1\rangle$ kg·m/s, **(b)** $\langle 0.04, 0.04, -0.1\rangle$ kg·m/s, **(c)** $\langle 1, 1, -2.5\rangle$ m/s
10 6, 6

CHAPTER

4

Contact Interactions

OBJECTIVES

After studying this chapter, you should be able to

- Using the ball–spring model, explain in words how an inanimate object can exert a force on an object that touches it.
- Mathematically relate the density of a solid and Young's modulus to the stiffness of an interatomic bond modeled as a spring, and to the speed of sound.
- Analyze systems subjected to tension, compression, friction, and buoyancy forces in terms of the derivative (instantaneous) form of the Momentum Principle.
- Mathematically describe the motion of an object that interacts with a spring, both analytically and computationally (iteratively).
- Determine the speed of sound in a solid in terms of the microscopic and macroscopic properties of the solid.

4.1 BEYOND POINT PARTICLES

In the first three chapters we have predicted the motion of a variety of systems, from protons to stars, by modeling them as point particles. However, when we consider the interactions of objects that are in physical contact with each other, such a model is no longer adequate. Objects that touch actually deform each other, stretching or compressing each other. We can't use a point particle model to understand contact interactions, because a point particle has no internal structure and can't be deformed. Therefore we'll need a more detailed model of the internal structure of solid objects. To frame our study of contact interactions and the structure of solids, we'll consider the following anecdote about Tarzan swinging from a vine.

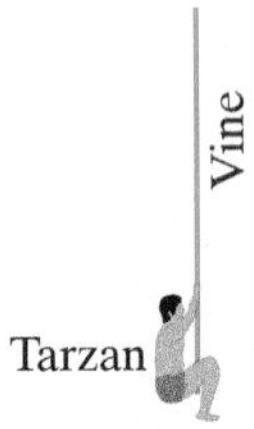

Figure 4.1 Tarzan hangs motionless from a vine, which does not break.

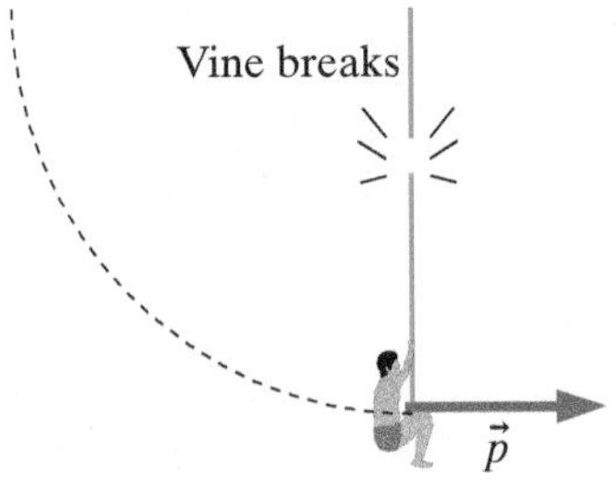

Figure 4.2 The vine breaks midway through Tarzan's swing.

Tarzan wants to use a vine to swing across a river. To make sure the vine is strong enough to support him, he tests it by hanging motionless on the vine for several minutes (Figure 4.1). The vine passes this test, so Tarzan grabs the vine and swings out over the river. He is annoyed and perplexed when the vine breaks midway through the swing (Figure 4.2), and he ends up drenched and shivering in the middle of the cold river, to the great amusement of the onlooking apes.

Why did the vine break while Tarzan was swinging on it, but not while he hung motionless from it? Apparently what happens to the vine is different in these two situations. To understand this, we need to understand how objects that are in contact exert forces on each other; this is the topic of the current chapter. The issue of how much force the vine must exert when Tarzan swings on it, as opposed to hanging motionless, will be discussed in Chapter 5.

4.2 THE BALL–SPRING MODEL OF A SOLID

The great 20th-century American physicist Richard Feynman said:

> "If, in some cataclysm, all of scientific knowledge were to be destroyed, and only one sentence passed on to the next generations of creatures, what statement would contain the most information in the fewest words? I believe it is the atomic hypothesis (or the atomic fact, or whatever you wish to call it) that all things are made of atoms—little particles that move around in perpetual motion, attracting each other when they are a little distance apart, but repelling upon being squeezed into one another. In that one sentence, you will see, there is an enormous amount of information about the world, if just a little imagination and thinking are applied." (*The Feynman Lectures on Physics* by R. P. Feynman, R. B. Leighton, & M. Sands, 1965; Palo Alto, Cal.: Addison-Wesley.)

In the quote above, Feynman summarized the basic properties of atoms and of interatomic forces. The main properties of atoms that will be important to us in this chapter are these:

- All matter consists of atoms, whose typical radius is about 1×10^{-10} m.
- Atoms attract each other when they are close to each other but not too close.
- Atoms repel each other when they get too close to each other.
- Atoms in solids, liquids, and gases keep moving even at very low temperatures.

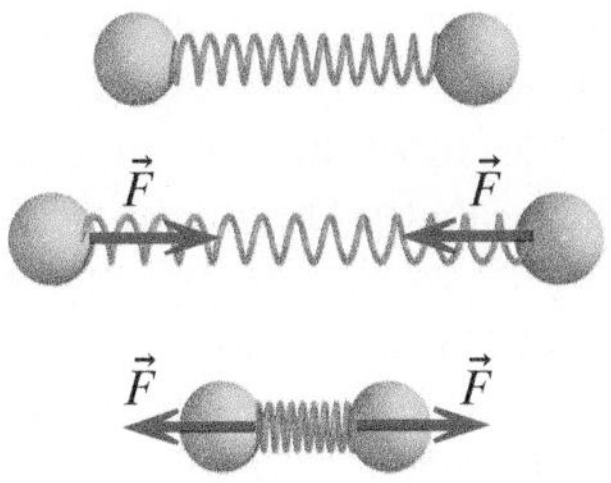

Figure 4.3 Two balls connected by a spring. Top: A relaxed spring exerts no forces on the balls. Middle: A stretched spring exerts forces to bring balls closer together. Bottom: A compressed spring exerts forces to move balls apart.

These properties were established during a century of intense study of atoms by physicists and chemists, using a wide variety of experimental techniques.

> QUESTION What properties of a block of aluminum, observable without special equipment, support the claims that "atoms attract each other when they are not too close to each other" and "atoms repel each other when they get too close to each other"?

Since the block of metal does not spontaneously fall apart or evaporate, the atoms in the block must be attracting each other. It is very difficult to compress the block, so the atoms must resist an attempt to push them closer together. Evidently the atoms in a block of aluminum are normally at just the right distance from each other, not too close and not too far away. This just-right distance is called the "equilibrium" distance between atoms.

A Chemical Bond Is Like a Spring

Two atoms linked by a chemical bond behave in a manner very similar to two macroscopic balls attached to the ends of a very low mass spring (Figure 4.3). The ball–spring system is a good model for the atomic system. In this model:

- Each ball in the model represents a massive atomic nucleus, surrounded by the inner electrons of the atom. Almost all the mass of an atom is concentrated in the tiny nucleus.
- The spring in the model represents the chemical bond, which is due to the shared outer electrons of both atoms.

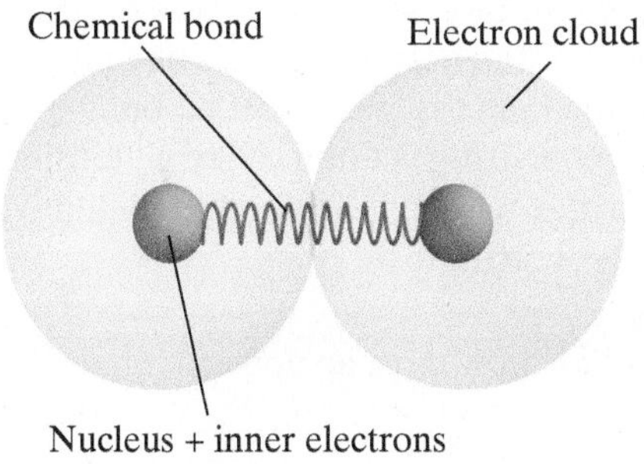

Figure 4.4 The ball–spring model for two atoms connected by a chemical bond (represented by a spring), superimposed on a space-filling model in which the entire electron clouds of the atoms are shown.

The microscopic atomic system behaves much like the macroscopic model system, as long as the stretch or compression is small, as it is in ordinary processes. If the atoms are moved farther apart, they experience forces resisting the separation. If the atoms are pushed closer together, they experience forces resisting the compression. In Figure 4.4 the ball–spring model of two bonded atoms is superimposed on a space-filling representation showing the full electron cloud of each atom.

Figure 4.5 STM images of two surfaces of a silicon crystal. (Images courtesy of Randall Feenstra, IBM Corp.)

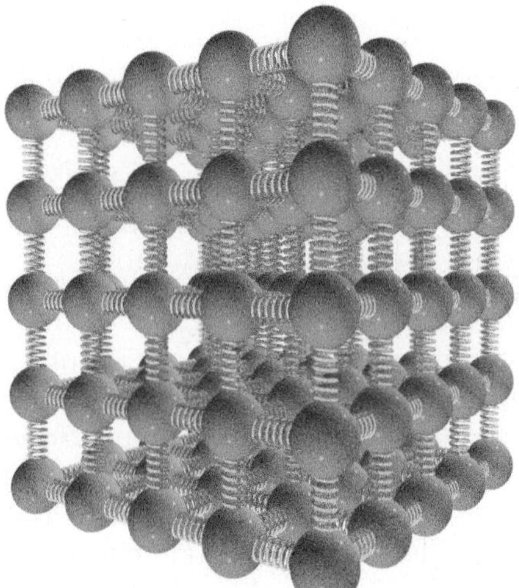

Figure 4.6 A simple model of a solid: tiny balls in constant motion, connected by springs. This figure shows only a small section of a solid object, which has many more atoms than are depicted here.

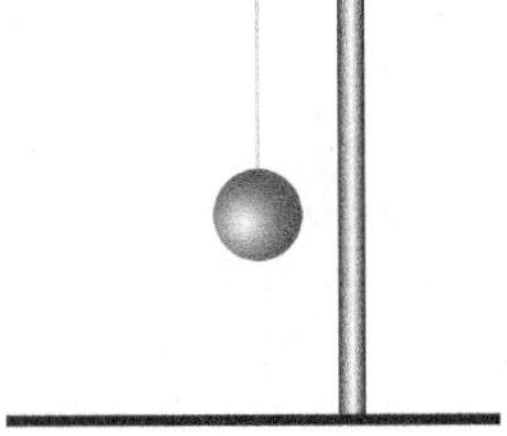

Figure 4.7 A heavy ball hangs motionless at the end of a thin wire.

A Ball–Spring Model for a Solid Object

A solid object contains many atoms, not just two. We know from x-ray studies and from images produced by scanning tunneling microscopes (STM) and atomic force microscopes (AFM) that many solid objects, such as metals, are crystals composed of regular arrays of atoms, as shown in the STM images of silicon surfaces in Figure 4.5. In these images, the spheres indicate the entire electron cloud associated with each atom, not just the nucleus and inner electrons represented by the balls in our model.

These STM images show only one surface of a solid, but inside the solid, atoms are arranged in 3D patterns, like the balls and springs in the model solid shown in Figure 4.6. Solids in which the atoms are arranged in regular "lattices" like those shown in Figure 4.6 are called crystals; these include metals, quartz, diamond, ice, and table salt (NaCl), but not most organic solids such as plastic or wood.

The lattice shown in Figure 4.6 is the simplest kind of crystal, and is called a "cubic" lattice because the atoms in the crystal are located at the corners of adjacent cubes. More complex lattice arrangements are possible; for example, in one common variant called "body-centered cubic" there is an additional atom at the center of each cube. In Figure 4.5 the lower image shows a hexagonal arrangement of atoms. Although most crystals are more complex than the simple cubic lattice, we will use this simple cubic ball-and-spring model because it incorporates all the important features we need.

In a solid at room temperature, the atoms are in motion, continually oscillating around their equilibrium positions. If we heat the solid, these atomic oscillations become more vigorous. One of the most important outcomes of research on the atomic nature of matter was the realization that for many materials, the temperature of an object, as measured by an ordinary thermometer, is just an indicator of the average energy of the atoms; the higher the temperature, the more vigorous the atomic motion. Figure 4.6 shows atoms in a model solid frozen in their equilibrium positions.

In later chapters we will see that the ball–spring model of a solid will also enable us to answer questions about energy (Chapter 6), energy dissipation and temperature changes (Chapter 7), and the irreversibility of certain processes (Chapter 11). By refining and extending the model in Chapter 14, we will also be able to model the electrical properties of metals and other conductors.

4.3 TENSION FORCES

By applying the Momentum Principle we can show that an object like a wire, string, or vine can exert a force on an object attached to it. Figure 4.7 shows a heavy iron ball hanging motionless on the end of a wire. The y momentum of the ball is not changing, so the y component of the net force on the ball must be zero. Since the Earth is pulling down on the ball, the wire must be pulling up. The force exerted by an object like a wire or a string is often called a "tension" force, or sometimes just "the tension in the wire," and usually labeled $\vec{F}_T$. A tension force always acts along the wire or string.

QUESTION If the mass of the ball is 1 kg, how large an upward force does the wire exert on the ball?

System: Ball
Surroundings: Earth, wire
Free-body diagram: Figure 4.8

Choose a value for Δt: 10 s (since the ball remains at rest indefinitely, we could choose any value of Δt other than zero)

As reflected in the free-body diagram, the only significant forces exerted on the ball by the surroundings are in the $\pm y$ direction, so we can focus on the y components of force and momentum.

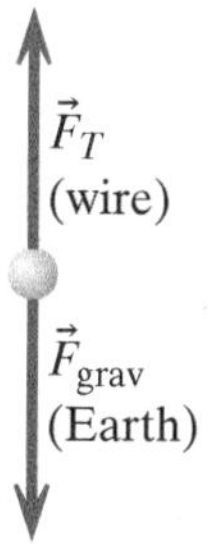

Figure 4.8 Free-body diagram for the ball.

$$\Delta p_y = (F_T - mg)(10\,\text{s})$$
$$0 = (F_T - mg)(10\,\text{s})$$
$$F_T = mg = (1\,\text{kg})(9.8\,\text{N/kg}) = 9.8\,\text{N}$$

QUESTION If the mass of the ball were 2 kg, how large an upward force would the wire exert on the ball?

$$F_T = (2\,\text{kg})(9.8\,\text{N/kg}) = 19.6\,\text{N}$$

This simple result has an interesting implication. Evidently the magnitude of the tension force F_T exerted by the wire depends on the mass of the ball.

QUESTION How does an inanimate object like a wire "know" how large a force to exert in a given situation?

We have already seen that the force exerted by a spring depends on how much the spring is stretched. A wire can be thought of as a very stiff spring. When a weight is hung on the end of a wire, the wire does stretch, although the stretch is usually very small. The ball–spring model of a solid helps us understand what happens when a wire stretches.

Microscopic View: Tension in a Wire

We can model a wire as a chain of balls and springs (atoms connected by spring-like chemical bonds). For simplicity, we'll consider a wire that is only one atom thick.

When the wire lies on a table, each of the springs (bonds) is relaxed. When the wire hangs vertically with nothing attached to it, the bonds stretch a tiny amount, just to support the weight of the atoms below them. However, when a massive object is hung from the wire, the spring-like bonds between atoms stretch significantly, because each bond must support the weight of everything below it (Figure 4.9). If we make the approximation that the total mass of all the atoms in the wire is negligible compared to the mass of the hanging object, then we can say that each bond in the wire is stretched by approximately the same amount, or equivalently, that the tension is the same throughout the wire.

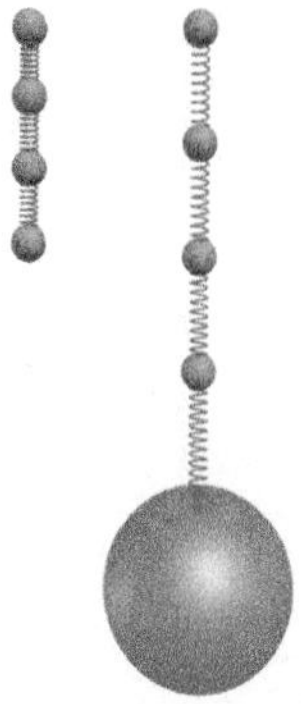

Figure 4.9 The interatomic bonds in a wire stretch when a heavy mass is hung from the wire. In this idealized diagram the wire is only one atom thick, and the stretch of each bond is greatly exaggerated.

QUESTION How does the wire manage to exert a larger upward force when a more massive object is hung from it?

The more massive object stretches the interatomic bonds more than a less massive object. Of course, there is a limit to how much the wire can stretch. A sufficiently heavy weight will break the wire.

4.4 LENGTH OF AN INTERATOMIC BOND

Our goal in the next sections is to determine the stiffness of an interatomic bond, considered as a spring. The first step is to determine the length of an interatomic bond in a particular material. Bond lengths will be slightly different in different materials (for example, aluminum vs. lead), depending on the size of the atoms. We will determine the interatomic bond length in solid copper.

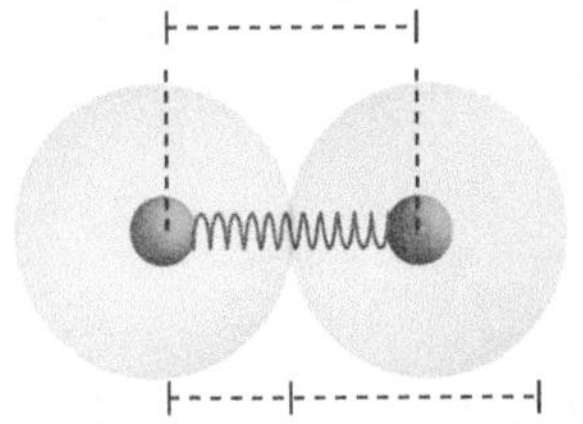

Figure 4.10 The length of an interatomic bond is defined as the center-to-center distance between adjacent atoms. This is the same as the diameter of an atom (including the full electron cloud).

We will define the length of one interatomic bond as the center-to-center distance between two adjacent atoms (Figure 4.10). If we consider the space-filling model of a solid, this distance is equal to twice the radius of an atom, or the diameter of an atom.

The diameter of an atom in a solid is one of the important properties of matter that plays a role in interactions. We can calculate atomic diameters for crystals of particular elements by using the measured density of the material in kilograms per cubic meter and Avogadro's number (the number of atoms in one mole of the material).

QUESTION Does the density of a block of aluminum depend on the dimensions of the block? Does it depend on the mass of the block?

Density does not depend on the size, shape, or mass of an object. Density is a property of the material itself, and the ratio of mass to volume will always be the same for objects made of the same solid material. The density of a wide range of materials has been measured and may easily be looked up in reference materials.

EXAMPLE **Diameter of a Copper Atom (Length of a Bond in Copper)**

One mole of copper (6.02×10^{23} atoms) has a mass of 64 g (see the periodic table on the inside front cover). The density of copper is 8.94 g/cm^3. What is the approximate diameter, in meters, of a copper atom in solid copper?

Solution Often densities are given in grams per cubic centimeter rather than in kilograms per cubic meter. We often need to convert these densities to SI units, like this:

$$\left(8.94\frac{\text{g}}{\text{cm}^3}\right)\frac{(1\,\text{kg})}{(1\times 10^3\,\text{g})}\frac{(1\times 10^2\,\text{cm})^3}{(1\,\text{m})^3} = 8.94\times 10^3\,\frac{\text{kg}}{\text{m}^3} \quad \text{(SI units)}$$

Figure 4.11 shows a cube that contains $5\times 5\times 5 = 125$ atoms. The number of atoms on one side of the cube is $\sqrt[3]{125} = 5$.

In a cube that is 1 m on each side, how many copper atoms are there?

$$(8.94\times 10^3\,\text{kg})\left(\frac{1\,\text{mol}}{0.064\,\text{kg}}\right)\frac{(6.02\times 10^{23}\,\text{atoms})}{(1\,\text{mol})} = 8.41\times 10^{28}\,\text{atoms}$$

Along one edge of the cube, which is 1 m long, there are

$$\sqrt[3]{8.41\times 10^{28}} = 4.38\times 10^9\,\text{atoms}$$

A row of 4.38×10^9 atoms is 1 m long, so the diameter of one atom is

$$d = \left(\frac{1\,\text{m}}{4.38\times 10^9\,\text{atoms}}\right) = 2.28\times 10^{-10}\,\text{m}$$

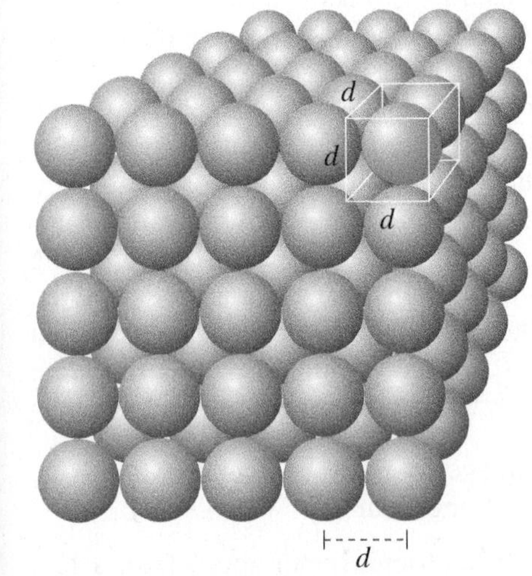

Figure 4.11 A simple cubic arrangement of atoms. The volume of space associated with each atom is a tiny cube d by d by d.

In a solid block of copper, the length of a bond between two adjacent copper atoms is 2.28×10^{-10} m.

An alternative approach to this calculation involves a "micro" view of density. Since density is independent of the amount of an object, the density of one copper atom ought to be the same as the density of a large block of copper. Figure 4.11 shows that the volume of space taken up by each atom is a tiny cube d on a side, where d is the diameter of an atom in the solid and is also the distance from the center of one atom to the center of a neighboring atom. The micro density is the mass of one atom divided by the volume of the tiny cube associated with each atom. This micro density is the same as the macro density, mass per volume:

$$\text{density} = \frac{\text{mass of } 8.41 \times 10^{28} \text{ atoms}}{\text{volume of } 8.41 \times 10^{28} \text{ atoms}} = \frac{\text{mass of 1 atom}}{\text{volume of 1 atom}} = \frac{m_a}{d^3}$$

The mass of one atom can be determined using the mass of one mole and the knowledge that one mole contains 6.02×10^{23} atoms (Avogadro's number):

$$m_a = \frac{\text{mass of one mole}}{6.02 \times 10^{23} \text{ atoms/mole}}$$

Knowing the macroscopic density and the mass of one atom, you can solve for the unknown atomic diameter d. It might be tempting to put these concepts together into an equation of the form $d = \ldots$, but try to avoid the temptation. It is much more memorable and much safer to think through these physical relationships each time you make the calculation, to avoid serious mistakes. Just set the macroscopic and microscopic densities equal to each other and solve for d.

Few solid elements actually have cubic lattices, because a cubic lattice is rather unstable. However, assuming that all crystalline solids have cubic lattices greatly simplifies our model, and gives adequately accurate results for our purposes.

The diameters of the smallest and largest atoms differ by only about a factor of 8. Most metallic elements have similar radii, on the order of 1.5×10^{-10} m. It is useful to remember that the radius of an "average" atom is on the order of 1×10^{-10} m.

Checkpoint 1 The density of aluminum is 2.7 g/cm^3 and the density of lead is 11.4 g/cm^3. What is the approximate diameter of an aluminum atom (length of a bond) in solid aluminum? What is the approximate diameter of a lead atom (length of a bond) in solid lead?

4.5 THE STIFFNESS OF AN INTERATOMIC BOND

Knowing the length of an interatomic bond in solid copper (and the diameter of a copper atom), we can now use experimental data to find the stiffness of the interatomic bond, considered as a spring. You have probably had some experience with ordinary macroscopic springs. A Slinky, for example, is a soft spring; its spring stiffness is small—around 1 N/m. The stiffness of the spring on a pogo stick is much larger—around 5000 N/m. It is difficult to measure the stiffness of an interatomic bond directly, but we can analyze data from macroscopic (large-scale) experiments to determine this quantity.

The basic idea is to hang heavy masses on a long wire and measure the stretch of the wire; then, by figuring out how many interatomic "springs" there are in the wire, we can determine the spring constant of a single interatomic bond. To do this, we need to be able to relate the stiffness of an object composed of many springs to the stiffness of each individual spring. We will consider the wire to be composed of many parallel long chains of atoms connected by springs (Figure 4.12). We will need to determine how the stiffness of one of the short springs (bonds) is related to the stiffness of an entire chain of springs, and how the stiffness of the entire wire is related to the stiffness of one chain of springs.

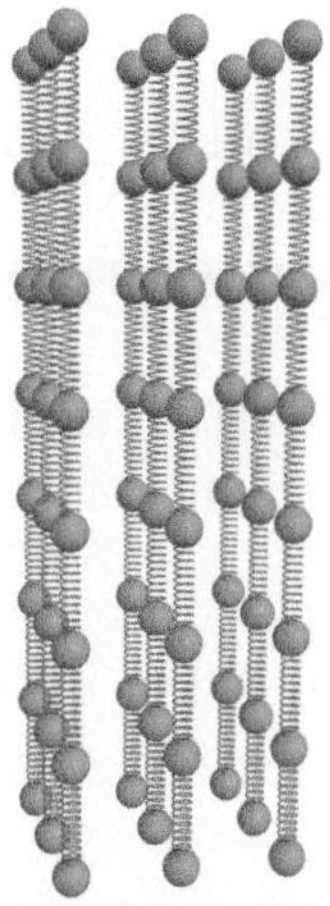

Figure 4.12 A solid wire considered as many parallel long chains of atoms connected by springs (interatomic bonds). For clarity, the horizontal bonds are not shown; they aren't significant when stretching the wire.

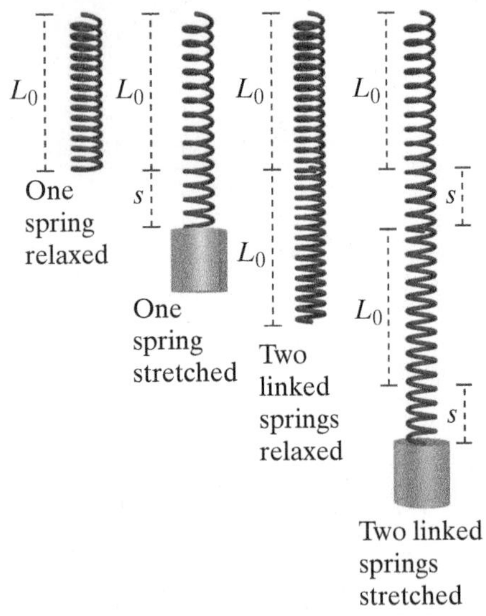

Figure 4.13 Two identical springs linked end to end stretch twice as much as one spring when the same force is applied. The combined spring therefore is only half as stiff as the individual springs.

Two Springs Linked End-to-End (in Series)

Suppose that we have a long, low-mass spring whose spring stiffness is 100 N/m. If we hang a block whose weight is 100 N (about a 10 kg mass) on the end of the spring, the spring will stretch 1 m. (This is a big stretch, so the relaxed spring must be several meters long; we are using simple numbers here to make our reasoning easier.)

> QUESTION We make a longer spring by linking two identical springs end-to-end (in "series"). Each shorter spring has a stiffness of 100 N/m. What will the stiffness of the longer spring be?

When we hang the 100 N block on the end of the longer spring, this longer spring will stretch 2 m, because each of the individual springs stretches 1 m (Figure 4.13). We can use the Momentum Principle to get the stiffness of this longer spring. When the block hangs motionless, its momentum is not changing. Choosing a Δt of 10 s, we find this:

System: Block
Surroundings: Spring, Earth
Free-body diagram: Figure 4.14

$$\begin{aligned}\Delta p_y &= F_{\text{net},y}\Delta t\\ 0 &= (k_s s - mg)(10\,\text{s})\\ k_s s &= mg\\ k_s(2\,\text{m}) &= 100\,\text{N}\\ k_s &= 50\,\text{N/m}\end{aligned}$$

A long spring made of two identical springs linked end-to-end is only half as stiff as each of the shorter springs.

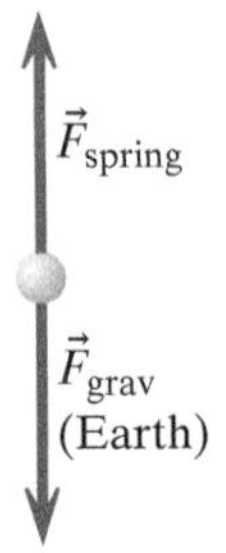

Figure 4.14 Free-body diagram for hanging block.

> **Checkpoint 2** If a chain of 20 identical short springs linked end-to-end has a stiffness of 40 N/m, what is the stiffness of one short spring?

Two Springs in Parallel (Side-by-Side)

> QUESTION Suppose that we support a 100 N weight with two identical springs side by side (Figure 4.15). If the spring stiffness of each spring is 100 N/m, how much will each spring stretch? What is the effective stiffness of the two-spring system?

Informally, we see that each spring supports only 50 N, so each spring will stretch only 0.5 m. You can work out the formal solution yourself using the Momentum Principle, remembering to include two separate upward spring forces.

We can think of the two springs as a single, wider spring. The effective spring stiffness of this "double-width" spring is:

$$k_{s,\text{effective}} = \frac{100\,\text{N}}{0.5\,\text{m}} = 200\,\text{N/m}$$

Two springs side by side are effectively twice as stiff as a single spring.

Figure 4.15 Two springs side by side supporting a block.

> **Checkpoint 3** Nine identical springs are placed side by side (in parallel, as in Figure 4.15), and connected to a large massive block. The stiffness of the nine-spring combination is 2700 N/m. What is the stiffness of one of the individual springs?

Cross-Sectional Area

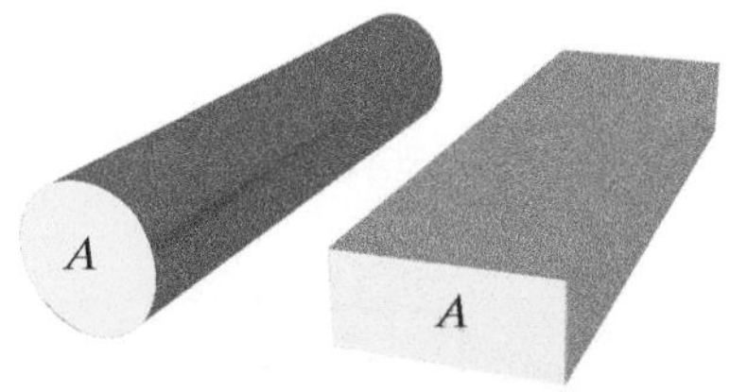

Figure 4.16 The cross-sectional area of a cylinder is the area of a circle. The cross-sectional area of a rectangular solid is the area of a rectangle.

The cross-sectional area of an object is the area of a flat surface made by slicing through the object (Figure 4.16). A cylindrical object like a round pencil has a circular cross-sectional area (imagine sawing crosswise through a pencil). A cylindrical pencil 10 cm long and 0.5 cm in diameter has a cross-sectional area of

$$A = \pi(0.0025\,\text{m})^2 = 1.96 \times 10^{-5}\,\text{m}^2$$

Note that the length of the object is irrelevant.

A long object with four flat sides, like a board, has a cross-sectional area that is rectangular, or perhaps square. A wooden board 7 m long, 5 cm wide, and 3 cm high, has a cross-sectional area of

$$A = (0.05\,\text{m})(0.03\,\text{m}) = 1.5 \times 10^{-3}\,\text{m}^2$$

EXAMPLE

Interatomic Bond Stiffness in Copper

A copper wire is 2 m long. The wire has a square cross section (that is, it has four flat sides—it is not round). Each side of the wire is 1 mm in width. Making sure the wire is straight, you hang a 10 kg mass on the end of the wire. Careful measurement shows that the wire is now 1.67 mm longer. From these measurements, determine the stiffness of one interatomic bond in copper.

Solution

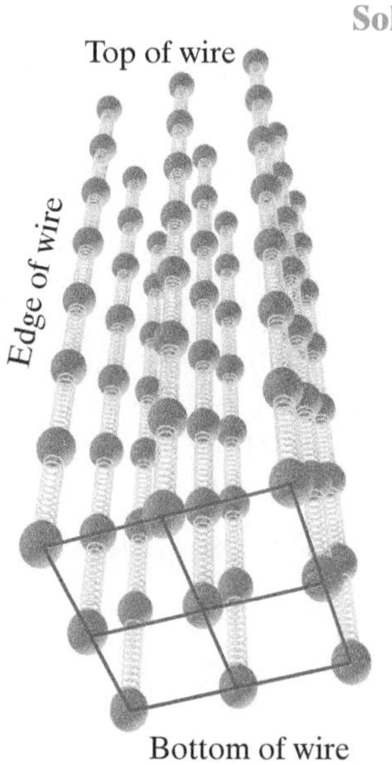

Figure 4.17 A wire modeled as an assembly of side-by-side chains of balls and springs, viewed from the bottom. Colored lines connect the bottom layer of atoms in the wire (only nine ball–spring chains are shown here).

1. What is the spring stiffness $k_{s,\text{wire}}$ of the entire wire, considered as a single macroscopic (large scale), very stiff spring?
   System: Mass
   Surroundings: Earth, wire
   Momentum Principle:

$$\Delta p_y = 0 = (k_{s,\text{wire}}s - mg)\Delta t$$

$$k_{s,\text{wire}} = \frac{mg}{s} = \frac{(10\,\text{kg})(9.8\,\text{N/kg})}{(1.67 \times 10^{-3}\,\text{m})} = 5.87 \times 10^4\,\text{N/m}$$

2. How many side-by-side atomic chains (long springs) are there in this wire? This is the same as the number of atoms on the bottom surface of the copper wire (see Figure 4.17; the horizontal bonds are not shown, as they aren't stretched significantly).
   Cross-sectional area of wire:

$$A_{\text{wire}} = (1 \times 10^{-3}\,\text{m})^2 = 1 \times 10^{-6}\,\text{m}^2$$

Cross-sectional area of one copper atom (even though an atom itself is spherical, each atom occupies a cubical space in the crystal):

$$A_{1\,\text{atom}} \approx (2.28 \times 10^{-10}\,\text{m})^2 = 5.20 \times 10^{-20}\,\text{m}^2$$

Number of side-by-side atomic chains in the wire:

$$N_{\text{chains}} = \frac{A_{\text{wire}}}{A_{1\,\text{atom}}} = 1.92 \times 10^{13}$$

3. How many interatomic bonds are there in one atomic chain running the length of the wire?

$$N_{\text{bonds in 1 chain}} = \frac{L_{\text{wire}}}{d} = \frac{2\,\text{m}}{2.28 \times 10^{-10}\,\text{m}} = 8.77 \times 10^9$$

4. What is the stiffness $k_{s,i}$ of a single interatomic spring?
Using our results about springs linked end to end and springs arrayed side by side:

$$k_{s,\text{wire}} = \frac{(k_{s,i})(N_{\text{chains}})}{N_{\text{bonds in 1 chain}}} \quad \text{(solve for } k_{s,i}\text{)}$$

$$k_{s,i} = \frac{(5.87 \times 10^4\ \text{N/m})(8.77 \times 10^9)}{(1.92 \times 10^{13})} = 26.8\ \text{N/m}$$

An interatomic bond in copper is stiffer than a Slinky, but less stiff than a pogo stick. The stiffness of a single interatomic bond is very much smaller than the stiffness of the entire wire (which we found to be around 6×10^4 N/m).

Checkpoint 4 The 2 m copper wire with square cross section of 1 mm by 1 mm stretched 1.67 mm when it supported a 10 kg mass. Cut a length of this wire 0.2 m long and hang a 10 kg mass from it. How much will this short wire stretch?

4.6 STRESS, STRAIN, AND YOUNG'S MODULUS

QUESTION Suppose that we had used a different copper wire, which was 3 m long, and had a circular cross section, with a diameter of 0.9 mm. Would our result for the interatomic bond stiffness have been different?

The result should not be different, because the interatomic bond stiffness is a property of the material (solid copper). The macroscopic dimensions of the wire will not change the intrinsic properties of copper.

You will not find tables of interatomic spring stiffness for different solid materials in reference libraries. Instead, what is published is a macroscopic quantity called Young's modulus. Like density and interatomic spring stiffness, Young's modulus is a property of a particular material (for example, copper) and is independent of the shape or size of a particular object made of that material. Young's modulus is a macroscopic measure of the "stretchability" of a solid material. It relates the fractional change in length of an object to the force per square meter of cross-sectional area applied to the object.

Strain

The longer a wire is, the more atomic bonds it has along its length, and the more it will stretch when a force is applied to it. In order to define a quantity (Young's modulus) that is a property of the material and doesn't depend on the shape of the wire, we need to take into account not only the measured stretch, but also the original length of the wire. If the wire has a length L, we call the amount of stretch of the wire ΔL (a small increment in the length, what we have called s in a spring), and the fractional stretch $\Delta L/L$ is called the "strain."

DEFINITION OF STRAIN

$$\text{strain} = \frac{\Delta L}{L}$$

Stress

As we saw above, all the chains of atoms help hold up the weight attached to the wire. In order to define a quantity that doesn't depend on the thickness

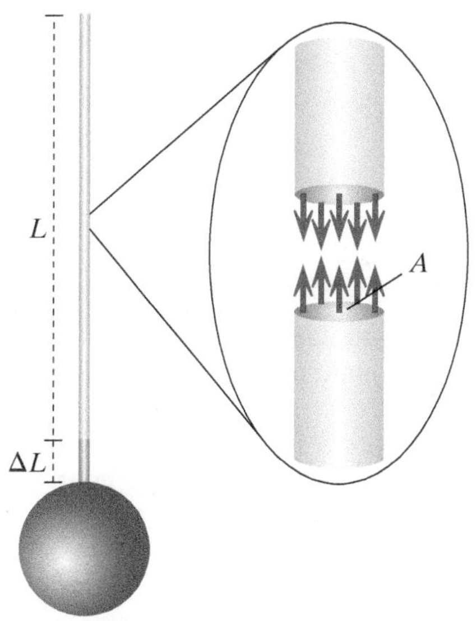

Figure 4.18 The heavy ball stretches the wire an amount ΔL. A is the cross-sectional area—the area of a slice through the wire, perpendicular to its length.

of the wire, we need to take into account not only the tension force F_T, but also the cross-sectional area of the wire A (Figure 4.18). The tension force per unit area is called the "stress."

DEFINITION OF STRESS

$$\text{stress} = \frac{F_T}{A}$$

Young's Modulus

As long as the strain isn't too big, the strain is proportional to the stress. At the atomic level, the stress F_T/A can be related to the force that each long chain of atomic bonds must exert, and the strain $\Delta L/L$ can be related to the stretch of the interatomic bond. The ratio of stress to strain is a property of the material and differs for steel, aluminum, and so on, but doesn't depend on the length or thickness of the wire we use in our measurement of Young's modulus.

DEFINITION OF YOUNG'S MODULUS

$$Y = \frac{\text{stress}}{\text{strain}} = \frac{\left(\frac{F_T}{A}\right)}{\left(\frac{\Delta L}{L}\right)}$$

Young's modulus is the ratio of stress to strain for a particular material. Young's modulus is a property of the material, and does not depend on the size or shape of an object. The stiffer the material, the larger is Young's modulus.

Using Young's modulus, a quantitative form of the relation between stress and strain in a wire is written as follows:

$$\frac{F_T}{A} = Y\frac{\Delta L}{L}$$

Note the similarity to the spring force, $F_T = k_s s \ (= k_s \Delta L)$.

You may have the opportunity to measure Young's modulus in a laboratory experiment, and to use your measurements to determine the interatomic spring stiffness for a particular material.

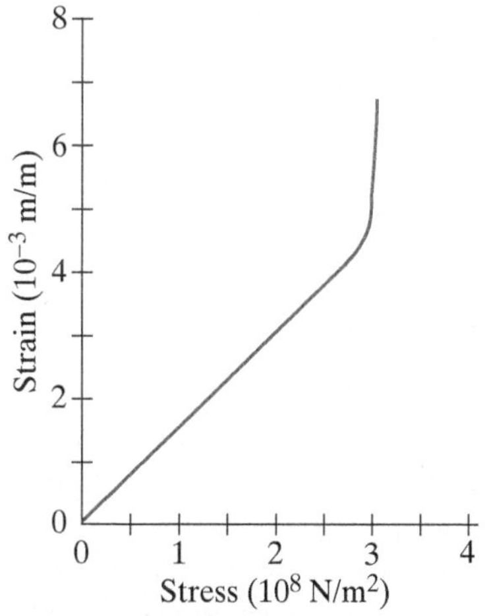

Figure 4.19 Strain vs. stress for a particular aluminum alloy.

Limit of Applicability of Young's Modulus

If you apply too large a stress, the wire "yields" (suddenly stretches a great deal) or breaks, and the proportionality of stress and strain is no longer even approximately true. Figure 4.19 shows a graph of the strain $\Delta L/L$ that results as a function of the applied stress F_T/A for a particular aluminum alloy.

You see that for moderate stress the resulting strain is proportional to the applied stress: double the stress, double the strain. However, once you reach the yield stress, a further slight increase in the stress leads to a very large increase in the length of the metal. This can be quite dramatic: as you add more weights to the end of a wire the wire gets slightly longer, then suddenly it starts to lengthen a lot, very rapidly, and then the wire breaks. In materials reference manuals, stress is usually plotted against strain, so that the slope of the line is equal to Young's modulus, stress divided by strain. We plot strain vs. stress here to emphasize that stress is the *cause* (the independent variable) and strain the *effect* (the dependent variable).

Checkpoint 5 You hang a 5 kg mass from the end of a rod that is 1.7 m by 2 mm by 2 mm. You observe that the rod's length increases to 1.74 m. **(a)** What is the strain? **(b)** What is the stress? **(c)** What is Young's modulus for this material?

Relating Young's Modulus to Interatomic Spring Stiffness

From macroscopic measurements of Young's modulus we can calculate an approximate spring stiffness $k_{s,i}$ for the interatomic bond by again making a "macro-micro" connection.

Consider a single stretched interatomic bond in a wire that is being stretched. Let d represent both the relaxed length of an interatomic bond and the diameter of one atom. The cross-sectional area occupied by one atom is d^2 (Figure 4.20) (recall that even though an atom is approximately spherical, it occupies a cube of space in the crystal lattice).

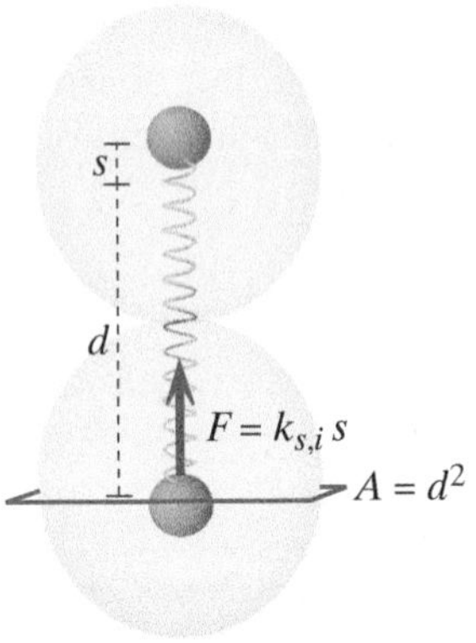

Figure 4.20 A single stretched bond in a solid wire. The effective cross-sectional area of one atom is d^2, and the force acting on that portion is $k_{s,i}s$.

QUESTION What is the magnitude of the stress on one atom in terms of the interatomic force $k_{s,i}s$?

If the stretch of the interatomic bond is s, the atomic stress (force per unit area) is $\dfrac{k_{s,i}s}{d^2}$.

QUESTION In terms of the stretch s of the interatomic "spring," what is the strain for one atom (change of height of the atom, divided by its normal height)?

The strain for one atom is s/d, because its height was originally d, and the change in its height is s. Therefore we can express Young's modulus in terms of atomic quantities:

YOUNG'S MODULUS IN TERMS OF ATOMIC QUANTITIES

$$Y = \frac{(k_{s,i}s/d^2)}{(s/d)} = \frac{k_{s,i}}{d}$$

where $k_{s,i}$ is the stiffness of an interatomic bond in a solid, and d is the length of an interatomic bond (and the diameter of an atom).

If you know Young's modulus for some metal, you can calculate the effective stiffness of the interatomic bond, modeled as a spring: $k_{s,i} = Yd$, where d is the atomic diameter in the metal. This is another example of an important theme, that of relating macroscopic properties to microscopic (atomic-level) properties.

Tarzan and the Vine Revisited

We now know that interatomic bonds in the vine stretch when Tarzan hangs from the vine. It is not yet clear why the vine breaks when Tarzan swings on it but not when he hangs motionless. Apparently the vine must stretch more when he swings—perhaps it must apply a larger force? In Chapter 5 we will learn how to apply the Momentum Principle to Tarzan's situation.

4.7 COMPRESSION (NORMAL) FORCES

Figure 4.21 A brick lies on a table.

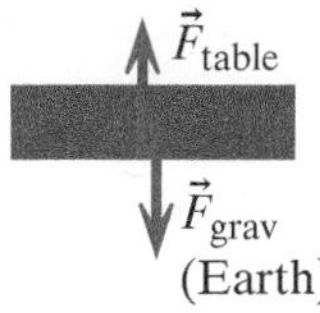

Figure 4.22 Forces acting on the brick.

Figure 4.23 A schematic view of a portion of a brick resting on a portion of a table (scale greatly exaggerated). Interatomic "springs" in the contact region are compressed by the weight of atoms above.

A closely related situation is that of a brick lying at rest on a metal table (Figure 4.21). In this situation the table is compressed, while in the previous discussion the wire was stretched.

Take the brick as the system of interest (Figure 4.22). The Earth pulls down on the brick of mass M, and the magnitude of the gravitational force is Mg.

Since the brick remains at rest, the vertical component of the net force must be zero:

$$0 = +F_{\text{Table}} - Mg$$

We conclude that the force F_{Table} exerted by the table is equal to the weight Mg of the brick, an unsurprising result.

How does the metal table exert a force on the brick? If the brick were extremely heavy, we might not be surprised to observe the table sagging under its weight. Even if the table does not sag visibly, the molecules in the top layers of the table are pushed down by molecules in the bottom layer of the brick. The table actually deforms somewhat—its interatomic "springs" are compressed. (Of course, atoms in the brick are compressed somewhat, too.)

Atoms in the table are squeezed closer than their normal equilibrium positions, compressing the spring-like interatomic bonds. Figure 4.23 schematically shows the contact region between the brick and the table.

It would be appropriate to call the upward force of the table F_C for "compression force," but it is common practice to label this kind of force as a "normal" force. The word is used in the mathematical sense, meaning "perpendicular to," because the direction of the force is perpendicular to the surface of the table. Usually "normal" forces are labeled as F_N.

Note: Neither normal force nor tension force is a different kind of fundamental force, like the gravitational force or electric force. These are merely descriptive names for particular kinds of interatomic electric interactions.

4.8 FRICTION

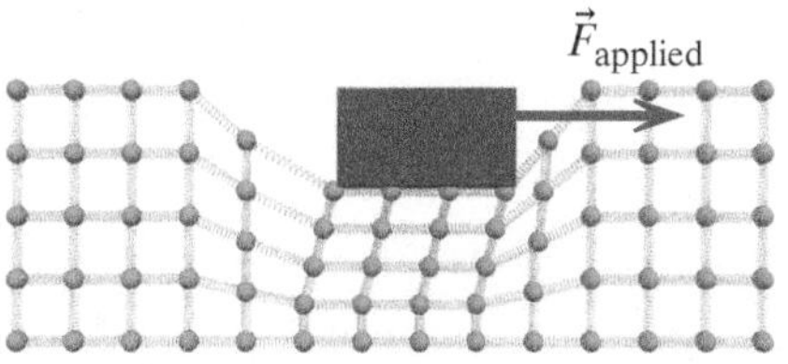

Figure 4.24 When you apply a force on the brick, it runs into atoms of the table, compressing interatomic bonds in front and stretching them behind (scale greatly exaggerated). Similar deformations occur in the block but for clarity are not shown.

Friction is another phenomenon that involves deformable objects; to understand it we again need to consider the internal structure of solid matter. As you've just seen, a brick sitting on a table sinks into the table, compressing the interatomic bonds, and the table pushes upward on the brick. If you then drag the brick across the table as in Figure 4.24, the brick continually runs into previously uncompressed regions of the table, compressing bonds ahead of the brick and stretching bonds behind, so we can consider the force of the table as the sum of two perpendicular components. The component that is parallel to the velocity (and opposite to the velocity in this case), is called a "frictional force," as shown in Figure 4.25. The full free-body diagram for the brick is shown in Figure 4.26. After the brick has passed by, the atoms tend to go back to their equilibrium positions.

However, the disturbance of the atoms away from their equilibrium positions leaves the atoms vibrating about their equilibrium positions, in both the table and in the block. As we will discuss in detail in later chapters, this increased vibrational motion or "internal energy" is associated with higher temperature. Dragging the brick across the table makes the temperature of the table and brick rise. The increase in internal energy means that you must continually apply a force and expend energy to do the dragging. If you stop pulling, the brick will slow down and stop as its motional ("kinetic") energy is dissipated away into internal energy in the form of agitation of atoms throughout the objects. We say that friction is a "dissipative" process.

Evidently for a full understanding of the block sliding on the table, we can't simply consider the block and table to be point particles. They are in fact multiparticle systems, and friction is very much a multiparticle phenomenon.

Friction is a complex process, and this picture of friction at the atomic level, based on the ball–spring model of solids, is necessarily simplified. In Chapter 9 we will discuss friction in more detail, after we have studied energy.

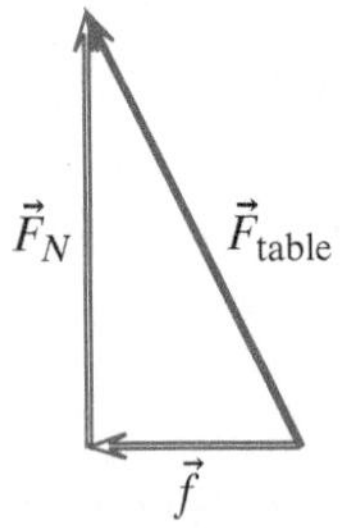

Figure 4.25 If the brick is pulled to the right, the table now exerts a force upward and to the left. Here the force $\vec{F}_{\text{table}}$ is expressed as the sum of two perpendicular components: the upward "normal" force $\vec{F}_N$ and the sideways "friction" force $\vec{f}$.

> **QUESTION** To keep the brick moving at a constant speed, you have to keep pulling. It's tempting to conclude that a constant force is needed to keep an object moving at a constant speed, which sounds like a violation of Newton's first law of motion and of the Momentum Principle. What's going on here?

It is the net force that governs the motion of the brick, not just the force that you apply. Since the net force is zero, the brick moves with constant speed. You are pulling with a force whose magnitude is equal to that of the parallel component of the force of the table, so the net force $\vec{F}_{\text{net}}$ is zero.

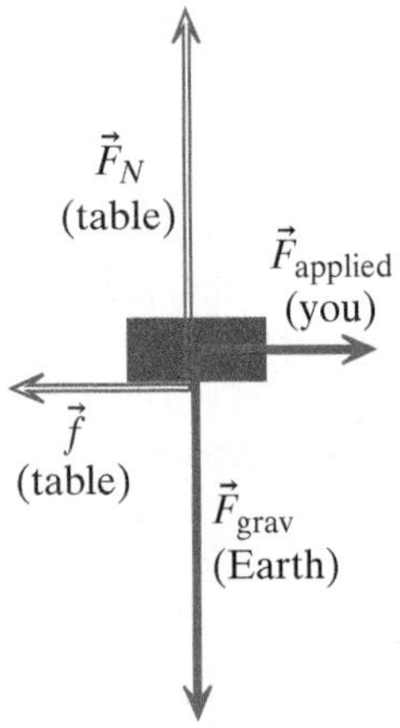

Figure 4.26 The free-body diagram for the brick now includes the components of the force by the table, as well as the force by you and the force by the Earth.

It is often the case that the direction of the friction force is opposite to the velocity, tending to slow down the object. However, this is not always true. For example, if you drop a box onto a moving conveyor belt, bonds at the back of the box are compressed and the friction force pushes the box forward with increasing speed (Figure 4.27). Eventually the box reaches the speed of the moving belt, at which time the friction force drops to zero and the box moves with constant speed, subject to a zero net force.

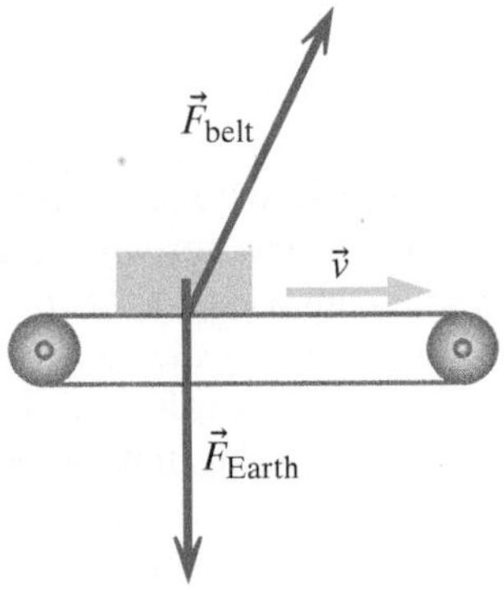

Figure 4.27 Place a box on a moving conveyor belt, and the horizontal component of $\vec{F}_{\text{belt}}$ accelerates the box.

When a sprinter speeds up, she pushes her feet backwards against the ground, and the horizontal component of the force of the ground on her feet points forward, accelerating her. On a very slick surface with almost no friction she can't accelerate.

Sliding Friction

When one object slides on another, the component of force exerted by one object on the other has a component parallel (or antiparallel) to the motion with the following properties:

THE SLIDING FRICTION FORCE

$$f_{\text{friction}} \approx \mu_k F_N$$

> F_N is the normal force, the perpendicular component of the force that is squeezing the two objects into each other. μ_k (lowercase Greek mu) is the "coefficient of kinetic friction," which depends on the nature of the two materials in contact, with values typically in the range of 0 to 1. The sliding friction force does not depend on the speed of the object.

The more you squeeze the two objects together, the deeper their interpenetration and the more difficult it is to slide sideways, which is why the force is proportional to F_N.

It is interesting that the friction force does not depend on the speed. This is in contrast to air resistance, another dissipative force. Air resistance is larger at higher speeds, as you can easily verify by putting your hand out of the window of a moving car and noticing how much greater the air resistance force is on your hand at high speed than it is at low speed.

$f_{\text{friction}} \approx \mu_k F_N$ is only approximate and does not have the fundamental character of the gravitational and electric forces. Nevertheless, it is a good approximation to the effects of friction when one object slides on another.

QUESTION If you are pulling a brick at a constant speed, what happens if you increase your force to be bigger than $\mu_k F_N$?

If while the brick is sliding you apply a force bigger than $\mu_k F_N$, the net force becomes nonzero, so the momentum of the brick will increase.

Static Friction

QUESTION What will happen if you exert a horizontal force *less* than $\mu_k F_N$ on a brick lying on a table?

In this case, although the brick doesn't slide, there is a sideways compression of the atomic bonds, so there is a parallel component of the force of the table on the brick, equal to the force you're applying. This situation is called "static friction," where "static" means "not moving." The static friction force can have any value that is less than $\mu_s F_N$, where μ_s is called the "coefficient of static friction." As you apply a bigger and bigger force, but less than $\mu_s F_N$, the bonds are compressed more but the object doesn't slide.

For some pairs of materials it may be necessary to exert a force somewhat bigger than $\mu_k F_N$ in order to get the object to start to move, but once the object has started to slide, you only need a force of $\mu_k F_N$ to keep it moving. In such situations we say that the coefficient of static friction μ_s is larger than the one for sliding friction μ_k (k for kinetic, meaning with motion).

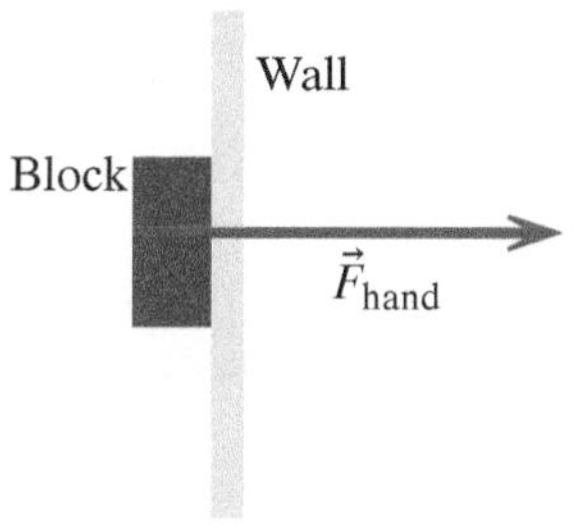

Figure 4.28 You press a block against a wall.

STATIC FRICTION

$$f_{friction} \leq \mu_s F_N$$

μ_s is the "coefficient of static friction." The parallel component $f_{friction}$ of the force squeezing the two objects together can be less than $\mu_s F_N$. The maximum static friction force is $\mu_s F_N$.

When μ_s is larger than μ_k, there can be "stick-slip" motion. You pull on a string attached to the brick and apply a large force $\mu_s F_N$ to start the motion, then the friction force decreases to $\mu_k F_N$, the net force becomes nonzero, and the brick accelerates, making the string go slack if you don't accelerate your hand. Now that the string has gone slack, the brick slows down and may stop, and you have to start over, hence the stick-slip motion.

EXAMPLE **Holding a Block Against a Wall**

You hold a 3 kg metal block against a wall by applying a horizontal force of 40 N, as shown in Figure 4.28. The coefficient of friction for the metal–wall pair of materials is 0.6 for both static friction and sliding friction. Does the block slip down the wall?

Solution System: Block

Surroundings: Earth, wall, hand

Free-body diagram: Figure 4.29

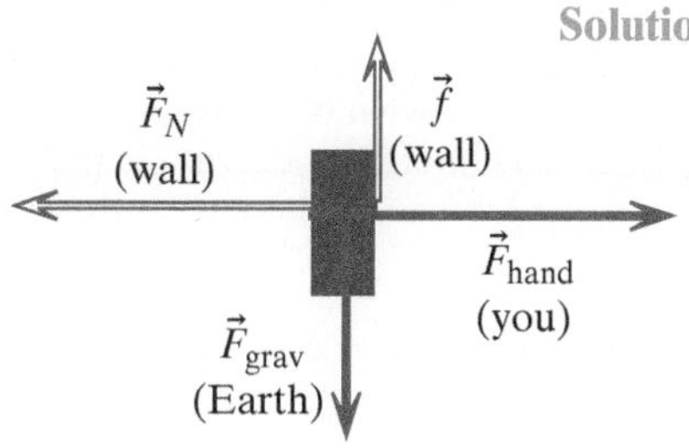

Figure 4.29 Free-body diagram for a block pressed against a wall.

$$x\text{: } \Delta p_x = (F_{hand} - F_N)\Delta t = 0$$
$$y\text{: } \Delta p_y = (\mu_k F_N - mg)\Delta t, \text{ assuming it slides}$$

Combining these two equations, we have

$$\Delta p_y = (\mu_k F_{hand} - mg)\Delta t = (0.6(40\,\text{N}) - (3\,\text{kg})(9.8\,\text{N/kg}))\Delta t$$
$$\Delta p_y = (-5.4\,\text{N})\Delta t$$

Since there is a nonzero impulse in the $-y$ direction, the block will slip downward with increasing speed.

To prevent the block from sliding downward, you would have to apply a force such that $F_{net,y} = 0$, in which case $\mu_s F_{hand} = mg$, so $F_{hand} = (3\,kg)(9.8\,N/kg)/0.6 = 49\,N$. Note that F_N isn't always equal to mg; here it is equal to the force of your hand.

If you apply a force greater than 49 N, how big is the static friction force f? If the block doesn't move, the y component of the net force $F_{net,y}$ is zero, so the static friction force will always be mg, no matter how hard you press the block against the wall, with $f_{friction} = mg < \mu_s F_N$.

EXAMPLE **Sliding to a Stop**

You take the same 3 kg metal block and slide it along the floor, where the coefficient of friction is only 0.4. You release the block with an initial velocity of $\langle 6,0,0\rangle$ m/s. How long will it take for the block to come to a stop? How far does the block move?

Solution System: Block
Surroundings: Earth, floor
Free-body diagram: Figure 4.30
Initial state: $v_x = 6$ m/s
Final state: $v_x = 0$

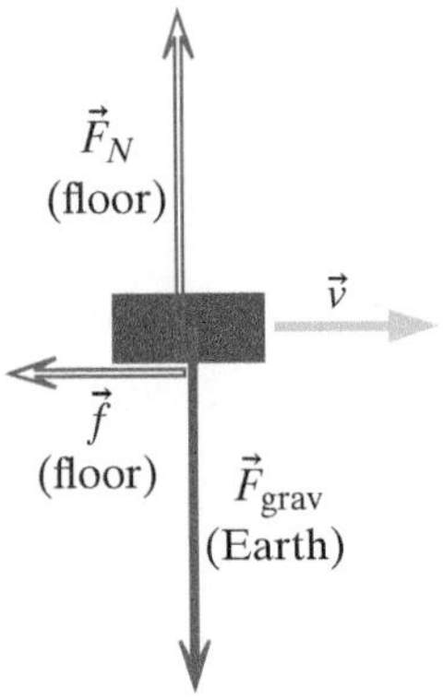

Figure 4.30 F_N is the normal component of the force of the floor and $f = \mu_k F_N$ is the parallel component.

$$x: \Delta p_x = -\mu_k F_N \Delta t$$
$$y: \Delta p_y = (F_N - mg)\Delta t = 0$$

Combining these two equations and writing $p_x = mv_x$, we have

$$\Delta(mv_x) = -\mu_k mg\Delta t$$
$$\Delta v_x = -\mu_k g\Delta t$$
$$\Delta t = \frac{0 - v_{xi}}{-\mu_k g} = \frac{v_{xi}}{\mu_k g}$$
$$\Delta t = \frac{6\,m/s}{0.4(9.8\,N/kg)} = 1.53\,s$$

Since the net force was constant, $v_{x,avg} = (v_{xi} + v_{xf})/2$, so

$$\Delta x/\Delta t = ((6+0)/2)\,m/s = 3\,m/s$$
$$\Delta x = (3\,m/s)(1.53\,s) = 4.5\,m$$

Checkpoint 6 When you apply a horizontal force of 80 N to a block, the block moves across the floor at a constant speed, with $\mu_k = \mu_s$. **(a)** What happens if instead you apply a force of 60 N? **(b)** When you apply a force of 60 N, what is the magnitude of the horizontal component of the force that the floor exerts on the block? **(c)** What happens if instead you exert a force of 100 N? **(d)** When you apply a force of 100 N, what is the magnitude of the horizontal component of the force that the floor exerts on the block?

4.9 SPEED OF SOUND IN A SOLID AND INTERATOMIC BOND STIFFNESS

When we have considered objects that exert tension forces, such as vines, strings, and wires, we have made the (very good) approximation that the stretch

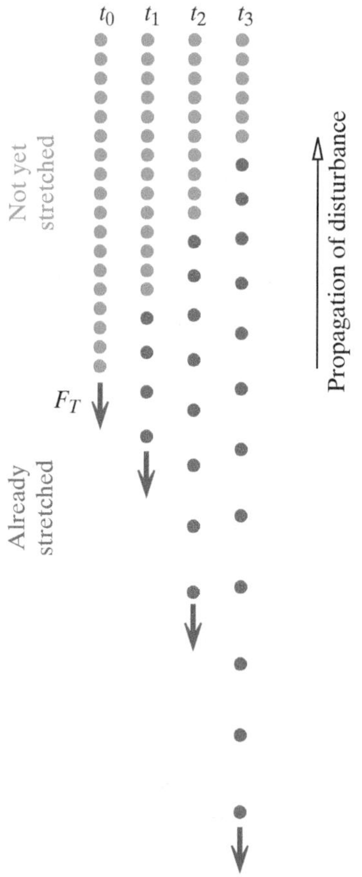

Figure 4.31 At time t_0 Tarzan starts pulling downward on the vine with a force F_T. The "disturbance" in the vine propagates upward. At successive times t_1, t_2, and t_3, more and more of the interatomic bonds in the vine are stretched. Atoms in the stretched region are indicated by colored dots. The amount of stretch is greatly exaggerated here.

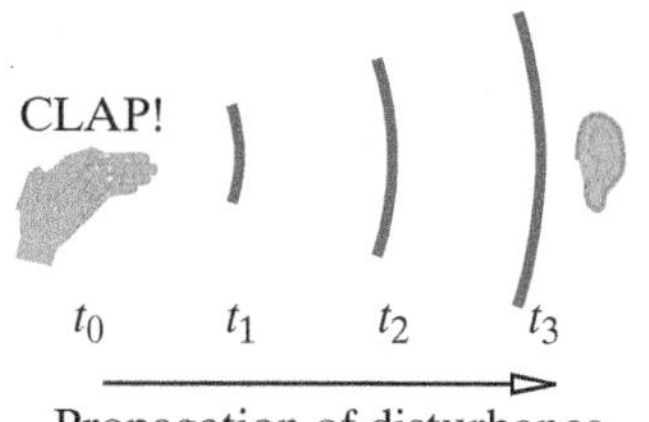

Figure 4.32 Clap your hands—sound travels to your ear as a pulse of air compression.

of the interatomic bonds is nearly the same throughout the length of the object. Likewise, we assume that inside objects exerting compression forces, the interatomic bonds are all compressed nearly the same amount.

The Propagation of Stretch or Compression

The equal stretch or compression of bonds does not, however, occur instantaneously. When Tarzan first starts pulling down on the end of a vine, he very slightly lengthens the interatomic bonds between neighboring atoms in the nearby section of the vine (Figure 4.31). As a result of their downward displacement, these atoms stretch the bonds of their neighbors, and very quickly this new interatomic bond-stretching propagates upward, all the way to the other end of the vine, and the whole vine is then in tension.

It is the boundary between the already stretched region of the vine and the newly stretched region that moves upward, along the length of the vine. Individual atoms move extremely short distances, though for clarity in the diagram we have enormously exaggerated the interatomic stretches and the distance any individual atom moves. The process is the same for compression.

This process is fast, but it is not instantaneous. The rate of propagation of the boundary between the stretched and unstretched regions is called the "speed of sound" in the material. There are various ways to measure this speed. One of the simplest methods is to place a microphone at one end of a metal bar, strike the other end of the bar with a hammer, and use a computer interface to measure how long it takes for the disturbance to travel from one end of the bar to the other.

QUESTION What does this have to do with sound?

What we call "sound" is a disturbance in a material. We are most familiar with sound traveling through air. When you clap your hands, a pulse of sound propagates through the air to your ear (Figure 4.32). This pulse starts as a momentary compression of the air (an increase in the air density) made by your hands. This compression is passed on to neighboring regions of the air, and the air near the hands relapses to its original density.

Eventually your eardrum is hit by the pulse and moves in response to the sudden increase in density (and pressure). The motion of the eardrum is detected by your inner ear and passed on to the brain, which interprets the signal as meaning that you clapped your hands. Note that no individual air molecule moves from near your hands to your ear. Rather it is the disturbance that moves from one place to another. The propagation of a disturbance is called a wave.

An Iterative Model of Sound Propagation in a Metal

We will model a copper rod as a single horizontal chain of one hundred copper atoms, each with mass $(64 \times 10^{-3}\ \text{kg})/(6.02 \times 10^{23})$. The initial distance between any two atoms is the length of the interatomic bond in solid copper (2.28×10^{-10} m, see Section 4.4), and the stiffness of each interatomic spring is 26.8 N/m (see Section 4.5). Initially each atom is at rest. The calculation is organized in a way similar to other iterative calculations we have done; the only difference is that we must update the position and momentum of 100 objects after each time step. An outline of the calculation is this:

Displace the leftmost atom by a small amount
Choose a very short time step ($\Delta t \approx 1 \times 10^{-14}$ s)
Set elapsed time = 0
Then, repeatedly:

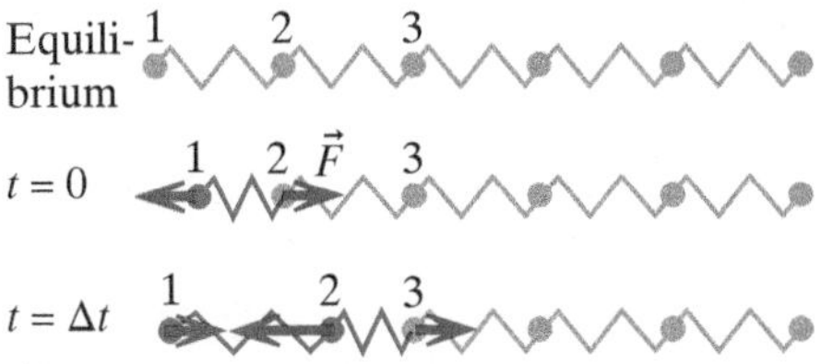

Figure 4.33 Top: A chain of atoms and bonds in their equilibrium position. At $t = 0$ one atom is displaced to the right, starting a wave that moves along the chain.

- From the current positions of the atoms, find the stretch of every spring.
- Find the net force on each atom, due to the two springs attached to it.
- Calculate the change in momentum of each atom.
- Calculate the new momentum of each atom.
- Update the position of each atom.
- Add Δt to the elapsed time.

Stop the calculation when the rightmost atom is displaced. Divide the distance traveled (100 atoms)(2.28×10^{-10} m/atom) by elapsed time to get the speed at which the disturbance traveled down the chain.

Figure 4.33 shows a visualization of the first step in this process. At the top is a chain of atoms and bonds in their equilibrium positions. At time $t = 0$ atom 1 is displaced to the right, compressing a bond, which then exerts a force on atom 2 (arrows denote the net forces on each atom). After applying the Momentum Principle to all atoms, and updating positions of all atoms over a time step Δt, both atoms 1 and 2 are displaced from equilibrium; the second bond is compressed, and the first is now stretched. Iterative updates continue, involving progressively more atoms after each time step.

Visualizing the propagation of the disturbance can be done in various ways. A particularly useful visualization is to plot the horizontal displacements (y_n) vertically, at the locations of the horizontal equilibrium locations (x_n). An example is shown in Figure 4.34, where you see that a displacement to the left is plotted below the axis, and a displacement to the right is plotted above the axis.

It is clearly desirable to program a computer to do this calculation. When we carry out this calculation for a copper bar, using the values given above, we obtain a value for the speed of sound in copper of about 3750 m/s. Experimentally measured values given in engineering and science reference materials are around 3800 m/s, so our simple model gives a surprisingly good result.

Although predictions based on iterative calculations can assure us that our understanding of the underlying nature of a phenomenon is correct, it is also useful to have a simple algebraic expression for a central relationship such as that between interatomic bond stiffness and speed of sound. In order to be able to do this, we will need to derive an analytical solution for the oscillations of a single mass and spring. This will give us added insight into the propagation of a disturbance along a chain of masses and springs.

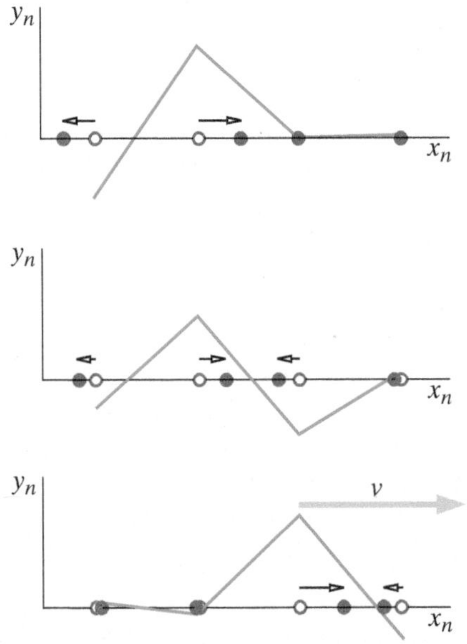

Figure 4.34 Visualizing the propagation by plotting the displacements y_n away from the equilibrium positions x_n.

4.10 DERIVATIVE FORM OF THE MOMENTUM PRINCIPLE

The forms of the Momentum Principle we have used so far,

$$\Delta\vec{p} = \vec{F}_{\text{net}}\,\Delta t \quad \text{and} \quad \vec{p}_f = \vec{p}_i + \vec{F}_{\text{net}}\,\Delta t$$

are particularly useful when we know the momentum at a particular time and want to predict what the momentum will be at a later time. We use these forms in repetitive computer calculations to predict future motion.

In solving problems involving objects whose momentum was not changing at all, such as a ball hanging motionless on the end of a wire, we saw that although we had to choose a value for Δt, it didn't matter what value we chose, as long as it was greater than zero. In such problems, we could also have used a different form of the Momentum Principle, which we obtain by dividing by the time interval Δt:

$$\frac{\Delta\vec{p}}{\Delta t} = \vec{F}_{\text{net}}$$

If we let Δt get smaller and smaller, the ratio $\Delta\vec{p}/\Delta t$ becomes the time derivative of $\vec{p}$:

$$\lim_{\Delta t \to 0} \frac{\Delta\vec{p}}{\Delta t} = \frac{d\vec{p}}{dt}$$

The derivative of a vector is itself a vector:

$$\frac{d\vec{p}}{dt} = \left\langle \frac{dp_x}{dt}, \frac{dp_y}{dt}, \frac{dp_z}{dt} \right\rangle$$

We have previously encountered other vector derivatives: the instantaneous velocity $\vec{v} = d\vec{r}/dt$, and the acceleration $\vec{a} = d\vec{v}/dt$. Like any other vector, the derivative of a vector has both a magnitude and a direction.

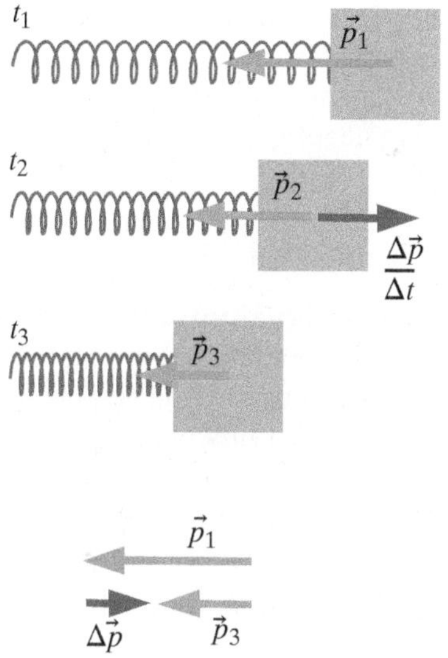

Figure 4.35 A block attached to a spring moves to the left, slowing down. To find $d\vec{p}/dt$ at time t_2, we take the difference of the two values of momentum at times t_1 and t_3 (at equal times before and after t_2), and divide by $\Delta t = t_3 - t_1$. As we take Δt to be smaller and smaller, $\Delta\vec{p}/\Delta t$ approaches $d\vec{p}/dt$. In the bottom diagram the collinear arrows $\vec{p}_1$ and $\vec{p}_3$ have been offset slightly to make the graphical subtraction clear.

QUESTION For a given physical system, such as the mass–spring system in Figure 4.35, how can we find $d\vec{p}/dt$?

Operationally, to find $d\vec{p}/dt$ at a particular instant we apply the definition of a derivative, as shown in Figure 4.35:

1. Find $\vec{p}$ at equal time intervals before and after the time of interest (t_2 in Figure 4.35).
2. Find $\Delta\vec{p} = \vec{p}_3 - \vec{p}_1$ (the bottom diagram in Figure 4.35).
3. Divide $\Delta\vec{p}$ by $\Delta t = t_3 - t_1$ to get $\Delta\vec{p}/\Delta t$ (whose direction is the same as the direction of $\Delta\vec{p}$).
4. Let Δt approach 0 by taking t_1 and t_3 closer to t_2. As $\Delta t \to 0$, $\Delta\vec{p}/\Delta t$ approaches $d\vec{p}/dt$.

We can now write the Momentum Principle in its derivative form:

THE MOMENTUM PRINCIPLE (DERIVATIVE FORM)

$$\frac{d\vec{p}}{dt} = \vec{F}_{\text{net}}$$

In words: The instantaneous time rate of change of the momentum of an object is equal to the net force acting on the object. In calculus terms, we can say that the derivative of the momentum with respect to time is equal to the net force acting on the object.

This form of the Momentum Principle is useful when we know something about the rate of change of the momentum at a particular instant. Knowing the rate of change of momentum, we can use this form of the Momentum Principle to deduce the net force acting on the object, which is numerically equal to the rate of change of momentum. Knowing the net force, we may be able to figure out particular contributions to the net force.

QUESTION What does the derivative form of the Momentum Principle tell us about the relationship between the direction of $d\vec{p}/dt$ and the direction of $\vec{F}_{\text{net}}$?

If two vectors are equal, both their magnitudes and their directions are the same. Consider the mass–spring system in Figure 4.35. The block is slowing down because the spring is being compressed, and consequently exerts a force to the right. The direction of this force is in fact the same as the direction of $d\vec{p}/dt$.

Checkpoint 7 At a certain instant the momentum of an object is changing at a rate of $\langle 0,0,4\rangle$ kg·m/s/s. **(a)** At this instant, what is the net force on the object? **(b)** What, if anything, can you conclude about the direction of the object's momentum at this instant?

4.11 ANALYTICAL SOLUTION: SPRING–MASS SYSTEM

In Chapter 2 we applied the finite-time form of the Momentum Principle iteratively to predict the motion of a mass attached to a spring. Now, using the derivative form of the Momentum Principle, we can derive an analytical (algebraic) solution for the motion of an ideal mass–spring system. By "ideal" we mean that the spring can stretch or compress nearly indefinitely, that the mass of the spring is nearly zero, that the spring never gets hot or deforms, and that there is neither friction nor air resistance.

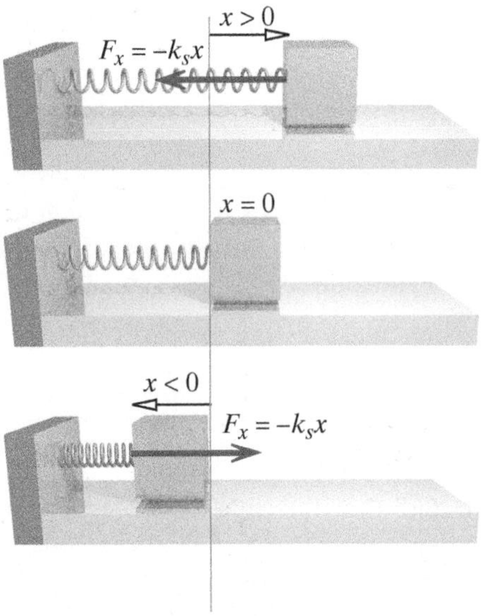

Figure 4.36 A block connected by a spring to a wall slides back and forth with little friction. The x component of the force that the spring exerts on the block is equal to $-k_s x$.

Applying the Momentum Principle to a Spring–Mass System

Consider a block of mass m connected to a spring whose stiffness is k_s (that is, the force exerted by the spring is $k_s s$ when the spring is stretched or compressed an amount s). The other end of the spring is attached to the wall (Figure 4.36). Our goal is to derive an equation for the position x as a function of time t,

$$x(t) = \ldots$$

which predicts the position of the block at any time in the future.

We will assume that:

- The block slides with almost no friction on an air table, where it is supported on a cushion of air.
- The mass of the spring is negligible compared to the mass of the block.
- Air resistance is negligible.

We can simplify our calculations by choosing to measure the position of the block with respect to an origin chosen to be at the equilibrium position, the location of the block when the spring is relaxed, so that the stretch s of the spring is equal to the position x of the block:

$$L = L_0 + x$$
$$s = L - L_0 = (L_0 + x) - L_0 = x$$

Therefore the x component of the force on the block by the spring is $F_x = -k_s x$, as shown in Figure 4.36.

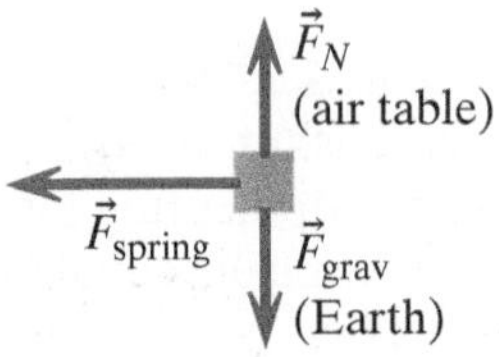

Figure 4.37 Free-body diagram.

QUESTION What would be the equation for the spring force in terms of the position of the block if the origin were located at the wall, and the relaxed length of the spring is L_0?

In that case we would write $F_x = -k_s(x - L_0)$, because what matters in this expression is the stretch (change in length of the spring), which is $(x - L_0)$. A check on this is the fact that when the block is at location $x = L_0$, the stretch $(x - L_0)$ is zero.

System: Block
Surroundings: Spring, Earth, air table
Free-body diagram: Figure 4.37

Momentum Principle:

$$d\vec{p}/dt = \vec{F}_{\text{net}}$$

Since the block oscillates along the x axis, we know that the y component of the momentum of the block is zero and remains zero. Therefore we can deduce that the y component of the net force on the block must be zero. Given this, we can focus only on the x components of force and momentum.

Because $v_x = dx/dt$:

$$\frac{dp_x}{dt} = \frac{d(mv_x)}{dt} = m\frac{d^2x}{dt^2}$$

From the Momentum Principle:

$$\frac{dp_x}{dt} = -k_s x$$

$$m\frac{d^2x}{dt^2} = -k_s x$$

If we ignore the constants and signs for a moment, this equation (which is a "differential equation") says that if we know x as a function of time, taking the second derivative of that function will give us back the original function.

QUESTION What function appears in its own second derivative?

If we differentiate a cosine function twice, we get a cosine again. Further, since the block oscillates back and forth, cosine sounds like a reasonable guess. Guessing a function isn't as peculiar as it may seem. This is actually one of the standard approaches to trying to solve a differential equation.

However, $x(t) = \cos(t)$ doesn't look quite right. The cosine function has values between 1 and -1, but the block may not move 1 m. Let's multiply by a constant A to change the maximum displacement to a reasonable value.

Now we have $x(t) = A\cos(t)$, but that still isn't quite right, because the period of the oscillations might not be 2π s. We can add another constant, ω (lowercase Greek omega), to adjust the period of the oscillations, which gives us the equation

$$x(t) = A\cos(\omega t)$$

To see if this expression for $x(t)$ is in fact a solution, we will try substituting this expression for x back into the Momentum Principle differential equation:

$$m\frac{d^2}{dt^2}(A\cos(\omega t)) = -k_s A\cos(\omega t)$$

$$m\frac{d}{dt}(-A\omega\sin(\omega t)) = -k_s A\cos(\omega t)$$

$$-Am\omega^2\cos(\omega t) = -k_s A\cos(\omega t)$$

$$m\omega^2 = k_s$$

$$\omega = \sqrt{\frac{k_s}{m}}$$

We find that $x(t) = A\cos(\omega t)$ is an analytical solution for the position of the mass as a function of time, if we choose ω to have the value $\sqrt{k_s/m}$. This is an interesting result, because it says that the period of the oscillations depends on the physical properties of the mass and of the spring.

ANALYTICAL SOLUTION: IDEAL SPRING–MASS SYSTEM

$$x(t) = A\cos(\omega t), \text{ with } \omega = \sqrt{\frac{k_s}{m}}$$

A (the "amplitude") is a constant that depends on the initial conditions. A is equal to the maximum stretch of the spring during an oscillation. ω is called the "angular frequency" and has units of radians/s.

Often we write simply $x = A\cos(\omega t)$, since the time dependence of x is made clear by the right-hand side of the equation. The constant A is called the "amplitude" of the motion. It is equal to the maximum stretch of the spring. If the spring stretches or compresses a maximum of 5 cm, then A will have the value 0.05 m. The constant ω in $x = A\cos(\omega t)$ is called the "angular frequency" and is measured in radians per second (ω is lowercase Greek omega). When the argument ωt of the cosine increases by 2π rad, the motion repeats (one complete cycle), so if we call the "period" T (the time required for one complete cycle, Figure 4.38), we have $\omega T = 2\pi$. One other quantity often used to describe oscillating systems is the "frequency" f, which is the number of complete cycles per second: $f =$ (1 cycle) per T seconds. Cycles per second are also called "hertz."

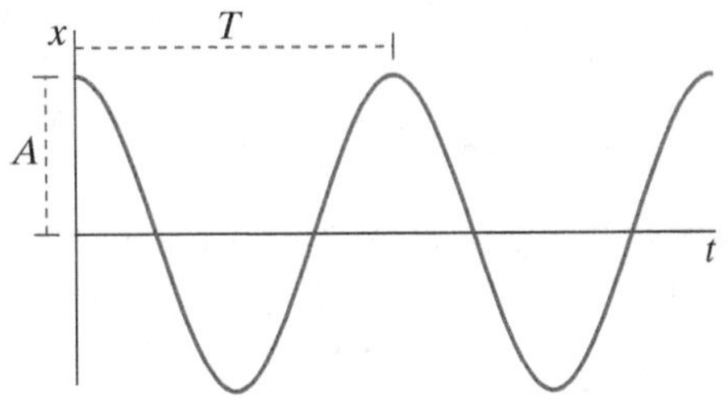

Figure 4.38 Amplitude A and period T, shown on a plot of x vs. t for the mass.

ANGULAR FREQUENCY, PERIOD, AND FREQUENCY

$$\text{Angular frequency } \omega = \frac{2\pi}{T} \quad \text{(radians per second)}$$

$$\text{Period } T = \frac{2\pi}{\omega} \quad \text{(seconds)}$$

$$\text{Frequency } f = \frac{1}{T} = \frac{\omega}{2\pi} \quad \text{cycles per second or hertz}$$

QUESTION According to the analytical solution for the motion of an ideal mass–spring system, how does the amplitude of the motion change over time?

In the solution the amplitude A is a constant, and does not decrease over time. This makes sense because in an idealized system there is no friction. This idealized system will continue to oscillate forever with a constant amplitude.

QUESTION According to the analytical solution, if you decreased the amplitude of the oscillations (for example, by stretching the spring less initially), how would the period of the motion change?

From the analytical solution, we see that the angular frequency (and therefore the period) does not depend on the amplitude at all! It depends only on the spring stiffness and the mass of the block. Such an idealized spring–mass system is called a "harmonic oscillator," and its oscillatory motion is called "harmonic motion." The special features of a harmonic oscillator are:

HARMONIC OSCILLATOR

- The position of the mass as a function of time is given by a cosine function.
- The period (round-trip time) is independent of the amplitude of the oscillations.

The fact that the period of a harmonic oscillator does not depend on the amplitude is surprising. With a larger amplitude, the mass travels farther in one period. However, it also reaches higher speeds, and for an ideal mass–spring system, these two effects exactly cancel.

Not all oscillating systems are harmonic. If you drop a rubber ball and let it bounce, as the bounce amplitude decreases, the period decreases too: the time between bounces gets shorter and shorter as the height of the bounces gets lower and lower. The bouncing ball is an example of an "anharmonic oscillator." "Anharmonic" simply means "not harmonic."

QUESTION Does the analytical solution above for a horizontal oscillator apply to a mass–spring system hanging vertically?

Yes. Although we need to take the gravitational force into account for a vertical system, we can show that this analytical solution is also valid for a vertical mass–spring system, in which the gravitational force on the mass acts parallel or antiparallel to the force on the mass by the spring. The details of this analysis are given in Section 4.16.

EXAMPLE **Period Depends on Mass and Stiffness**

Suppose that the period of a particular ideal mass–spring system is 2 s. **(a)** What would be the period of the system if the mass were tripled? **(b)** What would be the period of the system with the original mass, but with a spring that was four times as stiff?

Solution

(a) Triple the mass:

$$T_{\text{original}} = \frac{2\pi}{\sqrt{\dfrac{k_s}{m}}} = 2\pi\sqrt{\frac{m}{k_s}}$$

$$T_{\text{new}} = 2\pi\sqrt{\frac{3m}{k_s}} = \sqrt{3}\,T_{\text{original}}$$

$$= (1.73)(2\text{ s}) = 3.46\text{ s}$$

(b) Quadruple the spring stiffness:

$$T_{\text{new}} = 2\pi\sqrt{\frac{m}{4k_s}} = \frac{1}{\sqrt{4}}T_{\text{original}}$$

$$= 0.5(2\text{ s}) = 1\text{ s}$$

EXAMPLE **Mass–Spring Oscillation (Analytical Solution)**

A 50 g mass is attached to a low-mass horizontal spring whose relaxed length is 25 cm and whose stiffness is 4 N/m. The block slides on a very slippery, low-friction surface. You stretch the spring so it is 30 cm long, and release the block. **(a)** What is the amplitude of the oscillations? **(b)** How long will it take the block to make one full oscillation? **(c)** What would be the amplitude of the oscillations if the initial stretched length of the spring were 35 cm instead of 30 cm? **(d)** How long would it take the block to make one full oscillation if the initial stretched length of the spring were 35 cm instead of 30 cm? **(e)** How many oscillations will the block make in 10 s? **(f)** The initial length of the stretched spring is 35 cm. What will be the location of the block 3.15 s after you release it?

Solution System: Mass
Surroundings: Spring, Earth, table

QUESTION Are the assumptions for an ideal mass–spring system satisfied?

Yes. The mass of the spring, friction, and air resistance are all negligible, so we can use the analytical solution $x(t) = A\cos(\omega t)$.

(a) $A = s_i = 0.3\,\text{m} - 0.25\,\text{m} = 0.05\,\text{m}$
(b) Time for one round-trip $= T$ (period)

$$\omega = \sqrt{\frac{k_s}{m}} = \sqrt{\frac{4\,\text{N/m}}{0.05\,\text{kg}}} = 8.94\,\text{s}^{-1}$$

$$T = \frac{2\pi}{\omega} = \frac{2\pi}{8.94\,\text{s}^{-1}} = 0.70\,\text{s}$$

(c) $A = s_i = 0.35\,\text{m} - 0.25\,\text{m} = 0.10\,\text{m}$
(d) $T = 0.70\,\text{s}$
(e)

$$f = \frac{1}{T} = \frac{1}{0.70\,\text{s}} = 1.43\ \text{cycles/s}$$

$$\text{number of cycles} = ft = (1.43\,\text{cycles/s})(10\,\text{s}) = 14.3$$

(f)

$$x(t) = A\cos(\omega t)$$

$$x(3.15\,\text{s}) = (0.10\,\text{m})\cos\left((8.94\,\text{s}^{-1})(3.15\,\text{s})\right) = -0.0994\,\text{m}$$

Checkpoint 8 You have a rubber band whose relaxed length is 8.5 cm. You hang a coffee cup, whose mass is 330 g, from the rubber band, which stretches to a length of 14 cm. Consider the rubber band to be a single spring. **(a)** What is the stiffness of the rubber band? **(b)** If you start this mass–spring system oscillating, how long does one round-trip take?

4.12 ANALYTICAL VS. ITERATIVE SOLUTIONS

In Chapter 2 we modeled the motion of a block on a vertical spring by iteratively applying the Momentum Principle and updating position. In the previous section of the current chapter, by solving a differential equation involving the derivative form of the Momentum Principle, we obtained an analytical solution for the position of an ideal 1D mass–spring system.

QUESTION What are the advantages of the analytical solution?

Our analytical solution for the position of an ideal harmonic oscillator as a function of time allows us to see important general features of the behavior of the ideal system:

- The period does not depend on the amplitude.
- Increasing the mass increases the period.
- Increasing the spring stiffness decreases the period.

Although we could have learned these things by experimenting with different values of mass, spring stiffness, or initial stretch in an iterative solution, it is particularly easy to see this relationship from examining the analytical solution, as shown in the example entitled "Period Depends on Mass and Stiffness."

QUESTION What are the advantages of the iterative (computational) solution?

Although we were able to obtain an analytical solution for an ideal 1D system, it is not straightforward to extend this solution to the general case of a 3D mass–spring system that may be moving in a variety of ways. Recall from Chapter 2 that if $\vec{r}_0$ is the location of the fixed end of the spring, then we can calculate the 3D spring force in VPython this way:

$\vec{L} = \vec{r} - \vec{r}_0$	`L = block.pos - spring.pos`
$\hat{L} = \vec{L}/\lvert\vec{L}\rvert$	`Lhat = L/mag(L)`
$s = \lvert\vec{L}\rvert - L0$	`s = mag(L) - L0`
$\vec{F} = -k_s \cdot s \cdot \hat{L}$	`F_spring = -ks * s * Lhat`

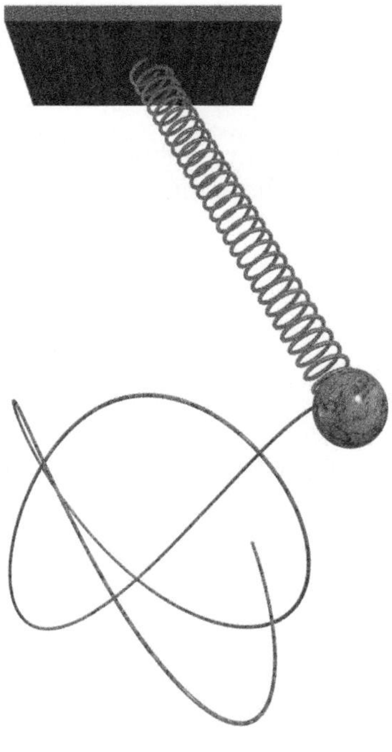

Figure 4.39 A snapshot from a computational model of a mass–spring system oscillating in 3D.

This force calculation forms the core of a 3D computational model of the system that is easy to refine and extend, in order to model more realistic systems. For example:

- Since we use the full 3D vector force in our computational model, by varying the initial conditions we can get 3D motion (Figure 4.39).
- We can easily add the effects of sliding friction.
- We can easily add the effects of air resistance.
- We can easily modify the spring force to model nonideal springs (for example, a spring whose force depended on $k_s s^3$ instead of $k_s s$).
- We can analyze the behavior of a spring whose mass is not negligible, such as a Slinky, by modeling it as a chain of tiny masses connected by ideal springs (Figure 4.40).

QUESTION Is it always possible to find an analytical solution?

Figure 4.40 A massive spring may be modeled as a chain of point masses connected by massless springs.

For many systems the differential equation corresponding to the derivative form of the Momentum Principle has a mathematical form that cannot be solved analytically and must be solved numerically. The power of iterative application of the Momentum Principle is that it can be used even in those situations where no analytical solution seems possible.

For example, $x = A\cos(\omega t)$ *cannot* solve the following "nonlinear" differential equation for any value of ω:

$$\frac{dp_x}{dt} = m\frac{d^2x}{dt^2} = -k_s x^3$$

There may or may not exist an analytical solution of this differential equation in terms of well-known mathematical functions, but *any* differential equation can be solved numerically, just as we did for the spring–mass equation. Given x and p at some instant in time, we can calculate the force ($-k_s x^3$) and determine the new values of x and p at a time Δt later, then repeat as often as necessary to reach the time of interest.

Note that either an analytical or an iterative prediction of the future applies only to our *model* of the actual real-world situation. If our model is a good approximation to the real world, our prediction will be a good approximation to what will actually happen, but even an analytical prediction cannot be exact because of the practical impossibility of knowing exactly the initial conditions and the net force due to all the other objects in the Universe.

Of course, for both types of solution—iterative or analytical—we need to know the initial conditions (position and momentum of the mass) in order to predict the position of the system at some future time. In the previous example, the mass was initially at rest, and the spring was stretched. It is also possible to start an oscillator by placing it at its equilibrium position and giving it a kick—in this case its initial stretch is zero, but its initial momentum is nonzero. In an iterative solution, this is easy to take into account. A discussion at the end of this chapter explains how this is taken into account in the analytical solution.

4.13 ANALYTICAL EXPRESSION FOR SPEED OF SOUND

Under ordinary conditions, the speed of sound in air is about 340 m/s. If your friend stands 340 m away and yells, you will hear the sound a second later. The speed of sound in a solid object is much faster than it is in air. In aluminum, the speed of sound is about 4800 m/s. In lead, which is a much softer material than aluminum, the speed of sound is only about 1200 m/s. There are two significant differences between lead and aluminum at the atomic level. First, the mass of a lead atom is about eight times larger than the mass of an aluminum atom. Second, the interatomic bond stiffness in aluminum is greater than the interatomic bond stiffness in lead (see Problem P36). Since the propagation of a disturbance through a metal rod involves displacements of atoms, it seems reasonable that both of these factors should play a role in determining the speed with which the disturbance propagates. Our goal in the following discussion is to use the ball–spring model of a solid to predict this relationship.

Qualitative Guess

Think of a solid metal rod as a network of atoms (balls) connected by springs (interatomic bonds) as in Figure 4.41.

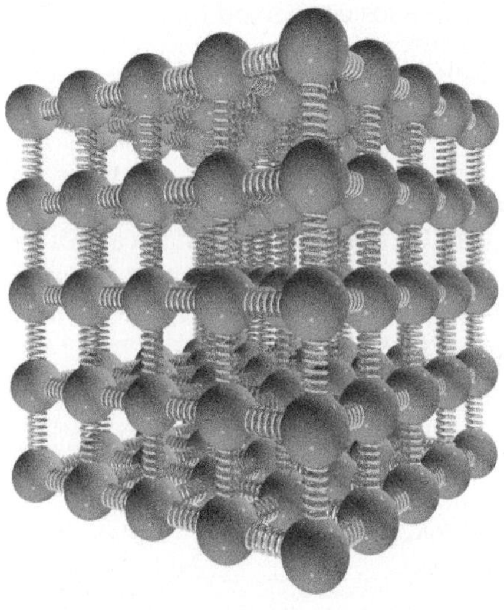

Figure 4.41 A model of a metal: atoms (balls) connected by springs (interatomic bonds).

> QUESTION How should the speed of sound depend on m_a, the mass of an atom in a material? (Think of hitting one end of a metal rod with a hammer.)

A given applied force produces the same change of momentum for a massive atom like lead or a lighter atom like aluminum. However, the change in speed of the lead atom will be less than the change in speed of the aluminum atom. This suggests that the greater the mass of the atoms, the slower a sound pulse (disturbance) might propagate through a rod.

> QUESTION How should the speed of sound depend on the stiffness of the interatomic bonds in a material?

The stiffer the bond, the greater the force the "spring" will exert on a neighboring atom in response to the same displacement. Stiffer interatomic bonds should make a sound pulse propagate faster through a rod.

These initial qualitative guesses don't tell us details of the mathematical relationships involved: for example, whether speed of sound is proportional to $1/m_a, 1/m_a^2, 1/\sqrt{m_a}$, or some other factor. We need to apply our understanding of mass–spring systems to refine these guesses.

Dimensional Analysis

We need to construct a mathematical expression that has the correct units, or "dimensions." Speed has units of m/s, so we need an equation involving interatomic bond stiffness $k_{s,i}$ and atomic mass m, ending up with meters in the numerator and seconds in the denominator:

$$v = \ldots$$

The angular frequency $\omega = \sqrt{k_{s,i}/m_a}$ and has units of s^{-1}, so ω might appear in the numerator of the expression:

$$v = \omega(?)$$

QUESTION Is it reasonable that v should be directly proportional to ω?

Yes. The larger the angular frequency ω is, the faster an atom oscillates back and forth, compressing and stretching intermolecular bonds, which affect neighboring atoms.

QUESTION What other quantity should go in the expression?

Since we need to end up with m/s, we need to multiply by something that has units of meters. ω is related to the rate at which one atom hits its neighbor, propagating a change through a distance of one interatomic spacing. Multiplying by the interatomic bond length d might be appropriate.

$$v = \omega d$$

This informal reasoning is an example of "dimensional analysis," in which we identify meaningful quantities in the situation and combine them to guess an expression for some other quantity of interest. While this can be a fruitful procedure, we shouldn't be surprised if we find that the speed of sound is actually given by $v = 2\omega d$, or $v = \omega d/\pi$, for example. Moreover, note that atomic oscillations in the x direction involve two atomic "springs" on each side of an atom, not one, so we may have omitted a factor of 2.

Nevertheless, we can guess that $\left(\sqrt{k_{s,i}/m_a}\right) d$ is proportional to the actual speed of sound, with the correct dependence on $k_{s,i}$ and m_a. In a section on the "wave equation" in Supplement S3 there is a derivation showing that this is in fact the correct result.

SPEED OF SOUND IN A SOLID

$$v = \omega d = \sqrt{\frac{k_{s,i}}{m_a}} d$$

where v is the speed of sound (m/s), $k_{s,i}$ is the stiffness of the interatomic bond, m_a is the mass of one atom, and d is the length of the interatomic bond.

EXAMPLE **Speed of Sound in Copper**

We found (Section 4.5) that the stiffness of the interatomic bond in solid copper is about 26.8 N/m, and that the interatomic bond length in solid copper is about 2.28×10^{-10} m (Section 4.4). There are 64 g of copper in one mole. Use these data to calculate the speed of sound in solid copper.

Solution

$$m_a = \frac{0.064\,\text{kg}}{6.02 \times 10^{23}\,\text{atoms}} = 1.06 \times 10^{-25}\,\text{kg/atom}$$

$$v = \sqrt{\frac{26.8\,\text{N/m}}{1.06 \times 10^{-25}\,\text{kg}}}(2.28 \times 10^{-10}\,\text{m}) = 3620\,\text{m/s}$$

This agrees well with experimentally measured values of around 3800 m/s.

Speed of Sound in Different Materials

In aluminum, which is very stiff (large k_s) and has a small atomic mass m_a, the speed of sound is very high, about 5000 m/s. If a wire is half a meter long, it takes about

$$\frac{0.5\,\text{m}}{5000\,\text{m/s}} = 1 \times 10^{-4}\,\text{s} \quad \text{(a tenth of a millisecond)}$$

from the time you pull on one end for the other end to become tense. On the other hand, lead is quite soft (small k_s) and has a large atomic mass m_a, and the speed of sound in lead is only about 1200 m/s, much lower than in aluminum. (The speed of sound in solids is much higher than the speed of sound in air, which is about 340 m/s. Because air molecules are usually far from one another, the speed of sound in air is determined by the average speed of the air molecules, not by interatomic forces.)

More on Waves

Additional material on waves is provided in Supplement S3, from the point of view of the "wave equation," a partial differential equation that describes many kinds of wave motion.

Checkpoint 9 A certain metal with atomic mass 2×10^{-25} kg has an interatomic bond with length 2.1×10^{-10} m and stiffness 40 N/m. What is the speed of sound in a rod made of this metal?

4.14 CONTACT FORCES DUE TO GASES

Buoyancy

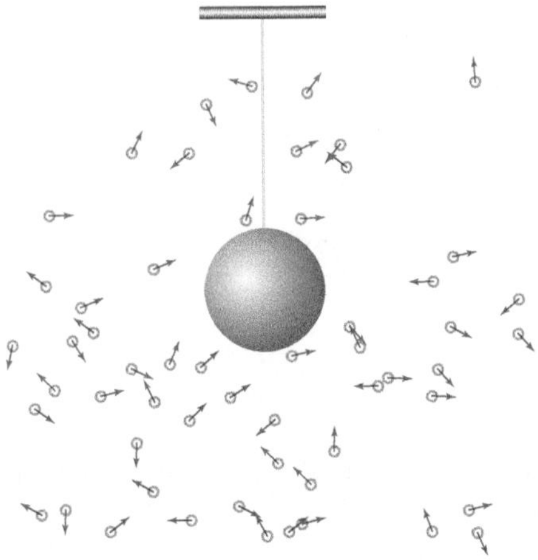

Figure 4.42 Air molecules in motion near a hanging ball. The density of the air beneath the ball is slightly greater than the air density above the ball, so there is a small upward force on the ball exerted by the air.

Until now we have considered only contact forces due to solids. However, gases can also exert contact forces. In everyday circumstances, the most significant of these is the force that the air exerts on objects—even stationary objects. Consider the surrounding air that is in contact with a ball hanging from a wire, as shown in Figure 4.42. You might think that the air has no effect on the ball, especially since the ball is not moving, so we don't need to worry about air resistance. To be certain, let's consider this issue at the microscopic level.

Air is a gas, composed of about 80% nitrogen and 20% oxygen, with small amounts of other gases, especially argon (about 1%) and water vapor (which varies with the humidity). The molecules in the air are continuously in motion, moving in random directions. Molecules in the air near the ball will frequently bump into the ball, exerting a small force on it. If there is no wind, the motion of the gas molecules is random, so in a short time period there should be about as many molecules colliding with the right-hand side of the ball as with the left-hand side, and there should be on average no sideways force on the ball.

However, the situation is not quite the same for collisions with the top and bottom of the ball. The directions of molecular velocities are still random, but there are more air molecules per cubic meter below the ball than there are above it. The density of the Earth's atmosphere decreases as one moves upward away from the Earth's surface, until finally at a sufficient distance (approximately 50 km) there is essentially no air left. You may have observed this variation yourself if you have ever gone from sea level to a location several thousand meters higher; you may feel light-headed if you exercise at high altitude because the density of oxygen is significantly lower.

This variation in air density occurs over kilometers; can it possibly be important over a distance of a few centimeters? Interestingly enough, the surprising answer is yes! There is actually a "buoyant force" upward on the ball, because the number of air molecules hitting the bottom of the ball per second is slightly greater than the number of molecules hitting the top of the ball per second.

A Macroscopic View of Buoyancy

How can we determine the magnitude of the buoyant force on the ball due to the air? Let's consider a macroscopic viewpoint. The following discussion, based on the Momentum Principle, applies equally well to the buoyant force exerted on an object by any fluid (air, water, other liquids, other gases). The result we will obtain is called the Archimedes principle, for the Greek thinker who first understood it.

Imagine a box filled with air (or water). Consider a ball-shaped region that a ball will eventually occupy, at a moment when the ball is not yet there, so the ball-shaped region is filled with the fluid. Let's choose this ball-shaped mass of fluid as our system, and consider what forces must be acting on it (Figure 4.43). This may seem odd, but it is a perfectly legitimate choice of system.

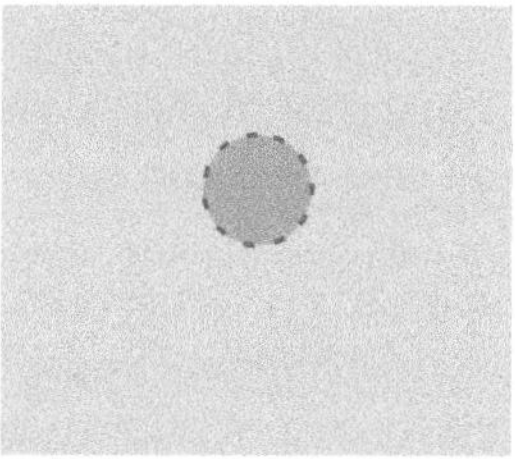

Figure 4.43 A system consisting of a ball-shaped spherical region of fluid, surrounded by more fluid.

QUESTION Is the momentum of the indicated sphere of fluid changing? What must be the net force on this sphere?

The momentum of the indicated sphere of fluid is not changing. The Earth exerts a downward gravitational force on it, but the sphere does not sink. Because its momentum isn't changing, the net force on the sphere must be zero, so there must be an upward force that balances the gravitational force.

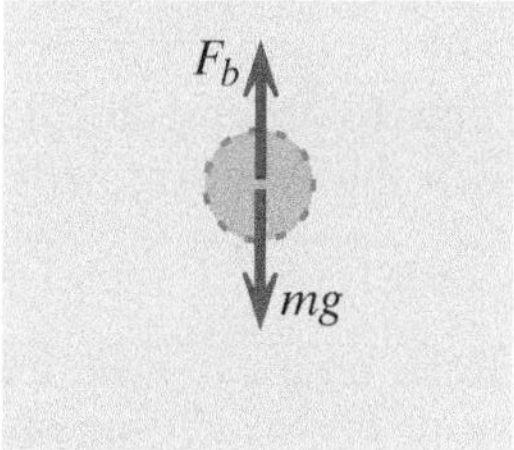

Figure 4.44 Forces on the sphere of fluid. Here m is the mass of the fluid (air or water) in the sphere.

QUESTION What exerts this upward force on the sphere?

The only object in contact with this sphere is the rest of the fluid, so it must be that the rest of the fluid exerts an upward force F_b ("buoyant force") whose magnitude is mg, where m is the mass of the sphere of fluid (Figure 4.44).

Now remove the sphere of fluid and replace it with the ball. It must be that initially the rest of the fluid still exerts the same upward force F_b of magnitude mg (Figure 4.45). This force may not be large enough to counteract the downward gravitational force on the ball; in the case of a ball hanging in air it clearly is not, since if the ball were not supported by a wire it would begin accelerating downward.

The upward force is called a "buoyant" force but is simply an interatomic contact force due to fluid molecules striking atoms in the surface of the ball. The key point is that despite the complexity of the interatomic interactions, the net effect is simply an upward force whose magnitude is the mass of an equivalent volume of fluid, times g.

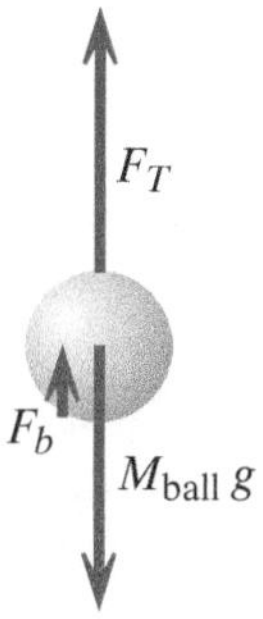

Figure 4.45 A ball hanging from a wire, surrounded by air. The tension force is actually less than the weight of the ball, because the buoyant force F_b due to the air is also in the upward direction.

When Can the Buoyant Force Be Neglected?

Often it is reasonable to neglect buoyant forces in air (though usually not in water). To see why, let's compare the buoyant force to the gravitational force F_g on a ball hanging in air, where we write V for the volume of the ball:

$$\frac{F_b}{F_g} = \frac{m_{\text{air}}\, g}{M_{\text{ball}}\, g} = \frac{m_{\text{air}}}{M_{\text{ball}}} = \frac{m_{\text{air}}/V}{M_{\text{ball}}/V} = \frac{\rho_{\text{air}}}{\rho_{\text{ball}}}$$

The ratio of the forces is equal to the ratio of ρ_{air} (the density of air) to ρ_{ball} (the density of the ball; ρ is the lowercase Greek letter rho). The density of air at STP (standard temperature and pressure, 0°C and 1 atmosphere) can be calculated if we remember from chemistry that one mole of a gas at STP occupies 22.4 liters. The molecular mass of N_2 is 28 and that of O_2 is 32, which gives an average molecular mass of about 29 for air (which is about 80% nitrogen and 20% oxygen):

$$\rho_{\text{air}} \approx \frac{29\ \text{g/mol}}{22.4 \times 10^3\ \text{cm}^3/\text{mol}} \approx 1.3 \times 10^{-3}\ \text{g/cm}^3$$

What is the density of the ball? We can estimate it by noting that the density of water is 1 g/cm^3, and that most solids are more dense than water. Using 1 g/cm^3 as the density of the ball, we find that $F_b/F_g \approx 1.3 \times 10^{-3}/1$.

The buoyant force $m_{\text{air}}g$ of air on a solid object is therefore very small compared to the gravitational force mg. In many cases we can safely choose to neglect it. Since the upward force exerted by the air on the ball is small compared to the gravitational force or the tension force exerted by the wire, we can choose to neglect it in an analysis of the forces on a hanging ball. However, it is not always appropriate to neglect the buoyant force; in accurate analytical chemistry it is necessary to correct for buoyancy when weighing a sample of liquid or low-density solid on an analytical balance.

A final note: The buoyant force being equal to $m_{\text{air}}g$ is really a time average, because the buoyant force is due to random collisions with air molecules and therefore has an intermittent character. However, the collision rate is so high that the force seems nearly continuous and constant. If, however, the ball is of microscopic size, it may not be possible to ignore the intermittent nature of the collisions. In a microscope it is possible to observe small particles being jostled about by the random collisions with the water molecules. This effect is called "Brownian motion."

Checkpoint 10 Calculate the buoyant force in air on a kilogram of iron (whose density is about 8 g/cm^3). Compare with the weight mg of this much iron.

Pressure

QUESTION Why is it difficult to remove a suction cup from a surface?

Figure 4.46 A one-square-meter column of air extending upward through the entire atmosphere.

There is an interesting case in which the effect of the air is very significant and cannot be neglected. Suppose that an object lies on a table and the surfaces of the table and object are so smooth that upon squeezing them together all air is expelled. Such a situation occurs when a suction cup is pressed down onto a smooth surface. Now there are no longer any air molecules striking the bottom surface of the suction cup, while air molecules continue to strike the top surface. The force of the air on the top surface is quite large, about 1×10^5 newtons on each square meter at sea level!

This large force per unit area is called the "pressure" P of the air. The pressure can be thought of as due to the weight of a one-square-meter column of the entire atmosphere, extending upward many kilometers (although with diminishing density), as illustrated in Figure 4.46.

DEFINITION OF PRESSURE

Pressure $P = F/A$ (force per unit area)

Atmospheric pressure at sea level is about $1 \times 10^5\ \text{N/m}^2$. In older units this is approximately 15 lb/in^2.

A Constant-Density Model of the Atmosphere

If the atmosphere had a constant density ρ like that near the surface (1.3 g/cm^3), we could calculate the height h of the atmosphere in the following way. Choose the column of air as the system. The volume of a column of air of height h and cross-sectional area A is Ah, and its mass is $M_{air} = \rho(Ah)$. Because the center of mass of the column is at rest, the Momentum Principle tells us that the upward force of the ground on the bottom of the air column must equal the downward gravitational force on the total mass of air in the column. By the reciprocity of interatomic electric forces (air molecules hitting the ground), the force of the air on the ground must also be equal to the weight of the air in the column. Therefore the atmospheric pressure on the ground is the weight $M_{air}g$ divided by the cross-sectional area: $P = M_{air}g/A = \rho gh$.

The height of a constant-density atmosphere, $h = P/(\rho g)$, would be about 8000 m (about 5 mi), about the height of Mount Everest. In reality air density is not constant but decreases as you go higher, so there is still some low-density air at much higher altitudes than this.

This macroscopic picture, taking the whole column of air as the system, is a useful one, but remember that the actual force exerted on a square meter at sea level is due to air molecules hitting an area, and they strike surfaces at all angles. Every square meter of the walls and ceiling of a room at sea level, as well as the floor, is subjected to the same rate of molecular bombardments. The momentum change per second $\Delta\vec{p}/\Delta t$ of all the molecules colliding with one square meter is equal in magnitude to the force of 1×10^5 N.

Checkpoint 11 Do the calculation and verify that the height of a constant-density atmosphere would be about 8000 m (8 km).

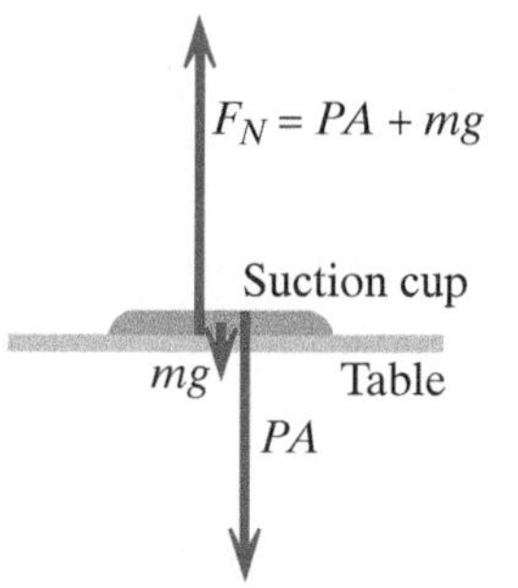

Figure 4.47 The weight of the suction cup is very small compared to the downward force by the air or the normal force by the table.

Suction

A free-body diagram for a suction cup looks like Figure 4.47, where PA represents the force of the air on the area A of the top surface of the suction cup. The normal force exerted upward by the table and the downward force of the air are huge compared to the weight of the suction cup mg.

The normal force $F_N = PA + mg \approx PA$, since the force of the air is so much larger than the weight mg of the suction cup. For example, a rubber suction cup with a diameter of 4 cm has an area of $\pi(0.02\,\text{m})^2 = 1.3 \times 10^{-3}\,\text{m}^2$, so the force of the air on the suction cup is

$$PA = (1 \times 10^5\,\text{N/m}^2)(1.3 \times 10^{-3}\,\text{m}^2) = 130\,\text{N}$$

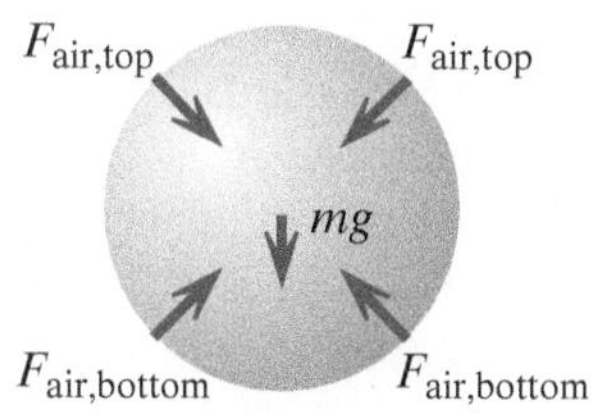

Figure 4.48 A ball hanging in air is subject to large compression forces due to the surrounding air.

whereas its own weight mg might be about 0.01 kg (10 g) times 9.8 N/kg, or only about 0.1 N. Note that a force of 130 N is the weight of an object whose mass is about 13 kg, or about 29 lb, which is why it isn't easy to pull even a small suction cup off a flat surface.

Every solid object in air, such as a hanging ball, is subjected to large compression forces by the pressure of the surrounding air (Figure 4.48). The upward buoyant force is the tiny difference between these huge forces. A hollow object must be very strong to support this compression, if air is removed from the inside. There is no problem, however, if air can flow into and out of the container, because then it is just the solid walls that are compressed, since the air exerts comparable pressure on the inner and outer surfaces.

All of this is even more relevant under water, because the density of water is about 1000 times greater than the density of air, and the compression forces are about 1000 times as strong. A hollow submersible vehicle that dives very deep must have an extremely strong hull. When a scuba diver descends to greater depth where the water pressure is larger, the breathing apparatus automatically increases the pressure of the air supplied to the diver to equal the increased pressure of the outside water at the new depth, to prevent the lungs from being crushed.

4.15 *ACCELERATION

In previous studies you may have seen a simplified, approximate form of the Momentum Principle, written $F = ma$. This equation is valid under the following conditions:

- Only one force acts on the system.
- Motion occurs only in one dimension.
- The mass of the system is constant.
- The speed of the system is much less than the speed of light.

This equation can be derived from the derivative form of the Momentum Principle, using the low speed approximation $\vec{p} \approx m\vec{v}$. If we take the approximate derivative of $\vec{p}$ with respect to time, we get:

$$\frac{d\vec{p}}{dt} \approx \frac{d(m\vec{v})}{dt} \quad \text{(assuming } v \ll c\text{)}$$

$$\frac{d(m\vec{v})}{dt} = \left(\frac{dm}{dt}\right)\vec{v} + m\frac{d\vec{v}}{dt} \quad \text{(product rule for derivatives)}$$

$$\frac{d(m\vec{v})}{dt} = 0 + m\frac{d\vec{v}}{dt} \quad \text{(assuming system mass is constant)}$$

As noted in Chapter 1, the rate of change of velocity with respect to time is called acceleration, and is usually denoted by the symbol $\vec{a}$. Given this, we can write the approximate rate of change of momentum as

$$\frac{d\vec{p}}{dt} \approx m\vec{a}$$

and an approximate form of the Momentum Principle, usually referred to as Newton's second law, as

$$m\vec{a} \approx \vec{F}_{\text{net}} \quad \text{(nonrelativistic form; constant mass)}$$

A highly simplified scalar form of this equation, $F = ma$, involves a single force and one-dimensional motion. Although $F = ma$ is casually referred to as Newton's second law, Newton himself worked with the Momentum Principle in the forms emphasized in this textbook, $d\vec{p}/dt = \vec{F}_{\text{net}}$ and $\Delta\vec{p} = \vec{F}_{\text{net}}\Delta t$.

The definition of acceleration is useful, because it is occasionally important to know the acceleration of an object (the rate at which its velocity is changing). However, we will generally find that the full form of the Momentum Principle, $d\vec{p}/dt = \vec{F}_{\text{net}}$, is more appropriate for our analyses.

- $d\vec{p}/dt = \vec{F}_{\text{net}}$ is relativistically correct (correct at any speed), and we will sometimes deal with fast-moving particles.
- $d\vec{p}/dt = \vec{F}_{\text{net}}$ involves momentum, which is a conserved quantity; we need to work with momentum when we analyze situations such as collisions, as we did in Chapter 3.
- $d\vec{p}/dt = \vec{F}_{\text{net}}$ is correct even if the mass of an object is not constant, such as a rocket including its exhaust gases ejecting out the back.
- $d\vec{p}/dt = \vec{F}_{\text{net}}$ is a vector principle, and contains information about directions. The arrows over the symbols are extremely important; they remind us that there are really three separate component equations, for x, y, and z.
- $d\vec{p}/dt = \vec{F}_{\text{net}}$ reminds us that we have to add up all vector forces to give the net force, so we write $\vec{F}_{\text{net}}$, not just F.

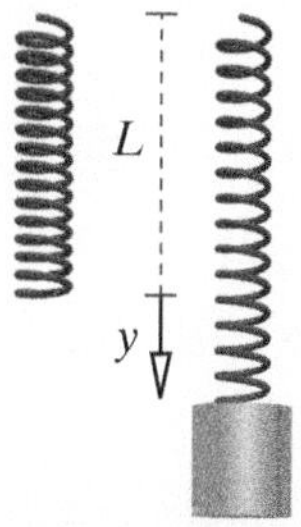

Figure 4.49 A vertical spring–mass system minimizes sliding friction.

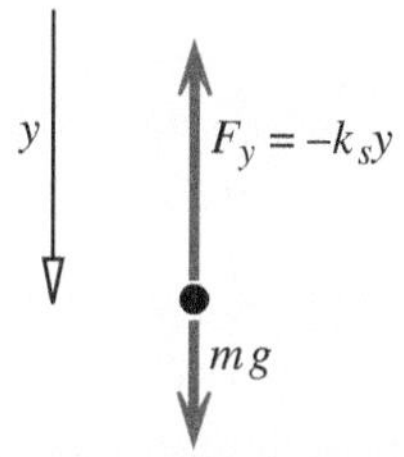

Figure 4.50 Free-body diagram for the hanging block.

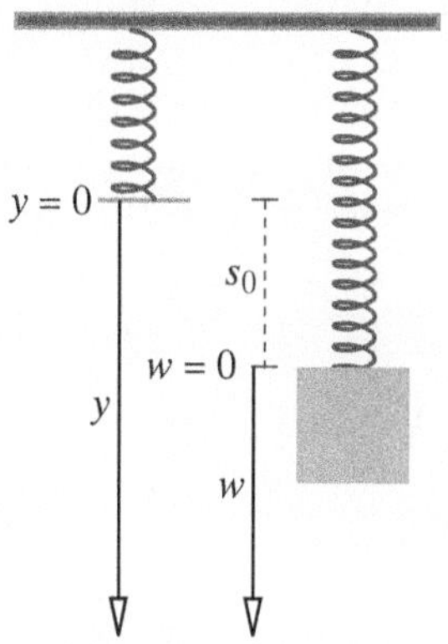

Figure 4.51 Measure w from the place where the block can hang motionless.

4.16 *A VERTICAL SPRING–MASS SYSTEM

To reduce friction, it is easier to experiment with a mass oscillating vertically up and down on a spring (Figure 4.49) rather than sliding on an air table. We will show that we expect the same period whether the spring is vertical or horizontal.

We choose a coordinate system with the positive y axis pointing downward (Figure 4.49). We'll measure y downward from the location at which the end of the relaxed spring would be (the relaxed length of the spring is L).

Neglecting air resistance and gravitational interactions with objects other than the Earth, the free-body diagram of the forces acting on the hanging block is shown in Figure 4.50.

We then have the following:

$$\frac{dp_y}{dt} = -k_s y + mg$$

We choose to analyze only simple vertical motion, with no bouncing side to side, so we know this:

$$p_x = 0 \text{ at all times, so } \frac{dp_x}{dt} = 0$$

We can reduce the p_y equation to a familiar form if we measure a new coordinate, w, from the equilibrium position, when the block is hanging motionless, rather than from the unstretched spring position (Figure 4.51). When the mass hangs motionless from the spring, the spring is stretched an amount $k_s s_0 = mg$, so $s_0 = mg/k_s$. We'll measure w downward from the end of the spring, so

$$y = (s_0 + w) = \frac{mg}{k_s} + w$$

and therefore

$$\frac{dp_y}{dt} = -k_s\left(\frac{mg}{k_s} + w\right) + mg = -k_s w$$

This says that if we measure from the end of the spring at its equilibrium location, mg/k_s below the bottom of the unstretched spring, the equation for the motion of the mass is the same as in a horizontal spring–mass system, whose motion we already know. This means that the analytical solution we obtained for the horizontal case is also the solution for the vertical case (substituting y for x in the appropriate places).

The moving mass will oscillate up and down around this equilibrium position, as you can observe if you experiment with a vertical mass–spring system.

4.17 *GENERAL SOLUTION FOR THE MASS–SPRING SYSTEM

It is easy to show that the function $x = A\cos(\omega t + \phi)$ is a general solution for a spring–mass system with arbitrary k_s, m, and initial conditions, where $\omega = \sqrt{k_s/m}$, and where the amplitude A and the phase shift ϕ (measured in radians and denoted by the Greek letter phi) are constants determined by the initial conditions, as is explained below. The phase shift essentially shifts the starting point of the oscillation. For example, if $\phi = \pi/2$, the cosine function actually represents a negative sine function (Figure 4.52):

$$x = A\cos\left(\omega t + \frac{\pi}{2}\right) = -A\sin(\omega t)$$

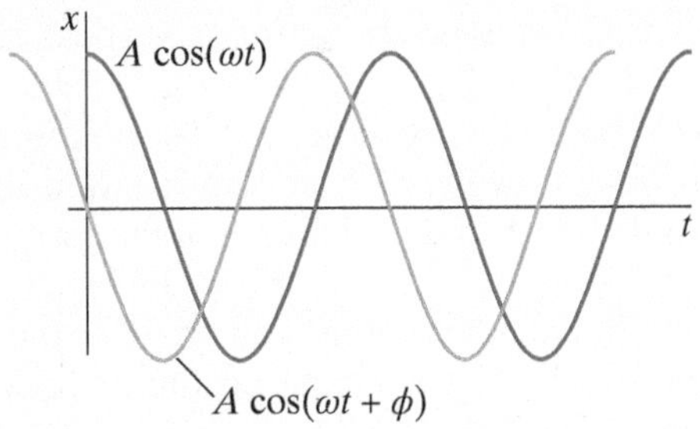

Figure 4.52 Phase shift ϕ.

Determining the Amplitude and Phase Shift

The amplitude A and phase shift ϕ can be determined from the initial values at time t_0 of position x_0 and velocity v_0:

$$x_0 = A\cos\left(\sqrt{\frac{k_s}{m}}t_0 + \phi\right)$$

$$v_0 = -\sqrt{\frac{k_s}{m}}A\sin\left(\sqrt{\frac{k_s}{m}}t_0 + \phi\right)$$

Using the important trigonometric identity $\sin^2\theta + \cos^2\theta = 1$ (which is the Pythagorean theorem applied to a triangle whose hypotenuse has length 1), we can obtain an equation that can be solved for the amplitude A:

$$x_0^2 + \frac{m}{k_s}v_0^2 = A^2, \text{ so we have } A = \sqrt{x_0^2 + \frac{m}{k_s}v_0^2}$$

For example, if the initial velocity is zero, the amplitude is simply equal to the initial stretch.

The value for A can be plugged back into either the x_0 equation or the v_0 equation to obtain the phase shift ϕ. Alternatively, by dividing the x_0 equation by the v_0 equation we obtain an equation that can be solved for the phase shift ϕ:

$$\frac{v_0}{x_0} = -\sqrt{\frac{k_s}{m}}\tan\left(\sqrt{\frac{k_s}{m}}t_0 + \phi\right)$$

$$\sqrt{\frac{k_s}{m}}t_0 + \phi = \arctan\left(-\sqrt{\frac{m}{k_s}}\frac{v_0}{x_0}\right)$$

$$\phi = \arctan\left(-\sqrt{\frac{m}{k_s}}\frac{v_0}{x_0}\right) - \sqrt{\frac{k_s}{m}}t_0$$

Checkpoint 12 Suppose that the system is not oscillating and you strike it with a hammer, applying a large force F for a *very* short time Δt. What is the initial speed v_0? Explain briefly. What will be the amplitude (the maximum displacement away from the equilibrium position)? What is the phase shift ϕ? (Does this make sense?)

Comment: Two Constants for a Second-Order Differential Equation

Note that whether we specify the starting conditions in terms of x_0 and v_0 or in terms of the amplitude A and the phase shift ϕ, two constants are required. You can see why two constants are needed by looking at the start of a numerical integration. You need an initial position x_0 in order to be able to calculate the initial force and impulse, so that you can use the Momentum Principle to update the initial momentum p_0, from which you get the new velocity that lets you update the position; then repeat.

We have been working with what is called a "second-order differential equation." That is, the highest derivative in the equation is a second derivative—the acceleration $dv/dt = d(dx/dt)/dt$. A second-order differential equation normally requires two constants to specify the starting conditions completely.

Linear Systems

We saw that doubling the amplitude of a spring–mass system simply multiplies the solution by a factor of two without changing the time dependence of the motion. This is an extremely important property of systems described by "linear" differential equations (equations in which the variables appear only to the first power, not squared, for example). More formally, in the Momentum Principle for a spring–mass system, let x be replaced by the quantity Rx, where R is a constant:

$$m\frac{d}{dt}\left(\frac{d(Rx)}{dt}\right) = -k_s(Rx)$$

$$Rm\frac{d}{dt}\left(\frac{d(x)}{dt}\right) = -Rk_s(x)$$

The R's cancel, leaving us with the original equation. Therefore starting with a different initial stretch will simply scale up the vertical axis of the graph of x vs. t. Also, since the motion is periodic, a change in the initial velocity corresponds to starting the motion at a different time. For example, instead of starting with an initial stretch and zero velocity, you could start with zero stretch and some initial velocity, and this would correspond to a point on the original graph where the function crosses the time axis.

In contrast, consider a differential equation in which there is an x^3, and try scaling by R:

$$m\frac{d}{dt}\left(\frac{d(Rx)}{dt}\right) = -b(Rx)^3$$

$$Rm\frac{d}{dt}\left(\frac{d(x)}{dt}\right) = -R^3bx^3$$

In this case the R's do *not* cancel. The solutions to this "nonlinear" equation do not scale up in the simple way that solutions to a linear equation do.

SUMMARY

Contact forces:

- A solid object may be modeled as a 3D lattice of balls (atoms) connected by springs (interatomic bonds).
- Based on this model, and given the density and atomic weight of an element, one can find the length of an interatomic bond in a solid metal.
- From macroscopic measurements of stress and strain, one can determine the stiffness of an interatomic bond in a solid material.
- Solid objects exert contact forces because the interatomic bonds in the solid are stretched (a tension force) or compressed (a normal force).
- Friction forces arise when one object slides on another.
- The speed of sound in a solid is the speed at which a disturbance propagates through the material.

Definitions:

For a long object like a rod, which is stretched or compressed:

$$\text{strain} = \frac{\Delta L}{L}$$

$$\text{stress} = \frac{F_T}{A}$$

L is the original length of the wire, ΔL is the change in length, F_T is the force applied to stretch or compress the object, and A is the cross-sectional area of the object.

$$\text{Young's modulus: } Y = \frac{\text{stress}}{\text{strain}} = \frac{\left(\frac{F_T}{A}\right)}{\left(\frac{\Delta L}{L}\right)}$$

Macro/micro connections:

Young's modulus in terms of microscopic quantities is

$$Y = \frac{k_{s,i}}{d}$$

Y is Young's modulus, $k_{s,i}$ is the stiffness of the interatomic bond, and d is the length of the interatomic bond (distance between adjacent nuclei).

$$\text{Speed of sound in a solid: } v = \sqrt{\frac{k_{s,i}}{m_a}}\, d$$

$k_{s,i}$ is the stiffness of the interatomic bond, d is the length of the interatomic bond (distance between adjacent nuclei), m_a is the mass of one atom.

Momentum Principle: derivative form

$$\frac{d\vec{p}}{dt} = \vec{F}_{\text{net}}$$

The instantaneous time rate of change of the momentum of an object is equal to the net force acting on the object.

A simplified version of the Momentum Principle is $\vec{F}_{\text{net}} = m\vec{a}$, valid for $v \ll c$ and constant mass.

Analytical solution for mass–spring oscillations

$$x = A\cos(\omega t), \text{ with } \omega = \sqrt{\frac{k_s}{m}}$$

assuming that the spring's mass is negligible, and there is no friction and no air resistance.

A (the "amplitude") is a constant equal to the maximum stretch of the spring during an oscillation. ω is the angular frequency, and its units are radians/s. k_s is the spring stiffness, and m is the mass of the object.

The period T is the time for one complete cycle, in seconds (Figure 4.53).

$$T = \frac{2\pi}{\omega}$$

The frequency f is the number of complete cycles per second.

$$f = \frac{1}{T} = \frac{\omega}{2\pi}$$

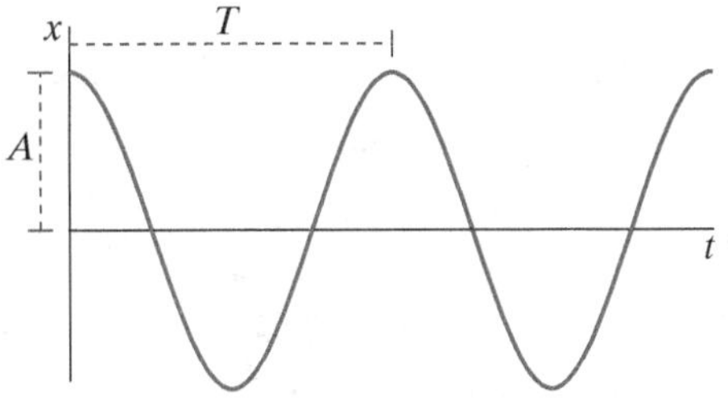

Figure 4.53

Contact forces due to gases

Upward buoyancy force = $m_{\text{fluid}}g$, where m_{fluid} is the mass of an equivalent volume of fluid.

Pressure is force per unit area; air pressure at sea level is about 1×10^5 N/m^2.

QUESTIONS

Q1 Why are two balls connected by a spring a good model for two atoms connected by a chemical bond? **(a)** If the two atoms get farther apart than the equilibrium interatomic distance, they attract each other. **(b)** Each spring represents a real microscopic coiled metal wire that connects two adjacent atoms. If the two atoms get closer together than the equilibrium interatomic distance, they repel each other. **(c)** The magnitude of the force one atom exerts on another is proportional to the stretch or compression of the bond between them.

Q2 Approximately what is the radius of a copper atom? Is it 1×10^{-15} m, 1×10^{-12} m, 1×10^{-10} m, 1×10^{-8} m, or 1×10^{-6} m?

Q3 **(a)** A climber whose mass is 55 kg hangs motionless from a rope. What is the tension in the rope? **(b)** Later, a different climber whose mass is 88 kg hangs from the same rope. Now what is the tension in the rope? **(c)** Compare the physical state of the rope when it supports the heavier climber to the state of the rope when it supports the lighter climber. Which statements about the physical state of the rope are true? Check all that apply. (1) Because the same rope is used, the tension in the rope must be the same in both cases. (2) The interatomic bonds in the rope are stretched more when the rope supports the heavier climber than when the rope supports the lighter climber. (3) The rope is slightly longer when it supports the heavier climber than when it supports the lighter climber.

Q4 You hang a 10 kg mass from a copper wire, and the wire stretches by 8 mm. **(a)** If you suspend the same mass from two copper wires, identical to the original wire, what happens? **(b)** If you suspend the same mass from a copper wire with half the cross-sectional area but the same length as the original wire, what happens? **(c)** If you suspend the same mass from a copper wire with the same cross-sectional area but twice the length of the original wire, what happens?

Q5 You hang a mass M from a spring, which stretches an amount s_1. Then you cut the spring in half, and hang a mass M from one half. How much does the half-spring stretch? **(a)** $2s_1$, **(b)** s_1, **(c)** $s_1/2$

Q6 A spring has stiffness k_s. You cut the spring in half. What is the stiffness of the half-spring? **(a)** $2k_s$, **(b)** k_s, **(c)** $k_s/2$

Q7 Lead is much softer than aluminum, and can be more easily deformed or pulled into a wire. What difference between the two materials best explains this? **(a)** Pb and Al atoms have different sizes. **(b)** Pb and Al atoms have different masses. **(c)** The stiffness of the interatomic bonds is different in Pb and Al.

Q8 Two wires are made of the same kind of metal. Wire A has a diameter of 2.4 mm and is initially 2.8 m long. You hang a 8 kg mass from wire A, measure the amount of stretch, and determine Young's modulus to be $Y_A = 1.37 \times 10^{12}\,\text{N/m}^2$.

Wire B, which is made of the same kind of metal as wire A, has the same length as wire A but twice the diameter. You hang the same 8 kg mass from wire B, measure the amount of stretch, and determine Young's modulus, Y_B.

Which one of the following is true? **1.** $Y_B = Y_A$, **2.** $Y_B > Y_A$, or **3.** $Y_B < Y_A$

Q9 Two wires with equal lengths are made of pure copper. The diameter of wire A is twice the diameter of wire B. When 6 kg masses are hung on the wires, wire B stretches more than wire A. You make careful measurements and compute Young's modulus for both wires. What do you find? **(a)** $Y_A > Y_B$, **(b)** $Y_A = Y_B$, **(c)** $Y_A < Y_B$

Q10 Suppose you attempt to pick up a very heavy object. Before you tried to pick it up, the object was sitting still—its momentum was not changing. You pull very hard, but do not succeed in moving the object. Is this a violation of the Momentum Principle? How can you be exerting a large force on the object without causing a change in its momentum? What does change when you apply this force?

Q11 **(a)** In outer space, a rod is pushed to the right by a constant force F (Figure 4.54). Describe the pattern of interatomic distances along the rod. Include a specific comparison of the situation at locations A, B, and C. Explain briefly in terms of fundamental principles.

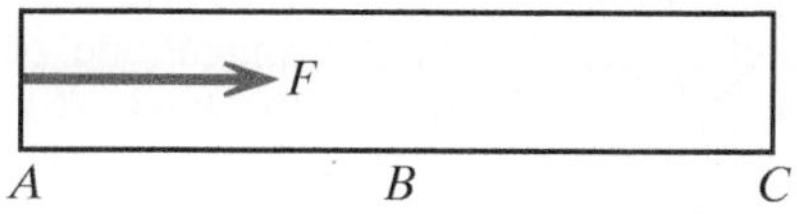

Figure 4.54

Hint: Consider the motion of an individual atom inside the rod, and various locations along the rod.

(b) After the rod in part (a) reaches a speed v, the object that had been exerting the force on the rod is removed. Describe the subsequent motion of the rod and the pattern of interatomic distances inside the rod. Include a specific comparison of the situation at locations A, B, and C. Explain briefly.

Q12 Bob is pushing a box across the floor at a constant speed of 1 m/s, applying a horizontal force whose magnitude is 20 N. Alice is pushing an identical box across the floor at a constant speed of 2 m/s, applying a horizontal force. **(a)** What is the magnitude of the force that Alice is applying to the box? **(b)** With the two boxes starting from rest, explain qualitatively what Alice and Bob did to get their boxes moving at different constant speeds.

Q13 In a spring–mass oscillator, when is the magnitude of momentum of the mass largest: when the magnitude of the net force acting on the mass is largest, or when the magnitude of the net force acting on the mass is smallest?

Q14 For a vertical spring–mass oscillator that is moving up and down, which of the following statements are true? (More than one statement may be true.) **(a)** At the lowest point in the oscillation, the momentum is zero. **(b)** At the lowest point in the oscillation, the rate of change of the momentum is zero. **(c)** At the lowest point in the oscillation, $mg = k_s s$. **(d)** At the lowest point in the oscillation, $mg > k_s s$. **(e)** At the lowest point in the oscillation, $mg < k_s s$.

Q15 The period of a particular spring–mass oscillator is 1 s when the amplitude is 5 cm. **(a)** What would be the period if we doubled the mass? **(b)** What would be the period if we replaced the original spring with a spring that is twice as stiff (keeping the original mass)? **(c)** What would be the period if we cut the original spring in half and use just one of the pieces (keeping the original mass)? **(d)** What would be the period if we increased the amplitude of the original system to 10 cm, so that the total distance traveled in one period is twice as large? **(e)** What would be the period if we took the original system to a massive planet where $g = 25$ N/kg?

Q16 How should you start the system going at $t = 0$ in order for the motion to be $A\cos(\omega t)$? How should you start the system going at $t = 0$ in order for the motion to be $A\sin(\omega t)$?

Q17 Describe two examples of oscillating systems that are *not* harmonic oscillators.

Q18 Two rods are both made of pure titanium. The diameter of rod A is twice the diameter of rod B, but the lengths of the rods are equal. You tap on one end of each rod with a hammer and measure how long it takes the disturbance to travel to the other end of the rod. In which rod did it take longer? **(a)** Rod A, **(b)** Rod B, **(c)** The times were equal.

Q19 A particular spring–mass oscillator oscillates with period T. Write out the general equation for the period of such an oscillator to use as a guide when answering the following questions. **(a)** If you double the mass but keep the stiffness the same, by what numerical factor does the period change? (That is, if the original period was T and the new period is bT, what is b?) **(b)** If, instead, you double the spring stiffness but keep the mass the same, what is the factor b? **(c)** If, instead, you double the mass and also double the spring stiffness, what is the factor b? **(d)** If, instead, you double the amplitude (keeping the original mass and spring stiffness), what is the factor b?

Q20 Uranium-238 (U^{238}) has three more neutrons than uranium-235 (U^{235}). Compared to the speed of sound in a bar of U^{235}, is the speed of sound in a bar of U^{238} higher, lower, or the same? Explain your choice, including justification for assumptions you make.

PROBLEMS

Section 4.4

•**P21** If in a certain material whose atoms are in a cubic array the interatomic distance is 1.7×10^{-10} m and the mass of one atom is 8.2×10^{-26} kg, what would be the density of this material?

•**P22** A block of one mole of a certain material whose atoms are in a cubic array has dimensions of 5 cm by 4 cm by 0.5 cm. What is the interatomic distance?

•**P23** The diameter of a copper atom is approximately 2.28×10^{-10} m. The mass of one mole of copper is 64 g. Assume that the atoms are arranged in a simple cubic array. Remember to convert to SI units. **(a)** What is the mass of one copper atom, in kg? **(b)** How many copper atoms are there in a cubical block of copper that is 4.6 cm on each side? **(c)** What is the mass of the cubical block of copper, in kg?

•**P24** One mole of tungsten (6.02×10^{23} atoms) has a mass of 184 g, as shown in the periodic table on the inside front cover of the textbook. The density of tungsten is $19.3\ \text{g/cm}^3$. What is the approximate diameter of a tungsten atom (length of a bond) in a solid block of the material? Make the simplifying assumption that the atoms are arranged in a cubic array.

Section 4.5

•**P25** If a chain of 50 identical short springs linked end to end has a stiffness of 270 N/m, what is the stiffness of one short spring?

•**P26** A certain spring has stiffness 190 N/m. The spring is then cut into two equal lengths. What is the stiffness of one of these half-length springs?

•**P27** Forty-five identical springs are placed side by side (in parallel) and connected to a large massive block. The stiffness of the 45-spring combination is 20,250 N/m. What is the stiffness of one of the individual springs?

•**P28** A certain spring has stiffness 140 N/m. The spring is then cut into two equal lengths. What is the stiffness of one of these half-length springs?

•**P29** Five identical springs, each with stiffness 390 N/m, are attached in parallel (that is, side by side) to hold up a heavy weight. If these springs were replaced by an equivalent single spring, what should be the stiffness of this single spring?

••**P30** A hanging titanium wire with diameter 2 mm (2×10^{-3} m) is initially 3 m long. When a 5 kg mass is hung from it, the wire stretches an amount 0.4035 mm, and when a 10 kg mass is hung from it, the wire stretches an amount 0.807 mm. A mole of titanium has a mass of 48 grams, and its density is $4.51\ \text{grams/cm}^3$. Find the approximate value of the effective spring stiffness of the interatomic force, and explain your analysis.

••**P31** A hanging copper wire with diameter 1.4 mm (1.4×10^{-3} m) is initially 0.95 m long. When a 36 kg mass is hung from it, the wire stretches an amount 1.83 mm, and when a 72 kg mass is hung from it, the wire stretches an amount 3.66 mm. A mole of copper has a mass of 63 g, and its density is $9\ \text{g/cm}^3$. Find the approximate value of the effective spring stiffness of the interatomic force.

••**P32** One mole of tungsten (6.02×10^{23} atoms) has a mass of 184 g, and its density is $19.3\ \text{g/cm}^3$, so the center-to-center distance between atoms is 2.51×10^{-10} m. You have a long thin bar of tungsten, 2.5 m long, with a square cross section, 0.15 cm on a side. You hang the rod vertically and attach a 415 kg mass to the bottom, and you observe that the bar becomes 1.26 cm longer. From these measurements, it is possible to determine the stiffness of one interatomic bond in tungsten. **(a)** What is the spring stiffness of the entire wire, considered as a single macroscopic (large scale), very stiff spring? **(b)** How many side-by-side atomic chains (long springs) are there in this wire (Figure 4.55)? This is the same as the number of atoms on the bottom surface of the tungsten wire. Note that the cross-sectional area of one tungsten atom is $(2.51 \times 10^{-10})^2\ \text{m}^2$.

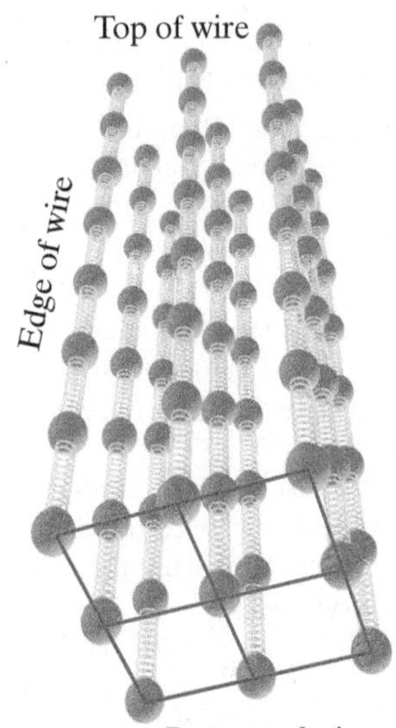

Figure 4.55

(c) How many interatomic bonds are there in one atomic chain running the length of the wire? **(d)** What is the stiffness of a single interatomic "spring"?

••**P33** A hanging iron wire with diameter 0.08 cm is initially 2.5 m long. When a 52 kg mass is hung from it, the wire stretches an amount 1.27 cm. A mole of iron has a mass of 56 g, and its density is $7.87\ \text{g/cm}^3$. **(a)** What is the length of an interatomic bond in iron (diameter of one atom)? **(b)** Find the approximate value of the effective spring stiffness of one interatomic bond in iron.

Section 4.6

•**P34** Steel is very stiff, and Young's modulus for steel is unusually large, $2 \times 10^{11}\ \text{N/m}^2$. A cube of steel 28 cm on a side supports a load of 85 kg that has the same horizontal cross section as the steel cube. **(a)** What is the magnitude of the normal force that the steel cube exerts on the load? **(b)** What is the compression of the steel cube? That is, what is the small change in height of the steel cube due to the load it supports? Give your answer as a positive number. The compression of a wide, stiff support can be extremely small.

•**P35** Suppose that you are going to measure Young's modulus for three rods by measuring their stretch when they are suspended vertically and weights are hung from them (Figure 4.56).

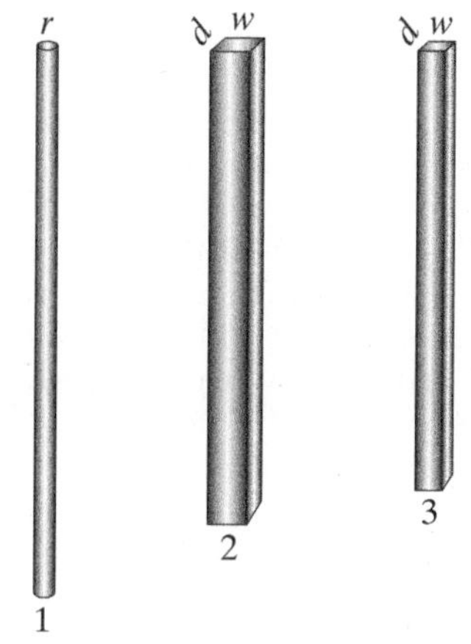

Figure 4.56

Rod 1 is 2.7 m long and cylindrical with radius 4 mm (1 mm is 0.001 m). Rod 2 is 3.2 m long by 12 mm wide by 6 mm deep. Rod 3 is 3 m long by 6 mm wide by 6 mm deep. The definition of Young's modulus, $Y = (F/A)/(\Delta L/L)$, includes the quantity A, the cross-sectional area. **(a)** What is the cross-sectional area of rod 1? **(b)** What is the cross-sectional area of rod 2? **(c)** What is the cross-sectional area of rod 3?

••**P36** Young's modulus for aluminum is 6.2×10^{10} N/m^2. The density of aluminum is 2.7 g/cm^3, and the mass of one mole (6.02×10^{23} atoms) is 27 g. If we model the interactions of neighboring aluminum atoms as though they were connected by springs, determine the approximate spring constant of such a spring. Repeat this analysis for lead: Young's modulus for lead is 1.6×10^{10} N/m^2, the density of lead is 11.4 g/cm^3, and the mass of one mole is 207 g. Make a note of these results, which we will use for various purposes later on. Note that aluminum is a rather stiff material, whereas lead is quite soft.

••**P37** Suppose that we hang a heavy ball with a mass of 10 kg (about 22 lb) from a steel wire 3 m long that is 3 mm in diameter. Steel is very stiff, and Young's modulus for steel is unusually large, 2×10^{11} N/m^2. Calculate the stretch ΔL of the steel wire. This calculation shows why in many cases it is a very good approximation to pretend that the wire doesn't stretch at all ("ideal nonextensible wire").

••**P38** You hang a heavy ball with a mass of 14 kg from a gold wire 2.5 m long that is 2 mm in diameter. You measure the stretch of the wire, and find that the wire stretched 0.00139 m. **(a)** Calculate Young's modulus for the wire. **(b)** The atomic mass of gold is 197 g/mole, and the density of gold is 19.3 g/cm^3. Calculate the interatomic spring stiffness for gold.

••**P39** A hanging wire made of an alloy of iron with diameter 0.09 cm is initially 2.2 m long. When a 66 kg mass is hung from it, the wire stretches an amount 1.12 cm. A mole of iron has a mass of 56 g, and its density is 7.87 g/cm^3. Based on these experimental measurements, what is Young's modulus for this alloy of iron?

•••**P40** A certain coiled wire with uneven windings has the property that to stretch it an amount s from its relaxed length requires a force that is given by $F = bs^3$, so its behavior is different from a normal spring. You suspend this device vertically, and its unstretched length is 25 cm. **(a)** You hang a mass of 18 g from the device, and you observe that the length is now 29 cm. What is b, including units? **(b)** Which of the following were needed in your analysis in part (a)? (1) The Momentum Principle, (2) The fact that the gravitational force acting on an object near the Earth's surface is approximately mg, (3) The force law for an ordinary spring ($F = k_s s$), (4) The rate of change of momentum being zero **(c)** Next you take hold of the hanging 18 g mass and throw it straight downward, releasing it when the length of the device is 33 cm and the speed of the mass is 5 m/s. After a very short time, 0.0001 s later, what is the stretch of the device, and what was the change in the speed of the mass (including the correct sign of the change) during this short time interval? It helps enormously to draw a diagram showing the forces that act on the mass after it leaves your hand.

Section 4.7

••**P41** Two blocks of mass m_1 and m_3, connected by a rod of mass m_2, are sitting on a low-friction surface, and you push to the left on the right block (mass m_1) with a constant force of magnitude F (Figure 4.57).

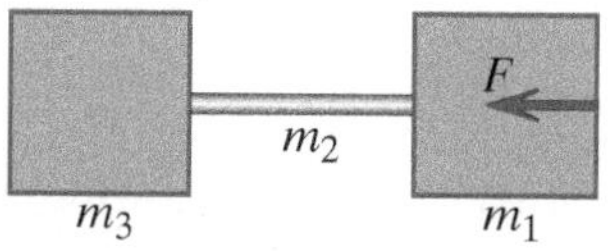

Figure 4.57

(a) What is the acceleration dv_x/dt of the blocks? **(b)** What is the vector force $\vec{F}_{\text{net }3}$ exerted by the rod on the block of mass m_3? ($|\vec{F}_{\text{net }3}|$ is approximately equal to the compression force in the rod near its left end.) What is the vector force $\vec{F}_{\text{net }1}$ exerted by the rod on the block of mass m_1? ($|\vec{F}_{\text{net }1}|$ is approximately equal to the compression force in the rod near its right end.) **(c)** Suppose that instead of pushing on the right block (mass m_1), you pull to the left on the left block (mass m_3) with a constant force of magnitude F. Draw a diagram illustrating this situation. Now what is the vector force $\vec{F}_{\text{net }3}$ exerted by the rod on the block of mass m_3?

Section 4.8

•**P42** A 5 kg box with initial speed 4 m/s slides across the floor and comes to a stop after 0.7 s. **(a)** What is the coefficient of kinetic friction? **(b)** How far does the box move? **(c)** You put a 3 kg block in the box, so the total mass is now 8 kg, and you launch this heavier box with an initial speed of 4 m/s. How long does it take to stop?

•**P43** A 3 kg block measures 5 cm by 10 cm by 20 cm. When it slides on a 10 cm by 20 cm face, it moves with constant speed when pulled horizontally by a force whose magnitude is 3 N. How big a horizontal force must be applied to pull it with constant speed if it slides on a 5 cm by 20 cm face?

•**P44** A 15 kg box sits on a table. The coefficient of static friction μ_s between table and box is 0.3, and the coefficient of kinetic friction μ_k is 0.2. **(a)** What is the force required to start the box moving? **(b)** What is the force required to keep it moving at constant speed? **(c)** What is the force required to maintain an acceleration of 2 m/s/s?

•**P45** A 20 kg box is being pushed across the floor by a constant force $\langle 90, 0, 0 \rangle$ N. The coefficient of kinetic friction for the table and box is 0.25. At $t = 5$ s the box is at location $\langle 8, 2, -1 \rangle$ m, traveling with velocity $\langle 3, 0, 0 \rangle$ m/s. What is its position and velocity at $t = 5.6$ s?

•**P46** You drag a block across a table while a friend presses down on the block. The coefficient of friction between the table and the block is 0.6. The vertical component of the force exerted by the table on the block is 190 N. How big is the horizontal component of the force exerted by the table on the block?

••P47 For this problem you will need measurements of the position vs. time of a block sliding on a table, starting with some initial velocity, slowing down, and coming to rest. If you do not have an appropriate laboratory setup for making these measurements, your instructor will provide you with such data. Analyze these data to determine the coefficient of friction and to see how well they support the assertion that the force of sliding friction is essentially independent of the speed of sliding.

•••P48 It is sometimes claimed that friction forces always slow an object down, but this is not true. If you place a box of mass 8 kg on a moving horizontal conveyor belt, the friction force of the belt acting on the bottom of the box speeds up the box. At first there is some slipping, until the speed of the box catches up to the speed of the belt, which is 5 m/s. The coefficient of kinetic friction between box and belt is 0.6. **(a)** How much time does it take for the box to reach this final speed? **(b)** What is the distance (relative to the floor) that the box moves before reaching the final speed of 5 m/s?

Section 4.10

•••P49 A chain of length L and mass M is suspended vertically by one end with the bottom end just above a table. The chain is released and falls, and the links do not rebound off the table, but they spread out so that the top link falls very nearly the full distance L. Just before the instant when the entire chain has fallen onto the table, how much force does the table exert on the chain? Assume that the chain links have negligible interaction with each other as the chain drops, and make the approximation that there is a very large number of links. *Hint*: Consider the instantaneous rate of change of momentum of the chain as the last link hits the table.

Section 4.11

•P50 A ball whose mass is 1.4 kg is suspended from a spring whose stiffness is 4 N/m. The ball oscillates up and down with an amplitude of 14 cm. **(a)** What is the angular frequency ω? **(b)** What is the frequency? **(c)** What is the period? **(d)** Suppose this apparatus were taken to the Moon, where the strength of the gravitational field is only 1/6 of that on Earth. What would be the period on the Moon? (Consider carefully how the period depends on properties of the system; look at the equation.)

•P51 A mass of 2.2 kg is connected to a horizontal spring whose stiffness is 8 N/m. When the spring is relaxed, $x = 0$. The spring is stretched so that the initial value of $x = +0.18$ m. The mass is released from rest at time $t = 0$. Remember that when the argument of a trigonometric function is in radians, on a calculator you have to switch the calculator to radians or convert the radians to degrees. Predict the position x when $t = 1.15$ s.

•P52 A bouncing ball is an example of an anharmonic oscillator. If you quadruple the maximum height, what happens to the period? (Assume that the ball keeps returning almost to the same height.)

••P53 Here on Earth you hang a mass from a vertical spring and start it oscillating with amplitude 1.7 cm. You observe that it takes 2.1 s to make one round-trip. You construct another vertical oscillator with a mass 6 times as heavy and a spring 10 times as stiff. You take it to a planet where $g_{\text{planet}} = 6.8$ N/kg. You start it oscillating with amplitude 3.3 cm. How long does it take for the mass to make one round-trip?

••P54 In the approximation that the Earth is a sphere of uniform density, it can be shown that the gravitational force it exerts on a mass m inside the Earth at a distance r from the center is $mg(r/R)$, where R is the radius of the Earth. (Note that at the surface, the force is indeed mg, and at the center it is zero). Suppose that there were a hole drilled along a diameter straight through the Earth, and the air were pumped out of the hole. If an object is released from one end of the hole, how long will it take to reach the other side of the Earth? Include a numerical result.

••P55 A spring suspended vertically is 18 cm long. When you suspend a 30 g weight from the spring, at rest, the spring is 22 cm long. Next you pull down on the weight so the spring is 23 cm long and you release the weight from rest. What is the period of oscillation?

••P56 It was found that a 20 g mass hanging from a particular spring had an oscillation period of 1.2 s. **(a)** When two 20 g masses are hung from this spring, what would you predict for the period in seconds? Explain briefly.

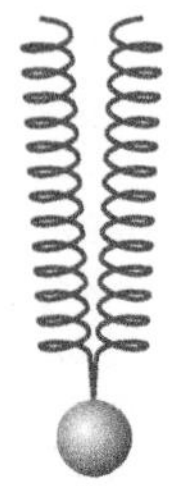

Figure 4.58

(b) When one 20 g mass is supported by two of these vertical, parallel springs (Figure 4.58), what would you predict for the period in seconds? Explain briefly. **(c)** Suppose that you cut one spring into two equal lengths, and you hang one 20 g mass from this half spring. What would you predict for the period in seconds? Explain briefly. **(d)** Suppose that you take a single (full-length) spring and a single 20 g mass to the Moon and watch the system oscillate vertically there. Will the period you observe on the Moon be longer, shorter, or the same as the period you measured on Earth? (The gravitational field strength on the Moon is about one-sixth that on the Earth.) Explain briefly.

••P57 A vertical mass–spring oscillator has an amplitude of 0.06 m and a period of 0.4 s. **(a)** What is the maximum speed of the mass? **(b)** What is the maximum acceleration of the mass?

••P58 In Problem P36 you can find the effective spring stiffness corresponding to the interatomic force for aluminum and lead. Let's assume for the moment that, *very* roughly, other atoms have similar values. **(a)** What is the (very) approximate frequency f for the vibration of H_2, a hydrogen molecule? **(b)** What is the (very) approximate frequency f for the vibration of O_2, an oxygen molecule? **(c)** What is the approximate vibration frequency f of the molecule D_2, both of whose atoms are deuterium atoms (that is, each nucleus has one proton and one neutron)? **(d)** Explain why the *ratio* of the deuterium frequency to the hydrogen frequency is quite accurate, even though you have estimated each of these quantities very approximately, and the effective spring stiffness is normally expected to be significantly different for different atoms. (*Hint:* What interaction is modeled by the effective "spring"?)

••**P59** Find a spring (or a rubber band) and one or more masses with which to study the motion of a mass hanging from a spring (Figure 4.59). **(a)** Measure the value of the spring stiffness k_s of the spring in N/m. The magnitude of a spring force is $k_s s$, where s is the stretch of the spring (change from the unstretched length). Explain briefly how you measured k_s. Include in your report the unstretched length of the spring. **(b)** Measure the period (the round-trip time) of a mass hanging from the vertical spring. Report the mass that you use, and the amplitude of the oscillation. Amplitude is the maximum displacement, plus or minus, from the equilibrium position (the position where the mass can hang motionless). **(c)** With twice the amplitude that you reported in part (b), measure the period again. Since the mass has to move twice as far, one might expect the period to lengthen. Does it?

Figure 4.59

There are unavoidable fluctuations in starting and stopping the timing, but you can minimize the error this contributes by timing many complete cycles so that the starting and stopping fluctuations are a small fraction of the total time measured.

It is good practice to count out loud *starting from zero, not one*. To count five cycles, say out loud "Zero, one, two, three, four, five." If on the other hand you say "One, two, three, four, five," you have actually counted only four cycles, not five.

Since the motion is periodic, you can start (say "zero") at any point in the motion. It is best to start and stop when the mass is moving fast past some marker near the equilibrium point, because it is difficult to estimate the exact time when the mass reaches the very top or the very bottom, because it is moving slowly at those turnaround points. Be sure to measure full round-trip cycles, not the half-cycles between returns to the equilibrium point (but going in the opposite direction).

•••**P60** An object of mass m is attached by two stretched springs (stiffness k_s and relaxed length L_0) to rigid walls, as shown in Figure 4.60. The springs are initially stretched by an amount $(L - L_0)$. When the object is displaced to the right and released, it oscillates horizontally. Starting from the Momentum Principle, find a function of the displacement x of the object and the time t that describes the oscillatory motion. **(a)** What is the period of the motion? **(b)** If L_0 were shorter (so that the springs are initially stretched more), would the period be larger, smaller, or the same?

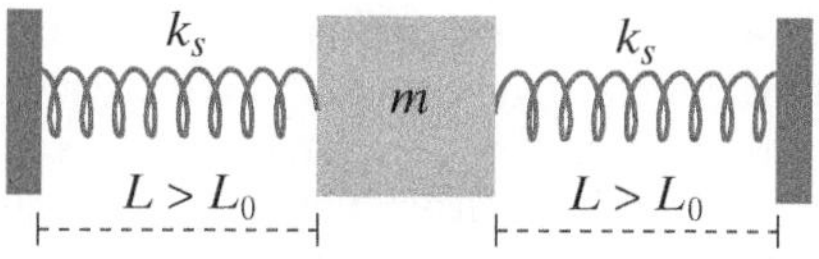

Figure 4.60

•••**P61** A simple pendulum (Figure 4.61) consists of a small mass of mass m swinging at the end of a low-mass string of length L (in contrast to a pendulum whose mass is distributed, such as a rod swinging from one end). **(a)** Show that the tangential momentum p of the small mass along the arc to the right (increasing θ) obeys the following equation, where s is the arc length $L\theta$:

$$\frac{dp}{dt} = -mg\sin\theta = -mg\sin\left(\frac{s}{L}\right)$$

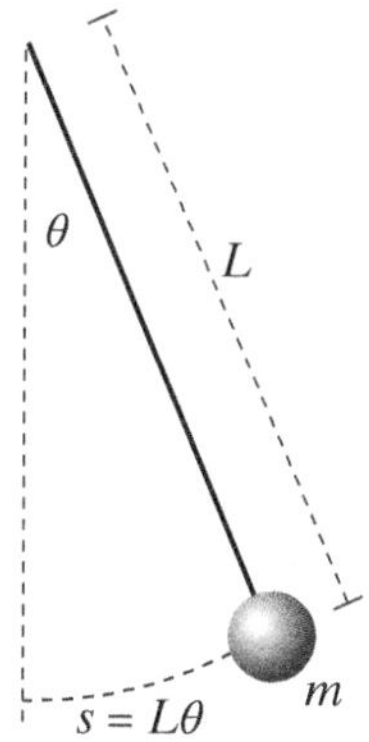

Figure 4.61

(b) What form does this equation take for small-amplitude swings? (Note that if θ is measured in radians, $\sin\theta \approx \theta$ for small angles. For example, even for an angle as large as $\theta = 30°$, $\theta = \pi/6 = 0.524$ radians, which is close to $\sin(30°) = 0.5$, and the approximation gets better for smaller angles.) **(c)** Compare the form of the approximate equation you obtained in part (b) with the form of the Momentum Principle for a spring–mass system. By comparing these equations for different systems, determine the period of a simple pendulum for small-amplitude swings. (Note that any time you can approximate the Momentum Principle for some system by an equation that looks like the equation for a mass on a spring, you know that the system will move approximately sinusoidally.) **(d)** Make a simple pendulum and predict its period, then measure the period. Do this for a long pendulum and for a short pendulum. Report your experimental data and results, with comparisons to your theoretical predictions.

Section 4.13

•**P62** Two metal rods are made of different elements. The interatomic spring stiffness of element A is three times larger than the interatomic spring stiffness for element B. The mass of an atom of element A is three times greater than the mass of an atom of element B. The atomic diameters are approximately the same for A and B. What is the ratio of the speed of sound in rod A to the speed of sound in rod B?

••**P63** You hang a heavy ball with a mass of 41 kg from a silver rod 2.6 m long by 1.5 mm by 3.1 mm. You measure the stretch of the rod, and find that the rod stretched 0.002898 m. Using these experimental data, what value of Young's modulus do you get? The atomic mass of silver is 108 g/mole, and the density of silver is 10.5 g/cm^3. Using this information along with the measured value of Young's modulus, calculate the speed of sound in silver.

••**P64** One mole of nickel (6.02×10^{23} atoms) has a mass of 59 g, and its density is 8.9 g/cm^3. You have a bar of nickel 2.5 m long, with a square cross section, 2 mm on a side. You hang the rod vertically and attach a 40 kg mass to the bottom, and you observe that the bar becomes 1.2 mm longer. Next you remove the 40 kg mass, place the rod horizontally, and strike one end with a hammer. How much time T will elapse before a microphone at the other end of the bar will detect a disturbance?

Section 4.14

•P65 It is hard to imagine that there can be enough air between a book and a table so that there is a net upward (buoyant) force on the book despite the large downward force on the top of the book. About how many air molecules are there between a textbook and a table, if there is an average distance of about 0.01 mm between the uneven surfaces of the book and table?

•P66 Calculate the buoyant force in air on a kilogram of lead (whose density is about 11 g/cm^3). The density of air is 1.3×10^{-3} g/cm^3. Remember that the buoyant force is equal to the weight of a volume of air that is equal to the volume of the object. (Compare with the weight mg of this much lead.)

•P67 Air pressure at the surface of a fresh water lake near sea level is about 1×10^5 N/m^2. At approximately what depth below the surface does a diver experience a pressure of 3×10^5 N/m^2? How would this be different in sea water, which has higher density than fresh water?

••P68 Here are two examples of floating objects: **(a)** A block of wood 20 cm long by 10 cm wide by 6 cm high has a density of 0.7 g/cm^3 and floats in water (whose density is 1.0 g/cm^3). How far below the surface of the water is the bottom of the block? Explain your reasoning. **(b)** An advertising blimp consists of a gas bag, the supporting structure for the gas bag, and the gondola hung from the bottom, with its cabin, engines, and propellers. The gas bag is about 30 m long with a diameter of about 10 m, and it is filled with helium (density about 4 g per 22.4 l under these conditions). Estimate the total mass of the blimp, including the helium.

COMPUTATIONAL PROBLEMS

More detailed and extended versions of some of these computational modeling problems may be found in the lab activities included in the *Matter & Interactions, 4th Edition*, resources for instructors.

••P69 Write an iterative computational model to predict and display the motion of a block of mass 60 g that sits on top of a vertical spring of stiffness 8 N/m and relaxed length 20 cm (Figure 4.62). **(a)** Use initial conditions that represent the block at rest on top of the spring, which has been pushed down until its total length is 10 cm. These values are the same as those used in an example in Chapter 2, Section 2.6, so you can use that example to check your work. **(b)** Add a graph of the y component of the block's position vs. time. What is the period of the block's motion? (It will be easiest to determine this if you stop the graph after two or three complete cycles. By holding down the mouse button while dragging over a VPython graph you will get crosshairs and a numerical readout of the coordinates.) **(c)** Double the stiffness of the spring. How does that affect the period of the motion? **(d)** Restore the stiffness of the spring to its original value, but double the mass. How does this affect the period of the motion?

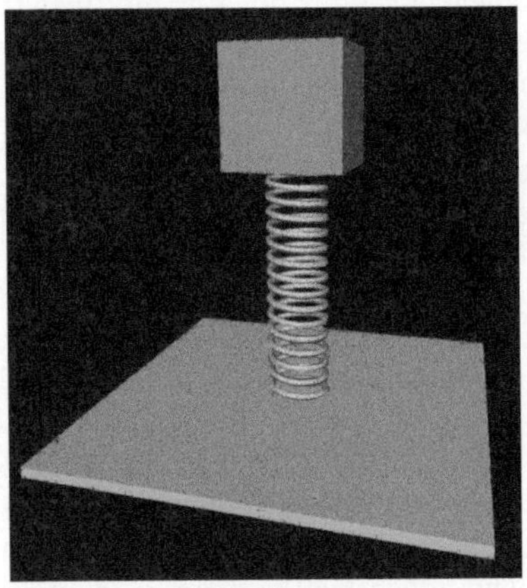

Figure 4.62

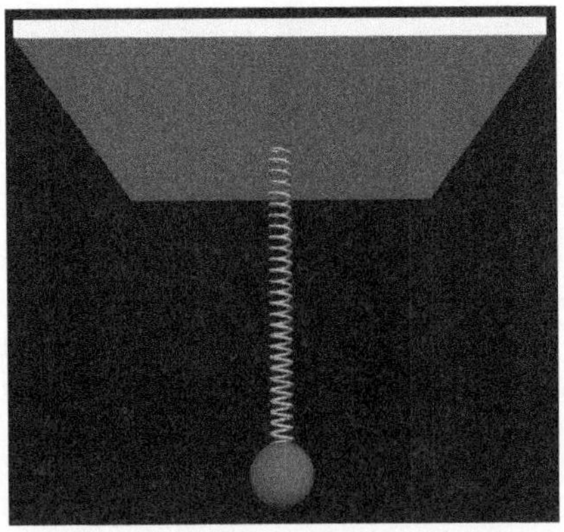

Figure 4.63

••P70 A ball of mass 20 g hangs from a spring whose relaxed length is 20 cm and whose stiffness is 0.9 N/m (Figure 4.63). **(a)** Write a VPython program to model the motion of this spring. **(b)** Add a graph of the y component of the ball's position vs. time. From the graph, determine the period of the ball's motion. (It will be easiest to determine this if you stop the graph after two or three complete cycles. By holding down the mouse button while dragging over a VPython graph you will get crosshairs and a numerical readout of the coordinates.) **(c)** How does doubling the mass affect the period of the oscillations? **(d)** How does keeping the mass unchanged but doubling the spring stiffness affect the period of the oscillations? **(e)** Make the ball leave a trail behind it as it moves. Find initial conditions that result in oscillations in all three dimensions. Rotate the display to make sure the motion is not planar.

••P71 Create a VPython program to model the motion of a 0.03 kg mass connected to two horizontal springs whose stiffness is 4 N/m, and whose relaxed length is 0.5 m, as shown in Figure 4.64. Neglect the effects of gravity (perhaps the system is in outer space). **(a)** Add a graph of x vs. t for the mass, and determine the period of the oscillator from this graph. It will be easiest to do this if you stop the graph after two or three complete cycles. If you hold down the mouse button while dragging over a VPython graph you will get crosshairs and a numerical readout of the coordinates. **(b)** Using the period you determined above, calculate the effective spring stiffness of this two-spring system. Is this the same as the spring stiffness you assigned to each spring in your model? Explain. **(c)** How does changing the amplitude of the oscillations affect the period?

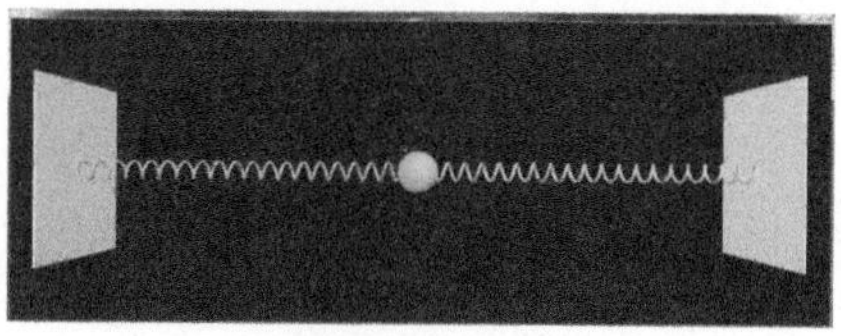

Figure 4.64

••P72 On a space station in outer space a horizontal chain of two identical 3 kg masses and three identical springs of relaxed length 14 m and stiffness 50 N/m is suspended between two walls, as shown in Figure 4.65. Write a VPython program to model the motion of this system. **(a)** Start the motion by displacing the leftmost mass to the left, and observe the motion of the system. **(b)** Add a graph of x vs. t for each mass, making the graphs different colors so you can tell which is which. Is this system a simple harmonic oscillator? **(c)** Experiment with different initial conditions. Make a screen shot of a graph and report the initial conditions that led to this motion. **(d)** Find a set of initial conditions that results in simple sinusoidal motions of both masses. What are the periods of these motions? **(e)** (Optional) The pattern of motion you found in part (d) is called a "normal mode" of the system. This particular system has two such normal modes. Can you find initial conditions that produce the second? **(f)** (Optional) It is interesting to graph p.x for each mass as well as x. To do this you will need to make a separate gdisplay (graph window) for each graph. See the VPython help to learn how to do this. Explain how the plots of p.x and x are related.

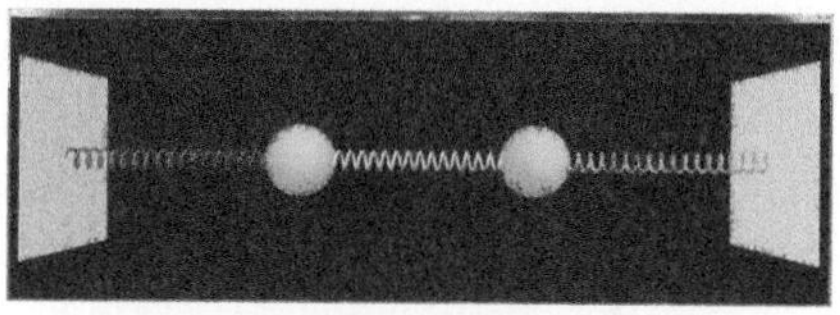

Figure 4.65

••P73 Create a computational model of a system like the one shown in Figure 4.66, in which a ball of mass 30 g is connected by six identical springs of stiffness 30 N/m and relaxed length 50 cm to six fixed walls. **(a)** Find initial conditions that make the system oscillate only along the x axis. **(b)** Find initial conditions that make the system oscillate in such a way that the ball doesn't travel in a straight line. Have the ball leave a trail to check your results. **(c)** Is this system a simple harmonic oscillator? Explain how you determined this.

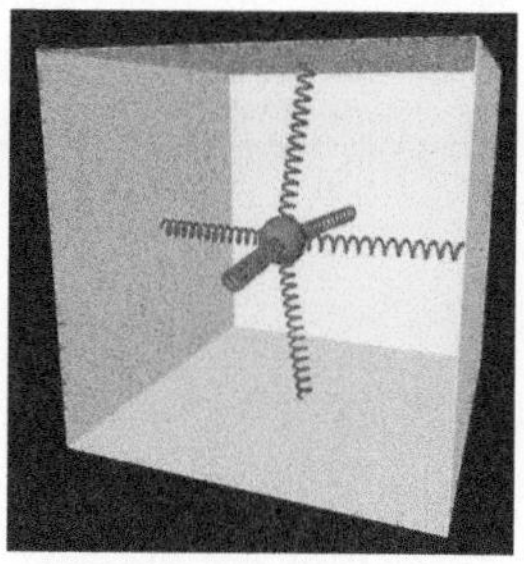

Figure 4.66

••P74 An object of mass m slides with negligible friction on a table. The object is connected to a post by a spring whose stiffness is k_s and whose relaxed length is L_0. Make a computational model of the situation and investigate the various kinds of motions that result from choosing different initial conditions and different values of k_s and L_0. Leave a trail behind the moving object to help in visualizing the motion (Figure 4.67). The program that generated Figure 4.67 used $m = 0.2$ kg, $k_s = 2$ N/m, and $L_0 = 0.16$ m.

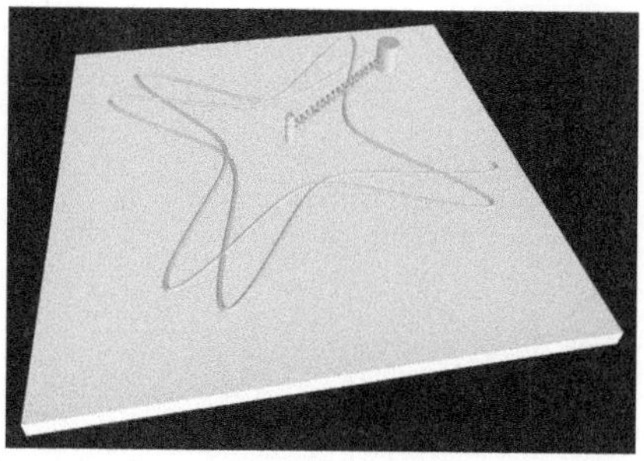

Figure 4.67

••P75 A peculiar spring-like device exerts a force whose magnitude is $2 \times 10^9 \times s^7$ N, where s is the stretch in meters. The relaxed length is 0.1 m and the initial stretch is 0.02 m. The device is attached along the x axis to an object of mass 0.1 kg that is free to move with negligible friction on a horizontal surface. Write a program to model this device. A time step Δt of about 0.01 s is a reasonable choice. **(a)** Release the system from rest (the initial speed is zero) and display the motion as a function of time for the first 20 seconds. **(b)** Display on the same graph both the stretch s and the x component of velocity, using different colors. **(c)** How does the motion differ from that of an ordinary spring–mass system? **(d)** How does changing the initial stretch to 0.03 m affect the period of the system? Is this a harmonic oscillator?

•••P76 As explained in Problem P61, a simple pendulum consisting of a small mass of mass m swinging at the end of a low-mass string of length L behaves like a harmonic oscillator when the amplitude of the swings is small. For large-amplitude swings, the small-angle approximation discussed in Problem P61 is not valid, but you can use an iterative model to calculate the motion. For a mass on a string, the largest possible amplitude is 90°, but if the string is replaced by a lightweight rod, the amplitude can be as large as 180° (standing upside-down, with the mass at the top). It turns out that such a pendulum can be modeled by the same equation as the equation derived in part (a) of Problem P61, but the concepts of torque and angular momentum are required to prove it. **(a)** Assume that the equation is valid, and write a computational model to predict the motion of the pendulum. Think of s and p as being like x and p_x so that your program is basically a one-dimensional calculation, with a force $-mg\sin(s/L)$ instead of $-k_s s$. **(b)** Plot the position s and momentum p as a function of time for a pendulum whose amplitude is nearly 180°. **(c)** How does the period of this system depend on the amplitude? Is it a harmonic oscillator?

ANSWERS TO CHECKPOINTS

1 2.6×10^{-10} m; 3.1×10^{-10} m
2 800 N/m
3 300 N/m
4 0.167 mm
5 (a) 0.0235, **(b)** 1.225×10^7 N/m^2, **(c)** 5.21×10^8 N/m^2
6 (a) The box would not move, **(b)** 60 N, **(c)** The box would accelerate, **(d)** 80 N
7 (a) $\langle 4, 0, 0 \rangle$ N, **(b)** Nothing. $d\vec{p}/dt$ gives no information about the instantaneous value of $\vec{p}$.
8 (a) 58.8 N/m, **(b)** 0.47 s
9 2970 m/s
10 $F_{\text{buoyant}} = 1.6 \times 10^{-3}$ N; $\dfrac{F_{\text{buoyant}}}{F_{\text{weight of iron}}} = 1.6 \times 10^{-4}$
12 $v_0 = \dfrac{F}{m}\Delta t$; $A = v_0\sqrt{\dfrac{m}{k}}$; 90° ($\pi/2$ radians), corresponding to a sine rather than a cosine (start at $x = 0$).

CHAPTER

5

Determining Forces from Motion

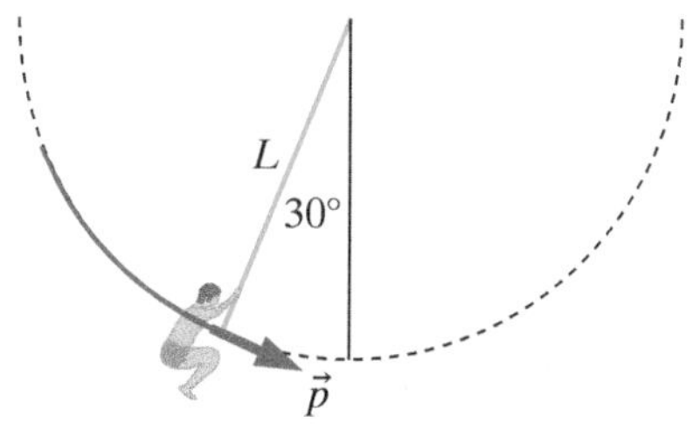

OBJECTIVES

After studying this chapter, you should be able to

- Identify systematically all the forces acting on a system.
- Determine the values of unknown 3D forces acting on a system whose motion is known.
- Analyze curving motion mathematically, and relate parallel and perpendicular components of $d\vec{p}/dt$ to the net force acting on the system.

5.1 UNKNOWN FORCES

In previous chapters we have applied the Momentum Principle to predict the motion of systems that are affected by a known net force. In this chapter we will do the opposite: knowing the motion of a system, we will deduce the net force acting on the system. In some cases, knowing the net force, and also knowing the values of some, but not all, of the individual forces contributing to the net force, we will be able to determine the forces whose values are unknown.

One important application of this procedure is to deduce the value of a contact force that we cannot calculate directly. In Chapter 4, what was the force exerted by the vine on Tarzan when he swung on it, just before it broke (Figure 5.1)? What is the force of the seat on an airplane passenger when the plane makes a sudden change of direction, and why is it different from the passenger's weight?

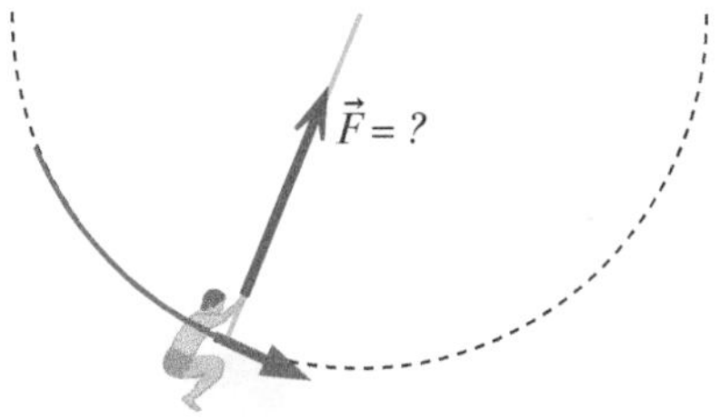

Figure 5.1 The force exerted by the vine on Tarzan is an unknown force.

5.2 IDENTIFYING ALL FORCES

It is surprisingly important to list all forces on a system explicitly when beginning a problem. Once we start doing mathematical operations, it is easy to forget a force, or to include it twice, so it is essential to have a written record for reference. Enumerating forces on a system is a little like following the rules of a game:

RULES FOR IDENTIFYING FORCES ON A SYSTEM

- Identify every force due to an *object* in the surroundings that interacts at a distance (gravitationally or electrically) with the system (*distance forces*).
- Identify every force due to an *object* in the surroundings that touches the system (*contact forces*).
- Draw a free-body diagram in which each force is labeled with the name of the object in the surroundings that is causing the force.

QUESTION What is the point of listing objects in the surroundings? Why should we say "the Earth" instead of "gravity"?

We know that an interaction involves two objects, so each force on the system must be due to an object in the surroundings. Naming the object is a way of avoiding double counting, because sometimes forces have more than one standard name. For example, if we included both "the gravitational force" and "the weight of the ball" we would be counting the same force twice. Identifying each force by the name of the interacting object helps avoid counting a force twice, or missing a force. If you can't name an *object* in the surroundings that is responsible for the force, there is no such force on the system as long as we use an inertial (nonaccelerating, nonrotating) reference frame, which we will always do in this textbook. If you find yourself identifying objects named "tension," "normal," "gravity," or "centripetal," you are in danger of double-counting forces, or including nonexistent forces.

Making Free-Body Diagrams

A free-body diagram can be very simple, as shown in Figure 5.2:

- If the system is modeled as a point particle, represent it by a dot.
- Represent each force by an arrow. If you don't know the magnitude of the force, that is okay—draw the arrow with an arbitrary length.
- Label each force. A label has two parts: a symbol that can be used in equations, and the name of the object in the surroundings that is responsible for the force.

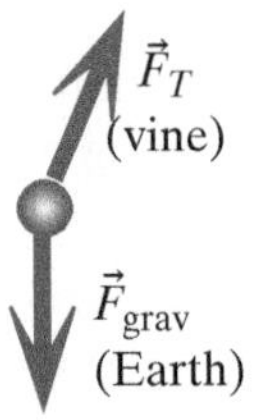

Figure 5.2 Free-body diagram for Tarzan at the instant shown in Figure 5.1. Tarzan is modeled as a point particle, and is represented by a dot. Each force is labeled with a symbol for use in equations and with the name of the interacting object in the surroundings. It is common to use the symbol $\vec{F}_T$ for a tension force. Since we do not yet know relative magnitudes, we have drawn the vine force with an arbitrary length.

If you would like to include more information in your free-body diagram, such as magnitudes of forces, direction of momentum, or values of angles or distances, that is fine. However, a simple free-body diagram, recording forces and their corresponding objects, is often quite adequate.

QUESTION In a free-body diagram should we show forces on objects in the surroundings?

No. We construct a free-body diagram showing forces on the chosen system to help us apply the Momentum Principle correctly to that system. Only forces on the system itself affect the momentum of the system.

QUESTION In Figure 5.2, why don't we include a force due to Tarzan's motion?

Force is a quantitative measure of interaction with an object in the surroundings. Motion isn't an object. The system's motion at this instant is certainly important, and it will be taken into account when we apply the Momentum Principle (remember that $\vec{p}_{future} = \vec{p}_{now} + \vec{F}_{net}\,\Delta t$).

> **Checkpoint 1** In each of the following cases identify all objects in the surroundings that exert forces on the system, and draw a free-body diagram for the system. Assume that air resistance is negligible. **(a)** You hit a baseball with a bat. Choose the baseball as the system, and consider the instant of contact with the bat. **(b)** You are playing with a yo-yo. Choose the yo-yo as the system. As the yo-yo moves downward, you pull up on the string.

5.3 DETERMINING UNKNOWN FORCES

Once we have identified all the objects that exert a force on a system, we may discover that we cannot calculate all of the forces directly. This is especially true for contact forces: we do not have an equation for calculating the force exerted by a table on a coffee cup or by a string on a yo-yo. We will need to use the Momentum Principle to deduce the values of unknown forces, by following this procedure:

FINDING UNKNOWN FORCES

- Explicitly choose a system, and stick to this choice.
- Systematically identify all the forces on the system due to objects in the surroundings, and draw and label a free-body diagram.
- Observe the rate at which the system's momentum changes ($d\vec{p}/dt$).
- Use the Momentum Principle ($d\vec{p}/dt = \vec{F}_{\text{net}}$) to deduce the net force on the system.
- Solve for the unknown contributions to the net force.

This procedure will work only if we know $d\vec{p}/dt$, and if we are careful to identify *all* of the contributions to the net force. It is important to be very clear and explicit about the choice of system, and to include all other objects in the surroundings. The free-body diagram is the basis for correctly accounting for all contributions to $\vec{F}_{\text{net}}$. The following example illustrates this process with a very simple example for which you already know the answer.

EXAMPLE **A Hanging Ball**

A 1 kg ball hangs motionless at the end of a wire (Figure 5.3). The top of the wire is attached to an iron support. How large an upward force does the wire exert on the ball?

Solution Here we use the convenient and widely used convention that F_T means the magnitude of the vector $\vec{F}_T$, for example,

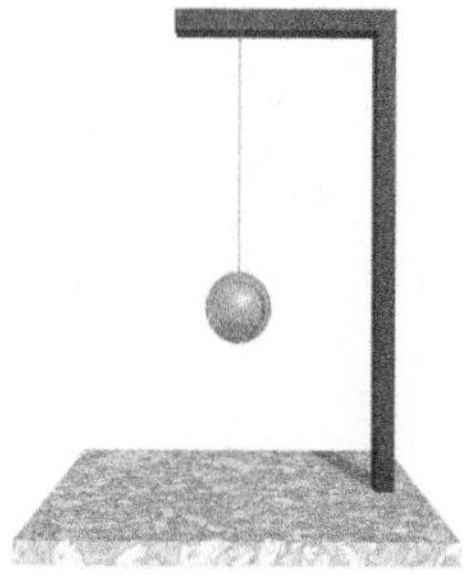

Figure 5.3 A heavy ball hangs motionless at the end of a thin wire.

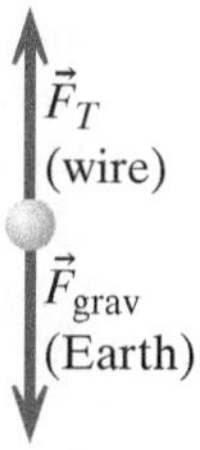

Figure 5.4 Free-body diagram for the hanging ball.

System: Ball
Surroundings: Earth, wire
Free-body diagram: Figure 5.4
Momentum Principle, *y* component:

$$dp_y/dt = F_{\text{net},y}$$
$$0 = F_T - F_{\text{grav}}$$
$$0 = F_T - 1\,\text{kg}\cdot 9.8\,\text{N/kg}$$
$$F_T = 9.8\,\text{N}$$

You almost certainly knew before going through these careful steps that the tension in the wire would be equal to the weight of the ball. The reason for going through this analysis without omitting any steps is that this same procedure can be used in all situations, including situations much more complicated than this one.

QUESTION Why didn't we include the iron support in the surroundings? Why doesn't the free-body diagram show a force due to the support?

Because the support doesn't touch the chosen system (the ball), the interatomic bonds in the ball are not stretched or compressed by the support. Only two kinds of forces can act on the ball: forces involving interaction at a distance and contact forces. To exert a contact force, the atoms of one object must be in contact with the atoms of another object, which is not the case here. The support does exert a very tiny gravitational force on the ball (acting over a distance), but this force is too small to have a noticeable effect.

It is of course important for the iron support to be present, but the support does not interact directly with the ball. The support exerts a force on the wire, but the wire is not the system of interest; it does get taken into account indirectly, since the wire would not be under tension if it were not supported by something.

QUESTION What if we had chosen the ball plus the wire as the system?

Compare the following example to the previous one.

EXAMPLE **Ball and Wire as the System (A Hanging Ball)**

Again consider Figure 5.3, but choose the ball and the wire as the system. Find the tension force exerted by the support on the ball–wire system.

Solution System: Ball and wire, modeled as a point particle
Surroundings: Earth
Free-body diagram: Figure 5.5

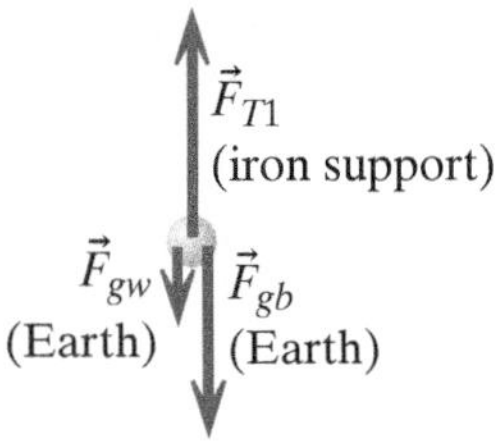

Figure 5.5 The free-body diagram for the system consisting of the ball plus the wire, modeled as a point particle. The symbol $\vec{F}_{gb}$ represents the gravitational force on the ball by the Earth, and the symbol $\vec{F}_{gw}$ represents the gravitational force on the wire by the Earth. The gravitational force of the Earth on the wire is exaggerated for clarity.

$$\frac{dp_y}{dt} = F_{T1} - F_{gb} - F_{gw}$$

$$0 = F_{T1} - m_b g - m_w g$$

$$F_{T1} = (m_b + m_w)g$$

where m_b and m_w are the masses of the ball and the wire, respectively. Note that the tension force by the support on the wire is slightly larger than the tension force by the wire on the ball, which we calculated in the previous example.

QUESTION Is the tension in the wire uniform throughout the wire?

Actually, the tension in the top of the wire is a little greater than the tension at the bottom of the wire, because the top of the wire supports not only the ball, but the rest of the wire as well. Of course if the mass of the wire is small compared to the mass of the ball, it is a good approximation to say that the interatomic bonds in the wire are stretched almost as much at the bottom as at the top.

5.4 UNIFORM MOTION

The simplest situation in which we can deduce the values of unknown forces is that of uniform motion. The momentum of a system in uniform motion is constant and does not change with time. Recall that a special case of uniform motion is a situation in which an object is at rest and remains at rest; this situation is called "static equilibrium." (Problems in which the system is at rest are often called "statics" problems. If you are studying engineering you may take an entire course on this topic.) In a typical problem involving uniform motion you will be asked to deduce the unknown magnitude of a force on the system by applying the Momentum Principle.

The underlying physics is the same for all systems in uniform motion (including those at rest):

QUESTION If the system's velocity is constant, what is $\frac{d\vec{p}}{dt}$?

The momentum of the system isn't changing, so the rate of change of the momentum is zero:

$$\frac{d\vec{p}}{dt} = \vec{0} = \langle 0,0,0 \rangle$$

QUESTION What does this imply about the net force acting on the system?

From the derivative form of the Momentum Principle, $d\vec{p}/dt = \vec{F}_{\text{net}}$, we deduce that the net force acting on a system that never moves must be zero:

$$\vec{F}_{\text{net}} = \vec{0} = \langle 0,0,0 \rangle$$

That's really all there is to it: knowing that the net force must be zero may allow us to deduce the values of some of the contributions to the net force. The previous example of a ball hanging from a wire is a simple 1D uniform motion problem.

Forces Exerted by Strings

Many uniform motion problems involve strings or wires. Because a string or wire is flexible, the tension force exerted by a string is always in the direction of the string. A string cannot exert a force perpendicular to the string. Also, of course, a string cannot push—it can only pull. It is common to refer to the tension force exerted by a string as "the tension in the string," and to use the symbol $\vec{F}_T$ to represent a tension force.

Component Equations

If a situation involves forces in two or three dimensions, it is frequently easiest to write one equation for each dimension:

$$\frac{dp_x}{dt} = F_{\text{net,x}} \quad \text{and} \quad \frac{dp_y}{dt} = F_{\text{net,y}} \quad \text{and} \quad \frac{dp_z}{dt} = F_{\text{net,z}}$$

Typically we know something about the directions of unknown forces, but we don't know their magnitudes. In such cases, we will start by expressing the components of each unknown force in terms of their magnitudes. For example, suppose that we know that a string exerts a force $\vec{F}_1$ in the $-x$ direction on a system, but we don't know the magnitude of the force. Using the notation convention that $F_1 = |\vec{F}_1|$, we can write the force as

$$\vec{F}_1 = F_1 \langle -1,0,0 \rangle = \langle -F_1,0,0 \rangle$$

so the x component of the momentum principle becomes

$$\frac{dp_x}{dt} = -F_1 + \ldots$$

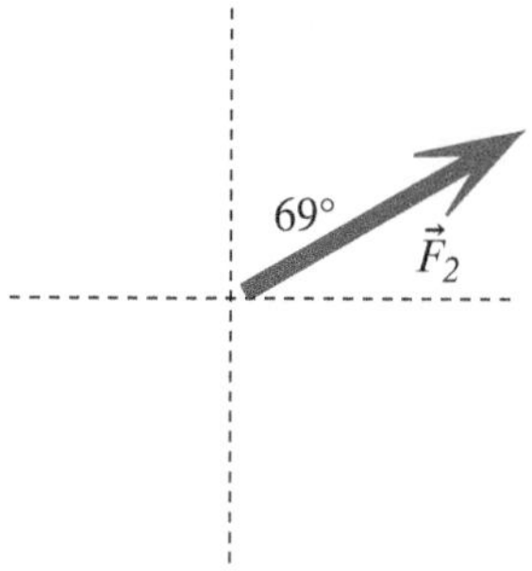

Figure 5.6 A force $\vec{F}_2$ is applied in the xy plane at an angle of 69° to the $+y$ axis.

If an unknown force is at some angle to the axes, the force will have more than one nonzero component. For example, suppose a spring pulls on a system at an angle of 69° to the $+y$ axis, as shown in Figure 5.6. If we call this unknown force $\vec{F}_2$, then by using direction cosines as we did in Chapter 1 we have this:

$$\vec{F}_2 = F_2 \langle \cos 21°, \cos 69°, \cos 90° \rangle$$

and the Momentum Principle can be written in component form as

$$\frac{dp_x}{dt} = F_2 \cos 21° + \ldots$$
$$\frac{dp_y}{dt} = F_2 \cos 69° + \ldots$$
$$\frac{dp_z}{dt} = 0 + \ldots$$

In the following examples, to simplify notation, we use the convention of writing F_1 instead of $|\vec{F}_1|$, etc. When doing this we need to be careful of signs, which must match the directions of forces.

EXAMPLE **Forces in Two Directions**

A metal block is suspended from a spring, which supports some, but not all, of the block's weight, so that the block rests lightly on a wooden board (Figure 5.7). The block is also pulled to the right by a spring and to the left by a copper wire. The vertical spring has a stiffness of 40 N/m and is stretched 3 cm beyond its relaxed length. The horizontal spring has a stiffness of 60 N/m and is stretched 1.5 cm. The mass of the block is 300 grams. **(a)** What is the tension force exerted on the block by the copper wire? **(b)** What is the normal force exerted on the block by the wooden board? **(c)** What force would the wooden board exert on the block if the vertical spring were removed?

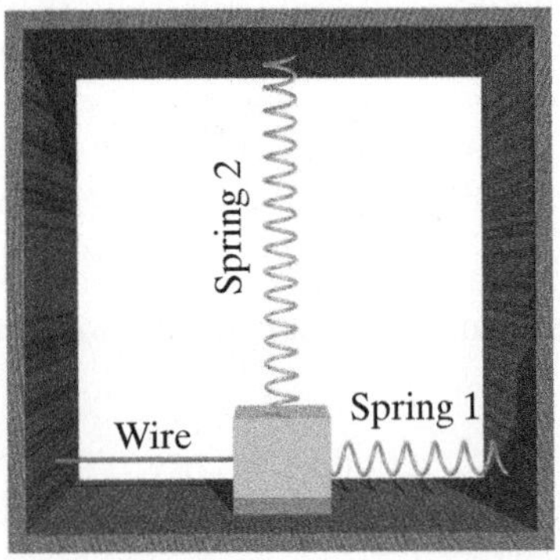

Figure 5.7 A block is partially supported by a vertical spring and rests lightly on a wooden surface. Another spring pulls the block from the right, and a wire pulls it from the left.

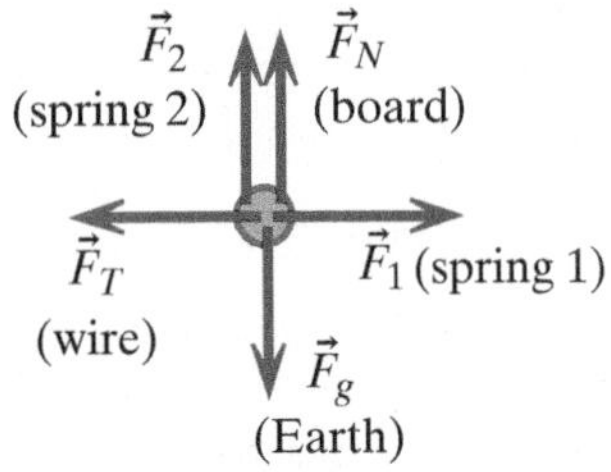

Figure 5.8 Free-body diagram for the block. Lengths of arrows representing unknown forces are arbitrary.

Solution System: Block
Surroundings: Spring 1, spring 2, board, wire, Earth
Free-body diagram: Figure 5.8

x components:

$$\frac{dp_x}{dt} = F_{\text{net},x}$$
$$0 = F_1 + (-F_T)$$
$$0 = (60\text{ N/m})(0.015\text{ m}) - F_T$$
$$F_T = 0.9\text{ N}$$

(a) The rod exerts a force of $\langle -0.9, 0, 0\rangle$ N on the block.
y components:

$$\frac{dp_y}{dt} = F_{\text{net},y}$$
$$0 = F_2 + F_N - F_g$$
$$0 = (40\text{ N/m})(0.03\text{ m}) + F_N - (0.3\text{ kg})(9.8\text{ N/kg})$$
$$F_N = 1.74\text{ N}$$

(b) The wood surface exerts a force of $\langle 0, 1.74, 0\rangle$ N on the block.
(c) If the vertical spring were removed the table would have to support the full weight of the block, which is 2.94 N, so the normal force would be $\langle 0, 2.94, 0\rangle$ N.

In the previous example every force was directed along the x, y, or z axis. In most problems, however, this isn't the case. Typically such problems involve relating angles to unit vectors in order to get the x, y, and z components of forces, as in the next example.

EXAMPLE **Pulling at an Angle**

A 3 kg block hangs from string 1 (Figure 5.9). Then you pull the block to the side by pulling horizontally with a second string (string 2). **(a)** What is the tension in string 2 in order for the block to hang motionless with string 1 at an angle of 33° to the vertical, as shown in Figure 5.9? **(b)** In this situation, what is the tension in string 1 (the force on the block by the string)? **(c)** In which situation are the interatomic bonds in string 1 stretched more: block hanging straight down, suspended from string 1, or block hanging at an angle, pulled on by string 1 and string 2?

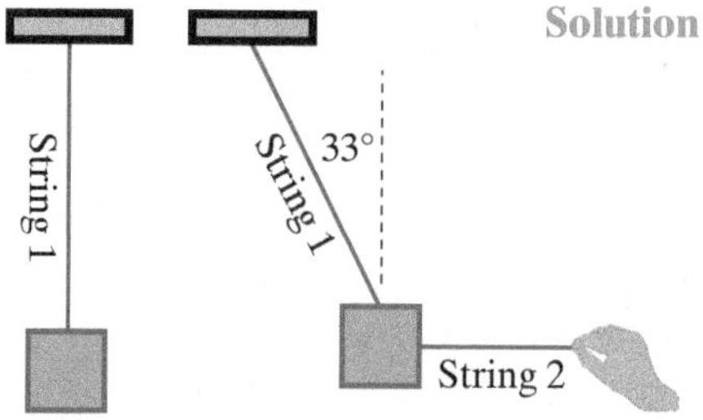

Figure 5.9 Left: A block hangs from a string. Right: Then you pull the block to the side with another string.

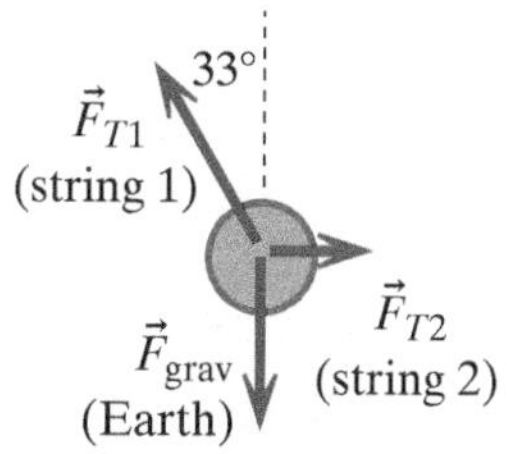

Figure 5.10 Free-body diagram.

Solution System: Block
Surroundings: String 1, string 2, and Earth
Free-body diagram: Figure 5.10

x components:

$$\frac{dp_x}{dt} = F_{\text{net},x}$$
$$0 = F_{T2} + F_{T1,x}$$

y components:

$$\frac{dp_y}{dt} = F_{\text{net},y}$$
$$0 = -mg + F_{T1,y}$$
$$0 = -(3\text{ kg})(9.8\text{N/kg}) + F_{T1,y}$$
$$0 = -29.4\text{ N} + F_{T1,y}$$

It may look as if we have two equations but three unknowns ($F_{T1,x}$, $F_{T1,y}$, and F_{T2}), which would make it impossible to solve the problem. However, we can write $F_{T1,x}$ and $F_{T1,y}$ in terms of F_{T1} by using direction cosines:

$$\hat{F}_{T1} = \langle \cos(90° + 33°), \cos 33°, 0\rangle$$
$$\vec{F}_{T1} = F_{T1}\langle \cos 123°, \cos 33°, 0\rangle$$

Now we have two equations in two unknowns:

$$0 = F_{T2} + F_{T1}\cos 123°$$
$$0 = -29.4\text{ N} + F_{T1}\cos 33°$$

and we can solve for the unknown force magnitudes (parts (a) and (b)):

$$F_{T1} = \frac{29.4\text{ N}}{\cos 33°} = 35.1\text{ N}$$
$$F_{T2} = -F_{T1}(\cos 123°) = -(35.1\text{ N})(-0.545) = 19.1\text{ N}$$

(c) When the block hangs straight down, the force exerted by string 1 must be equal in magnitude to the force by the Earth on the block, which is 29.4 N. Evidently the tension in string 1 is greater (35.1 N) when the block is pulled to the side than when the block hangs straight down. The interatomic bonds in string 1 must be stretched more when the block is pulled to the side.

It is interesting that in different situations the tension force exerted by an object such as a string or a wire may be different, associated with different amounts of stretch of interatomic bonds.

Check: The units are correct (N). Special cases: Consider other angles.

Call the angle between the string and the vertical θ (in the example above, $\theta = 33°$). We can rewrite the y component equation algebraically:

$$F_{T1} = \frac{mg}{\cos\theta}$$

If $\theta = 0°$, the block hangs vertically, and $F_{T1} = mg$, as we expect.
If $\theta = 90°$, string 2 is horizontal, and $F_{T1} = \infty$, which is impossible. This actually does make sense, since if both strings are horizontal, there is no y component to support the block—an impossible situation.

EXAMPLE **Tension and Young's Modulus (Pulling at an Angle)**

Suppose that we replace string 1 in the preceding example (Pulling at an Angle) with a thin steel wire whose length is 40 cm and thickness is 0.2 mm. How much will the wire stretch when the block is pulled to the side?

Solution Young's modulus for steel is approximately $2 \times 10^{11}\ \mathrm{N/m^2}$, though different kinds of steel differ in their stiffness.

$$Y = \frac{F/A}{\Delta L/L}$$

$$\Delta L = \frac{F/A}{Y/L} = \frac{(35.1\ \mathrm{N})/[\pi((0.2 \times 10^{-3}\ \mathrm{m})/2)^2]}{(2 \times 10^{11}\ \mathrm{N/m^2})/(0.4\ \mathrm{m})}$$

$$\Delta L = 2.2 \times 10^{-3}\ \mathrm{m} = 2.2\ \mathrm{mm}$$

EXAMPLE **Forces on a String in Tension (Pulling at an Angle)**

The horizontal string (string 2) in the previous example (Pulling at an Angle, Figure 5.9) applied a 19.1 N force to the right on the block. What was the force that your hand applied to the other end of that string?

Solution System: String 2 (Figure 5.9)
Surroundings: Block, hand (neglect Earth because the string mass is small)
Free-body diagram: Figure 5.11

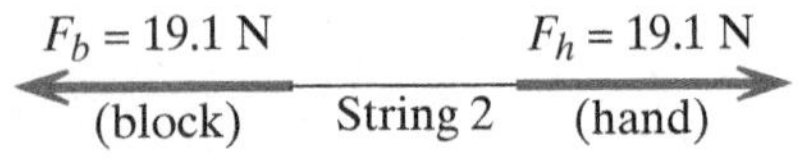

Figure 5.11 Free-body diagram for String 2 (Figure 5.9).

Choose the string as the system of interest, so the surroundings consist of the block, the Earth, and your hand. We can neglect the gravitational force on string 2, because in order for this force to be comparable to the tension, the string would have to have a mass of nearly 2 kg.

By reciprocity of (interatomic) electric forces, the force F_b that the block exerts to the left on the string is equal to F_T, the force that the string exerts on the block, so it has a magnitude of 19.1 N.

$$F_b = F_T$$

The momentum of string 2 is not changing, so we deduce that the force F_h exerted by your hand on the string must also have a magnitude of 19.1 N to the right on the string (Figure 5.11).

$$F_h = F_T$$

Evidently the forces at each end of a low-mass string in tension must be nearly equal in magnitude, and equal to the tension in the string.

EXAMPLE **Three Wires**

A 350 kg load is suspended from a vertical wire that is tied to two aluminum wires of radius 1.2 mm, as shown in Figure 5.12. Young's modulus for this aluminum alloy is $6.9 \times 10^{10}\ \mathrm{N/m^2}$. **(a)** Calculate the tension in each wire (that is, the magnitude of the tension force exerted by each wire). **(b)** What is the fractional stretch (the strain) of each wire?

Solution **(a)** This problem requires two steps, each with a different choice of system. In the first step we choose the load as the system, and find the tension force exerted on the load by the vertical wire. This is exactly the same problem as

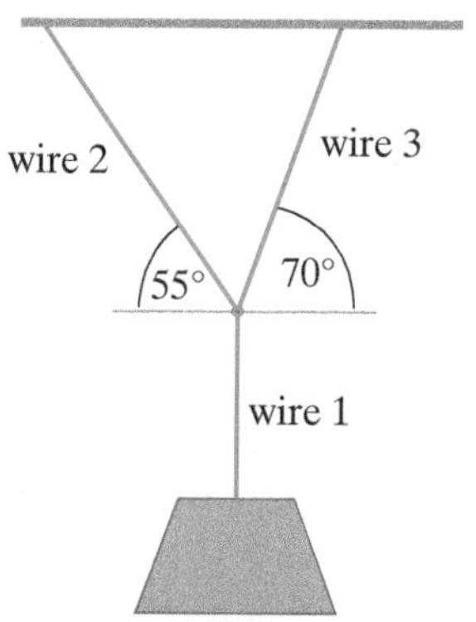

Figure 5.12 A load hangs from an aluminum wire that is tied to two other aluminum wires.

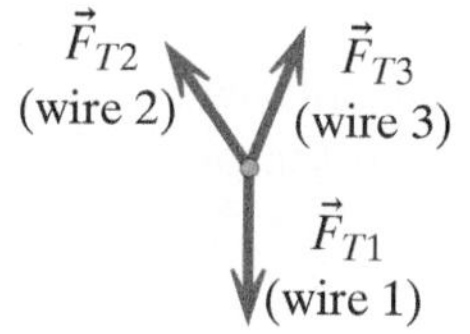

Figure 5.13 Free-body diagram for the knot.

the one we solved in the example entitled A Hanging Ball. We find that the magnitude of the tension in wire 1 is equal to the weight of the ball:

$$F_{T1} = (350\,\text{kg})(9.8\,\text{N/kg}) = 3430\,\text{N}$$

The choice of system for the second step may seem counterintuitive: we choose the knot connecting the three wires as the system. This allows us to relate the known force $\vec{F}_{T1}$ to the two unknown forces by the other wires.

System: Knot
Surroundings: Wire 1, wire 2, wire 3
Free-body diagram: Figure 5.13

Write the forces as vectors, using direction cosines:

$$\vec{F}_{T2} = F_{T2}\langle \cos 125°, \cos 35°, 0\rangle$$
$$\vec{F}_{T3} = F_{T3}\langle \cos 70°, \cos 20°, 0\rangle$$
$$\vec{F}_{T1} = \langle 0, -3430, 0\rangle\ \text{N}$$

Separate components:

$$\frac{dp_x}{dt} = F_{\text{net},x}$$
$$0 = F_{T2}\cos 125° + F_{T3}\cos 70° + 0$$
$$\frac{dp_y}{dt} = F_{\text{net},y}$$
$$0 = F_{T2}\cos 35° + F_{T3}\cos 20° - 3430\ \text{N}$$

We now have two equations in two unknowns (F_{T2} and F_{T3}). Solving for the unknown force magnitudes, we find that

$$F_{T2} = 1431\ \text{N}$$
$$F_{T3} = 2402\ \text{N}$$

Check: The units are correct (N). We can plug these results back into the momentum equations to verify that they satisfy those equations.

(b) Fractional stretch:

$$Y = \frac{F/A}{\Delta L/L}$$
$$A = \pi(1.2\times 10^{-3}\,\text{m})^2 = 4.52\times 10^{-6}\ \text{m}^2$$

$$\text{Wire 1:}\quad \Delta L/L = \frac{\left(\dfrac{1431\ \text{N}}{4.52\times 10^{-6}\ \text{m}^2}\right)}{6.9\times 10^{10}\,\text{N/m}^2} = 4.6\times 10^{-3}\quad \text{or about } 0.5\%$$

$$\text{Wire 2:}\quad \Delta L/L = \frac{\left(\dfrac{2402\ \text{N}}{4.52\times 10^{-6}\ \text{m}^2}\right)}{6.9\times 10^{10}\,\text{N/m}^2} = 7.7\times 10^{-3}\quad \text{or about } 0.8\%$$

Nonzero Velocity

Both systems at rest and systems moving with nonzero constant velocity exhibit uniform motion. The physical analysis process is the same in both cases.

EXAMPLE **Uniform Motion**

You place a 4 kg crate on the floor of an elevator that is temporarily stopped. Then the elevator starts moving, briefly speeds up, then moves upward at a constant speed of 2 m/s for many seconds (it is in a very tall building). While the elevator is moving upward at constant speed, what is the magnitude of the force exerted by the elevator floor on the crate?

Solution System: Crate
Surroundings: Earth, elevator floor
Free-body diagram: Figure 5.14
Momentum Principle:

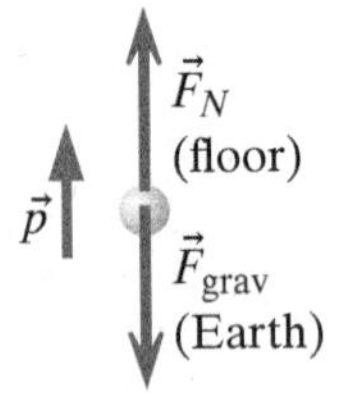

Figure 5.14 Free-body diagram for the crate. The momentum of the crate is also shown on the diagram.

$$\frac{dp_y}{dt} = F_N - mg$$
$$0 = F_N - (4\,\text{kg})(9.8\,\text{N/kg})$$
$$F_N = 39.2\ \text{N}$$

QUESTION How can the crate keep moving upward if the net force on it is zero?

Let's apply the Momentum Principle at a specific instant, when the crate (and elevator) are moving at constant speed. The momentum of the crate a short time Δt in the future depends on two things:

$$\vec{p}_{\text{future}} = \vec{p}_{\text{now}} + \vec{F}_{\text{net,now}}\Delta t$$

The current momentum of the crate, $\vec{p}_{\text{now}}$, is in the upward direction. If the net force on the crate is zero, then we can conclude that $\vec{p}_{\text{future}}$ will be equal in magnitude and direction to $\vec{p}_{\text{now}}$; the speed of the crate will not change. A nonzero net force would change the momentum of the crate, causing it either to speed up or to slow down, depending on the direction of that net force. For example, during the brief time when the elevator was speeding up, the crate's momentum was increasing, and we can deduce that during that time the net force on the crate must have been nonzero.

It is interesting to think about this situation in terms of the principle of relativity. We have analyzed the system from the viewpoint of an observer standing on the ground, who sees the elevator and crate moving at constant velocity, and who concludes that the net force on the crate must be zero. Suppose, however, that there is a video camera inside the elevator. From the frame of reference of the camera, the elevator and crate appear motionless. Looking at the video, an observer would also conclude that the net force on the crate was zero, because the momentum of the crate was not changing. The Momentum Principle is valid in both frames of reference.

EXAMPLE **Pulling a Sled**

You pull a loaded sled whose mass is 40 kg at constant speed in the x direction with a rope at an angle of 35°. The coefficient of kinetic friction between the snow and the sled runners is 0.2. What is the tension in the rope?

Solution System: Sled
Surroundings: Earth, snow, rope
Free-body diagram: Figure 5.15
Write out the vector components of $\vec{F}_T$:

$$\vec{F}_T = F_T\langle\cos 35°, \cos(90° - 35°), 0\rangle$$

The momentum of the sled is constant, so:

$$\frac{dp_x}{dt} = F_{\text{net},x}$$
$$0 = F_T \cos 35° - \mu_k F_N$$
$$\frac{dp_y}{dt} = F_{\text{net},y}$$
$$0 = F_N + F_T \cos(90° - 35°) - mg$$

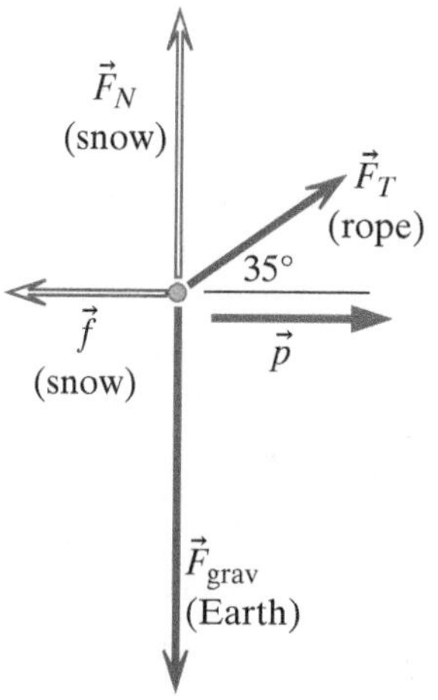

Figure 5.15 Free-body diagram for a sled pulled at constant speed. The sled's momentum is also shown.

Combine these two equations:

$$F_T \cos 35° - \mu_k (mg - F_T \cos 55°) = 0$$
$$F_T (\cos 35° + \mu_k \cos 55°) = \mu_k mg$$
$$F_T = \frac{\mu_k mg}{\cos 35° + \mu_k \cos 55°}$$
$$F_T = \frac{0.2(40\text{ kg})(9.8\text{ N/kg})}{0.819 + 0.2(0.574)} = 84.0\text{ N}$$

Pulling at an angle rather than horizontally makes the force squeezing the sled and snow together less than mg, which it would be if you pulled horizontally. Essentially you've partly lifted the sled out of the snow, which reduces the vertical component of the force of the snow, F_N, and therefore reduces the horizontal component of the force of the snow, the friction force $\mu_k F_N$.

Checkpoint 2 An object is moving with constant momentum $\langle 10, -12, -8 \rangle$ kg·m/s. What is the rate of change of momentum $d\vec{p}/dt$? What is the net force acting on the object? The object is acted on by three objects in the surroundings. Two of the forces are $\langle 100, 50, -60 \rangle$ N and $\langle -130, 40, -70 \rangle$ N. What is the third force?

Torque and Angular Momentum

Modeling a system as a point particle, and applying the Momentum Principle, can allow us to predict motion (when forces are known) or to find unknown forces (when motion is known). However, there are situations in which such an analysis, though valid, does not tell us everything we need to know. For example, consider two children sitting motionless on a seesaw (Figure 5.16). Choose as the system the two children and the board. For this choice of system, the objects in the surroundings that exert significant (external) forces on the system are the Earth (pulling down) and the support pivot (pushing up). The momentum of the system isn't changing, so the Momentum Principle correctly tells us that the net force on the system must be zero, and the support pivot must be pushing up with a force equal to the weight of the children plus the board.

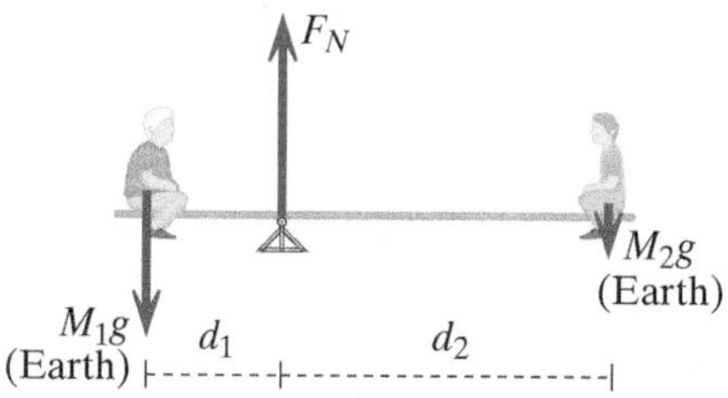

Figure 5.16 Two children sit on a seesaw in equilibrium.

However, the Momentum Principle alone doesn't tell us where the children have to sit in order to achieve balance and equilibrium. In order to learn more, we will need to model the system as an extended object (not a point particle) and to invoke the Angular Momentum Principle, which we will encounter in Chapter 11. By modeling the system as an extended object, we see that the force by the Earth on a child or on the unevenly balanced board exerts a twist (the technical term is "torque") about the pivot that tends to make the board turn. Torque has a magnitude and a direction, so it can be expressed as a vector. In order for the rate of rotation to be zero (or any other constant) the torques associated with the forces on the two children and the board must add up to zero. Torque is defined as force times lever arm, and a nonzero net *torque* causes changes in a quantity called angular momentum, just as a nonzero net *force* causes changes in the ordinary momentum we have been studying. The Angular Momentum Principle quantitatively relates the net torque on a system to the rate of change of the system's rotation.

In the present chapter we will continue to consider most objects (except things like wires, vines, and springs, which can stretch or compress) as point particles, which cannot deform or rotate.

5.5 CHANGING MOMENTUM

In the preceding sections we focused on systems whose momentum is constant. In such situations we know immediately that the value of $d\vec{p}/dt$ is zero, and that therefore the net force must also be zero. In this section we deal with systems consisting of several objects, in which the momentum may be changing. The motion of the center of mass is determined by the net force acting on the system, but we may want to know the forces that objects inside the system exert on each other.

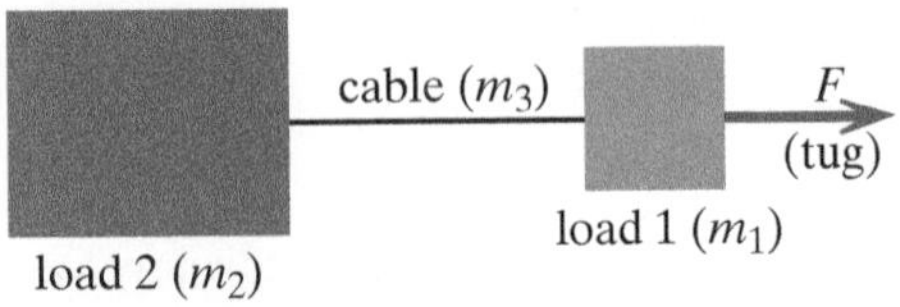

Figure 5.17 A space tug pulls two loads connected by a steel cable.

Consider the following example, which is typical of a class of problems called "dynamics" problems, in which it is often necessary to analyze different choices of system in order to determine some unknown forces, based on the motion. If you are studying engineering you may take an entire course on "dynamics."

EXAMPLE **Two Loads Connected by a Steel Cable**

You're an asteroid miner and your space tug is heading in the $+x$ direction toward a space station, pulling two loads connected by a steel cable. Your tug applies a force of $F = 10000$ N (Figure 5.17). The mass of load 1 is $m_1 = 500$ kg, the mass of load 2 is $m_2 = 2000$ kg, and the mass of the steel cable is $m_3 = 30$ kg. **(a)** What is the x component of the acceleration of the center of mass of the system? **(b)** What is the x component of the acceleration of load 1? Of load 2? Of the cable? **(c)** What is the tension in the cable? **(d)** Young's modulus for the steel used to make the cable is 2×10^{11} N/m^2. The cable is 5 m long and the diameter of the cable is 8 mm. How much is the cable stretched?

Solution **(a)** Choose the two loads and the cable as the system (Figure 5.18).
The Momentum Principle:

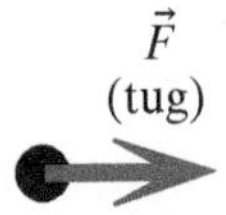

Figure 5.18 Free-body diagram for the two loads.

$$\frac{dp_{\text{cm},x}}{dt} = F_{\text{net,x}}$$

$$(m_1 + m_2 + m_3)\frac{v_{\text{cm},x}}{dt} = 10000 \text{ N}$$

$$a_{\text{cm},x} = \frac{10000 \text{ N}}{2530 \text{ kg}} = 3.95 \text{ m/s}^2$$

(b) As long as the cable remains taut, all of the parts of this multi-object system move together. Because the three objects have the same velocity but different masses, they have different momenta but the same acceleration, which is also the acceleration of the center of mass. This makes it convenient to analyze the motion in terms of acceleration.

(c) For the inclusive system of the two loads and the cable, the tension in the cable is internal to the system and doesn't contribute to the net force. To learn about the cable, we need to choose a different system to analyze, a system for which the cable is part of the surroundings. Let's choose load 2 as the system

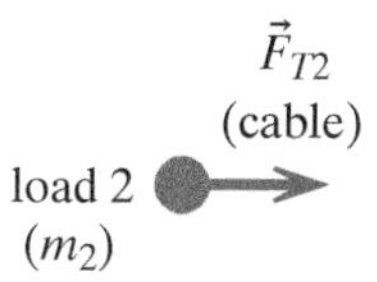

Figure 5.19 Free-body diagram for load 2.

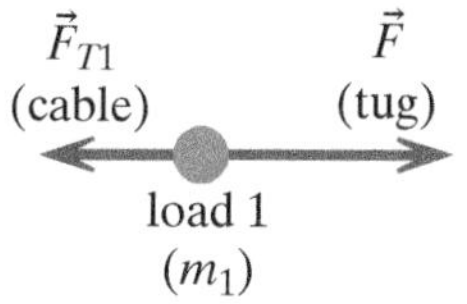

Figure 5.20 Free-body diagram for load 1.

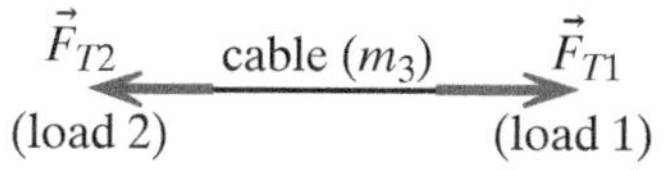

Figure 5.21 Free-body diagram for the cable.

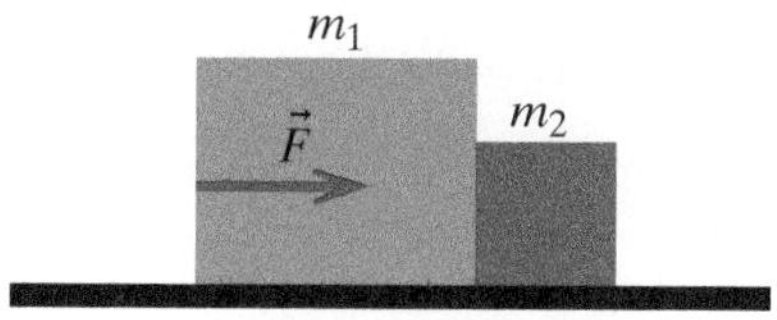

Figure 5.22 Push two blocks across a low-friction surface.

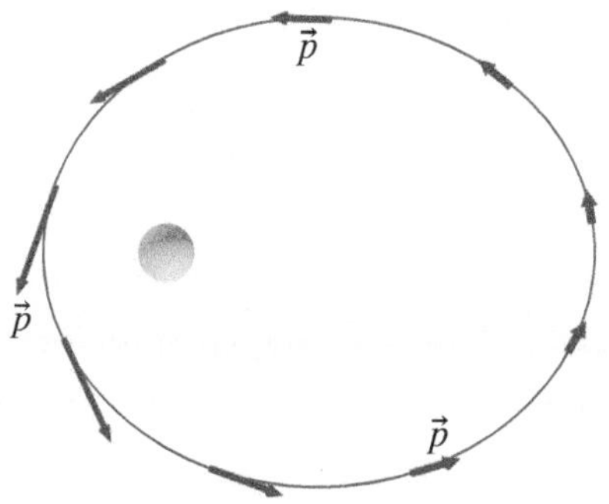

Figure 5.23 The momentum of a comet (arrows) changes in magnitude and direction as it orbits a star.

(Figure 5.19) and let F_{T2} represent the magnitude of the unknown force that the cable exerts on load 2:

$$\frac{dp_{2,x}}{dt} = F_{\text{net,x}}$$

$$m_2\frac{dv_{2,x}}{dt} = F_{T2}$$

$$F_{T2} = (2000 \text{ kg})(3.95 \text{ m/s}^2) = 7905 \text{ N}$$

Do we get the same answer if we choose load 1 as the system? Let F_{T1} represent the magnitude of the unknown force that the cable exerts on load 1 (Figure 5.20):

$$\frac{dp_{1,x}}{dt} = F_{\text{net,x}}$$

$$m_1\frac{dv_{1,x}}{dt} = F - F_{T1}$$

$$F_{T1} = 10000 \text{ N} - (500 \text{ kg})(3.95 \text{ m/s}^2) = 8024 \text{ N}$$

We see that the tension at the two ends of the cable differs by $(8024 - 7905) = 119$ N. The reason for this small difference is that we didn't neglect the mass of the cable. If you approximate the cable's mass as zero, you will find the same tension at both ends. A check on our calculations is to choose the cable as the system (Figure 5.21):

$$\frac{dp_{3,x}}{dt} = F_{\text{net,x}}$$

$$m_3\frac{dv_{3,x}}{dt} = F_{T1} - F_{T2}$$

$$F_{T1} - F_{T2} = (30 \text{ kg})(3.95 \text{ m/s}^2) = 119 \text{ N}$$

(d) Because the tension in the cable is somewhat greater on the right than on the left, the interatomic bonds are stretched more on the right than the left. To calculate the total stretch, an adequate approach is to use the average tension $(7905+8024)/2=7965$ N:

$$Y = \frac{F/A}{\Delta L/L}$$

$$\Delta L = \frac{FL}{YA} = \frac{(7965 \text{ N})(5 \text{ m})}{(2\times 10^{11} \text{N/m}^2)(\pi(4\times 10^{-3})^2 \text{ m}^2)}$$

$$\Delta L = 3.96\times 10^{-3} \text{ m} = 3.96 \text{ mm}$$

The key idea seen in this example is that by choosing different systems to which we apply the Momentum Principle, we obtain information about otherwise unknown forces acting among multiple objects inside the full system.

Checkpoint 3 You push two blocks across a low-friction surface, applying a known force $\vec{F}$ (Figure 5.22). **(a)** What system would you choose in order to determine the acceleration of the center of mass? **(b)** After determining the center-of-mass acceleration, what system would you choose in order to determine the magnitude of the force that block m_1 applies to block m_2?

5.6 FORCE AND CURVING MOTION

In some situations in which the momentum of a system is not constant, such as the case of an object moving along a curving path with varying speed (Figure 5.23), not only is $d\vec{p}/dt$ nonzero, but both $d\vec{p}/dt$ and $\vec{F}_{\text{net}}$ may be varying in magnitude and direction.

When analyzing the motion of systems moving in 2D or 3D with changing direction of momentum, it is often useful to divide both the $\vec{F}_{\text{net}}$ and $d\vec{p}/dt$ into two parts: a part parallel to the instantaneous momentum of the system and a part perpendicular to the momentum. Using this strategy, we will focus for the remainder of the chapter on understanding curving motion. We will find the values of unknown forces for such systems as a comet orbiting a star, Tarzan swinging on a vine, and a roller coaster looping the loop.

Parallel and Perpendicular Parts of Force

QUESTION What makes a moving object's path curve?

We know that a moving object whose path curves must be experiencing a nonzero net force due to objects in the surroundings, since

$$\vec{p}_{\text{future}} = \vec{p}_{\text{now}} + \vec{F}_{\text{net,now}}\Delta t$$

Figure 5.24 A force applied parallel to the momentum of a rolling ball.

and if $\vec{F}_{\text{net,now}}$ is zero, $\vec{p}_{\text{future}}$ will be the same as $\vec{p}_{\text{now}}$; the system's momentum will not change. But not all forces result in curving paths.

To begin, let's consider what forces *do not* cause an object's path to curve. Suppose that a soccer ball is rolling in a straight line on a smooth surface. You kick the rolling ball gently from behind, so the force you exert is in the direction of the ball's momentum (Fig 5.24).

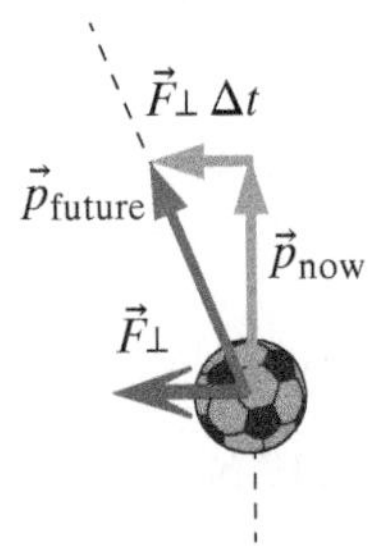

Figure 5.25 A kick perpendicular to the ball's momentum adds a sideways component to the ball's existing momentum.

QUESTION Will a force in the direction of motion cause the ball's momentum to change direction?

Clearly not. A force exerted parallel to the ball's momentum will speed the ball up, or slow it down or even reverse its direction if the force is opposite to $\hat{p}$, but it will not change the line along which the ball is moving. To make the ball turn, part of the force applied to the ball must be in a direction that is not along the line of the ball's motion.

The most extreme case of a force "not along the line of the ball's motion" would be a force that is perpendicular to the ball's momentum (Fig 5.25). Such a force would certainly change $\hat{p}$, the direction of the ball's momentum.

QUESTION Suppose that you kick the rolling soccer ball in a direction exactly perpendicular to its momentum. What will be the new direction of the ball's motion?

This depends on the relative magnitudes of the impulse you apply ($\vec{F}_{\perp}\Delta t$) and the ball's current momentum ($\vec{p}_{\text{now}}$), as shown in Figure 5.25. If the impulse you apply is relatively small, the ball won't turn very much. (Δt is the time your foot is in contact with the ball.)

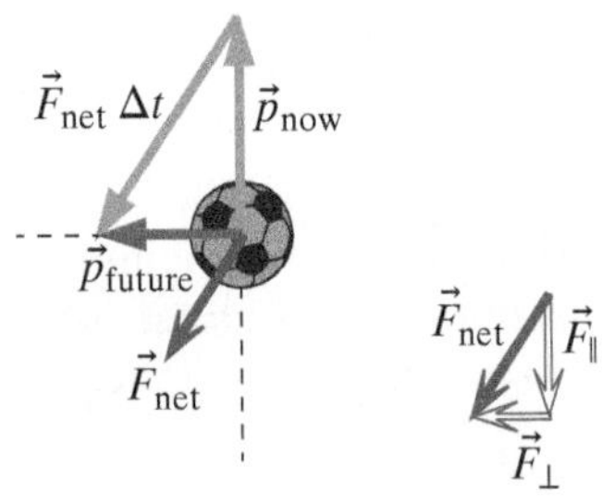

Figure 5.26 To make a 90° turn, you must kick backward at an angle. $\vec{F}_{||}\Delta t$ reduces the forward component of momentum to zero, and $\vec{F}_{\perp}\Delta t$ imparts a perpendicular component of momentum to the ball.

QUESTION If you want the future momentum of the ball to be in a direction perpendicular to its momentum now, in what direction should you kick the ball?

To force the rolling ball to turn 90° to the left, your kick would actually need to be directed partly backwards as well as to the left: both $\vec{F}_{||}$ and $\vec{F}_{\perp}$ would need to be nonzero, and $\vec{F}_{||}\Delta t$ would need to be equal in magnitude and opposite in direction to $\vec{p}_{\text{now}}$, as shown in Figure 5.26.

QUESTION Why can't you just kick exactly in the direction you want the ball to go?

We need to think about both $\vec{p}_{\text{now}}$ and the impulse $\vec{F}_{\text{net,now}}\Delta t$.

$$\vec{p}_{\text{future}} = \vec{p}_{\text{now}} + \vec{F}_{\text{net,now}}\Delta t$$

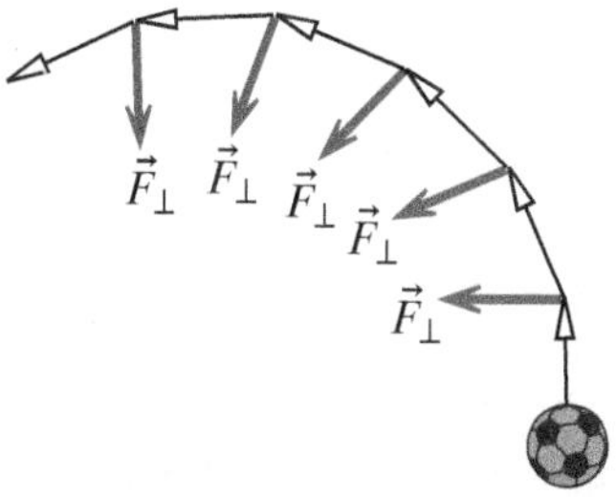

Figure 5.27 If you give a ball repeated big kicks perpendicular to its current heading, you can make the ball move in a roughly circular path.

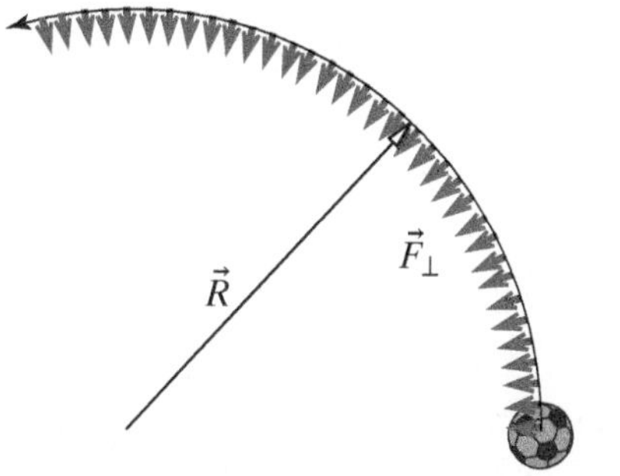

Figure 5.28 With a large number of small perpendicular kicks you can make a ball move along a path that is nearly indistinguishable from a circular path.

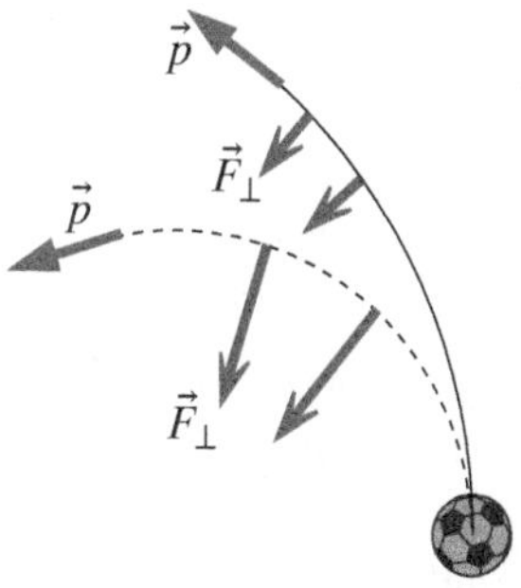

Figure 5.29 If you continually apply a bigger perpendicular force, you make the object turn more sharply, so the circle has a smaller radius.

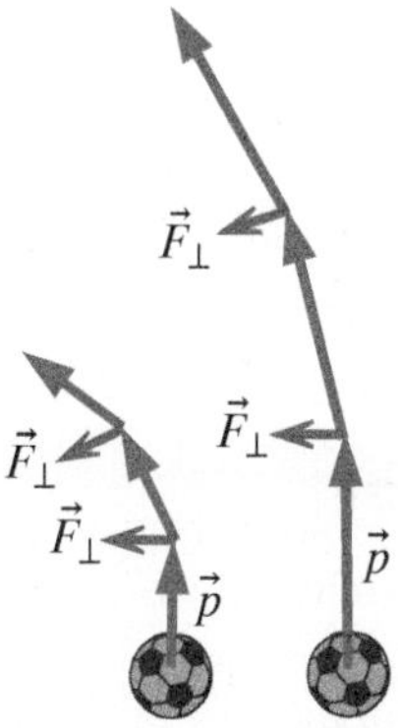

Figure 5.30 If the object's momentum is greater, the change due to the same perpendicular impulse is proportionally smaller, and the radius of curvature is larger.

If $\vec{p}_{\text{now}}$, the current momentum of the ball, is zero, then an impulse applied in the direction you want the ball to go would work. In the case discussed above, though, $\vec{p}_{\text{now}}$ is not zero, so we need to figure out what impulse would add the correct amount to $\vec{p}_{\text{now}}$ to get the desired $\vec{p}_{\text{future}}$ (Figure 5.26).

To express this idea generally, we note that any vector can be written as the sum of two other vectors that are perpendicular to each other. As shown in Figure 5.26, we can write the force $\vec{F}$ exerted on a moving system as the sum of two other forces: a force parallel to $\vec{p}$, called $\vec{F}_{\parallel}$, and a force perpendicular to $\vec{p}$, called $\vec{F}_\perp$:

$$\vec{F} = \vec{F}_{\parallel} + \vec{F}_\perp$$

We can write the Momentum Principle in a way that reflects this:

$$\Delta\vec{p} = (\vec{F}_{\parallel} + \vec{F}_\perp)\Delta t = \vec{F}_{\parallel}\Delta t + \vec{F}_\perp\Delta t$$

Circular Motion

Suppose that you deliver repeated kicks, at equal time intervals, to a rolling ball, each perpendicular to $\hat{p}$ at the current instant (Figure 5.27). If each perpendicular kick delivers the same impulse you can make the ball move along a roughly circular path, always turning to the left from its previous direction.

> QUESTION Suppose that instead of giving a series of infrequent big kicks you give a large number of small sideways kicks (with equal short times between these small kicks). What would the motion of the ball look like?

The ball would move along a path that is nearly indistinguishable from a circular path, as seen in Figure 5.28. We see that in order to make a ball move along a circular path, we have to continually push inward on the ball. If we always push perpendicular to the motion (and therefore toward the center of the circle), there is no parallel part of the force and so no change in the magnitude of the momentum, in which case the ball moves along a circular path with no change in speed, just a continuous change of direction.

> QUESTION Suppose that you kick just as often, but each of your kicks applies a bigger force than before. Would the radius of the circular path be larger or smaller?

Bigger kicks change the direction more, making sharper turns, so the radius of the circular path would be smaller (Figure 5.29). The larger the impulse, the larger $\vec{F}_\perp\Delta t$ is compared to $\vec{p}_{\text{now}}$, and the more the path will curve. Smaller impulses produce smaller changes; in fact, if you apply a very small sideways force, the motion will be nearly a straight line, which corresponds to an extremely large radius of curvature. In the limit of infinitesimal force, the radius will be infinite: the object moves in a straight line in the absence of a sideways force.

> QUESTION How does the speed of the ball affect the curvature of the path due to a perpendicular force of constant magnitude?

For a given impulse $\vec{F}_\perp\Delta t$, consider two different values of $\vec{p}_{\text{now}}$. The same impulse will produce a proportionally smaller change to a large initial momentum, so the higher the speed, the less a given force will make the path curve (Figure 5.30).

> QUESTION What would happen if you stopped kicking the ball? Would it continue to move in a circle?

The ball would move in a straight line, in the direction of its current momentum. Curving motion requires the continuous application of a force that is at least in part perpendicular to the instantaneous momentum of the object.

Calculating $\vec{F}_{\parallel}$ and $\vec{F}_{\perp}$

We have seen that the parallel part of a force, $\vec{F}_{\parallel}$, is associated with changes in the magnitude of the object's momentum (changes in $|\vec{p}|$) and the perpendicular part of the net force, $\vec{F}_{\text{net}\perp}$, is associated with changes in the direction of motion of the object (changes in $\hat{p}$). How do we calculate these vectors?

In Figure 5.31 the force $\vec{F}$ is expressed as the sum of two vectors: $\vec{F}_{\parallel}$ in the direction of $\hat{p}$ and $\vec{F}_{\perp}$ in the direction perpendicular to $\hat{p}$, in the plane defined by $\vec{F}$ and $\vec{p}$. Using direction cosines, we can see from the diagram that the part of $\vec{F}$ parallel to $\hat{p}$ is $|\vec{F}|\cos\theta$. This is the part of the force in the direction of $\hat{p}$, so:

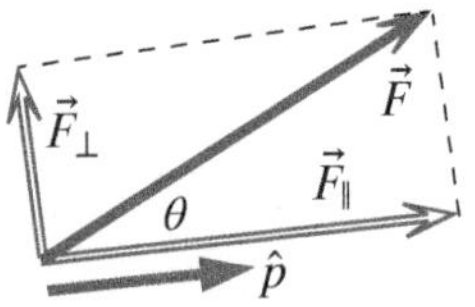

Figure 5.31 $\vec{F}$ can be expressed as the sum of $\vec{F}_{\parallel}$ and $\vec{F}_{\perp}$. We refer to the angle between $\vec{F}$ and $\hat{p}$ as θ.

$$\vec{F}_{\parallel} = |\vec{F}|\cos\theta\hat{p}$$

Since $\vec{F} = \vec{F}_{\parallel} + \vec{F}_{\perp}$ we can find $\vec{F}_{\perp}$ by subtraction:

$$\vec{F}_{\perp} = \vec{F} - \vec{F}_{\parallel}$$

EXAMPLE **Parallel Part of a Force**

A ball that is rolling in the $+z$ direction is kicked with a force of 50 N in a direction 33° from the $-z$ axis (Figure 5.32). Find $\vec{F}_{\parallel}$, the part of the force parallel to the ball's momentum. How will $\vec{F}_{\parallel}$ affect the magnitude of the ball's momentum?

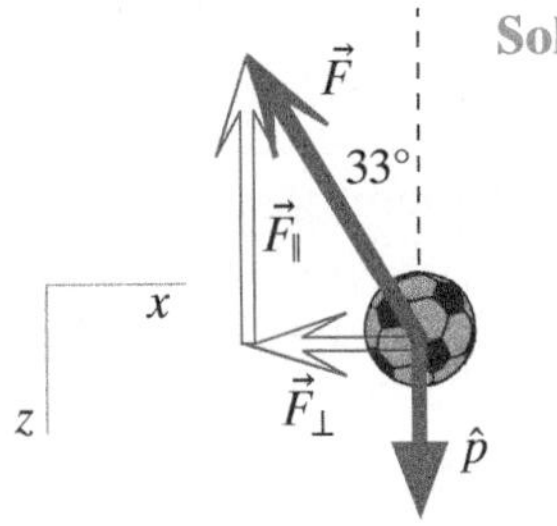

Figure 5.32 Top view, looking down on a ball rolling in the $+z$ direction. The ball is given a kick at an angle of 33° to the $-z$ axis.

Solution $\hat{p} = \langle 0,0,1\rangle$. The angle between $\vec{F}$ and $\hat{p}$ is $(180-33)° = 147°$.

$$\vec{F}_{\parallel} = (50\,\text{N})\cos(147°)\hat{p}$$
$$\vec{F}_{\parallel} = (-41.9\,\text{N})\hat{p} = \langle 0,0,-41.9\rangle\ \text{N}$$

The cosine of 147°, the angle between $\vec{F}$ and $\hat{p}$, is a negative number, so $\vec{F}_{\parallel}$ is opposite to $\hat{p}$. The magnitude of the ball's momentum will decrease.

QUESTION What would $\vec{F}_{\parallel}$ be if the angle between $\vec{F}$ and $\hat{p}$ were 90°?

The cosine of 90° is 0, so in this case $\vec{F}_{\parallel} = \vec{0}$, as it should.

The Vector Dot Product

The quantity $|\vec{F}|\cos\theta$ is an example of a more general mathematical operation called a vector dot product. For any two vectors $\vec{A}$ and $\vec{B}$, the dot product, which is written $\vec{A}\bullet\vec{B}$, is defined to be equal to $\vec{A}\bullet\vec{B} = |\vec{A}||\vec{B}|\cos\theta$, where θ is the angle between the two vectors (Figure 5.33):

$$\vec{A}\bullet\vec{B} = \vec{A}\bullet\vec{B} = |\vec{A}||\vec{B}|\cos\theta$$

The result of a vector dot product is a scalar, which may be positive or negative, depending on the angle θ.

We can write $\vec{F}_{\parallel}$ in terms of a vector dot product:

$$\vec{F}_{\parallel} = \left(\vec{F}\bullet\hat{p}\right)\hat{p} = |\vec{F}||\hat{p}|\cos\theta\hat{p}$$

Since $|\hat{p}| = 1$, the result is $|\vec{F}|\cos\theta\hat{p}$.

An alternative way of evaluating the vector dot product $\vec{A} \bullet \vec{B}$ uses the components of the vectors $\vec{A}$ and $\vec{B}$:

$$\vec{A} \bullet \vec{B} = A_x B_x + A_y B_y + A_z B_z$$

This result can be understood by expressing the two vectors in terms of the standard unit vectors $\hat{\imath} = \langle 1,0,0 \rangle$, $\hat{\jmath} = \langle 0,1,0 \rangle$, and $\hat{k} = \langle 0,0,1 \rangle$, and then multiplying out the product using the rules of ordinary algebra:

$$\langle A_x, A_y, A_z \rangle = (A_x\hat{\imath} + A_y\hat{\jmath} + A_z\hat{k})$$
$$\langle B_x, B_y, B_z \rangle = (B_x\hat{\imath} + B_y\hat{\jmath} + B_z\hat{k})$$
$$(A_x\hat{\imath} + A_y\hat{\jmath} + A_z\hat{k}) \bullet (B_x\hat{\imath} + B_y\hat{\jmath} + B_z\hat{k}) = A_x\hat{\imath} \bullet B_x\hat{\imath} + A_x\hat{\imath} \bullet B_y\hat{\jmath} + A_x\hat{\imath} \bullet B_z\hat{k} + \ldots$$

The dot product of any unit vector with itself is 1, because the angle between the vectors is 0°, and $\cos(0°) = 1$. Therefore:

$$\hat{\imath} \bullet \hat{\imath} = 1 \quad \text{and} \quad \hat{\jmath} \bullet \hat{\jmath} = 1 \quad \text{and} \quad \hat{k} \bullet \hat{k} = 1$$

In contrast, the dot product of any of these unit vectors with a different unit vector is zero, because the angles between them are 90°, and the cosine of 90° is zero. So

$$\hat{\imath} \bullet \hat{\jmath} = 0 \quad \text{and} \quad \hat{\jmath} \bullet \hat{k} = 0 \quad \text{and} \quad \hat{k} \bullet \hat{\imath} = 0$$

Therefore, any term such as $A_x\hat{\imath} \bullet B_y\hat{\jmath}$ is zero, and the only nonzero terms are those involving $\hat{\imath} \bullet \hat{\imath}$, $\hat{\jmath} \bullet \hat{\jmath}$, or $\hat{k} \bullet \hat{k}$:

$$(A_x\hat{\imath} + A_y\hat{\jmath} + A_z\hat{k}) \bullet (B_x\hat{\imath} + B_y\hat{\jmath} + B_z\hat{k}) = A_x B_x + A_y B_y + A_z B_z$$

Since in many circumstances we know the components of 3D vectors rather than knowing the angles between them (which are often not easy to calculate in 3D), the sum form of the dot product can be very useful.

Figure 5.33 Two vectors $\vec{A}$ and $\vec{B}$, with an angle θ between them.

VECTOR DOT PRODUCT

Given two vectors $\vec{A} = \langle A_x, A_y, A_z \rangle$ and $\vec{B} = \langle B_x, B_y, B_z \rangle$, the vector dot product is defined as

$$\vec{A} \bullet \vec{B} = |\vec{A}||\vec{B}|\cos\theta = A_x B_x + A_y B_y + A_z B_z$$

The angle θ is the angle between the two vectors; $\theta \leq 180°$ (Figure 5.33). The dot product is sometimes also called the "scalar product," because its result is a scalar.

EXAMPLE **$\vec{F}_{\parallel}$ in a Planetary Orbit**

A planet orbits a star, following an elliptical path. At an instant when the momentum of the planet is $\langle 3 \times 10^{29}, -6.0 \times 10^{28}, 0 \rangle$ kg·m/s, the force on the planet by the star is $\langle -4 \times 10^{22}, 1 \times 10^{23}, 0 \rangle$ N (Figure 5.34). Find the parts of the gravitational force parallel to and perpendicular to the momentum of the planet. How will these forces affect the momentum of the planet in the next small time interval?

Solution

$$\hat{p} = \frac{\langle 3\times10^{29}, -6\times10^{28}, 0\rangle \text{ kg}\cdot\text{m/s}}{\sqrt{(3\times10^{29})^2 + (-6\times10^{28})^2} \text{ kg}\cdot\text{m/s}}$$
$$= \langle 0.98, -0.20, 0\rangle$$
$$\vec{F}\bullet\hat{p} = (-4\times10^{22})(0.98) + (1\times10^{23})(-0.20) + (0)(0) \text{ N}$$
$$= -5.9\times10^{22} \text{ N}$$
$$\vec{F}_{\parallel} = (-5.9\times10^{22} \text{ N})\langle 0.98, -0.20, 0\rangle$$
$$= \langle -5.8\times10^{22}, 1.2\times10^{22}, 0\rangle \text{ N}$$
$$\vec{F}_{\perp} = \langle -4\times10^{22}, 1\times10^{23}, 0\rangle \text{ N} - \langle -5.8\times10^{22}, 1.2\times10^{22}, 0\rangle \text{ N}$$
$$= \langle 1.8\times10^{22}, 8.8\times10^{23}, 0\rangle \text{ N}$$

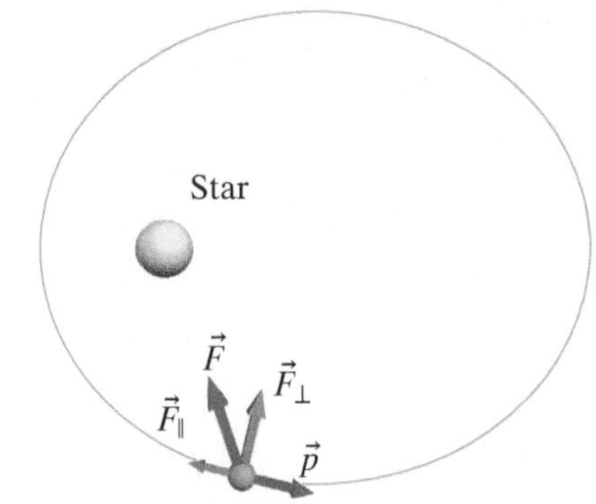

Figure 5.34 $\vec{F}$, $\vec{F}_{\parallel}$, and $\vec{F}_{\perp}$ at a particular instant for a planet orbiting a star.

Because the direction of $\vec{F}_{\parallel}$ is opposite to the direction of $\hat{p}$, the momentum of the planet will decrease during the next small time interval. Since $\vec{F}_{\perp}$ is not zero, the direction of the planet's momentum will also change.

In a computational model of curving motion, it can often be quite informative to display the parallel and perpendicular parts of the net force, since one determines changes in speed and the other changes in direction. The dot function in VPython provides a convenient way to calculate a dot product, and the norm function calculates a unit vector in the direction of a given vector. Assuming the current momentum and net force are the vectors `p` and `F`, one could use these VPython statements to calculate the force parts:

```
phat = norm(p)    # create unit vector from p
Fpara = dot(F, phat)*phat
Fperp = F - Fpara
```

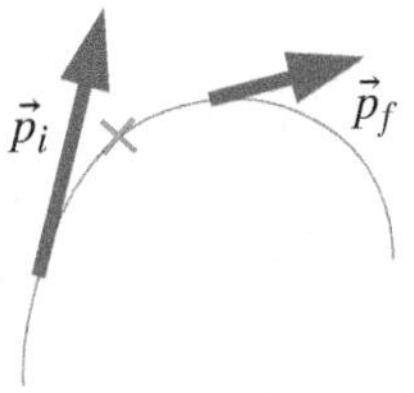

Figure 5.35 A system moves along the path shown. The arrows shown represent the momentum of the system at equal times before and after the time of interest, when the object is at the green x.

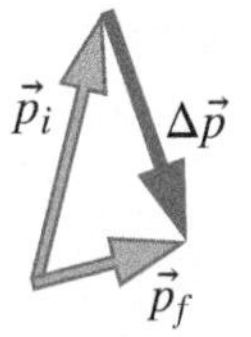

Figure 5.36 Graphically determine $\Delta\vec{p}$.

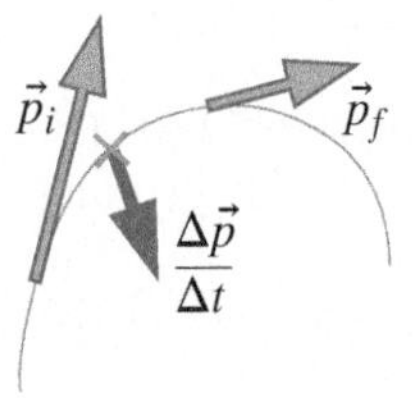

Figure 5.37 Divide $\Delta\vec{p}$ by Δt to get $\Delta\vec{p}/\Delta t$. Draw an arrow representing $\Delta\vec{p}/\Delta t$ at the location of interest. The smaller Δt is, the closer the result is to $d\vec{p}/dt$.

5.7 $d\vec{p}/dt$ FOR CURVING MOTION

It is not difficult to move from thinking about finite time steps to thinking about the instantaneous rate of change of momentum, $d\vec{p}/dt$, for curving motion. In Chapter 4 we used a three-step procedure to find $d\vec{p}/dt$ for a mass oscillating on a spring. We can use the same procedure, which is based on the definition of a derivative, to find $d\vec{p}/dt$ for an object whose path curves. Because we need to know the direction of $d\vec{p}/dt$ we often do this graphically, as in Figures 5.35–5.37.

FINDING $d\vec{p}/dt$ GRAPHICALLY

1. Draw two arrows representing $\vec{p}$ at equal times before and after the time of interest (Figure 5.35).
2. Find $\Delta\vec{p} = \vec{p}_f - \vec{p}_i$ (Figure 5.36).
3. Divide $\Delta\vec{p}$ by Δt to get $\Delta\vec{p}/\Delta t$ (Figure 5.37). The smaller Δt is, the closer the result is to $d\vec{p}/dt$.

QUESTION For curving motion, does $d\vec{p}/dt$ always point toward the inside of the curve?

Yes. Refer to Figures 5.27–5.30, in which we saw how a force perpendicular to motion caused an object's path to curve in the direction of that force. $d\vec{p}/dt$ is in the same direction as $\vec{F}_{\text{net}}$, and must point at least partly toward the inside of the curve, although it may also have a part parallel to $\vec{p}$ if the speed is changing as well.

Parallel and Perpendicular Parts of $d\vec{p}/dt$

We have seen that at any instant the net force on an object can be expressed as the sum of two parts: one that is parallel to the object's momentum and one that is perpendicular to the system's momentum. The same is true for the vector $d\vec{p}/dt$:

$$\frac{d\vec{p}}{dt} = \frac{d\vec{p}}{dt}_{\parallel} + \frac{d\vec{p}}{dt}_{\perp}$$

We can see this mathematically by "factoring" momentum into the product of the magnitude of the momentum (a scalar) and a unit vector in the direction of the momentum, and then taking the derivative, by applying the product rule. The resulting expression has two pieces, which correspond to the two ways in which momentum can change, in magnitude and in direction. (We use the notation convention that $p = |\vec{p}|$ to make the following discussion easier to follow.)

$$\vec{p} = p\hat{p}$$

$$\frac{d\vec{p}}{dt} = \frac{dp}{dt}\hat{p} + p\frac{d\hat{p}}{dt}$$

This may look unfamiliar, but we can make sense of it by considering each term separately.

> **QUESTION** Which term in the expression above involves a change in magnitude of momentum (change in speed)?

$\frac{dp}{dt}\hat{p}$ is nonzero if speed is changing, and zero otherwise.

> **QUESTION** Which term involves a change in direction?

$p\frac{d\hat{p}}{dt}$ is nonzero if direction is changing, and zero otherwise.

We can relate the parallel and perpendicular parts of $d\vec{p}/dt$ to the parallel and perpendicular parts of the net force responsible for changing the momentum of a system:

PARALLEL AND PERPENDICULAR

$$\frac{d\vec{p}}{dt}_{\parallel} = \frac{dp}{dt}\hat{p} = \vec{F}_{\text{net}\,\parallel}$$

$$\frac{d\vec{p}}{dt}_{\perp} = p\frac{d\hat{p}}{dt} = \vec{F}_{\text{net}\,\perp}$$

Note that this division is useful only for smooth, continuous curving motion. At turning points or other times when $\hat{p}$ changes abruptly, $d\hat{p}/dt$ is undefined.

This way of splitting the Momentum Principle is useful in finding the values of unknown forces that cause curving motion. If we can find the value of either part of $d\vec{p}/dt$, this will allow us to deduce the value of the corresponding part of the net force. We can show that, if we know an object's momentum and the instantaneous radius of curvature of its curving path, we can calculate $d\vec{p}/dt_{\perp}$.

The Kissing Circle

Motion along a circular path is easily described by giving the radius of the circle. However, we need a way to describe the instantaneous curvature of an object's path when that path is not a circle.

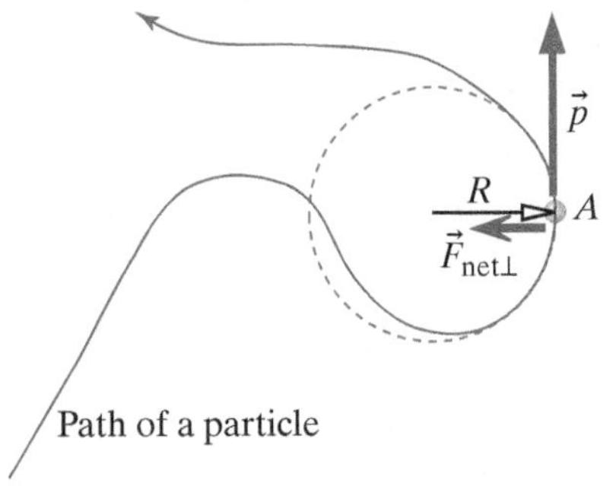

Figure 5.38 The "kissing circle" (dashed line) fits the trajectory at A as smoothly as possible.

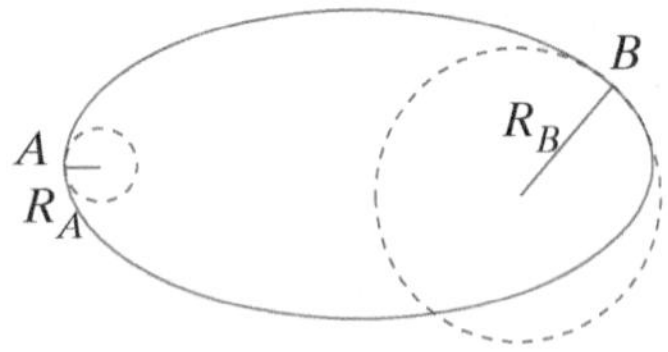

Figure 5.39 The radius of the kissing circle (dashed lines) can be different at different locations along a path, such as the elliptical path shown.

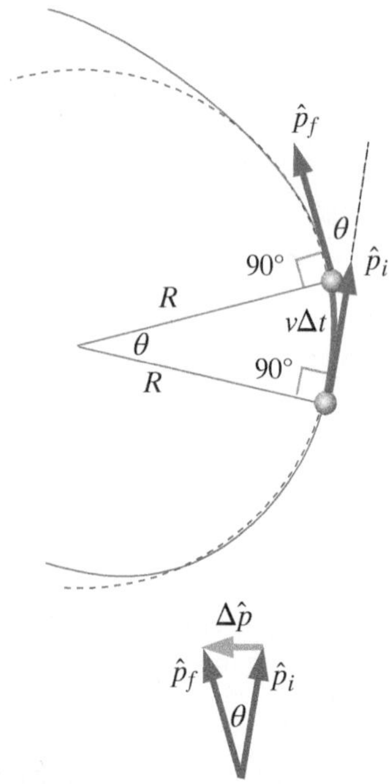

Figure 5.40 In a short time Δt, the particle moves a distance $v\Delta t$, through a small angle θ along a curve (red) whose radius of curvature is R. The angle θ, whose size is exaggerated in this diagram, is also the angle between $\hat{p}_i$ and $\hat{p}_f$. The kissing circle is indicated by a dashed line.

Consider motion along the curving path shown in Figure 5.38. A dashed circle has been drawn inside the path, tangent to the path at location A. This dashed circle of radius R is called the "osculatory" or "kissing" circle, because it just "kisses" the curving path at location A, fitting into the trajectory as smoothly as possible, with the circle and the trajectory sharing the same tangent ($\vec{p}$ is tangent to the path) and the same radius of curvature R at location A.

From the preceding discussion in which we considered applying a sequence of impulses, we see that both the perpendicular component of the net force, $\vec{F}_{net\perp}$, and the part of $d\vec{p}/dt$ perpendicular to the instantaneous momentum point toward the center of the kissing circle. If this force were bigger, the radius of the kissing circle would be smaller (a sharper turn). If on the other hand the momentum were bigger, the force would make a smaller change in direction in the same amount of time, and the kissing circle would be bigger. If the curvature of an object's path varies, then the radius of the kissing circle will be different at different locations along the path (Figure 5.39).

Calculating $d\vec{p}/dt_{\perp}$

We can start by using geometry to derive an expression for $d\hat{p}/dt$, the rate at which the direction of momentum changes. In Figure 5.40 we see a particle that in a short time Δt moves a distance $v\Delta t$, through a small angle θ along a curve whose radius of curvature is R. The small angle θ (measured in radians) is the arc length $v\Delta t$ divided by the radius R of the kissing circle: $\theta = v\Delta t/R$. (We've exaggerated the size of θ in order to be able to see it easily on the diagram.)

Because $\hat{p}$ is perpendicular to the radius of the kissing circle, as the particle moves through the small angle θ the vector $\hat{p}$ rotates through the same angle θ. At the bottom of Figure 5.40 we place the initial and final unit vectors tail to tail and show the change, $\Delta\hat{p}$. In this triangle, in the limit of a small angle θ, the angle θ is the length $|\Delta\hat{p}|$ divided by the length $|\hat{p}_i|$. Therefore

$$\theta = \frac{|\Delta\hat{p}|}{|\hat{p}_i|} = \frac{v\Delta t}{R}$$

Since $|\hat{p}_i| = 1$, we can rearrange to get this result:

$$\left|\frac{\Delta\hat{p}}{\Delta t}\right| = \frac{v}{R}$$

In the limit as Δt becomes infinitesimally small, we have the derivative:

$$\lim_{\Delta t \to 0}\left|\frac{\Delta\hat{p}}{\Delta t}\right| = \left|\frac{d\hat{p}}{dt}\right| = \frac{v}{R} \quad \text{on a smoothly curving path}$$

Check units: $\hat{p}$ is dimensionless, so $|d\hat{p}/dt|$ has units of $1/s$. Since v/R also has units of $1/s$, the units match.

RATE OF CHANGE OF DIRECTION

$$\left|\frac{d\hat{p}}{dt}\right| = \frac{v}{R}$$

R is the radius of the kissing circle. The direction of $d\hat{p}/dt$ is perpendicular to the particle's momentum, toward the center of the kissing circle. This result is valid *only* for a smoothly continuous curving motion, not for times when $\hat{p}$ changes abruptly, as in a collision or at a turning point.

We now have an expression for $|d\vec{p}/dt_\perp|$, the magnitude of the perpendicular part of $d\vec{p}/dt$:

MAGNITUDE OF $d\vec{p}/dt_\perp$

$$\left|\frac{d\vec{p}}{dt}_\perp\right| = p\left|\frac{d\hat{p}}{dt}\right| = p\left(\frac{v}{R}\right) = \frac{\gamma m v^2}{R}$$

At speeds much less than the speed of light:

$$\left|\frac{d\vec{p}}{dt}_\perp\right| \approx \frac{mv^2}{R}$$

The direction of $d\vec{p}/dt_\perp$ is toward the center of the kissing circle. R is the radius of the kissing circle. This result is valid *only* for a smoothly continuous curving motion, not for times when $\hat{p}$ changes abruptly, as in a collision or at a turning point.

If you examine the preceding derivation you'll see that it also applies to the rate of change of any unit vector that changes its direction at the same rate as the momentum unit vector, such as $\hat{v}$. Therefore we can write a similar expression for the perpendicular part of the acceleration $\vec{a} = d\vec{v}/dt$:

$$|\vec{a}_\perp| = \left|\left(\frac{d\vec{v}}{dt}\right)_\perp\right| = v\left|\frac{d\hat{v}}{dt}\right| = v\left(\frac{v}{R}\right) = \frac{v^2}{R}$$

Again, this result is valid *only* for a smoothly continuous curving motion, not for times when $\hat{v}$ changes abruptly, as in a collision or at a turning point.

EXAMPLE **Finding $\vec{F}_{\text{net}\perp}$**

Suppose that a 3 kg object is moving with a speed of 4 m/s and at a certain instant the radius of curvature of the kissing circle (the best fit to the curving motion at that point) is 2 m. What are the magnitude and direction of $\vec{F}_{\text{net}\perp}$?

Solution We don't know anything about the surroundings of the moving object, so we cannot calculate $\vec{F}_{\text{net}}$ directly. However, we can calculate $|d\vec{p}/dt_\perp|$, and use this to deduce the value of $|\vec{F}_{\text{net}\perp}|$:

$$\left|\frac{d\vec{p}}{dt}_\perp\right| = p\left(\frac{v}{R}\right) = (3\,\text{kg}\cdot 4\,\text{m/s})\frac{4\,\text{m/s}}{2\,\text{m}} = 24\,\text{kg}\cdot\text{m/s}^2$$

Therefore $|\vec{F}_{\text{net}\perp}| = 24\,\text{N}$

The direction of $(d\vec{p}/dt)_\perp$ is toward the center of the kissing circle, so the direction of $\vec{F}_{\text{net}\perp}$ is also toward the center of the kissing circle.

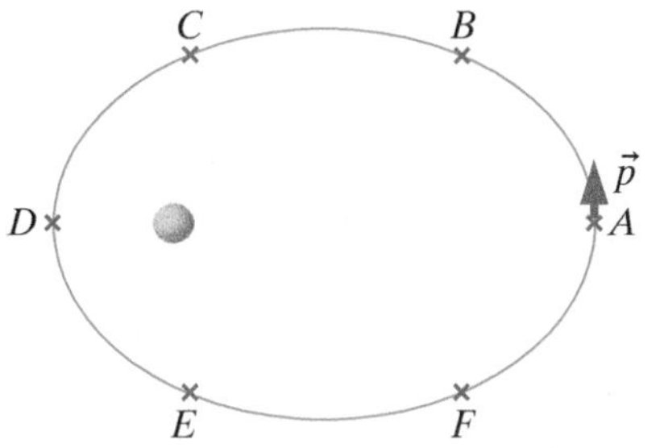

Figure 5.41 The elliptical orbit of a comet around a star. An arrow represents the momentum of the comet when it is at location A.

Checkpoint 4 A comet orbits a star in an elliptical orbit, as shown in Figure 5.41. The momentum of the comet at location A is shown in the diagram. At the instant the comet passes each location labeled A, B, C, D, E, and F, answer the following questions about the net force on the comet and the rate of change of the momentum of the comet: **(a)** Draw an arrow representing the direction and relative magnitude of the gravitational force on the comet by the star. **(b)** Is $\vec{F}_{\text{net}\perp}$ zero or nonzero? **(c)** Is $\vec{F}_{\text{net}\parallel}$ zero or nonzero? **(d)** Is $d|\vec{p}|/dt$ positive, negative, or zero? **(e)** Is $d\hat{p}/dt$ zero or nonzero?

"Centrifugal" and "Centripetal"

You may sometimes hear the terms "centrifugal" and "centripetal" in the context of discussions of forces and curving motion. These terms have simple meanings: "centrifugal" means "away from the center" and "centripetal" means "toward the center."

We have seen that both $d\vec{p}/dt_\perp$ and $\vec{F}_{\text{net}\,\perp}$ point toward the center of the kissing circle, so we could correctly refer to $\vec{F}_{\text{net}\,\perp}$ as a centripetal force, and $d\vec{p}/dt_\perp$ as a centripetal rate of change. However, there doesn't seem to be any advantage in using such terms.

> QUESTION Is there a "centrifugal" force involved in curving motion?

No. As long as we stay in our standard inertial reference frame, the force that causes an object's path to curve is directed toward the inside of the curve, not the outside. Probably it is the kinesthetic sensations that people have when riding in vehicles that turn that lead to confusion about this issue. For a discussion of how the kinesthetic sensations we experience are related to a physical analysis of motion, see Section 5.9. To avoid confusion, remember that each force included in a free-body diagram must be associated with an object in the surroundings.

Turning Points: System Momentarily at Rest

We have noted above that the quantity $d\hat{p}/dt$ is undefined at moments when an object's momentum changes abruptly, such as turning points where the direction of $\hat{p}$ is reversed. How do we analyze turning points, when an object is momentarily motionless?

EXAMPLE

Tarzan Swinging Back and Forth

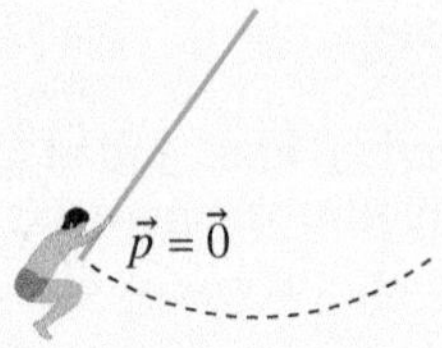

Figure 5.42 Tarzan is momentarily at rest.

Tarzan hangs from a vine, swinging back and forth. At the moment when he reaches his maximum displacement from the vertical, as indicated in Figure 5.42, Tarzan is momentarily at rest (a "turning point"). At this instant, is the rate of change of Tarzan's momentum zero or nonzero? If nonzero, what is the direction of $d\vec{p}/dt$? At this moment, is the net force on Tarzan zero or nonzero?

Solution

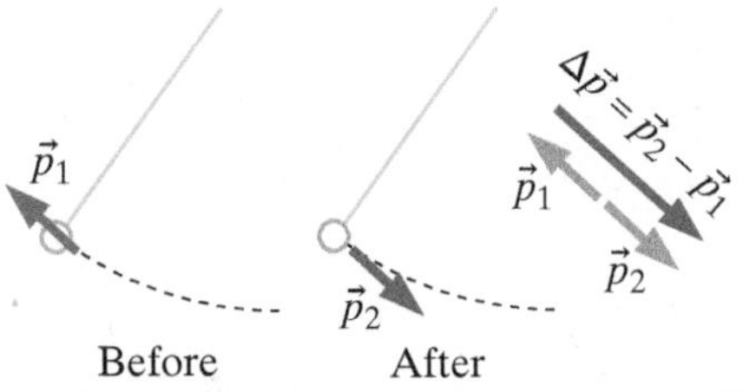

Figure 5.43 Tarzan's momentum an instant before and an instant just after he is momentarily at rest, and $\Delta\vec{p}$.

We use the procedure for finding $d\vec{p}/dt$ shown in Figure 5.43. Both immediately before this moment and immediately after this moment, Tarzan's momentum is nonzero; this is a turning point. The directions of his momenta at these two instants are shown in Figure 5.43, as is $\Delta\vec{p} = \vec{p}_2 - \vec{p}_1$. Therefore, at the instant Tarzan is motionless, $d\vec{p}/dt$ is nonzero, and its direction is the direction of $\Delta\vec{p}$, as shown in Figure 5.44.

Note that it is important to consider a very short time interval in order to get the direction of $d\vec{p}/dt$. The direction of $\Delta\vec{p}$ would be different for a longer time interval (and not the same as the direction of $d\vec{p}/dt$).

To decide if the net force on Tarzan is zero or not, we can reason from the update form of the Momentum Principle:

$$\vec{p}_{\text{future}} = \vec{p}_{\text{now}} + \vec{F}_{\text{net,now}}\Delta t$$

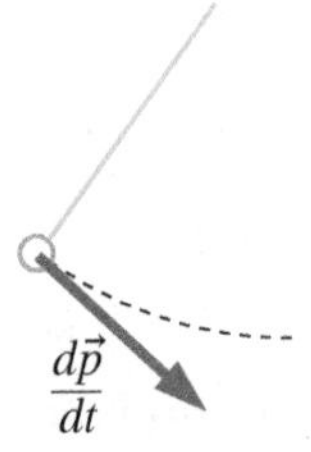

Figure 5.44 The direction of $d\vec{p}/dt$ at the instant Tarzan is motionless.

Pick the moment when the system is at rest as "now," and a short time Δt later as "future." We know that $\vec{p}_{\text{now}}$ is zero but $\vec{p}_{\text{future}}$ will not be zero, so we deduce that $\vec{F}_{\text{net,now}}$ is nonzero.

We get the same answer reasoning from the derivative form of the Momentum Principle:

$$d\vec{p}/dt = \vec{F}_{\text{net}}$$

Since at a turning point $d\vec{p}/dt \neq \vec{0}$, then $\vec{F}_{net}$ must also be nonzero. The direction of $\vec{F}_{net}$ must be the same as the direction of $d\vec{p}/dt$.

QUESTION At a particular instant a block hanging from a spring is at rest. What is the net force on the block?

Without more information, we can't answer this question. If the block is at rest and remains at rest, this is a case of uniform motion, and the net force on the system is zero. However, if the block is only momentarily at rest, then the rate of change of its momentum is not zero, and the net force on the block is not zero.

- For a system that remains at rest, $d\vec{p}/dt = \vec{0}$.
- For a system that is only momentarily at rest, $d\vec{p}/dt \neq \vec{0}$.

5.8 UNKNOWN FORCES: CURVING MOTION

If we know the mass and speed of an object moving along a curving path whose instantaneous radius is known, we can calculate $|d\vec{p}/dt_{\perp}|$, and use this to find the magnitude of $\vec{F}_{net\,\perp}$. We know that at any instant the direction of $\vec{F}_{net\,\perp}$ is toward the center of the kissing circle.

In each problem we also need to think about whether or not $|d\vec{p}/dt_{\parallel}|$ is zero or nonzero at the particular instant of interest. If it is zero, then $|\vec{F}_{net\,\parallel}| = 0$, and $\vec{F}_{net} = \vec{F}_{net\,\perp}$; if not, we need more information in order to calculate $\vec{F}_{net\,\parallel}$.

EXAMPLE **A Child on a Merry-Go-Round**

A child whose mass is 30 kg sits on a merry-go-round at a distance of 3 m from the center (Figure 5.45). The merry-go-round makes one revolution every 8 s. What are the magnitude and direction of the net force acting on the child?

Solution System: Child
Surroundings: Earth, seat

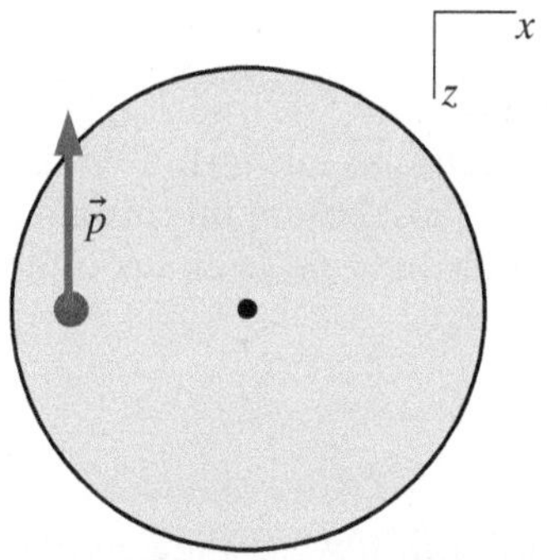

Figure 5.45 Top view of a child (red dot) on a merry-go-round that is rotating clockwise. At the instant shown the child's momentum is in the $-z$ direction.

Parallel: Because the merry-go-round revolves at a constant rate, the speed of the child, and the magnitude of the child's momentum, are constant. Therefore we know that:

$$\frac{d\vec{p}}{dt}_{\parallel} = \vec{0} \quad \text{and therefore} \quad \vec{F}_{net\,\parallel} = \vec{0}$$

Perpendicular: The direction of the child's momentum is changing continuously, so:

$$\left|\frac{d\vec{p}}{dt}_{\perp}\right| = \frac{mv^2}{R} = |\vec{F}_{net\,\perp}|$$

What is the speed v? In 8 s the child travels once around the circumference $2\pi R$, so the speed is $v = 2\pi(3\,\text{m})/(8\,\text{s}) = 2.36\,\text{m/s}$. So

$$|\vec{F}_{net\,\perp}| = \frac{(30\,\text{kg})(2.36\,\text{m/s})^2}{3\,\text{m}} = 55.5\,\text{N}$$

Since $\vec{F}_{net\,\perp}$ is the only nonzero part of the net force, we know that this is also the magnitude of the net force, which at any instant is directed toward the center of the circle. In the instant shown in Figure 5.45, this force would be in the $+x$ direction.

QUESTION What object in the surroundings exerts this force on the child?

We listed only two objects in the surroundings that exert significant forces on the child. The Earth exerts a downward force, so that isn't it. The seat exerts a force on the child that is partly upward, balancing the Earth's force, and partly toward the center of the circle. This part of the force is a friction force—if the seat were extremely slippery, the child would slide off it instead of moving around in a circle.

Tension Forces and Curving Motion

We now have all the tools we need to answer the question about Tarzan and the vine posed at the beginning of Chapter 4: Why does the vine break when Tarzan swings on it, but not when he hangs motionless? In the following problem, we can calculate $d\vec{p}/dt$, but the magnitude of the tension force on Tarzan by the vine is unknown. We can find the unknown $|\vec{F}_{\text{vine}}|$ by applying the Momentum Principle.

EXAMPLE **Tarzan and the Broken Vine**

Tarzan, whose mass is 90 kg, wants to use a vine to swing across a river. To make sure the 8-m-long vine is strong enough to support him, he tests it by hanging motionless on the vine for several minutes. The vine passes this test, so Tarzan grabs the vine and swings out over the river. He is annoyed and perplexed when the vine breaks midway through the swing, when his speed is 12 m/s (Figure 5.46), and he ends up drenched and shivering in the middle of the cold river, to the great amusement of the onlooking apes. Why did the vine break while Tarzan was swinging on it, but not while he hung motionless from it?

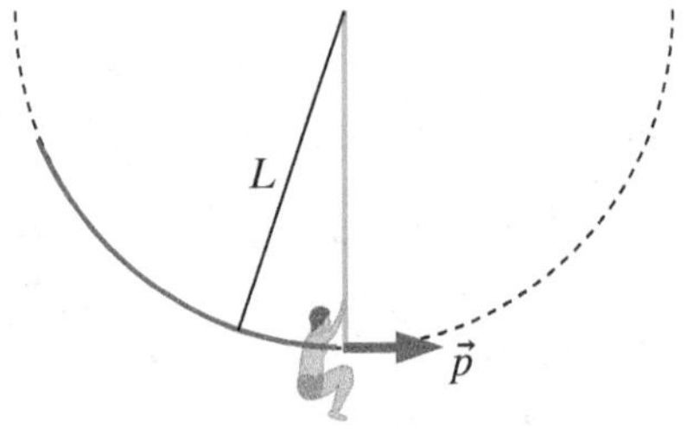

Figure 5.46 Tarzan at the bottom of his swing, the instant before the vine breaks. The colored line indicates his path up to this moment. The radius of the kissing circle (dashed line) is L, the length of the vine. At this moment Tarzan's speed is at a maximum.

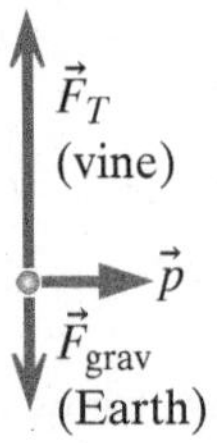

Figure 5.47 Free-body diagram.

Solution System: Tarzan
Surroundings: Vine, Earth (neglect air)
Free-body diagram: Figure 5.47

Parallel: At the instant shown in Figure 5.46, we know that the only forces acting on Tarzan are in the $+y$ or $-y$ direction (Figure 5.47), so $F_{\text{net}\|} = 0$. This implies that $|d\vec{p}/dt_{\|}| = 0$. This makes sense because at this instant $|\vec{p}|$ is going through a maximum: Tarzan's speed is increasing just before this instant (as he descends), and would decrease just afterward (as he went back up, if the vine did not break). So

$$\frac{d\vec{p}}{dt}_{\|} = \vec{F}_{\text{net}\,\|} = \vec{0}$$

Perpendicular: The direction of $d\vec{p}/dt_{\perp}$ is upward, toward the center of the kissing circle.

$$\left|\frac{d\vec{p}}{dt}_{\perp}\right| = \left|\vec{F}_{\text{net}\,\perp}\right|$$

$$\frac{mv^2}{L} = F_T - mg$$

$$\frac{(90\,\text{kg})(12\,\text{m/s})^2}{(8\,\text{m})} = F_T - (90\,\text{kg})(9.8\,\text{N/kg})$$

$$F_T = 2502\text{ N}$$

QUESTION When Tarzan hung motionless on the vine, what was the tension in the vine?

$$\frac{dp_y}{dt} = F_T - mg$$
$$0 = F_T - (90\,\text{kg})(9.8\,\text{N/kg})$$
$$F_T = 882\,\text{N}$$

The tension force exerted by the vine was nearly three times as great when Tarzan was midway through his swing as it was when he hung motionless. The interatomic bonds in the vine were stretched three times as much while he swung—evidently too much!

QUESTION Why must the vine exert a greater force when Tarzan's momentum is changing than when he hangs motionless?

It is not simply that Tarzan is moving—zero net force would be required if Tarzan's momentum were constant. However, to turn Tarzan's momentum from horizontal to upward requires an upward net force. The extra tension in the vine is necessary to change the direction of Tarzan's momentum. To swing upward Tarzan has to pull down harder on the vine, and therefore the vine stretches more, pulling up more on Tarzan.

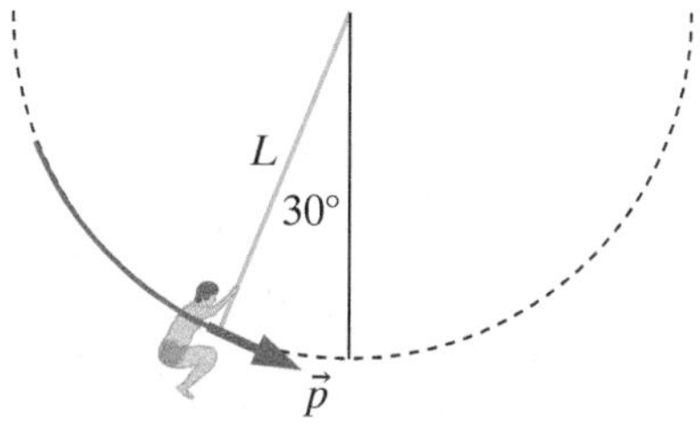

Figure 5.48 Tarzan swinging downward, with a speed of 11.1 m/s.

In the preceding example we picked a special instant when $|d\vec{p}/dt_{\parallel}|$ was zero. How would we go about finding the tension in the vine at some other instant?

EXAMPLE **Magnitude and Direction of Momentum Changing**

Consider the instant shown in Figure 5.48, before the vine breaks, when the angle between the vine and the vertical is 30°. At this instant Tarzan's speed is 11.1 m/s, and both the magnitude and direction of his momentum are changing. As before, Tarzan's mass is 90 kg, and the vine is 8 m long. **(a)** What is the tension in the vine at this moment? **(b)** What is the rate of change of the magnitude of Tarzan's momentum?

Solution System: Tarzan
Surroundings: Earth, vine (ignore the air)
Free-body diagram: Figure 5.49

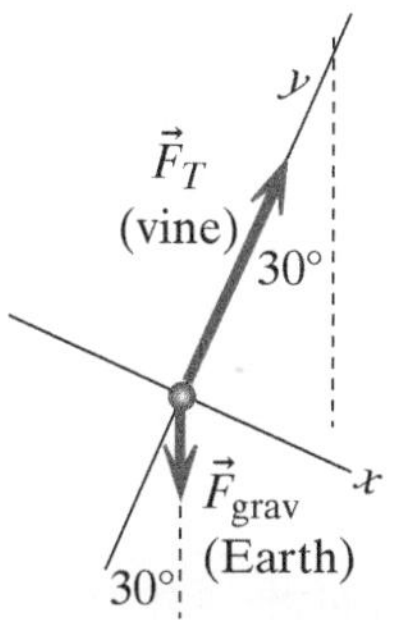

Figure 5.49 Free-body diagram. We have chosen to align the x axis with $\hat{p}$, and the y axis perpendicular to $\hat{p}$.

Here we have chosen to orient our coordinate axes so that the x axis is aligned with $\hat{p}$. Therefore the y axis is perpendicular to $\hat{p}$, and is aligned with the vine. This is done for convenience; we would get the same answers for the magnitudes of the quantities if we had chosen a different orientation for the axes, but the unit vectors $\hat{p}$ and $\hat{F}_T$ would be different.

(a) Perpendicular (y):

$$\left|\frac{d\vec{p}}{dt}_{\perp}\right| = \left|\vec{F}_{\text{net}\perp}\right|$$
$$\frac{mv^2}{L} = F_T + mg\cos(180° - 30°)$$
$$\frac{(90\,\text{kg})(11.1\,\text{m/s})^2}{(8\,\text{m})} = F_T + (90\,\text{kg})(9.8\,\text{N/kg})\cos 150°$$
$$F_T = 2150\,\text{N}$$
$$\vec{F}_T = \langle 0, 2150, 0\rangle\,\text{N}$$

(b) Parallel (x):

$$\left|\frac{d\vec{p}}{dt}_{\parallel}\right| = \left|\vec{F}_{\text{net}\,\parallel}\right|$$
$$\frac{d|\vec{p}|}{dt} = mg\cos(90-30)^\circ$$
$$= (90\,\text{kg})(9.8\,\text{N/kg})\cos 60^\circ$$
$$\frac{d|\vec{p}|}{dt} = 441\,\text{kg}\cdot\text{m/s}^2$$
$$\frac{d\vec{p}}{dt}_{\parallel} = \langle 441,0,0\rangle\ \text{kg}\cdot\text{m/s}^2$$

The tension in the vine is greater than when Tarzan was hanging motionless, but not as large as it will be at the bottom of the swing.

EXAMPLE **Swinging a Bucket in a Vertical Plane**

You swing a bucket of water in a vertical plane at the end of a rope, fast enough that the water stays in the bucket. The center of mass of the bucket and water moves approximately in a circle of radius 1.2 m. The speed of the bucket is faster at the bottom and slower at the top. The bucket with the water has a mass of 3 kg. At the instant when the rope is horizontal, the speed of the bucket is 9 m/s. What is the tension in the rope when the rope is horizontal? What is the rate of change of the speed at this moment?

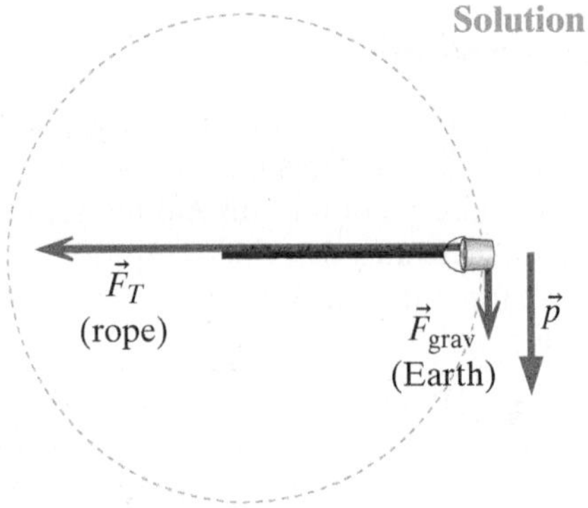

Figure 5.50 Swinging a bucket of water in a vertical plane.

Solution System: Bucket with water
Surroundings: Earth, rope
Free-body diagram: Figure 5.50

Perpendicular:

$$\left|\frac{d\vec{p}}{dt}_{\perp}\right| = \left|\vec{F}_{\text{net}\,\perp}\right|$$
$$\frac{mv^2}{R} = F_T$$
$$\frac{(3\,\text{kg})(9\,\text{m/s})^2}{1.2\,\text{m}} = F_T$$
$$F_T = 202.5\,\text{N}$$

If the bucket simply hung motionless from the rope, the tension in the rope would be only $mg = (3\,\text{kg})(9.8\,\text{N/kg}) = 29.4\,\text{N}$, much less than 202.5 N. However, as we saw with Tarzan and the vine, during curving motion the tension in the vine might be greater than mg.

Parallel:

$$\left|\frac{d\vec{p}}{dt}_{\parallel}\right| = \left|\vec{F}_{\text{net}\,\parallel}\right|$$
$$m\frac{d|\vec{v}|}{dt} = mg$$
$$\frac{d|\vec{v}|}{dt} = |\vec{a}| = g = 9.8\,\text{m/s}^2$$

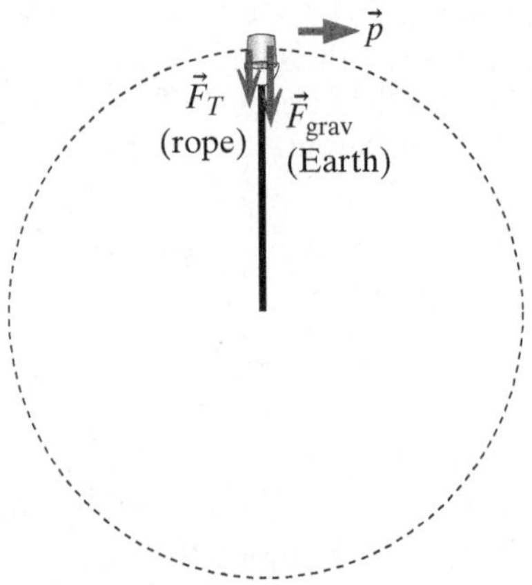

Figure 5.51 The bucket just barely makes it over the top with very little tension in the rope.

Note that this is only the case when the gravitational force is parallel to the momentum of the bucket. At other points in the swing, dv/dt will be different.

QUESTION How fast must the bucket be moving at the top of the swing?

There is a minimum speed necessary for the bucket to make it over the top without the rope going slack. In Figure 5.51 we see that as long as the rope is not slack there are two downward forces acting on the bucket, due to the rope and the Earth. If the bucket just barely makes it over the top of the circular loop with the rope just barely taut, the force of the rope is negligible compared to the force of the Earth, so we have $mv^2/R = mg$, or $v = \sqrt{gR} = \sqrt{(9.8\,\text{N/kg})(1.2\,\text{m})} = 3.4\,\text{m/s}$. Any speed at the top that is less than 3.4 m/s will mean that the rope has gone slack, and the motion won't be a circle any more.

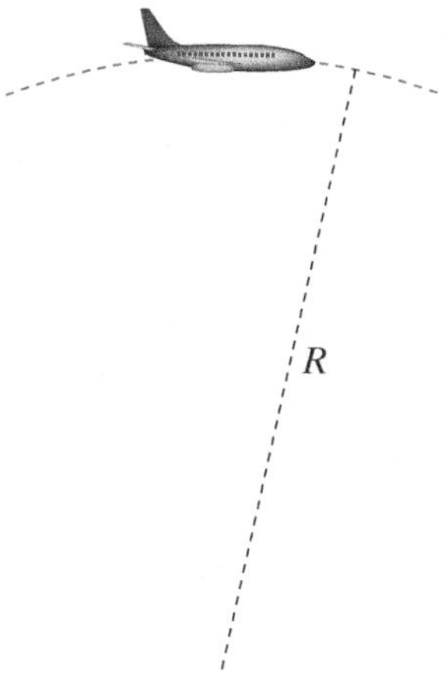

Figure 5.52 An airplane goes over the top of its path with speed v. The kissing circle has radius R at the top of the path.

QUESTION Why doesn't the water fall out of the bucket when the bucket is upside down?

As long as the speed at the top is at least 3.4 m/s the water will not fall out of the bucket, because its motion in a circle is perfectly compatible with the Momentum Principle. Of course the water actually is falling, moving downward and to the right, but you keep pulling inward on the bucket by pulling on the rope, and you're continually "catching" the water as it falls.

Compression Forces and Curving Motion

Just as we have deduced the value of unknown tension forces, we can use the known motion of a system to deduce the value of an unknown compression force, such as the force that a seat exerts on a person sitting on it.

EXAMPLE **Compression Force**

A passenger of mass M rides in an airplane flying along a curved path at constant speed v as shown in Figure 5.52, where the radius of the kissing circle at the top of the path is R. At the instant shown, when the airplane is at the top of its arc, how large is the compression force that the seat bottom exerts on the passenger? Compare this force to the force the seat exerts on the passenger when the airplane moves in a straight line at constant speed.

Solution System: Passenger
Surroundings: Seat, Earth
Free-body diagram: Figure 5.53

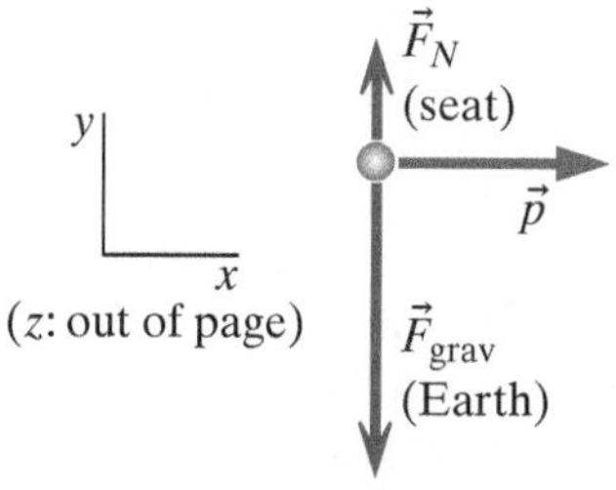

Figure 5.53 Free-body diagram.

Parallel: Because the passenger travels at constant speed, $|\vec{p}|$ is constant. Therefore we know that:

$$\frac{d\vec{p}}{dt}_{\parallel} = \vec{0} \quad \text{and therefore} \quad \vec{F}_{\text{net}\,\parallel} = \vec{0}$$

Perpendicular: We need to be careful of signs in this part. The direction of $d\hat{p}/dt$ is the $-y$ direction. Working with y components, we have:

$$\frac{dp_y}{dt} = F_{\text{net},y}$$

$$\frac{-Mv^2}{R} = F_N - Mg$$

$$F_N = Mg - \frac{Mv^2}{R}$$

QUESTION Why isn't $F_N = Mg$?

The minus sign in the result is important. The upward force of the seat F_N on the passenger's bottom is *less* than the passenger's weight Mg. The interatomic bonds in the seat are compressed less than they would be if the airplane's momentum were constant. The seat is being pulled down from under the passenger.

> QUESTION Since $F_N = Mg - Mv^2/R$, if the airplane's speed v is big enough, the force exerted by the seat could be zero. Does that make any sense?

Yes, that's the case where the airplane changes direction so quickly that the seat is essentially jerked out from under the passenger, and there is no longer contact between the seat and the passenger. No contact, no force of the seat on the passenger. You may have experienced a similar situation on a roller coaster going over a hill.

5.9 KINESTHETIC SENSATIONS

EXAMPLE **A Turning Car**

Suppose that you're riding as the passenger in a convertible with left-hand drive (American or continental European), so you are sitting on the right (Figure 5.54). Assume that you have foolishly failed to fasten your seat belt. The driver makes a sudden right turn. Before studying physics you would probably have said that you are "thrown to the left," and in fact you do end up closer to the driver. But there's something wrong with this "thrown to the left" idea. What object in the surroundings exerted a force to the left to make you move to the left?

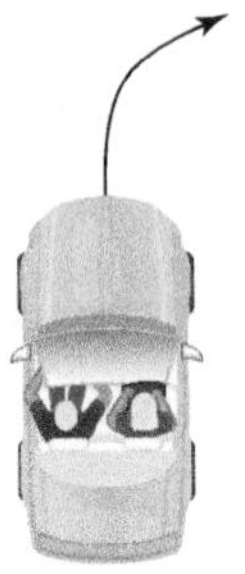

Figure 5.54 You are a passenger in a car that makes a sudden right turn.

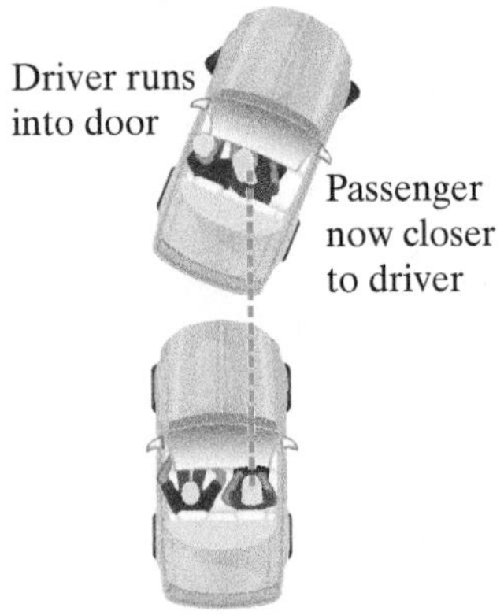

Figure 5.55 The passenger moves straight ahead and is now closer to the driver, who was thrown to the right by the door pushing the driver to the right.

Solution Don't confuse your kinesthetic perceptions with the physical analysis. There is no object that exerts an outward (leftward) force. To understand what's happening, watch the turning car from a fixed vantage point above the convertible, and observe carefully what happens to the passenger. In the absence of forces to change the passenger's direction of motion, the passenger keeps moving ahead, in a straight line at the same speed as before (Figure 5.55). It is the car that is yanked to the right, out from under the passenger. The driver moves closer to the passenger; it isn't that the passenger moves toward the driver.

The driver may also feel "thrown to the left" against the door. However, it is the door that runs into the driver, forcing the driver to the *right*, not to the left.

There exists an advanced technique for using a noninertial rotating reference frame fixed to the car, in which case it is necessary to introduce additional "pseudoforces" not associated with objects in the surroundings, such as a centrifugal force. In this textbook we will always carry out our analyses in an inertial reference frame, a frame that is not accelerating, including not rotating. In an inertial reference frame there is no centrifugal force, and all forces are associated with objects in the surroundings.

In the example of the airplane traveling through an arc at constant speed (Section 5.8), what does the situation look and feel like to the passenger? From the passenger's point of view it is natural to think that the passenger has been thrown upward, away from the seat. Actually, though, the seat has been yanked out from under the passenger, in which case the passenger is falling toward the Earth (being no longer supported by the seat), but is changing velocity less rapidly than the airplane and so the ceiling comes closer to the passenger.

In the extreme case, if the airplane's maneuver is very fast, it may seem as though the passenger is "thrown up against the cabin ceiling." In actual fact, however, it is the cabin ceiling that is yanked down and hits the passenger!

You can see the value of wearing a seat belt, to prevent losing contact with the seat. When the airplane yanks the seat downward, the seat belt yanks the passenger downward. This may be uncomfortable, but it is a lot better than having the cabin ceiling hit you hard in the head.

> QUESTION How would this analysis change if the airplane goes through the bottom of a curve rather than the top (Figure 5.56)?

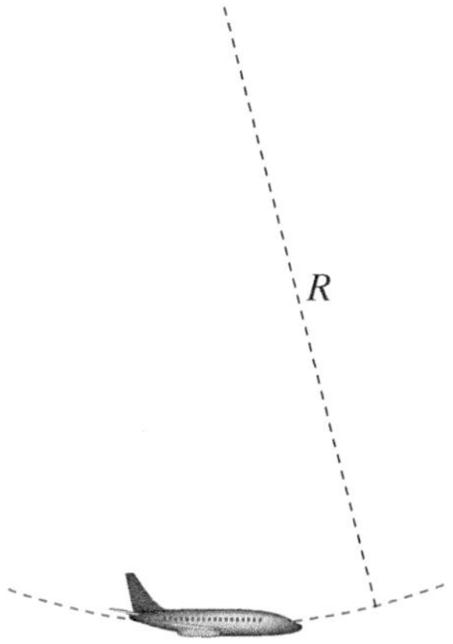

Figure 5.56 An airplane goes through the bottom of its path.

Now the center of the kissing circle is above the airplane, so $d\vec{p}/dt$ points upward. That means that the net force must also point upward.

> QUESTION Which is larger, the force of the seat on the passenger or the gravitational force of the Earth on the passenger?

The force of the seat is greater than Mg, since the net force is upward. The passenger sinks deeper into the seat. Or to put it another way, the airplane yanks the seat up harder against the passenger's bottom, squeezing the seat against the passenger.

We Perceive Contact Forces

As you sit reading this text, you feel a sensation that you associate with weight. If someone suddenly yanks the chair out from under you, you feel "weightless," which feels funny and odd. It also feels scary, because you know from experience that whenever you lose the support of objects under you (chair, airplane seat, floor, mountain path), something bad is about to happen.

In reality, what we perceive as "weight" is actually not the gravitational force at all but rather the forces of atoms in the chair (or airplane seat or floor or mountain path) on atoms in your skin. You have nerve endings that sense the compression of interatomic bonds in your skin, and you interpret this as evidence for gravity acting on you. However, if you were placed in a spaceship that was accelerating, going faster and faster, its floor would squeeze against you in a way that would fool your nerve endings and brain into thinking you were subjected to gravity, even if there are no stars or planets nearby.

If you lose contact with the seat in an airplane that is going rapidly over the top of a curving path, your nerve endings no longer feel any contact forces, and you feel "weightless." Yet this is a moment when the only force acting on you is in fact the Earth's gravitational force. Weightlessness near the Earth paradoxically is associated with being subject only to your weight Mg.

Contact Forces Internal to the Body

It isn't just nerve endings in your skin that give you the illusion of "weight." As you sit here reading, your internal organs press upward on other organs above them, making the net force zero. You feel these internal contact forces. If you are suddenly weightless, a main reason for feeling funny is that the forces one organ exerts on another inside your body are suddenly gone. Your inner ear is also affected, which can produce dizziness.

To train astronauts, NASA has a cargo plane that is deliberately flown in a curving motion over the top, so that people in the padded interior lose contact with the floor and seem to float freely (in actual fact, they are accelerating toward the Earth due to the Mg forces acting on them, but the plane is also deliberately accelerating toward the Earth rather than flying level, so the people don't touch the walls and appear to be floating). The brain is crying out for the usual comforting signals from the nerve endings and not getting them. This airplane is sometimes called the "vomit comet" in honor of the effects it can have on trainees.

5.10 MORE COMPLEX PROBLEMS

Problems that ask more than "What is the magnitude of $\vec{F}_{\text{net}\perp}$?" are more complex to solve. However, typically the solution emerges if we simply begin the analysis from the Momentum Principle, exactly as we have done in all the preceding problems. Even if it is not obvious at first where the desired answer will come from, this kind of "forward reasoning", which is typical in physics problem solving, will almost always lead to a solution.

EXAMPLE **An Amusement Park Ride**

There is an amusement park ride that some people love and others hate in which a bunch of people stand against the wall of a cylindrical room of radius R, and the room starts to rotate at higher and higher speed (Figure 5.57). The surface of the wall is designed to maximize friction between the person and the wall, so it is fuzzy or sticky, not slick. When a certain critical speed is reached, the floor drops away, leaving the people stuck against the wall as they whirl around at constant speed. How fast must the ride go to keep a person from falling down?

Solution System: Person
Surroundings: Wall, Earth (neglect air resistance)
Free-body diagram: Figure 5.58

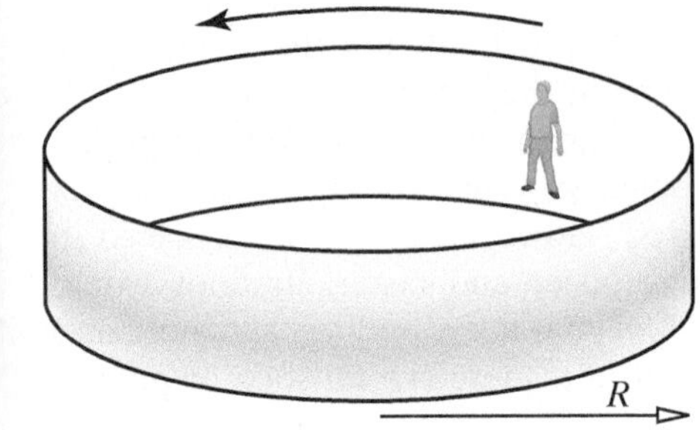

Figure 5.57 An amusement park ride.

At the instant shown in Figure 5.58 the person is in contact with the right wall, with momentum directed into the page ($-z$).

QUESTION Why doesn't the person slide down the wall?

We know the Earth pulls down on the person, so we deduce that there must be an upward force that is equal in magnitude to the downward gravitational force. The only object in contact with the person is the wall, so this force must be exerted by the wall. The free-body diagram reflects this deduction. In the diagram the force due to the wall has been separated into two parts: a normal part $\vec{F}_N$ that turns the direction of the person's momentum and a frictional part $\vec{f}$ that keeps the person from sliding downward.

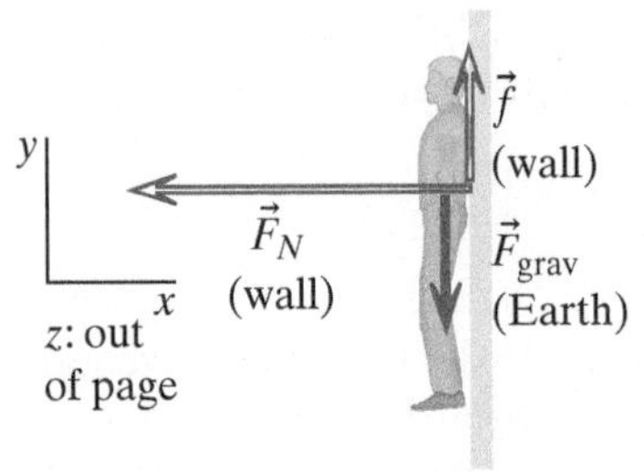

Figure 5.58 Free-body diagram.

Parallel (z): No change in speed, so $\dfrac{d\vec{p}}{dt}_{\parallel} = \vec{0}$.

There are two perpendicular directions, x and y. In the x direction the momentum of the person is changing:

$$\left|\frac{dp_x}{dt}\right| = |-F_N|$$

$$\frac{mv^2}{R} = F_N$$

This implies that the faster the ride goes, the larger the inward force (the normal force F_N) of the wall.

In the y direction there is no change in momentum, so

$$\frac{dp_y}{dt} = f - mg$$
$$0 = f - mg$$
$$f = mg$$

Recall from Chapter 4 that $\mu_s F_N \geq f$: a static friction force f (a friction force on an object that is not moving) has a maximum value of $\mu_s F_N$, where μ_s is the coefficient of static friction, and depends on the materials that are in contact with each other. If a force greater than $\mu_s F_N$ is applied to the object, it will begin to slide. In the case of the person on the ride, as long as $\mu_s F_N \geq mg$, the person will be supported. So the requirement is that

$$\mu_s F_N \geq mg$$
$$\mu_s \frac{mv^2}{R} \geq mg$$
$$\mu_s \geq \frac{gR}{v^2}$$

This predicts that the lower the speed v, the greater the coefficient of friction must be to hold the person up on the wall, which makes sense. Conversely, if the ride spins very fast, the wall doesn't have to have a very large coefficient of friction to hold the people against the wall.

EXAMPLE **Earth Satellite**

How long does it take a satellite in a circular orbit of radius R to make one complete revolution about the Earth?

Solution

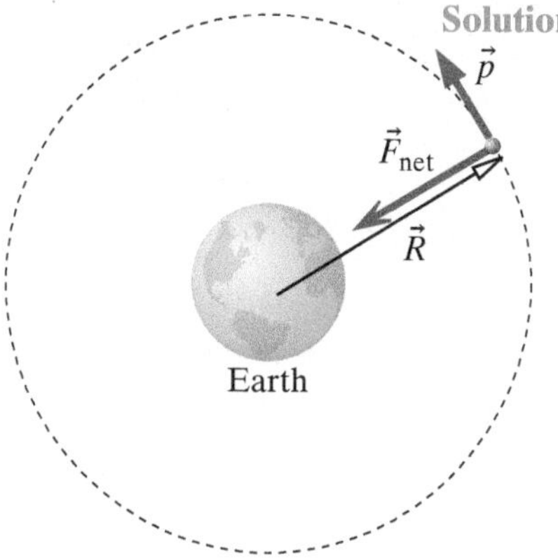

Figure 5.59 A satellite in a circular orbit around the Earth.

System: Satellite
Surroundings: Earth
Free-body diagram: Figure 5.59

Parallel:

Right-hand side of equation: $F_{net\parallel} = 0$

Therefore, left-hand side is $\frac{d|\vec{p}|}{dt} = 0$ (constant speed)

Perpendicular component (toward center):

Left-hand side of equation: $\left|\left(\frac{d\vec{p}}{dt}\right)_\perp\right| = \left(\frac{v}{R}\right)\gamma mv \approx \frac{mv^2}{R}$

Right-hand side of equation: $F_{net\perp} = \frac{GMm}{R^2}$

These two very different things, one the rate of change of the satellite's momentum and the other the gravitational attraction between the Earth and the satellite, are related to each other by the Momentum Principle to be numerically equal:

$$\frac{mv^2}{R} = \frac{GMm}{R^2}$$
$$v^2 = \frac{GM}{R}$$
$$v = \sqrt{\frac{GM}{R}}$$

The satellite goes a distance of the circumference $2\pi R$ in a time T, called the "period" of the orbit (and we say that the repetitive motion is "periodic"). So we have

$$\frac{2\pi R}{T} = \sqrt{\frac{GM}{R}}$$

$$T = \frac{2\pi R}{\sqrt{\frac{GM}{R}}} = \frac{2\pi R^{3/2}}{\sqrt{GM}}$$

Since the period is proportional to $R^{3/2}$, the smaller the radius of the orbit, the shorter the period T. The shortest period for an Earth satellite is for one in which the satellite orbits just high enough that air resistance isn't significant. This minimum height is about 50 km (5×10^4 m), so the radius of the orbit of a "near-Earth" satellite is $R_{\text{Earth}} = 6.4 \times 10^6$ m plus 5×10^4 m, which is not very different from the radius of the Earth. The period for a near-Earth satellite is therefore

$$T = \frac{2\pi(6.4 \times 10^6\,\text{m})^{3/2}}{\sqrt{(6.7 \times 10^{-11}\,\text{N}\cdot\text{m}^2/\text{kg}^2)(6 \times 10^{24}\,\text{kg})}} = 5074\,\text{s} = 85\,\text{min}$$

The speed of a near-Earth satellite is $v = 2\pi R/T = 2\pi(6.4 \times 10^6\,\text{m})/(5074\,\text{s}) = 7925\,\text{m/s} \approx 8$ km/s. A chemical rocket must accelerate from rest on the launch pad to 8 km/s in order to go into a minimal near-Earth orbit.

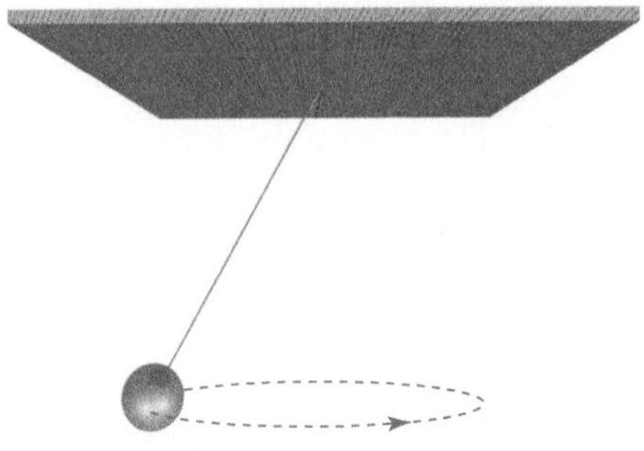

Figure 5.60 A circular pendulum, moving in the xz plane.

EXAMPLE **Circular Pendulum**

A ball is suspended from a string, and after being given a push, moves along a horizontal circular path, as shown in Figure 5.60. You measure the length of the string L and the angle θ, shown in Figure 5.61. You also time the motion, and find that it takes T s for the ball to make one complete circular trip. From these measurements, determine the gravitational constant g. (Newton used a large circular pendulum to measure g with considerable accuracy.)

Solution System: Ball

Surroundings: String, Earth (neglect air resistance)

Free-body diagram: Figure 5.61.

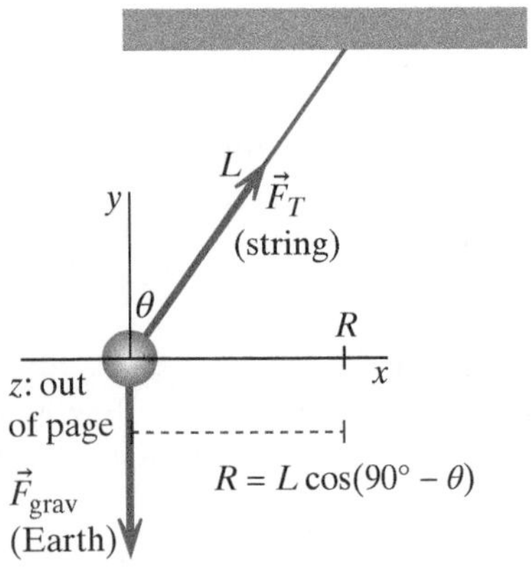

Figure 5.61 The circular pendulum at an instant when the string is in the xy plane, and the ball is moving in the $+z$ direction.

$$d\vec{p}/dt = \vec{F}_{\text{net}}$$

$$\hat{F}_T = \langle \cos(90° - \theta), \cos\theta, 0 \rangle$$

$$\vec{F}_T = F_T\langle \cos(90° - \theta), \cos\theta, 0 \rangle$$

$$\vec{F}_{\text{net}} = \langle F_T\cos(90° - \theta), (F_T\cos\theta - mg), 0 \rangle$$

Parallel component $(+z)$:

$$\frac{dp_z}{dt} = F_{\text{net},z} = 0$$

Perpendicular components:

$$y: \frac{dp_y}{dt} = F_T\cos\theta - mg$$

$$0 = F_T\cos\theta - mg$$

$$F_T = \frac{mg}{\cos\theta}$$

$$x: \frac{dp_x}{dt} = F_T \cos(90° - \theta)$$
$$\frac{mv^2}{R} = F_T \sin\theta$$
$$\frac{mv^2}{L\sin\theta} = \left(\frac{mg}{\cos\theta}\right)\sin\theta$$
$$g = \frac{v^2\cos\theta}{L(\sin\theta)^2}$$

Measuring the speed v, length L, and angle θ allows us to determine g. v can be found by measuring T, the time to go around once, since $v = (2\pi R)/T$, where $R = L\sin\theta$.

Special case: If $\theta = 0$ the string hangs vertically and there is no motion, so the tension in the string ought to be mg, which is consistent with the equation for F_T that we found above.

Note that $\vec{F}_{\text{net}}$ and $d\vec{p}/dt$ are not the same thing. In the example above, we knew the x and y components of $d\vec{p}/dt$ (the effect), and from this deduced the x and y components of $\vec{F}_{\text{net}}$ (the cause). In the z direction, it was the other way around: we knew that there were no forces in this direction, so we deduced that the z component of momentum was not changing.

There could in principle be many possible motions of this system. The ball could swing back and forth in a plane; this is the motion of a "simple pendulum." The ball could go around in a path something like an ellipse, in which case the ball's height would vary, just as with simple pendulum motion. The ball could move so violently that the string goes slack for part of the time, with abrupt changes of momentum every time the string suddenly goes taut. If the string can stretch noticeably, there can be a sizable oscillation superimposed on the swinging motion.

All of these motions are difficult to analyze. In principle, we could use a computer to predict the general motion of the ball hanging from the string if we had an equation for the tension force in the string, as a function of its length. Essentially, we would need the effective spring stiffness for this stiff "spring." However, the string may be so stiff that even a tiny stretch implies a huge tension (a nearly inextensible string). The observable length of the string is nearly constant, but the tension force that the string applies to the ball can vary a great deal. This makes it challenging to do a computer numerical integration, because a tiny error in position makes a huge change in the force. Special techniques have been developed to handle such situations.

SUMMARY

Determining unknown forces

Make a clear choice of the system; everything else is the surroundings.

Identify objects in the surroundings that can exert either a distant or contact force on the chosen system.

Draw the forces on a free-body diagram and include the name of the object responsible.

Apply the Momentum Principle to the chosen system.

Use what is known about $d\vec{p}/dt$, such as $d\vec{p}/dt = \vec{0}$ in the case of a system at rest or in uniform motion, or what can be obtained from the radius of the kissing circle.

Solve for the unknown forces. For a system containing several objects, it may be necessary to make a different choice of system in order to be able to determine forces between these objects.

Finding the rate of change of momentum graphically

1. Draw two arrows representing $\vec{p}$ at equal time intervals before and after the time of interest (Figure 5.35).
2. Find $\Delta\vec{p} = \vec{p}_f - \vec{p}_i$ (Figure 5.36).

3. Divide $\Delta\vec{p}$ by Δt to get $\Delta\vec{p}/\Delta t$ (Figure 5.37). The smaller Δt, the closer the result is to $d\vec{p}/dt$.

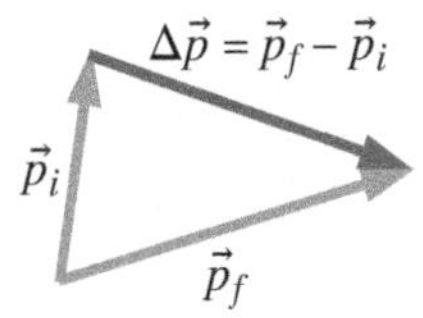

Figure 5.62

Momentarily at rest vs. uniform motion and equilibrium

For a system in equilibrium $d\vec{p}/dt = \vec{0}$.

For a system in uniform motion $d\vec{p}/dt = \vec{0}$.

For a system that is momentarily at rest, $d\vec{p}/dt \neq \vec{0}$.

The Momentum Principle: Parallel and perpendicular components

$$\frac{dp}{dt}\hat{p} = \vec{F}_{\parallel} = (\vec{F}_{\text{net}} \bullet \hat{p})\hat{p} \quad \text{and} \quad p\frac{d\hat{p}}{dt} = \vec{F}_{\perp} = \vec{F}_{\text{net}} - \vec{F}_{\parallel}$$

Valid only for a moving object, because $\hat{p}$ and $d\hat{p}/dt$ are undefined when $\vec{p} = \vec{0}$.

Component of net force parallel to momentum changes magnitude of momentum.

Component of net force perpendicular to momentum changes direction of momentum.

Effect of perpendicular component of net force on a particle moving along a curving path

$$p\left|\frac{d\hat{p}}{dt}\right| = p\frac{v}{R} = |\vec{F}_{\text{net}\perp}|$$

R is the radius of the kissing circle. The component of net force perpendicular to the particle's momentum changes its direction but not the magnitude of its momentum (speed). Must be a smoothly curving motion; not valid if there is an abrupt change of direction.

The dot product

$$\vec{A} \bullet \vec{B} = |\vec{A}||\vec{B}|\cos\theta = A_xB_x + A_yB_y + A_zB_z$$

QUESTIONS

Q1 You are riding in the passenger seat of an American car, sitting on the right side of the front seat. The car makes a sharp left turn. You feel yourself thrown to the right and your right side hits the right door. Is there a force that pushes you to the right? What object exerts that force? What really happens? Draw a diagram to illustrate and clarify your analysis.

Q2 A student said, "When the Moon goes around the Earth, there is an inward force due to the Moon and an outward force due to centrifugal force, so the net force on the Moon is zero." Give two or more physics reasons why this is wrong.

Q3 A space shuttle is in a circular orbit near the Earth. An astronaut floats in the middle of the shuttle, not touching the walls. On a diagram, draw and label **(a)** the momentum $\vec{p}_1$ of the astronaut at this instant, **(b)** all of the forces (if any) acting on the astronaut at this instant, **(c)** the momentum $\vec{p}_2$ of the astronaut a short time Δt later, and **(d)** the momentum change (if any) $\Delta\vec{p}$ in this time interval. **(e)** Why does the astronaut seem to "float" in the shuttle?

It is ironic that we say the astronaut is "weightless" despite the fact that the only force acting on the astronaut is the astronaut's weight (that is, the gravitational force of the Earth on the astronaut).

Q4 Tarzan swings back and forth on a vine. At the microscopic level, why is the tension force on Tarzan by the vine greater than it would be if he were hanging motionless?

Q5 Tarzan swings from a vine. When he is at the bottom of his swing, as shown in Figure 5.63, which is larger in magnitude: the force by the Earth on Tarzan, the force by the vine (a tension force) on Tarzan, or neither (same magnitude)? Explain how you know this.

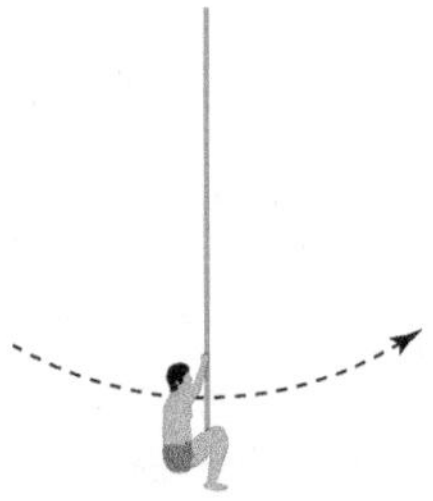

Figure 5.63

PROBLEMS

Section 5.4

•P6 A rope is attached to a block, as shown in Figure 5.64. The rope pulls on the block with a force of 210 N, at an angle of $\theta = 23°$ to the horizontal (this force is equal to the tension in the rope).

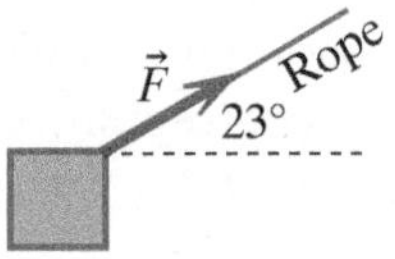

Figure 5.64

(a) What is the x component of the force on the block due to the rope? **(b)** What is the y component of the force on the block due to the rope?

••P7 A box of mass 40 kg hangs motionless from two ropes, as shown in Figure 5.65. The angle is 38 degrees. Choose the box as the system. The x axis runs to the right, the y axis runs up, and the z axis is out of the page.

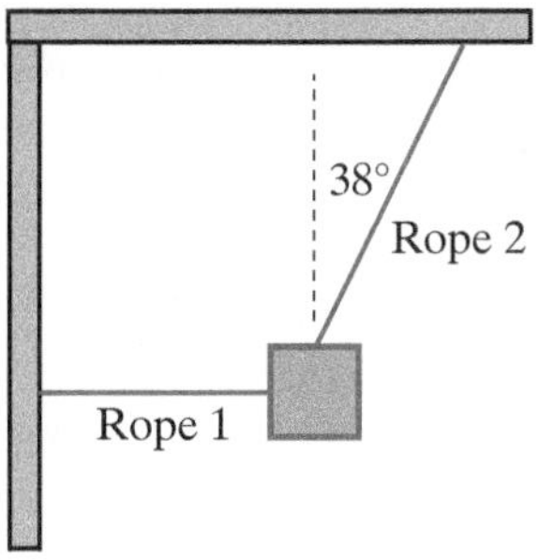

Figure 5.65

(a) Draw a free-body diagram for the box. **(b)** Is $d\vec{p}/dt$ of the box zero or nonzero? **(c)** What is the y component of the gravitational force acting on the block? (A component can be positive or negative). **(d)** What is the y component of the force on the block due to rope 2? **(e)** What is the magnitude of $\vec{F}_2$? **(f)** What is the x component of the force on the block due to rope 2? **(g)** What is the x component of the force on the block due to rope 1?

••P8 A helicopter flies to the right (in the $+x$ direction) at a constant speed of 12 m/s, parallel to the surface of the ocean. A 900 kg package of supplies is suspended below the helicopter by a cable as shown in Figure 5.66: the package is also traveling to the right in a straight line, at a constant speed of 12 m/s. The pilot is concerned about whether or not the cable, whose breaking strength is listed at 9300 N, is strong enough to support this package under these circumstances. **(a)** Choose the package as the system, and draw a free-body diagram. **(b)** What is the magnitude of the tension in the cable supporting the package? **(c)** Write the force exerted on the package by the cable as a vector. **(d)** What is the magnitude of the force exerted by the air on the package? **(e)** Write the force on the package by the air as a vector. **(f)** Is the cable in danger of breaking?

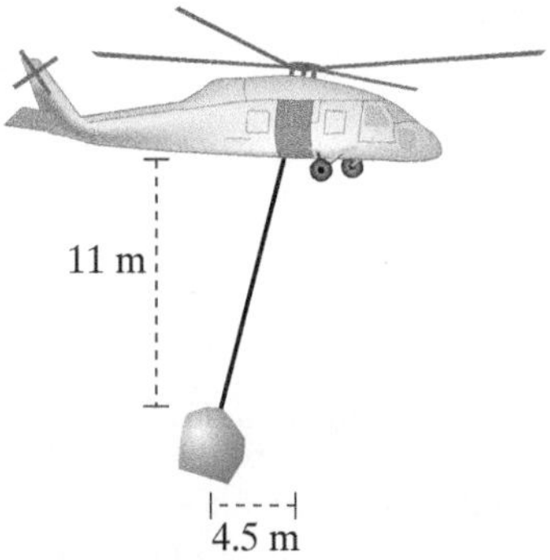

Figure 5.66

••P9 You pull with a force of 255 N on a rope that is attached to a block of mass 30 kg, and the block slides across the floor at a constant speed of 1.1 m/s. The rope makes an angle $\theta = 40°$ with the horizontal. Both the force and the velocity of the block are in the xy plane. See Figure 5.67.

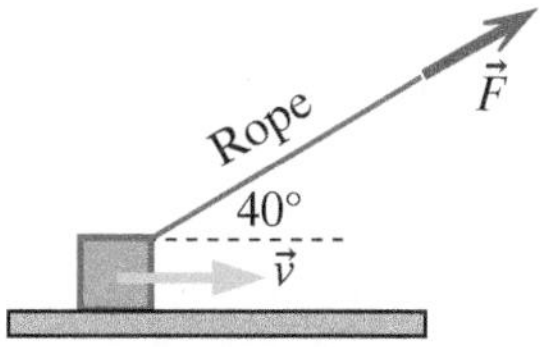

Figure 5.67

(a) Express the tension force exerted by the rope on the block as a vector. **(b)** Express the force exerted by the floor on the block as a vector.

••P10 A ball of mass 450 g hangs from a spring whose stiffness is 110 N/m. A string is attached to the ball and you are pulling the string to the right, so that the ball hangs motionless, as shown in Figure 5.68. In this situation the spring is stretched, and its length is 15 cm. What would be the relaxed length of the spring, if it were detached from the ball and laid on a table?

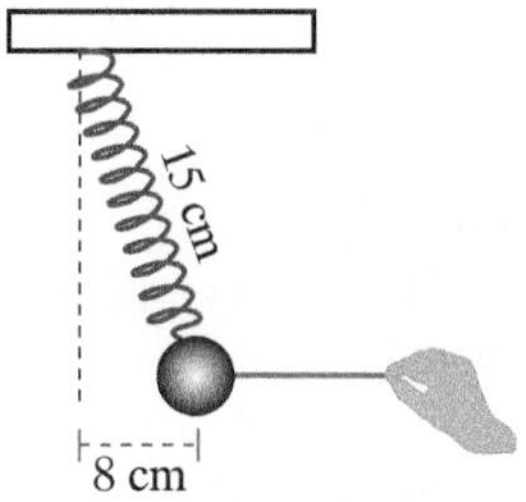

Figure 5.68

••P11 An 800 kg load is suspended as shown in Figure 5.69. **(a)** Calculate the tension in all three wires (that is, the magnitude of the tension force exerted by each of these wires). **(b)** These wires are made of a material whose value for Young's modulus is $1.3 \times 10^{11}\ \mathrm{N/m^2}$. The diameter of the wires is 1.1 mm. What is the strain (fractional stretch) in each wire?

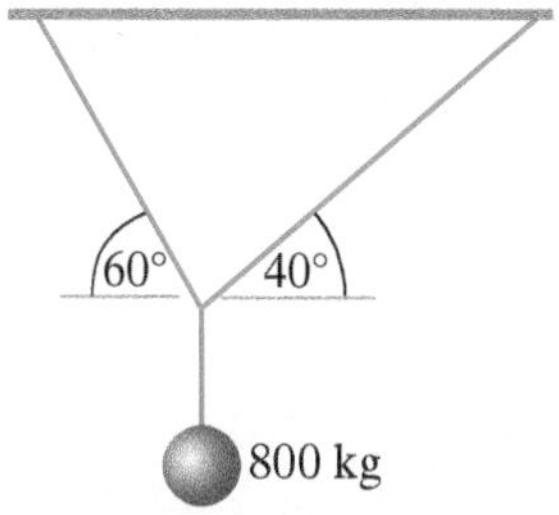

Figure 5.69

•••P12 A cylindrical steel rod with diameter 0.7 mm is 2 m long when lying horizontally (Figure 5.70). The density of this particular steel is 7.85 g/cm^3, and Young's modulus is $2 \times 10^{11}\ \mathrm{N/m^2}$. **(a)** You now hang the rod vertically from a support and attach a 60 kg mass to the bottom end. How much does the rod stretch? **(b)** Next you remove the 60 kg mass so the rod hangs under its own weight. Describe qualitatively how the stretch of the interatomic bonds depends on position along the rod. **(c)** Calculate how much the rod stretches under its own weight. (This requires the use of integral calculus, because the tension and strain vary along the rod.) **(d)** If the rod were twice as thick, how much would it stretch?

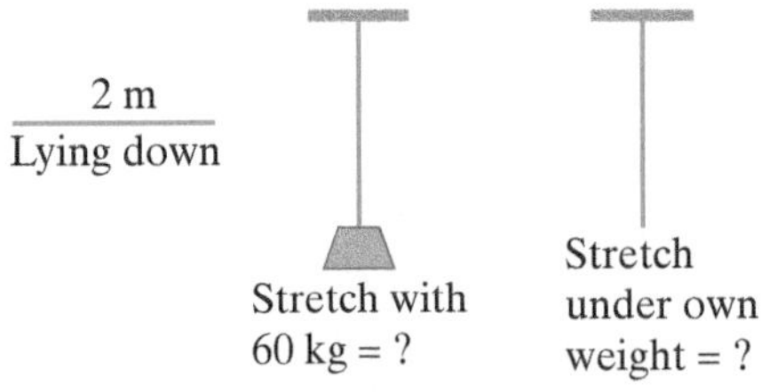

Figure 5.70

Section 5.5

••P13 An elevator is accelerating upward at a rate of 3.2 m/s^2. A block of mass 30 kg hangs by a low-mass rope from the ceiling, and another block of mass 80 kg hangs by a low-mass rope from the upper block. **(a)** What are the tensions in the upper and lower ropes? **(b)** What are the tensions in the upper and lower ropes when the elevator accelerates downward at a rate of 3.2 m/s^2?

••P14 A 10 kg block is placed on top of a 35 kg block (Figure 5.71). A force of 350 N is applied to the right on the lower block, and the upper block slips on the lower block (accelerating less than the lower block). The coefficient of kinetic friction between the upper block and the lower block is 0.2, and the coefficient of kinetic friction between the lower block and the floor is 0.7. **(a)** What is the acceleration of the upper block? **(b)** What is the acceleration of the lower block? **(c)** How big would the coefficient of static friction between the upper and lower block have to be so that the upper block would not slip on the lower block?

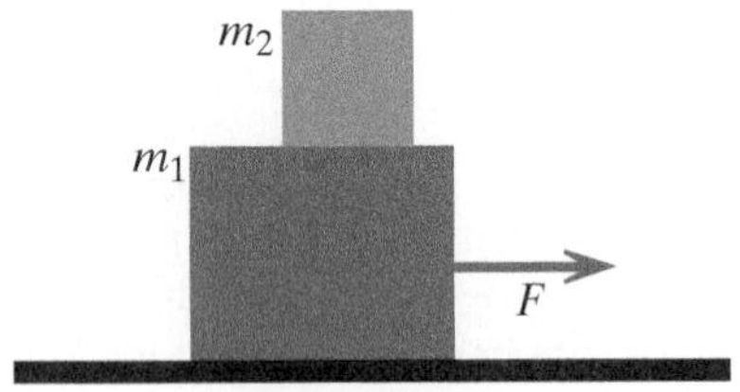

Figure 5.71

••P15 In Figure 5.72 $m_1 = 12$ kg and $m_2 = 5$ kg. The kinetic coefficient of friction between m_1 and the floor is 0.3 and that between m_2 and the floor is 0.5. You push with a force of magnitude $F = 110$ N. **(a)** What is the acceleration of the center of mass? **(b)** What is the magnitude of the force that m_1 exerts on m_2?

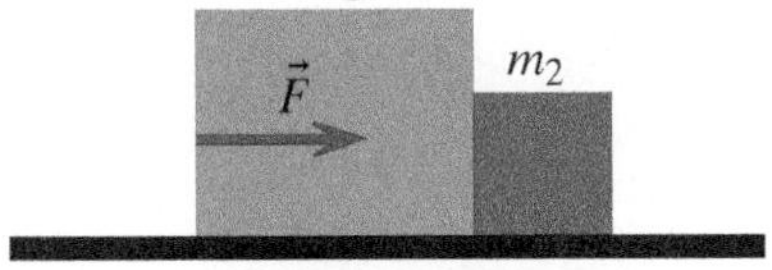

Figure 5.72

Section 5.6

•P16 The radius of a merry-go-round is 11 m, and it takes 12 s to go around once. What is the speed of an atom in the outer rim?

•P17 If the radius of a merry-go-round is 5 m, and it takes 14 s to go around once, what is the speed of an atom at the outer rim? What is the direction of the velocity of this atom: toward the center, away from the center, or tangential?

••P18 At a particular instant the magnitude of the momentum of a planet is 2.3×10^{29} kg·m/s, and the force exerted on it by the star it is orbiting is 8.9×10^{22} N. The angle between the planet's momentum and the gravitational force exerted by the star is 123°. **(a)** What is the parallel component of the force on the planet by the star? **(b)** What will the magnitude of the planet's momentum be after 9 h?

••P19 The angle between the gravitational force on a planet by a star and the momentum of the planet is 61° at a particular instant. At this instant the magnitude of the planet's momentum is 3.1×10^{29} kg·m/s, and the magnitude of the gravitational force on the planet is 1.8×10^{23} N. **(a)** What is the parallel component of the force on the planet by the star? **(b)** What will be the magnitude of the planet's momentum after 8 h?

••P20 A planet orbits a star in an elliptical orbit. At a particular instant the momentum of the planet is $\langle -2.6 \times 10^{29}, -1.0 \times 10^{29}, 0\rangle$ kg·m/s, and the force on the planet by the star is $\langle -2.5 \times 10^{22}, -1.4 \times 10^{23}, 0\rangle$ N. Find $\vec{F}_{\parallel}$ and $\vec{F}_{\perp}$.

••P21 A planet of mass 6×10^{24} kg orbits a star in a highly elliptical orbit. At a particular instant the velocity of the planet is $\langle 4.5 \times 10^4, -1.7 \times 10^4, 0\rangle$ m/s, and the force on the planet by the star is $\langle 1.5 \times 10^{22}, 1.9 \times 10^{23}, 0\rangle$ N. Find $\vec{F}_{\parallel}$ and $\vec{F}_{\perp}$.

Section 5.7

•P22 An object moving at a constant speed of 23 m/s is making a turn with a radius of curvature of 4 m (this is the radius of the kissing circle). The object's momentum has a magnitude of 78 kg·m/s. What is the magnitude of the rate of change of the momentum? What is the magnitude of the net force?

•P23 The radius of a merry-go-round is 7 m, and it takes 12 s to make a complete revolution. **(a)** What is the speed of an atom on the outer rim? **(b)** What is the direction of the momentum of this atom? **(c)** What is the direction of the rate of change of the momentum of this atom?

•P24 A proton moving in a magnetic field follows the curving path shown in Figure 5.73. The dashed circle is the kissing circle tangent to the path when the proton is at location A. The proton is traveling at a constant speed of 7.0×10^5 m/s, and the radius of the kissing circle is 0.08 m. The mass of a proton is 1.7×10^{-27} kg. Refer to the directional arrows shown at the right in Figure 5.73 when answering the questions below.

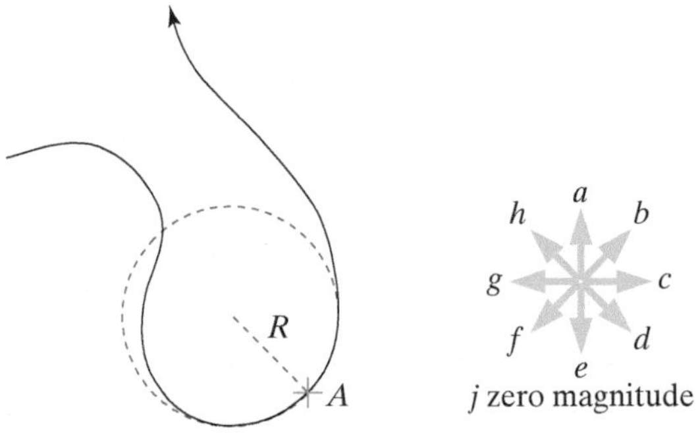

Figure 5.73

(a) When the proton is at location A, what are the magnitude and direction of $(d|\vec{p}|/dt)\hat{p}$, the parallel component of $d\vec{p}/dt$? **(b)** When the proton is at location A, what are the magnitude and direction of $|\vec{p}|d\hat{p}/dt$, the perpendicular component of $d\vec{p}/dt$?

•P25 A proton moving in a magnetic field follows the curving path shown in Figure 5.74, traveling at constant speed in the direction shown. The dashed circle is the kissing circle tangent to the path when the proton is at location A. Refer to the directional arrows shown at the right in Figure 5.74 when answering the questions below.

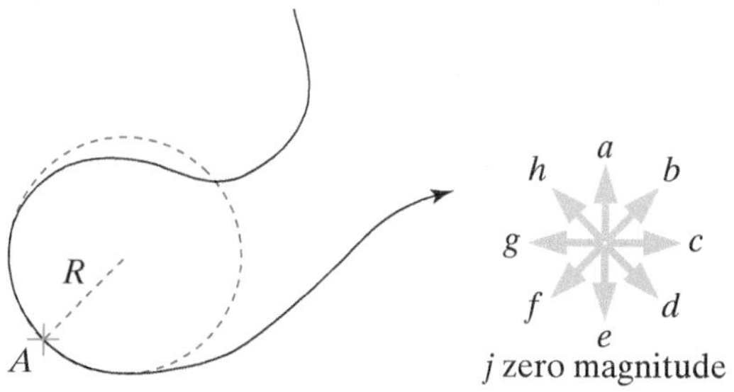

Figure 5.74

(a) When the proton is at location A, what is the direction of the proton's momentum? **(b)** When the proton is at location A, what is the direction of $d\vec{p}/dt$? **(c)** The mass of a proton is 1.7×10^{-27} kg. The proton is traveling at a constant speed of 6.0×10^5 m/s, and the radius of the kissing circle is 0.07 m. What is the magnitude of $d\vec{p}/dt$ of the proton?

••P26 A particle moving at nearly the speed of light ($v \approx c$) passes through a region where it is subjected to a magnetic force of constant magnitude that is always perpendicular to the momentum and has a magnitude of 2×10^{-10} N. As a result, the particle moves along a circular arc with a radius of 8 m. What is the magnitude of the momentum of this particle?

Section 5.8

•P27 A child of mass 40 kg sits on a wooden horse on a carousel. The wooden horse is 5 m from the center of the carousel, which completes one revolution every 90 s. What is $(d|\vec{p}|/dt)\hat{p}$ for the child, both magnitude and direction? What is $|\vec{p}|(d\hat{p}/dt)$ for the child? What is the net force acting on the child? What objects in the surroundings exert this force?

•P28 A child of mass 35 kg sits on a wooden horse on a carousel. The wooden horse is 3.3 m from the center of the carousel, which rotates at a constant rate and completes one revolution every 5.2 s. **(a)** What are the magnitude and direction (tangential in direction of velocity, tangential in the opposite direction of the velocity, radial outward, radial inward) of $(d|\vec{p}|/dt)\hat{p}$, the parallel component of $d\vec{p}/dt$ for the child? **(b)** What are the magnitude and direction of $|\vec{p}|d\hat{p}/dt$, the perpendicular component of $d\vec{p}/dt$ for the child? **(c)** What are the magnitude and direction of the net force acting on the child? **(d)** What objects in the surroundings contribute to this horizontal net force acting on the child? (There are also vertical forces, but these cancel each other if the horse doesn't move up and down.)

•P29 A 30 kg child rides on a playground merry-go-round, 1.4 m from the center. The merry-go-round makes one complete revolution every 5 s. How large is the net force on the child? In what direction does the net force act?

•P30 The orbit of the Earth around the Sun is approximately circular, and takes one year to complete. The Earth's mass is 6×10^{24} kg, and the distance from the Earth to the Sun is 1.5×10^{11} m. What is $(d|\vec{p}|/dt)\hat{p}$ of the Earth? What is $|\vec{p}|(d\hat{p}/dt)$ of the Earth? What is the magnitude of the gravitational force the Sun (mass 2×10^{30} kg) exerts on the Earth? What is the direction of this force?

••P31 You swing a bucket full of water in a vertical circle at the end of a rope. The mass of the bucket plus the water is 3.5 kg. The center of mass of the bucket plus the water moves in a circle of radius 1.3 m. At the instant that the bucket is at the top of the circle, the speed of the bucket is 4 m/s. What is the tension in the rope at this instant?

••P32 In outer space two identical spheres are connected by a taut steel cable, and the whole apparatus rotates about its center. The mass of each sphere is 60 kg. The distance between centers of the spheres is 3.2 m. At a particular instant the velocity of one of the spheres is $\langle 0,5,0 \rangle$ m/s and the velocity of the other sphere is $\langle 0,-5,0 \rangle$ m/s. What is the tension in the cable?

••P33 In the dark in outer space, you observe a glowing ball of known mass 2 kg moving in the xy plane at constant speed in a circle of radius 6.5 m, with the center of the circle at the origin ($\langle 0,0,0 \rangle$ m). You can't see what's making it move in a circle. At time $t = 0$ the ball is at location $\langle -6.5,0,0 \rangle$ m and has velocity $\langle 0,40,0 \rangle$ m/s.

On your own paper draw a diagram of the situation showing the circle and showing the position and velocity of the ball at time $t = 0$. The diagram will help you analyze the situation. Use letters *a–j* (Figure 5.75) to answer questions about directions ($+x$ to the right, $+y$ up).

Figure 5.75

At time $t = 0$: **(a)** What is the direction of the vector $\vec{p}$? **(b)** What are the magnitude and direction of $(d|\vec{p}|/dt)\hat{p}$, the parallel component of $d\vec{p}/dt$? **(c)** What are the magnitude and direction of $|\vec{p}|d\hat{p}/dt$, the perpendicular component of $d\vec{p}/dt$? **(d)** Even though you can't see what's causing the motion, what can you conclude must be the direction of the vector $\vec{F}_{net}$? **(e)** Even though you can't see what's causing the motion, what can you conclude must be the vector $\vec{F}_{net}$? **(f)** You learn that at time $t = 0$, two forces act on the ball, and that at this instant one of these forces is $\vec{F}_1 = \langle 196,-369,0 \rangle$ N. What is the other force?

••P34 You're driving a vehicle of mass 1350 kg and you need to make a turn on a flat road. The radius of curvature of the turn is 76 m. The coefficient of static friction and the coefficient of kinetic friction are both 0.25. **(a)** What is the fastest speed you can drive and still make it around the turn? Invent symbols for the various quantities and solve algebraically before plugging in numbers. **(b)** Which of the following statements are true about this situation? (1) The net force is nonzero and points away from the center of the kissing circle. (2) The rate of change of the momentum is nonzero and points away from the center of the kissing circle. (3) The rate of change of the momentum is nonzero and points toward the center of the kissing circle. (4) The momentum points toward the center of the kissing circle. (5) The centrifugal force balances the force of the road, so the net force is zero. (6) The net force is nonzero and points toward the center of the kissing circle. **(c)** Look at your algebraic analysis and answer the following question. Suppose that your vehicle had a mass five times as big (6750 kg). Now what is the fastest speed you can drive and still make it around the turn? **(d)** Look at your algebraic analysis and answer the following question. Suppose that you have the original 1350 kg vehicle but the turn has a radius twice as large (152 m). What is the fastest speed you can drive and still make it around the turn? This problem shows why high-speed curves on freeways have very large radii of curvature, but low-speed entrance and exit ramps can have smaller radii of curvature.

••P35 What is the minimum speed v that a roller coaster car must have in order to make it around an inside loop and just barely lose contact with the track at the top of the loop (see Figure 5.76)? The center of the car moves along a circular arc of radius R. Include a carefully labeled force diagram. State briefly what approximations you make. Design a plausible roller coaster loop, including numerical values for v and R.

Figure 5.76

••P36 In outer space a rock of mass 4 kg is attached to a long spring and swung at constant speed in a circle of radius 9 m. The spring exerts a force of constant magnitude 760 N. **(a)** What is the speed of the rock? **(b)** What is the direction of the spring force? **(c)** The relaxed length of the spring is 8.7 m. What is the stiffness of this spring?

••P37 A child of mass 26 kg swings at the end of an elastic cord. At the bottom of the swing, the child's velocity is horizontal, and the speed is 12 m/s. At this instant the cord is 4.30 m long. **(a)** At this instant, what is the parallel component of the rate of change of the child's momentum? **(b)** At this instant, what is the perpendicular component of the rate of change of the child's momentum? **(c)** At this instant, what is the *net* force acting on the child? **(d)** What is the magnitude of the force that the elastic cord exerts on the child? (It helps to draw a diagram of the forces.) **(e)** The relaxed length of the elastic cord is 4.22 m. What is the stiffness of the cord?

••P38 An engineer whose mass is 70 kg holds onto the outer rim of a rotating space station whose radius is 14 m and which takes 30 s to make one complete rotation. What is the magnitude of the force the engineer has to exert in order to hold on? What is the magnitude of the net force acting on the engineer?

••P39 In June 1997 the NEAR spacecraft ("Near Earth Asteroid Rendezvous"; see http://near.jhuapl.edu/), on its way to photograph the asteroid Eros, passed near the asteroid Mathilde. After passing Mathilde, on several occasions rocket propellant was expelled to adjust the spacecraft's momentum in order to follow a path that would approach the asteroid Eros, the final destination for the mission. After getting close to Eros, further small adjustments made the momentum just right to give a circular orbit of radius 45 km (45×10^3 m) around the asteroid. So much propellant had been used that the final mass of the spacecraft while in circular orbit around Eros was only 500 kg. The spacecraft took 1.04 days to make one complete circular orbit around Eros. Calculate what the mass of Eros must be.

••P40 A Ferris wheel is a vertical, circular amusement ride. Riders sit on seats that swivel to remain horizontal as the wheel turns. The wheel has a radius R and rotates at a constant rate, going around once in a time T. At the bottom of the ride, what are the magnitude and direction of the force exerted by the seat on a rider of mass m? Include a diagram of the forces on the rider.

••P41 A block with mass 0.4 kg is connected by a spring of relaxed length 0.15 m to a post at the center of a low-friction table. You pull the block straight away from the post and release it, and you observe that the period of oscillation is 0.6 s. Next you stretch the spring to a length of 0.28 m and give the block an initial speed v perpendicular to the spring, choosing v so that the motion is a circle with the post at the center. What is this speed v?

••P42 When a particle with electric charge q moves with speed v in a plane perpendicular to a magnetic field B, there is a magnetic force at right angles to the motion with magnitude qvB, and the particle moves in a circle of radius r (see Figure 5.77). This equation for the magnetic force is correct even if the speed is comparable to the speed of light. Show that

$$p = \frac{mv}{\sqrt{1-(|\vec{v}|/c)^2}} = qBr$$

even if v is comparable to c.

This result is used to measure relativistic momentum: if the charge q is known, we can determine the momentum of a particle by observing the radius of a circular trajectory in a known magnetic field.

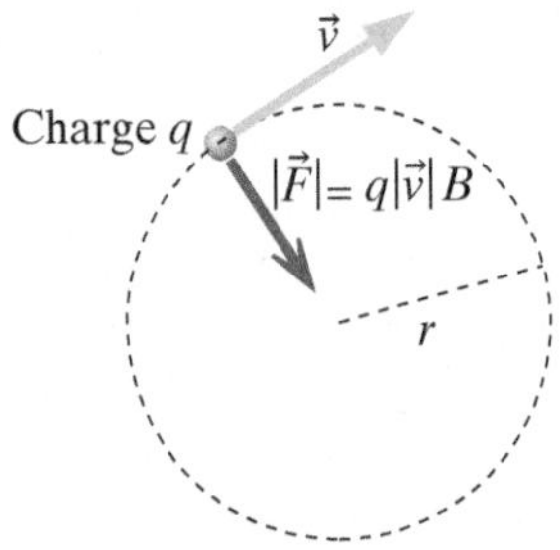

Figure 5.77

••P43 A ball of unknown mass m is attached to a spring. In outer space, far from other objects, you hold the other end of the spring and swing the ball around in a circle of radius 1.5 m at constant speed. **(a)** You time the motion and observe that going around 10 times takes 6.88 s. What is the speed of the ball? **(b)** Is the momentum of the ball changing or not? How can you tell? **(c)** If the momentum is changing, what interaction is causing it to change? If the momentum is not changing, why isn't it? **(d)** The relaxed length of the spring is 1.2 m, and its stiffness is 1000 N/m. While you are swinging the ball, since the radius of the circle is 1.5 m, the length of the spring is also 1.5 m. What is the magnitude of the force that the spring exerts on the ball? **(e)** What is the mass m of the ball?

••P44 A sports car (and its occupants) of mass M is moving over the rounded top of a hill of radius R. At the instant when the car is at the very top of the hill, the car has a speed v. You can safely neglect air resistance. **(a)** Taking the sports car as the system of interest, what object(s) exert nonnegligible forces on this system? **(b)** At the instant when the car is at the very top of the hill, draw a diagram showing the system as a dot, with force vectors whose tails are at the location of the dot. Label the force vectors (that is, give them algebraic names). Try to make the lengths of the force vectors be proportional to the magnitudes of the forces. **(c)** Starting from the Momentum Principle, calculate the force exerted by the road on the car. **(d)** Under what conditions will the force exerted by the road on the car be zero? Explain.

••P45 A small block of mass m is attached to a spring with stiffness k_s and relaxed length L. The other end of the spring is fastened to a fixed point on a low-friction table. The block slides on the table in a circular path of radius $R > L$. How long does it take for the block to go around once?

••P46 A person of mass 70 kg rides on a Ferris wheel whose radius is 4 m. The person's speed is constant at 0.3 m/s. The person's location is shown by a dot in Figure 5.78. **(a)** What is the magnitude of the rate of change of the momentum of the person at the instant shown?

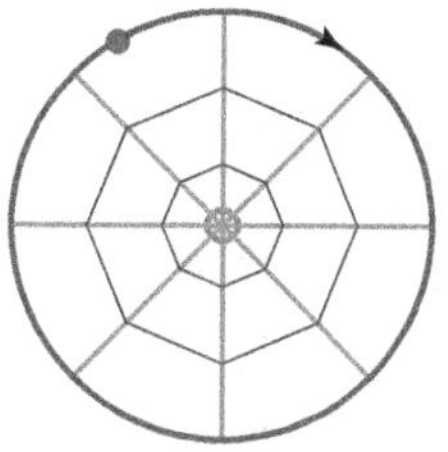

Figure 5.78

(b) What is the direction of the rate of change of momentum of the person at the instant shown? **(c)** What is the magnitude of the *net* force acting on the person at the instant shown? Draw the *net* force vector on the diagram at this instant, with the tail of the vector on the person.

••P47 The planets in our Solar System have orbits around the Sun that are nearly circular, and $v \ll c$. Calculate the period T (a year—the time required to go around the Sun once) for a planet whose orbit radius is r. This is the relationship discovered by Kepler and explained by Newton. (It can be shown by advanced techniques that this result also applies to elliptical orbits if you replace r by the semimajor axis, which is half the longer, major axis of the ellipse.) Use this analytical solution for circular motion to predict the Earth's orbital speed, using the data for Sun and Earth on the inside back cover of the textbook.

•••P48 In the 1970s the astronomer Vera Rubin made observations of distant galaxies that she interpreted as indicating that perhaps 90% of the mass in a galaxy is invisible to us ("dark matter"). She measured the speed with which stars orbit the center of a galaxy, as a function of the distance of the stars from the center. The orbital speed was determined by measuring the Doppler shift of the light from the stars, an effect that makes light shift toward the red end of the spectrum ("red shift") if the star has a velocity component away from us, and makes light shift toward the blue end of the spectrum if the star has a velocity component toward us.

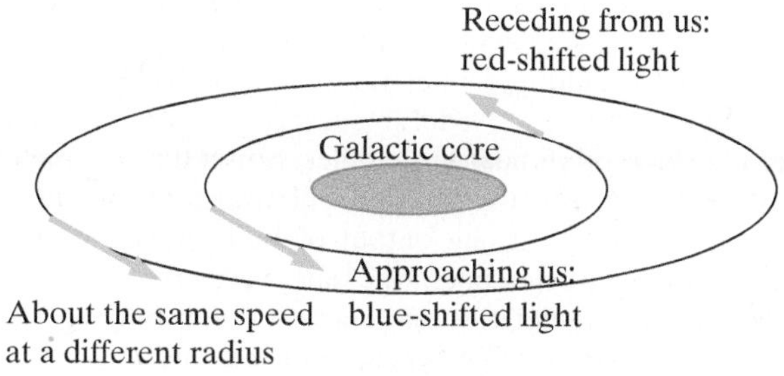

Figure 5.79

She found that for stars farther out from the center of the galaxy, the orbital speed of the star hardly changes with distance from the center of the galaxy, as is indicated in Figure 5.79. The visible components of the galaxy (stars, and illuminated clouds of dust) are most dense at the center of the galaxy and thin out rapidly as you move away from the center, so most of the visible mass is near the center. **(a)** Predict the speed v of a star going around the center of a galaxy in a circular orbit, as a function of the star's distance r from the center of the galaxy, assuming that almost all of the galaxy's mass M is concentrated at the center. **(b)** Construct a logical argument as to why Rubin concluded that much of the mass of a galaxy is not visible to us. Reason from principles discussed in this chapter, and your analysis of part (a). Explain your reasoning. You need to address the following issues: (i) Rubin's observations are not consistent with your prediction in (a). (ii) Most of the *visible* matter is in the center of the galaxy. (iii) Your prediction in (a) assumed that most of the mass is at the center.

This issue has not yet been resolved, and is still a current topic of astrophysics research. Here is a discussion by Rubin of her work: "Dark Matter in Spiral Galaxies" by Vera C. Rubin, *Scientific American*, June 1983 (96–108). You can find several graphs of the rotation curves for spiral galaxies on page 101 of this article.

Section 5.9

••P49 The Ferris wheel in Figure 5.80 is a vertical, circular amusement ride with radius 10 m. Riders sit on seats that swivel to remain horizontal. The Ferris wheel rotates at a constant rate, going around once in 10.5 s. Consider a rider whose mass is 56 kg.

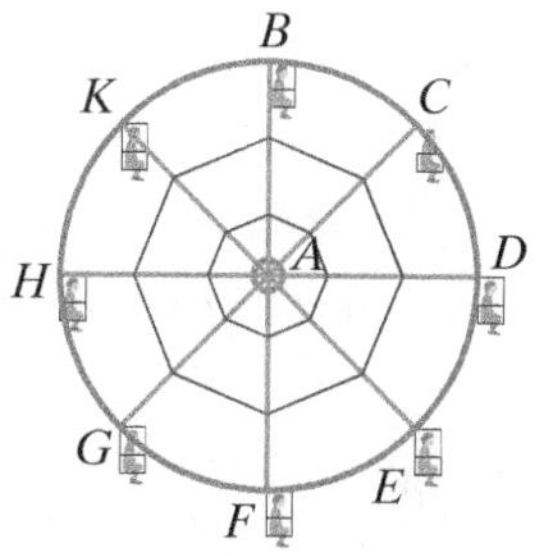

Figure 5.80

(a) At the bottom of the ride, what is the rate of change of the rider's momentum? **(b)** At the bottom of the ride, what is the vector gravitational force exerted by the Earth on the rider? **(c)** At the bottom of the ride, what is the vector force exerted by the seat on the rider? **(d)** Next consider the situation at the top of the ride. At the top of the ride, what is the rate of change of the rider's momentum? **(e)** At the top of the ride, what is the vector gravitational force exerted by the Earth on the rider? **(f)** At the top of the ride, what is the vector force exerted by the seat on the rider?

A rider *feels* heavier if the electric, interatomic contact force of the seat on the rider is larger than the rider's weight mg (and the rider sinks more deeply into the seat cushion). A rider *feels* lighter if the contact force of the seat is smaller than the rider's weight (and the rider does not sink as far into the seat cushion). **(g)** Does a rider *feel* heavier or lighter at the bottom of a Ferris wheel ride? **(h)** Does a rider *feel* heavier or lighter at the top of a Ferris wheel ride?

••P50 By weight we usually mean the gravitational force exerted on an object by the Earth. However, when you sit in a chair your own perception of your own weight is based on the contact force the chair exerts upward on your rear end rather than on the gravitational force. The smaller this contact force is, the less weight you perceive, and if the contact force is zero, you feel peculiar and "weightless" (an odd word to describe a situation when the only force acting on you is the gravitational force exerted by the Earth!). Also, in this condition pressure on your inner ear is released, which affects your sense of balance, and your internal organs no longer press on each other, all of which contributes to the odd sensation in your stomach. **(a)** How fast must a roller coaster car go over the top of a circular arc for you to feel "weightless"? The center of the car moves along a circular arc of radius R (see Figure 5.81). Include a carefully labeled force diagram.

Figure 5.81

(b) How fast must a roller coaster car go through a circular dip for you to feel three times as "heavy" as usual, due to the upward force of the seat on your bottom being three times as large as usual? The center of the car moves along a circular arc of radius R (see Figure 5.82). Include a carefully labeled force diagram.

Figure 5.82

Section 5.10

•P51 A circular pendulum of length 1.1 m goes around at an angle of 28 degrees to the vertical. Predict the speed of the mass at the end of the string. Also predict the period, the time it takes to go around once. Remember that the radius of the circle is the length of the string times the sine of the angle that the string makes to the vertical.

••P52 Use a circular pendulum to determine g. You can increase the accuracy of the time it takes to go around once by timing N revolutions and then dividing by N. This minimizes errors contributed by inaccuracies in starting and stopping the clock. It is wise to start counting from zero (0, 1, 2, 3, 4, 5) rather than starting from 1 (1, 2, 3, 4, 5 represents only four revolutions, not five). It also improves accuracy if you start and stop timing at a well-defined event, such as when the mass crosses in front of an easily visible mark. This was the method used by Newton to get an accurate value of g. Newton was not only a brilliant theorist but also an excellent experimentalist. For a circular pendulum he built a large triangular wooden frame mounted on a vertical shaft, and he pushed this around and around while making sure that the string of the circular pendulum stayed parallel to the slanting side of the triangle.

••P53 **(a)** Many communication satellites are placed in a circular orbit around the Earth at a radius where the period (the time to go around the Earth once) is 24 h. If the satellite is above some point on the equator, it stays above that point as the Earth rotates, so that as viewed from the rotating Earth the satellite appears to be motionless. That is why you see dish antennas pointing at a fixed point in space. Calculate the radius of the orbit of such a "synchronous" satellite. Explain your calculation in detail. **(b)** Electromagnetic radiation including light and radio waves travels at a speed of 3×10^8 m/s. If a phone call is routed through a synchronous satellite to someone not very far from you on the ground, what is the minimum delay between saying something and getting a response? Explain. Include in your explanation a diagram of the situation. **(c)** Some human-made satellites are placed in "near-Earth" orbit, just high enough to be above almost all of the atmosphere. Calculate how long it takes for such a satellite to go around the Earth once, and explain any approximations you make. **(d)** Calculate the orbital speed for a near-Earth orbit, which must be provided by the launch rocket. (The advantages of near-Earth communications satellites include making the signal delay unnoticeable, but with the disadvantage of having to track the satellites actively and having to use many satellites to ensure that at least one is always visible over a particular region.) **(e)** When the first two astronauts landed on the Moon, a third astronaut remained in an orbiter in circular orbit near the Moon's surface. During half of every complete orbit, the orbiter was behind the Moon and out of radio contact with the Earth. On each orbit, how long was the time when radio contact was lost?

••P54 There is no general analytical solution for the motion of a three-body gravitational system. However, there do exist analytical solutions for very special initial conditions. Figure 5.83 shows three stars, each of mass m, which move in the plane of the page along a circle of radius r. Calculate how long this system takes to make one complete revolution. (In many cases three-body orbits are not stable: any slight perturbation leads to a breakup of the orbit.)

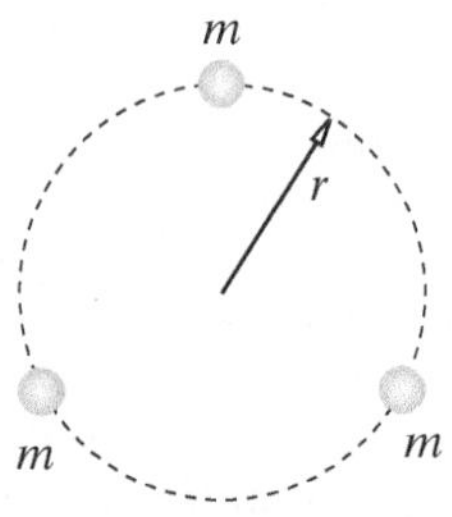

Figure 5.83

••P55 Remarkable data indicate the presence of a massive black hole at the center of our Milky Way galaxy. The W. M. Keck 10-m-diameter telescopes in Hawaii were used by Andrea Ghez and her colleagues to observe infrared light coming directly through the dust surrounding the central region of our galaxy (visible light is multiply-scattered by the dust, blocking a direct view). Stars were observed for several consecutive years to determine their orbits near a motionless center that is completely invisible ("black") in the infrared but whose precise location is known due to its strong output of radio waves, which are observed by radio telescopes. The data were used to show that the object at the center must have a mass that is huge compared to the mass of our own Sun, whose mass is a "mere" 2×10^{30} kg. Figure 5.84 shows positions from 1995 to 2012 of one of the stars, called S0-20, orbiting around the galactic center. The orbit appears nearly circular with the radius shown.

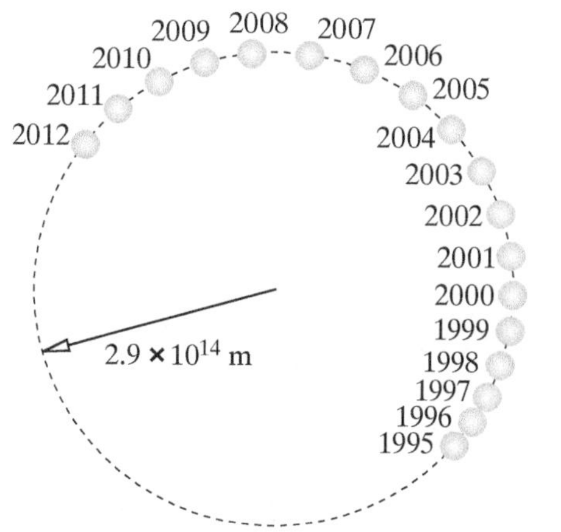

Figure 5.84 Positions on the sky of the star S0-20 near the center of our galaxy.

(a) Using the positions and times shown in Figure 5.84, what is the approximate speed of this star in m/s? Also express the speed as a fraction of the speed of light. **(b)** This is an extraordinarily high speed for a macroscopic object. Is it reasonable to approximate the star's momentum as mv? (Some other stars near the galactic center with highly elliptical orbits move even faster when they are closest to the center.) **(c)** Based on these data, estimate the mass of the massive black hole about which this star is orbiting. (You can see in Figure 5.84 that the speed appears greater in 2007–2008 than in 1995–1996. You're looking at a projection of an elliptical orbit, which only happens to look circular as viewed from Earth, and the black hole is not at the center of this circle. Therefore your result will be approximate and will differ from that obtained by Ghez and colleagues, who carefully made a consistent fit of the position data for seven different stars.) **(d)** How many of our Suns does this represent?

It is now thought that most galaxies have such a black hole at their centers, as a result of long periods of mass accumulation. When many bodies orbit each other, sometimes in an interaction an object happens to acquire enough speed to escape from the group, and the remaining objects are left closer together. Some simulations show that over time, as much as half the mass may be ejected, with agglomeration of the remaining mass. This could be part of the mechanism for forming massive black holes.

For more information, search the web for Andrea Ghez. You may see the term "arc seconds," which is an angular measure of how far apart objects appear in the sky, and "parsecs," which is a distance equal to 3.3 light-years (a light-year is the distance light goes in one year).

•••P56 You put a 10 kg object on a bathroom scale at the North Pole, and the scale reads exactly 10 kg (actually, it measures the force F_N that the scale exerts on the object, but displays a reading in kg). At the North Pole you are 6357 km from the center of the Earth. At the equator, the scale reads a different value due to two effects: (1) The Earth bulges out at the equator (due to its rotation), and you are 6378 km from the center of the Earth. (2) You are moving in a circular path due to the rotation of the Earth (one rotation every 24 hours). Taking into account ***both*** of these effects, what does the scale read at the equator?

COMPUTATIONAL PROBLEMS

More detailed and extended versions of some computational modeling problems may be found in the lab activities included in the *Matter & Interactions, 4th Edition,* resources for instructors.

••P57 Start with the program you wrote to model the motion of a spacecraft near the Earth (Chapter 3). Choose initial conditions that produce an elliptical orbit. **(a)** Create an arrow to represent the net force on the spacecraft, and place the tail of the arrow at the position of the spacecraft. Inside the calculation loop, update the arrow's position every time you update the craft's position. **(b)** Update the axis of the arrow inside the loop, so it always points in the direction of the net force. Find a scale factor that gives the arrow an appropriate length when the craft is near the Earth, and allows the arrow to be visible when the craft is far from the Earth. **(c)** Now create two additional arrows to represent $\vec{F}_{\parallel}$ and $\vec{F}_{\perp}$, the parallel and perpendicular parts of the net force on the spacecraft. In VPython you can take the dot product of two vectors, $\vec{A}$ and $\vec{B}$, like this: `C = dot(A,B)`. Make your program display $\vec{F}_{\text{net}}$, $\vec{F}_{\parallel}$, and $\vec{F}_{\perp}$ as the craft orbits the Earth. **(d)** In what part of the orbit does $\vec{F}_{\parallel}$ point in the same direction as $\vec{p}$? What effect does it have on the craft's momentum? **(e)** In what part of the orbit does $\vec{F}_{\parallel}$ point in the opposite direction to $\vec{p}$? What effect does it have on the craft's momentum? **(f)** Are there any locations in the orbit where $\vec{F}_{\parallel} = 0$? If so, what are they?

••P58 Start with the program you wrote to model the 3D motion of a mass hanging from a spring (Chapter 4). **(a)** Create an arrow to represent the net force on the mass, and place the tail of the arrow at the position of the mass. Update both the position and axis of the arrow inside the calculation loop, using an appropriate scale factor, so the arrow always shows the net force on the mass. **(b)** Now create two additional arrows to represent $\vec{F}_{\parallel}$ and $\vec{F}_{\perp}$, the parallel and perpendicular parts of the net force on the mass. Make your program display $\vec{F}_{\text{net}}$, $\vec{F}_{\parallel}$, and $\vec{F}_{\perp}$ as the mass oscillates. **(c)** Find initial conditions for which $\vec{F}_{\parallel}$ is zero and remains zero. Explain. **(d)** Find initial conditions for which $\vec{F}_{\perp}$ is zero and remains zero. Explain. **(e)** Find initial conditions for which both $\vec{F}_{\parallel}$ and $\vec{F}_{\perp}$ are nonzero most of the time. Explain.

•••P59 Start with the program you wrote to model the motion of a spacecraft near the Earth. Use initial conditions that produce an elliptical orbit. Modify the program so the kissing circle is continuously displayed as the craft orbits the Earth. You may want to use a `ring` object in VPython to display the circle.

ANSWERS TO CHECKPOINTS

1 **(a)** Baseball: left diagram in Figure 5.85, **(b)** Yo-yo: right diagram in Figure 5.85

Figure 5.85 Left diagram: baseball. Right diagram: yo-yo.

2 $\vec{0}$; $\vec{0}$; $\langle 30, -90, 130 \rangle$ N

3 **(a)** Both blocks. **(b)** Either one of the blocks; note that the magnitude of the force that block m_2 exerts on m_1 is the same as the magnitude of the force that block m_1 exerts on block m_2.

4 **(a)** Figure 5.86 shows the forces:

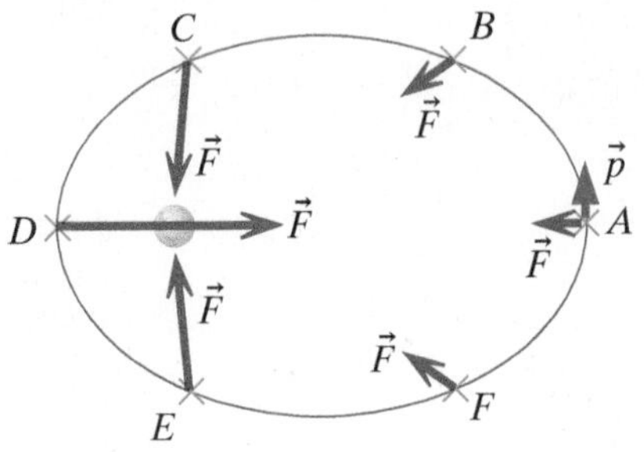

Figure 5.86

A: **(b)** zero, **(c)** nonzero, **(d)** zero, **(e)** changing; *B*: **(b)** nonzero, **(c)** nonzero, **(d)** positive, **(e)** changing; *C*: **(b)** nonzero, **(c)** nonzero, **(d)** positive, **(e)** changing; *D*: **(b)** zero, **(c)** nonzero, **(d)** zero, **(e)** changing; *E*: **(b)** nonzero, **(c)** nonzero, **(d)** negative, **(e)** changing; *F*: **(b)** nonzero, **(c)** nonzero, **(d)** negative, **(e)** changing

CHAPTER

6

The Energy Principle

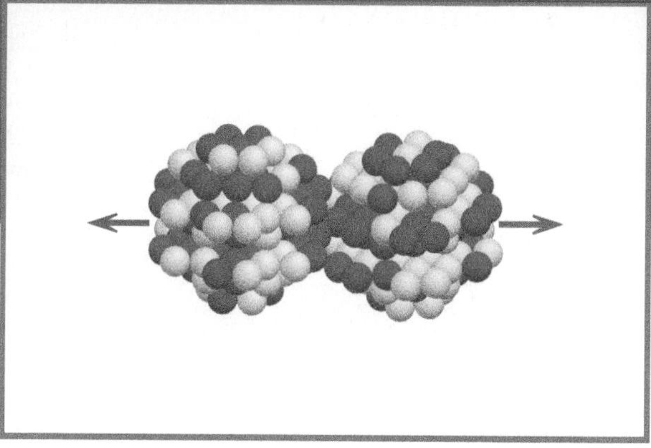

OBJECTIVES

After studying this chapter, you should be able to

- Calculate the total energy (rest energy + kinetic energy) of a single particle system.
- Calculate the total energy of a multiparticle system (rest energy, kinetic energy, and gravitational and electric potential energy).
- Mathematically relate changes in energy of a system to the work done by the surroundings.
- Analyze in detail processes involving changes in potential energy, kinetic energy, and rest energy.
- Construct and interpret graphs of multiparticle system energy as a function of separation.

6.1 THE ENERGY PRINCIPLE

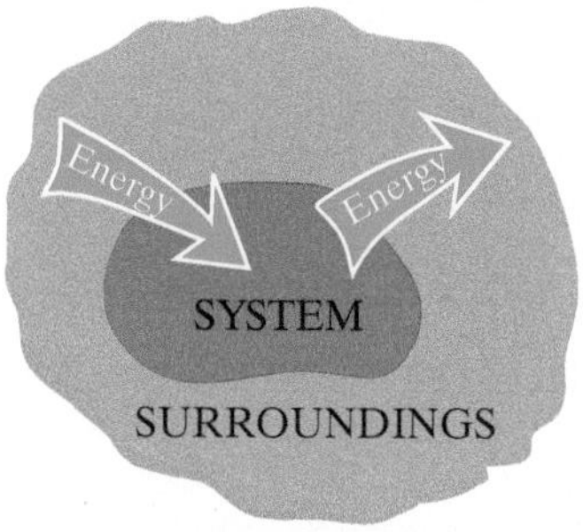

Figure 6.1 Energy may flow between system and surroundings.

You know a lot about the flow of energy from daily life. You ingest chemical energy in the form of food, and you use up this chemical energy when you are active. You put electrical energy into a toaster, and the toaster raises the temperature of the bread. You compress a spring, and the mechanical energy stored in the spring can launch a ball.

Energy cannot be created or destroyed, but it can change form. The only way for a system to gain or lose energy is if the surroundings lose or gain the same amount of energy (Figure 6.1). Therefore the Energy Principle can be stated most generally in terms of conservation of energy:

CONSERVATION OF ENERGY

$$\Delta E_{\text{system}} + \Delta E_{\text{surroundings}} = 0$$

In a detailed energy analysis of a system, it is often useful to express the energy principle in terms of inputs of energy from the surroundings:

THE ENERGY PRINCIPLE

$$\Delta E_{\text{system}} = (\text{Energy inputs from surroundings})$$

The energy inputs from the surroundings may be positive or negative, corresponding to flows of energy from surroundings to system or from system to surroundings, respectively.

There are a number of mechanisms by which energy may be transferred between system and surroundings. In this chapter, we will focus on "work," the result of a force acting through a distance. Other mechanisms will be discussed in subsequent chapters.

The validity of the Energy Principle has been verified through a very wide variety of observations and experiments, involving large and small objects, moving slowly or at speeds near the speed of light, and even undergoing nuclear reactions that change the identity of the objects. It is a summary of the way energy flows in the real world.

The Energy Principle is a fundamental principle because:

- It applies to every possible system, no matter how large or small it is (from clusters of galaxies to subatomic particles) or how fast it is moving.
- It is true for any kind of interaction (gravitational, electromagnetic, strong, weak).
- It relates an effect (change in energy of a system) to a cause (an interaction with the surroundings).

6.2 ENERGY OF A SINGLE PARTICLE

The simplest possible system consists of a single point particle. "A point particle" might refer to an electron, but we can also model a baseball or even a planet as a point particle if during the process of interest there are no significant changes internal to the system such as changes of shape or rotation or vibration or temperature.

In the current chapter we will discuss energy and changes in energy for systems of point particles. We will defer the consideration of systems that include extended objects that undergo internal changes, and hence cannot be modeled as point particles, until Chapter 7.

A single particle can have only two kinds of energy: the energy associated with its rest mass, called "rest energy," and the energy associated with its motion, called "kinetic energy." Both of these are combined into a single compact expression, first expressed by Einstein in 1905:

ENERGY OF A SINGLE PARTICLE

$$E_{\text{particle}} = \gamma mc^2$$

This equation defines particle energy. m is the mass of the particle. As usual $\gamma = \dfrac{1}{\sqrt{1 - \left(\frac{v}{c}\right)^2}}$, where c is the speed of light. The unit of energy is the joule, abbreviated J. One $\text{J} = \text{one kg} \cdot (\text{m/s})^2$.

Later in this chapter we'll see that this definition of particle energy is consistent with the definition of a particle's momentum, $\vec{p} = \gamma m\vec{v}$.

QUESTION What are the similarities and differences between the definition of particle energy and the definition of particle momentum?

Similarities:

- Both contain the same factor γ.
- Both are proportional to the mass m.

Differences:

- Particle energy is a scalar.
- Particle momentum is a vector.

Rest Energy

QUESTION If a particle is at rest, what is its energy?

If $v = 0$ then $\gamma = 1$, and we find the famous equation $E = mc^2$. This has the striking interpretation that a particle has energy even when it is sitting still, which was a revolutionary idea when Einstein proposed it. We call mc^2 the "rest energy," the amount of energy a particle has when it is sitting still.

REST ENERGY

$$E_{\text{rest}} = mc^2$$

The "rest energy" is the energy of a particle at rest.

Even more strikingly, Einstein realized that this means that the mass of an object at rest is its energy content divided by a constant: $m = E_{\text{rest}}/c^2$. For example, a hot object, with more internal energy, has very slightly more mass than a cold object. In a real sense, mass and energy are the same thing, although for historical reasons we use different units for them: mass in kilograms, energy in joules, with a constant factor of c^2 difference.

Note another difference between energy and momentum: A particle at rest has zero momentum but it has nonzero energy, its rest energy.

Kinetic Energy

QUESTION If a particle is moving with speed v, is its energy greater than mc^2?

If $v > 0$, $\sqrt{1-(v/c)^2}$ is less than 1, so $\gamma = 1/\sqrt{1-(v/c)^2}$ is greater than 1, and the particle energy γmc^2 increases with increasing speed. As a consequence of its motion, the particle has additional, "kinetic" energy K:

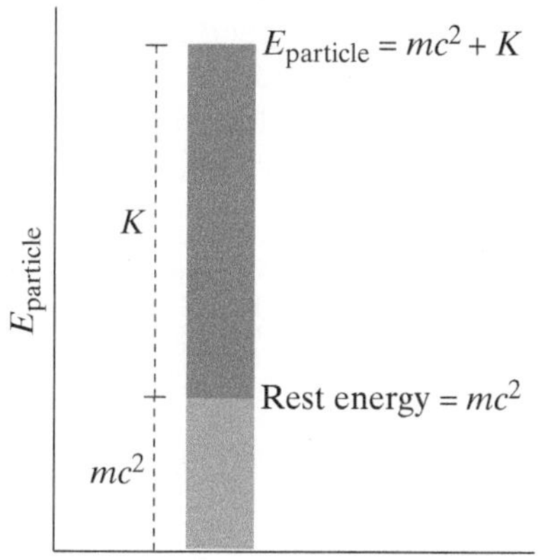

Figure 6.2 Particle energy is equal to rest energy plus kinetic energy K. This particle has high speed, and its kinetic energy is large compared to its rest energy.

KINETIC ENERGY K OF A PARTICLE

$$K = \gamma mc^2 - mc^2$$

The kinetic energy K of a particle is the energy a moving particle has in addition to its rest energy. It is often useful to turn this equation around:

$$E_{\text{particle}} = mc^2 + K$$

If a particle is at rest, its particle energy is just its rest energy mc^2. If the particle moves, it has not only its rest energy mc^2 but also an additional amount of energy, which we call the kinetic (motional) energy K (Figure 6.2).

EXAMPLE **Energy of a Fast-Moving Proton**

A proton in a particle accelerator has a speed of 2.91×10^8 m/s. **(a)** What is the energy of the proton? **(b)** What is the rest energy of the proton? **(c)** What is the kinetic energy of the moving proton?

Solution **(a)** Moving proton:

$$\gamma = \frac{1}{\sqrt{1-\left(\dfrac{2.91\times 10^8}{3\times 10^8}\right)^2}} = 4.11$$

$$E = (4.11)(1.7\times 10^{-27}\text{ kg})(3\times 10^8\text{ m/s})^2$$

$$E = 6.288\times 10^{-10}\text{ J}$$

(b) Proton at rest:

$$E_{\text{rest}} = mc^2$$
$$E_{\text{rest}} = (1.7 \times 10^{-27}\ \text{kg})(3 \times 10^8\ \text{m/s})^2$$
$$E_{\text{rest}} = 1.53 \times 10^{-10}\ \text{J}$$

(c) Kinetic energy of proton:

$$K = E - E_{\text{rest}}$$
$$K = 6.288 \times 10^{-10}\ \text{J} - 1.53 \times 10^{-10}\ \text{J} = 4.76 \times 10^{-10}\ \text{J}$$

In this example, the kinetic energy of the fast-moving proton turns out to be about three times as large as the rest energy of the proton—about the ratio shown in Figure 6.2.

QUESTION Is there a limit to the amount of energy a particle can have?

Delivering more and more energy to a particle increases its energy more and more, and there is no theoretical limit to how much energy a particle can acquire. As the speed approaches the speed of light $c = 3 \times 10^8$ m/s, the denominator $\sqrt{1-(v/c)^2}$ becomes very small, and the energy becomes very large.

However, even though you can increase the energy, it becomes very difficult to increase the speed v, because as you near the speed of light a tiny speed increase would require a huge energy increase. There is effectively a universal speed limit: a particle can't reach the speed of light c because an infinite amount of energy input would be required to achieve this.

Checkpoint 1 An electron has mass 9×10^{-31} kg. If the electron's speed is 0.988c (that is, $v/c = 0.988$), what is its particle energy? What is its rest energy? What is its kinetic energy?

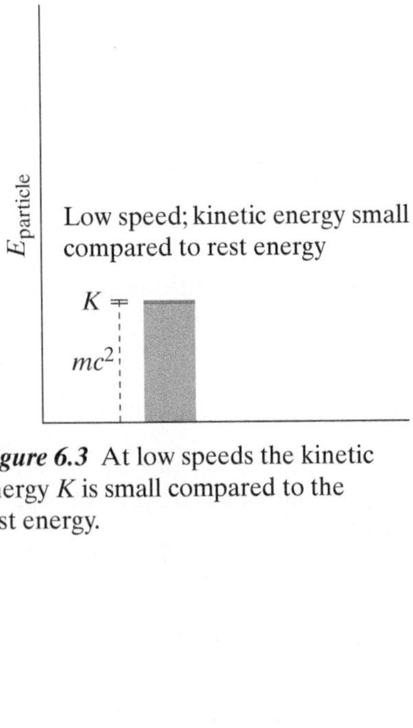

Figure 6.3 At low speeds the kinetic energy K is small compared to the rest energy.

Approximate Kinetic Energy at Low Speeds

If the speed v of a particle is small compared to the speed of light ($v \ll c$), we can find a simple approximate expression for the kinetic energy of the particle. In such a case, as shown in Figure 6.3, the kinetic energy is very small compared to the rest energy, and we can derive an approximate expression for the kinetic energy at low speeds by using the binomial expansion. (See Section 6.19 at the end of this chapter for a more detailed explanation.)

$$\frac{mc^2}{\sqrt{1-(v/c)^2}} = mc^2\left[1 + \frac{1}{2}\left(\frac{v}{c}\right)^2 + \frac{3}{8}\left(\frac{v}{c}\right)^4 + \frac{5}{16}\left(\frac{v}{c}\right)^6 + \cdots\right]$$
$$\frac{mc^2}{\sqrt{1-(v/c)^2}} \approx mc^2 + \frac{1}{2}mv^2 \quad \text{if } v/c \ll 1$$

Since we define kinetic energy through the relation $E_{\text{particle}} = mc^2 + K$, at low speeds the kinetic energy is approximately the following:

APPROXIMATE KINETIC ENERGY OF A PARTICLE

$$K \approx \frac{1}{2}mv^2 \quad (\text{if } v \ll c)$$

Approximate kinetic energy can also be written in terms of approximate magnitude of momentum.

$$K \approx \frac{1}{2}mv^2 \approx \frac{1}{2}m\left(\frac{p}{m}\right)^2 = \frac{p^2}{2m} \quad (\text{if } v \ll c)$$

EXAMPLE **Approximate Kinetic Energy**

The solar wind consists of charged particles streaming away from the Sun. The speed of a proton in the solar wind can be as high as 1×10^6 m/s. What is the kinetic energy of such a proton?

Solution

QUESTION Is it appropriate to use the approximate expression for kinetic energy in this case?

$$\left(\frac{v}{c}\right)^2 = \left(\frac{1 \times 10^6 \text{ m/s}}{3 \times 10^8 \text{ m/s}}\right)^2 = (3.33 \times 10^{-3})^2 = 1.11 \times 10^{-5}$$

so $\gamma \approx 1$.
It is appropriate to use the approximate expression for K.

$$K \approx \frac{1}{2}mv^2 = \frac{1}{2}(1.7 \times 10^{-27}\text{ kg})(1 \times 10^6\text{ m/s})^2 = 8.5 \times 10^{-16}\text{ J}$$

The kinetic energy of this relatively slow-moving proton is much less (nearly a million times less) than its rest energy, which was calculated in the previous example.

EXAMPLE **Pull a Block**

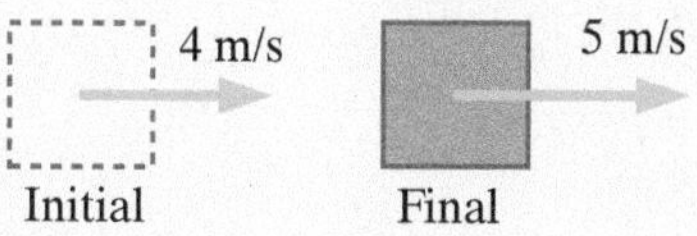

Figure 6.4 You pull a block, changing its speed from 4 m/s to 5 m/s.

If you pull on a 3 kg block and change its speed from 4 m/s to 5 m/s (Figure 6.4), what is the minimum change of chemical (food) energy in you? Give magnitude, sign, and units. (Your temperature rises when you do this, involving an increase in thermal energy, which we'll study in the next chapter, so there is more actual chemical cost than the minimum amount you calculate.)

Solution The speed is tiny compared to c, so the change in kinetic energy of the block is

$$\Delta K = \frac{1}{2}mv_f^2 - \frac{1}{2}mv_i^2 = \frac{1}{2}(3\text{ kg})[(5\text{ m/s})^2 - (4\text{ m/s})^2]$$

$$\Delta K = 13.5\text{ J}$$

Because you increased the kinetic energy of the block by 13.5 J, and your temperature also rose a bit, the change of your store of chemical (food) energy is at least −13.5 J.

Checkpoint 2 An automobile traveling on a highway has an average kinetic energy of 1.1×10^5 J. Its mass is 1.5×10^3 kg; what is its average speed? Convert your answer to miles per hour to see whether it makes sense. If you could use all of the mc^2 rest energy of some amount of fuel to provide the car with its kinetic energy of 1.1×10^5 J, what mass of fuel would you need?

Relativistic Energy and Momentum

Because both relativistic energy and momentum involve the factor

$$\gamma = \frac{1}{\sqrt{1-\left(\frac{v}{c}\right)^2}}$$

it seems likely that they can be related in some way. As you can show in Problem P8, the relationship turns out to be the following, where p represents $|\vec{p}|$:

RELATIVISTIC ENERGY AND MOMENTUM OF A PARTICLE

$$E^2 - (pc)^2 = (mc^2)^2$$

where $E = \gamma mc^2$; $m =$ particle mass.

This relation is true in all reference frames, for all particle speeds. Both energy and momentum will change if a phenomenon is viewed in a different reference frame, but this relation will still be true. The quantity $E^2 - (pc)^2$ is therefore called an "invariant" quantity—a quantity that does not vary when the reference frame is changed.

This relation turns out to be $E^2 - (pc)^2 = 0$ for those particles such as photons that always travel at the speed of light. Such particles have zero mass, and can never be at rest. The relation for so-called "massless" particles reduces to $E^2 = (pc)^2$, or $E = pc$. This relation is also approximately true for any particle that is traveling at very high speed, because in that case mc^2 is small compared to E. Neutrinos have extremely small mass and travel at nearly the speed of light.

The relation $E^2 - (pc)^2 = (mc^2)^2$ can be used to find another expression for kinetic energy, valid for both high and low speeds:

$$E^2 - (mc^2)^2 = (pc)^2$$

$$(E + mc^2)(E - mc^2) = p^2c^2$$

$$(\gamma mc^2 + mc^2)(K) = p^2c^2$$

$$K = \frac{p^2}{(\gamma+1)m}$$

At low speeds, $\gamma \approx 1$, and we find as before that $K \approx p^2/(2m)$.

EXAMPLE **Proton Momentum and Energy**

A proton has a mass of 1.7×10^{-27} kg. If its total energy is three times its rest energy, what is the magnitude of its momentum?

Solution

$$mc^2 = (1.7 \times 10^{-27}\text{ kg})(3 \times 10^8\text{ m/s})^2 = 1.5 \times 10^{-10}\text{ J}$$

$$E_{\text{particle}} = mc^2 + K = 3mc^2 \text{ and } (pc)^2 = (E_{\text{particle}})^2 - (mc^2)^2$$

$$pc = \sqrt{(E_{\text{particle}})^2 - (mc^2)^2} = \sqrt{(3mc^2)^2 - (mc^2)^2}$$

$$= \sqrt{8(mc^2)^2} = \sqrt{8}(mc^2)$$

$$p = \sqrt{8}(mc) = \sqrt{8}(1.7 \times 10^{-27}\text{ kg})(3 \times 10^8\text{ m/s}) = 1.4 \times 10^{-18}\text{ kg}\cdot\text{m/s}$$

6.3 WORK: MECHANICAL ENERGY TRANSFER

We are interested in changes in the energy of a system that are due to the interaction of the system with objects in its surroundings. Our goal is to figure out how to describe such changes quantitatively.

Displacement $\Delta\vec{r}$ Instead of Time Interval Δt

The Momentum Principle relates a change in momentum to the net impulse applied to a system. Impulse is the product of force and time.

$$\text{(change of momentum)} = \text{(force)(time)}$$

We might ask what property of a system changes due to forces acting not for a time but through a distance, something like this:

$$\text{(change of ?)} = \text{(force)(distance)}$$

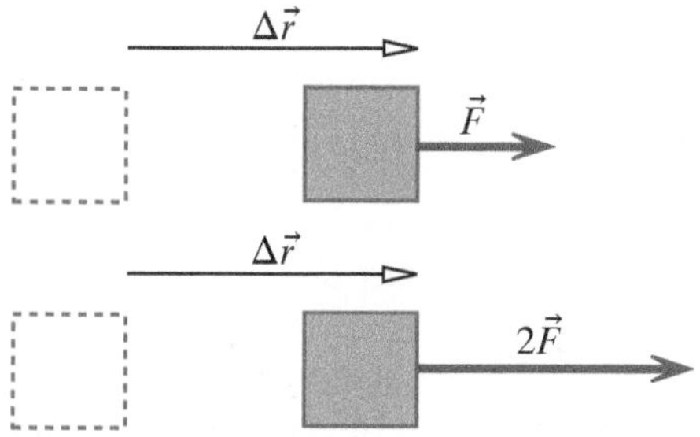

Figure 6.5 Pulling a block over a nearly frictionless surface with a force twice as big (through the same distance) means you expend twice as much chemical energy.

The quantity that changes turns out to be the energy of the system, γmc^2 in the case of a single particle. The quantity involving force times distance is called "work."

In Section 6.6 we will show that this is correct. For now we'll assume this to be true, and after explaining how to calculate work we'll use the Energy Principle for a particle to analyze several kinds of processes.

QUESTION What is the effect of increasing the applied force?

Suppose that you apply a force with constant magnitude F in a constant direction to a block that was initially at rest on a nearly frictionless surface, pulling it through a displacement Δr (Figure 6.5). The block moves faster and faster, acquiring a certain amount of kinetic energy. If you were to pull twice as hard ($2F$) through the same displacement Δr, the block would presumably acquire twice as much kinetic energy (and you would expend twice as much chemical energy). Note that the rest energy mc^2 of the block doesn't change, just the kinetic energy.

QUESTION What is the effect of increasing the displacement?

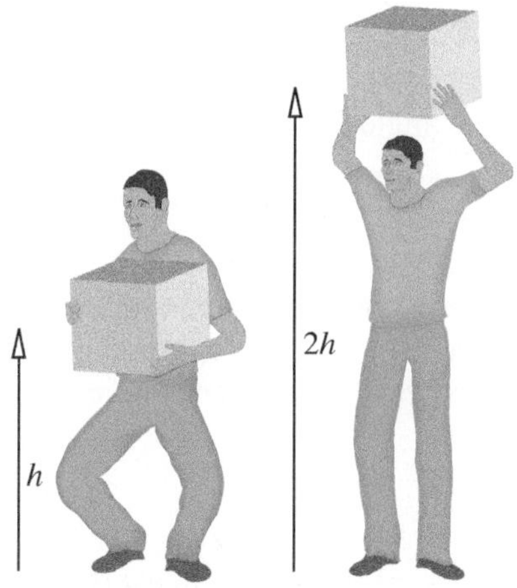

Figure 6.6 Lifting a box twice as high (with the same force) means you expend twice as much chemical energy.

If you lift a heavy box to a height h above the ground, you will have expended a certain amount of chemical energy (Figure 6.6). If you lift the box twice as high (a height $2h$ above the ground), you will presumably have had to expend twice as much energy.

Work

Because the amount of energy expended appears to be proportional both to the magnitude of the force exerted and to the distance through which an object is moved, we'll tentatively define the amount of energy transfer between you and the object as the product of force times displacement, $F\Delta r$. This type of mechanical energy transfer from the surroundings into the system is called "work." Here we use the word "work" as a technical term in the context of physics. This technical meaning is only loosely connected with everyday uses of the word.

The unit of work is newton · meter. From the Momentum Principle we know that a newton is the same as a kg · (m/s)/s, so a newton · meter is the same as a $\text{kg}\cdot\text{m}^2/\text{s}^2$, which is a joule (same units as mc^2). Work has the same units as energy, as it should if we expect the change in energy of a system to be equal to the work.

QUESTION Is doubling the displacement the same as doubling the time?

No, it isn't. If you pull a block through twice the displacement, you do twice as much work, and the block acquires twice as much kinetic energy. However, because the block is moving faster in the second half of the doubled displacement, the time interval is less than doubled (and the impulse is less than doubled), so the momentum change is not doubled.

Note carefully: work and kinetic energy change are proportional to displacement, whereas impulse and momentum change are proportional to time interval. In symbols: work $F\Delta r$ is not the same as impulse $F\Delta t$.

Work in 3D

When calculating the work done on a system by an applied force, we are interested in the effect of the component of the force that is parallel to the displacement of the system. This sounds familiar—we saw in Chapter 5 that it was useful to divide an applied force into two parts: $\vec{F}_{\parallel}$, the component of a force parallel to a system's momentum, which can change the magnitude of the momentum (and hence the speed), and $\vec{F}_{\perp}$, the perpendicular component of the force, which can change the direction of momentum but not the magnitude. For example, in the circular orbit of a planet around a star, the planet's speed is constant, since the force by the star is always perpendicular to the planet's momentum.

In calculating work we see a similar pattern: the part of an applied force that is parallel to $\Delta\vec{r}$, the displacement, does nonzero work on the system, but the perpendicular part of the force does not contribute to changing the system's energy. In order to calculate the work done on a system by a force $\vec{F}$, we want to use the parallel component of $\vec{F}$. This suggests using a dot product:

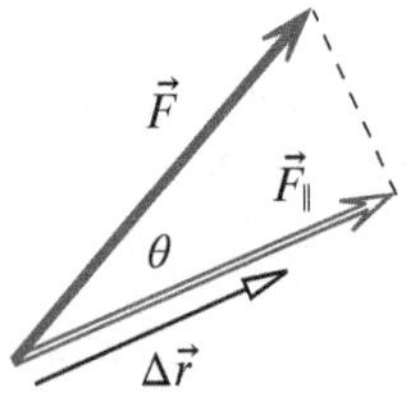

Figure 6.7 The part of a force that is parallel to the displacement does work on a system.

WORK DONE BY A CONSTANT FORCE

$$
\begin{aligned}
W &= \vec{F} \bullet \Delta\vec{r} \\
&= \left|\vec{F}\right| |\Delta\vec{r}| \cos\theta \\
&= F_x\Delta x + F_y\Delta y + F_z\Delta z
\end{aligned}
$$

The displacement $\Delta\vec{r} = \langle \Delta x, \Delta y, \Delta z \rangle$. The angle θ is the angle between $\vec{F}$ and $\Delta\vec{r}$, as shown in Figure 6.7.

As we'll discuss in more detail in a moment, work can be positive or negative. If you push or pull in the direction of displacement, you do positive work and you increase the energy of the system. If you push or pull in the direction opposite to the displacement, you do negative work and you decrease the energy of the system.

EXAMPLE **Pull a Block**

You pull a block 1.5 m across a table in the $-x$ direction while exerting a force of $\langle -0.3, 0.25, 0 \rangle$ N (Figure 6.8). How much work do you do on the block?

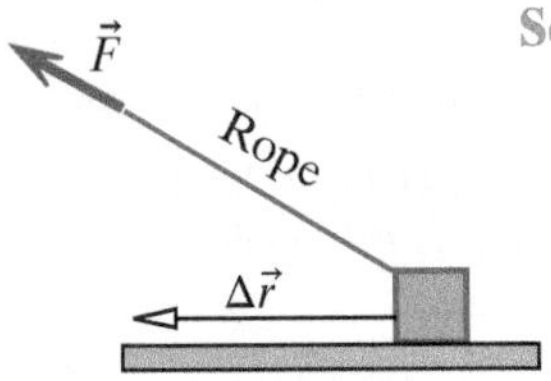

Figure 6.8 You pull a block across a table.

Solution

$$W = \vec{F} \bullet \Delta\vec{r} = \langle -0.3, 0.25, 0 \rangle \text{ N} \bullet \langle -1.5, 0, 0 \rangle \text{ m}$$
$$W = (-0.3\text{ N})(-1.5\text{ m}) + (0.25\text{ N})(0) + 0 \cdot 0 = 0.45\text{ J}$$

Checkpoint 3 A paper airplane flies from position $\langle 6, 10, -3 \rangle$ m to $\langle -12, 2, -9 \rangle$ m. The net force acting on it during this flight, due to the Earth and the air, is nearly constant at $\langle -0.03, -0.04, -0.09 \rangle$ N. What is the total work done on the paper airplane by the Earth and the air?

Positive and Negative Work

The Energy Principle states that the change in energy of a system (in this case, a single particle) is equal to the energy input from work done by the surroundings.

$$\Delta E_{\text{sys}} = W_{\text{surr}}$$

Since the kinetic energy of a particle might increase or decrease, it must be the case that the work done by the surroundings can be positive or negative.

QUESTION What is the meaning of negative work?

Suppose that you want to slow down a moving object. You push in a direction opposite to the object's motion, and although the object keeps moving in the original direction, it gradually slows down. The object's kinetic energy decreased, so you must have done negative work on the object. Evidently a force acting in a direction opposite to the displacement of a system does negative work.

It is important to determine the correct sign of the work done on a system, because increasing energy is associated with positive work, and decreasing energy is associated with negative work.

EXAMPLE **Positive and Negative Work**

Consider four simple situations that bring out the main issues. In Figure 6.9 a box is shown moving 2 m to the right or left, acted on by a force of magnitude 3 N to the right or left. In each case, calculate the work done on the box by the applied force, and state whether the box speeded up or slowed down.

Solution Since the forces and displacements are all in the $+x$ or $-x$ direction, in each case $W = F_x\Delta x + F_y\Delta y + F_z\Delta z$ reduces to $W = F_x\Delta x$.

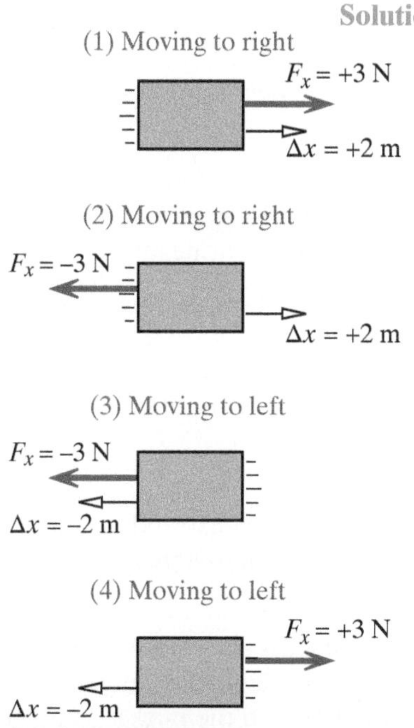

Figure 6.9 Force and displacement in various directions.

Case 1: Force and displacement both in $+x$ direction:

$$W = F_x\Delta x = (+3\,\text{N})(+2\,\text{m}) = +6\,\text{J}$$

The work done is positive, so the kinetic energy of the box increases. The box speeds up.

Case 2: Force in $-x$ direction and displacement in $+x$ direction:

$$W = F_x\Delta x = (-3\,\text{N})(+2\,\text{m}) = -6\,\text{J}$$

The work done is negative, so the kinetic energy of the box decreases. The box slows down.

Case 3: Force and displacement both in $-x$ direction:

$$W = F_x\Delta x = (-3\,\text{N})(-2\,\text{m}) = +6\,\text{J}$$

The work done is positive, so the kinetic energy of the box increases. The box speeds up.

Case 4: Force in $+x$ direction and displacement in $-x$ direction:

$$W = F_x\Delta x = (+3\,\text{N})(-2\,\text{m}) = -6\,\text{J}$$

The work done is negative, so the kinetic energy of the box decreases. The box slows down.

From these four cases we can infer that work is positive if the force and the displacement are in the same direction, and work is negative if force is opposite to displacement.

QUESTION Is it possible to determine whether work is positive or negative simply by looking at the direction of the force alone?

No, we need to know both the direction of the force and of the displacement. A force in the $-x$ direction can do positive work if the displacement of the system is also in the $-x$ direction.

EXAMPLE **Throw a Ball Straight Up**

You throw a 250 g ball straight up, and it rises 8 m into the air (Figure 6.10). Neglecting air resistance, is the work done on the ball positive or negative? How much work is done by the gravitational force?

Solution System: Ball
Surroundings: Earth

Physical reasoning: The ball slows down on its way up, so its kinetic energy is decreasing. Therefore the work must be negative. Calculate work:

$$F_y = -mg = -(0.25\,\text{kg})(+9.8\,\text{N/kg}) = -2.45\,\text{N}$$
$$W = F_y\Delta y = (-2.45\,\text{N})(8\,\text{m}) = -19.6\,\text{J}$$

Negative work is done on the ball, and the ball slows down, losing some kinetic energy.

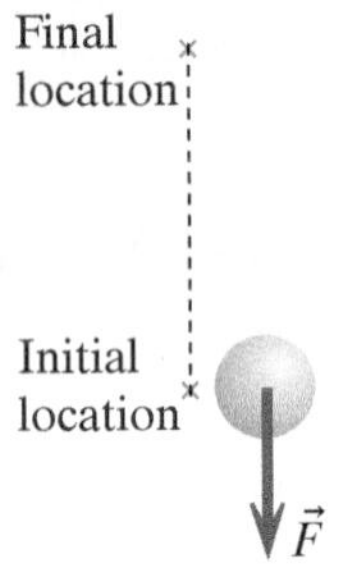

Figure 6.10 A ball travels straight up.

Zero Work

In Figure 6.11 a puck is sliding with low friction to the right on ice, being sped up by a hockey stick pushing on it. There are three different objects in the surroundings that are interacting with the puck. The hockey stick pushes horizontally, doing positive work to speed up the puck; the Earth pulls down, and the ice pushes up, supporting the puck.

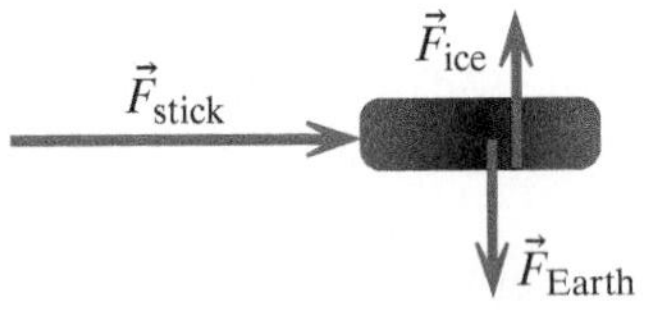

Figure 6.11 A hockey puck slides to the right with low friction on ice, sped up by a hockey stick pushing on the puck.

QUESTION How much work is done by the force exerted downward by the Earth? Does this force speed up or slow down the puck? Is the work done by the Earth positive or negative?

The work done by the Earth's gravitational force is zero. It doesn't make the puck speed up or slow down. We can check this:

$$W = F_x\Delta x + F_y\Delta y + F_z\Delta z = (0)\Delta x + (-mg)(0) + (0)(0) = 0$$

A force perpendicular to the motion does zero work.

QUESTION How much work is done by the force exerted upward by the ice? Does this force speed up or slow down the puck? Is the work done by the ice positive or negative?

Again, the work done by the force exerted by the ice is zero. It doesn't make the puck speed up or slow down (remember that we're neglecting any horizontal, frictional component of the force that the ice exerts on the puck).

A particularly striking case of zero work is circular motion at constant speed, such as the Moon going around the Earth in a (nearly) circular orbit under the influence of the Earth's gravitational force, which always acts perpendicular to the motion if the orbit is a circle (Figure 6.12).

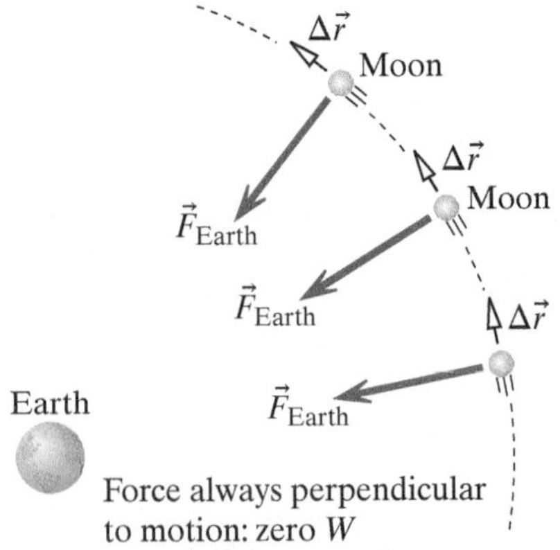

Figure 6.12 The gravitational force of the Earth on the Moon in a (nearly) circular orbit is always (nearly) perpendicular to the motion, so the work is zero.

QUESTION Is the Moon speeding up or slowing down? Therefore what can you say about the work done on the Moon by the Earth's gravitational attraction? Is this consistent with the definition of work?

The Moon's speed and kinetic energy aren't changing (and its rest energy isn't changing), so its energy isn't changing. Therefore the work done should be zero. This makes sense, because the force is always perpendicular to the motion.

> QUESTION Then what does the Earth's gravitational force do, if it doesn't do any work?

It changes the momentum of the Moon. The momentum is a vector quantity, and change of direction represents a change of the vector momentum. A force is required to change the momentum of the Moon, otherwise it would move in a straight line, not a circle.

Again we see important differences between energy and momentum, and between work and impulse. There is no change in the Moon's energy, since that depends on the *magnitude* of the velocity, but there is a change of the Moon's momentum, since that depends on the *direction* of the velocity.

Figure 6.13 The floor exerts an upward force through zero distance and does zero work.

EXAMPLE **Work Done on a Jumper by the Floor**

Suppose a person crouches down and jumps straight up (Figure 6.13). How much work is done by the force of the floor on the feet of the jumper?

Solution Consider the jumper as the system. The surroundings are the Earth, the floor, and the air (but we'll neglect the small air resistance force). As the jumper rises, the Earth pulls down, doing negative work.

> QUESTION What is the displacement through which the force exerted by the floor acts?

The distance that the atoms in the jumper's shoes move while they are in contact with the atoms of the floor is nearly zero (especially if the floor is quite rigid). Therefore the displacement through which the force acts is zero. This means that the floor does zero work!

> QUESTION If the contact force by the floor does no work, what *does* this force do?

The force of the floor contributes to the net force that determines the change in the jumper's momentum, but the floor force does no work. The only significant work done by the surroundings is the negative work done by the force of the Earth (correspondingly, the chemical energy of the jumper decreases).

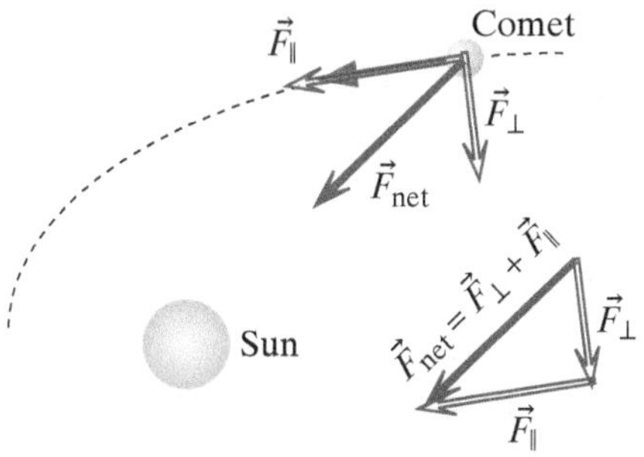

Figure 6.14 The parallel part of the force by the Sun does work, changing the comet's kinetic energy. The perpendicular part does zero work but changes the direction of the comet's momentum.

Of course, in many situations the forces applied to a system have parts both parallel and perpendicular to $\Delta\vec{r}$, as shown in Figure 6.14. $\vec{F}_\parallel$ does either positive or negative work, and $\vec{F}_\perp$ does zero work, but changes the direction of the system's momentum. Figure 6.15 summarizes the relationship between the relative directions of $\vec{F}$ and $\Delta\vec{r}$ and the sign of the work done.

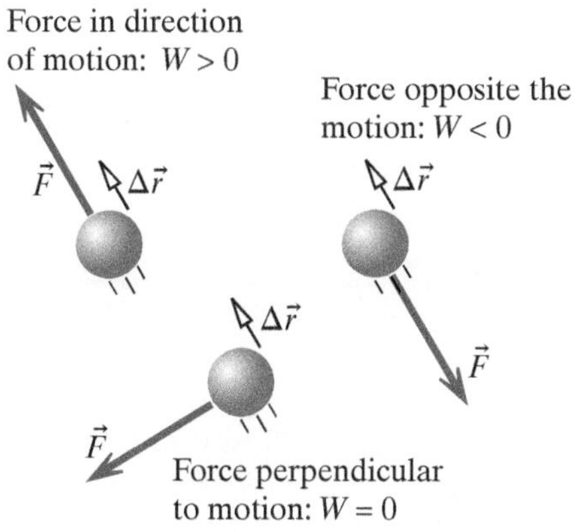

Figure 6.15 Force in direction of motion, opposite the motion, and perpendicular to motion.

SIGN OF WORK

- A force in the direction of motion does positive work.
- A force opposite to the direction of motion does negative work.
- A force perpendicular to the direction of motion does zero work.
- A force that acts through zero displacement does zero work.

Checkpoint 4 In each of the following cases state whether the work done by the specified force is positive, negative, or zero. Also state whether the kinetic energy of the object in question increases, decreases, or remains the same. **(a)** A ball is moving upward, acted on by a downward gravitational force. **(b)** A ball is falling downward, acted on by a downward gravitational force. **(c)** A car is moving rapidly to the left, and Superman exerts a force on it to the right to slow it down, backing up to the left as he pushes to the right. **(d)** You throw a ball downward. Consider the force exerted by your hand on the ball while they are in contact. **(e)** In a 6 month period the Earth travels halfway around its nearly circular orbit of the Sun. Consider the gravitational force exerted on the Earth by the Sun during this period.

Work Done by a Nonconstant Force

If a force changes in magnitude or direction during a process, we can't calculate the work simply by multiplying a constant force times the net displacement. The nonconstant force acts on the object along a path, and we split the path into small increments $\Delta\vec{r}$ (Figure 6.16). This is similar to the way in which we took small time steps Δt when computing the trajectories of objects.

If our increments along the path are sufficiently small, the force is approximately constant in magnitude and direction within one increment. We can write the total work done as a sum along the path:

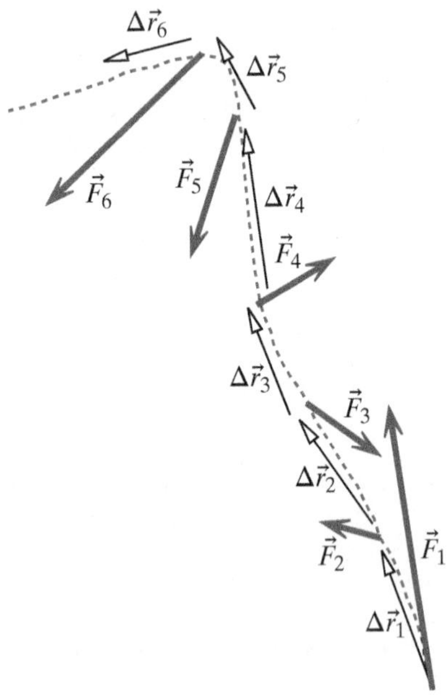

Figure 6.16 A varying force is exerted on an object moving along a curved path.

$$W = \vec{F}_1 \bullet \Delta\vec{r}_1 + \vec{F}_2 \bullet \Delta\vec{r}_2 + \vec{F}_3 \bullet \Delta\vec{r}_3 + \cdots \quad \text{(small steps)}$$

More compactly, we can use the symbol Σ (Greek capital sigma) to mean "sum":

$$W = \Sigma(\vec{F} \bullet \Delta\vec{r})$$

Checkpoint 5 You push a heavy crate out of a carpeted room and down a hallway with a waxed linoleum floor. While pushing the crate 2.3 m out of the room you exert a force of 30 N; while pushing it 8 m down the hallway you exert a force of 15 N. How much work do you do in all?

We have refined our definition of work to allow us to calculate the work done by varying forces. This more complete definition of work allows us to apply the Energy Principle in situations where forces vary.

Work as an Integral

If we go to the limit where the increments are infinitesimal ($\Delta\vec{r} \to d\vec{r}$), the sum used for calculating work turns into a sum of an infinite number of infinitesimal contributions. Such a summation is called a "definite integral." It is written as a distorted "S" for "sum," with an indication of the initial position "*i*" and the final position "*f*":

$$W = \int_i^f \vec{F} \bullet d\vec{r}$$

If you have already studied integration in a calculus course, you already know some techniques for evaluating definite integrals such as this; if not, you will learn them soon in calculus. However, in many real-world cases it is not possible to find an analytical form for evaluating an integral, in which case one uses numerical integration, approximating the infinite sum by a finite sum, with finite increments $\Delta\vec{r}$.

GENERAL DEFINITION OF WORK

$$W \equiv \int_i^f \vec{F} \bullet d\vec{r}$$

This general definition applies to any force, whether constant or not. The integral must be evaluated over the path along which the force is applied. The integral may be approximated computationally by the finite sum $\Sigma(\vec{F} \bullet \Delta\vec{r})$.

Later in this chapter we'll see that in some cases the work integral has the same value no matter what path is followed between the initial and final positions. This is called "path independence."

EXAMPLE **Work Done by a Spring**

A ball traveling horizontally runs into a horizontal spring, whose stiffness is 100 N/m (Figure 6.17). As the spring is compressed, the ball slows down. If the spring is compressed until its length is 20 cm shorter than its relaxed length, how much work does the spring do on the ball?

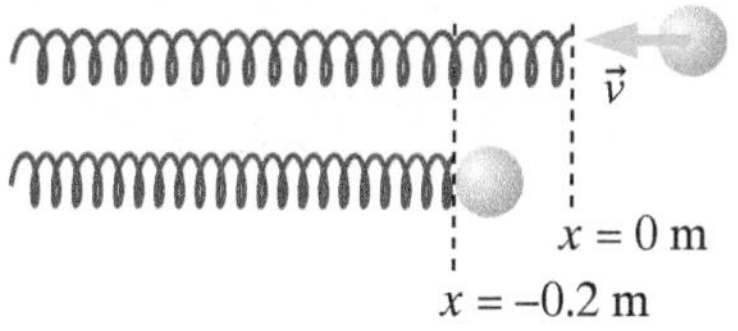

Figure 6.17 A ball runs into a horizontal spring.

Solution Initial state: $s = 0$

Final state: $s = -0.2\,\text{m}$

Choose the origin at the end of the relaxed spring, so $F_x = -k_s x$.

$$W = \int_0^{-0.2} (-100\,\text{N/m}) x\,dx$$

$$W = (-100\,\text{N/m}) \frac{1}{2} x^2 \Big|_{0\,\text{m}}^{-0.2\,\text{m}}$$

$$W = (-100\,\text{N/m}) \left(\frac{1}{2}(-0.2)^2 - 0 \right) = -2\,\text{N}\cdot\text{m}$$

The spring does -2 J of work on the ball.

Check: Sign ok—negative work decreases kinetic energy of ball. Units ok ($\text{N}\cdot\text{m} = \text{J}$).

6.4 WORK AND ENERGY

The Energy Principle involves a difference in the energy of a system in its final state and the energy in its initial state. We will refer to this as the "difference form" of the Energy Principle.

THE ENERGY PRINCIPLE: DIFFERENCE FORM

$$\Delta E_{\text{sys}} = W_{\text{surr}} + \text{other inputs}$$

$\Delta E_{\text{sys}} = E_{\text{sys},f} - E_{\text{sys},i}$, so we can also write the Energy Principle this way:

THE ENERGY PRINCIPLE: UPDATE FORM

$$E_{\text{sys},f} = E_{\text{sys},i} + W_{\text{surr}} + \text{other inputs}$$

Sometimes it's easier to think about the situation using the difference form of the Energy Principle, and sometimes the update form is more appropriate. Use whichever form you're comfortable with, but make sure you understand both forms.

A simple and effective procedure for solving problems involving energy is this:

SOLVING PROBLEMS INVOLVING ENERGY

- Write out all terms in the Energy Principle.
- Identify terms that don't change and therefore cancel.
- Substitute in all numerical values you know.
- Solve for the unknown quantity.

EXAMPLE **Pushing a Box in Space**

You're outside a spacecraft, pushing on a heavy box whose mass is $m =$ 2000 kg. You exert a force $\vec{F} = \langle 300, 500, 0\rangle$ N while the box moves through a displacement $\Delta\vec{r} = \langle 0.1, -0.3, 0.2\rangle$ m. Initially the box had a speed $|\vec{v}_i| =$ 0.7 m/s. **(a)** How much work do you do? **(b)** What is the final kinetic energy of the box? **(c)** What is the final speed of the box? **(d)** What is the direction of the final velocity of the box?

Solution System: Box (considered to be a particle)
Surroundings: You
Initial state: When you start pushing the box
Final state: When the box has been displaced the amount $\Delta\vec{r}$

$$
\begin{aligned}
E_f &= E_i + W \\
(\cancel{mc^2} + K_f) &= (\cancel{mc^2} + K_i) + W \\
K_f &= K_i + W \\
K_f &= K_i + (F_x\Delta x + F_y\Delta y + F_z\Delta z) \\
K_f &= K_i + (300\,\text{N})(0.1\,\text{m}) + (500\,\text{N})(-0.3\,\text{m}) + (0\,\text{N})(0.2\,\text{m}) \\
K_f &= K_i + (-120\,\text{J}) \\
K_f &= \left(\frac{1}{2}mv_i^2\right) + (-120\,\text{J}) \\
K_f &= \frac{1}{2}(2000\,\text{kg})(0.7\,\text{m/s})^2 + (-120\,\text{J}) \\
K_f &= (490\,\text{J}) + (-120\,\text{J}) = 370\,\text{J}
\end{aligned}
$$

Final speed:

$$
\begin{aligned}
\frac{1}{2}mv_f^2 &= K_f \\
v_f &= \sqrt{\frac{2K_f}{m}} \\
v_f &= \sqrt{\frac{2(370\,\text{J})}{2000\,\text{kg}}} = 0.608\,\text{m/s}
\end{aligned}
$$

(a) $W = -120$ J
(b) $K_f = 370$ J. This is less than K_i, as it should be since you did negative work on the system.
(c) $v_f = 0.608$ m/s
(d) We don't have enough information to answer this.

We answered all the questions except for (d), which asked for the direction of the final velocity of the box. We can't answer that question! The Energy

Principle is a scalar principle, and only the magnitude of the velocity appears in the particle energy, not the direction. In order to figure out the direction of the final velocity we would have to use the Momentum Principle, which is a vector principle. However, note that it was not stated how long the process took, so we can't apply the Momentum Principle directly. We would have to determine the approximate Δt from the known displacement and initial speed. (The actual trajectory is a curve; we've approximated the small displacement by a straight line.)

Note that we knew that the rest energy of the box did not change (it did not, for example, turn into solid gold), so we were able to eliminate mc^2 from both sides of the energy equation.

EXAMPLE

Throwing a Ball

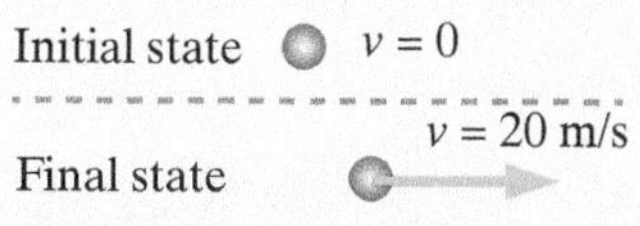

Figure 6.18 Throwing a ball.

You hold a ball of mass 0.5 kg at rest in your hand, then throw it forward, underhand, so that as it leaves your hand its speed is 20 m/s. How much work did you do on the ball? Note that you do not know how big a force you applied, nor do you know the distance through which you applied this force.

Solution

System: Ball
Surroundings: You (we can neglect the gravitational effect of the Earth in this short process)

Initial state: At the start of the throw, when the ball's speed is zero
Final state: When the ball just leaves your hand at 20 m/s (Figure 6.18)

$$E_f = E_i + W$$
$$(mc^2 + K_f) = (mc^2 + K_i) + W$$
$$K_f = K_i + W$$
$$\frac{1}{2}mv_f^2 = \frac{1}{2}mv_i^2 + W$$
$$\frac{1}{2}(0.5\,\text{kg})(20\,\text{m/s})^2 = \frac{1}{2}m(0)^2 + W$$
$$100\,\text{J} = W$$

There is no change of identity, so no change in rest energy; the mc^2 terms cancel. The energy of the ball increased; it sped up from 0 to 20 m/s. You did positive work, pushing the ball to speed it up.

Check: Work is positive; this makes sense because the final speed is larger than the initial speed. The units are correct.

We were able to calculate the work you did on the ball even though we didn't know how large a force you applied or how large was the displacement of your hand in the throwing motion. This is one aspect of the power of the Energy Principle—that simply by determining the change in energy of a system you know how much energy input there was to that system.

QUESTION In the previous example, what was the change in the energy of the surroundings? Can you say what kind of energy it was in the surroundings that changed?

Energy is a conserved quantity: $\Delta E_{\text{system}} + \Delta E_{\text{surroundings}} = 0$. The energy of the system (the ball) increased by 100 J. Evidently the change in energy of the surroundings was −100 J. This decrease of energy in the surroundings corresponds to a decrease in the chemical energy stored in your body, coming from food you consumed. In throwing the ball you converted some stored chemical energy into kinetic energy of the ball.

Actually, you need to expend more than 100 J of chemical energy (an energy decrease), because in addition to increasing the kinetic energy of the ball by 100 J, you also raised your body temperature a bit (thermal energy increase; more on this in Chapter 7). Your muscles aren't 100% efficient; it takes more than 100 J of chemical energy to give the ball 100 J of kinetic energy.

The net energy change of the surroundings is –100 J, which could consist (say) of –180 J of chemical energy change and +80 J of thermal energy rise in your body.

Checkpoint 6 In the example above, suppose that you wanted the final speed of the ball to be twice as big (40 m/s). How much work would you have to do?

EXAMPLE **The Energy Principle at High Speed**

An electron (mass 9×10^{-31} kg) in a particle accelerator is acted on by a constant electric force $\langle 5 \times 10^{-13}, 0, 0\rangle$ N. (This is much greater than the force of gravity, which is only $mg \approx 1 \times 10^{-29}$ N, so we can neglect the effect of the Earth on the electron.) The initial velocity of the electron is $\langle 0.995c, 0, 0\rangle$, where c as usual is the speed of light. The electron moves through a displacement of $\langle 5, 0, 0\rangle$ m. What is its final speed, expressed as a fraction of the speed of light?

Solution System: Electron
Surroundings: Accelerator (which applies the electric force); neglect the gravitational effect of the Earth in this short process.

Initial state: Beginning of given displacement
Final state: End of given displacement

The unknown quantity, final speed, is part of γ_f. It's easiest to find γ_f first, then solve for v_f. Because the electron's speed is near the speed of light, we can't use an approximate expression for K. We need to use the exact, relativistic equation for particle energy.

$$E_f = E_i + W$$

$$\gamma_f mc^2 = \gamma_i mc^2 + W$$

$$\gamma_f mc^2 = \gamma_i mc^2 + (F_x\Delta x + F_y\Delta y + F_z\Delta z)$$

$$\gamma_f mc^2 = \gamma_i mc^2 + (5 \times 10^{-13}\,\text{N})(5\,\text{m}) + (0\,\text{N})(0\,\text{m}) + (0\,\text{N})(0\,\text{m})$$

$$\gamma_f mc^2 = \gamma_i mc^2 + (2.5 \times 10^{-12}\,\text{J})$$

$$\gamma_f mc^2 = \frac{1}{\sqrt{1-(0.995c/c)^2}}mc^2 + (2.5 \times 10^{-12}\,\text{J})$$

$$\gamma_f mc^2 = 10.0mc^2 + (2.5 \times 10^{-12}\,\text{J})$$

$$\gamma_f = \frac{10.0mc^2 + (2.5 \times 10^{-12}\,\text{J})}{mc^2}$$

$$\gamma_f = 10.0 + \frac{2.5 \times 10^{-12}\,\text{N}}{(9 \times 10^{-31}\,\text{kg})(3 \times 10^8\,\text{m/s})^2} = 40.9$$

$$\sqrt{1-(v_f/c)^2} = \frac{1}{40.9} = 2.44 \times 10^{-2}$$
$$1-(v_f/c)^2 = (2.44 \times 10^{-2})^2 = 5.98 \times 10^{-4}$$
$$(v_f/c)^2 = 1 - 5.98 \times 10^{-4} = 0.9994$$
$$v_f/c = \sqrt{0.9994} = 0.9997$$
$$v_f = 0.9997c$$

Check: The final speed is greater than the initial speed, consistent with positive work. The units look right; in particular, the values for γ were dimensionless, as they should be.

The speed increased only slightly, from $0.995c$ to $0.9997c$, even though the energy changed a lot, from $10.0mc^2$ to $40.9mc^2$. The electron's energy increased by about a factor of 4, but the electron's speed increased very little. When a particle is traveling at nearly the speed of light, adding a lot of energy to the particle doesn't change its speed very much.

Checkpoint 7 In the preceding example, at the final speed, $0.9997c$, what was the particle energy as a multiple of the rest energy mc^2? (That is, if it was twice mc^2, write $2mc^2$.) What was the kinetic energy as a multiple of mc^2? Was the kinetic energy large or small compared to the rest energy? At low speeds, is the kinetic energy large or small compared to the rest energy?

6.5 CHANGE OF REST ENERGY

As we saw in the preceding examples, if a particle doesn't change its identity, both the initial and the final energies of the system include mc^2 rest-energy terms and these cancel out in the energy equation:

$$\cancel{mc^2} + K_f = \cancel{mc^2} + K_i + W$$

This is an important and common special case, but it is not the only case. In some processes the mass of a particle can change because the identity of the particle changes.

Neutron decay is a good example to illustrate the issues. A free neutron (one not bound into a nucleus) is unstable, with an average lifetime of about 15 min. As shown in Figure 6.19, it decays into a proton, an electron, and a nearly massless antineutrino, which travels at nearly the speed of light. All three of these particles have kinetic energy. That is, they have energy above and beyond their rest energy; the energy of the antineutrino is nearly all kinetic, because it has almost no rest energy. This is a situation with a change of particle identity, and a change in particle rest energy.

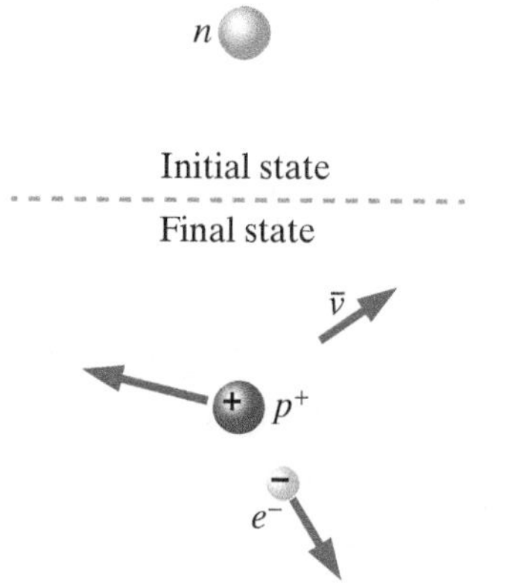

Figure 6.19 A stationary neutron (n) decays into a proton (p^+), an electron (e^-), and a nearly massless antineutrino ($\bar{\nu}$) in a reaction that can be written $n \rightarrow p^+ + e^- + \bar{\nu}$. This is an example of a process in which there is a change of identity and therefore a change in rest energy.

An Electron Volt (eV) Is a Unit of Energy

The mass of a neutron is 1.6749×10^{-27} kg, so its rest energy mc^2 is about 1.51×10^{-10} J. Because the rest energy of a single particle is such a small number of joules, it is usually expressed in different units of a million electron volts, abbreviated MeV. One electron volt is the amount of energy an electron acquires when moving through an electric potential difference of one volt. (Electric potential difference is introduced in a later chapter; an ordinary flashlight battery has an electric potential difference across it of 1.5 V.)

ELECTRON VOLT (eV) AND MeV

$$1\text{ eV} = 1.6 \times 10^{-19}\text{ J}$$

eV is the abbreviation for electron volt. 1 MeV $= 1 \times 10^6$ eV

The rest energy of a neutron is 939.6 MeV, and the rest energy of a proton is 938.3 MeV—slightly less than that of a neutron. An electron is much less massive; its rest energy is 0.511 MeV.

Scientists studying elementary particles often use the unit MeV/c^2 as a unit of mass, because dividing energy by c^2 gives mass. In these units the mass of a proton is 938.3 MeV/c^2. The mass of a neutron is 939.6 MeV/c^2. The mass of the electron is 0.511 MeV/c^2. A neutrino or antineutrino is nearly massless—almost all of its energy is kinetic energy.

EXAMPLE **Neutron Decay**

In the decay of a free neutron at rest (Figure 6.19), how much kinetic energy (in MeV) do the decay products (proton, electron, and antineutrino) share? Why can't a proton decay into a neutron?

Solution System: All the particles, both before and after the decay (as long as we include all particles in the system, it is okay if some of them change identity during the process)
Surroundings: Nothing (the Earth can be neglected, since the process is so fast that work done by gravitational forces is negligible)

Initial state: Neutron at rest
Final state: Proton (p), electron (e), and antineutrino ($\overline{\nu}$), far apart

Energy Principle:

$$E_f = E_i + W$$
$$(m_P c^2 + K_P) + (m_e c^2 + K_e) + (K_{\overline{\nu}}) = (m_N c^2 + K_N) + W$$
$$(938.3\text{ MeV} + K_P) + (0.511\text{ MeV} + K_e) + (K_{\overline{\nu}}) = (939.6\text{ MeV} + 0) + 0$$
$$(K_P + K_e + K_{\overline{\nu}}) + 938.8\text{ MeV} = 939.6\text{ MeV}$$
$$(K_P + K_e + K_{\overline{\nu}}) = (939.6 - 938.8)\text{ MeV}$$
$$(K_P + K_e + K_{\overline{\nu}}) = 0.8\text{ MeV}$$

Here the neutron undergoes a change of identity, so there is a change in rest energy, and the mc^2 terms do not cancel. Because the sum of the rest energies of the three decay particles is less than the rest energy of the parent neutron, there is energy left over that shows up as kinetic energy shared among the three decay particles. How this kinetic energy is divided up among the three particles varies from one neutron decay to the next, governed by probabilities that can be calculated using the science of quantum mechanics and the properties of the "weak interaction" that is responsible for neutron decay. We found the total kinetic energy of all the particles in the final state, but we don't know the separate kinetic energies of the individual particles.

QUESTION Why can't a proton decay into a neutron?

A proton can't decay into a neutron, because the neutron has more rest energy than a proton. In such a decay the kinetic energy of the decay products would turn out to be a negative number, which is impossible. Kinetic energy is always positive.

In a chemical reaction such as carbon combining with oxygen in the fire of a coal-burning power plant, the resulting kinetic energy (which is responsible for heating water to make steam) is on the order of 1 eV per molecule. Nuclear reactions produce vastly more kinetic energy, in the case of neutron decay almost a million times more per nucleus (1 MeV/1 eV is one million). This is why some power plants use nuclear rather than chemical reactions.

Checkpoint 8 In the preceding example, what fraction of the original neutron's rest energy was converted into kinetic energy?

Intelligent "Plug and Chug"

Look back at the last four worked-out examples (pushing a box, throwing a ball, energy at high speed, and neutron decay) and you'll see that the same procedure was used to solve each one. Applying the Energy Principle is basically an intelligent sort of "plug and chug" procedure.

- Specify system and surroundings.
- Specify the initial and final states.
- Write out the Energy Principle in detail for this system.
- Use the given information to evaluate all the terms you can.
- Solve for the unknown quantity, which may be inside one of the remaining terms.
- Check that signs and values make physical sense.

The four preceding examples were quite different, but the same procedure worked in each case. The central idea is to let the terms in the Energy Principle tell you what to calculate, and then go ahead and calculate them.

Often students have the impression that physics consists of "knowing what formula to use" for a particular type of problem, with a different formula for every problem. But look again at the last four examples. There weren't four "formulas" giving the four answers. Rather there was one logical procedure that worked in every case, based on the fundamental Energy Principle that applies to all situations.

In following this procedure, you still need to think, but let your thinking be organized by the principle itself. You need to decide whether to try using the Momentum Principle or the Energy Principle, you need to choose a system and initial and final states intelligently, and so on. However, the basic framework never changes.

QUESTION How do you decide whether to use the Momentum Principle or the Energy Principle?

Think about what quantities are given, and what you must find: Work involves a distance and affects the energy, whereas impulse involves a time and affects the momentum. Momentum is a vector, and involves information about directions; energy is a scalar, and has no directional information.

You should never find yourself saying, "I don't have any idea even where to start on this problem!"

If you do find yourself saying this, take a deep breath, choose a system, identify the objects in the surroundings that exert forces on the system, and write down the Momentum Principle or the Energy Principle for this situation. If you try using the Energy Principle, choose initial and final states, and flesh out the various terms in the Energy Principle for the particular situation at hand, then solve for the unknown quantities.

Physics can be easier than most people think, if you go with the flow. If on the other hand you try to invent a new and different approach for every new problem you encounter, or search futilely for some "formula," you'll waste huge amounts of time and get very frustrated.

6.6 PROOF OF THE ENERGY PRINCIPLE FOR A PARTICLE

We have asserted that the following is true for a particle:

$$\Delta E = W_{\text{surr}}$$

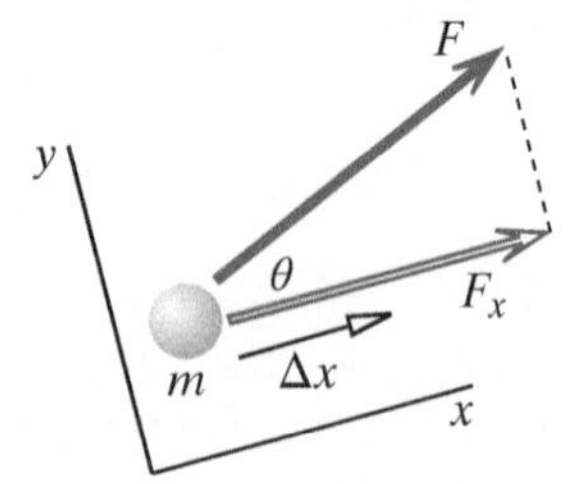

Figure 6.20 A particle of mass m. We choose the x axis to be in the direction of the motion. The net force acting on the particle has a component F_x in the direction of motion.

To prove this, we need to show that $\Delta E = W_{\text{surr}}$ is true given our definitions of particle energy and work. Consider a single pointlike particle of mass m. For convenience in the reasoning we are about to do, choose coordinate axes so that the x axis is in the direction of the motion at this instant, so the displacement in the next short time interval is simply Δx (Figure 6.20). There is a net force $\vec{F}$ (magnitude F) acting at an angle θ to the displacement. The physical results we will obtain do not depend on our choice of coordinate axes; we are free to choose axes that simplify our calculations.

The component of the net force that is parallel to the displacement, F_x, affects the speed and hence the kinetic energy of the particle. The perpendicular component F_y merely changes the direction of the momentum, and does not change speed or energy.

In the very short displacement Δx, the energy of the particle E_{particle} changes by a small amount, and this is supposed to be equal to the work done on the particle:

$$\Delta E = F_x \Delta x$$

According to the Momentum Principle $F_x = \dfrac{\Delta p_x}{\Delta t}$, so

$$\Delta E = \left(\frac{\Delta p_x}{\Delta t}\right)\Delta x$$

Dividing by Δx, we have this:

$$\frac{\Delta E}{\Delta x} = \frac{\Delta p_x}{\Delta t}$$

In the limit as $\Delta t \to 0$, this equation relates the spatial rate of change of energy to the time rate of change of momentum:

$$\frac{dE}{dx} = \frac{dp_x}{dt}$$

In words, this says that the change in energy of the particle per unit distance (in the direction of the motion) is equal to the change in the particle's momentum in that direction per unit time. This equation reflects the fact that a change in position (Δx or dx) is associated with a change in energy, whereas a change in time (Δt or dt) is associated with a change in momentum. If we can show that this equation is true if $E = \gamma mc^2$, we will have proven that $\Delta E = W$ is correct.

Proof for Low Speeds

In the special case that the speed of a particle is small compared to the speed of light, the proof is simple. We need to show that if particle energy is given by the approximate equation:

$$E \approx mc^2 + \frac{1}{2}mv^2 \quad (v \ll c)$$

then the equation derived above:

$$\frac{dE}{dx} = \frac{dp_x}{dt}$$

is satisfied.

We chose the x axis to be in the direction of the motion, so the particle energy is:

$$E \approx mc^2 + \frac{1}{2}mv_x^2 \quad (v \ll c)$$

Left-hand side of equation:

$$\frac{dE}{dx} = \frac{d}{dx}\left(\frac{1}{2}mv_x^2\right)$$
$$= \frac{1}{2}m\frac{d}{dx}(v_x^2) = mv_x\frac{dv_x}{dx} = m\frac{dx}{dt}\frac{dv_x}{dx} = m\frac{dv_x}{dt}$$

Right-hand side of equation (since mc^2 is constant, its derivative is zero):

$$\frac{dp_x}{dt} = \frac{d}{dt}(mv_x) = m\frac{dv_x}{dt} \quad (v \ll c)$$

So since the left-hand and right-hand sides of the equation are equal,

$$\frac{dE}{dx} = \frac{dp_x}{dt}$$

is indeed true with our definitions of energy and work. We have proved that at least at low speed the Energy Principle for a particle $\Delta E = W$ is correct.

General Proof (Relativistically Correct)

The proof that the relativistically correct expression $E = \gamma mc^2$ satisfies the relation

$$\frac{dE}{dx} = \frac{dp_x}{dt}$$

is given in Section 6.20, at the end of the chapter. It follows the same reasoning as was used to prove the low-speed case but is algebraically more complex.

6.7 POTENTIAL ENERGY IN MULTIPARTICLE SYSTEMS

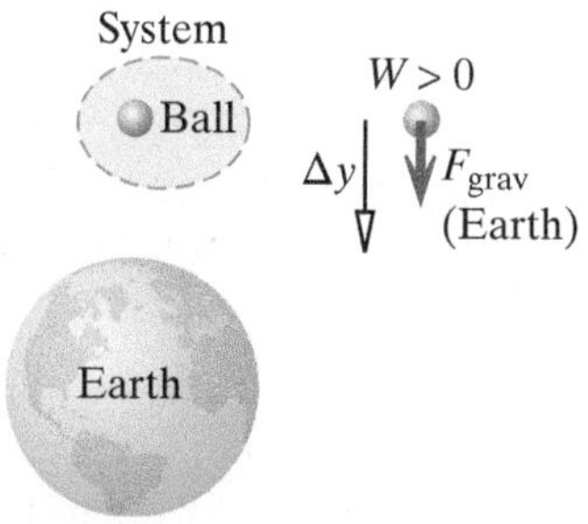

Figure 6.21 Ball alone: Earth is part of the surroundings and does positive work on the ball, increasing the kinetic energy of the ball.

Up to this point, our discussions of energy have been confined to systems consisting of a single particle (or systems that we could consider to be equivalent to a point particle because there were no internal changes). However, we saw in the case of momentum that it was often productive to choose a system containing two or more objects. We'll look at the effect of choice of system on the energy analysis of a simple situation: a ball falls from rest.

If the ball alone is the system, the Earth is part of the surroundings (Figure 6.21). The kinetic energy of the system (ball) increases, due to positive work done by the Earth on the system. The gravitational force acts in the same direction as the displacement of the ball, so the work done by the Earth on the ball is positive.

EXAMPLE **Ball and Earth: Ball Alone as the System**

In Figure 6.22 a ball of mass 100 g is released from rest 7 m above the ground. When the ball has fallen to 4 m above the ground, what is its kinetic energy K_f?

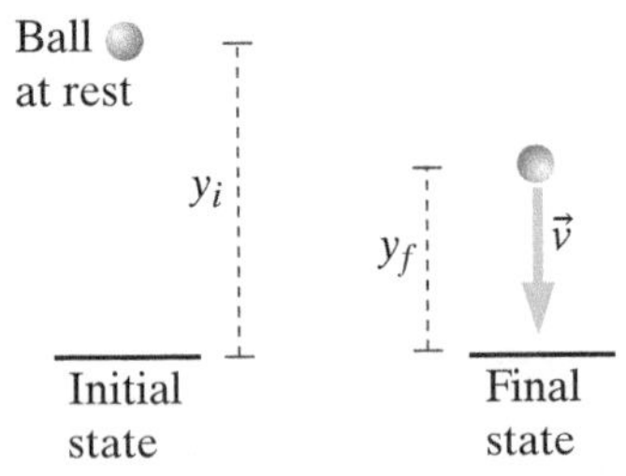

Figure 6.22 A ball falls from rest, and at a lower location its kinetic energy has increased.

Solution System: Ball
Surroundings: Earth
Initial State: Ball at rest 7 m above ground
Final State: Ball falling, 4 m above ground

$$\Delta(\cancel{m_{\text{ball}}c^2}) + \Delta K_{\text{ball}} = W_{\text{by Earth}}$$
$$\Delta K_{\text{ball}} = F_y \Delta y = -mg\Delta y$$
$$K_f - 0 = -(0.1\text{ kg})(9.8\text{ N/kg})(-3\text{ m}) = 2.94\text{ J}$$
$$K_f = 2.94\text{ J}$$

The Earth did an amount of work +2.94 J on the ball, and the kinetic energy of the ball increased from zero to 2.94 J.

QUESTION What if we choose the ball plus the Earth as the system (Figure 6.23)?

In this case there is nothing significant in the surroundings, so:

$$\Delta K_{\text{sys}} = W_{\text{surr}}$$
$$\Delta K_{\text{ball}} + \Delta K_{\text{Earth}} = 0 \quad \text{But this can't be right!}$$

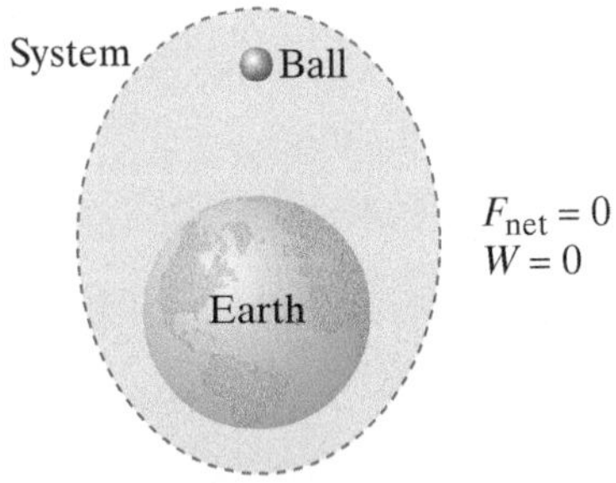

Figure 6.23 As the ball falls no work is done on the system of (Ball + Earth) because there are no significant interactions with the surroundings.

We know that the kinetic energy of the ball increased, and that the kinetic energy of the Earth also increased a tiny amount as it was drawn toward the ball. But the surroundings did no work! The change in kinetic energy of the system is not equal to the work done by the surroundings.

Before deciding that the Energy Principle has been violated, we should consider the possibility that we have overlooked a kind of energy that is present in systems containing more than one interacting object. In fact, this is the case.

Potential Energy Belongs to Pairs of Interacting Objects

In any system containing two or more interacting particles, such as compressed or stretched springs, galaxies of stars interacting gravitationally, or atoms in which the protons and electrons interact electrically, there is energy associated with the interactions between pairs of particles inside the system. This interaction energy is not the same as the rest energies or kinetic energies of the individual particles. We call this pairwise interaction energy "potential energy." Traditionally potential energy is represented by the symbol U.

Since the system of (Ball + Earth) contains two interacting objects, its energy is really this:

$$E_{\text{sys}} = (m_{\text{ball}}c^2 + K_{\text{ball}}) + (m_{\text{Earth}}c^2 + K_{\text{Earth}}) + U_{\text{ball-Earth}}$$

A change in potential energy is associated with a change in the separation of interacting objects. We can think of separation changes as change in the shape of the multiparticle system, such as the ball and Earth moving closer together, a spring stretching or compressing, or an electron moving away from a proton.

In the case above, as the ball and the Earth get closer together, the kinetic energy of the system increases, but we will see that the potential energy (interaction energy) decreases. The net change in the energy of the system is in fact zero, which is consistent with the fact that no work was done on the system by the surroundings:

$$\Delta K_{\text{ball}} + \Delta K_{\text{Earth}} + \Delta U_{\text{ball-Earth}} = 0$$

QUESTION How do we calculate changes in potential energy?

When we take the ball + Earth as the system, the force acting on the ball is exerted by another object inside the system. We call such a force an "internal force," which does "internal work." A simple rearrangement of the Energy Principle for the system of the ball alone leads to a useful way to account for energy in the ball + Earth system:

$$\Delta K_{\text{ball}} = W_{\text{by Earth}}$$

Rewrite this by moving the work to the left side of the equation:

$$\Delta K_{\text{ball}} + (-W_{\text{by Earth}}) = 0$$

For a system of more than one object, we define "the change of potential energy ΔU" to be the negative of the internal work: $\Delta U = -W_{\text{by Earth}}$. For the system of the ball + Earth, $\Delta U = -(F_y \Delta y) \approx -(-mg)\Delta y = \Delta(mgy)$. Actually, because we approximated the gravitational force by a constant mg, this is an approximate result valid only for height changes near the Earth's surface that are small compared to the radius of the Earth.

APPROXIMATE ΔU NEAR EARTH'S SURFACE

$$\Delta U_{\text{grav}} \approx \Delta(mgy)$$

for a system consisting of (object + Earth) near the surface of the Earth

Writing the Energy Principle for the ball + Earth system as

$$\Delta K_{\text{ball}} + \Delta K_{\text{Earth}} + \Delta U_{\text{ball-Earth}} = 0$$

has the advantage of putting on the left side of the equation quantities associated with the system (kinetic energy of the ball, interaction energy of the ball + Earth), and leaving on the right side of the equation only those quantities associated with the surroundings (in this case, nothing).

EXAMPLE **Ball and Earth: Ball + Earth as the System**

In Figure 6.22 a ball of mass 100 g is 7 m above the ground, initially at rest ($K_i = 0$). When the ball is 4 m above the ground, what is its kinetic energy K_f? Choose the ball + Earth as the system.

Solution System: Ball + Earth
Surroundings: Nothing significant

$$\begin{aligned}
\Delta(\cancel{m_{\text{Earth}}c^2}) + \Delta K_{\text{Earth}} + \Delta(\cancel{m_{\text{ball}}c^2}) + \Delta K_{\text{ball}} + \Delta U &= 0 \\
0 + (K_{\text{ball,f}} - 0) + \Delta(mgy) &= 0 \\
K_{\text{ball,f}} + (0.1\ \text{kg})(9.8\ \text{N/kg})(-3\ \text{m}) &= 0 \\
K_{\text{ball,f}} + (-2.94\ \text{J}) &= 0 \\
K_{\text{ball,f}} &= 2.94\ \text{J}
\end{aligned}$$

In the second step we assume that $\Delta K_{\text{Earth}} \approx 0$. This makes sense because although the Earth is attracted to the ball, the displacement of the Earth is negligibly small, so essentially zero work is done on the Earth.

When we chose the ball alone as the system, the Earth was in the surroundings and did +2.94 J of work on the ball. When we choose the ball + Earth as the system, no work is done on the system, but the potential energy of the ball–earth pair of interacting objects decreases; $\Delta U = -2.94$ J. We of course get the same result for the final kinetic energy of the ball no matter which system we choose, but the viewpoint is rather different in the two cases.

QUESTION Why didn't we have to include potential energy when we chose the ball alone as the system?

A system consisting of a single object has no potential energy. Potential energy is associated with the interactions of pairs of objects inside the system, and there are no such pairs of objects in a one-particle system.

Interaction Energy in a Multiparticle System

Having illustrated the basic idea of potential energy in a situation where the internal force was constant (mg), we now offer a more general analysis. Consider a system of three particles that exert forces on each other and that are also acted upon by objects in the surroundings, outside this system, as shown in Figure 6.24. For example, these could be three charged particles that exert electric forces on each other but also have springs connecting them to objects in the surroundings. The forces that are exerted by other particles in the system are called "internal" forces, and are denoted here by lowercase f's. The forces exerted by objects in the surroundings are called "external" forces, and are denoted here by uppercase F's.

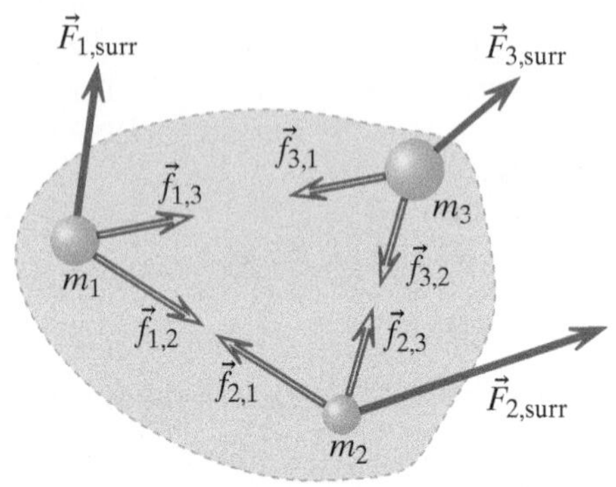

Figure 6.24 A system consisting of three particles. They exert "internal" forces on each other (lowercase f's), and additional "external" forces (black arrows, uppercase F's) are exerted by objects in the surroundings, which are not shown.

The work done by these internal and external forces on each particle changes the particle energy E_{particle} of that particle. Using the symbol W_{int} for work done by internal forces, and W_{surr} for work done by external forces, we can write the Energy Principle for each particle:

$$\Delta E_1 = (\vec{f}_{1,2} + \vec{f}_{1,3} + \vec{F}_{1,\text{surr}}) \bullet \Delta \vec{r}_1 = W_{1,\text{int}} + W_{1,\text{surr}}$$
$$\Delta E_2 = (\vec{f}_{2,1} + \vec{f}_{2,3} + \vec{F}_{2,\text{surr}}) \bullet \Delta \vec{r}_2 = W_{2,\text{int}} + W_{2,\text{surr}}$$
$$\Delta E_3 = (\vec{f}_{3,1} + \vec{f}_{3,2} + \vec{F}_{3,\text{surr}}) \bullet \Delta \vec{r}_3 = W_{3,\text{int}} + W_{3,\text{surr}}$$

Adding up the changes in the particle energy for each of the three particles (the left sides of the equations) and the work done on each particle (the right sides of the equations), we can write this:

$$\Delta(E_1 + E_2 + E_3) = W_{\text{int}} + W_{\text{surr}}$$

where W_{int} is the sum of the work done by internal forces on each of the particles, and W_{surr} is the sum of the work done by external forces on each of the particles. Added together, these two terms equal the sum of the changes in the individual particle energies. Note that the work done by each individual external force involves the displacement of that force (equivalently, the displacement of the particle on which that particular force acts).

Separating Internal and External Energy Terms

As we saw with the falling ball, it can be useful to have all terms relating to energy changes internal to the system on one side of the equation and terms relating to objects in the surroundings on the other side. As we did with the falling ball, we can do this by making a simple algebraic rearrangement, moving the work done on the particles by the internal forces, W_{int}, to the other side of the equation:

$$\Delta(E_1 + E_2 + E_3) + (-W_{\text{int}}) = W_{\text{surr}}$$

We relabel the term $(-W_{\text{int}})$ as a change of potential energy U, energy associated with the interactions of the particles with each other.

CHANGE IN POTENTIAL ENERGY

$$\Delta U = -W_{\text{int}}$$

The change in potential energy (ΔU) of a system is defined to be the negative of the work done by forces internal to the system.

The left side of the Energy Principle equation for a multiparticle system now represents all changes in the energy of the system, and the right side represents energy inputs from the surroundings (work). It is customary to drop the subscript and write W_{surr} simply as W, with the understanding that the work W is mechanical energy transfer from the surroundings to the system, with a resulting change in the energy of the system.

Because the "internal work" term involves the forces exerted on each particle by each other particle in the system, the potential energy U is really the sum of the interaction energies of all pairs of particles in the system:

$$U = U_{12} + U_{23} + U_{31}$$

In general, we can now extend our definition of energy to a system of interacting particles:

ENERGY OF A SYSTEM OF INTERACTING PARTICLES

$$E_{\text{sys}} = (m_1c^2 + K_1) + (m_2c^2 + K_2) + (m_3c^2 + K_3) + \cdots$$
$$+ U_{12} + U_{23} + U_{13} + \cdots$$

And we can write the Energy Principle in a way that explicitly includes this interaction energy:

ENERGY PRINCIPLE FOR A MULTIPARTICLE SYSTEM

$$\Delta(E_1 + E_2 + E_3 + \cdots) + \Delta(U_{12} + U_{13} + U_{23} + \cdots) = W$$

$E_1 = m_1c^2 + K_1$ is the particle energy of object 1, etc.

Checkpoint 9 If a system contains four particles, how many potential energy pairs U_{12}, etc., are there? List them.

Multiparticle Systems vs. Single-Particle Systems

The Energy Principle is fundamentally the same for a multiparticle system as it is for a single-particle system. To the left of the equals sign are quantities that are internal to the system, while to the right of the equals sign are the external effects on the system due to objects in the surroundings. The interpretation, as indicated in (Figure 6.25), is:

(Change in energy of a system) = (Energy inputs from the surroundings)

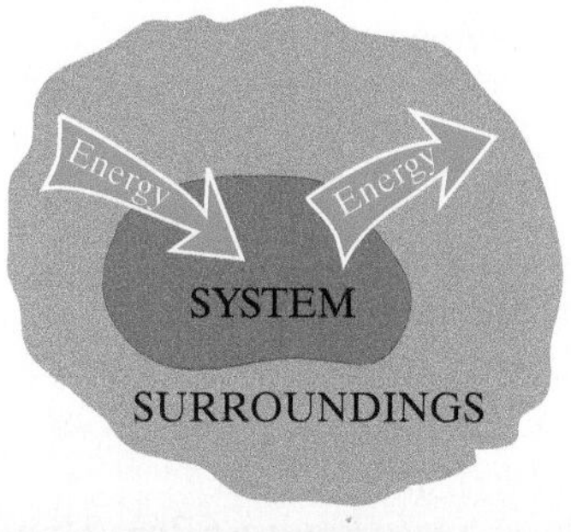

Figure 6.25 Change of system energy is equal to energy inputs from the surroundings, which can be either positive or negative. Work done by the surroundings is one form of energy input.

The difference is in the kinds of energy found inside each kind of system:

- A single-particle system has only particle energy (rest energy and kinetic energy).
- A multiparticle system has particle energy and potential energy (interaction energy).

Work done by the surroundings, which may be positive or negative, is one form of energy input to a system. Work can affect both particle energy and potential energy inside a multiparticle system.

You get to choose how to define the system you want to analyze, but once you have specified the portion of the Universe you're considering to be the system of interest, you must be consistent about what energy flows into or out of the system (work done by the surroundings), versus what energy changes occur inside your system (possible changes in particle energies or potential energy).

The Experimental Basis for the Energy Principle for Multiparticle Systems

We were able to make a mathematical proof of the Energy Principle for a single particle, starting from the Momentum Principle. For a complex, multiparticle system, the more general Energy Principle ultimately is based on over a hundred years of experiments starting in the early 1800s. Slowly scientists learned how to identify flows of energy in many different kinds of systems, and once the energy was accounted for, they found that conservation of energy, $\Delta E_{\text{sys}} + \Delta E_{\text{surr}} = 0$, was always true, not only in physics but also in chemistry and biology.

The Discovery of Neutrinos

■ One of the recent experiments that has helped to determine the mass of neutrinos is called MINOS, and involved shooting a beam of neutrinos underground from Fermilab, near Chicago, to a detector deep in a mine in northern Minnesota. To learn more about both MINOS and the history of the discovery of neutrinos, visit the MINOS Experiment website.

Neutrinos were discovered because of the Energy Principle. In the early part of the 20th century, scientists found that apparently energy was not conserved in nuclear reactions called "beta decays," in which certain radioactive nuclei spontaneously emit an electron, and a neutron in the nucleus changes into a proton. (Early in the history of atomic physics, energetic electrons were called "beta rays.") It was expected that the emitted electrons would all have exactly the same energy, corresponding to the change in mc^2 of the nucleus, but instead it was observed that the emitted electrons had a wide range of energies. In many cases the energies didn't add up. Was the Energy Principle in fact not true? The Swiss physicist Wolfgang Pauli made the bold proposal that an unseen particle, later named the "neutrino," was also emitted, with energy shared between it and the electron.

Pauli was so convinced of the universal correctness of energy conservation that he was willing to postulate an unseen particle rather than discard energy conservation. It took many years before experiments were sensitive enough to detect the elusive neutrinos, but eventually Pauli's analysis was confirmed. For many years it was thought that neutrinos had only kinetic energy, with zero rest energy (and zero mass). However, recent careful experiments have verified theoretical predictions that neutrinos should have a very tiny mass.

6.8 GRAVITATIONAL POTENTIAL ENERGY

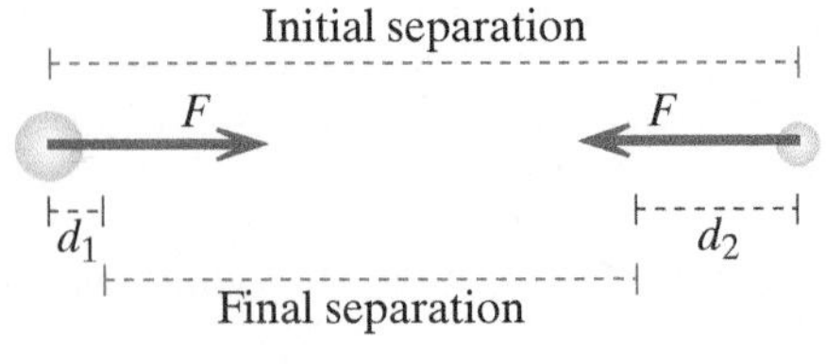

Figure 6.26 Two stars move toward each other during a short time interval. The internal work for the two-star system is F times the change in the separation distance of this pair of objects. The displacements are shown overly large for clarity.

We found that for a system of a ball falling near the surface of the Earth, where the force of the Earth was a constant mg, the change of gravitational potential energy for the ball + Earth system was $\Delta U \approx \Delta(mgy)$. Next we will find an expression for gravitational potential energy in the more general case where the force depends on distance, $\vec{F}_{\text{grav}} = -(GMm/r^2)\hat{r}$.

A big star and a small star are initially at rest (Figure 6.26). They start to move toward each other along the x axis due to the gravitational forces they exert on each other, with the same magnitude F (reciprocity). In a short time the big star moves a distance d_1 and the small star moves a distance d_2 (the displacements are shown overly large for clarity). The increase in the magnitude of the momentum of each star is the same, because the magnitude of the impulse $F\Delta t$ is the same, so the small star reaches a higher speed ($v \approx p/m$) and moves farther: $d_2 > d_1$, as indicated in Figure 6.26. Therefore more work is done on the small star.

The impulse (and momentum change) have the same magnitude for both stars because both forces act for the same time interval Δt. However, the work done is different because the two forces act through different distances. This is an important difference between impulse and work, and therefore between the Momentum Principle and the Energy Principle.

Consider the system consisting of both stars. The internal work W_{int} is the work done by forces internal to the system (not exerted by objects in the surroundings):

$$W_{\text{int}} = Fd_1 + Fd_2 = F(d_1 + d_2)$$

Change of potential energy is defined as the negative of the internal work:

$$\Delta U = -W_{\text{int}} = -F(d_1 + d_2)$$

Since ΔU for the two stars is negative, the potential energy decreased, which would imply that the kinetic energy increased, which is indeed what happened.

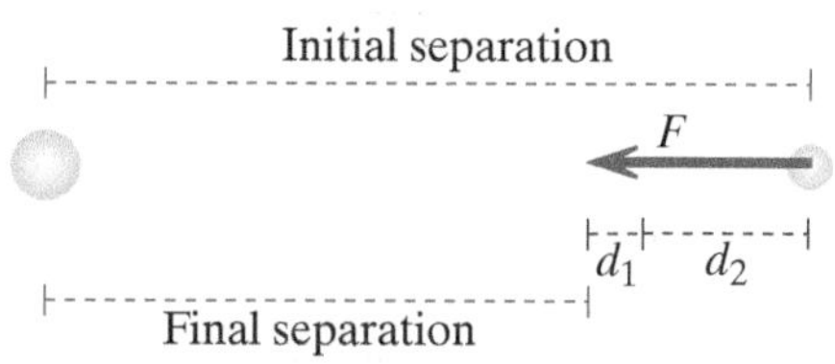

Figure 6.27 We can calculate internal work as though one of the stars were stationary and the other moved.

Change in Separation of Two Objects

The stars got closer to each other; their separation decreased by an amount $(d_1 + d_2)$. We can calculate internal work by multiplying the force one object exerts on another by the change in the separation between the two objects. The individual positions of the objects don't matter. The way the algebra works out, the internal work is calculated as though a single force F acted through a distance equal to the change in the separation of the pair, as in Figure 6.27.

Next we'll prove this more generally. Again consider a system of two interacting stars. We define the relative position vector $\vec{r}$ to be the difference between the positions of two stars, as shown in Figure 6.28. When the stars move a little closer together, star 1 does work on star 2, and star 2 does work on star 1. Since both stars are inside the system, the change in gravitational potential energy of the system is the negative of the sum of these work terms. Assume that the changes in position are small enough that we can consider the forces to be constant:

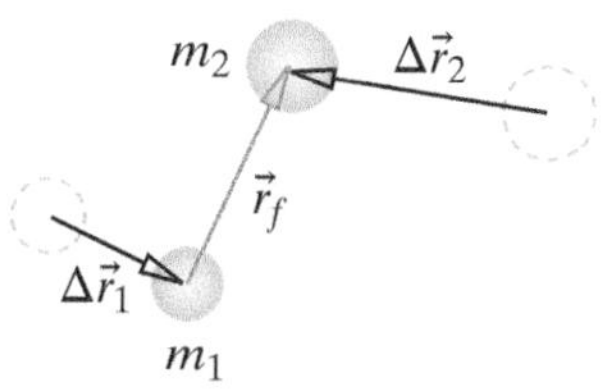

Figure 6.28 Two stars whose relative positions are changing (in this case, the stars move closer together).

$$\Delta U = -W_{\text{int}} = -(\vec{f}_{1,2} \bullet \Delta\vec{r}_1 + \vec{f}_{2,1} \bullet \Delta\vec{r}_2)$$
$$\vec{f}_{1,2} = -\vec{f}_{2,1}, \text{ so } \Delta U = -\vec{f}_{2,1} \bullet (\Delta\vec{r}_2 - \Delta\vec{r}_1)$$

The term $(\Delta\vec{r}_2 - \Delta\vec{r}_1)$ involves the change of position of each star (Figure 6.28). We can rewrite this in terms of $\Delta\vec{r}$, the change in relative position of the two stars:

$$\begin{aligned}\Delta\vec{r}_2 - \Delta\vec{r}_1 &= (\vec{r}_{2f} - \vec{r}_{2i}) - (\vec{r}_{1f} - \vec{r}_{1i})\\ &= (\vec{r}_{2f} - \vec{r}_{1f}) - (\vec{r}_{2i} - \vec{r}_{1i})\\ &= \vec{r}_f - \vec{r}_i\\ &= \Delta\vec{r}\end{aligned}$$

where

$$\vec{r}_i = \vec{r}_{2i} - \vec{r}_{1i}$$
$$\vec{r}_f = \vec{r}_{2f} - \vec{r}_{1f}$$

Using this result we can rewrite the change in potential energy:

$$\Delta U = -\vec{f}_{2,1} \bullet \Delta\vec{r}$$

$\vec{r}$ is the position of star 2 relative to star 1. Therefore we can calculate the amount of change of potential energy, ΔU, in terms of the *relative displacements* or separation between *pairs of particles*, and any change of potential energy is associated with a change of "shape," with pairs of particles getting closer together or farther apart. The algebra works out in such a way that the force we use in calculating ΔU is just one of the pairs of forces, not both. The result is the same if the interaction is an electric interaction instead of a gravitational interaction, since the reasoning above depended only on reciprocity of forces, and not on any other details of the force.

We write U_{12} for the potential energy of the pair consisting of particles 1 and 2, U_{23} for pair 2 and 3, and so on. If a system consists of two particles, there are three energy terms: two particle energies E_1 and E_2, and one potential energy term, U_{12}, since there is only one pair of particles. (Of course the particle energies can be further subdivided into rest energy plus kinetic energy.)

If a system consists of three particles, there are six energy terms: three particle energies E_1, E_2, and E_3, and three potential energy terms, U_{12}, U_{23}, and U_{31}, since there are three pairs of particles. Change of potential energy is associated with change of distance between pairs of particles.

Force Is the Negative Gradient of U

Since it is the relative displacement that matters, even if both objects move we can pretend that object 1 remains stationary, and just calculate the work done on object 2. The force is in a line with $\vec{r}$, so we can evaluate the dot product and write the following, where F_r is the component of the gravitational force on object 2 in the direction of $\vec{r}$, and r is the distance from object 1 to object 2 (Figure 6.29):

$$dU = -F_r dr$$

Therefore if we already knew the expression for potential energy, we could calculate the associated force like this: $F_r = -dU/dr$. A rate of change of a quantity with respect to position such as this is called a "gradient." For example, if a hill is said to have "a 7% grade," this means that it rises 7 m vertically for every 100 m horizontally—the tangent of the angle of the hill (the slope) is 0.07 (Figure 6.30).

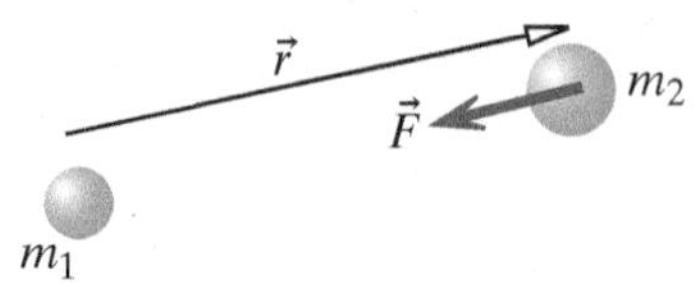

Figure 6.29 The r component of the gravitational force, F_r.

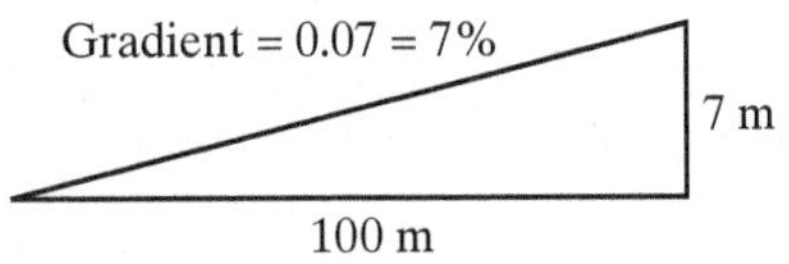

Figure 6.30 The gradient is the rate of change of y with respect to x.

FORCE IS THE NEGATIVE GRADIENT OF POTENTIAL ENERGY

$$F_r = -\frac{dU}{dr}$$

All that remains to be done to find an expression for gravitational energy is to think of an expression whose (negative) derivative is the component of the gravitational force acting on m_2 in the direction of $\vec{r}$. From Figure 6.29 we see that this component of the gravitational force is negative: $F_r = -G(m_1 m_2/r^2)$, where r is the center-to-center distance. The minus sign reflects the fact that the attractive gravitational force on object 2 points back toward object 1.

QUESTION Can you think of a function of r whose (negative) derivative with respect to r has this value of F_r?

The derivative of r^n with respect to r is nr^{n-1}. We have a r^{-2} factor in the gravitational force, so $n = -1$, and the "antiderivative" of r^{-2} (the function whose derivative gives r^{-2}) must be $-r^{-1}$.

QUESTION Therefore, what is the equation for gravitational energy?

Being careful about the signs, we find that $U = -Gm_1m_2/r$. To check, we can differentiate this expression with respect to r to get the gravitational force.

QUESTION What about the minus sign on U? Is it correct?

Yes, because the potential energy decreases as the particles get closer together. When stars fall toward each other, their kinetic energies increase and the pair-wise potential energies must decrease to more and more negative values. Figure 6.31 shows that although U is negative, it does increase with increasing separation r, which means that you have to do work to move two gravitationally attracting objects farther apart. It may seem odd that potential energy is negative, but it makes physical sense. We'll examine how this negative quantity contributes to the total energy of a multiparticle system in Section 6.12.

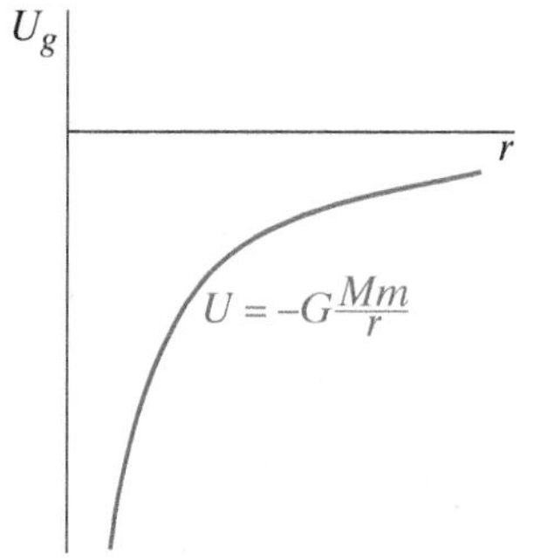

Figure 6.31 Gravitational potential energy is negative. It increases with increasing separation r.

In summary,

GRAVITATIONAL POTENTIAL ENERGY

$$U = -G\frac{m_1m_2}{r}$$

r is the center-to-center separation of m_1 and m_2.

QUESTION Could there be an additional constant in this equation?

It is true that the gradient of $U = (-Gm_1m_2/r + \text{constant})$ would also give us the correct gravitational force, because the derivative of a constant is zero. However, as we will soon discuss in more detail, when the objects are very far apart the total energy of the system must be equal just to the sum of the particle energies, and the potential energy must be zero.

QUESTION When the particle separation r is very large, what is the value of $-Gm_1m_2/r$?

When r is very large, $-Gm_1m_2/r$ is nearly zero. Therefore, for the potential energy to be zero when the separation is very large, the value of the constant in $U = (-Gm_1m_2/r + \text{constant})$ must be zero, in order for U to be zero at large separations. The pair-wise gravitational potential energy must be simply $U = -Gm_1m_2/r$, with no added constant.

EXAMPLE **A Meteor Falls Toward the Earth**

In February 2013 a large meteor, whose mass has been estimated to be 1.2×10^7 kg, fell to Earth near Chelyabinsk, Russia. (This meteor exploded spectacularly at a height of about 30 km, doing significant damage to objects on the ground.) Consider a meteor of the same mass falling toward the Earth. Choose the Earth plus the meteor as the system. As the meteor falls from a distance of 1×10^8 m from the center of the Earth to 1×10^7 m, what is the change in the kinetic energy of the meteor? Explain the signs of the changes in kinetic and potential energy of the system.

Solution System: Earth and meteor
Surroundings: Nothing significant (because the distance of the objects from the Sun doesn't change much)

Initial state: Center to center distance of 1×10^8 m
Final state: Center to center distance of 1×10^7 m

To get a sense of the magnitudes and signs of the gravitational potential energy, we will calculate the value of U_{grav} explicitly for each state instead of making algebraic simplifications.

$$U_i = -G\frac{M_E m_m}{r_i} = -6.7\times10^{-11}\,\text{Nm}^2/\text{kg}^2\frac{(6\times10^{24}\,\text{kg})(1.2\times10^7\,\text{kg})}{1\times10^8\,\text{m}}$$
$$= -4.82\times10^{13}\,\text{J}$$
$$U_f = -G\frac{M_E m_m}{r_i} = -6.7\times10^{-11}\,\text{Nm}^2/\text{kg}^2\frac{(6\times10^{24}\,\text{kg})(1.2\times10^7\,\text{kg})}{1\times10^7\,\text{m}}$$
$$= -4.82\times10^{14}\,\text{J}$$

Both the initial and final values of the gravitational potential energy are negative. U_f, the value when the meteor is closer to Earth, is a larger negative number, so $U_f - U_i < 0$. This is consistent with Figure 6.31, which shows that as the separation between two objects decreases, the gravitational potential energy of the two-object system becomes more negative.

Applying the Energy Principle, we find that:

$$\Delta K_{\text{meteor}} + (U_f - U_i) = 0$$
$$\Delta K_{\text{meteor}} + (-4.82\times10^{14}\,\text{J} - -4.82\times10^{13}\,\text{J}) = 0$$
$$\Delta K_{\text{meteor}} + -4.34\times10^{14}\,\text{J} = 0$$
$$\Delta K_{\text{meteor}} = 4.34\times10^{14}\,\text{J}$$

EXAMPLE

A Robot Spacecraft Leaves an Asteroid

A robot spacecraft lands on an asteroid, picks up a sample, and blasts off to return to Earth; its total mass is 1500 kg. When it is 200 km (2×10^5 m) from the center of the asteroid, its speed is 5.0 m/s, and the rockets are turned off. At the moment when it has coasted to a distance 500 km (5×10^5 m) from the center of the asteroid, its speed has decreased to 4.1 m/s. **(a)** Draw and label a diagram showing your choice of initial and final states. **(b)** Calculate the mass of the asteroid.

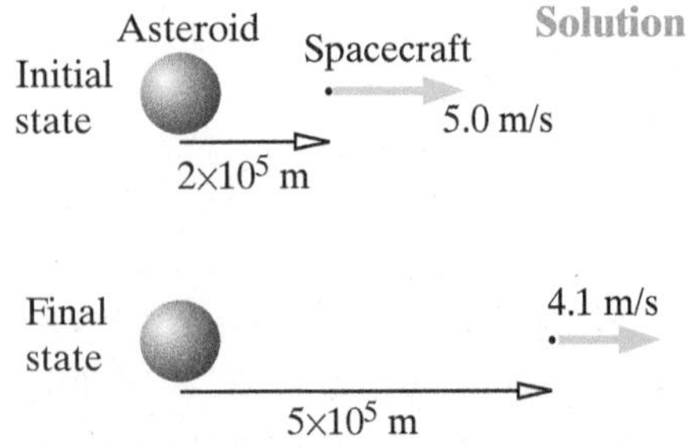

Figure 6.32 Initial and final states.

Solution

(a) Figure 6.32 shows the situation. It is important to keep in mind that r is measured center to center in $U = -Gm_1m_2/r$.

(b) We'll use M for the mass of the asteroid and m for the mass of the spacecraft.

System: Asteroid and spacecraft

Surroundings: Nothing significant

Initial state: Rockets off, 5.0 m/s, 200 km from center of asteroid, asteroid at rest

Final state: Speed 4.1 m/s, 500 km from center of asteroid

$$E_f = E_i + W$$
$$\cancel{Mc^2} + \cancel{K_{M,f}} + \cancel{mc^2} + K_{m,f} + U_f = \cancel{Mc^2} + \cancel{K_{M,f}} + \cancel{mc^2} + K_{m,i} + U_i + W$$
$$K_{m,f} + U_f = K_{m,i} + U_i + 0$$
$$\frac{1}{2}\cancel{m}v_f^2 + \left(-G\frac{M\cancel{m}}{r_f}\right) = \frac{1}{2}\cancel{m}v_i^2 + \left(-G\frac{M\cancel{m}}{r_i}\right)$$
$$\frac{1}{2}v_f^2 + \left(-G\frac{M}{r_f}\right) = \frac{1}{2}v_i^2 + \left(-G\frac{M}{r_i}\right)$$
$$GM\left(\frac{1}{r_i} - \frac{1}{r_f}\right) = \frac{1}{2}(v_i^2 - v_f^2)$$

$$M = \frac{\frac{1}{2}(v_i^2 - v_f^2)}{G(1/r_i - 1/r_f)}$$

$$M = \frac{\frac{1}{2}(5.0^2 - 4.1^2)\,\text{m/s}^2}{(6.7\times10^{-11}\,\text{N}\cdot\text{m}^2/\text{kg}^2)\left(\frac{1}{2\times10^5\,\text{m}} - \frac{1}{5\times10^5\,\text{m}}\right)} = 2\times10^{16}\,\text{kg}$$

By choosing both the asteroid and the spacecraft as the system, we didn't have to calculate work, since there are no significant external forces. We solved the problem algebraically before putting in numbers, and found that m canceled—the answer doesn't depend on the mass of the spacecraft.

QUESTION Could we have used the Momentum Principle instead of the Energy Principle to solve this problem?

We could not have used the Momentum Principle to carry out an accurate analysis without writing an iterative computer program because we didn't know how much time the process took. The velocity was changing at a nonconstant rate, so knowing the distance wouldn't give us the time, since we didn't know the average velocity. We were able to use the Energy Principle because we knew something about distances.

QUESTION We made the approximation that K_M, the kinetic energy of the asteroid, did not change. Was this reasonable? Don't the spacecraft and asteroid exert equal magnitude forces on each other?

It is true that the spacecraft pulls just as hard on the asteroid as the asteroid pulls on the spacecraft (reciprocity of gravitational forces). However, although the forces act during equal time intervals, they do not act through equal displacements, because the speeds of the objects are different.

We can use Conservation of Momentum for the system of asteroid + spacecraft to get the actual final speed of the asteroid (V_A). We'll assume all velocities have only x components, so the following equations are for the x component of the system's momentum:

$$(MV_A - 0) + (mv_f - mv_i) = 0$$

$$V_A = \frac{m}{M}(v_i - v_f)$$

$$= \frac{1.5\times10^3\,\text{kg}}{2\times10^{16}\,\text{kg}}(5.0 - 4.1)\,\text{m/s}$$

$$= 6.75\times10^{-14}\,\text{m/s}$$

Since the asteroid's speed is so low, its displacement during the interval of interest is tiny, and therefore the work done on it by the spacecraft is negligible. The change in kinetic energy of the asteroid is:

$$\Delta K_A = \frac{1}{2}(2\times10^{16}\,\text{kg})(6.75\times10^{-14}\,\text{m/s})^2 - 0$$

$$= 4.6\times10^{-11}\,\text{J}$$

In contrast, the kinetic energy change of the spacecraft was

$$\Delta K_s = \frac{1}{2}(1.5\times10^3\,\text{kg})(4.1^2 - 5.0^2)(\text{m/s})^2 = -6143\,\text{J}$$

More Than Two Interacting Objects

We have been applying energy considerations to two-particle systems. A homework problem in Chapter 3 involves predicting the motion of the *Ranger 7* spacecraft as it travels from the Earth to the Moon (where it crash lands). In our simple model for this voyage there are three "particles": the spacecraft with mass m, the Earth with mass M_{Earth}, and the Moon with mass M_{Moon}.

QUESTION How many interaction pairs are there in this three-particle system?

In a three-particle system there are three interaction pairs: U_{12}, U_{13}, and U_{23}. We can write the gravitational potential energy like this:

$$U = \left[-G\frac{M_{\text{Earth}}m}{r_{\text{to Earth}}}\right] + \left[-G\frac{M_{\text{Moon}}m}{r_{\text{to Moon}}}\right] + \left[-G\frac{M_{\text{Earth}}M_{\text{Moon}}}{r_{\text{Earth to Moon}}}\right]$$

In our simple model, the Earth and Moon are fixed in space, so the Earth–Moon term of this expression doesn't change. Also, there is no change of identity of the three particles, so their masses do not change. Therefore the energy equation for the system can be written like this, where v is the speed of the spacecraft, the only moving particle in the model system (and $v \ll c$):

$$\Delta\left(\frac{1}{2}mv^2\right) + \Delta\left[-G\frac{M_{\text{Earth}}m}{r_{\text{to Earth}}}\right] + \Delta\left[-G\frac{M_{\text{Moon}}m}{r_{\text{to Moon}}}\right] = 0$$

The change in the energy of the system is zero because no external work is done on the system (we're ignoring the effects of other objects, including the Sun). We can rearrange the terms and write the energy equation in terms of initial and final energies, which must be equal:

$$\frac{1}{2}mv_f^2 + \left[-G\frac{M_{\text{Earth}}m}{r_{\text{to Earth},f}}\right] + \left[-G\frac{M_{\text{Moon}}m}{r_{\text{to Moon},f}}\right] =$$
$$\frac{1}{2}mv_i^2 + \left[-G\frac{M_{\text{Earth}}m}{r_{\text{to Earth},i}}\right] + \left[-G\frac{M_{\text{Moon}}m}{r_{\text{to Moon},i}}\right]$$

That is, the final (kinetic plus potential) energy of the system is equal to the initial (kinetic plus potential) energy of the system, because no external work is done on the system (and the masses don't change; no change of identity).

Gravitational Potential Energy Near Earth's Surface

In Section 6.7 we found that for motion near the surface of the Earth, $\Delta U \approx \Delta(mgy)$. However, in Section 6.8 we found that the general expression $\Delta U_g = \Delta(-Gm_1m_2/r)$ describes a change in gravitational potential energy.

QUESTION How are $\Delta(mgy)$ and $\Delta(-Gm_1m_2/r)$ related?

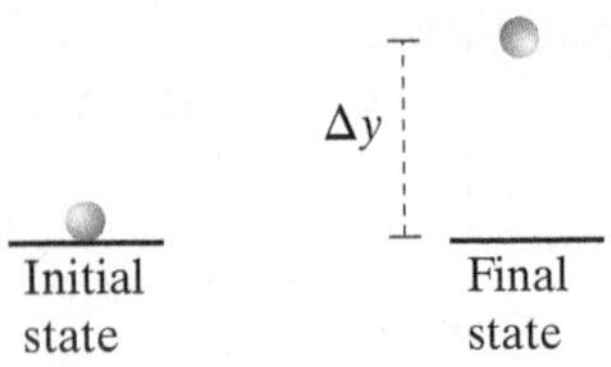

Figure 6.33 An object is lifted from the surface of the Earth to a distance Δy above the surface.

The expression $\Delta(mgy)$ is an approximate expression valid only near the Earth's surface. We can derive it from the full expression for gravitational potential energy. Consider a system of the Earth plus an object of mass m located on the surface of the Earth. Suppose you move the object from the surface of the Earth to a location a small distance Δy above the Earth's surface (Figure 6.33). If R_E is the radius of the Earth, then the change in potential energy of the system is:

$$\Delta U_g = \left(-G\frac{M_E m}{R_E+\Delta y}\right) - \left(-G\frac{M_E m}{R_E}\right)$$

$$\Delta U_g = (-GM_E m)\frac{R_E-(R_E+\Delta y)}{R_E(R_E+\Delta y)}$$

$$\Delta U_g = (-GM_E m)\frac{-\Delta y}{R_E^2+R_E\Delta y}$$

$$(R_E^2+R_E\Delta y) \approx R_E^2 \text{ since } \Delta y \ll R_E, \text{ so}$$

$$\Delta U_g \approx G\frac{M_E}{R_E^2} m\Delta y$$

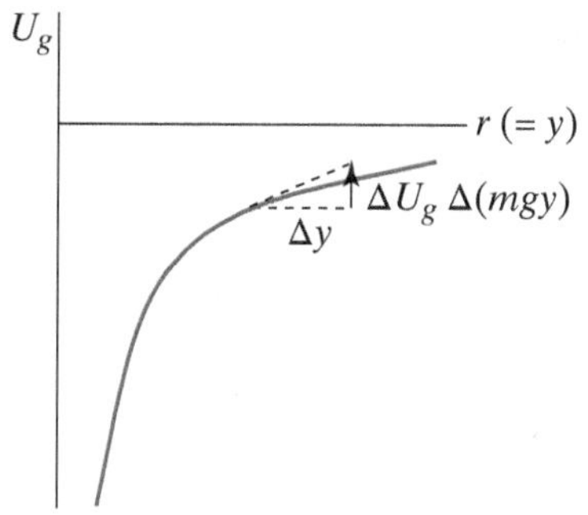

Figure 6.34 A small increase Δy in height y corresponds to a small increase in the (negative) gravitational energy.

In Chapter 3 we found that $G\frac{M_E}{R_E^2} = g$, so

$$\Delta U_g \approx gm\Delta y \quad \text{near the Earth's surface}$$

As expected, the sign of ΔU_g is correct; as you go up (farther from the center of the Earth), the change $\Delta(mgy)$ is positive; if you go down, it is negative (Figure 6.34).

QUESTION Is mgy the actual potential energy of the (Earth + object) system?

No. The actual potential energy of the system in a given state is $(-GM_E m/R_E + mgy)$, which is a negative quantity, as we will see in the following example.

EXAMPLE **ΔU Near the Earth**

A 500 kg horse stands on the ground. **(a)** What is the potential energy of a system consisting of the Earth plus the horse? **(b)** Suppose the horse jumps over a fence. What is the potential energy of the system when the horse is 1.5 m above the ground?

Solution **(a)** The actual potential energy of the system in the initial state is

$$U_g = -G\frac{M_E m_{\text{horse}}}{R_E}$$

$$U_g = -6.7\times10^{-11}\,\text{Nm}^2/\text{kg}^2\frac{(6\times10^{24}\,\text{kg})(5\times10^2\,\text{kg})}{6.4\times10^6\,\text{m}}$$

$$U_g = -3.15\times10^{10}\,\text{J}$$

(b) The *change* of the system's potential energy is

$$mg\Delta y = (500\,\text{kg})(9.8\,\text{N/kg})(1.5\,\text{m}) = 7350\,\text{J}$$

so the new potential energy of the system is

$$U_{gf} = -3.15\times10^{10}\,\text{J} + 7350\,\text{J}$$

To indicate that the actual potential energy of a system contains a large negative constant term we could write the energy equation for systems near Earth's surface like this:

$$K_f + (mgy_f + -GmM_E/R_E) = K_i + (mgy_i + -GmM_E/R_E) + W$$

However, it is more convenient to omit the constant, since it will not contribute to a change, and write

$$K_f + mgy_f = K_i + mgy_i + W$$

QUESTION In what situations is the approximation $\Delta U_g \approx \Delta mgy$ a good approximation?

As long as Δy is small compared to the radius of the Earth (6400 km), this is a good approximation. The gravitational force is inversely proportional to the square of the distance from the center of the Earth, and decreases as distance from Earth increases. However, if y doesn't change much, neither does gravity. If you go up 0.1 km (100 m, the length of a football field), the gravitational force decreases by a tiny factor:

$$\left(\frac{6400\text{ km}}{6400.1\text{ km}}\right)^2 \approx 0.99997$$

EXAMPLE **A Falling Ball**

At a certain instant a ball is falling with a speed of 6 m/s, and its position is $y = 35$ m above the Earth's surface. How fast is the ball falling when it has fallen to a position $y = 20$ m above the Earth's surface, assuming that we can neglect air resistance?

Solution System: Ball + Earth
Surroundings: Nothing significant (neglect air resistance)

Initial state: Speed 6 m/s, 35 m above ground
Final state: 20 m above ground

Energy Principle:

$$K_f + U_f = K_i + U_i + W$$

$$\frac{1}{2}mv_f^2 + mgy_f = \frac{1}{2}mv_i^2 + mgy_i + 0$$

$$\frac{1}{2}mv_f^2 = \frac{1}{2}mv_i^2 - (mgy_f - mgy_i)$$

$$v_f^2 = v_i^2 - 2g(y_f - y_i)$$

$$v_f = \sqrt{v_i^2 - 2g(y_f - y_i)}$$

$$v_f = \sqrt{(6\text{ m/s})^2 - 2(9.8\text{ N/kg})[(20\text{ m}) - (35\text{ m})]} = 18.2\text{ m/s}$$

There is nothing in our equations to specify whether the ball is initially heading upward or downward. In fact, it doesn't matter. In either case, if we neglect air resistance, the speed will be 18.2 m/s when the ball reaches a height of 20 m above the ground. (If it was initially headed upward, it will again have a speed of 6 m/s when it returns to a height of 35 m.)

Checkpoint 10 Suppose that in this situation we measure y from 25 m above the surface, so that $y_{\text{initial}} = +10$ m, and $y_{\text{final}} = -5$ m. Why do we get the same value for the final speed?

Including Horizontal Motion

Moving vertically up or down involves a change in gravitational energy approximately equal to $\Delta(mgy)$. What about sideways horizontal movements? There is negligible change in the gravitational energy of the Universe when an object is moved a short distance horizontally (that is, tangent to the Earth's surface). Consider moving a block along a surface with it rolling very easily on low-friction wheels, or a hockey puck sliding easily along the ice. You hardly have to use any energy at all to move the block or puck horizontally, as long as there is little friction.

What if a block is moved horizontally, but at a constant height of 3 m above the surface? You could support the block with a tall cart, and again there would be little effort involved in moving the block horizontally. Or you could pile up dirt 3 m deep and place a slick surface on top, and then you could easily move the block horizontally. We conclude that horizontal position doesn't affect the gravitational energy. (Again, we're considering movements over distances small compared with the radius of the Earth, so we can ignore the Earth's curvature.)

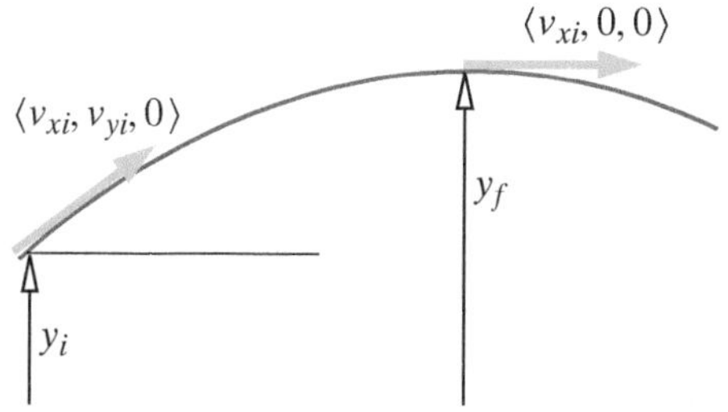

Figure 6.35 Path of a ball thrown at an angle to the horizontal.

Checkpoint 11 Suppose that you throw a ball at an angle to the horizontal, and just after it leaves your hand at a height y_i its velocity is $\langle v_{xi}, v_{yi}, 0\rangle$ (Figure 6.35). Assuming that we can neglect air resistance, at the top of its trajectory, when it is momentarily traveling horizontally, its velocity is $\langle v_{xi}, 0, 0\rangle$. What is the height y_f at the top of the trajectory, in terms of the other known quantities? Use the Energy Principle.

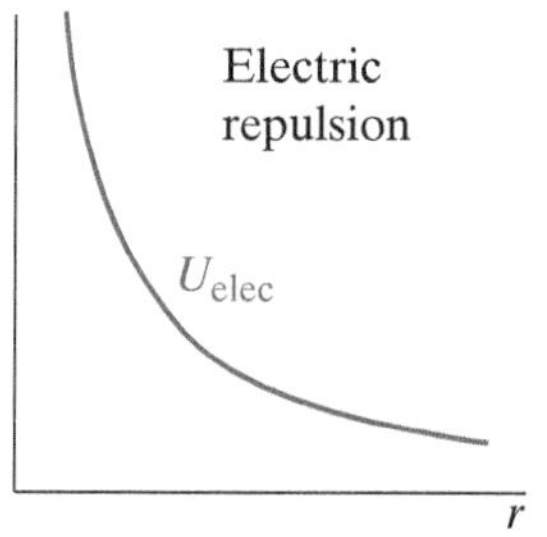

Figure 6.36 Electric potential energy is positive when like charged objects repel each other.

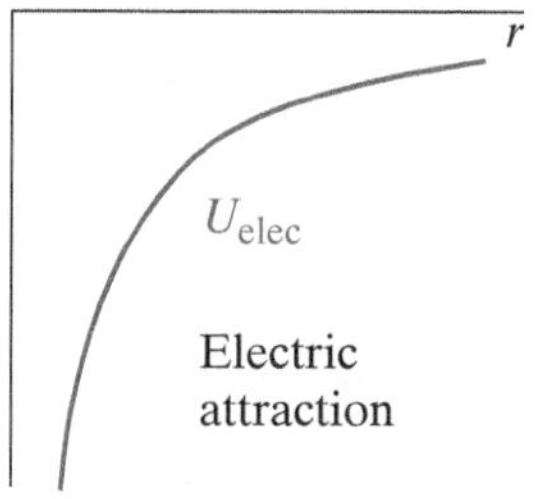

Figure 6.37 Electric potential energy is negative (like gravitational potential energy) when objects attract.

6.9 ELECTRIC POTENTIAL ENERGY

In Chapter 3 we saw that the electric force law for two charged particles is similar to the gravitational force law:

$$F_{\text{elec}} = \frac{1}{4\pi\varepsilon_0}\frac{|q_1 q_2|}{r^2} \quad \text{where } \frac{1}{4\pi\varepsilon_0} = 9\times 10^9 \frac{\text{N}\cdot\text{m}^2}{\text{C}^2}$$

The constant $1/(4\pi\varepsilon_0)$ is read as "one over four pi epsilon-zero." The quantities q_1 and q_2 represent the amount of charge on the two particles (measured in coulombs), and r is the distance between the two particles. Because this is an inverse square force, like gravitation, the electric potential energy has an equation very similar to that for gravitational potential energy:

ELECTRIC POTENTIAL ENERGY U_{elec}

$$U_{\text{elec}} = \frac{1}{4\pi\varepsilon_0}\frac{q_1 q_2}{r}$$

There is no minus sign in the expression for electric potential energy, because when the electric interaction is attractive, like gravitation, q_1 and q_2 have different signs, and the minus sign shows up automatically. However, if the two charges have the same sign, the interaction is repulsive, and there is a plus sign in the expression for the electric energy (Figure 6.36). If the two charges have opposite signs, the interaction is attractive, and there is a minus sign in the expression for electric energy (Figure 6.37).

EXAMPLE **Ionization Energy**

A certain amount of energy is required to remove an outer electron from a neutral atom, leaving behind a positive "ion" that is missing one electron. The charge of the ion is $+e$ (where $e = 1.6\times 10^{-19}$ C); the electron's charge is of course $-e$. The minimum energy to remove one outer electron is called the "ionization" energy. A typical atom has a radius of about 1×10^{-10} m. Estimate a typical ionization energy in joules, then convert your result to electron volts (eV).

Solution System: Two particles, the outer electron with charge $-e$, and the remaining part of the atom, the ion with charge $+e$.

Surroundings: An object that exerts an external force to remove the electron from the atom.

Initial state: The electron is bound to the atom, at a distance r_{atom} from the center.
Final state: The two particles are very far from each other, at rest.

We don't know the initial K of the electron, so we will assume it is zero, to get an approximate result. We will assume that the energy input to the system is in the form of work.

$$E_f = E_i + W$$
$$K_{\text{ion},f} + K_{e,f} + \cancel{m_i c^2} + \cancel{m_e c^2} + U_f = K_{\text{ion},i} + K_{e,i} + \cancel{m_i c^2} + \cancel{m_e c^2} + U_i + W$$
$$0 + 0 + U_f = 0 + 0 + U_i + W$$
$$W = U_f - U_i$$

$$W = \frac{1}{4\pi\varepsilon_0}\frac{q_1 q_2}{r_\infty} - \frac{1}{4\pi\varepsilon_0}\frac{q_1 q_2}{r_{\text{atom}}}$$
$$W = -\frac{1}{4\pi\varepsilon_0}\frac{(+e)(-e)}{r_{\text{atom}}}$$
$$W = \left(9\times10^9\,\frac{\text{N}\cdot\text{m}^2}{\text{C}^2}\right)\frac{(1.6\times10^{-19}\,\text{C})^2}{(1\times10^{-10}\,\text{m})}$$
$$W = 2.3\times10^{-18}\,\text{J}$$
$$W = (2.3\times10^{-18}\,\text{J})\left(\frac{1\,\text{eV}}{1.6\times10^{-19}\,\text{J}}\right) = 14\,\text{eV}$$

This is an estimate because we don't know the electron's initial kinetic energy K_i, but typical ionization energies are indeed a few electron volts.

Checkpoint 12 Two protons are hurled straight at each other, each with a kinetic energy of 0.1 MeV. You are asked to calculate the separation between the protons when they finally come to a stop. **(a)** Write out the Energy Principle for this system, using the update form and including all relevant terms. **(b)** Which term of the equation contains the unknown quantity?

6.10 PLOTTING ENERGY vs. SEPARATION

A graphical representation of energy can be exceptionally helpful in reasoning about a process. In Figure 6.38 we show an energy graph for the spacecraft leaving the asteroid, the process analyzed in Section 6.8 (the graph continues beyond the separation $r_f = 500\,\text{km}$). We plot the pair-wise gravitational potential energy U, the spacecraft's kinetic energy K, and the sum $K + U$, as a function of the separation distance r between the center of the asteroid and the spacecraft. We omit the rest energies, which are not changing during the motion.

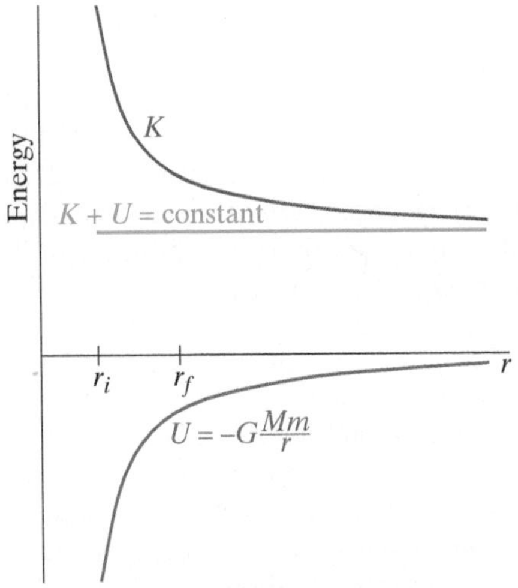

Figure 6.38 Energy vs. the separation distance between the asteroid and the spacecraft. (We omit the rest energies, which aren't changing.)

Important features of the graph are these: The gravitational potential energy U is negative because the interaction is attractive, not repulsive, and the potential energy increases with increasing separation r. As the potential energy increases, the kinetic energy K of the spacecraft decreases (the gravitational force exerted on the spacecraft by the asteroid is slowing it down). Because there is negligible work done on the combined system of asteroid plus spacecraft, $K_f + U_f = K_i + U_i$, so $K + U$ has a constant value at all times, and the graph of $K + U$ is a straight horizontal line on the graph, representing a constant positive value.

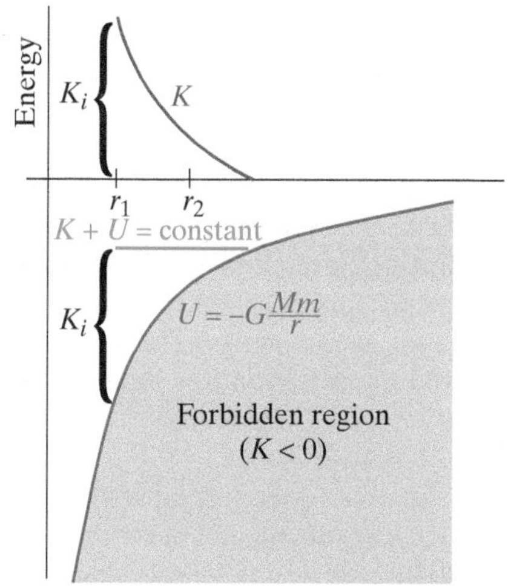

Figure 6.39 Energy vs. the separation distance between the asteroid and the spacecraft. (We omit the rest energies, which aren't changing.) At r_1 the potential energy is U_1 and the kinetic energy is K_1.

As the spacecraft gets farther and farther from the asteroid, you can see from the graph that its kinetic energy K will continue to fall, and the graph of K will approach the $K+U$ line at large separation. When the two objects are very far apart, U is nearly zero, and $K+U$ is just K.

This is an example of an "unbound" system, because the separation between spacecraft and asteroid will increase without bound. A gravitationally unbound system has $K+U$ greater than zero.

As an example of a "bound" system, again consider the spacecraft and the asteroid but this time give the spacecraft a smaller initial velocity, headed away from the asteroid. As a function of the center-to-center asteroid-to-spacecraft separation r, we again plot a graph of the potential energy $-GMm/r$ (Figure 6.39) and a graph of kinetic energy K. As the spacecraft moves away from the asteroid, its kinetic energy K falls as the pair-wise potential energy U rises. You can see on the graph that K falls to zero, and it cannot go negative, because kinetic energy is always a positive quantity. What happens next? The spacecraft momentarily comes to a stop (a turning point), then turns around and speeds up toward the asteroid; eventually it will crash onto the asteroid's surface.

We add to the graph a horizontal line whose height above or below the axis represents the total kinetic plus potential energy of the two-object system. The line on the graph for $K+U$ is horizontal because no external work is done on this isolated system, and there is no change of the masses (no change of identity), so $K+U$ is constant during the motion.

QUESTION Is $K+U$ positive or negative for the system shown in Figure 6.39?

The horizontal line representing the constant value of $K+U$ is below the axis representing $U=0$, so $K+U$ is negative. This corresponds to a "bound" state of the spacecraft and asteroid; the spacecraft and asteroid cannot get completely away from each other.

The vertical distance from the U curve up to the $K+U$ line is equal to K, the total kinetic energy of the spacecraft and asteroid. At all separations between the spacecraft and the asteroid, the height of the $K+U$ line above the U curve tells you the total kinetic energy. Notice that when K falls to zero, $K+U=U$ and the right end of the $K+U$ line touches the U graph.

The shaded area in Figure 6.39 represents a forbidden region, where the kinetic energy K would be negative, no matter what the value of $K+U$ (more about this in the following discussion).

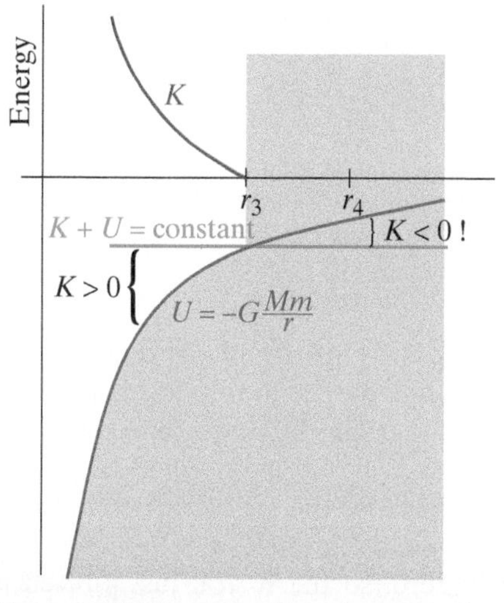

Figure 6.40 The asteroid and spacecraft cannot get farther away from each other than r_3; at r_4 the kinetic energy would be negative. (We omit the rest energies, which aren't changing.)

Limits on Possible Motion

The energy graph shows at a glance the limits on the motion for a given energy. In Figure 6.40, the asteroid and spacecraft with $K+U<0$ cannot get farther away from each other than r_3, a bound that is set by the potential-energy curve. At a larger separation r_4, the kinetic energy would have to be negative, which is "classically forbidden" (that is, impossible according to the laws of classical, prequantum mechanics). Kinetic energy cannot be negative. At low speeds, $K \approx \frac{1}{2}mv^2$, which is always positive. At high speeds, $K=\gamma mc^2 - mc^2$, which is always positive, because γ is always greater than 1.

If the system has the total $K+U$ shown ($K+U<0$), the separation of the asteroid and spacecraft can never be larger than r_3, and we say that this is a "bound" system. An energy graph shows at a glance the range of possible positions achievable during the motion, for a given total $K+U$. (We should mention, however, that in the world of atoms where a full analysis requires quantum mechanics, in some cases a system can "tunnel through" a region that is classically forbidden!)

An energy graph can show at a glance whether a system is bound or unbound. An example of a bound system is a planet in circular or elliptical orbit around a star: the planet cannot escape. Another example is an electron bound to an atom.

An example of an unbound system is the Earth plus a spacecraft whose initial speed is great enough that the spacecraft will get away from the Earth and never come back. For another example, if an amount of energy greater than or equal to the ionization energy is supplied to an atom, an electron can become unbound and escape from the atom. A gravitationally or electrically unbound system has total $K + U \geq 0$, which is easy to see on an energy graph, where it is clear that arbitrarily large separations are possible. Only when $K + U \geq 0$ can the system become separated by large distances, because K is never negative, and U goes to zero at large separations. (The following Checkpoint shows a different kind of potential energy situation, where there can be a bound state even when $K + U \geq 0$.)

Checkpoint 13 (a) In Figure 6.39, as the asteroid–spacecraft separation increases from r_1 to r_2 does the kinetic energy of the asteroid–spacecraft system increase, decrease, or remain constant? **(b)** In the energy graph for a two-object system shown in Figure 6.41, consider the various energy states indicated. Which of these values for the energy $K + U$ (A, B, or C) represents a bound state? Which represents an unbound state (the particle can escape)? Which represents a bound state with enough energy to be unbound but with a barrier that (classically) prevents escape?

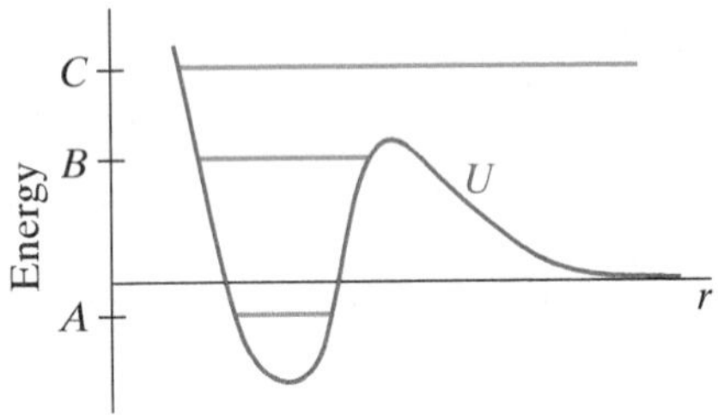

Figure 6.41 An energy graph for a two-object system. Which states are bound?

Drawing Energy Graphs

It is important to be able to draw (create) energy graphs as well as to be able to read them. Here is a practical scheme:

- Draw U vs. r for the particular interaction (gravitational, electric).
- At some r where you happen to know K, plot the point (r, K).
- Add that value of K to the value U has at the same separation r.
- Plot $K + U$ at that r, then draw a horizontal line through the point.
- At some other r, find a K that when added to U at that r gives $K + U$.
- Given these two points on the K graph, sketch the behavior of K vs. r.

EXAMPLE **Making an Energy Graph**

Let's follow through these steps for the energy graph for the spacecraft leaving the asteroid, which is shown in Figure 6.42.

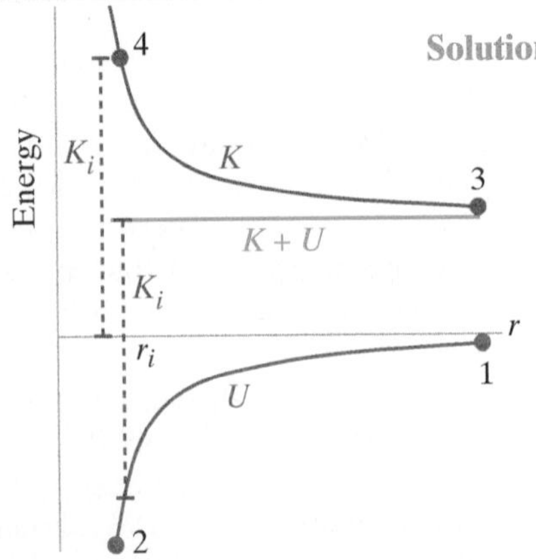

Figure 6.42 Energy vs. the separation distance between the asteroid and the spacecraft.

Solution

- We know the shape of U for gravity; it is $-GMm/r$. If r is small, U is a large negative number, so we plot point 1 on the graph. If r is large, U is nearly zero, so we plot point 2 on the graph. Now we can sketch the behavior of U vs. r.
- We know that K is nonzero at large r, because the spacecraft escapes, so at large r, we plot a positive K (point 3 on the graph).
- At that distant location, U is nearly zero, so the K you just plotted is nearly the same as $K + U$. Draw a horizontal line representing $K + U$ for all separations.
- At the initial separation r_i, determine graphically the value K_i that when added to the value of U_i at r_i yields the known value of $K + U$ (this is the height of the $K + U$ line about the U value at r_i). This is point 4 on the graph.
- Sketch the behavior of K vs. r. Note that its shape is the mirror image of U vs. r.

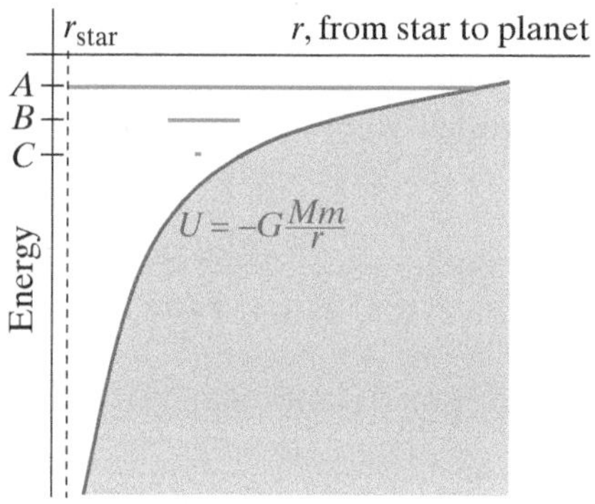

Figure 6.43 What kind of motion is represented by each of these situations?

Checkpoint 14 In Figure 6.43, what kind of motion is represented by the situation with $K + U = A$? B? C? Think about the range of r in each situation. For example, C represents a circular orbit (constant r).

Escape Speed

An analysis based on the Energy Principle will allow us to answer the following question: With what minimum speed v must a spacecraft leave the surface of an airless planet (no air resistance) of mass M and radius R in order that it can coast away without ever coming back (Figure 6.44)? Assume that there are no other objects nearby.

If you did the Moon voyage problem in Chapter 3 you found that there was a minimum initial speed required to reach the Moon. With an initial speed smaller than this, the spacecraft fell back to Earth without reaching the Moon. Your determination of the minimum initial speed was done by trial and error. However, we can now use energy relationships to calculate such minimum speeds directly.

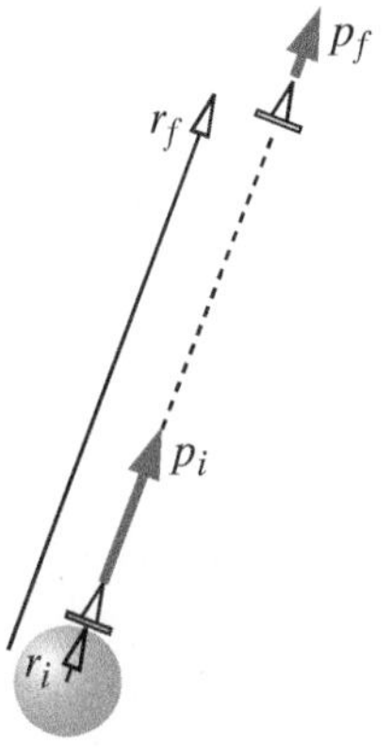

Figure 6.44 A spacecraft is launched with momentum p_i from near an airless planet's surface; it has momentum p_f when it has traveled far away.

As is often the case, an energy diagram can be very helpful. Consider Figure 6.45. Motions with $K + U = A$, B, C, or D all start from the surface of an airless planet, with a separation R between the center of the planet and the spacecraft. (Remember that the gravitational force outside a uniform sphere is exactly the same as it would be if all the mass collapsed to the center of the sphere.)

QUESTION Which of these motions starts with the largest kinetic energy K? Which starts with the least K?

Each of the horizontal lines represents motion with $K + U$ constant. Evidently A is the highest value of $K + U$, and therefore involves the largest initial kinetic energy. D is the lowest value of $K + U$, and starts with the least kinetic energy.

QUESTION For which of these motions does the spacecraft "escape" and never come back?

For motions with $K + U = A$, B, or C, no matter how far from the planet the spacecraft gets, there is still some kinetic energy, so the spacecraft will never return. This is another example of an unbound state being associated with a positive value of $K + U$.

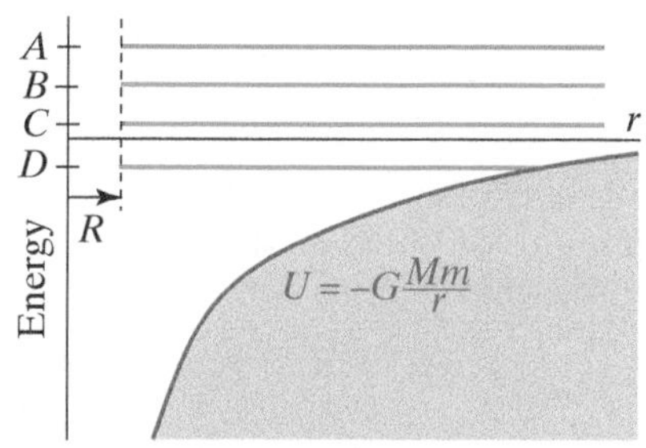

Figure 6.45 What is the minimum K required to escape?

In contrast, with motion $K + U = D$ the spacecraft will reach a maximum separation from the planet, at which point its kinetic energy has fallen to zero. The spacecraft will fall back to the planet.

QUESTION How would you describe the motion that requires the least possible initial kinetic energy to achieve escape?

Evidently the least costly escape is to have $K = 0$ when the spacecraft has reached a distance very far from the planet. At a large separation, $U = 0$, so $K + U = 0$. This corresponds on the diagram to a horizontal line lying on the axis. Because $K + U$ doesn't change (no external work on the planet–spacecraft system), it must also be true that $K + U = 0$ at the start of the motion:

MINIMAL CONDITION FOR ESCAPE

$$K + U = 0$$

The minimal initial kinetic plus potential energy of the system composed of planet plus spacecraft is this, assuming that the kinetic energy of the planet is negligible:

$$K_i + U_i = \frac{1}{2}mv_{\text{esc}}^2 + \left(-G\frac{Mm}{R}\right) = 0 \quad (\text{for } v \ll c)$$

This lets us calculate the minimum speed, which is called the "escape speed" v_{esc}, for a planet with mass M and radius R. Though we speak of escape speed, it is wrong to say that the spacecraft has "escaped from gravity." As long as the spacecraft is a finite distance from the planet, it feels a finite gravitational attraction to the planet.

If you give the spacecraft more than the minimum kinetic energy, it not only escapes but also has nonzero kinetic energy when it is far away (for example, see motions A and B in Figure 6.45).

It is interesting to note that the initial direction of motion doesn't matter. An energy analysis is insensitive to direction.

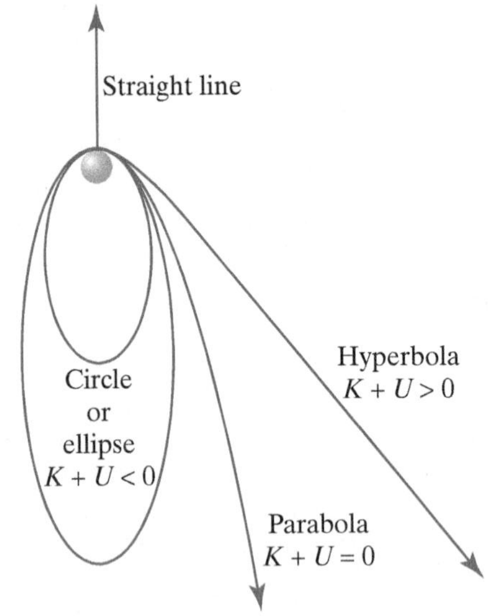

Figure 6.46 Possible orbital trajectories.

Bound vs. Unbound States

If the initial velocity points directly away from the planet, the path is a straight line. If the initial velocity points in some other direction, the path can be shown to be a parabola if the initial speed is exactly equal to escape speed and a hyperbola for higher speeds.

If the speed is below escape speed, there is no escape. The various possible orbits (Figure 6.46) can be classified on the basis of their total kinetic plus potential energy, with initial speed v_i:

$$\frac{1}{2}mv_i^2 + \left(-G\frac{Mm}{R}\right) > 0: \quad \text{escape with } v_\infty > 0\text{; straight line or hyperbola}$$

$$\frac{1}{2}mv_i^2 + \left(-G\frac{Mm}{R}\right) = 0: \quad \text{escape with } v_\infty = 0\text{; straight line or parabola}$$

$$\frac{1}{2}mv_i^2 + \left(-G\frac{Mm}{R}\right) < 0: \quad \text{no escape; straight line, circular, or elliptical}$$

This is a specific example of an important general principle. If the kinetic plus potential energy is negative, the system is bound: the objects cannot become widely separated, because that would require that the kinetic energy become negative, which is impossible (classically). If the kinetic plus potential energy is positive, the system is unbound or free (unless the particle is in a "trap" as in Figure 6.41): the objects can become widely separated, with net kinetic energy (and zero potential energy).

BOUND AND UNBOUND STATES

If $K + U < 0$, the system is in a bound state.

If $K + U \geq 0$, the system is unbound (if not in a "trap").

EXAMPLE **Escape Speed**

What is escape speed from Earth?

Solution System: Earth + spacecraft
Surroundings: Nothing significant (ignore Sun and Moon)

Initial state: At surface of Earth, with speed
Final state: Spacecraft extremely far away from Earth, nearly at rest
Energy Principle (negligible change in kinetic energy of Earth; no work by surroundings):

$$K_{s,f} + \cancel{K_{E,f}} + U_f = K_{s,i} + \cancel{K_{E,i}} + U_i + W$$

$$0 + 0 = \frac{1}{2}mv_i^2 + \left(-G\frac{Mm}{r_i}\right) + 0$$

$$v_i = \sqrt{\frac{2GM}{r_i}}$$

$$v_{esc} = \sqrt{\frac{2(6.7 \times 10^{-11}\,\text{N} \cdot \text{m}^2/\,\text{kg}^2)(6 \times 10^{24}\,\text{kg})}{(6.4 \times 10^6\,\text{m})}} = 1.12 \times 10^4\ \text{m/s}$$

Checkpoint 15 Turn the argument around. If an object falls to Earth starting from rest a great distance away, what is the speed with which it will hit the upper atmosphere? (Actually, a comet or asteroid coming from a long distance away might well have an even larger speed, due to its interaction with the Sun.) Small objects vaporize as they plunge through the atmosphere, but a very large object can penetrate and hit the ground at very high speed. Such a massive impact is thought to have killed off the dinosaurs (see *T. Rex and the Crater of Doom*, Walter Alvarez, Princeton University Press, 1997).

6.11 GENERAL PROPERTIES OF POTENTIAL ENERGY

There are some general properties of potential energy that are true not only for gravitational interactions but also for other kinds of interactions as well, including electric interactions. Here is a list of important properties:

- Potential energy depends on the separation between pairs of particles, not on their individual positions. (As we saw with gravitational interactions, this depends on the reciprocity of gravitational and electric forces.)
- Potential energy must approach zero as the separation between particles becomes very large, as we will soon show.
- If an interaction is attractive, potential energy becomes negative as the distance between particles decreases. (We saw this in the case of gravitational interactions.)
- If an interaction is repulsive, potential energy becomes positive as the distance between particles decreases.

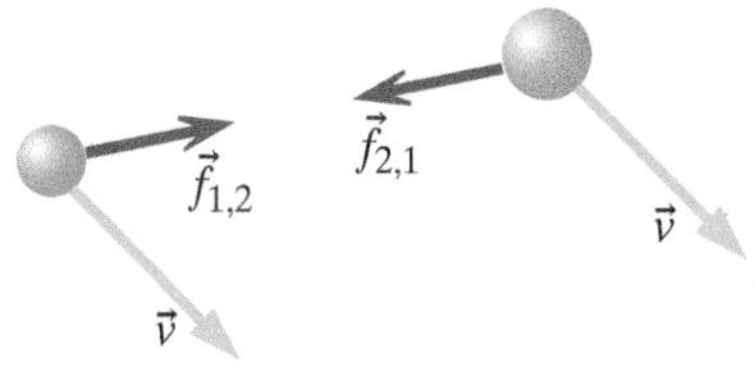

Figure 6.47 The two particles move together, with the same velocity.

Let's look in a bit more detail at the consequences of the fact that potential energy depends on the separation between the interacting objects.

QUESTION Suppose that two particles are part of a rigid system and move together, with the same velocity (Figure 6.47). In that case, what is $\Delta\vec{r}_{21}$ for this two-particle system? What is ΔU_{12}?

In this case the relative positions don't change, $\Delta\vec{r}_{21}$ is zero, and there is no change in the pair-wise potential energy.

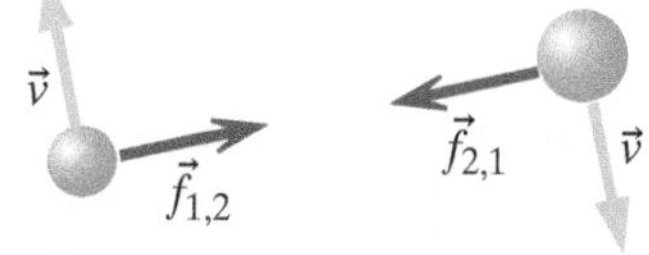

Figure 6.48 The two particles rotate around each other.

QUESTION Suppose instead that the two particles rotate around each other (Figure 6.48). What can you say about ΔU_{12}?

This is a bit trickier, because if there is rotation, $\Delta\vec{r}_{21}$ isn't zero, because $\vec{r}_{21}$ is now a rotating vector. In rotation, however, $\Delta\vec{r}_{21}$ is perpendicular to $\vec{r}_{21}$ and to the force $\vec{f}_{2,1}$, so the dot product $\vec{f}_{2,1} \bullet \Delta\vec{r}_{21}$ is zero. So again there is no change in the pair-wise potential energy.

QUESTION What can you conclude about potential energy for a rigid system?

Evidently U is constant for a rigid system, one whose shape doesn't change. For the potential energy to change, there must be a change of shape or size. Conversely, a change of shape or size is an indicator that U may have changed.

$U \rightarrow 0$ When the Particles Are Very Far Apart

Consider a system consisting of two identical stars, which attract each other gravitationally. If the stars are at rest, very, very far apart—almost infinitely far apart, so that the force one exerts on the other is nearly zero—then the total energy of the two-star system must be simply the sum of their rest energies:

$$E_{sys} = mc^2 + mc^2$$

This implies that the potential energy associated with the interaction of these two stars must be zero if the stars are "infinitely" far apart.

More generally, the total energy of a multiparticle system is given by the expression $E_{system} = (E_1 + E_2 + \cdots) + U$, where $E_1 = \gamma m_1 c^2$ is the particle energy of particle 1, and so on. When the particles are very far apart, we must define U to be zero, so that the total energy of these far-apart particles is simply the sum of the particle energies:

POTENTIAL ENERGY OF A PAIR OF PARTICLES THAT ARE VERY FAR APART

$$U \rightarrow 0 \quad \text{as} \quad r \rightarrow \infty$$

Attractive Interaction: U Becomes Negative as r Decreases

If the particles start to come closer together, we find that U starts to become negative if the interactions are attractive, while U starts to become positive if the interactions are repulsive. To see this, first consider an isolated system consisting of two stars that are very far apart, initially at rest, so the initial energy of the system is just the sum of the rest energies of the two stars. At first slowly, then more and more rapidly, their kinetic energies increase due to their mutual gravitational attractions (Figure 6.49).

Figure 6.49 Initially two stars are very far apart, at rest, but they attract each other.

QUESTION U was initially zero, so what must be the sign of U when the stars have acquired significant kinetic energies?

If the system is isolated, there are no external forces and hence no external work, so E_{sys} does not change. If the individual particle energies increase (so the total kinetic energy increases), the pair-wise potential energy must decrease. Because U was defined to be zero initially, U must become negative, as we found for gravitational potential energy. A graph of the energies is shown in Figure 6.50, omitting the very large rest energies of the stars. Since the rest energies don't change, $K + U$ is a constant.

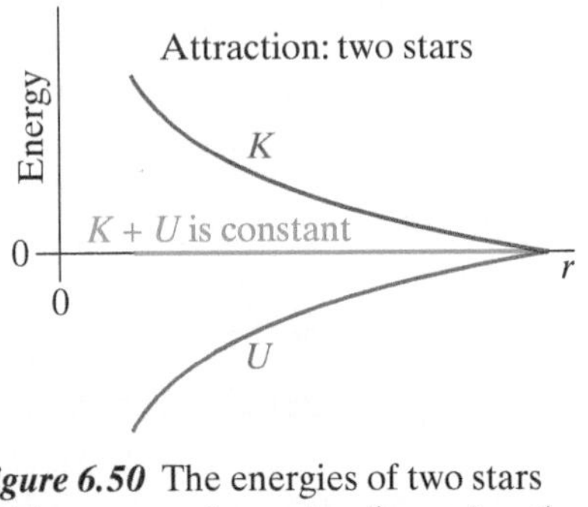

Figure 6.50 The energies of two stars (with rest energies omitted) as a function of separation r. U must be zero at large r. As r decreases, K increases and U decreases. The total energy of the two-star system is constant, and $K + U$ is constant.

Does this negative sign make sense in terms of the definition of potential energy change, $\Delta U = -W_{int}$? The attractive gravitational forces that the stars exert on each other certainly do a positive amount of work on each star, so the potential energy change must indeed be negative.

Repulsive Interaction: U Becomes Positive as r Decreases

Next consider an isolated system consisting of two protons that are very far apart, and suppose that initially they have high speeds, heading inward (Figure 6.51). Their mutual electric repulsions make the protons slow down (their electric repulsion is enormously larger than their gravitational attraction).

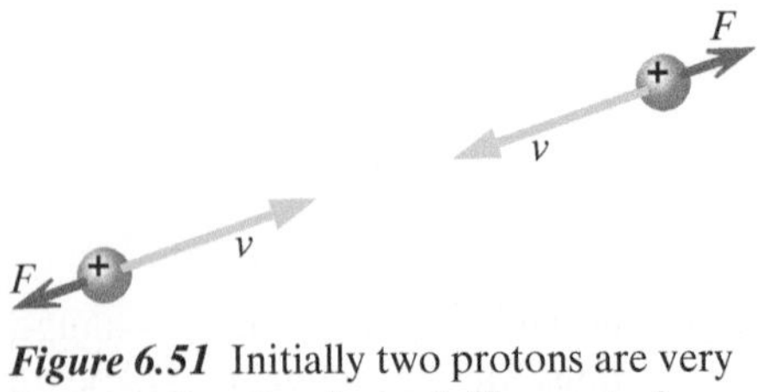

Figure 6.51 Initially two protons are very far apart, heading inward. The mutual repulsions slow the protons down.

QUESTION U was initially zero, so what must be the sign of U when the protons have slowed down somewhat?

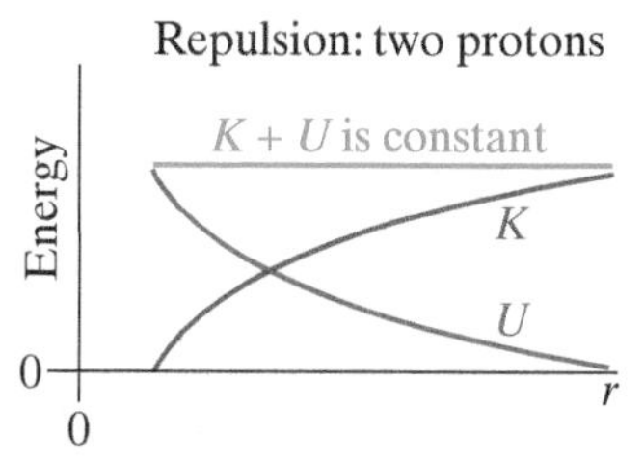

Figure 6.52 The energies of two protons as a function of separation (with rest energies omitted). U must be zero at large r. As r decreases, K decreases and U increases. The total energy of the two-proton system is constant, and $K+U$ is constant.

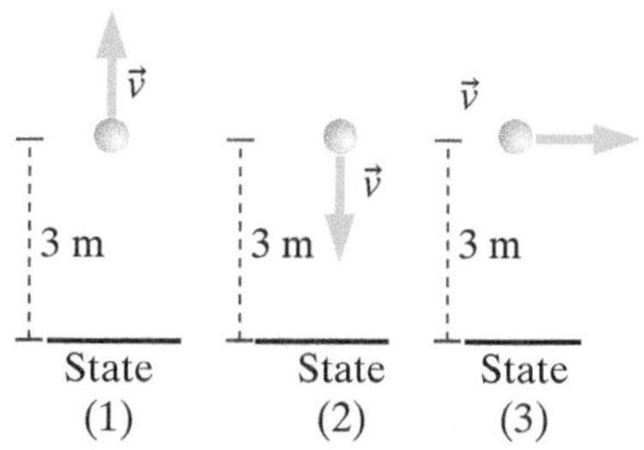

Figure 6.53 The potential energy of the system of (ball + Earth) is the same in each of these states, since it depends only on the separation of the ball and the Earth.

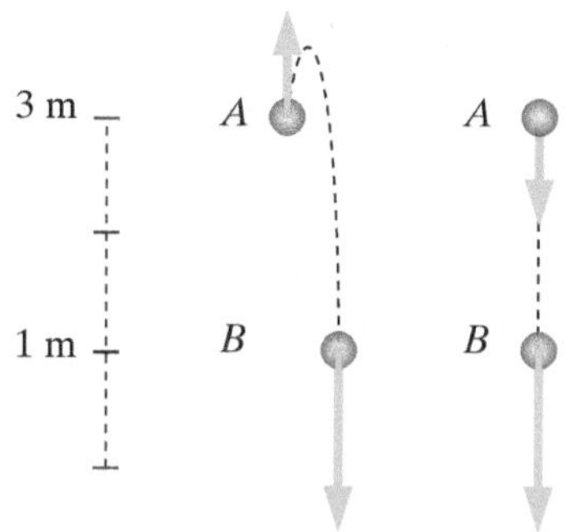

Figure 6.54 ΔU for the system of (ball + Earth) does not depend on the path taken to get from initial to final state.

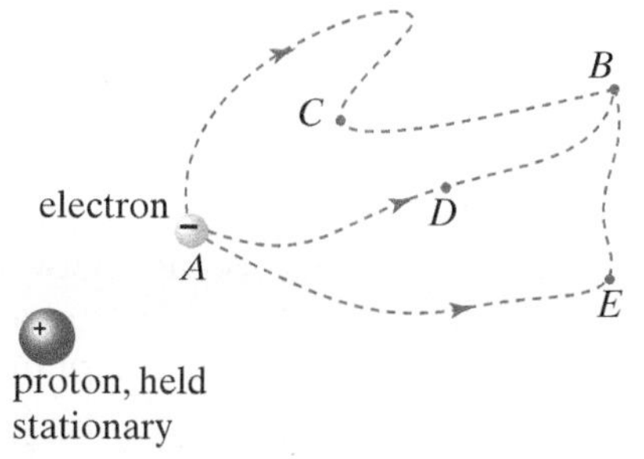

Figure 6.55 A proton is held stationary while an electron is moved from location A to location B along different paths.

Again, there is no external work done on this isolated multiparticle system, so E_{sys} does not change. The particle energies decrease, so the potential energy must increase; U becomes positive (Figure 6.52). Also, the repulsive electric forces do negative work on each other, so $\Delta U = -W_{int}$ is positive. The graph in Figure 6.52 omits the rest energies of the protons. Since the rest energies don't change, $K+U$ is a constant.

Note that the change in potential energy for the two stars or the two protons was associated with a change in shape (change in configuration) of the two-particle system. If there is no change in shape or size, there is no change in potential energy.

Path Independence

One of the most interesting things about the expressions for different kinds of potential energy is what is *not* in these equations: history. Since the potential energy of two interacting point particles depends only on the relative positions of the particles, it does not depend on how the particles got to those positions. If a tennis ball is 3 m above the surface of the Earth, it doesn't matter how the tennis ball got there (Figure 6.53). It could be moving up, down, or horizontally—only the relative positions of the ball and the Earth matter. This may seem obvious and unremarkable, but it has some important implications.

One consequence is that since U depends only on separation, any "trip" between two locations, or separations, for a two-particle system, results in the same ΔU. For the system of (ball + Earth) in Figure 6.54, the difference between the potential energy of the system in the initial state shown (3 m above the ground) and the final state in which the ball is only 1 m above the ground does not depend on whether the ball goes straight down or goes up, stops, and descends. The fact that the difference in potential energy of a multiparticle system depends only on the states chosen, and not on the path taken by the particles to get to those states, is called "path independence."

$$U_B - U_A = \text{constant}$$
$$\text{independent of path taken between } A \text{ and } B$$

A special case of a trip between two states is a round trip, in which the initial state is the same as the final state. The potential energy change associated with a round trip must be zero, since

$$\Delta U_{\text{round trip}} = U_A - U_A = 0$$

A second interesting consequence is that the dependence of potential energy on configuration, not history, allows us to make deductions about the amount of work required to change the state of a system no matter how that change happens. This is easiest to see if we consider processes in which the kinetic energy of the system does not change. Suppose that a proton remains fixed at the origin while we move an electron from location A to location B at constant speed (Figure 6.55). The electron is attracted to the proton, so we need to do positive work to move it farther from the proton.

To move the electron along the path that goes through location D we will have to do positive work to separate the particles. In contrast, on part of the path that goes through location C the electron is moving back toward the proton, and on that path segment we will have to do negative work to keep it from speeding up. In the end, though, all the positive and negative work we do to move the electron along any path from A to B will add up to the same amount. We can see this from the Energy Principle:

$$K_{e,f}+K_{p,f}+U_B=K_{e,i}+K_{p,i}+U_A+W$$
$$\cancel{K_{e,f}}+0+U_B=\cancel{K_{e,i}}+0+U_A+W$$
$$U_B-U_A=W$$

It also follows from this that the total amount of work required to move the electron from A to B along any path, and then back to A along any other path, without changing its kinetic energy, must add up to zero:

$$W_{\text{round trip}}=0$$

What About Friction?

We've defined potential energy for systems of interacting point particles. Point particles are a remarkably good model for many physical situations, but not for situations in which there is friction.

> **QUESTION** Why isn't the point particle model appropriate when there is friction?

As you know from everyday experience, when there is significant friction between two moving objects, the objects can get hot. A temperature change indicates that the amount of energy inside an object has changed. We'll see in Chapter 9 that friction results from the deformation of extended objects that are in contact with each other. However, we can't model such objects as point particles; since a point particle, by definition, has no internal structure, it can't deform or get hot.

Further, the work done by friction is clearly not independent of the path taken; the longer the path, the hotter the objects get, regardless of direction. We call forces like friction and air resistance "dissipative forces," and there isn't a way to define anything like potential energy for such interactions. We'll consider dissipative interactions further in Chapter 7.

You may have heard friction described as a "nonconservative" force, an older term used to denote interactions in which a potential energy cannot be defined. This isn't a very informative term, because friction itself is simply a consequence of the electric interactions of electrons and protons in two extended objects in contact with each other. The real issue is the fact that these electric forces lead to deformations of the objects, and that energy can flow into the vibrations of the atoms inside the objects.

6.12 THE MASS OF A MULTIPARTICLE SYSTEM

A surprising prediction of Einstein's Special Theory of Relativity (1905) was that mass and energy are closely related. This prediction has been confirmed in many different kinds of experiments. The most dramatic example of mass–energy equivalence is the annihilation reaction that occurs between matter and antimatter, as in the case of the annihilation of an electron and a positron (a positive electron) into two photons (high-energy quanta of light denoted by lowercase Greek gamma, γ):

$$e^- + e^+ \rightarrow \gamma + \gamma$$

Before the reaction, an electron and a positron can be almost at rest far from each other, a situation we might be inclined to describe as one in which there is mass but seemingly no energy. After the reaction there is light (in the form

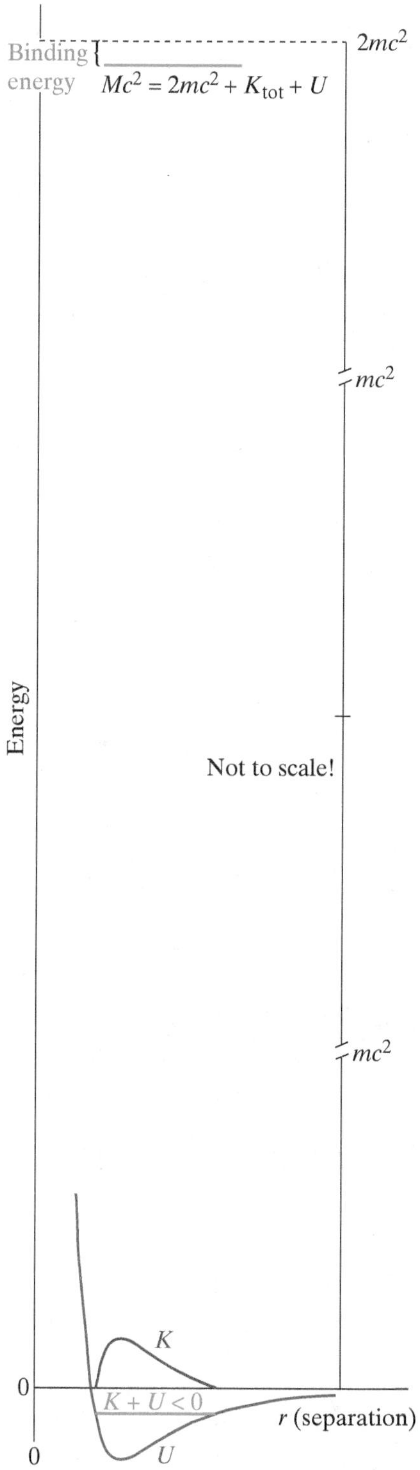

Figure 6.56 A plot of rest energy, kinetic energy, potential energy, and total energy (Mc^2) for O_2. Note that the total energy is less than the sum of the rest energies of the two individual atoms.

of two photons), a situation that we would naturally describe as pure energy but seemingly no mass. However, from the point of view of Special Relativity, the mass of the electron–positron system *is* energy, and the energy of the two-photon system *is* mass.

The equivalence of mass and energy is of course visible in Einstein's equation for the energy of a particle:

$$E = \gamma mc^2 = mc^2 + K \quad (\textit{definition} \text{ of kinetic energy } K)$$

In this equation m is the mass of the particle and is a constant. The particle has rest energy $E_{\text{rest}} = mc^2$ even if it is not moving. If it is moving, it has additional, kinetic energy K.

What is the mass of a multiparticle system consisting of interacting particles? Consider the simple case where the multiparticle system is overall at rest. The individual particles are moving around, but the system as a whole is not moving (more technically, the total momentum of the system is zero, and its center of mass is at rest). An example would be a box containing air molecules that are moving around rapidly in random directions, but the box isn't going anywhere. Another example is a diatomic molecule such as O_2 that is vibrating but has no overall translational motion.

It is natural to suppose that the total mass of such a system is just the sum of the masses of the particles: $M = m_1 + m_2 + \cdots$. However, this is not quite right. Rather, as predicted by the theory of relativity, the rest energy of the multiparticle system is Mc^2, and therefore its mass is this:

$$M = \frac{E_{\text{system}}}{c^2} = \frac{m_1c^2 + K_1 + m_2c^2 + K_2 + \cdots + U}{c^2}$$

$$M = (m_1 + m_2 + \cdots) + \left(\frac{K_1 + K_2 + \cdots + U}{c^2}\right)$$

The mass M of the system may be greater or less than the sum of the individual particle masses, because the final term in this equation may be positive or negative.

In most everyday situations the kinetic energies of the individual particles are very small compared to their rest energies (that is, the speeds of the particles are small compared to the speed of light), and the interaction energy U (the sum of all the pair-wise interactions) is also small compared to the rest energies. In that case, the mass of the multiparticle system is very nearly equal to the sum of the individual masses. However, this is really only an approximation, and it is often not a good approximation in nuclear and particle physics.

The Total Energy of a Multiparticle System

Consider an oxygen molecule (O_2) at rest. Instead of plotting only the kinetic and potential energy for this two-atom system, let's include the rest energy of the atoms, too, as we attempt to show in Figure 6.56. We can't draw the full vertical scale, because the rest energies are enormous compared to the kinetic and potential energies. In order to show the rest energy to the same scale as K and U, we would need a page stretching from the Earth to the Moon!

The total energy of the system (indicated by the green horizontal line near the top of the graph) is the sum of the rest energies of the two atoms, the potential energy U (which is negative), and K_{tot}, the total kinetic energy of the two atoms. For the bound state drawn on the graph you can see that the rest energy of the two-atom O_2 system is very slightly less than the sum of the individual atomic rest energies, because $K + U < 0$.

Binding Energy

When the diatomic molecule O_2 forms, about 5 eV of energy are given off (for example, by emitting photons, which carry energy). Correspondingly, it takes an input of about 5 eV of energy to break O_2 apart into two oxygen atoms. This is called the "binding energy" of the molecule and is shown at the top of Figure 6.56.

Two Ways of Thinking About the Energy of the System

Given these aspects of particle energy, there are two rather different ways to think about the energy of a multiparticle system:

- The energy of a multiparticle system consists of the individual particle energies of the particles that make up the system, plus their pair-wise interaction energies.
- A multiparticle system itself has energy, rather like a particle, and if the system is at rest (its center of mass is not moving; its net momentum is zero) its energy E is simply Mc^2, where M is the mass of the system.

The Mass of the System

The mass M may be greater or smaller than the sum of the masses of the individual particles, depending on whether the total $K+U$ of the particles is positive or negative. The graph in Figure 6.56 indicates that in the case of the O_2 molecule the quantity Mc^2 (where M is the mass of the whole molecule) is less than $2mc^2$, and the difference is the binding energy of the molecule. From the graph we can see that:

$$Mc^2 = 2mc^2 + K + U < 2mc^2$$

We analyzed the O_2 molecule in terms of the masses of the individual oxygen atoms and the molecular binding energy. We could make a more detailed analysis by considering the energy terms that contribute to the mass of each oxygen atom. The oxygen atom is itself a multiparticle system consisting of protons and neutrons bound together in the nucleus, plus eight electrons. These particles have rest energy and kinetic energy, and there is potential energy associated with pairs of particles. One could even go further, recognizing that the protons and neutrons are made of quarks and gluons. However, if we're just interested in the binding energy of the O_2 molecule, we can start with the rest mass of each oxygen atom, because all of the internal complexity of each atom is summarized by the value of the atom's mass.

Mass of a Hot Object

This way of thinking about mass leads us to conclude that a hot block has a greater mass than a cold block. As we'll discuss in more detail in the next chapter, a hot metal block has more thermal energy than a cold metal block, with more kinetic energy of the atoms and more potential energy in the interactions with neighboring atoms (represented as springs in our simple model of a solid). Although the effect is much too small to measure directly, the equivalence of mass and energy embodied in the special theory of relativity, $M = E/c^2$, leads to the conclusion that a hot block is actually more massive than a cold block! The same impulse $\vec{F}\Delta t$ applied to the two blocks should change the velocity of the cold block very slightly more than the velocity of the hot block.

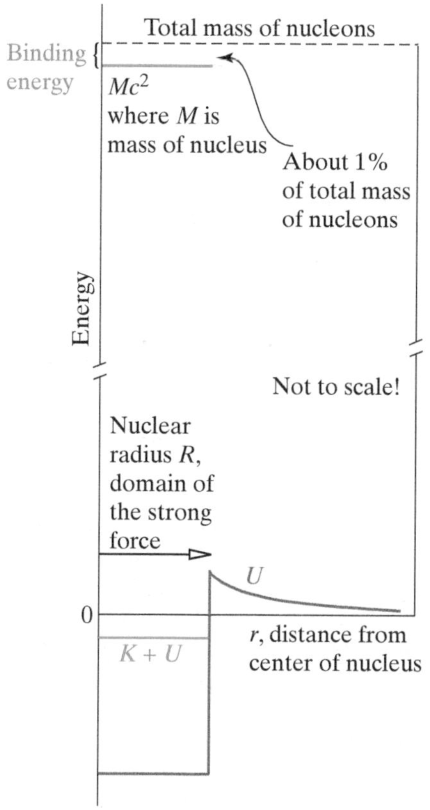

Figure 6.57 An energy diagram for a proton in an iron nucleus, not to scale. The mass of the nucleus is about 1% less than the sum of the masses of the individual nucleons (protons and neutrons).

Checkpoint 16 Calculate the approximate ratio of the binding energy of O_2 (about 5 eV) to the rest energy of O_2. Most oxygen nuclei contain 8 protons and 8 neutrons, and the rest energy of a proton or neutron is about 940 MeV. Do you think you could use a laboratory scale to detect the difference in mass between a mole of molecular oxygen (O_2) and two moles of atomic oxygen?

Nuclear Binding

In chemical reactions it is nearly impossible to measure the change of mass directly, but in nuclear interactions the changes are easily measurable. Consider the nuclear binding energy of an iron nucleus, which contains 26 protons and 30 neutrons, a total of 56 nucleons. When the protons and neutrons bind to each other through the "strong" (nuclear) force, the resulting iron nucleus has a mass that is about 1% smaller than the total mass of its individual constituents when separated. There is a sizable negative potential energy U associated with the strong force, which makes the multiparticle system's rest energy measurably smaller than the sum of the individual rest energies.

The average binding energy per nucleon is largest in iron. Lighter nuclei (hydrogen through manganese) are less tightly bound and can in principle fuse together to form the more tightly bound iron, with the release of energy. In fact, fusion reactions in stars build up from hydrogen to heavier and heavier nuclei, stopping at iron (heavier elements are made in supernova explosions).

Heavier nuclei (cobalt through uranium and beyond) are also less tightly bound than iron and can in principle fission to form the more tightly bound iron. However, only a few nuclei such as uranium and plutonium do actually spontaneously fission.

Figure 6.57 shows an energy diagram (not to scale) for one of the protons in an iron nucleus, subjected to the forces of all the other protons and neutrons. There is a deep potential well corresponding to the strong or nuclear force, which is essentially the same for protons and neutrons. This potential well has a very steep side corresponding to the very strong nuclear force (force is negative gradient of potential energy). The well is so steep it can be described approximately as a "square well." This deep well extends only out to the radius R of the nucleus, because the nuclear force has a very short range. A proton (but not a neutron) in addition experiences the repulsive electric force of the other protons (the electric U curve outside the nucleus). The total mass of the nucleus is about 1% less than the sum of the masses of the individual nucleons (protons and neutrons).

EXAMPLE **Nuclear Binding Energy**

The rest energy of an iron nucleus (26 protons and 30 neutrons) is 52,107 MeV. Show that the average binding energy per nucleon is about 1% of the rest energy of a nucleon, which is easily measurable. The rest energy of a proton is 938.3 MeV; of a neutron, 939.6 MeV.

Solution As shown in Figure 6.57, the binding energy is essentially the difference between the rest energy of the multiparticle system (the nucleus) and the sum of the rest energies of its constituent particles.

$$\begin{aligned}\text{binding energy} &= ((26)(938.3)\text{ MeV} + (30)(939.6)\text{ MeV}) - 52107\text{ MeV}\\ &= 494.8\text{ MeV}\end{aligned}$$

$$\text{energy per nucleon} = \frac{494.8\text{ MeV}}{26+30} = 8.8\text{ MeV}$$

This is about 1% of the rest energy of one nucleon.

Checkpoint 17 The deuteron, the nucleus of the deuterium atom ("heavy" hydrogen), consists of a proton and a neutron. It is observed experimentally that a high-energy photon ("gamma ray") with a minimum energy of 2.2 MeV can break up the deuteron into a free proton and a free neutron; this process is called "photo-dissociation." About what fraction of the deuteron rest energy corresponds to its binding energy? The result shows that the deuteron is very lightly bound compared to the iron nucleus.

Nuclear Fission

The nuclear fission of uranium leads to an easily measured change in masses. The average binding energy per nucleon in uranium is about 7.8 MeV, so the uranium nucleus is less tightly bound than an iron nucleus, whose binding is 8.8 MeV per nucleon. This suggests the possibility of obtaining energy from the fission of a uranium nucleus, corresponding to roughly 0.1% (0.7 MeV per nucleon) of its rest energy (about 940 MeV per nucleon).

In other words, the "mass deficiency" of a medium-mass nucleus is about 1%, with a variation of about a tenth of that. Even though there is only a small difference in average binding energy between iron and uranium, the energy involved is vast compared to the energy of a chemical reaction, which only involves the outer electrons of an atom, not the nucleus.

A uranium-235 nucleus contains 92 protons and 143 neutrons. The 92 protons exert strong electric repulsions on each other, but the large number of neutrons spreads the protons throughout a larger volume and reduces the electric repulsions. The neutrons and the protons also attract neighboring protons with strong nuclear forces. (Low-mass nuclei don't have an overabundance of neutrons; for example, ordinary carbon has 6 protons and 6 neutrons, and iron has only 30 neutrons with its 26 protons.)

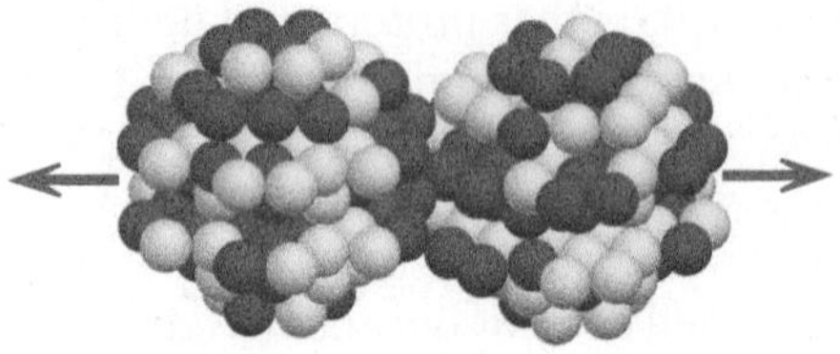

Figure 6.58 A U-235 nucleus in the process of fissioning.

Uranium-235 has an unusual property that few other nuclei have. If you hit a U-235 nucleus with a slowly moving neutron, the nucleus may deform into a dumbbell-like form and then split apart because the electric repulsion of the protons can now overcome the short-range attraction of the nucleons for each other (Figure 6.58). (In addition to the two main fission fragments there are typically one or more free neutrons in the final state.)

The fission fragments quickly acquire very high speed, and they leave the original atom's 92 electrons behind. These highly charged nuclei storm through the block of uranium metal, tearing up the material and violently heating the metal. In a nuclear power reactor the reaction is controlled in such a way that the metal merely gets hot enough to boil water. With many uranium-235 nuclei fissioning, the metal can boil water in a nuclear power plant to produce steam and power a turbine to drive an electric generator.

The masses of the two fission-fragment nuclei total less than the mass of the uranium-235 nucleus. The mass difference Δm is only about one-tenth of one percent of the original mass, but this mass change is easily measurable, and c^2 times this mass change is indeed found to be equal to the observed kinetic energy of the fission fragments. This 0.1% change is about ten million times as big an effect as the mass changes that occur in chemical reactions, which only involve rearrangements of the outer electrons in the atoms, and do not affect the nucleus.

Nuclear Waste

The problem of radioactive waste from nuclear power plants comes from the fact that the fission products are nuclei with a large overabundance of neutrons, and they emit energetic radiations. For example, uranium can fission into two palladium nuclei, each with 46 protons and 71 or 72 neutrons, whereas ordinary

$$n \rightarrow p^+ + e^- + \overline{\nu}$$

Figure 6.59 A neutron in a neutron-rich nucleus can change into a proton with emission of an electron and an antineutrino.

$$N^* \rightarrow N + \gamma$$

Figure 6.60 A nucleus in an excited state (called N^*) can drop to a lower energy state and emit a gamma ray (a high-energy photon).

stable palladium has only 60 neutrons (see the inside front cover of this book). A neutron in such a neutron-rich nucleus can change into a proton (Figure 6.59) with the emission of an energetic electron and an antineutrino (the antineutrinos almost never interact with matter, so they aren't dangerous). This process is called "beta decay" (energetic electrons were initially called "beta rays"). After a beta decay the new nucleus, in which a neutron has changed into a proton, may be in an "excited state" and emit a gamma ray (a high-energy photon, see Figure 6.60). Heavy elements such as plutonium also emit alpha particles (the nuclei of helium atoms) and there may be heavy elements mixed in with the rest of the nuclear waste.

When energetic electrons or gamma rays or alpha particles from radioactive nuclei go through matter, energy is absorbed. If the radiation goes through living tissue there can be damage to cells, and to the genetic information in the DNA in cells, through ionization of the biological molecules. The range of the radiation is relatively short and it can be stopped by a layer of concrete or other shielding. The big concern is that radioactive wastes might leak out of storage facilities, spread through the ground water, and get into the water supply. Drinking contaminated water would bring radioactive nuclei into the body, close to cells, where damage could occur.

A great deal of engineering and scientific study is under way to try to find a secure storage scheme that would prevent leakage for hundreds or even thousands of years, because some of the radioactive nuclei have very long "half-lives" (the time it takes for half of the nuclei to decay).

6.13 REFLECTION: WHY ENERGY?

It is appropriate at this point to reflect on why we introduced the topic of energy. Energy methods often make it possible to analyze some aspect of a phenomenon with much less effort than is required by direct application of the Momentum Principle. A striking example is the ability to calculate escape speed, or the impact speed of the *Ranger* spacecraft on the Moon, without carrying out a lengthy numerical integration.

Limits on the Possible

Another powerful capability of energy analyses is that they can easily predict whether some process can occur or not. For example, if there is insufficient energy, the rocket can't escape from the planet, or the atom cannot be ionized, or the biochemical reaction cannot proceed, or a proton can't decay into a neutron.

Less Detail

However, energy analyses give less detailed information than is given by a step-by-step numerical integration of the Momentum Principle. In particular, energy analyses tell us nothing about how fast a process will proceed; time does not appear in the statement of conservation of energy. Also, since the Energy Principle is a scalar principle, it doesn't tell us about change in the direction of motion.

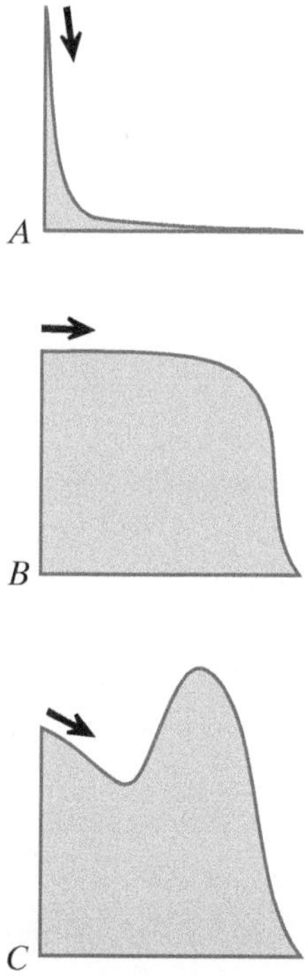

Figure 6.61 The final speed of the snowboarder will be the same at the bottom of each hill.

Here is an example that illustrates the strengths and weaknesses of energy analyses. Consider snowboarding down the three hills shown in Figure 6.61, starting from rest. If we can neglect friction, the final speed at the bottom should be the same in all three cases, because the final kinetic energy should be equal to *mgh*, the amount by which the gravitational energy decreases.

> QUESTION But how will the amount of *time* compare for the ride down the hill?

On the first hill you get going very fast very quickly, and the trip is short. On the second hill you move very slowly for a long time before dropping to the bottom, and the total trip takes a long time. On the third hill you never get to the bottom, despite the fact that the change in potential energy would be negative, because you would have to pass through a region that is energetically forbidden. So energy methods are powerful, but at the cost of losing some information, including the amount of time required for a process.

6.14 IDENTIFYING INITIAL AND FINAL STATES

Some processes involve several identifiable states, not just two. Any two states can be chosen as initial and final states; the rationale for making the choice depends on what unknown quantity you are trying to determine. In this section we offer two examples of situations involving more than two states, one simple case and one more complex process.

EXAMPLE **Throwing a Ball: Three States**

Throw a metal ball with mass 0.3 kg straight up with initial speed 10 m/s; air resistance is negligible. Consider these three states (Figure 6.62): (1) $y_1 = 7\,\text{m}$, $v_{y1} = 10\,\text{m/s}$, (2) $y_2 = ?$, $v_{y2} = 0$ (maximum height; a turning point in the motion), (3) $y_3 = 4$ m, $v_{y3} = ?$ (after coming back down, to a lower level). Calculate the maximum height y_2 and the final speed, $|v_{y3}|$.

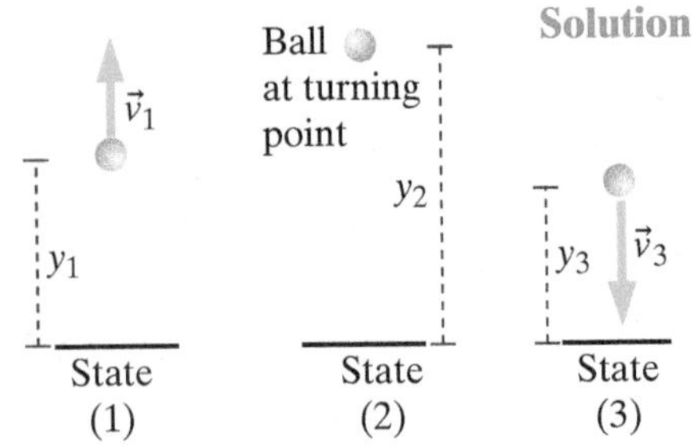

Figure 6.62 Throw a ball straight up. Find the maximum height and the final y component of the velocity.

Solution System: Ball + Earth

Initial state: 1
Final state: 2 (because the unknown is y_2)

$$K_2 + U_2 = K_1 + U_1 + W$$
$$0 + mgy_2 = K_1 + mgy_1 + 0$$
$$y_2 = y_1 + \frac{\frac{1}{2}mv_{y1}^2}{mg}$$
$$y_2 = 7\,\text{m} + \frac{\frac{1}{2}(10\,\text{m/s})^2}{9.8\,\text{N/kg}}$$
$$y_2 = 12.1\,\text{m}$$

Initial state: 2
Final state: 3 (because the unknown is v_3)

$$K_3 + U_3 = K_2 + U_2 + W$$
$$K_3 + mgy_3 = 0 + mgy_2 + 0$$
$$\frac{1}{2}mv_{y3}^2 = mg(y_2 - y_3)$$

$$v_{y3} = \pm\sqrt{2g(y_2 - y_3)}$$
$$|v_{y3}| = \sqrt{2(9.8\,\text{N/kg})(12.1 - 4)\,\text{m}}$$
$$|v_{y3}| = 12.6\,\text{m/s}$$

(See below for a discussion of the ± issue.)

QUESTION Could we have used states 1 and 3 instead of 2 and 3 to find the final speed?

Yes. We should get the same result for the final speed:

Initial state: 1
Final state: 3

$$K_3 + U_3 = K_1 + U_1 + W$$
$$K_3 + mgy_3 = K_1 + mgy_1 + 0$$
$$\frac{1}{2}mv_{y3}^2 = \frac{1}{2}mv_{y1}^2 + mg(y_1 - y_3)$$
$$v_{y3} = \pm\sqrt{v_{y1}^2 + 2g(y_1 - y_3)}$$
$$|v_{y3}| = \sqrt{(10\,\text{m/s})^2 + 2(9.8\,\text{N/kg})(7-4)\,\text{m}}$$
$$|v_{y3}| = 12.6\,\text{m/s}$$

QUESTION Aren't there two possible solutions to the quadratic that gives v_{y3}?

Yes. Unlike the Momentum Principle, the Energy Principle doesn't give us information about the direction of the velocity. In taking the square root above, v_{y3} could be either +12.6 m/s or −12.6 m/s. Which is correct? Both!

On the way down, $v_{y3} = -12.6$ m, which corresponds to the situation we analyzed. However, the mathematics claims that another possible value is $v_{y3} = +12.6$ m. This solution corresponds to the past, before the time of state 1. If the ball was thrown upward from the ground, v_y was +12.6 m/s as it rose past the height we've called $y_3 = 4$ m. It kept on rising, and v_y was +10 m/s as it rose past the height we've called $y_1 = 7$ m. Whenever you take a square root in a physics equation, you need to use physical sense to decide which root is the one you want.

QUESTION How do we decide which two states to use?

When using the Energy Principle, you get to choose which two states to use, just as you get to choose the system. The energy of one of the states must involve the unknown quantity whose value you want to find—for example, a distance or a speed. The choice of the other state is often a matter of convenience. Selecting different pairs of initial and final states often allows us to answer different questions about the process.

EXAMPLE **Reaction Between an Alpha Particle and a Carbon Nucleus**

If an energetic alpha particle (a helium nucleus: two protons and two neutrons) collides with a carbon nucleus (containing six protons and six neutrons), a nuclear reaction can occur, forming an oxygen nucleus whose rest energy is higher than an ordinary oxygen nucleus (an excited state, a concept discussed in Chapter 8). After a short time, the excited oxygen nucleus can emit its excess energy as a photon (a particle of light) called a "gamma ray" whose energy is 10.352 MeV. A photon (γ) has no rest energy, only kinetic energy. The emission of the photon reduces the rest energy of the nucleus to that of an ordinary oxygen nucleus. The oxygen nucleus recoils with momentum equal and opposite to the photon's momentum, but negligible kinetic energy.

This reaction won't occur unless the original particles have enough kinetic energy. Let's find out how much kinetic energy they must have. To simplify the analysis we will consider a situation in which the alpha particle and carbon nucleus start with equal magnitudes of momenta p_1, so the total momentum of the system is zero (Figure 6.63). The speeds involved are nonrelativistic.

Masses of the ground (not excited) states of selected nuclei, where $1\ \text{u} = 1.6605 \times 10^{-27}$ kg, are:

$$\text{He-4 } (2\text{p}+2\text{n}) = 4.00040868\ \text{u}$$
$$\text{C-12 } (6\text{p}+6\text{n}) = 11.99670852\ \text{u}$$
$$\text{O-16 } (8\text{p}+8\text{n}) = 15.99052636\ \text{u}$$

Figure 6.63 State 1: Alpha and C far apart, moving.

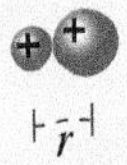

Figure 6.64 State 2: Alpha and C touch.

Figure 6.65 State 3: Excited O* nucleus.

Figure 6.66 State 4: O nucleus and photon.

A proton or neutron has a radius of roughly 1×10^{-15} m, and a nucleus is a tightly packed collection of nucleons. Experiments show that the radius of a nucleus containing N nucleons is approximately $(1.3 \times 10^{-15}\ \text{m}) \times N^{1/3}$.

(a) What is the minimum magnitude of momentum p_1 that the alpha particle must have in order for this sequence of phenomena to occur?
(b) Since both particles are positively charged, they repel each other. If the particles do not have sufficient kinetic energy to start with, they will not get close enough to touch, and the nuclear reaction will not occur. Find the minimum initial momentum p_2 each particle must have to get close enough to touch.
(c) Compare your answers to parts (a) and (b) to determine what momentum the particles must have initially to make this reaction occur.

Solution

In this process there are four identifiable states:

- State 1: Alpha particle and carbon nucleus moving toward each other (Figure 6.63)
- State 2: Alpha particle and carbon nucleus just touch (Figure 6.64)
- State 3: Excited oxygen nucleus at rest (extra rest energy) (Figure 6.65)
- State 4: Oxygen nucleus and photon, heading in opposite directions (Figure 6.66)

(a) Find p_1.

System: All particles in existence (alpha + C, or excited O, or O + photon)
Surroundings: No external objects

Initial state: State 1, Figure 6.63, because it involves the unknown momentum
Final state: State 4, Figure 6.66, because particle energies are known

$$E_f = E_i + W$$
$$(m_O c^2 + K_O) + K_\gamma + U_f = (m_\alpha c^2 + K_\alpha) + (m_C c^2 + K_C) + U_i$$
$$m_O c^2 + 0 + K_\gamma + 0 = m_\alpha c^2 + m_C c^2 + \frac{p_1^2}{2m_\alpha} + \frac{p_1^2}{2m_C} + 0$$
$$p_1 = \sqrt{\frac{(m_O - m_\alpha - m_C)c^2 + K_\gamma}{\dfrac{1}{2m_\alpha} + \dfrac{1}{2m_C}}}$$

$U_f = 0$ because the photon has no charge. $K_O \approx 0$ since there is negligible recoil kinetic energy for this comparatively massive particle. $U_i = 0$ because particles are initially far apart.

We need to use many significant figures when solving for p_1, because the mass change is very small:

$$(m_O - m_\alpha - m_C) = (15.99052636 - 4.00040868 - 11.99670852)\ \text{u} = -0.00659084\ \text{u}$$

$$(-0.00659084\ \text{u})(1.6605\times10^{-27}\frac{\text{kg}}{\text{u}})(3\times10^{8}\frac{\text{m}}{\text{s}})^2 = -9.8497\times10^{-13}\ \text{J}$$

$$K_\gamma = (10.352\times10^{6}\text{eV})(1.60\times10^{-19}\text{eV/J}) = 16.5632\times10^{-13}\ \text{J}$$

$$p_1 = \sqrt{\frac{(16.5632\times10^{-13} - 9.8497\times10^{-13})\ \text{J}}{\frac{1}{2}\left(\frac{1}{4.000}+\frac{1}{11.997}\right)\left(\frac{1}{1.6605\times10^{-27}\ \text{kg}}\right)}}$$

$$p_1 = 8.18\times10^{-20}\ \text{kg}\cdot\text{m/s}$$

(b) Find p_2.
System: alpha particle and carbon nucleus.
Surroundings: No external objects.

Initial state: State 1, Figure 6.63, because it involves the unknown momentum
Final state: State 2, Figure 6.64, because this is the final state of interest

$$E_f = E_i + W$$

$$\cancel{m_a c^2} + K_{a,f} + \cancel{m_C c^2} + K_{C,f} + U_f = \cancel{m_a c^2} + K_{a,i} + \cancel{m_C c^2} + K_{C,i} + U_i + W$$

$$0 + 0 + \frac{1}{4\pi\varepsilon_0}\frac{(2e)(6e)}{r_f} = \frac{p_2^2}{2m_a} + \frac{p_2^2}{2m_C} + 0 + 0$$

$$p_2 = \sqrt{\frac{\frac{1}{4\pi\varepsilon_0}\frac{12e^2}{r_f}}{\frac{1}{2}\left(\frac{1}{m_a}+\frac{1}{m_C}\right)}}$$

There is no change of rest energies yet, so the rest energies cancel. In the final state the particles are at rest, just touching (remember that we want to know the minimum kinetic energy required to do this).

Final center-to-center distance:

$$r_f = (1.3\times10^{-15}\ \text{m})(4^{1/3} + 12^{1/3}) = 5.04\times10^{-15}\ \text{m}$$

$$p_2 = \sqrt{\frac{\left(9\times10^{9}\frac{\text{N}\cdot\text{m}^2}{\text{C}^2}\right)\frac{12(1.6\times10^{-19}\text{C})^2}{(5.04\times10^{-15}\ \text{m})}}{\frac{1}{2}\left(\frac{1}{4}+\frac{1}{12}\right)\left(\frac{1}{1.66\times10^{-27}\ \text{kg}}\right)}}$$

$$p_2 = 7.39\times10^{-20}\ \text{kg}\cdot\text{m/s}$$

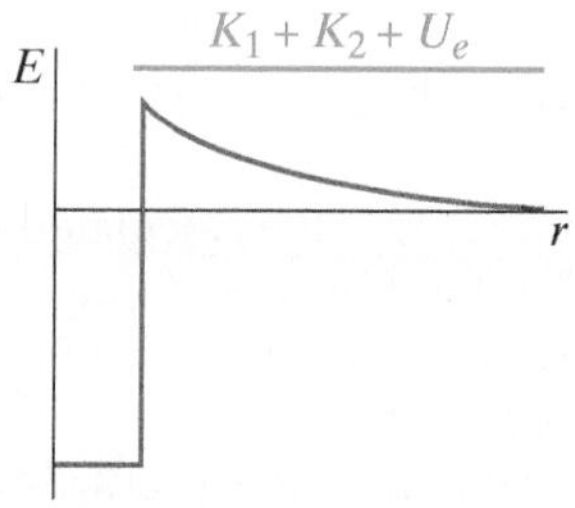

Figure 6.67 The initial kinetic energies are a bit greater than what is needed to overcome the electric potential energy barrier.

(c) The momentum required to touch is less than the momentum required to initiate the reaction.

The potential energy graph for this process is shown in Figure 6.67. The electric potential energy rises (due to a repulsive interaction) as the particles get closer together, up to the point at which they touch. After this, the attractive strong (nuclear) force dominates the interaction, and the electric repulsion between the particles is insignificant.

This kind of potential energy curve is sometimes described as having a "Coulomb barrier." This means that $K + U$ total must be larger than the maximum value of U_{elec}.

We found that the incoming momentum p_i required to create the excited oxygen nucleus is greater than the momentum p_2 required just to overcome the electric potential energy barrier and bring the alpha particle and carbon nucleus in contact. It is necessary to bring nuclei into contact for a reaction to proceed, since the nuclear force has a very short range. However, in this reaction just making contact is insufficient. A bit more energy is needed in order to produce the larger-mass oxygen nucleus.

We never did use state 3 in our analysis. However, if we needed to find the mass of the excited O* nucleus, we would use this state as either the final or initial state.

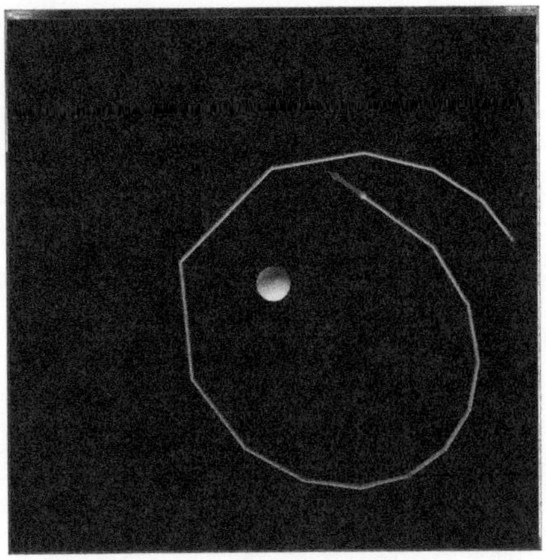

Figure 6.68 An elliptical orbit computed with a time step of two weeks.

6.15 ENERGY IN COMPUTATIONAL MODELS

In this chapter we have analyzed a wide variety of systems, involving objects as diverse as alpha particles, asteroids, and horses. We have been able to draw conclusions such as "When the objects are this far apart they will have that much kinetic energy," or "These objects do not have enough kinetic energy to get that far apart." However, we have not predicted the detailed trajectory of any moving object. We have put limits on what can happen, but we have not made detailed predictions of motion.

Because the Energy Principle involves neither direction nor time, it does not tell us which way objects are moving, or how long it will take them to get to a particular separation. However, energy can still play an important role in iterative computational models.

Figure 6.69 Energy versus time. Kinetic energy (cyan line), gravitational potential energy (green line), and the sum of $K+U$ (yellow line), as a function of time for the elliptical orbit shown in Figure 6.68 (Δt of two weeks).

Monitoring Step Size

We have seen that if we use a time step size that is too large in an iterative model, we predict trajectories that don't look right. In such computational models, tracking the kinetic and potential energy of such a system can provide important quantitative information about the performance of the model.

Figure 6.68 shows the display from a computational model of a planet orbiting a star. This model iteratively invoked the Momentum Principle, using a Δt of two weeks. The trajectory is jagged and the orbit does not close. Figure 6.69 shows plots of K_{planet}, U_g, and $(K_{\text{planet}} + U_g)$ as a function of time, computed and plotted by the same program.

> **QUESTION** What is wrong with the energy plots shown in Figure 6.69?

Figure 6.70 An elliptical orbit computed with a time step of one day.

The graph of $K + U$ in Figure 6.69 shows an apparent violation of the Energy Principle. In the model there are no objects in the surroundings, and therefore the sum of $K + U$ for the system should be constant. The fact that $K + U$ is not constant tells us that our calculations are insufficiently accurate. This is important information that we can get even without knowing what the motion of the system should really look like.

When we watch the running program, we can see that the calculated sum of $K + U$ varies most when the planet is close to the star. Since the force is largest when the planet is closest, this is the region in which the assumption of constant force over the time Δt is most problematic for a long time step.

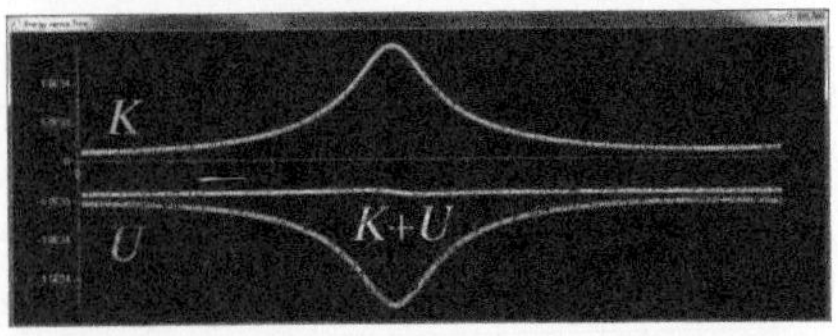

Figure 6.71 Kinetic energy (cyan line), gravitational potential energy (green line), and the sum of $K+U$ (yellow line), for an elliptical orbit computed with a time step of one day.

To confirm the hypothesis that too large a value for Δt is producing large errors in our calculations, we can try reducing the time step and re-running the program. Figure 6.70 shows the orbit predicted by the same model, using a Δt of one day instead of two weeks, and Figure 6.71 shows the associated plots of energy versus time. Although the trajectory now appears smooth, and the orbit closes, there is still a small wiggle in the graph of the sum of $K + U$. It is necessary to reduce the time step even further to get $K+U$ to remain constant, as shown in Figure 6.72, in which a time step of one hour was used.

Evidently monitoring the energy of a system in a computational model can provide a sensitive check on the accuracy of the computation.

Figure 6.72 Kinetic energy (cyan line), gravitational potential energy (green line), and the sum of $K+U$ (yellow line), for an elliptical orbit computed with a time step of one hour.

Energy Graphs in VPython

Adding graphs of energy is straightforward in VPython. In the partial program below, line numbers have been added to make it easy to refer to particular lines. In addition to the usual calculations done in a program modeling motion, which are indicated by comments, these things are done:

- Line 02 imports the graph module.
- Lines 04–06 create and assign names to three gcurve objects, one for each plot desired.
- Each time through the loop, K and U are calculated (lines 13–14).
- Each time through the loop, the statements in lines 15–17 are executed to add a point to the graph of each quantity of interest.

```
01 from visual import *
02 from visual.graph import *
03 ## create some objects ...
04 kineticenergy = gcurve(color=color.cyan)
05 potentialenergy = gcurve(color=color.green)
06 kplusu = gcurve(color=color.yellow)
07 ## set initial values...
08 while t < 60*60*24*365:
09     rate(100)
10     ## calculate forces...
11     ## update momentum...
12     ## update position...
13     K = ## calculate kinetic energy
14     U = ## calculate potential energy
15     kineticenergy.plot( pos=(t,K) )
16     potentialenergy.plot( pos=(t,U) )
17     kplusu.plot( pos=(t,K+U) )
18     t = t + delta_t
```

6.16 *A PUZZLE

The principle of conservation of energy states that when a system's energy increases, the energy of the surroundings decreases by the same amount. For example, take a block as the system of interest. If you push the block away from you and raise its kinetic energy, your store of chemical energy decreases by the same amount. The flow of energy from you (the surroundings) to the block takes the form of mechanical energy transfer we call "work." We can demonstrate that there is a puzzle concerning the energy in the surroundings of a star whose kinetic energy increases due to an interaction with another star.

Consider two stars of equal mass initially at rest, far apart (Figure 6.73). The two stars do work on each other:

Figure 6.73 Two stars with equal mass attract each other, starting from rest.

$$\Delta E_1 = W_{\text{on }1} \quad \text{star 1 as system}$$

$$\Delta E_2 = W_{\text{on }2} \quad \text{star 2 as system}$$

If we add these two energy equations for the two stars, we have this:

$$\Delta E_1 + \Delta E_2 = W_{\text{on }1} + W_{\text{on }2} \quad \text{(sum of two energy equations)}$$

Moving $(W_{\text{on}\,1} + W_{\text{on}\,2})$ to the left side of the equation and renaming the negative of this quantity ΔU, we have this familiar result:

$$\Delta E_1 + \Delta E_2 + \Delta U = 0 \quad \text{(energy equation for two stars)}$$

Our interpretation of this equation is that the gravitational potential energy U of the two-star system (calculated from the work on the individual particles) decreases as the particle energies increase. This point of view is valid, yields physically correct results, and is the view that we will take in this introductory textbook.

However, it is not difficult to show that there is a puzzle concerning energy flow between one of the stars and its surroundings. Consider star 1 as the system. Since it is a single object, there is no potential energy U, no pair-wise interaction energy. The energy of star 1 increases due to work done on it: $\Delta E_1 = E_{\text{on}\,1}$. There are very strong reasons to believe in energy conservation, so we expect that when star 1 gains an amount of energy ΔE_1 the surroundings of star 1 should lose that much energy.

> QUESTION Part of the surroundings is star 2. Does star 2 lose energy?

No! Star 2 doesn't lose energy, it *gains* energy, and of the same amount $\Delta E_2 = \Delta E_1$, due to the symmetry of the situation, with the two stars having equal mass.

What else is there in the surroundings that could experience a change in energy? The only other matter is star 2, but its energy change is inconsistent with the principle of energy conservation. Evidently our model of the world is incomplete. To be able to analyze this situation fully, we need to introduce the abstract idea of a "field." Fields, which are the subject of later chapters of this textbook, are associated with objects having mass (gravitational fields) or charge (electric and magnetic fields), and fields extend throughout all space. It is energy stored in the gravitational field that accounts for the energy in the surroundings of star 1. Fields can also have momentum, as we will see in the chapters that deal with electric and magnetic interactions.

Despite the fact that we haven't yet defined what a field is, we can determine indirectly how much the field energy must change. Since the energy in the surroundings of star 1 changes by an amount $-\Delta E_1$, and the energy of star 2 increases by an amount $+\Delta E_1$, it must be that the energy stored in the gravitational field changes by an amount $-2\Delta E_1$. It is an indication of the power of the principle of conservation of energy that we can calculate the energy change of an entity that we know nothing about! This energy change is precisely the amount of change in the potential energy of the two-star system: $\Delta U = -(\Delta E_1 + \Delta E_2) = -2\Delta E_1$.

Because the potential energy U correctly accounts for the energy changes, we don't have to consider field energy in our calculations. Our analyses in terms of U that focus on particles and their interactions, not their surroundings, are correct. However, the difficulty in accounting for the energy in the surroundings of a star suggests that eventually we will need to include fields in our models of the world.

6.17 *GRADIENT OF POTENTIAL ENERGY

Because the (negative) gradient of U is equal to the force, the force is the (negative) slope of a graph of potential energy. This means that you can mentally read the force right off a potential energy graph (Figure 6.74). Where the curve is steep, the force is large (large gradient), and where the curve is

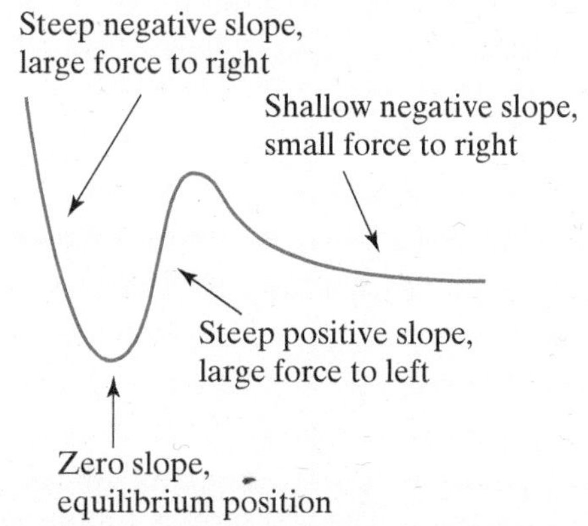

Figure 6.74 The negative of the slope (gradient) of a graph of potential energy vs. separation gives the force.

shallow, the force is small. Where the slope of the curve is horizontal, the force is zero, and this represents a possible equilibrium position for the system. The component of the force is given by the negative gradient, so a negative slope corresponds to a force to the right, while a positive slope corresponds to a force to the left.

Actually, in determining the x component of the force, we differentiate with respect to x while holding y and z constant. This kind of derivative is called a "partial derivative" and has a special symbol to emphasize that everything else is held constant:

$$F_x = -\frac{\partial U}{\partial x}$$

To get all three components of the force we take partial derivatives of the potential energy with respect to x, y, and z:

FORCE IS NEGATIVE GRADIENT OF POTENTIAL ENERGY

$$F_x = -\frac{\partial U}{\partial x}, F_y = -\frac{\partial U}{\partial y}, F_z = -\frac{\partial U}{\partial z}$$

$$\vec{F} = \left\langle -\frac{\partial U}{\partial x}, -\frac{\partial U}{\partial y}, -\frac{\partial U}{\partial z} \right\rangle$$

A standard mathematical notation for this relation is $\vec{F} = -\vec{\nabla} U$.

Checkpoint 18 With y pointing upward from the center of the Earth, find the y component of the gravitational force from $-\partial U/\partial y$, where $U = -GMm/y$. Find the x component of the force from $-\partial U/\partial x$.

6.18 *INTEGRALS AND ANTIDERIVATIVES

In determining the expression for gravitational potential energy, we started from these complementary expressions:

$$\Delta U_g = -F_r \Delta r \quad \text{and} \quad F_r = -\frac{dU_g}{dr}$$

We calculated the change in potential energy from the negative of the internal work:

$$U_{s,f} - U_{s,i} = \Delta\left(-G\frac{m_1 m_2}{r}\right) = -\Sigma F_r \Delta r$$

However, if we had let Δr approach zero, the finite sum would have turned into an integral:

$$\Delta\left(-G\frac{m_1 m_2}{r}\right) = -\int_i^f F_r dr = \int_i^f G\frac{m_1 m_2}{r^2} dr$$

You can see that a definite integral (the infinitesimal version of a finite sum) can be evaluated in terms of the antiderivative. The antiderivative of Gm_1m_2/r^2 is $-Gm_1m_2/r$, and the change in the value of the antiderivative turns out to be equal to the definite integral (the infinite sum of infinitesimal elements).

The reason why this works this way can be traced back to the fact that the force is the derivative of the energy, so the energy must be the antiderivative of the force, yet the energy is also the sum of lots of $F_r\Delta r$'s. Hence the sum is given by the change in the antiderivative.

When the quantity to be integrated (here, the force) is a simple function, it may be easy to find the antiderivative. However, in many practical cases there is no known simple function that is the antiderivative, in which case an integral must be evaluated by numerical integration, as a finite sum.

6.19 *APPROXIMATION FOR KINETIC ENERGY

An approximation for the kinetic energy of a particle at low speeds can be obtained through use of the mathematics of the "binomial expansion":

$$(1+\varepsilon)^n = 1 + n\varepsilon + \frac{n(n-1)}{2}\varepsilon^2 + \frac{n(n-1)(n-2)}{3\cdot 2}\varepsilon^3 + \cdots$$

This expression comes from working through what happens when you multiply the quantity $(1+\varepsilon)$ times itself n times. The first term is 1 multiplied by itself n times (1^n); the next term is made up of all those products, n of them, in which ε appears only once [$n\varepsilon$]; and so on.

Now for the important point that makes the binomial expansion useful in obtaining approximate results. If ε is a number that is very small compared to 1, each term in the expansion is much smaller than the preceding term:

$$n\varepsilon \text{ is much smaller than } 1, \frac{n(n-1)}{2}\varepsilon^2 \text{ is much smaller than } n\varepsilon, \text{ etc.}$$

Therefore we can get a good approximation if we keep just the first few terms and ignore the rest.

Although we have developed the binomial expansion for the case where n is an integer, it can be shown that the expression is valid even for noninteger values of n. In particular,

$$\frac{1}{\sqrt{1-\varepsilon}} = (1-\varepsilon)^{-1/2}$$
$$= 1 + \left(-\frac{1}{2}\right)(-\varepsilon) + \frac{(-\frac{1}{2})(-\frac{3}{2})}{2}(-\varepsilon)^2 + \frac{(-\frac{1}{2})(-\frac{3}{2})(-\frac{5}{2})}{6}(-\varepsilon)^3 + \cdots$$

If v/c is small compared to 1, then $(v/c)^2$ is even smaller compared to 1. For example, 1/10 is small compared to one, but $(1/10)^2 = 1/100$ is very small indeed. Therefore we can write this for low speeds:

$$\frac{mc^2}{\sqrt{1-(v/c)^2}} = mc^2\left[1 + \frac{1}{2}\left(\frac{v}{c}\right)^2 + \frac{3}{8}\left(\frac{v}{c}\right)^4 + \frac{5}{16}\left(\frac{v}{c}\right)^6 + \cdots\right]$$
$$\frac{mc^2}{\sqrt{1-(v/c)^2}} \approx mc^2 + \frac{1}{2}mv^2$$

The last term can be rewritten in terms of momentum by inserting $v \approx p/m$ into $K \approx \frac{1}{2}mv^2$:

$$K \approx \frac{1}{2}mv^2 \approx \frac{1}{2}m\left(\frac{p}{m}\right)^2 = \frac{p^2}{2m}$$

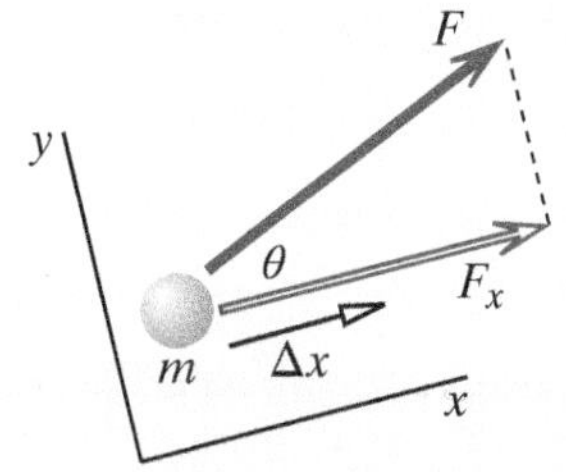

Figure 6.75 Choose the x axis to be in the direction of the motion. The net force acting on the particle has a component F_x in the direction of the motion.

6.20 *FINDING THE EXPRESSION FOR PARTICLE ENERGY

If we choose our axes so that x is in the direction of the motion at this instant (Figure 6.75) we have

$$\Delta E = F_x \Delta x = \left(\frac{\Delta p_x}{\Delta t}\right)\Delta x$$

$$\text{so } \frac{\Delta E}{\Delta x} = \frac{\Delta p_x}{\Delta t}$$

for the change in energy of a point particle acted on by a net force. In the limit of small displacement and short time interval, the energy E should satisfy the following relationship between two derivatives:

$$\frac{dE}{dx} = \frac{dp_x}{dt}$$

We will show that if we choose the following function for E, the relationship between the two derivatives will be satisfied:

$$E = \frac{mc^2}{\sqrt{1-(v/c)^2}}$$

The proof proceeds by calculating dE/dx, calculating dp_x/dt, and showing that the two derivatives are equal to each other. First, here is dE/dx:

$$\frac{dE}{dx} = \frac{d}{dx}\left(\frac{mc^2}{\sqrt{1-(v/c)^2}}\right)$$

$$= \frac{mc^2}{(1-(v/c)^2)^{3/2}}\left(-\frac{1}{2}\right)\left(-\frac{2v}{c^2}\right)\frac{dv}{dx}$$

$$\frac{dE}{dx} = \frac{mv}{(1-(v/c)^2)^{3/2}}\frac{dv}{dx}$$

Next we calculate dp_x/dt, keeping in mind that $v_x = v$, because we chose our axes so that x is in the direction of the motion at this instant:

$$\frac{dp_x}{dt} = \frac{d}{dt}\left(\frac{mv}{\sqrt{1-(v/c)^2}}\right)$$

$$= \frac{m}{\sqrt{1-(v/c)^2}}\frac{dv}{dt} + \frac{mv}{(1-(v/c)^2)^{3/2}}\left(-\frac{1}{2}\right)\left(-\frac{2v}{c^2}\right)\frac{dv}{dt}$$

$$\frac{dp_x}{dt} = \frac{m}{(1-(v/c)^2)^{3/2}}\left[\left(1-\frac{v^2}{c^2}\right)+\frac{v^2}{c^2}\right]\frac{dv}{dt}$$

$$= \frac{m}{(1-(v/c)^2)^{3/2}}\frac{dv}{dt}$$

This is almost in the same form as the derivative dE/dx. Now note this:

$$\frac{dv}{dt} = \left(\frac{dv}{dt}\right)\left(\frac{dx}{dx}\right)$$

$$= \left(\frac{dv}{dx}\right)\left(\frac{dx}{dt}\right) = \frac{dv}{dx}v \quad \text{because } v = \frac{dx}{dt}$$

Substituting this result for dv/dt into the derivative dp_x/dt, we find that this derivative is indeed equal to the derivative dE/dx if E is defined as

$$E = \frac{mc^2}{\sqrt{1-(v/c)^2}}$$

Hence the relativistic expression for energy does follow from the relativistic expression for momentum.

It might seem that we could define the particle energy with an additive constant, which would drop out when we took the derivative dE/dx. However, when one considers a variety of particle reactions, it is found that to account for energy in a consistent way requires that the additive constant be zero.

This approach to solving a "differential equation" (an equation containing derivatives), by proposing a solution and showing that it satisfies the equation, is a legitimate technique for solving differential equations, and in fact is used quite frequently.

6.21 *FINDING AN ANGLE FROM THE DOT PRODUCT

The alternative forms of the dot product offer a way to determine the angle between any two given vectors $\vec{a} = \langle a,b,c\rangle$ and $\vec{q} = \langle q,r,s\rangle$.

$$\vec{a}\bullet\vec{q} = (aq+br+cs) \quad \text{and also} \quad \vec{a}\bullet\vec{q} = |\vec{a}||\vec{q}|\cos\theta$$

So $\cos\theta = \dfrac{(aq+br+cs)}{|\vec{a}||\vec{q}|}$, from which θ can be determined.

SUMMARY

The unit of energy is the joule, abbreviated J.

Energy of a single particle

The energy of a particle with mass m is:

$$E_{\text{particle}} \equiv \gamma mc^2 \quad \text{where } \gamma = \frac{1}{\sqrt{1-(v/c)^2}}$$

Particle energy has two parts: rest energy and kinetic energy:

$$E_{\text{particle}} = mc^2 + K$$

The rest energy is the energy of a particle at rest.

$$E_{\text{rest}} = mc^2$$

The kinetic energy K of a particle is the energy a moving particle has in addition to its rest energy.

$$K = \gamma mc^2 - E_{\text{rest}}$$

For a particle whose speed is low compared to the speed of light, kinetic energy is approximately:

$$K \approx \frac{1}{2}mv^2 = \frac{p^2}{2m} \quad (\text{if } v \ll c)$$

For a particle with any speed:

$$E^2 - (pc)^2 = (mc^2)^2$$

For a massless particle this reduces to

$$E = pc$$

The Energy Principle

$$\Delta E_{\text{sys}} = W_{\text{surr}}$$

The change in energy of a system is equal to the work done by the surroundings.

Work

Work W_F done by a single constant force $\vec{F}$:

$$W_F = \vec{F}\bullet\Delta\vec{r} = F_x\Delta r_x + F_y\Delta r_y + F_z\Delta r_z$$
$$= F\Delta r\cos\theta$$

Work done by a varying force:

$$W_F = \Sigma\vec{F}\bullet\Delta\vec{r} \quad \text{or} \quad W = \int_i^f \vec{F}\bullet d\vec{r}$$

Work can be positive, negative, or zero. The sign of work depends on the relative directions of the force and the displacement of the point of application of the force.

Conservation of energy

$$\Delta E_{\text{sys}} + \Delta E_{\text{surr}} = 0$$

Potential energy
Multiparticle system energy consists of particle energy and potential energy:

$$E_{\text{sys}} \equiv (K_1 + m_1c^2) + (K_2 + m_2c^2) + \cdots + (U_{12} + \cdots)$$

Potential energy U is the energy of interaction of a pair of objects, and change in U is the negative of work done by internal forces:

$$\Delta U = -W_{\text{int}}$$

Force is the negative gradient of potential energy:

$$F_x = -\frac{dU}{dx}$$

Potential energy U depends on the separation r between two objects.
Potential energy U goes to zero as objects get very far apart:

$$U \to 0 \quad \text{as } r \to \infty$$

For an attractive interaction, U becomes negative as separation decreases.

For a repulsive interaction, U becomes positive as separation decreases.
The potential energy change between two states is independent of the path taken. The change in potential energy in a round trip is zero.

Gravitational potential energy

$$U = -G\frac{m_1m_2}{r}$$

$$G = 6.7 \times 10^{-11}\text{N} \cdot \text{m}^2/\text{kg}^2$$

m_1 and m_2 are the masses of the two interacting objects, and r is the center-to-center distance between them.

$$\Delta U \approx \Delta(mgy) \quad \text{near Earth's surface}$$

Electric potential energy

$$U_{\text{elec}} = \frac{1}{4\pi\varepsilon_0}\frac{q_1q_2}{r}$$

$$\frac{1}{4\pi\varepsilon_0} = 9 \times 10^9 \text{ N} \cdot \text{m}^2/\text{C}^2$$

q_1 and q_2 are the charges of the two interacting particles, in coulombs, and r is the center-to-center distance between them.

Energy graphs are highly useful tools.

Nuclear binding
Heavy nuclei can fission, light nuclei can fuse, with energy released in both cases.

$$\text{Electron volt: } 1 \text{ eV} = 1.6 \times 10^{-19} \text{ J}$$

QUESTIONS

Q1 Show that the units of $p^2/(2m)$ and mc^2 are indeed joules. (Note that 1 N is 1 kg · (m/s)^2.)

Q2 Give brief explanations for your answers to each of the following questions: **(a)** You hold a 1 kg book in your hand for 1 min. How much work do you do on the book? **(b)** In a circular pendulum, how much work is done by the string on the mass in one revolution? **(c)** For a mass oscillating horizontally on a spring, how much work is done by the spring on the mass in one complete cycle? In a half cycle?

Q3 You pull a block of mass m across a frictionless table with a constant force. You also pull with an equal constant force a block of larger mass M. The blocks are initially at rest. If you pull the blocks through the same distance, which block has the greater kinetic energy, and which block has the greater momentum? If instead you pull the blocks for the same amount of time, which block has the greater kinetic energy, and which block has the greater momentum?

Q4 One often hears the statement, "Nuclear energy production is fundamentally different from chemical energy production (such as burning of coal) because the nuclear case involves a change of mass." Critique this statement. Discuss the similarities and differences of the two kinds of energy production.

Q5 Figure 6.76 shows the path of a comet orbiting a star.

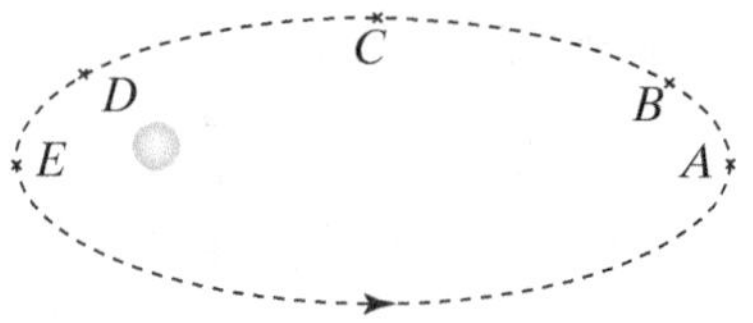

Figure 6.76

(a) Rank-order the locations on the path in terms of the magnitude of the comet's momentum at each location, starting with the location at which the magnitude of the momentum is the largest. **(b)** Rank-order the locations on the path in terms of the comet's kinetic energy at each location, starting with the location at which the kinetic energy is the largest. **(c)** Rank-order the locations on the path in terms of the potential energy of the system at each location, largest first. **(d)** Rank-order the locations on the path in terms of the sum of the kinetic energy and the potential energy of the system at each location, largest first.

Q6 Figure 6.77 is a graph of the energy of a system of a planet interacting with a star. The gravitational potential energy U_g is shown as the thick curve, and plotted along the vertical axis are various values of $K+U_g$.

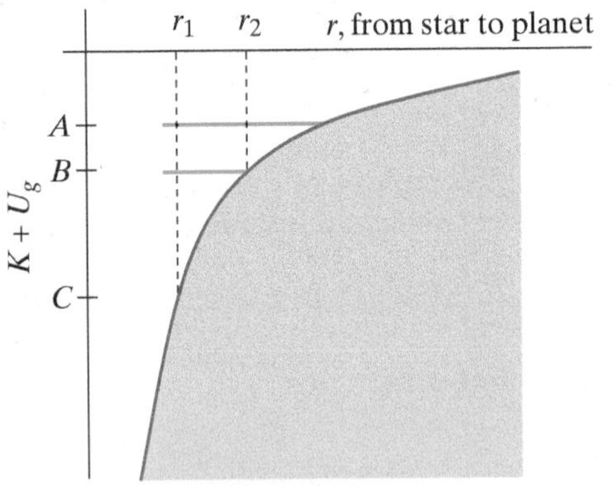

Figure 6.77

Suppose that $K+U_g$ of the system is A. Which of the following statements are true? **(a)** The potential energy of the system decreases as the planet moves from r_1 to r_2. **(b)** When the separation between the two bodies is r_2, the kinetic energy of the system is $(A-B)$. **(c)** The system is a bound system; the planet can never escape. **(d)** The planet will escape. **(e)** When the separation between the two bodies is r_2, the kinetic energy of the system is $(B-C)$. **(f)** The kinetic energy of the system is greater when the distance between the star and planet is r_1 than when the distance between the two bodies is r_2.

Suppose instead that $K+U_g$ of the system is B. Which of the following statements are true? **(a)** When the separation between the planet and star is r_2, the kinetic energy of the system is zero. **(b)** The planet and star cannot get farther apart than r_2. **(c)** This is not a bound system; the planet can escape. **(d)** When the separation between the planet and star is r_2, the potential energy of the system is zero.

Q7 A particle moves inside a circular glass tube under the influence of a tangential force of constant magnitude F (Figure 6.78). Explain why we cannot associate a potential energy with this force. How is this situation different from the case of a block on the end of a string, which is swung in a circle?

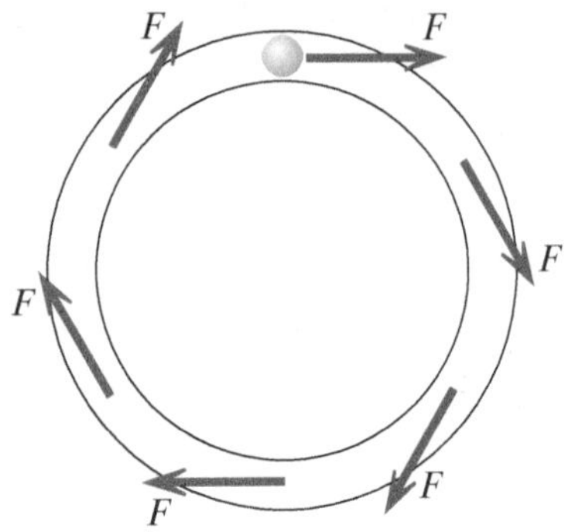

Figure 6.78

Q8 Show the validity of the relation $E^2_{\text{particle}} - (pc)^2 = (mc^2)^2$ when $m \neq 0$, by making these substitutions:

$$E_{\text{particle}} = \frac{mc^2}{\sqrt{1-(v/c)^2}} \quad \text{and} \quad p = \frac{mv}{\sqrt{1-(v/c)^2}}$$

PROBLEMS

Section 6.2

•P9 It is not very difficult to accelerate an electron to a speed that is 99% of the speed of light, because the electron has such a very small mass. What is the ratio of the kinetic energy K to the rest energy mc^2 in this case? In the definition of what we mean by kinetic energy K, $E = mc^2 + K$, you must use the full relativistic expression for E, because v/c is not small compared to 1.

•P10 A pitcher can throw a baseball at about 100 mi/h (about 44 m/s). What is the ratio of the kinetic energy to the rest energy mc^2? (Can you use $K \approx \frac{1}{2}mv^2$?)

•P11 What is the speed of an electron whose total energy is equal to the total energy of a proton that is at rest? What is the kinetic energy of this electron?

•P12 The point of this question is to compare rest energy and kinetic energy at low speeds. A baseball is moving at a speed of 17 m/s. Its mass is 145 g (0.145 kg). **(a)** What is its rest energy? **(b)** Is it okay to calculate its kinetic energy using the expression $\frac{1}{2}mv^2$? **(c)** What is its kinetic energy? **(d)** Which is true? A. The kinetic energy is approximately equal to the rest energy. B. The kinetic energy is much bigger than the rest energy. C. The kinetic energy is much smaller than the rest energy.

•P13 The point of this question is to compare rest energy and kinetic energy at high speeds. An alpha particle (a helium nucleus) is moving at a speed of 0.9993 times the speed of light. Its mass is 6.40×10^{-27} kg. **(a)** What is its rest energy? **(b)** Is it okay to calculate its kinetic energy using the expression $\frac{1}{2}mv^2$? **(c)** What is its kinetic energy? **(d)** Which is true? A. The kinetic energy is approximately equal to the rest energy. B. The kinetic energy is much bigger than the rest energy. C. The kinetic energy is much smaller than the rest energy.

•P14 A runner whose mass is 60 kg runs in the $+x$ direction at a speed of 7 m/s. **(a)** What is the kinetic energy of the runner? **(b)** The runner turns around and runs in the $-x$ direction at the same speed. Now what is the kinetic energy of the runner?

•P15 A baseball of mass 144 g has a velocity of $\langle 22, 23, -11 \rangle$ m/s. What is the kinetic energy of the baseball?

•P16 One mole of helium atoms has a mass of 4 grams. If a helium atom in a balloon has a kinetic energy of 1.437×10^{-21} J, what is the speed of the helium atom? (The speed is much lower than the speed of light.)

•P17 You throw a ball of mass 160 g upward (Figure 6.79). When the ball is 2 m above the ground, headed upward (the initial state), its speed is 19 m/s. Later, when the ball is again 2 m above the ground, this time headed downward (the final state), its speed is 19 m/s. What is the change in the kinetic energy of the ball from initial to final state?

Figure 6.79

•P18 A fan cart of mass 0.8 kg initially has a velocity of $\langle 0.9, 0, 0\rangle$ m/s. Then the fan is turned on, and the air exerts a constant force of $\langle -0.3, 0, 0\rangle$ N on the cart for 1.5 s. **(a)** What is the change in momentum of the fan cart over this 1.5 s interval? **(b)** What is the change in kinetic energy of the fan cart over this 1.5 s interval?

Section 6.3

•P19 You push a crate 3 m across the floor with a 40 N force whose direction is 30° below the horizontal. How much work do you do?

•P20 You pull your little sister across a flat snowy field on a sled. Your sister plus the sled have a mass of 20 kg. The rope is at an angle of 35 degrees to the ground. You pull a distance of 50 m with a force of 30 N. How much work do you do?

•P21 A 2-kg ball rolls off a 30-m-high cliff, and lands 25 m from the base of the cliff. Express the displacement and the gravitational force in terms of vectors and calculate the work done by the gravitational force. Note that the gravitational force is $\langle 0, -mg, 0\rangle$, where g is a *positive* number (+9.8 N/kg).

•P22 A boat is coasting toward a dock you're standing on, and as it comes toward you, you push back on it with a force of 300 N. As you do this, you back up a distance of 2 m. How much work do you do on the boat?

•P23 A jar of honey with a mass of 0.5 kg is nudged off the kitchen counter and falls 1 m to the floor. What force acts on the jar during its fall? How much work is done by this force?

•P24 An object that is originally at location $\langle -17, 0, 0\rangle$ m moves to location $\langle -25, 0, 0\rangle$ m as shown in Figure 6.80. While it is moving it is acted on by a constant force of $\langle 22, 0, 0\rangle$ N.

Figure 6.80

(a) How much work is done on the object by this force? **(b)** Does the kinetic energy of the object increase or decrease?

Figure 6.81

(c) A different object moves from location $\langle -25, 0, 0\rangle$ m to location $\langle -17, 0, 0\rangle$ m, as shown in Figure 6.81. While it is moving it is acted on by a constant force of $\langle 22, 0, 0\rangle$ N. How much work is done on the second object by this force? **(d)** Does the kinetic energy of the object increase or decrease?

•P25 A constant force $\langle 23, -12, 32\rangle$ N acts through a displacement $\langle 0.12, 0.31, -0.24\rangle$ m. How much work does this force do?

•P26 One end of a spring whose spring constant is 20 N/m is attached to the wall, and you pull on the other end, stretching it from its equilibrium length of 0.2 m to a length of 0.3 m. Estimate the work done by dividing the stretching process into two stages and using the average force you exert to calculate work done during each stage.

•P27 An electron traveling through a curving wire in an electric circuit experiences a constant force of 5×10^{-19} N, always in the direction of its motion through the wire. How much work is done on the electron by this force as it travels through 0.5 m of the wire?

•P28 You bring a boat toward the dock by pulling on a rope with a force of 130 N through a distance of 6 m. **(a)** How much work do you do? (Include the appropriate sign.) **(b)** Then you slow the boat down by pushing against it with a force of 40 N, opposite to the boat's movement of 5 m. How much work do you do? (Include the appropriate sign.) **(c)** What is the total amount of work that you do?

••P29 You push a box out of a carpeted room and along a hallway with a waxed linoleum floor. While pushing the crate 2 m out of the room you exert a force of 34 N; while pushing it 6 m along the hallway you exert a force of 13 N. To slow it down you exert a force of 40 N through a distance of 2 m, opposite to the motion. How much work do you do in all?

Section 6.4

•P30 A ball of mass 0.7 kg falls downward, as shown in Figure 6.82. Initially you observe it to be 4.5 m above the ground. After a short time it is just about to hit the ground.

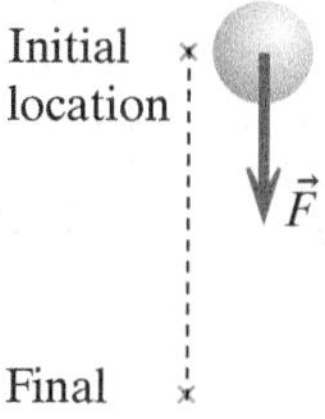

Figure 6.82

(a) During this interval how much work was done on the ball by the gravitational force? **(b)** Does the kinetic energy of the ball increase or decrease?

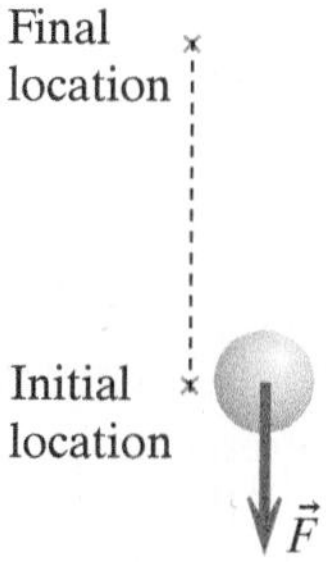

Figure 6.83

(c) The ball hits the ground and bounces back upward, as shown in Figure 6.83. After a short time it is 4.5 m above the ground again. During this second interval (between leaving the ground and reaching a height of 4.5 m) how much work was done on the ball by the gravitational force? **(d)** Does the kinetic energy of the ball increase or decrease?

•P31 A lithium nucleus has mass 5.1×10^{-27} kg. If its speed is 0.984c (that is, $|\vec{v}|/c = 0.984$), what are the values of the particle energy, the rest energy, and the kinetic energy?

Next an electric force acts on the lithium nucleus and does 4.7×10^{-9} J of work on the particle. Now what are the values of the particle energy, the rest energy, and the kinetic energy?

•P32 A space probe in outer space has a mass of 111 kg, and it is traveling at a speed of 29 m/s. When it is at location $\langle 445, 535, -350\rangle$ m it begins firing two booster rockets. The rockets exert constant forces of $\langle 90, 150, 195\rangle$ N and $\langle 90, -90, -585\rangle$ N, respectively. The rockets fire until the spacecraft reaches location $\langle 449, 541, -354\rangle$ m. Now what is its speed? There is negligible mass loss due to the rocket exhaust.

••P33 An object with mass 100 kg moved in outer space. When it was at location $\langle 9, -24, -4\rangle$ its speed was 3.5 m/s. A single constant force $\langle 250, 400, -170\rangle$ N acted on the object while the object moved from location $\langle 9, -24, -4\rangle$ m to location $\langle 15, -17, -8\rangle$ m. Then a different single constant force $\langle 140, 250, 150\rangle$ N acted on the object while the object moved from location $\langle 15, -17, -8\rangle$ m to location $\langle 19, -24, -3\rangle$ m. What is the speed of the object at this final location?

••P34 Outside the space shuttle, you and a friend pull on two ropes to dock a satellite whose mass is 700 kg. The satellite is initially at position $\langle 3.5, -1, 2.4\rangle$ m and has a speed of 4 m/s. You exert a force $\langle -400, 310, -250\rangle$ N. When the satellite reaches the position $\langle 7.1, 3.2, 1.2\rangle$ m, its speed is 4.01 m/s. How much work did your friend do?

••P35 A crate with a mass of 100 kg glides through a space station with a speed of 3.5 m/s. An astronaut speeds it up by pushing on it from behind with a force of 180 N, continually pushing with this force through a distance of 6 m. The astronaut moves around to the front of the crate and slows the crate down by pushing backward with a force of 170 N, backing up through a distance of 4 m. After these two maneuvers, what is the speed of the crate?

••P36 An object with mass 120 kg moved in outer space from location $\langle 11, -24, -5\rangle$ to location $\langle 17, -28, -10\rangle$ m. A single constant force $\langle 250, 440, -220\rangle$ N acted on the object while the object moved. Its final speed was 7.9 m/s. What was the speed of the object at the initial location?

••P37 An object with mass 120 kg moved in outer space. When it was at location $\langle 7, -20, -8\rangle$ its speed was 12 m/s. A single constant force $\langle 250, 490, -160\rangle$ N acted on the object while the object moved to location $\langle 10, -29, -13\rangle$ m. What is the speed of the object at this final location?

••P38 Jack and Jill are maneuvering a 3000 kg boat near a dock. Initially the boat's position is $\langle 2, 0, 3\rangle$m and its speed is 1.3 m/s. As the boat moves to position $\langle 4, 0, 2\rangle$m, Jack exerts a force $\langle -400, 0, 200\rangle$ N and Jill exerts a force $\langle 150, 0, 300\rangle$ N. **(a)** How much work does Jack do? **(b)** How much work does Jill do? **(c)** Without doing any calculations, say what is the angle between the (vector) force that Jill exerts and the (vector) velocity of the boat. Explain briefly how you know this. **(d)** Assuming that we can neglect the work done by the water on the boat, what is the final speed of the boat?

••P39 An electron traveling at a speed $0.99c$ encounters a region where there is a constant electric force directed opposite to its momentum. After traveling 3 m in this region, the electron's speed was observed to decrease to $0.93c$. What was the magnitude of the electric force acting on the electron?

••P40 A mass of 0.12 kg hangs from a vertical spring in the lab room. You grab hold of the mass and throw it vertically downward. The speed of the mass just after leaving your hand is 3.40 m/s. **(a)** While the mass moved downward a distance of 0.07 m, how much work was done on the mass by the Earth? **(b)** The speed of the mass has decreased to 2.85 m/s. How much work was done on the mass by the spring?

••P41 An electron (mass 9×10^{-31} kg) is traveling at a speed of $0.91c$ in an electron accelerator. An electric force of 1.6×10^{-13} N is applied in the direction of motion while the electron travels a distance of 2 m. You need to find the new speed of the electron. Which of the following steps must be included in your solution to this problem? **(a)** Calculate the initial particle energy $\gamma_i mc^2$ of the electron. **(b)** Calculate the final particle energy $\gamma_f mc^2$ of the electron. **(c)** Determine how much time it takes to move this distance. **(d)** Use the expression $\frac{1}{2}m|\vec{v}|^2$ to find the kinetic energy of the electron. **(e)** Calculate the net work done on the electron. **(f)** Use the final energy of the electron to find its final speed.

What is the new speed of the electron as a fraction of c?

••P42 SLAC, the Stanford Linear Accelerator Center, located at Stanford University in Palo Alto, California, accelerates electrons through a vacuum tube 2 mi long (it can be seen from an overpass of the Junipero Serra freeway that goes right over the accelerator). Electrons that are initially at rest are subjected to a continuous force of 2×10^{-12} N along the entire length of two mi (1 mi is 1.6 km) and reach speeds very near the speed of light. A similar analysis in a previous chapter required numerical integration, but with the new techniques of this chapter you can analyze the motion analytically. **(a)** Calculate the final energy, momentum, and speed of the electron. **(b)** Calculate the approximate time required to go the 2 mi distance.

Section 6.5

•P43 In positron-emission tomography (PET) used in medical research and diagnosis, compounds containing unstable nuclei that emit positrons are introduced into the brain, destined for a site of interest in the brain. When a positron is emitted, it goes only a short distance before coming nearly to rest. It forms a bound state with an electron, called "positronium," which is rather similar to a hydrogen atom. The binding energy of positronium is very small compared to the rest energy of an electron. After a short time the positron and electron annihilate. In the annihilation, the positron and the electron disappear, and all of their rest energy goes into two photons (particles of light) that have zero mass; all their energy is kinetic energy. These high-energy photons, called "gamma rays," are emitted at nearly 180° to each other. What energy of gamma ray (in MeV, million electron volts) should each of the detectors be made sensitive to? (The mass of an electron or positron is 9×10^{-31} kg.)

••P44 A nucleus whose mass is 3.499612×10^{-25} kg undergoes spontaneous alpha decay. The original nucleus disappears and there appear two new particles: a He-4 nucleus of mass 6.640678×10^{-27} kg (an "alpha particle" consisting of two protons and two neutrons) and a new nucleus of mass

3.433132×10^{-25} kg (note that the new nucleus has less mass than the original nucleus, and it has two fewer protons and two fewer neutrons). **(a)** When the alpha particle has moved far away from the new nucleus (so the electric interactions are negligible), what is the combined kinetic energy of the alpha particle and new nucleus? **(b)** How many electron volts is this? In contrast to this nuclear reaction, chemical reactions typically involve only a few eV.

••P45 In a location in outer space far from all other objects, a nucleus whose mass is 3.894028×10^{-25} kg and that is initially at rest undergoes spontaneous alpha decay. The original nucleus disappears, and two new particles appear: a He-4 nucleus of mass 6.640678×10^{-27} kg (an alpha particle consisting of two protons and two neutrons) and a new nucleus of mass 3.827555×10^{-25} kg. These new particles move far away from each other, because they repel each other electrically (both are positively charged).

Because the calculations involve the small difference of (comparatively) large numbers, you need to keep seven significant figures in your calculations, and you need to use the more accurate value for the speed of light, 2.99792e8 m/s.

Choose all particles as the system. Initial state: Original nucleus, at rest. Final state: Alpha particle + new nucleus, far from each other. **(a)** What is the rest energy of the original nucleus? *Give seven significant figures.* **(b)** What is the sum of the rest energies of the alpha particle and the new nucleus? *Give seven significant figures.* **(c)** Did the portion of the total energy of the system contributed by rest energy increase or decrease? **(d)** What is the sum of the kinetic energies of the alpha particle and the new nucleus?

••P46 A proton (1.6726×10^{-27} kg) and a neutron (1.6749×10^{-27} kg) at rest combine to form a deuteron, the nucleus of deuterium or "heavy hydrogen." In this process, a gamma ray (high-energy photon) is emitted, and its energy is measured to be 2.2 MeV (2.2×10^6 eV). **(a)** Keeping all five significant figures, what is the mass of the deuteron? Assume that you can neglect the small kinetic energy of the recoiling deuteron. **(b)** Momentum must be conserved, so the deuteron must recoil with momentum equal and opposite to the momentum of the gamma ray. Calculate approximately the kinetic energy of the recoiling deuteron and show that it is indeed small compared to the energy of the gamma ray.

••P47 Many heavy nuclei undergo spontaneous "alpha decay," in which the original nucleus emits an alpha particle (a helium nucleus containing two protons and two neutrons), leaving behind a "daughter" nucleus that has two fewer protons and two fewer neutrons than the original nucleus. Consider a radium-220 nucleus that is at rest before it decays to radon-216 by alpha decay.

The mass of the radium-220 nucleus is 219.96274 u (unified atomic mass units) where $1\text{ u} = 1.6603 \times 10^{-27}$ kg (approximately the mass of one nucleon).

The mass of a radon-216 nucleus is 215.95308 u, and the mass of an alpha particle is 4.00151 u. Radium has 88 protons, radon has 86 protons, and an alpha particle has 2 protons. **(a)** Make a diagram of the final state of the radon-216 nucleus and the alpha particle when they are far apart, showing the momenta of each particle to the same relative scale. Explain why you drew the lengths of the momentum vectors the way you did. **(b)** Calculate the final kinetic energy of the alpha particle. For the moment, assume that its speed is small compared to the speed of light. **(c)** Calculate the final kinetic energy of the radon-216 nucleus. **(d)** Show that the nonrelativistic approximation was reasonable.

•••P48 A nucleus whose mass is 3.917268×10^{-25} kg undergoes spontaneous alpha decay. The original nucleus disappears and there appear two new particles: a He-4 nucleus of mass 6.640678×10^{-27} kg (an alpha particle consisting of two protons and two neutrons) and a new nucleus of mass 3.850768×10^{-25} kg. (Note that the new nucleus has less mass than the original nucleus, and it has two fewer protons and two fewer neutrons.) **(a)** What is the total kinetic energy of the alpha particle and the new nucleus? **(b)** Use the Conservation of Momentum in order to determine the kinetic energy of the alpha particle and the kinetic energy of the new nucleus.

Section 6.8

•P49 You throw a ball straight up, and it reaches a height of 20 m above your hand before falling back down. What was the speed of the ball just after it left your hand?

•P50 A 1 kg block rests on the Earth's surface. How much energy is required to move the block very far from the Earth, ending up at rest again?

•P51 An object with mass 7 kg moves from a location $\langle 22, 43, -41 \rangle$ m near the Earth's surface to location $\langle -27, 11, 46 \rangle$ m. What is the change in the potential energy of the system consisting of the object plus the Earth?

•P52 The radius of the Moon is 1750 km, and its mass is 7×10^{22} kg. What would be the escape speed from an isolated Moon? Why was a small rocket adequate to lift the lunar astronauts back up from the surface of the Moon?

•P53 Use energy conservation to find the approximate final speed of a basketball dropped from a height of 2 m (roughly the height of a professional basketball player). Why don't you need to know the mass of the basketball?

•P54 Under certain conditions the interaction between a "polar" molecule such as HCl located at the origin and an ion located along the x axis can be described by a potential energy $U = -b/x^2$, where b is a constant. What is F_x, the x component of the force on the ion? What is F_y, the y component of the force on the ion?

•P55 **(a)** A 0.5 kg teddy bear is nudged off a window sill and falls 2 m to the ground. What is its kinetic energy at the instant it hits the ground? What is its speed? What assumptions or approximations did you make in this calculation? **(b)** A 1.0 kg flowerpot is nudged off a window sill and falls 2 m to the ground. What is its kinetic energy at the instant it hits the ground? What is its speed? How do the speed and kinetic energy compare to that of the teddy bear in part (a)?

•P56 You throw a ball of mass 1.2 kg straight up. You observe that it takes 3.1 s to go up and down, returning to your hand. Assuming we can neglect air resistance, the time it takes to go up to the top is half the total time, 1.55 s. Note that at the top the momentum is momentarily zero, as it changes from heading upward to heading downward. **(a)** Use the Momentum Principle to determine the speed that the ball had just after it left your hand. **(b)** Use the Energy Principle to determine the maximum height above your hand reached by the ball.

•P57 Suppose that a pitcher can throw a ball straight up at 100 mi/h (about 45 m/s). Use energy conservation to calculate how high the baseball goes. Explain your work. Actually, a pitcher can't attain this high a speed when throwing straight up, so your result will be an overestimate of what a human can do; air resistance also reduces the achievable height.

••P58 The radius of Mars (from the center to just above the atmosphere) is 3400 km (3400×10^3 m), and its mass is 0.6×10^{24} kg. An object is launched straight up from just above the atmosphere of Mars. **(a)** What initial speed is needed so that when the object is far from Mars its final speed is 1000 m/s? **(b)** What initial speed is needed so that when the object is far from Mars its final speed is 0 m/s? (This is called the escape speed.)

••P59 The radius of an airless planet is 2000 km (2×10^6 m), and its mass is 1.2×10^{23} kg. An object is launched straight up from the surface. **(a)** What initial speed is needed so that when the object is far from the planet its final speed is 900 m/s? **(b)** What initial speed is needed so that when the object is far from the planet its final speed is 0 m/s? (This is called the escape speed.)

••P60 The escape speed from an asteroid whose radius is 10 km is only 10 m/s. If you throw a rock away from the asteroid at a speed of 20 m/s, what will be its final speed?

••P61 The escape speed from a very small asteroid is only 24 m/s. If you throw a rock away from the asteroid at a speed of 35 m/s, what will be its final speed?

••P62 Calculate the speed of a satellite in a circular orbit near the Earth (just above the atmosphere). If the mass of the satellite is 200 kg, what is the minimum energy required to move the satellite from this near-Earth orbit to very far away from the Earth?

••P63 A spacecraft is coasting toward Mars. The mass of Mars is 6.4×10^{23} kg and its radius is 3400 km (3.4×10^6 m). When the spacecraft is 7000 km (7×10^6 m) from the center of Mars, the spacecraft's speed is 3000 m/s. Later, when the spacecraft is 4000 km (4×10^6 m) from the center of Mars, what is its speed? Assume that the effects of Mars's two tiny moons, the other planets, and the Sun are negligible. Precision is required to land on Mars, so make an accurate calculation, not a rough, approximate calculation.

••P64 A comet is in an elliptical orbit around the Sun. Its closest approach to the Sun is a distance of 4×10^{10} m (inside the orbit of Mercury), at which point its speed is 8.17×10^4 m/s. Its farthest distance from the Sun is far beyond the orbit of Pluto. What is its speed when it is 6×10^{12} m from the Sun? (This is the approximate distance of Pluto from the Sun.)

••P65 An electron is traveling at a speed of $0.95c$ in an electron accelerator. An electric force of 1.6×10^{-13} N is applied in the direction of motion while the electron travels a distance of 2 m. What is the new speed of the electron?

••P66 You stand on a spherical asteroid of uniform density whose mass is 2×10^{16} kg and whose radius is 10 km (10^4 m). These are typical values for small asteroids, although some asteroids have been found to have much lower average density and are thought to be loose agglomerations of shattered rocks. **(a)** How fast do you have to throw the rock so that it never comes back to the asteroid and ends up traveling at a speed of 3 m/s when it is very far away? **(b)** Sketch graphs of the kinetic energy of the rock, the gravitational potential energy of the rock plus asteroid, and their sum, as a function of separation (distance from center of asteroid to rock). Label the graphs clearly.

•••P67 In the rough approximation that the density of the Earth is uniform throughout its interior, the gravitational field strength (force per unit mass) inside the Earth at a distance r from the center is gr/R, where R is the radius of the Earth. (In actual fact, the outer layers of rock have lower density than the inner core of molten iron.) Using the uniform-density approximation, calculate the amount of energy required to move a mass m from the center of the Earth to the surface. Compare with the amount of energy required to move the mass from the surface of the Earth to a great distance away.

••P68 This problem is closely related to the spectacular impact of the comet Shoemaker-Levy with Jupiter in July 1994:

http://www.jpl.nasa.gov/sl9/sl9.html

A rock far outside our Solar System is initially moving very slowly relative to the Sun, in the plane of Jupiter's orbit around the Sun. The rock falls toward the Sun, but on its way to the Sun it collides with Jupiter. Calculate the rock's speed just before colliding with Jupiter. Explain your calculation and any approximations that you make.

$M_{\text{Sun}} = 2 \times 10^{30}$ kg, $M_{\text{Jupiter}} = 2 \times 10^{27}$ kg

Distance, Sun to Jupiter $= 8 \times 10^{11}$ m

Radius of Jupiter $= 1.4 \times 10^8$ m

•••P69 A pendulum (see Figure 6.84) consists of a very light but stiff rod of length L hanging from a nearly frictionless axle, with a mass m at the end of the rod. **(a)** Calculate the gravitational potential energy as a function of the angle θ, measured from the vertical. **(b)** Sketch the potential energy as a function of the angle θ, for angles from $-210°$ to $+210°$. **(c)** Let $s = L\theta$ = the arc length away from the bottom of the arc. Calculate the tangential component of the force on the mass by taking the (negative) gradient of the energy with respect to s. Does your result make sense? **(d)** Suppose that you hit the stationary hanging mass so it has an initial speed v_i. What is the minimum initial speed needed for the pendulum to go over the top ($\theta = 180°$)? On your sketch of the potential energy (part b), draw and label energy levels for the case in which the initial speed is less than, equal to, or greater than this critical initial speed.

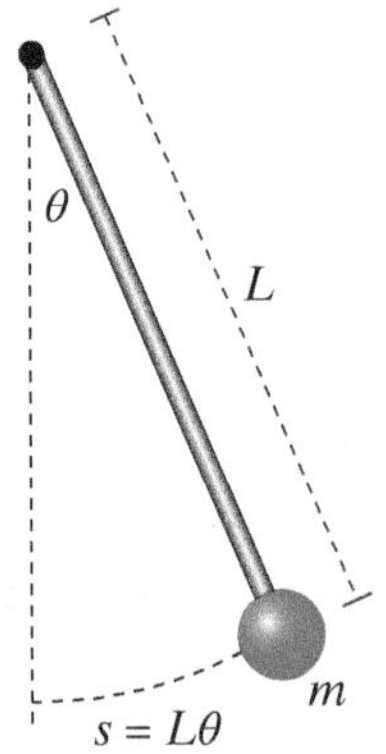

Figure 6.84

Section 6.9

•P70 (a) Two protons are a distance 4×10^{-9} m apart. What is the electric potential energy of the system consisting of the two protons? If the two protons move closer together, will the electric potential energy of the system increase, decrease, or remain the same? **(b)** A proton and an electron are a distance 4×10^{-9} m apart. What is the electric potential energy of the system consisting of the proton and the electron? If the proton and the electron move closer together, will the electric potential energy of the system increase, decrease, or remain the same? **(c)** Which of the following statements are true? A. In some situations charged particles released from rest would move in a direction that increases electric potential energy, but not in other situations. B. If released from rest two protons would move closer together, increasing the potential energy of the system. C. If any two charged particles are released from rest, they will spontaneously move in the direction in which the potential energy of the system will be decreased.

••P71 **(a)** A particle with mass M and charge $+e$ and its antiparticle (same mass M, charge $-e$) are initially at rest, far from each other. They attract each other and move toward each other. Make a graph of energy terms vs. separation distance r between the two particles. Label the various energies involved in this process. Include the rest energy of the particles, assuming that the other energy terms are comparable to the rest energy. **(b)** When the particle and antiparticle collide, they annihilate and produce two new particles, one with mass m (much smaller than M) and charge $+e$, and its antiparticle (same mass m, charge $-e$). When these two particles have moved far away from each other, how fast are they going? Is this speed large or small compared to c? **(c)** Now take the specific case of a proton and antiproton colliding to form a positive and negative pion. Each pion has a mass of 2.5×10^{-28} kg. When the pions have moved far away, how fast are they going? **(d)** How far apart must the two pions be (in meters) for their electric potential energy to be negligible compared to their kinetic energy? Be explicit and quantitative about your criterion and your result.

••P72 Four protons, each with mass M and charge $+e$, are initially held at the corners of a square that is d on a side. They are then released from rest. What is the speed of each proton when the protons are very far apart?

Section 6.11

•P73 There are three different ways to get from location A to location E in Figure 6.85. Along path 1, you take an elevator directly from location A straight up to location E. Along path 2, you walk from A to B, climb a rope from B to C, then walk from C to E. Along path 3, you walk from location A to D, then climb a ramp up to location E. The following questions focus on the work done on a 80 kg person by the Earth while following each of these paths.

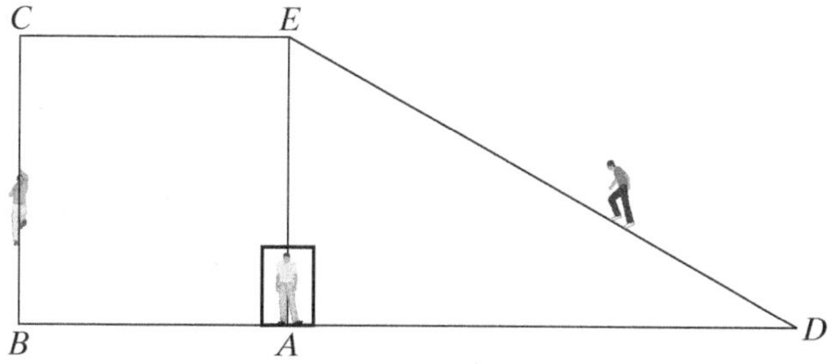

Figure 6.85

Taking the origin at location A, the coordinates of the other locations are: B $\langle 0,-3,0\rangle$; C $\langle -3,13,0\rangle$; D $\langle 33,0,0\rangle$; E $\langle 0,13,0\rangle$. **(a)** Path 1 (A to E): What is the displacement vector $\Delta\vec{r}$ for the path from A to E on the elevator? **(b)** What is the gravitational force acting on the person along path 1, as a vector? **(c)** What is the work done by the gravitational force of the Earth on the 80 kg person during the elevator ride from A to E? **(d)** Path 2 (A to B to C to E): What is the work done by the gravitational force of the Earth on the person as he walks from A to B? **(e)** What is the work done by the gravitational force of the Earth on the person as he climbs from B to C? **(f)** What is the work done by the gravitational force of the Earth on the person as he walks from C to E? **(g)** What is the total work done by the gravitational force of the Earth on the person as the person goes from A to E along path 2? **(h)** Path 3 (A to D to E): What is the work done by the gravitational force of the Earth on the person as he walks from A to D? **(i)** What is the work done by the gravitational force of the Earth on the person as he climbs from D to E? **(j)** What is the total work done by the gravitational force of the Earth on the person as the person goes from A to E along path 3? **(k)** You calculated the work done by the gravitational force of the Earth on the person along three different paths. How do these quantities compare with each other?

•P74 Refer to Figure 6.86. Calculate the change in electric energy along two different paths in moving charge q away from charge Q from A to B along a radial path, then to C along a circle centered on Q, then to D along a radial path. Also calculate the change in energy in going directly from A to D along a circle centered on Q. Specifically, what are $U_B - U_A$, $U_C - U_B$, $U_D - U_C$, and their sum? What is $U_D - U_A$? Also, calculate the round-trip difference in the electric energy when moving charge q along the path from A to B to C to D to A.

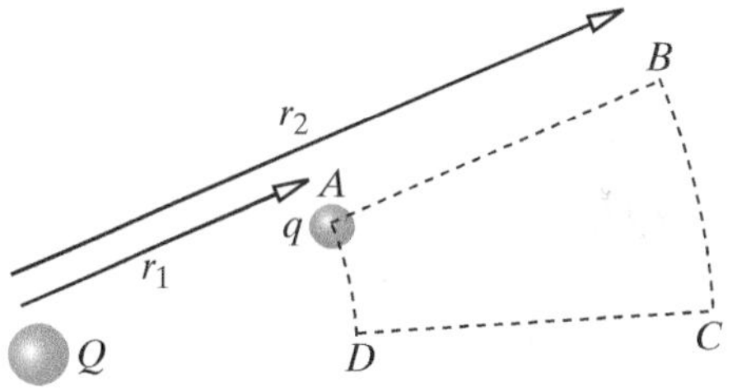

Figure 6.86

Section 6.12

••P75 One of the possible fission modes of Uranium-235 involves nearly equal fragments, palladium nuclei with $Q_1 = Q_2 = 46e$. The masses of the two palladium nuclei add up to less than the mass of the original nucleus. (In addition to the two main fission fragments there are typically one or more free neutrons in the final state; in your analysis make the simplifying assumption that there are no free neutrons, just two palladium nuclei.)

The mass of the U-236 nucleus is 235.996 u (unified atomic mass units), and the mass of each Pd-118 nucleus is 117.894 u, where 1 u $= 1.7 \times 10^{-27}$ kg (approximately the mass of one nucleon). Although in most problems you solve in this course it is adequate to use values of constants rounded to two or three significant figures, in this problem you must keep at least six significant figures throughout your calculation. Problems involving mass changes require many significant figures because the changes in mass are small compared to the total mass. **(a)** Calculate the final speed v, when the palladium nuclei have moved far apart (due to their mutual electric repulsion). Is this speed small enough that $p^2/(2m)$ is an adequate approximation for the kinetic energy of one of the palladium nuclei? (It is all right to go ahead and make the nonrelativistic assumption first, but you then must check that the calculated v is indeed small compared to c.) **(b)** Using energy considerations, calculate the distance between centers of the palladium nuclei just after fission, when they are starting from rest. **(c)** A proton or neutron has a radius of roughly 1×10^{-15} m, and a nucleus is a tightly packed collection of nucleons. Experiments show that the radius of a nucleus containing N nucleons is approximately $(1.3 \times 10^{-15}\text{ m}) \times N^{1/3}$. What is the approximate radius of a palladium nucleus? Draw a sketch of the two palladium nuclei in part (b), and label the distances you calculated in parts (b) and (c). If the two palladium nuclei are nearly touching, this would be consistent with our model of fission, in which the uranium nucleus fissions into two pieces that are nearly at rest. **(d)** The kinetic energy of the fast-moving daughter nuclei is eventually absorbed in the surrounding material and raises the temperature of that material. In a fission power plant, this thermal energy is used to boil water to drive a steam turbine and generate electricity. If a mole of uranium undergoes this fission reaction, how much kinetic energy is generated? For comparison, only around 1×10^6 J are obtained from burning a mole of gasoline, which is why energy from fission is of great interest, if some way can be found to deal with the waste products.

••P76 A proton (^{1}H) and a deuteron (^{2}H, "heavy" hydrogen) start out far apart. An experimental apparatus shoots them toward each other (with equal and opposite momenta). If they get close enough to make actual contact with each other, they can react to form a helium-3 nucleus and a gamma ray (a high-energy photon, which has kinetic energy but zero rest energy):

$$^1\text{H} + {}^2\text{H} \rightarrow {}^3\text{He} + \gamma$$

This is one of the thermonuclear or fusion reactions that takes place inside a star such as our Sun.

The mass of the proton is 1.0073 u (unified atomic mass unit, 1.7×10^{-27} kg), the mass of the deuteron is 2.0136 u, the mass of the helium-3 nucleus is 3.0155 u, and the gamma ray is massless. Although in most problems you solve in this course it is adequate to use values of constants rounded to two or three significant figures, in this problem you must keep at least six significant figures throughout your calculation. Problems involving mass changes require many significant figures because the changes in mass are small compared to the total mass. **(a)** The strong interaction has a very short range and is essentially a contact interaction. For this fusion reaction to take place, the proton and deuteron have to come close enough together to touch. The approximate radius of a proton or neutron is about 1×10^{-15} m. What is the approximate initial total kinetic energy of the proton and deuteron required for the fusion reaction to proceed, in joules and electron volts ($1\text{ eV} = 1.6 \times 10^{-19}$ J)? **(b)** Given the initial conditions found in part (a), what is the kinetic energy of the ^{3}He plus the energy of the gamma ray, in joules and in electron volts? **(c)** The net energy released is the kinetic energy of the ^{3}He plus the energy of the gamma ray found in part (b), minus the energy input that you calculated in part (a). What is the net energy release, in joules and in electron volts? Note that you do get back the energy investment made in part (a). **(d)** Kinetic energy can be used to drive motors and do other useful things. If a mole of hydrogen and a mole of deuterium underwent this fusion reaction, how much kinetic energy would be generated? (For comparison, around 1×10^6 J are obtained from burning a mole of gasoline.) **(e)** Which of the following potential energy curves (1–4) in Figure 6.87 is a reasonable representation of the interaction in this fusion reaction? Why?

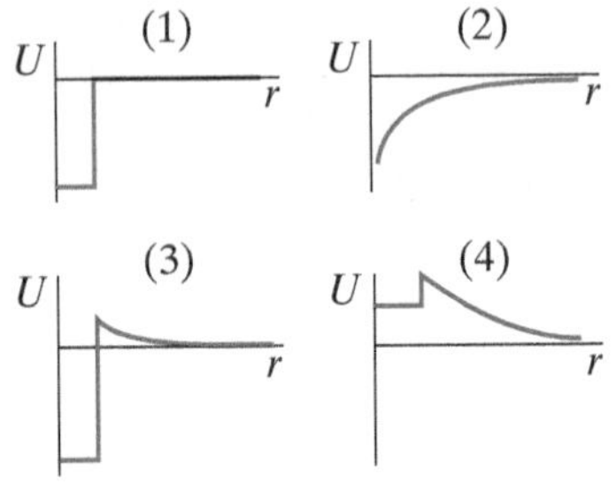

Figure 6.87

As we will study later, the average kinetic energy of a gas molecule is $\frac{3}{2}k_BT$, where k is the "Boltzmann constant," 1.4×10^{-23} J/K, and T is the absolute or Kelvin temperature, measured from absolute zero (so that the freezing point of water is 273 K). The approximate temperature required for the fusion reaction to proceed is very high. This high temperature, required because of the electric repulsion barrier to the reaction, is the main reason why it has been so difficult to make progress toward thermonuclear power generation. Sufficiently high temperatures are found in the interior of the Sun, where fusion reactions take place.

COMPUTATIONAL PROBLEMS

More detailed and extended versions of some computational modeling problems may be found in the lab activities included in the *Matter & Interactions, 4th Edition*, resources for instructors.

••P77 Start with the program you wrote to model the motion of a spacecraft near the Earth, in Chapter 3 (Problem P65). Add the statements necessary to plot graphs of K, U, and $(K+U)$ versus time, like the graphs in Figure 6.71. Use initial conditions that produce an elliptical orbit. **(a)** Refer to the shapes of the plots to explain the flow of energy within the system. **(b)** Use the plot of $K+U$ versus time produced by the program to determine the maximum reasonable value of Δt for this model. **(c)** Predict the shape each plot should have for a perfectly circular orbit. **(d)** Use your energy plots to find (by trial and error) the initial speed of the spacecraft that produces a circular orbit, and explain what information from the plots you used.

••P78 As in Problem P77, start with the program you wrote to model the motion of a spacecraft near the Earth in Chapter 3 (Problem P64). Add the statements necessary to plot graphs of graphs of K, U, and $(K+U)$ versus r, the separation between the spacecraft and the Earth. Use initial conditions that produce an elliptical orbit. In VPython, when you create a `gcurve` object, it can be helpful to add a moving dot to clarify what is happening in the plot, like this:

```
ugraph = gcurve(color=color.green, dot=True)
```

(a) Refer to the shapes of the plots to explain the flow of energy within the system. **(b)** What effect does changing the initial speed of the spacecraft have on these three plots? **(c)** Predict the shape each plot should have for a perfectly circular orbit. **(d)** Use your energy plots to find (by trial and error) the initial speed of the spacecraft that produces a circular orbit, and explain what information from the plots you used. **(e)** Use your energy plots to help you find the escape speed for this system. **(f)** Describe your plot of $K+U$ when the spacecraft starts with this escape speed.

••P79 Start with the program you wrote to model the motion of a spacecraft near the Earth and a stationary Moon, in Chapter 3 (Problem P66). Add the statements necessary to plot graphs of K, U, and $(K+U)$ versus time, like the graphs in Figure 6.71. Give the spacecraft an initial velocity of $\langle 0, 3.27\times 10^3, 0\rangle$ m/s. **(a)** Refer to the shapes of the plots to explain the flow of energy within the system. **(b)** Use the plot of $K+U$ versus time produced by the program to determine the maximum reasonable value of Δt for this model. **(c)** What changes in the energy graphs when you vary the initial speed?

••P80 As in Problem P79, start with the program you wrote to model the motion of a spacecraft near the Earth and a stationary Moon, in Chapter 3 (Problem P66). Add the statements necessary to plot graphs of graphs of K, U, and $(K+U)$ versus r, the separation between the spacecraft and the Earth. Give the spacecraft an initial velocity of $\langle 0, 3.27\times 10^3, 0\rangle$ m/s. In VPython, when you create a `gcurve` object, it can be helpful to add a moving dot to clarify what is happening in the plot, like this:

```
ugraph = gcurve(color=color.green, dot=True)
```

(a) Why are the shapes of the graphs of U and K versus separation different for this three-body system than they were for the two-body (spacecraft + Earth) system? **(b)** Refer to the shapes of the plots to explain the flow of energy within the system. **(c)** What effect does changing the initial speed of the spacecraft have on these three plots?

••P81 Start with the program you wrote in Problem P71 in Chapter 3 to model the Ranger 7 mission to the Moon. (If you have not already written this program, do it now.) **(a)** Add a calculation of the work done by the gravitational forces of the Earth and the Moon to your analysis of sending a spacecraft to the Moon. You need to approximate the work by adding up the amount of work done by gravitational forces along each step of the path:

$$W = \Sigma\vec{F}\bullet\Delta\vec{r} = \vec{F}_1\bullet\Delta\vec{r}_1 + \vec{F}_2\bullet\Delta\vec{r}_2 + \vec{F}_3\bullet\Delta\vec{r}_3 + \cdots$$

(b) Compare the numerical value of the work with the change in the kinetic energy (final kinetic energy just before crashing on the Moon, minus initial kinetic energy when released above the Earth's atmosphere). **(c)** Modify your program to make graphs of two quantities: the kinetic energy of the spacecraft, and the work done by the Earth and the Moon on the spacecraft, as a function of time. **(d)** Find a way to determine the maximum value of Δt you can use without introducing significant errors into your computation.

••P82 Energy conservation is a powerful check on the accuracy of a numerical integration. Modify the program for the Chapter 3 problem on the Ranger 7 mission to the Moon to plot graphs of kinetic energy, of gravitational potential energy, and of the sum of the kinetic energy and the gravitational potential energy, vs. position. Does the kinetic plus potential energy remain constant? What if you vary the step size (which varies the accuracy of the numerical integration)? Vary the launch speed, and explain the effect that this has on your graphs.

••P83 Use energy conservation to calculate analytically (that is, without doing a numerical integration) the final speed of the spacecraft just before it hits the Moon. Include the gravitational effect of the Moon. Use a launch speed of 1.3×10^4 m/s. Modify your program for the Chapter 3 problem on the Ranger 7 mission to the Moon to print out the speed of the spacecraft when it hits the surface of the Moon, and compare this value to your analytical result. What questions that could be addressed in the numerical integration are you *not* able to answer by doing this energy calculation?

ANSWERS TO CHECKPOINTS

1 Particle energy = 5.2×10^{-13} J, rest energy = 8.1×10^{-14} J; kinetic energy = 4.4×10^{-13} J
2 12 m/s; 27 miles per hour; 1.2×10^{-12} kg (an extremely small amount of mass!)
3 1.4 J
4 **(a)** decreases, **(b)** increases, **(c)** decreases, **(d)** increases, **(e)** stays the same
5 189 J
6 400 J
7 40.8; 39.8; very large; very small
8 8.5×10^{-4}, approximately 0.1%
9 Six pairs: $U_{12}+U_{13}+U_{14}+U_{23}+U_{24}+U_{34}$
10 Same ΔU
11 $\frac{1}{2}m(v_{xi}^2) - \frac{1}{2}m(v_{xi}^2+v_{yi}^2) + mg(y_f - y_i) = 0$, so $y_f = y_i + v_{yi}^2/(2g)$
12 **(a)** $\dfrac{1}{4\pi\varepsilon_0}\dfrac{e^2}{r} = (0.2\times 10^6\text{ eV})(1.6\times 10^{-19}\text{ J/eV})$, **(b)** the electric potential energy term
13 **(a)** decrease, **(b)** A is bound; C is unbound; B is trapped. In a quantum system in state B there is some probability of "tunneling" through the barrier and getting out of the trap!
14 A: go out in a straight line and stop, then fall back; B: elliptical (motion with varying separation); C: nearly circular (nearly constant separation)
15 1.12×10^4 m/s
16 About 1.7×10^{-10}; accuracy of laboratory scale is far below what would be needed to detect this
17 About 0.1%
18 $-GMm/y^2$; 0

CHAPTER

7

Internal Energy

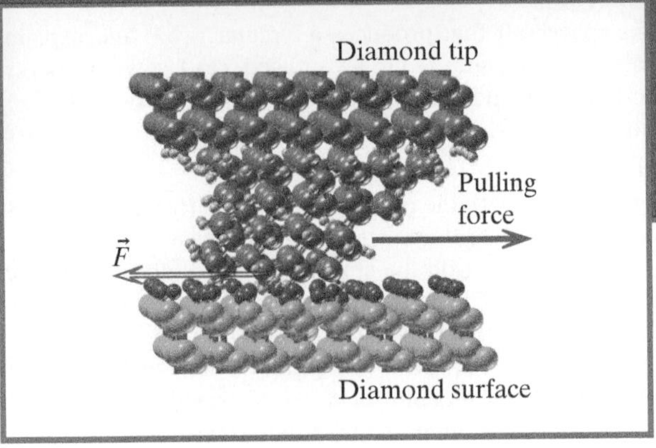

OBJECTIVES

After studying this chapter, you should be able to

- Explain changes in the internal energy of a solid object at a microscopic level in terms of the ball–spring model.
- Calculate internal energy changes in extended objects.
- Mathematically relate temperature change, energy transfer due to a temperature difference, and specific heat.
- Calculate and use the potential energy of a system containing a spring.
- Create a computational model of a system in which there is dissipation (for example, air resistance or sliding friction).

7.1 EXTENDED OBJECTS

In Chapter 4 we considered contact forces between extended objects—objects that can be stretched, compressed, and deformed, and that therefore cannot be modeled as point particles. We found that at the microscopic level a ball–spring model of an extended solid object allowed us to understand the origin of tension and compression forces, and to predict quantities such as the speed of sound in an object composed of a particular metal.

When considering energy changes in systems consisting of extended objects, we again find that it is not possible to model such systems as point particles that have no internal structure. We know that, as a result of interactions with their surroundings, macroscopic objects can get hotter or colder. We will see that a change in the temperature of an object is correlated with the amount of energy inside the object. In order to understand the nature of internal energy, we will return to the ball–spring model of the microscopic structure of a solid object (Figure 7.1).

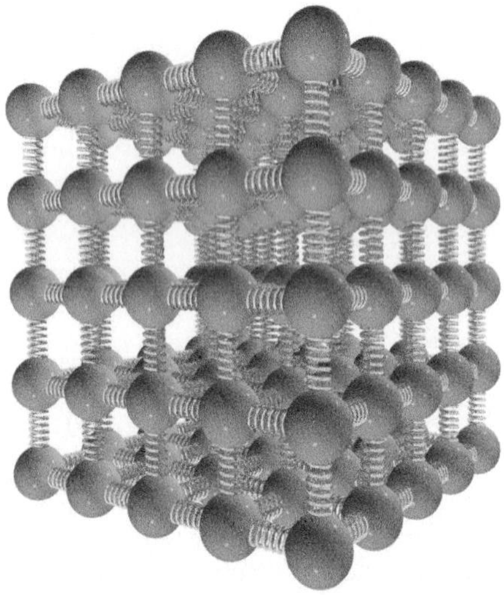

Figure 7.1 The ball–spring model of a solid object.

Recall that in this model, the balls represent atoms, and the springs represent the spring-like covalent chemical bonds between the atoms. Because the atoms inside a solid object can move, they can have kinetic energy. Since work is required to compress or stretch a spring, it is plausible that potential energy can be associated with springs. In order to think about internal energy in the context of the ball–spring model of a solid, we need to find an expression for the potential energy associated with a spring, which is discussed in the following section.

7.2 POTENTIAL ENERGY OF MACROSCOPIC SPRINGS

A real spring has a nonzero mass and can be damaged in various ways: it can be deformed or broken, the metal can fatigue, and so on. There is a limit to how much a real spring can be stretched (it will eventually break) or compressed (the coils will eventually touch each other, turning the spring into a stiff hollow rod). Initially, to reduce the complexity of our model, we will consider an idealized version of a spring, which we will call an ideal spring.

An Ideal Spring

An idealized spring has none of the problems of a real spring. There are no limits on the stretch s in the ideal spring force equation (here we're calling the axis of the spring the s axis):

$$F_s = -k_s s$$

Despite its limitations, an ideal spring proves to be a useful model for both real macroscopic springs and for spring-like interatomic bonds.

To determine the potential energy associated with the spring force, recall that the (negative) gradient of potential energy is equal to the associated force. So to find the spring potential energy U_s we need a solution to this equation:

$$-\frac{dU_s}{ds} = -k_s s$$

which simplifies to

$$\frac{dU_s}{ds} = k_s s$$

QUESTION Can you think of a function of s whose derivative with respect to s is $k_s s$?

The derivative with respect to s of $\frac{1}{2}k_s s^2$ is $k_s s$, and the negative gradient of $\frac{1}{2}k_s s^2$ is the force $-k_s s$.

IDEAL SPRING POTENTIAL ENERGY

$$U_s = \frac{1}{2}k_s s^2$$

s is stretch, measured from the equilibrium point.

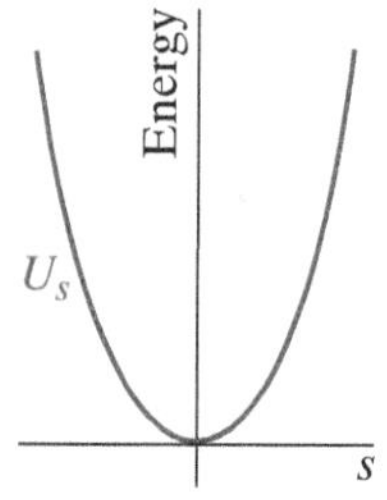

Figure 7.2 Potential energy as a function of stretch s for an ideal spring.

The function $\frac{1}{2}k_s s^2$ is an upturned parabola (Figure 7.2). Note that the stretch s appears squared in the expression for spring energy. This reflects the fact that either a lengthening (positive stretch) or a compression (negative stretch) of the spring involves increased potential energy.

Peculiar Features of the Ideal Spring Potential Energy

This ideal spring potential energy curve has some peculiar features:

- As the absolute value of stretch becomes large, U becomes infinite, instead of approaching zero. (This curve is sometimes called an "infinite potential energy well.")
- For a system of a mass and ideal spring, all states are bound states; there are no unbound states.

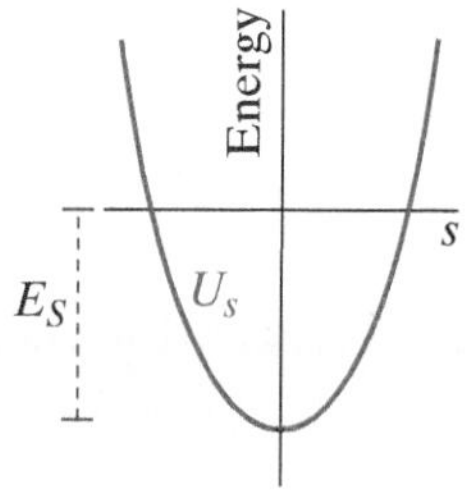

Figure 7.3 Subtracting a constant E_S shifts the ideal spring potential energy curve downward.

As a result of these peculiarities, this ideal potential energy function doesn't fully represent the nature of a macroscopic spring you can hold in your hand, or of an interatomic bond.

One possible modification to the equation would be to subtract a constant E_S. Such a function would still satisfy the gradient equation above, but it would allow for U to be negative in places, as in Figure 7.3.

$$U_s = \frac{1}{2}k_s s^2 - E_S$$

For the ideal spring, the constant is conventionally set to zero, but the negative offset may be needed when analyzing real systems.

Real Springs

We are primarily interested in helical (spiral) springs. A modest force applied to a helical spring gives a substantial change in the length of the spring, even though the total length of the coiled wire hardly changes. A large stretch of the spring with a small force implies a much smaller spring stiffness than the wire would have when straightened out.

If you stretch a helical spring far enough, it straightens out into a long straight wire, and it suddenly becomes very difficult to lengthen the wire any further. If you continue to stretch the straight wire, eventually it yields and then breaks. If you compress a helical spring so far that its coils run into each other, it suddenly gets extremely hard to compress the spring any further. Therefore, the linear relation between stretch and force, $F_s = -k_s s$, is valid only over a limited region of stretch, although this linear range is much wider than for the microscopic "spring."

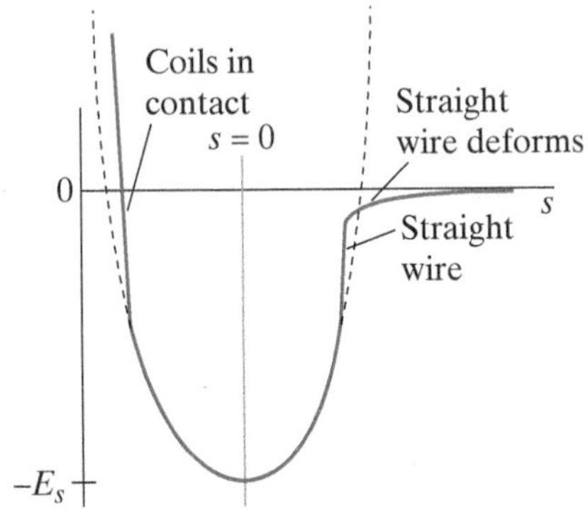

Figure 7.4 The dashed curve represents the spring potential energy

$$U_s = \frac{1}{2}k_s s^2 - E_S$$

within the linear range where $F_s = -k_s s$. The colored solid curve represents a real macroscopic spring.

In many situations the entire phenomenon of interest takes place within the valid range of the linear force approximation. One can speak of an "infinite" potential energy well as shown as the dashed curve in Figure 7.4, which is an idealization of the real situation. Within the linear range where $F_s = -k_s s$ is the negative gradient of $\frac{1}{2}k_s s^2$, we approximate the real curve by a parabola. In our calculations with macroscopic springs we're almost always interested only in *changes* in the potential energy, so we shift the origin to the bottom of the well by an amount $-E_s$ and measure $U_s = \frac{1}{2}k_s s^2$ from there.

In the blue curve shown in Figure 7.4 (which is still an idealization of the real situation) there are very steep sides of the potential energy well, corresponding to the large forces required to compress a spring whose coils are in contact and the large forces required to stretch the spring after it has straightened out into a straight wire. Since force is the negative gradient of the potential energy, a steep slope corresponds to a large force. The region marked "Straight wire deforms" in Figure 7.4 refers to the fact that a wire may "yield" (lengthen with little additional force required) or break.

Energy in an Oscillating Spring–Mass System

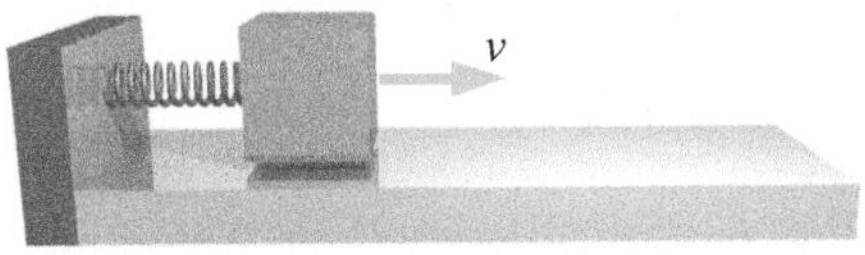

Figure 7.5 As the compressed spring expands the kinetic energy of the block increases, and spring potential energy decreases.

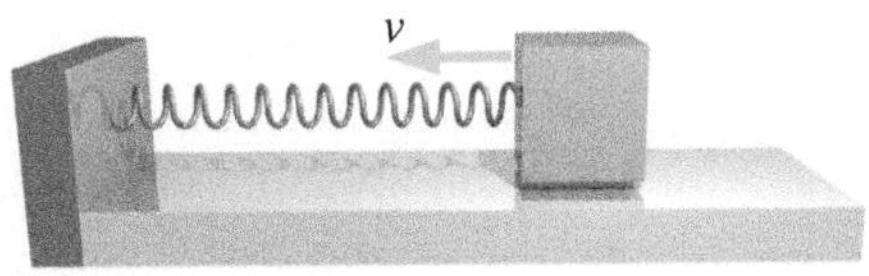

Figure 7.6 As the stretched spring contracts, the kinetic energy of the block increases and spring potential energy decreases.

In an oscillating spring–mass system such as the one discussed in Chapter 4, the energy of the system is continually changing from kinetic energy to potential energy, and back again.

Consider a block on a low-friction surface, connected by a helical spring to a wall. If you compress the spring and then let go, the block acquires kinetic energy (Figure 7.5), with a loss of spring potential energy. If you stretch the spring and then let go, the block also acquires kinetic energy, again with a loss of spring potential energy (Figure 7.6).

QUESTION At the moment that the spring is released, what is the kinetic energy of the spring–mass system? What is the total energy of the system?

Since the block is released from rest, its kinetic energy is zero. The energy of the system is equal to the energy stored in the spring, $\frac{1}{2}k_s s^2$ (omitting the constant rest energy and constant $-E_S$).

QUESTION At what point in the oscillation is the kinetic energy of the system highest? At that moment, what is the potential energy of the system?

When the contracting spring reaches its equilibrium length, the energy stored in the spring is a minimum, $\frac{1}{2}k_s s^2 = 0$. The maximum possible amount of spring potential energy has now been converted to kinetic energy, so this is the instant when the kinetic energy of the system is highest.

QUESTION At what point in the oscillation will the kinetic energy of the system be lowest?

At either turning point of the oscillation—spring fully compressed or spring fully extended—the instantaneous speed of the mass is zero. The kinetic energy of the system is lowest here (zero in fact), and all the energy has been momentarily converted back into spring potential energy. Since the absolute value of the stretch is highest at these locations, the potential energy $\frac{1}{2}k_s s^2$ is also highest.

Does the Wall Do Work on the System?

For the system of spring + block, the surroundings include the Earth, the low-friction table, and the wall. Both the table and the wall touch the system and therefore exert contact forces on it.

The table and Earth exert forces perpendicular to the block's motion, so they do zero work on the system (assuming negligible friction).

QUESTION The force by the wall on the spring is parallel to the block's motion. Does it do nonzero work on the system?

This force does zero work, because the point at which the force acts (the left end of the spring) undergoes zero displacement. So

$$F_{\text{wall},x}\Delta x = 0 \quad \text{because } \Delta x = 0 \text{ for the left end of the spring}$$

This force does have an effect on the momentum of the system, however, because the time interval over which it acts is not zero. The momentum of the mass–spring system is changed by the force on the spring by the wall. (Think about what would happen if the wall were not there.) This is another example of the difference between the Momentum Principle (all forces act for the same time interval) and the Energy Principle (different forces may act over different distances).

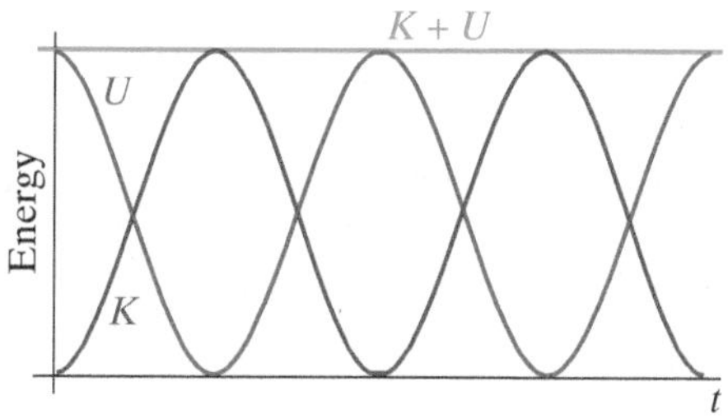

Figure 7.7 Kinetic energy (red), potential energy (blue), and the sum of kinetic + potential energy (green) for an oscillating spring–mass system.

Flow of Energy in the Mass–Spring System

Since the surroundings do no work on the system, the total energy of the system must be constant (in the absence of friction), and energy flows from potential to kinetic and back again, as shown in Figure 7.7.

Here we can see a nice example of "path independence," which was discussed in Chapter 6. Consider the round trip starting with the block moving at maximum speed as it passes the equilibrium point, where the potential energy is zero, moving in the $+x$ direction. The block slows down due to the spring force that acts in the $-x$ direction, momentarily comes to a stop, then moves with increasing speed in the $-x$ direction due to the spring force, which is still in the $-x$ direction.

The block passes the equilibrium point with the same kinetic energy it had when it was previously at this location, but now moving in the $-x$ direction. The round trip of the spring–mass system, from equilibrium configuration and return, with no change in the kinetic energy, required zero work on our part (path independence). The key to this is that the spring force and associated potential energy are purely functions of position. At any location you choose on the x axis, the spring force is the same in magnitude and direction when the block is moving to the right at that location and when it is returning to the left at that location.

If there were significant friction, the frictional force would be in the $-x$ direction outward bound but in the $+x$ direction during the return, unlike the spring force which is always in the $-x$ direction during the entire round trip. The frictional force depends on the direction of motion, the velocity, whereas the spring force depends solely on the position.

Checkpoint 1 During one complete oscillation of a mass on a spring (one period), what is the change in potential energy of the mass + spring system, in the absence of friction?

EXAMPLE **Energy in a Spring–Mass System**

A mass of 0.2 kg is attached to a horizontal spring whose stiffness is 12 N/m. Friction is negligible. At $t=0$ the spring has a stretch of 3 cm and the mass has a speed of 0.5 m/s. **(a)** What is the amplitude (maximum stretch) of the oscillation? **(b)** What is the maximum speed of the block?

Solution System: Mass and spring
Surroundings: Earth, table, wall (neglect friction, air)

(a) The stretch is maximum when the speed of the mass is zero, so:
Initial state: $s_i = 3$ cm, $v_i = 0.5$ m/s
Final state: $v_f = 0$

$$K_f + U_f = K_i + U_i + W$$
$$0 + \frac{1}{2}k_s s_f^2 = \frac{1}{2}mv_i^2 + \frac{1}{2}k_s s_i^2 + 0$$
$$\frac{1}{2}k_s s_f^2 = \frac{1}{2}(0.2\text{ kg})(0.5\text{ m/s})^2 + \frac{1}{2}(12\text{ N/m})(0.03\text{ m})^2$$
$$\frac{1}{2}k_s s_f^2 = 0.03\text{ J}$$
$$s_f = \sqrt{2(0.03\text{ J})/(12\text{ N/m})} = 0.07\text{ m}$$
$$s_f = 0.07\text{ m}$$

(b) The speed is maximum when the stretch is zero, so:
Initial state: $s_i = 3$ cm, $v_i = 0.5$ m/s
Final state: $s_f = 0$

$$K_f + U_f = K_i + U_i + W$$
$$K_f + 0 = 0.03\text{ J} + 0$$
$$\frac{1}{2}mv_f^2 = 0.03\text{ J}$$
$$v_{max} = \sqrt{2(0.03\text{ J})/(0.2\text{ kg})} = 0.55\text{ m/s}$$

Assuming negligible energy dissipation by friction or air resistance, the energy of the system never changes. If we can find $K+U$ for one state, we can always use this as the initial state in any calculation.

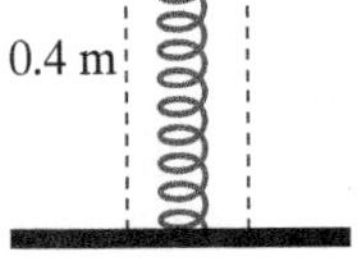

Figure 7.8 A block falls onto a vertical spring, compresses the spring, then rebounds upward.

EXAMPLE **A Rebounding Block**

A metal block of mass 3 kg is moving downward with speed 2 m/s when the bottom of the block is 0.8 m above the floor (Figure 7.8). When the bottom of the block is 0.4 m above the floor, it strikes the top of a relaxed vertical spring 0.4 m in length. The stiffness of the spring is 2000 N/m. **(a)** The block continues downward, compressing the spring. When the bottom of the block is 0.3 m above the floor, what is its speed? **(b)** The block eventually heads back upward, loses contact with the spring, and continues upward. What is the maximum height reached by the bottom of the block above the floor? **(c)** What approximations did you make?

Solution **(a)** Find the speed of the block when it is 0.3 m above the floor.
System: Earth, block, spring
Surroundings: Nothing significant
Initial state: Block 0.8 m above floor, moving downward, $v_i = 2$ m/s, spring relaxed
Final state: Block 0.3 m above floor, spring compressed

$$K_f + U_{s,f} + U_{g,f} = K_i + U_{s,i} + U_{g,i} + W$$

$$\frac{1}{2}mv_f^2 + \frac{1}{2}k_s s_f^2 + mgy_f = \frac{1}{2}mv_i^2 + 0 + mgy_i + 0$$

$$v_f^2 = v_i^2 + 2g(y_i - y_f) - \frac{k_s}{m}s_f^2$$

$$v_f = \sqrt{(2\ \text{m/s})^2 + 2(9.8\ \text{N/kg})(0.5\ \text{m}) - \frac{2000\ \text{N/m}}{3\ \text{kg}}(0.1\ \text{m})^2}$$

$$v_f = 2.7\ \text{m/s}$$

(b) Find the maximum height.
Initial state: Same as in part (a)
Final state: Block at highest point, $v_f = 0$, spring relaxed

$$K_f + U_{s,f} + U_{g,f} = K_i + U_{s,i} + U_{g,i} + W$$

$$0 + 0 + mgy_f = \frac{1}{2}mv_i^2 + 0 + mgy_i + 0$$

$$y_f = y_i + \frac{v_i^2}{2g} = (0.8\ \text{m}) + \frac{(2\ \text{m/s})^2}{2(9.8\ \text{N/kg})} = 1.0\ \text{m}$$

(c) Approximations: Air resistance and dissipation in the spring are negligible. $U_g \approx mgy$ near Earth's surface. ΔK_{Earth} is negligible.

Both Earth and spring are part of the system, so there are two potential energy terms, one for (block + spring) and one for (block + Earth). Since the mass of the spring is negligible, there is not a (spring + Earth) potential energy term. In part (b) we could equally well have used the final state from part (a) as the initial state.

Checkpoint 2 How many joules of energy can you store in a spring whose stiffness is 0.6 N/m, by starting from a relaxed spring and stretching it 20 cm?

Nonhelical Macroscopic Springs

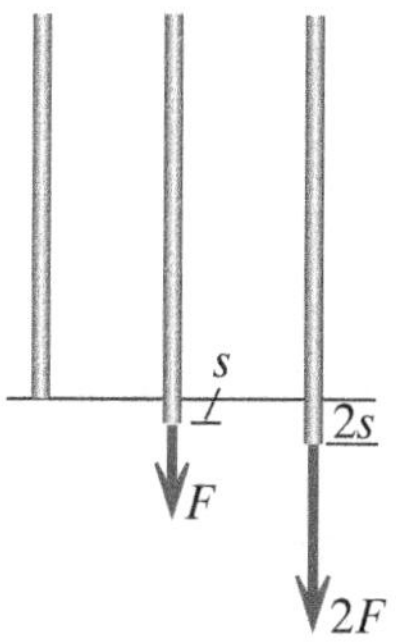

Figure 7.9 A bar of metal can be treated as a spring. For a limited range, double the force gives double the stretch.

We saw in Chapter 4 that a bar of metal can be treated as a spring, since stretching or compressing the bar stretches or compresses the interatomic bonds (Figure 7.9). Double the force produces double the change in length, as long as we're in the range where the spring approximation to the interatomic force is adequate. This is usually described for a bar in terms of Young's modulus Y, which we studied in Chapter 4, where the tension force F_T (per unit area) is proportional to the stretch ΔL (per unit length):

$$\frac{F_T}{A} = Y\frac{\Delta L}{L}$$

Since ΔL, like s, represents a change in length, and F_T is a spring-like force (tension force), this equation can be rewritten to parallel the spring force equation:

$$F_T = \left(\frac{YA}{L}\right)|\Delta L| \quad \text{is like} \quad F = (k_s)|s|$$

The effective "spring stiffness" of the metal bar is YA/L.

If you stretch the bar too much, two things can make the bar stop behaving in a spring-like manner. You might exceed the interatomic stretch for which $k_s s$ is a good approximation to the magnitude of the interatomic force. Also, the regular array of atoms may be disrupted by dislocations of the crystal structure, leading to large-scale slippage of crystal planes. Suddenly the bar "yields" and grows very much longer with little applied force.

A straight object is not usually used as a longitudinal spring but may be used in a bending mode. For example, when a diving board bends under a load, atomic bonds in the upper part of the board are stretched, and atomic bonds in the lower part are compressed (Figure 7.10). The net effect is a spring-like behavior of the diving board, with the amount of bend proportional to the applied force.

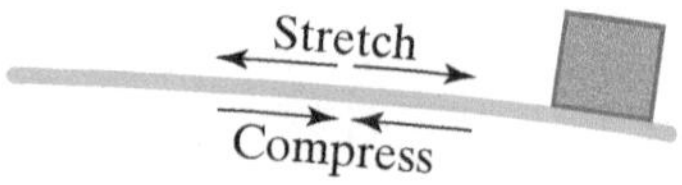

Figure 7.10 The bending of a diving board stretches and compresses interatomic bonds.

7.3 POTENTIAL ENERGY OF A PAIR OF NEUTRAL ATOMS

In Chapter 6 we found an equation for electric potential energy for pairs of charged particles. In Chapter 4 we studied the spring-like forces between electrically *neutral* atoms (a neutral object has equal amounts of positive and negative charge, so it has zero net charge). These interatomic forces are the superposition of electric attractions and repulsions among the positively charged protons and negatively charged electrons of which the atoms are made.

Figure 7.11 Two neutral atoms interact very little when far apart, attract each other at intermediate distances, and repel each other at very short distances.

When two electrically neutral atoms are far from each other (Figure 7.11), they exert almost no force on each other because the attractions between unlike charges (electrons in atom one with protons in atom two, and vice versa) are nearly equal to the repulsions between like charges. However, when the atoms come quite near each other, the electron clouds distort in such a way that the two atoms attract each other. Molecules and solid objects made of two or more atoms are bound together by these electric forces.

However, if you try to push two atoms even closer together, you eventually encounter a rapidly increasing repulsion. The attraction at a small distance and the repulsion when very close are due ultimately to the superposition of the electric forces of the various protons and electrons in the atoms, with the probable locations of these particles governed by the laws of quantum mechanics.

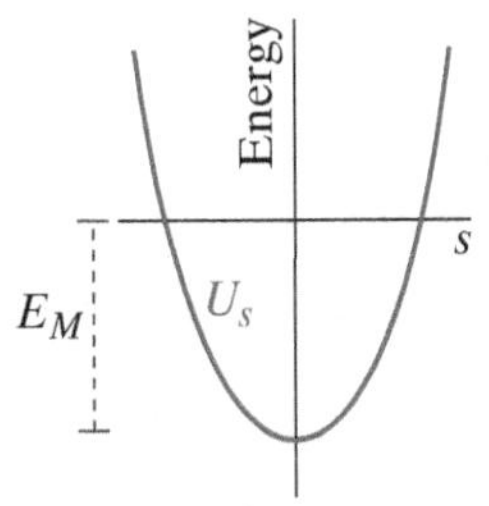

Figure 7.12 The interatomic potential energy is approximately spring-like if the bond is not stretched or compressed too much.

Potential Energy for Spring-like Interatomic Bonds

Since an interatomic bond behaves like a spring as long as it is not stretched or compressed excessively, we can guess that the equation for interatomic potential energy has the same form as the potential energy for an ideal spring (Figure 7.12):

APPROXIMATE INTERATOMIC POTENTIAL ENERGY

$$U = \frac{1}{2}k_s s^2 - E_M$$

What About Large Separations?

Of course, the potential energy of a real interatomic bond can't become infinitely large at large separation—it needs to approach zero. A more realistic potential energy function for the interatomic bond is called the Morse potential energy and is shown in Figure 7.13. The Morse potential energy is described by this equation:

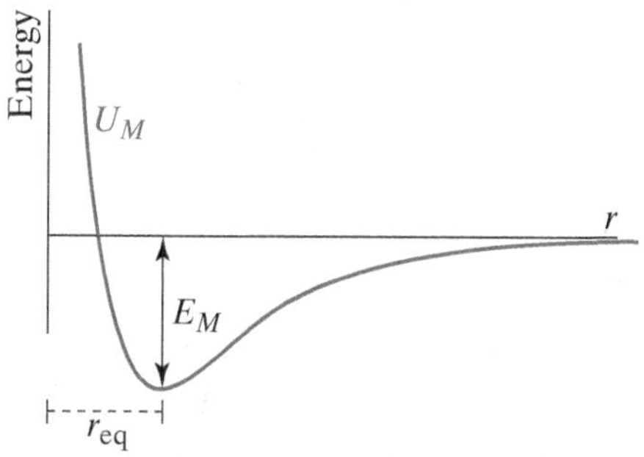

Figure 7.13 The Morse potential energy function U_M, representing the potential energy of two interacting neutral atoms as a function of separation.

$$U_M = E_M[1 - e^{-\alpha(r - r_{eq})}]^2 - E_M$$

The Morse approximation is designed to model the basic behavior of the interatomic force and is in approximate agreement with experimental measurements of the interatomic force.

The parameter E_M gives the depth of the potential energy at its minimum, and the parameter r_{eq} (on the order of 1×10^{-10} m) is the distance at which the potential energy curve is a minimum. The parameter α is adjusted to give the curve the right width. When the interatomic separation r is very large, U_M goes to zero, as it should.

The depth of the potential energy well, E_M, is typically a few electron volts (recall that 1 eV is equal to 1.6×10^{-19} J). For example, the potential energy well for oxygen, O_2, is about 5 eV deep. (As you saw in Chapter 6, a typical ionization energy is also in the range of a few electron volts.) However, for the noble gases such as helium and argon the well is so shallow that these atoms don't form stable diatomic molecules at room temperature, because collisions easily provide enough energy to break the molecules apart.

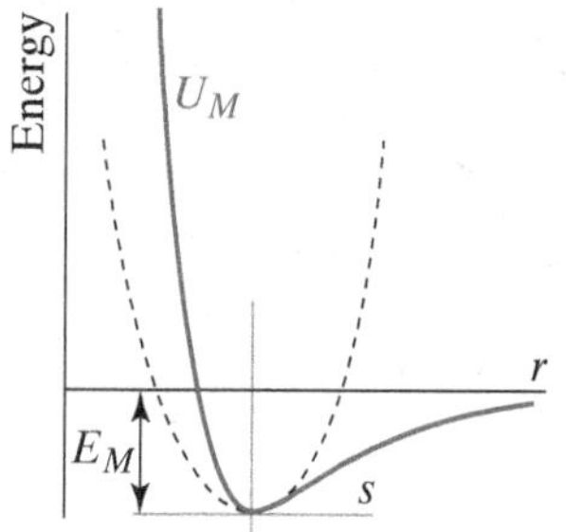

Figure 7.14 The bottom of the interatomic potential energy curve may be approximated by a parabola.

QUESTION What is the physical significance of the minimum of the curve?

At the minimum of the potential energy curve, the slope is zero, so the force between the two atoms is zero ($F = -dU/dr$). This is the position of stable equilibrium.

QUESTION What is the direction of the force on an atom at a distance smaller than the equilibrium distance?

The left side of the curve has a large negative slope, so the force $F = -dU/dr$ will be large and to the right (in the $+r$ direction), away from the other atom. This makes sense; when the atoms are too close together they push each other apart.

Likewise, if the atom moves to the right (farther away) there should be an attractive restoring force to the left. The slope of this region of the curve is positive, so the sign of the force $F = -dU/dr$ is negative, meaning its direction is to the left, as expected.

As indicated by the dashed curve in Figure 7.14, the bottom of the interatomic potential energy curve can be approximated by the spring-like potential energy function:

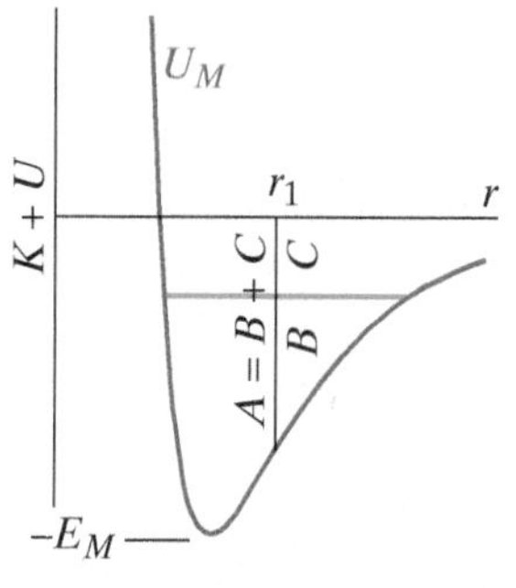

Figure 7.15 A horizontal line represents constant kinetic plus potential energy.

$$U = \frac{1}{2}k_s s^2 - E_M$$

Because we can make this approximate fit, we know that for ordinary small oscillations around the equilibrium point the two atoms will act as though they are connected by a spring with a $-k_s s$ force. This is the rationale for modeling a solid as a collection of balls connected by springs.

Bound and Unbound States with Interatomic Potential Energy

In Figure 7.15, a horizontal line represents motion in which the sum of the kinetic energies and the potential energy of a two-atom system does not change. A horizontal line corresponds to a possible state of the system when it is isolated from external forces.

Checkpoint 3 At separation r_1 in Figure 7.15, what is the physical significance of the quantity A? Of the quantity B? Of the quantity C? In Figure 7.16, which of the states are bound states of a two-atom system? Which are unbound states?

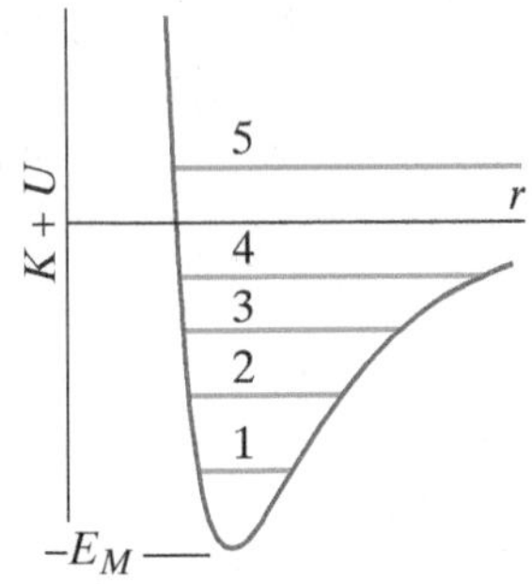

Figure 7.16 Which are bound states? Which are unbound states?

7.4 INTERNAL ENERGY

Consider two massive balls connected by a spring. If we model such an object as a point particle, in addition to its rest energy it can have only kinetic energy, which is associated with the motion of the center of mass of the object. However, by modeling the object as a point particle, we ignore the fact that in addition to the motion of the center of mass of the system, the balls could be oscillating back and forth, compressing and stretching the spring, and that the entire object could be rotating. We refer to the energy associated with the vibration and rotation of the system as "internal energy," and it is possible for the system to have such internal energy because it is not a point particle, and has some internal structure.

For example, Figure 7.17 shows two identical objects each consisting of two balls connected by identical springs, one relaxed, and the other compressed (and kept compressed by a string with negligible mass tied around it). If these two objects move at the same speed v, they have the same kinetic energy $K = \frac{1}{2}Mv^2$, where M is the total mass, but the object with the compressed spring has more internal energy, associated with the increase in spring potential energy.

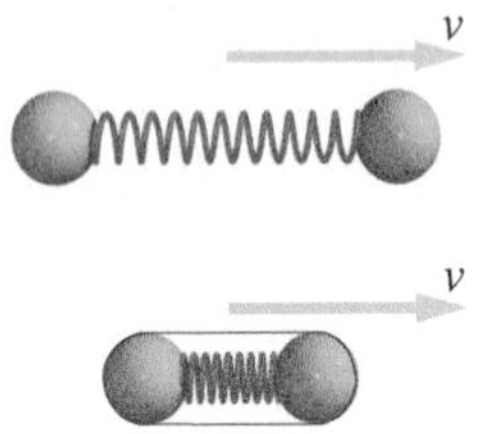

Figure 7.17 The object with the compressed spring has more "internal energy." It is kept compressed by a string of negligible mass.

It is often useful to split the total internal energy into various categories of interest to us in a particular situation. For example, if an object rotates about its center of mass, this internal energy is called "rotational energy" (Figure 7.18). If two balls connected by a spring oscillate relative to the center of mass of the system, this internal energy is called "vibrational energy" (Figure 7.19). If an object is stretched or compressed, there is an associated internal energy change, as in Figure 7.17. When you eat, you increase your internal energy in the form of more "chemical energy." If the temperature of an object increases, there is an increase in the internal energy that is called "thermal energy."

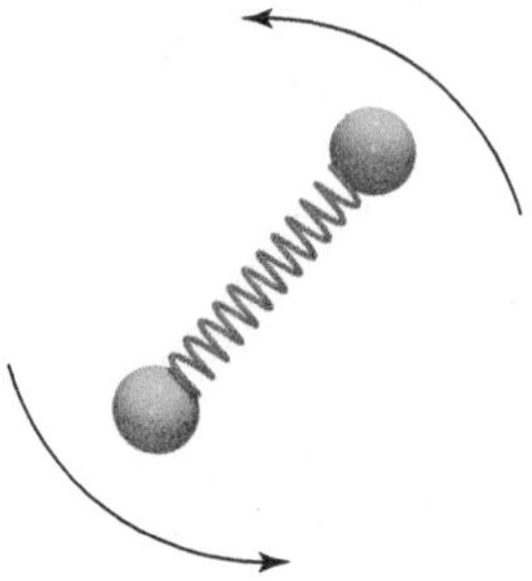

Figure 7.18 More rotation means more internal energy in the form of "rotational energy."

INTERNAL ENERGY

$$\text{Internal energy} = E_{\text{thermal}} + E_{\text{rotational}} + E_{\text{vibrational}} + E_{\text{chemical}} + \cdots$$

It is important to note that internal energy, thermal energy, rotational energy, vibrational energy, chemical energy, nuclear energy, and so on are not new kinds of energy. The energy of a multiparticle system has only three fundamental components:

- rest energy
- kinetic energy
- potential energy

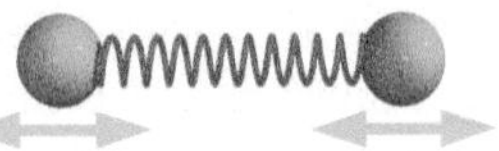

Figure 7.19 More vibration means more internal energy in the form of "vibrational energy."

We classify certain kinds of energy as internal energy mostly as a matter of convenience in analyzing energy changes in a complicated system. In fact, from a fundamental perspective, all of these internal energy increases can be thought of as contributing to an increase in the mass of the system: $M_{\text{sys}} = E_{\text{sys}}/c^2$. However, for many practical macroscopic calculations it makes sense to focus just on the changes in the thermal energy, or vibrational energy, or rotational energy, or deformational energy, or chemical energy, because these changes are extremely small compared to the huge total rest energy Mc^2.

Internal Energy at the Microscopic Level

At the microscopic level, a solid object is composed of atoms that are held together by spring-like chemical bonds. By considering the ball–spring model of a solid, we can see that an extended object can have internal kinetic energy associated with the motion of its component atoms and internal potential energy associated with the compression and stretch of interatomic bonds. Since all of the atoms in the solid are interconnected, this internal energy can flow throughout the system, and it ends up being distributed randomly among the many atoms and spring-like bonds. We call this energy "thermal energy." As the name suggests, this randomized energy at the microscopic level is associated with the temperature of an object—the greater the amount of thermal energy, the more internal motion there is, and the higher the temperature of the object.

Changes in the temperature of a macroscopic object are associated only with changes in its randomized thermal energy, and not with macroscopic rotations or vibrations. For example, if we increase the spin rate of a wheel that is spinning with negligible friction and air resistance, the rotational energy of the wheel increases, but the temperature of the wheel doesn't change. However, in some processes macroscopic kinetic energy may be converted to thermal energy. If two objects collide and stick to each other, some of their initial kinetic energy may be converted into random motions at the atomic level, with a corresponding increase in the temperature of the objects.

We can measure macroscopic spring potential energy and kinetic energy with simple measurements that don't require a microscope. However, when the temperature of a solid object increases, the increased energy inside the solid is not visible to the naked eye, and we face the problem of how to evaluate the increase in the thermal energy.

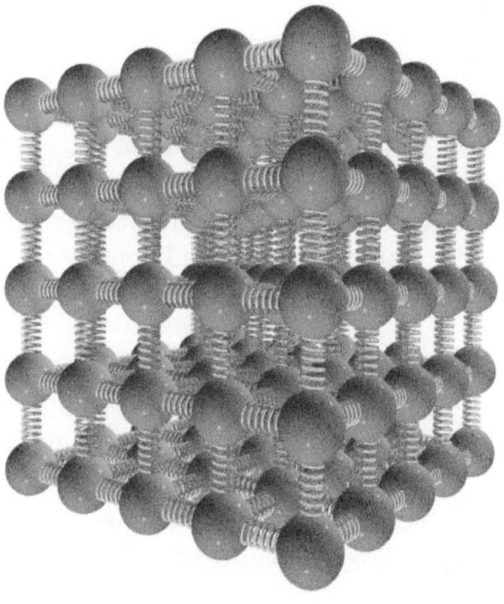

Figure 7.20 In our model solid we must account for the kinetic energy of every ball and the spring energy of every spring.

Microscopic Kinetic and Spring Potential Energy

The increased microscopic energy in a solid is in two forms. First, the atoms on average are moving around faster, with increased kinetic energy $\frac{1}{2}mv^2$. Second, there is increased spring potential energy $\frac{1}{2}k_s s^2$ in the interatomic bonds ("springs" in our model of solids), where k_s is the stiffness of an interatomic bond modeled as a spring (Figure 7.20).

In principle the internal energy in a solid could be evaluated by simultaneously measuring at some instant the momenta of *all* of the atoms, and the stretches or compressions of *all* of the springs (that is, changes in atomic positions away from their equilibrium positions, together with knowledge of the spring stiffness). Since there are about 1×10^{23} atoms (and even more "springs") in a macroscopic object, it is in practice impossible to evaluate the energy this way.

We might try to make microscopic measurements of just one atom (and its attached "springs"), assume that this is the average energy of each atom, and then multiply by the number of atoms there are in the object to get the total internal energy. That doesn't work either, because at a given instant we may choose an atom that is momentarily sitting motionless at its equilibrium position (zero $K + U$), but an instant later will be displaced away from its equilibrium position and moving rapidly. Energy keeps getting passed back and forth among the many atoms and springs, and measuring the energy of just one atom and its attached springs at one instant doesn't give us the correct average energy we would need to determine the total energy of the object.

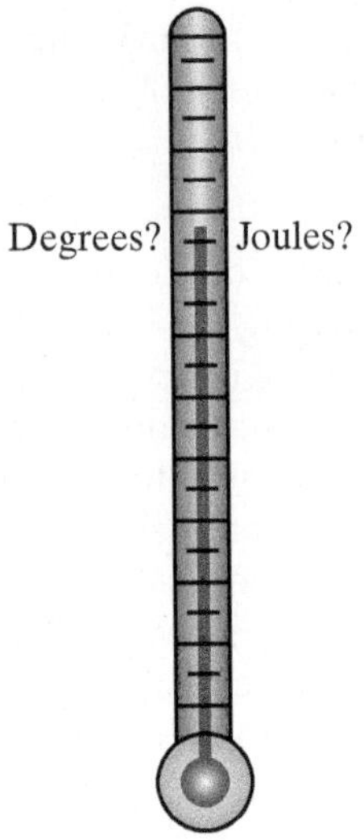

Figure 7.21 A thermometer might be calibrated in joules instead of in degrees.

Temperature

In the 1800s it was realized that for many systems the temperature is essentially a measure of the average random energy of the atoms in the system. Therefore we can use a thermometer to measure thermal energy. A familiar example is the mercury thermometer (silver colored) or the alcohol thermometer (tinted red for visibility). In these thermometers, a very thin column of liquid expands more than the glass when heated. By tradition, marks on the thermometer are placed to represent "degrees of temperature," but we could just as well place marks to indicate the average energy of the molecules in the thermometer, in joules (Figure 7.21). One kelvin (or one Celsius degree of temperature) is equivalent to an average molecular energy of about 1×10^{-23} J.

This is not the whole story. In Chapter 12 we will make more precise the relationship between temperature and thermal energy, and we will see that temperature is more directly related to a quantity called "entropy" than it is to energy. Nevertheless, for most ordinary systems at room temperature, the temperature of a system is a good measure of the average thermal energy of its atoms.

How does a thermometer work? When two objects are in good contact with each other, the average kinetic energies of the molecules in the two objects slowly come to be equal. At a location where two objects touch each other, molecules collide with each other, and the faster molecules on average lose kinetic energy to the slower molecules in a collision. Once the average kinetic energies of molecules in both objects have come to be equal, additional collisions on average make no further change.

To measure the thermal energy of an object, we place a thermometer in contact with the object. We wait for the temperatures (average molecular kinetic energies) of the object and thermometer to equilibrate, then we read the thermometer. A useful thermometer must have a relatively small mass compared to the object of interest, so that attaching the thermometer doesn't add or subtract much energy to or from the object. Another advantage of a small thermometer is that it will reach thermal equilibrium quicker than a large thermometer will.

How was the calibration between one kelvin (or one Celsius degree) and energy established? In one of a series of classic experiments performed by Joule in the 1840s, a paddle wheel in water was turned by a falling weight. The work done by the Earth turned the paddle wheel, which stirred the water. When the paddle wheel stopped, the increase in the internal energy of the water was equal to the work done by the Earth, assuming little rise in internal energy of the paddle wheel and container, and little energy transfer to the surroundings.

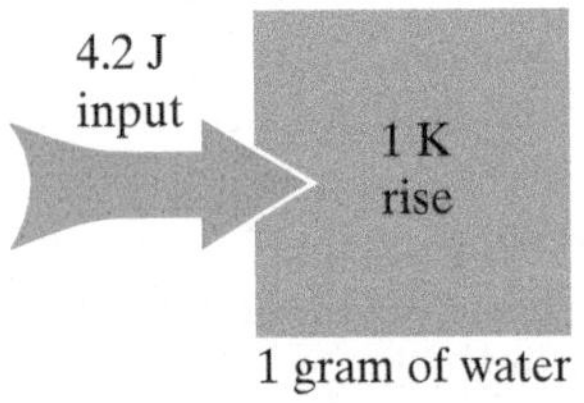

Figure 7.22 Energy input of 4.2 J into a gram of water raises the temperature by 1 K. We say that the specific heat of water is 4.2 J/K/g.

It was found for water that the energy required to raise the temperature of one gram of water by one kelvin (1 K) is 4.2 J. The "heat capacity" of an object is the amount of energy required to raise its temperature one kelvin. The "specific heat" is the heat capacity on a per-gram or per-mole or per-atom basis, and is a property of the material. The specific heat of water is 4.2 J/K/g (Figure 7.22). Other materials have different specific heats. We will study an atomic theory of specific heat in a later chapter.

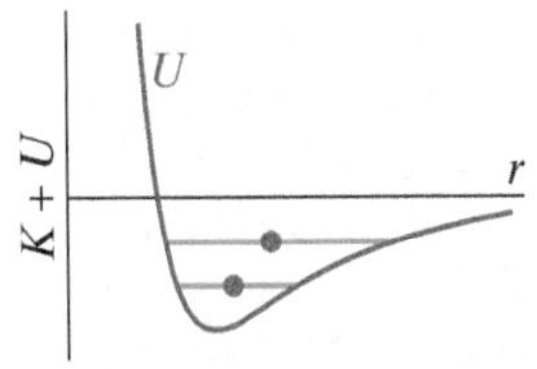

Figure 7.23 As $K + U$ (horizontal lines) increases, the average separation of the two bound atoms also increases.

Volume and Temperature

A liquid thermometer works because an ordinary liquid has a larger volume at higher temperature. Why does this happen?

For large oscillations of atoms whose interaction is described by a potential energy such as that shown in Figure 7.23, the average stretch increases. Because the actual potential energy curve is not really symmetric around the equilibrium point, with increasing energy the average interatomic bond length gets slightly longer. This is why an object typically expands at higher temperature. Higher temperature implies larger amplitude of oscillations and higher energy, and there is a slight shift in the center of the oscillations at higher energy.

In the interior of a solid the potential energy curve for an atom is symmetric, not asymmetric, because the potential energy is associated with forces exerted by atoms to the left and to the right. However, since the bond lengthening can start at the surface and propagate into the interior, a solid ultimately expands at higher temperatures.

Other Kinds of Thermometers

There are many other kinds of thermometers. All materials change in some way when they get hotter or colder, and some of these effects are the basis for useful temperature indicators. Liquid crystals are increasingly used in thermometers. In a liquid crystal there is some molecular order over long distances in the liquid, unlike a true liquid in which there is hardly any long-range order. In one form of liquid crystal thermometer, helical molecules remain roughly parallel to each other, and the distance between coils in the helix changes with temperature. Because the distance between coils is comparable to the wavelength of visible light, there is an effect on the way light is reflected by or transmitted through the liquid crystal. The result is that a change in temperature affects the appearance of the material.

Calculations Using Specific Heat

Joule found that it takes 4.2 J of energy input to a gram of water to raise its temperature 1 K. If such experiments are done on other materials, the temperature rise is found to be different. For example, the specific heat of ethanol is found to be 2.4 J/K/g, and the specific heat of copper is only 0.4 J/K/g. One of the many unusual and useful properties of water is that it has a very large specific heat on a per-gram basis, which means that it is difficult to change its temperature. Here is the meaning of specific heat on a per-gram basis, where $\Delta E_{\text{thermal}}$ is the rise in the internal energy of the system, in the form of increased atomic kinetic and potential energy:

SPECIFIC HEAT *C* ON A PER-GRAM BASIS

$$C = \frac{\Delta E_{\text{thermal}}}{m\Delta T} \quad (m \text{ in grams})$$

In principle, it should be possible to predict the specific heat of a material if something is known about its atomic structure, since a temperature rise is a measure of increased energy at the atomic level. Comparisons of calculations with experimental values of the specific heat data are a good test of our understanding of atomic models of solids, liquids, and gases. In later chapters you will learn how to predict the specific heat of a solid (based on the ball-and-spring model of a solid) and of a gas.

If the specific heat of a material is known, the amount of energy transfer into an object can be determined by measuring the temperature rise.

EXAMPLE **Energy Input Raises the Temperature**

You stir 12 kg of water vigorously, doing 36,000 J of work. If the container is well insulated (so that all of your energy input goes into increasing the energy of the water), what temperature rise would you expect?

Solution

$$\Delta E_{sys} = W$$
$$mC\Delta T = W$$
$$\Delta T = \frac{W}{mC} = \left(\frac{36{,}000\ \text{J}}{12{,}000\ \text{g}}\right)\left(\frac{1}{4.2\ \text{J/K/g}}\right) = 0.7\ \text{K}$$

Including units helps avoid making calculational mistakes. Note that a lot of work is required to achieve a small temperature change.

EXAMPLE **Thermal Equilibrium**

A 300 g block of aluminum at temperature 500 K is placed on a 650 g block of iron at temperature 350 K in an insulated enclosure. At these temperatures the specific heat of aluminum is approximately 1.0 J/K/g, and the specific heat of iron is approximately 0.42 J/K/g. Within a few minutes the two metal blocks reach the same common temperature T_f. Calculate T_f.

Solution System: The two blocks
Surroundings: Due to the insulation, no objects exchange energy with the chosen system
Initial state: Different temperatures, not in contact
Final state: Blocks have come to thermal equilibrium, same T_f
Energy Principle: $\Delta E_{Al} + \Delta E_{Fe} = 0$ (The total energy of the two blocks does not change, because there is no energy transfer from or to the surroundings.)

$$m_1C_1(T_f - T_{1i}) + m_2C_2(T_f - T_{2i}) = 0$$

$$(300\,\text{g})(1.0\ \text{J/K/g})(T_f - 500) + (650\ \text{g})(0.42\ \text{J/K/g})(T_f - 350) = 0$$

Solving for the final temperature, we find $T_f = 429$ K.

In words, what happens in this process is that the aluminum temperature falls from 500 K to 429 K, and the iron temperature rises from 350 K to 429 K. The thermal energy decrease in the aluminum is equal to the thermal energy increase in the iron.

What Part of Internal Energy is "Thermal"?

What we mean by a change of "thermal" energy is that part of the internal energy that is associated with a temperature change. When two cars hit head-on and stop dead, their lost macroscopic kinetic energies show up as increased internal energy. A thermometer will register a temperature increase in the metal parts, and we say there is increased thermal energy $\Delta E_{thermal} = mC\Delta T$. This "random" energy does not account for all of the increased internal energy. In such collisions some of the internal energy rise is in the form of locked-in potential energy in compressed interatomic bonds, a "nonrandom" form of internal energy.

Suppose you compress a spring, then hold it compressed with a rubber band. Simply sliding the rubber band off the spring will release the energy stored in the spring. Similarly, one can get back some of the stored energy in deformed metal by heating the metal to free up the locked-in "springs."

In many situations it isn't possible to say how much of the internal energy is thermal, but if the heat capacity is known, we can use a thermometer to measure a *change* in the thermal energy.

Checkpoint 4 300 g of water whose temperature is 25°C are added to a thin glass containing 800 g of water at 20°C (about room temperature). What is the final temperature of the water? What simplifying assumptions did you have to make in order to determine your approximate result?

7.5 ENERGY TRANSFER DUE TO A TEMPERATURE DIFFERENCE

Work is mechanical energy transfer into or out of a system involving forces acting through macroscopic displacements. When a hot object is placed in contact with a cold object, energy is transferred from the hot object to the cold one, but there are no macroscopic forces or displacements, so we don't refer to this kind of energy transfer as work.

Instead, we speak of "energy transfer Q due to a temperature difference." At the microscopic level there is actual work; when a hot block is placed in contact with a cold block (Figure 7.24), at the interface the atoms in the two blocks collide with each other, and do work on each other. The atoms in the hot block have greater average kinetic energy than the atoms in the cold block, so in an individual collision it is likely that a fast-moving atom in the hot block loses energy to a slow-moving atom in the cold block.

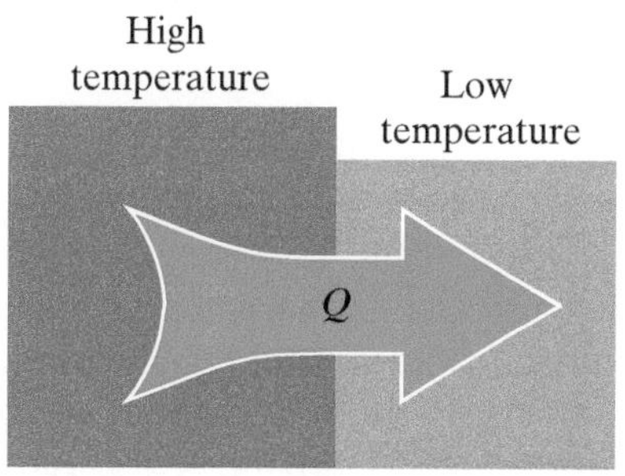

Figure 7.24 A hot block in contact with a cold block.

It can happen that the atom in the hot block happens to be moving slowly and gains energy when it is hit by an atom in the cold object that happens to be moving fast, but this is less likely. On average there is energy flow ("microscopic work") from the hot block to the cold block. These energy changes will diffuse throughout the blocks. If left to themselves, the two blocks will eventually come to the same temperature, intermediate between their two initial temperatures.

Q = ENERGY TRANSFER DUE TO A TEMPERATURE DIFFERENCE

The symbol Q represents an amount of energy that flows from the surroundings into a system, due to a temperature difference between the system and the surroundings. One can also call this process "microscopic work." Q can be positive or negative.

If we choose the cold block as our system of interest, this system gains energy (at the expense of the surroundings, the hot block). At the atomic level, this energy increase is the result of work done on the atoms at the surface of the cold block. Usually we are unable to observe these atomic-level interactions directly. We can infer the amount of energy transfer into the system by observing the temperature rise of that system, if we know the specific heat of the material.

It is common practice to denote energy transfer due to a temperature difference (microscopic work) by the letter Q and work (macroscopic, mechanical energy transfer) by the letter W. The energy equation for an "open" (not isolated) system is then

$$\Delta E_{\text{sys}} = Q + W$$

That is, energy transfer due to a temperature difference and mechanical energy transfers into or out of the system produce a change in the energy of the system. The energy of the system includes particle energy and potential energy, at the microscopic or macroscopic level.

Sign of Q

Like W, Q can be negative. If there is energy transfer out of a system, we give Q a negative sign, because the change in the energy of the system is a decrease. See Figures 7.25 and 7.26. Typically such transfer would be due to the system having a higher temperature than the surroundings. Similarly, if the system does work on its surroundings rather than the other way around, the sign of W is negative. (Some older textbooks on thermal physics reverse the sign of W, counting energy outputs from the system as positive.)

It is thought that for the Universe as a whole Q and W are zero, and the energy of the Universe does not change. However, even a small system may be effectively closed, with its energy unchanging, if it is thermally well insulated (to prevent the flow of energy due to a temperature difference) and if there are no other energy transfers from the surroundings.

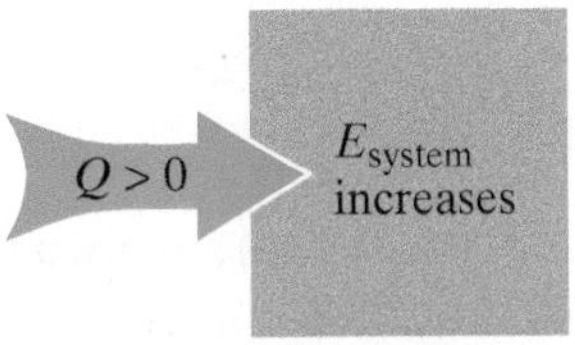

Figure 7.25 If energy flows from the surroundings into the system, Q is positive.

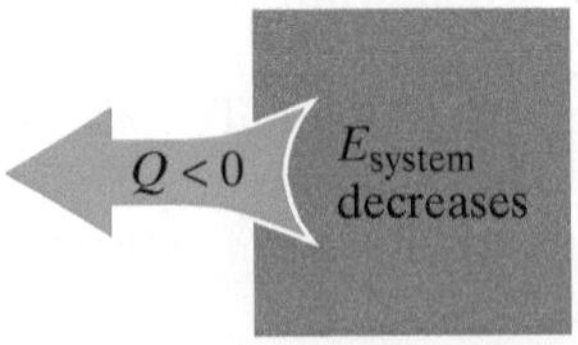

Figure 7.26 If energy flows from the system into the surroundings, Q is negative.

Terminology: The Word *Heat*

The technical meaning of the word "heat" in science is not the same as its everyday meaning, just as the scientific meaning of the word "work" is different from its everyday meaning. In science the word "heat" is reserved to refer to energy transfer across a system boundary due to a temperature difference (microscopic work); this quantity is represented by the symbol Q. One cannot say "there is heat in the object"; instead, one says there is thermal energy $E_{thermal}$ in the object. The amount of energy in the object might increase due to heat Q (microscopic work). However, even professional scientists sometimes slip in their usage of the word heat!

To avoid confusion, we will avoid the use of the word heat as a noun. Instead, we will speak of "energy transfer Q due to a temperature difference" between system and surroundings, and change of "thermal energy" inside the system $\Delta E_{thermal}$ (in the form of increased atomic kinetic energy and potential energy).

Checkpoint 5 Suppose that you warm up 500 g of water (half a liter, or about a pint) on a stove while doing 5×10^4 J of work on the water with an electric beater. The temperature of the water is observed to rise from 20°C to 80°C. What was the change in the thermal energy of the water? Taking the water as the system, how much transfer of energy Q due to a temperature difference was there across the system boundary? What was the energy change of the surroundings?

Other Kinds of Energy Transfers

Work W and energy transfer Q due to a temperature difference are very common kinds of energy transfers into or out of an open system, but there are several other kinds of energy transfer. Here is a list, together with examples of each type.

- W, mechanical work $\vec{F} \bullet \Delta\vec{r}$, in which a force acts through a macroscopic distance: Throw a ball, increasing the ball's kinetic energy; compress a macroscopic spring.
- Q, energy transfer due to a temperature difference between system and surroundings: Boil water over a hot flame; a warm house loses energy to the cold outside air.
- Matter transfer: Put gasoline into a car; natural gas is piped into the home.
- Mechanical waves: Sound enters a microphone and moves a plate to generate electric signals; ocean waves drive an electric generator.
- Electricity: Electrical current enters the house and is used to run appliances; electrical current stores energy in the batteries of an electric car.
- Electromagnetic radiation, including visible light, infrared light, gamma rays, x-rays, and ultraviolet light: Sunlight absorbed by the Earth raises the temperature (and the thermal energy) of the daylight side of the Earth. We will usually account for this kind of energy transfer in terms of photons

(packets of electromagnetic energy) crossing the system boundary, to be discussed in a later chapter.

A general form of the Energy Principle is this:

THE ENERGY PRINCIPLE

$$\Delta E_{sys} = W + Q + \text{other energy transfers}$$

Other energy transfers into a system may include matter transfer, mechanical waves, electric current, and electromagnetic radiation.

For historical reasons, many textbooks dealing with thermal physics define work W as the work done *by* a system rather than work done *on* a system, so the sign of W is changed. We define W as the work done on the system to be consistent with other, non-thermal uses of the Energy Principle.

Checkpoint 6 An electric hot plate raises its own internal energy and the internal energy of a cup of water by 8000 J, and there is at the same time 1000 J transferred to the cooler air (that is, $Q = -1000\,\text{J}$). How much energy was transferred to the hot plate in the form of electricity?

The Steady State

A system is in a *steady state* if its energy does not change despite the fact that there are energy transfers between the system and the surroundings. In the steady state, the rates of energy flow into and out of the system must be equal.

EXAMPLE **An Electric Heater**

An electric heater receives an energy input of 5000 J of electric energy. During this time the heating element is maintained at a constant high temperature. What is $\Delta E_{thermal}$ for the heater, and what is Q, the energy transfer between the heater's hot heating element and the cooler air?

Solution System: Heater
Surroundings: Electric wires, air
Energy Principle: $\Delta E_{sys} = W + Q + \text{other energy transfers}$

Since there is no internal energy change, $\Delta E_{thermal} = \Delta E_{sys} = 0$. This is a steady-state situation.

$$\Delta E_{sys} = W + Q + \text{electric energy input}$$
$$0 = 0 + Q + 5000\,\text{J}$$
$$Q = -5000\,\text{J}$$

In this situation the sign of Q is negative, reflecting the fact that energy is transferred from the heating element to the air. We describe this situation as a steady state, not equilibrium, because there are energy transfers into and out of the system. However, these energy transfers, plus and minus, add up to zero, so there is no change internally.

If the heater were in an insulating enclosure and no energy could flow into the air, as electric energy continually flowed into the heater the temperature of the heating element would keep increasing until the element melted.

QUESTION Is Q the same as $\Delta E_{thermal}$?

No. This is an important distinction. Q represents an energy transfer between system and surroundings due to a temperature difference. $\Delta E_{thermal}$, the change of the thermal energy of a system, can differ from Q, because there may be other positive or negative energy inputs to the system. In the previous example of the electric heater, the thermal or internal energy of the heating element did not change ($\Delta E_{sys} = 0$, steady state), but $Q = -5000$ J.

Checkpoint 7 In a certain time interval, natural gas with energy content of 10,000 J was piped into a house during a winter day. In the same time interval sunshine coming through the windows delivered 1000 J of energy into the house. The temperature of the house didn't change. What was $\Delta E_{thermal}$ of the house, and what was Q, the energy transfer between the house and the outside air?

7.6 POWER: ENERGY PER UNIT TIME

In technical usage, the word "power" is defined to mean "energy per unit time." From the point of view of energy usage, it makes no difference whether you take a minute or a month to climb a flight of stairs, yet from a practical point of view you certainly do notice the *rate* of energy usage, called power. The units of power are joules per second, or watts (honoring James Watt, the developer of the first efficient steam engine).

POWER

Energy per unit time (J/s or W)

We can construct an equation for the instantaneous power associated with the work done by a force, through the following reasoning. If a force $\vec{F}$ does an amount of work $\vec{F}\bullet\Delta\vec{r}$ in a time Δt, then:

$$\text{power} = \frac{\vec{F}\bullet\Delta\vec{r}}{\Delta t} \rightarrow \vec{F}\bullet\frac{d\vec{r}}{dt}$$

INSTANTANEOUS POWER

$$\text{power} = \vec{F}\bullet\vec{v}$$

EXAMPLE **Light Bulb Power**

How much energy is required to run a 100 W light bulb for an hour?

Solution

$$(100\text{ J/s})(1\text{ h})\left(60\frac{\text{min}}{\text{h}}\right)\left(60\frac{\text{s}}{\text{min}}\right) = 3.6\times10^5\text{ J}$$

Checkpoint 8 A vehicle with a mass of 1000 kg has an engine whose maximum power output is 50 kW (about 67 hp; 1 hp is 746 W). At a speed of 20 m/s (about 45 mi/h) and maximum power, determine the maximum acceleration by calculating the force that is acting.

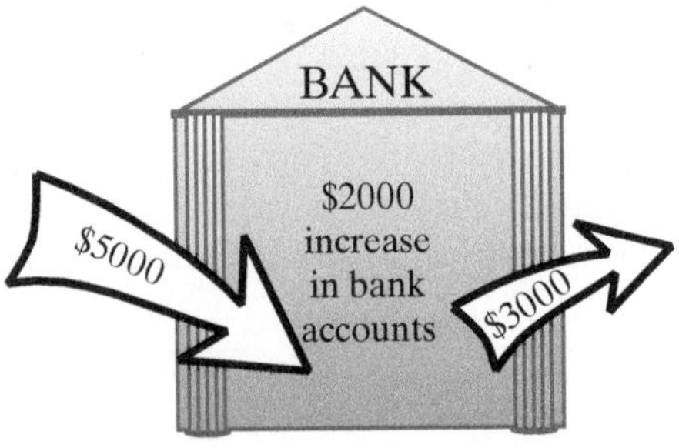

Figure 7.27 A bank is an open system.

7.7 OPEN AND CLOSED SYSTEMS

We have already had some experience with the difference between a closed system and an open system. Because of the importance of this distinction, in this section we will go more deeply into the issues. Figure 7.27 shows transfers of money into and out of a bank during a particular time period, and the corresponding change in accounts inside the bank.

> QUESTION We have shown a situation in which more money is transferred into the bank than is transferred out. During these transactions, does the total amount of money inside the bank remain unchanged?

In Figure 7.27 \$5000 came in and \$3000 went out, for a net gain of \$2000. We say that the bank is an "open system"—a portion of the world open to transfers in and out, and therefore subject to changes in its internal amount of money. During times when the bank does not permit transfers, the bank is temporarily a "closed system" and its total internal amount of money is unchanged, although there may be changes in the form of money inside the bank such as transfers between checking accounts and savings accounts. Now consider Figure 7.28, which shows energy transfers during a particular time period in the winter, into and out of a house that is heated by gas, and the corresponding change in the energy inside the house. In this situation more energy comes into the house than goes out, and as a result the temperature inside the house rises.

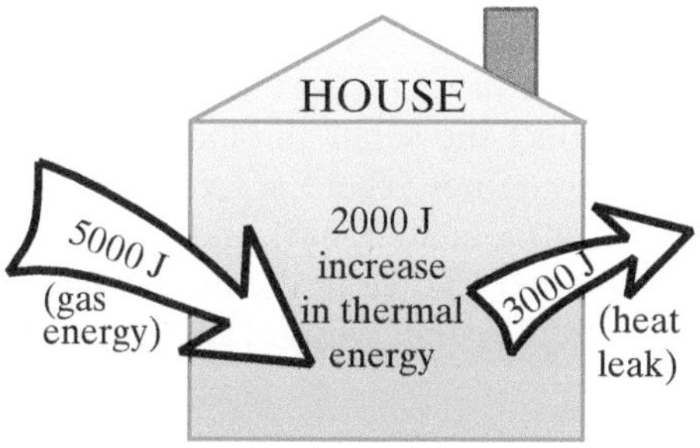

Figure 7.28 A house is an open system.

> QUESTION During these energy transactions, the total amount of energy in the whole Universe is unchanged, but is the amount of energy inside the house unchanged?

The amount of energy inside the house increased by 2000 J, and the energy of the surroundings decreased by 2000 J. We say that the house is an "open system"—a portion of the Universe open to energy transfers, and therefore subject to changes in its internal amount of energy. If the house could be perfectly insulated against heat leakage (and solar radiation) and the gas (and electricity) turned off, the house would be a "closed system" with respect to energy and its total internal amount of energy would be unchanged, although there may be changes in the form of energy inside the house, such as the family dog converting chemical energy into kinetic energy when chasing its tail.

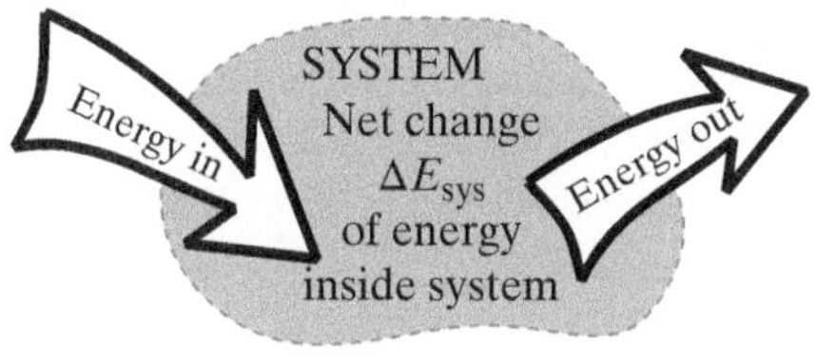

Figure 7.29 The change of energy of the system is equal to energy transferred into the system across the system boundary minus energy transferred out of the system across the system boundary.

These considerations lead to the following scheme for keeping track of energy. Choose some portion of the Universe and mentally surround it by a dashed line marking the boundary (Figure 7.29). Then the energy transfer into the system across the system boundary minus the energy transfer out of the system across the system boundary is equal to the change of the energy inside the system.

The change in energy *inside* the system, ΔE_{sys}, can be either a positive or a negative amount of energy. For example, take an electric car as the system of interest:

> Charge the battery: ΔE_{sys} is positive (more energy stored in the battery).
>
> Do work on the car by pushing it: ΔE_{sys} is positive (more kinetic energy).
>
> Run the headlights (radiate electromagnetic energy): ΔE_{sys} is negative (less energy stored in the battery).

The car is an open system whose total amount of energy changes due to energy transfers into or out of the system: the energy of an *open* system is *not* constant.

In contrast, the total energy of the Universe remains constant, because there are compensating energy changes in the surroundings of the car: the electric company lost a store of chemical energy by running its generators to charge the car, you used up some chemical energy to push the car, and the light from the headlights is absorbed by the surroundings and leads to a rise in temperature of the surroundings. The measurements that we are able to make confirm the premise that the Universe is a closed system whose total amount of energy never changes, although the form of the energy may change.

For any closed system, *inflow* = *outflow* = 0, so $\Delta E_{sys} = 0$.
The energy of a closed system does not change.

The Universe as a whole is the most important example of a closed system, but it is often the case that a portion of the Universe can be considered to be a closed system, at least approximately. For example, put hot water and ice cubes into a very well insulated container. During the short time that the ice melts and the water gets somewhat cooler, we can neglect the small amount of energy leakage through the walls of the insulated container. There is an increase in the energy of the ice cubes (which changed from solid to liquid), and a decrease in the energy of the hot water, but negligible net change in the energy inside the container, which is approximately a closed system whose total energy is (approximately) unchanged.

7.8 THE CHOICE OF SYSTEM AFFECTS ENERGY ACCOUNTING

In Section 6.7 we analyzed a falling ball in two ways. When we chose the ball alone as the system, there was no potential energy, but the surroundings did work on the ball. Alternatively, when we chose the ball plus the Earth as the system, no external work was done, but the potential energy of the system changed as the ball fell. The form of the energy equation depended on our choice of system, but the physical results were consistent.

Let's explore this idea further in the case of a complicated system in which internal energy is important. We will analyze the same problem three times, using three different choices of system. As before, we will find that the form of the Energy Principle will be different in each case, but the results of each analysis are consistent.

Figure 7.30 A woman lifts a barbell a distance h, exerting a force F. In the final state the barbell is moving upward with speed v.

The Woman, the Barbell, and the Earth

Consider a somewhat complicated situation: a woman applying a constant force F to lift a barbell of mass m from rest through a distance h, at which point the barbell is not only higher above the Earth but also has acquired some speed v (Figure 7.30). We're considering a time when the barbell is still headed upward, before the woman has brought the barbell to a stop above her.

For convenience in the following discussion, let E_w represent the energy of the woman, with the following energy terms:

$$\begin{aligned} E_w = {} & \text{internal energy of woman} + \\ & \text{kinetic energy of woman (moving arm)} + \\ & \text{gravitational energy of woman and Earth} \end{aligned}$$

System: Woman + Barbell + Earth

System: Woman + barbell + Earth (Figure 7.31)
Surroundings: Nothing significant
Initial state: Barbell at rest
Final state: Barbell has moved upward a distance h, and has speed v

$$E_f = E_i + W$$

$$K_f + U_f + E_{w,f} = K_i + U_i + E_{w,i} + W$$

$$\frac{1}{2}mv^2 + mgh + E_{w,f} = 0 + 0 + E_{w,i} + 0$$

$$\Delta E_w = -\left(\frac{1}{2}mv^2 + mgh\right)$$

Figure 7.31 System: woman + barbell + Earth.

Because the woman was part of the chosen system, we were able to calculate the energy term ΔE_w, which represents changes in energy associated with the woman. Her internal energy change involves both a decrease in the chemical energy stored in her body and an increase in her thermal energy. The woman must expend more energy than the rest of the system gets, since she not only has to supply energy to the rest of the system but also has to supply her own energy increases: increased kinetic energy and gravitational potential energy associated with her moving, raised arm, and increased thermal energy, because her temperature rises a bit when she exerts herself.

System: Barbell

System: Barbell (Figure 7.32)
Surroundings: Woman and Earth
Initial state: Barbell at rest
Final state: Barbell has moved upward a distance h, and has speed v

Figure 7.32 System: barbell only.

$$E_f = E_i + W$$
$$K_f = K_i + W_{\text{woman}} + W_{\text{Earth}}$$
$$\frac{1}{2}mv^2 = 0 + Fh - mgh$$

QUESTION Why didn't we include potential energy in the energy equation?

Since the chosen system consists of a single object (the barbell), there are no interactions between objects within the system, and hence no potential energy.

QUESTION Which is bigger, F or mg? Why?

Since the barbell's upward momentum increases, the force F exerted by the woman must be larger than mg, since the net force in the upward direction is $F - mg$.

By choosing the barbell alone as the system, we were able to calculate the change in kinetic energy of the barbell—a quantity we could not find when we chose all three objects as the system.

QUESTION Can we relate ΔE_w to the force exerted by the woman?

We can compare our results for system 1 (woman + barbell + Earth) and system 2 (barbell only).

$$\Delta E_w = -\left(\frac{1}{2}mv^2 + mgh\right)$$
$$Fh = \frac{1}{2}mv^2 + mgh$$
$$\text{So } \Delta E_w = -Fh$$

The sign makes sense: we had concluded earlier that her energy change must be negative.

System: Barbell + Earth

System: Barbell + Earth
Surroundings: Woman
Initial state: Barbell at rest

Figure 7.33 System: barbell + Earth.

Final state: Barbell has moved upward a distance h, and has speed v

$$E_f = E_i + W$$
$$K_f + U_f = K_i + U_i + W_{\text{woman}}$$
$$\frac{1}{2}mv^2 + mgh = 0 + 0 + Fh$$

If we choose the system to consist of the barbell plus the Earth (Figure 7.33), this system changes configuration as the woman pushes the two pieces of the system apart. The result we get with this choice of system is consistent with our previous results, even though in the previous case (barbell alone) the term $-mgh$ corresponded to work done by the surroundings, whereas when we chose the Earth plus the barbell we found a potential energy change of $+mgh$.

QUESTION Does the woman do any work on the Earth? Why or why not?

The woman's feet push down on the Earth but do negligible work, because there is essentially no displacement there (the Earth hardly moves). Work done by a force is calculated by taking into account the displacement of the point where the force is applied. In this case, the point of application hardly moves, so the work done by the force is negligible.

Looking back over the equations obtained for different choices of system, you can see how energy terms that represent transfers across the system boundary for one choice of system become changes of energy *inside* the system for a different choice of system. Comparing equations for different choices of system can be useful in determining an unknown quantity (such as the energy change in the woman, $-Fh$, in the example you just worked through). Also, analyzing a process for more than one choice of system is a good check on your calculations and your understanding.

Checkpoint 9 Consider a harmonic oscillator (mass on a spring without friction). Taking the mass alone to be the system, how much work is done on the system as the spring of stiffness k_s contracts from its maximum stretch A to its relaxed length? What is the change in kinetic energy of the system during this motion? For what choice of system does energy remain constant during this motion?

7.9 THE CHOICE OF REFERENCE FRAME AFFECTS ENERGY ACCOUNTING

We have just seen how the choice of system affects energy accounting. The Energy Principle is valid independent of our choice of system, but the terms that appear in the energy equation are different for different choices of system. In a similar way, the Energy Principle is independent of our choice of (inertial) reference frame, but the terms that appear in the energy equation are different for different choices of reference frame. It is instructive to look at the specific example of a compressed spring launching a block, as seen from two different reference frames.

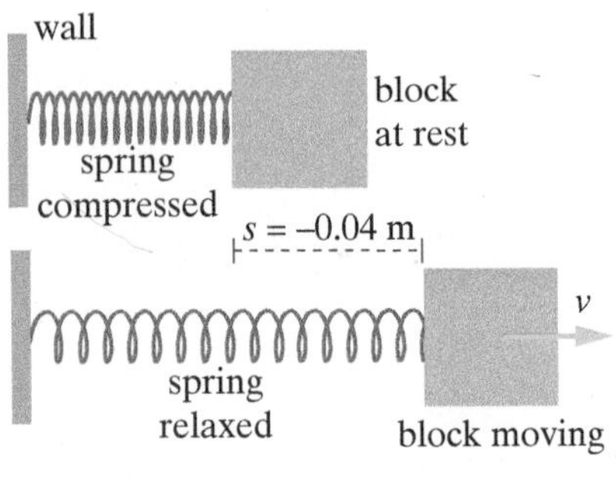

Figure 7.34 A block at rest is launched by a compressed spring.

Consider a spring with stiffness 400 N/m attached on the left to a wall, with a block of mass 1 kg held at rest against the spring (Figure 7.34). The spring is compressed, with $s = -0.04$ m. You release the block from rest, and the spring expands, accelerating the block on a low-friction surface. Choose the spring plus the block as the system.

A Stationary Reference Frame

First, let's view the system from a stationary reference frame, as usual. Assume that you, the observer, are standing motionless beside the wall.

QUESTION How much work is done on the spring–block system?

The upward force of the surface and the downward force of the Earth do no work, because there is no displacement in the y direction. Similarly, the wall does no work, because the force that the wall exerts on the left end of the spring acts through zero distance and does no work ($F_{\text{wall}}\Delta x = 0$). The wall force does increase the momentum of the system by providing an impulse $F_{\text{wall}}\Delta t$, but it does not affect the energy of the system. Here is the Energy Principle, with the initial state being just after release and the final state being when the spring reaches its relaxed length:

$$K_f + U_f = K_i + U_i$$
$$\frac{1}{2}mv^2 + 0 = 0 + \frac{1}{2}k_s s^2$$
$$\frac{1}{2}(1\text{ kg})v^2 = \frac{1}{2}(400\text{ N/m})(0.04^2\text{ m}^2) = 0.32\text{ J}$$
$$v = 0.8\text{ m/s}$$

A Moving Reference Frame

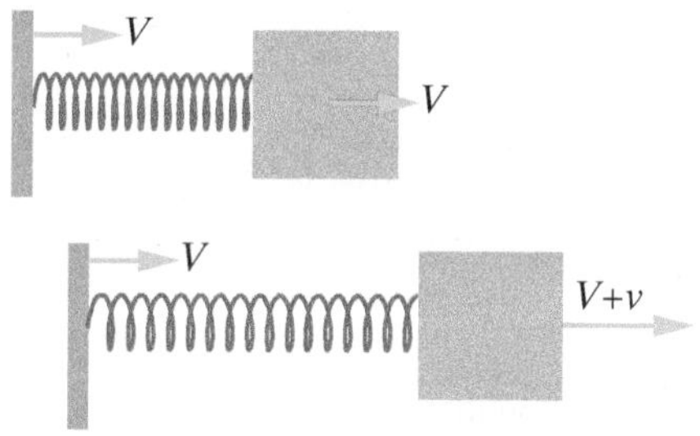

Figure 7.35 In a reference frame moving to the left with speed V, the wall moves to the right with speed V, and the initial speed of the block is also V. At a later moment the block is moving faster than the wall, with some speed $V+v$.

Now let's view the same system from a moving reference frame. Suppose you (the observer) move to the left with constant speed $V = 10$ m/s. In this altered perspective, the wall has a constant velocity of 10 m/s to the right, and the block's initial velocity is also 10 m/s to the right (Figure 7.35). You would see the system the same way if you stood still and the apparatus moved to the right. We again choose the spring plus the block as the system.

QUESTION In this moving reference frame, is the work done by the wall on the left end of the spring negative, zero, or positive?

In the original stationary reference frame, with the wall always at rest, the wall force did no work because the point of application of the wall force did not move; there was zero displacement. Now, however, the wall is moving at a speed of 10 m/s, and during the expansion of the spring the wall force acts through a significant distance to the right, doing positive work W on the spring.

In this moving reference frame the initial speed of the block is not zero but $V = 10$ m/s, and, using the results of the previous analysis, we know that its final speed is $V + v = 10.8$ m/s, because it is moving 0.8 m/s faster to the right than the wall, which is moving at 10 m/s to the right. We can calculate the work done by the wall in this reference frame:

$$K_f + U_f = K_i + U_i + W$$
$$\frac{1}{2}m(V+v)^2 + 0 = \frac{1}{2}mV^2 + \frac{1}{2}k_s s^2 + W$$
$$\frac{1}{2}(1\text{ kg})(10.8\text{ m/s})^2 = \frac{1}{2}(1\text{ kg})(10\text{ m/s})^2 + \frac{1}{2}(400\text{ N/m})(0.04^2\text{ m}^2) + W$$
$$58.32\text{ J} = 50\text{ J} + 0.32\text{ J} + W$$
$$W = 8\text{ J}$$

Notice that the initial and final potential energies of the spring were the same in both reference frames because spring potential energy only depends on the stretch of the spring, which is the same in both reference frames.

Checking Our Assumptions

In our calculation of the work done by the wall in the moving reference frame, we assumed that the Energy Principle functions correctly in both reference frames. To check this assumption, we can integrate $F_x dx$ to calculate the work done by the wall directly, in the moving reference frame, without using results obtained from the stationary reference frame. If the work is indeed 8 J, our assumption will be justified.

The force exerted by the wall on the spring is $-k_s s$. During the process of interest the stretch of the spring depends on time, and can be expressed as

$$s = (-0.04\text{ m})\cos\left(\sqrt{\frac{k_s}{m}}t\right)$$

Initially the spring is compressed: $s = -0.04$ m. At the end of the process $s = 0$, because the spring has reached its relaxed length. The duration of this process is one-quarter of the total oscillation period T of the spring–mass system, which is

$$T = 2\pi\sqrt{\frac{m}{k_s}} = 2\pi\sqrt{\frac{1\text{ kg}}{400\text{ N/m}}} = 0.314\text{ s}$$

During each time interval dt the wall moves a distance $dx = Vdt$. So the work done by the wall on the spring–mass system is:

$$\begin{aligned}
W &= \int_{x_i}^{x_f} F_x dx \\
&= \int_0^{T/4} -k_s s V dt \\
&= \int_0^{0.314/4} (-400\text{ N/m})(-0.04\text{ m})\cos\left(\sqrt{\frac{400\text{ N/m}}{1\text{ kg}}}t\right)(10\text{ m/s})dt \\
&= 8\text{ J}
\end{aligned}$$

This direct calculation of the work done by the wall agrees with our previous result.

In changing from the original reference frame to the moving one, the work done by the wall changed from 0 to 8 J, the initial kinetic energy of the block changed from 0 to 50 J, and the final kinetic energy changed from 0.32 J to 58.32 J. All of these individual energy values changed a lot, but the Energy Principle is still valid. This is an example of the relativity principle, that physics principles are valid in all uniformly moving reference frames.

7.10 ENERGY DISSIPATION

The total energy of the Universe does not change, but useful energy is often "dissipated" into forms less useful to us. Push a chair across the floor, and some of the work that you do goes into raising the temperature of the floor and the chair, rather than into increasing the macroscopic kinetic energy of the chair. Throw a ball up into the air, and some of the initial energy is dissipated into increased microscopic energy of the air. Sliding friction, air resistance, viscous friction—all of these phenomena are examples of energy dissipation discussed in the rest of this chapter.

Air Resistance

Air resistance isn't a major factor when you drop a metal ball a short distance. A video sequence of a falling metal ball shows that the ball moves faster and faster as it falls (an effect that is hard to observe by eye alone). The time between adjacent frames in Figure 7.36 is 1/15 s, and the increasing distances between heights of the ball in adjacent frames show that the speed is getting faster and faster. (Also note the increasing blur due to faster motion of the ball.) The gravitational attraction of the Earth acting on the ball makes the momentum of the ball continually increase. A curve is drawn along the tops of the ball images. The major visible marks on the vertical meter sticks are 10 cm apart. In 7/15 s the ball falls about 1 m.

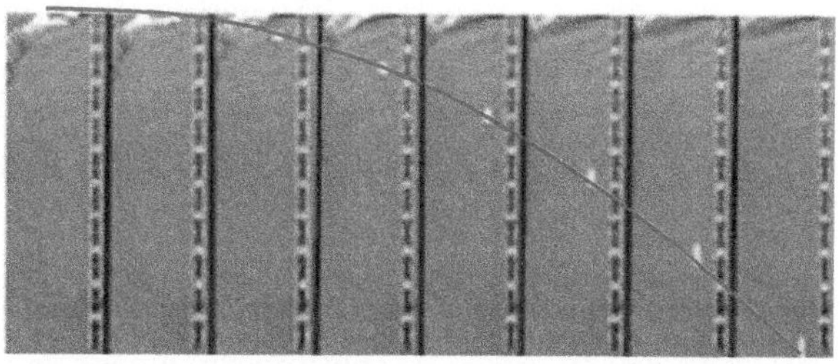

Figure 7.36 A sequence of video frames of a falling metal ball, which travels farther in each frame, reflecting the increase in its speed. (Image courtesy of Stacy Benson)

In contrast, if you drop a flat-bottomed paper coffee filter, a video sequence of the falling filter (Figure 7.37) shows that the filter's speed does increase at first, but instead of continuing to gain speed it quickly reaches a constant speed despite the gravitational force acting on it.

In the video sequence shown in Figure 7.37, the time between adjacent frames is again 1/15 s. In 16/15 s the filter falls about 1 m (the dense ball took only 7/15 s to fall that far). A curve is drawn through the centers of the coffee filter images. The nearly straight line in the later part of the motion indicates motion at constant speed.

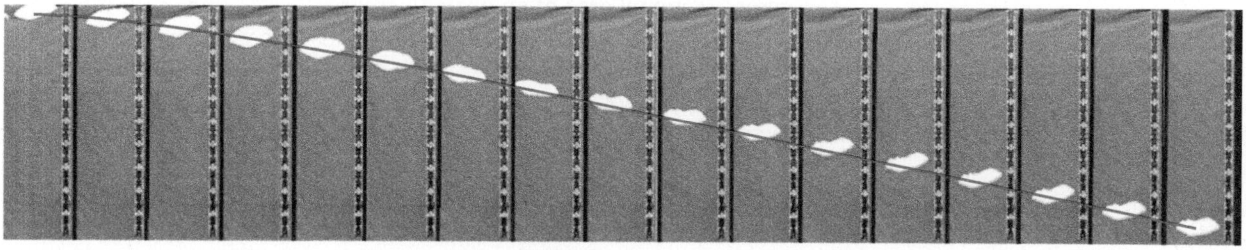

Figure 7.37 A sequence of video frames of a falling coffee filter, which speeds up briefly, then falls at constant speed. (Image courtesy of Stacy Benson)

Dependence of Air Resistance on Speed

The restraining effect of the air is called "air resistance" or "drag." You have probably observed that it is harder to move something quickly through a fluid such as water or air than to move it slowly. This suggests that air resistance acting on a falling object might depend on the speed of the object.

QUESTION If the air resistance force on a falling object increases as the speed of the object increases, how will this affect the net force on the object?

Since the downward gravitational force on the falling object does not change, while the upward air resistance force increases as the object speeds up, the net downward force on the object should get smaller and smaller, and eventually reach zero (Figure 7.38).

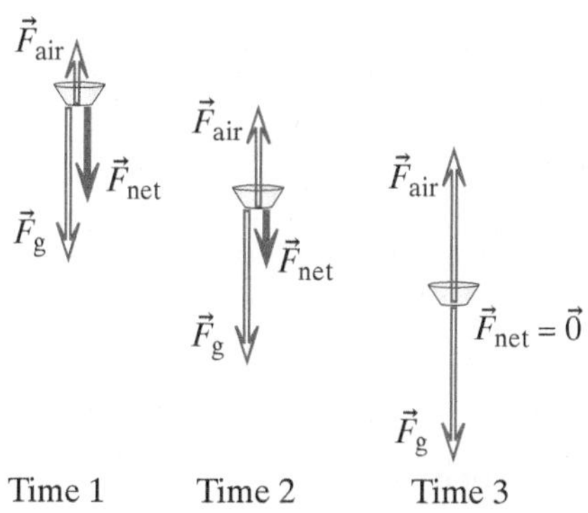

Figure 7.38 Forces acting on a falling coffee filter at three times during its fall.

QUESTION If the net force on the falling coffee filter becomes zero, why doesn't the filter just float motionless in the air?

Consider the Momentum Principle:

$$\vec{p}_{\text{future}} = \vec{p}_{\text{now}} + \vec{F}_{\text{net}}\,\Delta t$$

At the instant when $\vec{F}_{\text{net}}$ becomes zero, the object is falling, so $\vec{p}_{\text{now}}$ is nonzero. Therefore a short time Δt in the future, the momentum $\vec{p}_{\text{future}}$ of the coffee filter will be equal to $\vec{p}_{\text{now}} + \vec{0}$, and the filter will continue to fall at constant speed.

Experiments show that the magnitude of the air resistance force is approximately proportional to the square of an object's speed. You can do a simple experiment to confirm this approximate dependence (see Problem P40).

Terminal Speed

A sky diver initially speeds up due to the gravitational force acting downward, but the sky diver's speed doesn't keep getting bigger and bigger, because the upward air resistance force increases with speed. Eventually the sky diver reaches a "terminal speed" and falls thereafter at constant speed despite the gravitational force. This terminal speed for falling humans is so high (about 60 m/s, or about 135 mi/h) that hitting the ground without a parachute is normally fatal, although there have been cases of people hitting deep snow at terminal speed and surviving.

Checkpoint 10 A sky diver whose mass is 90 kg is falling at a terminal speed of 60 m/s. What is the magnitude of the force of the air on the sky diver?

Dependence of Air Resistance on Cross-Sectional Area

It is easy to observe that the effect of air resistance increases as the cross-sectional area of an object increases—a sky diver falls much more slowly with an open parachute than with a closed parachute. A low-density paper coffee filter has a large cross-sectional area, so air resistance has a large effect on a falling coffee filter. In contrast, a high-density metal ball has a small cross-sectional area, but a large gravitational force acts on it, and air resistance may be negligible in comparison with gravity.

Dependence of Air Resistance on Shape

QUESTION From your own experience, would you expect larger air resistance (at a particular speed) for a pointed object or a blunt object?

As you might expect, the air resistance force depends not only on density, cross-sectional area, and speed, but also on the shape of the object. There is a larger air resistance force on a blunt object such as a coffee filter than on a ball with the same cross-sectional area. Streamlined objects such as arrows have reduced air resistance.

Dependence of Air Resistance on Air Density

QUESTION If you could double the density of the air through which a coffee filter falls, what change in the air resistance force would you expect?

You would probably expect a larger air resistance force in denser air. In fact, measurements of a variety of ordinary-sized objects moving through air or other fluids show that air resistance is proportional to the density ρ of the air.

For example, there is less air resistance at higher altitudes, where the air is less dense. Balls travel farther in Denver, Colorado, a mile above sea level, than they do in New York City.

An Empirical Equation for Air Resistance

An equation describing the air resistance force on a moving object must incorporate all the effects we have listed above: dependence on speed v, cross-sectional area A, shape of the object, and density of the air ρ. The shape effect is captured by a parameter denoted by the symbol C that reflects the sharpness or bluntness of the object. The parameter C is called the "drag coefficient," and is typically determined empirically. Typically $0.3 \leq C \leq 1.0$; blunter objects have higher values of C. The direction of the force is $(-\hat{v})$, opposite to the velocity.

APPROXIMATE AIR RESISTANCE FORCE (EMPIRICAL)

$$\vec{F}_{\text{air}} \approx -\frac{1}{2}C\rho A v^2 \hat{v}$$

for blunt objects at ordinary speeds. A is the cross-sectional area of the object; ρ is the density of the air; C reflects the bluntness of the object.

How important is the effect of air resistance in the everyday world? In Chapter 2 we analyzed the motion of a ball thrown through the air. However, we neglected air resistance, which can have a sizable effect. In Problem P47 you are asked to include the air resistance force given above. You should find that a baseball thrown at high speed by a professional baseball pitcher goes only about half as far in air as it would go in a vacuum.

However, even when we include air resistance in our predictions of the motion of a baseball, we still have ignored a force that can have a major effect on the trajectory. If a ball has spin, there is an effect of fluid flow around the ball that raises the air pressure on the side where the rotational motion is in the same direction as the ball's velocity, and lowers the air pressure on the other side, where the rotational motion is in the opposite direction to the velocity. If the force points to the side, the ball curves; this is an important effect in baseball pitching. If the force points upward (due to "backspin," as in Figure 7.39), it lifts the ball and extends the range. If the force points downward (due to "topspin") it decreases the range. The calculations involved in modeling this effect are quite complex, and involve "fluid dynamics" (the study of fluid flow).

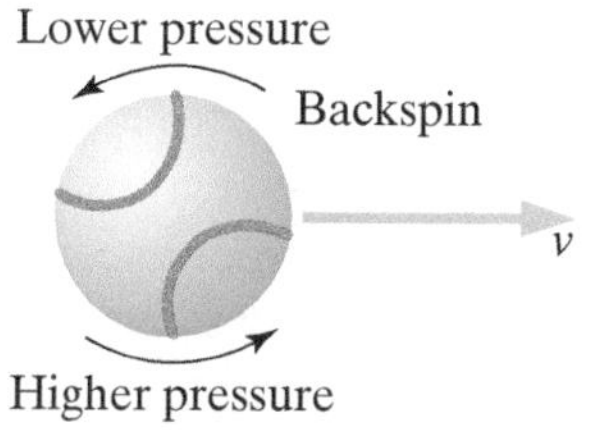

Figure 7.39 A spinning ball experiences additional forces.

QUESTION Is air resistance a fundamental force like the gravitational and electric forces?

No, air resistance is a result of the electric interactions between the molecules on the surface of a moving object and gas molecules in the air. Air resistance is the average result of a huge number of momentary contact electric interactions of air molecules hitting atoms on the surfaces of the falling object, and the average net air resistance is very difficult to calculate from fundamental principles at the molecular level.

Checkpoint 11 A coffee filter of mass 1.4 g dropped from a height of 2 m reaches the ground with a speed of 0.8 m/s. How much kinetic energy K_{air} did the air molecules gain from the falling coffee filter?

Mechanism of Air Resistance

We would like to understand the details of the interaction of the air with a falling object. In particular, we'd like to find a reason that the air resistance would depend on the square of the speed of the object, the cross-sectional area of the object, and the density of the air.

Let's consider the interactions of the air and object from a microscopic viewpoint. When you hold a coffee filter stationary, air molecules hit it nearly equally from above and below, as seen in Figure 7.40. (Actually there are slightly more collisions per second on the lower surface than on the upper surface, leading to a small upward buoyant force that is very small compared to air resistance.) However, when the filter is moving downward due to the gravitational attraction of the Earth, the bottom side of the filter is running into the air molecules, while the top side is moving away from the air molecules.

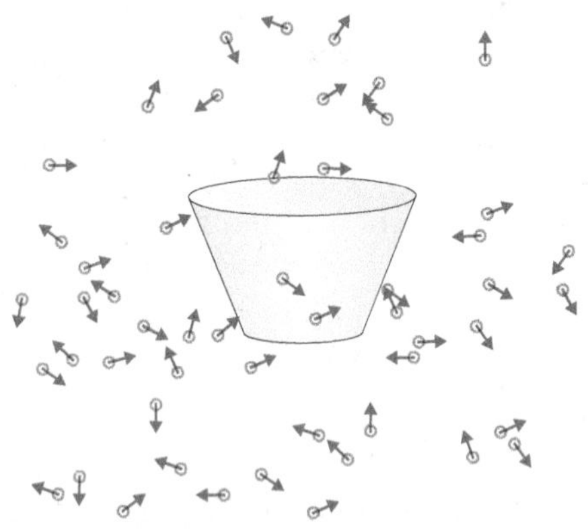

Figure 7.40 Air molecules collide with a falling coffee filter.

On average, the bottom side will have an increased number of collisions per second with air molecules, and with greater impact. On the top side there will be a reduced number of collisions per second, and with less impact. There is a net upward push on the filter. This air resistance force increases with increasing downward speed of the filter, because higher speed increases the rate and impact of collisions on the bottom and decreases the rate and impact of collisions on the top.

Eventually, when the downward speed of the coffee filter has become big enough, the net upward push by the air molecules becomes as large as the downward gravitational pull of the Earth. From then on the filter falls at a constant speed (terminal speed).

While this model captures the qualitative aspects of the situation, a quantitative analysis that predicts the v^2 dependence of the force is beyond the scope of this introductory course. We can accept the approximate force law as expressing a summary of experimental data.

Checkpoint 12 If you let a mass at the end of a string start swinging, at first the maximum swing decreases rather quickly, but once the swing has become small it takes a long time for further significant decrease to occur. Try it! Explain this simple observation.

Viscous Friction

Very small particles such as dust or droplets of fog falling slowly through air, or small objects moving through a thick liquid such as honey, experience a friction force that is proportional to v rather than v^2. This is called "viscous" friction.

The dependence of the friction force on v (for viscous friction) or v^2 (for air resistance) is approximately valid only in specific situations. The expressions for friction forces do not have the wide applicability of those for the gravitational and electric forces. However, even approximate expressions for friction forces allow us to make much better predictions than we could make if we ignored friction completely.

Viscous Friction in a Mass–Spring System

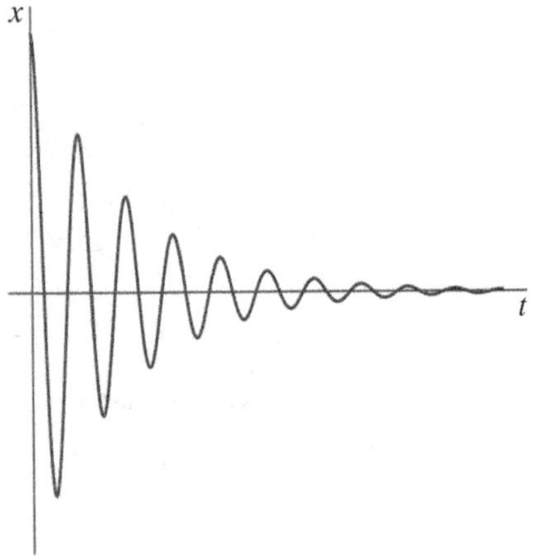

Figure 7.41 Position vs. time for an oscillator with viscous friction.

One of the most obvious things about the motion of a real macroscopic spring–mass system is that the amplitude gets smaller and smaller with time. Our model of a spring–mass system that we developed in Chapter 4 is too simple: we should include the effects of friction. See Problem P45 on numerical integration of a spring–mass system with friction.

In the case of viscous friction, where the friction force is $\vec{f} = -c\vec{v}$ with c constant, it is possible with elaborate math (or by guessing the solution and plugging it into the Momentum Principle for the system) to find an analytical solution for the position x as a function of time.

$$x(t) = A\,e^{-(c/2m)t}\cos(\omega t)$$

The amplitude dies away exponentially with time (Figure 7.41). The angular frequency ω of this damped oscillator is

$$\omega = \sqrt{\frac{k_s}{m} - \left(\frac{c}{2m}\right)^2}$$

which is approximately $\sqrt{k_s/m}$ for low friction.

There is no such simple analytical solution if the friction force is independent of the speed (sliding friction) or proportional to v^2 (air resistance for large blunt objects), but such frictional forces are easily handled in a computational model.

Dissipation with Sliding Friction

In Chapter 4 we saw that when a block slides on a table, the table exerts a force on the block associated with compression of the interatomic bonds in the table and block. This force has two components: $\vec{F}_N$ normal to the surface, and $\vec{f}$ parallel to the surface, called the frictional force. Atoms of the block run into atoms in the table, and the resulting interactions stretch and compress interatomic bonds in the objects.

The frictional force is nearly independent of the speed of the block, and $f \approx \mu_k F_N$, where μ_k is the kinetic (motional) coefficient of friction.

If the block does not slide across the table, the frictional force is just big enough to make the net force be zero. In this case, $f \leq \mu_s F_N$, where μ_s is called the coefficient of static friction.

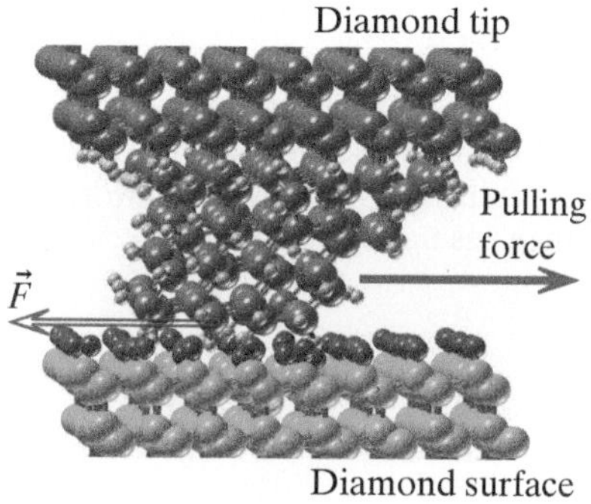

Figure 7.42 A computational model of sliding friction between a diamond tip and diamond surface. The vertical forces are not shown. (© U.S. Government. Courtesy of Professor Judith A. Harrison, U.S. Naval Academy.)

Figure 7.42 provides an atomic-level picture of friction. This image is a single frame from a molecular dynamics computer simulation of sliding friction between two objects, created by Judith A. Harrison and coworkers at the United States Naval Academy in Annapolis, Maryland. The computations are similar to the numerical integrations you have done, but they involve a large number of objects (atoms). The forces between the atoms are modeled by spring-like forces.

In this image, a diamond tip is dragged to the right across a diamond surface. The large spheres represent carbon atoms, and the small ones are hydrogen atoms on the carbon surfaces. Note that the projecting tip has caught on the lower surface, causing interatomic bonds in the two objects to deform. The stretching and compression of interatomic bonds is responsible for the horizontal component of the force exerted on the moving object by the lower object. The collisions between neighboring atoms lead to an increase in the motion of the atoms, with an increase in the internal energy. Some of the work is dissipated throughout the objects.

Dissipation and Path Independence

In Chapter 6 we saw that the work required to change the configuration of gravitationally or electrically interacting objects that could be considered to be point particles was independent of the path along which the work was done. We cannot consider a block sliding with friction to be a point particle because its internal energy changes.

Consider sliding a heavy block in a round trip on an uphill and downhill surface where there is considerable friction. The Energy Principle for the round-trip movement of the block contains a term corresponding to a rise in the internal energy of the block:

$$\Delta K + \Delta U_{\text{grav}} + \Delta E_{\text{internal}} = W$$
$$0 + 0 + \Delta E_{\text{internal}} = W$$

The longer the round-trip path taken, the more increase in internal energy. The total work W is nonzero and is greater for longer paths. Path independence does not apply. Some energy is dissipated/shared among the many atoms of the block.

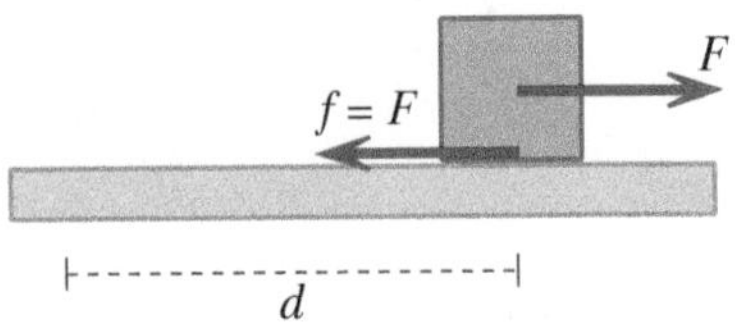

Figure 7.43 Forces in the x direction on a block pulled with constant speed across a table. (Forces in the y direction are not shown.)

A Paradox Involving Friction

If we are careful in our energy bookkeeping for a system involving motion with friction, we encounter a paradox that cannot easily be resolved. Consider the situation shown in Figure 7.43.

You exert a constant force F, and pull a block a distance d across a table. Because there is friction between the block and the table, the block travels at constant speed. Since the momentum of the block does not change, we conclude correctly that the net force on the block is zero, and that therefore the magnitude of the friction force must also be F.

In terms of energy inputs, you have done an amount of work Fd on the block. The friction force, acting in the opposite direction, has apparently done an amount of work $-Fd$. The net work done on the block therefore would seem to be zero. This is consistent with the fact that the kinetic energy of the block did not seem to change.

It is not, however, consistent with the fact that both the block and the table get hotter as they rub together! If the net work done on the block was zero, where did that increased internal energy come from?

This is not a "trick question," but a genuine puzzle. We will return to this issue in a later chapter, when we will deal with it in enough detail to resolve the paradox.

Checkpoint 13 You move a block slowly to the right a distance of 0.2 m on a table. You apply a constant force of 20 N. How much work do you do? You then move the block slowly back to the starting point; how much work do you do in this second move? What change has there been in the internal energy of the block and table?

7.11 ENERGY DISSIPATION IN COMPUTATIONAL MODELS

We saw in the previous chapter that although we cannot predict the detailed motion of a system directly by iteratively applying the Energy Principle (because neither direction nor time appears in the Energy Principle), it can be extremely useful to monitor and display the kinetic and potential energy of a system whose motion we predict by the iterative application of the Momentum Principle. This is especially true when we include dissipation in our models in order to produce a more realistic picture of the behavior of everyday systems. Such a model enables us to answer questions such as how rapidly energy dissipation occurs, whether energy is dissipated more rapidly at the beginning or end of a process, and so on. Incorporating dissipation into a computational model of motion requires that we translate our algebraic expressions for forces such as air resistance, sliding friction, or viscous friction into computational program statements.

In the previous section we saw that the magnitude of the air resistance force depended on the shape of an object, its speed, and the density of the air. If we wanted to model the motion of a baseball, including the effects of air resistance, we would need values not only for the mass of the baseball, but also its cross-sectional area A (which you can calculate given data for a standard baseball), and also a value for the drag coefficient C, which you can look up (0.3 is a reasonable value). You would also need the density of the air ρ at the altitude of interest, in SI units. Given all this, the calculation of the net force inside your loop might look like this:

```
while ball.pos.y > ground.pos.y:
    rate(100)
    F_grav = m_ball * vector(0, -g, 0)
    v = p_ball / m_ball
```

```
vhat = v / mag(v)
F_air = 0.5*C*rho_air*A_ball*mag(v)**2*(-vhat)
F_net = F_grav + F_air

## update momentum and update position...
```

If you wanted to model the damping of the motion of an object moving through a viscous liquid, the net force would have to include a viscous friction force proportional to the speed, and proportional to a damping coefficient, which is usually symbolized by lowercase c (not to be confused with the speed of light).

```
F_viscous = c * (-p_object/m_object)
```

Modeling dissipation due to sliding friction can be a little tricky. As we saw in Chapter 4, the sliding friction force is in a direction opposite to an object's velocity, but is independent of speed, depending only on how much two objects compress each other (represented in the equation by F_N, the magnitude of the normal force):

$$f_{\text{friction}} \approx \mu_k F_N$$

Calculating the sliding friction force in VPython is not difficult:

```
F_sliding = mu_k * F_N * (-p/mag(p))
```

The tricky part is sensing when the moving object has come nearly to a stop, and exiting the computational loop so the friction force doesn't start the object moving again!

QUESTION Why is it difficult to detect when a moving object has come exactly to a stop in a computational model?

There are two reasons. The first is that the time step Δt is finite; it is unlikely that the object's speed will be zero exactly at the end of a time step, when we can test to see if v is zero. The second reason, which was discussed in Chapter 2, is that the internal representations of floating-point numbers in computers and calculators are necessarily imprecise, and there is always a small but finite round-off error. So instead of testing to see if v is exactly zero, we need to test to see if v is very close to zero. Such a test in a computational loop might look like this:

```
while True:
   rate(100)
   F_sliding = mu_k * F_N * (-p/mag(p))
   ## calculate net force
   ## update momentum and update position...
   if mag(p_object / m_object) < 0.001:
      break
```

In VPython (as in many computational languages), a logical test comparing the value of a variable to a target value can be done using an `if` statement, as in the program fragment above. If the result of the comparison is true, then the statements indented below the `if` statement will be executed; if the result is false, then they will be skipped. The test is at the end of the loop to make it possible to start the system with $v = 0$.

The `break` command terminates the computational loop. Because it is indented below the `if` statement in the program fragment above, the `break` command is executed only if the test is true—that is, only if $v < 0.001$ m/s.

In the Computational Problems section at the end of this chapter, you will find several problems involving adding dissipative forces to models of motion.

7.12 *RESONANCE

Imagine pushing somebody in a swing. If you time your pushes to match the natural frequency of the swing, you can build up an increasingly large amplitude of the swing. However, larger amplitude means higher speed and a higher rate of energy dissipation.

> **QUESTION** You keep pushing with constant (small) amplitude, and the swing eventually reaches a "steady state" with constant (large) amplitude. What determines the size of this steady-state amplitude?

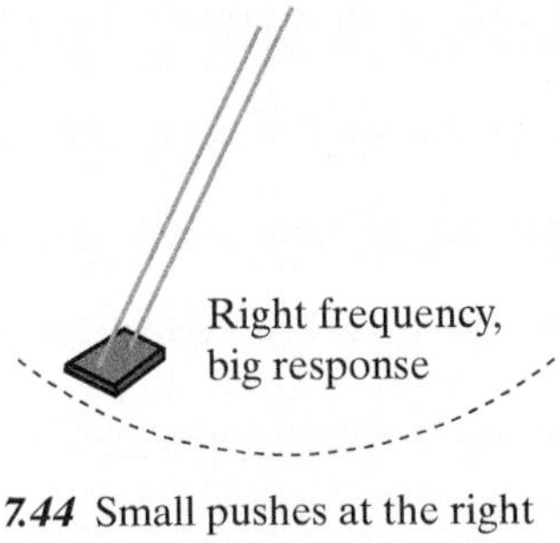

Figure 7.44 Small pushes at the right times lead to a big response.

The steady state is established when the energy dissipated through air resistance on each cycle has grown to be equal to the energy input you make on each cycle. The large steady-state oscillation needs only little pushes at the right times from you to make up for energy dissipation, and the amplitude of the oscillation can be much larger than the amplitude of the pushes (Figure 7.44).

On the other hand, if you deliberately push at the wrong times, not matching the natural frequency of the swing, the amplitude of the swing won't continually build up (Figure 7.45). To get a large response from an oscillating system, inputs should be made at the natural, free-oscillation frequency of that system. This important feature of driven oscillating systems is called "resonance."

Figure 7.45 Pushes at the wrong times lead to a small response.

*Analytical Treatment of Resonance

Problem P50 is a numerical integration of a sinusoidally driven spring–mass oscillator, as shown in Figure 7.46.

This problem lets you study the buildup of the steady state as well as the steady state itself. For the steady-state portion of the motion, it is possible to obtain an analytical solution, if the energy dissipation is due to viscous friction. Here is the Momentum Principle for the driven oscillator in the diagram, with viscous friction force $-cv$:

$$\frac{dp}{dt} = -k_s[x - D\sin(\omega_D t)] - cv$$

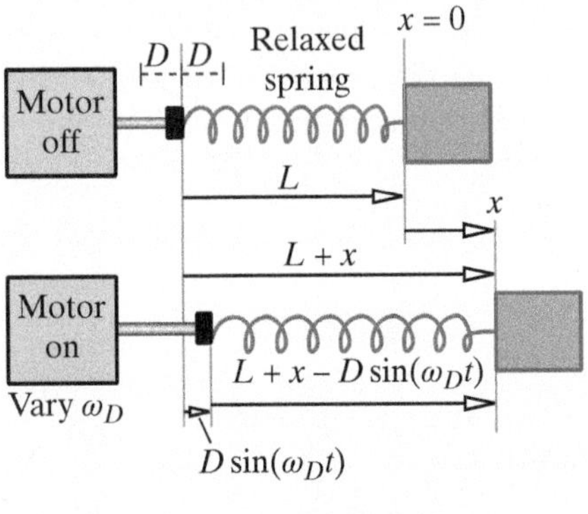

Figure 7.46 A sinusoidally driven oscillator.

where x is the position of the mass, D is the amplitude of the sinusoidal motion, and ω_D is the variable angular frequency of the driven end of the spring. For more details on how this equation was obtained, see the discussion in Problem P50.

In the steady state x must be a sinusoidal function, with amplitude A and phase shift ϕ depending strongly on the driving angular frequency ω_D:

$$x = A\sin(\omega_D t + \phi)$$

By taking derivatives of this expression to obtain $v = dx/dt$ and dv/dt as a function of time t and plugging these derivatives into the Momentum Principle, it can be shown that this expression for x as a function of time is a solution in the steady state if the amplitude A and phase shift ϕ have values determined by the following expressions, where $\omega_F^2 = k_s/m$:

$$A = \frac{\omega_F^2}{\sqrt{(\omega_F^2 - \omega_D^2)^2 + \left(\frac{c}{m}\omega_D\right)^2}}D$$

$$\cos\phi = \frac{(\omega_F^2 - \omega_D^2)}{\sqrt{(\omega_F^2 - \omega_D^2)^2 + \left(\frac{c}{m}\omega_D\right)^2}}$$

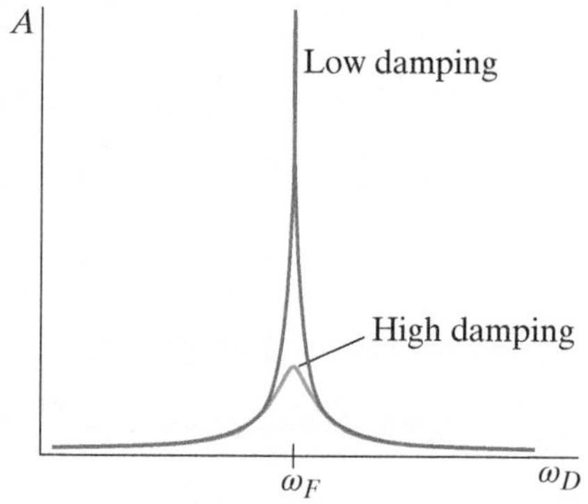

Figure 7.47 Amplitude as a function of driving frequency. The damping in the lower curve is five times greater than in the upper curve.

The detailed proof is rather complicated, so we omit it. Instead, in the exercises that follow we give you the opportunity to see that these expressions for amplitude and phase shift are in agreement with your experimental observations and the computer modeling of Problem P50. Figure 7.47 is a graph of the amplitude A as a function of the driving angular frequency ω_D for two different values of the viscous friction parameter c.

Checkpoint 14 (a) Using the equation for the amplitude A, show that if the viscous friction is small, the amplitude is large when ω_D is approximately equal to ω_F. Using the equation involving the phase shift ϕ, show that the phase shift is approximately $0°$ for very low driving frequency ω_D, approximately $180°$ for very high driving frequency ω_D, and $90°$ at resonance, consistent with your experiment. (b) Show that with small viscous friction, the amplitude A drops to $1/\sqrt{2}$ of the peak amplitude when the driving angular frequency differs from resonance by this amount:

$$|\omega_F - \omega_D| \approx \frac{c}{2m}\omega_F$$

(*Hint:* Note that near resonance $\omega_D \approx \omega_F$, so $\omega_F + \omega_D \approx 2\omega_F$.)

Given these results, how does the width of the resonance peak depend on the amount of friction? What would the resonance curve look like if there were very little friction?

Resonance in Other Systems

We have discussed resonance in the context of a particular system—a mass on a spring driven by a motor. We study resonance because it is an important phenomenon in a great variety of situations.

When you choose a radio or television station, you adjust the parameters of an electrically oscillating circuit so that the circuit has a narrow resonance at the chosen station's main frequency, and other stations with significantly different frequencies have little effect on the circuit.

Magnetic resonance imaging (MRI) is based on the phenomenon of nuclear magnetic resonance (NMR) in which nuclei acting like toy tops precess in a strong magnetic field. The precessing nuclei are significantly affected by radio waves only at the precession frequency. Because the precession frequency depends slightly on what kind of atoms are nearby, resonance occurs for different radio frequencies, which makes it possible to identify different kinds of tissue and produce remarkably detailed images.

SUMMARY

Potential energy of an ideal spring–mass system:

$$U_s = \frac{1}{2}k_s s^2$$

Approximate interatomic potential energy:

$$U = \frac{1}{2}k_s s^2 - E_M$$

A better approximation to interatomic potential energy:

$$U_M = E_M[1 - e^{-\alpha(r-r_{eq})}]^2 - E_M$$

Thermal energy: related to the temperature of the object
Specific heat: energy per gram to raise temperature 1 K

$$\Delta E_{\text{thermal}} = mC\Delta T$$

$$C = \frac{\Delta E_{\text{thermal}}}{m\Delta T} \quad (C \text{ in J/K/g}, m \text{ in grams})$$

Energy transfer due to a temperature difference, Q
Other energy transfers: work W, matter transfer, mechanical waves, electricity, electromagnetic radiation

Power $= \dfrac{\text{energy}}{\text{time}}$ (J/s = watts); also calculable as $\vec{F} \bullet \vec{v}$
Open system: energy flows into and/or out of system.
Closed system: no energy flows into or out of system.
The choice of system (open or closed) affects energy accounting, as does the choice of reference frame.
Dissipation of energy
Terminal speed
Air resistance: collisions between air molecules and moving object
Air resistance is approximately proportional to speed squared, and in a direction opposite to the velocity:

$$\vec{F}_{\text{air}} \approx -\frac{1}{2} C\rho A v^2 \hat{v}$$

For small particles ("viscous" friction), fluid friction is approximately proportional to v.

Sliding friction force due to forming and breaking of bonds between two solid objects
Impossibility of describing friction in terms of point particles and potential energy
*Analytical solution for driven spring–mass system: $x = A\sin(\omega_D t + \phi)$, where $\omega_F = \sqrt{k_s/m}$; viscous friction cv, driven at ω_D:

$$A = \frac{\omega_F^2}{\sqrt{(\omega_F^2 - \omega_D^2)^2 + \left(\frac{c}{m}\omega_D\right)^2}} D$$

$$\cos\phi = \frac{(\omega_F^2 - \omega_D^2)}{\sqrt{(\omega_F^2 - \omega_D^2)^2 + \left(\frac{c}{m}\omega_D\right)^2}}$$

Resonance: sensitivity of driven oscillator to driving frequency

QUESTIONS

Q1 Write an equation for the total energy of a system consisting of a mass suspended vertically from a spring, and include the Earth in the system. Place the origin for gravitational energy at the equilibrium position of the mass, and show that the changes in energy of a vertical spring–mass system are the same as the changes in energy of a horizontal spring–mass system.

Q2 Substance A has a large specific heat (on a per gram basis), while substance B has a smaller specific heat. If the same amount of energy is put into a 100 g block of each substance, and if both blocks were initially at the same temperature, which one will now have the higher temperature?

Q3 An oil company included in its advertising the following phrase: "Energy—not just a force, it's power!" In technical usage, what are the differences among the terms energy, force, and power?

Q4 Electricity is billed in kilowatt-hours. Is this energy or power? How much is one kilowatt-hour in standard physics units? (The typical cost of one kilowatt-hour is 5 to 10 cents.)

Q5 State which of the following are open systems with respect to energy, and which are closed: a car, a person, an insulated picnic chest, the Universe, the Earth. Explain why.

Q6 When a falling object reaches terminal speed, its kinetic energy reaches a constant value. However, the gravitational energy of the system consisting of object plus Earth continues to decrease. Does this violate the principle of conservation of energy? Explain why or why not.

Q7 You lift a heavy box. We'll consider this process for different choices of system and surroundings. **(a)** Choose the box as the system of interest. What objects in the surroundings exert significant forces on this system? **(b)** Choose you and the box as the system of interest. What objects in the surroundings exert significant forces on this system? **(c)** Choose you, the box, and the Earth as the system of interest. What objects in the surroundings exert significant forces on this system?

Q8 Consider the process of a woman lifting a barbell discussed in Section 7.8. Analyze the energy changes in this process, choosing the woman alone as the system. What quantities can be calculated with this choice of system?

Q9 A block of mass m is projected straight upward by a strong spring whose stiffness is k_s. When the block is a height y_1 above the floor, it is traveling upward at speed v_1, and the spring is compressed an amount s_1. A short time later the block is at height y_2, traveling upward at speed v_2, and the spring is compressed an amount s_2. Assume that thermal transfer of energy (microscopic work) Q between the block and the air is negligible. For each of the following choices of system, write the Energy Principle in the update form $E_f = E_i + W$. **(a)** The block, the spring, and the Earth; **(b)** the block plus spring; **(c)** the block alone.

Q10 A horse whose mass is M gallops at constant speed v up a long hill whose vertical height is h, taking an amount of time t to reach the top. The horse's hooves do not slip on the rocky ground, so the work done by the force of the ground on the hooves is zero (no displacement of the force). When the horse started running, its temperature rose quickly to a point at which from then on, heat transferred from the horse to the air keeps the horse's temperature constant.

(a) First consider the horse as the system of interest. In the initial state the horse is already moving at speed v. In the final state the horse is at the top of the hill, still moving at speed v. Write out the Energy Principle $\Delta E_{\text{sys}} = W + Q$ for the system of the horse alone. The terms on the left-hand side should include only energy changes for the system, while the terms on the right-hand side should relate to the surroundings (everything else). Which of the terms are equal to 0? Which of the terms should go on the system (left) side (ΔE_{sys})? Which of the terms should go on the surroundings (right) side?

(b) Next consider the Universe as the system of interest. Write out the energy principle for this system. Remember that terms on the left-hand side should include all energy changes for the system, while the terms on the right-hand side should relate to the surroundings. Which of the terms are equal to 0? Which of the terms should go on the system (left) side (ΔE_{sys})? Which of the terms should go on the surroundings (right) side?

Q11 Describe a situation in which it would be appropriate to neglect the effects of air resistance.

Q12 Describe a situation in which neglecting the effects of air resistance would lead to significantly wrong predictions.

Q13 Throw a ball straight up and catch it on the way down, at the same height. Taking into account air resistance, does the ball take longer to go up or to come down? Why?

Q14 You drag a block with constant speed v across a table with friction. Explain in detail what you have to do in order to change to a constant speed of $2v$ on the same surface. (That is, the puzzle is to explain how it is possible to drag a block with sliding friction at different constant speeds.)

Q15 Figure 7.48 is a portion of a graph of energy terms vs. time for a mass on a spring, subject to air resistance. Identify and label the three curves as to what kind of energy each represents. Explain briefly how you determined which curve represented which kind of energy.

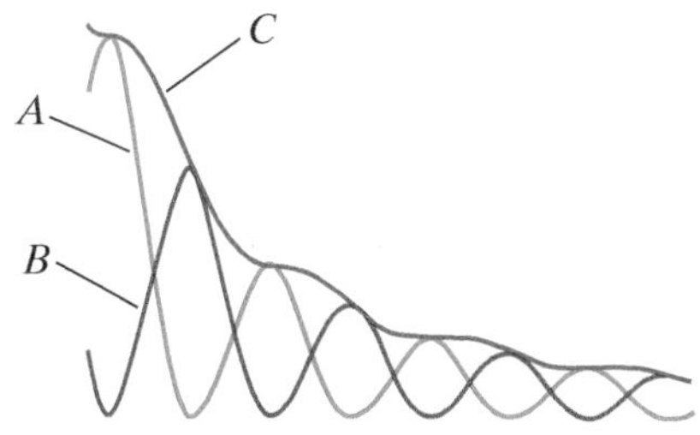

Figure 7.48

PROBLEMS

Section 7.2

•P16 A spring has a relaxed length of 6 cm and a stiffness of 100 N/m. How much work must you do to change its length from 5 cm to 9 cm?

•P17 A horizontal spring with stiffness 0.5 N/m has a relaxed length of 15 cm. A mass of 20 g is attached and you stretch the spring to a total length of 25 cm. The mass is then released from rest and moves with little friction. What is the speed of the mass at the moment when the spring returns to its relaxed length of 15 cm?

•P18 A spring whose stiffness is 800 N/m has a relaxed length of 0.66 m. If the length of the spring changes from 0.55 m to 0.96 m, what is the change in the potential energy of the spring?

••P19 A spring with stiffness k_s and relaxed length L stands vertically on a table. You hold a mass M just barely touching the top of the spring. **(a)** You *very slowly* let the mass down onto the spring a certain distance, and when you let go, the mass doesn't move. How much did the spring compress? How much work did *you* do? **(b)** Now you again hold the mass just barely touching the top of the spring, and then let go. What is the maximum compression of the spring? State what approximations and simplifying assumptions you made. **(c)** Next you push the mass down on the spring so that the spring is compressed an amount s, then let go, and the mass starts moving upward and goes quite high. When the mass is a height of $2L$ above the table, what is its speed?

••P20 A horizontal spring–mass system has low friction, spring stiffness 200 N/m, and mass 0.4 kg. The system is released with an initial compression of the spring of 10 cm and an initial speed of the mass of 3 m/s. **(a)** What is the maximum stretch during the motion? **(b)** What is the maximum speed during the motion? **(c)** Now suppose that there is energy dissipation of 0.01 J per cycle of the spring–mass system. What is the average power input in watts required to maintain a steady oscillation?

••P21 A package of mass 9 kg sits on an airless asteroid of mass 8.0×10^{20} kg and radius 8.7×10^5 m. We want to launch the package in such a way that it will never come back, and when it is very far from the asteroid it will be traveling with speed 226 m/s. We have a large and powerful spring whose stiffness is 2.8×10^5 N/m. How much must we compress the spring?

••P22 A mass of 0.3 kg hangs motionless from a vertical spring whose length is 0.8 m and whose unstretched length is 0.65 m. Next the mass is pulled down so the spring has a length of 0.9 m and is given an initial speed upward of 1.2 m/s. What is the maximum length of the spring during the following motion? What approximations or simplifying assumptions did you make?

•••P23 A relaxed spring of length 0.15 m stands vertically on the floor; its stiffness is 1000 N/m. You release a block of mass 0.4 kg from rest, with the bottom of the block 0.8 m above the floor and straight above the spring. How long is the spring when the block comes momentarily to rest on the compressed spring?

•••P24 Design a "bungee jump" apparatus for adults. A bungee jumper falls from a high platform with two elastic cords tied to the ankles. The jumper falls freely for a while, with the cords slack. Then the jumper falls an additional distance with the cords increasingly tense. You have cords that are 10 m long, and these cords stretch in the jump an additional 24 m for a jumper whose mass is 80 kg, the heaviest adult you will allow to use your bungee jump (heavier customers would hit the ground). You can neglect air resistance. **(a)** Make a series of five simple diagrams, like a comic strip, showing the platform, the jumper, and the two cords at various times in the fall and the rebound. On each diagram, draw and label vectors representing the forces acting on the jumper, and the jumper's velocity. Make the relative lengths of the vectors reflect their relative magnitudes. **(b)** At what instant is there the greatest tension in the cords? How do you know? **(c)** What is the jumper's speed at this instant? **(d)** Is the jumper's momentum changing at this instant or not? (That is, is $d\vec{p}/dt$ nonzero or zero?) Explain briefly. **(e)** Focus on this instant, and use the principles of this chapter to determine the spring stiffness k_s for each cord. Explain your analysis. **(f)** What is the maximum tension that each cord must support without breaking? **(g)** What is the maximum acceleration (in g's) that the jumper experiences? What is the direction of this maximum acceleration? **(h)** State clearly what approximations and estimates you have made in your design.

Section 7.3

••P25 Figure 7.49 is a potential energy curve for the interaction of two neutral atoms. The two-atom system is in a vibrational state indicated by the green horizontal line.

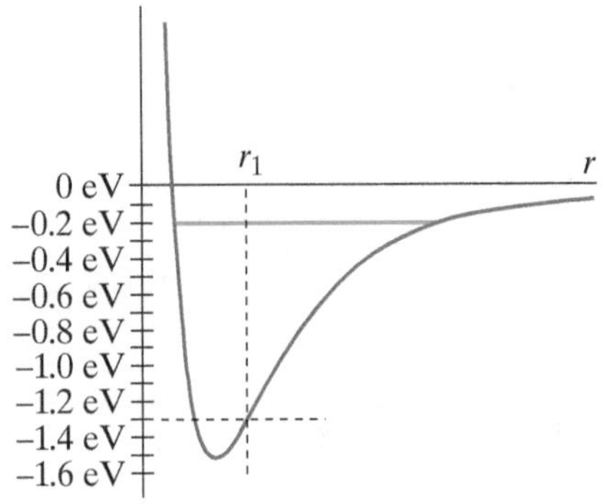

Figure 7.49

(a) At $r = r_1$, what are the approximate values of the kinetic energy K, the potential energy U, and the quantity $K + U$? **(b)** What minimum energy must be supplied to cause these two atoms to separate? **(c)** In some cases, when r is large, the interatomic potential energy can be expressed approximately as $U = -a/r^6$. For large r, what is the algebraic form of the magnitude of the force the two atoms exert on each other in this case?

Section 7.4

•P26 Niagara Falls is about 50 m high. What is the temperature rise in kelvins of the water from just before to just after it hits the rocks at the bottom of the falls, assuming negligible air resistance during the fall and that the water doesn't rebound but just splats onto the rock? It is helpful (but not essential) to consider a 1 g drop of water.

••P27 400 g of boiling water (temperature 100°C, specific heat 4.2 J/K/g) are poured into an aluminum pan whose mass is 600 g and initial temperature 20°C (the specific heat of aluminum is 0.9 J/K/g). After a short time, what is the temperature of the water? Explain. What simplifying assumptions did you have to make?

••P28 Here are questions about human diet. **(a)** A typical candy bar provides 280 calories (one "food" or "large" calorie is equal to 4.2×10^3 J). How many candy bars would you have to eat to replace the chemical energy you expend doing 100 sit-ups? Explain your work, including any approximations or assumptions you make. (In a sit-up, you go from lying on your back to sitting up.) **(b)** How many days of a diet of 2000 large calories are equivalent to the gravitational energy difference for you between sea level and the top of Mount Everest, 8848 m above sea level? (However, the body is not anywhere near 100% efficient in converting chemical energy into change in altitude. Also note that this is in addition to your basal metabolism.)

••P29 You observe someone pulling a block of mass 43 kg across a low-friction surface. While they pull a distance of 3 m in the direction of motion, the speed of the block changes from 5 m/s to 7 m/s. Calculate the magnitude of the force exerted by the person on the block. What was the change in internal energy (chemical energy plus thermal energy) of the person pulling the block?

Section 7.5

••P30 You place into an insulated container a 1.5 kg block of aluminum at a temperature of 45°C in contact with a 2.1 kg block of copper at a temperature of 18°C. The specific heat of aluminum is 0.91 J/g and the specific heat of copper is 0.39 J/g. What is the final temperature of the two blocks?

••P31 180 g of boiling water (temperature 100°C, heat capacity 4.2 J/K/g) are poured into an aluminum pan whose mass is 1050 g and initial temperature 26°C (the heat capacity of aluminum is 0.9 J/K/g). **(a)** After a short time, what is the temperature of the water? **(b)** What simplifying assumptions did you have to make? (1) The thermal energy of the water doesn't change. (2) The thermal energy of the aluminum doesn't change. (3) Energy transfer between the system (water plus pan) and the surroundings was negligible during this time. (4) The heat capacities for both water and aluminum hardly change with temperature in this temperature range. **(c)** Next you place the pan on a hot electric stove. While the stove is heating the pan, you use a beater to stir the water, doing 29,541 J of work, and the temperature of the water and pan increases to 86.9°C. How much energy transfer due to a temperature difference was there from the stove into the system consisting of the water plus the pan?

Section 7.6

•P32 A certain motor is capable of doing 3000 J of work in 11 s. What is the power output of this motor?

•P33 In the Niagara Falls hydroelectric generating plant, the energy of falling water is converted into electricity. The height of the falls is about 50 m. Assuming that the energy conversion is highly efficient, approximately how much energy is obtained from one kilogram of falling water? Therefore, approximately how many kilograms of water must go through the generators every second to produce a megawatt of power (1×10^6 W)?

••P34 Here are questions dealing with human power. **(a)** If you follow a diet of 2000 food calories per day (2000 kC), what is your average rate of energy consumption in watts (power input)? (A food or "large" calorie is a unit of energy equal to 4.2×10^3 J; a regular or "small" calorie is equal to 4.2 J.) Compare with the power input to a table lamp. **(b)** You can produce much higher power for short periods. Make appropriate measurements as you run up some stairs, and report your measurements. Use these measurements to estimate your power output (this is in addition to your basal metabolism—the power needed when resting). Compare with a horsepower (which is about 750 W) or a toaster (which is about 1000 W).

•••P35 Humans have about 60 ml (60 cm^3) of blood per kilogram of body mass, and blood makes a complete circuit in about 20 s, to keep tissues supplied with oxygen. Make a crude estimate of the *additional* power output of your heart (in watts) when you are standing compared with when you are lying down. Note that we're not asking you to estimate the power output of your heart when you are lying down, just the change in power when you are standing up. You will have to estimate the values of some of the relevant parameters. Because we're only looking for a crude estimate, try to construct a model that is as simple as possible. Describe the approximations and estimates you made.

Section 7.7

••P36 During three hours one winter afternoon, when the outside temperature was 0°C (32°F), a house heated by electricity was kept at 20°C (68°F) with the expenditure of

45 kWh (kilowatt·hours) of electric energy. What was the average energy leakage in joules per second through the walls of the house to the environment (the outside air and ground)?

The rate at which energy is transferred between two systems is often proportional to their temperature difference. Assuming this to hold in this case, if the house temperature had been kept at 25°C (77°F), how many kWh of electricity would have been consumed?

Section 7.9

•••P37 A man sits with his back against the back of a chair, and he pushes a block of mass $m = 2$ kg straight forward on a table in front of him, with a constant force $F = 30$ N, moving the block a distance $d = 0.3$ m. The block starts from rest and slides on a low-friction surface. **(a)** How much work does the man do on the block? **(b)** What is the final kinetic energy K of the block? **(c)** What is the final speed v of the block? **(d)** How much time Δt does this process take? **(e)** Consider the system of the man plus the block: how much work does the chair do on the man? **(f)** What is the internal energy change of the man?

Now suppose that the man is sitting on a train that is moving in a straight line with speed $V = 15$ m/s, and you are standing on the ground as the train goes by, moving to your right. From your perspective (that is, in your reference frame), answer the following questions: **(g)** What is the initial speed v_i of the block? **(h)** What is the final speed v_f of the block? **(i)** What is the initial kinetic energy K_i of the block? **(j)** What is the final kinetic energy K_f of the block? **(k)** What is the change in kinetic energy $\Delta K = K_f - K_i$, and how does this compare with the change in kinetic energy in the man's reference frame? **(l)** How far does the block move (Δx)? **(m)** How much work does the man do on the block, and how does this compare with the work done by the man in his reference frame, and with ΔK in your reference frame? **(n)** How far does the chair move? **(o)** Consider the system of the man plus the block: how much work does the chair do on the man, and how does this compare with the work done by the chair in the man's reference frame? **(p)** What is the internal energy change of the man, and how does this compare with the internal energy change in his reference frame?

Section 7.10

•P38 A coffee filter of mass 1.8 g dropped from a height of 4 m reaches the ground with a speed of 0.8 m/s. How much kinetic energy K_{air} did the air molecules gain from the falling coffee filter? Start from the Energy Principle, and choose as the system the coffee filter, the Earth, and the air.

•P39 You are standing at the top of a 50 m cliff. You throw a rock in the horizontal direction with speed 10 m/s. If you neglect air resistance, where would you predict it would hit on the flat plain below? Is your prediction too large or too small as a result of neglecting air resistance?

••P40 A simple experiment can allow you to determine approximately the dependence of the air resistance force on the speed of a falling object. The basic logic of the experiment is this: if we can determine the air resistance force on two objects of the same shape as they fall at different speeds, we can figure out a mathematical relationship between the air resistance force and speed. For example, if the air resistance force is eight times as large when the speed is only twice as large, then we may conclude that $F \propto v^3$. With only a timer and a ruler, it is possible to measure the terminal speed of a falling object. A simple way to increase the terminal speed of an object is to retain its shape but make it more massive. This will increase the gravitational force on the object, and therefore in order to get to a speed at which $\vec{F}_{net} = \vec{0}$ and the object falls at a constant terminal speed, the air resistance force will need to be larger. If you use a paper object (Figure 7.50) as the falling object, you can make it more massive without changing its shape simply by stacking two or more of these objects together.

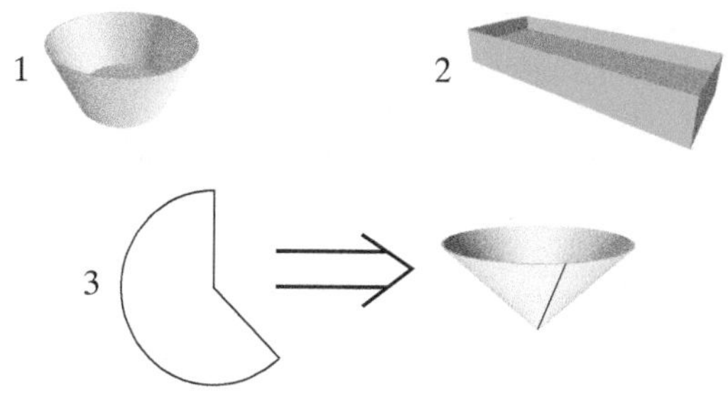

Figure 7.50 A flat-bottomed coffee filter, a folded dollar bill, or a paper cone can be used in Problem P40. The mass of a dollar bill is 1 g.

(a) Measure the terminal speed of a single falling paper object. Drop the object from as high a starting position as possible. Start timing when you think the object has reached terminal speed (think of a way to test this). Do this several times and average your results. **(b)** At terminal speed, what is the value of the air resistance force on a single falling object? If you don't have the equipment to measure the mass of a filter, leave your answer in terms of m. **(c)** Repeat this measurement for stacks of different numbers of objects of the same shape; one of the measurements should be with a stack of four objects. Qualitatively, as the mass of the falling object increases, what happens to the terminal speed? Explain why. **(d)** Now determine the mathematical form of the dependence of the air resistance force on speed, using your measured terminal speeds. (There are many ways to do this.)

•P41 A box with its contents has a total mass of 36 kg. It is dropped from a very high building. **(a)** After reaching terminal speed, what is the magnitude of the air resistance force acting upward on the falling box? **(b)** The box survived the fall and is returned to the top of the building. More objects are put into the box, and the box with its contents now has a total mass of 71 kg. The box is dropped, and it reaches a higher terminal speed than before. After reaching terminal speed, what is the magnitude of the air resistance force acting upward on the falling box? (The fact that the heavier object reaches a higher terminal speed shows that the air resistance force increases with increasing speed.)

••P42 You drop a single coffee filter of mass 1.7 g from a very tall building, and it takes 52 s to reach the ground. In a small fraction of that time the coffee filter reached terminal speed. **(a)** What was the upward force of the air resistance while the coffee filter was falling at terminal speed? **(b)** Next you drop a stack of five of these coffee filters. What was the upward force of the air resistance while this stack of coffee filters was falling at terminal speed? **(c)** Again assuming that the stack reaches terminal speed very quickly, about how long will the stack of coffee filters take to hit the ground? (*Hint:* Consider the relation between speed and the force of air resistance.)

Section 7.12

••P43 You can observe the main effects of resonance with very simple experiments. Hold a spring vertically with a mass suspended at the other end, and observe the frequency of "free" oscillations with your hand kept still. Then stop the oscillations, and move your hand *extremely* slowly up and down in a kind of slow sinusoidal motion. You will see that the mass moves up and down with the same very low frequency. **(a)** How does the amplitude (plus or minus displacement from the center location) of the mass compare with the amplitude of your hand? (Notice that the phase shift of the oscillation is 0°; the mass moves up when your hand moves up.) **(b)** Next move your hand up and down at a significantly higher frequency than the free-oscillation frequency. How does the amplitude of the mass compare to the amplitude of your hand? (Notice that the phase shift of the oscillation is 180°; the mass moves down when your hand moves up.) **(c)** Finally, move your hand up and down at the free-oscillation frequency. How does the amplitude of the mass compare with the amplitude of your hand? (It is hard to observe, but the phase shift of the oscillation is 90°; the mass is at the midpoint of its travel when your hand is at its maximum height.) **(d)** Change the system in some way so as to increase the air resistance significantly. For example, attach a piece of paper to increase drag. At the free-oscillation frequency, how does this affect the size of the response? A strong dependence of the amplitude and phase shift of the system to the driving frequency is called resonance.

COMPUTATIONAL PROBLEMS

More detailed and extended versions of some computational modeling problems may be found in the lab activities included in the *Matter & Interactions* resources for instructors.

••P44 Energy graphs with springs: Start with the program you wrote to model the motion of a mass hanging from a spring (Chapter 4, Problem P70). Add the statements necessary to plot graphs of K, U_{spring}, and $K + U_{\text{spring}}$ vs. time. **(a)** Use initial conditions that produce 1D vertical oscillations. Is the sum $K + U_{\text{spring}}$ constant over time? Should it be? Explain. **(b)** Use your graphs to explain the flow of energy within this system. If you need to plot an additional quantity in order to support your explanation, do so. **(c)** Now experiment with different initial conditions that produce 2D or 3D oscillations. What do your graphs show about energy flow within this system?

••P45 Air resistance: Start with the program you wrote to model the motion of a mass hanging from a spring (Chapter 4, Problem P70). Add the statements necessary to plot graphs of K, U, and $K + U$ vs. time. Use initial conditions that produce 1D vertical oscillations. **(a)** Make sure that the plots generated by your program make sense. Have you included all potential energy terms? **(b)** Add an air resistance force to your model. You may wish to modify the hanging mass to be a flat disk. Choose an approximate value for the drag coefficient C in your model, and adjust this value until the model behaves appropriately for a system experiencing the effects of air resistance. **(c)** Consider the energy graphs produced by the model. Is the rate of energy dissipation constant? If not, at what point(s) in the oscillation cycle is this rate largest? Explain this on physical grounds.

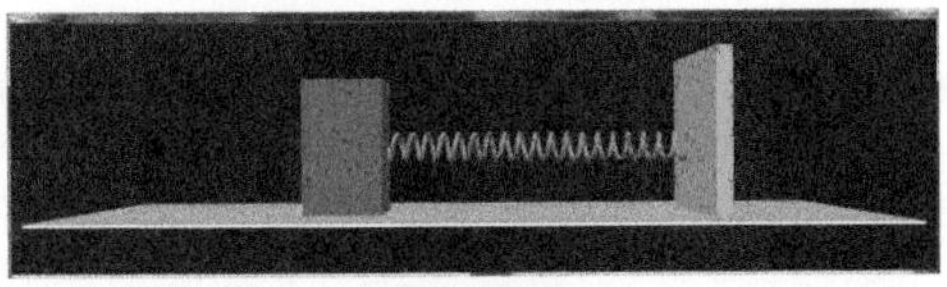

Figure 7.51

••P46 Sliding and viscous friction: Write a program to model a horizontal mass–spring system, like the one shown in Figure 7.51. Give the spring a relaxed length of 0.9 m and a stiffness of 1.5 N/m. Give the block a mass of 0.02 kg. **(a)** Compare the period of the oscillations of your model system (without any friction) to the period you would expect, to make sure your model behaves correctly. **(b)** Add graphs of K, U, and $K + U$. Again, check to make sure these make physical sense, in the absence of friction. **(c)** Now add a viscous friction force to the model. Adjust the viscous friction coefficient to make the behavior of the system clear. (A reasonable starting value is about 0.03). **(d)** Is the rate of energy loss constant over time? Use the energy graphs produced by your program to explain physically why the rate of energy loss varies as observed. **(e)** Replace the viscous friction force with a sliding friction force. A reasonable value for the coefficient in this model is about 0.15. **(f)** Compare the shape of the graph of $K + U$ for sliding friction to the shape of the $K + U$ graph for viscous friction. Explain the difference in the shapes in physical terms (consider the mathematical expressions for these forces).

••P47 Effect of air resistance on a baseball: Write a computational model to predict the motion of a baseball that is hit at a speed of 44 m/s (100 mi/h) at an angle of 45° to the horizontal. A baseball has a mass of 155 g and a diameter of 7 cm. The drag coefficient C for a baseball is about 0.35, and the density of air at sea level is about 1.3×10^{-3} g/cm^3, or 1.3 kg/m^3. **(a)** In your initial model, neglect air resistance. How far does the ball go? Is this distance reasonable? (A baseball field is about 400 ft from home plate to the fence in center field. An outfielder cannot throw a baseball in the air from the fence to home plate.) **(b)** Plot a graph of $K + U_g$ vs. time for the system of baseball plus Earth. **(c)** Now add air resistance to your model. How far does the ball go? **(d)** Compare the graphs of $K + U_g$ with and without air resistance. **(e)** In Denver, a mile above sea level, the air is about 83% as dense as the air at sea level. Including the effect of air resistance, use your computer model to predict the trajectory and range of a baseball thrown in Denver. How does the predicted range compare with the predicted range at sea level?

••P48 Determining a drag coefficient: Write a VPython program to model the motion of a paper object falling straight down, including the effects of air resistance. If you did the experiments in Problem P40, use your own data. Otherwise, you can use these data from an actual experiment that used coffee filters: filter mass = 8.52×10^{-3} kg, outer radius = 5.8×10^{-2} m, initial height = 2.03 m, drop time for one filter = 1.44 s, drop time for a stack of four coffee filters = 0.88 s. This experiment was performed in Santa Fe, New Mexico (altitude 2100 m), where the air density is only 1.045 kg/m^3; the air density at an altitude of H meters is $e^{-H/8500}$ times the air density at sea level (see Chapter 12).

(a) Add a plot of v vs. t to your program. Set the air resistance force to zero and use this plot to check that your program is working correctly. What should this plot look like without air resistance? **(b)** Now add the air resistance force to your program. What is the initial slope of your graph of speed vs. time? (By dragging the mouse across a VPython graph you can accurately read coordinates off the graph.) What should be the initial slope? Why? **(c)** Adjust the drag coefficient C until the drop time predicted by your model matches your experimental data for the case of the smallest mass. (Run your loop until the y coordinate of the moving object reaches zero, then print the computed time.) **(d)** What terminal speed does your model predict for this case? **(e)** Using the value of C that you determined with the smallest mass, what drop time does your model predict for an object of four times the mass but the same cross section? Does this agree with your measurement? **(f)** Using the value of C that you determined with the smallest mass, what terminal speed does your model predict for an object of four times the smallest mass but the same cross section? Is this what you expected? Why or why not?

••P49 Terminal speed for a falling sky diver has been measured to be about 60 m/s. Write a computational model to determine how far (in meters) and how long (in seconds) a sky diver falls before reaching terminal speed. (You will need to think about the meaning of "terminal speed" and how to test for it.) Plot graphs of speed vs. time and position vs. time.

•••P50 You can study resonance in a driven oscillator in detail by modifying the program you wrote for Problem P46. Let one end of the spring be moved back and forth sinusoidally by a motor, with a motion given by $D\sin(\omega_D t)$ (see Figure 7.52). Here D is the amplitude of the motor motion and ω_D is the angular frequency of the motor, which can be varied. (The free-oscillation angular frequency $\omega_F = \sqrt{k_s/m}$ of the spring–mass system has a fixed value, determined by the spring stiffness k_s and the mass m.)

We need an expression for the stretch s of the spring in order to be able to calculate the force $-k_s s$ of the spring on the mass. We have to do a bit of geometry to figure out the length of the spring when the mass is displaced a distance x from the equilibrium position, and the motor has moved the other end of the spring by an amount $D\sin(\omega_D t)$. In the figure, the spring gets longer when the mass moves to the right ($+x$), but the spring gets shorter when the motor moves to the right ($-D\sin(\omega_D t)$). The new length of the spring is $L + x - D\sin(\omega_D t)$, and the net stretch of the spring is this quantity minus the unstretched length L, yielding ($s = x - D\sin(\omega_D t)$). A check that this is the correct expression for the net stretch of the spring is that if the motor moves to the right the same distance as the mass moves to the right, the spring will have zero stretch. Replace x in your computer computation with the quantity $[x - D\sin(\omega_D t)]$.

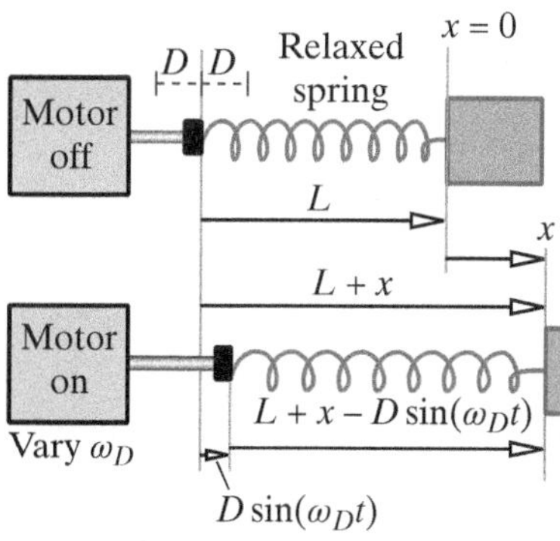

Figure 7.52

Use viscous damping (friction proportional to v). The damping should be small. That is, without the motor driving the system ($D = 0$), the mass should oscillate for many cycles. Set ω_D to $0.9\omega_F$ (that is, 0.9 times the free-oscillation angular frequency, $\omega_F = \sqrt{k_s/m}$), and let x_0 be 0. Graph position x vs. time for enough cycles to show that there is a transient buildup to a "steady state." In the steady state the energy dissipation per cycle has grown to exactly equal the energy input per cycle.

Vary ω_D in the range from 0 to $2.0\omega_F$, with closely spaced values in the neighborhood of $1.0\omega_F$. Have the computer plot the position of the motor end as well as the position of the mass as a function of time. For each of these driving frequencies, record for later use the steady-state amplitude and the phase shift (the difference in angle between the sinusoids for the motor and for the mass). If a harmonic oscillator is lightly damped, it has a large response only for driving frequencies near its own free-oscillation frequency.

Is the steady-state angular frequency equal to ω_F or ω_D for these various values of ω_D? (Note that during the transient buildup to the steady state the frequency is not well-defined, because the motion of the mass isn't a simple sinusoid.)

Sketch graphs of the steady-state amplitude and phase shift vs. ω_D. Note that when $\omega_D = \omega_F$ the amplitude of the mass can be much larger than D, just as you observed with your hand-driven spring–mass system. Also note the interesting variation of the phase shift as you go from low to high driving frequencies—does this agree with the phase shifts you observed with your hand-driven spring–mass system?

Repeat the analysis with viscous friction twice as large. What happens to the resonance curve (the graph of steady-state amplitude vs. angular frequency ω_D)?

ANSWERS TO CHECKPOINTS

1 No change; same stretch
2 0.012 J
3 A is $|U|$; B is K; C is $|K+U|$; 1–4 are bound; 5 is unbound.
4 21.36°C; neglect the material of the thin glass.
5 1.26×10^5 J; 7.6×10^4 J; -1.26×10^5 J
6 9000 J
7 0, since no temperature change; $-11{,}000$ J
8 2.5 m/s^2
9 $\frac{1}{2}k_sA^2$; $\frac{1}{2}k_sA^2$; spring plus mass
10 (90 kg)(9.8 N/kg) = 882 N; the net force must be zero.
11 2.7×10^{-2} J
12 Air resistance is proportional to v^2, so the energy dissipation rate is higher at higher speeds.
13 +4 J; another +4 J; +8 J
14 Denominator gets very small when $\omega_D = \omega_F$.

Low frequency:

$$\cos\phi \approx \frac{(\omega_F^2)}{\sqrt{(\omega_F^2)^2 + \left(\frac{c}{m}\omega_D\right)^2}} \approx 1 \text{ so } \phi \approx 0^\circ$$

High frequency:

$$\cos\phi \approx \frac{(-\omega_D^2)}{\sqrt{(-\omega_D^2)^2 + \left(\frac{c}{m}\omega_D\right)^2}} \approx -1,\ \phi \approx 180^\circ$$

Resonance frequency, $\cos\phi = 0$, so $\phi \approx 90^\circ$

14 When $\omega_D = \omega_F$, denominator $= \frac{c}{m}\omega_D$.

Denominator $= \sqrt{2}\frac{c}{m}\omega_D$ when $|(\omega_F^2 - \omega_D^2)| = \frac{c}{m}\omega_D$.
However, since near resonance $\omega_D \approx \omega_F$, we have

$$(\omega_F^2 - \omega_D^2) = (\omega_F + \omega_D)(\omega_F - \omega_D)$$
$$\approx 2\omega_D(\omega_F - \omega_D)$$

Hence when $|\omega_F - \omega_D| \approx \frac{c}{2m}\omega_D \approx \frac{c}{2m}\omega_F$, the amplitude is down by a factor of $1/\sqrt{2}$. Larger friction, wider resonance peak. If friction is very small, the resonance peak is a very narrow spike.

CHAPTER

8

Energy Quantization

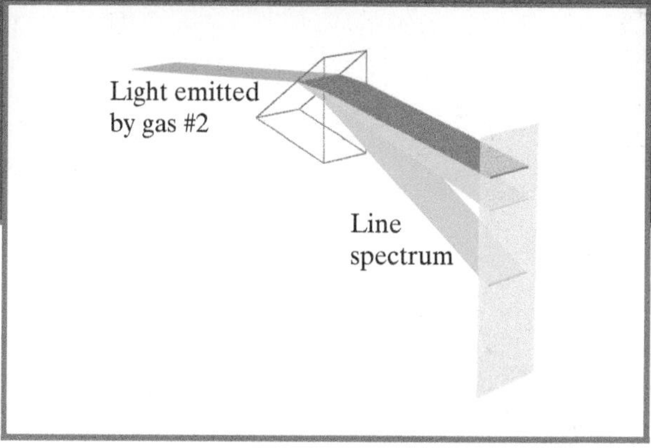

OBJECTIVES

After studying this chapter, you should be able to

- Calculate the energy of a photon emitted or absorbed in a transition between given energy levels.
- Given an emission spectrum, construct a possible energy-level diagram for a system.
- Calculate the energy levels of a quantized harmonic oscillator from its microscopic properties (stiffness, mass).

The Energy Principle applies to microscopic as well as macroscopic systems. In this chapter we will apply the Energy Principle to systems such as atoms and molecules, in both gas and solid form. In particular, an understanding of the energy of vibrations of atoms in a solid lattice and in diatomic molecules will provide an important foundation for our study of entropy in Chapter 12.

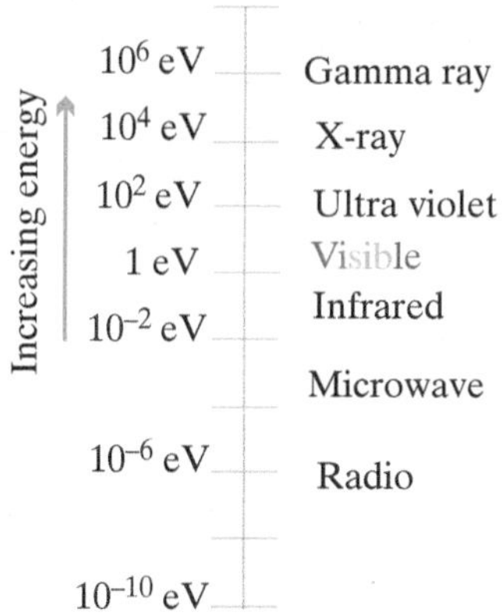

Figure 8.1 The electromagnetic spectrum. See Figure 8.2 for an expanded view of the portion of this spectrum that is visible to the human eye.

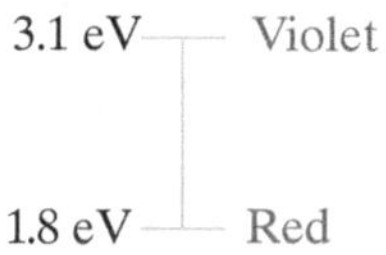

Figure 8.2 An expanded view of the portion of the electromagnetic spectrum that is visible to the human eye.

8.1 PHOTONS

One of the things that we will consider in this chapter is the interaction of light and matter. For the purposes of this chapter, we are interested only in the fact that light is a form of energy—we will defer considerations of the optical properties of light until later. In a contemporary model of light, a beam of light is composed of many particles of light, called "photons." A photon is a particle that has only kinetic energy—its rest energy is zero. A photon can never be at rest; it may be thought of as a packet of energy, traveling at the speed of light. An atom can gain energy by absorbing a photon; if this happens, all of the photon's energy is transferred to the object, and the photon disappears. Conversely, an object can lose energy by emitting a photon with a particular amount of energy. The general term applied to photons is "electromagnetic radiation" (or simply "light"), and in later chapters on electricity and magnetism we will go more deeply into the electric and magnetic aspects of electromagnetic radiation, and we will also discuss its wave-like aspects.

Photons also have momentum, and the Momentum Principle applies to absorption and emission of photons. In Chapter 6 we saw that for any particle $E^2 - (pc)^2 = (mc^2)^2$. Since the photon's mass is zero, we have $E^2 = (pc)^2$, and the magnitude of the momentum of a photon whose energy is E is $p = E/c$ (this result is approximately true for any particle traveling so fast that mc^2 is negligible compared to E).

Historically, photons in different energy ranges have been given different names, usually because the instruments needed to detect and measure electromagnetic radiation depend on the photon energies involved. The entire range of possible photon energies is called the electromagnetic spectrum

(Figure 8.1). Only photons whose energies fall between about 1.8 eV (red light) and 3.1 eV (violet light) can be detected by the human eye (Figure 8.2), so photons in this energy range are referred to as visible light. Colloquially we often refer to all photons as "light," whether they are high-energy gamma rays or low-energy radio waves.

Although earlier scientists, including Isaac Newton, had speculated that light might be made up of discrete particles, it was not until early in the 20th century that clear evidence for the particle nature of light was obtained. As in the case of atoms, the noncontinuous, lumpy nature of light was not noticed for a long time because the amount of energy carried by one photon in an ordinary beam of light is extremely small compared to the total energy of the beam. It has now become routine to detect individual photons in a very weak beam of light and to measure the energy of an individual photon. Observations of the interaction of photons and ordinary matter have provided key insights into the internal structure of molecules, atoms, and nuclei.

One can separate a beam of visible light into its component energies by passing a narrow beam of light through a prism. You have probably seen the continuous rainbow spectrum produced when ordinary white light goes through a prism, as in Figure 8.3. Your eye reacts differently to photons of different energies, which is why you see a colorful spectrum.

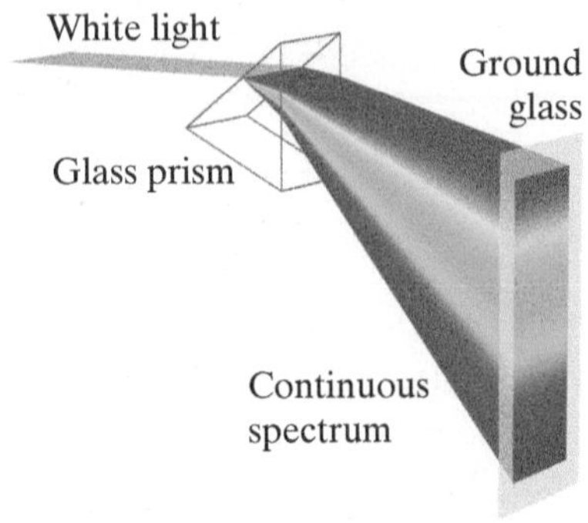

Figure 8.3 A prism separates a narrow beam of white light into a continuous rainbow of colors. The color of visible light is correlated with the energy of the photons; a red photon has less energy than a blue photon.

Checkpoint 1 The photon energy for green light lies between the values for red and violet light. What is the approximate energy of the photons in green light? The intensity of sunlight above the Earth's atmosphere is about 1400 W (J/s) per square meter. That is, when sunlight hits perpendicular to a square meter of area, about 1400 W of energy can be absorbed. Using the photon energy of green light, about how many photons per second strike an area of one square meter? (This is why the lumpiness of light was not noticed for so long.)

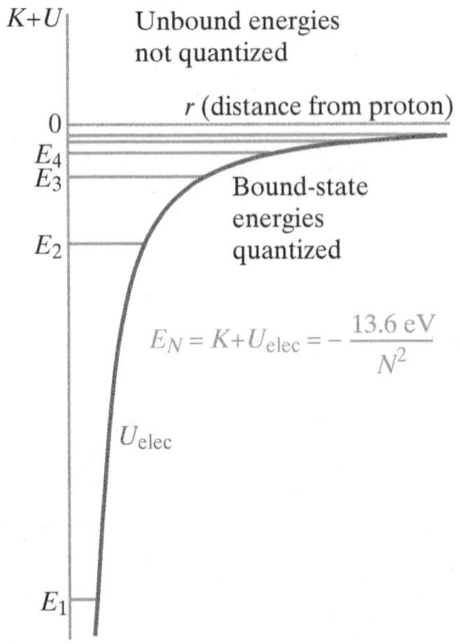

Figure 8.4 Electronic energy levels for a hydrogen atom. The potential energy curve describes the potential energy of the electron–proton system as a function of separation. Each colored horizontal line corresponds to a stable value of the internal energy of the atom.

8.2 ELECTRONIC ENERGY LEVELS

One of the most exciting discoveries of the early 20th century was that for the internal energy of an atom to be stable, there can be only certain discrete values of the internal energy, called energy "levels." An example of discrete energy levels is found in the "electronic" energy associated with the motion of electrons around the nucleus of an atom. You probably know from your study of chemistry that electronic energy levels in an atom are discrete—the only stable bound states of a system of nucleus plus electrons are those with certain specific values of $K+U$. The electronic energy levels in atomic hydrogen are shown in Figure 8.4 (atomic hydrogen is a single H atom, composed of one proton and one electron, not the usual diatomic molecule H_2). The term "quantized" is often used colloquially as a synonym for "discrete." We will use it this way, although the technical meanings of the two words are not quite identical.

Figure 8.4 shows the familiar electric potential energy curve for the proton–electron system (an atom of hydrogen). The horizontal lines represent the possible, stable bound energy states for the hydrogen atom; stable bound states with other values of $(K+U)$ are not observed. Unbound states $(K+U \geq 0)$, however, are not quantized.

The electric potential energy of an electron and a proton bound together in a hydrogen atom is negative for all separations:

$$U_{elec} = \frac{1}{4\pi\varepsilon_0}\frac{q_1 q_2}{r} = \frac{1}{4\pi\varepsilon_0}\frac{(+e)(-e)}{r} = -\frac{1}{4\pi\varepsilon_0}\frac{e^2}{r}$$

A quantum mechanical calculation of the energy levels of a system involves solving a differential equation called the Schrödinger equation, a form of the Energy Principle that can be applied to atomic phenomena. This equation includes both a potential energy term and a kinetic energy term. Solving this equation for the hydrogen atom predicts that the discrete energy levels for bound states of hydrogen (ignoring the rest energies of the proton and the electron) will be these:

ELECTRONIC ENERGY LEVELS OF A HYDROGEN ATOM

$$E_N = K + U_{elec} = -\frac{13.6\ \text{eV}}{N^2}, \quad N = 1,2,3,\text{etc.}$$

This arrangement of hydrogen energy levels has been verified by observations of the interaction of light and hydrogen, as we will see.

These energy levels denote different bound states of the electron + proton system. Recall that a bound state has a negative value of $K + U$, since a potential energy of zero is associated with a very large separation between the electron and the proton. We have seen that bound gravitational orbits are also associated with negative values of $K + U$. As you can see from the energy diagram (Figure 8.4), in higher-energy states the average location of the electron is farther from the proton.

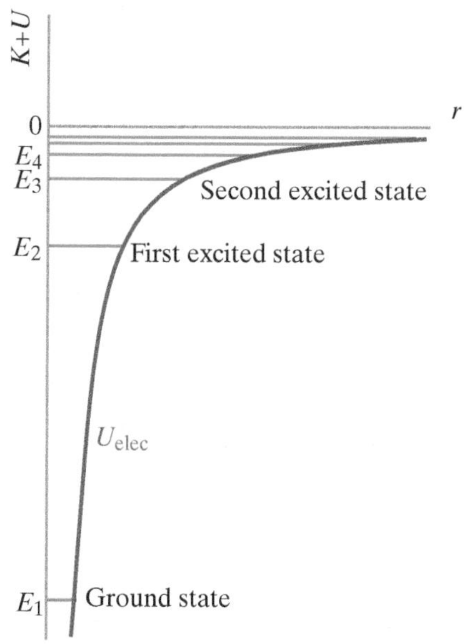

Figure 8.5 The lowest stable energy state is called the ground state. States with higher energy are called excited states.

The Ground State and Excited States

In the energy level diagram shown in Figure 8.5 there is a lowest-energy state, called the "ground state." This is the lowest possible internal energy that a single hydrogen atom can have: $K + U$ cannot be less than $-13.6\,\text{eV}$. Stable energy states whose energy is higher than the energy of the ground state are called "excited states."

A note on potentially confusing terminology: the ground state of the hydrogen atom is called E_1, corresponding to a value of $N = 1$ in the equation above. This means that for this system, E_2 (corresponding to $N = 2$) is called the "first excited state," because it is the first state above the ground state. Similarly, E_3 ($N = 3$) is the second excited state, and so on.

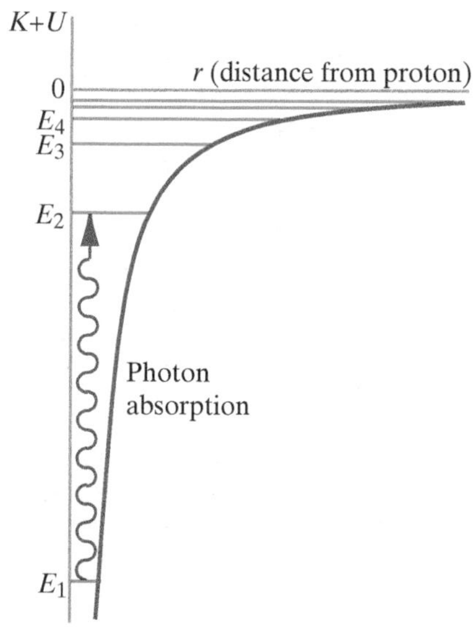

Figure 8.6 An atom can gain energy by absorbing a photon whose energy corresponds to the difference between two energy levels. By convention a squiggly line with an arrowhead indicates a photon. The upward arrow indicates a transition to a higher energy level.

Raising the Internal Energy of an Atom

If an atom is in its ground state, how can it gain enough energy to get to one of its excited states? We will consider two important mechanisms for this process:

- An atom (or molecule) can gain energy by absorbing a photon.
- An atom (or molecule) can gain energy through a collision with a particle such as an energetic electron, or another fast-moving atom.

Photon Absorption

In order for an atom to gain energy by absorbing a photon, the energy of the photon must match the difference in energy between two energy states of the atom (Figure 8.6). When a photon is absorbed by an atom, nothing is left over—the photon no longer exists, and the atom has more energy. We can describe the absorption of a photon by a hydrogen atom in this compact way (the asterisk is used to denote an atom with higher internal energy):

$$\text{H} + \text{photon} \rightarrow \text{H}^*$$

Applying the Energy Principle to this process, and choosing all particles as the system, we can write this equation:

$$E_{H^*,f} = E_{H,i} + K_{photon}$$

EXAMPLE **A Hydrogen Atom Absorbs a Photon**

A hydrogen atom is originally in its ground (lowest) electronic energy state ($N = 1$). It absorbs a photon, which raises the hydrogen atom's energy to the next energy level ($N = 2$). What was the energy of the photon, in eV?

Solution System: All particles
Surroundings: Nothing significant

$$E_f = E_i + K_{photon}$$

$$\left(-\frac{13.6\text{ eV}}{2^2}\right) = \left(-\frac{13.6\text{ eV}}{1^2}\right) + K_{photon}$$

$$K_{photon} = 10.2\text{ eV}$$

It is not possible to add less than 10.2 eV to a hydrogen atom in its lowest energy level. Any smaller amount fails to be absorbed because there is no stable energy level between the ground state and the first excited state.

Collisional Excitation

If a fast-moving particle collides with an atom, and if the kinetic energy of the particle is equal to or greater than the difference between two energy levels of the atom, some of the particle's kinetic energy can be converted to internal energy of the atom, raising it to an excited state. Any excess kinetic energy is retained by the particle. For example, a beam of fast-moving electrons may be used to excite atoms in a gas to a higher energy state. If we consider the system of a moving electron plus the atom:

$$E_{H^*,f} + K_{e,f} = E_{H,i} + K_{e,i}$$

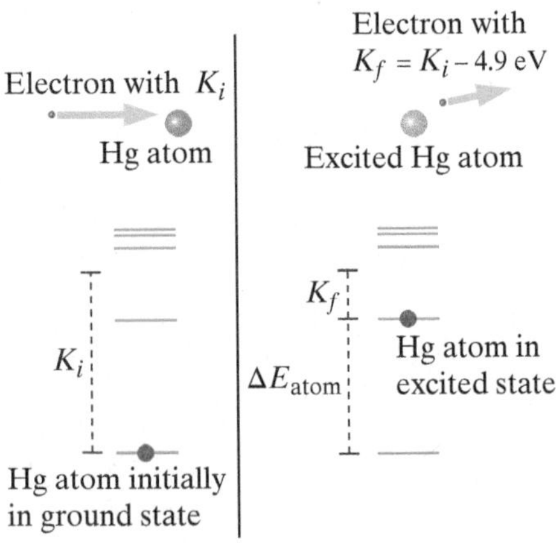

Figure 8.7 An electron with initial kinetic energy of more than 4.9 eV can excite a mercury atom (Hg) from the ground state to the first excited state. The electron loses 4.9 eV of kinetic energy.

Because electrons are (negatively) charged electrically, they interact strongly with atoms, and an electron that passes near an atom may excite the atom from the ground state to an excited state whose internal energy is ΔE_{atom} higher, with a corresponding loss of kinetic energy of the electron. Unlike photon absorption, which almost always occurs only if the photon energy exactly matches the difference between energy levels, an electron can give up some kinetic energy and have some kinetic energy left over. The process is illustrated in Figure 8.7 for an electron colliding with a mercury atom, whose first excited state is 4.9 eV above the ground state.

QUESTION For example, suppose an electron with kinetic energy of 6 eV collides with a mercury atom that is in the ground state. What can happen?

The first excited state of mercury is 4.9 eV above the ground state. This 6 eV electron has enough kinetic energy to excite the mercury atom to the first excited state, giving up 4.9 eV of energy. The electron moves away with 1.1 eV of kinetic energy remaining. The mercury atom will quickly drop back down to the ground state, emitting a 4.9 eV photon. If a beam containing a large number of energetic electrons passes through a gas of mercury vapor, there will be continual collisional excitation and continual photon emission. However, as soon as you turn off the electron beam, the mercury will stop emitting photons.

> QUESTION What happens if you send a beam of electrons with kinetic energy of 4 eV through the mercury vapor?

4 eV isn't enough energy to go from the ground state to the first excited state in mercury, so the electrons can pass right through the gas without changing the electronic energy states of the atoms. (An electron may collide with an atom, but because the mass of an electron is several thousand times smaller than the mass of a mercury atom, the electron will bounce off the atom like a Ping-Pong ball bouncing off a bowling ball—the kinetic energy of the atom will not change significantly.)

The German physicists James Franck and Gustav Hertz performed a collisional excitation experiment in 1914 that provided early dramatic evidence for discrete energy levels in atoms. Electrons with varying amounts of initial kinetic energy were sent through a gas of mercury vapor, and the number of electrons per second making it through the gas was measured. When the initial kinetic energy of the electrons was less than 4.9 eV, many electrons reached the detector. However, when the initial kinetic energy reached 4.9 eV, few electrons reached the detector. The Franck–Hertz experiment was correctly interpreted to mean that the first excited state of a mercury atom is 4.9 eV above the ground state (Figure 8.7), and an electron with kinetic energy of 4.9 eV can excite the atom, leaving the electron with almost no kinetic energy, so the number of electrons that made it through the gas became very small.

> QUESTION What happens if the initial kinetic energy of the electron is 9.8 eV?

In this case the electron could excite one mercury atom and still have 4.9 eV of kinetic energy available to excite a second mercury atom. In the Franck–Hertz experiment it was observed that the number of electrons reaching the detector again dropped sharply when the initial kinetic energy of the electrons was increased past 9.8 eV.

Every element has a different set of discrete energy levels. The first excited state in mercury is 4.9 eV above the ground state, whereas the first excited state in hydrogen is 10.2 eV above the ground state. These differences make it possible to identify elements from emission or absorption spectra, or measurements of the Franck–Hertz type.

> QUESTION What happens if you send a beam of electrons with kinetic energy of 25 eV through a gas of atomic hydrogen?

Because the ground state of hydrogen has $K + U = -13.6$ eV, only 13.6 eV is needed to raise a hydrogen atom to any of its excited states, or even to break the atom apart into a proton and an electron. Therefore a beam containing lots of 25 eV electrons passing through the atomic hydrogen gas will keep many excited states populated, and you'll observe many different photon energies being emitted as atoms drop from excited states to lower excited states or to the ground state.

EXAMPLE **An Electron Collides with a Hydrogen Atom**

An electron with kinetic energy 12.5 eV collides with a hydrogen atom that is initially in its ground state. After the collision the hydrogen atom is its second excited state ($N = 3$). **(a)** What is the kinetic energy of the electron after the collision? **(b)** Would it have been possible for this electron to excite the hydrogen atom to its third excited state instead?

Solution **(a)** System: Electron and H atom
Surroundings: Nothing

$$E_{\mathrm{H}^*,f} + K_{\mathrm{e},f} = E_{\mathrm{H},i} + K_{\mathrm{e},i}$$
$$\left(-\frac{13.6\ \mathrm{eV}}{3^2}\right) + K_{\mathrm{e},f} = \left(-\frac{13.6\ \mathrm{eV}}{1^2}\right) + 12.5\ \mathrm{eV}$$
$$K_{\mathrm{e},f} = 0.411\ \mathrm{eV}$$

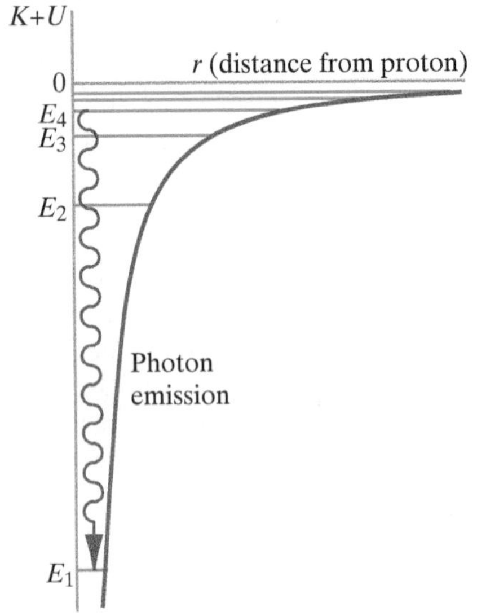

Figure 8.8 An atom can drop to a lower energy level by emitting a photon whose energy corresponds to the difference between the energies of the levels.

(b) The energy difference between the ground state and the third excited state ($N = 4$) is

$$\Delta E = \left(-\frac{13.6\ \mathrm{eV}}{4^2}\right) - \left(-\frac{13.6\ \mathrm{eV}}{1^2}\right)$$
$$= 12.75\ \mathrm{eV}$$

The initial kinetic energy of the electron is less than 12.75 eV, so it is not possible for the collision to excite the atom to the third excited state.

Checkpoint 2 The first excited state of a mercury atom is 4.9 eV above the ground state. A moving electron collides with a mercury atom, and excites the mercury atom to its first excited state. Immediately after the collision the kinetic energy of the electron is 0.3 eV. What was the kinetic energy of the electron just before the collision?

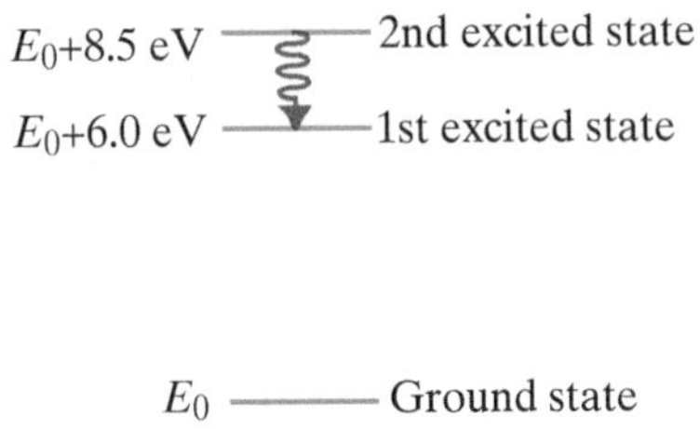

Figure 8.9 Photon emission from an atom.

Decreasing the Internal Energy of an Atom

Just as an atom or molecule can absorb a photon, ending up in an excited state (with more internal energy), it can lose energy by emitting a photon (Figure 8.8), which comes into existence specifically to carry away a particular amount of energy. For a hydrogen atom, emission of a photon can be described this way:

$$\mathrm{H}^* \rightarrow \mathrm{H} + \text{photon}$$
$$E_{\mathrm{H},f} + K_{\text{photon}} = E_{\mathrm{H}^*,i}$$

EXAMPLE **Photon Emission from an Atom**

Different atoms have different quantized energy levels. Consider a particular atom (not hydrogen) whose first excited state is 6.0 eV above the ground state, and the second excited state is 8.5 eV above the ground state (Figure 8.9). What is the energy of a photon emitted when such an atom drops from the second excited state to the first excited state?

Solution System: All particles
Surroundings: Nothing

$$E_f = E_i$$
$$(E_0 + 6.0\ \mathrm{eV}) + K_{\text{photon}} = (E_0 + 8.5\ \mathrm{eV})$$
$$K_{\text{photon}} = 2.5\ \mathrm{eV}$$

The energy of the emitted photon (2.5 eV) is the difference between the energies of the second and third energy levels.

Figure 8.10 shows a portion of a neon sign, in which neon atoms in a clear glass tube have gained energy through collisions with energetic electrons, and are emitting orange photons whose energy is characteristic of neon.

Figure 8.10 Collisional excitation of neon gas in a neon sign leads to emission of photons with the characteristic orange color. (Ruth Chabay/Bruce Sherwood)

Excited States Have Limited Lifetimes

If an isolated hydrogen atom is initially in a high energy level, over time it will spontaneously emit photons and drop to lower and lower states, eventually ending up in the lowest energy level, which is called the "ground state." Once an atom has dropped to the ground state, it cannot lose any more energy, or emit photons, because there is no lower energy level than the ground state.

Excited states are not in fact completely stable and eventually transition to a lower energy state. When a hydrogen atom drops from a higher energy level E_4 to a lower energy level E_1 it emits a photon with the associated energy $E_4 - E_1$ (Figure 8.8). In the energy diagrams we worked with for gravitational orbits, the orbit was stable and the energy was constant, but in an atom in a (nearly) stable state a photon can be emitted and the atom's energy drop to a lower level. When a change from one energy level to another happens, the change is instantaneous and takes no time, which is why these transitions are often called "quantum jumps" or "quantum leaps."

Another distinctive feature of photon emission from an excited state is that there is no way to predict exactly when it will happen. The probability that a hydrogen atom in the first excited state will remain in that state a time t after its formation is an exponential function of t, $e^{-t/\tau}$, where τ is approximately 2 ns (2 nanoseconds = 2×10^{-9} s). The quantity τ is called the "mean lifetime," the average time the excited state will persist. At the time $t = \tau$, the probability of the excited state still existing is $e^{-1} = 1/e = 0.368$. That is, when $t = \tau$, there is only a 36.8% probability that the atom is still in the first excited state (Figure 8.11). Most other excited states of hydrogen also have mean lifetimes of a few nanoseconds. Quantum mechanics can be used to predict the value of the mean lifetime τ, but when the emission from an excited state will actually occur is probabilistic and cannot be predicted. This is very different from the situation in our everyday macroscopic world.

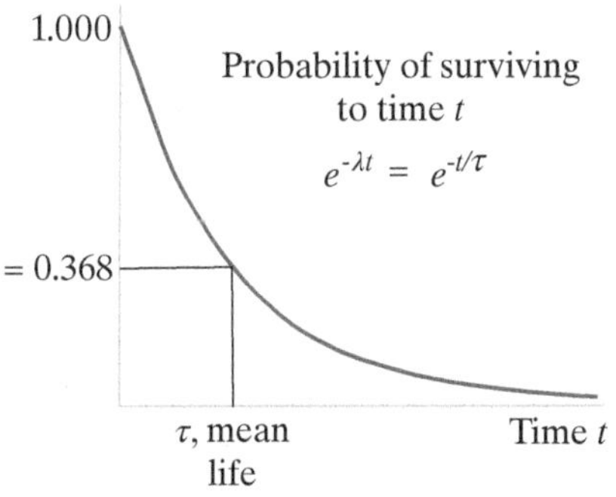

Figure 8.11 The probability of an excited atom remaining excited without emitting a photon, as a function of time.

Observations Typically Involve Many Atoms

It is much easier to observe the behavior of many atoms interacting with many photons than to observe a single atom absorbing or emitting a single photon. For example, a tube full of a gas (such as a fluorescent light tube) contains many atoms, which can be collisionally excited by a beam of electrons, and which emit many easily observable photons as they transition back down to the ground state. Instead of considering a single atom, therefore, another way of thinking about the probabilistic nature of photon emission is to consider a collection of 100 hydrogen atoms, all of them initially in the first excited state.

Different atoms in this collection will emit photons at different times. After a time $t = \tau$, there will on average be only $100e^{-1} \approx 37$ atoms still in the first excited state; the other 63 atoms will have dropped to the ground state with the emission of 63 photons of energy 10.2 eV. Because emission comes at random times, at $t = \tau$ there might still be 40 atoms in the excited state, or 35, but the average number in repeated experiments will be $100/e$. (For more details, see Section 8.8.)

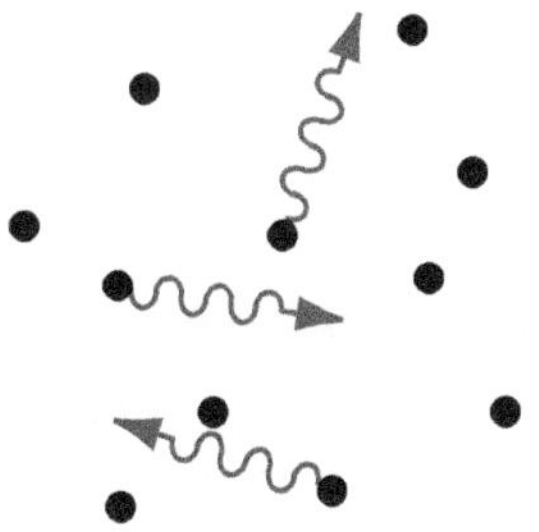

Figure 8.12 In a collection of atoms originally in the first excited state, many emitted photons will escape.

When one of the original 100 atoms emits a 10.2 eV photon, it is possible that an atom that is already in the ground state might absorb that photon and jump back up to the first excited state. However, in hydrogen gas at ordinary densities the atoms are rather far apart, so most photons will escape from a glass container of excited hydrogen atoms (Figure 8.12). Moreover, just because a photon is headed toward an atom in the ground state doesn't necessarily mean that it will interact with that atom, as this kind of interaction is also probabilistic.

EXAMPLE

Emission from Hydrogen Atoms

Consider a collection of many hydrogen atoms. Initially each individual atom is in one of the lowest three energy states, $N=1$ (the ground state), $N=2$ (the first excited state), or $N=3$ (the second excited state). What will be the energies of photons emitted by this collection of hydrogen atoms?

Solution There are three possibilities for atoms in the $N=3$ state: transitions from $N=3$ to $N=2$, from $N=2$ to $N=1$, or from $N=3$ to $N=1$ (Figure 8.13). Any atoms that are in the ground state ($N=1$) cannot emit a photon. Any atoms in the $N=2$ state can only emit a 10.2 eV photon. The photon energies for the three transitions are these:

$$N=3 \text{ to } N=2: \quad \left(-\frac{13.6\text{ eV}}{3^2}\right)-\left(-\frac{13.6\text{ eV}}{2^2}\right)=1.9\text{ eV}$$

$$N=2 \text{ to } N=1: \quad \left(-\frac{13.6\text{ eV}}{2^2}\right)-\left(-\frac{13.6\text{ eV}}{1^2}\right)=10.2\text{ eV}$$

$$N=3 \text{ to } N=1: \quad \left(-\frac{13.6\text{ eV}}{3^2}\right)-\left(-\frac{13.6\text{ eV}}{1^2}\right)=12.1\text{ eV}$$

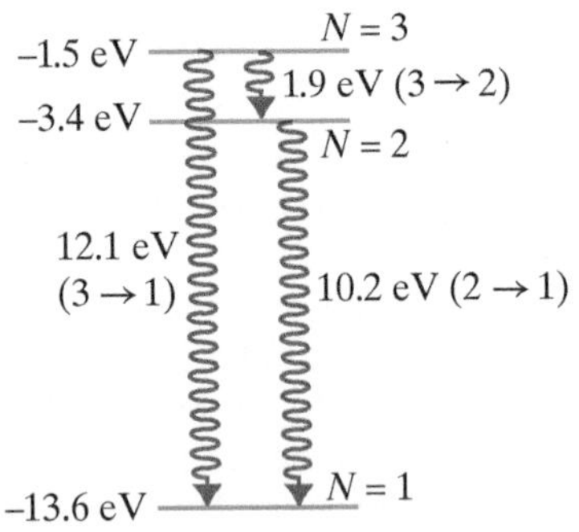

Figure 8.13 Photon emissions involving the ground state and first two excited states of a hydrogen atom. Only the $K+U$ energy levels are shown; for simplicity one often omits the potential energy curve.

Here again probability plays a distinctive quantum mechanical role. If a hydrogen atom is in the $N=3$ state, it can spontaneously (at an unpredictable time) change instantaneously to the ground state ($N=1$) with the emission of a 12.1 eV photon, or it can change instantaneously to the $N=2$ state with the emission of a 1.9 eV photon, followed shortly thereafter by the emission of a photon of 10.2 eV as the state changes instantaneously to the ground state.

Which of these events will happen? Will the atom emit one photon of 12.1 eV or two photons of 1.9 eV and 10.2 eV? With quantum mechanics it is only possible to calculate the probabilities of which event will happen; it is not possible to predict which event will in fact happen. This is very different from the situation in pre-1900 "classical mechanics."

Figure 8.14 illustrates this striking difference between classical mechanics and quantum mechanics. A ball falls to Earth in a continuous way, starting to move downward at a known time, and taking a predictable amount of time to reach the ground; the change in the Earth + ball system is continuous. In contrast, consider a hydrogen atom that starts out in the $N=3$ state at a known time. It is impossible to predict when the atom will drop to a lower state, and when it does drop, it does so instantaneously, a "quantum jump." There is no continuous change from the $N=3$ state to a lower state (with photon emission).

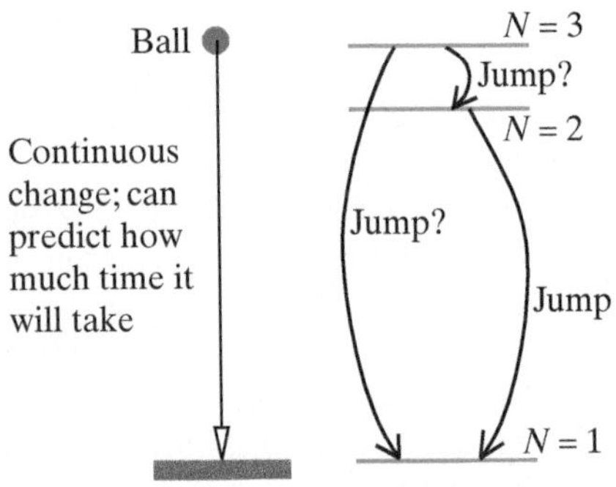

Figure 8.14 Classical mechanics vs. quantum mechanics: a ball falls in a continuous way; an atom changes state instantaneously, at unpredictable times.

Moreover, it is impossible to predict whether this particular atom will make an instantaneous quantum jump from the $N=3$ state directly to the ground state (with the emission of a 12.1 eV photon) or whether it will get to the ground state in two jumps, stopping briefly in the $N=2$ state. Quantum mechanics can determine the probability of jumping directly to the ground state but cannot predict whether this particular atom will do so.

Checkpoint 3 How many different photon energies would emerge from a collection of hydrogen atoms that occupy the lowest four energy states ($N = 1,2,3,4$)? (You need not calculate the energies of these transitions.)

Emission Spectra

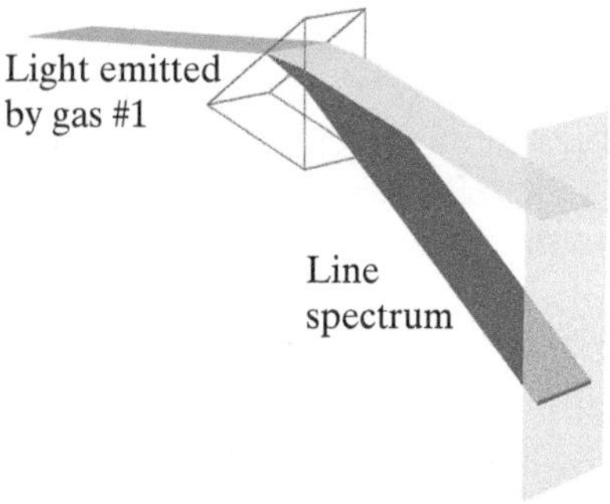

Figure 8.15 The light emitted by atoms of an excited gas.

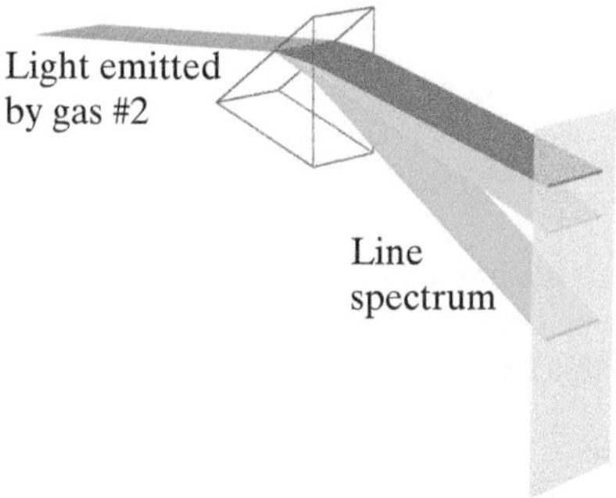

Figure 8.16 The light emitted by atoms of a second kind of excited gas.

Suppose we have a glass tube containing a gas, through which it is possible to run a beam of electrons that can transfer some of their kinetic energy to the atoms through collisions. These atoms remain in excited states for a few nanoseconds, then emit photons and return to the ground state, where they can repeat the process by gaining energy from collisions, and so on. Some of the photons emitted in this process may have energies in the visible range of the spectrum, so if we direct the emitted light through a slit and onto a prism, we will see a "line spectrum" like the one in Figure 8.15.

The light emitted by an excited gas is called an "emission spectrum," and the bright lines of light of particular energies in the spectrum are called "spectral lines" due to the line-like shape of the light coming through a slit in the apparatus. The energies of the photons emitted will depend on the particular gas, because the energy levels of every atom are different. Figure 8.16 shows an emission spectrum from a gas that is different from the one in Figure 8.15. The energy of the photons in any spectral line correspond to the differences in energy between two electronic energy levels of the particular atom emitting the photons.

The observed energies of light emitted by excited atomic hydrogen are consistent with the energy level differences predicted by quantum mechanics. Other atoms have different quantized energy levels and emit different photon energies, so a spectrum can be used to identify the emitting atoms by the pattern of emitted photon energies.

Absorption Spectra

Suppose that we have a container of hydrogen atoms, all in their ground state.

QUESTION Could we observe an emission spectrum from this gas?

No. If all the atoms are in the ground state, there is no lower energy state for them to go to, so they cannot lose energy by emitting photons. However, such a collection of atoms could gain energy by absorbing photons. If a beam of light composed of photons with a continuous distribution of energies (white light) is directed at a transparent container of gas, some of the photons in the beam will be absorbed by gas atoms.

QUESTION What should we expect to see if we pass a beam of white light through a transparent container of gas atoms in their ground state?

Only photons whose energy is the difference of atomic energy levels will be strongly absorbed, while other photons can pass right through the material. Energy absorption is by far most likely to occur if the photon energy corresponds exactly to the difference between two energy levels of the atom. What we will observe in this case is essentially the opposite of what we see in an emission spectrum. If we inspect the beam of light after it has passed through the container of gas, we will observe photons that were not absorbed, but photons whose energies match differences between energy levels of the gas will be missing because they were absorbed by the gas. In this case, the spectral lines are dark lines (missing energies) on a light background, instead of the bright lines on a dark background observed in an emission spectrum. This is called an "absorption spectrum" (Figure 8.17).

Photon absorption may be followed almost immediately by emission from the excited state to a lower state, so at any instant there are few atoms in excited states that could absorb photons. The result is that in dark-line absorption spectra only absorption from the ground state is observed, not from higher-level states.

The atoms that have jumped to higher energy levels through photon absorption will eventually drop to lower energy levels with the emission of photons. However, these photons are emitted in all directions, not just in the direction of the original beam of light, so the light that has passed through the material will have little intensity at photon energies corresponding to the strongly absorbed energies (Figure 8.18).

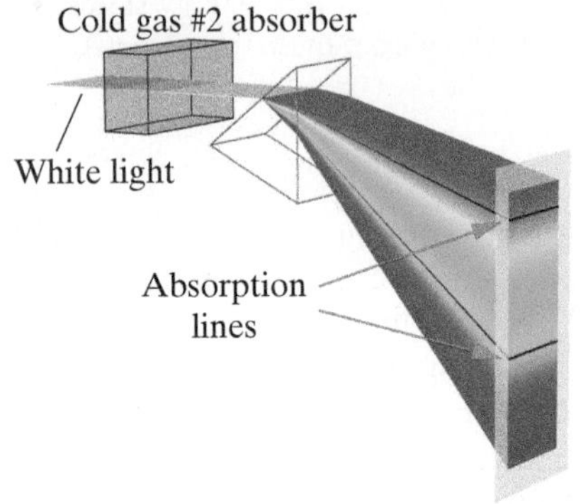

Figure 8.17 In an absorption spectrum, a beam of photons of many energies is directed through a gas. Although most photons pass through, photons matching the differences between energy levels are absorbed by gas atoms, leaving dark lines in the spectrum.

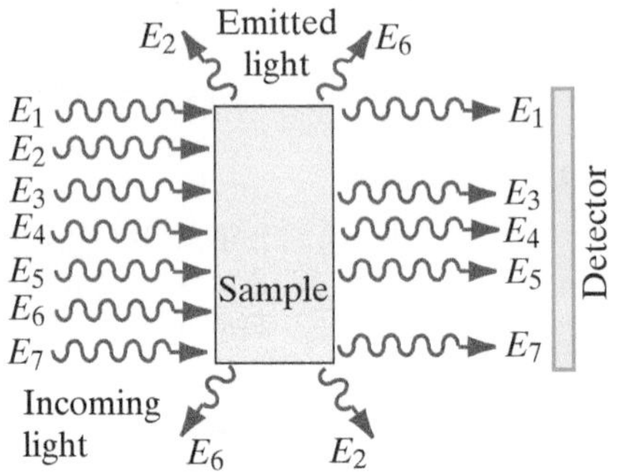

Figure 8.18 Measuring an absorption spectrum. A sample of a material is placed in a transparent container. Some energies of incoming light are absorbed. The outgoing light is depleted of the absorption energies (in this case E_2 and E_6).

QUESTION The gas that produced the three line emission spectrum in Figure 8.16 is the same gas that produced the two line absorption spectrum in Figure 8.17. Why do we observe three emission lines but only two absorption lines?

In an absorption spectrum the gas starts out in the ground state, so only transitions from the ground state to an excited state are possible. Evidently the middle line in the emission spectrum corresponds to a transition between two excited states, and won't appear in an absorption spectrum.

EXAMPLE **Why Is Hydrogen Gas Transparent?**

At room temperature in a gas of atomic hydrogen almost all of the atoms are in the ground state. Why might you expect the gas to be completely transparent to visible light?

Solution As we saw in an earlier example, the energy required to go from the ground state of atomic hydrogen to the first excited state is 10.2 eV, but visible light consists of photons whose energies lie between 1.8 eV and 3.1 eV. So visible light cannot be absorbed by ground-state hydrogen atoms. (In a collision between a photon of visible light and a hydrogen atom, the atom can acquire some kinetic energy, but no internal energy change is possible.)

Checkpoint 4 Suppose you had a collection of hypothetical quantum objects whose individual energy levels were −4.0 eV, −2.3 eV, and −1.6 eV. If nearly all of the individual objects were in the ground state, what would be the energies of dark spectral lines in an absorption spectrum if visible white light (1.8 to 3.1 eV) passes through the material?

EXAMPLE

Determining the Energy Levels from Photon Energies

You observe photon emissions from a collection of quantum objects, each of which is known to have just four quantized energy levels. The collection is continually bombarded by an electron beam, and you detect emitted photons using a detector sensitive to photons in the energy range from 2.5 eV to 30 eV. With this detector, you observe photons emitted with energies of 3 eV, 6 eV, 8 eV, and 9 eV, but no other energies.

(a) It is known that the ground-state energy of each of these objects is −10 eV. Propose two possible arrangements of energy levels that are consistent with the experimental observations. Explain in detail, using diagrams.

(b) You obtain a second detector that is sensitive to photon energies in the energy range from 0.1 eV to 2.5 eV. What additional photon energies do you observe to be emitted? Explain briefly.

(c) You turn off the electron beam so that essentially all the objects are in the ground state. Then you send a beam of photons with a wide range of energies through the material. Using both detectors to determine the absorption spectrum, what are the photon energies of the dark lines? How can this information be used to choose between your two proposed energy level schemes? Explain briefly.

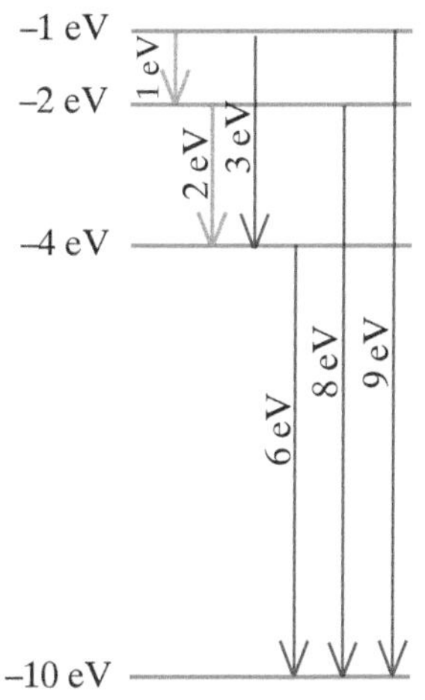

Figure 8.19 A possible energy-level scheme that fits the emission data. Lighter arrows indicate photons not detectable by this detector.

Solution System: Everything—some unknown object and a photon
Surroundings: Nothing significant
Initial state: Object in excited state (no photon)
Final state: Object in lower energy state, plus photon
$E_{\text{obj},f} + K_{\text{photon}} = E_{\text{obj},i}$

The energy of any emitted photon must be equal to the difference in energy between an excited state and a lower-energy state.

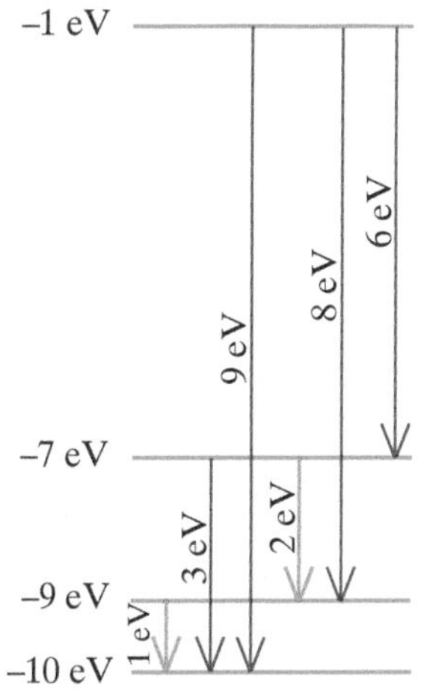

Figure 8.20 Another possible energy level scheme that fits the emission data.

(a) Figure 8.19 shows a possible energy-level scheme that fits the emission data. The highest photon energy (9 eV) tells us the difference between the highest and lowest states. The rest is trial and error, checking that the correct photon energies would be produced. The arrows are labeled with the energies of photons emitted in transitions between the two states indicated. The 1 eV and 2 eV emissions would not be detected by a detector that is sensitive to the photon energy range 2.5 eV to 30 eV.

Figure 8.20 shows another possible scheme that fits the emission data. This is somewhat less likely, because it is more common to find the energy-level spacing decreasing at higher energies, as shown in Figure 8.19.

(b) With this detector we observe the 1 eV and 2 eV photon emissions.

(c) With the electron beam turned off, almost all of the objects are in the ground state. If the energy levels are as shown in Figure 8.19, the dark lines will be at 6 eV, 8 eV, and 9 eV, corresponding to absorption in the ground state. If the energy levels are as shown in Figure 8.20, the dark lines will be at 1 eV, 3 eV, and 9 eV. Therefore the absorption spectrum allows us to distinguish between the two possible energy level schemes.

Note that since this was an energy analysis, we didn't need to know what kind of object this was, or whether the energy levels were electronic energy levels or some other kind.

Figure 8.20 is an example of an energy-level diagram, with photon emissions indicated by arrows that start on the upper level and point downward, with arrowheads on the lower level. Photon absorptions are indicated by arrows pointing up from the ground state to an upper level.

Checkpoint 5 The energy levels of a particular quantum object are −8 eV, −3 eV, and −2 eV. If a collection of these objects is bombarded by an electron beam so that there are some objects in each excited state, what are the energies of the photons that will be emitted?

Selection Rules

For simplicity, we've been assuming that photon emissions can occur from any discrete energy level to a lower level. Real atomic and molecular systems may be more complex, and some photon emissions are "forbidden." In a later chapter we will study angular momentum, a quantity that is related to rotational motion and is a conserved quantity, like momentum and energy. Photon emission and absorption, like all other physical processes, must conserve angular momentum, and as a result transitions between two particular energy levels may be forbidden. Such constraints are called "selection rules" and govern photon emission and absorption (collisional excitation is not constrained by selection rules). Selection rules are beyond the scope of this introductory course, but you will learn about them if you take a later course in quantum mechanics. For the purposes of this introduction to quantized energy, we will ignore selection rules and assume that all energetically possible transitions are allowed.

We also ignore questions of how probable an allowed transition is, which requires full quantum mechanics to calculate. Suppose that an object is in the second excited state, from which it could drop either to the first excited state or directly to the ground state. One of these two transitions could be much more probable than the other one, even if the low-probability transition isn't forbidden by a selection rule.

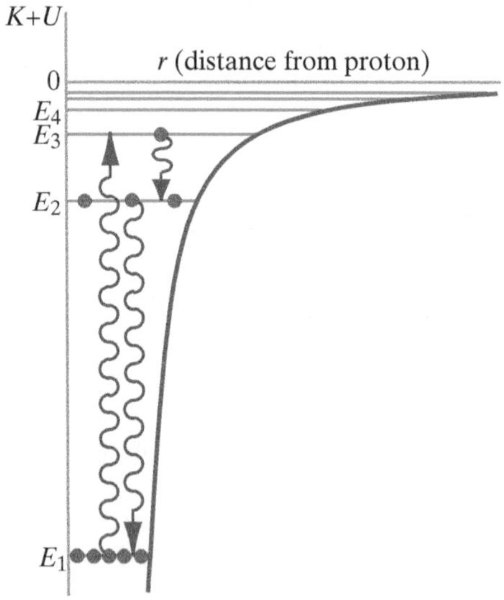

Figure 8.21 A dot represents the energy state of one hydrogen atom in a group of atoms. Energy is exchanged with the surroundings through photon emission and absorption.

8.3 THE EFFECT OF TEMPERATURE

When we observe atoms we are usually observing a collection of atoms, not an isolated single atom. Each atom in the collection (the "sample") interacts with other atoms and perhaps with a container. If an atom absorbs radiation coming from elsewhere, or absorbs energy from being jostled by a neighboring atom (collisional excitation), it may jump to a higher energy level, and eventually emit photons and drop again to a lower level (Figure 8.21 shows emissions as downward transitions and absorptions as upward transitions for a collection of hydrogen atoms). In thermal equilibrium with its surroundings, an atom's energy continually goes up and down, but with an average energy determined by the temperature of the collection of atoms.

In Figures 8.21 and 8.22 each dot represents one hydrogen atom in a container of atomic hydrogen. If three dots are on a line, this means three atoms are in this particular state. Each hydrogen atom has the same discrete energy levels, and at any particular instant some of the atoms will be in the ground state, some in the first excited state, and so on.

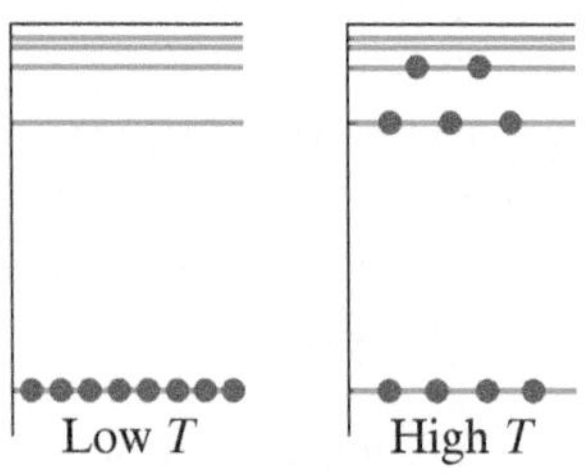

Figure 8.22 At a low temperature (system on the left) all atoms are in low-energy states. At a high temperature (system on the right) some atoms are in higher-energy states, though the largest fraction will still be in the ground state.

For a particular temperature, at any instant there is a certain probability that one particular atom will be in state 1, 2, 3, and so on. At low temperature almost all atoms are in the ground state (Figure 8.22). At high temperature the various atoms are in various states, from the ground state up to a high energy level, though the largest fraction will still be in the ground state. As we will see in a later chapter on statistical mechanics, in a collection of atoms the fraction of the atoms in a state whose energy is E above the ground state is proportional to the "Boltzmann factor" e^{-E/k_BT}, where k_B is the Boltzmann constant (1.38×10^{-23} J/K) and T is the absolute temperature in kelvins (K), where 0 °C is 273.15 K.

QUESTION Can you observe an emission spectrum if the material is very cold? Can you observe an absorption spectrum if the material is very cold?

In very cold matter the atoms will normally be in the ground state and will not emit photons. However, an absorption spectrum will show gaps (dark lines) corresponding to absorption of photons of energies appropriate to lift an atom from its ground state to an excited state. In order to observe an absorption spectrum you would of course need to use a beam of photons whose energies are sufficiently high to raise the system energy at least to the first excited state above the ground state.

Cold Samples

In a sample at room temperature (about 300 K), almost all atoms are in their electronic ground state. For example, if the first excited state is 1 eV above the ground state, the Boltzmann factor is

$$e^{-E/k_BT} = e^{-\frac{1.6\times10^{-19}\,\mathrm{J}}{(1.4\times10^{-23}\,\mathrm{J/K})(300\,\mathrm{K})}} = e^{-38} = 3\times10^{-17}$$

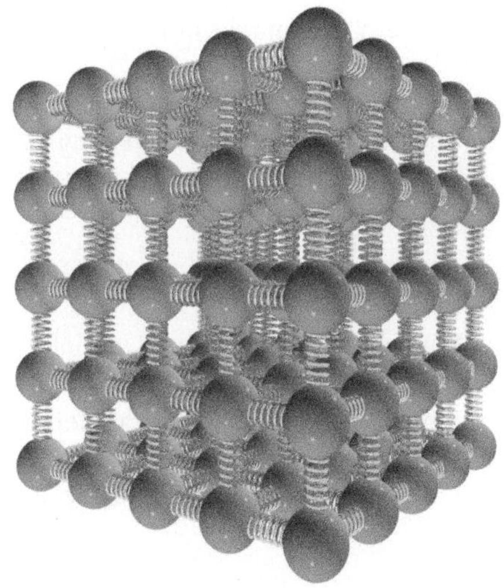

Figure 8.23 Model of a solid as a network of balls and springs.

This extremely tiny number represents the fact that at room temperature an insignificant fraction of the atoms will be excited thermally to 1 eV above the ground state. Therefore, it is reasonable to assume that essentially all the atoms in an ordinary room-temperature absorption measurement start out in the electronic ground state.

On the other hand, the surface temperature of the Sun is about 5000 K, and with $E = 1$ eV the Boltzmann factor is $e^{-E/k_BT} = e^{-2.3} = 0.1$, so a sizable fraction of atoms can be in an excited state 1 eV above the ground state, due to thermal excitation.

8.4 VIBRATIONAL ENERGY LEVELS

For concreteness we have described energy quantization for the electronic states of atomic hydrogen. Many other kinds of systems have quantized energy. We will next discuss vibrational energy quantization in atomic oscillators.

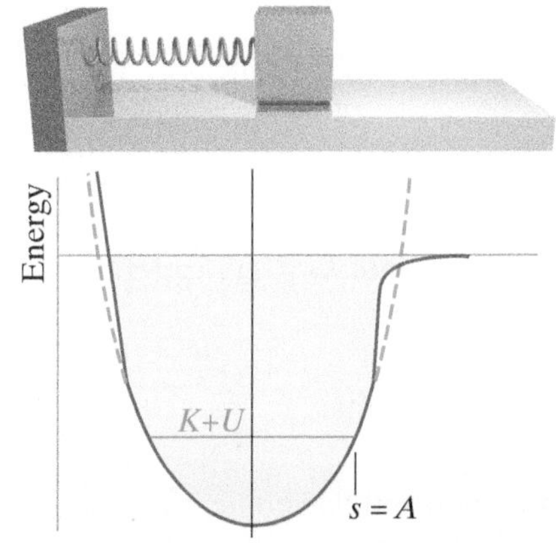

Figure 8.24 A classical harmonic oscillator (spring–mass system) can vibrate with any amplitude, and hence can have any energy. $K+U$ of the system could have any value in the shaded region.

Vibrational Energy in a Classical Oscillator

We have modeled a solid as a network of classical harmonic oscillators, balls and springs (Figure 8.23). Although this classical model explains many aspects of the behavior of solids, the predictions it makes regarding the thermal properties of solids do not correspond to experimental measurements at low temperatures. The classical model also fails to explain some aspects of the interaction of light with solid matter. To explain these phenomena we need to make our model of a solid more sophisticated, by incorporating the quantum nature of these atomic oscillators into our model.

The vibrational energy of a classical (nonquantum) spring–mass system can have *any* value within the colored region in Figure 8.24. Equivalently, the amplitude A (maximum stretch) can have *any* value within the allowed range. We can express the energy as $E = \frac{1}{2}k_sA^2$, because at the maximum stretch A the speed is zero, so the kinetic energy is zero, and the energy $K+U$ is just the potential energy at that instant. Classically, the amplitude A can have continuous rather than quantized values.

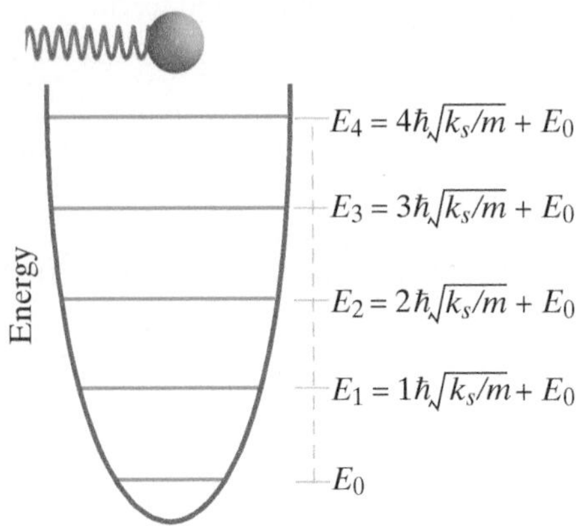

Figure 8.25 A quantum harmonic oscillator can have only certain energies, indicated by the horizontal lines on the graph. This idealized potential energy well represents the situation near the bottom of a real well (which is at a negative energy).

Vibrational Energy in a Quantum Mechanical Oscillator

Historically, the first hint that energy might be quantized came from an attempt by the German physicist Max Planck in 1900 to explain some puzzling features of the spectrum of light emitted through a small hole in a hot furnace (so-called "blackbody radiation"). Planck found that he could correctly predict the properties of blackbody radiation if he assumed that transfers of energy in the form of electromagnetic radiation were quantized—that is, an energy transfer could take place only in fixed amounts, rather than having a continuous range of possible values.

Later, with the full development of quantum mechanics, it was recognized that one could model an atom in the solid wall of the furnace as a tiny spring–mass system whose energy levels are discrete (Figure 8.25). Classically, the energy of a mass on a spring is related to the amplitude of its oscillation ($E = \frac{1}{2}k_sA^2$), so saying that the stable or nearly stable energy states of a spring–mass oscillator are discrete is equivalent to saying that only certain amplitudes correspond to these states. To see this, note that we have

$$K + U = \frac{1}{2}\frac{p^2}{m} + \frac{1}{2}k_s s^2 = \frac{1}{2}k_s A^2$$

since when the stretch s is equal to the amplitude A (the maximum stretch), the momentum is zero.

By solving the quantum mechanical Schrödinger equation for a system involving an atom and a chemical bond, one can show that the spacing ΔE between the quantized energies has the following value:

$$\Delta E = \hbar\sqrt{\frac{k_s}{m}} \quad \text{or} \quad \Delta E = \hbar\omega_0, \quad \text{where } \omega_0 = \sqrt{\frac{k_s}{m}}$$

In this case, the quantity $k_s = k_{s,i}$ is the interatomic spring stiffness, which we estimated in Chapter 4, $m = m_a$ is the mass of one atom, and $\omega_0 = \sqrt{k_s/m}$ is the angular frequency of the oscillator (in radians per second), as defined in Chapter 4. The quantity $\hbar$, pronounced "h-bar," is equal to h, "Planck's constant," divided by 2π. Planck's constant h has an astoundingly small value:

$$h = 6.6 \times 10^{-34}\ \text{J}\cdot\text{s}$$

The energy levels for an atomic harmonic oscillator are these:

VIBRATIONAL ENERGY LEVELS

$$E_N = N\hbar\omega_0 + E_0,\ N = 0, 1, 2, \ldots, \quad \text{where } \omega_0 = \sqrt{k_s/m}$$

In other words, the quantized energies have the following values:

$$E = E_0,\ 1\hbar\omega_0 + E_0,\ 2\hbar\omega_0 + E_0,\ 3\hbar\omega_0 + E_0, \ldots.$$

For an atom and a spring-like interatomic bond: $k_s = k_{s,i}$ (interatomic spring stiffness) and $m = m_a$ (mass of an atom)

$$\hbar = \frac{h}{2\pi} = 1.05 \times 10^{-34}\ \text{J}\cdot\text{s ("h-bar")}$$

It is an unusual and special feature of the quantized harmonic oscillator that the spacing between adjacent energy levels is the same for all levels ("uniform spacing"). This is quite different from the uneven spacing of energy levels in the hydrogen atom. These energy levels are predicted by quantum mechanics, and experiments confirm this prediction.

It is important to understand that in both the quantum and the classical picture, the angular frequency $\omega_0 = \sqrt{k_s/m}$ of a particular oscillator is fixed and is the same whether the energy of the oscillator is large or small. A useful way to think about the situation is to say that the frequency is determined by the parameters k_s and m, while the energy is determined by the initial conditions (or equivalently by the amplitude A). In the quantum world, it is as though the amplitude can have only certain specific values.

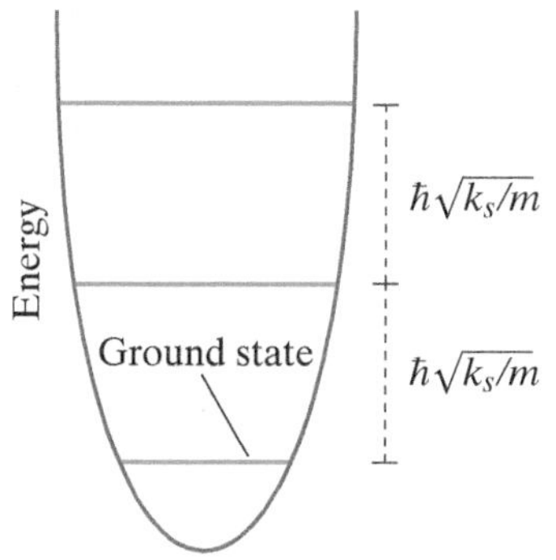

Figure 8.26 A quantum oscillator with a large k_s has widely spaced energy levels.

Checkpoint 6 Suppose that a collection of quantum harmonic oscillators occupies the lowest four energy levels, and the spacing between levels is 0.4 eV. What is the complete emission spectrum for this system? That is, what photon energies will appear in the emissions? Include all energies, whether or not they fall in the visible region of the electromagnetic spectrum.

Spring Stiffness k_s

The energy quantum depends on the stiffness k_s of the "spring." A stiffer spring (larger value of k_s, Figure 8.26) means a steeper, narrower potential curve and larger spacing between the allowed energies. A smaller value of k_s (Figure 8.27) means a shallower, wider potential curve and smaller spacing between the allowed energies.

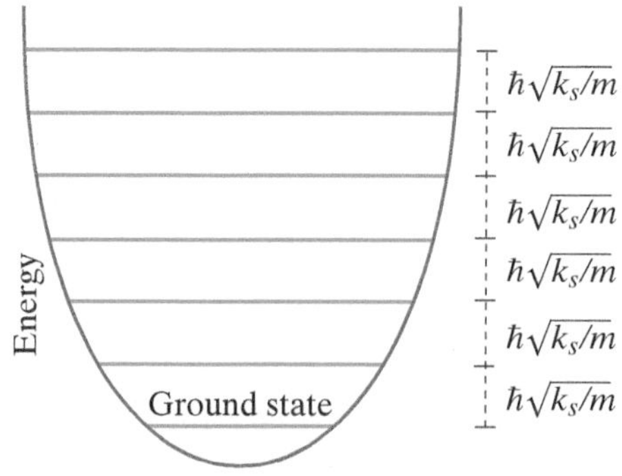

Figure 8.27 A quantum oscillator with a small k_s has closely spaced energy levels.

Checkpoint 7 You may have measured the properties of a simple spring–mass system in the lab. Suppose that you found $k_s = 0.7$ N/m and $m = 0.02$ kg, and you observed an oscillation with an amplitude of 0.2 m. What is the approximate value of N, the "quantum number" for this oscillator? (That is, how many levels above the ground state is this oscillator?)

The Ground State of a Quantum Oscillator

The lowest possible energy level E_0 is not zero relative to the bottom of the potential curve, but is $\frac{1}{2}(\hbar\sqrt{k_s/m})$ above the bottom of the well (Figures 8.26 and 8.27). This is closely related to the "Heisenberg uncertainty principle." When applied to the harmonic oscillator, this principle implies that if you knew that the mass on the spring were exactly at the center of the potential, its speed would be completely undetermined, whereas if you knew that the speed of the mass were exactly zero, the position of the mass could be literally anywhere! The minimum energy $\frac{1}{2}(\hbar\sqrt{k_s/m})$ above the bottom of the well can be viewed as a kind of a compromise, with neither the speed nor the position fixed at zero. The minimum energy is called the ground state of the system.

In our work with the quantized oscillator we will measure energy from the ground state rather than from the bottom of the potential well, so the offset from the bottom of the well won't come into our calculations.

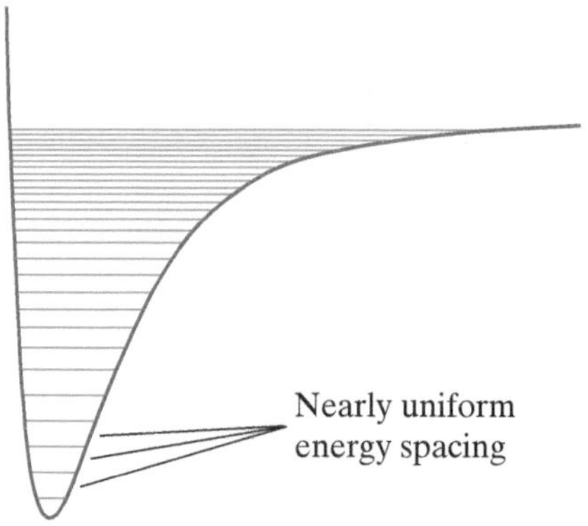

Figure 8.28 Energy levels for the interatomic potential energy, which describes the interaction of two neighboring atoms. This diagram is not to scale; the number of vibrational energy levels in the uniform region may be very large.

Energy Levels for the Interatomic Potential Energy

If we model the atoms in a solid as simple harmonic oscillators, whose potential energy curve is a parabola, then the energy levels of these systems are uniformly spaced. However, if we use a more realistic interatomic potential energy such as is shown in Figure 8.28, the energy levels of the system follow a more complex pattern. The lowest energy levels are nearly evenly spaced, corresponding to the fact that for small amplitude the bottom of the potential energy curve can be fit rather well by a curve of the form $\frac{1}{2}k_s s^2$. However, for large energies the energy spacing is quite small. Of course there are no bound states for energies greater than zero in Figure 8.28 (where zero is the value of the system energy at very large separations). The spacing of the vibrational energy levels of a diatomic molecule (Figure 8.29) typically corresponds to photons in the infrared region of the spectrum (see Problem P26).

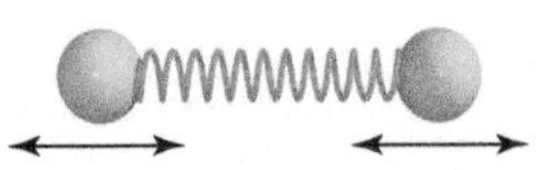

Figure 8.29 The vibrational energy levels of a diatomic molecule are discrete.

Energy Spacing in Various Systems

Most quantum objects have energy-level spacings that decrease as you go to higher levels. The harmonic oscillator potential energy is unusual in leading to uniform energy-level spacing. Even less common is the situation where the energy-level spacing actually increases with higher energy. One such exception is the so-called "square well" with very steep sides and a flat bottom. The energy levels in a square well actually get farther and farther apart for higher energies.

8.5 ROTATIONAL ENERGY LEVELS

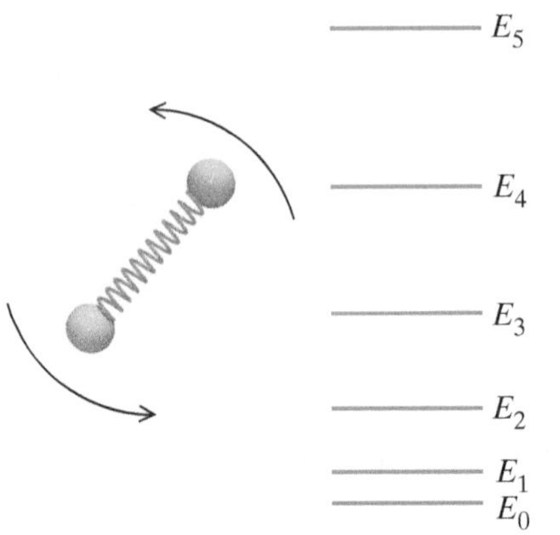

Figure 8.30 Rotational kinetic energy levels of a diatomic molecule.

A molecule in the gas phase has not only electronic energy and vibrational energy but rotational energy as well. Like electronic and vibrational energy levels, rotational energy levels are also discrete—only certain rotational energies are permitted. As we will see in a later chapter, the angular momentum of a molecule is quantized, and this quantization gives rise to discrete rotational energy levels. In this case all the energy of rotation is kinetic energy—there is no potential energy involved. Rotational energy levels for a diatomic molecule such as N_2 are shown in Figure 8.30. Note that the spacing between levels actually increases with increasing energy, which is the opposite of the trend seen in many atomic systems. Rotational energy levels are typically close together; transitions between these levels usually involve photons in the microwave region of the electromagnetic spectrum. Microwave ovens excite rotational energies of water molecules in food. These excited molecules then give up their energy to the food, making the food hotter.

Case Study: A Diatomic Molecule

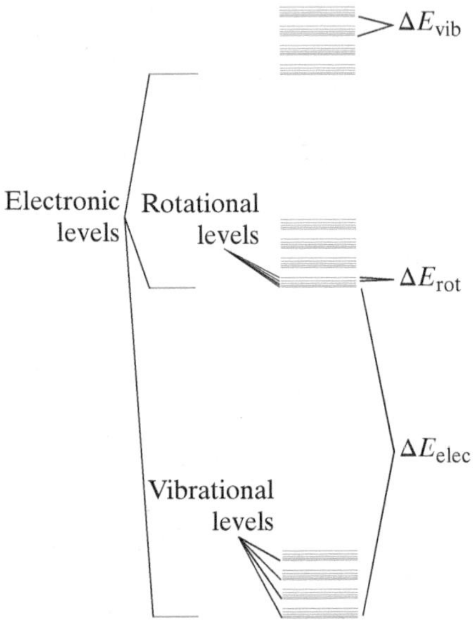

Figure 8.31 The energy bands of a diatomic molecule (not to scale). This energy-level structure gives rise to a band spectrum of emitted photons.

Analyzing a diatomic molecule such as N_2 provides a good review of three important types of discrete energy levels: electronic, vibrational, and rotational.

There are different configurations of the electron clouds in a diatomic molecule that correspond to discrete electronic energy levels. The spacings of the low-lying molecular electronic energy levels are typically in the range of 1 eV or more, like the spacings of the low-lying energy levels in the hydrogen atom.

For each major configuration of the electron clouds, many different vibrational energy states are possible, with energy spacings of the order of 1×10^{-2} eV (see Problem P26).

For each major configuration of the electron clouds and each vibrational energy state, many different rotational kinetic energy states are possible, with energy spacings of the order of 1×10^{-4} eV, as we will see in a later chapter.

These three different families of energy levels, with their very different energy-level spacings, give rise to energy "bands" and a distinctive "band" spectrum of emitted photons. See Figure 8.31, which is not to scale due to the large differences in level spacing for the three kinds of quantized energy levels. An emitted photon can have the energy of an electronic transition plus a smaller-energy vibrational transition plus an even smaller-energy rotational transition. The various possible energy differences give rise to "bands" in the observed emission spectrum.

8.6 OTHER ENERGY LEVELS

It is not only electronic, vibrational, and rotational energies in atoms and molecules that are quantized. Small objects such as nuclei and hadrons (particles made of quarks) also have discrete energy levels.

Nuclear Energy Levels

Just as atoms have discrete, stable or nearly stable states corresponding to allowed configurations of the electron cloud, so nuclei have discrete states corresponding to allowed configurations of the nucleons (protons and neutrons). An excited nucleus can drop to a lower state with the emission of a photon. The energy spacing of nuclear levels is very large, of the order of 1×10^6 eV (1 MeV), so the emitted photons have energies in the range of millions of electron volts. Such energetic photons are called "gamma rays."

Checkpoint 8 A gamma ray with energy greater than or equal to 2.2 MeV can dissociate a deuteron (the nucleus of deuterium, heavy hydrogen) into its constituents, a proton and a neutron. What is the nuclear binding energy of the deuteron?

Hadronic Energy Levels

Atoms have energy levels associated with the configurations of electrons in the atom, and nuclei have energy levels associated with the configurations of nucleons in the nucleus. Hadrons are particles made of quarks. In a simple model we represent baryons (particles like protons) as a combination of three quarks, and mesons (particles like pions) as a combination of a quark and an antiquark. Different configurations of quarks correspond to different energy levels of the multiquark system. These different configurations are the different hadrons we observe, and the corresponding energy levels are the observed rest energies of these particles. The energy spacing is of the order of hundreds of MeV, or about 1×10^8 eV. (Leptons, which include electrons, muons, and neutrinos, are not made of quarks.)

An example of a hadron is the Δ^+ particle, which has a rest energy of 1232 MeV. It can be considered to be an excited state of the three-quark system whose ground state is the proton, whose rest energy is 938 MeV. The Δ^+ particle can decay into a proton with the emission of a very high energy photon. However, this "electromagnetic" decay occurs in only about 0.5% of the decays. In most cases the Δ^+ decays into a proton plus a pion, a strong-interaction decay that has a much higher rate, corresponding to the much stronger interaction.

An excited electronic state of a hydrogen atom is still called hydrogen, but for historical reasons we give excited quark states distinctive names, such as Δ^+.

Checkpoint 9 What is the approximate energy of the photon emitted when the Δ^+ decays electromagnetically?

Type of state	Typical energy level spacing
hadronic	10^8 eV
nuclear	10^6 eV
electronic (atoms, molecules)	1 eV
vibrational (molecules)	10^{-2} eV
rotational (molecules)	10^{-4} eV

Figure 8.32 Typical energy level spacings in various quantum objects.

8.7 COMPARISON OF ENERGY-LEVEL SPACINGS

Figure 8.32 is a summary of typical energy scales for various kinds of objects. It is important to distinguish among hadronic, nuclear, electronic, vibrational, rotational, and other kinds of quantized energy situations. We can generalize the main issues to any kind of generic quantum object. The key features of all these different kinds of objects are these:

- discrete energy levels
- discrete emissions whose energies equal differences in energy levels

Some of the homework problems at the end of the chapter ask you to analyze generic objects, independent of the details of what gave rise to the particular discrete energy levels.

8.8 *RANDOM EMISSION TIME

Suppose we initially have an atom in its first excited state. In a short time interval dt there is a probability λdt that this atom will emit a photon and fall to the ground state, where λ is the average emission rate for this kind of atom. If we initially have $N = N_0$ atoms each in their first excited state, in a short time interval dt there is a probability of $N\lambda dt$ atoms emitting photons and falling to the ground state, so that the change in the number of excited atoms is $dN = -N\lambda dt$ or $dN/dt = -N\lambda$, where the negative sign indicates that N decreases. The equation shows that emissions are proportional to the remaining number of excited atoms, which makes sense. This differential equation can be solved:

$$\frac{dN}{dt} = -N\lambda$$

$$\frac{dN}{N} = -\lambda dt$$

$$\int \frac{dN}{N} = \int -\lambda dt$$

$$\ln N - \ln N_0 = -\lambda t$$

$$\ln N = \ln N_0 - \lambda t$$

$$N = N_0 e^{-\lambda t}$$

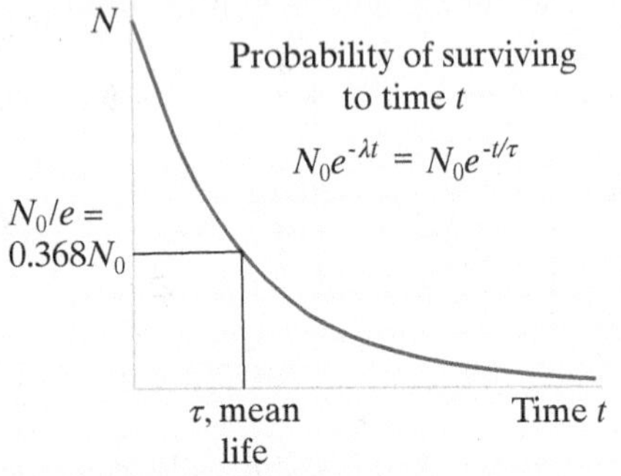

Figure 8.33 The average number of initially excited atoms remaining excited without emitting a photon, as a function of time.

The final equation is the result of equating e raised to the expression on the left to e raised to the expression on the right, where $e^{\ln x} = x$. The result shows that, on average, we expect the number of atoms remaining in the excited state to decrease exponentially with time t. The number remaining after a time $\tau = 1/\lambda$ is $1/e$ times the original number, and we often write $N = e^{-t/\tau}$, where τ is called the "mean lifetime" of the excited state (Figure 8.33). These same equations are used to describe radioactive decay, in which the number of radioactive nuclei decreases exponentially with time.

Problems P29 and P30 illustrate the full probabilistic character of emission, and the fact that the smooth exponential behavior shown in Figure 8.33 is seen only if the number of atoms is very large.

8.9 *CASE STUDY: HOW A LASER WORKS

A laser (Light Amplification by Stimulated Emission of Radiation) offers an interesting case study of the interaction of light and atomic matter. We will give a basic introduction to lasers, but we can only scratch the surface of a vast topic that continues to be an active field of research and development. Laser action depends on a process called "stimulated emission" and on creating an "inverted population" of quantum states, which can lead to a chain reaction and produce a special kind of light, called "coherent" light.

Stimulated Emission

If an atom can be placed into an excited state—that is, a state of higher energy than the ground state—normally it drops quickly to a lower-energy state and emits a photon in the process. For example, in a neon sign high-energy electrons collide with ground-state neon atoms and raise them to an excited state. An excited neon atom quickly drops back to the ground state, emitting a photon whose energy is the difference in energy between the two quantized energy levels, and you see the characteristic orange color corresponding to these photons.

We call this emission process "spontaneous emission." Spontaneous emission happens at an unpredictable time. All that quantum mechanics can predict is the *probability* that the excited atom will drop to the ground state in the next small time interval. It is truly not possible to predict the exact time of the emission, just as it is impossible to predict exactly when a free neutron will decay into a proton, electron, and antineutrino.

A collection of excited atoms such as those in a neon sign emit spontaneously, at random times. In the full quantum-mechanical description of light, light has both the properties of a particle (the photon picture) and the properties of a wave; see Section 8.10. Emissions at random times have random phases in a wave description: sine, cosine, and all phases between a sine and a cosine. We say that such light is "incoherent."

"Stimulated emission" of light is a different process that produces coherent light. Consider an atom in an energy level E above the ground state. A photon of this same energy E can interact with the atom in a remarkable way, causing the atom to drop immediately to the ground state with the emission of a second, "clone" photon. Where there had been one photon of energy E, there are now two photons of exactly the same energy E, and in a wave description they have exactly the same phase (Figure 8.34). That is, the two waves are both sines, or both cosines, or both with exactly the same phase somewhere between a sine and a cosine. Such light is called coherent light, and it has valuable properties for such applications as making holograms.

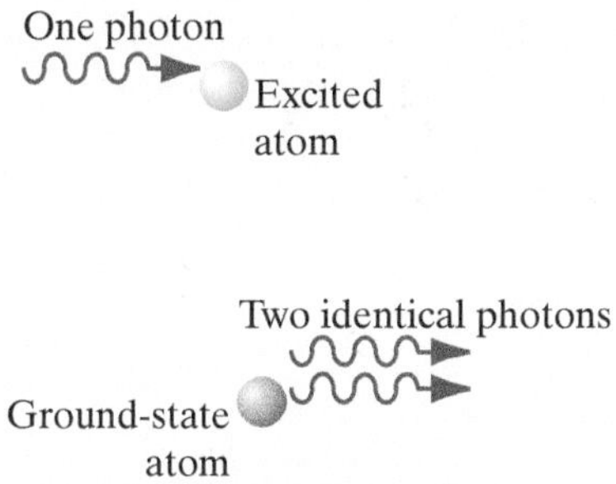

Figure 8.34 Stimulated emission produces two identical photons from one.

Chain Reaction

The phenomenon of stimulated emission can be exploited to create a chain reaction. Start with one photon of the right energy E to match the energy difference between the ground state and an excited state of the atoms in the system. If that photon causes stimulated emission from an excited atom, there are now two identical photons, both capable of triggering stimulated emission of other excited atoms, which yields four identical photons, then 8, 16, 32, 64, 128, 256, 512, 1024, and so on.

However, the chain reaction won't proceed if all of the photons escape from the system, so some kind of containment mechanism is required. In a common type of gas laser the gas is in a long tube with mirrors at both ends (Figure 8.35). Photons that happen to be going in the direction of the tube are continually reflected back and forth through the tube, which increases the probability that they will interact with excited atoms and cause stimulated emission. Since the stimulated-emission photons are clones, they too are headed in the direction of the tube and they too are likely to interact. One of the mirrors is deliberately made to be an imperfect reflector, so a fraction of the photons leaks out and provides a coherent beam of light. Unfortunately, the stimulated-emission photons also have just the right energy E to be absorbed by an atom in the ground state, in

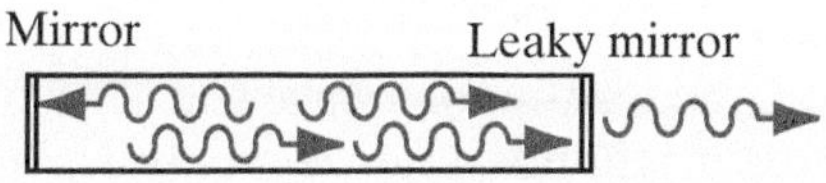

Figure 8.35 A gas laser with mirrors at the ends. Photons leak through one of the mirrors.

which case they can no longer contribute to the chain reaction. For this reason we need an "inverted population," with few atoms in the ground state to absorb the photons and stop the chain reaction. We discuss this issue next.

Inverted Population

As we have discussed earlier in this chapter, in a normal collection of atoms more atoms are in the ground state than in any other state, and the number of atoms decreases in each successively higher state. In fact, for a collection of atoms in thermal equilibrium the fraction of atoms in a particular energy state decreases exponentially with the energy above the ground state, as we will see in a later chapter.

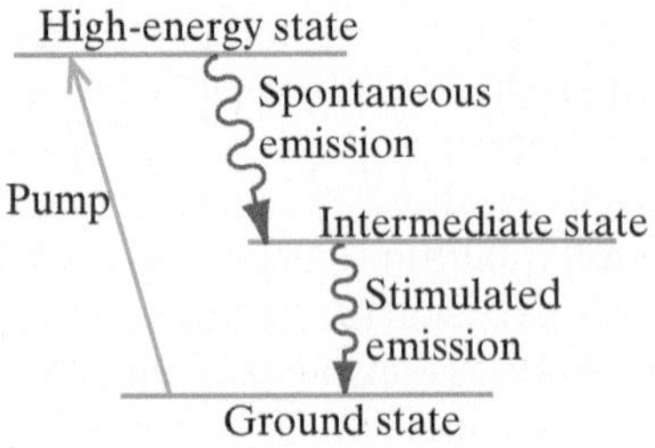

Figure 8.36 A three-state scheme for sustaining an inverted population.

By clever means it is possible to invert this population scheme, so that there are more atoms in an excited state than in the ground state (this is not a state of thermal equilibrium). An inverted population is crucial to laser action, so that stimulated emission with cloning of photons will dominate over simple absorption of photons by atoms in the ground state.

There are many different schemes for creating an inverted population. For example, the scheme used in the common helium–neon laser is to "pump" atoms to a high energy level (Figure 8.36). There are diverse energy input schemes, including electric discharges and powerful flashes of ordinary light. Atoms in this high energy level rapidly drop to an intermediate energy level, higher than the ground state. This intermediate state happens to have a rather long lifetime for spontaneous emission, so that it is possible to sustain an inverted population, with more atoms in the intermediate state than in the ground state. Stimulated emission drives these excited atoms down to the ground state, from which they are again pumped up to the high-energy state. Spontaneous emission from the high-energy state to the intermediate state maintains the inverted population.

8.10 *WAVELENGTH OF LIGHT

In this chapter we have emphasized the particle model of light, in terms of photons and their energy. However, as we have hinted earlier, some aspects of the behavior of light (of any energy) can best be understood by considering light to be a wave. For completeness we will mention a connection between the particle model and the wave model of light. Light consisting of photons whose energy is E can also be treated as a wave, and the energy E and wavelength λ are related as follows:

ENERGY AND WAVELENGTH OF LIGHT

$$E_{\text{photon}} = \frac{hc}{\lambda_{\text{light}}}$$

where h is Planck's constant.

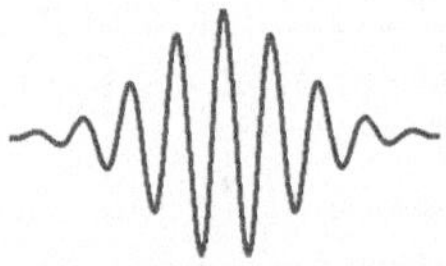
Figure 8.37 A wave packet is a traveling wave of finite spatial extent.

A model for light that incorporates both the features of a wave and a localized particle is a "wave packet," a wave whose magnitude is nonzero only in a small region of space (Figure 8.37). We will return to this topic in a later chapter on waves and particles.

SUMMARY

Light consists of particles called photons with energy and momentum but no rest mass, so $E = K$. Their speed is always 3×10^8 m/s.

Stable or nearly stable quantum systems can have only certain bound-state energies (unbound states have a continuous range of energies).

Three important processes involving energy-level change are collisional excitation, photon emission, and photon absorption.

The temperature is a measure of the average energy of a collection of objects, so high temperature is associated with significant probability of finding particles in any of many states, from the ground state up to a high energy level. Very low temperature is associated with little probability of the particles being in any state other than the ground state.

Examples were given of electronic, vibrational, and rotational energy levels, and of nuclear and hadronic energy levels.

Discrete hydrogen atom energy levels

$E_N = -(13.6 \text{ eV})/N^2$, where N is a nonzero positive integer (1,2,3...)

Discrete harmonic oscillator energy levels

$$E_N = N\hbar\omega_0 + E_0, \quad N = 0,1,2,\ldots, \quad \text{where } \omega_0 = \sqrt{k_s/m}$$

Planck's constant $h = 6.6 \times 10^{-34}$ J·s

$$\hbar = \frac{h}{2\pi} = 1.05 \times 10^{-34} \text{ J}\cdot\text{s}$$

$1 \text{ eV} = 1.6 \times 10^{-19}$ J

Photon energy and wavelength: $E_{\text{photon}} = \dfrac{hc}{\lambda_{\text{light}}}$

QUESTIONS

Q1 Match the description of a process with the corresponding arrow in Figure 8.38: **(a)** Absorption of a photon whose energy is $E_1 - E_0$. **(b)** Absorption from an excited state (a rare event at ordinary temperatures). **(c)** Emission of a photon whose energy is $E_3 - E_1$. **(d)** Emission of a photon whose energy is $E_2 - E_0$. **(e)** In drawing arrows to represent energy transitions, which of the following statements are correct? (1) It doesn't matter in which direction you draw the arrow as long as it connects the initial and final states. (2) For emission, the arrow points down. (3) For absorption, the arrow points up. (4) The tail of the arrow is drawn on the initial state. (5) The head of the arrow is drawn on the final state. (6) It is not necessary to draw an arrowhead.

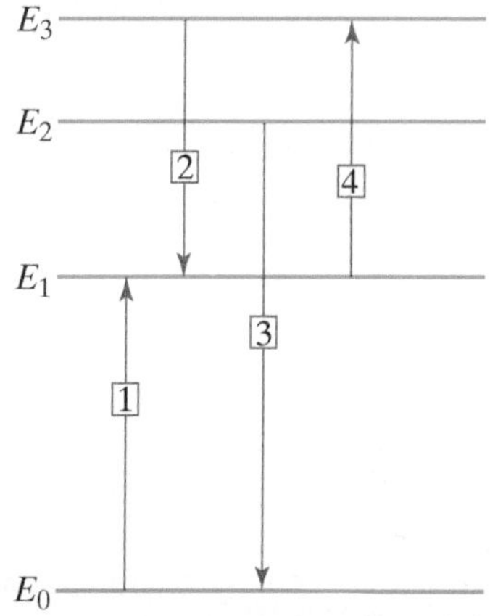

Figure 8.38

Q2 When starlight passes through a cold cloud of hydrogen gas, some hydrogen atoms absorb energy, then reradiate it in all directions. As a result, a spectrum of the star shows dark absorption lines at the energies for which less energy from the star reaches us. How does the spectrum of dark absorption lines for very cold hydrogen differ from the spectrum of bright emission lines from very hot hydrogen?

Q3 Which energy diagram in Figure 8.39 is appropriate for each of the following situations? **(a)** Hadronic (such as Δ^+). **(b)** Vibrational states of a diatomic molecule such as O_2. **(c)** Idealized quantized spring–mass oscillator. **(d)** Nuclear (such as the nucleus of a carbon atom). **(e)** Electronic, vibrational, and rotational states of a diatomic molecule such as O_2. **(f)** Rotational states of a diatomic molecule such as O_2. **(g)** Electronic states of a single atom such as hydrogen.

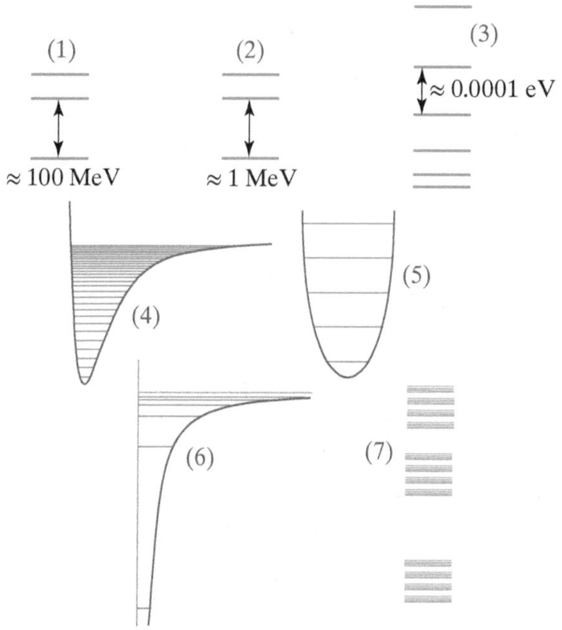

Figure 8.39

Q4 If you try to increase the energy of a quantum harmonic oscillator by adding an amount of energy $\frac{1}{2}\hbar\sqrt{k_s/m}$, the energy doesn't increase. Why not?

Q5 If you double the amplitude, what happens to the frequency in a classical (nonquantum) harmonic oscillator? In a quantum harmonic oscillator?

Q6 What is the energy of the photon emitted by a harmonic oscillator with stiffness k_s and mass m when it drops from energy level 5 to energy level 2?

Q7 Summarize the differences and similarities between different energy levels in a quantum oscillator. Specifically, for the first two levels in Figure 8.26, compare the angular frequency $\sqrt{k_s/m}$, the amplitude A, and the kinetic energy K at the same value of s. (In a full quantum-mechanical analysis the concepts of angular frequency and amplitude require reinterpretation. Nevertheless there remain elements of the classical picture. For example, larger amplitude corresponds to a higher probability of observing a large stretch.)

PROBLEMS

Section 8.1

•P8 A certain laser outputs pure red light (photon energy 1.8 eV) with power 700 mW (0.7 W). How many photons per second does this laser emit?

Section 8.2

•P9 How much energy in electron volts is required to ionize a hydrogen atom (that is, remove the electron from the proton), if initially the atom is in the state $N = 2$? (Remember that $N = 1$ if the atom is in the lowest energy level.)

•P10 The mean lifetime of a certain excited atomic state is 5 ns. What is the probability of the atom staying in this state for 10 ns or more?

•P11 At $t = 0$ all of the atoms in a collection of 10000 atoms are in an excited state whose lifetime is 25 ns. Approximately how many atoms will still be in this excited state at $t = 12$ ns?

•P12 $N = 1$ is the lowest electronic energy state for a hydrogen atom. **(a)** If a hydrogen atom is in state $N = 4$, what is $K + U$ for this atom (in eV)? **(b)** The hydrogen atom makes a transition to state $N = 2$. Now what is $K + U$ in electron volts for this atom? **(c)** What is the energy (in eV) of the photon emitted in the transition from level $N = 4$ to $N = 2$? **(d)** Which of the arrows in Figure 8.40 represents this transition?

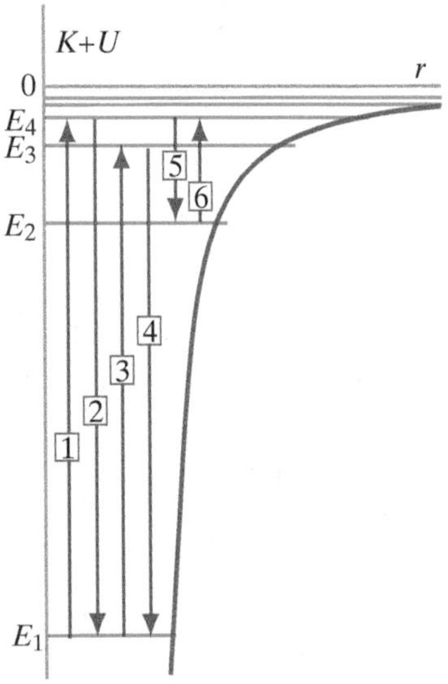

Figure 8.40

•P13 The Franck–Hertz experiment involved shooting electrons into a low-density gas of mercury atoms and observing discrete amounts of kinetic energy loss by the electrons. Suppose that instead a similar experiment is done with a very cold gas of atomic hydrogen, so that all of the hydrogen atoms are initially in the ground state. If the kinetic energy of an electron is 11.6 eV just before it collides with a hydrogen atom, how much kinetic energy will the electron have just after it collides with and excites the hydrogen atom?

••P14 Hydrogen atoms: **(a)** What is the minimum kinetic energy in electron volts that an electron must have to be able to ionize a hydrogen atom that is in its ground state (that is, remove the electron from being bound to the proton)? **(b)** If electrons of energy 12.8 eV are incident on a gas of hydrogen atoms in their ground state, what are the energies of the photons that can be emitted by the excited gas? **(c)** If instead of electrons, photons of all energies between 0 and 12.8 eV are incident on a gas of hydrogen atoms in the ground state, what are the energies at which the photons are absorbed?

••P15 Suppose we have reason to suspect that a certain quantum object has only three quantum states. When we excite such an object we observe that it emits electromagnetic radiation of three different energies: 2.48 eV (green), 1.91 eV (orange), and 0.57 eV (infrared). **(a)** Propose two possible energy-level schemes for this system. **(b)** Explain how to use an absorption measurement to distinguish between the two proposed schemes.

••P16 Predict how many emission lines will be seen by a human in the visible spectrum of atomic hydrogen. Give the energies of the emitted photons, and specify the energy levels involved in the transitions that are responsible for these lines.

••P17 Assume that a hypothetical object has just four quantum states, with the following energies:

−1.0 eV (third excited state)
−1.8 eV (second excited state)
−2.9 eV (first excited state)
−4.8 eV (ground state)

(a) Suppose that material containing many such objects is hit with a beam of energetic electrons, which ensures that there are always some objects in all of these states. What are the six energies of photons that could be strongly emitted by the material? (In actual quantum objects there are often "selection rules" that forbid certain emissions even though there is enough energy; assume that there are no such restrictions here.) List the photon emission energies. **(b)** Next, suppose that the beam of electrons is shut off so that all of the objects are in the ground state almost all the time. If electromagnetic radiation with a wide range of energies is passed through the material, what will be the three energies of photons corresponding to missing ("dark") lines in the spectrum? Remember that there is hardly any absorption from excited states, because emission from an excited state happens very quickly, so there is never a significant number of objects in an excited state. Assume that the detector is sensitive to a wide range of photon energies, not just energies in the visible region. List the dark-line energies.

••**P18** Energy graphs: **(a)** Figure 8.41 shows a graph of potential energy vs. interatomic distance for a particular molecule. What is the direction of the associated force at location A? At location B? At location C? Rank the magnitude of the force at locations A, B, and C. (That is, which is greatest, which is smallest, and are any of these equal to each other?) For the energy level shown on the graph, draw a line whose height is the kinetic energy when the system is at location D.

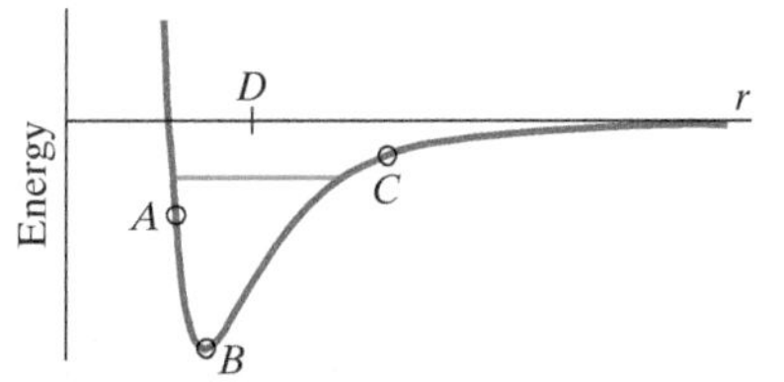

Figure 8.41

(b) Figure 8.42 shows all of the quantized energies (bound states) for one of these molecules. The energy for each state is given on the graph, in electron volts (1 eV $= 1.6 \times 10^{-19}$ J). How much energy is required to break a molecule apart, if it is initially in the ground state? (Note that the final state must be an unbound state; the unbound states are not quantized.)

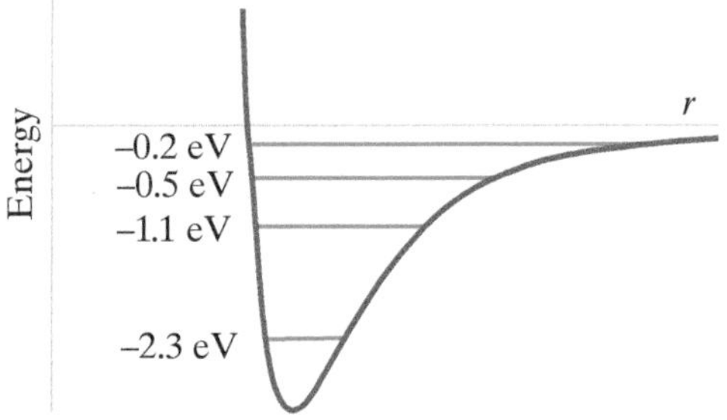

Figure 8.42

(c) At high enough temperatures, in a collection of these molecules there will be at all times some molecules in each of these states, and light will be emitted. What are the energies in electron volts of the emitted light? **(d)** The "inertial" mass of the molecule is the mass that appears in Newton's second law, and it determines how much acceleration will result from applying a given force. Compare the inertial mass of a molecule in the ground state and the inertial mass of a molecule in an excited state 10 eV above the ground state. If there is a difference, briefly explain why and calculate the difference. If there isn't a difference, briefly explain why not.

••**P19** A bottle contains a gas with atoms whose lowest four energy levels are -12 eV, -6 eV, -3 eV, and -2 eV. Electrons run through the bottle and excite the atoms so that at all times there are large numbers of atoms in each of these four energy levels, but there are no atoms in higher energy levels. List the energies of the photons that will be emitted by the gas.

Next, the electron beam is turned off, and all the atoms are in the ground state. Light containing a continuous spectrum of photon energies from 0.5 eV to 15 eV shines through the bottle. A photon detector on the other side of the bottle shows that some photon energies are depleted in the spectrum ("dark lines"). What are the energies of the missing photons?

••**P20** Suppose we have reason to suspect that a certain quantum object has only three quantum states. When we excite a collection of such objects we observe that they emit electromagnetic radiation of three different energies: 0.3 eV (infrared), 2.0 eV (visible), and 2.3 eV (visible). **(a)** Draw a possible energy-level diagram for one of the quantum objects, which has three bound states. On the diagram, indicate the transitions corresponding to the emitted photons, and check that the possible transitions produce the observed photons and no others. The energy $K + U$ of the ground state is -4 eV. Label the energies of each level ($K + U$, which is *negative*). **(b)** The material is now cooled down to a very low temperature, and the photon detector stops detecting photon emissions. Next, a beam of light with a continuous range of energies from infrared through ultraviolet shines on the material, and the photon detector observes the beam of light after it passes through the material. What photon energies in this beam of light are observed to be significantly reduced in intensity ("dark absorption lines")? Remember that there is hardly any absorption from excited states, because emission from an excited state happens very quickly, so there is never a significant number of objects in an excited state. **(c)** There exists another possible set of energy levels for these objects that produces the same photon emission spectrum. On an alternative energy-level diagram, *different from the one you drew in part (a)*, indicate the transitions corresponding to the emitted photons, and check that the possible transitions produce the observed photons and no others. Label the energies of each level ($K + U$, which is *negative*). **(d)** For your second proposed energy-level scheme, what photon energies would be observed to be significantly reduced in intensity in an absorption experiment ("dark absorption lines")? (Given the differences from part (b), you can see that an absorption measurement can be used to tell which of your two energy-level schemes is correct.)

••**P21** Assume that a hypothetical object has just four quantum states, with the energies shown in Figure 8.43.

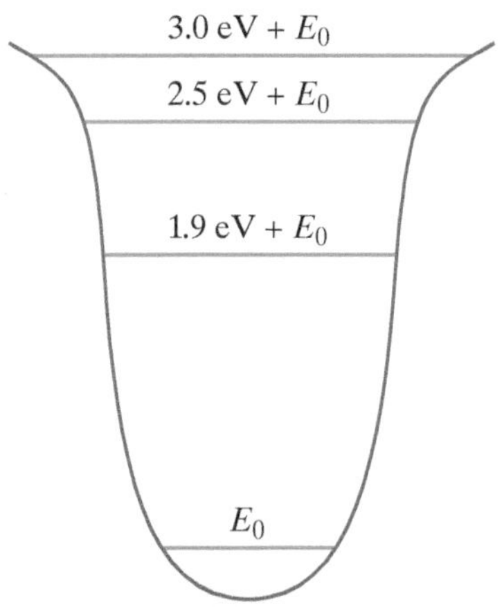

Figure 8.43

(a) Suppose that the temperature is high enough that in a material containing many such objects, at any instant some objects are found in all of these states. What are all the energies of photons that could be strongly emitted by the material?

(In actual quantum objects there are often "selection rules" that forbid certain emissions even though there is enough energy; assume that there are no such restrictions here.) **(b)** If the temperature is very low and electromagnetic radiation with a wide range of energies is passed through the material, what will be the energies of photons corresponding to missing ("dark") lines in the spectrum? (Assume that the detector is sensitive to a wide range of photon energies, not just energies in the visible region.)

Section 8.3

••P22 A certain material is kept at very low temperature. It is observed that when photons with energies between 0.2 and 0.9 eV strike the material, only photons of 0.4 eV and 0.7 eV are absorbed. Next, the material is warmed up so that it starts to emit photons. When it has been warmed up enough that 0.7 eV photons begin to be emitted, what other photon energies are also observed to be emitted by the material? Explain briefly.

••P23 Some material consisting of a collection of microscopic objects is kept at a high temperature. A photon detector capable of detecting photon energies from infrared through ultraviolet observes photons emitted with energies of 0.3 eV, 0.5 eV, 0.8 eV, 2.0 eV, 2.5 eV, and 2.8 eV. These are the only photon energies observed. **(a)** Draw and label a possible energy-level diagram for one of the microscopic objects, which has four bound states. On the diagram, indicate the transitions corresponding to the emitted photons. Explain briefly. **(b)** Would a spring–mass model be a good model for these microscopic objects? Why or why not? **(c)** The material is now cooled down to a very low temperature, and the photon detector stops detecting photon emissions. Next, a beam of light with a continuous range of energies from infrared through ultraviolet shines on the material, and the photon detector observes the beam of light after it passes through the material. What photon energies in this beam of light are observed to be significantly reduced in intensity ("dark absorption lines")? Explain briefly.

Section 8.4

•P24 For a certain diatomic molecule, the lowest-energy photon observed in the vibrational spectrum is 0.17 eV. What is the energy of a photon emitted in a transition from the 5th excited vibrational energy level to the 2nd excited vibrational energy level, assuming no change in the rotational energy?

••P25 Consider a microscopic spring–mass system whose spring stiffness is 50 N/m, and the mass is 4×10^{-26} kg. **(a)** What is the smallest amount of vibrational energy that can be added to this system? **(b)** What is the difference in mass (if any) of the microscopic oscillator between being in the ground state and being in the first excited state? **(c)** In a collection of these microscopic oscillators, the temperature is high enough that the ground state and the first three excited states are occupied. What are possible energies of photons emitted by these oscillators?

••P26 Molecular vibrational energy levels: **(a)** A HCl molecule can be considered to be a quantized harmonic oscillator, with quantized vibrational energy levels that are evenly spaced. Make a rough estimate of this uniform energy spacing in electron volts (where $1\text{ eV} = 1.6 \times 10^{-19}$ J). You will need to make some rough estimates of atomic properties based on prior work. For comparison with the spacing of these vibrational energy states, note that the spacing between quantized energy levels for "electronic" states such as in atomic hydrogen is of the order of several electron volts. **(b)** List several photon energies that would be emitted if a number of these vibrational energy levels were occupied due to collisional excitation. To what region of the spectrum (x-ray, visible, microwave, etc.) do these photons belong? (See Figure 8.1 at the beginning of the chapter.)

••P27 A hot bar of iron glows a dull red. Using our simple ball-spring model of a solid (Figure 8.23), answer the following questions, explaining in detail the processes involved. You will need to make some rough estimates of atomic properties based on prior work. **(a)** What is the approximate energy of the lowest-energy spectral emission line? Give a numerical value. **(b)** What is the approximate energy of the highest-energy spectral emission line? Give a numerical value. **(c)** What is the quantum number of the highest-energy occupied state? **(d)** Predict the energies of two other lines in the emission spectrum of the glowing iron bar. (*Note*: Our simple model is too simple—the actual spectrum is more complicated. However, this simple analysis gets at some important aspects of the phenomenon.)

COMPUTATIONAL PROBLEMS

The VPython programs used in these computational problems are available on the Wiley student website for this textbook.

•P28 There is a "random" module for Python that contains a function (also named "random") that generates a random number, which we can use to model photon emission at random times. Here is a little program that graphs and prints successive values generated by this function (the graph is shown in Figure 8.44):

```
from visual import *
from visual.graph import *
from random import random
gg = gcurve(color=color.yellow)
n = 0
while n < 20:
    rate(10)
    a = random()
    print(a)
    gg.plot(pos=(n,a))
    n= n + 1
```

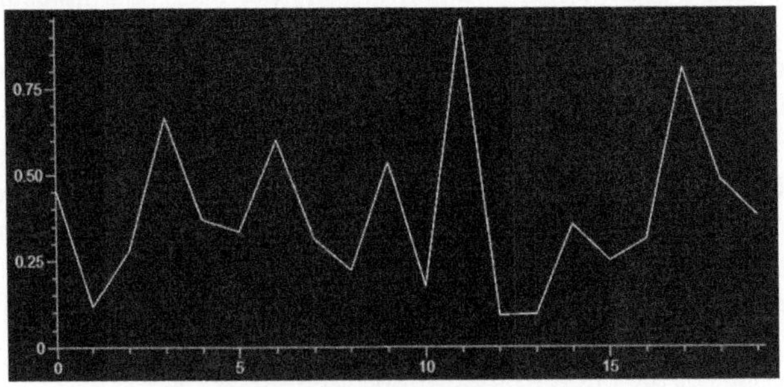

Figure 8.44

(a) Run this program several times, looking at the graph produced each time. What is being plotted in this graph? Is the graph the same each time you run the program? **(b)** What is the maximum random number generated by the `random()` function? **(c)** What is the minimum number generated by the `random()` function?

••P29 Refer to Problem P28 for an explanation of the `random()` function. The VPython program below models random photon emission in a collection of excited atoms. Read through the program, then answer the questions below. **(a)** In which line of code is it decided whether a particular atom will emit or not? How is a random number used in this decision? **(b)** What is the probability that a given atom will emit a photon in one nanosecond? **(c)** In which line of code is the count of excited atoms decreased after an atom emits a photon? **(d)** Start with 5 excited atoms, and run the program 10 times. In your observations, what is the longest time it takes for every atom to emit a photon? What is the shortest time? **(e)** Increase the number of atoms until the results of every run look the same. Approximately how many atoms are required? (You may wish to use a larger value for `rate()`). **(f)** With 10,000 atoms, drag the mouse across the graph and find a vertical bar whose height is $10000/e = 10000*0.368$. What is the value of t at this location? This value is called the "mean lifetime" and can be shown to be equal to the reciprocal of the emission rate (emissions per second, which is `P/dt`). How does your mean lifetime compare with `dt/P`)?

```
from visual import *
from visual.graph import *
from random import random

Natoms = 5
# P is the probability for an atom to emit
# during a time interval dt
P = 0.1
dt = 0.2  # ns
t = 0
tmax = 5*dt/P # 5 mean lifetimes
# Create a bar graph
gdisplay(xtitle='t, ns',
         ytitle='Atoms in excited state')
excited = gvbars(color=color.yellow,
delta=dt/2)

while t < tmax:
    rate(10)
    # Show number of excited atoms remaining
    excited.plot(pos=(t,Natoms))
    emissions = 0
    atom = 0
    while atom < Natoms:
        if random() < P: # emits?
            # count emissions in this dt
            emissions = emissions + 1
        atom = atom + 1
    Natoms = Natoms - emissions
    t = t + dt
```

Figure 8.45 shows sample output of this program.

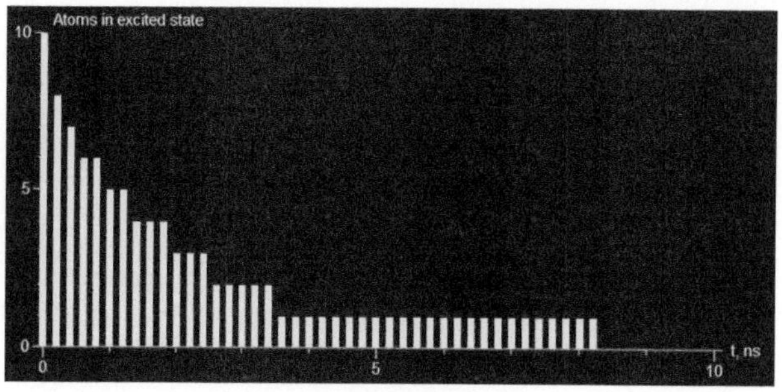

Figure 8.45

••P30 Refer to Problem P28 for an explanation of the `random()` function. The VPython program below models the random emission of red and green photons from a collection of atoms in their second excited state. These atoms can either emit a green photon and drop to the ground state, or they can emit a red photon and drop to the first excited state. Read the program carefully before running it, then answer the questions below. **(a)** What quantity is plotted in each of the graphs generated by the program? **(b)** Which is higher: the probability of emission of a red photon or a green photon? **(c)** Run the program repeatedly. Approximately, how much does the number of green emissions vary in repeated trials? **(d)** It can be shown that statistically one expects that the number of green emissions in repeated trials lies in the range $N \pm \sqrt{N}$, where N is the average number (which we expect to be `Pgreen*Natoms`). Do 30 or more trials and determine the experimental average number N and the fraction of trials in which the number of green emissions is within the range $N \pm \sqrt{N}$.

```
from visual import *
from visual.graph import *
from random import random

# Start with 100 atoms in an excited state
# (try larger or smaller numbers)

Natoms = 100
# P is the probability for an atom to emit
# during a time interval dt
P = 0.1
# Pgreen is the probability that when an
# atom emits, it emits a green photon
Pgreen = 0.3
dt = 0.2  # ns
t = 0
tmax = 5*dt/P # 5 mean lifetimes
# Create bar graphs
gdisplay(xtitle='t, ns', xmax=tmax,
         ytitle='Emissions of green photons')
greeng = gvbars(color=color.green, delta=dt/2)
gdisplay(y=400, xtitle='t, ns', xmax=tmax,
        ytitle='Emissions of red photons')
redg = gvbars(color=color.red, delta=dt/2)
greens = reds = 0
```

```
while t < tmax:
    rate(100)
    atom = 0
    g = r = 0
    while atom < Natoms:
        if random() < P: # emits?
            if random() < Pgreen: # green?
                g = g + 1
            else:              # emits red
                r = r + 1
        atom = atom + 1
    greeng.plot(pos=(t,g))
    redg.plot(pos=(t,r))
    greens = greens + g
    reds = reds + r
    Natoms = Natoms - (g + r)
    t = t + dt

print(greens,'green emissions,',
      reds,'red emissions')
```

Figure 8.46 shows sample output of this program.

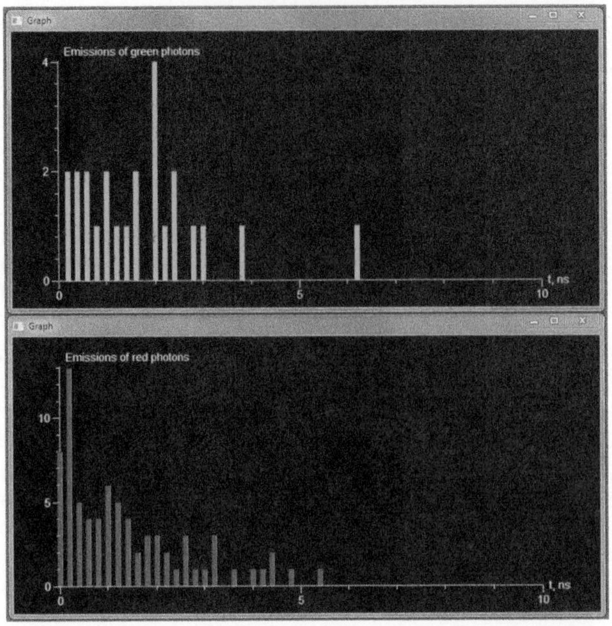

Figure 8.46 The vertical scales in the two graphs are different, due to the different probabilities of green and red emissions.

ANSWERS TO CHECKPOINTS

1 About 2.5 eV; about 3.5×10^{21} per second per square meter
2 5.2 eV
3 Six different photon energies, corresponding to transitions from 4 to 3, 4 to 2, 4 to 1, 3 to 2, 3 to 1, and 2 to 1
4 Just one dark line at 2.4 eV
5 1 eV, 5 eV, and 6 eV
6 0.4 eV (three transitions), 0.8 eV (two transitions), and 1.2 eV (one transition)
7 $N \approx 2 \times 10^{31}$. A system in a very high quantum state behaves like a classical system.
8 2.2 MeV
9 About 294 MeV

CHAPTER

9

Translational, Rotational, and Vibrational Energy

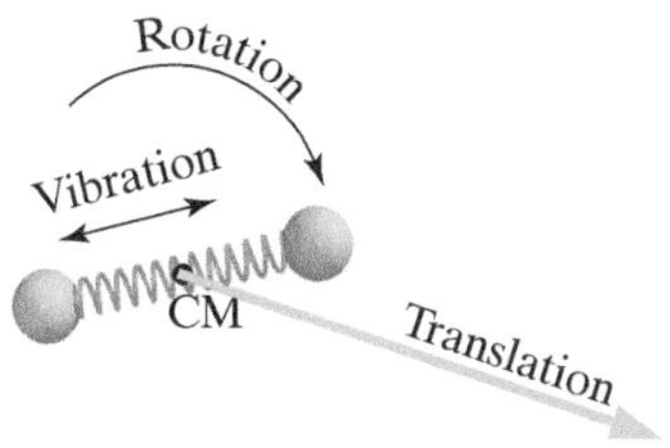

OBJECTIVES

After studying this chapter, you should be able to

- Analyze the kinetic energy of a multiparticle system in terms of translational kinetic energy and kinetic energy relative to the center of mass.
- Combine the results of modeling a multiparticle system as a point particle and as an extended system to obtain information about particular forms of internal energy.

Suppose you throw a flying disc (such as a Frisbee®) of mass 140 g in such a way that it leaves your hand with a speed of 15 m/s. What is the kinetic energy of the disc?

Although this sounds like a straightforward question, we actually don't have enough information to answer it, because the motion of the disc is complicated—it is traveling through the air, but it is also spinning around its center. The kinetic energy associated with the motion of the center of mass of the object is called "translational kinetic energy," because in mathematics a "translation" is a function that moves every point in an object a constant distance in a specified direction. In addition to its translational kinetic energy, the spinning disc also has rotational kinetic energy, associated with the rotation of the atoms of the disc around the disc's center. A different kind of object, such as two balls connected by a spring, might be vibrating as it travels through the air—it would have vibrational kinetic energy as well.

This chapter will focus on determining how a system's kinetic energy is divided between translational kinetic energy and energy associated with motion relative to the center of mass of a system, such as rotation. Along the way we will find that by modeling a system two different ways—as a point particle, and as an extended system—we can predict how much of the work done on a system will go into different kinds of energy. This analysis method will allow us to solve some very complex problems in a very simple way.

9.1 SEPARATION OF MULTIPARTICLE SYSTEM ENERGY

In many situations we find that the energy of a multiparticle system can be analyzed simply, by dividing the kinetic energy of the system into two parts: "translational kinetic energy" K_{trans}, kinetic energy associated with the motion of the center of mass, and K_{rel}, the kinetic energy relative to the center of mass.

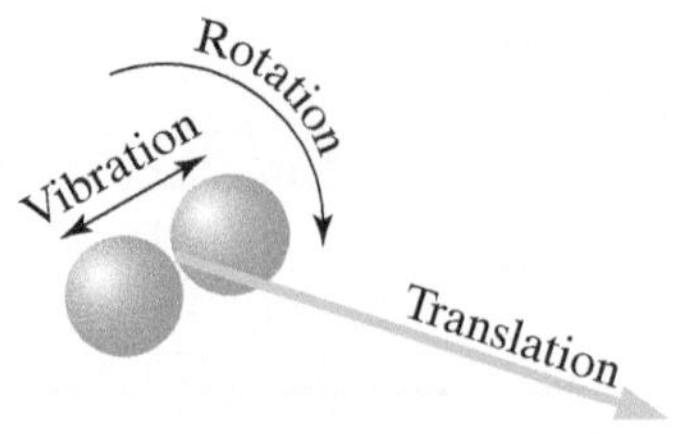

Figure 9.1 A diatomic molecule such as O_2 can vibrate and rotate while also translating.

K_{rel} includes the kinetic energy associated with rotation of the object and the kinetic energy associated with vibration of the object.

MULTIPARTICLE KINETIC ENERGY

$$K_{tot} = K_{trans} + K_{rel}$$

where $K_{rel} = K_{vib} + K_{rot}$.

For example, the kinetic energy of a rotating, vibrating diatomic molecule such as oxygen (O_2) is the sum of the kinetic energy that the moving molecule would have if it were not rotating or vibrating, plus the additional kinetic energy of rotation around the center of mass and kinetic energy of vibration relative to the center of mass (Figure 9.1).

This split of the total kinetic energy into different parts sounds plausible. A formal derivation of this result is given in Section 9.5 at the end of this chapter.

Translational Kinetic Energy

In physics and mathematics, the word "translate" means "move from one location to another." Translational motion is the ordinary kind of motion we have studied most so far. The translational motion of a macroscopic object is described by $\vec{v}_{CM}$, the velocity of the object's center of mass. We use the term "translate" here in order to distinguish between translational motion, in which the center of mass of a system moves from one location to another, and other kinds of motion such as rotation and vibration, in which parts of a system move relative to the center of mass.

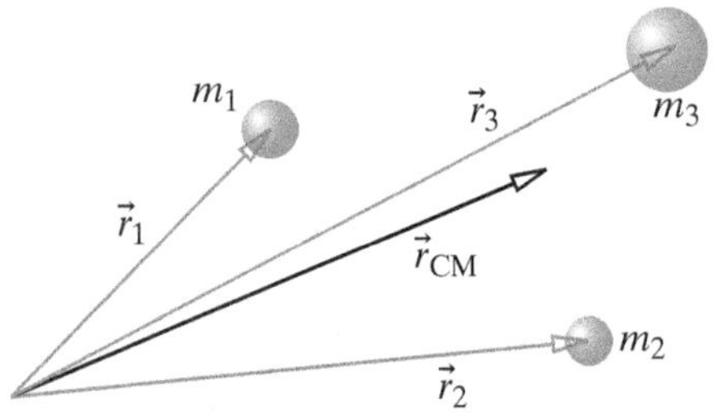

Figure 9.2 Location of the center of mass of a multiparticle system.

In Section 3.11 we defined the center of mass of a multiparticle system to be the weighted average of the positions of all atoms making up an object (Figure 9.2):

$$\vec{r}_{CM} = \frac{m_1\vec{r}_1 + m_2\vec{r}_2 + m_3\vec{r}_3 + \cdots}{M_{total}}$$

If each of the objects is itself a large extended object (Figure 9.3), we can calculate the center of mass by pretending that we shrink each object to a point at its own center of mass:

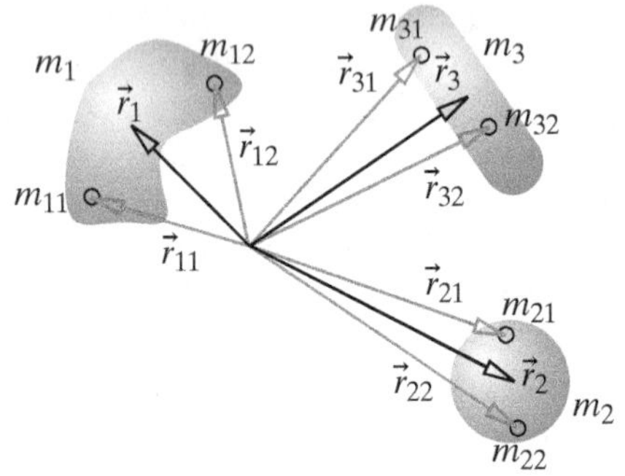

Figure 9.3 Finding the center of mass of a system consisting of several large objects.

$$\vec{r}_{CM} = \frac{(m_{11}\vec{r}_{11} + m_{12}\vec{r}_{12} + \cdots) + (m_{21}\vec{r}_{21} + m_{22}\vec{r}_{22} + \cdots) + \cdots}{(m_{11} + m_{12} + \cdots) + (m_{21} + m_{22} + \cdots) + \cdots}$$

$$\vec{r}_{CM} = \frac{m_1\vec{r}_1 + m_2\vec{r}_2 + m_3\vec{r}_3 + \cdots}{M_{total}}$$

This looks just like the equation for finding the center of mass of a bunch of atoms, except that the $\vec{r}$ vectors refer to the centers of mass of large objects.

By differentiating the expression for $\vec{r}_{CM}$ with respect to time, we found that the velocity of the center of mass is

$$\vec{v}_{CM} = \frac{m_1\vec{v}_1 + m_2\vec{v}_2 + m_3\vec{v}_3 + \cdots}{M_{total}}$$

and that therefore if $v \ll c$,

$$\vec{p}_{sys} \approx M_{total}\vec{v}_{CM}$$

Using the center-of-mass concept, we define the "translational kinetic energy" of a system as kinetic energy associated with translational motion, which does not include kinetic energy associated with rotation or vibration:

TRANSLATIONAL KINETIC ENERGY

$$K_{trans} = \frac{1}{2}M_{total}v_{CM}^2 = \frac{p_{sys}^2}{2M_{total}}$$

for an object with speed that is small compared to the speed of light ($v \ll c$).

EXAMPLE **Translational Kinetic Energy of a Frisbee®**

What is the translational kinetic energy of a spinning 140 g Frisbee that is traveling at a speed of 15 m/s? Is this the total kinetic energy of the Frisbee?

Solution

$$K_{trans} = \frac{1}{2}(0.14\,\text{kg})(15\,\text{m/s})^2 = 15.8\,\text{J}$$

This does not include the kinetic energy associated with the rotation of the disk.

Vibrational Energy

One form of energy that is internal to a system is vibrational energy, both elastic and kinetic. We already know how to calculate vibrational energy. Consider an oxygen molecule (O_2) that has no translational motion (that is, the center of mass is stationary), but is vibrating (Figure 9.4), with elastic energy and kinetic energy continually interchanging, but with the sum of the two energies remaining constant, like the mass and spring you modeled in earlier chapters.

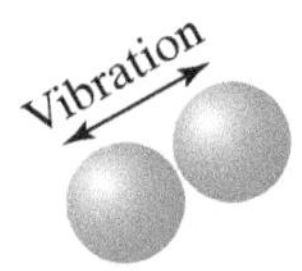

Figure 9.4 A vibrating oxygen molecule whose center of mass is at rest.

VIBRATIONAL ENERGY

$$E_{vib} = K_{vib} + U_{spring}$$

If in addition the oxygen molecule is translating (Figure 9.5), the total energy of the molecule is the sum of the translational kinetic energy, the vibrational kinetic energy (in terms of velocities of the two atoms relative to the center of mass), and the elastic energy of the "spring" holding them together, and the rest energies of the constituent atoms:

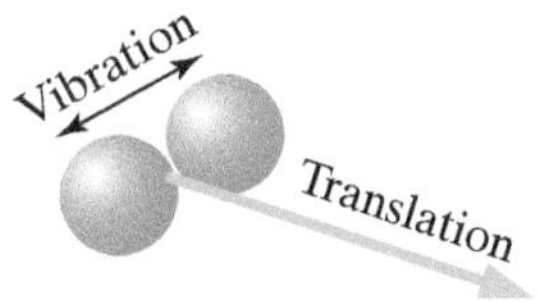

Figure 9.5 A translating, vibrating oxygen molecule.

$$E_{tot} = \frac{1}{2}(2m)v_{CM}^2 + K_{vib} + \frac{1}{2}k_s s^2 + 2mc^2$$

Note that the elastic energy $\frac{1}{2}k_s s^2$ is clearly unaffected by the translational motion of the center of mass, because it depends only on the stretch of the distance between the two atoms.

The same reasoning applies to the motion of a hot block of metal. In addition to its translational kinetic energy $\frac{1}{2}Mv_{CM}^2$ there is vibrational kinetic and elastic energy of the atoms around their equilibrium positions, which we call thermal energy, with more thermal energy if the block's temperature is higher.

Rotational Kinetic Energy

Just as a vibrating object has kinetic energy associated with vibration, even if its center of mass is at rest, so a rotating object has kinetic energy associated with rotation, even if its center of mass is at rest.

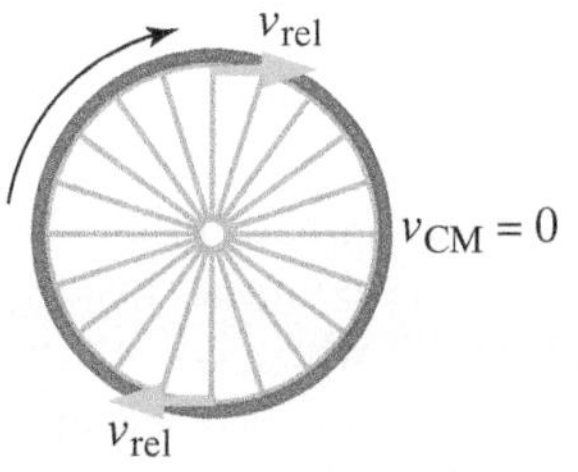

Figure 9.6 A bicycle wheel spinning on its axle, which is at rest. Atoms in the rim have a speed v_{rel} relative to the center of mass.

As an example, suppose that you spin a bicycle wheel on its axis, and hold the axle stationary, as in Figure 9.6. The spinning wheel has some kinetic energy K_{rel}, which in this case we will call K_{rot}, because the atoms are moving in rotational motion relative to the center of mass. Almost all of the mass M of the wheel is in the rim, so if the rim is traveling at a (tangential) speed v_{rel}, the kinetic energy of the wheel is approximately $K_{rot} \approx \frac{1}{2}Mv_{rel}^2$. (This isn't exact, because there is some mass in the spokes, and the atoms in the spokes are moving at speeds smaller than v_{rel}.)

In Section 9.2 we will see how to calculate the rotational kinetic energy of other kinds of rotating objects.

Rotation, Vibration, and Translation

The most general motion of an oxygen molecule involves rotation, vibration, and translation. It is easy to show that the internal (nontranslational) kinetic energy separates cleanly into rotational and vibrational contributions. Consider one atom with momentum $\vec{p}$ relative to the center of mass, and resolve its momentum into "tangential" and "radial" components (Figure 9.7). By "radial" component we mean a component in the direction of the line from the center of mass to the particle. By "tangential" component we mean a component perpendicular to that line. If the particle is in circular motion around the center of mass, the radial component is zero, so tangential momentum is associated with rotation. Radial momentum is associated with vibration.

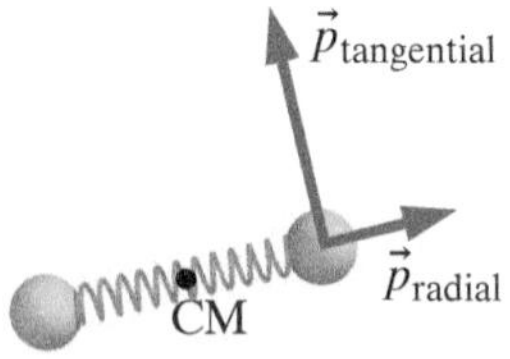

Figure 9.7 Radial and tangential components of momentum of a single atom, relative to the center of mass of an object.

Find the kinetic energy in terms of these two momentum components:

$$\frac{p^2}{2m} = \frac{(p^2_{tangential} + p^2_{radial})}{2m} = K_{rot} + K_{vib}$$

Therefore the kinetic energy relative to the center of mass splits into two terms, rotational energy and vibrational energy:

$$K_{rel} = K_{rot} + K_{vib}$$

The total energy of a translating, rotating, vibrating oxygen molecule can be written like this:

$$E_{tot} = K_{trans} + K_{rot} + K_{vib} + \frac{1}{2}k_s s^2 + 2mc^2$$

If the object can fly apart into separate pieces, we might call the kinetic energy relative to the center of mass $K_{explosion}$! For example, a fireworks rocket has translational kinetic energy that doesn't change from just before to just after its explosion, and the pieces move at high speeds relative to the center of mass of the rocket. These large kinetic energies relative to the center of mass come from the chemical energy used up in the explosion.

Vibrational and rotational kinetic energy are common types of K_{rel} but they aren't the only ones. For example, a person walking around on a moving train contributes relative kinetic energy to the total kinetic energy of the train and its passengers.

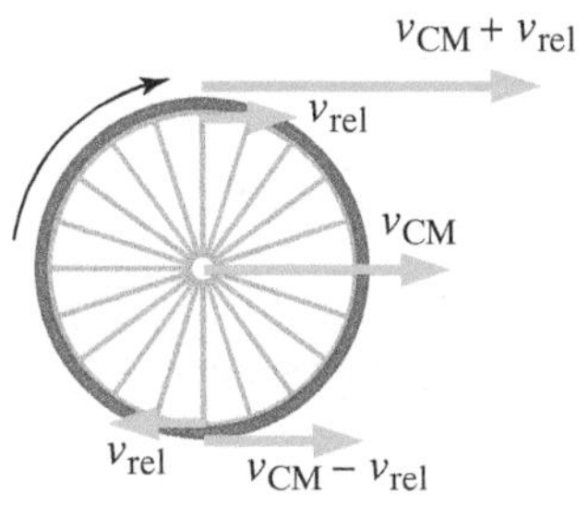

Figure 9.8 The velocity of an atom in the rim is the vector sum of the center-of-mass velocity and the velocity relative to the center of mass.

Why Separating Kinetic Energy Is Useful

If we do not separate kinetic energy into $K_{trans} + K_{rel}$, adding up the kinetic energies of all the particles in a system may be a very complex task. Consider the bicycle wheel shown in Figure 9.8. The fact that this wheel is both translating to the right, and also spinning on its axis, means that particles at different locations in the wheel have quite different speeds. For example, an atom at the bottom of the wheel is actually moving at a speed less than v_{CM}, while an atom at the top of the wheel is moving at a speed greater than v_{CM}. In order to calculate the kinetic energy of the whole system, we would have to figure out the instantaneous speed of each constituent particle!

Gravitational Energy of a Multiparticle System

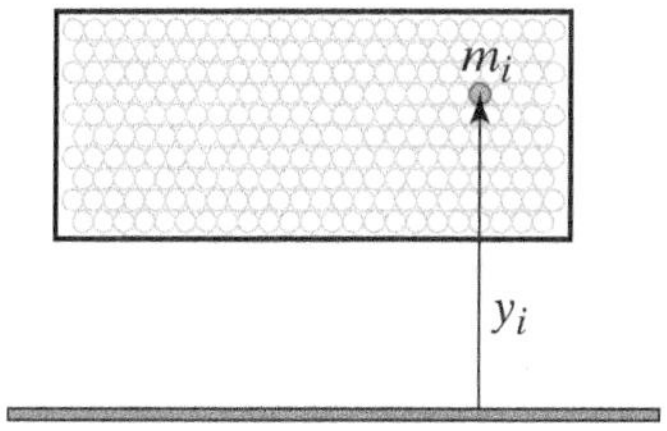

Figure 9.9 The i-th atom in the block is a distance y_i from the Earth's surface.

In previous chapters we have calculated the gravitational potential energy of a system consisting of the Earth and a multiparticle object simply by treating the object as if all its mass were concentrated at a single point (its center of mass). Why is this correct?

Consider the gravitational energy of a block of mass M and the Earth, near the Earth's surface. The gravitational energy associated with the interaction of the Earth with the i-th atom in the block is $m_i g y_i$ (Figure 9.9).

We can calculate the total gravitational energy by adding up the gravitational energy associated with the interaction of each of the N atoms with the Earth:

$$U_g = m_1 g y_1 + m_2 g y_2 + m_3 g y_3 + \cdots = g(m_1 y_1 + m_2 y_2 + m_3 y_3 + \cdots)$$

However, the last expression in parentheses appears in the calculation of the y component of the location of the center of mass:

$$M y_{CM} = m_1 y_1 + m_2 y_2 + m_3 y_3 + \cdots$$

Therefore for a block made up of many atoms, near the Earth's surface, we have this:

GRAVITATIONAL ENERGY U_g

$$U_g = M g y_{CM} \quad \text{near the Earth's surface}$$

In other words, the gravitational energy of a block of material and the Earth, near the Earth's surface, can be evaluated as though the block were a tiny particle of mass M located at the center of mass of the block.

This simple equation is not valid if the object is so large that the strength of the gravitational field is significantly different at different locations in the object. In that case mg would not be a good approximation for the gravitational force, nor mgy a good approximation for the gravitational energy associated with each atom.

> QUESTION In a calculation of gravitational energy in the voyage of a spacecraft to the Moon, is it a valid approximation to treat the spacecraft as if all its mass were concentrated at one point?

This is a good approximation, because the spacecraft is very small compared to the distances to the Earth and Moon, so the gravitational force on each atom is nearly the same. Also, if the spacecraft can be well approximated by a sphere of constant density, or nested spherical shells of constant density, it turns out that the gravitational force acts on it as though all the mass were concentrated at the center of mass.

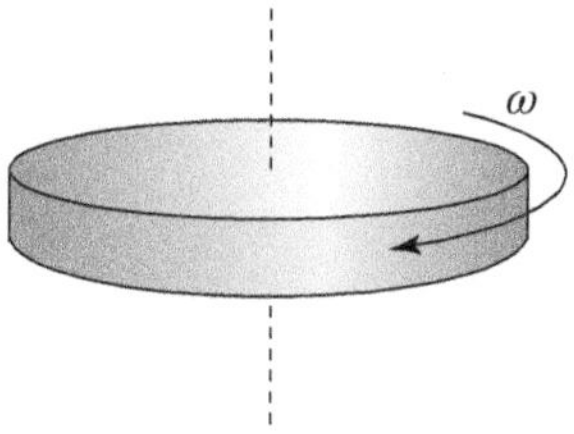

Figure 9.10 A rigid disk rotating about the y axis with angular speed ω.

9.2 ROTATIONAL KINETIC ENERGY

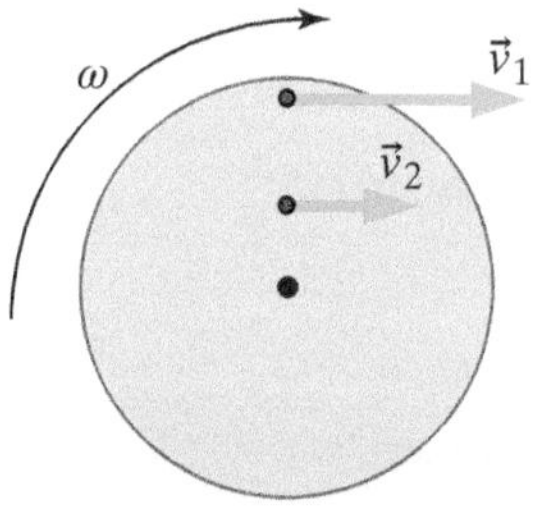

Figure 9.11 A top view of a disk rotating about an axis. A particle near the edge has a higher speed than a particle nearer to the axis. In each case, $v_i = \omega r_i$.

A common instance of rotational motion is that in which a rigid system is rotating on an axis, as shown in Figure 9.10. In this situation all the atoms in the system share the same "angular speed" in radians per second. However, they have different linear speeds in meters per second, depending on their distances from the axis, because an atom near the edge must travel farther in one revolution (Figure 9.11).

Angular speed, normally denoted by ω (lowercase Greek omega), is a measure of how fast something is rotating. If an object makes one complete turn of 360 degrees (2π radians) in a time T, we say that its angular speed is $\omega = 2\pi/T$ radians per second. The time T is called the period.

PERIOD AND ANGULAR SPEED

$$\omega = \frac{2\pi}{T}$$

The period T is the time it takes for a rotating object to make one complete revolution. The angular speed ω has units of radians per second.

An atom rotating at a constant rate around an axis at a distance r from the axis goes a distance of the circumference $2\pi r$ in the period T, so the speed v of the atom can be expressed in terms of the angular speed:

$$v = \frac{2\pi r}{T} = \left(\frac{2\pi}{T}\right) r = \omega r$$

We measure angular speed ω in the "natural" units of radians per second rather than degrees per second, because the fundamental geometrical relationship between angle and arc length (arc length $= r\theta$) is valid only if the angle θ is measured in radians.

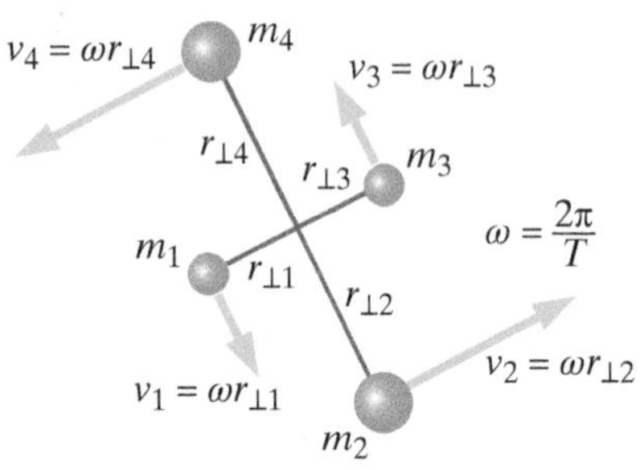

Figure 9.12 A case of rigid rotation about an axis with angular speed ω. Note the different speeds, with $v_i = \omega r_i$.

Moment of Inertia

The kinetic energy of the system shown in Figure 9.12, which is a rotating rigid object, is $\frac{1}{2}m_1v_1^2 + \frac{1}{2}m_2v_2^2 + \frac{1}{2}m_3v_3^2 + \cdots$. Since the speed of each mass is $v = \omega r_\perp$ (where $r_\perp$ is the perpendicular distance of the mass from the axis of rotation), we can write the rotational kinetic energy like this:

$$K_{\text{rot}} = \frac{1}{2}\left[m_1(\omega r_{\perp 1})^2 + m_2(\omega r_{\perp 2})^2 + m_3(\omega r_{\perp 3})^2 + \cdots\right]$$
$$= \frac{1}{2}\left[m_1 r_{\perp 1}^2 + m_2 r_{\perp 2}^2 + m_3 r_{\perp 3}^2 + \cdots\right]\omega^2$$

The quantity in brackets is called the "moment of inertia" and is usually denoted by the letter I:

MOMENT OF INERTIA

$$I = m_1 r_{\perp 1}^2 + m_2 r_{\perp 2}^2 + m_3 r_{\perp 3}^2 + m_4 r_{\perp 4}^2 + \cdots$$

The units of moment of inertia are $\text{kg} \cdot \text{m}^2$.

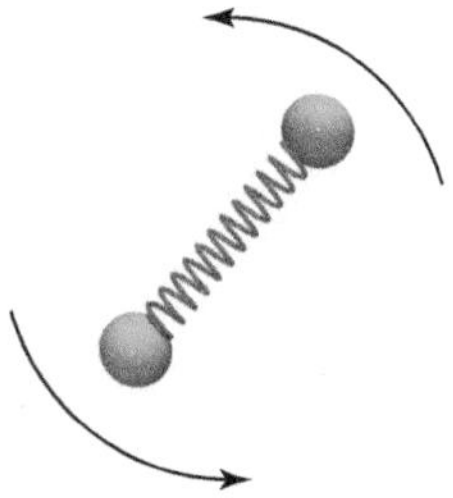

Figure 9.13 A rotating diatomic molecule.

Using this definition, we have a compact expression for rotational kinetic energy:

ROTATIONAL KINETIC ENERGY

$$K_{\text{rot}} = \frac{1}{2}I\omega^2$$

EXAMPLE **The Moment of Inertia of a Diatomic Molecule**

What is the moment of inertia of a diatomic nitrogen molecule N_2 around its center of mass (Figure 9.13)? The mass of a nitrogen atom is 2.3×10^{-26} kg and the average distance between nuclei is 1.5×10^{-10} m. Use the definition of moment of inertia carefully.

Solution For two masses, $I = m_1 r_{\perp 1}^2 + m_2 r_{\perp 2}^2$. The distance between masses is d, so the distance of each object from the center of mass is $r_{\perp 1} = r_{\perp 2} = d/2$. Therefore

$$I = M(d/2)^2 + M(d/2)^2 = 2M(d/2)^2$$

$$I = 2 \cdot (2.3 \times 10^{-26}\,\text{kg})(0.75 \times 10^{-10}\text{m})^2$$

$$I = 2.6 \times 10^{-46}\,\text{kg} \cdot \text{m}^2$$

EXAMPLE **The Moment of Inertia of a Bicycle Wheel**

A bicycle wheel has almost all its mass M located in the outer rim at radius R (Figure 9.14). What is the moment of inertia of the bicycle wheel about its center of mass? (*Hint:* It's helpful to think of dividing the wheel into the atoms it is made of and think about how much each atom contributes to the moment of inertia.)

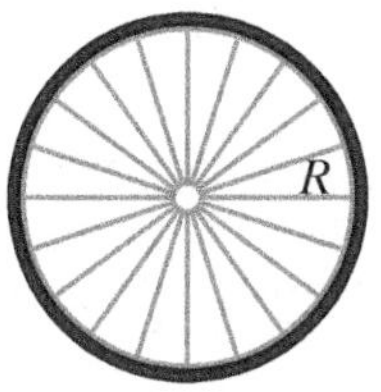

Figure 9.14 Most of the mass of a bicycle wheel is in the rim.

Solution Let m represent the mass of one atom in the rim. The moment of inertia is

$$I = m_1 r_{\perp 1}^2 + m_2 r_{\perp 2}^2 + m_3 r_{\perp 3}^2 + m_4 r_{\perp 4}^2 + \cdots$$

$$I = m_1 R^2 + m_2 R^2 + m_3 R^2 + m_4 R^2 + \cdots$$

$$I = [m_1 + m_2 + m_3 + m_4 + \cdots] R^2$$

$$I = MR^2$$

We've assumed that the mass of the spokes is negligible compared to the mass of the rim, so that the total mass is just the mass of the atoms in the rim.

EXAMPLE **Rotational Kinetic Energy and Work**

In Figure 9.15 a wheel is mounted on a stationary axle, which is nearly frictionless so that the wheel turns freely. The wheel has an inner ring with mass 5 kg and radius 10 cm and an outer ring with mass 2 kg and radius 25 cm; the spokes have negligible mass. A string with negligible mass is wrapped around the outer ring and you pull on it, increasing the rotational speed of the wheel. During the time that the wheel's rotation changes from 4 revolutions per second to 7 revolutions per second, how much work do you do?

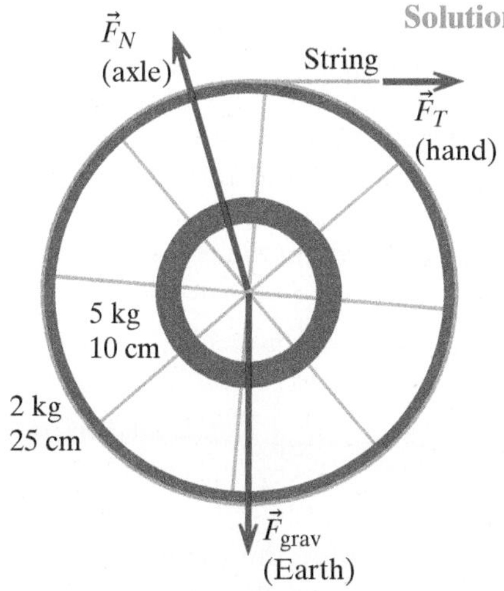

Figure 9.15 A rotating wheel with two rings.

Solution System: Wheel and string
Surroundings: Your hand, axle, Earth
Free-body diagram: Figure 9.15
Principle: Energy Principle

$$E_f = E_i + W$$

$$\frac{1}{2} I \omega_f^2 = \frac{1}{2} I \omega_i^2 + W$$

$$W = \frac{1}{2} I \left(\omega_f^2 - \omega_i^2 \right)$$

What is the moment of inertia $I = m_1 r_{\perp 1}^2 + m_2 r_{\perp 2}^2 + m_3 r_{\perp 3}^2 + m_4 r_{\perp 4}^2 + \cdots$? Group this sum into a part that includes just the atoms of the inner ring and another part that includes just the atoms of the outer ring:

$$I = (m_1 r_{\perp 1}^2 + m_2 r_{\perp 2}^2 + \cdots)_{\text{inner}} + (m_1 r_{\perp 1}^2 + m_2 r_{\perp 2}^2 + \cdots)_{\text{outer}}$$

$$I = I_{\text{inner}} + I_{\text{outer}}$$

This is a general result: The moment of inertia of a composite object is the sum of the moments of inertia of the individual pieces, because we have to add up all the contributions of all the atoms. We already determined that the moment of inertia of a ring is MR^2 (all the atoms are at the same perpendicular distance R from the center), so the moment of inertia of this wheel is

$$I = M_{\text{inner}}R^2_{\text{inner}} + M_{\text{outer}}R^2_{\text{outer}}$$
$$I = (5\,\text{kg})(.1\text{m})^2 + (2\,\text{kg})(.25\text{m})^2 = (0.050 + 0.125)\,\text{kg}\cdot\text{m}^2 = 0.175\,\text{kg}\cdot\text{m}^2$$

We need to convert revolutions per second into radians per second:

$$\omega_i = \left(4\frac{\text{rev}}{\text{s}}\right)\left(\frac{2\pi\ \text{rad}}{\text{rev}}\right) = 25.1\ \text{rad/s}$$
$$\omega_f = \left(7\frac{\text{rev}}{\text{s}}\right)\left(\frac{2\pi\ \text{rad}}{\text{rev}}\right) = 44.0\ \text{rad/s}$$

You, the Earth, and the axle all exert forces on the system. How much work does the Earth do? Zero, because the center of mass of the wheel doesn't move. How much work does the axle do? If there is negligible friction between the axle and the wheel, the axle does no work, because there is no displacement of the axle's force. Therefore only you do work, and the work that you do is

$$W = \frac{1}{2}(0.175\,\text{kg}\cdot\text{m}^2)(44.0^2 - 25.1^2)\frac{\text{rad}^2}{\text{s}^2} = 114\ \text{J}$$

Although the outer ring has a smaller mass than the inner ring, the outer ring has a larger moment of inertia because its radius is 2.5 times as big, so the factor R^2 is 6.25 times as big. Applying the Momentum Principle to the wheel, we see that the three forces must add up to zero, because the momentum $M\vec{v}_{\text{CM}}$ of the wheel isn't changing.

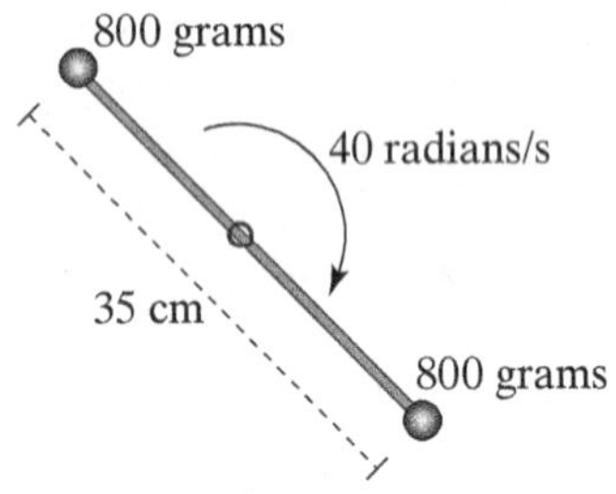

Figure 9.16 A rotating barbell.

Checkpoint 1 A barbell spins around a pivot at its center (Figure 9.16). The barbell consists of two small balls, each with mass 800 g, at the ends of a very low mass rod whose length is 35 cm. The barbell spins with angular speed 40 rad/s. Calculate K_{rot}.

Calculating Moment of Inertia

We've seen how to calculate the moment of inertia in a few simple special cases. We will calculate the moment of inertia of a long thin rod about its center to illustrate a more general technique. Consider a rod of mass M and length L whose thickness is small compared to its length, and that rotates around an axis perpendicular to the page (Figure 9.17). The density of the rod is the same everywhere ("uniform density"); every centimeter along the rod has the same mass. We need to evaluate the sum

$$I = m_1 r^2_{\perp 1} + m_2 r^2_{\perp 2} + m_3 r^2_{\perp 3} + m_4 r^2_{\perp 4} + \cdots$$

for all the atoms in the rod.

Divide into Small Slices As shown in Figure 9.17, we've chosen the origin of our axes to be at the center of the rod, and we've oriented the x axis to lie along the rod. We mentally divide the rod into N small slices of equal length $\Delta x = L/N$. Each slice has a mass $\Delta M = M/N$, because the rod's density is the same everywhere. We need to add up all the contributions to I made by all the slices.

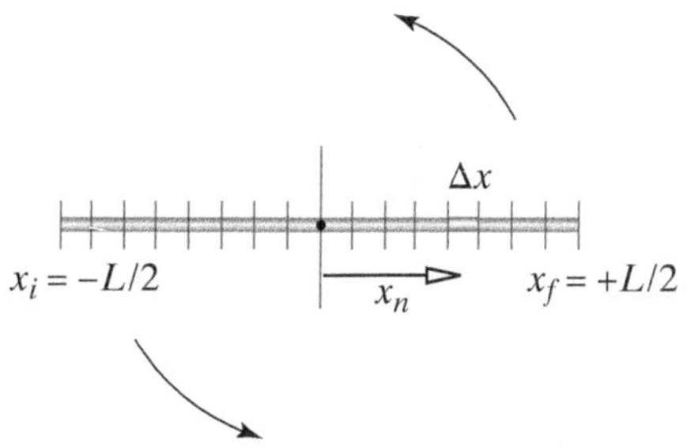

Figure 9.17 A uniform-density rod of length L rotates around an axis perpendicular to the page. We divide it into short slices and choose the origin to be at the center of the rod, with the x axis lying along the rod.

The Mass of One Slice Concentrate on one representative slice, whose center in Figure 9.17 is located at $\langle x_n, 0, 0\rangle$ and whose mass is $\Delta M = M/N$. Since $\Delta x = L/N$,

$$N = \frac{L}{\Delta x}$$

$$\Delta M = \frac{M}{N} = \frac{M}{L/\Delta x} = M\frac{\Delta x}{L}$$

Another way to see this is that $\Delta x/L$ is the fraction of the rod contained in this slice of length Δx, so the mass of this slice should be $M(\Delta x/L)$.

The Contribution of One Slice Now we're able to write down an expression for the contribution ΔI to the moment of inertia I of the rod that is made by this one representative slice of the rod. We make the approximation that all the atoms of this slice are approximately the same perpendicular distance $r_\perp \approx x_n$ from the center of the rod:

$$\Delta I = (\Delta M)\,x_n^2 = \left(M\frac{\Delta x}{L}\right)x_n^2 = \frac{M}{L}x_n^2\Delta x$$

Adding Up the Contributions We need to add up all such contributions by all such slices. One way to do this is numerically, by dividing the rod into, say, $N = 50$ slices, evaluating $(M/L)x_n^2\Delta x$ for each slice, and adding up all the contributions ($\sum$ means "summation"):

$$I = \sum_{n=1}^{N}\Delta I = \frac{M}{L}\sum_{n=1}^{N}x_n^2\Delta x$$

The Finite Sum Becomes a Definite Integral In some cases, including this case of a long thin rod, it is possible to use integral calculus to evaluate the summation. In the limit of large N (which means small Δx), the sum becomes an integral (the $\int$ symbol is a distorted S for Summation). The finite slice length Δx turns into the infinitesimal quantity dx, and the sum from 1 to N turns into a definite integral with lower and upper limits on x (x_i and x_f):

$$I = \frac{M}{L}\lim_{N\to\infty}\sum_{n=1}^{N}x_n^2\Delta x = \frac{M}{L}\int_{x_i}^{x_f}x^2dx$$

The Limits of Integration Since we put the origin at the center of the rod in Figure 9.17, the lower limit x_i is $-L/2$ and the upper limit x_f is $+L/2$:

$$I = \frac{M}{L}\int_{-L/2}^{+L/2}x^2dx = \left(\frac{1}{12}\right)ML^2$$

Because it is easy to make a mistake in this kind of calculation, it is important to check whether the results are reasonable. First, the units are correct, since $(\frac{1}{12})ML^2$ has units of $\text{kg}\cdot\text{m}^2$, which are units of moment of inertia. Second, if we double the mass M, I doubles, which makes sense, and if we double the length L, I gets four times bigger, which also makes sense given the factors of $r_\perp^2$ that appear in the definition of moment of inertia.

QUESTION Why is this expression valid only for a *thin* rod?

We made the approximation that all the atoms in one slice had the same perpendicular distance x_n from the center of the rod. If the rod is thick, this isn't a valid approximation, even in the limit of infinitesimally short slices, because now those short slices are thick.

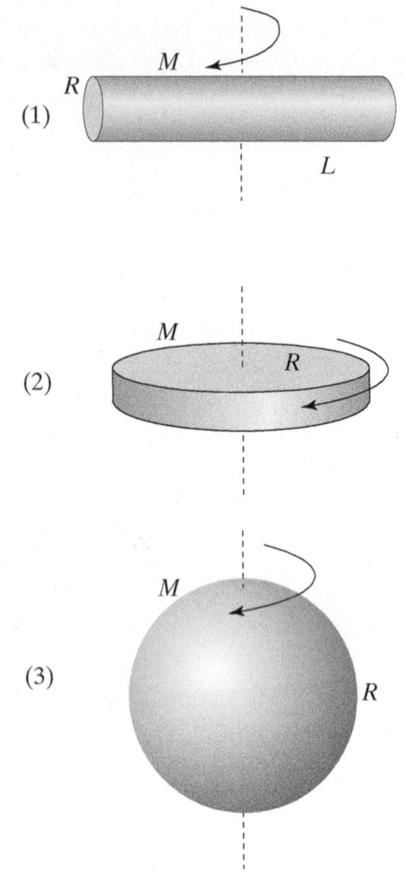

Figure 9.18 Some common uniform-density solids whose moments of inertia are known.

Checkpoint 2 A thin uniform-density rod whose mass is 1.2 kg and whose length is 0.7 m rotates around an axis perpendicular to the rod, with angular speed 50 radians/s. Its center moves with a speed of 8 m/s. **(a)** What is its rotational kinetic energy? **(b)** What is its total kinetic energy?

Moments of Inertia of Simple Shapes

By applying integral calculus as we did above in the case of a long thin uniform-density rod, it is possible to calculate the moments of inertia of some other simple shapes. Some useful results for objects of uniform density (see Figure 9.18) are as follows:

(1) For a cylinder of length L, radius R, rotating about an axis perpendicular to the cylinder, through center of cylinder:

$$I_{\text{cylinder}} = \frac{1}{12}ML^2 + \frac{1}{4}MR^2$$

(2) For a disk or cylinder rotating about its axis:

$$I_{\text{disk}} = I_{\text{cylinder}} = \frac{1}{2}MR^2$$

(3) For a sphere rotating around an axis passing through the center of the sphere:

$$I_{\text{sphere}} = \frac{2}{5}MR^2$$

A thin rod is like a cylinder with radius $R \approx 0$, in which case the moment of inertia for rotation about an axis perpendicular to the rod reduces to $\frac{1}{12}ML^2$.

Because only the perpendicular distances of atoms from the axis matter ($r_\perp$), the moment of inertia for rotation about the axis of a long cylinder has exactly the same form as that of a disk, $I_{\text{disk}} = \frac{1}{2}MR^2$. It doesn't matter how long the cylinder is (or to put it another way, how thick the disk is). See Figure 9.19.

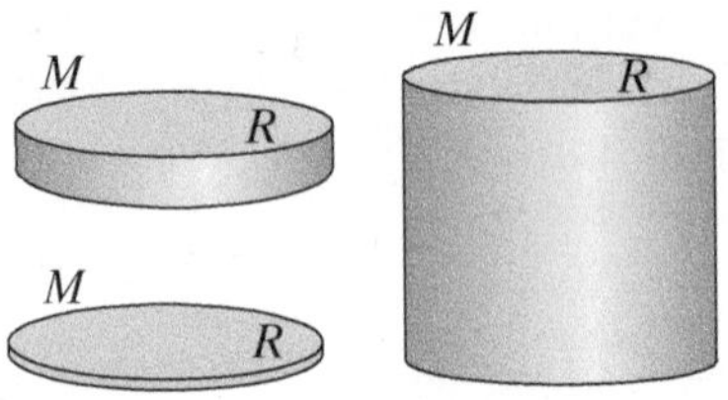

Figure 9.19 These disks and cylinders all have moment of inertia $\frac{1}{2}MR^2$ about their axes (but if they have the same density, they have different masses).

Checkpoint 3 A uniform-density sphere whose mass is 10 kg and radius is 0.4 m makes one complete rotation every 0.2 s. What is the rotational kinetic energy of the sphere?

Rigid Rotation about a Point Not the Center of Mass

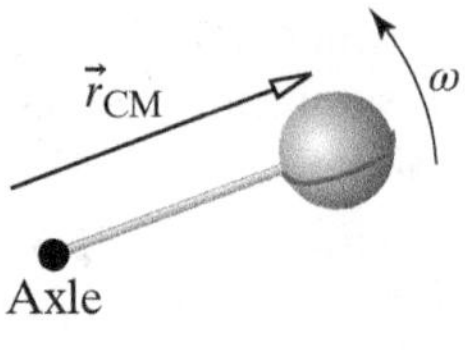

Figure 9.20 An object with known I_{CM} is connected to a low-mass rod and rotates around an axle. The red stripe on the object rotates at the same rate ω as the rod.

In Figure 9.20 a rigid object with known moment of inertia I_{CM} about its center of mass is connected to a low-mass rod and rotates about an axle that is a distance r_{CM} from the center of mass. What is the kinetic energy of this rotating object?

We know that in general $K_{\text{tot}} = K_{\text{trans}} + K_{\text{rel}}$, and in this case K_{rel} is just K_{rot}, since there is no vibration or explosion. The red line painted on the object in Figure 9.20 rotates with the same angular speed ω as the rod, which rotates around the axle; both the line and the object make one revolution in the same amount of time, so the kinetic energy that is relative to the (moving) center of mass is, as usual, $K_{\text{rot}} = \frac{1}{2}I_{\text{CM}}\omega^2$.

QUESTION What is K_{trans}, the kinetic energy associated with the motion of the center of mass?

The center of mass of the object rotates about the axle with angular speed ω at a distance r_{CM} from the axle, so the speed of the center of mass (in meters per second) is $v_{CM} = \omega r_{CM}$. Therefore we have

$$K_{trans} = \frac{1}{2}Mv_{CM}^2 = \frac{1}{2}M(\omega r_{CM})^2$$

$$K_{trans} = \frac{1}{2}(Mr_{CM}^2)\omega^2$$

$$K_{tot} = \frac{1}{2}(Mr_{CM}^2 + I_{CM})\omega^2$$

We see that when the center of mass rotates around a fixed point, K_{trans} brings into the total kinetic energy an additional term $\frac{1}{2}(Mr_{CM}^2)\omega^2$. We can summarize this result by saying that the total kinetic energy is $\frac{1}{2}I\omega^2$, where $I = Mr_{CM}^2 + I_{CM}$. This result is called the parallel axis theorem.

RIGID ROTATION ABOUT A POINT NOT THE CENTER OF MASS

$$K_{tot} = K_{trans} + K_{rot} = \tfrac{1}{2}(Mr_{CM}^2 + I_{CM})\omega^2$$

This is valid only for a rigid body, one whose rotation rate is the same as the rotation rate about the center of mass.

EXAMPLE **A Rod Rotating Not Around Its Center**

A thin rod with mass 140 g and 60 cm long rotates at an angular speed of 25 rad/s about an axle that is 20 cm from one end of the rod (Figure 9.21). What is its kinetic energy?

Solution There is translational kinetic energy because the center of mass is moving, and there is rotational kinetic energy because there is rotation about the center of mass. From what we just showed, we have this ($r_{CM} = 0.1$ m):

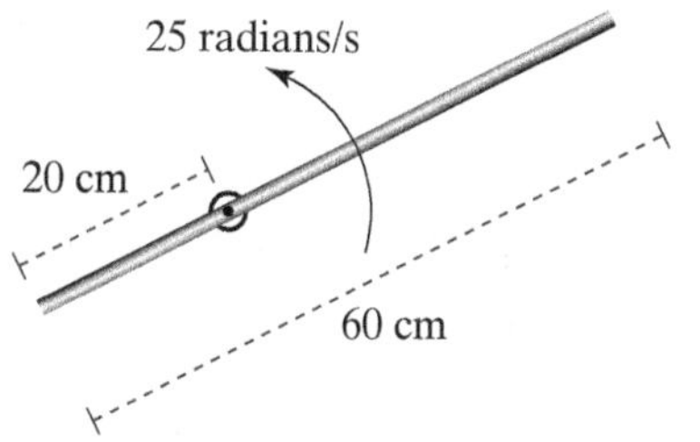

Figure 9.21 A rod rotates around an axle that is not located at the center of the rod.

$$K_{tot} = K_{trans} + K_{rot} = \frac{1}{2}(Mr_{CM}^2 + I_{CM})\omega^2$$

$$K_{tot} = \frac{1}{2}\left(Mr_{CM}^2 + \frac{1}{12}ML^2\right)\omega^2 = \frac{1}{2}M\left(r_{CM}^2 + \frac{1}{12}L^2\right)\omega^2$$

$$K_{tot} = \frac{1}{2}(.140\,\text{kg})\left((0.1\text{ m})^2 + \frac{1}{12}(0.6\text{ m})^2\right)(25\text{ rad/s})^2$$

$$K_{tot} = 1.75\text{ J}$$

Checkpoint 4 A solid uniform-density sphere is tied to a rope and moves in a circle with speed v. The distance from the center of the circle to the center of the sphere is d, the mass of the sphere is M, and the radius of the sphere is R. **(a)** What is the angular speed ω? **(b)** What is the rotational kinetic energy of the sphere? **(c)** What is the total kinetic energy of the sphere?

9.3 COMPARING TWO MODELS OF A SYSTEM

In analyzing the motion and energy of multiparticle systems, we encounter situations in which it would be extremely convenient to be able to calculate only the change in the translational kinetic energy K_{trans} of the system. Perhaps this might be all we are interested in, or perhaps we might wish to subtract this energy change from the change of the total energy of the system to find out how much energy has gone into vibration or rotation. Fortunately, it is quite simple to calculate ΔK_{trans} by itself.

Modeling a System as a Point Particle

Because $d\vec{p}_{\text{sys}}/dt = \vec{F}_{\text{net}}$, the center of mass of a multiparticle system moves exactly like a simple point particle whose mass is the mass of the entire system, under the influence of the net external force applied to the entire system. Imagine that you could crush the system that is modeled as an extended object (the "extended system") down to a very tiny ball. Next, apply to this system, modeled as a point particle (the "point particle system"), forces with the same magnitudes and directions as those that acted on the extended system, but applied directly to the point particle rather than at their original locations. What would the motion of this system modeled as a point particle look like? It would look exactly like the motion of the center of mass of the extended system (Figure 9.22).

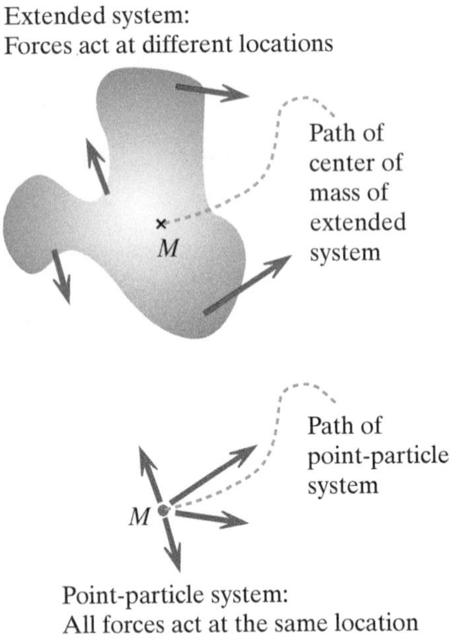

Figure 9.22 The path of the system modeled as a point particle is exactly the same as the path of the center of mass of the system modeled as an extended object.

These two different systems have the same total mass M, and both are acted on by the same net force, so the two paths are exactly the same. The difference is that the system when modeled as an extended object may rotate and stretch and vibrate due to the effects of the forces acting at different locations on the extended object. In contrast, the system when modeled as a point particle has no rotational motion, no vibrational motion, no internal energy of any kind. All of the forces act at the location of the point particle, and these forces don't stretch or rotate it. The only energy the point particle system can have is translational kinetic energy, and this is exactly the same as the translational kinetic energy of the extended system:

$$K_{\text{trans}} = \frac{p_{\text{sys}}^2}{2M} = \frac{1}{2}Mv_{\text{CM}}^2 \quad \text{(extended system or point particle system)}$$

An analysis of the point particle system gives the translational kinetic energy of the extended system. This is the reason it is useful to model an extended system as a point particle.

WORK AND ENERGY FOR A POINT PARTICLE

$$\Delta K_{\text{trans}} = \vec{F}_{\text{net}} \bullet \Delta\vec{r}_{\text{CM}}$$

This assumes that $\vec{F}_{\text{net}}$ is constant during the displacement $\Delta\vec{r}_{\text{CM}}$. If this is not the case, one must calculate the work as an integral of $\vec{F}_{\text{net}} \bullet d\vec{r}_{\text{CM}}$, analytically or numerically.

In Chapter 6 we showed that starting from the Momentum Principle for a single point particle we could derive the result that the change in the kinetic energy of the particle is equal to the work done on the particle by the net force. By modeling the system as a point particle, we assume that the net force is applied at the center-of-mass point, doing work to increase the kinetic energy of the point particle. Although this equation involves energy, it is actually derived from the Momentum Principle. A formal derivation may be found at the end of this chapter.

Modeling a System as an Extended Object

Because an extended object can rotate, vibrate, and change shape, not every part of the system necessarily moves the same distance as the center of mass moves. We need to consider this in calculating the work done on a system modeled as an extended object, because it matters where each force is applied.

To find the total work done on the system we need to consider separately the displacement of every point $\vec{r}_i$ where a force $\vec{F}_i$ is applied:

WORK AND ENERGY FOR AN EXTENDED SYSTEM

$$\Delta E_{\text{sys}} = \sum \vec{F}_i \bullet \Delta\vec{r}_i$$

Again, this assumes that each force $\vec{F}_i$ is constant during the displacement $\Delta\vec{r}_i$. If this is not the case, one must calculate the work of each force as an integral of $\vec{F}_i \bullet d\vec{r}_i$, analytically or numerically.

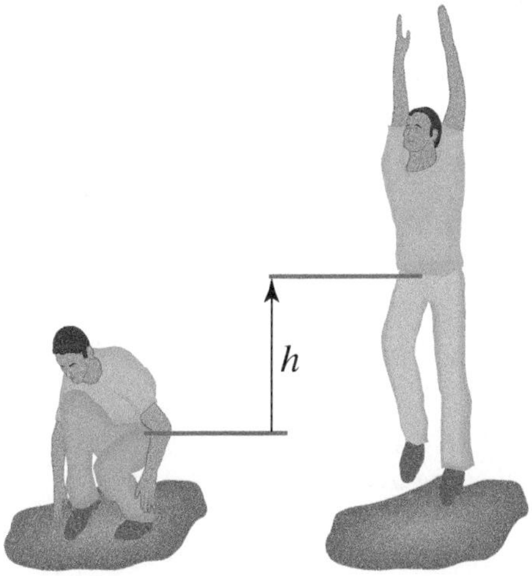

Figure 9.23 As the jumper leaves the floor, the center of mass has risen h. The system changes shape.

A Jumper Modeled as a Point Particle

As an example of how to exploit the point particle model, we'll analyze jumping (Figure 9.23). A jumper crouches down and then jumps straight upward. We want to calculate how fast the jumper's center of mass is moving at the instant that the feet leave the floor, v_{CM}. Let h be the distance through which the center of mass rises in this process. Note that there is a change of shape of the system.

Let F_N be the electric contact force that the atoms in the floor exert on the bottom of your foot, and let M be the jumper's total mass. We don't know exactly how the floor force F_N varies with time, but to get an idea of what happens let's make the crude approximation that it is constant as long as the feet are in contact with the floor (of course it falls suddenly to zero as the foot leaves the floor). Consider the system modeled as a point particle in Figure 9.24, which does not change shape.

QUESTION What is the net force on the jumper?

As indicated in Figure 9.24 the net force in the upward direction is $(F_N - Mg)$. Therefore the energy equation for the point particle system is

Figure 9.24 Free-body diagram for a jumper's body, modeled as a point particle. This system does not change shape.

$$\Delta K_{\text{trans}} = \vec{F}_{\text{net}} \bullet \Delta\vec{r}_{\text{CM}}$$

$$\frac{1}{2}Mv_{\text{CM}}^2 - 0 = (F_N - Mg)h$$

The work done on the point particle system involves the net force multiplied by the displacement of the center-of-mass point. As we will see, this is not the same as the energy equation for the system when modeled as an extended object, consisting of the jumper. The reason F_N shows up in the point particle energy equation is that it contributes to the net force and impulse, which affects the momentum (which is related to the motion of the center of mass).

A Jumper Modeled as an Extended Object

By modeling a jumper as a point particle, we were able to find out the change in the person's translational kinetic energy.

QUESTION What are some of the energy changes that occur during the jump that are not represented in the energy equation for the system when modeled as a point particle?

There is change of kinetic energy of the moving arms and legs relative to the center of mass, decrease of chemical energy, and increase of thermal energy (the jumper's temperature rises somewhat).

QUESTION Is the work done on an extended system the same as the work done on a point particle?

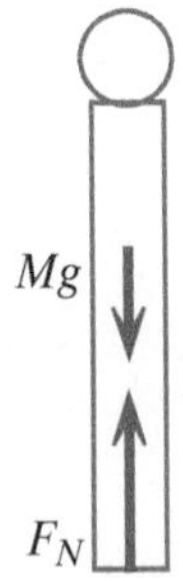

Figure 9.25 Free-body diagram for a jumper's body, modeled as an extended system. This system changes shape.

No. When calculating the work done on an extended system we must take into account the fact that the individual forces acting on the system are applied at different locations, as shown in the free-body diagram in Figure 9.25. Typically some of these forces act through very different distances than the displacement of the center of mass of the system, because different parts of the system move different distances.

The energy equation for the extended system (the jumper) looks something like the following, ignoring Q, transfer of energy due to a temperature difference between the jumper and the surrounding air:

$$\Delta E_{\text{sys}} = \sum \vec{F}_i \bullet \Delta \vec{r}_i$$
$$\Delta K_{\text{trans}} + \Delta K_{\text{rel}} + \Delta E_{\text{internal}} = -Mg \cdot h + F_N \cdot 0$$

QUESTION Why does the force F_N exerted by the floor on the jumper's foot do zero work?

The floor force F_N is applied to the bottom of the foot, and the contact point does not move during the entire process leading to lift-off, so there is no work done by the floor force. The definition of the work done by a force involves (the parallel component of) the displacement of the point of contact, at the place where the force is applied to the object. No displacement, no work done on the system. Of course, F_N does affect the momentum of the system, by contributing to the net impulse.

ΔK_{rel} includes the rotation of the upper and lower legs and the swinging of the arms. $\Delta E_{\text{internal}}$ includes both the increase of thermal energy in the jumper's body due to the exertion, and the decrease of chemical energy stored in the jumper's body.

None of these terms appear in the energy equation for the point particle system, because the point particle equation deals just with the motion of the mathematical center of mass point, which doesn't rotate or stretch or get hot. A crucial difference is that the extended system changes shape (legs unbend, etc.), but the point particle system doesn't.

Combining the Two Models

Replace ΔK_{trans} in the energy equation for the extended system with the value of ΔK_{trans} we found for the point particle system:

$$\Delta K_{\text{trans}} + \Delta K_{\text{rel}} + \Delta E_{\text{internal}} = -Mgh$$
$$(F_N - Mg)h + \Delta K_{\text{rel}} + \Delta E_{\text{internal}} = -Mgh$$
$$\Delta K_{\text{rel}} + \Delta E_{\text{internal}} = -F_N h$$

QUESTION What are the differences between the energy equations for the system modeled as a point particle and as an extended system?

Most energy terms do not appear at all in the point particle equation, and the work that appears in the energy equation for the extended system is not the same as the work that appears in the energy equation for the point particle system.

Stretching a Spring

A simple situation—stretching a spring—illustrates why the work done on a system modeled as a point particle can be different from the work done on an extended system. Suppose that you pull on one end of a spring with a force $\vec{F}_L$ to the left, making a short displacement $\Delta\vec{r}_L$ to the left, and you pull on the other end with an equal but opposite force $\vec{F}_R$ and equal but opposite

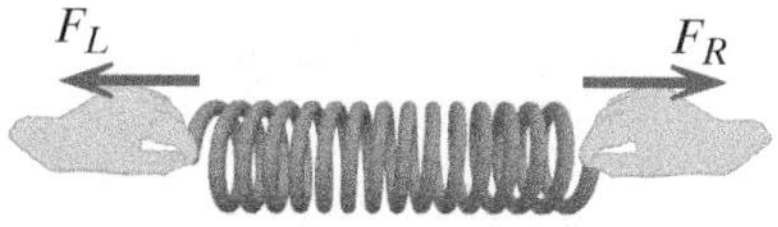

Figure 9.26 You pull on both ends of a spring with equal and opposite forces. The system (the spring) changes shape.

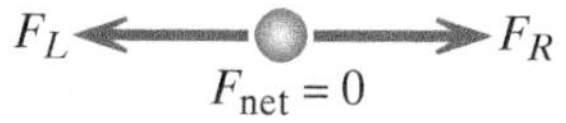

Figure 9.27 Free-body diagram for a spring pulled from each end, modeled as a point particle.

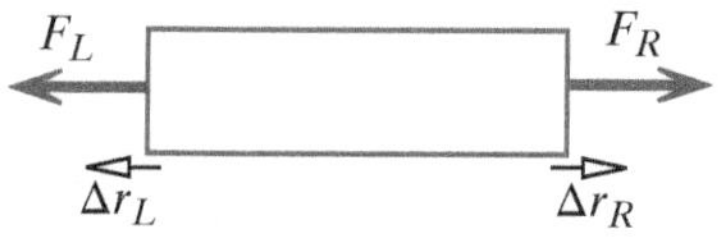

Figure 9.28 Free-body diagram for a spring modeled as an extended object.

short displacement $\Delta\vec{r}_R$ (Figure 9.26). The system (the spring) changes shape. Our analysis should yield the expected result: the spring should not gain translational kinetic energy!

A Spring Modeled as a Point Particle

To begin, we will model the system as a point particle, located at the center of mass of the spring (Figure 9.27). This point particle system cannot change shape.

QUESTION What is the net force $\vec{F}_{net}$? How far does the center of mass move? How much work is done on the point particle system? What is the change in K_{trans}?

$$\vec{F}_{net} = \vec{F}_L + \vec{F}_R = \vec{0}$$
$$\Delta\vec{r}_{CM} = \vec{0}$$
$$\Delta K_{trans} = \vec{0} \bullet \vec{0} = 0$$

This result is consistent with what we would expect: zero work is done, and the translational kinetic energy of the system does not change.

A Spring Modeled as an Extended Object

Now we will model the spring as a deformable extended system (Figure 9.28).

QUESTION Does your right hand do any work on the spring? Positive or negative? Does your left hand do any work on the spring? Positive or negative? Is there any change in the energy of the spring?

$$\Delta E_{sys} = \sum \vec{F}_i \bullet \Delta\vec{r}_i$$
$$\Delta K_{trans} + \Delta U_{spring} = \vec{F}_L \bullet \Delta\vec{r}_L + \vec{F}_R \bullet \Delta\vec{r}_R$$
$$\Delta K_{trans} + \Delta U_{spring} = F_L \Delta r_L + F_R \Delta r_R$$

Both hands do positive work, so the energy of the system increases. Combining the models, we find that

$$0 + \Delta U_{spring} = F_L \Delta r_L + F_R \Delta r_R$$
$$\Delta\left(\frac{1}{2}k_s s^2\right) = F_L \Delta r_L + F_R \Delta r_R$$

Checkpoint 5 A runner whose mass is 50 kg accelerates from a stop to a speed of 10 m/s in 3 s. (A good sprinter can run 100 m in about 10 s, with an average speed of 10 m/s.) **(a)** What is the average horizontal component of the force that the ground exerts on the runner's shoes? **(b)** How much displacement is there of the force that acts on the sole of the runner's shoes, assuming that there is no slipping? Therefore, how much work is done on the extended system (the runner) by the force you calculated in the previous exercise? How much work is done on the point particle system by this force? **(c)** The kinetic energy of the runner increases—what kind of energy decreases? By how much?

Analyzing Point Particle and Extended Systems

We have seen that we can often extract significant information about the energy distribution within a system by comparing the results of two models: a model of the system as a point particle, and a model of the system as an extended object. Modeling the system as a point particle gives us information about the translational kinetic energy K_{trans} of the system—the only kind of energy a point particle can have other than rest energy. Knowing K_{trans} can allow us to resolve other kinds of energy when we model the system as an extended object. The basic approach is to:

(a) Model the system as a point particle.

- Assume that the entire mass of the system is concentrated into a point at the location of the system's center of mass.
- Assume that all forces applied to the system act at this point.
- Find the displacement of the center of mass.
- Calculate the net work done on the point particle.
- Find ΔK_{trans}, the change in the system's translational energy.

(b) Model the system as an extended object.

- Draw a diagram showing the point of application of each force acting on the system. Typically some of these forces will not act at the center of mass of the system.
- Identify the distance through which each of these forces acts on the system. Typically these forces will act through different distances, especially if the object changes shape.
- Separately calculate the work done on the system by each force, acting through the appropriate distance.
- Apply the Energy Principle to find ΔE, the change in the extended system's energy.

(c) Find $\Delta E - \Delta K_{trans}$. This is the change in kinds of energy other than translational energy for the extended system.

The following examples apply this method to answer questions about changes in internal energy and relative kinetic energy.

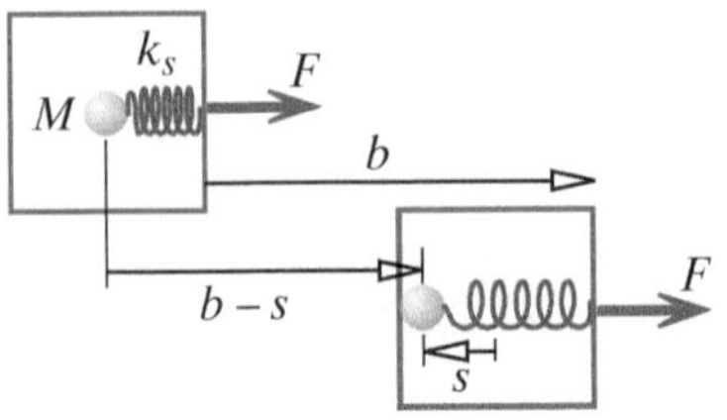

Figure 9.29 A constant force acts on a thin box containing a spring and a ball of clay. The box moves a distance b, but the ball (center of mass of the system) moves only a distance $b-s$. Tails of arrows representing forces are drawn at the point at which they act.

EXAMPLE

A Box Containing a Spring

A thin box in outer space contains a large ball of clay of mass M, connected to an initially relaxed spring of stiffness k_s (Figure 9.29). The mass of the box and spring are negligible compared to M. The apparatus is initially at rest. Then a force of constant magnitude F is applied to the box. When the box has moved a distance b, the clay makes contact with the left side of the box and sticks there, with the spring stretched an amount s. See the diagram for distances. **(a)** Immediately after the clay sticks to the box, how fast is the box moving? **(b)** What is the increase in internal energy of the clay? Assume that the process takes place so quickly that there isn't time for any significant energy transfer Q due to a temperature difference between the (extended) system and surroundings.

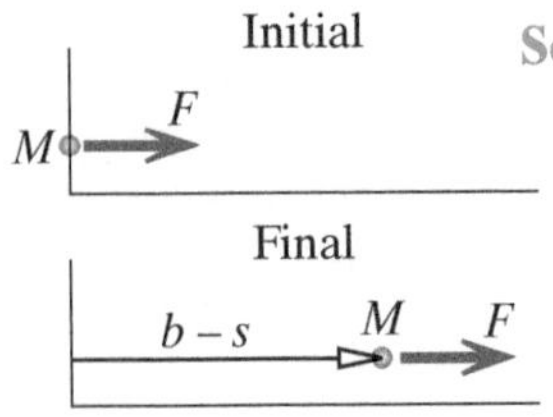

Figure 9.30 The point particle system for the apparatus. Tails of arrows representing forces are attached to the point particle, which moves the same distance as the center of mass of the extended system.

Solution **(a)** Point particle system

System: Point particle of mass M

Surroundings: External object exerting force F

Initial state: Point particle at rest

Final state: Point particle moving with (unknown) speed v

Free-body diagram: Figure 9.30

Energy Principle (point particle only has K_{trans}):

$$\Delta K_{trans} = F\Delta x_{CM}$$

$$\Delta x_{CM} = (b-s) \quad \text{(from the diagram)}$$

$$\frac{1}{2}Mv^2 - 0 = F(b-s)$$
$$v = \sqrt{\frac{2F(b-s)}{M}}$$

(b) Extended system
System: Mass, box, and spring
Surroundings: External object exerting force F
Initial state: System at rest
Final state: Box moving with (known) speed v (same as clay)
Free-body diagram: Figure 9.29
Energy Principle:

$$\Delta E_{\text{sys}} = W$$
$$\Delta(K_{\text{trans}} + U_{\text{spring}} + E_{\text{internal}}) = Fb$$
$$\Delta K_{\text{trans}} = F(b-s) \quad \text{(from part (a))}$$
$$F(b-s) + \frac{1}{2}k_s s^2 + \Delta E_{\text{internal}} = Fb$$
$$\Delta E_{\text{internal}} = Fs - \frac{1}{2}k_s s^2$$

In the energy equation for the point particle system there is nothing about the spring or internal energy, because the point particle system doesn't stretch or get hot. In Figure 9.30 the point at which the force vector is applied moves a distance $(b-s)$, moving with the point particle, so the work done on the point particle system is $F(b-s)$.

In the extended system, the point at which the force vector is applied moves a longer distance b, so the work done on the extended system is Fb. More work is done, and the extra work goes into spring potential energy and internal energy of the clay.

Good labeled diagrams of the two choices of system are extremely important, in order to determine the distances through which forces act in the two cases. Review the procedure at the beginning of the section and see how these steps were implemented in the example given above.

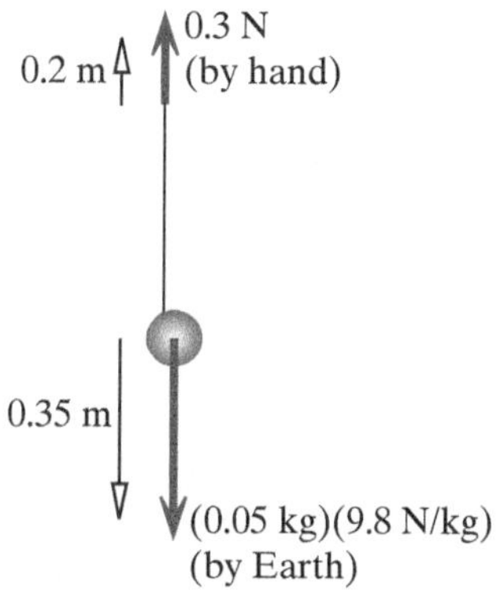

Figure 9.31 You pull up with a force of magnitude 0.3 N through a distance of 0.2 m on the string of a yo-yo while the yo-yo moves downward a distance 0.35 m.

EXAMPLE **A Yo-yo**

You're playing with a yo-yo whose mass is 0.05 kg attached to a low-mass string (Figure 9.31). You pull up on the string with a force of magnitude 0.3 N, and your hand moves up a distance of 0.2 m. During this time the mass falls a distance 0.35 m and some of the string reels off the yo-yo's axle. **(a)** What is the change in the translational kinetic energy of the yo-yo? **(b)** What is the change in the rotational kinetic energy of the yo-yo, which spins faster?

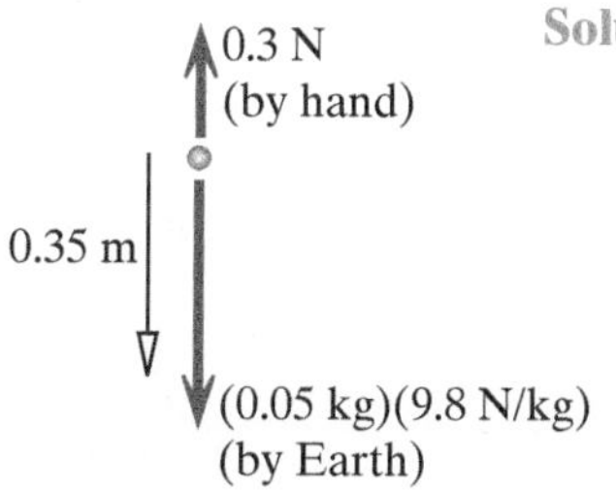

Figure 9.32 The point particle system for the yo-yo. Tails of arrows representing forces are attached to the point particle, which moves the same distance as the center of mass of the extended system.

Solution **(a)** Point particle system
System: Point particle
Surroundings: Earth and hand
Initial state: Point particle with initial translational kinetic energy
Final state: Point particle with final translational kinetic energy
Free-body diagram: Figure 9.32
Energy Principle (point particle only has K_{trans}):

$$\Delta K_{\text{trans}} = F_{\text{net},y}\Delta y_{\text{CM}}$$
$$\Delta K_{\text{trans}} = [0.3\text{ N} - (0.05\text{ kg})(9.8\text{ N/kg})](-0.35\text{ m}) = 0.0665\text{ J}$$

(b) Extended system
System: Mass and string
Surroundings: Earth and hand

Initial state: Initial rotational and translational kinetic energy
Final state: Final rotational and translational kinetic energy
Free-body diagram: Figure 9.31
Energy Principle:

$$\begin{aligned}
\Delta E_{\text{sys}} &= W_{\text{hand}} + W_{\text{Earth}} \\
\Delta K_{\text{trans}} + \Delta K_{\text{rot}} &= (0.3\text{ N})(0.2\text{ m}) + [-(0.05\text{ kg})(9.8\text{ N/kg})](-0.35\text{ m}) \\
&= 0.2315\text{ J} \\
\Delta K_{\text{trans}} &= 0.0665\text{ J} \quad \text{(from part (a))} \\
0.0665\text{ J} + \Delta K_{\text{rot}} &= 0.2315\text{ J} \\
\Delta K_{\text{rot}} &= 0.165\text{ J}
\end{aligned}$$

The key issue is that the distance through which the force of your hand moves is different in the extended system and the point particle system. In the extended system, your hand applies an upward force of magnitude 0.3 N through an upward displacement of 0.2 m, doing a positive amount of work, 0.06 J. However, in the "crushed" system, where all the mass is located at the center of mass of the extended system and all the force vectors are moved to that point, the upward-pointing force of your hand is applied to the point particle as the point particle drops a distance of 0.35 m. Therefore the work done by this force on the point particle is negative: $(0.3\text{ N})(-0.35\text{ m}) = -0.105\text{ J}$. The work done by the Earth's force is $(0.05\text{ kg})(9.8\text{ N/kg})](0.35\text{ m}) = +0.1715\text{ J}$ for both systems, because this force is applied to the center of mass of the extended system, which is also the center of mass of the crushed system. The diagrams labeled with distances are critical to calculating the work correctly for the two different systems.

It is important to understand that the point particle or "crushed" system is a different model of the situation than is the extended-system model. The forces acting on the point particle model are equal in magnitude and direction to the forces acting on the extended-system model, but applied directly to the point particle, so the displacements through which these forces act are different from the displacements of the forces acting on the extended system.

In this yo-yo example the difference between the two models is quite striking, because the work done by the force of your hand on the extended system is positive, but the work done by this force on the point particle system is negative. Recall that the point particle equation was derived from the Momentum Principle; if we think about momentum, we note that the force of the hand acts to oppose the downward motion of the center of mass of the yo-yo.

It is interesting to note that the change of the rotational kinetic energy, 0.165 J, is the same as the work that you would do to increase the purely rotational kinetic energy of a yo-yo whose axle was fixed in space. The distance $(0.2+0.35)\text{ m}=0.55\text{ m}$ is the length of the string that you would reel off the axle, and the force of your hand would do an amount of work $(0.3\text{ N})(0.55\text{ m})$ $=0.165\text{ J}$, which is ΔK_{rot}.

We chose to analyze this situation with specific numerical values in order to make the sign issues particularly clear. It is instructive to analyze the problem again symbolically, calling the upward force of your hand F, the distance you pull up d, the mass m, and the distance the yo-yo falls h.

EXAMPLE **An Ice Skater Pushes Off from a Wall**

Consider a woman on ice skates who pushes on a wall with a nearly constant contact force to the left of magnitude F_N (normal to the wall in Figure 9.33). By the reciprocity of electric interatomic forces, the wall exerts a force of magnitude F_N to the right on the skater. As a result, she moves backward with increasing speed. Her center of mass is marked on each frame of the sequence. Look closely at Figure 9.33 and you'll see that as the skater straightens her arms, the location in her body of the center of mass shifts a little toward her outstretched hands and arms; it is not a point fixed at some place in her body. What can we learn from analyzing the **(a)** point particle and **(b)** extended versions of the system of the skater?

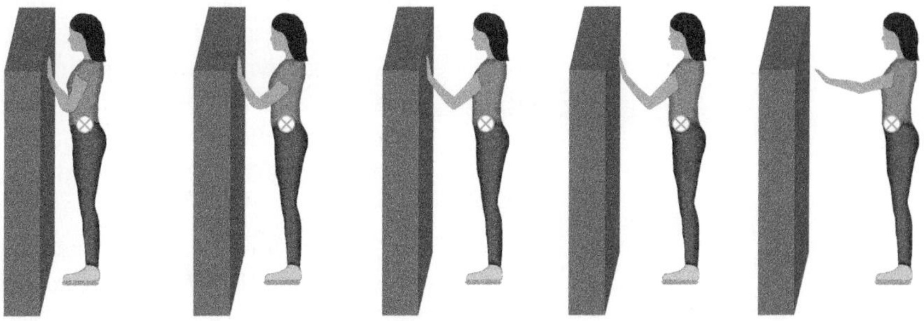

Figure 9.33

Solution Let's make the approximation that the change in the position of the center of mass in the skater's body is negligible. This is equivalent to neglecting the mass of the skater's arms and hands. Her center of mass moves to the right a distance d, which is approximately the length of her arms.

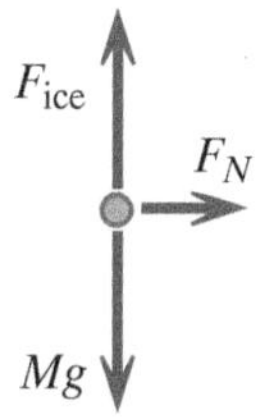

Figure 9.34 Free-body diagram for the point particle system.

(a) Point particle system
System: Point particle of mass M
Surroundings: Wall, ice, Earth
Initial state: Point particle at rest
Final state: Point particle moving with (unknown) speed v
Free-body diagram: Figure 9.34
Energy Principle (point particle only has K_{trans}):

$$\Delta K_{trans} = Fd$$

$$\frac{1}{2}Mv^2 - 0 = Fd$$

$$v = \sqrt{\frac{2Fd}{M}}$$

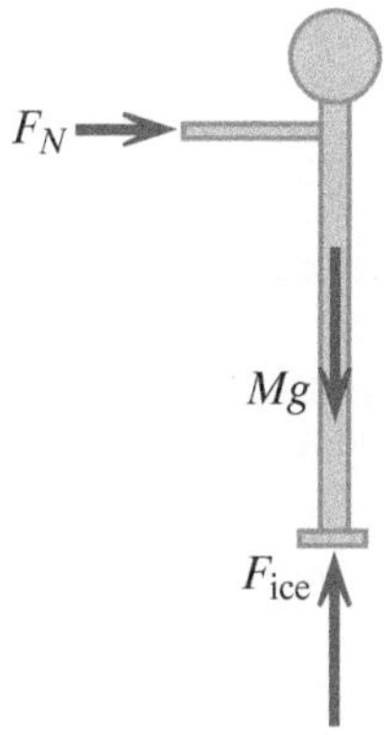

Figure 9.35 Free-body diagram for the extended system.

(b) Extended system
System: The skater
Surroundings: Wall, ice, Earth
Initial state: System at rest
Final state: Skater moving with speed v, with internal energy change
Free-body diagram: Figure 9.35
Energy Principle:

$$\Delta E_{sys} = W$$

$$\Delta(K_{trans} + E_{internal}) = F(0)$$

$$\Delta K_{trans} = Fd \quad \text{(from part (a))}$$

$$Fd + \Delta E_{internal} = 0$$

$$\Delta E_{internal} = -Fd$$

From the analysis of the point particle system we are able to determine the final speed of the skater. From the analysis of the extended system we are able to determine the internal energy change of the skater, which is negative, corresponding to a decrease in stored chemical energy (and a smaller increase in her thermal energy associated with a higher body temperature).

EXAMPLE **Pull on Two Hockey Pucks**

Tie a string to the center of a hockey puck, and wrap another string around the outside of a second hockey puck, as shown in Figure 9.36. Then two people pull on the two strings so that both strings have the same tension F_T. It may be surprising that the center of mass of both pucks moves the same distance, in the same direction, but remember that the x component of the Momentum Principle for speeds small compared to the speed of light can be written as $M\Delta v_{\text{CM},x} = F_{\text{net},x}\Delta t$. Here the net force is the same for both pucks, so the motion of the center of mass is the same for both pucks.

As shown in the figure, in a time Δt hand 1 has pulled a distance Δx_{CM} while hand 2 has pulled the bottom puck an additional distance d, because some string has unrolled from the puck. What can we learn from analyzing the point particle and extended versions of the system for each puck?

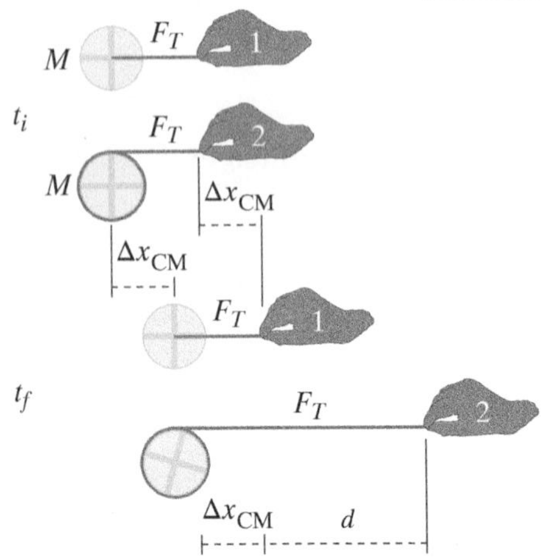

Figure 9.36 One puck is pulled by a string attached to its center. The other puck is pulled by a string wrapped around its edge, which unrolls a distance d as the puck is pulled.

Solution For either puck the point particle version of the system of puck plus string yields the result $\Delta K_{\text{trans}} = F\Delta x_{\text{CM}}$, and the same result applies to the case of the extended system of the puck and string pulled by hand 1.

What remains to be done is to analyze the extended version of the system consisting of the puck plus string that is pulled by hand 2:

$$\begin{aligned}
\Delta E_{\text{sys}} &= W \\
\Delta(K_{\text{trans}} + K_{\text{rot}}) &= F(\Delta x_{\text{CM}} + d) \\
\Delta K_{\text{trans}} &= F\Delta x_{\text{CM}} \quad \text{(from the point particle analysis)} \\
F\Delta x_{\text{CM}} + \Delta K_{\text{rot}} &= F(\Delta x_{\text{CM}} + d) \\
\Delta K_{\text{rot}} &= Fd
\end{aligned}$$

This makes sense: hand 2 pulls through a distance d farther than does hand 1, and this extra work Fd is equal to the increase in rotational kinetic energy.

A Reflection

If you look back over the various examples in this section, you can see that the Energy Principle for the point particle system and the Energy Principle for the extended system differ if any of the forces acting on the extended system acts through a displacement that is different from the displacement of the center of mass point. Or to put it the other way, the Energy Principle will be the same in both analyses only if all the atoms in the system have the same displacements (which will necessarily be the same as the displacement of the center of mass).

A change of shape or configuration, including deformation, or rotation with forces exerted at some location that isn't the center of mass, are examples of situations where you can expect to find that the Energy Principle for the two analyses will give different information, and where you can expect to find changes in the internal energy.

9.4 MODELING FRICTION IN DETAIL

In Chapter 7 we described a seeming paradox involving friction—that when you drag a block at constant speed across a table it seems that no work is done on the block, since you do an amount of work Fd, and the table apparently does an amount of work $-Fd$. Yet the block's temperature increases, indicating an increase in its internal energy.

In this section we'll see that the paradox arose from mistakenly treating the sliding block only as a point particle, which is inadequate for fully analyzing the energy aspects of the situation. Evidently there must be some kind of deformation of the block—some change of shape—and the friction force that the table exerts on the block must act through a displacement that is different from the displacement of the center of mass.

Sliding Block: Point Particle System

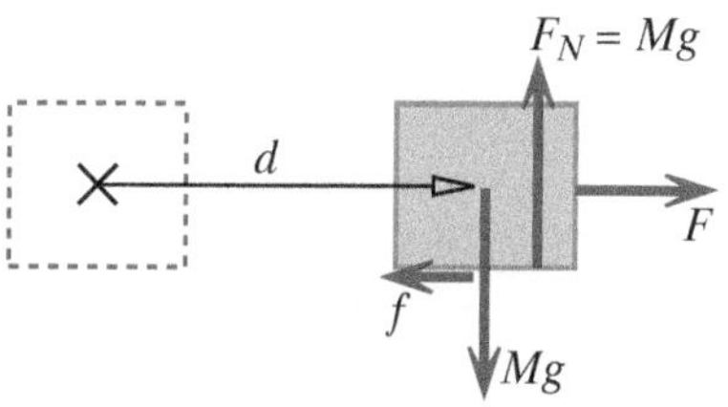

Figure 9.37 The center of mass of the block moves a distance d under the influence of a net force $F - f$.

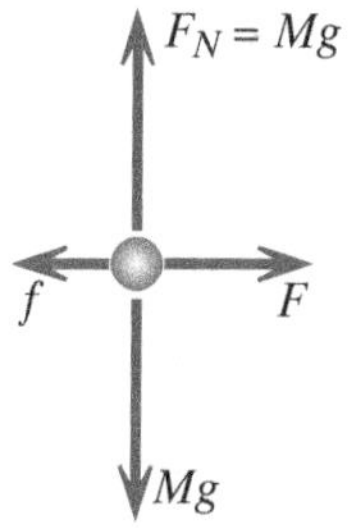

Figure 9.38 The block considered as a point particle.

A key issue that we will address is the question of how much work is done on the block by the friction force f exerted by the table. One of the new tools is the energy equation for the point particle system (Figure 9.37 and Figure 9.38). You apply a force of magnitude F to the right and the table exerts a force of magnitude f to the left. (Pretend for the moment that you don't know the magnitude of the friction force f.) The change in the kinetic energy of the fictitious center-of-mass point particle is given by the product of the net force $(F - f)$ times the displacement d of the center-of-mass point:

$$\Delta K_{\text{trans}} = (F - f)d$$

Since the speed of the block is constant, $\Delta K_{\text{trans}} = 0$, so $F - f = 0$, which shows that the friction force is equal and opposite to the force you apply $f = F$. The internal energy does not appear in this energy equation for the point particle system, because this equation deals merely with the motion of the (mathematical) center of mass, not with the many kinds of energy in the extended system.

We also could have obtained the result $f = F$ directly from the Momentum Principle: $dp_{\text{sys}}/dt = F - f = 0$. The Momentum Principle is closely related to the energy equation for the point particle system, because both involve the net force acting on a system and the motion of the center of mass.

Sliding Block: Extended System

The energy equation for the extended system of the block can be written as follows, where Fd is the work done by you, and W_{fric} is the work done by the table:

$$\Delta K_{\text{trans}} + \Delta E_{\text{internal}} = Fd + W_{\text{fric}}$$

where $\Delta E_{\text{internal}}$ is the rise in the internal energy of the block. We're assuming that the process takes a short enough time that there is negligible transfer of energy Q due to a temperature difference between the system of the block and the surroundings (the table), because energy transfer due to a temperature difference between table and block is a relatively slow process. (To remove such energy transfer as a possible complication, we could consider a block sliding not on a table but on an identical block. Then the symmetry of the situation is such that there cannot be any net transfer of energy due to a temperature difference into or out of the upper block. This tactic—reducing the complexity of a model—is a useful approach to complex problems.)

Since the speed of the center of mass v_{CM} does not change, $\Delta K_{\text{trans}} = 0$, and the energy equation reduces to

$$\Delta E_{\text{internal}} = Fd + W_{\text{fric}}$$

If we conclude that the friction force does an amount of work $-Fd$, then we would have to conclude that the internal energy of the block doesn't change, which is absurd. The block definitely gets hotter, indicating an increase in its internal energy. We need to find a way around the entirely plausible but apparently incorrect conclusion that the friction force does an amount of work $-Fd$.

> **QUESTION** Considering the sign of the change of internal energy, must the magnitude of the work done by the friction force be greater or less than Fd?

Since the internal energy change $\Delta E_{\text{internal}}$ is surely positive (the block gets hotter, not colder), the magnitude of the work done by the friction force must be less than Fd. Since the friction force is definitely equal to F, the friction force must act through some effective distance d_{eff} *that must be less than* d!

Evidently the energy equation for the extended system has the following form, with d_{eff} less than d:

$$\Delta E_{\text{internal}} = Fd - Fd_{\text{eff}}$$

A Physical Model for Dry Friction

Figure 9.39 A block slides on a small number of protruding teeth (vertical scale exaggerated).

How can the effective distance through which the friction force does work be less than the distance through which the block moves? We can understand this by looking at a microscopic picture of what happens on the surfaces in contact.

When a metal block slides on a metal surface, the block is supported by as few as three protruding "teeth," called "asperities" in the scientific literature on friction (Figure 9.39). The very high load per unit area on these teeth makes the material partially melt and flow, and high local temperatures produced during sliding lead to adhesion (welding) in the contact regions. The frictional force divided by the tiny contact area corresponds to the large "shear" (sideways) stress required to break these welds. This shearing of contact welds is the dominant friction mechanism for a dry metal sliding on the same metal.

Because the tooth tips can become stronger than the bulk metal due to a process called "work-hardening" (which introduces dislocations into the otherwise regular geometrical arrangements of atoms), shearing often occurs in the weaker regions of the teeth, away from the tip. This is a major effect when the two objects are made of the same material, and chunks of metal can break off and embed in the other surface. Nevertheless, this wear will be ignored in the further discussion. It is in any case a symmetrical effect for identical blocks. If the metal surfaces have oxide coatings, this can reduce the shear stress required to break the temporary weld (which reduces the friction force) and can prevent the breaking off of chunks of metal, if the oxide contact area is the weakest section.

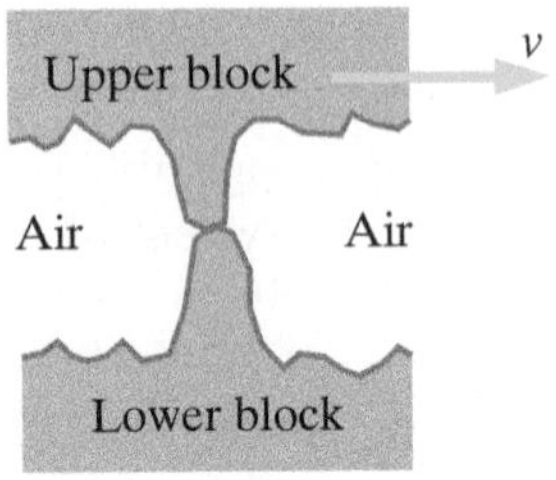

Figure 9.40 Two projecting teeth of identical blocks, which have temporarily adhered to each other. (Exaggerated vertical scale.)

This is the model of dry friction developed in a classic treatise on friction, *The Friction and Lubrication of Solids*, Part 1 and Part II, by F. P. Bowden and D. Tabor (Oxford University Press, 1950 and 1964). The physics and chemistry of friction continues to be an active field of research, because the effects of friction can be quite complex, and there is high practical interest in controlling friction. A more recent textbook is *Friction, Wear, Lubrication*, by K. C. Ludema (CRC Press, 1996).

Having briefly reviewed a basic model of dry friction, we proceed to use this model to calculate the work done by frictional forces exerted at the contact points. The key issue is that the surface is deformable, which leads to differences in the energy equation for the extended system compared with the energy equation for the point particle system.

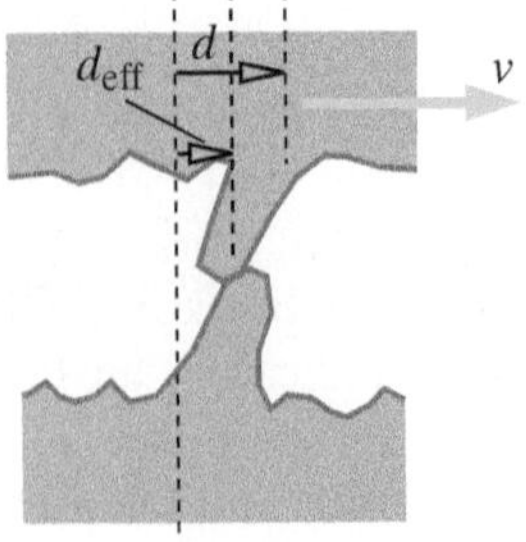

Figure 9.41 The top block has moved a distance d to the right, but the point of contact has moved only d_{eff}.

Actual Work Done by Friction Forces

We'll consider a microscopic picture of the contact region between two identical blocks. Figure 9.40 shows in a schematic way two teeth that have temporarily adhered to each other. The vertical scale has been greatly exaggerated for clarity—machined surfaces have much gentler slopes.

As the top block is dragged to the right, the teeth continue to stick together for a while. Both teeth must deform as a result; the top tooth is stretched backward, and the bottom tooth is stretched forward. In Figure 9.41 we see that when the top block has moved a distance d to the right, the point where the teeth touch each other has moved a distance of only d_{eff}, because the bottom block hasn't moved. (For two identical blocks made of the same material, on average the two teeth will bend the same amounts, and d_{eff} will be equal to $d/2$.)

> *This is the key issue:* The point of contact where the friction force is applied moves a shorter distance (d_{eff}) than the block itself moves (d). This means that the work done by the friction force is less than one might expect.

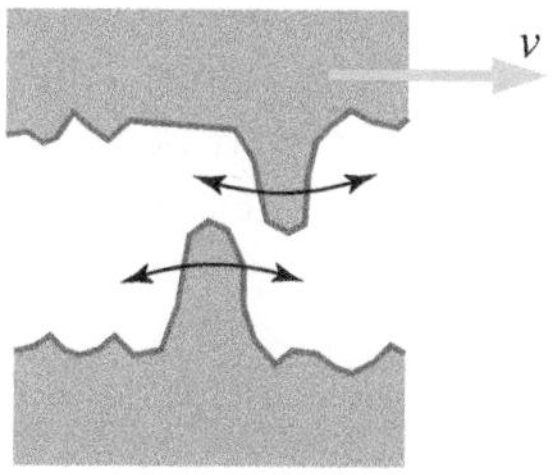

Figure 9.42 After the temporary weld breaks, the top tooth catches up to the rest of the block. Now no force acts on the teeth. Both teeth vibrate.

Eventually the weld breaks, and the top tooth snaps forward, so all the atoms in the tooth now catch up with the rest of the top block, but during this part of their journey, there is no external force acting on them (Figure 9.42).

The time average of the contact forces is indeed F (as determined by the fact that the net force must be zero, $F - F$), but the effective displacement d_{eff} at the point of contact of the frictional force is less than the displacement of the center of mass of the upper block. For two identical blocks we expect to find that $d_{\text{eff}} = d/2$. For other combinations of two materials in contact, all we know is that d_{eff} must be less than d but not necessarily equal to $d/2$.

Internal Energy of the Blocks

Once the weld has broken, the teeth can vibrate, as we saw in Figure 9.42. The vibration of the tooth belonging to the top block and thermal conduction upward from the hot tip into the main body of the block contribute to the increase in the internal energy of the upper block. Similarly, vibration and thermal conduction in the tooth belonging to the bottom block end up as increase in the internal energy of the bottom block.

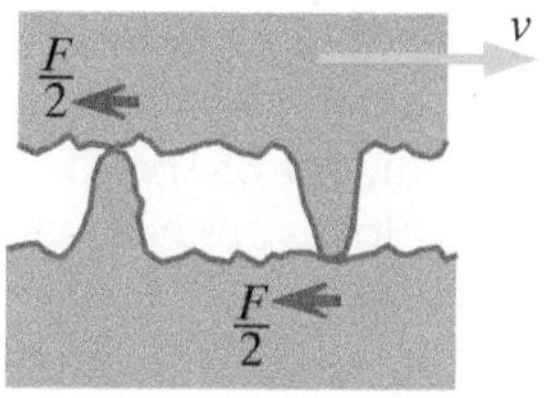

Figure 9.43 A more representative picture of contact between two surfaces, with projecting teeth in contact with flat areas.

A More Physical Picture

Having two long teeth in contact is rather unphysical. A better picture is the one shown in Figure 9.43 (again with a greatly exaggerated vertical scale), where we show one of the top block's longest teeth in contact with the average surface level of the bottom block, and one of the bottom block's longest teeth similarly in contact with the average surface level of the top block. The two teeth shown here are to be taken as representative of time and space averages of the sliding friction. Take the special case of identical blocks. By symmetry, the frictional force F exerted on the top block is divided on the average into two forces each of magnitude $F/2$ at the ends of the two sets of long teeth.

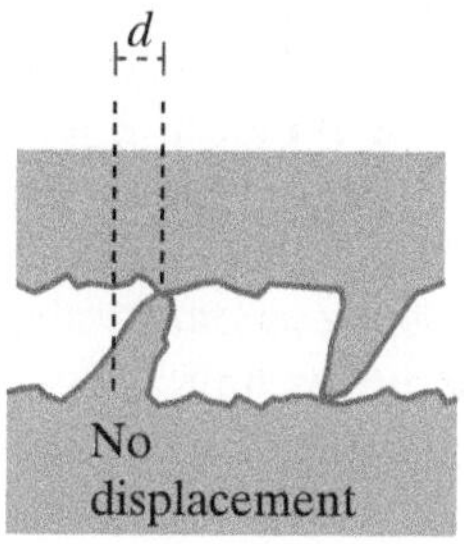

Figure 9.44 During this time interval the upper contact moves a distance d, while the lower contact does not move at all.

In Figure 9.44 we see that when the top block moves a distance d to the right the upper contact also moves a distance d, whereas the lower contact does not move at all. The frictional work is therefore $(-F/2)(0) + (-F/2)(d)$, which is $-Fd/2 = -Fd_{\text{eff}}$, so d_{eff} is equal to $d/2$ in the special case of two identical blocks.

Summary of Dry Friction

The fundamental reason why the friction force can act through a distance less than d is that the block is deformable. All atoms in the top block eventually move the same distance d, but because of the stick/slip contact between the blocks the friction force F acts only for a portion of the displacement ($d/2$ in the special case of identical blocks). The point of contact for the friction force does not move in the same way as the center of mass. Note again that the energy equation for the point particle system is not the same as the energy equation for the extended system if the system changes shape.

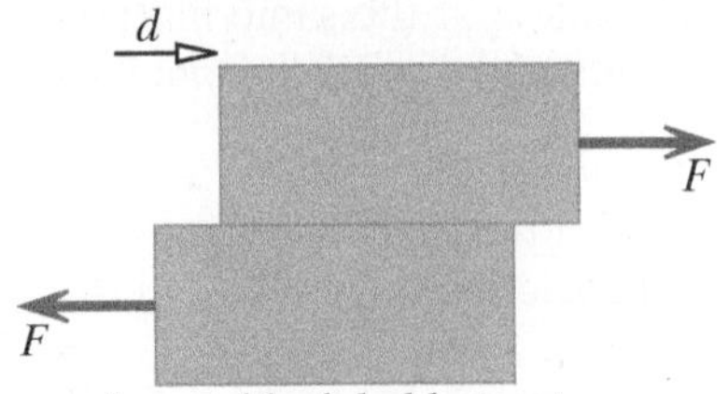

Figure 9.45 The lower block is held at rest, while the upper block slides a distance *d*. Forces in the vertical direction are not shown.

A Model-Independent Calculation of the Effective Distance

In the special case of a block sliding on an identical block we can calculate d_{eff}, independent of the particular model of the surfaces. Consider a system consisting of both blocks together (Figure 9.45). The bottom block is held stationary by applying a force to the left (to prevent it from being dragged to the right). This constraining force acts through no distance (the bottom block stands still), so this force does no work. The only work done on the two-block system is done by you, of magnitude Fd. So the energy equation for the extended two-block system is this:

$$\Delta E_{\text{internal},1} + \Delta E_{\text{internal},2} = Fd$$

Since the two blocks are identical, half of this increased internal energy shows up in the upper block ($Fd/2$), and half in the lower block ($Fd/2$). We can plug this into our earlier energy equation for the upper block:

$$\Delta E_{\text{internal},1} = \frac{Fd}{2} = Fd - Fd_{\text{eff}}$$

QUESTION Calculate d_{eff}.

We find that the effective distance through which the friction force acts is half the distance d that the block moves: $d_{\text{eff}} = d/2$. This is in agreement with the model of dry friction with bending teeth that we examined earlier, but our result is a general one that applies to any kind of model of friction surfaces for two symmetrical blocks. Bear in mind that if the blocks are not made of the same material, d_{eff} need not be equal to $d/2$.

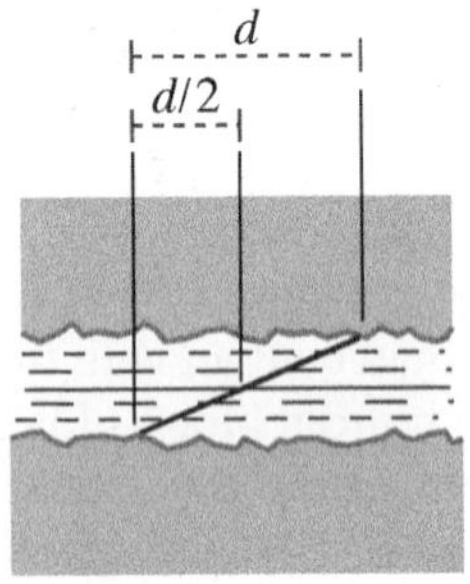

Figure 9.46 Two objects separated by a film of oil, which is modeled as a stack of fluid layers that can move relative to each other.

Lubricated Friction

It is interesting to see that the model-independent result for two identical blocks, $d_{\text{eff}} = d/2$, is also consistent with the case of lubricated friction. We separate the identical two blocks with a film of viscous lubricating oil, so that the two blocks do not make direct contact, as shown in Figure 9.46.

It is a property of simple fluid flow that fluid layers immediately adjacent to the blocks are constrained to share the motion of the blocks. Also, for a common type of flow called "laminar" flow, the displacement profile in the oil is linear. In particular, at the midplane the fluid moves half as far as the top block moves. As a result, as the top block is pulled a distance d to the right, the top layer of the oil is dragged along and moves a distance d to the right, the bottom layer of oil doesn't move, and the layer in the midplane (halfway between the blocks) moves a distance $d/2$.

If we take as the symmetrical systems of interest the top block with the upper half of the oil and the bottom block with the lower half of the oil, we see that the shear force between the two systems (at the midplane in the oil) acts through a distance that is again half the displacement of the top block: $d_{\text{eff}} = d/2$. (Of course the magnitude of the friction force is much reduced by the lubrication, and the applied force F must be much smaller if the velocity is to be constant.)

This discussion is based on an article "Work and heat transfer in the presence of sliding friction," by B. Sherwood and W. Bernard, *American Journal of Physics*, volume 52, number 11, Nov. 1984, pages 1001–1007, which in turn draws on an earlier article "Real work and pseudowork," by B. Sherwood, *American Journal of Physics*, volume 51, number 7, July 1983, pages 597–602.

9.5 *DERIVATION: KINETIC ENERGY OF A MULTIPARTICLE SYSTEM

In this section we derive the important result that $K_{tot} = K_{trans} + K_{rel}$, where $K_{trans} = \frac{1}{2}Mv_{CM}^2$ (for $v_{CM} \ll c$) and K_{rel} is the kinetic energy relative to the center of mass. This result sounds entirely plausible, but the formal proof is rather difficult.

As in the case of calculating the gravitational energy of a multiparticle object, the derivation of the kinetic energy of a multiparticle system hinges on the definition of the center of mass point of a collection of atoms. The (vector) location of the i-th atom of the object can be expressed as the sum of two vectors, one from the origin to the center of mass ($\vec{r}_{CM}$) plus another from the center of mass to the i-th atom ($\vec{r}_{i,CM}$): $\vec{r}_i = \vec{r}_{CM} + \vec{r}_{i,CM}$ (Figure 9.47).

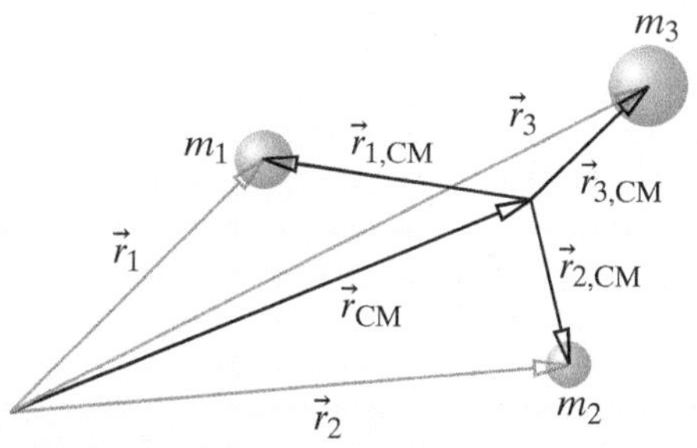

Figure 9.47 Center of mass of a multiparticle system.

The kinetic energy $K_i = \frac{1}{2}mv_i^2$ of the i-th atom is this, where we write $\vec{v}_{i,CM} = d\vec{r}_{i,CM}/dt$, the velocity of the i-th particle relative to the center of mass:

$$K_i = \frac{1}{2}m_i\left|\frac{d}{dt}(\vec{r}_{CM} + \vec{r}_{i,CM})\right|^2 = \frac{1}{2}m_i\left|\vec{v}_{CM} + \vec{v}_{i,CM}\right|^2$$

This can be expanded, using the fact that the magnitude squared of a vector can be written as a vector dot product, $|\vec{C}|^2 = \vec{C} \bullet \vec{C} = C^2\cos(0) = C^2$:

$$K_i = \frac{1}{2}m_i[\vec{v}_{CM} + \vec{v}_{i,CM}] \bullet [\vec{v}_{CM} + \vec{v}_{i,CM}] = \frac{1}{2}m_i v_{CM}^2 + 2\frac{1}{2}m_i\vec{v}_{CM} \bullet \vec{v}_{i,CM} + \frac{1}{2}m_i v_{i,CM}^2$$

Now that we have the kinetic energy of the i-th atom in terms of its velocity $\vec{v}_{i,CM}$ relative to the center of mass, we need to add up the total kinetic energy of all the atoms. The total kinetic energy K_{tot} can be written in the following compact way, where Σ (Greek capital sigma) means "sum," and the sum goes from $i = 1$ through $i = N$:

$$K_{tot} = \underbrace{\sum_{i=1}^{N}\frac{1}{2}m_i v_{CM}^2}_{\textbf{First term}} + \underbrace{\sum_{i=1}^{N}2\left(\frac{1}{2}m_i\right)\vec{v}_{CM} \bullet \vec{v}_{i,CM}}_{\textbf{Second term}} + \underbrace{\sum_{i=1}^{N}\frac{1}{2}m_i v_{i,CM}^2}_{\textbf{Third term}}$$

First Term

The first term in this summation turns out to be the kinetic energy of a single particle of mass M, moving at the speed of the center of mass:

$$\sum_{i=1}^{N}\frac{1}{2}m_i v_{CM}^2 = \frac{1}{2}v_{CM}^2\left(\sum_{i=1}^{N}m_i\right) = \frac{1}{2}v_{CM}^2(m_1 + m_2 + m_3 + \cdots) = \frac{1}{2}Mv_{CM}^2$$

Second Term

The second term can be shown to be zero:

$$\sum_{i=1}^{N}2\left(\frac{1}{2}m_i\right)\vec{v}_{CM} \bullet \vec{v}_{i,CM} = \vec{v}_{CM} \bullet \sum_{i=1}^{N}m_i\vec{v}_{i,CM} = \vec{v}_{CM} \bullet \frac{d}{dt}\left(\sum_{i=1}^{N}m_i\vec{r}_{i,CM}\right) = 0$$

This result follows from the way that we calculate the location of the center of mass of a system:

$$\vec{r}_{\text{CM}} = \frac{m_1\vec{r}_1 + m_2\vec{r}_2 + m_3\vec{r}_3 + \cdots}{m_1 + m_2 + m_3 + \cdots} = \frac{\sum_{i=1}^{N} m_i\vec{r}_i}{M}$$

We are measuring the location $\vec{r}_{i,\text{CM}}$ of the i-th atom relative to the center of mass, and since the distance from the center of mass to the center of mass is of course zero, we have

$$\vec{0} = \frac{\sum_{i=1}^{N} m_i\vec{r}_{i,\text{CM}}}{M} \quad \text{and therefore} \quad \sum_{i=1}^{N} m_i\vec{r}_{i,\text{CM}} = \vec{0}$$

Third Term

The third term by definition is the kinetic energy of the atoms relative to the (possibly moving) center of mass:

$$\sum_{i=1}^{N} \frac{1}{2} m_i v_{i,\text{CM}}^2 = \frac{1}{2} m_1 v_{1,\text{CM}}^2 + \frac{1}{2} m_2 v_{2,\text{CM}}^2 + \cdots = K_{\text{rel}}$$

Putting the three pieces together, we find that the total kinetic energy splits into two parts: a term associated with the overall motion of the center of mass plus the kinetic energy relative to the center of mass,

$$K_{\text{tot}} = K_{\text{trans}} + K_{\text{rel}}$$

where $K_{\text{trans}} = \frac{1}{2} M v_{\text{CM}}^2 = \frac{p_{\text{sys}}^2}{2M}$.

9.6 *DERIVATION: THE POINT PARTICLE ENERGY EQUATION

In this chapter we showed that the energy equation for the point particle version of the system follows from the fact that the motion of the center of mass is just like that of a point particle with the total mass of the extended system and subjected to the *net* force acting on the extended system. Here we give a more formal derivation of this important result.

Start from the x component of the Momentum Principle for a multiparticle system whose center of mass is moving at nonrelativistic speed:

$$\frac{dp_{\text{sys},x}}{dt} = M_{\text{tot}} \frac{dv_{\text{CM},x}}{dt} = F_{\text{net},x}$$

Integrate through the x displacement of the center of mass:

$$M_{\text{tot}} \int_i^f \frac{dv_{\text{CM},x}}{dt} dx_{\text{CM}} = \int_i^f F_{\text{net},x} dx_{\text{CM}}$$

Switch dv and dx:

$$M_{\text{tot}} \int_i^f \frac{dx_{\text{CM}}}{dt} dv_{\text{CM},x} = \int_i^f F_{\text{net},x}\, dx_{\text{CM}}$$

However, dx_{CM}/dt is the x component of the center of mass velocity:

$$M_{\text{tot}} \int_i^f v_{\text{CM},x}\, dv_{\text{CM},x} = \int_i^f F_{\text{net},x}\, dx_{\text{CM}}$$

The integral on the left can be carried out:

$$M_{\text{tot}} \left[\frac{1}{2} v_{\text{CM},x}^2\right]_i^f = \int_i^f F_{\text{net},x}\, dx_{\text{CM}}$$

$$\Delta\left[\frac{1}{2} M_{\text{tot}} v_{\text{CM},x}^2\right] = \int_i^f F_{\text{net},x}\, dx_{\text{CM}}$$

We could repeat exactly the same argument for the y and z motions:

$$\Delta\left[\frac{1}{2} M_{\text{tot}} v_{\text{CM},y}^2\right] = \int_i^f F_{\text{net},y}\, dy_{\text{CM}}$$

$$\Delta\left[\frac{1}{2} M_{\text{tot}} v_{\text{CM},z}^2\right] = \int_i^f F_{\text{net},z}\, dz_{\text{CM}}$$

Note that

$$\frac{1}{2} M_{\text{tot}} v_{\text{CM},x}^2 + \frac{1}{2} M_{\text{tot}} v_{\text{CM},y}^2 + \frac{1}{2} M_{\text{tot}} v_{\text{CM},z}^2 = \frac{1}{2} M_{\text{tot}} v_{\text{CM}}^2$$

Moreover,

$$F_{\text{net},x}\, dx_{\text{CM}} + F_{\text{net},y}\, dy_{\text{CM}} + F_{\text{net},z}\, dz_{\text{CM}} = \vec{F}_{\text{net}} \bullet d\vec{r}_{\text{CM}}$$

Adding the three equations together (the sum of the left sides is equal to the sum of the right sides), we have the following equation for the translational kinetic energy of the system:

$$\Delta\left[\frac{1}{2} M_{\text{tot}} v_{\text{CM}}^2\right] = \int_i^f \vec{F}_{\text{net}} \bullet d\vec{r}_{\text{CM}}$$

In words: the change in the translational kinetic energy of a system is equal to the integral of the *net* force acting through the displacement of the center of mass point.

The derivation that we have just carried out shows that although this equation looks like an energy equation, it is actually closely related to the Momentum Principle from which it was derived. The common element is the *net* force. Note that there are three separately valid equations, for x, y, and z, which indicates that we're really dealing with momentum (a vector) rather than energy (a scalar).

In contrast, the actual energy equation for the extended system involves the work done by each individual force through the displacement of the point of application of that force. If the system deforms or rotates, these displacements of the individual forces need not be the same as the displacement of the center of mass.

SUMMARY

Center of mass

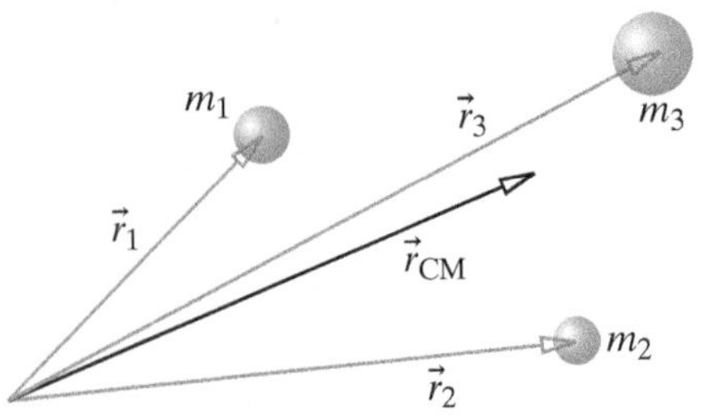

Figure 9.48

$$\vec{r}_{CM} \equiv \frac{m_1\vec{r}_1 + m_2\vec{r}_2 + m_3\vec{r}_3 + \cdots}{m_1 + m_2 + m_3 + \cdots}$$

Problem-solving techniques

Combined use of the energy equations for two different models, the extended system and the point particle system, in analyzing complex phenomena involving multiparticle systems.

Point particle system: Information on change of translational kinetic energy; $\Delta K_{trans} = \vec{F}_{net} \bullet \Delta\vec{r}_{CM}$

Extended system: Information on change of total energy; $\Delta E_{sys} = \sum \vec{F}_i \bullet \Delta\vec{r}_i$

The Momentum Principle for multiparticle systems, derived from Newton's second and third laws of motion, was extended by defining the center of mass:

$$\frac{d\vec{p}_{sys}}{dt} = \vec{F}_{net}$$

$\vec{p}_{sys} \approx M\vec{v}_{CM}$ (if $v \ll c$, and constant mass).

Gravitational energy of multiparticle systems plus Earth, near the Earth's surface:

$$U_g = Mgy_{CM}$$

Kinetic energy of multiparticle systems:

$$K_{tot} = K_{trans} + K_{rel}$$

where $K_{trans} = \dfrac{p_{sys}^2}{2M} = \dfrac{1}{2}Mv_{CM}^2$.

Kinetic energy relative to center of mass can be split into two terms:

$$K_{rel} = K_{rot} + K_{vib}$$

Moment of inertia about an axis of rotation:

$$I = m_1 r_{\perp 1}^2 + m_2 r_{\perp 2}^2 + m_3 r_{\perp 3}^2 + \cdots$$

Kinetic energy relative to center of mass:

$$K_{rot} = \frac{1}{2}I\omega^2$$

Moments of inertia:

$$I_{disk} = I_{cylinder} = \frac{1}{2}MR^2$$

about center of disk, rotating around axis of cylinder

$$I_{sphere} = \frac{2}{5}MR^2$$

for axis passing through center of sphere.

Uniform solid cylinder of length L, radius R, about axis perpendicular to cylinder, through center of cylinder:

$$I = \frac{1}{12}ML^2 + \frac{1}{4}MR^2$$

For a rigid object, $K_{tot} = K_{trans} + K_{rot} = \frac{1}{2}(Mr_{CM}^2 + I_{CM})\omega^2$

Point particle system:

$$K_{trans} = \Delta\left(\frac{1}{2}Mv_{CM}^2\right) = \int_i^f \vec{F}_{net,ext} \bullet d\vec{r}_{CM}$$

Sliding friction can deform the contact surfaces, with the result that the frictional force may act through a distance that is different from the distance through which the center-of-mass point moves. As with any deformable system, the energy equations for the extended system and for the point particle system differ (though both are correct).

QUESTIONS

Q1 Discuss qualitatively the motion of the atoms in a block of steel that falls onto another steel block. Why and how do large-scale vibrations damp out?

Q2 Can you give an example of a system that has no atoms located at its center of mass?

Q3 The Momentum Principle for multiparticle systems would seem to say that the center of mass of a system moves just as though the system were a point particle. This led Chris to ask, "Wouldn't that mean that a piece of paper ought to fall in the same way as a small metal ball, if they have the same mass?" Explain carefully to Chris the resolution of this puzzle.

Q4 Consider the voyage to the Moon that you studied in Chapter 3. Would it make any difference, even a very tiny difference, whether the spacecraft is long or short, if the mass is the same? Explain briefly.

Q5 Under what conditions does the energy equation for the point particle system differ from the energy equation for the extended system? Give two examples of such a situation. Give one example of a situation where the two equations look exactly alike.

Q6 Consider the acceleration of a car on dry pavement, if there is no slipping. The axle moves at speed v, and the outside of the tire moves at speed v relative to the axle. The instantaneous velocity of the bottom of the tire is zero. How much work is done by the force exerted on the tire by the road? What is the source of the energy that increases the car's translational kinetic energy?

Q7 Two people with different masses but equal speeds slide toward each other with little friction on ice with their arms extended straight out to the side (so each has the shape of a "T"). Her right hand meets his right hand, they hold hands and spin 90°, then release their holds and slide away. Make a rough sketch of the path of the center of mass of the system consisting of the two people, and explain briefly. (It helps to mark equal time intervals along the paths of the two people and of their center of mass.)

PROBLEMS

Section 9.1

•P8 Three uniform-density spheres are positioned as follows:

- A 3 kg sphere is centered at $\langle 10, 20, -5\rangle$ m.
- A 5 kg sphere is centered at $\langle 4, -15, 8\rangle$ m.
- A 6 kg sphere is centered at $\langle -7, 10, 9\rangle$ m.

What is the location of the center of mass of this three-sphere system?

•P9 Relative to an origin at the center of the Earth, where is the center of mass of the Earth–Moon system? The mass of the Earth is 6×10^{24} kg, the mass of the Moon is 7×10^{22} kg, and the distance from the center of the Earth to the center of the Moon is 4×10^{8} m. The radius of the Earth is 6400 km. One can show that the Earth and Moon orbit each other around this center of mass.

•P10 A meter stick whose mass is 300 g lies on ice (Figure 9.49). You pull at one end of the meter stick, at right angles to the stick, with a force of 6 N. The ensuing motion of the meter stick is quite complicated, but what are the initial magnitude and direction of the rate of change of the momentum of the stick, dp_{sys}/dt, when you first apply the force? What is the magnitude of the initial acceleration of the center of the stick?

Figure 9.49

•P11 Determine the location of the center of mass of an L-shaped object whose thin vertical and horizontal members have the same length L and the same mass M. Use the formal definition to find the x and y coordinates, and check your result by doing the calculation with respect to two different origins, one in the lower left corner at the intersection of the horizontal and vertical members and one at the right end of the horizontal member.

••P12 A man whose mass is 80 kg and a woman whose mass is 50 kg sit at opposite ends of a canoe 5 m long, whose mass is 30 kg. **(a)** Relative to the man, where is the center of mass of the system consisting of man, woman, and canoe? (*Hint:* Choose a specific coordinate system with a specific origin.) **(b)** Suppose that the man moves quickly to the center of the canoe and sits down there. How far does the canoe move in the water? Explain your work and your assumptions.

•P13 If an object has a moment of inertia 19 kg · m^2 and rotates with an angular speed of 70 rad/s, what is its rotational kinetic energy?

•P14 A group of particles of total mass 35 kg has a total kinetic energy of 340 J. The kinetic energy relative to the center of mass is 85 J. What is the speed of the center of mass?

••P15 By calculating numerical quantities for a multiparticle system, one can get a concrete sense of the meaning of the relationships $\vec{p}_{sys} = M_{tot}\vec{v}_{CM}$ and $K_{tot} = K_{trans} + K_{rel}$. Consider an object consisting of two balls connected by a spring, whose stiffness is 400 N/m. The object has been thrown through the air and is rotating and vibrating as it moves. At a particular instant the spring is stretched 0.3 m, and the two balls at the ends of the spring have the following masses and velocities:

1: 5 kg, $\langle 8, 14, 0\rangle$ m/s
2: 3 kg, $\langle -5, 9, 0\rangle$ m/s

(a) For this system, calculate $\vec{p}_{sys}$. **(b)** Calculate $\vec{v}_{CM}$. **(c)** Calculate K_{tot}. **(d)** Calculate K_{trans}. **(e)** Calculate K_{rel}. **(f)** Here is a way to check your result for K_{rel}. The velocity of a particle relative to the center of mass is calculated by subtracting $\vec{v}_{CM}$ from the particle's velocity. To take a simple example, if you're riding in a car that's moving with $v_{CM,x} = 20$ m/s, and you throw a ball with $v_{rel,x} = 35$ m/s, relative to the car, a bystander on the ground sees the ball moving with $v_x = 55$ m/s. So $\vec{v} = \vec{v}_{CM} + \vec{v}_{rel}$, and therefore we have $\vec{v}_{rel} = \vec{v} - \vec{v}_{CM}$. Calculate $\vec{v}_{rel} = \vec{v} - \vec{v}_{CM}$ for each mass and calculate the corresponding K_{rel}. Compare with the result you obtained in part (e).

••P16 Consider a system consisting of three particles:

$m_1 = 2$ kg, $\vec{v}_1 = \langle 8, -6, 15\rangle$ m/s
$m_2 = 6$ kg, $\vec{v}_2 = \langle -12, 9, -6\rangle$ m/s
$m_3 = 4$ kg, $\vec{v}_3 = \langle -24, 34, 23\rangle$ m/s

What is K_{rel}, the kinetic energy of this system relative to the center of mass?

••P17 Binary stars: **(a)** About half of the visible "stars" are actually binary star systems, two stars that orbit each other with no other objects nearby (the small, dim stars called "red dwarfs" however tend to be single). Describe the motion of the center of mass of a binary star system. Briefly explain your reasoning. **(b)** For a particular binary star system, telescopic observations repeated over many years show that one of the stars (whose unknown mass we'll call M_1) has a circular orbit with radius $R_1 = 6\times10^{11}$ m, while the other star (whose unknown mass we'll call M_2) has a circular orbit of radius $R_2 = 9\times10^{11}$ m about the same point. Make a sketch of the orbits, and show the positions of the two stars on these orbits at some instant. Label the two stars as to which is which, and label their orbital radii. Indicate on your sketch the location of the center of mass of the system, and explain how you know its location, using the concepts and results of this chapter. **(c)** This double-star system is observed to complete one revolution in 40 years. What are the masses of the two stars? (For comparison, the distance from Sun to Earth is about 1.5×10^{11} m, and the mass of the Sun is about 2×10^{30} kg.) This method is often used to determine the masses of stars. The mass of a star largely determines many of the other properties of a star, which is why astrophysicists need a method for measuring the mass.

Section 9.2

•P18 If an object's rotational kinetic energy is 50 J and it rotates with an angular speed of 12 rad/s, what is its moment of inertia?

•P19 A uniform-density disk of mass 13 kg, thickness 0.5 m, and radius 0.2 m makes one complete rotation every 0.6 s. What is the rotational kinetic energy of the disk?

•P20 A sphere of uniform density with mass 22 kg and radius 0.7 m is spinning, making one complete revolution every 0.5 s. The center of mass of the sphere has a speed of 4 m/s. **(a)** What is the rotational kinetic energy of the sphere? **(b)** What is the total kinetic energy of the sphere?

•P21 A uniform-density disk whose mass is 10 kg and radius is 0.4 m makes one complete rotation every 0.2 s. What is the rotational kinetic energy of the disk?

•P22 A cylindrical rod of uniform density is located with its center at the origin, and its axis along the x axis. It rotates about its center in the xy plane, making one revolution every 0.03 s. The rod has a radius of 0.08 m, length of 0.7 m, and mass of 5 kg. It makes one revolution every 0.03 s. What is the rotational kinetic energy of the rod?

•P23 A uniform-density 6 kg disk of radius 0.3 m is mounted on a nearly frictionless axle. Initially it is not spinning. A string is wrapped tightly around the disk, and you pull on the string with a constant force of 25 N through a distance of 0.6 m. Now what is the angular speed?

••P24 A high diver tucks himself into a ball and spins rapidly (Figure 9.50). Make estimates of the relevant parameters and calculate K_{rot}, the rotational kinetic energy of the diver.

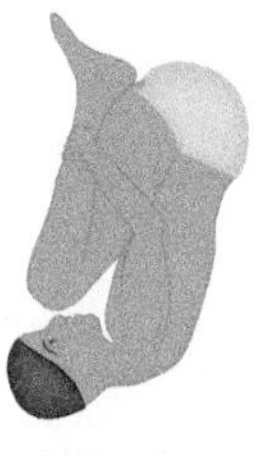

Figure 9.50

••P25 The Earth is 1.5×10^{11} m from the Sun and takes a year to make one complete orbit. It rotates on its own axis once per day. It can be treated approximately as a uniform-density sphere of mass 6×10^{24} kg and radius 6.4×10^{6} m (actually, its center has higher density than the rest of the planet, and the Earth bulges out a bit at the equator). Using this crude approximation, calculate the following: **(a)** What is v_{CM}? **(b)** What is K_{trans}? **(c)** What is ω, the angular speed of rotation around its own axis? **(d)** What is K_{rot}? **(e)** What is K_{tot}?

••P26 Show that the moment of inertia of a disk of mass M and radius R is $\frac{1}{2}MR^2$. Divide the disk into narrow rings, each of radius r and width dr. The contribution to I by one of these rings is simply r^2dm, where dm is the amount of mass contained in that particular ring. The mass of any ring is the total mass times the fraction of the total area occupied by the area of the ring. The area of this ring is approximately $2\pi r dr$. Use integral calculus to add up all the contributions.

Section 9.3

•P27 You pull straight up on the string of a yo-yo with a force 0.235 N, and while your hand is moving up a distance 0.18 m, the yo-yo moves down a distance 0.70 m. The mass of the yo-yo is 0.025 kg, and it was initially moving downward with speed 0.5 m/s and angular speed 124 rad/s. **(a)** What is the increase in the translational kinetic energy of the yo-yo? **(b)** What is the new speed of the yo-yo? **(c)** What is the increase in the rotational kinetic energy of the yo-yo? **(d)** The yo-yo is approximately a uniform-density disk of radius 0.02 m. What is the new angular speed of the yo-yo?

••P28 A string is wrapped around a disk of mass 2.1 kg (its density is not necessarily uniform). Starting from rest, you pull the string with a constant force of 9 N along a nearly frictionless surface. At the instant when the center of the disk has moved a distance 0.11 m, your hand has moved a distance of 0.28 m (Figure 9.51).

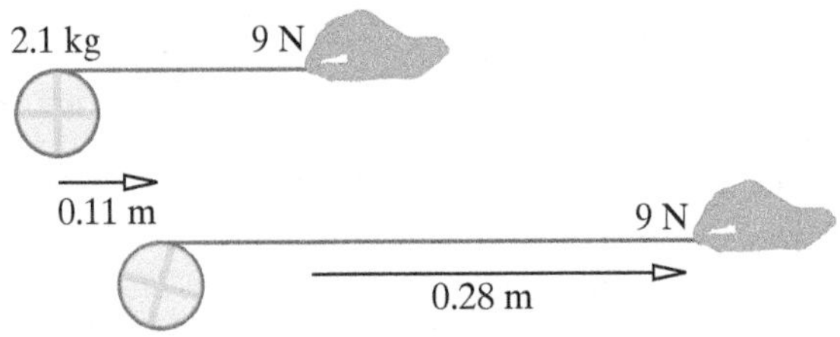

Figure 9.51

(a) At this instant, what is the speed of the center of mass of the disk? **(b)** At this instant, how much rotational kinetic energy does the disk have relative to its center of mass? **(c)** At this instant, the angular speed of the disk is 7.5 rad/s. What is the moment of inertia of the disk?

••P29 A chain of metal links with total mass $M = 7$ kg is coiled up in a tight ball on a low-friction table (Figure 9.52). You pull on a link at one end of the chain with a constant force $F = 50$ N. Eventually the chain straightens out to its full length $L = 2.6$ m, and you keep pulling until you have pulled your end of the chain a total distance $d = 4.5$ m.

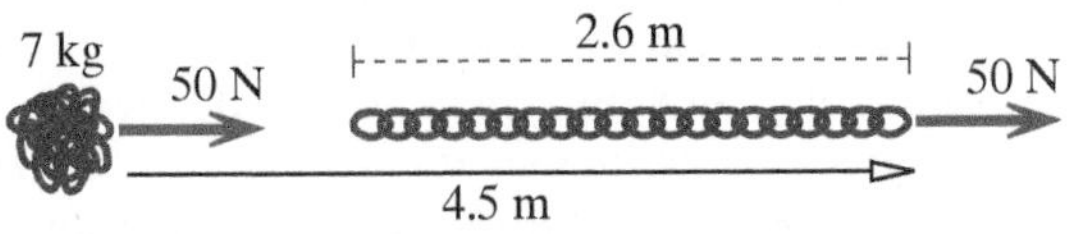

Figure 9.52

(a) Consider the point particle system. What is the speed of the chain at this instant? **(b)** Consider the extended system. What is the change in energy of the chain? **(c)** In straightening out, the links of the chain bang against each other, and their temperature rises. Assume that the process is so fast that there is insufficient time for significant transfer of energy from the chain to the table due to the temperature difference, and ignore the small amount of energy radiated away as sound produced in the collisions among the links. Calculate the increase in internal energy of the chain.

••P30 Tarzan, whose mass is 100 kg, is hanging at rest from a tree limb. Then he lets go and falls to the ground. Just before he lets go, his center of mass is at a height 2.9 m above the ground and the bottom of his dangling feet are at a height 2.1 above the ground. When he first hits the ground he has dropped a distance 2.1, so his center of mass is $(2.9 - 2.1)$ above the ground. Then his knees bend and he ends up at rest in a crouched position with his center of mass a height 0.5 above the ground. **(a)** Consider the point particle system. What is the speed v at the instant just before Tarzan's feet touch the ground? **(b)** Consider the extended system. What is the net change in internal energy for Tarzan from just before his feet touch the ground to when he is in the crouched position?

••P31 Here is an experiment on jumping up you can do. **(a)** Crouch down and jump straight up, as high as you can. Estimate the location of your center of mass, and measure its height at three stages in this process: in the crouch, at lift-off, and at the top of the jump. Report your measurements. You may need to have a friend help you make the measurements. **(b)** Analyze this process as fully as possible, using all the theoretical tools now available to you, especially the concepts in this chapter. Include a calculation of the average force of the floor on your feet, the change in your internal energy, and the approximate time of contact from the beginning of the jump to lift-off. Be sure to explain clearly what approximations and simplifying assumptions you made in modeling the process.

••P32 Consider an ice skater who pushes away from a wall. **(a)** Estimate the speed she can achieve just after pushing away from the wall. Then estimate the average acceleration during this process. How many g's is this? (That is, what fraction or multiple of 9.8 m/s^2 is your estimate?) Be sure to explain clearly what approximations and simplifying assumptions you made in modeling the process. **(b)** For this process, choose the woman as the system of interest and discuss the energy transfers, and the changes in the various forms of energy. Estimate the amount of each of these, including the correct signs.

••P33 A hoop of mass M and radius R rolls without slipping down a hill, as shown in Figure 9.53. The lack of slipping means that when the center of mass of the hoop has speed v, the tangential speed of the hoop relative to the center of mass is also equal to v_{CM}, since in that case the instantaneous speed is zero for the part of the hoop that is in contact with the ground $(v - v = 0)$. Therefore, the angular speed of the rotating hoop is $\omega = v_{CM}/R$.

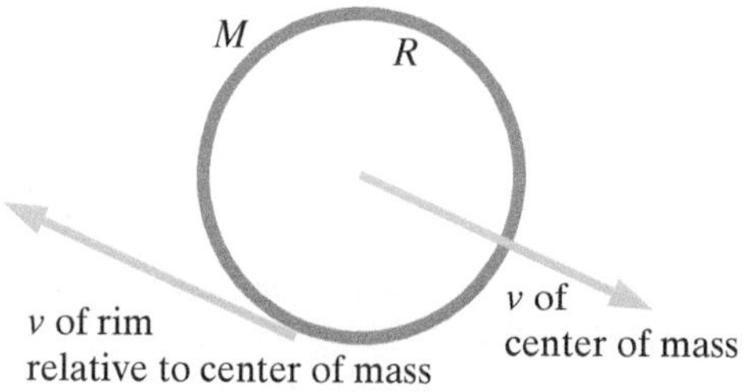

Figure 9.53

(a) The initial speed of the hoop is v_i, and the hill has a height h. What is the speed v_f at the bottom of the hill? **(b)** Replace the hoop with a bicycle wheel whose rim has mass M and whose hub has mass m, as shown in Figure 9.54. The spokes have negligible mass. What would be the bicycle wheel's speed at the bottom of the hill?

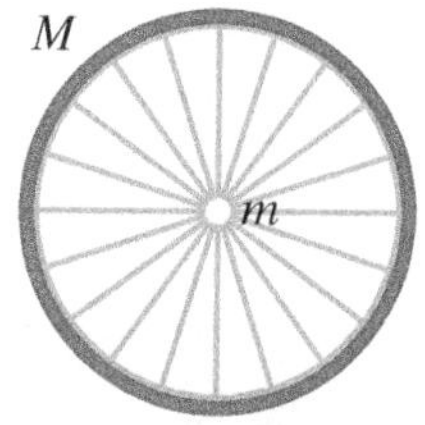

Figure 9.54

••P34 A sphere or cylinder of mass M, radius R, and moment of inertia I rolls without slipping down a hill of height h, starting from rest. As explained in Problem P33, if there is no slipping $\omega = v_{CM}/R$. **(a)** In terms of the given variables (M, R, I, and h), what is v_{CM} at the bottom of the hill? **(b)** If the object is a thin hollow cylinder, what is v_{CM} at the bottom of the hill? **(c)** If the object is a uniform-density solid cylinder, what is v_{CM} at the bottom of the hill? **(d)** If the object is a uniform-density sphere, what is v_{CM} at the bottom of the hill? An interesting experiment that you can perform is to roll various objects down an inclined board and see how much time each one takes to reach the bottom.

••P35 Two disks are initially at rest, each of mass M, connected by a string between their centers, as shown in Figure 9.55. The disks slide on low-friction ice as the center of the string is pulled by a string with a constant force F through a distance d. The disks collide and stick together, having moved a distance b horizontally.

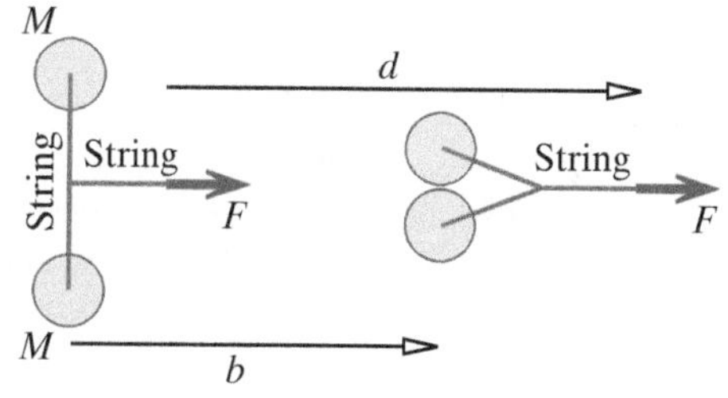

Figure 9.55

(a) What is the final speed of the stuck-together disks? **(b)** When the disks collide and stick together, their temperature rises. Calculate the increase in internal energy of the disks, assuming that the process is so fast that there is insufficient time for there to be much transfer of energy to the ice due to a temperature difference. (Also ignore the small amount of energy radiated away as sound produced in the collisions between the disks.)

••P36 You hang by your hands from a tree limb that is a height L above the ground, with your center of mass a height h above the ground and your feet a height d above the ground, as shown in Figure 9.56. You then let yourself fall. You absorb the shock by bending your knees, ending up momentarily at rest in a crouched position with your center of mass a height b above the ground. Your mass is M. You will need to draw labeled physics diagrams for the various stages in the process.

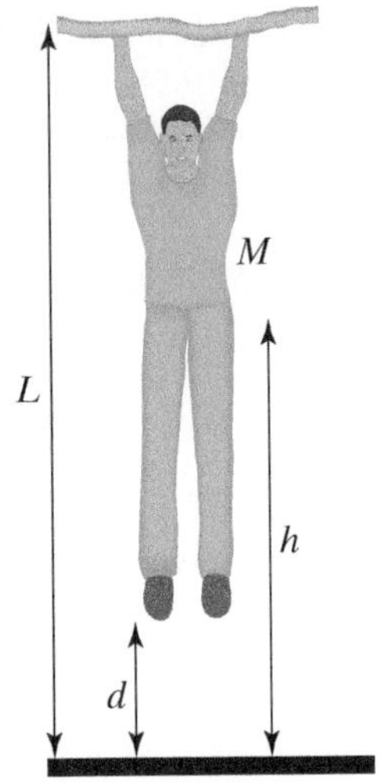

Figure 9.56

(a) What is the net internal energy change ΔE_{int} in your body (chemical plus thermal)? **(b)** What is your speed v at the instant your feet first touch the ground? **(c)** What is the approximate average force F exerted by the ground on your feet during the time when your knees are bending? **(d)** How much work is done by this force F?

••**P37** A box and its contents have a total mass M. A string passes through a hole in the box (Figure 9.57), and you pull on the string with a constant force F (this is in outer space—there are no other forces acting).

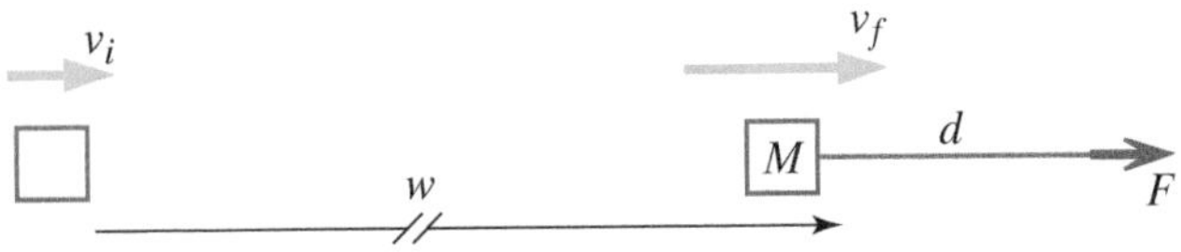

Figure 9.57

(a) Initially the speed of the box was v_i. After the box had moved a long distance w, your hand had moved an additional distance d (a total distance of $w + d$), because additional string of length d came out of the box. What is now the speed v_f of the box? **(b)** If we could have looked inside the box, we would have seen that the string was wound around a hub that turns on an axle with negligible friction, as shown in Figure 9.58. Three masses, each of mass m, are attached to the hub at a distance r from the axle. Initially the angular speed relative to the axle was ω_i. In terms of the given quantities, what is the final angular speed relative to the axis, ω_f?

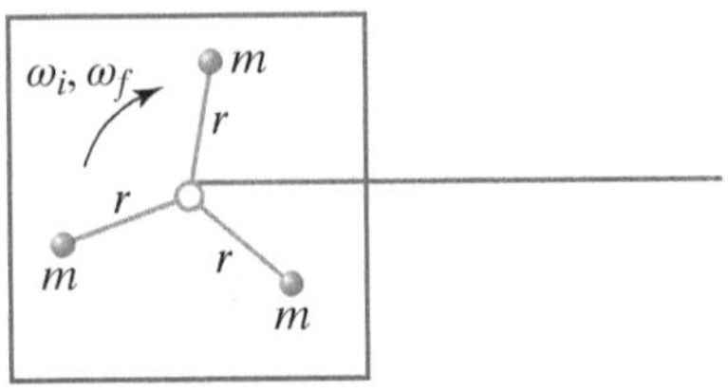

Figure 9.58

••**P38** Two identical 0.4 kg blocks (labeled 1 and 2) are initially at rest on a nearly frictionless surface, connected by an unstretched spring, as shown in the upper portion of Figure 9.59.

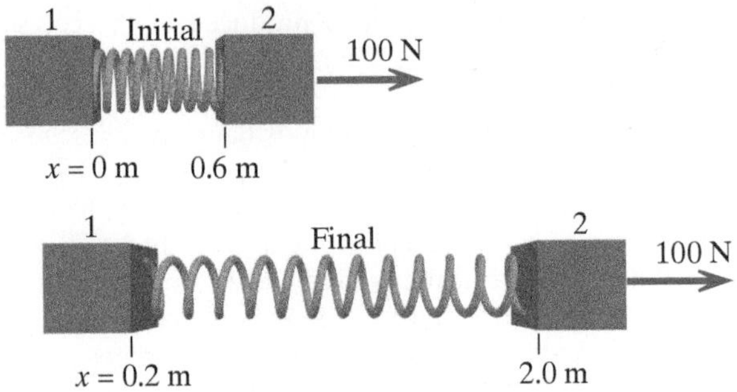

Figure 9.59

Then a constant force of 100 N to the right is applied to block 2, and at a later time the blocks are in the new positions shown in the lower portion of Figure 9.59. At this final time, the system is moving to the right and also vibrating, and the spring is stretched. **(a)** The following questions apply to the system modeled as a point particle. (i) What is the initial location of the point particle? (ii) How far does the point particle move? (iii) How much work was done on the particle? (iv) What is the change in translational kinetic energy of this system? **(b)** The following questions apply to the system modeled as an extended object. (1) How much work is done on the right-hand block? (2) How much work is done on the left-hand block? (3) What is the change of the total energy of this system? **(c)** Combine the results of both models to answer the following questions. (1) Assuming that the object does not get hot, what is the final value of $K_{vib} + U_{spring}$ for the extended system? (2) If the spring stiffness is 50 N/m, what is the final value of the vibrational kinetic energy?

••**P39** You hold up an object that consists of two blocks at rest, each of mass $M = 5$ kg, connected by a low-mass spring. Then you suddenly start applying a larger upward force of constant magnitude $F = 167$ N (which is greater than $2Mg$). Figure 9.60 shows the situation some time later, when the blocks have moved upward, and the spring stretch has increased.

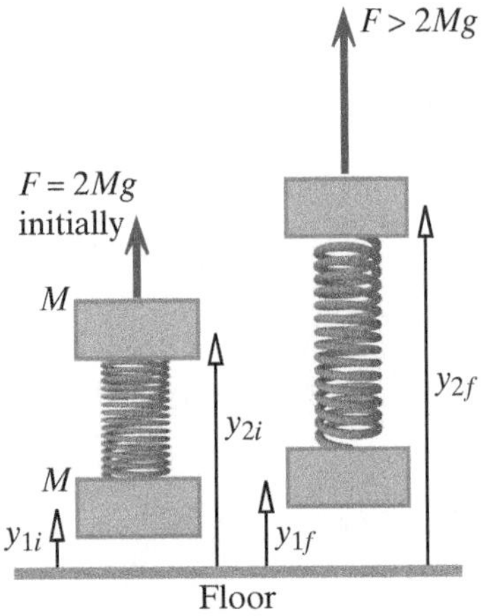

Figure 9.60

The heights of the centers of the two blocks are as follows:

Initial and final positions of block 1: $y_{1i} = 0.3$ m, $y_{1f} = 0.5$ m
Initial and final positions of block 2: $y_{2i} = 0.7$ m, $y_{2f} = 1.2$ m

It helps to show these heights on a diagram. Note that the initial center of mass of the two blocks is $(y_{1i} + y_{2i})/2$, and the final center of mass of the two blocks is $(y_{1f} + y_{2f})/2$. **(a)** Consider the point particle system corresponding to the two blocks and the spring. Calculate the increase in the total translational kinetic energy of the two blocks. It is important to draw a diagram showing all of the forces that are acting, and through what distance each force acts. **(b)** Consider the extended system corresponding to the two blocks and the spring. Calculate the increase of $(K_{vib} + U_s)$, the vibrational kinetic energy of the two blocks (their kinetic energy relative to the center of mass) plus the potential energy of the spring. It is important to draw a diagram showing all of the forces that are acting, and through what distance each force acts.

••**P40** A box contains machinery that can rotate. The total mass of the box plus the machinery is 7 kg. A string wound around the machinery comes out through a small hole in the top of the box. Initially the box sits on the ground, and the machinery inside is not rotating (left side of Figure 9.61). Then you pull upward on the string with a force of constant magnitude 130 N. At an instant when you have pulled 0.6 m of string out of the box (indicated on the right side of Figure 9.61), the box has risen a distance of 0.2 m and the machinery inside is rotating.

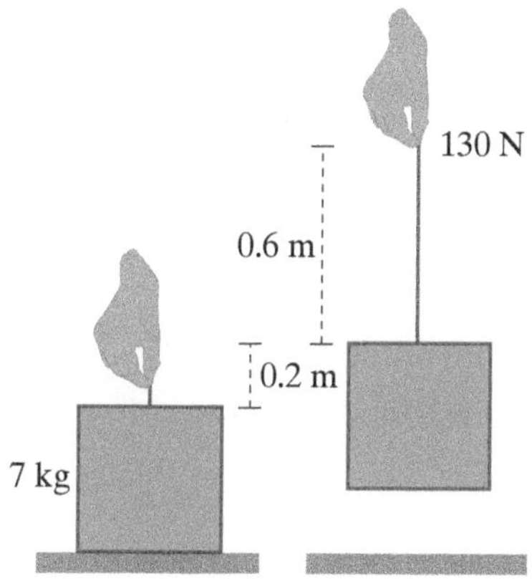

Figure 9.61

POINT PARTICLE SYSTEM (a) List all the forms of energy that change for the point particle system during this process. **(b)** What is the y component of the displacement of the point particle system during this process? **(c)** What is the y component of the net force acting on the point particle system during this process? **(d)** What is the distance through which the net force acts on the point particle system? **(e)** How much work is done on the point particle system during this process? **(f)** What is the speed of the box at the instant shown in the right side of Figure 9.61? **(g)** Why is it not possible to find the rotational kinetic energy of the machinery inside the box by considering only the point particle system?

EXTENDED SYSTEM (h) The extended system consists of the box, the machinery inside the box, and the string. List all the forms of energy that change for the extended system during this process. **(i)** What is the translational kinetic energy of the extended system, at the instant shown in the right side of Figure 9.61? **(j)** What is the distance through which the gravitational force acts on the extended system? **(k)** How much work is done on the system by the gravitational force? **(l)** What is the distance through which your hand moves? **(m)** How much work do you do on the extended system? **(n)** At the instant shown in the right side of Figure 9.61, what is the total kinetic energy of the extended system? **(o)** What is the rotational kinetic energy of the machinery inside the box?

••**P41** String is wrapped around an object of mass M and moment of inertia I (the density of the object is not uniform). With your hand you pull the string straight up with some constant force F such that the center of the object does not move up or down, but the object spins faster and faster (Figure 9.62). This is like a yo-yo; nothing but the vertical string touches the object.

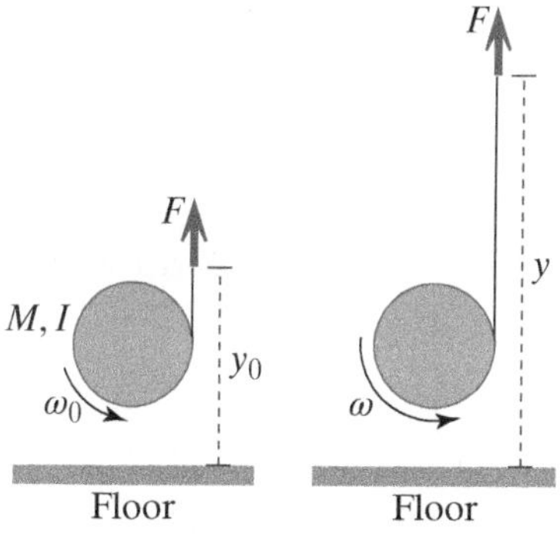

Figure 9.62

When your hand is a height y_0 above the floor, the object has an angular speed ω_0. When your hand has risen to a height y above the floor, what is the angular speed ω of the object?

Your result should not contain F or the (unknown) radius of the object. Explain the physics principles you are using.

••**P42** String is wrapped around an object of mass 1.2 kg and moment of inertia 0.0015 kg · m^2 (the density of the object is not uniform). With your hand you pull the string straight up with some constant force F such that the center of the object does not move up or down, but the object spins faster and faster (Figure 9.62). This is like a yo-yo; nothing but the vertical string touches the object. When your hand is a height $y_0 = 0.25$ m above the floor, the object has an angular speed $\omega_0 = 12$ rad/s. When your hand has risen to a height $y = 0.35$ m above the floor, what is the angular speed ω of the object? Your answer must be numeric and not contain the symbol F.

••**P43** A string is wrapped around a uniform disk of mass M and radius R. Attached to the disk are four low-mass rods of radius b, each with a small mass m at the end (Figure 9.63).

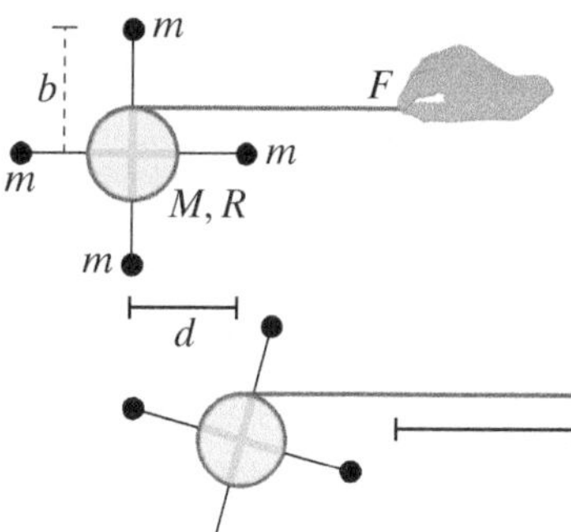

Figure 9.63

The apparatus is initially at rest on a nearly frictionless surface. Then you pull the string with a constant force F. At the instant when the center of the disk has moved a distance d, an additional length w of string has unwound off the disk. **(a)** At this instant, what is the speed of the center of the apparatus? Explain your approach. **(b)** At this instant, what is the angular speed of the apparatus? Explain your approach.

••**P44** A string is wrapped around a uniform disk of mass $M = 1.2$ kg and radius $R = 0.11$ m (Figure 9.63). Attached to the disk are four low-mass rods of radius $b = 0.14$ m, each with a small mass $m = 0.4$ kg at the end. The device is initially at rest on a nearly frictionless surface. Then you pull the string with a constant force $F = 21$ N. At the instant that the center of the disk has moved a distance $d = 0.026$ m, an additional length $w = 0.092$ m of string has unwound off the disk. **(a)** At this instant, what is the speed of the center of the apparatus? Explain your approach. **(b)** At this instant, what is the angular speed of the apparatus? Explain your approach.

••**P45** A rod of length L and negligible mass is attached to a uniform disk of mass M and radius R (Figure 9.64). A string is wrapped around the disk, and you pull on the string with a constant force F. Two small balls each of mass m slide along the rod with negligible friction. The apparatus starts from rest, and when the center of the disk has moved a distance d, a length of string s has come off the disk, and the balls have collided with the ends of the rod and stuck there. The apparatus slides on a nearly frictionless table. Here is a view from above:

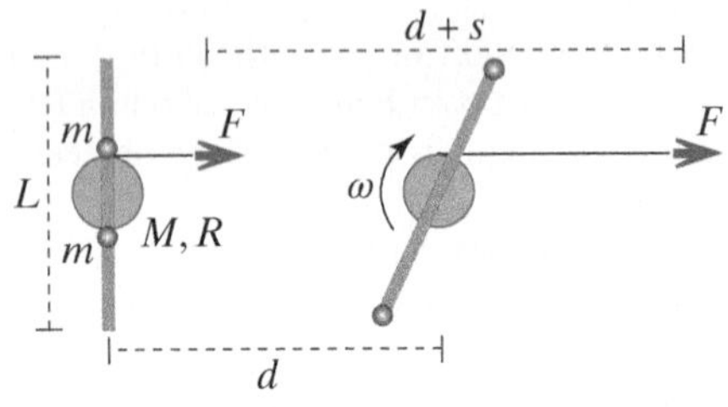

Figure 9.64

(a) At this instant, what is the speed v of the center of the disk? **(b)** At this instant the angular speed of the disk is ω. How much internal energy change has there been?

Section 9.4

•••**P46** It is sometimes claimed that friction forces always slow an object down, but this is not true. If you place a box of mass M on a moving horizontal conveyor belt, the friction force of the belt acting on the bottom of the box speeds up the box. At first there is some slipping, until the speed of the box catches up to the speed v of the belt. The coefficient of friction between box and belt is μ. **(a)** What is the distance d (relative to the floor) that the box moves before reaching the final speed v? Use energy arguments, and explain your reasoning carefully. **(b)** How much time does it take for the box to reach its final speed? **(c)** The belt and box of course get hot. Is the effective distance through which the friction force acts on the box greater than or less than d? Give as quantitative an argument as possible. You can assume that the process is quick enough that you can neglect transfer of energy Q due to a temperature difference between the belt and the box. Do not attempt to use the *results* of the friction analysis in this chapter; rather, apply the *methods* of that analysis to this different situation. **(d)** Explain the result of part (c) qualitatively from a microscopic point of view, including physics diagrams.

ANSWERS TO CHECKPOINTS

1 39.2 J
2 (a) 61.3 J; **(b)** 99.7 J
3 316 J
4 (a) v/d; **(b)** $\frac{1}{2}(\frac{2}{5}MR^2)(v/d)^2$; **(c)** $\frac{1}{2}(Md^2 + \frac{2}{5}MR^2)(v/d)^2$

5 (a) 167 N; **(b)** zero displacement, so no work done on the extended system, but on the point particle system there is 2500 J of work done; **(c)** internal energy (especially chemical energy) decreases by 2500 J

CHAPTER

10

Collisions

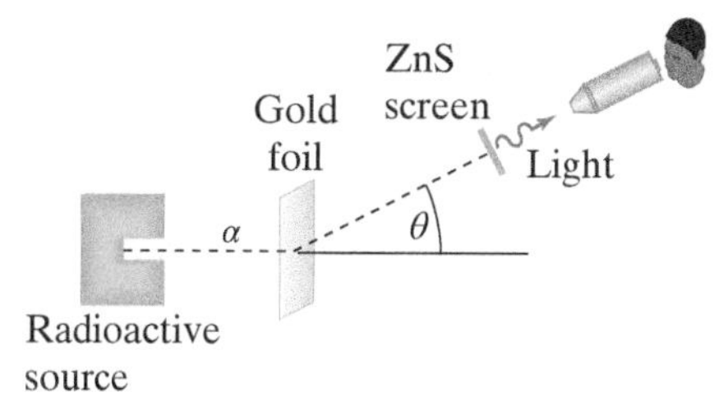

OBJECTIVES

After studying this chapter, you should be able to

- Compare the information provided by an analytical treatment of a collision to the information obtained from an iterative computational model of the process.
- Distinguish between elastic and inelastic collisions.
- Analyze both elastic and inelastic collisions using a combination of the Energy Principle and the Momentum Principle.
- Analyze collisions in multiple reference frames.

10.1 COLLISIONS

Collisions are common events. Neutral molecules in a gas frequently collide with each other—the interatomic electric force is effectively short range, so there is little interaction until the molecules come nearly into contact. Similarly, collisions between macroscopic objects are quite common: before and after a ball hits a wall it has little interaction with the wall, and the contact between ball and wall lasts a very short time.

Colliding objects need not come into actual contact during the collision, if they interact via long-range forces. Two magnets can repel each other without touching, yet we can treat the interaction between fast-moving magnets as a "collision." A collision is any process in which there is little interaction before and after a short time interval, and large interactions during that short time interval. Since the electric force acts at a distance, a moving electron can collide with a moving proton without physical contact, as shown in Figure 10.1.

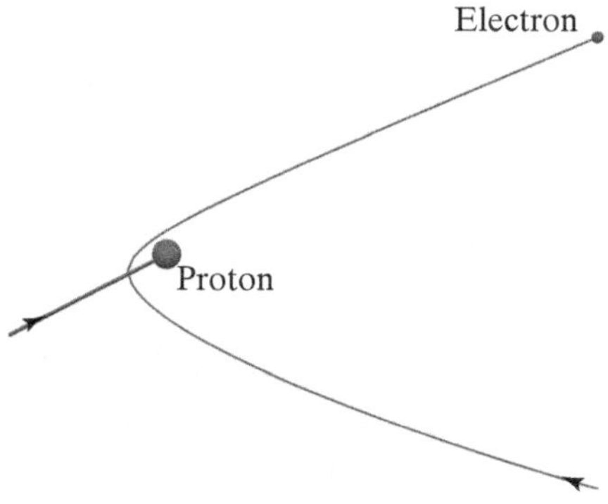

Figure 10.1 A collision between moving charged particles, such as a proton and an electron, need not involve actual contact.

In Chapters 2 and 3 we applied the Momentum Principle to collisions, interactions involving large forces acting for a short time. Many kinds of collisions cannot be fully analyzed without also using the Energy Principle, and this chapter deals with such situations. We will also see that depending on the information we have and on what we want to know about a collision, an analytical solution may not be possible at all, and we will need to model the process computationally.

Choice of System: Both Objects

In analyzing a collision, we typically choose both colliding objects as the system. With this choice of system, the large forces between the colliding objects are internal forces. Because these internal forces are very large—much larger than forces due to objects in the surroundings—the effect of external forces is negligible during the collision. Since a collision occurs during a brief time, energy flow due to any temperature differences between system and surroundings is negligible also. Because we can neglect interactions with the surroundings, both the total momentum and the total energy of the system must remain constant during the collision, although both momentum and energy can flow between the objects in the system.

$$\Delta\vec{p}_{\rm sys} + \cancel{\Delta\vec{p}_{\rm surr}} = 0$$
$$\Delta\vec{p}_{\rm sys} = 0$$
$$\vec{p}_f = \vec{p}_i$$

$$\Delta E_{\rm sys} + \cancel{\Delta E_{\rm surr}} = 0$$
$$\Delta E_{\rm sys} = 0$$
$$E_f = E_i$$

External Forces Are Negligible

In Chapter 2 we examined the forces and interaction times involved in two collisions. When a hockey stick struck a hockey puck hard enough that the stick broke, the force exerted by the stick on the puck was about 1000 N, and the interaction lasted only about 0.005 s. (For comparison, the gravitational force by the Earth on a 160 g hockey puck is only about 1.6 N.) We also analyzed a head-on collision between two running students; we found the interaction time to be about 0.02 s, and the force exerted by one student on the other to be about 21000 N—35 times the weight of one student. Such brief, intense interactions are characteristic of collisions.

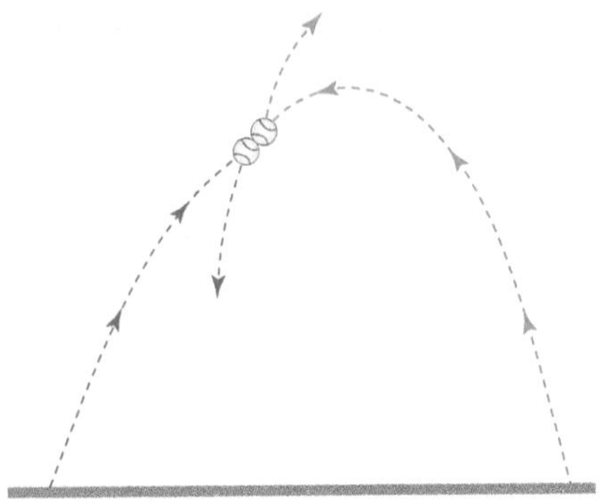

Figure 10.2 Midair collision between two baseballs.

These characteristics justify an approach to the analysis of collisions that proves fruitful. Up until the collision, and after the collision, we do need to take into account external forces, which affect the total momentum of the system. For example, consider the collision of two baseballs in midair, shown in Figure 10.2. We can tell from the curving trajectories of the balls that the gravitational force by the Earth had a significant effect on the balls before and after the collision. However, during the very short time of the collision, the Earth's gravitational force had a negligible effect, and there was little time for energy transfer from the balls to the air due to a temperature difference, so both the total momentum and total energy of the system were nearly constant during the impact.

Limitations of This Approach

In the following sections we'll see that applying the Momentum Principle and the Energy Principle will allow us to analyze several interesting types of collisions using an analytical (algebraic) approach. However, not all problems can be solved this way. We'll explore the power and the limitations of this approach in a later section.

10.2 ELASTIC AND INELASTIC COLLISIONS

Often we want to focus on the changes in translational kinetic energy that occur during a collision between two objects. We will use the term "internal energy" to refer to the other kinds of energy objects may have—for example, electronic energy, vibrational energy, rotational energy, or the randomized thermal energy of a solid object whose atoms and chemical bonds are vibrating.

Elastic Collisions: Internal Energy Doesn't Change

We call a collision "elastic" if there is no change in internal energy of the interacting objects: no thermal energy rise (associated with a temperature increase), no springs newly compressed, no lasting deformations, no new rotations or vibrations, and so on.

ELASTIC COLLISION

In an elastic collision, the internal energy of the objects in the system does not change: $\Delta E_{\text{int}} = 0$. Since no kinetic energy is converted to E_{int}, in an elastic collision $K_f = K_i$. In any collision $\vec{p}_f = \vec{p}_i$, because external forces are negligible.

When atomic systems with quantized energies collide, the collision will be perfectly elastic if there is insufficient energy available to raise the systems to excited quantum states (Figure 10.3). For example, as discussed in Chapter 8, the energy of the first excited electronic state of a mercury atom is 4.9 eV above the ground state. If an electron whose kinetic energy is only 3 eV collides with a ground-state mercury atom, the collision will be elastic, because there is not enough energy available to change the internal energy of the mercury atom.

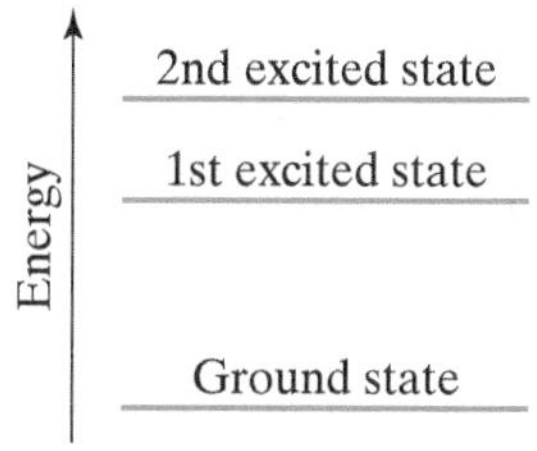

Figure 10.3 Energy levels in an atomic system. If the energy input is too small to reach an excited state, the energy cannot be absorbed.

With macroscopic systems there are no perfectly elastic collisions, because there is always some dissipation, but many macroscopic collisions are very nearly elastic. For example, collisions between billiard balls or steel balls can be nearly elastic.

Carts used to demonstrate collisions in physics lectures often contain magnets to repel each other without actually making contact, and such collisions are nearly elastic. Collisions between carts that interact through soft springs may also be nearly elastic.

Inelastic Collisions: Internal Energy Does Change

We call a collision "inelastic" if it isn't elastic; that is, there is some change in the internal energy of the colliding objects. They get hot, or deform, or rotate, or vibrate, and so on.

INELASTIC COLLISION

In an inelastic collision, the internal energy of the objects in the system changes: $\Delta E_{\text{int}} \neq 0$. Since some kinetic energy is converted to internal energy, in an inelastic collision $K_f \neq K_i$. In any collision $\vec{p}_f = \vec{p}_i$, because external forces are negligible.

As stated above, collisions of macroscopic objects are always at least a bit inelastic, but in many common situations the collision may be nearly elastic. When atomic systems collide, one or more of the colliding objects may undergo a change of quantized energy level, and we call that an inelastic collision. For example, if an electron with kinetic energy 5 eV collides with a mercury atom in its ground state, the mercury atom may be excited to a higher electronic state 4.9 eV above the ground state, leaving only 0.1 eV of kinetic energy in the system.

Maximally Inelastic Collisions ("Sticking Collisions")

The most extreme inelastic collision is the "maximally inelastic" collision, in which there is maximum dissipation. That doesn't necessarily mean that the objects stop dead, because momentum must be conserved. It can be shown that a collision in which the objects stick together is a maximally inelastic collision: the only remaining kinetic energy is present only because the total momentum can't change.

The simplest example of a maximally inelastic collision is that of two objects that have equal and opposite momenta, so that before the collision the total momentum of the combined system is zero. The momentum afterward must also be zero, and if they are stuck together that means that they must be at rest.

Identifying Inelastic Collisions

When deciding whether or not it is reasonable to treat a collision as approximately elastic, one can look for indications of inelasticity, such as:

- Objects are stuck together after the collision.
- An object is deformed after the collision.
- Objects are hotter after the collision.
- There is more vibration or rotation after the collision.
- An object is in an excited state after the collision.

10.3 A HEAD-ON COLLISION OF EQUAL MASSES

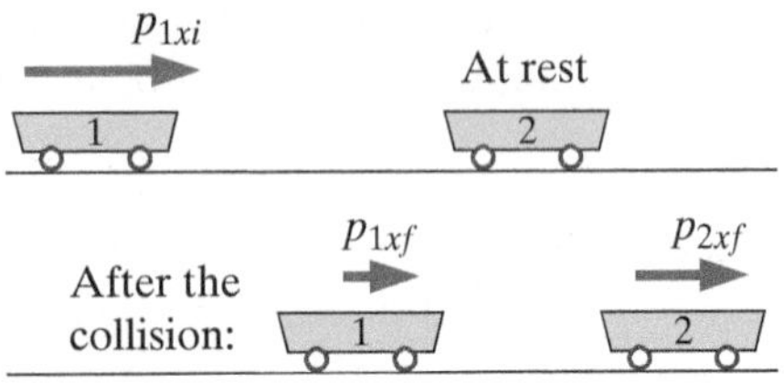

Figure 10.4 Cart 1 runs into stationary cart 2. The carts have equal masses. Friction and air resistance are negligible.

We will start by considering simple 1-D (head-on) collisions in order to see clearly how to apply both the Energy Principle and the Momentum Principle. Consider a head-on collision between two carts rolling or sliding on a track with low friction (Figure 10.4). Similar situations are billiard balls or hockey pucks or vehicles hitting each other head-on. Cart 1 with mass m moves to the right with x momentum p_{1xi}, and it runs into cart 2, which has the same mass m and is initially sitting still.

Taking the two carts as the system, and neglecting the small frictional force exerted by the track, all that the Momentum Principle can tell us is that after the collision the total final x momentum $p_{1xf}+p_{2xf}$ must equal the initial total x momentum p_{1xi}. By itself, the Momentum Principle doesn't tell us how this momentum will be divided between the two carts. We need to consider conservation of energy as well as conservation of momentum in order to predict what will happen.

Before the collision, nonzero energy terms included the kinetic energy of cart 1, K_{1i}, and the internal energies of both carts. After the collision, there is internal energy of both carts and kinetic energy of both carts, $K_{1f}+K_{2f}$.

> QUESTION What types of energy may have changed as a result of the collision?

K_1+K_2 might change, and internal energy may have changed. The temperature of the carts might be a bit higher now, with some of the initial kinetic energy having been dissipated into thermal energy. How much energy dissipation occurs depends on the details of the contact between the two interacting carts.

EXAMPLE **Elastic Collision of Two Identical Carts**

Cart 1 collides with stationary cart 2, which is identical (Figure 10.4). Suppose that the collision is (nearly) elastic, as it will be if the carts repel each other magnetically or interact through soft springs. In this case there is no change of internal energy. What are the final momenta of the two carts?

Solution Since the y and z components of momentum don't change, we can work with only x components.

System: Both carts
Surroundings: Earth, track, air (neglect friction and air resistance)

Initial situation: Just before collision
Final situation: Just after collision

Momentum Principle (negligible effect of surroundings; only x components change):

$$p_{xf} = p_{xi} + F_{\text{net},x}\Delta t$$
$$p_{1xf} + p_{2xf} = p_{1xi} + 0$$

Energy Principle (negligible effect of surroundings; negligible change in internal energy of the carts):

$$E_f = E_i + W + Q$$
$$(K_{1f} + \cancel{E_{\text{int}1f}}) + (K_{2f} + \cancel{E_{\text{int}2f}}) = (K_{1i} + \cancel{E_{\text{int}1i}}) + (K_{2i} + \cancel{E_{\text{int}2i}}) + 0 + 0$$
$$K_{1f} + K_{2f} = K_{1i} + 0$$

Since $K = p^2/2m$, we can combine the momentum and energy equations:

$$\frac{p_{1xf}^2}{2m} + \frac{p_{2xf}^2}{2m} = \frac{(p_{1xf} + p_{2xf})^2}{2m}$$
$$p_{1xf}^2 + p_{2xf}^2 = p_{1xf}^2 + 2p_{1xf}p_{2xf} + p_{2xf}^2$$
$$2p_{1xf}p_{2xf} = 0$$

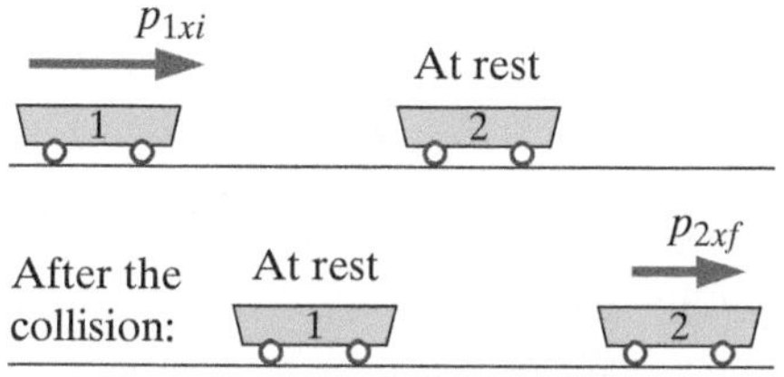

Figure 10.5 If the collision is elastic, cart 1 stops, and cart 2 has all the momentum cart 1 used to have.

There are two possible solutions to this equation. The term $p_{1xf}p_{2xf}$ can be zero if $p_{1xf} = 0$ or if $p_{2xf} = 0$.

If $p_{1xf} = 0$, the physical situation is that cart 1 came to a complete stop. In that case we see from the momentum equation $p_{1xf} + p_{2xf} = p_{2xf} = p_{1xi}$ that cart 2 now has the same momentum that cart 1 used to have. There has been a complete transfer of momentum from cart 1 to cart 2. There has also been a complete transfer of kinetic energy from cart 1 to cart 2. See Figure 10.5.

If $p_{2xf} = 0$, what is the physical situation? In that case $p_{1xf} + p_{2xf} = p_{1xf} = p_{1xi}$, and cart 1 just keeps going, missing cart 2! Of course that won't happen if the carts are on the same narrow track, but the algebra doesn't know that.

It is not possible for both final momenta to be zero, since the total final momentum of the system must equal the nonzero total initial momentum of the system.

EXAMPLE

Maximally Inelastic Collision of Two Identical Carts

Consider the opposite extreme—a maximally inelastic collision of the two identical carts, one initially at rest. That means the carts stick together (perhaps they have sticky material on their ends), and each has the same final momentum $p_{1xf} = p_{2xf}$ (Figure 10.6). **(a)** Find the final momentum, final speed, and final kinetic energy of the carts in terms of their initial values. **(b)** What is the change in internal energy of the two carts?

Solution Since the y and z components of momentum don't change, we can work with only x components.

System: Both carts
Surroundings: Earth, track, air (neglect friction and air resistance)

(a) Momentum Principle (x components):

$$p_{1xf}+p_{2xf}=p_{1xi}$$
$$2p_{1xf}=p_{1xi}$$
$$p_{1xf}=\frac{1}{2}p_{1xi}$$

The final speed of the stuck-together carts is half the initial speed:

$$v_f=\frac{1}{2}v_i.$$

Final translational kinetic energy:

$$(K_{1f}+K_{2f})=2\left(\frac{1}{2}mv_f^2\right)$$
$$(K_{1f}+K_{2f})=2\left(\frac{1}{2}m\left(\frac{1}{2}v_i\right)^2\right)=\frac{1}{4}mv_i^2$$
$$(K_{1f}+K_{2f})=\frac{K_{1i}}{2}$$

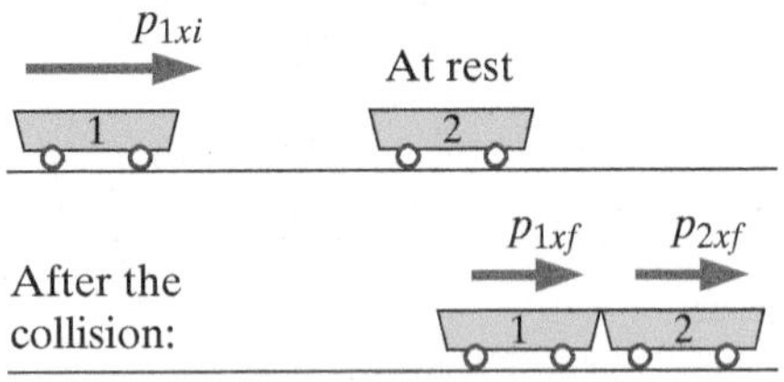

Figure 10.6 In a maximally inelastic collision, cart 1 sticks to cart 2, so the carts have the same final momentum.

(b) Energy Principle:

$$K_{1f}+K_{2f}+E_{\text{int},f}=K_{1i}+E_{\text{int},i}$$
$$E_{\text{int},f}-E_{\text{int},i}=K_{1i}-(K_{1f}+K_{2f})$$
$$\Delta E_{\text{int}}=K_{1i}-\frac{K_{1i}}{2}$$
$$\Delta E_{\text{int}}=\frac{K_{1i}}{2}$$

The Momentum Principle is still valid even though the collision is inelastic. Fundamental principles apply in all situations. The final kinetic energy of the system is only half of the original kinetic energy, which means that the other half of the original kinetic energy has been dissipated into increased internal energy ΔE_{int} of the two carts.

In both cases (elastic collision and inelastic collision) we needed to apply both the Momentum Principle and the Energy Principle to find all the unknown quantities. One equation alone was not enough.

We have shown the two extremes of energy conservation in the head-on collision of the moving cart with the stationary cart:

- If the collision is elastic, cart 1 stops and cart 2 moves with the speed cart 1 used to have.
- If the collision is maximally inelastic, the carts stick together and move with half the original speed. Half of the original kinetic energy is dissipated into increased internal energy.

Between these two extremes there can be an inelastic but not maximally inelastic collision, in which some amount of the original kinetic energy (less than half) is dissipated into increased internal energy.

Checkpoint 1 A 6 kg mass traveling at speed 10 m/s strikes a stationary 6 kg mass head-on, and the two masses stick together. **(a)** What was the initial total kinetic energy? **(b)** What is the final speed? **(c)** What is the final total kinetic energy? **(d)** What was the increase in internal energy of the two masses?

10.4 HEAD-ON COLLISIONS BETWEEN UNEQUAL MASSES

Suppose that a Ping-Pong ball hits a stationary bowling ball head-on, nearly elastically. Imagine that this takes place in outer space or floating in an orbiting spacecraft, so there is no friction—no significant external forces act on the combined system during the collision.

QUESTION What do you expect? How will the Ping-Pong ball move after hitting the bowling ball? How will the bowling ball move after the collision?

You'll see the Ping-Pong ball bounce straight back with very little change of speed (Figure 10.7), and we'll show this is consistent with conservation of momentum and conservation of energy. Less obvious is that the bowling ball does move, although very slowly.

EXAMPLE **A Ping-Pong Ball Hits a Stationary Bowling Ball Head-On**

In an orbiting spacecraft a Ping-Pong ball of mass m (object 1) traveling in the $+x$ direction with initial momentum $\vec{p}_{1i}$ hits a stationary bowling ball of mass M (object 2) head on, as shown in Figure 10.7. What are the **(a)** momentum, **(b)** speed, and **(c)** kinetic energy of each object after the collision? Assume little change in the speed of the Ping-Pong ball, and assume that the collision is elastic.

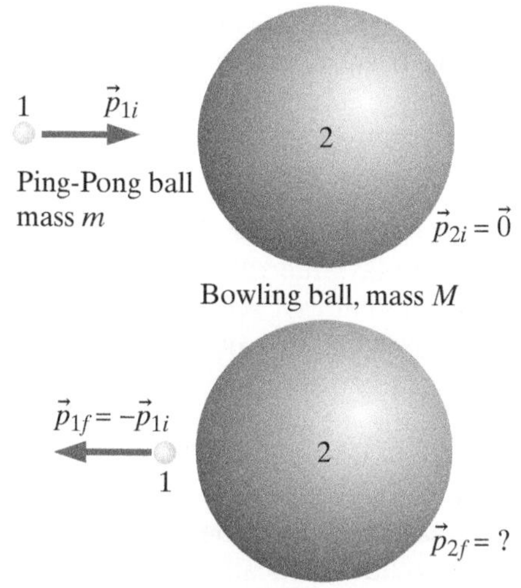

Figure 10.7 A Ping-Pong ball hits a bowling ball head-on and bounces off elastically.

Solution System: Ping-Pong ball and bowling ball

Surroundings: Nothing that exerts significant forces

Momentum Principle:

$$\vec{p}_{1f}+\vec{p}_{2f}=\vec{p}_{1i}+\vec{p}_{2i}$$

Assume that the speed of the Ping-Pong ball does not change significantly in the collision, so $\vec{p}_{1f}\approx-\vec{p}_{1i}$.

$$-\vec{p}_{1i}+\vec{p}_{2f}=\vec{p}_{1i}$$
$$\vec{p}_{2f}=2\vec{p}_{1i}$$

(a) The final momentum of the bowling ball is twice the initial momentum of the Ping-Pong ball. It may be surprising that the bowling ball ends up with about twice the momentum of the Ping-Pong ball. One way to understand this is that the final momentum of the Ping-Pong ball is approximately $-\vec{p}_{1i}$, so the change in the Ping-Pong ball's momentum is approximately

$$-\vec{p}_{1i}-\vec{p}_{1i}=-2\vec{p}_{1i}$$

The Ping-Pong ball's speed hardly changed, but its momentum changed a great deal. Because momentum is a vector, a change of direction is just as much a change as a change of magnitude. This big change is of course due to the interatomic electric contact forces exerted on the Ping-Pong ball by the bowling ball. By reciprocity, the same magnitude of interatomic contact forces are exerted by the Ping-Pong ball on the bowling ball, which undergoes a momentum change of $+2\vec{p}_{1i}$.

(b) Final speed of bowling ball:

$$v_{2f}\approx\frac{p_{2f}}{M}=\frac{2p_{1i}}{M}=\frac{2mv_{1i}}{M}=2\left(\frac{m}{M}\right)v_{1i}$$

This is a very small speed since $m\ll M$. For example, if the mass of the bowling ball is about 5 kg, and the 2 g Ping-Pong ball is initially traveling at 10 m/s, the final speed of the bowling ball will be 0.008 m/s.

(c) Kinetic energies:

$$K_{2f} = \frac{(2p_{1i})^2}{2M} \quad \text{and} \quad K_{1f} = \frac{p_{1i}^2}{2m}$$

Because the mass of the bowling ball is much larger than the mass of the Ping-Pong ball, the kinetic energy of the bowling ball is much smaller than the kinetic energy of the Ping-Pong ball. The kinetic energy of the 2 g Ping-Pong ball, traveling at 10 m/s, is about 0.1 J, while the 5 kg bowling ball has acquired a kinetic energy of 1.6×10^{-4} J—nearly 1000 times less.

Although the kinetic energy of the slowly moving bowling ball is very small, it is not zero, which means that the Ping-Pong ball does lose a little speed in the collision. That's why our assumption that the Ping-Pong ball's speed hardly changes is an approximation.

We made the plausible assumption that the Ping-Pong ball bounced back with nearly its original speed, and showed that this is consistent with momentum conservation if the bowling ball gets twice the momentum of the Ping-Pong ball, and also consistent with energy conservation, because the bowling ball acquires very little kinetic energy. With a lot of algebra, it is also possible to solve the momentum and energy equations exactly for the unknown final momentum of the Ping-Pong ball, and when one makes the approximation that the mass of the bowling ball is very much larger than the mass of the Ping-Pong ball, the results are those given above (see Problem P15).

When a Ping-Pong ball hits a stationary bowling ball head-on, elastically:

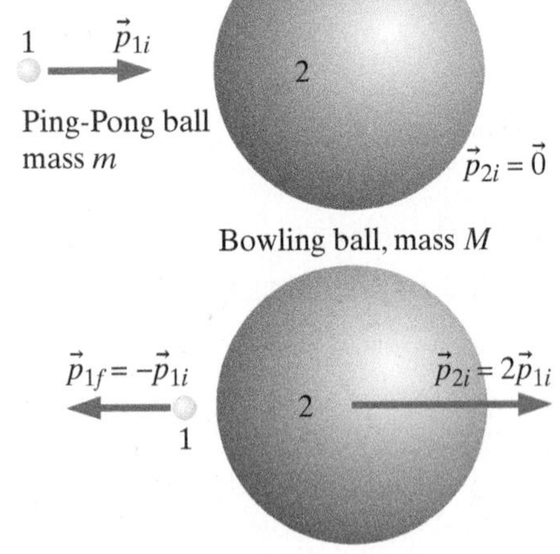

Figure 10.8 The bowling ball gets twice the momentum of the Ping-Pong ball (but its speed and kinetic energy are very small).

- The Ping-Pong ball bounces back with almost the same speed.
- The Ping-Pong ball has a large change of momentum $-2\vec{p}_{1i}$.
- The bowling ball has a large change of momentum $+2\vec{p}_{1i}$ (Figure 10.8).
- The bowling ball's final speed is very small, because it has large mass but comparable momentum.
- The bowling ball's final kinetic energy is very small, because it has large mass but comparable momentum.

Similar situations include a ball bouncing off a wall, or falling and bouncing off the ground, in which case the entire Earth plays the role of the bowling ball. The Earth gets twice the momentum of the ball but an extremely small speed and extremely small kinetic energy. If a bowling ball sits on a table and is hit by a Ping-Pong ball, and the table exerts a large enough frictional force to prevent the bowling bowl from moving, the bowling ball is essentially part of the Earth.

Although we've looked at the very special case of a head-on collision, it is generally true that when low-mass projectiles collide elastically with stationary large masses, the low-mass projectile bounces off with little change of speed and kinetic energy but may undergo a large change of momentum. The large-mass target can pick up lots of momentum, but little speed or kinetic energy.

Checkpoint 2 We can use our results for head-on elastic collisions to analyze the recoil of the Earth when a ball bounces off a wall embedded in the Earth. Suppose that a professional baseball pitcher hurls a baseball ($m = 155$ g) with a speed of 100 mi/h ($v_1 = 44$ m/s) at a wall that is securely anchored to the Earth, and the ball bounces back with little loss of kinetic energy. **(a)** What is the approximate recoil speed of the Earth ($M = 6 \times 10^{24}$ kg)? **(b)** calculate the approximate recoil kinetic energy of the Earth and compare it to the kinetic energy of the baseball. The Earth gets lots of momentum (twice the momentum of the baseball) but very little kinetic energy.

10.5 FRAME OF REFERENCE

A frame of reference is a perspective from which a system is observed. A frame of reference may be stationary or it may move at constant velocity. (If the velocity of a reference frame is not constant, the frame is "noninertial," and the laws of physics as we have stated them do not apply. We'll stick to inertial frames in this textbook.) In Chapter 7 we saw that our choice of reference frame affected energy accounting. It is sometimes significantly simpler to analyze collisions from a moving reference frame.

As a simple example, imagine that you are standing on a street corner and a bus passes you. In your stationary frame of reference, a passenger sitting on the bus is moving with the same velocity as the bus. However, the bus driver may observe the passenger from a moving reference frame "anchored" to the bus. In the moving bus reference frame, the driver sees the passenger sitting still.

QUESTION How does the velocity of an object depend on the frame of reference we choose?

Suppose that the bus mentioned in the previous paragraph is heading in the $+x$ direction at a speed of 14 m/s. A passenger on the bus walks toward the front of the bus at a speed of 1.2 m/s. To the bus driver, the velocity of the passenger is $\langle 1.2, 0, 0\rangle$ m/s. However, as you stand on the street corner, you see the passenger moving past you slightly faster than the sitting passengers on the bus. You observe the walking passenger's velocity to be $\langle 15.2, 0, 0\rangle$ m/s, the sum of the velocity of the bus ($\langle 14, 0, 0\rangle$ m/s) and his velocity relative to the bus ($\langle 1.2, 0, 0\rangle$ m/s).

Evidently, the velocity $\vec{v}$ of an object viewed from a stationary reference frame is the sum of the velocity $\vec{v}_{\text{frame}}$ of a moving reference frame plus the velocity $\vec{v}\,'$ ("v-prime") of the object as viewed from the moving frame.

VELOCITY IN A MOVING REFERENCE FRAME

$$\vec{v} = \vec{v}\,' + \vec{v}_{\text{frame}}$$

where $\vec{v}$ is the velocity in a stationary frame, $\vec{v}_{\text{frame}}$ is the velocity of the moving reference frame, and $\vec{v}\,'$ is the velocity observed in the moving frame.

Changing frames of reference is a very powerful technique that is discussed in more depth later in this chapter. A change of reference frame often makes a problem much simpler to analyze, or reduces the problem to one you've already analyzed.

Bowling Ball Hits Ping-Pong Ball: Change of Reference Frame

In the previous example we analyzed a collision in which a moving Ping-Pong ball hit a stationary bowling ball head-on. What happens when a bowling ball of large mass M moving to the right with speed V hits elastically head-on a stationary Ping-Pong ball of small mass m? What is the final speed of the Ping-Pong ball?

We can perform a change of reference frame that reduces this problem to the previous one. Suppose that this experiment is being done in a spacecraft. Jack and Jill are two crew members in the spacecraft. Jack hangs onto a wall, remaining stationary, and watches the bowling ball go by (Figure 10.9), while Jill launches herself away from the wall, moving to the right with speed V, and glides along beside the bowling ball.

Figure 10.9 In a stationary reference frame, Jack sees a bowling ball with speed V move to the right toward a stationary Ping-Pong ball.

According to the principle of relativity, in any reference frame in uniform motion one can apply physics principles and make correct predictions, even though the details of the calculations look different from the details in some other reference frame in uniform motion. If the walls are dark, and the Ping-Pong ball and bowling ball glow in ultraviolet light, Jill truly cannot tell whether she is moving and the Ping-Pong ball is stationary relative to the walls, or whether she is stationary and the Ping-Pong ball is moving relative to the walls. This is at the heart of the principle of relativity.

QUESTION What is the speed of the bowling ball as measured by Jill?

For Jill, the bowling ball isn't moving. It is always the same distance from her. In Jill's reference frame, the bowling ball is at rest, and the Ping-Pong ball is moving toward her with speed V (Figure 10.10).

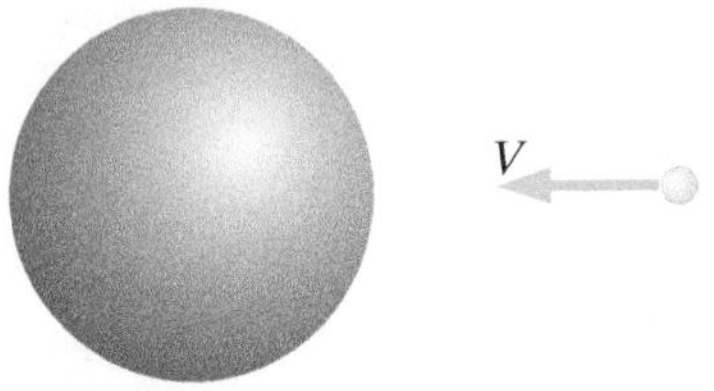

Figure 10.10 Moving with the bowling ball, at speed V, Jill sees the bowling ball at rest and the Ping-Pong ball moving toward her with speed V.

QUESTION What do Jack and Jill see during and after the collision?

Jill sees the moving Ping-Pong ball bounce off the stationary bowling ball. She observes the Ping-Pong ball traveling to the right with speed $\approx V$ (Figure 10.11). To Jill the process looks exactly the same as in the example in the previous section (A Ping-Pong Ball Hits a Stationary Bowling Ball Head-On).

Jack, however, sees the moving bowling ball hit the stationary Ping-Pong ball. After the collision he sees the bowling ball still moving to the right with speed $\approx V$, but he sees the Ping-Pong ball moving to the right with speed $\approx 2V$ (Figure 10.12).

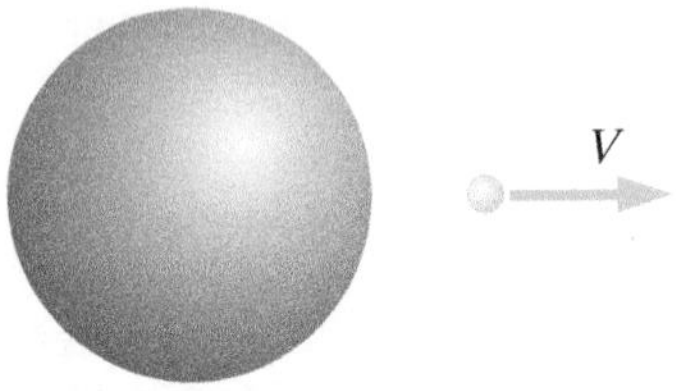

Figure 10.11 Jill (in the moving frame of reference) sees the moving Ping-Pong ball hit the bowling ball and bounce off. After the collision, she sees the Ping-Pong ball moving to the right with nearly its original speed V.

QUESTION How does this emerge from changing reference frames?

We need to apply the equation $\vec{v} = \vec{v}\,' + \vec{v}_{\text{frame}}$ to the velocities of both balls: $\vec{v}_B$ (bowling ball) and $\vec{v}_P$ (Ping-Pong ball). After the collision:

$$\begin{aligned}\vec{v}_B &= \vec{v}_B{}' + \vec{v}_{\text{frame}}\\ &= \langle 0,0,0\rangle + \langle V,0,0\rangle\\ &= \langle V,0,0\rangle\\ \vec{v}_P &= \vec{v}_P{}' + \vec{v}_{\text{frame}}\\ &= \langle V,0,0\rangle + \langle V,0,0\rangle\\ &= \langle 2V,0,0\rangle\end{aligned}$$

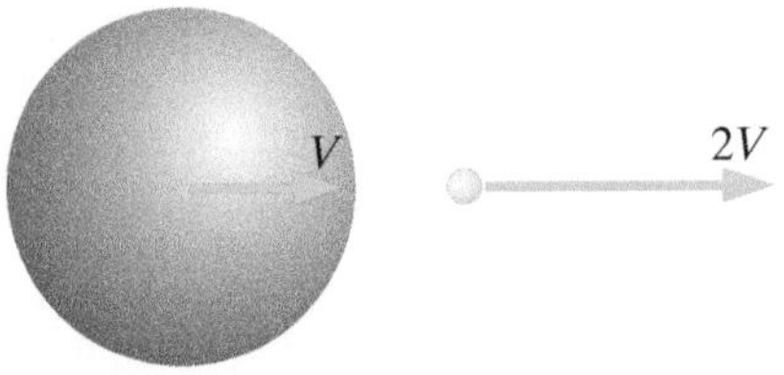

Figure 10.12 In Jack's stationary reference frame, the Ping-Pong ball's speed after the collision is nearly $2V$.

By transforming into a moving reference frame and back again, we were able to show that when a moving bowling ball hits a stationary Ping-Pong ball head-on, elastically, the Ping-Pong ball ends up moving with nearly twice the speed of the bowling ball. The change in speed of the bowling ball is insignificant. It just plows straight ahead, having lost very little kinetic energy to the Ping-Pong ball. The small amount of momentum lost by the bowling ball was gained by the Ping-Pong ball ($2mV$).

Checkpoint 3 A uranium atom traveling at speed 4×10^4 m/s collides elastically with a stationary hydrogen molecule, head-on. What is the approximate final speed of the hydrogen molecule?

10.6 SCATTERING: COLLISIONS IN 2D AND 3D

Scattering experiments are used to study the structure and behavior of atoms, nuclei, and other small particles. In a scattering experiment, a beam of particles collides with other particles. In atomic or nuclear collisions, we can't observe in detail the curving trajectories inside the tiny interaction region. We only observe the trajectories before and after the collision, when the particles are far apart and their mutual interaction is very weak, so they are traveling in nearly straight lines. An example of scattering is the collision of an alpha particle (helium nucleus) with the nucleus of a gold atom.

Momentum and Energy Equations

An alpha particle collides with a gold nucleus, and is deflected through some final angle θ. The gold nucleus recoils at some other final angle ϕ. The final angles are measured relative to the initial direction of the alpha particle (Figures 10.13 and 10.14). We say that the alpha particle "scattered" through an angle θ.

We take the "before" time early enough and the "after" time late enough that the two particles are sufficiently far away from each other that we can neglect their electric potential energy. As for conservation of momentum, we must write two equations, for the x and y components of momentum conservation.

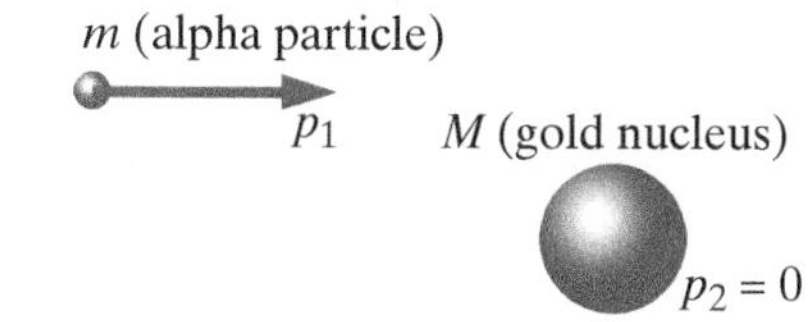

Figure 10.13 Before the collision, the alpha particle (mass m) approaches the gold nucleus (mass M).

QUESTION What are the momentum and the energy conservation equations for this system?

All the speeds are small compared to the speed of light. Choose x in the direction of the incoming alpha particle, and y at right angles to it. Consider ϕ to be a positive angle. Write component equations in terms of the magnitudes of the momenta, p_1, p_3, and p_4; the gold nucleus is initially at rest.

$$p_{xi} = p_{xf}$$
$$p_1 = p_3 \cos\theta + p_4 \cos\phi$$

$$p_{yi} = p_{yf}$$
$$0 = p_3 \cos(90° - \theta) + p_4 \cos(90° + \phi)$$

$$K_f = K_i$$
$$\frac{p_1^2}{2m} = \frac{p_3^2}{2m} + \frac{p_4^2}{2M}$$

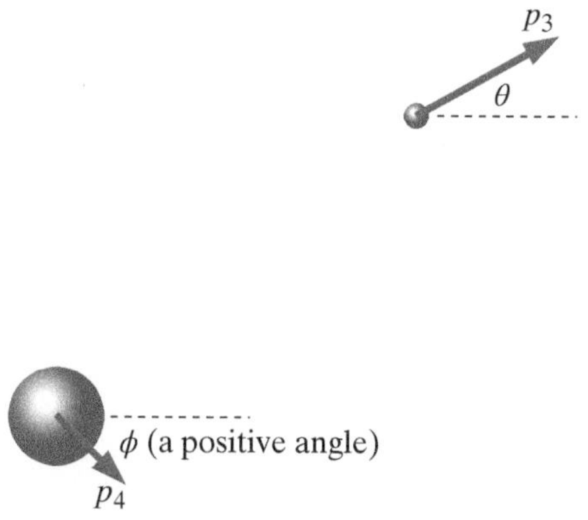

Figure 10.14 After the collision, when the particles are far apart and moving with nearly uniform velocities. Final angles are measured relative to the initial direction of the alpha particle.

Note that in the equations above, direction cosines are used to express vector components in terms of the magnitudes of momenta and the angles to the x axis. For example, the final momentum of the gold nucleus is $\vec{p}_4 = |\vec{p}_4|\langle \cos\phi, \cos(90° + \phi), 0\rangle$.

QUESTION Which quantities in these three equations are unknown? Can we solve for the unknown quantities?

We know the masses of the alpha particle and the gold nucleus, and we know the initial kinetic energy of the alpha particle (so we know its speed and magnitude of momentum). You should have identified the two final momenta and the two final angles as unknown quantities (p_3, p_4, θ, and ϕ, all of which represent positive numbers). But if there are four unknowns and only three equations, how can we solve for the unknowns? The simple answer is, we can't!

Nevertheless, there are many situations in which these three equations can be exploited. For example, if we measure the final direction of motion of the alpha particle, the angle θ, we can solve for the other three unknown quantities and thereby predict the final momenta p_3 and p_4 of the two particles and the direction of motion ϕ of the gold nucleus, even without measuring all of these quantities.

It is worth mentioning that in experimental particle physics the momentum is often measured in a very direct way. Charged particles curve in a magnetic field due to magnetic forces that act on moving charges, and the radius of curvature is a direct measure of the momentum of the particle, if you know how much charge it has, which you do know if you know what kind of particle it is. Usually the momentum of a particle can be measured more accurately than the velocity, though once you have measured the momentum you can deduce the velocity if you know what kind of particle it is, since in that case you know its mass.

EXAMPLE

Elastic Collision: Identical Particles, One Initially at Rest

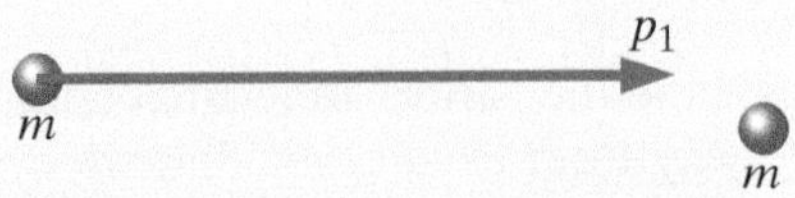

Figure 10.15 Before a collision of two identical particles, one initially at rest.

An interesting special case of an elastic collision that can be analyzed this way is one involving particles of equal mass, such as an elastic collision between two alpha particles, or two gold nuclei, or two billiard balls (whose collision is nearly elastic), in which one of the two objects (the "target") is initially at rest (Figure 10.15 and Figure 10.16). In this special case, show that the angle A must be 90° if the speeds are small compared to the speed of light.

Solution

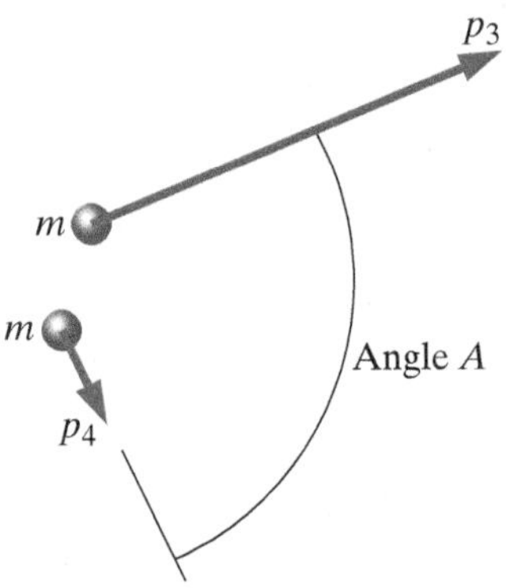

Figure 10.16 After the collision of two identical particles, one initially at rest.

Because the vector momentum of the system is constant, we can write $\vec{p}_1 = \vec{p}_3 + \vec{p}_4$, then calculate p_1^2 by expanding the dot product:

$$p_1^2 = \vec{p}_1 \bullet \vec{p}_1 = (\vec{p}_3 + \vec{p}_4) \bullet (\vec{p}_3 + \vec{p}_4) = p_3^2 + p_4^2 + 2p_3p_4 \cos A$$

Divide by $2m$:

$$\frac{p_1^2}{2m} = \frac{p_3^2}{2m} + \frac{p_4^2}{2m} + \frac{2p_3p_4 \cos A}{2m}$$

Since the kinetic energy K at low speed can be written as $p^2/2m$,

$$K_1 = K_3 + K_4 + \frac{2p_3p_4 \cos A}{2m}$$

However, energy conservation for this elastic collision says that $K_1 = K_3 + K_4$, so we have

$$\cos A = 0 \text{ or } A = 90°$$

In this special situation—an elastic collision at low speeds between equal particles, one of which is initially at rest—the angle between the final velocities must be 90° as long as neither p_3 nor p_4 is zero, in which case the equation reduces to $0 \cos A = 0$, and $\cos A$ is indeterminate. If one of the final momenta is indeed zero, the other particle must have all the momentum of the initial particle. This can happen either because the incoming particle missed the target entirely and kept going in the original direction, or because the incoming particle hit the target dead center, in which case the only consistent solution of the equations is for the incoming particle to come to rest and the target particle to move forward with the initial momentum, as we saw in the head-on collision of two carts earlier in the chapter.

Impact Parameter

The distance between centers perpendicular to the incoming velocity is called the "impact parameter" and is often denoted by b. A head-on collision has an impact parameter of zero. Here in Figures 10.17–10.20 are possible elastic collisions between two billiard balls, for various impact parameters:

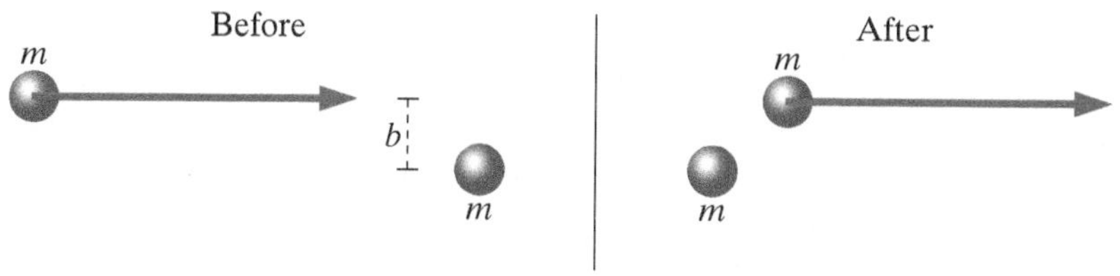

Figure 10.17 If the impact parameter, b, is too large, there is no interaction at all between two billiard balls.

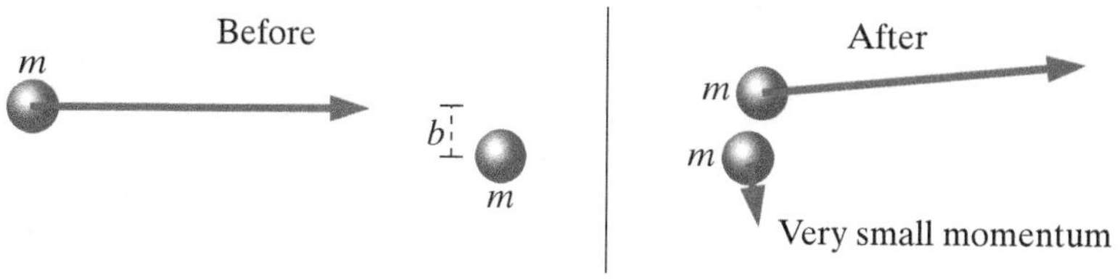

Figure 10.18 A grazing impact results in small-angle scattering.

Smaller-impact parameters lead to larger scattering unless $b = 0$:

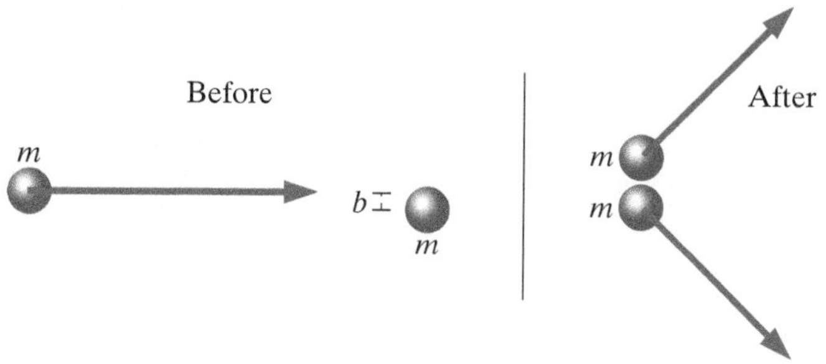

Figure 10.19 A medium-impact parameter results in symmetric scattering.

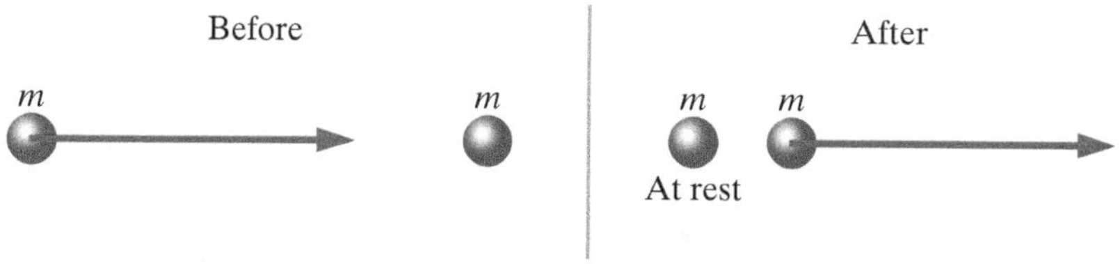

Figure 10.20 In a head-on collision, the impact parameter $b = 0$. In this case all the momentum is transferred to the target.

The smaller the impact parameter, the more severe is the collision, and the larger the deflection angle of the incoming particle (larger "scattering"), except for a head-on collision, where if the masses are equal the incoming ball stops dead and the target ball gets the entire momentum.

10.7 DISCOVERING THE NUCLEUS INSIDE ATOMS

A famous experiment involving collisions led to the discovery of the nucleus inside the atom. It was carried out in 1911 by a group in England led by the New Zealander Ernest Rutherford. In the experiment, high-speed alpha particles (now known to be helium nuclei, consisting of two protons and two neutrons) were shot at a thin gold foil (whose nuclei consist of 79 protons and 118 neutrons).

When Rutherford and his coworkers decided to perform a scattering experiment in order to probe the internal structure of matter, they were familiar with the relationships of momenta in collisions that we have just studied, and they reasoned that by shooting microscopic particles at a thin layer of metal, they might be able to learn something about the microscopic structure of the metal by looking at patterns of scattering.

In the Rutherford experiment, the velocity of the alpha particles was not high enough to allow them to make actual contact with the nucleus. It is nonetheless appropriate to think of the process as a collision, because there are sizable electric forces acting between the alpha particle and the gold nucleus for a very short time. We call any process of this kind a collision, even if there is no direct contact but only gravitational, electric, or magnetic interactions at a distance. Figure 10.21 shows what the trajectories look like. We say that the alpha particle projectile is "scattered" (deflected) by the interaction with the initially stationary gold nucleus target, and such experiments are called "scattering" experiments.

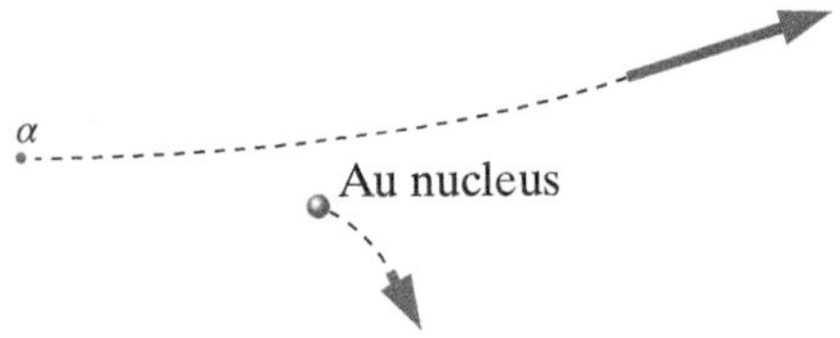

Figure 10.21 A collision that is not head-on.

Elastic Collisions in the Rutherford Experiment

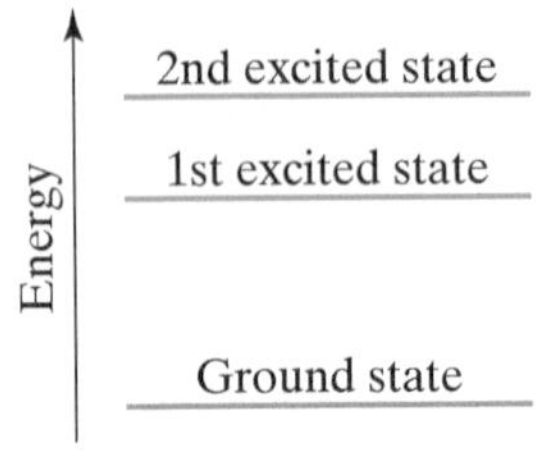

Figure 10.22 Energy levels in a nucleus. The potential energy curve is not shown.

The conditions of this experiment were such that there was no change in the internal energy of either the alpha particle or the gold nucleus (elastic collision). There do exist "excited" states of nuclei, in which there is a change in the configuration of the nucleons corresponding to a higher energy than the ground state, the normal configuration of the nucleus. The internal energy of the nucleus is quantized, and there are no stable or semi-stable states with energy intermediate between these quantized energies (Figure 10.22). In the Rutherford experiment, the interaction did not raise the gold nucleus or the alpha particle to an excited state; both nuclei remained in their ground states. This happy accident made it much easier for Rutherford to analyze the experiment.

Electric Potential Energy in the Rutherford Experiment

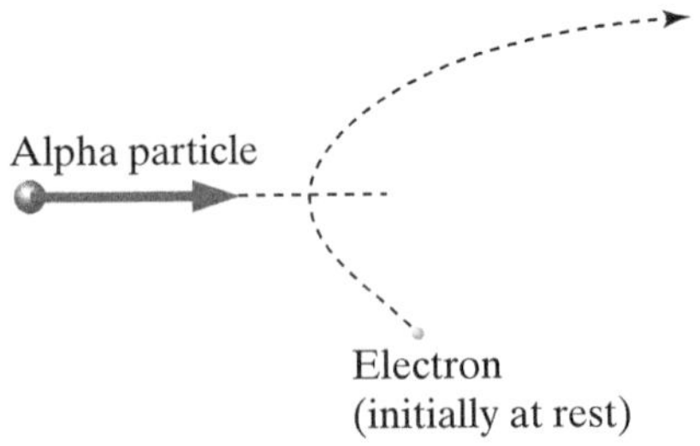

Figure 10.23 The alpha particle sweeps away the low-mass electrons and plows straight on ("bowling ball hits Ping-Pong ball").

What about the electric potential energy associated with the electric forces? Long before the collision the two nuclei were far from each other, and long after the collision they are again far from each other. Like gravitational energy, electric energy is proportional to $1/r$, where r is the distance between the two charged objects. Therefore, from long before the collision to long after the collision the change in electric energy would be very nearly zero ($\Delta U_{\text{el}} \approx 0$) if the alpha particle and gold nucleus are completely isolated from other particles. However, what about all the electrons surrounding a gold nucleus?

As the alpha particle plows through the gold foil it knocks low-mass electrons away (Figure 10.23), much as a moving bowling ball knocks Ping-Pong balls away (the alpha particle does of course lose a little energy in each such interaction). Therefore we can start our analysis when the alpha particle is deep inside the atom, with the electrons of the gold atom mostly far outside the region of interest. An atomic diameter is about 1×10^{-10} m, whereas the diameter of a gold nucleus is much smaller, less than 1×10^{-14} m, so we might start our analysis of the collision when the particles are about 1×10^{-12} m apart, a distance that is very small compared to an atom and very large compared to a nucleus. At this distance the electric potential energy is

$$U_{\text{elec}} = \frac{1}{4\pi\varepsilon_0}\frac{(2e)(79e)}{r} = \left(9 \times 10^9 \frac{\text{N}\cdot\text{m}^2}{\text{C}^2}\right)\frac{(2)(79)(1.6 \times 10^{-19}\,\text{C})^2}{1 \times 10^{-12}\,\text{m}}$$

$$U_{\text{elec}} = (3.6 \times 10^{-14}\,\text{J})\frac{1\,\text{eV}}{1.6 \times 10^{-19}\,\text{J}} = 0.2 \times 10^6\,\text{eV} = 0.2\,\text{MeV}$$

The radioactive source used by the Rutherford group provided alpha particles whose kinetic energy was 10 MeV, so if we take as the initial state a separation of 1×10^{-12} m the electric energy is only 2% of the kinetic energy. If we also take as the final state a separation of 1×10^{-12} m we have $\Delta U_{\mathrm{elec}}=0$. Evidently the only significant energy changes in this experiment are changes in the kinetic energies of the alpha particle and the gold nucleus.

The gold nuclei in the gold foil are about 2×10^{-10} apart, the approximate diameter of a typical atom in a metal, so when an alpha particle is 1×10^{-12} m from a gold nucleus, it is comparatively very far from any other gold nucleus. This explains why Rutherford could analyze the experiment as though isolated alpha particles scattered off isolated gold nuclei; electrons and other gold nuclei were far away.

The Plum Pudding Model of an Atom

For about a decade, the prevailing model of an atom had been what was called a "plum pudding" model proposed by J. J. Thomson. Electrons were discovered in 1897 by Thomson, using a cathode ray tube, a device similar to an old-style television picture tube. Thomson was able to show that the "cathode rays" that he and others had observed in electric gas discharges were in fact particles that were negatively charged, by the already existing convention for labeling electrical charges. This was the first conclusive demonstration of the existence of what we now call "electrons," one of the major constituents of all atoms.

Since these electrons were extracted from ordinary matter, which is normally not electrically charged, Thomson's discovery implied that matter also contained something having a positive charge. Because the electrons apparently had little mass, it seemed likely that most of the mass in ordinary matter was charged positively. Thomson proposed what was called a plum pudding model of atoms: a positively charged substance of uniform, rather low density, with tiny electrons distributed like raisins throughout it. Rutherford's group expected to gather evidence supporting this model from their collision experiments. Rutherford was extremely surprised by the results of these experiments, which led to a radical revision of the model.

The Rutherford Experiment

A new source of fast-moving particles was provided through the discoveries of natural radioactivity by the French scientists Henri Becquerel and Marie and Pierre Curie. These newly discovered particles included high-speed, massive, positively charged "alpha particles," which we now know to be the nuclei of helium atoms, consisting of two protons and two neutrons bound together. Alpha particles are emitted in the radioactive decay of heavy elements. Alpha particles were convenient microscopic projectiles, and Rutherford's group decided to shoot alpha particles at a very thin piece of gold foil and observe where they reappeared.

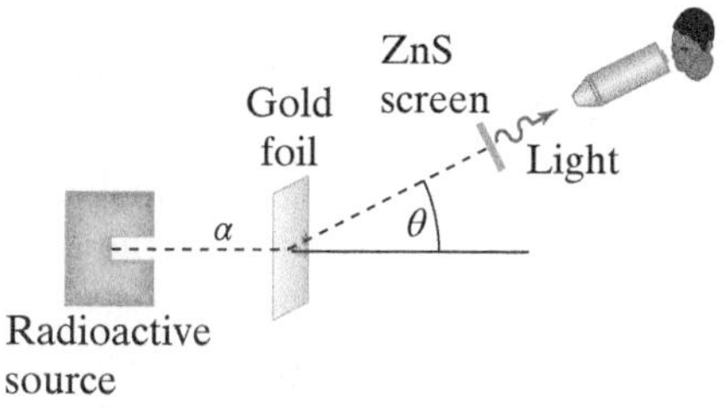

Figure 10.24 Schematic diagram of the Rutherford experiment. A radioactive source emits alpha particles, which are directed at a thin piece of gold foil. Particles scattered by gold nuclei in the gold foil hit a screen coated with ZnS, which fluoresces, giving off light when hit by an energetic particle. The light was observed by the experimenter.

> QUESTION Based on the plum pudding model, what did Rutherford and his colleagues expect to see in their collision experiments?

The researchers expected the alpha particles to be deflected only very slightly by interactions with the low-mass electrons and with the low-density "pudding" of positively charged matter in the gold atoms. They could detect a deflected ("scattered") alpha particle by a pulse of light that was emitted when the particle struck a fluorescent material (zinc sulfide). They moved their detector around at various angles to the incoming beam of alpha particles and measured what fraction of the alpha particles were scattered through various angles (Figure 10.24).

Rutherford and his coworkers were astounded to observe that an alpha particle, which is much more massive than an electron, sometimes bounced straight backward, as though it had encountered an even more massive yet very tiny positive particle inside the gold foil, rather than being slightly deflected in its passage through the "pudding" of positive charge. In other words, the experimenters thought they were throwing bowling balls at Ping-Pong balls, and they were amazed to see what they thought were bowling balls come bouncing back.

"It was quite the most incredible event that has ever happened to me in my life," Rutherford said. "It was almost as incredible as if you fired a 15-inch [artillery] shell at a piece of tissue paper and it came back and hit you."

He continued, "On consideration, I realized that this scattering backward must be the result of a single collision, and when I made the calculations I saw that it was impossible to get anything of that order of magnitude unless you took a system in which the greater part of the mass of the atom was concentrated in a minute nucleus. It was then that I had the idea of an atom with a minute massive center, carrying a charge." (See Figure 10.25.)

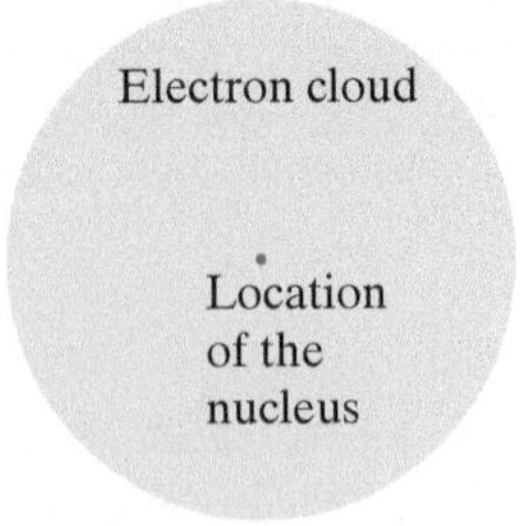

Figure 10.25 This diagram of an atom is not to scale; the radius of the nucleus is around 1×10^{-15} m, which is only 1/100,000 the radius of the electron cloud (about 1×10^{-10} m), so the nucleus would not be visible at all on this scale.

Computer Modeling of the Rutherford Experiment

We can make a computational model of the Rutherford experiment; see Problems P34 and P35. Remember that Coulomb's law for electric forces is very similar to Newton's law of gravitation. The electric force $\vec{F}_{elec}$ between two small charged particles carrying amounts of charge q_1 and q_2, and located a distance r apart, is

$$\vec{F}_{elec} = \frac{1}{4\pi\varepsilon_0}\frac{q_1 q_2}{r^2}\hat{r}$$

The charges q_1 and q_2 are measured in units of coulombs (C), and the proportionality factor is measured to be $1/(4\pi\varepsilon_0) = 9 \times 10^9\ \text{N}\cdot\text{m}^2/\text{C}^2$. The electric charge of the proton is $e = +1.6 \times 10^{-19}$ C, and the electric charge of the electron is $-e = -1.6 \times 10^{-19}$ C. In the Rutherford experiment the charge on the alpha particle is $2e$ (2 protons and 2 neutrons), and the charge on the gold nucleus is $79e$ (79 protons and 118 neutrons).

10.8 DISTRIBUTION OF SCATTERING ANGLES

In general, we have three equations describing an elastic collision—two momentum equations (in the x and y directions) and one energy equation, but there are four unknown quantities (the outgoing speeds and directions for each particle). The concept of impact parameter provides an additional quantity, and you saw earlier how the impact parameter determines the scattering angle θ of the incoming particle, from which you can predict the other final values (p_3, p_4, and ϕ).

In a macroscopic situation, such as shooting billiard balls, you can choose an impact parameter simply by carefully aiming at a point a perpendicular distance b from the center of the target. Unfortunately, in experiments where you fire alpha particles or other subatomic projectiles at nuclei, you can't aim at a specific impact parameter because the target is so incredibly small. All you can do is fire lots of projectiles at the subatomic target and expect a probabilistic distribution of impact parameters.

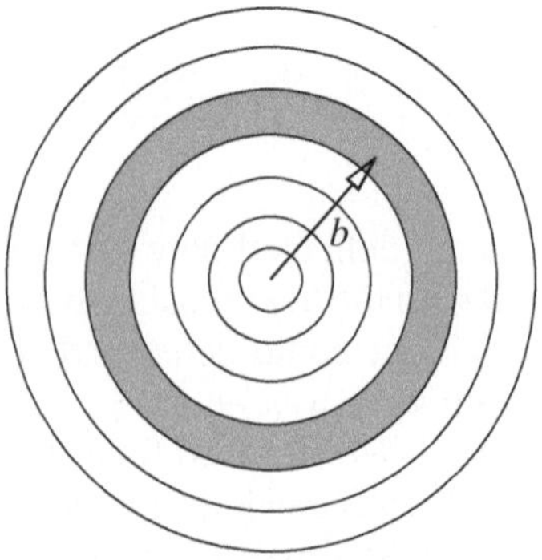

Figure 10.26 The probability of hitting a particular ring is proportional to the area of the ring.

When we fire a projectile at this microscopic target, the probability of hitting one particular ring (whose radius corresponds to the impact parameter) is proportional to the area of that ring (Figure 10.26). If you calculate the range of scattering angles that corresponds to hitting somewhere within a particular ring, you can predict the angular distribution of scattering angles that will be observed.

Cross Section

If a target particle such as a gold nucleus has a radius R, we say that its "geometrical cross section" is πR^2. A high-speed alpha particle that hits within this cross-sectional area will interact strongly with the gold nucleus. Quantum-mechanical effects can make the effective target size bigger or smaller than the geometrical cross section, and the effective target size is called the "cross section" for interactions of this kind. Cross sections are measured experimentally by observing how often projectiles are deflected by an interaction, instead of passing straight through the material without interacting. Cross sections have units of square meters.

A "differential" cross section for scattering within a particular range of angles is the area in square meters of a ring bounded by impact parameters corresponding to the given angular range.

Conservation Laws vs. Details of the Interaction

We used general principles to determine the final momenta of two colliding particles (conservation of momentum, conservation of energy). We didn't have to say what kind of force was involved. However, doesn't it make a difference whether the interaction is electric or nuclear? Yes, the type of interaction determines the *distribution* of scattering angles. For example, in what fraction of the collisions is the scattering angle between 0 and 10 degrees? Between 10 and 20 degrees? Between 20 and 30 degrees? These fractions are different for different forces.

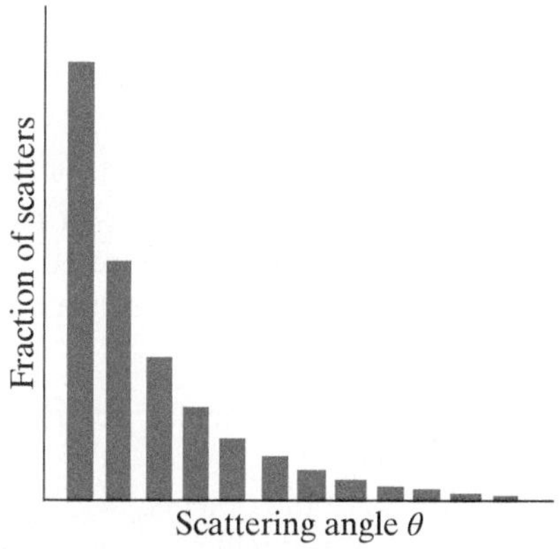

Figure 10.27 For an electric interaction, the fraction of scatters at different scattering angles falls rapidly with angle.

To put it another way, if the scattering angle is 20 degrees, the momenta are determined. The question is, how often is the scattering angle in fact near 20 degrees? Measurement of the distribution of scattering angles in collisions at modern particle accelerators is one of the ways used to study the nature of the strong and weak interactions.

Rutherford used Coulomb's law of electric force to make a detailed prediction of the distribution of scattering angles, assuming that the positive nucleus was pointlike and very massive. That is, he predicted what fraction of the alpha particles would be scattered by less than 10 degrees, by an angle between 10 and 20 degrees, by an angle between 20 and 30 degrees, and so on. The observed distribution of scattering angles agreed with the predicted distribution (Figure 10.27 shows a portion of this distribution).

The calculation that Rutherford did involves a lot of complicated geometry, algebra, and probability (with a random distribution of impact parameters), so we won't go through it. The main features of Figure 10.27 are easily understood, however. Because the gold nucleus is very small, most of the time the alpha particle will miss the nucleus by a large distance. With a large impact parameter the electric force is very small, and the deflection (scattering angle) is very small. For that reason small scattering angles are common. Large scattering angles occur only for nearly head-on collisions (small impact parameter), and these events are rare.

Since Rutherford's prediction assumed a pointlike nucleus, the observations were consistent with the nucleus being very small. Also, the good fit between prediction and observation strengthened the belief that the interaction was purely electric in nature (the strong interaction, which was unknown at the time, played no role because it is a very short-range interaction, and the alpha particle was not going fast enough to overcome the electric forces and contact the gold nucleus).

If the alpha particle has high enough energy, it can overcome the electric repulsion and make contact with the gold nucleus. In that case the strong interaction comes into play, and the angular distribution changes. The radius of the gold nucleus can be determined by finding out how much energy an alpha particle has to have in order to obtain an angular distribution that deviates from what is expected due just to the electric interaction.

> QUESTION For a particular scattering angle, can we use momentum and energy conservation to determine the unknown angle and momenta, even if the alpha particle makes contact with the gold nucleus?

Yes. These conservation principles are completely general and apply to all types of interactions. However, they cannot by themselves predict the angular distribution—that is, how often a particular range of scattering angles will occur. To do that you have to have a mathematical description for the specific interaction, such as the $1/r^2$ force law for electric interactions.

10.9 COMPUTATIONAL AND ANALYTICAL APPROACHES

In the previous sections we have analyzed collisions analytically (algebraically) by choosing both objects as the system, neglecting external forces, and applying Conservation of Momentum and Conservation of Energy to determine the outcome of the process. It is also possible to model collisions computationally, by doing exactly what we have done before in modeling interacting objects such as a spacecraft and a planet or a mass and a spring—using a known force law, and updating position and momentum iteratively. These approaches are complementary; depending on what information we know, and what questions we are trying to answer, one approach may be preferable to the other.

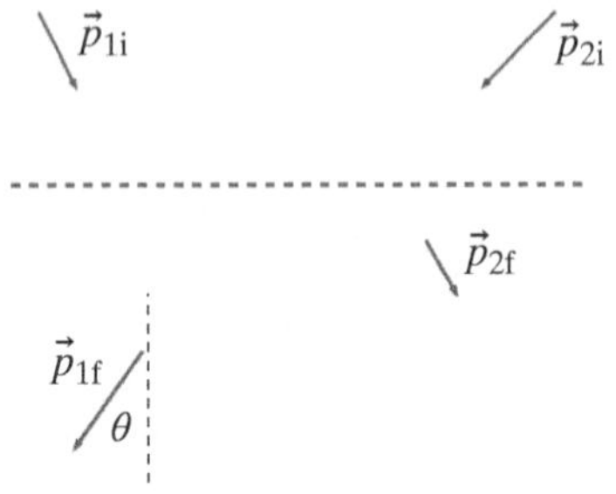

Figure 10.28 Using momentum and energy conservation let's predict the final momenta of the objects, but only if we have partial information about the outcome (for example, an experimental measurement of the direction of one of the final momenta, indicated here by the angle θ).

Using Momentum and Energy Conservation

This approach is useful in analyzing the results of elastic collisions in which there are experimental observations of the outcome of the collision–at least the unit vector $\hat{p}$ in the direction of one of the final momenta (Figure 10.28). Except in special cases (such as collisions in 1D) this approach isn't useful in predicting full outcomes.

- This approach does not require knowledge of the particular force law involved in the interaction.
- This approach does not predict the full trajectories of the objects.
- As we observed in Section 10.6, in general (2D or 3D) it will not be possible to predict the final momenta of both objects without additional information, such as experimental observations of some of the final quantities.

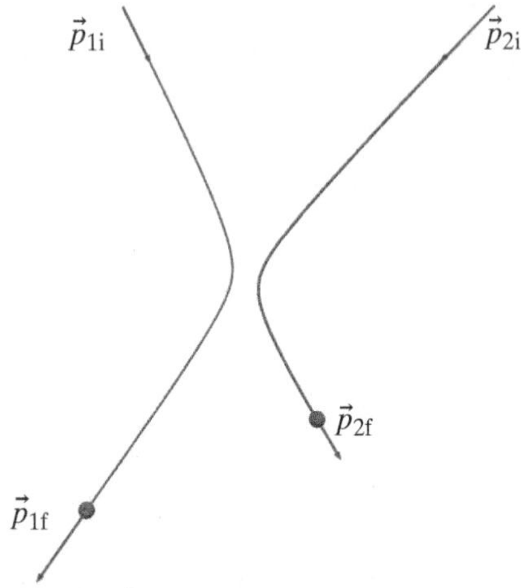

Figure 10.29 A computational approach predicts the full trajectories of the objects, if we know their initial positions and the appropriate force law.

Calculating Full Trajectories

This is a predictive approach for elastic collisions involving a known force law—it allows the prediction of full trajectories (Figure 10.29).

- This approach requires knowledge of initial positions as well as initial momenta.
- This approach requires knowledge of the force law involved in the interaction.
- This approach allows us to predict the full trajectories of the objects.
- This approach is always possible if we know forces, initial positions, and momenta.

Analyzing Inelastic Collisions

It is difficult to predict ahead of time how much initial kinetic energy will end up as internal energy after an inelastic collision. Most analyses of inelastic collisions depend on having detailed information about final momenta. For example, if both the magnitude and direction of $\vec{p}_{1f}$ are measured experimentally, it is then possible to use momentum conservation to determine $\vec{p}_{2f}$, and hence to determine the final kinetic energies. Given initial and final kinetic energies, we can determine the change in total internal energy of the system.

10.10 RELATIVISTIC MOMENTUM AND ENERGY

Rutherford's original experiments were carried out with alpha particles whose kinetic energy was about 10 MeV (10 million electron-volts), with a corresponding speed that was small compared to the speed of light. In order to probe the nucleus more deeply, physicists later built special machines, called "particle accelerators," which use electric forces to accelerate charged particles to very much higher energies, with speeds that may be as much as 99.9999% of the speed of light. For such high speeds we cannot use the low-speed approximate expressions for momentum and energy.

Physicists analyzing collisions produced by high-energy accelerators routinely do calculations based on the relativistic energy and momentum expressions. Consider a collision in which a particle of mass m_1 and magnitude of momentum p_1 strikes a stationary particle of mass m_2 (Figure 10.30). There is a change of identity, with the production of two new particles of mass m_3 and m_4, with momenta p_3 and p_4, at angles θ and ϕ to the horizontal (Figure 10.31).

Figure 10.30 Before a high-energy collision.

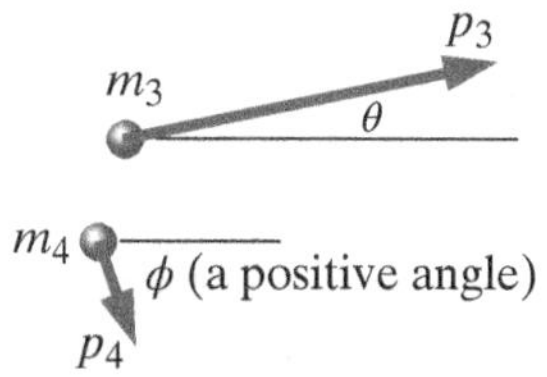

Figure 10.31 After the collision.

We will write down relativistically correct equations for momentum conservation and energy conservation, in terms of momentum p and energy E, knowing that we can determine E from p, since $E^2 - (pc)^2 = (mc^2)^2$:

$$p_1 = p_3 \cos\theta + p_4 \cos\phi, \text{ } x \text{ momentum conservation}$$

$$0 = p_3 \cos(90° - \theta) - p_4 \cos(90° - \phi), \text{ } y \text{ momentum conservation}$$

$$E_1 + m_2c^2 = E_3 + E_4, \text{ energy conservation (scalar)}$$

The energy equation is equivalent to

$$\sqrt{(p_1c)^2 + (m_1c^2)^2} + m_2c^2 = \sqrt{(p_3c)^2 + (m_3c^2)^2} + \sqrt{(p_4c)^2 + (m_4c^2)^2}$$

QUESTION Suppose that the initial momentum p_1 of the projectile is known, and all four masses are known (because we were able to identify the particles). How many unknown quantities are there, considering that we can write the energy equation in terms of p?

There are four unknown quantities, the magnitudes of the momenta and the directions of the two outgoing particles (p_3, p_4, θ, and ϕ), but only three equations. A typical situation in a high-energy particle experiment is that one of these unknown quantities is measured, and momentum and energy conservation are used to calculate the other three quantities.

As in the Rutherford experiment, measurements of the outgoing particle momenta alone do not tell us anything about the detailed nature of the interaction involved in the collision, because they are determined by general momentum and energy conservation laws. The identity changes of the particles tell us a lot, and the distribution of scattering angles also tells us a lot.

Photons Have Momentum

The relation $E^2 - (pc)^2 = (mc^2)^2$ is valid even for particles with zero mass, so that a photon with energy E has momentum of magnitude $p = E/c$, which is verified by experiment. One cannot write $E = \gamma mc^2$ or $p = \gamma mv$ for a photon, because when $v = c$, γ is infinite, and γm (infinity times zero) is undefined. Nevertheless, both the energy and the momentum of a photon are well-defined.

EXAMPLE **Photodissociation of the Deuteron**

A photon of energy E is absorbed by a stationary deuteron. As a result, the deuteron breaks up into a proton and a neutron. The photon energy is sufficiently low that the proton and neutron have speeds small compared to the speed of light, but there is a change of rest mass. The proton is observed to move in a direction at an angle θ to the direction of the incoming photon. Write equations that could be solved to determine the unknown magnitudes of the proton momentum p_p and neutron momentum p_n and the positive angle α of the neutron to the direction of the incoming photon. Let the precise masses of the particles be M_d, M_p, and M_n for the deuteron, proton, and neutron. Explain clearly on what principles your equations are based. Provide a physics diagram of the reaction and label the angles. Do not attempt to solve the equations.

Solution System: All of the particles (Figure 10.32)
Surroundings: Nothing significant

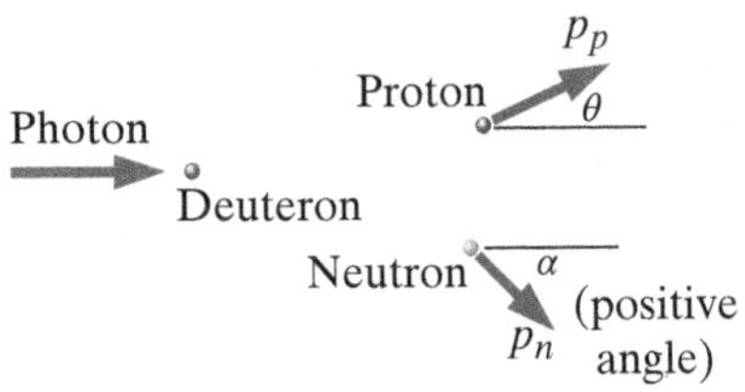

Figure 10.32 Physics diagram for photodissociation of deuteron. The angle θ was measured.

Momentum conservation (Figure 10.32): $\vec{p}_{\text{photon}} = \vec{p}_p + \vec{p}_n$

$$\langle E/c, 0, 0\rangle = \langle p_p\cos\theta + p_n\cos\alpha, p_p\cos(90^\circ - \theta) - p_n\cos(90^\circ - \alpha), 0\rangle$$

This is equivalent to two component equations:

$$E/c = p_p\cos\theta + p_n\cos\alpha$$

$$0 = p_p\cos(90^\circ - \theta) - p_n\cos(90^\circ - \alpha)$$

Energy conservation (small proton and neutron speeds):

$$E + M_dc^2 = \left(M_pc^2 + \frac{p_p^2}{2M_p}\right) + \left(M_nc^2 + \frac{p_n^2}{2M_n}\right)$$

We have three equations (two momentum component equations and one energy equation) with three unknowns: p_p, p_n, and α (θ was measured). In principle we can solve for the unknown quantities, though the algebra is messy.

We could also have written the energy equation relativistically, using the relationship $E^2 - (pc)^2 = (mc^2)^2$, which gives $E = \sqrt{(pc)^2 + (mc^2)^2}$:

$$E + M_dc^2 = \sqrt{(p_pc)^2 + (M_pc^2)^2} + \sqrt{(p_nc)^2 + (M_nc^2)^2}$$

Identifying a Particle

A variation on the relativistic collision analysis can be used to identify one of the outgoing particles. Suppose that we measure p_3 and θ, and we know m_3 but not m_4. We can use our three equations with the three unknowns (m_4, p_4, and ϕ) to solve for the unknown mass m_4. This is a very powerful technique when we are unable to observe the particle directly. For example, neutrinos don't have electromagnetic or strong interactions and therefore typically fail to register in ordinary detectors. However, if we measure the other outgoing particle we can determine the properties of the unseen particle.

The Discovery of the Neutrino

A specific example of such analyses was the original motivation for proposing the existence of the neutrino. The neutron (either free or when bound into some nuclei) was observed to decay with the creation of a proton and an electron, both charged particles that are easy to detect. However, the amount of energy and momentum carried by the proton and electron varied from one decay to the next, which made no sense:

$$n \to p^+ + e^-, \text{ violating momentum and energy conservation?}$$

It was proposed that there was an unseen third particle, called the neutrino (or antineutrino in neutron decay), which carried energy and momentum that would salvage the conservation laws:

$$n \to p^+ + e^- + \bar{\nu}, \text{ conserving momentum and energy}$$

It was possible to solve for the properties of the unseen particle by the techniques outlined above, and it was found that the neutrino must have nearly zero mass and must travel at nearly the speed of light. (Recent experiments show that the neutrino has a very small nonzero mass and must travel at slightly less than the speed of light.)

For many years the neutrino was just a theoretical construct that balanced the accounts, but eventually the neutrino was detected directly. The neutrino is very hard to detect because it does not have electromagnetic or strong interactions, only weak interactions, so the most probable thing for it to do is to pass right through any detector you place in its way. In fact, neutrinos emitted in nuclear reactions in the Sun mostly pass right through the entire Earth! However, with extremely intense neutrino beams produced in nuclear reactors or accelerators, or with gigantic underground detectors used to identify neutrinos coming from the Sun, it is possible to have a measurable rate of neutrino reactions.

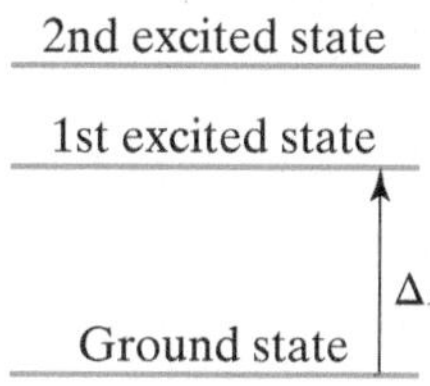

Figure 10.33 Energy levels for a gold nucleus. A sufficiently energetic alpha particle may have enough energy to raise the nucleus above its ground state.

10.11 INELASTIC COLLISIONS AND QUANTIZED ENERGY

The Rutherford experiment involved elastic collisions, because the internal energies of the gold nucleus and the alpha particle didn't change during the interaction. When there is an internal energy change, or the production of additional particles or the emission of light, we call the collision "inelastic."

If we shoot alpha particles at a gold nucleus with higher energy than the energy of alpha particles in the Rutherford experiment, we may be able to excite the gold nucleus to a higher quantum energy level, an amount ΔE above the ground state of the gold nucleus (Figure 10.33). Such a state corresponds to a different configuration of the nucleons inside the gold nucleus.

Suppose that you know the incoming momentum p_1 of the alpha particle, and you are able to measure its outgoing momentum p_3 and scattering angle θ (Figure 10.34).

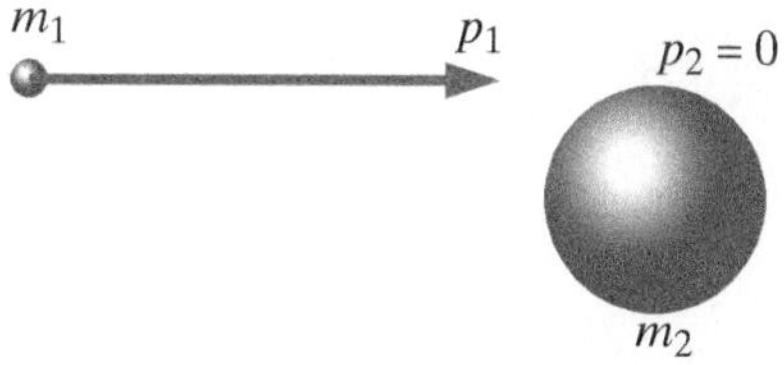

Figure 10.34 You measure the incoming and outgoing momenta of the alpha particle and deduce something about the energy levels in the gold nucleus.

> QUESTION What is the form of the energy conservation and momentum conservation equations in this case? Don't assume that the speeds are small compared to the speed of light. Note that the rest energy of the recoiling gold nucleus is $M_4c^2 + \Delta E$ because it is in an excited state.

Here are the relevant equations:

$$p_1 = p_3 \cos\theta + p_4 \cos\phi,$$

where p_4 and ϕ refer to the gold nucleus, which is not observed in the experiment

$$0 = p_3 \cos(90^\circ - \theta) - p_4 \cos(90^\circ - \phi)$$

$$E_1 + Mc^2 = E_3 + E_4, \text{ where } E_4 = \sqrt{(p_4c)^2 + (M_4c^2 + \Delta E)^2}$$

QUESTION How many unknown quantities are there in the energy and momentum equations? Is there enough information to be able to determine the internal energy difference ΔE? (This is one of the methods used to study the energy levels of nuclei. This energy level determination can be verified by measuring the energy of a gamma-ray photon emitted when the nucleus drops back to the ground state.)

There are three unknowns (p_4, ϕ, and ΔE), and there are three equations (energy, x and y momentum), so we could determine ΔE.

Change of Identity

Instead of merely exciting the target particle to a higher internal energy state, suppose that there is a change of identity in which the target particle turns into some other particle. An example is the following particle reaction, which was used in the 1950s to discover the Δ^+ particle of mass m_Δ, which is a particle more massive than a proton:

$$\pi^- + p^+ \rightarrow \pi^- + \Delta^+$$

A beam of negative pions (π^-) of known mass m_π and known kinetic energy, produced by a particle accelerator, is directed at a container of hydrogen gas, and the targets for the collisions are the protons of mass m_p that are the nuclei of the hydrogen atoms. Sometimes an incoming pion interacts through the strong interaction with a proton to produce a Δ^+ particle. The momentum of the scattered pion can be measured by measuring the radius of curvature of its curving trajectory in a magnetic field.

QUESTION What is the form of the energy conservation and momentum conservation equations in this case? (Don't assume that the speeds are small compared to the speed of light.)

Here are the relativistically correct equations:

$$p_1 = p_3 \cos\theta + p_4 \cos\phi$$
$$0 = p_3 \cos(90° - \theta) - p_4 \cos(90° - \phi)$$
$$E_1 + m_p c^2 = E_\pi + E_\Delta, \text{ where } E^2 - (pc)^2 = (mc^2)^2$$

QUESTION How many unknown quantities are there in the energy and momentum equations? Is there enough information to be able to determine the mass of the Δ^+ particle?

There are three unknowns (p_4, ϕ, and E_Δ) and three equations, so it is possible to determine the mass of the Δ^+ particle.

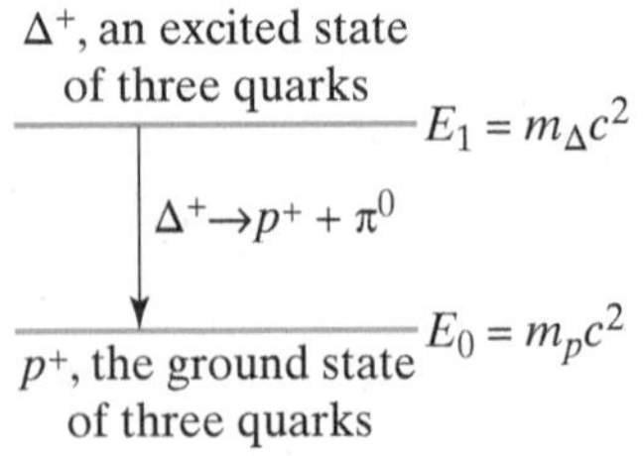

Figure 10.35 The Δ^+ particle can be considered to be an excited state of three quarks, with the proton representing the ground state. A decay to the ground state can occur with the emission of a neutral pion (π^0).

What We Learn from Inelastic Collisions

The analysis for finding the mass change in the preceding exercises is essentially the same as the analysis for determining the internal energy change of an excited gold nucleus. In fact, we can say that the Δ^+ particle is an excited state of three quarks, with the proton representing the ground state of three quarks (Figure 10.35). The excited state (the Δ^+ particle) can decay to the ground state (the proton) with the emission of a neutral pion (π^0), which is a quark–antiquark pair. The extra kinetic energy is equal to $(m_\Delta c^2 - m_p c^2)$. Scattering experiments have been an important tool for discovering new particles (or new excited states of quark systems).

10.12 COLLISIONS IN OTHER REFERENCE FRAMES

In the case of a bowling ball colliding head-on with a Ping-Pong ball, we saw that it can be useful to carry out the analysis in a moving reference frame. In this section we will explore this further by analyzing a two-dimensional sticking collision in two different reference frames.

A Sticking Collision

We'll consider an inelastic collision in which a truck and a car collide in an icy intersection and stick together (Figure 10.36). In the figure we have shown velocity vectors rather than momentum vectors; a typical truck would have much more momentum than the car, given the truck's much greater mass.

Let M be the mass of the truck, m the mass of the car, and θ the angle of their mutual final velocity, as shown in the diagram. Also let v_1 be the initial speed of the truck, v_2 the initial speed of the car, and v_3 the final speed when they're stuck together and sliding on the ice.

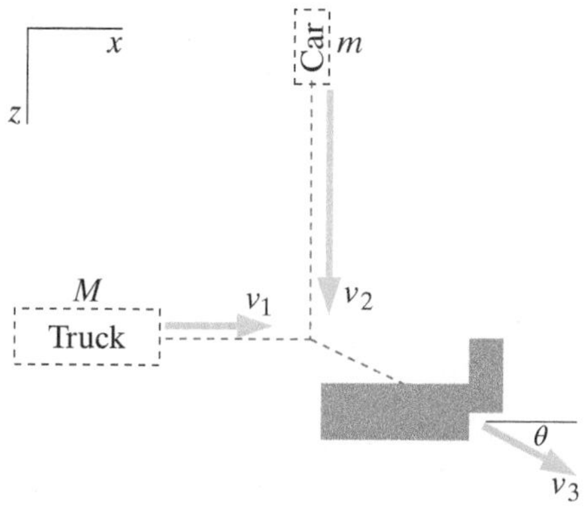

Figure 10.36 A sticking collision between a car and a truck on an icy road. This is a top view, looking down on the xz plane from a location on the $+y$ axis.

QUESTION How does the fact that the intersection is icy affect our ability to apply conservation of momentum to the truck–car collision?

Friction with the road is small if the intersection is icy, so we can presumably ignore road friction and air resistance during the short time of the actual collision. Therefore the total momentum of the truck + car system should be (approximately) unchanged. Momentum conservation is really three equations, one each for the x, y, and z directions. In the y direction (vertically upward) the ice simply supports the weights of the vehicles, so the net force is zero, and the zero y component of momentum remains zero throughout the collision.

QUESTION In terms of the named quantities, write two equations for the x and z components of the momentum conservation equation, equating momentum components before the collision to momentum components after the vehicles stick together.

The unit vector $\hat{v}_3$ giving the final direction of motion is:

$$\hat{v}_3 = \langle \cos\theta, 0, \cos(90^\circ - \theta)\rangle = \langle \cos\theta, 0, \sin\theta\rangle$$

so

$$\vec{v}_3 = v_3\langle \cos\theta, 0, \sin\theta\rangle = \langle v_3\cos\theta, 0, v_3\sin\theta\rangle$$

The momentum conservation equation is:

$$\langle Mv_1, 0, 0\rangle + \langle 0, 0, mv_2\rangle = \langle (M+m)v_3\cos\theta, 0, (M+m)v_3\sin\theta\rangle$$

Separating this equation into x and z components:

$$x:\quad Mv_1 = (M+m)v_3\cos\theta$$
$$z:\quad mv_2 = (M+m)v_3\sin\theta$$

Assume that we know the initial speeds of the truck and car (v_1 and v_2). Then we have two equations with two unknowns (v_3 and θ), and we can solve for the final speed and angle. To find θ, divide the z equation by the x equation:

$$\frac{(M+m)v_3\sin\theta}{(M+m)v_3\cos\theta} = \frac{mv_2}{Mv_1} = \tan\theta$$

so

$$\theta = \arctan\left(\frac{mv_2}{Mv_1}\right)$$

To find v_3, square the equations and add them:

$$(Mv_1)^2 + (mv_2)^2 = [(M+m)v_3]^2[(\sin\theta)^2 + (\cos\theta)^2]$$
$$\sqrt{(Mv_1)^2 + (mv_2)^2} = (M+m)v_3$$
$$v_3 = \frac{\sqrt{M^2v_1^2 + m^2v_2^2}}{M+m}$$

We used the trig identity $\sin^2\theta + \cos^2\theta = 1$, which corresponds to the Pythagorean theorem applied to a triangle whose hypotenuse has length 1.

Energy in a Sticking Collision

A sticking collision is unusual in that momentum conservation alone gives us enough equations (two) to be able to solve for the unknown final speed and unknown direction. If the car and truck bounce off each other instead of sticking, there are four unknowns instead of two, because there are two final velocities, each with a speed and a direction to be calculated. In that case momentum conservation alone doesn't provide enough equations to be able to analyze the situation fully.

There is an important issue associated with energy in a sticking collision such as this one. Using our result for the final speed, we can write the kinetic energy after the collision:

$$\frac{1}{2}(M+m)v_3^2 = \frac{1}{2}(M+m)\left[\frac{(Mv_1)^2 + (mv_2)^2}{(M+m)^2}\right]$$

After minor rearrangements we get the following:

$$\frac{1}{2}(M+m)v_3^2 = \left(\frac{M}{M+m}\right)\left(\frac{1}{2}Mv_1^2\right) + \left(\frac{m}{M+m}\right)\left(\frac{1}{2}mv_2^2\right)$$

Since $M/(M+m)$ is less than 1, and $m/(M+m)$ is also less than 1, the final kinetic energy is less than the initial kinetic energy.

QUESTION Where is the "missing" kinetic energy?

Part of the energy was radiated away as sound waves in the air, produced when the vehicles collided with a bang. Some of the energy went into raising the temperature of the vehicles, so that their thermal energy is now higher than before (larger random motions of the atoms).

Some of the energy went into deforming the metal bodies of the truck and car. A bent fender has a higher internal energy, associated with a change in the configuration of the atoms. Sometimes you can get such energy back (as when you compress a spring, changing the configuration of the atoms in the spring, and later let it push back, returning to its original configuration), and we speak of "elastic" energy. Sometimes, however, you can't get the energy back, because the configuration change gets "locked in."

A sticking collision is an example of an "inelastic" collision, referring to the effective "loss" of kinetic energy into forms that are inaccessible.

Checkpoint 4 Car 1 headed north and car 2 headed west collide. They stick together and leave skid marks on the pavement, which show that car 1 was deflected 30° (so car 2 was deflected 60°). What can you conclude about the cars before the collision?

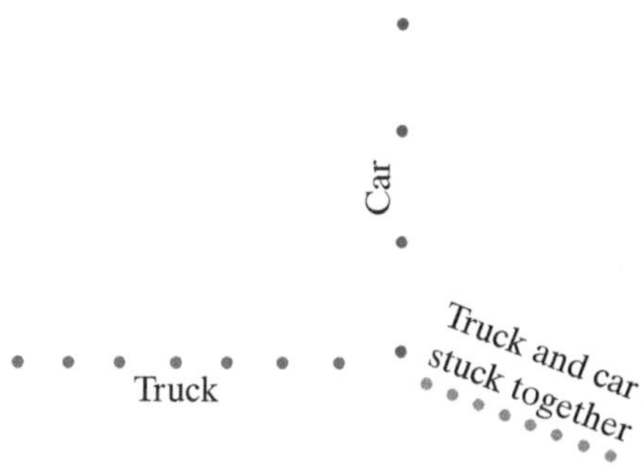

Figure 10.37 View of collision from hovering (stationary) helicopter.

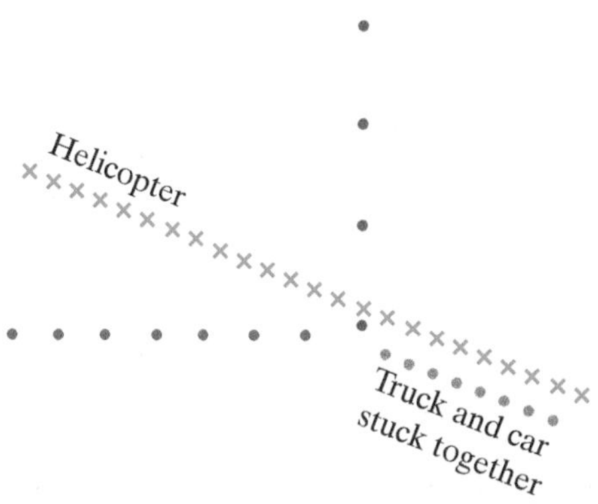

Figure 10.38 Path of moving helicopter that produces the simple view seen in Figure 10.39.

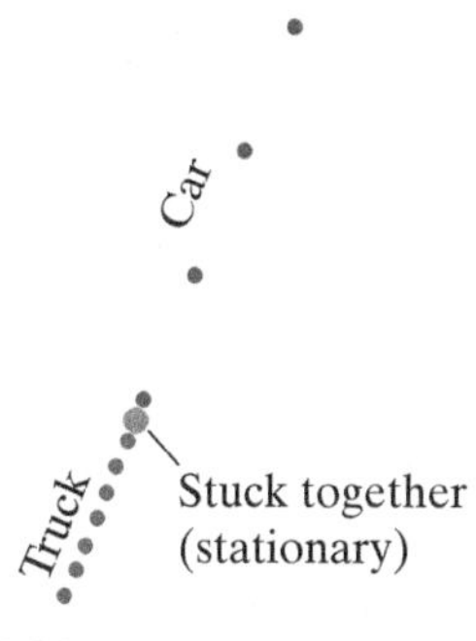

Figure 10.39 View of collision from helicopter moving with velocity of stuck-together wreck.

The Collision in a Different Reference Frame

We can show that a truck–car collision process looks a lot simpler from the vantage point of a helicopter flying over the scene with the final velocity of the truck + car combined system. This will lead to new insights. Suppose the accident takes place at night, and from the helicopter all you can see are lights mounted on top of the two vehicles. Figure 10.37 shows a multiple exposure of the scene that you would see from a helicopter that is hovering over the intersection.

Suppose instead that your helicopter flies over the intersection with the same velocity that the stuck-together truck and car share after their collision, as shown in Figure 10.38. The view thus obtained is in an important sense "simpler," as shown in Figure 10.39. From the moving observation platform you see the truck and car run into each other and then stand still (since you and they now have the same velocity). In particular, in this (moving) reference frame it is obvious that there is a loss of kinetic energy into other forms of energy, because after the collision the kinetic energy is clearly zero. In contrast, we had to go through a fair amount of algebra in the nonmoving reference frame in order to establish the loss of kinetic energy.

The reason that we imagine making these observations at night is so that you won't be distracted by the moving background. If you actually view this collision from a moving helicopter in daylight, there is a strong tendency to compensate mentally for your motion and to put yourself (mentally) into the reference frame of the road, in which case you could miss the important fact that *relative to you*, the car and truck are stationary after the collision.

You may have had an experience related to this situation. You're sitting in a bus or train, waiting to leave. You think you've started forward, but then you realize that you were fooled by a neighboring bus or train moving backward. For a moment you were mentally in the reference frame of the moving vehicle.

Total Momentum Is Zero in the Center-of-Mass Frame

Viewing a collision from a moving reference frame can provide insights into all kinds of collisions, not just sticking collisions. The key point that leads to a useful generalization is that the total momentum of the truck + car system after the collision is zero when the vehicles are motionless relative to your helicopter.

> QUESTION In this moving reference frame, if the total momentum is zero after the collision, what was the total momentum before the collision? Why?

The helicopter is moving at constant velocity (constant speed and direction), and we know that the Momentum Principle is valid in this uniformly moving reference frame. Thus, the principle of momentum conservation still holds, and the initial total momentum of the car + truck system must be zero if the final total momentum is zero. That's what makes the process look so simple: the total momentum is zero before and after the collision in the moving reference frame.

In viewing the scene from the moving helicopter, we have subtracted from each of the velocities an unchanging, constant vector velocity, the velocity $\vec{v}_h$ of the helicopter, and this subtraction makes no change in the Momentum Principle for the car (or truck):

$$\frac{d[m(\vec{v}-\vec{v}_h)]}{dt} = \frac{d[m\vec{v}]}{dt} = \vec{F}_{\text{net}}$$

The net force is the same in both reference frames because the contact electric interaction forces depend on interatomic distances, which are the same in both reference frames.

Evidently we gain insight into a collision by viewing the collision in a moving reference frame where the total momentum is zero. How do we determine the appropriate velocity for such a reference frame? Let this velocity be $\vec{v}_{fr}$, which is to be subtracted from all the original velocities to obtain the velocities as viewed in the moving reference frame. We want the total momentum to add up to zero in the moving frame:

$$\vec{p}_{sys,new\ frame} = m_1(\vec{v}_1 - \vec{v}_{fr}) + m_2(\vec{v}_2 - \vec{v}_{fr}) = \vec{0}$$
$$(m_1 + m_2)\vec{v}_{fr} = m_1\vec{v}_1 + m_2\vec{v}_2$$
$$\vec{v}_{fr} = \frac{m_1\vec{v}_1 + m_2\vec{v}_2}{m_1 + m_2}$$

We have solved for a particular velocity for the moving reference frame that makes the total momentum in that frame be zero—but look at the form of this equation.

QUESTION Does this equation look familiar?

Evidently $\vec{v}_{fr} = \vec{v}_{CM}$, the velocity of the center of mass of the system, since

$$\vec{r}_{CM} = \frac{m_1\vec{r}_1 + m_2\vec{r}_2}{m_1 + m_2}$$

We have obtained an important result. In a reference frame moving with the center of mass of a system, the total momentum of the system is zero. If no external forces act, the total momentum is unchanged, and the total momentum is zero before and after a collision.

We can also express the frame velocity directly in terms of the total momentum:

$$\vec{v}_{CM} = \frac{m_1\vec{v}_1 + m_2\vec{v}_2}{m_1 + m_2} = \frac{\vec{p}_{sys}}{M_{tot}}$$

This provides another way to calculate the frame velocity and shows why this is often called the "center-of-momentum" velocity.

Let's reanalyze the car–truck collision from this new point of view. Again let M be the mass of the truck, m the mass of the car, and θ the angle of their mutual final velocity, as shown in Figure 10.40. Also let v_1 be the initial speed of the truck, v_2 the initial speed of the car, and v_3 the final speed when they're stuck together and sliding on the ice.

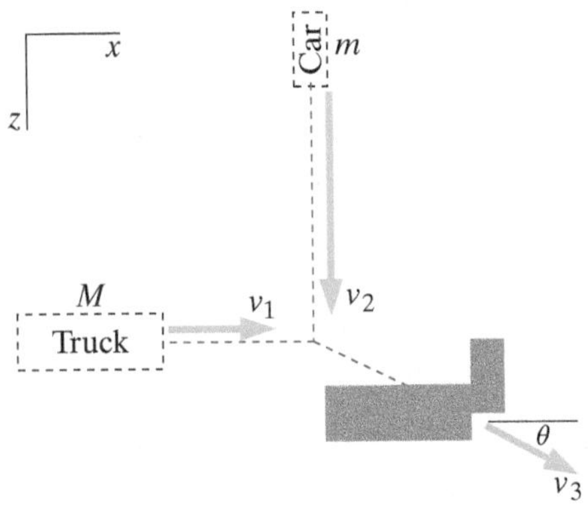

Figure 10.40 Collision of car and truck.

QUESTION What are the x and z components of $\vec{v}_{CM}$ in terms of v_1 and v_2?

$$\vec{v}_{CM} = \left\langle \frac{Mv_1}{M+m}, 0, \frac{mv_2}{M+m} \right\rangle$$

QUESTION In the moving center-of-momentum reference frame, what are the x and z components of the car's velocity before the collision? Remember that you find the car's velocity (or its components) in the center-of-momentum frame by subtracting the center-of-momentum velocity (or its components) from that of the car. (That is, the view from the helicopter is obtained by subtracting the helicopter's velocity from the velocities of the car and truck.)

$$\left\langle -\frac{Mv_1}{M+m}, 0, v_2 - \frac{mv_2}{M+m} \right\rangle = \left\langle -\frac{Mv_1}{M+m}, 0, \frac{Mv_2}{M+m} \right\rangle$$

QUESTION In the moving center-of-momentum reference frame, what are the x and z components of the truck's velocity before the collision? Show that this velocity is in a direction opposite to the velocity of the car, and that the total momentum is zero.

$$\left\langle v_1 - \frac{Mv_1}{M+m}, 0, -\frac{mv_2}{M+m} \right\rangle = \left\langle \frac{mv_1}{M+m}, 0, -\frac{mv_2}{M+m} \right\rangle$$

The x and z components of velocity are in the opposite direction to those of the car (multiply the car's velocity by $-m/M$ and you have the truck's velocity). Also, the total momentum is zero in the center-of-momentum frame:

$$\left\langle \left[m\left(-\frac{Mv_1}{M+m}\right) + M\left(\frac{mv_1}{M+m}\right)\right], 0, \left[m\left(\frac{Mv_2}{M+m}\right) + M\left(-\frac{mv_2}{M+m}\right)\right] \right\rangle = \vec{0}$$

QUESTION In the moving center-of-momentum reference frame, what is the speed of the car after the collision? Of the truck? Why?

Both are zero; in the center-of-mass frame the total momentum is zero, and with the car and truck stuck together the only way for this to be true is for them to be at rest. Because of this, the kinetic energy loss to other kinds of energy is equal to the total kinetic energy in the center-of-momentum frame before the collision.

QUESTION Now, after the collision, transform back to the (nonmoving) reference frame of the intersection, by adding back in the $\vec{v}_{CM}$ that had been subtracted. What are the speed and direction of the car after the collision? Of the truck? Compare with the results obtained in our earlier analysis of this collision.

$$\vec{v}_3 = \left\langle \frac{Mv_1}{M+m}, 0, \frac{mv_2}{M+m} \right\rangle$$

which yields
$$v_3 = \frac{\sqrt{M^2v_1^2 + m^2v_2^2}}{M+m}; \quad \theta = \arctan\left(\frac{mv_2}{Mv_1}\right)$$

These are the same results we obtained in our earlier analysis.

Energy in the Two Reference Frames

As far as energy is concerned, note that the kinetic energy of the car + truck system is, as usual,

$$K_{tot} = K_{trans} + K_{rel}$$

where $K_{trans} = \dfrac{p_{sys}^2}{2M_{tot}}$. In the original reference frame, $K_{trans} > 0$ but is the same before and after the collision, since p_{sys} doesn't change. In the center-of-momentum reference frame, $K_{trans} = 0$ both before and after the collision, because $p_{sys} = 0$ in the center-of-momentum reference frame.

In either reference frame, the kinetic energy relative to the center of mass, K_{rel}, decreases to zero in the collision, because the car and truck after the collision are traveling with the center of mass. It is ΔK_{rel} that represents the loss of kinetic energy in this inelastic collision, and this quantity is the same in both reference frames. The increase in internal energy is relatively easy to calculate in the center-of-momentum reference frame, as it is simply

$$\Delta E_{int} = -\Delta K_{rel} = 0 - (K_{truck,i} + K_{car,i})$$

Also note that distances between atoms are the same in both reference frames, so potential energy associated with bending of metal parts is the same in both frames.

Checkpoint 5 A 1000 kg car moving east at 30 m/s runs head-on into a 3000 kg truck moving west at 20 m/s. The vehicles stick together. Use the concept of the center-of-momentum frame to determine how much kinetic energy is lost.

SUMMARY

New concepts
Elastic and inelastic collisions
Scattering, and scattering distributions
Impact parameter (0 if head-on)
Reference frame transformations:

$$\vec{v} = \vec{v}\,' + \vec{v}_{\text{frame}}$$

The center-of-mass reference frame

$$\vec{v}_{\text{cm}} = \frac{m_1\vec{v}_1 + m_2\vec{v}_2}{m_1 + m_2} = \frac{\vec{p}_{\text{sys}}}{M_{\text{tot}}}$$

Simple results for head-on elastic collisions between Ping-Pong balls and bowling balls

Evidence for the nuclear model of an atom

Problem-solving techniques
Conservation of momentum and conservation of energy applied to collisions
Simplification of calculations in the center-of-mass reference frame
Computational models can predict full trajectories of colliding objects. Applying conservation of momentum and energy can predict only final momenta, and in most cases additional information is required for such an analysis.

QUESTIONS

Q1 Under what conditions is the momentum of a system constant? Can the x component of momentum be constant even if the y component is changing? In what circumstances? Give an example of such behavior.

Q2 In a collision between an electron and a hydrogen atom, why is it useful to select both objects as the system? Pick all that apply: (1) The total momentum of the system does not change during the collision. (2) The sum of the final kinetic energies must equal the sum of the initial kinetic energies for a two-object system. (3) The kinetic energy of a two-object system is nearly zero. (4) The forces the objects exert on each other are internal to the system and don't change the total momentum of the system. (5) During the time interval from just before to just after the collision, external forces are negligible.

Q3 Two asteroids in outer space collide and stick together. The mass of each asteroid, and the velocity of each asteroid before the impact, are known. To find the momentum of the stuck-together asteroids after the impact, what approach would be useful? (1) Use the Energy Principle. (2) Use the Momentum Principle. (3) It depends on whether or not the speed of the asteroids was near the speed of light. (4) Use the relationship among velocity, displacement, and time. (5) It depends on whether the collision was elastic or inelastic.

Q4 You know that a collision must be "elastic" if: (1) The colliding objects stick together. (2) The colliding objects are stretchy or squishy. (3) The sum of the final kinetic energies equals the sum of the initial kinetic energies. (4) There is no change in the internal energies of the objects (thermal energy, vibrational energy, etc.). (5) The momentum of the two-object system doesn't change.

Q5 What happens to the velocities of the two objects when a high-mass object hits a low-mass object head-on? When a low-mass object hits a high-mass object head-on?

Q6 It has been proposed to propel spacecraft through the Solar System with a large sail that is struck by photons from the Sun. **(a)** Which would be more effective, a black sail that absorbs photons or a shiny sail that reflects photons back toward the Sun? Explain briefly. **(b)** Suppose that N photons hit a shiny sail per second, perpendicular to the sail. Each photon has energy E. What is the force on the sail? Explain briefly.

Q7 In an elastic collision involving known masses and initial momenta, how many unknown quantities are there after the collision? How many equations are there? In a sticking collision involving known masses and initial momenta, how many unknown quantities are there after the collision? Explain how you can determine the amount of kinetic energy change.

Q8 What properties of the alpha particle and the gold nucleus in the original Rutherford experiment were responsible for the collisions being elastic collisions?

Q9 Give an example of what we can learn about matter through the use of momentum and energy conservation applied to scattering experiments. Explain what it is that we cannot learn this way, for which we need to measure the distribution of scattering angles.

Q10 What is it about analyzing collisions in the center-of-mass frame that simplifies the calculations?

Q11 In order to close a door, you throw an object at the door. Which would be more effective in closing the door, a 50 g tennis ball or a 50 g lump of sticky clay? Explain clearly what physics principles you used to draw your conclusion.

Q12 Consider a head-on collision between two objects. Object 1, which has mass m_1, is initially in motion, and collides head-on with object 2, which has mass m_2 and is initially at rest. Which of the following statements about the collision are true? (1) $\vec{p}_{1,\text{initial}} = \vec{p}_{1,\text{final}} + \vec{p}_{2,\text{final}}$. (2) $|\vec{p}_{1,\text{final}}| < |\vec{p}_{1,\text{initial}}|$. (3) If $m_2 \gg m_1$, then $|\Delta\vec{p}_1| > |\Delta\vec{p}_2|$. (4) If $m_1 \gg m_2$, then the final speed of object 2 is less than the initial speed of object 1. (5) If $m_2 \gg m_1$, then the final speed of object 1 is greater than the final speed of object 2.

Q13 In a nuclear fission reactor, each fission of a uranium nucleus is accompanied by the emission of one or more high-speed neutrons, which travel through the surrounding material. If one of these neutrons is captured in another uranium nucleus, it can trigger fission, which produces more fast neutrons, which could make possible a chain reaction (Figure 10.41).

However, fast neutrons have low probability of capture and usually scatter off uranium nuclei without triggering fission. In order to sustain a chain reaction, the fast neutrons must be slowed down in some material, called a "moderator." For reasons having to do with the details of nuclear physics, slow neutrons have a high probability of being captured by uranium nuclei.

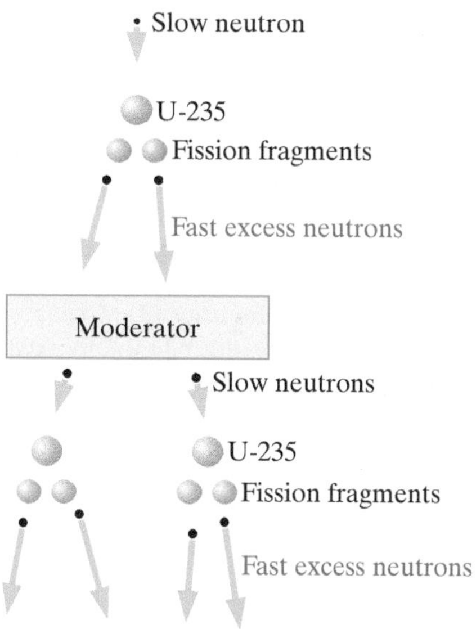

Figure 10.41

A slow neutron induces fission of U-235, with the emission of additional (fast) neutrons. The moderator is some material that slows down the fast neutrons, enabling a chain reaction.

In the following analyses, remember that neutrons have almost no interaction with electrons. Neutrons do, however, interact strongly with nuclei, either by scattering or by being captured and made part of the nucleus. Therefore you should think about neutrons interacting with nuclei (through the strong force), not with entire atoms. **(a)** Based on what you now know about collisions, explain why fast neutrons moving through a block of uranium experience little change in speed. **(b)** Explain why carbon should be a much better moderator of fast neutrons than uranium. **(c)** Should water be a better or worse moderator of fast neutrons than carbon? Explain briefly.

Background: The first fission reactor was constructed in 1941 in a squash court under the stands of Stagg Field at the University of Chicago by a team led by the physicist Enrico Fermi. The moderator consisted of blocks of graphite, a form of carbon. The graphite had to be exceptionally pure because certain kinds of impurities have nuclei that capture neutrons with high probability, removing them from contributing to the chain reaction. Many reactors use ordinary "light" water as a moderator, though sometimes a proton captures a neutron and forms a stable deuterium nucleus, in which case the neutron is lost to contributing to the chain reaction. Heavy water, D_2O, in which the hydrogen atoms are replaced by deuterium atoms, actually works better as a moderator than light water, because the probability of a deuterium nucleus capturing a neutron to form tritium is quite small.

PROBLEMS

Section 10.1

•**P14** A ball whose mass is 0.2 kg hits the floor with a speed of 8 m/s and rebounds upward with a speed of 7 m/s. The ball was in contact with the floor for 0.5 ms (0.5×10^{-3} s). **(a)** What was the average magnitude of the force exerted on the ball by the floor? **(b)** Calculate the magnitude of the gravitational force that the Earth exerts on the ball. **(c)** In a collision, for a brief time there are forces between the colliding objects that are much greater than external forces. Compare the magnitudes of the forces found in parts (a) and (b).

Section 10.3

••**P15** A projectile of mass m_1 moving with speed v_1 in the $+x$ direction strikes a stationary target of mass m_2 head-on. The collision is elastic. Use the Momentum Principle and the Energy Principle to determine the final velocities of the projectile and target, making no approximations concerning the masses. After obtaining your results, see what your equations would predict if $m_1 \gg m_2$, or if $m_2 \gg m_1$. Verify that these predictions are in agreement with the analysis in this chapter of the Ping-Pong ball hitting the bowling ball, and of the bowling ball hitting the Ping-Pong ball.

Section 10.6

•**P16** Object A has mass $m_A = 7$ kg and initial momentum $\vec{p}_{A,i} = \langle 17, -5, 0 \rangle$ kg·m/s, just before it strikes object B, which has mass $m_B = 11$ kg. Object B has initial momentum $\vec{p}_{B,i} = \langle 4, 6, 0 \rangle$ kg·m/s. After the collision, object A is observed to have final momentum $\vec{p}_{A,f} = \langle 13, 3, 0 \rangle$ kg·m/s. In the following questions, "initial" refers to values before the collisions, and "final" refers to values after the collision. Consider a system consisting of both objects A and B. Calculate the following quantities: **(a)** The total initial momentum of this system. **(b)** The final momentum of object B. **(c)** The initial kinetic energy of object A. **(d)** The initial kinetic energy of object B. **(e)** The final kinetic energy of object A. **(f)** The final kinetic energy of object B. **(g)** The total initial kinetic energy of the system. **(h)** The total final kinetic energy of the system. **(i)** The increase of internal energy of the two objects. **(j)** What assumption did you make about Q (energy flow from surroundings into the system due to a temperature difference)?

•**P17** In outer space a rock whose mass is 3 kg and whose velocity was $\langle 3900, -2900, 3000 \rangle$ m/s struck a rock with mass 13 kg and velocity $\langle 220, -260, 300 \rangle$ m/s. After the collision, the 3 kg rock's velocity is $\langle 3500, -2300, 3500 \rangle$ m/s. **(a)** What is the final velocity of the 13 kg rock? **(b)** What is the change in the internal energy of the rocks? **(c)** Which of the following statements about Q (transfer of energy into the system because of a temperature difference between system and surroundings) are correct? (1) $Q \approx 0$ because the duration of the collision was very short. (2) $Q = \Delta E_{\text{thermal}}$ of the rocks. (3) $Q \approx 0$ because there are no significant objects in the surroundings. (4) $Q = \Delta K$ of the rocks.

••**P18** A spring has an unstretched length of 0.32 m. A block with mass 0.2 kg is hung at rest from the spring, and the spring becomes 0.4 m long (Figure 10.42). Next the spring is stretched to a length of 0.43 m and the block is released from rest. Air resistance is negligible.

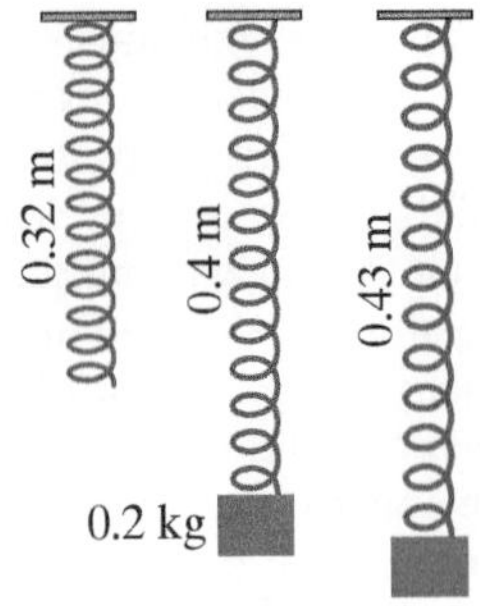

Figure 10.42

(a) How long does it take for the block to return to where it was released? **(b)** Next the block is again positioned at rest, hanging from the spring (0.4 m long) as shown in Figure 10.43. A bullet of mass 0.003 kg traveling at a speed of 200 m/s straight upward buries itself in the block, which then reaches a maximum height h above its original position. What is the speed of the block immediately after the bullet hits? **(c)** Now write an equation that could be used to determine how high the block goes after being hit by the bullet (a height h), but you need not actually solve for h.

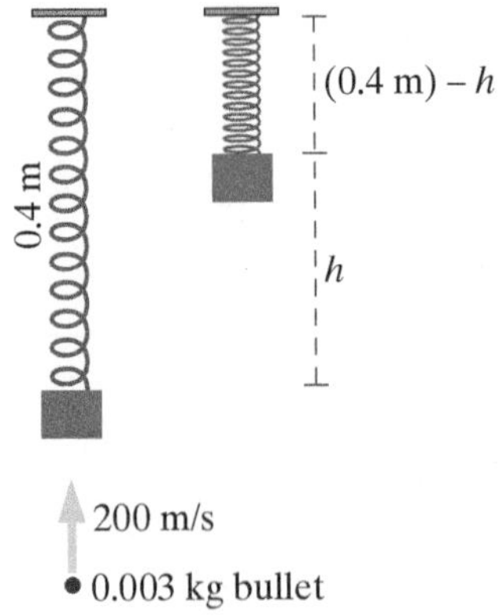

Figure 10.43

•P19 In outer space, rock 1 whose mass is 5 kg and whose velocity was $\langle 3300, -3100, 3400\rangle$ m/s struck rock 2, which was at rest. After the collision, rock 1's velocity is $\langle 2800, -2400, 3700\rangle$ m/s. **(a)** What is the final momentum of rock 2? **(b)** Before the collision, what was the kinetic energy of rock 1? **(c)** Before the collision, what was the kinetic energy of rock 2? **(d)** After the collision, what is the kinetic energy of rock 1? **(e)** Suppose that the collision was elastic (that is, there was no change in kinetic energy and therefore no change in thermal or other internal energy of the rocks). In that case, after the collision, what is the kinetic energy of rock 2? **(f)** On the other hand, suppose that in the collision some of the kinetic energy is converted into thermal energy of the two rocks, where $\Delta E_{\text{thermal},1} + \Delta E_{\text{thermal},2} = 7.16 \times 10^6$ J. What is the final kinetic energy of rock 2? **(g)** In this case (some of the kinetic energy being converted to thermal energy), what was the transfer of energy Q (microscopic work) from the surroundings into the two-rock system during the collision? (Remember that Q represents energy transfer due to a temperature difference between a system and its surroundings.)

•P20 A bullet of mass 0.102 kg traveling horizontally at a speed of 300 m/s embeds itself in a block of mass 3.5 kg that is sitting at rest on a nearly frictionless surface. **(a)** What is the speed of the block after the bullet embeds itself in the block? **(b)** Calculate the kinetic energy of the bullet plus the block before the collision. **(c)** Calculate the kinetic energy of the bullet plus the block after the collision. **(d)** Was this collision elastic or inelastic? **(e)** Calculate the rise in internal energy of the bullet plus block as a result of the collision.

••P21 A car of mass 2300 kg collides with a truck of mass 4300 kg, and just after the collision the car and truck slide along, stuck together, with no rotation. The car's velocity just before the collision was $\langle 38, 0, 0\rangle$ m/s, and the truck's velocity just before the collision was $\langle -16, 0, 27\rangle$ m/s. **(a)** Your first task is to determine the velocity of the stuck-together car and truck just after the collision. What system and principle should you use? (1) Energy Principle (2) Car plus truck (3) Momentum Principle (4) Car alone (5) Truck alone **(b)** What is the velocity of the stuck-together car and truck just after the collision? **(c)** In your analysis in part (b), why can you neglect the effect of the force of the road on the car and truck? **(d)** What is the increase in internal energy of the car and truck (thermal energy and deformation)? **(e)** Is this collision elastic or inelastic?

Section 10.7

••P22 A gold nucleus contains 197 nucleons (79 protons and 118 neutrons) packed tightly against each other. A single nucleon (proton or neutron) has a radius of about 1×10^{-15} m. Remember that the volume of a sphere is $\frac{4}{3}\pi r^3$. **(a)** Calculate the approximate radius of the gold nucleus. **(b)** Calculate the approximate radius of the alpha particle, which consists of 4 nucleons, 2 protons and 2 neutrons. **(c)** What kinetic energy must alpha particles have in order to make contact with a gold nucleus?

Rutherford correctly predicted the angular distribution for 10-MeV (kinetic energy) alpha particles colliding with gold nuclei. He was lucky: if the alpha particle had been able to touch the gold nucleus, the strong interaction would have been involved and the angular distribution would have deviated from that predicted by Rutherford, which was based solely on electric interactions.

•••P23 An alpha particle (a helium nucleus, containing 2 protons and 2 neutrons) starts out with kinetic energy of 10 MeV (10×10^6 eV), and heads in the $+x$ direction straight toward a gold nucleus (containing 79 protons and 118 neutrons). The particles are initially far apart, and the gold nucleus is initially at rest. Assuming that all speeds are small compared to the speed of light, answer the following questions about the collision. **(a)** What is the final momentum of the alpha particle, long after it interacts with the gold nucleus? **(b)** What is the final momentum of the gold nucleus, long after it interacts with the alpha particle? **(c)** What is the final kinetic energy of the alpha particle? **(d)** What is the final kinetic energy of the gold nucleus? **(e)** Assuming that the movement of the gold nucleus is negligible, calculate how close the alpha particle will get to the gold nucleus in this head-on collision.

Section 10.10

••P24 A Fe-57 nucleus is at rest and in its first excited state, 14.4 keV above the ground state (14.4×10^3 eV, where 1 eV = 1.6×10^{-19} J). The nucleus then decays to the ground state with the emission of a gamma ray (a high-energy photon). **(a)** What is the recoil speed of the nucleus? **(b)** Calculate the slight difference in eV between the gamma-ray energy and the 14.4 keV difference between the initial and final nuclear states. **(c)** The "Mössbauer effect" is the name given to a related phenomenon discovered by Rudolf Mössbauer in 1957, for which he received the 1961 Nobel Prize for physics. If the Fe-57 nucleus is in a solid block of iron, occasionally when the nucleus emits a gamma ray the entire solid recoils as one object. This can happen due to the fact that neighboring atoms and nuclei are connected by the electric interatomic force. In this case, repeat the calculation of part (b) and compare with your previous result. Explain briefly.

••**P25** There is an unstable particle called the "sigma-minus" (Σ^-), which can decay into a neutron and a negative pion (π^-): $\Sigma^- \to n + \pi^-$. The mass of the Σ^- is 1196 MeV/c^2, the mass of the neutron is 939 MeV/c^2, and the mass of the π^- is 140 MeV/c^2. Write equations that could be used to calculate the momentum and energy of the neutron and the pion. You do not need to solve the equations, which would involve some messy algebra. However, be clear in showing that you have enough equations that you could in principle solve for the unknown quantities in your equations.

It is advantageous to write the equations not in terms of v but rather in terms of E and p; remember that $E^2 - (pc)^2 = (mc^2)^2$.

••**P26** A beam of high-energy π^- (negative pions) is shot at a flask of liquid hydrogen, and sometimes a pion interacts through the strong interaction with a proton in the hydrogen, in the reaction $\pi^- + p^+ \to \pi^- + X^+$, where X^+ is a positively charged particle of unknown mass.

The incoming pion momentum is 3 GeV/c (1 GeV = 1000 MeV = 1×10^9 electron-volts). The pion is scattered through 40°, and its momentum is measured to be 1510 MeV/c (this is done by observing the radius of curvature of its circular trajectory in a magnetic field). A pion has a rest energy of 140 MeV, and a proton has a rest energy of 938 MeV.

What is the rest mass of the unknown X^+ particle, in MeV/c^2? Explain your work carefully.

It is advantageous to write the equations not in terms of v but rather in terms of E and p; remember that $E^2 - (pc)^2 = (mc^2)^2$.

••**P27** A particle of mass m, moving at speed $v = (4/5)c$, collides with an identical particle that is at rest. The two particles react to produce a new particle of mass M and nothing else. **(a)** What is the speed V of the composite particle? **(b)** What is its mass M?

••**P28** A charged pion ($m_\pi c^2 = 139.6$ MeV) at rest decays into a muon ($m_\mu = 105.7$ MeV) and a neutrino (whose mass is very nearly zero). Find the kinetic energy of the muon and the kinetic energy of the neutrino, in MeV (1 MeV = 1×10^6 eV).

••**P29** A hydrogen atom is at rest, in the first excited state, when it emits a photon of energy 10.2 eV. **(a)** What is the speed of the ground-state hydrogen atom when it recoils due to the photon emission? Remember that the magnitude of the momentum of a photon of energy E is $p = E/c$. Make the initial assumption that the kinetic energy of the recoiling atom is negligible compared to the photon energy. **(b)** Calculate the kinetic energy of the recoiling atom. Is this kinetic energy indeed negligible compared to the photon energy?

•••**P30** At the PEP II facility at the Stanford Linear Accelerator Center (SLAC) in California and at the KEKB facility in Japan, electrons with momentum 9.03 GeV/c were made to collide head-on with positrons whose momentum is 3.10 GeV/c (1 GeV = 10^9 eV); see Figure 10.44. That is, pc for the electron is 9.03 GeV and pc for the positron is 3.10 GeV. The values of pc and the corresponding energies are so large with respect to the electron or positron rest energy (0.5 MeV = 0.0005 GeV) that for the purposes of this analysis you may, if you wish, safely consider the electron and positron to be massless.

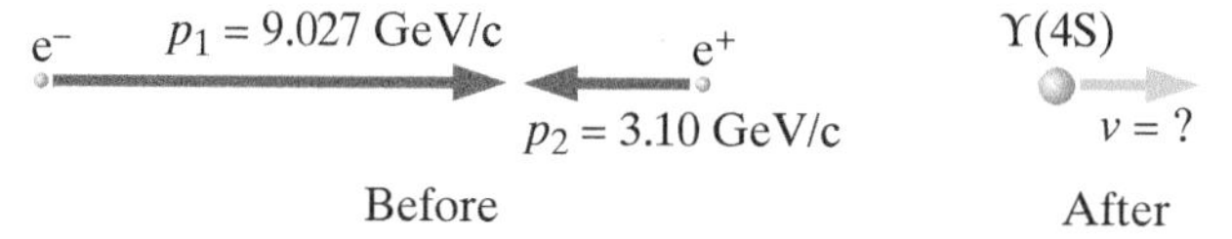

Figure 10.44

(a) The electron–positron collision produces in an intermediate state a particle called the Υ(4S) ("Upsilon 4S"), in the reaction $e^- + e^+ \to \Upsilon(4S)$. Show that the rest energy of the Υ(4S) is 10.58 GeV. **(b)** What is the speed of the Υ(4S) produced in the collision? **(c)** The Υ(4S) decays almost immediately into two "B" mesons: $\Upsilon(4S) \to B^0 + \overline{B^0}$. The $\overline{B^0}$ is the antiparticle of the B^0 and has the same rest energy Mc^2 = 5.28 GeV as the B^0. Consider the case in which both "B" mesons are emitted at the same angle θ to the direction of the moving Υ(4S), as shown in Figure 10.45. Calculate this angle θ.

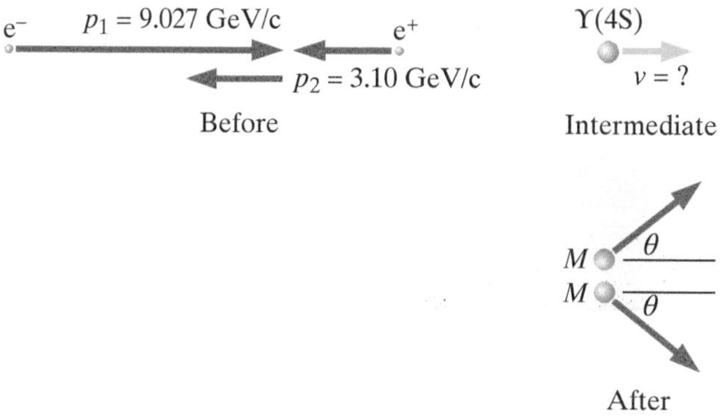

Figure 10.45

•••**P31** In a reference frame where a Δ^+ particle is at rest, it decays into a proton and a high-energy photon (a "gamma ray"): $\Delta^+ \to p^+ + \gamma$. The mass of the Δ^+ particle is 1232 MeV/c^2 and the mass of the proton is 938 MeV/c^2 (1 MeV = 1×10^6 eV). Calculate the energy of the gamma ray and the speed of the proton.

Section 10.12

••**P32** Redo Problem P21, this time using the concept of the center-of-momentum reference frame.

••**P33** Redo the analysis of the Rutherford experiment, this time using the concept of the center-of-momentum reference frame. Let m = the mass of the alpha particle and M=the mass of the gold nucleus. Consider the specific case of the alpha particle rebounding straight back. The incoming alpha particle has a momentum p_1, the outgoing alpha particle has a momentum p_3, and the gold nucleus picks up a momentum p_4. **(a)** Determine the velocity of the center of momentum of the system. **(b)** Transform the initial momenta to that frame (by subtracting the center-of-momentum velocity from the original velocities). **(c)** Show that if the momenta in the center-of-momentum frame simply turn around (180°), with no change in their magnitudes, both momentum and energy conservation are satisfied, whereas no other possibility satisfies both conservation principles. (Try drawing some other momentum diagrams.) **(d)** After the collision, transform back to the original reference frame (by adding the center-of-momentum velocity to the velocities of the particles in the center-of-mass frame). Although using the center-of-momentum frame may be conceptually more difficult, the algebra for solving for the final speeds is much simpler.

COMPUTATIONAL PROBLEMS

More detailed and extended versions of some of these computational modeling problems may be found in the lab activities included in the *Matter & Interactions*, 4th Edition, resources for instructors.

••P34 The following code is the skeleton of a computational model of a collision between a single moving alpha particle and a single gold nucleus which is initially stationary. Some pieces of the physical model are missing. An alpha particle consists of 2 protons and 2 neutrons. A gold nucleus contains 79 protons and 118 neutrons.

```
from visual import *
from visual.graph import *
scene.width = 1024
scene.height = 600
q_e = 1.6e-19
m_p = 1.7e-27
oofpez = 9e9
m_Au = (79+118) * m_p
m_Alpha = (2+2) * m_p
qAu = 2 * q_e
qAlpha = 79 * q_e
deltat = 1e-23
Au = sphere(pos=vector(0,0,0), radius=4e-15,
            color=color.yellow, make_trail=True)
Alpha = sphere(pos=vector(-1e-13,5e-15,0),
               radius=1e-15, color=color.magenta,
               make_trail=True)
p_Au = m_Au*vector(0,0,0)
p_Alpha = vector(1.043e-19,0,0)
t = 0
while t<1.3e-20:
    rate(100)
    Alpha.pos = Alpha.pos+(p_Alpha/m_Alpha) *
deltat
    t = t + deltat
```

(a) Read the code, and predict what will happen when the unmodified skeleton program is run. Then run the program to check your prediction. **(b)** Modify the program so that it represents a reasonable physical model of this interaction. **(c)** Does the gold nucleus move during these collisions? Should it? You can see the trail left by the gold nucleus better if you make this object slightly transparent by adding this line of code before the loop: `Au.opacity = 0.7` **(d)** What is the value of the impact parameter b in this program? Experiment with different impact parameters, and report what you observe. **(e)** After the loop, add to your program a calculation of the scattering angle θ (the angle between the final and initial momenta of the alpha particle). The VPython function for the dot product is `dot(A,B)`. The function `acos(D)` returns the angle (in radians) whose cosine is D. Check to make sure the angle you calculate makes sense in terms of what you observe on the screen. **(f)** Find values of the impact parameter that lead to the following scattering angles: (1) 90°, (2) 168°, (3) 38°, (4) 13°.

•P35 Start with the program you wrote for Problem P34. Add graphs to display the values of the x and y components of the momentum of each particle, and the sums of all x components and of all y components, as a function of time. Is momentum conserved during the collision? What should you change in your program if you find that momentum is not conserved? By adding the following code before the loop you can create two different graphing windows, and plot all curves simultaneously:

```
gdx = gdisplay(x=0,y=600, width=500,
               title='p_x')
p_Au_x_graph = gcurve(color=color.yellow)
p_Alpha_x_graph = gcurve(color=color.magenta)
## other gcurves if needed
gdy = gdisplay(x=500,y=600,width=500,title='p_y')
p_Au_y_graph = gcurve(color=color.yellow)
## other gcurves if needed
```

•••P36 The *Voyager 1* spacecraft was launched in 1977 and has recently left the Solar System and entered interstellar space. No existing rocket is powerful enough to launch a spacecraft with enough speed to coast that far against the Sun's gravitational pull, so a "gravity boost" was obtained from Jupiter (and another from Saturn). The basic idea is that it was possible to arrange *Voyager*'s trajectory in such a way that in a brief noncontact collision with Jupiter the spacecraft gained momentum, with a corresponding loss of the giant planet's momentum, which due to Jupiter's huge mass corresponded to a tiny change in Jupiter's speed.

Construct a computational model of this maneuver. The collision takes only a few days, so it is okay to approximate Jupiter's motion around the Sun with a straight line (Jupiter takes 12 Earth years to orbit the Sun). Jupiter's mass is 1.9×10^{27} kg. Give your spacecraft the same mass as the *Voyager* spacecraft (722 kg). Jupiter's radius is 7×10^7 m and it is 7.8×10^{11} m from the Sun, with an orbital speed of 1.3×10^4 m/s. **(a)** Place Jupiter at the location $\langle 0,0,0 \rangle$ m, with a velocity of $\langle -1.3 \times 10^4,0,0 \rangle$ m/s. Choose a starting location for the spacecraft at location $\langle x_i,-25R,0 \rangle$, where R is the radius of Jupiter. You will need to find a value for x_i that leads to the spacecraft passing close to the right side of Jupiter. Let the initial velocity of the spacecraft be $\langle 0,1 \times 10^4,0 \rangle$ m/s. (From this perspective you are looking down on the North pole of Jupiter, with the outward bound spacecraft heading toward Jupiter.) **(b)** Exit the computational loop when the spacecraft is as far from Jupiter as it was at the start. (Put this test at the end of your loop, so the loop doesn't stop immediately.) Print the change in the speed, $v_f - v_i$, and the time in Earth days for the process. Modify your starting position x_i until you find that the speed change is approximately 1.6×10^4 m/s. **(c)** Include a graph of the speed of the spacecraft as a function of time. **(d)** Find a different value of x_i for the spacecraft that results in the spacecraft losing a significant amount of speed instead of gaining speed.

Suggestions: If you precede a `gcurve` statement with `gdisplay(y=600)` the graph will be placed conveniently beneath your graphics window. If you specify `scene.autocenter = True` the camera will continually shift to point at the center of the scene, which is useful in this situation.

•••**P37** Start with the program you wrote to model a spacecraft's interaction with Jupiter (Problem P36). Use initial conditions that result in the spacecraft gaining speed. **(a)** Change to a center-of-mass reference frame by subtracting Jupiter's velocity from the initial velocities of the spacecraft and Jupiter. (Because Jupiter is so massive, its velocity is almost exactly the velocity of the center of mass of the combined system.) What do you observe when you re-run your program? **(b)** What is the speed change of the spacecraft in the center-of-mass reference frame?

ANSWERS TO CHECKPOINTS

1 (a) 300 J; **(b)** 5 m/s; **(c)** 150 J; **(d)** 150 J
2 (a) 2.3×10^{-24} m/s (!), so it is an excellent approximation to say that the Earth doesn't recoil. **(b)** 1.6×10^{-23} J; $K_{\text{baseball}} = 150$ J
3 8×10^4 m/s
4 Car 1's momentum was $\sqrt{3}$ greater than the momentum of car 2.
5 We outline the entire solution.

First find $v_{\text{cm},x}$:

$$v_{\text{cm},x} = \frac{(1000\,\text{kg})(30\,\text{m/s}) + (3000\,\text{kg})(-20\,\text{m/s})}{(4000\,\text{kg})} = -7.5\,\text{m/s}$$

Transform to center-of-momentum frame:

$$v_{\text{car},x} = (30\,\text{m/s}) - (-7.5\,\text{m/s}) = 37.5\,\text{m/s}$$
$$v_{\text{truck},x} = (-20\,\text{m/s}) - (-7.5\,\text{m/s}) = -12.5\,\text{m/s}$$

Check to make sure that the momenta are equal and opposite in this frame:

$$p_{\text{car},x} = (1000\,\text{kg})(37.5\,\text{m/s}) = 37500\,\text{kg}\cdot\text{m/s}$$
$$p_{\text{truck},x} = (3000\,\text{kg})(-12.5\,\text{m/s}) = -37500\,\text{kg}\cdot\text{m/s}$$

In this frame the kinetic energy before the collision is this:

$$K = \frac{(37500\,\text{kg}\cdot\text{m/s})^2}{2(1000\,\text{kg})} + \frac{(37500\,\text{kg}\cdot\text{m/s})^2}{2(3000\,\text{kg})} = 9.4 \times 10^5\,\text{J}$$

After the collision, the velocities are zero in this frame, so kinetic energy in this frame goes to zero, and there is an increase in the internal energy of the mangled car and truck of amount 9.4×10^5 J. There must be the same internal energy increase in the original reference frame, so the kinetic energy lost in the original reference frame was 9.4×10^5 J.

CHAPTER

11

Angular Momentum

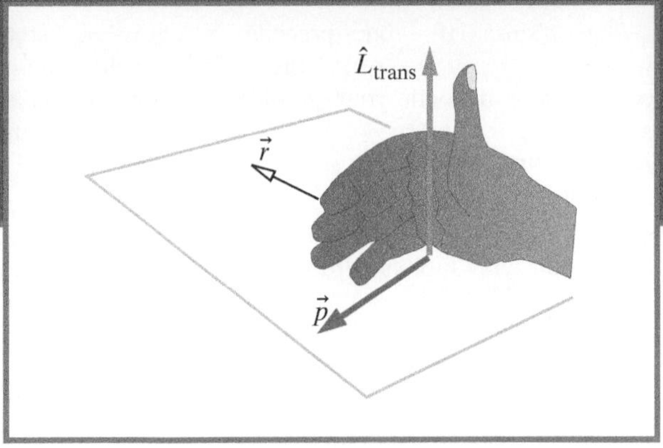

OBJECTIVES

After studying this chapter, you should be able to

- Calculate both translational and rotational angular momentum in 3D.
- Apply the Angular Momentum Principle to predict the new angular momentum of a system subject to a torque.
- Apply Conservation of Angular Momentum to a system consisting of more than one particle.
- Predict the position of a rotating object.

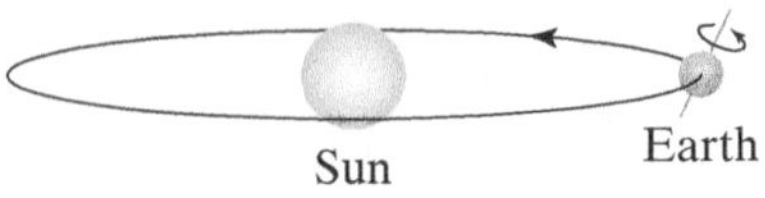

Figure 11.1 The Earth orbits the Sun, and it also spins about its own axis (which is tilted with respect to the orbital plane). The sizes of the Earth and Sun are exaggerated in this diagram.

Angular momentum is a measure of rotational motion. Translational (or "orbital") angular momentum describes motion such as the orbit of the Earth around the Sun. Rotational (or "spin") angular momentum describes motion such as the revolution of the Earth around its own axis (Figure 11.1). Just as the Momentum Principle relates a change in momentum to the net force on a system, the Angular Momentum Principle relates a change in angular momentum to the net torque, or twist, applied to a system.

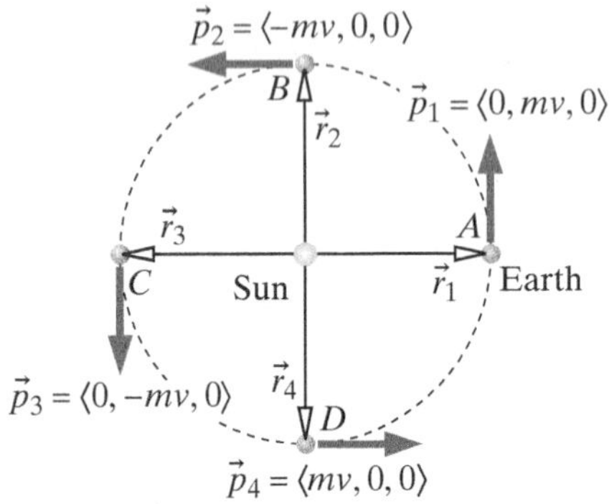

Figure 11.2 The Earth goes around the Sun in a nearly circular orbit. The position vector and the momentum vector change continuously.

11.1 TRANSLATIONAL ANGULAR MOMENTUM

As the Earth goes around the Sun in a nearly circular orbit, the position vector (from Sun to Earth) is constantly changing in direction, and the momentum of the Earth is also constantly changing in direction, as shown in Figure 11.2. Both the position vector and the momentum vector continually change, though their magnitudes are (nearly) constant. In Figure 11.2, as the Earth goes from A to B in three months (one quarter of a year), the Earth's momentum changes from $\langle 0, mv, 0\rangle$ to $\langle -mv, 0, 0\rangle$, and the change in the momentum is $\Delta\vec{p} = \langle -mv, 0, 0\rangle - \langle 0, mv, 0\rangle = \langle -mv, -mv, 0\rangle$. This change in the Earth's momentum is of course due to the gravitational force exerted on the Earth by the Sun.

Nevertheless, it feels like something is constant here. For example, the angle between the position vector and the momentum vector is always 90° for this circular orbit. Also, the direction of rotation of the circular motion stays constant (counterclockwise in Figure 11.2). The quantity that is constant is $\vec{L}_{\text{trans,Sun}}$, the translational angular momentum of the Earth relative to the Sun.

Magnitude of Translational Angular Momentum

The magnitude of the Earth's orbital or translational angular momentum, relative to the center of the Sun, involves $|\vec{r}\,|$, $|\vec{p}|$, and the angle θ between the two. It is defined like this:

$$\left|\vec{L}_{\text{trans,Sun}}\right| = |\vec{r}\,|\,|\vec{p}|\sin\theta$$

EXAMPLE **Earth's Translational Angular Momentum**

Calculate the magnitude of the Earth's translational (orbital) angular momentum relative to the Sun when the Earth is at location A and when the Earth is at location B as shown in Figure 11.2. The mass of the Earth is 6×10^{24} kg and its distance from the Sun is 1.5×10^{11} m (see the table of values at the end of this book).

Solution The Earth makes one complete orbit of the Sun in 1 y, so its average speed is:

$$v = \frac{2\pi(1.5 \times 10^{11}\ \text{m})}{(365)(24)(60)(60)\ \text{s}} = 3.0 \times 10^4\ \text{m/s}$$

At location A

$$\vec{p} = \langle 0, 6 \times 10^{24}\ \text{kg} \cdot 3.0 \times 10^4\ \text{m/s}, 0 \rangle = \langle 0, 1.8 \times 10^{29}, 0 \rangle\ \text{kg} \cdot \text{m/s}$$

$$|\vec{p}| = 1.8 \times 10^{29}\ \text{kg} \cdot \text{m/s}$$

$$\left|\vec{L}_{\text{trans,Sun}}\right| = (1.5 \times 10^{11}\ \text{m})(1.8 \times 10^{29}\ \text{kg} \cdot \text{m/s}) \sin 90^\circ$$

$$= 2.7 \times 10^{40}\ \text{kg} \cdot \text{m}^2/\text{s}$$

At location B, $|\vec{r}|$, $|\vec{p}|$, and θ are the same as they were at location A, so $\left|\vec{L}_{\text{trans,Sun}}\right|$ also has the same value it had at location A.

Figure 11.3 The right-hand rule.

Direction of Translational Angular Momentum

The magnitude of the translational angular momentum doesn't tell us everything we need to know about the orbital motion of the Earth. To describe the motion indicated in Figure 11.2 we could say something like this:

> If you stand at a location on the $+z$ axis and look toward the origin, the Earth appears to be moving counterclockwise in the xy plane.

However, this is annoyingly verbose and not useful mathematically; for example, one can't add "counterclockwise in the xy plane" to "clockwise in the yz plane." Instead, we specify the direction this way:

- The plane in which both $\vec{r}$ and $\vec{p}$ lie can be indicated by a unit vector perpendicular to that plane.
- The direction of motion within the plane (clockwise or counterclockwise) can be indicated by establishing a "right-hand" rule for this unit vector.
- Right-hand rule: Curl the fingers of your right hand in the direction of the rotational motion in the plane and extend your thumb, as shown in Figure 11.3. The unit vector representing the direction of the angular momentum is defined to point in the direction of your thumb.
- If the rotational motion is counterclockwise, the unit vector (and your right thumb) point out of the plane, as in Figure 11.3. If the rotational motion is clockwise, the unit vector (and your right thumb) points into the plane, as in Figure 11.4.

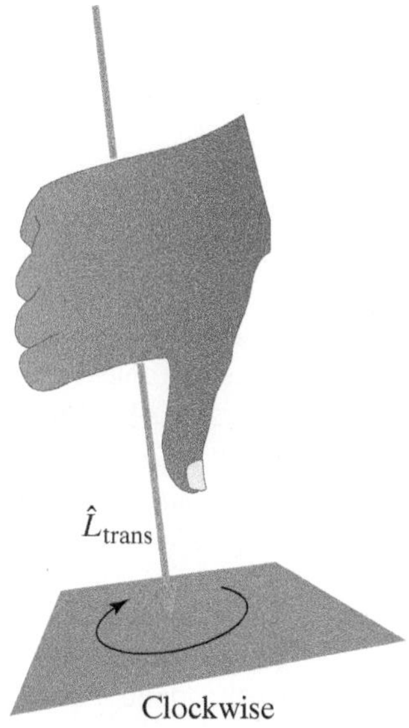

Figure 11.4 The right-hand rule.

EXAMPLE **Direction of Translational Angular Momentum**

Assuming the usual coordinate system, what is the direction of the Earth's translational angular momentum, as indicated in Figure 11.2?

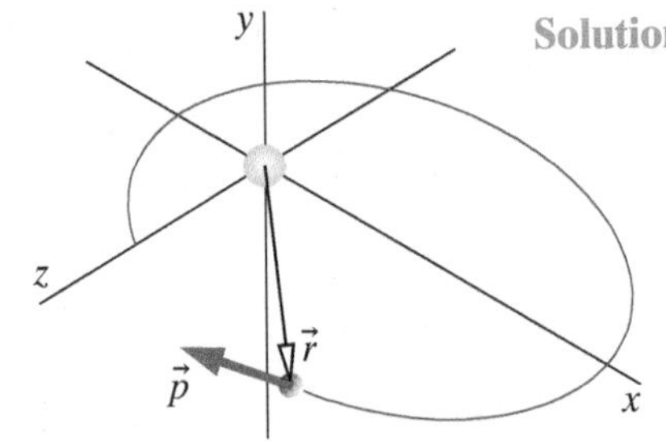

Figure 11.5 A comet orbits the Sun. The momentum vector and relative position vector lie in the xz plane.

Solution Applying the right-hand rule we find that the Earth's translational angular momentum is in the $+z$ direction: out of the page toward you.

Angular Momentum as a Vector

Putting together the results from the previous two examples, we can express the translational angular momentum of the Earth with respect to the Sun as a vector:

$$\vec{L}_{\text{trans,Sun}} = (2.7 \times 10^{40}\ \text{kg} \cdot \text{m}^2/\text{s}) \cdot \langle 0,0,1 \rangle = \langle 0,0,2.7 \times 10^{40} \rangle\ \text{kg} \cdot \text{m}^2/\text{s}$$

Checkpoint 1 What is the direction of the orbital (translational) angular momentum of the comet shown in Figure 11.5, relative to the Sun?

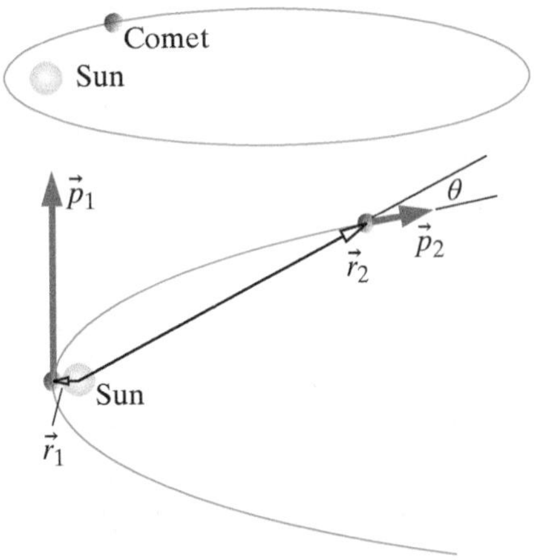

Figure 11.6 The orbit of Halley's comet, shown here in the xy plane. The top figure shows the full orbit. The bottom figure shows a magnified view of the portion of the orbit nearest the Sun.

An Elliptical Orbit: Halley's Comet

The previous exercise raises an interesting question: What if $|\vec{r}|$, $|\vec{p}|$, and θ are not constant (as they were in a circular orbit)? Is the angular momentum, relative to the Sun, of a comet in an elliptical orbit around the Sun constant, or does it change? During this orbit, all three parameters ($|\vec{r}|$, $|\vec{p}|$, and θ) change continuously.

Halley's comet orbits the Sun with a period of 75.3 y, and will next be visible on Earth in the year 2061. Figure 11.6 displays two views of the orbit, here shown in the xy plane. From the upper diagram it is clear that the orbit of the comet is highly elliptical. The lower diagram shows a magnified view of the portion of the orbit nearest to the Sun, and displays the momentum vectors and position vectors for the comet at two different points in its orbit. From available data we can calculate the angular momentum of Halley's comet relative to the Sun at two different locations.

EXAMPLE **Angular Momentum of Halley's Comet**

The highly elliptical orbit of Halley's comet is shown in Figure 11.6. When the comet is closest to the Sun, at the location specified by the position vector $\vec{r}_1$ ("perihelion"), it is 8.77×10^{10} m from the Sun, and its speed is 5.46×10^4 m/s. When the comet is at the location specified by the position vector $\vec{r}_2$, its speed is 1.32×10^4 m/s. At that location the distance between the comet and the Sun is 1.19×10^{12} m, and the angle θ is 17.81°. The mass of the comet is estimated to be 2.2×10^{14} kg. Calculate the translational (orbital) angular momentum of the comet, relative to the Sun, at both locations.

Solution Direction: At both locations, the direction of the translational angular momentum of the comet is in the $-z$ direction (into the page); determined by using the right-hand rule.

At location 1:

$$\left|\vec{L}_{\text{trans,Sun}}\right| = (8.77 \times 10^{10}\ \text{m})(2.2 \times 10^{14}\ \text{kg})(5.46 \times 10^4\ \text{m/s}) \sin 90°$$
$$= 1.1 \times 10^{30}\ \text{kg} \cdot \text{m}^2/\text{s}$$
$$\vec{L}_{\text{trans,Sun}} = \langle 0,0,-1.1 \times 10^{30} \rangle\ \text{kg} \cdot \text{m}^2/\text{s}$$

At location 2:

$$\left|\vec{L}_{\text{trans,Sun}}\right| = (1.19\times10^{12}\text{ m})(2.2\times10^{14}\text{ kg})(1.32\times10^{4}\text{ m/s})\sin 17.81°$$
$$= 1.1\times10^{30}\text{ kg}\cdot\text{m}^2\text{/s}$$
$$\vec{L}_{\text{trans,Sun}} = \langle 0, 0, -1.1\times10^{30}\rangle\text{ kg}\cdot\text{m}^2\text{/s}$$

Even in this highly elliptical orbit, the comet's translational angular momentum is constant throughout the orbit, despite the fact that its position, its momentum, and the angle between them change continuously. It was observations like this that first suggested that angular momentum might be a conserved quantity, and should not change if there is no twist exerted on the system by the surroundings.

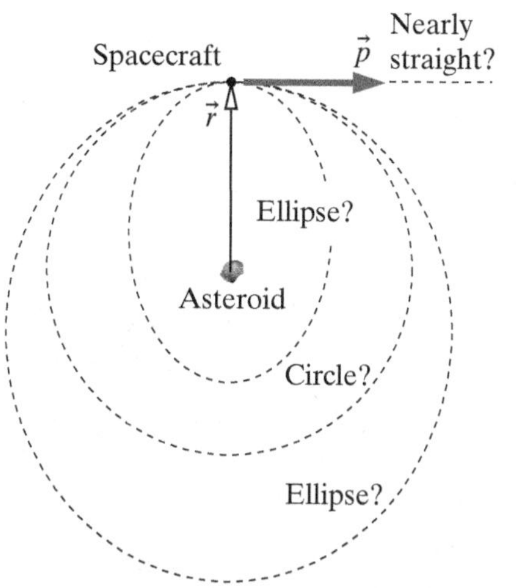

Figure 11.7 A snapshot of a spacecraft passing an asteroid. Its future motion might be a circle or an ellipse, or even a nearly straight line if the asteroid's mass is small.

A Straight Path

In calculating translational (orbital) angular momentum, we use the position and momentum of an object at a particular instant. We do not necessarily have any information about how the object's path will curve in the future or whether it has curved in the past. Figure 11.7 illustrates this point by showing many possible paths of a spacecraft passing an asteroid. It is possible that the spacecraft will be nearly unaffected by the asteroid, and will travel nearly in a straight line, but we can still calculate its orbital angular momentum at the instant shown.

Later in this chapter we will see why it may be physically meaningful to attribute angular momentum to an object that moves in a straight line.

Choosing a Location

Translational or orbital angular momentum is defined relative to a particular location. If we choose a location A, the magnitude of the angular momentum includes the distance r_A to the moving object. The translational angular momentum relative to some other location B would have a different magnitude because r_B would be a different distance. We will see later that it is usually advantageous to choose a location around which there is little or no twist applied by objects in the surroundings.

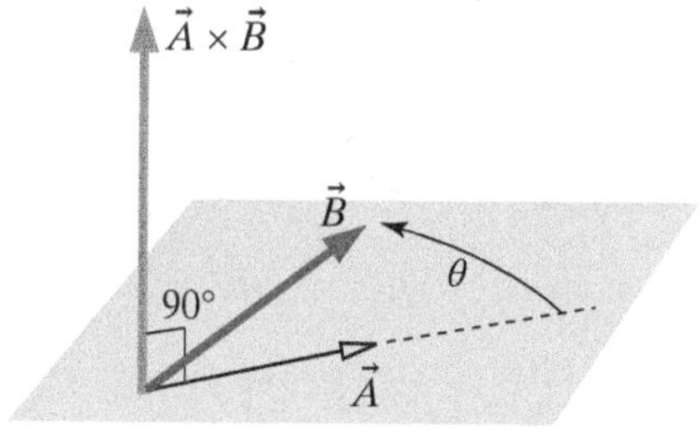

Figure 11.8 The cross product produces a vector that is perpendicular to the two original vectors.

The Vector Cross Product

The formal definition of translational angular momentum involves a vector cross product. In the mathematics of vectors, the cross product is an operation that combines two vectors to produce a third vector that is perpendicular to the plane defined by the original vectors (Figure 11.8). The cross product of two vectors $\vec{A}$ and $\vec{B}$ is written as $\vec{A}\times\vec{B}$.

We already know one way of evaluating a vector cross product, by finding the magnitude and direction separately. The quantity

$$\left|\vec{L}_{\text{trans,Sun}}\right| = |\vec{r}\,|\,|\vec{p}|\sin\theta$$

is the magnitude of the cross product $\vec{r}\times\vec{p}$:

$$|\vec{r}\times\vec{p}| = \left|\vec{L}_{\text{trans,Sun}}\right| = |\vec{r}\,|\,|\vec{p}|\sin\theta$$

where θ is the angle between the two vectors when they are placed tail to tail. θ is always less than or equal to 180°. The direction of $\vec{r}\times\vec{p}$ is given by a right-hand rule:

- Draw the vectors $\vec{r}$ and $\vec{p}$ so their tails are in the same location.

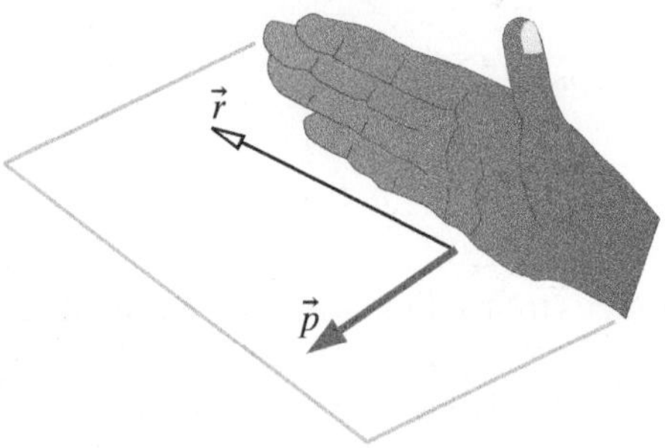

Figure 11.9 With the tails of $\vec{r}$ and $\vec{p}$ together, point your fingers along $\vec{r}$.

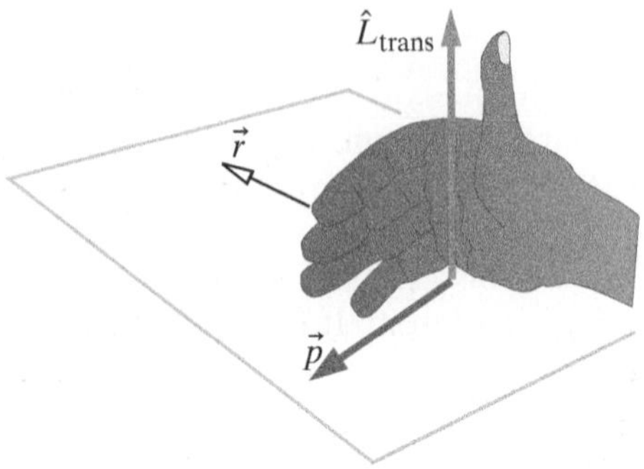

Figure 11.10 Fold your fingers toward $\vec{p}$. Your extended thumb points in the direction of $\vec{r} \times \vec{p}$.

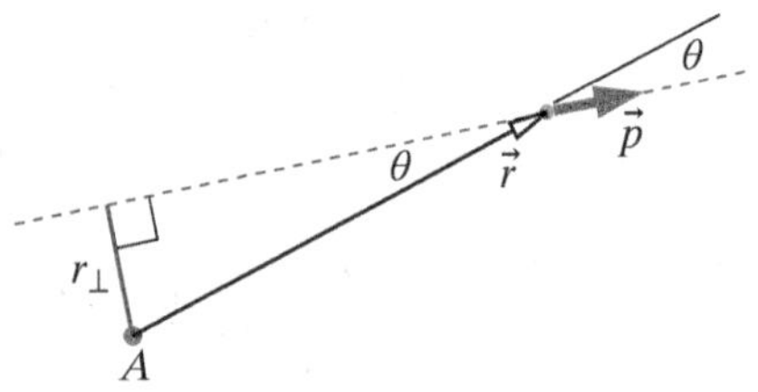

Figure 11.11 $r_\perp = r\sin\theta$ is the perpendicular distance from the tail of $\vec{r}$ to a line collinear with $\vec{p}$.

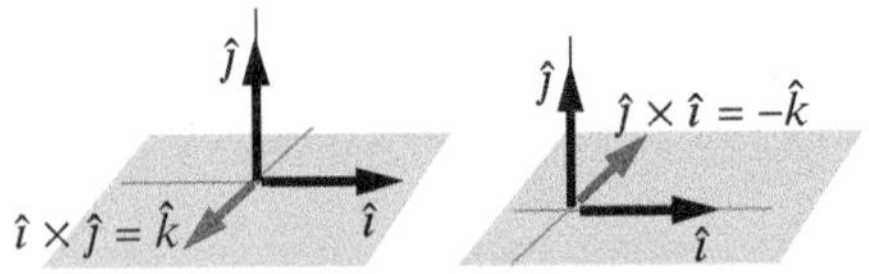

Figure 11.12 Cross products of unit vectors.

- Point the fingers of your right hand in the direction of $\vec{r}$ (Figure 11.9), then fold them through the angle θ toward $\vec{p}$ (Figure 11.10). You may need to turn your hand over to be able to do this.
- Stick your thumb out. Your thumb points in the direction of the angular momentum vector.

The Perpendicular Distance $r_\perp$

The quantity $r\sin\theta$ has a geometrical meaning. As shown in Figure 11.11, it is the component of $\vec{r}$ perpendicular to $\vec{p}$. If you draw a line perpendicular to $\vec{p}$, through the point about which $\vec{L}$ is to be calculated, $r_\perp = r\sin\theta$ is the distance along this line from the tail of $\vec{r}$ to the line along which the momentum or velocity points. $r_\perp$ is closely related to the impact parameter introduced in the previous chapter on collisions.

PERPENDICULAR DISTANCE $r_\perp$

$$r_\perp = r\sin\theta$$

Vector Components in the Cross Product

Given the components of two vectors, it is possible to evaluate the cross product in terms of unit vectors along the three axes (Figure 11.12). Recall that $\hat{\imath} = \langle 1,0,0 \rangle$, and so on.

First, note that $\hat{\imath} \times \hat{\imath} = 0$, $\hat{\jmath} \times \hat{\jmath} = 0$, and $\hat{k} \times \hat{k} = 0$, since when we cross a vector with itself the angle between the two vectors is zero, and $\sin 0° = 0$.

Second, $\hat{\imath} \times \hat{\jmath} = \hat{k}$, since the angle is 90° and the right-hand rule gives a result in the $+z$ direction (out of the page; Figure 11.12). On the other hand, $\hat{\jmath} \times \hat{\imath} = -\hat{k}$, because the right-hand rule gives a result in the $-z$ direction (into the page). Similarly, $\hat{\jmath} \times \hat{k} = \hat{\imath}$, $\hat{k} \times \hat{\jmath} = -\hat{\imath}$, $\hat{k} \times \hat{\imath} = \hat{\jmath}$, and $\hat{\imath} \times \hat{k} = -\hat{\jmath}$.

If $\vec{A}$ and $\vec{B}$ lie in the xy plane, we can use these results for the unit vectors to calculate the cross product, which will be in the $+z$ or $-z$ direction:

$$\begin{aligned}\vec{A} \times \vec{B} &= (A_x\hat{\imath} + A_y\hat{\jmath}) \times (B_x\hat{\imath} + B_y\hat{\jmath}) \\ &= (A_xB_x)\hat{\imath} \times \hat{\imath} + (A_yB_y)\hat{\jmath} \times \hat{\jmath} + (A_xB_y)\hat{\imath} \times \hat{\jmath} + (A_yB_x)\hat{\jmath} \times \hat{\imath} \\ &= 0 + 0 + A_xB_y\hat{k} - A_yB_x\hat{k} \\ &= (A_xB_y - A_yB_x)\hat{k}\end{aligned}$$

Similar results hold for components of the cross product in the x and y directions, so that in general we have the following result:

CROSS PRODUCT $\vec{A} \times \vec{B}$

$$\vec{A} \times \vec{B} = \langle A_yB_z - A_zB_y, A_zB_x - A_xB_z, A_xB_y - A_yB_x \rangle$$

$$|\vec{A} \times \vec{B}| = |\vec{A}||\vec{B}|\sin\theta$$

The direction of the vector is given by the right-hand rule.

For example,

$$\begin{aligned}\langle 2,3,4 \rangle \times \langle 5,6,7 \rangle &= \langle 3\cdot 7 - 4\cdot 6, 4\cdot 5 - 2\cdot 7, 2\cdot 6 - 3\cdot 5 \rangle \\ &= \langle -3, 6, -3 \rangle\end{aligned}$$

This approach to calculating a cross product is particularly useful in computer calculations. Note the cyclic nature of the subscripts: $x{:}yz, y{:}zx, z{:}xy$.

If you have studied determinants, you may recognize that the components of $\vec{A} \times \vec{B}$ can be thought of as the "minors" of a 3 by 3 determinant in which $\vec{A}$ and $\vec{B}$ are the second and third rows, and the first row is $\hat{\imath}$, $\hat{\jmath}$, $\hat{k}$.

Applying this general formalism to translational angular momentum, we can define the angular momentum of a particle, relative to location A, as:

TRANSLATIONAL ANGULAR MOMENTUM RELATIVE TO LOCATION A

$$\vec{L}_{\text{trans,A}} = \vec{r}_A \times \vec{p}$$

$\vec{r}_A$ extends from location A to the object.

The vector dot product $\vec{r}_A \times \vec{p}$ may be evaluated directly:

$$\vec{L}_{\text{trans,A}} = \langle r_y p_z - r_z p_y, r_z p_x - r_x p_z, r_x p_y - r_y p_x \rangle$$

Alternatively, magnitude and direction may be found separately:

$$\left|\vec{L}_{\text{trans,A}}\right| = |\vec{r}\,||\vec{p}|\sin\theta$$

The direction of $\vec{L}_{\text{trans,A}}$ may be found using the right-hand rule.

Note that this formal definition allows us to calculate the translational (orbital) angular momentum of an object at any instant, even if we do not know the direction of the path it will follow in the future.

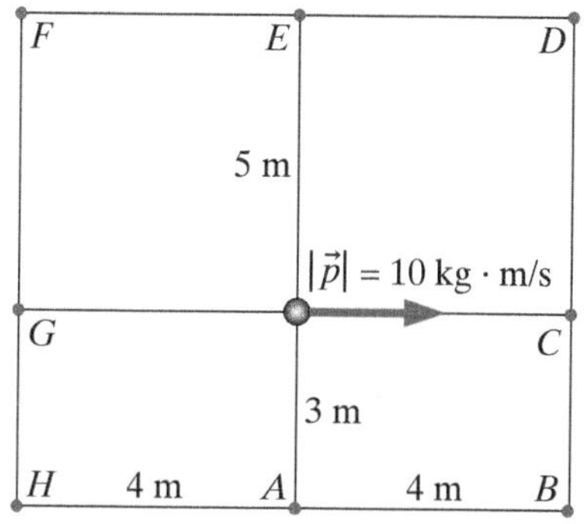

Figure 11.13 Determine the magnitude and direction of the angular momentum of the particle relative to various locations. All points lie in the xy plane.

Two-Dimensional Projections

Because it is more difficult to sketch a situation in three dimensions, whenever possible we will work with two-dimensional projections onto a plane. If $\vec{A}$ and $\vec{B}$ lie in the xy plane, the cross product vector $\vec{A} \times \vec{B}$ points in the $+z$ direction (out of the page, ⊙) or in the $-z$ direction (into the page, ⊗). The symbol ⊙ is supposed to suggest the tip of an arrow pointing toward you (angular momentum pointing toward you). The symbol ⊗ is supposed to suggest the feathers on the end of an arrow that is heading away from you (angular momentum pointing away from you).

EXAMPLE **Calculating Angular Momentum**

Use three different methods to calculate the angular momentum of the particle in Figure 11.13 relative to location B. How does this value compare to the angular momentum of the same particle relative to location A and to location C?

Solution Method 1: Vector cross product

$$\vec{r}_B = \langle -4, 3, 0 \rangle \text{ m}$$
$$\vec{p} = \langle 10, 0, 0 \rangle \text{ kg} \cdot \text{m/s}$$
$$\vec{L}_B = \vec{r}_B \times \vec{p}$$
$$= \langle r_y p_z - r_z p_y, r_z p_x - r_x p_z, r_x p_y - r_y p_x \rangle$$
$$= \langle (3 \cdot 0 - 0 \cdot 0), (0 \cdot 10 - -4 \cdot 0), (-4 \cdot 0 - 3 \cdot 10) \rangle$$
$$= \langle 0, 0, -30 \rangle \text{ kg} \cdot \text{m}^2/\text{s}$$

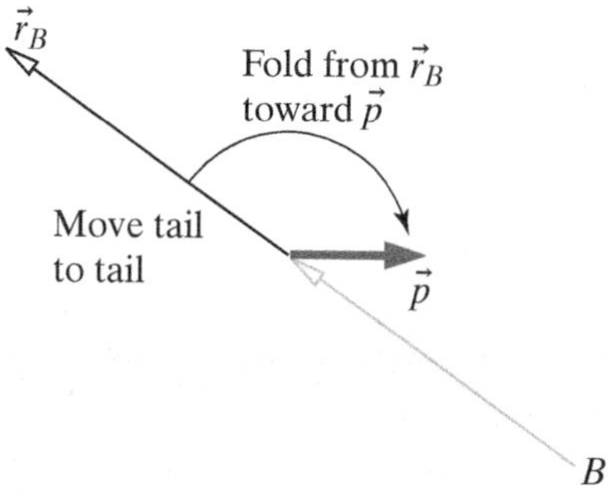

Figure 11.14 Place $\vec{r}_B$ and $\vec{p}$ tail to tail, then fold your right fingers from $\vec{r}_B$ toward $\vec{p}$.

Method 2: Direction and magnitude using $|\vec{r}_B||\vec{p}|\sin\theta$

In Figure 11.14 we have placed $\vec{r}_B$ and $\vec{p}$ tail to tail. The vectors $\vec{r}_B$ and $\vec{p}$ define a plane, which is the xy plane (the plane of the paper). Now fold the fingers of your right hand from $\vec{r}_B$ toward $\vec{p}$ (Figure 11.14). To be able to do this, your thumb must point into the page (⊗), so the angular momentum is in the $-z$ direction (into the plane of the paper). Note too that you fold your fingers clockwise, which is associated with angular momentum into the page.

$$|\vec{L}_B| = |\vec{r}_B||\vec{p}|\sin\theta$$
$$|\vec{r}_B| = \sqrt{3^2+4^2} = 5\text{ m}$$
$$|\vec{p}| = 10\text{ kg}\cdot\text{m/s}$$
$$\sin\theta = 3/5 \quad \text{(See Figure 11.15)}$$
$$|\vec{L}_B| = (5)(10)(3/5) = 30\text{ kg}\cdot\text{m}^2\text{/s}$$
$$\vec{L}_B = \langle 0, 0, -30\rangle\text{ kg}\cdot\text{m}^2\text{/s}$$

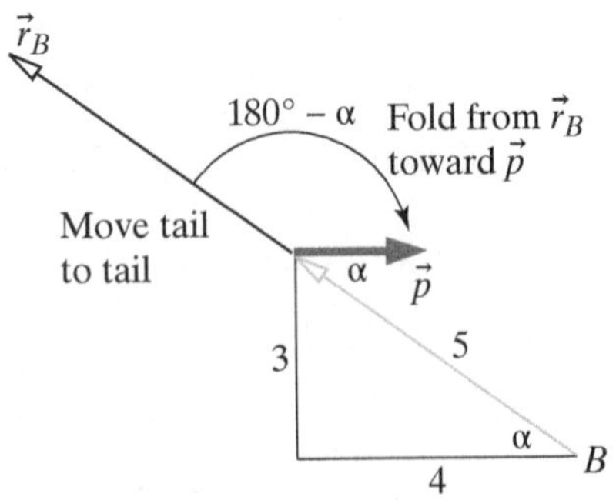

Figure 11.15 Place $\vec{r}_B$ and $\vec{p}$ tail to tail, then rotate $\vec{r}_B$ toward $\vec{p}$. The sine of this angle is $\sin(180° - \alpha) = \sin\alpha = 3/5$.

Method 3: Direction and magnitude using $r_\perp$

Determine the direction using the right-hand rule, as in Method 2 above.

Figure 11.15 shows that $r_\perp = 3$ m. Therefore

$$|\vec{L}_B| = (3)(10) = 30\text{ kg}\cdot\text{m}^2\text{/s}$$

By examining Figure 11.13 we can see that the translational angular momentum relative to location A is exactly the same as the angular momentum relative to location B, because the perpendicular distance $r_\perp$ is the same (3 m).

About location C, $r_\perp$ is zero because the momentum points straight toward C (the impact parameter is zero). Therefore $|\vec{L}_C| = 0$, and the direction of $\vec{L}_C$ is undefined.

Complete the following checkpoint before going on, because you need to be able to use the concept of cross product quickly and easily in later discussions. One way to check your work is to try more than one method to find the angular momentum.

Checkpoint 2 Determine both the direction and magnitude of the angular momentum of the particle in Figure 11.13, relative to the locations D, E, F, G, and H. We've already analyzed the angular momentum relative to A, B, and C in the example given above. Notice how the magnitude and direction of the angular momentum relative to the different locations differ in magnitude and direction.

11.2 ROTATIONAL ANGULAR MOMENTUM

We've just seen how to calculate the (translational) angular momentum of a point particle. However, an extended object spinning about its own axis (like the Earth) can't be modeled as a point particle. By modeling such an object as a collection of point particles, and adding up their translational angular momentum, we will see that there is an easy way to calculate the angular momentum of a system such as a spinning bicycle wheel (Figure 11.16).

In Chapter 9 we found that it is useful to express the kinetic energy of a multiparticle system as $K = K_{\text{trans}} + K_{\text{rel}}$. We'll see that it is similarly useful to express the angular momentum of a multiparticle system as a translational part $\vec{L}_{\text{trans},A}$ associated with the motion of the center of mass relative to a location A, plus a rotational part $\vec{L}_{\text{rot}}$ that represents the angular momentum relative to the center of mass, which we'll focus on in this section.

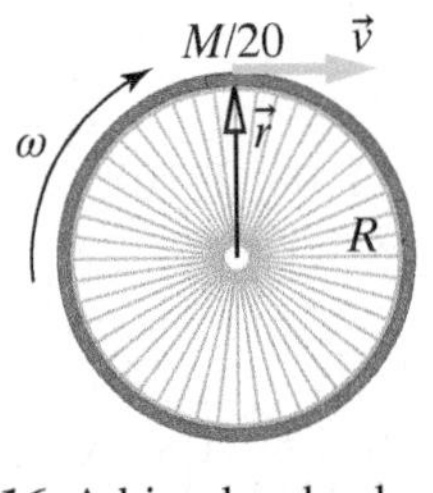

Figure 11.16 A bicycle wheel spinning with angular speed ω about a stationary axle. Almost all the mass of the wheel is in the rim.

Consider a bicycle wheel spinning about its center of mass with angular speed ω (Figure 11.16) about a stationary axle. Most of the mass is in the rim of the wheel, so we will neglect the low-mass spokes and the hub. Imagine dividing the rim into 20 equal-sized pieces, each with mass $M/20$, and consider each piece to be an object orbiting the axis. The speed of one piece is

$$v = \omega R$$

so the magnitude of the translational angular momentum of one piece, relative to the center of mass, is:

$$\left|\vec{L}_{\text{CM, one piece}}\right| = R(M/20)v\sin 90^\circ$$
$$= R(M/20)\omega R = (M/20)R^2\omega$$

By the right-hand rule, the direction of the translational angular momentum of each piece is into the page (the $-z$ direction), so to get the magnitude of the angular momentum of the entire wheel, we can add the contribution of each piece:

$$\left|\vec{L}_{\text{CM}}\right| = 20\cdot(M/20)R^2\omega$$
$$= MR^2\omega$$

Because every piece of the wheel is the same distance from the center of mass, the moment of inertia of the hoop-like wheel is $I = MR^2$. We can therefore rewrite the angular momentum of the wheel as

$$\left|\vec{L}_{\text{CM}}\right| = I\omega$$

The direction of the angular momentum of the entire wheel is $-z$ (into the page). Note that the result would have been the same if we had divided the wheel into 1000 pieces, or into its component atoms.

To convert the result to a vector, we define the angular velocity $\vec{\omega}$ to be a vector with magnitude ω and direction given by a variant of the right-hand rule (Figure 11.17):

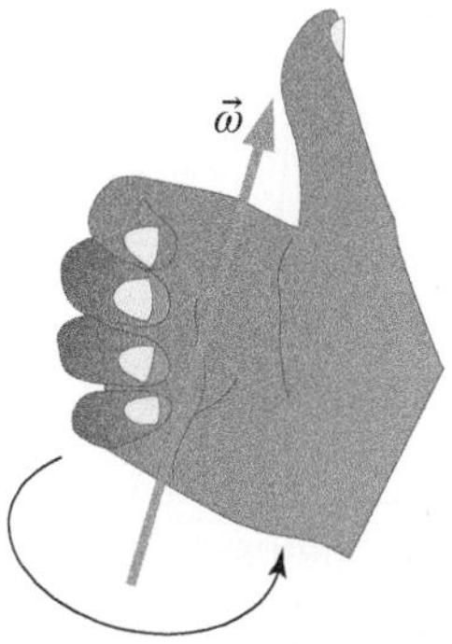

Figure 11.17 Finding the direction of the angular velocity vector. Curl your fingers in the direction of rotation; your thumb sticks out in the direction of $\vec{\omega}$.

ANGULAR VELOCITY VECTOR

$$\vec{\omega}$$

Magnitude: radians/second.
Direction: see Figure 11.17.

We call the angular momentum of a system relative to its center of mass the "rotational" angular momentum $\vec{L}_{\text{rot}}$. For the bicycle wheel we found that $\vec{L}_{\text{rot}} = I\vec{\omega}$, where $I = MR^2$.

EXAMPLE **Rotational Angular Momentum of a Bicycle Wheel**

A bicycle wheel has a mass of 0.8 kg and a radius of 32 cm. If the wheel rotates in the xz plane, spinning clockwise when viewed from the $+y$ axis, and making one full revolution in 0.75 s, what is the rotational angular momentum of the wheel?

Solution The direction of $\vec{\omega}$ is $-y$.

$$I = MR^2 = (0.8\,\text{kg})(0.32\text{ m})^2 = 0.082\text{ kg}\cdot\text{m}^2$$
$$\omega = \frac{2\pi}{0.75\,\text{s}} = 8.38\text{ s}^{-1}$$
$$\left|\vec{L}_{\text{rot}}\right| = (0.082\text{ kg}\cdot\text{m}^2)(8.38\text{ s}^{-1}) = 0.69\,\text{kg}\cdot\text{m}^2/\text{s}$$
$$\vec{L}_{\text{rot}} = \langle 0, -0.69, 0\rangle\text{ kg}\cdot\text{m}^2/\text{s}$$

$\vec{L}_{\text{rot}}$ in Terms of $\vec{\omega}$

The result that $\vec{L}_{\text{rot}} = I\vec{\omega}$ is a general one, and is true for any object whose atoms all share the same angular velocity $\vec{\omega}$ in radians per second but with

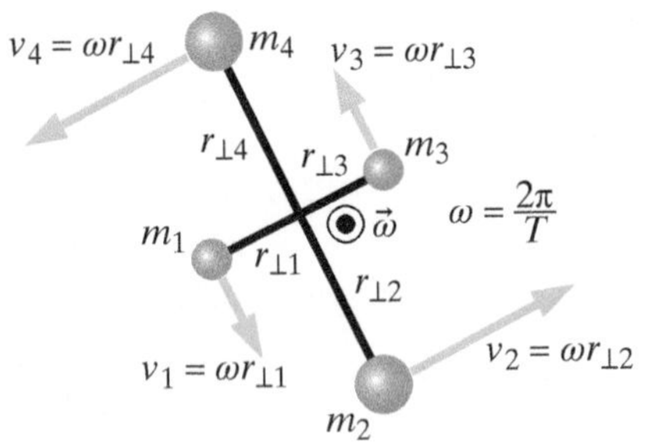

Figure 11.18 A case of rotation about an axis with all particles sharing the same angular speed ω. Note the different linear speeds, with $v_i = \omega r_{\perp i}$. $\vec{\omega}$ points out of the page.

different linear speeds in meters per second, depending on their distances from the axis (Figure 11.18). The angular velocity vector $\vec{\omega}$ points out of the page in Figure 11.18.

The magnitude of the angular momentum of each piece is this, where $r_\perp$ is the perpendicular distance of a piece from the axis:

$$|\vec{r} \times \vec{p}| = r_\perp m v = r_\perp m(\omega r_\perp) = m r_\perp^2 \omega$$

So the magnitude of the rotational angular momentum is this:

$$|\vec{L}_{\text{rot}}| = [m_1 r_{\perp 1}^2 + m_2 r_{\perp 2}^2 + m_3 r_{\perp 3}^2 + m_4 r_{\perp 4}^2]\omega$$

The quantity in brackets is the moment of inertia:

$$I = m_1 r_{\perp 1}^2 + m_2 r_{\perp 2}^2 + m_3 r_{\perp 3}^2 + m_4 r_{\perp 4}^2 + \cdots$$

Since the rotational angular momentum points in the direction of $\vec{\omega}$, we have the following for the angular momentum relative to the center of mass, even if the center of mass is moving in some complicated way:

ROTATIONAL ANGULAR MOMENTUM

$$\vec{L}_{\text{rot}} = I\vec{\omega}$$

for a system, such as a rigid object, whose components all rotate around its center of mass with the same angular velocity $\vec{\omega}$.

This way of evaluating rotational angular momentum is often easier to use than the more basic expression in terms of cross products $\vec{r} \times \vec{p}$, especially because the moments of inertia for common objects can be looked up in reference books. For example, we saw in Chapter 9 that the moment of inertia about an axis passing through the center of a uniform-density solid disk of radius R is $\frac{1}{2}MR^2$, and the moment of inertia about an axis passing through the center of a uniform-density solid sphere of radius R is $\frac{2}{5}MR^2$.

Since $L_{\text{rot}} = I\omega$, we can express K_{rot} in terms of the rotational angular momentum:

K_{rot} and L_{rot}

$$K_{\text{rot}} = \frac{1}{2}I\omega^2 = \frac{1}{2}\frac{(I\omega)^2}{I} = \frac{L_{\text{rot}}^2}{2I}$$

In this discussion we have considered only simple, symmetric rotating systems. The optional Section 11.13 deals with more complicated situations.

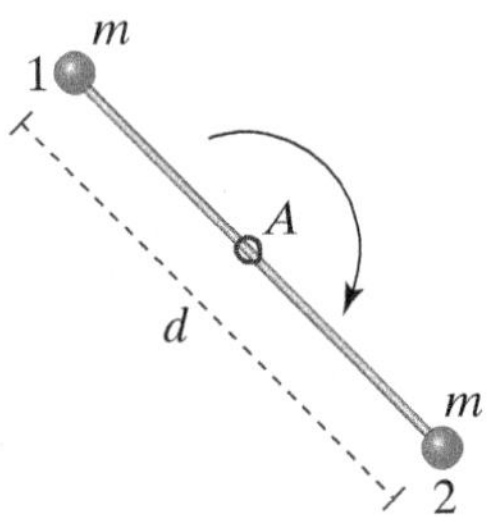

Figure 11.19 A rotating barbell.

Checkpoint 3 A barbell spins around a pivot at its center at A (Figure 11.19). The barbell consists of two small balls, each with mass $m = 0.4$ kg, at the ends of a very low mass rod of length $d = 0.6$ m. The barbell spins clockwise with angular speed $\omega_0 = 20$ radians/s. **(a)** Consider the two balls separately, and calculate $\vec{L}_{\text{trans},1,A}$ and $\vec{L}_{\text{trans},2,A}$ (both direction and magnitude in each case). **(b)** Calculate $\vec{L}_{\text{tot},A} = \vec{L}_{\text{trans},1,A} + \vec{L}_{\text{trans},2,A}$ (both direction and magnitude). **(c)** Next, consider the two balls together and calculate I for the barbell. **(d)** What is the direction of the angular velocity $\vec{\omega}_0$? **(e)** Calculate $\vec{L}_{\text{rot}} = I\vec{\omega}_0$ (both direction and magnitude). **(f)** How does $\vec{L}_{\text{rot}}$ compare to $\vec{L}_{\text{tot},A}$? The point is that the form $I\omega$ is just a convenient way of calculating the (rotational) angular momentum of a multiparticle system. In principle one can always calculate the angular momentum simply by adding up the individual angular momenta of all the particles. **(g)** Calculate K_{rot}.

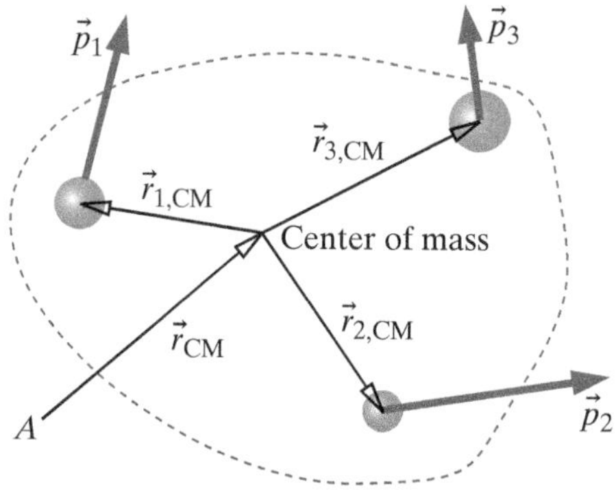

Figure 11.20 Position vectors relative to the center of mass.

11.3 TOTAL ANGULAR MOMENTUM

Kinetic energy can be usefully expressed as $K_{tot} = K_{trans} + K_{rel}$, where "rel" means "relative to the center of mass." Similarly, angular momentum can be usefully expressed as $\vec{L}_{tot,A} = \vec{L}_{trans,A} + \vec{L}_{rel}$, where we refer to the portion that is relative to the center of mass as rotational angular momentum, $\vec{L}_{rot}$.

We will calculate the total angular momentum relative to location A of a three-mass system (Figure 11.20) and prove that $\vec{L}_{tot,A} = \vec{L}_{trans,A} + \vec{L}_{rot}$, where the system need not be rigid. We specify the positions of each mass relative to location A by going from A to the center of mass (vector $\vec{r}_{CM}$) and then to the individual masses. The location A need not be in the same plane as the masses, nor are the momentum vectors necessarily in the plane of the masses.

The position of mass m_1 relative to location A is $\vec{r}_1 = \vec{r}_{CM} + \vec{r}_{1,CM}$, the position of mass m_2 is $\vec{r}_2 = \vec{r}_{CM} + \vec{r}_{2,CM}$, and the position of mass m_3 is $\vec{r}_3 = \vec{r}_{CM} + \vec{r}_{3,CM}$ (Figure 11.21). Using these vectors we calculate the total angular momentum $\vec{L}_A$ relative to location A:

$$\begin{aligned}\vec{L}_A &= (\vec{r}_{CM} + \vec{r}_{1,CM}) \times \vec{p}_1 + (\vec{r}_{CM} + \vec{r}_{2,CM}) \times \vec{p}_2 + (\vec{r}_{CM} + \vec{r}_{3,CM}) \times \vec{p}_3 \\ &= [\vec{r}_{CM} \times (\vec{p}_1 + \vec{p}_2 + \vec{p}_3)] + [\vec{r}_{1,CM} \times \vec{p}_1 + \vec{r}_{2,CM} \times \vec{p}_2 + \vec{r}_{3,CM} \times \vec{p}_3]\end{aligned}$$

Since $(\vec{p}_1 + \vec{p}_2 + \vec{p}_3)$ is the total momentum $\vec{P}_{tot}$, we have the following:

$$\vec{L}_A = \underbrace{[\vec{r}_{CM} \times \vec{P}_{tot}]}_{\text{translational}} + \underbrace{[\vec{r}_{1,CM} \times \vec{p}_1 + \vec{r}_{2,CM} \times \vec{p}_2 + \vec{r}_{3,CM} \times \vec{p}_3]}_{\text{rotational}}$$

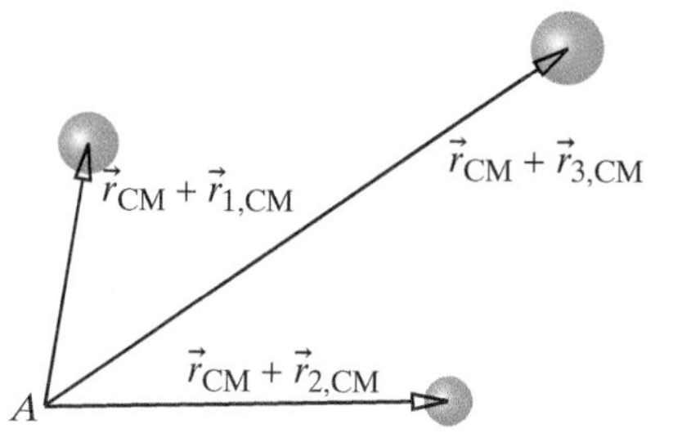

Figure 11.21 Position vectors relative to location A.

The first term in the total angular momentum, $\vec{r}_{CM} \times \vec{P}_{tot}$, is the angular momentum the system would have if all the mass were concentrated at the center of mass. It is the angular momentum that a "collapsed" point-particle version of the system would have. This is the translational (or orbital) angular momentum $\vec{L}_{trans,A}$, which is already familiar.

The second term in the total angular momentum,

$$\vec{r}_{1,CM} \times \vec{p}_1 + \vec{r}_{2,CM} \times \vec{p}_2 + \vec{r}_{3,CM} \times \vec{p}_3$$

is the angular momentum of the system relative to the center of mass of the system, which we call the rotational angular momentum $\vec{L}_{rot}$. An example of rotational angular momentum is the angular momentum due to the rotation of the Earth on its axis. This term is also called the "spin" angular momentum. In the previous section we saw that $\vec{L}_{rot}$ can be written as $I\vec{\omega}$ if the object is rigid, with all particles sharing the same ω.

In Figure 11.22 we show both the translational (orbital) angular momentum of the Earth relative to the location of the Sun, and the rotational (spin) angular momentum of the Earth relative to the Earth's center of mass. The total angular momentum of the Earth relative to the Sun is the vector sum of the Earth's translational and rotational angular momenta.

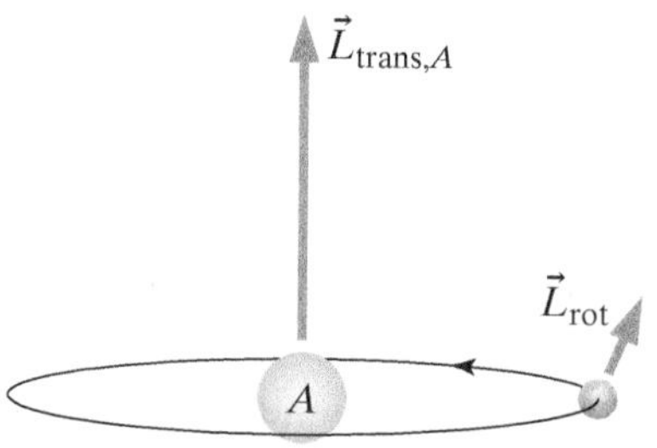

Figure 11.22 Translational angular momentum of the Earth (relative to the location of the Sun), and rotational angular momentum of the Earth (relative to the Earth's center of mass). At this instant the *linear* momentum $\vec{p}$ of the Earth is into the page. The Earth's axis is tilted 23° away from a perpendicular to the plane of the orbit, so the two angular momenta are not parallel.

TOTAL ANGULAR MOMENTUM

$$\vec{L}_A = \vec{L}_{trans,A} + \vec{L}_{rot}$$

Translational: $\vec{L}_{trans,A} = \vec{r}_{CM,A} \times \vec{P}_{tot}$

Rotational: $\vec{L}_{rot} = \vec{r}_{1,CM} \times \vec{p}_1 + \vec{r}_{2,CM} \times \vec{p}_2 + \cdots$

The translational angular momentum differs for different choices of the location A, so we need the "A" subscript on the translational angular momentum and the total angular momentum to remind us of this dependence. The rotational angular momentum doesn't need a subscript because it is calculated relative to the center of mass, and this calculation is unaffected by our choice of the point A and by the motion of the center of mass, which may even be accelerating.

For an ideal point particle, or the "collapsed" point-particle version of a multiparticle system, the translational angular momentum is the only kind of angular momentum there is, just as translational kinetic energy is the only kind of kinetic energy a point particle can have.

In Chapter 9 we saw that a rigid object that rotates about an axle that is a distance r_{CM} from the center of mass of the object has translational plus rotational kinetic energy equal to $\frac{1}{2}I\omega^2$, where $I = Mr_{\text{CM}}^2 + I_{\text{CM}}$ (this is also referred to as the "parallel axis theorem" for moment of inertia). It is easy to show that the angular momentum of a rotating rigid object is similar: the translational plus rotational angular momentum is $(Mr_{\text{CM}}^2 + I_{\text{CM}})\vec{\omega}$.

Checkpoint 4 Pinocchio rides a horse on a merry-go-round turning counterclockwise as viewed from above, with his long nose always pointing forward, in the direction of his velocity. Is Pinocchio's translational angular momentum relative to the center of the merry-go-round zero or nonzero? If nonzero, what is its direction? Is his rotational angular momentum zero or nonzero? If nonzero, what is its direction?

11.4 TORQUE

The (linear) momentum of a system can be changed by a force exerted by the surroundings.

QUESTION What can change angular momentum?

The angular momentum of a rotating or orbiting object can be changed by a twist—a force applied off-center. Such a twist is called a "torque," a word of Latin origin that simply means "twist." Torque is usually symbolized by τ, the lowercase Greek letter tau. The effectiveness of the torque depends on where the force is applied, so the definition of torque includes not only the magnitude and direction of the force, but also the position at which the force acts, relative to some location which we must specify.

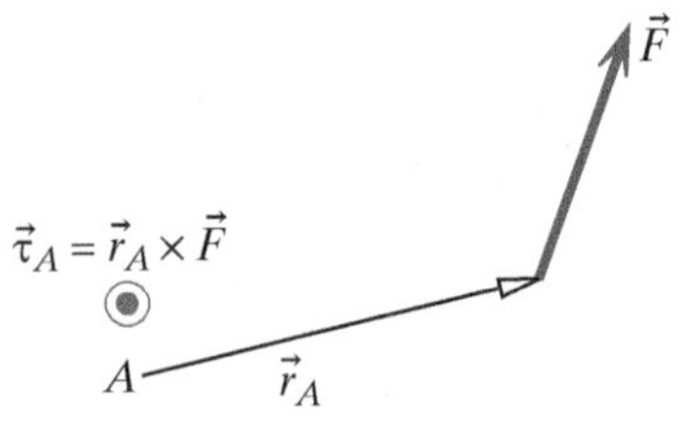

Figure 11.23 Torque relative to location A is defined as $\vec{\tau}_A = \vec{r}_A \times \vec{F}$.

DEFINITION OF TORQUE

$$\vec{\tau}_A = \vec{r}_A \times \vec{F}$$

Torque due to a force $\vec{F}$ is defined about a particular location, A. $\vec{r}_A$ is a vector from location A to the point of application of the force (Figure 11.23). The units of torque are $\text{N} \cdot \text{m}$.

Applying a torque, or twist, to a system changes the angular momentum of the system. Just as we call the vector sum of all the forces acting on a system $\vec{F}_{\text{net}}$, we call the vector sum of all the torques acting on a system $\vec{\tau}_{\text{net},A}$. Torque is a vector because it has both magnitude and direction. It is calculated with a cross product because it depends on both the direction of the applied force and the point of application.

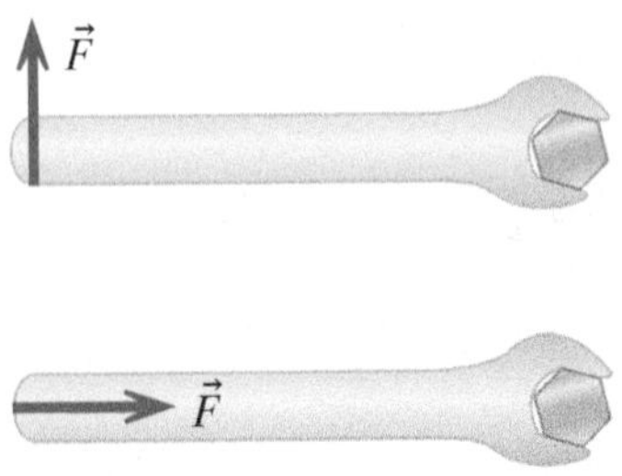

Figure 11.24 Twist a nut, or push straight at it?

Direction of the Force

QUESTION In Figure 11.24, which way of pushing on a wrench is more effective in twisting the nut?

You can see that a force parallel to the handle is completely ineffective in twisting the nut. Evidently it is only the perpendicular component of a force that is effective in twisting an object.

Point of Application of the Force

QUESTION In Figure 11.25, where should you push on the wrench—at the end of the wrench or near the nut?

The farther away from the nut we push, the more effective we are in twisting the nut. The advantage of a long wrench is to provide more leverage in twisting.

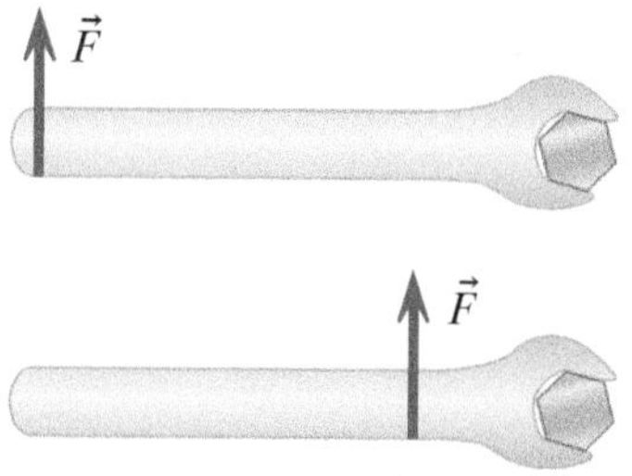

Figure 11.25 Push at the end, or near the nut?

Magnitude of Torque

Suppose that we apply a force at an angle θ to the radius (Figure 11.26). Evidently the component of the force that is effective in twisting is $F\sin\theta$, the perpendicular component. Moreover, the bigger the lever arm (r_A, the distance from the nut at location A), the more effective we'll be. To capture both effects, we define the magnitude of torque (which means twist) exerted by a force $\vec{F}$ relative to a location A as follows:

MAGNITUDE OF TORQUE RELATIVE TO LOCATION A

$$\tau = r_A F \sin\theta$$

For a purely perpendicular force, $\theta = 90°$, $\sin\theta = 1$, and the torque is $r_A F$. For a force that is parallel to the lever arm, $\theta = 0°$, $\sin\theta = 0$, and the torque is zero (Figure 11.24).

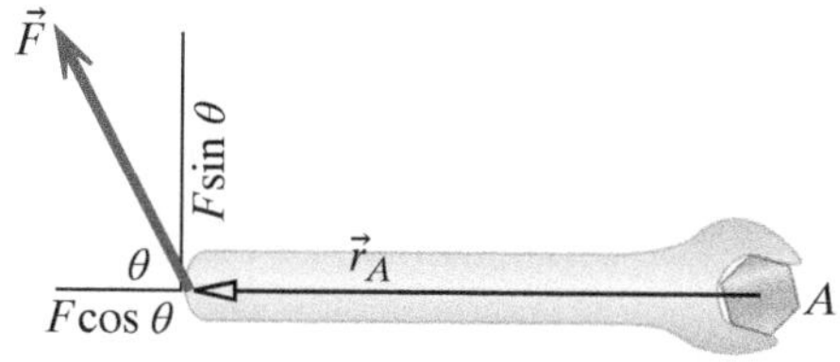

Figure 11.26 Push on the wrench at an angle.

Direction of Torque

The direction of the torque is given by the cross product and associated right-hand rule.

QUESTION In Figure 11.26, is the torque vector into the page or out of the page? Use the right-hand rule.

Move the $\vec{r}$ and $\vec{F}$ vectors tail to tail (Figure 11.27), and align your hand so you can fold the fingers of your right hand from $\vec{r}$ toward $\vec{F}$. The only way you can do this results in your extended right thumb pointing into the page. If you try to keep your extended right thumb pointing out of the page, you will find that you can't fold your fingers toward the direction of the force. To put it another way, folding your fingers from the direction of $\vec{r}$ toward the direction of $\vec{F}$ is a clockwise rotation, which means the angular momentum points into the page.

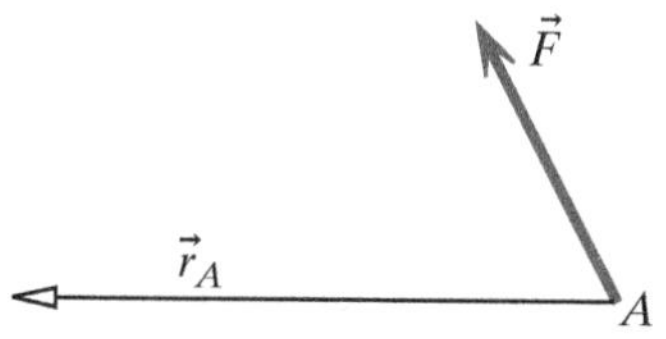

Figure 11.27 To find the direction of the cross product, move the vectors tail to tail.

Most nuts and bolts have what are called "right-handed threads." In Figure 11.26 a right-handed nut will advance in the direction of the torque vector, into the page. If you reverse the direction of the force, so the torque vector points out of the page, a right-handed nut will advance out of the page. (Exceptions include left-handed threads on gas cylinders containing explosive gases; the unexpected behavior alerts users to danger.)

The fact that a physical right-handed nut advances in the direction of the torque vector is another justification for drawing torque cross-product vectors along an axis of rotation, perpendicular to the plane of the motion.

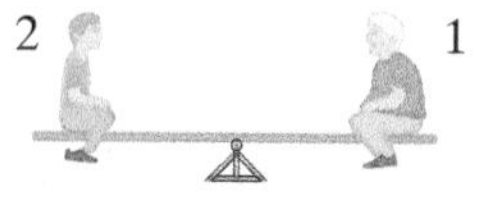

Figure 11.28 Two children on a seesaw.

EXAMPLE **Torque on a Seesaw**

Two children sit on a seesaw, each at a distance of 1.2 m from the pivot point (Figure 11.28). The mass of child 1 is 30 kg, and the mass of child 2 is 20 kg. Assume that the mass of the seesaw itself is small. What is the net torque on the system of children plus seesaw, about the pivot point?

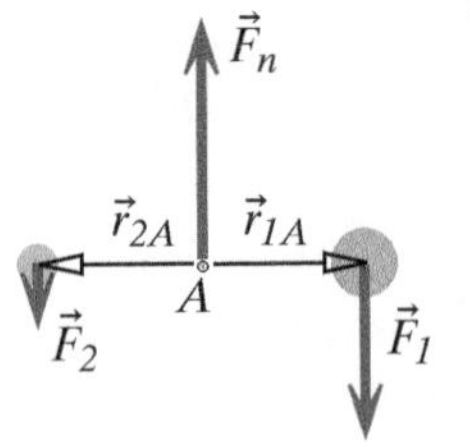

Figure 11.29 Forces on the see-saw and children.

Solution The diagram in Figure 11.29 shows all the forces acting on the system. We'll use the standard coordinate system, and we'll choose the location of the pivot as location A. To get the net torque we'll need to calculate all the torques about this location and add them up.

The torque due to child 1 is:

$$\begin{aligned}\vec{\tau}_{1,A} &= \vec{r}_{1,A} \times \vec{F}_1 \\ &= \langle 1.2, 0, 0\rangle \times \langle 0, -(30\,\text{kg})(9.8\text{N/kg}), 0\rangle \\ &= \langle 0, 0, -352.8\rangle\ \text{N}\cdot\text{m}\end{aligned}$$

Alternatively, if we put the vectors $\vec{r}_{1,A}$ and $\vec{F}_1$ tail to tail and apply the right-hand rule, we find that $\vec{\tau}_{1,A}$ is into the page ($-z$). Its magnitude is:

$$\begin{aligned}|\vec{\tau}_{1,A}| &= (1.2\text{ m})(30\,\text{kg})(9.8\text{N/kg})\sin 90^\circ \\ &= 352.8\,\text{N}\cdot\text{m}\end{aligned}$$

Combining these results for magnitude and direction, we again find that

$$\vec{\tau}_{1,A} = \langle 0, 0, -352.8\rangle\ \text{N}\cdot\text{m}$$

The torque due to child 2 is:

$$\begin{aligned}\vec{\tau}_{2,A} &= \vec{r}_{2,A} \times \vec{F}_2 \\ &= \langle -1.2, 0, 0\rangle \times \langle 0, -(20\,\text{kg})(9.8\text{N/kg}), 0\rangle \\ &= \langle 0, 0, 235.2\rangle\ \text{N}\cdot\text{m}\end{aligned}$$

The normal force $\vec{F}_n$ is important in holding up the seesaw. Its contribution to the net torque about A is:

$$\begin{aligned}\vec{\tau}_{n,A} &= \vec{r}_{n,A} \times \vec{F}_n \\ &= \langle 0, 0, 0\rangle\ \text{m} \times \langle 0, (50\,\text{kg})(9.8\text{N/kg}), 0\rangle \\ &= \langle 0, 0, 0\rangle\ \text{N}\cdot\text{m}\end{aligned}$$

The torque due to the normal force is zero because the force acts at location A, so it can't twist the seesaw. The net torque is:

$$\begin{aligned}\vec{\tau}_{\text{net},A} &= \vec{\tau}_{n,A} + \vec{\tau}_{1,A} + \vec{\tau}_{2,A} \\ &= \langle 0, 0, 0\rangle\ \text{N}\cdot\text{m} + \langle 0, 0, -352.8\rangle\ \text{N}\cdot\text{m} + \langle 0, 0, 235.2\rangle\ \text{N}\cdot\text{m} \\ &= \langle 0, 0, -117.6\rangle\ \text{N}\cdot\text{m}\end{aligned}$$

Checkpoint 5 In Figure 11.26, if $r_A = 3$ m, $F = 4$ N, and $\theta = 30^\circ$, what is the magnitude of the torque about location A, including units? If the force in Figure 11.26 were perpendicular to $\vec{r}_A$ but gave the same torque as before, what would be its magnitude?

11.5 THE ANGULAR MOMENTUM PRINCIPLE

The Angular Momentum Principle relates a change in the angular momentum of a system to the net torque on the system due to the surroundings. Like the Momentum Principle, it can be written in several different forms.

THE ANGULAR MOMENTUM PRINCIPLE FOR A PARTICLE

$$\vec{L}_{f,A} = \vec{L}_{i,A} + \vec{\tau}_{\text{net},A}\Delta t \quad \text{(update form)}$$

$$\frac{d\vec{L}_A}{dt} = \vec{\tau}_{\text{net},A} \quad \text{(derivative form)}$$

$$\Delta\vec{L}_{A,\text{system}} + \Delta\vec{L}_{A,\text{surroundings}} = 0 \quad \text{(conservation form)}$$

In each case, angular momentum and torque are defined relative to a location A.

These statements of the Angular Momentum Principle apply to a system modeled as a single particle. We'll see a little later how to extend these equations to apply to an extended system.

By considering the derivative form of the Angular Momentum Principle, we can show that it is consistent with the Momentum Principle. For a single particle, the Momentum Principle says that the time derivative of the momentum is the net force: $d\vec{p}/dt = \vec{F}_{\text{net}}$. In the following derivation we will use these facts:

$$\frac{d\vec{r}_A}{dt} = \vec{v}, \qquad \vec{p} = \gamma m\vec{v}, \qquad \vec{v}\times\vec{v} = \vec{0}, \qquad \frac{d\vec{p}}{dt} = \vec{F}_{\text{net}}, \qquad \vec{r}_A \times \vec{F}_{\text{net}} = \vec{\tau}_{\text{net},A}$$

We'll start with the definition of angular momentum of a particle, and take its time derivative. Then we'll apply the product rule for derivatives:

$$\begin{aligned}
\vec{L}_A &= \vec{r}_A \times \vec{p} \\
\frac{d\vec{L}_A}{dt} &= \frac{d(\vec{r}_A \times \vec{p})}{dt} \\
&= \left(\frac{d\vec{r}_A}{dt}\times\vec{p}\right) + \left(\vec{r}_A \times \frac{d\vec{p}}{dt}\right) \\
&= (\vec{v}\times\gamma m\vec{v}) + \left(\vec{r}_A \times \frac{d\vec{p}}{dt}\right) \\
&= \vec{0} + \left(\vec{r}_A \times \frac{d\vec{p}}{dt}\right) \\
&= \vec{0} + \vec{r}_A \times \vec{F}_{\text{net}} \\
\frac{d\vec{L}_A}{dt} &= \vec{\tau}_{\text{net},A}
\end{aligned}$$

EXAMPLE **Momentum Change for a Seesaw**

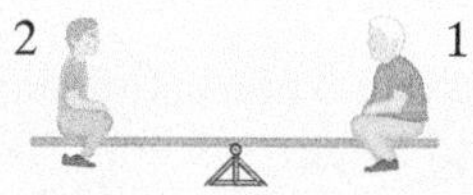

Figure 11.30 Two children on a seesaw.

In the previous example we calculated the net torque on the system of a seesaw plus two children (Figure 11.30). Apply the Angular Momentum Principle to calculate the angular momentum of the system at a time 0.4 s after it is released from rest. Will the seesaw be rotating clockwise or counterclockwise?

Solution We found the net torque on the seesaw to be $\langle 0,0,-117.6\rangle$ N·m.

$$\begin{aligned}
\vec{L}_{f,A} &= \vec{L}_{i,A} + \vec{\tau}_{\text{net},A}\Delta t \\
&= \vec{0} + \langle 0,0,-117.6\rangle\ \text{N}\cdot\text{m}(0.4\ \text{s}) \\
&= \langle 0,0,-47.0\rangle\ \text{kg m}^2/\text{s}
\end{aligned}$$

Since $\vec{L}_A$ is in the $-z$ direction, the seesaw will rotate clockwise (that is, child 1 will go down and child 2 will go up). If you point your right thumb in the direction of $\vec{L}_A$, your fingers curl in the direction of rotation.

Three Fundamental Principles

There are parallels between the three fundamental principles we have encountered so far: the Momentum Principle, the Energy Principle, and the Angular Momentum Principle. Each principle relates a change in a property of the system (momentum, energy, or angular momentum) to interactions with objects in the surroundings:

$$\Delta\vec{p} = \vec{F}_{\text{net}}\,\Delta t$$
$$\Delta E = W + Q$$
$$\Delta\vec{L}_A = \vec{\tau}_{\text{net},A}\,\Delta t$$

Momentum, energy, and angular momentum are all conserved quantities:

$$\Delta\vec{p}_{\text{sys}} + \Delta\vec{p}_{\text{surr}} = \vec{0}$$
$$\Delta E_{\text{sys}} + \Delta E_{\text{surr}} = 0$$
$$\Delta\vec{L}_{\text{sys},A} + \Delta\vec{L}_{\text{surr},A} = \vec{0}$$

All three of these fundamental principles apply to any choice of system and surroundings. In complicated situations it may be necessary to use all three principles in order to predict or explain the behavior of a system.

11.6 MULTIPARTICLE SYSTEMS

The Angular Momentum Principle has already given us new insight into the motion of a single particle acted on by a torque due to a single force. The equation becomes a really powerful tool when we extend it to apply to multiparticle systems acted on by multiple torques. For example, we will be able to understand the counterintuitive behavior of spinning tops and gyroscopes.

The derivation of the multiparticle version of the Angular Momentum Principle follows closely the derivation in Chapter 3 of a multiparticle version of the Momentum Principle. The basic idea is that, due to reciprocity of forces, torques internal to a multiparticle system cancel, and only torques due to objects in the surroundings can change the system's angular momentum.

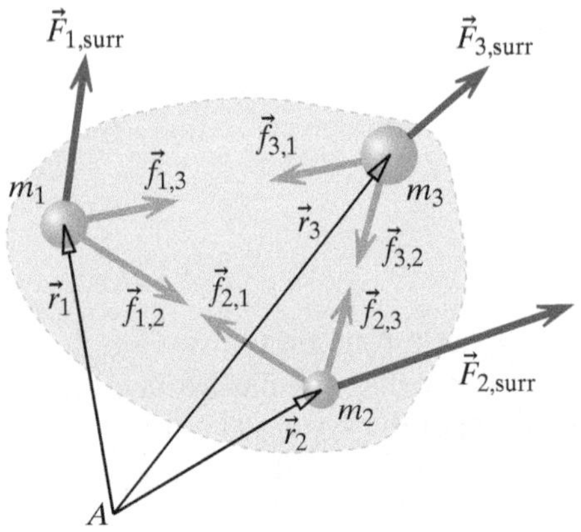

Figure 11.31 The Angular Momentum Principle in a multiparticle system.

To be as concrete as possible, we'll consider a system consisting of just three particles. It is easy to see how this generalizes to larger systems, including a block consisting of an astronomically huge number of atoms.

In Figure 11.31 we show all of the forces acting on each particle, where the lowercase $\vec{f}$'s are internal forces, and the uppercase $\vec{F}$'s are external forces applied by objects in the surroundings, such as the Earth that applies a gravitational attraction. We write the Angular Momentum Principle applied to each of the three particles. We measure angular momenta and torques relative to location A, but to avoid clutter we don't write the subscript A in any of these equations until the end. ($\vec{L}_1$ is the angular momentum of m_1 relative to A.)

$$\frac{d\vec{L}_1}{dt} = \vec{r}_1 \times (\vec{F}_{1,\text{surr}} + \vec{f}_{1,2} + \vec{f}_{1,3})$$
$$\frac{d\vec{L}_2}{dt} = \vec{r}_2 \times (\vec{F}_{2,\text{surr}} + \vec{f}_{2,1} + \vec{f}_{2,3})$$
$$\frac{d\vec{L}_3}{dt} = \vec{r}_3 \times (\vec{F}_{3,\text{surr}} + \vec{f}_{3,1} + \vec{f}_{3,2})$$

Nothing new so far—but now we add up these three equations. That is, we create a new equation by adding up all the terms on the left sides of the three equations, adding up all the terms on the right sides, and setting them equal to each other. In doing so, we take into account the fact that many of these terms cancel. By the reciprocity of electric and gravitational forces (Newton's third law), we have

$$\vec{f}_{1,2} = -\vec{f}_{2,1}$$
$$\vec{f}_{1,3} = -\vec{f}_{3,1}$$
$$\vec{f}_{2,3} = -\vec{f}_{3,2}$$

Therefore the torques of the internal forces cancel. For example, consider this piece of the sum:

$$\vec{r}_1 \times \vec{f}_{1,2} + \vec{r}_2 \times \vec{f}_{2,1} = \vec{r}_1 \times \vec{f}_{1,2} - \vec{r}_2 \times \vec{f}_{1,2} = (\vec{r}_1 - \vec{r}_2) \times \vec{f}_{1,2}$$

The vector $\vec{r}_1 - \vec{r}_2$ points from m_2 toward m_1. Because electric and gravitational forces act along the line connecting the particles, the angle between $\vec{r}_1 - \vec{r}_2$ and $\vec{f}_{1,2}$ is either 0° or 180°, and $\sin\theta$ is zero. Consequently all of the torques associated with the internal forces cancel, and all that remains is this:

$$\frac{d(\vec{L}_1 + \vec{L}_2 + \vec{L}_3)}{dt} = \vec{r}_1 \times \vec{F}_{1,\text{surr}} + \vec{r}_2 \times \vec{F}_{2,\text{surr}} + \vec{r}_3 \times \vec{F}_{3,\text{surr}}$$

The right side of this equation represents the net torque due to external forces, $\vec{\tau}_{\text{net},A}$. The great importance of this equation is that the reciprocity of gravitational and electric forces causes all the torques due to internal forces to cancel. We rewrite our result like this:

THE ANGULAR MOMENTUM PRINCIPLE FOR A MULTIPARTICLE SYSTEM

$$\frac{d\vec{L}_{\text{tot},A}}{dt} = \vec{\tau}_{\text{net},A}$$

or

$$\Delta\vec{L}_{\text{tot},A} = \vec{\tau}_{\text{net},A}\Delta t$$

In words, the rate of change of the "total angular momentum" of a system relative to a location A, $\vec{L}_{\text{tot},A} = \vec{L}_{1,A} + \vec{L}_{2,A} + \vec{L}_{3,A} + \cdots$, is equal to the net torque due to forces exerted on that system by objects in the surroundings, relative to location A. (Or, the change in the angular momentum is equal to the angular impulse.)

Often it pays to choose location A to be at the place where the center of mass happens to be at that instant. In that case, the Angular Momentum Principle simplifies:

$$\frac{d\vec{L}_{\text{CM}}}{dt} = \frac{d}{dt}[(\vec{r}_{\text{CM,CM}} \times \vec{P}_{\text{tot}}) + \vec{L}_{\text{rot}}] = \frac{d\vec{L}_{\text{rot}}}{dt}$$

since $\vec{r}_{\text{CM,CM}} = \vec{0}$. That is, the position of the center of mass, relative to the center of mass itself, is of course a zero vector. Therefore we have the following important and useful special case for the angular momentum relative to the center of mass:

THE ANGULAR MOMENTUM PRINCIPLE RELATIVE TO THE CENTER OF MASS

$$\frac{d\vec{L}_{\text{rot}}}{dt} = \vec{\tau}_{\text{net,CM}}$$

or

$$\Delta\vec{L}_{\text{rot}} = \vec{\tau}_{\text{net,CM}}\Delta t$$

11.7 SYSTEMS WITH ZERO TORQUE

In a situation in which there is zero external torque, the angular momentum of a system does not change, even if the moment of inertia of the system does change, due to a change in the shape or configuration of the system. Athletes like figure skaters, divers, and dancers can use this fact to change the rotation of their bodies.

Figure 11.32 The skater spins slowly with leg and arms stretched out.

A Spinning Skater

You may have seen an ice skater spin vertically on the tip of one skate, with her arms and one leg outstretched, then pull her leg in and bring her arms to a vertical position above her head. She then spins much faster (Figures 11.32 and 11.33).

What's going on? There is some frictional force of the ice on the tip of the skate, but this force is applied so close to the axis of rotation that the torque is small.

QUESTION During the short time when the skater quickly changes her configuration, what can you say about the rotational angular momentum? Why?

Small torque implies small change in angular momentum per unit time, so in a short time the angular momentum of the skater will hardly change.

QUESTION If the rotational angular momentum hardly changes, how can the skater spin faster?

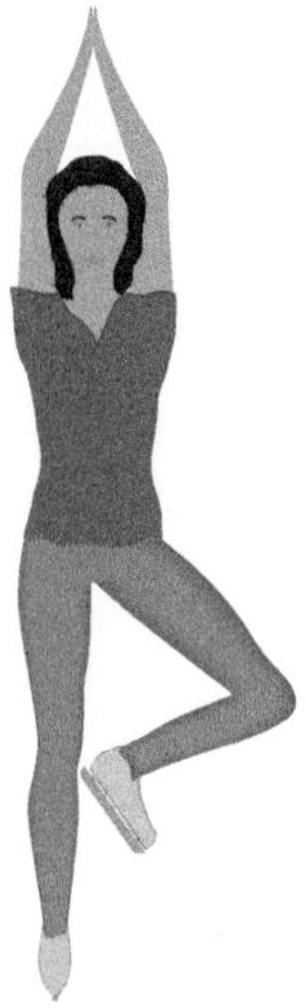

Figure 11.33 The skater pulls in her leg and arms and spins much faster.

When she changes her configuration, the moment of inertia decreases, because some parts of her body are now closer to the axis of rotation. For the rotational angular momentum not to change, the angular velocity must increase to compensate for the decreased moment of inertia: $I_1\omega_1 = I_2\omega_2$, and therefore $\omega_2 = I_1\omega_1/I_2$.

Now you may quite legitimately be puzzled about how this actually works! *Why* does moving her arms and leg closer to the spin axis increase her angular speed? At one level of discussion, you can close your eyes (and maybe hold your nose) and say, "Well, we did all that general analysis of the effect of torques on the angular momentum of multiparticle systems, so if that's what we get when we apply these general principles, I guess that's that." At another level, though, it would be very nice to get a better sense of the detailed mechanisms involved in this odd phenomenon.

One approach is to analyze the changes in energy involved in this skating maneuver. The angular momentum L_{rot} of the skater doesn't change, so as her moment of inertia decreases, the kinetic energy of the skater $\frac{1}{2}I\omega^2 = L_{\text{rot}}^2/(2I)$ actually increases.

QUESTION Where does this increased energy come from?

Evidently the skater has to expend chemical energy in order to increase her kinetic energy. In fact, at high spin rates it takes a noticeable effort to pull her arms and leg toward the spin axis. This is even more dramatic when holding heavy weights.

A popular physics demonstration is to sit on a rotating stool holding a dumbbell in each hand. Start spinning slowly with the arms held out, then pull your hands in toward your chest. It requires considerable effort to pull the dumbbells in. After all, the dumbbells would tend to move in straight lines, and you must exert a radial force just to keep turning them in a circle (though in that case you do no mechanical work, because the motion of the dumbbells has no radial component in the direction of the force you apply). To move the dumbbells into a smaller radius requires applying an even larger force that does work on the dumbbells, because there is now a radial component of the motion, in the direction of the radial force you exert.

Figure 11.34 A diver's moment of inertia is large when his body is extended.

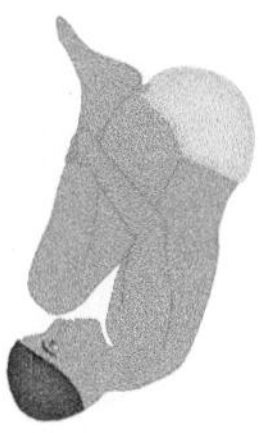

Figure 11.35 A diver's moment of inertia is smaller when he is curled into a ball.

A High Dive

You may have seen a skilled diver leap off a high board, tuck himself into a tight ball (holding onto his ankles), and rotate quite fast for a few turns. Then he straightens out and enters the water like a knife, hardly ruffling the surface (Figure 11.34).

> QUESTION What can you say about the diver's rotational angular momentum while he is in the air? Why?

We can probably neglect air resistance, because the diver is very far from reaching terminal speed. Although air resistance is distributed over the entire body, and there is a gravitational force acting on every atom in the body, torques to the left and right of the center of mass tend to cancel each other out (this cancellation is exact for the gravitational forces). Therefore there is negligible torque about the center of mass, and the rotational angular momentum $\vec{L}_{rot}$ does not change.

> QUESTION Early in the dive the diver is spinning rapidly (Figure 11.35). If rotational angular momentum is constant during the dive, where did that rapid spin come from?

When the diver jumps off the board, he must thrust with his feet in such a way that the force of the board on his feet has a sizable lever arm about his center of mass, so that by the time his feet lose contact with the board, the diver already has acquired a sizable rotational angular momentum.

> QUESTION When the diver pulls out of the tucked position, why does he stop spinning rapidly?

There is a large increase in the moment of inertia, because many atoms are now much farther from the center of mass than they were when the diver was in the tucked position. Larger moment of inertia implies smaller angular speed, since the magnitude of the rotational angular momentum $I\omega$ does not change.

> QUESTION Can his body go straight from then on?

His body can't really go completely straight, because he still has rotational angular momentum. However, his angular speed may be so small that you hardly notice it, especially in comparison with the very rapid spin in the preceding tucked position. Moreover, his body could approximately follow the curving path of his center of mass, with the body rotating just enough to stay tangent to the trajectory. This enhances the illusion of straight motion. The most important aspect for good form is to arrange for the body to rotate into the vertical position at the time of entering the water, so as to make little splash.

See Problem P52, which asks you to analyze a diver's motion.

EXAMPLE **A Satellite with Solar Panels**

A satellite has four low-mass solar panels sticking out (Figure 11.36). The satellite can be considered to be approximately a uniform solid sphere. Originally it is traveling to the right with speed v and rotating clockwise with angular speed ω_i.

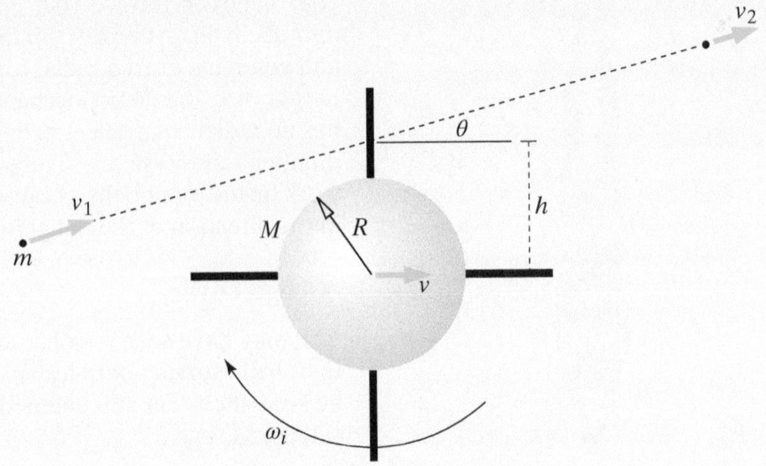

Figure 11.36 A satellite is struck by a tiny meteor.

A tiny meteor traveling at high speed v_1 rips through one of the solar panels and continues in the same direction but at reduced speed v_2. Afterward calculate the v_x and v_y components of the center-of-mass velocity of the satellite and its angular velocity ω_f (magnitude and direction). Additional data are provided on the diagram.

Solution What fundamental principles are useful starting points? The Momentum Principle and the Angular Momentum Principle. The Energy Principle would not be a useful starting point, because we don't know how much internal energy increase is associated with the meteor ripping through the panel. Take the satellite and meteor together as the system of interest, in which case the momentum and the angular momentum will remain constant, there being no external forces or torques.

System: Both the satellite and the meteor
Surroundings: Nothing during the collision

Initial State: Just before collision
Final state: Just after collision

Momentum Principle:
x components:

$$p_{xf} = p_{xi}$$
$$Mv_x + mv_2 \cos\theta = Mv + mv_1 \cos\theta$$
$$v_x = v + \frac{m}{M}(v_1 - v_2)\cos\theta$$

y components:

$$p_{yf} = p_{yi}$$
$$Mv_y + mv_2 \cos(90^\circ - \theta) = m_1 v_1 \cos(90^\circ - \theta)$$
$$v_y = \frac{m}{M}(v_1 - v_2)\sin\theta$$

Angular Momentum Principle, about the location in space where the center of the satellite was initially:

Component into the page ($-z$):

$$I\omega_i + hmv_1 \sin(90^\circ - \theta) = I\omega_f + hmv_2 \sin(90^\circ - \theta)$$

$$I = \frac{2}{5}MR^2$$

$$\omega_f = \omega_i + \frac{hm}{\left(\frac{2}{5}MR^2\right)}(v_1 - v_2)\cos\theta, \quad \text{clockwise, into the page}$$

Choosing times just before and just after the collision makes it particularly easy to see how to calculate the translational angular momentum of the meteor, since it is directly above the center of the satellite at those instants.

A Child Jumps onto a Playground Ride

A playground ride consists of a disk mounted on a low-friction axle (Figure 11.37). A child runs on a line tangential to the disk and jumps onto the outer edge of the disk. The disk spins, but at what rate? How many revolutions per minute, or radians per second?

Neither the Momentum Principle nor the Energy Principle is sufficient to analyze this collision. If we know the speed of the child just after jumping onto the disk, we can use the Momentum Principle to determine the impulse exerted by the axle on the disk, which prevents the disk from moving sideways, but we need to know the spin rate in order to be able to calculate the new speed of the child. The Energy Principle can be used to determine how much internal energy is produced in the inelastic sticking collision of the child and the disk, but only if we know the spin rate. We need another principle that will yield the spin rate: the Angular Momentum Principle.

It is important to see that even though the child is running in a straight line in Figure 11.37, the child does have angular momentum about the center of the disk, because the disk is caused to rotate when hit off-center by the child.

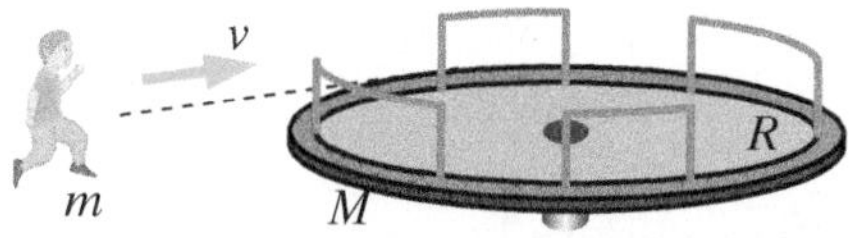

Figure 11.37 A child jumps onto a playground ride, and the disk spins. What is the rate of spin?

EXAMPLE

A Child Jumps onto a Playground Ride

A playground ride consists of a uniform-density disk of mass 300 kg and radius 2 m mounted on a low-friction axle (Figure 11.37). Starting from a distance of 5 m from the disk, a child of mass 40 kg runs at 3 m/s on a line tangential to the disk and jumps onto the outer edge of the disk. If the disk was initially at rest, how fast does it rotate just after the collision?

Solution

System: Disk plus child
Surroundings: Earth, ground, axle
Initial state: Child 5 m away, running at 3 m/s; disk not rotating
Final state: Just after collision; child now stuck to the disk

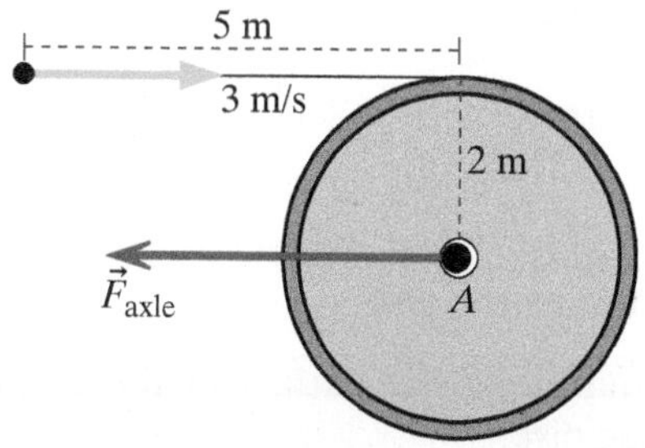

Figure 11.38 The free-body diagram for the system of disk plus child. The axle is in the surroundings and exerts a large impulsive force to the left when the child rams the disk against the axle.

Free-body diagram: Figure 11.38. The axle exerts a large impulsive force at the time of impact in response to the child ramming the disk against the axle; this force prevents the disk from moving to the right, and it changes the total momentum of the combined system.

Principle: The Angular Momentum Principle $\vec{L}_{A,f} = \vec{L}_{A,i} + \vec{\tau}_{\text{net}}\Delta t$

We can choose any point A about which to calculate torque and angular momentum. To simplify the analysis, we choose A to be at the location of the axle, so that the large impulsive force of the axle points toward A and therefore exerts no torque. The force of the Earth is downward and the force of the ground is upward, with equal magnitude, so the net torque on our chosen system is zero, and $\vec{L}_{A,f} = \vec{L}_{A,i}$. We consider just the component of angular momentum that is perpendicular to the xz plane of the motion, $L_{A,y}$.

The initial angular momentum of the system consists solely of the child's angular momentum, since the disk isn't rotating. Here is the magnitude:

$$\begin{aligned}|L_{A,y,i}| &= |\vec{r}_A||\vec{p}|\sin\theta = (|\vec{r}_A|\sin\theta)|\vec{p}|\\ &= (2\text{ m})(40\text{ kg})(3\text{ m/s})\\ &= 240\text{ kg}\cdot\text{m}^2/\text{s}\end{aligned}$$

Here we used the fact that in Figure 11.39 ($|\vec{r}_A|\sin\phi$) is the side opposite the angle ϕ, and that $\sin\phi = \sin(180° - \theta) = \sin\theta$.

Another way to calculate the magnitude of the initial angular momentum for this motion in the xz plane is to use the general expression for components of angular momentum, which gives the same result and is a check on our work:

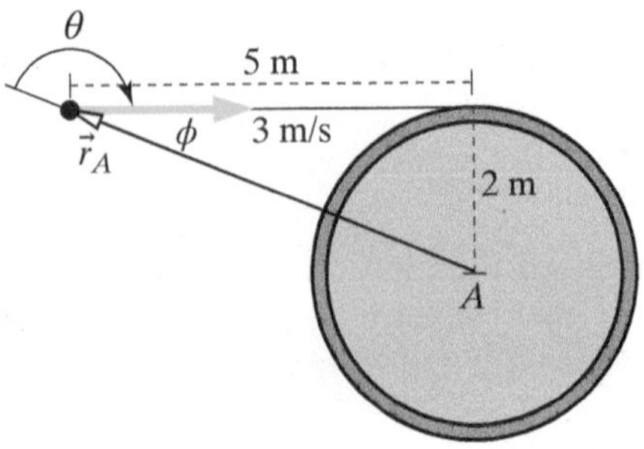

Figure 11.39 Applying the definition $L_A = r_A p \sin\theta$.

$$|L_{A,y,i}| = |zp_x - xp_z| = |zp_x| = (2\text{ m})(40\text{ kg})(3\text{ m/s}) = 240\text{ kg}\cdot\text{m}^2/\text{s}$$

The final angular momentum of the system consists of spin (rotational) angular momentum of the disk plus orbital (translational) angular momentum of the child on the disk. If the disk rotates with angular speed ω, the child's speed is $\omega r = \omega(2\text{ m})$.

$$\begin{aligned}|L_{A,y,f}| &= |L_{A,y,i}| = 240\text{ kg}\cdot\text{m}^2/\text{s}\\ &= I\omega + Rmv = \frac{1}{2}MR^2\omega + Rm(\omega R)\\ &= \frac{1}{2}(300\text{ kg})(2\text{ m})^2\omega + (40\text{ kg})(2\text{ m})^2\omega\\ \omega &= \frac{240\text{ kg}\cdot\text{m}^2/\text{s}}{\frac{1}{2}(300\text{ kg})(2\text{ m})^2 + (40\text{ kg})(2\text{ m})^2}\\ \omega &= 0.316\text{ rad/s}\end{aligned}$$

Before the collision there is only straight-line motion of the child, who nevertheless has (translational) angular momentum relative to the axle. There is obviously angular momentum after the collision, so it makes sense that there had to have been angular momentum before the collision, even though the child was running in a straight line.

Because the axle exerts a large impulsive force to the left, the Momentum Principle predicts that the child's speed should be smaller just after the collision (the rotating disk has zero linear momentum, because its center of mass doesn't move). In fact, the child's speed just after the collision is ωR, which is only 0.63 m/s.

Concerning the Energy Principle, this is an inelastic collision. The kinetic energy decreases and the internal energy increases; the child's feet and the disk get warmer.

Note that the perpendicular distance from A to the line of the child's momentum is $r_\perp = 2$ m, and $|L_{A,y}| = r_\perp|\vec{p}| = 240\text{ kg}\cdot\text{m}^2/\text{s}$.

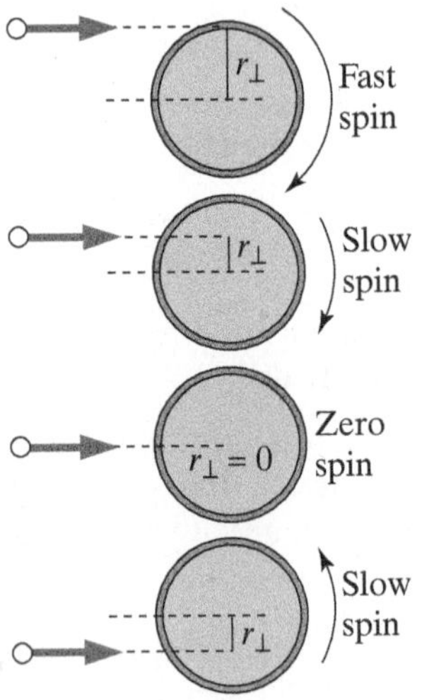

Figure 11.40 How much the disk spins, and in what direction, depends on the impact parameter $r_\perp$.

QUESTION What was the magnitude of the (translational) angular momentum of the child just before hitting the disk?

It was $240\text{ kg}\cdot\text{m}^2/\text{s}$, because $r_\perp = 2$ m all along the path, and the child's momentum $\vec{p}$ didn't change until hitting the disk.

In Figure 11.40 we see that how much the disk spins, and in what direction, depends on the impact parameter—how far off center the child hits the disk.

QUESTION Wait—doesn't the child have rotational angular momentum as well as translational angular momentum?

Yes. We've skipped over a subtle point in calculating the final angular momentum of the system consisting of the disk plus the child. After the collision the child is stuck to the disk and has not only orbital (translational) angular momentum $Rmv = Rm(R\omega) = mR^2\omega$ but also spin (rotational) angular momentum. As the disk rotates about the axle at a rate ω, the child also rotates about the child's own center of mass at a rate ω. If the child's moment of inertia about the child's center of mass is I_{CM}, there is an additional contribution to the final angular momentum of amount $I_{CM}\omega$.

We can make an approximate model of the child as a cylinder whose radius is about $r = 15$ cm, in which case the magnitude of the child's translational plus rotational angular momentum is

$$mR^2\omega + \frac{1}{2}mr^2\omega = m(2^2 + 0.5(0.15)^2)\omega$$

The quantity $(2^2 + 0.5(0.15)^2) = 4 + .011 \approx 4$, which is why in this case it makes only a very slight difference whether we include the child's rotational angular momentum around the child's own center of mass.

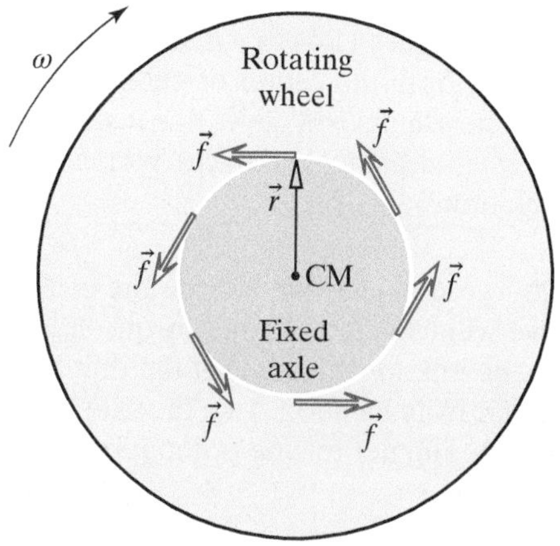

Figure 11.41 A wheel rotates on a fixed axle. There are frictional forces exerted by the axle on the inner surface of the wheel (other forces are omitted).

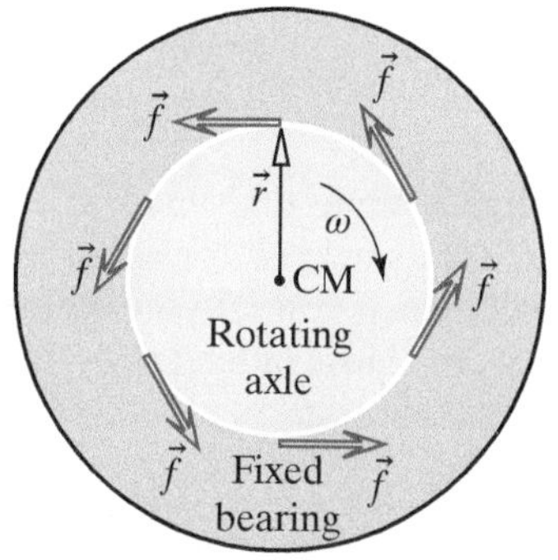

Figure 11.42 An axle rotates inside a fixed bearing. There are frictional forces exerted by the bearing on the outer surface of the axle (other forces are omitted).

Frictional Torque

In the real world, friction is important. After the child in the preceding example has jumped onto the disk and made it rotate, the rotation rate will slowly decrease due to friction between the disk and the axle. Figure 11.41 shows a wheel that rotates on a fixed axle. There are frictional forces exerted by the axle on the inner surface of the wheel (other forces are omitted from the diagram). The frictional forces exert torques $(\vec{r} \times \vec{f})$ about the center of mass of the wheel, which make the angular speed ω of the wheel steadily decrease.

Figure 11.42 shows the opposite case: an axle rotates inside a bearing, which is fixed in place. For example, the axle of a wagon may pass through bearings attached to the bottom of the wagon, and there are wheels rigidly attached to each end of the axle. There are frictional forces exerted by the bearing on the outer surface of the axle (other forces are omitted from the diagram). We again see that frictional forces exert torques $(\vec{r} \times \vec{f})$ about the center of the axle, which make the angular speed ω of the axle steadily decrease.

There are three situations in which it may be a good approximation to neglect frictional torques as we will often do in this chapter:

- If the coefficient of friction μ between the axle and the wheel is small, the frictional force may be negligible, so the frictional torque may be negligible. In rotating machinery μ is often reduced with lubricating oil.
- In Figure 11.41, if the axle and the inside surface of the wheel have small radii, the distance $|\vec{r}|$ from the center of mass to the point of contact between axle and wheel is small, and the frictional torque $\vec{r} \times \vec{f}$ is small. For example, in a magnetic compass the needle sits on a sharp point, which has a very small radius, so any frictional force has a very small lever arm and torque. When an ice skater spins on the tip of the skate, the frictional torque may be very small because the lever arm $|\vec{r}|$ is very small. Although a thin axle has the advantage of reducing frictional torque on a wheel, there may be practical limits imposed by the requirement that the axle be able to support the wheel or the weight of a vehicle.
- In a rotational collision like that of the child jumping on the playground ride, during the brief time interval of the collision any frictional angular impulse $\vec{\tau}\Delta t$ of the axle is small because Δt is small.

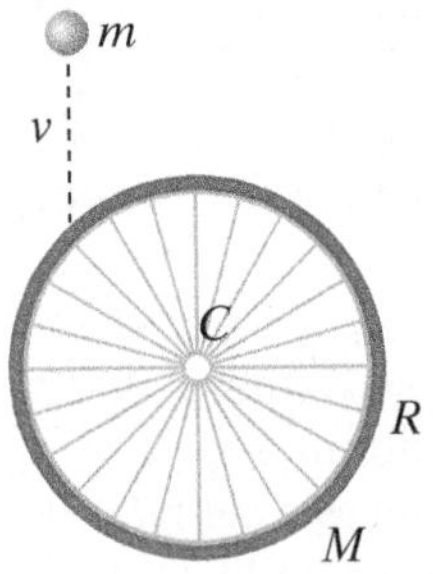

Figure 11.43 A lump of clay falls onto a wheel mounted on a frictionless axle.

Checkpoint 6 A stationary bicycle wheel of radius R is mounted in the vertical plane on a horizontal low-friction axle (Figure 11.43). The wheel has mass M, all concentrated in the rim (the spokes have negligible mass). A lump of clay with mass m falls and sticks to the outer edge of the wheel at the location shown. Just before the impact the clay has a speed v. **(a)** Just before the impact, what is the angular momentum of the combined system of wheel plus clay about the center C? **(b)** Just after the impact, what is the angular momentum of the combined system of wheel plus clay about the center C, in terms of the angular speed of the wheel? **(c)** Just after the impact, what are the magnitude and direction of the angular velocity of the wheel? **(d)** Qualitatively, what happens to the linear momentum of the combined system? Why?

EXAMPLE **A Physical Pendulum**

The name "physical pendulum" is given to hanging objects that swing back and forth and cannot be modeled as a point particle at the end of a string. An example is the object of mass M shown in Figure 11.44 that is supported by a low-friction axle located a distance r_{CM} from the center of mass. The moment of inertia about the object's center of mass is I_{cm}. At this instant the line from the axle to the center of mass is at an angle θ to the vertical. Calculate the period of small-angle oscillations.

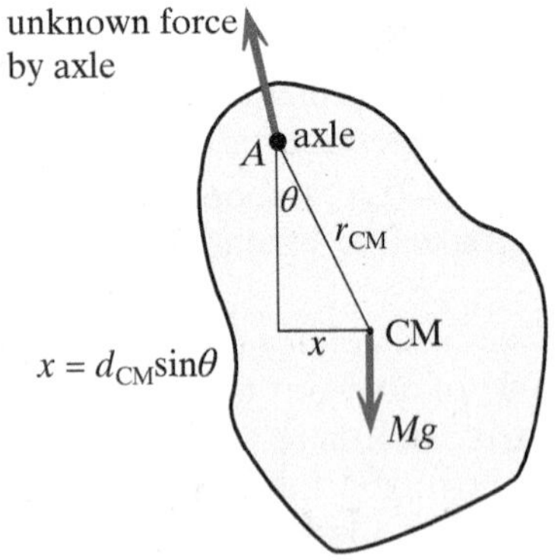

Figure 11.44 A physical pendulum consisting of an object hanging from a low-friction axle.

Solution Apply the Angular Momentum Principle to the object. We choose the center of the axle for location A, about which we will calculate the net torque $\vec{\tau}_{net,A}$. The advantage of this choice for A is that the unknown force that the thin axle exerts on the object has a nearly zero distance from location A, so the axle force contributes almost nothing to the net torque. The net torque is due just to the gravitational force acting at the center of mass.

System: the hanging object

Surroundings: axle, Earth

Angular Momentum Principle: We'll use the parallel axis theorem, that the moment of inertia about a location A a distance r_{CM} from the center of mass is equal to the moment of inertia about the center of mass plus Mr_{CM}^2. We write the z component of the Angular Momentum Principle:

$$I_A \frac{d\omega_z}{dt} = -Mgr_{CM}\sin\theta$$

where $I_A = I_{CM} + Mr_{CM}^2$.

It is clear that the object will oscillate. If the angle θ is small, it can be shown that it is a good approximation to say that $\sin\theta \approx \theta$ for small angles, where the angle is measured in radians. For example, if you set your calculator to radians you'll find that sin (0.01) is 0.0099998, very close indeed to 0.01. Since $\omega_z = d\theta/dt$, $d\omega_z/dt = d^2\theta/dt^2$, so to a good approximation

$$I_A \frac{d^2\theta}{dt^2} = -Mgr_{CM}\theta$$

QUESTION Does this differential equation remind you of a similar differential equation we encountered in an earlier chapter?

This differential equation has the same form as the differential equation for the spring–mass system:

$$m\frac{d^2x}{dt^2} = -k_s x$$

We know that the solution for the spring–mass equation is of the form

$$x = \cos\left(\sqrt{\frac{k_s}{m}}t\right)$$

Comparing the differential equation for the physical pendulum and that for the spring–mass system, we see that the following is a solution for the physical pendulum, for small-angle oscillations:

$$\theta = \cos\left(\sqrt{\frac{Mgr_{CM}}{I_A}}\,t\right)$$

For the spring–mass system we could calculate the period T from the fact that $2\pi/T = \sqrt{k_s/m}$. For the physical pendulum we have $2\pi/T = \sqrt{Mgr_{CM}/I_A}$. Hence the period is $T = 2\pi\sqrt{I_A/Mgr_{CM}}$.

If we needed to know the force that the axle exerts on the hanging object we would need to use the Momentum Principle, which involves the net force.

QUESTION An important special case is the "simple pendulum" where there is a small mass M at the end of a string of length L. What is the period of small-angle oscillations of a simple pendulum?

The moment of inertia of the small point-like object is $I_{CM} + ML^2$, or simply ML^2 since the moment of inertia about the center of mass of the point-like object is nearly zero. Hence $T = 2\pi\sqrt{ML^2/(MgL)} = 2\pi\sqrt{L/g}$.

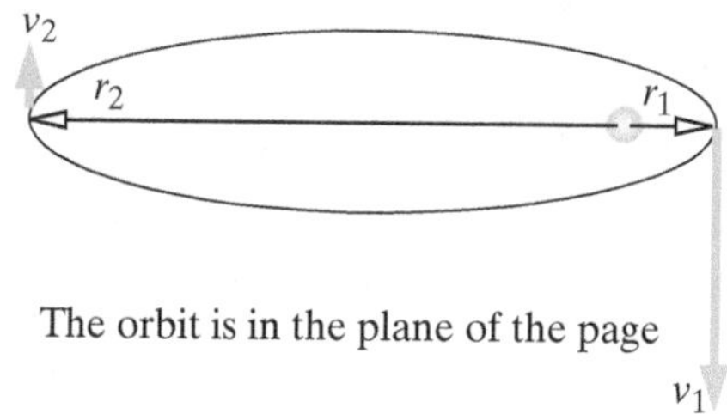

Figure 11.45 A long elliptical orbit of a comet around the Sun.

Comets and the Angular Momentum Principle

If the net torque around a location A is zero, the angular momentum about that location doesn't change. There may be forces acting, causing changes in the linear momentum, but if these forces don't exert any torques, the rate of change of angular momentum is zero, in which case angular momentum is constant.

Comet orbits provide a nice example. Most comets have very long elliptical orbits around the Sun (Figure 11.45). In our earlier discussion of Halley's comet (Section 11.1) we saw that the translational angular momentum of the comet relative to the Sun had the same value at different times along the orbit. We can now understand this observation in terms of the Angular Momentum Principle.

EXAMPLE **Torque on a Comet**

Relative to the center of the Sun, explain why the torque exerted by the Sun's gravitational force on the comet is zero at every point along the orbit (Figure 11.45). What does that say about the angular momentum of the comet relative to the center of the Sun? What if we choose location B instead of location A as the reference point?

Solution The gravitational force on the comet always points directly at the Sun, at location A (Figure 11.46). In $|\vec{\tau}_A| = |r_\perp \times \vec{F}_{grav}|$, the lever arm $r_\perp$ is always zero, so there is no torque relative to location A that can change the angular momentum of the comet. Therefore the angular momentum of the comet, relative to the location of the Sun, cannot change.

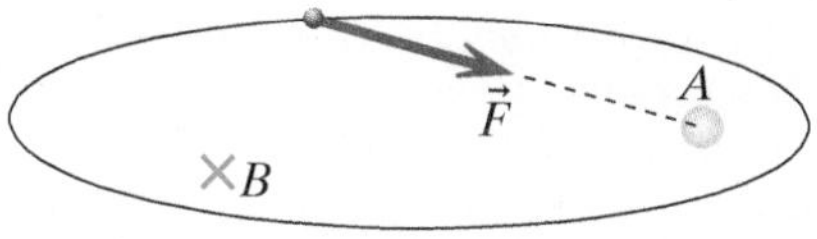

Figure 11.46 The gravitational force acting on the comet points directly at the center of the Sun (location A).

About location B in Figure 11.46, the torque acting on the comet is *not* zero, and the angular momentum of the comet does *not* stay constant. If on the other hand you consider the system consisting of both Sun and comet, the gravitational forces that the two objects exert on each other make torques about location B, or any location, that vectorially add up to zero, and the total angular momentum of the combined system will stay constant. The Sun actually moves a little due to the force that the comet exerts on it, so the Sun's angular momentum relative to location B changes, but its change is the opposite of the change of the angular momentum relative to B of the comet.

QUESTION When a comet is closest to the Sun, its speed is v_1 and its distance from the Sun is r_1. What is the comet's speed when it is farthest from the Sun (a distance r_2)?

The comet's angular momentum about the center of the Sun is constant, because the torque on the comet is zero. Therefore:

$$L_{2,A} = L_{1,A}$$
$$r_2 v_2 = r_1 v_1$$
$$v_2 = v_1 \frac{r_1}{r_2}$$

This equation tells us that the speed of a comet at its farthest point can be very small, since r_1/r_2 is a small quantity. Some comets go far beyond Pluto and spend most of their time there, because they're traveling very slowly. We see them when they come near the Sun, but only for a few months, because they're now traveling fast. The result $r_2 v_2 = r_1 v_1$ at the closest and farthest points in the orbit is correct no matter what kind of "central" force is involved. For any force that acts along a line connecting two objects, there is no torque about a point at the center of one of the objects, so angular momentum is constant about that location.

Kepler and Elliptical Orbits

Historically, the behavior of comets was described before either the Momentum Principle or the Angular Momentum Principle was discovered. Based on careful, accurate naked-eye measurements made by Tycho Brahe before the invention of the telescope, Johannes Kepler in 1609 announced his discovery that the planets follow elliptical orbits around the Sun. Kepler also stated that he had found that "a radius vector joining any planet to the Sun sweeps out equal areas in equal lengths of time." This is equivalent to conservation of angular momentum, as can be seen with the aid of Figure 11.47. The area swept out in a time Δt is the area of a triangle whose base is $v\Delta t$ and whose altitude is $r\sin\theta$. This area is $\frac{1}{2}(rv\sin\theta)\Delta t$, which is proportional to what we now call the angular momentum $rmv\sin\theta$.

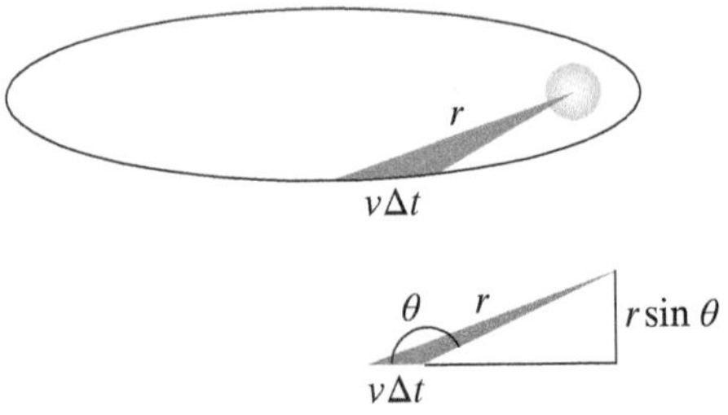

Figure 11.47 "Equal area in equal time."

In 1618 Kepler announced his discovery that the square of the time it takes a planet to go around the Sun is proportional to the cube of the mean distance from the Sun (a result you could derive in Chapter 5 for the simpler case of circular orbits). Later in the 1600s Newton explained all three of these discoveries as derivable from his second law of motion (the Momentum Principle) plus his universal law of gravitation. Kepler's insights provided important tests for Newton's theories.

Checkpoint 7 Because the Earth is nearly perfectly spherical, gravitational forces act on it effectively through its center. Explain why the Earth's axis points at the North Star all year long. Also explain why the Earth's rotation speed stays the same throughout the year (one rotation per 24 h). In your analysis, does it matter that the Earth is going around the Sun?

In actual fact, the Earth is not perfectly spherical. It bulges out a bit at the equator, and tides tend to pile up water at one side of the ocean. As a result, there are small torques exerted on the Earth by other bodies, mainly the Sun and the Moon. Over many thousands of years there are changes in what portion of sky the Earth's axis points toward (change of direction of rotational angular momentum), and changes in the length of a day (change of magnitude of rotational angular momentum). See Section 11.12.

11.8 SYSTEMS WITH NONZERO TORQUES

Of course, the angular momentum of a system is not always constant. External torques (due to objects in the surroundings) can change the angular momentum of a system.

EXAMPLE

A Meter Stick on the Ice

Consider a meter stick whose mass is 300 g and that lies on ice (in Figure 11.48 we're looking down on the meter stick). You pull at one end of the meter stick, at right angles to the stick, with a force of 6 N. Assume that friction with the ice is negligible. What is the rate of change of the center-of-mass speed v_{CM}? What is the rate of change of the angular speed ω? (In Chapter 9 we showed that the moment of inertia I around the center of mass of a uniform rod of mass M and length L is $ML^2/12$; $L = 1\,\text{m}$ here.)

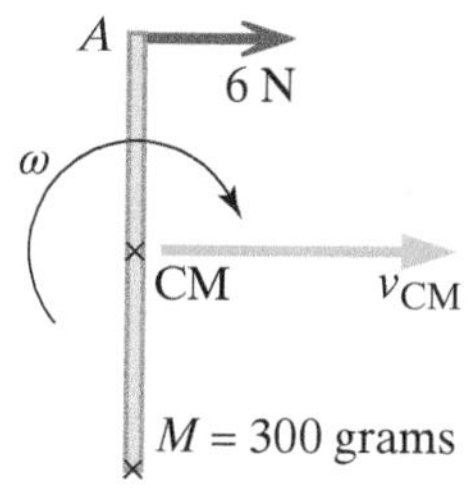

Figure 11.48 Pull one end of a meter stick that is lying on ice (negligible friction).

Solution

System: Stick
Surroundings: Your hand (pulling); ice (negligible effect)

Momentum Principle:
$d\vec{P}/dt = d(m\vec{v}_{CM})/dt = \vec{F}_{net}$
$dv_{CM}/dt = (6\text{N})/(0.3\,\text{kg}) = 20\,\text{m/s}^2$

Angular Momentum Principle about center of mass:
$d\vec{L}_{rot}/dt = \vec{\tau}_{net,CM}$

Component into page ($-z$ direction):
$I d\omega/dt = (0.5\,\text{m})(6\text{N})\sin 90° = 3\,\text{N}\cdot\text{m}$
$d\omega/dt = (3\,\text{N}\cdot\text{m})/[(0.3\,\text{kg}\cdot\text{m}^2)/12] = 120\,\text{rad/s}^2$

In vector terms, $d\vec{\omega}/dt$ points into the page, corresponding to the fact that the angular velocity points into the page and is increasing.

Alternative Analysis—Taking Torques Around the End of the Stick

It is interesting to re-analyze the motion of the meter stick by calculating torque and angular momentum about the end of the stick where the force is applied (location A in Figure 11.48), rather than about the center of mass. To be cautious and correct, we should say that we are taking torques about a location fixed in the ice next to the place where the end of the stick is momentarily located. This is a fixed location, not tied to the moving stick.

System: Stick
Surroundings: Your hand (pulling); ice (negligible effect)

Momentum Principle unchanged, which gives this:
$dv_{CM}/dt = (6\text{N})/(0.3\,\text{kg}) = 20\,\text{m/s}^2$

Angular Momentum Principle about location A (there is zero torque about location A, because there is zero distance from A to where the force is applied.):
$d\vec{L}_A/dt = \vec{\tau}_{net,A}$

Component into page ($-z$ direction):

$$\frac{d}{dt}\left[\frac{ML^2}{12}\omega - \left(\frac{L}{2}\right)Mv_{CM}\right] = 0$$

$$\frac{d\omega}{dt} = \frac{6}{L}\frac{dv_{CM}}{dt} = \frac{6}{(1\,\text{m})}(20\,\text{m/s}^2) = 120\ \text{rad/s}^2$$

This agrees with our analysis in which we took torques about the center of mass of the stick, but the details of the calculation are rather different. It is a good check on an analysis involving the Angular Momentum Principle to do the problem for two different choices of the location about which to calculate the net torque.

A Puck with String Wound Around It

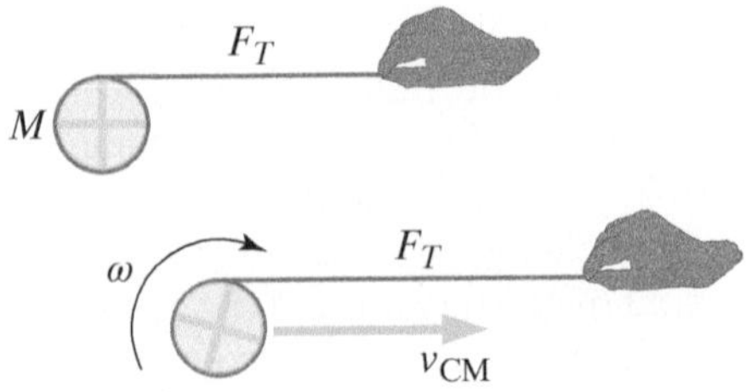

Figure 11.49 Wrap a string around a hockey puck, then pull it along the ice with negligible friction, with a constant tension F_T.

Wrap a string around the outside of a hockey puck. Then pull on the string with a constant tension F_T (Figure 11.49). The puck has mass M and radius R. Assume that friction with the ice rink is negligible. Evidently $dv_{CM}/dt = F_T/M$.

> **QUESTION** The moment of inertia of a uniform solid puck of mass M and radius R is $MR^2/2$. What is the initial rate of change $d\omega/dt$ of the angular speed ω?

The torque about the center of mass is RF_T, into the page (down into the ice). The rotational angular momentum is $(MR^2/2)\omega$, into the page (down into the ice). Therefore $(MR^2/2)d\omega/dt = RF_T$, and we have $d\omega/dt = (2F_T)/(MR)$.

> **Checkpoint 8** Redo the analysis, calculating torque and angular momentum relative to a fixed location in the ice anywhere underneath the string (similar to the analysis of the meter stick around one end). Show that the two analyses of the puck are consistent with each other.

Equilibrium

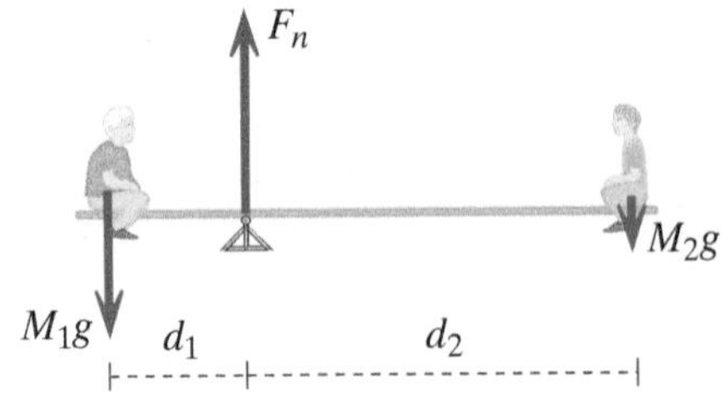

Figure 11.50 A seesaw in equilibrium.

If the net torque on a system is zero, its angular momentum is constant. Conversely, if we know that the angular momentum is not changing, we can conclude that the net torque must be zero. For example, if a system is in equilibrium, not only must the net force on the system be zero; the net torque must also be zero. This allows us to make conclusions about the individual torques whose vector sum is zero.

EXAMPLE **Two People on a Seesaw**

Figure 11.50 shows two persons on a seesaw. The person on the left has mass $M_1 = 90$ kg and sits at a distance $d_1 = 1.2$ m from the nearly frictionless axle. The person on the right has mass $M_2 = 40$ kg. What is the upward force that the axle must exert? Where must the person on the right sit in order that the seesaw not rotate? The seesaw itself has negligible mass.

Solution System: The two persons and the seesaw

Surroundings: Earth and axle

Principle: The Momentum Principle, $\frac{d\vec{p}_{sys}}{dt} = \vec{F}_{net}$

Since the system is not moving, the momentum of the system isn't changing, so $d\vec{p}_{sys}/dt = \vec{0}$, and therefore the net force $\vec{F}_{net} = \vec{0}$. Consider the y forces:

$$F_N - M_1g - M_2g = 0$$

$$F_N = (M_1 + M_2)g = ((90 + 40)\ \text{kg})(9.8\,\text{N/kg}) = 1274\,\text{N}$$

It isn't surprising to find that the axle must exert an upward force equal to the combined weights of the two persons. Next we use a different principle to determine where the person on the right should sit.

Principle: The Angular Momentum Principle, $\dfrac{d\vec{L}_A}{dt} = \vec{\tau}_{\text{net}}$

Since the angular momentum isn't changing, $d\vec{L}_A/dt = \vec{0}$ about any point whatsoever, and therefore $\vec{\tau}_{\text{net}} = \vec{0}$ about any point whatsoever. It is convenient to choose point A to be at the location of the axle, because then the upward force of the axle exerts no torque about point A. Consider the z component of the net torque:

$$M_1 g d_1 - M_2 g d_2 = 0$$

$$d_2 = \frac{M_1}{M_2} d_1 = \frac{90\,\text{kg}}{40\,\text{kg}}(1.2\text{ m}) = 2.7\text{ m}$$

This analysis would also be valid if the seesaw were rotating at a constant angular velocity, because it would still be the case that the angular momentum would not be changing, and the net torque would be zero.

This is an example of a category of "statics" problems in which it is necessary to consider both the Momentum Principle and the Angular Momentum Principle in order to carry out a complete analysis.

Checkpoint 9 Write a different torque equation around a location fixed in space where the person on the left is sitting, and show that it is in fact equivalent to the torque equation around the axle, when you take the force equation into consideration. It is often convenient to choose your fixed location so that some of the forces create no torque around that location and therefore don't appear in the Angular Momentum Principle.

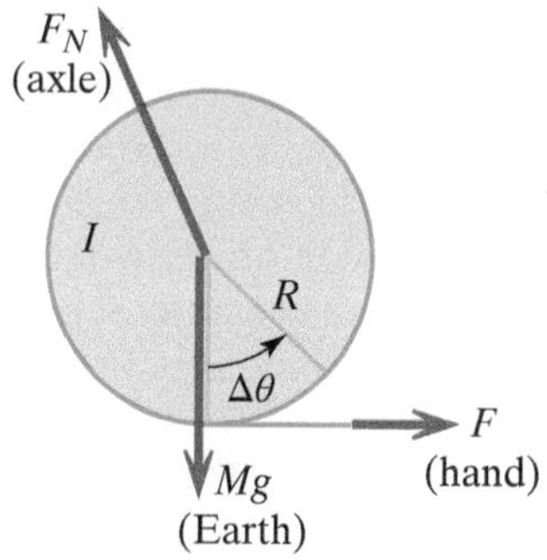

Figure 11.51 A wheel rotates on a low-friction axle, pulled by a string wrapped around the edge.

11.9 PREDICTING POSITIONS WHEN THERE IS ROTATION

In order to model the motion of a rotating system, we need to be able to predict its position. This usually means predicting the angle through which the object will rotate. If we know this angle, we can also figure out how far a given point has moved.

EXAMPLE **Predicting Position with Rotation**

A wheel of radius R and moment of inertia I is mounted on a low-friction axle (Figure 11.51). A string is wrapped around the edge, and you pull on it with a force F. At a certain time the angular speed is ω_i. **(a)** After a time interval Δt, what is the angular speed ω_f? **(b)** How far did your hand move during this time interval?

Solution System: Wheel plus string
Surroundings: Earth, axle, hand
Free-body diagram: Figure 11.51

Principle: The Momentum Principle tells us that the net force must be zero, because the center of mass does not move. To obtain information about rotation, we need to use the Angular Momentum Principle. We'll use the update form, about the center of mass:

$$\vec{L}_{\text{rot},f} = \vec{L}_{\text{rot},i} + \vec{\tau}_{\text{net}}\Delta t$$

(a) Calculate the final angular speed (only the force of the hand exerts a nonzero torque about the axle):

$$I\omega_f = I\omega_i + RF\Delta t$$
$$\omega_f = \omega_i + \frac{RF\Delta t}{I}$$

(b) To calculate how far your hand moved, first we will determine through how large an angle the wheel turned during the time interval Δt. Suppose that there is a blue line drawn radially from the center of the wheel to the edge, as shown in Figure 11.51. Let θ be the angle that this blue line makes with the y axis. This angle changes at an instantaneous rate of $\omega = d\theta/dt$, in radians per second. The change in the angle θ is

$$\Delta\theta = \omega_{\text{avg}}\Delta t$$

Because the angular speed ω changes at a constant rate RF/I, the average angular speed is

$$\omega_{\text{avg}} = \frac{\omega_i + \omega_f}{2}$$

Therefore the change in the angle of the blue line is

$$\Delta\theta = \frac{\omega_i + \omega_f}{2}\Delta t$$

Since we calculated ω_f from the Angular Momentum Principle, now we know through how big an angle the wheel turned. What remains to be calculated is how far your hand moved. If the angle through which the wheel rotates is 2π radians (360°), an amount of string $2\pi R$ will come off the edge of the wheel, since the circumference of the wheel is $2\pi R$. More generally, if the angle through which the wheel rotates is $\Delta\theta$ (in radians), the amount of string that comes off is $R\Delta\theta$.

The length of string that comes off the wheel is equal to the distance your hand moves:

$$\Delta x_{\text{hand}} = R\frac{\omega_i + \omega_f}{2}\Delta t$$

The rate of change of the angular velocity $d\vec{\omega}/dt$ is often called the "angular acceleration" $\vec{\alpha}$, and the Angular Momentum Principle applied to pure rotation of an object whose moment of inertia is not changing is written as $I\vec{\alpha} = \vec{\tau}_{\text{net}}$. In this form, the equation looks very similar to $M\vec{a} = \vec{F}_{\text{net}}$. Just as a constant acceleration $\vec{a}$ means that the average velocity is $(\vec{v}_i + \vec{v}_f)/2$, so also if the angular acceleration $\vec{\alpha}$ is constant the average angular velocity is $(\vec{\omega}_i + \vec{\omega}_f)/2$.

We could also determine the distance the hand moves by using the Energy Principle in combination with our result for ω_f, which we obtained using the Angular Momentum Principle:

$$\frac{1}{2}I\omega_f^2 = \frac{1}{2}I\omega_i^2 + F\Delta x_{\text{hand}}$$

Checkpoint 10 A uniform-density wheel of mass 6 kg and radius 0.3 m rotates on a low-friction axle. Starting from rest, a string wrapped around the edge exerts a constant force of 15 N for 0.6 s. **(a)** What is the final angular speed? **(b)** What is the average angular speed? **(c)** Through how big an angle did the wheel turn? **(d)** How much string came off the wheel?

11.10 COMPUTATION AND ANGULAR MOMENTUM

Computationally, modeling the motion of a rotating object subject to a nonzero net torque is very similar to modeling the motion of a moving object subject to a nonzero net force. After choosing a location A and specifying initial conditions, we iteratively update torque, angular momentum, and position (angle).

Repeat

- Calculate the net torque $\vec{\tau}_{\text{net},A}$ acting on the system.
- Update the angular momentum of the system: $\vec{L}_f = \vec{L}_i + \vec{\tau}_{\text{net},A}\Delta t$.
- Update the position: $\theta_f = \theta_i + \omega\Delta t$.

In VPython, an object can be rotated by specifying the angle through which to rotate, the axis of rotation, and the origin about which to rotate:

```
wheel.rotate(angle=dtheta, axis=vector(0,1,0),
             origin=wheel.pos)
```

The angle `dtheta` is a signed scalar. Assuming the angular momentum and the axis of rotation are along the same line, we can use a vector dot product to find the magnitude and sign of $\vec{\omega}$, and use this to find `dtheta`:

```
omega_scalar = dot(omega, norm(axis_of_rotation))
dtheta = omega_scalar * deltat
```

In VPython, the `cross` function calculates vector cross products:

```
torque = cross(r_A, F)
```

The calculational loop in a program modeling the motion of a wheel subject to a torque might look something like this:

```
axis_of_rotation = vector(0,1,0)
while True:
    rate(100)
    torque = cross(r_a,F)
    L = L + torque * deltat
    omega = L/I
    omega_scalar = dot(omega, norm(axis_of_rotation))
    dtheta = omega_scalar * deltat
    wheel.rotate(angle=dtheta, axis=axis_of_rotation,
                 origin=wheel.pos)
```

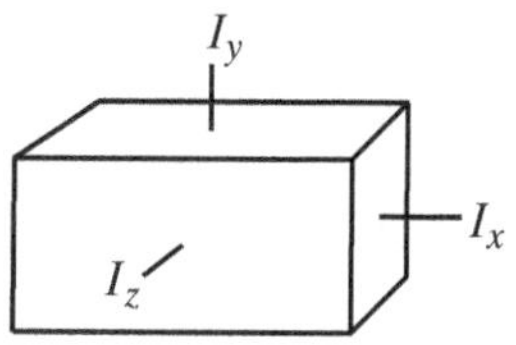

Figure 11.52

The approach outlined above is valid for situations in which the angular momentum and the axis of rotation lie along the same line. However, as discussed in the optional Section 11.13, the angular momentum vector $\vec{L}$ need not be in the same direction as the angular velocity vector $\vec{\omega}$. For example, the box shown in Figure 11.52 has three different moments of inertia about the three axes, and for rotations about some arbitrary axis $\vec{L}$ may not be in the direction of $\vec{\omega}$.

This complexity is beyond the scope of this textbook. In the computational modeling problems at the end of this chapter, we limit the complexity to situations where the net torque $\vec{\tau}_{\text{net}}$, the angular momentum $\vec{L}$, and the angular velocity $\vec{\omega}$ all lie along the same axis (the torque and angular momentum may be in opposite directions along that axis, for example if the torque is slowing down the rotation).

11.11 ANGULAR MOMENTUM QUANTIZATION

Many elementary particles have rotational angular momentum. Electrons bound in an atom may have translational angular momentum relative to the nucleus as well as their own rotational angular momentum. Atoms as a whole may have angular momentum, as do many nuclei. The surprising thing about angular momentum at the atomic and subatomic level is that in stable energy states the angular momentum is quantized:

THE ANGULAR MOMENTUM QUANTUM

$$\hbar = \frac{h}{2\pi} = 1.05 \times 10^{-34}\ \text{J} \cdot \text{s}$$

The constant h is "Planck's constant," 6.63×10^{-34} J · s. The constant $\hbar$ is called "h-bar." Whenever you measure a vector component of the angular momentum (that is, along the x or y or z axis), you get either a half-integer or integer multiple of this quantum of angular momentum.

Checkpoint 11 Show that $\hbar$ and angular momentum have the same units.

The Bohr Model of the Hydrogen Atom

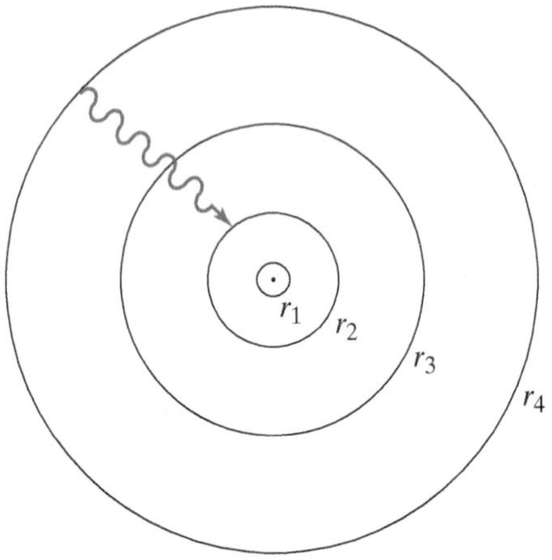

Figure 11.53 In the Bohr model of the hydrogen atom, electrons can be in only those circular orbits for which angular momentum is a multiple of $\hbar$. In a transition between allowed orbits, a photon is emitted.

As discussed in the previous chapter, in 1911 Rutherford and his group discovered the nucleus in atoms. Stimulated by this discovery, in 1913 the Danish physicist Niels Bohr made a bold conjecture that a hydrogen atom could be modeled as an electron going around a proton in circular orbits, but only in those orbits whose translational angular momentum is an integer multiple of $\hbar$. Consequently, these orbits would have only certain radii, and only certain values of energy. The differences in energies between these quantized orbits match the observed energies of photons emitted by atomic hydrogen (Figure 11.53).

Bohr's basic hypothesis—that the angular momentum of electrons in an atom is quantized—has proven to be an important insight, and has been retained as a fundamental tenet of the quantum mechanical model of an atom. As this model has been refined, a probabilistic view of the motion of electrons has been adopted. In more sophisticated models electrons do not have precise trajectories, but only a "probability density"—a probability of being found at a particular location. These more complex quantum mechanical models explain a much wider range of atomic and molecular phenomena than does Bohr's original model. Additionally, the actual quantization rules in a hydrogen atom are more complex than those that are assumed in the Bohr model. For example, the translational angular momentum in the ground state ($N = 1$) is actually zero, not $\hbar$ as the Bohr model predicts, and for the next higher state ($N = 2$), the z component of translational angular momentum can be either 0 or $\hbar$. The photon itself has angular momentum $\hbar$, so this must also be taken into account in photon emissions or absorptions.

The simple Bohr model does, however, predict correctly the allowed electronic energy levels for atomic hydrogen (as determined from the emission and absorption spectra of hydrogen atoms). We will work with this model to see how quantization of angular momentum leads to the prediction of electronic energy levels in a hydrogen atom.

Allowed Radii of Electron Orbits

Because the proton has much more mass than an electron (about 2000 times as much), we'll make the approximation that the proton is at rest, with the electron in a circular orbit around the proton.

QUESTION If the electron momentum is p, and the radius of the circular orbit is r, what is the translational angular momentum?

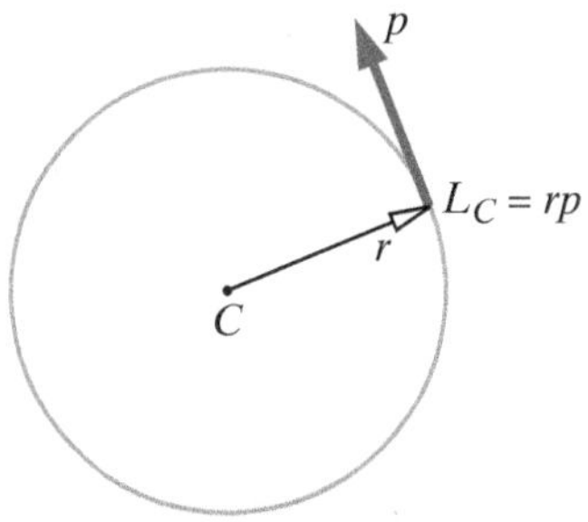

Figure 11.54 The angular momentum of the electron in a circular Bohr orbit relative to the proton. L_C is out of the page.

The translational angular momentum of the electron relative to the location of the proton is $L_{\text{trans},C} = rp\sin(90°) = rp$ (Figure 11.54). Bohr proposed that the only possible states of the hydrogen atom are those where the electron is in a circular orbit whose translational angular momentum is an integer multiple of $\hbar$:

BOHR: ANGULAR MOMENTUM IS QUANTIZED

$$|\vec{L}_{\text{trans},C}| = rp = N\hbar,$$

where N is an integer $(1, 2, 3, \ldots)$

The Electric Force and a Circular Orbit

QUESTION What is the magnitude of the electric force that the proton exerts on the electron?

$$F_{\text{el}} = \frac{1}{4\pi\varepsilon_0}\frac{e^2}{r^2}$$

where $+e$ is the electric charge on the proton ($-e$ for the electron).

QUESTION Use the Momentum Principle to relate this force to the circular motion of the electron.

In a circular orbit at constant speed we know that the rate of change of the momentum is $(v/r)p$, so we have in the approximation that $v \ll c$, where m is the mass of the electron,

$$\left(\frac{v}{r}\right)p = \frac{mv^2}{r} = F = \frac{1}{4\pi\varepsilon_0}\frac{e^2}{r^2}$$

Solving for r

We are looking for the allowed values of the orbit radius r, so we look for ways to express v in terms of r. Bohr's angular momentum condition gives us v in terms of r:

$$|\vec{L}_{\text{trans},C}| = rp = N\hbar \quad \text{leads to} \quad v = \frac{p}{m} = \frac{N\hbar}{mr}$$

Putting this result for r into the result obtained from the Momentum Principle, $mv^2/r = F$, we have

$$\frac{m}{r}\left(\frac{N\hbar}{mr}\right)^2 = \frac{1}{4\pi\varepsilon_0}\frac{e^2}{r^2}$$

Solving the latter relation for r, we obtain the following result (Figure 11.55):

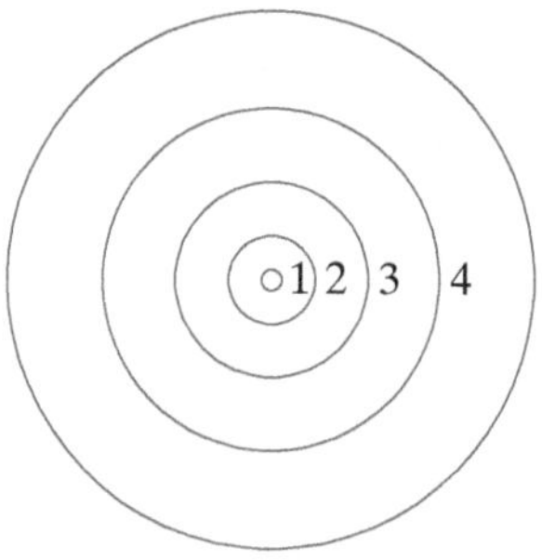

Figure 11.55 Bohr radii for $N = 1$ through 5. The nucleus is not shown.

ALLOWED BOHR RADII FOR ELECTRON ORBITS

$$r = N^2\frac{\hbar^2}{\left(\frac{1}{4\pi\varepsilon_0}\right)e^2 m}$$

Let's evaluate this result numerically, so that later we can compare with measurements of the spectrum of light emitted by excited atomic hydrogen. Using

$\hbar = 1.05 \times 10^{-34}$ J · s,
$1/4\pi\varepsilon_0 = 8.99 \times 10^9$ N · m^2/C^2,
$e = 1.60 \times 10^{-19}$ C,
and the mass of the electron $m = 9.11 \times 10^{-31}$,

we find this:

$$r = N^2(0.53 \times 10^{-10}\,\text{m})$$

This is a striking result. The simple Bohr model predicts that the smallest permissible electron radius ($N = 1$) is 0.53×10^{-10} m, and atoms are in fact observed to have radii of approximately this size.

Energy for a Circular Orbit

We cannot observe directly the radius of the orbit of an electron in an atom. What we can observe are photons emitted by excited atoms, which tell us the differences in energies between various energy levels of the atom. Using the Bohr model we can predict the energies of these energy levels, and see whether the differences between these levels match the energies of photons observed in an atomic spectrum.

Given a set of possible values for r, we can calculate the possible values for energy of the electron, starting from this observation:

$$\text{The result } mv^2/r = F \quad \text{leads to} \quad mv^2 = \frac{1}{4\pi\varepsilon_0}\frac{e^2}{r}.$$

QUESTION Now write an expression for the kinetic energy plus electric potential energy of the hydrogen atom, using the Bohr model.

Kinetic energy:

$$K = \frac{1}{2}mv^2 = \frac{1}{2}\left(\frac{1}{4\pi\varepsilon_0}\frac{e^2}{r}\right) \quad \text{(from the previous equation)}$$

Electric potential energy:

$$U_{\text{el}} = -\frac{1}{4\pi\varepsilon_0}\frac{e^2}{r}$$

Therefore the energy is this (omitting the rest energies):

$$E = K + U_{\text{el}} = -\frac{1}{2}\left(\frac{1}{4\pi\varepsilon_0}\frac{e^2}{r}\right)$$

According to the Bohr model only certain radii will actually occur. Inserting our previous expression for r, we get:

BOHR MODEL ENERGY LEVELS

$$E = -\frac{\left(\frac{1}{4\pi\varepsilon_0}\right)^2 e^4 m}{2N^2\hbar^2}$$

Evaluating this expression we get

$$E = -\frac{2.17 \times 10^{-18}\,\text{J}}{N^2}$$

Converting to electron-volts ($1\,\text{eV} = 1.6 \times 10^{-19}$ J) we have (Figure 11.56)

$$E = -\frac{13.6\,\text{eV}}{N^2}, \quad N = 1, 2, 3, \ldots$$

This prediction for the quantized energies of atomic hydrogen agrees well with the observed electronic spectrum of hydrogen. We quoted this result for the quantized energy levels in Chapter 8. The energies are negative, corresponding to bound states.

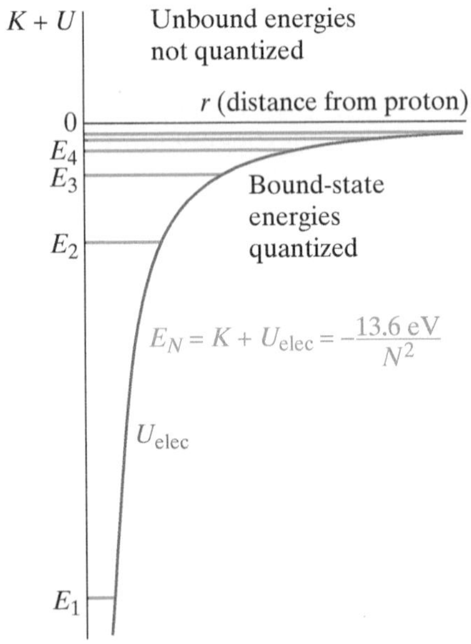

Figure 11.56 Bohr model prediction of electronic energy levels for a hydrogen atom. The potential energy curve for the electron–proton system is also shown.

Photons Emitted by Atomic Hydrogen

Bohr proposed that electromagnetic radiation would be emitted when there was a sudden change from a higher energy level to a lower one. We saw in Chapter 8 that the Bohr energy expression does correctly predict the energies of photons emitted by excited atomic hydrogen.

> QUESTION In the Bohr model, does photon emission correspond to the radius of the circular orbit getting larger or smaller? Does the quantum number N increase or decrease?

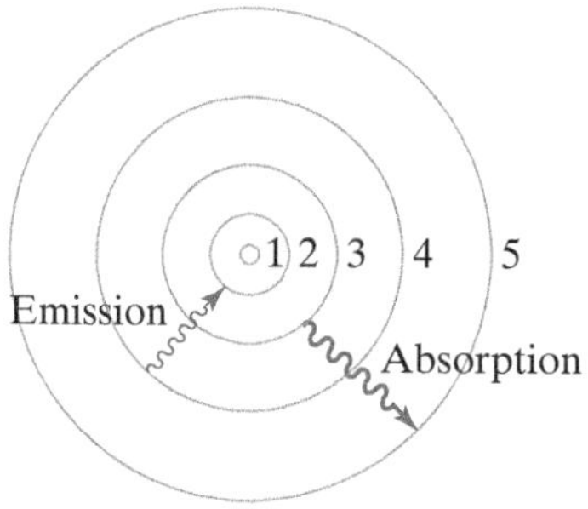

Figure 11.57 A photon is emitted in a transition from a higher to a lower energy level. A photon is absorbed in a transition from a lower to a higher energy level.

The kinetic plus potential energy is negative, corresponding to a bound state, so a higher energy is one that is less negative and therefore corresponds to a larger radius (since the energy is proportional to $-1/r$). Also, higher energy corresponds to a larger value of N, which is also associated with a larger radius. Therefore photon emission is associated with a decrease in r and a decrease in N (Figure 11.57). Photon absorption involves an increase in r and N.

When Bohr explained his model to Rutherford, Rutherford immediately realized that atoms display probabilistic behavior. For example, in the Bohr model an atom in the second excited state can either drop to the first excited state (with emission of a photon of energy 1.9 eV) or directly to the ground state (with emission of a photon of energy 12.4 eV). Both of these photon energies are observed in the spectrum of atomic hydrogen, so evidently some fraction of the time one process occurs, and the rest of the time the other process occurs. This is radically different from the deterministic behavior familiar in classical mechanics.

Particle Spin

We know that many elementary particles have rotational angular momentum because of observations of the interaction of these particles with other particles having nonzero angular momentum; in these interactions total angular momentum is constant.

For example, the electron, muon, and neutrino are said to have spin $1/2$, because measurement of a component of their rotational angular momentum always yields $\pm(1/2)\hbar$. Every quark has spin $1/2$, and particles built out of three quarks, such as protons and neutrons, necessarily have half-integral spin. The spin of the proton and neutron is $1/2$, as though two of the quarks have their spins opposed ($\Uparrow\Downarrow\Uparrow$), with no translational angular momentum of the quarks. Some short-lived three-quark particles have spin $3/2$. One way to get this is for all the quarks to have their spins aligned ($\Uparrow\Uparrow\Uparrow$), or there can be some translational angular momentum of the quarks. While rotational angular momentum can be half-integral, translational angular momentum is always integral (0, 1, 2, etc.).

Particles made out of one quark and one antiquark, called mesons, have integral spin. For example, the pion spin is zero (quark and antiquark spins opposed, $\Uparrow\Downarrow$), and the spin of the "rho" meson is 1 (quark and antiquark spins aligned, $\Uparrow\Uparrow$).

Presumably the angular momentum of a macroscopic object such as a baseball is also quantized, but the angular momentum of a baseball is huge compared to Planck's constant, and you don't notice the quantization, which is on an exceedingly fine scale (1×10^{-34} J · s!). Only at the atomic level are the effects of quantization really evident.

Particles, Nuclei, Atoms, and Molecules

Quantum mechanics predicts that the x, y, or z component of the angular momentum (L_x, L_y, or L_z) can only be an integer or half-integer multiple of $\hbar$, whereas the square magnitude of the angular momentum has the quantized values $l(l+1)\hbar^2$, where l has integer or half-integer values, depending on whether a component has integer or half-integer values:

QUANTIZED VALUES OF L^2

$$L^2 = l(l+1)\hbar^2$$

where l is integer or half-integer.

There is a very deep connection between angular momentum and the statistical behavior of particles. Particles such as the electron and the proton that have half-integral spin are called "fermions" and exhibit the peculiar behavior that two fermions cannot be in the exact same quantum state. This is why only two electrons can be in the lowest energy state of an atom, with their spins opposed. Their spin directions are forbidden to be the same, because then their energy and angular momentum would be exactly the same, which is forbidden for fermions. This prohibition is called the "Pauli exclusion principle."

On the other hand, particles with integral spin, called "bosons," are not subject to the Pauli exclusion principle, and there is no limit to the number of bosons that can be in the same energy state. In fact, there is a special state of matter called a "Bose–Einstein condensate" in which very large numbers of particles end up in exactly the same quantum state. This state of matter was predicted long before it was actually created and observed in 1995.

All nuclei that have an even number of protons and an even number of neutrons ("even-even" nuclei) have a total angular momentum of zero in their nuclear ground state. Examples include carbon-12 (6 protons and 6 neutrons) and oxygen-16 (8 protons and 8 neutrons). The spins of protons and neutrons in even-even nuclei are paired up in the lowest-energy state of a nucleus with each other in such a way as to produce zero net angular momentum. This would be an exceedingly unlikely outcome if angular momentum weren't quantized.

Angular momentum quantization plays a major role in determining the structure of atoms and the nature of the chemical periodic table. The two lowest-energy-state electrons in an atom always have zero translational angular momentum and zero rotational angular momentum (because the two rotational angular momenta are always oppositely aligned and add up exactly to zero).

In a diatomic gas molecule such as oxygen (O_2), the kinetic energy associated with rotation of the molecule is of course $L_{\text{rot}}^2/(2I)$. However, the rotational angular momentum L_{rot} is quantized, so the rotational energies of oxygen are quantized. The phenomenon of angular momentum quantization affects the specific heat of diatomic gases at low temperatures, as will be discussed in Chapter 12.

11.12 *GYROSCOPES

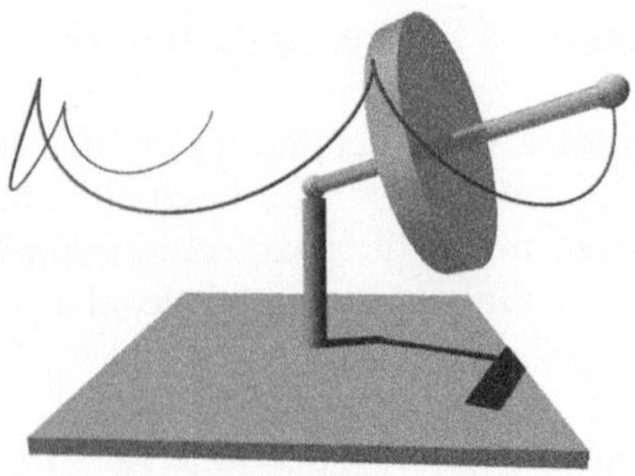

Figure 11.58 A gyroscope can exhibit both "nutation" and "precession."

A gyroscope is a fascinating device, and its unusual properties have been exploited to stabilize ships and spacecraft. Its behavior provides a good model and analogy for some important aspects of the quantum mechanical behavior of atoms and nuclei. Magnetic resonance imaging (MRI) is based on the gyroscope-like behavior of spinning nuclei.

A gyroscope has a spinning disk mounted on an axle. If you place one end of the axle on a vertical support, you may observe a very complex motion (Figure 11.58). The gyroscope rises and falls ("nutation") as it revolves around the support ("precession").

One fruitful approach in modeling a complex physical system is to try to analyze the simplest motion that the system is capable of. For example, we can observe a pure precession of a gyroscope, with no nutation. It is possible with care to start the gyroscope with special initial conditions so that it merely precesses, without bobbing up and down. We'll try to analyze this special kind of gyroscope motion. In fact, we'll start with a particularly simple form of precession, with the rotational axis horizontal (Figure 11.59).

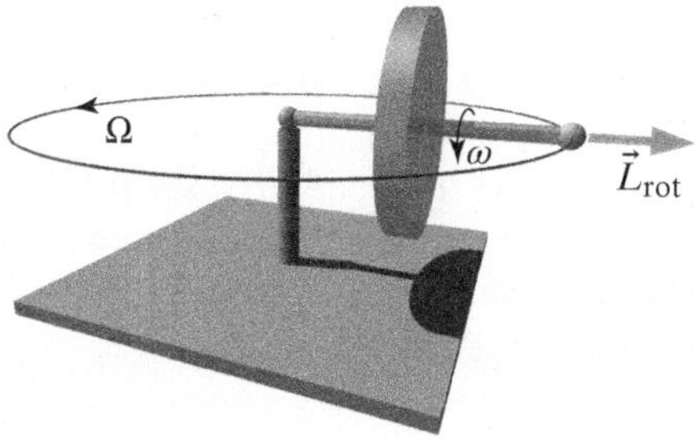

Figure 11.59 A gyroscope that is precessing horizontally, with no nutation.

We define some variables to make it easier to describe and discuss the situation. The gyroscope disk rotates around its own axis with some spin angular speed that we'll call ω, and the disk has a moment of inertia I. The rotational angular momentum $\vec{L}_{rot}$ of the disk always points horizontally but is continually changing direction as the gyroscope precesses (if we neglect friction, the angular speed ω is constant). The gyroscope revolves around the support with angular speed Ω (uppercase Greek omega), constant in magnitude and direction (if we neglect friction). Note that $v_{CM} = \Omega r$, so $\Omega = (v_{CM}/r)$, where r is the distance from the support to the center of mass.

In Chapter 5 we derived the expression for the time rate of change of the unit momentum vector $\hat{p}$, which had the magnitude (v/r), which is the same as angular speed ω. In general, the magnitude of the time rate of change of any rotating unit vector $\hat{X}$ is $|d\hat{X}/dt| = \omega$, and the magnitude of the rate of change of a rotating vector $\vec{X} = |\vec{X}|\hat{X}$ whose magnitude isn't changing is $\omega|\vec{X}|$. Since the angular speed of the gyroscope's precession is Ω, the magnitude of the rate of change of the (linear) momentum P of the gyroscope is ΩP, and the magnitude of the rate of change of the angular momentum is ΩL.

> **QUESTION** What forces act on the gyroscope? What relationships are there among these forces?

The support pushes up with a force we'll call F_N, and the Earth pulls down with a force Mg through the center of mass (Figure 11.60). Since the center of mass stays at the same height all the time in the case of simple precession, the vertical component of the net force must be zero. Therefore $F_N = Mg$.

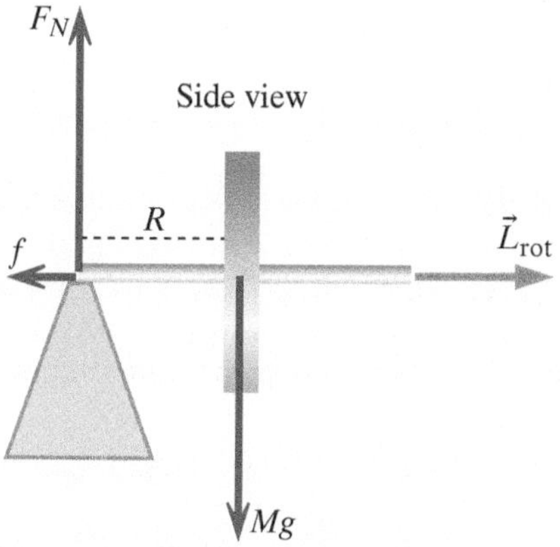

Figure 11.60 Side view showing external forces acting on the gyroscope and the rotational angular momentum vector.

This isn't quite the whole story, though. Note that the center of mass of the gyroscope is moving in a circle, which requires that there be a radially inward force to make the momentum vector rotate (with angular speed Ω). The only object that can exert this force is the support. There must be a small horizontal frictional force f such that $|d\vec{P}/dt| = \Omega P = f$, where $P = Mv_{CM} = M(\Omega R)$; R is the radius of the circle traveled by the center of mass. This force, $f = MR\Omega^2$, is small if the precession rate Ω is small.

It is easy to observe with a toy gyroscope that as the spin of the disk slows down due to friction (smaller ω), the precession actually speeds up (larger Ω). If we could predict the precession speed for a given spin, we would largely understand this simple example of gyroscope motion. What can we use to attempt a prediction?

> **QUESTION** What is the magnitude of the rotational angular momentum of the gyroscope? What is the magnitude of the translational angular momentum of the gyroscope around the support?

We said that the rotating disk has moment of inertia I and angular speed ω, so the rotational angular momentum is simply $L_{rot} = I\omega$. The magnitude of the translational angular momentum is this:

$$L_{support} = |\vec{R} \times \vec{P}| = RP$$

Evidently the magnitudes of both the rotational and translational angular momenta are constant, not changing with time. However, what about the directions of these angular momenta? Do they change with time?

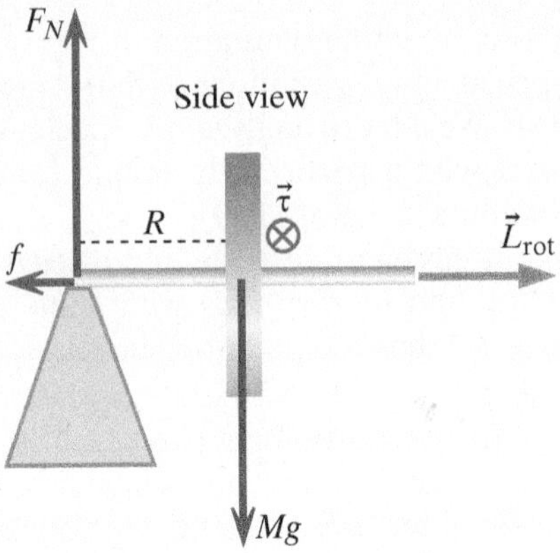

Figure 11.61 Side view showing external forces acting on the gyroscope and the rotational angular momentum vector. There is a torque around the center of mass, into the page, due to the F_N force.

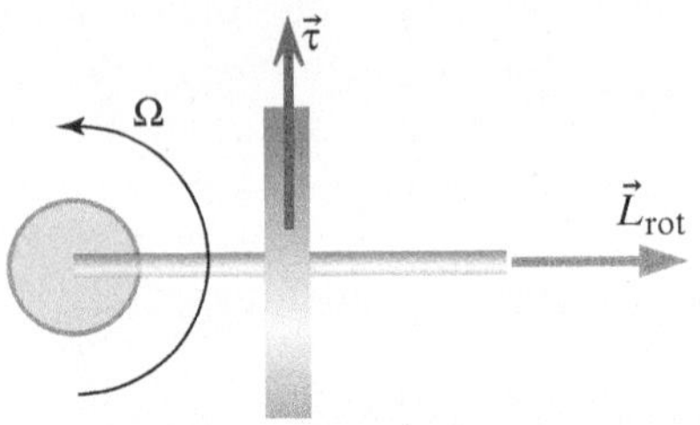

Figure 11.62 View from above, showing the torque vector and the rotational angular momentum vector.

QUESTION How does the direction of the translational angular momentum change with time? How does the direction of the rotational angular momentum change with time?

Neither the magnitude nor the direction of the translational angular momentum changes (neglecting friction): it has constant magnitude and points vertically upward at all times. However, the direction of the rotational angular momentum is constantly changing as the gyroscope precesses. We need to calculate the rate at which the rotational angular momentum vector is changing, $d\vec{L}_{rot}/dt$. Since the rotational angular momentum vector is a rotating vector, which rotates with the precession angular speed Ω, we have this:

$$\left|\frac{d\vec{L}_{rot}}{dt}\right| = \Omega L_{rot}$$

QUESTION Therefore, what is the Angular Momentum Principle for the gyroscope?

The remaining element we need for the Angular Momentum Principle is the torque that acts around the center of mass, at the center of the spinning disk. Look again at the force diagram (Figure 11.61), and you see that the magnitude of the torque about the center of mass is $RF_N = RMg$, and the direction of the torque is into the page. Seen from above, the torque points at a right angle to the rotational angular momentum (Figure 11.62). Therefore the Angular Momentum Principle yields

$$\left|\frac{d\vec{L}_{rot}}{dt}\right| = \Omega L_{rot} = \tau_{CM}$$

where $L_{rot} = I\omega$, and $\tau_{CM} = RMg$. Solving for the precession angular speed Ω, we have

$$\Omega = \frac{\tau_{CM}}{L_{rot}} = \frac{RMg}{I\omega}$$

This is a surprisingly simple result for such a complicated system, and it agrees at least qualitatively with observations. A smaller spin ω is associated with a larger precession rate Ω. Conversely, a gyroscope that spins very fast precesses very slowly. If possible, your instructor will provide an opportunity to test the theory quantitatively by observing an actual gyroscope whose spin and precession rates can be measured.

Reflection

Reflect on how we arrived at this result. Look over the steps involved. The key point is that the rotational angular momentum varies in direction without varying in magnitude, and the magnitude of the rate of change of the rotational angular momentum is simply Ω times L_{rot}. This rate of change of the rotational angular momentum is equal to the net torque acting on the system (around the center of mass).

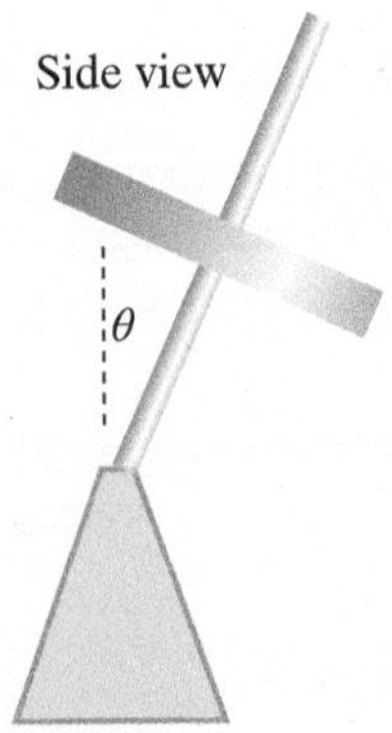

Figure 11.63 What is the precession rate when the spin axis is at an angle θ to the vertical?

Checkpoint 12 To complete this reflection, determine the relationship between ω and Ω for the case of pure precession, but with the spin axis at an arbitrary angle θ to the vertical (Figure 11.63; $\theta = 90°$ is the case of horizontal precession we treated). If you have the opportunity, see whether this relationship holds for a real gyroscope.

Uses of Gyroscopes

The precession we calculated is the result of a nonzero torque acting around the center of mass. Through clever mounting of the gyroscope it is possible to make this torque vanishingly small, in which case the axis does not precess but always maintains the same direction, which has been useful in navigation. The satellite-based Global Positioning System (GPS) can tell you where your airplane is; gyroscopically based instruments can tell you the orientation of your airplane at this known location.

Large gyroscopes have been used to stabilize ships against rolling in the sea. A gyroscope with very high rotational angular momentum (large moment of inertia and large angular speed) is hard to turn quickly, and this can provide mechanical stability. Gyroscopes are also used to stabilize the orientation of spacecraft and satellites.

Friction

We deliberately neglected friction in the calculation. Friction has two main effects on a gyroscope. Friction in the spin axis makes ω decrease with time, which leads to an increased precession rate Ω. Friction on the top of the support slows down the precession, which means that Ω is no longer equal to RMg/L_{rot} as it should be for pure precession. As a result of these effects, what you observe with a gyroscope that starts out in pure precession is that the spin axis eventually starts to tip lower, accompanied by faster precession. The gyroscope starts out with a stately, dignified, slow precession, but toward the end gives the impression of motion that is more and more frantic.

Magnetic Resonance Imaging (MRI)

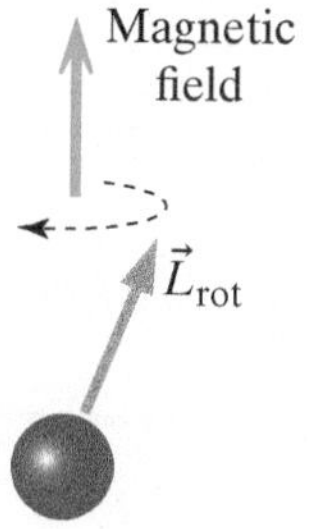

Figure 11.64 Precession of proton spin in an external magnetic field.

Electrically charged atomic and subatomic objects that have angular momentum act like little bar magnets with north and south poles along the angular momentum axis (we say that they have "magnetic moments"). It is as though the spinning charged particle constituted little loops of current, and current loops produce a magnetic field. Even some electrically neutral particles such as the neutron have magnetic moments, because they are built out of electrically charged quarks that have magnetic moments.

In the presence of an applied magnetic field (typically produced by other current-carrying coils of wire), a bar magnet tends to twist to line up with the applied magnetic field. That is, a magnetic field exerts a torque on a bar magnet. A well-known example is that a compass needle aligns with the Earth's magnetic field, which is useful for navigation. (The compass needle is magnetized and is itself a little bar magnet.)

Since a magnetic field applies a torque to a bar magnet, or to atomic or subatomic particles that have angular momentum, such particles precess in the presence of an applied magnetic field (Figure 11.64). This phenomenon is exploited in a technique important in physics, chemistry, and biology, called NMR—nuclear magnetic resonance. A particularly useful application of NMR is in magnetic resonance imaging (MRI).

In MRI, a large, steady, uniform magnetic field created by large direct-current-carrying coils of wire surrounding the patient makes protons in the person's body precess. Many of the most common nuclei in the body are even-even nuclei (in particular, the most common isotopes of carbon and oxygen) and have zero angular momentum and therefore no magnetic moment, and these do not precess. However, hydrogen nuclei (protons) are very common in the body, have rotational angular momentum, and do precess. The stronger the magnetic field, the larger the torque acting on the proton nuclear magnets, and the faster their precession. (Note that for a gyroscope, $\Omega = \tau/L_{rot}$; the precession rate is proportional to the applied torque τ.)

With the protons precessing in the presence of the large steady magnetic field, a small high-frequency (time-varying) magnetic field tuned to match exactly the precession frequency of the protons will flip the spins upside down. This is called a "resonance" phenomenon. The high-frequency signal is turned off, and the protons revert back to being aligned with the steady magnetic field. The act of flipping back emits radiation (at the precession frequency) that can be detected by coils connected to a receiver, and the strength of the signal indicates how many protons were affected.

That's the basic physical mechanism, but this by itself would not yield spatial detail about the interior of the body. The trick is to superimpose on the large steady magnetic field a small nonuniform magnetic field, which has the effect of establishing slightly different torques on protons at different locations in the body. As a result, when the protons flip back into alignment they radiate signals whose frequencies indicate their locations. A computer algorithm calculates how much of each frequency is present in the signal, and this indicates how many protons are at each location. In this way a very detailed image is built up of the interior of the body.

Precession of Spin Axes in Astronomy

As we mentioned earlier, the Earth is subject to small torques due to gravitational forces of the Sun and Moon acting on its nonspherical, "oblate" shape. There is a torque on the Earth's equatorial bulge, perpendicular to the spin rotation axis, and this causes the Earth's axis to precess very slowly, once around about every 26,000 years, so that the "North Star" hasn't always been the star we call Polaris. This effect is called the "precession of the equinoxes" because it leads to a change in what month the spring and fall equinoxes occur. To see why there is such an effect, consider Figure 11.65.

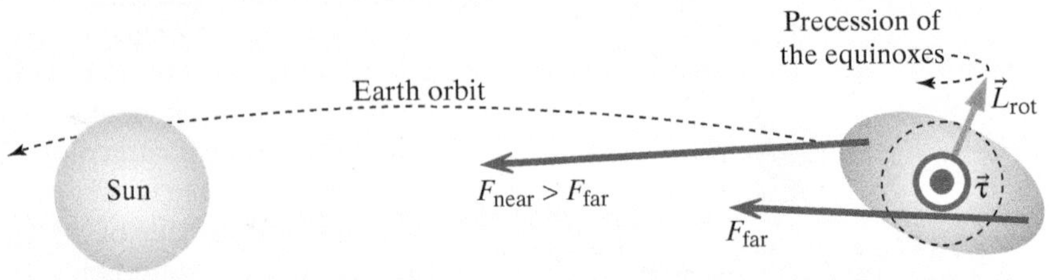

Figure 11.65 Forces on the Earth due to the Sun. The diagram is not to scale; the bulges of the Earth are greatly exaggerated. The net torque about the center of mass of the Earth is out of the page.

The Sun (or Moon) exerts a slightly larger force on the closer bulge due to the $1/r^2$ dependence of the force, so there is a nonzero net torque around the center of mass of the Earth, pointing toward you, out of the page. This makes the rotational angular momentum $\vec{L}_{rot}$ of the Earth precess in a "retrograde" way—that is, in the opposite direction to most rotations in our Solar System. The direction of the axis of the Earth slowly changes due to the torque. (The size of the equatorial bulge, and the differences in F_{near} and F_{far}, have been greatly exaggerated in Figure 11.65.)

Since the Earth's axis is tipped 23° away from perpendicular to the Earth–Sun orbit, the precession of the axis makes a big change in the location of the North Star. Thirteen thousand years from now the North Star will be a star that is 46° away from Polaris in the night sky.

The Moon going around the Earth is a kind of gyroscope. The Earth–Moon orbit is inclined a few degrees to the plane of the Earth–Sun orbit, and the Sun exerts a nonzero torque perpendicular to the angular momentum of the Earth–Moon system. To see why, consider a simpler case in which the Earth and Moon have equal mass, and you see the same effect as with the equatorial bulge of the Earth (Figure 11.66).

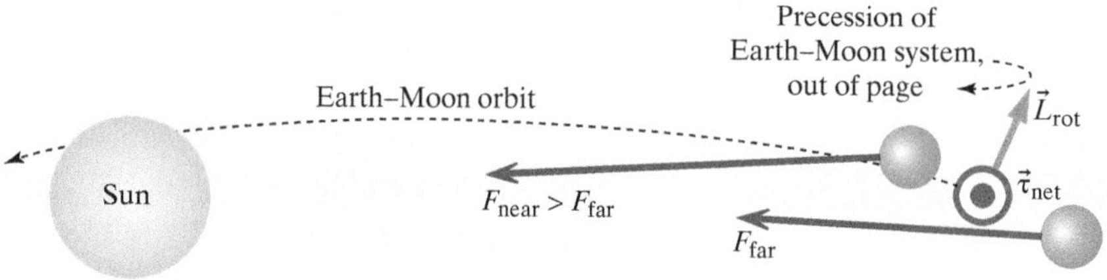

Figure 11.66 Forces on the Earth–Moon system due to the Sun. The diagram is not to scale; the Earth and Moon are shown as if their masses were equal. The net torque on the Earth–Moon system about its center of mass is out of the page.

This torque varies in magnitude during a month, but the averaged effect is that the Earth–Moon system precesses once around in about 18 y (this precession is also "retrograde"). This has an effect on the timing of eclipses, which can occur only when the Moon in its orbit is passing through the plane of the Earth–Sun orbit.

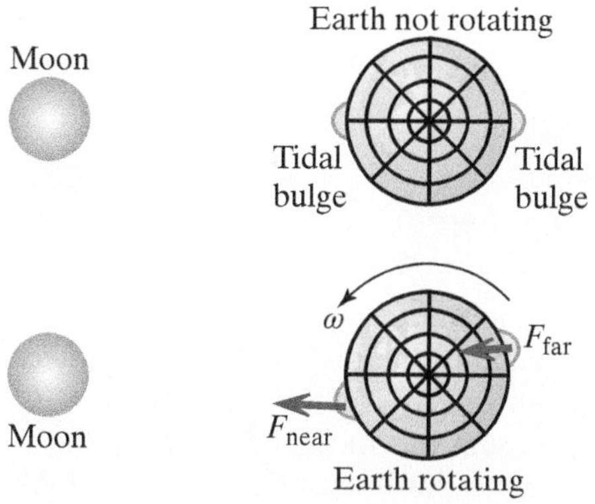

Figure 11.67 Tidal torques due to the Moon, viewed from a point above the North Pole. In the top image the Earth is not rotating; in the bottom image it is rotating. (Not to scale; the tidal effects are greatly exaggerated.)

Tidal Torques

If the Earth did not rotate, the Moon would create tides in the oceans that would pile up in line with the Moon. However, the rotating Earth drags these tidal bulges so that they are no longer in line with the Moon, and the Moon exerts a small torque (Figure 11.67). This torque is directed along the axis and acts to slow down the spin rate, so that the day is getting longer than 24 hours. This effect is called "tidal friction." This interaction with the Moon has the effect that as the rotational angular momentum of the Earth decreases, the translational angular momentum of the Earth–Moon system increases (conservation of total angular momentum), with the result that the Earth and Moon are getting farther apart. (These tidal forces have a small net tangential component that acts on both Earth and Moon to increase their translational angular momenta.)

Something like this presumably happened in the past to the Moon, when it was molten. The rotational angular momentum of the Moon decreased due to tidal torques exerted by the Earth, which are much larger than the tidal torques that the low-mass Moon exerts on the Earth. When the Moon's spin angular speed had decreased to be the same as its translational angular speed (currently about 2π rad or 360° per month), the process terminated. As a result, now the Moon always displays nearly the same face to the Earth.

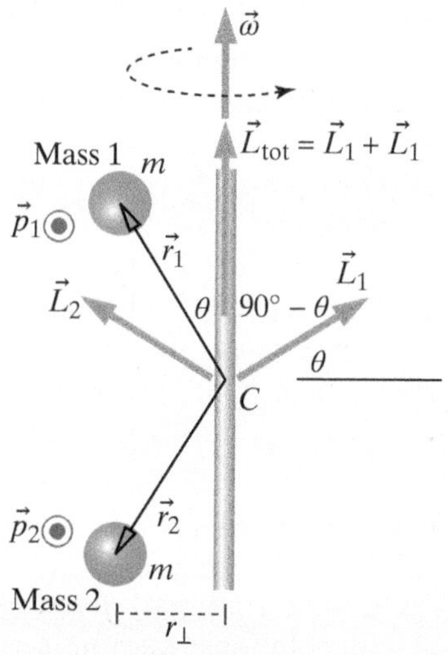

Figure 11.68 The off-axis components of the angular momenta of the two masses cancel. Their total angular momentum is along the axis of rotation.(The two masses at this instant are in the *xy* plane.)

11.13 *MORE ON MOMENT OF INERTIA

In Figure 11.18 all the masses lie in a plane that was perpendicular to the axis of rotation. What if they don't all lie in this plane? We'll show that for simple rigid systems the moment of inertia is still calculated by adding up terms like $mr_\perp^2\omega$, where the perpendicular distances are measured from the axis of rotation.

In Figure 11.68 two equal masses are mounted to a rotating shaft, and we want to determine the angular momentum relative to the center of the device, marked C. We show the directions of the angular momentum $\vec{L}_C = \vec{r}_C \times \vec{p}$ of each mass. You can see that the individual contributions have components that are not along the shaft. However, these components cancel for pairs of masses above and below each other. The effect is that the total angular momentum vector lies along the shaft, and can be calculated by adding up $mr_\perp^2\omega$ for each mass, where $r_\perp$ is the perpendicular distance from the axis.

Here is the proof: the magnitude of the angular momentum of mass 1 about location C is $|\vec{L}_1| = r_1 m v_1 = r_1 m \omega r_\perp$. The y component of this angular momentum is $|\vec{L}_1|\cos(90° - \theta) = r_1 m\omega r_\perp \sin\theta = m\omega r_\perp^2$ since $r_1 \sin\theta = r_\perp$. Similarly, the y component of the angular momentum of mass 2 about location C is also $m\omega r_\perp^2$. Therefore the total angular momentum of the two masses is in the y direction and is given by

$$L_{C,y} = mr_\perp^2\omega + mr_\perp^2\omega = (mr_\perp^2 + mr_\perp^2)\omega = I\omega$$

since $$I = mr_\perp^2 + mr_\perp^2$$

For example, the moment of inertia of a uniform-density thin disk rotating on its axle is $I_{\text{disk}} = \frac{1}{2}MR^2$, where M is the mass of the disk and R is its radius. Because only the distances of atoms from the axle matter, the moment of inertia of a long cylinder has exactly the same form, $I_{\text{disk}} = \frac{1}{2}MR^2$, and it doesn't matter how long the cylinder is (or to put it another way, how thick the disk is).

In this introductory treatment of angular momentum we will deal with simple, symmetrical rotating systems for which the moment of inertia can be calculated simply as $I = m_1 r_{\perp 1}^2 + m_2 r_{\perp 2}^2 + m_3 r_{\perp 3}^2 + m_4 r_{\perp 4}^2 + \cdots$. For a more complicated situation, see the following discussion.

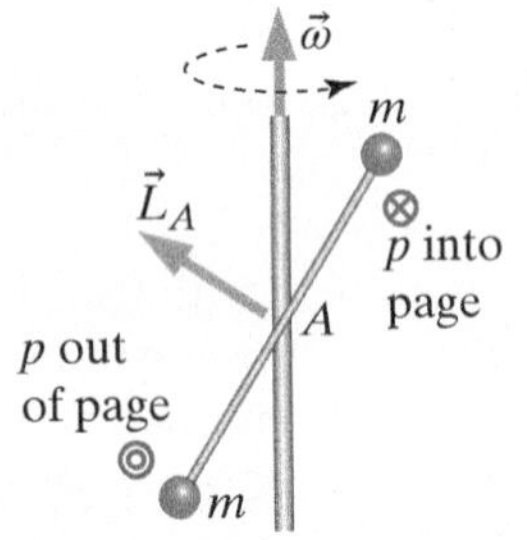

Figure 11.69 If the object lacks axial symmetry, the angular momentum and angular velocity may have different directions!

Lack of Symmetry

In more advanced courses you may study nonsymmetric rotational situations like the one in Figure 11.69. The angular velocity points along the axis of rotation, but the (translational) angular momentum of each mass is $\vec{r}_A \times \vec{p}$, so the total angular momentum $\vec{L}_A$ does *not* point along the axis! Moreover, the angular momentum vector continually changes direction. This change of angular momentum requires a nonzero torque, which is applied to the axle by the bearings in which the axle rotates. The effect of this "dynamic imbalance" is to cause severe wear on the axle and bearings. Car tires must be carefully balanced to prevent this.

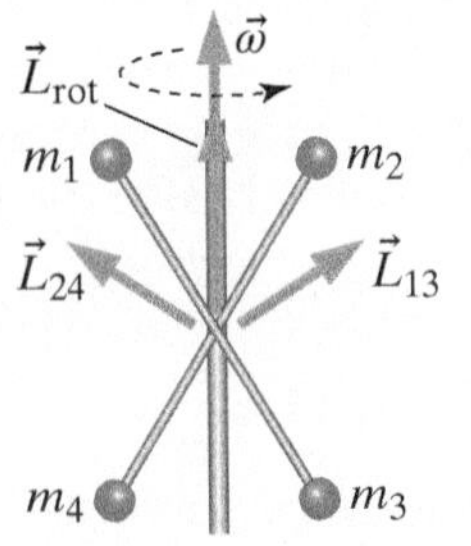

Figure 11.70 If the object is symmetric, the angular momentum is in the direction of the angular velocity.

A symmetric object like the one shown in Figure 11.70 has an angular momentum that does point along the axis, because the perpendicular components of the angular momenta cancel each other. This is the simpler kind of situation we have dealt with in this chapter. In the symmetric case shown in Figure 11.70 it is true that $I = m_1 r_{\perp 1}^2 + m_2 r_{\perp 2}^2 + m_3 r_{\perp 3}^2 + m_4 r_{\perp 4}^2 + \cdots$ about the axis of rotation, and the simple expressions $\vec{L}_{\text{rot}} = I\vec{\omega}$ and $K_{\text{rot}} = \frac{1}{2}I\omega^2$ are valid.

We calculated moment of inertia as a sum (integral) over $r_\perp^2$, the perpendicular distance to an axis. To deal with rotational motion in general requires expressing the moment of inertia as a "tensor," a 3 by 3 array of numbers representing integrals over $x^2, xy, xz, y^2, yx, yz, z^2, zx$, and zy. Matrix multiplication of this tensor times the angular velocity vector $\vec{\omega}$ yields an angular momentum vector that need not point in the same direction as $\vec{\omega}$. This concept is beyond the scope of this textbook.

SUMMARY

The Angular Momentum Principle for a system:

$$\frac{d\vec{L}_{\text{tot},A}}{dt} = \vec{\tau}_{\text{net},A} \quad \text{or} \quad \Delta\vec{L}_{\text{tot},A} = \vec{\tau}_{\text{net},A}\Delta$$

Angular momentum of a particle about location A (see Figure 11.71):

$$\vec{L}_A = \vec{r}_A \times \vec{p}$$
$$L_A = r_\perp p = r_A p \sin\theta$$

The direction is given by the right-hand rule.

$$\vec{L}_A = \langle (yp_z - zp_y), (zp_x - xp_z), (xp_y - yp_x) \rangle$$

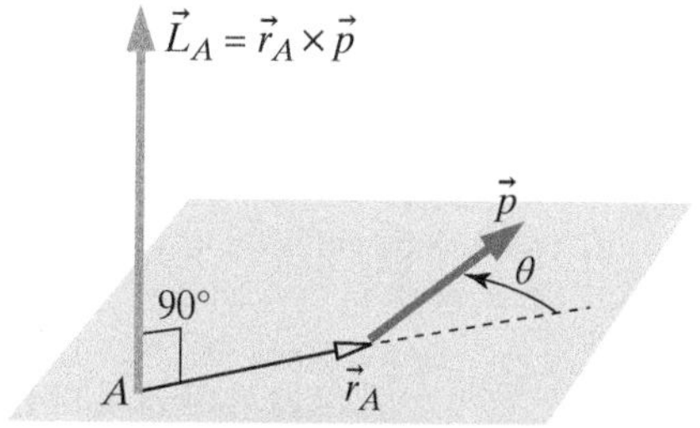

Figure 11.71

Torque is a measure of the twist imparted by a force about location A:

$$\vec{\tau}_A = \vec{r}_A \times \vec{F}$$

Angular momentum for a multiparticle system about some location A can be divided into translational and rotational angular momentum:

$$\vec{L}_A = \vec{L}_{\text{trans},A} + \vec{L}_{\text{rot}}$$
$$\vec{L}_{\text{trans},A} = \vec{r}_{\text{CM},A} \times \vec{P}_{\text{tot}}$$
$$\vec{L}_{\text{rot}} = \vec{r}_{1,\text{CM}} \times \vec{p}_1 + \vec{r}_{2,\text{CM}} \times \vec{p}_2 + \cdots$$

(See Figure 11.72.)

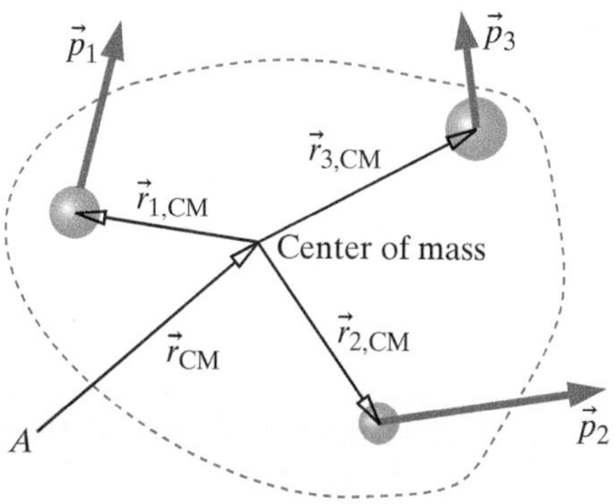

Figure 11.72

Rotational angular momentum in terms of moment of inertia:

$$\vec{L}_{\text{rot}} = I\vec{\omega}$$

Rate of change of rotational angular momentum:

$$\frac{d\vec{L}_{\text{rot}}}{dt} = \vec{\tau}_{\text{net,CM}}$$

Kinetic energy relative to center of mass in terms of moment of inertia and rotational angular momentum:

$$K_{\text{rot}} = \frac{1}{2}I\omega^2 = \frac{L_{\text{rot}}^2}{2I}$$

Moments of inertia:

$$I_{\text{disk}} = I_{\text{cylinder}} = \frac{1}{2}MR^2$$

about center of disk, rotating around axis of disk.

$$I_{\text{sphere}} = \frac{2}{5}MR^2$$

for axis passing through center of sphere.

$$I_{\text{cylinder}} = \frac{1}{12}ML^2 + \frac{1}{4}MR^2$$

for a uniform solid cylinder of length L, radius R, about axis perpendicular to cylinder, through center of cylinder.

$$I_{\text{cylinder}} \approx \frac{1}{12}ML^2$$

if the cylinder is very thin (R is very small).

Kinetic energy and angular momentum for a rigid object rotating about an axle that does not go through the center of mass:

$$K_{\text{tot}} = K_{\text{trans}} + K_{\text{rot}} = \frac{1}{2}(Mr_{\text{CM}}^2 + I_{\text{CM}})\omega^2$$
$$\vec{L}_A = \vec{L}_{\text{trans},A} + \vec{L}_{\text{rot}} = (Mr_{\text{CM}}^2 + I_{\text{CM}})\omega$$

Predicting positions when there is rotation:

$$\Delta\theta = \omega_{\text{avg}}\Delta t$$

Quantized angular momentum in microscopic systems:

A component of angular momentum is an integer or half-integer multiple of $\hbar$.

The square magnitude of angular momentum has quantized values

$$L^2 = l(l+1)\hbar^2$$

where l has integer or half-integer values.

Bohr model of hydrogen:

$$E = -\frac{13.6\,\text{eV}}{N^2}, \quad N = 1, 2, 3, \ldots$$

Angular speed of precession for a gyroscope:

$$\Omega = \frac{\tau}{L_{\text{rot}}} = \frac{RMg}{I\omega}$$

QUESTIONS

Q1 Give an example of a situation in which an object is traveling in a straight line, yet has nonzero angular momentum.

Q2 What are the units of moment of inertia? Of angular speed ω? Of angular momentum? Of linear momentum?

Q3 Under what circumstances is angular momentum constant? Give an example of a situation in which the x component of angular momentum is constant, but the y component isn't.

Q4 Give examples of translational angular momentum and rotational angular momentum in our Solar System.

Q5 Give an example of a physical situation in which the angular momentum is zero yet the translational and rotational angular momenta are both nonzero.

Q6 Under what conditions is the torque about some location equal to zero?

Q7 Make a sketch showing a situation in which the torque due to a single force about some location is 20 N · m in the positive z direction, whereas about another location the torque is 10 N · m in the negative z direction.

Q8 What is required for the angular momentum of a system to be constant? **(a)** zero net torque, **(b)** zero impulse, **(c)** no energy transfers, **(d)** zero net force

Q9 Consider a rotating star far from other objects. Its rate of spin stays constant, and its axis of rotation keeps pointing in the same direction. Why?

Q10 A device consists of eight balls, each of mass M, attached to the ends of low-mass spokes of length L, so the radius of rotation of the balls is $L/2$. The device is mounted in the vertical plane, as shown in Figure 11.73. The axle is held up by supports that are not shown, and the wheel is free to rotate on the nearly frictionless axle. A lump of clay with mass m falls and sticks to one of the balls at the location shown, when the spoke attached to that ball is at 45° to the horizontal. Just before the impact the clay has a speed v, and the wheel is rotating counterclockwise with angular speed ω.

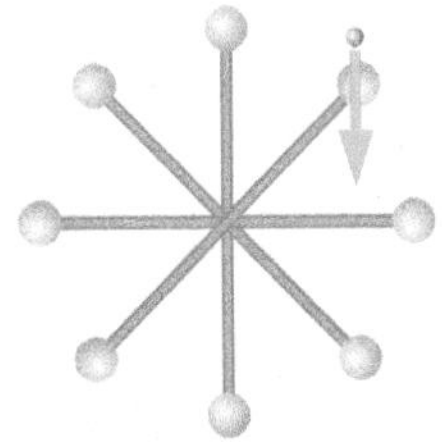

Figure 11.73

(a) Which of the following statements are true about the device and the clay, for angular momentum relative to the axle of the device? (1) The angular momentum of the device + clay just after the collision is equal to the angular momentum of the device + clay just before the collision. (2) The angular momentum of the falling clay is zero because the clay is moving in a straight line. (3) Just before the collision, the angular momentum of the wheel is 0. (4) The angular momentum of the device is the sum of the angular momenta of all eight balls. (5) The angular momentum of the device is the same before and after the collision. **(b)** Just before the impact, what is the (vector) angular momentum of the combined system of device plus clay about the center C? (As usual, x is to the right, y is up, and z is out of the screen, toward you.) **(c)** Just after the impact, what is the angular momentum of the combined system of device plus clay about the center C? **(d)** Just after the impact, what is the (vector) angular velocity of the device? **(e)** Qualitatively, what happens to the total linear momentum of the combined system? Why? (1) Some of the linear momentum is changed into energy. (2) Some of the linear momentum is changed into angular momentum. (3) There is no change because linear momentum is always conserved. (4) The downward linear momentum decreases because the axle exerts an upward force. **(f)** Qualitatively, what happens to the total kinetic energy of the combined system? Why? (1) Some of the kinetic energy is changed into linear momentum. (2) Some of the kinetic energy is changed into angular momentum. (3) The total kinetic energy decreases because there is an increase of internal energy in this inelastic collision. (4) There is no change because kinetic energy is always conserved.

Q11 A rod rotates in the vertical plane around a horizontal axle. A wheel is free to rotate on the rod, as shown in Figure 11.74. A vertical stripe is painted on the wheel. As the rod rotates clockwise, the vertical stripe on the wheel remains vertical. Is the translational angular momentum of the wheel relative to location A zero or nonzero? If nonzero, what is its direction? Is the rotational angular momentum of the wheel zero or nonzero? If nonzero, what is its direction? Consider a similar system, but with the wheel welded to the rod (not free to turn). As the rod rotates clockwise, does the stripe on the wheel remain vertical? Is the translational angular momentum of the wheel relative to location A zero or nonzero? If nonzero, what is its direction? Is the rotational angular momentum of the wheel zero or nonzero? If nonzero, what is its direction?

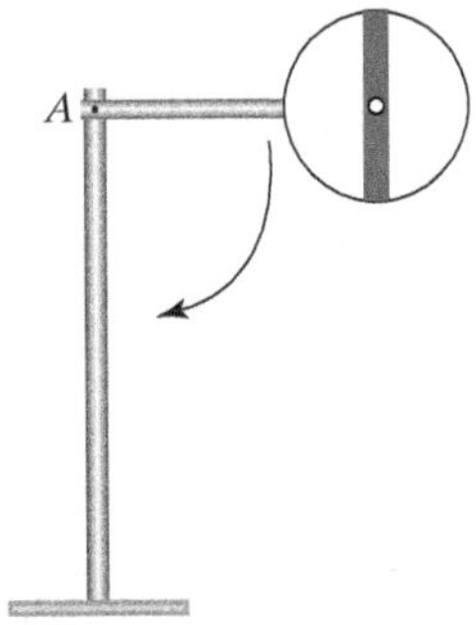

Figure 11.74 A rod rotating in the vertical plane with a wheel attached.

Q12 What features of the Bohr model of hydrogen are consistent with the later, full quantum mechanical analysis? What features of the Bohr model had to be abandoned?

Calculate the tension in the string as the yo-yo falls. Note that when the center of the yo-yo moves down a distance y, the yo-yo turns through an angle y/r, which in turn means that the angular speed ω is equal to v_{CM}/r. The moment of inertia of a uniform disk is $\frac{1}{2}MR^2$.

••P63 A string is wrapped around a uniform disk of mass $M = 1.2$ kg and radius $R = 0.11$ m (Figure 11.109). Attached to the disk are four low-mass rods of radius $b = 0.14$ m, each with a small mass $m = 0.4$ kg at the end. The device is initially at rest on a nearly frictionless surface. Then you pull the string with a constant force $F = 21$ N for a time of 0.2 s. Now what is the angular speed of the apparatus?

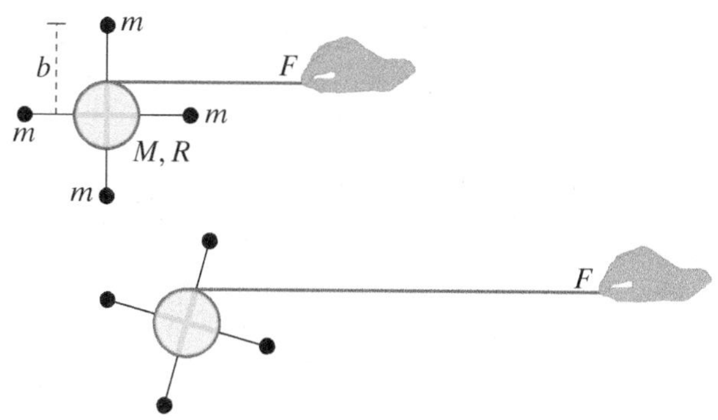

Figure 11.109

•••P64 A small solid rubber ball of radius r hits a rough horizontal floor such that its speed v just before striking the floor at location A makes an angle of 60° with the horizontal and also has back spin with angular speed ω. It is observed that the ball repeatedly bounces from A to B, then from B back to A, etc. Assuming perfectly elastic impact determine **(a)** the required magnitude of ω of the back spin in terms of v and r, and **(b)** the minimum magnitude of co-efficient of static friction μ_s to enable this motion. *Hint:* Notice that the direction of $\vec{\omega}$ flips in each collision.

Section 11.9

•P65 A wheel rotates at a rate of 200 rpm (revolutions per minute). **(a)** What is the angular speed in radians per second? **(b)** What is the angle in radians through which the wheel rotates in 5 s? **(c)** What is this angle in degrees?

•P66 A wheel is mounted on a stationary axle that lies along the z axis. The center of the wheel is at $\langle 0,0,0\rangle$ m. It is rotating with a constant angular speed of 30 rad/s. At time $t = 4.20$ s a red dot painted on the rim of the wheel is located at $\langle 0.25,0,0\rangle$ m. At time $t = 4.26$ s, what is the location of the red dot?

•P67 A rotating wheel accelerates at a constant rate from an angular speed of 24 rad/s to 36 rad/s in a time interval of 3 s. **(a)** What is the angular acceleration in rad/s/s? **(b)** What is the average angular speed? **(c)** What is the angle in radians through which the wheel rotates? **(d)** What is this angle in degrees?

•P68 A string is wrapped around a wheel of radius 25 cm mounted on a stationary axle. The wheel is initially not rotating. You pull the string with a constant force through a distance of 40 cm. What is the angle in radians and degrees through which the wheel rotates?

••P69 A wheel with radius 30 cm is rotating at a rate of 10 rev/s. **(a)** What is the angular speed in radians per second? **(b)** In a time interval of 6 s, what is the angle in radians through which the wheel rotates? **(c)** At $t = 10$ s the angular speed begins to increase at a rate of 1.4 rad/s/s. At $t = 15$ s, what is the angular speed in radians per second? **(d)** Through what angle in radians did the wheel rotate during the time between $t = 10$ s and $t = 15$ s? **(e)** If the wheel rolls along the ground without slipping, the instantaneous velocity of the atoms of the object that are momentarily in contact with the ground is zero. This zero-velocity condition implies that $v_{CM} = \omega R$, where ω is the angular speed of the object, since the instantaneous speed of the contact point is $v_{CM} - \omega R$. During the time between $t =$ 10 s and $t = 15$ s, how far did the center of the wheel move, in meters?

••P70 If you did not already do Problem P59, do it now. Also calculate the angle through which the disk turns, in radians and degrees.

••P71 If you did not already do Problem P60, do it now. Also calculate numerically the angle through which the yo-yo turns, in radians and degrees.

••P72 If you did not already do Problem P63, do it now. Also calculate numerically the angle through which the apparatus turns, in radians and degrees.

•••P73 Two balls of equal radius and mass, free to roll on a horizontal plane, are separated by a distance L large compared to their radius. One ball is solid, the other hollow with a thickness small compared to its radius. They are attracted by an electric force. How far will the solid ball roll before it collides with the hollow ball?

•••P74 An amusing trick is to press a finger down on a marble on a horizontal table top, in such a way that the marble is projected along the table with an initial linear speed v and an initial backward rotational speed ω about a horizontal axis perpendicular to v. The coefficient of sliding friction between marble and top is constant. The marble has radius R. **(a)** If the marble slides to a complete stop, what was ω in terms of v and R? **(b)** If the marble skids to a stop and then starts returning toward its initial position, with a final constant speed of $(3/7)v$, what was ω in terms of v and R? *Hint for part (b):* When the marble rolls without slipping, the relationship between speed and angular speed is $v = \omega R$.

Section 11.11

•P75 According to the Bohr model of the hydrogen atom, what is the magnitude of the translational angular momentum of the electron (relative to the location of the proton) when the atom is in the 2nd excited state above the ground state ($N = 3$)?

••P76 Calculate rotational energy levels and photon emission energies for a carbon monoxide molecule (CO). **(a)** Calculate the energies of the quantized rotational energy levels for the CO molecule. Estimate any quantities you need. See the discussion of diatomic molecules in the section on quantized angular momentum; the parameter l has values $0, 1, 2, 3 \ldots$. **(b)** Describe the emission spectrum for electromagnetic radiation emitted in transitions among the rotational CO energy levels. Include a calculation of the lowest-energy emission in electron volts (1.6×10^{-19} J). **(c)** It is transitions among "electronic" states of atoms that produce visible light, with photon energies on the order of a couple of electron-volts. Each electronic energy level has quantized rotational and vibrational (harmonic oscillator) energy sublevels. Explain why this leads to a visible spectrum that contains "bands" rather than individual energies.

••P77 Review the derivation of the Bohr model of the hydrogen atom and apply this reasoning to predict the energy levels of ionized helium He^+ (a helium atom with only one electron, and a nucleus containing two protons and two neutrons). What are the energies in eV of the ground state and the first excited state? What is the energy of a photon emitted in a transition from the first excited state to the ground state? How do these results differ from those for a hydrogen atom?

••P78 The Bohr model correctly predicts the main energy levels not only for atomic hydrogen but also for other "one-electron" atoms where all but one of the atomic electrons has been removed, such as in He^+ (one electron removed) or Li^{++} (two electrons removed). **(a)** Predict the energy levels in eV for a system consisting of a nucleus containing Z protons and just one electron. You need not recapitulate the entire derivation for the Bohr model, but do explain the changes you have to make to take into account the factor Z. **(b)** The negative muon (μ^-) behaves like a heavy electron, with the same charge as the electron but with a mass 207 times as large as the electron mass. As a moving μ^- comes to rest in matter, it tends to knock electrons out of atoms and settle down onto a nucleus to form a "one-muon" atom. For a system consisting of a lead nucleus (Pb^{208} has 82 protons and 126 neutrons) and just one negative muon, predict the energy in eV of a photon emitted in a transition from the first excited state to the ground state. The high-energy photons emitted by transitions between energy levels in such "muonic atoms" are easily observed in experiments with muons. **(c)** Calculate the radius of the smallest Bohr orbit for a μ^- bound to a lead nucleus (Pb^{208} has 82 protons and 126 neutrons). Compare with the approximate radius of the lead nucleus (remember that the radius of a proton or neutron is about 1×10^{-15} m, and the nucleons are packed closely together in the nucleus).

Comments: This analysis in terms of the simple Bohr model hints at the result of a full quantum-mechanical analysis, which shows that in the ground state of the lead–muon system there is a rather high probability for finding the muon *inside* the lead nucleus. Nothing in quantum mechanics forbids this penetration, especially since the muon does not participate in the strong interaction. Electrons in an atom can also be found inside the nucleus, but the probability is very low, because on average the electrons are very far from the nucleus, unlike the muon.

The eventual fate of the μ^- in a muonic atom is that it either decays into an electron, neutrino, and antineutrino, or it reacts through the weak interaction with a proton in the nucleus to produce a neutron and a neutrino. This "muon capture" reaction is more likely if the probability is high for the muon to be found inside the nucleus, as is the case with heavy nuclei such as lead.

••P79 The nucleus dysprosium-160 (containing 160 nucleons) acts like a spinning object with quantized angular momentum, $L^2 = l(l+1)\hbar^2$, and for this nucleus it turns out that l must be an even integer (0,2,4...). When a Dy-160 nucleus drops from the $l = 2$ state to the $l = 0$ state, it emits an 87 keV photon (87×10^3 eV). **(a)** What is the moment of inertia of the Dy-160 nucleus? **(b)** Given your result from part (a), find the approximate radius of the Dy-160 nucleus, assuming it is spherical. (In fact, these and similar experimental observations have shown that some nuclei are not quite spherical.) **(c)** The radius of a (spherical) nucleus is given approximately by $(1.3 \times 10^{-15}\,\text{m})A^{1/3}$, where A is the total number of protons and neutrons. Compare this prediction with your result in part (b).

Section 11.12

•P80 Two gyroscopes are made exactly alike except that the spinning disk in one is made of low-density aluminum, whereas the disk in the other is made of high-density lead. If they have the same spin angular speeds and the same torque is applied to both, which gyroscope precesses faster?

••P81 This problem requires that you have a toy gyroscope available. The purpose of this problem is to make as concrete as possible the unusual motions of a gyroscope and their analysis in terms of fundamental principles. In all of the following studies, the effects are most dramatic if you give the gyroscope as large a spin angular speed as possible. **(a)** Hold the spinning gyroscope firmly in your hand, and try to rotate the spin axis quickly to point in a new direction. Explain qualitatively why this feels "funny." Also explain why you *don't* feel anything odd when you move the spinning gyroscope in any direction *without* changing the direction of the spin axis. **(b)** Support one end of the spinning gyroscope (on a pedestal or in an open loop of the string) so that the gyroscope precesses *counterclockwise* as seen from above. Explain this counterclockwise precession direction; include sketches of top and side views of the gyroscope. **(c)** Again support one end of the spinning gyroscope so that the gyroscope precesses *clockwise* as seen from above. Explain this clockwise precession direction; include sketches of top and side views of the gyroscope.

••P82 This problem requires that you have a toy gyroscope available. The purpose of this problem is to make as concrete as possible the unusual motions of a gyroscope and their analysis in terms of fundamental principles. In all of the following studies, the effects are most dramatic if you give the gyroscope as large a spin angular speed as possible. **(a)** If you knew the spin angular speed of your gyroscope, you could predict the precession rate. Invent an appropriate experimental technique and determine the spin angular speed approximately. Explain your experimental method and your calculations. Then predict the corresponding precession rate, and compare with your measurement of the precession rate. You will have to measure and estimate some properties of the gyroscope and how it is constructed. **(b)** Make a *quick* measurement of the precession rate with the spin axis horizontal, then make another quick measurement of the precession rate with the spin axis nearly vertical. (If you make quick measurements, friction on the spin axis doesn't have much time to change the spin angular speed.) Repeat, this time with the spin axis initially nearly vertical, then horizontal. Making all four of these measurements gives you some indication of how much the spin unavoidably changes due to friction while you are quickly changing the angle. What do you conclude about the dependence of the precession rate on the angle, assuming the same spin rate at these different angles? What is the theoretical prediction for the dependence of the precession rate on angle (for the same spin rate)?

••P83 The axis of a gyroscope is tilted at an angle of 30° to the vertical (Figure 11.110). The rotor has a radius of 15 cm, mass 3 kg, moment of inertia $0.06\ \text{kg}\cdot\text{m}^2$, and spins on its axis at 30 rad/s. It is supported in a cage (not shown) in such a way that without an added weight it does not precess. Then a mass of 0.2 kg is hung from the axis at a distance of 18 cm from the center of the rotor.

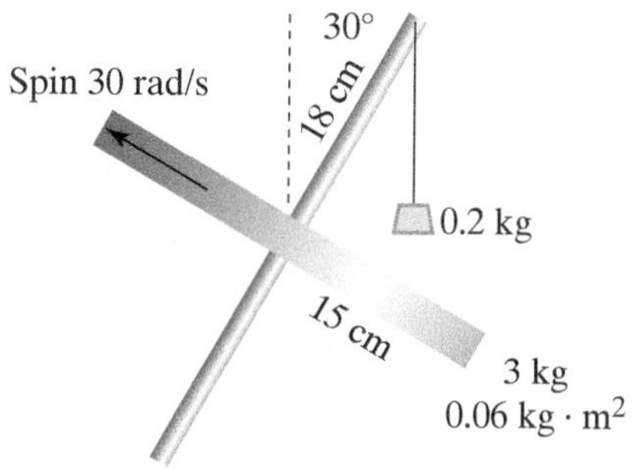

Figure 11.110

(a) Viewed from above, does the gyroscope precess in a (1) clockwise or a (2) counterclockwise direction? That is, does the top end of the axis move (1) out of the page or (2) into the page in the next instant? Explain your reasoning. **(b)** How long does it take for the gyroscope to make one complete precession?

••P84 A bicycle wheel with a heavy rim is mounted on a lightweight axle, and one end of the axle rests on top of a post. The wheel is observed to precess in the horizontal plane. With the spin direction shown in Figure 11.111, does the wheel precess clockwise or counterclockwise? Explain in detail, including appropriate diagrams.

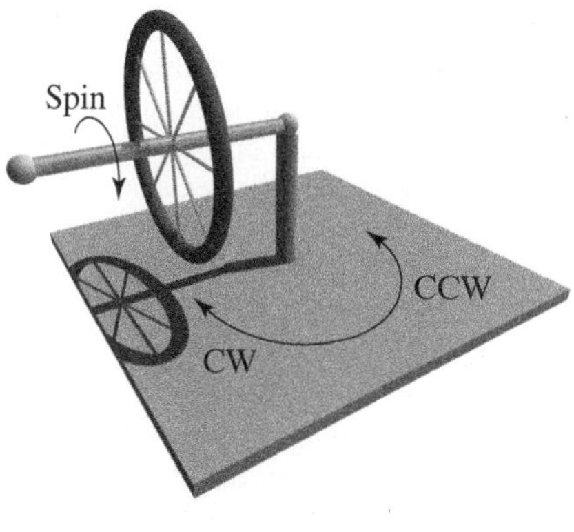

Figure 11.111

••P85 A solid wood top spins at high speed on the floor, with a spin direction shown in Figure 11.112.

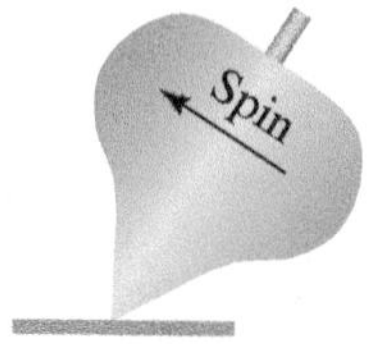

Figure 11.112

(a) Using appropriately labeled diagrams, explain the direction of motion of the top (you do not need to explain the magnitude). **(b)** How would the motion change if the top had a higher spin rate? Explain briefly. **(c)** If the top were made of solid steel instead of wood, explain how this would affect the motion (for the same spin rate).

COMPUTATIONAL PROBLEMS

•P86 A rod on a low-friction axle is initially at rest when a constant torque begins to be applied. **(a)** Complete the skeleton program shown below and make sure that it runs properly. See Section 11.10 for suggestions. After exiting the loop, print the values of the final angle `theta` and the final angular speed `mag(omega)`. Given the fact that the torque is constant, what should be the value of the ratio `mag(omega)/theta`? Is this true for your program? **(b)** Make graphs of `theta` vs. time and `mag(omega)` vs. t. What should these graphs look like? Do they? **(c)** Change the torque to oscillate: $\tau_{\text{net},z} = 3\cos(5t)$. Graph `theta` vs. time and measure the period. Is it consistent with the angular frequency of 5 rad/s of the torque? **(d)** Plot on the same graph the rotational kinetic energy of the rod vs. time and the magnitude of the torque as a function of time. Is the rotational kinetic energy large when the torque is large?

```
from visual import *
from visual.graph import *

M = 2
Lrod = 1
R = 0.1
Laxle = 4*R
I = (1/12)*M*Lrod**2 + (1/4)*M*R**2

rod = cylinder(pos=vector(-1,0,0),
               radius=R, color=color.orange,
               axis=vector(Lrod,0,0))
axle = cylinder(pos=vector(-1+Lrod/2,0,-Laxle/2),
                radius=R/6, color=color.red,
                axis=vector(0,0,4*R))

L = vector(0,0,0) # angular momentum
deltat = 0.0001   # for accuracy in later parts
t = 0
theta = 0
dtheta = 0

while t < 2:
    rate(10000)
    torque = vector(0,0,2) # constant torque
    # Apply Angular Momentum Principle
    # Update angle and rod position
    t = t + deltat
```

•P87 Start with the program for part **(c)** of Problem P86. Add a constant force $\vec{F} = \langle 0.1, 0, 0\rangle$ N applied to the axle. Let the while loop run to 7 s. Describe what you see.

••P88 Start with the program from Chapter 3 Problem P68 to model the motion of a planet going around a fixed star. In this problem you will build on that program. **(a)** Use initial conditions that produce an elliptical orbit. At each step calculate the translational angular momentum $\vec{L}_{\text{trans},A}$ of the planet with respect to a location A chosen to be in the orbital plane but outside the orbit. Display this in two ways (i and ii below), and briefly describe in words what you observe: (i) Display $\vec{L}_{trans,A}$ as an arrow with its tail at location A, throughout the orbit. Since the magnitude of $\vec{L}_{trans,A}$ is quite different from the magnitudes of the distances involved, you will need to scale the arrow by some factor to fit it on the screen. (ii) Graph the component of $\vec{L}_{trans,A}$ perpendicular to the orbital plane as a function of time. **(b)** Repeat part (a), but this time choose a different location B at the center of the fixed star, and calculate and display $\vec{L}_{trans,B}$ relative to that location B. As in part (a), display $\vec{L}_{trans,B}$ as an arrow (scaled appropriately), and also graph the component of $\vec{L}_{trans,B}$ perpendicular to the orbital plane, as a function of time, and briefly describe in words what you observe. **(c)** Choose a location C which is not in the orbital plane, and calculate and display $\vec{L}_{trans,C}$ as an arrow throughout the orbit. (You do not need to make a graph.) Briefly describe in words what you observe.

••P89 Figure 11.113 shows a glass disk mounted on a low-friction axle and held stationary. The center of the disk is at the origin, $\langle 0,0,0 \rangle$ m. A peg in the disk is connected by two identical stretched springs to a wall. The disk is released from rest. **(a)** Run the skeleton program shown below and rotate the camera to understand the apparatus. Then complete the program so that the disk rotates. **(b)** Make a graph of the angle `theta` and determine the period. Is this system a harmonic oscillator? Try varying the initial angle and/or the relaxed length of the spring. **(c)** Show on a second graph the kinetic energy of the disk, the spring energy, and their sum, as a function of time. What should the sum look like? Try a smaller time step. (The setting `material=materials.rough` for the disk makes it easier to see its rotation; this may not have a visible effect if your computer graphics card is not up to date.)

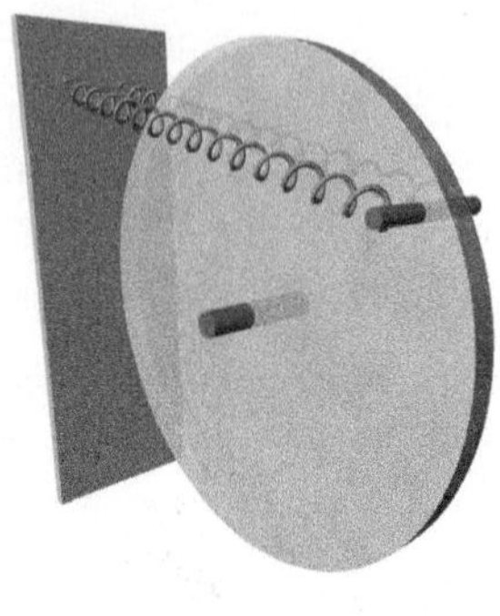

Figure 11.113

```
from visual import *
from visual.graph import *
M = 2              # mass of uniform-density disk
R = 0.2            # radius of disk
thick = 0.02       # thickness of disk
I = M*R**2         # moment of inertia of disk
r = 0.9*R          # peg in disk at this radius
Lpeg = 5*thick # length of peg
wall = box(pos=vector(-1.2*R,0,0),
         size=(0.01,2.4*R,0.8*R),
         color=color.green)
# Center of disk is at <0,0,0>:
disk = cylinder(pos=vector(0,0,-thick/2),
          radius=R, axis=vector(0,0,thick),
          color=color.white, opacity=0.7,
          material=materials.rough)
axle = cylinder(pos=vector(0,0,-Lpeg/2),
          radius=0.05*R,color=color.red,
          axis=vector(0,0,Lpeg))
# Place peg in the disk on x axis:
peg = cylinder(pos=(r,0,-Lpeg/2),
          radius=0.03*R, color=color.red,
          axis=vector(0,0,Lpeg))
# Rotate to initial position:
theta = pi/6      # radians; CCW from x axis
peg.rotate(angle=theta, axis=axle.axis,
          origin=axle.pos)
rspring = 0.05*R  # spring radius
# Front spring:
springF = helix(pos=(wall.x,r,1.5*thick),
          radius=0.05*R, color=color.orange,
          coils=15, thickness=0.4*rspring)
# Back spring:
springB = helix(pos=(wall.x,r,-1.5*thick),
          radius=0.05*R, color=color.orange,
          coils=15, thickness=0.4*rspring)
# Attach springs to peg:
end = peg.pos+vector(0,0,Lpeg/2+springF.pos.z)
springB.axis = springF.axis = end - springF.pos


t = 0
deltat = 0.01
dtheta = 0
ks = 1.5           # stiffness of each spring
L0 = 0.26          # relaxed length of each spring
L = vector(0,0,0) # initial angular momentum


while True:
    rate(100)
    # Calculate spring force F acting on peg
    # Calculate torque due to springs
    # Update angular momentum L
    # Calculate angular velocity omega
    # Calculate dtheta and theta
    # Update disk and peg positions:
    disk.rotate(angle=dtheta, axis=axle.axis,
             origin=axle.pos)
    peg.rotate(angle=dtheta, axis=axle.axis,
             origin=axle.pos)
    # Update spring lengths:
    end = peg.pos+vector(0,0,Lpeg/2+springF.pos.z)
    springF.axis = end - springF.pos
    springB.axis = springF.axis
```

•••P90 An electric "dipole" consists of an object that is positively charged at one end and negatively charged at the other. A molecular example is an HCl molecule in which the H end is positive and the Cl end is negative. In Figure 11.114 is a snapshot of a positively charged sphere on the left interacting with a dipole on the right in outer space; the two charges on the right are stuck together. The spheres are uniformly charged spheres of radius $R = 1$ mm which interact electrically as though they were point charges concentrated at the centers. Each sphere has uniform density with a mass of 1×10^{-6} kg. Each of the red spheres has an electric charge of $+2 \times 10^{-10}$ C and the negative sphere's charge is -2×10^{-10} C.

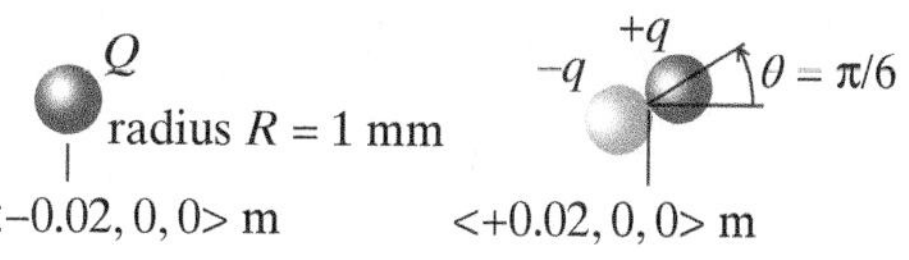

Figure 11.114

The charge on the left is fixed in position at $\langle -0.02, 0, 0\rangle$ m. The center of mass of the dipole is initially located at $\vec{r}_{CM} = \langle 0.02, 0, 0\rangle$ m, with the positive charge at $\vec{r}_{CM} + \vec{r}$ and the negative charge at $\vec{r}_{CM} - \vec{r}$, where $\vec{r} = \langle R\cos\theta, R\sin\theta, 0\rangle$ and $\theta = \pi/6$ rad. The dipole is free to rotate about an axle at its center of mass, but the axle is fixed and cannot move. **(a)** Create a computational model of the situation. Note in calculating the moment of inertia of the rotating dipole that each of the dipole charges has both translational and rotational angular momenta. We suggest using a time step of 0.001 s for accuracy, with a large value in the rate statement for display speed. Use the Angular Momentum Principle to predict and display the motion of the dipole, starting from rest (no rotation). Include a graph of θ vs. time. What is the period of the oscillation? **(b)** Determine the period for starting values of θ of 0.01π and 0.02π. Is this oscillator a harmonic oscillator for small oscillations (that is, is the period independent of the amplitude)? **(c)** Is the period the same if you increase the amplitude to $\theta = \pi/2$? **(d)** Measure the period for a small amplitude with quadruple the charge on the left. What is the ratio of this period to the small-amplitude period you measured in part (b)? How does this effect compare with quadrupling the spring stiffness for a spring–mass oscillator?

•••P91 This is a continuation of problem P90, in which we allow the center of mass of the dipole to move freely. Restore the charge on the left to $Q = +2 \times 10^{-10}$ C. **(a)** Use the Momentum Principle and the Angular Momentum Principle to predict and display the motion of the translating, rotating dipole. Stop the animation when the distance between the charge on the left and the center of mass of the dipole is less than 0.02 m. **(b)** Make a graph that shows as a function of time K_{trans}, K_{rot}, U_{elec}, and the total $E = K_{trans} + K_{rot} + U_{elec}$. (We can ignore the constant electric potential energy term associated with the dipole pair of charges; the distance between them does not change.) What does the Energy Principle predict for the graph of E vs. t? Is your graph consistent with the Energy Principle? **(c)** Let the charge on the left move freely. Make sure your graph of E vs. t still makes sense.

In this problem we've seen a situation where we needed to use all three fundamental mechanics principles: The Momentum Principle, the Energy Principle, and the Angular Momentum Principle.

•••P92 Model the motion of a meter stick suspended from one end on a low-friction axle. Do not make the small-angle approximation but allow the meter stick to swing with large angles. Plot on the same graph both θ and the z component of $\vec{\omega}$ vs. time. Try starting from rest at various initial angles, including nearly straight up (which would be $\theta_i = \pi$ radians). Is this a harmonic oscillator? Is it a harmonic oscillator for small angles?

ANSWERS TO CHECKPOINTS

1 $-y$
2 D: 50 kg · m²/s out of page; E: 50 kg · m²/s out of page; F: 50 kg · m²/s out of page; G: 0; H: 30 kg · m²/s into page
3 (a) $L_1 = 0.72$ kg · m²/s and $L_2 = 0.72$ kg · m²/s, both into page; **(b)** $L_{rot} = 1.44$ kg · m²/s into page; **(c)** $I = 0.072$ kg · m²; **(d)** $\vec{\omega}_0$ into page; **(e)** $I\omega_0 = 1.44$ kg · m²/s into page; **(f)** They're the same; **(g)** $K_{rot} = 14.4$ J
4 Nonzero; up (toward the sky); nonzero; up
5 6 N · m; 2 N
6 (a) $R\cos(45°)mv$ out of page; **(b)** $(M+m)R^2\omega$ out of page; **(c)** $R\cos(45°)mv/[(M+m)R^2]$ out of page; **(d)** The linear momentum of the clay decreases because the axle exerts an impulsive force upward (the wheel always has zero linear momentum because its center of mass doesn't move).
7 No torques around center of mass means no change in rotational angular momentum, so rotational angular momentum stays constant in magnitude (which determines length of day) and direction (which determines what "North Star" the axis points at). Doesn't matter that Earth is going around Sun; rotational angular momentum is affected solely by torque around center of mass.
8 $dL_z/dt = \frac{d}{dt}[RMv_{CM} - \frac{1}{2}MR^2\omega] = 0$, since there is no torque about a point under the string. Differentiating, we get $d\omega/dt = (2/R)(dv_{CM}/dt) = 2F_T/(MR)$, as before.
9 $dL_z/dt = d_1F_N - (d_1+d_2)M_2g = 0$ ($+z$ out of page), and since $F_N = M_1g + M_2g$, this is equivalent to $d_1M_1g - d_2M_2g$, as before.
10 (a) 10 rad/s; **(b)** 5 rad/s; **(c)** 3 rad; **(d)** 0.9 m
12 Same as in horizontal case: $\Omega = RMg/(I\omega)$

CHAPTER

12

Entropy: Limits on the Possible

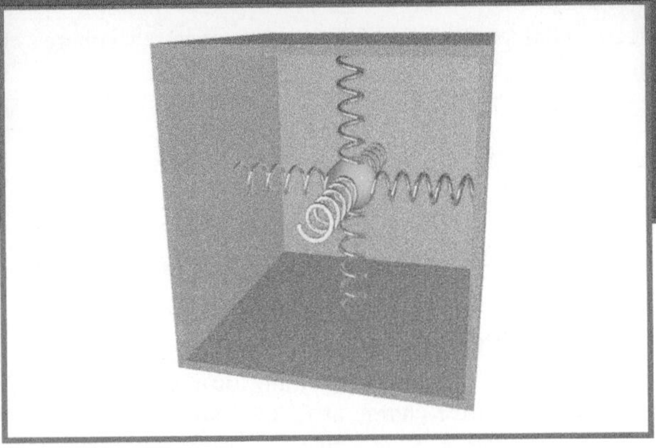

OBJECTIVES

After studying this chapter, you should be able to

- Calculate the entropy of a solid object, based on the Einstein model of a solid
- Calculate the absolute temperature of a solid object as a function of its energy
- Calculate the specific heat of a solid object as a function of temperature
- Calculate the probability of finding a microscopic system in a particular excited state

12.1 IRREVERSIBILITY

A pen lying on a table doesn't suddenly jump upward, despite the fact that there is plenty of energy in the table in the form of microscopic kinetic and potential energy. Why doesn't this happen? It wouldn't be a violation of the Energy Principle for energy to flow from the table into the pen. In fact, if instead of a pen you place a single atom on a table at room temperature, on average it will get so much energy from the table that it could jump thousands of meters upward (if there were no air or other obstacles in the way). This chapter deals with a statistical analysis of microscopic energy that puts limits on what is possible.

Much of this chapter is based on an article by Thomas A. Moore and Daniel V. Schroeder, "A different approach to introducing statistical mechanics," *American Journal of Physics*, vol. 65, pp. 26–36 (January 1997).

What Is Temperature?

We have repeatedly encountered a connection between the temperature of an object and the average motion of the atoms that make up the object, but our understanding of the meaning and role of temperature is incomplete. What is the quantitative relationship between temperature and microscopic energy?

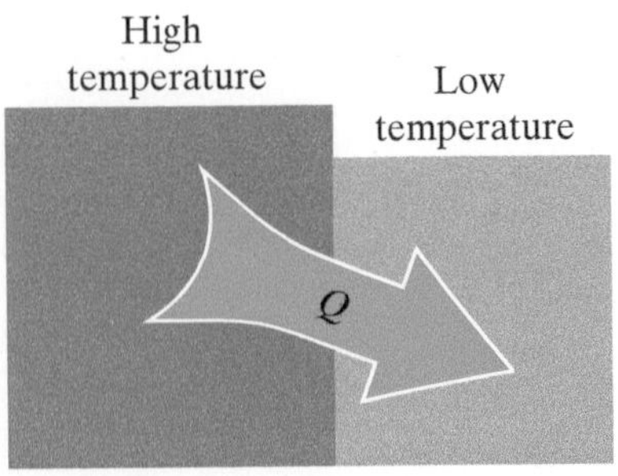

Figure 12.1 We observe that energy flows from a hotter object to a cooler one.

Why Does Energy Flow from Hot to Cold?

When two blocks of different temperatures are placed in contact with each other, we observe that energy flows from the hotter block to the colder block (Figure 12.1). We can understand this in a rough way. The average energy of atoms in the hotter block is greater than the average energy of atoms in the colder block. In the interface region, where atoms of the two blocks are in contact with each other, it seems more likely that an atom in the hotter block

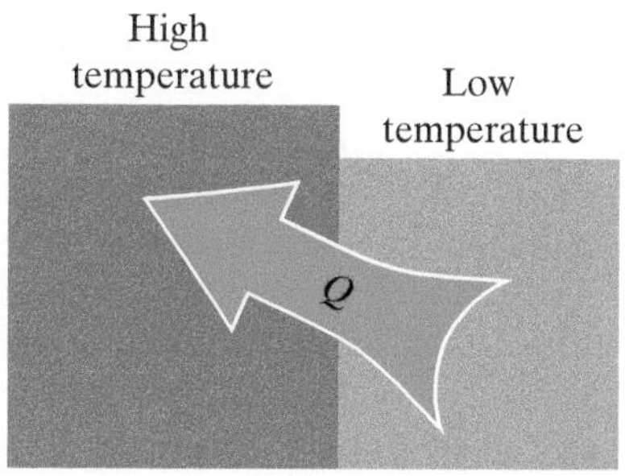

Figure 12.2 Could energy be transferred from the cooler object to the hotter one?

will lose energy rather than gain energy when it collides with an atom in the colder block.

Couldn't energy flow the other way? After all, sometimes an individual atom in the hotter block might happen to have a lot less than its average energy when it collides with an individual atom in the colder block that happens to have a lot more than its average energy. In that case energy would be transferred from the colder block to the hotter block (Figure 12.2).

This "uphill" movement of energy seems unlikely, but how unlikely is it? Should we occasionally observe a hot block get hotter when we put it in contact with a cold block? If you put an ice cube in your drink, could your drink get warmer and the ice cube get colder? This would not violate conservation of energy, because the total energy of the two objects would still be the same. So what physical principle would be violated?

Why Do Things Run Down?

A related question, and a very deep one, has to do with the "reversibility" of processes. The fundamental physical interactions seem to be completely "reversible" in the following technical, physics sense: Make a movie of an alpha particle scattering off a gold nucleus, then run the movie forward or reversed. Both views look entirely reasonable. In contrast, make a movie of a ball bouncing on the floor with decreasing height on each bounce, then run the movie forward or reversed. The reversed movie looks silly: you see a ball bouncing higher and higher with each bounce! Would such a motion violate conservation of energy? Not necessarily. It could be that the atoms in the floor happen on average to give energy to the ball, and the floor is getting colder as a result of this continuing loss of energy. So what physical principle is violated by a process represented by the reversed movie?

Statistical Models

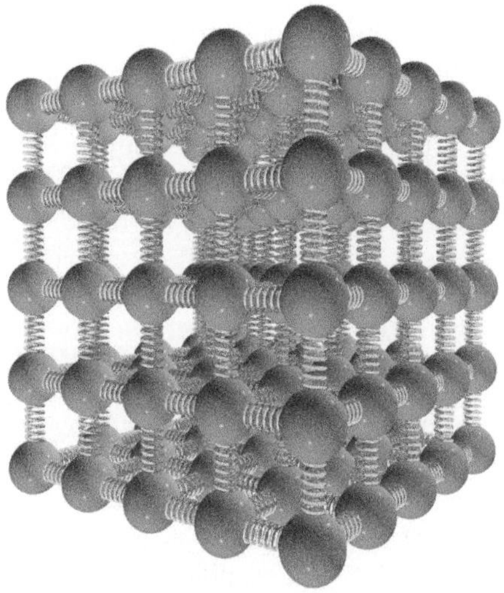

Figure 12.3 The familiar model of a solid as massive balls connected by springs.

This chapter addresses these questions by applying probabilistic and statistical ideas to the behavior of systems. This subject is called "statistical mechanics." We will first apply statistical mechanics to a simple model of a solid, and we'll find that we can go surprisingly far with just a few new concepts. The main concepts that we will develop for the specific case of a solid apply in general to a wide variety of systems.

How will we know whether our statistical model explains anything? Our criterion for understanding is whether the predictions of our microscopic model agree with measurements of macroscopic systems, such as measurements of heat capacity (the amount of energy required to increase the temperature of an object by one degree).

12.2 THE EINSTEIN MODEL OF A SOLID

We have noted in previous chapters that the interactions between atoms in a solid are electric in nature, but that a detailed model of them involves quantum mechanics. The interatomic potential energy function encountered in Chapter 7 provides a reasonably accurate description of these interactions. Because of the similarity of this function to the potential energy curve for a harmonic oscillator (a mass on a spring), in previous chapters we have modeled a solid as a large number of tiny masses (the atoms) connected to their neighbors by springs (the interatomic bonds), as shown in Figure 12.3. This model has allowed us to understand qualitatively how solids interact with other objects. We would now like to use a modified version of this model to ask detailed quantitative questions about the distribution of energy in a solid.

In this chapter we focus on calculating how probable a particular distribution of speeds or energies in a solid would be. We do this for a solid because it happens to be easier to calculate probabilities for a simple model of a solid than it is for a gas or a liquid. The atoms in a solid are nearly fixed in position, so we don't have to consider how likely different spatial arrangements might be (unlike the situation for a gas). Reasoning about the energy distribution in a solid will also enable us to draw conclusions about the transfer of thermal energy from one solid object to another.

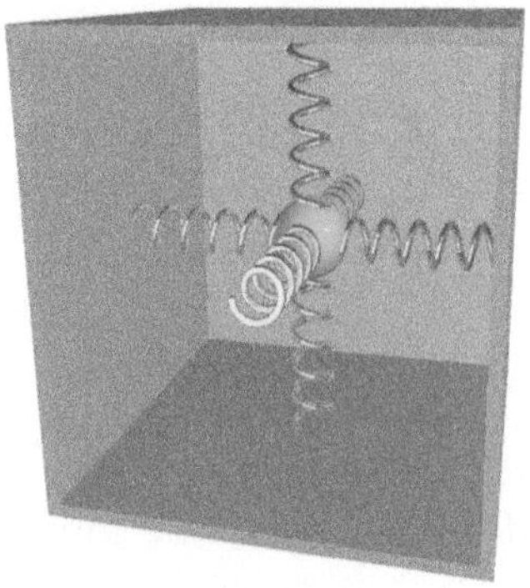

Figure 12.4 A single atom can oscillate in three dimensions. We simplify our model of a solid by assuming that each atom moves independently of the surrounding atoms—in effect, that it is connected to rigid walls instead of moving atoms.

Assume that Atoms Move Independently

Because our goal is to calculate the probability of particular distributions of energy among atoms, we can simplify our model of a solid even further. We can think of each atom as moving independently of its neighbors, as though it were connected to rigid walls rather than to other atoms (Figure 12.4). Of course in a real solid energy is exchanged with neighboring atoms. However, to address our current questions we do not need to worry about the mechanism of energy exchange between atoms, since we will focus on calculating probabilities for various distributions of energy among the atoms of the solid.

This is a different kind of model from those we have so far constructed, because this very simple model would not allow us to predict in detail the dynamic motion of each atom in the solid, nor to ask how long it would take for energy to flow from one end of a solid to the other. In fact, there could not be propagation of sound in such a solid, with its "rigid walls." The model will not shed light on the details of a process, but it does allow us to ask questions about initial and final states. Since at the moment the questions we want to address involve initial and final states, this model is useful, and it is mathematically much simpler than the connected-atoms model.

Einstein proposed this simple model in 1907, and he found that some basic properties of solids such as specific heat could in fact be understood using this model. This model also allows us to understand in detail the statistical nature of energy transfer between a hot object and a cold object, and why two objects come to "thermal equilibrium" (the same final temperature). This simple model of a solid will help us gain a more sophisticated and powerful understanding of the meaning of temperature.

A Three-Dimensional Oscillator—Three One-Dimensional Oscillators

We will consider each atom in a solid to be connected by springs to immovable walls. We will model each isolated atom as a three-dimensional spring–mass system, with $\vec{s}$ representing the three-dimensional vector displacement away from a fixed equilibrium position. Since $p^2 = p_x^2 + p_y^2 + p_z^2$ and $s^2 = s_x^2 + s_y^2 + s_z^2$, we can write the energy of a three-dimensional classical oscillator as the sum of three parts, corresponding to the x, y, and z oscillations:

$$K_{\text{vib}} + U_s = \left(\frac{p_x^2}{2m} + \frac{1}{2}k_s s_x^2\right) + \left(\frac{p_y^2}{2m} + \frac{1}{2}k_s s_y^2\right) + \left(\frac{p_z^2}{2m} + \frac{1}{2}k_s s_z^2\right)$$

Recall from Chapter 8 that a quantum oscillator (a quantum mechanical "ball and spring") has evenly spaced energy levels. A complete quantum mechanical analysis of an oscillator that is free to oscillate in three dimensions rather than just one dimension leads to the conclusion that the motion of a three-dimensional oscillator can be separated into x, y, and z

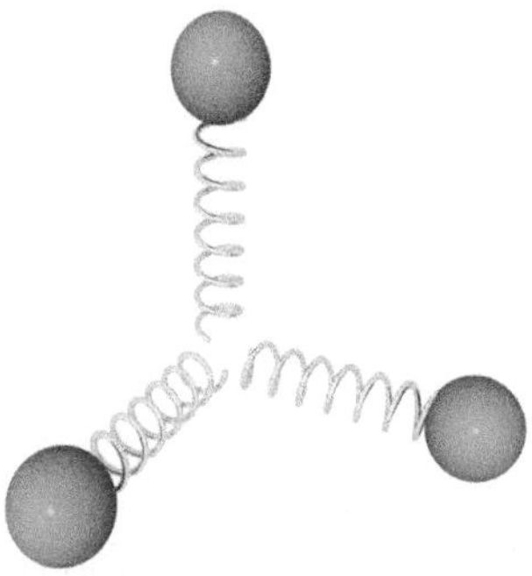

Figure 12.5 Since we are considering the atoms to be independent, in our model we can replace a single three-dimensional oscillator (one atom) by three independent one-dimensional oscillators.

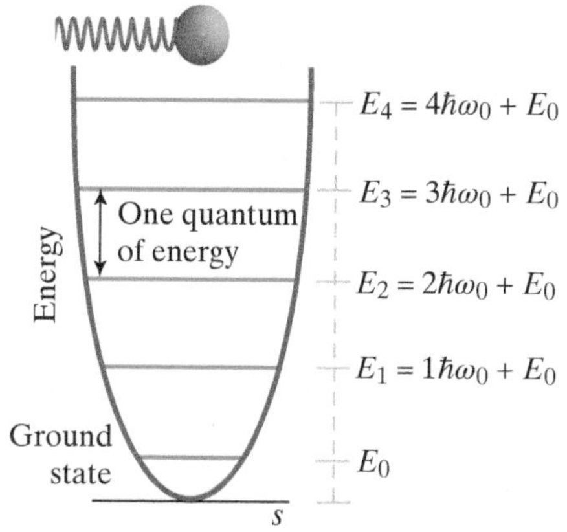

Figure 12.6 The energy levels of a one-dimensional atomic oscillator.

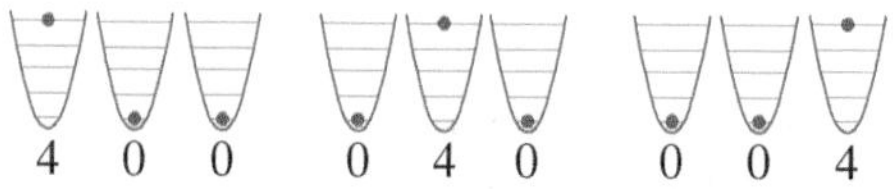

Figure 12.7 Three ways of distributing four quanta of vibrational energy among three one-dimensional oscillators.

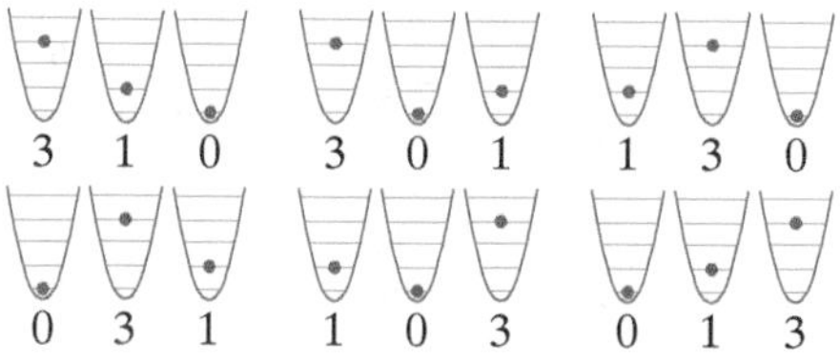

Figure 12.8 Six more ways of distributing four quanta of vibrational energy among three one-dimensional oscillators.

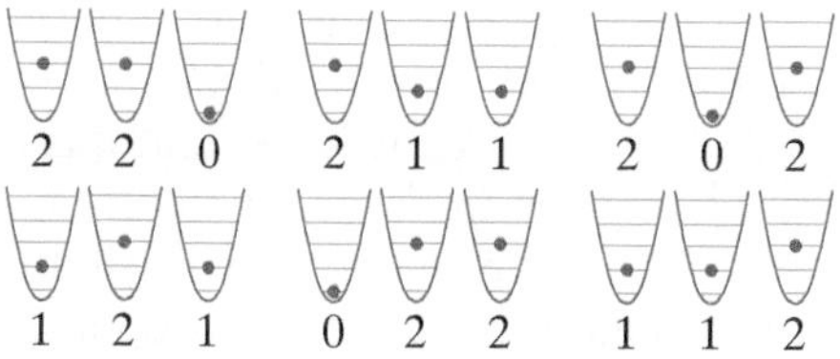

Figure 12.9 Still more ways of distributing four quanta of vibrational energy among three one-dimensional oscillators.

components, each of which has the same energy level structure as the familiar one-dimensional oscillator. This is mathematically equivalent to replacing each three-dimensional oscillator (which we're using to model an atom) with three ordinary one-dimensional oscillators, which we will do to simplify our model. We will think of a block as containing N one-dimensional oscillators, corresponding to $N/3$ atoms (Figure 12.5).

In Chapter 8 we discussed energy levels in atomic spring–mass systems. As predicted by quantum mechanics, the stable energy levels of a one-dimensional quantum oscillator are integer multiples of one "quantum" of energy $\hbar\omega_0$, where $\omega_0 = \sqrt{k_{s,i}/m_a}$ above the ground state (Figure 12.6). Here $\hbar = h/(2\pi) = 1.05 \times 10^{-34}$ J · s, $k_{s,i}$ is the interatomic spring stiffness, and m_a is the mass of the atom. The ground state of the quantum oscillator is not at the bottom of the potential energy well; $E_0 = \frac{1}{2}\hbar\omega_0$. This offset doesn't matter for what we are going to do; it just sets the baseline energy level, and we will measure energies starting from the ground state.

Distributing Energy Among Objects

When two identical blocks are brought into contact, their total energy is shared among all the oscillators (atoms) in both blocks. It seems plausible that the most likely outcome would be that the total thermal energy be shared equally between identical blocks. However, might there be some probability that the first block would have more energy than the other, or even have all of it? With a large number of atoms the probability of even a small deviation from the most probable distribution turns out to be hugely unlikely. In order to look at this question in detail, we need to find a way to calculate the probability of various possible distributions of the energy. If we can find the most probable energy distribution, we can predict the eventual equilibrium distribution of energy between two objects brought into thermal contact with each other.

Distributing Energy: A Single Atom Inside a Solid

Consider a single atom inside a solid, whose energy we model in terms of the energy of three one-dimensional oscillators (x, y, z), neglecting interactions with the neighbors. Each of these three oscillators can have 0, 1, 2, or more "quanta" of vibrational energy ($\hbar\omega_0$) added to its ground state.

Suppose that the total vibrational energy added to the atom is four quanta, and we ask how we might distribute this energy among the three oscillators. We will soon see that enumerating all the possible ways of distributing the energy among the oscillators leads to a deeper understanding of the statistical behavior of a solid. We could give all the energy to the first one-dimensional oscillator and none to the others, or all to the second, or all to the third, as shown in Figure 12.7.

We could also give three quanta to one oscillator, and give the remaining quantum to one of the others, as shown in Figure 12.8.

Or we could give two quanta to one oscillator, and distribute the other two to the others, as shown in Figure 12.9.

That's it—there aren't any more ways to distribute the energy. (Check to make sure that you can't think of any other arrangements.) By explicitly listing all the possible arrangements, we see that there are 15 different ways that the four energy quanta could be distributed among the three one-dimensional oscillators.

In each of these 15 cases the total energy of the three-oscillator atom is exactly the same. When we make macroscopic measurements of the energy of a block, we don't know and we usually don't care exactly how the energy is distributed among the atoms that make up the block, because the internal energy of the block is the same for all of these distributions. However, the number of different arrangements influences interactions with other objects, as we will see.

Microstates and Macrostates

In the context of the Einstein model of a solid, the term "macrostate" refers to the total amount of energy in the system. For every macrostate (for example, a total of four quanta of energy in the system), there are a number of possible "microstates." Each microstate reflects a different way of distributing the total energy among all the oscillators in the system. Figure 12.10 illustrates the 15 different microstates (possible ways of arranging energy) corresponding to a macrostate of a system of three oscillators with a total of four quanta of energy. These same microstates are shown pictorially in Figures 12.7–12.9.

1 1 2	1 3 0	0 1 3
4 0 0	1 2 1	2 0 2
3 1 0	0 4 0	2 1 1
2 2 0	0 2 2	1 0 3
0 3 1	3 0 1	0 0 4

Figure 12.10 The 15 microstates corresponding to a macrostate of a system of three oscillators with four quanta of energy. According to the fundamental assumption of statistical mechanics, each of these microstates is equally probable.

THE FUNDAMENTAL ASSUMPTION OF STATISTICAL MECHANICS

The fundamental assumption of statistical mechanics is that, over time, an isolated system in a given macrostate is equally likely to be found in any of its possible microstates.

For example, we isolate three oscillators (corresponding to one atom), containing a total energy of four quanta, which is a macrostate (we haven't said which oscillators have how many quanta). The fundamental assumption of statistical mechanics implies that if we repeatedly observe the detailed arrangement of the four quanta among the three oscillators (the microstate), over time we would find on the average that each of these microstates would occur 1/15th of the times that we looked. This fundamental assumption is plausible, but ultimately it is justified by the fact that deductions based on this assumption do agree with experimental observations.

In principle you could prepare a system in one particular microstate and hope it would stay that way, but if there is the slightest bit of interaction among the pieces of the system, or with the surroundings, over time the fundamental assumption of statistical mechanics claims that all microstates will occur with equal probability. The Einstein model is simplified in that it doesn't provide a mechanism for energy to transfer from one atom to another (a change of microstate) or even from one oscillator to another within the same atom. The Einstein model is useful for counting microstates, but for other purposes it needs to be supplemented by the notion that some slight interaction does occur.

Figure 12.11 Two small blocks in thermal contact.

Distributing Energy: Two Interacting Atoms

We can now consider the case of two blocks in thermal contact (Figure 12.11). The importance of counting arrangements ("microstates") can be seen when we have systems of oscillators interacting with each other. Consider the smallest possible "blocks," two neighboring atoms (a total of six one-dimensional oscillators), and suppose four quanta of vibrational energy, $4\hbar\omega_0$, are distributed among these six oscillators. We already know that there are 15 ways to distribute the four quanta among the three one-dimensional oscillators of the first atom, and in that case there is only one way to distribute zero quanta to the second atom. Similarly, in Figure 12.12 we see that there are 15 ways to distribute the four quanta among the three one-dimensional oscillators of the second atom, and in that case there is only one way to distribute zero quanta to the first atom.

Atom 1	Atom 2	# ways
4	0	$15 \cdot 1$
0	4	$1 \cdot 15$

Figure 12.12 If we give all four quanta to one atom or the other, there are 30 ways of distributing four quanta of energy between the two atoms (that is, among six independent oscillators).

However, since the two atoms are in contact with each other, the four quanta could also be shared between the atoms. For example, the first atom might have only three quanta, with the second atom having the other quantum.

QUESTION How many ways are there to distribute three quanta among three one-dimensional oscillators? Try to list all the possibilities, as we did for the case of four quanta.

You should have found 10 ways, which we could list as 300 (that is, three quanta on the first oscillator, zero on the second, and zero on the third), 030, 003, 201, 210, 021, 120, 012, 102, and 111.

QUESTION For each of these arrangements, how many ways are there to distribute the one remaining quantum among the other three one-dimensional oscillators?

There are just three ways to arrange one quantum among the other three oscillators: 100, 010, and 001.

The product of these two numbers is the total number of ways of distributing the four quanta in such a way that there are three quanta on the first atom and one on the other. This product is $10 \cdot 3 = 30$ ways. These results are summarized in Figure 12.13.

Atom 1	Atom 2	# ways
3	1	$10 \cdot 3$
1	3	$3 \cdot 10$

Figure 12.13 We find 60 more ways to distribute four quanta among two atoms (six oscillators) if we give three quanta to one atom and one quantum to the other.

Now let's give just two quanta to the first atom (leaving two quanta for the other atom).

QUESTION How many ways are there to distribute two quanta among three one-dimensional oscillators?

You should find that there are six ways: 200, 020, 002, 101, 110, and 011.

QUESTION For each of these arrangements, how many ways are there to distribute the two remaining quanta among the other three one-dimensional oscillators?

Clearly there are also six ways to arrange the remaining two quanta among the other three oscillators. The total number of ways of distributing the four quanta in such a way that there are two quanta on the first atom and two on the other is $6 \cdot 6 = 36$ (Figure 12.14).

Atom 1	Atom 2	# ways
2	2	$6 \cdot 6$

Figure 12.14 We find 36 ways of distributing four quanta of energy by giving two quanta to each atom.

Now we have carried out enough calculations to be able to make a table of all the ways of distributing four quanta between two atoms (that is, among six one-dimensional oscillators). In this table, q_1 is the number of quanta (0 to 4) assigned to the first atom, and q_2 is the number of quanta assigned to the second (with $q_1 + q_2$ necessarily equal to 4). In Figure 12.15 we show the number of ways to arrange the quanta among the first three one-dimensional oscillators, among the second three oscillators, and the product, which is the total number of ways of distributing q_1 and q_2 quanta.

q_1	q_2	# ways 1	# ways 2	# ways 1 · # ways 2
0	4	1	15	15
1	3	3	10	30
2	2	6	6	36
3	1	10	3	30
4	0	15	1	15

Figure 12.15 Summary of the 126 different ways to distribute four quanta of vibrational energy between two atoms (each consisting of three independent oscillators).

Figure 12.16 is a graph of the total number of ways of distributing the quanta vs. q_1. Remember that the basic assumption of statistical mechanics is that all of these 126 different arrangements ($15 + 30 + 36 + 30 + 15 = 126$ microstates) with a total energy of four quanta are equally probable.

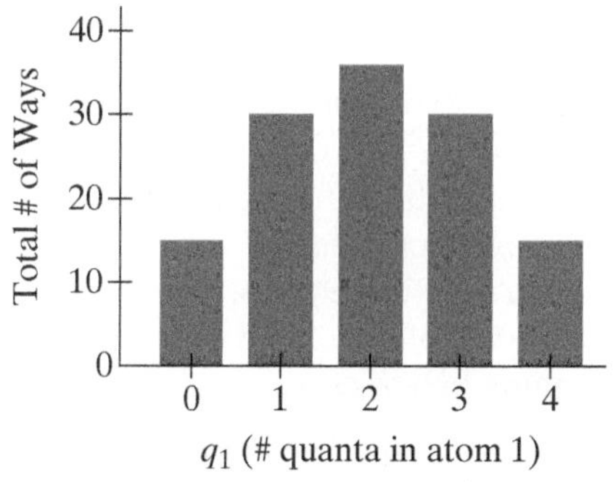

Figure 12.16 Bar graph showing the total number of ways of distributing four quanta of vibrational energy between two atoms (six one-dimensional oscillators).

QUESTION Therefore, if two atoms share four quanta, what is the most likely division of the energy between the two? What is the probability that at some instant the energy is equally divided? What is the probability that at some instant the first atom has all the energy?

Evidently the most probable division is 2 and 2, since there are 36 ways for this to be done, whereas all other divisions have fewer ways for this to happen. The probability for a 2-2 division is 36/126, or 29%, so even though this is the most probable division, it happens less than one-third of the time. The probability that the first atom has all the energy is the number of ways for this to happen (15) divided by 126: 15/126 = 12%.

This is a microcosm of what happens when two identical blocks are brought into contact: the most probable division of the thermal energy is that it is shared equally. In order to understand in detail how this works in a real macroscopic system, we need to consider large numbers of atoms and large numbers of quanta. For numbers of atoms or quanta even slightly larger than the few we've been considering, it becomes practically impossible to figure out the number of arrangements of quanta by explicitly listing all the possibilities as we did up to now. We need an equation for calculating the number of arrangements.

Checkpoint 1 For practice in counting microstates, determine how many ways there are to arrange three quanta among four one-dimensional oscillators. (This would represent one-and-a-third atoms, so this doesn't make physical sense.)

One System or Many?

The results for distributing four quanta among two atoms (six oscillators) can be thought of in two complementary ways. You can say, "I will make frequent observations of my isolated two-atom system, and I expect that in 29% of these observations I'll find the energy split evenly (2-2)." Alternatively, you can set up 100 of these isolated two-atom systems and say, "Whenever I look at my 100 systems, I expect that 29 of them will have the energy split evenly (2-2)." Sometimes the "one system, many observations" view is particularly helpful, and sometimes the "many systems, one observation" view is the more useful way to think about the statistical nature of a phenomenon.

An Equation for the Number of Arrangements of Quanta

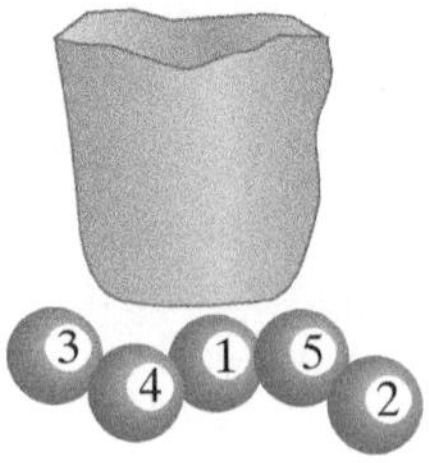

Figure 12.17 Five numbered billiard balls are taken one at a time from a bag. How many different number sequences are possible?

Clearly it would be not only tedious but also impractical to keep on counting states as we did above for systems involving very large numbers of atoms. We need an equation for calculating how many different ways a set of objects can be arranged. We can develop this equation in a general way, using concepts of probability, and then apply it to solids made up of atomic oscillators.

Suppose you have five billiard balls in a bag, numbered 1 through 5. You draw them out of the bag, one at a time, and record the sequence of numbers, such as 34152 (Figure 12.17). Then you put them back in the bag and repeat. How many different number sequences are possible?

- The first ball might be any one of the five balls.
- For each of these five possibilities, there are four possibilities for the second ball, or $5 \cdot 4 = 20$ choices so far.
- For each of the $5 \cdot 4 = 20$ choices of the first two balls, there are three different possible choices for the third ball, or $5 \cdot 4 \cdot 3$ choices so far.
- For each of the $5 \cdot 4 \cdot 3 = 60$ choices of the first three balls, there are two possible choices for the fourth ball, for a total of $5 \cdot 4 \cdot 3 \cdot 2 \cdot 1 = 120$ possibilities, since there is only one remaining ball to choose.

Evidently there are 120 different "permutations" of the five integers. It would be exceedingly tedious to list all these different arrangements, but we have a simple equation to calculate how many there are: 5!, which is the standard notation for "5 factorial," meaning $5 \cdot 4 \cdot 3 \cdot 2 \cdot 1$. Fortunately, for analyzing a solid all we care about is how many arrangements (microstates) there are for a particular total energy, not the details about how much of the energy is assigned to which oscillators. So all we need from an equation is the number of arrangements.

Checkpoint 2 We can easily check this factorial equation for the case of three balls. Explicitly list all possible permutations of the numbers 1, 2, and 3, and verify that there are indeed $3! = 3 \cdot 2 \cdot 1 = 6$ possible arrangements.

Making Fewer Distinctions

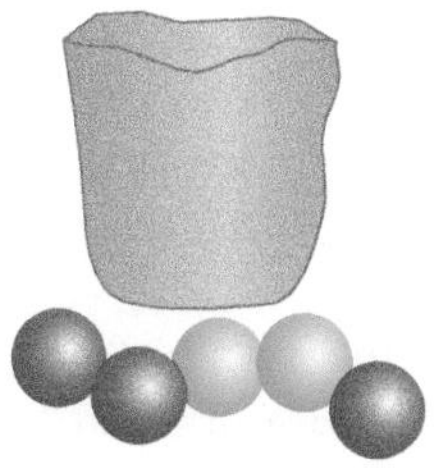

Figure 12.18 Five billiard balls are taken one at a time from a bag. Three are red and two are green. How many different color sequences are possible?

Now suppose that of the five balls, three are red (R) and two are green (G), and we ask how many different arrangements of the colors are possible, such as RRGGR shown in Figure 12.18. We know that there are $5! = 120$ different arrangements of the numbered balls, but if we're only interested in the color sequence, the numerical order of the red balls is irrelevant, as is the numerical order of the green balls.

There are $3! = 6$ permutations of the red balls among each other, and $2! = 2$ permutations of the green balls among each other. Therefore there are many fewer than 120 distinctively different color sequences, and we need to correct for this by dividing by the extra permutations:

$$\text{number of color sequences} = \frac{5!}{3!2!} = \frac{120}{(6)(2)} = 10$$

Checkpoint 3 Check this result by listing all the different ways of ordering three R's and two G's.

Arranging Quanta Among Oscillators

• • | • | •

Figure 12.19 A representation of a situation where the first oscillator has 2 quanta, the second oscillator has 1 quantum, and the third oscillator has 1 quantum.

By an appropriate choice of visual representation, we can convert our problem of calculating the number of arrangements of q quanta among N one-dimensional oscillators ($N/3$ atoms) into the problem we just solved, for which we have an equation. Consider again the specific case of $q = 4$ quanta distributed among $N = 3$ one-dimensional oscillators. We'll represent a quantum of energy by the symbol •, and a (fictitious) boundary between oscillators by a vertical bar **|**. Figure 12.19 is a picture of a particular situation where the first oscillator has 2 quanta, the second oscillator has 1 quantum, and the third oscillator has 1 quantum; together with the boundaries between oscillators we have 6 objects arranged in a particular sequence.

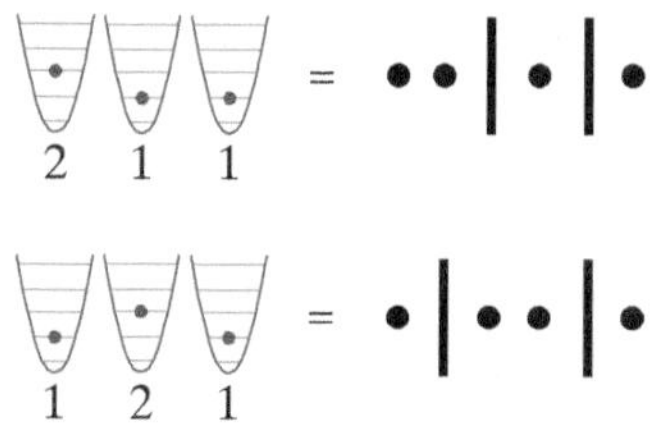

Figure 12.20 Representations of two situations for the number of quanta in 3 oscillators.

This sequence represents a total energy of 4 quanta and 2 boundaries (a total of 6 things, if we consider a boundary to be a "thing"). Figure 12.20 illustrates two such sequences. Notice that we need $N-1 = 2$ vertical bars to be able to indicate which oscillators have how much energy. How many such sequences like this are there? Rearranging quanta and boundaries is equivalent to moving quanta between oscillators. In this pictorial form, the problem is like having 4 red balls and 2 green ones. Therefore we can write an equation for the number of different arrangements (number of different microstates):

$$\frac{6!}{4!2!} = 15 \text{ ways of arranging 4 quanta among 3 oscillators}$$

This agrees with our earlier calculations. Generalizing to arbitrary numbers of quanta distributed among arbitrary numbers of oscillators, we have this important result for the number of microstates, written as Ω (Greek capital omega):

NUMBER OF WAYS Ω TO ARRANGE q QUANTA AMONG N ONE-DIMENSIONAL OSCILLATORS

$$\Omega = \frac{(q+N-1)!}{q!(N-1)!}$$

Checkpoint 4 Verify that this equation gives the correct number of ways to arrange 0, 1, 2, 3, or 4 quanta among 3 one-dimensional oscillators, given in earlier tables (1, 3, 6, 10, 15).

Very Big Numbers

As you increase q and/or N, this expression gets very big very fast. For example, the number of ways to distribute 100 quanta among 300 oscillators (100 atoms) is about 1.7×10^{96}, which is 17 followed by 95 zeros!

In striking contrast, there is only one way to arrange to have all 100 energy quanta be placed on just one particular oscillator out of all the 300 oscillators. Although this arrangement would satisfy the requirement that the total energy of the system be 100 quanta, this is extremely improbable. The fundamental assumption of statistical mechanics is that in an isolated system, over time, all microstates are equally probable, so the odds that the actual microstate is the one with all the energy given to just one particular oscillator, of your choice, is 1 in 1.7×10^{96}, which makes this unlikely event essentially impossible.

A typical macroscopic object such as a block of ordinary size contains 1×10^{23} or more atoms, not a mere hundred, and the number of quanta is even larger. How likely is it that all the energy will be found concentrated on one particular atom? The mind boggles at the astronomically huge odds against this ever happening. Could it happen? Yes. Is any human ever likely to observe it? No.

12.3 THERMAL EQUILIBRIUM OF BLOCKS IN CONTACT

We now have the tools necessary for analyzing in some detail what will happen when two blocks are brought into contact and approach thermal equilibrium. We'll choose two blocks made of the same material, so a quantum of energy is the same for the atomic oscillators in both blocks (same atomic mass m; same interatomic forces, so same "spring stiffness" k_s). We'll make the analysis somewhat general by choosing blocks of different sizes. One block contains N_1 one-dimensional oscillators ($N_1/3$ atoms) and initially contains q_1 quanta of energy, and the other block contains N_2 one-dimensional oscillators ($N_2/3$ atoms) and initially contains q_2 quanta of energy.

QUESTION We want to treat the simple case where the total energy $q_1 + q_2$ of the two-block system remains fixed at all times. What simplifying assumption should we make about the situation?

During the entire process we assume that there is little energy transferred into or out of the surrounding air or the supports for the blocks, so that the total energy of the two blocks doesn't change. However, the number of quanta in each block, q_1 or q_2, need not stay fixed, since energy can flow back and forth between the two blocks.

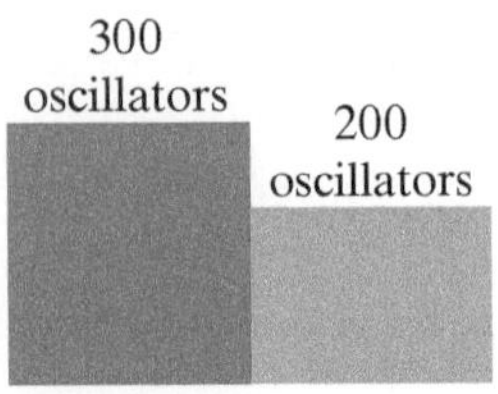

Figure 12.21 A total of 100 quanta of vibrational energy are available to be distributed between two systems, one consisting of 300 oscillators (100 atoms), the other consisting of 200 oscillators (about 67 atoms).

Consider a very concrete example. Suppose that $N_1 = 300$ (100 atoms), $N_2 = 200$ (about 67 atoms; or choose 201 oscillators if you wish to be exact), and the total energy distributed throughout the two blocks corresponds to $q_1 + q_2 =$ 100 quanta (Figure 12.21).

We use a computer to calculate the number of ways that q_1 quanta can be distributed among the $N_1 = 300$ oscillators of the first block (using the equation we developed earlier), and we multiply this number times the number of ways that $q_2 = (100 - q_1)$ quanta can be distributed among the $N_2 = 200$ oscillators of the second block.

The product of these two calculations is the number of ways that we can arrange the 100 quanta so that the first block has q_1 quanta and the second block has $q_2 = (100 - q_1)$ quanta, for a total energy shared between the two blocks of 100 quanta. We have the computer do this calculation for $q_1 = 0$, 1, 2, 3,... 99, 100 quanta, which corresponds to $q_2 = 100$, 99, 98,... 1, 0 quanta. The following table shows the first few results, where the number of microstates is denoted by Ω (Greek uppercase omega):

q_1	$q_2 = (100-q_1)$	$\Omega_1 = \dfrac{(q_1+300-1)!}{q_1!(300-1)!}$	$\Omega_2 = \dfrac{(q_2+200-1)!}{q_2!(200-1)!}$	**Total # of Ways** $\Omega_1\Omega_2$
0	100	1	2.772 E+81	2.772 E+81
1	99	300	9.271 E+80	2.781 E+83
2	98	4.515 E+04	3.080 E+80	1.391 E+85
3	97	4.545 E+06	1.016 E+80	4.619 E+86
4	96	3.443 E+08	3.331 E+79	1.147 E+88
	...	...	...	...

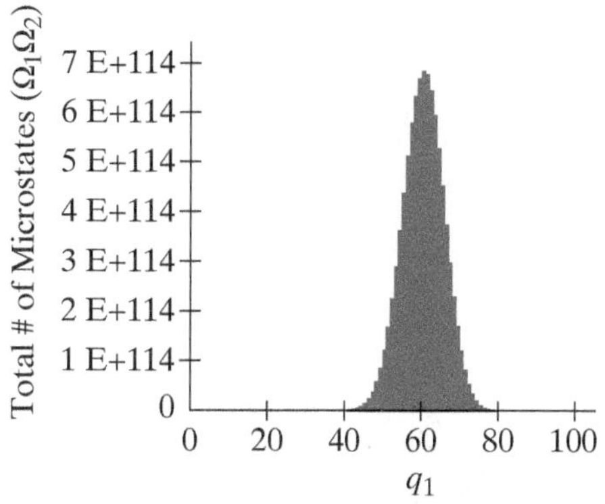

Figure 12.22 The number of ways of distributing 100 quanta of vibrational energy between two blocks having 300 and 200 oscillators, respectively. q_1 is the number of quanta in the first (larger) block.

We see in the last column that there is an enormous number of ways of arranging 100 quanta among 500 oscillators, and that this number is growing rapidly in the table. In Figure 12.22 the possible number of arrangements of quanta is plotted on the y axis, and q_1, the number of quanta assigned to the first block, is plotted on the x axis.

The most probable arrangement (indicated by the highest point of the peak on the graph) is that 60 of the 100 quanta will be found in the first block, which contains 300 oscillators out of the total of 500 oscillators. This seems reasonable, since it does seem most probable that the energy would be distributed uniformly throughout the two blocks (since they're made of the same material), and that would give 3/5 (60%) of the energy to the first block. It is gratifying that our statistical analysis leads to this plausible result.

It appears from the width of the peak shown in the graph that we shouldn't be too surprised if occasionally we would find that the first block contains anywhere between 40 to 80 of the 100 quanta, but it appears that it is very unlikely to find fewer than 40 or more than 80 of the quanta in the first block.

Relatively speaking, how likely is it for *none* of the energy to be in the first block? In that case, $q_1 = 0$, and from the table we see that the number of ways to arrange the 100 quanta this way is 2.772 E+81 (2.772×10^{81}). That's an awfully big number, but how big is it compared to the most probable arrangement, where $q_1 = 60$, for which there are about 7×10^{114} ways according to the graph? Evidently it is less likely by about a factor of 10^{33} (!) to find the energy split 0-100 rather than 60-40. Is it possible according to the laws of physics for none of the energy to be in the first block? Yes. Is it likely that we would ever observe such an unusual distribution? Most emphatically not!

Note that the number of ways to arrange the quanta 0-100 (2.772×10^{81}) isn't actually visible on the graph, because it is 1×10^{33} times smaller than the peak of the graph. Values of q_1 outside the 40 to 80 range are invisible on the graph, because they are relatively so very small compared to the peak.

At this point it is very useful to do Problem P61, in which these ideas are made very concrete.

Width of the Distribution

Compared to the graph in Figure 12.16 of our earliest calculation (6 oscillators and 4 quanta), the peak shown in Figure 12.22 occupies a much narrower range

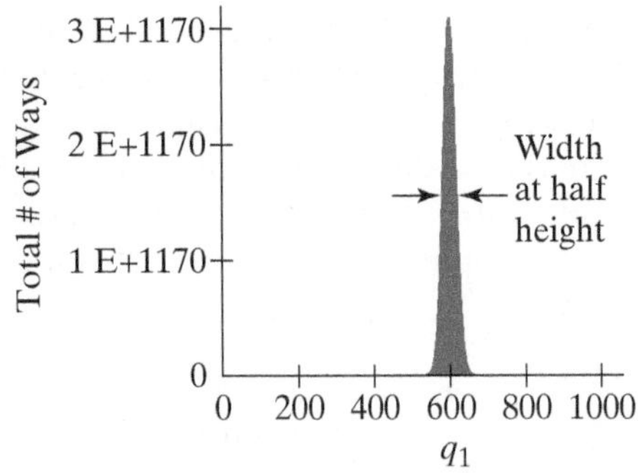

Figure 12.23 The number of ways of distributing 1000 quanta between two blocks containing 3000 and 2000 oscillators, respectively.

of the graph, reflecting the sharply decreased relative probability of observing extreme distributions of the energy.

For further comparison, Figure 12.23 is a graph of the number of ways of distributing 1000 quanta among two blocks containing 3000 and 2000 oscillators. This peak is much narrower than the peak in the previous graph (where there were only 100 quanta distributed among only 300 and 200 oscillators). This is a general trend. The larger the number of quanta and oscillators, the narrower is the peak around the most probable distribution.

The fractional width of the peak is the width of the peak at half height divided by the value of q_1 that gives the maximum probability (in the present case, the width divided by 600). It can be shown that the fractional width of the peak is proportional to $1/\sqrt{q}$ or to $1/\sqrt{N}$, whichever is larger. That is, if you quadruple the number of quanta or the number of oscillators, the fractional width of the peak decreases by a factor of 2.

> QUESTION Consider two blocks that are of ordinary macroscopic size, containing 3×10^{23} atoms and 2×10^{23} atoms, and many quanta per atom. Qualitatively, what would you expect about the width of the peak, if you could calculate it? How likely is it that you would ever observe a significant fluctuation away from the most probable 60-40 split?

You would expect the peak to be extremely narrow, and hence the probability of significant fluctuations would be very low.

The Most Probable Is the Only Real Possibility

These considerations show that in the world of macroscopic objects such as ordinary-sized blocks, the most probable arrangement is essentially the *only* arrangement that is ever observed. That is why you do not see a block suddenly leap up from the table when all of the thermal energy of the table floods into the block. On the other hand, at the microscopic level we should not be surprised if the energy of one of the atoms in a block varies a lot, since the most probable distribution includes many different arrangements of the quanta, with varying numbers of quanta on one particular atom.

Entropy and Equilibrium

We now have a good statistical description of the thermal equilibrium of two blocks in contact. At equilibrium, energy is distributed between the two blocks in the most probable manner, based on having the largest number of ways of achieving this distribution (largest number of microstates for the macrostate of given total energy). We have been studying very small systems consisting of a hundred or more atoms, and we found that in small systems there is some significant probability of finding the energy distributed somewhat differently than the most probable 60-40 division. (Objects consisting of only a few hundred atoms are called "nanoparticles" and are currently the subject of intense research, because many of their properties and behaviors are intermediate between those of atoms and those of large-scale objects.) For large macroscopic objects containing 10^{20} atoms or more, the most probable distribution of the energy is essentially the only energy distribution we will ever observe, because the probability of distributions that are only slightly different is very small.

Next we will study more deeply the details of why a particular thermal equilibrium becomes established. Suppose for example that the first block (a nanoparticle with 300 oscillators, or 100 atoms) starts out with 90 quanta and the second (200 oscillators, or about 67 atoms) with only 10 quanta (Figure 12.24). When we put them together, we expect this to shift toward a

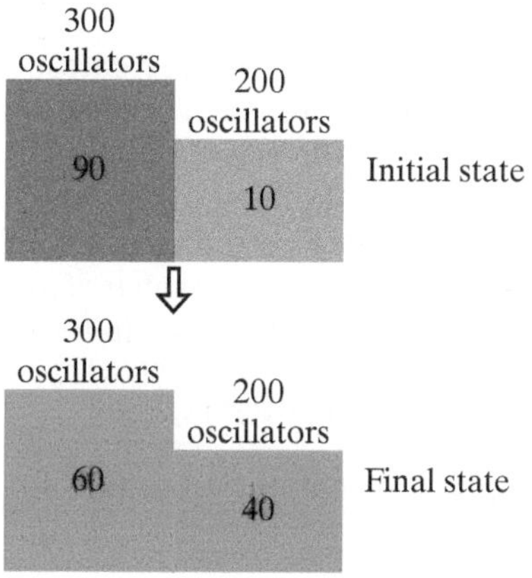

Figure 12.24 If the initial energy distribution between two systems in thermal contact is not the most probable energy distribution, energy will be exchanged until the most probable distribution is reached.

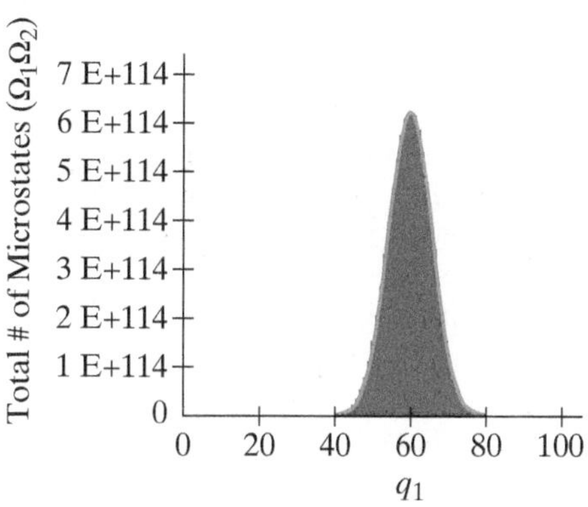

Figure 12.25 Ways of distributing 100 quanta of vibrational energy between a system of 300 oscillators and a system of 200 oscillators.

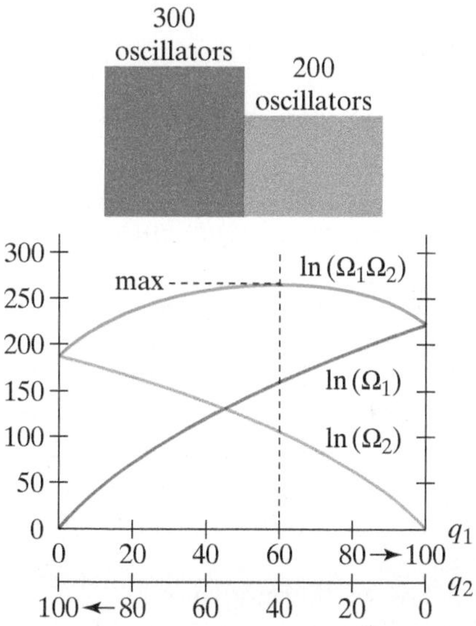

Figure 12.26 A logarithmic plot of the ways of distributing 100 quanta of vibrational energy between two systems, plotted against q_1, the number of quanta in object 1.

60-40 distribution (Figure 12.25). Studying the details of why this shift occurs will lead us to a deep understanding of the concepts of temperature and something called "entropy."

Since we start from a nonequilibrium energy distribution, it would be nice if we could make some kind of graph where we could see the total number of ways to arrange the quanta even when we are far from equilibrium. As we have seen, the graph of the total number of microstates is so strongly peaked that we don't see anything outside the peak. For example, on the graph in Figure 12.25 where we can see 1×10^{114} we can't see the relatively much smaller value of 1×10^{81}. A way to get around this problem is to plot the logarithm of the data, which makes the data visible across the whole range of energy distributions.

In statistical mechanics it is standard practice to use the base-*e* or "natural" logarithm ("ln"). For the case of distributing 100 quanta among 300 and 200 one-dimensional oscillators, which we studied before, Figure 12.26 shows a plot of the natural logarithms of the number of ways Ω_1 to arrange q_1 quanta in block 1, the number of ways Ω_2 to arrange $q_2 = (100 - q_1)$ quanta in block 2, and the total number of ways $\Omega_1\Omega_2$ to arrange the 100 quanta among the 500 oscillators. These quantities are plotted against q_1, of course running left to right, 0 to 100. You can think of q_2 as running right to left.

Because we are considering discrete microstates, each of these curves is really a set of 100 dots, but we connect the dots and make continuous curves. This is particularly appropriate when we deal with macroscopic objects, where an increment of one quantum along the *x* axis is practically an infinitesimal fraction of the axis.

By taking (natural) logarithms, we have converted a highly-peaked graph into a slowly varying one. To make sure you understand how this graph is related to the peaked graph, do the following:

Checkpoint 5 For this situation, we calculated earlier that there is just one way to arrange 0 quanta in block 1, and 2.772×10^{81} ways to arrange 100 quanta in block 2. Take the natural logarithm of this number and verify that all three curves on the graph make sense to you at $q_1 = 0$ ($q_2 = 100$). Note that at the right end of the graph, at $q_1 = 100$, the curve for $\ln(\Omega_1)$ goes higher than $\ln(2.772 \times 10^{81})$, because block 1 has a larger number of oscillators among which to distribute the 100 quanta (300 oscillators compared with 200).

Definition of Entropy

We will deal repeatedly with the natural logarithm of the number of ways to arrange energy among a group of atoms (the number of microstates corresponding to a particular macrostate of specified energy). This quantity, $\ln(\Omega)$, when multiplied by the Boltzmann constant k_B, is called the "entropy" of the object and is denoted by the letter "S" (the triple equal sign means "is defined as"):

DEFINITION OF ENTROPY *S*

$$S = k_B \ln \Omega$$

The Boltzmann constant is $k_B = 1.38 \times 10^{-23}$ J/K. (This is a simplified definition that is adequate for our purposes.)

Because the Boltzmann constant is $k_B = 1.38 \times 10^{-23}$ J/K, and the logarithm has no units, entropy has units of joules per kelvin. The Boltzmann constant is included in the definition for consistency with an older, macroscopically based definition of entropy.

Since $k_B \ln(\Omega_1\Omega_2) = k_B \ln(\Omega_1) + k_B \ln(\Omega_2)$ (this is a property of logarithms), the entropy S of the two-block system is equal to the entropy S_1

of the first block plus the entropy S_2 of the second block. A consequence of defining entropy in terms of a logarithm is that we can consider entropy as describing a property of a system, and when there is more than one object in a system we get the total entropy simply by adding up the individual entropies, $S_{tot} = S_1 + S_2$, as is the case with energy ($E_{tot} = E_1 + E_2$).

Figure 12.27 is a plot of the entropies of each block, and their sum. Since both $k_B \ln(\Omega_1\Omega_2)$ and $\Omega_1\Omega_2$ go through a maximum at the same value of q_1, we can state the condition for thermal equilibrium of the two blocks, in terms of entropy:

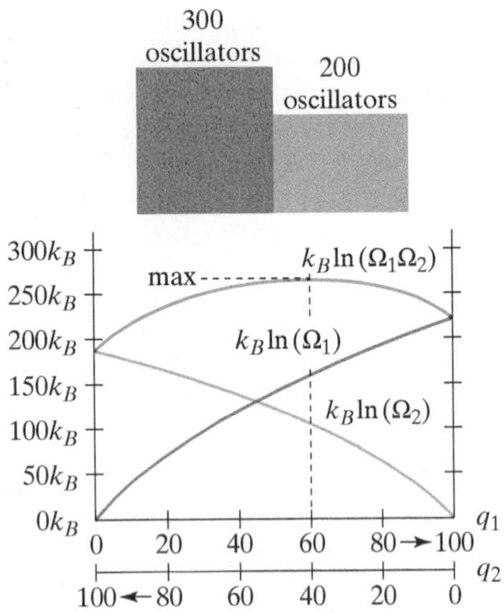

Figure 12.27 A plot of the entropies S_1, S_2, and $S_1 + S_2$ plotted against q_1, the number of quanta in object 1. Entropy is $k_B \ln \Omega$.

CONDITION FOR THERMAL EQUILIBRIUM

In equilibrium the most probable energy distribution is the one that maximizes the total entropy $S_{tot} = S_1 + S_2$ of two objects in thermal contact.

QUESTION In Figure 12.27, what is the physical significance of the fact that the top curve goes through a maximum for $q_1 = 60$?

The most probable energy distribution is a 60/40 sharing, corresponding to the 300/200 sizes of the two objects.

QUESTION In Figure 12.27, what is the physical significance of the point where the two lower curves cross each other?

Nothing! This point isn't anything special. It is the point where the entropy in one object is equal to the entropy in the other object (and equal to half the entropy of the combined system). It turns out that this doesn't have any real physical significance.

At this point it is very useful to do Problem P62, to see in detail how to produce Figure 12.26 or Figure 12.27.

12.4 THE SECOND LAW OF THERMODYNAMICS

The "first law of thermodynamics" is another name for the Energy Principle $\Delta E = W + Q$, which by now should be very familiar to you. The "second law of thermodynamics" is, however, something new. It can be stated in a number of equivalent forms, but there is a particularly useful formulation in terms of entropy:

THE SECOND LAW OF THERMODYNAMICS

If a closed system is not in equilibrium, the most probable consequence is that the entropy of the system will increase.

In other words, a closed system will tend toward maximum entropy.

As a specific example of the second law of thermodynamics, consider our two blocks. The most probable energy distribution is the one for which the total entropy is a maximum, and if initially the energy distribution is something else, it is highly likely that the entropy will increase. For nanoparticles there can be significant fluctuations away from this state of maximum entropy (with accompanying decrease in total entropy), but for ordinary-sized systems these fluctuations are extremely small, and for practical purposes the entropy of a closed macroscopic system never decreases.

Loosely, one can restate the second law of thermodynamics like this: "A closed system tends toward increasing disorder." This is overly vague without precise definitions of "order" and "disorder," and it is the definition of entropy as $k_B \ln \Omega$ that provides the needed precision. A closed system tends toward increasing entropy, which is a measure of the number of microstates corresponding to a particular macrostate of given energy.

Irreversibility

Any process in which the entropy of the Universe increases is "irreversible" in the technical, physics sense: a reversed movie of the process looks odd. However, any process in which the entropy of the Universe doesn't change is in principle "reversible," and there do exist processes that are approximately or nearly reversible. For example, a steel ball bearing that bounces vertically on a steel plate rebounds almost to the same height from which it was dropped, and this represents a (nearly) reversible process, because the system returns (nearly) to its original state.

However, if we watch for a while, the ball bearing returns to lower and lower heights and eventually settles down to rest on the plate. If the ball bearing kept returning to its original height, this would not be a violation of energy conservation. It is the second law of thermodynamics that says that we cannot expect the process to be completely reversible. There are a larger number of ways to share the energy between the ball and the plate than the number of ways for the ball to keep all the energy.

Suppose we make a movie of the ball bearing bouncing lower and lower and coming to rest on the plate. Then we run it backwards, and our friends see the ball starting to bounce, and bouncing higher and higher (Figure 12.28). This needn't violate energy conservation: energy in the plate could be flowing into the ball. This reversed movie certainly looks odd, though, presumably because we have an instinctive sense, based on lots of experience, that the entropy of the Universe increases rather than decreases.

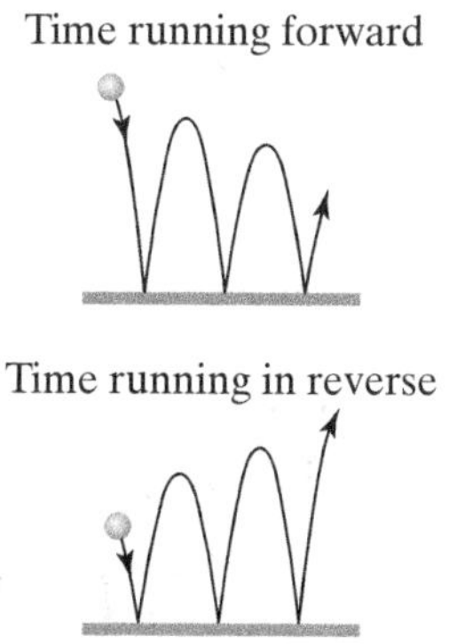

Figure 12.28 If the height of a bouncing ball increased with time instead of decreasing with time, this would not necessarily violate conservation of energy.

REVERSIBLE AND IRREVERSIBLE PROCESSES

Reversible process: $\Delta S_{\text{sys}} + \Delta S_{\text{surroundings}} = 0$

Irreversible process: $\Delta S_{\text{sys}} + \Delta S_{\text{surroundings}} > 0$

Checkpoint 6 You see a movie in which a shallow puddle of water coalesces into a perfectly cubical ice cube. How do you know the movie is being played backwards? Otherwise, what physical principle would be violated?

12.5 WHAT IS TEMPERATURE?

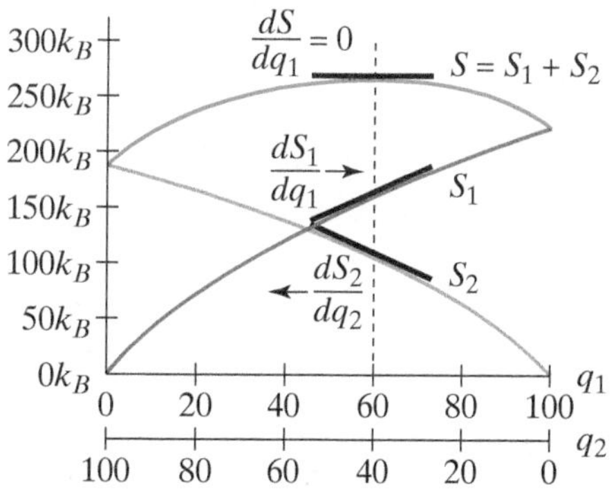

Figure 12.29 Entropy vs. number of quanta of energy in system 1.

Until now, we have loosely associated temperature with the average energy of a molecule. The concept of entropy makes possible a deeper connection between our macroscopic measurements of temperature and a fundamental, atomic, statistical view of matter and energy. We will develop a statistically based definition of temperature.

Consider again our two blocks. On the graph in Figure 12.29 we plot entropy $S = k_B \ln \Omega$. Remember that we are approximating a set of dots by a smooth curve, since the dots are very close together (tiny energy spacing). At the maximum of the total entropy $S = S_1 + S_2$, the total entropy curve is horizontal (slope = 0), and the following relationship is true at equilibrium:

$$\frac{dS}{dq_1} = \frac{dS_1}{dq_1} + \frac{dS_2}{dq_1} = 0$$

Since $q_2 = (100 - q_1)$, this leads to

$$\frac{dS_1}{dq_1} - \frac{dS_2}{dq_2} = 0 \quad \text{(at equilibrium)}$$

dS_2/dq_2 is the slope moving from right to left and is positive, since q_2 and S_2 increase from right to left. Therefore we see that these slopes are the same at equilibrium:

$$\frac{dS_1}{dq_1} = \frac{dS_2}{dq_2} \quad \text{(at equilibrium; maximum total entropy)}$$

dS_1/dq_1 is a measure of the state of the first block, and dS_2/dq_2 is a measure of the state of the second block.

QUESTION What physical property can these derivatives represent? What physical property of the blocks is the same when the two blocks reach thermal equilibrium?

The blocks reach the same temperature. It must be that the derivative of the entropy with respect to the energy is somehow related to temperature.

To see what this relationship is, suppose that when the two blocks were initially brought into contact, the first block contained $q_1 = 90$ quanta and the second block only $q_2 = 100 - 90 = 10$ quanta, as illustrated in Figure 12.30. Note that in Figure 12.30 the initial slope of the entropy curve for block 2 is steeper than the initial slope of the entropy curve for block 1. (Remember: dS_2/dq_2 is the slope moving from right to left and is positive, since q_2 and S_2 increase from right to left.) Therefore, if we remove one quantum of energy from block 1 and give it to block 2, we'll increase the entropy in block 2 more than we'll decrease the entropy in block 1.

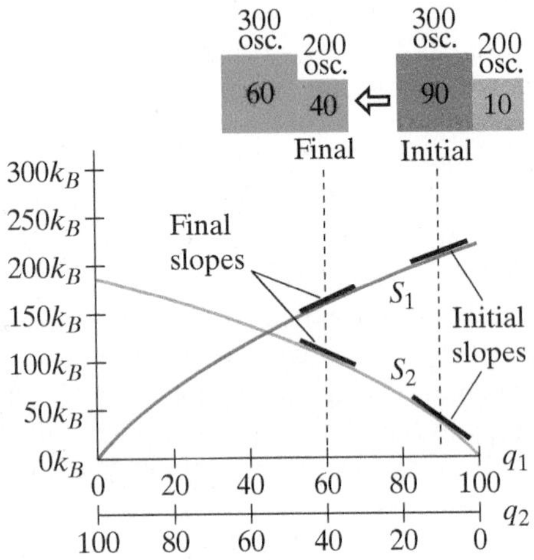

Figure 12.30 Location of the initial and final states of the two-block system on a plot of entropy vs. number of energy quanta in system 1.

QUESTION Will this result in a net increase or a net decrease in the total entropy S? (See Figure 12.30.)

Since the entropy of block 2 increases more than the entropy of block 1 decreases, there is a net increase of the entropy of the two blocks together. We already established that the state of the two blocks will evolve toward greater total entropy, not less total entropy.

QUESTION Consider which block gives up energy, and which block takes it up. Which block is initially at a higher temperature?

Evidently block 1 must be at the higher temperature, because on average it will give energy to block 2 in order that the total entropy increase.

QUESTION Does a steeper or a less steep slope of entropy vs. energy correspond to higher temperature?

Initially, block 1 has the higher temperature, and S_1 vs. q_1 has the smaller slope. Initially, block 2 has the lower temperature, and S_2 vs. q_2 has the larger slope.

QUESTION Therefore, which of the following looks like a better guess for relating temperature to entropy: $T_1 \propto dS_1/dq_1$ or $1/T_1 \propto dS_1/dq_1$?

Since block 1 has the higher temperature and the smaller slope (smaller dS/dq), it seems possible that $1/T_1 \propto dS_1/dq_1$.

Definition of Temperature

For these reasons, we *define* temperature in terms of dS/dq. However, the magnitude of one quantum of energy varies for different systems, so we can't compare different systems if we use q. So instead of q we use E_{int}, the internal energy above the ground state of a system measured in joules. For a group of oscillators, E_{int} is the number of quanta q times the energy per quantum: $E_{int} = q\hbar\sqrt{k_s/m}$ for a system of harmonic oscillators, where k_s is the spring stiffness.

DEFINITION OF TEMPERATURE T

$$\frac{1}{T} = \frac{dS}{dE_{\text{int}}}$$

The Boltzmann constant k_B in the definition $S = k_B \ln \Omega$ makes the units come out right. With the internal energy E_{int} of a system measured in joules (J), and entropy S measured in joules per kelvin (J/K), temperature T is measured in kelvins (K). (Remember that we're approximating a set of closely spaced dots by a curve, so taking a derivative makes sense.)

This is a highly sophisticated and abstract way of defining temperature. How does this relate to the temperature that is measured by an ordinary thermometer? In Supplement 1 we show that the temperature defined by $1/T = dS/dE_{\text{int}}$ is the same as the "absolute" temperature, which appears for example in the ideal gas law $PV = N_{\text{moles}}RT$, which you have probably encountered in previous chemistry or physics studies. On this temperature scale, ice melts at a temperature of +273.15 K (0 °C), and water boils at +373.15 K (100 °C). To put it another way, absolute zero Kelvin is at −273.15 °C. For our purposes, it will be adequate to say ice melts at 273 K. Room temperature is about 20 °C or about 293 K.

To recapitulate, the greater the dependence of the entropy on the energy for an object (the steeper the slope of S vs. E_{int}), the more "eager" the object is to take in energy to contribute to increasing the total entropy of the total system. It will do this if it can take energy from another object that has a smaller dependence of entropy on energy (smaller slope), since the second object's entropy will decrease less than the first object's entropy will increase. An object with a large dS/dE_{int} has a low temperature, since it is low-temperature objects that take in energy from high-temperature objects.

Conversely, the smaller the dependence of the entropy on the energy, the less reluctant the object is to give up energy and decrease its own entropy, if it can give the energy to an object with a greater dependence of entropy on energy. An object with a small dS/dE_{int} has a high temperature, since it is high-temperature objects that donate energy to low-temperature objects.

At this point it is useful to do Problem P63, in which you calculate the temperature of an object as a function of the amount of energy in the object.

W, Q, and Entropy

More formally, we should write $1/T = \partial S/\partial E_{\text{int}}$, involving a "partial derivative," which means that we hold everything but S and E constant when we take the derivative. In particular, we hold the volume constant, which means that we do no work on the system to compress it or stretch it, so we should write $1/T = (\partial S/\partial E_{\text{int}})_V$, where the subscript V means "take the (partial) derivative holding the volume constant."

One might ask whether there is any entropy change of the surroundings associated with energy exchange in the form of work. The answer turns out to be no for processes that are reversible (there do exist irreversible processes in which work can be associated with an increase in entropy). Energy transfer caused by a temperature difference between objects in contact, Q, is "disorganized" energy transfer and is associated with entropy change. Work is "organized" energy transfer and in many situations doesn't affect the entropy. It is beyond the scope of this textbook to prove this rigorously, but we can illustrate the basic issues in terms of an Einstein solid.

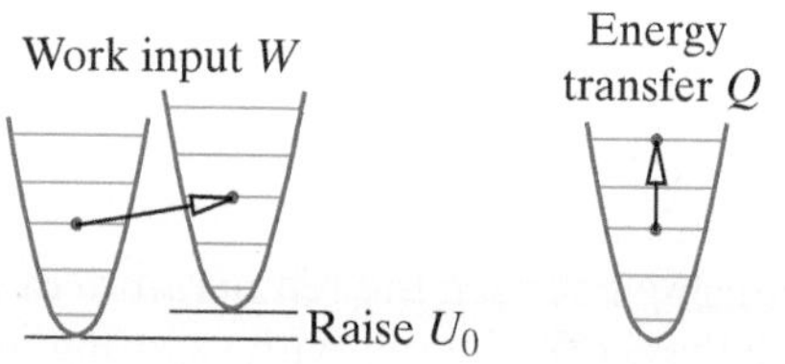

Figure 12.31 Mechanical work (compression or stretching of an oscillator system) compared to energy transfer due to a temperature difference to the same system.

If you mechanically squeeze or compress a block, doing work on it to raise its energy, it can be shown that you shift the energy levels upward without changing which state an oscillator is in, as shown in Figure 12.31. The "spring" stiffnesses don't change, but there is more energy stored in the "springs." On the other hand, if there is transfer of energy into the system due to a temperature difference, Q, there is no change in the energy level, but an oscillator will jump to a higher level. This jump from one level to another affects the probability calculations and the entropy, which is the natural log of the number of ways to arrange the energy quanta.

Roughly speaking, transfer of energy Q due to a temperature difference alters which state the system is in and affects the entropy. Work W alters the energy levels without changing which level the system is in, and doesn't affect the entropy.

Transfer of energy into a system due to a temperature difference, of amount Q, with $\Delta E_{int} = Q$, leads to an entropy change of that system. Since $1/T = \Delta S/\Delta E_{int} = \Delta S/Q$, we conclude that the entropy change associated with an input Q is $\Delta S = Q/T$ in the case that T doesn't change very much. If Q is large, causing a large temperature change in a system, we must add up (integrate) the contributions to the entropy change.

To repeat, there do exist irreversible processes in which work can be associated with an increase in entropy (for example, in rubbing two blocks together, with friction), so entropy change due to Q is only part of the story.

Checkpoint 7 There was transfer of energy of 5000 J into a system due to a temperature difference, and the entropy increased by 10 J/K. What was the approximate temperature of the system, assuming that the temperature didn't change very much?

Can the Entropy of an Object Decrease?

QUESTION We analyzed two blocks that initially had a 90-10 energy distribution. In the process of changing to a 60-40 distribution, was it possible for the entropy of *one* of the blocks to decrease significantly? Is this a violation of the second law of thermodynamics?

The entropy of block 1 decreased, but the entropy of block 2 increased even more. We believe that the Universe is a closed system, and that the total entropy of the Universe continually increases and never decreases. However, some portions of the Universe may experience a decrease of entropy, as long as there is at least as much increase elsewhere.

Irreversibility Again

We started with a 90-10 distribution of the energy but found the most probable final distribution to be 60-40. This change is effectively irreversible. The 60-40 distribution is so enormously more probable than the 90-10 arrangement that an observer will essentially *never* see a 90-10 distribution spontaneously recur. Closed macroscopic systems (such as two blocks inside an insulating box) always evolve toward the most probable arrangement, and stay there (or so close that you can hardly tell the difference).

12.6 SPECIFIC HEAT OF A SOLID

In Problem P63 you calculated the relationship between the temperature of a block and its energy. How can we compare these calculations with experiment, when we don't have a good way to measure the total energy in a solid block? It is, however, possible to measure *changes* in energy and temperature. Suppose that we add a known amount of energy ΔE to a system, and we measure the resulting rise in temperature ΔT. The ratio of the internal energy change to the temperature rise of an object is called the heat capacity of that object. If we express this on a per-gram or per-mole or per-atom basis, it is called "specific heat." We will work with specific heat on a per-atom basis:

$$C = \frac{\Delta E_{\text{atom}}}{\Delta T}$$

where

$$\Delta E_{\text{atom}} = \frac{\Delta E_{\text{system}}}{N_{\text{atoms}}}$$

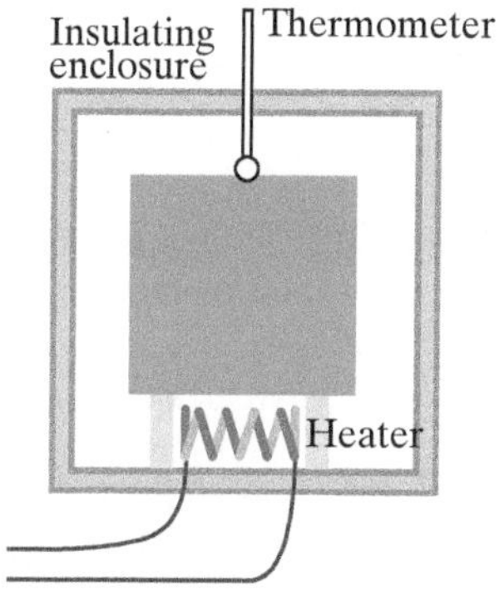

Figure 12.32 Schematic diagram of an apparatus for measuring the heat capacity of a solid.

Different materials have different specific heats. For example, at room temperature the specific heat of water (on a per-gram basis) is 4.2 J/K/g, while the specific heat of iron at room temperature is 0.84 J/K/g. Specific heat is an important property of a material for two reasons. It has practical consequences in science and engineering because it, along with the thermal conductivity, determines the thermal interactions that the material has with other objects. Also, specific heat is an experimentally measurable quantity that we can compare with theoretical calculations, to test the validity of our models.

How would we measure the heat capacity of a block? One way is to enclose the block inside an insulating box and use an electric heater inside the box to raise the temperature of the block (Figure 12.32). The electric power (in watts) times the amount of time the current runs (in seconds) gives us the input energy ΔE (in joules), which is equal to the internal energy change ΔE_{int}, and we can use a thermometer to measure the temperature rise ΔT of the block. The specific heat (on a per-atom basis) is $\Delta E_{\text{atom}}/\Delta T = (\Delta E_{\text{block}}/\Delta T)/N_{\text{atoms}}$. The heater itself should have a small mass compared to the sample of material whose heat capacity we're measuring, so that the heater isn't a significant part of the material that is being warmed up. Problem P32 illustrates another way to measure specific heat, on a per-gram basis.

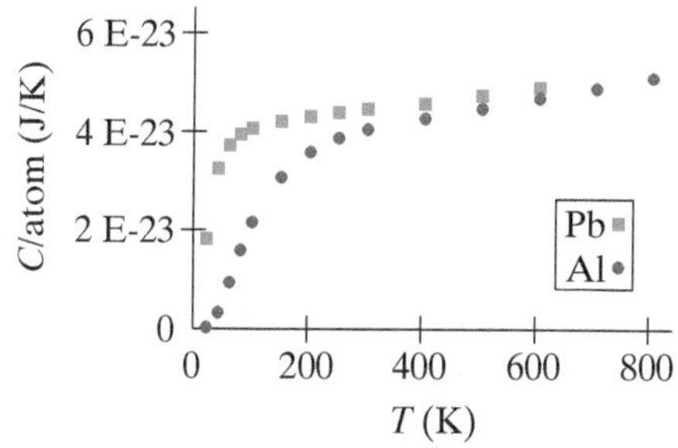

Figure 12.33 Measured heat capacities of lead and aluminum, shown on a per-atom basis.

Experimental Results for Specific Heat

Measurements of the specific heat of aluminum (circles) and lead (squares) are displayed in Figure 12.33 (the numerical data are given in a table accompanying Problem P64). They are plotted as heat capacity (J/K) per atom to facilitate comparison. Note that at low temperatures the specific heat of both substances changes dramatically with temperature. It is not initially obvious why this should be the case, and in fact this temperature dependence was not predicted by classical theory developed in the 1800s. Note also that at room temperature and above, the specific heat (on a per-atom basis) of both materials is about the same.

Will our simple statistical model of a solid be able to predict the temperature dependence of the specific heat of a solid? In Problem P63 you determined the relationship between energy and temperature for a block (in fact, you did the calculation for two different blocks). This calculation can easily be extended to predict the heat capacity per atom, which is $\Delta E_{\text{atom}}/\Delta T$, where E_{atom} is the average energy per atom.

Effective Spring Stiffness

In Chapter 4 we determined the interatomic spring stiffness $k_{s,i}$ of aluminum and lead by relating it to the macroscopic stress–strain relationship (Young's modulus). In Problem P64 you will use the Einstein independent-oscillator model of a solid to predict the specific heat of aluminum and lead. We need to use a value for $k_{s,i}$ in this model, in order to convert the number of quanta to a value for energy in joules. Presumably the spring stiffness $k_{s,i}$ that we use in $\hbar\sqrt{k_{s,i}/m_a}$ should be related to the spring stiffness $k_{s,i}$ determined from Young's modulus. However, we shouldn't expect our

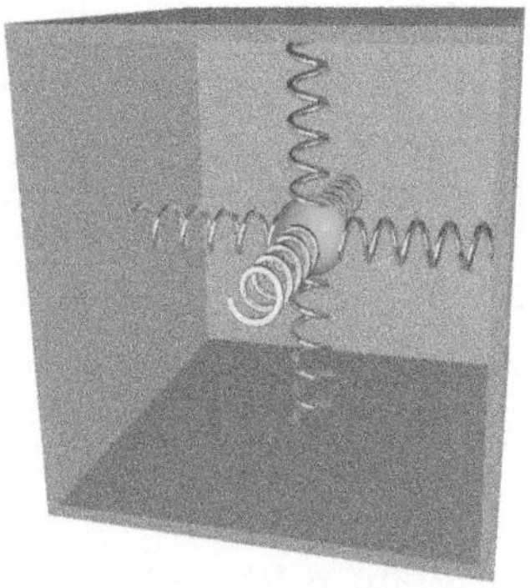

Figure 12.34 In our simple model of a solid, each atom experiences forces from two interatomic "springs." Each of these springs is half as long as an interatomic bond.

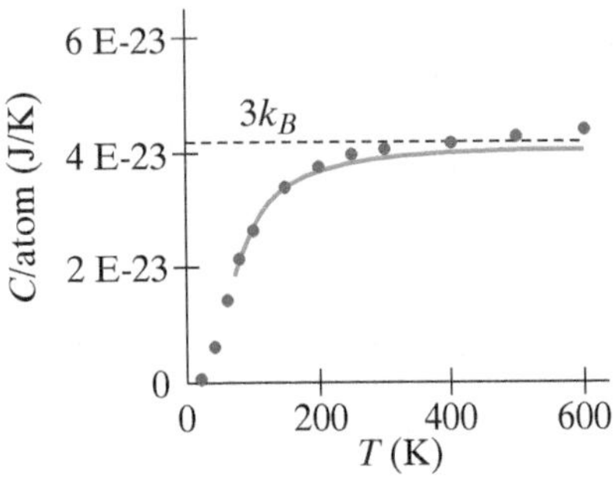

Figure 12.35 Computed specific heat of copper, compared with experimental values.

predictions to agree exactly with experimental data, because our model ignores the fact that the atoms are not actually isolated from each other.

Moreover, the effective spring stiffness for oscillations in the x, y, or z directions can be expected to be larger than the value of $k_{s,i}$ estimated in Chapter 4 for two reasons.

First, each atom is attached to two "springs" (Figure 12.34). When an atomic core is displaced a distance s_x from its equilibrium position, the spring to the left and the spring to the right each exert a force of $k_{s,i}s_x$, so the combined force is $2k_{s,i}s_x$. This implies that the "effective spring stiffness" for oscillations would be $2k_{s,i}$.

Second, in this simple model we divide the solid into "cubes" surrounding each independent atomic oscillator (Figure 12.34), so each of the two springs to the left and right of an atomic core is effectively half the length of a full interatomic spring, and each half-spring would have a spring stiffness of $2k_{s,i}$.

These considerations suggest that the effective spring stiffness for thermal oscillations in the context of the Einstein model might be a factor of 4 larger than the spring stiffness we determined in Chapter 4 from stretching a wire.

We emphasize that the Einstein model is clearly deficient, since an atom is not actually enclosed in rigid walls. However, having chosen this model we follow through in a consistent way, including using a spring stiffness k_s appropriate to the model. There are more sophisticated models for determining specific heat that fit experimental data even better than the Einstein model does, but the Einstein model captures the main features.

Specific Heat as a Function of Temperature

Figure 12.35 displays the result of a calculation of the specific heat of copper on a per-atom basis. The colored dots represent actual experimental data, and the solid line represents the computed values. An outline of the calculation that produced this curve is as follows:

- Use the mass of a copper atom and the effective interatomic spring stiffness for copper, calculated in Chapter 4 and modified by the factor discussed above, to find the value of one quantum of vibrational energy for this system.
- For a small particle of solid copper containing a given number of atoms, calculate the entropy of the system for increasing values of energy added to the system, in amounts of one quantum (as in Problem P62).
- Calculate the change in entropy due to the addition of each quantum of energy, and use the definition of temperature ($1/T \approx \Delta S/\Delta E$) to find the temperature of the particle for each value of energy (as in Problem P63).
- Calculate the change in temperature of the particle due to the addition of each quantum of energy, to find the specific heat ($C \approx \Delta E/\Delta T$), and plot values of C (per atom) versus T (in kelvins) as in Problem P64.

Our simple model of a solid as a collection of independent harmonic oscillators does a surprisingly good job of fitting experimental data over a wide range of temperatures. If we were able to include more oscillators in our calculations (by using double-precision arithmetic to handle larger numbers, or by using mathematical techniques such as Stirling's approximation for factorials), we could extend our predictions to lower temperatures.

The deviation of the experimental data from the prediction at very high temperatures suggests that at these temperatures a simple harmonic oscillator is not a good model of the atoms in this solid. At high quantum levels, the harmonic oscillator potential energy curve is a poor approximation to the actual potential energy curve describing interatomic interactions (see Chapter 7).

Note an interesting aspect of the graph: at room temperatures, the predicted specific heat on a per-atom basis approaches a constant value of $3k_B$ (three times the Boltzmann constant, 1.38×10^{-23} J/K). This value agrees quite well with the measured specific heat of a variety of substances at ordinary temperatures.

At this point it is important to do Problem P64, in which you compute the specific heat of aluminum and lead, and compare with actual experimental data.

Energy Quantization and Specific Heat

The key difference between our model of solids (the Einstein model) and earlier classical models that did not predict a temperature dependence of specific heat lies in the discrete energy levels of atomic oscillators. Statistical mechanics was originally developed in the 1800s, before the beginning of quantum theory. The classical theory did predict that at high temperatures the energy of each atom in a solid would be $3k_BT$ (k_BT per nonquantized one-dimensional oscillator), so that the heat capacity per atom would be $3k_B$, and this is discussed later in this chapter.

If you examine your calculations, you'll find that the high-temperature limit (heat capacity per atom $= 3k_B$) corresponds to temperatures high enough that k_BT is significantly larger than one quantum of energy. In a situation where the average energy per one-dimensional oscillator is about k_BT and is large compared to one quantum of energy, the quantization of the energy doesn't make much difference in the analysis. That is, if energy quanta are small compared to the energies of interest, mathematical analysis can be carried out adequately by considering energy to be continuous rather than discrete. In such cases pre-quantum and quantum calculations will give the same results, as is the case here at high temperatures.

On the other hand, at low temperatures the average energy per one-dimensional oscillator is comparable to or smaller than one quantum of energy, and the continuous, nonquantum calculations are not valid. The classical theory provided no explanation for the discovery that as materials were cooled down to very low temperatures, the specific heat decreased with decreasing temperature.

In 1907 Einstein carried through the analysis we have just done and predicted the curve we have just plotted. The good agreement with both low-temperature *and* high-temperature measurements of specific heat was strong additional evidence for the hypothesis that the energy levels of oscillators are indeed discrete. (At extremely low temperatures the model of a solid must be refined, taking into account the electrons in the metal, among other things, to achieve full agreement between theory and experiment.)

The fact that the specific heat for all materials decreases at low temperatures has practical consequences. For example, it makes it difficult to cool a sample to a very low temperature. Cooling a sample depends essentially on putting the sample in contact with a "sink," a large object that is already at a lower temperature, so that there is transfer of energy Q due to the temperature difference out of the sample into the sink, lowering the temperature of the sample and not raising the temperature of the sink very much. At very low temperatures, however, the sink has a low specific heat, so this is difficult to achieve.

Which of Our Results Are General?

We have analyzed simple models of solid matter. Nevertheless, the basic conclusions are quite general. For example, if our two model blocks were made of different materials, so that the energy quanta were of different size in the two blocks, this would complicate the procedures for evaluating the number of ways Ω to arrange the energy, but the basic conclusion would remain—that the entropy will increase to a maximum.

EXAMPLE **A Lead Nanoparticle**

In Chapter 4, from Young's modulus for lead we found that the effective interatomic "spring" stiffness was about 5 N/m. **(a)** For a nanoparticle consisting of 3 lead atoms, what is the approximate temperature when there are 5 quanta of energy in the nanoparticle? **(b)** What is the approximate specific heat (per atom) at this temperature? Compare with the approximate high-temperature specific heat (per atom) for lead.

Solution We'll use the Einstein model of a solid, though with only three atoms our conclusions will be very approximate.

(a) We model the three atoms as 9 independent quantized oscillators. One quantum of energy in one of these oscillators is this many joules:

q	Ω	$\ln \Omega$
4	$\frac{12!}{4!8!} = 495$	6.20
5	$\frac{13!}{5!8!} = 1287$	7.16
6	$\frac{14!}{6!8!} = 3003$	8.01

Figure 12.36 Calculation of entropy ($S = k_B \ln \Omega$).

$$\hbar\omega_0 = \hbar\sqrt{k_s/m} = (1.05 \times 10^{-34}\,\text{J}\cdot\text{s})\sqrt{\frac{4(5\,\text{N/m})}{(207 \times 1.7 \times 10^{-27}\,\text{kg})}}$$

$$= 7.92 \times 10^{-22}\,\text{J}$$

We need to calculate the number of ways to arrange q quanta in the neighborhood of $q = 5$, and the associated entropy (Figure 12.36).

Since $1/T = \partial S/\partial E$, $T \approx \Delta E/\Delta S$. Take differences from $q = 4$ to $q = 6$, in order to approximate the slope at $q = 5$; energy increase is 2 quanta:

$$T \approx \frac{N\hbar\omega_0}{k_B \Delta(\ln \Omega)} = \frac{2(7.92 \times 10^{-22}\,\text{J})}{(1.38 \times 10^{-23}\,\text{J/K})(8.01 - 6.20)} \approx 63.7\,\text{K}$$

(b) Find T_1 at the midpoint of the 4 to 5 interval, T_2 at the midpoint of the 5 to 6 interval; energy increase from T_1 to T_2 is one quantum:

$$T_1 \approx \frac{(7.92 \times 10^{-22}\,\text{J})}{(1.38 \times 10^{-23}\,\text{J/K})(7.16 - 6.20)} \approx 60.1\,\text{K}$$

$$T_2 \approx \frac{(7.92 \times 10^{-22}\,\text{J})}{(1.38 \times 10^{-23}\,\text{J/K})(8.01 - 7.16)} \approx 67.7\,\text{K}$$

As expected, these two temperatures bracket the temperature of 63.7 K at $q = 5$. We want the specific heat on a per-atom basis, and there are 3 atoms:

$$C_{\text{per atom}} = \frac{1}{3}\frac{\Delta E}{\Delta T} \approx \frac{1}{3}\frac{(7.92 \times 10^{-22}\,\text{J})}{(67.7 - 60.1)\,\text{K}} \approx 3.44 \times 10^{-23}\,\text{J/K}$$

The approximate high-temperature specific heat for a solid is $3k_B$ per atom, which is $3(1.38 \times 10^{-23}\,\text{J/K}) = 4.14 \times 10^{-23}\,\text{J/K}$. This suggests that at a temperature of about 63.7 K, lead is not quite at the high-temperature limit.

The actual experimental value for lead at 63.7 K (interpolation of the data accompanying Problem P64) is 22.6 J/K/mole, which on a per-atom basis is

$$(22.6\,\text{J/K/mole})\left(\frac{1\,\text{mole}}{6 \times 10^{23}\,\text{atoms}}\right) = 3.8 \times 10^{-23}\,\text{J/K}$$

Our three-atom calculation is not a very accurate model of a macroscopic block of lead, in part because with only three atoms it is not a very good model to say that each atom is connected by spring-like bonds to six neighboring atoms. Despite the failings of our simple model, our result is rather close to the actual value.

12.7 COMPUTATIONAL MODELS

Calculating entropy, energy, and specific heat by hand is possible but tedious. In order to be able to vary the number of atoms in the objects, their elemental composition, and the amount of energy available to the system, it is easier to use a computer.

The structure of the programs we will write to do statistical calculations is very similar to that of the programs we wrote to predict motion. Constants and initial values are set at the beginning of the program, and then calculations are done iteratively inside a while loop. The two most important differences are these:

- In models of motion, time is the variable that increases each time the loop repeats. In the statistical calculations you will do, the amount of energy in the system (or in a particular part of the system) is the variable that increases.
- In models of motion, the calculations done inside the loop involve forces and fundamental principles, and predict motion. In statistical models, calculations inside the loop involve calculating the number of ways of arranging energy among a specific number of independent quantum oscillators (three for each atom).

Combinations, Logs, and Bar Graphs in VPython

To get the `factorial` and `combin` functions, and be able to graph results, insert the following statements at the beginning of your program. These statements *must* be entered in the order shown:

```
from visual import*
from visual.graph import*
from visual.factorial import*
```

You will need to use the following functions:

- `combin()`. This function takes two arguments:

  `combin(a,b)` $= \dfrac{a!}{b!(a-b)!}$

  If we want to calculate the number of ways to arrange q quanta of energy among N oscillators, we can let $a = q + N - 1$ and $b = q$. Then $(a - b) = (N - 1)$, and

  `combin(q+N-1,q)` $= \dfrac{(q+N-1)!}{q!(N-1)!}$

- `log()`. In Python, the natural logarithm $ln(x)$ is written `log(x)`. The function to find the base 10 logarithm is `log10(x)`.

- `gvbars()`: To make a vertical bar graph, create a `gvbars` object. The parameter `delta` determines the width of a single bar:

  ```
  mybars = gvbars(color=color.red, delta=0.3)
  ```

  This code displays a vertical bar at location x:

  ```
  mybars.plot(pos=(x,height))
  ```

In addition, VPython does have a `factorial()` function, but the largest number whose factorial it can compute is 170. It is safer simply to use the `combin()` function, which does not have this limitation. `b = factorial(9)` will set b to $9! = 9 \cdot 8 \cdot 7 \cdot 6 \cdot 5 \cdot 4 \cdot 3 \cdot 2 \cdot 1 = 362880$. As is standard, `factorial(0)` is 1.

Computational Explorations

In Problems P61–P64 at the end of this chapter, you will apply the ideas discussed in this chapter to create computational models of nanoparticles—solid particles containing up to a few hundred atoms. Problems P61–P63 involve modeling two nanoparticles in thermal contact with each other.

The main part of such a program, referring to two nanoparticles labeled 1 and 2, might look like this:

```
Ntotal = 6 # total number of oscillators
N1 = 3 # number of oscillators in object 1
N2 = Ntotal-N1 # number of oscillators in object 2
qtotal = 4 # total quanta of energy available
q1 = 0 # initially 0 quanta in object 1

while q1 <= qtotal: # for each possible value of energy in 1
    q2 = ?? # number of quanta of energy in 2
    # Calculate number of ways to arrange q1 quanta in 1:
    ways1 = ?
    # Calculate number of ways to arrange q2 quanta in 2:
    ways2 = ?
    # Plot number of ways to arrange energy in both:
    waygraph.plot( pos=(q1,??) )
    q1 = q1+1
```

Since $\Delta S/\Delta E \approx 1/T$, to calculate one temperature we need two values of S. One way to do this inside a calculation loop is to "look ahead," like this:

```
while q1 < qtotal:
    # For each delta_S you need two values of S
    # so use q1, and q1a=(q1+1), etc.
    q2 = qtotal-q1
    q1a = q1+1
    q2a = ?
    ways1 = ?
    ways2 = ?
    ways1a = ?
    ways2a = ?
    # etc.
```

In Problem P64, you will modify your model to consider a single nanoparticle and will use the definitions of entropy, temperature, and specific heat to predict the specific heat of two different real substances (aluminum and lead) as a function of temperature. You'll compare the predictions of your model with published experimental data for these quantities.

12.8 THE BOLTZMANN DISTRIBUTION

How does the density of Earth's atmosphere change as altitude increases? How does the distribution of the speeds of molecules in a gas change as the temperature of the gas increases? So far we have mainly been concerned with the thermal equilibrium of two blocks, and how thermal equilibrium arises as that particular distribution of energy between the two blocks that has (by far) the largest number of ways to arrange the quanta. What can we say about the probability of observing a particular amount of energy associated with one particular atom or oscillator? Addressing this question will lead us to the "Boltzmann distribution," which provides insight into the behavior of a very wide variety of physical, chemical, and biological phenomena.

A Constant-Temperature Reservoir

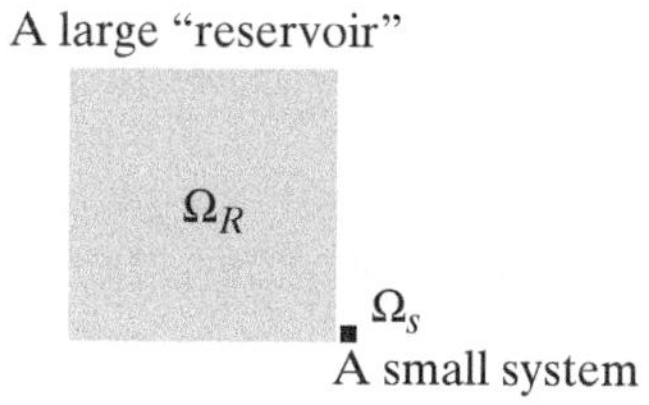

Figure 12.37 A small system in contact with a very large system (a "reservoir").

Consider a large system in contact with a small system, as shown in Figure 12.37. The two systems are isolated from their surroundings and share a fixed amount of energy E_{tot}. The large system's energy and temperature cannot change very much, so we call the large system a nearly constant-temperature "reservoir," which keeps the small system always at nearly the same temperature.

The entropy of the total system (reservoir plus small system) will increase rapidly if energy is transferred from the small system to the reservoir, because in the large system that energy can be distributed among a very much larger number of microstates than were available in the small system. Since we expect the total entropy to increase, we expect that most of the time, most of the total energy E_{tot} will be found in the reservoir, and little energy in the small system. We will show how to make this idea quantitative.

Let $\Omega_{\text{res}}(E_{\text{res}})$ be the number of microstates in the reservoir when the amount of energy in the reservoir is E_{res}. Similarly, let $\Omega(E)$ be the number of microstates in the small system when it has an amount of energy E, where E is much less than E_{res}, because the reservoir is so big. The total energy of the combined system is $E_{\text{tot}} = E_{\text{res}} + E$, which is a fixed number because the two systems are isolated from their surroundings. The total number of microstates for the combined system corresponding to E_{tot} is $\Omega_{\text{tot}}(E_{\text{tot}})$, which is also a fixed number.

Probability of a Particular Division of Energy

The number of ways of arranging E_{res} in the reservoir and E in the small system is $\Omega_{\text{res}}(E_{\text{res}})\Omega(E)$, while the total number of ways of arranging E_{tot} in the combined system is $\Omega_{\text{tot}}(E_{\text{tot}})$. Therefore the probability $P(E)$ of finding the energy split between the reservoir and the small system so that there is energy E in the small system is this:

$$P(E) = \frac{\Omega_{\text{res}}(E_{\text{res}})\Omega(E)}{\Omega_{\text{tot}}(E_{\text{tot}})}$$

The most probable value of the energy E to be found in the small system is zero, because the more energy E that is taken away from the big system and put into the small system, the fewer the microstates in the big system (without a comparable increase in the small system), which would mean a decrease in the total entropy.

We're interested in how fast $P(E)$ decreases as we move more and more of the total energy into the small system. We take logarithms and multiply by k_B to express everything in terms of entropy, then see how the expression varies with E:

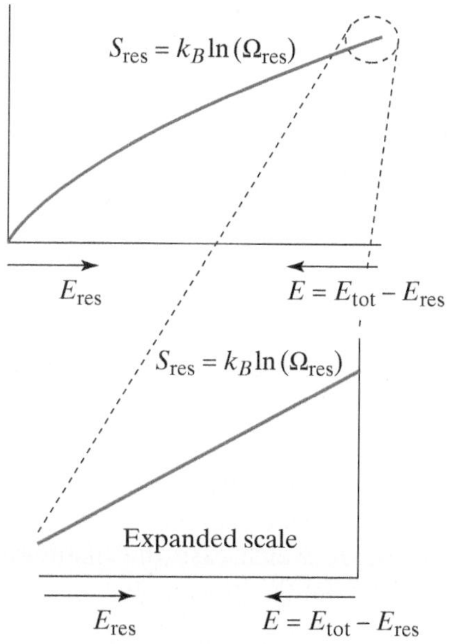

Figure 12.38 Entropy (in units of k_B) as a function of E_{res}, for the larger object.

$$k_B \ln P(E) = k_B \ln(\Omega_{\text{res}}(E_{\text{res}})) + k_B \ln(\Omega(E)) - k_B \ln(\Omega_{\text{tot}}(E_{\text{tot}}))$$

Consider the term $k_B \ln(\Omega_{\text{res}}(E_{\text{res}}))$, which is the entropy $S_{\text{res}}(E_{\text{res}})$ of the reservoir when the energy in the reservoir is $E_{\text{res}} = E_{\text{tot}} - E$ due to some energy E having gone into the small system. In Figure 12.38 we show the familiar calculation of the entropy of the reservoir as a function of the energy in the reservoir. On an expanded scale we show a portion of a plot of the entropy vs. E_{res} near the maximum possible value of E_{res}, which is the total energy E_{tot} of the combined system.

We need to evaluate the entropy of the reservoir for small values of $E = (E_{tot} - E_{res})$. We can obtain this by the following argument. The nearly straight line on the graph has a slope dS_{res}/dE_{res} and goes through $S_{res}(E_{tot})$, where the energy in the reservoir is E_{tot}. Therefore in this region $S_{res}(E_{res})$ can be represented by

$$S_{res}(E_{res}) = S_{res}(E_{tot}) - \frac{dS_{res}}{dE_{res}}E$$

The entropy of the reservoir decreases as more energy E is shifted into the small system.

However, $dS_{res}/dE_{res} = 1/T$, where T is the temperature of the reservoir, and the fact that the slope is nearly constant in this region reflects the fact that removing a small amount of energy from this large object hardly changes its temperature. The large object is a nearly constant-temperature reservoir, which keeps the small object at that temperature.

Making the substitution $dS_{res}/dE_{res} = 1/T$, we have

$$S_{res}(E_{res}) = k_B \ln(\Omega_{res}(E_{res})) = S_{res}(E_{tot}) - \frac{E}{T}$$

Substitute this into the equation for $k_B \ln P(E)$ and then divide by k_B:

$$\ln P(E) = \frac{S_{res}(E_{tot})}{k_B} - \frac{E}{k_B T} + \ln(\Omega(E)) - \ln(\Omega_{tot}(E_{tot}))$$

Note that the terms that don't involve E are all constants because E_{tot} is constant. Take the exponential of this equation—that is, raise e to both sides of the equation:

$$e^{\ln P(E)} = e^{\text{constant}} e^{\ln(\Omega(E))} e^{-E/k_B T}$$

Since e raised to the natural logarithm of some quantity is that quantity, we have the following, where A is some constant:

$$P(E) = A\Omega(E)e^{-E/k_B T}$$

We have calculated the probability of finding a small amount of energy E in a small system that is in contact with a large reservoir. This is called the Boltzmann distribution in honor of the Austrian physicist who developed statistical mechanics in the 19th century:

THE BOLTZMANN DISTRIBUTION

The probability of finding energy E in a small system in contact with a large reservoir is proportional to

$$\Omega(E)e^{-E/k_B T}$$

The exponential part, $e^{-E/k_B T}$, is called the "Boltzmann factor." $\Omega(E)$ is the number of microstates corresponding to energy E.

This result is very general and applies to a very wide variety of phenomena. In many situations the number of microstates $\Omega(E)$ changes much more slowly with energy than does the Boltzmann factor, in which case the qualitative behavior of the system can be determined just from the exponential.

QUESTION What is the most likely value of E, the energy to be found in a microscopic system that is in thermal equilibrium with a large system, if the Boltzmann factor is the important factor? Are you likely to find an energy that is much larger than $k_B T$?

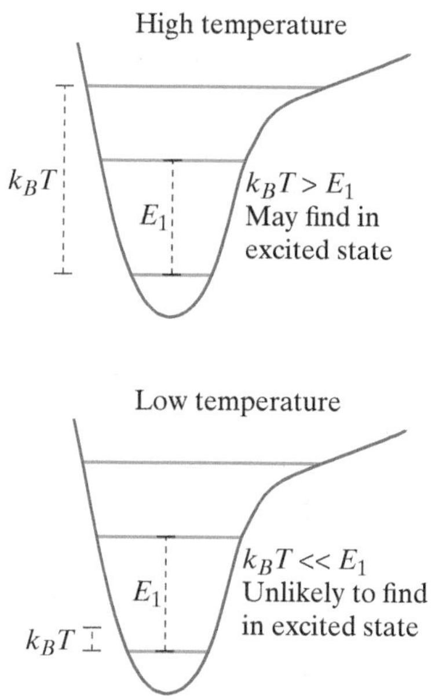

Figure 12.39 At a low temperature it is unlikely to find a system in an excited state, but at a high temperature the likelihood increases.

Since the exponent is negative, the exponential factor is largest for $E = 0$, so you're most likely to find the system in its ground state. This agrees with our expectation that taking energy out of the reservoir and putting it in the small system will reduce the number of total microstates and represent a decrease in entropy for the combined system, which is unlikely.

Since e^{-1} is 0.37, the probability of the energy being much larger than k_BT is rather small.

If the temperature is so low that k_BT is small compared to the energy E_1 of the first excited state, you are unlikely ever to find the system in one of its excited states. The system is thermally inert, because the probability of taking in *any* amount of energy is very small. In contrast, if the temperature is high enough that $k_BT > E_1$, then you will sometimes find the system in one of its excited states, although it is still true that the system is most likely to be found in its ground state (Figure 12.39).

The Boltzmann distribution has far-reaching consequences. For example, chemical and biochemical reaction rates typically depend strongly on temperature because with higher temperature the reactants are moving faster and may be found in excited states. Physical reaction rates are also affected. At the very high temperatures found in the interior of our Sun, kinetic energies are high enough to overcome the electric repulsion between nuclei and to allow the nuclei to come in contact, so that they can undergo thermonuclear fusion reactions. A gas becomes a plasma if the temperature is so high that k_BT is comparable to the ionization energy.

Checkpoint 8 At room temperature, show that $k_BT \approx \frac{1}{40}$ eV. It is useful to memorize this result, because it tells a lot about what phenomena are likely to occur at room temperature.

Many Observations vs. Many Systems

If a microscopic system (such as a single oscillator) is in contact with a large system, the probability of finding a particular amount of energy in the microscopic system is governed by the Boltzmann distribution.

One can think about this in two ways. The first is to imagine measuring the energy in this particular system repeatedly, over time, and recording the results. The fraction of the results that indicate a particular energy is predicted by the Boltzmann distribution.

A second way to think about this is to imagine assembling a large number of identical systems, each in contact with a large system, and taking one "snapshot" in which the energy of each system is recorded simultaneously. The fraction of systems with a particular energy is predicted by the Boltzmann distribution.

The second approach also gives us a way to predict the distribution of energy in a single large system. For example, consider a system consisting of all the air molecules in a room. Imagine that each molecule is identified by a letter: A, B, C ... One could consider molecule A to be a microscopic system in contact with the large system consisting of all other molecules (B, C, ...). However, one could also consider B to be the microscopic system, in contact with all other molecules (A, C, ...). We can use the Boltzmann distribution to predict the fraction of all molecules in the room that have a particular energy.

These two views of the Boltzmann distribution, one microscopic system observed repeatedly or a large number of microscopic systems observed once, complement each other. Sometimes one view is more helpful, sometimes the other.

12.9 THE BOLTZMANN DISTRIBUTION IN A GAS

The Boltzmann distribution applies to any kind of system—not just a solid. As a major application of the Boltzmann distribution, we will study a gas consisting of molecules that don't interact much with each other. Examples are the so-called "ideal gas" (with no interactions at all), and any real gas at sufficiently low density that the molecules seldom come near each other.

In order to apply the Boltzmann distribution, we need an equation for the energy of a molecule in the gas. We will omit rest energy, and we will also omit nuclear energy and electronic energy from our total, because at ordinary temperatures there is not enough energy available in the surroundings to raise the molecule to an excited nuclear state or an excited electronic state. Also, for simplicity, instead of writing ΔE_{vib} to represent an amount of vibrational energy above the ground vibrational state, we will simply write E_{vib}.

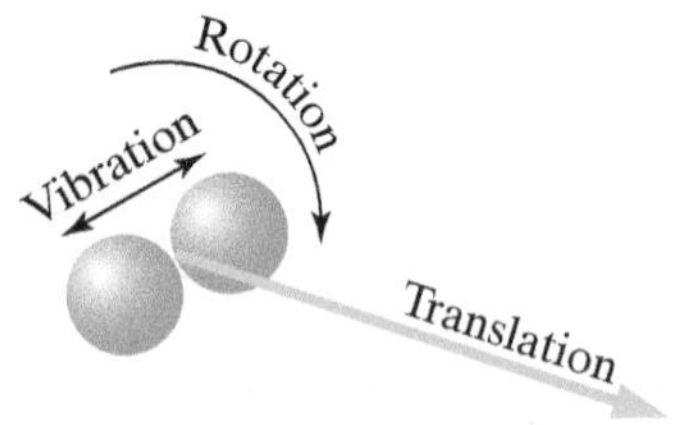

Figure 12.40 Energy of an oxygen molecule.

The energy of a single gas molecule in the gravitational field near the Earth's surface (excluding rest energy, nuclear energy, and electronic energy) is this:

$$K_{\text{trans}} + E_{\text{vib}} + E_{\text{rot}} + Mgy_{\text{CM}}$$

where E_{vib} and E_{rot} are the vibrational and rotational energies relative to the center of mass (if the molecule contains more than one atom, as in Figure 12.40) and y_{CM} is the height of the molecule's center of mass above the Earth's surface. The mass of the molecule is M. For brevity we will speak of "the energy of the gas molecule," but of course the gravitational potential energy really applies to the system of gas molecule plus Earth.

> QUESTION For this expression for the energy to be accurate, why must the gas be an "ideal" gas (or a real gas at low density)?

In a real gas at somewhat high density we may not be able to neglect the potential energy associated with the intermolecular forces, in which case this expression for the energy would not be adequate.

If the temperature T is the same everywhere in the (ideal) gas, the probability that a particular molecule will have a certain amount of energy is proportional to

$$\Omega(E)e^{-(K_{\text{trans}}+E_{\text{vib}}+E_{\text{rot}}+Mgy_{\text{CM}})/k_BT}$$

We mentally divide the gas into two systems: one particular molecule of interest and all the rest of the molecules. These two systems are in thermal equilibrium with each other, because the gas molecules are continually colliding with each other and can share energy. The energy for the one particular molecule is then expected to follow the Boltzmann distribution.

To avoid excess subscripts, in the following discussion we will simply write v for the center-of-mass speed v_{CM} and y for the center-of-mass height y_{CM}.

Separating the Various Factors

It is useful to group the terms of the Boltzmann factor like this:

$$\left[e^{-K_{\text{trans}}/k_BT}\right]\left[e^{-Mgy/k_BT}\right]\left[e^{-E_{\text{vib}}/k_BT}\right]\left[e^{-E_{\text{rot}}/k_BT}\right]$$

The first bracket is associated with the distribution of velocities, the second with the distribution of positions, the third with the distribution of vibrational energy, and the fourth with the distribution of rotational energy. We will discuss each of these individually.

Height Distribution in a Gas

A striking property of Earth's atmosphere is that in high mountains the air density and pressure are significantly lower than at sea level. The air density is so low on top of Mount Everest that most climbers carry oxygen tanks. Can we explain this? In order to get at the main issue, the variation of density with height, we make the rough approximation that the temperature is constant—the same in the mountains as at sea level.

> **QUESTION** How bad an approximation is it to consider the temperature to be the same at all altitudes?

Even in high mountains the temperature is typically above –29 °C (–20 °F, which is 244 K, and this is only 17% lower than room temperature of 293 K (20 °C or 68 °F). So maybe this approximation isn't too bad.

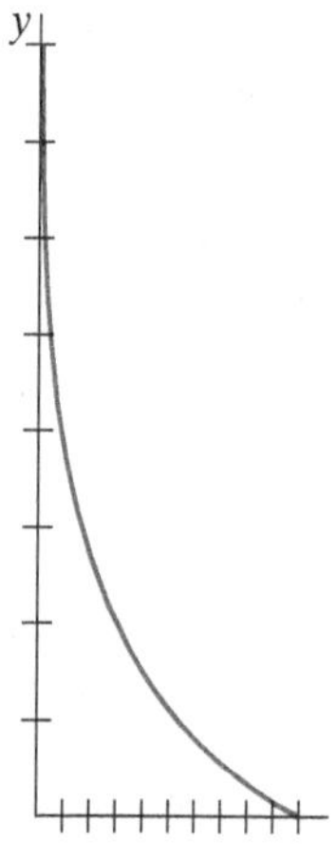

Figure 12.41 Number density vs. height in a constant-temperature atmosphere.

We speak of the probability that the x component of the gas molecule's position lies between some x and $x+dx$, and similarly for y and z. Here dx is considered to be a very short distance, small compared to the size of the container but large compared to the size of a molecule.

Focus just on that part of the Boltzmann distribution that deals with position, where the distribution is proportional to the probability of finding one particular molecule between x and $x+dx$, y and $y+dy$, and z and $z+dz$:

$$e^{-Mgy/k_BT}dx\,dy\,dz$$

Evidently there's nothing very interesting about x and z (directions parallel to the ground). However, in the vertical direction there is an exponential fall-off with increasing height for the probability of finding a particular molecule at height y (Figure 12.41).

Looked at another way, our exponential equation tells us how the number density of the atmosphere depends on height, because in telling us about the behavior of one representative molecule, the equation also tells us something about all the molecules.

We shifted gears from thinking about one molecule to thinking about many. Think again about one single air molecule. Suppose that you place it on a table, which is at room temperature. It is not a very unusual event for a thermally agitated atom in the table to hit our air molecule hard enough to send it kilometers high into the air! Actually, our particular molecule will very soon run into another air molecule and not make it very high in one great leap, but on average we do find lots of air molecules very high above sea level rather than finding them all lying on the ground.

Notice again that k_BT is the important factor in understanding the statistical behavior of matter. Here it sets the scale for the variation with height of the atmosphere's number density.

At a height where the gravitational potential energy $U_g = mgy$ is equal to k_BT, the number density has dropped by a factor of $e^{-1} = 1/e = 0.37$. Dry air at sea level is 78% nitrogen (one mole $N_2 = 28$ grams), 21% oxygen (one mole $O_2 = 32$ grams), about 1% argon, and 0.03% CO_2. An average mass of 29 grams per mole is good enough for most calculational purposes.

Distribution of Velocities in a Gas

Next we look at the distribution of molecular velocities. When the gas is confined inside a finite container the energy levels in terms of momentum and height are discrete, but under almost all conditions the size of the energy quantum is so small compared to k_BT that it is appropriate to take a nonquantum approach for these variables.

We speak of the probability that the gas molecule has an x component of velocity within the range between some v_x and $v_x + dv_x$, and similarly for v_y

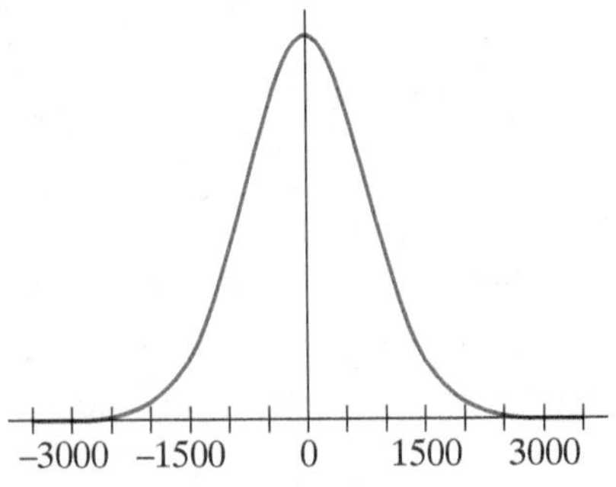

Figure 12.42 Distribution of the x component of velocity for helium at room temperature.

and v_z. Here dv_x is considered to be a very small amount, small compared to the average speed of the molecules.

Since we're explicitly interested in v_{CM}, it will be useful to express translational kinetic energy as $\frac{1}{2}Mv_{CM}^2$ rather than $p^2/(2M)$ in this discussion. Since $v^2 = v_x^2 + v_y^2 + v_z^2$, we can write the equation for the distribution of velocity in a gas as follows (remember that we are simply writing v for the center-of-mass speed v_{CM}):

$$\left[e^{-\frac{1}{2}Mv_x^2/k_BT}dv_x\right]\left[e^{-\frac{1}{2}Mv_y^2/k_BT}dv_y\right]\left[e^{-\frac{1}{2}Mv_z^2/k_BT}dv_z\right]$$

The distribution for each velocity component is a bell-shaped curve, called a "Gaussian." In Figure 12.42 we show the distribution of the x component of velocity for helium at room temperature.

QUESTION Judging from this graph, what is the average value for v_x? What are the average values of v_y and v_z? Why are these results plausible?

Evidently the average value for the x component of the velocity is zero. This is reassuring, because a molecule is as likely to be headed to the right as it is to be headed to the left. Similarly, the average values for the y and z components of velocity are zero.

It would not be particularly surprising to find a gas molecule with an x component of velocity such that $\frac{1}{2}Mv_x^2 \approx k_BT$. On the other hand, it would be very surprising to find a gas molecule with a value of v_x many times larger. As usual, k_BT sets the scale for thermal phenomena.

Distribution of Speeds in a Gas

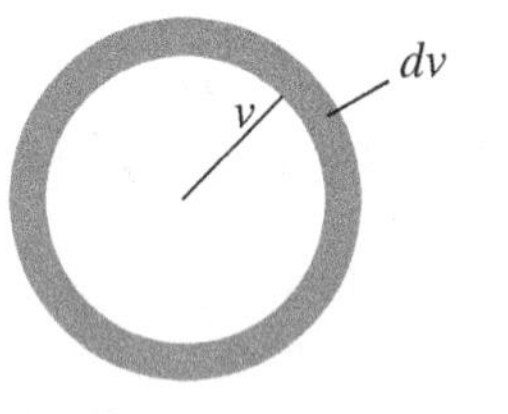

Figure 12.43 The volume of a shell with radius v and thickness dv is $4\pi v^2 dv$.

By converting from rectangular coordinates to spherical coordinates, the velocity-component distribution can be converted into a speed distribution. We won't go into the details, but the main idea is that the volume element in rectangular coordinates in "velocity space" $dv_x dv_y dv_z$ turns into the volume of a spherical shell with surface area $4\pi v^2$, and thickness dv, as shown in Figure 12.43.

Transferring to spherical coordinates, we have the following:

$$e^{-\frac{1}{2}Mv^2/k_BT}dv_x dv_y dv_z = e^{-\frac{1}{2}Mv^2/k_BT}4\pi v^2 dv$$

The only thing missing is a "normalization" factor in front to make the integral over all speeds from 0 to infinity be equal to 1.0 (since our one molecule must have a speed somewhere in that range). Here is the Maxwell–Boltzmann distribution for a low-density gas:

MAXWELL–BOLTZMANN SPEED DISTRIBUTION (LOW-DENSITY GAS)

$$P(v) = 4\pi\left(\frac{M}{2\pi k_BT}\right)^{3/2} v^2 e^{-\frac{1}{2}Mv^2/k_BT}$$

The probability that a molecule of a gas has a center-of-mass speed within the range v to $v+dv$ is given by $P(v)dv$.

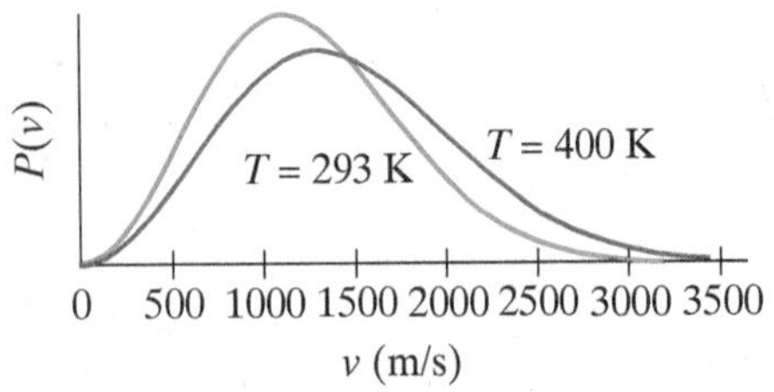

Figure 12.44 Distribution of speeds of helium atoms at two temperatures.

The distribution function for helium is shown for two different temperatures in Figure 12.44. You can see that the average speed of a helium atom is predicted to be about 1200 m/s at room temperature (20 °C, which is 293 K). The prediction was first made by Maxwell, in the mid-1800s, long before the development of quantum mechanics. In retrospect, this worked because the quantum levels of the translational kinetic energy are so close together as to be almost a continuum, so the classical model is a good approximation.

The average number of helium atoms in a container that have speeds in the range of 415.4 m/s to 415.7 m/s can be calculated by evaluating Maxwell's equation with $v = 415.4$ m/s and multiplying by $dv = 0.3$ m/s, which gives the probability of finding one molecule in this speed range. If there are N atoms in the container, multiplying by N gives the total number of molecules likely to be in that speed range at any given instant. The area of the vertical slice shown in Figure 12.45 represents the fraction of helium atoms that are likely to have speeds between 500 m/s and 600 m/s. At higher temperatures the distribution shifts to higher speeds (Figure 12.44).

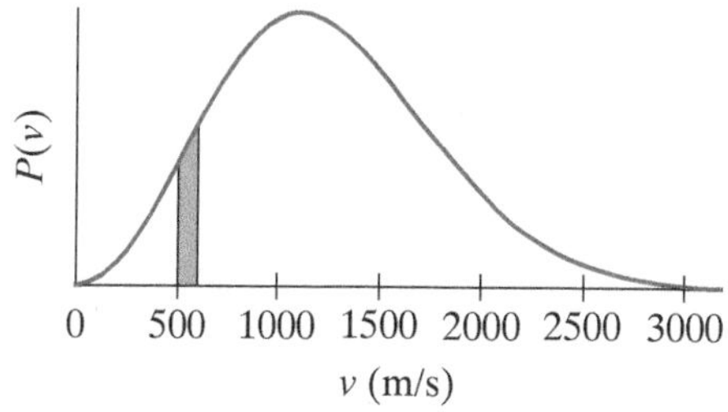

Figure 12.45 Fraction of helium atoms at 293 K with speeds between 500 and 600 m/s.

Measuring the Distribution of Speeds in a Gas

The actual distribution of speeds of molecules in any gas can be measured by an ingenious experiment. Make a tiny hole in a container of gas and let the molecules leak out into a vacuum (that is, a region from which the air is continually pumped out). The moving molecules run through collimating holes and then through a slot in a rapidly rotating drum and strike a row of devices at the other side of the drum that can detect gas molecules (Figure 12.46). Such measurements confirm the Maxwell prediction.

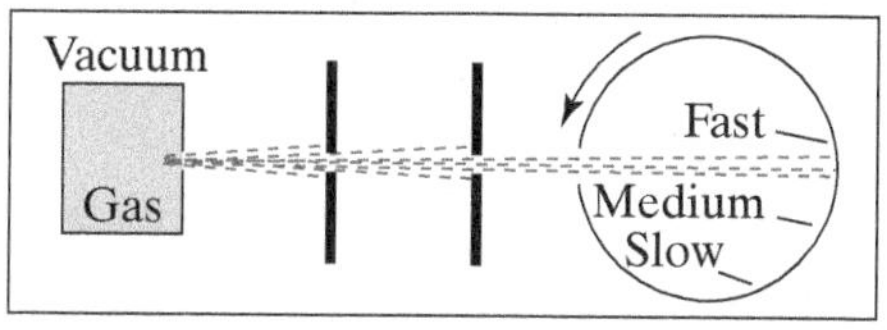

Figure 12.46 Apparatus for measuring the distribution of speeds of gas molecules.

Fast molecules strike soonest, followed by medium-speed molecules, and finally slow molecules, so molecules with different speeds strike at different locations along the rotating drum. The number of molecules striking various locations along the drum is a direct indication of the distribution of speeds of molecules in the gas. (A correction has to be applied to these data to obtain the speed distribution inside the gas, because high-speed molecules emerge through the hole more frequently than low-speed molecules do, even if their numbers are the same in the interior of the gas. Also, the apparatus measures the distribution of v_x rather than v.)

How might we display the results of this experiment? We could present the data as a graph (Figure 12.47) of the fraction $\Delta N/N$ of the N helium atoms that have speeds between 0 and 500 m/s, between 500 and 1000 m/s, between 1000 and 1500 m/s, and so on.

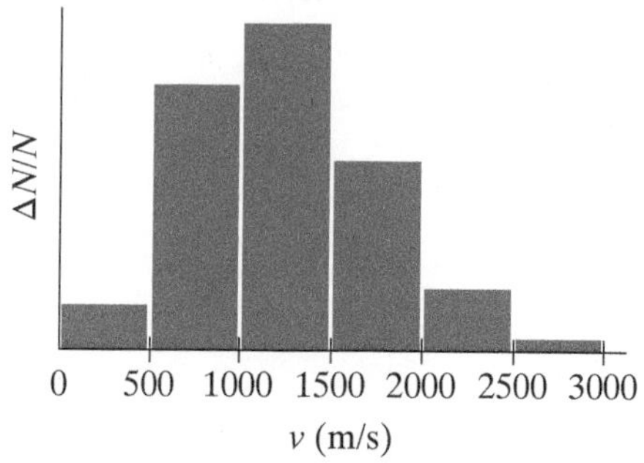

Figure 12.47 Distribution showing fraction of atoms with speeds in a given range.

Average Translational Kinetic Energy in a Gas

For a gas confined inside a container, quantum mechanics predicts that the kinetic energy $\frac{1}{2}Mv^2$ is quantized, but if the container is of ordinary macroscopic size, the energy quanta are extremely small compared to the average kinetic energy, even at very low temperatures. Most gases liquefy before the temperature drops so low as to invalidate the pre-quantum analysis for the velocity distribution.

Whenever the energy has a term containing a square of a position or momentum component, such as x^2 or p_x^2, pre-quantum theory predicts an associated average energy of $\frac{1}{2}k_BT$. This result is valid in quantum mechanics for high temperatures, where k_BT is large compared to the quantum energy spacing. This result follows from using an integral to take the average value of a quadratic term when the distribution is a Gaussian involving that quadratic term. That is, using integral tables (or integrating by parts), you can show that the average value of a quadratic term is $\frac{1}{2}k_BT$ (a bar over a quantity means "average value"):

$$\overline{w^2} = \frac{\displaystyle\int_0^\infty w^2 e^{-w^2/k_BT}dw}{\displaystyle\int_0^\infty e^{-w^2/k_BT}dw} = \frac{1}{2}k_BT$$

The average value is obtained by weighting the values of w^2 by the probability of finding that value; the denominator takes care of normalizing the distribution properly.

As an example, when we did quantum-based calculations for the Einstein solid, we found that at high temperatures the specific heat on a per-atom basis was $3k_B$, which implies an average energy of $3k_BT$. We modeled an atomic oscillator as three one-dimensional oscillators, each having an average energy of $\frac{1}{2}k_BT$ corresponding to kinetic energy $p^2/(2M)$ and another $\frac{1}{2}k_BT$ for the potential energy $\frac{1}{2}k_sx^2$. This is a total of six quadratic terms, implying an average energy per atom of $6(\frac{1}{2}k_BT)$, or $3k_BT$. We therefore expect $3k_BT$ the heat capacity per atom in a solid at high temperature to be $3k_B$, which is indeed what is observed.

HIGH-TEMPERATURE AVERAGE ENERGY

If $k_BT \gg$ the energy quantum, the average energy

associated with a quadratic energy term is $\dfrac{1}{2}k_BT$.

The number of quadratic terms in the expression for the energy is often called the "degrees of freedom."

QUESTION What is the average translational kinetic energy $\frac{1}{2}M\overline{v^2}$ in terms of k_BT? What is the associated contribution to the specific heat of the gas?

Since there are three quadratic terms in the translational kinetic energy, $K_{\text{trans}} = \frac{1}{2}mv_x^2 + \frac{1}{2}mv_y^2 + \frac{1}{2}mv_z^2$, the average value is three times $\frac{1}{2}k_BT$. We have the following important result:

AVERAGE K_{trans} FOR AN IDEAL GAS

$$\overline{K}_{\text{trans}} = \frac{1}{2}M\overline{v^2} = \frac{3}{2}k_BT$$

The contribution to the specific heat for a gas is $\frac{3}{2}k_B$.

Root-Mean-Square Speed

The square root of $\overline{v^2}$ is called the "root-mean-square" or "rms" speed:

$$v_{\text{rms}} = \sqrt{\overline{v^2}}$$

With this definition of v_{rms}, we write $\frac{1}{2}m\overline{v^2} = \frac{1}{2}mv_{\text{rms}}^2 = \frac{3}{2}k_BT$.

Average Speed vs. RMS Speed

The way you find the average speed of molecules in a gas is to weight the speed by the number of molecules that have that speed, and divide by the total number of molecules:

$$\overline{v} = \frac{N_1v_1 + N_2v_2 + \text{etc.}}{N} = \frac{N_1}{N}v_1 + \frac{N_2}{N}v_2 + \cdots$$

For a continuous distribution of speeds this turns into an integral, where the weighting factors are given by the Maxwell speed distribution, which gives the probability that a molecule has a center-of-mass speed in the range v to $v + dv$. Using integral tables one finds this:

$$\overline{v} = \int_0^\infty 4\pi\left(\frac{m}{2\pi k_BT}\right)^{3/2} e^{-\frac{1}{2}mv^2/k_BT}(v)v^2dv = \sqrt{\frac{8}{3\pi}}\sqrt{\frac{3k_BT}{m}} = 0.92v_{\text{rms}}$$

This shows that the average speed is smaller than the rms speed ($0.92v_{\text{rms}}$). This same calculational scheme can be used to determine any average. For example, if you calculate the average value of v^2 by this method, you find $\overline{v^2} = 3k_B T/m$, as expected.

Earlier we commented that the average speed $\overline{v}$ of helium atoms at room temperature is about 1200 m/s, but you can calculate that v_{rms} at room temperature is 1350 m/s. This is an example of the fact that the rms speed is higher than the average speed.

Checkpoint 9 The rms speed is somewhat higher than the average speed due to the averaging of squared speeds. Calculate the average of the numbers 1, 2, 3, and 4, then calculate the rms average (the square root of the average of their squares), and show that the rms average is larger than the simple average. Squaring gives extra weight to larger contributions.

Application: Retaining a Gas in the Atmosphere

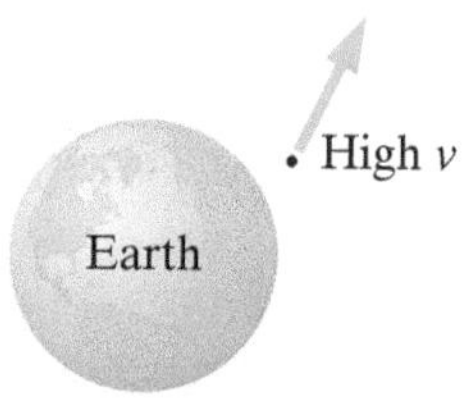

Figure 12.48 Low-mass atoms or molecules in our atmosphere may have speeds high enough to escape from the Earth entirely.

Since $v_{\text{rms}} = \sqrt{3k_B(T/m)}$, helium atoms typically travel much faster than nitrogen molecules in our atmosphere, due to the small mass of the helium atoms. Some few helium atoms will be going much faster than the average and may attain a high enough speed to escape from the Earth entirely (escape velocity from the Earth is about 1.1×10^4 m/s). This leads to a continuous leakage of high-speed helium atoms and other low-mass species such as hydrogen molecules (Figure 12.48). Other processes may also contribute to the flow of helium away from the Earth.

Where do we obtain helium for party balloons and low-temperature refrigeration and scientific experiments? It is extracted from natural gas when the gas is pumped out of the ground. Heavy radioactive elements such as uranium in the Earth's crust emit alpha particles (helium nuclei), which capture electrons and become helium atoms. These atoms are trapped with natural gas in underground cavities. When the helium-bearing natural gas is pumped to the surface, the helium is extracted (at some cost).

QUESTION Why doesn't the Moon have any atmosphere at all?

The Moon's gravitational field is so weak that escape speed is quite small, and all common gases can escape, not just hydrogen and helium.

Application: Speed of Sound

Sound waves in a gas consist of propagation of variations in density (Figure 12.49), and the fundamental mechanism for this kind of wave propagation involves collisions between neighboring molecules, whose speeds are proportional to v_{rms} (and roughly comparable to the speed of sound). For example, compare the v_{rms} for nitrogen, 510 m/s, with the speed of sound in air (which is mostly nitrogen) at 293 K, which is measured to be 344 m/s. (You may know the approximate rule that a 1-s delay between lightning and thunder indicates a distance of about 1000 ft, which is about 300 m.)

Figure 12.49 Periodic variations in gas density make a sound wave, which travels through a gas with speed v (the speed of sound).

Checkpoint 10 Should the speed of sound in air increase or decrease with increasing temperature? What percentage change would result from doubling the absolute temperature? (This effect is readily observed by measuring the speed of sound in a gas as a function of absolute temperature. The excellent agreement between theory and experiment provides additional evidence for our understanding of gases.)

Vibrational Energy in a Diatomic Gas Molecule

We have treated the distribution of velocity and position. Next we discuss the distribution of vibrational energy, with the Boltzmann factor

$$e^{-E_{\text{vib}}/k_B T}$$

For a monatomic gas such as helium, there is no vibrational energy term. However, for a diatomic molecule such as N_2 or HCl, the vibrational energy is

$$E_{\text{vib}} = \frac{p_1^2}{2m_1} + \frac{p_2^2}{2m_2} + \frac{1}{2}k_s s^2$$

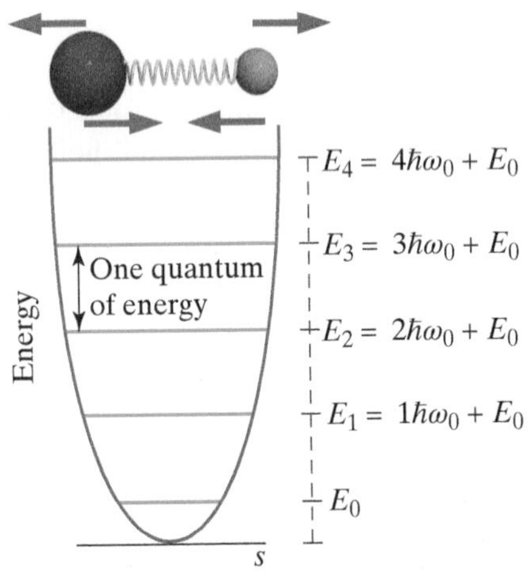

Figure 12.50 The quantized vibrational energy levels of a diatomic oscillator.

where k_s is the effective spring stiffness corresponding to the interatomic electric force (not to be confused with k_B, the Boltzmann constant), and s is the stretch of the interatomic bond. This equation for the vibrational energy is essentially the same as the equation for a one-dimensional spring–mass oscillator, which we have studied in detail (Figure 12.50).

Since $p_1 = p_2$ for momenta p_1 and p_2 relative to the center of mass, we can write

$$\frac{p_2^2}{2m_2} = \frac{p_1^2}{2m_2} = \left(\frac{m_1}{m_2}\right)\frac{p_1^2}{2m_1}$$

$$\frac{p_1^2}{2m_2} + \left(\frac{m_1}{m_2}\right)\frac{p_1^2}{2m_1} + \frac{1}{2}k_s s^2 = \left(1 + \frac{m_1}{m_2}\right)\frac{p_1^2}{2m_1} + \frac{1}{2}k_s s^2$$

This equation has a form exactly like that for a spring–mass oscillator, with a different mass.

Earlier we modeled a solid as a large number of isolated three-dimensional atomic oscillators (each corresponding to three one-dimensional oscillators, because there are spring-like interatomic forces on an atom from neighboring atoms in all directions). This is an overly simplified model of a solid, because in a solid the atoms interact with each other. For example, if an atom moves to the left this affects the atoms to the right and to the left.

In a low-density gas, however, the vibrational oscillators really are nearly independent of each other, because the gas molecules aren't even in contact with each other except when they happen to collide. So the analysis we carried out for the Einstein model of a solid applies even better to the vibrational portion of the energy in a real gas than it does to a real solid.

Among the results that apply immediately are that the specific heat associated with the vibrational energy of one oscillator is k_B at high temperatures. The specific heat decreases at very low temperatures, where $k_B T$ is small compared to the energy spacing of the quantized oscillator energies. Just as it was a surprise when the specific heat of metals was found to decrease at low temperatures, there was a similar surprise in the measurements of the specific heat of diatomic gases at low temperatures, because the contribution of the vibrational motion vanished.

QUESTION There are two quadratic terms in the vibrational energy (kinetic and spring). Therefore, at high temperatures what is the average vibrational energy in terms of $k_B T$? What is the associated contribution to the specific heat of the gas?

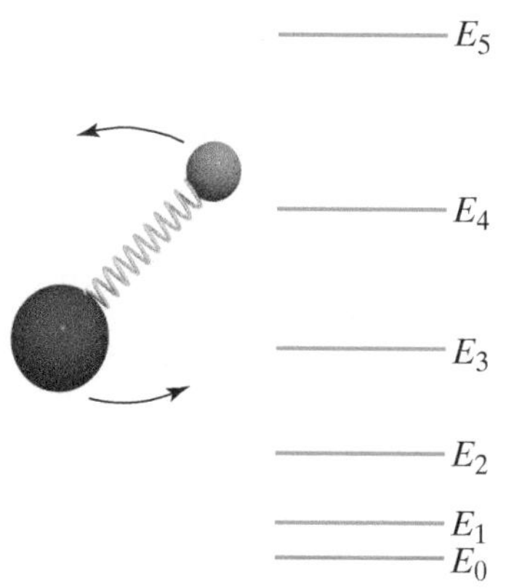

Figure 12.51 Rotational kinetic energy levels of a diatomic molecule.

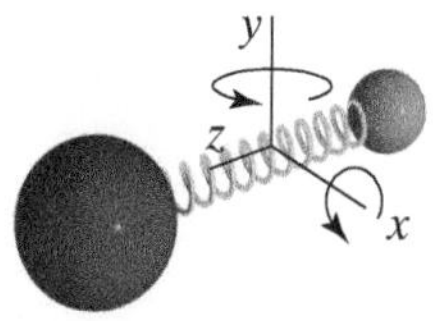

Figure 12.52 A diatomic molecule can rotate around the x or the y axis. We say there are two rotational degrees of freedom.

At high temperatures the vibrational motion of a diatomic molecule (a one-dimensional oscillator) is expected to have an average energy that is approximately $2(\frac{1}{2}k_BT) = k_BT$, and contributes k_B to the specific heat.

Rotational Energy in a Diatomic Gas Molecule

Finally we consider the rotational-energy portion of the distribution, for which the Boltzmann factor is

$$e^{-E_{\text{rot}}/k_BT}$$

For a monatomic gas there is no rotational term, but for a diatomic gas there is rotational kinetic energy associated with rotation of the "dumbbell" consisting of the two nuclei. The rotational kinetic energy is this (Figure 12.51):

$$E_{\text{rot}} = \frac{1}{2}I\omega_x^2 + \frac{1}{2}I\omega_y^2 = \frac{L_{\text{rot},x}^2}{2I} + \frac{L_{\text{rot},y}^2}{2I}$$

This corresponds to rotational angular momenta $L_{\text{rot},x}$ and $L_{\text{rot},y}$ about the x axis and about the y axis. Note that only the nuclei contribute significantly to the rotational kinetic energy, because the electrons have much less mass.

Rotation about the z axis connecting the two nuclei is irrelevant, for a somewhat subtle reason. The angular momentum is quantized, which leads to energy quantization. For rotations around the z axis (Figure 12.52) the energy is $L_{\text{rot},z}^2/(2I)$, where $L_{\text{rot},z}$ is the z component of the rotational angular momentum. Since I about the z axis for the tiny nuclei is extremely small compared with the moment of inertia about the x and y axes of the diatomic molecule, the rotational energy of the associated first excited state is enormous, and this state is not excited at ordinary temperatures.

Since there are two quadratic energy terms associated with rotation, we conclude that at high temperatures the rotational motion of a diatomic molecule has an average energy that is approximately $2(\frac{1}{2}k_BT) = k_BT$, and contributes k_B to the specific heat.

The spacing between quantized rotational energies for a diatomic molecule is even smaller than the vibrational energy quantum. As a result, the gas must be cooled to a very low temperature before the pre-quantum results become invalid. Many gases liquefy before this low-temperature regime is reached.

Specific Heat of a Gas

We are now in a position to discuss the specific heat of a gas as a function of temperature. This property is important in calculating the thermal interactions of a gas. Historically, measurements of the specific heat of gases were also important in testing theories of statistical mechanics. We'll concentrate on calculating C_V, the specific heat at constant volume (meaning no mechanical work is done on the gas). The associated experiment would be to add a known amount of energy to a gas in a rigid container and measure the temperature rise of the gas.

Consider the average energy of a diatomic molecule such as N_2 or HCl, consisting of the translational kinetic energy associated with the motion of the center of mass, plus the vibrational energy, plus the rotational energy, plus the gravitational energy of the molecule and the Earth:

$$\left[\frac{1}{2}m\overline{v_x^2} + \frac{1}{2}m\overline{v_y^2} + \frac{1}{2}m\overline{v_z^2}\right] + \left[\left(1+\frac{m_1}{m_2}\right)\frac{\overline{p_1^2}}{2m_1} + \frac{1}{2}k_s\overline{s^2}\right] + \left[\frac{\overline{L_{\text{rot},x}^2}}{2I} + \frac{\overline{L_{\text{rot},y}^2}}{2I}\right] + Mg\overline{y}_{\text{CM}}$$

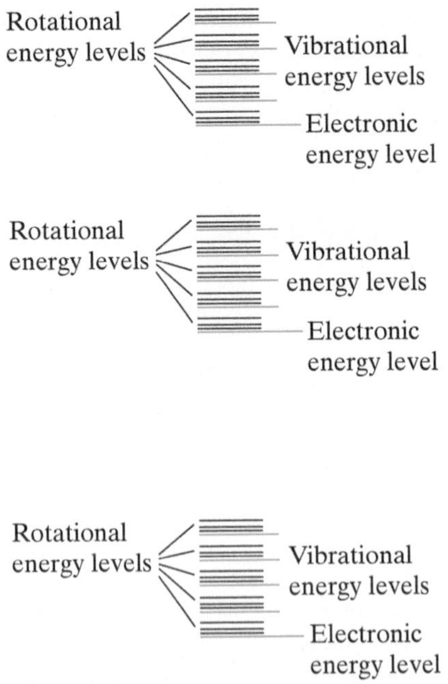

Figure 12.53 Electronic, vibrational, and rotational energy levels for a diatomic molecule.

QUESTION At high temperature, what should be the specific heat at constant volume?

There are seven quadratic terms in the expression for the energy, so if the temperature is very high, we expect an average energy of $7(\frac{1}{2}k_BT)$ and a specific heat at constant volume $C_V = \frac{7}{2}k_B$.

The discrete energy levels for a diatomic molecule include electronic, vibrational, and rotational energy levels (Figure 12.53). The electronic energy levels correspond to particular configurations of the electron clouds, and the spacing between these levels is typically 1 eV or more. Since at room temperature k_BT is about $\frac{1}{40}$ eV, the electronic levels do not contribute to the specific heat at ordinary temperatures. The vibrational energy spacing is much smaller, and the rotational energy spacing is smaller still. A diatomic gas at high temperature has a band-like spectrum, since transitions between electronic levels may be accompanied by one or more vibrational or rotational quanta.

What About *Mgy*?

What about the Mgy_{CM} term in the energy? Remember that we are trying to predict the specific heat of a gas at constant volume (so that no mechanical work is done on the system). We could measure C_V by having a known amount of energy transfer Q due to a temperature difference into a closed container of gas and observing the temperature rise of the gas. During this process there is an extremely small increase in y_{CM} of the gas in the container due to the temperature dependence of the height distribution, so the associated Mgy contribution to the specific heat is negligible.

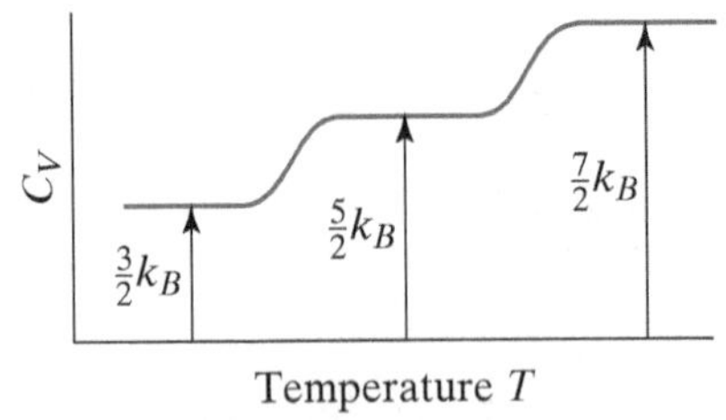

Figure 12.54 Specific heat C_V of a diatomic gas vs. temperature.

Specific Heat vs. Temperature

Figure 12.54 is a graph of the specific heat (at constant volume) of a diatomic gas as a function of temperature. In the following Checkpoint, see whether you can explain these values of the specific heat in terms of our analyses of the various contributions to the energy of a diatomic gas molecule.

Remember that k_BT is about $\frac{1}{40}$ eV at room temperature.

Checkpoint 11 What is the specific heat, on a per-molecule basis, at constant volume, of a diatomic gas at the following temperatures? **(a)** At high temperatures? **(b)** At a temperature where k_BT is small compared to the energy of the first excited vibrational state but larger than the energy of the first excited rotational state? **(c)** At a temperature low enough that k_BT is small compared to the energy of the first excited rotational state? (Some gases liquify above this temperature.) **(d)** For which gas, H_2 or D_2, does the specific heat become $\frac{3}{2}k_B$ at a higher temperature?

SUMMARY

The fundamental assumption of statistical mechanics is that, over time, an isolated system in a given macrostate is equally likely to be found in any of its possible microstates.

Two blocks in thermal contact evolve to that division of energy that has associated with it the largest number of ways $\Omega = \Omega_1\Omega_2$ of arranging the total energy among the atoms (largest number of microstates for a macrostate of given energy).

Second law of thermodynamics: The entropy of a closed system never decreases. Only in a reversible process does the entropy of a closed system stay constant.

Einstein solid: Each atom is modeled as three one-dimensional quantized oscillators.

Number of ways to arrange q quanta of energy among N one-dimensional oscillators:

$$\Omega = \frac{(q+N-1)!}{q!(N-1)!} \quad \text{(number of microstates)}$$

Entropy: $S \equiv k_B \ln \Omega$, where $k_B = 1.38 \times 10^{-23}$ J/K

Temperature: $\frac{1}{T} = \frac{\partial S}{\partial E}$

A small flow of energy Q due to a temperature difference into a system raises entropy by $\Delta S = \frac{Q}{T}$.

Specific heat per atom: $C = \frac{\Delta E_{atom}}{\Delta T}$, where

$$\Delta E_{atom} = \frac{\Delta E_{system}}{N_{atoms}}$$

The probability of finding energy E in a small system in contact with a large reservoir is proportional to

$$\Omega(E)e^{-E/k_BT}$$

$\Omega(E)$ is the number of microstates corresponding to energy E.

For each energy term involving a quadratic term such as x^2 or p_x^2 ("degree of freedom"), if the average energy is large compared to k_BT the average energy is $\frac{1}{2}k_BT$ and the contribution to the specific heat is $\frac{1}{2}k_B$. This contribution decreases at low temperatures. At high temperatures, specific heat on a per-atom basis in a solid $\approx 3k_B$.

Speed distribution for a gas:

$$P(v) = 4\pi \left(\frac{M}{2\pi k_BT}\right)^{3/2} e^{-\frac{1}{2}mv^2/k_BT}v^2$$

QUESTIONS

Q1 Energy conservation for two blocks in contact with each other is satisfied if all the energy is in one block and none in the other. Would you expect to observe this distribution in practice? Why or why not?

Q2 What is the advantage of plotting the (natural) logarithm of the number of ways of arranging the energy among the many atoms (natural logarithm of the number of microstates)?

Q3 Which has a higher temperature: a system whose entropy changes rapidly with increasing energy or one whose entropy changes slowly with increasing energy?

Q4 Explain why it is a disadvantage for some purposes that the specific heat of all materials decreases at low temperatures.

Q5 Two blocks with different temperatures had entropies of 10 J/K and 35 J/K before they were brought in contact. What can you say about the entropy of the combined system after the two came in contact with each other?

Q6 Can the entropy of a system decrease? What must be true about the entropy of the surroundings in this case?

Q7 Consider two blocks of copper. Block A contains 600 atoms and initially has a total of 20 quanta of energy. Block B contains 400 atoms and initially has 80 quanta of energy. The two blocks are placed in contact with each other, inside an insulated container (so no thermal energy can be exchanged with the surroundings). After waiting for a long time (for example, an hour), which of the following would you expect to be true? (1) Approximately 60 quanta of energy are in block A, and approximately 40 quanta of energy are in block B. (2) Approximately 50 quanta of energy are in block A, and approximately 50 quanta of energy are in block B. (3) The entropy of block A is equal to the entropy of block B. (4) The temperature of block A and the temperature of block B are equal.

Q8 Since $\Delta T = \Delta E/C$, what will happen at low temperatures to the temperature of a sink when some energy ΔE is transferred to it from the sample? Why is this unfortunate?

Q9 Which has more internal energy at room temperature—a mole of helium or a mole of air?

Q10 Which of the following are true about the Boltzmann factor $e^{-\Delta E/k_BT}$, where ΔE is the energy above the ground state? (1) $e^{-\Delta E/k_BT}$ is small when ΔE is large. (2) $e^{-\Delta E/k_BT}$ is small at low temperature T. (3) $e^{-\Delta E/k_BT}$ tends to zero as T gets larger and larger. (4) Even at high temperature, when many energy levels are excited, the ground state ($\Delta E = 0$) is the most populated state.

Q11 Explain qualitatively the basis for the Boltzmann distribution. Never mind the details of the math for the moment. Focus on the trade-offs involved with giving energy to a single oscillator vs. giving that energy to a large object.

Q12 Many chemical reactions proceed at rates that depend on the temperature. Discuss this from the point of view of the Boltzmann distribution.

Q13 A gas is made up of diatomic molecules. At temperature T_1, the ratio of the number of molecules in vibrational energy state 2 to the number of molecules in the ground state is measured, and found to be 0.35. The difference in energy between state 2 and the ground state is ΔE. **(a)** Which of the following conclusions is correct? (1) $\Delta E \approx k_BT_1$, (2) $\Delta E \ll k_BT_1$, (3) $\Delta E \gg k_BT_1$ **(b)** At a different temperature T_2, the ratio is found to be 8×10^{-5}. Which of the following is true? (1) $\Delta E \approx k_BT_2$, (2) $\Delta E \ll k_BT_2$, (3) $\Delta E \gg k_BT_2$.

Q14 How does the speed of sound in a gas change when you raise the temperature from 0 °C to 20 °C? Explain briefly.

Q15 This question follows the entire chain of reasoning involved in determining the specific heat of an Einstein solid. Start with two metal blocks, one consisting of one mole of aluminum (27 g) and the other of one mole of lead (207 g), both initially at a temperature very near absolute zero (0 K). From measurements of Young's modulus one finds that the effective stiffness of the interatomic bond modeled as a spring is 16 N/m for aluminum and 5 N/m for lead. **(a)** Is the number of quantized oscillators in the aluminum block greater, smaller, or the same as the number in the lead block? **(b)** What is the initial entropy of each block? **(c)** In which metal is the energy spacing of the quantized harmonic oscillators larger? **(d)** If we add 1 J of energy to each block, which metal now has the larger number of energy quanta? **(e)** In which block is the number of possible ways of arranging this 1 J of energy greater? **(f)** Which block now has the larger entropy? **(g)** Which block experienced a greater entropy change? **(h)** Which block experienced the larger temperature change? **(i)** Which metal has the larger specific heat at low temperatures? **(j)** Does your conclusion agree with the actual data given in Figure 12.33? (The numerical data are given in a table accompanying Problem P64.)

PROBLEMS

Section 12.2

•P16 List explicitly all the ways to arrange 2 quanta among 4 one-dimensional oscillators.

•P17 How many different ways are there to get 5 heads in 10 throws of a true coin? How many different ways are there to get no heads in 10 throws of a true coin?

•P18 How many different ways are there to arrange 4 quanta among 3 atoms in a solid?

•P19 A carbon nanoparticle (very small particle) contains 6000 carbon atoms. According to the Einstein model of a solid, how many oscillators are in this block?

•P20 In order to calculate the number of ways of arranging a given amount of energy in a tiny block of copper, the block is modeled as containing 8.7×10^5 independent oscillators. How many atoms are in the copper block?

•P21 In Chapter 4 you determined the stiffness of the interatomic "spring" (chemical bond) between atoms in a block of lead to be 5 N/m, based on the value of Young's modulus for lead. Since in our model each atom is connected to two springs, each half the length of the interatomic bond, the effective "interatomic spring stiffness" for an oscillator is 4×5 N/m = 20 N/m. The mass of one mole of lead is 207 g (0.207 kg). What is the energy, in joules, of one quantum of energy for an atomic oscillator in a block of lead?

•P22 Consider an object containing 6 one-dimensional oscillators (this object could represent a model of 2 atoms in an Einstein solid). There are 4 quanta of vibrational energy in the object. **(a)** How many microstates are there, all with the same energy? **(b)** If you examined a collection of 48,000 objects of this kind, each containing 4 quanta of energy, about how many of these objects would you expect to find in the microstate 000004?

•P23 Suppose that you look once every second at a system with 300 oscillators and 100 energy quanta, to see whether your favorite oscillator happens to have all the energy (all 100 quanta) at the instant you look. You expect that just once out of 1.7×10^{96} times you will find all of the energy concentrated on your favorite oscillator. On the average, about how many years will you have to wait? Compare this to the age of the Universe, which is thought to be about 1×10^{10} years. (1 y $\approx \pi \times 10^7$ s.)

••P24 The reasoning developed for counting microstates applies to many other situations involving probability. For example, if you flip a coin 5 times, how many different sequences of 3 heads and 2 tails are possible? Answer: 10 different sequences, such as HTHHT or TTHHH. In contrast, how many different sequences of 5 heads and 0 tails are possible? Obviously only one, HHHHH, and our equation gives $5!/[5!0!] = 1$, using the standard definition that 0! is defined to equal 1.

If the coin is equally likely on a single throw to come up heads or tails, any specific sequence like HTHHT or HHHHH is equally likely. However, there is only one way to get HHHHH, while there are 10 ways to get 3 heads and 2 tails, so this is 10 times more probable than getting all heads.

Use the expression $5!/[N!(5-N)!]$ to calculate the number of ways to get 0 heads, 1 head, 2 heads, 3 heads, 4 heads, or 5 heads in a sequence of 5 coin tosses. Make a graph of the number of ways vs. the number of heads.

Section 12.3

•P25 Object A and object B are two identical microscopic objects. Figure 12.55 below shows the number of ways to arrange energy in one of these objects, as a function of the amount of energy in the object.

E, J	0.8 E–20	1.0 E–20	1.2 E–20	1.4 E–20
# ways	37	60	90	122

Figure 12.55

(a) When there are 1.0×10^{-20} J of energy in object A, what is the entropy of this object? **(b)** When there are 1.4×10^{-20} J of energy in object B, what is the entropy of this object? **(c)** Now the two objects are placed in contact with each other. At this moment, before there is time for any energy flow between the objects, what is the entropy of the combined system of objects A and B?

••P26 For a certain metal the stiffness of the interatomic bond and the mass of one atom are such that the spacing of the quantum oscillator energy levels is 1.5×10^{-23} J. A nanoparticle of this metal consisting of 10 atoms has a total thermal energy of 18×10^{-23} J. **(a)** What is the entropy of this nanoparticle? **(b)** The temperature of the nanoparticle is 87 K. Next we add 18×10^{-23} J to the nanoparticle. By how much does the entropy increase?

Section 12.5

•P27 A block of copper at a temperature of 50 °C is placed in contact with a block of aluminum at a temperature of 45 °C in an insulated container. As a result of a transfer of 2500 J of energy from the copper to the aluminum, the final equilibrium temperature of the two blocks is 48 °C. **(a)** What is the approximate change in the entropy of the aluminum block? **(b)** What is the approximate change in the entropy of the copper block? **(c)** What is the approximate change in the entropy of the Universe? **(d)** What is the change in the energy of the Universe?

•P28 It takes about 335 J to melt one gram of ice. During the melting, the temperature stays constant. Which has higher entropy, a gram of liquid water at 0 °C or a gram of ice at 0 °C? Does this make sense? How large is the entropy difference?

•**P29** Suppose that the entropy of a certain substance (not an Einstein solid) is given by $S = a\sqrt{E}$, where a is a constant. What is the energy E as a function of the temperature T?

••**P30** A nanoparticle consisting of four iron atoms (object 1) initially has 1 quantum of energy. It is brought into contact with a nanoparticle consisting of two iron atoms (object 2), which initially has 2 quanta of energy. The mass of one mole of iron is 56 g. **(a)** Using the Einstein model of a solid, calculate and plot $\ln\Omega_1$ vs. q_1 (the number of quanta in object 1), $\ln\Omega_2$ vs. q_1, and $\ln\Omega_{total}$ vs. q_1 (put all three plots on the same graph). Show your work and explain briefly. **(b)** Calculate the approximate temperature of the objects at equilibrium. State what assumptions or approximations you made.

Section 12.6

•**P31** Suppose that the entropy of a certain substance (not an Einstein solid) is given by $S = a\sqrt{E}$, where a is a constant. What is the specific heat C as a function of the temperature T?

••**P32** A 100-g block of metal at a temperature of 20 °C is placed into an insulated container with 400 g of water at a temperature of 0 °C. The temperature of the metal and water ends up at 2 °C. What is the specific heat of this metal, per gram? Start from the Energy Principle. The specific heat of water is 4.2 J/K/g.

••**P33** A nanoparticle containing 6 atoms can be modeled approximately as an Einstein solid of 18 independent oscillators. The evenly spaced energy levels of each oscillator are 4×10^{-21} J apart. **(a)** When the nanoparticle's energy is in the range $5 \times 4 \times 10^{-21}$ J to $6 \times 4 \times 10^{-21}$ J, what is the approximate temperature? (In order to keep precision for calculating the specific heat, give the result to the nearest tenth of a kelvin.) **(b)** When the nanoparticle's energy is in the range $8 \times 4 \times 10^{-21}$ J to $9 \times 4 \times 10^{-21}$ J, what is the approximate temperature? (In order to keep precision for calculating the specific heat, give the result to the nearest tenth of a degree.) **(c)** When the nanoparticle's energy is in the range $5 \times 4 \times 10^{-21}$ J to $9 \times 4 \times 10^{-21}$ J, what is the approximate heat capacity per atom? Note that between parts (a) and (b) the average energy increased from 5.5 quanta to 8.5 quanta. As a check, compare your result with the high temperature limit of $3k_B$.

••**P34** The entropy S of a certain object (not an Einstein solid) is the following function of the internal energy E: $S = bE^{1/2}$, where b is a constant. **(a)** Determine the internal energy of this object as a function of the temperature. **(b)** What is the specific heat of this object as a function of the temperature?

••**P35** The interatomic spring stiffness for tungsten is determined from Young's modulus measurements to be 90 N/m. The mass of one mole of tungsten is 0.185 kg. If we model a block of tungsten as a collection of atomic "oscillators" (masses on springs), note that since each oscillator is attached to two "springs," and each "spring" is half the length of the interatomic bond, the effective interatomic spring stiffness for one of these oscillators is 4 times the calculated value given above.

Use these precise values for the constants: $\hbar = 1.0546 \times 10^{-34}$ J·s (Planck's constant divided by 2π), Avogadro's number $= 6.0221 \times 10^{23}$ molecules/mole, $k_B = 1.3807 \times 10^{-23}$ J/K (the Boltzmann constant). **(a)** What is one quantum of energy for one of these atomic oscillators? **(b)** Figure 12.56 contains the number of ways to arrange a given number of quanta of energy in a particular block of tungsten. Fill in the blanks to complete the table, including calculating the temperature of the block. The energy E is measured from the ground state. Nothing goes in the shaded boxes. Be sure to give the temperature to the nearest 0.1 kelvin. **(c)** There are about 60 atoms in this object. What is the heat capacity on a per-atom basis? (Note that at high temperatures the heat capacity on a per-atom basis approaches the classical limit of $3k_B = 4.2 \times 10^{-23}$ J/K/atom.)

q	# ways	E, J	S, J/K	ΔE, J	ΔS, J/K	T, K
20	4.91 E26					
21	4.44 E27					
22	3.85 E28					

Figure 12.56

••**P36** A 50-g block of copper (one mole has a mass of 63.5 g) at a temperature of 35 °C is put in contact with a 100-g block of aluminum (molar mass 27 g) at a temperature of 20 °C. The blocks are inside an insulated enclosure, with little contact with the walls. At these temperatures, the high-temperature limit is valid for the specific heat. Calculate the final temperature of the two blocks. Do NOT look up the specific heats of aluminum and copper; you should be able to figure them out on your own.

••**P37** Young's modulus for copper is measured by stretching a copper wire to be about 1.2×10^{11} N/m^2. The density of copper is about 9 g/cm^3, and the mass of a mole is 63.5 g. Starting from a very low temperature, use these data to estimate roughly the temperature T at which we expect the specific heat for copper to approach $3k_B$. Compare your estimate with the data shown on a graph in this chapter.

••**P38** Figure 12.57 shows a one-dimensional row of 5 microscopic objects each of mass 4×10^{-26} kg, connected by forces that can be modeled by springs of stiffness 15 N/m. These objects can move only along the x axis.

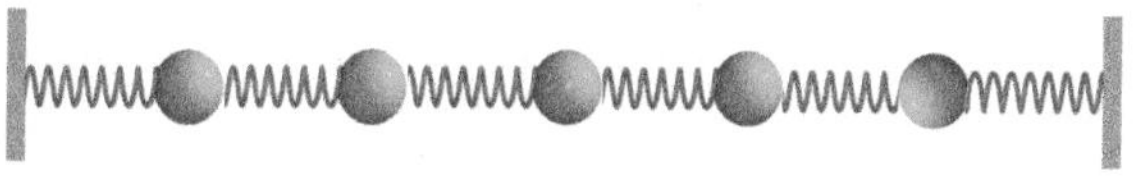

Figure 12.57

(a) Using the Einstein model, calculate the approximate entropy of this system for total energy of 0, 1, 2, 3, 4, and 5 quanta. Think carefully about what the Einstein model is, and apply those concepts to this one-dimensional situation. **(b)** Calculate the approximate temperature of the system when the total energy is 4 quanta. **(c)** Calculate the approximate specific heat on a per-object basis when the total energy is 4 quanta. **(d)** If the temperature is raised very high, what is the approximate specific heat on a per-object basis? Give a numerical value and compare with your result in part (c).

••**P39** In an insulated container a 100-W electric heating element of small mass warms up a 300-g sample of copper for 6 s. The initial temperature of the copper was 20 °C (room temperature). Predict the final temperature of the copper, using the $3k_B$ specific heat per atom.

••P40 The goal of this experiment is to understand, in a concrete way, what specific heat is and how it can be measured. You will need a microwave oven, a styrofoam coffee cup, and a clock or watch.

In the range of temperature where water is a liquid (0 C to 100 °C), it is approximately true that it takes 4.2 J of energy (1 calorie) to raise the temperature of 1 g of water through 1 Kelvin. To measure this specific heat of water, we need some way to raise a known mass of water from a known initial temperature to a final temperature that can also be measured, while we keep track of the energy supplied to the water. One way to do this, as discussed in the text, is to put water in a well-insulated container within which a heater, whose power output is known, warms up the water. In this experiment, instead of a well-insulated box with a heater, we will use microwave power, which preferentially warms up water by exciting rotational modes of the water molecules, as opposed to burners or heaters that warm up water in a pan by first warming up the pan.

Use a styrofoam coffee cup of known volume in which water can be warmed up. The density of water is 1 gram/cm^3.

> It is a good idea not to fill the cup completely full, because this makes it more likely to spill.

One method of recording the initial temperature of the water is to get water from the faucet and wait for it to equilibrate with room temperature (which can either be read off a thermostat or estimated based on past experience). After waiting about a half hour for this to happen, place the cup in the microwave oven and turn on the oven at maximum power. The cup needs to be watched as it warms up, so that when the water starts to boil, the elapsed time can be noted accurately.

> BE CAREFUL! A styrofoam cup full of hot liquid can buckle if you hold it near the rim. Hold the cup near the bottom. If the cup is full, do not attempt to move the cup while the water is hot. A spill can cause a painful burn.

On the back of the microwave oven (or inside the front door), there is usually a sticker with specifications that says "Output Power= ...Watts" which can be used to calculate the energy supplied. If there is no indication, use a typical value of 600 W for a standard microwave oven. Using all the quantities measured above and knowing the temperature interval over which you have warmed up the water, you can calculate the specific heat of water. **(a)** Show and explain all your data and calculations, and compare with the accepted value for water (4.2 J/K/g). **(b)** Discuss why your result might be expected to differ from the accepted value. For each effect that you consider, state whether this effect would lead to a result that is larger or smaller than the accepted value.

••P41 A box contains a uniform disk of mass M and radius R that is pivoted on a low-friction axle through its center (Figure 12.58). A block of mass m is pressed against the disk by a spring, so that the block acts like a brake, making the disk hard to turn. The box and the spring have negligible mass. A string is wrapped around the disk (out of the way of the brake) and passes through a hole in the box. A force of constant magnitude F acts on the end of the string. The motion takes place in outer space. At time t_i the speed of the box is v_i, and the rotational speed of the disk is ω_i. At time t_f the box has moved a distance x, and the end of the string has moved a longer distance d, as shown.

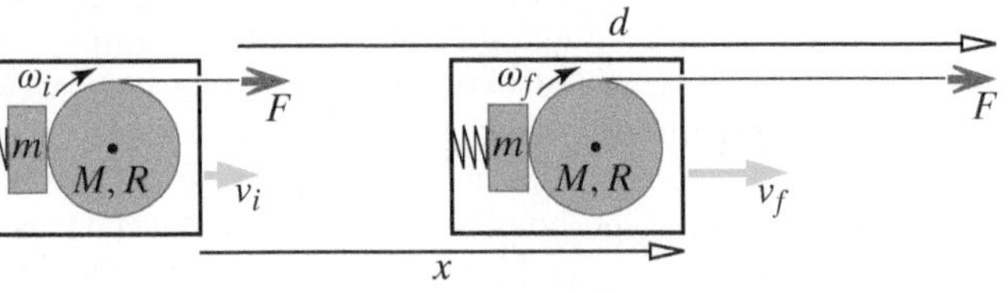

Figure 12.58

(a) At time t_f, what is the speed v_f of the box? **(b)** During this process, the brake exerts a tangential friction force of magnitude f. At time t_f, what is the angular speed ω_f of the disk? **(c)** At time t_f, assume that you know (from part b) the rotational speed ω_f of the disk. From time t_i to time t_f, what is the increase in thermal energy of the apparatus? **(d)** Suppose that the increase in thermal energy in part (c) is 8×10^4 J. The disk and brake are made of iron, and their total mass is 1.2 kg. At time t_i their temperature was 350 K. At time t_f, what is their approximate temperature?

Section 12.8

•P42 Consider the exponential function e^{-x}. Evaluate this function for $x = 1$, 10,000, and 0.01.

•P43 At room temperature (293 K), calculate $k_B T$ in joules and eV.

•P44 A microscopic oscillator has its first and second excited states 0.05 eV and 0.10 eV above the ground-state energy. Calculate the Boltzmann factor for the ground state, first excited state, and second excited state, at room temperature.

Section 12.9

•P45 Suppose that you put one air molecule on your desk, so it is in thermal equilibrium with the desk at room temperature. Suppose that there is no atmosphere to get in the way of this one molecule bouncing up and down on the desk. Calculate the typical height that the air molecule will be above your desk, so that $Mgy \approx k_B T$.

•P46 Approximately what fraction of the sea-level air density is found at the top of Mount Everest, a height of 8848 m above sea level?

•P47 Calculate v_{rms} for a helium atom and for a nitrogen molecule (N_2; molecular mass 28 g per mole) in the room you're in (whose temperature is probably about 293 K).

•P48 Calculate the escape speed from the Moon and compare with typical speeds of gas molecules. The mass of the Moon is 7×10^{22} kg, and its radius is 1.75×10^6 m.

•P49 Sketch and label graphs of specific heat vs. temperature for hydrogen gas (H_2) and oxygen gas (O_2), using the same temperature scale. Explain briefly.

•P50 Marbles of mass $M = 10$ g are lying on the floor. They are of course in thermal equilibrium with their surroundings. What is a typical height above the floor for one of these marbles? That is, for what value of y is $Mgy \approx k_B T$?

•P51 Viruses of mass $M = 2 \times 10^{-20}$ kg are lying on the floor at room temperature (about 20 °C = 293 K). They are of course in thermal equilibrium with their surroundings. What is a typical height above the floor for one of these viruses? That is, for what value of y is $Mgy \approx k_B T$?

•**P52** The temperature of the surface of a certain star is 7000 K. Most hydrogen atoms at the surface of the star are in the electronic ground state. What is the approximate fraction of the hydrogen atoms that are in the first excited state (and therefore could emit a photon)? The energy of the first excited state above the ground state is $(-13.6/2^2 eV) - (-13.6 eV) = 10.2 eV = 1.632 \times 10^{-18}$ J.

(In this estimate we are ignoring the fact that there may be several excited states with the same energies—for example, the 2s and 2p states in hydrogen—because this makes only a small difference in the answer.)

•**P53** Figure 12.59 shows the distribution of speeds of atoms in a particular gas at a particular temperature. Approximately what is the average speed? Is the rms (root-mean-square) speed bigger or smaller than this? Approximately what fraction of the molecules have speeds greater than 1000 m/s?

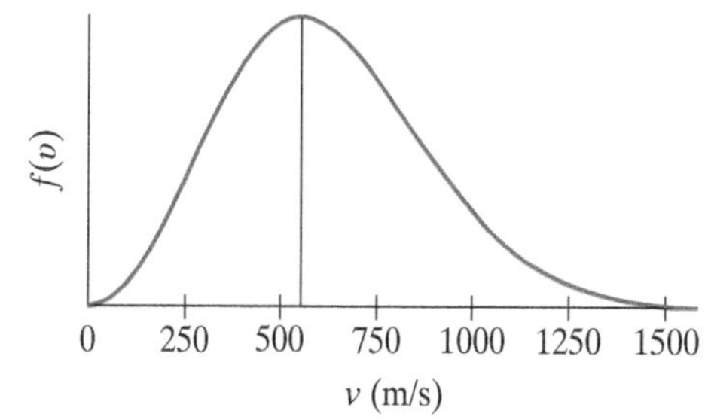

Figure 12.59

••**P54** Calculate the temperature rise of a gas. **(a)** You have a bottle containing a mole of a monatomic gas such as helium or neon. You warm up this monatomic gas with an electrical heater, which inputs $Q = 580$ J of energy. How much does the temperature of the gas increase? **(b)** You have a bottle containing a mole of a diatomic gas such as nitrogen (N_2) or oxygen (O_2). The initial temperature is in the range where many rotational energy levels are excited but essentially no vibrational energy levels are excited. You warm up this diatomic gas with an electrical heater, which inputs $Q = 580$ J of energy. How much does the temperature of the gas increase? **(c)** You have a bottle containing a mole of a diatomic gas such as nitrogen (N_2) or oxygen (O_2). The initial temperature is in the range where not only many rotational energy levels are excited but also many vibrational energy levels are excited. You warm up this diatomic gas with an electrical heater, which inputs $Q = 580$ J of energy. How much does the temperature of the gas increase?

••**P55** In studying a voyage to the Moon in Chapter 3 we somewhat arbitrarily started at a height of 50 km above the surface of the Earth. **(a)** At this altitude, what is the density of the air as a fraction of the density at sea level? **(b)** Approximately how many air molecules are there in one cubic centimeter at this altitude? **(c)** At what altitude is air density one-millionth (1×10^{-6}) that at sea level?

••**P56** At sufficiently high temperatures, the thermal speeds of gas molecules may be high enough that collisions may ionize a molecule (that is, remove an outer electron). An ionized gas in which each molecule has lost an electron is called a "plasma." Determine approximately the temperature at which air becomes a plasma.

••**P57** 100 J of energy transfer due to a temperature difference are given to air in a 50-L rigid container and to helium in a 50-L rigid container, both initially at STP (standard temperature and pressure). **(a)** Which gas experiences a greater temperature rise? **(b)** What is the temperature rise of the helium gas?

•••**P58** It is possible to estimate some properties of a diatomic molecule from the temperature dependence of the specific heat. **(a)** Below about 80 K the specific heat at constant volume for hydrogen gas (H_2) is $\frac{3}{2}k_B$ per molecule, but at higher temperatures the specific heat increases to $\frac{5}{2}k_B$ per molecule due to contributions from rotational energy states. Use these observations to estimate the distance between the hydrogen nuclei in an H_2 molecule. **(b)** At about 2000 K the specific heat at constant volume for hydrogen gas (H_2) increases to $\frac{7}{2}k_B$ per molecule due to contributions from vibrational energy states. Use these observations to estimate the stiffness of the "spring" that approximately represents the interatomic force.

•••**P59** In 1988, telescopes viewed Pluto as it crossed in front of a distant star. As the star emerged from behind the planet, light from the star was slightly dimmed as it went through Pluto's atmosphere. The observations indicated that the atmospheric density at a height of 50 km above the surface of Pluto is about one-third the density at the surface. The mass of Pluto is known to be about 1.5×10^{22} kg and its radius is about 1200 km. Spectroscopic data indicate that the atmosphere is mostly nitrogen (N_2). Estimate the temperature of Pluto's atmosphere. State what approximations and/or simplifying assumptions you made.

•••**P60** Buckminsterfullerene, C_{60}, is a large molecule consisting of 60 carbon atoms connected to form a hollow sphere. The diameter of a C_{60} molecule is about 7×10^{-10} m. It has been hypothesized that C_{60} molecules might be found in clouds of interstellar dust, which often contain interesting chemical compounds. The temperature of an interstellar dust cloud may be very low, around 3 K. Suppose you are planning to try to detect the presence of C_{60} in such a cold dust cloud by detecting photons emitted when molecules undergo transitions from one rotational energy state to another.

Approximately, what is the highest-numbered rotational level from which you would expect to observe emissions? Rotational levels are $l = 0, 1, 2, 3, \ldots$.

COMPUTATIONAL PROBLEMS

More detailed and extended versions of some computational modeling problems may be found in the lab activities included in the *Matter & Interactions, 4th Edition*, resources for instructors.

Working through Problems P61–P64 can be very informative and can help make the models discussed in this chapter concrete and clear.

••**P61** Write a program to calculate the number of ways to arrange energy in an Einstein solid. **(a)** Model a system consisting of two atoms (three oscillators each), among which 4 quanta of energy are to be distributed. Write a program to display a histogram showing the total number of possible microstates of the two-atom system vs. the number of quanta assigned to atom

1. Compare your histogram to the one shown in Figure 12.16 (you should get the same distribution). **(b)** Model a system consisting of two solid blocks, block 1 containing 300 oscillators and block 2 containing 200 oscillators. Find the possible distributions of 100 quanta among these blocks, and plot number of microstates vs. number of quanta assigned to block 1. Compare your histogram to the one shown in Figure 12.22. Determine the distribution of quanta for which the probability is half as large as the most probable 60-40 distribution. **(c)** Do a series of calculations distributing 100 quanta between two blocks whose total number of oscillators is 500, but whose relative number of atoms varies. For example, consider equal numbers of oscillators, and ratios of 2:1, 5:1, and so on. Describe your observations.

••P62 Start with your solution to Problem P61. For the same system of two blocks, with $N_1 = 300$ oscillators and $N_2 = 200$ oscillators, plot $\ln(\Omega_1)$, $\ln(\Omega_2)$, and $\ln(\Omega_1\Omega_2)$, for q_1 running from 0 to 100 quanta. Your graph should look like the one in Figure 12.26. Determine the maximum value of $\ln(\Omega_1\Omega_2)$ and the value of q_1 where this maximum occurs. What is the significance of this value of q_1?

••P63 Modify your program from Problem P62 to plot the temperature of block 1 in kelvins as a function of the number of quanta q_1 present in the first block. On the same graph, plot the temperature of block 2 in kelvins as a function of q_1 (of course, $q_2 = q_{\text{tot}} - q_1$).

In order to plot the temperature in kelvins, you must determine the values of ΔE and ΔS that correspond to a one-quantum change in energy. Consider the model we are using. The energy of one quantum, in joules, is $\Delta E = \hbar\sqrt{k_{s,i}/m_a}$. The increment in entropy corresponding to this increment in energy is $\Delta S = k_B\Delta(\ln\Omega)$. Assume that the blocks are made of aluminum. Based on Young's modulus measurements the interatomic spring constant $k_{s,i}$ for Al is approximately 16 N/m. In Section 12.6 we show that the effective $k_{s,i}$ for oscillations in the Einstein solid is expected to be about 4 times the value obtained from measuring Young's modulus.

What is the significance of the value of q_1 (and of q_2) where the temperature curves for the two blocks cross (the temperatures are equal)?

••P64 Modify your analysis of Problem P63 to determine the specific heat as a function of temperature for a single block of metal. In order to see all of the important effects, consider a single block of 35 atoms (105 oscillators) with up to 300 quanta of energy. Note that in this analysis you are calculating quantities for a *single* block, not two blocks in contact.

To make specific comparison with experimental data, consider the cases of aluminum (Al) and lead (Pb). For each metal, plot the theoretical specific heat C per atom vs. $T(K)$, with dots showing the actual experimental data given in the following table, which you should convert to the same per-atom basis as your theoretical calculations. Adjust the interatomic spring stiffness k_s until your calculations approximately fit the experimental data.

T	C, Al	C, Pb	T	C, Al	C, Pb
20	0.23	11.01	150	18.52	25.27
40	2.09	19.57	200	21.58	25.87
60	5.77	22.43	250	23.25	26.36
80	9.65	23.69	300	24.32	26.82
100	13.04	24.43	400	25.61	27.45

T is in kelvin; C is J/K/mole.

What value of k_s gives a good fit? Based on Young's modulus, estimated values of the interatomic spring stiffnesses are 16 N/m for Al and 5 N/m for lead. However, recall that these values may need modification since the "springs" in the Einstein model are half as long, and there are two of them per oscillator.

Show from your graph that the high-temperature limit of the specific heat is about $3k_B$ per atom. The rise above this limit at high temperatures may be due to the fact that the assumed uniform spacing of the quantized oscillator energy levels isn't a good approximation for highly excited states. See Chapter 8.

••P65 For some examples of your choice, demonstrate by carrying out actual computer calculations that the "square-root" rule holds true for the fractional width of the peak representing the most probable arrangements of the energy. *Warning:* Check to see what is the largest number you can use in your computations; some programs or programming environments won't handle numbers bigger than about 1×10^{307}, for example, and larger numbers are treated as "infinite."

••P66 Create a computational model of one atom of the Einstein solid as shown in Figure 12.4. Let the atom move under the influence of the six spring-like forces that act on the atom. Use the classical Momentum Principle, although the actual behavior is governed by quantum mechanics, with quantized energies. **(a)** Experiment with different initial displacements of the atom away from the equilibrium position. **(b)** In many materials the effective stiffness of the interatomic bonds is different in the x, y, and z directions; see how making the stiffnesses different affects the motion.

ANSWERS TO CHECKPOINTS

1 20
2 123, 132, 213, 231, 312, 321
3 RRRGG, RRGRG, RRGGR, RGRRG, RGRGR, RGGRR, GGRRR, GRGRR, GRRGR, GRRRG
4 See Figure 12.15.
6 This looks *very* improbable; the second law of thermodynamics would be violated by this process—the entropy of the Universe is decreasing.
7 500 K
9 2.50; 2.74
10 Increase, because the average speed of the air molecules increases; 40% (factor of $\sqrt{2}$)
11 (a) $\frac{7}{2}k_B$; (b) $\frac{5}{2}k_B$; (c) $\frac{3}{2}k_B$; (d) transition to $\frac{3}{2}k_B$ occurs at a lower temperature for deuterium because the rotational energy level spacing $L^2/(2I)$ is smaller for deuterium (larger nuclear mass means larger moment of inertia, since the internuclear distance is about the same; $L^2 = l(l+1)\hbar^2$ is the same for the rotational energy levels of hydrogen and deuterium molecules)

CHAPTER

13

Electric Field

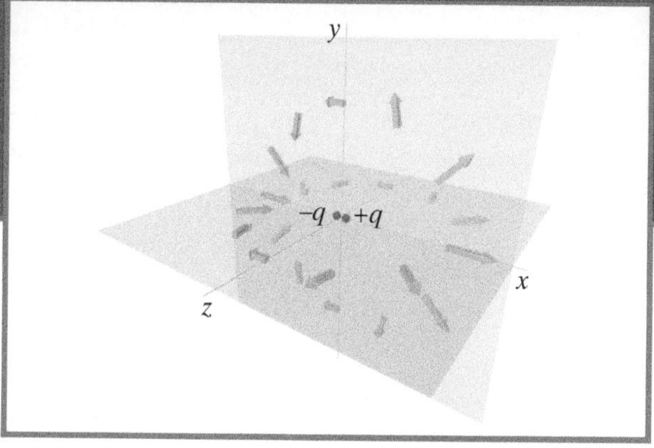

OBJECTIVES

After studying this chapter you should be able to

- Mathematically relate electric field and force.
- Calculate the 3D electric field at a particular location due to a collection of point charges.
- Explain the approximations made in deriving expressions for the electric field of a dipole, and use these approximate expressions appropriately.
- Graphically represent the magnitude and direction of the electric field of a dipole with arrows, at locations in a plane containing the dipole.
- Create a computational model to compute and display the electric field of a collection of point charges in 3D, and predict the motion of a charged particle that interacts with this field.

13.1 NEW CONCEPTS

Two important new ideas will form the core of our study of electric and magnetic interactions. The first is the concept of electric and magnetic fields. This concept is more abstract than the concept of force, which we used extensively in our study of modern mechanics. The reason we want to incorporate the idea of "field" into our models of the world is that this concept turns out to be a very powerful one, which allows us to explain and predict important phenomena that would otherwise be inaccessible to us.

The second important idea is a more sophisticated and complex model of matter. In our previous study of mechanics and thermal physics it was usually adequate to model a solid as an array of electrically neutral microscopic masses (atoms) connected by springs (chemical bonds). As we consider electric and magnetic interactions in more depth, we will find that we need to consider the individual charged particles—electrons and nuclei—that make up ordinary matter.

The material in this chapter lays the foundation for all succeeding chapters, so it is worth taking time to understand it thoroughly. In addition, if you did not use volume 1 of this textbook in your previous study of physics, you should work through the summary from Chapter 1 on 3D vectors, vector notation, and computational modeling available at no charge on the student website, `www.wiley.com/college/chabay`.

13.2 ELECTRIC CHARGE AND FORCE

In this section we briefly review concepts familiar to you from your previous studies.

Point Charges

There are two kinds of electric charge, which are called positive and negative. Particles with like charges repel each other (two positive or two negative particles); particles with unlike charges (positive and negative) attract each other. By "point particle" we mean an object whose radius is very small compared to the distance between it and all other objects of interest, so we can treat the object as if all its charge and mass were concentrated at a single mathematical point. Small particles such as protons and electrons can almost always be considered to be point particles.

The Coulomb Force Law for Point Particles

The electromagnetic interaction is one of the four fundamental physical interactions (see Chapter 3, The Fundamental Interactions). The electric force law, called Coulomb's law, describes the magnitude of the electric force between two point-like electrically charged particles:

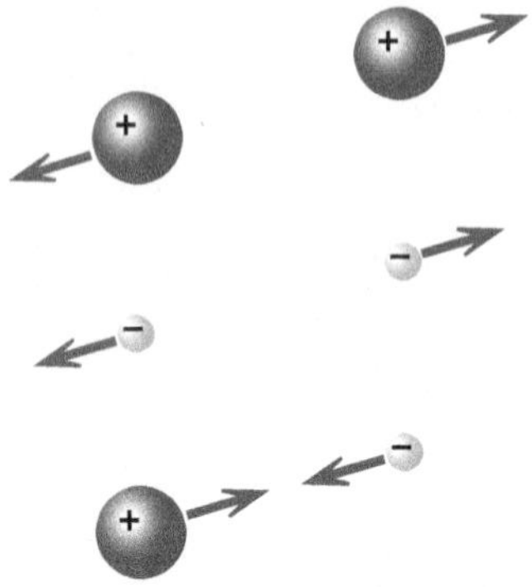

Figure 13.1 Two protons repel each other; two electrons repel each other; a proton and an electron attract each other.

$$|\vec{F}| = F = \frac{1}{4\pi\varepsilon_0}\frac{|Q_1 Q_2|}{r^2}$$

where Q_1 and Q_2 are the magnitudes of the electric charge of objects 1 and 2, and r is the distance between the objects. As indicated in Figure 13.1:

- The electric force acts along a line between two point-like objects.
- Like charges repel; unlike charges attract.
- Two charged objects interact even if they are some distance apart.

Units and Constants

The SI unit of electric charge is the coulomb, abbreviated C. The charge of one proton is 1.6×10^{-19} C in SI units. The constant e is often used to represent this amount of positive charge: $e = +1.6 \times 10^{-19}$ C. An electron has a charge of $-e$.

The constant $1/4\pi\varepsilon_0$ has the value $9 \times 10^9\ \mathrm{N \cdot m^2/C^2}$. We write this constant in this way, instead of using the letter k, for two reasons. Since the letter k is often used for other quantities, we avoid confusion by not using it here. Also, the constant $\varepsilon_0 = 8.85 \times 10^{-12}\ \mathrm{C^2/N \cdot m^2}$ will appear by itself in important situations.

Charged Particles

There are many microscopic particles, some of which are electrically charged and hence interact with each other through the electric interaction. The characteristics of some charged particles are shown in the table below.

Particle	Mass	Charge	Radius
Electron	9×10^{-31} kg	$-e\ (-1.6 \times 10^{-19}\ \mathrm{C})$	? (too small to measure)
Positron	9×10^{-31} kg	$+e\ (+1.6 \times 10^{-19}\ \mathrm{C})$	?
Proton	1.7×10^{-27} kg	$+e$	$\sim 1 \times 10^{-15}$ m
Antiproton	1.7×10^{-27} kg	$-e$	$\sim 1 \times 10^{-15}$ m
Muon	1.88×10^{-28} kg	$+e\ (\mu^+)$ or $-e\ (\mu^-)$	?
Pion	2.48×10^{-28} kg	$+e\ (\pi^+)$ or $-e\ (\pi^-)$	$\sim 1 \times 10^{-15}$ m

This is not an exhaustive list; you may also have learned about other charged particles, such as the W and Δ particles, and other particles that are short-lived and not commonly encountered in everyday circumstances. Positrons and antiprotons are *antimatter*. A positron is an antielectron; if a positron and an electron encounter each other they will annihilate, releasing all their energy as high-energy photons. Antimatter is therefore not found in ordinary matter.

Ordinary matter is composed of protons, electrons, and neutrons (which have about the same size and mass as protons but are uncharged). However, some other charged particles do play a role in everyday processes, as we will see, for example, in our study of sparks in a later chapter.

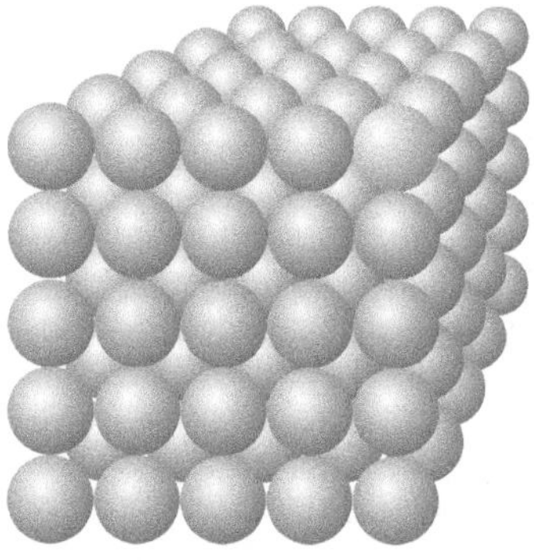

Figure 13.2 A metal lattice. The radius of a single atom is on the order of 1×10^{-10} m, and in a cube that is 1 cm on a side, there are on the order of 1×10^{23} atoms!

Size and Structure of Atoms

As you have learned in chemistry or from other sources, matter is made of tiny atoms. In a solid metal, atoms are arranged in a regular three-dimensional array, called a lattice (Figure 13.2). A cubic centimeter of solid metal in which atoms are packed right next to each other contains around 1×10^{23} atoms, which is an astronomically large number.

A neutral atom has equal numbers of protons and electrons. The protons and neutrons are all found in the nucleus at the center of the atom. The electrons are spread out in a cloud surrounding the nucleus.

Figure 13.3 The electron cloud of an iron atom. The radius of the electron cloud is approximately 1×10^{-10} m. On this scale the tiny nucleus, located at the center of the cloud, would not actually be visible.

Each atom consists of a cloud of electrons continually in motion around a central "nucleus" made of protons and neutrons. If we imagine taking a snapshot of an iron atom, with a nucleus of 26 protons and 30 neutrons surrounded by 26 moving electrons, it might look something like Figure 13.3. In this figure the nucleus is hardly visible, because it is much smaller than the electron cloud, whose radius is on the order of 1×10^{-10} m.

The nucleus of the iron atom, depicted in Figure 13.4, has a radius of roughly 4×10^{-15} m, about 25,000 times smaller than the tiny electron cloud. If an iron atom were the size of a football field, the nucleus would have a radius of only 4 mm! Yet almost all of the mass of an atom is in the nucleus, because the mass of a proton or neutron is about 2000 times larger than the mass of an electron.

Checkpoint 1 What is the approximate radius of the electron cloud of a typical atom? Which of the following charged particles are constituents of ordinary matter? Protons, positrons, electrons, antiprotons, muons

Figure 13.4 The nucleus of a iron atom contains 26 protons and 30 neutrons. Its radius is approximately 4×10^{-15} m.

13.3 THE CONCEPT OF "ELECTRIC FIELD"

Consider the following thought experiment. Having evacuated the air from the room (to avoid collisions with air molecules), you hold a proton in front of you and release it. There are no other objects nearby. You observe that the proton begins to move downward, picking up speed (accelerating) at the rate of 9.8 m/s each second (Figure 13.5). Recall that at speeds much less than the speed of light,

$$\frac{d\vec{p}}{dt} \approx m\frac{d\vec{v}}{dt} = m\vec{a}$$

Figure 13.5 Acceleration of a proton.

QUESTION What do you think is responsible for this change in the velocity of the proton?

You probably inferred that the gravitational interaction of the Earth and the proton caused the downward acceleration of the proton. That is a reasonable explanation for this observation.

Now suppose that at a later time, you release another proton in the same location. This time you observe that the proton begins to move to the right, picking up speed at a rate of 1×10^{11} m/s^2 (Figure 13.6).

Figure 13.6 Acceleration of another proton at a later time.

QUESTION What might be responsible for this change in the velocity of the proton?

This acceleration cannot be due to a gravitational interaction of the proton with the Earth, since the magnitude of the effect is too large, and the direction is not appropriate. Could it be due to a gravitational interaction with a nearby black hole? No, because if there were a black hole very nearby we would not be here to observe anything! Could it be an interaction via the strong (nuclear) force? No, because the strong force is a very short range force, and there are no other objects near enough.

It is, however, plausible that the interaction causing the acceleration of the proton could be an electric interaction, since electric interactions can have large effects and can occur over rather large distances.

QUESTION What charged objects might be responsible for this interaction, and where might they be?

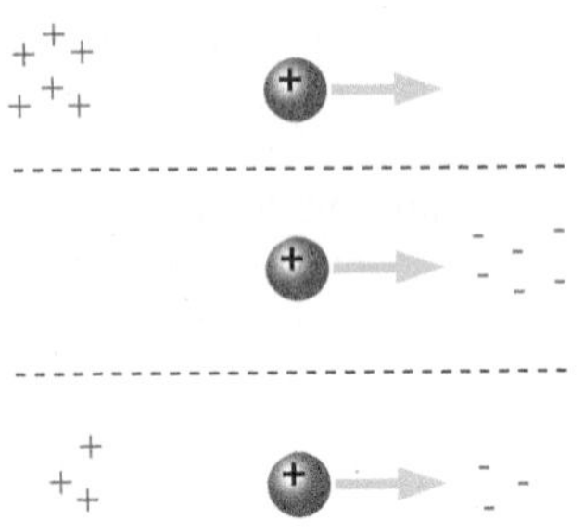

Figure 13.7 Three possible arrangements of charged particles that might be responsible for the observed high acceleration of a proton.

There are many possible configurations of charged particles that might produce the observed effect. As indicated in Figure 13.7, there might be positively charged objects to your left. Alternatively, there could be negative charges to your right. Perhaps there are both—you can't draw a definite conclusion from a single observation. There are many possible arrangements of charges in space that could produce the observed effect.

(If, however, you made several observations of the proton over some time period, you might note a change in the proton's acceleration. For example, if you noticed that the acceleration of the proton increased as it moved to the right, you might suspect that there was a negative charge to your right, which would have an increasingly large effect on the proton as it moved closer to the charge.)

QUESTION On the basis of your observations so far, can you predict what you would observe if you released an electron instead of a proton at the same location (Figure 13.8)?

Figure 13.8 A (negatively charged) electron.

An electron would accelerate to the left rather than to the right, since it has a negative charge. Its acceleration would be greater than 1×10^{11} m/s^2, because the mass of the electron is much less than the mass of the proton.

QUESTION What would happen if you placed an alpha particle (a helium nucleus, with charge $+2e$) at the observation location (Figure 13.9)?

Figure 13.9 An alpha particle with two (positively charged) protons and two (uncharged) neutrons.

Since the alpha particle has a positive charge, it would accelerate to the right as the proton did. However, because the charge of the alpha particle is twice the charge of the proton, and its mass is about four times that of a proton, the magnitude of its acceleration would be one-half of 1×10^{11} m/s^2.

QUESTION If you released a neutron at the observation location, what would you expect to observe (Figure 13.10)?

Figure 13.10 An (uncharged) neutron.

Since the neutron has no electric charge, it would experience no electric force, and should simply accelerate toward the Earth at 9.8 m/s^2.

QUESTION Finally, suppose we do not put any particle at the observation location. Is there anything there?

Since we know that if we were to put a charged particle at that location, it would experience a force, it seems that in a certain sense there is something there, waiting for a charged particle to interact with. This "virtual force" is called "electric field." The electric field created by a charge is present throughout space at all times, whether or not there is another charge around to feel its effects. The electric field created by a charge penetrates through matter. The field permeates the neighboring space, lying in wait to affect anything brought into its web of interaction.

We define the electric field at a location by the following equation:

DEFINITION OF ELECTRIC FIELD

$$\vec{F}_2 = q_2\vec{E}_1$$

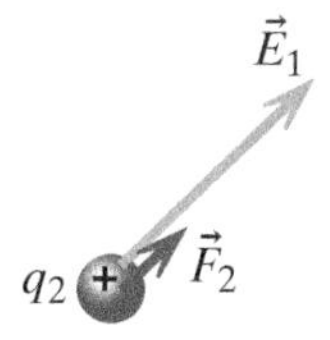

Figure 13.11 A particle with charge q_2 experiences a force $\vec{F}_2$ because of its interaction with the electric field $\vec{E}_1$ produced by all other charged particles in the vicinity.

This equation says that the force on particle 2 is determined by the charge of particle 2 and by the electric field $\vec{E}_1$ made by all other charged particles in the vicinity, as shown in Figure 13.11.

How might we measure the electric field at a given location in space? We can't see it, but we can measure it indirectly. If we place a charge at that location in space, we can measure the force on the charge due to its interaction with the electric field at that location. We can determine the magnitude and direction of $\vec{E}$ by measuring a force on a known charge q:

$$\vec{E} = \vec{F}/q$$

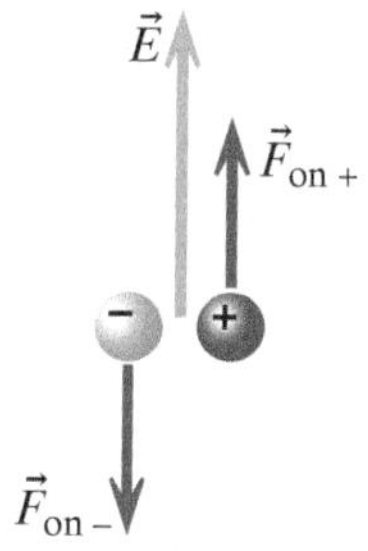

Figure 13.12 A positive charge experiences a force in the direction of the electric field at its location, while a negative charge experiences a force in the opposite direction.

Electric field therefore has units of newtons per coulomb. Note that it doesn't matter whether the affected charge q is positive or negative. If $\vec{E}$ points north, a positive charge would experience a force to the north, while a negative charge would experience a force to the south (Figure 13.12); dividing this force by a negative q would still yield a vector $\vec{E}$ pointing north. Though we will usually write $\vec{E}$ for the value of the electric field at a particular location and time, this is really shorthand for $\vec{E}(x,y,z,t)$, since the electric field has a value at all locations and times.

Drawing Fields: Tail of Arrow at Observation Location

In diagrams we will use arrows to represent the electric field (a vector) at a specific observation location. By convention, we will almost always draw such an arrow with its tail at the location where the field is measured (the "observation location"). The length of the arrow is proportional to the magnitude of the field.

EXAMPLE **Electric Force and Electric Field**

The charge of an alpha particle (a helium nucleus, consisting of two protons and two neutrons) is $2e = 2(1.6\times10^{-19}\,\text{C})$. An alpha particle at a particular location experiences a force of $\langle 0, -9.6\times10^{-17}, 0\rangle$ N. What is the electric field at that location? If the alpha particle were removed and an electron were placed at that location, what force would the electron experience?

Solution To find the field we divide the force on the alpha particle by the charge of the alpha particle:

$$\vec{E} = \frac{\vec{F}}{q}$$

$$\vec{E} = \frac{\langle 0, -9.6\times10^{-17}, 0\rangle\,\text{N}}{2(1.6\times10^{-19}\,\text{C})} = \langle 0, -300, 0\rangle\,\text{N/C}$$

To find the force on the electron, we multiply the electric field by the charge on the electron:

$$\vec{F} = (-e)\vec{E}_1$$
$$\vec{F} = (-1.6\times10^{-19}\ \text{C})\langle 0,-300,0\rangle\ \text{N/C} = \langle 0, 4.8\times10^{-17}, 0\rangle\ \text{N}$$

Since the electron's charge is negative, the force on the electron is in a direction opposite to the direction of the electric field. Note that we did not need to know the source of the electric field in order to calculate this force.

Checkpoint 2 What force would a proton experience if placed at the same location (see the preceding example)? What force would a neutron at the same location experience?

No "Self-Force"

It is important to note that a point charge is not affected by its *own* electric field. A point charge does not exert a force on itself! In the next section we will see that mathematically this is reassuring, because at the location of a point charge its own electric field would be infinite ($1/0^2$). Physically, this makes sense too, since after all, the charge can't start itself moving, nor is there any way to decide in what direction it should go.

When we use the electric field concept with point charges we always talk about a charge q_1 (the *source* charge) making a field $\vec{E}_1$, and a different charge q_2 in a different place being affected by that field with a force $\vec{F}_{\text{on } q_2} = q_2\vec{E}_1$.

The Physical Concept of "Field"

The word "field" has a special meaning in mathematical physics. A field is a physical quantity that has a value at every location in space. Its value at every location can be a scalar or a vector.

For example, the temperature in a room is a scalar field. At every location in the room, the temperature has a value, which we could write as $T(x,y,z)$, or as $T(x,y,z,t)$ if it were changing with time. The air flow in the room is a vector field. At every location in the room, air flows in a particular direction with a particular speed. Electric field is a vector field; at every location in space surrounding a charge the electric field has a magnitude and a direction.

QUESTION Think of another example of a quantity that is a field.

Figure 13.13 Gravitational field of the Earth, in a plane cutting through the center of the Earth.

The field concept is also used with gravitation. Instead of saying that the Earth exerts a force on a falling object, we can say that the mass of the Earth creates a "gravitational field" surrounding the Earth, and any object near the Earth is acted upon by the gravitational field at that location (Figure 13.13). Gravitational field has units of newtons per kilogram. At a location near the Earth's surface we can say that there is a gravitational field $\vec{g}$ pointing downward (that is, toward the center of the Earth) of magnitude $g = 9.8$ newtons per kilogram.

A falling object of mass $m = 10$ kg experiences a gravitational force of magnitude $mg = (10\ \text{kg})(9.8\ \text{N/kg}) = 98$ newtons. If the object is moving at a speed slow compared to the speed of light, we can calculate its acceleration, $a = F/m = mg/m = 9.8\ \text{m/s}^2$.

Of course, farther away from the surface of the Earth, the gravitational field of the Earth is smaller in magnitude, just as the electric field of a charge decreases in magnitude as the distance from the charge increases.

The field concept also applies to magnetism. As we will see in a later chapter, electric currents, including electric currents inside the Earth, create "magnetic fields" that affect compass needles.

QUESTION Find an example of a quantity that is *not* a field.

In contrast, many quantities are not fields. The position and velocity at one instant of a car going around a race track are not fields. The distance between the Earth and the Sun is not a field. The rate of change of the momentum of the Moon is not a field.

13.4 THE ELECTRIC FIELD OF A POINT CHARGE

By asking what expression for $\vec{E}_1$ satisfies the equation $\vec{F}_2 = q_2\vec{E}_1$ for the force on one point charge by another, we can find an algebraic expression for the electric field at a location in space called the "observation location"—the location where we detect or measure the field—due to a charged particle q_1 (the "source charge") at the source location. The electric field at the observation location (marked by an "×" in Figure 13.14) is:

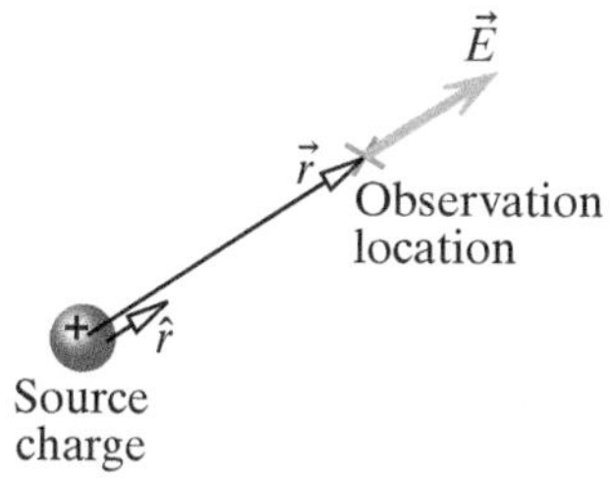

Figure 13.14 The relative position vector $\vec{r}$ extends from the source charge to the observation location, marked by "×". $\hat{r}$ is a unit vector with the same direction as the vector $\vec{r}$. Here the source charge is positive, so the electric field points away from the source charge.

ELECTRIC FIELD OF A POINT CHARGE

$$\vec{E}_1 = \frac{1}{4\pi\varepsilon_0}\frac{q_1}{|\vec{r}|^2}\hat{r}$$

where q_1 is the source charge, $\vec{r}$ is the relative position vector giving the position of the observation location relative to the source charge, and $\hat{r}$ is the unit vector in the direction of $\vec{r}$. $|\vec{r}|$, the magnitude of $\vec{r}$, is the distance from the source location to the observation location.

$$\vec{E}_1 = \frac{1}{4\pi\varepsilon_0}\frac{q_1}{|\vec{r}|^2}\hat{r}$$

Figure 13.15 The direction of the electric field depends on both the unit vector $\hat{r}$ and the sign of the source charge.

This equation for electric field of a point charge may look complex, but we can make sense of it by taking it apart and looking at one piece at a time.

Direction of Electric Field

The direction of the electric field at the observation location depends on both the direction of $\hat{r}$ and the sign of the source charge q_1, which may be positive or negative (Figure 13.15).

If the source charge q_1 is positive, the quantity $q_1\hat{r}$ is in the direction of $\hat{r}$, pointing away from the source charge (Figure 13.14).

If the source charge q_1 is negative, the direction of $q_1\hat{r}$ is in the direction of $-\hat{r}$, pointing toward the source charge (Figure 13.16).

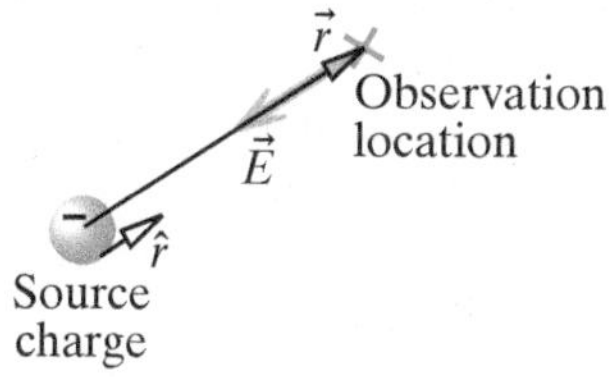

Figure 13.16 Here the source charge is negative, so the electric field points toward the source charge, in the direction of $-\hat{r}$.

Distance Dependence of Electric Field

As in the gravitational force law and the electric force law, the square of the distance appears in the denominator (Figure 13.17). This means that the electric field of a point charge depends very strongly on distance. For example, if you double the distance between the source location and the observation location, the only thing that changes is the denominator, which gets four times bigger, so the electric field is only 1/4 as large as before.

$$\vec{E}_1 = \frac{1}{4\pi\varepsilon_0}\frac{q_1}{|\vec{r}|^2}\hat{r}$$

Figure 13.17 The magnitude of the electric field is inversely proportional to the square of the distance from source charge to observation location.

QUESTION If you move the source charge five times farther away from the observation location, how does the electric field change?

The magnitude of the electric field decreases by a factor of 25.

$$\vec{E}_1 = \frac{1}{4\pi\varepsilon_0}\frac{q_1}{|\vec{r}|^2}\hat{r}$$

Figure 13.18 The magnitude of the electric field is proportional to the magnitude of the source charge.

Magnitude of Charge

Because q_1, the charge of the source particle, appears in the numerator (Figure 13.18), the larger the charge of the source particle, the larger the magnitude of the electric field at any location in space.

> QUESTION At a particular observation location, the magnitude of the electric field due to a point charge is found to be 10 N/C. If the source charge were replaced by a different particle whose charge was seven times larger, how would this change the electric field at the observation location?

The electric field would now have a magnitude of 70 N/C.

The Constant $\frac{1}{4\pi\varepsilon_0} = 9 \times 10^9\ \mathbf{N \cdot m^2/C^2}$

This may seem an odd way to write a single constant; this is done because it is based on the constant $\varepsilon_0 = 8.85 \times 10^{-12}\,\mathrm{C^2/N \cdot m^2}$, which is in a sense a more fundamental constant, as we will see in a later chapter.

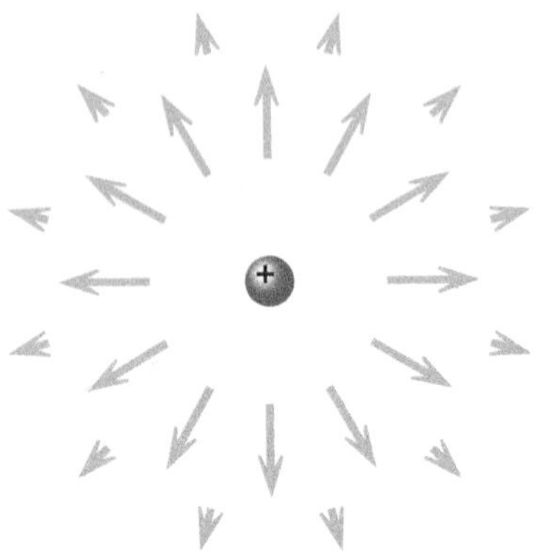

Figure 13.19 Electric field of a positive point charge, at various observation locations in a plane containing the charge.

Magnitude of Electric Field

The scalar quantity

$$\frac{1}{4\pi\varepsilon_0}\frac{q_1}{|\vec{r}|^2}$$

is *not* necessarily equal to the magnitude of the electric field. Because the charge q_1 can be negative or positive, the expression above may be negative or positive. The magnitude of a vector is a positive quantity. To get the magnitude of a field it is necessary to take the absolute value of the signed scalar quantity above. (Remember that the sign of q_1 helps determine the direction of the field.)

Patterns of Electric Field Near Point Charges

The electric field of a point charge is spherically symmetric. That is, the magnitude of the field at any location a distance r from the source charge is the same, and at each location the field points either toward the source charge or away from it, depending on the sign of the source charge. Figure 13.19 shows the electric field due to a positive source charge, at various locations in a plane containing the charge. The tail of each arrow is placed at the location where the electric field was measured. Notice that at every observation location the electric field points radially away from the positive source charge, and that the magnitude of the field decreases rapidly with distance from the charge.

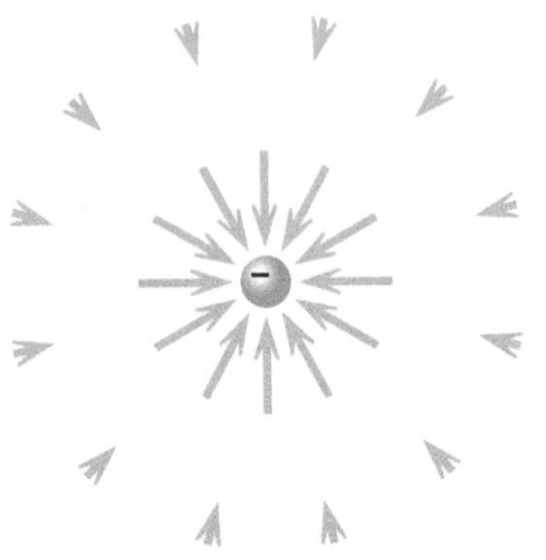

Figure 13.20 The electric field of a negative point charge, at various observation locations in a plane containing the charge.

Figure 13.20 shows the electric field due to a negative source charge. At each observation location, the electric field points radially inward toward the negative source charge (in the direction of $-\hat{r}$), so a positively charged particle would experience a force toward the negative charge. The magnitude of the field decreases rapidly with distance from the charge. We will frequently use two-dimensional diagrams because they are easy to draw, but it is important to remember the 3D character of electric fields in space.

EXAMPLE **Field of a Particle**

A particle with charge $+2\,\mathrm{nC}$ (a nanocoulomb is $1 \times 10^{-9}\,\mathrm{C}$) is located at the origin. What is the electric field due to this particle at a location $\langle -0.2, -0.2, -0.2 \rangle$ m?

Solution

$$\vec{r} = \langle\text{observation location}\rangle - \langle\text{source location}\rangle$$
$$= \langle -0.2, -0.2, -0.2\rangle\ \text{m} - \langle 0,0,0\rangle\text{m}$$
$$\vec{r} = \langle -0.2, -0.2, -0.2\rangle\text{m}$$
$$|\vec{r}| = \sqrt{(-0.2)^2 + (-0.2)^2 + (-0.2)^2}\text{m} = 0.35\ \text{m}$$
$$\hat{r} = \frac{\vec{r}}{|\vec{r}|} = \frac{\langle -0.2, -0.2, -0.2\rangle\ \text{m}}{0.35\ \text{m}} = \langle -0.57, -0.57, -0.57\rangle$$
$$\frac{1}{4\pi\varepsilon_0}\frac{q}{r^2} = \left(9\times 10^9\,\frac{\text{N}\,\text{m}^2}{\text{C}^2}\right)\left(\frac{2\times 10^{-9}\,\text{C}}{0.35^2\ \text{m}^2}\right) = 147\,\frac{\text{N}}{\text{C}}$$

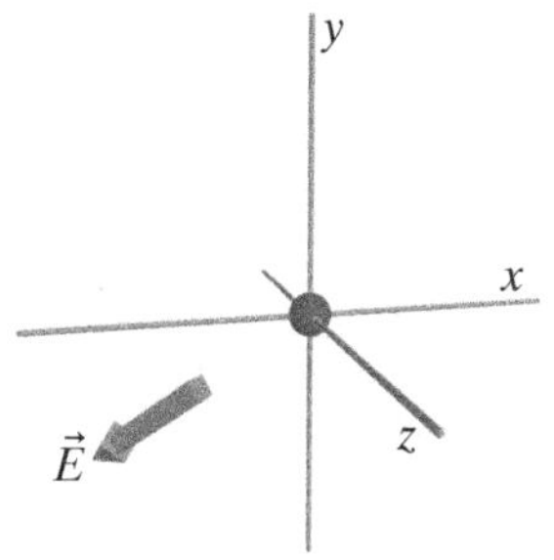

Figure 13.21 The electric field calculated in the example. This orientation of coordinate axes ($+x$ to the right, $+y$ up, and $+z$ out of the page, toward you) is the standard orientation we will use in this book.

$$\vec{E} = \frac{1}{4\pi\varepsilon_0}\frac{q}{r^2}\hat{r}$$
$$= \left(147\,\frac{\text{N}}{\text{C}}\right)\langle -0.57, -0.57, -0.57\rangle$$
$$= \langle -84, -84, -84\rangle\,\frac{\text{N}}{\text{C}}$$

The electric field points away from the positively charged particle, as shown in Figure 13.21.

EXAMPLE **Source Location**

The electric field at location $\langle -0.13, 0.14, 0\rangle$ m is found to be $\langle 6.48\times 10^3, -8.64\times 10^3, 0\rangle$ N/C. The only charged particle in the surroundings has charge $-3\,\text{nC}$. What is the location of this particle?

Solution Draw a diagram (Figure 13.22) to find the relative position of the source and the observation location. Remember that the electric field of a negatively charged particle points toward the particle. Find the magnitude of the electric field, and use this to determine the distance from the source to the observation location:

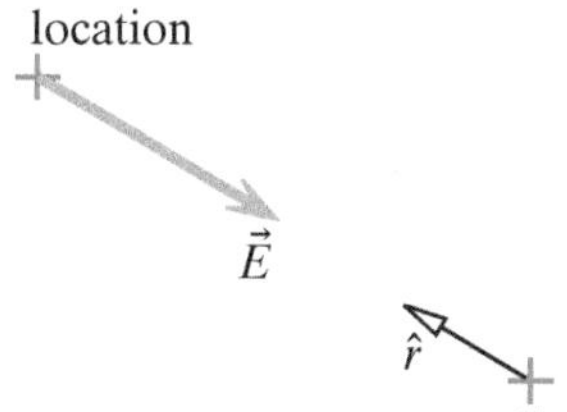

Figure 13.22 Draw this diagram, using the fact that the electric field due to a negative charge points toward the source charge.

$$|\vec{E}| = \sqrt{(6.48\times 10^3)^2 + (-8.64\times 10^3)^2}\,\text{N/C} = 1.08\times 10^4\,\text{N/C}$$
$$|\vec{E}| = \frac{1}{4\pi\varepsilon_0}\frac{q}{|\vec{r}|^2}$$
$$|\vec{r}| = \sqrt{\frac{1}{4\pi\varepsilon_0}\frac{q}{|\vec{E}|}}$$
$$= \sqrt{9\times 10^9\,\text{Nm}^2/\text{C}^2\,\frac{3\times 10^{-9}\,\text{C}}{1.08\times 10^4\,\text{N/C}}}$$
$$= 0.05\,\text{m}$$

Find the unit vector $\hat{r}$. From the diagram, note that in this case (because the source charge is negative) $\hat{r} = -\hat{E}$:

$$\hat{E} = \frac{\vec{E}}{|\vec{E}|}$$
$$= \frac{\langle 6.48\times 10^3, -8.64\times 10^3, 0\rangle\,\text{N/C}}{1.08\times 10^4\,\text{N/C}}$$
$$= \langle 0.6, -0.8, 0\rangle$$
$$\hat{r} = -\hat{E} = \langle -0.6, 0.8, 0\rangle$$

Find $\vec{r}$, and use this to find the source location, where location 2 is the observation location and location 1 is the source location:

$$\begin{aligned}\vec{r} &= |\vec{r}|\,\hat{r}\\ &= (0.05\,\text{m})\langle -0.6, 0.8, 0\rangle = \langle -0.03, 0.04, 0\rangle\ \text{m}\\ \vec{r} &= \vec{r}_2 - \vec{r}_1\\ \vec{r}_1 &= \vec{r}_2 - \vec{r} = \langle -0.13, 0.14, 0\rangle - \langle -0.03, 0.04, 0\rangle\\ &= \langle -0.10, 0.10, 0\rangle\ \text{m}\end{aligned}$$

Checkpoint 3 A particle with charge +1 nC (a nanocoulomb is 1×10^{-9} C) is located at the origin. What is the electric field due to this particle at a location $\langle 0.1, 0, 0\rangle$ m?

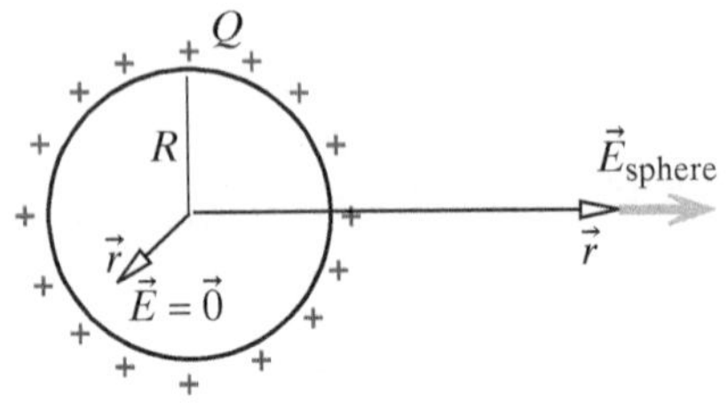

Figure 13.23 The electric field of a uniformly charged sphere, at locations outside and inside the sphere.

The Electric Field of a Uniformly Charged Sphere

In Chapter 15 we will apply the superposition principle to calculate the electric field of macroscopic objects with charge spread out over their surfaces. One of the results of such a calculation is sufficiently useful that we present it now, without going through the calculations involved.

As shown in Figure 13.23, a spherical object of radius R with charge Q uniformly spread out over its surface produces an electric field with the following distance dependence, if $\vec{r}$ is a vector from the center of the sphere to the observation location:

$$\vec{E}_{\text{sphere}} = \frac{1}{4\pi\varepsilon_0}\frac{Q}{r^2}\hat{r} \quad \text{for } r > R \text{ (outside the sphere)}$$

$$\vec{E}_{\text{sphere}} = \vec{0} \quad \text{for } r < R \text{ (inside the sphere)}$$

In other words, at locations outside the charged sphere, the electric field due to the sphere is the same as if all its charge were located at its center. Inside the sphere, the electric field due to all the charge on the surface of the sphere adds up to zero!

Informally, we may say that a uniformly charged sphere "acts like" a point charge, at locations outside the sphere. This is not only because the electric field due to the sphere is the same as the electric field of a point charge located at the center of the sphere, but also because the charged sphere responds to applied electric fields (due to other charges) in the same way as would a point charge located at its center.

Checkpoint 4 A sphere with radius 1 cm has a charge of 5×10^{-9} C spread out uniformly over its surface. What is the magnitude of the electric field due to the sphere at a location 4 cm from the center of the sphere?

13.5 SUPERPOSITION OF ELECTRIC FIELDS

An important property of electric field is a consequence of the superposition principle for electric forces: the net electric field due to two or more charges is the vector sum of each field due to each individual charge. Superposition as a vector sum holds true for forces in general, so it also holds true for fields, because field is force per unit charge.

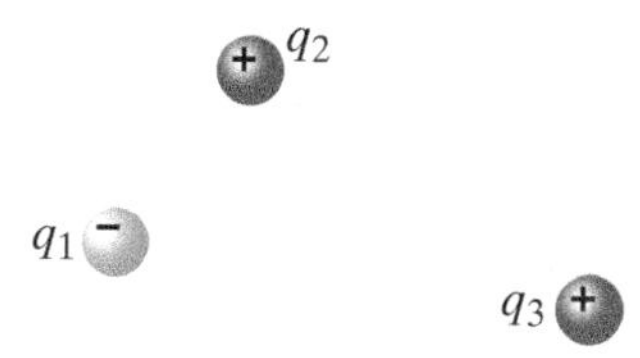

Figure 13.24 Three interacting charged particles.

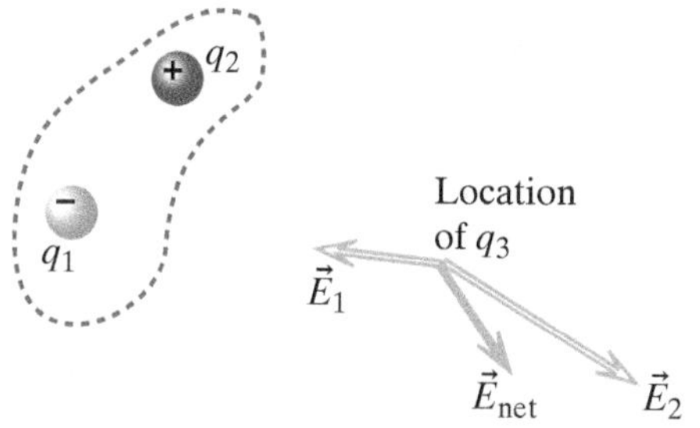

Figure 13.25 Electric field at the location of q_3.

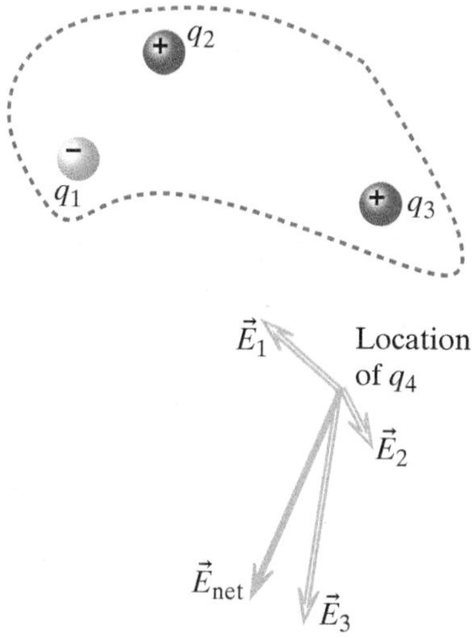

Figure 13.26 Electric field at the location of q_4.

We state the superposition principle here in terms of electric field, in a form intended to emphasize an important aspect of the principle:

THE SUPERPOSITION PRINCIPLE

The net electric field at a location in space is the vector sum of the individual electric fields contributed by all charged particles located elsewhere.

The electric field contributed by a charged particle is unaffected by the presence of other charged particles.

This may seem obvious, but it actually is a bit subtle. The presence of other particles does not affect the fundamental electric interaction between each pair of particles! The interaction doesn't get "used up"; the interaction between one particle and another is unaffected by the presence of a third particle.

A significant consequence of the superposition principle is that matter cannot "block" electric fields; electric fields "penetrate" through matter.

Applying the Superposition Principle

Consider three charges q_1, q_2, and q_3 that interact with each other, as shown in Figure 13.24. We could calculate the net force that q_1 and q_2 exert on q_3 by using Coulomb's law and vector addition.

However, let's use the field concept to find the force on q_3 due to the other two charges. We pretend that q_3 isn't there, and we find the electric field $\vec{E}_{net}$ due to q_1 and q_2, at the location where q_3 would be. We can then calculate the force on q_3 as $q_3\vec{E}_{net}$. In Figure 13.25 the scheme is shown graphically. $\vec{E}_{net}$ is the vector sum of $\vec{E}_1$ (due to q_1) and $\vec{E}_2$ (due to q_2), at the location of q_3.

Advantage of Electric Field in Calculations

We could just as easily have calculated the force on q_3 by using Coulomb's law directly, without using the more abstract concept of electric field. What advantage is there to using electric field? One of the advantages is that once you have calculated the field $\vec{E}_{net}$ due to q_1 and q_2, you can quickly calculate the force $Q\vec{E}_{net}$ on any amount of charge Q placed at that location, not just the force on q_3. The "force per unit charge" idea simplifies these kinds of calculations.

If we are interested in the effect that all three charges would have on a fourth charge q_4 placed at some location, we would include all three charges as the source of the electric field at that location, but exclude q_4, as shown in Figure 13.26.

Knowing the electric field $\vec{E}_{net}$ at a particular location, we know what force would act on a charge q_4 placed at that location: it would simply be $q_4\vec{E}_{net}$.

EXAMPLE **Electric Field and Force Due to Two Charges**

A small object with charge $Q_1 = 6\,\text{nC}$ is located at the origin. A second small object with charge $Q_2 = -5\,\text{nC}$ is located at $\langle 0.05, 0.08, 0\rangle$ m. What is the net electric field at location A $\langle -0.04, 0.08, 0\rangle$ m due to Q_1 and Q_2? If a small object with a charge of $Q_3 = -3\,\text{nC}$ were placed at location A, what would be the force on this object?

Solution Diagram: Figure 13.27

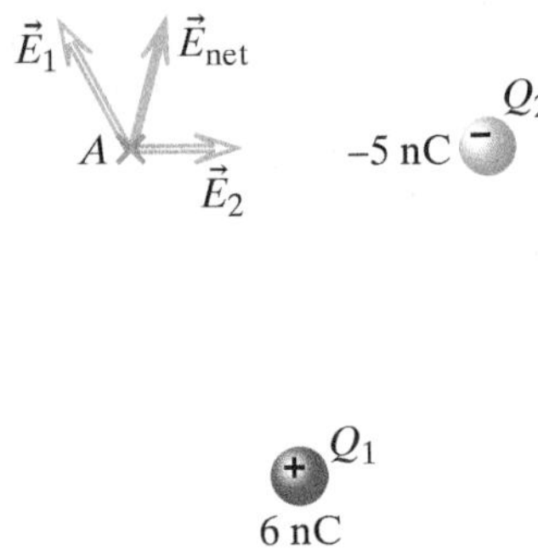

Figure 13.27 Two point charges make an electric field at location A.

$$\vec{r}_1 = \langle -0.04, 0.08, 0 \rangle \text{ m} - \langle 0,0,0 \rangle \text{ m} = \langle -0.04, 0.08, 0 \rangle \text{ m}$$

$$|\vec{r}_1| = \sqrt{(-0.04)^2 + (0.08)^2} = 0.0894 \text{ m}$$

$$\hat{r}_1 = \frac{\vec{r}_1}{|\vec{r}_1|} = \frac{\langle -0.04, 0.08, 0 \rangle \text{ m}}{(0.0894 \text{ m})} = \langle -0.447, 0.894, 0 \rangle$$

$$\vec{E}_1 = \frac{1}{4\pi\varepsilon_0} \frac{Q_1}{|\vec{r}_1|^2} \hat{r}_1$$

$$= \left(9 \times 10^9 \frac{\text{Nm}^2}{\text{C}^2}\right) \frac{(6 \times 10^{-9} \text{ C})}{(0.0894 \text{ m})^2} \langle -0.447, 0.894, 0 \rangle$$

$$\vec{E}_1 = \langle -3.02 \times 10^3, 6.04 \times 10^3, 0 \rangle \text{ N/C}$$

$$\vec{r}_2 = \langle -0.04, 0.08, 0 \rangle \text{ m} - \langle 0.05, 0.08, 0 \rangle \text{ m} = \langle -0.09, 0, 0 \rangle \text{ m}$$

$$|\vec{r}_2| = \sqrt{(-0.09)^2} = 0.09 \text{ m}$$

$$\hat{r}_2 = \frac{\vec{r}_2}{|\vec{r}_2|} = \frac{\langle -0.09, 0, 0 \rangle \text{ m}}{(0.09 \text{ m})} = \langle -1, 0, 0 \rangle$$

$$\vec{E}_2 = \frac{1}{4\pi\varepsilon_0} \frac{Q_2}{|\vec{r}_2|^2} \hat{r}_2$$

$$= \left(9 \times 10^9 \frac{\text{Nm}^2}{\text{C}^2}\right) \frac{(-5 \times 10^{-9} \text{ C})}{(0.09 \text{ m})^2} \langle -1, 0, 0 \rangle$$

$$\vec{E}_2 = \langle 5.56 \times 10^3, 0, 0 \rangle \text{ N/C}$$

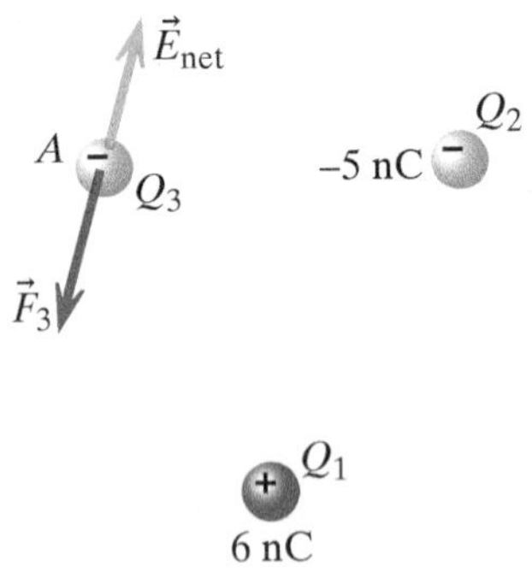

Figure 13.28 When a negative charge is placed at location A, it experiences a force in a direction opposite to the net field at that location.

$$\vec{E}_{net} = \vec{E}_1 + \vec{E}_2$$

$$= \langle -3.02 \times 10^3, 6.04 \times 10^3, 0 \rangle \text{ N/C} + \langle 5.56 \times 10^3, 0, 0 \rangle \text{ N/C}$$

$$\vec{E}_{net} = \langle 2.54 \times 10^3, 6.04 \times 10^3, 0 \rangle \text{ N/C}$$

In Figure 13.28 the charge Q_3 has been placed at the observation location.

$$\vec{F}_3 = Q_3 \vec{E}_{net}$$

$$= (-3 \times 10^{-9} \text{ C}) \langle 2.54 \times 10^3, 6.04 \times 10^3, 0 \rangle \text{ N/C}$$

$$= \langle -7.51 \times 10^{-6}, -1.81 \times 10^{-5}, 0 \rangle \text{ N}$$

The individual contributions and the net field are shown in Figure 13.27. When Q_3 is placed at location A, it experiences a force in the direction shown in Figure 13.28. The force on Q_3 is in a direction opposite to the direction of the net electric field at A, because Q_3 is negative.

13.6 THE ELECTRIC FIELD OF A DIPOLE

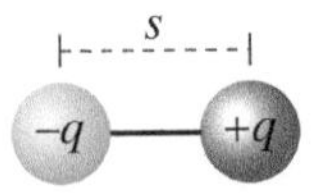

Figure 13.29 An electric dipole.

Neutral matter contains both positive and negative charges (protons and electrons). The simplest piece of neutral matter that we can analyze in detail is an "electric dipole" consisting of two equally but oppositely charged point-like objects, separated by a distance s, as shown in Figure 13.29. Dipoles occur frequently in nature, so this analysis has practical applications. For example, a single molecule of HCl is an electric dipole: the H end is somewhat positive and the Cl end is somewhat negative. Moreover, the analysis of a single dipole provides a basis for analyzing more complicated forms of matter.

By applying the superposition principle we can calculate the electric field due to a dipole at any location in space. We will be particularly interested in the electric field in two important locations: at a location along the axis of the dipole and at a location along the perpendicular bisector of the axis.

Along the Dipole Axis

Figure 13.30 shows a horizontal dipole with charge $+q$ and $-q$, and separation s (here q is a positive number, so $+q$ is a positive charge, and $-q$ is a negative charge). Its center is at the origin. We will apply the superposition principle to find the electric field $\vec{E}_1$ at an observation location on the axis of the dipole (in this case, the x axis), due to the dipole.

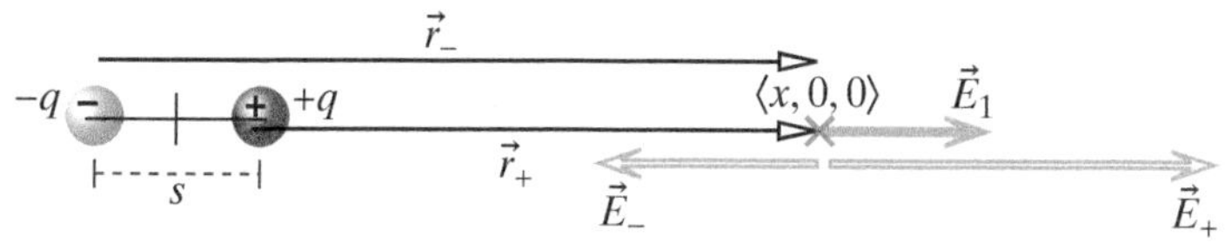

Figure 13.30 Electric field of a dipole at a location on the dipole axis (in the case shown, this is the x axis).

For the orientation shown, we can see that the net field $\vec{E}_1$ will be in the $+x$ direction, because at this location the field $\vec{E}_+$ of the closer $+q$ charge is larger than the field $\vec{E}_-$ of the more distant $-q$ charge. Note that if the dipole were oriented with its negative end on the right, $\vec{E}_1$ would point in the opposite direction. For the orientation shown, the electric field $\vec{E}_+$ due only to the positive end of the dipole axis is:

$$\vec{r}_+ = \langle x,0,0\rangle - \left\langle \frac{s}{2},0,0\right\rangle = \left\langle x-\frac{s}{2},0,0\right\rangle$$

$$|\vec{r}_+| = \sqrt{\left(x-\frac{s}{2}\right)^2 + 0^2 + 0^2} = \left(x-\frac{s}{2}\right)$$

$$\hat{r}_+ = \frac{\left\langle x-\frac{s}{2},0,0\right\rangle}{\left(x-\frac{s}{2}\right)} = \langle 1,0,0\rangle$$

$$\vec{E}_+ = \frac{1}{4\pi\varepsilon_0}\frac{q}{\left(x-\frac{s}{2}\right)^2}\langle 1,0,0\rangle = \left\langle \frac{1}{4\pi\varepsilon_0}\frac{q}{\left(x-\frac{s}{2}\right)^2},0,0\right\rangle$$

Similarly, the electric field $\vec{E}_-$ due only to the negative end of the dipole is:

$$\vec{r}_- = \langle x,0,0\rangle - \left\langle \frac{-s}{2},0,0\right\rangle = \left\langle x+\frac{s}{2},0,0\right\rangle$$

$$|\vec{r}_-| = x+\frac{s}{2}$$

$$\hat{r}_- = \langle 1,0,0\rangle$$

$$\vec{E}_- = \frac{1}{4\pi\varepsilon_0}\frac{-q}{\left(x+\frac{s}{2}\right)^2}\langle 1,0,0\rangle = \left\langle \frac{1}{4\pi\varepsilon_0}\frac{-q}{\left(x+\frac{s}{2}\right)^2},0,0\right\rangle$$

We add the two contributions to get the net field at the observation location:

$$\vec{E}_1 = \vec{E}_+ + \vec{E}_- = \left\langle \left(\frac{1}{4\pi\varepsilon_0}\frac{q}{\left(x-\frac{s}{2}\right)^2} + \frac{1}{4\pi\varepsilon_0}\frac{-q}{\left(x+\frac{s}{2}\right)^2}\right),0,0\right\rangle$$

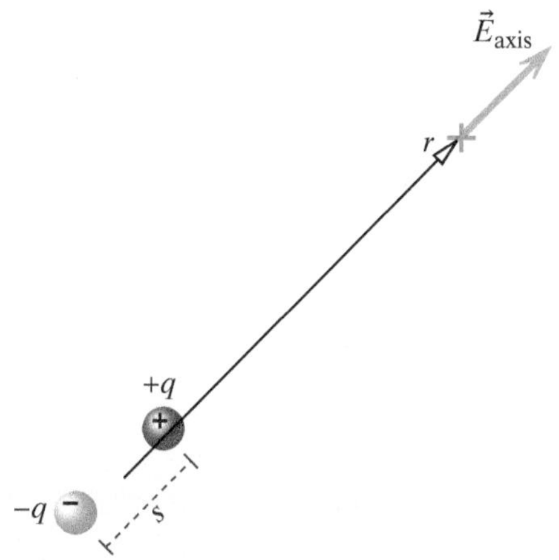

Figure 13.31 The electric field due to a dipole, at a location on the axis of the dipole, for a dipole with an arbitrary orientation.

Because the dipole could be oriented along any axis (see Figure 13.31), we can make this result more general by changing the x to an r, representing the distance from the center of the dipole to an observation location on the axis of the dipole, however the dipole may be oriented. Because the direction of the electric field at a location on the dipole axis depends on the orientation of the dipole, it is more useful to write an algebraic expression for the magnitude of the dipole, and to figure out the direction by looking at the particular situation.

$$|\vec{E}_{\text{on axis}}| = \left| \frac{1}{4\pi\varepsilon_0}\frac{q}{\left(r-\frac{s}{2}\right)^2} + \frac{1}{4\pi\varepsilon_0}\frac{-q}{\left(r+\frac{s}{2}\right)^2} \right|$$

$$= \frac{1}{4\pi\varepsilon_0} q \left(\frac{1}{\left(r-\frac{s}{2}\right)^2} - \frac{1}{\left(r+\frac{s}{2}\right)^2} \right)$$

$$= \frac{1}{4\pi\varepsilon_0} \frac{2qsr}{\left(r-\frac{s}{2}\right)^2 \left(r+\frac{s}{2}\right)^2}$$

QUESTION If the dipole in Figure 13.31 were rotated through an angle of 180°, so the negative end was closer to the observation location, what would be the direction of the electric field at the observation location?

If the dipole were rotated, the electric field would now point toward the dipole.

Approximation: Far from the Dipole

This is still a fairly messy expression, and it does not give us much information about how fast the magnitude of the field decreases as we get farther away from the dipole. Let's try to simplify it. If we are far from the dipole, as is normally the case when we interact with molecular dipoles, then the dipole separation s is very small compared to the distance r from the dipole, so:

$$\text{If } r \gg s, \text{ then } \left(r-\frac{s}{2}\right)^2 \approx \left(r+\frac{s}{2}\right)^2 \approx r^2$$

Using this approximation, we can simplify the expression:

$$|\vec{E}_{\text{on axis}}| \approx \frac{1}{4\pi\varepsilon_0}\frac{2qsr}{r^4} = \frac{1}{4\pi\varepsilon_0}\frac{2qs}{r^3}$$

ELECTRIC FIELD OF A DIPOLE, ON THE DIPOLE AXIS

$$|\vec{E}_{\text{axis}}| \approx \frac{1}{4\pi\varepsilon_0}\frac{2qs}{r^3} \quad \text{at a location on the dipole axis, if } r \gg s$$

q is the magnitude of one of the charges.
s is the separation between the charges.
r is the distance from the center of the dipole to the observation location.
The direction is determined by looking at the particular situation.

Since a dipole can have any orientation, we give only the magnitude of the field. For a particular dipole you will need to determine the direction of the field at a particular observation location by looking at the orientation of the dipole.

Is This Result Reasonable?

QUESTION Should we have expected the magnitude of the electric field of the dipole to be proportional to $1/r^2$?

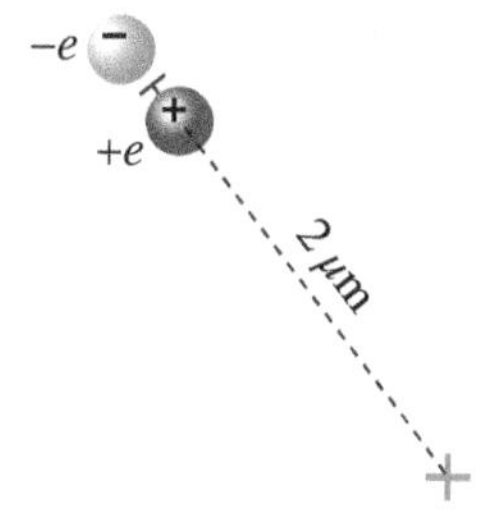

Figure 13.32 An HCl molecule.

The electric field of a single point charge is proportional to $1/r^2$, but when we add the electric fields contributed by more than one charge, the result may have quite a different distance dependence.

QUESTION Does our result have the correct units?

The units of electric field come out to N/C, as they should.

$$\left(\frac{\text{N}\cdot\text{m}^2}{\text{C}^2}\right)\frac{(\text{C})(\text{m})}{(\text{m}^3)}=\frac{\text{N}}{\text{C}}$$

EXAMPLE **HCl Molecule**

An HCl molecule is a dipole, which can be considered to be a particle of charge $+e$ (a H^+ ion) and a particle of charge $-e$ (a Cl^- ion) separated by a distance of about 1×10^{-10} m. What are the magnitude and direction of the electric field due to the HCl molecule at the location indicated in Figure 13.32, which is 2 μm (2×10^{-6} m) from the center of the molecule, on the molecular axis?

Solution

$$|\vec{E}|\approx\frac{1}{4\pi\varepsilon_0}\frac{2qs}{r^3}=\left(9\times10^9\,\frac{\text{N}\cdot\text{m}^2}{\text{C}^2}\right)\frac{2(1.6\times10^{-19}\,\text{C})(1\times10^{-10}\,\text{m})}{(2\times10^{-6}\,\text{m})^3}$$

$$=3.6\times10^{-2}\,\frac{\text{N}}{\text{C}}$$

Because the positive end of the dipole is closer to the observation location than the negative end, the electric field points away from the dipole, along the dipole axis.

Along the Perpendicular Axis

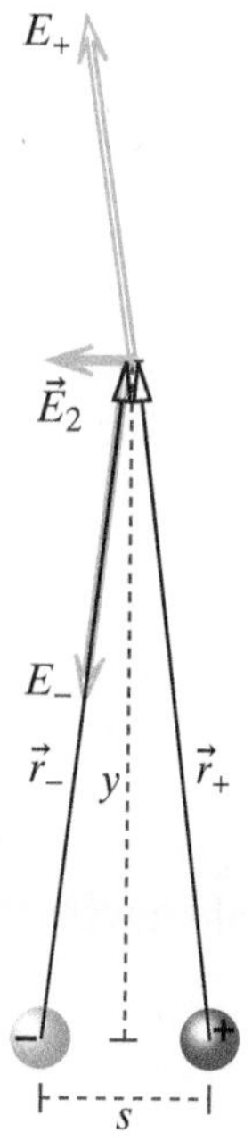

Figure 13.33 Electric field of a dipole at a location on the perpendicular axis.

Next consider the electric field $\vec{E}_2$ at an observation location on the axis perpendicular to the axis of the dipole. For a dipole aligned as shown in Figure 13.33, this is the y axis. First we need to find $\vec{r}_+$ and $\vec{r}_-$, the vectors from each source charge to the observation location.

$$\vec{r}_+=\langle 0,y,0\rangle-\left\langle\frac{s}{2},0,0\right\rangle=\left\langle-\frac{s}{2},y,0\right\rangle$$

$$\vec{r}_-=\langle 0,y,0\rangle-\left\langle-\frac{s}{2},0,0\right\rangle=\left\langle\frac{s}{2},y,0\right\rangle$$

The magnitudes of $\vec{r}_+$ and $\vec{r}_-$ are equal, because each charge is the same distance from the observation location.

$$|\vec{r}_+|=\sqrt{\left(-\frac{s}{2}\right)^2+y^2+0^2}=\sqrt{\left(\frac{s}{2}\right)^2+y^2}$$

$$|\vec{r}_-|=\sqrt{\left(\frac{s}{2}\right)^2+y^2}$$

$$\hat{r}_+=\frac{\left\langle-\frac{s}{2},y,0\right\rangle}{\left[\left(\frac{s}{2}\right)^2+y^2\right]^{1/2}}\quad\text{and}\quad\hat{r}_-=\frac{\left\langle\frac{s}{2},y,0\right\rangle}{\left[\left(\frac{s}{2}\right)^2+y^2\right]^{1/2}}$$

$$\vec{E}_+ = \frac{1}{4\pi\varepsilon_0}\frac{q}{\left[\left(\frac{s}{2}\right)^2 + y^2\right]}\frac{\left\langle -\frac{s}{2}, y, 0\right\rangle}{\left[\left(\frac{s}{2}\right)^2 + y^2\right]^{1/2}}$$

$$\vec{E}_- = \frac{1}{4\pi\varepsilon_0}\frac{-q}{\left[\left(\frac{s}{2}\right)^2 + y^2\right]}\frac{\left\langle \frac{s}{2}, y, 0\right\rangle}{\left[\left(\frac{s}{2}\right)^2 + y^2\right]^{1/2}}$$

The net field $\vec{E}_2$ is the vector sum of $\vec{E}_+$ and $\vec{E}_-$:

$$\vec{E}_2 = \vec{E}_+ + \vec{E}_- = \frac{1}{4\pi\varepsilon_0}\frac{qs}{\left[\left(\frac{s}{2}\right)^2 + y^2\right]^{3/2}}\langle -1, 0, 0\rangle$$

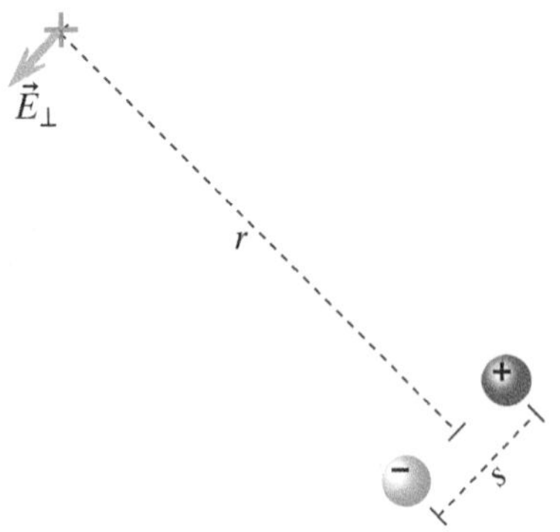

Figure 13.34 The electric field of a dipole at a location on the axis perpendicular to the dipole.

The y components of the two fields cancel, and the net field $\vec{E}_2$ is horizontal to the *left*, as shown in Figure 13.33, which may come as a bit of a surprise.

Again, since in general a dipole could have any orientation (for example, see Figure 13.34), it is appropriate to replace the y with an r representing the distance from the center of the dipole to a location on the axis perpendicular to the dipole, and to give only the magnitude of the electric field at this location.

$$|\vec{E}_\perp| = \frac{1}{4\pi\varepsilon_0}\frac{qs}{\left[\left(\frac{s}{2}\right)^2 + r^2\right]^{3/2}}$$

If we are far from the dipole, as is normally the case when we interact with molecular dipoles, the magnitude of the field can be further simplified. You should be able to show that in this important case ($r \gg s$), by making an appropriate approximation, we obtain:

ELECTRIC FIELD OF A DIPOLE, ON THE ⊥ AXIS

$$|\vec{E}_\perp| \approx \frac{1}{4\pi\varepsilon_0}\frac{qs}{r^3} \quad \text{on a location on the perpendicular axis, if } r \gg s$$

q is the magnitude of one of the charges.
s is the separation between the charges.
r is the distance from the center of the dipole to the observation location.
The direction is determined by looking at the particular situation.

Note that the magnitude of the electric field at a location along the y axis is half of the magnitude of the electric field at a location the same distance away on the x axis.

Along the Other Perpendicular Axis

QUESTION What would be the electric field of a dipole at a location on the other perpendicular axis?

By looking at the symmetry of the system, we can conclude by inspection that the electric field along both perpendicular axes should have exactly the same distance dependence.

Other Locations

By applying the superposition principle we can calculate the electric field of a dipole at any location, though we may not get a simple algebraic expression.

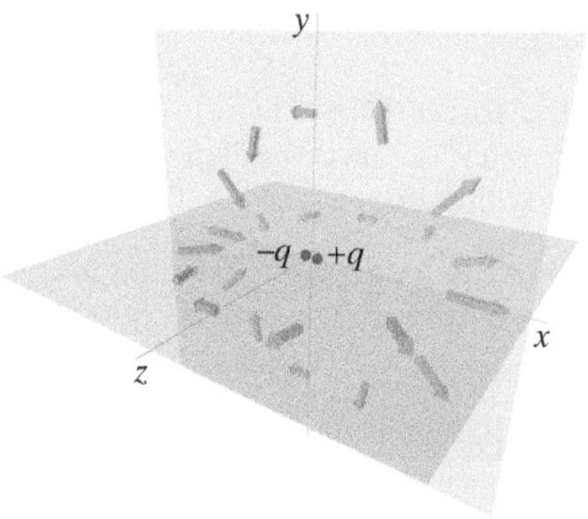

Figure 13.35 Electric field of a dipole at locations in two planes. Included are locations on the dipole axis and the perpendicular axis, as well as other locations.

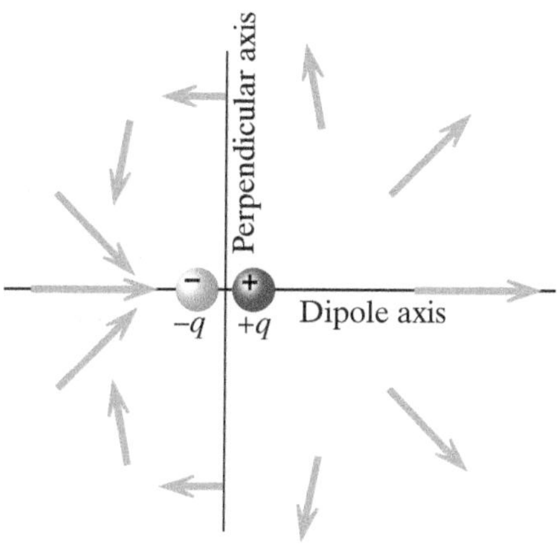

Figure 13.36 Electric field of a dipole, at locations in a plane containing the dipole.

Figure 13.35 shows the electric field of a dipole at locations in two perpendicular planes containing the dipole, calculated numerically. Figure 13.36 shows only locations in the vertical plane, displaying the symmetry of the field pattern. Problem P68 is a computer exercise that allows you to visualize the electric field of a dipole at any location.

The Importance of Approximations

By considering the common situation where $r \gg s$, we were able to come up with a simple algebraic expression for the electric field of a dipole at locations along the x and y axes, which gives us a clear insight into the distance dependence of the dipole field. If we had not made this simplification, we would have been left with a messy expression, and it would not have been easy to see that the electric field of a dipole falls off like $1/r^3$. (This turns out to be true for locations not on the axes as well, but the algebra required to demonstrate this is more complicated.)

QUESTION How would we calculate the electric field of a dipole at a location very near the dipole, where $r \approx s$?

Since the approximation we made in the calculations above is no longer valid, we would have to go back to adding the fields of the two charges exactly, as we did in the first step of our work above.

Checkpoint 5 A dipole is located at the origin, and is composed of charged particles with charge $+e$ and $-e$, separated by a distance 2×10^{-10} m along the x axis. **(a)** Calculate the magnitude of the electric field due to this dipole at location $\langle 0, 2 \times 10^{-8}, 0 \rangle$ m. **(b)** Calculate the magnitude of the electric field due to this dipole at location $\langle 2 \times 10^{-8}, 0, 0 \rangle$ m.

Interaction of a Point Charge and a Dipole

Since we have analytical expressions for the electric field of a dipole at locations along the x, y, or z axes, we can calculate the force exerted on a point charge that interacts with the field of the dipole. Figure 13.37 shows a dipole acting on a point charge $+Q$ that is a distance $d \gg s$ from the center of the dipole.

EXAMPLE **Force of a Dipole on a Point Charge**

What are the direction and the magnitude of the force exerted by the dipole on the point charge shown in Figure 13.37?

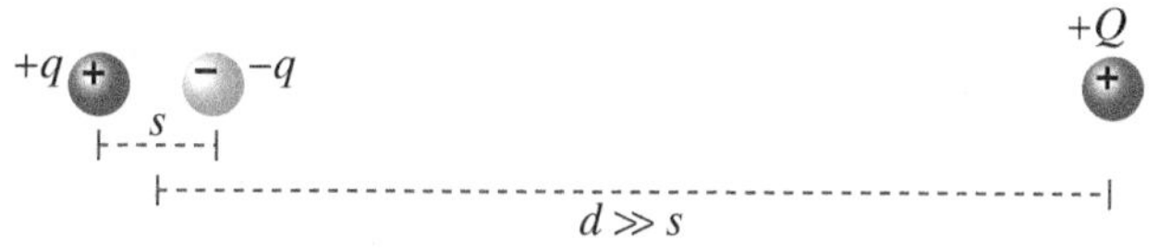

Figure 13.37 A dipole and a point charge.

Solution The electric field due to the dipole at the location of the point charge points in the $-x$ direction (Figure 13.39). The dipole exerts a force on Q toward the dipole:

$$\vec{F} = Q\vec{E}_{\text{dipole}} = Q\left\langle -\frac{1}{4\pi\varepsilon_0}\frac{2qs}{d^3}, 0, 0 \right\rangle$$

QUESTION Does the direction of this electric force make sense?

Yes. The negative end of the dipole is closer to the location of interest, so its contribution to the net electric field is larger than the contribution from the positive end of the dipole.

QUESTION What are the magnitude and direction of the force exerted on the dipole by the point charge?

By the principle of reciprocity of electric forces (Newton's third law), the force exerted by Q on the dipole must be equal in magnitude and opposite in direction to the force on the point charge by the dipole.

QUESTION How is this possible? Since the particles making up the dipole have equal and opposite charges, why isn't the net force on the dipole zero?

The key lies in the distance dependence of the electric field of the point charge. Although the two ends of the dipole are very close together, the positive end is very slightly farther away from the point charge. Hence, the magnitude of the point charge's electric field is slightly less at this location, and the resulting repulsive force on the positive end of the dipole is slightly less than the attractive force on the negative end of the dipole. The net force is small, but attractive (Figure 13.38).

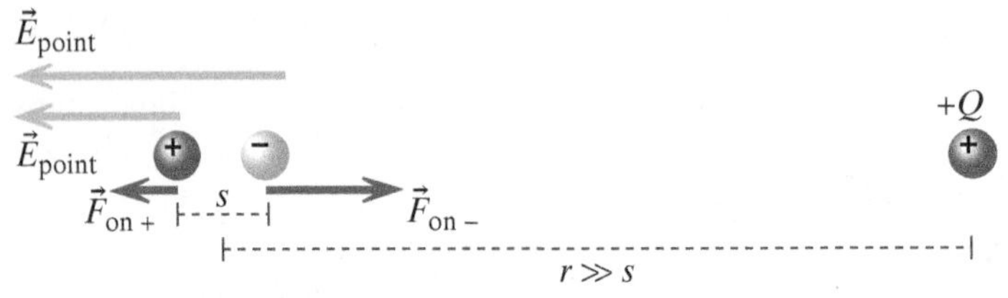

Figure 13.38 The electric field of the point charge is slightly larger in magnitude at the location of the negative end of the dipole. Hence, the net force on the dipole is to the right. The arrows representing field and force vectors are offset slightly from the location of the charges for clarity.

Checkpoint 6 Consider the situation in Figure 13.39. **(a)** If we double the distance d, by what factor is the force on the point charge due to the dipole reduced? **(b)** How would the magnitude of the force change if the point charge had a charge of $+3Q$? **(c)** If the charge of the point charge were $-2Q$, how would the force change?

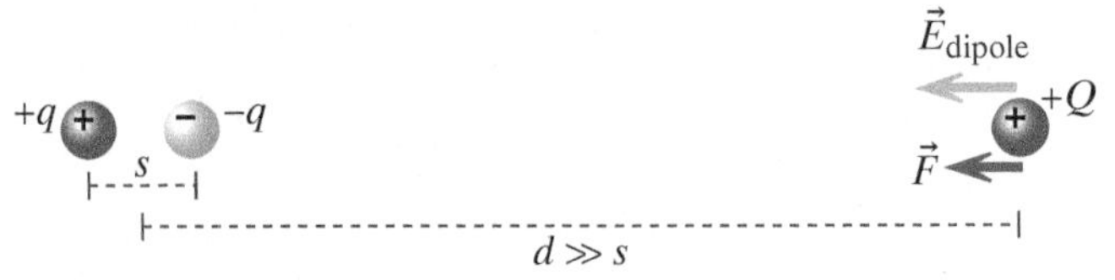

Figure 13.39 The electric field of the dipole at the location of the point charge, and the force on the point charge due to this field.

Electric Dipole Moment

The electric field of a dipole is proportional to the product qs, called the "electric dipole moment" and denoted by p.

$$p = qs \quad \text{(electric dipole moment)}$$

Far away from a dipole, the same electric field could be due to a small q and a large s, or a large q and a small s. The only thing that matters is the product qs. Moreover, this is the quantity that is measurable for molecules such as HCl and H_2O that are permanent dipoles, not the individual values of q (the amount of charge on an end) and s (the separation of the charges).

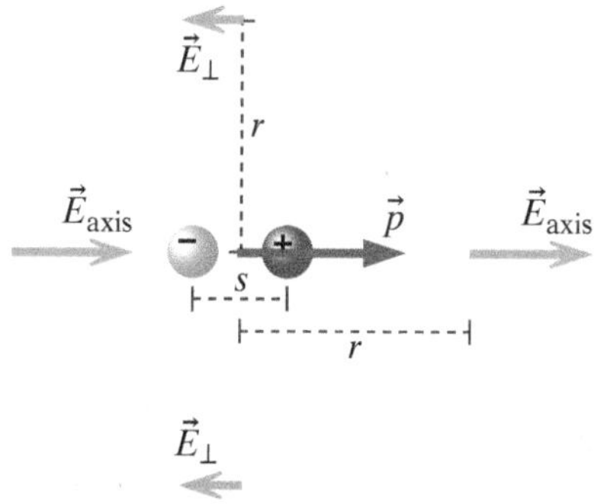

Figure 13.40 Electric field of a dipole on the axis and perpendicular axis, and the electric dipole moment vector.

We can rewrite our expressions for the electric field of a dipole in terms of the dipole moment p:

$$|\vec{E}_{\text{axis}}| \approx \frac{1}{4\pi\varepsilon_0}\frac{2p}{r^3} \quad \text{at a location on the dipole axis, if } r \gg s$$

$$|\vec{E}_{\perp}| \approx \frac{1}{4\pi\varepsilon_0}\frac{p}{r^3} \quad \text{at a location on the perpendicular axis, if } r \gg s$$

Dipole Moment as a Vector

The dipole moment can be defined as a vector $\vec{p}$ that points from the negative charge to the positive charge, with a magnitude $p = qs$ (Figure 13.40). Note that the electric field along the axis of the dipole (outside the dipole) points in the same direction as the dipole moment $\vec{p}$, which is one of the useful properties of the vector $\vec{p}$.

In Chapter 17 we will see a similar pattern of magnetic field around a magnetic dipole characterized by a magnetic dipole moment that is a vector.

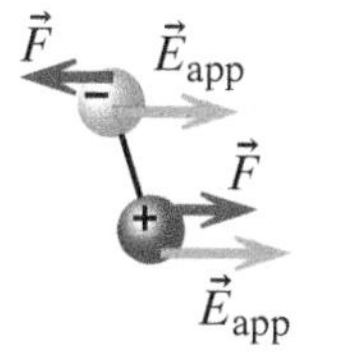

Figure 13.41 A dipole in a uniform electric field would experience no net force. It would experience a torque about its center of mass (in this example $\vec{\tau}_{CM}$ is out of the page, and the dipole would begin to rotate counterclockwise).

A Dipole in a Uniform Electric Field

We will see in Chapter 15 that it is possible to arrange a collection of point charges in a configuration that produces an electric field nearly uniform in magnitude and direction within a particular region. What would be the force on a dipole in such a uniform electric field?

The net force on the dipole would be zero. We see in Figure 13.41 that the forces on the two ends would be equal in magnitude and opposite in direction. However, there would be a torque on the dipole about its center of mass, and the dipole would begin to rotate.

Checkpoint 7 What would be the equilibrium position of the dipole shown in Figure 13.41? In this equilibrium position, compare the orientation of the vector dipole moment with the direction of the applied electric field.

EXAMPLE **Dipole and Charged Ball**

A dipole consisting of point charges $+4\,\text{nC}$ and $-4\,\text{nC}$ separated by a distance of 2 mm is centered at location $\langle 0.03, 0.15, 0\rangle$ m as shown in Figure 13.42 (the size of the dipole is exaggerated in the diagram). A hollow plastic ball with radius 3 cm and charge $-0.2\,\text{nC}$ distributed uniformly over its surface is centered at location $\langle 0.11, 0.15, 0\rangle$ m. What is the net electric field at location C, $\langle 0.03, 0.04, 0\rangle$ m? ($1\,\text{nC} = 1$ nanocoulomb $= 1\times10^{-9}\,\text{C}$.)

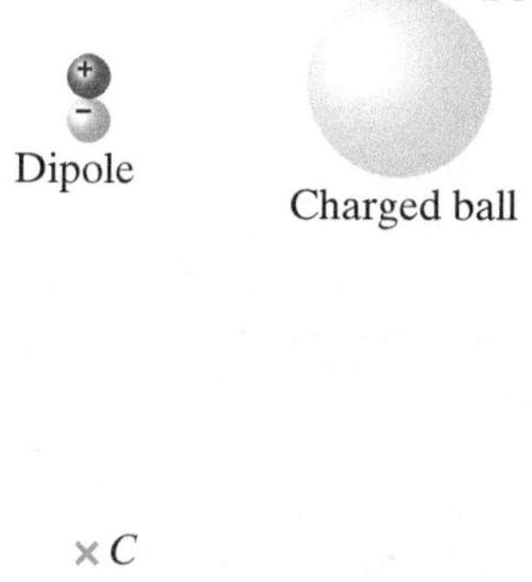

Figure 13.42 A dipole and a charged ball. Dipole size is exaggerated (not to scale).

Solution Apply the superposition principle:

$$\vec{E}_{\text{net}} = \vec{E}_{\text{dipole}} + \vec{E}_{\text{ball}}$$

The distance from the center of the dipole to the observation location is

$$|\vec{r}_d| = |\langle 0.03, 0.04, 0\rangle - \langle 0.03, 0.15, 0\rangle| = 0.11\text{ m}$$

Because location C is reasonably far from the dipole, along the dipole axis, we can approximate the magnitude of the dipole field as:

$$|\vec{E}| \approx \frac{1}{4\pi\varepsilon_0}\frac{2qs}{r^3} = \left(9\times10^9\,\frac{\text{N}\cdot\text{m}^2}{\text{C}^2}\right)\frac{2(4\times10^{-9}\,\text{C})(2\times10^{-3}\,\text{m})}{(0.11\text{ m})^3} = 108\,\frac{\text{N}}{\text{C}}$$

By looking at the diagram we see that the direction of the dipole field is $\langle 0,1,0\rangle$ (along the positive y axis, because the negative end is closer to location C). The electric field at location C due to the dipole is:

$$\vec{E}_{\text{dipole}} = \langle 0,108,0\rangle\ \frac{\text{N}}{\text{C}}$$

The electric field due to the ball is the same as if the ball were a point charge located at its center. The vector $\vec{r}$ from the center of ball to location C is:

$$\vec{r} = \langle 0.03,0.04,0\rangle\ \text{m} - \langle 0.11,0.15,0\rangle\ \text{m} = \langle -0.08,-0.11,0\rangle\ \text{m}$$

The magnitude of $\vec{r}$ is $|\vec{r}| = \sqrt{(-0.08)^2+(-0.11)^2+0^2}\ \text{m} = 0.136\ \text{m}$, so the unit vector $\hat{r}$ is:

$$\hat{r} = \frac{\vec{r}}{|\vec{r}|} = \frac{\langle -0.08,-0.11,0\rangle\ \text{m}}{0.136\ \text{m}} = \langle -0.588,-0.809,0\rangle$$

The electric field due to the charged ball is then

$$\vec{E}_{\text{ball}} = \frac{1}{4\pi\varepsilon_0}\frac{Q}{|\vec{r}|^2}\hat{r} = \left(9\times 10^9\ \frac{\text{N}\cdot\text{m}^2}{\text{C}^2}\right)\frac{(-2\times 10^{-10}\,\text{C})}{(0.136\ \text{m})^2}\langle -0.588,-0.809,0\rangle$$
$$= \langle 57.2,78.7,0\rangle\ \frac{\text{N}}{\text{C}}$$

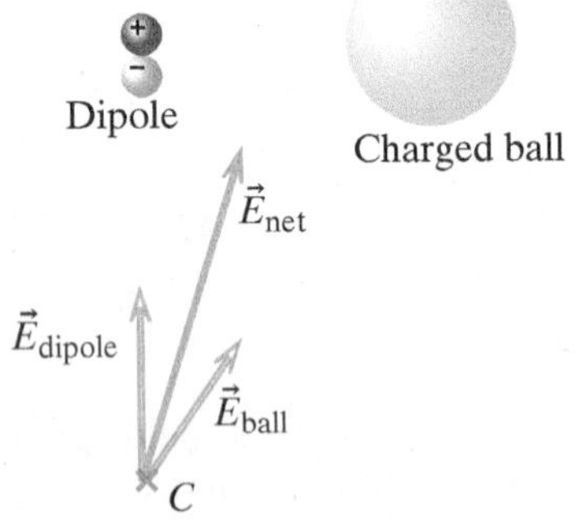

Figure 13.43 Contributions of dipole and ball, and net electric field at location C.

The net electric field is:

$$\vec{E}_{\text{net}} = \vec{E}_{\text{dipole}} + \vec{E}_{\text{ball}}$$
$$= \langle 0,108,0\rangle\ \frac{\text{N}}{\text{C}} + \langle 57.2,78.7,0\rangle\ \frac{\text{N}}{\text{C}}$$
$$= \langle 57.2,187,0\rangle\ \frac{\text{N}}{\text{C}}$$

As shown in Figure 13.43, both the x and y components of $\vec{E}_{\text{ball}}$ are positive, which is correct, since the ball is negatively charged. Arrows representing each contribution to the electric field, and the net electric field, are shown in Figure 13.43.

13.7 CHOICE OF SYSTEM

In our previous study of mechanics, when we analyzed multiparticle systems in terms of momentum or energy we had to decide what objects to include in our "system" and what objects to consider as external to the system. Similarly, when we use the field concept to calculate a force rather than using Coulomb's law directly, we split the Universe into two parts:

- the charges that are the sources of the field, and
- the charge that is affected by that field.

This is particularly useful when the sources of the field are numerous and fixed in position. We calculate the electric field due to such sources throughout a given region of space, and then we can predict what would happen to a charged particle that enters that region. (It is of course true that the moving charged particle exerts forces on all the other charges, but often we're not interested in those forces, or the other charges are not free to move.)

For example, in an oscilloscope, charges on metal plates are the sources of an electric field that affects the trajectory of single electrons (Figure 13.44). The electric field has nearly the same magnitude and direction everywhere within a box (this can be achieved with a suitable configuration of positive charges below and negative charges above). The trajectory of an electron outside this box, where the field is nearly zero, is a straight line. Upon entering the box (through a small hole) the electron experiences a downward force (it has negative charge, hence the force on it is opposite to the field) and has a trajectory as shown.

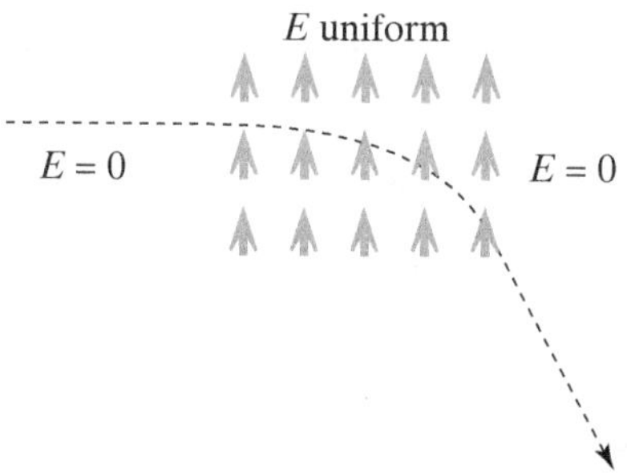

Figure 13.44 Path of an electron through a region of uniform electric field.

Checkpoint 8 If the uniform upward-pointing electric field depicted in Figure 13.44 has a magnitude of 5000 N/C, what is the magnitude of the force on the electron while it is in the box? If a different particle experiences a force of 1.6×10^{-15} N when passing through this region, what is the charge of the particle?

13.8 IS ELECTRIC FIELD REAL?

Reflecting on what we have done so far, we can give two reasons for why we introduced the concept of electric field:

1. Once we know the electric field $\vec{E}$ at some location, we know the electric force $\vec{F} = q\vec{E}$ acting on *any* charge q that we place there.
2. In terms of electric field, we can describe the electric properties of matter, independent of how the applied electric field is produced.

Later we'll see that if the electric field in air exceeds about 3×10^6 N/C, air becomes a conductor, no matter how this electric field is produced. This is an example of the value of the electric field concept for parameterizing the behavior of matter. It would be very awkward to describe the properties of matter if we didn't split the problem into two parts using the field concept, in which we talk about the field made by other charges and the effect that a field has on a particular kind of matter.

Even with these advantages of the field concept, it may seem that it doesn't matter whether we think of electric field as merely a calculational convenience or as something real. However, Einstein's special theory of relativity, which has been experimentally verified in a wide variety of situations, rules out instantaneous action at a distance and implies that the field concept is necessary, not merely convenient, as we will show next.

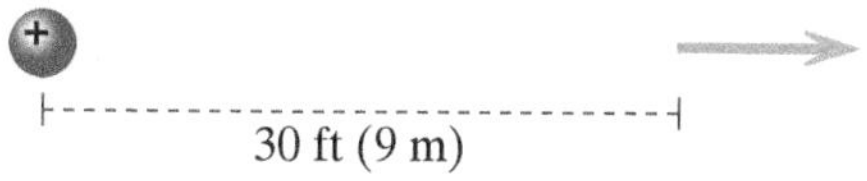

Figure 13.45 Electric field of a point charge.

Retardation: Electric Field and the Speed of Light

Special relativity predicts that nothing can move faster than the speed of light, not even information, and no one has ever observed a violation of this prediction. The speed of light is $c = 3 \times 10^8$ m/s, or about 1 ft (30 cm) per nanosecond, where a nanosecond is 1×10^{-9} s.

Consider a charge that is 30 ft away (about 9 m), as shown in Figure 13.45. It takes light 30 nanoseconds (30 ns) to travel from the charge to you.

Figure 13.46 At $t = 0$ charge is moved, but the electric field at the observation location does not change for 30 ns.

Suppose that the charge is suddenly moved to a new location, as shown in Figure 13.46. You cannot observe change in the electric field until 30 ns have elapsed! If the field could change instantaneously, that would provide a mechanism for sending signals faster than the speed of light.

For a while (30 ns), the electric field has some kind of reality and existence independent of the original source of the field. If you place a positive charge at your observation location during this 30-ns interval, you will see it accelerate to the right, in the direction of the old electric field that is still valid.

Figure 13.47 At $t = 30$ ns, the field at the observation location finally changes.

At the end of the 30-ns delay, the field finally changes to correspond to the place to which the charge was moved 30 ns ago, as shown in Figure 13.47. You *see* the new position of the charge at the same instant that you notice the change in the electric field, because of course light itself travels at the speed of light.

What If the Source Charges Vanish?

An even more dramatic example concerns the electric field made by a remote electric dipole. Suppose that an electron and a positron are separated by a small distance s, making a dipole, and creating an electric field throughout space. The electron and positron may suddenly come together and react, annihilating each other and releasing a large amount of energy in the form of high-energy photons (gamma rays), which have no electric charge:

$$e^- + e^+ \to \gamma + \gamma$$

At a time shortly after the annihilation has occurred, there are no longer any charged particles in the vicinity. However, if you monitor the electric field in this region, you will keep detecting the (former) dipole's field for a while even though the dipole no longer exists! If you are a distance r away from the original location of the dipole, you will still detect the dipole's electric field for a time r/c, which is the same as the time it would take light to travel from that location to your position.

Coulomb's Law Is Correct Only at Low Speeds

This interesting behavior, called "retardation," means that Coulomb's law is not completely correct, since the equation

$$\vec{F} = \frac{1}{4\pi\varepsilon_0}\frac{q_1 q_2}{r^2}\hat{r}$$

doesn't contain time t or speed of light c. Coulomb's law is an approximation that is quite accurate as long as charges are moving slowly. However, when charges move with speeds that are a significant fraction of the speed of light, Coulomb's law is not adequate. Similarly, the equation for the electric field of a point charge,

$$\vec{E} = \frac{1}{4\pi\varepsilon_0}\frac{q}{r^2}\hat{r}$$

is only an approximation, valid if the charge is moving at a speed that is small compared with the speed of light.

Later we will consider the fields of moving charges, which include magnetic fields as well as electric fields. We will also consider the fields of *accelerated* charges, which produce a very special kind of electric field—a component of the electromagnetic field that makes up radio and television waves and light.

Evidently the electric field isn't just a calculational device. Space can be altered by the presence of an electric field, even when the source charges that made the field are gone.

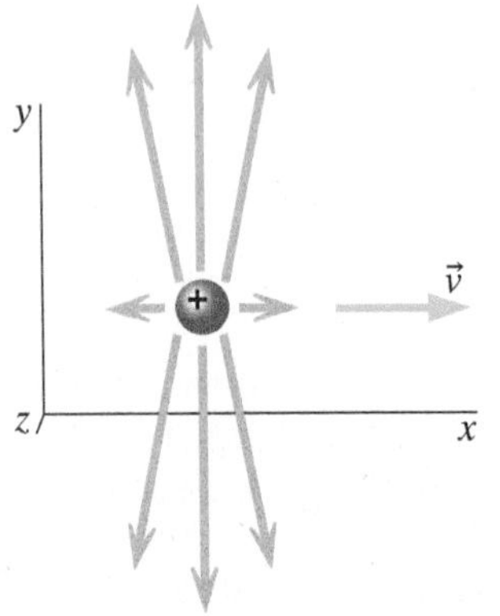

Figure 13.48 The pattern of electric field of a positively charged particle moving at high speed in the $+x$ direction is no longer spherically symmetric.

Relativistic Electric Field

It turns out that for a charged particle moving at high speed in the $+x$ direction, the x component of electric field at any location is unchanged. However, a factor of

$$\gamma = \frac{1}{\sqrt{1 - v^2/c^2}}$$

appears in the y and z components of electric field, and they are much larger, as shown in Figure 13.48. The fields of relativistic particles are discussed further in Chapter 20.

13.9 COMPUTATIONAL MODELING OF ELECTRIC FIELDS

■ In this section it is assumed that you have done at least some of the computational modeling activities in Chapters 1–12. If computational modeling in VPython is new to you, you will need to work through the computational part of the Chapter 1 supplement that is available on the Wiley student website before reading this section or doing the computational problems at the end of the chapter.

One of the goals of computational modeling in this and subsequent chapters is simply to calculate and visualize patterns of electric and magnetic fields in 3D space. Recognizing these patterns and understanding how they relate to source charge distributions will be important throughout the rest of this book.

The calculation of the electric field of a single point charge at a single observation location is similar to the calculation of gravitational force we have done before. In writing a program to tell a computer how to calculate an electric field, we use exactly the same organization that we use when doing the calculation with a calculator. Let's assume that our program creates an object named `particle_1` at some appropriate location in 3D space, and sets the variable `q1` equal to the charge of `particle_1`. The program also identifies the observation location by creating a vector called `r_obs`. Let's also assume that the program has defined a constant `oofpez` (standing for One-Over-Four-Pi-Epsilon-Zero) equal to $1/(4\pi\epsilon_0)$. A translation of the mathematical expressions (shown in red) into code (shown in blue) for calculating the electric field due to `particle_1` at the location `r_obs` into VPython code might look like this:

$\vec{r} = \vec{r}_{\text{obs}} - \vec{r}_1$
```
r = r_obs - particle_1.pos
```

$\hat{r} = \vec{r}/|\vec{r}|$
```
rhat = r / mag(r)
```

$\vec{E} = \left(\dfrac{1}{4\pi\epsilon_0}\dfrac{q_1}{|\vec{r}|^2}\right)\hat{r}$
```
E = (oofpez * q1 / mag(r)**2) * rhat
```

Instead of just printing the value of the electric field at a location, we typically want to use an arrow to display the field. Assuming that the program has defined an appropriate scale factor `sf` to scale the arrow so both the arrow and the sphere representing the source charge can be seen at the same time, we could use the following statement to do this (by convention, arrows representing electric field are orange throughout this book):

```
ea = arrow(pos=r_obs, color=color.orange, axis=sf*E)
```

Figure 13.49 Display generated by a VPython program to calculate the electric field due to one charge at a single observation location.

QUESTION Figure 13.49 shows the display from a VPython program that calculates and displays the electric field at a single observation location, due to a single source charge. What is the sign of the source charge?

Since the electric field at the observation location points toward the source charge, the source charge must be negative.

QUESTION In the program used to generate Figure 13.49 the distance between the source charge and the observation location was 1.4×10^{-8} m, and the magnitude of the electric field was $|\vec{E}| = 7.2 \times 10^6$ N/C. Approximately what was the value of the scale factor `sf` used to adjust the length of the arrow?

The arrow is about half as long as the distance between source and observation location, so the scale factor must have been about $0.7 \times 10^{-8}/7.2 \times 10^6 \approx 1 \times 10^{-15}$.

Finding the Net Field Due to Several Charges

In Chapter 15 we will concentrate on finding the electric field at one or more observation locations due to a large number of point charges. In that chapter we will introduce a new computational concept called a "list" that will facilitate this process. In the current chapter, we will stick to applying the computational concepts we have already covered. (If you already know how to use Python lists, you are welcome to use them in the problems at the end of the chapter.)

In Section 13.6 we learned how to calculate the electric field of a dipole at a particular location by adding the fields contributed by the two charges in the dipole. We were able to find analytical expressions for the magnitude of the electric field at any location on the axis of the dipole, or on the perpendicular axis. Finding an analytical expression for the magnitude and direction of the field of a dipole at observation locations not on an axis is much more difficult, but this is an easy problem to solve computationally.

If we call the two objects making up the dipole `particle_1` and `particle_2`, we can find the net field at an observation location `r_obs` this way:

```
E_net = vector(0,0,0)  ## initialize E_net

r = r_obs - particle_1.pos
rhat = r / mag(r)
E = (oofpez * q1 / mag(r)**2) * rhat
E_net = E_net + E

r = r_obs - particle_2.pos
rhat = r / mag(r)
E = (oofpez * q2 / mag(r)**2) * rhat
E_net = E_net + E

ea = arrow(pos=r_obs, axis=sf*E_net, color=color.orange)
```

In the code segment above, notice that the variable names `r` and `E` were reused. This is okay, because after we update `E_net` we no longer need the values stored in these variables, and we can set them to new values. Reusing variable names makes it possible to put calculations like these inside `while` loops, which will allow us to avoid repeating lines of code as is done in the segment above.

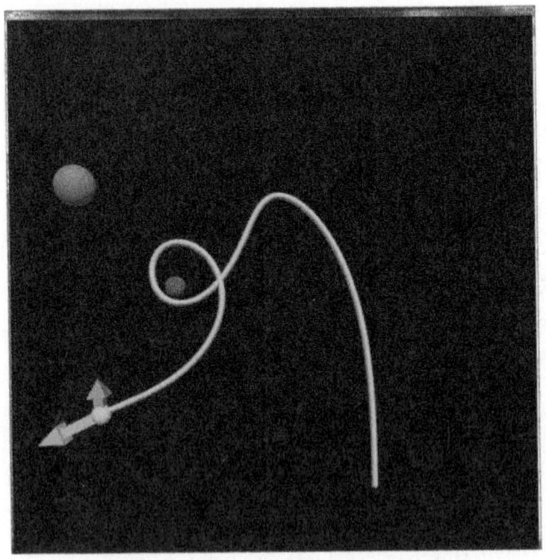

Figure 13.50 Display generated by a VPython program to predict the motion of a charged particle that is affected by the electric field due to two unequal source charges. The orange arrow represents the net electric field at the current location of the moving particle. The green arrow represents the particle's momentum.

QUESTION In the code above, why does the following instruction appear twice?

```
E_net = E_net + E
```

This is an "update" instruction of the kind we've used with momentum and position. We incrementally update the value of `E_net` by sequentially adding to it the electric field of each particle.

Predicting Particle Motion

We can predict the motion of charged particles by iteratively updating momentum and position, as illustrated in Figure 13.50. Such a computational model differs from what we have done previously only in that we must calculate the net electric field at the particle's location in order to get the net force on the particle. The basic approach is this:

Repeat:
- Calculate the net electric field $\vec{E}_{\text{net}}$ at the location of the moving particle.
- Calculate the net force $\vec{F}_{\text{net}} = q\vec{E}_{\text{net}}$ on the moving particle.
- Update the momentum of the particle: $\vec{p}_f = \vec{p}_i + \vec{F}_{\text{net}}\,\Delta t$.
- Update the position: $\vec{r}_f = \vec{r}_i + \vec{v}_{\text{avg}}\,\Delta t$.

As we did in Chapters 2–11, we use the approximation $\vec{v}_{\text{avg}} \approx \vec{p}_f/m$ as long as the speed of the particle is small compared to the speed of light.

The computational loop in a program that moves a charged particle named `mp` (for "moving particle") with momentum `p_mp`, charge `q_mp`, and mass `m_mp`, in a region near a single point charge, might look like this:

```
while t < tmax:
    rate(100)
    r = mp.pos - source.pos
    rhat = r / mag(r)
    E = (oofpez * q_source / mag(r)**2) * rhat
    F = q_mp * E
    p_mp = p_mp + F * deltat
    mp.pos = mp.pos + (p_mp/m_mp) * deltat
    t = t + deltat
```

QUESTION Why is the calculation of the electric field done inside the computational loop?

Any quantity that changes in magnitude or direction must be recalculated in each time step. The moving particle's position changes during each iteration, and the electric field at its new position is different from the electric field at its previous position. We need the new value of the electric field to find the new force on the particle.

Many Observation Locations

The electric field of even a single point charge has a value at every location in 3D space. Visualizing these patterns in 3D is straightforward in VPython. Here is a small program that calculates and visualizes the electric field of a single point charge at several locations on a circle. The display generated by the program is shown in Figure 13.51.

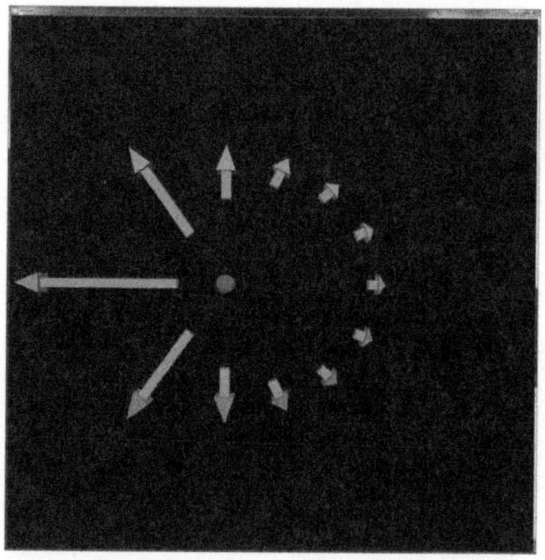

Figure 13.51 Display generated by the VPython program shown in the text at the right.

```
from visual import *
oofpez = 9e9 ## One-Over-Four-Pi-Epsilon-Zero
qe = 1.6e-19
sf = 3e-16   ## scale factor
source = sphere(pos=vector(-0.5e-8,0,0), radius=1e-9,
                color=color.red)
q_source = qe
thetamax = 2*pi
dtheta = pi/6
R = 1e-8  ## radius of circle

theta = 0
while theta < thetamax:
    rate(500)
    r_obs = R * vector(cos(theta),cos(pi/2-theta),0)
    r = r_obs - source.pos
    rhat = r/mag(r)
    E = (oofpez * q_source / mag(r)**2) * rhat
    arrow(pos=r_obs, color=color.orange, axis=sf*E,
          shaftwidth = source.radius)
    theta = theta + dtheta
```

The extra argument in the constructor for the arrows is needed to force VPython to make the shafts of the arrows all the same size. By default VPython scales the width of the arrow shaft to the length of the arrow, but this can make it difficult to compare arrows.

QUESTION What is the physical significance of the variable `theta` used in the `while` loop, and how is it related to the observation locations?

`theta` is the angle between the $+x$ axis and the relative position vector $\vec{r}$. Direction cosines are used to construct the unit vector $\hat{r}$, which is then multiplied by the radius of the desired circle to give the observation location.

QUESTION Why aren't the lengths of the orange arrows in Figure 13.51 all the same?

The distance from the source charge to the observation locations varies, so the magnitude of the electric field also varies. The center of the circle of observation locations is not at the location of the source charge.

SUMMARY

At a location where there is an electric field $\vec{E}_1$, a point particle with charge q_2 experiences a force $\vec{F}_2 = q_2\vec{E}_1$ (Figure 13.52).

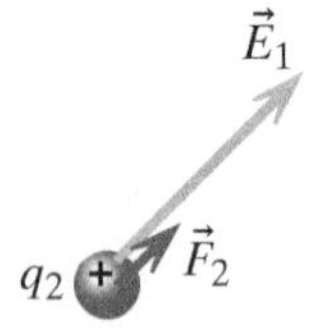

Figure 13.52

A point particle with charge q_1 makes an electric field at all locations in space (Figure 13.53) except its own location:

$$\vec{E}_1 = \frac{1}{4\pi\varepsilon_0}\frac{q_1}{|\vec{r}|^2}\hat{r}$$

where q_1 is the source charge, $\vec{r}$ is the relative position vector giving the position of the observation location relative to the source charge, $\hat{r}$ is the unit vector in the direction of $\vec{r}$, and $|\vec{r}|$, the magnitude of $\vec{r}$, is the distance from the source location to the observation location.

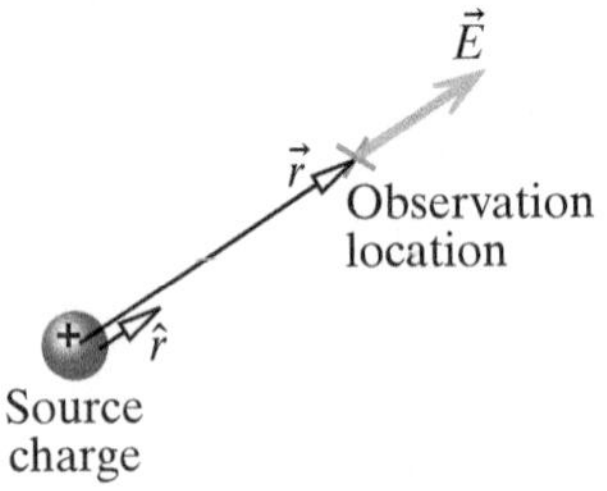

Figure 13.53

A point charge is not affected by its own electric field.

The source location is the location of the charge(s) that create an electric field.

The observation location is the location at which the electric field is detected (Figure 13.54).

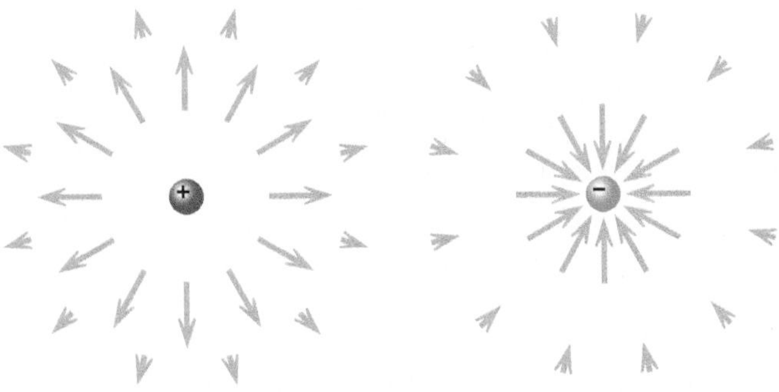

Figure 13.54 Electric field at selected observation locations in a plane containing a positive or negative point charge.

The Superposition Principle

The net electric field at a particular location due to two or more charges is the vector sum of each field due to each individual charge. Each individual contribution is unaffected by the presence of the other charges. Thus, matter cannot "block" electric fields.

Retardation

Changes in electric fields (due to changes in source charge distribution) propagate through space at the speed of light. Therefore the electric field at a distant point does not change instantaneously when the source charge distribution changes.

Electric Dipole

An electric dipole consists of two point particles having equal and opposite charges $+q$ and $-q$, separated by a distance s (Figure 13.55). The electric dipole moment of a dipole is defined as $p \equiv qs$. The electric dipole moment vector $\vec{p}$ points from the negative charge toward the positive charge.

Electric field at a location on the dipole axis:

$$|\vec{E}_{\text{axis}}| \approx \frac{1}{4\pi\varepsilon_0}\frac{2qs}{r^3} \quad \text{for } r \gg s$$

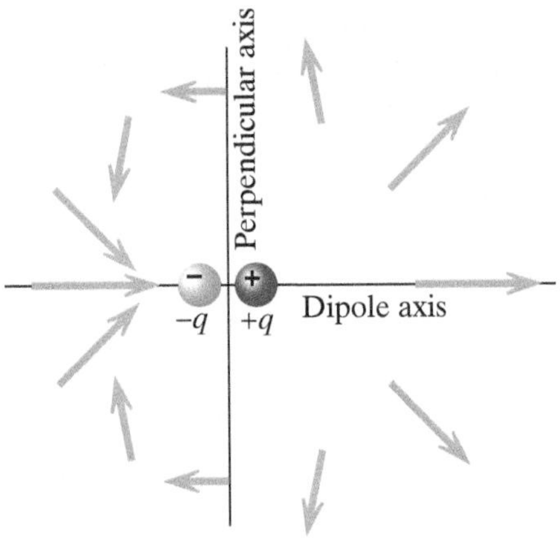

Figure 13.55 The pattern of electric field around an electric dipole.

The direction is determined by looking at the particular situation.

Electric field at a location on the perpendicular axis:

$$|\vec{E}_{\perp}| \approx \frac{1}{4\pi\varepsilon_0}\frac{qs}{r^3} \quad \text{for } r \gg s$$

The direction is determined by looking at the particular situation.

For a spherical object with a total charge Q distributed uniformly over its surface (a "uniform spherical shell of charge"):

at a location inside the shell, $|\vec{E}| = 0$

at a location outside the shell, $\vec{E} = \frac{1}{4\pi\varepsilon_0}\frac{Q}{|\vec{r}|^2}\hat{r}$

($\vec{r}$ extends from the center of sphere to the observation location; Q is the total charge on the sphere).

Constants:

$$\frac{1}{4\pi\varepsilon_0} = 9\times 10^9\,\frac{\mathrm{N}\cdot\mathrm{m}^2}{\mathrm{C}^2} \quad \text{and} \quad \varepsilon_0 = 8.85\times 10^{-12}\,\frac{\mathrm{C}^2}{\mathrm{N}\cdot\mathrm{m}^2}$$

QUESTIONS

Q1 What is the relationship between the terms "field" and "force"? What are their units?

Q2 You are the captain of a spaceship. You need to measure the electric field at a specified location P in space outside your ship. You send a crew member outside with a meter stick, a stopwatch, and a small ball of known mass M and net charge $+Q$ (held by insulating strings while being carried). **(a)** Write down the instructions you will give to the crew member, explaining what observations to make. **(b)** Explain how you will analyze the data that the crew member brings you to determine the magnitude and direction of the electric field at location P.

Q3 Criticize the following statement: "A proton can never be at rest, because it makes a very large electric field near itself that accelerates it."

Q4 Draw a diagram showing two separated point charges placed in such a way that the electric field is zero somewhere, and indicate that position. Explain your reasons.

Q5 At location A there is an electric field in the direction shown by the orange arrow in Figure 13.56. This electric field is due to charged particles that are not shown in the diagram. **(a)** If a proton is placed at location A, which of the arrows (a–h) best indicates the direction of the electric force on the proton? **(b)** If the proton is removed and an electron is placed at location A, which of the arrows in Figure 13.56 best indicates the direction of the electric force on the electron?

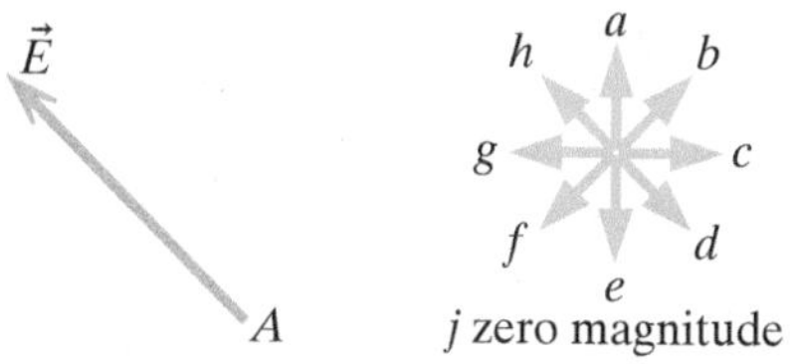

Figure 13.56

Q6 We found that the force exerted on a distant charged object by a dipole is given by

$$F_{\text{on } Q \text{ by dipole}} \approx Q\left(\frac{1}{4\pi\varepsilon_0}\frac{2qs}{r^3}\right)$$

In this equation, what is the meaning of the symbols q, Q, s, and r?

Q7 At a given instant in time, three charged objects are located near each other, as shown in Figure 13.57. Explain why the equation

$$F_{\text{on } Q \text{ by dipole}} \approx Q\left(\frac{1}{4\pi\varepsilon_0}\frac{2qs}{r^3}\right)$$

cannot be used to calculate the electric force on the ball of charge $+Q$.

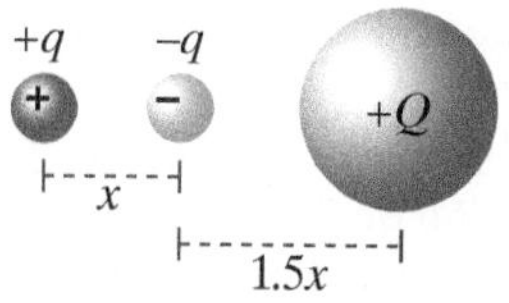

Figure 13.57

Q8 Where could you place one positive charge and one negative charge to produce the pattern of electric field shown in Figure 13.58? (As usual, each electric field vector is drawn with its tail at the location where the electric field was measured.) Briefly explain your choices.

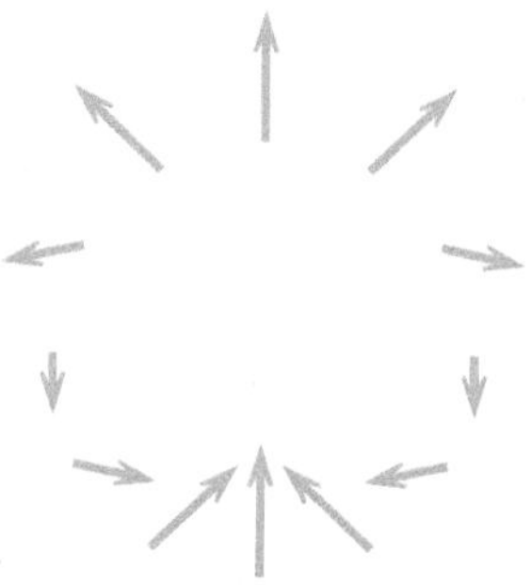

Figure 13.58

Q9 Consider Figure 13.59. Assume that the dipole is fixed in position. **(a)** What is the direction of the electric field at location A due to the dipole? **(b)** At location B? **(c)** If an electron were placed at location A, in which direction would it begin to move? **(d)** If a proton were placed at location B, in which direction would it begin to move? **(e)** Now suppose that an electron is placed at location A and held there, while the dipole is free to move. When the dipole is released, in what direction will it begin to move?

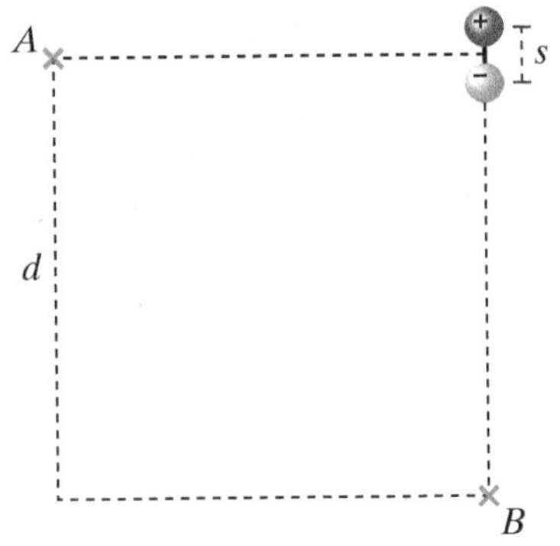

Figure 13.59

Q10 Which of these statements about a dipole are correct? Select all that are true. (1) At a distance d from a dipole, where $d \gg s$ (the separation between the charges), the magnitude of the electric field due to the dipole is proportional to $1/d^3$. (2) A dipole consists of two particles whose charges are equal in magnitude but opposite in sign. (3) The net electric field due to a dipole is zero, since the contribution of the negative charge cancels out the contribution of the positive charge. (4) At a distance d from a dipole, where $d \gg s$ (the separation between the charges), the magnitude of the electric field due to the dipole is proportional to $1/d^2$. (5) The electric field at any location in space, due to a dipole, is the vector sum of the electric field due to the positive charge and the electric field due to the negative charge.

Q11 If we triple the distance d, by what factor is the force on the point charge due to the dipole in Figure 13.60 reduced? (Note that the factor is smaller than one if the force is reduced and larger than one if the force is increased.)

Q12 If the charge of the point charge in Figure 13.60 were $-9Q$ (instead of Q): **(a)** By what factor would the magnitude of the force on the point charge due to the dipole change? Express your answer as the ratio (magnitude of new force / magnitude of Fv). **(b)** Would the direction of the force change?

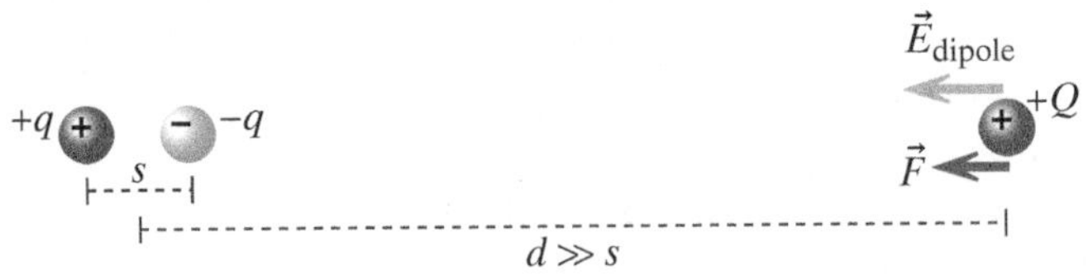

Figure 13.60

Q13 The distance between the dipole and the point charge in Figure 13.60 is d. If the distance between them were changed to $0.5d$, by what factor would the force on the point charge due to the dipole change? Express your answer as the ratio (magnitude of new force / magnitude of Fv).

Q14 Draw a diagram like the one in Figure 13.61.

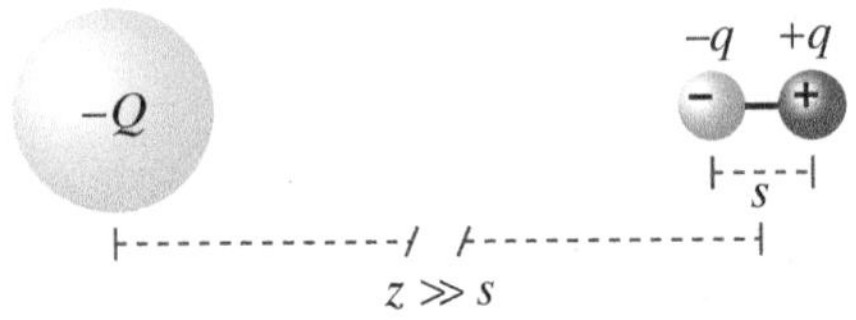

Figure 13.61

On your diagram, draw vectors showing: **(a)** the electric field of the dipole at the location of the negatively charged ball, **(b)** the net force on the ball due to the dipole, **(c)** the electric field of the ball at the center of the dipole, **(d)** the net force on the dipole due to the ball.

Q15 If the distance between the ball and the dipole in Figure 13.61 were doubled, what change would there be in the force on the ball due to the dipole?

PROBLEMS

Section 13.3

•P16 An electron in a region in which there is an electric field experiences a force of magnitude 3.8×10^{-16} N. What is the magnitude of the electric field at the location of the electron?

•P17 The electric field at a particular location is measured to be $\langle 0, -280, 0 \rangle$ N/C. What force would a positron experience if placed at this particular location?

•P18 An electron in a region in which there is an electric field experiences a force of magnitude 3.7×10^{-16} N. What is the magnitude of the electric field at the location of the electron?

•P19 If the particle in Figure 13.62 is a proton, and the electric field $\vec{E}_1$ has the value $\langle 2 \times 10^4, 2 \times 10^4, 0 \rangle$ N/C, what is the force $\vec{F}_2$ on the proton?

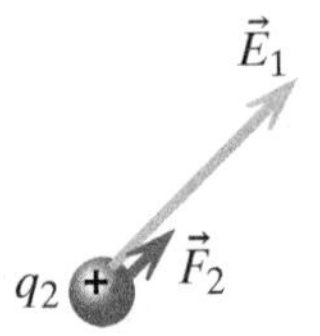

Figure 13.62

•P20 An electron in a region in which there is an electric field experiences a force of $\langle 8.0 \times 10^{-17}, -3.2 \times 10^{-16}, -4.8 \times 10^{-16} \rangle$ N. What is the electric field at the location of the electron?

••P21 In the region shown in Figure 13.63 there is an electric field due to a point charge located at the center of the dashed circle. The arrows indicate the magnitude and direction of the electric field at the locations shown.

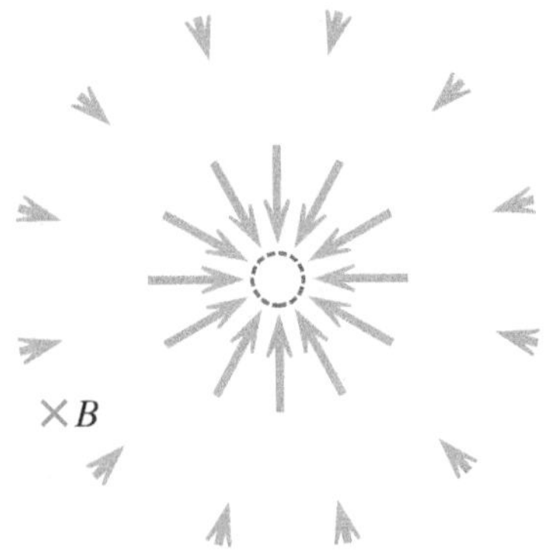

Figure 13.63

(a) What is the sign of the source charge? **(b)** Now a particle whose charge is -7×10^{-9} C is placed at location B. What is the direction of the electric force on the -7×10^{-9} C charge? **(c)** The electric field at location B has the value $\langle 2000, 2000, 0\rangle$ N/C. What is the unit vector in the direction of $\vec{E}$ at this location? **(d)** What is the electric force on the -7×10^{-9} C charge? **(e)** What is the unit vector in the direction of this electric force?

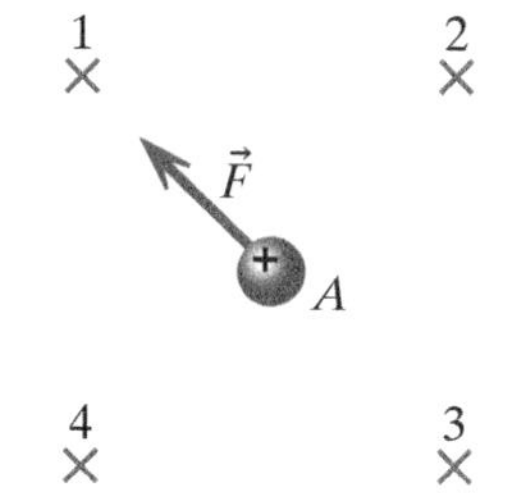

Figure 13.64

••**P22** In the region shown in Figure 13.64 there is an electric field due to charged objects not shown in the diagram. A tiny glass ball with a charge of 5×10^{-9} C placed at location A experiences a force of $\langle 4 \times 10^{-5}, -4 \times 10^{5}, 0\rangle$ N, as shown in the figure. **(a)** Which arrow in Figure 13.65 best indicates the direction of the electric field at location A? **(b)** What is the electric field at location A? **(c)** What is the magnitude of this electric field? **(d)** Now the glass ball is moved very far away. A tiny plastic ball with charge -6×10^{-9} C is placed at location A. Which arrow in Figure 13.65 best indicates the direction of the electric force on the negatively charged plastic ball? **(e)** What is the force on the negative plastic ball? **(f)** You discover that the source of the electric field at location A is a negatively charged particle. Which of the numbered locations in Figure 13.64 shows the location of this negatively charged particle, relative to location A?

Figure 13.65

••**P23** An electron is observed to accelerate in the $+z$ direction with an acceleration of 1.6×10^{16} m/s^2. Explain how to use the definition of electric field to determine the electric field at this location, and give the direction and magnitude of the field.

••**P24** An object falling in a vacuum near a planet has a charge of -4×10^{-8} C and a mass of 0.3 kg. In this region of space there is an electric field $\langle 2 \times 10^{7}, 0, 0\rangle$ N/C and a gravitational field $\langle 0, 5, 0\rangle$ N/kg. What is the net force acting on the object?

••**P25** A proton is observed to have an instantaneous acceleration of 9×10^{11} m/s^2. What is the magnitude of the electric field at the proton's location?

Section 13.4

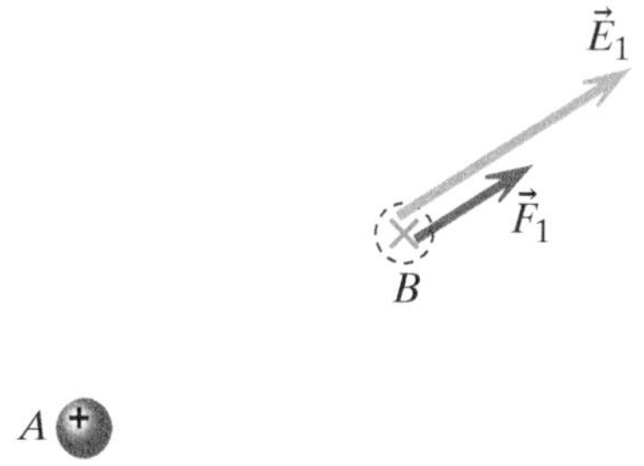

Figure 13.66

•**P26** In Figure 13.66 a proton at location A makes an electric field $\vec{E}_1$ at location B. A different proton, placed at location B, experiences a force $\vec{F}_1$.

Now the proton at B is removed and replaced by a lithium nucleus, containing three protons and four neutrons. **(a)** Now what is the value of the electric field at location B due to the proton? **(b)** What is the force on the lithium nucleus? **(c)** The lithium nucleus is removed, and an electron is placed at location B. Now what is the value of the electric field at location B due to the proton? **(d)** What is the magnitude of the force on the electron? **(e)** Which arrow in Figure 13.65 best indicates the direction of the force on the electron due to the electric field?

•**P27** You want to calculate the electric field at location $\langle 0.5, -0.1, -0.5\rangle$ m, due to a particle with charge +9 nC located at $\langle -0.6, -0.7, -0.2\rangle$ m. **(a)** What is the source location? **(b)** What is the observation location? **(c)** What is the vector $\vec{r}$ that points from the source location to the observation location? **(d)** What is $|\vec{r}|$? **(e)** What is the vector $\hat{r}$? **(f)** What is the value of $\dfrac{1}{4\pi\varepsilon_0}\dfrac{q}{|\vec{r}|^2}$? **(g)** Finally, what is the electric field, expressed as a vector?

•**P28** A particle with charge +5 nC (a nanocoulomb is 1×10^{-9} C) is located at the origin. What is the electric field due to this particle at a location $\langle 0.4, 0, 0\rangle$ m?

•**P29** What is the electric field at a location $\langle -0.1, -0.1, 0\rangle$ m, due to a particle with charge +4 nC located at the origin?

•**P30** In a hydrogen atom in its ground state, the electron is on average a distance of about 0.5×10^{-10} m from the proton. What is the magnitude of the electric field due to the proton at this distance from the proton?

•**P31** A sphere with radius 1 cm has a charge of 2×10^{-9} C spread out uniformly over its surface. What is the magnitude of the electric field due to the sphere at a location 4 cm from the center of the sphere?

•**P32** A sphere with radius 2 cm is placed at a location near a point charge. The sphere has a charge of -9×10^{-10} C spread uniformly over its surface. The electric field due to the point charge has a magnitude of 470 N/C at the center of the sphere. What is the magnitude of the force on the sphere due to the point charge?

•**P33** What are the magnitude and direction of the electric field $\vec{E}$ at location $\langle 20, 0, 0\rangle$ cm if there is a negative point charge of 1 nC (1×10^{-9} C) at location $\langle 40, 0, 0\rangle$ cm? Include units.

•**P34** A sphere with radius 2 cm is placed at a location near a point charge. The sphere has a charge of -8×10^{-10} C spread uniformly over its surface. The electric field due to the point charge has a magnitude of 500 N/C at the center of the sphere. What is the magnitude of the force on the sphere due to the point charge?

••**P35** An electron is located at $\langle 0.8, 0.7, -0.8\rangle$ m. You need to find the electric field at location $\langle 0.5, 1, -0.5\rangle$ m, due to the electron. **(a)** What is the source location? **(b)** What is the observation location? **(c)** What is the vector $\vec{r}$? **(d)** What is $|\vec{r}|$? **(e)** What is the vector $\hat{r}$? **(f)** What is the value of $\frac{1}{4\pi\varepsilon_0}\frac{q}{|\vec{r}|^2}$? **(g)** Finally, what is the electric field at the observation location, expressed as a vector?

••**P36** A charged particle located at the origin creates an electric field of $\langle -1.2\times 10^3, 0, 0\rangle$ N/C at a location $\langle 0.12, 0, 0\rangle$ m. What is the particle's charge?

••**P37** At a particular location in the room there is an electric field = $\langle 1000, 0, 0\rangle$ N/C. Where would you place a single negative point particle of charge 1 μC in order to produce this electric field?

••**P38** The electric field at a location C points north, and the magnitude is 1×10^6 N/C. Give numerical answers to the following questions: **(a)** Where relative to C should you place a single proton to produce this field? **(b)** Where relative to C should you place a single electron to produce this field? **(c)** Where should you place a proton and an electron, at equal distances from C, to produce this field?

••**P39** You want to create an electric field = $\langle 0, 4104, 0\rangle$ N/C at location $\langle 0, 0, 0\rangle$. **(a)** Where would you place a proton to produce this field at the origin? **(b)** Instead of a proton, where would you place an electron to produce this field at the origin? (*Hint:* This problem will be much easier if you draw a diagram.)

••**P40** A π^- ("pi-minus") particle, which has charge $-e$, is at location $\langle 7\times 10^{-9}, -4\times 10^{-9}, -5\times 10^{-9}\rangle$ m. **(a)** What is the electric field at location $\langle -5\times 10^{-9}, 5\times 10^{-9}, 4\times 10^{-9}\rangle$ m, due to the π^- particle? **(b)** At a particular moment an antiproton (same mass as the proton, charge $-e$) is at the observation location. At this moment what is the force on the antiproton, due to the π^-?

••**P41** What is the electric field at a location $\vec{b} = \langle -0.1, -0.1, 0\rangle$ m, due to a particle with charge +3 nC located at the origin?

••**P42** At a particular location in the room there is an electric field $\vec{E} = \langle 1000, 0, 0\rangle$ N/C. Figure out where to place a single positive point particle, and how much charge it should have, in order to produce this electric field (there are many possible answers!). Do the same for a single negatively charged point particle. Be sure to draw diagrams to explain the geometry of the situation.

••**P43** Where must an electron be to create an electric field of $\langle 0, 160, 0\rangle$ N/C at a location in space? Calculate its displacement from the observation location and show its location on a diagram.

••**P44** The electric field at a location C points west, and the magnitude is 2×10^6 N/C. Give numerical answers to the following questions: **(a)** Where relative to C should you place a single proton to produce this field? **(b)** Where relative to C should you place a single electron to produce this field? **(c)** Where should you place a proton and an electron, at equal distances from C, to produce this field?

••**P45** A lithium nucleus consisting of three protons and four neutrons accelerates to the right due to electric forces, and the initial magnitude of the acceleration is 3×10^{13} m/s/s. **(a)** What is the direction of the electric field that acts on the lithium nucleus? **(b)** What is the magnitude of the electric field that acts on the lithium nucleus? Be quantitative (that is, give a number). **(c)** If this acceleration is due solely to a single helium nucleus (two protons and two neutrons), where is the helium nucleus initially located? Be quantitative (that is, give a number).

Section 13.5

••**P46** **(a)** On a clear and carefully drawn diagram, place a helium nucleus (consisting of two protons and two neutrons) and a proton in such a way that the electric field due to these charges is zero at a location marked ×, a distance 1×10^{-10} m from the helium nucleus. Explain briefly but carefully, and use diagrams to help in the explanation. Be quantitative about the relative distances. **(b)** On a clear and carefully drawn diagram, place a helium nucleus and an electron in such a way that the electric field due to these charges is zero at a location marked ×. Explain briefly but carefully, and use diagrams to help in the explanation. Be quantitative about the relative distances.

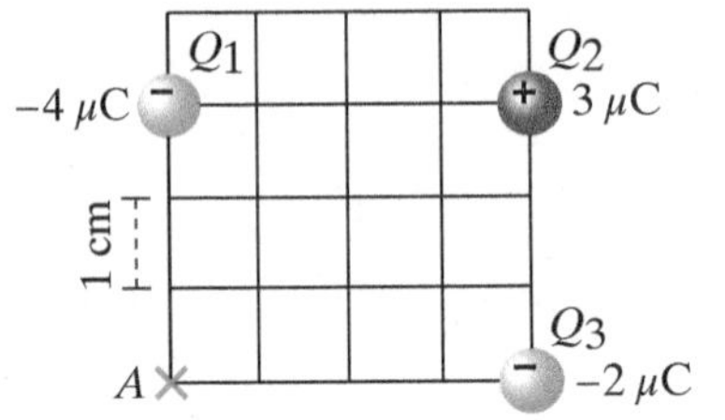

Figure 13.67

••**P47** At a particular moment, three charged particles are located as shown in Figure 13.67. $Q_1 = -4\,\mu$C, $Q_2 = +3\,\mu$C, and $Q_3 = -2\,\mu$C. Your answers to the following questions should be vectors. (Recall that $1\,\mu\text{C} = 1\times 10^{-6}$ C.) **(a)** Find the electric field at the location of Q_3, due to Q_1. **(b)** Find the electric field at the location of Q_3, due to Q_2. **(c)** Find the net electric field at the location of Q_3. **(d)** Find the net force on Q_3. **(e)** Find the electric field at location A due to Q_1. **(f)** Find the electric field at location A due to Q_2. **(g)** Find the electric field at location A due to Q_3. **(h)** What is the net electric field at location A? **(i)** If a particle with charge −3 nC were placed at location A, what would be the force on this particle?

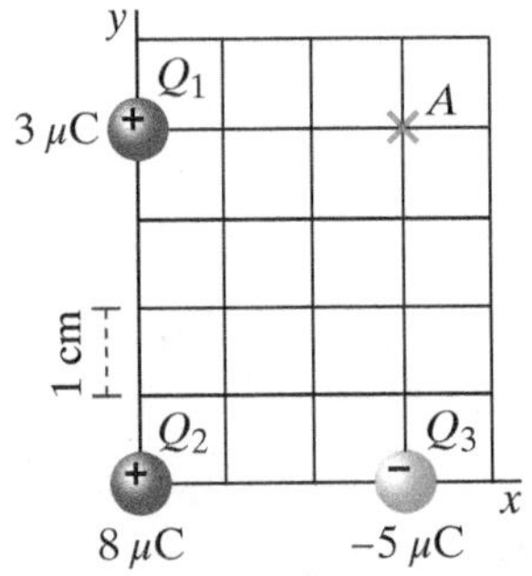

Figure 13.68

••**P48** At a particular moment, one negative and two positive charges are located as shown in Figure 13.68. $Q_1 = +3\,\mu$C, $Q_2 = +8\,\mu$C, and $Q_3 = -5\,\mu$C. Your answers to each part of this problem should be vectors. (Recall that $1\,\mu\text{C} = 1\times 10^{-6}$ C.) **(a)** Find the electric field at the location of Q_1, due to Q_2 and Q_3. **(b)** Use the electric field you calculated in part (a) to find the force on Q_1. **(c)** Find the electric field at location A, due to all three charges. **(d)** An alpha particle (He^{2+}, containing two protons and two neutrons) is released from rest at location A.

Use your answer from the previous part to determine the initial acceleration of the alpha particle.

••**P49** An Fe^{3+} ion is located 400 nm (400×10^{-9} m, about 4000 atomic diameters) from a Cl^- ion, as shown in Figure 13.69 (ions not shown to scale).

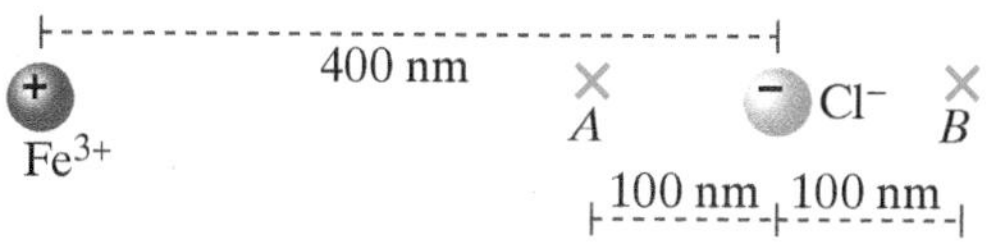

Figure 13.69

(a) Determine the magnitude and direction of the electric field $\vec{E}_A$ at location A, 100 nm to the left of the Cl^- ion. **(b)** Determine the magnitude and direction of the electric field $\vec{E}_B$ at location B, 100 nm to the right of the Cl^- ion. **(c)** If an electron is placed at location A, what are the magnitude and direction of the force on the electron?

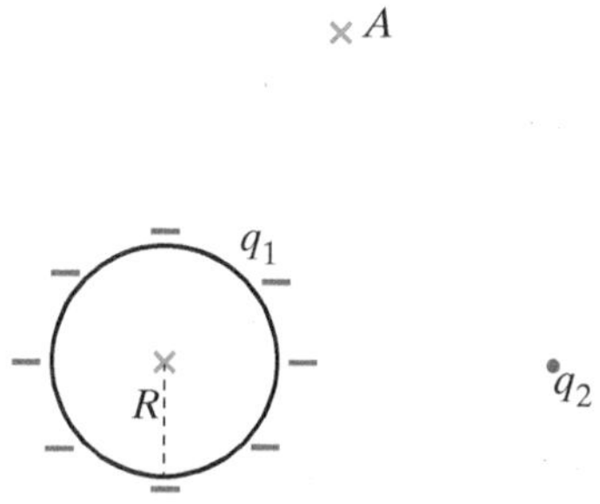

Figure 13.70

••**P50** A hollow ball with radius $R = 2$ cm has a charge of $q_1 = -3$ nC spread uniformly over its surface, as shown in Figure 13.70. The center of the ball is at $\langle -3,0,0 \rangle$ cm. A point charge of $q_2 = 5$ nC is located at $\langle 4,0,0 \rangle$ cm. **(a)** What is the net electric field at location A, when $\vec{r}_A = \langle 0,6,0 \rangle$ cm? **(b)** Draw an arrow representing the net electric field at that location. Make sure that the arrow you drew makes sense.

••**P51** Three nested hollow spheres have the same center. The innermost sphere has a radius of 2 cm and carries a uniformly distributed charge of 6 nC (1 nC $= 1 \times 10^{-9}$ C). The middle sphere has a radius of 5 cm and carries a uniformly distributed charge of −4 nC. The outermost sphere has a radius of 10 cm and carries a uniformly distributed charge of 8 nC. **(a)** What is the magnitude of the electric field at a distance of 1 cm from the center? **(b)** What is the magnitude of the electric field at a distance of 4 cm from the center? **(c)** What is the magnitude of the electric field at a distance of 9 cm from the center?

Section 13.6

•**P52** A dipole is located at the origin and is composed of charged particles with charge $+e$ and $-e$, separated by a distance 6×10^{-10} m along the x axis. The charge $+e$ is on the $+x$ axis. Calculate the electric field due to this dipole at a location $\langle 0, -5 \times 10^{-8}, 0 \rangle$ m.

•**P53** A dipole is located at the origin and is composed of charged particles with charge $+e$ and $-e$, separated by a distance 2×10^{-10} m along the x axis. Calculate the magnitude of the electric field due to this dipole at a location $\langle 0, 3 \times 10^{-8}, 0 \rangle$ m.

•**P54** The dipole moment of the HF (hydrogen fluoride) molecule has been measured to be 6.3×10^{-30} C · m. If we model the dipole as having charges of $+e$ and $-e$ separated by a distance s, what is s? Is this plausible?

••**P55** A dipole is located at the origin and is composed of charged particles with charge $+e$ and $-e$, separated by a distance 6×10^{-10} m along the y axis. The $+e$ charge is on the $-y$ axis. Calculate the force on a proton due to this dipole at a location $\langle 0, 4 \times 10^{-8}, 0 \rangle$ m.

••**P56** A dipole is located at the origin and is composed of charged particles with charge $+2e$ and $-2e$, separated by a distance 2×10^{-10} m along the y axis. The $+2e$ charge is on the $+y$ axis. Calculate the force on a proton at a location $\langle 0, 0, 3 \times 10^{-8} \rangle$ m due to this dipole.

•**P57** A dipole consists of two charges +6 nC and −6 nC, held apart by a rod of length 3 mm, as shown in Figure 13.71. **(a)** What is the magnitude of the electric field due to the dipole at location A, 5 cm from the center of the dipole? **(b)** What is the magnitude of the electric field due to the dipole at location B, 5 cm from the center of the dipole?

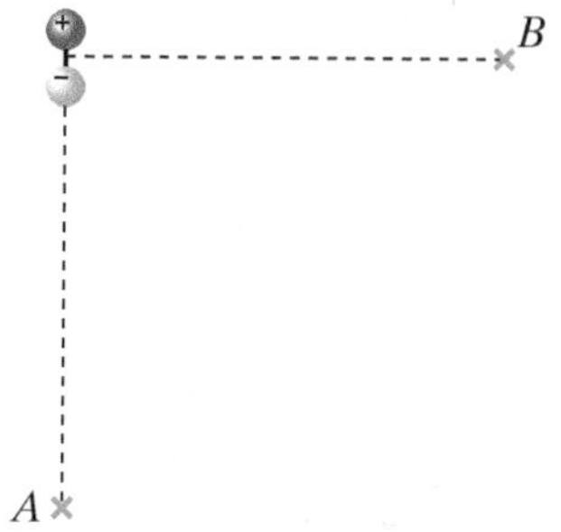

Figure 13.71

••**P58** A dipole is centered at the origin and is composed of charged particles with charge $+2e$ and $-2e$, separated by a distance 7×10^{-10} m along the y axis. The $+2e$ charge is on the $-y$ axis, and the $-2e$ charge is on the $+y$ axis. **(a)** A proton is located at $\langle 0, 3 \times 10^{-8}, 0 \rangle$ m. What is the force on the proton due to the dipole? **(b)** An electron is located at $\langle -3 \times 10^{-8}, 0, 0 \rangle$ m. What is the force on the electron due to the dipole? (*Hint:* Make a diagram. One approach is to calculate magnitudes, and get directions from your diagram.)

••**P59** Two dipoles are oriented as shown in Figure 13.72. Each dipole consists of two charges $+q$ and $-q$, held apart by a rod of length s, and the center of each dipole is a distance d from location A. If $q = 2$ nC, $s = 1$ mm, and $d = 8$ cm, what is the electric field at location A? (*Hint:* Draw a diagram and show the direction of each dipole's contribution to the electric field on the diagram.)

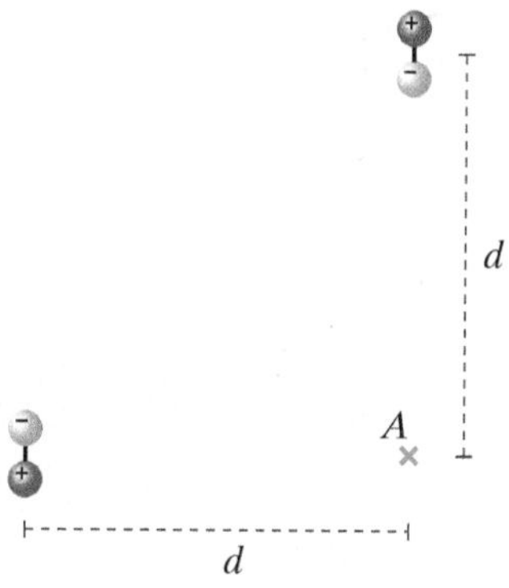

Figure 13.72

••P60 Two dipoles are oriented as shown in Figure 13.73. Each dipole consists of charges held apart by a short rod (not shown to scale). What is the electric field at location A? Start by drawing a diagram that shows the direction of each dipole's contribution to the electric field at location A.

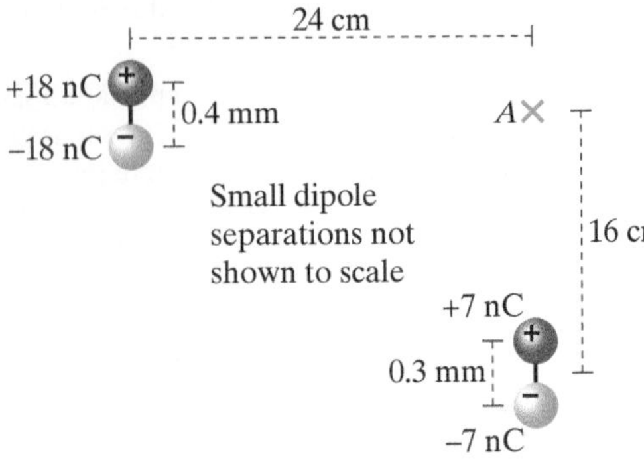

Figure 13.73

••P61 A charge of +1 nC (1×10^{-9} C) and a dipole with charges $+q$ and $-q$ separated by 0.3 mm contribute a net field at location A that is zero, as shown in Figure 13.74.

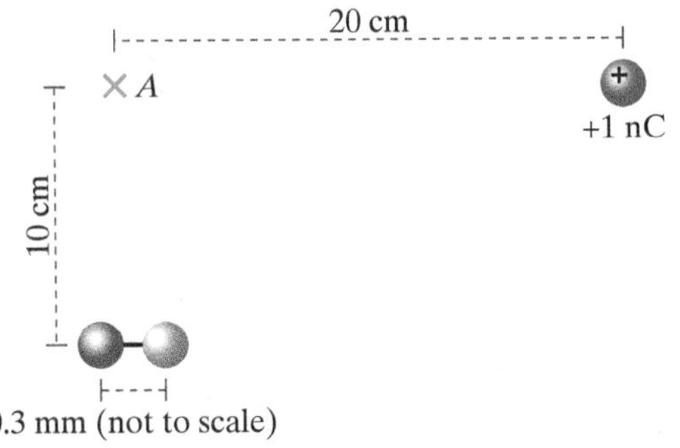

Figure 13.74

(a) Which end of the dipole is positively charged? **(b)** How large is the charge q?

•••P62 A water molecule is asymmetrical, with one end positively charged and the other negatively charged. It has a dipole moment whose magnitude is measured to be 6.2×10^{-30} C·m. If the dipole moment is oriented perpendicular to an electric field whose magnitude is 4×10^5 N/m, what is the magnitude of the torque on the water molecule? Also, show that the vector torque is equal to $\vec{p} \times \vec{E}$, where $\vec{p}$ is the dipole moment.

•••P63 Two identical permanent dipoles, each consisting of charges $+q$ and $-q$ separated by a distance s, are aligned along the x axis, a distance r from each other, where $r \gg s$ (Figure 13.75). Show all of the steps in your work, and briefly explain each step. **(a)** Draw a diagram showing all individual forces acting on each particle, and draw heavier vectors showing the net force on each dipole. **(b)** Show that the magnitude of the net force exerted on one dipole by the other dipole is this:

$$F \approx \frac{1}{4\pi\varepsilon_0}\frac{6q^2s^2}{r^4}$$

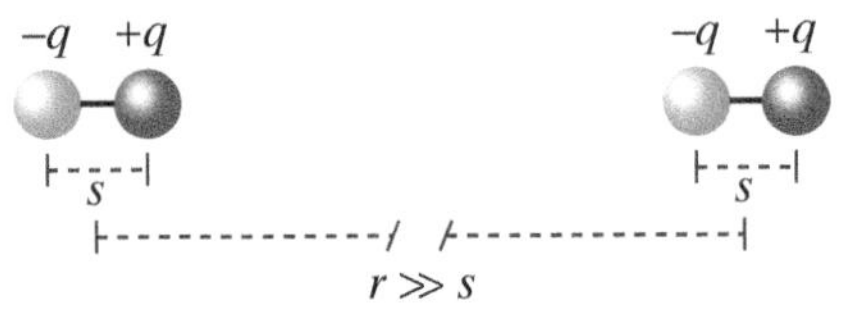

Figure 13.75

Section 13.8

••P64 You make repeated measurements of the electric field $\vec{E}$ due to a distant charge, and you find it is constant in magnitude and direction. At time $t = 0$ your partner moves the charge. The electric field doesn't change for a while, but at time $t = 45$ ns you observe a sudden change. How far away was the charge originally?

COMPUTATIONAL PROBLEMS

If you did not learn how to create 3D computational models using VPython in your previous physics course, you should work through the introductory materials from Chapter 1 that are available at `www.wiley.com/college/chabay`.

To install the free 3D programming environment VPython, go to vpython.org and (carefully) follow the instructions for your operating system (Windows, MacOS, or Linux). Note the instructions given there on how to zoom and rotate the "camera" when viewing a 3D scene you have created.

To use an `arrow` object to visualize an electric field, it is usually necessary to scale the length of the arrow in order to make it fit on the screen with the objects that produce or experience that field. Watch VPython Instructional Video 5: Scalefactors, at `vpython.org/video05.html` to learn how to do this. In general it may be helpful to print the magnitude of an electric field using `print(mag(E))` in order to decide on an appropriate scale factor.

More detailed and extended versions of some of these computational modeling problems may be found in the lab activities included in the *Matter & Interactions, 4th Edition*, resources for instructors.

•P65 In Section 13.9 there is a program to calculate the electric field of a single point charge at multiple observation locations. **(a)** Study this program and make sure you can explain every line of code. **(b)** Modify the program so that the magnitude of the electric field at each observation location is the same. (There is more than one way to do this.) **(c)** Add an additional circle of observation locations in the yz plane, centered on the point charge.

••P66 A particle with a charge of +3 nC is located at $\langle -0.04, 0, 0 \rangle$ m. **(a)** Calculate the electric field at location $\langle -0.04, 0, 0.05 \rangle$ m due to this particle, and create an arrow to visualize the field at the observation location. Try a scale factor of about 2×10^{-6}. **(b)** Add an arrow representing the electric field at location $\langle -0.04, 0, -0.05 \rangle$ due to this particle. **(c)** Add two more arrows, each representing the electric field at a location 0.05 m from the particle in the $\pm y$ direction. **(d)** Add two more arrows, each representing the electric field at a location 0.05 m from the particle in the $\pm x$ direction.

••**P67** Three charged spheres lie in the xz plane. Sphere a has a charge of $+2$ nC, and is located at $\langle -0.03, 0, 0.03\rangle$ m. Sphere b has a charge of $+4$ nC, and is located at $\langle 0.03, 0, 0.03\rangle$ m. Sphere c has a charge of -2 nC, and is located at $\langle 0, 0, -0.011\rangle$ m. Write a program to calculate and display (using an arrow) the electric field of these three charged spheres at location $\langle 0, 0.25, 0\rangle$ m.

••**P68** The following code creates two objects representing a dipole oriented along the y axis. **(a)** Extend the program to calculate and display (using arrows) the electric field due to the dipole at 12 equally spaced observation locations located on a circle of radius 0.5 nm in the xy plane, centered on the dipole. **(b)** Add a second circle of arrows representing the electric field at observation locations in the yz plane.

```
from visual import *
scene.width = scene.height = 800
oofpez = 9e9
qe = 1.6e-19
sf = 5e-20
source_01 = sphere(pos=vector(0,0.1e-9,0),
                   color=color.red,
                   radius=0.5e-10)
q_01 = +qe
source_02 = sphere(pos=vector(0,-0.1e-9,0),
                   color=color.blue,
                   radius=0.5e-10)
q_02 = -qe
```

••**P69** The following skeleton program creates objects representing a stationary source charge and a moving antiproton. **(a)** Complete the program so that the antiproton is affected by the net electric field at its current location. **(b)** Add two arrows, and use them to visualize the momentum of the antiproton, and the electric field at the location of the antiproton. These arrows should move with the antiproton.

```
from visual import *
scene.width = scene.height = 760
scene.range = 3e-8
oofpez = 9e9
qe = 1.6e-19
source = sphere(pos=vector(-0.5e-8,0,0),
                radius=5e-10,
                color=color.red)
source_q = qe
##antiproton
ap = sphere(pos=vector(0.5e-8,-1e-8,0),
            radius=5e-10,
            color=color.cyan,
            make_trail = True)
ap_q = -qe
ap_m = 1.7e-27
ap_p = ap_m * vector(0,5e3,0)
deltat = 1e-15
t=0

while t < 8e-12:
    rate(500)
    ## add your code here
    ap.pos = ap.pos + (ap_p/ap_m) * deltat
    t = t + deltat
```

•••**P70** Start with the program you wrote in Problem P68 to calculate and display the electric field of a dipole. **(a)** Place a proton at location $\langle 0.3\times 10^{-9}, 0, 0\rangle$ m, and release it from rest. Compute and display the trajectory of the proton as it moves under the influence of the electric field of the dipole. You may wish to start with $\Delta t = 1\times 10^{-17}$ s. **(b)** Simultaneously compute and plot a graph showing the potential energy U, kinetic energy K, and $(K+U)$ vs. time for the entire system (dipole + proton). Your graph will be more useful as a computational diagnostic tool if you do not include the potential energy associated with the interaction of the pair of charges making up the dipole, which does not change. **(c)** Explain the shape of the K and U graphs.

•••**P71** Write a computer program to calculate and plot a graph of the magnitude of the electric field of the dipole from Problem P68 at locations on the y axis as a function of distance from the center of the dipole. Vary y from 0.2 nm (0.2×10^{-9} m) to 0.5 nm from the center of the dipole. Do the calculation two different ways, and put both plots (in different colors) on the same axes: **(a)** Calculate the electric field exactly as the superposition of the fields due to the individual charges. **(b)** Calculate the electric field using the approximate equation for the dipole field derived in Section 13.6. **(c)** Comment on the validity of the approximate equation. How close to the dipole (compared to s, the dipole separation) do you have to get before the approximate equation no longer gives good results? What is your criterion for "good results"?

ANSWERS TO CHECKPOINTS

1 $\approx 1\times 10^{-10}$ m; protons and electrons
2 $\langle 0, -4.8\times 10^{-17}, 0\rangle$ N; $\langle 0,0,0\rangle$ N
3 $\langle 900, 0, 0\rangle$ N/C
4 2.8×10^{4} N/C
5 **(a)** 3.6×10^{4} N/C; **(b)** 7.2×10^{4} N/C
6 **(a)** 1/8; **(b)** three times bigger; **(c)** two times bigger, opposite direction
7 Its electric dipole moment would point horizontally, in the same direction as the applied electric field, with the + charge to the right.
8 3.2×10^{-19} C

CHAPTER

14

Electric Fields and Matter

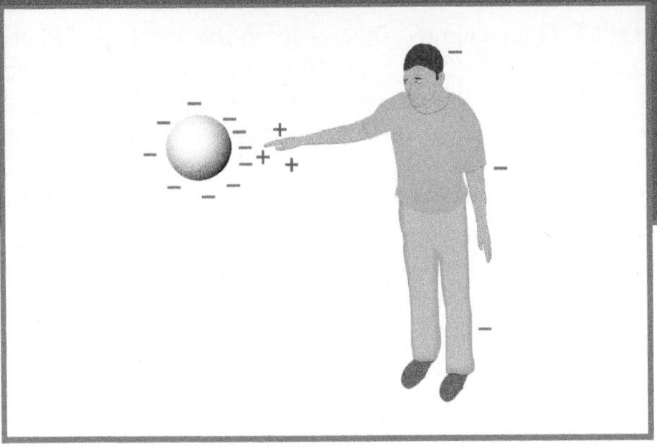

OBJECTIVES

After studying this chapter you should be able to

- Explain the difference between a conductor and an insulator, and graphically depict the polarization of each when subjected to an external electric field.
- Explain why a charged object is attracted to a neutral object.
- Calculate the drift speed of a charged particle in a conductor.
- Explain why the net electric field inside a conductor at equilibrium must be zero.
- Mathematically relate polarizability, induced dipole moment, and applied electric field.

14.1 CHARGED PARTICLES IN MATTER

Since ordinary matter is composed of charged particles, electric fields can affect matter. In order to understand the effect of electric fields on matter, in this chapter we will extend our microscopic model of matter to include the fact that matter contains charged particles: protons and electrons.

Net Charge

DEFINITION OF NET CHARGE

> The net charge of an object is the sum of the charges of all of its constituent particles. An object with a net charge of zero is called "neutral." An object with a nonzero net charge (either positive or negative) is called "charged."

Elementary particles such as protons and electrons are electrically charged. If a proton and an electron combine to form a hydrogen atom, however, the hydrogen atom is electrically "neutral"—its net charge is the sum of the charges of its constituent particles, which in this case is zero:

$$(+e)+(-e)=0$$

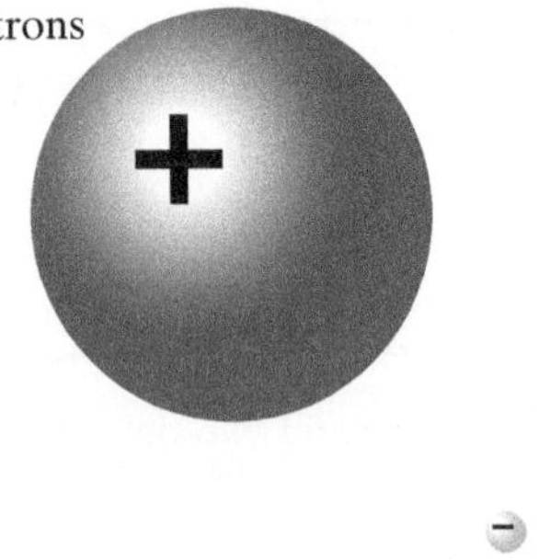

Figure 14.1 A sodium ion Na^+ consists of a sodium atom that has lost an electron.

A sodium atom has 11 protons in its nucleus and 11 electrons surrounding the nucleus, so it has a net charge of zero and is electrically neutral. However, a sodium atom can lose an electron, becoming a sodium ion, Na^+.

QUESTION What is the net charge of a sodium ion, Na^+ (Figure 14.1)?

A sodium ion has 11 protons and 10 electrons, so its net charge is

$$(+11e)+(-10e)=+e=+1.6\times 10^{-19}\,\text{C}$$

Ordinary matter is electrically neutral. However, it is possible to remove or add charged particles, giving an object a nonzero net charge.

QUESTION Is a dipole electrically neutral?

A dipole is neutral, because the sum of its constituent charges is zero:

$$(+q)+(-q)=0$$

Evidently even a neutral object can make a nonzero electric field in the surrounding space.

Conservation of Charge

In an extremely wide variety of experiments, no one has ever observed a change in the net charge of the universe. These results are summarized by the fundamental principle called "conservation of charge": if the net charge of a system changes, the net charge of the surroundings must change by the opposite amount. For example, if your comb acquires negative charge, your hair acquires an equal amount of positive charge:

CONSERVATION OF CHARGE

The net charge of a system plus its surroundings cannot change.

Conservation of charge is a fundamental principle, because it applies to every system in every situation.

Consider the annihilation reaction between an electron and a positron:

$$e^- + e^+ \rightarrow \gamma + \gamma$$

QUESTION In this reaction an electron and a positron are destroyed, creating two high-energy photons (called gamma rays). Does this reaction violate the principle of conservation of charge?

No. The net charge of the system (electron plus positron) was initially zero; the charge of the two photons is also zero. Even though charged particles were destroyed, the net charge of the system did not change.

Checkpoint 1 A carbon atom is composed of 6 protons, 6 neutrons, and 6 electrons. What is the net charge of this atom? A neutral chlorine atom contains 17 protons and 17 electrons. When a chlorine atom gains one extra electron, it becomes a chloride ion. What is the net charge of a chloride ion?

Conductors and Insulators

All materials are made of atoms that contain electrons and protons. However, at the microscopic level there can be differences in structure that lead to very different behavior when macroscopic objects are exposed to electric fields. In this chapter we will examine two different kinds of materials: conductors and insulators. (There are other classes of technologically important materials, such as semiconductors and superconductors, which we will discuss briefly in later chapters.)

Conductors: Some materials contain charged particles that can move easily through the material. These materials are called conductors. Most metals, such as copper, silver, iron, aluminum, and gold, are excellent conductors because they contain highly mobile electrons, as we will discuss later. Aqueous salt solutions are also conductors, because there are mobile positive and negative ions. An example is ordinary salt water, containing positive sodium ions (Na^+) and negative chloride ions (Cl^-).

Insulators: In many materials the electrons are tightly bound to the atoms, and there are no charged particles that can move through the material. These materials are called insulators because they can electrically "insulate" one charged object from another. You are familiar with many insulating materials, such as rubber, plastics, wood, paper, and glass. A related experiment is EXP11.

DEFINITION OF "CONDUCTOR" AND "INSULATOR"

A conductor contains mobile charged particles that can move throughout the material.
An insulator has no mobile charged particles.

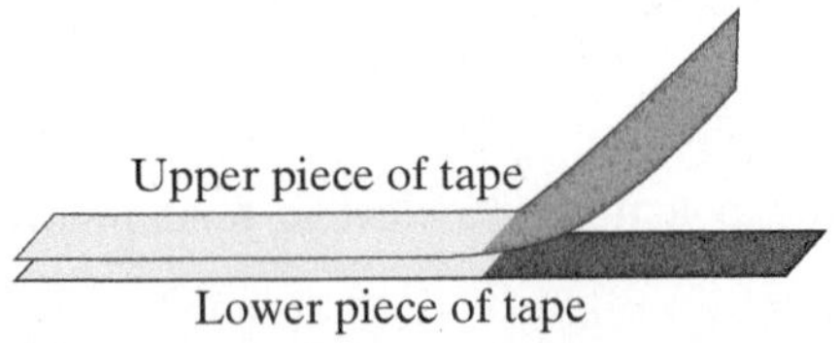

Figure 14.2 Pulling a piece of invisible tape off of another piece charges the tapes electrically. See Experiment EXP1.

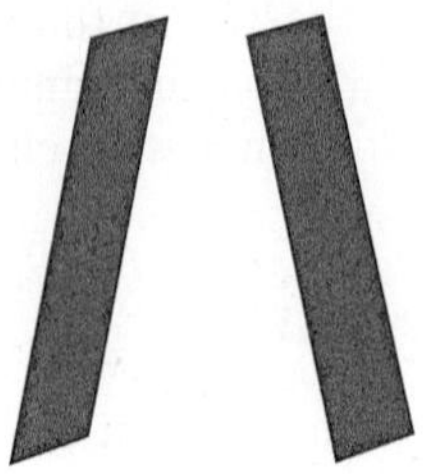

Figure 14.3 Two pieces of invisible tape with like charges repel each other.

14.2 HOW OBJECTS BECOME CHARGED

Many of the interactions we observe in our everyday lives are electric in nature. By being systematic in observing the behavior of simple systems and by thinking carefully in our analysis of this behavior, we can uncover some deep questions about the interaction of ordinary matter with electric fields.

In the section "Basic Experiments" at the end of this chapter are some simple experiments that you can do to make the issues vivid. All you need is a roll of "invisible" tape such as Scotch® brand Magic™ Tape or a generic brand of frosted tape (Figure 14.2). When you pull a long piece of invisible tape off a roll, it often curls up or sticks to your hand, and this is due to electric interactions between the electrically charged tape and your hand. We encourage you to do these simple experiments as you study this chapter.

In experimenting with charged objects such as invisible tape we find the following:

- Objects can have a charge that is positive, negative, or zero.
- Like charges repel (Figure 14.3), unlike charges attract.
- The electric force
  - acts along a line between the charges,
  - decreases rapidly as the distance between the charges increases, and
  - is proportional to the amounts of both charges.

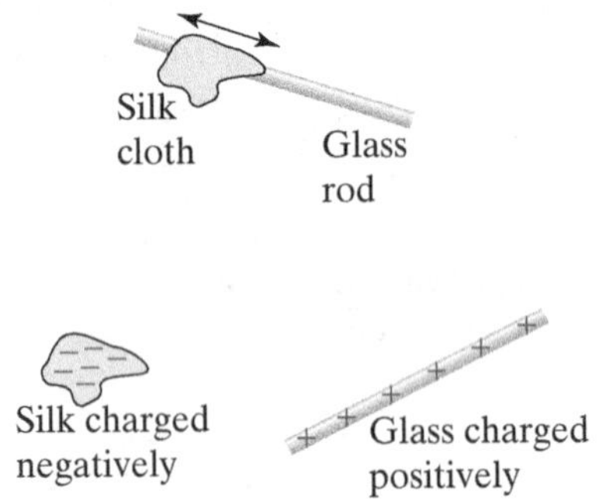

Figure 14.4 Rub silk on glass and the glass becomes positively charged.

Charging by Contact

A charged object that has a net negative charge has more electrons than protons. A positively charged object has fewer electrons than protons.

QUESTION Normally most ordinary objects appear to be neutral. How can they become significantly charged?

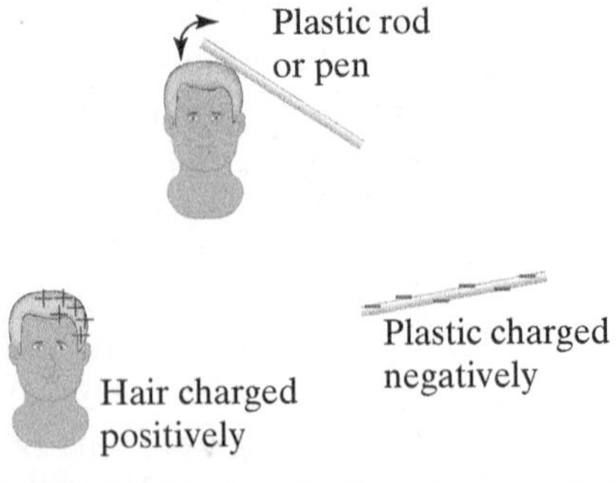

Figure 14.5 Rub a plastic pen on your hair or shirt and the pen becomes negatively charged.

It has been known for a long time that if you rub a glass rod with silk, the glass rod becomes positively charged and the silk negatively charged (Figure 14.4). If you rub a clear plastic object such as a pen through your hair (or with fur, wool, or even cotton), the plastic ends up having a negative charge and so repels electrons (Figure 14.5). A similar process occurs when you separate one piece of invisible tape from another. Many objects acquire a nonzero net charge through contact with other objects. See the related experiments EXP1–EXP4.

There are a variety of possible mechanisms for charging by contact. Large organic molecules in the plastic or your hair could break at their weakest bond in such a way that negative ions (negatively charged fragments) are deposited on the plastic and/or positive ions (positively charged fragments) are deposited on your hair. Electrons can move from one object to another, especially if the objects are metals, as we will discuss later. It is known that rubbing is not essential to transferring charge from one object to another: mere contact can be sufficient. However, rubbing produces many points of contact, which may facilitate transfer.

Protons Are Not Removed from Nuclei

One thing is certain: you cannot remove bare nuclei from inside the surface atoms or remove protons from inside the nuclei of the surface atoms by rubbing. The amount of energy required to do this would be enormous. Removing protons would amount to transmuting one element into another! The nucleus is buried deep inside the atom, and the protons are bound tightly in the nucleus. A much smaller force is required to remove one electron from an atom, or to break a chemical bond and transfer an entire ion to another object. Therefore the only things that can be transferred by contact are positive or negative ions, or electrons.

We can make an approximate comparison of the energy required to charge an object by different mechanisms: breaking a bond, removing an electron from an atom, or removing a proton from a nucleus. We saw in Chapter 8 (Energy Quantization) that the energy required to ionize a hydrogen atom (that is, to move the electron very far away from the nucleus) was about 14 eV (recall that $1\text{ eV} = 1.6 \times 10^{-19}\text{ J}$), so we can estimate the energy required to remove an electron from any atom as several eV. Similarly, the energy required to break one of the oxygen–hydrogen bonds in water is about 450 kJ/mol, which is 4.6 eV per molecular bond.

In contrast, as we saw in Chapter 6 (The Energy Principle), an input of 2.2 MeV (2.2×10^6 eV) is required to break apart the nucleus of a deuterium atom, which consists of one proton and one neutron, so we can estimate the energy required to remove a proton from a nucleus as millions of electron-volts.

Mechanism	**Energy Required**
Break chemical bond	several eV
Remove electron	several eV
Remove proton from nucleus	several *million* eV

It is clear that either breaking bonds or removing single electrons is a possible mechanism for charging a macroscopic object by contact, but removing protons from the atomic nuclei is not.

The Location of Charge Transfer

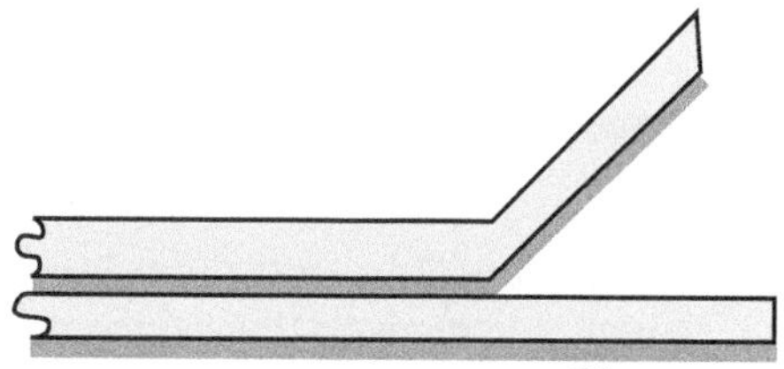

Figure 14.6 Two initially neutral tapes being pulled apart and becoming charged. Where are the regions with excess + and − charges?

Suppose that two neutral strips of invisible tape are stuck together, then pulled quickly apart, as shown in Figure 14.6. Both tapes will become charged as a result of this process. (The sign of the charge on each tape depends on the chemical composition of the tape and the glue, which varies from brand to brand.)

QUESTION If the upper tape ends up being negatively charged after the tapes are pulled apart, what is the charge of the lower tape?

The principle of conservation of charge requires that the net charge of this isolated system remain unchanged. Since the tapes started out neutral, the lower tape must now be positively charged, and the absolute value of the charge of each tape must be the same. For example, if the charge of the upper tape is -1 nC, the charge of the lower tape must be $+1$ nC ($1\text{ nC} = 1 \times 10^{-9}\text{ C}$).

QUESTION After the tapes are separated, where are the excess charges on the negative upper tape located? Where are the excess charges on the positive lower tape located?

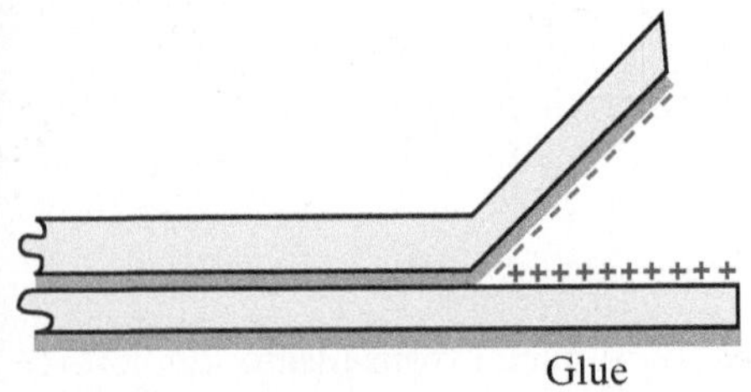

Figure 14.7 In this case the upper tape becomes negatively charged, so there must be excess electrons or negative ions on the part of its lower surface that is no longer touching the lower tape. The corresponding part of the lower tape has a deficiency of electrons, so there are now positive ions on it.

The tapes interact only along the bottom surface of the upper tape and the top surface of the lower tape, and because plastic is a good insulator the charges cannot move to a different location. Any excess charges must end up on one of these surfaces, as shown in Figure 14.7. The parts of the tapes that have been separated now have an excess or a deficiency of negative charge. Electrons or negative ions could have moved from the lower tape to the upper tape, or positive ions could have moved from the upper tape to the lower tape.

How Much Charge Is on a Charged Object?

Objects like invisible tapes, plastic pens, Ping-Pong balls, balloons, and glass rods can be electrically charged by rubbing with an appropriate material, ripping apart, or other similar contact. It would be useful to know approximately how much excess charge is on an ordinary small object like these when it is charged. Even knowing an approximate order of magnitude would be useful—is it closer to 10 C, 0.1 C, or 1×10^{-10} C?

EXAMPLE **Approximate Amount of Charge on a Piece of Tape**

We can make a rough estimate of the amount of excess charge on a piece of tape by doing a simple experiment, illustrated in Figure 14.8. Prepare a pair of oppositely charged invisible tapes by pulling apart a pair of initially neutral tapes, so that the magnitude of charge on the two tapes is the same. If we suspend the positively charged tape between two books and lower the negatively charged tape toward it, when the negative tape gets close enough we will see that the positive tape is lifted upward. If we know the mass of the tape, we can estimate the electric force on the tape, and from this and the distance between the tapes, estimate the charge on the tape.

Solution Momentum Principle:

$$\frac{dp_y}{dt} = F_{\text{electric}} - F_{\text{grav}}$$

$$0 = Q_B E_{A,y} - mg$$

In Figure 14.8, estimate the electric field due to A by approximating A as a point charge (a very rough approximation):

$$E_{A,y} \approx \frac{1}{4\pi\varepsilon_0}\frac{Q_A}{d^2}$$

so

$$Q_B \frac{1}{4\pi\varepsilon_0}\frac{Q_A}{d^2} \approx mg$$

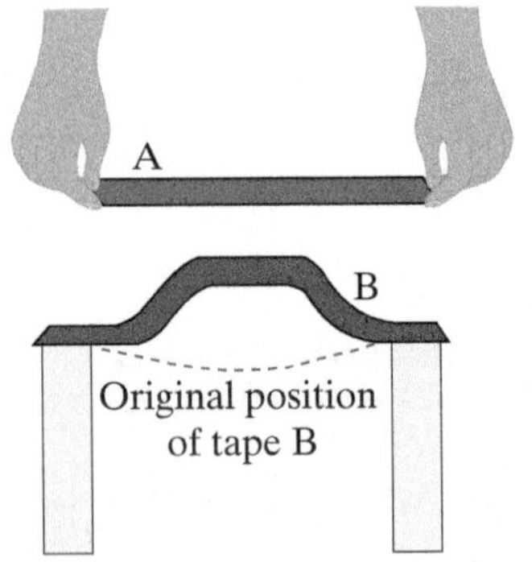

Figure 14.8 Tapes A and B are oppositely charged. When A gets sufficiently close, the electrical force on B is sufficient to lift it up.

If the 20-cm-long piece of tape has a mass of about 0.15 g, and the bottom tape starts to be lifted when the top tape is about 2.5 cm away, then, since the magnitude of the charge on the tapes is the same,

$$Q \approx \sqrt{\frac{(1.5 \times 10^{-4}\ \text{kg})(9.8\ \text{N/kg})(0.025\ \text{m})^2}{\left(9 \times 10^9 \dfrac{\text{N}\cdot\text{m}^2}{\text{C}^2}\right)}} \approx 1 \times 10^{-8}\,\text{C} = 10\ \text{nC}$$

This is a rough estimate, but it turns out to be a reasonable one—a small object charged by rubbing usually has a charge on the order of 10 nC. In Chapter 15 we'll see how to calculate the electric field of an object that is not point-like but has charge spread over its entire surface. Using these techniques, we find that the electric field at a location very close to the surface of the charged tape is around 2×10^5 N/C. Since the electric field required to ionize the air itself is about 3×10^6 N/C, the field near the surface of the charged tape is pretty large.

Experiment EXP6 guides you through a more careful version of this experiment. Your results may be somewhat different, depending on the design of the experiment (there are other possible geometries), the length of your tapes, and other factors.

Fraction of Surface Atoms with Excess Charge

Assuming that the net charge on the negative tape is distributed uniformly over its surface, we can estimate the fraction of those surface atoms that have gained an excess electron or negative ion.

EXAMPLE **Fraction of Surface Atoms with Excess Charge**

What fraction of the atoms on the charged surface of the tape have gained or lost charge? (Assume that an atom gains or loses at most one electron charge.)

Solution Surface area of a 1-cm-wide, 20-cm-long tape:

$$A = (0.2\ \text{m})(0.01\ \text{m}) = 2 \times 10^{-3}\ \text{m}^2$$

Approximate cross-sectional area occupied by an atom whose radius is approximately 1×10^{-10} m:

$$A_{\text{atom}} \approx (2 \times 10^{-10}\ \text{m})^2 = 4 \times 10^{-20}\ \text{m}^2$$

Number of atoms on the surface:

$$\frac{A}{A_{\text{atom}}} = \frac{2 \times 10^{-3}\ \text{m}^2}{4 \times 10^{-20}\ \text{m}^2} = 5 \times 10^{16}\ \text{atoms}$$

Number of excess electrons (or ions):

$$\frac{10 \times 10^{-9}\ \text{C}}{1.6 \times 10^{-19}\ \text{C}} = 6.25 \times 10^{10}$$

Fraction of surface atoms with excess charge:

$$\frac{6.25 \times 10^{10}}{5 \times 10^{16}} \approx 1 \times 10^{-6}$$

Thus only about one in a million atoms on the surface of the tape has acquired an excess electron or lost an electron—a small fraction.

At the macroscopic level, the charge on the surface of the tape may appear to be distributed quite uniformly. At the atomic level, though, we see that the atoms with excess charge are sprinkled quite sparsely over the surface.

14.3 POLARIZATION OF ATOMS

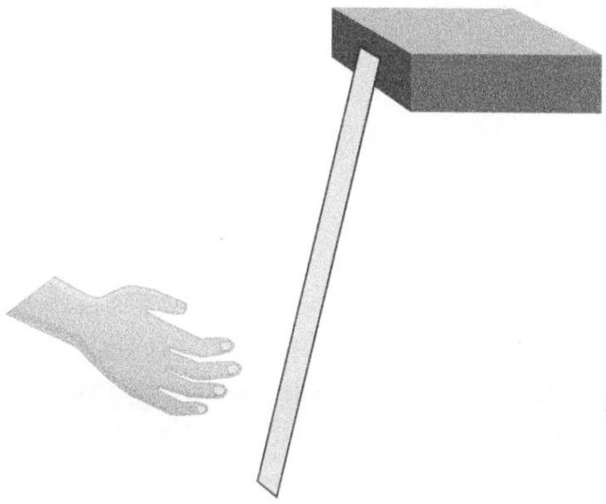

Figure 14.9 A charged tape, whether positive or negative, is attracted to your neutral hand.

A positively charged object, like a positive piece of invisible tape, is attracted not only to a negative tape, but also to your hand, your desk, your book, and every other nearby neutral object (Figure 14.9). The same is true for a negatively charged object, such as a negative piece of invisible tape. Since

$$\vec{F} = q\vec{E}$$

this implies that your neutral hand must make a nonzero electric field. It is not obvious how or why this can happen.

QUESTION Why are charged objects attracted to neutral objects?

The attraction of both positively and negatively charged invisible tape to your hand, and to many other neutral objects, is deeply mysterious. The net charge of a neutral object is zero, so your neutral hand should not make an electric field that could act on a charged tape, nor should your neutral hand experience a force due to the electric field made by a charged tape. Nothing in our statement of the properties of electric interactions allows us to explain this attraction!

The Structure of an Atom

An external charge can cause a shift in the position of the charges that make up a neutral atom or molecule. To see this clearly we need to look in more detail at the structure of atoms. We'll consider a hydrogen atom because it is the simplest atom, but the effects we discuss occur with other atoms as well.

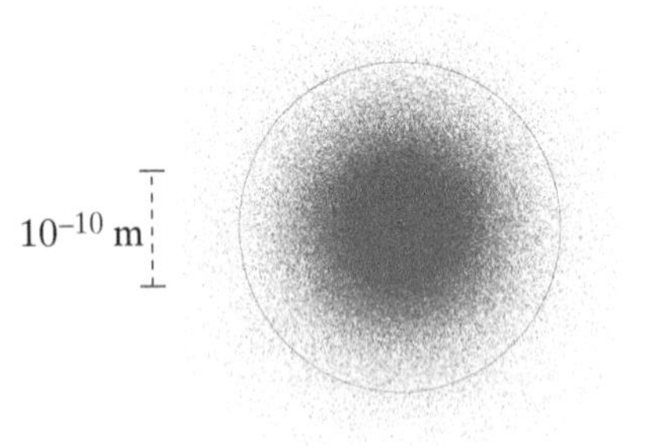

Figure 14.10 A quantum-mechanical view of a hydrogen atom. The picture is a two-dimensional slice through a three-dimensional spherical distribution. Each dot represents the location of the electron at the time of a multiple-exposure photo. The tiny nucleus is shown as a red dot at the center of the electron cloud (actual nuclear radius is only about 1×10^{-15} m). A circle is drawn through regions of constant density.

In Figure 14.10 we show a special kind of picture of a hydrogen atom, based on quantum mechanics, the theory that describes the detailed structure of atoms. A hydrogen atom consists of an electron and a nucleus normally consisting of one proton (and no neutrons). The lightweight electron doesn't follow a well-defined orbit around the heavy nucleus the way the Earth does around the Sun. Rather, there is only a probability for finding the electron in any particular place.

Figure 14.10 shows this probability graphically. You can think of the picture as a multiple exposure. For each exposure, the position of the electron at that time is shown as a dot. Because the electron is most likely to be found near the nucleus, that part of the multiple exposure is so dark you can't see the individual dots. The electron is seldom found a long way from the nucleus, so as you get farther and farther from the nucleus the density of dots gets less and less.

We call this probability distribution the "electron cloud." In hydrogen the cloud consists of just one electron, but in other atoms the electron cloud is made up of many electrons. The average location of the electron is in the center, at the same location as the nucleus. You're just as likely to find the electron to the right of the nucleus as to the left of the nucleus.

It is impossible to show the nucleus accurately on this scale. Although the mass of a proton is 2000 times the mass of an electron, the radius of the proton, about 1×10^{-15} m, is only about 1/100,000 as big as the radius of the electron cloud, which is itself only about 1×10^{-10} m! We used an oversize red dot to mark the position of the tiny nucleus in Figure 14.10.

In the following exercise, remember that in the previous chapter we pointed out that the electric field produced by a uniformly distributed sphere of charge, outside the sphere, is the same as though all the charge were located at the center of the sphere (this will be discussed in more detail in Chapter 15).

Checkpoint 2 A student asked, "Since the positive nucleus of the atom is hidden inside a negative electron cloud, why doesn't all matter appear to be negatively charged?" Explain to the student the flaw in this reasoning.

Polarization of Atoms

If the electron cloud in an atom could be considered to be spherically uniform and always centered on the nucleus, a neutral atom would have no interaction with an external charge. If the electron cloud were centered on the nucleus, the electric field produced by the N electrons would exactly cancel the field produced by the N protons. However, the electron cloud doesn't always stay centered, as we'll see next.

In an atom the electron cloud is not rigidly connected to the nucleus. The electron cloud and the nucleus can move relative to each other. If an external charge is nearby, it creates an electric field, which exerts forces on the electron cloud and on the nucleus. Under the influence of this "applied" electric field the electron cloud and the nucleus shift position relative to each other.

Figure 14.11 shows the probability distribution or electron cloud for hydrogen when there is an external positive charge located somewhere to the left of the hydrogen atom. You can see that the cloud has been distorted, because the positive charge attracts the electron to the left and repels the nucleus to the right.

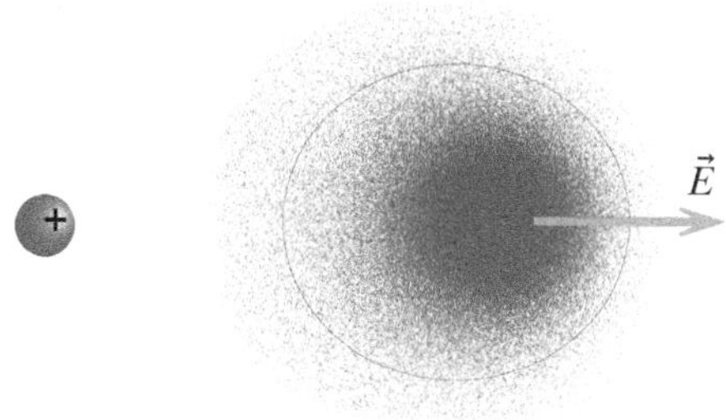

Figure 14.11 A positive charge makes an electric field that shifts the electron cloud of the hydrogen atom to the left (and shifts the hydrogen nucleus to the right). An ellipse is drawn through regions of constant density. It is now more probable that the electron will be found to the left of the nucleus than to the right.

Average Location of the Electron

The average location of the electron is now not at the center where the nucleus is located, but is displaced somewhat to the left of the nucleus. That is, each time you take a snapshot, you're more likely to find the electron to the left of the nucleus than to the right of the nucleus.

The hydrogen atom isn't immediately torn apart, because the attraction between the nucleus and the electron is stronger than the forces exerted by the distant external charge. However, if the external charge gets *very* close the hydrogen atom may break up or react with the external charge. If the external charge were a proton, it could combine with the hydrogen atom to form ionized molecular hydrogen ($H_2{}^+$).

You can see in Figure 14.11 that the outer regions of the cloud are affected the most by the external charge. This is because in the outer regions the electron is farther from the nucleus and can be influenced more by the external positive charge. In an atom containing several electrons, the outer electrons are affected the most. The picture is deliberately exaggerated to show the effect: unless the polarization is caused by charges only a few atomic diameters away, the shift in the electron cloud is normally too small to represent accurately in a drawing.

An atom is said to be "polarized" when its electron cloud has been shifted by the influence of an external charge so that the electron cloud is not centered on the nucleus.

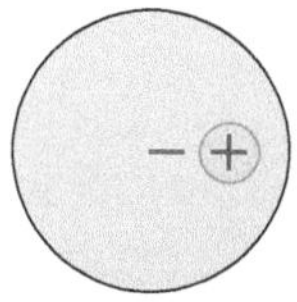

Figure 14.12 We can approximate a polarized atom as a roughly spherical electron cloud whose center is displaced from the positive nucleus.

Diagrams of Polarized Atoms or Molecules

For most purposes we can approximate the charge distribution of the polarized atom as consisting of an approximately spherical negative cloud whose center is displaced from the positive nucleus (Figure 14.12). A uniform spherical charge distribution acts as if it were a point charge located at the center of the sphere, both in the sense that it makes an electric field outside the sphere identical to the electric field of a point charge and that it responds to applied fields as though it were a point charge. It is therefore reasonable to model a polarized atom as a dipole, consisting of two opposite point charges separated by a small distance.

To simplify drawing a polarized atom or molecule and to emphasize its most important aspects, we will usually represent it as an exaggeratedly elongated blob, with + and − at the ends (Figure 14.13).

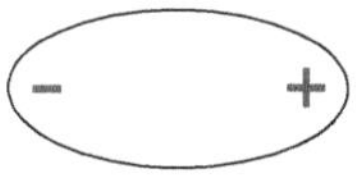

Figure 14.13 A simplified representation of a polarized atom or molecule.

Induced Dipoles Are Created by Applied Electric Fields

Figure 14.12 shows quite clearly that a polarized atom or molecule is a dipole, since there are two opposite charges separated by a distance. However, the polarized atom or molecule is not a permanent dipole. If the applied electric field is removed (for example, by removing the charges making that field), the electron cloud will shift back to its original position, and there will no longer be any charge separation. We call the polarized atom or molecule an "induced" dipole, because the dipole was induced (caused) to form by the presence of an applied electric field.

An "induced dipole" is created when a neutral object is polarized by an applied electric field. The induced dipole will vanish if the applied field is removed.

A "permanent dipole" consists of two opposite charges separated by a fixed distance, such as HCl or H_2O molecules, or the dipole that can be constructed out of + and − tapes (Experiment EXP8).

Polarizability

It has been found experimentally that for almost all materials, the amount of polarization induced (that is, the dipole moment $\vec{p}$ of the polarized atoms or molecules) is directly proportional to the magnitude of the applied electric field. This result can be written like this:

$$\vec{p} = \alpha \vec{E}$$

The constant α is called the "polarizability" of a particular material. The polarizability of many materials has been measured experimentally, and these experimental values may be found in reference volumes. Recall that the magnitude of the electric dipole moment of a dipole is

$$|\vec{p}| = qs$$

and the direction is from the negative charge toward the positive charge (see Chapter 13).

Checkpoint 3 A typical atomic polarizability is $1 \times 10^{-40}\ \text{C}\cdot\text{m}/(\text{N/C})$. If the q in $p = qs$ is equal to the proton charge e, what charge separation s could you produce in a typical atom by applying a large field of $3 \times 10^6\ \text{N/C}$, which is large enough to cause a spark in air?

A Neutral Atom and a Point Charge

In the previous chapter we found that since the electric field of a dipole was proportional to $1/r^3$, the force exerted by a dipole on a point charge was also proportional to $1/r^3$. Because of the reciprocity of the electric force, the force on the dipole by the point charge was therefore also proportional to $1/r^3$. Let us extend this analysis by considering the case of a point charge q_1 and a neutral atom.

Even though the entire process happens very quickly, it is instructive to analyze it as if it occurred in several steps. (Of course, the process is not instantaneous, since information about changes in electric field takes a finite time to propagate to distant locations.)

Step 1: At the location of the atom there is an electric field $\vec{E}_1$ due to the point charge (Figure 14.14). This electric field affects both the nucleus and the electron cloud, both of which, due to their spherical symmetry, can be modeled as point charges. The force on the electron cloud and the force on the nucleus are in opposite directions. Since the electron cloud and the nucleus can move relative to each other, they shift in opposite directions, until a new equilibrium position is reached.

The atom is now polarized, with dipole moment $\vec{p}_2 = \alpha\vec{E}_1$ proportional to the applied electric field $\vec{E}_1$.

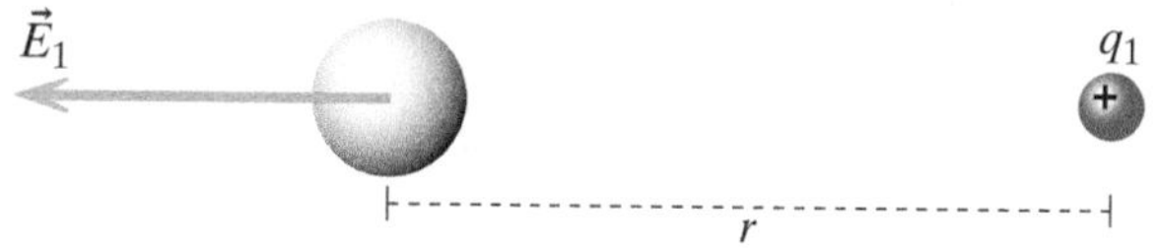

Figure 14.14 At the location of the atom there is an electric field $\vec{E}_1$ due to the point charge.

Step 2: The polarized atom now has a dipole moment $p = q_2 s$. The atom, which is now an induced dipole, makes an electric field $\vec{E}_2$ at the location of the point charge (Figure 14.15). We can write an expression for the magnitude of $\vec{E}_2$:

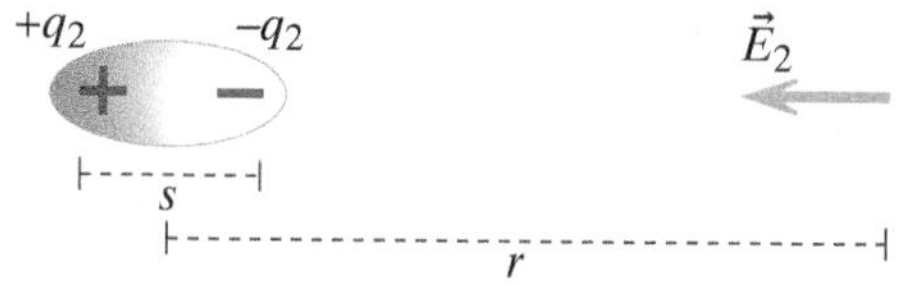

Figure 14.15 The polarized atom makes an electric field $\vec{E}_2$ at the location of the point charge.

$$|\vec{E}_2| = E_2 = \frac{1}{4\pi\varepsilon_0}\frac{2p}{r^3} = \frac{1}{4\pi\varepsilon_0}\frac{2\alpha E_1}{r^3}$$

Since we know $\vec{E}_1$, the electric field of the point charge at the location of the dipole, we can put that into our equation:

$$E_2 = \frac{1}{4\pi\varepsilon_0}\frac{2\alpha}{r^3}E_1 = \left(\frac{1}{4\pi\varepsilon_0}\frac{2\alpha}{r^3}\right)\left(\frac{1}{4\pi\varepsilon_0}\frac{q_1}{r^2}\right)$$
$$= \left(\frac{1}{4\pi\varepsilon_0}\right)^2\left(\frac{2\alpha q_1}{r^5}\right)$$

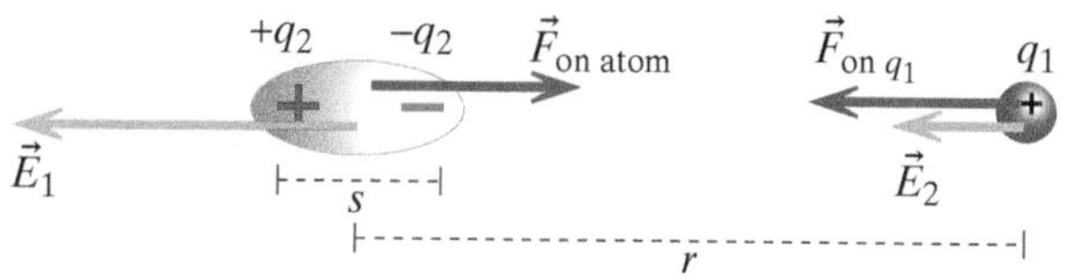

Figure 14.16 The force on the point charge due to the electric field of the polarized atom is equal in magnitude to the force on the polarized atom due to the electric field of the point charge.

QUESTION **Step 3:** What is the force exerted on the point charge by the induced dipole (Figure 14.16)?

$$\vec{F}_1 = q_1\vec{E}_2 = q_1\left(\frac{1}{4\pi\varepsilon_0}\right)^2\left(\frac{2\alpha q_1}{r^5}\right)\hat{r} = \left(\frac{1}{4\pi\varepsilon_0}\right)^2\left(\frac{2\alpha q_1^2}{r^5}\right)\hat{r}$$

We find that the force on the point charge by the polarized atom is proportional to $1/r^5$.

QUESTION What is the force on the neutral atom by the point charge?

Because of the reciprocity of the electric interaction (Newton's third law), the force on the neutral atom by the point charge is equal in magnitude and opposite in direction to the force on the point charge by the neutral atom:

$$\vec{F}_2 = -\vec{F}_1 = -\left(\frac{1}{4\pi\varepsilon_0}\right)^2\left(\frac{2\alpha q_1^2}{r^5}\right)\hat{r}$$

so the force on a (polarized) neutral atom by a point charge also is proportional to $1/r^5$.

EXAMPLE **A Proton Attracts a Helium Atom**

A proton and a helium atom are at rest, 20 nm apart. The polarizability of a helium atom is $\alpha = 2.3 \times 10^{-41}$ C·m/(N/C). **(a)** What is the initial acceleration of the helium atom? **(b)** What would be the initial acceleration of the helium atom if it were 100 nm from the proton?

Solution **(a)** The mass of a helium atom is

$$m = \frac{4 \times 10^{-3}\ \text{kg/mole}}{6.02 \times 10^{23}\ \text{atoms/mole}} = 6.64 \times 10^{-27}\ \text{kg}$$

The initial acceleration of the helium atom is

$$a = \frac{F}{m} = \left(\frac{1}{4\pi\varepsilon_0}\right)^2 \left(\frac{2\alpha q_1^2}{mr^5}\right)$$
$$= \left(9 \times 10^9 \frac{\text{N}\cdot\text{m}^2}{\text{C}^2}\right)^2 \left(\frac{2(2.3 \times 10^{-41}\ \text{C}\cdot\text{m/(N/C)})(1.6 \times 10^{-19}\ \text{C})^2}{(6.64 \times 10^{-27}\ \text{kg})(20 \times 10^{-9}\ \text{m})^5}\right)$$
$$= 4.5 \times 10^6\ \text{m/s}^2$$

This is an extremely high acceleration compared to $g = 9.8\ \text{m/s}^2$.

(b) Since the force is proportional to $1/r^5$, increasing the distance r by a factor of 5 means that the force and acceleration are reduced by a large factor of $5^5 = 3125$. The acceleration is now only $(4.5 \times 10^6\ \text{m/s}^2)/5^5 = 1400\ \text{m/s}^2$.

Checkpoint 4 If the distance between a neutral atom and a point charge is doubled, by what factor does the force on the atom by the point charge change?

Interaction of Charged Objects and Neutral Matter

We are now in a position to explain why both positively and negatively charged objects (such as + and − tapes) are strongly attracted to neutral matter.

QUESTION Try to explain in detail what happens when a positively charged tape is brought near your hand. This is a complex process; consider all the interactions involved.

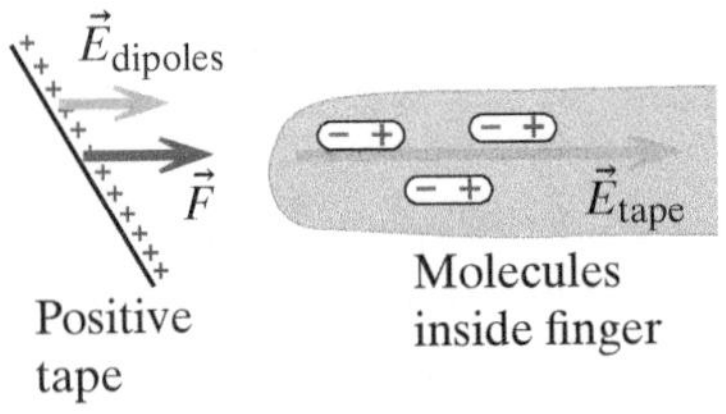

Figure 14.17 The electric field of a positively charged tape polarizes your finger. The induced dipoles in your finger create an electric field at the location of the tape, which attracts the tape.

In considering the interactions of fields and matter, the following scheme is useful. (1) Identify any sources of electric fields. (2) Identify any charges at other locations that can be affected by these fields. (3) Redistribution of the affected charges may create an electric field at the location of the original source charges: are they affected?

The positively charged tape makes an electric field, which points away from the tape. This electric field is present inside your hand, and affects atoms, molecules, and ions inside your hand. Figure 14.17 shows the polarization caused inside your finger by the electric field of the tape. The induced dipoles in your finger create an electric field at the location of the tape, which attracts the tape. You should be able to construct a diagram like the one in Figure 14.17 illustrating what happens when a negatively charged tape interacts with your finger.

You may have noticed that the attraction between your neutral hand and a hanging charged invisible tape changes much more rapidly with distance ($1/r^5$) than does the interaction between two charged tapes ($1/r^2$). A related experiment is EXP8.

Checkpoint 5 Explain in detail, including diagrams, what happens when a negatively charged tape is brought near your finger.

Determining the Charge of an Object

Suppose that you have a negatively charged tape hanging from the desk, and you rub a wooden pencil on a wool sweater and bring it near the tape.

QUESTION If the tape swings toward the pencil, does this show that the pencil had been charged positively by rubbing it on the wool?

Not necessarily. Even if the pencil is uncharged, the charged tape will polarize the pencil and be attracted by the induced dipoles.

QUESTION Can a charged object repel a neutral object? Why or why not? Draw diagrams to help you make your point.

Polarization always brings the unlike-sign charge closer, yielding a net attraction. Repulsion of an induced dipole can't happen. Therefore repulsion is the better test of whether an object is charged.

Electric Field Penetrates Intervening Matter

The superposition principle states that the presence of matter does not affect the electric field produced by a charged object. Intervening matter does not "screen" or "shield" the electric field, just as your desk does not "screen" or "shield" your book from the gravitational field of the Earth. See Figure 14.18; a related experiment is EXP10.

You may have already observed one case of electric field passing through intervening matter. You see the same interaction between a charged invisible tape and your hand, or another tape, when approaching either side of a hanging tape, despite the charges being on just one side of the tape. The charges are initially on either the slick side or the sticky side (depending on whether it is the upper or lower tape in a pair of tapes), and one can show that the charges can't move through the tape to the other side (Experiment EXP11).

Figure 14.18 Two like charged tapes repel each other even if another object such as a piece of paper is placed between them. Since the tapes are also attracted to the paper, the net effect may be small.

Intervening Matter and Superposition

The fact that an electric field acts through intervening matter is another example of the superposition principle. It is true that the repulsion between two like-charged pieces of tape is weaker when a piece of paper is in the way (Experiment EXP10), but when viewed in terms of the superposition principle this reduction is not due to the paper partially "blocking" the field of the other tape. Rather, we say that the net field is due to the superposition of two fields: the *same* field that you would have had without the paper intervening, plus another field due to the induced dipoles in the paper.

At this time we can't prove that this view is correct and that there is no blocking of electric field. However, we will find repeatedly that the superposition principle makes the right predictions for a broad range of phenomena and offers a simpler explanation than any kind of hypothetical blocking effect.

14.4 POLARIZATION OF INSULATORS

In insulators, all of the electrons are firmly bound to the atoms or molecules making up the material. We have seen that an individual atom or molecule can be polarized by an applied electric field, producing an induced dipole of atomic or molecular dimensions. The electrons in an atom or molecule of an insulator shift position slightly, but remain bound to the molecule—no charged particles can move more than about one atomic diameter, or 1×10^{-10} m (most move much less than this distance; see Checkpoint 3).

In Figure 14.19 we show a solid block of insulating material, each of whose molecules has been polarized by an applied electric field (that is, an electric field made by external charges—in this case a single positive charge). The molecules are of course not shown to scale! This is an example of "induced polarization"—the electric field has induced the normally unpolarized insulator to become polarized. In each molecule the electrons have moved a very short distance, and the molecules themselves are not free to move. However, the net effect can be very large because there are many molecules in the insulator to be affected. Note that the polarized molecules align with the electric field that is polarizing them, and that the stronger the electric field the larger the "stretch" of the induced dipole.

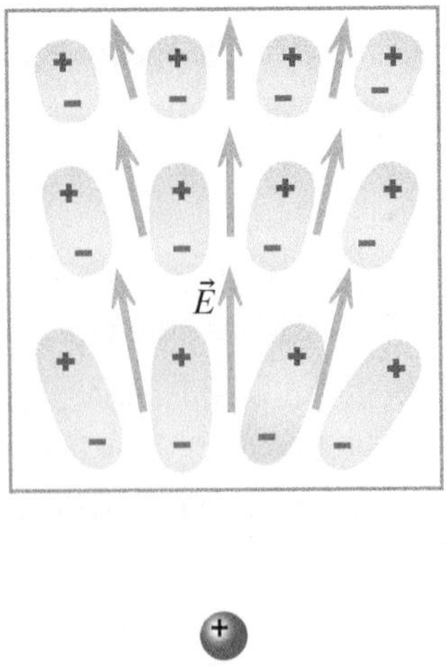

Figure 14.19 A block of insulating material (plastic, glass, etc.) polarized by an applied electric field. The molecules are not shown to scale!

Polarization Happens Very Rapidly

Because the electron cloud is displaced only a tiny distance when an atom or molecule polarizes, this process happens extremely rapidly. The process can take much less than a nanosecond to complete.

Diagrams Showing Polarization of Insulators

In diagrams of insulators we show polarized molecules exaggerated in size, to indicate that individual molecules in an insulator polarize, but the electrons remain bound to the molecule. We show the extent of polarization by the degree to which the molecule is "stretched." Keep this diagrammatic convention in mind, and compare it to diagrams of polarized conductors in the following sections.

Charge on or in an Insulator

Since there are no mobile charged particles in an insulator, excess charges stay where they are. Excess charge can be located in the interior of an insulator, or can be bound to a particular spot on the surface without spreading out along that surface (Figure 14.20).

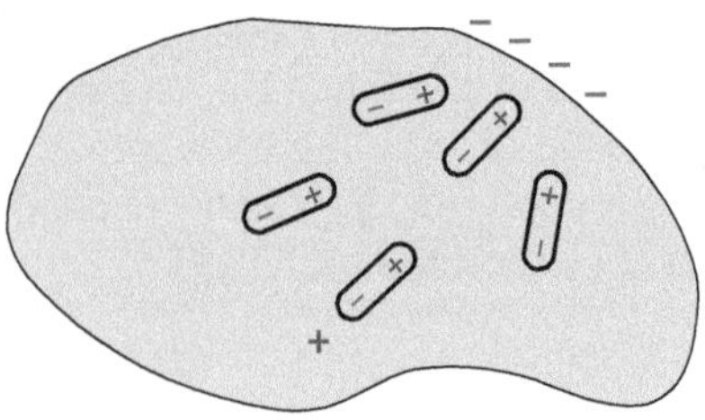

Figure 14.20 In an insulator, charge can occur in patches on the surface, and there can be excess charge inside.

Low-Density Approximation

When atoms or molecules in an insulator polarize in response to an electric field created by external charged objects, the polarized molecules themselves make electric fields that affect neighboring molecules. Because the effect of polarized molecules on each other is typically small compared to the effect of the original applied field, we will neglect this when discussing polarized insulators.

In formal terms, when an electric field E_{applied} is applied to a dense material (a solid or a liquid), the induced dipole moment of one of the atoms or molecules in the material isn't simply $p = \alpha E_{\text{applied}}$, but is really $p = \alpha|\vec{E}_{\text{applied}} + \vec{E}_{\text{dipoles}}|$, where E_{dipoles} is the additional electric field at the location of one of the molecules, due to all the other induced dipoles in the material. In this text we make the simplifying assumption of low density and assume that E_{dipoles} is small compared to E_{applied}. This is good enough for our purposes, but accurate measurements of polarizability must take this effect into account.

14.5 POLARIZATION OF CONDUCTORS

As we stated earlier, a conductor has some kind of charged particles that can move freely throughout the material. In contrast to an insulator, where electrons and nuclei can move only very small distances (around 1×10^{-10} m,

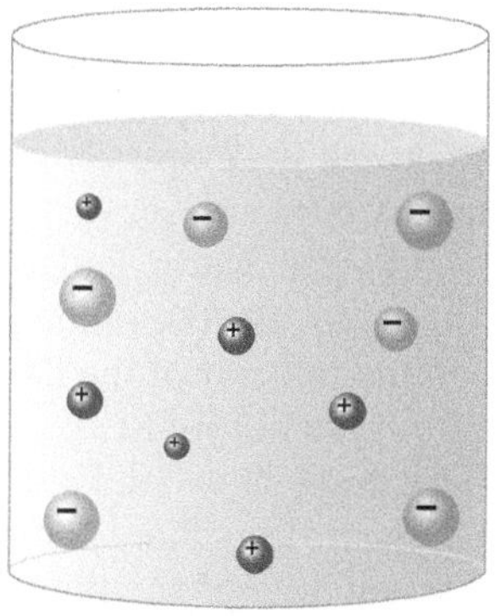

Figure 14.21 A beaker containing an ionic solution (salt water).

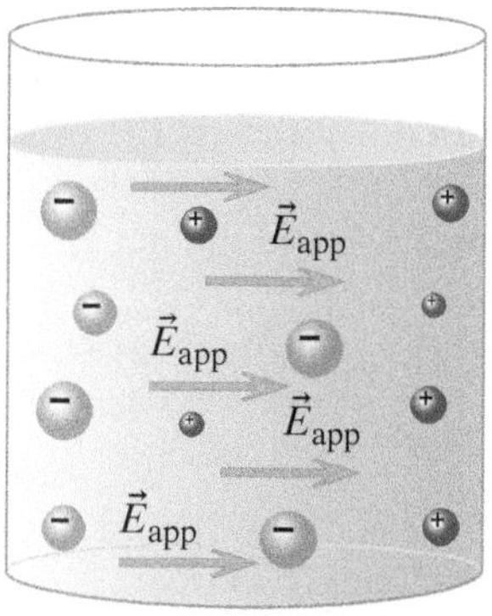

Figure 14.22 Under the influence of an applied electric field (labeled $\vec{E}_{app}$), the liquid polarizes. There is a slight excess ion concentration at the two sides of the beaker. The electric field due to the redistributed ions is not shown.

or much less), the charged particles in a conductor are free to move large distances.

Ionic Solutions

Ionic solutions are conductors, such as a solution of sodium chloride (table salt) in water. In salt water, the mobile charged particles are Na^+ ions and Cl^- ions (Figure 14.21; there are also very small concentrations of H^+ and OH^- ions, which are not shown).

QUESTION What happens when an electric field is created in the region of the beaker?

When an electric field is applied to a conductor, the mobile charged particles begin to move in the direction of the force exerted on them by the field. However, as the charges move, they begin to pile up in one location, creating a concentration of charge that itself creates an electric field in the region occupied by the remaining mobile charges. The net electric field in the region is the superposition of the applied (external) field and the electric field created by the relocated charges in the material. Figure 14.22 is a diagram of the polarization that occurs in the salt water. The ions (charged atoms or molecules) are in constant motion, so the actual situation isn't simple. Moreover, the interior of the liquid is full of positive and negative ions; there's just a slight excess concentration of ions near the sides of the beaker.

Drift Speed and Applied Electric Field

If a beaker of salt water is placed in a region where there is an electric field (due to charges outside the beaker), a sodium ion or a chloride ion will experience an electric force, and will begin to move in the direction of the force. However, even if the force remains constant, the ion will not keep accelerating, because it will collide with water molecules or with other ions. In effect, there is a kind of friction at the microscopic level.

To keep the ions in a salt solution moving at a constant speed, a constant electric field must be applied to the solution. The speed at which mobile charges (in this case, sodium or chloride ions) move through a conductor is called the *drift speed*. Drift speed is directly proportional to the net electric field at the location of the charge. The proportionality constant is called the *mobility* of the mobile charges.

DRIFT SPEED

$$\bar{v} = uE_{net}$$

$\bar{v}$ is the average drift speed of a mobile charge.

u is the mobility of the charge. The units of mobility are $\frac{\text{m/s}}{\text{N/C}}$.

E_{net} is the magnitude of the net electric field at the location of the mobile charge.

As implied by this equation, if the net electric field at the location of a mobile charge in a conductor is zero, the charge will stop moving.

Checkpoint 6 An electric field of magnitude 190 N/C is applied to a solution containing chloride ions. The mobility of chloride ions in solution is 7.91×10^{-8} (m/s)/(N/C). What is the average drift speed of the chloride ions in the solution?

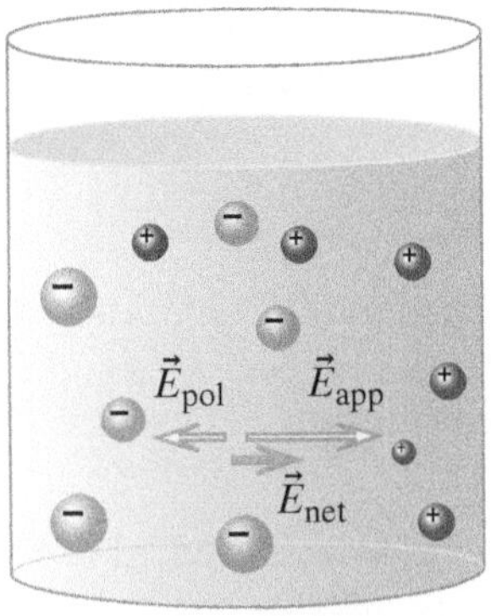

Figure 14.23 An intermediate stage in the polarization process, before polarization is complete.

The Polarization Process in an Ionic Solution

Polarization occurs very rapidly, but it is not instantaneous. Let's "slow down time" so we can talk about the process of polarization; we'll operate on a time scale of attoseconds (1×10^{-18} seconds!). To simplify our analysis, we'll imagine that we are able temporarily to "freeze" the ions in the salt water, and to release them after we have brought charges nearby to apply an electric field.

Consider the net electric field at a location in the interior of the liquid, at a time attoseconds after polarization has begun, but long before the process has finished. The electric field E_{app} due to external charges is shown in Figure 14.23, and also a smaller electric field E_{pol} due to the polarization charges present at this time. The net electric field at a location in the middle of the liquid is now smaller than it was before polarization began.

QUESTION Will the polarization of the salt water increase beyond what it is now?

At this instant the net electric field in the solution still has magnitude greater than zero, so ions in the solution will still experience forces in the direction to increase the polarization. Because $E_{net} \neq 0$, the drift speed $\bar{v}$ of the ions is not zero. More ions will pile up at the sides of the beaker, and the net electric field in the interior will be further weakened.

Eventually the conductor will reach equilibrium on the microscopic level. Equilibrium at the microscopic level means that there is no net motion of mobile charges in any direction:

EQUILIBRIUM INSIDE A CONDUCTOR

When a conductor is in equilibrium at the microscopic level:

$$\bar{v} = 0$$

The average drift speed of the mobile charges inside the conductor is zero. There is no net flow of charges in any direction.

QUESTION How weak does the net electric field inside the conductor get? In the final state of equilibrium (when there is no further increase in polarization), how big is the net electric field in the interior of the liquid?

An Example of a "Proof by Contradiction"

You may have correctly deduced that in the final state the net electric field in the conductor goes to zero at equilibrium. A rigorous way to reason about this using formal logic is to construct a "proof by contradiction." In a proof by contradiction, we assume the opposite of what we want to prove, then, making valid logical deductions from this assumption, show that we reach a conclusion that is impossible or contradictory. We therefore conclude that the original assumption was wrong, and its opposite must be true.

1. Assume that in equilibrium the magnitude of the net electric field in the interior of an ionic solution is nonzero.
2. Since $E_{net} > 0$, all mobile ions in the solution will experience a nonzero force. Since $u > 0$, the average drift speed $\bar{v} > 0$, and all ions move in the direction of the force.
3. Since $\bar{v} > 0$, there is a net flow of charges. Therefore the system cannot be in equilibrium, because by definition in equilibrium $\bar{v} = 0$ and there is no net flow of charges. This result, that $\bar{v} > 0$, contradicts our original assumption (point 1 above) that the ionic solution is in equilibrium.

4. Because we have reached a contradiction, we must conclude that the original assumption (that the net electric field in the solution may be nonzero in equilibrium) is wrong. Thus, we conclude that the net electric field in an ionic solution in equilibrium must be zero.

This reasoning holds true for any conductor, including not only ionic solutions, but solid metal objects as well.

Superposition

Note that the electric field inside the liquid is zero, not because of any "blocking" of fields due to external charges, but by the superposition of two effects: the effect of the external charges and the effect of the polarization charges. This is another example of the superposition principle in action.

It is not true that the net electric field in a solution is zero at *all* times. While the ionic solution is in the process of polarizing, it is not in equilibrium; there is a nonzero electric field, and hence a nonzero force on an ion in the liquid, as you saw above. If electrodes are placed in the ionic solution and connected to a battery, the battery prevents the system from reaching equilibrium. In such a case (no equilibrium), there can be a field continuously acting on ions inside the liquid, resulting in continuous shifting of the ions through the liquid, constituting an electric current.

Since there are a very large number of ions in the solution, none of them has to move very far during the polarization process. Even a tiny shift leads to the buildup of an electric field large enough to cancel out the applied electric field.

Polarization of Salt Water in the Body

Your own body consists mainly of salt water, including the blood and the insides of cells. Look again at the diagrams in which you focused on the way an external charge polarizes individual molecules inside your finger. An additional effect is the polarization of the salt water inside your finger. As shown in Figure 14.22, there will be a shift of Na^+ and Cl^- ions in the blood and tissues. This shift may be a larger effect than the molecular polarization. It is a bit unsettling to realize that a charged tape or comb messes with the inside of your body!

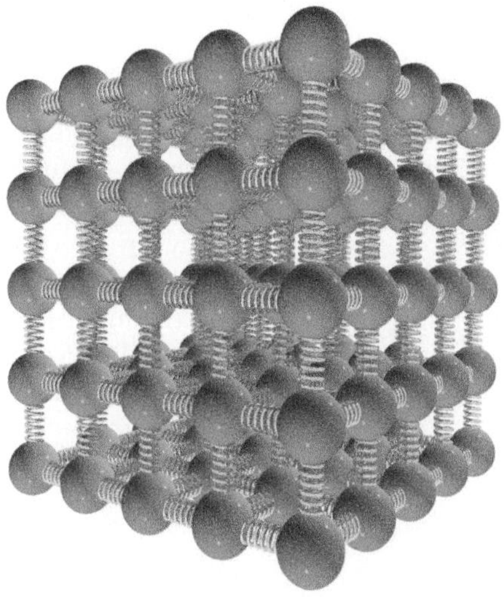

Figure 14.24 The ball-and-spring model of a solid.

14.6 CHARGE MOTION IN METALS

You probably know that metals are very good electrical conductors. In almost all metals, the mobile charged particles are electrons.

The Mobile Electron Sea

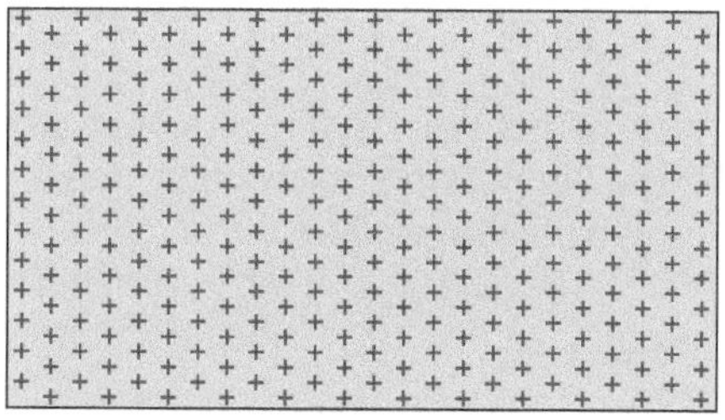

Figure 14.25 A 2D slice of an unpolarized metal: uniform mobile-electron sea (blue), positive atomic cores (red "+" symbols).

The atoms in a solid piece of metal are arranged in a regular 3D geometric array, called a "lattice" (Figure 14.24). The inner electrons of each metal atom are bound to the nucleus. Some of the outer electrons participate in chemical bonds between atoms (the "springs" in the ball-and-spring model of a solid). However, some of the outer electrons (usually one electron per atom) join a "sea" of mobile electrons that are free to move throughout the entire macroscopic piece of solid metal (Figure 14.25). In a sense, the entire hunk of metal is like one giant molecule, in which some of the electrons are spread out over the entire crystal. The electrons are not completely free; they are bound to the metal as a whole and are difficult to remove from the metal. (For example, electrons do not drip out when you shake a piece of metal!) Metals are excellent conductors because of the presence of these mobile electrons.

No Net Interaction between Mobile Electrons

Although the roaming electrons repel each other strongly, this repulsion between electrons is neutralized on the average by the attractions exerted by the positive atomic cores (a "core" is a neutral atom minus its roaming electron, so it has a charge of $+e$). The effect is that on average, the net electric field inside a piece of metal in equilibrium is zero.

Because of this, in some ways the mobile electrons look like an ideal gas: they move in a region free from the electric field, so they appear not to interact with each other or with the atomic cores. In fact, in some simple models of electron motion the mobile electron sea is treated as an ideal gas.

The Drude Model

In a simple classical model of electron motion (called the "Drude model" after the physicist who first proposed it), a mobile electron in the metal, under the influence of the electric field inside the metal, does accelerate and gain energy, but then it loses that energy by colliding with the lattice of atomic cores, which is vibrating because of its own thermal energy and acquires more thermal energy due to the collisions of the electrons with the lattice. After a collision, an electron again gets accelerated, and again collides. This process is what makes the metal filament in a light bulb get hot. Figure 14.26 shows a graph of this start–stop motion for a single electron.

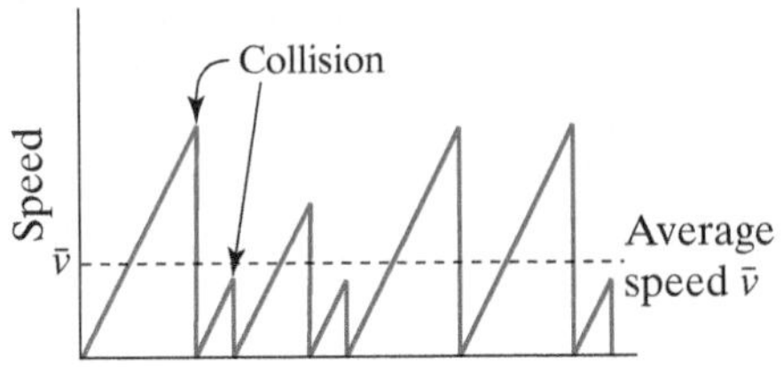

Figure 14.26 A mobile electron speeds up under the influence of the electric field inside the wire, then collides with an atomic core and loses energy.

Drift Speed and Electron Mobility

The average speed of an electron in this start–stop motion is called the "drift" speed $\bar{v}$, and we say that the electron "drifts" through the metal. Actually, the slow drift motion is superimposed on high-speed motion of the electrons in all directions inside the metal, much as the wind is a slow drift motion superimposed on the high-speed motion of air molecules in all directions. A full treatment of electrons in a metal, including the reason for the high-speed motion in all directions, requires quantum mechanics, but the simple classical Drude model allows us to understand most of the important aspects of circuits on a microscopic level.

We express the momentum principle

$$\frac{\Delta\vec{p}}{\Delta t} = \vec{F}_{\text{net}}$$

in a form involving finite time steps for momentum in the direction of $\vec{F}_{\text{net}}$,

$$\Delta p = F_{\text{net}}\,\Delta t = eE_{\text{net}}\,\Delta t$$

where E_{net} is the magnitude of the net electric field inside the wire, and Δt is the time between collisions. If we make the simplifying assumption that the electron loses all its momentum during each collision, we have

$$\Delta p = p - 0 = eE_{\text{net}}\,\Delta t$$

The speed of the electron (of mass m_e) at the time of collision turns out to be small compared to the speed of light, so we have

$$v = \frac{p}{m_e} = \frac{eE_{\text{net}}\,\Delta t}{m_e}$$

However, the time between collisions is not the same for all electrons. Some experience longer times between collisions, some shorter times. To get an average, drift speed $\bar{v}$ for all electrons at a particular instant, we need the average time $\overline{\Delta t}$ between collisions:

$$\bar{v} = \frac{eE_{\text{net}}\overline{\Delta t}}{m_e}$$

$\overline{\Delta t}$, the average time between collisions of the electrons with the atomic cores, is determined by the high-speed random motion of the electrons and by the temperature of the metal. (At a higher temperature the thermal motion of the atomic cores is greater, and the average time between collisions is reduced, leading to a smaller drift speed for the same field E.)

QUESTION Is drift speed directly proportional to the magnitude of the electric field?

Assuming that increasing the electric field does not result in a significant change in temperature, then doubling the electric field E doubles the drift speed $\bar{v}$ attained in that time; hence the drift speed is directly proportional to the electric field. As mentioned earlier, the proportionality factor is called the electron "mobility" and is denoted by u (or by μ in some books).

$$\bar{v} = uE_{\text{net}}$$

Evidently,

$$u = \frac{e}{m_e}\overline{\Delta t}$$

Different metals have different electron mobilities. The higher the mobility, the higher the drift speed for a given electric field. The direction of the drift velocity of a mobile (negatively charged) electron is opposite to the direction of the electric field.

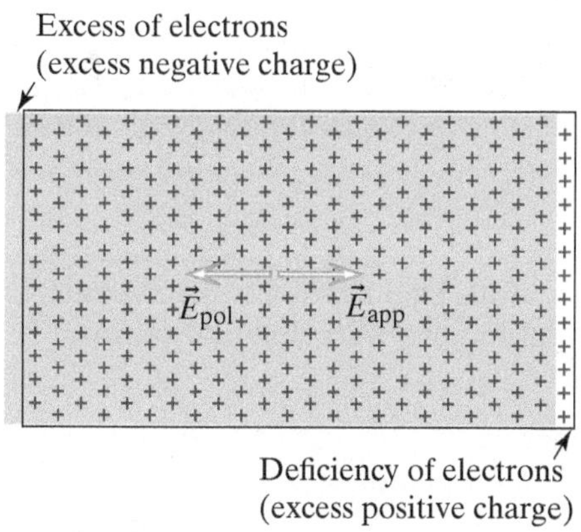

Figure 14.27 Polarized metal: mobile electron sea shifted left relative to the positive atomic cores, under the influence of an applied electric field. There is an excess of electrons on the left side and a deficiency of electrons on the right. These charges contribute to the net electric field inside the metal.

Checkpoint 7 The mobility of the mobile electrons in copper is 4.5×10^{-3} (m/s)/ (N/C). How large an electric field would be required to give the mobile electrons in a block of copper a drift speed of 1×10^{-3} m/s?

Polarization Happens Very Quickly

When an electric field (due to some external charges) is applied to a metal, the metal polarizes. We can describe the polarization of a metal as shifting the entire mobile electron sea relative to the fixed positive cores. In Figure 14.27, in response to an applied field, electrons have piled up on the left, creating a very thin negatively charged layer near the surface. There is a corresponding deficiency of electrons on the right, creating a very thin positively charged layer near the surface.

The shift in the electron sea is extremely small, much less than an atomic diameter! (See Checkpoint 3.) It is not necessary for electrons at one end of the block to move to the other end. Just displace the entire electron sea slightly and you have lots of excess electrons on one surface. Because the electron sea has to move only a tiny distance, this displacement can happen very rapidly—it can take much less than a nanosecond.

Diagrams Showing Polarization of Metals

Figure 14.28 shows the polarization of a neutral metal block caused by a positive point charge to the left of the block, as calculated in a computational model of the polarization process. Blue represents negative charge and red represents positive charge. The block is still neutral, but negative charge (excess electron density) has accumulated on the surface of the left side of the block, leaving positive charge (deficiency of electrons) on the right side of the block. Note that excess charge tends to concentrate along the edges and on the corners of the block. As expected, the charge on the block's surface is more dense where the polarizing field is greater, closer to the point charge.

The computational model used to generate this image does quantitatively what we did qualitatively in reasoning about the polarization of ionic solutions and metals. In Chapter 18, after introducing additional concepts that are involved, we will describe this algorithm in more detail.

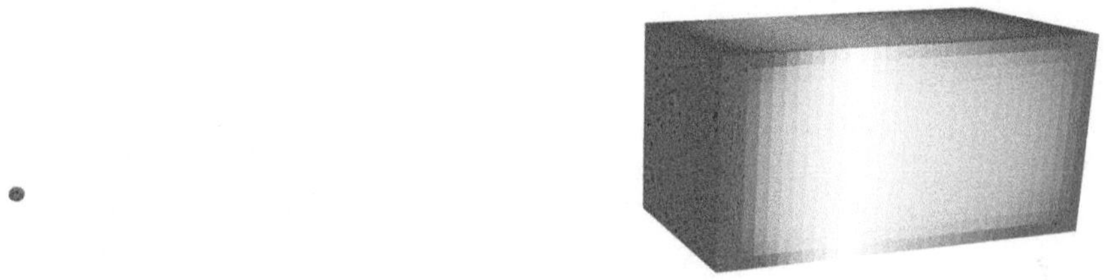

Figure 14.28 The polarization of a neutral block of metal by a positive point charge. Shades of blue represent negative charge; red represents positive charge.

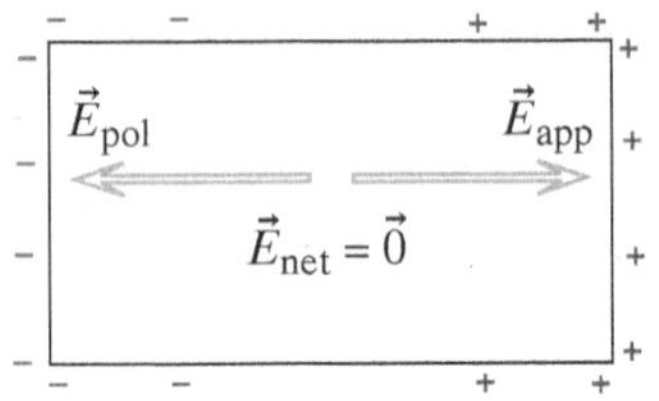

Figure 14.29 A simpler way to depict the polarization of a polarized metal. Excess charges are drawn outside the boundary lines, to indicate that they are on the surface.

In Figure 14.29 we depict a polarized metal in a simplified way that is both easy to draw and easy to interpret at a glance.

- We show − and + signs outside the surfaces to indicate which surfaces have thin layers of negative charge (electron excess) or positive charge (electron deficiency), as a result of shifts in the mobile electron sea. To avoid ambiguity we draw + and − signs just outside the surface of a metal object to indicate that the excess charge (excess or deficiency of electrons) is on the surface of the object.
- We do not show the positive atomic cores and the mobile electron sea inside the metal, because the interior is all neutral. The diagram is much easier to interpret if we do not clutter up neutral regions with charges that must be counted to see whether they balance.

Compare these conventions to the convention we used earlier to show how the individual atoms or molecules polarize in an insulator.

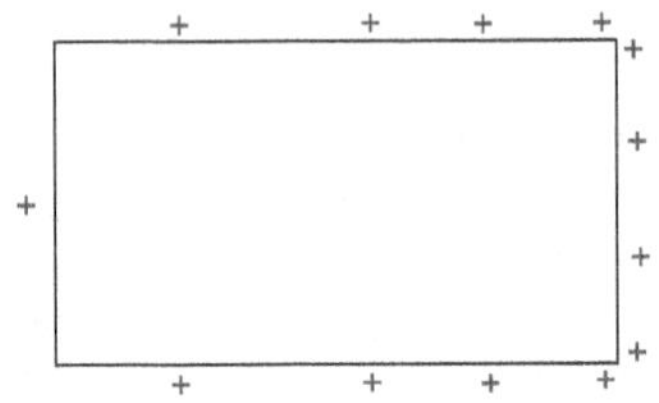

Figure 14.30 This metal block is both charged and polarized.

Polarized and/or Charged

Take care to use technical terms precisely. The metal block shown in Figure 14.29 is *polarized*. It is *not* charged; its net charge is still zero.

On the other hand, a charged object can also be polarized. The positively charged metal block depicted schematically in Figure 14.30 is also polarized. "Polarized" and "charged" are not synonyms.

Net Electric Field in a Metal Goes to Zero in Equilibrium

The reasoning process that we went through when considering the polarization of ionic solutions applies equally well to metals, or to any conductor. We used proof by contradiction to demonstrate that in equilibrium the net electric field inside a conductor must be zero (because if it were not zero, mobile charged particles would move under the influence of the field, and the system would not be in equilibrium).

It is intriguing that it is possible for mobile charges to rearrange themselves in such a way that the net electric field is zero not just at one single location, but also at every location inside the metal. It would be a very difficult problem for us to calculate exactly where to place charged particles to make a net field of zero inside a metal object, but in fact the many mobile charges do rearrange in just such a way as to accomplish this. It can be shown that it is because of the $1/r^2$ distance dependence of the electric field that this is possible—if the exponent were not exactly 2.0, the world would be quite different.

When equilibrium is reached in a metal, things are essentially unchanged in the interior of the metal. There is no excess charge—we still have a uniform sea of electrons filling the space around the positive atomic core. The net electric field inside the metal, which is the sum of the applied field and the field due to the charge buildup on the edges of the metal, is still zero.

$$\vec{E}_{\text{net}} = \vec{E}_{\text{app}} + \vec{E}_{\text{pol}} = 0 \quad \text{in a conductor at equilibrium}$$

At the surfaces there is some excess charge, so we can represent a polarized metal as having thin layers of charge on its surfaces but being unpolarized in the interior, unlike an insulator.

The shifting of the mobile electron sea in metals is a much larger effect than occurs in insulators, where the polarization is limited by the fact that all the electrons, including the outermost ones, are bound to the atoms, unlike the situation in metals. A polarized insulator is a collection of tiny (molecule-sized) dipoles, whereas a polarized metal forms one giant dipole.

E Is Not Always Zero Inside a Metal

Do not overgeneralize our previous conclusions. It is not true that the net electric field in a metal is zero at all times. While the metal is in the process of polarizing, the metal is not in equilibrium, and there is a nonzero electric field inside the metal, creating a nonzero force on electrons in the electron sea, as you saw above. In an electric circuit, the battery prevents the system from reaching equilibrium. In such a nonequilibrium situation, there can be an electric field inside the metal, and hence a force continuously acting on electrons in the mobile electron sea, resulting in continuous shifting of the electron sea around the closed circuit, constituting an electric current.

Excess Charges on Conductors

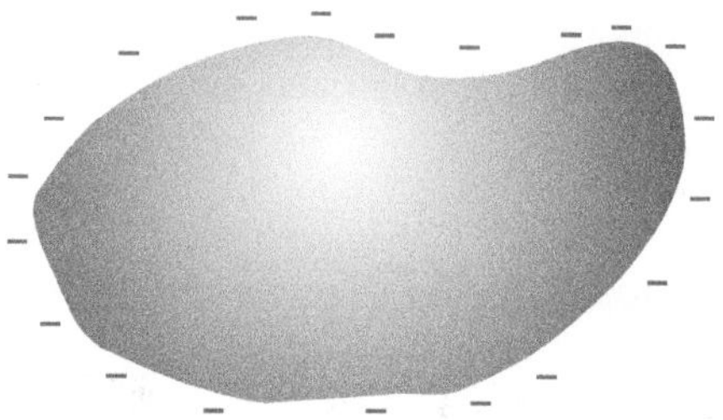

Figure 14.31 In a metal, charge is spread all over the surface (not necessarily uniformly), and there is no excess charge inside.

Another important property of metals (and of the $1/r^2$ property of the electric interaction, as we will see when we study Gauss's law in a later chapter) is that any excess charges on a piece of metal, or any conductor, are always found on an outer or inner surface. This makes intuitive sense, since any excess charges will repel each other and will end up as far apart as possible—on the surface of the conductor. Any multiatom region in the interior of the conductor has a net charge of zero. Moreover, the mutual repulsion among any excess charges makes the mobile electron sea redistribute itself in such a way that charge appears almost immediately all over the surface (Figure 14.31).

Here is a summary of the behavior of conductors vs. insulators:

	Conductor	**Insulator**
Mobile charges	Yes	No
Polarization	Entire sea of mobile charges moves	Individual atoms or molecules polarize
Equilibrium	$\vec{E}_{\text{net}} = \vec{0}$ inside	$\vec{E}_{\text{net}}$ nonzero inside
Excess charge	Spread over surface	In patches anywhere

EXAMPLE

Plastic and Metal Rods

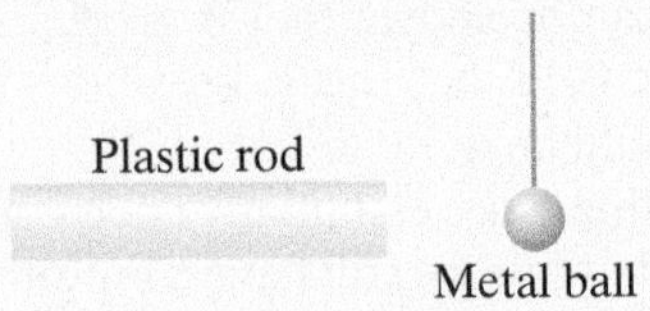

Figure 14.32 The plastic rod and ball are initially uncharged.

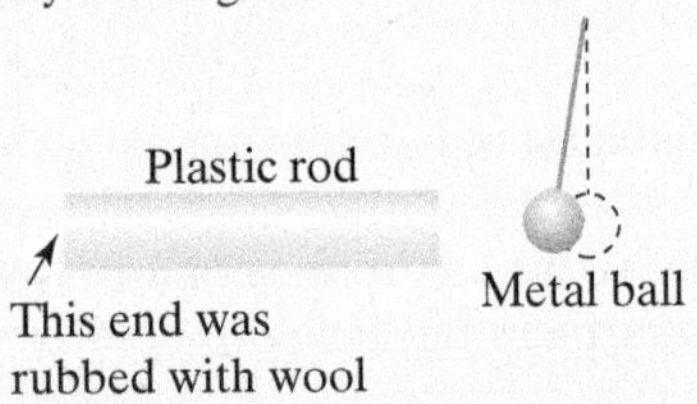

Figure 14.33 The left end of the plastic rod was rubbed with wool and became negatively charged.

A lightweight (conducting) metal ball hangs from a thread, to the right of an (insulating) plastic rod. Both are initially uncharged (Figure 14.32).

(a) You rub the left end of the plastic rod with wool, depositing charged molecular fragments whose total (negative) charge is that of 1×10^9 electrons. You observe that the ball moves toward the rod, as shown in Figure 14.33.

Explain. Show all excess charged particles, polarization, and so on, clearly in a diagram. Make it clear whether charged particles that you show are on the surface of an object or inside it.

(b) You perform a similar experiment with a (conducting) metal rod. You touch the left end of the rod with a charged metal object, depositing 1×10^9 excess electrons on the left end. You then remove the object. You see the ball deflect more than it did with the plastic rod in part (a), as shown in Figure 14.34.

Explain. Show all excess charged particles, polarization, and so on, clearly in a diagram. Make it clear whether charged particles that you show are on the surface of an object or inside it.

Solution

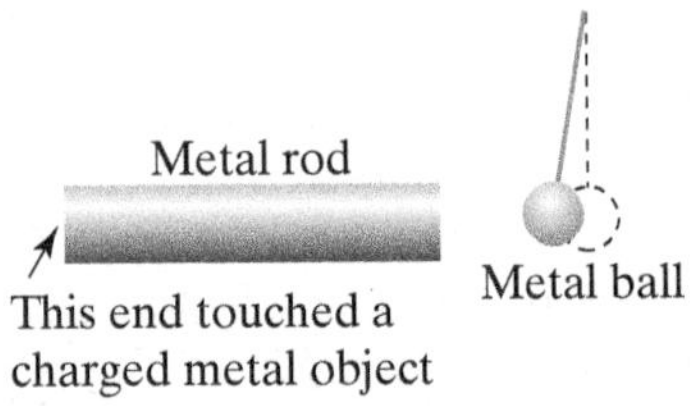

Figure 14.34 Replace the plastic rod with a neutral metal rod, then touch the left end with a negatively charged metal object. The ball deflects more than it did with the plastic rod.

(a) The plastic rod is an insulator, so the excess charge remains on the left-end surface of the rod. This charge polarizes the molecules inside the rod. The original charge plus the polarized molecules make a field that polarizes the neutral metal ball, as shown in Figure 14.35. The excess charge on the polarized metal ball is on the surface of the ball, because the ball is a conductor. The interior of the ball is neutral.

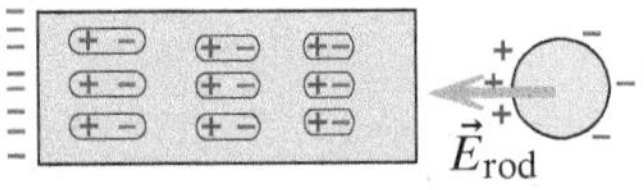

Figure 14.35

The field due to the plastic rod (charge on end plus polarized molecules) exerts a net force to the left on the polarized metal ball (Figure 14.36).

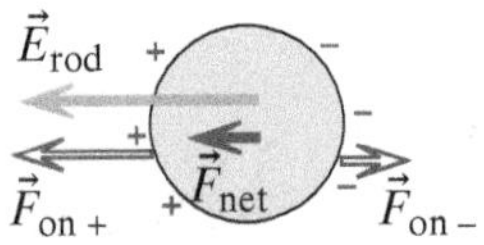

Figure 14.36

(b) The excess negative charge spreads all over the surface of the metal rod and ball, which are conductors. This excess negative charge polarizes the metal ball. The polarized metal ball in turn polarizes the negatively charged metal rod somewhat, as shown in Figure 14.37.

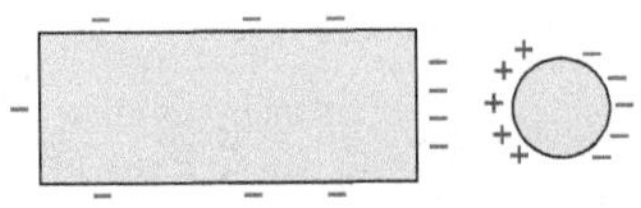

Figure 14.37

Polarization of the ball is greater in this case, because more of the original charge is closer to the ball (Figure 14.38). The net force on the ball is greater in this case than it was with the plastic rod.

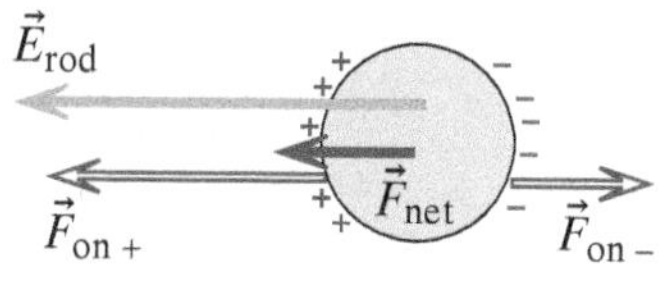

Figure 14.38

EXAMPLE **A Ball and a Wire**

The center of a small spherical metal ball of radius R, carrying a negative charge $-Q$, is located a distance r from the center of a short, thin, neutral copper wire of length L (Figure 14.39). The ball and the wire are held in position by threads that are not shown. If $R = 5\,\text{mm}$, $Q = 1 \times 10^{-9}\,\text{C}$, $r = 10\,\text{cm}$, and $L = 4\,\text{mm}$, calculate the force that the ball exerts on the wire.

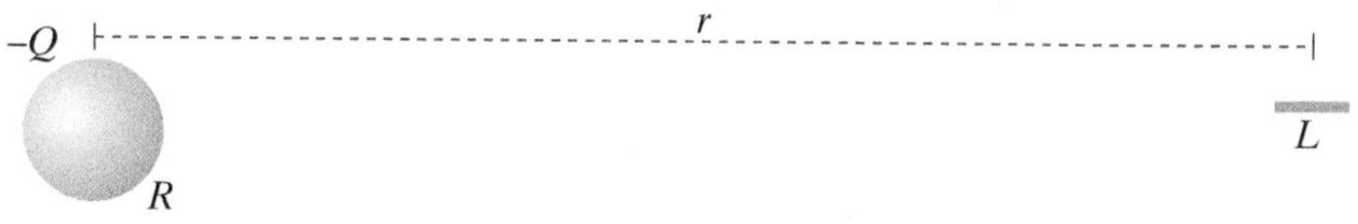

Figure 14.39

Solution The ball makes a field that polarizes the wire as shown in Figure 14.40. The polarized wire in turn makes a field that polarizes the ball, but let's assume that we can neglect this tiny effect, so we can model the ball as a point charge. The polarized wire will be attracted by the ball, and the ball will be attracted by the polarized wire.

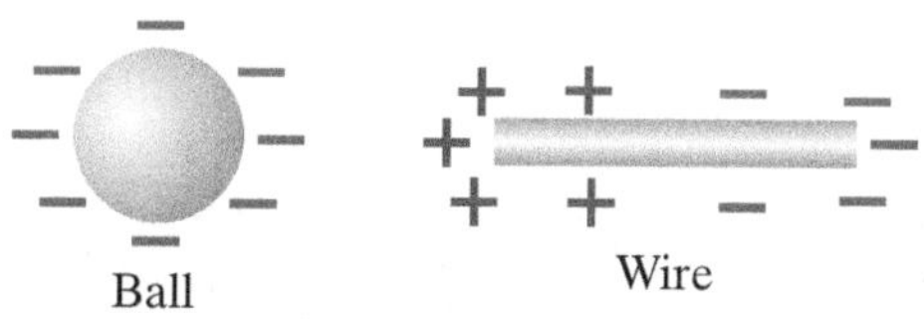

Figure 14.40 The ball polarizes the wire. Neglect the effect of the wire on the ball.

$F_{\text{on ball}} = F_{\text{on wire}}$ (reciprocity of electric forces)

$F_{\text{on ball}} = Q_{\text{ball}} E_{\text{wire}}$ (field made by polarized wire affects ball)

$E_{\text{wire}} \approx \frac{1}{4\pi\varepsilon_0}\frac{2qL}{r^3}$ (model wire as dipole with charges $+q$ and $-q$ at ends)

We need to find the charge q on one end of the wire. We know that at any location inside the metal wire, $\vec{E}_{\text{net}} = 0$ in equilibrium (Figure 14.41). Consider a location in the center of the wire, and model the wire as though there were $+q$ and $-q$ on the ends, a distance L apart, ignoring the small amount of charge on the rest of the wire (here we can't use dipole equations; we're between the charges).

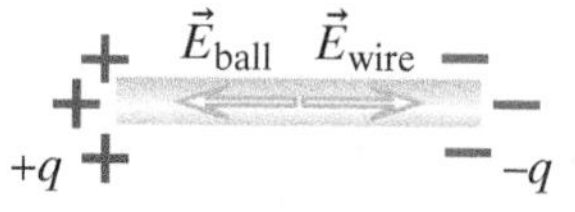

Figure 14.41 The net electric field inside the wire must be zero in equilibrium.

$$E_{\text{net},x} = E_{\text{ball},x} + E_{+\text{end},x} + E_{-\text{end},x}$$

$$0 = -\frac{1}{4\pi\varepsilon_0}\frac{Q}{r^2} + \frac{1}{4\pi\varepsilon_0}\frac{q}{(L/2)^2} + \frac{1}{4\pi\varepsilon_0}\frac{q}{(L/2)^2}$$

$$q \approx \frac{Q}{8}\left(\frac{L}{r}\right)^2$$

At a location outside the wire we model the wire as a dipole, so the force on the ball (which is equal in magnitude to the force on the wire) is this:

$$F \approx Q\left(\frac{1}{4\pi\varepsilon_0}\frac{2qL}{r^3}\right) = \frac{1}{4\pi\varepsilon_0}\frac{Q^2}{4}\left(\frac{L}{r}\right)^2\frac{L}{r^3} = \frac{1}{4\pi\varepsilon_0}\frac{Q^2L^3}{4r^5}$$

Not too surprisingly, we find a force proportional to $1/r^5$.

Now we can calculate numerical values:

$$q \approx \frac{(1\times10^{-9}\,\text{C})}{8}\left(\frac{4\times10^{-3}\,\text{m}}{0.1\,\text{m}}\right)^2 = 2\times10^{-13}\,\text{C}$$

which is a very small charge. This justifies our assumption that the polarized wire won't polarize the ball to any significant extent. The force is tiny:

$$F = \left(9\times10^9\frac{\text{N}\cdot\text{m}^2}{\text{C}^2}\right)\frac{(1\times10^{-9}\,\text{C})^2(4\times10^{-3}\,\text{m})^3}{4(0.1\,\text{m})^5} = 1.4\times10^{-11}\,\text{N}$$

If we double r, there is $1/4$ as much q and $1/32$ as much force. If we double Q, there is 2 times as much q and 4 times as much force.

Figure 14.42 This is a cross section of the metal ball.

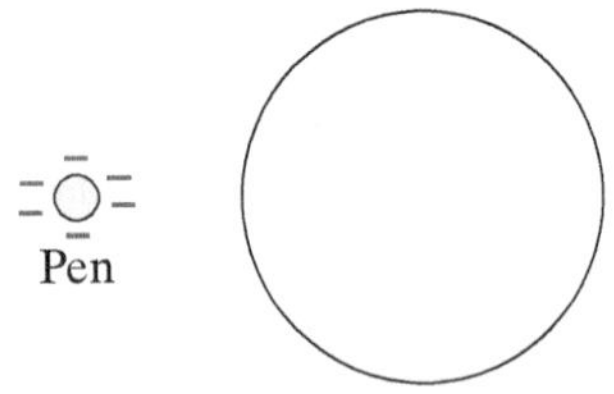

Figure 14.43 This is a cross section of the metal or plastic cylinder.

Checkpoint 8 (a) An object can be both charged *and* polarized. On a negatively charged metal ball, the charge is spread uniformly all over the surface (Figure 14.42). If a positive charge is brought near, the charged ball will polarize. If any of the following quantities is zero, state this explicitly. (1) Draw the approximate final charge distribution on the ball. (2) At the center, draw the electric field due to the external positive charge. (3) At the center, draw the electric field due to the charge on the surface of the ball. (4) At the center, draw the net electric field.

(b) Next, consider a negatively charged plastic pen that is brought near a neutral solid metal cylinder (Figure 14.43). If any of the following quantities is zero, state this explicitly. (1) Show the approximate charge distribution for the metal cylinder. (2) Draw a vector representing the net force exerted by the pen on the metal cylinder, and explain your force vector briefly but completely, including all relevant interactions. (3) At the center, draw the electric field due to the external negative charge. (4) At the center, draw the electric field due to the charge on the surface of the ball. (5) At the center, draw the net electric field.

(c) Replace the solid metal cylinder with a solid plastic cylinder. (1) Show the approximate charge distribution for the plastic cylinder. (2) Draw a vector representing the net force exerted by the pen on the plastic cylinder. (3) Explain your force vector briefly but completely, including all relevant interactions.

14.7 CHARGE TRANSFER

We have seen that objects made of insulators can often acquire a nonzero net charge if they are rubbed by another insulator. You can charge initially neutral pieces of invisible tape by stripping them off other pieces of tape. You can charge an initially neutral pen by rubbing it on your hair. In both these cases, some kind of charged particle is added to or removed from a surface that was originally neutral. In Section 14.2 we discussed possible mechanisms for the transfer of charge between one insulating object and another, including the transfer of positive or negative ions and the transfer of electrons.

Since a conducting object contains mobile charged particles, the process of charging or discharging a conductor involves a flow of charged particles from one conductor to another. For example, electrons from the mobile electron sea in a metal object can move into the mobile electron sea of a different metal object if the objects come into contact with each other.

Although you may not have previously thought of yourself as a conductor, your own body plays an interesting role in some kinds of charging or discharging phenomena. In the following discussion you will see why.

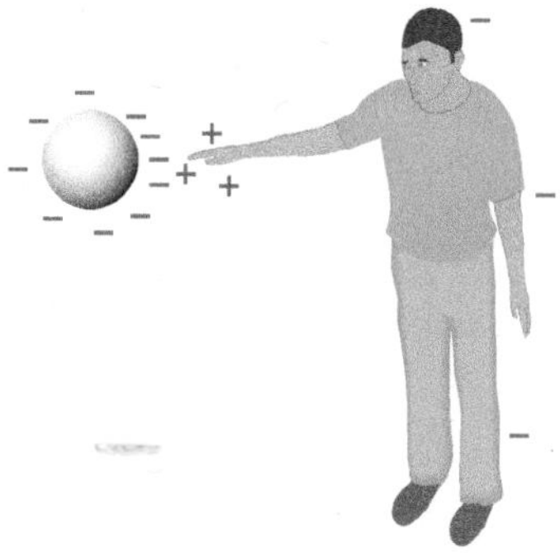

Figure 14.44 The metal is charged, and the person is uncharged but slightly polarized.

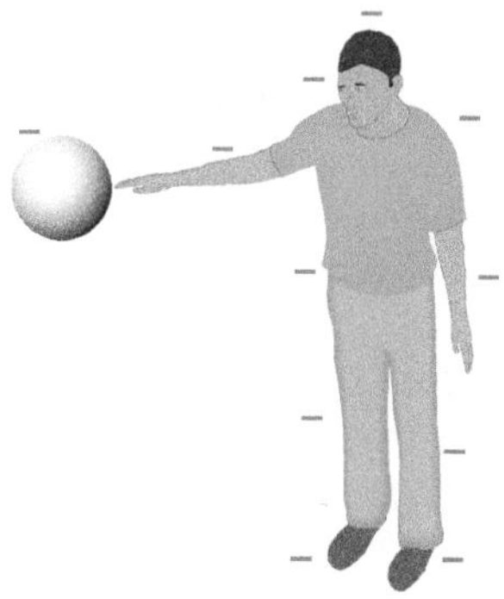

Figure 14.45 The net negative charge is distributed over a much larger area, nearly neutralizing the metal.

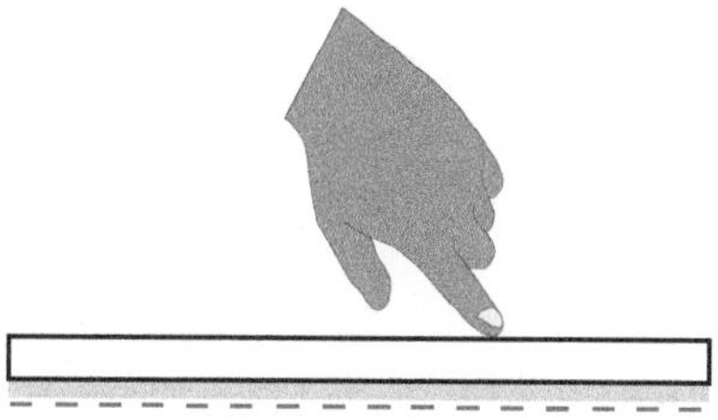

Figure 14.46 You run your finger along the slick side of the tape, and the tape seems to become neutralized.

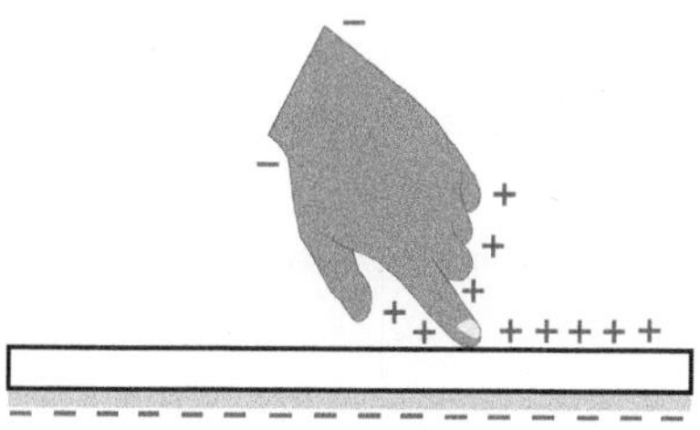

Figure 14.47 Positive ions from the salt solution on your skin are attracted to the negatively charged tape.

Discharging by Contact

If you exercise on a hot day, you sweat, and your body becomes covered with a layer of salt water. Even in a cool place, when you are not moving, there is usually a thin layer of salt water covering your skin. As we saw earlier, salt water is a conductor, so you have a conducting film all over the surface of your skin.

When you approach a negatively charged surface, your body polarizes as shown in Figure 14.44. The polarization includes not only atomic or molecular polarization but also polarization of the blood and sweat, which are salt solutions.

When you touch the charged object, the negatively charged object attracts positive Na^+ ions in the film of salt water on the skin. The Na^+ ions pick up an electron, partially neutralizing the excess negative charge of the object (Figure 14.45). The body acquires a net negative charge. (The Na atom can react with the water to form NaOH and hydrogen!) In the case of a small piece of metal, on which charge is free to redistribute itself, this process nearly neutralizes the metal, because the original net amount of charge is now spread out over the much larger area of metal plus human body.

Similarly, a positive metal surface would attract negative Cl^- ions from your skin, which give up an electron to the metal. The body acquires a net positive charge. (Chlorine can be emitted in tiny quantities!)

Grounding

Touching a small charged object is a pretty effective way to discharge the object, even though you're wearing shoes with insulating soles. An even better way to discharge a conducting object is to "ground" it by making a good connection to the earth or ground (typically through a water pipe that goes into the ground). Earth is a rather good conductor due to the presence of water containing ions. Grounding spreads charge throughout a huge region, neutralizing an object essentially completely.

Discharging an Insulator

You can easily discharge a charged metal foil by briefly touching it anywhere, because it is a conductor. It is more difficult to discharge a charged strip of invisible tape, which is an insulator and does not allow charge to move through the tape.

To discharge a charged piece of invisible tape, it turns out that one may simply rub one's fingers across the slick side of the tape, as shown in Figure 14.46. After doing this, one finds that the tape no longer interacts with neutral objects like your hand and appears to be neutral itself.

> QUESTION How is it possible to discharge a tape by rubbing the slick side even when it was the sticky side that got charged? Tape is an insulator, so charges can't move through the tape.

There are mobile charges on your skin. Positive ions from the salt solution on the skin are attracted to the negatively charged tape and are deposited on its slick surface, so the tape becomes neutral (net charge becomes zero) as shown in Figure 14.47. The + charges on the top and the – charges on the bottom actually make the tape into a dipole (and there are induced dipoles inside the tape), but these dipoles exert much weaker forces on other objects than the negatively charged tape did. Thus the tape acts like ordinary neutral matter. A related experiment is EXP12.

Charging by Induction

It is possible to make use of the polarizability of a conductor to make it acquire a net charge, without actually touching a charged object. The process, called "charging by induction," is illustrated in the sequence in Figure 14.48. In this example, a piece of neutral aluminum foil hangs from a neutral insulating tape. You charge a plastic pen by rubbing it on wool, and bring it near the left side of the neutral foil. A related experiment is EXP13.

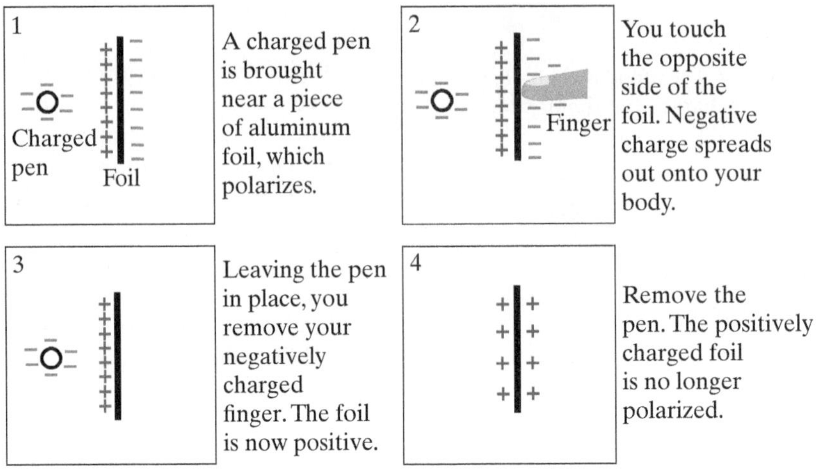

Figure 14.48

Effect of Humidity on Tapes

Isolated atoms are always symmetrical and unpolarized unless an external charge shifts the electron cloud and makes an induced dipole. However, some molecules are permanently polarized even in the absence of an external charge, and this leads to important physical and chemical effects.

For example, water molecules are permanently polarized. The water molecule (H_2O) is not spherically symmetrical but has both hydrogen atoms off to one side of the oxygen atom. In Figure 14.49 the δ^+ and δ^- symbols are used to indicate that slight shifts of the electron clouds to the right leave the right side of the molecule a bit negative and the left side a bit positive, so the water molecule is a permanent dipole.

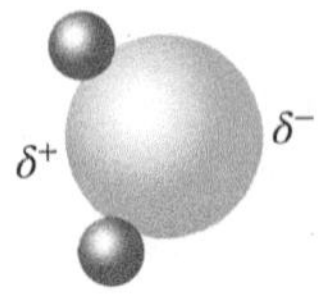

Figure 14.49 A water molecule is a permanent dipole.

Many of water's unusual chemical and physical properties are due to this structure. In particular, the charged ends can bind to ions, which is why many chemicals dissolve well in water.

When water molecules in the air strike a surface they sometimes become attached to the surface, probably because the charged ends bind to the surface. A film of water builds up on all surfaces. Pure water is a very poor conductor but does contain small amounts of mobile H^+ and OH^- ions. More important, the water dissolves surface contaminants such as salt, and the impure water provides an effective path for charges to spread onto neighboring objects. After a while a charged surface loses its original charge, so experiments with charged objects work better when the humidity is low. A related experiment is EXP14.

14.8 PRACTICAL ISSUES IN MEASURING ELECTRIC FIELD

There are practical issues in measuring the electric field by placing a charge somewhere and measuring the force on it, if the charge affected by the field significantly alters the original distribution of source charges. Consider a negatively charged metal sphere (Figure 14.50). The electric field due to the sphere points radially inward.

If we place a particle with very little charge q near this charged sphere (a single proton, for example), it hardly alters the distribution of charge on the metal sphere. The small force on the small charge is $q\vec{E}$, where $\vec{E}$ is the electric field we calculated in the absence of the small additional charge q.

However, if we place a particle with a big charge Q near the sphere, the sphere polarizes to a significant extent (Figure 14.51). We show the electric field due solely to the new charge distribution on the sphere (we don't show the large additional contribution to the net electric field due to Q). Clearly, the force on Q is not simply Q times the *original* $\vec{E}$, but Q times a significantly larger field.

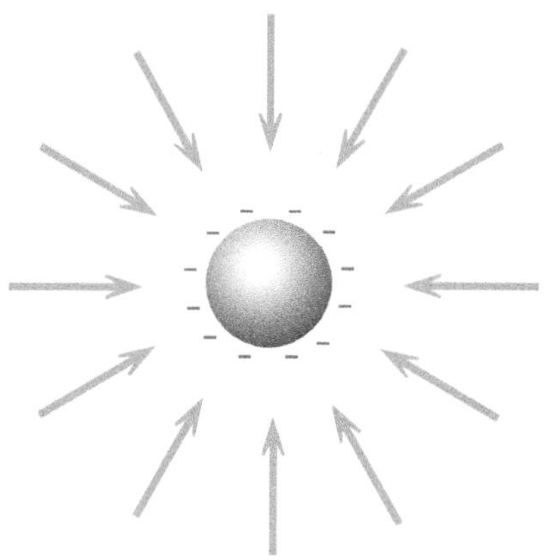

Figure 14.50 Electric field of a uniformly charged metal sphere.

With these effects in mind, we need to qualify our previous method for measuring electric field, in which we measure the force exerted on a charge q and determine the force per unit charge:

$$\vec{E} = \vec{F}/q$$

This procedure is valid only if q is small enough not to disturb significantly the arrangement of other charges that create $\vec{E}$.

Since no object can have a charge smaller than e (the charge of a proton), sometimes it is not possible to find a charge small enough that it doesn't disturb the arrangement of source charges. In this case, we can't measure the electric field without changing the field!

On the other hand, if we know the locations of the source charges, we can calculate the electric field at a location, by applying the superposition principle and adding up the contributions of all the point charges that are the sources of the field:

$$\vec{E}_{\text{net}} = \frac{1}{4\pi\varepsilon_0}\left(\frac{q_1}{r_1^2}\hat{r}_1 + \frac{q_2}{r_2^2}\hat{r}_2 + \frac{q_3}{r_3^2}\hat{r}_3 + \cdots\right)$$

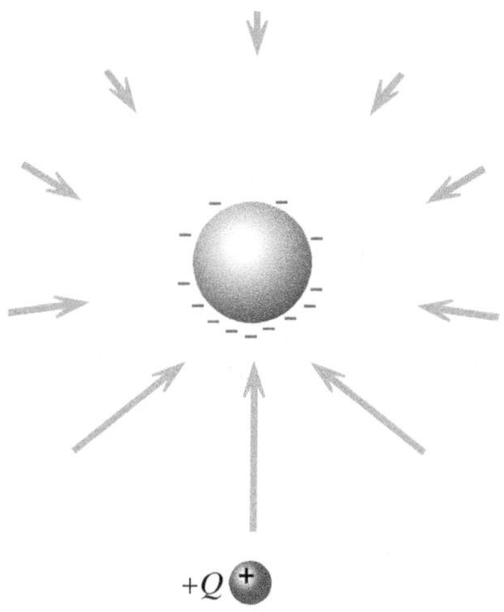

Figure 14.51 Electric field due to the polarized sphere.

If even the smallest possible charge e would disturb this arrangement of source charges, we can't use the calculated field to predict the force that would act on a charge placed at this location. However, we could use the calculated field to predict the polarization of a neutral atom placed at that location, because a neutral atom, even if (slightly) polarized, would disturb the existing arrangement of source charges much less than a charged object would. To turn this around, placing a neutral atom at a location and observing its polarization could offer a less intrusive way to measure an electric field at that location.

Another way to improve the measurement would be to measure $\vec{F}/q$ for a positive q and also measure $\vec{F}/q$ for a negative q, and average the results. A negative charge would polarize the sphere in Figure 14.50 in such a way as to reduce rather than increase the value of $\vec{F}/q$, by pushing the negative charges on the sphere farther away.

SUMMARY

Net Charge
The net charge of an object is the sum of the charges of all of its constituent particles. An object with a net charge of zero is called "neutral."

Conservation of Charge
The net charge of a system plus the net charge of its surroundings cannot change.

Polarization of an atom or molecule produces an induced dipole

$$p = qs = \alpha E_{\text{applied}}$$

(where p is the dipole moment and α is the atomic polarizability).

Insulator
An insulator has no mobile charges.

Conductor
A conductor contains mobile charges that can move through the material.

$$\bar{v} = uE_{\text{net}}$$

The average drift speed $\bar{v}$ of a mobile charged particle in a conductor is directly proportional to the magnitude of the net electric field inside the material. The proportionality constant u is called the "mobility" and has units of (m/s)/(N/C). Different materials have different mobilities.

Metal
A metal is a conductor. It has a mobile electron sea, spread throughout the object, like an ideal gas.

	Conductor	**Insulator**
Mobile charges	Yes	No
Polarization	Entire sea of mobile charges moves	Individual atoms or molecules polarize
Equilibrium	$\vec{E}_{\text{net}} = \vec{0}$ inside	$\vec{E}_{\text{net}}$ nonzero inside
Excess charge	Spread over surface	In patches anywhere

Force between a point charge and a neutral atom is proportional to $1/r^5$:

$$F_{\text{on pt}} = q_{\text{pt}}E_{\text{ind. dipole}} = \left(\frac{1}{4\pi\varepsilon_0}\right)^2\left(\frac{2\alpha q_1^2}{r^5}\right)$$

EXPERIMENTS

Is Invisible Tape Electrically Charged?
When you pull a long piece of invisible tape off a roll, it often curls up or sticks to your hand. We will do some simple experiments with invisible tape.

Obtain a roll of invisible tape, such as Scotch® brand Magic™ Tape or a generic brand. It must be the kind of tape that almost disappears when you smooth it down on a surface, not ordinary cellophane tape.

Our first task is to determine whether or not a piece of invisible tape might be electrically charged.

> QUESTION How can we decide whether a piece of invisible tape is electrically charged?

If an object has a net electric charge, it should create an electric field in the surrounding space. Another charged object placed nearby should therefore experience an electric force. If we observe a change in an object's momentum, we can conclude that a force acts on the object.

We know that the electric field of a point charge has these characteristics:

- The magnitude of $\vec{E}$ is proportional to the amount of charge.
- The magnitude of $\vec{E}$ decreases with distance from the charge.
- The direction of $\vec{E}$ is directly away from or toward the source charge.

Therefore, since $\vec{F}_2 = q_2\vec{E}_1$, the electric force on object 2 should have the same properties. In addition, we should be able to observe both attraction and repulsion, since charges of different sign will be affected differently by a particular field.

We will observe the interactions of two pieces of invisible tape, and see whether they meet the criteria listed above.

Preparing a U Tape
Use a strip of tape about 20 cm long (about 8 in, about as long as this paper is wide). Shorter pieces are not flexible enough, and longer pieces are difficult to handle. Fold under one end of the strip to make a nonsticky handle, as shown in Figure 14.52.

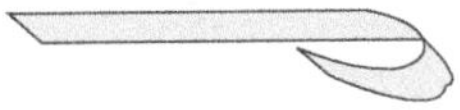

Figure 14.52

HOW TO PREPARE A U TAPE

- Stick a strip of tape with a handle down onto a smooth flat surface such as a desk. This is a "base" tape.
- Smooth this base tape down with your thumb or fingertips. This base tape provides a standard surface to work from. (Without this base tape, you get different effects on different kinds of surfaces.)
- Stick another tape with a handle down on top of the base tape, as shown in Figure 14.53.

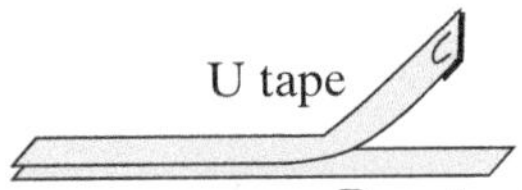

Figure 14.53

- Smooth the upper tape down well with your thumb or fingertips.
- Write U (for Upper) on the handle of the upper tape.
- With a quick motion, pull the U tape up and off the base tape, leaving the base tape stuck to the desk.
- Hang the U tape vertically from the edge of the desk, and bring your hand near the hanging tape, as shown in Figure 14.54.

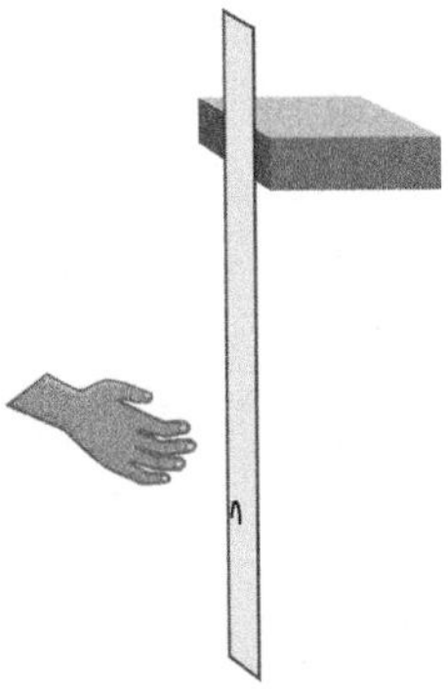

Figure 14.54

- If the tape is in good condition and the room is not too humid, you should find that there is an attraction between the hanging strip of tape and your hand when you get close to the tape. If there is no attraction, remake the U tape.

Experimental technique: try to handle the tapes only by their ends while you are doing an experiment.

EXP1 Interaction of Two U Tapes

(a) If U tapes are electrically charged, how would you expect two U tapes to interact with each other? Would you expect them to repel each other, attract each other, or not to interact at all? Make a prediction, and briefly state a reason.

(b) Make two U ("upper") tapes by following the procedure detailed above. Make sure that both tapes interact with your hand. Hang one on the edge of a desk. Bring the second U tape near the hanging U tape. Since the hanging tape is attracted to your hands, try to keep your hands out of the way. For example, you might approach the vertically hanging tape with the other tape oriented horizontally, held by two hands at its ends. What happens?

You should have seen the two U tapes repel each other. If you did not observe repulsion, try remaking the U tapes (or making new ones, both from the same roll of invisible tape). It is important to see this effect before continuing further.

Making a Tape Not Interact

You may have already discovered that if you handle a U tape too much, it no longer repels another U tape. Next we will learn a systematic way for making this happen.

- Make sure that you have an active U tape, which is attracted to your hand.
- Holding onto the bottom of the U tape, slowly rub your fingers or thumb back and forth along the *slick* side of the tape (Figure 14.55).
- You should find that the U tape no longer interacts with your hand. If it still does, repeat the process.

This is a little odd; if the U tape was originally electrically charged, the charges would presumably have been on its sticky side. However, by running a finger along the other side (the slick side) we have apparently "neutralized" it—it now appears uncharged. It will be a while before we can explain this peculiar effect, but now we have a useful way to neutralize a U tape.

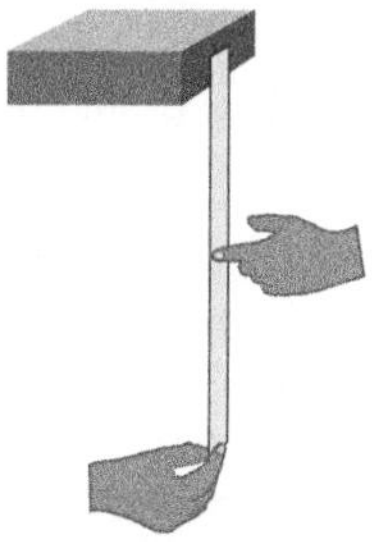

Figure 14.55

EXP2 Is This an Electric Interaction?

To decide whether the interaction between two U tapes (Figure 14.56) is or is not an electric interaction, we will see whether it obeys the criteria for an electric interaction. (As is done throughout the scientific community, it is important to compare your results to the results of other experimenters.)

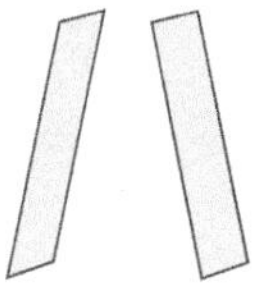

Figure 14.56

(a) Does the force act along a line connecting the two tapes? Think of a way to determine whether or not the force between two tapes acts along a line drawn from one object to the other, and do the experiment. What did you find? (What would you see if this were not the case?)

(b) Does the force decrease rapidly as the distance between the tapes increases? How can you determine this?

(c) Is the force proportional to the amounts of both charges? Design and carry out an experiment to test this. One way to vary the amount of charge on a tape is to neutralize part of one of the tapes, by running your finger along the length of the slick side of the tape, being careful that your finger touches only a portion of the width of the tape. What do you observe?

The real world is messy! You may have noted several difficulties in making your measurements. For example, both of the tapes are attracted to your hand, and also repel each other. If you tried to use a ruler to measure the distance between the tapes, you might have found that the tapes are attracted to the ruler, too.

Unlike Charges

So far we have observed that two U tapes repel each other, that the force acts along a line between the tapes, that the strength of the repulsion decreases as the tapes get farther away from each other, and that the strength of the interaction depends on the amount of charge on the tape. These observations are consistent with the hypothesis that the U tapes are electrically charged and that all U tapes have like electric charge.

QUESTION How could you prepare a tape that might have an electric charge unlike the charge of a U tape? Think of a plan before reading further.

Perhaps you reasoned along these lines: We don't know how the U tape became charged, but if the tapes started out neutral, maybe the U tape pulled some charged particles off of the lower tape (or vice versa). So now the lower tape should have an equal amount of charge, of the opposite sign.

Making an L Tape

Here is a reproducible procedure for making an L tape, whose charge is unlike the charge of a U tape:

HOW TO PREPARE AN L TAPE

- Stick a strip of tape with a handle down onto a base tape, smooth this tape down thoroughly with your thumb or fingertips, and write L (for Lower) on the handle of this tape.
- Stick another tape with a handle down on top of the L tape, and write U (for Upper) on the handle of this tape. Smooth the upper tape down well with your thumb or fingertips.

- You now have three layers of tape on the desk: a base tape, an L tape, and a U tape (Figure 14.57).

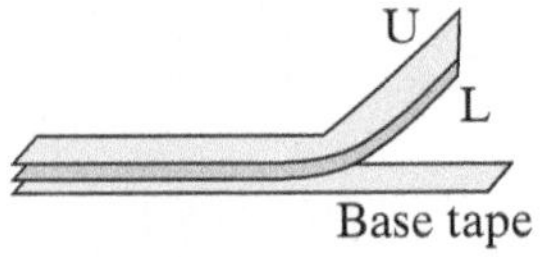

Figure 14.57

- Slowly lift the L tape off the base tape, bringing the U tape along with it (and leaving the bottom base tape stuck to the desk). Hang the double layer of tape vertically from the edge of the desk and see whether there is attraction between it and your hand (Figure 14.58). If so, get rid of these interactions (hold the bottom of the tape and slowly rub the slick side with your fingers or thumb).

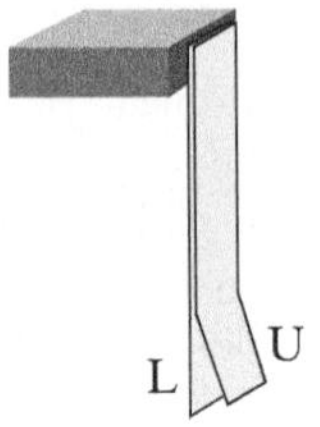

Figure 14.58

- Check that the tape pair is no longer attracted to your hand. This is important!
- Hold onto the bottom tab of the L tape and quickly pull the U tape up and off (Figure 14.59). Hang the U tape vertically from the edge of the desk, not too close to the L tape!

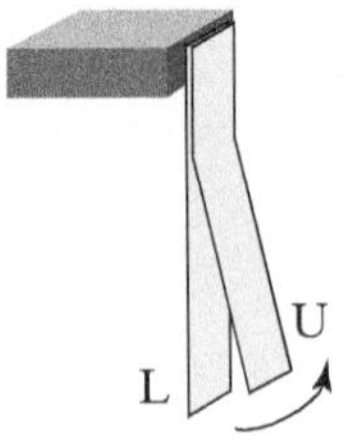

Figure 14.59

Repeating exactly the same procedure, make another pair of tapes so that you have at least two U tapes and two L tapes. Before separating the tapes from each other, always remember to make sure that the tapes are not attracted to your hand.

> QUESTION An important step in preparing an L tape is to neutralize the L/U tape pair before separating the two tapes. Considering the principle of charge conservation, why is this step important? What can go wrong if this step is omitted?
>
> The principle of conservation of charge states that if the pair has a total charge of zero before separation, the two tapes will have a total charge of zero after separation: one tape will have a charge of $+q$ and the other a charge of $-q$. However, if the total charge before separation is nonzero and positive (say), the separated tapes could both have positive charge, as long as their individual charges add up to the original amount.

EXP3 Observations of L and U Tapes
You should now have two L tapes and two U tapes. Make sure that both the U tapes and the L tapes are active (attracted to your hand).

(a) If an L tape is indeed electrically charged, and its charge is unlike the charge on a U tape, what interaction would you *predict* between an L tape and a U tape?
(b) What interaction do you *observe* between an L tape and a U tape?
(c) What interaction would you *predict* between two L tapes?
(d) What interaction do you *observe* between two L tapes?
(e) Is the pattern of interactions consistent with the statement: "Like charges repel; unlike charges attract"?

A U Tape and an L Tape: Distance Dependence of Attraction

If U and L tapes are electrically charged, then we would expect the strength of the attractive interaction to decrease as the distance between the tapes increases. Make the same sort of observations you made with two U tapes.

EXP4 Distance Dependence of Force between U and L Tapes
Move a U tape very slowly toward a hanging L tape. Observe the deflections of the tapes from the vertical, at several distances (for example, the distance at which you first see attraction, half that distance, etc.). The deflections of the tapes away from the vertical is a measure of the strength of the interaction.

(a) Does the force decrease rapidly as the distance between the tapes increases?
(b) Why is this measurement more difficult with a U and an L tape than with two U tapes?

Summary and Conclusions: U and L Tapes

Let's summarize the observations and try to conclude, at least tentatively, whether U and L tapes are electrically charged. Presumably you have observed the following:

- Objects can have a charge that is positive, negative, or zero.
- Like charges repel, unlike charges attract.
- The electric force
  - acts along a line between the charges,
  - decreases rapidly as the distance between the charges increases, and
  - is proportional to the amounts of both charges.

Our observations of U and L tapes seem to be consistent with a description of the electric interactions between charged objects. We tentatively conclude that U and L tapes are electrically charged, and have unlike charges.

How a Plastic Comb or Pen Becomes Charged

Charged objects, such as invisible tape, are negatively charged if they have more electrons than protons, and positively charged if they have fewer electrons than protons. Are U tapes positively or negatively charged? How can we tell? Charging an object in a standard manner gives us a "litmus test."

It is known that if you rub a glass rod with silk, the glass rod becomes positively charged and the silk negatively charged. Likewise, if you rub a clear plastic object such as a pen through your hair (or with fur, wool, or even cotton), the plastic ends up having a negative charge and so repels electrons. A similar process occurs when you separate one tape from another.

See Section 14.2 for a discussion of how objects become charged through rubbing or contact.

EXP5 Determining the Charge on U and L Tapes
Prepare a U tape and an L tape, and hang them from your desk. Test them with your hand to make sure they are both charged. Rub a plastic pen or comb on your hair (clear plastic seems to charge best), or on a piece of cotton or wool, and bring it close to each tape. You should observe that one of the tapes is repelled by the pen, and one is attracted to it.

Knowing that the plastic is negatively charged, what can you conclude about the sign of the electric charge on U tapes? On L tapes?

These results may be reversed if you try a different brand of "invisible tape." Be sure to compare your results with those of other students. Make sure you all agree on the assignment of "+" and "−" labels to your tapes. (If other groups are using different brands of tape, you may disagree on whether U tapes or L tapes are positive, but the electric interactions between your + tapes and their + tapes should be repulsive!)

In any of your experiments, did you find any objects, other than tapes or a charged comb or pen, that repelled a U or L tape? If so, those objects must have been charged. List these objects and whether the charge was + or −.

Amount of Charge on a Tape
We have concluded that U tapes and L tapes are electrically charged, but we have no idea how much charge is on one of the charged tapes—we don't even know an approximate order of magnitude for this quantity. Even a rough measurement of the amount of charge on a tape would be useful, because it would give us a feel for the amount of charge there might be on an ordinary object that is observed to interact electrically with other objects. Therefore the following experiment is an important one.

EXP6 Amount of Excess Charge on a Tape
In this problem you will design and carry out an experiment to determine the approximate number of excess electron charges on the surface of a negatively charged tape.

Initial Estimates
Since we do not know what order of magnitude to expect for our answer, it is important to put upper and lower bounds on reasonable answers.

(a) What is the smallest amount of excess charge that a tape could possibly have?

(b) What is the largest amount of excess charge a tape could possibly have?

Design and Perform an Experiment
A centimeter ruler is printed on the inside back cover of this textbook. A piece of half-inch-wide (1.2 cm) invisible tape, 20 cm long (8 in), has a mass of about 0.16 g.

(c) Make a clear and understandable diagram of your experimental setup, indicating each quantity you measured. Report all measurements you made.

Analyze the Results

(d) Clearly present your physical analysis of your data. Make an appropriate diagram, labeling all vector quantities. Reason from fundamental physics principles. Explicitly report any simplifying assumptions or approximations you have made in your analysis. Report two quantities:

- The amount of charge on a tape, in coulombs
- The number of excess electrons to which this charge corresponds

Present your analysis clearly. Your reasoning must be clear to a reader.

(e) Estimate whether the true amount of excess charge is larger or smaller than the value you calculated from your experimental data. Explain your reasoning briefly.

Is This a Lot of Charge?

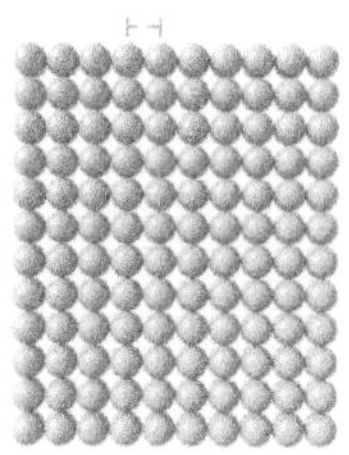

Figure 14.60

(f) What fraction of the molecules on the surface of the tape have gained an extra electronic charge? To estimate this, you may assume that molecules in the tape are arranged in a square array, as indicated in Figure 14.60, and that the diameter of a molecule in the tape is about 3×10^{-10} m. Does your answer suggest that it is a common event or a rare event for a molecule to gain an extra electron?

(g) If the electric field at a location in air exceeds 3×10^6 N/C, the air will become ionized and a spark will be triggered. In Chapter 15 we will see that the electric field in a region very close to a uniformly charged disk or plate depends approximately only on the charge Q per unit area A:

$$E = \frac{1}{e\varepsilon_0}\left(\frac{Q}{A}\right)$$

Use this model (or make a different but justifiable simplifying assumption) to calculate the magnitude of the electric field at a location in the air very close to your tape (less than 1 mm from the surface of the tape). How does it compare to the electric field needed to trigger a spark in the air?

Interaction of Charges and Neutral Matter
We have focused on the interactions of U and L tapes with other U and L tapes. Let's look more broadly at the interactions of charged tapes with other objects.

EXP7 Interactions of U and L Tapes with Other Objects
Which other objects (paper, metal, plastic, etc.) have an attractive interaction with a hanging U or L tape, and which objects have a repulsive interaction? Which objects have no interaction? Record the objects you try and the interactions observed.

The attraction of both U and L tapes to your hand, and to many other objects, is deeply mysterious. The net charge of a neutral object is 0, so your neutral hand should not make an electric field that could act on a charged tape; nor should your neutral hand experience a force due to the electric field made by a charged tape. Nothing in our statement of the properties of electric interactions allows us to explain this attraction!

Observing Interactions with Dipoles

You can also use charged tapes to observe the behavior of a dipole.

> QUESTION Consider the forces that a positive charge Q exerts on the charges making up a dipole (Figure 14.61) and describe the main features of the resulting motion of the dipole.

Figure 14.61

There is a twist (torque) that tends to align the dipole along the line connecting the charge Q and the center of the dipole, with the negative end of the dipole closer to the single positive charge. The dipole has a nonzero net force acting on it that makes it move toward the positive charge.

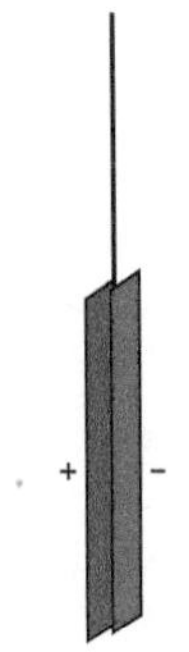

Figure 14.62

EXP8 An Electric "Compass"

Make a tall dipole and observe the motion. Take a + tape and a − tape and stick them together, overlapping them only enough to hold them together. Avoid discharging the tapes with too much handling. Hang the combination from a thread or a hair, as in Figure 14.62.

Now approach the tapes with a charged object and admire how sensitive a charge detector you have made! Slowly move the charged object all around the dipole and observe how the dipole tracks the object.

If you draw an appropriately labeled arrow on the tape, you have an electric "compass" that points in the direction of electric field.

EXP9 Observing Attraction of Like-Charged Objects (!)

Because of the very rapid $1/r^5$ increase in the attraction to neutral matter at short distances, it is sometimes the case that at short distances the attractive effect can actually overcome the repulsion between like-charged objects. Specifically, it could be that

$$N\left(\frac{1}{4\pi\varepsilon_0}\right)^2\frac{2\alpha q_1^2}{r^5} > \frac{1}{4\pi\varepsilon_0}\frac{q_1 q_2}{r^2}$$

where q_1 and q_2 are the excess charges on the surfaces of the two tapes, and N is the total number of neutral atoms in tape 2 (since each neutral atom participates in the attraction.) As can be seen by dividing the inequality by q_1, you can enhance the effect by making q_1 be significantly larger than q_2, so you might wish to partially discharge one of the tapes.

Try it! Hold one of the U tapes horizontally, with its slick side facing away from you and toward the slick side of the hanging tape. Move toward the hanging tape and check that the hanging tape is repelled as you approach. Then move close enough so that the tapes touch each other (a partner may have to hold the bottom of the hanging tape in order to be able to get very close).

You may be able to detect some slight attraction when the tapes are very close together or touch each other, despite the fact that the tapes repel at longer distances. Do you see such an effect? The effect is quite easy to see in the interaction between a highly charged Van de Graaff generator and a charged tape.

EXP10 Interaction through a Piece of Paper

Have a partner hold a piece of paper close to, but not touching, a hanging U tape. Bring another U tape toward the hanging tape from the other side of the paper, holding both ends of this tape so that it can't swing (Figure 14.63).

Figure 14.63

Can you observe repulsion occurring right through the intervening paper? This is difficult, because the paper attracts the hanging tape, which masks the repulsion due to the other tape. You can heighten the sensitivity of the experiment by moving the tape rhythmically toward and away from the hanging tape, as though you were pushing a swing. This lets you build up a sizable swing in the hanging tape even though the repulsive force is quite small, because you are adding up lots of small interactions. Using rhythmic movements, are you able to observe repulsion through the intervening paper?

The effect is especially hard to observe if you have weak repulsion due to high humidity. The farther away you can detect repulsion, the better, because the competing attraction falls off rapidly with distance. Under good conditions of low humidity, when tapes remain strongly charged and repulsion is observable with the tapes quite far apart from each other, it is possible to see repulsion with the paper in place, showing that electric field does go right through intervening matter. You have seen evidence of this when you observed attraction between a tape and your hand even when you approached the slick side of the tape.

EXP11 Is Tape a Conductor or an Insulator?
Prepare a hanging tape that has the top half charged and the bottom half uncharged. After a second or two, check to see whether the bottom half of the tape has become charged.

Based on this observation, is tape a conductor or an insulator? That is, are the charges free or bound? Explain fully and rigorously how your observations justify your conclusion. (*Hint:* Draw a diagram showing what effect the charges on the upper half of the tape have on each other and on charges inside the tape, and reason through what will happen if any of these charges are free to move.)

EXP12 Discharging a Tape
The previous exercise suggests that a key element in neutralizing a tape is the salt solution on the surface of your finger. Design one or more experiments you can do to confirm or reject this explanation of discharging a tape.

EXP13 Charging by Induction
Hang a short piece of aluminum foil (about the width of the tape and half as long as your thumb) from a tape, with another piece of tape added to the bottom of the foil as a handle (Figure 14.64).

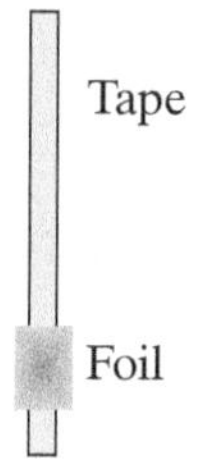

Figure 14.64

Now carry out the following operations *exactly* as specified:

1. Make sure that the tape and foil are uncharged (touch the foil, and rub the slick side of the tape).
2. Have a partner hold onto the bottom tape to keep the foil from moving.
3. Bring a charged plastic pen or comb very close to the foil, *but don't touch the foil with the plastic.*
4. *While holding the plastic near the foil,* tap the *back* of the foil with your finger.
5. Move your finger away from the foil, *then* move the plastic away from the foil.

You should find that the metal foil is now strongly charged. This process is called "charging by induction." Now touch the charged aluminum foil with your finger and observe that this discharges the foil, as predicted by our earlier discussion.

Complete a "comic strip" of diagrams (Figure 14.65) illustrating the charging by induction process you carried out. Make sure you have the sign of the charges right. In each diagram show charge distributions, polarization, movement of charges, and so on. For each frame, explain briefly what happens. Remember that excess charges on the metal foil can only be on the surface.

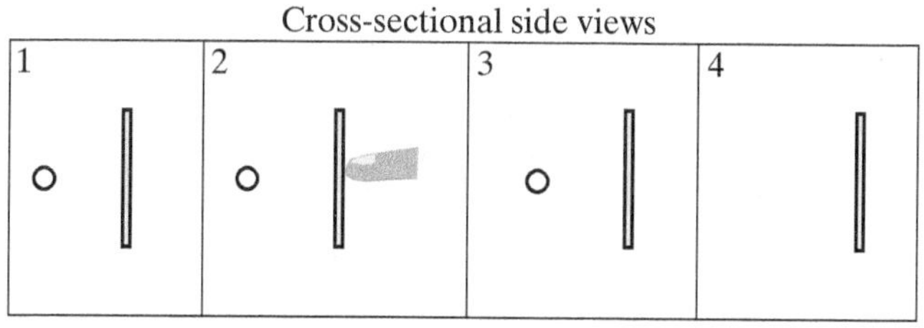

Figure 14.65

This process is called charging by induction because the entire piece of foil becomes an induced dipole when it is polarized by the external charge. Charging by induction makes it possible to charge a metal without touching the external charge to the metal.

Explain the process of discharging the metal foil by touching it, using the same kind of time-sequence "comic strip" diagrams you used previously. Illustrate the important aspects of each step in the process. Include any changes to your body as well as to the foil. Be precise in your use of words.

EXP14 A Water Film as a Conductor
Prepare a hanging tape that has the top half charged and the bottom half uncharged. Let it hang while you do other work, but check every few minutes to see what has happened in the two halves. What do you predict will happen to the state of charge in the two halves? What do you observe over a period of many minutes? (If the room is very dry or very wet you may not be able to see this effect.)

Suppose you were to breathe heavily through your mouth onto the slick and sticky sides of a short section in the middle of a long charged tape. Your breath is very moist. What do you predict you would find immediately afterward? Try the experiment—what do you observe? (Repeat if you see no effect.)

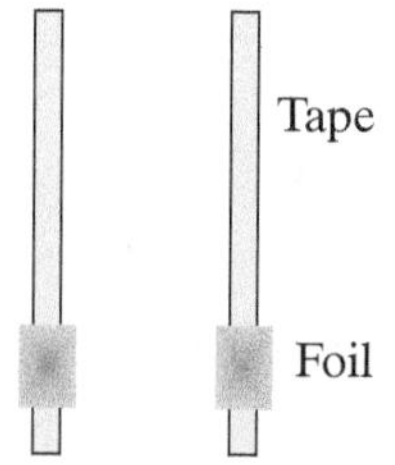

Figure 14.66

EXP15 Transferring Charge by Contact
Make two identical hanging foil arrangements (Figure 14.66), and charge one of the aluminum foils by induction. Discharge the other foil by touching it with your finger. Charge a tape or pen or comb, and note the approximate strength of the interaction between it and the charged foil.

Next make the two foils touch each other, being careful not to touch either foil with your fingers. Note the approximate strength of the interaction that there is now between the plastic and each foil.

Compared with the situation before the two foils touched, what sign of charge, and roughly how much, is there now on each foil? Discuss this fully with your partners, and convince yourselves that you understand the process. Make a written explanation, including appropriate diagrams.

What would you expect to happen if one piece of foil were much larger than the other?

QUESTIONS

Q1 Criticize the following statement: "Since an atom's electron cloud is spherical, the effect of the electrons cancels the effect of the nucleus, so a neutral atom can't interact with a charged object." ("Criticize" means to explain why the given statement is inadequate or incorrect, as well as to correct it.)

Q2 Criticize the following statement: "A positive charge attracts neutral plastic by polarizing the molecules and then attracting the negative side of the molecules." ("Criticize" means to explain why the given statement is inadequate or incorrect, as well as to correct it.)

Q3 Jill stuck a piece of invisible tape down onto another piece of tape. Then she yanked the upper tape off the lower tape, and she found that this upper tape strongly repelled other upper tapes and was charged positive. Jack ran his thumb along the slick (upper) side of the upper tape, and the tape no longer repelled other upper tapes. Jill and Jack explained this by saying that Jack rubbed some protons out of the carbon nuclei in the tape. Give a critique of their explanation. If Jill's and Jack's explanation is deficient, give a physically possible explanation for why the upper tape no longer repelled other upper tapes. Include explanatory diagrams.

Q4 Atom A is easier to polarize than atom B. Which atom, A or B, would experience a greater attraction to a point charge a distance r away? Explain your reasoning.

Q5 Is the following statement true or false? If true, what principle makes it true? If false, give a counterexample or say why. See Figure 14.67.

"The electric field E_{point} at the center of an induced dipole, due to the point charge, is equal in magnitude and opposite in direction to the electric field E_{dipole} at the location of the point charge, due to the induced dipole."

Figure 14.67

Q6 Explain briefly why the attraction between a point charge and a dipole has a different distance dependence for induced dipoles ($1/r^5$) than for permanent dipoles ($1/r^3$). (You need not explain either situation in full detail: just explain why there is this difference in their behavior.)

Q7 A large positive charge pulls on a distant electron. How does the *net* force on the electron change if a slab of glass is inserted between the large positive charge and the electron? Does the net force get bigger, smaller, or stay the same? Explain, using only labeled diagrams.

(Be sure to show *all* the forces on the electron before determining the net force on the electron, not just the force exerted by the large positive charge. Remember that the part of the net force on the electron contributed by the large positive charge does not change when the glass is inserted: the electric interaction extends through matter.)

Q8 Explain briefly why repulsion is a better test for the sign of a charged object than attraction is.

Q9 Carbon tetrachloride (CCl_4) is a liquid whose molecules are symmetrical and so are not permanent dipoles, unlike water molecules. Explain briefly how the effect of an external charge on a beaker of water (H_2O) differs from its effect on a beaker of CCl_4. (*Hint:* Consider the behavior of the permanent dipole you made out of U and L tapes.)

Q10 A positive charge is located between a neutral block of plastic and a neutral block of copper (Figure 14.68). Draw the approximate charge distribution for this situation.

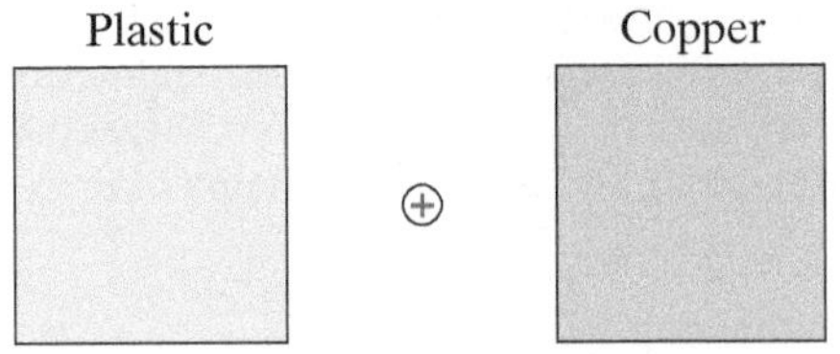

Figure 14.68

Q11 Make a table showing the major differences in the electric properties of plastic, salt water, and copper. Include diagrams showing polarization by an external charge.

Q12 Figure 14.69 shows a neutral, solid piece of metal placed near two point charges. Copy this diagram. **(a)** On your diagram, show the polarization of the piece of metal. **(b)** Then, at location A *inside* the solid piece of metal, carefully draw and label three vectors: (1) $\vec{E}_1$, the electric field due to $-q_1$; (2) $\vec{E}_2$, the electric field due to $+q_2$; (3) $\vec{E}_3$, the electric field due to all of the charges on the metal. **(c)** Explain briefly why you drew the vectors the way you did.

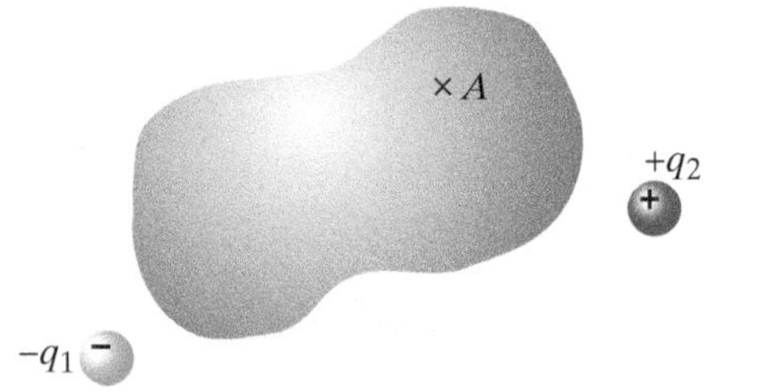

Figure 14.69

Q13 Two identical metal spheres are suspended from insulating threads. One is charged with excess electrons, and the other is neutral. When the two spheres are brought near each other, they swing toward each other and touch, then swing away from each other. **(a)** Explain in detail why both these swings happen. In your explanation, include clear diagrams showing charge distributions, including the final charge distribution. **(b)** Next the spheres are moved away from each other. Then a block of plastic is placed between them as shown in Figure 14.70. The original positions of the spheres are indicated, before the plastic is placed between them. Sketch the new positions of the spheres. Explain, including charge distributions on the spheres.

Figure 14.70

(c) Show the polarization of a molecule inside the plastic at points A, B, C, D, and E. Explain briefly.

Q14 A small glass ball is rubbed all over with a small silk cloth and acquires a charge of $+5\,\text{nC}$. The silk cloth and the glass ball are placed 30 cm apart.

Figure 14.71

(a) On a diagram like that shown in Figure 14.71, draw the electric field vectors qualitatively at the locations marked ×. Pay careful attention to directions and to relative magnitudes. Use dashed lines to explain your reasoning graphically, and draw the final electric field vectors with solid lines.

(b) Next, a neutral block of copper is placed between the silk and the glass.

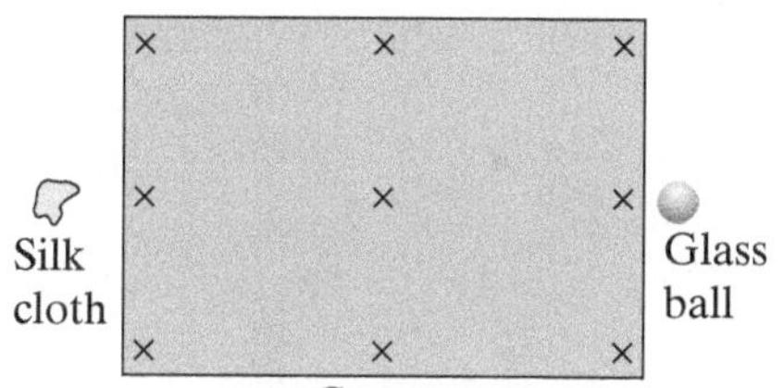

Figure 14.72

On a diagram like that shown in Figure 14.72, carefully show the approximate charge distribution for the copper block and the electric field vectors inside the copper at the locations marked ×.

(c) The copper block is replaced by a neutral block of plastic. Carefully show the approximate molecular polarization of the plastic block at the locations marked × in Figure 14.73.

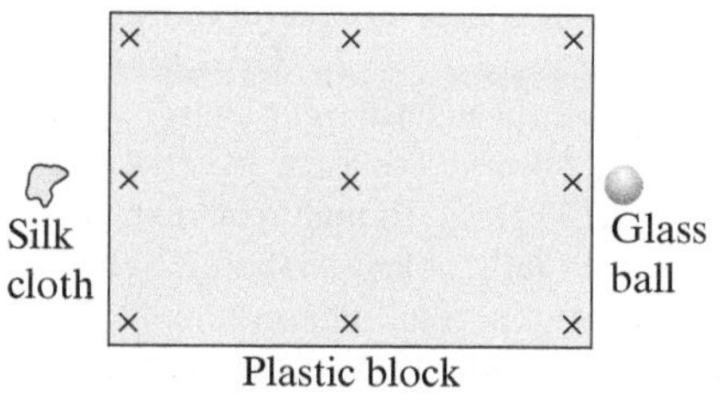

Figure 14.73

(d) Even if you have to state your result as an inequality, make as quantitative a statement as you can about the electric field at the location of the glass ball and the net force on the ball when the plastic block is in place compared to when there is no block. Explain briefly.

Q15 A student said, "When you touch a charged piece of metal, the metal is no longer charged: all the charge on the metal is neutralized." As a practical matter, this is nearly correct, but it isn't exactly right. What's wrong with saying that all the charge on the metal is neutralized?

Q16 You are wearing shoes with thick rubber soles. You briefly touch a negatively charged metal sphere. Afterward, the sphere seems to have little or no charge. Why? Explain in detail.

Q17 Criticize the following statement: "When you rub your finger along the slick side of a U tape, the excess charges flow onto your finger, and this discharges the tape." Draw diagrams illustrating a more plausible explanation.

Q18 Can you charge a piece of plastic by induction? Explain, using diagrams. Compare with the amount of charging obtained when you charge a piece of metal by induction.

Q19 As shown in Figure 14.74(a), an electroscope consists of a steel ball connected to a steel rod, with very thin gold foil leaves connected to the bottom of the rod (in good electric contact with the rod). The bottom of the electroscope is enclosed in a glass jar and held in place by a rubber stopper.

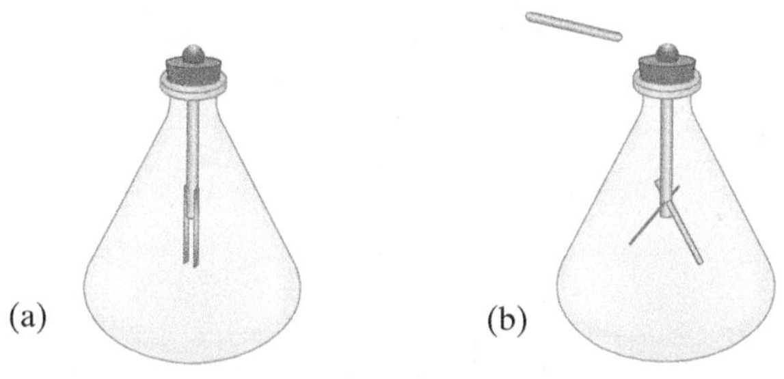

Figure 14.74

(a) The electroscope is brought near to but not touching a positively charged glass rod as shown in Figure 14.74(b). The foil leaves are observed to spread apart. Explain why in detail, using as many diagrams as necessary.

(b) The electroscope is moved far away from the glass rod and the steel ball is touched momentarily to a metal block. The foil leaves spread apart and stay spread apart when the electroscope is moved away from the block. As the electroscope is moved close to but not touching the positively charged glass rod, the foil leaves move closer together. Is the metal block positive, negative, or neutral? How do you know? Explain.

Q20 Two plastic balls are charged equally and positively and held in place by insulating threads, as shown in Figure 14.75.

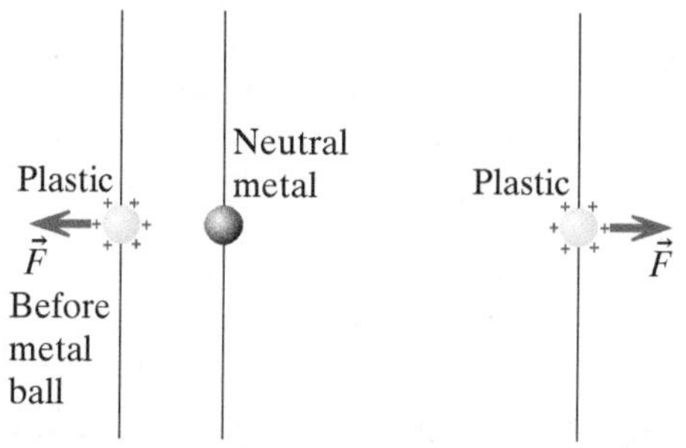

Figure 14.75

They repel each other with an electric force of magnitude F. Then an uncharged metal ball is held in place by insulating threads between the balls, closer to the left ball. State what change (if any) there is in the net electric force on the left ball and on the net electric force on the right ball. Show relevant force vectors. Also show the charge distribution on the metal ball. Explain briefly but completely.

Q21 You take two invisible tapes of some unknown brand, stick them together, and discharge the pair before pulling them apart and hanging them from the edge of your desk. When you bring an uncharged plastic pen within 10 cm of either the U tape or the L tape you see a slight attraction. Next you rub the pen through your hair, which is known to charge the pen negatively. Now you find that if you bring the charged pen within 8 cm of the L tape you see a slight repulsion, and if you bring the pen within 12 cm of the U tape you see a slight attraction. Briefly explain all of your observations.

Q22 You have three metal blocks marked A, B, and C, sitting on insulating stands. Block A is charged $+$, but blocks B and C are neutral (Figure 14.76).

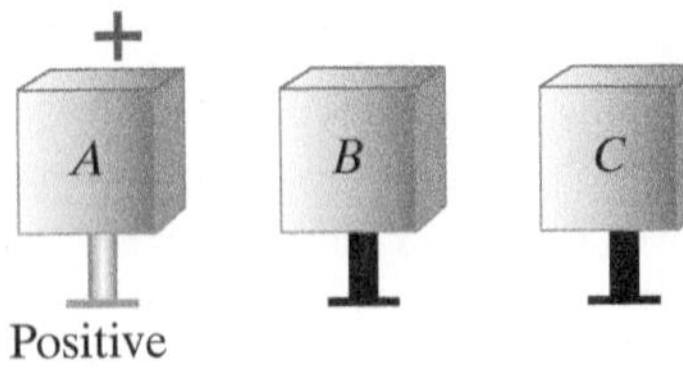

Figure 14.76

Without using any additional equipment and without altering the amount of charge on block A, explain how you could make block B be charged $+$ and block C be charged $-$. Explain your procedure in detail, including diagrams of the charge distributions at each step in the process.

Q23 You have two identical neutral metal spheres labeled A and B, mounted on insulating posts, and you have a plastic pen that charges negatively when you rub it on your hair (Figure 14.77).

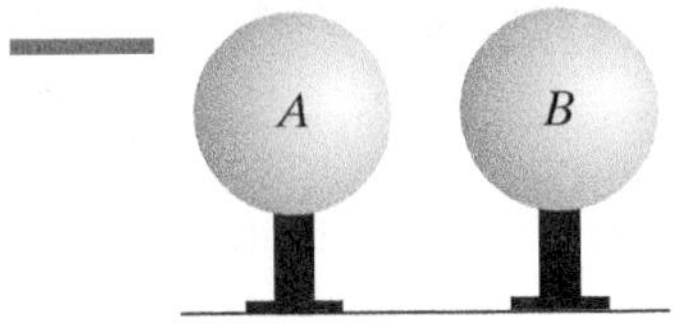

Figure 14.77

(a) ($+$ and $-$) Explain in detail, including diagrams, what operations you would carry out to give sphere A some positive charge and sphere B an equal amount of negative charge.
(b) ($+$ and $+$) Explain in detail, including diagrams, what operations you would carry out on the neutral spheres to give sphere A some positive charge and sphere B an equal amount of positive charge (the spheres are initially uncharged).

Q24 Here is a variant of "charging by induction." Place two uncharged metal objects so as to touch each other, one behind the other. Call them *front* object and *back* object. While you hold a charged comb in front of the *front* object, your partner moves away the *back* object (handling it through an insulator so as not to discharge it). Now you move the comb away. Explain this process. Use only labeled diagrams in your explanation (no prose!).

Q25 Metal sphere A is charged negatively and then brought near an uncharged metal sphere B (Figure 14.78). Both spheres rest on insulating supports, and the humidity is very low.

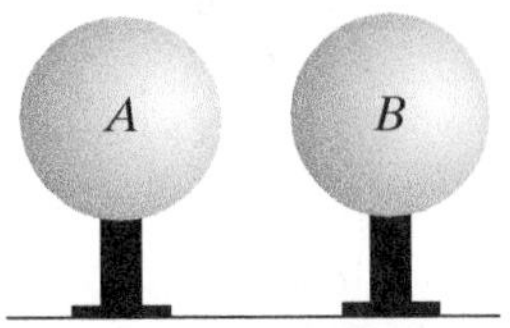

Figure 14.78

(a) Use $+$'s and $-$'s to show the approximate distribution of charges on the two spheres. (*Hint:* Think hard about *both* spheres, not just B.)

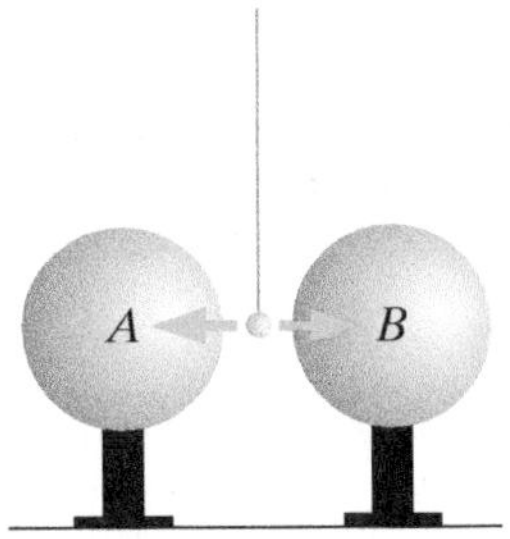

Figure 14.79

(b) A small, lightweight hollow metal ball, initially uncharged, is suspended from a string and hung between the two spheres (Figure 14.79). It is observed that the ball swings rapidly back and forth hitting one sphere and then the other. This goes on for 5 seconds, but then the ball stops swinging and hangs between the two spheres. Explain in detail, step by step, why the ball swings back and forth and why it finally stops swinging. Your explanation must include good physics diagrams.

Q26 Suppose that you try to measure the electric field $\vec{E}$ at a location by placing a charge Q_1 there and observing the force $\vec{F}_1$, so that you measure $E_1 = F_1/Q_1 = 1000$ N/C. Then you remove Q_1 and place a much larger charge $Q_2 = 30Q_1$ at the same location, and observe the force $\vec{F}_2$. This time you measure $E_2 = F_2/Q_2 = 1100$ N/C, though you expected to measure 1000 N/C again. What's going on here? Why didn't you get $E = 1000$ N/C in your second measurement? Sketch a possible situation that would lead to these measurements.

PROBLEMS

Section 14.1

•P27 You rub a plastic comb through your hair and it now carries a charge of -4×10^{-10} C. What is the charge on your hair?

•P28 Many heavy nuclei are "alpha emitters": they emit an alpha particle, which is the historical name for the nucleus of a helium atom, which contains two protons. For example, a thorium nucleus containing 90 protons is an alpha-emitter. What element does thorium turn into as a result of emitting an alpha particle?

Section 14.2

•P29 Which of the following could be reasonable explanations for how a piece of invisible tape gets charged? Select all that apply. (1) Protons are pulled out of nuclei in one tape and transferred to another tape. (2) Charged molecular fragments are broken off one tape and transferred to another. (3) Electrons are pulled out of molecules in one tape and transferred to another tape. (4) Neutrons are pulled out of nuclei in one tape and transferred to another tape.

•**P30** You rub a clear plastic pen with wool, and observe that a strip of invisible tape is attracted to the pen. Assuming that the pen has a net negative charge, which of the following could be true? Select all that apply. (1) The tape might be negatively charged. (2) The tape might be positively charged. (3) The tape might be uncharged. (4) There is not enough information to conclude anything.

•**P31** Which observation provides evidence that two objects have the same sign charge? (a) The two objects repel each other. (b) The two objects attract each other. (c) The two objects do not interact at all. (d) The strength of the interaction between the two objects depends on distance.

Section 14.3

•**P32** Which statements about a neutral atom are correct? Select all that apply. (1) A neutral atom is composed of both positively and negatively charged particles. (2) The positively charged particles in the nucleus are positrons. (3) The electrons are attracted to the positively charged nucleus. (4) Positively charged protons are located in the tiny, massive nucleus. (5) The radius of the electron cloud is twice as large as the radius of the nucleus. (6) The negatively charged electrons are spread out in a "cloud" around the nucleus.

•**P33** If the distance between a neutral atom and a point charge is tripled, by what factor does the force on the atom by the point charge change? Express your answer as a ratio: new force/original force.

•**P34** There is a region where an electric field points to the right, due to charged particles somewhere. A neutral carbon atom is placed inside this region. Draw a diagram of the situation, and use it to answer the following question: Which of the following statements are correct? Select all that apply. (1) Because the net charge of the carbon atom is zero, it cannot be affected by an electric field. (2) The electron cloud in the carbon atom shifts to the left. (3) The neutral carbon atom polarizes and becomes a dipole. (4) The nucleus of the carbon atom shifts to the left.

•**P35** A charged particle with charge q_1 is a distance r from a neutral atom, as shown in Figure 14.80.

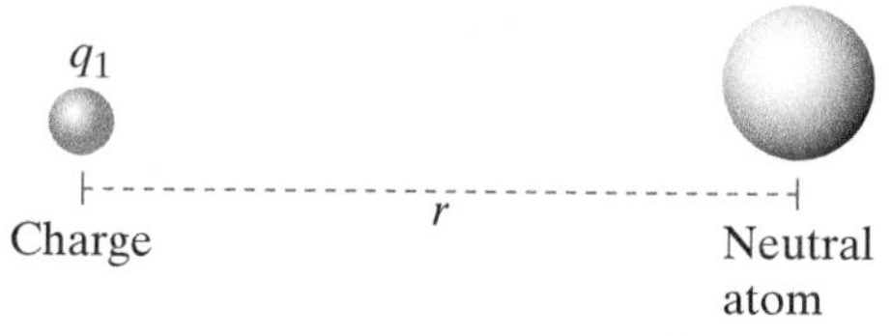

Figure 14.80

(1) If q_1 is negative, which diagram (1–10) in Figure 14.81 best shows the charge distribution in the neutral atom in this situation?

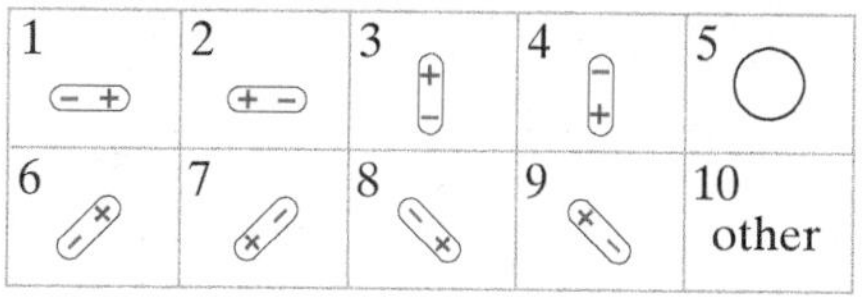

Figure 14.81

(2) Which of the arrows (*a*–*j*) in Figure 14.82 best indicates the direction of the electric field at the location of the charged particle, made by the polarized neutral atom?

Figure 14.82

(3) Which of the arrows (*a*–*j*) in Figure 14.82 best indicates the direction of the force on the charged particle, due to the polarized neutral atom? **(4)** Which of the arrows (*a*–*j*) in Figure 14.82 best indicates the direction of the force on the polarized neutral atom, due to the charged particle?

•**P36** A charged piece of invisible tape is brought near your hand, as shown in Figure 14.83. Your hand is initially neutral.

Figure 14.83

(a) If the tape is negatively charged, which of the diagrams 1–10 in Figure 14.81 best shows the polarization of a neutral molecule in your hand? **(b)** Which arrow in Figure 14.82 best indicates the direction (*a*–*j*) of the electric field at the location of the tape due to the large number of polarized molecules in your hand? **(c)** Which arrow in Figure 14.82 best indicates the direction (*a*–*j*) of the force on the tape due to the polarized molecules in your hand? **(d)** Which arrow in Figure 14.82 best indicates the direction (*a*–*j*) of the force on your hand due to the charged tape?

••**P37** An electron and a neutral carbon atom are initially 1×10^{-6} m apart (about 10 000 atomic diameters), and there are no other particles in the vicinity. The polarizability of a carbon atom has been measured to be $\alpha = 1.96 \times 10^{-40}\,\mathrm{C \cdot m/(N/C)}$. **(a)** Calculate the initial magnitude and direction of the acceleration of the electron. Explain your steps clearly. Pay particular attention to clearly defining your algebraic symbols. Don't put numbers into your calculation until the very end. **(b)** If the electron and carbon atom were initially twice as far apart, how much smaller would the initial acceleration of the electron be?

••**P38** In Problem P37, replace "electron" with "water molecule" and repeat the analysis. A water molecule has a permanent dipole moment whose magnitude is $6.2 \times 10^{-30}\,\mathrm{C \cdot m}$, which is much larger than the induced dipole for this situation. Assume that the dipole moment of the water molecule points toward the carbon atom.

•••**P39** In Figure 14.84 there is a permanent dipole on the left with dipole moment $\mu_1 = Qs_1$ and a neutral atom on the right with polarizability α, so that it becomes an induced dipole with dipole moment $\mu_2 = qs_2 = \alpha E_1$, where E_1 is the magnitude of the electric field produced by the permanent dipole. Show that the force the permanent dipole exerts on the neutral atom is

$$F \approx \left(\frac{1}{4\pi\varepsilon_0}\right)^2 \frac{12\alpha\mu_1^2}{r^7}$$

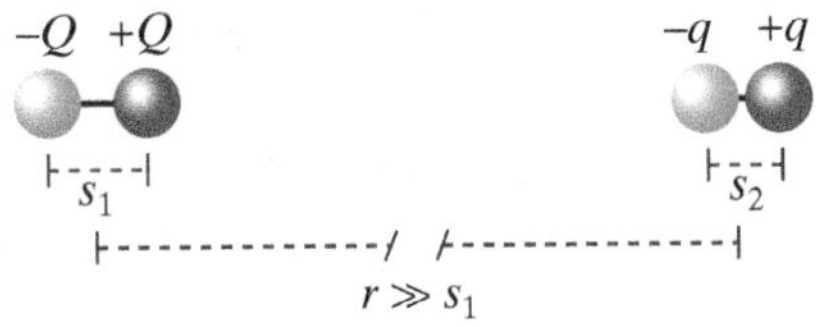

Figure 14.84

Hint: It is convenient to use the "binomial expansion" that you may have learned in calculus, that $(1+\epsilon)^n \approx 1+n\epsilon$ is $\epsilon \ll 1$. Note that n can be negative.

•••P40 Try rubbing a plastic pen through your hair, and you'll find that you can pick up a tiny scrap of paper when the pen is about one centimeter above the paper. From this simple experiment you can estimate how much an atom in the paper is polarized by the pen! You will need to make several assumptions and approximations. Hints may be found at the end of the chapter. **(a)** Suppose that the center of the outer electron cloud ($q = -4e$) of a carbon atom shifts a distance s when the atom is polarized by the pen. Calculate s algebraically in terms of the charge Q on the pen. **(b)** Assume that the pen carries about as much charge Q as we typically find on a piece of charged invisible tape. Evaluate s numerically. How does this compare with the size of an atom or a nucleus? **(c)** Calculate the polarizability α of a carbon atom. Compare your answer to the measured value of $1.96 \times 10^{-40}\ \mathrm{C \cdot m/(N/C)}$ (T. M. Miller and B. Bederson, "Atomic and molecular polarizabilities: a review of recent advances," *Advances in Atomic and Molecular Physics*, 13, 1–55, 1977). **(d)** Carefully list all assumptions and approximations you made.

Section 14.4

•P41 A solid plastic ball has negative charge uniformly spread over its surface. Which of the diagrams in Figure 14.85 best shows the polarization of molecules inside the ball?

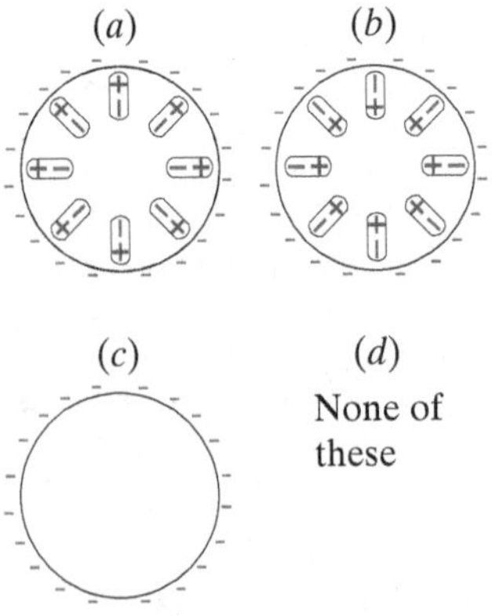

Figure 14.85

•P42 A dipole consisting of two oppositely charged balls connected by a wooden stick is located as shown in Figure 14.86. A block of plastic is located nearby, as shown. Locations B, C, and D all lie on a line perpendicular to the axis of the dipole, passing through the midpoint of the dipole.

Before selecting answers to the following questions, draw your own diagram of this situation, showing all the fields and charge distributions requested.

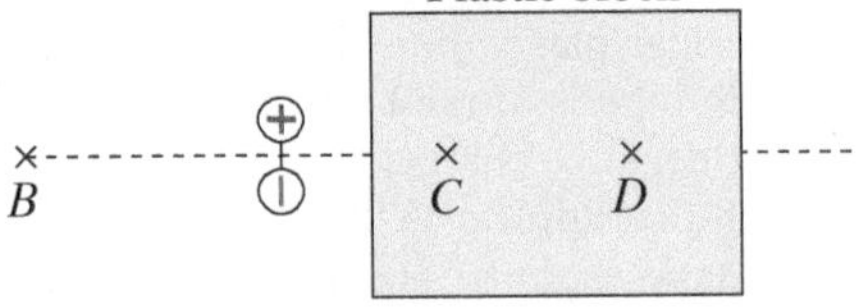

Figure 14.86

Answer the following questions by selecting either a direction (*a–j*) from Figure 14.82 or an orientation of a polarized molecule (1–10) from the diagrams in Figure 14.81. **(a)** Which of the arrows (*a–j*) best indicates the direction of the electric field at location C due only to the dipole? **(b)** Which of the arrows (*a–j*) best indicates the direction of the electric field at location D due only to the dipole? **(c)** Which of the diagrams (1–10) best indicates the polarization of a molecule of plastic at location C? **(d)** Which of the diagrams (1–10) best indicates the polarization of a molecule of plastic at location D? **(e)** Which of the following statements is correct? (1) A molecule located at C would not be polarized at all. (2) The polarization of a molecule located at D would be the same as the polarization of a molecule located at C. (3) A molecule located at D would be polarized more than a molecule located at C. (4) A molecule located at D would be polarized less than a molecule located at C. **(f)** Which of the arrows (*a–j*) best indicates the direction of the electric field at location B due only to the dipole? **(g)** Which of the arrows (*a–j*) best indicates the direction of the electric field at location B due only to the plastic block? The magnitude of the electric field at B due to the plastic is less than the magnitude of the electric field at B due to the dipole. **(h)** Which of the arrows (*a–j*) best indicates the direction of the net electric field at location B? **(i)** Which of the following statements is correct? (1) The electric field at B due only to the dipole would be larger if the plastic block were not there. (2) The electric field at B due only to the dipole would be the same if the plastic block were not there. (3) The electric field at B due only to the dipole would be smaller if the plastic block were not there. (4) The electric field at B due only to the dipole would be zero if the plastic block were not there.

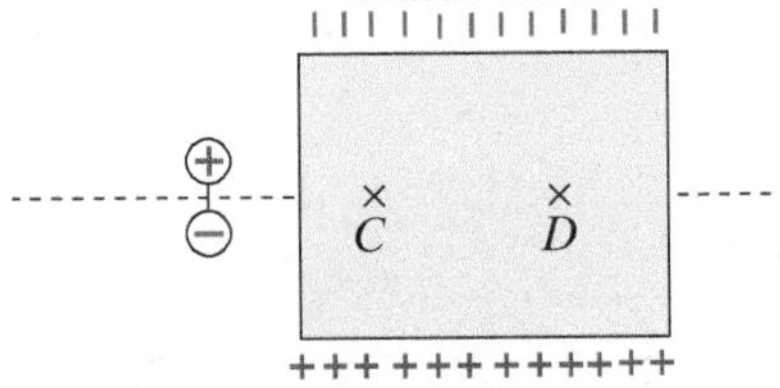

Figure 14.87

(j) Using the diagrammatic conventions discussed in the text, a student drew the diagram in Figure 14.87 to help answer the questions asked above. Which of the following statements about the student's diagram are true? Check all that apply. (1) The direction of polarization of the plastic block is wrong. (2) The diagram is correct; this is just a different way of drawing the polarization. (3) The diagram shows mobile charges; this is wrong because an insulator does not have mobile charged particles.

Section 14.5

•**P43** The mobility of Na^+ ions in water is 5.2×10^{-8} (m/s)/(N/C). If an electric field of 2400 N/C is maintained in the fluid, what is the drift speed of the sodium ions?

•**P44** An electric field is applied to a solution containing bromide ions. As a result, the ions move through the solution with an average drift speed of 3.7×10^{-7} m/s. The mobility of bromide ions in solution is 8.1×10^{-8} (m/s)/(N/C). What is the magnitude of the net electric field inside the solution?

Section 14.6

•**P45** Which of the following are true? Check all that apply. (1) If the net electric field at a particular location inside a piece of metal is zero, the metal is not in equilibrium. (2) The net electric field inside a block of metal is zero under all circumstances. (3) The net electric field at any location inside a block of copper is zero if the copper block is in equilibrium. (4) The electric field from an external charge cannot penetrate to the center of a block of iron. (5) In equilibrium, there is a net flow of mobile charged particles inside a conductor.

•**P46** Which of the following are true? Select all that apply. (1) In equilibrium, there is no net flow of mobile charged particles inside a conductor. (2) The electric field from an external charge cannot penetrate to the center of a block of iron. (3) The net electric field inside a block of aluminum is zero under all circumstances. (4) If the net electric field at a particular location inside a piece of metal is not zero, the metal is not in equilibrium. (5) The net electric field at any location inside a block of copper is zero if the copper block is in equilibrium.

•**P47** A negatively charged iron block is placed in a region where there is an electric field downward (in the $-y$ direction) due to charges not shown. Which of the diagrams (a–f) in Figure 14.88 best describes the charge distribution in and/or on the iron block?

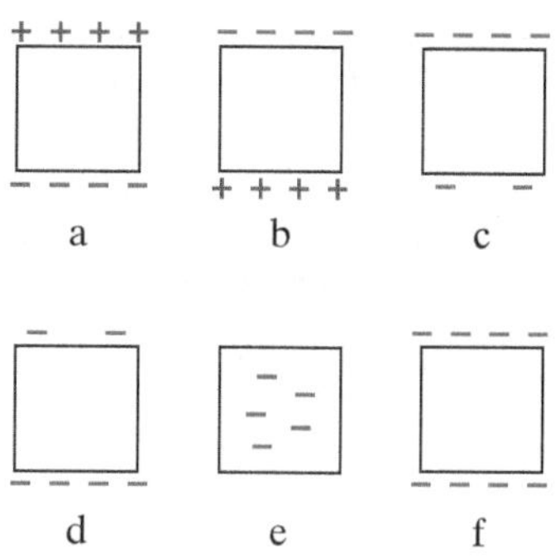

Figure 14.88

•**P48** Two small, negatively charged plastic spheres are placed near a neutral iron block, as shown in Figure 14.89. Which arrow (*a–j*) in Figure 14.89 best indicates the direction of the net electric field at location *A*?

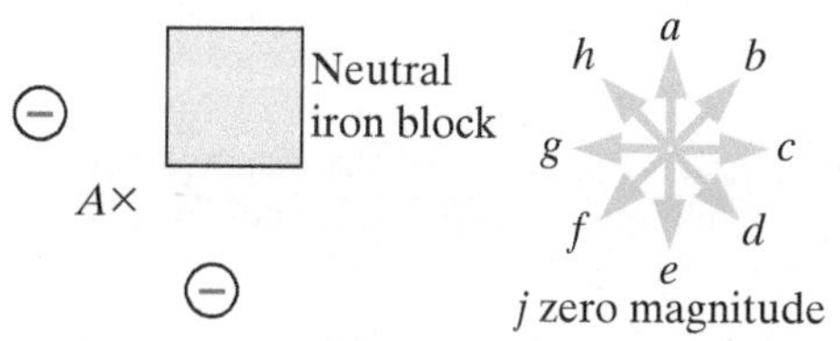

Figure 14.89

•**P49** A neutral copper block is polarized as shown in Figure 14.90, due to an electric field made by external charges (not shown). Which arrow (*a–j*) in Figure 14.90 best indicates the direction of the net electric field at location *B*, which is inside the copper block?

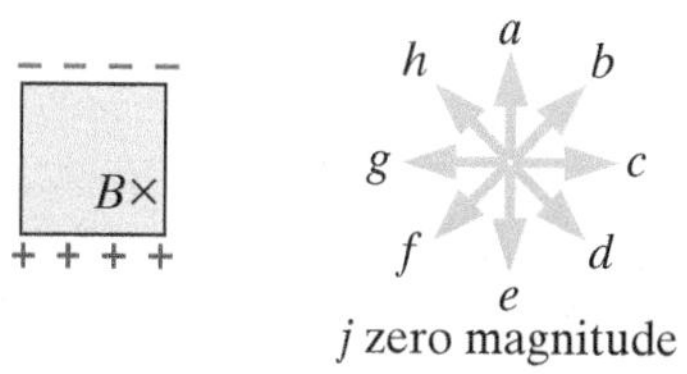

Figure 14.90

•**P50** **(a)** Which of the diagrams (A–F) in Figure 14.91 correctly displays the polarization of a *metal* sphere by an electric field that points to the left, using the conventions discussed in this chapter? **(b)** Which of the diagrams (A–F) in Figure 14.91 correctly displays the polarization of a *plastic* sphere by an electric field that points to the left, using the conventions discussed in this chapter?

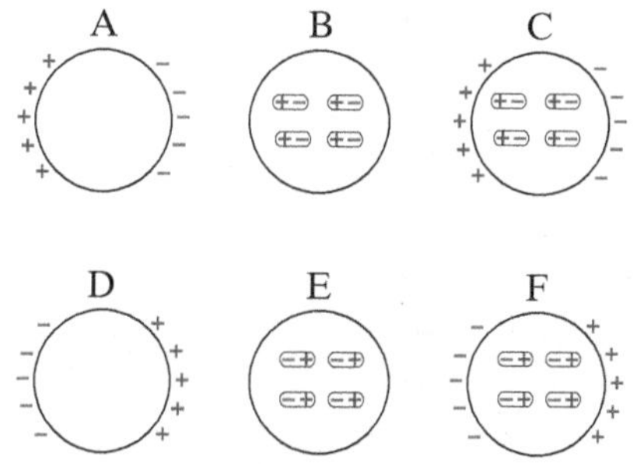

Figure 14.91

•**P51** You place a neutral block of nickel near a small glass sphere that has a charge of 2×10^{-8} C uniformly distributed over its surface, as shown in Figure 14.92.

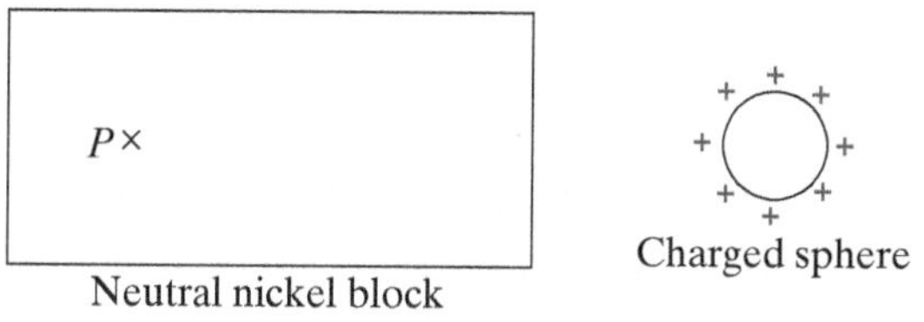

Figure 14.92

(a) About how long do you have to wait to make sure that the mobile electron sea inside the nickel block has reached equilibrium? (1) Less than a nanosecond (1×10^{-9} s), (2) Several hours, (3) About 1 s, (4) About 10 min **(b)** In equilibrium, what is the average drift speed of the mobile electrons inside the nickel block? (1) About 1×10^5 m/s, (2) About 1×10^{-5} m/s, (3) 0 m/s **(c)** In the equation $\bar{v} = uE$, what is the meaning of the symbol u? (1) The density of mobile electrons inside the metal, in electrons/m^3, (2) The mobility of an electron inside the metal, in (m/s)/(N/C), (3) The time it takes a block of metal to reach equilibrium, in seconds

•**P52** This question focuses on reasoning about equilibrium inside the nickel block shown in Figure 14.92. Start with these premises:

- The definition of equilibrium inside a conductor and
- The relationship between average drift speed and electric field in a conductor

to reason about which situations are *possible* inside the nickel block at equilibrium. Some of the situations listed below are possible, some are ruled out by one premise, and some are ruled out by two premises. If a situation is ruled out by two premises, choose both.

Case 1: $\overline{v} = 0$ and $E_{net} = 0$ (1) Possible, (2) Not possible by definition of equilibrium, (3) Not possible because $\overline{v} = uE_{net}$
Case 2: $\overline{v} = 0$ and $E_{net} > 0$ (1) Possible, (2) Not possible by definition of equilibrium, (3) Not possible because $\overline{v} = uE_{net}$
Case 3: $\overline{v} > 0$ and $E_{net} = 0$ (1) Possible, (2) Not possible by definition of equilibrium, (3) Not possible because $\overline{v} = uE_{net}$
Case 4: $\overline{v} > 0$ and $E_{net} > 0$ (1) Possible, (2) Not possible by definition of equilibrium, (3) Not possible because $\overline{v} = uE_{net}$
Now that you have considered each case, in equilibrium, which one is the only situation that is physically possible? (1) Case 1, (2) Case 2, (3) Case 3, (4) Case 4

•P53 A positively charged sphere is placed near a neutral block of nickel, as shown in Figure 14.92. **(a)** Which of the diagrams in Figure 14.93 best represents the equilibrium distribution of charge on the neutral nickel block?

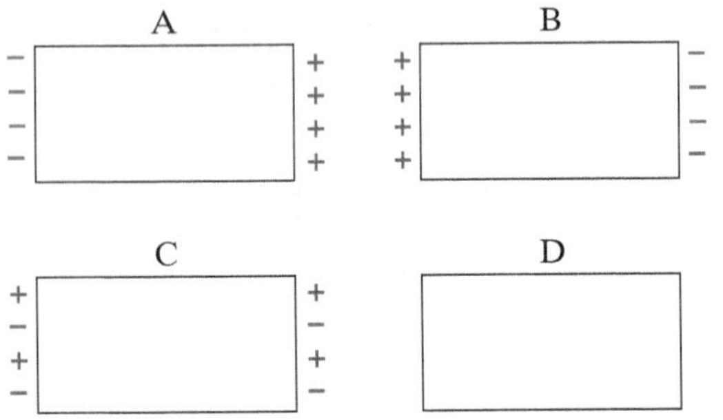

Figure 14.93

(b) At location *P* inside the nickel block the electric field due to the charged sphere is $\langle -625, 0, 0 \rangle$ N/C. At equilibrium, which of the following statements must be true? (1) It is not possible to determine the electric field at location *P* due only to charges on the surface of the nickel block. (2) The electric field at location *P* due only to charges on the surface of the nickel block is $\langle 0,0,0 \rangle$ N/C. (3) Because the net electric field at location *P* is $\langle 0,0,0 \rangle$ N/C, the field at *P* due only to charges on the surface of the polarized nickel block must be $\langle 625,0,0 \rangle$ N/C.

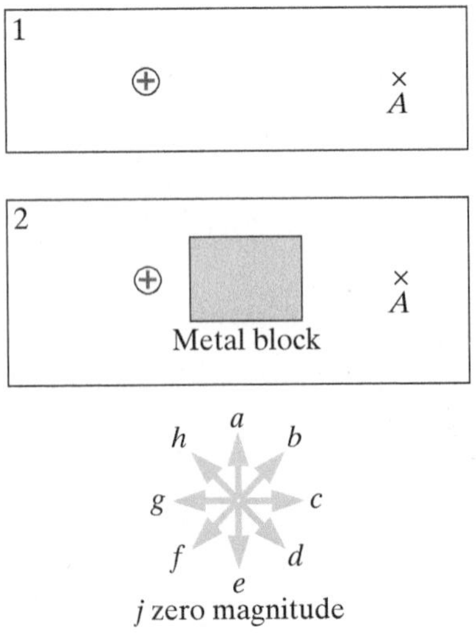

Figure 14.94

•P54 **(a)** The positively charged particle shown in diagram 1 in Figure 14.94 creates an electric field $\vec{E}_p$ at location A. Which of the arrows (*a–j*) in Figure 14.94 best indicates the direction of $\vec{E}_p$ at location A? **(b)** Now a block of metal is placed in the location shown in diagram 2 in Figure 14.94. Which of the arrows (*a–j*) in Figure 14.94 best indicates the direction of the electric field $\vec{E}_m$ at location A due only to the charges in and/or on the metal block? **(c)** $|\vec{E}_p|$ is greater than $|\vec{E}_m|$. With the metal block still in place, which of the arrows (*a–j*) in Figure 14.94 best indicates the direction of the net electric field at location A? **(d)** With the metal block still in place, which of the following statements about the magnitude of $\vec{E}_p$, the field due only to the charged particle, is correct? (1) $|\vec{E}_p|$ is less than it was originally, because the block is in the way. (2) $|\vec{E}_p|$ is the same as it was originally, without the block. (3) $|\vec{E}_p|$ is zero, because the electric field due to the particle can't go through the block. **(e)** With the metal block still in place, how does the magnitude of $\vec{E}_{net}$ at location A compare to the magnitude of $\vec{E}_p$? **(f)** Which of the arrows (*a–j*) in Figure 14.94 best indicates the direction of the net electric field at the center of the metal block (inside the metal)?

•P55 In a particular metal, the mobility of the mobile electrons is 0.0077 (m/s)/(N/C). At a particular moment the net electric field everywhere inside a cube of this metal is 0.053 N/C in the $+x$ direction. What is the average drift speed of the mobile electrons in the metal at this instant?

••P56 A neutral solid metal sphere of radius 0.1 m is at the origin, polarized by a point charge of 6×10^{-8} C at location $\langle -0.3, 0, 0 \rangle$ m. At location $\langle 0, 0.07, 0 \rangle$ m, what is the electric field contributed by the polarization charges on the surface of the metal sphere? How do you know?

••P57 A point charge of 3×10^{-9} C is located at the origin. **(a)** What is the magnitude of the electric field at location $\langle 0.2, 0, 0 \rangle$ m? **(b)** Next, a short, straight, thin copper wire 3 mm long is placed along the x axis with its center at location $\langle 0.1, 0, 0 \rangle$ m. What is the approximate change in the magnitude of the electric field at location $\langle 0.2, 0, 0 \rangle$ m? **(c)** Does the magnitude of the electric field at location $\langle 0.2, 0, 0 \rangle$ m increase or decrease as a result of placing the copper wire between this location and the point charge? **(d)** Does the copper metal block the electric field contributed by the point charge?

••P58 A metal ball with diameter of a half a centimeter and hanging from an insulating thread is charged up with 1×10^{10} excess electrons. An initially uncharged identical metal ball hanging from an insulating thread is brought in contact with the first ball, then moved away, and they hang so that the distance *between their centers* is 20 cm. **(a)** Calculate the electric force one ball exerts on the other, and state whether it is attractive or repulsive. If you have to make any simplifying assumptions, state them explicitly and justify them. **(b)** Now the balls are moved so that as they hang, the distance *between their centers* is only 5 cm. Naively one would expect the force that one ball exerts on the other to increase by a factor of $4^2 = 16$, but in real life the increase is a bit less than a factor of 16. Explain why, including a diagram. (Nothing but the distance between centers is changed—the charge on each ball is unchanged, and no other objects are around.)

••P59 A thin, hollow spherical plastic shell of radius R carries a uniformly distributed negative charge $-Q$. A slice through the plastic shell is shown in Figure 14.95.

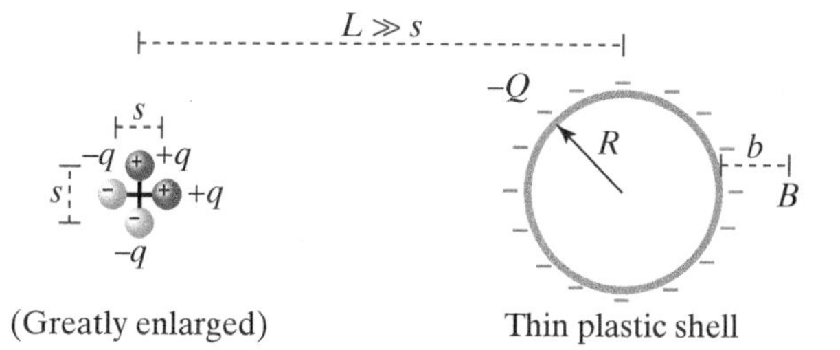

Figure 14.95

To the left of the spherical shell are four charges packed closely together as shown (the distance s is shown greatly enlarged for clarity). The distance from the center of the four charges to the center of the plastic shell is L, which is much larger than s ($L \gg s$). Remember that a uniformly charged sphere makes an electric field as though all the charge were concentrated at the center of the sphere. **(a)** Calculate the x and y components of the electric field at location B, a distance b to the right of the outer surface of the plastic shell. Explain briefly, including showing the electric field on a diagram. Your results should not contain any symbols other than the given quantities R, Q, q, s, L, and b (and fundamental constants). You need not simplify the final algebraic results except for taking into account the fact that $L \gg s$. **(b)** What simplifying assumption did you have to make in part (a)? **(c)** The plastic shell is removed and replaced by an uncharged metal ball, as in Figure 14.96.

At location A inside the metal ball, a distance b to the left of the outer surface of the ball, accurately draw and label the electric field $\vec{E}_{\text{ball}}$ due to the ball charges and the electric field $\vec{E}_4$ of the four charges. Explain briefly.

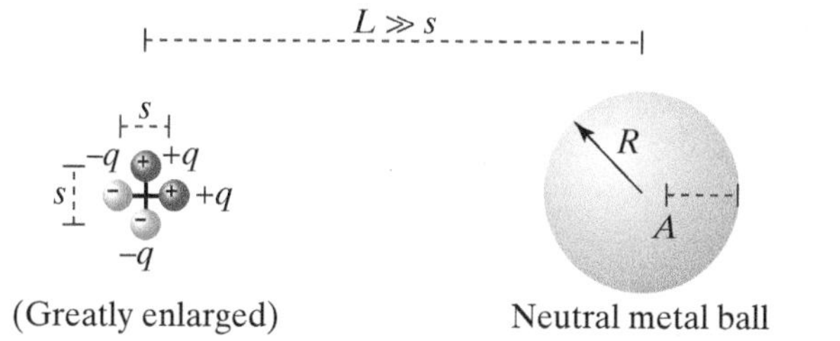

Figure 14.96

(d) Show the distribution of ball charges. **(e)** Calculate the x and y components of the net electric field at location A.

••**P60** A very thin spherical plastic shell of radius 15 cm carries a uniformly distributed negative charge of −8 nC (-8×10^{-9} C) on its outer surface (so it makes an electric field as though all the charge were concentrated at the center of the sphere). An uncharged solid metal block is placed nearby. The block is 10 cm thick, and it is 10 cm away from the surface of the sphere. See Figure 14.97. **(a)** Sketch the approximate charge distribution of the neutral solid metal block.

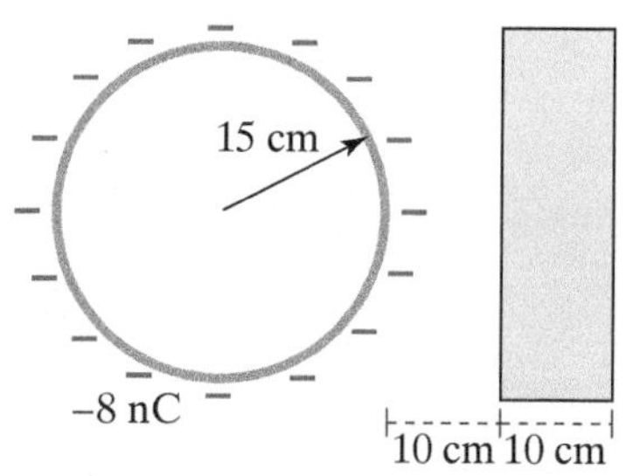

Figure 14.97

(b) Draw the electric field vector at the center of the metal block that is due solely to the charge distribution you sketched (that is, excluding the contributions of the sphere). **(c)** Calculate the magnitude of the electric field vector you drew. Explain briefly. If you must make any approximations, state what they are.

Section 14.7

•**P61** You run your finger along the slick side of a positively charged tape, and then observe that the tape is no longer attracted to your hand. Which of the following are not plausible explanations for this observation? Check all that apply. (1) Sodium ions (Na^+) from the salt water on your skin move onto the tape, leaving the tape with a zero (or very small) net charge. (2) Electrons from the mobile electron sea in your hand move onto the tape, leaving the tape with a zero (or very small) net charge. (3) Chloride ions (Cl^-) from the salt water on your skin move onto the tape, leaving the tape with a zero (or very small) net charge. (4) Protons are pulled out of the nuclei of atoms in the tape and move onto your finger.

•**P62** You observe that a negatively charged plastic pen repels a charged piece of invisible tape. You then observe that the same piece of tape is repelled when brought near a metal sphere. You are wearing rubber-soled shoes, and you touch the metal sphere with your hand. After you touch the metal sphere, you observe that the tape is attracted to the metal sphere. Which of the following statements could be true? Check all that apply. (1) Electrons from the sphere traveled through your body into the Earth. (2) Electrons from the sphere moved into the salt water on your skin, where they reacted with sodium ions. (3) After you touched it, the metal sphere was very nearly neutral. (4) Chloride ions from the salt water on your hand moved onto the sphere. (5) The excess negative charge from the sphere spread out all over your body. (6) Electrons from your hand moved onto the sphere. (7) Sodium ions from the salt water on your hand moved onto the sphere.

•**P63** Blocks A and B are identical metal blocks. Initially block A is neutral, and block B has a net charge of 5 nC. Using insulating handles, the blocks are moved so they touch each other. After touching for a few seconds, the blocks are separated (again using insulating handles). **(a)** What is the final charge of block A? **(b)** What happened while the blocks were in contact with each other? (1) Protons moved from block B to block A. (2) Positrons moved from block B to block A. (3) Electrons moved from block A to block B. (4) Both protons and electrons moved. (5) No charged particles moved.

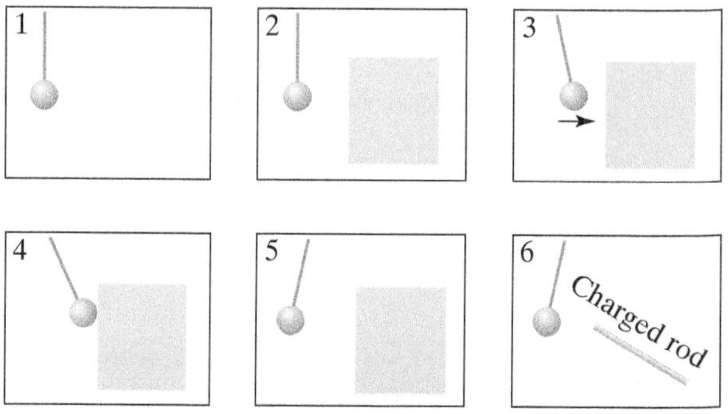

Figure 14.98

••**P64** The diagrams in Figure 14.98 show a sequence of events involving a small lightweight aluminum ball that is suspended from a cotton thread. In order to get enough information, you will need to read through the entire sequence of events described below before beginning to answer the questions. Before trying to select answers, you will need to draw your own diagrams showing the charge state of each object in each situation. **(a)** A small, lightweight aluminum ball hangs from a cotton thread. You touch the ball briefly with your fingers, then release it (Diagram 1 in Figure 14.98). Which of the diagrams in Figure 14.99 best shows the distribution of charge in and/or on the ball at this moment, using the diagrammatic conventions discussed in this chapter? **(b)** A block of metal that is known to be charged is now moved near the ball (Diagram 2 in Figure 14.98). The ball starts to swing toward the block of metal, as shown in Diagram 3 in Figure 14.98. Remember to read through the whole sequence before answering this question: Which of the diagrams in Figure 14.99 best shows the distribution of charge in and/or on the ball at this moment? **(c)** The ball briefly touches the charged metal block (Diagram 4 in Figure 14.98). Then the ball swings away from the block and hangs motionless at an angle, as shown in Diagram 5 in Figure 14.98. Which of the diagrams in Figure 14.99 best shows the distribution of charge in and/or on the ball at this moment? **(d)** Finally, the block is moved far away. A negatively charged rod is brought near the ball. The ball is repelled by the charged rod, as shown in Diagram 6 in Figure 14.98. Which of the diagrams in Figure 14.99 best shows the distribution of charge in and/or on the ball at this moment?

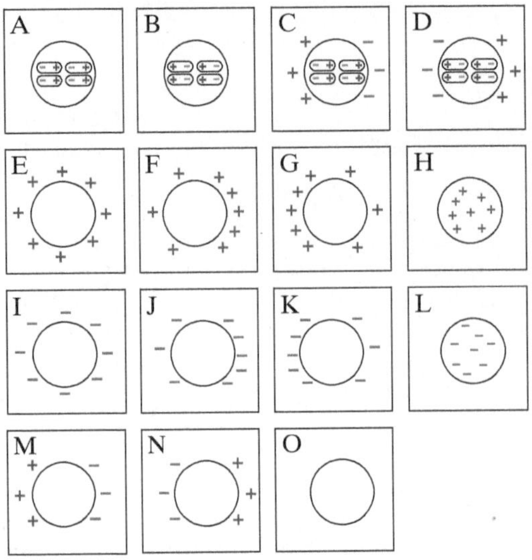

Figure 14.99

ANSWERS TO CHECKPOINTS

1 0; $-e = -1.6 \times 10^{-19}$ C

2 The student has forgotten to consider the superposition principle. Electric interactions go right through matter, so the effect of the positive nucleus is not blocked by the surrounding electron cloud. There are exactly as many protons in the nucleus as there are electrons, and normally the electron cloud is centered on the nucleus, so the net effect is zero.

3 About 2×10^{-15} m, about the diameter of a proton!

4 $2^{-5} = 1/32$

5 Negative tape is attracted to finger:

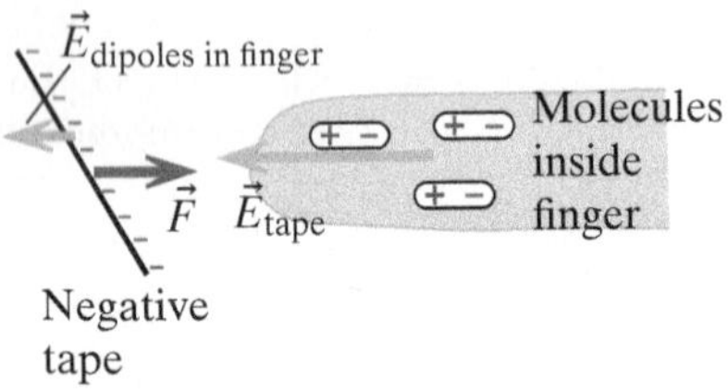

Figure 14.100

6 1.5×10^{-5} m/s

7 0.22 N/C

8 (a) Note shift of charge distribution; it is no longer uniform:

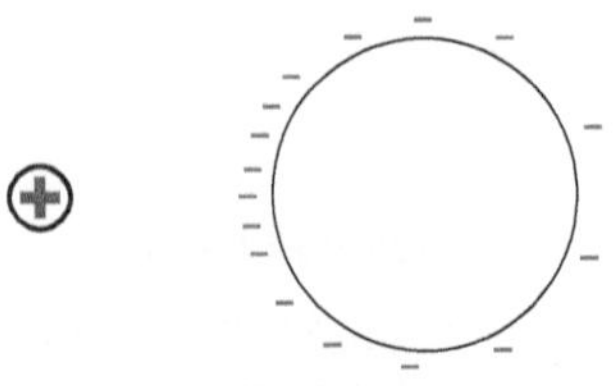

Figure 14.101

The net field is zero:

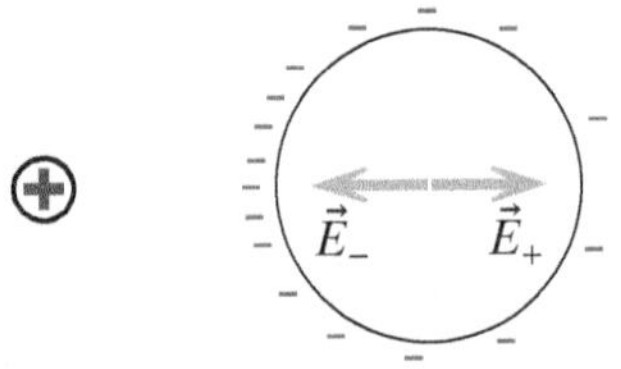

Figure 14.102

(b) Negative pen polarizes the neutral metal cylinder by shifting the electron sea; + charges are closer than − charges, so the pen exerts a net attraction on the cylinder.

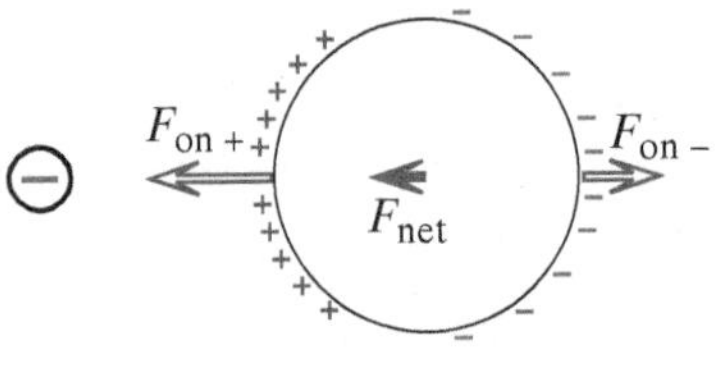

Figure 14.103

The net field is zero:

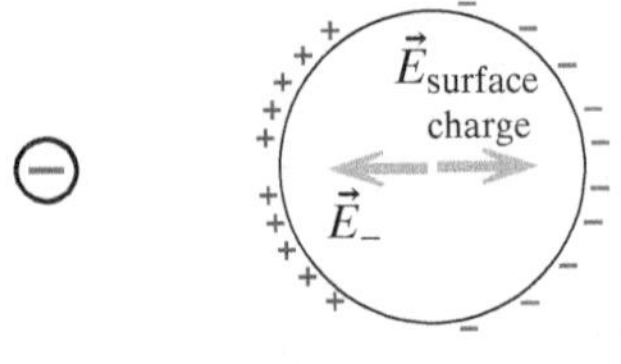

Figure 14.104

(c) Negative pen polarizes the neutral plastic cylinder by polarizing the molecules; + charges are closer than − charges, so the pen exerts a net attraction on the cylinder.

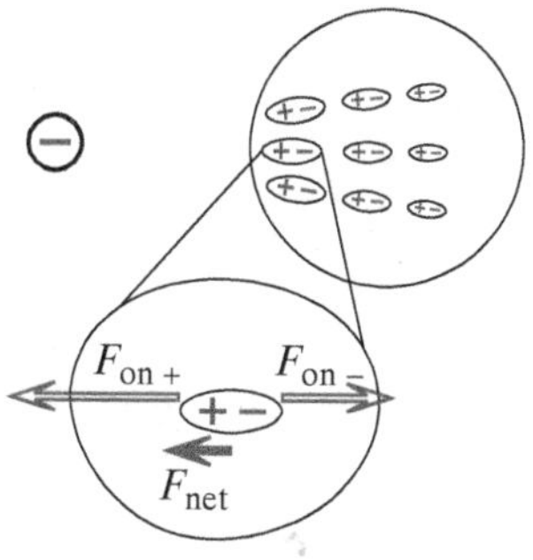

Figure 14.105

Hints for Problem P40, Polarizability of a carbon atom What must be the force on a single carbon atom in the paper at the moment the paper is lifted by the pen? You know how to calculate the force on a point charge due to a dipole. How does this relate to the force on the dipole by the point charge? In this problem, is there something you can model as a dipole and something else you can model as a point charge? Note that the dipole moment ($p = qs$) of a polarized atom or molecule is directly proportional to the applied electric field. In this case the charged pen is generating the applied electric field.

CHAPTER

15

Electric Field of Distributed Charges

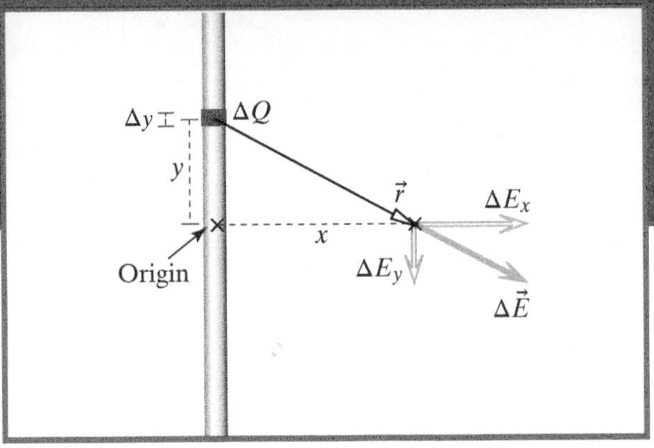

OBJECTIVES

After studying this chapter you should be able to

- Set up an integral to calculate the electric field at any location due to a continuous charge distribution.
- Correctly use approximate analytical expressions for the electric field of various charge distributions.
- Numerically (computationally) calculate the electric field at a single location due to a continuous charge distribution.
- Numerically (computationally) calculate the electric field at many observation locations due to a charge distribution.

In many real-world situations, charge is spread out over the surface of a macroscopic object. This chapter focuses on mathematical techniques for adding up the contributions to the electric field of very large numbers of point charges distributed over large areas. The most general technique is to divide the charge distribution into a large but finite number of pieces, approximate each piece by a point charge, and use a computer to add up the contributions ("numerical integration"). This chapter introduces an approach to setting up and carrying out such a computation. Such a calculation can be done by hand, but it is most easily done using a computer. Section 15.9 introduces a new computational concept that greatly facilitates the calculation and display of electric fields in 3D.

In a few special but important cases we can get an analytical solution by using an integral to add up the contributions. We are able to do this for some locations near a charged rod, ring, disk, capacitor, and sphere. A major advantage of the analytical approach is that the result allows us to see how the field varies with distance from the charge distribution, which is important in many applications. These results turn out to be useful because it is often possible to model ordinary objects as combinations of spheres, rods, rings, and disks, and thereby to estimate their electric fields.

Whether we solve a problem analytically or numerically, a key step in the process is coming up with an algebraic description of the location of an object or part of an object in 3D space. Drawing and labeling diagrams can save time and minimize frustration in this process.

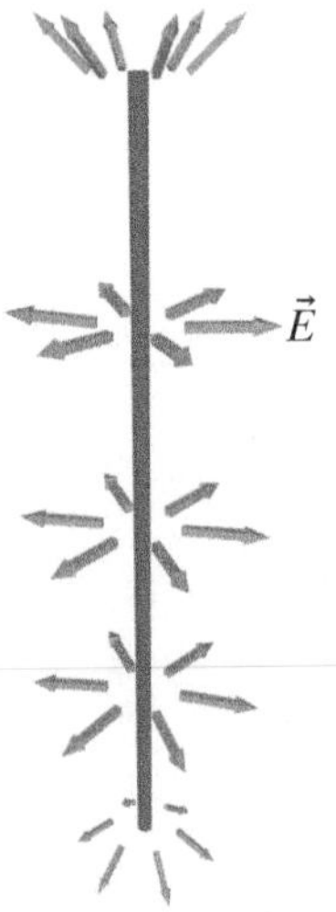

Figure 15.1 The electric field at locations near a positively charged rod. Near the middle of the rod the electric field vectors lie in a plane perpendicular to the rod.

15.1 A UNIFORMLY CHARGED THIN ROD

As an example of how to calculate the electric field due to large numbers of charges, we'll consider the electric field of a uniformly charged thin rod (Figure 15.1). The rod might, for example, be a glass rod that was rubbed all over with silk, giving it a nearly uniform distribution of positive charge on its surface. We'll consider a thin rod of length L and total positive charge Q. To illustrate the process, we will pick an observation location on the midplane of the rod (the plane perpendicular to the rod, passing through the midpoint of the rod); this simplifies the algebra we will need to do.

> QUESTION Before launching into a calculation, think about the pattern of electric field you would expect to observe around a rod. Viewed end-on, what would it look like?

The field of a point charge was spherically symmetric, since a point charge itself has spherical symmetry. Since a rod is shaped like a cylinder, we should expect cylindrical symmetry in the electric field of a uniformly charged rod. Near a positively charged rod, the electric field might look like Figure 15.1. If we took a slice perpendicular to the rod, near the middle of the rod, the pattern of field might look like Figure 15.2.

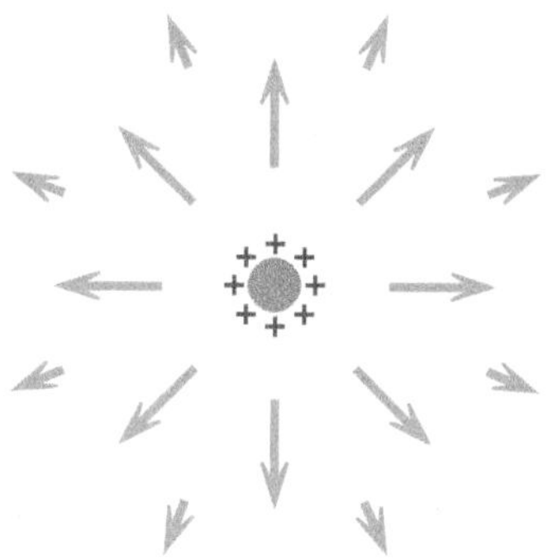

Figure 15.2 The electric field near a charged rod should be cylindrically symmetric. This is a representation of the field in a plane perpendicular to the rod, near the middle of the rod.

The process of finding the electric field due to charge distributed over a macroscopic object has four steps:

1. Divide the charged object into small pieces. Make a diagram and draw the electric field $\Delta\vec{E}$ contributed by one of the pieces.
2. Choose an origin and axes. Write an algebraic expression for the electric field $\Delta\vec{E}$ due to one piece.
3. Add up the contributions of all pieces, either numerically or symbolically.
4. Check that the result is physically correct.

We will carry out this process to find the electric field of a uniformly charged thin rod, at a location on the midplane.

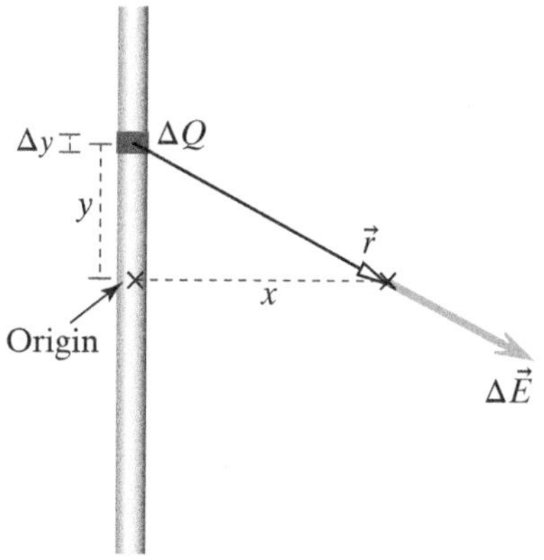

Figure 15.3 $\Delta\vec{E}$ is the contribution to the total field at location $\langle x,0,0\rangle$ made by a small piece of the rod of length Δy.

Step 1: Divide the Distribution into Pieces; Draw $\Delta\vec{E}$

To apply the superposition principle, we imagine cutting up the thin rod into very short sections each with positive charge ΔQ, as shown in Figure 15.3. The Greek capital letter delta (Δ) denotes a small portion of something or a change in something. Here ΔQ is a small portion of the total charge Q of the rod, which contributes $\Delta\vec{E}$ to the net field at the observation location. We can treat the piece ΔQ as though it were a point charge, which should be a fairly good approximation as long as its size is small compared to the distance to the observation location.

In Figure 15.3 we have picked a representative piece of the rod that is not at a "special" location (in this case, neither at an end nor at the middle), and drawn its contribution $\Delta\vec{E}$ to the net field at our chosen observation location.

Assumptions

We assume that the rod is so thin that we can ignore the thickness of the rod. We choose the piece of charge ΔQ to be small enough that it can be modeled as a point charge.

Step 2: Write an Expression for the Electric Field Due to One Piece

We will approximate each piece of the rod as a point charge. Therefore the expression for $\Delta\vec{E}$ will have the familiar form of the equation for the electric field of a point charge. The challenge is to write this expression in terms of our chosen origin and axes, and to figure out a general expression for the amount of charge on each piece.

In order to write an algebraic expression for the contribution to the electric field of one representative piece of the rod, we need to pick an origin and axes. It is possible to put the origin anywhere, but some choices make the algebra easier than others. Here we put the origin at the center of the rod; the x axis extends to the right, and the y axis extends up.

The location of one piece of the rod (shown in Figure 15.4) depends on y.

QUESTION What variables should remain in our answer?

The coordinates of the observation location should remain, but we will sum (integrate) over all pieces of the rod, so the coordinates of the piece of the rod should not remain. However, in the expression we want to integrate it is okay to have variables representing the location of the rod segment; they are called "integration variables" and will disappear after we do the sum.

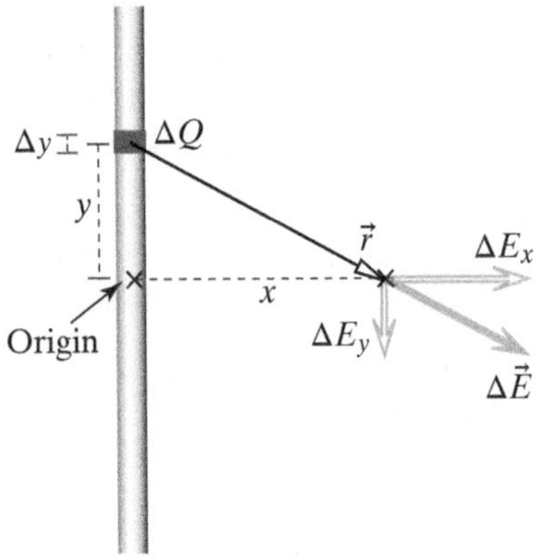

Figure 15.4 The components of $\Delta \vec{E}$ may be calculated by multiplying the scalar quantity by the components of $\hat{r}$.

A piece of the rod is a distance y from the origin. The length of the piece is Δy, a small increment in the integration variable y, shown in Figure 15.4. Our invented variable y will not appear in the final result. It is important to understand that if there is no Δ *something* in the expression (such as Δy), we can't do a sum or evaluate an integral.

The vector $\vec{r}$ points from the source (the representative little piece of charge ΔQ) to the observation location. We can read the components of $\vec{r}$ from the diagram in Figure 15.4:

$$\vec{r} = \langle \text{obs. loc.} \rangle - \langle \text{source} \rangle = \langle x,0,0 \rangle - \langle 0,y,0 \rangle$$
$$= \langle x,-y,0 \rangle$$

Note that $-y$, the y component of $\vec{r}$, is negative (as it should be).

$$|\vec{r}| = [x^2 + (-y)^2]^{1/2}$$
$$\hat{r} = \frac{\vec{r}}{|\vec{r}|} = \frac{\langle x,-y,0 \rangle}{[x^2 + (-y)^2]^{1/2}}$$

The scalar part of the field, due to this piece, is:

$$\frac{1}{4\pi\varepsilon_0}\frac{\Delta Q}{r^2} = \frac{1}{4\pi\varepsilon_0}\frac{\Delta Q}{[x^2 + (-y)^2]}$$

We can now write the vector $\Delta \vec{E}$ (Figure 15.3):

$$\Delta \vec{E} = \frac{1}{4\pi\varepsilon_0}\frac{\Delta Q}{[x^2 + (-y)^2]}\frac{\langle x,-y,0 \rangle}{[x^2 + (-y)^2]^{1/2}}$$
$$\Delta \vec{E} = \frac{1}{4\pi\varepsilon_0}\frac{\Delta Q}{(x^2 + y^2)^{3/2}}\langle x,-y,0 \rangle$$

The components of $\Delta \vec{E}$ (Figure 15.4) are then:

$$\Delta E_x = \frac{1}{4\pi\varepsilon_0}\frac{x\Delta Q}{(x^2 + y^2)^{3/2}}$$
$$\Delta E_y = \frac{1}{4\pi\varepsilon_0}\frac{-y\Delta Q}{(x^2 + y^2)^{3/2}}$$
$$\Delta E_z = 0$$

ΔQ and the Integration Variable

QUESTION Which things are constants and which things are variables in the expressions for ΔE_x and ΔE_y?

You should have identified y as a variable, since it differs for each small piece. We need to rewrite the expression in a form in which it is easy to add up all the ΔE_x's and ΔE_y's due to all the pieces of the rod. This means expressing everything in terms of one integration variable, related to the coordinates of the piece; in this case y is the integration variable.

In particular, we need to express the charge ΔQ in terms of the integration variable y. The rod is uniformly charged with a total charge Q (positive or negative), so the amount of charge on a section of length Δy is equal to

$$\Delta Q = \left(\frac{\Delta y}{L}\right) Q$$

since $\Delta y/L$ is the fraction of the whole rod represented by Δy. Alternatively, there is a linear charge density of Q/L in coulombs/meter, so on a length of Δy m there is an amount of charge $\Delta Q = (Q/L)\Delta y$. The important thing is to realize that ΔQ is a small fraction of the total charge Q, and that you must express ΔQ in terms of Δy.

Algebraic Expression for $\Delta\vec{E}$

Putting it all together, we have an expression for the x component:

$$\Delta E_x = \frac{1}{4\pi\varepsilon_0}\frac{Q}{L}\frac{x}{(x^2+y^2)^{3/2}}\Delta y$$

A similar approach yields an expression for the y component:

$$\Delta E_y = \frac{1}{4\pi\varepsilon_0}\frac{Q}{L}\frac{-y}{(x^2+y^2)^{3/2}}\Delta y$$

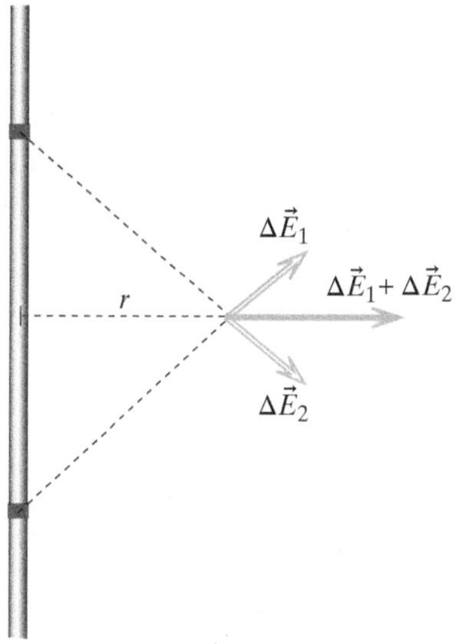

Figure 15.5 Along a line in the midplane only the x component of the electric field is nonzero, because the y contributions cancel.

A Note on the Symbol Δ

We have used the Greek letter Δ (capital delta) in two ways in setting up this problem. ΔE or ΔQ refers to a small contribution to a total quantity. Δy refers to a change in the integration variable y, which determines the location of the piece of the rod currently under consideration. You will need to use the symbol Δ in both these ways when setting up problems like this one.

Step 3: Add Up the Contributions of All the Pieces

Before performing the actual summation, we should think about what we expect the answer to be. Since we have chosen an observation location on the midplane of the rod, we should expect the y component of the final result to be zero, as indicated in Figure 15.5.

x components:

Each ΔQ contributes ΔE_x to the net field. If we number each piece 1, 2, 3, and so on, we have

$$\Delta E_x = \Delta E_{x_1} + \Delta E_{x_2} + \Delta E_{x_3} + \cdots$$

It is standard practice to write such sums in a more compact form using a Σ (the Greek capital sigma stands for "Summation"):

$$E_x = \sum \Delta E_x = \sum \frac{1}{4\pi\varepsilon_0}\frac{Q}{L}\frac{x}{(x^2+y^2)^{3/2}}\Delta y$$

y components:

$$E_y = \sum \Delta E_y = \sum \frac{1}{4\pi\varepsilon_0}\frac{Q}{L}\frac{-y}{(x^2+y^2)^{3/2}}\Delta y$$

At this point we have to decide how to add up these terms. Each term in this summation is different. Although we can choose to make all the pieces the same length Δy and take Δy out of the sum, y itself is different for every piece along the rod. How can we add up all the contributions of all the pieces?

Numerical Summation

One way to add up all these different contributions would be to divide the rod into 10 slices as shown in Figure 15.6, calculate the contribution of each of those 10 slices (using our equation with $\Delta y = L/10$), and add up these 10 numbers. Of course that would be only an approximate result, because each slice isn't really a point particle, but it might be good enough for many purposes. Let's do this calculation for the case of $L = 1$ m, $\Delta y = 0.1$ m, $Q = 1$ nC, and $x = 0.05$ m. For each slice we'll take y to be at the center of the slice, and we'll use a calculator to evaluate each term in the summation. The results are shown in Figure 15.7, including the sum for all pieces. As expected, we see that the sum of the y components is zero.

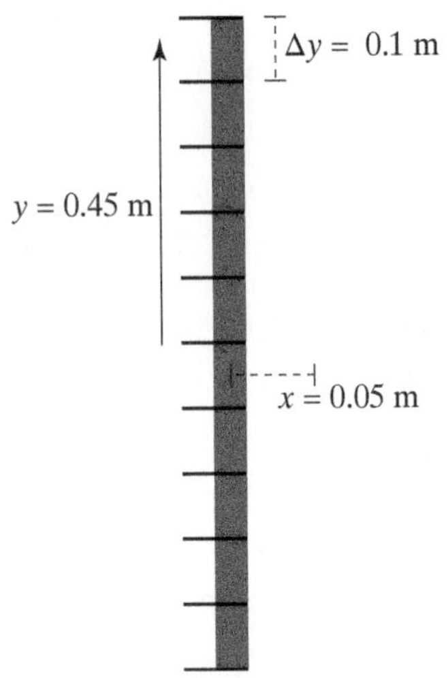

Figure 15.6 Division of the rod into 10 equal slices.

Accuracy

It would be possible to get a more accurate value by cutting the rod into more slices, but it would be tedious to do the calculations by hand. We wrote a little computer program to do these calculations for various numbers of slices, and the results of these calculations are shown in Figure 15.8.

Judging from the computations summarized in Figure 15.8, our calculation with just 10 slices was not very accurate (though it might be good enough for some purposes), while taking more than 50 slices makes almost no difference. The accuracy depends on how good an approximation it is to consider one slice as a point charge. With 50 slices, each slice Δy is 1/50th of a meter long (0.02 m), and the nearest slice is 0.05 m away from the observation location. Apparently the 0.02-m slices are adequately approximated by point charges at their centers, since using smaller 0.01-m slices (100 slices) gives practically the same result.

How do we know that our computer summations are correct? For 10 slices they agree with the calculator results. For larger and larger numbers of slices the computer results approach a constant value, which is expected. Later we will discuss additional ways to check such work.

y	E_x	E_y
+0.45	0.484	−4.363
+0.35	1.018	−7.127
+0.25	2.715	−13.577
+0.15	11.384	−34.153
+0.05	127.279	−127.279
−0.05	127.279	127.279
−0.15	11.384	34.153
−0.25	2.715	13.577
−0.35	1.018	7.127
−0.45	0.484	4.363
Sum	285.764	0.000

Figure 15.7 Contributions to $\vec{E}$.

Number of Slices	E_x
10	285.764
20	353.764
50	358.214
100	358.214

Figure 15.8 Effect of number of slices.

Summation as an Integral

A major disadvantage of adding up the contributions numerically is that we don't get an analytical (algebraic) form for the electric field. This means that we can't easily answer such questions as, "How does the electric field vary with r near a rod?" For a point charge we know that the field goes like $1/r^2$, and far from a permanent dipole it goes like $1/r^3$. Is there a way to do the summation to get an algebraic rather than a numerical answer for the rod? That is what integral calculus was invented to do.

The key idea of integral calculus applied to problems like ours is to imagine taking not 50 or 100 slices but an infinite number of infinitesimal slices. We let the number of slices $N = L/(\Delta y)$ increase without bound, and the corresponding slice length $\Delta y = L/N$ decreases without bound. We take the limit as Δy gets arbitrarily small:

$$E_x = \lim_{\Delta y \to 0} \frac{1}{4\pi\varepsilon_0} \frac{Q}{L} x \sum \frac{1}{(x^2 + y^2)^{3/2}} \Delta y$$

This limit is called a "definite integral" and is written like this:

$$E_x = \frac{1}{4\pi\varepsilon_0} \frac{Q}{L} x \int_{-L/2}^{+L/2} \frac{1}{(x^2 + y^2)^{3/2}} dy$$

The integral sign $\int$ is a distorted S standing for Summation, just as the Greek sigma Σ stands for summation. It is important that in this context you think of an integral as a sum of many contributions.

The integration variable y ranges from $y = -L/2$ (the bottom of the rod) to $y = +L/2$ (the top of the rod), so these are the limits on the definite integral. If you have drawn an appropriately labeled diagram, you should be able to read the correct limits of the integration variable off the diagram. This is another of the benefits of drawing a well-labeled physics diagram.

Our small length Δy has now changed into dy, an "infinitesimal" increment in y that *must* appear in the integrand. This is why Δy *must* appear in the algebraic expression for the contribution of one piece of the charge distribution.

Evaluating the Integral

Most of the physics in this problem went into setting up the integral. Evaluating the integral is simply mathematics. In some cases it is easy to evaluate the integral. In this particular case, the integral is not a very simple one, but if you would like to exercise your skill at integration, give it a try! Otherwise, it can be found in tables of integrals in mathematical handbooks and in some calculus textbooks. Some calculators can do such integrals, or one can use a tool such as Maple or Mathematica. Looking up the result in a table of integrals, we get the following:

$$E_x = \frac{1}{4\pi\varepsilon_0}\left(\frac{Q}{L}x\right)\left[\frac{y}{x^2\sqrt{x^2+y^2}}\right]_{-L/2}^{+L/2}$$

$$E_x = \frac{1}{4\pi\varepsilon_0}\left[\frac{Q}{x\sqrt{x^2+(L/2)^2}}\right]$$

Note that as we expected this result does not contain the integration variable y, which was simply a variable referring to the coordinates of one piece of the rod and was necessary in setting up the summation.

The y component of the electric field can also be found by integration. In this situation the result comes out to zero, as expected.

$$E_y = \frac{1}{4\pi\varepsilon_0}\frac{Q}{L}\int_{-L/2}^{+L/2}\frac{-y}{(x^2+y^2)^{3/2}}dy$$

$$E_y = 0$$

Replace x with r

Because the rod, and its associated electric field, are cylindrically symmetric, the axis we called the x axis could have been rotated around the rod by any angle, and we would have obtained the same answer. To indicate this, we replace x with r in our result (you may recognize this as converting to cylindrical coordinates):

ELECTRIC FIELD OF A UNIFORMLY CHARGED THIN ROD

$$E = \frac{1}{4\pi\varepsilon_0}\left[\frac{Q}{r\sqrt{r^2+(L/2)^2}}\right]$$

at a location a distance r from the midpoint of the rod along a line perpendicular to the rod. Q is the total charge on the rod, and L is the length of the rod. The direction is either radially away from or toward the rod (depending on the sign of Q). See Figure 15.9.

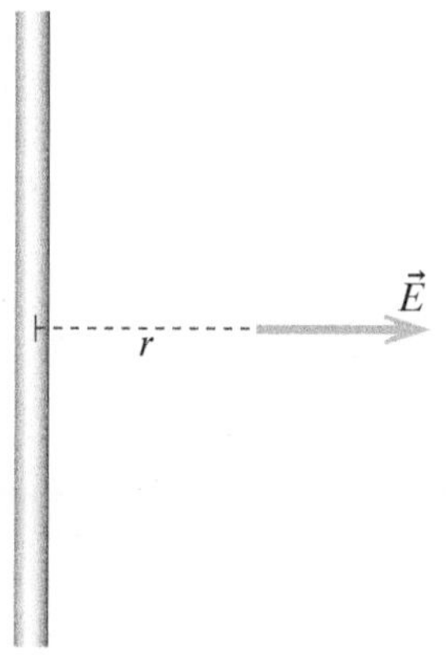

Figure 15.9 Electric field of a uniformly charged thin rod, near the midpoint of the rod.

Step 4: Check the Result

Because there are many opportunities to make mistakes in this procedure, it is extremely important to check the result in as many ways as you can. Different checks provide information about different kinds of possible errors.

Direction: First, is the direction qualitatively correct? We have the electric field pointing straight away from the midpoint of the rod, which is correct, given the symmetry of the situation. The vertical component of the electric field should indeed be zero.

Units: Second, do we have the right units? The units should be the same as the units of the expression for the electric field for a single point particle:

$$\frac{1}{4\pi\varepsilon_0}\frac{Q}{r^2}$$

We easily verify that our answer does have the right units, since

$$\frac{1}{r\sqrt{r^2+(L/2)^2}} \quad \text{has the same units as} \quad \frac{1}{r^2}$$

QUESTION Next, let's try a special case for which we already know the answer. If r is very much larger than L, the distant rod looks almost like a point charge, so the net field ought to look like the field of a point charge. Does it?

If $r \gg L$, then $\dfrac{1}{r\sqrt{r^2+(L/2)^2}} \approx \dfrac{1}{r\sqrt{r^2+0}} = \dfrac{1}{r^2}$.

Special case, $L \ll r$: For another special case, suppose that r is fairly near the rod, but the rod is so short as to look almost like a point charge. In that case $L \ll r$, and we again get $E \to \dfrac{1}{4\pi\varepsilon_0}\dfrac{Q}{r^2}$, as we should.

Compare to numerical calculation: Earlier we carried out a numerical calculation for the case of $Q = 1\times10^{-9}$ C, $L = 1$ m, and $r = 0.05$ m, and we found that if we used 50 or more slices the result was $E_x = 358.214$ N/C. The analytical solution gives

$$\begin{aligned} E &= \frac{1}{4\pi\varepsilon_0}\frac{Q}{r\sqrt{r^2+(L/2)^2}} \\ &= \frac{1}{4\pi\varepsilon_0}\left(\frac{1\times10^{-9}}{(0.05)\sqrt{0.05^2+0.5^2}}\right) \\ &= 358.213 \text{ N/C} \end{aligned}$$

which agrees very well with the numerically calculated value.

Checking a Numerical Solution

Similar techniques can be used to check a numerical integration done on a computer. For example, set the length of the rod to be very short or the distance to be very large, and the numerical integration should give a result equal to what you calculate by hand for a point charge.

Checkpoint 1 If the total charge on a thin rod of length 0.4 m is 2.5 nC, what is the magnitude of the electric field at a location 1 cm from the midpoint of the rod, perpendicular to the rod?

Special Case: A Very Long Rod

A very important special case, which we will refer to often, is the case of a rod that is very long (or alternatively a rod very close to the observation location). In either case $L \gg r$.

QUESTION Before reading further, try to show that if $L \gg r$, the electric field is approximately:

$$\frac{1}{4\pi\varepsilon_0}\frac{2(Q/L)}{r}$$

$$E = \frac{1}{4\pi\varepsilon_0}\left[\frac{Q}{r\sqrt{r^2+(L/2)^2}}\right]$$

$$\text{If } L \gg r, \text{ then } \frac{1}{r\sqrt{r^2+(L/2)^2}} \approx \frac{1}{r\sqrt{(L/2)^2}} = \frac{1}{r(L/2)}, \text{ so}$$

$$E \approx \frac{1}{4\pi\varepsilon_0}\frac{2(Q/L)}{r}$$

Figure 15.10 The electric field of a very long uniformly charged positive rod, shown at locations along a line very near the center of the rod. Only a small portion of the rod is visible.

This distance dependence is illustrated in Figure 15.10. This also holds for a short rod if we are very close, so that $L \gg r$.

APPROXIMATE ELECTRIC FIELD: VERY LONG ROD

$$E \approx \frac{1}{4\pi\varepsilon_0}\frac{2(Q/L)}{r} \quad \text{if } L \gg r$$

at a location a perpendicular distance $r \ll L$ from the rod, as long as the observation location is not too near the ends of the rod. Q is the total charge on the rod, and L is the length of the rod. The direction is either radially away from or toward the rod (depending on the sign of Q).

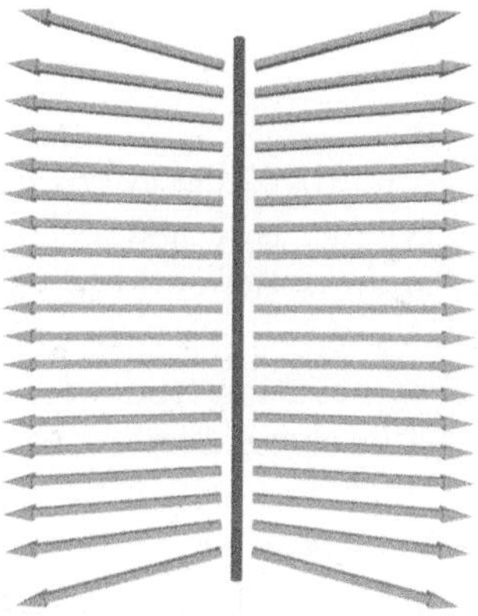

Figure 15.11 A numerical integration of the electric field near a uniformly positively charged rod.

Although we proved this result only for observation locations a perpendicular distance $r \ll L$ from the center of the rod, the result is actually a good approximation all along the rod as long as you're not too near the ends of the rod. Figure 15.11, which is the result of an accurate numerical integration, shows that $\vec{E}$ hardly varies along the central region of the rod. You can also see this in a three-dimensional numerical integration shown in Figure 15.12.

15.2 PROCEDURE FOR CALCULATING ELECTRIC FIELD

Here is a summary of the procedure for calculating electric field:

Step 1: Cut Up the Charge Distribution into Pieces and Draw $\Delta\vec{E}$

- Divide the charge distribution into pieces whose field is known. In particular, very small pieces can be approximated by point particles.
- Pick a representative piece, and at the location of interest draw a vector $\Delta\vec{E}$ showing the contribution to the electric field of this representative piece. Drawing this vector helps you figure out the direction of the net field at the location of interest.

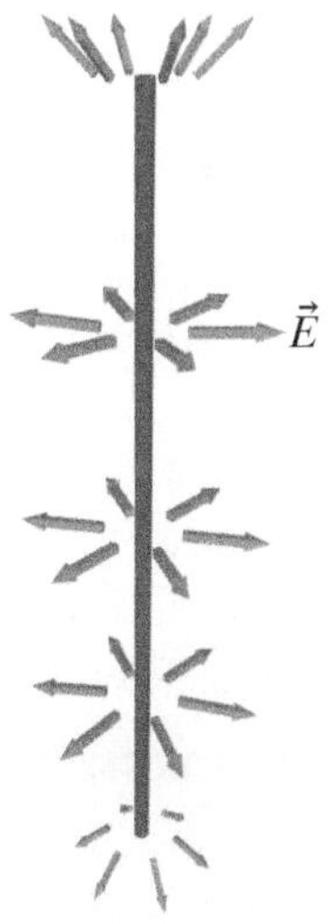

Figure 15.12 A numerical integration in three dimensions of the electric field of a uniformly positively charged rod.

Step 2: Write an Expression for the Electric Field Due to One Piece

- Pick an origin for your coordinate system, and show it on your diagram.
- Draw the vector $\vec{r}$ from the source piece to the observation location. Write algebraic expressions for $\vec{r}$ and $\hat{r}$ (a unit vector in the direction of $\vec{r}$).
- Write an algebraic expression for the magnitude $|\Delta\vec{E}|$ contributed by the representative piece. Multiply by $\hat{r}$ to get a vector $\Delta\vec{E}$, from which you can read the components ΔE_x, ΔE_y, and ΔE_z. Your expressions should contain one or more "integration variables" related to the coordinates of the piece.

- Write the amount of charge on the piece, Δq, in terms of your variables.
- If a representative piece is small in size, your algebraic expressions should include small increments of the integration variable. For example, if your integration variable is y, your expressions must be proportional to Δy.

Step 3: Add Up the Contributions of All the Pieces

- The net field is the sum of the contributions of all the pieces. To write the sum as a definite integral, you must include limits given by the range of the integration variable. If the integral can be done symbolically, do it. If not, choose a finite number of pieces and do the sum with a calculator or a computer.

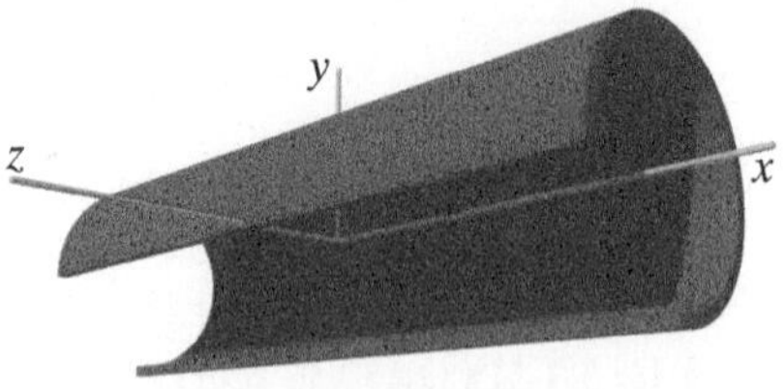

Figure 15.13 A charged hollow 3/4 cylinder. The observation location is at the center of the cylinder.

Step 4: Check the Result

- Check that the direction of the net field is qualitatively correct.
- Check the units of your result, which should be newtons per coulomb.
- Look at special cases. For example, if the net charge is nonzero, your result should reduce to the field of a point charge when you are very far away. For a numerical integration on a computer, check that the computation gives the correct numerical result for special cases that can be calculated by hand.

EXAMPLE **A Hollow Cylinder**

A hollow 3/4 cylinder (Figure 15.13) carries a positive charge Q spread uniformly over its surface. The radius of the cylinder is R, and its length is L, where $L \gg R$. Calculate the electric field at the center of the cylinder.

Solution

(1) Since we know the electric field near the center of a long rod, we could consider the cylinder to be made up of many thin charged rods, as shown in Figure 15.14. We will place the origin at the observation location, as shown. Each rod contributes $\Delta\vec{E}$ as shown in Figure 15.15.

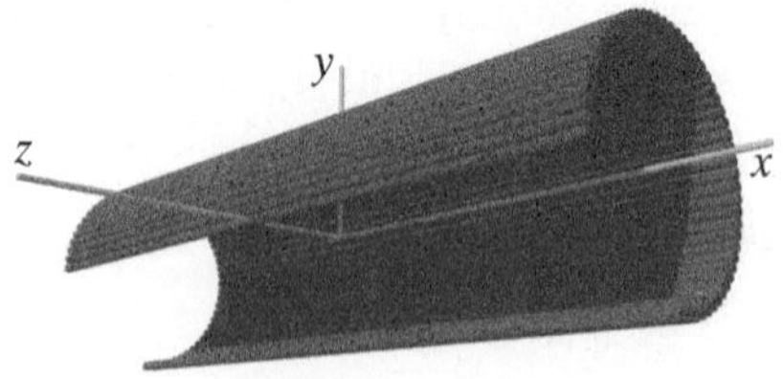

Figure 15.14 We can consider the cylinder to be composed of a large number of very thin charged rods.

(2) The location of the center of a particular rod can be described in terms of the angle θ, defined as shown in Figure 15.15.

$$\vec{r} = \langle \text{obs. location}\rangle - \langle\text{source}\rangle = \langle 0,0,0\rangle - \langle 0, R\sin\theta, R\cos\theta\rangle$$
$$= \langle 0, -R\sin\theta, -R\cos\theta\rangle$$
$$r = \sqrt{(-R\cos\theta)^2 + (-R\sin\theta)^2} = R$$
$$\hat{r} = \frac{\vec{r}}{r} = \langle 0, -\sin\theta, -\cos\theta\rangle$$

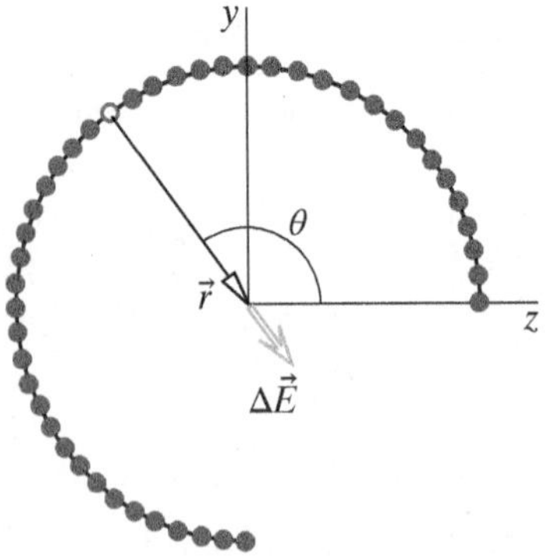

Figure 15.15 An end view of the cylinder, looking along the x axis in the $+x$ direction.

The amount of charge on one rod is given by:

$$dQ = Q\left(\frac{d\theta}{\theta_{\text{total}}}\right) = \frac{Q\,d\theta}{(3/2)\pi}$$

where $d\theta$ is the angular width of one rod, and $\theta_{\text{total}} = (3/2)\pi$ is the angular extent of the cylinder.

The contribution of this rod to the net electric field is:

$$\Delta\vec{E} = |\Delta\vec{E}|\hat{r} \approx \frac{1}{4\pi\varepsilon_0}\frac{2dQ}{L}\frac{1}{r}\langle 0, -\sin\theta, -\cos\theta\rangle$$
$$= \frac{1}{4\pi\varepsilon_0}\frac{2}{LR}\frac{Q\,d\theta}{(3/2)\pi}\langle 0, -\sin\theta, -\cos\theta\rangle$$

because in this case $r = R$; the center of every rod is the same distance R from the observation location.

(3) We need to consider the x, y, and z components of electric field separately. To add up the contributions of all of the rods, we can integrate:

$$E_x = \int_0^{\frac{3}{2}\pi} \frac{1}{4\pi\varepsilon_0}\frac{2Q}{LR(3/2)\pi} \cdot 0 d\theta = 0$$

$$E_y = \int_0^{\frac{3}{2}\pi} \frac{1}{4\pi\varepsilon_0}\frac{2Q}{LR(3/2)\pi}(-\sin\theta)d\theta = \frac{1}{4\pi\varepsilon_0}\frac{2Q}{LR(3/2)\pi}\cos\theta\Big|_0^{3\pi/2}$$
$$= -\frac{1}{4\pi\varepsilon_0}\frac{4Q}{3\pi LR}$$

$$E_z = \int_0^{\frac{3}{2}\pi} \frac{1}{4\pi\varepsilon_0}\frac{2Q}{LR(3/2)\pi}(-\cos\theta)d\theta = \frac{-1}{4\pi\varepsilon_0}\frac{2Q}{LR(3/2)\pi}\sin\theta\Big|_0^{3\pi/2}$$
$$= \frac{1}{4\pi\varepsilon_0}\frac{4Q}{3\pi LR}$$

so the net electric field is:

$$\vec{E} = \left\langle 0, -\frac{1}{4\pi\varepsilon_0}\frac{4Q}{3\pi LR}, \frac{1}{4\pi\varepsilon_0}\frac{4Q}{3\pi LR} \right\rangle$$

(4) The direction is reasonable, since the y component of $\vec{E}$ should be negative, and the z component should be positive.

Units are $(\text{N} \cdot \text{m}^2/\text{C}^2)(\text{C}/(\text{m} \cdot \text{m})) = \text{N/C}$, which is correct.

Special case: If this were a complete hollow cylinder, with $\theta_{\text{total}} = 2\pi$, the limits of integration would be 0 to 2π, and both E_y and E_z would be zero. The zero field makes sense, because there is just as much charge above the center as below, or to the $+z$ or $-z$ side of the center. Later when we introduce Gauss's law it will be possible to prove that inside a long uniformly charged cylindrical tube the contributions of all the charges cancel out, and everywhere inside the tube (not too near the ends) the electric field is zero, not just at the center.

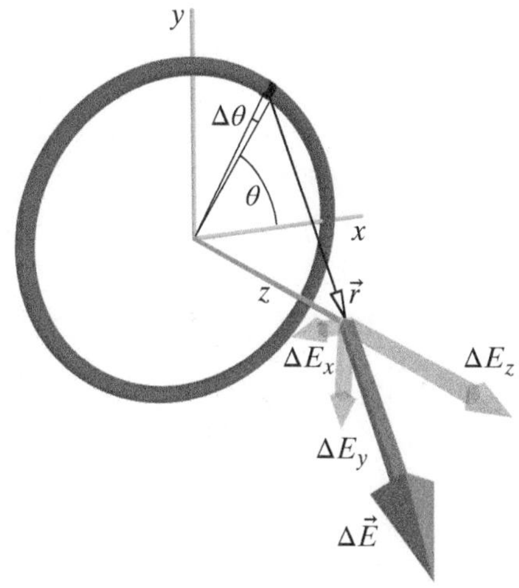

Figure 15.16 The contribution $\Delta\vec{E}$ to the total electric field of a charged ring, made by a segment of the ring of angular size $\Delta\theta$. We have chosen the center of the ring for the origin of our coordinate system. The x, y, and z components of $\Delta\vec{E}$ are indicated.

15.3 A UNIFORMLY CHARGED THIN RING

Next we'll calculate the electric field of a uniformly charged thin ring. We'll encounter rings of charge later when we study electric circuits. We'll also use the results for a ring to find the electric field of a disk later in this chapter.

Field of a Ring, at a Point on the Axis

We'll calculate the electric field due to a uniformly charged ring of radius R and total positive charge q. We'll do only the easiest case—the field at a location along the axis of the ring, which is a line going through the center and perpendicular to the ring. Finding the electric field at other locations is harder.

Step 1: Cut Up the Charge Distribution into Pieces and Draw $\Delta\vec{E}$

In Figure 15.16 a piece has angular length $\Delta\theta$.

Step 2: Write an Expression for the Electric Field Due to One Piece

Origin: Center of the ring. Axes shown in Figure 15.16.
Location of piece: Described by θ, where $\theta = 0$ is along the x axis.

Integration variable: θ

$$\vec{r} = \langle \text{obs. loc.} \rangle - \langle \text{source} \rangle = \langle 0,0,z \rangle - \langle R\cos\theta, R\sin\theta, 0 \rangle$$
$$= \langle -R\cos\theta, -R\sin\theta, z \rangle$$

$$|\vec{r}| = \sqrt{(-R\cos\theta)^2 + (-R\sin\theta)^2 + z^2} = (R^2+z^2)^{1/2}$$

$$\hat{r} = \frac{\vec{r}}{|\vec{r}|} = \frac{\langle -R\cos\theta, -R\sin\theta, z \rangle}{(R^2+z^2)^{1/2}}$$

$$\frac{1}{4\pi\varepsilon_0}\frac{\Delta q}{r^2} = \frac{1}{4\pi\varepsilon_0}\frac{\Delta q}{(R^2+z^2)}$$

$$\Delta q = q\left(\frac{\Delta\theta}{2\pi}\right) \quad \text{(There are } 2\pi \text{ radians in the complete ring.)}$$

$$\Delta\vec{E} = (\Delta E)\hat{r} = \left(\frac{1}{4\pi\varepsilon_0}\frac{q\left(\frac{\Delta\theta}{2\pi}\right)}{(R^2+z^2)}\right)\frac{\langle -R\cos\theta, -R\sin\theta, z \rangle}{(R^2+z^2)^{1/2}}$$

$$\Delta\vec{E} = \frac{1}{4\pi\varepsilon_0}\frac{q}{2\pi}\frac{\Delta\theta}{(R^2+z^2)^{3/2}}\langle -R\cos\theta, -R\sin\theta, z \rangle$$

Components: From the symmetry of the situation, we see that the x components will sum to zero, as will the y components. The z components will not cancel, so we need ΔE_z:

$$\Delta E_z = \frac{1}{4\pi\varepsilon_0}\frac{q}{2\pi}\frac{z}{(R^2+z^2)^{3/2}}\Delta\theta$$

$$dE_z = \frac{1}{4\pi\varepsilon_0}\frac{q}{2\pi}\frac{z}{(R^2+z^2)^{3/2}}d\theta$$

Step 3: Add Up the Contributions of All the Pieces

In this case we can evaluate the integral (sum) analytically:

$$E_z = \int_0^{2\pi} dE_z$$
$$= \int_0^{2\pi} \frac{1}{4\pi\varepsilon_0}\frac{q}{2\pi}\frac{z}{(R^2+z^2)^{3/2}}d\theta$$
$$= \frac{1}{4\pi\varepsilon_0}\frac{q}{2\pi}\frac{z}{(R^2+z^2)^{3/2}}\int_0^{2\pi} d\theta$$

ELECTRIC FIELD OF A UNIFORMLY CHARGED THIN RING

$$E = \frac{1}{4\pi\varepsilon_0}\frac{qz}{(R^2+z^2)^{3/2}}$$

along the axis, a distance z from the center of the ring, for a ring of radius R and charge q (Figure 15.17). The direction is parallel or antiparallel to the axis, depending on the sign of q.

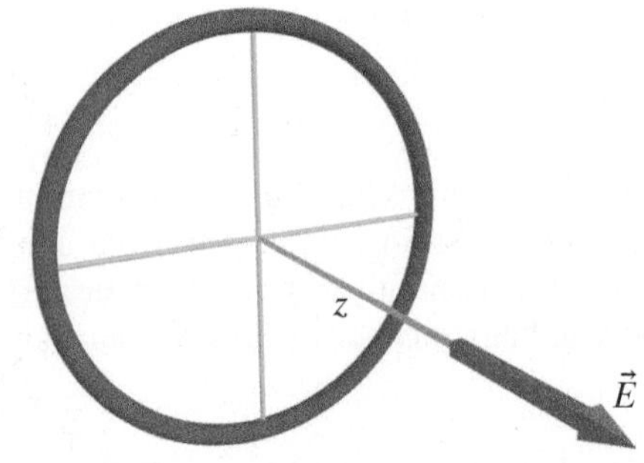

Figure 15.17 Electric field of a uniformly charged ring at a location on the z axis.

Step 4: Check the Result

Direction: Correct, by symmetry.

Units: $\left(\frac{\text{N}\cdot\text{m}^2}{\text{C}^2}\right)\left(\frac{\text{C}\cdot\text{m}}{(\text{m}^2)^{3/2}}\right) = \frac{\text{N}}{\text{C}}$, which is correct.

Special case: Exact center of the ring:

$$z = 0 \Rightarrow E = 0$$

This is correct, since all contributions to E will cancel at this location.

Special case: $z \gg R$:

$$(R^2 + z^2)^{3/2} \approx (z^2)^{3/2} = z^3, \text{ so } E \propto z/z^3 = 1/z^2$$

This is correct; at locations far from the ring, the ring should look like a point charge.

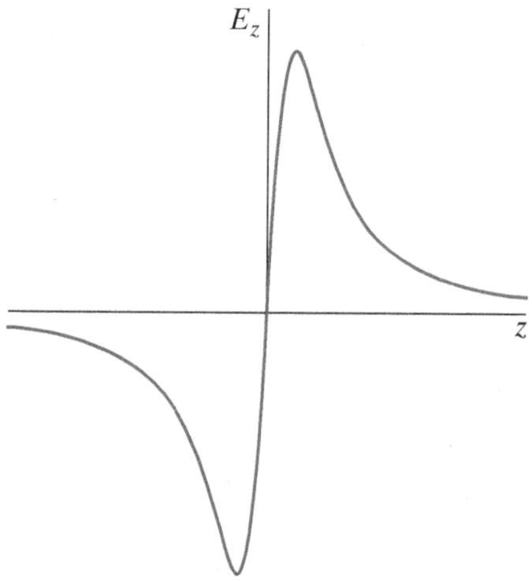

Figure 15.18 The z component of electric field of a uniformly charged ring along the z axis (perpendicular to the ring).

Distance Dependence of the Electric Field of a Ring

Since the field is zero at the center of the ring but falls off like $1/z^2$ far from the ring, the field must first increase, then decrease, with distance from the ring. With a graphing calculator you can easily show that a plot of E_z vs. z looks like Figure 15.18.

Note that we have calculated the field of a ring only along the axis perpendicular to the ring (Figure 15.19). As shown in Figure 15.20, the pattern of the field at other locations is much more complex.

Figure 15.20 is the result of a numerical integration, in which the ring was cut up into short segments and the summation was made at many observation locations. We calculated the field analytically along the z axis not just because this is one of the rare cases where an analytical solution is possible, but also because this result will be useful to us later, including in the next section.

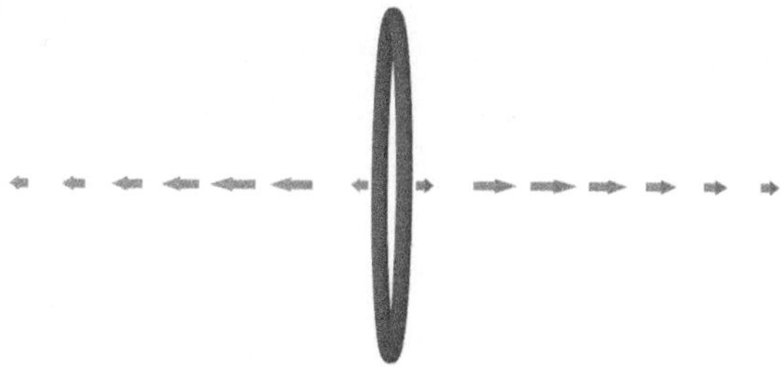

Figure 15.19 The electric field of a ring along the axis, varying with distance. As the graph in Figure 15.18 indicates, as the distance from the ring increases the magnitude of the field first increases, then decreases.

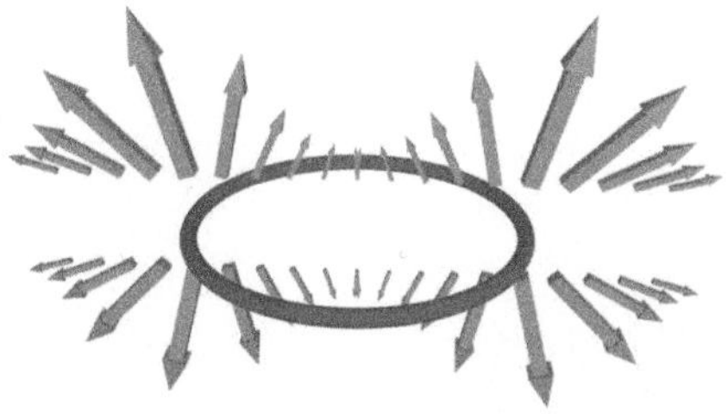

Figure 15.20 The electric field of a charged ring at locations across the ring, a short distance out of the plane of the ring.

Checkpoint 2 Two rings of radius 5 cm are 20 cm apart and concentric with a common horizontal axis. The ring on the left carries a uniformly distributed charge of +35 nC, and the ring on the right carries a uniformly distributed charge of −35 nC. **(a)** What are the magnitude and direction of the electric field on the axis, halfway between the two rings? **(b)** If a charge of −5 nC were placed midway between the rings, what would be the magnitude and direction of the force exerted on this charge by the rings? **(c)** What are the magnitude and direction of the electric field midway between the rings if both rings carry a charge of +35 nC?

15.4 A UNIFORMLY CHARGED DISK

One reason that a uniformly charged disk is important is that two oppositely charged metal disks can form a "capacitor," a device that is important in electric circuits. Before discussing metal disks, we will consider a glass disk that has been rubbed with silk in such a way as to deposit a uniform density of positive charge all over the surface.

Field Along the Axis of a Uniformly Charged Disk

The electric field of a uniformly charged disk of course varies in both magnitude and direction at observation locations near the disk, as illustrated in Figure 15.21, which shows the computed pattern of electric field at many locations near a uniformly charged disk (done by numerical integration, with the surface of the disk divided into small areas). Note, however, that near the center of the disk the field is quite uniform for a given distance from the disk. Even near the edge of the disk the magnitude of the electric field isn't very different, though the direction is no longer nearly perpendicular to the disk.

Again, we'll pick an easy location for an analytical solution—the field at any location along the axis of the disk, which is a line going through the center and perpendicular to the disk, as shown in Figure 15.22. This is a more useful choice than one might expect, since the field turns out to be nearly uniform in regions far from the edge of the disk as can be seen in Figure 15.21; our result will be applicable to a variety of situations.

We consider a disk of radius R, with a total charge Q uniformly distributed over the front surface of the disk.

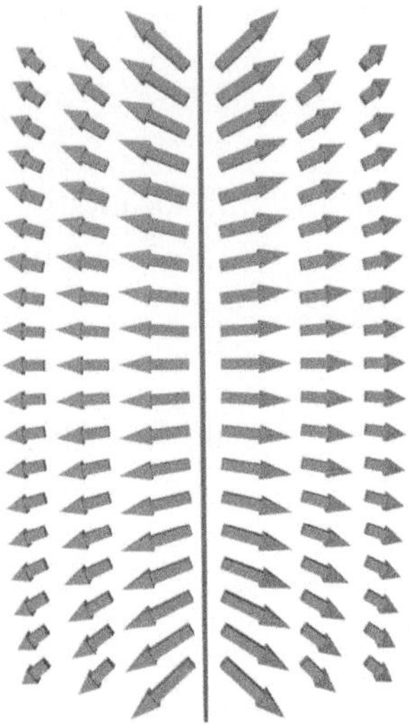

Figure 15.21 A uniformly charged disk viewed edge-on. The electric field is shown at locations across the diameter of the disk. Note the uniformity of the field near the central region of the disk for a given distance from the disk.

Step 1: Cut Up the Charge Distribution into Pieces; Draw $\Delta\vec{E}$

Use thin concentric rings as pieces, as shown in Figure 15.22, since we already know the electric field of a uniform ring. Approximate each ring as having some average radius r.

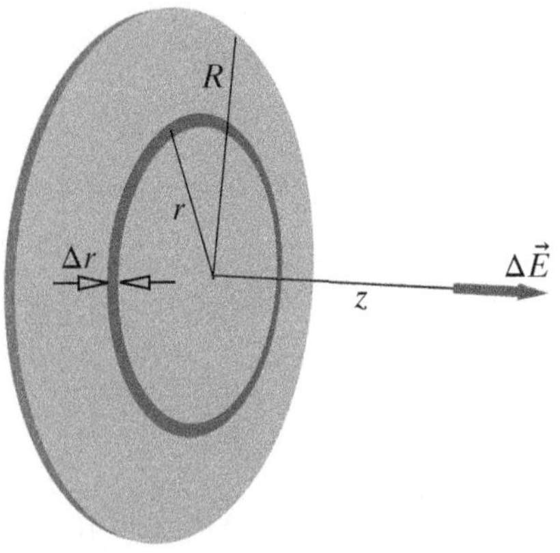

Figure 15.22 A ring of width Δr makes a contribution ΔE to the total electric field.

Step 2: Write an Expression for the $\Delta\vec{E}$ Due to One Piece

Origin: Center of ring.
Location of piece: Given by radius r of ring.
Integration variable: r

$$\vec{r} = \langle 0, 0, z\rangle$$
$$|\vec{r}| = z$$
$$\hat{r} = \frac{\vec{r}}{|\vec{r}|} = \langle 0, 0, 1\rangle$$
$$\Delta\vec{E} = \frac{1}{4\pi\varepsilon_0}\frac{(\Delta q)z}{(r^2+z^2)^{3/2}}\langle 0, 0, 1\rangle$$

Δq in terms of variables: (See Figure 15.23.)

$$\Delta q = Q\frac{(\text{area of ring})}{(\text{area of disk})} = Q\frac{2\pi r\Delta r}{\pi R^2}$$

$$\Delta\vec{E} = \frac{1}{4\pi\varepsilon_0}\frac{\left(Q\frac{2\pi r\Delta r}{\pi R^2}\right)z}{(r^2+z^2)^{3/2}}\langle 0, 0, 1\rangle$$

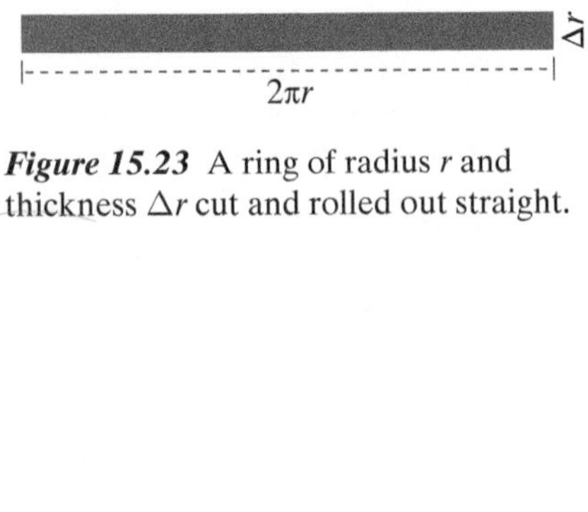

Figure 15.23 A ring of radius r and thickness Δr cut and rolled out straight.

Note that both Δq and r will be different for each piece.

Components to calculate: Only ΔE_z is nonzero.

$$\Delta E_z = \frac{1}{2\varepsilon_0}\left(\frac{Q}{\pi R^2}\right)\frac{z}{(r^2+z^2)^{3/2}}r\Delta r$$

or, for infinitesimally thin rings:

$$dE_z = \frac{1}{2\varepsilon_0}\left(\frac{Q}{\pi R^2}\right)\frac{z}{(r^2+z^2)^{3/2}}rdr$$

Step 3: Sum All Contributions

$$E_z = \int_0^R \frac{1}{2\varepsilon_0}\left(\frac{Q}{\pi R^2}\right)\frac{z}{(r^2+z^2)^{3/2}} r dr$$

Many of these quantities have the same values for different values of r, and these can be taken out of the integral as common factors:

$$E_z = \frac{1}{2\varepsilon_0}\left(\frac{Q}{\pi R^2}\right) z \int_0^R \frac{r}{(r^2+z^2)^{3/2}} dr$$

This particular integral can be done by a change of variables, letting $u = (r^2 + z^2)$. You can work it out yourself, look up the result in a table of integrals, or use a symbolic math package or calculator to evaluate it. The result is:

$$E_z = \frac{1}{2\varepsilon_0}\left(\frac{Q}{\pi R^2}\right)\left[1 - \frac{z}{(R^2+z^2)^{1/2}}\right]$$

for a uniformly charged disk of charge Q and radius R, at locations along the axis of the disk. This is often written in terms of the area A of the disk ($A = \pi R^2$):

$$E_z = \frac{(Q/A)}{2\varepsilon_0}\left[1 - \frac{z}{(R^2+z^2)^{1/2}}\right]$$

Step 4: Check

Direction: Away from the disk if Q is positive, as expected.
Special location: $0 \ll z \ll R$ (very close to the disk, but not touching it. See Figure 15.25.)

$$E \approx \frac{Q/A}{2\varepsilon_0}\left[1 - \frac{z}{R}\right]$$

If z/R is extremely small, $[1 - z/R]$ reduces to 1, and

$$E \approx \frac{Q/A}{2\varepsilon_0}$$

Interestingly, this field is nearly independent of distance! This result is approximately true near any large uniformly charged plate, not just a circular one.

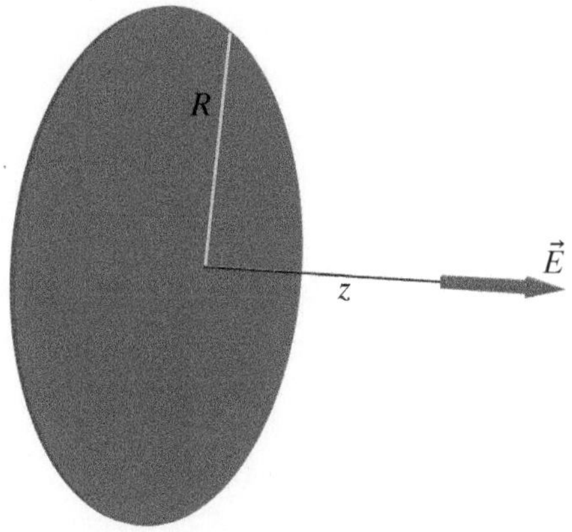

Figure 15.24 Electric field of a uniformly charged disk, at a location on the axis.

ELECTRIC FIELD OF A UNIFORMLY CHARGED DISK

$$E = \frac{(Q/A)}{2\varepsilon_0}\left[1 - \frac{z}{(R^2+z^2)^{1/2}}\right]$$

at a location along the axis, a distance z from the disk (Figure 15.24). The direction is perpendicular to the surface of the disk. Q is the total charge on the disk, A is the area of the surface of the disk, and R is the radius of the disk.

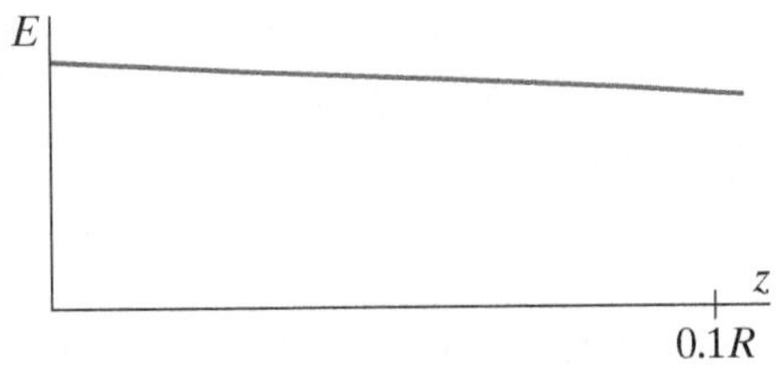

Figure 15.25 Magnitude of electric field along the axis of a disk, for $z < 0.1R$.

APPROXIMATE ELECTRIC FIELD OF A UNIFORMLY CHARGED DISK

$$E \approx \frac{Q/A}{2\varepsilon_0}\left[1 - \frac{z}{R}\right]$$

or

$$E \approx \frac{Q/A}{2\varepsilon_0}$$

at a location a perpendicular distance $z \ll R$ from the disk, as long as the observation location is not too near the edge of the disk (Figure 15.25). The direction is perpendicular to the surface of the disk. Q is the total charge on the disk, A is the area of the surface of the disk, and R is the radius of the disk.

Although we proved this result only for observation locations a perpendicular distance $z \ll R$ from the center of the disk, the result is actually a good approximation as long as you're not too near the edge of the disk, as can be seen in Figure 15.21, which is the result of an accurate numerical integration.

EXAMPLE **A Rod and a Disk**

In Figure 15.26 a thin plastic disk of radius 0.6 m is uniformly charged with $Q_{\text{disk}} = -3 \times 10^{-7}$ C and is attached to a thin glass rod of length 2.4 m that is uniformly charged with $Q_{\text{rod}} = 5 \times 10^{-8}$ C. The center of the rod and the center of disk are at the origin. The rod lies along the x axis and the disk lies in the yz plane. What is the (vector) electric field at location $\langle 0.02, 0.01, 0 \rangle$ m?

Solution According to the superposition principle $\vec{E}$ is the vector sum of the electric field $\vec{E}_{\text{disk}}$ contributed by the disk and the electric field $\vec{E}_{\text{rod}}$ contributed by the rod:

$$\vec{E} = \vec{E}_{\text{disk}} + \vec{E}_{\text{rod}}$$

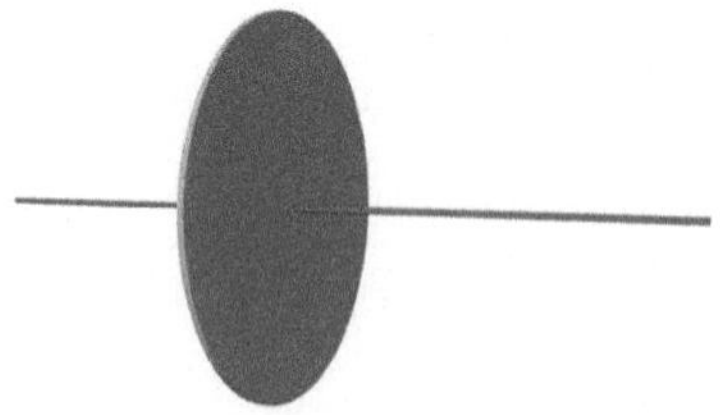

Figure 15.26 A thin positively charged glass rod (red) passes through a thin negatively charged circular plastic disk (blue).

Because the observation location is close to the disk (0.02 m $\ll$ 0.6 m) and far from the edge of the disk, we can approximate the field contributed by the disk:

$$\begin{aligned} E_{\text{disk}} &\approx \frac{Q/A}{2\varepsilon_0} \\ &= \frac{(|-3 \times 10^{-7}|\text{C})/(\pi(0.6\text{ m})^2)}{2(8.85 \times 10^{-12}\,\text{C}^2/\text{N}\cdot\text{m}^2)} \\ &= 1.49 \times 10^4\,\text{N/C} \end{aligned}$$

Because the disk is negatively charged, $\vec{E}_{\text{disk}}$ is in the $-x$ direction, so

$$\vec{E}_{\text{disk}} \approx \langle -1.49 \times 10^4, 0, 0 \rangle\,\text{N/C}$$

Because the observation location is close to the rod (0.01 m $\ll$ 2.4 m) and far from the ends of the rod, we can approximate the field contributed by the rod:

$$\begin{aligned} E_{\text{rod}} &\approx \frac{1}{4\pi\varepsilon_0}\frac{2Q/L}{r} \\ &= (9 \times 10^9\,\text{N}\cdot\text{m}^2/\text{C}^2)\frac{2(5 \times 10^{-8}\,\text{C})/2.4\text{ m}}{0.01\text{ m}} \\ &= 3.75 \times 10^4\,\text{N/C} \end{aligned}$$

Because the rod is positively charged, $\vec{E}_{\text{rod}}$ is in the $+y$ direction, so

$$\vec{E}_{\text{rod}} \approx \langle 0, 3.75 \times 10^4, 0 \rangle \text{ N/C}$$

Therefore the net electric field is

$$\begin{aligned}\vec{E} &= \vec{E}_{\text{disk}} + \vec{E}_{\text{rod}} \\ &\approx \langle -1.49 \times 10^4, 3.75 \times 10^4, 0 \rangle \text{ N/C}\end{aligned}$$

The disk does not "block" the contribution to the field by the left half of the rod, on the other side of the disk from the observation location. As the superposition principle stated, the field $\vec{E}_{\text{rod}}$ contributed by the rod is unaffected by the presence of the disk, and the field $\vec{E}_{\text{disk}}$ contributed by the disk is unaffected by the presence of the rod.

The disk and rod were "thin," so they didn't contain much matter to be polarized. If these objects were "thick," the net electric field might have a significant additional contribution from induced dipoles in the disk and rod.

Checkpoint 3 Suppose that the radius of a disk is $R = 20$ cm, and the total charge distributed uniformly all over the disk is $Q = 6 \times 10^{-6}$ C. Use the exact result to calculate the electric field 1 mm from the center of the disk, and also 3 mm from the center of the disk. Does the field decrease significantly?

Figure 15.27 The two charged metal plates of a capacitor are separated by a very small gap s (exaggerated in this figure).

15.5 TWO UNIFORMLY CHARGED DISKS: A CAPACITOR

Consider two uniformly charged metal disks placed very near each other (separation or gap distance s), carrying charges of $-Q$ and $+Q$ (Figure 15.27). This arrangement is called a capacitor, and we will work with such devices later in electric circuits.

A single charged metal disk cannot have a truly uniform charge density, because the mobile charges tend to push each other to the edge of the disk. Nevertheless, our results for disks made of insulating material are approximately correct for metal disks, especially in this two-disk configuration if the disks are very close together ($s \ll R$). Due to the attraction by the neighboring disk, almost all of the charge is nearly uniformly distributed on the inner surfaces of the disks, with very little charge on the outer surfaces of the disks.

We'll calculate the strength of the electric field at locations near the center of the disks, both inside and outside the capacitor.

Step 1: Cut Up the Charge Distribution into Pieces and Draw $\Delta\vec{E}$

We know the electric field made by a single uniformly charged disk, so we can use this as a "piece." Consider the locations labeled 1, 2, and 3 in Figure 15.28, which shows a side view of the region near the center of the disks, blown up so there is room to draw. The disks are very close together ($s \ll R$), and extend up and down beyond the boundaries of the drawing.

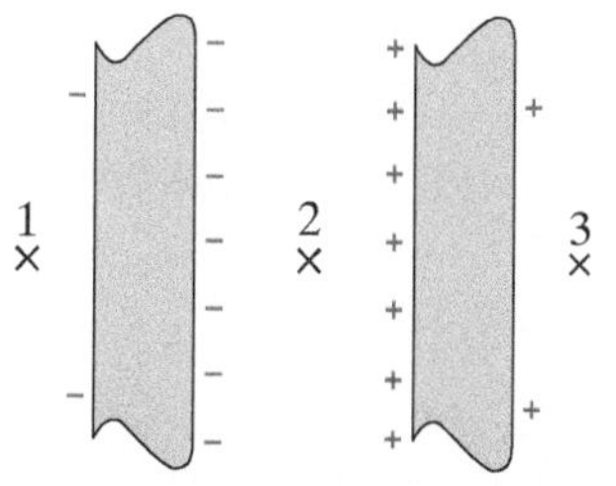

Figure 15.28 Expanded view of a region near the center of a capacitor. Only a small section of the plates is shown; the radius R of a plate is actually much larger than the distance s between plates.

QUESTION Before reading farther, predict the direction and relative magnitude (that is, whether the net field will be large or small) of $\vec{E}_{\text{net}}$ at locations 1, 2, and 3 in Figure 15.28.

At locations 1 and 3, the contributions of the two disks are in opposite directions, and nearly equal, so the net field is very small. However, the distances to the two disks are slightly different, so $E_+ \neq E_-$, and the net field is to the right at locations 1 and 3 (Figure 15.29). At location 2, both disks contribute $\Delta\vec{E}$ in the same direction, so $\vec{E}_{\text{net}}$ is large, and to the left.

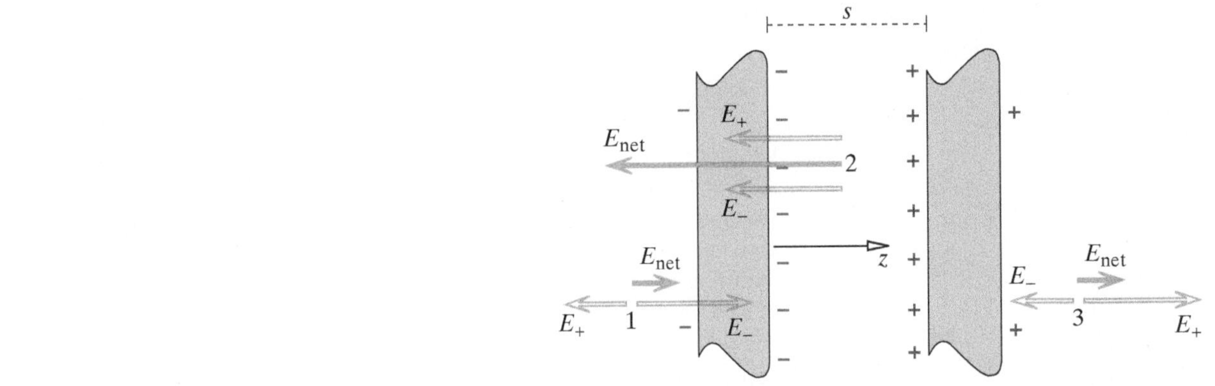

Figure 15.29 Electric field at three locations near the center of a capacitor. As shown, we choose an origin on the inner face of the left disk, with $+z$ running to the right.

Step 2: Write an Expression for the Electric Field Due to One Piece

We will write the expressions in terms of the horizontal coordinate z, so the expressions will be valid at all locations (1, 2, and 3).

Origin: At surface of left plate; z axis runs to right (Figure 15.29).
Location of a piece given by: Pieces are at $z = 0$ and $z = s$.
Distance from piece to observation location: z (for the left plate) or $s - z$ (for the right plate).
Δq in terms of variables: $+Q$ or $-Q$ (all the charge on a disk).
Assumptions: Uniform charge density on inner surface of each disk (ignore the small charge on the outer surfaces of the disks); ($s \ll R$); locations 1 and 3 are very near the disks ($z \ll R$).

$$A = \pi R^2 \quad \text{(the area of one disk)}$$

Contribution of each piece:

$$E_- \approx \frac{Q/A}{2\varepsilon_0}\left[1 - \frac{z}{R}\right] \text{ to the left}$$

$$E_+ \approx \frac{Q/A}{2\varepsilon_0}\left[1 - \frac{s-z}{R}\right] \text{ to the left}$$

Step 3: Add Up the Contributions of All the Pieces

Location 2: At a distance z from the negative plate (and a distance $s - z$ from the positive plate), the electric field is to the left, with magnitude given by the following:

$$E_2 \approx \frac{Q/A}{2\varepsilon_0}\left[1 - \frac{z}{R}\right] + \frac{Q/A}{2\varepsilon_0}\left[1 - \frac{s-z}{R}\right]$$

$$\approx \frac{Q/A}{2\varepsilon_0}\left[1 - \frac{z}{R} + 1 - \frac{s-z}{R}\right]$$

$$\approx \frac{Q/A}{\varepsilon_0}\left[1 - \frac{s/2}{R}\right]$$

$$\approx \frac{Q/A}{\varepsilon_0}$$

The electric field between the plates is essentially twice the field due to one plate. The magnitude hardly depends on z, so the field is remarkably

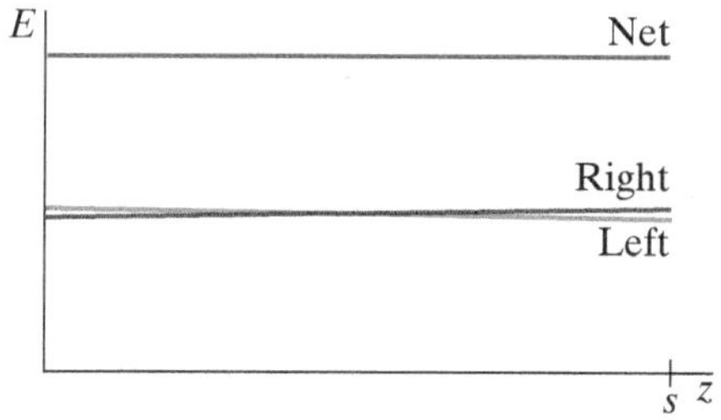

Figure 15.30 E at a location inside a capacitor, showing the contributions from each plate, and the net field, as a function of location inside the capacitor.

uniform. As you move away from the negative plate, the very slightly smaller contribution of the negative plate is nearly compensated by the very slightly larger contribution of the positive plate, as shown in Figure 15.30.

Location 3: $\vec{E}_3$ points to the right (see Figure 15.29) and its magnitude is the difference of the magnitudes of the fields of the two disks:

$$E_3 \approx \frac{Q/A}{2\varepsilon_0}\left[1-\frac{z-s}{R}\right]-\frac{Q/A}{2\varepsilon_0}\left[1-\frac{z}{R}\right]=\frac{Q/A}{2\varepsilon_0}\left(\frac{s}{R}\right)$$

Location 1: $\vec{E}_1$ also points to the right, and

$$E_1 \approx \frac{Q/A}{2\varepsilon_0}\left[1-\frac{z}{R}\right]-\frac{Q/A}{2\varepsilon_0}\left[1-\frac{z+s}{R}\right]=\frac{Q/A}{2\varepsilon_0}\left(\frac{s}{R}\right)$$

The electric field at locations outside the capacitor is called the "fringe field." The fringe field is very small compared to the field inside the capacitor. Inside the gap, the fields of the two disks are in the same direction, but outside the gap the fields of the two disks are in opposite directions.

QUESTION Calculate the ratio of the outside field to the inside field, and show that this ratio is very small if $s \ll R$.

The result is $s/(2R)$, which is small if $s \ll R$. This is the fringe field at a location outside the capacitor but very close to it. (If, however, you are very far away, so $z \gg R$, the capacitor looks like an electric dipole, and the fringe field falls off like $1/z^3$.)

Step 4: Check the Result

Units: Inside the capacitor

$$\frac{\mathrm{C/m^2}}{\mathrm{N\cdot m^2/C^2}}=\frac{\mathrm{N}}{\mathrm{C}}$$

Units: Fringe field

$$\left(\frac{\mathrm{C/m^2}}{\mathrm{N\cdot m^2/C^2}}\right)\left(\frac{\mathrm{m}}{\mathrm{m}}\right)=\frac{\mathrm{N}}{\mathrm{C}}$$

Applicability

Because we have considered locations far from the edges of the plates, our results apply not only to circular capacitors but also to capacitors with rectangular or other shapes of plates, as long as the same conditions are satisfied—the plate separation s must be much smaller than the width or height of a plate.

Although we cannot prove it at this time (we will need Gauss's law—see Chapter 21), it is also true that the magnitude and direction of the electric field are practically the same everywhere in the gap, not just near the center of the plates. Only if you get near the outer edges of the plates does the electric field vary much from its center value.

ELECTRIC FIELD OF A CAPACITOR

$$E \approx \frac{Q/A}{\varepsilon_0}$$

near the center of a two-plate capacitor (each plate has area A, one plate has charge $+Q$, other plate has charge $-Q$; separation s between plates is very small compared to the radius of a plate). The direction is perpendicular to the plates.

Fringe field (just outside the plates, near center of disk):

$$E_{\text{fringe}} \approx \frac{Q/A}{2\varepsilon_0}\left(\frac{s}{R}\right)$$

Checkpoint 4 If the magnitude of the electric field in air exceeds roughly 3×10^6 N/C, the air breaks down and a spark forms. For a two-disk capacitor of radius 50 cm with a gap of 1 mm, what is the maximum charge (plus and minus) that can be placed on the disks without a spark forming (which would permit charge to flow from one disk to the other)? Under these conditions, what is the strength of the fringe field just outside the center of the capacitor?

15.6 A SPHERICAL SHELL OF CHARGE

As mentioned in Chapter 13, a sphere with charge spread uniformly over its surface produces a surprisingly simple pattern of electric field. For brevity, we will use the phrase "a uniform spherical shell" to refer to "a sphere with charge spread uniformly over its surface," since the charge itself forms a thin, shell-like spherical layer. A spherical object of radius R with charge Q uniformly spread out over its surface produces an electric field with the following distance dependence, if $\vec{r}$ is a vector from the center of the sphere to the observation location (Figure 15.31):

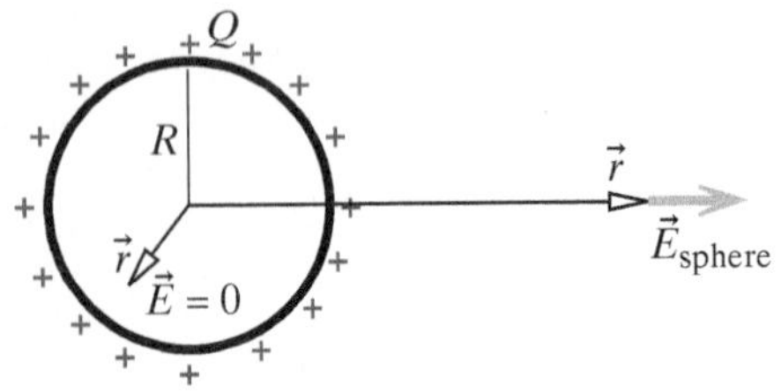

Figure 15.31 The electric field of a uniformly charged sphere (shown in cross section), at locations outside and inside the sphere.

ELECTRIC FIELD OF A UNIFORMLY CHARGED SPHERICAL SHELL

$$\vec{E}_{sphere} = \frac{1}{4\pi\varepsilon_0}\frac{Q}{r^2}\hat{r} \quad \text{for } r > R \text{ (outside the sphere)}$$

$$\vec{E}_{sphere} = 0 \quad \text{for } r < R \text{ (inside the sphere)}$$

Q is the total charge on the surface of the sphere, and R is the radius of the sphere.

This can be shown by setting up and evaluating an integral, as we did in the case of the rod, the ring, and the disk, following the procedure that we used in those cases. The geometry involved is somewhat complex, and the details of the process are given in Section 15.10. In later chapters we will see alternative ways to prove this, one involving Gauss's law and one involving electric potential.

In this section we will try to develop a qualitative understanding of how this surprising result can be true.

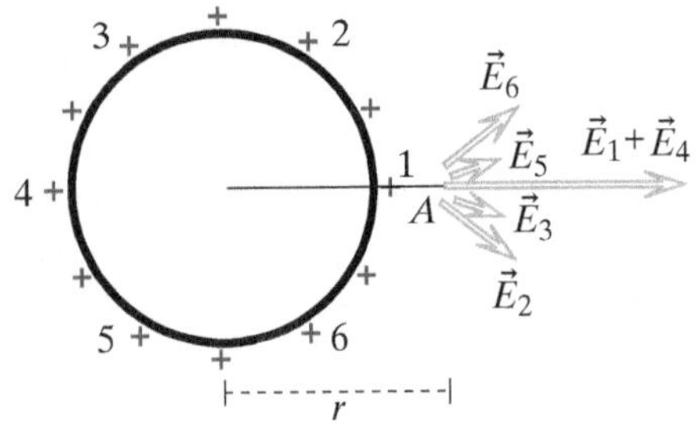

Figure 15.32 Contributions to the net electric field outside a charged sphere.

Outside a Uniform Spherical Shell of Charge

Let's divide the sphere into six regions, as shown in Figure 15.32. $\vec{E}_2$ means "the electric field due to the charges in region 2 of the spherical shell." As long as we are rather far from a region of distributed charge we can approximate the electric field of that region as being due to a point particle with the total charge of the region.

Direction

Consider the field at location A outside the shell, a distance r from the center of the sphere. By symmetry, the net electric field is horizontal and to the right, because the vertical component of the net field cancels for pairs of fields such as $\vec{E}_2$ and $\vec{E}_6$, or $\vec{E}_3$ and $\vec{E}_5$. It is clear that the net field (the superposition of the contributions of all the source charges) is radially outward from the center of the sphere.

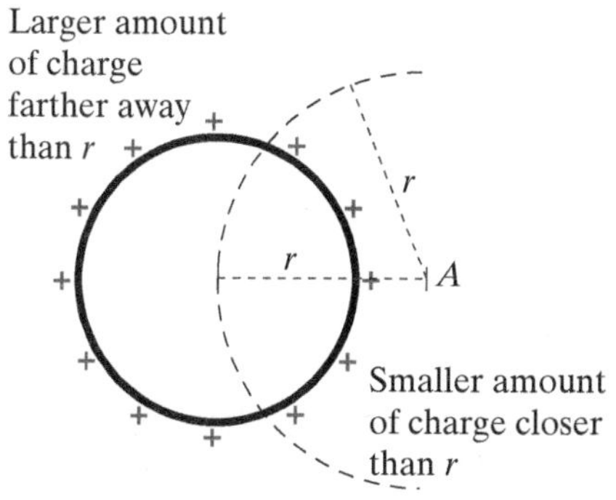

Figure 15.33 Some charge is closer than r to point A, but more of the charge is farther away.

Magnitude

It is somewhat surprising that the magnitude of the net field turns out to be

$$E = \frac{1}{4\pi\varepsilon_0}\frac{Q}{r^2}$$

where Q is the total charge on the sphere and r is the distance to location A from the center of the sphere (not from the surface of the sphere!).

In Figure 15.33, notice that the distance from location A to some regions is less than r, while the distance from location A to other regions is greater than r. Because of the $1/r^2$ distance dependence for point charges, a smaller amount of charge closer than r makes as big a contribution to the net field as the larger amount of charge that is farther away than r.

It is an extraordinary aspect of the $1/r^2$ behavior of Coulomb's law that the net effect of all these charges comes out to be simply equivalent to placing one point charge Q at the center of the sphere. This is not an obvious result!

Force on a Uniformly Charged Sphere

A uniformly charged spherical shell not only makes a field outside that looks as though it were made by a point charge, but the shell also reacts to outside charges as though it were a point charge.

Consider the interaction between a uniform spherical shell and an outside point charge. The shell exerts a force on the point charge as though both were point charges. By the reciprocity principle for electric forces (Newton's third law), the real point charge must exert an equal and opposite force on the shell, and so it exerts a force on the shell as though the shell were a point charge. Hence from the outside a uniform spherical shell looks just like a point charge, both as a source of electric field and when it reacts to external fields.

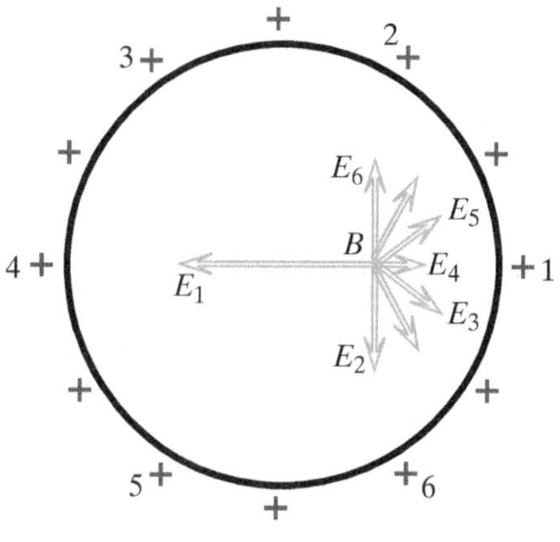

Figure 15.34 Contributions to the net electric field at a location inside a uniformly charged sphere add up to zero.

Inside a Uniform Spherical Shell of Charge

Now consider a location B inside the shell, as indicated in Figure 15.34. A small number of charges near location B (region 1) make a large contribution to the left, while a large number of charges far away from location B (regions 3, 4, and 5) each make small contributions to the right. Because the surface area of a portion of the sphere is proportional to r^2 while the electric field contributed by this region is proportional to $1/r^2$, these contributions exactly cancel each other. The electric field at location B due to charges on the surface of the sphere turns out to be exactly zero. Although intuitively plausible, this is not an obvious result.

Warning: E Is Not Always Zero Inside a Charged Sphere

Do not overgeneralize this result. Other charges in the Universe may make a nonzero electric field inside the shell. It is only the electric field contributed by charges uniformly distributed on the surface of the sphere that is zero inside the sphere.

Implications

The fact that the electric field due to uniform charges on the shell is zero at any location inside the shell has interesting implications. For example, if the uniformly charged shell is filled with plastic, the charges on the surface do not polarize the molecules in the plastic, because the surface charges contribute zero field. Even if the location of a molecule is very close to the charged surface of the sphere, E inside the sphere is still zero at that location. (Charges external to the sphere may contribute to a nonzero electric field inside the sphere, however.)

Since a solid metal sphere is a conductor, any excess charge on a metal sphere arranges itself on the outer surface. Inside the metal there is no excess charge, and there is no field, because inside a metal in equilibrium the electric field is zero. The field inside a metal sphere in equilibrium is zero even if external charges polarize the metal and make the charge on the outer surface have a nonuniform distribution.

What Is the Electric Field Right at the Surface?

The electric field is zero just inside the surface of a uniformly charged spherical shell, but the field is nonzero just outside the surface. You might wonder what the field is right at the surface. The physical reality is that the electric field is highly variable in direction and magnitude right at the surface, because the individual excess point charges (electrons and ions) on the surface are many atomic diameters away from each other. In Chapter 14 we found that typically only about one in a million atoms on a surface has an extra charge. It is only when you get some thousands of atomic diameters away from the surface (about 1×10^{-7} m) that it is appropriate to consider the surface charge as approximately a uniform, continuous sheet of charge. As long as you are this far away from the surface, the field outside the surface looks like the field of a point charge at the center of the sphere, and the field inside the surface is zero.

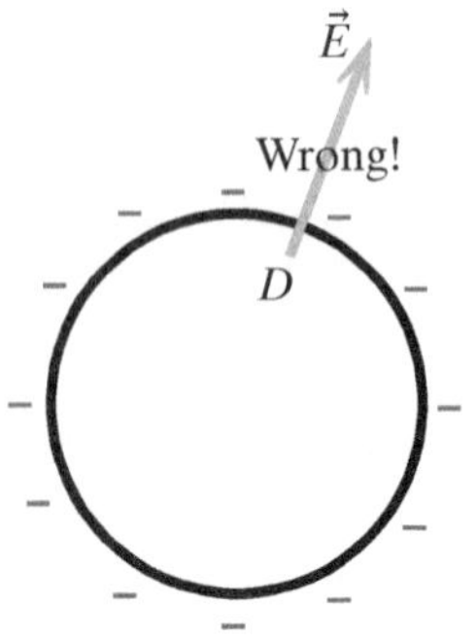

Figure 15.35 A student's incorrect prediction of the electric field inside a uniformly charged sphere.

Checkpoint 5 What is wrong with Figure 15.35 and this associated incorrect student explanation? "The electric field at location D inside the uniformly charged sphere points in the direction shown, because the charges closest to this location have the largest effect." (Spheres provide the most common exception to the normally useful rule that the nearest charges usually make the largest contribution to the electric field.)

15.7 A SOLID SPHERE CHARGED THROUGHOUT ITS VOLUME

Until this point, all of the charge distributions we have considered involved charge distributed over the surface of an object. There are some cases, however, in which charge can be distributed throughout an object. One such case is the nucleus of an atom, which is composed of protons and neutrons packed into a sphere of radius on the order of 10^{-15} m. We can treat the charge density inside the nucleus as approximately uniform. Another such case is the electron cloud in an atom, in which the negative charge of the electrons appears to be distributed throughout a spherical region of radius on the order of 10^{-10} m (although not necessarily uniformly).

We will consider a solid sphere with radius R, having charge Q uniformly distributed throughout its volume.

Cut Up the Charge Distribution into Pieces and Draw $\Delta\vec{E}$

We can model the sphere as a series of concentric spherical shells.

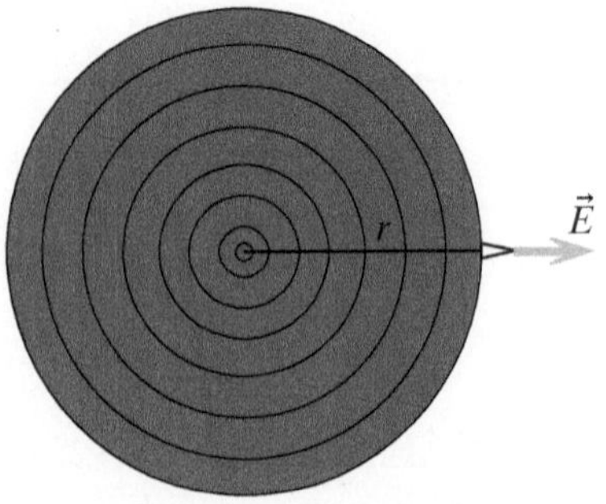

Figure 15.36 A charged solid sphere modeled as a series of concentric spherical shells, all charged.

Outside a Solid Sphere of Charge

At a location outside the sphere, and hence outside all the shells (Figure 15.36), each shell looks like a point charge at the center of the sphere. Hence, outside the sphere:

$$E = \frac{1}{4\pi\varepsilon_0}\frac{Q}{r^2} \quad \text{for } r > R$$

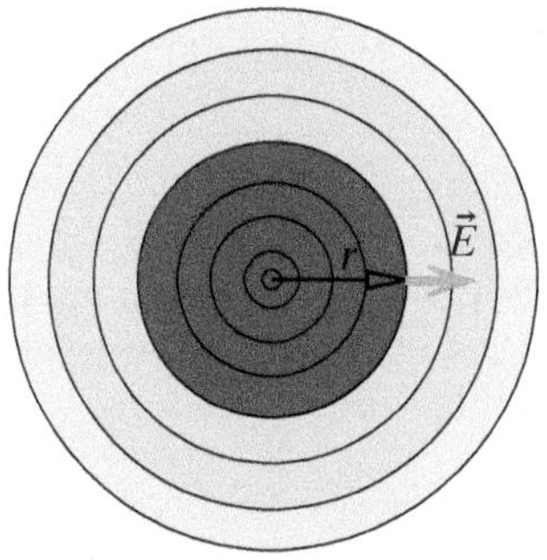

Figure 15.37 A location inside the gray shells, but outside the red shells. At this location only the inner shells make a nonzero contribution to the electric field.

Inside a Solid Sphere of Charge

At a location inside the sphere, we are inside some of the spherical shells (gray shells in Figure 15.37); the contribution of these shells to $\vec{E}_{\text{net}}$ is therefore zero.

We are outside the rest of the shells (red shells in Figure 15.37); each of these shells looks like a point charge at the center of the sphere. To get $\vec{E}_{\text{net}}$ at this location we must add the contributions of all the inner shells. Since all the inner shells together will contribute

$$E = \frac{1}{4\pi\varepsilon_0}\frac{\Delta Q}{r^2}$$

our only remaining task is to find ΔQ.

$$\Delta Q = Q\frac{(\text{volume of inner shells})}{(\text{volume of sphere})} = Q\frac{\frac{4}{3}\pi r^3}{\frac{4}{3}\pi R^3}$$

Assembling the result:

$$E = \frac{1}{4\pi\varepsilon_0}\frac{Q}{r^2}\frac{\frac{4}{3}\pi r^3}{\frac{4}{3}\pi R^3} = \frac{1}{4\pi\varepsilon_0}\frac{Q}{R^3}r \quad \text{for } r < R$$

so the electric field inside the sphere is directly proportional to r.

QUESTION Does this make sense?

Yes. While inside the sphere, as r increases, the amount of charge inside the observation location increases. The larger volume ($V \propto r^3$) more than compensates for the $1/r^2$ field dependence. By symmetry, $E = 0$ at $r = 0$.

In the special case that $r = R$, we have $E = \dfrac{1}{4\pi\varepsilon_0}\dfrac{Q}{R^2}$, which is correct.

15.8 INFINITESIMALS AND INTEGRALS IN SCIENCE

In pure mathematics, an infinitesimal quantity is conceived to be arbitrarily close to zero. In science, however, this may not make sense. For example, in the case of the uniformly charged thin rod it makes no physical sense to choose dz to be smaller than an atomic diameter. It would enormously complicate the summation if we had to take into account the detailed distribution of charged particles inside each atom. All that is required is that dz be small compared to the significant dimensions of the situation (r and L). We want each slice dz to be an adequate approximation of a point charge, and $dz = 0.02$ m may be good enough! We can consider the mathematical infinitesimal dz to be an idealization of what a scientist means by an infinitesimal quantity, which is a quantity that is “small enough” within the desired and possible precision of measurements and analysis.

Oddly, the result of doing an integral is approximate, not exact, because at the microscopic level the actual charge distribution is not in fact uniformly distributed. In Chapter 14 we found that on a charged piece of tape the individual electrons or ions were quite far from each other on an atomic scale. However, if you are many atomic diameters away from a “uniformly” charged rod, the mathematical integral gives an excellent approximation to the actual electric field.

There is another important sense in which our analysis is approximate. If a rod is an insulator, and has been charged by rubbing it with a cloth, the charge distribution will be only approximately uniform, even on a macroscopic scale, since charge transfer cannot be accurately controlled. If the rod is a metal, charge will tend to pile up at the ends rather than being uniformly distributed. Nevertheless, modeling a charged rod as a uniformly charged rod in practice can be a useful model. We just need to remember that our analytical solution is approximate, not exact.

15.9 3D NUMERICAL INTEGRATION WITH A COMPUTER

In Section 15.1 we found an algebraic expression for the electric field of a uniformly charged thin rod, at observation locations on the midplane of the rod. However, this expression is not valid for observation locations that are not near the midplane. We can easily write a computer program to calculate the electric field of the rod at any observation location.

A Thin Rod: One Observation Location

■ The use of Python lists to facilitate calculations involving many charges and observation locations is introduced in two VPython videos: VPython Video 7: Lists, Part 1 (http://vpython.org/video07.html), and VPython Video 8: Lists, Part 2 (http://vpython.org/video08.html). Watch these videos and do the challenge activities at the end of each video before reading this section.

We'll start by reproducing the numerical calculation for an observation location in the midplane, approximating the rod by a collection of 10 point charges. We did this before by hand, with results that are summarized in Figure 15.7. We'll start with this observation location so we can check our result by comparing it to the analytical result.

The following code creates a Python list of point charges represented by spheres, each corresponding to one slice of the rod (see the note about lists in the margin of this page). We'll use the values used to generate Figure 15.7: a total charge of 1 nC spread uniformly over a thin rod of length 1 m, and an observation location of $\vec{r}_{\text{obs}} = \langle 0.05, 0, 0 \rangle$ m.

```
from visual import *
scene.height = 800
oofpez = 9e9
Qtot = 1e-9
L = 1
N = 10
dy = L/N
## create a list of point charges
slices = [] ## an empty list
i = 0
y0 = -L/2 + dy/2  ## center of bottom slice
while i < N:
    a = sphere(pos=vector(0,y0+i*dy,0), radius=dy/2,
               color=color.red, q=Qtot/N)
    slices.append(a)  ## add sphere to list
    i = i + 1
```

QUESTION A sphere has only three attributes: `pos`, `radius`, and `color`. Why is an attribute `q` added in the instruction that creates a sphere named `a`?

As explained in VPython Video 7, in Python it is legal to make up new attributes for objects. This is done here for convenience: we need the charge of each slice, and it is easier to store this information in a new attribute than to make a separate list of the charges of every slice. In this case, we happen to know that each slice has the same charge, but in a more complex situation this might not be true (for example, if the rod were not uniformly charged), so this is a very convenient way of keeping track of the charge of each slice.

QUESTION Why is `y0` set to `-L/2 + dy/2`?

The thin rod is aligned along the y axis, and centered at the origin. The bottom end of the rod is at $\langle -L/2, 0, 0 \rangle$. However, the center of the bottom slice is not at the end of the rod. The height of a slice is `dy`, so the center of this slice must be a distance `dy/2` up from the bottom of the rod.

QUESTION Why is the y coordinate calculated as `y0+i*dy`?

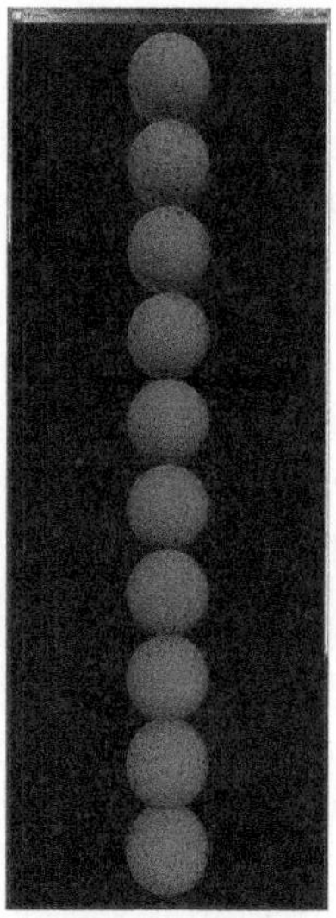

Figure 15.38 A uniformly charged rod modeled as ten point charges.

Before entering the loop we set `y0` to represent the *y* coordinate of the center of the bottom slice, and the first time through the loop `i` is zero, so `y0+i*dy` is equal to `y0`. The second time through the loop `i` is 1, so the *y* coordinate is `y0+dy`, which is the *y* coordinate of the next higher slice, and so on.

To make sure that the positions of the spheres representing the point charges in our model of the rod are correct, we can print their positions. Recall that the first element of a list is element 0. As explained in the VPython videos on lists, we can refer to the 0th object in a list named `slices` by the name `slices[0]`; if the object is a sphere, we can refer to its position attribute by adding `.pos` as usual (`slices[0].pos`). In the code below the variable `i` is used for the list index.

```
## print index and position
i = 0
while i < len(slices):
    print(i, slices[i].pos)
    i = i + 1
```

Figure 15.38 shows the 10 spheres created by the program. The output from the `print()` statement is shown below. These positions agree with the *y* values in Figure 15.7, although the order in which they are listed is different because we created the bottom sphere first.

```
0 <0, -0.45, 0>
1 <0, -0.35, 0>
2 <0, -0.25, 0>
3 <0, -0.15, 0>
4 <0, -0.05, 0>
5 <0, 0.05, 0>
6 <0, 0.15, 0>
7 <0, 0.25, 0>
8 <0, 0.35, 0>
9 <0, 0.45, 0>
```

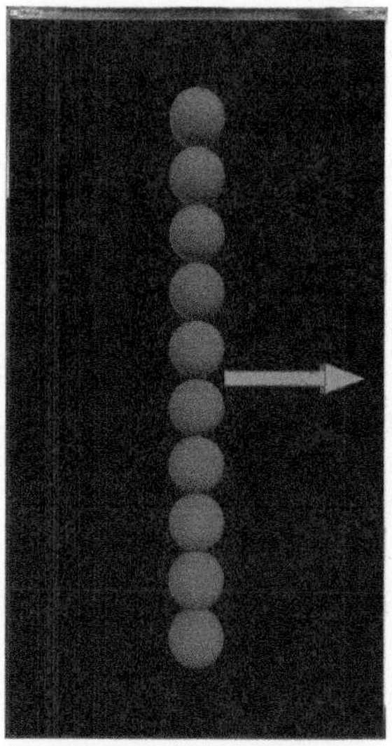

Figure 15.39 The electric field at a location in the midplane, calculated by modeling a uniformly charged rod as 10 point charges.

To calculate the net electric field at the observation location, we can write a loop to add up the contributions of each slice:

```
## calculate E at observation location
r_obs = vector(0.05,0,0)  ## to test calculation
E_net = vector(0,0,0)
i = 0
while i < N:
    rate(100)
    r = r_obs - slices[i].pos
    rhat = r/mag(r)
    E = (oofpez * slices[i].q / mag(r)**2) * rhat
    E_net = E_net + E
    i = i + 1
print(E_net)
```

The net field calculated by this program comes out to $\langle 285.764, 1.43885 \times 10^{-13}, 0\rangle$ N/C. In Figure 15.39 an arrow has been added to display the electric field at the observation location.

QUESTION Why isn't the *y* component of $\vec{E}_{\text{net}}$ zero, as expected?

There are slight rounding errors in any calculation. Since the calculated *y* component of the net field is 15 orders of magnitude smaller than the *x* component, it is sufficiently close to zero for our purposes. More seriously, the magnitude of the electric field given by this calculation is significantly smaller than the value obtained from the analytical expression, which is 385.213 N/C. As discussed in Section 15.1, increasing the number of point charges in the model of the rod will initially increase the accuracy of the calculation. However, at some point adding more slices does not produce a noticeable change in the result.

In Problem P63 you will improve the accuracy of this calculation by increasing the number of point charges used to model the rod, visualize the result with an arrow, and then use the resulting program to calculate and display the electric field due to the rod at locations that are not near the midplane.

There is a somewhat more elegant way to go through the elements of a list within a loop. In the following version of the code we've been discussing, the name `this_slice` is repeatedly recycled; during the first iteration it refers to the 0th element of the list, during the second iteration it refers to the 1st element of the list, and so on. Either version of the loop is acceptable—both accomplish the same thing.

```
## calculate E at observation location
r_obs = vector(0.05,0,0)   ## to test calculation
E_net = vector(0,0,0)
for this_slice in slices:
    rate(100)
    r = r_obs - this_slice.pos
    rhat = r/mag(r)
    E = (oofpez * this_slice.q / mag(r)**2) * rhat
    E_net = E_net + E
print(E_net)
```

A Thin Rod: Multiple Observation Locations

To see the 3D pattern of electric field associated with a distribution of source charges, we need to calculate and display the field at many observation locations. We can easily do this by making a list of observation locations.

As an example, let's create a list of arrows at observation locations lying on a circle around the rod. Initially the arrows will have zero length (although if we wanted to make sure their positions were correct, we could give them a nonzero axis, pointing in an arbitrary direction).

```
## create a list of arrows
observation = []
yobs = 0.4
dtheta = pi/6
theta = 0
R = 0.15 ## radius of circle
while theta < 2*pi:
   a = arrow(pos=vector(R*cos(theta), yobs, R*sin(theta)),
                color=color.orange,
                axis=vector(0,0,0))
    observation.append(a)
    theta = theta + dtheta
```

We now have two lists: a list of source charges and a list of arrows whose positions are observation locations. For each observation location, we need to add up the field contributed by all source charges. This means we need two nested loops, and therefore two levels of indentation. In the outer loop, which uses the index j, the next observation location is chosen. In the inner loop, which uses the index i, the field at the chosen observation location is calculated.

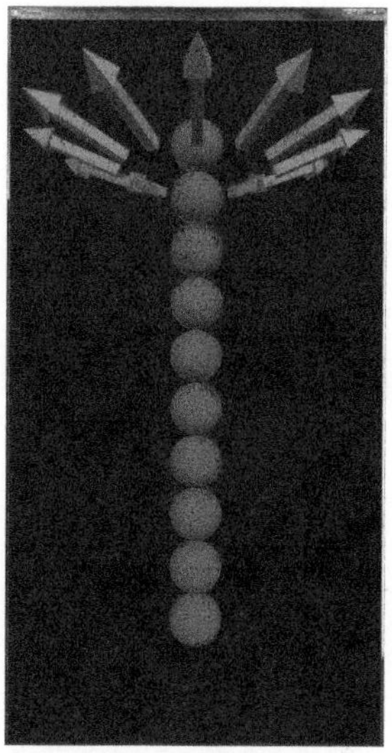

Figure 15.40 The calculated electric field at many observation locations.

```
## calculate E at observation location
sf = 0.002   ## arrow scale factor
j = 0
## outer loop
while j < len(observation):
    rate(500)
    earrow = observation[j]
    ## add E of all slices for this obs. loc.
    i = 0
    E_net = vector(0,0,0)
    ## inner loop
    while i < N:
        r = earrow.pos - slices[i].pos
        rhat = r/mag(r)
        E = (oofpez * slices[i].q / mag(r)**2) * rhat
        E_net = E_net + E
        i = i + 1
        ## end of inner loop
    earrow.axis = sf*E_net
    j = j + 1
    ## end of outer loop
```

The result of this calculation is shown in Figure 15.40. In Problems P66–P67 you will add additional observation locations, or create a different pattern of observation locations.

15.10 *INTEGRATING THE SPHERICAL SHELL

In this section, we'll sketch the proof that a uniformly charged spherical shell looks from the outside like a point charge but on the inside has a zero electric field. Two quite different proofs are given in Chapters 16 and 21.

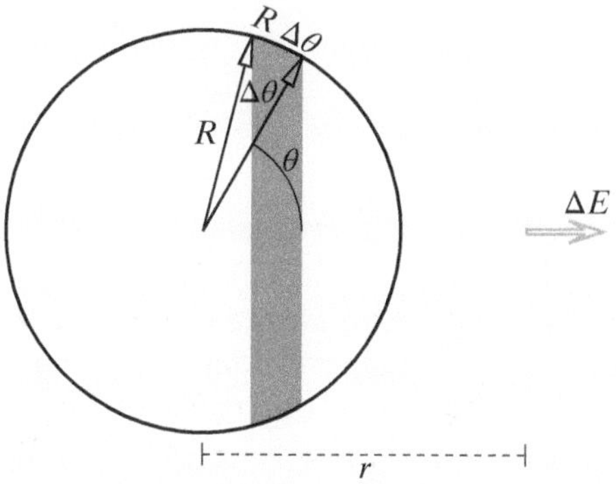

Figure 15.41 Divide the spherical shell (shown here in cross section) into rings, each carrying a (variable) amount of charge ΔQ.

Step 1: Divide the Sphere into Pieces

Divide the spherical shell into rings of charge, each delimited by the angle θ and the angle $\theta + \Delta\theta$, and carrying an amount of charge ΔQ (Figure 15.41). Each ring contributes ΔE at an observation point a distance r from the center of the spherical shell.

Step 2: Write an Expression for ΔE

Origin: At the center of the sphere.

Use spherical coordinates (r,θ,ϕ). As shown in Figure 15.42, θ is an angle measured from the r (horizontal) axis; ϕ refers to rotation about the r (horizontal) axis.

Location of one piece (ring): Given by angle θ.

Components: There is only a horizontal component.

Distance from center of one ring to observation location: $d = (r - R\cos\theta)$

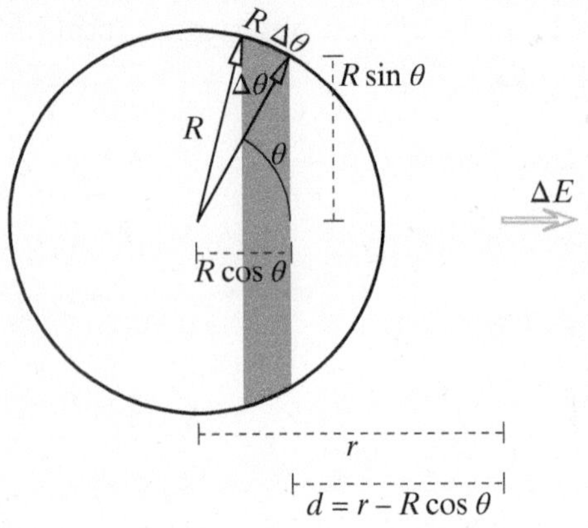

Figure 15.42 The sphere may be divided into ring-shaped segments.

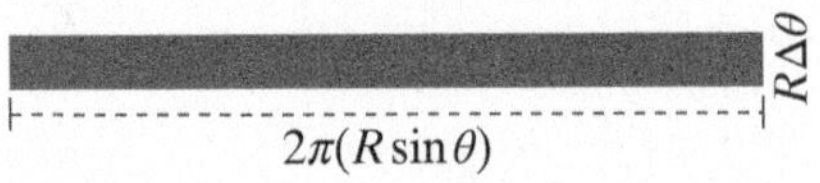

Figure 15.43 We can calculate the area of the ring of charge by "unrolling" it.

Amount of charge on each ring:

$$\Delta Q = Q\frac{\text{(surface area of ring)}}{\text{(surface area of sphere)}} = Q\left(\frac{2\pi(R\sin\theta)(R\Delta\theta)}{4\pi R^2}\right)$$

As before, we calculate the surface area ΔA of the ring by laying the ring out flat (Figure 15.43), noting that the radius of the ring is $R\sin\theta$ and its width is $R\Delta\theta$ (since arc length is radius times angle, with angle measured in radians).
Integration variable: θ
Magnitude of ΔE*:*

$$\Delta E = \frac{1}{4\pi\varepsilon_0}\frac{(\Delta Q)d}{(d^2+(R\sin\theta)^2)^{3/2}}$$
$$= \frac{1}{4\pi\varepsilon_0}\frac{(r-R\cos\theta)}{[(r-R\cos\theta)^2+(R\sin\theta)^2]^{3/2}}Q\left(\frac{2\pi(R\sin\theta)}{4\pi R^2}\right)(R\Delta\theta)$$

Step 3: Add Up Contributions of All the Pieces

$$E = \frac{1}{4\pi\varepsilon_0}\frac{Q}{2}\int_0^{\pi}\frac{(r-R\cos\theta)}{[(r-R\cos\theta)^2+(R\sin\theta)^2]^{3/2}}\sin\theta\, d\theta$$

The limits on the integral are determined by the fact that if we let θ range between 0 and π radians (180°), we add up rings that account for the entire surface of the spherical shell.

Evaluating the integral is simply math (in this case, it is rather difficult math). The results are:
Outside the shell ($r > R$):

$$E = \frac{1}{4\pi\varepsilon_0}\frac{Q}{r^2} \quad \text{for } r > R$$

as though all the charge were concentrated into a point at the center of the spherical shell.
Inside the shell ($r < R$): $E = 0$ for $r < R$.

SUMMARY

A procedure for calculating the electric field due to a distribution of electric charges, by applying the superposition principle.

Step 1: Divide the charge distribution into pieces and draw $\Delta\vec{E}$.
Choose pieces whose electric field you know.

Step 2: Write an expression for the electric field due to one piece.
This involves selecting an origin, finding $\vec{r}$ and $\hat{r}$, describing the location of one piece in terms of an integration variable, expressing Δq in terms of your variables, and writing expressions for ΔE_x, ΔE_y, and ΔE_z.

Step 3: Add up the contributions of all the pieces.
Sometimes the sum can be written as a definite integral that can be symbolically evaluated; otherwise you must calculate a finite sum numerically.

Step 4: Check the result.
Checks should include a qualitative check of field direction, a check of the units of the answer, and checks of special cases.

Useful results:

Rod
Electric field of a uniformly charged thin rod, at a location a distance r from the midpoint of the rod along a line perpendicular to the rod (Figure 15.44). Q is the total

charge on the rod, and L is the length of the rod. Direction: either radially away from or toward the rod (depending on the sign of Q).

$$E = \frac{1}{4\pi\varepsilon_0} \frac{Q}{r\sqrt{r^2 + (L/2)^2}}$$

Approximate electric field of a rod:

$$E \approx \frac{1}{4\pi\varepsilon_0} \frac{2(Q/L)}{r} \quad (\text{if } L \gg r)$$

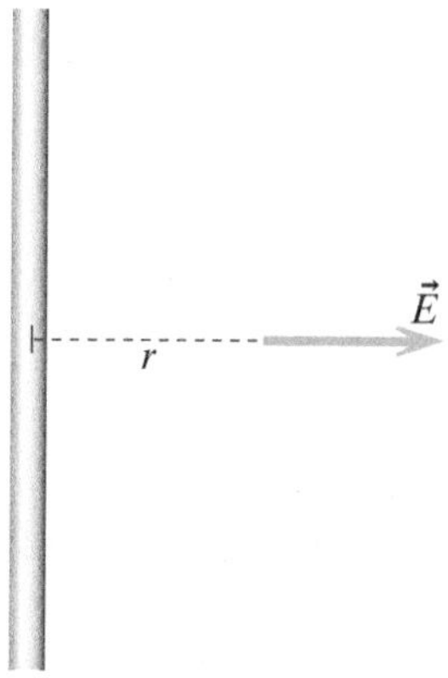

Figure 15.44

Ring

Electric field of a thin uniformly charged ring at a location along the axis, a distance z from the center of the ring, for a ring of radius R and charge q (Figure 15.45). Direction: parallel or antiparallel to the axis, depending on the sign of q.

$$E = \frac{1}{4\pi\varepsilon_0} \frac{qz}{(R^2 + z^2)^{3/2}}$$

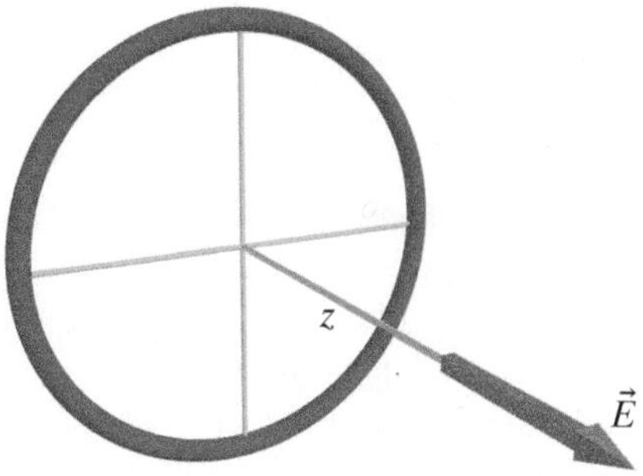

Figure 15.45

Disk

Electric field of a uniformly charged disk of radius R, a distance z from the center of the disk along a line perpendicular to the disk (Figure 15.46):

$$E = \frac{(Q/A)}{2\varepsilon_0} \left[1 - \frac{z}{(R^2 + z^2)^{1/2}}\right]$$

where $A = \pi R^2$.

Approximation: close to the disk ($z \ll R$):

$$E \approx \frac{Q/A}{2\varepsilon_0} \left[1 - \frac{z}{R}\right]$$

Approximation: extremely close to the disk ($z \ll R$):

$$E \approx \frac{Q/A}{2\varepsilon_0}$$

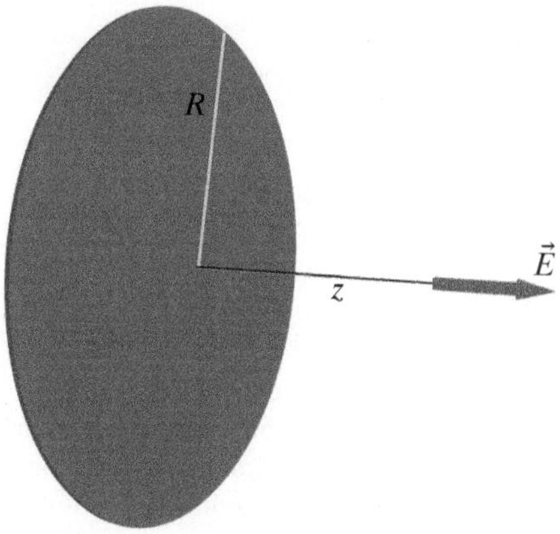

Figure 15.46

Capacitor

Electric field at a location inside a capacitor (Figure 15.47):

$$E \approx \frac{Q/A}{\varepsilon_0}$$

Near center of a two-plate capacitor (each plate has area A, one plate has charge $+Q$, other plate has charge $-Q$; separation between plates is very small compared to the radius of a plate). Direction: perpendicular to the plates. Fringe field just outside the plates:

$$E_{\text{fringe}} \approx \frac{Q/A}{2\varepsilon_0}\left(\frac{s}{R}\right)$$

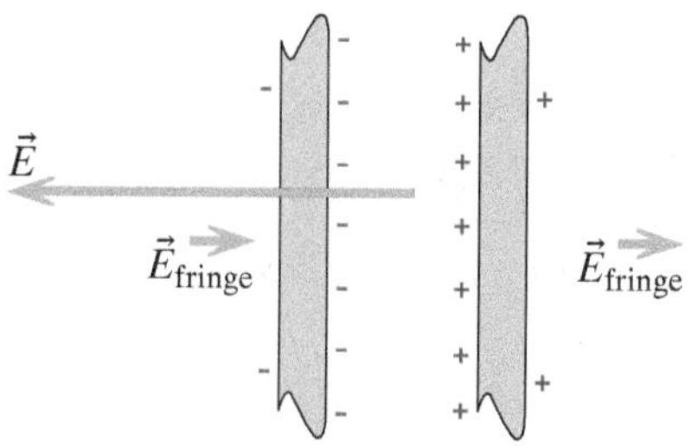

Figure 15.47

Spherical shell

Inside a uniformly charged thin spherical shell:

$$E = 0 \quad \text{due to charges on the shell}$$

Outside a uniformly charged thin spherical shell:

$$E = \frac{1}{4\pi\varepsilon_0}\frac{Q}{r^2} \quad \text{due to charges on the shell}$$

where r is the distance from the center of the sphere to the observation location (Figure 15.48):

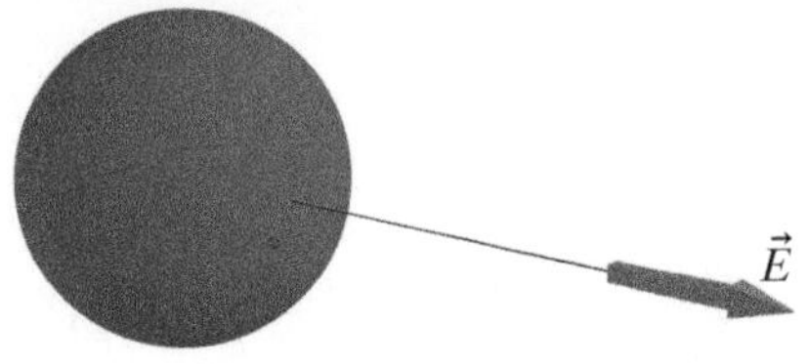

Figure 15.48

Sphere charged throughout its volume

Inside a solid sphere of radius R, charged throughout its volume, at a location a distance r from the center of the sphere:

$$E = \frac{1}{4\pi\varepsilon_0}\frac{Q}{R^3}r$$

Warning: Do not confuse this result with the previous one. The words "solid" and "shell" refer to the location of the charge (on the surface or throughout the object). If a plastic ball is solid but has charge uniformly distributed over its surface and none inside, the electric field inside the plastic, due to the charge on the surface, is zero—this is an example of a spherical shell of charge.

QUESTIONS

Q1 Consider setting up an integral to find an algebraic expression for the electric field of a uniformly charged rod of length L, at a location on the midplane. If we choose an origin at the center of the rod, what are the limits of integration?

Q2 A rod is 2.5 m long. Its charge is -2×10^{-7} C. The observation location is 4 cm from the rod, in the midplane. In the expression

$$E = \frac{1}{4\pi\varepsilon_0}\frac{Q}{r\sqrt{r^2+(L/2)^2}}$$

what is r in meters?

Q3 Graph the magnitude of the full expression for the field E of a rod along the midplane vs. r. Does E fall off monotonically (with distance)?

Q4 A rod with uniformly distributed charge 2×10^{-8} C is 50 cm long. We need to calculate E at a distance of 1 cm from the midpoint of the rod. Which equation for the electric field of a rod should we use? (1) Exact, (2) Approximate, (3) Either exact or approximate, (4) Neither—we have to do it numerically, (5) Neither—we need to integrate

Q5 A student claimed that the equation for the electric field outside a cube of edge length L, carrying a uniformly distributed charge Q, at a distance x from the center of the cube, was

$$E = \frac{Q}{\varepsilon_0}\frac{L}{x^{1/2}}$$

Explain how you know that this cannot be the right equation.

Q6 A student claimed that the equation for the electric field outside a cube of edge length L, carrying a uniformly distributed charge Q, at a distance x from the center of the cube, was

$$\frac{1}{4\pi\varepsilon_0}\frac{50QL}{x^3}$$

Explain how you know that this cannot be the right equation.

Q7 When calculating the electric field of an object with electric charge distributed approximately uniformly over its surface, what is the order in which you should do the following operations? (1) Check the direction and units. (2) Write an expression for the electric field due to one point-like piece of the object. (3) Divide up the object into small pieces of a shape whose field is known. (4) Sum the vector contributions of all the pieces.

Q8 Explain briefly how knowing the electric field of a ring helps in calculating the field of a disk.

Q9 Consider the algebraic expression for the electric field of a uniformly charged ring, at a location on the axis of the ring. Q is the charge on the entire ring, and ΔQ is the charge on one piece of the ring. $\Delta\theta$ is the angle subtended by one piece of the ring (or, alternatively, Δr is the arc length of one piece). What is ΔQ, expressed in terms of given constants and an integration variable? What are the integration limits?

Q10 The rod in Figure 15.49 carries a uniformly distributed positive charge. Which arrow (a–h) best represents the direction of the electric field at the observation location marked with a red X?

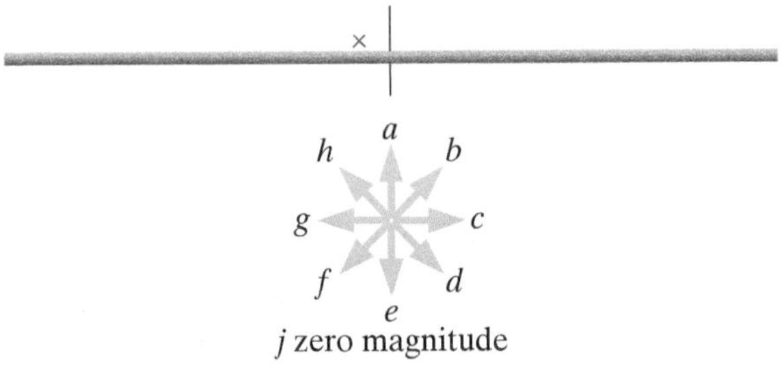

Figure 15.49

Q11 By thinking about the physical situation, predict the magnitude of the electric field at the center of a uniformly charged ring of radius R carrying a charge $+Q$. Then use the equation derived in the text to confirm this result.

Q12 Define "fringe field."

Q13 Explain qualitatively how it is possible for the electric field at locations near the center of a uniformly charged disk not to vary with distance away from the disk.

Q14 Coulomb's law says that electric field falls off like $1/z^2$. How can E for a uniformly charged disk depend on $[1 - z/R]$, or be independent of distance?

Q15 Two very thin circular plastic sheets are close to each other and carry equal and opposite uniform charge. Explain briefly why the field between the sheets is much larger than the field outside. Illustrate your argument on a diagram.

Q16 The electric field inside a capacitor is shown on the left in Figure 15.50. Which option (1–5) best represents the electric field at location A?

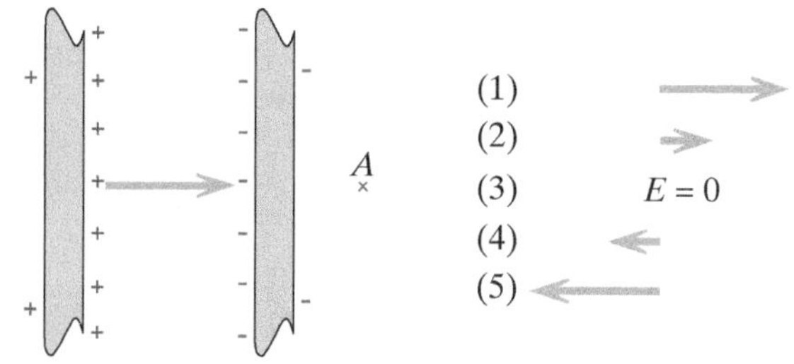

Figure 15.50

Q17 A solid spherical plastic ball was rubbed with wool in such a way that it acquired a uniform negative charge all over the surface. Make a sketch showing the polarization of molecules inside the plastic ball, and explain briefly.

Q18 A student said, "The electric field inside a uniformly charged sphere is always zero." Describe a situation where this is not true.

Q19 Give an example of a configuration of charges that yields an electric field or force whose magnitude varies approximately with distance as specified: (1) Field independent of distance, (2) Field proportional to $1/r$, (3) Field proportional to $1/r^2$, (4) Field proportional to $1/r^3$, (5) Force (not field) that is proportional to $1/r^5$.

PROBLEMS

Section 15.1

•P20 If the total charge on a uniformly charged rod of length 0.4 m is 2.2 nC, what is the magnitude of the electric field at a location 3 cm from the midpoint of the rod?

•P21 A thin rod lies on the x axis with one end at $-A$ and the other end at A, as shown in Figure 15.51. A charge of $-Q$ is spread uniformly over the surface of the rod. We want to set up an integral to find the electric field at location $\langle 0, y, 0\rangle$ due to the rod. Following the procedure discussed in this chapter, we have cut up the rod into small segments, each of which can be considered as a point charge. We have selected a typical piece, shown in red on the diagram.

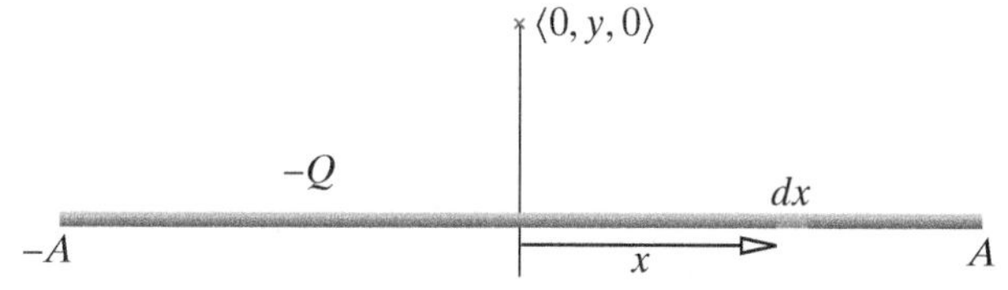

Figure 15.51

Answer using the variables x, y, dx, A, Q as appropriate. Remember that the rod has charge $-Q$.
(a) In terms of the symbolic quantities given above and on the diagram, what is the charge per unit length of the rod? **(b)** What is the amount of charge dQ on the small piece of length dx? **(c)** What is the vector from this source to the observation location? **(d)** What is the distance from this source to the observation location? **(e)** When we set up an integral to find the electric field at the observation location due to the entire rod, what will be the integration variable?

•P22 A plastic rod 1.7 m long is rubbed all over with wool, and acquires a charge of -2×10^{-8} C (Figure 15.52). We choose the center of the rod to be the origin of our coordinate system, with the x axis extending to the right, the y axis extending up, and the z axis out of the page. In order to calculate the electric field at location $A = \langle 0.7, 0, 0\rangle$ m, we divide the rod into eight pieces, and approximate each piece as a point charge located at the center of the piece.

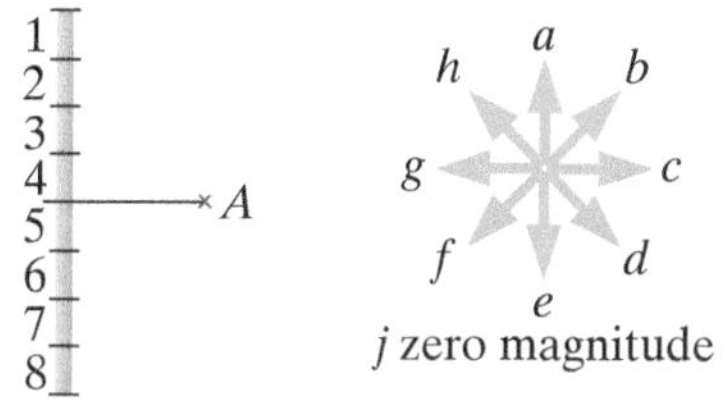

Figure 15.52

(a) What is the length of one of these pieces? **(b)** What is the location of the center of piece number 3? **(c)** How much charge is on piece number 3? (Remember that the charge is negative.) **(d)** Approximating piece 3 as a point charge, what is the electric field at location A due only to piece 3? **(e)** To get the net electric field at location A, we would need to calculate $\Delta\vec{E}$ due to each of the eight pieces, and add up these contributions. If we did that, which arrow (a–h) would best represent the direction of the net electric field at location A?

•P23 A clear plastic pen 12 cm long is rubbed all over with wool, and acquires a negative charge of -2 nC. You want to figure out the electric field a distance of 18 mm from the pen, near the middle of the pen. **(a)** You decide to model the pen as a rod consisting of a series of five segments, each of which you will consider to be approximately point-like. What is the length of each segment in meters? **(b)** What is the amount of charge ΔQ on each of the five segments? **(c)** In general, if the rod has a length L and total charge Q, and you divide the rod into N segments, what is the amount of charge ΔQ on each piece? **(d)** If the length of each segment is dL, write a symbolic expression for the number of pieces N in terms of the length of the rod L and the length of one piece dL. **(e)** Now write a symbolic expression for the amount of charge on each piece in terms of the length of the rod and the length of a small piece.

•P24 A thin glass rod of length 80 cm is rubbed all over with wool and acquires a charge of 60 nC, distributed uniformly over its surface. Calculate the magnitude of the electric field due to the rod at a location 7 cm from the midpoint of the rod. Do the calculation two ways, first using the exact equation for a rod of any length, and second using the approximate equation for a long rod.

•••P25 A water molecule is a permanent dipole with a known dipole moment p ($=qs$). There is a water molecule in the air a very short distance x from the midpoint of a long glass rod of length L carrying a uniformly distributed positive charge $+Q$ (Figure 15.53). The axis of the dipole is perpendicular to the rod. Note that $s \ll x \ll L$. (The charged rod induces an increase in the dipole moment, but the induced portion of the dipole moment is completely negligible compared to p. It is convenient to use the "binomial expansion" that you may have learned in calculus, that $(1+\epsilon)^n \approx 1+n\epsilon$ is $\epsilon \ll 1$. Note that n can be negative.)

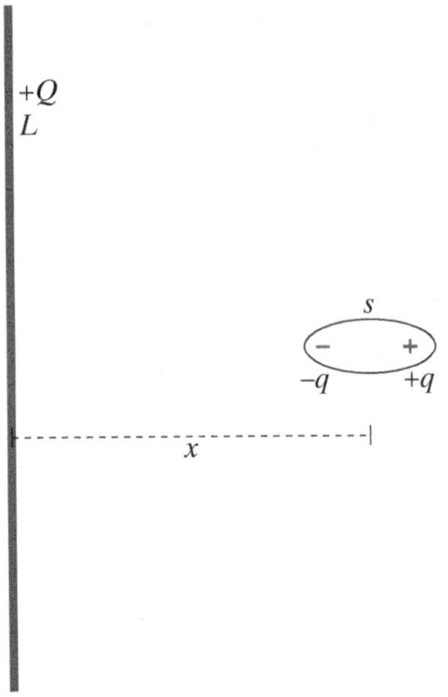

Figure 15.53

(a) Find the magnitude and direction of the electric force acting on the water molecule. Your final result for the magnitude of the force must be expressed in terms of Q, L, x, and p. You can use q and s in your calculations, but your final result must not include q or s, since it is only their product $p = qs$ that is measurable for a water molecule. Explain your work carefully, including appropriate diagrams. **(b)** If the electric field of the rod has the magnitude 1×10^6 N/C at the location of the water molecule, 1 cm from the rod, what is the magnitude of the horizontal component of the acceleration of the water molecule? The measured dipole moment for H_2O is 6.2×10^{-30} C · m, and the mass of one mole is 18 g (1+1+16). Be sure to check units in your calculation!

•••P26 An electrostatic dust precipitator that is installed in a factory smokestack includes a straight metal wire of length $L = 0.8$ m that is charged approximately uniformly with a total charge $Q = 0.4 \times 10^{-7}$ C. A speck of coal dust (which is mostly carbon) is near the wire, far from both ends of the wire; the distance from the wire to the speck is $d = 1.5$ cm. Carbon has an atomic mass of 12 (6 protons and 6 neutrons in the nucleus). A careful measurement of the polarizability of a carbon atom gives the value

$$\alpha = 1.96 \times 10^{-40} \frac{\text{C} \cdot \text{m}}{\text{N/C}}$$

(a) Calculate the initial acceleration of the speck of coal dust, neglecting gravity. Explain your steps clearly. Your answer must be expressed in terms of Q, L, d, and α. You can use other quantities in your calculations, but your final result must not include them. Don't put numbers into your calculation until the very end, but then show the numerical calculation that you carry out on your calculator. It is convenient to use the "binomial expansion" that you may have learned in calculus, that $(1+\epsilon)^n \approx 1+n\epsilon$ is $\epsilon \ll 1$. Note that n can be negative. **(b)** If the speck of coal dust were initially twice as far from the charged wire, how much smaller would be the initial acceleration of the speck?

Section 15.2

••P27 Consider a thin plastic rod bent into an arc of radius R and angle α (Figure 15.54). The rod carries a uniformly distributed negative charge $-Q$.

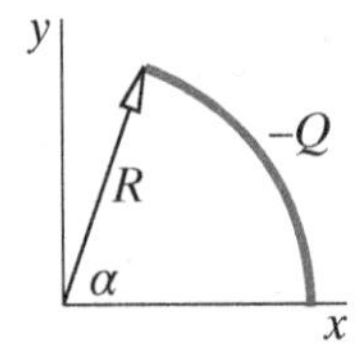

Figure 15.54

What are the components E_x and E_y of the electric field at the origin? Follow the standard four steps: **(a)** Use a diagram to explain how you will cut up the charged rod, and draw the $\Delta\vec{E}$ contributed by a representative piece. **(b)** Express algebraically the contribution each piece makes to the x and y components of the electric field. Be sure to show your integration variable and its origin on your drawing. (*Hint:* An arc of radius R and angle $\Delta\theta$ measured in radians has a length $R\Delta\theta$.) **(c)** Write the summation as an integral, and simplify the integral as much as possible. State explicitly the range of your integration variable. Evaluate the integral. **(d)** Show that your result is reasonable. Apply as many tests as you can think of.

••P28 A strip of invisible tape 0.12 m long by 0.013 m wide is charged uniformly with a total net charge of 3 nC (nano = 1×10^{-9}) and is suspended horizontally, so it lies along the x axis, with its center at the origin, as shown in Figure 15.55. Calculate the approximate electric field at location $\langle 0, 0.03, 0 \rangle$ m (location A) due to the strip of tape. Do this by dividing the strip into three equal sections, as shown in Figure 15.55, and approximating each section as a point charge.

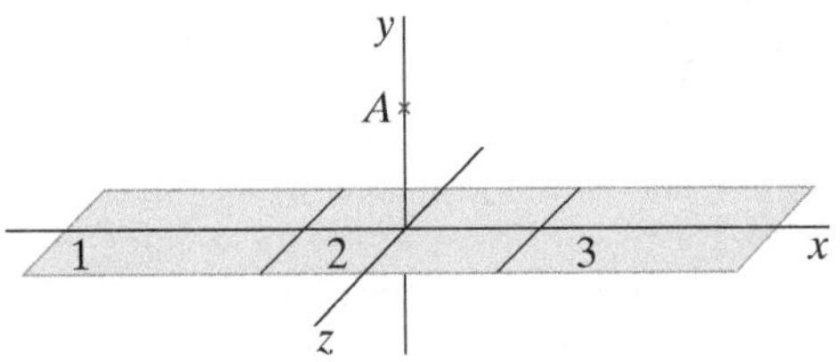

Figure 15.55

(a) What is the approximate electric field at A due to piece 1? **(b)** What is the approximate electric field at A due to piece 2? **(c)** What is the approximate electric field at A due to piece 3? **(d)** What is the approximate net electric field at A? **(e)** What could you do to improve the accuracy of your calculation?

••**P29** Consider a thin glass rod of length L lying along the x axis with one end at the origin. The rod carries a uniformly distributed positive charge Q.

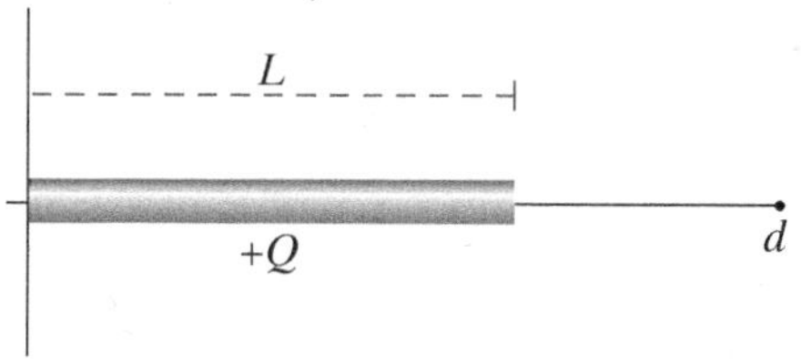

Figure 15.56

At a location $d > L$, on the x axis to the right of the rod in Figure 15.56, what is the electric field due to the rod? Follow the standard four steps. **(a)** Use a diagram to explain how you will cut up the charged rod, and draw the $\Delta\vec{E}$ contributed by a representative piece. **(b)** Express algebraically the contribution each piece makes to the electric field. Be sure to show your integration variable and its origin on your drawing. **(c)** Write the summation as an integral, and simplify the integral as much as possible. State explicitly the range of your integration variable. Evaluate the integral. **(d)** Show that your result is reasonable. Apply as many tests as you can think of.

••**P30** Consider a thin plastic rod bent into a semicircular arc of radius R with center at the origin (Figure 15.57). The rod carries a uniformly distributed negative charge $-Q$.

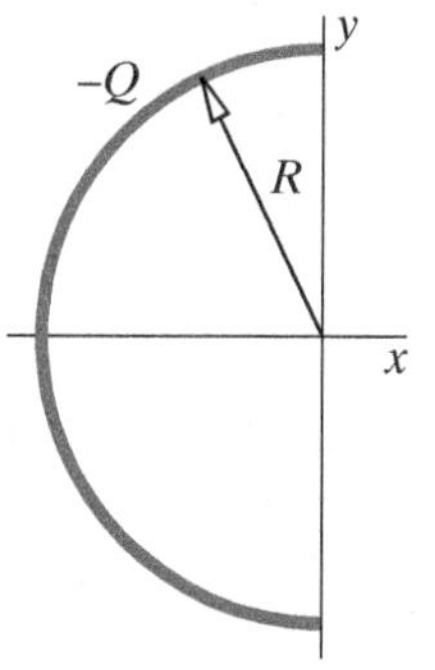

Figure 15.57

(a) Determine the electric field $\vec{E}$ at the origin contributed by the rod. Include carefully labeled diagrams, and be sure to check your result. **(b)** An ion with charge $-2e$ and mass M is placed at rest at the origin. After a *very* short time Δt the ion has moved only a very short distance but has acquired some momentum $\vec{p}$. Calculate $\vec{p}$.

Section 15.3

•**P31** Two rings of radius 5 cm are 24 cm apart and concentric with a common horizontal x axis. The ring on the left carries a uniformly distributed charge of +31 nC, and the ring on the right carries a uniformly distributed charge of −31 nC. **(a)** What are the magnitude and direction of the electric field on the x axis, halfway between the two rings? **(b)** If a charge of −9 nC were placed midway between the rings, what would be the force exerted on this charge by the rings?

•**P32** Two rings of radius 4 cm are 12 cm apart and concentric with a common horizontal x axis. The ring on the left carries a uniformly distributed charge of +40 nC, and the ring on the right carries a uniformly distributed charge of −40 nC. **(a)** What is the electric field due to the right ring at a location midway between the two rings? **(b)** What is the electric field due to the left ring at a location midway between the two rings? **(c)** What is the net electric field at a location midway between the two rings? **(d)** If a charge of −2 nC were placed midway between the rings, what would be the force exerted on this charge by the rings?

•**P33** Two rings of radius 2 cm are 20 cm apart and concentric with a common horizontal x axis. What is the magnitude of the electric field midway between the rings if both rings carry a charge of +35 nC?

•••**P34** A thin-walled hollow circular glass tube, open at both ends, has a radius R and length L. The axis of the tube lies along the x axis, with the left end at the origin (Figure 15.58). The outer sides are rubbed with silk and acquire a net positive charge Q distributed uniformly. Determine the electric field at a location on the x axis, a distance w from the origin. Carry out all steps, including checking your result. Explain each step. (You may have to refer to a table of integrals.)

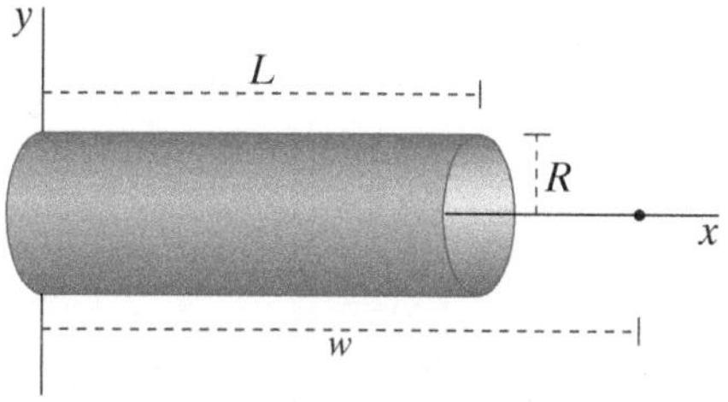

Figure 15.58

Section 15.4

•**P35** Suppose that the radius of a disk is 21 cm, and the total charge distributed uniformly all over the disk is 5.0×10^{-6} C. **(a)** Use the exact result to calculate the electric field 1 mm from the center of the disk. **(b)** Use the exact result to calculate the electric field 3 mm from the center of the disk. **(c)** Does the field decrease significantly?

•**P36** For a disk of radius $R = 20\,\text{cm}$ and $Q = 6 \times 10^{-6}\,\text{C}$, calculate the electric field 2 mm from the center of the disk using all three equations:

$$E = \frac{(Q/A)}{2\varepsilon_0}\left[1 - \frac{z}{(R^2+z^2)^{1/2}}\right],$$

$$E \approx \frac{Q/A}{2\varepsilon_0}\left[1 - \frac{z}{R}\right], \quad \text{and} \quad E \approx \frac{Q/A}{2\varepsilon_0}.$$

How good are the approximate equations at this distance? For the same disk, calculate E at a distance of 5 cm (50 mm) using all three equations. How good are the approximate equations at this distance?

•**P37** A disk of radius 16 cm has a total charge 4×10^{-6} C distributed uniformly all over the disk. **(a)** Using the exact equation, what is the electric field 1 mm from the center of the disk? **(b)** Using the same exact equation, find the electric field 3 mm from the center of the disk. **(c)** What is the percent difference between these two numbers?

•**P38** For a disk of radius 20 cm with uniformly distributed charge 7×10^{-6} C, calculate the magnitude of the electric field on the axis of the disk, 5 mm from the center of the disk, using each of the following equations:

(a) $E = \dfrac{(Q/A)}{2\varepsilon_0}\left[1 - \dfrac{z}{(R^2+z^2)^{1/2}}\right],$

(b) $E \approx \dfrac{Q/A}{2\varepsilon_0}\left[1-\frac{z}{R}\right]$,

(c) $E \approx \dfrac{Q/A}{2\varepsilon_0}$.

(d) How good are the approximate equations at this distance? **(e)** At what distance does the least accurate approximation for the electric field give a result that is closest to the most accurate: at a distance $R/2$, close to the disk, at a distance R, or far from the disk?

••**P39** A large, thin plastic disk with radius $R = 1.5$ m carries a uniformly distributed charge of $-Q = -3 \times 10^{-5}$ C as shown in Figure 15.59. A circular piece of aluminum foil is placed $d = 3$ mm from the disk, parallel to the disk. The foil has a radius of $r = 2$ cm and a thickness $t = 1$ mm.

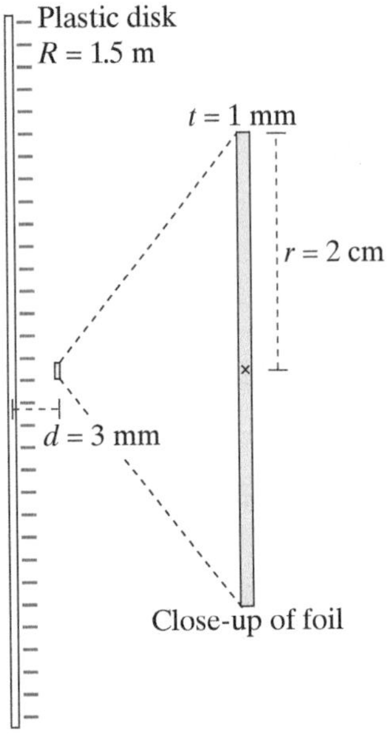

Figure 15.59

(a) Show the charge distribution on the close-up of the foil. **(b)** Calculate the magnitude and direction of the electric field at location × at the center of the foil, inside the foil. **(c)** Calculate the magnitude q of the charge on the left circular face of the foil.

••**P40** A thin circular sheet of glass of diameter 3 m is rubbed with a cloth on one surface and becomes charged uniformly. A chloride ion (a chlorine atom that has gained one extra electron) passes near the glass sheet. When the chloride ion is near the center of the sheet, at a location 0.8 mm from the sheet, it experiences an electric force of 5×10^{-15} N, toward the glass sheet. It will be useful to you to draw a diagram on paper, showing field vectors, force vectors, and charges, before answering the following questions about this situation.

Which of the following statements about this situation are correct? Select all that apply. (1) The electric field that acts on the chloride ion is due to the charge on the glass sheet and to the charge on the chloride ion. (2) The electric field of the glass sheet is equal to the electric field of the chloride ion. (3) The charged disk is the source of the electric field that causes the force on the chloride ion. (4) The net electric field at the location of the chloride ion is zero. (5) The force on the chloride ion is equal to the electric field of the glass sheet.

In addition to an exact equation for the electric field of a disk, the text derives two approximate equations. In the current situation we want an answer that is correct to three significant figures. Which of the following is correct? We should not use an approximation if we have enough information to do an exact calculation. (1) $R \gg z$, so it is adequate to use the most approximate equation here. (2) z is nearly equal to R, so we have to use the exact equation. (3) $z \ll R$, so we can't use an approximation.

How much charge is on the surface of the glass disk? Give the amount, including sign and correct units.

Section 15.5

•**P41** If the magnitude of the electric field in air exceeds roughly 3×10^6 N/C, the air breaks down and a spark forms. For a two-disk capacitor of radius 47 cm with a gap of 1 mm, what is the maximum charge (plus and minus) that can be placed on the disks without a spark forming (which would permit charge to flow from one disk to the other)?

•**P42** Consider a capacitor made of two rectangular metal plates of length L and width W, with a very small gap s between the plates. There is a charge $+Q$ on one plate and a charge $-Q$ on the other. Assume that the electric field is nearly uniform throughout the gap region and negligibly small outside. Calculate the attractive force that one plate exerts on the other. Remember that one of the plates doesn't exert a net force on itself.

•**P43** A capacitor consists of two large metal disks of radius 1.1 m placed parallel to each other, a distance of 1.2 mm apart. The capacitor is charged up to have an increasing amount of charge $+Q$ on one disk and $-Q$ on the other. At about what value of Q does a spark appear between the disks?

••**P44** If the magnitude of the electric field in air exceeds roughly 3×10^6 N/C, the air breaks down and a spark forms. For a two-disk capacitor of radius 51 cm with a gap of 2 mm, if the electric field inside is just high enough that a spark occurs, what is the strength of the fringe field just outside the center of the capacitor?

••**P45** In a cathode-ray tube, an electron travels in a vacuum and enters a region between two deflection plates where there is an upward electric field of magnitude 1×10^5 N/C (Figure 15.60).

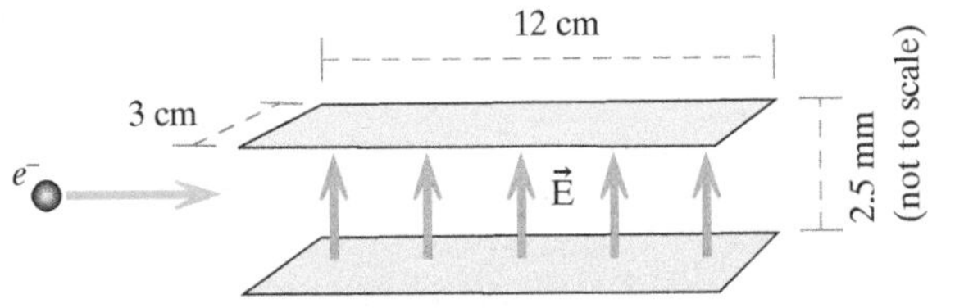

Figure 15.60

(a) Sketch the trajectory of the electron, continuing on well past the deflection plates (the electron is going fast enough that it does not strike the plates). **(b)** Calculate the acceleration of the electron while it is between the deflection plates. **(c)** The deflection plates measure 12 cm by 3 cm, and the gap between them is 2.5 mm. The plates are charged equally and oppositely. What are the magnitude and sign of the charge on the upper plate?

••**P46** In Figure 15.61 are two uniformly charged disks of radius R that are very close to each other (gap $\ll R$). The disk on the left has a charge of $-Q_{\text{left}}$ and the disk on the right has a charge of $+Q_{\text{right}}$ (Q_{right} is greater than Q_{left}). A uniformly charged thin rod of length L lies at the edge of the disks, parallel to the axis of the disks and centered on the gap. The rod has a charge of $+Q_{\text{rod}}$.

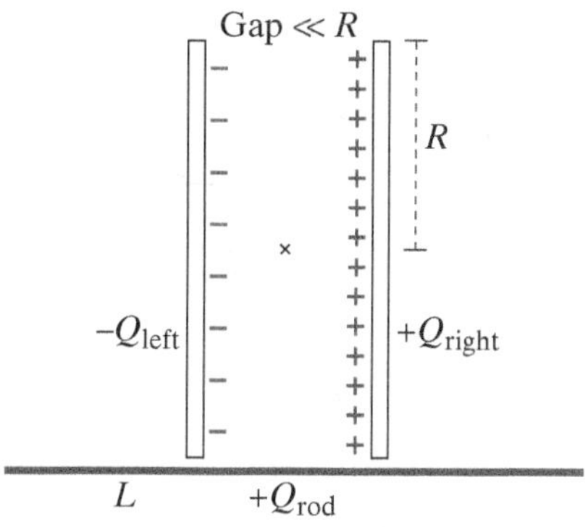

Figure 15.61

(a) Calculate the magnitude and direction of the electric field at the point marked × at the center of the gap region, and explain briefly, including showing the electric field on a diagram. Your results must not contain any symbols other than the given quantities R, Q_{left}, Q_{right}, L, and Q_{rod} (and fundamental constants), unless you define intermediate results in terms of the given quantities. **(b)** If an electron is placed at the center of the gap region, what are the magnitude and direction of the electric force that acts on the electron?

••P47 A capacitor made of two parallel uniformly charged circular metal disks carries a charge of $+Q$ and $-Q$ on the inner surfaces of the plates and very small amounts of charge $+q$ and $-q$ on the outer surfaces of the plates. Each plate has a radius R and thickness t, and the gap distance between the plates is s. How much charge q is on the outside surface of the positive disk, in terms of Q?

•••P48 This is a challenging problem that requires you to construct a model of a real physical situation, to make idealizations, approximations, and assumptions, and to work through a detailed analysis based on physical principles. Be sure to allow yourself enough time to think it through. You may need to measure, estimate, or look up various quantities.

A clear plastic ball-point pen is rubbed thoroughly with wool. The charged plastic pen is held above a small, uncharged disk-shaped piece of aluminum foil, smaller than a hole in a sheet of three-ring binder paper. **(a)** Make a clear physics diagram of the situation, showing charges, fields, forces, and distances. Refer to this physics diagram in your analysis. **(b)** Starting from fundamental physical principles, predict quantitatively how close you must move the pen to the foil in order to pick up the foil (that is, predict an actual numerical distance). State explicitly all approximations and assumptions that you make. **(c)** Try the experiment and compare your observation to your prediction.

Section 15.6

•P49 A solid metal ball of radius 1.5 cm bearing a charge of −17 nC is located near a solid plastic ball of radius 2 cm bearing a uniformly distributed charge of +7 nC (Figure 15.62) on its outer surface. The distance between the centers of the balls is 9 cm. **(a)** Show the approximate charge distribution in and on each ball. **(b)** What is the electric field at the center of the metal ball due only to the charges on the plastic ball? **(c)** What is the net electric field at the center of the metal ball? **(d)** What is the electric field at the center of the metal ball due only to the charges on the surface of the metal ball?

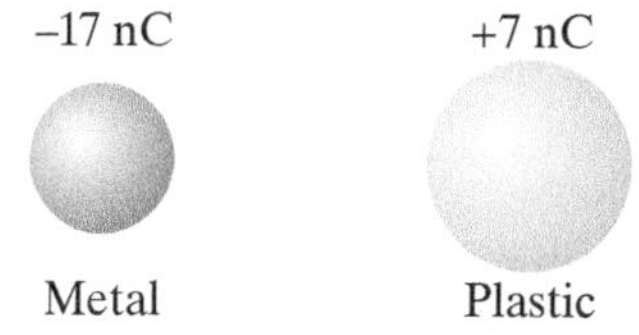

Figure 15.62

•P50 You stand at location A, a distance d from the origin, and hold a small charged ball. You find that the electric force on the ball is 0.08 N. You move to location B, a distance $2d$ from the origin, and find the electric force on the ball to be 0.04 N. What object located at the origin might be the source of the field? (1) A point charge, (2) A dipole, (3) A uniformly charged rod, (4) A uniformly charged ring, (5) A uniformly charged disk, (6) A capacitor, (7) A uniformly charged hollow sphere, (8) None of the above

If the force at B were 0.0799 N, what would be your answer? If the force at B were 0.01 N, what would be your answer? If the force at B were 0.02 N, what would be your answer?

•P51 A thin plastic spherical shell of radius 5 cm has a uniformly distributed charge of −25 nC on its outer surface. A concentric thin plastic spherical shell of radius 8 cm has a uniformly distributed charge of +64 nC on its outer surface. Find the magnitude and direction of the electric field at distances of 3 cm, 7 cm, and 10 cm from the center. See Figure 15.63.

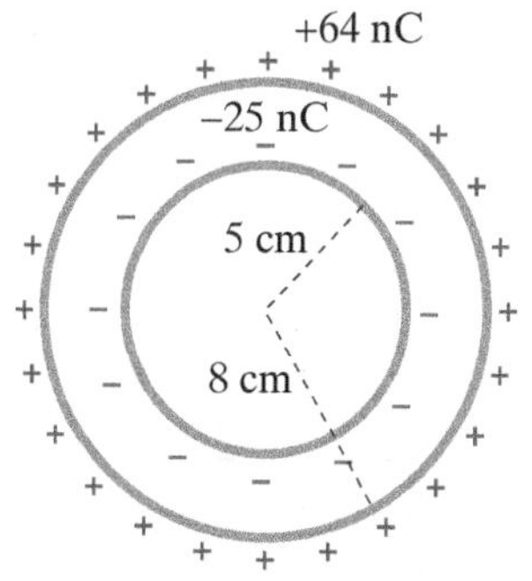

Figure 15.63

•P52 A hollow ball of radius 7 cm, made of very thin glass, is rubbed all over with a silk cloth and acquires a negative charge of -9×10^{-8} C that is uniformly distributed all over its surface. Location A in Figure 15.64 is inside the sphere, 1 cm from the surface. Location B in Figure 15.64 is outside the sphere, 2 cm from the surface. There are no other charged objects nearby.

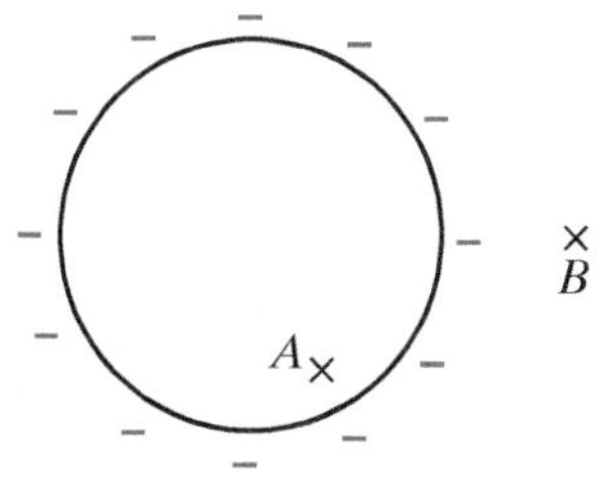

Figure 15.64

Which of the following statements about E_{ball}, the magnitude of the electric field due to the ball, are correct? Select all that apply. **(a)** At location A, E_{ball} is 0 N/C. **(b)** All of the charges on the surface of the sphere contribute to E_{ball} at location A. **(c)** A hydrogen atom at location A would polarize because it is close to the negative charges on the surface of the sphere. What is E_{ball} at location B?

••P53 A thin hollow spherical glass shell of radius 0.17 m carries a uniformly distributed positive charge $+6 \times 10^{-9}$ C, as shown in Figure 15.65. To the right of it is a horizontal permanent dipole with charges $+3 \times 10^{-11}$ and -3×10^{-11} C separated by a distance 2×10^{-5} m (the dipole is shown greatly enlarged for clarity). The dipole is fixed in position and is not free to rotate. The distance from the center of the glass shell to the center of the dipole is 0.6 m.

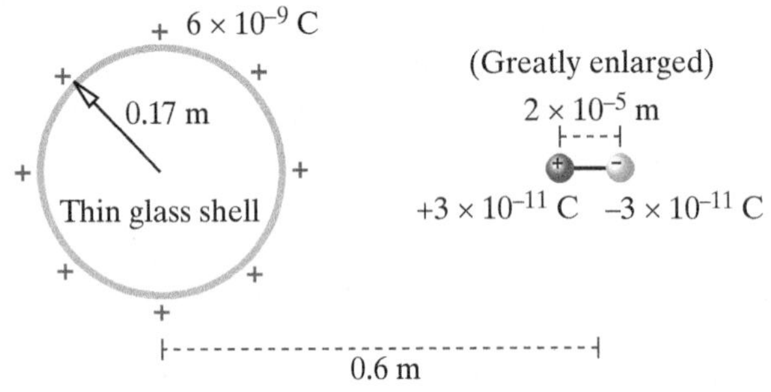

Figure 15.65

(a) Calculate the net electric field at the center of the glass shell. **(b)** If the sphere were a solid metal ball with a charge $+6 \times 10^{-9}$ C, what would be the net electric field at its center? **(c)** Draw the approximate charge distribution in and/or on the metal sphere.

••P54 Two large, thin, charged plastic circular plates each of radius R are placed a short distance s apart; s is much smaller than the dimensions of a plate (Figure 15.66). The right-hand plate has a positive charge of $+Q$ evenly distributed over its inner surface (Q is a positive number). The left-hand plate has a negative charge of $-2Q$ evenly distributed over its inner surface. A very thin plastic spherical shell of radius r is placed midway between the plates (and shown in cross section). It has a uniformly distributed positive charge of $+q$. You can ignore the contributions to the electric field due to the polarization of the thin plastic shell and the thin plastic plates.

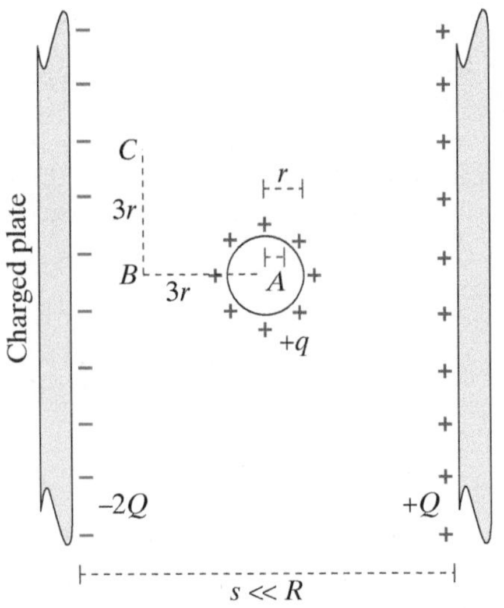

Figure 15.66

(a) Calculate the x and y components of the electric field at location A, a horizontal distance $r/2$ to the right of the center of the sphere. **(b)** Calculate the x and y components of the electric field at location B, a horizontal distance $3r$ to the left of the center of the sphere. **(c)** Calculate the x and y components of the electric field at location C, a horizontal distance $3r$ to the left and a vertical distance $3r$ above the center of the sphere.

••P55 A glass sphere carrying a uniformly distributed charge of $+Q$ is surrounded by an initially neutral spherical plastic shell (Figure 15.67).

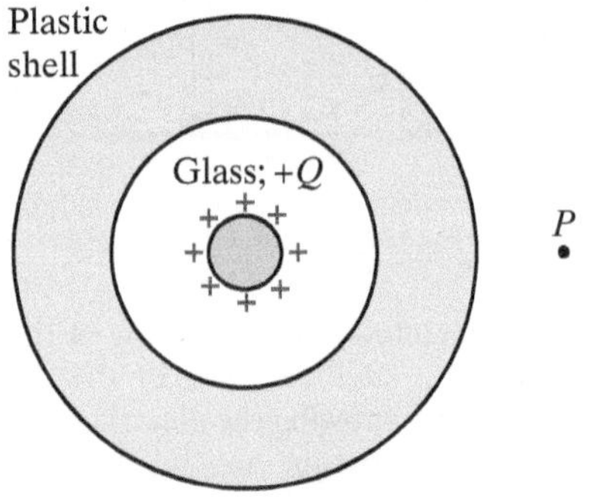

Figure 15.67

(a) Qualitatively, indicate the polarization of the plastic. **(b)** Qualitatively, indicate the polarization of the inner glass sphere. Explain briefly. **(c)** Is the electric field at location P outside the plastic shell larger, smaller, or the same as it would be if the plastic weren't there? Explain briefly. **(d)** Now suppose that the glass sphere carrying a uniform charge of $+Q$ is surrounded by an initially neutral metal shell (Figure 15.68). Qualitatively, indicate the polarization of the metal.

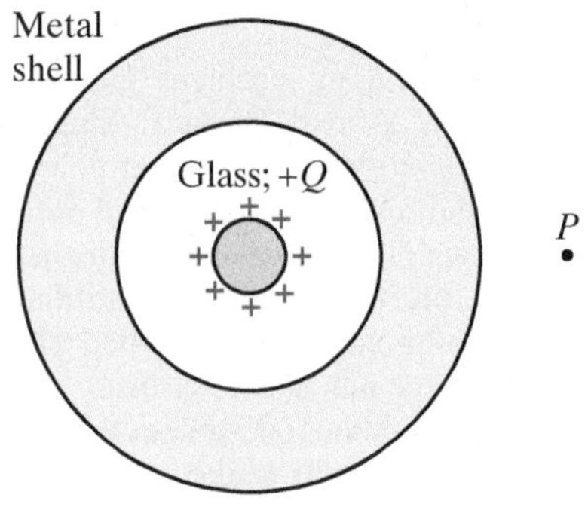

Figure 15.68

(e) Now be *quantitative* about the polarization of the metal sphere and prove your assertions. **(f)** Is the electric field at location P outside the metal shell larger, smaller, or the same as it would be if the metal shell weren't there? Explain briefly.

••P56 Breakdown field strength for air is roughly 3×10^6 N/C. If the electric field is greater than this value, the air becomes a conductor. **(a)** There is a limit to the amount of charge that you can put on a metal sphere in air. If you slightly exceed this limit, why would breakdown occur, and why would the breakdown occur very near the surface of the sphere, rather than somewhere else? **(b)** How much excess charge can you put on a metal sphere of 10-cm radius without causing breakdown in the neighboring air, which would discharge the sphere? **(c)** How much excess charge can you put on a metal sphere of only 1-mm radius? These results hint at the reason why a highly charged piece of metal tends to spark at places where the radius of curvature is small, or at places where there are sharp points.

••P57 Two thin plastic spherical shells (shown in cross section in Figure 15.69) are uniformly charged. The center of the larger sphere is at (0,0); it has a radius of 12 cm and a uniform positive charge of $+4 \times 10^{-9}$ C. The center of the smaller sphere is at (25 cm, 0); it has a radius of 3 cm and a uniform negative charge of -1×10^{-9} C. **(a)** What are the components $E_{A,x}$ and $E_{A,y}$ of the electric field $\Delta\vec{E}$ at location A (6 cm to the right of the center of the large sphere)? Neglect the small contribution of the polarized molecules in the plastic, because the shells are very thin and don't contain much matter. **(b)** What are the components $E_{B,x}$ and $E_{B,y}$ of the electric field at location B (15 cm above the center of the small sphere)? Neglect the small contribution of the polarized molecules in the plastic, because the shells are very thin and don't contain much matter. **(c)** What are the components F_x and F_y of the force on an electron placed at location B?

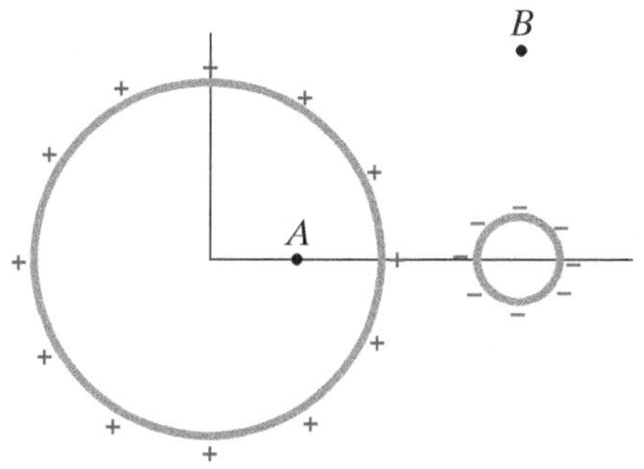

Figure 15.69

••P58 A solid plastic sphere of radius R_1 has a charge $-Q_1$ on its surface (Figure 15.70). A concentric spherical metal shell of inner radius R_2 and outer radius R_3 carries a charge Q_2 on the inner surface and a charge Q_3 on the outer surface. Q_1, Q_2, and Q_3 are positive numbers, and the total charge $Q_2 + Q_3$ on the metal shell is greater than Q_1.

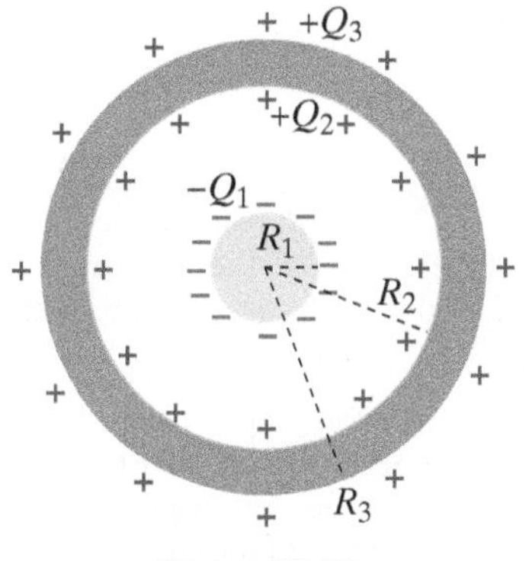

Figure 15.70

At an observation location a distance r from the center, determine the magnitude and direction of the electric field in the following regions, and explain briefly in each case. For parts (a)–(d), be sure to give both the direction and the magnitude of the electric field, and explain briefly: **(a)** $r < R_1$ (inside the plastic sphere), **(b)** $R_1 < r < R_2$ (in the air gap), **(c)** $R_2 < r < R_3$ (in the metal), **(d)** $r > R_3$ (outside the metal). **(e)** Suppose $-Q_1 = -5$ nC. What is Q_2? Explain fully on the basis of fundamental principles. **(f)** What can you say about the molecular polarization in the plastic? Explain briefly. Include a drawing if appropriate.

••P59 A small, thin, hollow spherical glass shell of radius R carries a uniformly distributed positive charge $+Q$. Below it is a horizontal permanent dipole with charges $+q$ and $-q$ separated by a distance s (s is shown greatly enlarged in Figure 15.71 for clarity). The dipole is fixed in position and is not free to rotate. The distance from the center of the glass shell to the center of the dipole is L.

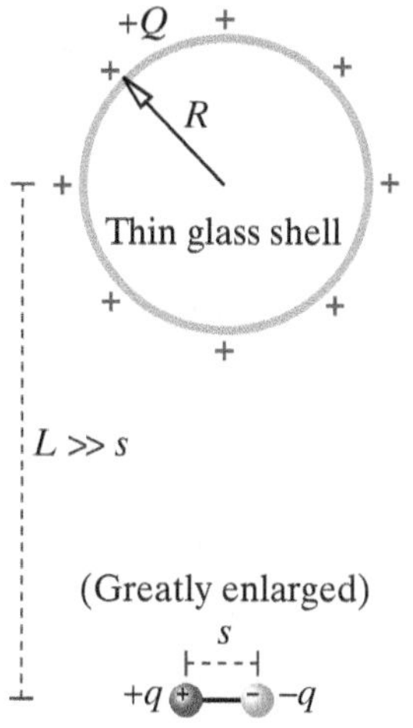

Figure 15.71

(a) Calculate the magnitude and direction of the electric field at the center of the glass shell, and explain briefly, including showing the electric field on the diagram. Your results must not contain any symbols other than the given quantities R, Q, q, s, and L (and fundamental constants), unless you define intermediate results in terms of the given quantities. What simplifying assumption do you have to make? **(b)** If the upper sphere were a solid metal ball with a charge $+Q$, what would be the magnitude and direction of the electric field at its center? Explain briefly. Show the distribution of charges everywhere, and at the center of the ball accurately draw and label the electric field due to the ball charges $\vec{E}_{\text{ball}}$ and the electric field $\vec{E}_{\text{dipole}}$ of the dipole.

••P60 A very thin glass rod 4 m long is rubbed all over with a silk cloth (Figure 15.72). It gains a uniformly distributed charge $+1.3 \times 10^{-6}$ C. Two small spherical rubber balloons of radius 1.2 cm are rubbed all over with wool. They each gain a uniformly distributed charge of -2×10^{-8} C. The balloons are near the midpoint of the glass rod, with their centers 3 cm from the rod. The balloons are 2 cm apart (4.4 cm between centers).

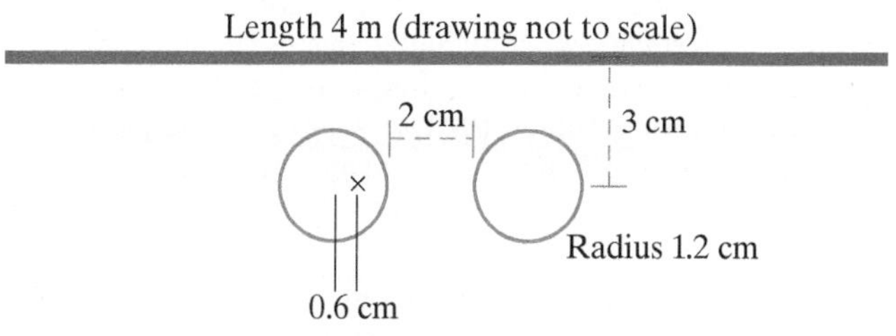

Figure 15.72

(a) Find the magnitude of the electric field at the location marked by the $\times$, 0.6 cm to the right of the center of the left balloon. Also calculate the angle the electric field makes with the horizontal. Show all your work, including showing vectors. State any approximations you are making. **(b)** Next, a neutral hydrogen atom is placed at that same location (marked by the $\times$). Draw a diagram showing the effect on the hydrogen atom while it is at that location. The polarizability of atomic hydrogen has been measured to be $\alpha = 7.4 \times 10^{-41}\ \text{C}\cdot\text{m}/(\text{N}/\text{C})$. What is the distance between the center of the proton and the center of the electron cloud in the hydrogen atom?

••P61 A very long thin wire of length L carries a total charge $+Q$, nearly uniformly distributed along the wire (Figure 15.73). The center of a small neutral metal ball of radius r is located a distance d from the wire, where $d \ll L$. Determine the approximate magnitude of the force that the rod exerts on the ball. Show labeled physics information on the diagram. Explain how you model the situation.

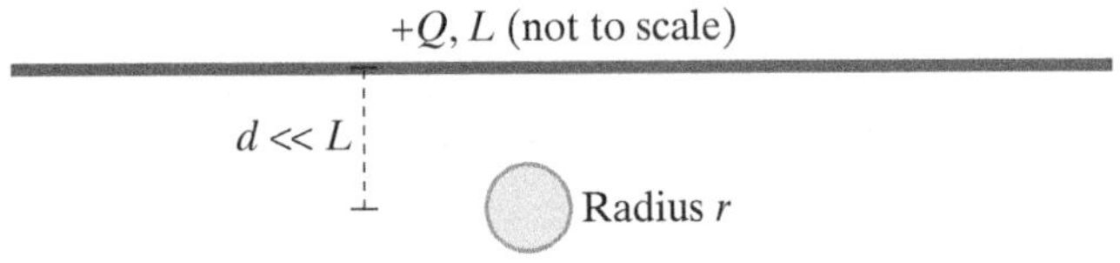

Figure 15.73

Section 15.7

•••P62 A simplified model of a hydrogen atom is that the electron cloud is a sphere of radius R with uniform charge density and total charge $-e$. (The actual charge density in the ground state is nonuniform.) See Figure 15.74.

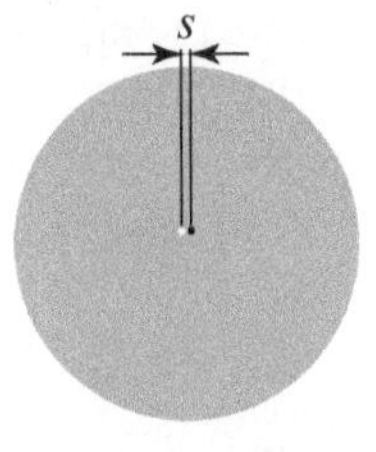

Figure 15.74

(a) For the uniform-density model, calculate the polarizability α of atomic hydrogen in terms of R. Consider the case where the magnitude E of the applied electric field is much smaller than the electric field required to ionize the atom. Suggestions for your analysis: Imagine that the hydrogen atom is inside a capacitor whose uniform field polarizes but does not accelerate the atom. Consider forces on the proton in the equilibrium situation, where the proton is displaced a distance s from the center of the electron cloud ($s \ll R$ in the diagram). **(b)** For a hydrogen atom, R can be taken as roughly 1×10^{-10} m (the Bohr model of the H atom gives $R = 0.5 \times 10^{-10}$ m). Calculate a numerical value for the polarizability α of atomic hydrogen. For comparison, the measured polarizability of a hydrogen atom is $\alpha = 7.4 \times 10^{-41}\ \text{C}\cdot\text{m}(\text{N}/\text{C})$; see the note below. **(c)** If the magnitude E of the applied electric field is 1×10^{6} N/C, use the measured value of α to calculate the shift s shown in Figure 15.74. **(d)** For some purposes it is useful to model an atom as though the nucleus and electron cloud were connected by a spring. Use the measured value of α to calculate the effective spring stiffness k_s for atomic hydrogen. For comparison, measurements of Young's modulus show that the effective spring stiffness of the *interatomic* force in solid aluminum is about 16 N/m. **(e)** If α were twice as large, what would k_s be?

Note: Quantum-mechanical calculations agree with the experimental measurement of α reported in T. M. Miller and B. Bederson, "Atomic and molecular polarizabilities: a review of recent advances," *Advances in Atomic and Molecular Physics* **13**, 1–55, 1977. They use cgs units, so their value is $1/(4\pi\varepsilon_0)$ greater than the value given here.

COMPUTATIONAL PROBLEMS

More detailed and extended versions of some of these computational modeling problems may be found in the lab activities included in the *Matter & Interactions* resources for instructors.

•P63 Start with the program in Section 15.9 that calculates the electric field of a uniformly charged rod at a single location on the midplane. **(a)** Improve the accuracy of the calculation by increasing the number of point charges used to model the rod. Determine experimentally the minimum number of point charges required to get an accurate answer, and explain your criterion for accuracy. **(b)** Use your improved program to calculate the electric field of the rod at location $\langle 0.05, -0.35, 0\rangle$ m, and display your result with an arrow. Report the value of the electric field you calculated.

•P64 Write a program to model a long thin rod of length 0.5 m with a net charge of -2 nC, lying on the x axis and centered at the origin. **(a)** Calculate the electric field of the rod at location $\langle -0.4, 0.3, 0.3\rangle$ m, and display the field with an arrrow. **(b)** Determine experimentally the minimum number of point charges required to give an accurate result, and explain how you determined this.

••P65 A strip of invisible tape 12 cm by 2 cm is charged uniformly with a total net charge of 4×10^{-8} and hangs vertically. Write a program to calculate and display both the tape and the electric field due to this strip of tape, at a location a perpendicular distance of 3 cm from the center of the tape. Use the same scale factor to display the arrow representing the electric field in all parts of this problem. **(a)** Divide the strip into three sections, each 4 cm high by 2 cm wide, and use numerical summation to calculate the magnitude of the electric field at a perpendicular distance of 3 cm from the center of the tape. **(b)** Divide the strip into eight sections, each 3 cm high by 1 cm wide, and use numerical summation to calculate the magnitude of the electric field at a perpendicular distance of 3 cm from the center of the tape. Compare the result to your previous result. **(c)** Divide the tape into a large number of small sections. Compare your result to your previous results.

•P66 Start with the program in Section 15.9 that calculates the electric field of a uniformly charged rod, modeled as 10 point charges, at multiple observation locations in a circle around the rod. **(a)** Add more arrows to the list named "observation," so that the electric field of the rod is displayed with 5 rings of arrows centered on the rod, at different locations. **(b)** Increase the number of point charges used to model the rod to increase the accuracy of the calculation. Explain your criteria for determining an appropriate number of point charges.

••P67 Write a computer program to compute and display the electric field of a thin rod of length 2 m, aligned along the z axis and centered at the origin. A total charge of -60 nC is uniformly distributed along the surface of the rod. Start by modeling the rod as a collection of 5 point charges. **(a)** Increase the number of point charges until you are satisfied with the accuracy of the calculation. Explain the criteria you used to make this decision. **(b)** Display the electric field at locations around the rod with arrows arranged in 6 rings centered on the rod, at different z locations along the rod. Scale the arrows appropriately to fit the display.

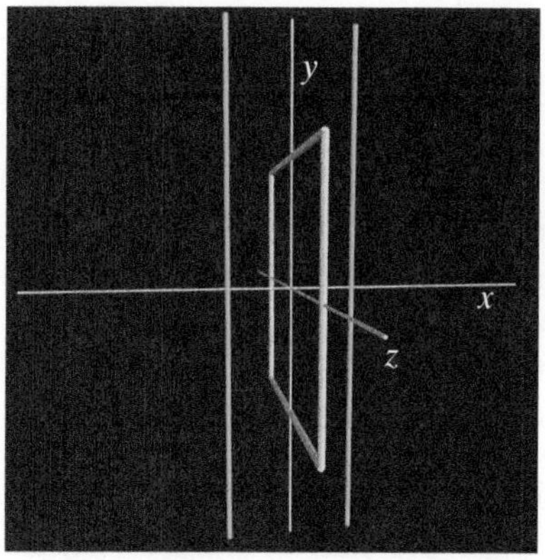

Figure 15.75

•••P68 Four thin rods, each of length 0.4 m, form a square loop in the yz plane, as shown in yellow in Figure 15.75. A charge of 3 nC is uniformly distributed over the surface of each rod (the net charge of the loop is 12 nC). Write a program to calculate and display the electric field of the loop at 12 observation locations: six that lie on a vertical line that crosses the x axis at $x = -0.1$ m, and six that lie on a vertical line that crosses the x axis at $x = +0.1$ m. These lines are shown in cyan in Figure 15.75.

•••P69 Model a uniformly charged ring as a set of point charges placed on a circle. Place the ring in the yz plane so that its axis is along the x axis. Give the ring a total charge of $Q = 50 \times 10^{-9}$ C and a radius of $R = 0.1$ m. **(a)** Start an electron from rest at location $\langle 0.15, 0, 0 \rangle$ m, and model its motion. In each step you will need to calculate the net electric field at the location of the electron. Describe the motion you observe. **(b)** Vary `deltat` and the number of point charges used to model the uniformly charged ring to make sure your calculation is accurate. **(c)** Experiment with different initial locations that do not lie on the axis of the ring. You may also wish to experiment with giving the electron some initial momentum. Make a screen shot of the most interesting trajectory you find. Figure 15.76 is an example of one such trajectory.

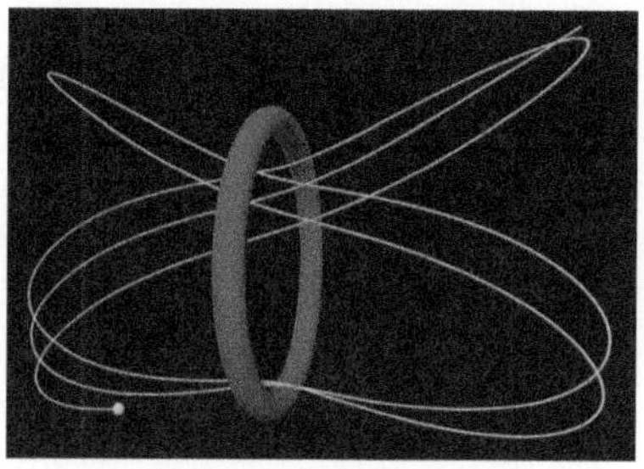

Figure 15.76

ANSWERS TO CHECKPOINTS

1 1.12×10^4 N/C
2 **(a)** 4.51×10^4 N/C to the right; **(b)** 2.25×10^{-4} N to the left; **(c)** zero
3 2.684×10^6 N/C; 2.657×10^6 N/C
4 About 2.1×10^{-5} C; about 3000 N/C
5 Although the other charges are much farther away, there are a lot more of them, and in a sphere the two effects exactly cancel each other. The electric field at location D is zero.

CHAPTER

16 Electric Potential

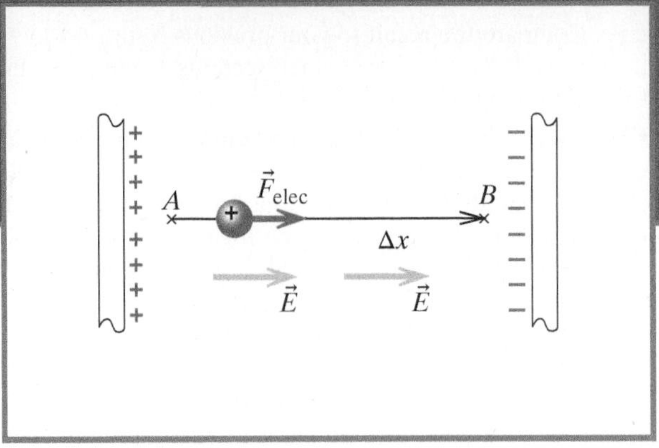

OBJECTIVES

After studying this chapter you should be able to

- Mathematically relate electric field, electric potential energy, and electric potential
- Calculate differences in electric potential in and around conductors and insulators

In analyzing mechanical systems, both at the macroscopic and the microscopic level, we found that it was frequently important to consider not only forces and momenta, but also work and energy, in trying to model the behavior of a physical system. Similarly, to complement our use of the concept of electric field and electric force, we need the concept of electric potential. Electric potential is defined as electric potential energy per unit charge.

The concept of electric potential is useful for some of the same reasons that the concept of electric field is useful. It allows us to reason about energy in a range of situations without having to worry about the details of some particular distribution of point charges. Electric potential has practical importance, in part because batteries and electric generators maintain a potential difference across themselves that is nearly independent of what is connected to them. The concept also provides significant theoretical power, enabling us to draw conclusions about a surprising range of issues, including, for example, what patterns of electric field in space are possible, and the magnitude and direction of the average electric field inside an insulator due to the polarization of molecules in the insulator.

16.1 A REVIEW OF POTENTIAL ENERGY

This section is a review of the concept of potential energy, which was first introduced in Chapter 6. If this concept is very familiar to you, you may wish to skip ahead to Section 16.2, in which we apply the concept to systems of interacting charged particles similar to those that we will typically consider in this volume.

Energy of a Single Particle

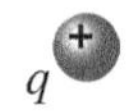

Figure 16.1 No electric potential energy is associated with a single charged particle.

The energy of a single particle with charge q_1 (Figure 16.1) consists solely of its particle energy. The particle energy is the sum of the rest energy mc^2 of the particle and its kinetic energy K.

$$\text{Particle energy} = mc^2 + K$$

Most of the processes we will study in this volume are low-energy processes that do not involve significant changes in rest energy. (In contrast, processes such as fission and fusion, in which atomic nuclei combine or split apart to form different elements, involve the release of very large amounts of energy. These are the subject of some of the applications in Chapter 6.)

Kinetic energy is energy associated with motion. The kinetic energy of a particle moving at low speed may be approximated as

APPROXIMATE KINETIC ENERGY OF A PARTICLE

$$K \approx \frac{1}{2}mv^2 \quad \text{for } v \ll c$$

Recall that energy is a scalar quantity, not a vector, so it depends only on the particle's speed but not the direction of its motion. (For a discussion of the relativistically correct form for particle energy and kinetic energy, see Chapter 6.)

A Reminder about the Symbol Δ

In this chapter we will be discussing changes in energy. Remember that the symbol Δ (uppercase Greek delta) indicates a change in a quantity, calculated by subtracting the initial value of the quantity from its final value.

$$\Delta K = K_{\text{final}} - K_{\text{initial}}$$

EXAMPLE **Accelerated Electron**

An electron is traveling at a speed of 6×10^4 m/s. After passing through a region in which there is an electric field, its speed is 2×10^4 m/s. (Note that these speeds are small compared to the speed of light.) What is the change in kinetic energy of the electron?

Solution

$$K_{\text{initial}} \approx \frac{1}{2}m_e v_i^2 = \frac{1}{2}(9 \times 10^{-31}\text{ kg})(6 \times 10^4\text{ m/s})^2 = 1.6 \times 10^{-21}\text{ J}$$

$$K_{\text{final}} \approx \frac{1}{2}m_e v_f^2 = \frac{1}{2}(9 \times 10^{-31}\text{ kg})(2 \times 10^4\text{ m/s})^2 = 1.8 \times 10^{-22}\text{ J}$$

$$\Delta K = 1.8 \times 10^{-22}\text{ J} - 1.6 \times 10^{-21}\text{ J} = -1.42 \times 10^{-21}\text{ J}$$

Since the electric force slowed down the electron, it makes sense that $K_{\text{final}} < K_{\text{initial}}$, the electron's kinetic energy decreased, and $\Delta K < 0$.

Checkpoint 1 A proton initially travels at a speed of 3000 m/s. After it passes through a region in which there is an electric field, the proton's speed is 5000 m/s. **(a)** What is the initial kinetic energy of the proton? **(b)** What is the final kinetic energy of the proton? **(c)** What is the change in kinetic energy of the proton?

A Single Particle Does Not Have Potential Energy

When analyzing changes in energy, we must choose a *system* consisting of one or more objects to analyze. In choosing the system, we must decide what objects to include in the system; all objects not included in the system are part of the surroundings. We are free to choose whatever system we find convenient or interesting—the laws of physics apply to any choice of system. However, the choice of system determines the forms of energy we need to consider.

One possible choice of system is to choose a single particle as the system and classify all other particles as part of the surroundings. In this case, the system can have only the two kinds of energy mentioned above: rest energy and kinetic energy. Potential energy is associated with interactions between particles in a system. A single particle does not have potential energy.

The kinetic energy of a single particle can be changed if positive or negative work is done on the particle by its surroundings.

CHANGE IN ENERGY OF A SINGLE-PARTICLE SYSTEM

$$\Delta K_{\text{sys}} = W_{\text{surr}}$$

Work

As discussed in Chapter 6, work is a scalar quantity, with units of joules. The work done on a particle is simply the product of the force exerted on the particle times the displacement through which the force acts. Because force is a vector and displacement is a vector, work is a sum:

$$W = F_x\Delta x + F_y\Delta y + F_z\Delta z$$

This sum can be written in a compact form as the "dot product" of two vectors.

$$\vec{A}\cdot\vec{B} = (A_xB_x + A_yB_y + A_zB_z) \quad \text{(a scalar quantity)}$$

so work can be written as

$$W = \vec{F}\cdot\Delta\vec{l} = \langle F_x, F_y, F_z\rangle\cdot\langle\Delta x, \Delta y, \Delta z\rangle = F_x\Delta x + F_y\Delta y + F_z\Delta z$$

where $\langle\Delta x, \Delta y, \Delta z\rangle = \Delta\vec{l}$ denotes the displacement of the particle.

It is occasionally useful to know another, equivalent, evaluation equation for a vector dot product:

$$\vec{A}\cdot\vec{B} = |\vec{A}||\vec{B}|\cos\theta$$

where θ is the angle between the two vectors.

EXAMPLE **Work Done by Electric Force**

A dust particle with charge 2×10^{-11}C moves from $\langle 0.1, -0.3, 0.4\rangle$ m to $\langle 0.2, -0.3, -0.2\rangle$ m. In this region there is an electric field of $\langle 2000, 0, 4000\rangle$ N/C. How much work is done on the dust particle by the electric force?

Solution

$$\Delta\vec{l} = \langle 0.2, -0.3, -0.2\rangle \text{ m} - \langle 0.1, -0.3, 0.4\rangle \text{ m} = \langle 0.1, 0, -0.6\rangle \text{ m}$$

$$W = [(2\times10^{-11})(2000)(0.1) + (2\times10^{-11})(0)(0) + (2\times10^{-11})(4000)(-0.6)]\text{ J}$$
$$= -4.4\times10^{-8}\text{ J}$$

The work done on the particle was negative, so the kinetic energy of the particle decreased.

Potential Energy Is Associated with Pairs of Interacting Objects

A different choice is to include two or more particles in the system. In a system consisting of two or more interacting particles, there can be energy associated with the interactions of particles inside the system. If the rest energy of the particles does not change (that is, if the identity of the particles does not change), then the kinetic energy of the system can be changed either by work done by forces exerted by objects in the surroundings (W_{surr}) or by work done by forces due to interactions between particles inside the system (internal work, W_{int}). (In general, there can also be other kinds of transfer of energy into the system, including transfers due to a temperature difference between the system and the surroundings (Q). In this volume, we will deal primarily with situations in which there are no energy transfers other than mechanical work W, and we will omit these other terms from the Energy Principle equation.)

$$\Delta K_{sys} = W_{surr} + W_{int}$$

We can rearrange this equation to put all terms associated only with particles inside the system on the left side of the equation:

$$\text{Energy change in system} = \text{external energy input}$$
$$\Delta K_{sys} + (-W_{int}) = W_{surr}$$

The term $-W_{int}$, the negative of the work done by internal forces, is called the change in the *potential energy* U of the system of particles:

DEFINITION: CHANGE IN POTENTIAL ENERGY OF A SYSTEM

$$\Delta U = -W_{int}$$

Therefore, the energy principle for a multiparticle system in which the particles do not change identity can be written:

ENERGY PRINCIPLE FOR A MULTIPARTICLE SYSTEM

$$\Delta K_{sys} + \Delta U = W_{surr}$$

for a situation in which $\Delta(mc^2) = 0$ and other energy transfers such as Q are zero.

The preceding equation, expressed in words, says that for a system of two or more particles, the change in kinetic energy of the system plus the change in the potential energy of the system is equal to the amount of work done on the system by external forces, assuming that there are no other (nonwork) kinds of energy transfers into the system.

16.2 SYSTEMS OF CHARGED OBJECTS

In most cases of interest in this volume, we will choose a system that includes all interacting charged objects. In some situations we will need to consider the effect of an external force, since a force may be required to move a charged particle in a direction opposite to the electric force acting on it. However, in many simple cases, we can neglect external forces. Although the Earth exerts a gravitational force on moving charged particles, this gravitational force is typically very small compared to the electric forces acting on the particles, and we can usually neglect it. In many cases, therefore, the energy principle for the systems in which we are interested will simplify to this:

$$\Delta K_{sys} + \Delta U = 0 \quad \text{(zero work done by surroundings)}$$

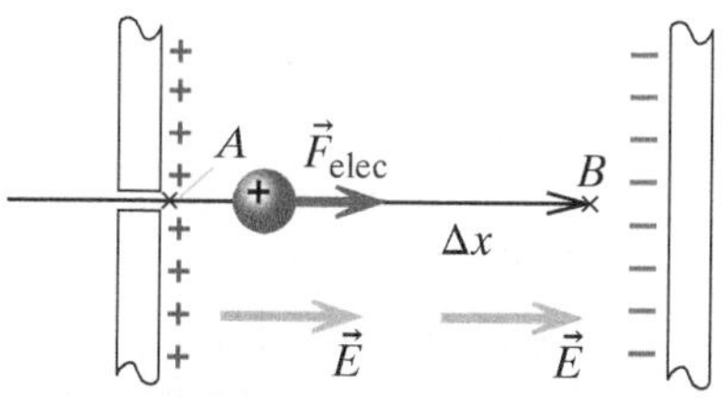

Figure 16.2 A proton traveling to the right enters a capacitor through a tiny hole in the left plate. Inside the capacitor it travels from location A to location B, in a region of nearly uniform electric field. The electric field outside the capacitor is negligible.

Decrease in *U*, Increase in *K*

As an example, consider a system consisting of two uniformly charged plates (a capacitor) and a single proton, as shown in Figure 16.2. Suppose that the proton is initially traveling to the right, and enters the capacitor through a small hole in the left plate. While the proton is inside the capacitor, an electric force (an internal force, due to the electric field inside the capacitor) acts on it, doing work on the proton.

The energy principle, applied to this system, is

$$\Delta K_{\text{proton}} + \Delta U_{\text{electric}} = 0$$

QUESTION Will the kinetic energy of the proton increase or decrease?

Because the force on the proton acts in the *same* direction as the displacement of the proton, the kinetic energy of the proton increases. When it reaches location B near the negative plate, the kinetic energy of the proton–capacitor system will have increased by an amount equal to the work done on the proton by the electric field inside the capacitor, which is due to all the charged particles on the plates of the capacitor. The speed of the proton will be greater.

QUESTION Will the electric potential energy of the system increase or decrease?

Since the kinetic energy will increase, the electric potential energy must decrease.

QUESTION How do we calculate the change in the electric potential energy of this system?

The change in the electric potential energy of this system is equal to the negative of the internal work: in this case, the work done on the proton by the uniform electric field inside the capacitor.

$$\begin{aligned}\Delta U_{\text{electric}} &= -W_{\text{int}} = -(F_x\Delta x + F_y\Delta y + F_z\Delta z)\\ &= -(eE_x\Delta x + eE_y\Delta y + eE_z\Delta z)\end{aligned}$$

In this case, since $\vec{E} = (E_x, 0, 0)$ and $\Delta\vec{l} = \langle \Delta x, 0, 0\rangle$,

$$\begin{aligned}\Delta U_{\text{electric}} &= -(eE_x\Delta x + 0 + 0)\\ &= -eE_x\Delta x\end{aligned}$$

From Figure 16.2 we can see that $\Delta x = x_f - x_i$ is a positive quantity, and E_x is also positive, so the change in the potential energy of the system $-eE_x\Delta x$ is negative, as predicted. For this system and this process, we find that

$$\Delta K = -\Delta U_{\text{electric}} > 0$$

and the kinetic energy of the proton increases.

EXAMPLE **ΔU and ΔK**

For concreteness, find ΔU and ΔK if the magnitude of the electric field inside the capacitor is 2×10^3 N/C, and the distance traveled by the proton is 4 mm.

Solution Since for the proton both the displacement and the force are in the x direction:

$$\Delta U_{\text{electric}} = -(1.6 \times 10^{-19}\,\text{C})(2 \times 10^3\,\text{N/C})(0.004\,\text{m}) = -1.3 \times 10^{-18}\,\text{J}$$

$$\Delta K = -\Delta U_{\text{electric}} = +1.3 \times 10^{-18}\,\text{J}$$

Since the potential energy of the system is less in the final state than it was initially, the kinetic energy of the system must have increased; the proton is moving faster when it reaches location B, as we would expect.

Increase in U, Decrease in K

Suppose that instead of a proton, an electron is initially traveling to the right, and enters the capacitor through the tiny hole in the left plate. The electron moves to location B, as shown in Figure 16.3.

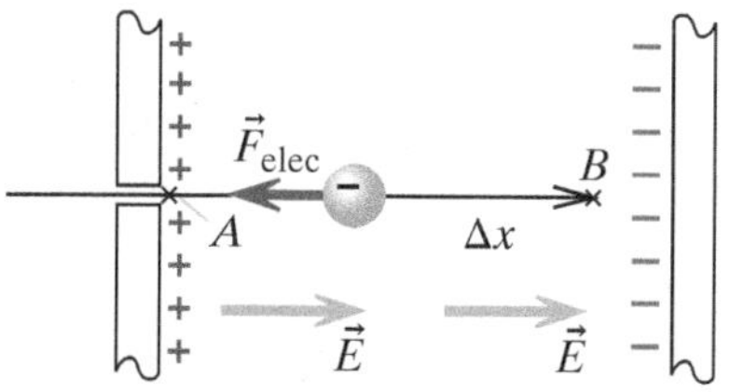

Figure 16.3 An electron traveling to the right enters a capacitor through a tiny hole in the left plate. Inside the capacitor it travels from location A to location B, in a region of nearly uniform electric field. The electric field outside the capacitor is negligible.

QUESTION Will the kinetic energy of the electron increase or decrease?

Because the electric force acts in a direction *opposite* to the motion of the electron, the kinetic energy of the electron will decrease as it moves toward location B. The speed of the electron will be lower at location B.

QUESTION Will the electric potential energy of the system increase or decrease?

As in the previous example, there are no significant energy inputs to the system from external sources. Therefore, applying the energy principle to this system and this process:

$$\Delta K_{\text{electron}} + \Delta U_{\text{electric}} = 0$$

Since kinetic energy will decrease, electric potential energy must increase.

QUESTION What is the change in the electric potential energy of this system?

The change in the electric potential energy of this system is equal to the negative of the internal work: in this case, the work done on the electron by the uniform electric field inside the capacitor. In the simple situation illustrated in Figure 16.3, the only nonzero component of the electric field is the x component, so

$$\begin{aligned}\Delta U_{\text{electric}} &= -(F_x\Delta x + 0 + 0)\\ &= -((-e)E_x\Delta x + 0 + 0)\\ &= eE_x\Delta x\end{aligned}$$

From Figure 16.3 we can see that $\Delta x = x_f - x_i$ is a positive quantity, and E_x is also positive, so the change in the potential energy of the system $eE_x\Delta x$ is positive. Since the electric force on the electron (due to the field inside the capacitor) is in a direction *opposite* to the displacement of the electron, the work done by this internal force is negative, and the change in potential energy is positive.

EXAMPLE **ΔU and ΔK**

As before, suppose that the magnitude of the electric field inside the capacitor is 2×10^3 N/C; the distance traveled by the electron is 4 mm.

Solution For the electron the direction of the force is opposite to the displacement:

$$\Delta U_{\text{electric}} = -(-1.6 \times 10^{-19}\text{C})(2 \times 10^3\ \text{N/C})(0.004\ \text{m}) = +1.3 \times 10^{-18}\ \text{J}$$

$$\Delta K = -\Delta U_{\text{electric}} = -1.3 \times 10^{-18}\ \text{J}$$

The kinetic energy of the electron decreases, and the potential energy of the system increases.

EXAMPLE **Proton Moving in a Uniform Electric Field**

A proton moves from location A to location B in a region of uniform electric field, as shown in Figure 16.4. If the magnitude of the electric field inside the capacitor shown is 190 N/C and the distance from location A to location B is 1.5 cm, what is the change in electric potential energy of the system during this process? What is the change in kinetic energy of the proton during this process? If the proton is initially at rest, what is its speed when it reaches location B?

Solution Initial location: A
Final location: B

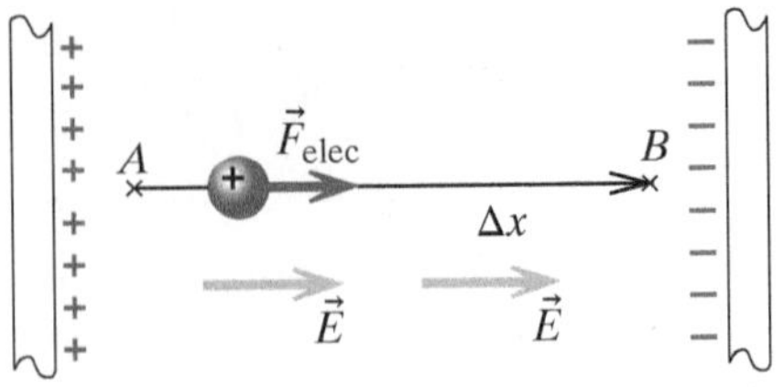

Figure 16.4 A proton moves from location A to location B through a region of uniform electric field.

Change in electric potential energy:

$$\Delta U = -W_{int} = -eE_x \Delta x$$
$$= -(1.6 \times 10^{-19}\,\text{C})(190\,\text{N/C})(0.015\,\text{m})$$
$$= -4.6 \times 10^{-19}\,\text{J}$$

Change in kinetic energy of proton:

$$\Delta K + \Delta U = 0$$
$$\Delta K = -\Delta U = +4.6 \times 10^{-19}\,\text{J}$$

Final speed of proton:

$$K_f - 0 = \frac{1}{2}mv_f^2 - 0$$
$$v_f = \sqrt{\frac{2K_f}{m}} = \sqrt{\frac{2(4.6 \times 10^{-19}\,\text{J})}{(1.7 \times 10^{-27}\,\text{kg})}} = 2.3 \times 10^4\,\frac{\text{m}}{\text{s}}$$

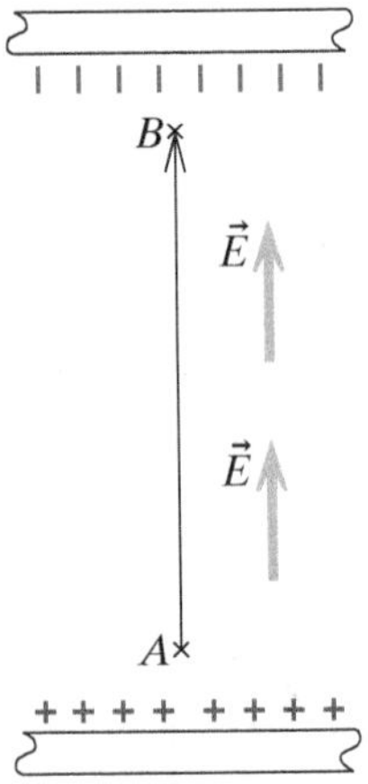

Figure 16.5 A particle moves from location A to location B in a region in which there is a uniform electric field.

Checkpoint 2 A proton moves from location A to location B in a region of uniform electric field, as shown in Figure 16.5. **(a)** If the magnitude of the electric field inside the capacitor in Figure 16.5 is 3500 N/C, and the distance between location A and location B is 3 mm, what is the change in electric potential energy of the system (proton + plates) during this process? **(b)** What is the change in the kinetic energy of the proton during this process? **(c)** If the proton is initially at rest, what is its speed when it reaches location B? **(d)** How do the answers to (a) and (b) change if the proton is replaced by an electron?

16.3 POTENTIAL DIFFERENCE IN A UNIFORM FIELD

When working with electric forces, we found it quite useful to be able to calculate the electric field at a particular location without having to worry about what kind of charged particle might come along to be affected by this field. When we know the electric field $\vec{E}_1$ at a location, to find the electric force on whatever particle happens to be at that location, we simply multiply $\vec{E}_1$ by the charge of the particle, q_2:

$$\vec{F}_2 = q_2\vec{E}_1$$

Likewise, when calculating changes in potential energy of a system, it would be useful to find a quantity that is independent of the charge of the particle that might move through a region. Then we could simply multiply this quantity by the charge of whatever particle happened to move through the region, to get the change in potential energy of the system. We would like to be able to write:

$$\Delta U = q\,(\text{something})$$

In the two cases discussed in the previous section (proton or electron moving in a uniform field), many things about the system were the same: the capacitor, the electric field in the region, and the path taken by the particle between the initial location A and final location B. Let's rewrite the change in electric potential energy in each case as the product of the charge of the displaced particle and another quantity:

If the particle is a proton: $\Delta U_{\text{electric}} = -eE_x\Delta x = (+e)(-E_x\Delta x)$

If the particle is an electron: $\Delta U_{\text{electric}} = +eE_x\Delta x = (-e)(-E_x\Delta x)$

The quantity that is the same in both cases is $(-E_x\Delta x)$. This quantity is called the "difference of the electric potential between locations A and B," and is given the symbol ΔV.

POTENTIAL DIFFERENCE

$$\Delta U_{\text{electric}} = q\Delta V \quad \text{and therefore} \quad \Delta V = \frac{\Delta U_{\text{electric}}}{q}$$

Like energy, electric potential is a scalar quantity. The dimensions of a difference in electric potential ΔV are joules/coulomb, or "difference of energy per unit charge." This is such an important quantity that it has its own name:

$$1\frac{\text{joule}}{\text{coulomb}} = 1 \text{ volt}$$

Units of Field and of Energy

Note that since $\Delta V = -(E_x\Delta x + E_y\Delta y + E_z\Delta z)$, and the units of ΔV are volts, then the units of electric field must be volts per meter. This is in fact equivalent to newtons per coulomb, and can be used interchangeably.

We can also now understand the origin of the energy unit "electron volt," or eV. One eV is equal to 1.6×10^{-19}J. If an electron moves through a potential difference of one volt, there is a change in the electric potential energy whose magnitude is

$$|\Delta U| = (e)(1 \text{ volt}) = (1.6 \times 10^{-19}\text{C})(1\,\text{J/C}) = 1.6 \times 10^{-19}\text{ J}$$

The electron volt is a convenient unit for measuring energies of atomic processes in physics and chemistry. Note that an electron volt is a unit of energy, not potential (potential is measured in volts, not electron volts).

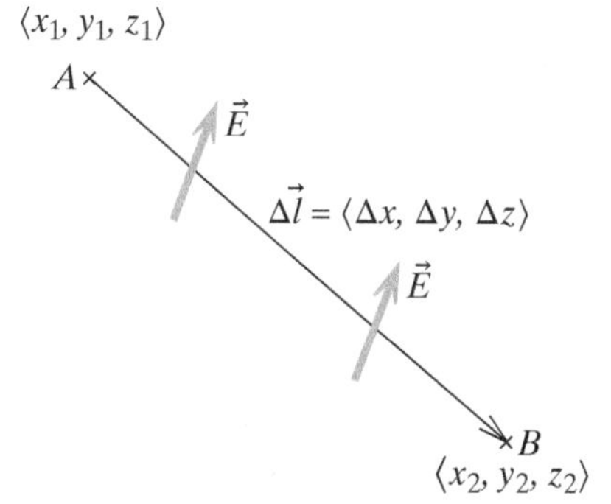

Figure 16.6 In general, $\Delta V = -(E_x\Delta x + E_y\Delta y + E_z\Delta z)$.

Path Not Parallel to Electric Field

The path between two locations does not have to be parallel to the electric field, as shown in Figure 16.6. The general three-dimensional expression for change in electric potential in a uniform field is:

$$\Delta V = -(E_x\Delta x + E_y\Delta y + E_z\Delta z)$$

Note that this quantity can be positive or negative, since each component of the electric field can be positive or negative, and each component of the displacement can also be positive or negative.

The quantity $(E_x\Delta x + E_y\Delta y + E_z\Delta z)$ can be written as the dot product of the electric field and the displacement vector $\Delta\vec{l} = \langle\Delta x, \Delta y, \Delta z\rangle$:

POTENTIAL DIFFERENCE IN A UNIFORM FIELD

$$\Delta V = -(E_x\Delta x + E_y\Delta y + E_z\Delta z)$$

or, using dot product notation:

$$\Delta V = -\vec{E}\bullet\Delta\vec{l} = -\langle E_x, E_y, E_z\rangle \bullet \langle\Delta x, \Delta y, \Delta z\rangle$$

EXAMPLE **Potential and Potential Energy**

Figure 16.7 shows two locations, A and B, in a region of uniform electric field. For a path starting at A and going to B, calculate the following quantities: **(a)** the difference in electric potential, **(b)** the potential energy change for the system when a proton moves along this path, and **(c)** the potential energy change for the system when an electron moves along this path.

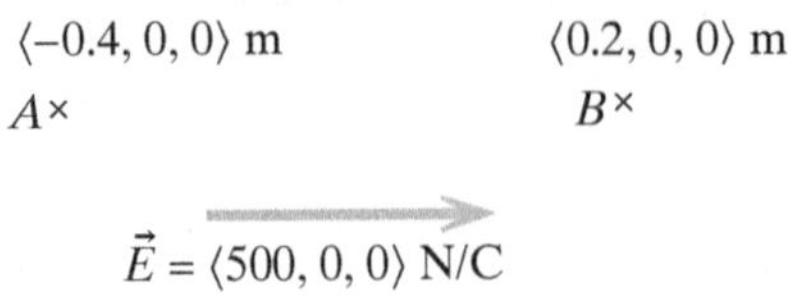

Figure 16.7 A region of uniform electric field.

Solution **(a)** Potential difference:
Initial location: A
Final location: B

$$\Delta\vec{l} = \langle 0.2,0,0\rangle \text{ m} - \langle -0.4,0,0\rangle \text{ m} = \langle 0.6,0,0\rangle \text{ m}$$

$$\Delta V = -(E_x\Delta x + E_y\Delta y + E_z\Delta z)$$

$$= -\left(500\frac{\text{N}}{\text{C}}\cdot 0.6\text{ m} + 0\cdot 0 + 0\cdot 0\right)$$

$$= -300\text{ V}$$

(b) For a proton:

$$\Delta U = (+e)(\Delta V) = (1.6\times 10^{-19}\text{C})(-300\text{ V}) = -4.8\times 10^{-17}\text{ J}$$

(c) For an electron:

$$\Delta U = (-e)(\Delta V) = (-1.6\times 10^{-19}\text{C})(-300\text{ V}) = 4.8\times 10^{-17}\text{ J}$$

EXAMPLE **Field and Potential**

Suppose that in a certain region of space (Figure 16.8) there is a nearly uniform electric field of magnitude 100 N/C (that is, the electric field is the same in magnitude and direction throughout this region). If you move 2 m at an angle of 30° to this electric field, what is the change in potential?

Solution

$$\Delta V \approx -\vec{E}\bullet\Delta\vec{l} = -E\Delta l\cos\theta$$

$$\Delta V = -(100\text{ N/C})(2\text{ m})\cos(30^\circ) = -173\text{ V}$$

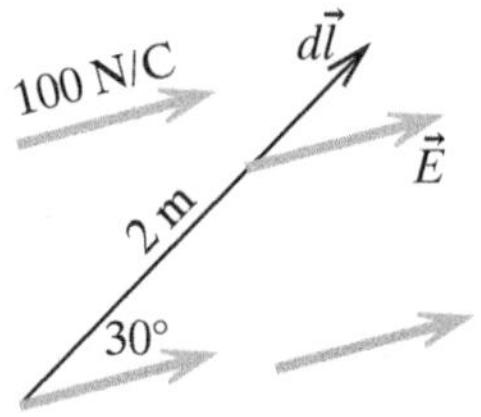

Figure 16.8 Moving at a 30° angle to the electric field.

Checkpoint 3 For a path starting at B and going to A (Figure 16.9), calculate **(a)** the change in electric potential, **(b)** the potential energy change for the system when a proton moves from B to A, and **(c)** the potential energy change for the system when an electron moves from B to A. For a path starting at B and going to C, calculate **(d)** the change in electric potential, **(e)** the potential energy change for the system when a proton moves from B to C, and **(f)** the potential energy change for the system when an electron moves from B to C.

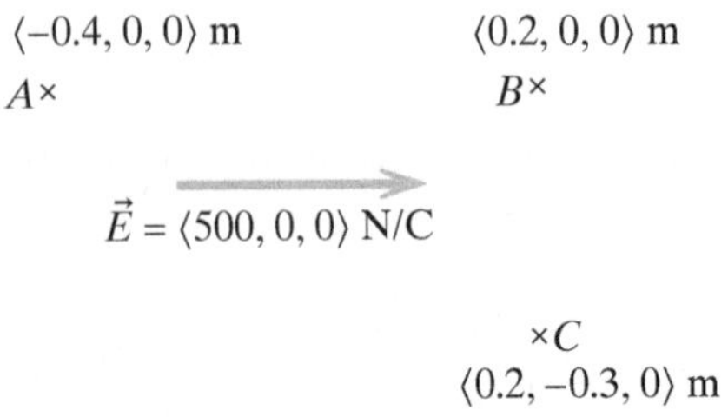

Figure 16.9 A region of uniform electric field (Checkpoint 3).

Calculating Field from Potential Difference

We have just seen how to calculate potential difference ΔV from the electric field $\vec{E}$. Conversely, we can calculate electric field from potential difference. Moving in the x direction, we have $\Delta V = -E_x\Delta x$, or

$$E_x = -\frac{\Delta V}{\Delta x}$$

EXAMPLE **Calculating Field from Potential Difference**

A capacitor has large plates that are 1 mm apart, and the potential difference from one plate to the other is 50 V. What is the magnitude of the electric field between the plates?

Solution Choose the x axis to be perpendicular to the plates. Then

$$|E| = \left| -\frac{50\text{ V}}{1 \times 10^{-3}\text{ m}} \right| = 5 \times 10^4\text{ V/m}$$

Gradient of Potential

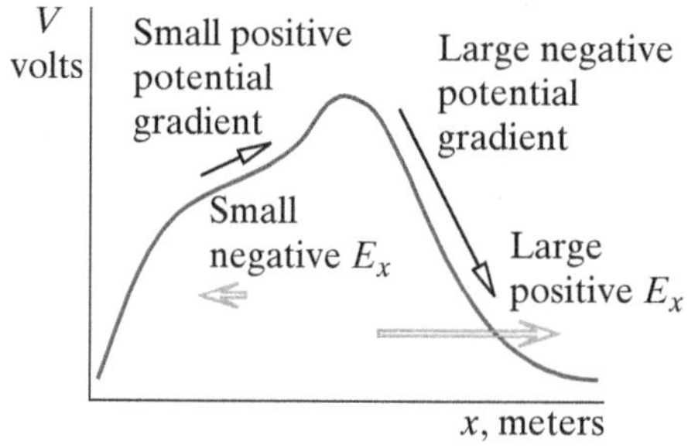

Figure 16.10 The negative slope of potential vs. x is the x component of the electric field.

In the limit of very short distance, E_x is given by a derivative:

$$E_x = -\lim_{\Delta x \to 0} \frac{\Delta V}{\Delta x} = -\frac{dV}{dx}$$

This is called the "gradient" of the potential, by analogy with the gradient of a hill, where a steep hill might have a "15% grade." Figure 16.10 emphasizes that a large gradient of potential (rapid change of potential with position) corresponds to a large electric field.

ELECTRIC FIELD IS THE NEGATIVE GRADIENT OF THE POTENTIAL

$$E_x = -\frac{dV}{dx}, \quad E_y = -\frac{dV}{dy}, \quad E_z = -\frac{dV}{dz}$$

More technically, the gradient should be written using partial derivatives, like this:

$$E_x = -\frac{\partial V}{\partial x}, \quad E_y = -\frac{\partial V}{\partial y}, \quad E_z = -\frac{\partial V}{\partial z}$$

The meaning of $\partial V/\partial x$ is simply "take the derivative of V with respect to x while holding y and z constant." The fact that $E_x = -\partial V/\partial x$ is closely related to the fact that (as we saw in Chapter 6 on energy) $F_x = -\partial U/\partial x$, because $E = F/q$ and $V = U/q$.

The gradient is often written in the form $\vec{E} = -\vec{\nabla}V$, where the inverted delta (nabla) represents taking the x, y, and z partial derivatives of the potential to find the x, y, and z components of the electric field.

Checkpoint 4 Suppose that the potential difference in going from location $\langle 2.00, 3.50, 4.00 \rangle$ m to location $\langle 2.00, 3.52, 4.00 \rangle$ m is 3 V. What is the approximate value of E_y in this region? Include the appropriate sign.

16.4 SIGN OF POTENTIAL DIFFERENCE

The potential difference ΔV can be positive or negative, and the sign is extremely important, since the sign determines whether a particular charged particle will gain or lose energy in moving from one place to another. An increase in potential energy $q\Delta V$ is associated with a decrease in kinetic energy, whereas a decrease in potential energy $q\Delta V$ is associated with an increase in kinetic energy.

Direction of Path Same as Direction of Electric Field: ΔV Is Negative

If the direction of the path from initial location to final location is the same as the direction of the electric field, the potential difference is negative. Mathematically, this is a result of the minus sign in the definition of potential difference.

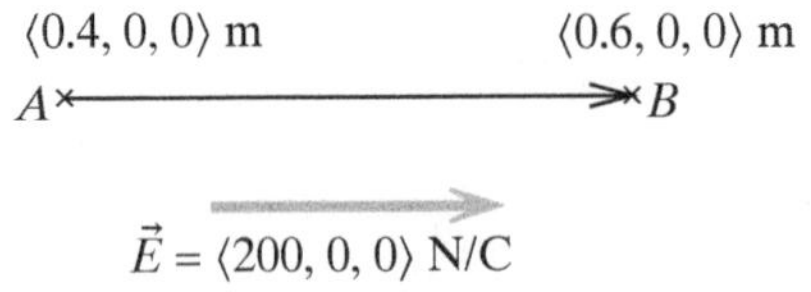

Figure 16.11 Direction of path same as direction of electric field.

In Figure 16.11, the path starting at A and ending at B travels in the same direction as the electric field in the region, so

$$\Delta x = (0.6 - 0.4)\text{ m} = 0.2\text{ m}$$

$$\Delta V = -E_x \Delta x = -\left(200\frac{\text{V}}{\text{m}}\right)(0.2\text{ m}) = -40\text{ V}$$

Direction of Path Opposite to Direction of Electric Field: ΔV Is Positive

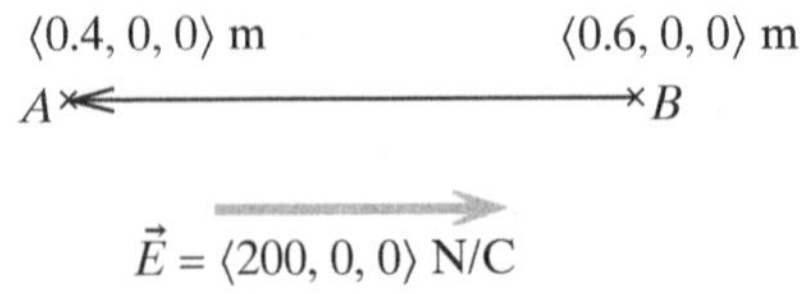

Figure 16.12 Direction of path opposite to direction of electric field.

On the other hand, in Figure 16.12, the path starting at B and ending at A travels opposite to the electric field in the region, and

$$\Delta V = -E_x \Delta x = -\left(200\frac{\text{V}}{\text{m}}\right)(-0.2\text{ m}) = +40\text{ V}$$

Path Perpendicular to the Electric Field

QUESTION What if a path is perpendicular to the electric field?

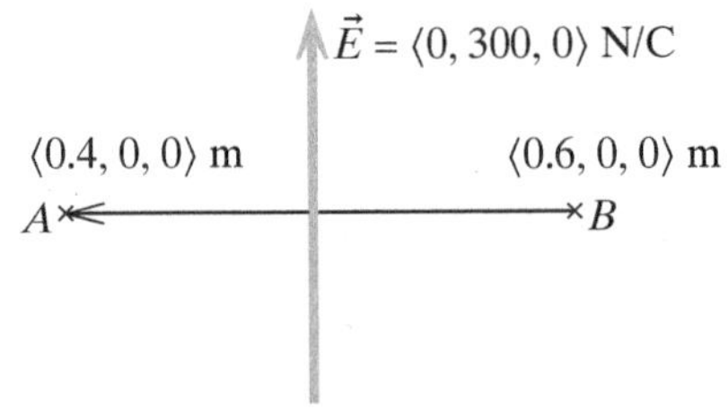

Figure 16.13 Path perpendicular to the direction of the electric field.

Consider the situation in Figure 16.13, where the path starting at B and ending at A is perpendicular to the electric field. Then

$$\Delta V = -(E_x\Delta x + E_y\Delta y) = -\left(\left(0\frac{\text{V}}{\text{m}}\right)(-0.2\text{ m}) + \left(300\frac{\text{V}}{\text{m}}\right)(0\text{ m})\right) = 0\text{ V}$$

Along a path whose direction is perpendicular to the electric field in the region, the potential does not change! Clearly, if a charged particle traveled along this path, an external force would be required to counteract the electric force. However, neither the external force nor the electric force would do any work on the particle, because the displacement in the direction of these forces would be zero. Thus, the kinetic energy of the particle would not change, and the potential energy of the system would not change.

Summary: Direction of the Path Relative to the Electric Field

It is easy to make sign errors in formal calculations using potential, and we need an independent check on the sign. The following rules are extremely useful:

SIGN OF ΔV AND DIRECTION OF PATH RELATIVE TO $\vec{E}$

Path going in the direction of $\vec{E}$: Potential is decreasing ($\Delta V < 0$)

Path going opposite to $\vec{E}$: Potential is increasing ($\Delta V > 0$)

Path perpendicular to $\vec{E}$: Potential does not change ($\Delta V = 0$)

Indicating the Path Direction

To show the direction of the path, we use the following notation. As usual, note that the symbol Δ indicates "final – initial."

$\Delta V = V_B - V_A$ indicates a path starting at A and ending at B.

$\Delta V = V_A - V_B$ indicates a path starting at B and ending at A.

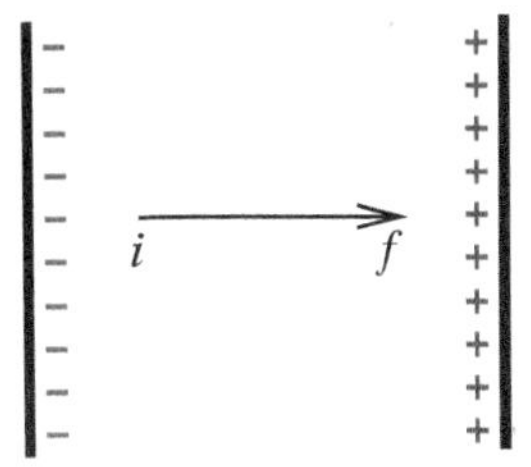

Figure 16.14 Checkpoint 5 (a).

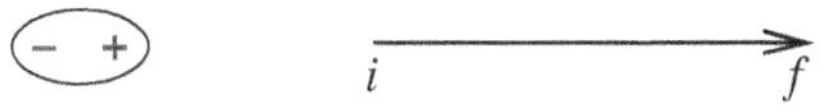

Figure 16.15 Checkpoint 5(b).

(We will see later that the symbol V_A actually has a meaning, and denotes the "potential at location A," which is defined as the potential difference along a path starting infinitely far away and traveling to location A.)

Because you can determine the sign of a potential difference on purely physical grounds, you should never get the wrong sign! In the following exercises you are asked to give the correct sign for the potential difference between two locations. Use the rule given above.

Checkpoint 5 **(a)** In Figure 16.14, what is the direction of the electric field? Is $\Delta V = V_f - V_i$ positive or negative? **(b)** In Figure 16.15, what is the direction of the electric field? Is $\Delta V = V_f - V_i$ positive or negative?

16.5 POTENTIAL DIFFERENCE IN A NONUNIFORM FIELD

Two Adjacent Regions with Different Fields

In the previous sections we considered only situations in which the electric field is uniform within a region. However, often the real world is more complex than this. We need to be able to calculate changes in electric potential and electric potential energy in regions of space in which the electric field is not uniform.

Consider a situation in which the path from the initial location to the final location of interest passes through two regions of uniform but different electric field. In region 1 the electric field is $\vec{E}_1$, and in region 2 the electric field is $\vec{E}_2$, as shown in Figure 16.16. What would be the change in electric potential energy of a particle traveling along a path from location A to location B? (Assume that there is a tiny hole in the middle plate, so a particle can pass through it.)

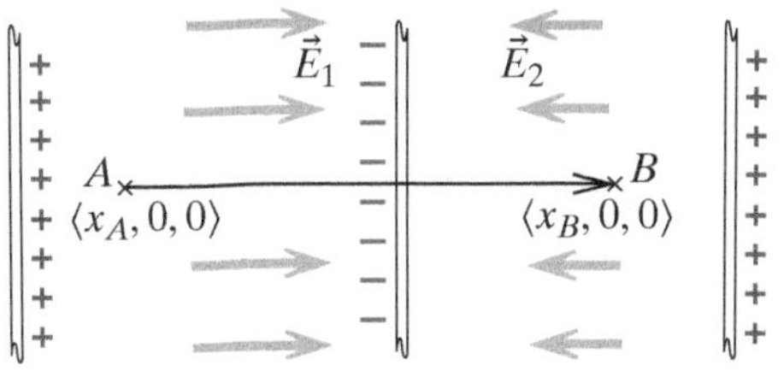

Figure 16.16 The path from location A to location B passes through two regions in which the field is different.

In order to calculate ΔV in this situation, we need to divide the path into two pieces. Each displacement vector $\Delta\vec{l}$ must be small enough that the electric field is uniform in the region through which it passes. Essentially, we have to invent a new point, which we can call C, on the boundary between the two regions, as shown in Figure 16.17.

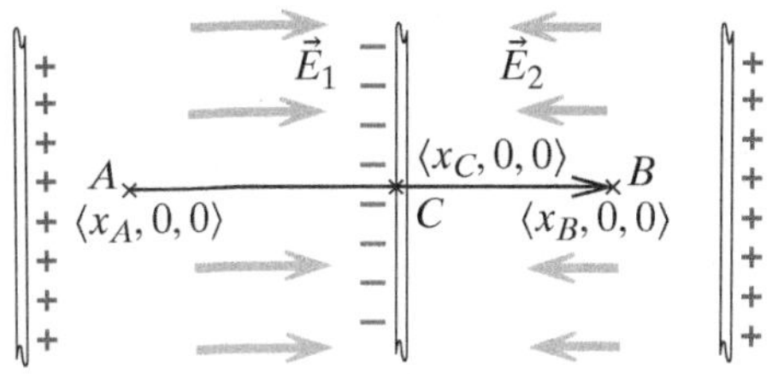

Figure 16.17 We divide the path into two segments by inventing a point C. Each segment of the path now passes through a region of uniform electric field.

Now we can calculate the difference in potential along the two segments of the path:

From location A to location C: $\Delta V_1 = -E_{1x}(x_C - x_A)$

From location C to location B: $\Delta V_2 = -E_{2x}(x_B - x_C)$

From location A to location B:

$$\Delta V = \Delta V_1 + \Delta V_2 = -E_{1x}(x_C - x_A) - E_{2x}(x_B - x_C)$$

In general, we need to add up all components (x, y, and z), so we can write the general equation as a sum:

POTENTIAL DIFFERENCE ACROSS SEVERAL REGIONS

$$\Delta V = \sum_i [(-E_{i,x}\Delta x_i) + (-E_{i,y}\Delta y_i) + (-E_{i,z}\Delta z_i)] = -\sum \vec{E} \bullet \Delta\vec{l}$$

EXAMPLE **Potential Difference Across Two Regions**

Suppose that from $x = 0$ to $x = 3$ the electric field is uniform and given by $\vec{E} = \langle 400, 0, 0\rangle$ N/C, and that from $x = 3$ to $x = 5$ the electric field is uniform and given by $\vec{E} = \langle 1000, 500, 0\rangle$ N/C (see Figure 16.18). What is the potential difference $\Delta V = V_C - V_A$?

Solution Split the path into two parts, A to B and B to C. In each part the electric field is constant in magnitude and direction. We have the following:

$$\Delta V = -\sum \vec{E} \bullet \Delta \vec{l}$$

$$\Delta V = -\langle 400, 0, 0 \rangle \frac{\text{V}}{\text{m}} \bullet \langle 3, 0, 0 \rangle \text{ m} - \langle 1000, 500, 0 \rangle \frac{\text{V}}{\text{m}} \bullet \langle 2, 0, 0 \rangle \text{ m}$$

$$\Delta V = (-1200 - 2000) \text{ V} = -3200 \text{ V}$$

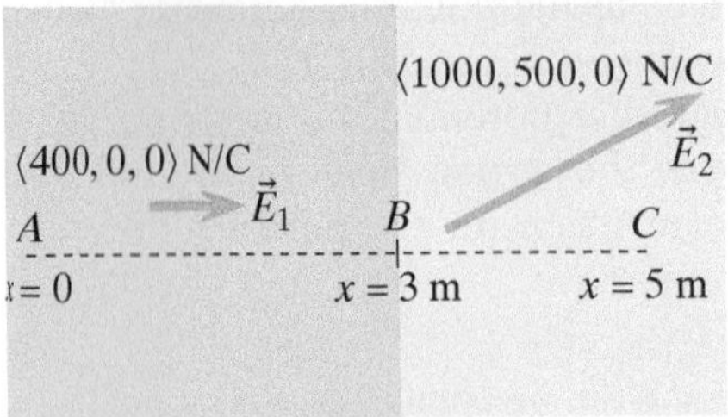

Figure 16.18 The electric field differs in the two regions.

> **Checkpoint 6** In the earlier Figure 16.17, location A is at $\langle 0.5, 0, 0 \rangle$ m, location C is $\langle 1.3, 0, 0 \rangle$ m, and location B is $\langle 1.7, 0, 0 \rangle$ m. $\vec{E}_1 = \langle 650, 0, 0 \rangle$ N/C and $\vec{E}_2 = \langle -350, 0, 0 \rangle$ N/C. Calculate the following quantities: **(a)** ΔV along a path going from A to B, and **(b)** ΔV along a path going from B to A.

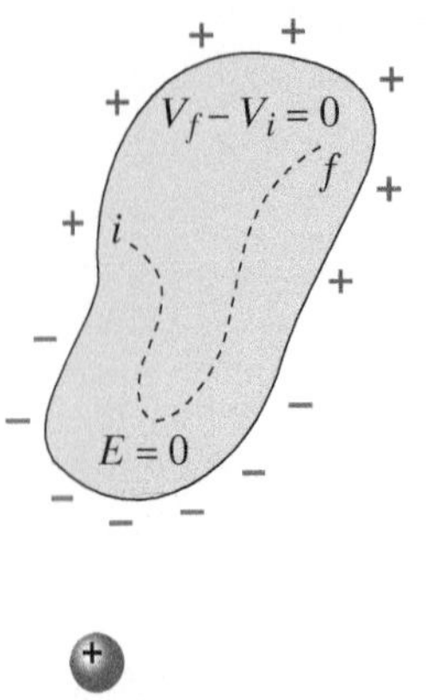

Figure 16.19 Potential difference is zero inside a metal in equilibrium.

A Conductor in Equilibrium

In Chapter 14 we proved that inside a conductor, such as a metal object, in equilibrium, the net electric field is zero (because otherwise mobile charges would shift until they contribute a field large enough to cancel the applied field). Therefore it must be the case that in equilibrium the net electric field is zero at all locations along any path through the conductor. This means that the potential difference is zero between two locations inside the metal (Figure 16.19).

In a conductor in equilibrium the difference in potential between any two locations is zero, because $E = 0$ inside the conductor.

$$\Delta V = -(E_x \Delta x + E_y \Delta y + E_z \Delta z) = 0, \quad \text{so } V = \text{constant}$$

We'll use this fact, in combination with the approach developed in this section, to examine a situation involving the potential difference across multiple regions, one of which is inside a metal object. Suppose that a capacitor with large plates and a small gap of 3 mm initially has a potential difference of 6 V from one plate to the other (Figure 16.20).

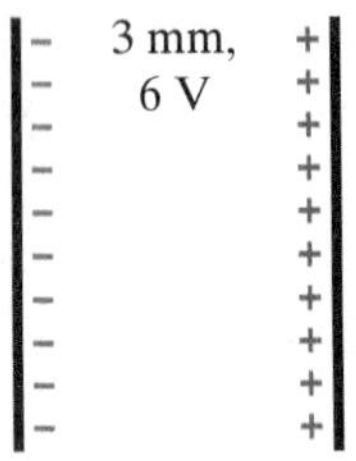

Figure 16.20 A capacitor with large plates and small gap.

> QUESTION What are the direction and magnitude of the electric field in the gap?

The direction of the electric field in the gap is toward the left, away from the positive plate and toward the negative plate. The magnitude can be found from noting that the potential difference $\Delta V = Es = 6$ V, so that $E = (6\text{ V})/(0.003\text{ m}) = 2000$ V/m.

> QUESTION Would inserting a metal slab into the gap change ΔV?

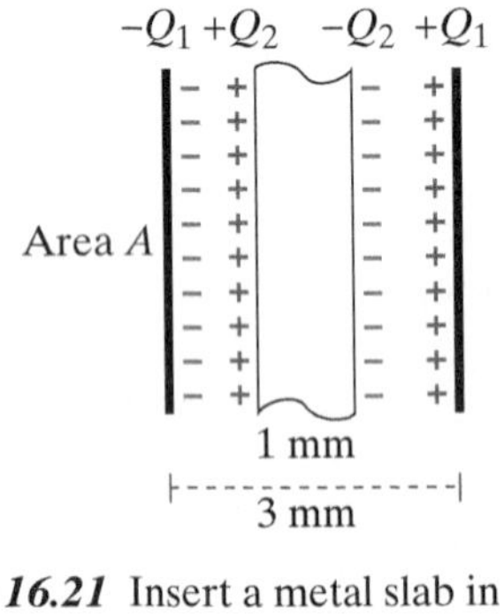

Figure 16.21 Insert a metal slab in the middle of the capacitor gap.

Suppose we insert into the center of the gap a 1-mm-thick metal slab with the same area as the capacitor plates, as shown in Figure 16.21. We are careful not to touch the charged capacitor plates as we insert the metal slab. We want to calculate the new potential difference between the two outer plates.

First we need to understand the new pattern of electric field that comes about when the metal slab is inserted. The charges on the outer plates are $-Q_1$ and $+Q_1$, both distributed approximately uniformly over a large plate area A. The metal slab of course polarizes and has charges of $+Q_2$ and $-Q_2$ on its surfaces, as indicated in Figure 16.21.

The electric field inside a capacitor is approximately $E = (Q/A)/\varepsilon_0$, where Q is the charge on one plate and A is the area of the plate, if the plate separation s is small compared to the size of the plates.

> QUESTION Since the electric field inside the metal slab must be zero, we can conclude that Q_2 is equal to Q_1. Why?

The plates and the surfaces of the slab have the same area A. The outer charges $-Q_1$ and $+Q_1$ produce an electric field in the metal slab $E_1 = (Q_1/A)/\varepsilon_0$ to the left. The inner charges $+Q_2$ and $-Q_2$ are arranged like the charges on a capacitor, so they produce an electric field in the metal slab $E_2 = (Q_2/A)/\varepsilon_0$ to the right. The sum of these two contributions must be zero, because the electric field inside a metal in equilibrium must be zero. Hence Q_2 is equal to Q_1.

As a result, the left pair of charges produces a field like that of a capacitor, and the right pair of charges also produces a field like that of a capacitor. The effect is that after inserting the metal slab, the electric field is still approximately 2000 V/m in the air gaps but is now zero inside the metal slab. (The fringe fields are small if the gap is small.)

> **QUESTION** Now that we know the electric field everywhere, what are the potential differences across each of the three regions between the plates (air gap, metal slab, air gap)?

The electric field in the air gap is essentially unchanged, so

$$\Delta V_{\text{left}} = \Delta V_{\text{right}} = (2000\ \text{V/m})(0.001\ \text{m}) = 2\ \text{V}$$

ΔV inside the metal slab must be zero, because $E = 0$ inside the slab. The potential difference between the plates of the capacitor is now:

$$\Delta V = 2\ \text{V} + 0\ \text{V} + 2\ \text{V} = 4\ \text{V}$$

not 6 V as it was originally. (Note that there is no such thing as "conservation of potential"; the potential difference changed when we inserted the slab.)

When you know that the electric field inside a metal must be zero, you know that the sum of all the contributions to that electric field must be zero. This is a powerful tool for reasoning about fields and charges in and on metals in equilibrium.

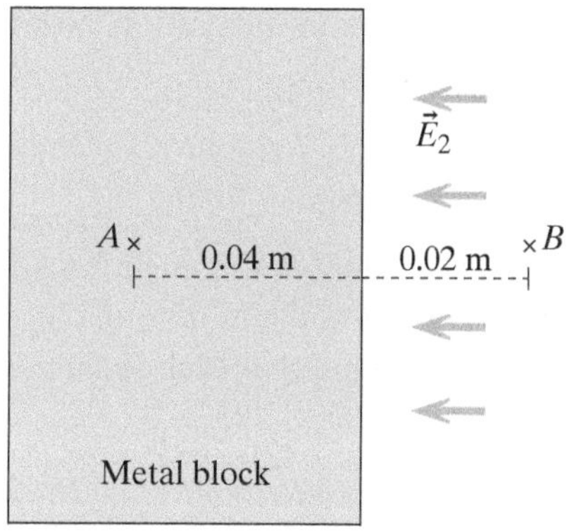

Figure 16.22 Location A is inside a charged metal block. Location B is outside the block.

> **Checkpoint 7** In Figure 16.22, location A is inside a charged metal block, and location B is outside the block. The metal block sits on an insulating surface and is not in contact with any other object. The electric field outside the block is $\vec{E}_2 = \langle -50, 0, 0\rangle$ V/m. **(a)** Calculate ΔV along a path going from A to B. **(b)** Calculate ΔV along a path going from B to A.

ΔV in a Region of Varying Electric Field

In some situations, the electric field in a region varies continuously. For example, the electric field due to a single point charge is different in magnitude and direction at every location in space. To find a change in electric potential between two locations in such a region, we need to take into account the varying electric field.

In the previous chapter we evaluated the electric field produced by distributed charges (rod, ring, disk, sphere) by adding up the contributions of pieces of the distribution. For example, to find the electric field due to a uniformly charged rod we divided the rod into short pieces, approximated the field of each short piece by the field of a point charge, and added up all these contributions either numerically or, where possible, analytically, by evaluating an integral. The more pieces, and the shorter each piece, the more accurate was the result, and in the case of an integral there were an infinite number of infinitesimally short pieces of the rod.

We can use a similar approach to evaluate electric potential differences between two locations in a region of varying electric field. The basic idea is to choose a path from the initial location to the final location, break that path into short pieces, evaluate ΔV for each piece, add up all these ΔV's, and check the result. Here are details of these steps:

Step 1: Choose a Path and Divide It into Pieces $\Delta\vec{l}$

- Choose a path from the initial location *i* to the final location *f*. Although the potential difference is independent of the path from *i* to *f*, we need to choose a specific path in order to carry out the calculation, and it needs to be a path along which we are able to calculate the electric field $\vec{E}$ at every location along the path.
- Divide the path into many short pieces, where each piece is represented as a vector $\Delta\vec{l}$.
- At a couple of locations somewhere near the middle of the path, sketch the electric field $\vec{E}$ to visualize and guide the calculation. (Remember that the result depends on $\vec{E}$ all along the path—don't focus on the endpoints.)

Step 2: Write an Expression for $\Delta V = -\vec{E} \bullet \Delta\vec{l}$ of One Piece

- Each piece of the path contributes an amount $\Delta V = -\vec{E} \bullet \Delta\vec{l}$ to the potential difference. $\vec{E}$ has a different value at each location on the path. To calculate each $\Delta V = -\vec{E} \bullet \Delta\vec{l}$ we will use the value of $\vec{E}$ midway along $\Delta\vec{l}$. If we are doing a numerical calculation, we (or a computer) must calculate $\vec{E}$ for each step. If we are setting up an expression that we can try to integrate analytically, our expression for $\vec{E}$ should depend on an integration variable related to position along the path.

Step 3: Add Up the Contributions of All the Pieces

- Add up the ΔV contributions to get $V_f - V_i$. The summation might be done numerically, or in some cases it may be possible to evaluate the summation as the integral $V_f - V_i = -\int_i^f \vec{E} \bullet d\vec{l}$, where we imagine that each short piece $\Delta\vec{l}$ becomes an infinitesimal length $d\vec{l}$.

Step 4: Check the Result

- Check that the result makes sense. Most importantly, does the overall sign make physical sense? Does the result have the correct units of volts (joules per coulomb)? Does the result give the right value in special cases where the correct result is known?

Potential Difference Near a Point Charge

To see how this scheme works, we'll carry out these steps to calculate the potential difference between two locations near a point charge of amount $Q = 3 \times 10^{-9}$ C. The initial location is a distance $r_i = 2 \times 10^{-3}$ m from the point charge, and the final location is a distance $r_f = 8 \times 10^{-3}$ m from the point charge. First we will solve the problem numerically; second we'll solve the problem analytically by integrating symbolically.

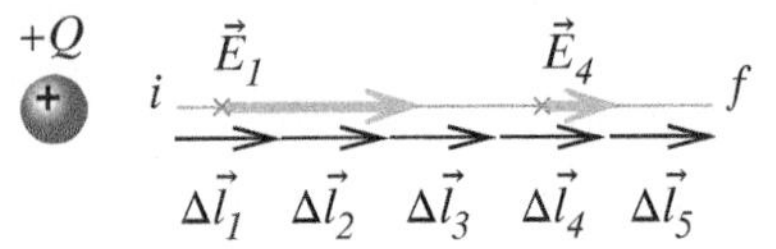

Figure 16.23 The path from *i* to *f* has been divided into five steps. The electric field is drawn at the center of the first and fourth steps. The arrows representing each $\Delta\vec{l}$ have been offset downward slightly from the path for clarity.

(1) Choose a Path and Cut it into Pieces $\Delta\vec{l}$: In Figure 16.23 we want to calculate the potential difference in going from location *i* to location *f*, that is, $V_f - V_i$. By sketching $\vec{E}$ at various places, we see that along the path $\vec{E}$ is parallel to the path. We'll divide the path into five equal lengths, each of length $(8 \times 10^{-3}\text{ m} - 2 \times 10^{-3}\text{ m})/5 = 1.2 \times 10^{-3}$ m, and we'll calculate $\Delta V = -\vec{E} \bullet \Delta\vec{l}$ numerically for each of the five pieces.

(2) Write an expression for $\Delta V = -\vec{E} \bullet \Delta\vec{l}$: The dot product is simple: $\Delta V = -E\Delta l \cos 0 = -E\Delta l$, where E is the usual field of a point charge. Therefore the potential difference of each piece is the following, where we can write $\Delta l = \Delta r$:

$$\Delta V = \frac{1}{4\pi\varepsilon_0}\frac{Q}{r^2}\Delta r$$

Although we've divided the path into five short pieces, within each piece the electric field does vary with distance from the point charge. In order to use something like the average value of E for a piece, we'll let r be the distance from the point charge to the middle of the piece. We could of course improve the accuracy of our calculation by dividing the path into more than five pieces.

r, m	ΔV, volts
$2.6E-3$	$-4.793E3$
$3.8E-3$	$-2.244E3$
$5.0E-3$	$-12960E3$
$6.2E-3$	$-8.429E2$
$7.4E-3$	$-5.917E2$
Sum	$-9.767E3$

Figure 16.24 Numerical integration for $V_f - V_i$.

(3) Add Up the Contributions of All the Pieces: In Figure 16.24 are the calculations in which we calculate the ΔV for each piece and add them up to find $V_f - V_i$.

It would be possible to get a more accurate value by cutting the path into more pieces, but it would be tedious to do the calculations by hand. We wrote a little computer program to do these calculations for various numbers of pieces, and the results of these calculations are shown in Figure 16.25.

# pieces	$V_f - V_i$, volts
5	$-9.767E3$
10	$-10.028E3$
15	$-10.081E3$
20	$-10.100E3$
100	$-10.124E3$

Figure 16.25 Effect of number of pieces.

(4) Check the Result: The first check to make is whether the sign of the potential difference makes physical sense. Referring back to Figure 16.23, we see that $\vec{E}$ is in the same direction as $\Delta\vec{l}$ along the path, so the sign of the potential difference along each piece, $-E\Delta l\cos 0 = -E\Delta l$, is negative, and the sum of all these ΔVs is negative. Note too that a positive charge placed at r_i would be repelled and would head "downhill" to lower potential, which means that the potential difference is negative. So the negative sign is correct. You should also check your own calculations to make sure that the units are correct.

Analytical Integration

In step 2 above we developed an expression for the potential difference along a short piece of the path. We expect that as we divide the path into more and more shorter and shorter pieces our calculation should be more accurate, and in fact in Figure 16.25 we see that for 15, 20, or 100 pieces there is very little change in the calculated potential difference (they're all about -10.1×10^3 V), which suggests that the true value is in fact about -10.1×10^3 V.

In step 3 (adding up all the contributions to the potential difference), let's go to the limit of an infinite number of pieces, each piece of infinitesimal length. This means that $\Delta r \rightarrow dr$ and the sum becomes an integral:

$$\begin{aligned} V_f - V_i &= -\int_{r_i}^{r_f} \frac{1}{4\pi\varepsilon_0}\frac{Q}{r^2}dr \\ &= -\frac{Q}{4\pi\epsilon_0}\left[-\frac{1}{r}\right]_{r_i}^{r_f} \\ &= \frac{1}{4\pi\varepsilon_0}\frac{Q}{r_f} - \frac{1}{4\pi\varepsilon_0}\frac{Q}{r_i} \end{aligned}$$

When we evaluate this expression for the specified values of Q, r_i, and r_f, we find that $V_f - V_i = -10.125 \times 10^3$ V. Compare with the numerical integration with 100 pieces, which gave $V_f - V_i = -10.124 \times 10^3$ V. Note that if r increases, as it does in this example, the quantity $1/r$ decreases and $1/r_f$ is smaller than $1/r_i$, so moving to larger r implies a negative potential difference.

Note the important result that near a point charge the potential difference along a radial path can be written like this:

$$V_f - V_i = \Delta\left(\frac{1}{4\pi\varepsilon_0}\frac{Q}{r}\right)$$

We'll say more about this later in the chapter.

EXAMPLE **Potential Difference Due to a Proton**

A proton is located at the origin. Location C is 1×10^{-10} m from the proton, and location D is 2×10^{-8} m from the proton, along a line radially outward. **(a)** What is the potential difference $V_D - V_C$? **(b)** How much work would be required to move an electron from location C to location D?

Solution **(a)** Potential difference:

$$\Delta V = -\int_C^D E_x dx = -\int_{1\times10^{-10}}^{2\times10^{-8}} \left(\frac{1}{4\pi\varepsilon_0}\frac{Q}{x^2}\right) dx$$

$$\Delta V = \frac{1}{4\pi\varepsilon_0}(1.6\times10^{-19}\text{C})\left(\frac{1}{2\times10^{-8}\text{ m}} - \frac{1}{1\times10^{-10}\text{ m}}\right)$$

$$= -14.3\text{ V}$$

(b) Work to move an electron: The least amount of work will be required if we don't change the kinetic energy of the proton.

$$\cancel{\Delta K} + \Delta U = W_{\text{ext}}$$

$$0 + (-e)(\Delta V) = (-1.6\times10^{-19}\text{C})(-14.3\text{ V}) = 2.3\times10^{-18}\text{ J}$$

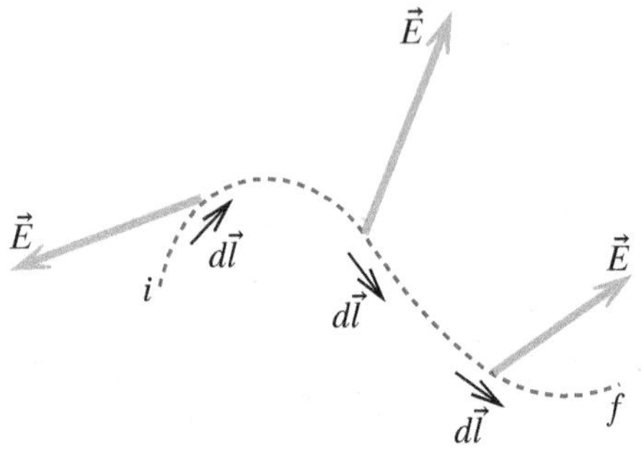

Figure 16.26 The potential difference for each step along this path is $-\vec{E}\bullet\Delta\vec{l}$. The potential difference $\Delta V = V_f - V_i$ is the sum of all these contributions, which becomes an integral for infinitesimal steps.

General Definition of Potential Difference

The most general definition of potential difference is the integral expression, which can be used in any situation (see Figure 16.26).

GENERAL DEFINITION OF POTENTIAL DIFFERENCE

$$\Delta V = -\int_i^f \vec{E}\bullet d\vec{l}$$

This general expression is correct for regions in which the field is uniform, as well as for regions in which the field varies. For example, in the situation shown in Figure 16.27, where the field is uniform,

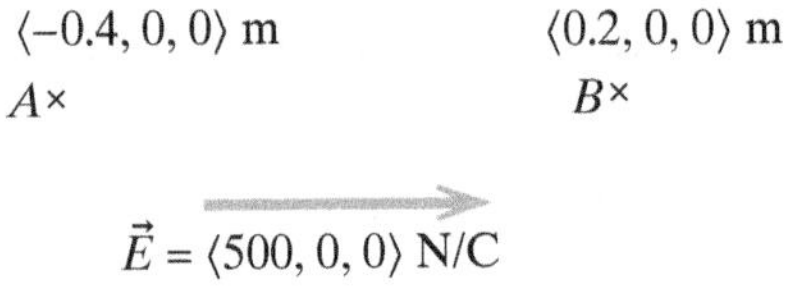

Figure 16.27 A region of uniform electric field.

$$\Delta V = -\int_i^f \vec{E}\bullet d\vec{l} = -\int_{-0.4}^{0.2}(500\text{ V/m})dx = -500\text{ V/m}\int_{-0.4}^{0.2}dx = -300\text{ V}$$

The result of the integration is the same as we got in the example involving uniform field in Section 16.3.

Complex Paths

Using $V_f - V_i = \Delta V = -\int_i^f \vec{E}\bullet d\vec{l}$ to calculate the potential difference along a path that is not straight, along which the electric field is varying in magnitude and direction (such as that shown in Figure 16.26), can be mathematically challenging. In many cases the best way to do this is to do it numerically, using a computer. In this text we will not ask you to do arbitrarily complex integrations of this kind analytically.

To establish the basic principles we will typically consider relatively simple situations. Often the electric field is uniform (same magnitude and direction) all along a straight path, which makes the calculation very simple. Another relatively simple situation is one in which the magnitude varies but not the direction along a straight path. A third simple situation that we will encounter is one in which we move along a curved path, but the electric field is always in the same direction as this curved path, and often constant in magnitude, which occurs in electric circuits. In each of these simple cases it can be quite easy to evaluate the potential difference, as we saw in the preceding exercises.

EXAMPLE

A Disk and a Spherical Shell

In Figure 16.28 a thin spherical shell made of plastic carries a uniformly distributed negative charge $-Q_1$ and a thin circular disk made of glass carries a uniformly distributed positive charge $+Q_2$. The radius R_1 of the plastic spherical shell is very small compared to the large radius R_2 of the glass disk. The distance from the surface of the spherical shell to the glass disk is d, and d is much smaller than R_2. Find the potential difference $V_2 - V_1$. Location 1 is at the center of the plastic sphere, and location 2 is just outside the glass disk. State what approximations or simplifying assumptions you make.

Solution Initial location: 1
Final location: 2
Path: Straight line from 1 to 2 (see Figure 16.29). $\vec{E}_{\text{net}} = \vec{E}_{\text{shell}} + \vec{E}_{\text{disk}}$ according to the Superposition Principle, so

$$-\int_1^2 \vec{E}_{\text{net}} \bullet d\vec{l} = -\int_1^2 \vec{E}_{\text{shell}} \bullet d\vec{l} - \int_1^2 \vec{E}_{\text{disk}} \bullet d\vec{l}$$

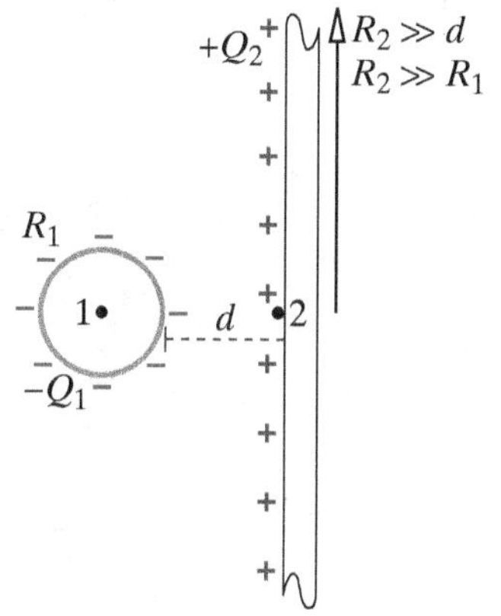

Figure 16.28 A uniformly charged plastic shell and a uniformly charged glass disk.

This allows us to find ΔV due to the shell and ΔV due to the disk separately, then add them. Alternatively, we could add the two electric field contributions, then integrate $\vec{E}_{\text{net}} \bullet d\vec{l}$ along a path from 1 to 2.

Simplifying assumption: Neglect the polarization of the plastic and glass, because there is little matter in the thin shell and thin disk, so the field of the polarized molecules is negligible compared to the contributions of $-Q_1$ and $+Q_2$.

ΔV due to shell: Put origin at center of shell.

Inside the shell: $V_{\text{surface of shell}} - V_1 = 0$ because $E_{\text{shell}} = 0$ inside the shell.

To the right of the shell:

$$\vec{E}_{\text{shell}} = \left\langle \frac{1}{4\pi\varepsilon_0}\frac{-Q_1}{x^2}, 0, 0 \right\rangle, \ \Delta\vec{l} = \langle dx, 0, 0 \rangle$$

$$V_2 - V_{\text{surface of shell}} = -\int_{R_1}^{R_1+d} \frac{1}{4\pi\varepsilon_0}\frac{-Q_1}{x^2}dx$$

$$= \frac{1}{4\pi\varepsilon_0}\left(\frac{-Q_1}{R_1+d} - \frac{-Q_1}{R_1}\right)$$

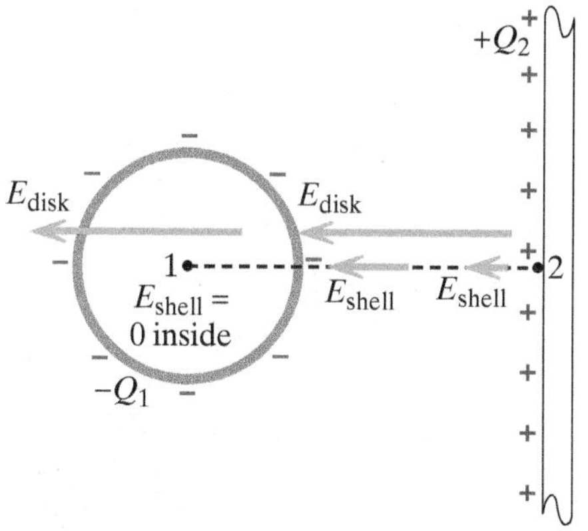

Figure 16.29 Electric field contributions of the shell and the disk, shown at locations along the path.

Check the sign: As we move toward the disk, we're moving opposite to the field, so the potential should increase. Result agrees, since $+Q_1/R_1$ is the larger term.

ΔV due to disk: an approximation is that since $d \ll R_2$ and $R_1 \ll R_2$,

$$\vec{E}_{\text{disk}} \approx \left\langle \frac{-Q_2/(\pi R_2^2)}{2\varepsilon_0}, 0, 0 \right\rangle$$

We could set up an integral, placing the origin at the center of the disk. However, since $\vec{E}_{\text{disk}}$ is approximately uniform, we recognize that the result of integrating will be

$$V_2 - V_1 = \frac{Q_2/(\pi R_2^2)}{2\varepsilon_0}(R_1 + d)$$

Check the sign: As we move toward the disk, we're moving opposite to the field, so the potential should increase. The result agrees.

ΔV due to both shell and disk:

$$V_2 - V_1 = \frac{1}{4\pi\varepsilon_0}\left(\frac{-Q_1}{R_1+d} - \frac{-Q_1}{R_1}\right) + \frac{Q_2/(\pi R_2^2)}{2\varepsilon_0}(R_1+d)$$

16.6 PATH INDEPENDENCE

In Chapter 6 we showed that potential energy differences depend only on the initial and final states of a system and are independent of path: different path, different process, but same change of state. Similarly, the potential difference $\Delta V = V_B - V_A$ between two locations A and B does not depend on the path taken between the locations.

To calculate the potential difference between two locations based on the electric field in the region, it is necessary to choose a path. We are free, however, to choose a path that simplifies our calculations.

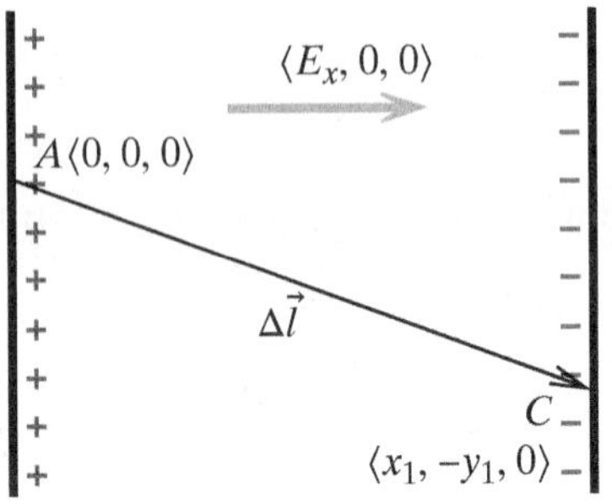

Figure 16.30 Moving at an angle to the electric field.

A Simple Example of Two Different Paths

To see how the potential difference between two locations can be the same along different paths, we'll consider two different paths through a capacitor. Suppose you move from the positive plate of a capacitor to the negative plate, moving at an angle to the electric field $\vec{E}$ (Figure 16.30).

QUESTION Calculate the potential difference in going from A to C: $\Delta V = V_C - V_A = ?$

Since the electric field is the same in magnitude and direction all along the path, we can write

$$\begin{aligned}\Delta V &= -(E_x\Delta x + E_y\Delta y + E_z\Delta z)\\ &= -E_x(x_1 - 0) + 0(-y_1 - 0) + 0\cdot 0 = -E_x x_1\end{aligned}$$

Suppose instead that we move along the two-part path shown in Figure 16.31.

QUESTION Now $\Delta V = V_C - V_A = ?$

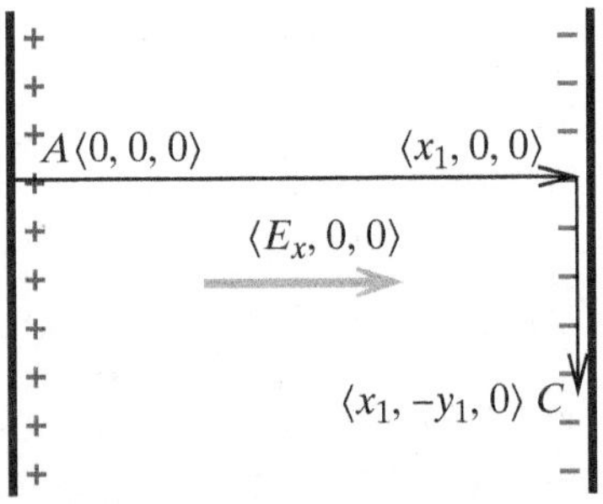

Figure 16.31 Moving at right angles to the electric field.

Along the path from A to B,

$$\begin{aligned}V_B - V_A &= -(E_x\Delta x + E_y\Delta y + E_z\Delta z)\\ &= -E_x(x_1 - 0) + 0\cdot 0 + 0\cdot 0 = -E_x x_1\end{aligned}$$

Along the path from B to C,

$$\begin{aligned}V_C - V_B &= -(E_x\Delta x + E_y\Delta y + E_z\Delta z)\\ &= -E_x(0) + 0(-y_1 - 0) + 0\cdot 0 = 0\end{aligned}$$

Therefore we again find that

$$V_C - V_A = (V_C - V_B) + (V_B - V_A) = -E_x x_1$$

which is what we found for the direct path from A to C.

QUESTION Does it make sense that $V_C - V_B = 0$?

Yes. Just as no work is required to move at right angles to a force, the electric potential does not change along a path perpendicular to the electric field. The potential difference doesn't depend on the angle of our path in Figure 16.30 because the sideways components of our steps don't affect ΔV.

Two Different Paths Near a Point Charge

Along a path straight away from a stationary charge Q we know this:

$$\Delta V = \frac{1}{4\pi\varepsilon_0}Q\left[\frac{1}{r_f} - \frac{1}{r_i}\right] \quad \text{near a positive or negative point charge } Q$$

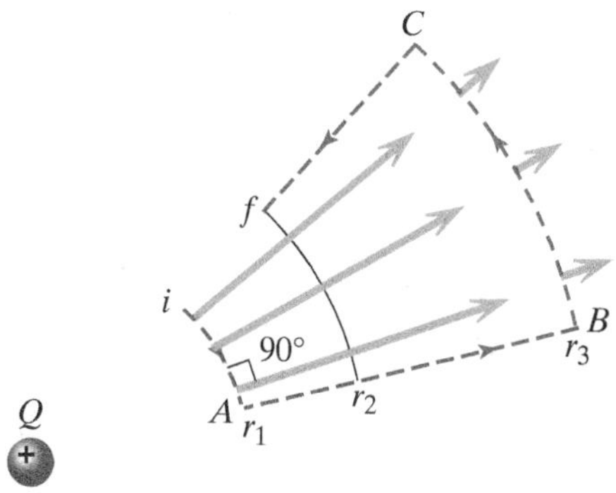

Figure 16.32 What is the potential difference along this complicated path?

It is instructive to work out the potential difference along a different path.

QUESTION In the situation shown in Figure 16.32, calculate the potential difference for each branch of the path from location i to A to B to C to location f, in terms of Q and the radii r_1, r_2, and r_3, then add these up to get the potential difference along this path. As a first step, draw the electric field at various locations along the path.

From i to A, $\vec{E} \perp \Delta\vec{l}$, so $\Delta V_1 = 0$.

From A to B, $\Delta V_2 = \frac{1}{4\pi\varepsilon_0}Q\left[\frac{1}{r_3} - \frac{1}{r_1}\right]$.

From B to C, $\vec{E} \perp \Delta\vec{l}$, so $\Delta V_3 = 0$.

From C to f, $\Delta V_4 = \frac{1}{4\pi\varepsilon_0}Q\left[\frac{1}{r_2} - \frac{1}{r_3}\right]$.

Adding up these ΔV's, we get:

$$\begin{aligned} V_f - V_i &= \Delta V_1 + \Delta V_2 + \Delta V_3 + \Delta V_4 \\ &= 0 + \frac{1}{4\pi\varepsilon_0}Q\left[\frac{1}{r_3} - \frac{1}{r_1}\right] + 0 + \frac{1}{4\pi\varepsilon_0}Q\left[\frac{1}{r_2} - \frac{1}{r_3}\right] \\ &= \frac{1}{4\pi\varepsilon_0}Q\left[\frac{1}{r_2} - \frac{1}{r_1}\right] \end{aligned}$$

This is exactly the same result that you would get by going directly in a straight line from i to f. We see here that it is only when $d\vec{l}$ has a radial component that there is a nonzero contribution to ΔV. Along the curved arcs the electric field is perpendicular to the path, so the arcs contribute zero to the potential difference. Going along the straight lines is equivalent to just going from r_1 to r_2. Evidently it doesn't matter what path we take. We say that potential difference is "path independent."

An Arbitrary Path Near a Point Charge

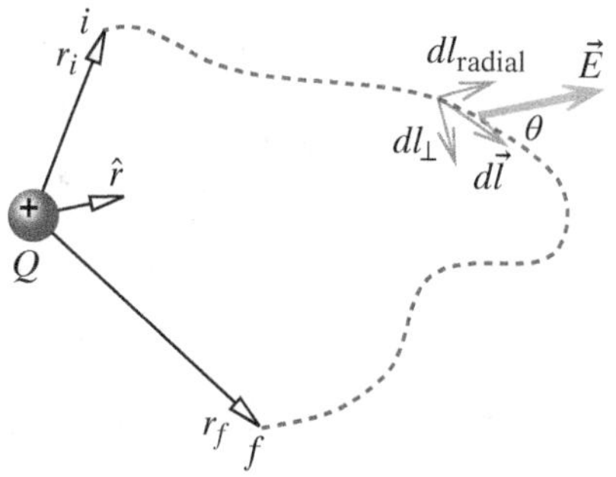

Figure 16.33 Calculating the potential difference along an arbitrary path.

We can see more generally how this works out for any path near a stationary point charge. Follow an arbitrary path from initial location i to final location f (Figure 16.33). At each location along the path draw the component dl_{radial} of $d\vec{l}$ in the direction of $\vec{E}$, pointing away from the charge Q.

At every location along the path we have

$$-\vec{E} \bullet d\vec{l} = -E\,dl\cos\theta$$

However, $dl\cos\theta$ is the radial component of $d\vec{l}$ and is equal to dl_{radial}, the change in distance from the charge. Therefore along this arbitrary path we are simply integrating $-E\,dl_{\text{radial}}$, so

$$\Delta V = \frac{1}{4\pi\varepsilon_0}Q\left[\frac{1}{r_f} - \frac{1}{r_i}\right]$$

All that matters are the initial and final distances from the charge, not the path we happen to follow. This is consistent with our earlier result for potential

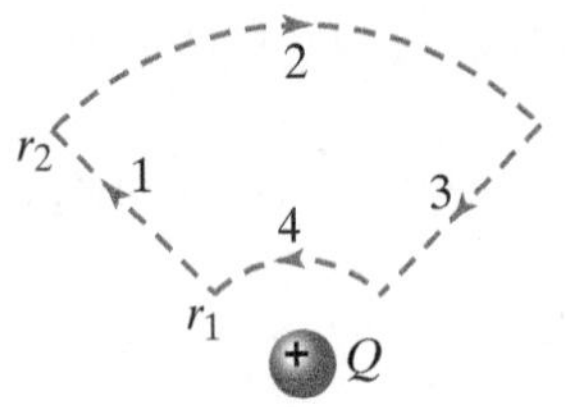

Figure 16.34 Checkpoint 8.

difference $\Delta V = (1/(4\pi\varepsilon_0))(Q/r)$ along a straight path between location i and location f. What is interesting is seeing how it worked out that the integral of $-\vec{E} \bullet d\vec{l}$ didn't depend on what path we took.

> **Checkpoint 8** Calculate the potential difference along the closed path consisting of two radial segments and two circular segments centered on the charge Q (Figure 16.34). Show that the four ΔV's add up to zero. It is helpful to draw electric field vectors at several locations on each path segment to help keep track of signs.

Round-Trip Potential Difference Is Zero

An important consequence of path independence is that the integral of $-\vec{E} \bullet d\vec{l}$ along a round-trip path gives a potential difference of zero, as we can now show (Figure 16.35). Move along path 1 (A-B-C), and then back to location A along path 2 (C-D-A).

> QUESTION Suppose that the potential difference along path 1 is -25 V. What would be the potential difference in going backward along path 2 (A-D-C)? Forward along path 2 (C-D-A)?

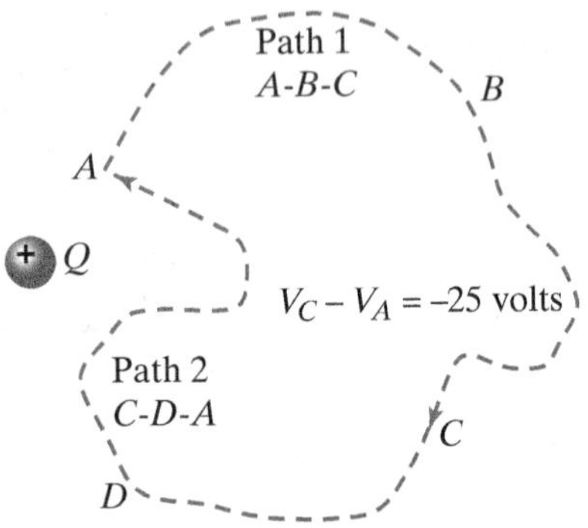

Figure 16.35 Round-trip potential difference on a path from A back to A again is zero.

Going backward along path 2 is the same as going forward along path 1, so the potential difference along the path (A-D-C) must be -25 V. Going in the opposite direction, along path 2 (C-D-A), reverses the sign of the parallel component of the electric field, so the potential difference along path 2 must be $+25$ V.

> QUESTION If we walk all the way around the closed path, A-B-C-D-A, what is the potential difference?

The round-trip potential difference is zero (-25 V plus $+25$ V). Note that this result is the direct consequence of the path independence of electric potential. It is also not surprising, since we do expect $V_A - V_A = 0$!

Potential difference due to a stationary point charge
is independent of the path.
Potential difference along a round trip is zero.

Since potential difference can be calculated as an integral involving the electric field, and the electric field is the sum of all the contributions of many stationary point charges, potential difference is the sum of the contributions of all the point charges. The potential difference integral is independent of the path for one point charge, and since all charge distributions are made up of atomic point charges, the round-trip potential difference must be zero for any configuration of stationary charges whatsoever.

PATH INDEPENDENCE

ΔV is always independent of the path taken between two locations.
For a round-trip path, ΔV is zero.

Conservation of Energy

The fact that $\Delta V = 0$ for a round trip along any path is fundamentally related to the principle of Conservation of Energy. We can use a proof by contradiction to show this connection.

Let's assume that it is possible to fasten down a collection of stationary charged particles in such a configuration that a round trip along the path shown in Figure 16.35 gives a potential difference greater than zero. If this is possible, then an electron could travel around this path many times, and $\Delta U = -e\Delta V$

of the system would be negative each time. Therefore the kinetic energy of the system would increase each time, despite the fact that no external energy inputs were made. We would have invented a perpetual motion machine! Because this outcome would violate the Energy Principle, we can conclude that our assumption (that it was possible for $\Delta V_{\text{round trip}}$ to be nonzero) must have been incorrect.

These are extremely important general results, and in the future we will often have occasion to refer to these properties of electric potential due to charges. Path independence is important because when we use potential we are reasoning about the relationship between initial and final states, independent of the intervening process. We are free to analyze the most convenient process that takes the system from the initial state to the final state.

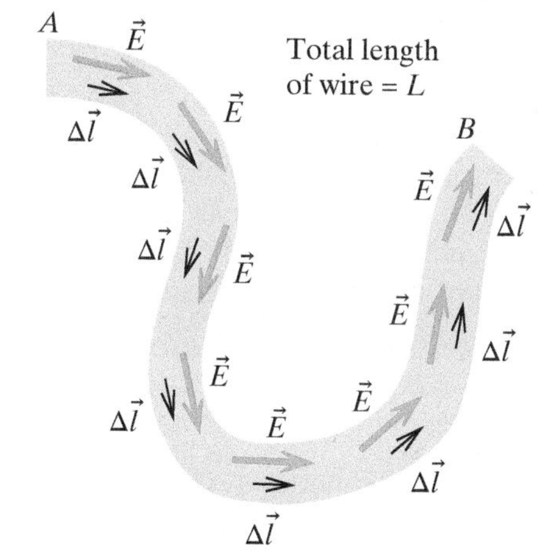

Figure 16.36 Inside a wire in a circuit, there can be an electric field that has uniform magnitude E but follows the direction of the wire. The wire's length is L.

Application: A Metal Not in Equilibrium

During the (extremely) brief time when a metal is polarizing due to an external charge, there is a nonzero electric field inside the metal. Before equilibrium is reached, the electric field in the metal is not zero.

As we will see in detail in a later chapter, in an electric circuit the battery maintains a nonzero electric field inside the wires, as a kind of continuing polarization process that doesn't lead to equilibrium. The electric field continually pushes the sea of mobile electrons through the wires. The electric field is not zero in this situation, which is not equilibrium.

QUESTION In Figure 16.36 there is a nonzero electric field of uniform magnitude E throughout the interior of a wire of length L, and the direction of the electric field follows the direction of the wire. What is the potential difference $V_B - V_A$?

In this case, since for each step $\vec{E} \| \Delta\vec{l}$, the potential difference is simply $-EL$, the sum of all the $-\vec{E} \bullet \Delta\vec{l} = -E\Delta l$ terms along the wire. Thus if a metal is not in equilibrium, there can be a nonzero potential difference inside the metal. However, in a circuit a thick copper wire may have such a small electric field in it that there is very little potential difference from one end to the other, in which case the potential is almost (but not quite) constant in the wire.

Checkpoint 9 In a circuit there is a copper wire 40 cm long with a potential difference from one end to the other of 0.01 V. What is the magnitude of the electric field inside the wire?

Impossible Patterns of Electric Field

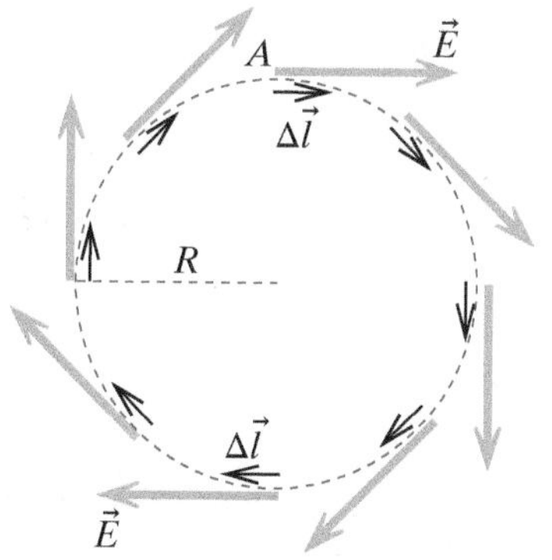

Figure 16.37 Is it possible to produce this pattern of electric field ($\vec{E}$ tangent to a circle at every point on the circle) with some arrangement of point charges?

We can use the fact that the potential difference over a round-trip path must be zero to reason about particular patterns of electric field. Surprisingly, we can actually conclude that certain patterns of electric field can never be created by a collection of stationary charged particles. For example, consider the pattern of electric field shown in Figure 16.37, where the electric field is tangent to the circle at every point on the circle.

QUESTION Try to think of an arrangement of stationary charges that could produce the pattern of electric field shown in Figure 16.37.

We can demonstrate very simply that it is impossible to produce this pattern of electric field by any arrangement of stationary point charges! Consider a path that starts at location A and goes clockwise around the circle, ending back at location A. Since at every location along the path $\vec{E}$ is parallel to $d\vec{l}$, the round-trip potential difference along this path is:

$$\Delta V = -\int \vec{E} \bullet d\vec{l} = -(2\pi R)E$$

However, we showed previously that for a single point charge, and hence for any assemblage of point charges, $\Delta V = 0$ for a round trip. Therefore this "curly" pattern of electric field must be *impossible* to produce by arranging any number of stationary point charges!

Important Limitation

In the next chapter we will study magnetic fields, which can be produced by moving charges. We will see in Chapter 22 on Faraday's Law that in the presence of time-varying magnetic fields the round-trip integral of the electric field can be nonzero. In this chapter we have assumed that there were no time-varying magnetic fields.

16.7 THE POTENTIAL AT ONE LOCATION

Usually we are interested in the potential difference between two locations; for example, $\Delta V = V_B - V_A$. However, it is sometimes useful to consider the potential at only one location.

The potential at location A is defined as the potential difference between a location infinitely far away from all charged particles, and the location of interest:

POTENTIAL AT ONE LOCATION

$$V_A = V_A - V_\infty$$

The potential at location A, sometimes referred to as "the potential relative to infinity," is the potential difference between location A and a location infinitely far away from all charged particles.

Of course, this equation makes sense only if V_∞, the potential at infinity, is zero, and by convention it is defined to be so. This is consistent with the fact that the potential energy of a system of two charged particles that are infinitely far from each other must be zero.

Since potential has a value at every location in space, potential (but not potential difference) is itself a field, but it is a scalar field, not a vector field.

Potential at a Location Near a Point Charge

As an example, we will find the potential at a location a distance r from a single point particle of charge q. To do this, we will find the potential difference between this location and a location infinitely far away.

$$V_r = V_r - V_\infty = -\int_\infty^r \frac{1}{4\pi\varepsilon_0}\frac{q}{x^2}dx = \frac{1}{4\pi\varepsilon_0}\frac{q}{x}\Big|_\infty^r = \frac{1}{4\pi\varepsilon_0}\frac{q}{r}$$

QUESTION How does the sign of this result depend on the sign of the charged particle?

If $q < 0$, $V_r < 0$, but if $q > 0$, $V_r > 0$. Figure 16.38 is a graph of potential as a function of distance from a positive or negative point charge.

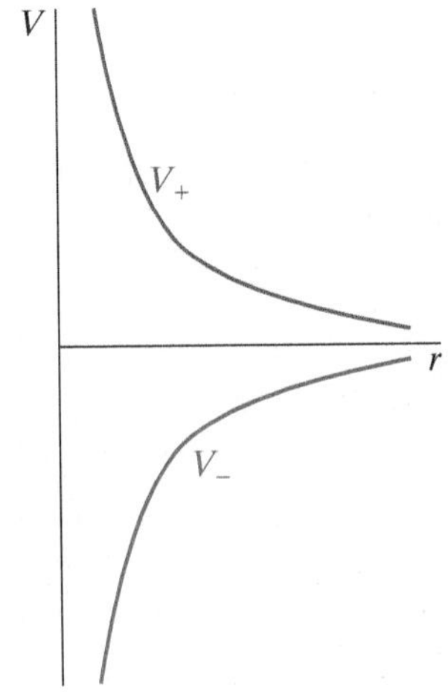

Figure 16.38 Potential as a function of distance from a positive (red) or negative (blue) point charge.

Potential and Potential Energy at One Location

If we know the value of the potential at location A, we can use this to find what the potential energy of the system would be if we placed a charged object at that location:

$$U_A = qV_A$$

Using our result for V we get the familiar expression for the electric potential energy of two particles with charges q_1 and q_2 separated by a distance r:

$$U = q_1\left(\frac{1}{4\pi\varepsilon_0}\frac{q_2}{r}\right) = \frac{1}{4\pi\varepsilon_0}\frac{q_1 q_2}{r}$$

EXAMPLE **Potential at a Location Near a Sphere**

A metal sphere of radius 3 mm has a charge of -2×10^{-9} C. What is the potential at location C, which is 5 mm from the center of the sphere?

Solution

$$V_C = V_C - V_\infty = \frac{1}{4\pi\varepsilon_0}Q\left(\frac{1}{x_C}-\frac{1}{\infty}\right)$$

$$= \left(9\times10^9\,\frac{\text{N}\cdot\text{m}^2}{\text{C}^2}\right)(-2\times10^{-9}\text{C})\left(\frac{1}{0.005\text{ m}}-\frac{1}{\infty}\right)$$

$$= -3600\text{ V}$$

We can calculate the potential due to a charge distribution in two ways: either by dividing the distribution into point-like pieces and adding the potential due to each piece, or by calculating $-\int \vec{E}\bullet d\vec{l}$ along a path from infinity to the location of interest.

EXAMPLE **Potential Along the Axis of a Ring**

Use superposition to calculate the potential at a location z from the center of a thin ring, along the axis, by adding the potential due to each piece of the ring. The ring carries a total charge Q. The total charge Q is made up of point charges, each of charge q, and each of these point charges is a distance $\sqrt{z^2+R^2}$ from the location of interest (Figure 16.39).

Solution The potential contributed by any one of these point charges q is

$$V = \frac{1}{4\pi\varepsilon_0}\frac{q}{\sqrt{z^2+R^2}}$$

for one point charge q.

Adding up the potential contributed by all of the point charges, we have the following result:

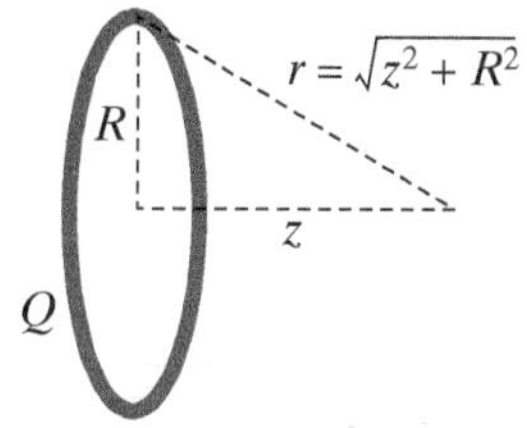

Figure 16.39 The potential due to a uniformly charged ring.

$$V_{\text{ring}} = \sum\frac{1}{4\pi\varepsilon_0}\frac{q}{\sqrt{z^2+R^2}}$$

$$V_{\text{ring}} = \frac{1}{4\pi\varepsilon_0}\frac{1}{\sqrt{z^2+R^2}}\sum q$$

$$V_{\text{ring}} = \frac{1}{4\pi\varepsilon_0}\frac{Q}{\sqrt{z^2+R^2}}$$

Integrating

This same result can be obtained by integrating the electric field of the ring along a path, using the expression for this field that we found in Chapter 15.

$$V(z) = -\int_\infty^z \frac{1}{4\pi\varepsilon_0}\frac{Qw}{(R^2+w^2)^{3/2}}dw$$

Conversely, from the result for the potential of the ring we can determine the electric field of the ring by differentiation, since $E_z = -\partial V/\partial z$ (see Problem P85).

Checkpoint 10 Show that if you are very far from the ring ($z \gg R$), the potential is approximately equal to that of a point charge. (This is to be expected, because if you are very far away, the ring appears to be nearly a point.)

Potential Inside a Conductor at Equilibrium

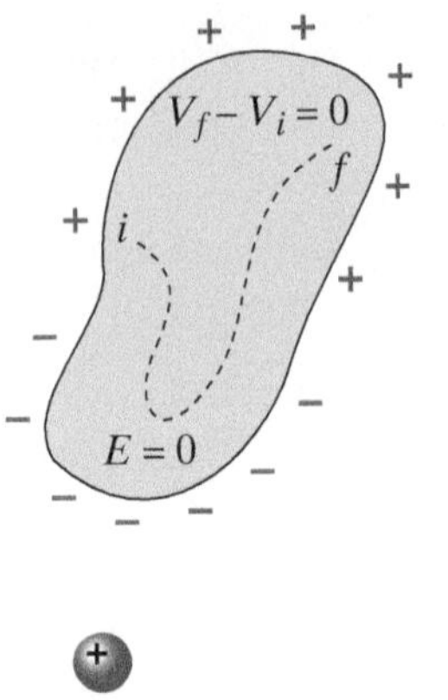

Figure 16.40 The potential is constant inside a metal in equilibrium.

In a previous section we noted that because the net electric field is zero inside a metal object in equilibrium, the potential difference between any two locations inside the metal must be zero. Since this is true, we can conclude that the potential at any location inside the metal must be the same as the potential at any other location (Figure 16.40).

In a conductor in equilibrium the potential is exactly the same everywhere inside the conductor, because $E = 0$ inside the conductor.

$$\Delta V = -(E_x \Delta x + E_y \Delta y + E_z \Delta z) = 0$$

so V is constant.

QUESTION Does this mean that the potential is zero at every location inside a metal in equilibrium?

No. The potential inside a metal object in equilibrium is constant (the same at every location), but it need not be zero. Charges on the surface of the metal object or on other objects may contribute to a nonzero but uniform potential inside the metal.

EXAMPLE **The Potential Inside a Charged Metal Sphere**

A solid metal sphere of radius R has a charge of Q uniformly distributed over its surface. What is the potential (relative to infinity) throughout the sphere?

Solution At any distance $r > R$, the potential relative to infinity due to the charged sphere is $\frac{1}{4\pi\varepsilon_0}\frac{Q}{r}$, so the potential relative to infinity just outside the surface of the sphere is $\frac{1}{4\pi\varepsilon_0}\frac{Q}{R}$. Inside the sphere the electric field is zero everywhere, so the potential difference between any location inside the sphere and the surface is zero. Therefore the potential inside the sphere is $\frac{1}{4\pi\varepsilon_0}\frac{Q}{R}$ everywhere.

This situation is a good example of the indirect relationship between field and potential. Inside the sphere the electric field is zero everywhere but the electric potential is nonzero everywhere.

Avoiding a Common Confusion

A common error is to assume that the electric field at a location determines the potential at a location. In fact, the electric field at location A has very little to do with the potential at location A! Similarly, the electric field at A and the electric field at B have very little to do with the potential difference between locations A and B; it is the electric field in the intervening region that determines ΔV. As we have just seen, the electric field due to a charged spherical shell is zero inside the shell, but the potential inside the shell due to the charges on the sphere is nonzero.

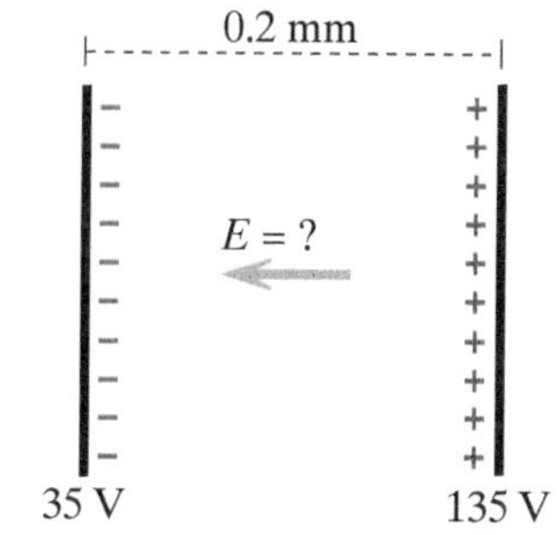

Figure 16.41 What is the magnitude of the electric field?

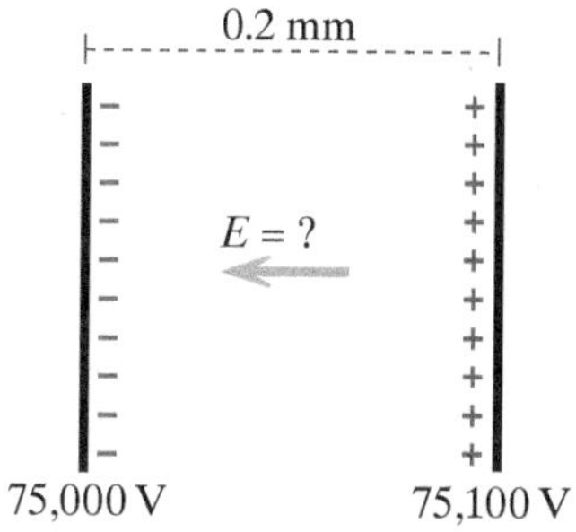

Figure 16.42 What is the magnitude of the electric field?

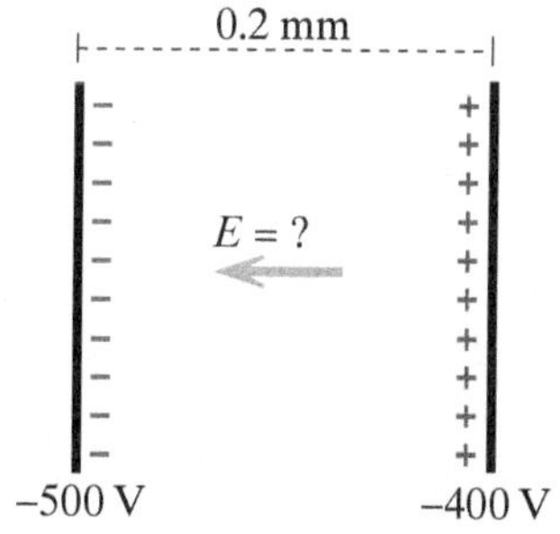

Figure 16.43 What is the magnitude of the electric field?

Shifting the Zero of Potential

As we saw in Chapter 6, electric potential energy (and therefore potential) must go to zero at large distances, to satisfy the requirements of relativity. Nevertheless, we are often interested only in differences of potential or potential energy, in which case we can choose a different zero location as a matter of convenience. This is exactly analogous to approximating change in gravitational potential energy as $\Delta(-GMm/r) \approx \Delta(mgy)$ near the surface of the Earth, and then choosing the zero of potential energy to be at or near the surface, rather than at infinity.

The electric field $\vec{E}$ is given by the gradient of the potential, not by the value of the potential. This means that only potential difference ΔV matters in determining the electric field, not the value V of the potential.

QUESTION Find the magnitude of the electric field in the 0.2-mm gap of a narrow-gap capacitor whose negative plate is at a potential of 35 V and whose positive plate is at a potential of 135 V (Figure 16.41).

The magnitude of the electric field in this case is

$$(100\text{ V})/(0.2 \times 10^{-3}\text{ m}) = 5 \times 10^5\text{ V/m}$$

The direction is from the positive plate toward the negative plate (direction of decreasing potential).

QUESTION Suppose instead that the potential of the negative plate is 75,000 V and the potential of the positive plate is 75,100 V (Figure 16.42). Now what is the electric field in the 0.2-mm gap?

This makes no difference in the calculation, because it is only the 100-V difference in the potential that counts, not the particular values of the potential.

QUESTION Now suppose that the potential of the negative plate is –500 V and the potential of the positive plate is –400 V (Figure 16.43). What is the electric field in the 0.2-mm gap?

The magnitude of the electric field is again 5×10^5 V/m. These calculations show that what counts in calculating the electric field is the gradient of potential or potential difference per unit distance, not the value of the potential. This is exactly analogous to the fact that it is no harder to walk up from the 75th floor to the 76th floor of a building than it is to walk up from the 1st floor to the 2nd floor. To put it another way, adding a constant V_0 to all the potential values changes nothing as far as potential difference and electric field are concerned:

$$(V_f + V_0) - (V_i + V_0) = V_f - V_i$$

Reflection: Potential and Potential Difference

We have seen that there are two different ways to find the potential at a particular location:

- Add up the contributions of all point charges at all other locations:

$$V_A = \sum \frac{1}{4\pi\varepsilon_0}\frac{q_i}{r_i}$$

- Travel along a path from a point very far away to the location of interest, adding up $-\vec{E} \bullet d\vec{l}$ at each step:

$$V_A = -\int_{\infty}^{A} \vec{E} \bullet d\vec{l}$$

One way to think of the meaning of the potential at location A is to consider how much work per unit charge you would have to do to move a charged particle from a location very far away to location A. This work would of course depend on the electric field in the region through which you moved the charge.

We have also seen that there are two ways to find the potential difference between two locations A and B:

- Subtract the potential at the initial location A from the potential at the final location B:

$$\Delta V = V_B - V_A$$

- Travel along a path from A to B, adding up $-\vec{E} \bullet d\vec{l}$ at each step:

$$\Delta V = V_B - V_A = -\int_A^B \vec{E} \bullet d\vec{l}$$

Again, note that the potential difference between two locations depends on the electric field in the region between the two locations. If you were to move a charge from location A to location B, the electric field in the intervening region would determine how much work per unit charge you would have to do.

16.8 COMPUTING POTENTIAL DIFFERENCES

A computer can be useful in calculating potential differences in two situations: first, if the locations and charges of the sources of electric field are known, and second, if we have information about the electric field in the region of interest—either an analytical expression or a sufficient number of measurements.

A Known Charge Distribution

If the distribution of source charges is known, the easiest approach is to model each source as a point charge or collection of point charges. From Section 16.7, we have the following results:

- The potential relative to infinity at location A, due to a single point charge q, is $\dfrac{1}{4\pi\epsilon_0}\dfrac{q}{r_A}$.
- The potential at location A due to many point charges is simply the sum of the individual contributions of each point charge.
- The potential difference $V_B - V_A$ is independent of the path taken from A to B.

A program to compute the potential difference $V_B - V_A$, therefore, has a relatively straightforward structure:

- Make a list of source charges and locations.
- Add up the contributions of each source charge to the potential at A.
- Add up the contributions of each source charge to the potential at B.
- Find $V_B - V_A$.

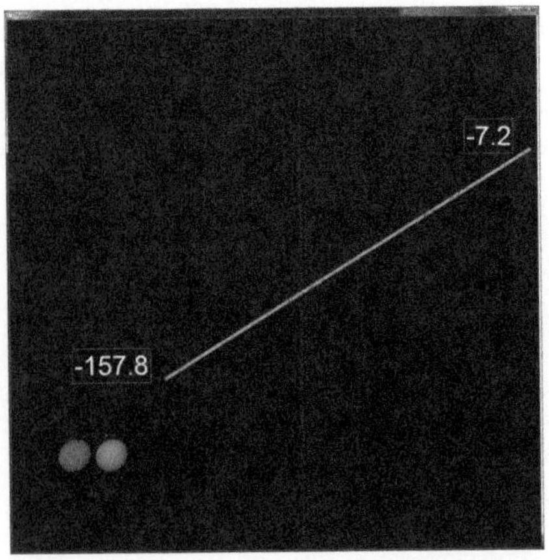

Figure 16.44 The display generated by a program to calculate the potential difference between two locations near a dipole, not on the dipole axis.

The program below does this calculation for a path near a dipole. Neither point A nor point B lies on the dipole axis. The display generated by the program (modified to use label objects to display the potential at each endpoint) is shown in Figure 16.44.

```
from visual import *
scene.width = scene.height = 800
oofpez = 9e9
```

```
s = 0.01
qd = 2e-9
## source charges
s1 = sphere(pos=(-s/2,0,0), radius=0.003,
            color=color.red, q=qd)
s2 = sphere(pos=(s/2,0,0), radius=0.003,
            color=color.blue, q=-qd)
sources = [s1,s2]
## path
A = vector(0.02,0.02,0)
B = vector(0.12,0.08,0)
path = [A,B]
curve(pos=path, color=color.green)  ## draw path
## potentials
VA = 0
VB = 0
i = 0
while i < len(sources):
    rate(500)
    r_A = A - sources[i].pos
    VA = VA + oofpez*sources[i].q/mag(r_A)
    r_B = B - sources[i].pos
    VB = VB + oofpez*sources[i].q/mag(r_B)
    i = i + 1
print('VA=', VA, 'VB=', VB, 'VB-VA=', VB-VA)
```

Problems P88–91 ask you to modify this program to calculate potential differences near other charge distributions such as rings and rods.

A Known Electric Field

It is sometimes the case that the spatial dependence of electric field in a region of space is known, even though we may not know exactly the location and value of the source charges associated with the field. In such a case, we can write a program to follow the procedure used in Section 16.5 to calculate a potential difference numerically. The structure of such a program is discussed in optional Section 16.13.

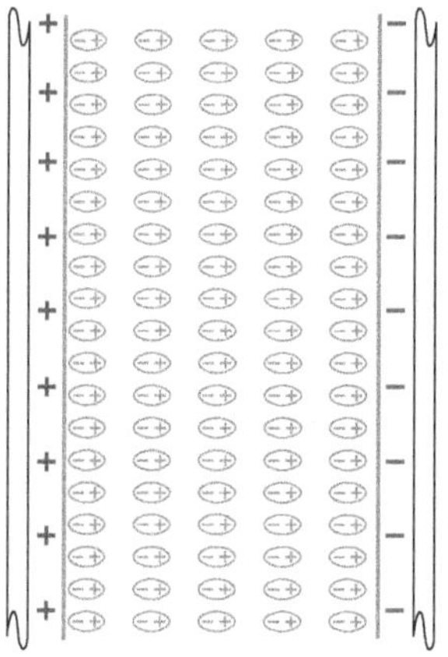

Figure 16.45 Polarization of molecules in a piece of plastic inside a capacitor.

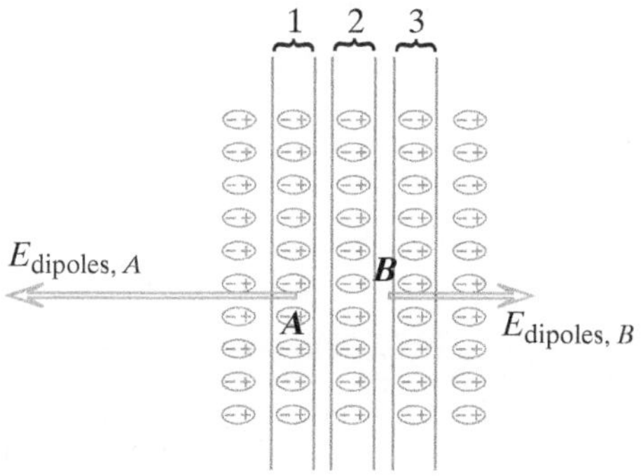

Figure 16.46 Electric field contributed solely by the induced dipoles in the plastic.

16.9 POTENTIAL DIFFERENCE IN AN INSULATOR

Reasoning from the definition of equilibrium and the existence of a sea of mobile electrons in a metal, we were able to conclude that in equilibrium the net electric field everywhere inside a metal is zero. Given this, we were able to conclude that the potential difference between any two locations inside a metal object in equilibrium must be zero.

The situation inside a polarized insulator is more complex. An applied field, such as the uniform field inside a capacitor (Figure 16.45), polarizes the molecules in the insulator. These polarized molecules themselves contribute to the net field inside the material. Presumably, however, the electric field inside the plastic due to the polarized molecules varies depending on the observation location.

For example, consider locations A and B inside the polarized plastic shown in Figure 16.46, where we show electric field contributed solely by the induced dipoles in the polarized plastic. If we consider columns of polarized molecules to be similar to capacitors (consisting of two vertical sheets of charge), then location A is inside "capacitor" 1, while location B is between "capacitors" 2 and 3. At location A the dominant contribution by the dipoles will be from the molecules in capacitor 1, and the electric field due to the dipoles will be large and to the left.

At location B, the fringe fields of the nearest two capacitors represent the dominant contributions by the dipoles, and the field due to the dipoles will point to the right and be smaller than the field due to the dipoles at A. If we follow a path from the left side of the plastic to the right side, $\vec{E}$ due to the dipoles will sometimes point in the direction of $d\vec{l}$ and sometimes opposite to the direction of $d\vec{l}$.

It would be useful to consider an average electric field due to the polarized molecules inside the plastic, but it isn't clear how we would calculate such a field. It is not even clear from simply inspecting the situation what the direction of this average field will be. We can construct a surprisingly simple argument based on potential to help us answer this question.

Round-Trip Potential Difference

Although the pattern of E_{dipoles} is complex inside the material, it is much less complex at locations outside the plastic. Intuitively, you may be able to see that at locations outside the plastic the electric field due to the polarized plastic will have a similar pattern to the electric field near a single dipole.

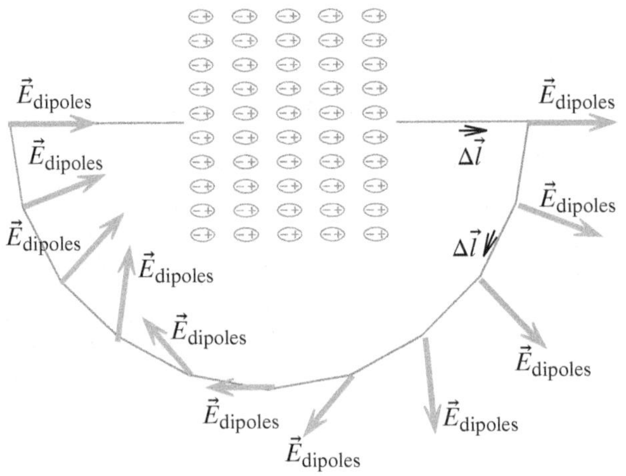

Figure 16.47 Electric field due solely to the induced dipoles in the plastic, along a path outside the plastic.

In Figure 16.47 the electric field due just to the polarized plastic is shown at locations along a path outside the plastic. If we travel clockwise around the path, it is clear from the diagram that $\vec{E} \bullet d\vec{l}$ is always a positive quantity: there is never a component of $\vec{E}$ opposite to $d\vec{l}$. Therefore $-\int \vec{E} \bullet d\vec{l} < 0$ along the path outside the plastic.

Since the molecular dipoles consist of stationary point charges, the round-trip path integral of the electric field (the round-trip potential difference) due to the molecular dipoles must sum to zero.

> QUESTION Explain why this implies that the average field inside the plastic must point to the left and cannot point to the right.

If the average field inside the plastic pointed to the right, the round-trip path integral of electric field in Figure 16.47 would be nonzero, which is impossible. A very general argument concerning the nature of potential lets us deduce the direction of the average field inside the plastic.

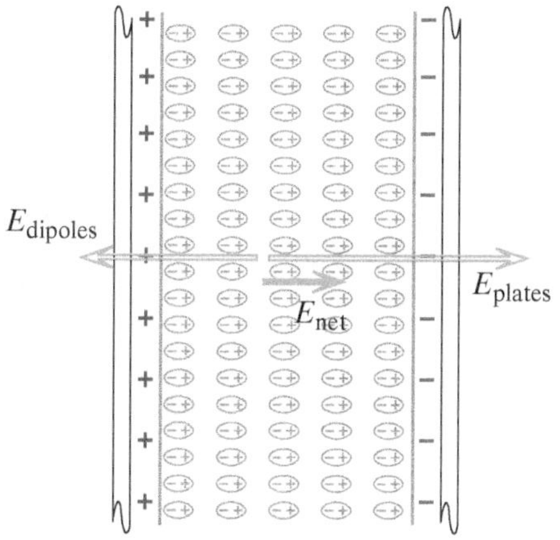

Figure 16.48 The net field inside the polarized insulator is smaller in magnitude than the applied field.

Material	K
Vacuum	1 (exactly)
Air	1.0006
Carbon tetrachloride	2.2
Typical plastic	5
Sodium chloride	6.1
Water	80 at 20 °C
Strontium titanate	310

Figure 16.49 Values of the dielectric constant of several materials.

Dielectric Constant and Net Field

The net electric field inside the plastic is the sum of the field due to the capacitor plates and the field due to the induced dipoles in the plastic, which, as we have just demonstrated, points in a direction opposite to the capacitor's field (Figure 16.48). The net field inside the plastic is therefore smaller than the field due to the capacitor alone. We can say informally that the electric field is "weakened" inside an insulator. Note carefully that the *net* field is in the same direction as the field made by the plates, but smaller in magnitude.

We define a constant K, called the "dielectric constant" (dielectric is another word for insulator), as the factor by which the net electric field is weakened inside an insulator:

$$\text{Inside an insulator:}\quad \vec{E}_{\text{insulator}} = \frac{\vec{E}_{\text{applied}}}{K}$$

For a capacitor, the applied field is $(Q/A)/\varepsilon_0$ and the net field inside a capacitor that is filled with an insulator is $(Q/A)/\varepsilon_0/K$. The dielectric constant is related to the atomic polarizability, but the details are beyond the scope of this introductory textbook. (The difficulty in making the connection is in accounting for the electric fields produced by the other polarized molecules.)

Note that the dielectric constant K is always bigger than 1, because polarization always weakens the net electric field inside the insulator. The easier it is to polarize a molecule, the bigger is the field-weakening effect and the bigger the value of K. Figure 16.49 gives representative values of the dielectric constant K for various insulators.

If there is no insulator in the gap, the potential difference across the capacitor is $|\Delta V| = Es$ (since $\vec{E}$ is uniform, and parallel to $d\vec{r}$). If the insulator fills the gap and we maintain the same charges $+Q$ and $-Q$ on the metal plates, both the electric field E and the potential difference ΔV are reduced by the same factor K:

$$\Delta V_{\text{insulator}} = \frac{\Delta V_{\text{vacuum}}}{K}$$

Since we can't actually get inside the insulator and measure the field there, we determine K for a particular insulator by measuring the effect on the potential difference ΔV between the plates for a fixed capacitor charge Q.

In summary, placing an insulator between the plates of a capacitor

- decreases electric field inside the insulator
- decreases potential difference across the insulator

If the insulator doesn't fill the gap, the electric field inside the insulator is reduced by the factor K, but the electric field at other places in the gap is hardly affected, because the electric field of the insulator in those places turns out to be small compared to the field made by the plates. (In our model of the insulator, consisting of capacitor-like sheets of plus and minus charges, the field made by the dipoles outside the insulator is a fringe field.) Outside the capacitor plates, however, the fringe field of the plates is small, and the small field contributed by the insulator significantly reduces the net field.

In Section 16.5 we examined the electric field and potential difference between the two metal plates of a capacitor, then inserted a metal slab. In the following Checkpoint we will start with the same device, but insert a glass slab.

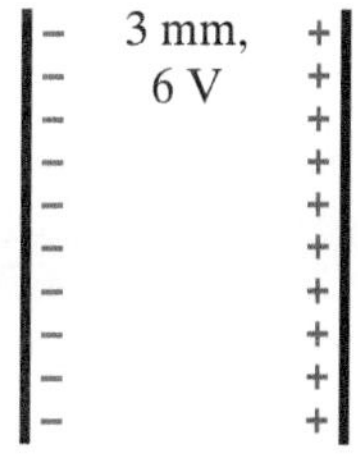

Figure 16.50 Capacitor before inserting a glass disk 1 mm in thickness between the plates.

Checkpoint 11 A capacitor with a 3-mm gap has a potential difference of 6 V (Figure 16.50). A disk of glass 1 mm thick, with area the same as the area of the metal plates, has a dielectric constant of 2.5. It is inserted in the middle of the gap between the metal plates. Now what is the potential difference of the two metal plates? (It helps to make a diagram showing the electric field along a path.)

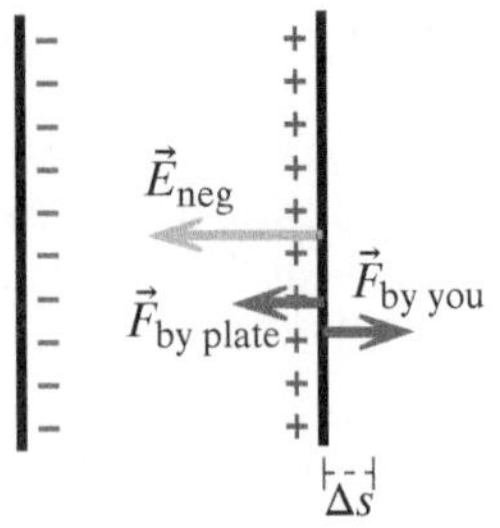

Figure 16.51 To move a capacitor plate you must exert a force equal to the force on the plate by the other plate.

16.10 ENERGY DENSITY AND ELECTRIC FIELD

Until now we have thought of energy (that is, electric potential energy) as associated with interacting charged particles. There is an alternative view, however, that considers energy to be stored in electric fields themselves. To see how this view works out quantitatively, we'll consider moving one plate of a capacitor.

The force that one capacitor plate exerts on the other (Figure 16.51) is equal to the charge Q on one plate times the field made by the other plate, which is half the total field in the gap:

$$E_{\text{one plate}} = \frac{(Q/A)}{2\varepsilon_0} \quad \text{(for very small gap)}$$

so the force on one plate is

$$F = Q\frac{(Q/A)}{2\varepsilon_0}$$

Suppose that you pull the positive plate of the capacitor slowly away from the negative plate, exerting a force only infinitesimally larger than the force exerted by the other plate. The work that you do in moving the plate a distance Δs goes into increasing the electric potential energy U:

$$\Delta U = W = Q\left(\frac{Q/A}{2\varepsilon_0}\right)\Delta s$$

We can rearrange this expression in the following way:

$$\Delta U = \frac{1}{2}\varepsilon_0\left(\frac{Q/A}{\varepsilon_0}\right)^2 A\,\Delta s$$

The expression in parentheses is the electric field E inside the capacitor. The quantity $A\,\Delta s$ is the change in the volume occupied by electric field inside the capacitor (length times width times increase in the plate separation). Therefore we can write

$$\frac{\Delta U}{\Delta(\text{volume})} = \frac{1}{2}\varepsilon_0 E^2$$

We ascribe an *energy density* (joules per cubic meter) to the electric field. By pulling the capacitor plates apart we increased the volume of space in which there is a sizable electric field. We say that the energy expended by us was converted into energy stored *in the electric field*.

This view is numerically equivalent to our earlier view of electric potential energy, but it turns out to be a more fundamental view. For example, the energy carried by electromagnetic radiation propagating through space far from any charges can best be expressed in terms of the energy density associated with the field.

FIELD ENERGY DENSITY

$$\frac{1}{2}\varepsilon_0 E^2 \text{ (units are J/m}^3\text{)}$$

The energy density (joules per cubic meter) is stored in a region where there is an electric field of magnitude E.

Although we derived the field energy density for the particular case of a capacitor, the result is general. Anywhere there is electric field, there is energy density given by this equation.

Checkpoint 12 The energy density inside a certain capacitor is 10 J/m^3. What is the magnitude of the electric field inside the capacitor? What is the energy density associated with an electric field of 3×10^6 V/m (large enough to initiate a spark)?

*An Electron and a Positron

In Chapter 6 we stated the principle of conservation of energy in this way: The change in energy of a system plus the change in energy of its surroundings must be zero. In the following example we will see that in order to understand energy conservation in quite a simple situation it is necessary to invoke the idea that energy is stored in fields.

Consider an electron and a positron that are released from rest some distance from each other (Figure 16.52). We will take the electron to be the system under consideration, so therefore the positron and everything else in the Universe are the "surroundings." Because of the attractive electric force between the particles, the electron accelerates toward the positron, gaining kinetic energy. By the principle of conservation of energy, the energy of the surroundings must therefore decrease. (Remember that the system can have no potential energy, since it consists of a single particle, and potential energy is a property of pairs of particles.)

Figure 16.52 An electron and a positron are released from rest, some distance apart. We choose the electron as the system and everything else (including the positron) as the surroundings.

QUESTION Does the energy of the positron decrease?

No, the energy of the positron also increases, since it accelerates toward the electron, gaining kinetic energy.

QUESTION Where is there a decrease of energy in the surroundings?

Evidently the energy stored in the fields surrounding the two particles must decrease. Clearly, the electric field at any location in space does change as the positions of the particles change. The electric field in the region between the particles gets larger, but the electric field everywhere else in space decreases (since E_{dipole} is proportional to s, the distance between the particles). It would be a somewhat daunting task to integrate E^2 over the volume of the Universe, with the additional complication that close to a charged particle E approaches infinity. However, we do not actually need to do this integral to figure out the change in energy of the electric field throughout space. Since

$$\Delta(\text{Field energy}) + \Delta K_{\text{positron}} + \Delta K_{\text{electron}} = 0$$

then

$$\Delta(\text{Field energy}) = -2(\Delta K_{\text{electron}})$$

In this example, the principle of conservation of energy leads us directly to the idea that energy must be stored in electric fields, since there is no other way to account for the decrease of energy in the surroundings.

If we had chosen the electron plus the positron as our system, we would have found that ΔU_{el} is equal to $-2(\Delta K_{\text{electron}})$. The change in potential energy for the two-particle system is the same as the change in the field energy. Evidently in a multiparticle system we can either consider a change in potential energy or a change in field energy (but not both); the quantities are equal.

The idea of energy stored in fields is a general one. It is not only electric fields that carry energy, but magnetic fields and gravitational fields as well.

16.11 *POTENTIAL OF DISTRIBUTED CHARGES

Potential Along the Axis of a Uniformly Charged Disk

Consider a disk of radius R (area $A = \pi R^2$) with charge Q uniformly distributed over its surface. To calculate the potential due to this disk we follow a procedure similar to the one we used in finding the electric field of a charge distribution. We divide the disk into rings as we did in finding the electric field of a disk in Chapter 15 (Figure 16.53).

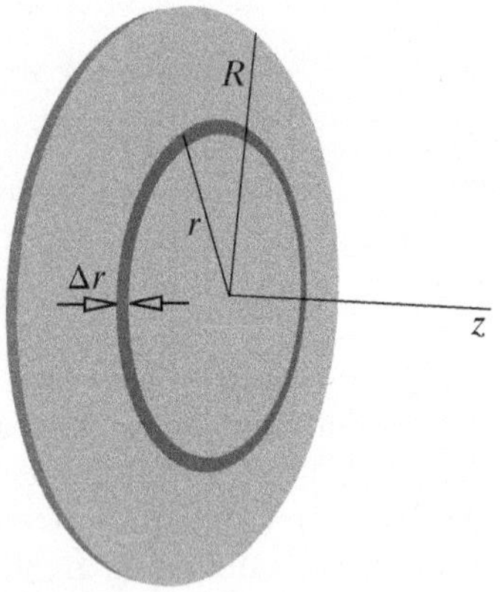

Figure 16.53 A ring of radius r and width Δr makes a contribution V_{ring} to the potential of the disk.

At location z along the axis of a disk, the potential contributed by one ring is given by the result found earlier:

$$V_{\text{ring}} = \frac{1}{4\pi\varepsilon_0}\frac{\Delta q}{(z^2+r^2)^{1/2}} = \frac{1}{4\pi\varepsilon_0}\frac{Q}{A}\frac{2\pi r\Delta r}{(z^2+r^2)^{1/2}}$$

We have used the result from the previous chapter for the charge on one of the rings, $\Delta q = (Q/A)2\pi r\Delta r$. Add up the contributions of all the rings:

$$V = \frac{1}{2\varepsilon_0}\frac{Q}{A}\int_0^R \frac{rdr}{(z^2+r^2)^{1/2}}$$

$$V = \frac{1}{2\varepsilon_0}\frac{Q}{A}\left[(z^2+r^2)^{1/2}\right]_0^R$$

$$V = \frac{1}{2\varepsilon_0}\frac{Q}{A}\left[(z^2+R^2)^{1/2} - z\right]$$

The units for this result check: meters in the bracket, and m^2 in the A term in the denominator, so the units are those of $Q/(\varepsilon_0 R)$, which are indeed units of potential.

We can find the electric field from the negative gradient of the potential. We want the z component of electric field, and taking the derivative of the potential we find the following:

$$E = -\frac{\Delta V}{\Delta z} \to -\frac{dV}{dz} = \frac{(Q/A)}{2\varepsilon_0}\left[1 - \frac{z}{(z^2+R^2)^{1/2}}\right]$$

This is the same result obtained in Chapter 15 by summing the contributions to the electric field.

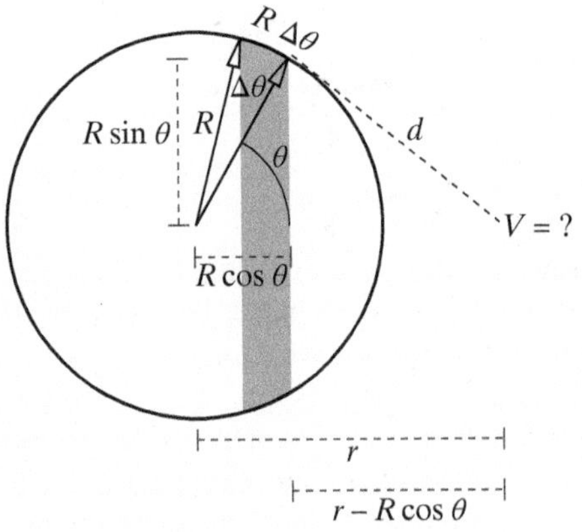

Figure 16.54 The sphere may be divided into ring-shaped segments.

Checkpoint 13 What is the potential at the center of a spherical shell of radius R that has a charge Q uniformly distributed over its surface? Don't do an integral of the electric field; just add up the contributions to the potential by the charges.

16.12 *INTEGRATING THE SPHERICAL SHELL

In this section, we'll use potential to prove that a uniformly charged spherical shell looks from the outside like a point charge but on the inside has a zero electric field. We did this by integrating the electric field directly in Chapter 15, and a very different kind of proof using Gauss's law is given in a later chapter.

Divide the spherical shell into rings of charge, each delimited by the angle θ and the angle $\theta+\Delta\theta$, and carrying an amount of charge ΔQ (Figure 16.54). The angle θ will be the integration variable used in summing up the contributions of the various rings.

Each ring contributes V_{ring} at an observation point a distance r from the center of the spherical shell. Each ring is a distance $(r - R\cos\theta)$ from the observation location (we will eventually choose $\Delta\theta$ so small that it makes no difference whether we measure to the edge of the ring or to the center of the ring). Therefore the distance d from the observation location to each charge on the ring is $\sqrt{[(r - R\cos\theta)^2 + (R\sin\theta)^2]}$:

$$V_{\text{ring}} = \frac{1}{4\pi\varepsilon_0}\frac{\Delta Q}{[(r - R\cos\theta)^2 + (R\sin\theta)^2]^{1/2}}$$

The total amount of charge on each ring is

$$\Delta Q = \left(\frac{Q}{4\pi R^2}\right)\Delta A$$

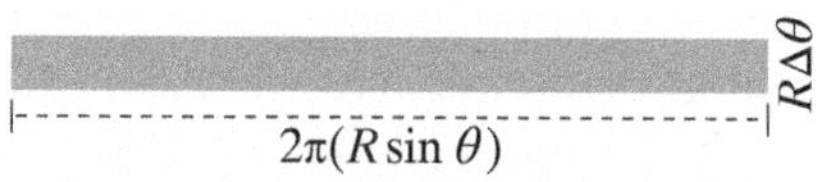

Figure 16.55 We can calculate the area of the ring of charge by "unrolling" it.

where ΔA is the surface area of the ring, since the total charge Q is spread uniformly all over the spherical surface, whose total area is $4\pi R^2$. To calculate the surface area ΔA of the ring, as before we lay the ring out flat (Figure 16.55), noting that the radius of the ring is $R\sin\theta$ and its width is $R\Delta\theta$ (since arc length is radius times angle, with angle measured in radians).

Putting these elements together, we have for the potential due to the ring

$$V_{\text{ring}} = \frac{1}{4\pi\varepsilon_0}\frac{1}{[(r - R\cos\theta)^2 + (R\sin\theta)^2]^{1/2}}\left(\frac{Q}{4\pi R^2}\right) \times 2\pi(R\sin\theta)R\Delta\theta$$

We can add up all these contributions in the form of a definite integral:

$$V = \frac{1}{4\pi\varepsilon_0}\frac{Q}{2}\int_0^{\pi}\frac{\sin\theta d\theta}{[(r - R\cos\theta)^2 + (R\sin\theta)^2]^{1/2}}$$

The limits on the integral are determined by the fact that if we let θ range between 0 and π rad (180°), we add up rings that account for the entire surface of the spherical shell.

A change of variables lets us evaluate this integral. Let

$$u = (r - R\cos\theta)^2 + (R\sin\theta)^2$$
$$u = r^2 - 2rR\cos\theta + (R\cos\theta)^2 + (R\sin\theta)^2$$

However, $(R\cos\theta)^2 + (R\sin\theta)^2 = R^2$, from the Pythagorean theorem. Therefore

$$u = r^2 - 2rR\cos\theta + R^2$$

$$du = 2rR\sin\theta d\theta \quad \text{and} \quad \sin\theta d\theta = \frac{1}{2rR}du$$

$$V = \frac{1}{4\pi\varepsilon_0}\frac{Q}{2}\frac{1}{2rR}\int_{\theta=0}^{\theta=\pi}\frac{du}{u^{1/2}}$$

$$V = \frac{1}{4\pi\varepsilon_0}\frac{Q}{2}\frac{1}{2rR}\left[2\sqrt{(r - R\cos\theta)^2 + (R\sin\theta)^2}\right]_0^{\pi}$$

$$V = \frac{1}{4\pi\varepsilon_0}\frac{Q}{2rR}\left[\sqrt{(r+R)^2} - \sqrt{(r-R)^2}\right]$$

$$V = \frac{1}{4\pi\varepsilon_0}\frac{Q}{2rR}[(r+R) \mp (r-R)]$$

This gives two results, depending on which sign is taken for the square root. Take the – sign, which turns out to correspond to a location outside the shell:

$$V = \frac{1}{4\pi\varepsilon_0}\frac{Q}{r} \quad \text{(outside the shell; } r > R\text{)}$$

This says that the potential outside a uniformly charged spherical shell is exactly the same as the potential due to a point charge located at the center of the shell, as though the shell were collapsed to a point. Since the electric field is the negative gradient of the potential, the electric field outside the shell is the same as if the shell were collapsed to its center.

If we take the + sign, we have

$$V = \frac{1}{4\pi\varepsilon_0}\frac{Q}{R} \quad \text{(inside the shell; } r < R\text{)}$$

This says that the potential inside the shell is constant, and equal to the potential just outside the shell. A constant potential means that the electric field inside the shell is zero. Note that for locations inside the spherical shell, some of the rings give a field to the left, and some give a field to the right. It turns out that these contributions exactly cancel each other.

It is the $1/r^2$ behavior of the electric field of point charges that leads to these unusual results. If the electric field of a point charge were not exactly proportional to $1/r^2$, the electric field outside a uniform spherical shell would not look like the field of a point charge at the center, and the electric field would not be zero inside the shell.

16.13 *NUMERICAL INTEGRATION ALONG A PATH

The key issue in writing a program to calculate $-\int_A^B \vec{E}\cdot\Delta\vec{l}$ is specifying the path. One way to do this is to find the vector $\vec{B}-\vec{A}$ from the initial to the final location, and then to divide this vector by the number of steps you want to take along the path to get $\Delta\vec{l}$. For example:

```
A = vector(0.15,0,0)
B = vector(0.20,0.35,0)
Nsteps = 10
deltal = (B-A)/Nsteps
```

Each iteration of the loop corresponds to one step along the path. For each step, we need to find $-\vec{E}\cdot\Delta\vec{l}$, using the value of $\vec{E}$ in the middle of the step. We then need to advance the position one step along the path. For example:

```
deltaV_BA = 0
L = vector(0.15,0,0)  ## initial position
i = 0
while i < Nsteps:
    rate(10) ## slow rate to watch the process
    E = ## calculate E at location L + deltal/2
    ## display E and deltal as arrows
    deltaV = -dot(E,deltal)
    deltaV_BA = deltaV_BA + deltaV
    L = L + deltal
    i = i + 1
```

If the full path is more complicated than a single straight line, it may be necessary to divide the path into several segments.

SUMMARY

Potential difference

$$\Delta V = V_f - V_i = -\int_i^f \vec{E} \bullet d\vec{l}$$

along a path from i to f

$$= -\int_i^f E_x dx - \int_i^f E_y dy - \int_i^f E_z dz$$

Units of electric potential are joules/coulomb, or volts.
Special case:
Regions of uniform electric field:

$$\Delta V = \sum(-\vec{E} \bullet \Delta\vec{l}) = \sum(-E_x\Delta x + -E_y\Delta y + -E_z\Delta z)$$

Change in electric potential energy for a particle of charge q moved through a potential difference ΔV:

$$\Delta U_{\text{electric}} = q\Delta V$$

Sign of ΔV and direction of path relative to $\vec{E}$:
Path going in the direction of $\vec{E}: \Delta V < 0$
Path going opposite to $\vec{E}: \Delta V > 0$
Path perpendicular to $\vec{E}: \Delta V = 0$
Point charge: $\Delta V = \Delta\left(\frac{1}{4\pi\varepsilon_0}\frac{Q}{r}\right)$
If the electric field is known along a path from location A to location B, the potential difference can be calculated as a path integral of the electric field:

Step 1: Choose a path and cut it into pieces $\Delta\vec{l}$.
Step 2: Write an expression for $\Delta V = -\vec{E} \bullet \Delta\vec{l}$ of one piece.
Step 3: Add up the contributions of all the pieces.
Step 4: Check the result, especially the sign.

If you can calculate the potential relative to infinity at locations A and B, it is not necessary to integrate the electric field along a path.

Path independence
Change in potential between two locations is independent of path.
For a round-trip path, $\Delta V = 0$.

Conductor in equilibrium
$\Delta V = -\int_i^f \vec{E} \bullet d\vec{l} = 0$ along any path, so potential is uniform (constant everywhere) throughout a metal in equilibrium.

Field is gradient of potential

$$E_x = \frac{\partial V}{\partial x}, \quad E_y = \frac{\partial V}{\partial y}, \quad E_z = \frac{\partial V}{\partial z}$$

Dielectric constant

Inside an insulator $\vec{E}_{\text{insulator}} = \frac{\vec{E}_{\text{applied}}}{K}$; K is the dielectric constant; $K > 1$, so

$$\Delta V_{\text{insulator}} = \frac{\Delta V_{\text{vacuum}}}{K}$$

In summary, placing an insulator between the plates of a capacitor

- decreases the electric field inside the insulator
- decreases the potential difference across the insulator

Field energy density

$$\frac{1}{2}\varepsilon_0 E^2 \quad (\text{units are J/m}^3)$$

QUESTIONS

Q1 Which statements about potential energy are correct? Select all that apply. (1) Potential energy can be either positive or negative. (2) Potential energy is a property of a pair of objects. (3) When two interacting charged objects get very far away from each other, their potential energy approaches zero. (4) A single particle does not have potential energy. (5) A single isolated object can have potential energy. (6) Potential energy is always a positive number. (7) The potential energy of a pair of interacting charged objects increases as the distance between them increases.

Q2 The graph in Figure 16.56 shows the electric potential energy for a system of two interacting objects, as a function of the distance between the objects. What system(s) might this graph represent?

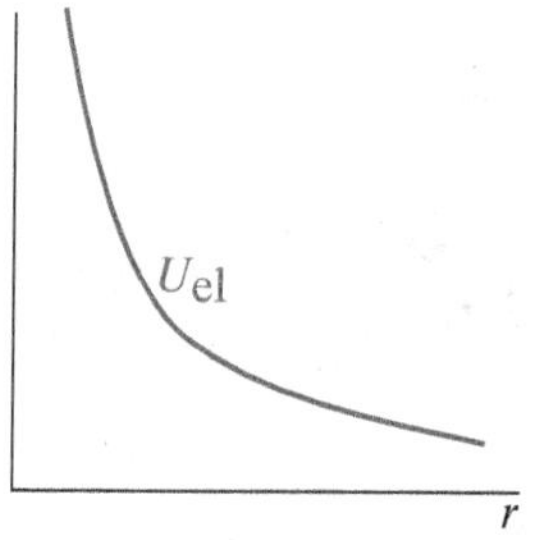

Figure 16.56

(1) Two protons, (2) Two sodium ions, (3) Two neutrons, (4) Two chloride ions, (5) Two electrons, (6) A proton and an electron, (7) A sodium ion and a chloride ion

Q3 The graph in Figure 16.57 shows the electric potential energy for a system of two interacting objects, as a function of the distance between the objects. What system(s) might this graph represent?

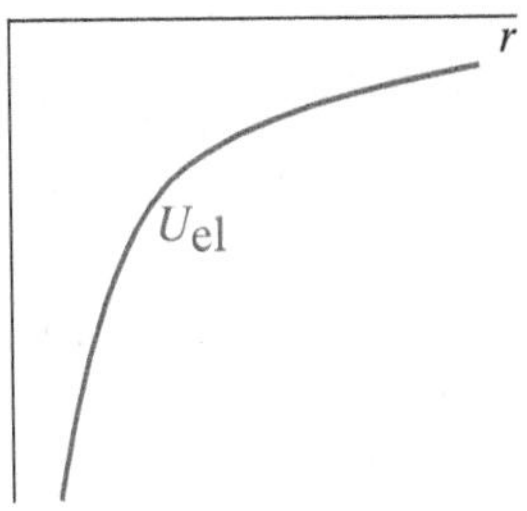

Figure 16.57

(1) Two protons, (2) Two sodium ions, (3) Two neutrons, (4) Two chloride ions, (5) Two electrons, (6) A proton and an electron, (7) A sodium ion and a chloride ion

Q4 What is the difference between electric potential energy and electric potential?

Q5 What are the units of electric potential energy, of electric potential, and of electric field?

Q6 In Figure 16.58, what is the direction of the electric field? Is $\Delta V = V_f - V_i$ positive or negative?

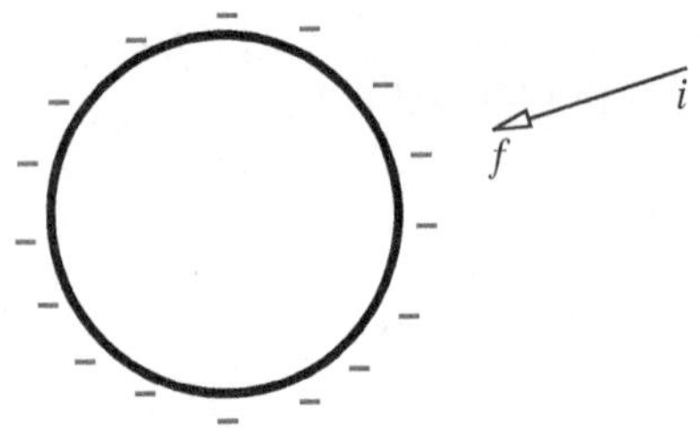

Figure 16.58

Q7 You travel along a path from location A to location B, moving in the same direction as the direction of the net electric field in that region. What is true of the potential difference $V_B - V_A$? (1) $V_B - V_A > 0$, (2) $V_B - V_A < 0$, (3) $V_B - V_A = 0$

Q8 You travel along a path from location A to location B, moving in a direction opposite to the direction of the net electric field in that region. What is true of the potential difference $V_B - V_A$? (1) $V_B - V_A > 0$, (2) $V_B - V_A < 0$, (3) $V_B - V_A = 0$

Q9 You travel along a path from location A to location B, moving in a direction perpendicular to the direction of the net electric field in that region. What is true of the potential difference $V_B - V_A$? (1) $V_B - V_A > 0$, (2) $V_B - V_A < 0$, (3) $V_B - V_A = 0$

Q10 Figure 16.59 shows several locations inside a capacitor. You need to calculate the potential difference $V_D - V_C$.

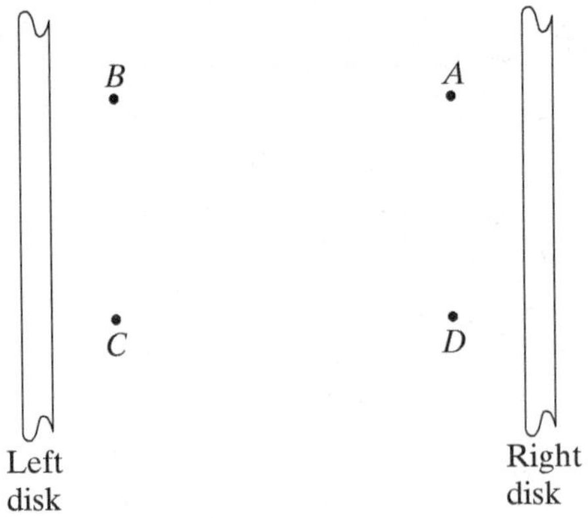

Figure 16.59

(a) What is the direction of the path ($+x$ or $-x$)? **(b)** If the charge on the right plate is negative and the charge on the left plate is positive, what is the sign of $V_D - V_C$?

Q11 Figure 16.60 shows a portion of a long, negatively charged rod. You need to calculate the potential difference $V_A - V_B$.

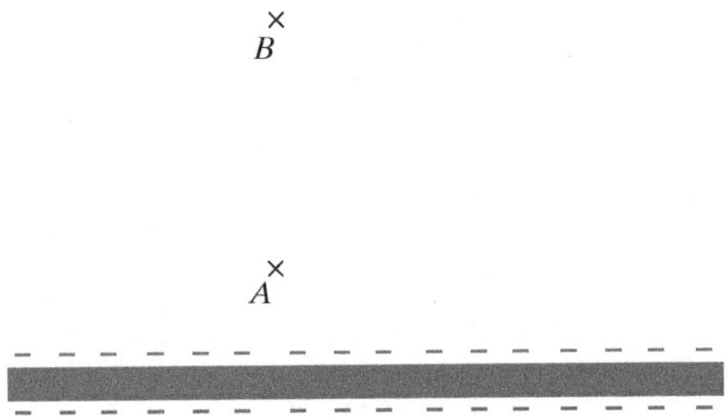

Figure 16.60

(a) What is the direction of the path ($+y$ or $-y$)? **(b)** What is the sign of $V_A - V_B$?

Q12 Location A is a distance d from a charged particle. Location B is a distance $2d$ from the particle. Which of the following statements are true? It may help to draw a diagram. (1) If the charge of the particle is negative, $V_B - V_A$ is negative. (2) If the charge of the particle is positive, $V_A < V_B$. (3) If $V_B < V_A$, we know that the particle must be positive. (4) $V_B < V_A$, regardless of the sign of the charge of the particle. (5) The sign of $(V_B - V_A)$ does not give us any information about the sign of the charge of the particle.

Q13 Locations $A = \langle a,0,0 \rangle$ and $B = \langle b,0,0 \rangle$ are on the $+x$ axis, as shown in Figure 16.61. Four possible expressions for the electric field along the x axis are given below. For each expression for the electric field, select the correct expression (1–8) for the potential difference $V_A - V_B$. In each case K is a numerical constant with appropriate units.

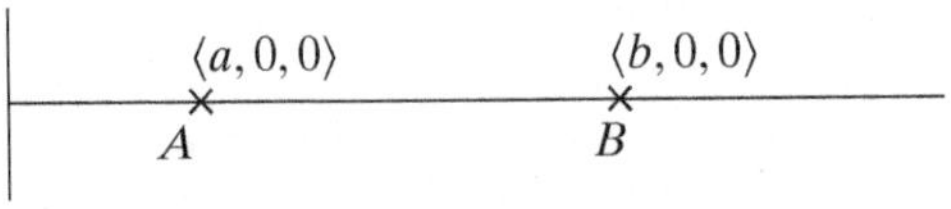

Figure 16.61

(a) $\vec{E} = \left\langle \frac{K}{x^2}, 0, 0 \right\rangle$, **(b)** $\vec{E} = \left\langle \frac{K}{x^3}, 0, 0 \right\rangle$, **(c)** $\vec{E} = \left\langle \frac{K}{x}, 0, 0 \right\rangle$,
(d) $\vec{E} = \langle Kx, 0, 0 \rangle$

1. $V_A - V_B = 0$
2. $V_A - V_B = K(a - b)$
3. $V_A - V_B = K\left(\frac{1}{a} - \frac{1}{b}\right)$
4. $V_A - V_B = K\left(\frac{1}{a^3}a - \frac{1}{b^3}b\right)$
5. $V_A - V_B = \frac{1}{2}K\left(b^2 - a^2\right)$
6. $V_A - V_B = K\ln\left(\frac{b}{a}\right)$
7. $V_A - V_B = K\left(a^2 - b^2\right)$
8. $V_A - V_B = \frac{1}{2}K\left(\frac{1}{a^2} - \frac{1}{b^2}\right)$

Q14 Figure 16.62 shows a pattern of electric field in which the electric field is horizontal throughout this region but is larger toward the top of the region and smaller toward the bottom. If this pattern of electric field can be produced by some arrangement of stationary charges, sketch such a distribution of charges. If this pattern of electric field cannot be produced by any arrangement of stationary charges, prove that it is impossible.

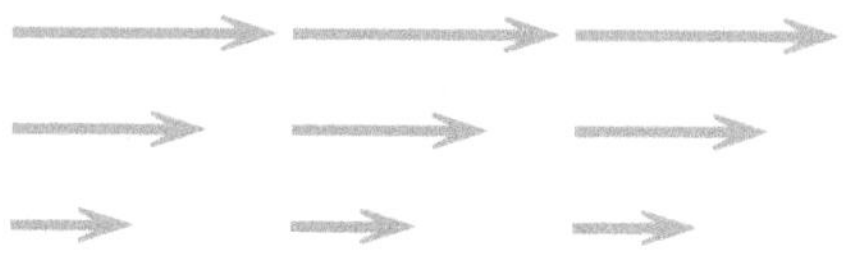

Figure 16.62

Q15 The graph in Figure 16.63 is a plot of electric potential versus distance from an object. Which of the following could be the object?
(1) A neutron, (2) A sodium ion (Na^+), (3) A chloride ion (Cl^-), (4) A proton, (5) An electron

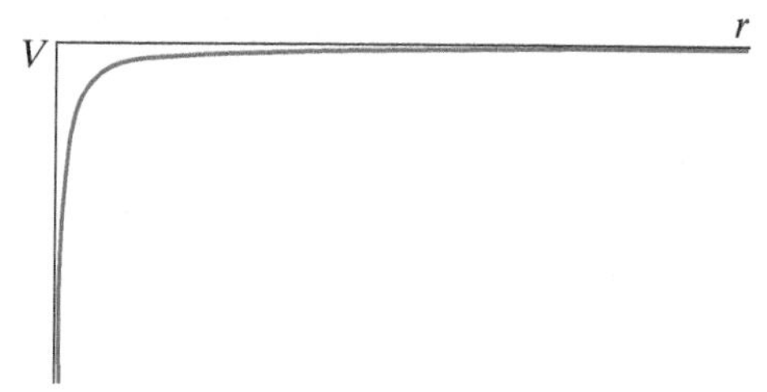

Figure 16.63

Q16 The graph in Figure 16.64 is a plot of electric potential versus distance from an object. Which of the following could be the object?

Figure 16.64

(1) A neutron, (2) A sodium ion (Na^+), (3) A chloride ion (Cl^-), (4) A proton, (5) An electron

Q17 A student said, "The electric field at the center of a charged spherical shell is zero, so the potential at that location is also zero." Explain to the student why this statement is incorrect.

Q18 For each of the following statements, say whether it is true or false and explain why it is true or false. Be complete in your explanation, but be brief. Pay particular attention to the distinction between potential V and potential difference ΔV. **(a)** The electric potential inside a metal in equilibrium is always zero. **(b)** If there is a constant large positive potential throughout a region, the electric field in that region is large. **(c)** If you get close enough to a negative point charge, the potential is negative, no matter what other charges are around. **(d)** Near a point charge, the potential difference between two points a distance L apart is $-EL$. **(e)** In a region where the electric field is varying, the potential difference between two points a distance L apart is $-(E_f - E_i)L$.

Q19 We discussed a method for measuring the dielectric constant by placing a slab of the material between the plates of a capacitor. Using this method, what would we get for the dielectric constant if we inserted a slab of metal (not quite touching the plates, of course)?

PROBLEMS

Section 16.1

•**P20** What is the kinetic energy of a proton that is traveling at a speed of 3725 m/s?

•**P21** If the kinetic energy of an electron is 4.4×10^{-18} J, what is the speed of the electron? You can use the approximate (nonrelativistic) equation here.

•**P22** An electron passes through a region in which there is an electric field, and while it is in the region its kinetic energy decreases by 4×10^{-17} J. Initially the kinetic energy of the electron was 4.5×10^{-17} J. What is the final speed of the electron? (You can use the approximate (nonrelativistic) equation here.)

•**P23** At a particular instant an electron is traveling with speed 6000 m/s. **(a)** What is the kinetic energy of the electron? **(b)** If a proton were traveling at the same speed (6000 m/s), what would be the kinetic energy of the proton?

•**P24** In Chapter 6 we saw that the electric potential energy of a system of two particles is given by the equation

$$U_{el} = \frac{1}{4\pi\varepsilon_0}\frac{q_1 q_2}{r}$$

(a) What is the electric potential energy of two protons separated by a distance of 9 nm (9×10^{-9} m)? **(b)** What is the electric potential energy of a proton and an electron separated by the same distance?

•**P25** A proton that initially is traveling at a speed of 300 m/s enters a region where there is an electric field. Under the influence of the electric field the proton slows down and comes to a stop. What is the change in kinetic energy of the proton?

Section 16.3

•**P26** You move from location i at $\langle 2,7,5\rangle$ m to location f at $\langle 5,6,12\rangle$ m. All along this path there is a nearly uniform electric field of $\langle 1000, 200, -510\rangle$ N/C. Calculate $V_f - V_i$, including sign and units.

•**P27** A capacitor with a gap of 1 mm has a potential difference from one plate to the other of 36 V. What is the magnitude of the electric field between the plates?

•**P28** An electron starts from rest in a vacuum, in a region of strong electric field. The electron moves through a potential difference of 35 V. **(a)** What is the kinetic energy of the electron in electron volts (eV)? **(b)** What would happen if the particle were a proton?

•**P29** Locations A, B, and C are in a region of uniform electric field, as shown in the diagram in Figure 16.65. Location A is

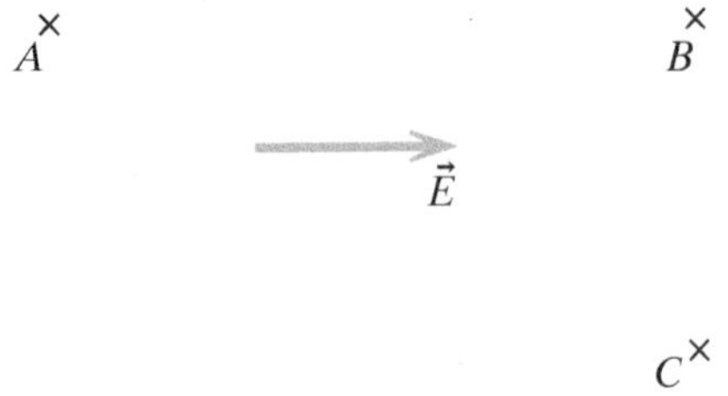

Figure 16.65

at $\langle -0.5,0,0\rangle$ m. Location B is at $\langle 0.5,0,0\rangle$ m. In the region the electric field has the value $\langle 750,0,0\rangle$ N/C. For a path starting at B and ending at A, calculate: **(a)** the displacement vector $\Delta\vec{l}$, **(b)** the change in electric potential, **(c)** the potential energy change for the system when a proton moves from B to A, **(d)** the potential energy change for the system when an electron moves from B to A.

•**P30** Locations A, B, and C are in a region of uniform electric field, as shown in Figure 16.65. Location A is at $\langle -0.5,0,0\rangle$ m. Location B is at $\langle 0.5,0,0\rangle$ m. In the region the electric field has the value $\langle 750,0,0\rangle$ N/C. **(a)** For a path starting at B and ending at C, calculate: (1) the displacement vector $\Delta\vec{l}$, (2) the change in electric potential, (3) the potential energy change for the system when a proton moves from B to C, (4) the potential energy change for the system when an electron moves from B to C, **(b)** Which of the following statements are true in this situation? Choose all that are correct. (1) the potential difference cannot be zero because the electric field is not zero along this path, (2) when a proton moves along this path, the electric force does zero net work on the proton, (3) $\Delta\vec{l}$ is perpendicular to $\vec{E}$.

•**P31** Locations A, B, and C are in a region of uniform electric field, as shown in Figure 16.66. Location A is at $\langle -0.3,0,0\rangle$ m. Location B is at $\langle 0.4,0,0\rangle$ m. In the region the electric field has the value $\langle -850,400,0\rangle$ N/C. For a path starting at A and ending at B, calculate:

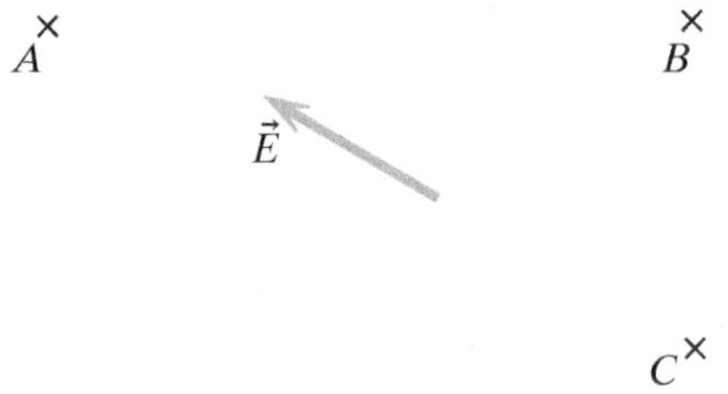

Figure 16.66

(a) the displacement vector $\Delta\vec{l}$, **(b)** the change in electric potential, **(c)** the potential energy change for the system when a proton moves from A to B, **(d)** the potential energy change for the system when an electron moves from A to B.

•**P32** Locations A, B, and C are in a region of uniform electric field, as shown in Figure 16.66. Location A is at $\langle -0.3,0,0\rangle$ m. Location B is at $\langle 0.4,0,0\rangle$ m. In the region the electric field has the value $\langle -850,400,0\rangle$ N/C. For a path starting at C and ending at A, calculate: **(a)** the displacement vector $\Delta\vec{l}$, **(b)** the change in electric potential, **(c)** the potential energy change for the system when a proton moves from C to A, **(d)** the potential energy change for the system when an electron moves from C to A.

•**P33** Locations A and B are in a region of uniform electric field, as shown in Figure 16.67. Along a path from B to A, the change in potential is -2200 V. The distance from A to B is 0.28 m. What is the magnitude of the electric field in this region?

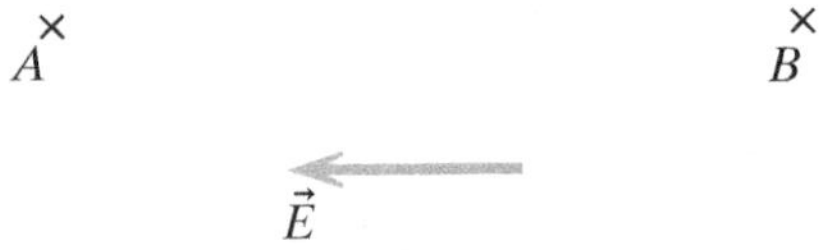

Figure 16.67

•**P34** An electron starts from rest in a vacuum, in a region of strong electric field. The electron moves through a potential difference of 44 V. **(a)** What is the kinetic energy of the electron in electron volts (eV)? **(b)** Which of the following statements would be true if the particle were a proton? Choose both if they are both correct. (1) The kinetic energy of the proton would be negative. (2) The proton would move in the opposite direction from the electron.

•**P35** If the electric field exceeds about 3×10^6 N/C in air, a spark occurs. Approximately, what is the absolute value of the maximum possible potential difference between the plates of a capacitor whose gap is 3 mm, without causing a spark in the air between them?

•**P36** A capacitor with a gap of 2 mm has a potential difference from one plate to the other of 30 V. What is the magnitude of the electric field between the plates?

•**P37** You move from location i at $\langle 2,5,4\rangle$ m to location f at $\langle 3,5,9\rangle$ m. All along this path there is a nearly uniform electric field whose value is $\langle 1000, 200, -500\rangle$ N/C. Calculate $\Delta V = V_f - V_i$, including sign and units.

•**P38** In a particular region there is a uniform electric field of $\langle -760, 380, 0\rangle$ V/m. Location A is $\langle 0.2, 0.1, 0\rangle$ m, location B is $\langle 0.7, 0.1, 0\rangle$, and location C is $\langle 0.7, -0.4, 0\rangle$ m. **(a)** What is the change in potential along a path from B to A? **(b)** What is the change in potential along a path from A to C? **(c)** An alpha particle (two protons and two neutrons) moves from A to C. What is the change in potential energy of the system (alpha + source charges)?

••**P39** The potential difference from one end of a 1-cm-long wire to the other in a circuit is $\Delta V = V_B - V_A = 1.5$ V, as shown in Figure 16.68. Which end of the wire is at the higher potential? What are the magnitude and direction of the electric field E inside the wire?

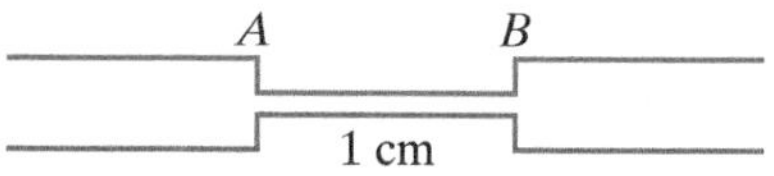

Figure 16.68

••**P40** In the cathode ray tube found in old television sets, which contains a vacuum, electrons are boiled out of a very hot metal filament placed near a negative metal plate. These electrons start out nearly at rest and are accelerated toward a positive metal plate. They pass through a hole in the positive plate on their way toward the picture screen, as shown in the diagram in Figure 16.69. If the high-voltage supply in the television set maintains a potential difference of 15,000 V between the two plates, what speed do the electrons reach?

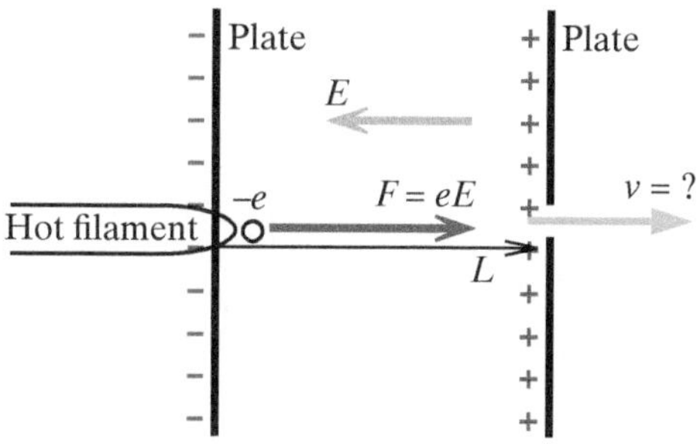

Figure 16.69

••**P41** Two very large disks of radius R are carrying uniformly distributed charges Q_A and Q_B. The plates are parallel and 0.1 mm apart, as shown in Figure 16.70. The potential difference between the plates is $V_B - V_A = -10$ V. **(a)** What is the direction of the electric field between the disks? **(b)** Invent values of Q_A, Q_B, and R that would make $V_B - V_A = -10$ V.

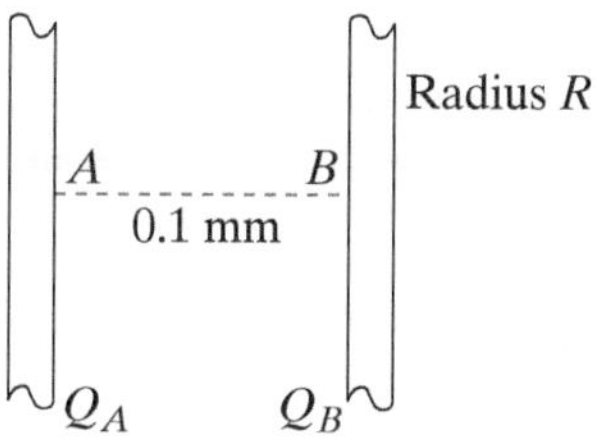

Figure 16.70

••**P42** A particular oscilloscope, shown in Figure 16.71, has an 18,000-V power supply for accelerating electrons to a speed adequate to make the front phosphor-coated screen glow when the electrons hit it. Once the electron has emerged from the accelerating region, it coasts through a vacuum at nearly constant speed.

One can apply a potential difference of plus or minus 40 V across the deflection plates to steer the electron beam up or down on the screen to paint a display (other deflection plates not shown in the diagram are used to steer the beam horizontally).

Each of the two deflection plates is a thin metal plate of length $L = 8$ cm and width (into the diagram) 4 cm. The distance between the deflection plates is $s = 3$ mm. The distance from the deflection plates to the screen is $d = 30$ cm.

When there is a 40-V potential difference between the deflection plates, what is the deflection y of the electron beam where it hits the screen? An approximate treatment is fine, but state your assumptions. As is usually the case, it pays to carry out all of your calculations algebraically and only evaluate the final algebraic result numerically. Note the exaggerated vertical scale: the deflection is actually small compared to the distance to the screen.

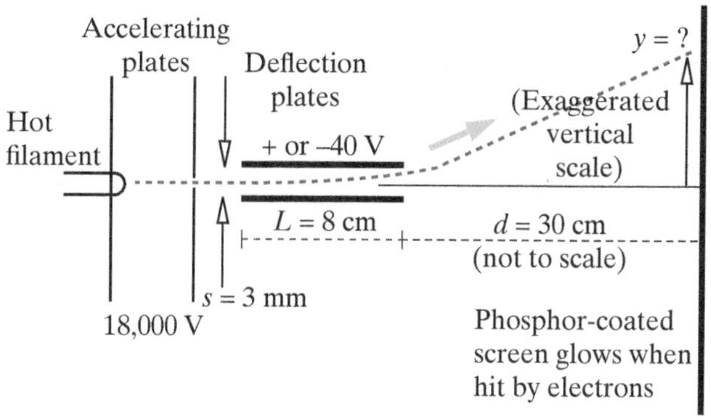

Figure 16.71

Section 16.5

•**P43** Location C is 0.02 m from a small sphere that has a charge of 4 nC uniformly distributed on its surface. Location D is 0.06 m from the sphere. What is the change in potential along a path from C to D?

•**P44** What is the change in potential ΔV in going from a location 3×10^{-10} m from a proton to a location 4×10^{-10} m from the proton?

•**P45** An electron is initially at rest. It is moved from a location 4×10^{-10} m from a proton to a location 6×10^{-10} m from the proton. What is the change in electric potential energy of the system of proton and electron?

•**P46** How much work is required to move a proton and an electron at rest a distance 3×10^{-8} m apart to be at rest a distance 7×10^{-8} m apart?

••**P47** As shown in Figure 16.72, three large, thin, uniformly charged plates are arranged so that there are two adjacent regions of uniform electric field. The origin is at the center of the central plate. Location A is $\langle -0.4, 0, 0\rangle$ m, and location B is $\langle 0.2, 0, 0\rangle$ m. The electric field $\vec{E}_1$ has the value $\langle 725, 0, 0\rangle$ V/m, and $\vec{E}_2$ is $\langle -425, 0, 0\rangle$ V/m.

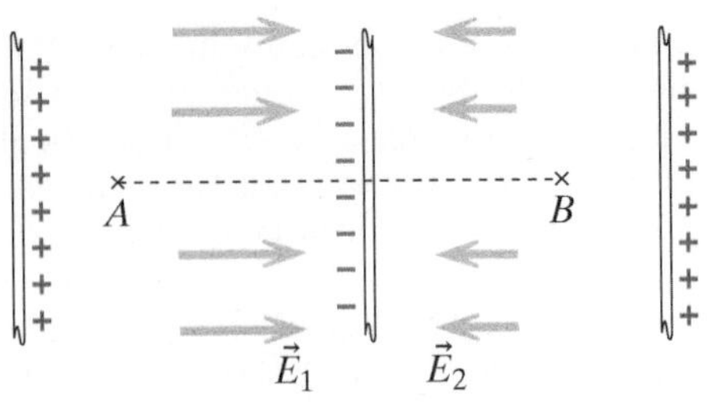

Figure 16.72

(a) Consider a path from A to B. Along this path, what is the change in electric potential? **(b)** What is the change in electric potential along a path from B to A? **(c)** There is a tiny hole in the central plate, so a moving particle can pass through the hole. If an electron moved from A to B along the path shown, what would be the change in its kinetic energy? **(d)** What is the minimum kinetic energy the electron must have at location A in order to ensure that it reaches location B?

••P48 If throughout a particular region of space the potential can be expressed as $V = 4xz + 2y - 5z$, what are the vector components of the electric field at location $\langle x,y,z \rangle$?

••P49 A dipole is oriented along the x axis. The dipole moment is p $(= qs)$. **(a)** Calculate exactly the potential V (relative to infinity) at a location $\langle x,0,0 \rangle$ on the x axis and at a location $\langle 0,y,0 \rangle$ on the y axis, by superposition of the individual $1/r$ contributions to the potential. **(b)** What are the approximate values of V at the locations in part (a) if these locations are far from the dipole? **(c)** Using the approximate results of part (b), calculate the gradient of the potential along the x axis, and show that the negative gradient is equal to the x component E_x of the electric field. **(d)** Along the y axis, $dV/dy = 0$. Why isn't this equal to the magnitude of the electric field E along the y axis?

••P50 A capacitor consists of two large metal disks placed a distance s apart. The radius of each disk is R $(R \gg s)$, and the thickness of each disk is t, as shown in Figure 16.73. The disk on the left has a net charge of $+Q$, and the disk on the right has a net charge of $-Q$. Calculate the potential difference $V_2 - V_1$, where location 1 is inside the left disk at its center, and location 2 is in the center of the air gap between the disks. Explain briefly.

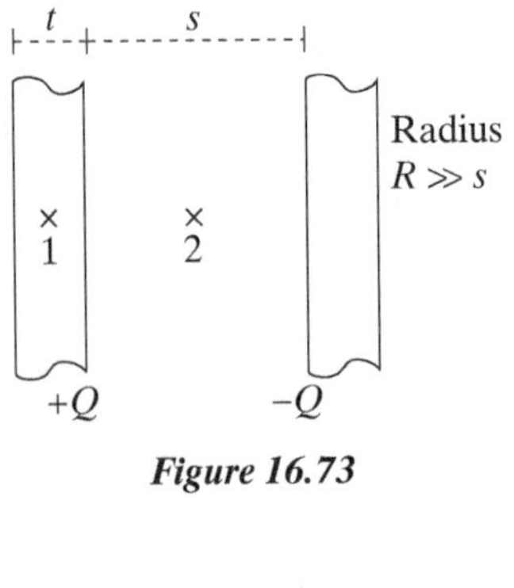

Figure 16.73

••P51 The diagram in Figure 16.74 shows three very large metal disks (seen edgewise), carrying charges as indicated. On each surface the charges are distributed approximately uniformly. Each disk has a very large radius R and a small thickness t. The distances between the disks are a and b, as shown; they also are small compared to R. Calculate $V_2 - V_1$, and explain your calculation briefly.

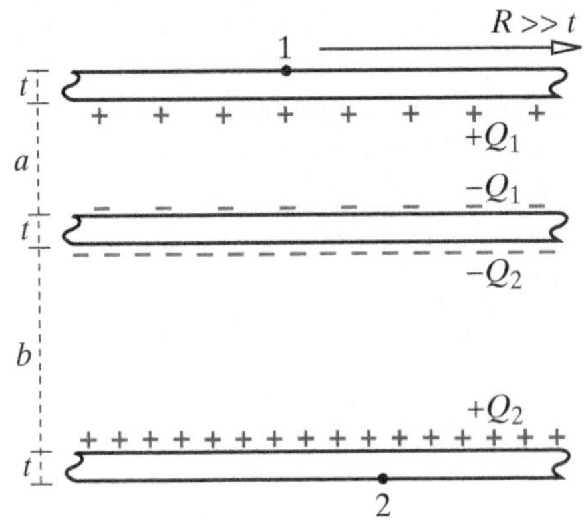

Figure 16.74

••P52 Three charged metal disks are arranged as shown in Figure 16.75 (cutaway view). The disks are held apart by insulating supports not shown in the diagram. Each disk has an area of 2.5 m^2 (this is the area of one flat surface of the disk). The charge $Q_1 = 5 \times 10^{-8}$ C and the charge $Q_2 = 4 \times 10^{-7}$C.

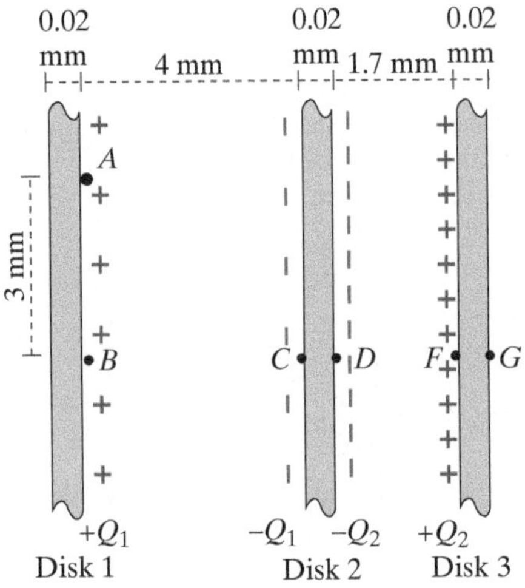

Figure 16.75

(a) What is the electric field (magnitude and direction) in the region between disks 1 and 2? **(b)** Which of the following statements are true? Choose all that apply. (1) Along a path from A to B, $\vec{E} \perp \Delta\vec{l}$. (2) $V_B - V_A = 0$. (3) $V_B - V_A = -\dfrac{Q_1/2.5}{\varepsilon_0} *$ (0.003) V. **(c)** To calculate $V_C - V_B$, where should the path start and where should it end? **(d)** Should $V_C - V_B$ be positive or negative? Why? (1) Positive, because $\Delta\vec{l}$ is opposite to the direction of $\vec{E}$. (2) Negative, because $\Delta\vec{l}$ is in the same direction as $\vec{E}$. (3) Zero, because $\Delta\vec{l} \perp \vec{E}$. **(e)** What is the potential difference $V_C - V_B$? **(f)** What is the potential difference $V_D - V_C$? **(g)** What is the potential difference $V_F - V_D$? **(h)** What is the potential difference $V_G - V_F$? **(i)** What is the potential difference $V_G - V_A$? **(j)** The charged disks have tiny holes that allow a particle to pass through them. An electron that is traveling at a fast speed approaches the plates from the left side. It travels along a path from A to G. Since no external work is done on the system of plates + electron, $\Delta K + \Delta U = W_{ext} = 0$. Consider the following states: initial, electron at location A; final, electron at location G. (1) What is the change in potential energy of the system? (2) What is the change in kinetic energy of the electron?

••**P53** The long rod shown in Figure 16.76 has length L and carries a uniform charge $-Q$. Calculate the potential difference $V_A - V_C$. All of the distances are small compared to L. Explain your work carefully.

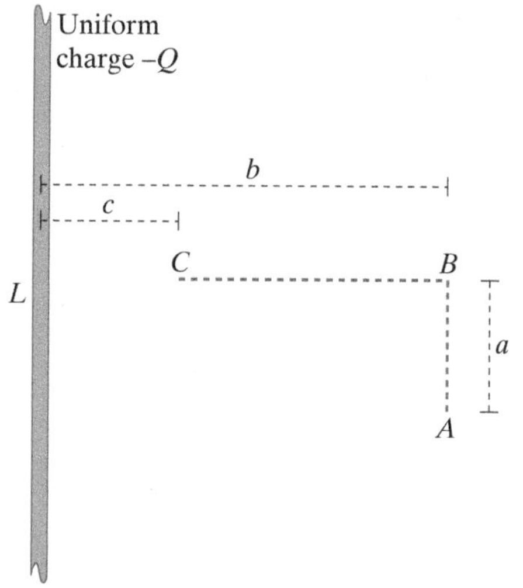

Figure 16.76

••**P54** Three very large charged metal plates are arranged as shown in Figure 16.77. The radius of each plate is 4 m, and each plate is $w = 0.05$ mm thick. The separation d_1 is 6 mm, and the separation d_2 is 2 mm. Each plate has a tiny hole in it, so it is possible for a small charged particle to pass through all the plates.

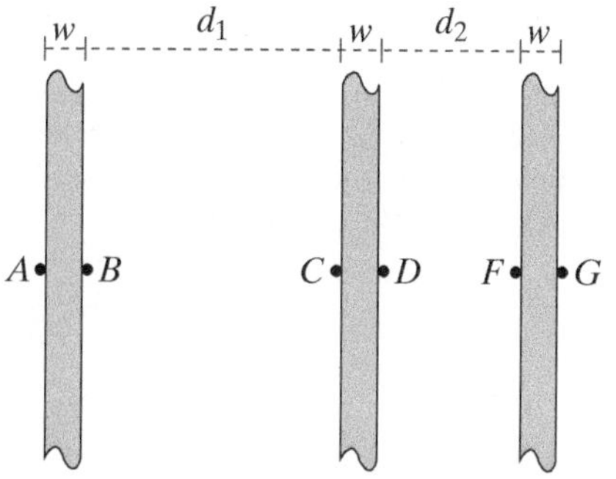

Figure 16.77

You are able to adjust the apparatus by varying the electric field in the region between location D and location F. You need to adjust this setting so that a fast-moving electron moving to the right, entering at location A, will have lost exactly 5.2×10^{-18} J of kinetic energy by the time it reaches location G. Using a voltmeter, you find that the potential difference $V_C - V_B = -16$ V. Based on this measurement, you adjust the electric field between D and F to the appropriate value. **(a)** Consider the system of (electron + plates). Neglecting the small amount of work done by the gravitational force on the electron, during this process (electron going from A to G), what is $\Delta K + \Delta U$? **(b)** What is the change in potential energy for the system during this process? **(c)** What is $V_G - V_A$? **(d)** What is $V_F - V_D$? **(e)** What is the electric field (magnitude and direction) in the region between locations D and F?

••**P55** In the Van de Graaff generator shown in Figure 16.78, a rubber belt carries electrons up through a small hole in a large hollow spherical metal shell. The electrons come off the upper part of the belt and drift through a wire to the outer surface of the metal shell, so that the metal shell acquires a sizable negative charge, approximately uniformly distributed over the sphere. At a time when the sphere has acquired a sizable charge $-Q$, approximately how much work must be done by the motor to move one more electron from the base (a distance h below the sphere) to the upper pulley (located a distance $R/2$ from the center of the hollow sphere)? Explain your work, and state explicitly what approximations you had to make.

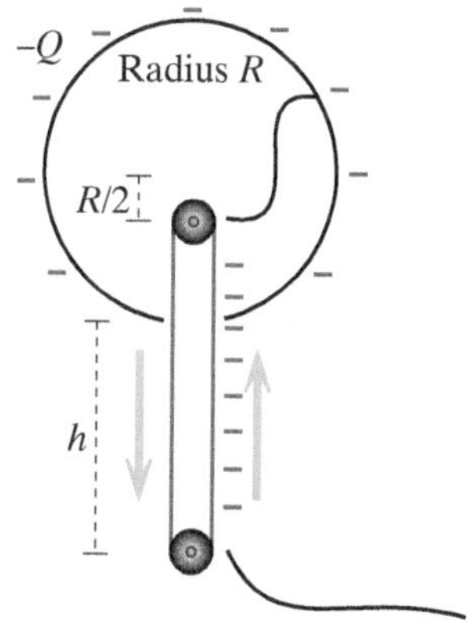

Figure 16.78

••**P56** A thin spherical shell made of plastic carries a uniformly distributed negative charge $-Q_1$. As shown in Figure 16.79, two large thin disks made of glass carry uniformly distributed positive and negative charges $+Q_2$ and $-Q_2$. The radius R_1 of the plastic spherical shell is very small compared to the radius R_2 of the glass disks. The distance from the center of the spherical shell to the positive disk is d, and d is much smaller than R_2. **(a)** Find the potential difference $V_2 - V_1$ in terms of the given quantities (Q_1, Q_2, R_1, R_2, and d). Point 1 is at the center of the plastic sphere, and point 2 is just outside the sphere. **(b)** Find the potential difference $V_3 - V_2$. Point 2 is just below the sphere, and point 3 is right beside the positive glass disk.

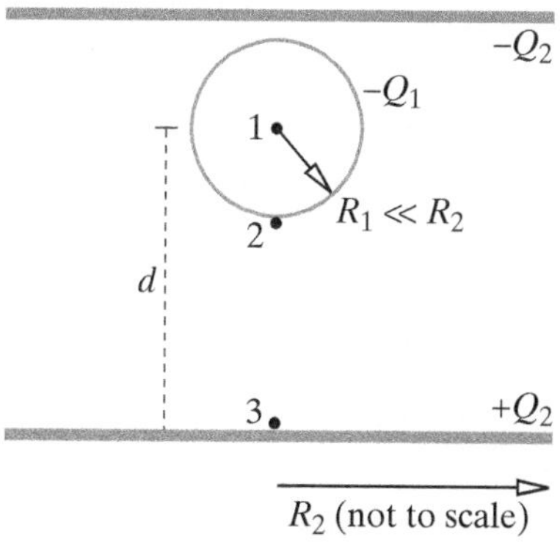

Figure 16.79

(c) Suppose that the plastic shell is replaced by a solid metal sphere with radius R_1 carrying charge $-Q_1$. State whether the absolute magnitudes of the potential differences would be greater than, less than, or the same as they were with the plastic shell in place. Explain briefly, including an appropriate diagram.

••P57 A thin spherical shell of radius R_1 made of plastic carries a uniformly distributed negative charge $-Q_1$. A thin spherical shell of radius R_2 made of glass carries a uniformly distributed positive charge $+Q_2$. The distance between centers is L, as shown in Figure 16.80. **(a)** Find the potential difference $V_B - V_A$. Location A is at the center of the glass sphere, and location B is just outside the glass sphere. **(b)** Find the potential difference $V_C - V_B$. Location B is just outside the glass sphere, and location C is a distance d to the right of B. **(c)** Suppose the glass shell is replaced by a solid metal sphere with radius R_2 carrying charge $+Q_2$. Would the magnitude of the potential difference $V_B - V_A$ be greater than, less than, or the same as it was with the glass shell in place? Explain briefly, including an appropriate physics diagram.

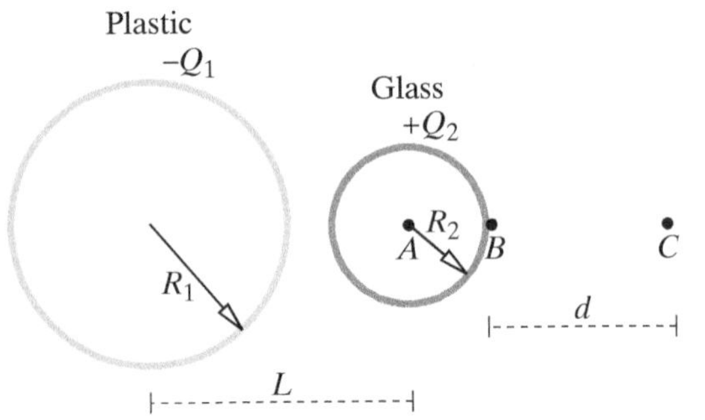

Figure 16.80

••P58 As shown in Figure 16.81, a solid metal sphere of radius r_1 has a charge $+Q$. It is surrounded by a concentric spherical metal shell with inner radius r_2 and outer radius r_3 that has a charge $-Q$ on its inner surface and $+Q$ on its outer surface. In the diagram, point A is located at a distance r_4 from the center of the spheres. Points B and C are inside the metal shell, very near the outer and inner surfaces, respectively. Point E is just inside the surface of the solid sphere. Point D is halfway between C and E. Point F is a distance $r_1/2$ from the center. **(a)** Is each of the following potential differences greater than zero, equal to zero, or less than zero? Briefly explain why in terms of the directions of the electric field and of the path: (1) $V_B - V_A$, (2) $V_C - V_B$, (3) $V_D - V_C$, (4) $V_F - V_E$. **(b)** Calculate V_F, the potential at location F. Explain your work.

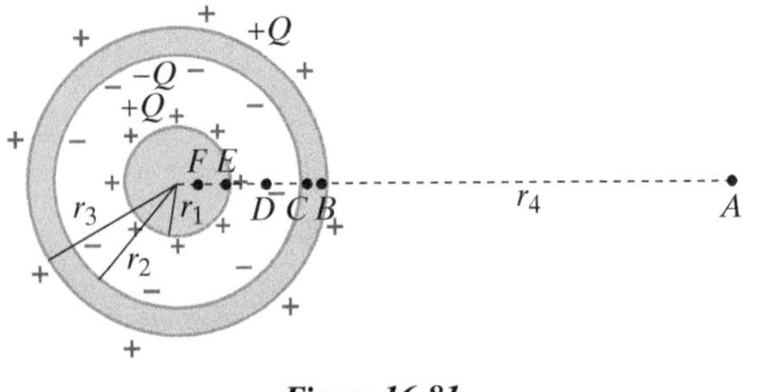

Figure 16.81

••P59 In a region with an uniform electric field, you measure a potential difference of 74 V from the origin to a position of $\langle 0,0,10\rangle$ m. Now we add a uniformly charged, thin spherical plastic shell centered at the origin. The spherical shell has a radius of 5 m and a charge of -3530 nC. Draw a diagram to help answer the following questions: **(a)** What is the potential difference from the origin to a position of $\langle 0,0,5\rangle$ m (at the surface of the spherical shell)? **(b)** What is the potential difference from the position of $\langle 0,0,5\rangle$ m to a position of $\langle 0,0,10\rangle$ m?

••P60 A dipole is centered at the origin, with its axis along the y axis, so that at locations on the y axis, the electric field due to the dipole is given by

$$\vec{E} = \left\langle 0, \frac{1}{4\pi\varepsilon_0}\frac{2qs}{y^3}, 0\right\rangle \frac{\text{V}}{\text{m}}$$

The charges making up the dipole are +3 nC and −3 nC, and the dipole separation is 2 mm (Figure 16.82). What is the potential difference along a path starting at location $\langle 0,0.03,0\rangle$ m and ending at location $\langle 0,0.04,0\rangle$ m?

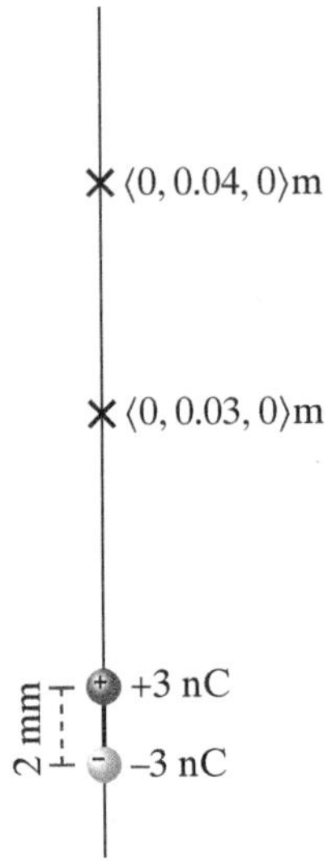

Figure 16.82

••P61 A thin spherical glass shell of radius R carries a uniformly distributed charge $+Q$, and a thin spherical plastic shell of radius R carries a uniformly distributed charge $-Q$. The surfaces of the spheres are a distance $L+2d$ from each other, and locations A and B are a distance d from the surfaces of the spheres, as shown in Figure 16.83. Calculate the potential difference $V_B - V_A$.

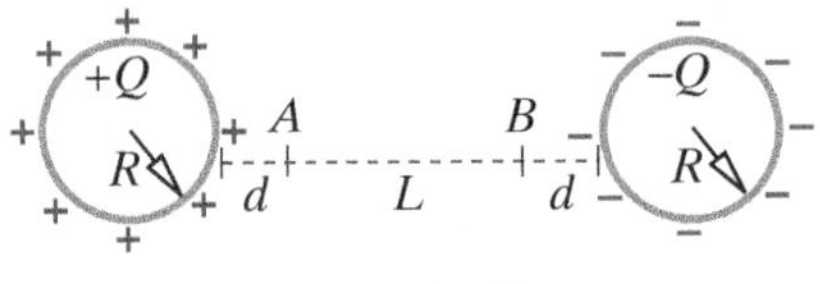

Figure 16.83

••P62 An HCl molecule (Figure 16.84) in the gas phase has an internuclear separation s. For the purposes of this problem, assume that

- The molecule can be considered as two point charges, H^+ and Cl^-, located at the centers of the nuclei (i.e., ignore polarization of one ion by the other, sharing of electrons, etc.).
- The molecule remains fixed in position.
- The distances a and b shown in the diagram are much greater than shown ($a > b \gg s$).

(a) Draw vectors showing the relative magnitude and direction of the electric field at each of the three points indicated by a square box. Your drawing should be accurate enough to show clearly whether one vector is equal in magnitude, greater in magnitude, or less in magnitude than another vector, but need not be more accurate than that. **(b)** What is the potential difference $\Delta V = V_B - V_A$ in terms of the given quantities? **(c)** An electron passes point A traveling along the x axis in the negative x direction, toward the Cl^-. Its kinetic energy is K_A. What is its kinetic energy at point B?

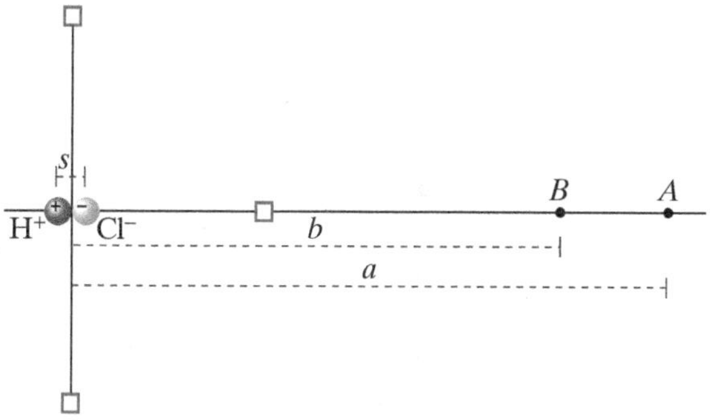

Figure 16.84

••**P63** A very long, thin glass rod of length $2R$ carries a uniformly distributed charge $+q$, as shown in Figure 16.85. A very large plastic disk of radius R, carrying a uniformly distributed charge $-Q$, is located a distance d from the rod, where $d \ll R$. Calculate the potential difference $V_B - V_A$ from the center of the surface of the disk (location A) to a location a distance h from the center of the disk (location B). If you have to make any approximations, state what they are.

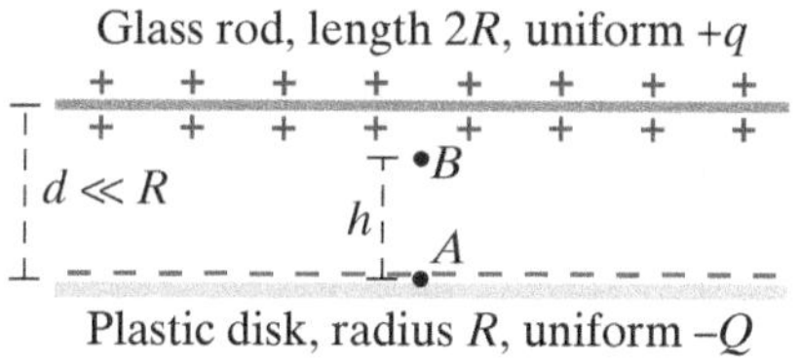

Figure 16.85

•••**P64** A long thin metal wire with radius r and length L is surrounded by a concentric long narrow metal tube of radius R, where $R \ll L$, as shown in Figure 16.86. Insulating spokes hold the wire in the center of the tube and prevent electrical contact between the wire and the tube. A variable power supply is connected to the device as shown. There is a charge $+Q$ on the inner wire and a charge $-Q$ on the outer tube. As we will see when we study Gauss's law in a later chapter, the electric field inside the tube is contributed solely by the wire, and the field outside the wire is the same as though the wire were infinitely thin; the outer tube does not contribute as long as we are not near the ends of the tube. **(a)** In terms of the charge Q, length L, inner radius r, and outer radius R, what is the potential difference $V_{\text{tube}} - V_{\text{wire}}$ between the inner wire and the outer tube? Explain, and include checks on your answer. **(b)** The power-supply voltage is slowly increased until you see a glow in the air very near the inner wire. Calculate this power-supply voltage (give a numerical value), and explain your calculation. The length $L = 80$ cm, the inner radius $r = 0.7$ mm, and the outer radius $R = 3$ cm. This device is called a "Geiger–Müller tube" and was one of the first electronic particle detectors. The voltage is set just below the threshold for making the air glow near the wire. A charged particle that passes near the center wire can trigger breakdown in the air, leading to a large current that can be easily measured.

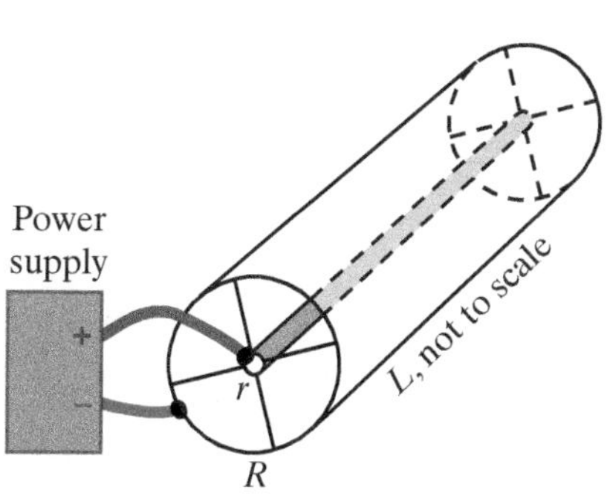

Figure 16.86

Section 16.6

•**P65** In the region of space depicted in Figure 16.87 there are several stationary charged objects that are not shown in the diagram. Along a path $A = B = C = D$ you measure the following potential differences: $V_B - V_A = 12$ V; $V_C - V_B = -5$ V; $V_D - V_C = -15$ V. What is the potential difference $V_A - V_D$?

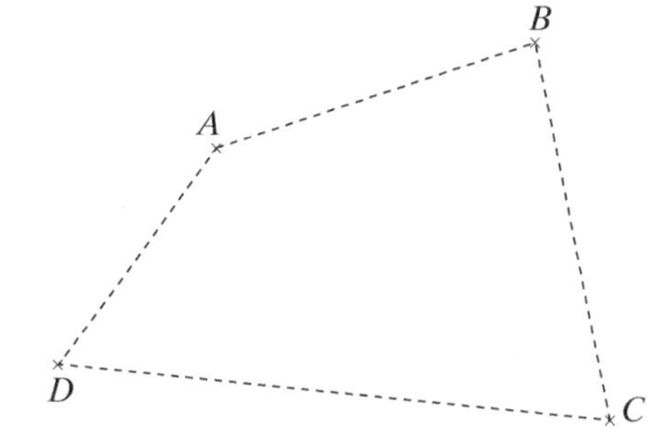

Figure 16.87

••**P66** A capacitor consists of two charged disks of radius R separated by a distance s, where $R \gg s$. The magnitude of the charge on each disk is Q. Consider points A, B, C, and D inside the capacitor, as shown in Figure 16.88. **(a)** Show that $\Delta V = V_C - V_A$ is the same for these paths by evaluating ΔV along each path: (1) Path 1: $A = B = C$, (2) Path 2: $A = C$, (3) Path 3: $A = D = B = C$. **(b)** If $Q = 43\ \mu$C, $R = 4.0$ m, $s_1 = 1.5$ mm, and $s_2 = 0.7$ mm, what is the value of $\Delta V = V_C - V_A$? **(c)** Choose two different paths from point A back to point A again, and show that $\Delta V = 0$ for a round trip along both of these paths.

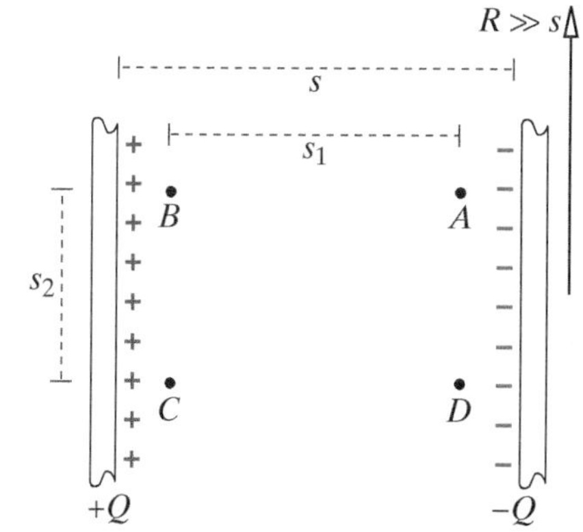

Figure 16.88

••**P67** A uniform spherical shell of charge $+Q$ is centered at point B, as shown in Figure 16.89. Show that $\Delta V = V_C - V_A$ is independent of path by calculating ΔV for each of these two paths (actually do the integrals): **(a)** Path 1: $A = B = C$ (along a straight line through the shell). **(b)** Path 2: $A = D = C$ (along a circular arc around point B).

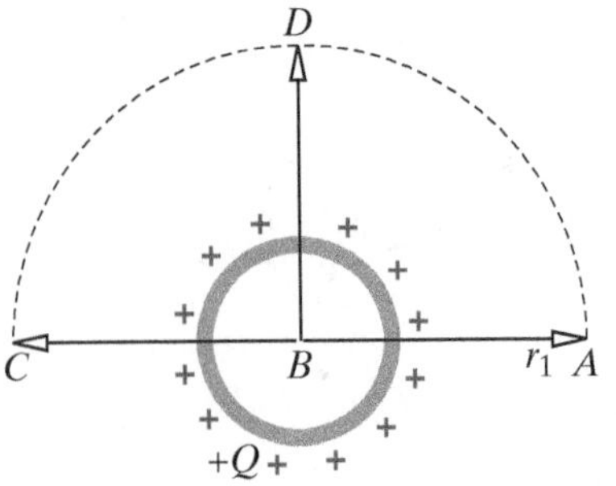

Figure 16.89

••P68 Four voltmeters are connected to a circuit as shown in Figure 16.90. As is usual with voltmeters, the reading on the voltmeter is positive if the negative lead (black wire, usually labeled COM) is connected to a location at lower potential, and the positive lead (red) is connected to a location at higher potential. The circuit contains two devices whose identity is unknown, and a rod (green) of length 9 cm made of a conducting material. At a particular moment, the readings observed on the voltmeters are: voltmeter A: −1.6 V; voltmeter B: −6 V; voltmeter C: −3.5 V. **(a)** At this moment, what is the reading on voltmeter D, both magnitude and sign? **(b)** What are the magnitude and direction of the electric field inside the rod?

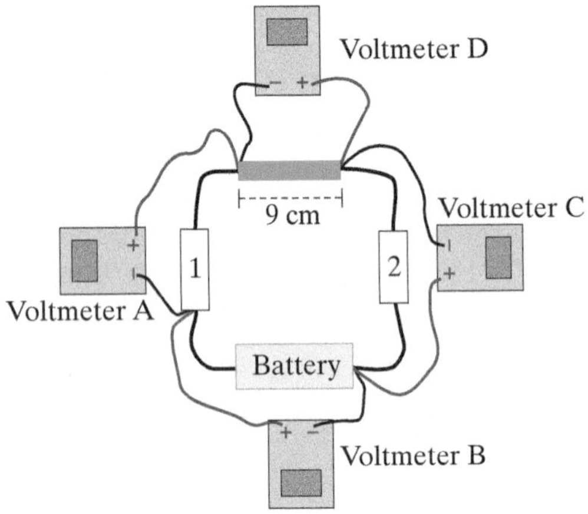

Figure 16.90

Section 16.7

•P69 A particle with charge $+q_1$ and a particle with charge $-q_2$ are located as shown in Figure 16.91. What is the potential (relative to infinity) at location A?

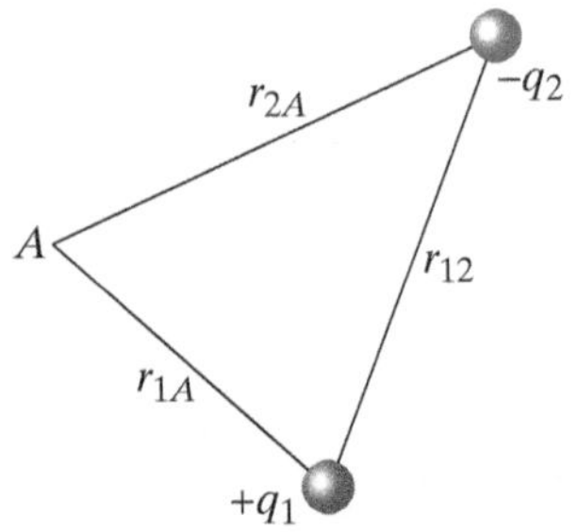

Figure 16.91

•P70 What is the electric potential at a location 2.5×10^{-9} m from an electron?

•P71 Calculate the potential at location A in Figure 16.92.

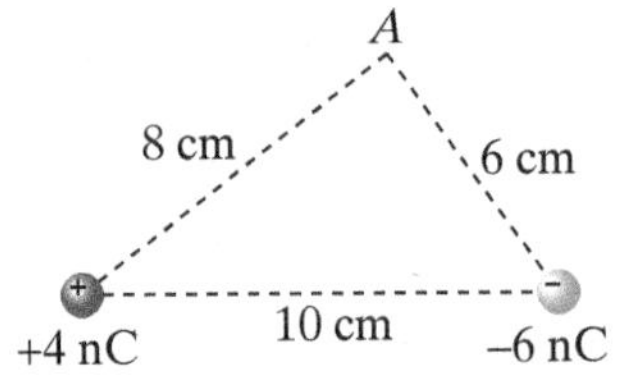

Figure 16.92

•P72 Two capacitors have the same size of plates and the same distance (2 mm) between plates. The potentials of the two plates in capacitor 1 are −10 V and +10 V. The potentials of the two plates in capacitor 2 are 350 V and 370 V. What is the electric field inside capacitor 1? Inside capacitor 2?

•P73 Suppose that the potential at $z = 3.15$ m is 37 V, and the potential at $z = 3.27$ m is −36 V. What is the approximate value of E_z in this region? Include the appropriate sign.

•P74 Four point charges, q_1, q_2, q_3, and q_4, are located at the corners of a square with side d, as shown in Figure 16.93. What is the potential (relative to infinity) at location A, at the center of the square? What is the potential at location B, in the middle of one side?

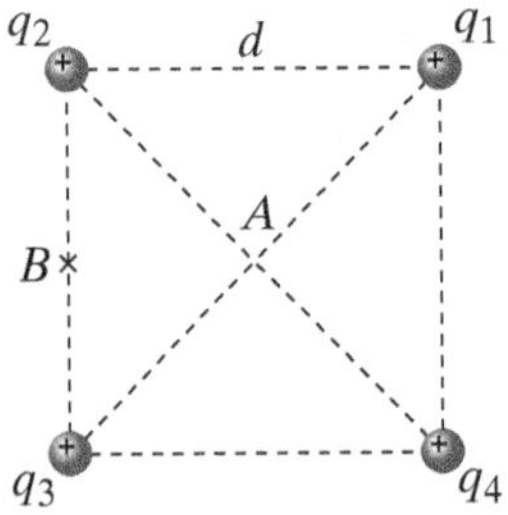

Figure 16.93

•P75 What is the potential (relative to infinity) at location B, a distance h from a ring of radius a with charge $-Q$, as shown in Figure 16.94?

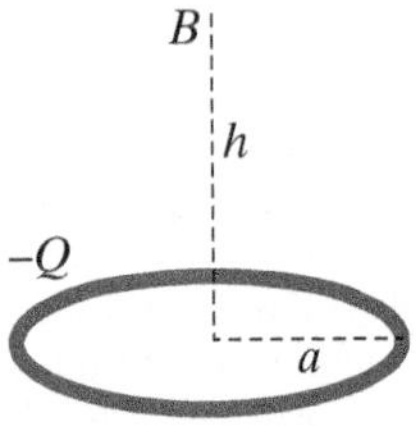

Figure 16.94

••P76 What is the maximum possible potential (relative to infinity) of a metal sphere of 10-cm radius in air? What is the maximum possible potential of a metal sphere of only 1-mm radius? These results hint at the reason why a highly charged piece of metal (with uniform potential throughout) tends to spark at places where the radius of curvature is small or at places where there are sharp points. Remember that the breakdown electric field strength for air is roughly 3×10^6 V/m.

••P77 A small solid glass ball of radius r is irradiated by a beam of positive ions and gains a charge $+q$ distributed uniformly throughout its volume. What is the potential (relative to infinity) at location A, inside the ball, a distance $r/2$ from the center of the ball?

••P78 A small metal sphere of radius r initially has a charge q_0. Then a long copper wire is connected from this small sphere to a distant, large, uncharged metal sphere of radius R. Calculate the final charge q on the small sphere and the final charge Q on the large sphere. You may neglect the small amount of charge on the wire. What other approximations did you make? (Think about potential ....)

••P79 A metal sphere of radius r_1 carries a positive charge of amount Q. A concentric spherical metal shell with inner radius r_2 and outer radius r_3 surrounds the inner sphere and carries a total positive charge of amount $4Q$, with some of this charge on the outer surface (at r_3) and some on the inner surface (at r_2). **(a)** How is the charge of $4Q$ distributed on the two surfaces of the outer shell? Prove this! **(b)** What is the potential (relative to infinity) just outside r_3, halfway between r_2 and r_3, just inside r_2, just outside r_1, and at the center?

••P80 A thin plastic spherical shell of radius a is rubbed all over with wool and gains a charge of $-Q$. What is the potential relative to infinity at location B, a distance $a/3$ from the center of the sphere, as shown in Figure 16.95?

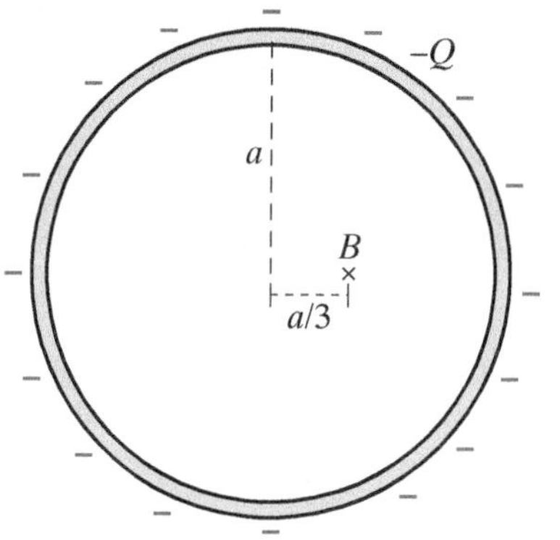

Figure 16.95

••P81 A solid plastic sphere of radius R has charge $-Q$ distributed uniformly over its surface. It is far from all other objects. **(a)** Sketch the molecular polarization in the interior of the sphere and explain briefly. **(b)** Calculate the potential relative to infinity at the center of the sphere.

•••P82 A nucleus contains Z protons that on average are uniformly distributed throughout a tiny sphere of radius R. **(a)** Calculate the potential (relative to infinity) at the center of the nucleus. Assume that there are no electrons or other charged particles in the vicinity of this bare nucleus. **(b)** Graph the potential as a function of distance from the center of the nucleus. **(c)** Suppose that in an accelerator experiment a positive pion is produced at rest at the center of a nucleus containing Z protons. The pion decays into a positive muon (essentially a heavy positron) and a neutrino. The muon has initial kinetic energy K_i. How much kinetic energy does the muon have by the time it has been repelled very far away from the nucleus? (The muon interacts with the nucleus only through Coulomb's law and is unaffected by nuclear forces. The massive nucleus hardly moves and gets negligible kinetic energy.) **(d)** If the nucleus is gold, with 79 protons, what is the numerical value of $K_f - K_i$ in electron volts?

Section 16.9

••P83 An isolated large-plate capacitor (not connected to anything) originally has a potential difference of 1000 V with an air gap of 2 mm. Then a plastic slab 1 mm thick, with dielectric constant 5, is inserted into the middle of the air gap as shown in Figure 16.96. Calculate the following potential differences, and explain your work.

$$V_1 - V_2 = ? \qquad V_2 - V_3 = ?$$
$$V_3 - V_4 = ? \qquad V_1 - V_4 = ?$$

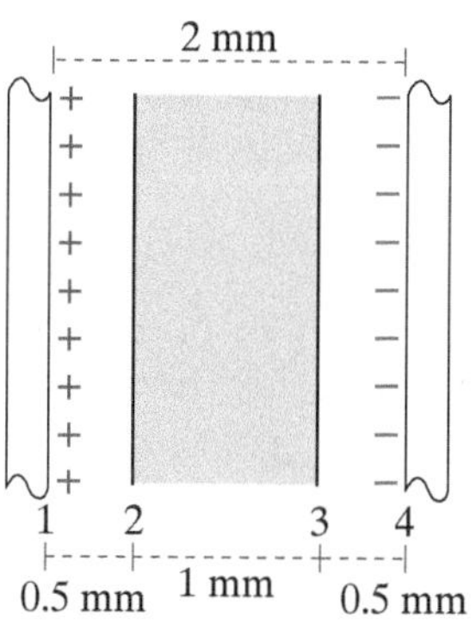

Figure 16.96

••P84 An isolated parallel-plate capacitor of area A_1 with an air gap of length s_1 is charged up to a potential difference ΔV_1. A second parallel-plate capacitor, initially uncharged, has an area A_2 and a gap of length s_2 filled with plastic whose dielectric constant is K. You connect a wire from the positive plate of the first capacitor to one of the plates of the second capacitor, and you connect another wire from the negative plate of the first capacitor to the other plate of the second capacitor. What is the final potential difference across the first capacitor?

Section 16.11

••P85 Refer to Section 16.7 for the potential relative to infinity along the axis of a ring of radius R carrying a charge Q. Let the axis of the ring be the z axis of the coordinate system, and determine E_z at any location z along the axis. (If the charge is nearly uniformly distributed around the ring, at these locations there is no E_x or E_y, due to the symmetry of the situation.) Compare with the result obtained in the previous chapter by integration.

••P86 A rod uniformly charged with charge $-q$ is bent into a semicircular arc of radius b, as shown in Figure 16.97. What is the potential relative to infinity at location A, at the center of the arc?

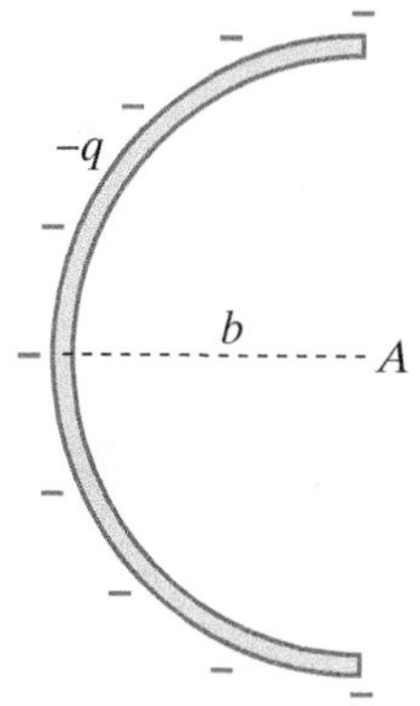

Figure 16.97

••**P87** We know the magnitude of the electric field of a dipole at a location on the x axis and at a location on the y axis, if we are far from the dipole. **(a)** Find $\Delta V = V_B - V_A$ along a line perpendicular to the axis of the dipole shown in Figure 16.98. Do it two ways: from the superposition of V due to the two charges and from the integral of the electric field. **(b)** Find $\Delta V = V_C - V_D$ along the axis of the dipole, as shown in Figure 16.99. Include the correct sign. Do it two ways: from the superposition of V due to the two charges and from the integral of the electric field. **(c)** What is the change in potential energy ΔU in moving an electron from D to C?

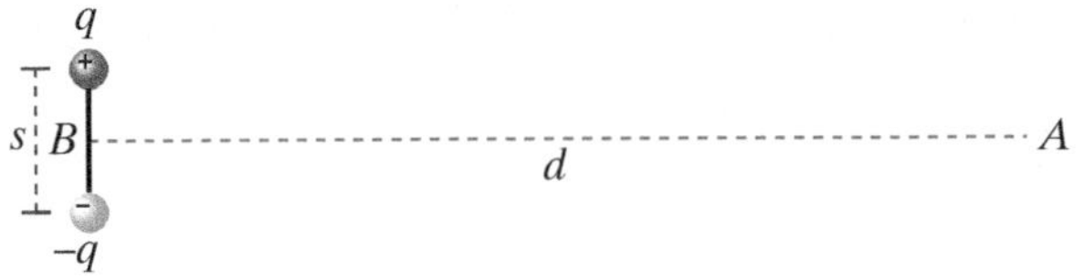

Figure 16.98

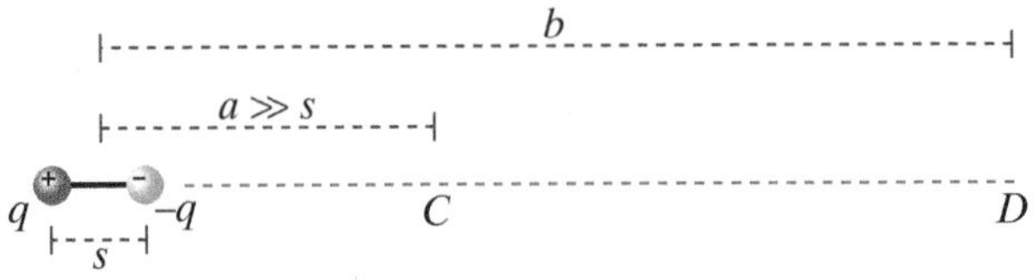

Figure 16.99

COMPUTATIONAL PROBLEMS

More detailed and extended versions of some computational modeling problems may be found in the lab activities included in the *Matter & Interactions* resources for instructors.

•**P88** Modify the program from Section 16.8 to calculate the potential difference $V_C - V_D$ near the dipole, where $C = \langle -0.03, -0.02, 0\rangle$ m and $D = \langle -0.05, -0.07, 0\rangle$ m.

••**P89** Replace the dipole in the program from Section 16.8 with a uniformly charged thin rod. The rod is centered at the origin, and its axis is aligned along the y axis. The rod is 0.05 m long, and its net charge is -2×10^{-8} C. Calculate the potential difference $V_B - V_A$, where $A = \langle 0.02, 0.02, 0\rangle$ m, and $B = \langle 0.12, 0.08, 0\rangle$ m. Explain how you decided how many point charges to use in modeling the rod.

••**P90** Replace the dipole in the program from Section 16.8 with a uniformly charged thin ring. The ring is centered at the origin, and its axis is aligned with the z axis. The radius of the ring is 0.04 m, and its net charge is -3×10^{-8} C. Calculate the potential difference $V_B - V_A$, where $A = \langle 0.02, 0, 0.02\rangle$ m, and $B = \langle 0.12, 0, 0.08\rangle$ m. Explain how you decided how many point charges to use in modeling the ring.

••**P91** Consider a uniformly charged cylinder 0.3 m long, with a radius of 1 mm, carrying a total charge of 4×10^{-8} C. The cylinder lies along the x axis, with its center at the origin. Calculate the potential difference along a path going from $\langle 0, 0.05, 0\rangle$ to $\langle 0.15, 0.05, 0\rangle$. Explain how you decided how many point charges to use in modeling the cylinder.

•••**P92** As in Problem P91, consider a uniformly charged cylinder 0.3 m long, with a radius of 1 mm, carrying a total charge of 4×10^{-8} C. The cylinder lies along the x axis, with its center at the origin. Compute the potential difference along a path from $\langle 0, 0.05, 0\rangle$ to $\langle 0.15, 0.05, 0\rangle$ by calculating and displaying (with arrows) $\vec{E}$ and $\Delta\vec{l}$ at many steps along the path and adding up the contributions $-\vec{E}\bullet\Delta\vec{l}$. The VPython function `dot(A,B)` gives the dot product of the two vectors `A` and `B`. See optional Section 16.13 for suggestions on how to structure the program.

ANSWERS TO CHECKPOINTS

1 (a) 7.65×10^{-21} J; **(b)** 2.12×10^{-20} J; **(c)** 1.36×10^{-20} J
2 (a) -1.68×10^{-18} J; **(b)** 1.68×10^{-18} J; **(c)** 4.45×10^{4} m/s; **(d)** 1.68×10^{-18} J; -1.68×10^{-18} J
3 (a) 300 V; **(b)** 4.8×10^{-17} J; **(c)** -4.8×10^{-17} J; **(d)** 0; **(e)** 0; **(f)** 0
4 −150 V/m
5 (a) To the left, positive; **(b)** To the right, negative
6 (a) −380 V; **(b)** +380 V
7 (a) +1 V; **(b)** −1 V
8 $\Delta V_1 = \frac{1}{4\pi\varepsilon_0}\left(\frac{Q}{r_2} - \frac{Q}{r_1}\right)$
$\Delta V_2 = 0$
$\Delta V_3 = \frac{1}{4\pi\varepsilon_0}\left(\frac{Q}{r_1} - \frac{Q}{r_2}\right)$
$\Delta V_4 = 0$
$\Delta V_{\text{round trip}} = 0$
9 0.025 V/m (a very small field)
10 If $z \gg R$, $\sqrt{z^2 + R^2} \approx z$, and $V \approx \frac{1}{4\pi\varepsilon_0}\frac{Q}{z}$.
11 4.8 V
12 1.5×10^{6} V/m; 39.8 J
13 Since each little charge dQ is a distance R from the center, the potential at the center is $\sum \frac{1}{4\pi\varepsilon_0}\frac{dQ}{R} = \frac{1}{4\pi\varepsilon_0}\frac{Q}{R}$.

CHAPTER

17

Magnetic Field

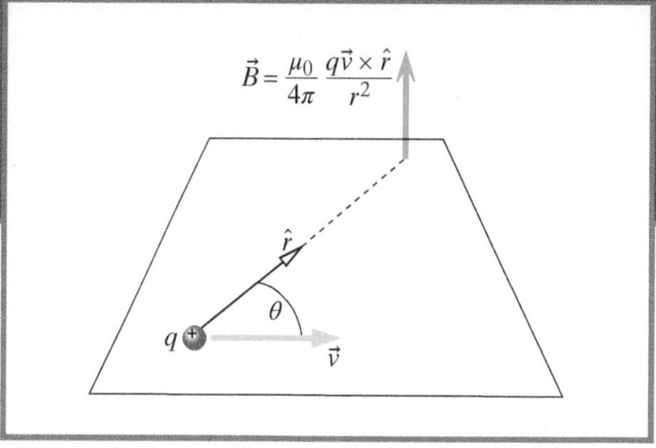

OBJECTIVES

After studying this chapter you should be able to

- Calculate the 3D magnetic field due to a single moving charge or to a current, and represent it graphically using arrows.
- Use a magnetic compass to determine the magnitude and direction of current in a wire.
- Relate the magnetic field of a bar magnet to its magnetic dipole moment.

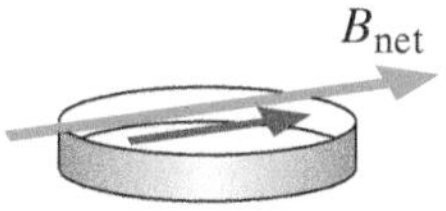

Figure 17.1 A compass needle points in the direction of the net magnetic field at its location.

Electric fields are not the only kind of field associated with charged particles. When a compass needle turns and points in a particular direction, we say that there is a "magnetic field" pointing in that direction, which forces the needle to line up with it (Figure 17.1). Initially we'll simply define magnetic field as whatever it is that is detected by a compass. The twist of a compass needle is an indicator of magnetic fields, just as the twist of a suspended electric dipole is an indicator of electric fields, as you saw in Chapter 14.

Magnetic fields are made by moving charges, so we will need to have some moving charges at hand. We will assemble some simple electric circuits in which currents can run steadily, providing a convenient source of moving charges. In this chapter we will study magnetic field, and we will also develop an atomic-level description of magnets. Later, in Chapter 20, we will study the forces that magnetic fields exert on moving charges.

There is a set of simple experiments related to the topics discussed in this chapter, which allow you to explore these phenomena yourself using equipment such as batteries, wires, flashlight bulbs, and a compass. See the Experiments section at the end of this chapter for further information.

17.1 ELECTRON CURRENT

Magnetic fields are associated with moving charges. A current in a wire provides a convenient source of moving charges and allows us to experiment with producing and detecting magnetic fields.

In equilibrium, there is no net motion of the sea of mobile electrons inside a metal. In the electric circuits we will construct in this chapter, we can arrange things so the electron sea does keep moving continuously. This continuous flow of electrons is called an "electron current," and it is an indication that the system is not in equilibrium. In order to be able to talk about what things affect electron current, we need a precise definition:

DEFINITION OF ELECTRON CURRENT

The electron current i is the number of electrons per second that enter a section of a conductor.

In an electric circuit with a steady current flowing, the electron current i is the same in every section of a wire of uniform thickness and composition (in the next chapter we'll see why). If we could count the number of electrons per second passing a particular point in a circuit (Figure 17.2), we could measure electron current directly. This is difficult to do, so we use indirect measurements to determine the magnitude of electron current in a wire. One such indirect measurement involves measuring the magnetic field created by the moving electrons.

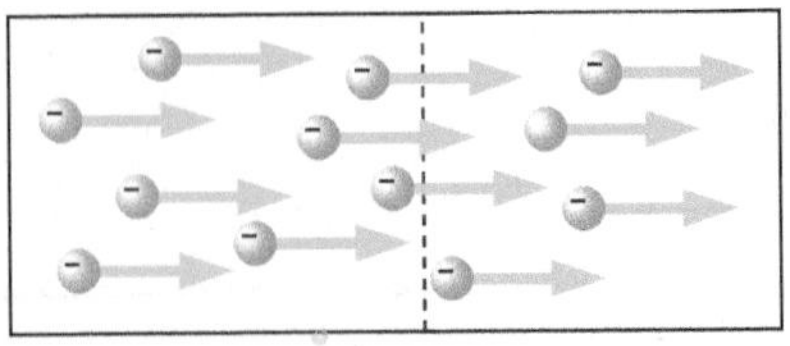

Figure 17.2 The electron current in this conductor is the number of electrons per second passing the dashed line.

As electrons drift through a wire, they collide with the atomic cores, and this "friction" heats the wire (and prevents the electrons from going faster and faster). Both the heating and the magnetic effects are proportional to the electron current—the number of electrons that enter the wire every second.

Checkpoint 1 If 1.8×10^{16} electrons enter a light bulb in 3 ms, what is the magnitude of the electron current at that point in the circuit? If the electron current at a particular location in a circuit is 9×10^{18} electrons/s, how many electrons pass that point in 10 min?

Simple Circuits

A simple electric circuit, involving a battery, wire, and a flashlight bulb, is a convenient means of producing a supply of moving electrons. We will refer to such circuits throughout this chapter. Despite the simplicity of the materials, the physics questions that can be investigated with such circuits are centrally important ones. If you have the appropriate equipment, you may wish to do these experiments yourself. See EXP1.

17.2 DETECTING MAGNETIC FIELDS

We can use a magnetic compass as a detector of magnetic fields. Just as the deflection of a charged piece of invisible tape or the twisting of a permanent electric dipole indicates the presence of an electric field, the twisting of a compass needle indicates the presence of a magnetic field. See EXP2.

QUESTION How can you be sure that a compass needle is not simply responding to electric fields?

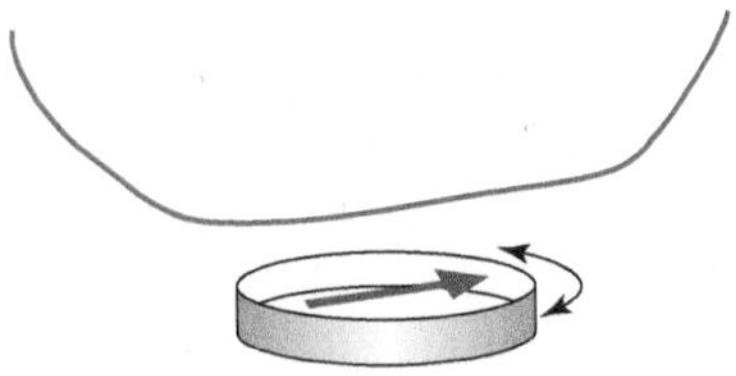

Figure 17.3 If a compass needle is originally pointing north, and a current-carrying wire aligned north–south is placed on the compass, the needle deflects. The deflection direction depends on whether the current runs northward or southward.

In observing the behavior of a compass, we find that:

- The compass needle is affected by the proximity of objects made of iron or steel, even if these objects are electrically neutral (and therefore attract both positively and negatively charged tapes). It is also affected by the presence of nickel or cobalt, though these are less readily available.
- The compass needle is unaffected by objects made of most other elements, including aluminum, copper, zinc, and carbon, whereas charged tapes interact with these objects.
- If it is not near objects made of iron, the compass needle points toward the Earth's magnetic north pole, whereas neither electrically charged objects nor electric dipoles do this.

If you bring a current-carrying wire near the compass, something very interesting happens: the compass needle is deflected while current is running in the wire, but not while the wire is disconnected from the battery. This effect was discovered by accident by the Danish scientist Oersted in 1820 while doing

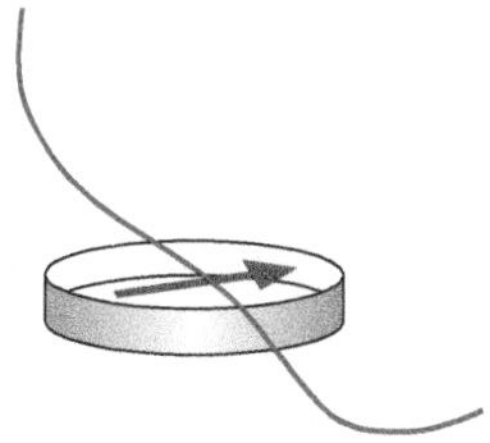

Figure 17.4 If a current-carrying wire aligned east–west is placed on a compass, the needle does not deflect.

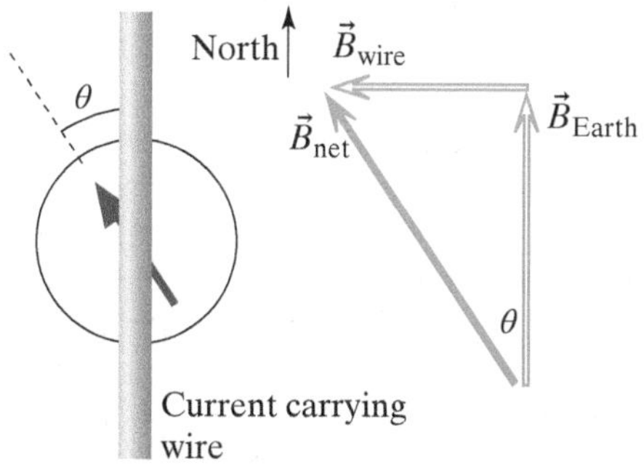

Figure 17.5 A compass needle points in the direction of the net magnetic field, which is the superposition of the magnetic field of the Earth and the magnetic field of the current-carrying wire that passes over the compass, aligned north–south.

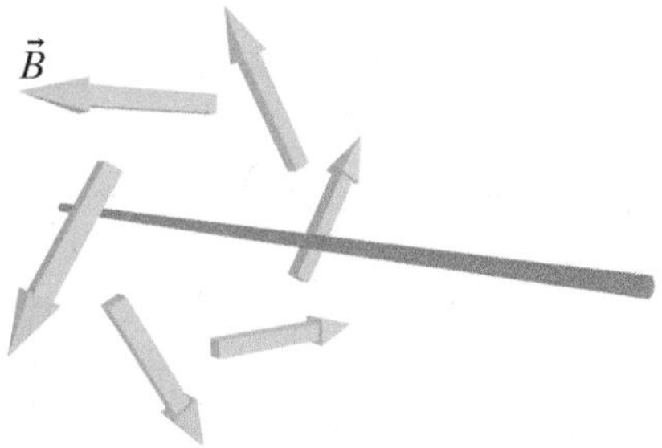

Figure 17.6 The magnetic field made by a long straight current-carrying wire. The magnetic field is perpendicular to the wire everywhere.

a lecture demonstration in a physics class. (He must have been surprised!) The phenomenon is often called "the Oersted effect." The biggest effect occurs if the wire points north–south (Figure 17.3); the deflection direction depends on whether the current runs northward or southward. The compass needle doesn't deflect if the wire points east–west (Figure 17.4).

From such experiments, one can draw the following conclusions:

- The magnitude of the magnetic field produced by a current of moving electrons depends on the amount of current.
- A wire with no current running in it produces no magnetic field.
- The magnetic field due to the current appears to be perpendicular to the direction of the current.
- The direction of the magnetic field due to the current under the wire is opposite to the direction of the magnetic field due to the current above the wire.

A Model for the Observations

The moving electrons in a wire create a magnetic field at various locations in space, including at the location of the compass. The vector sum of the Earth's magnetic field $\vec{B}_{Earth}$ plus the magnetic field $\vec{B}_{wire}$ of a current-carrying wire makes a net magnetic field $\vec{B}_{net}$ (Figure 17.5). Since a compass needle points in the direction of the net magnetic field at its location, the needle turns away from its original north–south direction to align with the net field. Because the magnetic field made by the wire is perpendicular to the wire and is in opposite directions above and below the wire, the pattern of magnetic field made by the wire must look like Figure 17.6.

It happens (conveniently) to be the case that the magnitudes of $\vec{B}_{Earth}$ and $\vec{B}_{wire}$ are similar. If B_{Earth} were much larger than B_{wire}, then we would see almost no response from the compass.

QUESTION What would we observe if B_{wire} were much larger than B_{Earth}?

If B_{wire} were much larger than B_{Earth}, then the compass deflection would be nearly 90°, since the net magnetic field would be primarily due to the magnetic field made by the current in the wire.

There is hardly any observable electric interaction of the current-carrying wire with nearby materials, because the wires have no net charge. (Actually, we'll see later that there are tiny amounts of excess charge on the surface, but too little to observe easily.)

EXAMPLE **Compass Deflection**

A current-carrying wire is oriented north–south and laid on top of a compass. At the location of the compass needle the magnetic field due to the wire points west and has a magnitude of 3×10^{-6} T (tesla). The horizontal component of Earth's magnetic field has a magnitude of about 2×10^{-5} T. What compass deflection will you observe?

Solution The compass needle will point in the direction of the net magnetic field, so from the diagram in Figure 17.7:

$$\theta = \tan^{-1}\left(\frac{B_{wire}}{B_{Earth}}\right) = \tan^{-1}\left(\frac{3 \times 10^{-6}\,\text{T}}{2 \times 10^{-5}\,\text{T}}\right) = 8.5° \text{ west}$$

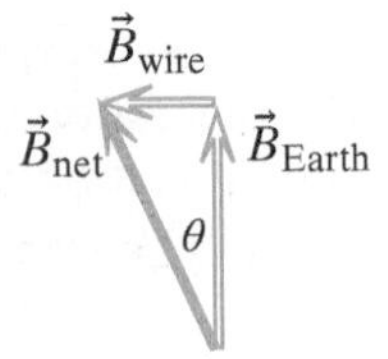

Figure 17.7 The compass points in the direction of the net magnetic field.

Checkpoint 2 A current-carrying wire oriented north–south and laid over a compass deflects the compass 15° east. What are the magnitude and direction of the magnetic field made by the current? The horizontal component of Earth's magnetic field is about 2×10^{-5} T.

17.3 BIOT–SAVART LAW: SINGLE MOVING CHARGE

Careful experimentation has shown that a stationary point charge makes an electric field given by this equation, called Coulomb's law:

$$\vec{E} = \frac{1}{4\pi\varepsilon_0}\frac{q}{r^2}\hat{r}$$

Similarly, careful experimentation shows that a moving point charge not only makes an electric field but also makes a magnetic field (Figure 17.8), which curls around the moving charge. This curly pattern is characteristic of magnetic fields. (In contrast, we saw in Chapter 16 that it is impossible to produce a curly electric field by arranging stationary point charges.) Magnetic field is measured in "teslas" and its magnitude and direction are given by the Biot–Savart law (pronounced bee-oh sah-VAR):

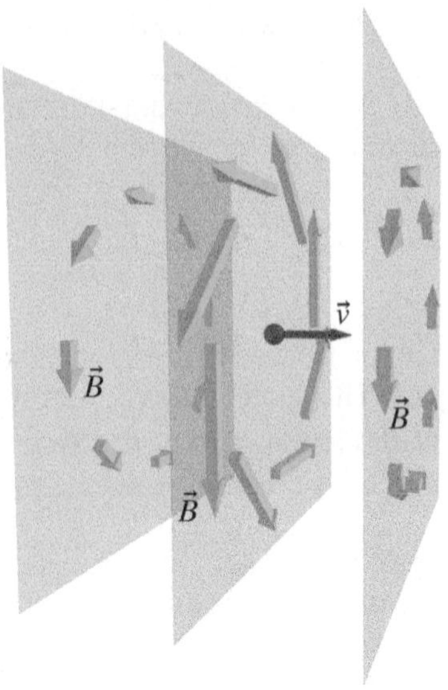

Figure 17.8 The magnetic field made by a moving positive charge, shown in three planes normal to $\vec{v}$.

THE BIOT–SAVART LAW FOR A SINGLE CHARGE

$$\vec{B} = \frac{\mu_0}{4\pi}\frac{q\vec{v} \times \hat{r}}{r^2}$$

where

$$\frac{\mu_0}{4\pi} = 1 \times 10^{-7}\,\frac{\text{T}\cdot\text{m}^2}{\text{C}\cdot\text{m/s}} \quad \textit{exactly}$$

$\vec{v}$ is the velocity of the point charge q, and $\hat{r}$ is a unit vector that points from the source charge toward the observation location.

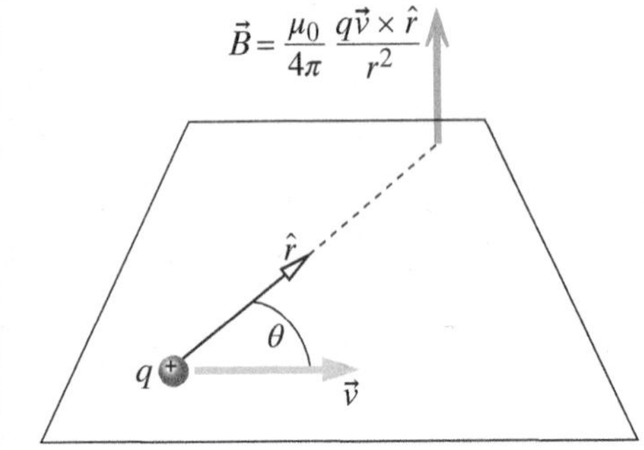

Figure 17.9 The Biot–Savart law involves a cross product, $\vec{v} \times \hat{r}$.

The Vector Cross Product

The Biot–Savart law involves a cross product, $\vec{v} \times \hat{r}$ (Figure 17.9). We will summarize the discussion of cross products found in Chapter 11 on angular momentum.

In the mathematics of vectors, the cross product is an operation that combines two vectors to produce a third vector that is perpendicular to the plane defined by the original vectors (Figure 17.10). The cross product of two vectors $\vec{A}$ and $\vec{B}$ is written as $\vec{A} \times \vec{B}$. Its magnitude is

$$|\vec{A} \times \vec{B}| = |\vec{A}||\vec{B}|\sin\theta$$

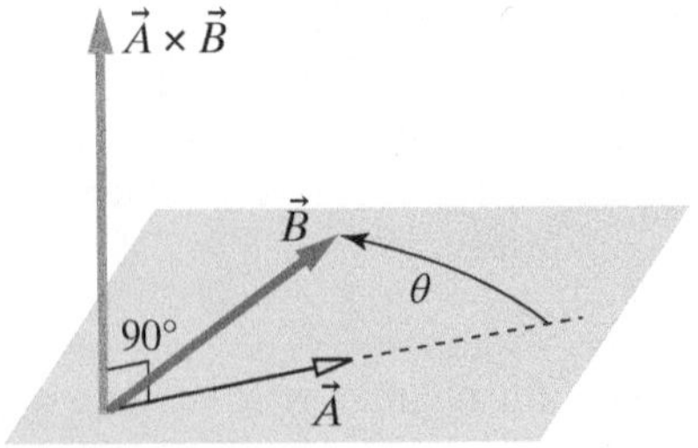

Figure 17.10 The cross product produces a vector that is perpendicular to the two original vectors.

where θ is the angle between the two vectors when they are placed tail to tail. θ is always less than or equal to 180°. The direction of $\vec{A} \times \vec{B}$ is given by a right-hand rule:

- Draw the vectors $\vec{A}$ and $\vec{B}$ so their tails are in the same location.
- Point the fingers of your right hand in the direction of $\vec{A}$ (Figure 17.11), then fold them through the angle θ toward $\vec{B}$ (Figure 17.12). You may need to turn your hand over to be able to do this.
- Stick your thumb out. Your thumb points in the direction of the cross product.

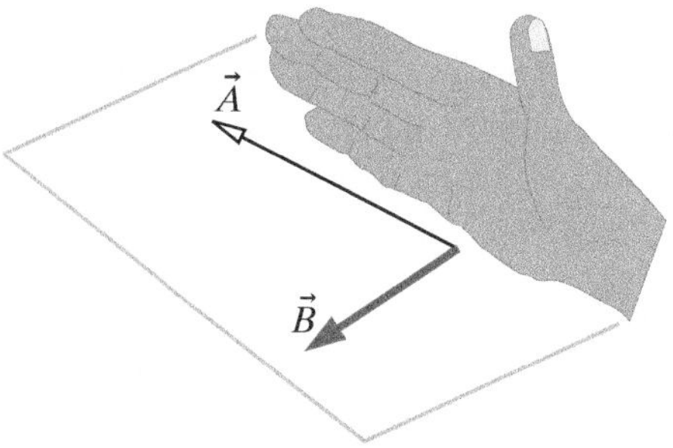

Figure 17.11 With the tails of $\vec{A}$ and $\vec{B}$ together, point your fingers along $\vec{A}$.

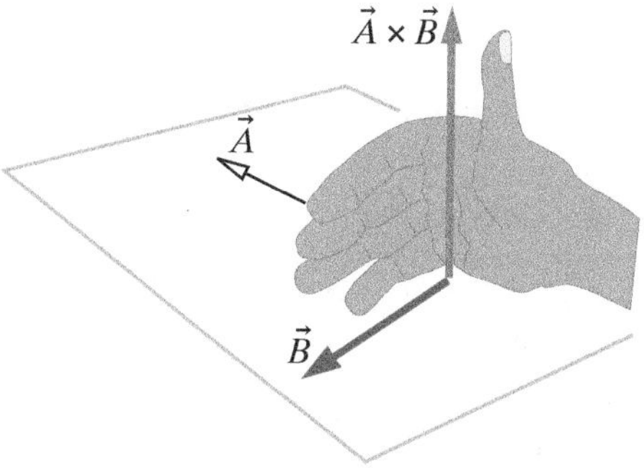

Figure 17.12 Fold your fingers toward $\vec{B}$. Your extended thumb points in the direction of $\vec{A}\times\vec{B}$.

In Figure 17.12, your fingers fold counterclockwise and $\vec{A}\times\vec{B}$ points out of the plane. On the other hand, if you were evaluating $\vec{B}\times\vec{A}$, in Figure 17.12 you would have had to turn your hand over, your fingers would have folded clockwise from $\vec{B}$ to $\vec{A}$, and your thumb would have pointed into the plane. In summary: fold counterclockwise and the cross product is out of the plane; fold clockwise and the cross product is into the plane.

Another way to determine the magnitude and direction of a cross product is to use the algebraic expression given below:

CROSS PRODUCT $\vec{A}\times\vec{B}$

$$\vec{A}\times\vec{B} = \langle A_yB_z - A_zB_y, A_zB_x - A_xB_z, A_xB_y - A_yB_x\rangle$$
$$|\vec{A}\times\vec{B}| = |\vec{A}||\vec{B}|\sin\theta$$

The direction of the vector is given by the right-hand rule.

For example,

$$\begin{aligned}\langle 2,3,4\rangle \times \langle 5,6,7\rangle &= \langle 3\cdot 7 - 4\cdot 6, 4\cdot 5 - 2\cdot 7, 2\cdot 6 - 3\cdot 5\rangle \\ &= \langle -3, 6, -3\rangle\end{aligned}$$

This approach to calculating a cross product is particularly useful in computer calculations. Note the cyclic nature of the subscripts: xyz, yzx, zxy.

If you have studied determinants, you may recognize that the components of $\vec{A}\times\vec{B}$ can be thought of as the "minors" of a 3 by 3 determinant in which $\vec{A}$ and $\vec{B}$ are the second and third rows, and the first row is $\hat{\imath} = \langle 1,0,0\rangle$, $\hat{\jmath} = \langle 0,1,0\rangle$, $\hat{k} = \langle 0,0,1\rangle$.

Two-Dimensional Projections

Because it is more difficult to sketch a situation in three dimensions, whenever possible we will work with two-dimensional projections onto a plane. If $\vec{A}$ and $\vec{B}$ lie in the xy plane, the cross-product vector $\vec{A}\times\vec{B}$ points in the $+z$ direction (out of the page, $\odot$) or in the $-z$ direction (into the page, $\otimes$). The symbol $\odot$ is supposed to suggest the tip of an arrow pointing toward you (cross product pointing toward you). The symbol $\otimes$ is supposed to suggest the feathers on the end of an arrow that is heading away from you (cross product pointing away from you).

Work through the following checkpoint before going on, to make sure you understand how to find the direction of $\vec{B}$ using the right-hand rule.

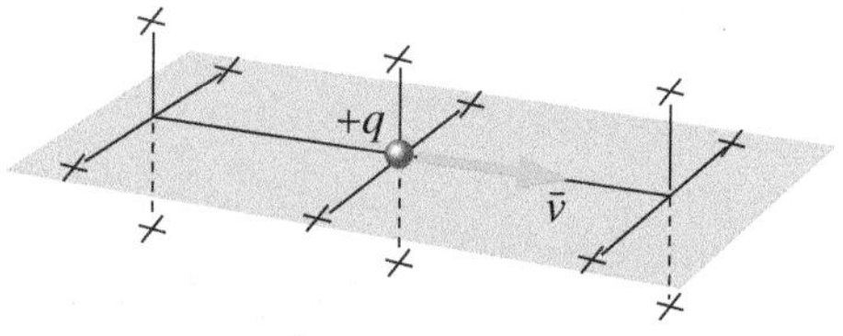

Figure 17.13

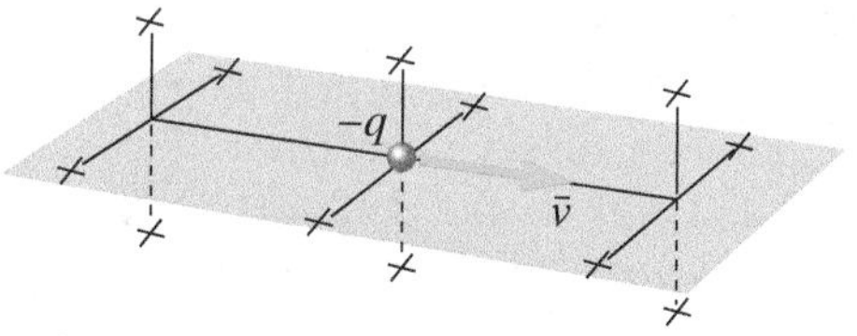

Figure 17.14

Checkpoint 3 At the locations marked with $\times$ in Figure 17.13, determine the direction of the magnetic-field vectors due to a positive charge $+q$ moving with a velocity $\vec{v}$. **(a)** For each observation location, draw the unit vector $\hat{r}$ *from* the charge *to* that location, then consider the cross product $\vec{v}\times\hat{r}$. Pay attention not only to the directions of the magnetic field but also to the relative magnitudes of the vectors. **(b)** If the charge is negative $(-q)$ as in Figure 17.14, how does this change the pattern of magnetic field?

Sign of the Moving Charge

Assume that the moving charged particles in a current-carrying wire are negatively charged and move in the direction shown in Figure 17.15.

QUESTION Use the Biot–Savart law and the right-hand rule to predict the direction of the magnetic field at several locations around the wire (Figure 17.15).

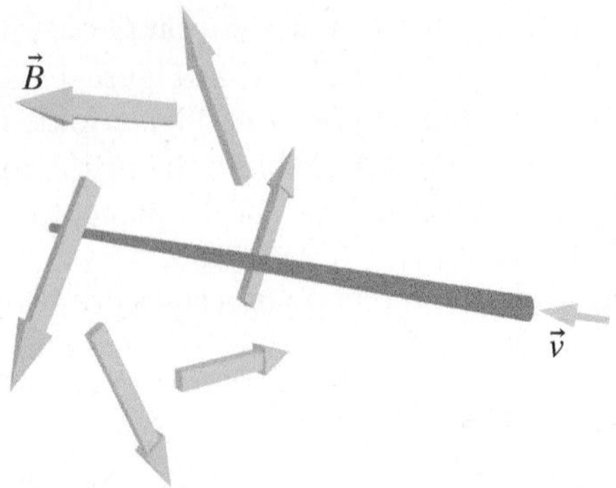

Figure 17.15 The magnetic field due to moving charges in the wire curls around the wire. Here the moving charges are assumed to be negative.

Convince yourself that $q\vec{v} \times \hat{r}$ does correctly give the direction and pattern of magnetic field near a wire; be sure to take into account the negative sign of q. From this curly pattern you can see that the direction of compass deflection will be different if the wire is above or below the compass, which is observed.

QUESTION Try the analysis again, but assume that the moving charges are positively charged particles moving in a direction opposite to that shown. Do you still predict the curly magnetic field as shown in Figure 17.15?

You should have concluded that for the purpose of predicting the direction of magnetic field, it does not matter whether we assume that the moving charges are negative and moving in a given direction or positive and moving in the opposite direction. It is common to describe current flow in terms of "conventional current," which means assuming that the moving charges are positive, even if this is known not to be the case in a particular situation.

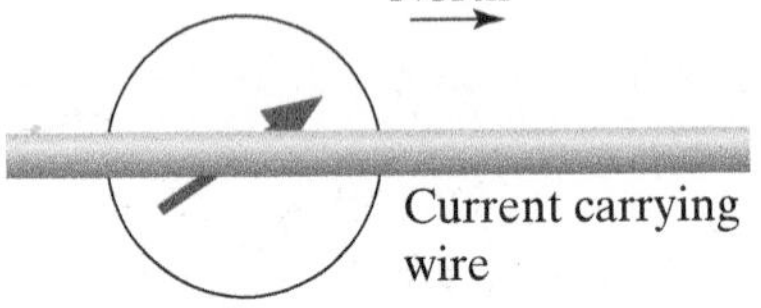

Figure 17.16 If the moving particles in the wire are electrons, in what direction are they moving? If we assume that the particles are positively charged, in which direction would they be moving?

Checkpoint 4 In Figure 17.16 a current-carrying wire lies on top of a compass. Judging from the deflection of the compass away from north, what is the direction of the electron current in the wire? If the current were due to the motion of positive charges, which way would the positive charges be moving?

17.4 RELATIVISTIC EFFECTS

Magnetic Field Depends on Your Frame of Reference

Electric fields are made by charges, whether at rest or moving. Magnetic fields seem to be made solely by moving charges—but consider the following odd "thought experiment."

Suppose that Jack sits in the classroom with a charged piece of invisible tape stuck to the edge of his desk. He can of course observe an electric field due to the charged tape, but he doesn't observe any magnetic field. His compass is unaffected by the charged tape, since those charges aren't moving:

$$\vec{B} = \frac{\mu_0}{4\pi}\frac{q\vec{v} \times \hat{r}}{r^2} = \vec{0} \quad \text{since } v = 0$$

Jill runs at high speed through the classroom past the charged tape, carrying her own very sensitive compass. She observes an electric field due to the charged tape. In addition, however, in her frame of reference the charged tape is moving, so she observes a small magnetic field due to the moving charges, which affects her compass! As far as Jack is concerned, the charged tape just makes an electric field, but apparently Jill sees a mixture of electric and magnetic fields made by what is for her a moving charged tape.

Up until now we have implied that electric fields and magnetic fields are fundamentally different, but this "thought experiment" shows that they are in fact closely related. Moreover, this connection raises questions about the Biot–Savart law:

$$\vec{B} = \frac{\mu_0}{4\pi}\frac{q\vec{v} \times \hat{r}}{r^2}$$

Just what velocity are we supposed to use in this equation? The velocity of the tape relative to Jack (which is zero) or the velocity of the tape relative to Jill (opposite her own velocity)?

The correct answer is that you use the velocities of the charges as you observe them in your frame of reference. Using these velocities in the

Biot–Savart law, you calculate the magnetic field; in your frame of reference you observe a magnetic field that agrees with your prediction. Observers in a different reference frame use the velocities observed in their frame to calculate the magnetic field using the Biot–Savart law, and in their frame of reference they observe a magnetic field that agrees with their prediction. You and they make different predictions and observe different magnetic fields, but both you and they find agreement between theory and experiments. Jack predicts that his compass won't deflect because the charged tape isn't moving, and that's what he observes. Jill predicts that her compass will deflect because she sees a moving charged tape, and she observes that her compass does deflect. (Actually, for Jill to see a large effect she would have to move at very high speed, because there isn't much charge on a charged tape.)

Evidently there is a deep connection between electric fields and magnetic fields, and this connection is made explicit in Einstein's special theory of relativity. Later in the text we will see further aspects of this connection.

Retardation

Remember that when you move a charge, the electric field of that charge at a distance from the charge doesn't change instantaneously. The electric field doesn't change until a time sufficient for light to reach the observation location, and you measure a change in the electric field at the same instant that you see the charge move.

The same retardation effect is observed with magnetic fields. If you suddenly change the current in a wire, the magnetic field at some distance from the source of the magnetic field stays the same until enough time has elapsed for light to reach the observation location. Thus magnetic field has some reality in its own right, independent of the moving charges that originally produced it.

The Biot–Savart law does not contain any reference to time, so it cannot be relativistically correct. Like Coulomb's law, the Biot–Savart law is only approximately correct and will give accurate results only if the speeds of the moving source charges are small compared to the speed of light.

17.5 ELECTRON CURRENT AND CONVENTIONAL CURRENT

An easy way to observe the magnetic fields made by moving charged particles is to initiate and sustain a current—a continual flow of charged particles in one direction. In order to do this, we need to find a way to produce and sustain an electric field inside a wire, since we know that if the electric field inside a metal becomes zero, the metal object will be in equilibrium, and no current will flow. How exactly it is possible to arrange charges in order to create such an electric field everywhere in a wire is the subject of Chapter 18. For the moment, we'll assume that we have somehow accomplished this by assembling a circuit from batteries, wires, and perhaps light bulbs, since evidently currents do flow in such circuits.

In order to apply the Biot–Savart law to predict the magnitude and direction of the magnetic field associated with this current, we need to know the number of moving charges making the field and how fast they are moving.

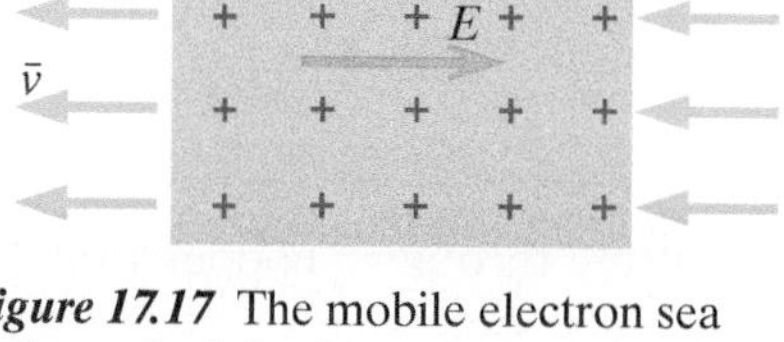

Figure 17.17 The mobile electron sea drifts to the left with a speed $\bar{v}$.

An Equation for Electron Current

Consider a section of a metal wire through which the mobile electron sea is continuously shifting under the influence of a a nonzero electric field, as indicated in Figure 17.17. There can be no excess charge anywhere inside the wire. To every mobile electron there corresponds a singly charged positive atomic core—an atom minus one electron that has been released to roam freely in the mobile-electron sea. Averaged over a few atomic diameters, the

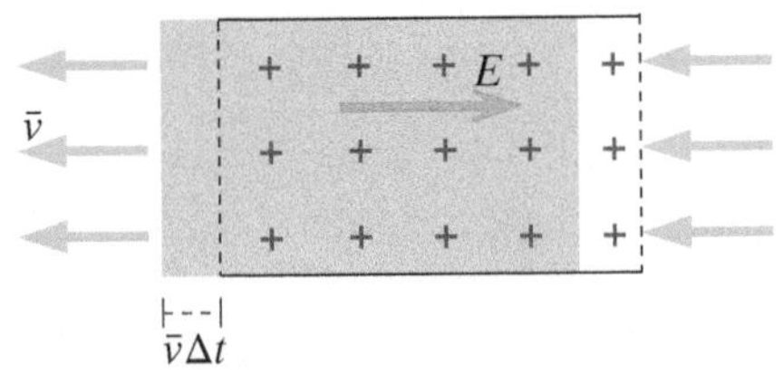

Figure 17.18 During a time Δt the electron sea shifts a distance $\bar{v}\Delta t$.

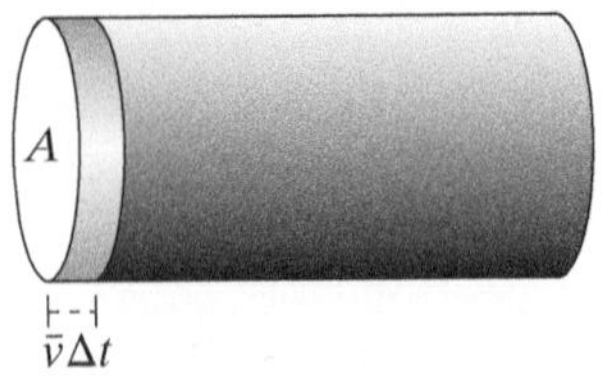

Figure 17.19 The volume of the electron sea that has flowed past a point in the wire during a time Δt is $A\bar{v}\Delta t$.

interior of the metal is neutral, and the repulsion between the mobile electrons is balanced out on the average by the attractions of the positive cores.

Suppose that in a metal wire the mobile electron sea as a whole has an average "drift speed" $\bar{v}$. Assume that the current is evenly distributed across the entire cross section of the wire (in a later chapter we will be able to show that this is true.) After a short time Δt, this section of the electron sea will drift to the left a distance $\bar{v}\Delta t$, as indicated in Figure 17.18.

If the cross-sectional area of this wire is A, the volume of the disk-shaped portion of the sea that has flowed out of the left end of this section of the wire (and into the adjoining section of wire) in the time Δt is $A\bar{v}\Delta t$ (Figure 17.19).

QUESTION If the density of mobile electrons (that is, the number of mobile electrons per unit volume) everywhere in the metal is n, how many electrons are in this disk?

$$n \times \text{volume} = n(A\bar{v}\Delta t) = nA\bar{v}\Delta t$$

The number of electrons in this disk is the number of electrons that have flowed out of this section of the wire in the short time Δt.

QUESTION What is the rate at which mobile electrons flow past the left end of this piece of the wire?

We divide by the time Δt to calculate the rate at which mobile electrons flow past the left end of this piece of the wire:

$$\frac{nA\bar{v}\Delta t}{\Delta t} = nA\bar{v}$$

This result, that the rate at which electrons pass some section of the wire is $nA\bar{v}$, is sufficiently important to be worth remembering (though it is even better to be able to re-do the simple derivation that gives this result).

ELECTRON CURRENT

The rate i at which electrons pass a section of a wire:

$$i = nA\bar{v}$$

n is the mobile electron density (the number of mobile electrons per unit volume), A is the cross-sectional area of the wire, and $\bar{v}$ is the average drift speed of the electrons. In SI units, the units of i are electrons per second.

In a circuit, electrons leave the negative end of the battery (the end that is marked –) and flow through the wire to the positive end of the battery. See EXP3.

EXAMPLE **Drift Speed of Electrons**

Suppose that $i = 3.4 \times 10^{18}$ electrons/s are drifting through a copper wire. (This is a typical value for a simple circuit.) The diameter of the wire is 1 mm, and the density of mobile electrons in copper is $8.4 \times 10^{28}\text{m}^{-3}$. **(a)** What is the drift speed of the electrons? **(b)** About how many minutes would it take for an electron traveling at this speed to drift from one end to the other end of a wire 30 cm long, about one foot?

Solution **(a)**

$$A = \pi\left(\frac{1\times10^{-3}\,\text{m}}{2}\right)^2 = 8\times10^{-7}\,\text{m}^2$$

$$\bar{v} = \frac{i}{nA} = \frac{3.4\times10^{18}\,\text{s}^{-1}}{(8.4\times10^{28}\text{m}^{-3})(8\times10^{-7}\,\text{m}^2)} = 5.06\times10^{-5}\,\text{m/s}$$

(b)

$$t = \frac{0.3\,\text{m}}{5.06\times10^{-5}\,\text{m/s}} = 5930\text{ s} = 99\text{ min}$$

This result is puzzling: If the drift speed is so slow, how can a lamp light up as soon as you turn it on? We'll come back to this in Chapter 18.

Conventional Current

In most metals, current consists of drifting electrons. However, there are a few materials in which the moving charges act as though they were positive "holes" in the sea of mobile electrons. The positive holes act in every way like real positive particles. For example, holes drift in the direction of the electric field, and they have a charge of $+e$.

Electron current moves from the negative end of a battery through a circuit to the positive end of the battery; hole current would go in the opposite direction. Given this information, consider the following question:

QUESTION From observations of the direction of magnetic field around a copper wire, can one tell whether the current in copper consists of electrons, or holes?

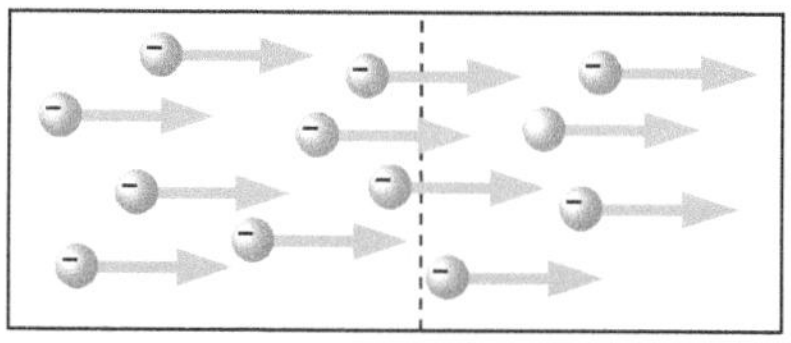

Figure 17.20 An electron current to the right moves negative charge from left to right.

Observations of the magnetic field due to a current-carrying wire are not sufficient to tell the difference, because if the sign of the moving charge is changed, the direction of the drift velocity is also changed: $(+e)(+\bar{v}) \rightarrow (-e)(-\bar{v})$ (Figures 17.20 and 17.21). Therefore the prediction of the Biot–Savart law is exactly the same in either case.

In copper, as in most metals, the current is due to moving electrons. The few metals that have hole current include aluminum and zinc. In the doped semiconductors important in electronics, p-type material (positive type) involves hole current, and n-type (negative type) involves electron current. In Chapter 20 we will study the Hall effect, which can be used to determine whether the current in a wire is due to electrons or to holes.

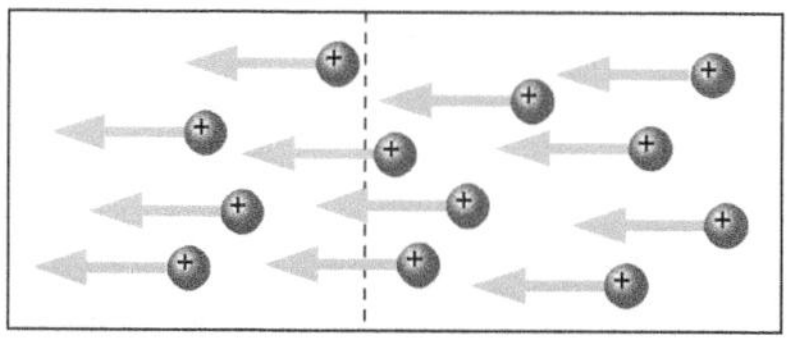

Figure 17.21 A conventional current to the left would move positive charge from right to left, which would have the same effect as moving negative charge from left to right, as in the previous figure.

Because most effects (other than the Hall effect) are the same for electron current and hole current, it is traditional to define "conventional current" to run in the direction of hole current, even if the actual current consists of moving electrons, in which case the conventional current runs in the opposite direction to the electrons (Figure 17.21). This simplifies calculations by eliminating the minus sign associated with electrons. (This sign ultimately goes back to a choice made by Benjamin Franklin when he arbitrarily assigned "positive" charge to be the charge we now know to be carried by protons.)

In addition, conventional current I is defined not as the number of holes passing some point per second but rather as the amount of charge (in coulombs) passing that point per second. This is the number of holes per second multiplied by the (positive) charge $|q|$ associated with one hole:

CONVENTIONAL CURRENT

$$I = |q|nA\bar{v} \quad \text{(coulombs per second, or amperes)}$$

1 C/s = 1 ampere (abbreviated A). The direction of conventional current is opposite to the direction of electron current. The moving charged particles are assumed to be positive. A battery is assumed to send conventional current out of the + end.

In a metal, $|q| = e$, but in an ionic solution the moving ions might have charges that are a multiple of e.

EXAMPLE **Electron Current and Conventional Current**

In the preceding example an electron current of $i = 3.4 \times 10^{18}$ electrons/s ran through a copper wire. What was the conventional current in the wire?

Solution

$$I = |q|i = (1.6 \times 10^{-19}\,\text{C})(3.4 \times 10^{18}\,\text{s}^{-1}) = 0.544\,\text{A}$$

Electric Field and Current

A nonzero electric field is necessary to maintain a current in a conductor. In Chapter 14 we saw that the average drift speed in a metal is $\bar{v} = uE$, where u is the electron mobility and E is the electric field inside the metal (which is not in equilibrium). Therefore, the conventional current in a wire is

$$I = |q|nA\bar{v} = |q|nA(uE)$$

If we know the magnitude of the electric field, we can predict what the current will be in the wire. In Chapter 18 we will see how to maintain a nonzero electric field inside a metal.

Checkpoint 5 The electron mobility in copper is 4.5×10^{-3} (m/s)/(N/C), and the density of mobile electrons is $8.4 \times 10^{28}\,\text{m}^{-3}$. In a copper wire 1 mm in diameter there is an electric field whose magnitude is 0.05 N/C. What is the conventional current in this wire? (Note that a small electric field drives a sizable current in copper, which is a good conductor.)

17.6 THE BIOT–SAVART LAW FOR CURRENTS

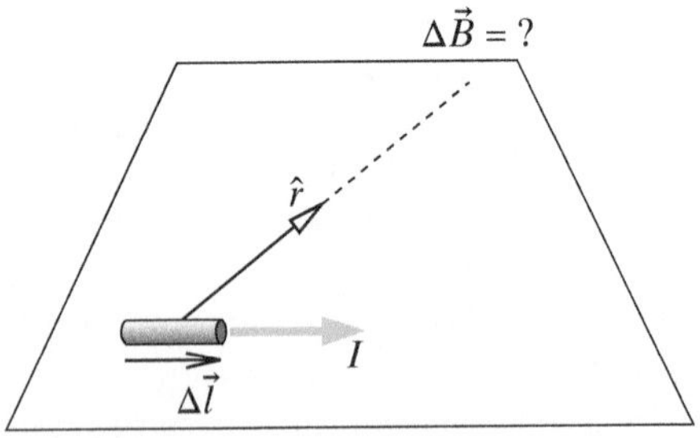

Figure 17.22 Magnetic field contributed by a short thin length Δl of current-carrying wire.

We don't often observe the magnetic field of a single moving charge. Usually we are interested in the magnetic field produced by a large number of charges moving through a wire in a circuit. The superposition principle is valid for magnetic fields, so we need to add up the contributions of the individual charges to the magnetic field.

Let's calculate the magnetic field due to a bunch of moving positive charges contained in a small thin wire of length Δl and cross-sectional area A (Figure 17.22). If there are n moving charges per unit volume, in this small volume $A\,\Delta l$ there are $nA\,\Delta l$ moving charges. We will measure the magnetic field at a location far enough away from this small volume that each moving charge produces approximately the same magnetic field at that location.

QUESTION Show that the net magnetic field of all the moving charges in this small volume, far from the volume, is

$$\Delta\vec{B} = \frac{\mu_0}{4\pi}\frac{I\Delta\vec{l} \times \hat{r}}{r^2}$$

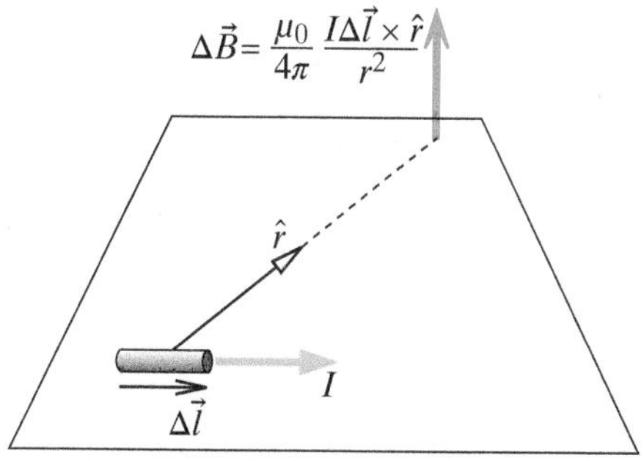

Figure 17.23 The Biot–Savart law for the magnetic field contributed by a short thin length of wire carrying conventional current I. The magnetic field due to this current segment is perpendicular to the plane defined by $\Delta\vec{l}$ and $\hat{r}$.

where $\Delta\vec{l}$ is a vector with magnitude Δl (the length of the segment of wire) pointing in the direction of the conventional current I.

This results from the fact that the volume $A\,\Delta l$ contains $n(A\,\Delta l)$ moving charges, so the magnetic field is $n(A\,\Delta l)$ times as large as the magnetic field of one of the charges. Each moving charge has a charge q, and the conventional current is $I = |q|(nA\,\bar{v})$. Collecting terms, we find the result given above.

We have an alternative form of the Biot–Savart law for the magnetic field contributed by a short thin length of current-carrying wire (Figure 17.23):

THE BIOT–SAVART LAW FOR A SHORT THIN LENGTH OF WIRE

$$\Delta\vec{B} = \frac{\mu_0}{4\pi}\frac{I\Delta\vec{l}\times\hat{r}}{r^2}$$

where $\frac{\mu_0}{4\pi} = 1\times10^{-7}\frac{\text{T}\cdot\text{m}^2}{\text{C}\cdot\text{m/s}}$ exactly.

$\Delta\vec{l}$ is a vector in the direction of the conventional current I, and whose magnitude Δl is the length of the segment of wire. $\hat{r}$ is a unit vector that points from the wire segment toward the observation location.

QUESTION Explain why this equation also gives the right results if the moving charges are (negative) electrons, as long as we interpret $\Delta\vec{l}$ as a vector in the direction of the conventional current.

The law works for moving electrons because although they have the opposite charge, they also move in the direction opposite to the direction of the conventional current. These two changes in sign cancel each other.

It is important to keep in mind that there is really only one Biot–Savart law, not two. Always remember that

$$\Delta\vec{B} = \frac{\mu_0}{4\pi}\frac{I\Delta\vec{l}\times\hat{r}}{r^2}$$

is simply the result of adding up the effects of many moving charges in a short thin length of wire, each of which contributes a magnetic field

$$\vec{B} = \frac{\mu_0}{4\pi}\frac{q\vec{v}\times\hat{r}}{r^2}$$

The key point is that there are $nA\,\Delta l$ electrons in a short length of wire, each moving with average speed $\bar{v}$, so that the sum of all the $q\bar{v}$ contributions is $(nA\,\Delta l)|q|\bar{v} = (|q|nA\bar{v})\Delta l = I\Delta l$.

17.7 THE MAGNETIC FIELD OF CURRENT DISTRIBUTIONS

In the next sections we will apply the Biot–Savart law to find the magnetic field of various distributions of currents in wires. We will use the same four-step approach we used to find the electric field of distributed charges in Chapter 15:

1. Cut up the current distribution into pieces; draw $\Delta\vec{B}$ for a representative piece.
2. Write an expression for the magnetic field due to one piece.
3. Add up the contributions of all the pieces.
4. Check the result.

Applying the Biot–Savart Law: A Long Straight Wire

Using the procedure outlined above, we can calculate the magnetic field near a long straight wire. This will make it possible to predict the compass needle deflection that you observe when you bring a wire near your compass, and you can compare your prediction with your experimental observation.

The long straight wire is one of the few cases that we can calculate completely by hand. Except for a few special cases, calculating the magnetic field due to distributed currents is often best done by computer. The basic concepts you would use in a computer program are the same as we will use here, but the computer would carry out the tedious arithmetic involved in the summation step of the procedure.

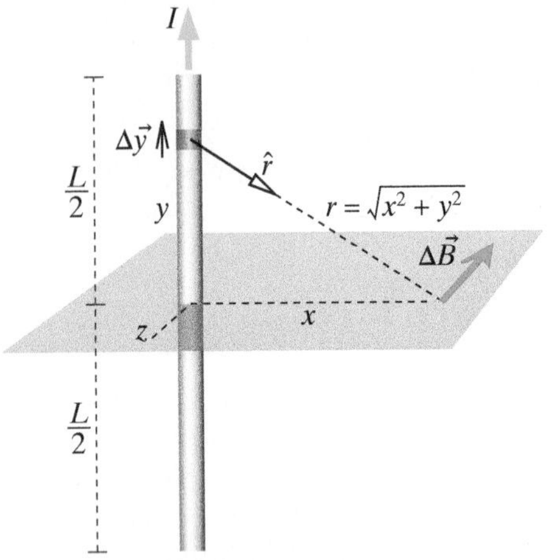

Figure 17.24 Cut the wire into short pieces and draw the magnetic field contributed by one of the pieces.

Step 1: Cut Up the Distribution into Pieces and Draw $\Delta\vec{B}$

We will consider a wire of length L (Figure 17.24). We cut up the wire into very short sections each of length Δy, where Δy is a small portion of the total length. We will calculate the magnetic field a perpendicular distance x from the center of the wire. Using the right-hand rule, we can draw the direction of $\Delta\vec{B}$ for this "piece" of current.

Step 2: Write an Expression for the Magnetic Field Due to One Piece

We will place the origin at the center of the wire.

$$\vec{r} = \langle x,0,0\rangle - \langle 0,y,0\rangle = \langle x,-y,0\rangle$$

$$|\vec{r}| = [x^2+(-y)^2]^{1/2} = [x^2+y^2]^{1/2}$$

$$\hat{r} = \frac{\vec{r}}{r} = \frac{\langle x,-y,0\rangle}{[x^2+y^2]^{1/2}}$$

Since the location of one piece is specified by y, this will be the integration variable. We can express $\Delta\vec{l}$ in terms of y:

$$\Delta\vec{l} = \Delta y\langle 0,1,0\rangle$$

$$\Delta\vec{B} = \frac{\mu_0}{4\pi}\frac{I\Delta\vec{l}\times\hat{r}}{(x^2+y^2)} = \frac{\mu_0}{4\pi}\frac{I\Delta y\langle 0,1,0\rangle}{(x^2+y^2)} \times \frac{\langle x,-y,0\rangle}{[x^2+y^2]^{1/2}}$$

Evaluating the cross product, we get:

$$\langle 0,1,0\rangle \times \langle x,-y,0\rangle = \langle 0,0,-x\rangle$$

Our final expression for $\Delta\vec{B}$ is:

$$\Delta\vec{B} = -\frac{\mu_0}{4\pi}\frac{Ix\Delta y}{(x^2+y^2)^{3/2}}\langle 0,0,1\rangle$$

Every piece of the straight wire contributes some magnetic field in the $-z$ direction, so we need to calculate only the z component.

Step 3: Add Up the Contributions of All the Pieces

By letting $\Delta y \to 0$, we can calculate the magnitude B by writing the sum as an integral:

$$B = \frac{\mu_0}{4\pi}Ix\int_{y=-L/2}^{y=+L/2}\frac{dy}{(x^2+y^2)^{3/2}}$$

Fortunately, this integral can be found in standard tables of integrals:

$$B = \frac{\mu_0}{4\pi} Ix \left[\frac{y}{x^2\sqrt{x^2+y^2}}\right]_{y=-L/2}^{y=+L/2}$$

$$B = \frac{\mu_0}{4\pi} Ix \left[\frac{L/2}{x^2\sqrt{x^2+(L/2)^2}} - \frac{-L/2}{x^2\sqrt{x^2+(L/2)^2}}\right]$$

$$B = \frac{\mu_0}{4\pi} \frac{LI}{x\sqrt{x^2+(L/2)^2}}$$

An extremely important case is that of a very long wire ($L \gg x$) or, equivalently, the magnetic field very near a short wire ($x \ll L$).

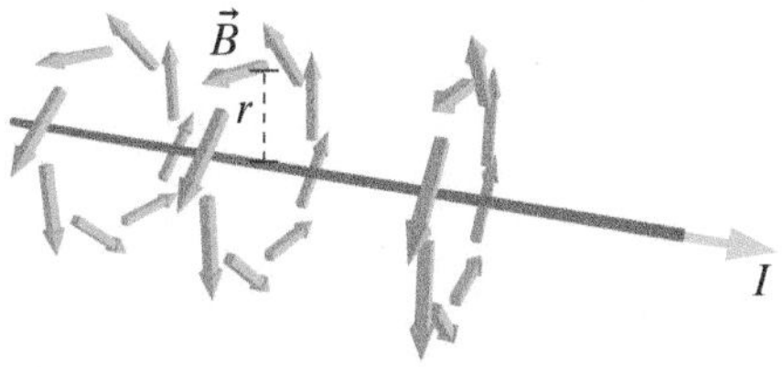

Figure 17.25 Magnetic field of a straight current-carrying wire, at selected locations a distance r radially outward from the wire.

QUESTION Show that you get

$$B \approx \frac{\mu_0}{4\pi} \frac{2I}{x} \quad \text{if } L \gg x$$

If $L \gg x$, $\sqrt{x^2+(L/2)^2} \approx L/2$, which leads to the stated result.

In summary, Figure 17.25 shows the pattern of magnetic field a distance r from the center of a straight wire of length L. Because of the symmetry of this field, it does not matter where we draw our x axis—the field will be the same all around the rod. To indicate this, we replace x in our equation with r, the radial distance from the rod.

Although the magnitude of the field is constant at constant r, the direction of the field is different at each angle, since the field curls around the wire. To express this as a vector equation we would have to include a cross product in the expression; it's simpler to get the direction using the right-hand rule.

MAGNETIC FIELD OF A STRAIGHT WIRE

$$B_{\text{wire}} = \frac{\mu_0}{4\pi} \frac{LI}{r\sqrt{r^2+(L/2)^2}}$$

for length L, conventional current I, a perpendicular distance r from the center of the wire.

$$B_{\text{wire}} \approx \frac{\mu_0}{4\pi} \frac{2I}{r} \quad \text{if } L \gg r$$

Historically, the result for a very long straight wire was first obtained by the French physicists Biot and Savart. Their names have come to be associated with the more fundamental principle (the Biot–Savart law) that leads to this result. See EXP4.

Step 4: Check the Result

Direction: In Figure 17.25 the right-hand rule is consistent with the diagram.
Units:

$$\left(\frac{\text{T}\cdot\text{m}}{\text{A}}\right)\frac{(\text{m}\cdot\text{A})}{\text{m}\cdot\text{m}} = \text{T}$$

Far away ($r \gg L$):

$$B_{\text{wire}} \approx \frac{\mu_0}{4\pi} \frac{LI}{r^2} = \frac{\mu_0}{4\pi} \frac{|I\Delta\vec{l} \times \hat{r}|}{r^2}$$

which is indeed the magnetic field contributed by a short wire of length Δl.

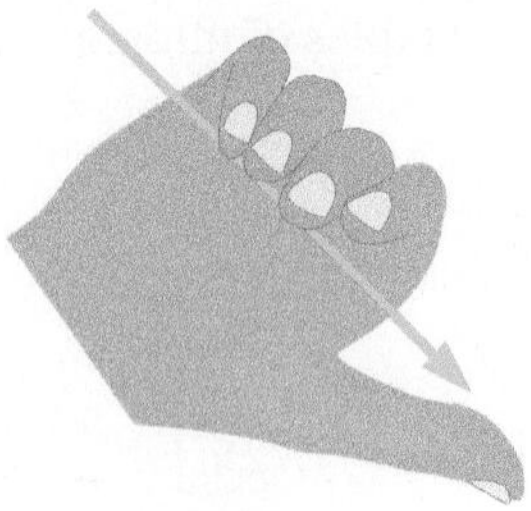

Figure 17.26 The thumb of the right hand points in the direction of conventional current flow I, and the fingers curl around in the direction of the magnetic field.

Another Right-Hand Rule

There is another right-hand rule that is often convenient to use with current-carrying wires. As you can see in Figure 17.26, if you grasp the wire in your right hand with your thumb pointing in the direction of conventional current, your fingers curl around in the direction of the magnetic field. Clearly this right-hand rule is consistent with the one we used for the Biot–Savart law.

17.8 A CIRCULAR LOOP OF WIRE

Next we'll calculate the magnetic field of a circular loop of wire that carries a conventional current I. We'll do only the easiest case—the magnetic field at any location along the axis of the loop, which is a line going through the center and perpendicular to the loop.

This calculation is important for two reasons. First, many scientific and technological applications of magnetism involve circular loops of current-carrying wire. Second, the calculation will also lead into an analysis of atomic current loops in your bar magnet. After calculating the magnetic field we will compare our predictions with experiments.

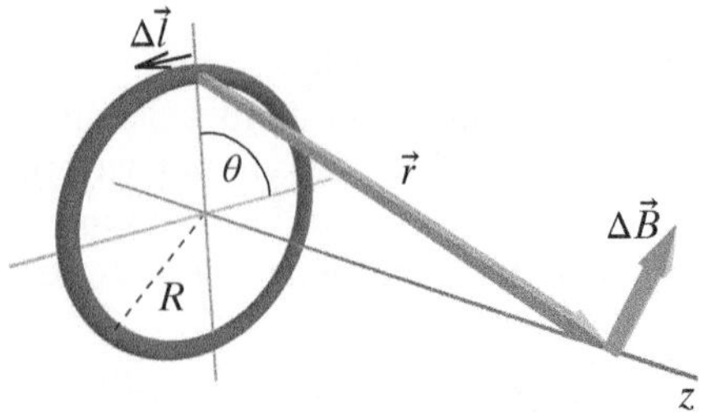

Figure 17.27 Cut the loop into short pieces and draw the magnetic field contributed by one of the pieces. Note that $d\vec{l}\perp\hat{r}$ for every piece of the loop. The loop is in the xy plane, with x to the right and y up. $\Delta\vec{B}$ is perpendicular to the plane defined by $\Delta\vec{l}$ and $\vec{r}$.

Step 1: Cut Up the Distribution into Pieces and Draw $\Delta\vec{B}$

See Figure 17.27. We cut up the loop into very short sections each of length Δl (a small portion of the total circumference $2\pi R$) (Figure 17.27; side view in Figure 17.28). Determine the direction of $\Delta\vec{B}$ with the right-hand rule. Note that the angle between $\Delta\vec{l}$ and $\hat{r}$ is 90° for every location on the ring.

Step 2: Write an Expression for the Magnetic Field Due to One Piece

To work out $\Delta\vec{B}$ for an arbitrary $\Delta\vec{l}$ will be algebraically messy, since

$$\Delta\vec{l} = \langle R\cos(\theta+d\theta), R\sin(\theta+d\theta), 0\rangle - \langle R\cos\theta, R\sin\theta, 0\rangle$$

and

$$\vec{r} = \langle 0,0,z\rangle - \langle R\cos\theta, R\sin\theta, 0\rangle$$

However, we can simplify the problem considerably by noticing the symmetry of the situation. Because of the circular symmetry of the ring, the ΔB_x and ΔB_y contributed by one piece will be canceled by the contributions of a piece located on the other side of the loop.

Furthermore, the ΔB_z contributed by each piece of the ring will be exactly the same. This allows us to select one piece for which ΔB_z is easy to calculate, and use this value for every piece in the sum. We will select the piece shown in Figure 17.27, located on the y axis. We choose the origin to be the center of the loop.

$$\vec{r} = \langle 0,0,z\rangle - \langle 0,R,0\rangle = \langle 0,-R,z\rangle$$

$$|\vec{r}| = [R^2+z^2]^{1/2}$$

$$\hat{r} = \frac{\vec{r}}{r} = \frac{\langle 0,-R,z\rangle}{[R^2+z^2]^{1/2}}$$

The location of the piece depends on θ, so this will be the integration variable.

$$\Delta\vec{l} = \langle -R\Delta\theta, 0, 0\rangle$$

$$\Delta\vec{B} = \frac{\mu_0}{4\pi}\frac{I\Delta\vec{l}\times\hat{r}}{r^2} = \frac{\mu_0}{4\pi}I\frac{\langle -R\Delta\theta,0,0\rangle\times\langle 0,-R,z\rangle}{[R^2+z^2]^{3/2}}$$

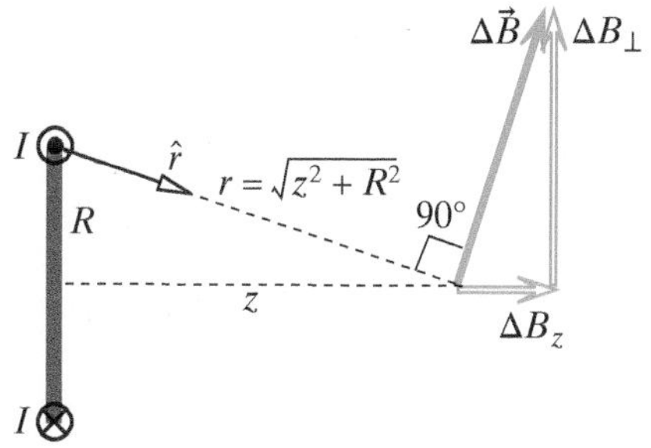

Figure 17.28 Side view of the current-carrying loop. Again note that $d\vec{l} \perp \hat{r}$ for every piece of the loop.

Evaluating the cross product, we get:

$$\langle -R\Delta\theta, 0, 0\rangle \times \langle 0, -R, z\rangle = \langle 0, zR\Delta\theta, R^2\Delta\theta\rangle$$

We need only the z component, since the others will add up to zero:

$$\Delta B_z = \frac{\mu_0}{4\pi}\frac{IR^2\Delta\theta}{[R^2+z^2]^{3/2}}$$

In Figure 17.28 we show the component ΔB_z along the axis and the component $\Delta B_\perp$ perpendicular to the axis.

Step 3: Add Up the Contributions of All the Pieces

We can express the sum as an integral, where θ, which specifies the location of a piece of the ring, runs all the way around the ring. Remember that ΔB_z contributed by each piece is the same.

$$B_z = \int_0^{2\pi} \frac{\mu_0}{4\pi}\frac{IR^2 d\theta}{(z^2+R^2)^{3/2}} = \frac{\mu_0}{4\pi}\frac{IR^2}{(z^2+R^2)^{3/2}}\int_0^{2\pi} d\theta$$

The integral $\int_0^{2\pi} d\theta$ is just 2π. Here is the result (Figure 17.29):

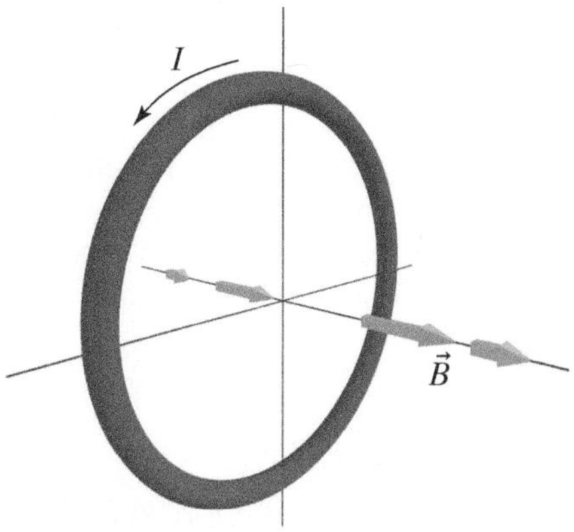

Figure 17.29 Magnetic field of a circular loop of current-carrying wire.

MAGNETIC FIELD OF A LOOP

$$B_{\text{loop}} = \frac{\mu_0}{4\pi}\frac{2\pi R^2 I}{(z^2+R^2)^{3/2}}$$

for a circular loop of radius R and conventional current I at a distance z from the center, along the axis. See EXP5.

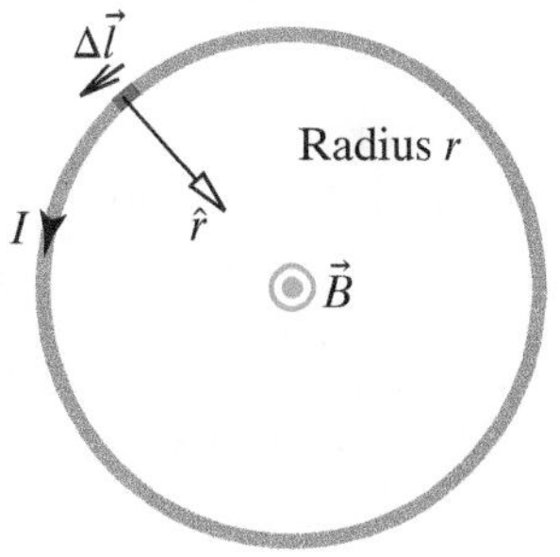

Figure 17.30 Magnetic field at the center of a loop of current-carrying wire.

Step 4: Check the Result

Units:

$$\left(\frac{\text{T}\cdot\text{m}}{\text{A}}\right)\frac{(\text{m}^2\cdot\text{A})}{(\text{m}^2)^{3/2}} = \text{T}$$

Direction: Check several pieces with the right-hand rule.
Special case: Center of the loop (Figure 17.30), where the magnetic field is especially easy to calculate from scratch. By the right-hand rule, $\Delta\vec{l} \times \hat{r}$ points out of the page for every piece of the loop, and its magnitude is simply Δl, since $\Delta\vec{l} \perp \hat{r}$. The magnitude of $\hat{r}$ is 1, and $\sin(90°) = 1$.

We have this:

$$B = \sum \frac{\mu_0}{4\pi}\frac{I\Delta l}{R^2} = \frac{\mu_0}{4\pi}\frac{I(2\pi R)}{R^2} = \frac{\mu_0}{4\pi}\frac{2\pi I}{R}$$

because the sum of all the Δl's is just $2\pi R$, the circumference of the ring.

QUESTION Show that the general equation for the magnetic field of a loop reduces to this result if you let $z = 0$, which is another kind of check.

Qualitative Features of the Magnetic Field of a Loop of Wire

It is interesting to see what the magnetic field is *far* from the loop, along the axis of the loop.

QUESTION Show that if z is very much larger than the radius R, the magnetic field is approximately equal to

$$\frac{\mu_0}{4\pi}\frac{2\pi R^2 I}{z^3}$$

The key to this result is that if $z \gg R$, $(z^2+R^2)^{3/2} \approx (z^2)^{3/2} = z^3$. We see that the magnetic field of a circular loop falls off like $1/z^3$.

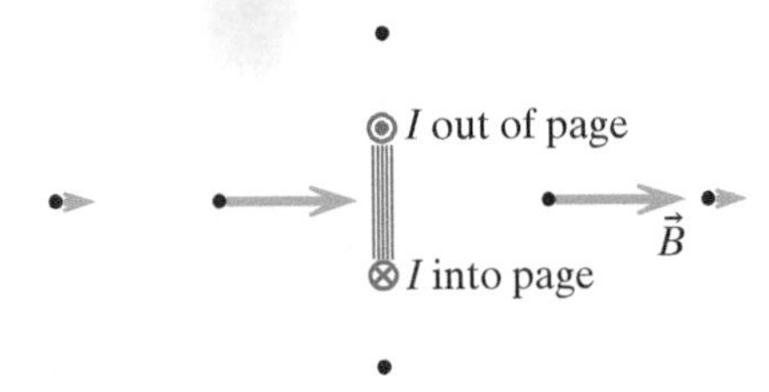

Figure 17.31 Predict the direction of the magnetic field of the coil, above and below the coil. The plane of the coil is perpendicular to the page.

QUESTION Figure 17.31 shows the pattern of magnetic field along the axis of a coil containing many loops. What is the direction of the magnetic field at the points above and below the coil? (*Hint:* Apply the Biot–Savart law qualitatively to the near and far halves of the circular loop.)

Far above the coil in Figure 17.31, the closer upper part of the coil contributes a larger magnetic field to the left than does the slightly farther away lower part of the coil to the right. A detailed calculation shows that this is also true a short distance above the coil (Figure 17.32). Compare the pattern of magnetic field with the pattern of *electric* field around an electric dipole.

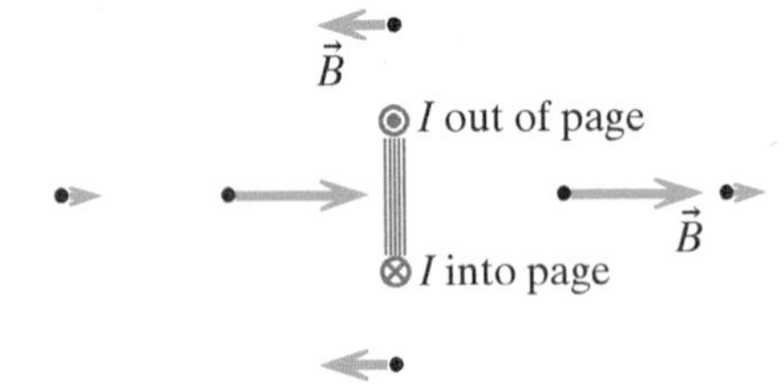

Figure 17.32 The magnetic field above and below the coil points in the opposite direction to the field along the axis.

Magnetic Field at Other Locations Outside the Loop

The magnetic field at other locations outside the loop is more difficult to calculate analytically, but the magnetic field has a characteristic dipole pattern as shown in Figure 17.33, which is the result of a computer calculation that added up all the contributions of many short sections of the loop.

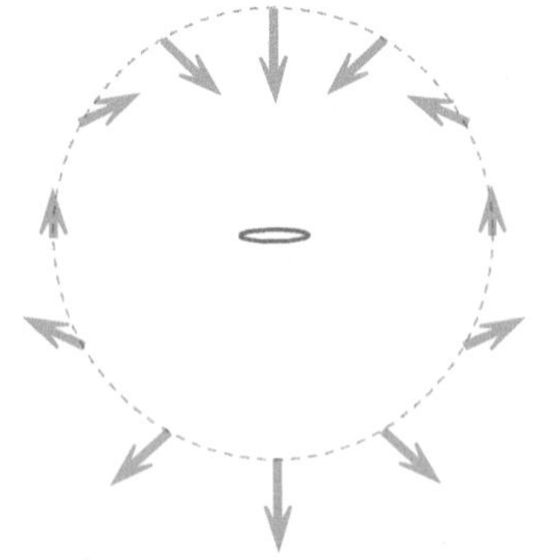

Figure 17.33 The magnetic field of a current loop (which lies in the xz plane, viewed edge-on), at locations outside the loop, in a plane containing the axis of the loop.

A Special Right-Hand Rule for Current Loops

There is another "right-hand rule" that is often used to get the direction of the magnetic field along the axis of a loop. Let the fingers of your right hand curl around in the direction of the conventional current, and your thumb will point in the direction of the magnetic field at any location on the axis.

QUESTION Try using this right-hand rule to determine the direction of the magnetic field at the indicated observation location in Figure 17.34.

You should find that the magnetic field points down. This right-hand rule should of course give the same result as applying the more general right-hand rule to the cross product $\Delta\vec{l}\times\hat{r}$ and adding up the contributions of the various parts of the loop, as called for by the Biot–Savart law.

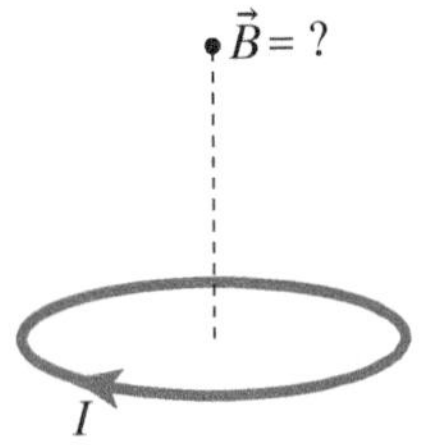

Figure 17.34 Curl the fingers of your right hand in the direction of the conventional current, and your thumb will point in the direction of the magnetic field.

QUESTION On the diagram, consider $\Delta\vec{l}\times\hat{r}$ for two short pieces of the loop, on opposite sides of the loop. Show that the two pieces together contribute a magnetic field in the downward direction above the loop.

Magnetic Field of Thin and Long Coils

A coil of wire with N turns like the one in Figure 17.35 produces a magnetic field that is approximately N times the magnetic field produced by one turn, as long as the coil is "thin"; that is, that all the turns are in nearly the same plane.

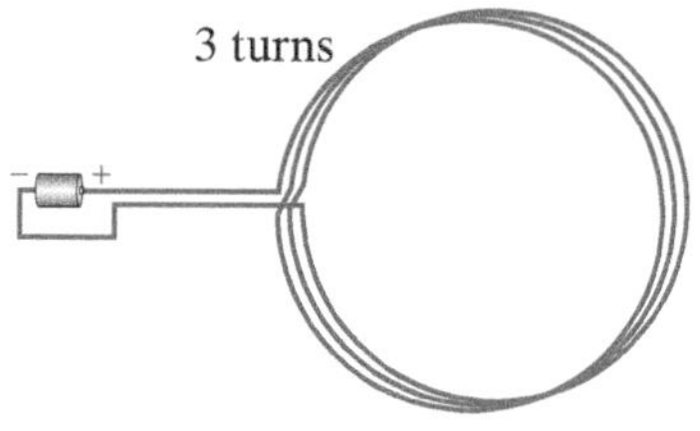

Figure 17.35 A thin coil containing three turns.

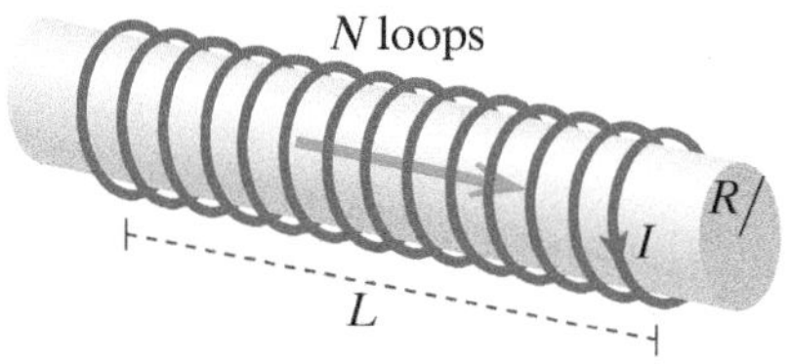

Figure 17.36 A solenoid is a long coil that produces a uniform magnetic field inside.

In contrast, a solenoid is a long coil made by winding a wire around a long tube (Figure 17.36). In Section 17.14 we show that the magnetic field along the centerline of a solenoid of length L has a surprisingly simple form, $B \approx \mu_0 NI/L$, which is nearly the same all along the centerline except near the ends. In Chapter 21 we'll see that the magnetic field is also nearly uniform across the cross section of the solenoid, so a solenoid is useful for producing a uniform magnetic field (constant in magnitude and direction) throughout the interior of the solenoid, not too near the ends.

17.9 COMPUTATION AND 3D VISUALIZATION

Although it is possible to understand something about patterns of electric field by considering only 2-D planar slices of space, this is not the case with magnetic fields. The fact that the vector cross-product occurs in the expression for calculating magnetic field is an indication of how three-dimensional patterns of magnetic field are.

It is easy to write a short VPython program that helps in visualizing both the spatial and temporal variation of the magnetic fields due to moving charged particles. We already have all of the computational tools we'll need to do this.

Recall from Chapter 11 on angular momentum that the VPython syntax for a vector cross product is

```
cross(v,rhat)
```

The organization of a program to calculate and display the magnetic field of a moving particle is straightforward:

- Set up a list of arrows at observation locations.
- Create a `while` loop to move a charged particle.
- For each time step Δt, loop through all observation locations, calculating the magnetic field at each one, and adjusting the axis of the arrow to represent the field.

Assuming that we have done the usual setup, including defining a constant `mzofp` (mu zero over four pi, equal to $\mu_0/4\pi$), and creating a list of arrows called `obs`, with their positions at the desired locations, the structure of the computational loop in such a program might look like this:

```
while t < tmax:
    source.pos = source.pos + v*deltat
    i = 0
    while i < len(obs):
        r = obs[i].pos - source.pos
        rhat = r/mag(r)
        B = mzofp * source.q * cross(v,rhat)/mag(r)**2
        obs[i].axis = sf*B
        i = i + 1
    t = t + deltat
```

Several of the problems at the end of the chapter involve visualizations of this kind, dealing with the magnetic field of a moving charge. A somewhat different approach is needed for calculating the magnetic field generated by current-carrying wires, where $q\vec{v} \times \hat{r}$ is replaced by $I\Delta\vec{l} \times \hat{r}$. The easiest way to approach these situations in the VPython context is to use a `curve` object, whose `pos` attribute is a list of points that are connected by straight lines. The ith vector piece $\Delta\vec{l}$ of a current-carrying wire represented as a `curve` object named `wire` is `dl = wire.pos[i+1] - wire.pos[i]`, and the center of the $\Delta\vec{l}$ vector is `wire.pos[i] + dl/2`. Make sure that the last value of `i` in your `while` loop references the next to the last element in the curve's position list, so that `wire.pos[i+1]` is a legal reference. Here is the basic idea:

```
# Create a curve object to represent a wire loop
wire = curve(radius=R/4, color=color.orange)
N = 40 # N dl's around the loop
dtheta = 2*pi/N
theta = 0
# Include one extra point for end of last dl
while theta <= 2*pi+dtheta/2:
    # Append another point to the curve
    wire.append(pos=R*vector(0, cos(theta), sin(theta)))
    theta = theta + dtheta

# Create a list named obs of observation locations
# ......

# Calculate B at all observation locations
iobs = 0
while iobs < len(obs):
    B = vector(0,0,0)
    i = 0
    while i < N:
        rate(1000)
        dl = wire.pos[i+1] - wire.pos[i]
        loc = wire.pos[i] + dl/2
        r = obs[iobs].pos-loc
        rhat = r/mag(r)
        B = B + mzofp*I*cross(dl,rhat)/mag(r)**2
        i = i + 1
    obs[iobs].axis = scalefactor*B
    iobs = iobs + 1
```

17.10 MAGNETIC DIPOLE MOMENT

Recall the equation for the electric field along the axis of an electric dipole, at a distance r far from the dipole:

$$E_{\text{axis}} \approx \frac{1}{4\pi\varepsilon_0}\frac{2p}{r^3}$$

where the electric dipole moment $p = qs$. Similarly, in the equation for the magnetic field along the axis of a current-carrying coil at a distance r far from the coil, we can write this:

$$B_{\text{axis}} \approx \frac{\mu_0}{4\pi}\frac{2\mu}{r^3}$$

where the magnetic dipole moment $\mu = IA$. (If there are N loops, $\mu = NIA$.) Here A is the area of the loop (πR^2 for circular loops). This equation for magnetic field is approximately valid even if the loop is not circular. The magnetic dipole moment $\vec{\mu}$ is considered to be a vector pointing in the direction of the magnetic field along the axis (Figure 17.37). This means that the direction of the magnetic dipole moment can be obtained by curling the fingers of your right hand in the direction of the conventional current, and your thumb points in the direction of the magnetic dipole moment.

We will see later that the concept of magnetic dipole moment also applies to magnets and provides a way of characterizing the strength of a magnet.

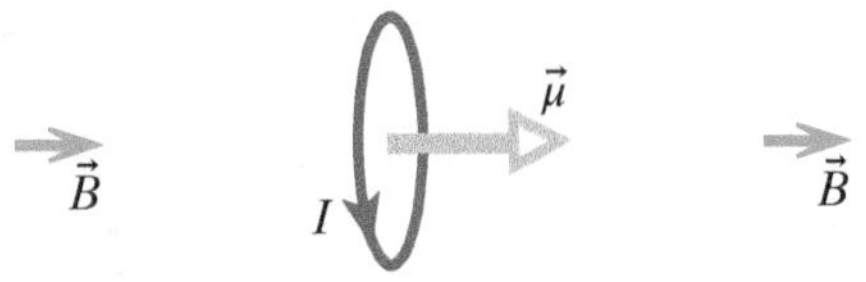

Figure 17.37 The magnetic dipole moment $\vec{\mu}$ is considered to be a vector pointing in the direction of the magnetic field along the axis.

Checkpoint 6 What is the magnetic dipole moment of a 3000-turn rectangular coil that measures 3 cm by 5 cm and carries a current of 2 A?

Twisting of a Magnetic Dipole

For reasons that we will discuss in a later chapter, the magnetic dipole moment vector $\vec{\mu}$ acts just like a compass needle. In an applied magnetic field, a current-carrying loop rotates so as to align the magnetic dipole moment $\vec{\mu}$ with the field. See EXP6.

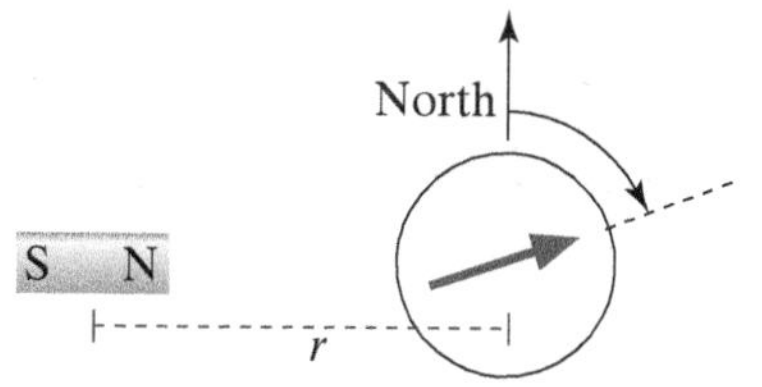

Figure 17.38 A bar magnet affects a compass.

17.11 THE MAGNETIC FIELD OF A BAR MAGNET

From its name, you might guess that a magnet also makes a magnetic field. We say that a magnetic field is present if we see a compass needle twist. If you bring a bar magnet near a compass, you see that a bar magnet does make a compass needle deflect (Figure 17.38). Two magnets interact with each other, attracting or repelling depending on their relative orientations.

Magnetic interactions have some similarity to electric interactions: one can observe both attraction and repulsion, the superposition principle is valid, and the interactions pass through matter. However, there are also some major differences. A bar magnet interacts only with iron or steel objects, whereas a charged invisible tape interacts with *all* objects. A magnet can be permanently magnetic, but a charged invisible tape or plastic pen loses its charge within a relatively short time. Two negatively charged objects always repel each other, whereas two magnets may repel or attract, depending on their orientations. The magnetic field around a current-carrying wire has a "curly" pattern that we don't see with electric fields. See EXP7 and EXP8.

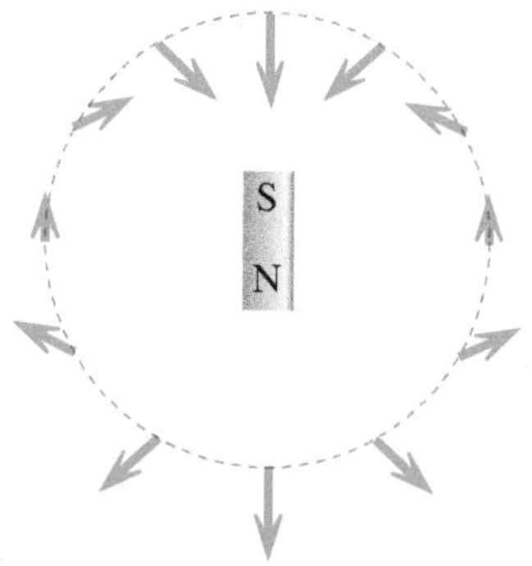

Figure 17.39 A bar magnet is a magnetic dipole. The field pattern is that of a dipole, and the magnitude of the field is proportional to $1/r^3$.

A Bar Magnet Is a Magnetic Dipole

The pattern of directions of magnetic field around a bar magnet is very similar to the pattern of directions of the magnetic field around a current loop, and to the pattern of the electric field around a permanent electric dipole seen in Chapter 13 (Figure 17.39). Also, along the axis of a magnet or a current loop or a permanent electric dipole, the magnetic or electric field varies like $1/r^3$ as can be seen in Experiment EXP9. Moreover, the magnitude of the field off to the side of an electric dipole or a magnet is half as large as it is at the same distance along the axis. Because of these strong similarities, a magnet is often called a "magnetic dipole." In Problem P59 you can use experimental data to determine the magnetic dipole moment of a bar magnet. See EXP9–EXP11.

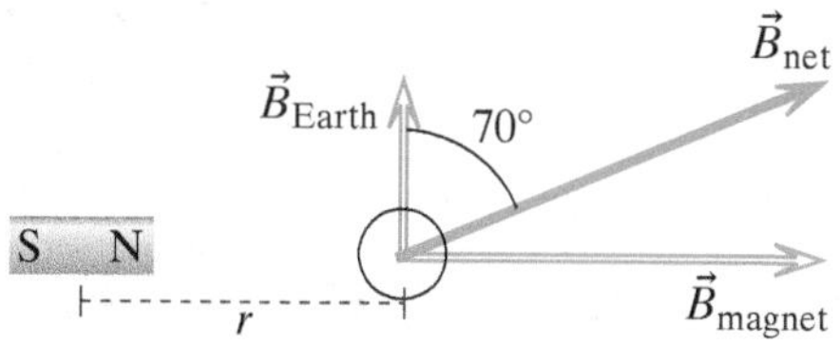

Figure 17.40 Determining the magnetic dipole moment of a bar magnet.

EXAMPLE **Determining the Magnetic Dipole Moment of a Magnet**

A compass originally points north. A bar magnet whose mass is 69.5 g is aligned east–west and placed near a compass as shown in Figure 17.40. When the distance between the center of the magnet and the center of the compass is 23.3 cm, the compass deflects 70°. What is the magnetic dipole moment of the bar magnet?

Solution

$$\vec{B}_{\text{net}} = \vec{B}_{\text{Earth}} + \vec{B}_{\text{magnet}}$$
$$B_{\text{magnet}} = B_{\text{Earth}} \tan 70°$$

$$B_{\text{magnet}} = (2 \times 10^{-5}\,\text{T}) \tan 70° = 5.5 \times 10^{-5}\,\text{T}$$

$$B_{\text{magnet}} = \frac{\mu_0}{4\pi} \frac{2\mu}{r^3}$$

$$\mu = \frac{B_m r^3}{2\frac{\mu_0}{4\pi}} = \frac{(5.5 \times 10^{-5}\,\text{T})(0.233\,\text{m})^3}{2\left(1 \times 10^{-7} \frac{\text{T} \cdot \text{m}}{\text{A}}\right)} = 3.5\,\text{A} \cdot \text{m}^2$$

The Magnetic Field of the Earth

The magnetic field of the Earth has a pattern that looks like that of a bar magnet (Figure 17.41). In the Northern Hemisphere, the Earth's magnetic field dips downward toward the Earth. A horizontally held compass is affected only by the horizontal component of the magnetic field, and this horizontal component gets smaller as you move closer to the magnetic poles.

For example, the horizontal component is smaller in Canada than it is in Mexico, despite the fact that the magnitude of the magnetic field is larger in Canada. The horizontal component is zero right at the magnetic poles, where the magnitude of the magnetic field is largest.

The Earth is a big magnet, but its type S pole is in the Arctic, and the N end of a compass needle points toward this type S pole.

The Earth's magnetic poles are not located exactly at the geographic poles. The S pole is located in northern Canada, about 1300 km (800 mi) from the geographic North Pole, and the N pole is on the Antarctic continent, about 1300 km from the geographic South Pole.

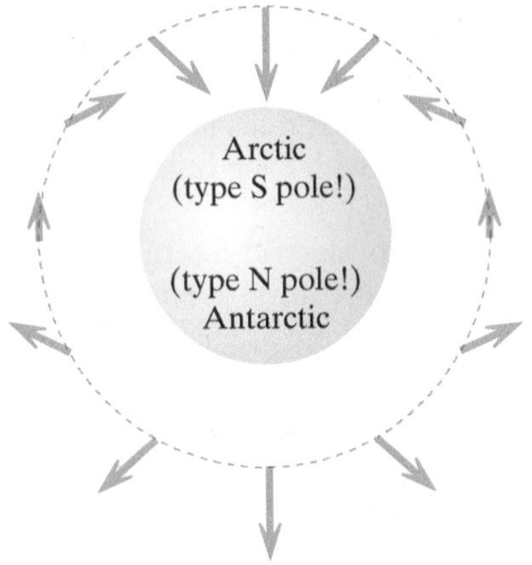

Figure 17.41 The Earth acts like a magnet. Its type N pole is in the Antarctic, and its type S pole is in the Arctic.

Dependence on Distance

The horizontal component of the Earth's magnetic field, which is the component that affects a horizontally held compass, is different at different latitudes, depending on the distance from the magnetic poles of the Earth. Here are measurements of the magnitude of the horizontal component of the Earth's magnetic field at a few selected locations in the United States:

Location	Horizontal Component of Earth's Magnetic Field
Maine	about 1.5×10^{-5} T
Much of the United States	about 2×10^{-5} T
Florida, Hawaii	about 3×10^{-5} T

Knowing the horizontal component of the Earth's magnetic field, we can use the deflection of the compass needle to measure the magnitude of the magnetic field of the magnet. We can then study both the distance dependence and the directional pattern of the magnetic field of the magnet.

Figure 17.42 No magnetic monopoles. This pattern of outward-going magnetic field has never been observed.

Magnetic Monopoles

There is one dramatic difference between electric and magnetic dipoles. The individual positive and negative electric charges (monopoles) making up an electric dipole can be separated from each other, and these point charges make outward-going or inward-going electric fields (which vary like $1/r^2$). One might expect a magnetic dipole to be made of positive and negative magnetic monopoles, but no one has ever found an individual magnetic monopole. Such a magnetic monopole would presumably make an outward-going or inward-going magnetic field (which would vary like $1/r^2$), but such a pattern of magnetic field has never been observed.

Figure 17.42 is an example of the sort of magnetic field pattern that many scientists have diligently looked for but not found. If you cut a magnet in two, you don't get two magnetic monopoles—you just get two magnets!

EXAMPLE

A Circuit in the Antarctic

In a research station in the Antarctic, a circuit containing a partial loop of wire, of radius 5 cm, lies on a table. A top view of the circuit (looking down on the table) is shown in Figure 17.43. A current of 5 A runs in the circuit in the direction shown, and other portions of the circuit are far away. You have a bar magnet with magnetic dipole moment 1.2 $A \cdot m^2$. How far above location A, at the center of the loop, would you have to hold the bar magnet, and in what orientation, so that the net magnetic field at location A would be zero? The magnitude of the Earth's magnetic field in the Antarctic is about 6×10^{-5} T.

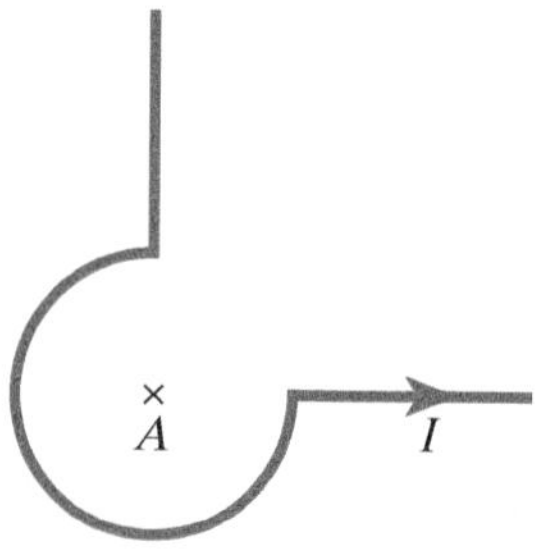

Figure 17.43 Top view of a circuit lying on a table (looking down at the table).

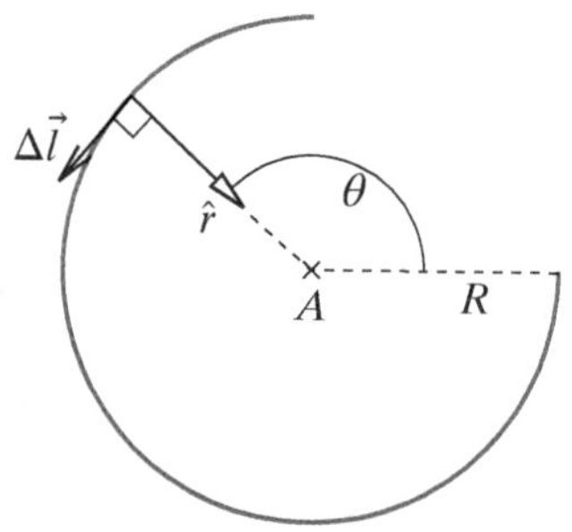

Figure 17.44 $d\vec{l} \perp \hat{r}$ for every step around the loop. The magnitude of r is constant and equal to the radius of the loop.

Solution

Orientation:

$\vec{B}_{\text{Earth}}$ points out of the page (out of the ground, in the Antarctic).

$\vec{B}_{\text{circuit}}$ points out of the page (out of the table).

The bar magnet must be held with its north pole downward (into page), to make a magnetic field into the ground.

Components of magnetic field out of page at location A:

$B_{\text{Earth}} = 6 \times 10^{-5}\,\text{T}$

$B_{\text{straight wires}} = 0$ (The straight wires contribute nothing to the magnetic field at A, because for each wire $d\vec{l} \times \hat{r} = 0$.)

B due to 3/4 loop: $\vec{B} = \int d\vec{B} = \int \frac{\mu_0}{4\pi} \frac{I d\vec{l} \times \hat{r}}{r^2}$

Origin: Center of the loop

$|\vec{r}| = R$ (constant)

At all locations $d\vec{l} \perp \hat{r}$, so $|d\vec{l} \times \hat{r}| = dl \sin 90^\circ = dl = R d\theta$, as shown in Figure 17.44. Therefore

$$|\vec{B}_{\text{loop}}| = \int_{\pi/2}^{2\pi} \frac{\mu_0}{4\pi} \frac{I(R d\theta)}{R^2} = \frac{\mu_0}{4\pi} \frac{IR}{R^2} \int_{\pi/2}^{2\pi} d\theta = \frac{\mu_0}{4\pi} \frac{I}{R} \theta \Big|_{\pi/2}^{2\pi} = \frac{\mu_0}{4\pi} \frac{3\pi I}{2R}$$

$$|\vec{B}_{\text{loop}}| = \left(1 \times 10^{-7} \frac{\text{T} \cdot \text{m}}{\text{A}}\right) \frac{(3\pi)(5\,\text{A})}{2(0.05\,\text{m})} = 4.7 \times 10^{-5}\,\text{T}$$

$$B_{\text{Earth}} + B_{3/4\ \text{loop}} = 6 \times 10^{-5}\,\text{T} + 4.7 \times 10^{-5}\,\text{T} = 1.06 \times 10^{-4}\,\text{T}$$

We need

$$B_{\text{magnet}} = 1.06 \times 10^{-4}\,\text{T} \approx \frac{\mu_0}{4\pi} \frac{2\mu}{z^3}$$

$$z = \left[\frac{\frac{\mu_0}{4\pi}(2\mu)}{B_{\text{magnet}}}\right]^{1/3} = \left[\frac{\left(1 \times 10^{-7} \frac{\text{T} \cdot \text{m}}{\text{A}}\right)(2)(1.2\,\text{A} \cdot \text{m}^2)}{(1.1 \times 10^{-4}\,\text{T})}\right]^{1/3} = 0.13\,\text{m}$$

Thus we hold the magnet with its center about 13 cm above the circuit, with its north pole toward the ground. (We model the magnet as a dipole, located at the center of the magnet.)

17.12 THE ATOMIC STRUCTURE OF MAGNETS

We have seen that the magnetic field of a circular current loop looks suspiciously similar to the magnetic field of a magnet. The pattern of directions of the magnetic field looks the same, and it is even the case that for both magnets and current loops the magnitude of the field at large distances along the axis falls off like $1/r^3$. Both magnets and current loops can be described in terms of a magnetic dipole moment $\vec{\mu}$.

A Single Atom Can Be a Magnetic Dipole

Inside every atom there are moving charged particles. In an atom in a magnetic material, these subatomic "currents" may produce tiny magnetic dipole moments that add up to a nonzero magnetic dipole moment for each atom. (In contrast, in a nonmagnetic material, these tiny magnetic dipole moments would add up to zero.)

There are three possibilities for the source of the subatomic currents that might give a single atom a nonzero magnetic dipole moment:

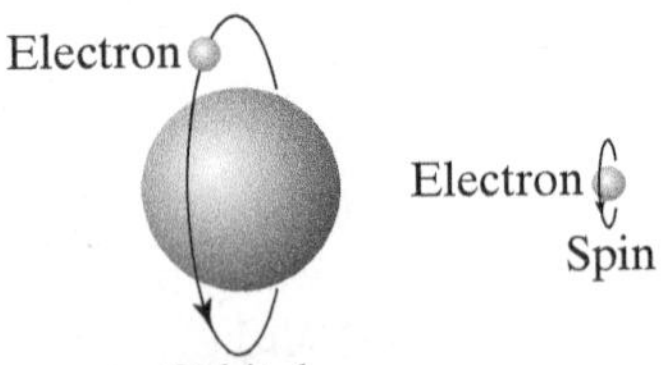

Figure 17.45 Electrons have spin as well as orbital motion that can contribute to the magnetic dipole moment.

- An electron orbiting around the nucleus (Figure 17.45)
- An electron spinning on its own axis (Figure 17.45)
- Rotational motion in the nucleus (protons and neutrons have spin about their own axes and can also orbit around inside the nucleus)

In all of these situations the object has angular momentum, and the magnetic dipole moment turns out to be proportional to the angular momentum L of the particle:

$$\mu = (\text{factor})L$$

We can use a simple model to estimate this proportionality factor.

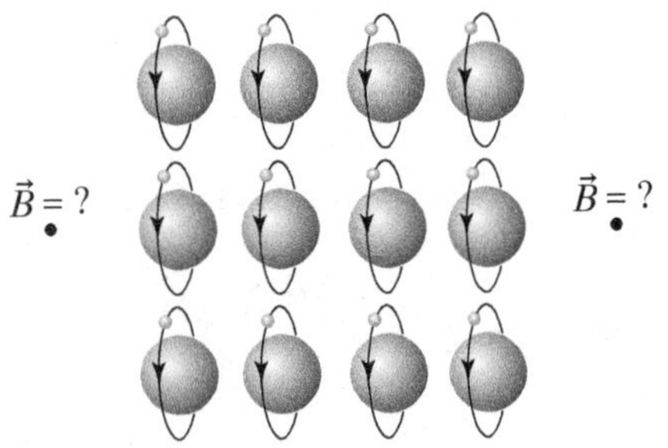

Figure 17.46 A simple model for a magnet might involve electrons in circular atomic orbits.

Magnetic Dipole Moment of an Orbiting Electron

It is easiest to estimate this proportionality factor by considering the magnetic dipole moment of an orbiting electron, using the simple Bohr model of an atom to get an order of magnitude estimate. As indicated in Figure 17.46, let's suppose that each atom has one unpaired electron that orbits the nucleus along a circular path, at constant speed. Each orbiting electron can be considered to be a tiny current loop that makes a dipole magnetic field.

> QUESTION In Figure 17.46, what is the direction of the resulting magnetic field to the left and right of the atoms, and of the magnetic dipole moment of one atom? (Remember that electrons are negative.)

The magnetic field due to each orbiting electron would point to the left, so the magnetic dipole moment of each atom, and of the whole object, would point to the left.

For a current loop, we know that

$$\mu = I(\pi R^2)$$

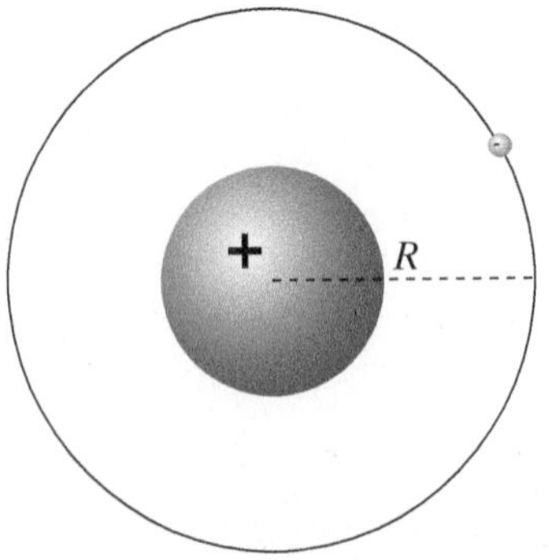

Figure 17.47 Simple model of an atom in which an outer electron orbits a positive inner core.

> QUESTION What would be the current I for one orbiting electron (Figure 17.47)?

The current I is a measure of how much charge passes one location per second. The charge on the electron is $-e$, and the time it takes the electron to go around once is $T = 2\pi R/v$, where v is the speed of the electron, so

$$I = \frac{e}{\left(\dfrac{2\pi R}{v}\right)} = \frac{ev}{2\pi R}$$

Therefore the magnetic dipole moment of a single orbiting electron would be

$$\mu = I(\pi R^2) = \left(\frac{ev}{2\pi R}\right)(\pi R^2) = \frac{1}{2}eRv$$

Relating Magnetic Dipole Moment to Angular Momentum

The magnitude of the angular momentum of an electron in a circular orbit is

$$|\vec{L}| = |\vec{r} \times \vec{p}|$$

In a circular orbit $\vec{r} \perp \vec{p}$, so this reduces to

$$L = Rp\sin 90° = Rmv \quad (v \ll c)$$

We can express the magnetic dipole moment of the orbiting electron in terms of angular momentum by multiplying by 1:

$$\mu = \left(\frac{m}{m}\right)\left(\frac{1}{2}eRv\right) = \frac{1}{2}\frac{e}{m}(Rmv) = \frac{1}{2}\frac{e}{m}L$$

Our estimate of the proportionality factor relating magnetic dipole moment to angular momentum for a particle in an atom is this:

$$(\text{factor}) \approx \frac{1}{2}\frac{e}{m}$$

We will assume that this expression for the proportionality factor is valid for orbiting electrons, electrons spinning on their axes, and nucleons (protons and neutrons) spinning on their axes. How big is this factor?

$$\text{For an electron: } \frac{1}{2}\frac{e}{m} = \frac{1}{2}\frac{(1.6\times10^{-19}\,\text{C})}{(9\times10^{-31}\,\text{kg})} = 8.9\times10^{10}\,\text{C/kg}$$

$$\text{For a nucleon: } \frac{1}{2}\frac{e}{m} = \frac{1}{2}\frac{(1.6\times10^{-19}\,\text{C})}{(1.7\times10^{-27}\,\text{kg})} = 4.7\times10^{7}\,\text{C/kg}$$

This factor is nearly 2000 times smaller for a nucleon. Angular momentum is quantized, with similar values for electrons and nucleons, so the magnetic dipole moment of a single atom comes almost entirely from orbiting electrons or electrons spinning on their own axes.

Angular Momentum of an Orbiting or Spinning Electron

You may recall from Chapter 11 that both translational angular momentum (like the orbital angular momentum of an electron) and rotational angular momentum (like the angular momentum of an electron spinning on its axis) are quantized in atoms; only certain discrete values are observed. For an electron orbiting a nucleus, the angular momentum has to be an integer multiple of Planck's constant h divided by 2π, written as $\hbar$:

$$L = N\hbar$$

where $N = 0, 1, 2, 3, \ldots$, and $\hbar = 1.05\times10^{-34}\,\text{J}\cdot\text{s}$.

The electron's spin angular momentum is always one-half of $\hbar$, but there is a compensating factor of 2 in its relation to the magnetic dipole moment. (If you are not familiar with angular momentum quantization, you may wish to read Section 17.13, which uses more familiar classical reasoning about circular motion to estimate the orbital angular momentum of an electron. The answer is the same.)

Assuming one quantum of angular momentum, we have the following:

$$L = \hbar \quad \text{so} \quad \mu \approx \frac{1}{2}\frac{e}{m}\hbar$$

Evaluating this expression, we find that:

$$\mu \approx \left(\frac{1}{2}\right)\frac{(1.6\times 10^{-19}\,\text{C})}{(9\times 10^{-31}\,\text{kg})}(1.05\times 10^{-34}\,\text{J}\cdot\text{s})$$

$$\mu \approx 1\times 10^{-23}\,\text{A}\cdot\text{m}^2 \text{ per atom}$$

Comparing with Experiment

If you have a bar magnet, a compass, and a meter stick, you can determine the magnetic dipole moment of the bar magnet experimentally. The procedure for doing this is detailed in Problem P59.

In an earlier example we found that a particular magnet whose mass was 69.5 g has a magnetic dipole moment of 3.5 $\text{A}\cdot\text{m}^2$. Assume that the magnetic dipole moment of this magnet is due to the fact that each atom has one unpaired electron contributing a magnetic dipole moment of $\mu \approx 1\times 10^{-23}\ \text{A}\cdot\text{m}^2$ (due either to orbital or spin angular momentum) and that the magnetic dipole moments of all the atoms are aligned in the same direction. What would be the predicted magnetic dipole moment of this bar magnet?

Estimated Magnetic Dipole Moment of a Bar Magnet

The bar magnet in an earlier example had a mass of 69.5 g. Assume that almost all of the atoms in this magnet are iron atoms. The mass of one mole of iron is 56 g.

$$N = \left(\frac{69.5\,\text{g}}{56\,\text{g/mol}}\right)(6.02\times 10^{23}\,\text{atoms/mol}) = 7.5\times 10^{23}\,\text{atoms}$$

$$\mu = N\mu_{\text{atom}} = (7.5\times 10^{23}\,\text{atoms})\left(1\times 10^{-23}\frac{\text{A}\cdot\text{m}^2}{\text{atom}}\right) = 7.5\,\text{A}\cdot\text{m}^2$$

QUESTION The estimated value of the magnetic dipole moment of this magnet is 7.5 $\text{A}\cdot\text{m}^2$, but the observed value is only 3.5 $\text{A}\cdot\text{m}^2$. Is this good agreement or poor agreement?

This is excellent agreement! The values differ by a factor of only about 2. The fact that the prediction and measurement are this close is evidence that the basic assumptions of our very simple model—that each atom in a magnet is itself a small dipole, and that the field produced by the magnet is the sum of the fields of all the atoms—are reasonable assumptions. If the predicted and measured values had differed by a factor of a thousand or a million, we would have had to re-examine our model. A related problem is Problem P59.

Simplified Models

We made a number of simplifying assumptions in our model of an atomic dipole. You probably know from chemistry courses that more sophisticated atomic models do not assume that electrons orbit the nucleus in well-defined circular paths. In these models, there is only a probability for finding the electron at a particular place. Moreover, many electrons in atoms have spherically symmetric probability distributions (s orbitals), and such a symmetric distribution has a zero average angular momentum. However, electrons in nonspherically symmetric probability distributions (p, d, or f orbitals) have nonzero angular momentum, and can contribute a nonzero magnetic dipole moment. Additionally, we assumed that in every atom there is just one electron that contributed a magnetic field, but in some materials two or more unpaired electrons may contribute. Moreover, the "spinning ball" model of the electron isn't really adequate, because the most sensitive experiments are consistent with the notion that an electron is a true point charge and has zero radius! It is nevertheless an experimental fact that an electron does make a magnetic field as though it were a spinning ball of charge, and it can be useful to think of it literally as a spinning ball.

Nonetheless, the fact that our simplified model gave a prediction with an appropriate order of magnitude shows the power of reasoning with simple models.

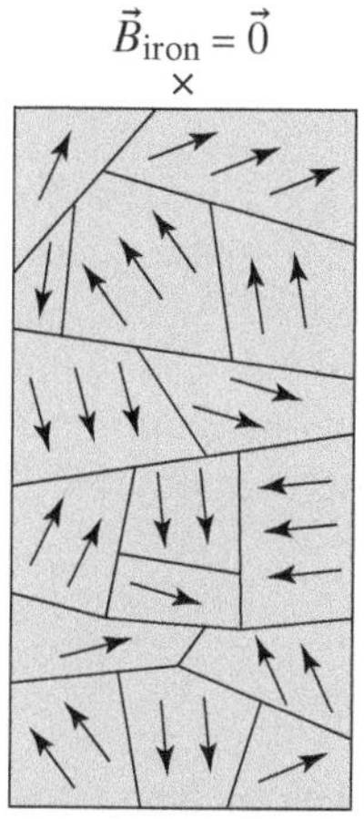

Figure 17.48 Disordered magnetic domains—the net magnetic field produced by the iron is nearly zero.

The Modern Theory of Magnets

Modern theories of the nature of magnets take into account not only the contribution of individual atoms in a solid magnet, but also the effects of their interactions with each other. These more complex theoretical treatments suggest that the magnetic dipole moments due to the rotational angular momentum of spinning electrons contribute most of the magnetic field of a magnet, although the orbital angular momentum of electrons does make some contribution.

As we saw above, because the factor e/m is so much smaller for a proton than for an electron, we can ignore nuclear contributions to the magnetic field of your bar magnet. Nuclear magnetic dipole moments do play an important role in the phenomenon of nuclear magnetic resonance (NMR) and in the technology of magnetic resonance imaging (MRI) used in medicine, which is based on NMR.

Alignment of the Atomic Magnetic Dipole Moments

In many materials, an atom has no net orbital or spin magnetism. In most materials whose individual atoms do make a magnetic field, the orbital and spin motions in different atoms don't line up with each other, so the net field of the many atoms in a piece of the material averages out to zero. However, in a few materials, notably iron, nickel, cobalt, and some alloys of these elements, the orbital and spin motions in neighboring atoms line up with each other and therefore produce a sizable magnetic field. These unusual materials are called "ferromagnetic." The reason for the alignment can be adequately discussed only within the framework of quantum mechanics; basically the alignment is due to electric interactions between the atoms, not to the much weaker magnetic interactions.

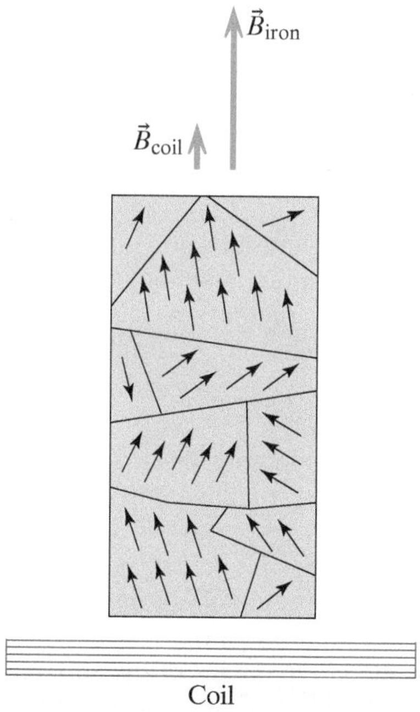

Figure 17.49 The magnetic field of a coil causes a partial ordering of the magnetic domains, which produces a significant nonzero magnetic field contributed by the iron.

Magnetic Domains

In an ordinary piece of iron that isn't a magnet, the iron is a patchwork of small regions, called "magnetic domains," within which the alignment of all the atomic magnetic dipole moments is nearly perfect. Normally, though, these domains are oriented in random directions, so the net effect is that this piece of iron doesn't produce a significant net magnetic field. Figure 17.48 shows a picture of the situation, in which the arrows indicate the magnetic dipole moments of the atoms.

If you use a coil to apply a large magnetic field to the iron, domains that happen to be nearly aligned with the applied field tend to grow at the expense of domains that have a different orientation (Figure 17.49). Also, with a sufficiently large applied magnetic field, the magnetic dipole moments of the atoms in a domain tend to rotate toward aligning with the applied field, just as a compass needle turns to align with an applied field. The result of both of these effects is to partially align the magnetic dipole moments of most of the atoms, which produces a magnetic field $\vec{B}_{\text{iron}}$ that may be much larger than the applied magnetic field. There is a kind of multiplier effect.

In the case of very pure iron, when you turn off the current in the coil, the piece of iron goes back nearly to its original disordered patchwork of domains and doesn't act like a magnet any more. In some other ferromagnetic materials, though, including some alloys of iron, when you remove the applied field the domains remain nearly aligned. In that case you end up with a permanent magnet, which is commonly made of the alloy alnico (alnico V contains 51% iron, 8% aluminum, 14% nickel, 24% cobalt, and 3% copper). Hitting a magnet with a hard blow may disorder the domains and make the metal no longer act like a magnet. Heating above a critical temperature also destroys the alignments.

Iron Inside a Coil

It is possible to observe this multiplier effect. A small magnetic field created by current-carrying loops of wire wrapped around a piece of iron can lead to a large observed magnetic field being contributed by the iron, due to the alignment of the magnetic dipole moments in the iron. Thanks to this multiplier effect, putting an iron core inside a current-carrying coil of wire makes a powerful "electromagnet" that can pick up iron objects. See EXP12 and EXP13.

Why Are There Multiple Domains?

Within one domain of a ferromagnetic material such as iron, strong electric forces between neighboring atoms make the atomic magnetic dipole moments line up with each other. Why don't all the atoms in a piece of iron spontaneously align with each other? Why does the piece divide into small magnetic domains with varying magnetic orientations in the absence of an applied magnetic field?

There is a simple way to get some insight into why this happens. Consider two possible patterns of magnetic domains in a bar of iron in the absence of an applied magnetic field, a single domain or two opposed domains (Figure 17.50).

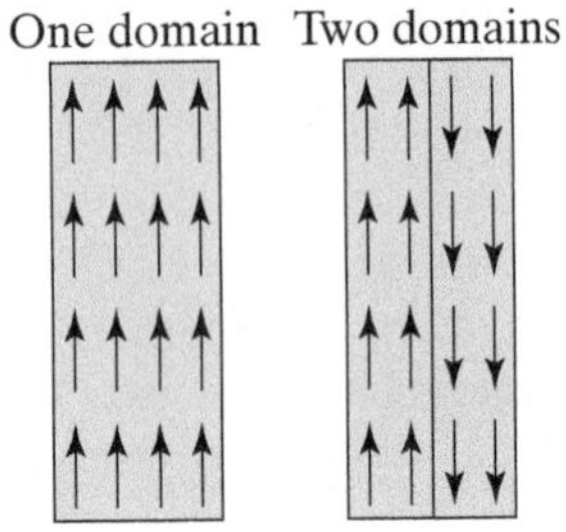

Figure 17.50 Possible arrangements—one magnetic domain or two.

You can simulate this situation by taking two permanently magnetized bar magnets and holding them in one of these two positions, as shown in Figure 17.51.

QUESTION Which of these two magnet configurations do you find to be more stable, the parallel or the antiparallel alignment? By analogy, which is more likely for a bar of iron: that it have just one domain or that it split into two domains in the absence of an applied magnetic field?

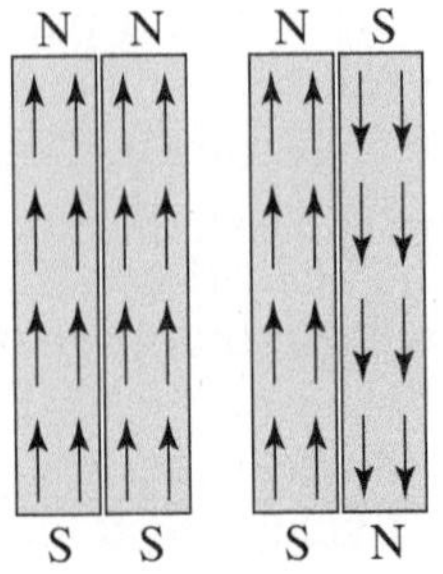

Figure 17.51 Arrange two bar magnets side by side, aligned parallel or antiparallel.

You will find that if you start with the parallel alignment the magnets tend to flip into the antiparallel alignment. The domain structure of iron can be a compromise between a tendency for neighboring atoms to have their magnetic dipole moments line up with each other (a strong but short-range electric interaction), and a competing tendency for magnetic dipole moments to flip each other (a weaker but longer-range magnetic interaction).

However, the net effect depends critically on the details of the geometric arrangement. Although two long bar magnets that are side by side tend to line up with their magnetic dipole moments in opposite directions, two thin disk-shaped magnets tend to line up with their magnetic dipole moments in the same direction. What happens inside a magnetic material depends on the details of the geometrical arrangement of the atoms and on the details of the electric interactions between neighboring atoms.

17.13 *ESTIMATE OF ORBITAL ANGULAR MOMENTUM OF AN ELECTRON IN AN ATOM

We can estimate the orbital angular momentum of an electron in an atom by using the simple Bohr model of an atom, in which an electron orbits a nucleus at constant speed in a circular orbit, and applying the Momentum Principle. In this model (Figure 17.52), an outer electron is held in a circular orbit of radius R by the electric attraction to the rest of the atom (which has a net charge of $+e$).

We want to find the magnitude of the angular momentum of the electron:

$$|\vec{L}| = |\vec{r} \times \vec{p}|$$

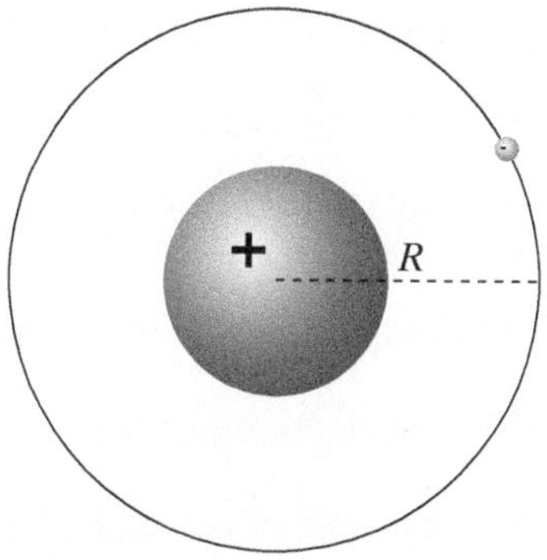

Figure 17.52 Simple model of an atom in which an outer electron orbits a positive inner core.

Since in a circular orbit $\vec{r} \perp \vec{p}$, this reduces to:

$$L = Rp = Rmv \quad (v \ll c)$$

We know the radius of an atom (about 1×10^{-10} m) and the mass of an electron (9×10^{-31} kg), so we need to find the speed of the electron.
System: Electron, charge $-e$
Surroundings: Nucleus and outer electrons, with charge $+e$
Momentum Principle:

$$\frac{d\vec{p}}{dt} = \vec{F}_{\text{net}}$$

Parallel component (constant speed, so the parallel component of momentum is not changing):

$$\frac{dp}{dt}\hat{p} = \vec{0}$$

Perpendicular component (electric force on electron is toward the nucleus, perpendicular to momentum):

$$p\left|\frac{d\hat{p}}{dt}\right| = F\perp$$

$$(mv)\left(\frac{v}{R}\right) = e\left(\frac{1}{4\pi\varepsilon_0}\frac{e}{R^2}\right)$$

$$\frac{mv^2}{R} = \frac{1}{4\pi\varepsilon_0}\frac{e^2}{R^2}$$

$$v = \sqrt{\frac{1}{4\pi\varepsilon_0}\frac{e^2}{Rm}}$$

Plugging in the electron charge and mass, and assuming an approximate atomic radius of 1×10^{-10} m, we get:

$$v = \sqrt{\frac{(9 \times 10^9\ \text{N}\cdot\text{m}^2/\text{C}^2)(1.6 \times 10^{-19}\ \text{C})^2}{(1 \times 10^{-10}\ \text{m})(9 \times 10^{-31}\ \text{kg})}}$$

$$v \approx 1.6 \times 10^6\ \text{m/s}$$

This gives a value for angular momentum of

$$L = Rmv = (1\times10^{-10}\ \text{m})(9\times10^{-31}\ \text{kg})(1.6\times10^{6}\ \text{m/s})$$
$$L \approx 1.4\times10^{-34}\,\text{kg}\cdot\text{m}^2/\text{s}^2$$

This is close to the quantum mechanical value of $\hbar = 1.05\times10^{-34}$ J · s. If we had reproduced Bohr's full calculation we would have obtained a slightly different value for R, and L would be exactly $\hbar$.

17.14 *MAGNETIC FIELD OF A SOLENOID

The analysis in this section offers one more example of how to apply the Biot–Savart law. It shows that the magnetic field is nearly uniform along the axis of a solenoid, far from the ends. A solenoid is a coil that is much longer than its radius.

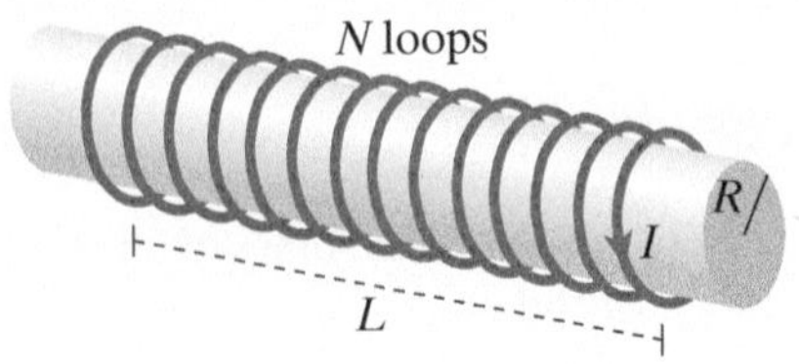

Figure 17.53 A solenoid of length L with N loops of radius R carrying current I.

Step 1: Cut Up the Distribution into Pieces and Draw $\Delta\vec{B}$

We consider a solenoid of length L that is made up of N circular loops wound tightly right next to each other, each of radius R (Figure 17.53). We consider each loop as one piece. The conventional current in the loops is I.

QUESTION Given what you know about individual current loops, what will be the direction of the magnetic field $\Delta\vec{B}$ contributed by each of the loops at *any* location along the axis of the solenoid in the figure above?

All of the contributions inside the solenoid will point to the right in the diagram.

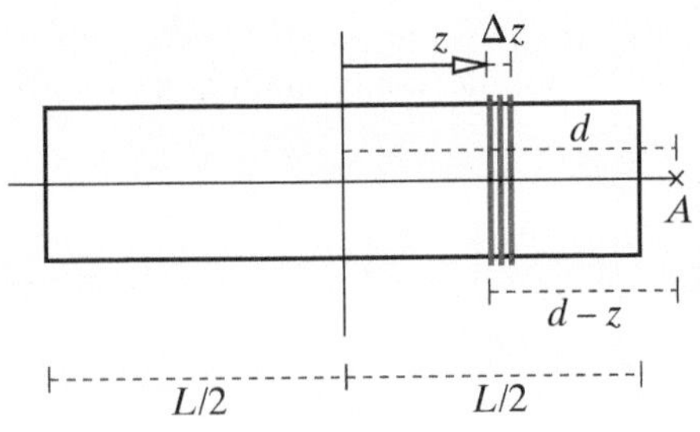

Figure 17.54 Defining an integration variable z for summing the contributions of the many loops.

Step 2: Write an Expression for the Magnetic Field Due to One Piece

Origin: Center of solenoid
Location of one piece: Given by z, so the integration variable is z; the distance from the loop to the observation location A is $d - z$ (Figure 17.54). Our result will be valid for any location along the axis, inside or outside the solenoid.

QUESTION How many closely packed loops are contained in a short length Δz of the solenoid?

There are N/L loops per meter, so the number of loops in a length Δz is $(N/L)\Delta z$. We know the magnetic field made by each loop along its axis.

QUESTION Show that the loops contained in the section Δz of the solenoid in Figure 17.54 contribute a magnetic field at the observation location in the z direction of this amount:

$$\Delta B_z = \frac{\mu_0}{4\pi}\frac{2\pi R^2 I}{[(d-z)^2+R^2]^{3/2}}\frac{N}{L}\Delta z$$

Step 3: Add Up the Contributions of All the Pieces

The net magnetic field lies along the axis and is the summation of all the ΔB_z contributed by all the loops:

$$B_z = \sum \Delta B_z = \sum \frac{\mu_0}{4\pi}\frac{2\pi R^2 IN}{L}\frac{\Delta z}{[(d-z)^2+R^2]^{3/2}}$$

Many of these quantities are the same for every piece and can be taken outside the summation as common multiplicative factors:

$$B_z = \frac{\mu_0}{4\pi}\frac{2\pi R^2 IN}{L}\sum \frac{\Delta z}{[(d-z)^2+R^2]^{3/2}}$$

This can be turned into an integral, with z ranging from $-L/2$ to $+L/2$:

$$B_z = \frac{\mu_0}{4\pi}\frac{2\pi R^2 IN}{L}\int_{z=-L/2}^{z=+L/2} \frac{dz}{[(d-z)^2+R^2]^{3/2}}$$

Let $u = d - z$, in which case $dz = -du$, and therefore the limits on the integration run from $u = d-(-L/2)$ to $u = d-(+L/2)$:

$$B_z = \frac{\mu_0}{4\pi}\frac{2\pi R^2 IN}{L}\int_{u=d+L/2}^{u=d-L/2} \frac{-du}{[u^2+R^2]^{3/2}}$$

$$= \frac{\mu_0}{4\pi}\frac{2\pi R^2 IN}{L}\int_{u=d-L/2}^{u=d+L/2} \frac{du}{[u^2+R^2]^{3/2}}$$

This integral can be found in standard tables of integrals:

$$B_z = \frac{\mu_0}{4\pi}\frac{2\pi R^2 IN}{L}\left[\frac{u}{R^2\sqrt{u^2+R^2}}\right]_{u=d-L/2}^{u=d+L/2}$$

$$B_z = \frac{\mu_0}{4\pi}\frac{2\pi IN}{L}\left[\frac{d+L/2}{\sqrt{(d+L/2)^2+R^2}} - \frac{d-L/2}{\sqrt{(d-L/2)^2+R^2}}\right]$$

In this expression, d is the distance from the center of the solenoid.

Figure 17.55 is a plot of this expression for B_z for a particular solenoid, as a function of the distance from the center of the solenoid. Notice that the magnetic field is nearly uniform inside the solenoid, as long as you are not too near the ends. The field is even more uniform for solenoids that are longer and/or thinner than this one.

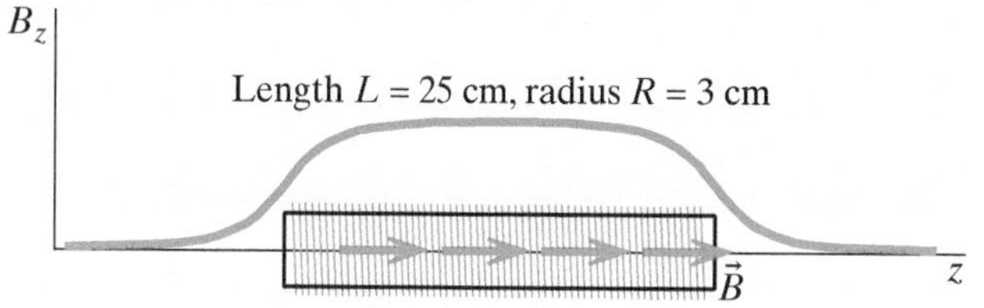

Figure 17.55

QUESTION If the radius R is small compared with the length L ($R \ll L$), show that at the center of the solenoid the magnetic field has a remarkably simple value:

$$B_z \approx \frac{\mu_0 NI}{L}$$

With $d = 0$ and $R \ll L$, the quantity in square brackets above reduces to

$$\frac{L/2}{\sqrt{(L/2)^2}} - \frac{-L/2}{\sqrt{(-L/2)^2}}$$

which is equal to 2, from which follows the simple form $B_z \approx \mu_0 NI/L$.

The function plotted above shows that this is also the approximate magnetic field inside the solenoid along much of the axis, not too close to the ends. Outside the solenoid, the magnetic field falls off rapidly. At a large distance the field falls off like $1/r^3$, since the solenoid then looks like a simple collection of current loops.

MAGNETIC FIELD INSIDE A LONG SOLENOID

$$B_z \approx \frac{\mu_0 NI}{L}$$

inside a long solenoid, with the radius of the solenoid $\ll L$.

The same result can be obtained with much less effort by using Ampere's law, which we will study in Chapter 21, where we will also show that the magnetic field is uniform across the cross section as well as along the centerline (not too near the ends).

Step 4: Check the Result

Does our final result make sense? In particular, do we have the right units? Comparing with the magnetic field for a single current element,

$$\frac{\mu_0}{4\pi}\frac{I\Delta\vec{l}\times\hat{r}}{r^2}$$

we easily verify that our answer does have the right units, since

$$\frac{NI}{L} \quad \text{has the same units as} \quad \frac{I\Delta l}{r^2}$$

(Remember that $\hat{r}$ is dimensionless.)

SUMMARY

The Biot–Savart Law

$$\vec{B} = \frac{\mu_0}{4\pi}\frac{q\vec{v}\times\hat{r}}{r^2} \quad \text{(single moving particle, Figure 17.56)}$$

$\hat{r}$ is a unit vector that points *from* the source charge *toward* the observation location.

$$\frac{\mu_0}{4\pi} = 1\times 10^{-7}\frac{\text{T}\cdot\text{m}^2}{\text{C}\cdot\text{m/s}} \quad \textit{exactly}$$

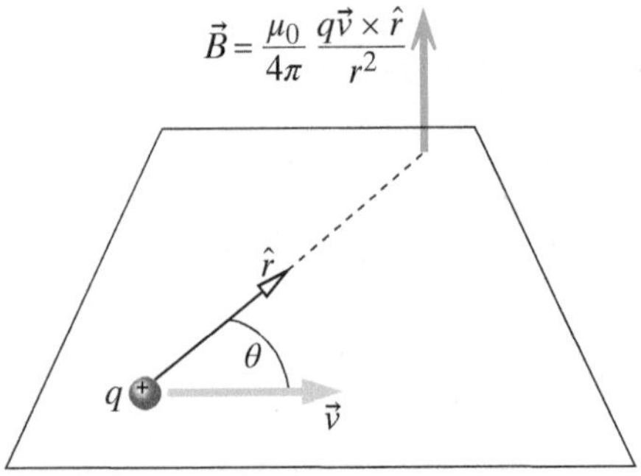

Figure 17.56

$$\Delta\vec{B} = \frac{\mu_0}{4\pi}\frac{I\Delta\vec{l}\times\hat{r}}{r^2} \quad \text{(current distribution, Figure 17.57)}$$

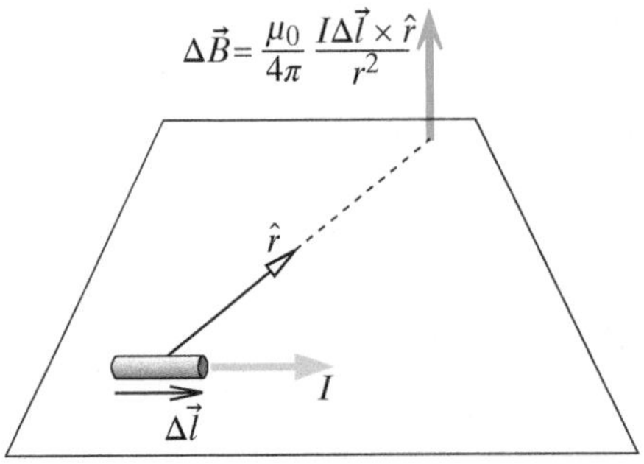

Figure 17.57

Electron current: $i = nA\bar{v}$
Conventional current: $I = |q|nA\bar{v}$
A magnetic field can be detected by a compass; the horizontal component of $B_{\text{Earth}} \approx 2\times 10^{-5}$ T in much of the continental United States.

Magnetic Field of a Wire
For a straight wire of length L and conventional current I, at a perpendicular distance r from the center of the wire (Figure 17.58):

$$B_{\text{wire}} = \frac{\mu_0}{4\pi}\frac{LI}{r\sqrt{r^2+(L/2)^2}} \approx \frac{\mu_0}{4\pi}\frac{2I}{r} \quad \text{for } r \ll L$$

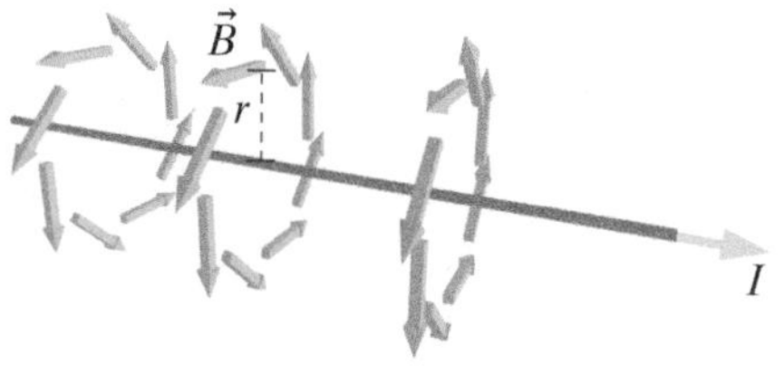

Figure 17.58

Magnetic Field of a Loop

For a circular loop of radius R and conventional current I, at a distance z from the center, along the axis (Figure 17.59):

$$B_{\text{loop}} = \frac{\mu_0}{4\pi}\frac{2\pi R^2 I}{(z^2+R^2)^{3/2}} \approx \frac{\mu_0}{4\pi}\frac{2\mu}{z^3} \quad \text{for } z \gg R$$

Magnetic dipole moment of a loop: $\mu = (\pi R^2)I$

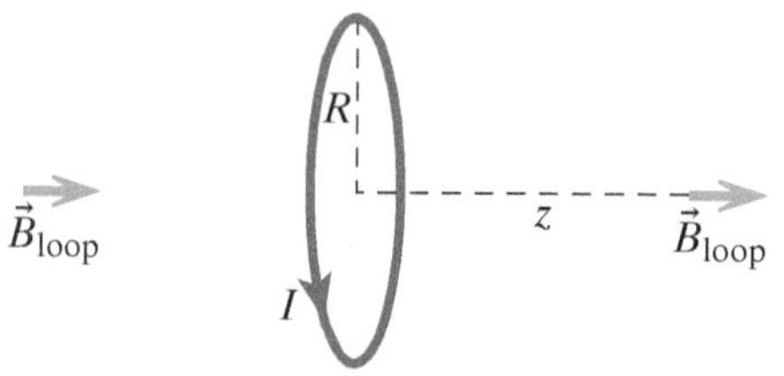

Figure 17.59

Magnetic Field inside a Long Solenoid

Inside a long solenoid (radius of solenoid $\ll L$)

$$B_z \approx \frac{\mu_0 NI}{L}$$

Note that a solenoid is very long compared to its radius—it is not the same thing as a coil composed of a few loops close together (Figure 17.60).

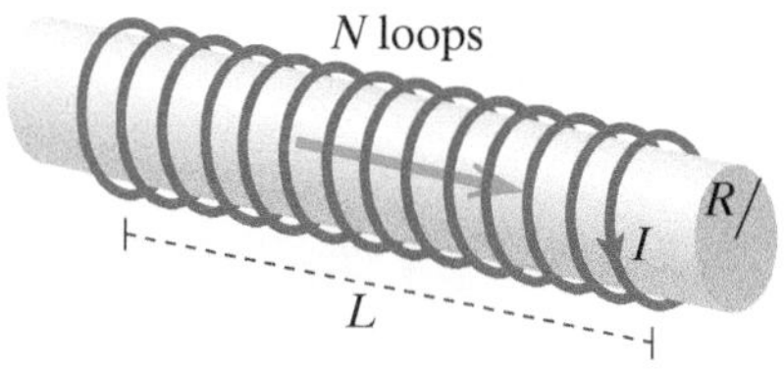

Figure 17.60

Atomic Model of Magnets

An atom can be a magnetic dipole, whose magnetic dipole moment is primarily due to the spin (rotational angular momentum) of unpaired electrons. The orbital angular momentum of electrons can also contribute.

The magnetic dipole moment of a bar magnet is the sum of the magnetic dipole moments of all its constituent atoms.

Ferromagnetic Material

It is organized into domains of aligned atomic magnetic dipole moments, and an applied field can orient these domains. Removal of the applied field may leave the material partially aligned, forming a permanent magnet.

EXPERIMENTS

Equipment

In the study of magnetic field you will need the following equipment:

- two flashlight batteries in a battery holder
- light bulbs of two kinds
- screw-in bulb sockets
- several short copper wires with clips on the ends ("clip leads" used as connecting wires)
- a long wire (about 2 m in length)
- a liquid-filled magnetic compass
- unmagnetized nails for Experiment EXP12

Simple Circuits

To observe the magnetic effects of electric currents, it is useful to construct simple circuits containing wires, light bulbs, and batteries. These are the simplest examples of systems in which we can observe the fundamental electric and magnetic properties of continuous electric currents.

The equipment needed for the experiments in this chapter is the following: two D cells (and it is useful to have a battery holder for them), flashlight bulbs (#48 and #14 if possible), sockets for these bulbs, insulated "hookup" wire, some clip leads (wires with alligator clips on the end), a liquid-filled compass (air-filled compasses typically don't work well because the needle tends to get stuck on its pivot), and a bar magnet (a magnet with a north end and a south end).

Suitable equipment may be available in a laboratory, or you can purchase experiment kit EM-8675 from http://www.pasco.com.

Light Bulbs and Sockets

The filament of a light bulb (the very thin metal wire that glows) is made of tungsten, a metal that does not melt until reaching a very high temperature. A glowing tungsten wire would rapidly oxidize and burn up in air, so there is a vacuum or an inert gas such as argon inside the bulb.

The thin tungsten filament in the bulb strongly resists the passage of electrons. When the electron sea is forced to move through the tungsten, the mobile electrons collide with the positive cores (nuclei plus inner electrons), and this "friction" makes the metal get hot and glow.

EXP1 Experiments with simple circuits

(a) Using one battery, some connecting wires (insulated wires with clips on the ends, not bare Nichrome wire), and a round bulb (#14) *but no socket*, make the bulb light up. (See Figure 17.61.) If the bulb glows with a steady light, this is a "steady state." Make a diagram showing how you connected the circuit.

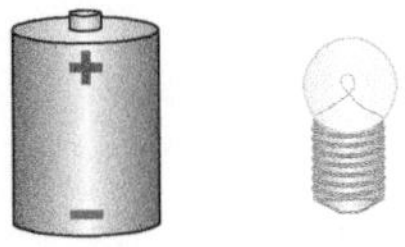

Figure 17.61

(b) Examine a light bulb carefully, and imagine slicing the bulb in half lengthwise. Figure 17.62 is a cutaway sketch:

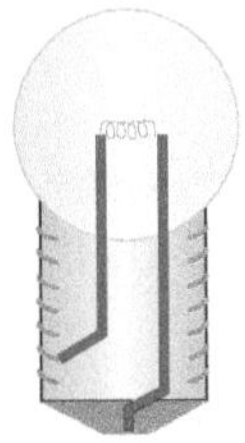

Figure 17.62

On a copy of the sketch, label the important parts and connections, indicating which parts of the bulb you think are metal conductors and which parts are insulators. Using a different color if available, show the conducting path that electric current follows through the bulb. What happens if you switch the connections to the battery? Does the light bulb still light?

(c) Closely examine the long #48 bulb and the round #14 bulb (Figure 17.63).

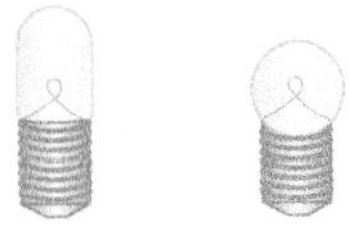

Figure 17.63 A long #48 bulb (left) and a round #14 bulb (right).

The tungsten filaments in both bulbs are about the same length, but perhaps you can see even with the naked eye that the filament in the long bulb is extremely thin—thinner than the filament in the round bulb. Would you guess it would be easier to push electric current through a thin filament (as in the long bulb) or through a thick filament (as in the round bulb)? Why? (At this point this is mostly just a guess, but we'll study this in detail later.)

(d) Examine a bulb *socket* (the receptacle into which a bulb is screwed), and imagine slicing the socket in half lengthwise. Make a cutaway sketch, and label the important parts and connections. Indicate which parts are metal conductors and which parts are insulators. On your sketch, trace the path (in a different color if available) along which electrons will move through the socket when there is a bulb in it. Screw a bulb into the socket and connect the socket to the battery with two connecting wires. Make sure the bulb lights.

(e) Connect a round bulb in a socket to two batteries in series (that is, one after the other) using connecting wires (Figure 17.64).

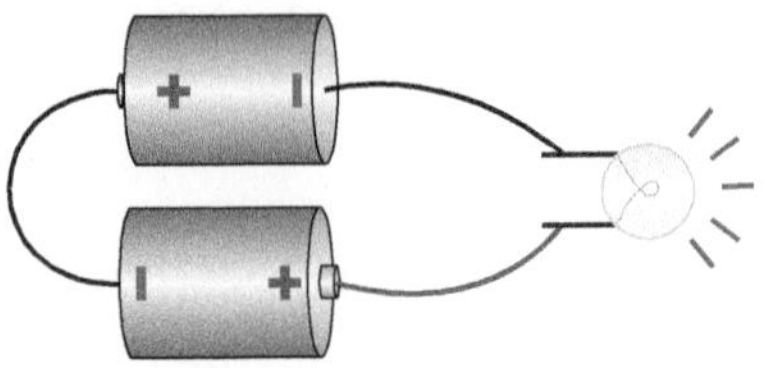

Figure 17.64 Two batteries and a round bulb in series. The socket is not shown.

To connect two batteries in series, put them in the battery holder in opposite directions, and connect them as shown (note that + is connected to −). Compare the brightness of the bulb with one battery and with two batteries in series. Because of this difference we'll usually use two batteries in series in our experiments.

EXP2 The magnetic effects of currents

Make a two-battery circuit with a round bulb in a socket. Place your magnetic compass on a flat surface under one of the wires as shown in Figure 17.65. Keep the compass away from steel objects, such as the steel-jacketed batteries and the alligator clips on the ends of your wires. If you are working on a steel table, you may need to put the compass on a thick book. (For flexibility in placement, you may find it useful to make a long wire by connecting two of your wires together.)

Connect the circuit so that the bulb glows, and do the following:

- Lift the wire up above the compass.
- Orient the wire to be horizontal and lined up with the compass needle. (Using a long wire may make it easier to do this.)
- Bring the aligned wire down onto the compass, as in Figure 17.65.

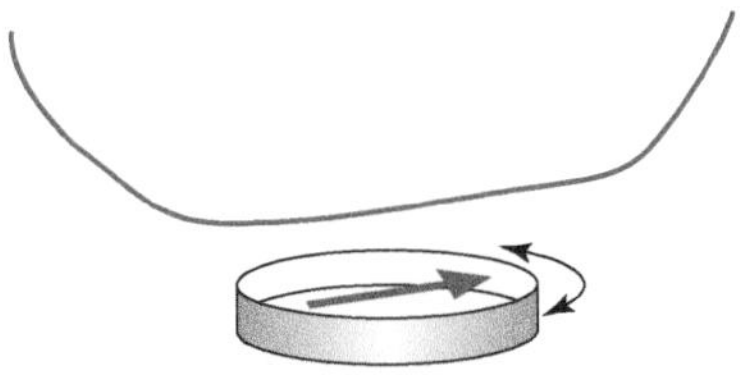

Figure 17.65 Wire aligned with the compass needle.

(a) What is the effect on the compass needle as you bring the wire down on top of the compass?

(b) What happens when the wire is initially aligned perpendicular instead of parallel to the needle as shown in Figure 17.66?

(c) Reverse the connections to the batteries, or reverse the direction of the wire over the compass, in order to force electrons through the circuit in the opposite direction. Again make the compass needle deflect. How is the deflection of the compass needle affected by changing the direction of the current?

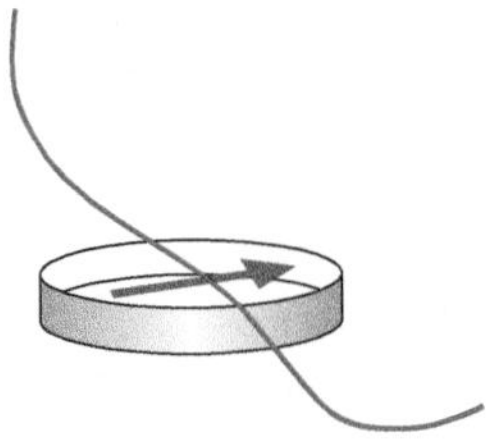

Figure 17.66 Wire perpendicular to the compass needle.

(d) Run the wire under the compass (Figure 17.67) instead of over the compass. What changes?

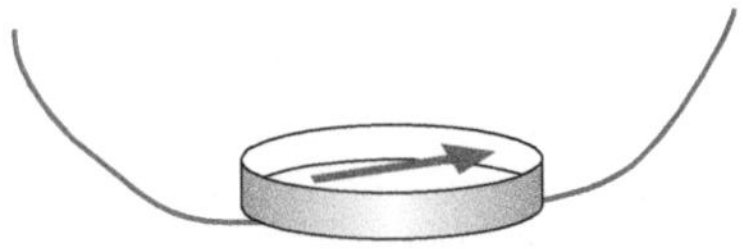

Figure 17.67 Wire under compass.

(e) To make sure that it is the current in the wire, and not the metal wire itself, that affects the compass, disconnect the batteries. When you bring the wire down on the compass, is there a deflection?

(f) Record the observed compass deflection in the following cases:

1. Two batteries and the (bright) round bulb
2. Two batteries and the (dim) long bulb
3. Two batteries, and just a long wire (no bulb). This is called a short circuit, and it puts a large drain on the batteries, so you should not leave this connected for many minutes.

The effect you have just observed was discovered by accident by the Danish scientist Oersted in 1820 while doing a lecture demonstration in a physics class. The phenomenon is often called "the Oersted effect." From your experiments, you should have drawn the following conclusions:

- The magnitude of the magnetic field produced by a current of moving electrons depends on the amount of current.
- A wire with no current running in it produces no magnetic field.
- The magnetic field due to the current appears to be perpendicular to the direction of the current.
- The direction of the magnetic field due to the current under the wire is opposite to the direction of the magnetic field due to the current above the wire.

EXP3 Electron current and battery
Use the right-hand rule to predict whether the compass needle should deflect to the left or right, assuming that electron current flows out of the negative end of the battery and into the positive end. Check to see that the needle does deflect in the predicted direction.

EXP4 The magnetic field of a long straight wire
Use clip leads to connect a wire about 2 m long to a battery, but don't make the final connection yet. Lay the wire down on the floor or a table, making a straight length as long as possible, heading north and south, as far from the steel-clad battery as possible.

- Hold the compass above the wire at a height where the compass deflection is 20° when you turn the current on and off. Mark on a vertically held piece of paper this height $r_{20°}$ of the compass needle above the wire, and record this height.
- Mark the paper at twice this height and use the second mark to place the compass at a height $2(r_{20°})$ above the paper, and record the compass deflection.

(a) Are your data consistent with the theoretical prediction that the magnitude of the magnetic field near a long straight wire is proportional to $1/r$, where r is the perpendicular distance to the wire?

(b) Using your value of r_{20}, determine the amount of current I in the wire.

Experimental Observations of a Coil of Wire
Next we will make measurements of a coil of wire consisting of multiple loops to make a larger magnetic field than a single loop would provide.

EXP5 The magnetic field of a coil of wire
Take an insulated wire about 2 m in length and wrap a coil of about $N = 20$ turns loosely around two fingers. Twist the ends together to help hold the coil together, and remove the coil from your fingers. Use clip leads to connect the coil to a battery.

By appropriate placement and orientation of your compass relative to your N-turn current loop, measure the experimental magnetic field direction and relative magnitude at the locations indicated. Remember that you must always position the compass in such a way that the coil's magnetic field is perpendicular to the Earth's magnetic field.

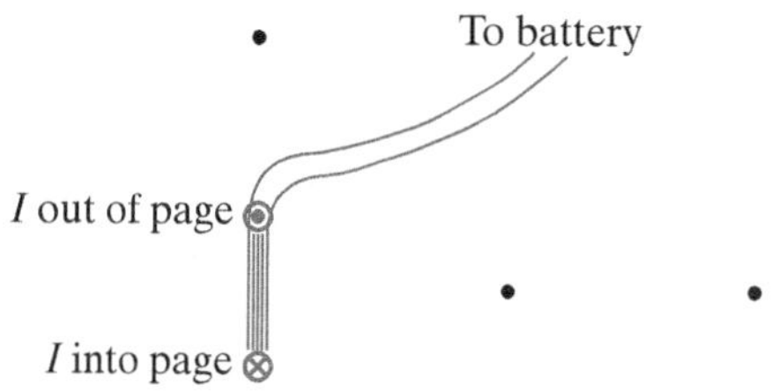

Figure 17.68

On a diagram like that in Figure 17.68, draw the magnetic field vectors at all marked locations. Record the number of turns N, the approximate radius of the coil R, the distances to the measurement locations, and the compass deflections at those locations.

(a) Do your measurements agree with the predicted pattern of magnetic field around a coil of wire?

(b) Determine the amount of current I in the wire.

EXP6 A coil is a compass
For reasons that we will discuss in a later chapter, the magnetic dipole moment vector $\vec{\mu}$ acts just like a compass needle. In an applied magnetic field, a current-carrying loop rotates so as to align the magnetic dipole moment $\vec{\mu}$ with the field.

It is hard to observe this twisting with your own hanging coil, because the rather stiff wires prevent the coil from freely rotating, unless you suspend the entire apparatus from a thread, batteries and all. Or do this:

If you have access to kitchen equipment, you may be able to float your batteries and coil in a bowl or large glass or aluminum pan, or on a block of wood. (Avoid steel containers!) Then you may be able to observe the axis of the coil line up with the Earth's magnetic field, just as though the magnetic dipole moment of the coil were a compass needle.

EXP7 Directions of the magnetic field of a bar magnet
From its name, you might guess that a magnet makes a magnetic field. We say that a magnetic field is present if we see a compass needle twist, so let's see whether your bar magnet does make your compass needle deflect.

Place your compass to the right of your bar magnet, with the magnet oriented in such a way as to make the compass deflect to the east as shown in the accompanying figure. Make a similar diagram in your notebook.

We define the direction of a magnetic field $\vec{B}$ at a particular location as the direction that the "north" end of a compass points to when placed at that location.

(a) Assuming that the bar magnet makes a magnetic field along the axis of the magnet, draw a vector in your notebook to show the direction of the compass needle and another arrow to show the direction of the magnetic field $\vec{B}_m$ of your bar magnet at the present location of the compass. For future reference, write "N" for "north" on the end of the magnet that the magnetic field points out of (attach tape to write on if necessary).

(b) Move the compass to the left of your magnet, above your magnet, and below your magnet, as shown in Figure 17.69. For each location, record the direction of the compass needle and also record on your diagram the direction of the magnetic field due to the bar magnet at that location.

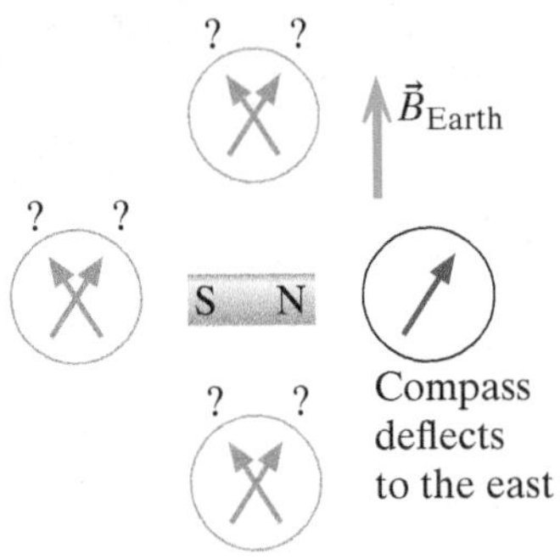

Figure 17.69

Place your compass at each of the locations shown above, relative to your bar magnet, and note the deflection of the compass needle. What is the direction of the magnet's magnetic field at these locations?

(c) Does this pattern of field look familiar? Where have we seen a similar pattern of field directions in space before?

(d) Suspend your magnet from a thread or a hair using a piece of tape, or float the magnet on a dish or plate in water. Note which end points toward the north. Evidently a compass consists simply of a magnet in the shape of a needle, mounted on a pivot.

EXP8 Magnets and matter

Obtain another magnet by working with a partner or finding a kitchen magnet. Make sure that you have labeled the north and south ends of both magnets.

(a) Describe briefly the interactions that the two magnets have with each other.

(b) You probably know that magnets don't interact strongly with anything but iron or steel objects (steel is mostly iron). Magnets also interact with nickel or cobalt objects, but these aren't so readily available. Check to see for yourself that your magnet doesn't interact with aluminum (aluminum foil), copper (a penny, or an electrical wire), or any nonmetal. Also notice that your magnet is strongly attracted to steel parts of your electricity kit.

(c) Check what happens with charged invisible tape, and see that although there is the usual attraction of a charged tape for any uncharged object, there doesn't seem to be any magnetic interaction, because either end of the magnet acts the same.

(d) Can you give any evidence that magnetic interactions pass through matter, as electric interactions do?

(e) Place two magnets in various positions near your compass to demonstrate to your satisfaction that the superposition principle holds for the magnetic fields of magnets; that is, the net field of the two magnets is the vector sum of the two fields. Give one example of your observation of the superposition principle for magnetic fields.

EXP9 Dependence on distance for a bar magnet

Again place the compass to the right of your bar magnet, oriented in such a way as to make the compass deflect 70° to the east. (See Figure 17.70.) *Do not work on a table that contains steel components*, which can interfere with the measurements. The best place to make the measurements is outdoors, away from iron and steel objects.

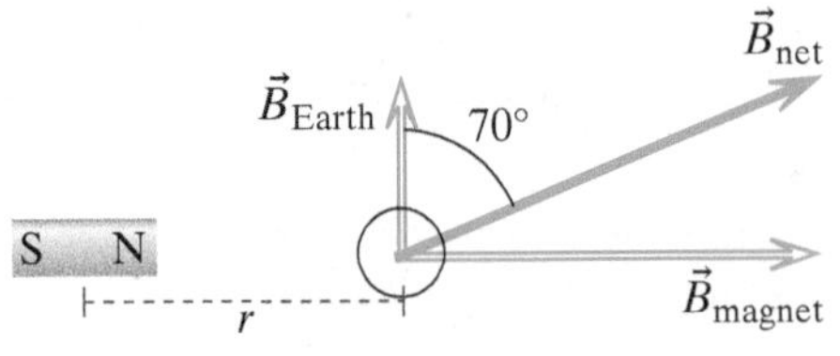

Figure 17.70

(a) At the location where you get a 70° deflection, measure the distance r from the center of the magnet to the center of the compass. A centimeter ruler is provided at the back of the book.

(b) Use vector analysis and the magnitude of the Earth's field to calculate the magnitude B_r of the magnetic field of the magnet at this distance r from the center of the magnet.

(c) Now move the magnet farther away by a factor of two; that is, place the magnet so that the distance from the center of the magnet to the center of the compass is $2r$. Record the new distance and the new compass deflection angle. Calculate the magnitude B_{2r} of the magnetic field of the magnet at this new distance from the magnet.

(d) By what ratio did the magnetic field of the magnet decrease when the distance from the magnet was doubled?

(e) The magnetic field of the magnet gets smaller with distance, and it is plausible to guess that the magnetic field of the magnet might vary as $1/r^n$, where n is initially unknown. According to your measurements, what is n? (You may wish to make measurements at some other distances to give further support to your analysis.)

(f) Extrapolating to a location near one end of your bar magnet, approximately how strong is the magnetic field in teslas near one end of your magnet? (Remember, measure r to the center of the magnet.)

(g) Place the compass north or south of the magnet, at the original distance r from the center of the magnet, and determine the magnitude of the magnetic field at this location. (See Figure 17.71.) How does this magnitude compare to your result in (b)?

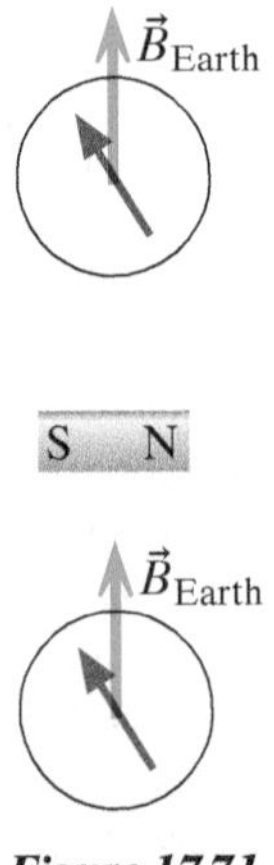

Figure 17.71

EXP10 Coils attract and repel like magnets

There is yet another way in which the behavior of a magnet and a current-carrying coil of wire is very similar. Hang your coil over the side of the table and bring your bar magnet near the hanging coil, along the axis of the coil. Reverse the magnet. Repeat from the other side of the coil. Is this behavior similar to the interaction you have observed between two bar magnets?

EXP11 Cutting a magnet in two
Without actually cutting a magnet into two pieces, you can simulate such an operation by putting two bar magnets together north end to south end as in Figure 17.72, then taking them apart.

Figure 17.72

Do you in fact find that the put-together longer magnet acts just like a single magnet, and that after pulling them apart the two pieces still act like magnets?

If you have the equipment available, you can magnetize a soft iron nail by stroking it with a magnet; then use a hacksaw to cut it in half and investigate the properties of each half.

EXP12 The magnetic multiplier effect
Note: You need unmagnetized nails for this experiment. If the nails have ever been near a magnet, they may have become strongly magnetized. Check this by bringing a nail near the compass, first one end, then the other. If the nail is unmagnetized, both ends of the nail will affect the compass equally. However, if the nail is strongly magnetized, the two ends of the nail will affect the compass quite differently.

If a nail is strongly magnetized, bring your bar magnet near it (but not touching the nail) to magnetize the nail in the opposite direction. Check again with the compass and repeat as necessary, trying to reduce the magnetization to zero. Do not proceed until you have at least three unmagnetized nails.

Now you can study the magnetic multiplier effect. Take an insulated wire about 2 m in length and wrap a coil of about 20 turns loosely around two fingers. Twist the ends together to help hold the coil together and remove it from your fingers. Connect the coil to a battery and position your compass to the east or west of the coil so that the compass needle deflects away from north by about 5°. Turn off the current, so the compass deflection goes to zero.

(a) Insert an unmagnetized iron nail into the coil and record the compass deflection (if any). Now start the current again and observe what happens to the compass deflection. Add additional nails and observe and record what happens to the compass deflection.

(b) Explain briefly why each nail produces a sizable increase in the compass deflection when you turn on the current.

(c) To check whether the nails are permanently magnetized after being inserted into the coil, turn off the current. If the compass still deflects significantly without any current in the coil, the nails have become strongly magnetized. Are they?

(d) Bring one of these nails near the compass, first one end and then the other. Is the nail slightly magnetized?

EXP13 An electromagnet
Next, construct an "electromagnet." Take an insulated wire about 2 m in length and wrap it tightly around an iron nail, all along its length. You can make multiple layers of windings, back and forth along the nail, to increase the effects, but leave the ends of the nail sticking out so that you can touch the nail to other objects. Twist the ends of the wire together to help hold the coil together. Connect the coil to a battery and bring it near another nail.

Can you pick up the other nail? (If the battery is not fresh, you may not get enough current to lift the other nail, but you should be able to lift a paper clip!) Why does the other nail fall when you cut the current? Is there any evidence that the electromagnet nail retains some magnetization?

QUESTIONS

Q1 Given that $\vec{v} = \langle v_x, v_y, v_z \rangle$ and $\hat{r} = \dfrac{\langle x, y, z \rangle}{\sqrt{x^2+y^2+z^2}}$, write out $\vec{v} \times \hat{r}$ as a vector.

Q2 The vectors $\vec{a}$ and $\vec{b}$ both lie in the xz plane. The vector $\vec{c}$ is equal to $\vec{a} \times \vec{b}$. Which of these statements about $\vec{c}$ must be true? (1) The x and y components of $\vec{c}$ are zero. (2) The y and z components of $\vec{c}$ are zero. (3) The z and x components of $\vec{c}$ are zero.

Q3 How does the magnetic field of a moving point charge fall off with distance at a given angle: like $1/r$, $1/r^2$, or $1/r^3$?

Q4 Describe the magnetic field made by a charge that is not moving.

Q5 A current-carrying wire is laid on a table, oriented north–south. Electrons in the wire are flowing north. **(a)** What is the direction of the magnetic field at a location directly underneath the wire, due only to these moving electrons? **(b)** If a compass is placed under the wire, in which direction will the needle deflect?

Q6 A current carrying wire is laid on a table, oriented north–south. Electrons in the wire are flowing south. **(a)** What is the direction of the magnetic field at a location directly underneath the wire, due only to these moving electrons? **(b)** If a compass is placed under the wire, in which direction will the needle deflect?

Q7 Consider the portion of a circuit in Figure 17.73. The wire rests on top of the two compasses. When no current is running, both compasses point north (direction shown by the gray arrows). When current runs in the circuit, the needle of compass 1 deflects as shown. In what direction will the needle of compass 2 point? Draw a sketch indicating its deflection.

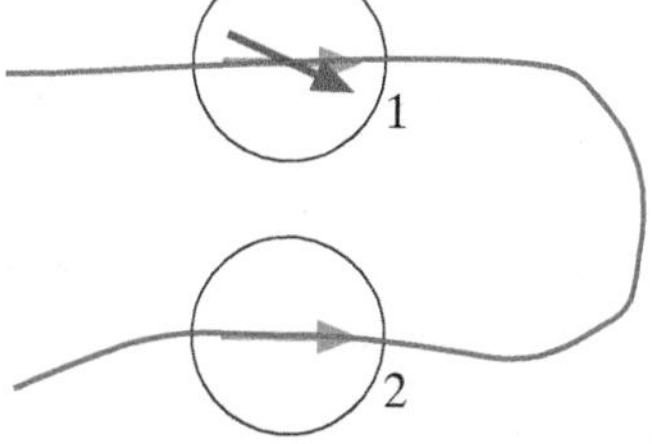

Figure 17.73

Q8 In a particular wire in a circuit, conventional current flows in the $+y$ direction. What is the direction of electron current in the wire?

Q9 In a particular wire in a circuit, electrons flow in the $+z$ direction. What is the direction of conventional current in the wire?

Q10 An iron bar magnet makes a pattern of magnetic field that looks just like the pattern of magnetic field outside a long current-carrying coil of wire. Are there currents in the iron? Explain briefly.

Q11 Suppose you have two alnico bar magnets, one with a mass of 100 g and one with a mass of a kg. At a distance of a meter from the center of either one, how would the magnetic field differ? Why?

Q12 Conventional current flows in a ring in the direction indicated in Figure 17.74. If you stand at location A, on the $+x$ axis, and look toward the ring, current flows clockwise. At each of the locations labeled by a letter, find the direction of the magnetic field at that location due to the current in the ring.

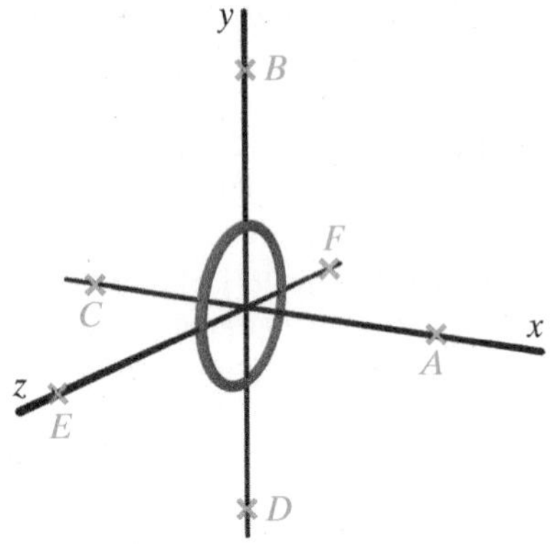

Figure 17.74

Q13 **(a)** How can you produce a magnetic field that is nearly uniform in a region? **(b)** How can you produce a magnetic field that falls off like $1/r$? **(c)** How can you produce a magnetic field that falls off like $1/r^2$? (Note that you *cannot* use just a short piece of current-carrying wire, because the other parts of the wire also contribute.) **(d)** How can you produce a magnetic field that falls off like $1/r^3$?

Q14 What is the direction of the magnetic field at the indicated locations inside and outside this current-carrying rectangular coil of wire in Figure 17.75? Explain briefly. (The direction of conventional current is shown.)

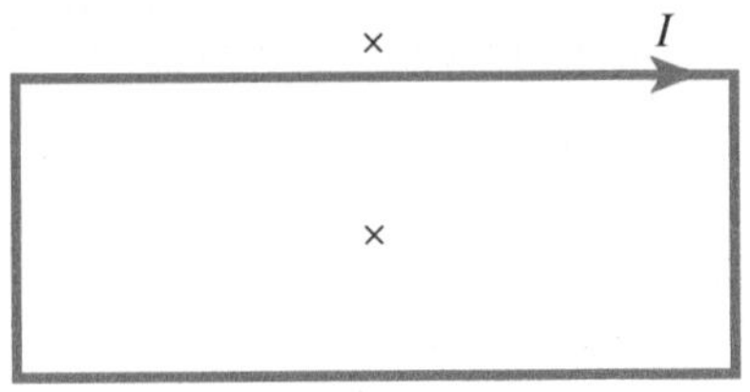

Figure 17.75

PROBLEMS

Section 17.3

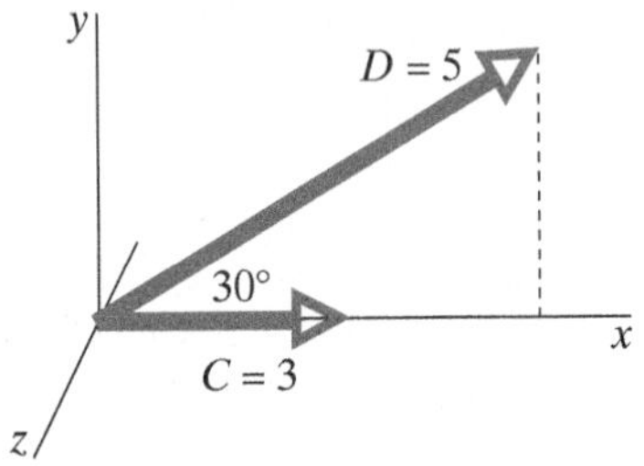

Figure 17.76

•P15 A vector $\vec{C}$ of magnitude 3 lies along the x axis, and a vector $\vec{D}$ of magnitude 5 lies in the xy plane, 30° from the x axis (Figure 17.76). What are the magnitude and direction of the cross product $\vec{C} \times \vec{D}$? What are the magnitude and direction of the cross product $\vec{D} \times \vec{C}$? Draw both vectors on a diagram.

•P16 If $\vec{v} = \langle 390, -480, 333 \rangle$ m/s and $\hat{r} = \langle 0.577, 0.577, -0.577 \rangle$ what is $\vec{v} \times \hat{r}$?

•P17 What is $\langle 2, 0, -5 \rangle \times \langle -5, 3, 5 \rangle$?

•P18 To get an idea of the size of magnetic fields at the atomic level, consider the magnitude of the magnetic field due to the electron in the simple Bohr model of the hydrogen atom. In the ground state the Bohr model predicts that the electron speed would be 2.2×10^6 m/s, and the distance from the proton would be 0.5×10^{-10} m. What is B at the location of the proton?

•P19 An electron is moving horizontally to the right with speed 4×10^6 m/s (Figure 17.77). Each of the indicated locations is 5 cm from the electron. What is the magnetic field due to this moving electron at the indicated locations? Give both magnitude and direction of the magnetic field at each location.

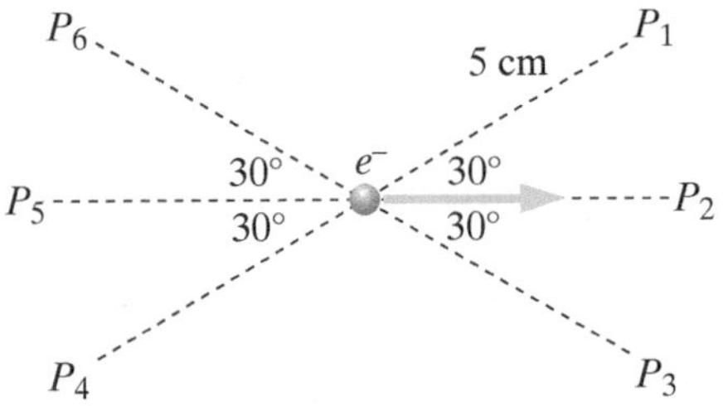

Figure 17.77

•P20 At a particular instant, a proton at the origin has velocity $\langle 2 \times 10^4, -2 \times 10^4, 0 \rangle$ m/s. You need to calculate the magnetic field at location $\langle 0.02, 0.06, 0 \rangle$ m due to the moving proton. **(a)** What is $\vec{r}$? **(b)** What is $\hat{r}$? **(c)** What is $\vec{v} \times \hat{r}$? **(d)** What is the magnetic field at the observation location due to the moving proton?

•**P21** In Figure 17.78 a proton is moving upward with speed 5×10^6 m/s. What is the magnetic field due to this moving proton at the indicated locations? Each location is a distance $r = 8$ cm from the proton. Give both magnitude and direction of the magnetic field at each location.

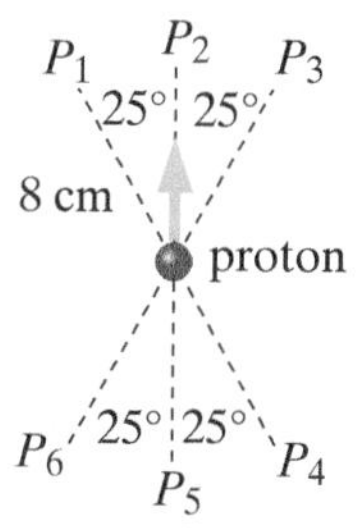

Figure 17.78

••**P22** At a particular instant a proton is at the origin, moving with velocity $\langle 3 \times 10^4, -2 \times 10^4, -7 \times 10^4 \rangle$ m/s. **(a)** At this instant, what is the electric field at location $\langle 4 \times 10^{-3}, 4 \times 10^{-3}, 3 \times 10^{-3} \rangle$ m, due to the proton? **(b)** At this instant, what is the magnetic field at the same location due to the proton?

••**P23** The electron in Figure 17.79 is traveling with a speed of 3×10^6 m/s.

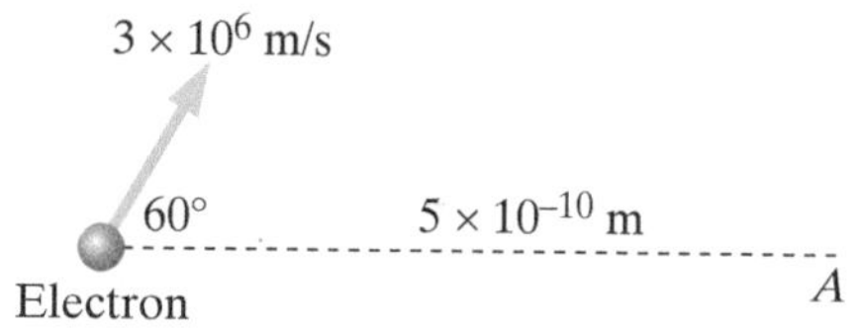

Figure 17.79

(a) At location A, what are the directions of the electric and magnetic fields contributed by the electron? **(b)** Calculate the magnitudes of the electric and magnetic fields at location A.

Section 17.6

•**P24** The electron current in a horizontal metal wire is 3.7×10^{18} electrons/s, and the electrons are moving to the right. What are the magnitude and direction of the conventional current?

•**P25** A conventional current of 12 A runs to the right in a horizontal metal wire. **(a)** How many electrons pass some point in the wire per second? **(b)** In what direction are the electrons moving? **(c)** The electron mobility in this metal is 2.1×10^{-4} (m/s)/(N/C) and the electric field in the wire is 0.15 N/C. What is the average drift speed of the electrons?

•**P26** You have used copper wires in your circuits. Let's calculate the mobile electron density n for copper. A mole of copper has a mass of 64 g (0.064 kg), and one mobile electron is released by each atom in metallic copper. The density of copper is about 9 g/cm^3, or 9×10^3 kg/m^3. Show that the number of mobile electrons per cubic meter in copper is 8.4×10^{28} m^{-3}.

•**P27** Calculate the mobile electron density for nickel. A mole of nickel has a mass of 59 g (0.059 kg), and one mobile electron is released by each atom in metallic nickel. The density of nickel is about 8.8 g/cm^3, or 8.8×10^3 kg/m^3.

Section 17.7

•**P28** In a circuit consisting of a long bulb and two flashlight batteries in series the conventional current is about 0.1 A. What is the magnetic field 5 mm from the wire? (This is about how far away the compass needle is when you place the wire on top of the compass.) Is this a big or a small field?

•**P29** A current-carrying wire oriented north–south and laid over a compass deflects the compass 10° to the east. **(a)** What is the magnitude of the magnetic field made by the current? The horizontal component of Earth's magnetic field is about 2×10^{-5} T. **(b)** In what direction does the electron current flow in the wire?

•**P30** A straight wire of length 0.62 m carries a conventional current of 0.8 A. What is the magnitude of the magnetic field made by the current at a location that is a perpendicular distance 2.9 cm from the center of the wire? Use both the exact equation and the approximate equation to calculate the field.

•**P31** A long straight wire carries a conventional current of 0.9 A. What is the approximate magnitude of the magnetic field at a location a perpendicular distance of 0.035 m from the wire due to the current in the wire?

•**P32** A battery is connected to a Nichrome wire and a conventional current of 0.3 A runs through the wire. The wire is laid out in the form of a rectangle 50 cm by 3 cm. **(a)** What is the approximate magnetic field at the center of the rectangle? Give the direction as well as the magnitude. **(b)** What approximations did you make?

•**P33** A wire through which a current is flowing lies along the x axis in Figure 17.80. Connecting wires that are not shown in the diagram connect the ends of the wire to batteries (which are also not shown). Electron current flows through the wire in the $-x$ direction, as indicated in the figure. To calculate the magnetic field at location A due to the current in the wire, we divide the wire into pieces, approximate each piece as a point charge moving in the direction of conventional current, and calculate the magnetic field at the observation location due only to this piece; then add the contributions of all pieces to get the net magnetic field.

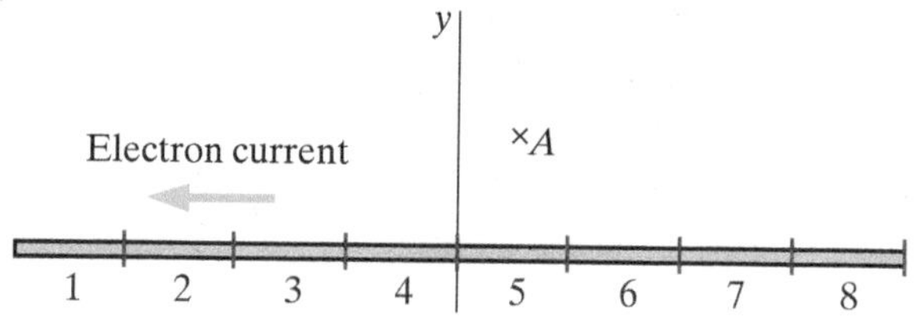

Figure 17.80

The wire is 1.3 m long, and is divided into eight pieces. The observation location A is located at $\langle 0.081, 0.178, 0 \rangle$ m. The conventional current running through the wire is 6.5 A. In this exercise you will calculate the magnetic field at the observation location due only to segment 4 of the wire. **(a)** What is the direction of conventional current in this wire? **(b)** How long is segment 4? **(c)** What is the magnitude of the vector $\Delta\vec{l}$ for segment 4? **(d)** What is the vector $\Delta\vec{l}$ for segment 4? **(e)** What is the location of the center of segment 4? **(f)** What is the vector $\vec{r}$ from source to observation location for segment 4? **(g)** What is the unit vector $\hat{r}$? **(h)** What is $\Delta\vec{l} \times \hat{r}$? **(i)** Calculate the magnetic field $\Delta\vec{B}$ at the observation location due only to the current in segment 4 of the wire.

••**P34** You place a long straight wire on top of your compass, and the wire is a height of 5 mm above the compass needle

(Figure 17.81). If the conventional current in the wire is $I = 0.2$ A and runs left to right as shown, calculate the approximate angle the needle deflects away from north and draw the position of the compass needle.

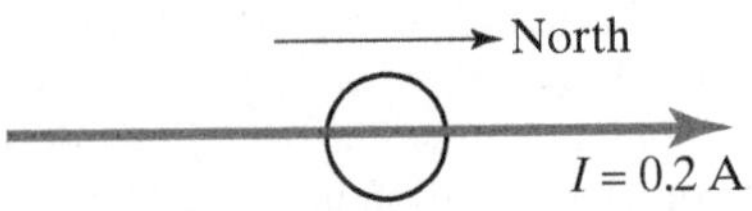

Figure 17.81

••P35 When you bring a current-carrying wire down onto the top of a compass, aligned with the original direction of the needle and 5 mm above the needle, the needle deflects by 10° (Figure 17.82).

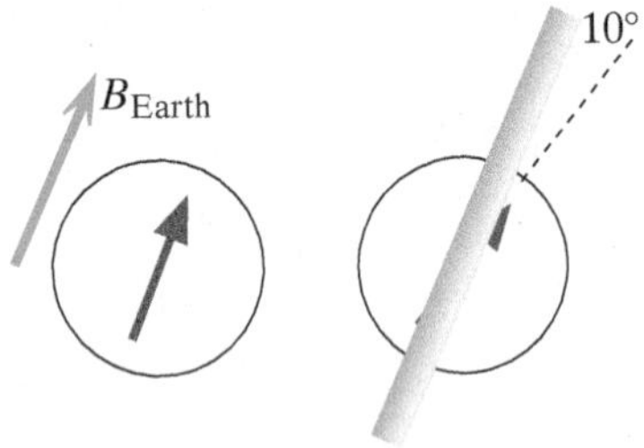

Figure 17.82

(a) Show on a diagram the direction of conventional current in the wire and the direction of the additional magnetic field made by the wire underneath the wire, where the compass needle is located. Explain briefly. **(b)** Calculate the amount of current flowing in the wire. The measurement was made at a location where the horizontal component of the Earth's magnetic field is $B_{\text{Earth}} \approx 2 \times 10^{-5}$ T.

••P36 A compass is placed inside a triangular coil of wire with three turns, as shown in Figure 17.83. Each side of the triangle has a length L. The compass is a perpendicular distance d from the center of each side of the triangle. The coil is in the xy plane; magnetic north (due to the Earth) is in the negative x direction. Conventional current runs in the coil as shown (clockwise, as viewed from a location on the $+z$ axis).

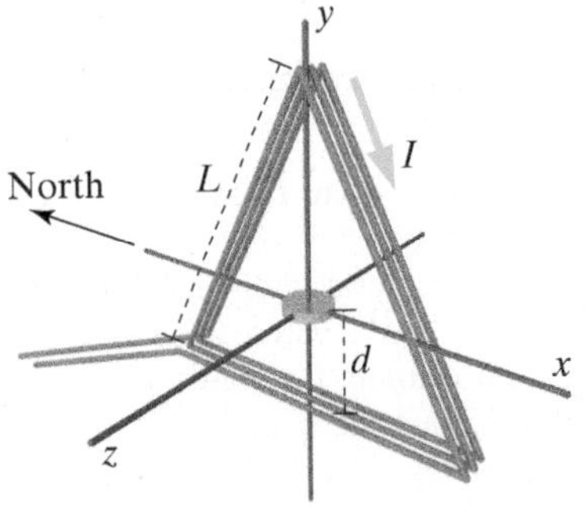

Figure 17.83

(a) While current is running through the coil, in what direction does the compass needle point? **(b)** Which of the following statements about how to model the coil are reasonable? List all that apply. (1) Since the coil is made from a single wire, we could model one turn of the coil as one long wire. (2) A circular loop is a poor model for one turn of the coil, because the distance from the wire to the compass varies so much. (3) One turn of the coil could be modeled as three short straight wires. **(c)** A current of 0.62 A runs through the wires. If $L = 8.3$ cm and $d = 2.4$ cm, what is the magnitude of the magnetic field at the location of the compass due to the coil? (Remember that the coil has three turns.) **(d)** What is the approximate magnitude of the compass deflection in degrees?

••P37 A thin wire is part of a complete electrical circuit that carries a current I. For this problem consider only the piece of wire of length d as shown in Figure 17.84. Answer the following questions based on this figure. **(a)** What are the magnitude and direction of the magnetic field due to the wire at Q (location $\langle -w,0,0\rangle$)? **(b)** Set up the integrals necessary to determine the x, y, and z components of the magnetic field at P (location $\langle -w,h,0\rangle$). The integrals must be in a form that can be evaluated (no cross products in the integrand), but you do not need to evaluate them. **(c)** What is the direction of the magnetic field at location P?

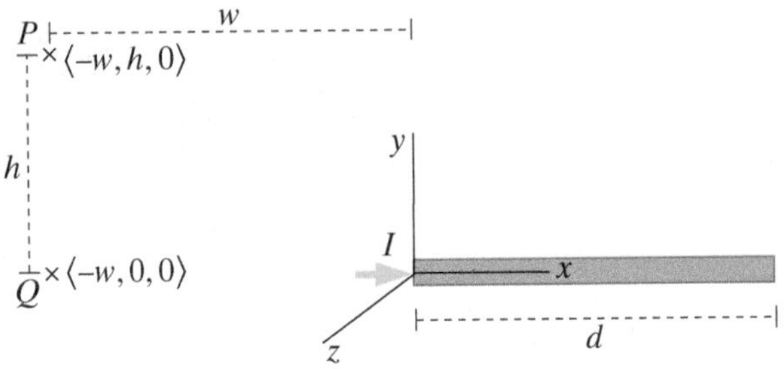

Figure 17.84

••P38 In Figure 17.85 a long current-carrying wire, oriented north–south, lies on a table (it is connected to batteries that are not shown). A compass lies on top of the wire, with the compass needle about 3 mm above the wire. With the current running, the compass deflects 11° to the west. At this location, the horizontal component of the Earth's magnetic field is about 2×10^{-5} T.

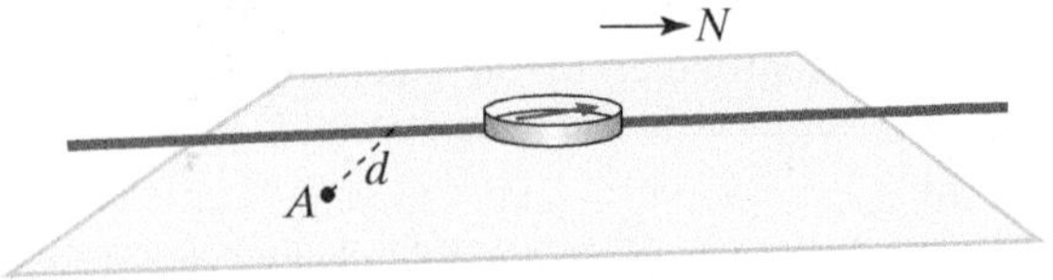

Figure 17.85

(a) What is the magnitude of the magnetic field at location A, on the tabletop, a distance 2.8 cm to the east of the wire, due only to the current in the wire? **(b)** What is the direction of the magnetic field at location A, due only to the current in the wire?

••P39 In Figure 17.86 a wire is connected to batteries (not shown) and current flows through the wire. The wire lies flat on a table, and you are looking down on it from above. The wire is laid on top of a compass, resting about 3 mm above the needle. The distance d between the straight wires is 4.5 cm. The compass needle deflects 17° from north, as shown. At this location the horizontal component of the Earth's magnetic field is 2×10^{-5} T.

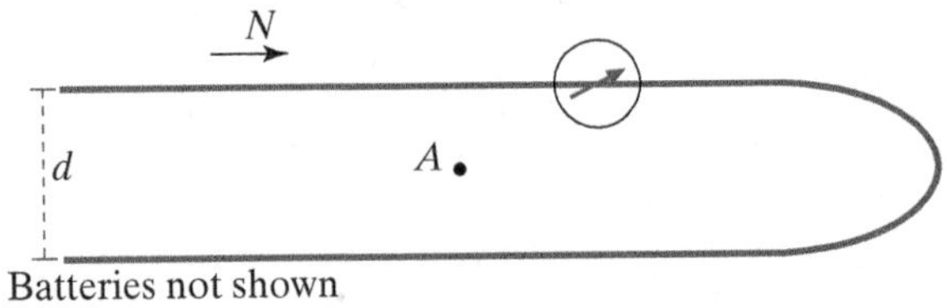

Figure 17.86

(a) What is the direction of the magnetic field at location A, midway between the straight wires, due to the current in the wire? **(b)** What is the approximate magnitude of the magnetic field at location A, due to the current in the wire? **(c)** What approximations or assumptions did you make in determining the magnetic field at location A?

••P40 At one time, concern was raised about the possible health effects of the small alternating (60-Hz) magnetic fields created by electric currents in houses and near power lines. In a house, most wires carry a maximum of 15 A (there are 15-A fuses that melt and break the circuit if this current is exceeded). The two wires in a home power cord are about 3 mm apart as shown in Figure 17.87, and at any instant they carry currents in opposite directions (both of which change direction 60 times per second).

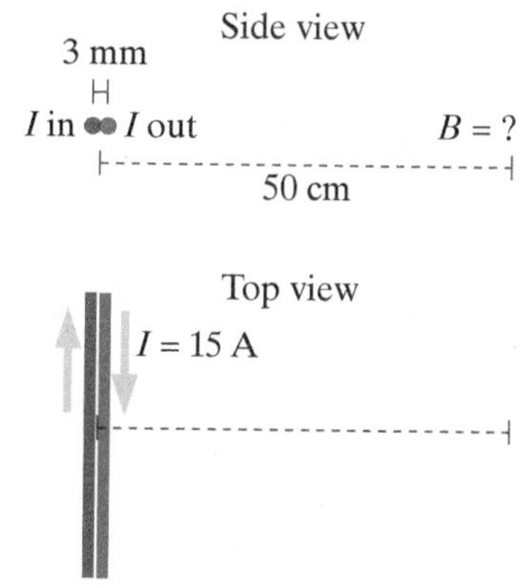

Figure 17.87

(a) Calculate the maximum magnitude of the alternating magnetic field, 50 cm away from the center of a long straight power cord that carries a current of 15 A. Both wires are at the same height as the observation location. **(b)** Explain briefly why twisting the pair of wires into a braid as shown in Figure 17.88 greatly decreases the magnetic field at the location discussed in (a), a generally useful technique.

Figure 17.88

(c) Optional: For two long parallel wires like those shown in part (a), a distance d apart and carrying conventional current I in opposite directions, find an *algebraic* equation for the magnetic field a perpendicular distance $r \gg d$ to the right of the wires.

(The magnitude of the field that you calculated in part (a) is very small compared to the Earth's magnetic field, but there were questions as to whether a very small alternating magnetic field might have health effects. After many detailed studies, the consensus of most scientists now seems to be that these small alternating magnetic fields are not a hazard after all.)

Section 17.8

•P41 **(a)** A loop of wire carries a conventional current of 0.8 A. The radius of the loop is 0.09 m. Calculate the magnitude of the magnetic field at a distance of 0.32 m from the center of the loop, along the axis of the loop. **(b)** What would be the magnitude of the magnetic field at the same location if there were 100 loops of wire in a coil instead of one loop?

•P42 A thin circular coil of wire of radius 5 cm consists of 100 turns of wire (Figure 17.89). If the conventional current in the wire is 4 A, what are the magnitude and direction of the magnetic field at the center of the coil? (The direction of conventional current is shown.)

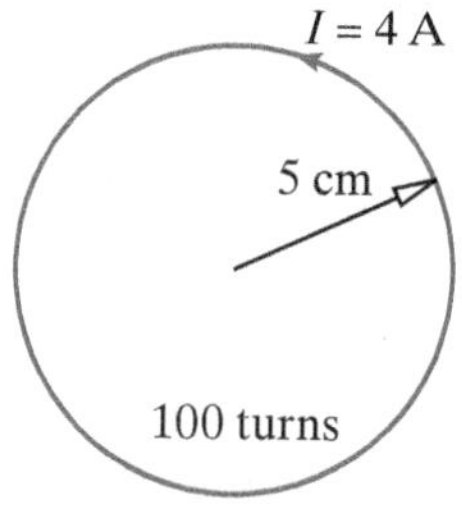

Figure 17.89

•P43 A conventional current of 6 A runs clockwise in a circular loop of wire in the xy plane, with center at the origin and with radius 0.08 m. Another circular loop of wire lies in the same plane, with its center at the origin and with radius 0.02 m. How much conventional current must run counterclockwise in this smaller loop in order for the magnetic field at the origin to be zero?

•P44 How much conventional current must you run in a solenoid with radius = 0.04 m and length = 0.35 m to produce a magnetic field inside the solenoid of 2×10^{-5} T, the approximate field of the Earth? The solenoid has 200 turns.

••P45 In Figure 17.90 two long wires lie very close together and carry a conventional current I as shown and each wire has a semicircular kink, one of radius R_1 and the other of radius R_2. Calculate the magnitude and direction of the magnetic field at the common center of the two semicircular arcs.

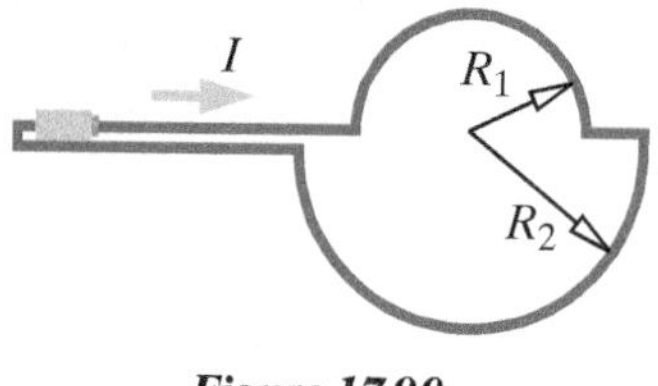

Figure 17.90

••P46 A very long wire carrying a conventional current of 3.5 A is straight except for a circular loop of radius 5.8 cm (Figure 17.91). Calculate the approximate magnitude and the direction of the magnetic field at the center of the loop.

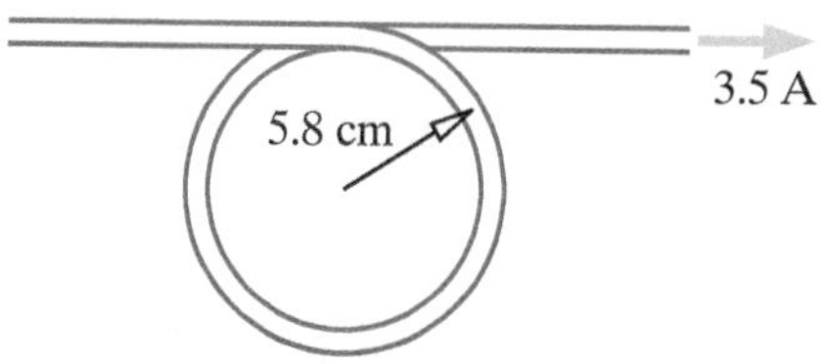

Figure 17.91

••P47 You can use measurements of the magnetic field of a coil to determine how much current your battery is supplying to the coil. Using your value of B (Experiment EXP5), determine the conventional current I through your coil. If this current is less than 3 A, you should replace the battery.

••P48 A thin circular coil of radius $r = 15$ cm contains $N = 3$ turns of Nichrome wire. A small compass is placed at the center of the coil, as shown in Figure 17.92. With the battery disconnected, the compass needle points to the right, in the plane of the coil. Assume that the horizontal component of the Earth's magnetic field is about $B_{\text{Earth}} \approx 2 \times 10^{-5}$ T.

When the battery is connected, a current of 0.25 A runs through the coil. Predict the deflection of the compass needle. If you have to make any approximations, state what they are. Is the deflection outward or inward as seen from above? What is the magnitude of the deflection?

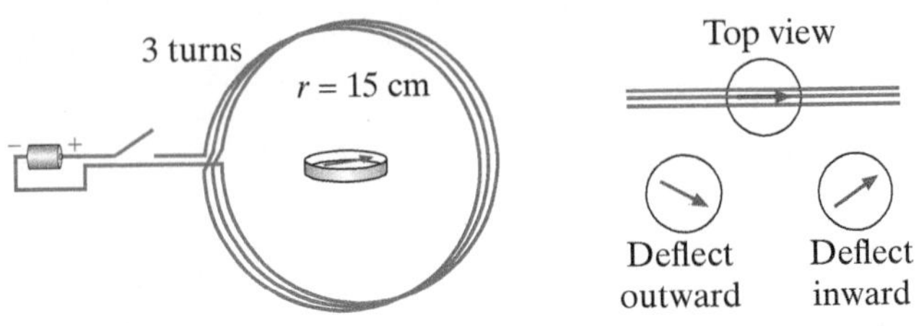

Figure 17.92

••P49 The circuit in Figure 17.93 consists of a battery and a Nichrome wire, through which runs a current I.

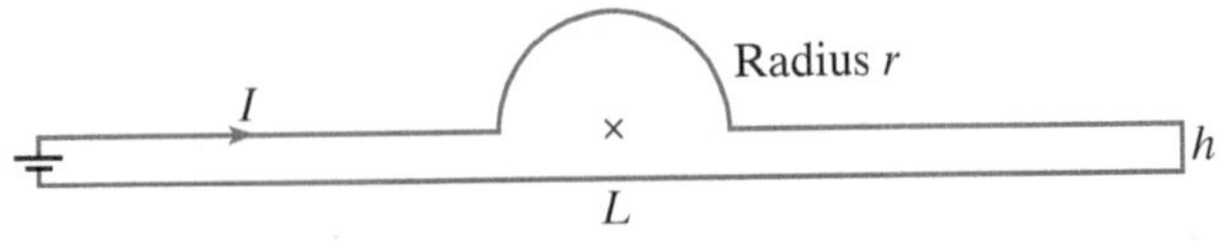

Figure 17.93

(a) At the location marked × (the center of the semicircle), what is the direction of the magnetic field? **(b)** At the location marked × (the center of the semicircle), what is the magnitude of the magnetic field? If you have to make any approximations, state what they are.

••P50 Two thin coils of radius 3 cm are 20 cm apart and concentric with a common axis (Figure 17.94). Both coils contain 10 turns of wire with a conventional current of 2 A that runs counterclockwise as viewed from the right side.

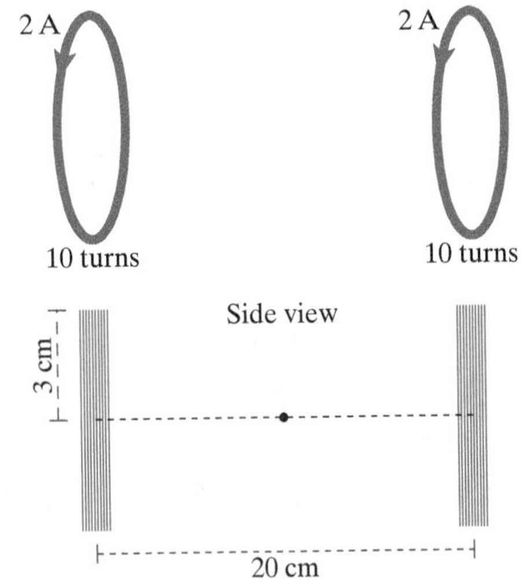

Figure 17.94

(a) What are the magnitude and direction of the magnetic field on the axis, halfway between the two loops, without making the approximation $z \gg r$? (For comparison, remember that the horizontal component of magnetic field in the United States is about 2×10^{-5} T). **(b)** In this situation, the observation location is not very far from either coil. How bad is it to make the $1/z^3$ approximation? That is, what percentage error results if you calculate the magnetic field using the approximate equation for a current loop instead of the exact equation? **(c)** What are the magnitude and direction of the magnetic field midway between the two coils if the current in the right loop is reversed to run clockwise?

••P51 A conventional current I runs in the direction shown in Figure 17.95. Determine the magnitude and direction of the magnetic field at point C, the center of the circular arcs.

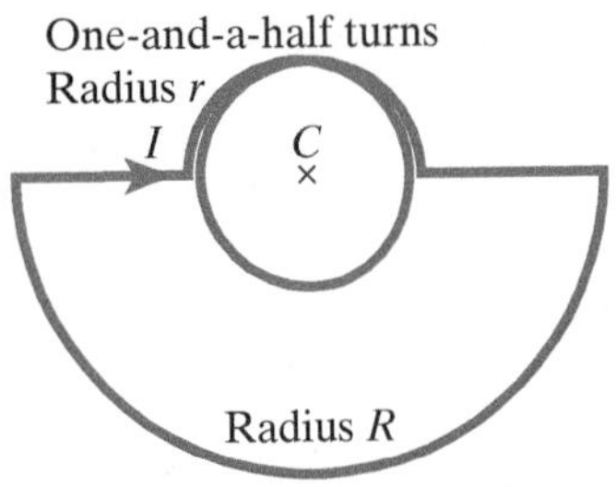

Figure 17.95

••P52 A square loop of wire with sides of length L and carrying a conventional current I lies in the yz plane with its center at the origin. **(a)** What is the magnitude of the magnetic field due to this loop at a location x on the x axis? **(b)** What is the approximate equation for the magnitude of the magnetic field at a location x that is much greater than L, in terms of the magnetic moment of the loop? **(c)** Is the result of part (b) consistent with our previously derived result for the magnetic field of a magnetic dipole?

Section 17.11

••P53 A compass originally points north; at this location the horizontal component of the Earth's magnetic field has a magnitude of 2×10^{-5} T. A bar magnet is aligned east–west, pointing at the center of the compass. When the center of the magnet is 0.25 m from the center of the compass, the compass deflects 70°. What is the magnetic dipole moment of the bar magnet?

••P54 In the region shown in Figure 17.96, the magnitude of the horizontal component of the Earth's magnetic field is about 2×10^{-5} T. Originally a compass placed at location A points north. Then a bar magnet is placed at the location shown in the diagram, with its center 16 cm from location A. With the magnet present, the compass needle points 60° west of north.

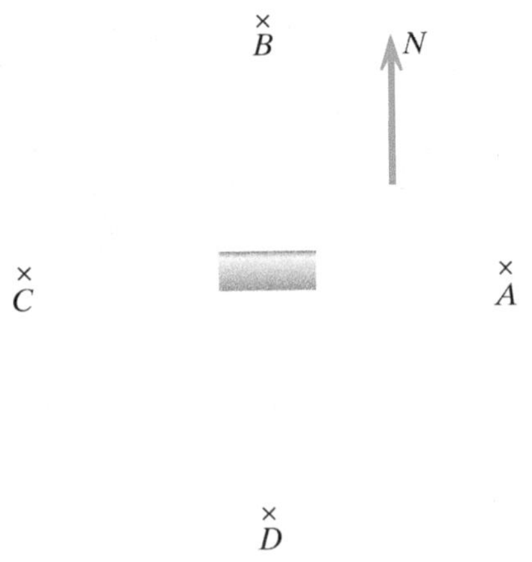

Figure 17.96

(a) Which pole of the magnet is nearer to location A? **(b)** What are the directions of the magnetic field at locations B, C, and D? **(c)** What is the magnetic dipole moment of the bar magnet?

••P55 A bar magnet with magnetic dipole moment $\langle 8,0,0\rangle$ A·m^2 is located at the origin. A second bar magnet with magnetic dipole moment $\langle \mu,0,0\rangle$ is located at $\langle 0.4,0.2,0\rangle$ m. Calculate the value of μ that makes the magnetic field at $\langle 0.4,0,0\rangle$ m be zero.

••P56 A bar magnet is aligned east–west, with its center 16 cm from the center of a compass (Figure 17.97). The compass is observed to deflect 50° away from north as shown, and the horizontal component of the Earth's magnetic field is known to be 2×10^{-5} T.

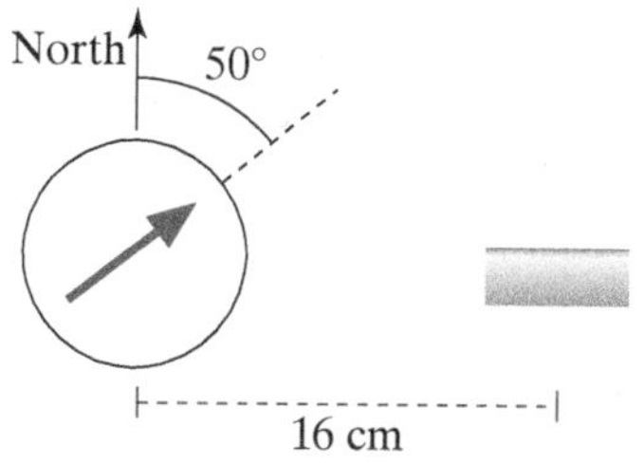

Figure 17.97

(a) Label the N and S poles of the bar magnet and explain your choice. **(b)** Determine the magnetic dipole moment of this bar magnet, including correct units.

••P57 A bar magnet with magnetic dipole moment 0.54 A·m^2 lies on the negative x axis, as shown in Figure 17.98. A compass is located at the origin. Magnetic north is in the negative z direction. Between the bar magnet and the compass is a coil of wire of radius 3.5 cm, connected to batteries not shown. The distance from the center of the coil to the center of the compass is 9.7 cm. The distance from the center of the bar magnet to the center of the compass is 22.5 cm. A steady current of 0.96 A runs through the coil. Conventional current runs clockwise in the coil when viewed from the location of the compass.

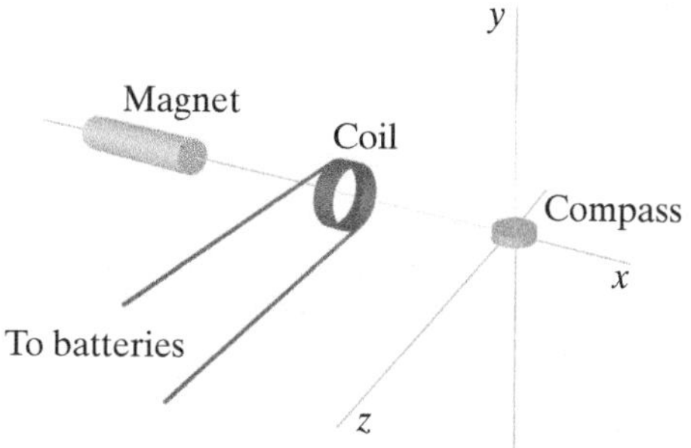

Figure 17.98

Despite the presence of the coil and the bar magnet, the compass points north. **(a)** Which end of the bar magnet is closer to the compass? **(b)** List all of the following statements that are correct: (1) At the location of the compass, the magnetic field due to the magnet is equal in magnitude to the magnetic field due to the coil. (2) It isn't necessary to take the magnetic field of the magnet into account, since the coil is in between the magnet and the compass. (3) Because not all parts of the magnet are the same distance from the compass, treating the magnet as a dipole located at the center of the magnet is an approximation. (4) The magnetic dipole moment of the coil is equal to the magnetic dipole moment of the compass. (5) At the location of the compass, the net magnetic field is equal to the magnetic field of the Earth. **(c)** How many turns of wire are in the coil?

Section 17.12

•P58 A particular alnico (aluminum, cobalt, nickel, and iron) bar magnet (magnet A) has a mass of 10 g. It produces a magnetic field of magnitude 6×10^{-5} T at a location 0.19 m from the center of the magnet, on the axis of the magnet. **(a)** Approximately what is the magnitude of the magnetic field of magnet A a distance of 0.38 m from the center of the magnet, along the same axis? **(b)** If you removed the original magnet and replaced it with a magnet made of the same material but with a mass of 50 g (magnet B), approximately what would be the magnetic field at a location 0.19 m from the center of the magnet, on the axis of the magnet?

••P59 The magnetic field along the axis of a bar magnet can be written like this:

$$B_{\text{axis}} \approx \frac{\mu_0}{4\pi}\frac{2\mu}{r^3}$$

(a) Use your own data from Experiment EXP9 to determine the magnetic dipole moment μ of your bar magnet. If you have mislaid those data, quickly repeat the measurements now. You will use the value of your magnetic dipole moment in later work. The magnetic dipole moment describes how strong a magnet is: the bigger the magnetic dipole moment, the bigger the magnetic field at some distance r. **(b)** Determine the mass of your magnet and thereby determine the number N of atoms in your magnet. Although your magnet is probably made of some alloy such as alnico V (51% iron, 8% aluminum, 14% nickel, 24% cobalt, and 3% copper), for simplicity assume it is made just of iron, which has a density of 8 g/cm^3 and an atomic mass of 56 (that is, 6×10^{23} atoms weigh a total of 56 g). Assuming that each of the atoms has a magnetic dipole moment with a value estimated in Section 17.12, see how well the atomic model for a magnet fits the measured value of the magnetic dipole moment of your bar magnet.

COMPUTATIONAL PROBLEMS

More detailed and extended versions of some computational modeling problems may be found in the lab activities included in the *Matter & Interactions, 4th Edition,* resources for instructors.

••P60 A proton moves with a constant velocity of 4×10^4 m/s in the $+x$ direction, along the x axis, starting at an initial location of $\langle -4\times 10^{-10},0,0\rangle$ m. **(a)** Write a program that calculates and displays the magnetic field of the moving proton at four observation locations in the yz plane (two on the y axis and two on the z axis), each a distance of 8×10^{-11} m from the x axis. Use cyan arrows to represent the magnetic field. **(b)** Does the magnetic field at one of your fixed observation locations change or stay constant? **(c)** What do you observe if you change the speed of the proton? **(d)** What do you observe if you change the proton to an antiproton?

••P61 Start with the program you wrote in Problem P60. Modify the program so that there are many observation locations, positioned on a cylindrical surface aligned with the x axis.

••P62 Start with the program you wrote in either Problem P60 or P61. Add to your display the electric field made by the proton at your fixed observation locations. Make the electric field arrows orange. Note that you need to scale the arrows representing electric field differently from the arrows used to represent magnetic field, since these quantities have very different magnitudes and units.

•••P63 Calculate and display (as arrows) the magnetic field made by a square current-carrying loop whose center is at the origin and which lies in the yz plane, with axis in the $+x$ direction. Each side of the square loop is 0.1 m long and the current in the loop is 3 A. Represent the square loop as a curve object with $N = 40$ positions. Show the field at many locations on a circle of radius 1 m in the xy plane centered on the origin, so that you can get a feel for the magnetic field at locations both along the axes and away from the axes. Print the values of the magnetic field where the circle intersects the x and y axes and check that they agree with the analytical expression for the magnetic field of this square loop when far from the loop. When you have the program working properly, determine the approximate minimum value of N that gives good agreement with the analytical solution.

•••P64 This problem is similar to Problem P63 but with a circular loop instead of a square loop, where the radius of the loop is 0.05 m and the current is again 3 A. Represent the circular loop as a curve object with $N = 40$ positions. Show the field at many locations on a circle of radius 1 m in the xy plane centered on the origin, so that you can get a feel for the magnetic field at locations both along the axes and away from the axes. Print the values of the magnetic field where the circle intersects the x and y axes and check that they agree with the analytical expression for the magnetic field of this circular loop when far from the loop. When you have the program working properly, determine the approximate minimum value of N that gives good agreement with the analytical solution.

•••P65 A solenoid of length $L = 0.5$ m and radius $R = 3$ cm is wound with $N = 50$ turns of wire carrying a current $I = 1$ A. Its center line lies on the x axis, with the origin at the center of the solenoid (Figure 17.99). This is a loosely wound solenoid, so it is not clear how uniform will be the magnetic field in its interior. To explore this, do the following:

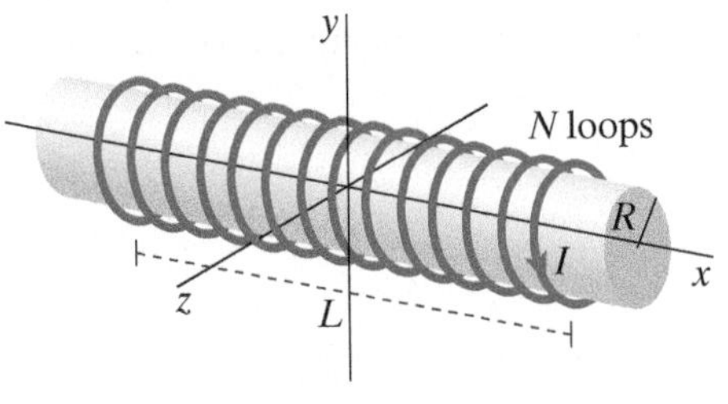

Figure 17.99

(a) Calculate and display magnetic field vectors at the locations in the xy plane, inside and outside of the solenoid, as shown in Figure 17.100. Do the pattern and direction of the magnetic field make sense? **(b)** Display the numerical value of the magnitude of the magnetic field at one location, the center of the solenoid. Also display the theoretical numerical value of the magnetic field, in the approximation that this is a very long solenoid ($B \approx \mu_0 NI/L$). **(c)** What is the minimum number of steps around one turn that are necessary to obtain good agreement between the theoretical value and your numerical integration? What is your criterion for good agreement? **(d)** Vary the number of turns of the helix. What is the minimum number of turns necessary to get an approximately uniform field inside the solenoid? **(e)** How do the magnitude and direction of the magnetic field outside the solenoid compare to the magnitude and direction of the magnetic field inside the solenoid?

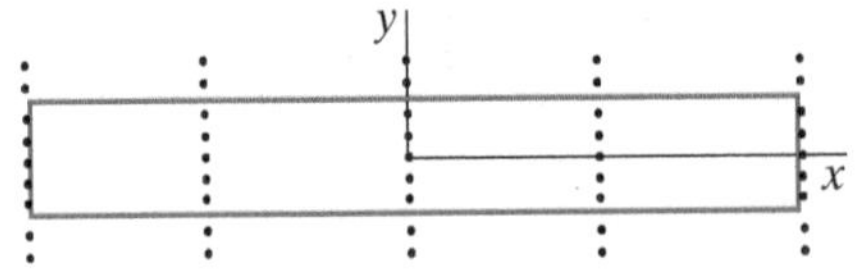

Figure 17.100

The code segment below creates a solenoid-shaped `curve`, and also creates a list of `arrows` at the specified observation locations. The `arrow` at the center of the array of observation locations is given the special name `center`.

```
from visual import *
scene.width = 800

L = 0.5             ## length of solenoid
R = 0.03            ## radius of solenoid
mzofp = 1e-7        ## mu0/4pi
I = 1.0             ## conventional current
Nturns = 75         ## number of turns
Nsteps = 20         ## number of line segments/turn

## a curve is a list of points
## start with an empty curve (no points)
solenoid = curve(color=color.yellow,
            radius = 0.001)
dx = L/(Nturns*Nsteps)
omega = 2*pi*Nturns/L
x = -L/2
## add points to the curve
while x < L/2+dx:
    solenoid.append(pos=vector(x,
                R*sin(x*omega),
                R*cos(x*omega)))
    x = x + dx

## a list of arrows
## at observation locations with
## arbitrary directions initially
Barrows = []      ## empty list
dx = L/4
## skip y = 0.03 which is inside wire
ylist = [-0.05, -0.04, -0.02, -0.01,
         0., 0.01, 0.02, 0.04, 0.05]
x = -L/2
while x < L/2 + dx:
    i = 0
    while i < len(ylist):
        aa = arrow(pos=vector(x,ylist[i],0),
```

```
                      axis=vector(0,0,0.005),
                      color=color.cyan,
                      shaftwidth=0.003)
            Barrows.append(aa)
            i = i + 1
        x = x + dx

## outer while loop: choose observation location
## Barrows[i]
## inner while loop: calculate field at
## Barrows[i].pos
## and scale axis of arrow appropriately
## ...
```

ANSWERS TO CHECKPOINTS

1 6×10^{18} electrons/s; 5.4×10^{21} electrons
2 5.4×10^{-6} T
3 **(a)** Figure 17.101; **(b)** Figure 17.102

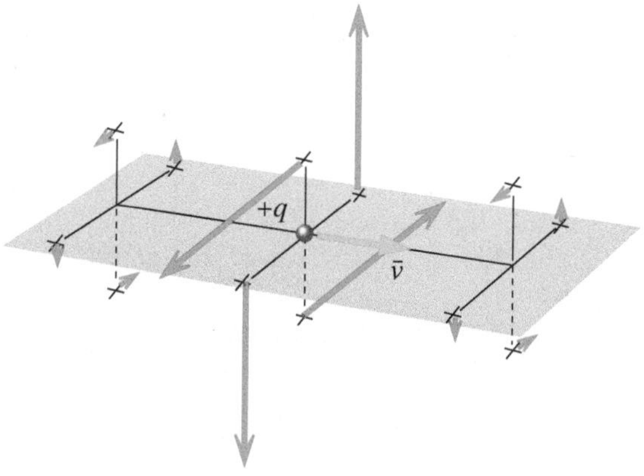

Figure 17.101

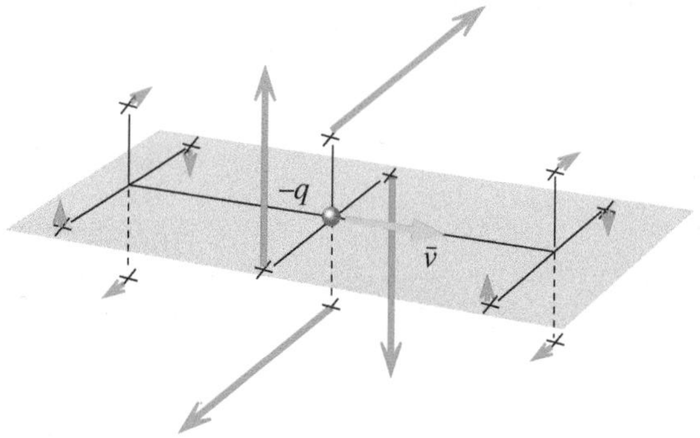

Figure 17.102

4 Electron current flows to the left, conventional current to the right.
5 2.4 A
6 9 $A \cdot m^2$

CHAPTER

18

Electric Field and Circuits

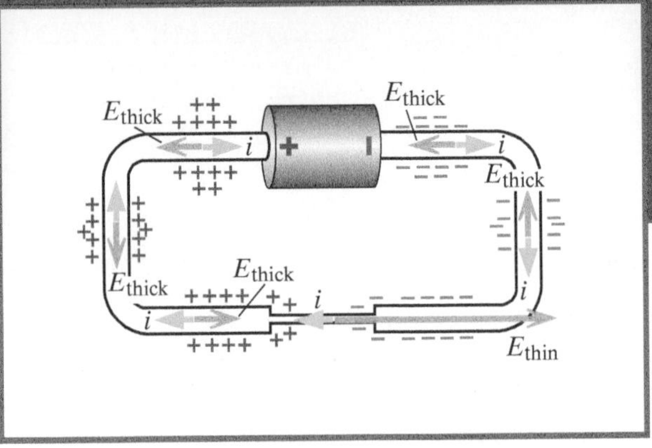

OBJECTIVES

After studying this chapter you should be able to

- Calculate the electric fields inside elements of a steady-state circuit, and depict these fields using arrows.
- Explain the role of charges on the surface of a circuit, and draw the approximate distribution of charges on the surface of the components of a simple steady-state circuit.
- Use the principles of conservation of energy and conservation of charge to analyze a circuit quantitatively, using microscopic quantities such as mobile charge density and charge mobility.

18.1 A CIRCUIT IS NOT IN EQUILIBRIUM

When an electric field is applied to a conductor, the mobile charges in the conductor experience forces, and begin to move in the direction of those forces. We have previously studied situations in which the flow of charges lasts for a very short time, until the buildup of charge at the surface of the conductor produces an electric field that is equal and opposite to the applied field, the net electric field inside the conductor becomes zero, and the system reaches equilibrium. In contrast, an electric circuit is a system that does not reach equilibrium. Despite the motion of charges, the net electric field inside the conductor does not go to zero, and charge flow continues for a long period in this nonequilibrium system.

Our goal in this chapter is to construct a microscopic model of what happens in an electric circuit, using familiar physics principles (the Energy Principle, the Momentum Principle), the concepts of electric field and potential, and the microscopic model of a metal that we developed in Chapter 14. This is a somewhat different view of electric circuits from the macroscopic view you may have seen in a previous physics course. We will be interested in questions such as:

- Are charges used up in a circuit?
- How is it possible to create and maintain a nonzero electric field inside a wire?
- What is the role of the battery in a circuit?

Looking ahead, once we understand the motion of charges in electric circuits, we will be able to explore magnetic forces. In the previous chapter we studied the magnetic fields made by moving charges. However, we did not discuss the effect of those magnetic fields on other charges. Just as moving charges make magnetic fields, only moving charges are affected by magnetic fields. One situation in which we can easily observe these effects is in electric circuits where currents are flowing. Before we can study magnetic forces in detail, we need to know more about the motion of charges in electric circuits.

18.2 CURRENT IN DIFFERENT PARTS OF A CIRCUIT

As a first step in constructing a microscopic model to explain current flow in a circuit, we need to address a crucial question, both experimentally and theoretically: What happens to the charges that flow through the circuit? Do they get used up? Is there a different amount of current in different parts of a series circuit, a circuit with a single path?

The Steady State

Once a circuit has been assembled, current keeps flowing at about the same rate for quite a long time. The brightness of a light bulb or the amount of compass deflection does not change noticeably over several minutes. In fact, since it takes quite a while for the batteries to run down under these circumstances, the current remains about the same for several hours. We say that these circuits are in a "steady state." The steady state is not the same as equilibrium.

EQUILIBRIUM VERSUS STEADY STATE

Equilibrium means that no current is flowing: $\overline{v} = 0$.

Steady state means that charges are moving ($\overline{v} \neq 0$), but their drift velocities at any location do not change with time, and there is no change in the deposits of excess charge anywhere.

Current in Different Parts of a Simple Circuit

One can measure the amount of current flowing in different parts of a circuit in the steady state by using compass deflections as an indicator of current, since the magnetic field made by a wire is proportional to the current in the wire.

Before making such a measurement, it is instructive to think about what we expect to see. Consider a circuit consisting of two batteries and a bulb as shown in Figure 18.1. Electron current flows from the negative terminal of the battery toward the bulb.

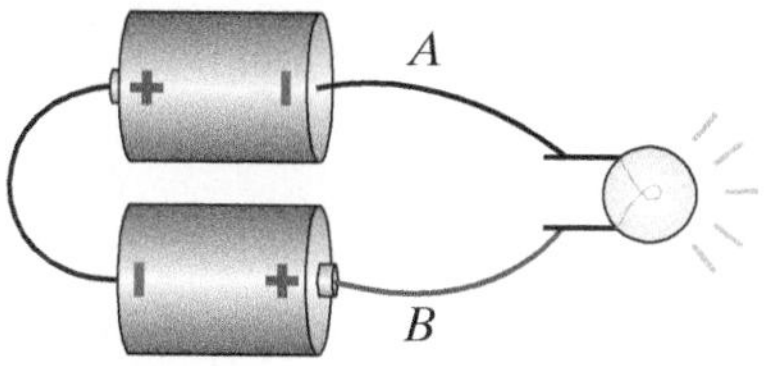

Figure 18.1 A circuit containing a bulb and two batteries in series.

QUESTION How would you expect the amount of electron current at location A to compare to the electron current at location B? Make a clear prediction—write it down. (It is ok to guess.)

Most people choose one of the following possibilities:

1. There should be no current at all at B, because all of the electron current coming from the negative end of the battery is used up in the bulb.
2. The current at B should be less than the current at A, because some of the current is used up to make the bulb give off light and heat.
3. The current should be the same at A and at B.

Before doing the experiment, let's consider these alternatives from a theoretical viewpoint, using what we know about electric interactions and the properties of metals.

Can Current Be Used Up in the Bulb?

It is reasonable to expect that something will be used up in the bulb in order to produce heat and light. Let's consider whether electric current can be the thing that is used up.

> QUESTION Current in a metal wire is simply the flow of electrons past a point. Can electrons be used up in the bulb?

Electrons alone cannot be destroyed, because this would violate the fundamental principle of conservation of charge. If negative particles were destroyed, the universe would become more and more positive!

We do know that electrons can react with positrons (positively charged anti-electrons), annihilating both particles and producing large amounts of energy. However, ordinary matter cannot coexist with antimatter for longer than a tiny fraction of a second, so there cannot be a supply of positrons in the circuit to annihilate electrons.

> QUESTION Could electrons just accumulate in the bulb, so that the current at B would be less than at A?

If electrons accumulated in the bulb, the bulb would become negatively charged. The negative charge of the bulb would become large enough to repel incoming electrons and stop the current. Since in the steady state current keeps flowing, this can't be happening. (Moreover, an increasingly negatively charged bulb would strongly attract nearby objects, which you don't observe.)

We have used proof by contradiction to rule out the possibility that electrons are annihilated or get stuck in the bulb, so we must conclude that current cannot be used up in a light bulb. Since we have ruled out both possibilities (1) and (2) by this argument, we expect that alternative (3) is correct: the current should be the same at A and at B.

Measuring Current All around a Circuit

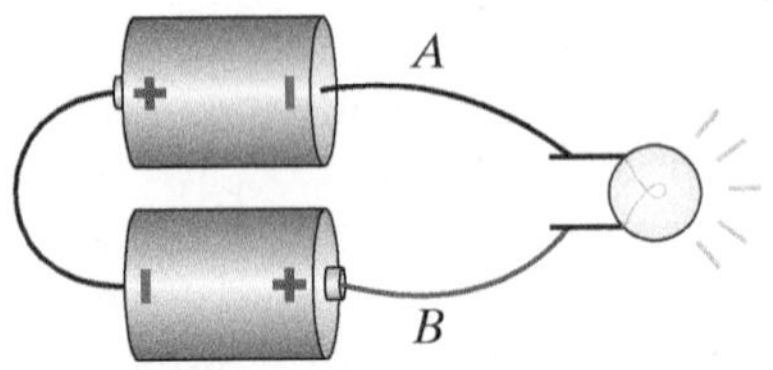

Figure 18.2 If we measure the current at A and at B, we should find these currents to be the same.

We have predicted on theoretical grounds that current cannot be used up in a circuit. If you make careful measurements, you should indeed find the same current everywhere in a series circuit (a circuit with no parallel branches, such as the circuit in Figure 18.2). To make these measurements yourself using simple equipment, do Experiment EXP1 at the end of this chapter.

What Is Used Up in the Light Bulb?

As can be observed in Experiment EXP1, the amount of current entering and leaving a lighted bulb in a steady state circuit is the same. You may still feel uneasy about all this. You might ask, "If electrons don't get used up in the bulb, what makes the light that we see? You can't get something for nothing!"

> QUESTION Rub your hands together as hard as you can and as fast as you can for several seconds. What change do you observe in your hands?

Your hands get hot, but they are certainly not used up! However, the feeling that something must be used up to light the bulb is entirely reasonable.

> QUESTION What does get used up to warm your hands or to make a light bulb hot enough to emit visible light?

In both cases energy is being "used up," or rather transformed from a form of stored energy into thermal energy (and energy in the form of emitted light, in the case of the bulb). In order to move your hands you must use up some of the stored chemical energy in your body. Similarly, to force electrons through the filament, some of the chemical energy stored in the battery must be used up.

Forcing electrons through the filament heats the metal with a kind of friction, not unlike the friction that heats your hands when you rub them. In both cases chemical energy is converted into thermal energy, but without destroying the objects that rub against each other.

Since energy flows out of the system (the circuit) in the form of heat and light, we will need to take this into account when we study the energetics of circuits later in this chapter.

Current at a Node

We can extend these considerations to more complicated arrangements of wires. In general we have the following important principle, based on charge conservation in the steady state (a node is a junction of two or more conductors in a circuit):

THE CURRENT NODE RULE

In the steady state, the electron current entering a node in a circuit is equal to the electron current leaving that node.

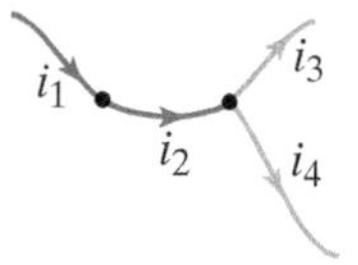

Figure 18.3 Electron current at different locations in part of a circuit.

This rule is not a fundamental principle, but rather a consequence of the fundamental principle of conservation of charge and the definition of steady state. It is also called "the Kirchhoff node rule."

EXAMPLE **Current at a Node**

Figure 18.3 depicts a portion of a circuit. Suppose that in the steady state $i_1 = 3 \times 10^{19}$ electrons/s and $i_4 = 7 \times 10^{19}$ electrons/s. What is i_3?

Solution

$$i_1 = i_2$$
$$i_2 = i_3 + i_4$$
$$3 \times 10^{19} \text{ electrons/s} = i_3 + 7 \times 10^{19} \text{ electrons/s}$$
$$i_3 = -4 \times 10^{19} \text{ electrons/s}$$

What does the minus sign mean physically in this answer? It means that we guessed wrong about the direction of i_3. The arrow representing i_3 shown on the diagram is incorrect, and electrons flow in the opposite direction—into the node instead of out of it. In circuit diagrams we use arrows to indicate our guesses as to the direction of electron or conventional current, and symbols such as i_3, but if i_3 turns out to be negative, it just means that we guessed wrong as to which way the current was running.

QUESTION Is the current always the same everywhere in a steady-state circuit?

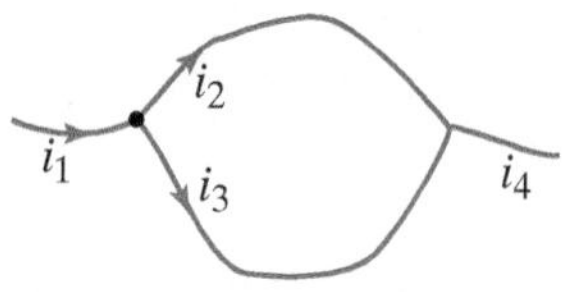

Figure 18.4 A portion of a circuit with two parallel branches.

No. Consider Figure 18.4. There are two parallel branches carrying currents i_2 and i_3. Assuming the direction of the arrows is correct, both i_2 and i_3 must be less than i_1; depending on the wires used, it may be the case that $i_2 \neq i_3$. We do know, however, that in this circuit $i_2 + i_3 = i_1$. One can of course express the node rule in terms of conventional current; we could write $I_2 + I_3 = I_1$.

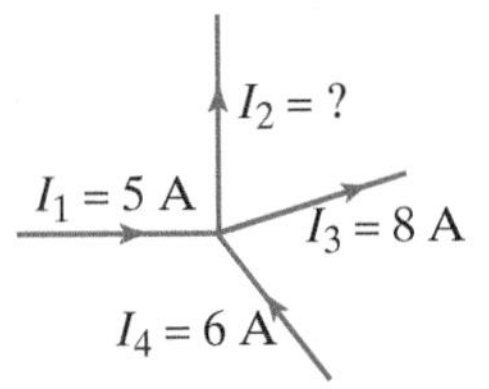

Figure 18.5 A node in a circuit.

Checkpoint 1 Figure 18.5 shows a portion of a steady-state circuit, with the magnitudes of the conventional currents given in amperes (A), and the directions given by arrows. Write the node equation for the circuit in Figure 18.5. What is the value of the outward-going current I_2? If I_4 were 1 A instead of 6 A, what would be the value of the outward-going current I_2? What is the meaning of the minus sign?

18.3 ELECTRIC FIELD AND CURRENT

In Chapter 14 we discussed the Drude model for the motion of mobile electrons in a metal and the relationship $\bar{v} = uE$: The average drift speed of the mobile electrons is proportional to the magnitude of the electric field. The proportionality constant u is called the electron mobility.

QUESTION According to the Momentum Principle, once an object is in motion, no force is required to keep it moving at constant speed. Given this, why is an electric field needed to keep a current flowing?

If the mobile electrons in a metal did not interact at all with the lattice of positively charged atomic cores, then once a current got started, it could flow forever. However, as we have observed, there is an interaction; the moving electrons do lose energy to the lattice, increasing the thermal motion of the atoms. We detect this by observing that the wire gets hot—in some cases hot enough to emit visible photons. Unless an electric field is present to increase the momentum of the mobile electrons repeatedly after collisions, their energy will quickly be dissipated (as energy can be dissipated by friction), and the current flow will stop.

Electrons Cannot Push Each Other through the Wire

QUESTION One might argue that in a circuit each mobile electron makes an electric field that affects the mobile electron to the left of it ("pushing" on its neighbor), thereby causing current to flow. Given our description of the mobile electron sea in a metal, what is wrong with this analysis?

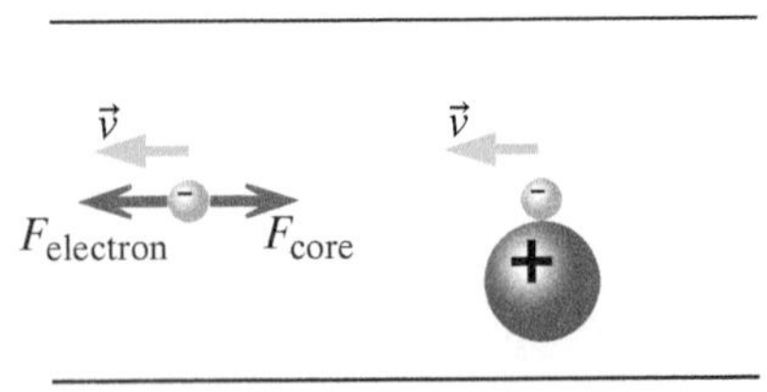

Figure 18.6 A mobile electron cannot push another mobile electron through a wire. The total force on a mobile electron due to the other mobile electrons and the positive cores is zero.

As we have stated previously, there can be no excess charge inside a conductor (and we'll see later that this also applies to the steady state as well as to equilibrium), so the number density of mobile electrons inside a metal wire must equal the number density of positive atomic cores—the inside of the wire is electrically neutral. In Figure 18.6 the repulsion of the mobile electron on the left by a mobile electron on the right is canceled by the attraction of a neighboring positive core, so the total force on a mobile electron due to the other mobile electrons and the positive cores is zero. The result is that the mobile electron sea behaves rather like an ideal gas, in which none of the particles interact with other particles. Electrons cannot continually push each other through the wire like peas pushed through a tube from one end of the tube. Rather, it must be other charges somewhere outside the wire that make an electric field throughout the wire that continually drives the electron current. At this point in our analysis it is not yet clear exactly where these charges are.

Magnitude of $\vec{E}$ in a Circuit

Because there are many mobile electrons in a metal wire, and because in most metals the electron mobility is quite high, the magnitude of the electric field needed to sustain a steady-state current in a wire is small.

EXAMPLE **Magnitude of $\vec{E}$ in a Circuit**

In the previous chapter we calculated the drift speed in a copper wire to be 5×10^{-5} m/s for a typical electron current. The mobility of mobile electrons in copper is $u = 4.5 \times 10^{-3}$ (m/s)/(N/C). Calculate the magnitude of the electric field inside the wire.

Solution

$$\bar{v} = uE$$

$$E = \frac{5 \times 10^{-5}\,\text{m/s}}{4.5 \times 10^{-3}\,\dfrac{\text{m/s}}{\text{N/C}}} = 0.011\,\text{N/C}$$

Because copper is a very good conductor, only a small electric field is needed to sustain a current in a copper wire.

Electric Field and Drift Speed in Different Elements of a Circuit

We confirmed by experiment that in the steady state, the current was the same everywhere in a series circuit (a circuit without parallel branches). However, we know that the electron current i depends on the cross-sectional area A of a wire, since $i = nA\bar{v}$. The cross-sectional area of the tungsten filament of a light bulb is much less than the cross-sectional area of the copper connecting wires in the circuit.

In order to simplify thinking about this situation, let's first consider a circuit in which a wire leads into another, thinner wire of the same material (Figure 18.7).

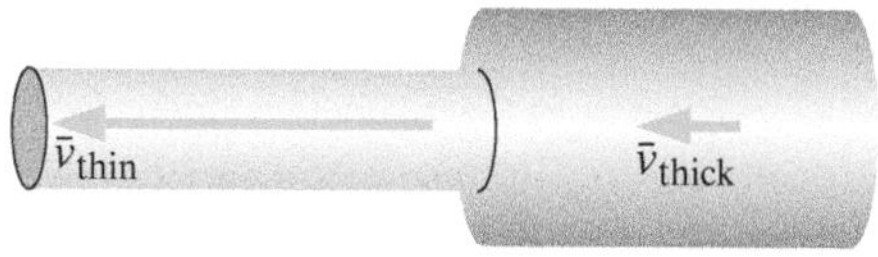

Figure 18.7 The drift speed of electrons in the thin and thick wires.

QUESTION How must the drift speed of mobile electrons in the thin wire compare to the drift speed of mobile electrons in the thick wire?

Since the current must be the same in the thick and thin wires, the electron drift speed must be greater in the thin wire.

$$nA_{thin}\bar{v}_{thin} = nA_{thick}\bar{v}_{thick}$$

and therefore

$$\bar{v}_{thin} = \frac{A_{thick}}{A_{thin}}\bar{v}_{thick}$$

Note that this is very different from "compressible" flows such as highway traffic. On a highway the density of cars (how close they are to each other) can vary. For example, when a highway narrows down from two lanes to one, cars usually go slower in the narrow region but are correspondingly closer together (higher density). Because the electron sea in a metal is nearly incompressible (its density cannot change significantly), in the steady state the electrons must go faster in the narrow region. The density n of mobile electrons in copper is the same in both wires.

QUESTION If the electron drift speed is greater in the thin wire, how must the electric field in the thin wire compare to the field in the thick wire?

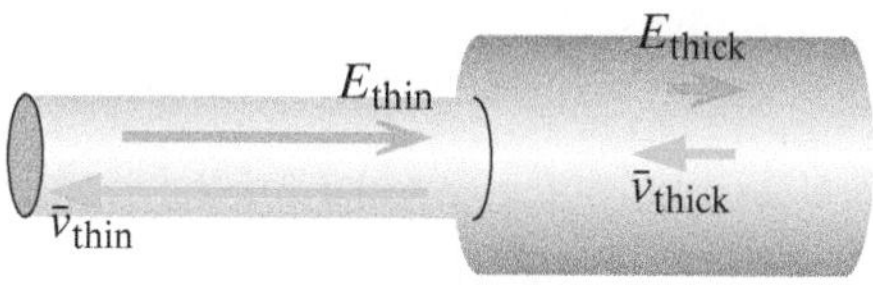

Figure 18.8 Electric field at locations inside the thin and thick wires.

Since the drift speed must be greater in the thinner wire, the electric field in the thinner wire must also be larger, since $\bar{v} = uE$ (Figure 18.8).

Checkpoint 2 Suppose that a wire leads into another, thinner wire of the same material that has only half the cross-sectional area. In the "steady state," the number of electrons per second flowing through a cross section of the thick wire must be equal to the number of electrons per second flowing through a cross section of the thin wire. If the drift speed $\bar{v}_1$ in the thick wire is 4×10^{-5} m/s, what is the drift speed $\bar{v}_2$ in the thinner wire? If the electric field E_1 in the thick wire is 9×10^{-3} N/C, what is the electric field E_2 in the thinner wire?

Direction of Electric Field in a Wire

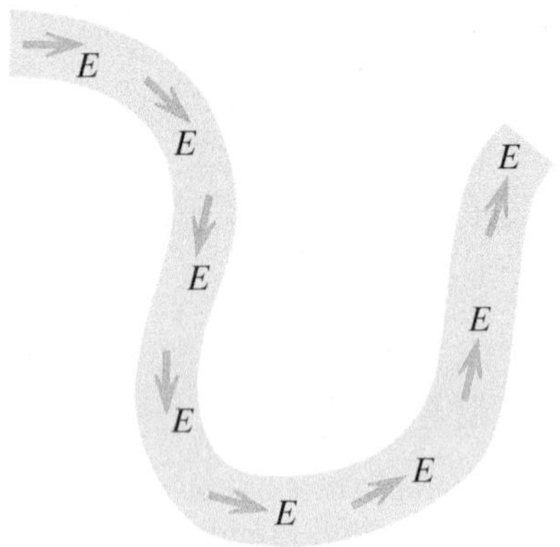

Figure 18.9 The electric field in a current-carrying wire must follow the wire.

According to our measurements of the magnetic field made by the moving charges in our circuits, the current is the same in every part of a series circuit. Since conventional current $I = |q|nA\bar{v} = |q|nAuE$, and since conventional current flows in the direction of $\vec{E}$, this must mean that E is the same in every part of a wire with uniform properties ($|q|$, n, A, and u), and that $\vec{E}$ must be parallel to the wire at every location—even if the wire twists and turns, as indicated in Figure 18.9.

Proof That Field and Current Inside the Wire Are Uniform

Not only must the electric field follow the wire, but it also must be uniform across a cross section of the wire. Suppose for a moment that this isn't true—that instead, at different locations in the cross section of a straight wire, the drift speed and therefore the electric field are different (Figure 18.10).

Consider the potential difference

$$\Delta V_{ABCDA} = -\int_A^A \vec{E} \bullet d\vec{l}$$

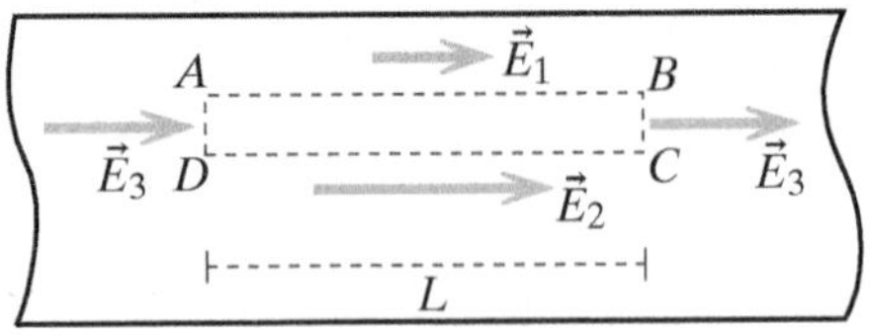

Figure 18.10 Could E vary at different locations in a uniform wire?

along the round-trip path $ABCDA$ in Figure 18.10 (from A to B to C to D and back to A).

QUESTION Calculate the contributions along each part of the path, $\Delta V_{AB}, \Delta V_{BC}, \Delta V_{CD}$, and ΔV_{DA}, and the sum ΔV_{ABCDA}. What can you conclude about E_1 and E_2?

The requirement that the round-trip potential difference be zero means that E_1 and E_2 have to be equal. Therefore the electric field must be uniform both along the length of the wire and also across the cross-sectional area of the wire (Figure 18.11). Since the drift speed is proportional to E, we find that the current is indeed uniformly distributed across the cross section. This result is true only for uniform cross section and uniform material, in the steady state. The current is not uniformly distributed across the cross section in the case of high-frequency (non-steady-state) alternating currents, because time-varying currents can create non-Coulomb forces, as we will see in a later chapter on Faraday's law.

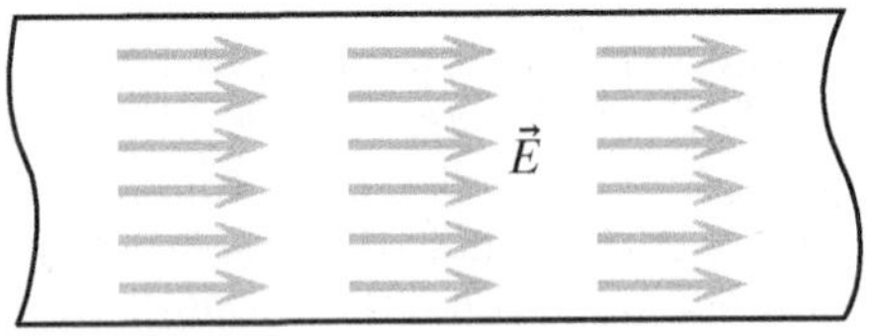

Figure 18.11 The electric field in a wire in a steady-state circuit must be uniform both along the wire and across the cross-sectional area of the wire.

18.4 WHAT CHARGES MAKE THE ELECTRIC FIELD INSIDE THE WIRES?

We have concluded that in a steady-state circuit:

- There must be a nonzero electric field in the wires.
- The magnitude of $\vec{E}$ must be the same throughout a wire of uniform cross section and material.
- The direction of the electric field at every location must be along the wire, since current flow follows the wire.

Consider a very simple circuit consisting of a bulb connected by long wires to a battery (Figure 18.12). We have concluded that there must be an electric field inside the metal bulb filament, forcing electrons through the filament, heating it so much that it glows. Since electric fields are produced by charges, there must be excess charges somewhere to produce the electric field inside a wire in a circuit. Where might these charges be? What distribution of charges could produce a pattern of electric field following the wire?

Figure 18.12 There is an electric field in the bulb filament. Is it charges in and on the battery that produce this field?

Are the Excess Charges Inside the Conductors?

The bulb filament and the wires are conductors, so there can be no excess charge in the interior of the bulb filament and wires. In these metal objects, there are equal numbers of mobile electrons and positive atomic cores. The excess charges cannot be inside the wires, so they must be somewhere else.

Are the Excess Charges on the Battery?

One might reasonably assume that the charges that make the electric field are in and on the battery, since that's the active element in the circuit. Perhaps there are + charges on the + end of the battery, and − charges on the − end? In that case the field made by the battery will be something like the field of a dipole.

> QUESTION Suppose the bulb is 10 cm from the battery, and shining brightly (Figure 18.12). Move the bulb to a distance of about 1 cm from the battery, allowing the wires to bend (Figure 18.13). Very roughly, about how much larger is the electric field made by the battery at this closer location?

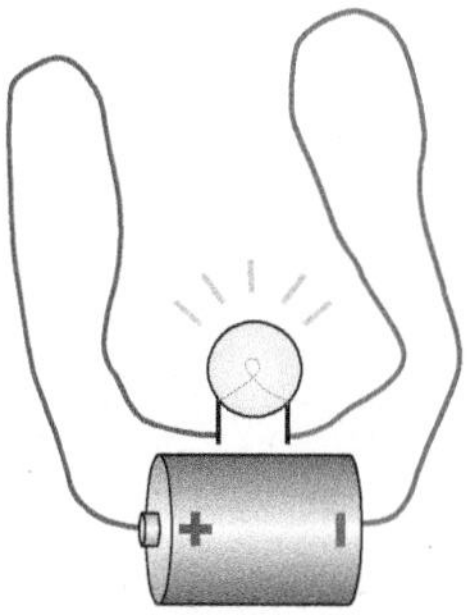

Figure 18.13 Bend the wires so the bulb is much closer to the battery. Presumably the electric field in the bulb should get much larger, but does it?

Although we're too close to the battery for the dipole-like electric field to be proportional to $1/r^3$, a rough estimate is that the electric field at 1 cm distance is about $1 \times 10^3 = 1000$ times as large as the field at 10 cm.

> QUESTION Now that the field made by the battery at the location of the bulb filament has increased by a factor of 1000 or so, what should happen to the electron current in the bulb and the brightness of the bulb?

For an ordinary metal, the electron current is proportional to the electric field: $i = nA\bar{v} = nAuE$. Therefore the electron current should increase by about a factor of 1000. The bulb should be enormously brighter than before! Yet this doesn't happen. As you have seen in working with a simple circuit, moving the bulb has no effect at all on the bulb brightness, or on the amount of current indicated by your compass.

Another test of the "charge on the battery" hypothesis also fails. To drive current through the bulb filament, the electric field must have a component parallel to the filament. Yet if you rotate the bulb, you don't see any change in the bulb brightness (Figure 18.14). If the charges responsible for making the electric field were solely in and on the battery, rotating the bulb ought to make a big difference in the brightness—the bulb might even go out.

Figure 18.14 Rotate the bulb. Does the brightness change, due to a change in the parallel component of electric field?

The fact that the bulb doesn't get brighter when moved toward the battery, and does not go out when rotated, proves that charges in and on the battery cannot be the only contributors to the electric field in the bulb filament. There must be other charges somewhere else (and these charges can't be inside the wires or bulb filament, either, where there is no net charge, averaged over a few atomic diameters). This is deeply puzzling.

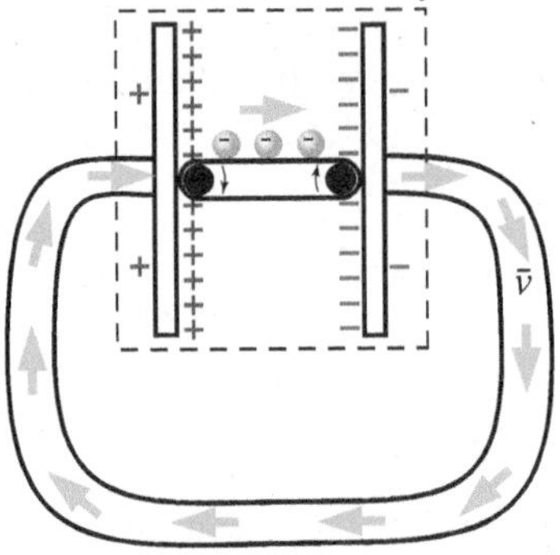

Figure 18.15 A "mechanical battery": a conveyor belt maintains a charge separation that drives a steady current in the wire.

A Mechanical Battery

To simplify our further analysis of circuits, including in particular our search for the location of the charges responsible for producing the electric field in a bulb filament, we introduce a simplified model of a battery, a "mechanical battery" that is much easier to understand in detail than a chemical battery yet behaves in a circuit in ways that are similar to a chemical battery.

Our mechanical battery has a conveyor belt driven by a motor or a hand crank that pulls electrons out of one plate, making it positive, and pushes them onto another plate, making it negative (Figure 18.15). This action replenishes electrons that leave the negative plate and move through the wire with drift speed $\bar{v}$, and it removes electrons that enter the positive plate after traveling through the wire. You may have seen a Van de Graaff generator, which has a motor-driven conveyor belt like this that pumps lots of charge onto a metal sphere.

This combination of charged plates plus the motor-driven conveyor belt acts very much like a chemical battery. The main difference is that here we imagine moving the charges by mechanical means, and this mechanism is easier to understand than the mechanism for charge transfer in a chemical battery. As long as the motor is able to maintain the charge separation across the two plates, we can have a steady-state current running in the wire.

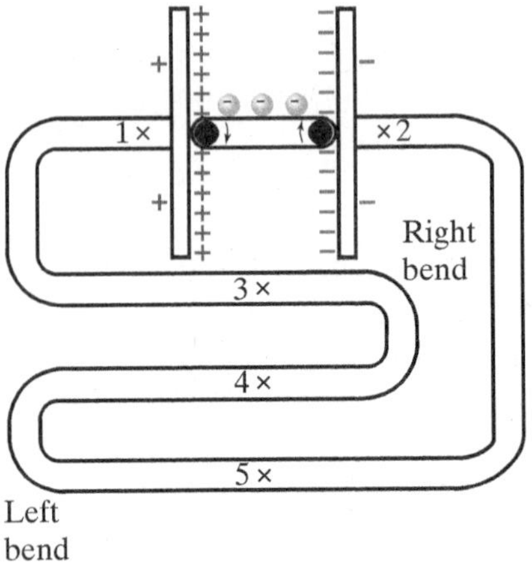

Figure 18.16 A circuit with a Nichrome wire and a mechanical battery.

Field Due to the Battery

We will now consider a simple circuit consisting of a mechanical battery and a resistive Nichrome wire that has some twists and turns in it. (The bare wire in the experiment kit mentioned in the experiment section of Chapter 17 is Nichrome; the heating elements in toasters are made of such wire.) You know from your own experience with a simple circuit that twists and turns do not seem to affect the amount of steady-state current, which seems odd. It will seem much odder in a moment!

QUESTION At the locations marked × on a copy of Figure 18.16 (locations 1 through 5), draw arrows representing the approximate electric field due solely to the charges on the metal plates of the mechanical battery. (The amount of charge on the belt is completely negligible compared to the charge on the plates.) Forget the presence of the wire for the moment.

QUESTION Next, in a different color or with a different-looking arrow, indicate the drift velocity of the mobile-electron sea at those locations, assuming that the drift velocity is due solely to the electric field made by the charges on the plates of the mechanical battery.

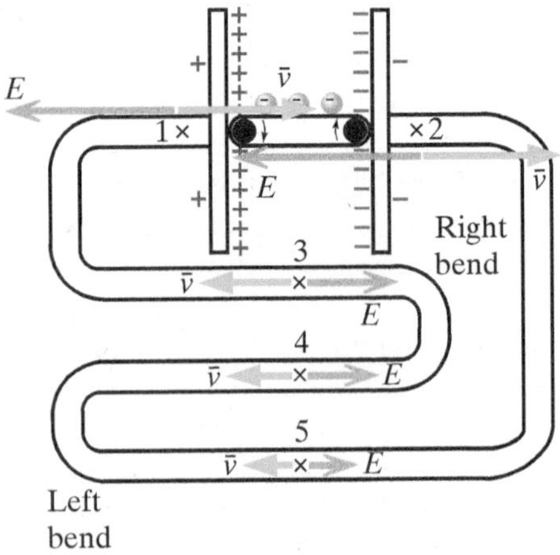

Figure 18.17 Electric field (orange) due solely to the mechanical battery, and electron drift velocities (green) that would result from this pattern of electric field. Some arrows are offset from the observation locations for clarity.

Good grief! Figure 18.17 shows that we've got the electron current running *upstream* at location 4 in the wire! That can't be right in the steady state (the situation in which charge distributions and currents are not changing).

We are forced to conclude that in the steady state there must be some other charges somewhere that contribute to the net electric field in such a way that the electric field points upstream everywhere (giving an electron drift velocity downstream everywhere).

Charge Buildup on the Surface of the Wire

Remember that the electrons in the mobile-electron sea inside the metal effectively don't interact with each other, because their mutual repulsions are on the average canceled by attraction to the positive atomic cores so they can't push each other through the wire. Some other charges must contribute to the electric field that is responsible for pushing the electrons through the wire.

Take a look at the left bend in the wire in Figure 18.17. If the only charges were on the plates, electron current would flow toward the left bend from both neighboring sections. Here we have a section of wire that has electron current flowing into it but not out.

QUESTION What effect will this have on this section of the wire? Be as specific as you can, bearing in mind that the wire is made of metal. Illustrate your thoughts on your copy of Figure 18.16.

If electrons are flowing into a section of wire from both ends, that section will gain a net negative charge. Since the wire is made of metal, electrons are free to move, and all excess charge will move to the surface of the wire. In Chapter 14 we asserted that excess charge goes to the surface of a metal. This assertion is plausible, because excess charges in the interior of a metal repel each other toward the surface.

More precisely, excess negative charge expands the mobile-electron sea so that some of the sea peeks out at the surface, giving a negative surface charge. Excess positive charge contracts the electron sea, allowing positive cores to peek out at the surface. A rigorous proof that all the excess charge goes to the surface of a metal requires Gauss's law, which we will study in Chapter 21.

Thus, negative charge accumulates on the surface of the left bend.

QUESTION What happens at the right bend in the wire? Illustrate your thoughts on a copy of Figure 18.16.

At the right bend electrons flow out of both ends, leaving a net positive charge, also on the surface of the wire.

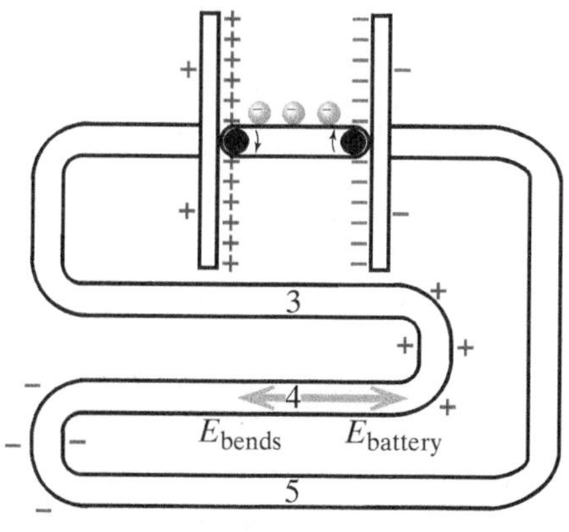

Figure 18.18 The electric field at location 4 is reduced by the contribution of the charges that have piled up on the bends.

QUESTION Focus your attention on location 4, a place where the electric field of the mechanical battery alone would drive current in the wrong direction. At location 4, what is the direction of the additional contribution to the electric field made by the excess + and − charges on the left and right bends?

The charges on the bends contribute an electric field to the left at location 4 that is opposite to the electric field of the mechanical battery. The "wrongly" directed net electric field is reduced in magnitude (Figure 18.18). However, as long as the net field at location 4 still points to the right (the "wrong" direction), more and more excess charge will pile up on the bends, and the electric field contributed by these charges will grow.

QUESTION How far will this pileup process go? What stops it?

The pileup will continue automatically until there is so much charge on the bends that the net field at 4 points to the left. Other charges must appear at various places on the surfaces of the wires in order that eventually the net field at locations 3, 4, and 5 all have the same magnitude. Only then will the electron current $i = nA\bar{v} = nAuE$ into a bend equal the electron current out of a bend, with no further change in the amount of charge on the surface of the bend.

We see examples of "feedback" in both the left bend and the right bend of the circuit. If the initial field is such that it drives current that is different in amount or direction from the steady-state current, surface charge automatically builds up in such a way as to alter the current to be more like the steady-state current. Moreover, once the steady state is established, there is "negative feedback": any deviation away from the steady state will produce a change in the surface charge that tends to restore the steady-state conditions.

We have deliberately emphasized the most dramatic aspect of this feedback, which is what happens at the bends in the wire. However, the resulting buildup of surface charge isn't limited to bends in the circuit. There must be positive and negative charge on the surface of the wires near the

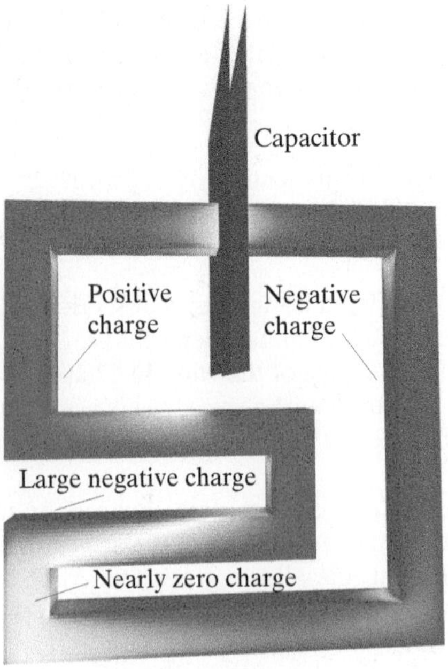

Figure 18.19 The distribution of surface charge on the twisty circuit discussed above, generated by a computational model. Red represents positive surface charge and blue represents negative surface charge. The intensity ("saturation") of the color is proportional to the amount of charge. Square wires were used to simplify the computation.

+ and − plates of the mechanical battery. The surface charge must arrange itself in such a way as to produce a pattern of electric field that follows the direction of the wire and has the same magnitude throughout the wire. Since the current is given by $i = nA\bar{v} = nAuE$, to have the same i everywhere in this series circuit we must have the same E everywhere inside the wires.

Figure 18.19 shows a detailed visualization of the distribution of surface charge, produced by a computational model to be described later, with a charged capacitor playing the role of the battery (we continually replenish the supply of charges on the capacitor plates). The calculations confirm the predictions we made using qualitative reasoning: there is an accumulation of positive charge on the right of the lower middle wire and of negative charge on the left. It is also evident that in this geometrically complicated circuit the distribution of surface charge is rather complex because some parts of the circuit polarize other parts. For example, note how the wire containing the location we've been considering is strongly polarized by the positive wire just above it and the negative wire just below it. There is however a general trend of the surface charge running from positive to negative around the circuit.

18.5 SURFACE CHARGE DISTRIBUTIONS

Despite the complex details of the pattern of charge on the twisty circuit discussed in the previous section, one thing stands out dramatically: the amount and sign of the charge on the surface of the wire varies. The general trend is clear: the wire to the left of the capacitor is strongly positive; the wire to the right is strongly negative, and there is only a small amount of charge in the middle of the circuit. This variation can be described as a gradient of surface charge. Mathematically, a gradient is a derivative describing the variation of some quantity in space. If we think of a circuit as a one-dimensional path, we can find the gradient of surface charge as we travel along the wire from one end to the other.

To see how a gradient of surface charge is necessary to produce a uniform field inside a wire in a steady-state circuit, let's start with a simple case. We'll consider a long straight wire in a circuit, and we'll look at a section of the middle of the wire, far from the battery. We'll model the surface of the wire as a series of rings of charge. In our simple model circuit each ring will be uniformly charged. In order to calculate the electric field inside the wire, we'll model each ring as a collection of point charges (Figure 18.20).

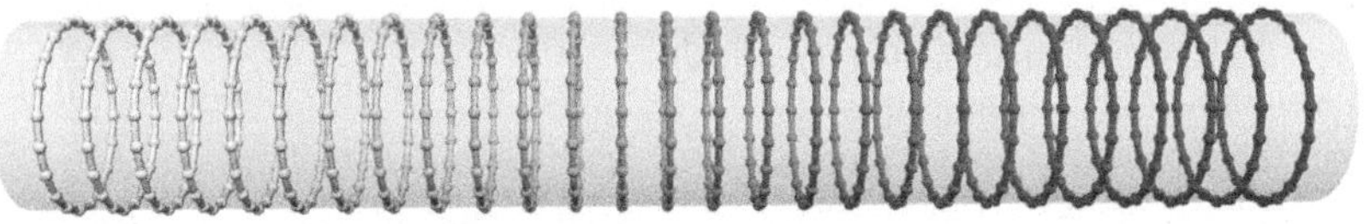

Figure 18.20 A model of the charge on the surface of a cylindrical wire as a series of charged rings, each approximated by a collection of point charges. Red is positive, blue is negative, and white is uncharged. The saturation of the color indicates the amount of charge. The surface charge on this wire changes linearly from slightly negative at the left end to strongly positive at the right end.

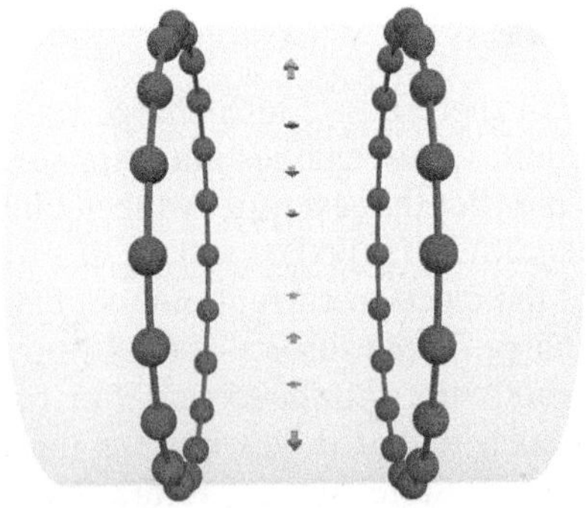

Figure 18.21 Two equally charged rings on the surface of a wire and their contributions to the net electric field (orange arrows) inside a wire in a circuit.

Two Equally Charged Rings

First, we'll consider two rings with equal charges (Figure 18.21). At a location on the axis, midway between the rings, the net electric field is zero. At locations not on the axis, the field is nonzero but very small, and does not point along the wire. No matter how large the charges on these rings are, they cannot create an electric field aligned with the wire.

QUESTION What is the gradient of charge along this two-ring wire?

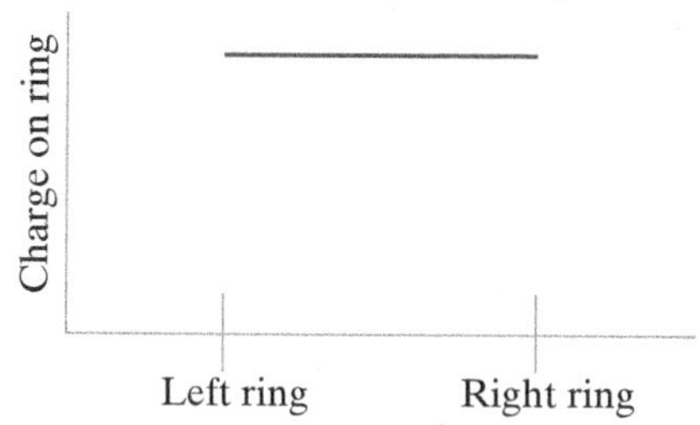

Figure 18.22 The gradient of charge due to the two equally charged rings shown in Figure 18.21 is zero (the gradient is the slope of the line).

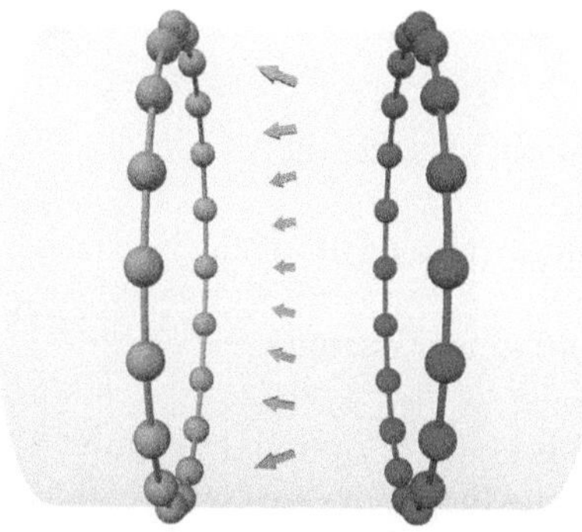

Figure 18.23 The electric field between two rings with different amounts of positive charge.

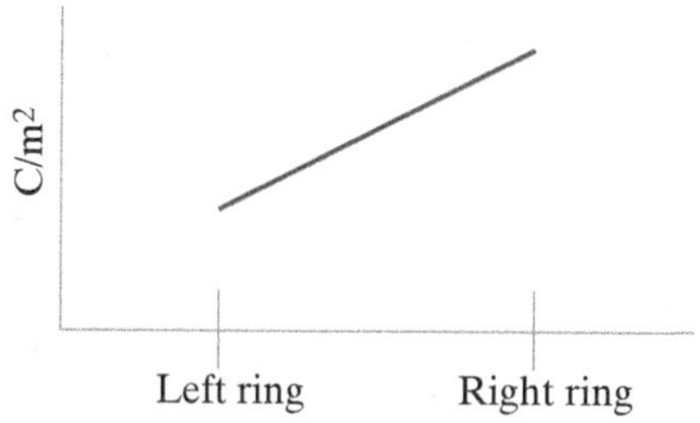

Figure 18.24 Between two rings with different charges, the charge gradient is nonzero.

If we plot a graph of ring charge vs. position, we see that the slope of this plot is zero—the charge does not vary from one ring to the next—so the gradient of surface charge is zero in this case (Figure 18.22). It is clear from this example that simply adding more equally charged rings will not produce a uniform electric field pointing along the wire. Let's try a different approach.

A Constant Gradient of Charge

The simplest configuration of charged rings that has a nonzero gradient of charge is two rings with different amounts of charge (Figure 18.23). This looks promising. Although both rings are positive, since one ring has more charge than the other the electric field between the rings is nonzero and points roughly along the axis. The charge gradient along the wire is not zero (Figure 18.24).

Let's extend our model of the charges on the surface of the wire, keeping the charge gradient constant, and extending it further along the wire in both directions (Figure 18.25).

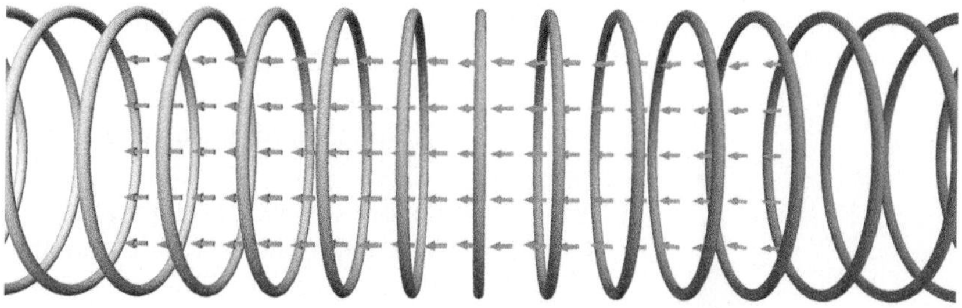

Figure 18.25 A closeup view of the electric field near the middle of the series of charged rings shown in Figure 18.20. To make it easier to see the orange arrows representing the electric field, the point charges making up the rings are not shown.

The result is surprisingly good—even though the charged rings are widely spaced, producing a "lumpy" distribution of charge instead of a smooth one, the electric field throughout the interior of the wire is remarkably uniform, both longitudinally and laterally, even near the surface of the wire. Evidently a constant gradient of charge density on the surface of a wire produces an electric field inside the wire that is uniform in magnitude along and across the wire and points along the wire.

Surface Charge Distribution on Circuits

Even the simplest circuit is more complex than a very long straight wire—it includes a battery, and the wire must bend around to connect to both ends of the battery. Therefore we should expect the distribution of surface charge on a circuit to be more complex than a simple constant gradient of charge. However, we might anticipate that:

- On long straight sections of wire that are far from other circuit elements there may be a nearly constant gradient of surface charge.
- The surface charge gradient will be largest in regions of the circuit in which the electric field is largest.
- In regions such as sharp bends in a wire, and locations where two different elements (wires of different diameter, or a wire and a battery) are connected, the distribution of surface charge may be quite complex.

In the following discussion, we show the results of a computational model that calculates the distribution of charge on the surface of several simple circuits.

> QUESTION Why don't we just measure the distribution of charge on the surface of a simple circuit, instead of calculating it?

Even when there is a large gradient of charge on the surface of a wire in a circuit, the total amount of charge is so small that it is very difficult to measure. Section 18.11 describes a demonstration in which surface charge can indeed be detected on a very high voltage circuit, but this isn't possible with an ordinary circuit.

How the Model Calculates Surface Charge Distributions

Basically, the computational model used to calculate surface charge uses this reasoning:

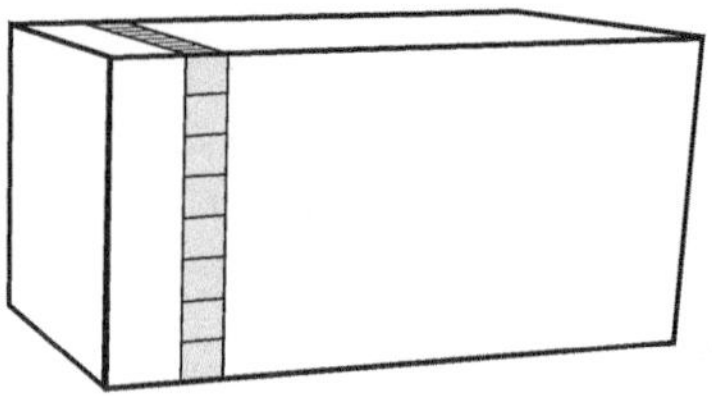

Figure 18.26 A (square) ring of charged tiles on the surface of a circuit wire.

- Start out with charge only on the battery (modeled as a capacitor whose charge is continually replenished).
- Use square wires to simplify the calculations. Divide the surface of the wires into many tiny square tiles (Figure 18.26).
- Now, repeatedly:
  - Calculate the electric field at the location of each tile due to all other charges on the circuit.
  - Use this field to calculate the flow of charge onto or off of each surface tile in the next time step Δt.
  - Repeat until the distribution of charge no longer changes.

A more detailed explanation of the model is given in optional Section 18.12. The computational techniques of implementing the model are complex and are beyond those taught in this book.

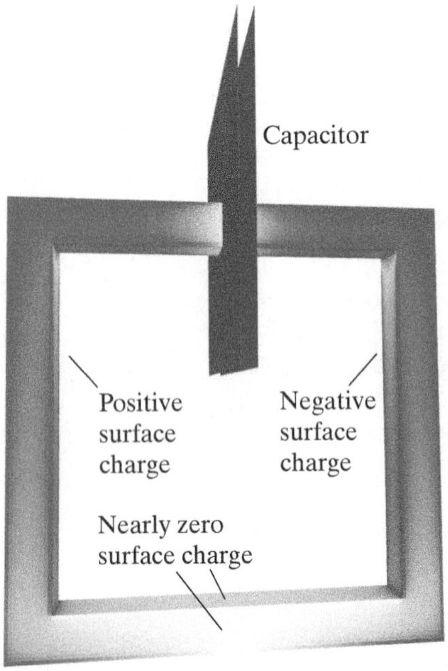

Figure 18.27 A model of a simple circuit. The resistive wire has a 6 mm by 6 mm square cross section. The circuit is 60 mm wide, and the capacitor gap is 2 mm, with a potential difference across the gap of 1.5 V.

A Single Wire Circuit

Figure 18.27 shows the calculated distribution of surface charge on a very simple circuit: one resistive wire and a battery. Starting at the positive left plate of the capacitor and following the circuit counterclockwise around to the negative capacitor plate, we find that the surface charge starts out strongly positive, becoming less and less positive and reaching zero in the middle of the bottom wire, then becoming increasingly negative as we continue on around the circuit.

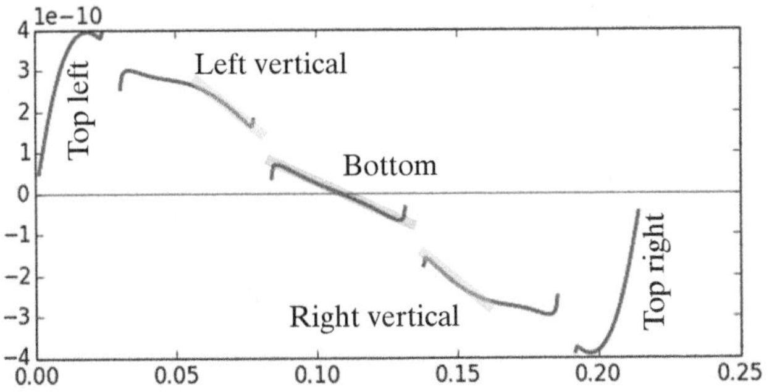

Figure 18.28 Surface charge density (C/m^2) on rings of charge along a path (meters) counterclockwise around the circuit of Figure 18.27, starting from the positive plate of the capacitor. We omit the charge situation at the square corners, where adjacent wires polarize each other. The charge gradient on the bottom wire, far from the battery, is constant (pink shading). The charge gradients on the bottom parts of the vertical wires (green shading) are also constant.

A graph of surface charge density vs. position in the circuit is shown in Figure 18.28. Surface charge density is calculated by adding up the charge on all the tiles in a single ring (Figure 18.26) and dividing by the total surface area of all those tiles. Figure 18.28 shows that there is a roughly constant gradient of surface charge density on those straight sections of wire that are farthest from the battery. The graph does not show the complications at the corners, which are in part due to the square geometry of the model circuit.

As can be seen from Figure 18.29, inside the wire the net electric field (shown in orange) produced by the capacitor and the surface charges is quite uniform, which is a check on the validity of the computational model. Figure 18.29 also shows the contributions to the net field from the battery and from charges on the surface of the wires. In the bottom wire, which is farthest from the battery, almost all of the electric field is due to the surface charge gradient on the wire.

Figure 18.29 The electric field at a number of locations around the circuit. The net electric field (orange), the field due to the capacitor (magenta), and the field due to all the surface charges (green) are shown. The fields are calculated along the centerline of the wire, but the arrows are displayed in front of the wire in order to make them visible. Note that to the left of the capacitor the field due to surface charge is actually opposite to the net field. At this location the large field due to the capacitor is opposed by the field due to the piled-up positive surface charge near the top left corner.

Electric Field Is Simple; Surface Charge Distributions Are Complicated

The main features of the surface charge distribution shown in Figure 18.27 are what we expected. We also expected some deviations from this pattern in special locations, and we do in fact see these. For example, in the inside corners you can see red puddles of positive surface charge inside blue negative regions, and blue puddles of negative surface charge inside red positive regions. These "wrong-sign" puddles are due to the polarizing effects of the capacitor field and the field of the neighboring wires. In these wrong-sign places one finds on the opposite side of the wire an extra amount of "correct-sign" charge; what counts is the total charge on a ring of tiles. Right next to the capacitor plates the charge density is reduced due to vertical components of the electric field contributed by the capacitor plates.

Although the graph of surface charge in Figure 18.28 shows that there is a roughly constant gradient of ring surface charge along the left, bottom, and right wires, Figure 18.29 shows that along the left and right wires the net field has large contributions from the capacitor, not just from the nearby surface charge gradient. Note too that the horizontal component of the field due to the surface charge means that there is sizable lateral polarization of these wires. In contrast, in the bottom wire, far from the capacitor, almost all of the field is due to the gradient of surface charge on the bottom wire.

If we make the circuit taller (Figure 18.30), the field of the capacitor is very small in the lower parts of the circuit, where we see truly constant gradients of surface charge (Figure 18.31).

Figure 18.30 In the lower parts of this taller version of the circuit the field contributed by the capacitor is very small, and there are constant surface charge gradients.

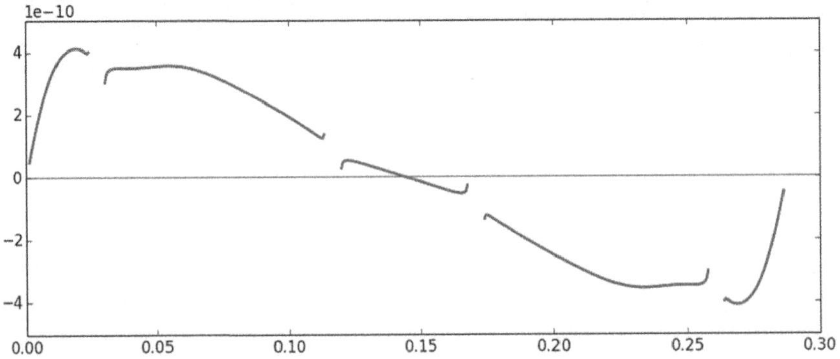

Figure 18.31 Surface charge along the circuit shown in Figure 18.30.

Although the surface charge distribution on a circuit is complex, and the details of the contributions of each charge to the net field are complex, the pattern of the net electric field inside the wires is very simple: the net field is uniform in magnitude and points along the wire (see the orange arrows in Figure 18.29), as we knew must be the case. We'll see that electric fields in circuits are easy to calculate, whereas it is very complicated to figure out where all the charges on the circuit are!

A Simplified Representation

In this chapter we will often draw approximate surface distributions like that in Figure 18.32. At selected locations we indicate high density of surface charge (C/m^2) with lots of +'s or −'s, and low density with few +'s or −'s. We emphasize that, as we've just seen, these charge distributions are not accurate in detail; our goal is to give a rough indication of the surface charge distribution, in a way that is easily written and read. Feedback in the circuit will tend to put some additional charge on the bends, the charges near location 2 will polarize the bottom wire, etc. We ignore the sideways polarization that occurs. The key point is that the surface charge near location 1 must be positive, and the surface charge near location 5 must be negative. When we connect a wire from one end of the battery to the other, there will be some kind of transition along the wire from + surface charge to − surface charge. There will even be some location at which there is zero surface charge (location 3 in Figure 18.32).

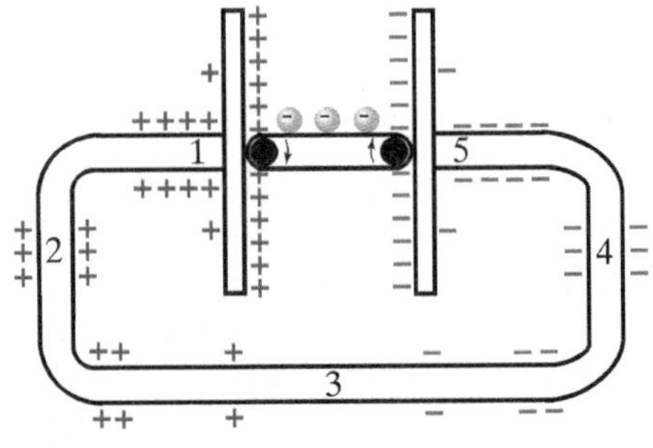

Figure 18.32 A geometrically simple circuit with approximate surface charge shown. The number of plus or minus signs indicates schematically the density of surface charge at that location. Charges are drawn only at sample locations, to make it possible to see differences in amount of charge.

Checkpoint 3 At locations 1, 2, 3, 4, 5 on a diagram like Figure 18.32, draw the net electric field. Also draw the resulting electron drift velocity. Pay close attention to direction and relative magnitudes.

A Circuit with a Thin Resistor

A slightly more complex circuit, including a thin resistor and thick connecting wires, is shown in Figure 18.33. In Section 18.3 we reasoned that in order for the current to be the same everywhere in a single-loop circuit, the drift speed of electrons must be larger in wires whose cross-sectional area is smaller.

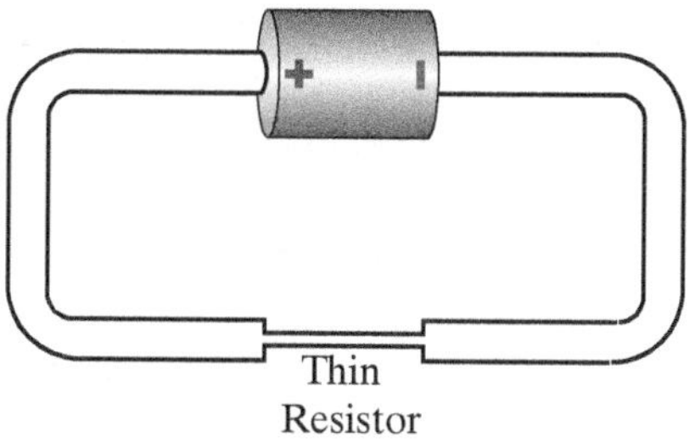

Figure 18.33 A circuit with a thin section acting as a resistor.

QUESTION What can we conclude about the gradient of surface charge across the thin resistor in Figure 18.33, compared to the gradient of surface charge on the thick connecting wires?

Two steps in reasoning are needed to answer this question. First, since the drift speed must be larger in the thin resistor, we deduce that the electric field inside the resistor must be larger than that inside the thick wires, since $\bar{v} = uE$. Second, we infer that there must be a larger surface charge gradient across the resistor, in order to produce a larger electric field.

We can see both of these features in Figures 18.34 and 18.35. In Figure 18.34 the thick wires (6 mm × 6 mm) have little charge gradient—the amount of charge per m^2 varies very little on these wires. This small charge gradient is associated with a small electric field inside these wires. In contrast, the charge on the thin wire (2 mm × 2 mm) varies rapidly from positive to negative across the wire. This large gradient is associated with a large electric field inside the resistor.

In the graph shown in Figure 18.35 we can see that there is an extra pile-up of surface charge at the entrance and exit of the thin resistor and a large gradient along the resistor (pink shading), both of which contribute to a large electric field in the resistor. In contrast, the surface charge gradient on the left and right vertical wires is nearly zero (green shading), contributing to a small electric field inside these wires.

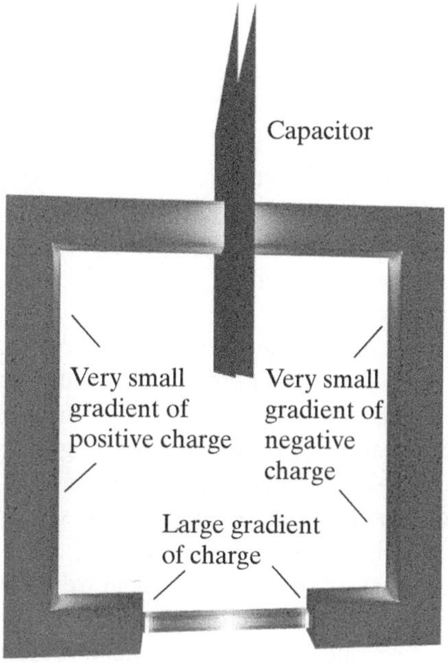

Figure 18.34 Computed surface charge on a circuit similar to that in Figure 18.33. Positive charge is red and negative charge is blue.

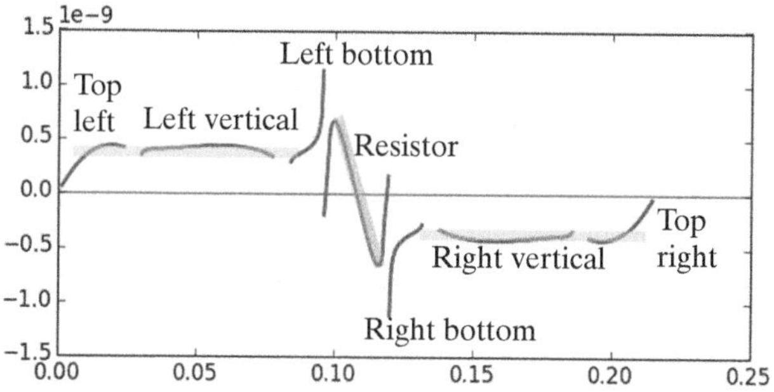

Figure 18.35 Surface charge density (C/m^2) on the circuit of Figure 18.34. The gradient of surface charge on the thick wires is very small (green shading), but the gradient of charge on the surface of the thin wire is large (pink shading).

In Figure 18.34 we again see complex charge patterns near junctions: there is wrong-sign charge density at each end of the resistor, due to polarization by the adjacent larger, strongly charged thick wires, which apply an electric field with large vertical components on the ends of the thin section.

Figure 18.36 shows a schematic version of the surface charge distribution shown in Figure 18.34. Gradients can be inferred by comparing the amount and sign of charge at neighboring locations. The large change in amount and sign of charge across the resistor indicates a large surface charge gradient.

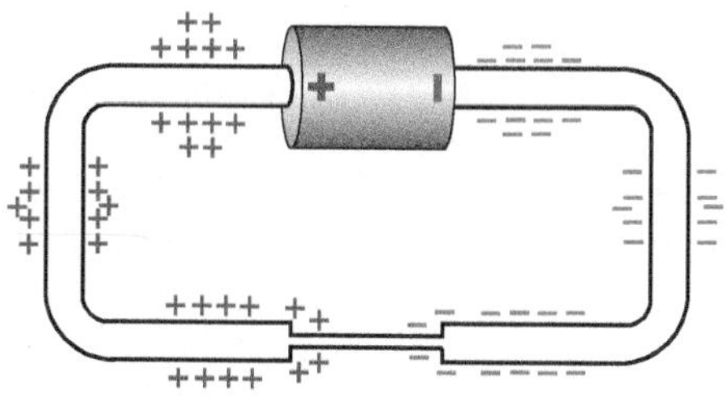

Figure 18.36 A schematic version of the surface charge distribution shown in Figure 18.34. The charge gradient across the thin resistor is large; the gradient on the thick wires is small.

Amount of Surface Charge

Despite the large number of electrons moving around in a circuit, you will find that the wires in one of your circuits will not repel a hanging charged tape. Try it! (The tape is of course attracted to neutral matter, which masks the tiny repulsion due to surface charges.) The large number of electrons moving inside the wires is balanced by an equally large number of positively-charged atomic cores, so the net charge inside the wire is zero and cannot repel charges by electric interaction. (The current does of course produce magnetic effects.)

The surface charges and the electric field they make are essential to driving and guiding the current through the interiors of the wires, but the amount of charge present on the wires in a circuit powered by two 1.5 V batteries is very small: in Figure 18.35 the maximum surface charge density is only about 1 nC per m^2. Little surface charge is needed in such a circuit; it is very easy to move electrons through metal, and very small electric fields are sufficient to drive sizable currents. We will see that in a typical circuit the magnitude of the electric field inside a copper wire is very small—typically much less than a volt per meter. Compare this with typical electrostatic fields, which can easily be large enough to create a spark (3×10^6 V/m). Evidently the amount of surface charge on a typical circuit is very small.

It is possible to run a circuit with a 10,000 V power supply and demonstrate how the surface charges can repel charged objects. The amount of surface charge is proportional to the voltage, and such circuits have much more surface charge than a 3 V circuit. A description of such a demonstration is given in Section 18.11.

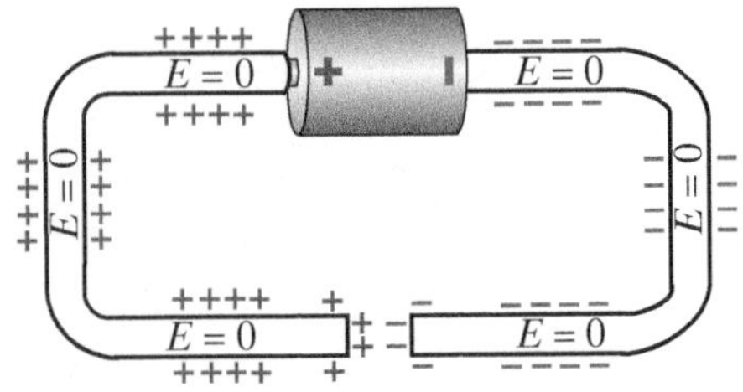

Figure 18.37 A circuit with a gap; approximate surface charge shown.

18.6 CONNECTING A CIRCUIT: THE INITIAL TRANSIENT

When you complete a circuit by making the final connection, feedback forces a rapid rearrangement of surface charges leading to the steady state. This period of adjustment before establishing the steady state is called the initial transient. How does this work, and how long does the transient take? We'll look now in closer detail at how the steady state is established.

Consider the circuit in Figure 18.37, containing one chemical battery and two Nichrome wires with a gap between them. The system is in equilibrium, so $E = 0$ everywhere inside the wires. An approximate surface-charge distribution is indicated, with one wire charged positive and the other charged negative, ignoring the exact details of how these charges are distributed. (In actual fact, there are larger concentrations of surface charge near the gap, due to strong mutual polarization in the gap region, so this is a crude diagram.)

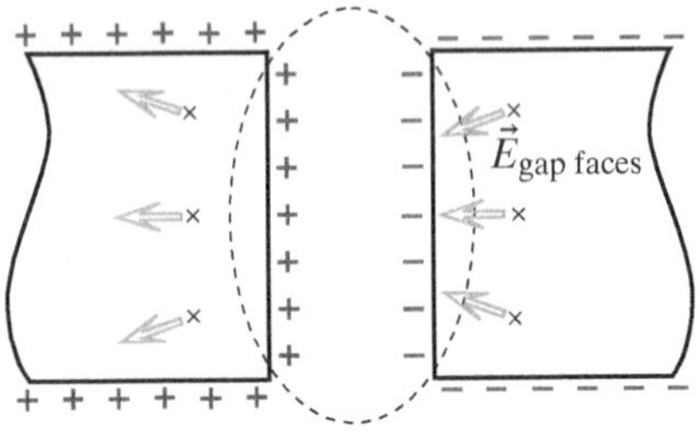

Figure 18.38 A close-up of the gap, showing the electric field contributed *solely* by the charges on the faces of the gap. At the locations marked ×, draw the electric field due to all *other* charges.

Let's look closely at the neighborhood of the gap. In Figure 18.38 we show the part of the electric field inside the wire that is contributed *solely* by the charges on the faces of the gap. You should convince yourself that these vectors are indeed drawn correctly.

> **QUESTION** On a similar diagram, sketch the part of the electric field that is due to all the *other* charges (on the ends of the battery and along the wires). Why is the field due to all the other charges exactly opposite to the field due to the gap charges?

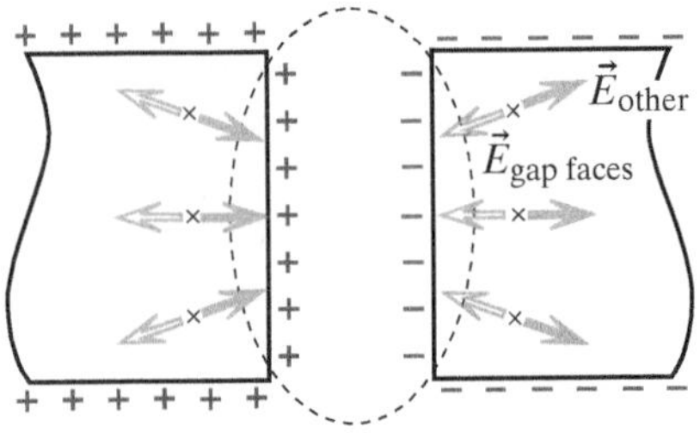

Figure 18.39 Field due to charges on the faces of the gap, and due to all others.

The net field inside the wire must be zero, because at this time the system is in equilibrium. Therefore the field due to all other charges must be exactly equal in magnitude and opposite in direction to the field made by the charges on the faces of the gap (Figure 18.39).

Let's complete the circuit by connecting the Nichrome wires together. The charges on the facing ends of the wires neutralize each other, leaving a tube of surface charge on the outside of the wire, as shown in Figure 18.40 in a cross section of the region near the connection point. At this instant this surface charge distribution has a big, unstable discontinuity in it—closely neighboring areas on the outside of the wire have positive and negative charges.

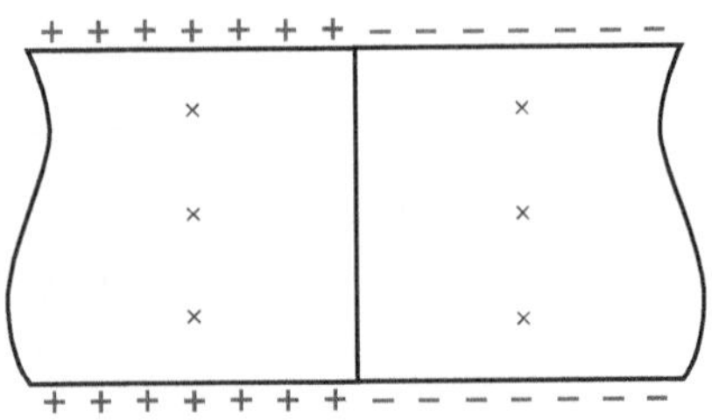

Figure 18.40 A cross section near the former gap, just after closing the gap.

> **QUESTION** In a diagram similar to Figure 18.40, use vectors to indicate the electric field at the locations marked × inside the wire just after closing the gap. (*Hint:* The gap charges are now gone.)

Immediately after closing the gap we've lost the contribution to the net field formerly made by the charges on the faces of the gap. Therefore the net field will look like the field of the "other" charges that we drew before.

Electrons in the mobile electron sea inside the metal will move due to these fields. Imagine that you could take a snapshot of the surface charge after the electrons inside the wire have moved just a little bit (that is, after a very short time). Electrons moving toward a positive region of the surface will make the surface charge in that region less positive, while electrons moving away from a negative region of the surface will make the surface charge in that region less negative.

> **QUESTION** On your diagram corresponding to Figure 18.40, where there is a change in the surface charge, write what that change is (for example, increase of positive charge, decrease of negative charge, etc.). (Remember that a + sign on the surface does not represent some exotic positive charge. Rather, a + sign indicates a region where there is a deficiency of electrons, leaving partially exposed the metal's positive atomic cores.)

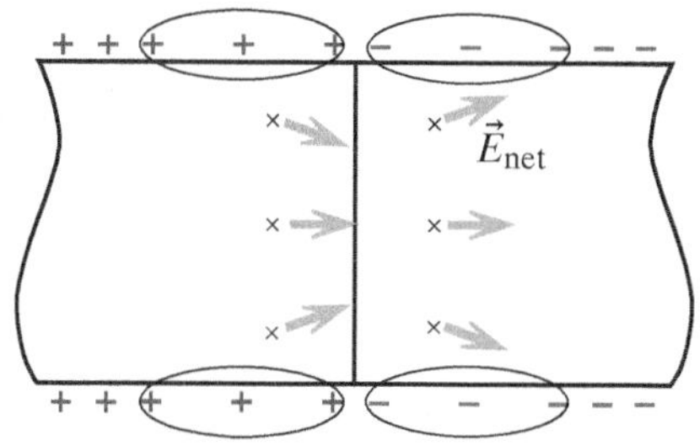

Figure 18.41 The dilution of the original surface charge. The circled regions at left become less positive, while the circled regions at right become less negative.

Figure 18.41 shows the new distribution of surface charge after a very short time (a fraction of a nanosecond!). The motion of electrons toward and away from the surface, driven by the new electric field, dilutes the surface charge near the location where the gap was, so that in the gap region the initial abrupt charge discontinuity turns into a more gradual change, indicated by placing fewer + or − symbols near the gap location. A nonzero electric field has come into being where the gap used to be, due to the + charges to the left and − charges to the right of that location.

In Figure 18.42 we see that the affected region, where there has been a rearrangement of the surface charge, grows rapidly outward from the gap location at approximately the speed of light, which is the speed with which the change in the distribution of charges is felt at places some distance away. At the instant sketched in Figure 18.42, the surface charges and the electric field at distant locations haven't been affected yet, because those regions haven't yet received the information that the gap has been closed! The electric field is still zero inside the wires except near the location of the gap.

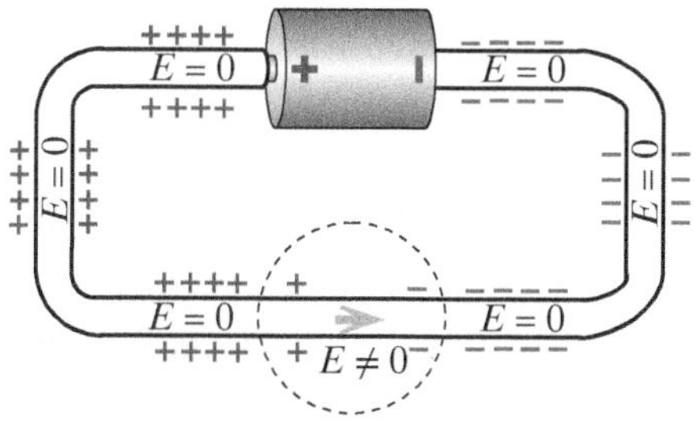

Figure 18.42 Parts of the circuit haven't yet been informed that the gap has been closed! The sphere enclosing the region of changed electric field expands at the speed of light.

It is the speed of light that determines the minimum time required to establish the steady state. Light travels 30 cm (about one foot) in a nanosecond (1×10^{-9} s). The time required to move the electrons toward or away from the surface is completely negligible, because the electrons only have to move an infinitesimal distance in order to establish a significant amount of surface charge, through slight expansion or contraction of the mobile-electron sea (compare with the molecular stretch associated with polarizing an insulator, which is less than the radius of a single proton!).

In just a few nanoseconds the rearrangement of the surface charges will extend all the way around the circuit. There is some wave-like sloshing around of the surface-charge distribution, which due to dissipation eventually settles down to the establishment of a "steady state," in which the surface charge and current no longer vary with time. In the steady state there is a variation of surface charge around the circuit, and the electric field has uniform magnitude throughout the inside of the wire (Figure 18.43).

In terms of the average drift speed $\bar{v}$ of the mobile charges:

- Equilibrium: $\bar{v} = 0$ (no current)
- Steady state: $\bar{v} =$ constant (constant, nonzero current)
- Initial transient: $\bar{v}$ changing (short process leading to the steady state)

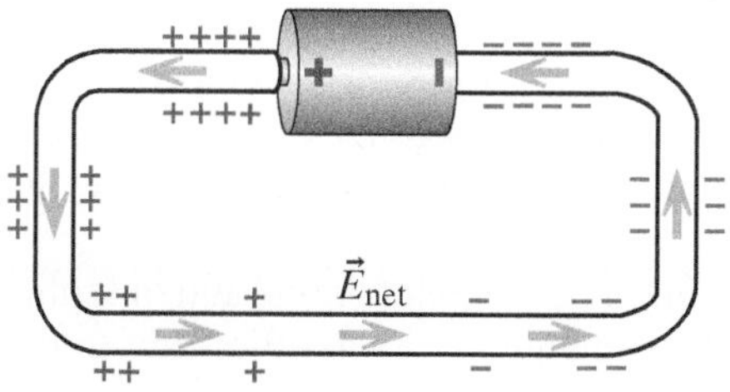

Figure 18.43 In the steady state the electric field has the same magnitude everywhere in the wire.

Why a Light Comes on Right Away

As discussed earlier, the drift speed in one of your battery-and-bulb circuits is only about 5×10^{-5} m/s. However, a change in the circuit, such as making or breaking a connection, propagates its effects very rapidly through the circuit, at approximately the speed of light. It is not necessary for charges to move physically to a distant place to cause a change at that place: a rearrangement of charges due to tiny movements produces a change at the speed of light in the electric field at a distance. It is important to keep in mind that the effects do take a nonzero amount of time to propagate (nanoseconds). The propagation is not instantaneous, but it is very fast.

When you turn on the light switch in your room (making the final connection in a circuit), it could take hours on the average for an electron in the switch to drift to the overhead light. However, the light comes on right away, because the rearrangement of surface charges in the circuit takes place at about the speed of light. The final steady state of the circuit is established in a few nanoseconds, after which the electron sea circulates slowly and majestically around and around, everywhere in the circuit, including through the switch and the wires and the light bulb. (Most lighting actually uses "alternating current," in which case the electron sea doesn't drift continuously but merely sloshes back and forth very short distances, everywhere in the circuit, 50 or 60 times per second.)

The electrons don't have to move from the switch to the light bulb; there are already plenty of electrons in the bulb filament. All that is required is to begin to establish the appropriate steady-state surface-charge distribution, very quickly, and then the electrons that are already in the filament start moving, propelled by the new nonzero electric field in the wire.

18.7 FEEDBACK

Feedback during the initial transient produces surface charge of the right amount to create the appropriate steady-state electric field. It also maintains these steady-state conditions, as we will show next.

Feedback Leads to Current Equalization

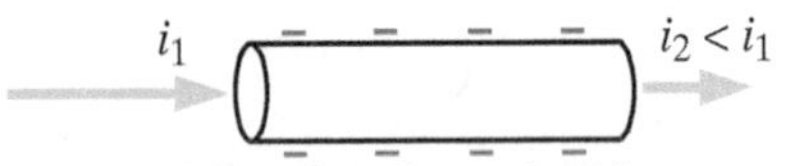

Negative charge buildup

Figure 18.44 What if more electron current enters than leaves a section of wire?

Consider a straight section of wire that at some moment has more incoming electron current than outgoing electron current, leading to a buildup of excess negative charge on the surface of this section of the wire (Figure 18.44).

This surface-charge buildup will tend to equalize the incoming and outgoing electron currents. The outer surface of this section of wire becomes more negative, and this negative surface charge tends to impede the incoming electrons and tends to speed up the outgoing electrons, thus making the incoming and outgoing electron currents $i = nA\bar{v}$ more nearly equal to each other. This feedback process will continue until the two currents are exactly equal to each other. At that point there will be no further change in the surface charge.

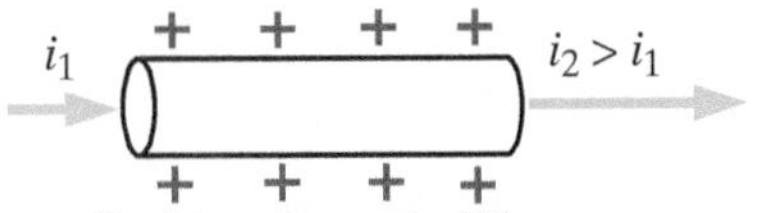

Positive charge buildup

Figure 18.45 What if less electron current enters than leaves a section of wire?

QUESTION If the incoming electron current is smaller than the outgoing current, what will happen on the outer surface of this section of the wire?

In this case (Figure 18.45) this section of wire becomes more positive, which speeds up the incoming electrons and slows down the outgoing electrons, so the incoming electron current increases and the outgoing electron current decreases. Again, the effect of this feedback mechanism is to equalize the current into and out of a section of wire.

Thus, whichever way there is an imbalance between incoming and outgoing current, the two currents will quickly become equal to each other due to the automatic feedback mechanism of changes in the surface charge. Earlier you saw that in the steady state the current everywhere in a series circuit is the same. Now you can see how the feedback mechanism creates and maintains this steady state.

Feedback Makes Current Follow the Wire

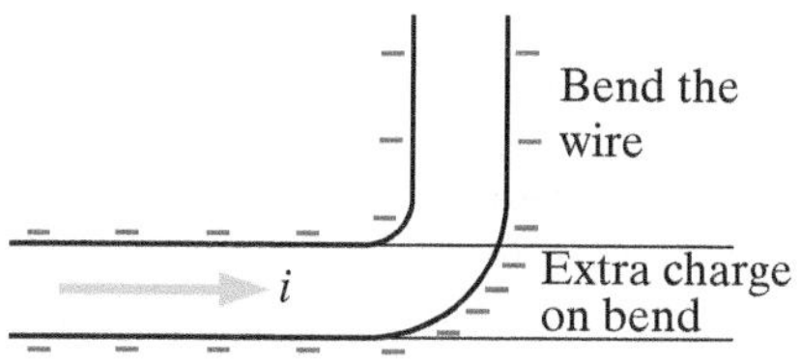

Figure 18.46 What happens if you bend a straight wire while current is running?

It is instructive to see what happens if you bend a straight wire while the current is running (Figure 18.46). The oncoming electrons don't immediately know what has happened and continue in their original direction, but as they run into the bend, electrons pile up on the outside of the bend, which tends to turn the oncoming electrons.

> QUESTION What determines how much charge piles up? (That is, what makes the pileup process come to a stop?)

The electrons pile up on the bend until there are enough there to repel oncoming electrons just enough to make them turn the corner (follow the wire) without running into the side of the wire. A calculation by Rosser indicates that a single extra electron on the bend is sufficient to turn a sizable current! (The required inward force is equal to mv^2/r, and the electron mass m and drift speed v are very small.) Not only does the feedback mechanism equalize incoming and outgoing current for any section of the wire, it also forces the electron current to follow the wire, no matter how the wire is bent or twisted.

For details of the calculation about the amount of charge needed to turn the current, see W. G. V. Rosser, "Magnitudes of surface charge distributions associated with electric current flow," *American Journal of Physics* **38** (1970), 265–266.

Summary of Feedback

We have learned that there must be some excess charge on the outside of the wire, all along the wire. Through the feedback mechanism, this surface charge will automatically arrange itself (or rearrange itself) in such a way as to ensure that the net electric field everywhere inside the wire points along the wire and has the appropriate magnitude to drive the appropriate amount of steady-state current. (The amount of this current is ultimately determined by the strength of the motor in our mechanical battery, and by the electron mobility u of the material that the wire is made of.)

This is very similar to the feedback we have seen in static electricity. If an external field is applied to an isolated metal block, charges automatically arrange themselves on the surface of the metal block in such a way as to make the electric field inside the metal block have a particular form: the electric field must be zero throughout the inside of the block in equilibrium.

To summarize:

> Feedback in a circuit leads to surface charges and steady-state current: $\vec{E}_{net} \neq \vec{0}$ inside a metal.
>
> Feedback in static electricity situations leads to equilibrium: $\vec{E}_{net} = \vec{0}$ inside a metal.

18.8 SURFACE CHARGE AND RESISTORS

We have used surface charges and electric fields to analyze a circuit made up of just a battery and a high-resistance Nichrome wire. In this section we will use the surface-charge model to analyze circuits involving "resistors"—sections of a circuit that resist the passage of electrons more than other parts of the circuit.

Narrow Resistors and Thick Wires

We will analyze a circuit similar to a circuit you assembled, consisting of a battery, copper wires, and a bulb. The very thin tungsten filament of the light bulb is an example of a "resistor"—a component that has a much higher

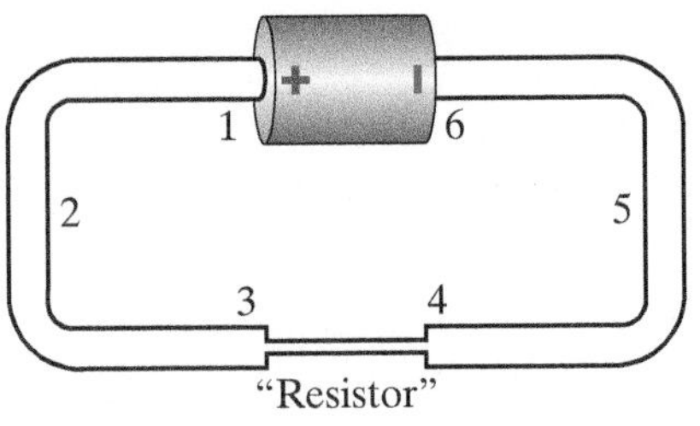

Figure 18.47 A simple circuit with a "resistor"—a thin section of the Nichrome wire.

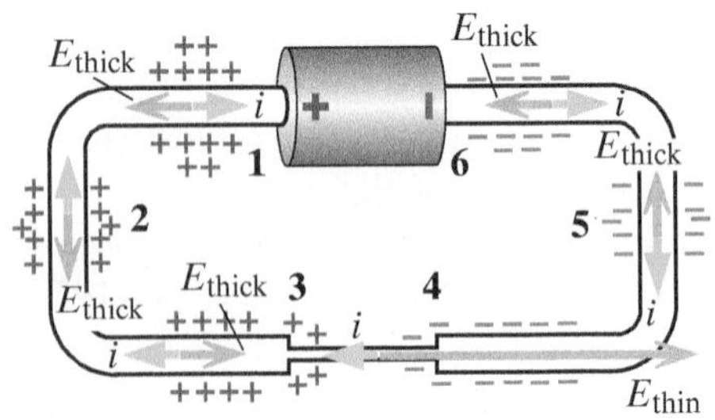

Figure 18.48 An approximate surface-charge distribution that is consistent with what we know about the electric field.

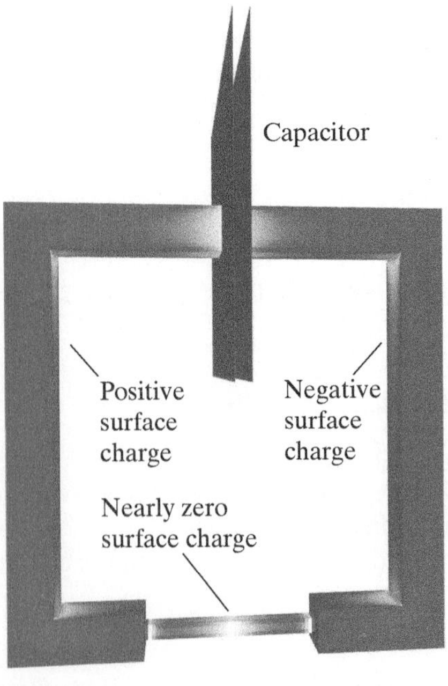

Figure 18.49 A detailed representation of the surface charge on a circuit with a thin resistor.

resistance to current flow than do the thick wires connecting the resistor to the battery.

For a simple example of a circuit containing a resistor, we'll again consider a battery and a high-resistance Nichrome wire, but we'll make part of the Nichrome wire very much thinner than the rest of the wire (Figure 18.47). We'll call the narrow section a "resistor," and there are two thick wires connecting the resistor to the battery.

Just after connecting the circuit but before the establishment of the steady-state surface charge, the electric field in the narrow resistor might temporarily be about the same as the electric field in the neighboring thick wires, leading to a comparable drift speed v in both sections. However, the number of electrons per second trying to enter the resistor is the large number $nA_{\text{thick}}\bar{v}$, whereas the number of electrons per second passing through the resistor is the small number $nA_{\text{thin}}\bar{v}$, so electrons pile up at location 3, the entrance to the resistor. (Remember that the mobile-electron density n is the same in the wires and the resistor, because they are both made of Nichrome in our example.)

Similarly, a deficiency of electrons, or + charge, accumulates at location 4, the exit from the resistor, because the number of electrons per second emerging from the resistor is small compared to the electron current in the thick wire that is removing electrons from the region.

> QUESTION Later, in the steady state, how does the electron current in the thick wires compare with the electron current in the resistor? Why?

Once the steady state has been established, the currents must have become the same. Reason by contradiction: if the currents weren't the same in the steady state, surface charge would be building up on the resistor, which would mean we hadn't reached the steady state yet. Charge piles up until the resulting increased electric field in the resistor gives a high enough drift speed to make an electron current that equals the electron current in the thick wires. Once these two currents are the same, there will be no more pileup.

> QUESTION In the steady state, how does the electric field E_{thin} in the resistor compare with the electric field in the thick wires? Why? On a diagram like Figure 18.47, draw the steady-state electric field in the resistor and in the neighboring thick wires, paying attention to relative magnitudes.

Since the steady-state current is the same in the thick and thin sections, we have the relation $nA_{\text{thick}}uE_{\text{thick}} = nA_{\text{thin}}uE_{\text{thin}}$. Therefore the electric field in the resistor must be quite a lot larger than the electric field in the thick sections. Quantitatively, $E_{\text{thin}} = (A_{\text{thick}}/A_{\text{thin}})E_{\text{thick}}$, since the mobile-electron density n and the mobility u are the same in the thick and thin wires (both made of Nichrome). The electric field has a uniform large magnitude throughout the resistor and a uniform small magnitude throughout the thick wires.

> QUESTION Now draw the approximate surface charge in the steady state. Be sure to make the distribution of surface charge consistent with the distribution of electric field that you drew already.

In the steady state the surface charge varies only a little along the thick wires, producing a small electric field E_{thick} and a small drift speed. Along the resistor there is a big variation from + at one end to − at the other, which makes a large field E_{thin} inside the thin resistor (Figure 18.48). This makes a large drift speed in the resistor (with the same electron current i as in the thick wires). We have an approximate charge distribution that is consistent with the distribution of electric field. Figure 18.49 again shows a detailed representation of the surface charge that you saw earlier, with a charged capacitor playing the role of the battery.

This resistor circuit offers another example of feedback in an electric circuit—an automatic adjustment leads to producing the electric fields that are needed in the steady state. Note that if the drift speed in the thick wires were momentarily to increase for any reason, extra charge would pile up at the resistor, which would tend to decrease the drift speed in the thick wires.

On the other hand, if the drift speed in the thick wires were to momentarily decrease, charge would drain through the resistor in such a way as to decrease the pileup of charge at the ends of the resistor, which would tend to increase the drift speed in the thick wires.

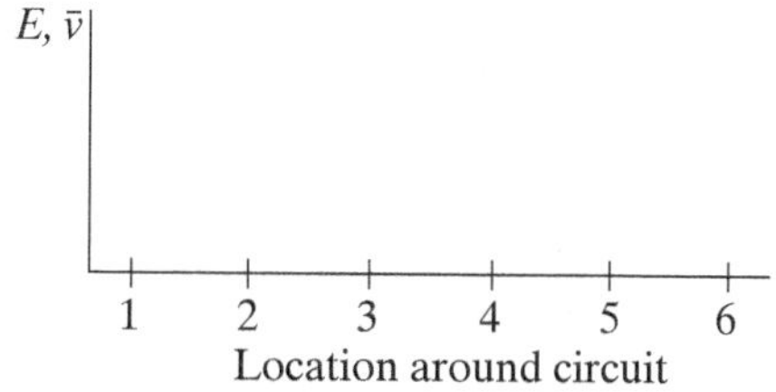

Figure 18.50 Make graphs of magnitude of electric field and of drift speed at locations around the circuit.

Checkpoint 4 Sketch a graph (Figure 18.50) of the magnitude of the electric field inside the wire in Figure 18.48 as a function of location around the circuit. In a different color, graph the drift speed of the electrons.

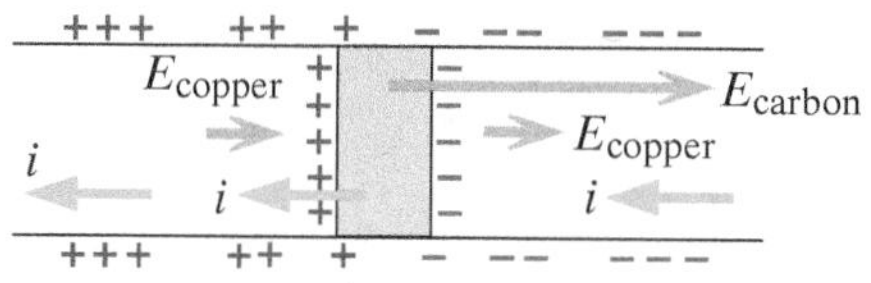

Figure 18.51 A portion of a circuit with a wide resistor.

A Good Analysis Strategy

The analysis in the preceding section is an example of a general strategy for understanding the relationship of charge and field in a circuit. Here is a summary of this strategy:

- Based on $i = nAuE$, draw the distribution of electric field in the wires.
- Next, draw an approximate charge distribution that is consistent with the distribution of electric field.

The reason for carrying out the analysis in this order is that the distribution of electric field is fairly simple and can be analyzed nearly exactly, whereas the details of the pattern of surface charge can be quite complicated.

What we know is that feedback will produce an arbitrarily complicated distribution of surface charge appropriate to the steady-state distribution of electric field. The pattern of electric field gives us quantitative information about the electron current; the pattern of surface charge gives us a clear sense of the mechanism that produces the pattern of electric field.

A Wide Resistor: Charges on the Interface

Another kind of resistor is wide rather than narrow but has low mobility. Consider copper wires leading up to a short piece of carbon having the same, wide cross section (Figure 18.51; we don't show the entire circuit). Carbon has a much smaller electron mobility u than copper, so the steady-state electric field has to be much larger in the carbon than in the copper in order to have the same current in both materials.

In the transient leading up to the steady state, electrons pile up not only on the outer surfaces of the copper and carbon, but also on the interfaces between the copper and the carbon, and these interface charges make a major contribution to the large electric field inside the carbon resistor.

Our assertion that "excess charge is always on the surface of a conductor" extends to this case, where the surface may be in contact with a different conducting material.

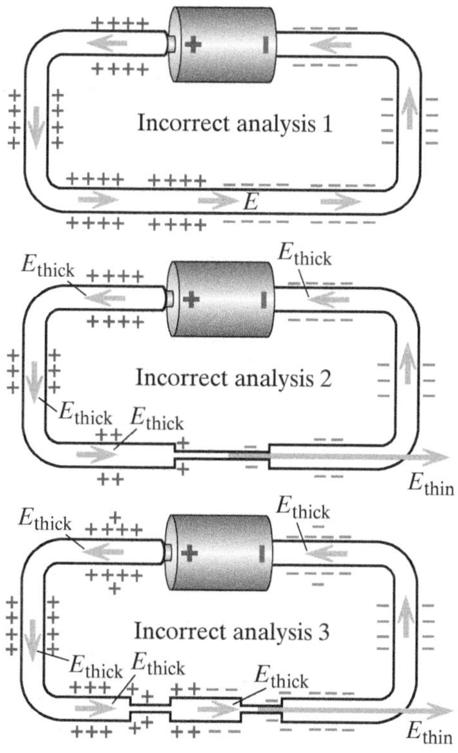

Figure 18.52 Circuits incorrectly analyzed: distribution of surface charge inconsistent with the pattern of electric field.

Checkpoint 5 The circuits in Figure 18.52 have been analyzed incorrectly; the distribution of surface charge is inconsistent with the distribution of electric field. Briefly identify what is inconsistent in each case.

18.9 ENERGY IN A CIRCUIT

Charge conservation plus the definition of a steady state led to the current node rule, which tells us that the current in a simple battery-and-bulb circuit must be the same everywhere in the circuit. But how many amperes of current flow? What determines the magnitude of the current? The Energy Principle provides the answer, plus the properties of batteries.

Consider a circuit containing various circuit elements such as resistors and batteries. We could discuss energy conservation in the circuit by considering one electron as it makes a complete trip around the circuit. The energy gained by the electron as it is transported across the mechanical battery is dissipated in collisions with atomic cores as it travels through the wire. Alternatively, we can analyze the energy per unit charge gained and lost in a trip around the circuit. We know that over any path the round-trip path integral of the electric field must be zero, as long as the current is steady or nearly steady. (This is not true in alternating current or AC circuits where there are time-varying currents, with time-varying magnetic fields, as we will see in Chapter 22 on Faraday's law.) If we follow a round-trip path through a circuit, through circuit element 1, circuit element 2, and so on, we find the following:

ENERGY CONSERVATION (THE LOOP RULE)

$\Delta V_1 + \Delta V_2 + \cdots = 0$ along any closed path in a circuit

This is valid for circuits with steady or slowly time-varying currents.

This is essentially the Energy Principle, but on a per-coulomb basis. Remember that electric potential difference is defined as potential energy difference per unit charge: $\Delta V = \Delta U/q$. The loop rule says that you can't gain energy in a round trip. In circuits this energy principle is often called the "Kirchhoff loop rule," but its validity is not confined to circuits.

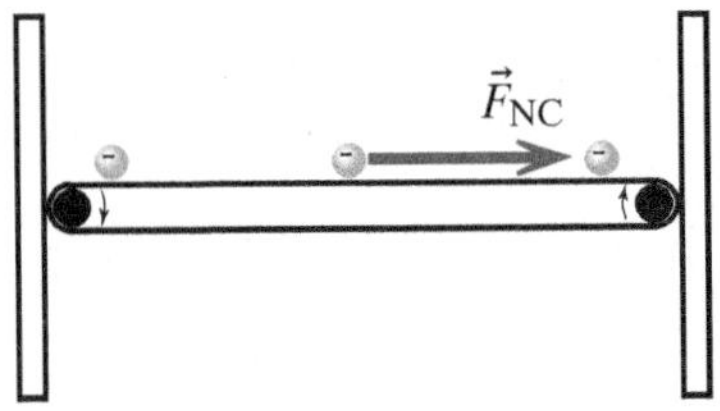

Figure 18.53 Start the motor and begin moving electrons from left to right.

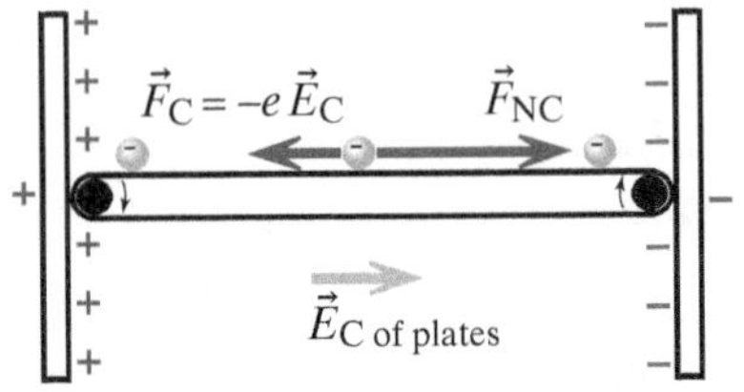

Figure 18.54 Charges build up on the plates and exert a "Coulomb" force on the electrons being transported.

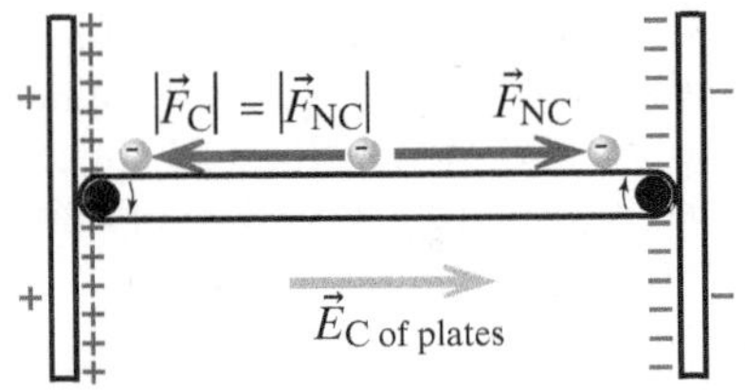

Figure 18.55 The Coulomb force grows to be as big as the non-Coulomb force.

Potential Difference Across a Battery

Along a wire or resistor of length L in which there is a uniform electric field of magnitude E, the magnitude of the potential difference is just EL. What determines how much potential difference the motor of a mechanical battery can maintain across the battery? Suppose you start with uncharged plates (Figure 18.53) and turn on the motor (with no wire connected to the plates), extracting electrons from the left-hand plate and transporting them to the right-hand plate. The conveyor belt, driven by the motor, exerts what we will call a "non-Coulomb" force $\vec{F}_{NC}$ on each electron. This being the only force, the belt accelerates electrons.

As the motor continues to transport electrons to the right, charge builds up on the plates. The charges on the plates exert a "Coulomb" force $F_C = eE_C$ on the electrons being transported, in the opposite direction to F_{NC}, the non-Coulomb force that the conveyor belt exerts (Figure 18.54). There is less acceleration of the electrons.

By "Coulomb force" and "Coulomb field" we mean the force and field due to point charges, as given by Coulomb's law. We use the term "non-Coulomb" to refer to other kinds of interactions. The Coulomb and non-Coulomb forces oppose each other inside the battery.

Eventually there is enough charge on the plates to make $F_C = eE_C = F_{NC}$. At that point the motor cannot pump any more charge, and the plates are charged up as much as they can be (Figure 18.55). We see that the function of a mechanical battery (or of a chemical battery) is to produce and maintain a potential difference by pulling electrons out of the positive plate and pushing them onto the negative plate. The amount of potential difference is limited by and determined by the strength of the motor in a mechanical battery (or by the nature of the chemical reactions in a chemical battery).

If the distance between the (large) plates of a mechanical battery is s, and the electric field E_C of the charged plates is uniform between the plates, then the potential difference across the battery is

$$|\Delta V_{\text{batt}}| = E_C s = \frac{F_{NC} s}{e}$$

The quantity $F_{NC}s/e$, the energy input per unit charge, is a property of the battery and is called the "emf" of the battery (pronounced "e" "m" "f"). Historically, emf was an abbreviation for "electromotive force," which is a bad name, since emf is not a force at all but energy per unit charge. We will avoid this terminology by just using the abbreviation emf. In a chemical battery, the emf is a measure of the chemical energy per unit charge expended by the battery in moving charge through the battery. The emf of a flashlight battery is about 1.5 V, and the associated charge separation is due to chemical reactions in which electrons are reactants or products.

The emf of a chemical battery turns out to be nearly constant, independent of how much current is running through the battery (as long as the current is not too large), and approximately independent of how long the battery has been used (though it does decline with time). It is the approximate constancy of the emf that makes this a useful concept, and that makes it useful to analyze circuits in terms of emf and the potential differences in the circuit elements attached to the battery.

It is important to keep in mind that although the units of emf are volts, the emf is not a potential difference. Potential difference is a path integral of the electric field made by real charges. The emf is the energy input per unit charge and might in principle be gravitational or nuclear in nature. For example, our mechanical battery could be run by a falling weight.

ROLE OF A BATTERY

A battery maintains a potential difference across the battery. This potential difference is numerically equal to the battery's emf for batteries supplying a small amount of current. Later we will discuss how the potential difference decreases if the current is large.

Checkpoint 6 If the plates of a mechanical battery are very large compared to the distance s between the plates, and the area of a plate is A, show that the amount of charge Q on one of the plates of an isolated battery is determined by the battery's emf: $Q = A\varepsilon_0(\text{emf})/s$. For an ordinary flashlight battery, emf $= 1.5$ V. Make a rough estimate of the amount of charge in coulombs on the end plate of a battery, whose length $s = 5$ cm and radius $r = 1$ cm. (This is a very rough estimate because 5 cm isn't small compared to 1 cm.) How does this compare to the charge on a U tape? Should you expect to be able to repel a U tape with one end of a battery?

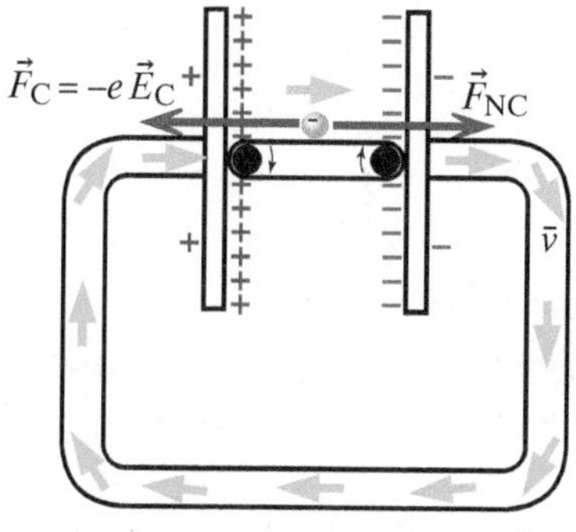

Figure 18.56 $F_C = F_{NC}$ if there is little internal resistance.

Internal Resistance

If we now connect a Nichrome wire from one plate of the mechanical battery to the other (Figure 18.56), we can achieve a steady state if the motor transports as many electrons per second from the + plate to the − plate as are flowing through the wire. The larger the current, the more electrons per second the motor must transport to keep the plates resupplied. If there is no resistance to the flow of charge inside the battery, we will have $F_C = eE_C = F_{NC}$.

In any real battery there is some resistance to the flow of charge through the battery, and this resistance is called the "internal resistance" of the battery.

If the mobility of an ion of charge e through the battery is u, the drift velocity through the battery will be proportional to the net force per unit charge acting on the ion: $\bar{v} = u(F_{NC}/e - E_C)$. Since the motor force F_{NC} is fixed, the maximum drift speed is obtained when $E_C = 0$, which means that there is no charge on the ends of the battery: the motor isn't keeping up, and $|\Delta V_{batt}| = 0$. Much of the time we will make the simplifying assumption that the mobility u of ions inside the battery is very high (low internal resistance), so that we get a sizable drift speed even if F_C $(= eE_C)$ is nearly as large as F_{NC}. Correspondingly, we assume that $|\Delta V_{batt}| \approx \text{emf}$.

In Chapter 19 we will show how to model a battery when the internal resistance is not negligible. For now we will consider "ideal" batteries for which the internal resistance is negligible, and $|\Delta V_{batt}| = \text{emf}$.

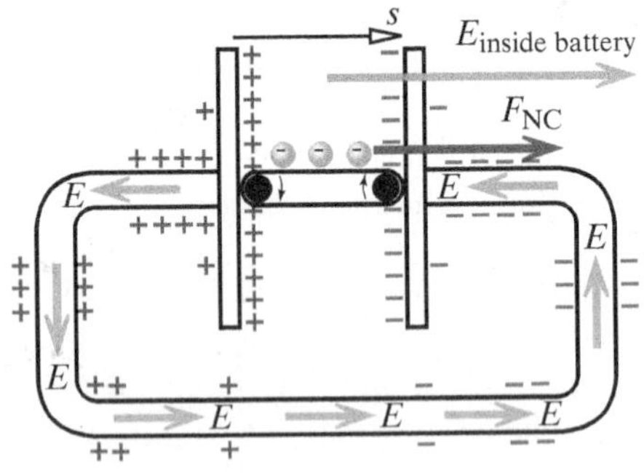

Figure 18.57 The electric field inside the mechanical battery is opposite to the electric field in the neighboring wires.

Field and Current in a Simple Circuit

In Figure 18.57, note that the electric field inside the mechanical battery points from the + end toward the – end, in the opposite direction to the electric field in the neighboring wires. The + end of a battery is at a higher potential than the – end.

If we start at the negative plate of the battery and go counterclockwise in Figure 18.57, there is a potential increase of + emf across the battery and a potential drop of $-EL$ along the wire of length L. The round-trip potential difference is zero:

$$\Delta V_{batt} + \Delta V_{wire} = 0$$

$$\text{emf} + (-EL) = 0$$

$$E = \frac{\text{emf}}{L}$$

We used the energy principle (round-trip potential difference is zero) to determine the magnitude of the electric field in the wire. If we know the properties of the wire (cross-sectional area A, electron density n, and electron mobility u), we can now determine the magnitude of the conventional current $I = enAuE$.

A Parallel Circuit: Two Different Paths

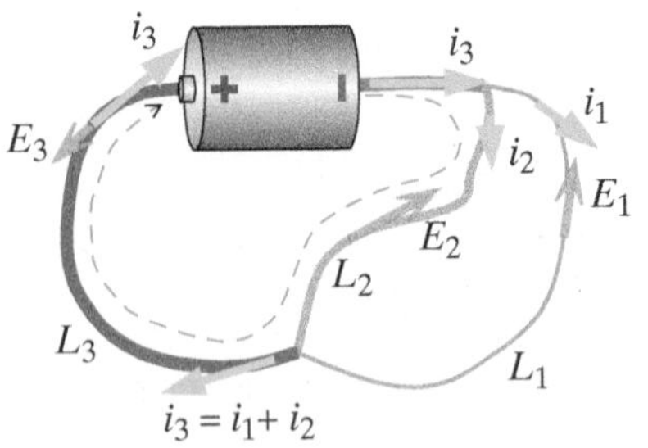

Figure 18.58 Follow the dashed path through wires L_2 and L_3.

Figure 18.58 shows a parallel circuit with more than one path for current to follow. Start at the negative end of the battery and follow the dashed path clockwise through wires L_2 and L_3, and then through the battery to the starting location. There is a potential rise $+E_2L_2$ along the part of the path of length L_2, a potential rise $+E_3L_3$ along the rest of the path of length L_3, then a potential drop through the battery of amount $-\text{emf}$, yielding $E_2L_2 + E_3L_3 + (-\text{emf}) = 0$ for the round-trip potential difference. If we instead follow the path through L_1 and L_3, we find a round-trip potential difference $E_1L_1 + E_3L_3 + (-\text{emf}) = 0$.

These two equations imply (correctly) that $E_1L_1 = E_2L_2$, which makes sense: because the two wires have the same starting and ending points, they have the same potential differences. Also notice in Figure 18.58 that $i_3 = i_1 + i_2$, due to the current node rule (charge conservation in the steady state).

Potential Difference Along Connecting Wires

In your own experiments you have probably noticed that in a circuit containing a light bulb or a Nichrome wire, the number or length of the connecting wires used has a negligible effect on the amount of current in the circuit. In a series circuit with thick copper wires and a light bulb, the electric field is large in the bulb and quite small in the thick copper wires, because the steady-state electron current $nA\bar{v} = nAuE$ must be the same in both. The electron mobility in cool copper is much higher than the electron mobility in hot tungsten, and typically the copper wires are much thicker than the tungsten bulb filament, so that not only is it the case that $E_{\text{wires}} \ll E_{\text{bulb}}$, but even for rather long wires $E_{\text{wires}}L_{\text{wires}} \ll E_{\text{bulb}}L_{\text{bulb}}$. Therefore, in the energy equation (emf $- E_{\text{bulb}}L_{\text{bulb}} - E_{\text{wires}}L_{\text{wires}} = 0$), the term $E_{\text{wires}}L_{\text{wires}}$ may be negligible; the work done by the battery goes mostly into energy dissipation in the bulb, with a small amount of energy dissipation in the thick copper wires.

Checkpoint 7 Explain briefly why the brightness of a bulb doesn't change noticeably when you use longer copper wires to connect it to the battery.

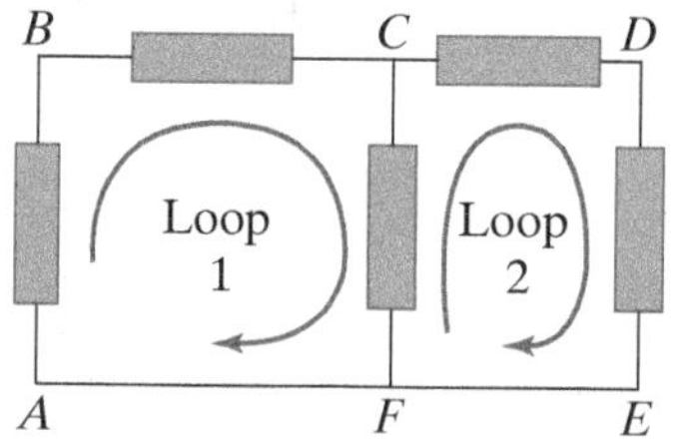

Figure 18.59 A circuit containing devices connected by thick copper wires.

General Use of the Loop Rule

The loop rule can be applied along any path through an arbitrarily complicated circuit. Consider a circuit made up of five circuit elements connected together by thick copper wires that have high electron mobility (Figure 18.59). The circuit elements could be resistors, batteries, capacitors, or even more complex devices such as semiconductor diodes or even refrigerators!

In Figure 18.60 we show the left portion of the complete circuit (loop 1). Start at location A and walk along the path labeled $ABCFA$. From A to B the potential difference is $\Delta V_1 = V_B - V_A$; from B to C the potential difference is $\Delta V_2 = V_C - V_B$; and from C to F the potential difference is $\Delta V_3 = V_F - V_C$. From F to A there is negligible potential difference $\Delta V_4 = V_A - V_F \approx 0$, because the thick wire has high electron mobility and large cross-sectional area, and so requires a negligibly small electric field to drive the current.

Any round-trip potential difference must be zero, and the loop rule gives this:

$$\Delta V_1 + \Delta V_2 + \Delta V_3 + \Delta V_4 = 0$$
$$(V_B - V_A) + (V_C - V_B) + (V_F - V_C) + (V_A - V_F) = 0$$

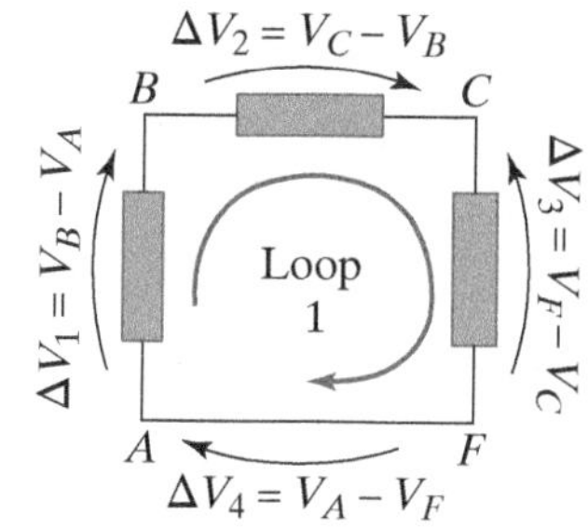

Figure 18.60 Walk around loop 1, following the path $ABCFA$.

We could write a similar equation corresponding to walking along loop 2 (path $FCDEF$). The idea is that you start anywhere you like on the circuit diagram and walk around a loop that brings you back to your starting location, adding up potential drops and rises as you go.

You can even walk along a round-trip path through the air, outside the circuit elements, but unless you know something about the electric field and potential differences outside the devices and connecting wires, this usually isn't very useful.

Checkpoint 8 In Figure 18.59, suppose that $V_C - V_F = 5$ V, and $V_D - V_E = 3.5$ V. What is the potential difference $V_C - V_D$? If the element between C and D is a battery, is the + end of the battery at C or at D?

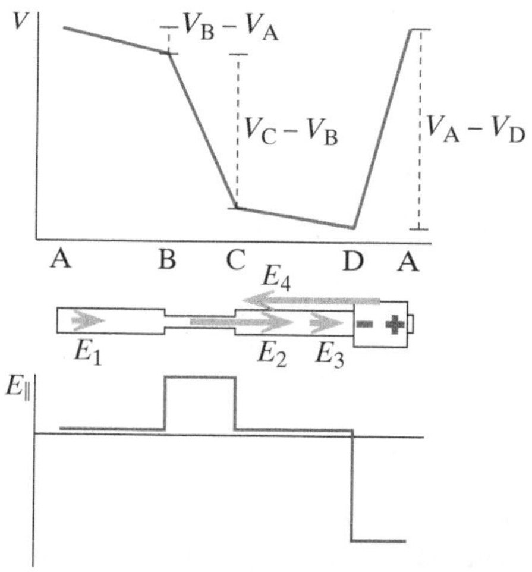

Figure 18.61 Potential rises and falls as you walk around a circuit. The electric field is the gradient of the potential. The connection from the + end of the battery to the left wire is not shown.

Potential Around a Circuit

At the top of Figure 18.61 is a graph of potential around the circuit shown in the middle, along the path ABCDA. Below the circuit is a graph of $E_{||}$, the component of the electric field in the direction that we walk, counterclockwise around the circuit. Inside the battery we are walking opposite to the direction of the electric field, so there is a potential rise equal in magnitude to the emf of the battery.

A big potential gradient (volts per meter) means a large field, so a steep slope on the potential graph corresponds to a large electric field (see bottom of Figure 18.61). Along the thick wires, the potential gradient is small and the field is small ($E = |\Delta V|/L$), yielding a small drift speed (wire AB, for instance).

Along the thin wire (BC) there is a large potential gradient corresponding to a large field and a high drift speed.

For this circuit with its two thick wires and thin resistor connected to a battery, all the potential drops (negative potential differences) plus the potential rise across the battery (numerically equal to the emf) add up to 0:

$$\Delta V_1 + \Delta V_2 + \Delta V_3 + \Delta V_{\text{battery}} = 0$$
$$(-E_1L_1) + (-E_2L_2) + (-E_3L_3) + \text{emf} = 0$$

Checkpoint 9 What would be the potential difference ΔV_2 across the thin resistor in Figure 18.61 (between locations B and C) if the battery emf is 1.5 V? Do you have enough information to determine the current I in the circuit? Assume that the electric field in the thick wires is very small (so that the potential differences along the thick wires are negligible).

18.10 APPLICATIONS OF THE THEORY

We now have two principles on which to base the analysis of circuits. One has to do with charge conservation:

THE CURRENT NODE RULE

In the steady state, for the many electrons flowing into and out of a node,

electron current: net $i_{\text{in}} = \text{net } i_{\text{out}}$, where $i = nAuE$

conventional current: net $I_{\text{in}} = \text{net } I_{\text{out}}$, where $I = |q|nAuE$

The second principle has to do with energy:

ENERGY CONSERVATION (THE LOOP RULE)

In the steady state, for any round-trip path,

$$\Delta V_1 + \Delta V_2 + \cdots = 0$$

The current node rule refers to many electrons moving together, whereas the energy equation is on a per-unit-charge basis.

We will apply these two principles to predict the behavior of simple circuits. These predictions can be tested by measuring the currents in the circuits.

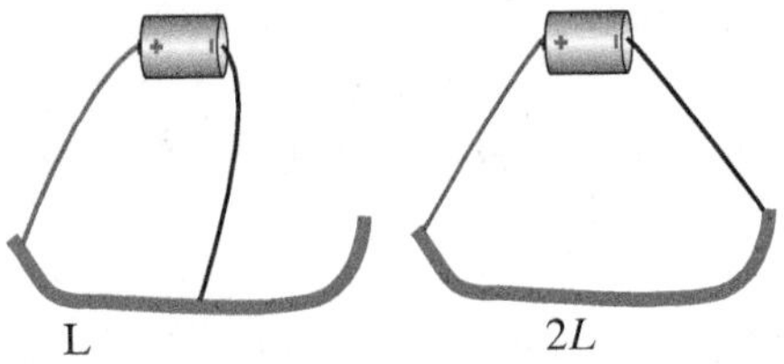

Figure 18.62 How does the current change when we double the length of a Nichrome wire?

Application: Doubling the Length of a Wire

In each of the two circuits shown in Figure 18.62, a Nichrome wire (which has low electron mobility) is connected to a battery by thick copper connecting wires. The circuits differ only in the length of the Nichrome wire that forms part of the circuit. Suppose that you measure the compass deflection produced by the current in each circuit.

QUESTION Based on our model for current in a circuit, if the current through a Nichrome wire of length L produces a compass deflection θ, what compass deflection would you expect from the current in the Nichrome wire of length $2L$?

Around the circuit loop emf $-EL = 0$, so if we double the length L, we'll get half as big an electric field (since the emf doesn't change). With half the field we get half the drift speed ($\bar{v} = uE$), half the electron current ($i = nAuE$), and half the conventional current ($I = |q|nAuE$). Therefore we expect half as large a magnetic field and half the compass deflection.

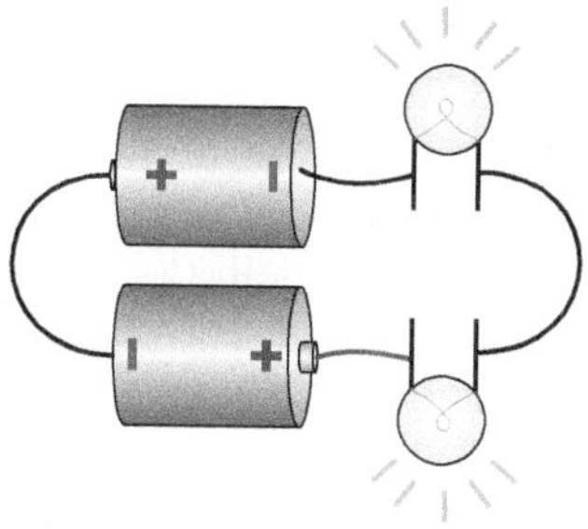

Figure 18.63 A circuit with two identical bulbs in series.

Note that placing two identical bulbs in series is analogous to doubling the length of the bulb filament. However, if you compare a one-bulb circuit to a two-bulb series circuit (Figure 18.63), you will probably find more than half the current in the two-bulb circuit. The issue here is that the electron mobility in very hot tungsten (several thousand degrees!) is less than the electron mobility in cooler tungsten. The greater thermal agitation of the atomic cores leads to larger "friction." Note that the bulbs in the two-bulb circuit emit redder light than the single bulb, indicating that their filaments are not as hot.

Application: Doubling the Cross-Sectional Area of a Wire

Figure 18.64 shows two circuits in which a Nichrome wire is connected to a battery by thick copper connecting wires. The circuits differ only in the cross-sectional area of the Nichrome wire: the thicker wire has twice the cross-sectional area of the thinner wire.

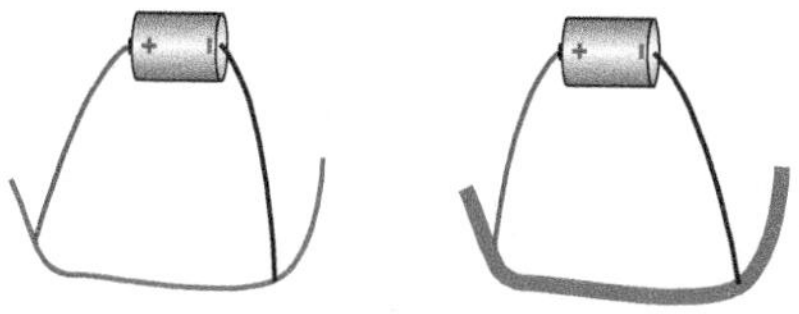

Figure 18.64 How does the current change when we double the cross-sectional area of a Nichrome wire?

QUESTION If the current in the circuit with the thin Nichrome wire is I, what compass deflection would you expect from the current in the circuit with the thick wire?

It seems plausible that doubling A should double the conventional current $I = |q|nAuE$. It is like having two wires connected to the battery, doubling the drain on the battery.

We can make this argument quantitative by considering energy. Since we have emf $- EL = 0$ around the loop, the electric field E in the wire does not change when we use a thicker wire, so the current in the second circuit should be $2I$. (This neglects internal resistance in the battery, which has the effect of limiting the amount of current that the battery can supply, so that for large cross-sectional areas and large currents, doubling a large cross-sectional area may not yield twice the current.)

Note that placing two bulbs in parallel, as shown in Figure 18.65, is analogous to doubling the cross-sectional area of a bulb filament.

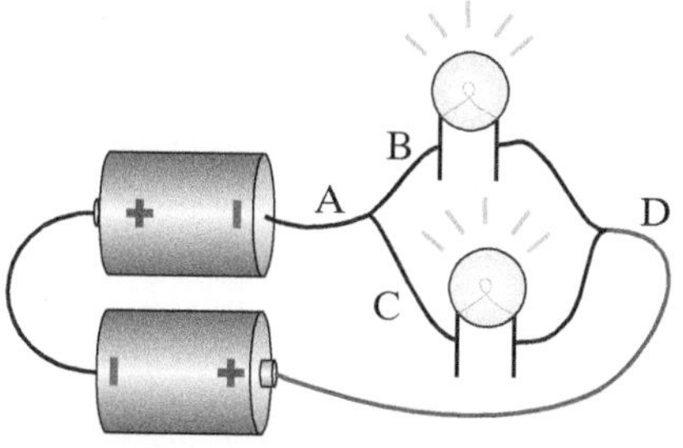

Figure 18.65 Placing two bulbs in parallel is analogous to doubling the cross-sectional area of a bulb filament.

Quantitative Measurements of Current with a Compass

QUESTION How should the current at location D in Figure 18.65 compare to the current in a circuit with a single bulb?

We should expect the current at D to be twice as large as the current in a circuit with only one bulb, for the reasons discussed above.

Experiments EXP1 through EXP5 are designed to let you explore the consequences of the node rule (charge conservation in the steady state) and the loop rule (energy conservation on a per-coulomb basis).

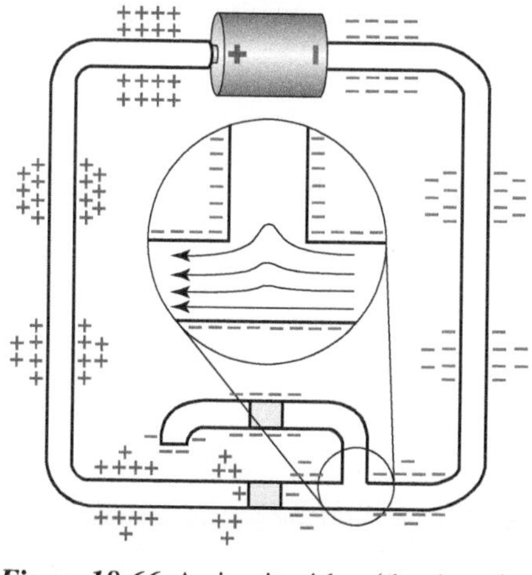

Figure 18.66 A circuit with a (dead-end) parallel branch.

How Does Current Know How to Divide between Parallel Resistors?

When there are parallel branches in a circuit, as in Experiment EXP5, there is a question of how the current "knows" how to divide between the branches.

In Figure 18.66 we show very approximately the steady state of a circuit containing wide carbon resistors in two parallel branches. During the initial transient just after connecting the battery, electrons flowed into both branches, but the surface of the dead-end branch became charged so negatively that no more electrons could enter. In the inset, note the deflection of the electron stream lines due to the lack of charge immediately above the branch point, inside the dead-end branch. The wire is effectively a bit wider as a result of adding the dead-end branch.

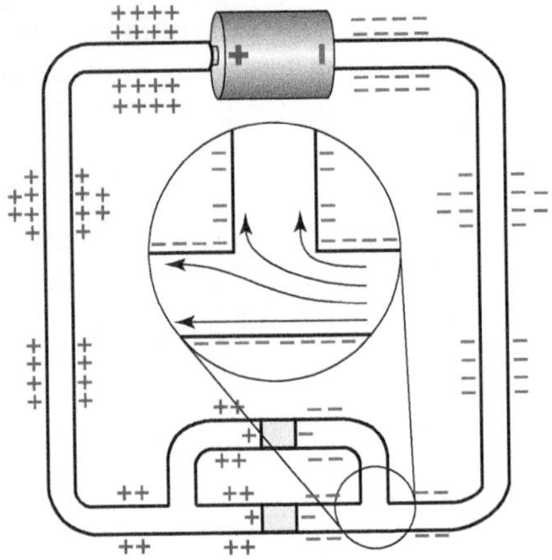

Figure 18.67 Complete the connection: the electron current splits.

Next we complete the parallel connection, which leads to a rearrangement of surface and interface charges (Figure 18.67). The inset shows how the new arrangement of surface charges guides and splits the oncoming electron current, with some of the electrons taking the upper branch and some the lower branch. Note that there is now a larger current through the battery and a larger gradient of surface charge along the wires to drive this larger current.

The steady-state current in each branch is determined by how well each branch conducts current. For example, if the mobility is low in one of the branches, surface charge will build up in such a way as to decrease the fraction of the current that goes down that branch. The extreme case is a dead end, where surface charge builds up until it prevents any current at all from going down the dead-end branch.

Application: Thin Filament Bulb and Thick Filament Bulb in Series

In the Experiments section of the previous chapter we described simple equipment that can be used to explore electric and magnetic phenomena. This equipment includes a thin-filament bulb (a #48 flashlight bulb) that has an elongated shape and a thick-filament bulb (a #14 flashlight bulb) that has a round shape. If you put a thin-filament bulb and a thick-filament bulb in series with each other (Figure 18.68), the thin-filament bulb lights up but the thick-filament bulb doesn't light up! What happened to the current that ran through the thin-filament bulb? We can explain this puzzling behavior. Consider that:

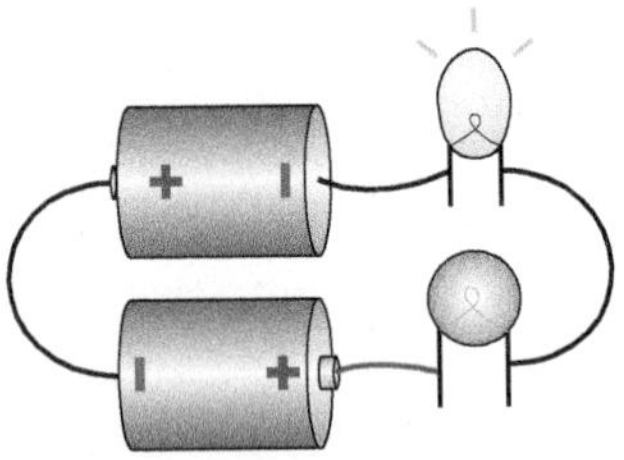

Figure 18.68 A thin-filament bulb and a thick-filament bulb in series—why doesn't the thick-filament bulb light up?

Both bulb filaments are made of tungsten.

The round, thick-filament bulb has large cross-sectional area A_r, length L_r, electric field E_r, and drift speed $\bar{v}_r$.

The long, thin-filament bulb has small cross-sectional area A_l, length L_l, electric field E_l, and drift speed $\bar{v}_l$.

Assume for the moment that the electron mobility is the same in both bulbs, even though in fact the glowing thin-filament bulb is very hot (which should mean low mobility) and the thick-filament bulb is apparently cool (since it isn't glowing, which should mean high mobility).

QUESTION Using the current node rule, write an equation relating the drift speeds $\bar{v}_r$ and $\bar{v}_l$.

In the steady state, the number of electrons per second passing through each bulb must be the same:

$$nA_r\bar{v}_r = nA_l\bar{v}_l$$

QUESTION Based on the equation you just wrote, write an equation relating the electric fields E_r and E_l, and state which is larger, E_r or E_l.

Assuming the same mobility in both filaments, since

$$\bar{v} = uE$$

we know that $nA_r uE_r = nA_l uE_l$, so

$$E_l = \frac{A_r E_r}{A_l}$$

Since $A_r > A_l, E_l > E_r$.

QUESTION Applying the loop rule, we find that $2\text{emf} = E_r L_r + E_l L_l$, which is approximately equal just to $E_l L_l$, since E_r is very small. Explain why the thick-filament bulb doesn't glow.

If the thick-filament bulb were alone in a circuit, it would be bright, and the electric field in the filament would be:

$$2\text{emf} = E_{r,\text{bright}}\,L$$

so

$$E_{r,\text{bright}} = \frac{2\text{emf}}{L}$$

In the two-bulb circuit (if the lengths of the filaments are the same) we have:

$$\begin{aligned} 2\text{emf} &= E_r L + E_l L = L\left(E_r + \frac{A_r E_r}{A_l}\right) \quad \text{(from previous exercise)} \\ &= LE_r(1 + A_r/A_l) \end{aligned}$$

so $$E_r = \frac{2\text{emf}}{L(1+A_r/A_l)} < \frac{2\text{emf}}{L}$$

Thus the electric field in the thick-filament bulb is significantly smaller than it is in the single-bulb circuit. We can estimate how much smaller by noting that one typically measures about 1/4 as much current in a circuit with a single thin-filament bulb as in a circuit with a single thick-filament bulb. This suggests that $A_r \approx 4A_l$, so the electric field in the thick-filament bulb has been reduced by a factor of 5. Evidently this field does not produce sufficient current through the filament to make it glow.

Moreover, since the thick-filament bulb isn't glowing, its temperature is low and its mobility is much larger than usual. This further reduces the E_r required to drive the same current that runs through the thin-filament bulb.

What Determines Bulb Brightness?

For each electron that moves through a wire of length L due to an electric field E, the electric potential energy of the system (electron plus wire) decreases by an amount $e\Delta V = eEL$. This loss of electric potential energy is equal to the gain in kinetic energy of the electron, which is dissipated in collisions with the atoms in the wire, with consequent increase in the thermal energy and temperature of the wire. As the wire's temperature increases, the rate of energy transfer to the surroundings increases in the form of the emission of light. The temperature of the wire stops rising once the temperature is high enough that the rate of emitted energy (as light) has become equal to the rate of energy dissipation in the wire.

In a time Δt the number of electrons entering (and leaving) the wire is $nA\bar{v}\Delta t = nAuE\Delta t$, so the energy dissipation per second (power in joules per second, or watts) is $(eEL)(nAuE) = enAuLE^2$. We see that the rate of energy dissipation is proportional to the square of E, or alternatively to the square of the current $I = enAuE$. In the previous example the field E in the thick-filament bulb was reduced by a factor of 5, so the rate of energy dissipation in the wire was reduced by a factor of 25. This in turn means that the rate of energy emitted as light was reduced by a factor of 25. The bulb is so dim that you don't see it glow.

Application: Two Batteries in Series

It should now be clear why two batteries in series can drive more current through a wire than one battery can: the potential difference across two batteries in series is 2 emf, and this doubles the electric field everywhere in the circuit. Double the electric field, double the drift speed, double the current (if the mobility doesn't decrease due to a higher temperature).

In Experiment EXP6 you can test this prediction with a Nichrome wire rather than a light bulb. When you add a second battery, a bulb filament undergoes a sizable temperature increase that changes the mobility significantly (you can see the color become whiter, which is an indication of the higher filament temperature). With a length of Nichrome wire the temperature stays close to room temperature, so the mobility is nearly the same with one or two batteries.

EXAMPLE

A Circuit with a Wide Wire

The circuit in Figure 18.69 contains a low-resistance mechanical battery that exerts a non-Coulomb force F_{NC} to move electrons through the battery (with negligible internal resistance). The end plates of the battery are very large compared to the distance $0.1L$ between the plates (plates not drawn to scale). Two thin Nichrome wires of length L and cross-sectional area A connect the battery to a thick Nichrome wire of length $0.6L$ and cross-sectional area $4A$. The electron mobility of the Nichrome is u, and there are n mobile electrons per cubic meter in the Nichrome.

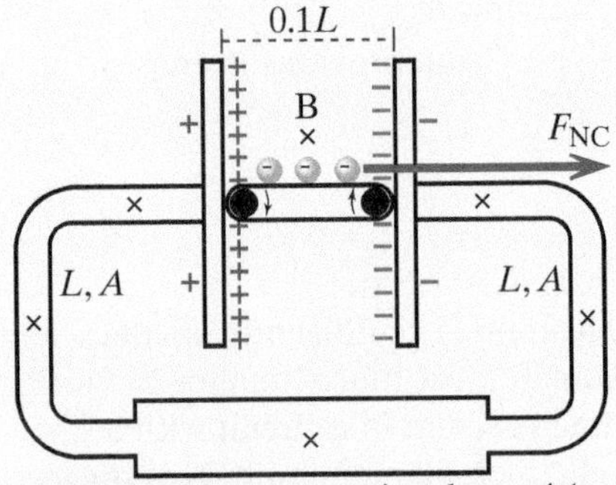

Length $0.6L$, cross-sectional area $4A$

Figure 18.69 Circuit with a wide wire.

(a) Show the electric field at the six locations marked with × (including location B between the plates). Pay attention to the relative magnitudes of the six vectors that you draw.

(b) Show the approximate distribution of charge on the surface of the Nichrome wires. Make sure that your distribution is compatible with the electric fields that you drew in part (a).

(c) Calculate the number of electrons that leave the battery every second, in terms of the given quantities L, A, n, u, and F_{NC} (and fundamental constants). Be sure to show all of your work.

(d) Calculate the magnitude of the electric field between the plates of the battery (location B).

Solution

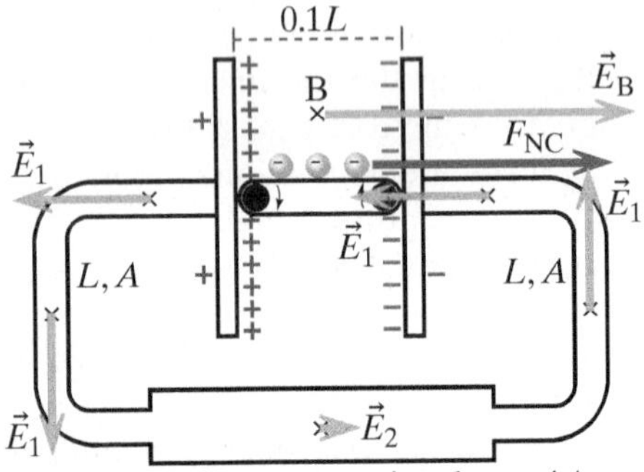

Length $0.6L$, cross-sectional area $4A$

Figure 18.70 Part (a), electric field at locations in the wire.

(a) By the current node rule, current in the thin and thick wires will be equal. Since $i = nAuE$, E will be larger in wires with thinner cross-sectional area. This suggests the pattern of electric field shown in the wires in Figure 18.70.

In order for the round-trip potential difference to be zero, the electric field between the plates must point to the right and be large compared to the electric field in the wires, as shown in Figure 18.70.

(b) A large gradient of surface charge is associated with a large electric field; a small gradient with a small field, as shown in Figure 18.71.

(c) Apply basic principles:
Current node rule: current in thin wire (1) = current in thick wire (2) gives

$$nAuE_1 = n(4A)uE_2$$

$\Delta V_{\text{round trip}} = 0$ (energy conservation), going counterclockwise from the negative plate of the mechanical battery:

$$\frac{F_{NC}(0.1L)}{e} - E_1L - E_2(0.6L) - E_1L = 0$$

From the current equation:

$$E_2 = \left(\frac{A}{4A}\right)E_1 = \frac{E_1}{4}$$

We now have two equations in two unknowns (E_1 and E_2). Substitute for E_2:

$$\frac{F_{NC}(0.1L)}{e} - 2(E_1L) - \left(\frac{E_1}{4}\right)0.6L = 0$$

$$E_1 = \frac{F_{NC}}{e(21.5)}$$

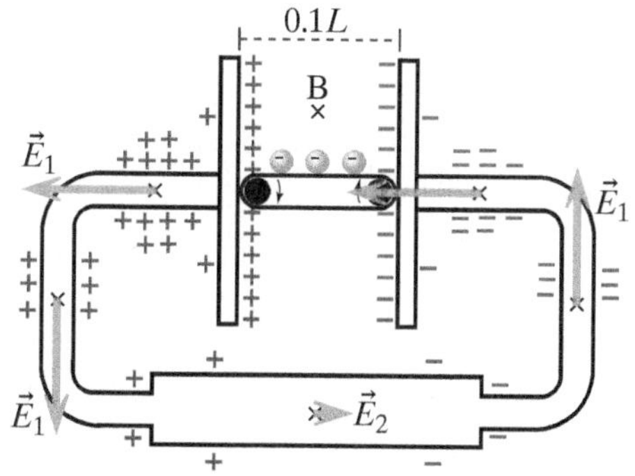

Figure 18.71 Part (b), an approximate surface-charge distribution.

Now we can solve for electron current (number of electrons/s leaving the battery):

$$i = nAuE_1 = nAu\left(\frac{F_{NC}}{e(21.5)}\right)$$

(d) The round-trip potential difference must be zero, so:

$$E_B(0.1L) - 2E_1L - \left(\frac{E_1}{4}\right)(0.6L) = 0$$

$$E_B = 21.5\left(\frac{F_{NC}}{e(21.5)}\right) = 21.5\,E_1$$

The electric field E_B inside the battery is 21.5 times as large as the electric field E_1.

18.11 DETECTING SURFACE CHARGE

The reason we can't detect the surface charge on the wires of the simple circuits we have been assembling is that the amount of charge on the wires is very small—very little charge is needed to make the small electric fields in the wires and bulb filaments of your circuits. However, it is possible to detect surface charge in a high-voltage circuit (around 10,000 V is required). The following interesting demonstration of surface charge was suggested by the science education group at the Weizmann Institute in Israel. Perhaps your instructor will be able to show this to you.

Warning: Working with high voltages is dangerous, so don't try this without an instructor present. Keep one hand behind your back when you are near a high-voltage circuit. Do not touch the bare wires!

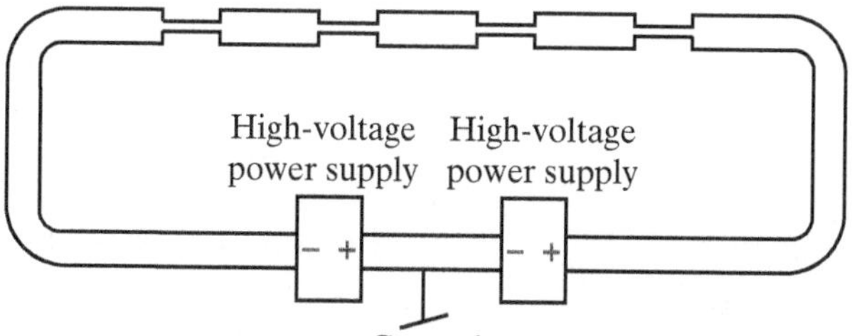

Figure 18.72 A high-voltage (low-current) circuit for observing surface charge.

In Figure 18.72 a chain of four identical high-resistance resistors is supported in air from Styrofoam columns, so that the resistor chain is far from other objects, including the table, to avoid unwanted polarization effects. The four resistors are represented in Figure 18.72 as thin wires. The resistors are connected by bare (uninsulated) thick wires to two high-voltage power supplies, which maintain one end of the resistor chain at a positive potential and the other end at a negative potential with respect to ground (the round hole in a three-prong socket). Note that you need to avoid touching the wires when the circuit is connected!

We model the resistors as very thin wires. The conventional current everywhere in this series circuit is $I = enA\bar{v} = enAuE$.

QUESTION Given the expression for the current, sketch the pattern of electric field around the circuit.

Evidently the electric field must be very much larger inside the thin resistors than inside the thick connecting wires, and the electric field must have the same magnitude inside each resistor. As usual, the pattern of electric field is simple.

The feedback mechanism will lead to an arbitrarily complicated distribution of surface charge that produces this simple pattern of field.

QUESTION Sketch a plausible approximate distribution of surface charge, consistent with the pattern of electric field.

Very roughly, the electric field inside the leftmost resistor (AB) is in large part due to a large negative surface-charge density at A and a medium surface-charge density at B, with a large gradient along the resistor (Figure 18.73). This is an approximate analysis, because the field in the resistor is due not just to the nearby charges but to all charges. The electric field must have the same magnitude in the next resistor (BC) and is largely due to the medium surface-charge density at B and the nearly zero surface-charge density at C. Similar remarks apply to the other two resistors.

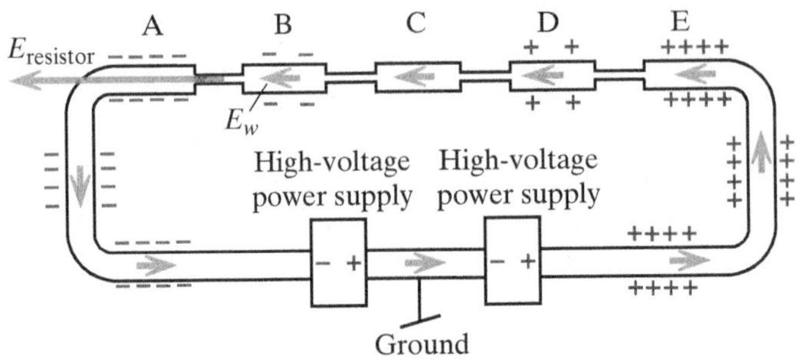

Figure 18.73 Electric field and surface charge for a high-voltage circuit.

Inside the thick connecting wires the electric field is very small, so we indicate a roughly uniform charge distribution along these wires (the details of the actual charge distribution, except for the sign of the charge, are presumably more complicated).

A neutral strip of aluminum foil is suspended by a hair from an insulating rod and brought close to the bare wire near location A.

QUESTION What do you expect will happen?

The strip is observed to be attracted to the wire (due to polarization of the strip by the surface charge on the wire) and then to jump away, due to being charged by contact with the bare wire and repelled by the surface charge at that location. The strip is found to be negatively charged, because it is repelled by a plastic pen rubbed through one's hair (a clear plastic pen is known to charge negatively). A similar effect is observed at E, but the strip is observed to be positively charged. A smaller effect is observed at B and D, where the surface-charge density is expected to be smaller. No effect is observed at point C, where by symmetry there is essentially no surface charge. These experimental observations confirm the validity of the approximate surface charge analysis.

The surface-charge density is proportional to the circuit voltage, and only at very high voltages is there enough charge to observe electrostatic repulsion in a mechanical system. It is difficult to observe repulsion by surface charge in a low-voltage circuit, because a charged object is of course attracted to neutral matter, and the attraction between a charged object and a circuit can mask the repulsion, if the surface charge on the circuit is small as it is in low-voltage circuits.

The object must be very lightweight so that one can observe the interaction at a sizable distance (a small piece of aluminum foil hanging from a thread works well). The problem is that at short distances there is competition between repulsion and the attraction that is due to polarization of neutral matter. The attractive part of the interaction falls off with distance much more rapidly than

$1/r^2$. As we saw in Chapter 14, the attraction between a point charge and a small neutral object is proportional to $1/r^5$, since the polarization of the neutral object is proportional to $1/r^2$, and the interaction between a point charge and a dipole is proportional to $1/r^3$.

18.12 *COMPUTATIONAL MODEL OF A CIRCUIT

The computational model used to calculate surface charge distributions can be applied to the polarization of a neutral metal block, as well as to a circuit. Consider the case of a positive point charge polarizing a neutral metal block (Figure 18.74). Imagine that at first there are no charges near the block, and then we suddenly bring the positive point charge into position to the left of the block.

As we saw in Chapter 14, at first there is a nonzero net field to the right inside the block, due solely to the point charge. This field drives mobile electrons to the left, and in a *very* short time Δt some excess electrons will appear to the left on the surface of the block. There will be a corresponding deficiency of electrons to the right. The time to create some nonzero surface charge is very short because there are plenty of mobile electrons everywhere inside the metal; the mobile electron sea has to shift only an extremely short distance in order for significant nonzero charges to build up on the surface.

Figure 18.74 A 4 mm × 4 mm × 8 mm metal block polarized by a positive point charge 12 mm to the left of the block. Shades of blue represent negative charge and shades of red represent positive charge.

These new surface charges, negative on the left and positive on the right, contribute an electric field inside the block that points to the left, so that the net field inside is reduced. In the next very short time interval, the (reduced) electric field, still to the right, will drive more electrons onto the left portion of the surface, with increased electron deficiency on the right. These charges will reduce the net field even more. Eventually there will be so much surface charge, negative on the left and positive on the right, that the sum of the field contributed by the point charge and the field contributed by the surface charge add up to zero.

In our computational model of this process we divide the surface into small square tiles, as though we were covering the surface with little postage stamps (in Figure 18.74 the tiles are 0.25 mm × 0.25 mm). In each step of an iterative procedure, we calculate the electric field just inside each tile based on the contributions of all the charged tiles and any external fixed charges. Knowing the field just inside a surface tile, we can determine how much charge will be added to or subtracted from that tile in the next short time interval.

The next iteration uses the new tile charges in the computation of the field near each tile. We graph the amount of charge on a representative tile and halt the iterations when we see that this tile's charge is no longer changing. We find that the electric field contributed by the final distribution of charged tiles, plus the field of the external charge in Figure 18.74, is as expected zero inside the metal block, which is a check on the procedure.

In order to get a more quantitative representation of the distribution of surface charge we consider the charge on (square) rings of tiles whose axes are the center line of the block as shown in Figure 18.75. Figure 18.76 is a graph of the computed surface charge density on such rings of charged tiles from left to right along the center line of the block. For each ring we plot the total charge of the ring divided by the surface area of the ring (the number of tiles times the area of each tile), which is the average surface charge on this ring. Note the approximately constant gradient of surface charge density except near the left and right faces of the block, where there are large buildups of charge.

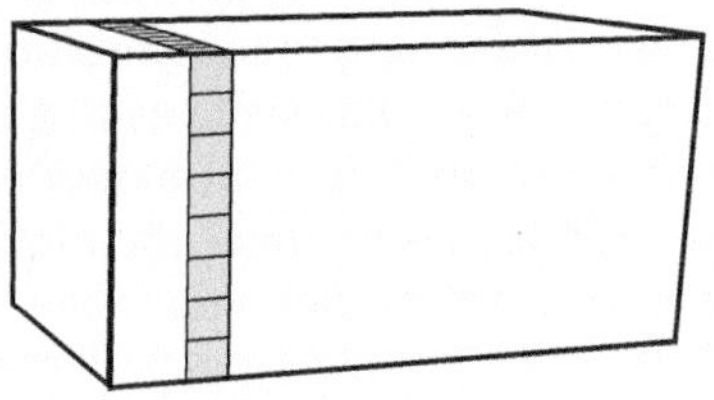

Figure 18.75 A (square) ring of charged tiles on the surface.

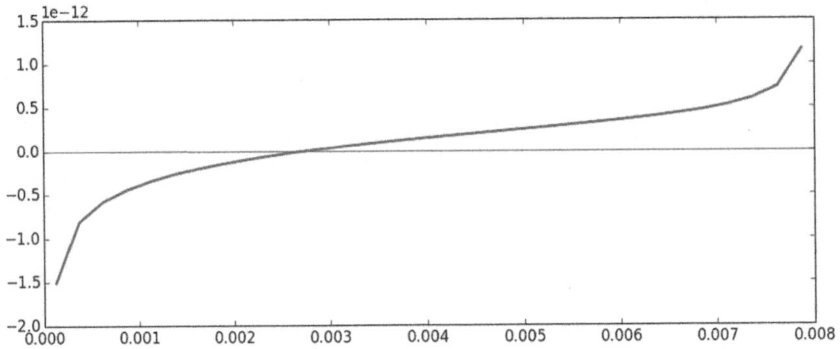

Figure 18.76 Surface charge density (C/m^2) of rings of charge surrounding a horizontal path (in meters) going left to right through the center of a polarized neutral metal block.

The computational model for determining the distribution of surface charge was implemented in VPython, exploiting the fast array processing provided by Python's NumPy module, which is imported by VPython. In preparation for the iterative loop that moves charge onto and off the surface tiles, we calculate for each tile the effect that every other tile would have if the other tile were a point charge of 1 C. To do this we calculate the field the other tile contributes just inside our tile, take the component E_{normal} along the outward-going unit vector that is normal to our tile, and multiply by $enAu\Delta t$ to get the change in our tile's charge in a time Δt.

After calculating this for every one of the N tiles, we have an N by N matrix of the effects of all tiles on all other tiles, on a per-coulomb basis. We add to these effects the effect of each tile's nearby field on itself, considering the tile to be a uniformly charged square plate. In a similar way, for each tile we compute what effect the fixed external charges will have.

Having prepared this information, each iteration of the iterative process consists of multiplying the current tile charges times their per-coulomb effects on all of the tile charges, and we add in the fixed effects of the external charges. After making a list of all the Δq's for all tiles, we update the charges on all the tiles. We graph the charge on a representative tile as a function of time and stop the iteration when we see that this charge is no longer changing. A conclusive test of the procedure is that when we calculate the electric field due to all of the computed tile charges and the external charges, the distribution of electric field is correct: zero inside a polarized metal block, nonzero and uniform inside circuit wires in the steady state.

Currents in the Interior

We need not concern ourselves with charge in the interior of the metal, thanks to a remarkable, special property of the $1/r^2$ electric force law that ensures that the interior of the metal will remain neutral at all times. Consider what

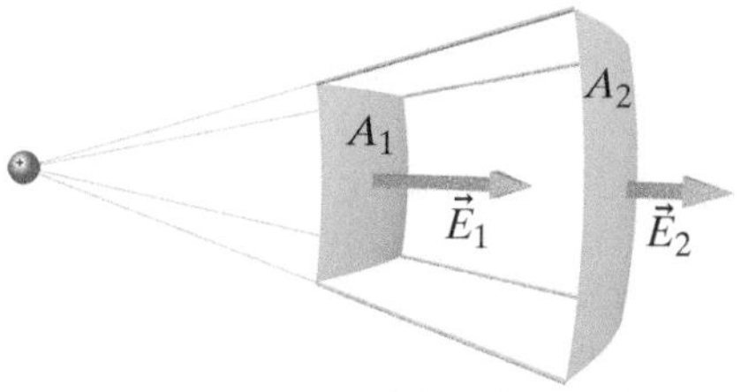

Figure 18.77 A volume similar to a truncated pyramid, located in the interior of the metal. The volume is bounded by two spherical faces centered on the point charge. A_1 is the area of the smaller face (nearer to the point charge), and A_2 is the area of the larger face (farther from the point charge). On each face the electric field due to the point charge is uniform.

happens initially to a (neutral) region in the interior of our metal block, chosen to have the form of a truncated pyramid, with the apex of the pyramid at the location of the point charge (Figure 18.77).

The field of the point charge drives an electron current to the left, out of the near face of the truncated pyramid, of magnitude $i_1 = nAv = nA_1uE_1$, where A_1 is the area of the left face and E_1 is the field on the near face. The field of the point charge also drives an electron current into the far face of the region, of magnitude $i_2 = nAv = nA_2uE_2$, where E_2 is the field on the far face.

On the far side of the region the field is smaller but the area is bigger: the area A is proportional to r^2 (where r is the distance to the point charge) while the electric field is proportional to $1/r^2$, so that $A_1E_1 = A_2E_2$, which means that $i_1 = nA_1uE_1$ has the same magnitude as $i_2 = nA_2uE_2$. Also, no electrons enter or leave the sides of the truncated cone because there is no component of the field that is perpendicular to the surface there. Therefore, in the next short time Δt, the outflow of electrons is exactly the same as the inflow, so this initially neutral interior region of the metal stays neutral. In each time interval Δt, the same number of electrons flow out of the region as flow into it, so the interior of the metal remains neutral at all times. The only changes in charge are on the surface of the metal.

In Chapter 21 we will see, using a property of electric fields known as Gauss's law, that this result is general, that the net charge contained in any region in the interior of the metal, of any shape, will remain zero at all times. As charge accumulates on the surface of the metal block, the additional fields contributed by these surface charges also have the property of leaving all regions of the interior with a net charge of zero. As a result, the computational model only has to calculate changes in the amount of charge of the surface tiles and does not need to track what happens in the interior of the metal.

An Interactive Presentation

On the Wiley student website for this textbook is a program that lets you inspect these surface charge distributions in 3D and explore by dragging the mouse the electric fields contributed by the fixed charges and by the surface charges inside and outside the metal.

SUMMARY

New Concepts

- In a uniform wire, a uniform electric field drives a uniform steady-state current.
- A gradient of charge on the surface of the wires makes electric fields inside the wires.
- There is an initial transient, lasting a minimum of a few nanoseconds, in which feedback establishes the steady-state distribution of surface charges and currents.
- A battery maintains a charge separation and therefore a potential difference. The emf of the battery is the energy input per unit charge required to maintain the charge separation.

Analysis Techniques

In analyzing electric circuits we apply two rules:
The current node rule:

In the steady state:
electron current: net i_{in} = net i_{out}, where

$$i = nAuE$$

conventional current: net I_{in} = net I_{out}, where

$$I = |q|nAuE$$

q is the charge of a mobile charged particle; n is the density of mobile charges; A is the cross-sectional area of the conducting object, u is the mobility of the charged particles.

Energy conservation (the loop rule):
In the steady state, for any round-trip path,

$$\Delta V_1 + \Delta V_2 + \cdots = 0$$

Results

Our model of the microscopic behavior of circuits provides answers to several questions about circuit behavior:

1. *Where are the charges that are the sources of the electric field inside the wires?* These charges are on the surfaces of the wires, and on and in the battery. The necessary charge distribution is created and maintained by feedback.
2. *How can the light come on instantaneously despite the very slow drift speed of the mobile charges?* Electric fields propagate at the speed of light. The electrons inside the bulb are affected by the electric field which is quickly (in a few nanoseconds) established in the bulb filament.
3. *What does a battery actually do in a circuit?* A battery does a fixed amount of work per unit charge (emf), maintaining a charge separation and potential difference across itself, and consequently maintaining a distribution of surface charge all over the circuit.

EXPERIMENTS

EXP1 Current in different parts of a circuit

Assemble a circuit containing a thick-filament bulb and two batteries in series, as shown in Figure 18.78.

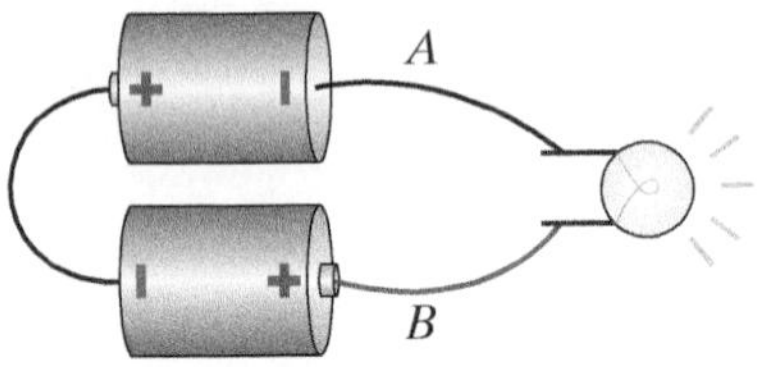

Figure 18.78

Use the following procedure to make careful measurements of the magnetic field produced by each wire. (Note that equal currents will produce equal magnetic fields, and hence equal compass deflections.)

- Disconnect one wire.
- Make sure the batteries and the steel alligator clips are not near the compass.
- Lay the wire in which you wish to measure current over the compass, carefully lining up the wire with the compass needle.
- Carefully reconnect the circuit without moving anything.
- Read the compass deflection to within 2° or better.

(a) Is the circuit in a steady state? Leave the compass under the wire at A, and observe it for 15–20 s. Does the deflection remain constant?

(b) Using the compass, measure the current at A and B, and record the observed compass deflections.

(c) Use this sensitive "null" test to compare the current in the two wires: run wires A and B side-by-side in opposite directions over the compass, then connect the circuit. The magnetic fields produced by the two wires are in opposite directions. If the currents are equal and the wires are very close together, the magnetic fields in the two wires will cancel each other and give zero deflection.

Making Careful Measurements of Currents

In the following experiments (Experiments EXP 16 through EXP5) we will use our model of circuits to predict the ratio of current in one circuit to current in another circuit, and then we will do experiments to test our predictions. In order to get good experimental results, we need to understand how to use a compass to make accurate measurements of current. There are two important issues to consider.

1. Is Magnetic Field Uniform All Along the Compass Needle?

The compass needle aligns with the net magnetic field at its location. As we have seen, the net magnetic field is the sum of the Earth's magnetic field and the magnetic field made by the current in the wire. In Chapter 17 we found that the magnitude of the magnetic field made by current in a long wire is inversely proportional to distance from the wire:

$$B = \frac{\mu_0}{4\pi}\frac{2I}{r}$$

If the distance from the wire to the center of the needle is significantly different from the distance from the wire to the ends of the needle, the magnetic field at the ends of the needle will be weaker than at the center. In this case the deflection of the needle will be smaller than it should be. If the wire lies on top of the compass, it is about 6 mm away from the center of the needle. If the needle deflects more than 15°, the ends of the needle are significantly farther away from the wire than the center is, as shown in Figure 18.79, and the deflection will not be proportional to the current.

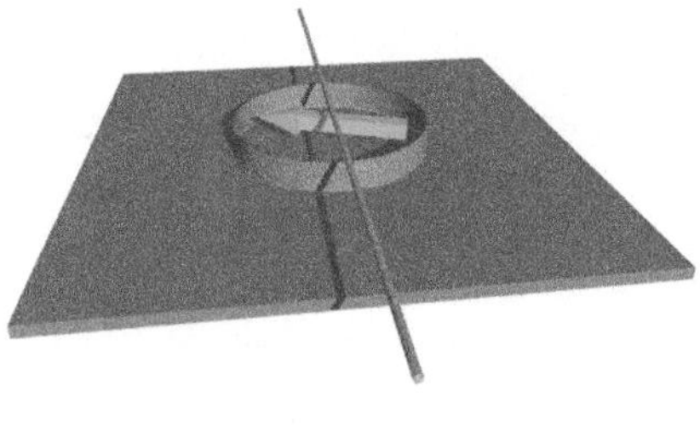

Figure 18.79

Elevating the Wire

By elevating the wire about 1 cm above the top of the compass, we minimize this effect. Now even if the needle deflects up to 40°, the ends and the center will still be approximately the same distance from the wire. An easy way to elevate the wire is to support it on two fingers, as shown in Figure 18.80.

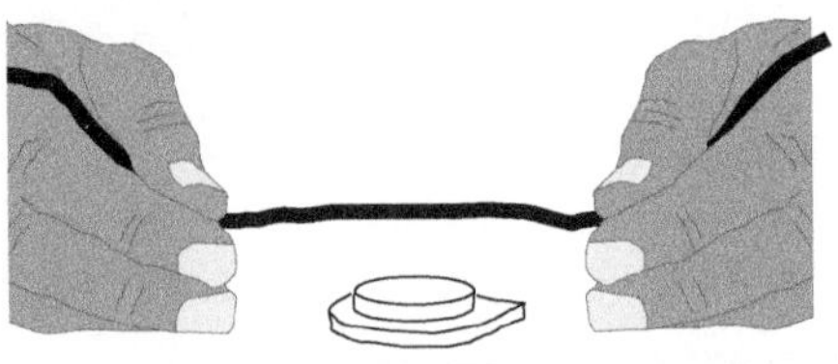

Figure 18.80

2. Is Current Proportional to Deflection Angle?

As can be seen from Figure 18.81, the tangent of the deflection angle is proportional to the magnetic field made by the current in the wire. For small deflection angles (measured in radians), $\theta \approx \tan\theta$. However, this approximation is not good for angles greater than about 0.3 rad (17°).

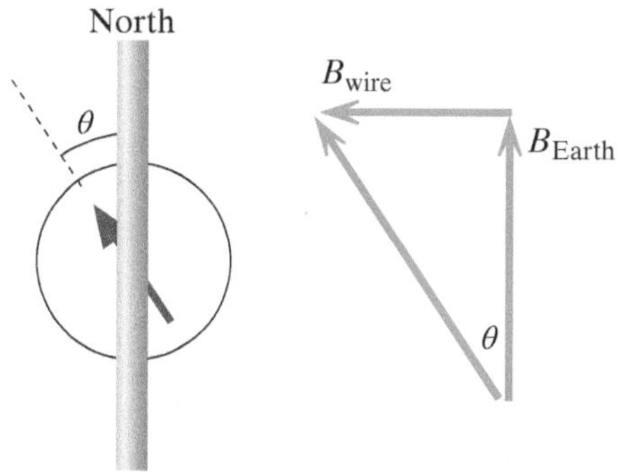

Figure 18.81 $B_{\text{wire}}/B_{\text{Earth}} = \tan\theta$

Tangent of Deflection Angle

- For deflection angles less than 15°, we can assume that the magnetic field, and hence the current in the wire, is proportional to the deflection angle.
- For deflection angles greater than 15°, we need to use the tangent of the deflection angle to calculate magnetic field (and therefore current).

EXP2 Effect of twice the length

Here is an experiment in which we compare the electron current in a Nichrome wire of length L to the current in a Nichrome wire of length $2L$.

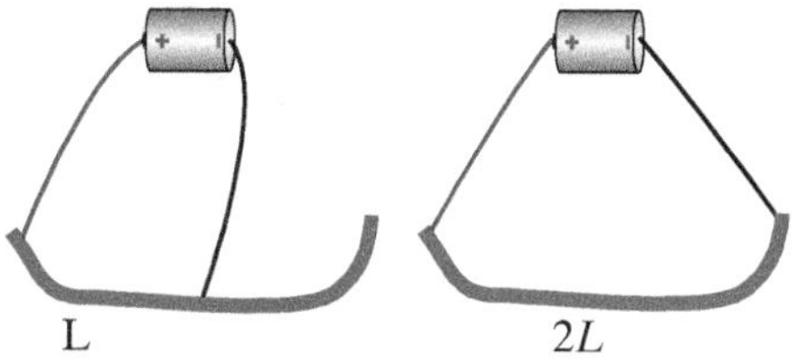

Figure 18.82

QUESTION What does our model for current in a circuit predict for the result of the experiment?

Around the circuit loop we have emf $- EL = 0$, so if we double the length L, we'll get half as big an electric field (since the emf doesn't change). With half the field we get half the drift speed ($\bar{v} = uE$), half the electron current ($i = nAuE$), and half the conventional current ($I = |q|nAuE$). Therefore we expect half as large a magnetic field and half the compass deflection.

Next make the measurement. Be careful when connecting Nichrome wire. If you run current through a short (< 10 cm) length of the wire it can get hot enough to burn you. (It may even get hot enough to glow.)

Following the directions given above, use the compass to compare the amount of current in the two different circuits shown in Figure 18.82:

(a) One battery (to keep the compass deflection small) and half the length of the thin Nichrome wire (about 20 cm). Don't cut the Nichrome wire, just connect clips to a portion half the length of the wire.

(b) One battery and the full length (about 40 cm) of the thin Nichrome wire.

Make careful, quantitative measurements of the angles, and record your measurements.

- Are you able to verify the prediction that the current is halved when you double the length of the Nichrome wire?
- Do you find that the current is unaffected if you bend the wire or add additional connecting wires?

EXP3 Current in a two-bulb series circuit

Two identical light bulbs in series (Figure 18.83) are essentially the same as one light bulb with twice as long a filament, since the copper connecting wires contribute little energy dissipation. From the preceding experiment with doubling the length of a Nichrome wire, we might expect to get half the current compared with just one bulb.

(a) Following the directions given above, use a compass to measure the current in the circuit with two thick filament bulbs in series shown in Figure 18.83, and record the observed compass deflections. Compare your observations to your prediction.

(b) Compare the compass deflection you observe in the two-bulb circuit with the compass deflection you observe in a one-bulb circuit. Record your measurements (by removing one bulb from the circuit).

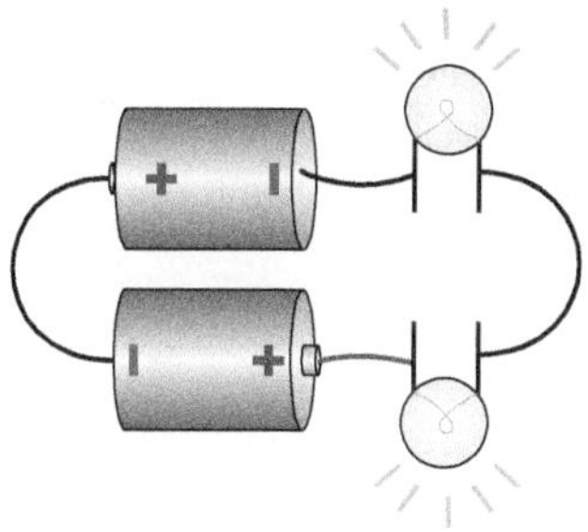

Figure 18.83

You presumably found that the two-bulb circuit had more than half the current of the one-bulb circuit (if necessary, repeat this measurement carefully). The issue here is that the electron mobility in very hot tungsten (several thousand degrees!) is less than the electron mobility in cooler tungsten. The greater thermal agitation of the atomic cores leads to larger "friction." Note that the bulbs in the two-bulb circuit emit redder light than the single bulb, indicating that their filaments are not as hot.

Explain why the current in the two-bulb series circuit is more than half the current in the one-bulb circuit. How is the electric field affected? How is the drift speed affected?

Why didn't we see this effect when we experimented with doubling the length of the Nichrome wire?

EXP4 Effect of doubling the cross-sectional area

Figure 18.84 illustrates an experiment in which we see what happens to the current in a Nichrome wire when we double its cross-sectional area.

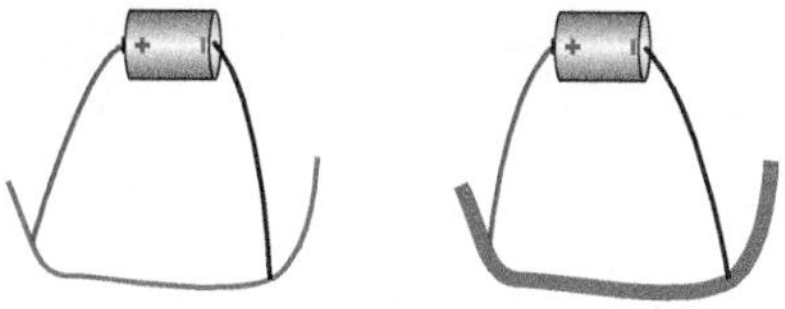

Figure 18.84

QUESTION Explain why the electron current in the wire should increase by a factor of two if the cross-sectional area of the wire doubles.

In Figure 18.84 it seems plausible that doubling A should double the conventional current $I = |q|nAuE$. It is like having two wires in parallel, doubling the drain on the battery.

We can make this argument quantitative by considering energy. Since we have emf $- EL = 0$ around the loop, the electric field E in the wire does not change when we use a thicker wire, so the current doubles. (This neglects internal resistance in the battery, which has the effect of limiting the amount of current that the battery can supply, so that for large cross-sectional areas and large currents, doubling a large cross-sectional area may not yield twice the current.)

Use the compass to compare the amount of current in two different circuits (Figure 18.84): one battery (to keep the compass deflection small) and the same lengths of thin and thick Nichrome wires.

Use only one battery. Connect the connecting wires about 30 cm apart on each Nichrome wire (don't cut the Nichrome wire). Following the directions given above, make careful, quantitative measurements of the angles.

The thick Nichrome wire has a cross section about twice the cross section of the thin Nichrome wire. Your instructor may give you a more precise ratio of the areas. What do you measure for the ratio of the current in the thick wire to the current in the thin wire? Does this agree with predictions?

Your own data are probably consistent with what more precise experiments confirm. These measurements imply that the drift speed of individual electrons is the same in the thick wire and in the thin wire (because the electric field is the same in the two circuits if the length of the wires is the same), and that the many-electron current is distributed uniformly over the cross section.

If the drift speed were not the same everywhere across the cross section of the wire, we couldn't simply write $nA\bar{v}$ for the electron current, using the same $\bar{v}$ for all the electrons passing that point in the circuit. In Section 18.3 we proved that the current is indeed uniform across the cross section.

EXP5 A parallel circuit

Connect two thick filament bulbs in "parallel" (that is, one beside the other), as shown in Figure 18.85.

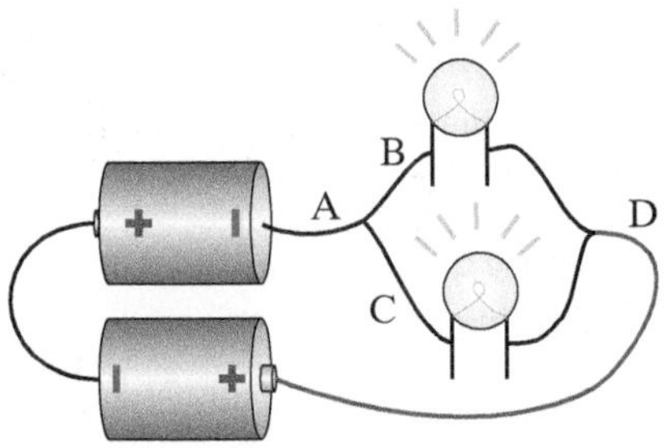

Figure 18.85

(a) How does the brightness of the thick-filament bulbs in this parallel circuit compare to the brightness of two thick-filament bulbs in series?

(b) Unscrew one of the bulbs in the parallel circuit. What effect does this have on the brightness of the remaining bulb?

(c) Based on these observations, what relationship would you expect to see between the compass deflections at locations A, B, C, and D, in this circuit? Record your prediction.

(d) Measure the compass deflections at locations A, B, C, and D, and record your measurements. Do your observations agree with your predictions in (c)?

In the preceding experiment you should again have seen that the current entering a location is equal to the current leaving the same location in the steady state. In particular, the current at A should be equal to the sum of the currents at B and C. If you did not observe this, you should repeat the experiment. We can also think of the two bulbs in parallel as equivalent to increasing the cross-sectional area of one of the bulb filaments.

We can analyze the situation quantitatively. If we follow the path through one of the bulbs, we have 2 emf $- EL = 0$, (2 emf for two batteries), and L is the length of the bulb filament (neglecting dissipation in the connecting wires). If we follow the path through the other thick filament bulb, we also have 2 emf $- EL = 0$. Thus the electric field $E = 2\ \text{emf}/L$ is the same in both bulb filaments, and it is the same as there would be with only one bulb present.

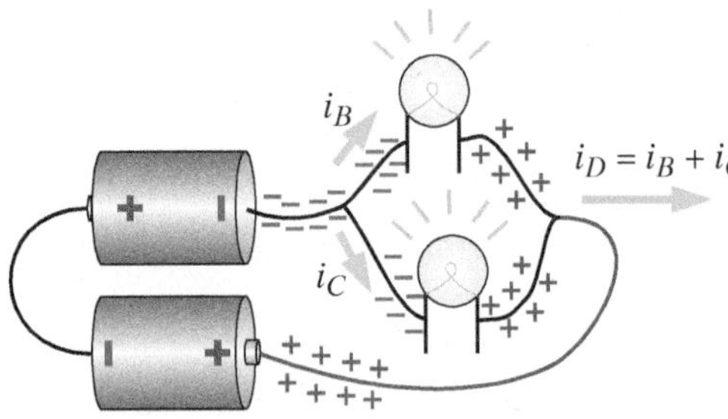

Figure 18.86

Figure 18.86 shows an approximate surface-charge distribution on the circuit. Each bulb is flanked by positive and negative charges, which make a large electric field in the bulb filament. Removing one of the bulbs doesn't change the surface-charge distribution very much, and the brightness of the remaining bulb hardly changes. The batteries of course have to deliver more current ($i_B + i_C$) to the two parallel bulbs than they do to one bulb.

EXP6 Effect of two batteries

Use the compass to compare the amount of current in two different circuits: one battery and a length of thin Nichrome wire, and two batteries and the same length of thin Nichrome wire. Following the directions given with Experiment EXP21, make careful, quantitative measurements of the angles.

Are you able to verify the prediction that the current doubles when you use two batteries? (If you wish, try three or more batteries with the same length of thin Nichrome wire—but don't use more than two batteries with your light bulbs, because it may burn them out.)

QUESTIONS

Q1 What is the most important general difference between a system in steady state and a system in equilibrium?

Q2 Describe the following attributes of a *metal wire* in steady state vs. equilibrium:

Metal Wire	Steady-State	Equilibrium
Location of excess charge		
Motion of mobile electrons		
E inside the metal wire		

Q3 How can there be a nonzero electric field inside a wire in a circuit? Isn't the electric field inside a metal always zero?

Q4 State your own theoretical *and* experimental objections to the following statement: In a circuit with two thick-filament bulbs in series, the bulb farther from the negative terminal of the battery will be dimmer, because some of the electron current is used up in the first bulb. Cite relevant experiments.

Q5 Since there is an electric field inside a wire in a circuit, why don't the mobile electrons in the wire accelerate continuously?

Q6 Why don't all mobile electrons in a metal have exactly the same speed?

Q7 There are very roughly the same number of iron atoms per m^3 as there are copper atoms per m^3, but copper is a much better conductor than iron. How does u_{iron} compare with u_{copper}?

Q8 In the few nanoseconds *before* the steady state is established in a circuit consisting of a battery, copper wires, and a single bulb, is the current the same everywhere in the circuit? Explain.

Q9 During the initial transient leading to the steady state, the electron current going into a bulb may be greater than the electron current leaving the bulb. Explain why and how these two currents come to be equal in the steady state.

Q10 At a typical drift speed of 5×10^{-5} m/s, an electron traveling at that speed would take about 100 min to travel through one of your connecting wires. Why, then, does the bulb light immediately when the connecting wire is attached to the battery?

Q11 Criticize the statement below on theoretical and experimental grounds. Be specific and precise. Refer to your own experiments, or describe any new experiments you perform: "A flashlight battery always puts out the same amount of current, no matter what is connected to it."

Q12 What is the difference between emf and electric potential difference?

Q13 Compare the direction of the average electric field inside a battery to the direction of the electric field in the wires and resistors of a circuit.

Q14 In the circuit shown in Figure 18.87, bulbs 1 and 2 are identical in mechanical construction (the filaments have the same length and the same cross-sectional area), but the filaments are made of different metals. The electron mobility in the metal used in bulb 2 is three times as large as the electron mobility in the metal used in bulb 1, but both metals have the same number of mobile electrons per cubic meter. The two bulbs are connected in series to two batteries with thick copper wires (like your connecting wires).

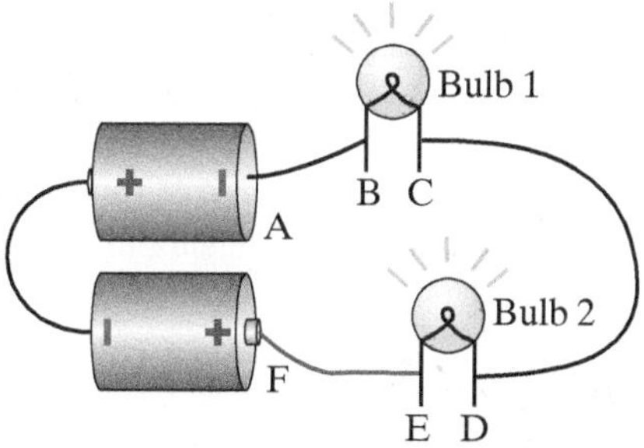

Figure 18.87

(a) In bulb 1, the electron current is i_1 and the electric field is E_1. In terms of these quantities, determine the corresponding quantities i_2 and E_2 for bulb 2, and explain your reasoning. **(b)** When bulb 2 is replaced by a wire, the electron current through bulb 1 is i_0 and the electric field in bulb 1 is E_0. How big is i_1 in terms of i_0? Explain your answer, including explicit mention of any approximations you must make. Do not use ohms or series-resistance equations in your explanation, unless you can show in detail how these concepts follow from the microscopic analysis introduced in this chapter. **(c)** Explain why the electric field inside the thick copper wires is very small. Also explain why this very small electric field is the same in all of the copper wires, if they all have the same cross-sectional area. **(d)** Figure 18.88 is a graph of the magnitude of the electric field at each location around the circuit when bulb 2 is replaced by a wire. Copy this graph and add to it, on the same scale, a graph of the magnitude of the electric field at each location around the circuit when both bulbs are in the circuit. The very small field in the copper wires has been shown much larger than it really is in order to give you room to show how that small field differs in the two circuits.

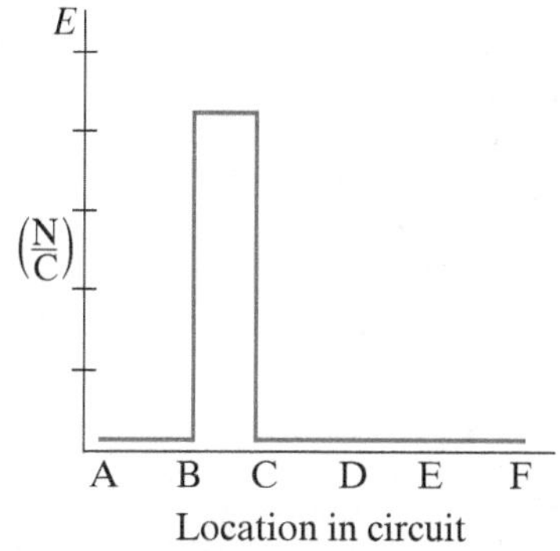

Figure 18.88

Q15 Some students intended to run a light bulb off two batteries in series in the usual way, but they accidentally hooked up one of the batteries backwards, as shown in Figure 18.89 (the bulb is shown as a thin filament).

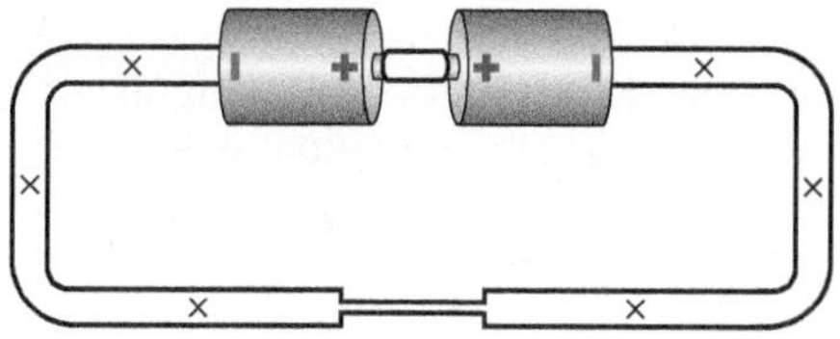

Figure 18.89

(a) Use +'s and −'s to show the approximate steady-state charge distribution along the wires and bulb. **(b)** Draw vectors for the electric field at the indicated locations inside the connecting wires and bulb. **(c)** Compare the brightness of the bulb in this circuit with the brightness the bulb would have had if one of the batteries hadn't been put in backwards. **(d)** Try the experiment to check your analysis. Does the bulb glow about as you predicted?

Q16 For each of the following experiments, state what effect you observed (how the current in the circuit was affected) and why, in terms of the relationships:

$$\text{electron current} = nA\bar{v} \text{ and } \bar{v} = uE$$

Experiment	Effect on Current (Be Quantitative)	Circle the Parameter(s) That Changed and Explain Briefly
Double the length of a Nichrome wire		$n\,A\,u\,E$
Double the cross-sectional area of a Nichrome wire		$n\,A\,u\,E$
Two identical bulbs in series compared to a single bulb		$n\,A\,u\,E$
Two batteries in series compared to a single battery		$n\,A\,u\,E$

Q17 In a table like the one shown, write an inequality comparing each quantity in the steady state for a narrow resistor and thick connecting wires, which are made of the same material as the resistor.

Electron Current in Resistor	<, =, or >	Electron Current in Thick Wires
n_R		n_W
A_R		A_W
u_R		u_W
E_R		E_W
v_R		v_W

Q18 A steady-state current flows through the Nichrome wire in the circuit shown in Figure 18.90.

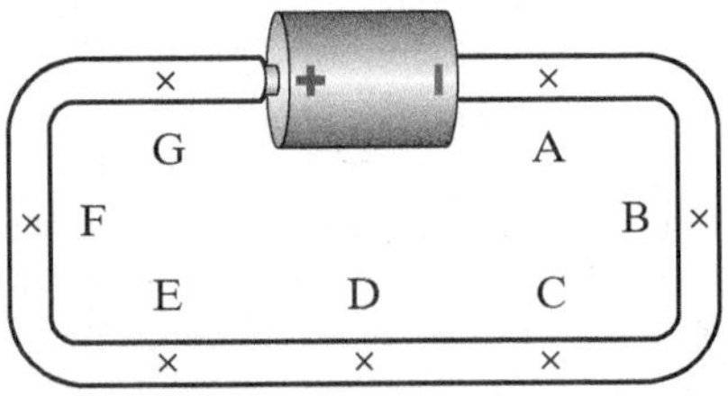

Figure 18.90

Before attempting to answer the following questions, draw a copy of this diagram. All of the locations indicated by letters are inside the wire. **(a)** On your diagram, show the electric field at the locations indicated, paying attention to relative magnitude. **(b)** Carefully draw pluses and minuses on your diagram to show the approximate surface charge distribution that produces the electric field you drew. Make your drawing show clearly the differences between regions of high surface charge density and regions of low surface-charge density. Use your diagram to determine which of the following statements about this circuit are true. (1) There is some excess negative charge on the surface of the wire near location B. (2) Inside the metal wire the magnitude of the electric field is zero. (3) The magnitude of the electric field is the same at locations G and C. (4) The electric field points to the left at location G. (5) There is no excess charge on the surface of the wire. (6) There is excess charge on the surface of the wire near the batteries but nowhere else. (7) The magnitude of the electric field inside the wire is larger at location G than at location C. (8) The electric field at location D points to the left. (9) Because the current is not changing, the circuit is in static equilibrium.

Q19 In the circuit shown in Figure 18.91, all of the wire is made of Nichrome, but one segment has a much smaller cross-sectional area. On a copy of this diagram, using the same scale for magnitude that you used in the previous question for Figure 18.90, show the steady-state electric field at the locations indicated, including in the thinner segment. Before attempting to answer these questions, draw a copy of this diagram. All of the locations indicated by letters are inside the wire.

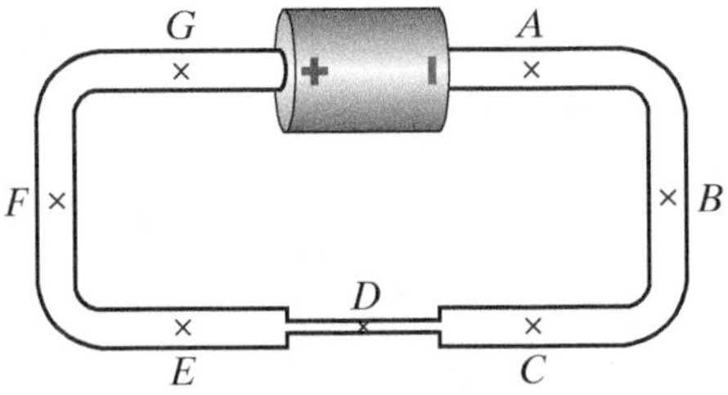

Figure 18.91

(a) On your diagram, show the electric field at the locations indicated, paying attention to relative magnitude. Use the same scale for magnitude as you did in the previous question. **(b)** Carefully draw pluses and minuses on your diagram to show the approximate surface charge distribution that produces the electric field you drew. Make your drawing show clearly the differences between regions of high surface charge density and regions of low surface-charge density. Use your diagram to determine which of the following statements about this circuit are true. (1) There is a large gradient of surface charge on the wire between locations C and E. (2) The electron current is the same at every location in this circuit. (3) Fewer electrons per second pass location E than location C. (4) The magnitude of the electric field at location G is smaller in this circuit than it was in the previous circuit (Figure 18.90). (5) The magnitude of the electric field is the same at every location in this circuit. (6) The magnitude of the electric field at location D is larger than the magnitude of the electric field at location G. (7) There is no surface charge at all on the wire near location G. (8) The electron current in this circuit is less than the electron current in the previous circuit (Figure 18.90).

PROBLEMS

Section 18.2

•P20 Which of the following statements about a metal wire *in equilibrium* are true? List all that apply. (1) There cannot be excess charges on the surface of the wire. (2) There is no net flow of mobile electrons inside the wire. (3) There are no excess charges in the interior of the wire. (4) There may be excess charges in the interior of the wire. (5) The net electric field everywhere inside the wire is zero. (6) There may be excess charges on the surface of the wire. (7) The interior of the metal wire is neutral. (8) There may be a constant flow of mobile electrons inside the wire. (9) The electric field inside the wire may be nonzero but uniform.

•P21 Which of the following statements about a metal wire *in the steady state* are true? List all that apply. (1) There is a constant flow of mobile electrons inside the wire. (2) The net electric field everywhere inside the wire is zero. (3) There are no excess charges in the interior of the wire. (4) There cannot be excess charges on the surface of the wire. (5) There is no net flow of mobile electrons inside the wire. (6) There may be excess charges on the surface of the wire. (7) There may be excess charges in the interior of the wire. (8) There may be a nonzero, uniform electric field inside the wire. (9) The interior of the metal wire is neutral.

•P22 Electron current $i = nA\bar{v} = nAuE$: **(a)** What are the units of electron current? **(b)** What is n? What are its units? **(c)** What is A? What are its units? **(d)** What is $\bar{v}$? What are its units? **(e)** What is u? What are its units?

•P23 In the circuit shown in Figure 18.92, what are the relationships among i_A, i_B, i_C, and i_D? How much current flows through the lower battery?

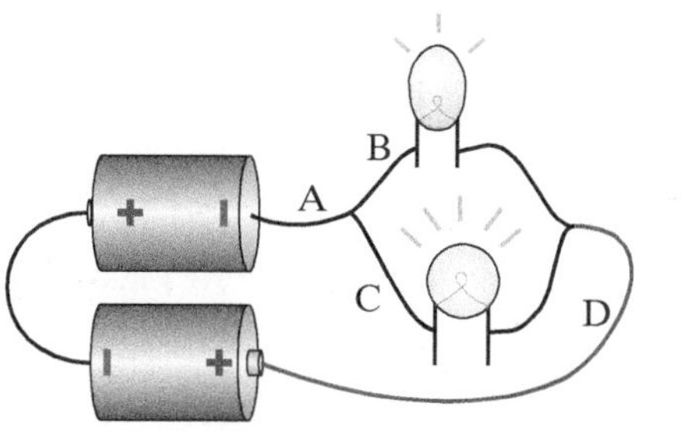

Figure 18.92

•P24 A steady-state current runs in the circuit shown in Figure 18.93. The narrow resistor and thick connecting wires are made of the same material.

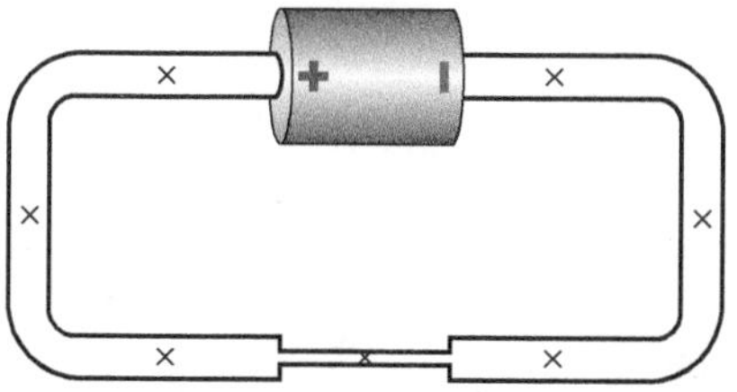

Figure 18.93

Which of the following quantities are greater in the thin resistor than in the thick wire? List all that apply: u, n, E, v, A, i, or none of these.

•P25 Consider the circuit fragment in Figure 18.94. **(a)** What is the absolute value of the outward-going conventional current I_2? **(b)** In this case, did we make the right guess about the direction of the conventional current I_2? **(c)** Suppose I_3 is 20 A; what is the absolute value of the outward-going conventional current I_2? **(d)** In this case, did we make the right guess about the direction of the conventional current I_2?

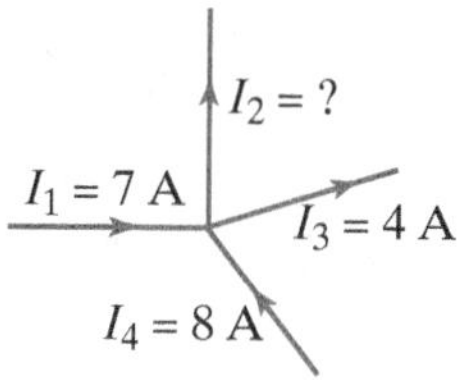

Figure 18.94

•P26 In the circuit in Figure 18.95 the narrow resistor is made of the same material as the thick connecting wires. In the steady state, which graph in Figure 18.96 correctly shows the magnitude of the electric field at locations around the circuit?

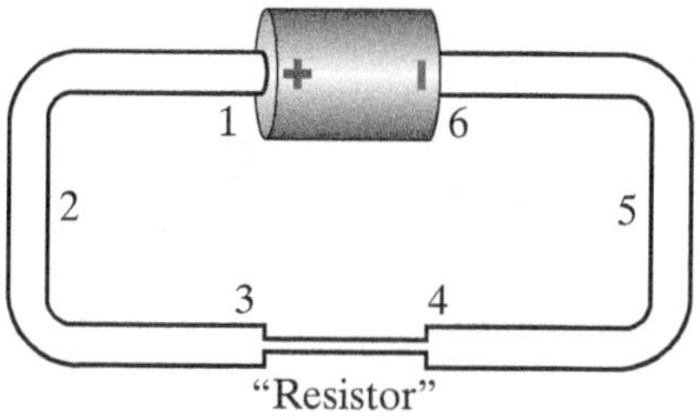

Figure 18.95

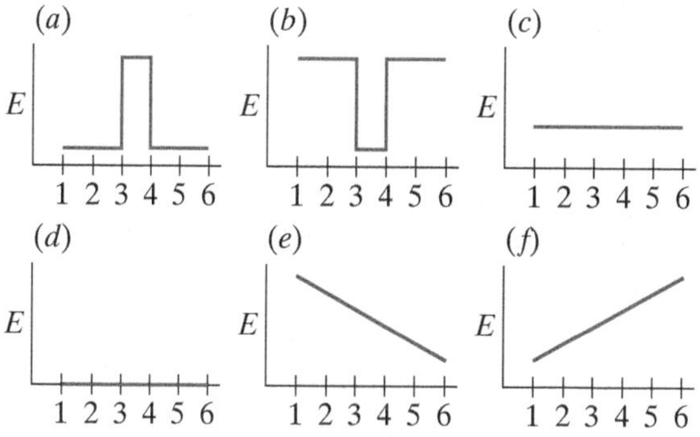

Figure 18.96

In the steady state, which graph in Figure 18.97 correctly shows the drift speed of the electrons at locations around the circuit?

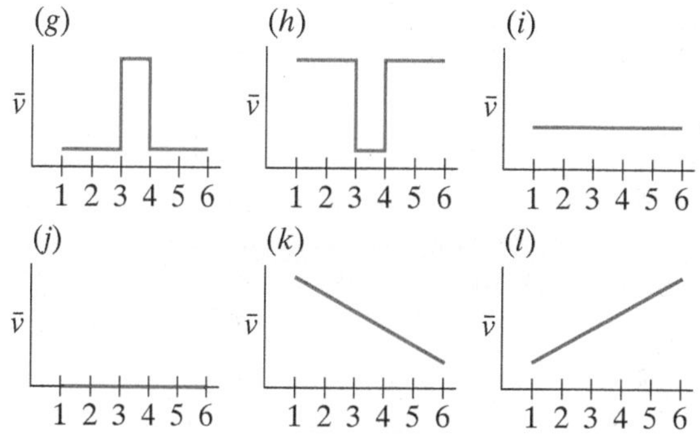

Figure 18.97

Section 18.3

•P27 The drift speed in a copper wire is 7×10^{-5} m/s for a typical electron current. Calculate the magnitude of the electric field E inside the copper wire. The mobility of mobile electrons in copper is 4.5×10^{-3} (m/s)/(N/C). (Note that though the electric field in the wire is very small, it is adequate to push a sizable electron current through the copper wire.)

•P28 Suppose that a wire leads into another, thinner wire of the same material that has only a third the cross-sectional area. In the steady state, the number of electrons per second flowing through the thick wire must be equal to the number of electrons per second flowing through the thin wire. If the drift speed $\bar{v}_1$ in the thick wire is 4×10^{-5} m/s, what is the drift speed $\bar{v}_2$ in the thinner wire?

•P29 In the circuit in Figure 18.98 a mechanical battery keeps a steady-state current running in a wire that has rather low electron mobility.

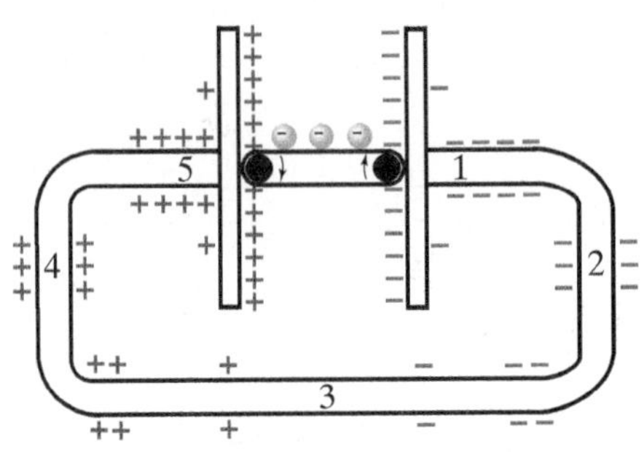

Figure 18.98

Which of the following statements about the circuit are true? List all that apply. (1) The electric field inside the wire varies in magnitude, depending on location. (2) At location 3 inside the wire the electric field points to the right. (3) The electric field is zero at all locations inside the metal wire. (4) At location 3 the electric field points to the left. (5) The magnitude of the electric field inside the wire is the same at all locations inside the wire. (6) Mobile electrons inside the wire push each other through the wire. (7) The nonzero electric field inside the wire is created by the moving electrons in the wire. (8) At every location inside the wire the direction of the electric field is parallel to the wire. (9) The nonzero electric field inside the wire is created by excess charges on the surface of the wire and in and on the mechanical battery.

•P30 Suppose that a wire leads into another, thinner wire of the same material that has only half the cross-sectional area. In the steady state, the number of electrons per second flowing through the thick wire must be equal to the number of electrons per second flowing through the thin wire. If the electric field E_1 in the thick wire is 1×10^{-2} N/C, what is the electric field E_2 in the thinner wire?

•P31 Suppose that wire A and wire B are made of different metals and are subjected to the same electric field in two *different* circuits. Wire B has 6 times the cross-sectional area, 1.3 times as many mobile electrons per cubic centimeter, and 4 times the mobility of wire A. In the steady state, 2×10^{18} electrons enter wire A every second. How many electrons enter wire B every second?

••P32 Inside a chemical battery it is not actually individual electrons that are transported from the + end to the − end. At the + end of the battery an "acceptor" molecule picks up an electron entering the battery, and at the − end a different "donor" molecule gives up an electron, which leaves the battery. Ions rather than electrons move between the two ends to transport the charge inside the battery.

When the supplies of acceptor and donor molecules are used up in a chemical battery, the battery is dead, because it can no longer accept or release electrons. The electron current in electrons per second, times the number of seconds of battery life, is equal to the number of donor (or acceptor) molecules in the battery.

A flashlight battery contains approximately half a mole of donor molecules. The electron current through a thick-filament bulb powered by two flashlight batteries in series is about 0.3 A. About how many hours will the batteries keep this bulb lit?

••P33 All of the wires in the circuit shown in Figure 18.99 are made of the same material, but one wire has a smaller radius than the other wires.

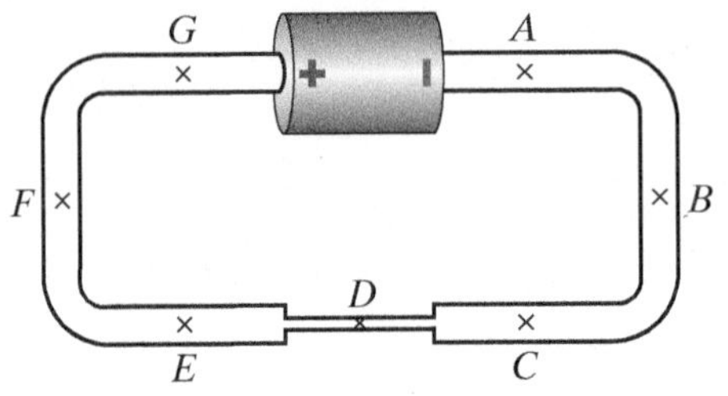

Figure 18.99

Which of the following statements are true of this circuit in the steady state? List all that apply. (1) The drift speed of electrons passing location D is greater than the drift speed of electrons passing location G. (2) The magnitude of the electric field is the same at each location labeled by a letter. (3) The electric field at location F points up. (4) The electric field at G is larger in magnitude than the electric field at C. (5) The number of electrons passing location B each second is the same as the number of electrons passing location D each second.

The radius of the thin wire is 0.22 mm, and the radius of the thick wire is 0.55 mm. There are 4×10^{28} mobile electrons per cubic meter of this material, and the electron mobility is 6×10^{-4} (m/s)/(V/m). If 6×10^{18} electrons pass location D each second, how many electrons pass location B each second? What is the magnitude of the electric field at location B?

••P34 Figure 18.100 is a top view of a portion of a circuit containing three identical light bulbs (the rest of the circuit including the batteries is not shown). The connecting wires are made of copper.

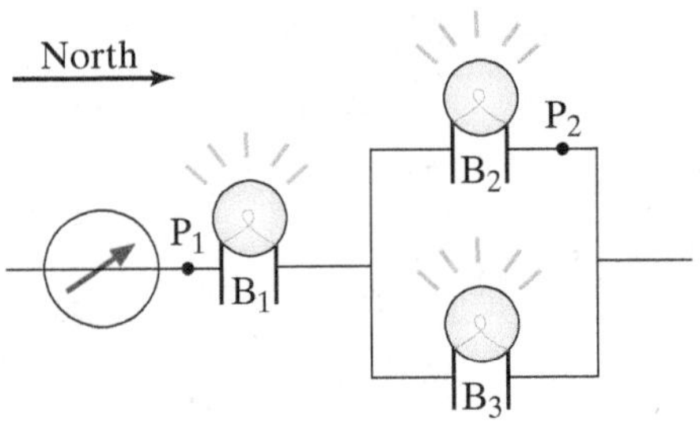

Figure 18.100

(a) The compass is placed on top of the wire, and it deflects 20° away from north as shown (the wire is underneath the compass). In what direction are the electrons moving at location P_1? How do you know? **(b)** In the steady state, 3×10^{18} electrons pass location P_1 every second. How many electrons pass location P_2 every second? Explain briefly. **(c)** Give the relative brightnesses of bulbs B_1, B_2, and B_3. Explain briefly. **(d)** There are 6.3×10^{28} mobile electrons per cubic meter in tungsten. The cross-sectional area of the tungsten filament in bulb B_1 is 0.01 mm^2 (which is 1×10^{-8} m^2). The electron mobility in hot tungsten is 1.2×10^{-4} (m/s)/(N/C). Calculate the electric field inside the tungsten filament in bulb B_1. Give both the direction and the magnitude of the electric field.

••P35 Consider the circuit containing three identical light bulbs shown in Figure 18.101. North is indicated in the diagram. Compasses are placed under the wires at locations A and B.
(a) The magnitude of the deflection of compass A is 13° away from north. In what direction does the needle point? Draw a sketch. **(b)** What is the magnitude of the deflection of the needle of compass B? In what direction does the needle point? Draw a sketch. Explain your reasoning. **(c)** In the steady state 1.5×10^{18} electrons per second enter bulb 1. There are 6.3×10^{28} mobile electrons per cubic meter in tungsten. The cross-sectional area of the tungsten filament in bulb 1 is 1×10^{-8} m^2. The electron mobility in hot tungsten is 1.2×10^{-4} (m/s)/(N/C). Calculate the electric field inside the tungsten filament in bulb 3. Give both the direction and the magnitude of the electric field.

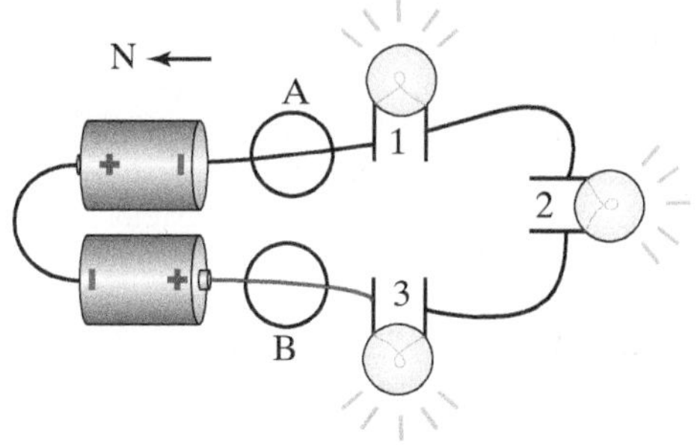

Figure 18.101

Section 18.10

Base your analyses in the following problems on the principles discussed in this chapter. If you wish to use equations derived elsewhere, you must justify them in terms of the microscopic analysis in terms of field that was introduced in this chapter.

•P36 **(a)** The emf of a particular flashlight battery is 1.7 V. If the battery is 4.5 cm long and the radius of the cylindrical battery is 1 cm, estimate roughly the amount of charge on the positive end plate of the battery. **(b)** Is this amount of charge sufficient to repel noticeably a positively charged piece of invisible tape?

•P37 A Nichrome wire 30 cm long and 0.25 mm in diameter is connected to a 1.5 V flashlight battery. What is the electric field inside the wire? Why don't you have to know how the wire is bent? How would your answer change if the wire diameter were 0.35 mm? (Note that the electric field in the wire is quite small compared to the electric field near a charged tape.)

•P38 Why does the brightness of a bulb not change noticeably when you use longer copper wires to connect it to the battery? (1) Very little energy is dissipated in the thick connecting wires. (2) The electric field in connecting wires is very small, so emf $\approx E_{\text{bulb}}L_{\text{bulb}}$. (3) Electric field in the connecting wires is zero, so emf $= E_{\text{bulb}}L_{\text{bulb}}$. (4) Current in the connecting wires is smaller than current in the bulb. (5) All the current is used up in the bulb, so the connecting wires don't matter.

•P39 A Nichrome wire 48 cm long and 0.25 mm in diameter is connected to a 1.6 V flashlight battery. What is the electric field inside the wire? Next, the wire is replaced by a different Nichrome wire with the same length, but diameter 0.20 mm. Now what is the electric field inside the wire?

•P40 In a circuit with one battery, connecting wires, and a 12 cm length of Nichrome wire, a compass deflection of 6° is observed. What compass deflection would you expect in a circuit containing two batteries in series, connecting wires, and a 36 cm length of thicker Nichrome wire (double the cross-sectional area of the thin piece)? Explain.

•P41 In Figure 18.102, suppose that $V_C - V_F = 8$ V, and $V_D - V_E = 4.5$ V.

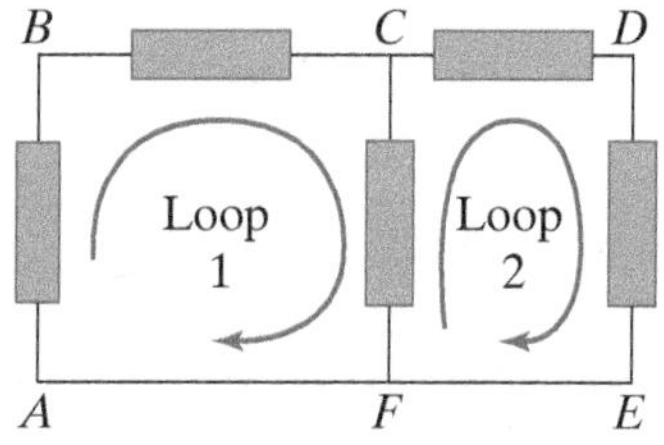

Figure 18.102

(a) What is the potential difference $V_C - V_D$? **(b)** If the element between C and D is a battery, is the + end of the battery at C or at D?

•P42 What would be the potential difference $V_C - V_B$ across the thin resistor in Figure 18.103 if the battery emf is 3.5 V? Assume that the electric field in the thick wires is very small (so that the potential differences along the thick wires are negligible). Do you have enough information to determine the current I in the circuit?

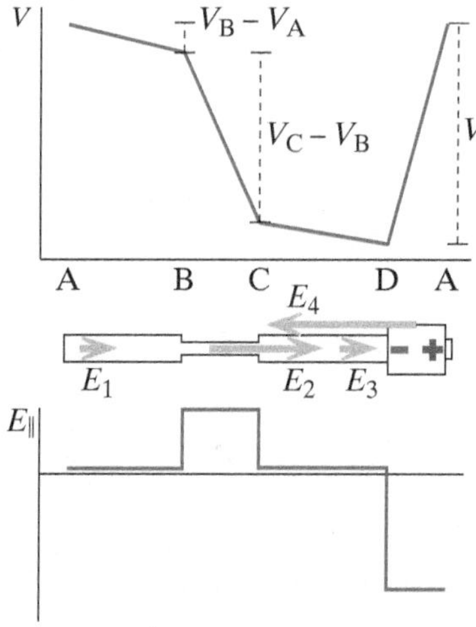

Figure 18.103

••P43 A circuit is constructed from two batteries and two wires, as shown in Figure 18.104. Each battery has an emf of 1.3 V. Each wire is 26 cm long and has a diameter of 7×10^{-4} m. The wires are made of a metal that has 7×10^{28} mobile electrons per cubic meter; the electron mobility is 5×10^{-5} (m/s)/(V/m). A steady current runs through the circuit. The locations marked by × and labeled by a letter are in the interior of the wire.

(a) Which of these statements about the electric field in the interior of the wires, at the locations marked by ×'s, are true? List all that apply. (1) The magnitude of the electric field at location G is larger than the magnitude of the electric field at location F.

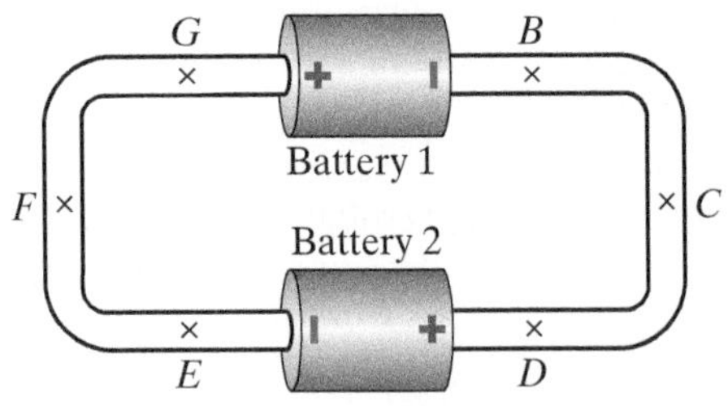

Figure 18.104

(2) At every marked location the magnitude of the electric field is the same. (3) At location B the electric field points to the left. **(b)** Write a correct energy conservation (round-trip potential difference) equation for this circuit, along a round-trip path starting at the negative end of battery 1 and traveling counterclockwise through the circuit (that is, traveling to the left through the battery, and continuing on around the circuit in the same direction). **(c)** What is the magnitude of the electric field at location B? **(d)** How many electrons per second enter the positive end of battery 2? **(e)** If the cross-sectional area of both wires were increased by a factor of 2, what would be the magnitude of the electric field at location B? **(f)** Which of the diagrams in Figure 18.105 best shows the approximate distribution of excess charge on the surface of the circuit?

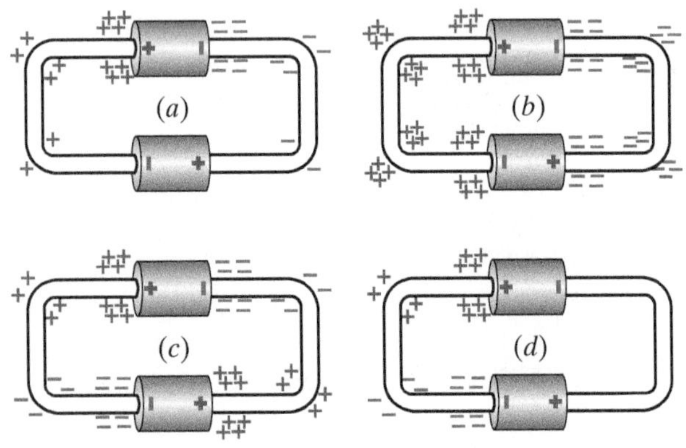

Figure 18.105

••P44 Three identical light bulbs are connected to two batteries as shown in Figure 18.106.

(a) To start the analysis of this circuit you must write energy conservation (loop) equations. Each equation must involve a round-trip path that begins and ends at the same location. Each segment of the path should go through a wire, a bulb, or a battery (not through the air). How many valid energy conservation (loop) equations is it possible to write for this circuit? **(b)** Which of the following equations are valid energy conservation (loop) equations for this circuit? E_1 refers to the electric field in bulb 1; L refers to the length of a bulb filament. Assume that the electric field in the connecting wires is small enough to neglect.

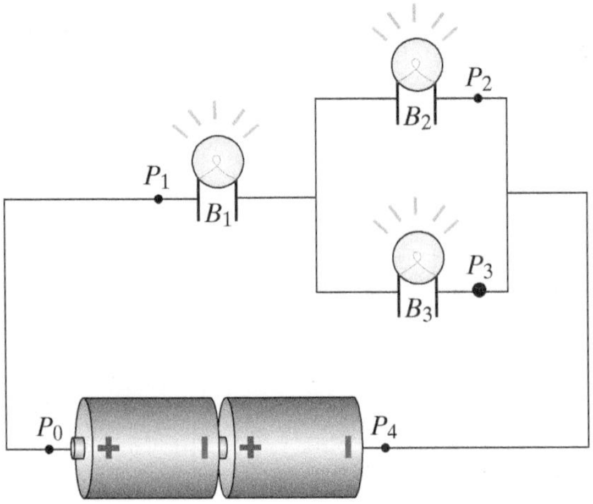

Figure 18.106

(1) $+E_2L - E_3L = 0$, (2) $E_1L - E_3L = 0$, (3) $+2\text{emf} - E_2L - E_3L = 0$, (4) $E_1L - E_2L = 0$, (5) $+2\text{emf} - E_1L - E_2L = 0$, (6) $+2\text{emf} - E_1L - E_3L = 0$, (7) $+2\text{emf} - E_1L - E_2L - E_3L = 0$. **(c)** It is also necessary to write charge conservation equations (node) equations. Each such equation must relate electron current flowing into a node to electron current flowing out of a node. Which of the following are valid charge conservation equations for this circuit? (1) $i_1 = i_3$, (2) $i_1 = i_2$, (3) $i_1 = i_2 + i_3$.

Each battery has an emf of 1.5 V. The length of the tungsten filament in each bulb is 0.008 m. The radius of the filament is 5×10^{-6} m (it is very thin!). The electron mobility of tungsten is 1.8×10^{-3} (m/s)/(V/m). Tungsten has 6×10^{28} mobile electrons per cubic meter. Since there are three unknown quantities, we need three equations relating these quantities. Use any two valid energy conservation equations and one valid charge conservation equation to solve for E_1, E_2, i_1, and i_2.

••P45 The circuit shown in Figure 18.107 consists of a single battery, whose emf is 1.8 V, and three wires made of the same material but having different cross-sectional areas. Each thick wire has cross-sectional area 1.4×10^{-6} m^2 and is 25 cm long. The thin wire has cross-sectional area 5.9×10^{-8} m^2 and is 6.1 cm long.

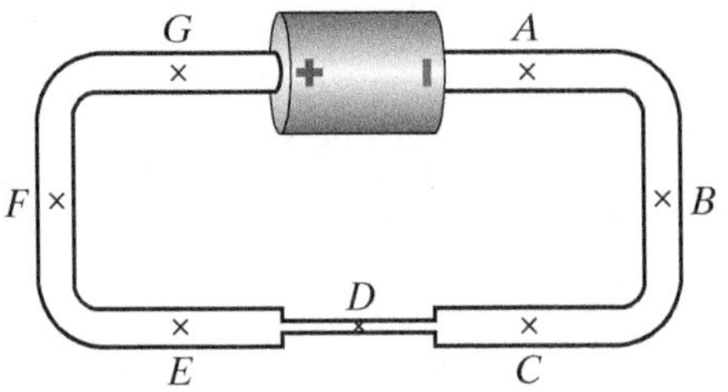

Figure 18.107

In this metal, the electron mobility is 5×10^{-4} (m/s)/(V/m), and there are 4×10^{28} mobile electrons/m^3.

(a) Which of the following statements about the circuit in the steady state are true? (1) At location B the electric field points toward the top of the page. (2) The magnitude of the electric field at locations F and C is the same. (3) The magnitude of the electric field at locations D and F is the same. (4) The electron current at location D is the same as the electron current at location F. **(b)** Write a correct energy conservation (loop) equation for this circuit, following a path that starts at the negative end of the battery and goes counterclockwise. **(c)** Write a correct charge conservation (node) equation for this circuit. **(d)** Use the appropriate equation(s), plus the equation relating electron current to electric field, to solve for the magnitudes E_D and E_F of the electric field at locations D and F. **(e)** Use the appropriate equation(s) to calculate the electron current at location D in the steady state.

••P46 In the circuit shown in Figure 18.108, two thick copper wires connect a 1.5 V battery to a Nichrome wire. Each thick connecting wire is 17 cm long and has a radius of 9 mm. Copper has 8.4×10^{28} mobile electrons per cubic meter and an electron mobility of 4.4×10^{-3} (m/s)/(V/m). The Nichrome wire is 8 cm long, and has a radius of 3 mm. Nichrome has 9×10^{28} mobile electrons per cubic meter and an electron mobility of 7×10^{-5} (m/s)/(V/m).

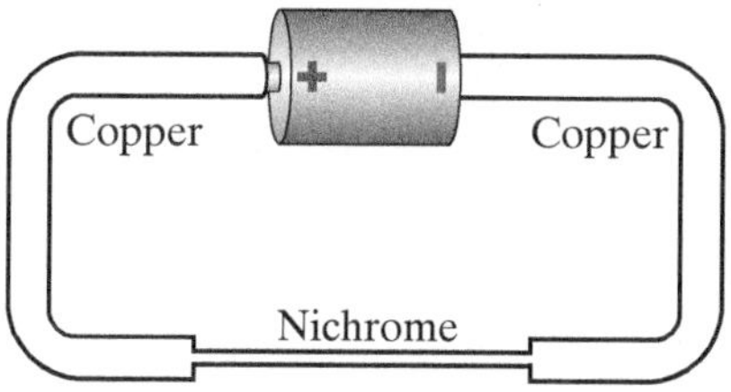

Figure 18.108

(a) What is the magnitude of the electric field in the thick copper wire? **(b)** What is the magnitude of the electric field in the thin Nichrome wire?

••P47 When a single thick-filament bulb of a particular kind and two batteries are connected in series, 3×10^{18} electrons pass through the bulb every second. When two batteries connected in series are connected to a single thin-filament bulb, with a filament made of the same material and the same length as that of the thick-filament bulb but smaller cross section, only 1.5×10^{18} electrons pass through the bulb every second.

(a) In the circuit shown in Figure 18.109, how many electrons per second flow through the thin-filament bulb? **(b)** What approximations or simplifying assumptions did you make? **(c)** Show approximately the surface charge on a diagram of the circuit.

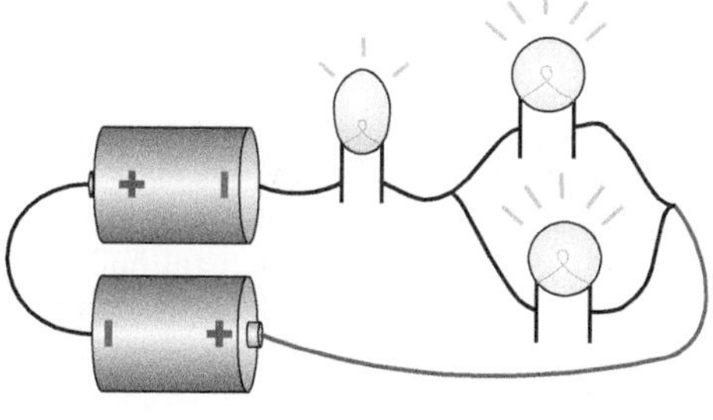

Figure 18.109

••P48 In the circuit shown in Figure 18.110, the two thick wires and the thin wire are made of Nichrome.

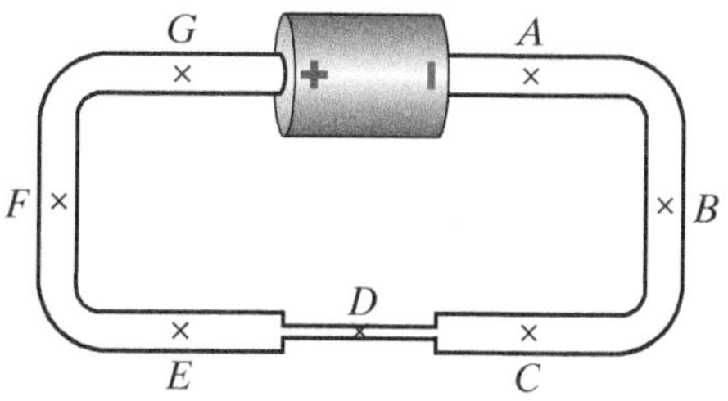

Figure 18.110

(a) Show the steady-state electric field at the locations indicated, including in the thin wire. **(b)** Carefully draw pluses and minuses on your own diagram to show the approximate surface-charge distribution in the steady state. Make your drawing show clearly the differences between regions of high surface-charge density and regions of low surface-charge density. **(c)** The emf of the battery is 1.5 V. In Nichrome there are $n = 9 \times 10^{28}$ mobile electrons per m^3, and the mobility of mobile electrons is $u = 7 \times 10^{-5}$ (m/s)/(N/C). Each thick wire has length $L_1 = 20\,\text{cm} = 0.2\,\text{m}$ and cross-sectional area $A_1 = 9 \times 10^{-8}\,\text{m}^2$. The thin wire has length $L_2 = 5\,\text{cm} = 0.05\,\text{m}$ and cross-sectional area $A_2 = 1.5 \times 10^{-8}\,\text{m}^2$. (The total length of the three wires is 45 cm.) In the steady state, calculate the number of electrons entering the thin wire every second. Do not make any approximations, and do not use Ohm's law or series-resistance equations. State briefly where each of your equations comes from.

••P49 Three identical thick-filament bulbs are in series as shown in Figure 18.111. Thick copper wires connect the bulbs. In the steady state, 3×10^{17} electrons leave the battery at location A every second.

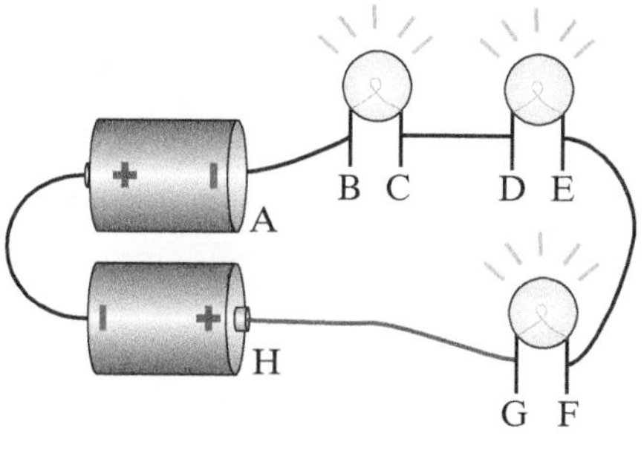

Figure 18.111

(a) How many electrons enter the second bulb at location D every second? If there is insufficient information to give a numerical answer, state how it compares with 3×10^{17}. Justify your answer carefully. **(b)** Next, the middle bulb (at DE) is replaced by a wire, as shown in Figure 18.112. Now how many electrons leave the batteries at location A every second? Explain clearly! If you have to make an approximation, state what it is. Do not use ohms or series-resistance equations in your explanation, unless you can show in detail how these concepts follow from the microscopic analysis introduced in this chapter.

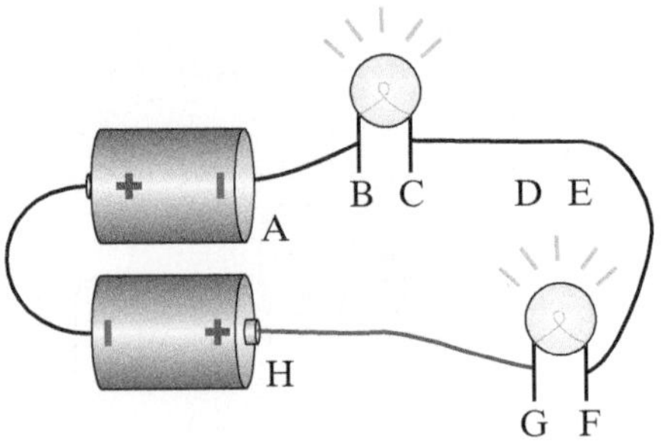

Figure 18.112

(c) Check your analysis by trying the experiment with a partner. Try to find three thick-filament bulbs that glow equally brightly when in series with each other, because bulb construction varies slightly in manufacturing. Remember to arrange the circuit so that the largest compass deflection is no more than 15°. Report the deflections that you observe. Does the experiment agree with your prediction? (If not, can you explain the discrepancy? Be specific. For example, if the current is larger than predicted, explain why it is larger than predicted.) **(d)** Finally, the last bulb (at FG) is replaced by a bulb identical in every way except that its filament has twice as large a cross-sectional area, as shown in Figure 18.113. Now how many electrons leave the batteries at location A every second? Explain clearly! If you have to make an approximation, state what it is.

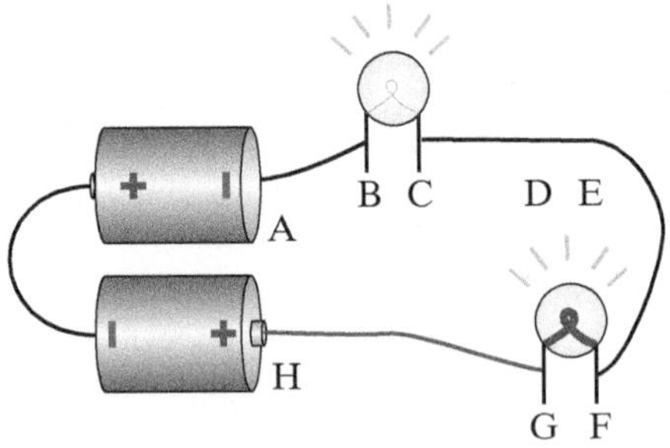

Figure 18.113

••P50 The following questions refer to the circuit shown in Figure 18.114, consisting of two flashlight batteries and two Nichrome wires of different lengths and different thicknesses as shown (corresponding roughly to your own thick and thin Nichrome wires).

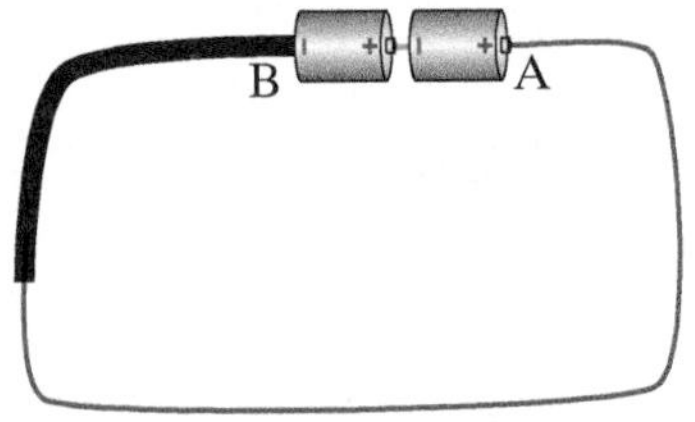

Figure 18.114

The thin wire is 50 cm long, and its diameter is 0.25 mm. The thick wire is 15 cm long, and its diameter is 0.35 mm. **(a)** The emf of each flashlight battery is 1.5 V. Determine the steady-state electric field inside each Nichrome wire. Remember that in the steady state you must satisfy both the current node rule and energy conservation. These two principles give you two equations for the two unknown fields. **(b)** The electron mobility in room-temperature Nichrome is about 7×10^{-5} (m/s)/(N/C). Show that it takes an electron 36 min to drift through the two Nichrome wires from location B to location A. **(c)** On the other hand, about how long did it take to establish the steady state when the circuit was first assembled? Give a very approximate numerical answer, not a precise one. **(d)** There are about 9×10^{28} mobile electrons per cubic meter in Nichrome. How many electrons cross the junction between the two wires every second?

•P51 A Nichrome wire 75 cm long and 0.25 mm in diameter is connected to a 1.7 V flashlight battery. **(a)** What is the electric field inside the wire? **(b)** Next, the Nichrome wire is replaced by a wire of the same length and diameter, and same mobile electron density but with electron mobility 4 times as large as that of Nichrome. Now what is the electric field inside the wire? **(c)** The electron current in the first circuit (Nichrome) is i_1. The electron current in the second circuit (wire with higher mobility) is i_2. Which of the following statements is true? (1) $i_2 = i_1$, (2) $i_2 < i_1$, (3) $i_2 > i_1$, (4) Not enough information is given to compare the two currents.

••P52 Two circuits are assembled using 1.5 V batteries, thick copper connecting wires, and thin-filament bulbs (Figure 18.115). Bulbs A, B, and C are identical thin-filament bulbs. Compasses are placed under the wires at the indicated locations. **(a)** On sketches of the circuits, draw the directions the compass needles will point, and indicate the approximate magnitude of the compass deflections in degrees. Note that the deflection is given at one location. If you do not have enough information to give a number, then indicate whether it will be greater than, less than, or equal to 5°. (Assume that the compasses are adequately far away from the steel-jacketed batteries.) **(b)** Briefly explain your reasoning about the magnitudes of the compass deflections. **(c)** On the same sketches of the circuits, show very approximately the distribution of surface charge. **(d)** The tungsten filament in each of the bulbs is 4 mm long with a radius of 6×10^{-6} m. Calculate the electric field inside each of the three bulbs, E_A, E_B, and E_C.

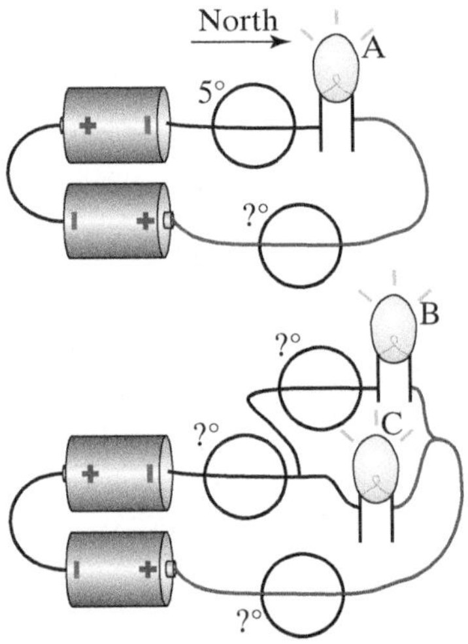

Figure 18.115

••P53 A circuit is assembled that contains a thin-filament bulb and a thick-filament bulb as shown in Figure 18.116, with four compasses placed underneath the wires (we're looking down on the circuit). When the thin-filament bulb is unscrewed from its socket, the compass on the left (next to the battery) deflects 15°. When the thin-filament bulb is screwed back in and the thick-filament bulb is unscrewed, the compass on the left deflects 4°.

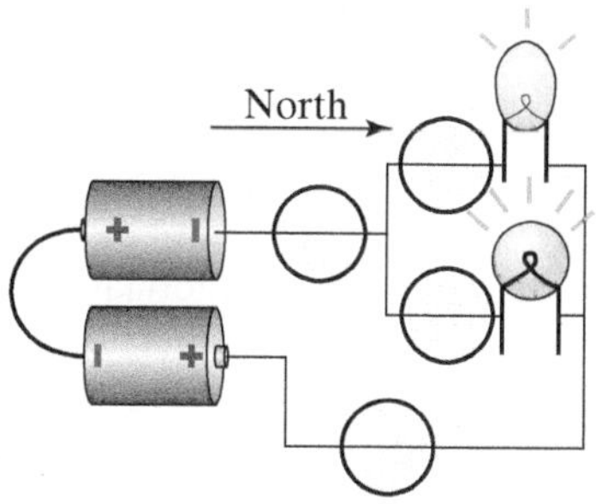

Figure 18.116

With both bulbs screwed into their sockets, draw the orientations of the needle on each compass, and write the compass deflection in degrees beside the compass. Explain briefly.

•••**P54** A solid metal sphere of radius R carries a uniform charge of $+Q$. Another solid metal sphere of radius r carries a uniform charge $-q$. The amount of charge is not enough to cause breakdown in air. The two spheres are very far apart (distance $\gg R$ and distance $\gg r$). At $t = 0$ a very thin wire of length L is connected to the two spheres (Figure 18.117). The mobility u of mobile electrons in this wire is very small, and the wire conducts electrons so poorly that it takes about an hour for the system to reach equilibrium. In a short time Δt (a few seconds) how many electrons leave the sphere of radius r? There are n mobile electrons per cubic meter in the wire, and the wire has a constant cross-sectional area A. Explain your work and any approximations you need to make.

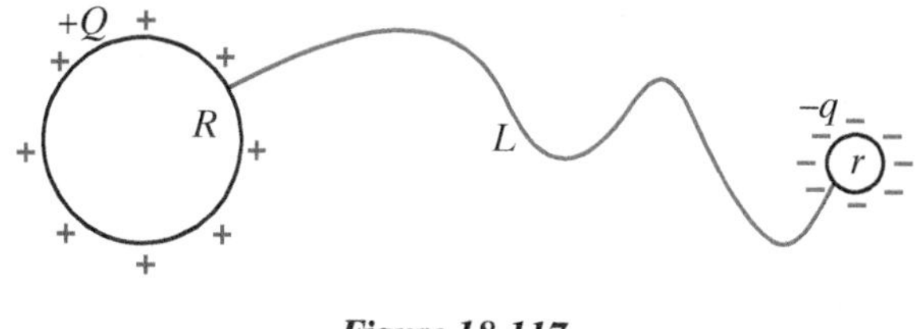

Figure 18.117

ANSWERS TO CHECKPOINTS

1 $I_1 + I_4 = I_2 + I_3$; $I_2 = 3$ A; $I_2 = -2$ A, where the minus sign means that the current is entering the node rather than leaving. It is not true that current must be the same in every part of a circuit, only that the current entering a node must equal the current leaving a node; here $i_1 = i_2 + i_3$.

2 $\overline{v_2} = 8 \times 10^{-5}$ m/s; $E_2 = 1.8 \times 10^{-2}$ N/C

3 See Figure 18.118.

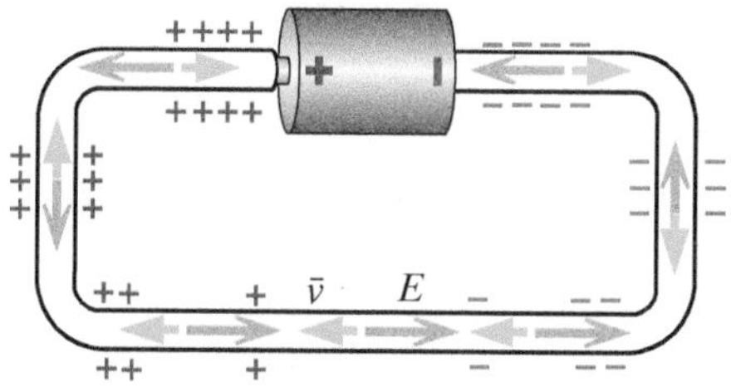

Figure 18.118

4 See Figure 18.119.

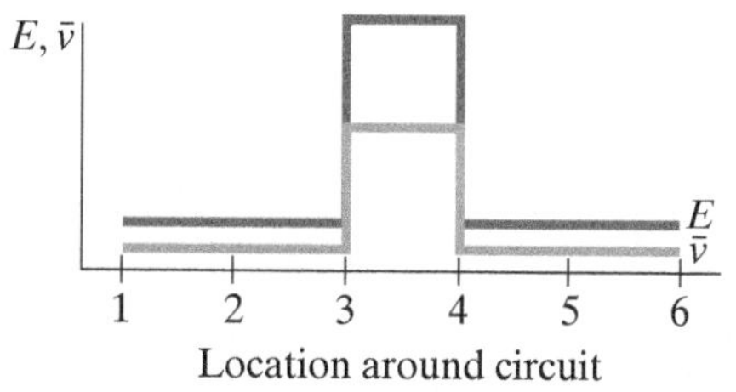

Figure 18.119

5 See Figure 18.120.

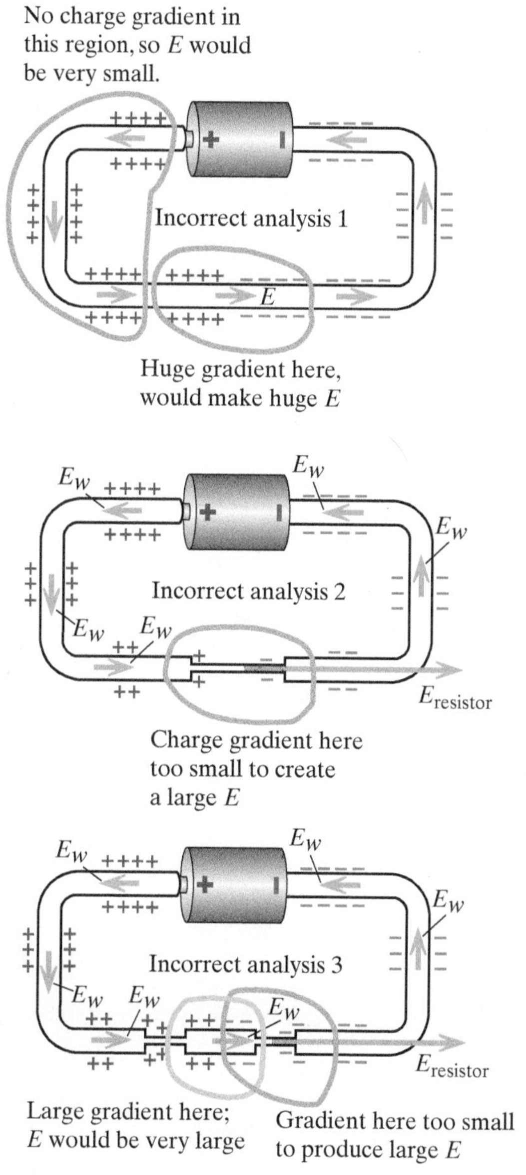

Figure 18.120

6 8.3×10^{-14} C. This is much less than the charge on a tape (about 1×10^{-8} C). One would not expect to observe repulsion, since the repulsive force would probably be smaller than the force attracting the tape to the battery (which had been polarized by the charged tape).

7 The electric field in the thick copper connecting wires is very small, so emf $\approx E_{\text{bulb}} L_{\text{bulb}}$ and the electric field in the bulb filament depends almost entirely on the length of the filament itself.

8 $+1.5$ V; $+$ end of battery is at C.

9 Approximately 1.5 V; not enough information to determine I.

CHAPTER

19

Circuit Elements

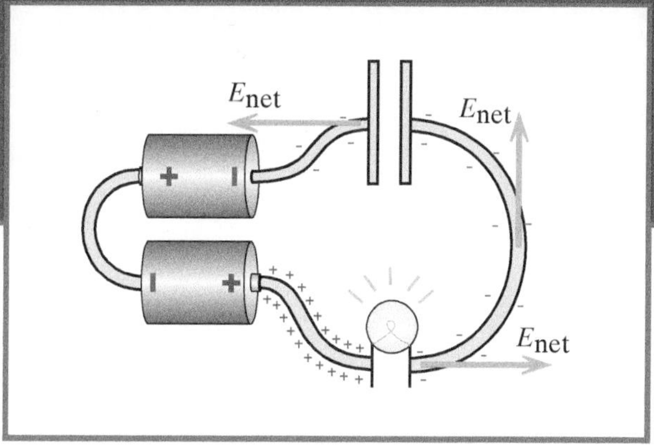

OBJECTIVES

After studying this chapter you should be able to

- Qualitatively and quantitatively explain the behavior of a circuit containing a capacitor.
- Analyze a circuit using macroscopic quantities such as resistance and capacitance.

19.1 CAPACITORS

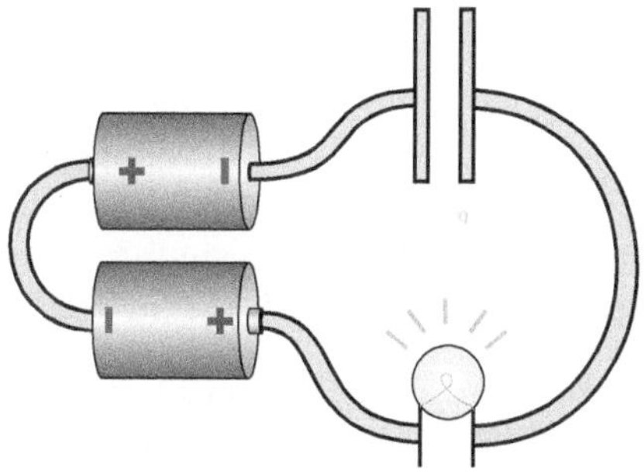

Figure 19.1 Charging a capacitor. Initially the bulb glows brightly, but with time the bulb dims and finally goes out.

Experiment EXP1 at the end of this chapter shows an interesting phenomenon. In the experiment you connect batteries in series with a certain kind of capacitor (1 farad, F) and a flashlight bulb (Figure 19.1), and what you observe in this "charging" process is that initially the bulb glows brightly, as brightly as if there were no capacitor in the circuit, despite the fact that no current can flow across the gap between the capacitor plates. As time goes on, however, the bulb gets dimmer and dimmer and eventually goes out after many seconds.

After the bulb has stopped glowing, if you remove the batteries and attach the bulb to the capacitor ("discharging," Figure 19.2), the bulb again glows brightly but then gets dimmer and dimmer and eventually stops glowing after many seconds.

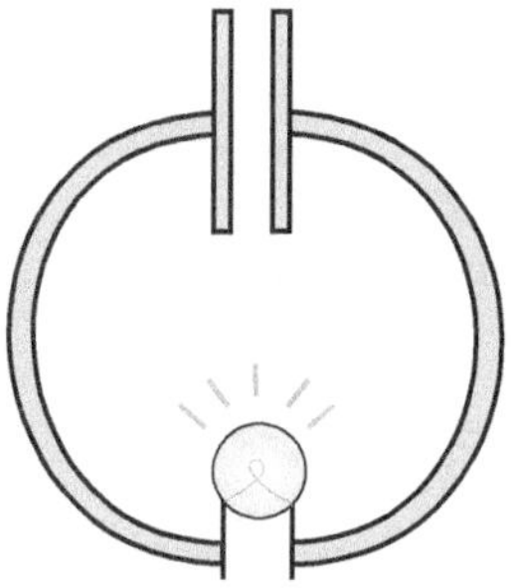

Figure 19.2 Discharging a capacitor. Initially the bulb glows brightly, but with time the bulb dims and finally goes out.

There are two different time scales for events in this circuit. The bulb lights instantly when a connection is made (surface charge rearrangement takes only nanoseconds), but the current in the circuit declines slowly over a period of many seconds—the two time scales differ by a factor of 1×10^{10}. We call the slow process a "quasi-steady state," in which the currents change slowly with time as the circuit slowly comes into equilibrium, with the current coming to a stop. (It should be mentioned that in high-frequency circuits with capacitors measured in picofarads, 1×10^{-12} F, there may be little difference in these two time scales. Also, surface charge rearrangement in a circuit as big as a national power grid will take far longer than a nanosecond.)

Capacitor Construction

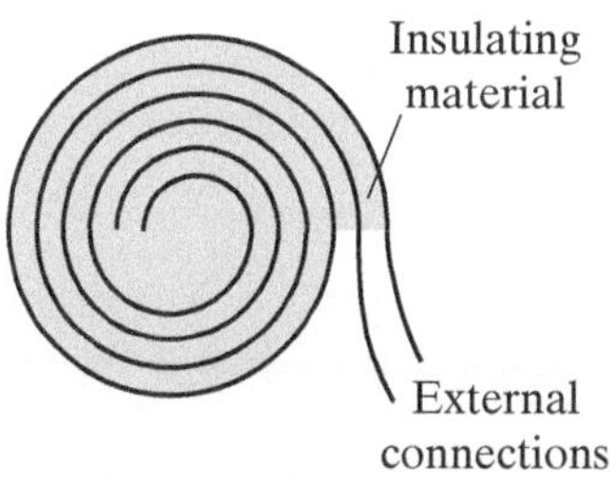

Figure 19.3 Typical capacitor construction—a coiled-up sandwich of metal foils separated by insulating material.

The capacitors we've discussed in previous chapters consisted of two very large flat plates separated by a gap. However, it wouldn't be very practical to try to use such a large device in a circuit. A capacitor of the kind used in circuits often consists of a sandwich of two strips of metal foil separated by insulating material and coiled up into a small package (Figure 19.3). Note that there is no conducting path through the capacitor, even though you are able to light a bulb in series with the capacitor. Charges cannot jump from one plate to the other.

Discharging a Capacitor

Using what we already know, we can explain the charging and discharging of a capacitor in terms of the fundamental properties of charge and field. First we'll discuss the discharging process, which is easier to understand. We'll consider a simple circuit like that shown in Figure 19.2.

Our reasoning will focus on the net electric field at a location inside the wire, and the effect it has on the electrons at that location in the wire. We'll think about it step by step, iteratively, choosing a time step that is long compared to the time required to discharge the capacitor (several seconds), but short compared to the time it takes surface charge to arrange itself on the wire and battery (nanoseconds). Figure 19.4 shows the pattern of electric field in the connecting wires 0.01 s after connecting the bulb to the charged capacitor, while the bulb is glowing brightly.

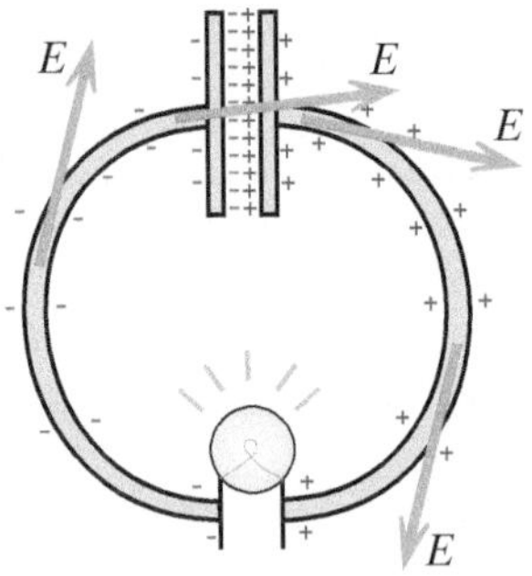

Figure 19.4 Charge and electric field 0.01 s after connecting the bulb to the charged capacitor. The bulb is bright.

Just outside the capacitor, we know that the electric field at a location inside the wire points away from the positive plate and toward the negative plate, due in part to the fringe field of the capacitor. After the quick transient (a matter of nanoseconds), the electron current everywhere in the connecting wires must be the same, and surface charge will build up to make a uniform electric field everywhere in the wires. There is a big field inside the bulb filament (not shown), across which is a big surface charge gradient, going from + to −.

> **QUESTION** In Figure 19.4, electron current will move opposite to the electric field. In the next short time interval, given the direction of electron current, what will happen to the amount of charge on the negative plate of the capacitor? On the positive plate?

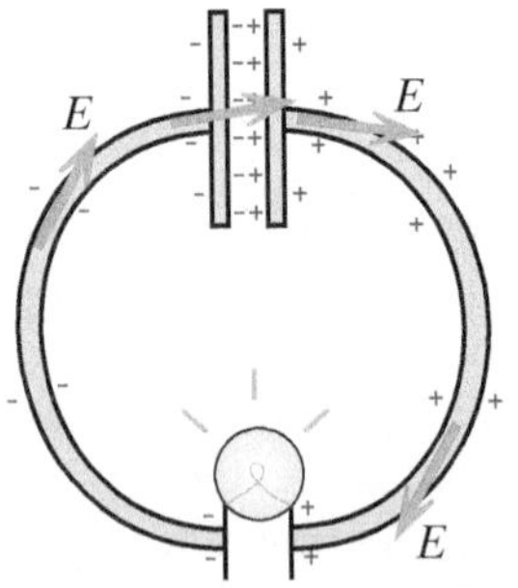

Figure 19.5 Charge and electric field 1 s after connecting the bulb to the charged capacitor. The bulb is less bright than before.

Since electron current moves away from the negative plate, the negative plate will lose some of its negative charge. Since electron current moves toward the positive plate, that plate will become less positively charged. Since the plates now have less charge on them, they contribute a smaller fringe field in the connecting wires, and the electric field everywhere in the circuit gets smaller, as seen in Figure 19.5. Smaller electric field means smaller electron current, so the bulb glows less brightly.

> **QUESTION** As the process continues, what will be the final state?

With time, the capacitor plates will lose more and more of their charge, so the electric field will get smaller and smaller, and the electron current will get smaller and smaller. The bulb will get dimmer and dimmer and finally stop glowing visibly when the current is so small that the bulb filament isn't hot enough to glow. Eventually the capacitor plates lose all their charge and the current stops completely (Figure 19.6). The circuit is now in equilibrium.

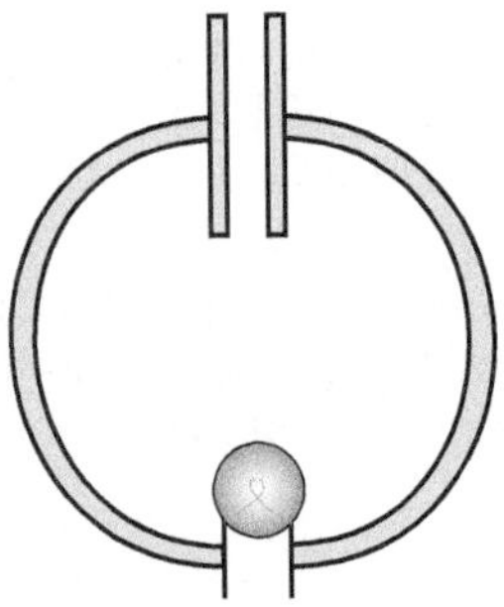

Figure 19.6 After many seconds there is no more charge on the capacitor plates, and the current is zero (equilibrium).

The dimming of the light bulb is gradual because the less charge there is on the capacitor plates, the smaller the fringe field and the smaller the current draining charge off the plates. The amount of charge that leaves the plates in the first second is larger than the amount that leaves in the second second, which in turn is larger than the amount that leaves in the third second. The bulb glows brightly for a short time but dimly for a long time.

> **QUESTION** Since the positive and negative charges on the plates of the capacitor are attracted to each other, why does charge leave the capacitor and go through the light bulb?

In the reasoning above we focused on the electric field in the wire close to but outside the plates. Initially, right next to the negative plate, the electric field in the wire is not zero and points toward the negative plate, as shown in Figure 19.4. Electrons in the wire at this location feel a force and move away from the negative plate; this will create a positively charged region inside the wire unless electrons flow off the negative plate into the wire. Electrons in the wire near the positive plate likewise move toward the positive plate, under the influence of the fringe electric field in the wire.

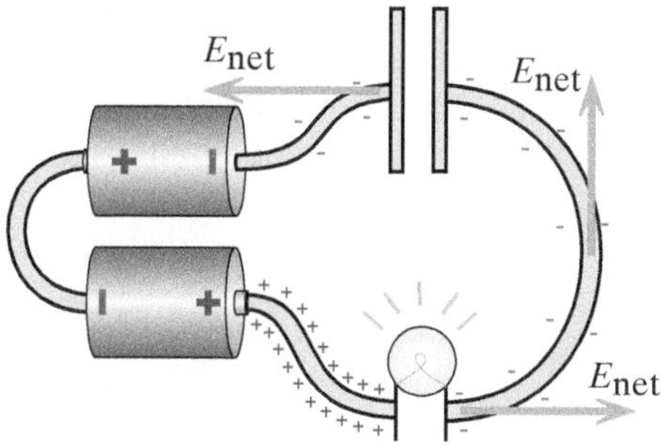

Figure 19.7 Charge and electric field 0.01 s after connecting the circuit. The bulb is as bright as in a noncapacitor circuit.

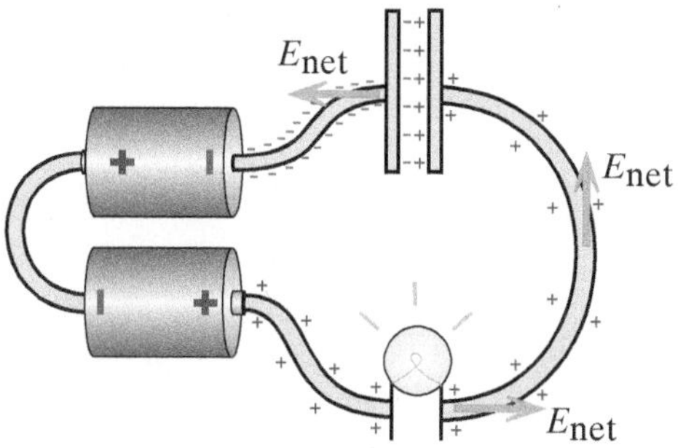

Figure 19.8 Charge and electric field 1 s after connecting the circuit. The bulb is less bright than before.

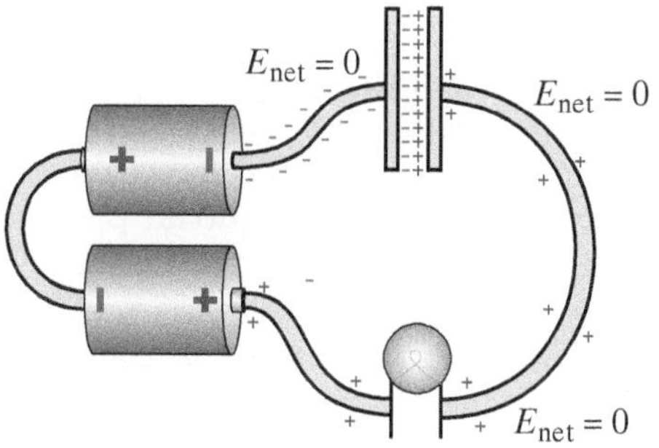

Figure 19.9 After many seconds the current is zero (equilibrium).

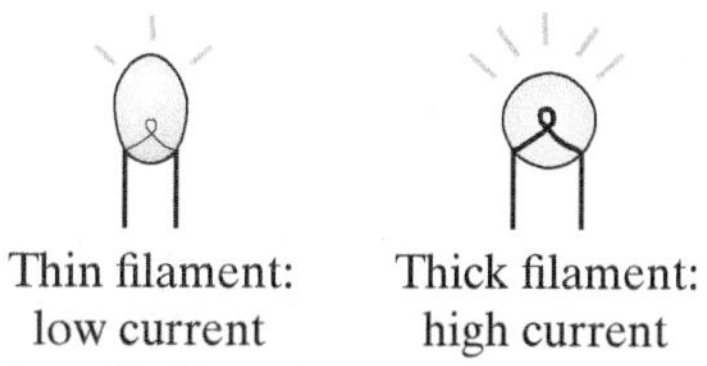

Figure 19.10 When a thin-filament bulb is connected to a battery, the current is small and the bulb is dim. With a thick-filament bulb the current is high and the bulb is bright.

Charging a Capacitor

We can use similar reasoning to explain the initial charging of the capacitor in a circuit containing a battery. As we did above, we will consider the electric field at a location inside the wire, and the effect this field has on electrons in the wire at that location. Again we will reason step by step, considering this as an iterative process.

We'll choose a time step of about 0.01 s, which is short compared to the total time it takes to charge the capacitor (a few seconds), but long compared to the time it takes the surface charges on the battery and wire to rearrange themselves (a few nanoseconds). Because the capacitor is initially uncharged, during the first time step the capacitor's contribution to the net field inside the wire is nearly zero (Figure 19.7).

> QUESTION Since the capacitor is initially uncharged, how bright will the bulb be initially, compared to a circuit containing just batteries and the bulb?

Evidently the initial current will be the same as it would be in a simple bulb circuit, and the bulb just as bright as usual (Figure 19.7). However, after several time steps, when current has been running in the wires for a second or more, a significant amount of + and − charge will have built up on the capacitor plates.

> QUESTION One second after the circuit is connected, will the fringe field of the capacitor affect the net electric field in the neighboring wire—will it increase or decrease the net field? What will happen to the current?

The fringe field of the capacitor acts in a direction opposite to the conventional current, thus decreasing the net field. As a result, the current will decrease (Figure 19.8). Eventually the current stops completely when there is enough charge on the capacitor plates to create a large enough fringe field to counteract the field made by all the other charges. The light stops glowing (Figure 19.9). We see that for both charging and discharging, a key to explaining the qualitative behavior of the circuit is the varying fringe field of the capacitor.

The initial establishment of surface charges takes a few nanoseconds, whereas it takes several seconds to reach equilibrium. During the several seconds while the bulb is glowing, the current is not constant, but it changes so slowly that it makes sense to describe the situation as a "quasi-steady state."

The Effect of the Resistor

You might wonder what effect the light bulb in the circuit has on the charging process and on the final state of the capacitor. Suppose that you have two light bulbs, one with a thick filament and the other with a thin filament (Figure 19.10). If you charge a capacitor through a thin-filament bulb, the bulb is initially dim (low current) and takes a long time to go out. In contrast, if you charge the capacitor through a thick-filament bulb, the bulb is initially bright (high current) but glows for only a short time.

QUESTION Should the final amount of charge on the capacitor plates depend on what kind of bulb is used during the charging process? Consider the following two arguments.

- Since charging the capacitor through the thin-filament bulb goes on much longer than charging through the thick-filament bulb, one might argue that the capacitor gets charged up a lot more when the thin-filament bulb is used.
- Alternatively, one might argue that since the thick-filament bulb glows much more brightly, the capacitor would get charged up a lot more even though the bulb goes out much quicker.

In the final state there is no current, so the only potential differences are across the battery and the capacitor. Therefore the potential difference across the capacitor is equal to that across the battery, no matter which bulb is in the circuit, and the charge on the plates is the same no matter which bulb we use. With the thin-filament bulb a small current runs for a long time, and with the thick-filament bulb a large current runs for a short time, but the total amount of charge is the same in either case. Experiment EXP2 is a test of the effect of the kind of bulb on the final amount of charge on the capacitor plates.

In Section 19.7 we will see how to predict the current and the amount of charge on the capacitor quantitatively at any time during the charging or discharging process.

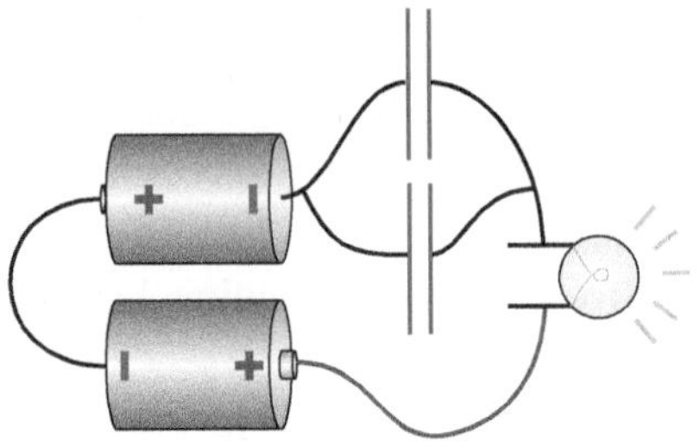

Figure 19.11 Two capacitors in parallel; Experiment EXP3.

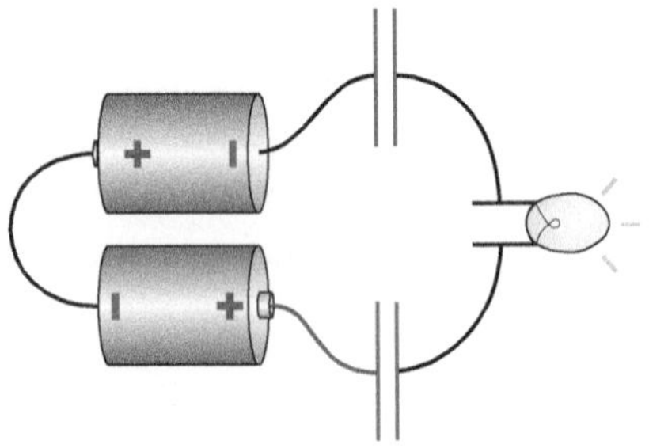

Figure 19.12 A glowing bulb completely isolated from the batteries; Experiment EXP4.

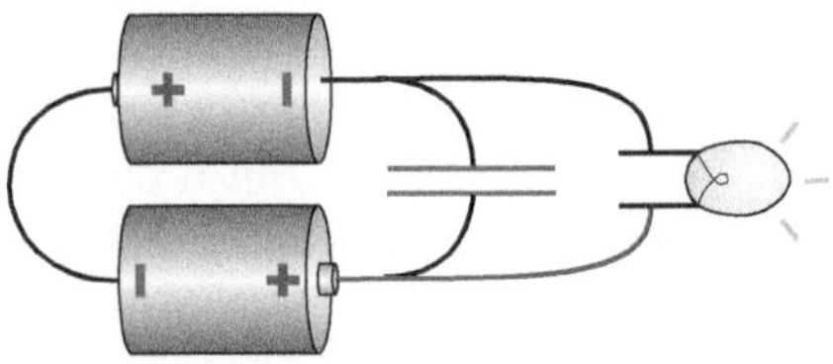

Figure 19.13 What capacitors are used for; Experiment EXP6.

Capacitors in Circuits

Capacitors are used in various ways in circuits, some of which are indicated in Figures 19.11–19.13.

Parallel capacitors: Two capacitors in parallel as shown in Figure 19.11 can be thought of as one capacitor with plates that have twice the area (see Checkpoint 1). Experiment EXP3 shows that the bulb glows twice as long with this arrangement.

An isolated bulb: In Experiment EXP4, shown in Figure 19.12, you can light a bulb despite the fact that there is no conducting path between the bulb and the battery! In the isolated section of the circuit, electrons flow from the plate on the right of the capacitor on the top through the bulb onto the plate on the right of the capacitor on the bottom, at the same time that there is current in the other wires.

Reversing a capacitor: In Experiment EXP5 you can see that reversing the capacitor in a circuit after charging it leads to an interesting effect. Temporarily, it's like putting an extra battery in the circuit.

Compensating for brief power interruptions: Experiment EXP6 (shown in Figure 19.13) helps understand how these capacitors can be useful. What you find if you intermittently break the connection to the batteries for fractions of a second is that the bulb stays lit despite being disconnected from the battery. The current is supplied from the capacitor, which had been charged by the batteries. We say that the capacitor is connected "in parallel" with the batteries, rather than in series.

The large-capacity capacitor used in these experiments is designed to be used in this way, connected in parallel to a power supply in a computer, to supply the computer circuits with current during momentary power outages. For a given voltage, these capacitors hold almost a million times more charge than ordinary capacitors!

More generally, capacitors placed in parallel are used in all kinds of electric circuits to even out rapid changes in voltage. For example, in a power supply, parallel capacitors are used to filter out high-frequency (AC) voltage changes that are riding on top of a constant (DC) voltage. You might say that a capacitor makes a circuit behave in a sluggish manner, being unable to change conditions rapidly.

Energy Stored in a Capacitor

Let's consider briefly the energy aspects of charging and discharging a capacitor through a light bulb.

> QUESTION In the charging phase, from a time just before the circuit is completed to the time when there is essentially no more current, state whether there is energy gain or loss in the battery, in the bulb, in the capacitor, and in the surroundings. (*Hint:* Think especially carefully about the initial and final states of the bulb.)

The battery loses energy, the bulb neither loses nor gains (it wasn't glowing originally, and it isn't glowing afterward either), the capacitor gains energy (enough to power a bulb later), and the surroundings gain energy (light and heat transfer from the bulb).

The Current Node Rule in a Capacitor Circuit

The current node rule in circuits results from charge conservation in the steady state. Figure 19.14 shows a simple application of the node rule to a node in a circuit where the incoming current I_1 supplies the outgoing currents I_2 and I_3, so that $\sum(I_{\text{entering}}) = I_1 - I_2 - I_3 = 0$.

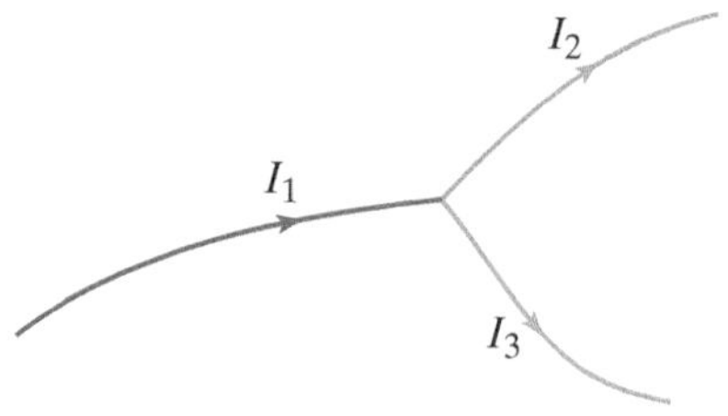

Figure 19.14 In the steady state it must be that $I_1 = I_2 + I_3$.

Figure 19.15 shows a more complex situation. Consider locations 1 and 2 as nodes. While the capacitor is being charged, there is current running onto one plate of the capacitor without any current coming off that plate (region 1 in Figure 19.15), and as a result the charge on the plate is changing. This is clearly not a steady state. Similarly, current comes off of the other plate (region 2 in Figure 19.15), which changes the charge on that plate.

However, if you look at the capacitor as a whole (region 3 in Figure 19.15), just as much current leaves one plate as enters the other plate. Otherwise the two plates together would acquire a nonzero charge that would quickly repel or attract electrons in such a way as to restore the net charge of the capacitor to be zero. So during charging or discharging of a capacitor the current node rule applies to the capacitor as a whole, even though it doesn't apply to just one plate of the capacitor.

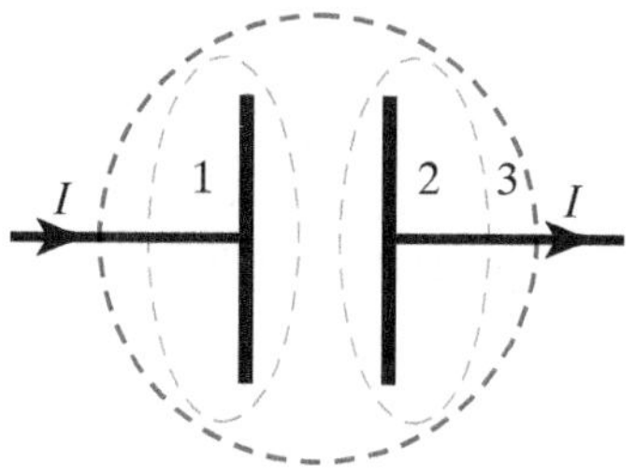

Figure 19.15 In a non-steady-state situation, the current node rule does not apply.

Capacitance

The capacitors you used in your circuit experiments may have had something like "0.47 F" or "1 F" marked on them, which would mean that they have a "capacitance" of about 0.5 or 1 farad (a unit honoring Michael Faraday). Let's see what is meant by this "capacitance" rating.

The more charge on the capacitor plates ($+Q$ and $-Q$), the bigger the electric field in the gap, and the greater the potential difference ΔV across the gap (Figure 19.16). Thus the absolute magnitude of the potential difference $|\Delta V|$ is proportional to the amount of positive charge Q on the positive plate, and conversely Q is proportional to $|\Delta V|$. We define capacitance C as this proportionality constant:

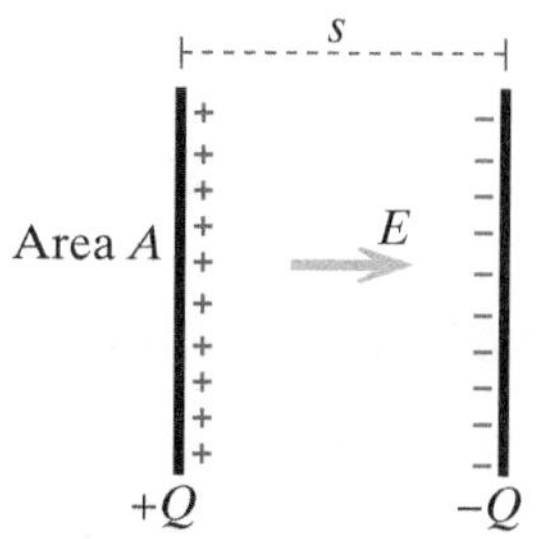

Figure 19.16 The more charge there is on the plates, the bigger the electric field and the bigger the potential difference.

DEFINITION OF CAPACITANCE

$$Q = C|\Delta V|$$

The units of capacitance are coulombs per volt, or farads.

QUESTION By using what you know about E and ΔV, show that $C = (\varepsilon_0 A)/s$ in a parallel-plate capacitor with plate area A and plate separation s.

Since the magnitude of the electric field inside the gap is

$$E = (Q/A)/\varepsilon_0$$

we can calculate the potential difference as

$$|\Delta V| = [(Q/A)/\varepsilon_0]s$$

which shows that the potential difference is proportional to the charge Q on one of the plates, as we expect. The capacitance is therefore

$$C = Q/|\Delta V| = (\varepsilon_0 A)/s \quad \text{(parallel-plate capacitor)}$$

Note that the capacitance depends on the geometry of the capacitor: the area of the plates and the separation between them.

The concept of capacitance is useful in the detailed analysis of circuits, and we will use it later to study capacitor circuits quantitatively. Bear in mind, however, that capacitance is just a derived quantity relating more fundamental quantities: charge and potential difference (and ultimately charge and field).

Reflection: Capacitors in Circuits

Circuits containing capacitors display a behavior that may be called a "quasi-steady state." The current in the capacitor circuit is approximately steady over a short time, but only approximately: the current slowly decreases and finally goes to zero—equilibrium is attained.

Actually, even an ordinary battery-and-bulb circuit is not in a true steady state, because when the chemicals in the battery are used up, the system comes into equilibrium (no current). There is an important difference, however. In a capacitor circuit the current continually decreases. In a battery-powered circuit, however, the current may stay nearly constant for many hours, then drop to zero in a much shorter time.

We were able to understand the behavior of a capacitor circuit in terms of the fringe field of the capacitor competing with the electric field of the other charges (battery charges and surface charges). The more charge on the capacitor, the larger the fringe field. When charging the capacitor, equilibrium is finally reached when there is so much charge on the capacitor that the fringe field is large enough to cancel the other contributions to the electric field.

We emphasized a qualitative analysis of capacitor circuits to get at the fundamental issues, and to illustrate the power of reasoning with charge and field. To analyze such circuits quantitatively we need to apply the energy principle in our analysis. Later in this chapter we do this, and we will find that we can quantitatively predict the rate of decrease of the current.

The design of a capacitor has an effect on the behavior of a capacitor circuit. By thinking about what happens in the first fraction of a second when the current decreases slightly, and how the state of the system at the end of this time interval will affect what happens next, we can predict qualitatively how capacitor circuits will behave. In Section 19.7 we will see how to make these predictions quantitative.

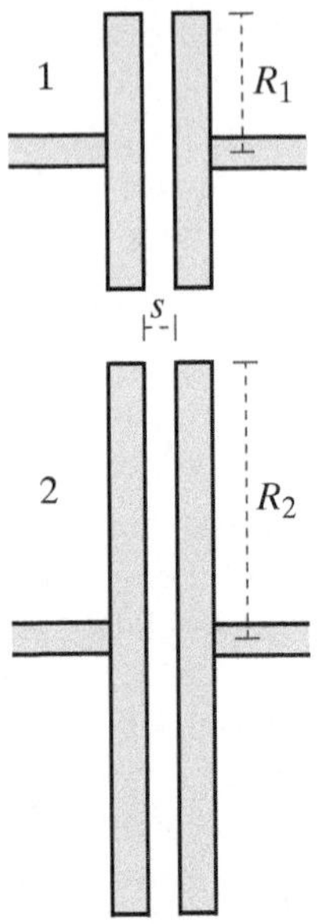

Figure 19.17 What is the effect of increasing the size of the capacitor plates?

Checkpoint 1 Consider two capacitors whose only difference is that the plates of capacitor number 2 are larger than those of capacitor number 1 (Figure 19.17). Neither capacitor has an insulating layer between the plates. They are placed in two different circuits having similar batteries and bulbs in series with the capacitor.

Show that in the first fraction of a second the current stays more nearly constant (decreases less rapidly) in the circuit with capacitor number 2. Explain your reasoning in detail. *Hint:* Show charges on the metal plates, and consider the electric fields they produce in the nearby wires. Remember that the fringe field near a plate outside a circular capacitor is approximately

$$\frac{Q/A}{\varepsilon_0}\left(\frac{s}{2R}\right)$$

A more extensive analysis shows that this trend holds true for the entire charging process: the larger capacitor ends up with more charge on its plates.

19.2 RESISTORS

Our study of electric circuits has been based on a microscopic picture of the motion of electrons inside the wires, due to electric fields produced by charges on and in the battery and on the surfaces of the wires. This microscopic picture gives us insight into the fundamental physical mechanisms of circuit behavior. However, it is not easy to measure electric field, surface charge, electron drift speed, or mobility directly.

On the other hand, it is easy to use relatively inexpensive commercially available meters to measure conventional current (rather than electron current), potential difference (rather than electric field), and "resistance" (rather than mobility). For practical purposes, it is useful to describe and analyze circuits in terms of these macroscopic quantities. We will link what we already know about circuits at the atomic level to a macroscopic description in terms of quantities that can be measured easily with standard meters.

To analyze the behavior of circuits in detail at the microscopic level, we have found it necessary to take into account both properties of particular materials and the geometry of particular circuit elements, such as thin and thick wires. It is sometimes useful to use a single number to summarize the properties of a particular object.

Conductivity: Combining the Properties of a Material

Conventional current is $I = |q|nA\bar{v} = |q|nAuE$, where $|q|$ is the absolute value of the charge of one of the mobile "charge carriers" (electrons or holes), n is the number density of the charge carriers, A is the cross-sectional area of the wire, u is the mobility of the charge carriers, and E is the electric field inside the wire that drives the mobile charges. This equation for conventional current mixes together three quite different kinds of quantities: $|q|$, n, and u are properties of the material, A describes the geometry of the material, and E is the electric field due to charges outside the material (and/or on its surface).

It is helpful to group the material properties together: $I = (|q|nu)AE$. Then if we divide both sides of the equation by the area A, we arrive at the following way of describing the situation:

$$J = \frac{I}{A} = (|q|nu)E = \sigma E$$

where J = current density (A/m^2) and σ = conductivity = $(|q|nu)$.

By dividing the current by the cross-sectional area, we obtain a "current density" J in amperes per square meter, which is independent of the geometry of the situation. The current density is proportional to the applied electric field, and the proportionality constant σ is called the "conductivity," which lumps together all the relevant properties of the material, $|q|nu$. (σ is a lowercase Greek sigma.)

We can consider $\vec{J}$ to be a vector pointing in the direction of the conventional current flow, which is in the direction of the applied electric field $\vec{E}$ (Figure 19.18):

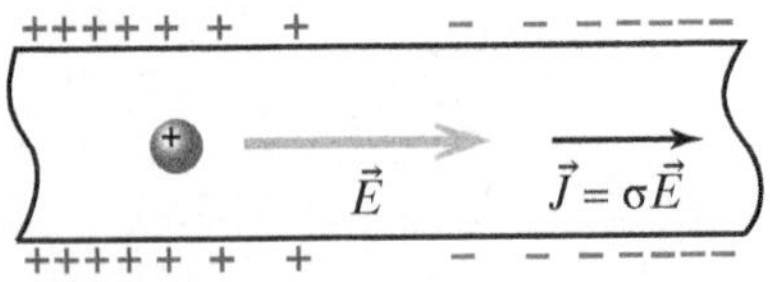

Figure 19.18 $\vec{J}$ is in the direction of the electric field, and in the direction of motion of a fictitious positive charge.

CURRENT DENSITY AND CONDUCTIVITY

$$\vec{J} = \sigma\vec{E}$$

The current density $J = (I/A)$; units are A/m^2.
The conductivity $\sigma = |q|nu$ depends on

- $|q|$, the absolute value of the charge on each carrier;
- n, the number of charge carriers per m^3; and
- u, the mobility of the charge carriers.

A higher conductivity means that a given applied electric field $\vec{E}$ gives larger current densities. The conductivity is a property of the material and has nothing to do with what shape the material has or with how large an electric field is applied to the material.

Conductivity with Two Kinds of Charge Carrier

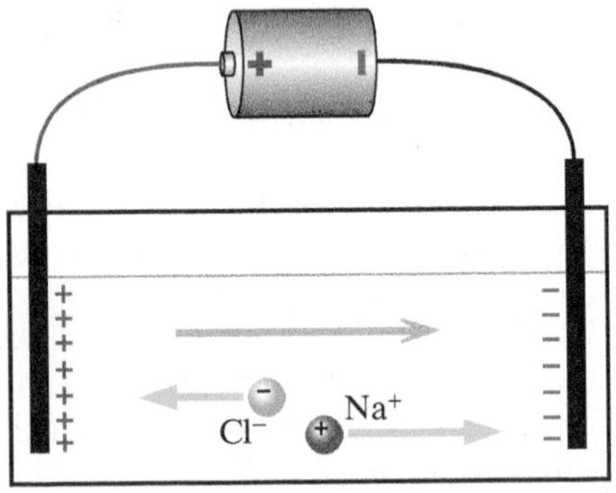

Figure 19.19 Both positive and negative charges move in salt water.

When electric forces are applied to salt water, both the sodium ions (Na^+) and the chloride ions (Cl^-) move and constitute an electric current. If the electric field is to the right in the water, the force on the Na^+ ions is to the right, but the force on the Cl^- ions is to the left (Figure 19.19).

The flow of Na^+ ions to the right constitutes a conventional current to the right, and the flow of Cl^- ions to the left also constitutes a conventional current to the right, because the motion of these negative ions has the effect of tending to make the region on the right less negative and hence more positive, with an effect similar to the flow of Na^+ ions to the right.

If the charge on the positive ion is q_1 (e in the case of Na^+ ions), the positive-ion contribution to the conventional current in coulombs per second is $q_1 n_1 A v_1$. If the charge on the negative ion is q_2 (the charge on a Cl^- ion is $q_2 = -e$), the negative-ion contribution to the conventional current in coulombs per second is $|q_2| n_2 A v_2$. We have to take the absolute value of the charge because the negative ions contribute conventional current in the same direction as do the positive ions.

Note that the mobilities (and therefore the drift speeds) may differ for the two kinds of ions. Larger ions usually have lower mobilities because it is more difficult to move them through the water.

The conventional current in amperes (coulombs per second) is the sum of the conventional current due to the positive charges and the conventional current in the same direction due to the negative charges:

$$I = |q_1| n_1 A v_1 + |q_2| n_2 A v_2$$

Note that both ionic currents contribute to the total current, and that the ionic charge q_1 or q_2 may be a multiple of e. For example, the charge of Zn^{++} is $2e$.

The current density $J = I/A = |q_1| n_1 u_1 E + |q_2| n_2 u_2 E = \sigma E$, so the conductivity in this situation where there are two kinds of charge carrier is

CONDUCTIVITY WITH TWO KINDS OF CHARGE CARRIERS

$$\sigma = |q_1| n_1 u_1 + |q_2| n_2 u_2$$

Resistance Combines Conductivity and Geometry

Often when we have compared two different electric circuits, we have said such things as, "The cross-sectional area goes up, the length decreases, and the mobility increases." It is practical to have a single quantity to describe a resistor that encapsulates both the properties of the material (mobility, mobile-carrier number density) and the geometrical properties of the particular object (cross-sectional area, length). This quantity is called "resistance."

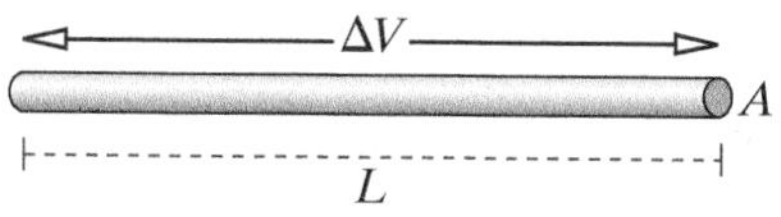

Figure 19.20 A resistor—for example, a Nichrome wire.

Consider the potential difference ΔV from one end to the other of a wire of constant cross-sectional area A, length L, and uniform composition (Figure 19.20). In the steady state, the electric field inside the wire has the same magnitude everywhere (otherwise the drift speed and current would vary) and is everywhere parallel to the wire (otherwise additional charge would be driven to the surface). Since potential difference is defined as

$$\Delta V = -\int_i^f \vec{E} \bullet d\vec{l}$$

we have $|\Delta V| = EL$. The conventional current density $J = I/A$ is the same everywhere on the cross-sectional area A of the wire, as was proven in the previous chapter (if E varied across the cross section, one could construct a path with a nonzero round-trip potential difference, and uniform E implies uniform v and therefore uniform J).

> QUESTION For some materials $J = \sigma E$. For such materials, write an equation for I in terms of σ and ΔV (consider I and ΔV to both have positive values).

Combining the two equations for ΔV and J, we find a relation between two macroscopic quantities, conventional current I and potential difference ΔV:

$$J = \sigma E$$

$$\frac{I}{A} = \sigma \frac{\Delta V}{L}$$

It is useful to rearrange this relation between current and potential difference in the following way:

$$I = \frac{\Delta V}{L/(\sigma A)} = \frac{\Delta V}{R}$$

where the resistance R is measured in volts per ampere or ohms (abbreviated Ω and named for Georg Ohm, the German physicist who first studied resistance quantitatively in the early 1800s). We have this definition:

DEFINITION OF RESISTANCE

$$R = \frac{L}{\sigma A}$$

Note that resistance depends on

- the properties of the material: $\sigma = |q|nu$
- the geometry of the resistor: L, A

These aspects are lumped into one quantity—resistance. We can make a large-resistance resistor in a variety of ways. The resistor could have a large length L, or a small cross-sectional area A, or a low conductivity σ. Sometimes resistance is expressed in terms of the "resistivity" $\rho = 1/\sigma$. In that case, $R = \rho L/A$.

We can of course rearrange $I = \Delta V/R$ and write $\Delta V = RI$. Usually, however, it is appropriate to think of the potential difference as causing a current, not of the current causing a potential difference, and the form $I = \Delta V/R$ reminds us that the current is the result of applying an electric field, with an associated potential difference.

Note that we haven't developed any new physical principles. We have simply rewritten $I = |q|nA\bar{v} = |q|nA(uE)$ in terms of potential difference, so that we have $I = \Delta V/R$.

Reflection: Connecting Macroscopic and Microscopic Viewpoints

We can relate the analysis we've just done to our earlier microscopic analysis of a current in a material with a single charge carrier, such as electrons in a metal:

$$\bar{v} = uE$$
$$I = |q|nA\bar{v} = |q|nA(uE)$$

Taking ΔV to be positive, $\Delta V = EL$ for a straight wire, so $E = \Delta V/L$.

$$I = |q|nA\left(u\frac{\Delta V}{L}\right) = \frac{\Delta V}{\left(\dfrac{L}{|q|nAu}\right)}$$

$I = \dfrac{\Delta V}{R}$, where $R = \dfrac{L}{\sigma A}$, with $\sigma = |q|nu$. You should derive this relationship yourself without referring to notes.

For some kinds of scientific work the microscopic picture in terms of electric field, mobility, and drift speed ($\bar{v} = uE$) is more useful. However, when you need to make contact with the technological world of voltmeters, ammeters, and ohmmeters, you also need to be able to describe a situation in terms of macroscopic quantities such as potential difference (voltage), resistance, and conventional current.

Here is a side-by-side comparison of the two views:

Microscopic view	Macroscopic view
$\bar{v} = uE$	$J = \sigma E$
$i = nA\bar{v} = nAuE$	$I = \lvert q\rvert nA\bar{v} = \left(\frac{1}{R}\right)\Delta V$

In other books you may see this equation written as $V = IR$. This can lead to confusion between potential at one location and the potential difference between two locations, so we always write $\Delta V = IR$.

Ohmic and Non-Ohmic Materials

As you may have seen in experiments with light bulbs, the mobility u of the tungsten in a light bulb filament varies with the temperature. For example, the current through one bulb is less than twice the current through two bulbs in series, because the mobility is lower at higher temperatures. Since conductivity and resistance both depend on mobility ($\sigma = |q|nu$ and $R = L/(\sigma A)$), both these quantities may change as the current through an object varies.

In some materials the conductivity σ is nearly constant, independent of the amount of current flowing through the resistor. We call such constant conductivity materials "ohmic." An ohmic material has the property that $I = \Delta V/R$, where the resistance $R = L/(\sigma A)$ is constant, independent of the amount of current. No material is truly ohmic, because conductivity depends somewhat on temperature, and higher currents tend to raise the temperature of the material. However, many materials can be considered to be approximately ohmic as long as the temperature doesn't change very much.

Resistors made of ohmic materials are called "ohmic" resistors. Doubling the length L of an ohmic resistor doubles the resistance R and cuts the current in half if the potential difference ΔV across the resistor is kept constant. Doubling the cross-sectional area A cuts the resistance in half and doubles the current if the potential difference ΔV across the resistor is kept constant.

Nichrome wires, and the carbon resistors used in electronic circuits, are nearly ohmic resistors, as long as not enough current runs through them to raise the temperature significantly. In metals the conductivity decreases somewhat at higher temperatures, and metals are only approximately ohmic. You can think of the electrons as having more frequent collisions with atoms that have increased random motion at higher temperatures. You may have seen experimental evidence for non-ohmic behavior in the previous chapter, where the current through two bulbs in series was more than half the current through one bulb.

Checkpoint 2 By actual measurement, the current through a thin-filament bulb when connected to two batteries in series (3 V) is about 100 mA; connected to one battery (1.5 V) the current is about 80 mA; and connected to a small voltage of only 50 mV the current is about 6 mA. (Different thin-filament bulbs may differ from these values somewhat.) Using the equation $I = \Delta V/R$, what is R for each of these cases? Is R a constant? Is a thin-filament bulb an ohmic resistor over this whole range of currents?

After calculating R, notice that from 0 to 3 V a thin-filament bulb is not ohmic, although in the range from 1.5 to 3 V the resistance changes by only about 50% and the bulb can be said to be very roughly ohmic.

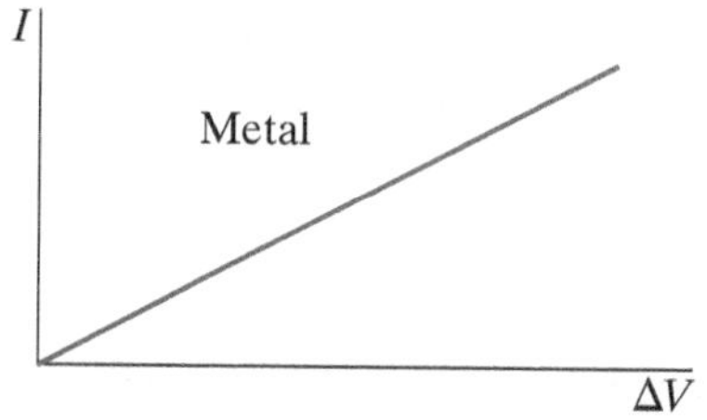

Figure 19.21 Mobile electrons, not bound to atoms—ohmic behavior.

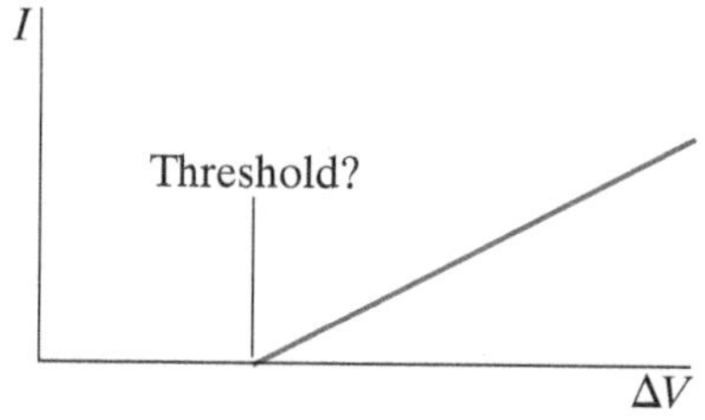

Figure 19.22 If electrons in a metal were bound to the atoms, the current might behave like this.

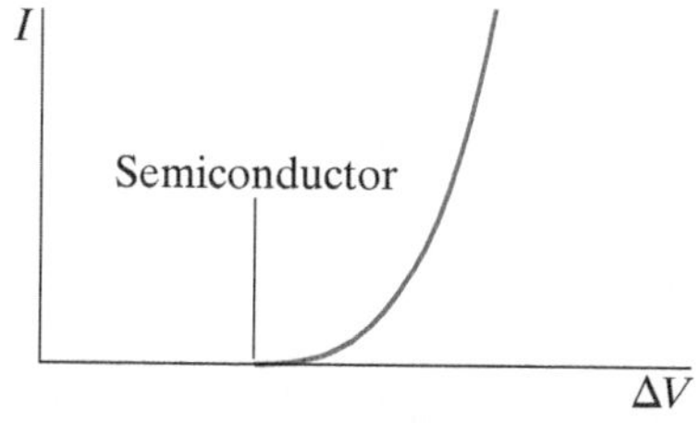

Figure 19.23 Behavior of a pure semiconductor such as silicon or germanium.

Semiconductors

In a metal the current is zero when the potential difference is zero, but the slightest potential difference produces some current, as shown in Figure 19.21. Such behavior implies that at least some of the electrons in a metal are truly free, not bound to the atoms. If they were all bound, we would have to apply some field just to free some up, and then more field to increase the current, as shown in Figure 19.22.

Pure silicon and germanium are non-ohmic in a spectacular and extremely useful way. In these "semiconductor" materials, the density of mobile electrons n and therefore the conductivity $\sigma = enu$ (where u is the mobile electron mobility) depend exponentially on the voltage. Applying a slight voltage produces almost no current. After applying enough voltage to get a small current, doubling this voltage may make the current go up by a factor of a hundred rather than by a factor of two! (See Figure 19.23.)

The key difference between a metal and a semiconductor lies in n, the number of available charge carriers per unit volume. In a metal, n is a fixed number (one mobile electron or hole per atomic core). In a semiconductor, at very low temperatures there are no mobile electrons or holes and the material acts nearly like an insulator for small applied fields. However, the electrons are not tightly bound to the atomic cores and it is possible to free up some electrons by applying a large enough electric field, so n is variable and increases with increasing potential difference.

Raising the temperature can also free up some electrons (or holes), and the effect of increasing n more than compensates for the higher rate of collisions with the atomic cores, so that for a semiconductor the resistance decreases with increasing temperature. This is also true for carbon: the resistance of a carbon resistor decreases somewhat with increasing temperature.

Other Non-Ohmic Circuit Elements

Not all resistors are ohmic, and we have seen that even those called ohmic are only approximately so if the temperature changes with current. Moreover, circuit elements other than resistors are definitely not ohmic.

Capacitors aren't ohmic. If you double the current flowing onto and off of the capacitor plates, the capacitor gets charged faster, but it is not the case that ΔV is proportional to I for a capacitor. The voltage is proportional to the charge on one of the plates rather than being proportional to the current: $|\Delta V| = Q/C$, where as usual Q is the charge on the positively charged plate of the capacitor. Therefore capacitors are not ohmic devices.

Batteries aren't ohmic. If you double the current through a battery (by changing the circuit elements attached to the battery), the battery voltage hardly changes at all. The battery voltage certainly doesn't double and will actually decrease slightly. It definitely is not the case that ΔV is proportional to I for a battery.

It is extremely important to apply $I = \Delta V/R$ *only* to resistors. This equation is not a fundamental physics principle with broad validity. It is just an expression of the fact that some materials are (approximately) ohmic.

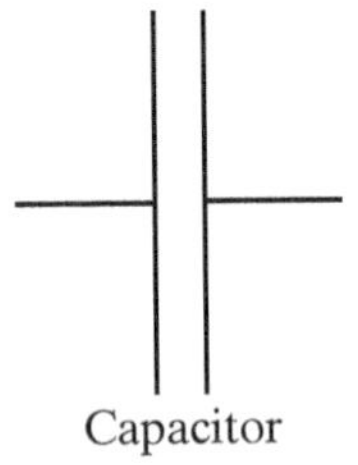

Figure 19.24 The conventional symbol for a capacitor.

19.3 CONVENTIONAL SYMBOLS AND TERMS

In conventional circuit diagrams wires are represented as straight lines, and stylized symbols are used to represent circuit elements. A capacitor is shown as though we had unrolled the coiled-up capacitor and made external connections to the center of the metal plates (Figure 19.24). The insulating material between the plates is usually not indicated.

Resistors are represented by jagged lines. The symbol for a battery is two parallel lines of unequal length. The longer line represents the positive end of the battery and the short line represents the negative end. Figure 19.25 shows the symbols for a resistor and a battery.

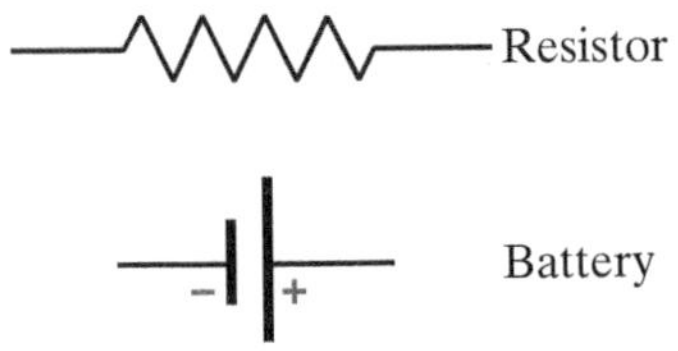

Figure 19.25 Conventional symbols for resistors and batteries.

Series Resistors

When circuit elements are connected along a single path with no branches, they are said to be in series. Figure 19.26 shows three ohmic resistors in series with an ideal battery (one with negligible internal resistance).

We can show that several ohmic resistors in series are equivalent to one ohmic resistor with a resistance that is the sum of the individual resistances. Consider the circuit of Figure 19.26.

The current I is the same everywhere in this series circuit.

> QUESTION Write the energy-conservation loop equation for this circuit, using the fact that $\Delta V = RI$ for an ohmic resistor.

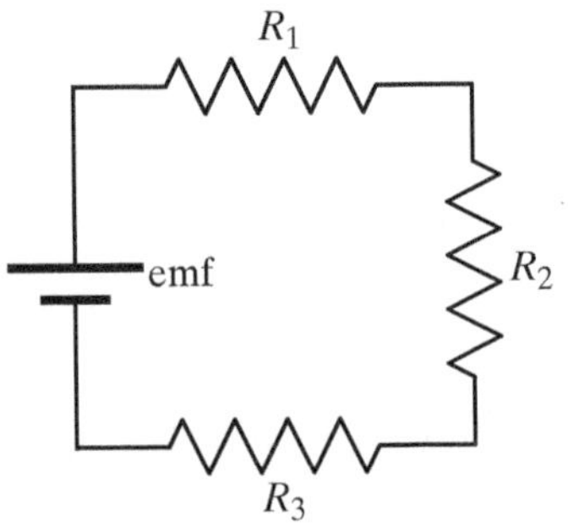

Figure 19.26 Three resistors in series with a battery.

We can rearrange the loop equation, $\text{emf} - R_1I - R_2I - R_3I = 0$, in this form:

$$\text{emf} = (R_1 + R_2 + R_3)I = R_{\text{equivalent}}I$$

Evidently a series of ohmic resistors acts like a single ohmic resistor, with a resistance equal to the sum of all the resistances:

$$R_{\text{equivalent}} = R_1 + R_2 + R_3$$

This is not an entirely obvious result! The proof depends on the following:

- the definition of resistance $R = L/(\sigma A)$;
- the assumption that the conductivity σ is independent of the current for the material in these resistors (ohmic resistors);

- the fact that the magnitude of the sum of all the resistor potential drops is equal to the potential rise across the battery, which is numerically equal to the emf of the battery; and
- the fact that in a series circuit the steady-state current I is the same in every element, due to charge conservation, and an unchanging charge distribution in the steady state.

Because $R = L/(\sigma A) = \rho L/A$, if the various resistors are all made of the same material and have the same cross-sectional area, note that the series-resistance equation is equivalent to just adding the various lengths: $L_{\text{equivalent}} = L_1 + L_2 + L_3$.

Parallel Resistors

Instead of connecting circuit elements in series, it is possible to connect them in parallel, creating several branches in a circuit. Figure 19.27 shows three resistors connected in parallel.

We can show that several ohmic resistors in parallel are equivalent to one ohmic resistor with a resistance that is the reciprocal of the sum of reciprocals of the individual resistances. Consider the circuit of Figure 19.27, with three different ohmic resistors in parallel.

The potential difference is the same across each resistor in this parallel circuit and is numerically equal to the emf of the battery.

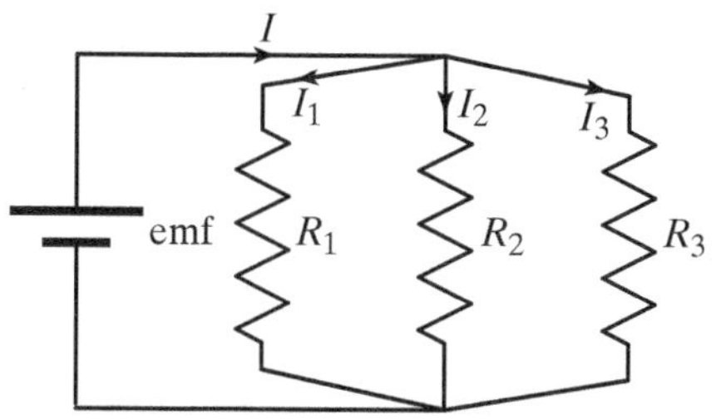

Figure 19.27 Three resistors in parallel.

QUESTION Write the current node equation for this circuit, using the fact that $I = \Delta V/R$ for an ohmic resistor:

We can rearrange the current equation, $I - \dfrac{\text{emf}}{R_1} - \dfrac{\text{emf}}{R_2} - \dfrac{\text{emf}}{R_3} = 0$, in this form:

$$I = \left(\frac{1}{R_1} + \frac{1}{R_2} + \frac{1}{R_3}\right)(\text{emf}) = \frac{\text{emf}}{R_{\text{equivalent}}}$$

Evidently ohmic resistors in parallel act like a single ohmic resistor, with a resistance whose reciprocal is equal to the sum of the reciprocals of the individual resistances:

$$\frac{1}{R_{\text{equivalent}}} = \frac{1}{R_1} + \frac{1}{R_2} + \frac{1}{R_3}$$

Because $1/R = \sigma A/L$, if the various resistors are all made of the same material and have the same length, note that the parallel-resistance equation is equivalent to just adding the various cross-sectional areas: $A_{\text{equivalent}} = A_1 + A_2 + A_3$. This corresponds to what you see if you connect two bulbs in parallel to a battery: there is twice as much current through the battery, as though you had one bulb with twice the cross-sectional area.

Checkpoint 3 When glowing, a thin-filament bulb has a resistance of about 30 Ω and a thick-filament bulb has a resistance of about 10 Ω. If they are in parallel, what is their equivalent resistance? How much current goes through two 1.5 V flashlight batteries in series if a thin-filament bulb and a thick-filament bulb are connected in parallel to the batteries?

19.4 WORK AND POWER IN A CIRCUIT

If you move a small amount of charge Δq from one location to another, the small amount of electric potential energy change ΔU_e is equal to the charge Δq times the potential difference $\Delta V, \Delta U_e = (\Delta q)(\Delta V)$, since electric potential is electric potential energy per unit charge (joules per coulomb, or volts). If this work is done in a short amount of time Δt, we can calculate the power, which is energy per unit time:

$$\text{Power} = \frac{\Delta U_e}{\Delta t} = \frac{(\Delta q)(\Delta V)}{\Delta t}$$

However, $\Delta q/\Delta t$ is the amount of charge moved per unit time, which is the current I. Since $\Delta q/\Delta t = I$, we have the following important equation:

POWER FOR ANY KIND OF CIRCUIT COMPONENT

$$\text{Power} = I\Delta V$$

You should be able to repeat this argument without referring to your notes. This is a very general result, applying to any kind of circuit component. Note that the power dissipated in the resistor is $I(RI) = RI^2$, which can also be written as $(\Delta V/R)\Delta V = (\Delta V)^2/R$.

Checkpoint 4 Power in a battery: In a certain circuit, a battery with an emf of 1.5 V generates a current of 3 A. What is the output power of the battery? Include appropriate units. (If on the other hand you were charging a rechargeable battery with a charging current of 3 A, this would be the input power to the battery.)

EXAMPLE **Two Different Bulbs in Series**

The circuit in Figure 19.28 consists of two flashlight batteries, a thick-filament bulb, a thin-filament bulb, and very low-resistance copper wires.

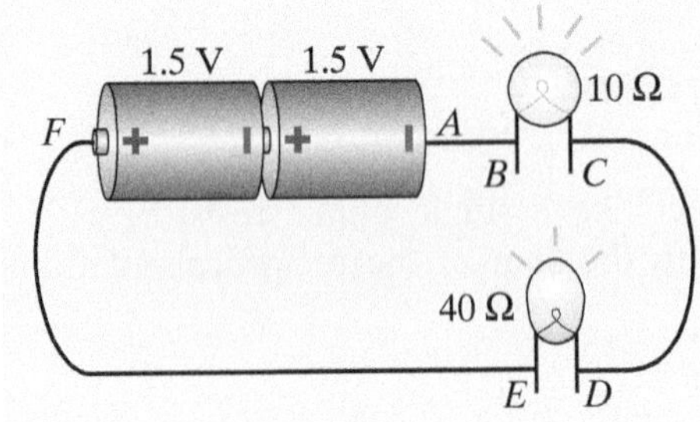

Figure 19.28 A circuit with a thin-filament bulb and a thick-filament bulb.

(a) Calculate the number of electrons that flow out of the battery every second at location A, in the steady state. Assume that in these steady-state conditions the resistance of the thick-filament bulb is 10 Ω and the resistance of the thin-filament bulb is 40 Ω.

(b) Show the approximate surface charge on the steady-state circuit.

(c) Draw an accurate graph of potential around the circuit, with the x axis running from A to F to A. Label the y axis with numerical values of the potential.

(d) The tungsten filament in the thin-filament bulb is 8 mm long and has a cross-sectional area of 2×10^{-10} m^2. How big is the electric field inside this metal filament?

(e) What is the power output of the battery?

Solution **(a)** Loop rule: $2\,\text{emf} - IR_{10} - IR_{40} = 0$

$$I = \frac{2\,\text{emf}}{(R_{10}+R_{40})} = \frac{2(1.5\,\text{V})}{(10\,\Omega + 40\,\Omega)} = 6\times10^{-2}\,\text{A}$$

$$i = \frac{I}{|q|} = \frac{6\times10^{-2}\,\text{A}}{1.6\times10^{-19}\,\text{C}} = 3.8\times10^{17}\,\text{electrons/s}$$

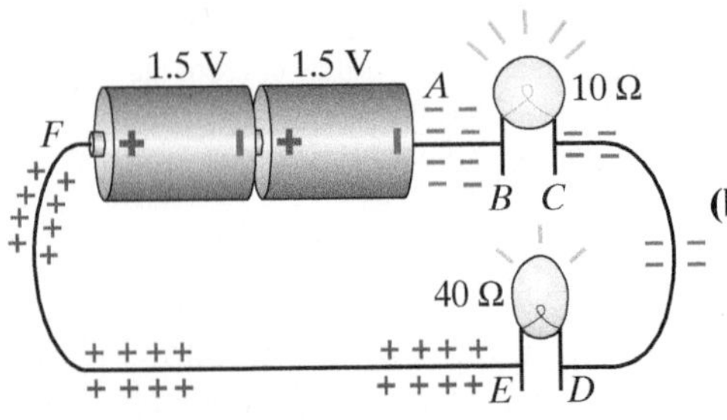

Figure 19.29 Approximate surface charge distribution on a circuit in steady state.

(b) See Figure 19.29. A bigger electric field is needed in the thin-filament bulb (because it has higher resistance), so the surface charge gradient across the thin-filament bulb is much bigger than the surface charge gradient across the thick-filament bulb. The surface charge gradient across the copper wires is extremely small, since only a very small electric field is needed in the wires. This gradient is too small to show up on this diagram, so the surface charge is shown as approximately uniform on the wires.

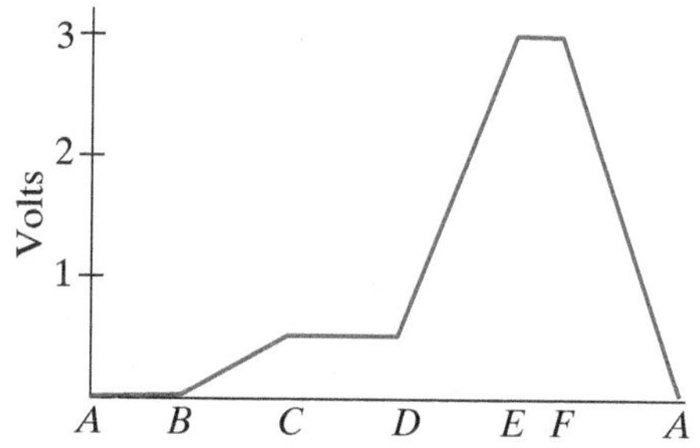

Figure 19.30 Potential (volts) vs. location in circuit. Distances (x axis) are not to scale.

(c) See Figure 19.30. Distances on the x axis are not to scale.

(d) The drop in potential across the thin-filament bulb is given by $\Delta V_{40} = R_{40}I$. The electric field in the filament of the thin-filament bulb can be found from the potential difference:

$$E = \frac{\Delta V}{L} = \frac{R_{40}I}{L} = \frac{(40\,\Omega)(6\times10^{-2}\,\text{A})}{(8\times10^{-3}\,\text{m})} = 300\,\frac{\text{V}}{\text{m}}$$

(e) For the battery:

$$P = I\Delta V = (6\times10^{-2}\,\text{A})(3\,\text{V}) = 0.18\,\text{W}$$

19.5 BATTERIES

We have been treating batteries as if they were ideal devices: that is, as if they always managed to maintain the same potential difference across themselves regardless of what they are connected to. Real batteries, however, don't quite succeed in maintaining a potential difference equal to the battery emf, due to "internal resistance"—the resistance of the battery itself to the passage of current. We will show that a real battery can be modeled as an ideal, resistance-less battery in series with the internal resistance of the battery.

Consider again our mechanical "battery" consisting of two large charged plates whose charge is continually replenished by transporting electrons on a conveyor belt from the + plate to the − plate. The electrons on the motor-driven belt are acted on both by the non-Coulomb force exerted by the motor, $\vec{F}_{NC}$, and by the Coulomb force exerted by the charges on the plates, $\vec{F}_C = -e\vec{E}_C$. These two forces oppose each other (Figure 19.31).

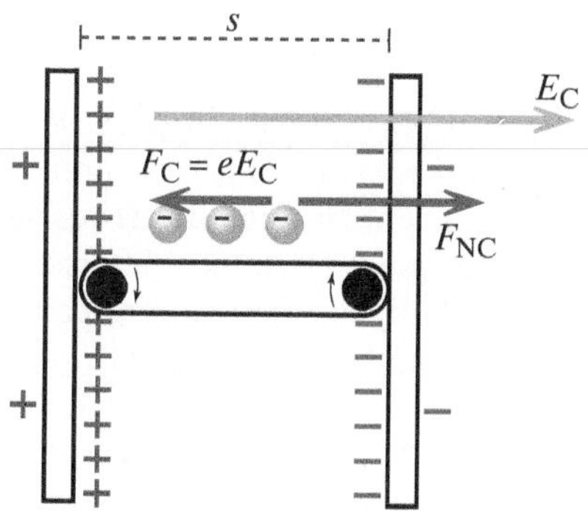

Figure 19.31 A mechanical battery with some friction in the mechanism.

In general, the non-Coulomb force F_{NC} has to be somewhat larger than the Coulomb force $F_C = eE_C$ because of finite mobility of the charges moving through the battery. In a chemical battery, this takes the familiar form of collisions, leading to an average drift speed. The drift speed of ions through the battery is proportional to the net force $(F_{NC} - eE_C)$. If σ is the conductivity for moving ions through the battery and A is the cross-sectional area of the battery, the current density $J = I/A$ is given not by the usual $J = \sigma E$ but by

$$J = \frac{I}{A} = \sigma \times (\text{force per unit charge}) = \sigma\left(\frac{F_{NC}}{e} - E_C\right)$$

Rearranging this equation, we have

$$E_C = \frac{F_{NC}}{e} - \frac{I/A}{\sigma}$$

If the (short) length of our mechanical battery is s, the electric field is nearly uniform, and the potential difference across the battery is $\Delta V = E_C s$, so

$$\Delta V = \frac{F_{NC}s}{e} - \frac{s}{\sigma A}I$$

Note, however, that $F_{NC}s/e$ is the work per unit charge or emf of the battery, and $s/(\sigma A)$ has the units of resistance. We call $s/(\sigma A)$ the "internal resistance" r_{int} of the battery, and finally we have

$$\Delta V_{\text{battery}} = \text{emf} - r_{int}I$$

This says: If there is very little internal resistance r_{int} to movement of ions through the battery (or very little current I), the potential difference ΔV across the battery is numerically nearly equal to the emf of the battery. Up until now we have made this assumption to simplify the discussion. However, if the current I through the battery is large enough, the term $r_{int}I$ may be sizable, and the potential difference across the battery (and therefore across the rest of the circuit) may be significantly less than the battery emf.

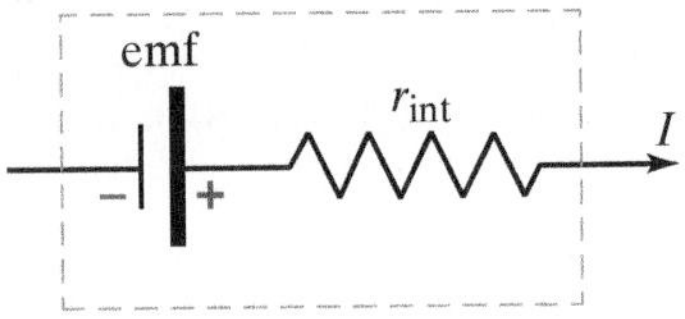

Figure 19.32 A real battery can be modeled as an ideal battery in series with an internal resistance.

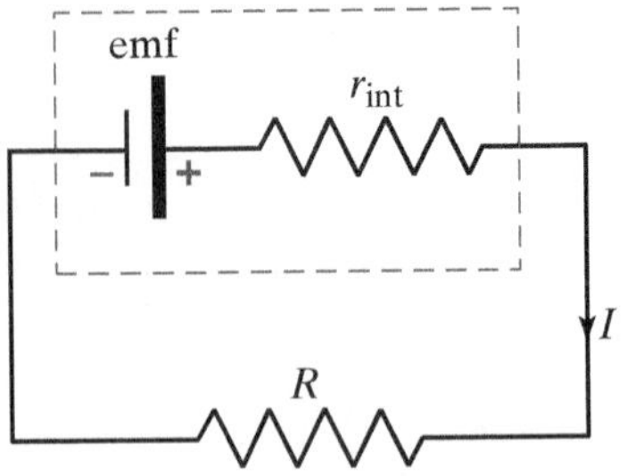

Figure 19.33 A nonideal battery connected to a resistor.

Since $\Delta V_{battery} = \text{emf} - r_{int}I$, a real (nonideal) battery can be modeled as an ideal battery with the internal resistance in series, so that the potential difference across the battery is less than the emf by an amount $r_{int}I$, the potential drop across the internal resistance (Figure 19.32).

If a resistor R is connected to a real, nonideal battery (Figure 19.33), energy conservation (round-trip potential difference = 0) gives

$$(\text{emf} - r_{int}I) - RI = 0$$

Solving for the current I, we have

$$I = \frac{\text{emf}}{R + r_{int}}$$

Let's see how this works out in practice. A fresh 1.5 V alkaline flashlight battery has an internal resistance r_{int} of about 0.25 Ω.

QUESTION What current would we get when we connect various resistors to the battery? Calculate the various values to get an idea of the effect:

Ideal Battery $r_{int} = 0$		**Real Battery $r_{int} = 0.25\ \Omega$**
100 Ω:	0.015 A	100 Ω:
10 Ω:	0.15 A	10 Ω:
1 Ω:	1.5 A	1 Ω:
0 Ω (short circuit):	infinite!	0 Ω (short circuit):

The currents with the real battery are 0.01496 A, 0.146 A, 1.2 A, and 6 A. There are several things to notice in this table: With a real battery, with nonzero internal resistance, you don't get 10 times the current through a 1 Ω resistor that you get through a 10 Ω resistor. Also, the internal resistance determines the maximum current that you can get out of a battery—this battery cannot deliver more than 6 A, no matter how small a resistor you attach to it. However, if the current is small, the behavior is nearly ideal.

An example of the role of a battery's internal resistance is seen in the requirements for a car battery, which must deliver a very large current to the starting motor. The battery must be built in such a way as to have a particularly low internal resistance. There are very large metal electrodes inside a car battery (large cross-sectional area A), which helps reduce the internal resistance to the transport of ions.

We should point out that the internal resistance r_{int} of a chemical battery is only approximately constant. As the battery is used more and more, the internal resistance increases. However, at any particular moment, the equation $\Delta V_{battery} = \text{emf} - r_{int}I$ does relate the potential difference across the battery to the current through the battery (using the values of emf and internal resistance valid at that moment). Experiments EXP7 to EXP9 let you explore internal resistance.

Checkpoint 5 What is the potential difference ΔV across a 1.5 V battery when it is short-circuited, if the internal resistance is 0.25 Ω and the resistance of the connecting wire is much less than 0.25 Ω? What is the average net electric field E inside the battery? For a mechanical battery, how much charge is on the battery plates?

This again shows the difference between the non-Coulomb emf of the battery, which does not change no matter what you connect to the battery, and the potential difference ΔVacross the battery, which depends on how much charge the battery is able to maintain on its ends. Only for an ideal battery with no internal resistance is ΔV exactly equal to the emf. For real batteries that are discharging, ΔV is always somewhat less than the emf (and if you are charging a real rechargeable battery, you have to apply a ΔV that is somewhat greater than the emf).

19.6 AMMETERS, VOLTMETERS, AND OHMMETERS

An important application of circuit theory is to explain the operation of ammeters, voltmeters, and ohmmeters used for measuring conventional currents, potential differences, and resistances. Modern digital "multimeters" combine these functions into one compact instrument, with a switch that lets you choose between measuring current, voltage, or resistance, and with a convenient digital readout, complete with sign. Here are the most important things to know about ammeters and voltmeters:

- An ammeter must be inserted into the circuit in series with the circuit element whose current you want to measure. An ammeter must have a very small resistance, so as not to alter significantly the circuit into which it has been inserted (Figure 19.34).
- A voltmeter must be placed in parallel with the circuit element along which you want to measure the potential difference. A voltmeter must have a very high resistance, so as not to create a significant alternative pathway for current, which would alter the circuit (Figure 19.35).

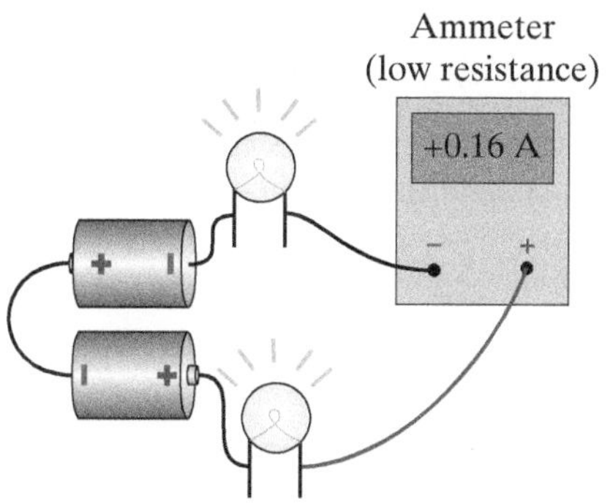

Figure 19.34 An ammeter must be inserted in series with the circuit element whose current you want to measure. The ammeter must have low resistance so as not to alter the circuit significantly.

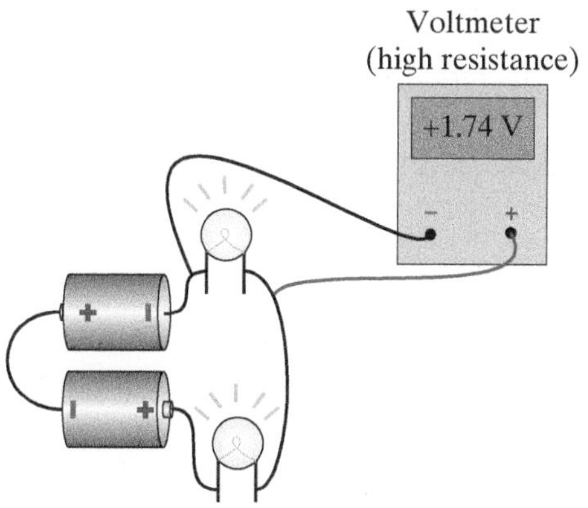

Figure 19.35 A voltmeter is placed in parallel with the circuit element along which you want to measure the potential difference. The voltmeter must have high resistance so as not to alter the circuit significantly by providing an alternative current pathway.

Using an Ammeter

An ammeter must be inserted into a circuit, and the current to be measured is brought into the + socket and out the − socket. If conventional current flows into the + socket, the ammeter displays a positive number. When conventional current flows into the − socket, the ammeter indicates a negative current reading.

AMMETER SIGN CONVENTION

If conventional current flows into the socket marked "+" an ammeter indicates a positive current.

(This is the opposite of the convention for batteries, since conventional current flows out of the + terminal of a battery.)

A Simple Ammeter

If you have been doing experiments with currents and compasses, you have been using a simple ammeter, in the form of a compass whose needle is deflected by a passing current. If you could run known amounts of current over the compass, you could calibrate this ammeter by determining the relationship between the known current and the observed deflection angle. This is a "noninvasive" ammeter, because you don't have to alter the circuit in order to be able to detect and measure the current in a wire.

A serious practical problem in using standard ammeters is that if you want to measure the current somewhere in an existing circuit, you must break the circuit at the place of interest and insert your ammeter. In order that this make as little change as possible in the current to be measured, ammeters are designed to have very low resistance, so that if you insert the ammeter it's as though you're just lengthening the wire a bit.

AN AMMETER SHOULD HAVE LOW RESISTANCE

An ammeter should have as small a resistance as possible, so as not to change the current that you are trying to measure.

QUESTION Suppose that you are trying to measure the current through the bulb in the circuit shown in Figure 19.36. What is wrong with the placement of the ammeter?

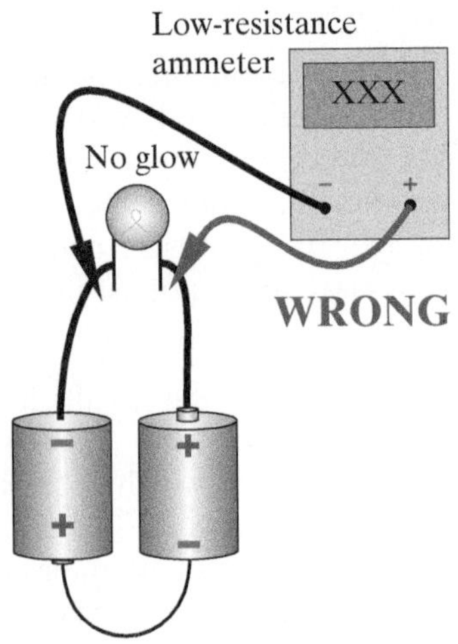

Figure 19.36 This ammeter is placed incorrectly in the circuit.

Connecting the ammeter in parallel to the bulb means that there is a very low-resistance path for the current (a "short circuit"). Most current will go through the ammeter, with very little current going through the bulb. The bulb no longer glows, and the large current measured by the ammeter has nothing to do with the tiny current that goes through the bulb but rather is a measure of the maximum current these batteries can deliver.

The fact that an ammeter must be inserted into a circuit may not seem like a problem with simple circuits, but it can be a big problem if the circuit of interest is built in such a way that you can't break in to insert your ammeter without damaging the circuit. Moreover, a commercial ammeter may blow a fuse or even be permanently damaged if used incorrectly as shown in Figure 19.36, because a current of several amperes will run through the very low-resistance ammeter, and many commercial ammeters cannot handle such large currents.

Voltmeters Measure Potential Difference

If we add a series resistor to an ammeter, in principle we can make a voltmeter to measure potential differences. Suppose that we attach an ammeter across a circuit element, through a resistor whose resistance R is very large (Figure 19.37).

> QUESTION If the measured current through the ammeter is I, what is the potential difference $V_A - V_B$ in this circuit, in terms of R and I?

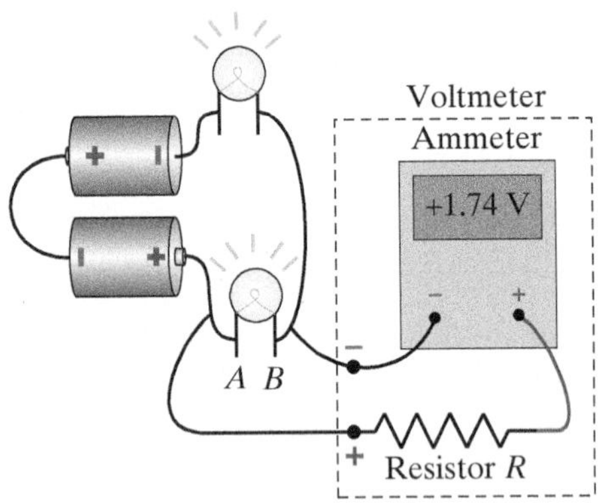

Figure 19.37 An ammeter with a resistor in series forms a voltmeter.

The potential difference along the resistor is $V_A - V_B = RI$. A small current I running through the resistor is measured by the ammeter, and the potential difference is RI. If the ammeter is relabeled in volts (RI), the device reads potential differences directly. A voltmeter can be thought of as an ammeter with a series resistance, all in one box (shown inside dashed lines in Figure 19.37).

Notice that a voltmeter has two leads and measures potential difference, not potential! Commercial voltmeters have their + and − sockets arranged as follows:

VOLTMETER SIGN CONVENTION

If the potential is higher at the socket marked "+" a voltmeter indicates a positive potential difference.

If the − socket is at the higher potential, the voltmeter displays a *negative* potential difference. The meter in Figure 19.37 will read positive.

An ammeter should have very small resistance so that inserting the ammeter into a circuit should disturb the circuit as little as possible. A voltmeter, however, should have as large a resistance as possible, so that placing the voltmeter in parallel with a circuit element will disturb the circuit as little as possible, with only a tiny current being deflected through the voltmeter.

A VOLTMETER SHOULD HAVE HIGH RESISTANCE

A voltmeter (used in parallel) should have very large resistance.
An ammeter (used in series) should have very small resistance.

There is a practical limit to the voltmeter resistance. The larger the resistance of the voltmeter, the more sensitive must be the ammeter part of the voltmeter. Commercial voltmeters often have resistances as high as 10 MΩ (10 megohms or 10×10^6 ohms), and the ammeter portion of the voltmeter is correspondingly sensitive. Even a commercial voltmeter with a resistance of 10 MΩ is useless for measuring potential differences in a circuit consisting of 100 MΩ resistors, because placing a 10 MΩ voltmeter in parallel with a 100 MΩ resistor drastically alters the circuit.

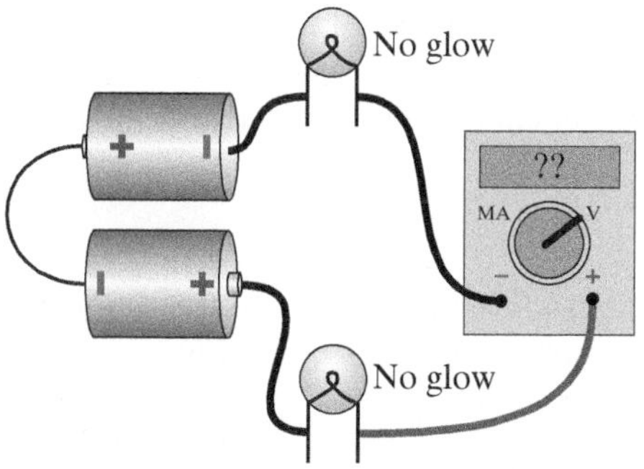

Figure 19.38 This voltmeter is placed incorrectly in the circuit.

Checkpoint 6 Two lab partners measured the current in a series circuit by placing a digital multimeter (set to read current in milliamperes) in series between the two thick-filament bulbs (Figure 19.38). Then they wanted to measure the voltage across a bulb, so they simply switched the multimeter to the V setting (volts). They were surprised that the bulbs stopped glowing. What did they do wrong? What did their voltmeter read?

Ohmmeters

Commercial multimeters can act not only as ammeters and voltmeters but also as ohmmeters to measure the resistance of an element that has been removed from a circuit. The ohmmeter section of the multimeter consists of an ammeter in series with a small voltage source (for example, 50 mV). When you connect an unknown resistor to the ohmmeter, the small voltage drives a small current through the resistor and ammeter. The ohmmeter is calibrated to display the applied voltage divided by the observed current, which is the resistance (assuming that the resistance of the ammeter is small compared to the resistance you are trying to measure). You can see that you have to remove the resistor of interest from its circuit before you can use the ohmmeter, because an ohmmeter is an active device with its own voltage source.

How Commercial Meters Are Really Constructed

We have seen that one could construct a voltmeter from a sensitive ammeter with a large resistor in series. Modern digital voltmeters read voltage directly, by timing how long it takes a known current to discharge a capacitor! A digital ammeter is basically a voltmeter with a very small resistance in parallel, and the voltmeter reads the very small potential difference across the small resistance. A digital ohmmeter drives a known current through the unknown resistor and measures the resulting potential difference. Independent of the construction details, it is important to understand that an ammeter has very low resistance, and a voltmeter has very high resistance.

19.7 QUANTITATIVE ANALYSIS OF AN RC CIRCUIT

We are now in a position to analyze quantitatively a series circuit containing an ideal battery with known emf, a resistor R, and a capacitor C—a so-called "RC" circuit. Recall that the potential difference across a capacitor is $\Delta V = Q/C$, where Q is the charge on the positive plate and C is the capacitance (Section 19.1; for a parallel-plate capacitor, $C = K\varepsilon_0 A/s$, where A is the area of one of the plates, s is the gap distance between the plates, and K is the dielectric constant of the material filling the gap, as discussed in Chapter 16). The energy equation for the RC circuit in Figure 19.39 is

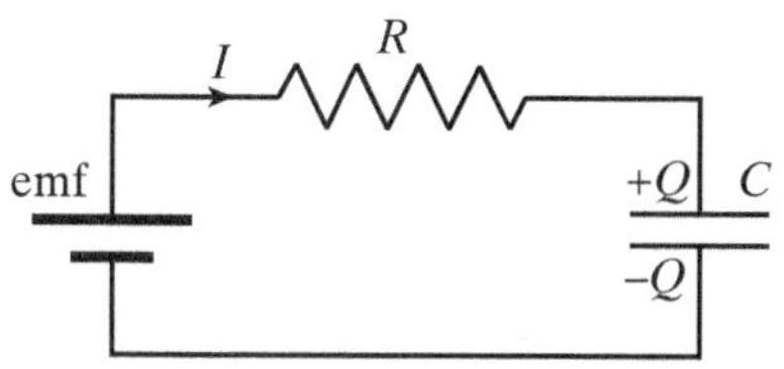

Figure 19.39 An RC circuit.

$$\Delta V_{\text{round trip}} = \text{emf} - RI - Q/C = 0$$

The Final State

The simplest situation to analyze is the final equilibrium state of the circuit, in which the current has dropped to zero, and the capacitor is fully charged. Then the term RI is zero, and the loop equation gives this:

$$\Delta V_{\text{round trip}} = \text{emf} - Q/C = 0$$
$$Q = \text{emf} \cdot C$$

We see that the final charge on the capacitor depends only on the emf and the capacitance. The resistor in the circuit determines how long it takes to reach equilibrium, but it has no effect on the final charge on the capacitor.

The Initial Situation

Recall what happens when a battery and thick-filament bulb are connected in series to a (discharged) capacitor. Immediately after making the final connection the light bulb is very bright (and then it gets dimmer). The initial brightness looks about the same as you get in a circuit containing only a thick-filament bulb without a capacitor.

Let's see how we can understand this quantitatively. Rearrange the energy equation to solve for I in terms of the other quantities:

$$I = \frac{\text{emf} - Q/C}{R}$$

> QUESTION Use this equation for I to explain why the initial brightness is the same as though the capacitor weren't there. (*Hint:* What is Q initially?)

The capacitor is initially uncharged, so initially $Q = 0$, and plugging this value for Q into the equation for I gives us $I = \text{emf}/R$. This is the same current we would have in a simple circuit without the capacitor. Having understood the initial situation, let's see what happens next.

The rate at which charge Q accumulates on the positive capacitor plate is equal to $I = dQ/dt$. To put it another way, in a time dt the plate receives an additional amount of charge $dQ = Idt$. We rewrite the energy equation:

$$I = \frac{dQ}{dt} = \frac{\text{emf} - Q/C}{R}$$

$$dQ = \left(\frac{\text{emf} - Q/C}{R}\right) dt$$

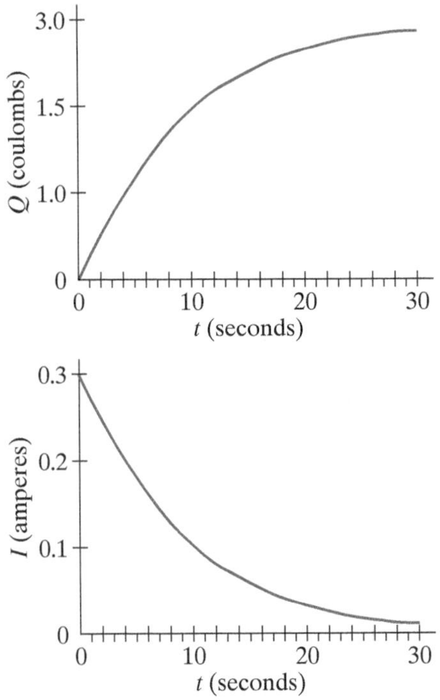

Figure 19.40 Charge and current as a function of time.

This is in a form suitable for doing a numerical integration. At each instant, we know Q (initially zero), so we can calculate the change dQ from this equation. Then $Q + dQ$ is the new value of Q, and we can use this new Q to calculate a new value of dQ, and so on.

Numerical Integration

Let's take a couple of steps in this numerical integration to see how things go. In the first time interval dt, $dQ = (\text{emf}/R)dt$, since the initial value, Q_0, is zero. The new value, Q_1, is

$$Q_1 = Q_0 + dQ = (\text{emf}/R)dt$$

Now the capacitor has a small nonzero charge Q_1. In the second time interval, we have

$$dQ = \left(\frac{\text{emf} - Q_1/C}{R}\right) dt$$

$$Q_2 = Q_1 + dQ$$

Because Q_1 is nonzero, the increase in charge dQ is smaller this time. As time goes on, and Q gets larger and larger, dQ for each additional time step will be smaller and smaller. Therefore a graph of Q vs. time will look like the upper graph of Figure 19.40. The graph is the result of a numerical integration using

$R = 10\ \Omega$ (approximately the resistance of a glowing thick-filament bulb) and $C = 1.0$ F, with two 1.5 V batteries in series.

The lower graph of Figure 19.40 is the current flowing into the resistor, calculated as $I = dQ/dt$. As Q increases, I decreases, because the fringe field of the capacitor increasingly opposes the current flow. In terms of potential, as Q increases, the potential difference across the capacitor increases, which means that the potential difference across the resistor must decrease, and therefore $I = \Delta V_{\text{resistor}}/R$ decreases.

Analytical Solution

The graph in Figure 19.40 looks like some kind of exponential function. Let's make a guess that the current is in fact given by

$$I = I_0 e^{-at}$$

where I_0 is the initial current (which we found earlier to be $I_0 = \text{emf}/R$), and a is an unknown constant. Consider this line of reasoning:

$$I = \frac{dQ}{dt} = \frac{\text{emf} - Q/C}{R} \quad \text{(energy equation in terms of } I \text{ and } Q\text{)}$$

$$\frac{dI}{dt} = -\frac{1}{RC}\frac{dQ}{dt} = -\frac{1}{RC}I \quad \text{(since emf}/R \text{ is constant)}$$

$$\frac{dI}{dt} = \frac{d}{dt}(I_0 e^{-at}) = -aI_0 e^{-at} = -aI = -\frac{1}{RC}I \quad \text{(using our guess for } I\text{)}$$

This last equation is true for all time if $a = 1/RC$, so we have a solution:

CURRENT IN A SERIES RC CIRCUIT

$$I = \left(\frac{\text{emf}}{R}\right) e^{-t/RC}$$

This solution has all the right properties. When $t = 0$, the exponential is equal to 1, and $I = \text{emf}/R$, so the initial brightness is indeed the same as if there were no capacitor. As t becomes very large, the exponential goes to 0, and the current goes to zero, which is what is observed in an RC circuit.

We used a very common and powerful method for finding an analytical solution for a differential equation (an equation involving derivatives): Guess the form of the solution (containing unknown constants such as a), plug this form and its derivatives into the differential equation, and determine what values the constants must have to satisfy the equation ($a = 1/RC$).

Since $I = dQ/dt$, $Q = \int_0^t I dt$, and it is easy to show this:

CHARGE IN A SERIES RC CIRCUIT

$$Q = C(\text{emf})\left[1 - e^{-t/RC}\right]$$

This solution has all the right properties. When $t = 0$, the exponential is equal to 1, and $Q = 0$ (no charge yet on the positive plate of the capacitor). As t becomes very large, the exponential goes to 0, and Q approaches C(emf). This is correct: $\Delta V_{\text{capacitor}} = Q/C = C(\text{emf})/C = \text{emf}$. When the current stops, the potential difference across the capacitor becomes equal to the potential difference across the battery.

QUESTION What is the numerical value of the final charge on one plate of the capacitor in this circuit, in terms of coulombs?

We find that $Q_F = C \times \text{emf} = (1.0\ \text{farad})(3\ \text{V}) = 3.0\ \text{C}$. That's a huge amount of charge! Compare, for example, with the amount of charge on a strip of invisible tape, which is of the order of 1×10^{-8} C.

Actually, our analytical solution of the differential equation is physically only an approximation, because as the current changes the resistor changes temperature and changes resistance. This is particularly significant when the resistor is a light bulb, since its resistance drops a great deal as the current decreases and it cools off.

The *RC* Time Constant

A rough measure of how long it takes to reach final equilibrium is the "time constant" RC. When the time $t = RC$, the factor $e^{-t/RC}$ has fallen from the value $e^0 = 1$ to the value

$$e^{-t/RC} = e^{-1} = \frac{1}{e} = \frac{1}{2.718} = 0.37$$

QUESTION Calculate the time constant for a circuit with $R = 10\ \Omega$ (approximately the resistance of a glowing thick-filament bulb) and $C = 1.0$ F. This is the time when the current has decreased from the original 0.30 A to 0.11 A (0.37 times 0.30 A).

The time constant $RC = (10\ \Omega)(1.0\ \text{F}) = 10$ s. The bulb will be very dim with this current. This is roughly consistent with what is observed with a thick-filament bulb and a 1 F capacitor, which loses most of its brightness in this amount of time. Figure 19.41 shows graphs of Q vs. t and I vs. t for the RC circuit with the thick-filament bulb, with the 10 s RC time constant indicated.

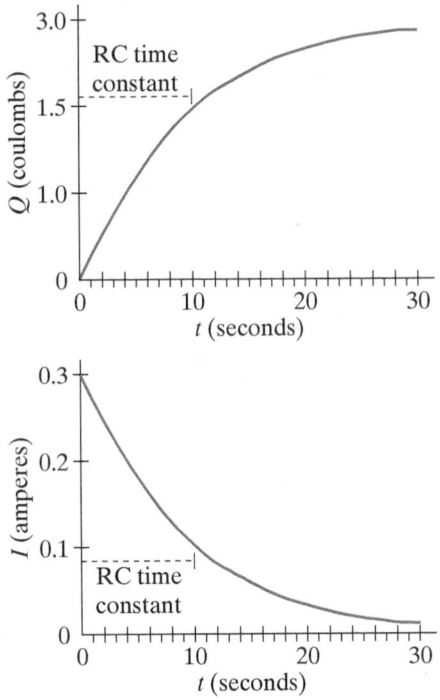

Figure 19.41 Charge and current as functions of time.

QUESTION Show that the power dissipated in the bulb at $t = RC$ is only 14% of the original power.

The power dissipated in the bulb is $I\Delta V = RI^2$, so a reduction in the current by a factor of 0.37 gives a reduction in the power by a factor of $(0.37)^2 = 0.136$.

QUESTION Calculate the time constant for a circuit in which you replace the thick-filament bulb with a thin-filament bulb (whose resistance R when glowing is typically about 30 Ω, which may vary from one bulb to another).

The time constant $RC = (30\ \Omega)(1\ \text{F}) = 30$ s, which agrees roughly with observations with a 1 F capacitor.

We now have a complete quantitative description of the behavior of an RC circuit. In experiments where one times how long it takes for the current to decrease to a point where the bulb no longer glows, this is not the same as timing how long it takes for the current to get smaller by a factor of e^{-1}.

19.8 REFLECTION: THE MACRO–MICRO CONNECTION

In this chapter we made connections between a microscopic and macroscopic view of resistance. In our original microscopic view, individual electrons acquire a drift speed proportional to the electric field, with $\bar{v} = uE$. This electric field is made by all the charges in and on the battery and on the surface of the conducting circuit elements. Electron current i is measured in number of electrons per second.

In the macroscopic view, the electric field made by all the charges produces a conventional current density $J = \sigma E$ and a conventional current $I = JA$ measured in coulombs per second (amperes). The integral of the electric field is potential difference, and any round-trip integral of the electric field due to point charges must equal zero, which represents energy conservation per unit charge. We also used a conventional-current version of the steady-state current node rule.

19.9 *WHAT ARE AC AND DC?

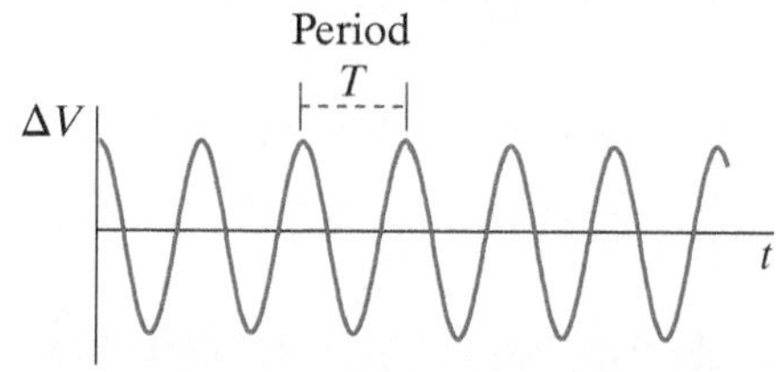

Figure 19.42 The voltage between two slots in a wall socket varies sinusoidally.

The steady constant current in a circuit consisting of a battery and a resistor is called "direct current" or DC. Home appliances such as a floor lamp use "alternating current" or AC. Two slots of the wall socket supply a sinusoidally varying voltage ("AC voltage") that drives a sinusoidally varying current in the lamp (Figure 19.42). (A third wall socket connection goes to ground for safety reasons.) In many countries including the United States the frequency is 60 Hz (60 hertz means 60 cycles of a sine wave every second, with a period $T = 1/60$ s). Many other countries, including those in Europe, use 50 Hz AC. For reasons discussed below, what is called "110 volts AC" is actually a voltage that varies sinusoidally between −155 volts and +155 volts.

Although current in an incandescent floor lamp rises and falls sinusoidally, the light doesn't seem to flicker for two reasons. First, humans have difficulty perceiving flickering that occurs 50 or 60 times per second. For example, movies are presented in the form of 24 still pictures per second, but you perceive a continuous image. Second, the temperature of an incandescent bulb filament cannot change rapidly; there isn't time in a fraction of a 60 Hz cycle for the filament to cool down. However, with a fluorescent lamp, where the light is produced nonthermally in sync with the applied voltage, you may see strobe-like images of your fingers if you wave your fingers rapidly back and forth while looking through your fingers at the lamp.

Very high voltages (100,000 V or more) are used in power transmission lines to reduce energy losses between a generating station and a remote city. If R is the resistance of the transmission line, you want the voltage drop RI along the line to be small, so you want the current I to be small. The power delivered to the line is $I\Delta V$, so for I to be small you need ΔV to be large. However, very high voltages would be extremely dangerous in the home. It is relatively cheap and efficient to convert high AC voltages to low AC voltages using "transformers" whose operation is based on Faraday's law, which we will study in a later chapter. Until quite recently, converting high DC voltages to low DC voltages was difficult and inefficient (though this is changing with modern semiconductor devices) so historically it made sense to use AC in distributing electrical power.

Electronic devices such as computers need low DC voltages. The small box between a wall socket and a laptop uses semiconductor elements to convert the AC voltage to 12 volts DC; similar AC to DC converters are built into other devices such as desktop computers and radios.

AC Power

The reason a sinusoid varying between −155 V and +155 V is called "110 volts AC" is that it delivers the same energy in one second as would a constant 110 V. Consider a circuit containing an AC power supply and a resistor, such as a floor lamp. With $\Delta V = \Delta V_{\max}\cos(\omega t)$, we'll integrate the instantaneous power over one cycle (with period T and $\omega = 2\pi/T$) and see how much energy is delivered to a resistor R in one cycle:

$$\begin{aligned}\text{energy} &= \int_0^T \frac{\Delta V^2}{R}dt \\ &= \frac{\Delta V_{\max}^2}{R}\int_0^T \cos(\omega t)^2 dt \\ &= \frac{\Delta V_{\max}^2}{R}\int_0^T \frac{1}{2}[1+\cos(2\omega t)]dt \quad \text{(using a trig identity)}\end{aligned}$$

$$\text{energy} = \frac{\Delta V_{max}^2}{R}\frac{1}{2}T$$

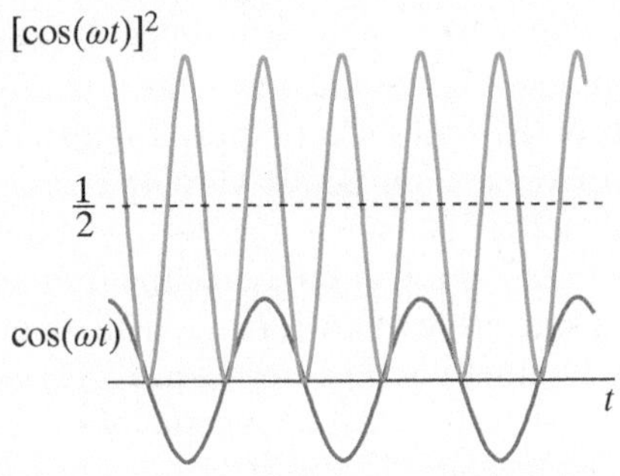

Figure 19.43 The average value of the square of a cosine is 1/2.

The average power is the energy delivered in time T divided by the time T:

$$\text{average power} = \frac{1}{2}\frac{\Delta V_{max}^2}{R}$$

The factor of 1/2 comes ultimately from the fact that the average value of the square of a cosine is 1/2 (Figure 19.43).

The "root mean square" (rms) voltage is defined as $\Delta V_{rms} = \sqrt{\frac{1}{2}}\Delta V_{max}$, in which case the average power can be written in a form that looks like the equation for DC power:

$$\text{average power} = \frac{\Delta V_{rms}^2}{R}$$

The "110 volts AC" available at the wall socket means $V_{rms} = 110$ V, which corresponds to $\Delta V_{max} = \sqrt{2}\Delta V_{rms} = 155$ V.

Phase Shift

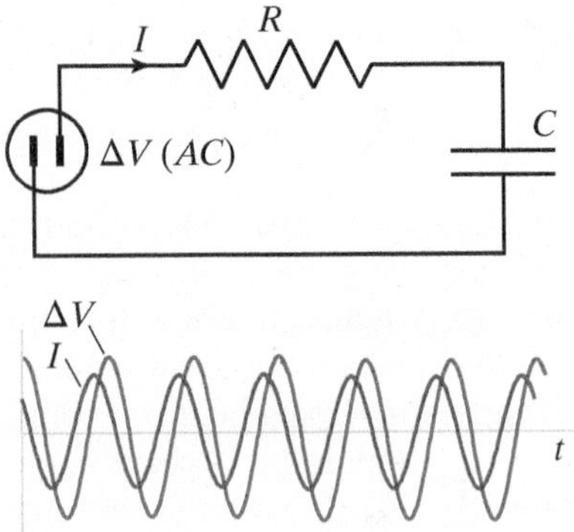

Figure 19.44 An AC voltage is applied to an RC circuit. The current in the circuit peaks earlier than does the applied ΔV.

An unusual aspect of AC circuits is that the current need not be in phase with the voltage. For example, if an AC voltage is maintained across an RC circuit, this applied voltage peaks later than the current (Figure 19.44). We say that the current in the circuit "leads" the applied voltage. The energy equation (loop rule) is

$$\Delta V_{max}\cos(\omega t) - RI - \frac{Q}{C} = 0$$

where $Q = \int I dt$, or $dQ/dt = I$; the rate at which charge increases on a plate is equal to the charge per second entering the plate, which is the current I. Differentiating the energy equation with respect to t we have

$$-\Delta V_{max}\omega\sin(\omega t) - R\frac{dI}{dt} - \frac{I}{C} = 0$$

By substituting into the energy equation it is possible to show that the current is

$$I = I_{max}\cos(\omega t + \phi)$$

where ϕ is called the "phase angle" and expresses the result that the current I peaks at a different time than the applied voltage. The values for ϕ and I_{max} can be obtained from the following relations, which emerge from the solution process:

$$\tan\phi = \frac{1}{RC\omega}$$

$$I_{max} = \frac{\Delta V_{max}}{\sqrt{R^2 + 1/(C\omega)^2}}$$

The frequency-dependent factor $\sqrt{R^2 + 1/(C\omega)^2}$ is called the "impedance" and plays the role of resistance in this circuit. If the capacitance C or the angular frequency ω is very large, the impedance is approximately equal to R, and $I = \Delta V_{max}/R$, just as in a simple resistor circuit. Also, in this case $\tan\phi \approx 0$ and $\phi \approx 0$, so the current is in phase with the applied voltage as it is for a DC circuit.

19.10 *ELECTRONS IN METALS

It is possible to show experimentally that the mobile charged particles in a metal are indeed electrons. For example, magnetic forces can lead to a transverse polarization of a current-carrying wire, and the sign of this polarization is different for positive and negative moving charges (this is called the "Hall effect," and we will study this in Chapter 20).

A different kind of experiment was carried out in 1916–1917 by Tolman and Stewart, which showed that the mobile charge carriers in a metal are indeed electrons, which was not known for certain before their experiment. We will describe a simpler experiment that is conceptually equivalent to the experiment they actually carried out.

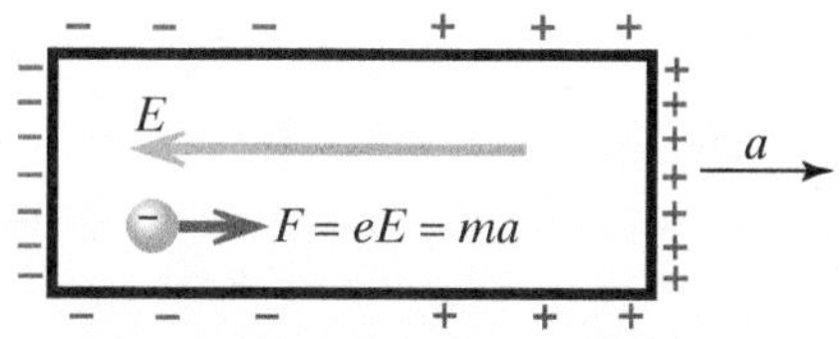

Figure 19.45 In an accelerated metal bar the mobile electrons are initially left behind, polarizing the bar.

Suppose that we accelerate a metal bar, which means that we accelerate the atomic cores, which are bound to each other (Figure 19.45).

On the average, the mobile electrons normally feel no force; the mutual repulsion of the mobile electrons is effectively canceled by the attraction to the atomic cores. Therefore as the bar and atomic cores accelerate, the mobile electrons are left behind. As a result, they pile up at the trailing end of the bar (they can't easily leave the bar). This polarizes the bar, with negative charge on the trailing surface and positive charge (deficiency of electrons) on the leading surface, as shown in Figure 19.45.

The surface charges produce an electric field E to the left, which means that there is a force eE to the right on electrons inside the metal (Figure 19.45). As the polarization increases, E increases until the electric force eE is equal to the mass of the electron times the acceleration of the metal bar, at which point the acceleration of the mobile electrons matches the acceleration of the bar, and then the polarization no longer increases.

The observed direction of the polarization (negative at the left end) led the experimenters to conclude that the mobile particles are indeed negatively charged. You should run through the argument to see that if the movable charges had been positive, the trailing end of the bar would have been positive instead of negative.

Although this experiment does not give separate values for the charge and the mass of the mobile particles, it does provide a measurement of the ratio q/m. There had already been measurements of the ratio e/m for electrons moving in a vacuum, in a device similar to an old television cathode-ray tube. The modern value for this ratio for the electron is about

$$\frac{e}{m} = \frac{(1.6 \times 10^{-19}\,\text{C})}{(9 \times 10^{-31}\,\text{kg})} = 1.7 \times 10^{11}\,\text{C/kg}$$

Tolman and Stewart measured the acceleration a, and they used a voltmeter to measure the potential difference ΔV from one end of the metal bar to the other. The potential was lower at the left end of the bar, which shows that the mobile charge carriers are negative. Moreover, the experimental value of q/m had the same value as that obtained for electrons in other experiments. This was very strong evidence that currents in metals consist of moving electrons.

Checkpoint 7 If the length of the bar is L, express q/m in terms of the quantities measured by Tolman and Stewart (a and ΔV).

19.11 *A COMPLICATED RESISTIVE CIRCUIT

We can use energy conservation (loop equations) and the current node rule to determine how much current would flow in each element of a complicated multiloop circuit made up of batteries and ohmic resistors (resistors for which the current is proportional to the potential difference). Analyzing complex

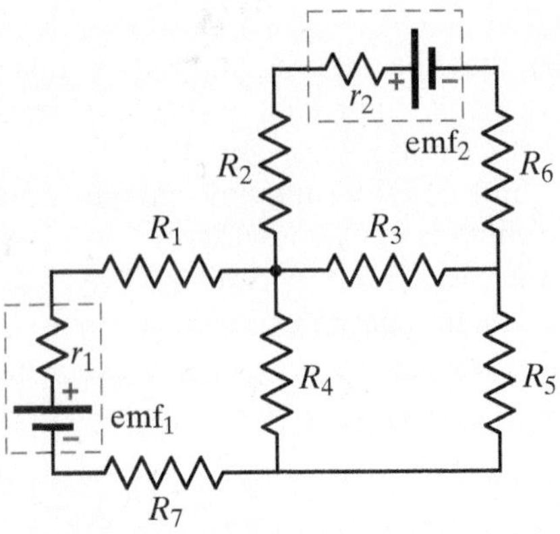

Figure 19.46 A circuit with unknown currents, which we will determine.

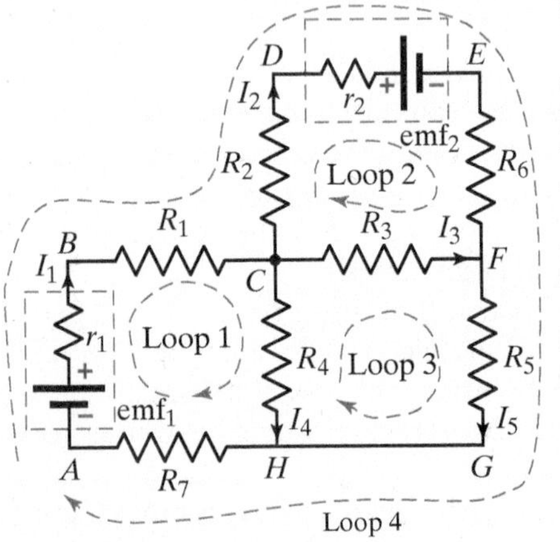

Figure 19.47 Name the unknown currents, draw paths along which to apply energy conservation.

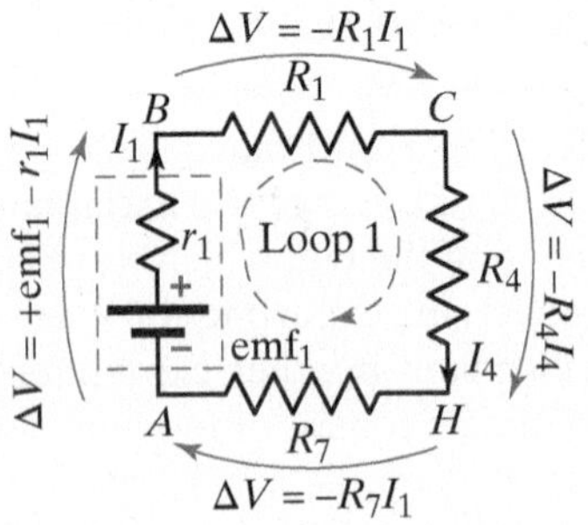

Figure 19.48 Apply energy conservation for a round-trip path, loop 1 ($ABCHA$).

circuits like the one discussed in this section is more a matter of technology than fundamental physics, so we won't analyze many such circuits. However, it is useful to analyze at least one such circuit as another illustration of the application of energy conservation and the current node rule. We will use the loop and node rules to generate as many independent equations as we have unknown currents, and then solve the system of equations. Assume that the emf's, battery internal resistances, and R's are known for the circuit shown in Figure 19.46, but the currents are unknown. We will determine the current in each part of this circuit. The first step in the solution is to assign names and directions to the unknown currents (I_1 through resistor R_1, I_2 through resistor R_2, etc.). We don't know what the actual directions of these unknown conventional currents are, so we guess, and we record our guesses on a diagram, as shown in Figure 19.47. We also draw possible paths to follow for writing loop equations for energy conservation.

We don't have to assign a different current variable to every branch. Consider conventional current I_1. On the circuit diagram we have assumed that it flows from A toward B. This also means that we have assumed the same amount of current I_1 goes from B to C through resistor R_1, and from H to A through resistor R_7. Similar comments apply to the other currents. These simplifications are trivial applications of the current node rule.

First let's use energy conservation for loop 1 (the path $ABCHA$). We can start anywhere around the loop; let's start at location A and walk around the loop, adding up the potential differences we encounter on our walk (Figure 19.48). The potential rises across the battery (the electric field inside the battery is opposite to the direction we're walking) and the potential falls along the resistors, since we are walking in the assumed direction of the conventional current, which runs in the direction of the electric field from higher potential to lower potential.

When we get back to our starting location at A, the net potential change along our walk should be zero, since potential difference in a round trip is zero (and $V_A - V_A = 0$):

$$\text{Loop 1:} \quad (\text{emf}_1 - r_1I_1) + (-R_1I_1) + (-R_4I_4) + (-R_7I_1) = 0$$

Note that in writing this loop rule we implicitly used the current node rule to show that the current is the same I_1 through the battery, r_1, R_1, and R_7. We also used the property of ohmic resistors to be able to relate the current through R_1 to the potential difference along R_1 (as $-R_1I_1$).

If I_1 turns out to be -1.3 A when we solve for the unknown currents, that will merely mean that we guessed the wrong direction and that I_1 actually runs from C to B (and that C is at a higher potential than B). Thus there is no real penalty for guessing the wrong current directions.

Next consider loop 2 (path $CDEFC$), starting at location C (Figure 19.49). When we get back to our starting location at C, the net potential change along our walk should be zero:

$$\text{Loop 2:} \quad (-R_2I_2) + (-\text{emf}_2 - r_2I_2) + (-R_6I_2) + (R_3I_3) = 0$$

Comments on loop 2: From D to E there is a potential drop (not a rise) across the battery numerically equal to ($-\text{emf}_2 - r_2I_2$), because we are traversing the battery from its high-potential + end to its low-potential – end, in the direction of the electric field inside the battery. (If we've guessed right about the direction of I_2, this battery is being charged rather than discharged.) From F to C there is a potential *rise* $+R_3I_3$ (not a drop) because we are heading upstream against the current I_3. That is, the assumed direction of I_3 (from C to F) implies that the potential at C is higher than the potential at F. Again,

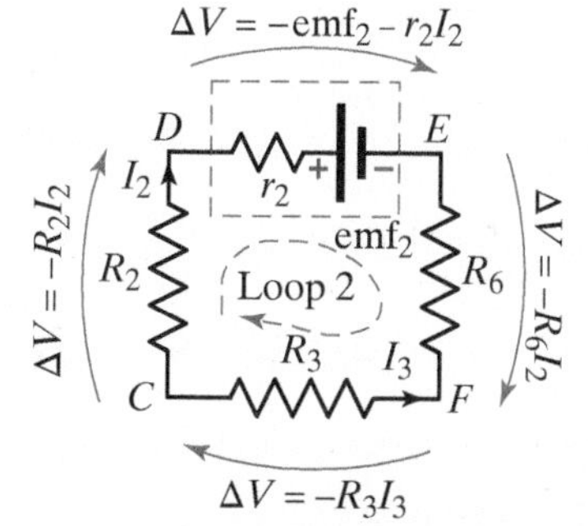

Figure 19.49 Apply energy conservation for a round-trip path, loop 2 (*CDEFC*).

if we have made the wrong assumption, all that will happen is that the final numerical answer for I_3 will have a negative sign.

QUESTION In the same manner, use energy conservation for loop 3 (path *HCFGH*) and for loop 4 (path *ABCDEFGHA*).

The equation for loop 4 around the outside of the circuit is the algebraic sum of the equations for the other three loops and is not an independent equation. Only three of these four loop equations are actually significant in solving for the unknown currents, although it doesn't matter which three we choose to use in the solution.

Loop 3: $(R_4I_4)+(-R_3I_3)+(-R_5I_5)=0$

Loop 4:

$$(\text{emf}_1 - r_1I_1)+(-R_1I_1)+(-R_2I_2)+(-\text{emf}_2 - r_2I_2)+(-R_6I_2)+(-R_5I_5)+(-R_7I_1)=0$$

QUESTION Now write current node rule equations for nodes *C*, *H*, and *F*, paying attention to signs (+ for incoming current, − for outgoing current).

Adding the equations for node *C* and node *F* gives an equation that is equivalent to the equation for node *H*. Only two of these three node equations are actually independent, and we can use any two of them in solving for the unknown currents.

Node *C*: $I_1 - I_2 - I_3 - I_4 = 0$

Node *F*: $I_2 + I_3 - I_5 = 0$

Node *H*: $I_4 + I_5 - I_1 = 0$

We have a total of five independent equations (three independent loop equations and two independent node equations).

It can be shown that in any resistive circuit there are as many independent loop equations as there are minimal-sized loops (that is, ignoring combined loops such as loop 4). The number of independent node equations can be determined by subtracting the number of independent loop equations from the number of independent currents.

Our five equations are sufficient to be able to solve for the five unknown currents in terms of the known emf's and known resistances. This solution step can involve extremely tedious algebraic manipulation and should usually be turned over to a symbolic manipulation package such as Maple or Mathematica. Of course in some very simple cases it may be easy to solve the equations, but in general the algebra can be quite forbidding.

Once the unknown currents have been determined (typically by using a computer program to solve the equations), we can know essentially everything there is to know about the circuit. For example, we can calculate the potential drop $\Delta V = -RI$ across each resistor, since we know R and I for each resistor. We can therefore also calculate the potential difference between any two locations in the circuit.

Checkpoint 8 Refer to Figure 19.47, and suppose that $\text{emf}_1 = 12\text{ V}$ and $R_1 = 20\,\Omega$, and that after we solved the equations we found that I_1 was 0.4 A. What is the potential difference $V_C - V_B$? How much power is dissipated in R_1? What is the power expenditure of battery 1 (some of which is dissipated in the battery itself due to internal resistance r_1)?

SUMMARY

New Concepts

In a capacitor circuit there is a slow transient leading to a final equilibrium state.

Current can run (for awhile) even if there is a gap in a circuit. The fringe field of the capacitor in the neighboring wires is critical for understanding the behavior of a capacitor circuit in terms of charge and electric field.

Conventional current I:

- is in the direction of motion of positive carriers
- is measured in coulombs per second

Conductivity $\sigma = |q|nu$ is a property of a material;

$$J \equiv \frac{I}{A} = \sigma E$$

Ohms and nonohmic resistors

Results

Charging and discharging times are larger with larger plates, smaller gap, or with an insulator in the gap.

Resistance $R = \frac{L}{\sigma A}$ incorporates the geometry of a resistor.

$I = \frac{\Delta V}{R}$ is the current through a resistor.

$\Delta V = RI$ is the potential difference across a resistor.

Power $= I\Delta V$ for any circuit element.

Potential difference across a real, nonideal battery (Figure 19.50):

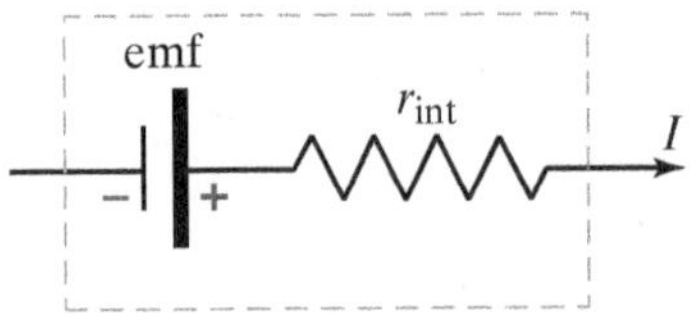

Figure 19.50 Equivalent nonideal battery.

Capacitors in circuits can be analyzed in terms of $\Delta V = Q/C$, where C is the capacitance, and the charge Q and current I vary exponentially with time, with a time constant RC.

Capacitance $C = Q/|\Delta V|$ depends on the geometry of the capacitor (Section 19.1; for a parallel-plate capacitor, $C = K\varepsilon_0 A/s$, where A is the area of one of the plates, s is the gap distance between the plates, and K is the dielectric constant of the material filling the gap).

EXPERIMENTS

Capacitors in Circuits

EXP1 Charging and discharging a capacitor

Find the capacitor in your experiment kit or lab (your instructor will tell you what it looks like).

> **Caution: Do not exceed the voltage rating marked on the capacitor—typically a few volts. (If the rating is 2.5 V, it is normally okay to use two ordinary 1.5 V batteries.)**

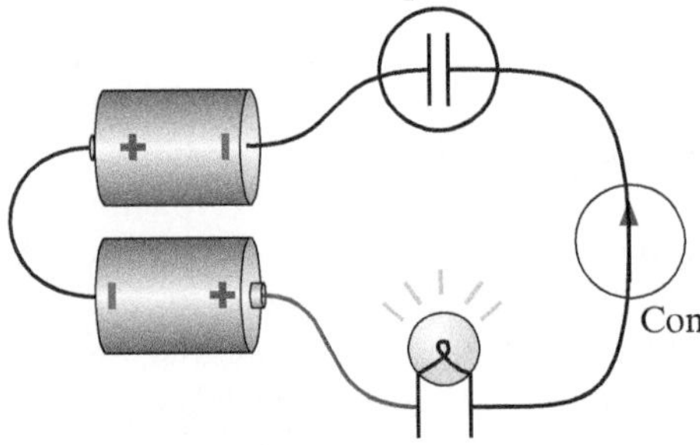

Figure 19.51 Charging a capacitor.

- To make sure that the capacitor is initially uncharged, start by connecting a wire across the capacitor for a couple of seconds in order to discharge the capacitor.
- Figure 19.51: Connect the discharged capacitor in series with a thick-filament bulb (#14), and run the wire over a compass. The process is called "charging the capacitor." What do you observe?
- After the bulb stopped glowing, was there still current flowing? (You may need to place the wire under the compass in order to see small deflections easily.)
- Figure 19.52: Remove the batteries and connect the bulb directly to the capacitor. The process is called "discharging the capacitor." What do you observe?

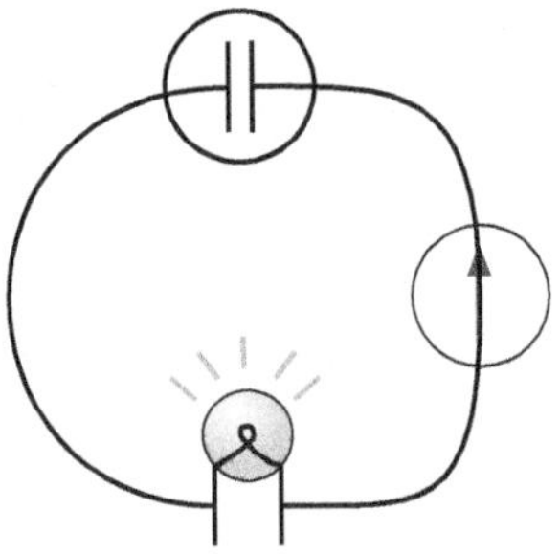

Figure 19.52 Discharging a capacitor.

- Repeat both stages of the experiment (charging and discharging), but this time use a thin-filament bulb (#48) with the two batteries and the capacitor. What difference does the choice of bulb make?
- Sketch graphs of current through the bulb vs. time for both kinds of bulbs. On your graphs label the time axis, and note which curve refers to which bulb. Pay special attention to the magnitude of the initial current (you may need to repeat the experiment to observe this).

You see that the time can be varied by varying the resistance. If you like, you can experiment with different lengths or thicknesses of your Nichrome wires, which provide a wide range of charging and discharging times.

EXP2 The effect of different bulbs on final charge
Design an experiment that will show how the final amount of charge on each plate of the capacitor depends on which bulb is involved in charging the capacitor. You might discuss your ideas with other students before carrying out your experiments. (If you wish, you can conduct your study using different lengths or thicknesses of Nichrome wire rather than two kinds of light bulbs.)

Be careful in your experiments. Before charging a capacitor be sure that it doesn't already have some charge on its plates: discharge it fully by connecting a wire across it for a couple seconds. Repeat any numerical measurements several times to make sure you know how reproducible the measurements are: 15.43 s and 15.67 s may not be significantly different!

Describe your experiments, results, and conclusions.

EXP3 Two capacitors in parallel
Work with another group so that you can try lighting a thick-filament bulb with two capacitors in parallel as shown in Figure 19.53. What do you predict for the charging time compared to a circuit with one capacitor? (Again, consider what happens in the first fraction of a second, and what effect this will have on the process in the next time interval.) What do you observe?

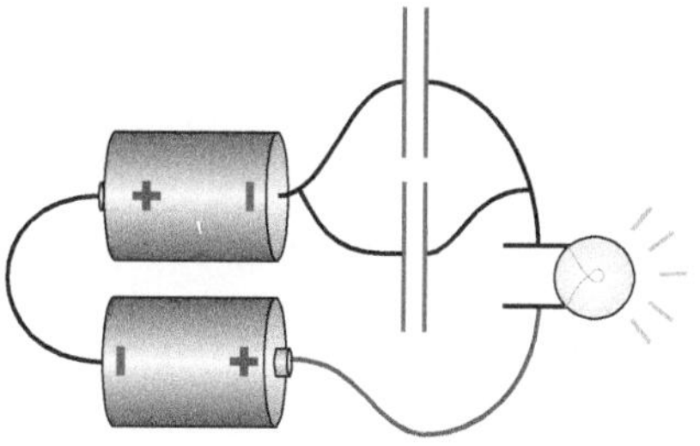

Figure 19.53

EXP4 Lighting an isolated bulb
Work with another group so that you can make a circuit with two capacitors as shown in Figure 19.54. Note that the thin-filament bulb is in a conducting "island" completely isolated from the batteries.

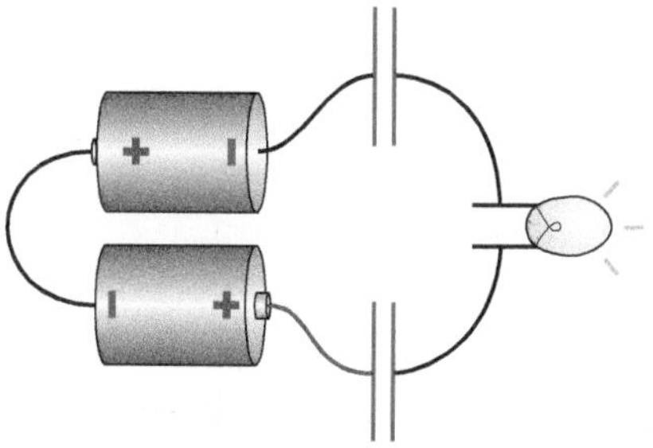

Figure 19.54

Before running the circuit, predict in detail what you expect will happen, and explain your reasoning. Try to work out agreement among the students in the two groups. Remember that during the few-nanosecond transient, the uncharged capacitors have no effect on the electric field.

Now construct the two-capacitor circuit and observe what happens. Discuss any discrepancies between the prediction you made and what you see. You might use compasses to verify your theory about the currents in the various parts of the circuit.

EXP5 Reversing the capacitor
Follow the following instructions carefully—otherwise you might damage the capacitor or the bulb.

Charge the capacitor using one battery and a thick-filament bulb (use only one battery!). Now that the capacitor is charged, what do you predict would happen if you reversed the connections to the capacitor, leaving all other connections intact, including the battery? Explain your reasoning, including an appropriate diagram.

Now try it. What do you observe? If the observation is different from your prediction, explain what was wrong with the prediction. Why is it important to use only one battery?

EXP6 The usefulness of a capacitor
Connect the capacitor and a thin-filament bulb in parallel rather than in series (Figure 19.55). Intermittently break the connection to the batteries for very brief fractions of a second. What happens to the brightness of the bulb during the very brief time that the batteries are disconnected, compared to what happens without a capacitor in the circuit? Explain.

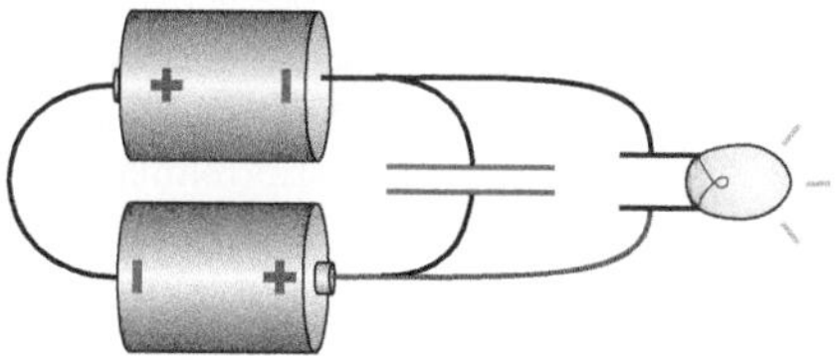

Figure 19.55

Internal Resistance of a Battery

EXP7 Two bulbs in parallel
Connect two thick-filament bulbs in parallel to two batteries in series. Unscrew one of the two bulbs. When you screw it back in, you may be able to see a slight change in the brightness of the other bulb. Why is this?

EXP8 Two batteries in parallel
Connect two batteries in parallel to a thin-filament bulb. You will see that the bulb has about the same brightness (dimness) as it has with just one battery, because the potential difference is essentially unchanged. What would you expect concerning how long the batteries would last, compared with just one battery? Why?

EXP9 The short-circuit current
Hold a copper connecting wire high enough above a compass that when the wire is connected by other connecting wires to short-circuit a single battery, you get about a 20° compass deflection. Try a different battery, trying not to move the wire. What compass deflections do you get for each battery? For the two batteries in series? For the two batteries in parallel? Are your measurements consistent with your predictions in the exercises above for the maximum current from batteries in series and in parallel?

QUESTIONS

Q1 How is the initial current through a bulb affected by putting a capacitor in series in the circuit? Explain briefly.

Q2 How is the charging time for a capacitor correlated with the initial current? That is, if the initial current is bigger, is the charging time longer, shorter, or the same? Explain briefly.

Q3 Consider two capacitors whose only difference is that the plates of capacitor number 2 are closer together than those of capacitor number 1 (Figure 19.56). Neither capacitor has an insulating layer between the plates. They are placed in two different circuits having similar batteries and bulbs in series with the capacitor.

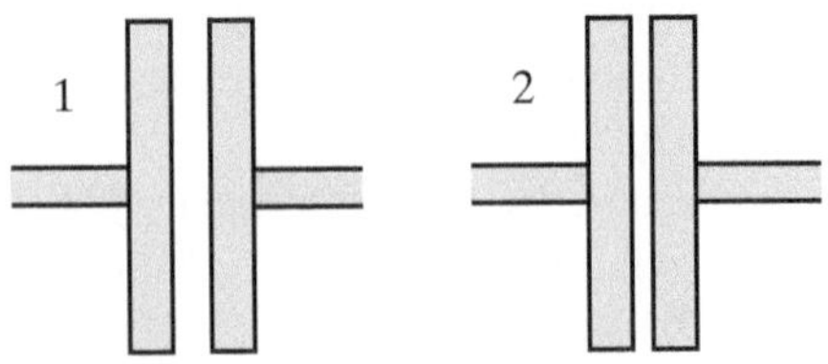

Figure 19.56

Show that in the first fraction of a second the current stays more nearly constant (decreases less rapidly) in the circuit with capacitor number 2. Explain your reasoning in detail. *Hint:* Show charges on the metal plates, and consider the electric fields they produce in the nearby wires. Remember that the fringe field near a plate outside a circular capacitor is approximately

$$\frac{Q/A}{\varepsilon_0}\left(\frac{s}{2R}\right)$$

A more extensive analysis shows that this trend holds true for the entire charging process: the capacitor with the narrower gap ends up with more charge on its plates.

Q4 The insulating layer between the plates of a capacitor not only holds the plates apart to prevent conducting contact but also has a big effect on charging. Consider two capacitors whose only difference is that capacitor number 1 has nothing between the plates, while capacitor number 2 has a layer of plastic in the gap (Figure 19.57). They are placed in two different circuits having similar batteries and bulbs in series with the capacitor.

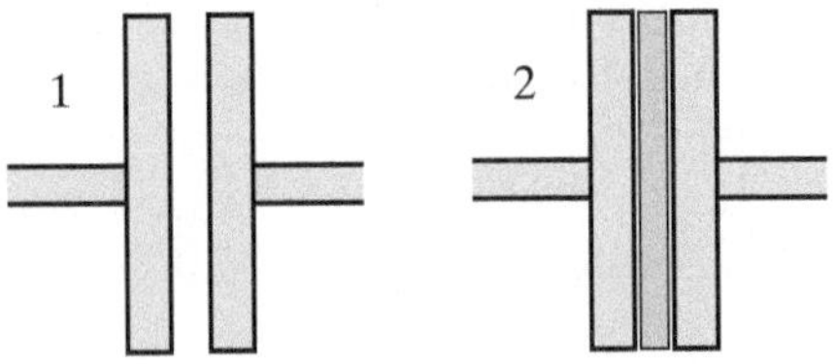

Figure 19.57

Show that in the first fraction of a second the current stays more nearly constant (decreases less rapidly) in the circuit with capacitor number 2. Explain your reasoning in detail. *Hint:* Consider the electric fields produced in the nearby wires by this plastic-filled capacitor.

Suppose that the plastic is replaced by a different plastic that polarizes more easily. In the same circuit, would this capacitor keep the current more nearly constant or less so than capacitor 2?

A more extensive analysis shows that this trend holds true for the entire charging process: the capacitor containing an easily polarized insulator ends up with more charge on its plates. The capacitor you have been using is filled with an insulator that polarizes extremely easily.

Q5 Suppose that instead of placing an insulating layer between the plates of the capacitor shown in Figure 19.57, you inserted a metal slab of the same thickness, just barely not touching the plates. In the same circuit, would this capacitor keep the current more nearly constant or less so than capacitor 2 in Question Q4? Explain why this is essentially equivalent to making a capacitor with a shorter distance between the plates.

Q6 How does the final (equilibrium) charge on the capacitor plates depend on the particular resistor (for example, the kind of bulb or the length of Nichrome wire) in the circuit during charging? Explain briefly.

Q7 How does the final (equilibrium) charge on the capacitor plates depend on the size of the capacitor plates? On the spacing between the capacitor plates? On the presence of a plastic slab between the plates?

Q8 Researchers have developed an experimental capacitor that takes about eight hours (!) to discharge through a thin-filament bulb. They propose using this in electric cars to provide a burst of energy for emergency situations (it would be charged at a slow rate during normal driving). Describe in general terms the key design elements of this extraordinary capacitor.

Q9 Give a complete but brief explanation for the behavior of the current during the discharging of a capacitor in a circuit consisting of a capacitor and light bulb. Include detailed diagrams. Explain, don't just describe.

Q10 Give a complete but brief explanation for the behavior of the current during the charging of a capacitor in a circuit consisting of batteries, bulb, and capacitor in series. Include detailed diagrams. Explain, don't just describe.

Q11 Figure 19.58 shows three circuits labeled A, B, and C. All the thin-filament bulbs, capacitors, and batteries are identical and are like the equipment you used in class. The capacitors are initially uncharged. In each circuit the batteries are connected for a short time T and then disconnected. The time T is only 10% of the total charging time through a single thin-filament bulb, so that the bulb brightness doesn't change much during the time T.

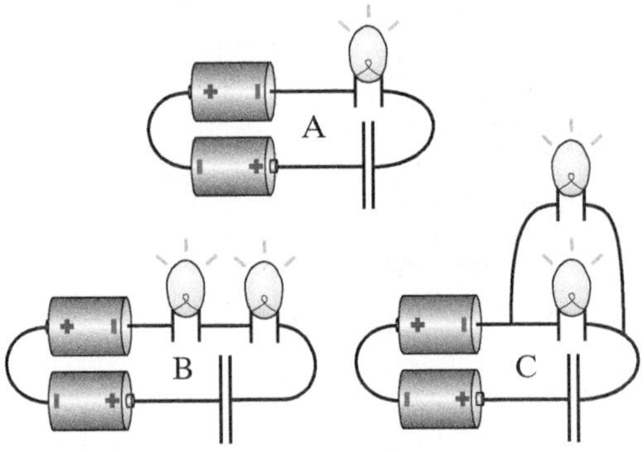

Figure 19.58

(a) In which circuit (A, B, or C) does the capacitor now have the most charge? Explain. **(b)** In which circuit (A, B, or C) does the capacitor now have the least charge? Explain. **(c)** Design and carry out experiments to check your answers. Describe your experiments and the numerical results. Before each experiment, connect a wire across the capacitor for a few seconds to fully discharge the capacitor. One way to compare the amount of charge stored in the capacitor during the time T is to finish charging it through a single thin-filament bulb, and see how much less time is required than when you start with a discharged capacitor.

Q12 The two circuits shown in Figure 19.59 have different capacitors but the same batteries and thin-filament bulbs. The capacitors in circuit 1 and circuit 2 are identical except that the capacitor in circuit 2 was constructed with its plates closer together. Both capacitors have air between their plates. The capacitors are initially uncharged. In each circuit the batteries are connected for a time that is short compared to the time required to reach equilibrium, and then they are disconnected. In which circuit (1 or 2) does the capacitor now have more charge? Explain your reasoning in detail.

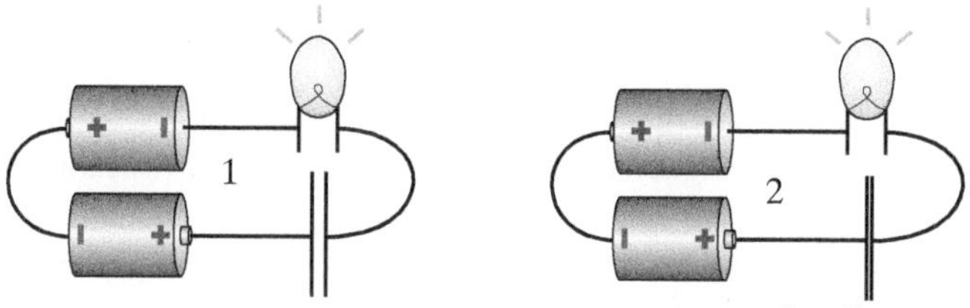

Figure 19.59

Q13 The circuit shown in Figure 19.60 is composed of two batteries, two identical thin-filament bulbs, and an initially uncharged capacitor. Three compasses are placed on the desktop underneath the wires, all pointing north before the circuit is closed. Then the gap in the wire is closed, and the compass on the left immediately shows about a 20° deflection.

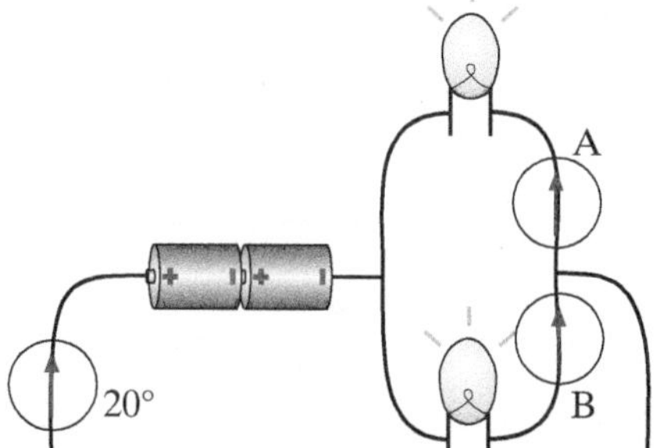

Figure 19.60

(a) On a diagram, draw the new needle positions on all three compasses at this time, and write the approximate angle of deflection beside the compasses labeled A and B. Explain carefully. **(b)** When this circuit was connected with one of the bulbs removed from its socket, the single bulb glowed for T seconds. In the two-bulb circuit shown in Figure 19.60, predict how long the two bulbs would glow (in terms of T). Explain carefully. **(c)** If possible, carry out the experiment, observing initial compass deflections and length of time of glow with one or two thin-filament bulbs. State your numerical results. Do your observations agree with your predictions?

Q14 The circuit shown in Figure 19.61 consists of two flashlight batteries, a large air-gap capacitor, and Nichrome wire. The circuit is allowed to run long enough that the capacitor is fully charged with $+Q$ and $-Q$ on the plates.

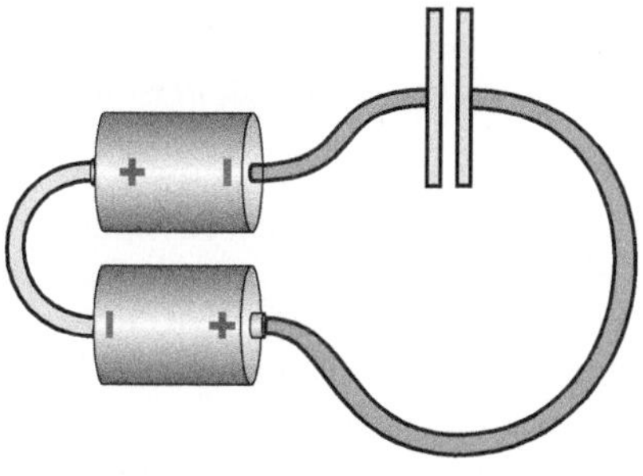

Figure 19.61

Next you push the two plates closer together (but the plates don't touch each other). Describe what happens, and explain why in terms of the fundamental concepts of charge and field. Include diagrams showing charge and field at several times.

Q15 The charge on an isolated capacitor does not change when a sheet of glass is inserted between the capacitor plates, and we find that the potential difference decreases (because the electric field inside the insulator is reduced by a factor of $1/K$). Suppose instead that the capacitor is connected to a battery, so that the battery tries to maintain a fixed potential difference across the capacitor. **(a)** A light bulb and an air-gap capacitor of capacitance C are connected in series to a battery with known emf. What is the final charge Q on the positive plate of the capacitor? **(b)** After fully charging the capacitor, a sheet of plastic whose dielectric constant is K is inserted into the capacitor and fills the gap. Does any current run through the light bulb? Why? What is the final charge on the positive plate of the capacitor?

Q16 An isolated capacitor consisting of two large metal disks separated by a very thin sheet of plastic initially has equal and opposite charges $+Q$ and $-Q$ on the two disks. Then we connect a long thin Nichrome wire to the capacitor as shown in Figure 19.62.

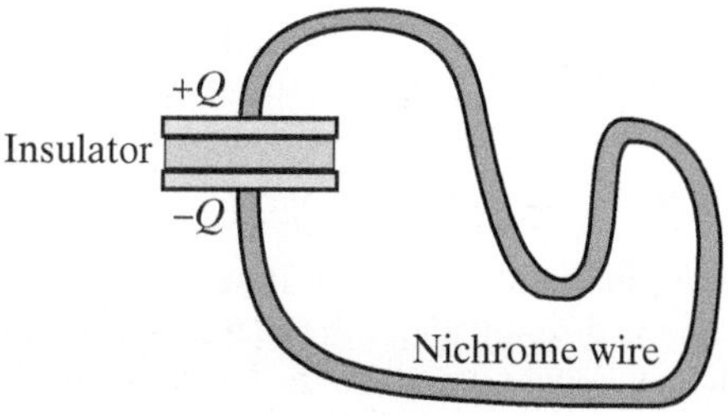

Figure 19.62

Describe and explain in detail what happens, including the first few nanoseconds, the approach to the final state, and the final state. Base your explanation on the fundamental concepts of charge and field; do not use the concept of potential in your explanation. To allow room for drawing charges and fields, make your diagram like Figure 19.62, with the thickness of the capacitor and the wire greatly exaggerated.

Q17 A capacitor with a slab of glass between the plates is connected to a battery by Nichrome wires and allowed to charge completely. Then the slab of glass is removed. Describe and explain what happens. Include diagrams. If you give a direction for a current, state whether you are describing electron current or conventional current.

Q18 A capacitor is connected to batteries by Nichrome wires and allowed to charge completely. Then the plates are suddenly moved farther apart. Describe what happens and explain in detail why it happens, based on fundamental physical principles. If you give a direction for a current, state whether you are describing electron current or conventional current. Include appropriate diagrams to support your explanation.

Q19 Two 1.5 V batteries in series were connected to a capacitor consisting of two very large metal disks placed very close to each other, so that the disks became charged (left disk positive). The batteries were then removed.

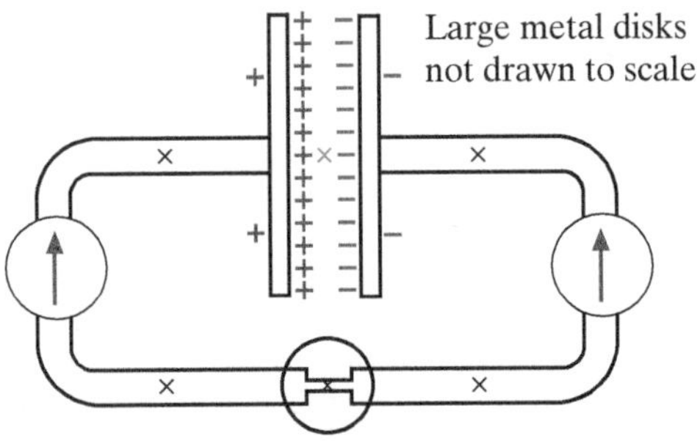

Figure 19.63

Next, thick copper wires and a thick-filament bulb were attached to the charged capacitor, as indicated in Figure 19.63. Two magnetic compasses lying on top of the copper wires initially pointed north. You find experimentally that when this thick-filament bulb is connected directly to the two batteries in series you observe a compass deflection of 15°. **(a)** Briefly describe in detail what you would observe about the bulb and both compasses during the next minute. You do not need to explain yet why this happens. Just describe in detail what you would see. **(b)** At a time that is 0.01 s after assembling this circuit, draw and label the electric field at the six locations marked × on the diagram. Pay attention to the relative magnitudes of your vectors. Make sure you have clearly labeled your electric field vectors (for example, E_{bulb}, etc.). **(c)** Sketch roughly a possible charge distribution on the diagram. If necessary for clarity, add comments on the diagram about the charge distribution. **(d)** The length of the left copper wire is L, the length of the right copper wire is also L, the length of the bulb filament is L_{bulb}, and the gap between the capacitor plates is s. The diameter of the copper wires is d, and the diameter of the bulb filament is d_{bulb}. The mobility in the copper is u, and the mobility in the hot tungsten is u_{bulb}. The number of mobile electrons per unit volume is approximately the same in the copper and the tungsten. At a time that is 0.01 s after assembling this circuit, write two valid physics equations involving the magnitudes of the electric fields, using the labels that you gave for the electric fields in part (b). What are the principles underlying these physics equations? **(e)** In part (a) you described what you would see happening with the bulb and compasses over the course of a minute. Now explain qualitatively but in detail why this happens. Be brief but be complete in your explanation.

Q20 Someone said, "Current takes the path of least resistance." What's wrong with this statement?

Q21 What are the units of conductivity σ, resistivity ρ, resistance R, and current density J?

Q22 Which of the following are ohmic resistors? For those that aren't, briefly state why they aren't. **(a)** Nichrome wire, **(b)** a thin filament light bulb, **(c)** a plastic rod, **(d)** salt water, **(e)** silicon (a semiconductor)

Q23 A desk lamp that plugs into a 120 V wall socket can use a 60 W or a 100 W light bulb. Which bulb has the larger resistance? Explain briefly.

Q24 **(a)** If the current through a battery is doubled, by what factor is the battery power increased? **(b)** If the current through a resistor is doubled, by what factor is the power dissipation increased? **(c)** Explain why these factors are the same or different (depending on what you find).

Q25 State whether the following statement is true or false, and briefly explain why: "In the two circuits shown in Figure 19.64, the battery output power is greater in circuit 2 because there is an additional resistor R_2 dissipating power."

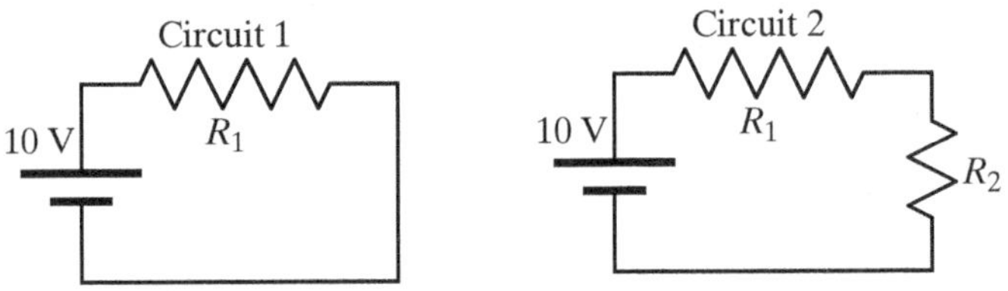

Figure 19.64

Q26 Why can birds perch on the bare wire of a 100,000 V power line without being electrocuted?

Q27 Should an ammeter have a low or high resistance? Why? Should a voltmeter have a low or high resistance? Why?

Q28 You are marooned on a desert island full of all kinds of standard electrical apparatus including a sensitive voltmeter, but you don't have an ammeter. Explain how you could use the voltmeter to measure currents.

PROBLEMS

Section 19.1

•**P29** The capacitor in Figure 19.65 is initially charged, then the circuit is connected. Which graph in Figure 19.66 best describes the current through the bulb as a function of time?

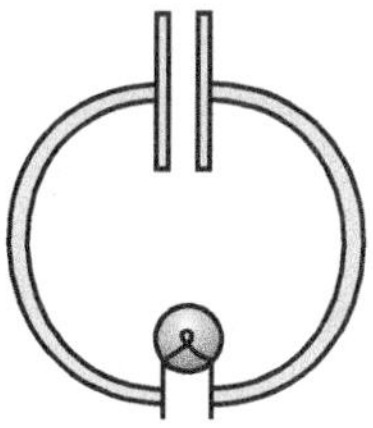

Figure 19.65

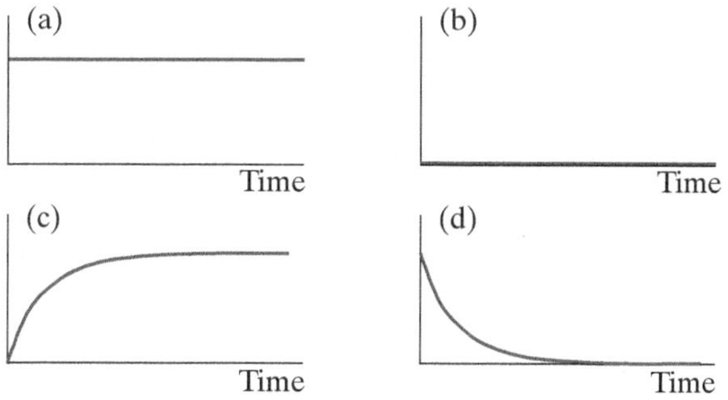

Figure 19.66

•**P30** The capacitor in Figure 19.67 is initially uncharged, then the circuit is connected. Which graph in Figure 19.66 best describes the current through the bulb as a function of time?

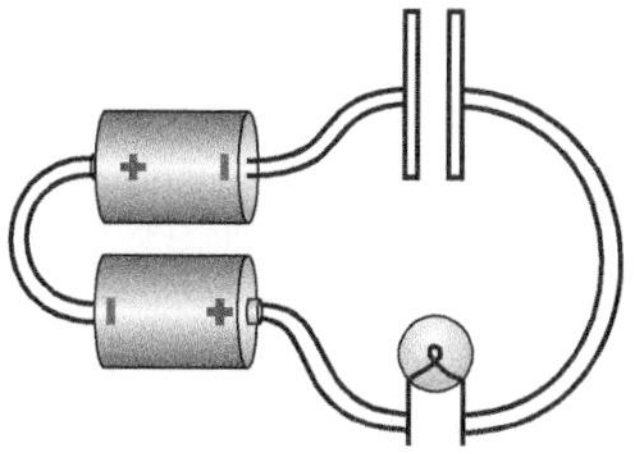

Figure 19.67

•**P31** The capacitor in Figure 19.67 is initially uncharged, then the circuit is connected. Which graph in Figure 19.66 best describes the absolute value of the charge on the left plate as a function of time?

•**P32** The capacitor in Figure 19.68 is initially uncharged, then the circuit is connected. Which graph in Figure 19.66 best describes the magnitude of the fringe field of the capacitor at location A (inside the connecting wire) as a function of time?

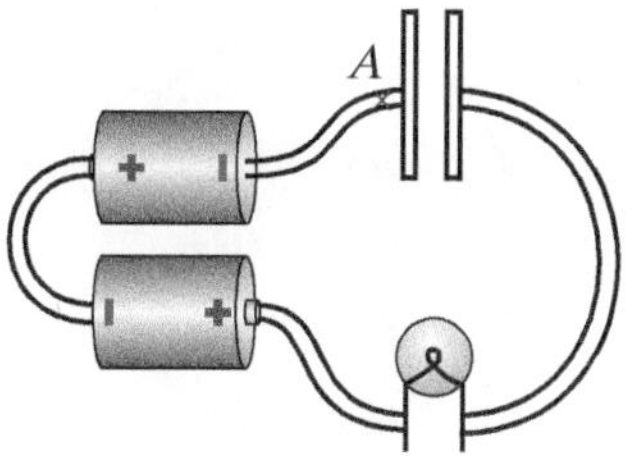

Figure 19.68

•**P33** The capacitor in Figure 19.68 is initially uncharged, then the circuit is connected. Which graph in Figure 19.66 best describes the magnitude of the net electric field at location A (inside the connecting wire) as a function of time?

•**P34** A particular capacitor is initially charged. Then a high-resistance Nichrome wire is connected between the plates of the capacitor, as shown in Figure 19.69. The needle of a compass placed under the wire deflects 20° to the east as soon as the connection is made. After 60 s the compass needle no longer deflects. **(a)** Which of the diagrams in Figure 19.69 best indicates the electron current at three locations in this circuit? (1) 0.01 s after the circuit is connected, (2) 15 s after the circuit is connected, (3) 120 s after the circuit is connected. **(b)** Which of the diagrams in Figure 19.70 best indicates the net electric field inside the wire at three locations in this circuit? (1) 0.01 s after the circuit is connected, (2) 15 s after the circuit is connected, (3) 120 s after the circuit is connected.

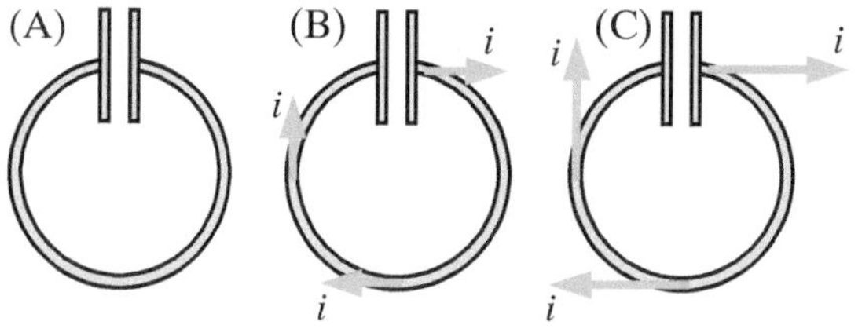

Figure 19.69

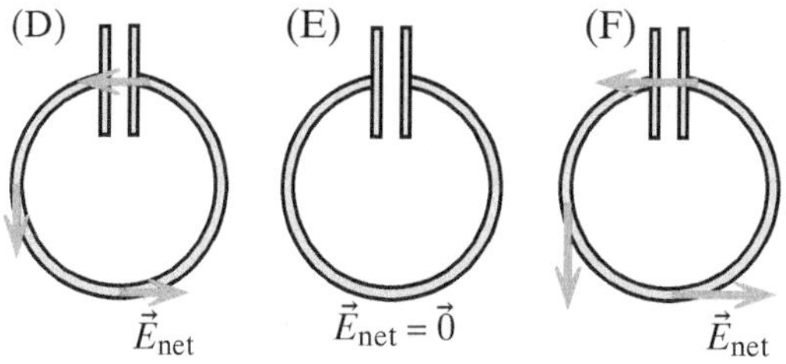

Figure 19.70

•**P35** Which of the following statements about the discharging of a capacitor through a light bulb are correct? Choose all that are true. (1) The fringe field of the capacitor decreases as the charge on the capacitor plates decreases. (2) Electrons flow across the gap between the plates of the capacitor, thus reducing the charge on the capacitor. (3) The electric field at a location inside the wire is due to charge on the surface of the wires and charge on the plates of the capacitor. (4) Electrons in the wires flow away from the negative plate toward the positive plate, reducing the charge on the plates.

•**P36** When a particular capacitor, which is initially uncharged, is connected to a battery and a small light bulb, the light bulb is initially bright but gradually gets dimmer, and after 45 s it goes out. The diagrams in Figure 19.71 show the electric field in the circuit and the surface charge distribution on the wires at three different times (0.01 s, 2 s, and 240 s) after the connection to the bulb is made. The diagrams in Figure 19.71 show the pattern of electric field in the wires and charge on the surface of the wires at three different times (0.01 s, 8 s, and 240 s) after the connection to the battery is made. Which of the diagrams best represents the state of the circuit at each time specified? **(a)** 0.01 s after the connection is made, **(b)** 8 s after the connection is made, **(c)** 240 s after the connection is made

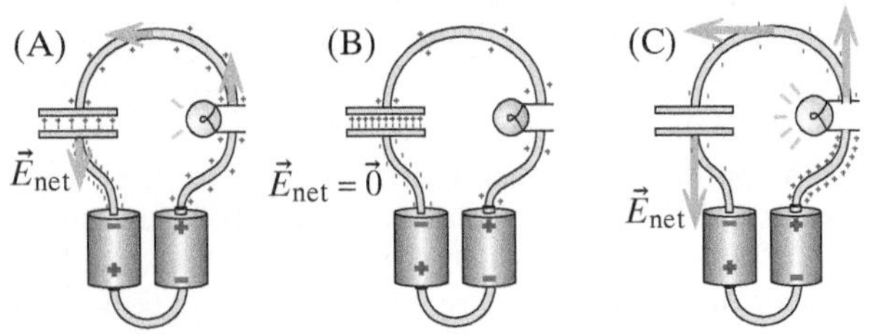

Figure 19.71

•**P37** In circuit 1 (Figure 19.72), an uncharged capacitor is connected in series with two batteries and one light bulb. Circuit 2 (Figure 19.72) contains two light bulbs identical to the bulb in circuit 1; in all other respects it is identical to circuit 1. In circuit 1, the light bulb stays lit for 25 s. The following questions refer to these circuits. You should draw diagrams representing the fields and charges in each circuit at the times mentioned, in order to answer the questions.

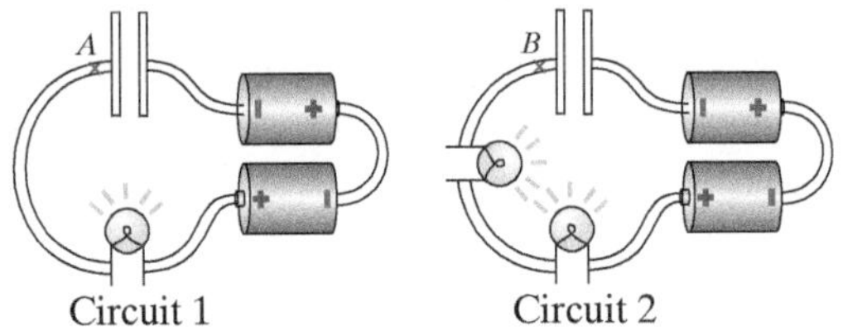

Figure 19.72

(a) One microsecond (1×10^{-6} s) after connecting both circuits, which of the following are true? Choose all that apply: (1) The net electric field at location A in circuit 1 is larger than the net electric field at location B in circuit 2. (2) At location A in circuit 1, electrons flow to the left. (3) At location A in circuit 1, the electric field due to charges on the surface of the wires and batteries points to the right. (4) In circuit 1 the potential difference across the capacitor plates is equal to the emf of the batteries. (5) The current in circuit 1 is larger than the current in circuit 2. **(b)** Two seconds after connecting both circuits, which of the following are true? Choose all that apply: (1) There is more charge on the plates of capacitor 1 than there is on the plates of capacitor 2. (2) There is negative charge on the right plate of the capacitor in circuit 1. (3) At location B in circuit 2 the net electric field points to the right. (4) At location B in circuit 2 the fringe field of the capacitor points to the right. (5) At location A in circuit 1 the fringe field of the capacitor points to the left. **(c)** Which of the graphs in Figure 19.73 represents the amount of charge on the positive plate of the capacitor in circuit 1 as a function of time? **(d)** Which of the graphs in Figure 19.73 represents the current in circuit 1 as a function of time?

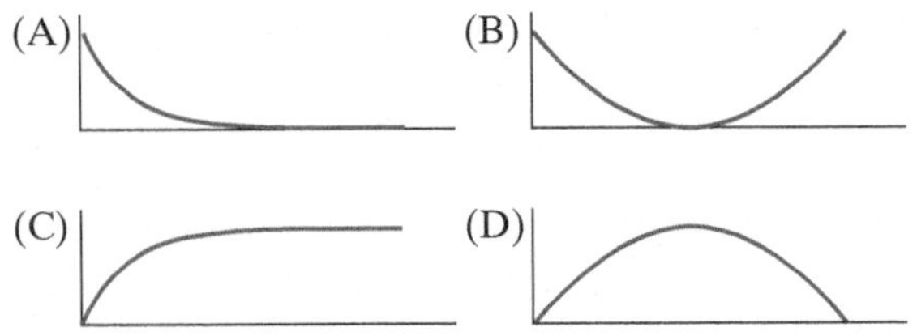

Figure 19.73

•**P38** A certain capacitor has rectangular plates 56 cm by 24 cm, and the gap width is 0.20 mm. What is its capacitance? We see that typical capacitances are very small when measured in farads. A 1 F capacitor is quite extraordinary! Apparently it has a very large area A (all wrapped up in a small package), and a very small gap s.

•**P39** A certain capacitor has rectangular plates 59 cm by 33 cm, and the gap width is 0.27 mm. If the gap is filled with a material whose dielectric constant is 2.9, what is the capacitance of this capacitor?

•**P40** Suppose that you charged a 2.5 F capacitor with two 1.5 V batteries. How much charge would be on each plate in the final state? How many excess electrons would be on the negative plate?

•**P41** You connect a 9 V battery to a capacitor consisting of two circular plates of radius 0.08 m separated by an air gap of 2 mm. What is the charge on the positive plate?

•**P42** Two circular plates of radius 0.12 m are separated by an air gap of 1.5 mm. The plates carry charge $+Q$ and $-Q$ where $Q = 3.6 \times 10^{-8}$ C. **(a)** What is the magnitude of the electric field in the gap? **(b)** What is the potential difference across the gap? **(c)** What is the capacitance of this capacitor?

•**P43** You may have noticed that while discharging a capacitor through a light bulb, the light bulb glows just about as brightly, and for just about as long, as it does while charging the same capacitor through the same bulb. Let E stand for the energy emitted by the light bulb (as light and heat) in the discharging phase, from just before the bulb is connected to the capacitor until the time when there is essentially no more current. In terms of $+E$ or $-E$, what was the energy change of the battery, capacitor, bulb, and surroundings during the charging phase, and during the discharging phase? One answer is already given in the following table:

	Charging	Discharging
Battery:		
Bulb:		
Capacitor:		
Surroundings:		$+E$

It is somewhat surprising that we can get this much information out of one simple observation.

••**P44** As shown in Figure 19.74, a spherical metal shell of radius r_1 has a charge Q (on its outer surface) and is surrounded by a concentric spherical metal shell of radius r_2 which has a charge $-Q$ (on its inner surface).

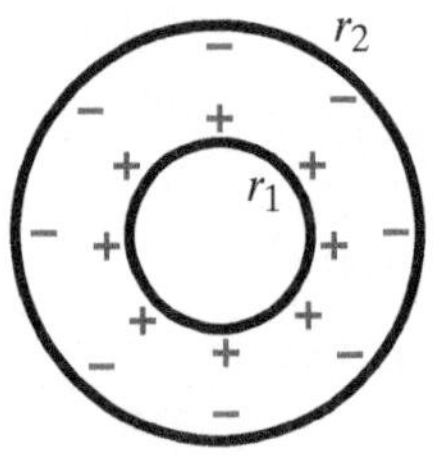

Figure 19.74

(a) Use the definition of capacitance: $Q = C|\Delta V|$ to find the capacitance of this spherical capacitor. **(b)** If the radii of the spherical shells r_1 and r_2 are large and nearly equal to each other, show that C can be written as $\varepsilon_0 A/s$ (which is also the equation for the capacitance of a parallel-plate capacitor) where $A = 4\pi r^2$ is the surface area of one of the spheres, and s is the small gap distance between them ($r_2 = r_1 + s$).

Section 19.2

•P45 In gold at room temperature, the mobility of mobile electrons is about 4.3×10^{-3} (m/s)/(V/m), and there are about 5.9×10^{28} mobile electrons per cubic meter. Calculate the conductivity of gold, including correct units.

•P46 Consider a silver wire with a cross-sectional area of 1 mm^2 carrying 0.3 A of current. The conductivity of silver is 6.3×10^7 (A/m^2)/(V/m). Calculate the magnitude of the electric field required to drive this current through the wire.

•P47 When a thin-filament light bulb is connected to two 1.5 V batteries in series, the current is 0.075 A. What is the resistance of the glowing thin-filament bulb?

•P48 In copper at room temperature, the mobility of mobile electrons is about 4.5×10^{-3} (m/s)/(V/m), and there are about 8×10^{28} mobile electrons per m^3. Calculate the conductivity σ and include the correct units. In actual practice, it is usually easier to measure the conductivity σ and deduce the mobility u from this measurement.

•P49 (a) A carbon resistor is 5 mm long and has a constant cross section of 0.2 mm^2. The conductivity of carbon at room temperature is $\sigma = 3 \times 10^4$ per ohm·m. In a circuit its potential at one end of the resistor is 12 V relative to ground, and at the other end the potential is 15 V. Calculate the resistance R and the current I. **(b)** A thin copper wire in this circuit is 5 mm long and has a constant cross section of 0.2 mm^2. The conductivity of copper at room temperature is $\sigma = 6 \times 10^7$ ohm^{-1}m^{-1}. The copper wire is in series with the carbon resistor, with one end connected to the 15 V end of the carbon resistor, and the current you calculated in part (a) runs through the carbon resistor wire. Calculate the resistance R of the copper wire and the potential $V_{\text{at end}}$ at the other end of the wire.

You can see that for most purposes a thick copper wire in a circuit would have practically a uniform potential. This is because the small drift speed in a thick, high-conductivity copper wire requires only a very small electric field, and the integral of this very small field creates a very small potential difference along the wire.

•P50 When a single thin-filament bulb is connected to a 1.5 V battery, the current through the battery is about 80 mA. If you add another thin-filament bulb in parallel, the battery current of course increases to 160 mA. Is the battery ohmic? That is, is the current through the battery proportional to the potential difference across the battery?

•P51 A certain ohmic resistor has a resistance of 40 Ω. A second resistor is made of the same material but is three times as long and has half the cross-sectional area. What is the resistance of the second resistor? What is the effective resistance of the two resistors in series?

•P52 In the circuit shown in Figure 19.75, the emf of the battery is 7.9 V. Resistor R_1 has a resistance of 23 Ω, and resistor R_2 has a resistance of 44 Ω. A steady current flows through the circuit. **(a)** What is the absolute value of the potential difference across R_1? **(b)** What is the conventional current through R_2?

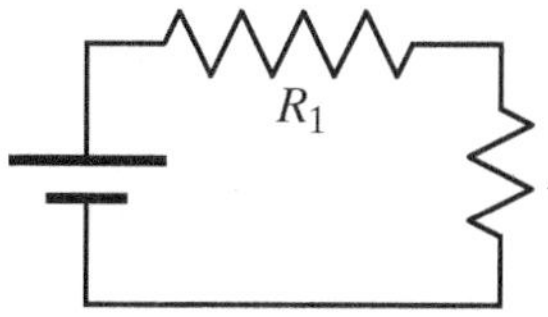

Figure 19.75

•P53 When a thick-filament bulb is connected to one flashlight battery, the current is 0.20 A. When you use two batteries in series, the current is not 0.40 A but only 0.33 A. Briefly explain this behavior.

•P54 In Figure 19.76 the resistance R_1 is 10 Ω, R_2 is 5 Ω, and R_3 is 20 Ω. If this combination of resistors were to be replaced by a single resistor with an equivalent resistance, what should that resistance be?

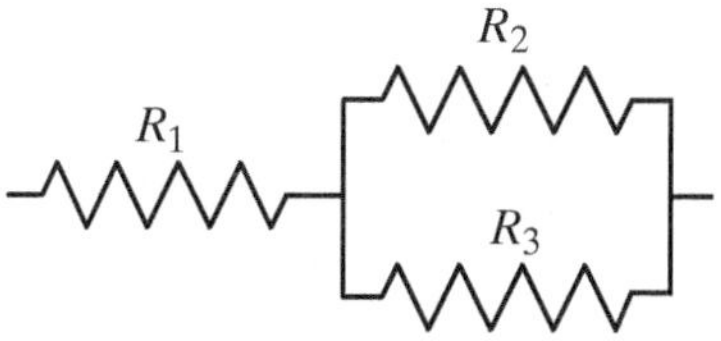

Figure 19.76

•P55 Suppose that you charge a 1 F capacitor in a circuit containing two 1.5 V batteries, so the final potential difference across the plates is 3 V. How much charge is on each plate? How many excess electrons are on the negative plate?

•P56 In the circuit shown in Figure 19.77 the emf of the battery is 7.4 V. Resistor R_1 has a resistance of 31 Ω , resistor R_2 has a resistance of 47 Ω, and resistor R_3 has a resistance of 52 Ω. A steady current flows through the circuit.

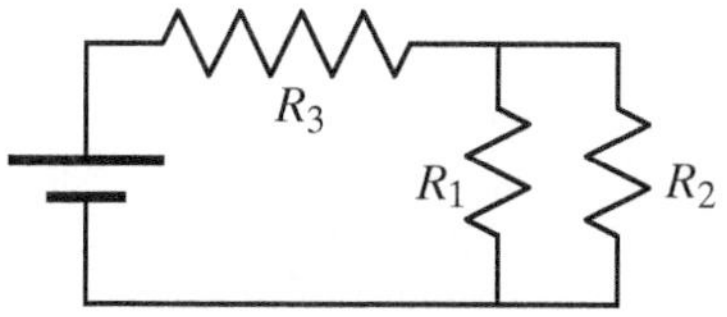

Figure 19.77

(a) What is the equivalent resistance of R_1 and R_2? **(b)** What is the equivalent resistance of all three resistors? **(c)** What is the conventional current through R_3?

•P57 Consider a copper wire with a cross-sectional area of 1 mm^2 (similar to your connecting wires) and carrying 0.3 A of current, which is about what you get in a circuit with a thick-filament bulb and two batteries in series. Calculate the strength of the very small electric field required to drive this current through the wire.

••P58 The conductivity of tungsten at room temperature, 1.8×10^7 (A/m^2)/(V/m), is significantly smaller than that of copper. At the very high temperature of a glowing light-bulb filament (nearly 3000 kelvins), the conductivity of tungsten is 18 times smaller than it is at room temperature. The tungsten filament of a thick-filament bulb has a radius of about 0.015 mm. Calculate the electric field required to drive 0.20 A of current through the glowing bulb and show that it is very large compared to the field in the connecting copper wires.

••P59 A circuit consists of a battery, whose emf is K, and five Nichrome wires, three thick and two thin as shown in

Figure 19.78. The thicknesses of the wires have been exaggerated in order to give you room to draw inside the wires. The internal resistance of the battery is negligible compared to the resistance of the wires. The voltmeter is not attached until part (e) of the problem.

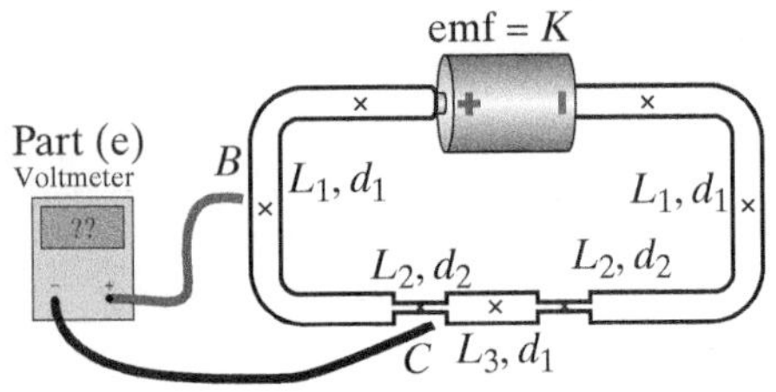

Figure 19.78

(a) Draw and label appropriately the electric field at the locations marked × inside the wires, paying attention to appropriate relative magnitudes of the vectors that you draw. **(b)** Show the approximate distribution of charges for this circuit. Make the important aspects of the charge distribution very clear in your drawing, supplementing your diagram if necessary with very brief written descriptions on the diagram. Make sure that parts (a) and (b) of this problem are consistent with each other. **(c)** Assume that you know the mobile-electron density n and the electron mobility u at room temperature for Nichrome. The lengths (L_1, L_2, L_3) and diameters (d_1, d_2) of the wires are given on the diagram. Calculate accurately the number of electrons that leave the negative end of the battery every second. Assume that no part of the circuit gets very hot. Express your result in terms of the given quantities (K, L_1, L_2, L_3, d_1, d_2, n, and u). Explain your work and identify the principles you are using. **(d)** In the case that $d_2 \ll d_1$, what is the *approximate* number of electrons that leave the negative end of the battery every second? **(e)** A voltmeter is attached to the circuit with its + lead connected to location B (halfway along the leftmost thick wire) and its – lead connected to location C (halfway along the leftmost thin wire). In the case that $d_2 \ll d_1$, what is the approximate voltage shown on the voltmeter, including sign? Express your result in terms of the given quantities (K, L_1, L_2, L_3, d_1, d_2, n, and, u).

••P60 Using thick connecting wires that are very good conductors, a Nichrome wire ("wire 1") of length L_1 and cross-sectional area A_1 is connected in series with a battery and an ammeter (this is circuit 1). The reading on the ammeter is I_1. Now the Nichrome wire is removed and replaced with a different wire ("wire 2"), which is 2.5 times as long and has 5.5 times the cross-sectional area of the original wire (this is circuit 2).

In the following questions, a subscript 1 refers to circuit 1, and a subscript 2 refers to circuit 2. It will be helpful to write out your solutions to the following questions algebraically before doing numerical calculations. (*Hint:* Think about what is the same in these two circuits.) **(a)** What is the value of I_2/I_1, the ratio of the conventional currents in the two circuits? **(b)** What is the value of R_2/R_1, the ratio of the resistances of the wires? **(c)** What is the value of E_2/E_1, the ratio of the electric fields inside the wires in the steady state?

•••P61 A long iron slab of width w and height h emerges from a furnace, as shown in Figure 19.79. Because the end of the slab near the furnace is hot and the other end is cold, the electron mobility increases significantly with the distance x.

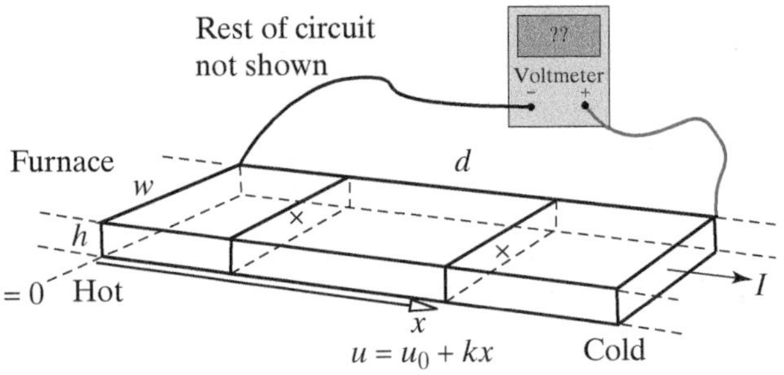

Figure 19.79

The electron mobility is $u = u_0 + kx$, where u_0 is the mobility of the iron at the hot end of the slab. There are n iron atoms per cubic meter, and each atom contributes one electron to the sea of mobile electrons (we can neglect the small thermal expansion of the iron). A steady-state conventional current runs through the slab from the hot end toward the cold end, and an ammeter (not shown) measures the current to have a magnitude I in amperes. A voltmeter is connected to two locations a distance d apart, as shown. **(a)** Show the electric field inside the slab at the two locations marked with ×. Pay attention to the relative magnitudes of the two vectors that you draw. **(b)** Explain why the magnitude of the electric field is different at these two locations. **(c)** At a distance x from the left voltmeter connection, what is the magnitude of the electric field in terms of x and the given quantities $w,h,d,u_0,k,n,$ and I (and fundamental constants)? **(d)** What is the sign of the potential difference displayed on the voltmeter? Explain briefly. **(e)** In terms of the given quantities $w,h,d,u_0,k,n,$ and I (and fundamental constants), what is the magnitude of the voltmeter reading? Check your work. **(f)** What is the resistance of this length of the iron slab?

Section 19.4

••P62 A battery with negligible internal resistance is connected to a resistor. The power produced in the battery and the power dissipated in the resistor are both P_1. Another resistor of the same kind is added, so the circuit consists of a battery and two resistors in series. **(a)** In terms of P_1, now how much power is dissipated in the first resistor? **(b)** In terms of P_1, now how much power is produced in the battery? **(c)** The circuit is rearranged so that the two resistors are in parallel rather than in series. In terms of P_1, now how much power is produced in the battery?

••P63 Two flashlight batteries in series power a circuit consisting of four ohmic resistors as shown in Figure 19.80, connected by copper wires with negligible resistance. The internal resistance of the batteries is negligible. Two magnetic compasses are placed underneath the wires, initially pointing northward before closing the circuit. A very-high-resistance voltmeter is connected as shown. The resistance R_1 is 20 Ω, R_2 is 20 Ω, R_3 is 30 Ω, and R_4 is 15 Ω.

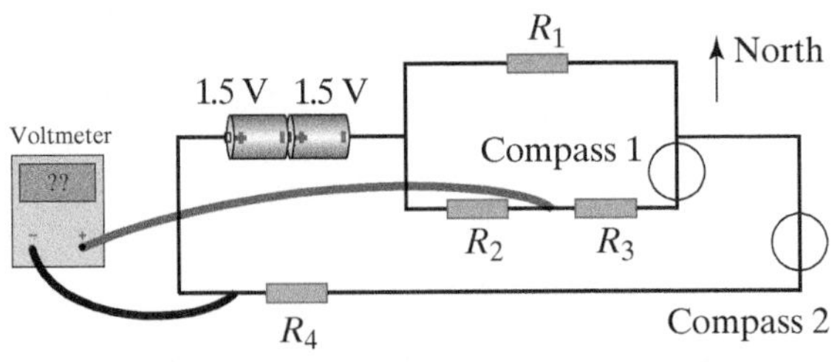

Figure 19.80

(a) Starting from fundamental principles, write equations that could be solved to determine the conventional current through each resistor, and show the directions of these currents on the diagram. Explain where each of your equations comes from. Do not use equations for parallel and series resistances (but you are free to use such equations to check your work if you like). If these equations are solved, we find that the current through R_1 is 0.073 A and the current through R_4 is 0.102 A. **(b)** Sketch the positions of the compass needles. Compass 1 deflects by 3 degrees. What is the deflection angle of compass 2? **(c)** The resistors are made of a material that has 8×10^{28} free electrons per cubic meter and a mobility of 3×10^{-5} (m/s)/(N/C). Find E_2, the magnitude of the electric field in resistor R_2, which is in the form of a short wire with a constant cross-sectional area of 6×10^{-10} m^2. **(d)** Is the field E_1 in resistor R_1 larger, smaller, or the same as E_2? **(e)** What is the length of resistor R_2? **(f)** What does the voltmeter read, including sign? **(g)** How much energy in joules is expended by the batteries in moving a singly charged ion from one end of one of the batteries to the other end of that same battery? **(h)** How much power output is there from one of the batteries?

••P64 Consider the two circuits depicted in Figure 19.81. In circuit 1, ohmic resistor R_1 dissipates 5 W; in circuit 2, ohmic resistor R_2 dissipates 20 W. The wires and batteries have negligible resistance. The circuits contain 10 V batteries.

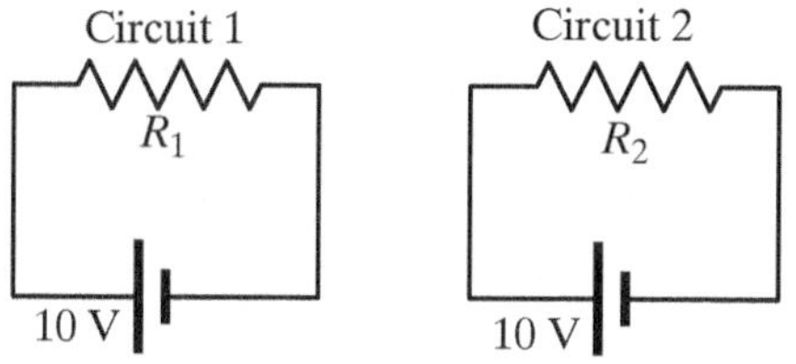

Figure 19.81

(a) What is the resistance of R_1 and of R_2? **(b)** Resistor R_1 is made of a very thin metal wire that is 3 mm long, with a diameter of 0.1 mm. What is the electric field inside this metal resistor? **(c)** The same resistors are used to construct circuit 3 (Figure 19.82), using the same 10 V battery as before. Make a complete accurate graph of potential versus location for circuit 3. Label the y axis numerically. Be sure to explain the significant features of your graph.

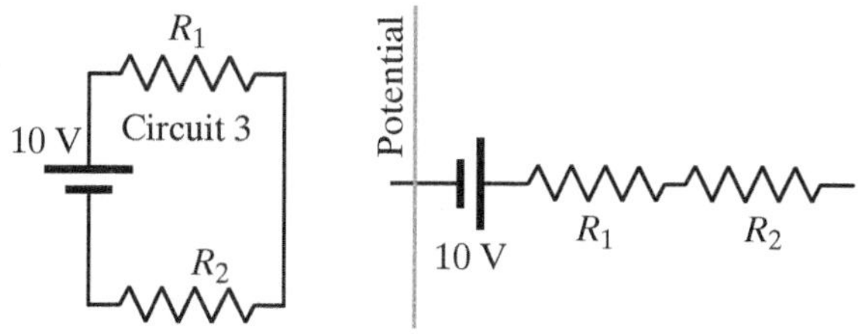

Figure 19.82

(d) On a copy of the diagram of circuit 3, place +'s and −'s to indicate the distribution of surface charge everywhere on the circuit. Assume that both resistors are far from the battery. **(e)** In circuit 3, calculate the number of electrons entering R_1 every second and the number of electrons entering R_2 every second. **(f)** What is the power output of the battery in circuit 3?

••P65 Work and energy with a capacitor: A capacitor with capacitance C has an amount of charge q on one of its plates, in which case the potential difference across the plates is $\Delta V = q/C$ (definition of capacitance). The work done to add a small amount of charge dq when charging the capacitor is $dq(\Delta V) = dq(q/C)$. Show by integration that the amount of work required to charge up the capacitor from no charge to a final charge Q is $\frac{1}{2}(Q^2/C)$. Since this is the amount of work required to charge the capacitor, it is also the amount of energy stored in the capacitor. Substituting $Q = C\Delta V$, we can also express the energy as $\frac{1}{2}C(\Delta V)^2$.

••P66 A circuit is made of two 1.5 V batteries and three light bulbs as shown in Figure 19.83.

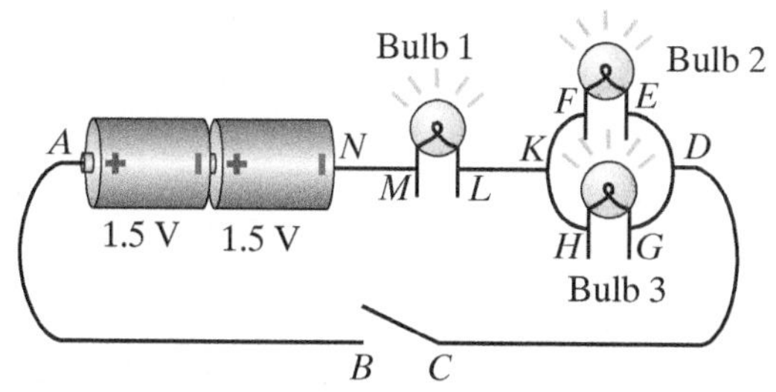

Figure 19.83

When the switch is closed and the bulbs are glowing, bulb 1 has a resistance of 10 Ω, bulb 2 has a resistance of 40 Ω, bulb 3 has a resistance of 30 Ω, and the copper connecting wires have negligible resistance. You can also neglect the internal resistance of the batteries. **(a)** Make a copy of the circuit diagram. With the switch open, indicate the approximate surface charge with +'s and −'s on your diagram. **(b)** With the switch open, find the potential differences $V_B - V_C$ and $V_D - V_K$. **(c)** After the switch is closed and the steady state is established, the currents through bulbs 1, 2, and 3 are I_1, I_2, and I_3, respectively. Write loop and node equations that could be solved to determine these three unknown currents, but do not solve the equations. Label on the diagram what current directions, loops, and nodes you are using, and explain which equation is which. In order to learn about the general approach, do not use equations for series and parallel resistance in this problem. (You can use these equations to check your work.) **(d)** In terms of the unknown currents I_1, I_2, and I_3, what is the potential difference $V_C - V_F$ (with the switch closed)? **(e)** In terms of the unknown currents I_1, I_2, and I_3, how much power is delivered by the batteries (with the switch closed)? **(f)** Solve your equations and give values for I_1, I_2, and I_3. **(g)** How many electrons leave the battery at location N every second? **(h)** What is the numerical value of $V_C - V_F$? **(i)** What is the numerical value of the power delivered by the batteries? **(j)** The tungsten filament in the 40 Ω bulb is 8 mm long and has a cross-sectional area of 2×10^{-10} m^2. How big is the electric field inside this metal filament?

Section 19.5

•P67 A certain 6 V battery delivers 12 A when short-circuited. How much current does the battery deliver when a 1 Ω resistor is connected to it?

••P68 **(a)** You short-circuit a 9 V battery by connecting a short wire from one end of the battery to the other end. If the current in the short circuit is measured to be 18 A, what is the internal resistance of the battery? **(b)** What is the power generated by the battery? **(c)** How much energy is dissipated in the internal resistance every second? (Remember that one watt is one joule per second.) **(d)** This same battery is now connected to a 10 Ω resistor. How much current flows through this resistor? **(e)** How much power is dissipated in the 10 Ω resistor? **(f)** The leads to a voltmeter are placed at the two ends of the battery of this circuit containing the 10 Ω resistor. What does the meter read?

••P69 A 9 V battery is connected to a 100 Ω resistor, and a voltmeter shows the potential difference across the battery to be 6.7 V. **(a)** What is the internal resistance of the battery? **(b)** If the resistor is replaced by a very low-resistance wire, what does the voltmeter read? **(c)** What is the current through this wire?

••P70 **(a)** Suppose that you connect two identical flashlight batteries in series to get a 3 V emf. Write a loop equation to show that the maximum (short-circuit) current these batteries can deliver is exactly the same as the maximum current from one battery. **(b)** Alternatively, suppose that you connect two identical flashlight batteries in parallel, which gives you just a 1.5 V emf. Write loop and node equations to show that the maximum (short-circuit) current these batteries can deliver is double that of one battery.

Section 19.6

•P71 You connect an ammeter to a 1.5 V battery whose internal resistance is 0.1 ohm. **(a)** What does the ammeter read? **(b)** Replace the ammeter with a voltmeter. What does the voltmeter read?

••P72 A 40-cm-long high-resistance wire with rectangular cross section 7 mm by 3 mm is connected to a 12 V battery through an ammeter, as shown in Figure 19.84. The resistance of the wire is 50 Ω. The resistance of the ammeter and the internal resistance of the battery can be considered to be negligibly small compared to the resistance of the wire.

Leads to a high-resistance voltmeter are connected as shown, with the − lead connected to the inner edge of the wire, at the top (location A), and the + lead connected to the outer edge of the wire, at the bottom (location C). The distance d along the wire between voltmeter connections is 5 cm.

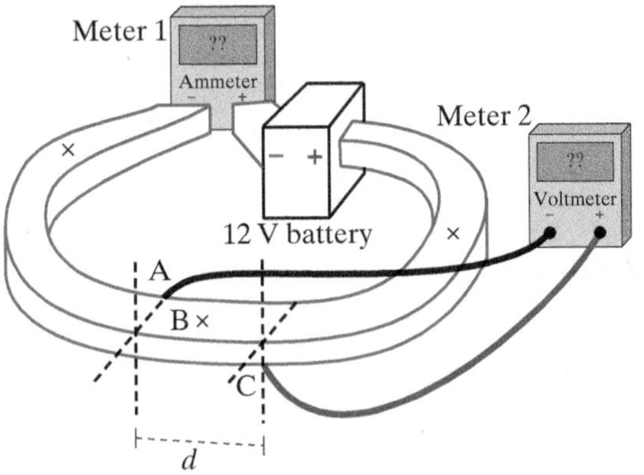

Figure 19.84

(a) On a copy of the diagram, show the approximate distribution of charge. **(b)** On a copy of the diagram, draw the electric field inside the wire at the 3 locations marked ×. **(c)** What is the magnitude of the electric field at location B? **(d)** What does the voltmeter read, both magnitude and sign? **(e)** What does the ammeter read, both magnitude and sign? **(f)** In a 60 s period, how many electrons are released from the − end of the battery? **(g)** There are 1.5×10^{26} free electrons per cubic meter in the wire. What is the drift speed v of the electrons in the wire? **(h)** What is the mobility u of the material that the wire is made of? **(i)** Switch meter 1 from being an ammeter to being a voltmeter. Now what do the two meters read? **(j)** The 12 V battery is removed from the circuit and both the ammeter and voltmeter are connected in parallel to the battery. The voltmeter reads 1.8 V, and the ammeter reads 20.4 A. What is the internal resistance of the battery?

••P73 Two resistors each with resistance of 4×10^6 Ω are connected in series to a 60 V power supply whose internal resistance is negligible. You connect a voltmeter across one of these resistors, and this voltmeter has an internal resistance of 1×10^6 ohms. What is the reading on the voltmeter?

Section 19.7

•P74 A 20 Ω resistor and a 2 F capacitor are in series with a 9 V battery. What is the initial current when the circuit is first assembled? What is the current after 50 s?

••P75 A resistor with resistance R and an air-gap capacitor of capacitance C are connected in series to a battery (whose strength is "emf"). **(a)** What is the final charge on the positive plate of the capacitor? **(b)** After fully charging the capacitor (so there is no current), a sheet of plastic whose dielectric constant is K is inserted into the capacitor and fills the gap. Explain why a current starts running in the circuit. You can base your explanation either on electric field or on electric potential, whichever you prefer. **(c)** What is the initial current through the resistor just after inserting the sheet of plastic? **(d)** What is the final charge on the positive plate of the capacitor after inserting the plastic?

••P76 The deflection plates in an oscilloscope are 10 cm by 2 cm with a gap distance of 1 mm. A 100 V potential difference is suddenly applied to the initially uncharged plates through a 1000 Ω resistor in series with the deflection plates. How long does it take for the potential difference between the deflection plates to reach 95 V?

••P77 A capacitor consists of two rectangular metal plates 3 m by 4 m, placed a distance 2.5 mm apart in air (Figure 19.85). The capacitor is connected to a 9 V power supply long enough to charge the capacitor fully, and then the battery is removed.

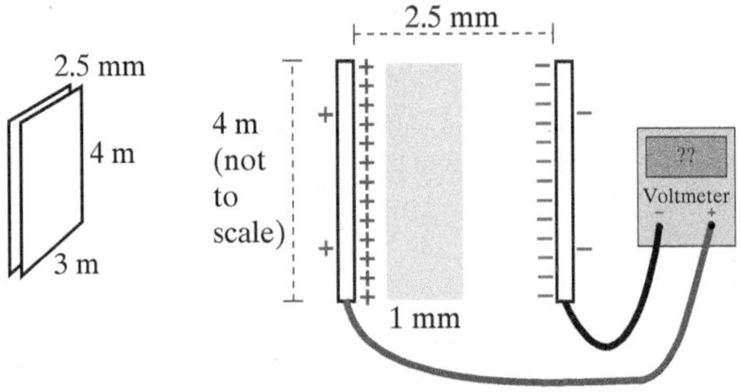

Figure 19.85

(a) Show that there will not be a spark in the air between the plates. **(b)** How much charge is on the positive plate of the capacitor?

With the battery still disconnected, you insert a slab of plastic 3 m by 4 m by 1 mm between the plates, next to the positive plate, as shown in Figure 19.85. This plastic has a dielectric constant of 5. **(c)** After inserting the plastic, you connect a voltmeter to the capacitor. What is the initial reading of the voltmeter? **(d)** The voltmeter has a resistance of 100 MΩ (1×10^8 Ω). What does the voltmeter read 3 s after being connected?

Section 19.11

•P78 For the circuit shown in Figure 19.86, which consists of batteries with known emf and ohmic resistors with known resistance, write the correct number of energy-conservation and current node rule equations that would be adequate to solve for the unknown currents, but do not solve the equations. Label nodes and currents on the diagram, and identify each equation (energy or current, and for which loop or node).

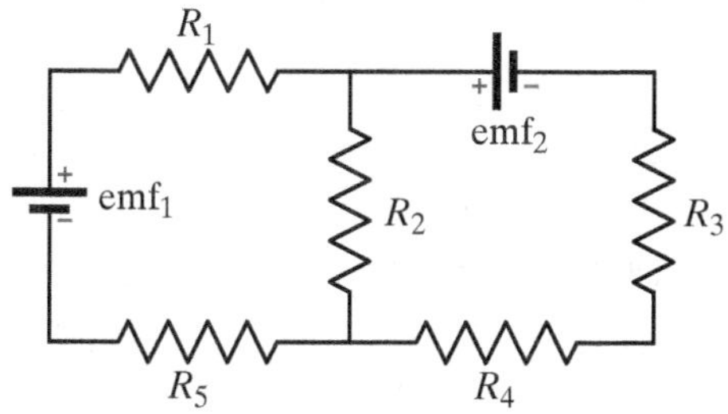

Figure 19.86

••P79 A circuit consists of two batteries (with negligible internal resistance), six ohmic resistors, and connecting wires that have negligible resistance. The resistance R_1 is 10 Ω, R_2 is 20 Ω, R_3 is 30 Ω, R_4 is 12 Ω, R_5 is 15 Ω, and R_6 is 20 Ω. Unknown currents I_1, I_2, I_3, I_4, I_5, and I_6 have their directions marked on the circuit diagram in Figure 19.87.

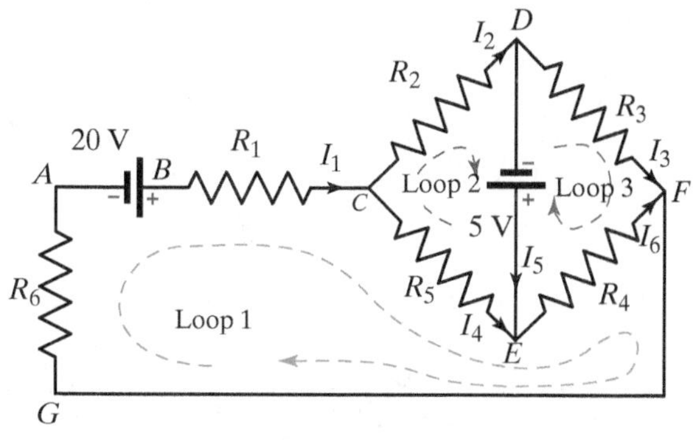

Figure 19.87

(a) Write down a set of equations that could be solved for the six unknown currents. Make sure you can explain how you got these equations. **(b)** When a correct set of equations is solved, the currents are as follows (to the nearest milliampere): $I_1 = 0.4394\,\text{A}$, $I_2 = 0.3312\,\text{A}$, $I_3 = 0.0065\,\text{A}$, $I_4 = 0.1082\,\text{A}$, $I_5 = 0.3247\,\text{A}$, $I_6 = 0.4329\,\text{A}$. Check your equations by substituting in these numbers. **(c)** Suppose that you connect the negative lead of a voltmeter to location G and the positive lead of the voltmeter to location C. What does the voltmeter read, including both magnitude and sign? **(d)** What is the power output of the 5 V battery? **(e)** Resistor R_4 is made of a very thin metal wire that is 3 mm long, with a diameter of 0.1 mm. What is the electric field inside this metal resistor?

••P80 A circuit contains two batteries (with negligible internal resistance) and five ohmic resistors (Figure 19.88). The connecting wires have negligible resistance. The letters A through H are shown to make it possible to refer to specific parts of the circuit.

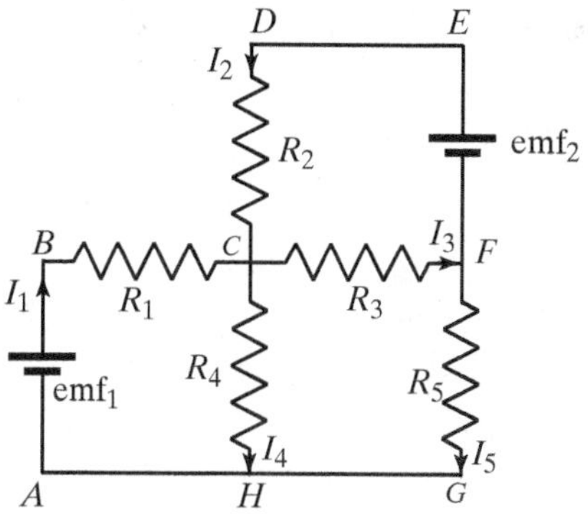

Figure 19.88

(a) Write all the equations necessary to solve for the unknown currents I_1, I_2, I_3, I_4, and I_5, whose directions are indicated on the circuit diagram. Do not solve the equations, but do explain very clearly what your equations are based on, and to what they refer.

Assume that a computer program has solved your equations in terms of the known battery voltages and known resistances, so that the currents I_1, I_2, I_3, I_4, and I_5 are known. **(b)** In terms of the known quantities, calculate $V_D - V_A$, and check that your sign makes sense. **(c)** In terms of the known quantities, calculate the power produced in battery number 2.

ANSWERS TO CHECKPOINTS

1 Consider the very first short time interval—say, 0.01 s. At the beginning of this interval both capacitors have very little charge. The electric field in the wires is due almost entirely to the battery and surface charge on the wires, and is the same in both cases. Thus the number of electrons flowing onto the left plate and off of the right plate is approximately the same in the first 0.01 s. At the end of this time interval, both capacitors have nearly the same small amount of charge, q. This is the situation illustrated in Figure 19.89.

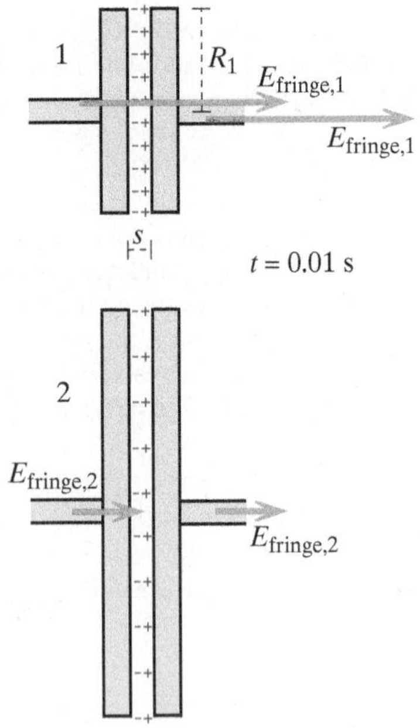

Figure 19.89

The fringe field of the smaller capacitor is this:

$$E_{\text{fringe},1} \approx \frac{q/A}{\varepsilon_0}\left(\frac{s}{2R_1}\right) = \frac{1}{\varepsilon_0}\frac{q}{\pi R_1^2}\left(\frac{s}{2R_1}\right)$$

The fringe field of the larger capacitor is smaller, since $R_2 > R_1$:

$$E_{\text{fringe},2} = \frac{1}{\varepsilon_0}\frac{q}{\pi R_2^2}\left(\frac{s}{2R_2}\right) < E_{\text{fringe},1}$$

The smaller fringe field of the large capacitor means a larger *net* field and larger drift speed, so the current flowing onto the large capacitor has decreased less than the current flowing onto the small capacitor. See further discussion in Section 19.7.

2 30 Ω; 18.8 Ω; 8.3 Ω; R is not constant, and the bulb filament is not ohmic for this range of currents

3 7.5 Ω; 0.4 A

4 4.5 W

5 nearly 0; nearly 0; nearly 0

6 A voltmeter has very high resistance, so the current is extremely small; the voltmeter reads 3 V ($+3-3=0$ round-trip potential difference).

7 $q/m = aL/\Delta V$

8 −8.0 V; 3.2 W; 4.8 W

CHAPTER

20

Magnetic Force

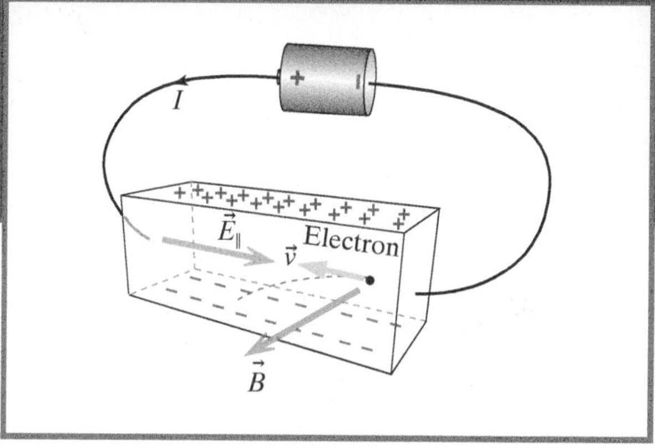

OBJECTIVES

After studying this chapter you should be able to

- Calculate the magnetic force on a moving charged particle, and on a current-carrying wire.
- Determine the sign of the mobile charges in a conductor using the Hall effect.
- Explain motional emf and calculate its magnitude.

Review: Magnetic Field of Moving Charge

Recall that the magnetic field associated with a moving charge can be calculated from the Biot–Savart law:

$$\vec{B} = \frac{\mu_0}{4\pi}\frac{q\vec{v}\times\hat{r}}{r^2}$$

From this law and the definition of current we derived an expression for the magnetic field contributed by a small current element:

$$\Delta\vec{B} = \frac{\mu_0}{4\pi}\frac{I\Delta\vec{l}\times\hat{r}}{r^2}$$

and we were able to integrate this form of the Biot–Savart law to predict the magnetic field of various configurations of currents, such as straight wires, loops, and solenoids.

In this chapter we will study the effects of magnetic fields. Often our analysis will require two steps: first, find the field at a location made by a moving charge or a current, and second, find the effect of this field on a different moving charge. Just as the electric field made by a charge does not affect the source charge itself, so the magnetic field made by a moving charge affects other moving charges, not the source charge.

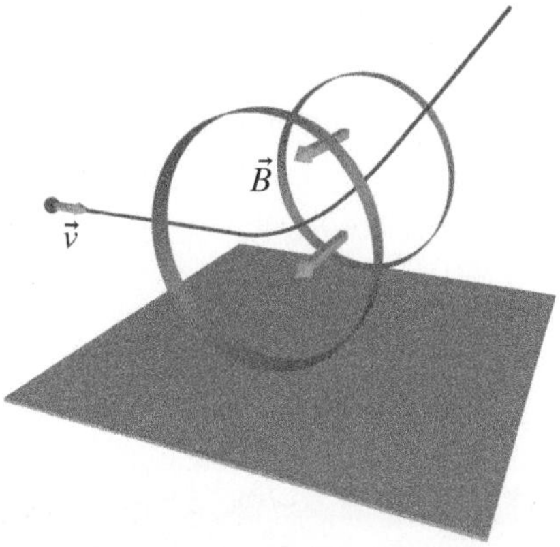

Figure 20.1 Magnetic deflection of a moving electron. Current-carrying coils are the source of the magnetic field.

20.1 MAGNETIC FORCE ON A MOVING CHARGE

In the cathode ray tube (CRT) displays that were formerly used in home television sets and computer monitors (not the flat-panel displays used today), electrons were accelerated to fairly high speeds by electric fields. The moving electrons were deflected vertically by a magnetic field. This magnetic field was horizontal and was produced by currents in coils positioned as shown in Figure 20.1 (not shown are other coils whose magnetic field deflected the electrons from side to side).

There is a 3D geometrical relationship between the direction of the magnetic force on a moving charge and

- the direction of the applied magnetic field,
- the direction of the velocity of the moving charge, and
- the sign of the moving charge.

Experiments in which these parameters are systematically varied show that the magnetic force on a moving charge is given by the following equation:

MAGNETIC FORCE

$$\vec{F}_{\text{magnetic}} = q\vec{v} \times \vec{B}$$

q is the charge (including sign) of the moving charge.
$\vec{v}$ is the velocity of the moving charge.
$\vec{B}$ is the applied magnetic field (in tesla).

Since $\vec{F}_{\text{magnetic}} = q\vec{v} \times \vec{B}$, we can see that a tesla has the units of F/qv:

$$T = \frac{\text{N}}{\text{C} \cdot \text{m/s}} = \frac{\text{N/C}}{\text{m/s}}$$

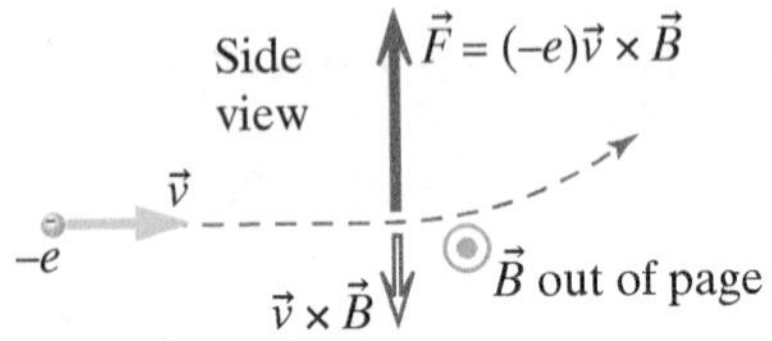

Figure 20.2 Side view of the magnetic deflection of the moving electron.

Use the right-hand rule for cross products to verify that the equation for magnetic force gives the observed direction of magnetic force in Figure 20.1 and in the side view shown in Figure 20.2. Remember that the electron charge $q = -e$, which means that the magnetic force on a moving electron is in the opposite direction to the direction of the cross product $\vec{v} \times \vec{B}$.

The equation $\vec{F} = q\vec{v} \times \vec{B}$ summarizes the experimental observations of the effect of a magnetic field on a moving charge.

> QUESTION What do these observations say about the effect that a magnetic field has on a charge that is not moving? How does this compare with the effect that an electric field has on a charge that is not moving?

If the charge is not moving, the magnetic field has no effect on it, whereas electric fields affect charges even if they are at rest.

Magnitude of the Magnetic Force

According to the equation $\vec{F}_{\text{magnetic}} = q\vec{v} \times \vec{B}$, there should be a magnetic force on a charged piece of invisible tape when you move it near a magnet. However, the magnitude of this magnetic force is extremely small compared to the familiar electric force acting between the tape and the magnet (due to the charged tape polarizing the metal).

The magnetic force on a single moving electron in a CRT is significant because the particle's acceleration $\vec{a} = d\vec{v}/dt$ is proportional to the ratio q/m:

$$\left|\frac{d\vec{p}}{dt}\right| = qvB\sin\theta$$

$$\left|\frac{d\vec{v}}{dt}\right| = \frac{q}{m}vB\sin\theta \quad (\text{for } v \ll c)$$

The q/m ratio is enormous for a single electron:

$$(1.6 \times 10^{-19}\ \text{C})/(9 \times 10^{-31}\ \text{kg}) \approx 1 \times 10^{11}\ \text{C/kg}$$

but q/m is very small for a charged tape:

$$(1 \times 10^{-8}\ \text{C})/(1 \times 10^{-4}\ \text{kg}) \approx 1 \times 10^{-4}\ \text{C/kg}$$

Magnetic forces on moving objects that have been charged by rubbing will always be tiny, because only an extremely small fraction of the atoms in the object have lost or gained an electronic charge. Moreover, the magnetic force is proportional to the speed of the moving particle, and the speed of electrons in a CRT is very high (it may be nearly 1×10^8 m/s).

Checkpoint 1 In Chapter 14 we found that a charged invisible tape has a net charge of about 1×10^{-8} C. You could move it past a magnet with a speed of about 10 m/sec. In Chapter 17, we saw that the magnetic field near the end of a small bar magnet could be around 0.1 T. Compare the approximate magnetic force on a moving invisible tape near a magnet with the approximate electric force that two charged tapes exert on each other when they are 10 cm apart.

Circular (or Helical) Motion in a Magnetic Field

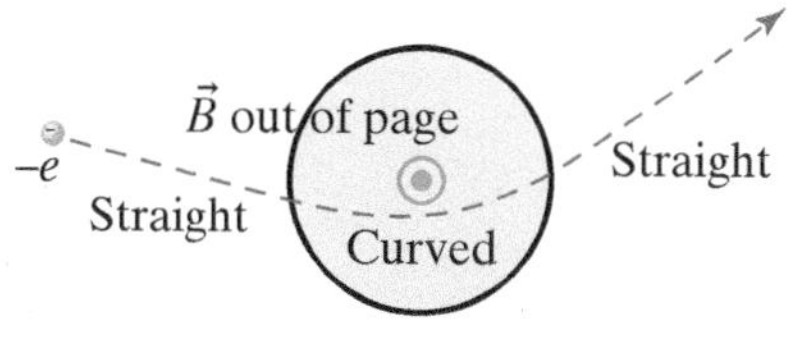

Figure 20.3 Electron trajectory through a limited region of magnetic field.

In a CRT the region of magnetic field that deflects the electrons is mainly confined to the region of the deflection coils, and the trajectory of an electron is nearly straight before entering the magnetic field region and after leaving it (Figure 20.3).

However, if there is a large enough region of uniform magnetic field, a charge can go around and around in a circle, and the curved portion of the electron trajectory in a CRT is a circular arc (if the magnetic field is fairly uniform). There are many other examples of circular orbits of charged particles in magnetic fields, including particle trajectories in particle accelerators, and the orbits in the magnetic field of the Earth of high-altitude electrons that produce the Northern Lights.

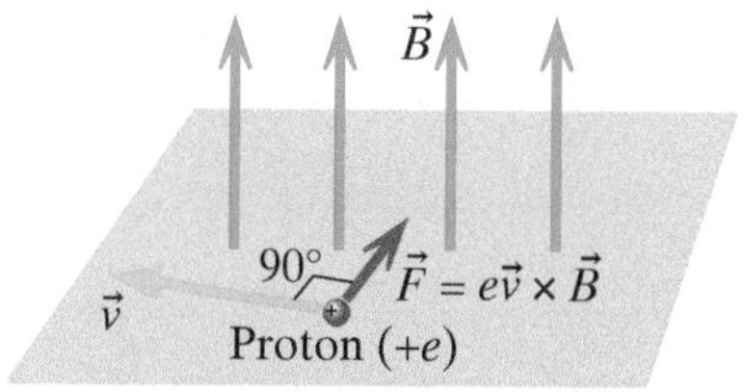

Figure 20.4 Circular motion of a proton in a region of uniform magnetic field.

In Figure 20.4 a proton moves horizontally with speed v in a large region of uniform vertically upward magnetic field (neglect the much smaller effects of gravity). The magnetic force $\vec{F} = q\vec{v} \times \vec{B}$ is perpendicular to the velocity $\vec{v}$, which means that the magnetic force will change the direction of the velocity without changing the magnitude of the velocity (the speed v), and no work is done by the magnetic force in this situation. If the motion is not horizontal, the component of velocity parallel to the magnetic field will not be changed, so the particle's path will be a helix (spiral) instead of a circle.

Circular Motion at Any Speed

This situation is similar to that of a rock going around in a circle at the end of a rope, or the Moon orbiting the Earth in a circular orbit. Consider the general case of a particle traveling at any speed (including near the speed of light). Recall that for a particle traveling in a circular orbit of radius R at constant speed, the rate of change of the particle's momentum points toward the center of the circle, and has this magnitude:

$$\left|\frac{d\vec{p}}{dt}\right| = p\left(\frac{v}{R}\right) \quad \text{where } p = \gamma m v = \frac{1}{\sqrt{1-(v/c)^2}} m v$$

The quantity v/R is the angular speed ω (omega), which has units of radians/s, so

$$p\left(\frac{v}{R}\right) = \omega p = \omega \gamma m v$$

The Momentum Principle tells us that the rate of change of the particle's momentum is equal to the net force acting on the particle, which in this case is the magnetic force $\vec{F} = q\vec{v} \times \vec{B}$, so

$$\omega \gamma m v = |q\vec{v} \times \vec{B}| = |q| v B \sin 90^\circ$$

$$\omega = \frac{|q|B}{\gamma m} \quad \text{valid at any speed}$$

Circular Motion at Low Speeds

An important special case is circular motion of a charged particle at speeds small compared to the speed of light c, in which case $\gamma \approx 1$ and the angular speed ω is (approximately) independent of the speed v:

$$\omega \approx \frac{|q|B}{m} \quad \text{if } v \ll c$$

An alternative derivation of this result starts from the nonrelativistic form of the Momentum Principle. If $v \ll c$, $\vec{p} \approx m\vec{v}$, so

$$\left|\frac{d\vec{p}}{dt}\right| = p\left(\frac{v}{R}\right) \approx \frac{v(mv)}{R} = \frac{mv^2}{R}$$

$$\frac{d\vec{p}}{dt} \approx m\frac{d\vec{v}}{dt} = \vec{F}_{\text{net}}$$

$$\frac{mv^2}{R} = |q|vB\sin 90°$$

$$m\omega = |q|B \quad \text{or} \quad \omega = \frac{|q|B}{m}$$

Time Required for a Circular Orbit

Let T = the time required for a charged particle to go around once in a complete circular orbit.

QUESTION Show that $T = 2\pi\frac{m}{|q|B}$ for nonrelativistic speeds ($v \ll c$).

Since by definition the angular speed ω is equal to 2π rad per the amount of time T required to go around once, $\omega = 2\pi/T$, and $\omega \approx |q|B/m$ if $v \ll c$, the result follows. This is an unusual result, because it is independent of the speed v and radius R. It says that the particle will take the same amount of time to go around in a circle in a uniform magnetic field, no matter how fast it is going (as long as $v \ll c$). The faster it goes, the bigger the radius of the circle, but the longer distance around the bigger circle is exactly compensated by the higher speed. This independence of the orbit time is the basis for the cyclotron, to be discussed in a moment.

Determining the Momentum of a Particle

In a region of known magnetic field, we can find the momentum of a particle of known charge (even one traveling at a speed close to the speed of light) by measuring the radius of curvature of its motion in a magnetic field.

$$\left|\frac{d\vec{p}}{dt}\right| = p\left(\frac{v}{R}\right) = |q|vB \quad \text{at any speed, so} \quad p = |q|BR$$

This result is used to measure momentum in high-energy particle experiments.

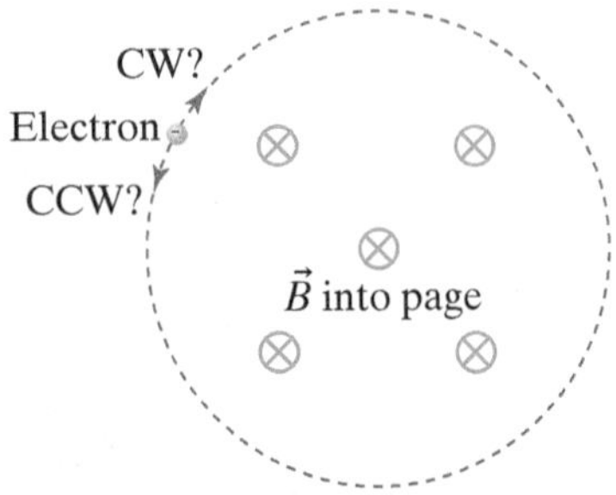

Figure 20.5 In which direction will the electron go in this uniform magnetic field (Checkpoint 2)?

Checkpoint 2 In Figure 20.5 there is a magnetic field going into the page, produced by a large current-carrying coil (not shown). Would an electron go in a clockwise or counterclockwise circle? Explain briefly. (*Hint:* Try both directions, and see which direction is consistent with the magnetic force.) Note that we're not asking what magnetic field is made by the moving electron, but what effect the coil's magnetic field has on the moving electron.

Application: A Cyclotron

One of the principal methods for studying the nucleus of an atom is to shoot high-velocity protons at the nucleus and observe reactions that take place. Because they repel each other electrically, in order to drive a proton into a nucleus we have to accelerate the proton through a potential difference of a million volts or more. For example, there is a million-volt potential difference from very far away (infinity) to the surface of a carbon nucleus ($q = 6e$), whose radius is about 1×10^{-14} m. A compact device for creating the equivalent of such large potential differences is the cyclotron. Ernest Lawrence invented the idea for a cyclotron in 1929 [see Richard Rhodes, *The Making of the Atomic Bomb,* Simon & Schuster (1986), pp. 145–148; this book contains an excellent history of early 20th-century physics].

A cyclotron consists of two D-shaped metal boxes (called "dees") placed in a strong magnetic field (Figure 20.6 and Figure 20.7). Protons can enter the dees through slits on the flat faces. Near the center of the cyclotron is a source of low-velocity ionized hydrogen (protons). Inside the metal dees the electric field is zero (because polarization of the metal box contributes an electric field inside the box that cancels any electric fields of external charges—see Chapter 21), and a proton follows a circular path in the magnetic field. As a proton goes from the left dee to the right dee, it receives a kick to the right due to a potential difference across the two dees. This increases the proton's momentum, so its path in the right dee has a larger radius of curvature.

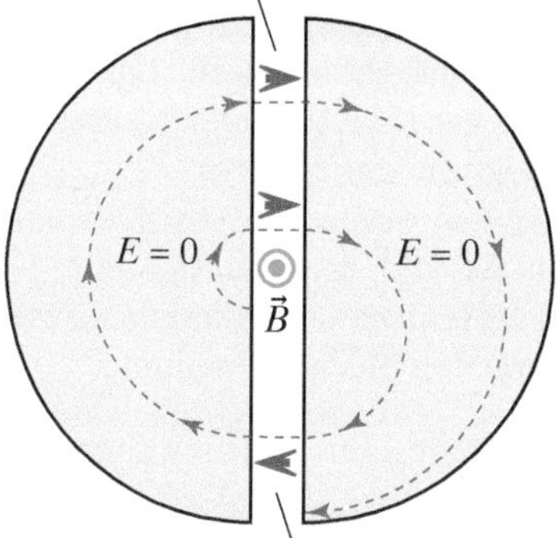

Figure 20.6 A cyclotron. Magnetic field is uniform in the $+y$ direction; the proton orbits in the x-z plane.

While the proton is inside the right dee (where the electric field is zero), we reverse the potential difference across the dees so that when the proton goes from the right dee to the left dee at the bottom of the drawing, it again receives a kick, this time to the left, resulting in a higher momentum and a larger radius of curvature in the left dee. By repeatedly kicking the proton through a modest potential difference, we can accelerate the proton up to very high momentum and energy that would otherwise require a practically unattainable potential difference of millions of volts. Problems P32 and P82 deal with the operation of a cyclotron.

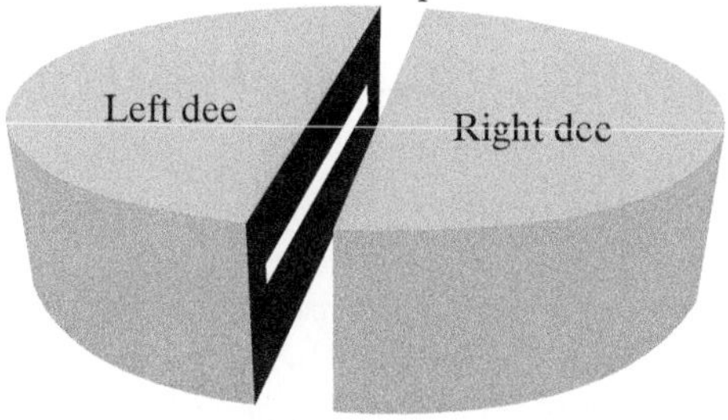

Figure 20.7 3D view of a cyclotron.

As we found earlier in this section, if $v \ll c$, the period of a proton's orbit does not change as its speed increases, so it is possible to apply a sinusoidally oscillating potential difference $\Delta V = \Delta V_0 \cos(\omega t)$ across the dees to accelerate the proton.

However, if the equivalent accelerating potential is more than about 100 million volts, as it is in modern accelerators, the speed v of a proton becomes a sizable fraction of the speed of light c, and relativistic effects are significant, so that

$$\omega = \frac{2\pi}{T} = \frac{|q|B}{\gamma m} \quad \text{depends on } v$$

At high speeds γ is no longer approximately equal to 1, so the period of the motion is no longer independent of the speed. Either the frequency of the accelerating voltage or the strength of the magnetic field in a cyclotron must be adjusted during the acceleration to compensate.

The largest modern circular accelerator is the Large Hadron Collider (LHC) at CERN (European Organization for Nuclear Research) in Geneva, Switzerland. In the final stage of acceleration, protons travel in a circle having a circumference of 27 km (17 mi)! The high-vacuum ring is buried underground and passes through parts of Switzerland and France. A smaller accelerator injects protons into this ring with a speed very close to the speed of light, $v = 0.999998c$, and in the large ring their speed gets even closer as they are accelerated to a speed of $v = 0.9999999c$. Although the speed hardly changes, the energy change is enormous, with $E = \gamma mc^2$ going from $480mc^2$ to $7500mc^2$. The magnetic field is increased during the acceleration to keep the protons at the same radius despite the large change in γ. At the maximum momentum, $B = \gamma mv/eR$ is 5.6 T, which is provided by electromagnets containing superconducting coils. (Actually, these magnets don't cover the entire circumference, and there are straight sections between the bending regions; the magnets reach 8.4 T.) The time to go around once is nearly constant as in a cyclotron, but for a completely different reason: the nearly constant speed is close to c. For more information, search online for LHC, or the campy and informative video, "Large Hadron Rap."

20.2 MAGNETIC FORCE ON A CURRENT-CARRYING WIRE

Often we are interested in the magnetic force exerted on a large number of charges moving through a wire in a circuit. Because the net force on a system (the wire) is the sum of all the forces acting on its individual components, adding up these forces will give the net force on the entire wire.

Let's calculate the magnetic force on a bunch of moving positive charges contained in a small volume with length Δl and cross-sectional area A (Figure 20.8). Their average drift velocity is $\bar{v}$. If there are n moving charges per unit volume, there are $nA\,\Delta l$ moving charges in this small volume.

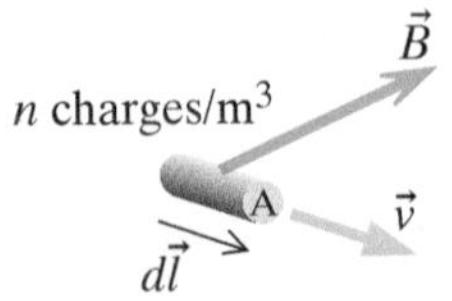

Figure 20.8 Calculate the magnetic force on a short current-carrying wire.

> **QUESTION** Show that the net magnetic force on all the moving charges in this short length of wire is
>
> $$\Delta\vec{F}_{\text{magnetic}} = I\Delta\vec{l} \times \vec{B}$$
>
> where $\Delta\vec{l}$ is a vector with magnitude Δl, pointing in the direction of the conventional current I.

Since there are $nA\,\Delta l$ moving charges affected by the magnetic field, the total force on this piece of wire is

$$(nA\Delta l)(q\bar{v} \times \vec{B}) = (qnA\bar{v})(\Delta\vec{l} \times \vec{B})$$

Note that $\bar{v} = |\bar{v}|\hat{v}$, so $\bar{v}(\Delta l) = |\bar{v}|\hat{v}(\Delta l) = |\bar{v}|(\hat{v}\Delta l) = |\bar{v}|\Delta\vec{l}$, and the conventional current is $I = qnA\bar{v}$, so we get

$$\Delta\vec{F}_{\text{magnetic}} = I\Delta\vec{l} \times \vec{B}$$

> **QUESTION** In most metal wires the moving charges are negative electrons traveling in the opposite direction. Explain briefly why the equation is valid even if the actual charge carriers are negative.

The electrons drift in a direction opposite to the conventional current, and their charge is negative. Therefore the quantity $q\bar{v}$ has two canceling minus signs, and the equation in terms of conventional current works fine, even if the actual charge carriers are negative electrons.

FORCE ON A SHORT LENGTH OF CURRENT-CARRYING WIRE

$$\Delta\vec{F}_{\text{magnetic}} = I\Delta\vec{l} \times \vec{B}$$

I is the conventional current; $|\Delta\vec{l}|$ is the length of the wire segment; $\Delta\vec{l}$ points in the direction of the conventional-current flow; $\vec{B}$ is the applied magnetic field.

This is not a new fundamental equation. It is merely the magnetic force added up for all the individual moving charges in a piece of wire. One way to keep this straight is to memorize the derivation. The key point is that there are $nA\,\Delta l$ electrons in a short length of wire, each moving with average speed $\bar{v}$, so that the sum of all the $|q|\bar{v}$'s is $(nA\,\Delta l)|q|\bar{v} = (|q|nA\bar{v})\Delta l = I\Delta l$.

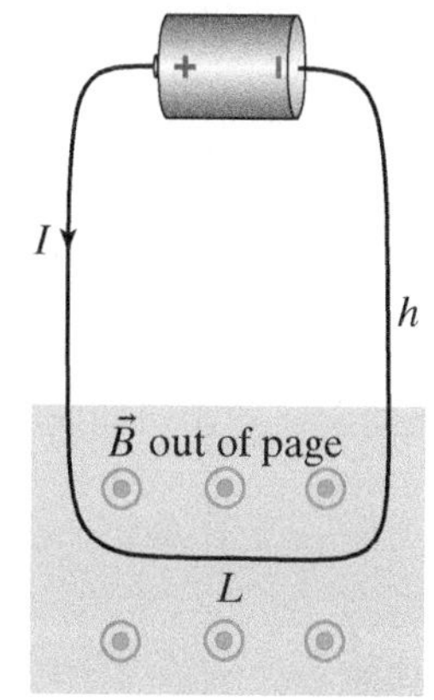

Figure 20.9 Magnetic forces on a loop of wire carrying conventional current I.

QUESTION Figure 20.9 shows a circuit consisting of a battery and a long wire, part of which is in a uniform magnetic field pointing out of the page. What is the direction of the force on those parts of the wire that are in the magnetic field?

Along each wire, the magnetic force points outward from the loop (toward the left along the left wire, down along the bottom wire, and toward the right along the right wire). Note that you get this result whether you consider positive charges drifting in the direction of the conventional current or electrons drifting in the opposite direction.

QUESTION Given those force directions, what is the net force on the wire in terms of the magnetic field B, current I, height h, and length L? (*Note:* Add up the contributions of lots of $\Delta\vec{l}$'s.)

The outward forces on the left and right wires cancel (though they tend to stretch the loop), and the net magnetic force is down. The magnitude is $(IL)B\sin 90° = IBL$.

QUESTION To get an idea of the numerical order of magnitude, suppose $B = 1$ T, $I = 10$ A, $h = 50$ cm, and $L = 10$ cm. What is the magnitude of the force?

The magnetic force depends only on L, not h, and $IBL = 1$ N, a rather small force (1 N is approximately the gravitational force on a small apple). Related experiments: EXP1–EXP3.

Keep in mind that there is really only one magnetic force expression, not two. Always remember that $\Delta\vec{F}_{\text{magnetic}} = I\Delta\vec{l} \times \vec{B}$ is simply the result of adding up the effects on many moving charges in a short length of wire, each of which experiences a magnetic force $\vec{F} = q\vec{v} \times \vec{B}$.

Application: Forces between Parallel Wires

An important application of the theory is the calculation of the magnetic force that parallel wires exert on each other, because the ampere is defined as the amount of conventional current in two very long parallel wires, a meter apart, which gives a force of one wire on the other of 2×10^{-7} N per meter of length, from which it follows that $\mu_0/4\pi$ is *exactly* equal to 1×10^{-7} T · m/A.

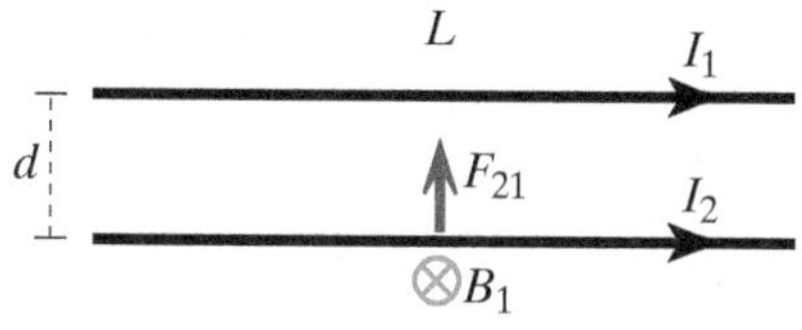

Figure 20.10 Parallel current-carrying wires exert magnetic forces on each other.

Figure 20.10 shows two very long parallel straight wires of length L, a distance d apart. The upper wire carries a conventional current I_1 and the lower wire carries a conventional current I_2 in the same direction. There are no magnets present.

Step 1, magnetic field due to upper wire: In order to calculate the force that the upper wire exerts on the lower wire, first we need to determine the magnitude and direction of the magnetic field $\vec{B}_1$ produced by the upper wire (carrying I_1) at the location of the lower wire. In Chapter 17 we calculated the magnetic field near a long straight wire. Using that result we have $B_1 \approx [\mu_0/(4\pi)][2I_1/d]$. The direction is into the page, using the right-hand rule.

Step 2, magnetic force on lower wire: Now we can calculate the magnitude of the magnetic force $\vec{F}_{21}$ exerted on the lower wire by the field produced by the upper wire:

$$F_{21} = I_2 L B_1 \sin 90° = I_2 L \left(\frac{\mu_0}{4\pi}\frac{2I_1}{d}\right)$$

Using the right-hand rule with $I\Delta\vec{l} \times \vec{B}$, the direction of the force is up, so that the lower wire is attracted to the upper wire.

Force on upper wire: Find the magnitude and direction of the magnetic force $\vec{F}_{12}$ exerted on the upper wire. This requires two steps:

> **QUESTION** First, what are the magnitude and direction of the magnetic field produced by the lower wire (carrying I_2) at the location of the upper wire?

We have $B_2 = [\mu_0/(4\pi)][2I_2/d]$. The direction is out of the page.

> **QUESTION** What are the magnitude and direction of the magnetic force $\vec{F}_{12}$ exerted on the upper wire by the field produced by the lower wire? Is Newton's third law obeyed (reciprocity of forces)? That is, is it the case that $\vec{F}_{12} = -\vec{F}_{21}$?

The magnitude of the force on the upper wire is this:

$$F_{12} = I_1 L B_2 \sin 90° = I_1 L \left(\frac{\mu_0}{4\pi}\frac{2I_2}{d}\right)$$

Using the right-hand rule with $I\Delta\vec{l} \times \vec{B}$, the direction of the force is down, so that the upper wire is attracted to the lower wire. The forces on the two wires have the same magnitude and opposite directions, so in this case the reciprocity of the magnetic forces holds. (However, this need not always be the case: see Question Q3 and Problem P30.)

Currents in Opposite Directions

> **QUESTION** If the two currents run in the same direction, we see that the magnetic interaction is an attraction. If the two currents run in opposite directions, is the magnetic interaction an attraction or a repulsion? Prove your assertion by showing the direction of a magnetic field and the force it exerts.

In this case you should find that two currents running in opposite directions repel each other. For electric forces we say that "likes repel, unlikes attract," but for the magnetic forces between two parallel current-carrying wires, we can say "likes attract, unlikes repel."

> **Checkpoint 3** To get a feel for the size of the effects, calculate the magnitude of the force per meter if $I_1 = I_2 = 10$ A and the distance between the wires is $d = 1$ cm.

20.3 COMBINING ELECTRIC AND MAGNETIC FORCES

If a charged particle is moving in a magnetic field, it is subject to a magnetic force $\vec{F}_m = q\vec{v} \times \vec{B}$. If there is an electric field in the region, the particle is also subject to an electric force $\vec{F}_e = q\vec{E}$.

$$\vec{F} = q\vec{E} + q\vec{v} \times \vec{B} \quad \text{net force due to electric and magnetic fields}$$

This net electric plus magnetic force is called the "Lorentz force." In one sense, the terms in this equation are a kind of definition of electric and magnetic fields in terms of the observable effects that these fields have on charges.

> **Checkpoint 4 (a)** Consider the situation in Figure 20.11, in which there is a uniform electric field in the x direction and a uniform magnetic field in the y direction. For each example of a proton at rest or moving in the x, y, or z direction, what is the direction of the net electric and magnetic force on the proton at this instant? **(b)** In each example in Figure 20.12, what is the direction of the net electric and magnetic force on an electron at this instant?

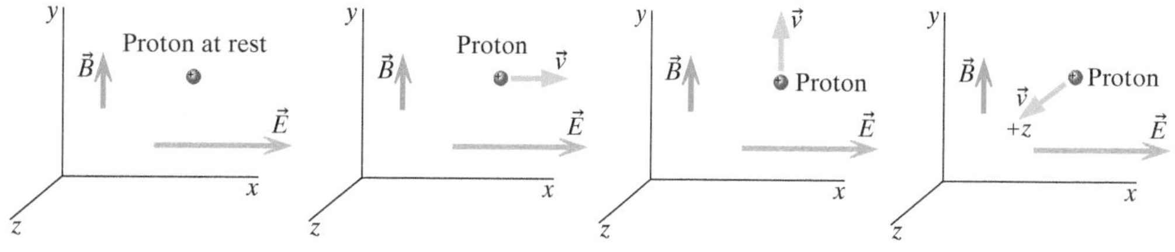

Figure 20.11 A proton in electric and magnetic fields.

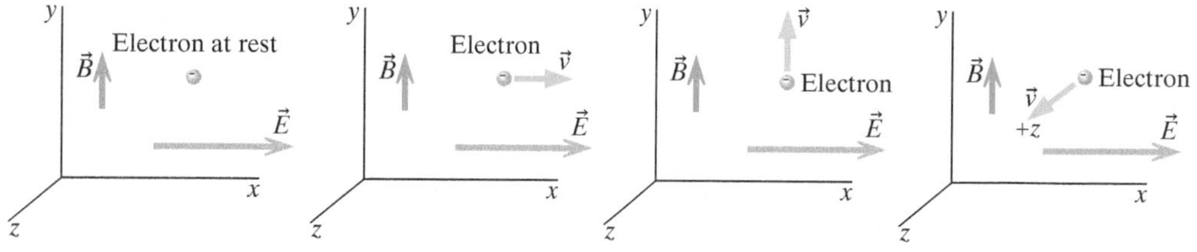

Figure 20.12 An electron in electric and magnetic fields.

Application: A Velocity Selector

It is possible to arrange electric and magnetic fields so that the combined electric and magnetic force $\vec{F} = q\vec{E} + q\vec{v} \times \vec{B}$ acting on a positive charge $+q$ moving with velocity $\vec{v}$ is zero.

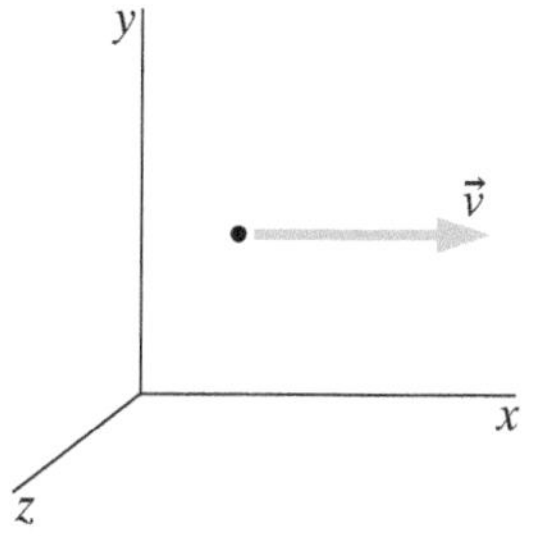

Figure 20.13 Sketch electric and magnetic fields in three-dimensional space so that the net electric and magnetic force on the particle is zero.

QUESTION With the velocity $\vec{v}$ in the x direction (Figure 20.13), try to sketch a three-dimensional diagram of an arrangement of $\vec{E}$ and $\vec{B}$ that could lead to a net electric and magnetic force of zero, in which case the particle will move with constant velocity through the region.

Several arrangements are possible, one of which is shown in Figure 20.14.

QUESTION If the magnetic field in this situation has a magnitude B, what must be the magnitude E of the electric field?

Since the magnetic and electric forces must be equal and opposite to produce a zero net force,

$$eE = evB \sin 90^\circ$$
$$E = vB$$

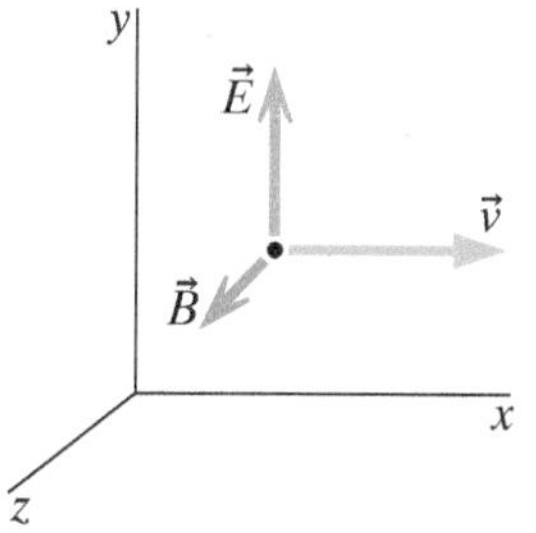

Figure 20.14 One configuration of fields that can allow a charged particle to travel in a straight line.

QUESTION Now suppose that a particle with the same charge q comes through this region, traveling in the same direction but at a slower speed. Explain briefly why it will *not* move in a straight line.

Now $qE > qvB$, so the net force isn't zero.

QUESTION If a particle with the correct velocity but with a *negative* charge $-q$ comes through, will it pass straight through or not? (Make a diagram.)

Yes. Both the electric force and the magnetic force are flipped 180°, and the net force is still zero.

Thus a region containing crossed electric and magnetic fields (that is, perpendicular to each other) forms a kind of velocity selector, allowing particles to pass straight through only if they have the chosen speed and direction. You might notice that the cross product $\vec{E} \times \vec{B}$ points in the direction of the velocity that can pass straight through.

EXAMPLE **Electron and Current Loop**

A two-turn circular loop of wire of radius R is fed a large conventional current I from long straight wires (Figure 20.15). A charge of magnitude $+Q$ is a distance $3R$ to the right of the center of the loop. An electron traveling upward with speed v passes through the center of the loop, and the net force on this electron at this instant is zero. The Earth's magnetic field is negligible in this region. What are the magnitude and direction of the current I? Assume that all other quantities are known.

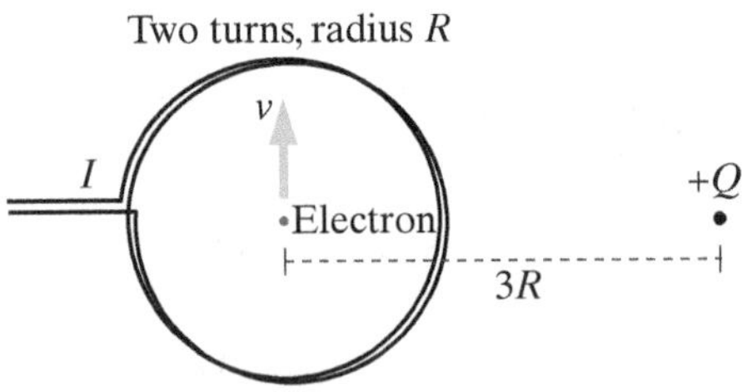

Figure 20.15 An electron inside two turns of current-carrying wire.

Solution The key idea here is that $\vec{F}_{\text{magnetic}} + \vec{F}_{\text{electric}} = 0$ at this instant.

Direction of current

At the location of the electron, the electric field due to the positive charge is in the $-x$ direction, so the electric force on the electron is in the $+x$ direction, as shown in Figure 20.16. Since at the instant specified the net force on the electron is zero, the magnetic force on the electron must be equal and opposite to the electric force. $\vec{F}_{\text{magnetic}}$ is in the $-x$ direction, but the electron's charge is negative, so $\vec{v} \times \vec{B}$ must be in the $+x$ direction, as shown in Figure 20.17. By trial and error with the right-hand rule (thumb in direction of $\vec{v} \times \vec{B}$, fingers initially pointing up), we find that $\vec{B}$ must be in the $+z$ direction. Applying the right-hand rule for a current loop, we find that conventional current must flow counterclockwise around the loops to make a magnetic field in this direction.

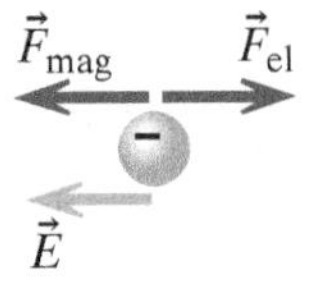

Figure 20.16 The directions of the electric and magnetic forces on the electron.

Magnitude of current

Since the net force on the electron is zero, we know that:

$$|\vec{F}_{\text{magnetic}}| = |\vec{F}_{\text{electric}}|$$

$$\frac{1}{4\pi\varepsilon_0}\frac{eQ}{(3R)^2} = evB$$

$$B = \frac{\dfrac{1}{4\pi\varepsilon_0}\dfrac{Q}{9R^2}}{v}$$

We found in an earlier chapter that the magnetic field of a loop is this:

$$B_{\text{loop}} = \frac{\mu_0}{4\pi}\frac{2\pi R^2 I}{(z^2 + R^2)^{3/2}}$$

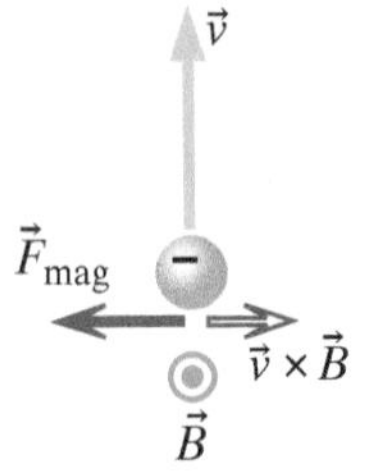

Figure 20.17 Finding the direction of the magnetic field.

At the center of the loops $z = 0$, so $B_{\text{loop}} = \dfrac{\mu_0}{4\pi}\dfrac{2\pi I}{R}$, and for two loops:

$$\frac{\dfrac{1}{4\pi\varepsilon_0}\dfrac{Q}{9R^2}}{v} = 2\frac{\mu_0}{4\pi}\frac{2\pi I}{R}$$

$$I = \frac{1}{\varepsilon_0\mu_0}\frac{Q}{36\pi R v}$$

This quantity can be shown to have units of C/s, or amperes.

Checkpoint 5 If an electron traveling at a velocity of $\langle 0, 0.9c, 0\rangle$ passes through a region where the magnetic field is $\langle 1, 0, 0\rangle$ T, what electric field would be necessary to keep the electron traveling in a straight line? Give $\vec{E}$ as a vector.

20.4 THE HALL EFFECT

Magnetic forces on moving charged particles inside a conductor can cause the conductor to polarize. This is the basis of the Hall effect.

The Initial Transient: Polarization Develops

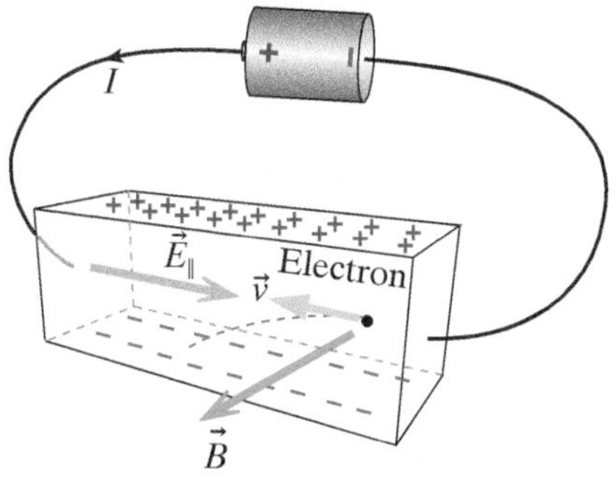

Figure 20.18 Magnetic deflection of electrons in a conducting bar. $\vec{E}_{\parallel}$ is the electric field due to the battery and surface charge on the bar (not shown), which drives the current. Polarization shown is due to magnetic deflection of electrons during the transient phase.

Mobile electrons moving inside a metal wire can be deflected to the surface of the wire by a magnetic field that is perpendicular to the wire. In Figure 20.18, large coils (not shown) apply a uniform magnetic field of magnitude B throughout the region of the circuit. As usual, mobile electrons move through the wire in a direction opposite to the electric field driving the current. We will call this electric field $\vec{E}_{\parallel}$, because it is parallel to the wire, as we saw in Chapter 18.

Verify for yourself that the moving electrons will be deflected toward the bottom surface of the wire (shown as a rectangular bar for simplicity). This will make the bottom surface negatively charged and the top surface positively charged (due to a deficiency of electrons). These charges are in addition to the usual surface charges, which for clarity we don't show on the diagram.

The Steady State: Magnetic Force Balanced by Electric Force

As time goes on, more and more electrons will pile up on the bottom of the wire. However, this piling up won't continue forever. When enough electrons have piled up on the bottom (with a corresponding depletion of electrons on the top surface), these extra surface charges will produce a new electric field that applies a vertical electric force to the moving electrons inside the wire.

> QUESTION What is the direction of this new electric field?

Inside the metal wire, the new electric field points down, pointing away from the positive charge on the top and toward the negative charge on the bottom. We'll call this the "transverse" electric field, $\vec{E}_{\perp}$.

> QUESTION When no more electrons pile up on the bottom, what is the magnitude $E_{\perp}$ of the transverse (perpendicular) electric field, in terms of the drift speed $\bar{v}$ of the free electrons and the magnetic field B? (*Hint:* If no more electrons are piling up, what do you know about the electric and magnetic forces on the moving electrons?)

Pileup will cease when the electric force grows to be as large as the magnetic force, at which time the net vertical force becomes zero, and the electrons move horizontally through the wire, with no vertical deflection. When this happens we know the vertical force on an electron is zero, so the magnetic force must equal the perpendicular electric force:

$$|\vec{F}_{\text{electric}\perp}| = |\vec{F}_{\text{magnetic}}|$$
$$eE_{\perp} = e\bar{v}B\sin 90°, \text{ so } E_{\perp} = \bar{v}B$$

Note the similarity between this analysis and the analysis of the velocity selector in Section 20.3.

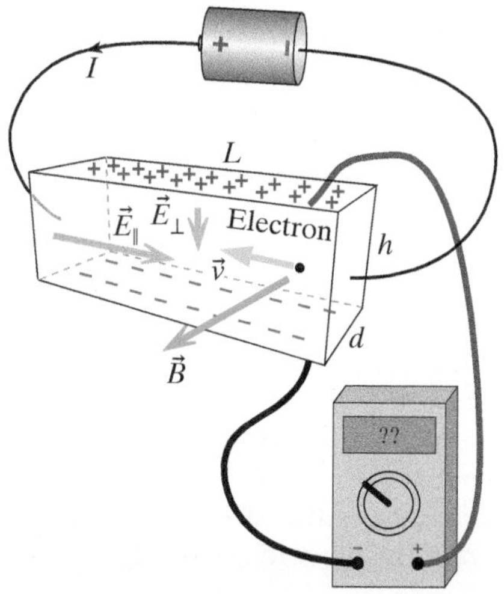

Figure 20.19 Measuring the transverse potential difference. The voltmeter leads are directly across from each other on the top and bottom surfaces.

> QUESTION Suppose that you attach the leads of a digital voltmeter to the top and bottom surfaces as shown in Figure 20.19. Determine the magnitude and sign of the voltage displayed on the voltmeter (the wire has height h, depth d, and length L; there is a uniform magnetic field of magnitude B).

Since the electric field points down inside the wire, the electric potential is higher on top. The electric field has the uniform value $E_{\perp} = \bar{v}B$ throughout the wire, so the voltmeter will read a positive value $+E_{\perp}h = +\bar{v}Bh$.

> QUESTION How big is this voltage if there is no magnetic field?

Since the potential difference is $\bar{v}Bh$, the voltage will be zero if $B = 0$.

The appearance of a sideways or transverse potential difference across a current-carrying wire in the presence of a magnetic field is called the "Hall effect," named for its discoverer.

Let's see approximately how big the Hall voltage is in a metal. Consider a bar of iron of length 5 cm, height 2 cm, and depth 0.5 mm, carrying a conventional current of 15 A to the right in the presence of a perpendicular magnetic field of magnitude 2 T, as shown in Figure 20.19. There are about 8.4×10^{28} mobile electrons per cubic meter in iron.

The magnetic force involves the average drift speed of the moving charge:

$$\bar{v} = \frac{I}{|q|nA} = \frac{15\text{ A}}{(1.6 \times 10^{-19}\text{ C})(8.4 \times 10^{28}\text{ m}^{-3})(0.02\text{ m})(5 \times 10^{-4}\text{ m})}$$
$$= 1.11 \times 10^{-4}\text{ m/s}$$

In the steady state the vertical force (the force perpendicular to current flow) on a moving charge is zero:

$$F_{\text{magnetic}} = F_{\text{electric}\perp} \quad \text{and therefore} \quad evB = eE_{\perp}$$
$$E_{\perp} = \bar{v}B = (1.11 \times 10^{-4}\text{ m/s})(2\text{ T}) = 2.22 \times 10^{-4}\text{ V/m}$$
$$\Delta V_{\text{Hall}} = E_{\perp}h = (2.22 \times 10^{-4}\text{ V/m})(0.02\text{ m}) = 4.5 \times 10^{-6}\text{ V}$$

This is a very small voltage, but it can be measured with a sensitive voltmeter whose leads are carefully aligned vertically across the bar.

Determining the Density of Mobile Charges

We can use measurements of the Hall voltage and the conventional current to determine the density of mobile charges in a metal.

$$\Delta V_{\text{Hall}} = E_{\perp}h = \bar{v}Bh = \left(\frac{I}{|q|nA}\right)(Bh)$$
$$n = \frac{I}{|q|\Delta V_{\text{Hall}}A}(Bh)$$

For some monovalent metals such as sodium, Hall effect measurements show n to be the same as the number of atoms per cubic meter, indicating that each atom gives up one electron into the free-electron sea. In other metals there may be more than one electron contributed per atom.

Determining the Sign of the Mobile Charges

The Hall effect offers a very special insight into the nature of the "charge carriers" (mobile charges) responsible for a current. Most electrical measurements come out the same whether electrons are moving inside the wire or positive particles (holes in a metal) are moving in the opposite direction. However, the Hall effect differs in the two cases.

Consider what would happen if *positive* particles (holes in the case of aluminum or zinc) were moving to the *right* in the wire as in Figure 20.20, instead of electrons moving to the left.

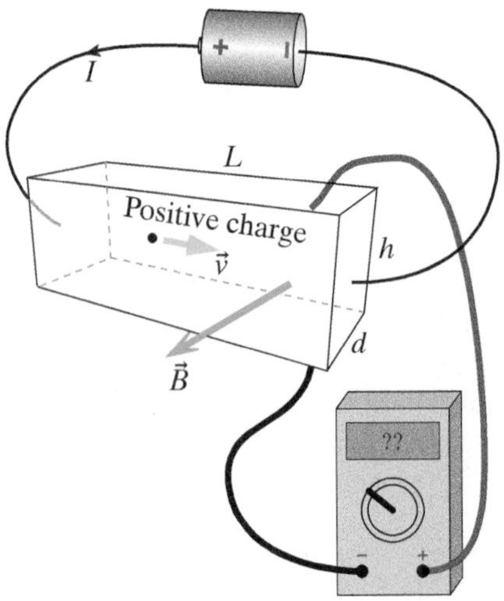

Figure 20.20 What happens if the charge carriers are positive? The voltmeter leads are directly across from each other on the top and bottom surfaces.

QUESTION On a diagram like Figure 20.20, show the charge pileup on the top and bottom surfaces and the electric field due to that pileup. What would be the sign of the voltage displayed on the voltmeter?

Now it is the bottom of the wire that has additional positive surface charge, rather than the top being positive as it was with negative charge carriers. The transverse electric field now points up, and the voltmeter reads a negative voltage.

HALL EFFECT AND THE SIGN OF CHARGE CARRIERS

By measuring the Hall effect for a particular material, we can determine the sign of the moving particles that make up the current.

In the steady state, the Hall effect is an example of the straight-line motion of charged particles through crossed electric and magnetic fields. Remember that the surface charges that accumulate due to the Hall effect are not the only charges on the wires; there are also the usual surface charges responsible for the longitudinal electric field in the wires. The distinction is that the ordinary surface charges produce electric fields inside the wire in the direction of the conventional current, whereas the Hall effect surface charges produce a transverse electric field $\vec{E}_{\perp}$, perpendicular to the current.

Checkpoint 6 Calculate the Hall voltage ΔV_{Hall} for the case of a ribbon of copper 5 mm high and 0.1 mm deep, carrying a current of 20 A in a magnetic field of 1 T. For copper, $q = -e$, and $n = 8.4 \times 10^{28}$ free electrons per cubic meter (one per atom).

This is a very small voltage, but it can be measured with a sensitive voltmeter. In semiconductors the density of charge carriers n may be quite small compared to a metal, in which case the Hall effect voltage can be quite large, since it is inversely proportional to n.

It is instructive to compare this transverse Hall voltage in copper with the voltage you would measure *along* the copper ribbon. Calculate the potential difference you would measure along 5 mm of the ribbon; the conductivity of copper is

$$\sigma = 6 \times 10^7 \frac{\text{A/m}^2}{\text{V/m}}$$

Since this voltage is much larger than the Hall voltage, experimenters must take special precautions to place their voltmeter leads exactly across from each other, to avoid a spurious reading due to the potential difference along the bar. One way to check for this problem is to remove the magnetic field, in which case the Hall voltage should go to zero.

Practical Applications of the Hall Effect

Measuring the Hall effect potential difference for a material can provide information about the sign of the mobile charges and about the number of them per unit volume. For example, Hall effect measurements have confirmed that in some metals and semiconductors, it is positive holes in the electron sea rather than the electrons themselves that are the dominant charge carriers. Hall effect measurements of mobile electron density for some monovalent metals such as sodium and copper show n to be the same as the number of atoms per cubic meter, which shows that each atom gives up one electron into the free-electron sea. In other metals there may be more than one electron contributed per atom.

Another practical use of the Hall effect turns the relationship around. There are commercial devices that measure the magnitude and direction of an unknown magnetic field by observing the Hall voltage in a material in which the sign and number density of the mobile charges are known.

Why Does a Current-Carrying Wire Move?

There is a subtle point about the magnetic force that a magnetic field exerts on a current-carrying wire. The magnetic force only acts on moving particles, which in the case of a copper wire are the drifting electrons. The stationary positive atomic cores do not experience a magnetic force—so why does the entire wire move?

The answer to this puzzle is closely related to the Hall effect. In Figure 20.21, consider again a current-carrying metal wire in a magnetic field, with a growing number of electrons piling up on the bottom surface (and a deficiency of electrons accumulating on the top surface). Again, for clarity we don't show the usual surface charges along the wire; we just show the extra surface charges due to the magnetic deflection of the electrons.

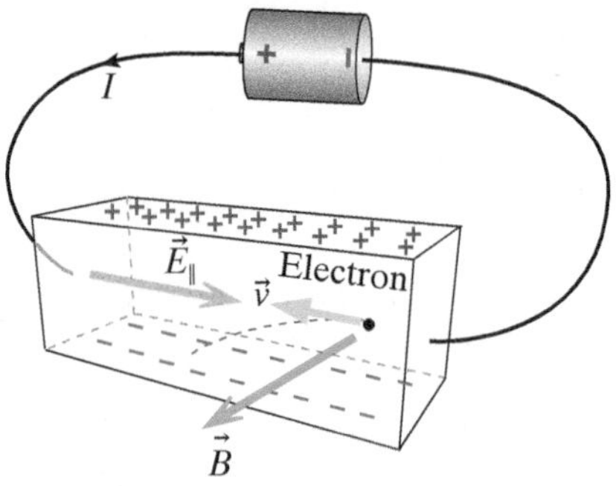

Figure 20.21 Deflected electrons drag the wire along too.

The extra surface charges exert an electric force $eE_{\perp}$ upward on the moving electrons, which in the steady state just balances the magnetic force evB. The extra surface charges also exert an electric force $eE_{\perp}$ on the positive atomic cores, and this is an unbalanced force, because the atomic cores initially are not moving and hence are not subject to a magnetic force.

> QUESTION What is the direction of this electric force exerted by the surface charges on the atomic cores?

Evidently the positively charged atomic cores are pulled downward. The electric force due to the Hall effect surface charges forces the wire as a whole to move in the direction of $\Delta\vec{F}_{\text{magnetic}} = I\Delta\vec{l} \times \vec{B}$, and the magnitude of the electric force is equal to this force. The overall effect is that in Figure 20.21 the atomic cores are dragged downward as the electrons are dragged downward. The motion of the wire is an electric side effect of the magnetic force on the moving electrons.

Another way of thinking about this is to say that the electrons are bound electrically to the metal wire—they can't simply fall out of the wire. Thus when the electrons are pushed down by the magnetic force, they drag the wire along with them. Note that ultimately it is the battery that supplies energy to the wire.

EXAMPLE

Determining Electrical Properties

A bar made out of a new conducting material is 9 cm long with a rectangular cross section 4 cm wide and 2 cm deep. The bar is inserted into a circuit with a 1.5 V battery and carries a constant current of 0.7 A (Figure 20.22). The resistance of the copper connecting wires and the ammeter, and the internal resistance of the battery, are all negligible compared to the resistance of the bar.

Using large coils not shown in the diagrams, a uniform magnetic field of 1.7 T is applied perpendicular to the bar (to the left, as shown in Figure 20.22). A voltmeter is connected across the bar, with the connections across the bar carefully placed directly across from each other as shown. The voltmeter reads −0.29 mV. Remember that a voltmeter reads positive if the positive lead is connected to a higher potential location than the negative lead. There is only one kind of mobile charge in the bar material. **(a)** What is the sign of the mobile charges in the bar, and which way do they move? Explain carefully, using diagrams to support your explanation. **(b)** How long does it take for a mobile charge to go from one end of the bar to the other? **(c)** What is the mobility u of the mobile charges? **(d)** If each mobile charge is singly charged ($|q| = e$), how many mobile charges are there in 1 m^3 of this material?

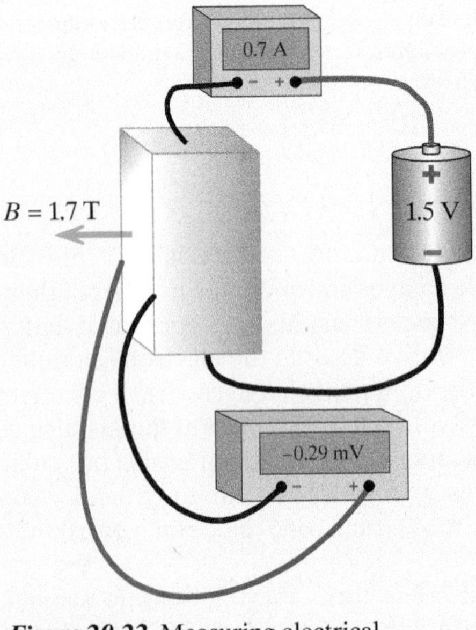

Figure 20.22 Measuring electrical properties of a new material.

Solution **(a)** This requires trial and error—we assume that the mobile charges have a particular sign, and see whether this would lead to the observed polarization. First we need to draw $\vec{E}_{\parallel}$, the field established by the batteries and surface charge on the bar, which drives the current in the bar (see Figure 20.23).

If the mobile charges are positive, they will move in the direction of $\vec{E}_{\parallel}$. As is shown in Figure 20.23, this would lead to positive charge buildup on the back of the bar (magnetic force into the page). The voltmeter would give a positive reading, because $\vec{E}_{\perp}$ would point out of the page. This is not what is observed, so the mobile charges must not be positive.

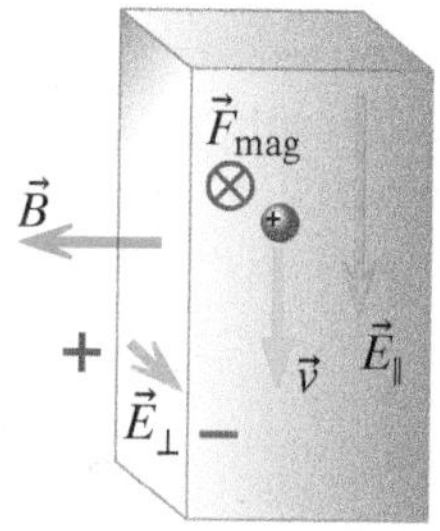

Figure 20.23 If the mobile charges are positive their velocity will be in the direction of $\vec{E}_{\parallel}$. The magnetic force will be into the page, so the back of the bar will become positive.

If the mobile charges are negative, they will move opposite to $\vec{E}_{\parallel}$, as shown in Figure 20.24. The magnetic force would again be into the page, making the back of the bar negative, and creating $\vec{E}_{\perp}$ pointing into the page. The voltmeter reading would be negative, as is observed. We can conclude that the mobile charges are negative.

(b) We need to find the average drift speed of the mobile charges, which we can find from the magnetic force on an electron. In the steady state, the transverse magnetic and electric forces balance:

$$eE_{\perp} = e\bar{v}B$$

$$\frac{\Delta V_{\perp}}{d} = \bar{v}B \quad \text{because} \quad E_{\perp} = \frac{\Delta V_{\perp}}{d}$$

$$\bar{v} = \frac{\Delta V_{\perp}}{Bd} = \frac{0.29 \times 10^{-3}\ \text{V}}{(1.7\ \text{T})(0.02\ \text{m})} = 8.5 \times 10^{-3}\ \text{m/s}$$

$$\Delta t = \frac{0.09\ \text{m}}{8.5 \times 10^{-3}\ \text{m/s}} = 10.6\ \text{s}$$

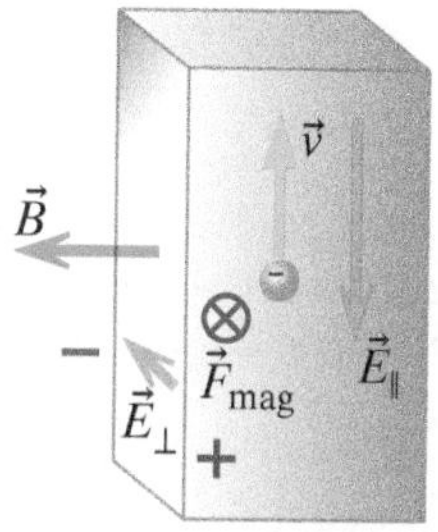

Figure 20.24 If the mobile charges are negative, their velocity will be opposite to $\vec{E}_{\parallel}$. The magnetic force will again be into the page, so the back of the bar will become negative.

(c) The drift speed of the mobile charges depends on the longitudinal electric field and on the charge mobility.

$$\bar{v} = uE_{\parallel}$$

$$E_{\parallel} = (1.5\ \text{V})/(0.09\ \text{m}) = 16.7\ \text{V/m}$$

$$u = \frac{\bar{v}}{E_{\parallel}} = \frac{(8.5 \times 10^{-3}\ \text{m/s})}{(16.7\ \text{V/m})} = 5 \times 10^{-4} (\text{m/s})/(\text{V/m})$$

(d) $I = enA\bar{v}$, so $n = I/(eA\bar{v})$.

$$n = \frac{(0.7\ \text{A})}{(1.6 \times 10^{-19}\ \text{C})(0.02\ \text{m})(0.04\ \text{m})(8.5 \times 10^{-3}\ \text{m/s})} = 6.4 \times 10^{23}\ \text{m}^{-3}$$

20.5 MOTIONAL EMF

A magnetic field exerts a force on a current-carrying wire. This effect can be reversed: moving a wire through a magnetic field makes a current run in the wire, which provides a way to generate electricity from mechanical work.

The Initial Transient: Polarization Develops

Consider a metal bar of length L that is moving through a region of uniform magnetic field, out of the page, with magnitude B (Figure 20.25). Because a mobile electron inside the moving metal bar is moving to the right, it experiences a magnetic force.

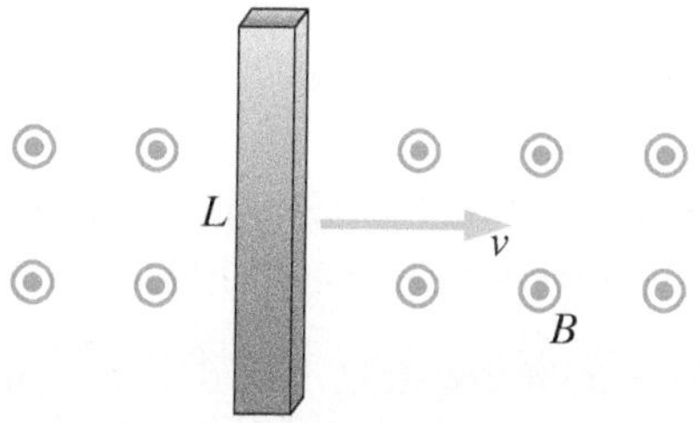

Figure 20.25 A metal bar moving at constant velocity through a region of uniform magnetic field into the page.

QUESTION If the bar is moving at a speed v, what are the magnitude and direction of the magnetic force on this electron?

The magnetic force $\vec{F} = (-e)\vec{v} \times \vec{B}$ points upward, and the magnitude of the force is $evB\sin(90°) = evB$. As a result of this magnetic force, the mobile-electron sea will shift upward.

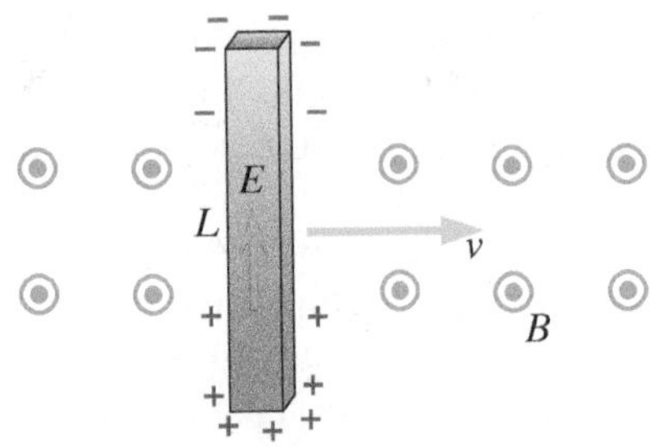

Figure 20.26 The metal bar is polarized as shown, with a downward Coulomb electric field inside the metal.

QUESTION What will be the resulting approximate charge distribution on the metal bar? What is the direction of the electric field $\vec{E}$ produced by this charge distribution inside the metal?

The magnetic force on the moving bar polarizes the bar so that it becomes negative at the top and positive at the bottom. As a result, there is an electric field $\vec{E}$ inside the metal that points upward (Figure 20.26). A mobile electron inside the metal is subject to an electric force due to this electric field as well as to the magnetic force.

The Steady State: Magnetic Force Balanced by Electric Force

QUESTION If the bar is kept moving at a constant speed v, how big will E be compared to B, in the steady state? Why? (*Hint:* Why doesn't the polarization keep increasing as more and more electrons pile up at one end of the bar?)

At first the only force on a mobile electron is the upward magnetic force, but the more the mobile-electron sea shifts upward, the larger the downward electric force $-e\vec{E}$ on an electron. Polarization continues until the electric field grows so large that the net upward force ($evB - eE$) is zero. Therefore in the steady state we have $E = vB$.

This situation is a bit odd. Here we have a nonzero electric field inside a metal in equilibrium (bar moving at constant speed, no accelerations). There is no current, because the electric force is balanced by a magnetic force: the net force on a mobile electron inside the metal is zero.

QUESTION What is the potential difference ΔV from one end of the bar to the other? Which end is at the higher potential?

Since the electric field $E = vB$, and v and B are the same throughout the bar, the electric field E is uniform, and the potential difference is simply $\Delta V = EL = vBL$, with the lower (positive) end of the bar at the higher potential (moving downward in the bar, we go against the direction of the electric field, which means we're moving toward higher potential).

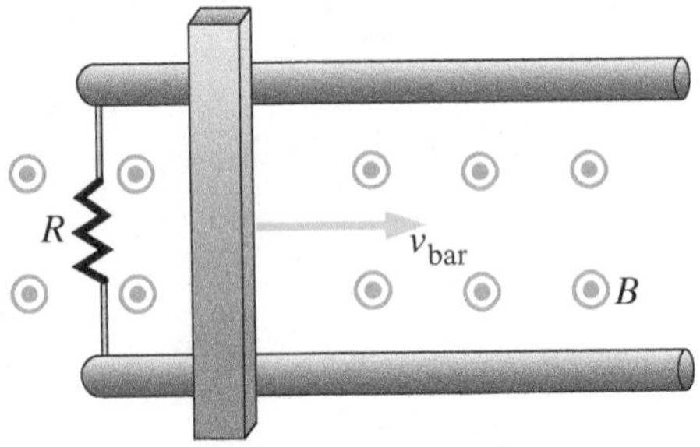

Figure 20.27 Pull the metal bar at constant speed along frictionless metal rails. A resistor is connected between the rails.

QUESTION Once the bar is completely polarized, how much force must we apply to the bar to keep it moving at a constant speed v?

In the steady state there is no current running in the bar, so there is no net force on the bar due to the magnetic field, and the polarized bar will coast along with no force required to keep it going.

Motional Emf Can Drive Current

QUESTION Does this polarized bar remind you of any object we've studied previously? What object or device have we seen before that is most similar to the polarized bar?

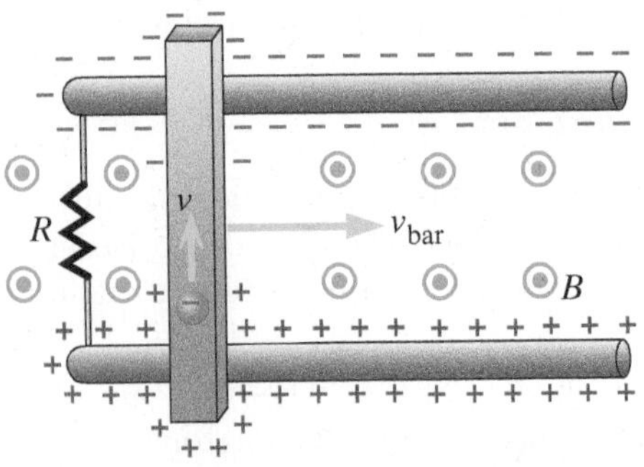

Figure 20.28 Approximate surface-charge distribution on the circuit.

This looks like a battery, because a charge separation is maintained from one end to the other by a non-Coulomb force, in this case a magnetic force. We can exploit this effect to generate an electric current.

Suppose that you pull the metal bar at a constant speed v_{bar} along frictionless metal rails through the uniform magnetic field (Figure 20.27). The moving metal bar and the metal rails have negligible resistance, but a resistor with resistance R is connected between the rails. A current will run in this circuit, driven by a "battery" whose role is played by the moving bar, and the phenomenon is called "motional emf." We show an approximate surface-charge distribution on the circuit (Figure 20.28). Electrons are driven the "wrong" way through the moving bar, moving toward the negative end of the bar.

To understand motional emf in detail, we'll start with the bar at rest, so there is no emf and no current, and we will reason step by step. To begin, you apply a force of magnitude F to the right to start the bar moving. Because the net electric force on a mobile electron in a metal is effectively zero, the first thing that happens when the bar starts moving to the right is that the mobile electrons are left behind (see the optional section "Electrons in Metals" near the end of Chapter 19 for another example of this). The bar almost instantaneously becomes polarized as shown in Figure 20.29. Now there is a nonzero electric force on a mobile electron, due to the bar having become polarized, so the electrons begin to move to the right. Another way of describing the situation is to say that you pull the lattice of positive ions to the right and the mobile electrons in turn are pulled to the right by electric interactions with the positive charges.

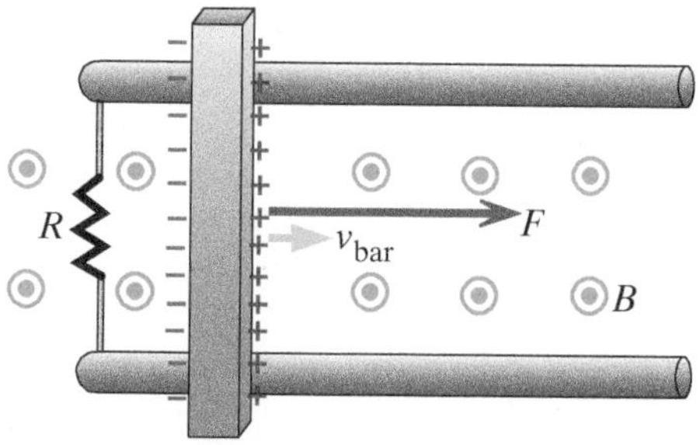

Figure 20.29 When the bar starts to move, initially the mobile electrons are left behind, so the bar polarizes left–right.

> QUESTION In the presence of the magnetic field out of the page ($+z$), as shown in Figure 20.27, what is the direction of the magnetic force on the mobile electrons that are starting to move to the right?

There will be a small upward magnetic force on a mobile electron, in the $+y$ direction.

As a result of this magnetic force, a mobile electron will acquire a small $+y$ component of velocity in addition to its component of velocity in the $+x$ direction. We'll write the vector velocity of the electron as $\vec{v}_{\text{electron}} = \langle v_{\text{bar}}, v, 0\rangle$, where v_{bar} is the speed of the bar and v is the vertical component, as shown in Figures 20.28 and 20.30.

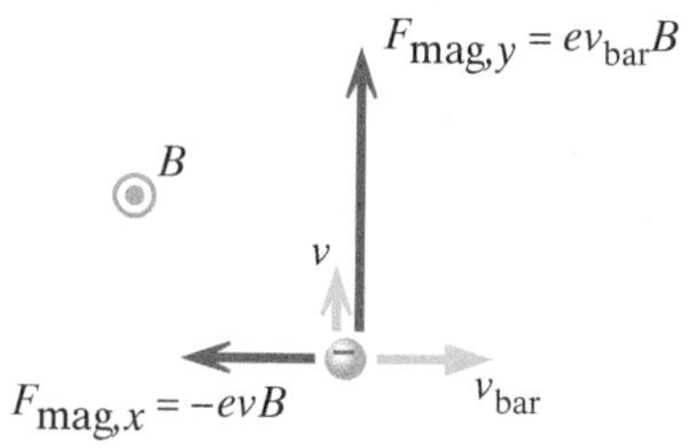

Figure 20.30 The x and y components of the magnetic force on a mobile electron in the bar.

> QUESTION At this instant, what are the components of the magnetic force on a mobile electron? The magnitude of the magnetic field is B and the electron charge is $-e$.

$\vec{F}_{\text{mag}} = -e\vec{v}_{\text{electron}} \times \vec{B} = \langle -evB, ev_{\text{bar}}B, 0\rangle$. The magnetic force on a mobile electron now has a component in the $-x$ direction as well as in the $+y$ direction (Figure 20.30). Both components are very small just after applying the force F to the bar, because the velocity of the mobile electron is small. Considering the bar as the system of interest, the x component of the net force on the bar is $F_{\text{net},x} = F - NevB$, where N is the number of mobile electrons in the bar.

As long as $F_{\text{net},x}$ is nonzero, the bar will continue to accelerate to the right. However, higher speed means that the vertical component $ev_{\text{bar}}B$ of the magnetic force is bigger, which increases the vertical component v of the mobile electron's velocity. The faster the bar moves, the bigger is v and the larger the force $NevB$ to the left on the bar, due to all the magnetic forces on the mobile electrons.

As time goes on, the net force acting on the bar gets smaller and smaller. Eventually the net force becomes zero, and the bar moves at constant speed with you pulling to the right and the $-x$ portion of the magnetic force pulling to the left. As you keep pulling, you do work to maintain a charge separation as shown in Figure 20.28, and a current runs through the resistor.

> QUESTION You can see that there is an electric field inside the bar, due to the charges maintained above and below the center of the bar. Why do the electrons move the "wrong" way through the bar, vertically upward?

The electron current through the resistor is continually depleting the charges on the ends of the moving bar, with the result that the electric field inside the bar is always slightly less than is needed to balance the magnetic force. The downward electric force eE is slightly less than the upward magnetic force evB, so the electrons move toward the negative end of the bar.

The magnetic force evB is a non-Coulomb force (F_{NC} in our earlier discussions of batteries). The work done by this force in moving an electron from one end of the bar to the other is

$$F_{NC}L = ev_{bar}BL$$

QUESTION The emf is the work done per unit charge in moving a charge from one end of a "battery" to the other, so what is the emf of this "battery"?

The non-Coulomb work per unit charge is $(ev_{bar}BL)/e = v_{bar}BL$. If the resistance of the bar is negligible, $\Delta V = \text{emf} = v_{bar}BL$. If the bar has some resistance r_{int}, this resistance is like the internal resistance of any other kind of battery, and $\Delta V = \text{emf} - r_{int}I$ (see the discussion of batteries with internal resistance in the section "Batteries" in Chapter 19).

It is interesting to see that although the vertical component of the magnetic force does positive work on a mobile electron, the horizontal component does negative work, and the total work is zero, which is consistent with the fact that the magnetic force is always perpendicular to the electron's velocity. The power input to the electron due to the magnetic force is $\vec{F}_{mag} \bullet \vec{v}_{electron}$, which is zero, as you can verify by taking the dot product of $\vec{F}_{mag} = \langle -evB, ev_{bar}B, 0\rangle$ and $\vec{v}_{electron} = \langle v_{bar}, v, 0\rangle$. We'll see in a moment that the work that drives the current is due to the force that you apply to the bar to keep it going.

The emf that is created by moving a wire in a magnetic field is called "motional emf." This effect is reminiscent of the Hall effect, because when you move a wire through a magnetic field, the magnetic force creates a polarization (and electric field) in a direction perpendicular to the motion.

QUESTION Does it look like the potential difference $\Delta V = -\oint \vec{E} \bullet d\vec{l}$ for a round trip around the circuit in Figure 20.31 could be equal to zero, as should be the case for the path integral of electric fields due to point charges? (In Figure 20.31, pay attention to the directions of the electric field in the moving bar, in the rails, and in the resistor.)

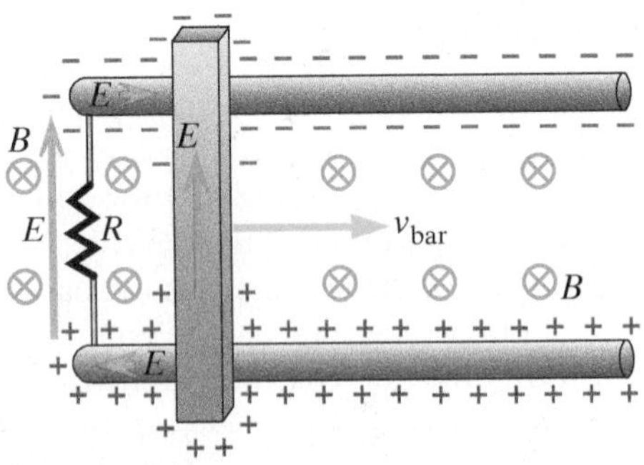

Figure 20.31 The pattern of electric field in the circuit driven by the moving bar.

Going around the loop clockwise from the bottom end of the moving bar, we get a small potential drop in the lower rail, a large potential drop in the resistor, a small drop in the upper rail, and a large potential increase in the moving bar. Therefore it does look plausible that the net ΔV for the round trip could be zero, which we know should be true.

This situation is quite similar to a circuit with a mechanical battery in it, where charges move through the mechanical battery in the "wrong" direction, due to non-Coulomb forces, while the round-trip potential difference is zero. Here the non-Coulomb force is the magnetic force.

Checkpoint 7 Can you light a flashlight bulb by moving a connecting wire rapidly past your bar magnet? Let's put in the numbers. You found that near your bar magnet the magnetic field was about 0.1 T, and the length of wire affected by the magnetic field might be about a centimeter. You might be able to move the wire at a speed of 10 m/s past the magnet. What is the emf you can produce? How does this compare with the emf of a flashlight battery?

Small Size of Motional Emf

From the preceding exercise dealing with your flashlight bulb you can see that it is hard to see the effects of motional emf with batteries and bulbs. To get a sizable emf you need larger magnetic field B extending over larger regions (L) with very fast motion (v_{bar}) of the wires through the magnetic field and many turns of wire, each contributing to the net emf.

Also, if you move a wire rapidly past your magnet, the emf doesn't last very long. Finally, a light bulb or compass isn't a very sensitive detector of small currents. Consequently, to study this motional emf you need somewhat sensitive lab equipment. Your instructor may arrange a demonstration of these effects or a formal laboratory exercise where you can observe the phenomena for yourself.

Forces on the Moving Bar and Energy Conservation

If the metal bar were to slide effortlessly along the rails at a constant speed v_{bar}, and the resistor were to continually warm the surroundings, we would be getting something for nothing. Thus there must be an energy input into the system consisting of the moving bar, the resistor circuit, and the magnet that provides the magnetic field. The magnetic force is internal to this system; the only external force is the force you continuously apply to the bar. You must exert a force on the bar to keep it moving at constant speed (Figure 20.32), because there is a horizontal component of the magnetic force on the moving bar, due to the conventional current I in the bar.

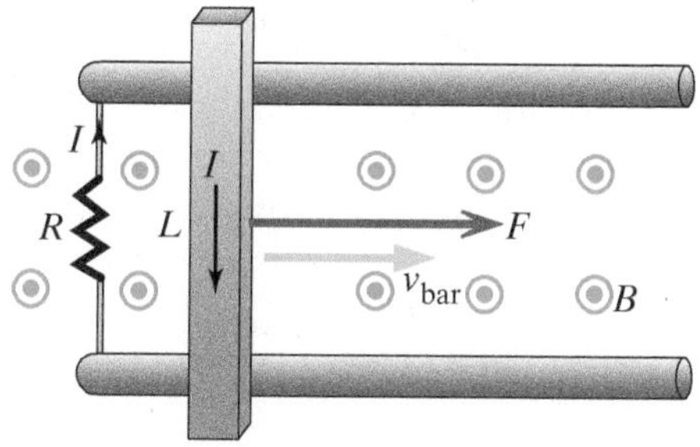

Figure 20.32 You have to keep pulling on the bar in order to keep current running in this circuit.

> **QUESTION** What is the direction of this magnetic force? What are the direction and magnitude of the force that we have to exert on the bar to keep it going at a constant speed?

Taking just the bar as the system for a moment, we've seen that there is a $-x$ component of magnetic force of magnitude $NevB$ opposing the force we apply. N is the number of mobile electrons in the bar, which is equal to the electron density n (electrons per unit volume) times the volume of the bar AL, where A is the cross-sectional area of the bar. Therefore the magnitude of the magnetic force opposing us is $NevB = (nAL)evB = (enAv)LB$, which is ILB, the usual expression for the force on a current-carrying wire with a magnetic field perpendicular to the wire.

In the steady state, in order for the bar to keep moving at constant speed, the x component of the net force on the bar must be zero, so we have to exert a force of magnitude ILB to the right.

In a short time Δt we move the bar a distance Δx. We do an amount of work $F\Delta x$ in a time Δt, so we have to supply energy at a rate of $F\Delta x/\Delta t = Fv_{bar}$.

> **QUESTION** Is this power input Fv_{bar} numerically equal to the power dissipated in the resistor?

We have power $= (ILB)v_{bar} = I(LBv_{bar}) = I(\text{emf})$, but $I(\text{emf})$ is equal to the power output of the battery that is input to the resistor. Moreover, $I = \text{emf}/R$, so the power dissipated in the resistor is $\text{emf}^2/R = RI^2$, as expected.

This is the basic principle of electric generators: mechanical power is converted into electric power. Commercially, the mechanical energy is supplied by falling water at power dams, or by expanding steam in a turbine (where the steam is heated by chemical or nuclear reactions—burning coal or oil or natural gas, or fissioning uranium). A practical generator has a rotating coil rather than a sliding bar, and we'll discuss practical generators later in this chapter.

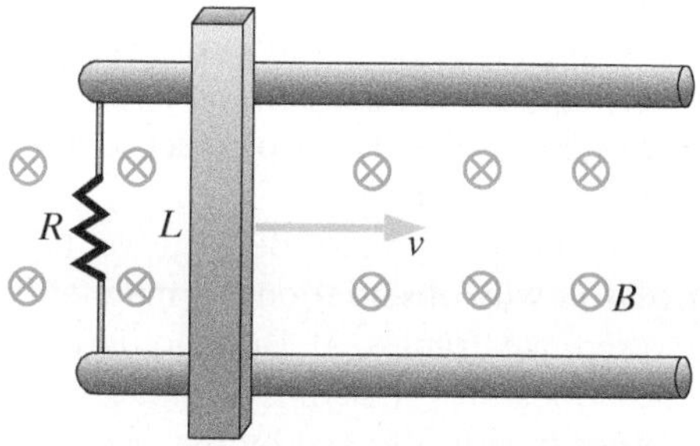

Figure 20.33 Pull the metal bar at constant speed along frictionless metal rails. A resistor R is connected between the rails.

> **Checkpoint 8** Here is a review of the basic issues: A metal rod of length L is dragged horizontally at constant speed v_{bar} on frictionless conducting rails through a region of uniform magnetic field into the page of magnitude B (Figure 20.33). There is a resistance R. What is the magnitude of the induced

emf? What is the direction of the conventional current I? What is the magnitude of the current I? What are the magnitude and direction of the force you have to apply to keep the rod moving at a constant speed v_{bar}?

EXAMPLE **A Metal Bar Moving Downward**

In Figure 20.34 a horizontal metal bar is moving downward in the plane of the page with electrical contact along two metal vertical bars that are a distance w apart. Throughout this region there is a uniform horizontal magnetic field with magnitude B out of the plane of the page, made by large coils that are not shown. At the instant that the moving bar has a speed v, what is the reading on the voltmeter? Give magnitude and sign, and explain briefly.

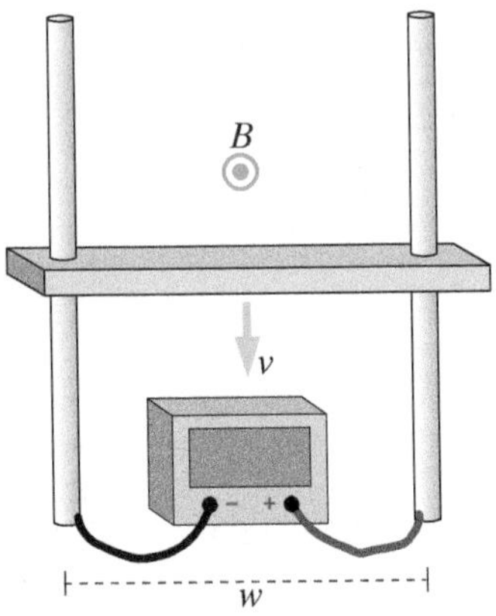

Figure 20.34 As the metal bar moves downward, what does the voltmeter read?

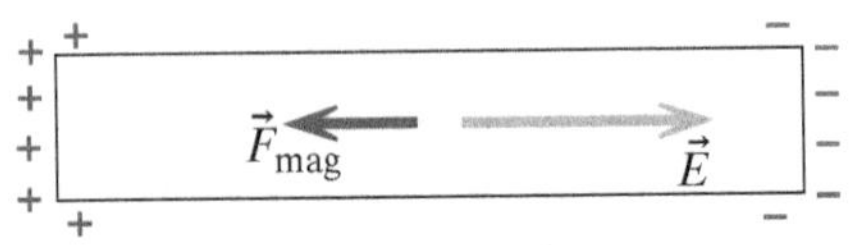

Figure 20.35 Magnetic force on a positive charge, resulting polarization of the bar, and electric field due to polarization.

Solution The charged particles in the bar are moving through a magnetic field, and therefore experience a magnetic force. It does not matter what sign we assume for the mobile charges, so we will assume they are positive because it simplifies keeping track of signs.

$\vec{v} \times \vec{B}$ is to the left, so the bar polarizes as shown in Figure 20.35.

The polarization takes an extremely short time (less than nanoseconds), after which for every mobile charge in the bar we have the following (note that the voltmeter has very high resistance, so there is almost no current):

$$\vec{F}_{\text{magnetic}} + \vec{F}_{\text{electric}} = 0$$
$$qvB\sin(90°) = qE$$
$$E = vB$$

Since E is uniform throughout the bar,

$$|\Delta V| = \left| -\int \vec{E} \cdot d\vec{l} \right| = vBw$$

The left end of the bar is at a higher potential than the right end. Since the high potential end is connected to the negative terminal of the voltmeter, the reading on the voltmeter will be negative. The reading on the voltmeter is therefore $-vBw$.

20.6 MAGNETIC FORCE IN A MOVING REFERENCE FRAME

The magnetic field produced by a moving charge depends on the velocity of the charge, and this leads to the curious result that a stationary Bob may observe a different magnetic field than a moving Alice sees. Suppose that Bob holds a stationary charge, which of course produces a pure electric field and no magnetic field. If Alice runs past Bob, she sees a mixture of electric and magnetic fields produced by what is for her a moving charge.

The magnetic force experienced by a moving charge also depends on the velocity of the charge, which is another reason why observations of magnetic effects may differ in different moving reference frames. If Bob holds two charged tapes that repel each other, he describes this by saying that one tape produces a purely electric field, and the other tape is affected by that electric field. Alice, however, as she moves past Bob says that one (moving) tape produces a mixture of electric and magnetic fields, and the other (moving) tape feels both electric and magnetic forces.

Two Moving Protons

We are now in a position to do a detailed calculation of these odd effects. We'll consider two protons initially traveling parallel to each other with the same speed v, a distance r apart, and we will calculate the magnitudes and directions of the electric and magnetic forces that proton 1 exerts on proton 2 at this moment (Figure 20.36). The electric force is easy. Proton 1 makes an electric field $\vec{E}_1$ at the location of proton 2, and proton 2 experiences an electric force $\vec{F}_{21,e} = q_2\vec{E}_1$:

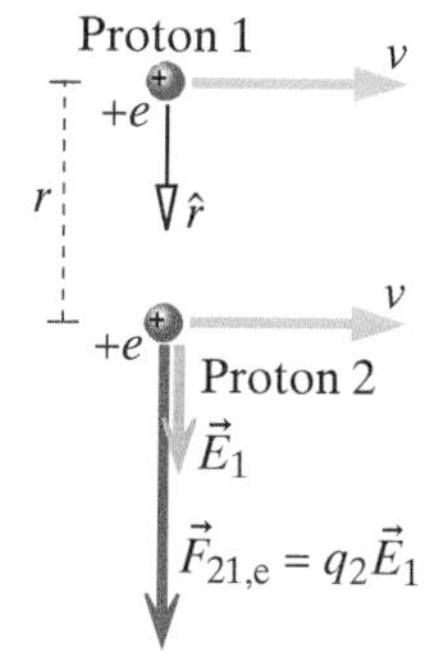

Figure 20.36 Proton 1 exerts an electric force on proton 2.

$$F_{21,e} = \frac{1}{4\pi\varepsilon_0}\frac{e^2}{r^2}, \quad \text{downward}$$

We're making an approximation when we use Coulomb's law to calculate the electric force. Due to relativistic effects, the electric field $\vec{E}_1$ made by the moving charge (proton 1) is somewhat different from the electric field made by a stationary charge. We'll say a bit more about this approximation later.

The magnetic force involves cross products both in the field and in the force. To determine the magnetic force $\vec{F}_{21,m}$ acting on proton 2, first we need to calculate the magnetic field $\vec{B}_1$ made by proton 1 at the location of proton 2. It helps enormously to draw the unit vector on the diagram (from source to observation point; that is, from proton 1 to proton 2).

QUESTION Give the direction and the magnitude of $\vec{B}_1 = \dfrac{\mu_0}{4\pi}\dfrac{q_1\vec{v}_1 \times \hat{r}}{r^2}$.

The magnetic field $\vec{B}_1$ is into the page at the location of proton 2, and its magnitude is $B_1 = [\mu_0/(4\pi)](ev/r^2)\sin 90°$.

QUESTION Now that you know the direction and magnitude of the magnetic field $\vec{B}_1$, determine the magnitude and direction of the magnetic force exerted on proton 2, $\vec{F}_{21,m} = q_2\vec{v}_2 \times \vec{B}_1$.

The direction is upward, and the magnitude is $F_{21,m} = \dfrac{\mu_0}{4\pi}\dfrac{e^2v^2}{r^2}$.

The Ratio of the Magnetic and Electric Forces

Figure 20.37 summarizes the situation. The magnetic force on proton 2 is opposite to the electric force, and the net force on proton 2 is reduced from what it would be for a stationary proton.

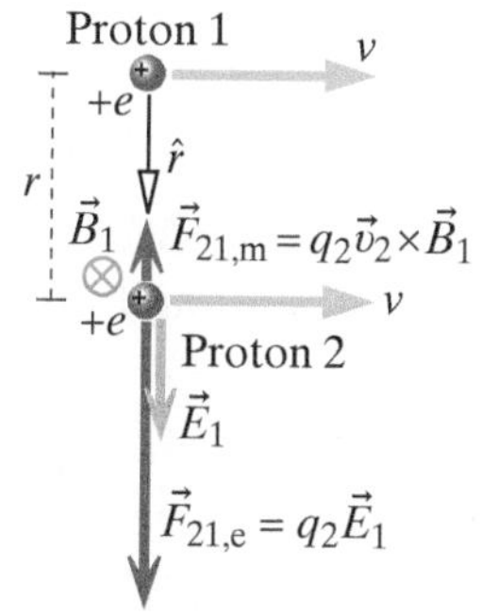

Figure 20.37 The magnetic force on proton 2 is opposite to the electric force.

QUESTION What is the ratio of the magnetic force to the electric force on proton 2 due to proton 1?

$$\frac{F_{21,m}}{F_{21,e}} = \frac{\left(\dfrac{\mu_0}{4\pi}\dfrac{e^2v^2}{r^2}\right)}{\left(\dfrac{1}{4\pi\varepsilon_0}\dfrac{e^2}{r^2}\right)}$$

$$= \frac{v^2}{\left(\dfrac{1}{4\pi\varepsilon_0}\Big/\dfrac{\mu_0}{4\pi}\right)} = \frac{v^2}{(9\times 10^9/1\times 10^{-7})} = \frac{v^2}{(3\times 10^8)^2} = \frac{v^2}{c^2}$$

This is a surprise! Why should the speed of light show up in a comparison between electric and magnetic forces? What does the speed of light have to do with this? There are some deep issues here. (We did not show units in the calculation above, but they do work out correctly.)

This result indicates that at ordinary speeds ($v \ll c$) the magnetic interaction between two charged particles is much weaker than the electric interaction. However, if v approaches the speed of light c, the magnetic force may be comparable to the electric force. Even though at low speeds magnetic forces are inherently weak compared to electric forces, magnetic forces are more important in many industrial applications, because the magnetic force on a current-carrying wire is typically much larger than an electric force on the wire, since the net charge of the wire is nearly zero (neglecting the small amount of surface charge).

QUESTION Now determine the full electric and magnetic force, $\vec{F} = q\vec{E} + q\vec{v} \times \vec{B}$, on proton 2 due to proton 1 (magnitude and direction).

The electric force is downward and the magnetic force is upward, so the net force on proton 2 is

$$\frac{1}{4\pi\varepsilon_0}\frac{e^2}{r^2} - \frac{\mu_0}{4\pi}\frac{e^2v^2}{r^2} = \frac{1}{4\pi\varepsilon_0}\frac{e^2}{r^2}\left(1 - \frac{v^2}{c^2}\right), \quad \text{downward}$$

QUESTION Describe the subsequent trajectories of the protons.

The protons repel each other, so their trajectories are curving lines moving apart (Figure 20.38).

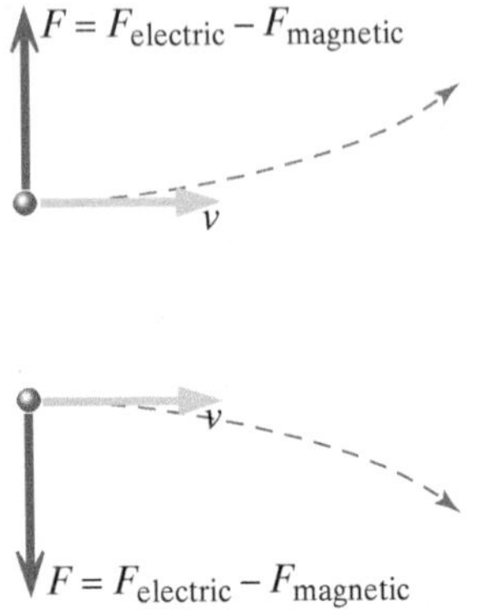

Figure 20.38 Bob is sitting still, and he sees the protons move along curving paths.

Checkpoint 9 As a numerical example, suppose that $v = 1500$ m/s, a very high speed typical of a molecule of hydrogen gas at room temperature. What is the numerical ratio of the magnetic force to the electric force for protons moving at this speed?

Alice and Bob and Einstein

The fact that the net force is smaller on the moving protons may seem innocuous, but it has startling consequences. Suppose that Bob is sitting still and sees the protons go by. The protons repel each other along curving paths, with a force somewhat weakened by the magnetic effects (Figure 20.38).

Suppose that Alice is running along beside the protons, with the same speed v that the protons have initially. She sees the two protons repel each other straight away from each other (Figure 20.39). The repulsion is purely electric, with no weakening of this electric repulsion (the magnetic field of one proton is zero at the location of the other proton, because $\vec{v} \times \hat{r} = 0$ along the line of motion of the proton).

Figure 20.39 Alice runs along beside the protons, at speed v; she sees the protons move along straight paths.

Let both Alice and Bob use accurate timers to measure how long it takes for the protons to hit the floor and ceiling of the room, starting from halfway between floor and ceiling.

QUESTION Given the difference in the force that accelerates the protons vertically, which observer will measure a shorter time for the protons to reach the floor and ceiling?

Alice sees a pure electric repulsion, while Bob sees the electric repulsion partially counteracted by an attractive magnetic force. To take a concrete example, suppose Bob's timer advances by 20 ns during this process. Alice's timer might advance by only 15 ns (if v is a large fraction of the speed of light), corresponding to her observation that the protons get to the floor and ceiling in less time due to the greater force of repulsion.

But that's crazy! Surely Alice and Bob wouldn't measure different amounts of time for the same process, would they? It was Einstein who thought very deeply about this and related paradoxes and concluded that, much as it violates common sense, it has to be the case that Alice's timer seems to run slower for Bob than his own timer!

This sounds utterly absurd, yet a great variety of measurements of this kind all agree with Einstein's radical proposal that time runs at different rates for different observers. One of the reasons that Einstein was driven to these strange notions was that by the time that he proposed his special theory of relativity (1905), it was clear that the existing theories of electricity and magnetism seemed to lead to paradoxes, as we have seen in the observations of Alice and Bob. In fact, the title of Einstein's 1905 paper on relativity was "On the electrodynamics of moving bodies."

The net force (electric minus magnetic) observed by Bob is always smaller than the purely electric force observed by Alice, and Alice and Bob disagree on how long it takes the protons to move apart, and on whether their timers are marking time correctly. Yet both Alice and Bob can correctly predict how much their own timers will advance during the time it takes for the protons to hit the floor and ceiling.

The details of our calculations are not correct, because relativistic aspects of the fields have not been taken into account. Both the electric field and the magnetic field of a fast-moving charge are different from the fields given by Coulomb's law and the Biot–Savart law for a slow-moving charge. However, both fields are modified by the same factor $1/\sqrt{1-v^2/c^2}$, so the ratio of the magnetic force to the electric force is equal to v^2/c^2 for all speeds. Alice and Bob will indeed observe different times for the same process. See Section 20.11 at the end of this chapter for a more detailed discussion of the nature of electric and magnetic fields in moving reference frames.

An Asymmetry in the Measurements

You may wonder about the asymmetry in the measurements that Alice and Bob make. If Bob thinks that Alice's timer runs slow, why doesn't Alice think that Bob's timer runs slow? After all, Alice should be able to take the point of view that she is standing still while Bob moves past her.

In reality, Bob cannot make his measurements unaided. Rather, Bob and his friend Fred must station themselves an appropriate distance apart in the room. Bob records the time (on Bob's timer) when the protons pass by him, and he also records the time that he sees on Alice's timer at that moment. Fred records the later time (on Fred's timer) when the protons reach him and hit the floor and ceiling, and he also records the time that he sees on Alice's timer at that moment. Two people, Bob and Fred, are needed to make these measurements. Otherwise Bob would have to apply corrections to his measurements, due to the time it takes for light from Alice's timer to reach him from the distant location.

After Fred has made his measurements, Bob and Fred compare notes and find that the times that they wrote down differ by 20 ns (they had previously synchronized their timers). Their two views of Alice's timer indicate that her timer advanced only 15 ns. Bob and Fred describe Alice's timer as running slow. The situation is different for Alice, because she looks at two *different* timers, Bob's and Fred's. The asymmetry lies in the fact that there are two observers sitting in the room (Bob and Fred), but only one observer moving through the room (Alice).

On the other hand, suppose that Alice sits at the front of a space ship moving past Bob, and her friend Sue sits further back in the space ship. Alice notes the time on her own timer and on Bob's timer when she passes Bob, and Sue does the same when she passes Bob. When Alice and Sue compare notes, they will describe Bob's timer as seeming to run slow.

20.7 MAGNETIC TORQUE

A magnetic field can cause a current-carrying coil of wire to twist. You may have seen this for yourself, if you were able to suspend a coil of wire and see it twist in the magnetic field of a bar magnet (or you may have floated a coil and seen it twist in the magnetic field of the Earth). Moreover, we have seen that a compass needle can be thought of as a collection of atomic current loops, and you have certainly seen a compass needle twist in a magnetic field.

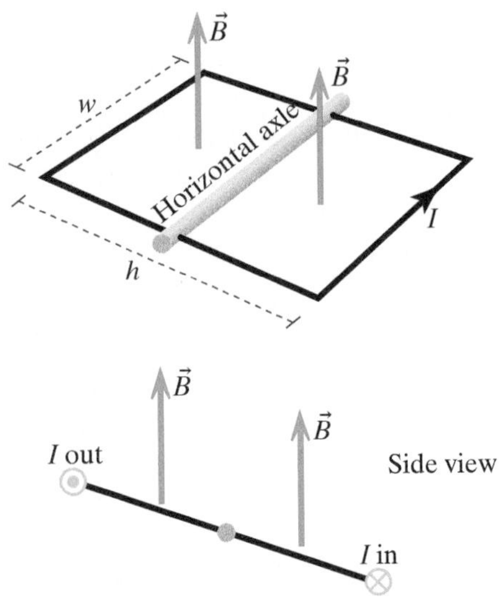

Figure 20.40 A current-carrying loop connected to an insulating horizontal axle is free to rotate.

> **QUESTION** In Figure 20.40, what is the direction of the magnetic force on each of the four sides of the loop?

Consider a rectangular current-carrying loop of wire in a uniform magnetic field (that is, the magnetic field has the same direction and magnitude throughout this region). The loop is free to rotate on an insulating horizontal axle (Figure 20.40).

On the sides of length h the magnetic force is horizontal and points outward, tending to stretch the loop. On the other sides, of length w, the magnetic forces are horizontal and tend to make the loop twist on the axle. This may be easiest to see in the side view shown in Figure 20.40.

> **QUESTION** In the side view, will the loop rotate clockwise (to the right) or counterclockwise (to the left)?

The magnetic forces act to twist the loop counterclockwise.

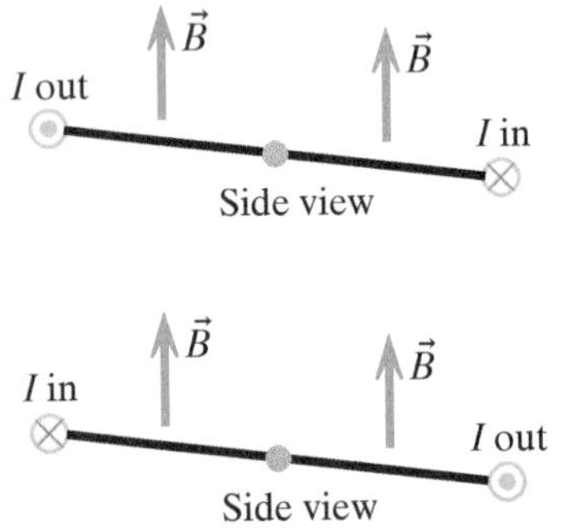

Figure 20.41 Which of these orientations is stable?

> **QUESTION** When the plane of the loop is perpendicular to the magnetic field, the magnetic forces don't exert any twist. However, one of these orientations is unstable: if you nudge it slightly, the loop will flip over. Which of the two orientations shown in side view in Figure 20.41 is stable, and which is unstable?

The situation on the top is stable: the magnetic forces twist the loop back toward the horizontal plane. The situation on the bottom is unstable: a small displacement away from the horizontal plane leads to magnetic forces that rotate it even farther out of the plane.

It is often easier to talk about the twist on a current-carrying loop of wire in terms of the "magnetic dipole moment" (see Chapter 17). The magnetic dipole moment $\vec{\mu}$ of a current-carrying loop of wire is defined as a vector pointing in the direction of the magnetic field that the loop makes along its axis, which is given by a right-hand rule (this magnetic field made by the loop is of course different from the applied magnetic field that makes the loop twist). The magnitude of the magnetic dipole moment is $\mu = IA$.

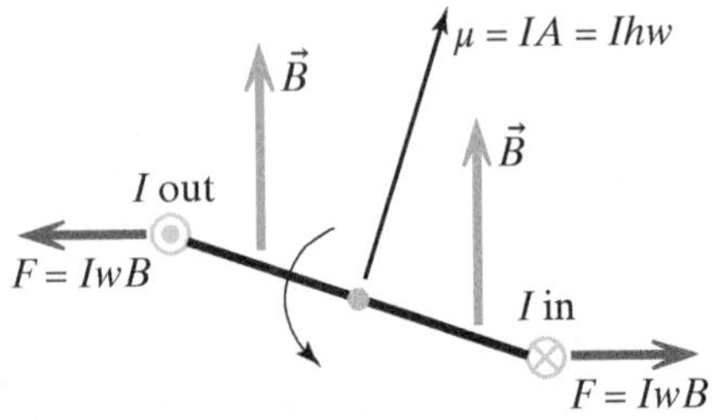

Figure 20.42 The magnetic dipole moment of the loop is a vector that tends to line up with the applied magnetic field.

Our rectangular loop has an area $A = wh$, so its magnetic dipole moment is $\mu = IA = Ihw$. In Figure 20.42, note that the coil tends to twist in a direction to make $\vec{\mu}$ line up with the applied magnetic field $\vec{B}$.

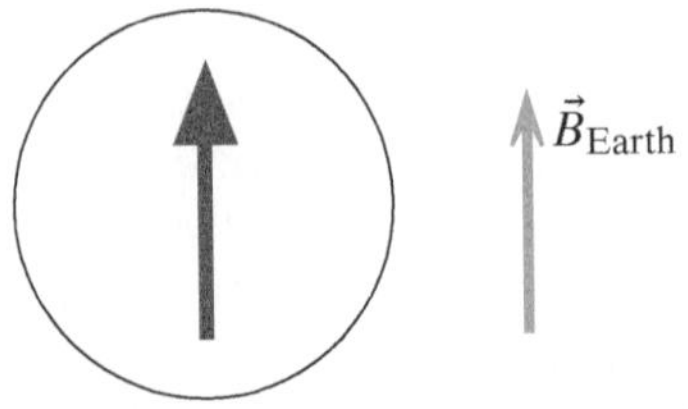

Figure 20.43 What is the direction of the atomic magnetic dipole moments inside this compass needle, which points north?

> **Checkpoint 10** Inside the magnetized compass needle shown in Figure 20.43, which points north, what is the direction of a typical atomic magnetic dipole moment $\vec{\mu}$? If the needle is moved away from north and then released, what will happen? Explain briefly.
>
> At last we have an explanation in terms of the fundamental principles of magnetism for the behavior of a compass needle, and why it always lines up with the applied magnetic field. (A practical compass has some friction to damp out the oscillations: without friction the needle would swing back and forth forever, like a frictionless pendulum. Many compasses are filled with liquid to damp out the oscillations.)

Quantitative Torque

The fact that a magnetic dipole twists in a magnetic field is very important. The torque provided by each of the magnetic forces around the axle is equal to the distance from the axle (the "lever arm") times the component of the force perpendicular to the lever arm. The twist applied to this rectangular loop is due to the magnetic forces on the w-lengths of the loop, as is seen in Figure 20.44, where the magnetic torque is out of the page (there is a clockwise twist).

The perpendicular component of each force is $F_\perp = IwB\sin\theta$, and the lever arm is $h/2$. (Note that the other component of F, and the forces on the h lengths of the loop, just tend to stretch the loop without causing a twist.) Each of the forces exerts a torque of $(F_\perp)(h/2)$, so the total torque is $2IwB\sin\theta(h/2) = (Iwh)B\sin\theta$.

Since $\mu = Iwh$, we can express the torque in the form $\vec{\mu}\times\vec{B}$, because this cross product has a magnitude $|\vec{\mu}||\vec{B}|\sin\theta = \mu B\sin\theta$. The direction of this torque vector is along the axle around which the loop rotates. If the thumb of your right hand points in the direction of $\vec{\mu}\times\vec{B}$, your fingers curl around in the direction that the coil will twist. In Figure 20.44, the torque vector is into the page, and the coil rotates clockwise.

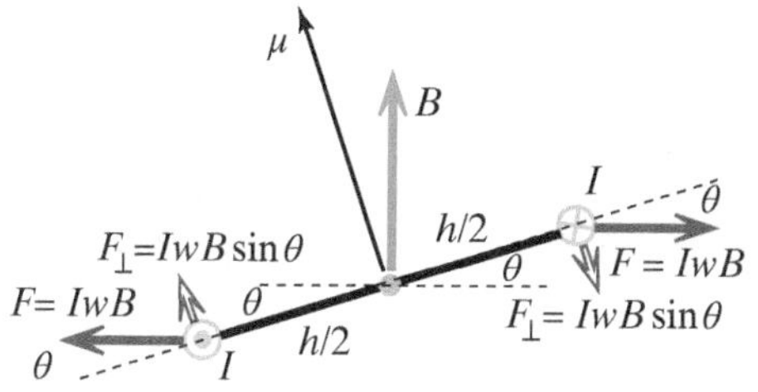

Figure 20.44 The magnetic forces on this rectangular current-carrying loop tend to twist it clockwise (there is a torque $\vec{\mu}\times\vec{B}$ into the page).

We showed that the torque is $\mu B\sin\theta$ for a rectangular loop, but a loop of any shape can be approximated by a set of current-carrying rectangles along whose adjacent sides the current cancels (Figure 20.45). This shows that the torque is $\mu B\sin\theta$ for any kind of loop.

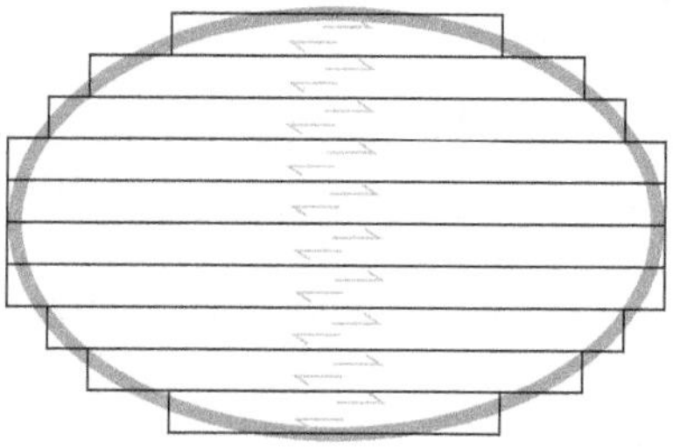

Figure 20.45 A loop of any shape can be approximated by a set of rectangular loops.

TORQUE ON A MAGNETIC DIPOLE

$$\vec{\tau} = \vec{\mu}\times\vec{B}$$

20.8 POTENTIAL ENERGY FOR A MAGNETIC DIPOLE

If allowed to pivot freely, a magnetic dipole turns toward alignment with the applied magnetic field. That means that an aligned magnetic dipole is associated with lower potential energy in the magnetic field, and the system tries to go to the lower potential-energy configuration. If we calculate how much work it takes to move the magnetic dipole out of alignment, we will have a measure of the increased potential energy associated with being out of alignment.

We consider the rectangular loop shown in Figure 20.46, and we calculate how much work we have to do to move it from an initial angle $\theta = \theta_i$ to a final angle $\theta = \theta_f$. We move slowly, without changing the kinetic energy of the loop. The loop is always very nearly in equilibrium.

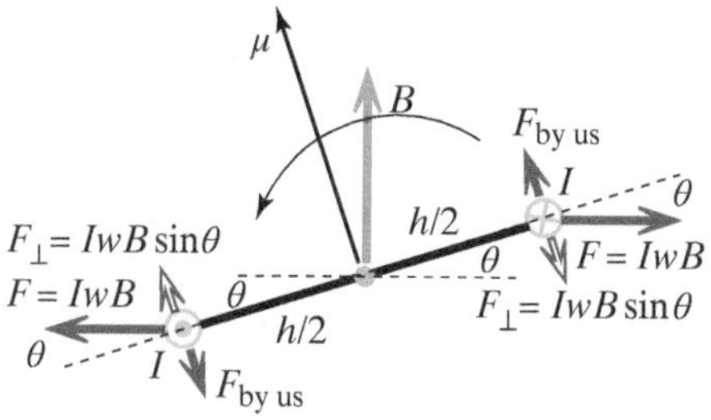

Figure 20.46 Calculating how much work is required to turn a current loop in a magnetic field.

As we rotate the loop through a small angle $d\theta$, both of the forces that we exert act through a distance $(h/2)d\theta$, because arc length is radius $(h/2)$ times angle ($d\theta$ measured in radians). If we force the loop to rotate from the initial angle $\theta = \theta_i$ to a final angle $\theta = \theta_f$, the amount of work we do by exerting our two forces must be calculated by an integral, because F isn't constant. The work that we do goes into changing the magnetic potential energy U_m of the system consisting of the magnetic dipole in the magnetic field:

$$\text{work} = \Delta U_m = \int_{\theta_i}^{\theta_f} 2IwB\sin\theta\left(\frac{h}{2}d\theta\right) = IwhB\int_{\theta_i}^{\theta_f}\sin\theta\, d\theta$$

$$\Delta U_m = IwhB[-\cos\theta]_{\theta_i}^{\theta_f} = -IwhB[\cos\theta_f - \cos\theta_i]$$

$$\Delta U_m = \Delta(-\mu B\cos\theta) \text{ since } \mu = IA = Iwh$$

It is customary to define the potential energy of a magnetic dipole in a magnetic field by choosing the zero of potential energy in such a way that one writes this:

$$U_m = -\mu B\cos\theta$$

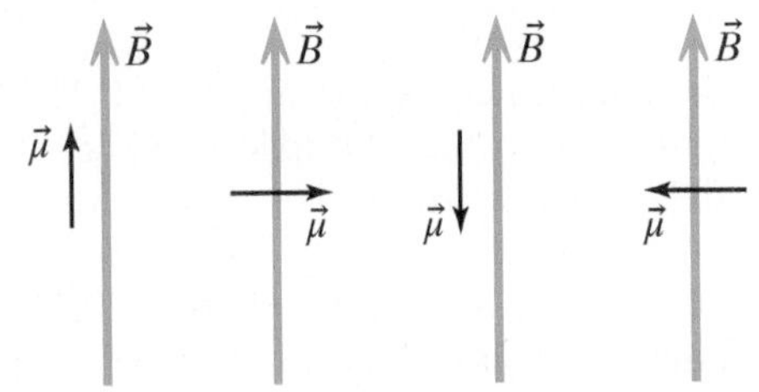

Figure 20.47 Which alignment represents the lowest potential energy? The highest?

This can also be written as a dot product:

THE POTENTIAL ENERGY FOR A MAGNETIC DIPOLE

$$U_m = -\vec{\mu} \bullet \vec{B}$$

This convention corresponds to calling the potential energy zero when $\theta = 90°$, since $\cos(90°) = 0$.

> QUESTION Given that $U_m = -\vec{\mu} \bullet \vec{B}$, which of the alignments in Figure 20.47 corresponds to the lowest possible potential energy? To the highest possible potential energy?

The lowest energy corresponds to $\vec{\mu}$ and $\vec{B}$ pointing in the same direction, so that $U_m = -\mu B \cos 0 = -\mu B$. This is also the most stable configuration; if you twist the magnetic dipole away from this position, it will be twisted back by the magnetic field. The highest energy corresponds to $\vec{\mu}$ and $\vec{B}$ pointing in opposite directions, so that $U_m = -\mu B \cos 180° = +\mu B$. This is an unstable configuration; a slight displacement will lead to a large twist. The other two configurations correspond to medium energy: $U_m = 0$ (because $\cos 90° = 0$).

This picture of the potential energy of a magnetic dipole in a magnetic field is very important in atomic and nuclear physics. In a magnetic field, atoms or nuclei can make transitions between higher and lower energy states that correspond to different orientations of the magnetic dipole moments of the atoms or nuclei in the magnetic field.

For example, in the magnetic imaging equipment now used in hospitals, the patient lies inside a large solenoid. In the presence of the magnetic field of the solenoid, nuclei in the patient's body have slightly different energies depending on whether the nuclear magnetic dipole moment of the nucleus is in the direction of or opposite to the applied field. Sensitive equipment detects transitions between these two slightly different energy levels.

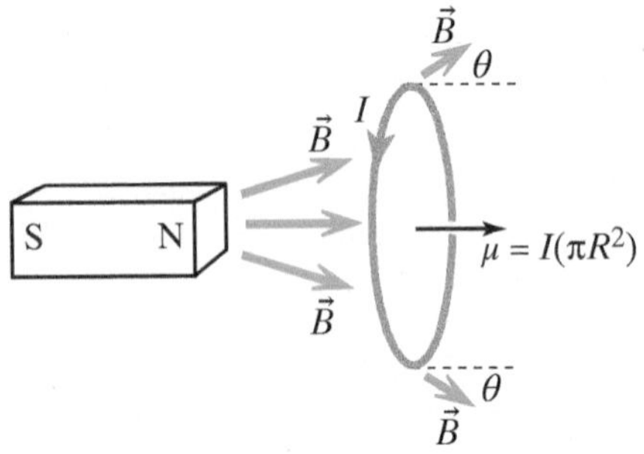

Figure 20.48 A bar magnet exerts a force on a current-carrying loop.

Checkpoint 11 What is the energy difference between the highest and lowest states?

Force on a Magnetic Dipole

We have successfully explained the twist of a magnetic dipole in a magnetic field. There remains the question of how to explain the net force on a magnetic dipole in a magnetic field. For example, how can we explain the attraction or repulsion between two magnets?

We can show that if a magnetic dipole is placed in a nonuniform magnetic field, it experiences a force. Figure 20.48 shows a circular current loop placed near the north pole of a bar magnet, along the axis of the magnet. The magnetic field of the bar magnet diverges with an angle θ as shown and so is nonuniform.

> QUESTION Show that the net force on the current loop is to the left, with magnitude $F = I(2\pi R)B\sin\theta = (2\mu B/R)\sin\theta$, where R is the radius of the circular loop.

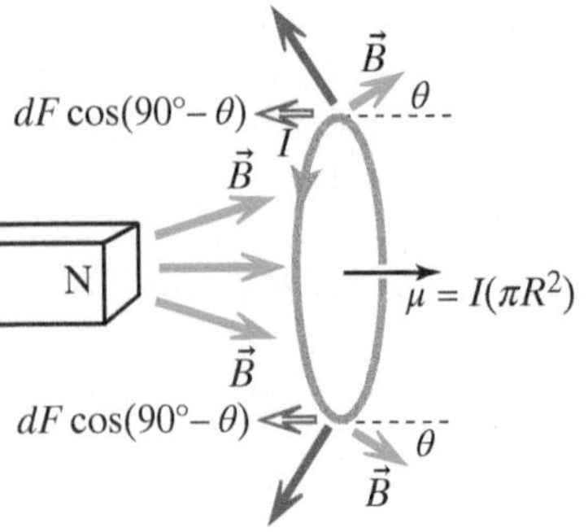

Figure 20.49 Force on a current loop by a bar magnet.

As shown in Figure 20.49, each segment of the ring experiences a magnetic force $d\vec{F} = I\Delta\vec{l} \times \vec{B}$. The horizontal component of the force is $dF\cos(90° - \theta) = dF\sin\theta$. The vertical components of the forces on two opposite pieces of the ring cancel, so the net force is in the $-x$ direction (toward the bar magnet).

$$F_{\text{net}} = IB\sin\theta \int dl = IB(2\pi R)\sin\theta$$

Since $\mu = IA = I(\pi R^2)$,

$$F_{\text{net}} = \frac{\mu(2B\sin\theta)}{R}$$

If we knew how the magnetic field spread out in angle (that is, how big θ is at a distance R off-axis), we could calculate the net force F. Without that information, we can nevertheless determine F by using an argument based on potential energy, which we will do in the next section.

Checkpoint 12 In what direction would the net force be if the bar magnet were turned around, or the current in the loop reversed? What can you say about the net force if the field were uniform ($\theta = 0$)?

Calculating the Force on a Magnetic Dipole

Using an energy argument, we can calculate the force on a magnetic dipole due to a nonuniform magnetic field. For example, we can calculate the force on one magnet due to another magnet.

If we move a magnetic dipole from a region of low magnetic field to a region of high magnetic field, the magnetic potential energy $U_m = -\vec{\mu} \bullet \vec{B}$ will change. If released from rest, the object will spontaneously move to the place where the potential energy is lower, picking up kinetic energy as it goes. If we prevent the kinetic energy from changing, by moving the object at a slow constant speed, we have to exert a force and do some work. By calculating how much work we have to do, we can determine how big a force we have to apply, and this is numerically equal to the magnetic force on the object. This will let us calculate how big a force one magnet exerts on another magnet.

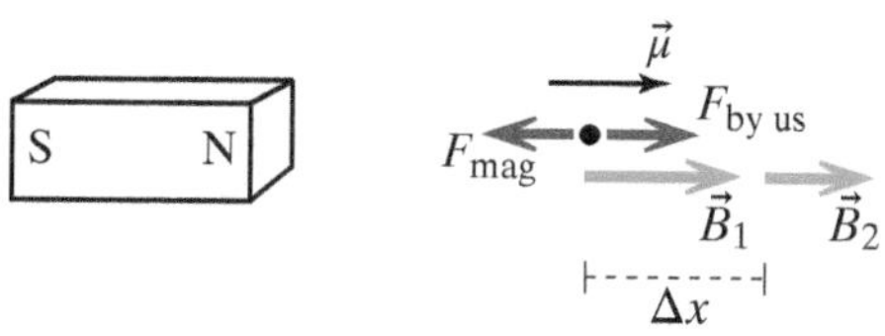

Figure 20.50 A bar magnet acts on a magnetic dipole.

Consider the situation shown in Figure 20.50. A magnetic dipole is aligned with the magnetic field made by a bar magnet, and the magnet exerts a force F_{magnetic} to the left on the magnetic dipole moment, as we saw in the previous section. We exert a force $F_{\text{by us}}$ to the right, infinitesimally larger than F_{magnetic}, and we move the magnetic dipole moment a short distance Δx to the right.

The magnet makes a nonuniform magnetic field (proportional to $1/x^3$ if we are far from the magnet), and we are moving the magnetic dipole from a larger magnetic field B_1 to a smaller magnetic field B_2. The magnetic potential energy increases from $-\mu B_1$ to the somewhat less negative $-\mu B_2$. Therefore we have to do an amount of work

$$F_{\text{by us}}\Delta x = \Delta U_m = (-\mu B_2) - (-\mu B_1) = -\mu(B_2 - B_1) = -\mu\Delta B$$

This is greater than zero because $B_2 < B_1$. The magnitude of the force we have to exert is this:

$$F_{\text{by us}} = \frac{\Delta U_m}{\Delta x} = -\mu\frac{\Delta B}{\Delta x} \to -\mu\frac{dB}{dx} \quad \text{(a positive value; } dB/dx < 0\text{)}$$

The force that the magnet exerts to the left on the magnetic dipole moment is numerically equal to the force that we exert. This magnitude is also correct in the case of repulsion. If the magnetic dipole moment is at some angle to the magnetic field, there is both a net force and a twist.

This is a special case of the general result that the force associated with a potential energy is the negative gradient of that potential energy:

$$F_x = -\frac{dU}{dx} \quad \text{(general result)}$$

$$F_x = -\frac{d(-\vec{\mu} \bullet \vec{B})}{dx} = \mu\frac{dB}{dx} \quad \text{(our specific case)}$$

In this specific case, the quantity $\vec{\mu} \bullet \vec{B}$ is decreasing in the $+x$ direction, so the equation predicts that the x component of the force on the magnetic dipole moment is in the $-x$ direction, which is correct.

Note that there is no force if the field is uniform ($dB/dx = 0$), in agreement with our previous results, though there is a twist. This is the situation with a compass needle affected by the magnetic field of the Earth, which hardly varies over short distances and therefore applies an extremely weak net force, but there is a twist due to equal and opposite forces that align the needle to point north.

We can now calculate the force that one bar magnet exerts on another. The magnetic field along the axis of one magnet, far from that magnet, is

$$B_1 \approx \frac{\mu_0}{4\pi}\frac{2\mu_1}{x^3}$$

where μ_1 is the total magnetic dipole moment of all the atomic magnetic dipole moments in the magnet.

Figure 20.51 The magnetic dipole moment μ_2 of the second magnet lies along the axis of μ_1 of the first magnet.

QUESTION Show that the attractive force that this magnet would exert on a magnetic dipole moment μ_2 lying along the axis of μ_1 (Figure 20.51) is

$$F \approx 3\mu_2\left(\frac{\mu_0}{4\pi}\frac{2\mu_1}{x^4}\right)$$

Just take the negative gradient of the magnetic potential energy to obtain this result. This says that two magnets should attract or repel each other with a force proportional to $1/x^4$.

Moreover, using our measurements of the magnetic dipole moments of our bar magnets in Chapter 17, we should be able to calculate how close one of our magnets must be to another magnet in order to pick it up. See Experiment EXP4.

Magnetic Force and Work

Because the basic magnetic force involves a cross product, the magnetic force $q\vec{v} \times \vec{B}$ on an individual moving charged particle acts perpendicular to the velocity of the particle (and its momentum). Therefore the magnetic force can change the direction of the particle's momentum but not its magnitude or its kinetic energy. Another way of seeing this is to note that the quantity $\vec{F} \cdot d\vec{r} = 0$, so no work is done on the particle.

However, applying a magnetic field can cause a compass needle to begin to rotate; the needle gains rotational kinetic energy. The directions of the magnetic torque and the needle's rotation are the same, so the work done is nonzero. When a bar magnet picks up another bar magnet, the second magnet accelerates upward, gaining kinetic energy. Again, the magnetic force and the displacement are in the same direction, so work is nonzero. Note that these are not cases of magnetic forces exerted on point particles.

Additionally, if a charged particle is bound to other particles by electric forces, the magnetic force may indirectly result in changes in kinetic energy. Recall that magnetic forces acting on a current of mobile electrons in a wire can indirectly move the stationary positive cores, because the electrons are bound to the positive cores by electric forces, and the battery that drives the current can supply energy.

The Stern–Gerlach Experiment

An important milestone in our understanding of atoms was the Stern–Gerlach experiment (1922). Silver was vaporized in an oven, and a beam of neutral silver atoms was defined by a series of slits (Figure 20.52). On theoretical grounds, each silver atom was expected to have angular momentum, with a magnetic dipole moment proportional to the angular momentum. The beam of neutral atoms passed through a strongly nonuniform magnetic field, and then struck a cold glass plate onto which the atoms condensed, leaving a visible silvery trace on the glass. (The entire apparatus was in a vacuum to avoid collisions with air molecules.)

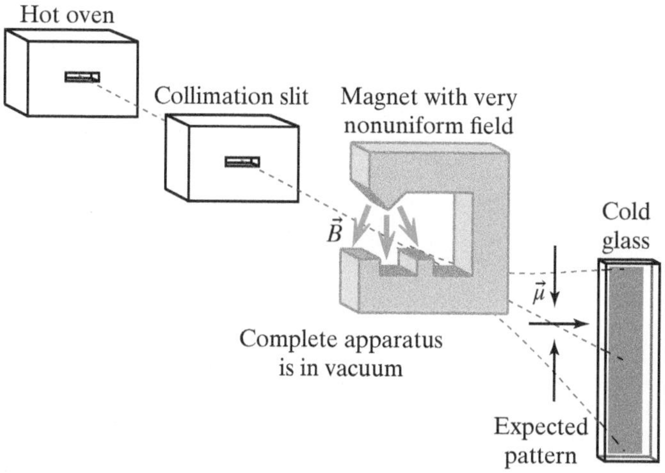

Figure 20.52 Classically, one expects a continuous pattern of silver deposited on the glass.

If the atoms have magnetic dipole moments, they should experience a vertical force, up or down in varying degrees depending on the orientation of their magnetic dipole moments. One would expect a random orientation of the magnetic dipole moments, so that there should be a vertical band of silver deposited on the glass. For example, if the magnetic dipole moment happens to be horizontal, the atom goes straight through without any deflection. (The motion is actually more complicated because the magnetic dipole moment oscillates back and forth in the magnetic field, like the motion around north of an undamped compass needle, unless it happens to point in the direction of the magnetic field. This does not affect the general conclusions, however.)

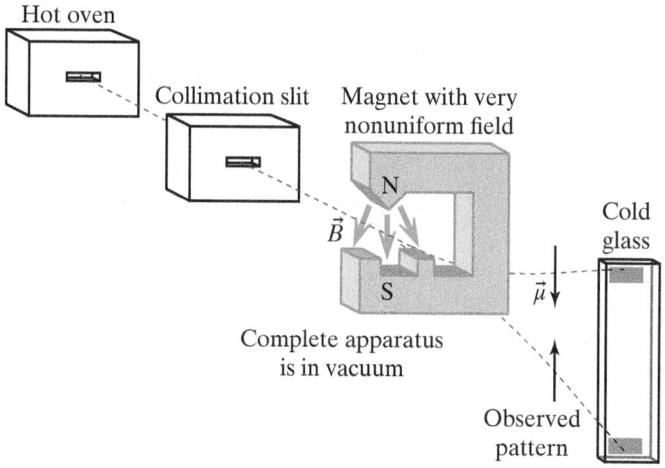

Figure 20.53 What one actually observes is a quantized pattern of silver deposited on the glass.

Instead of being spread out uniformly as one would expect, what Stern and Gerlach actually observed was that the atoms always hit either at the top or at the bottom of the expected region (Figure 20.53). This was interpreted as indicating that the orientation of the angular momentum of the atom is "quantized"—that whenever the atom is observed in this way we find that a component of angular momentum can have only certain quantized values. This distinctive quantization effect is typical of the behavior of atoms and subatomic particles.

20.9 MOTORS AND GENERATORS

Electric motors are the most important everyday application of the torque that a magnetic field exerts on a current-carrying coil. Our daily lives are greatly affected by electric motors in refrigerator compressors, water pumps, computer disk drives, elevators, automobile starter motors, and so on.

We have seen that a current-carrying coil twists to its most stable position in a magnetic field. How can we get the coil to turn continuously? The key is to make electrical connections to the coil in such a way that just as it is coming to its stable position, we reverse the direction of the current. Assuming that it is rotating fast enough at that moment, the coil will continue on around in the same direction to the new stable position rather than simply going backward. Then as the coil approaches the new stable position we again switch the direction of the current.

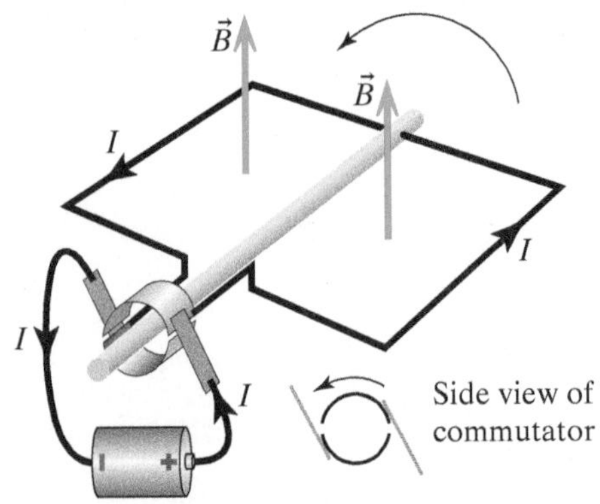

Figure 20.54 A split-ring commutator rotates with the coil, and as the coil reaches its most stable position the current in the coil reverses direction.

A simple way to achieve continuous rotational motion of the coil is with a "split-ring commutator" that automatically changes the direction of the current through the coil at just the right moment (Figure 20.54). Springy metal tabs or brushes make contact between the battery and the commutator. Whereas actual motors have a wide variety of designs, this simple single-loop motor illustrates the basic principle involved in motors driven by direct current (DC, as opposed to alternating current, AC).

Electric Generators

It would be inconvenient to build a generator in the form of a bar riding on rails, such as we considered earlier in this chapter. It is mechanically awkward to move a bar or loop back and forth in a linear motion, and it is difficult to avoid having a lot of friction between a bar and supporting rails. Commercial generators rotate a loop of wire in a magnetic field rather than moving a bar back and forth. This has the same effect of achieving a charge separation along part of the rotating loop, but rotary motion is much easier to arrange mechanically, and with relatively low friction on the axle.

Rotating Loop in an External Magnetic Field

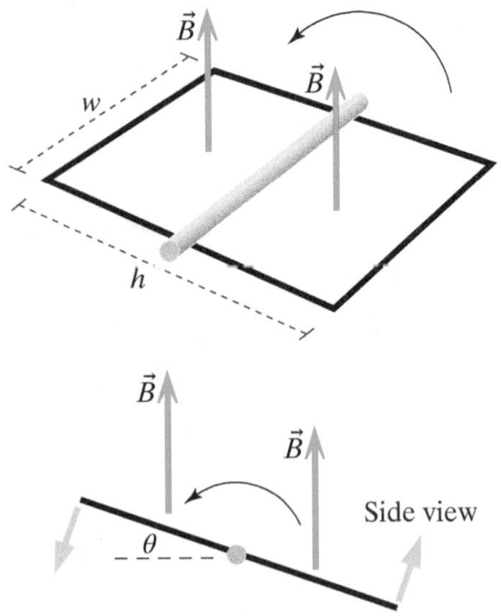

Figure 20.55 Rotating a loop in a magnetic field produces a current.

Suppose that we turn a crank to rotate a loop of dimensions $w \times h$ at a constant angular speed $\omega = d\theta/dt$ (in radians per second) inside a uniform magnetic field made by some large magnet or coils (Figure 20.55). The tangential speed of the left or right wire is $v = \omega(h/2)$, since the wire is a distance $h/2$ from the axle.

> QUESTION Look at the magnetic forces on conventional positive charges inside the wires. At the moment shown in Figure 20.55, in which direction will conventional current run in this loop—clockwise or counterclockwise?

The magnetic forces drive current out of the page on the right wire and into the page on the left wire. Therefore there will be a clockwise conventional current, as seen from above (Figure 20.56). (As far as the front and back wires are concerned, the magnetic forces merely polarize them, perpendicular to the wires.)

> QUESTION At a moment when the loop is at an angle θ to the horizontal, what is the magnitude of the magnetic force that acts on a charge carrier q in the right wire?

The magnetic force on a charge carrier q in the wire is $\vec{F} = q\vec{v} \times \vec{B}$, where $\vec{v}$ is the velocity of the wire. The magnitude of this force is $F = qvB\sin\theta$.

QUESTION At a moment when the loop is at an angle θ to the horizontal, what is the motional emf created along the right wire in Figure 20.55?

The magnetic force is uniform throughout the right wire at this instant, so the motional emf, non-Coulomb work per unit charge, is given by the magnetic force times the length w, divided by q:

$$\text{emf along left wire} = \frac{(qvB\sin\theta)w}{q} = vBw\sin\theta$$

The left wire contributes the same emf, acting in the same direction around the loop, so the emf around the loop is this:

$$\text{emf around loop} = 2vBw\sin\theta$$

Since $v = \omega(h/2)$ and $\sin\theta = \sin(\omega t)$, the emf can be written as

$$\text{emf around loop} = \omega B(hw)\sin(\omega t)$$

where hw is the area of the loop.

We can take current out of the loop through metal tabs pressed against rotating rings connected to the loop, as shown in Figure 20.56.

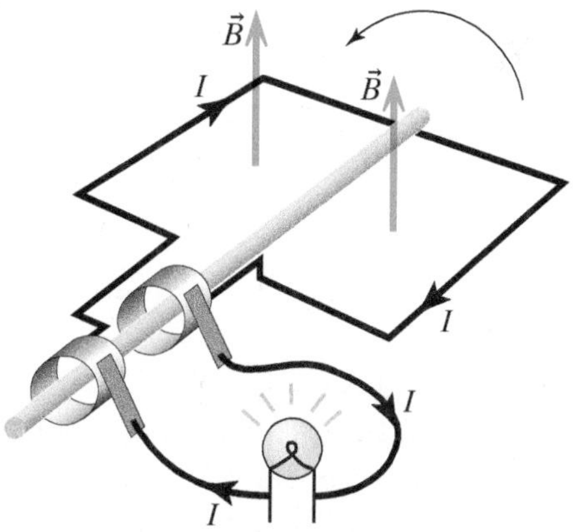

Figure 20.56 If you rotate the loop you can light the bulb, whose resistance is R.

QUESTION If a light bulb with resistance R is connected as shown in Figure 20.56, what is the magnitude of the current I in the loop at this instant, assuming that the loop itself has negligible resistance?

The current will be $I = \omega B(hw)\sin(\omega t)/R$. We see that if we rotate a loop at constant speed inside a magnetic field, we generate an emf and can drive current through a light bulb. This is the basic principle of the electric generators used to power our cities, farms, and factories. The only significant difference from this simple generator is that a commercial generator has many turns in the loop to get a bigger emf. Having N turns is equivalent to having N single-loop generators in series, so we get N times the single-loop emf.

The emf produced in this rotating loop is proportional to $\sin(\omega t)$, so it will drive current through a bulb first one way, then the other way, with a sinusoidal variation. Such currents are called alternating currents or AC, and this is what is supplied at wall sockets. This is in contrast to the steady one-directional current maintained by a battery (or a bar sliding on rails through a uniform magnetic field), which is called direct current or DC.

If the frequency of the alternating current is high enough, a bulb filament doesn't have time to cool off very much between peaks in the cycle, so that the light seems steady to the eye (and there is persistence of vision in the eye, as well). This is the situation with incandescent bulbs in reading lamps, where the frequency f is 60 Hz in some countries and 50 Hz in others (with the angular frequency $\omega = 2\pi f$ in radians per second).

Checkpoint 13 For what value of the angle $\theta = \omega t$ is the instantaneous emf a maximum? A minimum? Rotate a 40-turn rectangular loop that is 0.1 m by 0.2 m in a uniform magnetic field of 2 T, turning at a rate of 30 revolutions (2π rad) per second. What is the maximum emf?

Power Required to Turn a Generator

Presumably we have to exert ourselves to rotate the loop in the magnetic field—we're not going to light the bulb for nothing! Let's see how much power

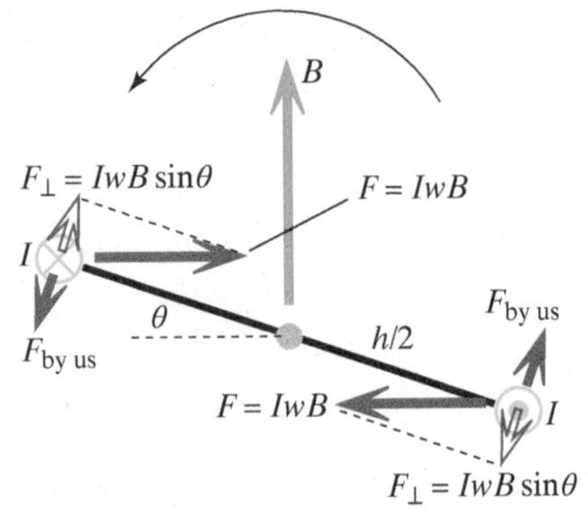

Figure 20.57 Calculating how much work is required to turn a current loop in a magnetic field.

we have to input to make the loop rotate. In the previous section we calculated the current I through an attached light bulb (as a function of the angle $\theta = \omega t$ of the loop to the horizontal). In order to turn the loop in Figure 20.57, we have to exert a varying force $F_{\text{by us}} = IwB\sin\theta$ at both ends of the loop, perpendicular to the loop, to balance the perpendicular component $F_\perp = IwB\sin\theta$ of the magnetic force on the loop.

When we rotate the loop through a small angle $\Delta\theta$, we move each end a short distance $(h/2)\Delta\theta$ and do a small amount of work

$$\Delta W = 2[IwB\sin\theta]\left[\frac{h}{2}\Delta\theta\right]$$

Divide this result by Δt and take the limit as Δt approaches zero to calculate the mechanical power dW/dt we supply:

$$\frac{dW}{dt} = 2[IwB\sin\theta]\left[\frac{h}{2}\frac{d\theta}{dt}\right] = I[Bwh\omega\sin\theta]$$

Given your calculation in the preceding section that $I = Bwh\omega\sin(\theta)/R$ when a bulb of resistance R is attached (where $\theta = \omega t$), we can write this:

$$\frac{dW}{dt} = I[RI] = RI^2$$

This is the power dissipated in the resistor at this instant. Thus we don't get something for nothing; the output power of the resistor that heats the room is equal to the input power that we supply. An electric "generator" might better be called a "converter," because it doesn't create electric power for free; it merely converts mechanical power into electric power.

20.10 *CASE STUDY: SPARKS IN AIR

We now have all the tools and concepts we need to discuss in detail a very complex electric phenomenon: the occurrence of sparks in air. You may occasionally see a spark when you flip a switch, unplug a power cord, or when you separate clothing that has stuck together in the dryer or pull a blanket away from a sheet in the winter. During a spark, charge is transferred from one object to another. How can such sparks occur in air?

Like other gases, air is an excellent insulator, consisting mostly of neutral nitrogen and oxygen molecules. (If air were a good conductor, a charged tape would very quickly become discharged.) There are no movable charged particles—all the electrons are bound to neutral gas molecules. During a spark, however, electrons are ripped out of molecules (ionization), leaving free electrons and positive N_2^+ and O_2^+ ions, all of which can move and transfer significant amounts of charge from one place to another through the air.

An ionized gas is called a plasma. Plasma physics is a domain of physics concerned with the behavior of plasmas, which are found in stars, fusion reactors, and sparks.

The goal of the following sections is to figure out a physically reasonable mechanism for the creation of sparks in air. You will be asked to estimate various physical quantities, in order to judge whether alternative models of sparks are physically reasonable. We will see that sometimes simple calculations make it possible to completely rule out particular models of complex phenomena!

To explain a spark, we must explain these aspects of the phenomenon:

How can electric charge move through air?
Why does a spark last only a short time?
Why is light given off?
And finally, how can the air become ionized?

A Long Chain of Reasoning

There is considerable interest in understanding sparks, since they are one of the most dramatic kinds of electric phenomena. In addition, the kind of explanation we will find for sparks provides a good example of an important kind of scientific reasoning, in which there is a long chain of reasoning. Each of the steps in the explanation is relatively simple; the complexity lies in linking many simple steps into one long, rigorous, logical chain.

How Can Electric Charge Move through Air? Two Models

We will explore two different possible models for how charge might travel through the air, which is normally not a conductor. We will first examine a seemingly plausible model for the situation but then show that this first model for conduction doesn't work. One of the powerful capabilities of science is the ability to rule out possible models or explanations, and we will illustrate this kind of reasoning here.

Suppose that we have two metal balls, one negatively charged and one positively charged. When the two balls are brought close together (but not touching), there is a spark between them. How and why does this happen?

Model 1: Electrons Jump between Balls

It is tempting to guess that during a spark electrons could simply jump from the negatively charged ball to the positively charged ball. Let's try to determine whether this is a reasonable model or not. One important issue is how far, on average, an electron could travel through air before colliding with a gas molecule and losing most of its energy. Since molecules in air are relatively far apart, compared to those in a solid or liquid, perhaps an electron could travel all the way from the negative ball to the positive ball without colliding with anything. We can determine this by estimating what is called the "mean free path" of a particle in air. You may recall a similar argument from Supplement S1 on gases; we review it here.

Mean Free Path

The mean free path is the distance that a particle can travel, on average, before colliding with another particle.

> QUESTION Qualitatively, what factors should the mean free path of a particle in a gas depend on?

You may have guessed that the mean free path d of a particle should be inversely proportional to the density of the gas surrounding it—the higher the gas density, the smaller the mean free path. You may also have guessed that d should be inversely proportional to the size of the gas molecules, or more precisely, since we are interested in collisions, the cross-sectional area of a gas molecule—the larger the molecule, the smaller the mean free path. Both of these factors will indeed prove to be important. If you guessed that the speed of the particle was a factor, think again; because we are not interested in the time between collisions, but only in the distance a particle can travel between collisions, speed does not matter.

It is not difficult to estimate the mean free path d of an electron in air, based on a simple geometrical insight. Draw a cylinder along the direction of motion of the electron, with length d and cross-sectional area A equal to the cross-sectional area of an air molecule (Figure 20.58). The geometrical significance of this cylinder is that if the path of the electron comes within one molecular radius of a molecule, the electron will hit the molecule, so any molecule whose center is inside the cylinder will be hit.

By definition, d is the average distance the electron will travel before colliding with a molecule, so the cylinder drawn in Figure 20.58 should on average contain about one molecule.

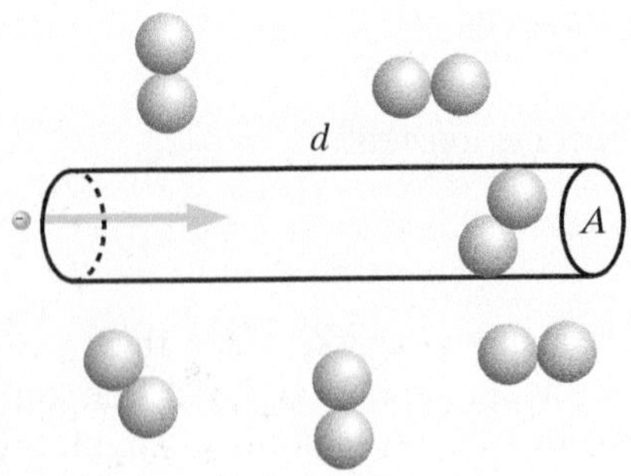

Figure 20.58 A cylindrical volume containing on average one air molecule.

QUESTION What is the volume of air that contains on average one molecule?

At standard temperature and pressure (one atmosphere at 0°C), one mole of gas occupies 22.4 L. One liter is one cubic decimeter—that is, one liter would fill a cube whose sides were 0.1 m long, and whose volume is $(0.1\text{ m})^3$.

$$\left(\frac{22.4\text{ L}}{1\text{ mol}}\right)\left(\frac{1\text{ mol}}{6.02\times10^{23}\text{ molecules}}\right)\left(\frac{(0.1\text{ m})^3}{\text{L}}\right)=3.72\times10^{-26}\text{ m}^3$$

QUESTION To guarantee that an electron traveling through a cylinder of volume $3.72\times10^{-26}\text{ m}^3$ will collide with a molecule, what should be the radius of the cylinder?

The radius of the cylinder should be the radius of a molecule. We can use the approximate value of 1.5×10^{-10} m for this radius.

QUESTION Given the radius and the volume of our hypothetical cylinder containing one molecule, what is the length d of the cylinder?

$$\text{volume}=\pi r^2 d$$

$$d=\frac{\text{volume}}{\pi r^2}=\frac{3.72\times10^{-26}\text{ m}^3}{\pi(1.5\times10^{-10}\text{ m})^2}\approx5\times10^{-7}\text{ m}$$

Thus we see that a free electron in air can travel approximately 5×10^{-7} m before colliding with an air molecule or ion.

QUESTION During a spark in air, the positive ball becomes less positive and the negative ball becomes less negative. Do electrons jump through the air from one ball to the other?

Apparently not! Since a free electron can travel only about 5×10^{-7} m before colliding with a gas molecule and losing much of its energy, this is not a possible mechanism. On the basis of this estimate, we can rule out Model 1 for charge transfer through air.

Model 2: Positive Ions and Electrons Move in Ionized Air

If many of the oxygen and nitrogen molecules in air became ionized, we would have a gas composed of charged particles, all of which are free to move. In other words, the air would now contain both free electrons and positive N_2^+ and O_2^+ ions, and would therefore be a conductor. Leaving aside for the moment the question of how air could become ionized, let's consider the mechanism of charge transfer in ionized air. Just as in a metal wire, there is now a "sea" of mobile charged particles.

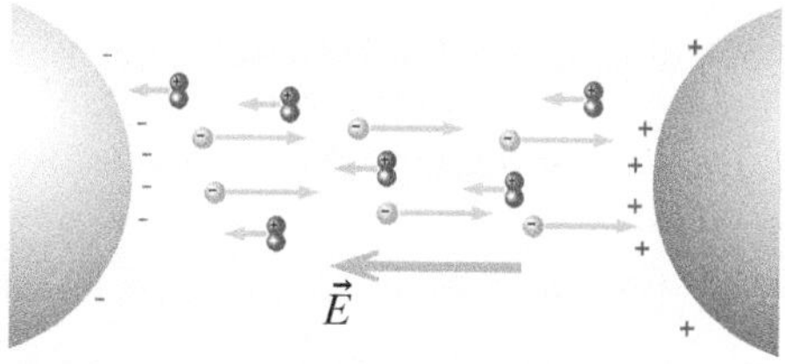

Figure 20.59 Free electrons and positively charged ions (O_2^+ or N_2^+) move through the ionized air between the balls.

QUESTION In which direction will the electrons in Figure 20.59 move? In which direction will the positive ions move?

All of the free electrons in the ionized air will move opposite to the electric field, toward the positively charged ball. The positive ion sea will be driven by the electric field toward the negatively charged ball.

QUESTION How would the negative ball become less negative? How would the positive ball become less positive?

As nearby positive ions come in contact with the negative ball, they can acquire electrons, making neutral molecules once again. The negative ball therefore becomes less negative. Free electrons near the positive ball can come in contact with the ball and join the free electron sea in the metal, making the ball less positive.

QUESTION Does any particle jump from one ball to the other?

No, since there are a very large number of charged particles near each ball, the seas of free electrons and positive ions need move only a very small distance to supply adequate charge to neutralize both balls.

Model 2 (ionized air) appears to be significantly better than Model 1 (jumping electrons), because in the ionized air model no particle travels farther than around one mean free path. We still need to answer the question of how the air might become ionized, though; we will return to this shortly.

Checkpoint 14 At high altitudes the air is less dense. If the density of air were half that at sea level, how would the mean free path of an electron in air change? If the radius of an air molecule were twice as large, how would the mean free path change?

Why Does a Spark Last Only a Short Time?

Assuming that the air can somehow become ionized (we haven't yet explained how), with plenty of mobile electrons and ions, what happens next? Electrons drift toward the positive ball, and the positive ions drift (much more slowly) toward the negative ball. Since the low-mass electrons move so much more quickly, the ionized air conducts in a manner that is rather similar to electric conduction in a metal. There is a "sea" of free electrons that drifts through the air. There is a kind of "start-stop" motion of an individual electron, in that it is accelerated until it collides with a molecule and comes nearly to a stop, then accelerates again. This accelerated "start-stop" motion has some average speed, called the "drift speed." Note the similarity to the Drude model of mobile electrons in a metal.

QUESTION What happens when an electron collides with the positive metal ball? Does the electric field in the air between the balls change?

The electron fills an electron deficiency in the positively charged metal ball, making the ball less positively charged. As a result, the magnitude of the electric field at all locations in the air between the spheres decreases slightly.

QUESTION What happens when an ion collides with the negative metal ball? What happens to the ion? Does the electric field in the air between the balls change?

The ion can pick up an excess electron from the negatively charged metal ball, and the ion becomes an ordinary neutral air molecule. The negatively charged ball is now less negatively charged. Again, as a result, the magnitude of the electric field at all locations in the air between the spheres decreases slightly.

We'll see later why it is that as long as the electric field is larger than about 3×10^6 V/m throughout a region of air, the air remains heavily ionized, but if the electric field becomes too small, the ions and electrons recombine with each other, and the gas is no longer a conductor.

QUESTION Why does the spark last only a very short time?

The excess charge on both metal balls is quickly reduced to an amount that is not sufficient to make a large enough electric field to keep the air heavily ionized, and the spark goes out. However, if you continually resupply the metal balls with charge by connecting the balls to a battery or power supply, you may be able to maintain a steady spark. A dramatic example is the very bright spark maintained between two carbon electrodes in a searchlight or commercial movie projector (a "carbon arc" light). A gentler example is the continuous glow seen in neon lights, which contain neon at low pressure.

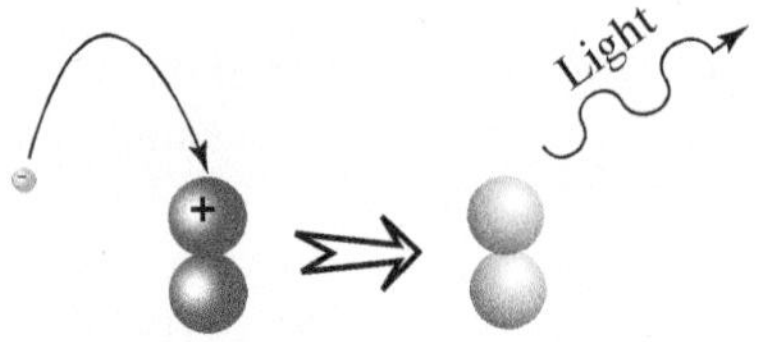

Figure 20.60 A free electron recombines with a positive ion to produce a neutral molecule. The excess energy is given off as visible light.

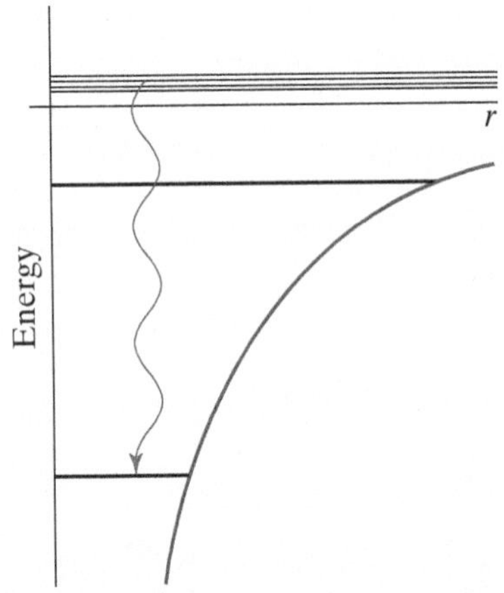

Figure 20.61 The transition of electron plus ion from an unbound (high-energy) state to a bound (lower-energy) state is accompanied by emission of a photon.

Why Is Light Given Off?

What causes the light that we see emitted by a spark (why are photons emitted)? Actually, the emission of light is a sort of side effect of the ionization of the air. Occasionally a free electron comes near enough to a positive ion (not necessarily the original ion) to be attracted and recombine with the ion to form a neutral molecule. When the free electron and the ion recombine (Figure 20.60), there is a transition from the high-energy unbound state (free electron and ion) to a low-energy bound state (Figure 20.61), with the emission of a photon.

It is important to understand that the emission of light goes on simultaneously with charge conduction through the ionized air. We will see later that as long as the applied electric field is big enough, neutral molecules are continually being ionized, at the same time that some free electrons and ions are recombining to make neutral molecules (with the emission of light). Although the light that we see is the most obvious aspect of a spark, the light can be considered to be essentially a mere side effect of conduction in a gas.

We have now seen the main aspects of sparks in ionized gas. Sparks are often referred to as "gas discharges," and a heavily ionized gas whose net charge is zero is called a "plasma." Plasma physics is an important area of contemporary research.

There remains the big question, how does the gas get ionized in the first place? That is the subject of the remaining sections.

How Does the Air Become Ionized? Two Models

It is observed experimentally that an electric field of about 3×10^6 V/m is sufficient to ionize air and make it a conductor. We will try to come up with a model of this process. To determine whether our model is a good one, we will compare the predictions of the model to experimental observations—in particular, the observation that the critical field value is $E_{crit} = 3 \times 10^6$ V/m.

As we did with models for conduction in the air, we will explore two different possible models for how the air can become ionized. First we'll examine a seemingly plausible model but then show that this model for ionization doesn't work. Again, it is an important and powerful aspect of science that it is often possible to rule out proposed explanations even if you don't know what the correct explanation is. After discarding the first model for ionization we'll discuss another model that explains many aspects of sparks.

Model 2A: A Strong Electric Field Pulls an Electron out of a Molecule

Since an atom or molecule is polarized by an electric field, we might suppose that an electric field of magnitude 3×10^6 V/m is large enough to pull an electron completely out of an air molecule, creating a positive ion and a free electron. Let's check this model.

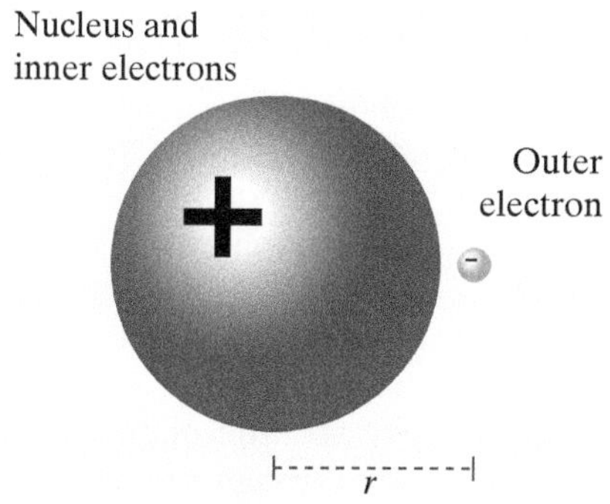

Figure 20.62 An outer electron interacts with the nucleus and inner electrons of the atom.

QUESTION Knowing what you know about atoms, estimate about how big an electric field would be needed to rip an outer electron out of a neutral atom.

If you could not do this calculation, think again for a moment before reading farther. You really do know enough to make this calculation. In particular, remember that the radius of an atom is about 1×10^{-10} m. For simplicity you may want to think about a hydrogen atom.

You know that a spherical charged object acts like a point charge (outside of the sphere). Think of an outer electron as being attracted by the rest of the atom, which has a charge $+e$ (Figure 20.62). The electric field made by this $+e$ charge that acts on the outer electron is

$$E = \frac{1}{4\pi\varepsilon_0}\frac{e}{r^2} = (9 \times 10^9\ \text{N} \cdot \text{m}^2/\text{C}^2)\frac{(1.6 \times 10^{-19}\ \text{C})}{(1 \times 10^{-10}\ \text{m})^2} = 1.4 \times 10^{11}\ \text{V/m}$$

where $r = 1 \times 10^{-10}$ m is the approximate radius of an atom. We would have to apply at least that big a field in order to pull an outer electron out of an atom.

The enormous difference between the experimentally observed field of 3×10^6 V/m and the value predicted by our model for the electric field necessary to ionize an atom means that we can rule out the possibility of direct atomic ionization by the applied electric field. This is an excellent example of one of the strengths of science: it is often possible to rule out a proposed model for a process, even if we can't figure out a better explanation.

Having ruled out direct atomic ionization by the applied electric field, what other process can we think of?

Model 2B: Fast-Moving Charged Particles Knock Electrons out of Atoms

It takes a dramatic event to ionize air, because it requires a very large force to rip electrons out of air molecules. A fast-moving charged particle that collides with an atom or molecule can knock out an electron, leaving a singly ionized ion behind. Where would such energetic charged particles come from? As it happens, there are fast-moving charged particles passing through your body at this very moment, ionizing some atoms and molecules in your body! Some of these charged particles are "muons" produced by cosmic rays in nuclear reactions at the top of the atmosphere. Others are electrons, positrons, or alpha particles (helium nuclei) emitted by radioactive isotopes present in trace quantities in materials in and around you.

This process is the cause of some potentially dangerous events. A neutral atom in your DNA that suddenly turns into an ion can cause biochemical havoc and genetic mutations. The passage of fast-moving charged particles through the tiny and very sensitive components in a computer chip can change a memory bit with potentially disastrous consequences. To guard against such disasters, both DNA and computer circuits have checks built into them to try to compensate for damage caused by the unavoidable passage of high-speed charged particles.

Invaders from Outer Space

"Cosmic rays," consisting mostly of very high energy protons coming from outside our Solar System, strike the nuclei of molecules in the Earth's upper atmosphere. The resulting nuclear reactions produce a spray of other particles. Many of the products of these reactions are particles such as protons, neutrons, and pions that interact very strongly with other nuclei in air molecules. For that reason these particles typically don't go very far through the upper atmosphere. It is mainly positive and negative muons (μ^+ and μ^-) and neutral neutrinos that are able to penetrate the atmosphere and reach the surface of the Earth.

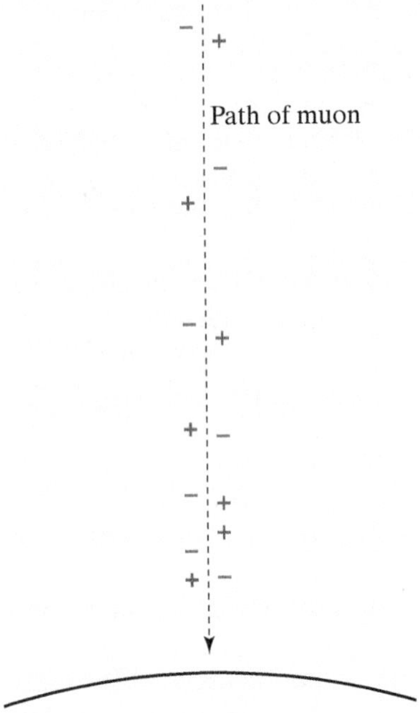

Figure 20.63 Ion trail left by a muon plunging down through the Earth's atmosphere.

Electrons, muons, and neutrinos do not interact with nuclei through the strong interaction. Electrons and muons, being charged, do interact electrically with atoms and nuclei. Electrons have so little mass that they undergo large deflections and don't travel very far through the air. Muons are essentially just massive electrons, behaving in almost every way just like electrons except for having a mass about 200 times the electron mass. Because of their large mass, the muons undergo small deflections and can travel long distances through the air.

The neutral neutrinos, lacking electric charge, don't have any electric interactions with matter, and like electrons and muons they have extremely weak nonelectric interactions with matter. The interactions of neutrinos with matter are so unimaginably weak that most of them go right through the entire Earth, coming out the other side with no change!

As the rapidly moving charged muons plunge downward through the air, they have enough energy to occasionally knock electrons out of air molecules, producing free electrons and positive ions (Figure 20.63). Each muon leaves a trail of slightly ionized air. Muons produced by cosmic rays are not the only cause of ionization in the air. Materials around you contain trace quantities of radioactive nuclei that can emit high-speed electrons, positrons, or alpha particles (helium nuclei, He^{2+}). These rapidly moving charged particles can also ionize the air.

If the ionization were extensive, air would be a good conductor at all times, but the number of free electrons and ions produced by cosmic-ray muons and natural radioactivity is much too small to make the air be a good conductor. However, in the next section we will see that the small amount of ionization produced in this way can act as a trigger for large-scale ionization.

It is surprising that the muons actually manage to reach the Earth's surface, because a positive or negative muon at rest has an average lifetime of about 2 μs before splitting up (decaying) into a positron (if μ^+) or electron (if μ^-), a neutrino, and an antineutrino. Even if moving at nearly the speed of light, a muon would be expected to travel only about $(3 \times 10^8\ \text{m/s})(2 \times 10^{-6}\ \text{s}) = 600$ m before decaying, which is far too short a distance to reach the ground from the top of the atmosphere where the muon was produced.

However, as we saw in Section 20.6, Einstein's special theory of relativity predicts that time elapses more slowly for a fast-moving object, so that in the reference frame of the muon it takes less than 2 μs to reach the ground, and many muons plunge through your room every minute.

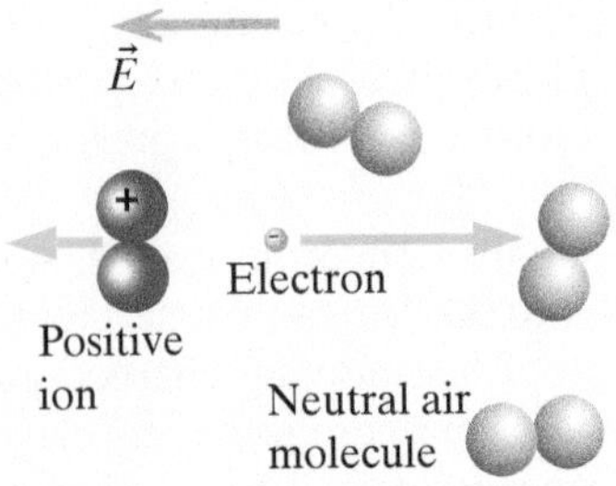

Figure 20.64 A single free electron is accelerated by the electric field, initiating a chain reaction.

A Chain Reaction

If there happens to be a single free electron in the air (due to muons or natural radioactivity), it can be accelerated by an applied electric field due to nearby charged objects such as the charged metal balls shown earlier (Figure 20.64).

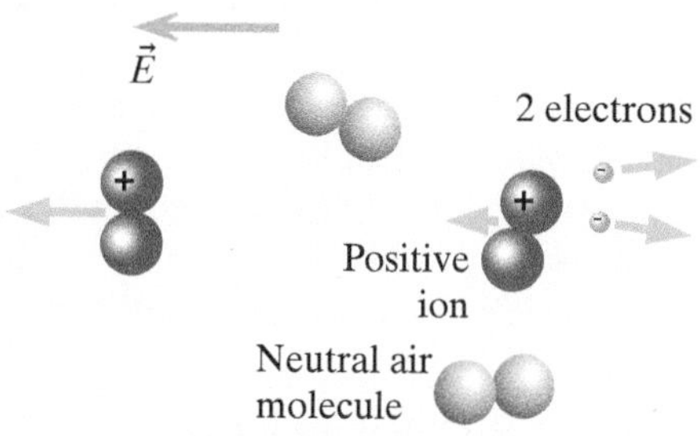

Figure 20.65 The first free electron ionizes an air molecule, producing a second free electron.

If the electron gets going fast enough before colliding with an air molecule, it can knock another electron out of that molecule, so now there are two free electrons and two ions (Figure 20.65).

These two free electrons can knock out two more electrons, so the number of free electrons grows rapidly: 2, 4, 8, 16, 32, 64, and so on. This is an avalanche, or "chain reaction." The air becomes significantly ionized with a large number of mobile charged particles and is now a rather good conductor (Figure 20.66). Note that the air is still neutral overall, just as the interior of a metal wire is neutral overall.

Of course, the positive ions created when molecules are ionized are also affected by the applied electric field, and also begin to move under the influence of the field. However, because their mass is much larger than the mass of an electron, these ions move much more slowly. Therefore it makes sense to focus on free electrons as the particles whose collisions ionize a molecule when they collide with it.

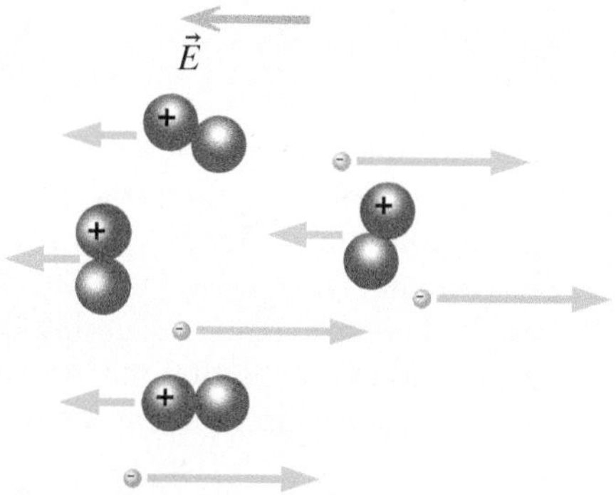

Figure 20.66 As the chain reaction progresses, more and more charged particles are freed up, and the air becomes a conductor.

Electric Field Required to Trigger a Chain Reaction

The few free electrons due to cosmic rays and radioactivity frequently bump into neutral air molecules, but these collisions won't start a chain reaction unless the electrons are traveling fast enough to rip electrons out of the molecules. Let's see whether we can predict from fundamental principles and the structure of matter the magnitude of the critical electric field E_{crit} that would be required to give a free electron a high enough kinetic energy to trigger a chain reaction.

> QUESTION What energy input is required to remove an outer electron from an oxygen or nitrogen molecule?

When a molecule is ionized, an outer electron is moved from its average location a distance R from a nucleus to a location far enough away that the field due to the positive ion is very small. The change in potential energy of the electron plus positive ion system is therefore:

$$\Delta U = (-e)\Delta V = (-e)(V_\infty - V_R)$$

As we saw in the previous discussion of Model 2A, near the ion the electric field due to the ion is very large compared to any applied field, so we only need to consider the potential difference due to the positive ion, which has a charge $+e$. Estimating the radius R of the ion to be about 1×10^{-10} m, we find that

$$\begin{aligned}(V_\infty - V_R) &= 0 - \frac{1}{4\pi\varepsilon_0}\frac{e}{r}\\ &= -\frac{1}{4\pi\varepsilon_0}\frac{e}{(1\times10^{-10}\ \text{m})}\\ &= -14.4\ \text{V}\end{aligned}$$

so $\Delta U = -1.6\times10^{-19}\text{C}\cdot(-14.4\text{V}) = 2.3\times10^{-18}\ \text{J}$

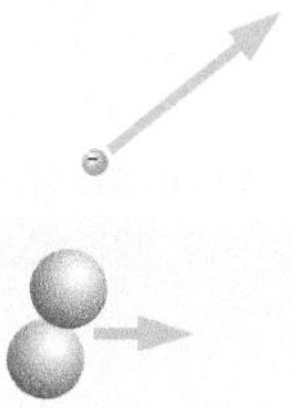

Figure 20.67 An electron with insufficient energy does not remove an electron from the air molecule.

This increase in the potential energy of the system must come from the kinetic energy of the fast-moving electron that collides with the molecule. (For simplicity we're ignoring the kinetic energy of the outer electron bound in the molecule, though it happens that 2.3×10^{-18} J is in fact close to the ionization energy for molecular nitrogen and molecular oxygen.) If the incoming electron's kinetic energy K is less than 2.3×10^{-18} J, it just bounces off or excites the air molecule to a higher quantum state without removing an electron (Figure 20.67). However, if K is equal to or greater than the ionization energy of an air molecule, after the collision there can be two electrons where there had been one (Figure 20.68).

Both of these electrons in Figure 20.68 can now be accelerated on average through a distance d, and both may ionize molecules, leading to the existence of 4 free electrons, then 8, 16, 32, 64, and so on—a chain reaction. (After a typical ionizing collision both electrons are moving rather slowly, so each time they start nearly from rest, and the collision energy is always about the same.)

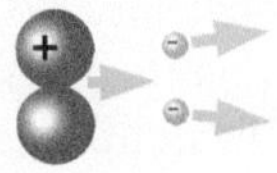

Figure 20.68 An electron with sufficient energy knocks another electron out of an air molecule during a collision.

QUESTION Could a free electron that starts from rest gain adequate kinetic energy in one mean free path to ionize a molecule?

Considering the system of free electron plus the sources of the applied field:

$$\Delta K + \Delta U = 0$$
$$(K_f - 0) + (-e)\Delta V = 0$$
$$K_f = -(-e)\Delta V$$
$$\Delta V = \frac{K_f}{e} = \frac{2.3 \times 10^{-18}\ \text{J}}{e} = +14.4\ \text{V}$$

On average a free electron will travel only a distance of one mean free path before colliding with a molecule (Figure 20.69). To gain adequate kinetic energy, the free electron must travel through a potential difference of +14.4 V while traveling this distance. Unlike the electron that is bound to the molecule, the free electron is affected only by the applied electric field. Assuming that the applied field is uniform over this distance, and assuming that the electron travels a distance of one mean free path, which we estimated earlier to be 5×10^{-7} m:

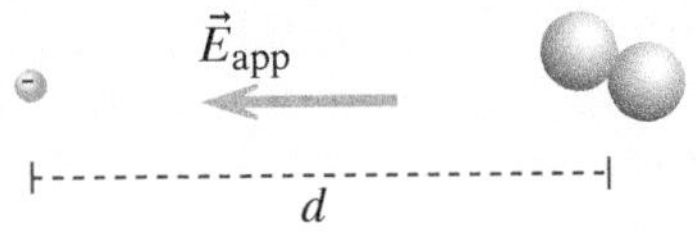

Figure 20.69 Mean free path d.

$$\Delta V = E_{app} \cdot d$$
$$+14.4\ \text{V} \approx E_{app} \cdot 5 \times 10^{-7}\ \text{m}$$
$$E_{app} \approx 2.9 \times 10^{7}\ \frac{\text{V}}{\text{m}}$$

Model 2 therefore predicts that the applied electric field required to initiate a spark must be approximately 3×10^7 V/m.

QUESTION Should we consider this good or poor agreement with the experimental value of 3×10^6 V/m?

For the previous model (direct ionization by the applied field), the prediction of the model was off by 5 orders of magnitude (1×10^5) from the experimental value. It is satisfying that with the chain reaction model we predict the right order of magnitude (we're off by a factor of 10 instead of being off by a factor of 1×10^5). This tells us that we now have a pretty good intuitive feeling for the issues.

To make an accurate prediction would involve difficult calculations because of the statistical nature of the process. In our simple calculations we assumed that every electron goes one mean free path and picks up the same amount of kinetic energy. However, occasionally an electron happens to go much farther than d, and a smaller electric field is sufficient to accelerate such an electron to a high enough energy to ionize the molecule that it finally hits. It also isn't necessary that every electron ionize an atom in order to create a chain reaction. That is why our simple calculation has overestimated the magnitude of electric field required to cause a spark.

Additional Tests of the Models

It is possible to make additional tests of the two models by comparing their predictions with other experimental observations, such as the effect on the critical field of changing the density of the gas. If a physical model explains a variety of different kinds of experimental results, this increases our confidence in the validity of the model.

> **QUESTION** If the density of air were half as large as it is at sea level, according to each model how would this affect the electric field required to ionize air, which is normally about 3×10^6 V/m?

Halving the density of the air would double the mean free path d. According to model 1 (electric field large enough to detach an electron) this would make no difference. According to model 2 (accelerated free electron collides with molecule), since the kinetic energy of the free electron (starting from rest) is $\Delta K = K_{\text{final}} - 0 = eEd$, doubling d would decrease E_{critical} by a factor of two. What is observed experimentally is that the required electric field is indeed proportional to the air density, supporting model 2.

An ordinary long fluorescent light tube or a neon sign exemplifies this. If the gas density inside the tube were not very low, it would be difficult to create a sufficiently high electric field throughout the tube to sustain a discharge (a continuous spark). Because the gas density in the tube is low, the electric field required is also low, making such lights practical. On the other end of the scale, gases at high pressures (hence high densities) are sometimes used as insulators.

> **Checkpoint 15** If you observed that under certain conditions an electric field three times as great as usual were required to ionize air, what might you conclude about the mean free path under these conditions? Approximately how large an electric field would be required to cause a spark in a gas at standard temperature and pressure if the ionization energy for the molecules of this gas is 1.5 times the ionization energy for air molecules, and the cross-sectional area of the molecules is 0.8 times the cross-sectional area of air molecules?

Drift Speed of Free Electrons in a Spark

We can now estimate the average "drift speed" of free electrons in a spark. Knowing that the kinetic energy acquired in one mean free path is 2.4×10^{-18} J, we can determine the typical final speed v_f of an electron when it hits a molecule:

$$\frac{1}{2}mv_f^2 \approx 2.3 \times 10^{-18}\text{ J}$$

$$v_f \approx \left(\frac{2(2.3 \times 10^{-18}\text{ J})}{9 \times 10^{-31}\text{ kg}}\right)^{1/2} = 2 \times 10^6\text{ m/s}$$

$$\bar{v} \approx \frac{1}{2}v_f = 1 \times 10^6\text{ m/s}$$

The statistical average speed is difficult to compute precisely, since there is a distribution of free paths. Evidently, however, the average drift speed is of the order of 1×10^6 m/s.

Propagation of Ionization

The magnitude of the electric field is not uniform in the region between two oppositely charged metal balls, and it is largest near the balls. (Do you see why?) Suppose that the field near the surface of the negative ball is bigger than 3×10^6 V/m, so that the air near the surface of this ball becomes strongly ionized. How can the ionization extend into other regions where E is smaller?

Positive ions are attracted toward the negative ball, and electrons are repelled into regions of the air where the electric field was less than 3×10^6 V/m. These electrons increase the electric field in the un-ionized region, and their large number facilitates a chain reaction, so the spark penetrates into the air farther and farther from the ball. It is possible to form an unbroken column of heavily ionized air leading from one ball to the other, even though initially the field was stronger than 3×10^6 V/m only near the negatively charged ball. Once the air is heavily ionized, an electric field significantly less than 3×10^6 V/m is sufficient to keep the spark going for a while. Recombination of electrons and ions along this column makes the column glow, and you see a spark connecting the two balls.

Something very similar happens in a lightning storm. A column of ionized air propagates from the negatively charged bottom of a cloud downward toward the ground. The speed of propagation from cloud to ground is somewhat slower than the average drift speed you calculated (about 1×10^6 m/s), because the propagation occurs in steps, with pauses between steps due to the statistical nature of triggering a chain reaction in the neighboring air. Once the column of ionized air reaches all the way to the ground, a large current runs through this ionized column. The visible lightning flash is of course due to the recombination of ions and electrons in this column. The large current heats the column explosively, and the rapidly expanding gas makes the noise heard as thunder.

20.11 *RELATIVISTIC FIELD TRANSFORMATIONS

Alice and Bob observed different electric and magnetic fields in their different reference frames, though they both could correctly predict the motion of the protons in their own reference frames, in terms of the fields they observe and the electric and magnetic forces.

The theory of relativity predicts how measurements of electric and magnetic fields transform when you change from one reference frame to another. We state these transform equations without proof. In the following, the "unprimed" quantities (E_x, etc.) are the fields that are observed in the lab frame (fixed to our laboratory). The "primed" quantities (E'_x, etc.) are the fields that are observed in a reference frame that is moving with speed v in the $+x$ direction relative to the lab frame (Figure 20.70).

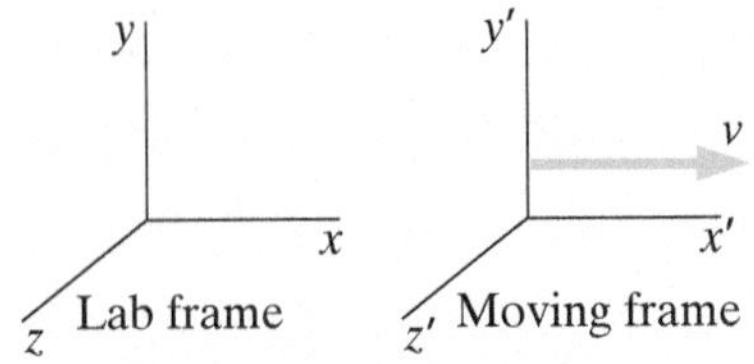

Figure 20.70 The moving reference frame moves with speed v in the $+x$ direction relative to the lab frame.

$$E'_x = E_x \qquad E'_y = \frac{(E_y - vB_z)}{\sqrt{1-v^2/c^2}} \qquad E'_z = \frac{(E_z + vB_y)}{\sqrt{1-v^2/c^2}}$$

$$B'_x = B_x \qquad B'_y = \frac{\left(B_y + \frac{v}{c^2}E_z\right)}{\sqrt{1-v^2/c^2}} \qquad B'_z = \frac{\left(B_z - \frac{v}{c^2}E_y\right)}{\sqrt{1-v^2/c^2}}$$

Magnetic Field of a Moving Charged Particle

Let's apply these transforms to a positively charged particle that is at rest in the lab frame. It of course makes only an electric field, yet if you run past the particle, you observe it in motion, and you observe not only an electric field but also a magnetic field.

In the lab frame the electric field of the stationary charge is given by:

$$\vec{E} = \frac{1}{4\pi\varepsilon_0}\frac{q}{r^2}\hat{r}$$

We would like to determine the fields made by this charge in the moving reference frame. In the moving reference frame the particle is seen to be moving with speed v in the $-x$ direction (Figure 20.71).

In the lab frame there is no magnetic field. However, straight above the moving particle, E_y is positive, so we have this in the moving reference frame:

$$B'_z = \frac{\left(B_z - \frac{v}{c^2}E_y\right)}{\sqrt{1-v^2/c^2}} = \frac{-\frac{v}{c^2}E_y}{\sqrt{1-v^2/c^2}}$$

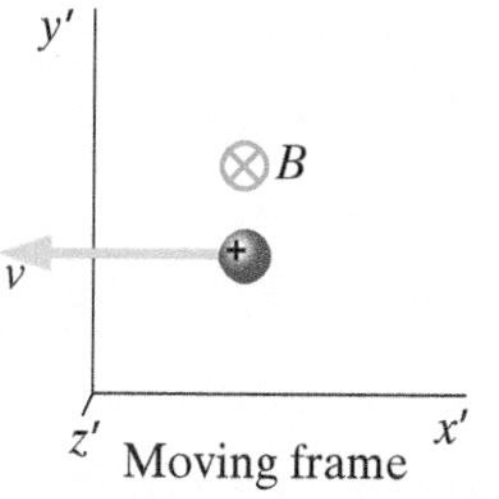

Figure 20.71 In the moving reference frame the particle moves with speed v in the $-x$ direction. A magnetic field appears in this frame.

At low speeds, this is approximately

$$B'_z = -\frac{v}{c^2}\left(\frac{1}{4\pi\varepsilon_0}\frac{q}{r^2}\right) \quad (v \ll c)$$

The magnetic field above the positively charged particle is in the $-z$ direction (Figure 20.71). This is consistent with the prediction of the Biot–Savart law for the direction of magnetic field above the charge. In an exercise at the end of this section you are asked to show that the transforms predict the usual curly magnetic field all the way around the moving charge.

Let's evaluate the magnitude of the magnetic field in the moving frame in terms of $\mu_0/4\pi$. In analyzing the electric and magnetic forces between two moving protons (the Alice and Bob episode), we found that

$$\frac{1}{\mu_0\varepsilon_0} = c^2 \quad \text{so} \quad \frac{1}{4\pi\varepsilon_0} = c^2\frac{\mu_0}{4\pi}$$

Therefore we can write

$$B'_z = -\frac{v}{c^2}\left(\frac{1}{4\pi\varepsilon_0}\frac{q}{r^2}\right) = -\frac{\mu_0}{4\pi}\frac{qv}{r^2}$$

This is exactly the magnitude of the magnetic field of the moving charge that is predicted by the Biot–Savart law, which is valid for $v \ll c$. This is partial verification that the field transforms are consistent with what you already know about electric and magnetic fields.

ELECTRIC AND MAGNETIC FIELDS ARE INTERRELATED

Electric fields and magnetic fields are closely related to each other, since a mere change of reference frame mixes the two together.

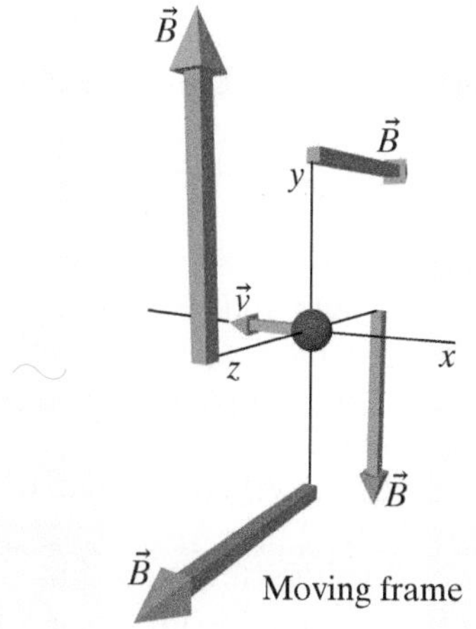

Figure 20.72 Use the field transforms to show that in the moving frame there is a curly magnetic field all the way around the moving positive charge.

In many situations one can consider magnetic fields to be a relativistic consequence of electric fields. For example, if you run past a stationary charge you will observe a magnetic field, but if you stand still you will not observe a magnetic field. However, it is not always the case that magnetic fields are merely the result of changing the reference frame. For example, a stationary electron or proton or neutron acts like a magnetic dipole and produces a magnetic field, and there is no inertial reference frame in which that magnetic field is zero.

Checkpoint 16 Use the field transforms to find the direction of the magnetic field in front of, behind, and below the positively charged particle in the moving reference frame, and therefore show that there is a curly magnetic field all the way around the moving charge, as expected (Figure 20.72).

Electric Field of a Rapidly Moving Charged Particle

If the speed of the moving reference frame is very large (very near c), the electric field of the particle is altered significantly in the moving frame. The magnitude of the electric field in front of and behind the particle in the moving frame is unchanged: $E'_x = E_x$. However, straight off to the side of the particle the electric field is much larger at high speeds, since we have

$$E'_y = \frac{(E_y + vB_z)}{\sqrt{1 - v^2/c^2}} = \frac{E_y}{\sqrt{1 - v^2/c^2}}$$

which is much greater than E_y if v approaches c. Figure 20.73 shows the general pattern of electric field in the moving frame, near the charge that is rapidly moving in this reference frame. The region of high field strength becomes a thinner and thinner pancake as the speed of the particle approaches the speed of light.

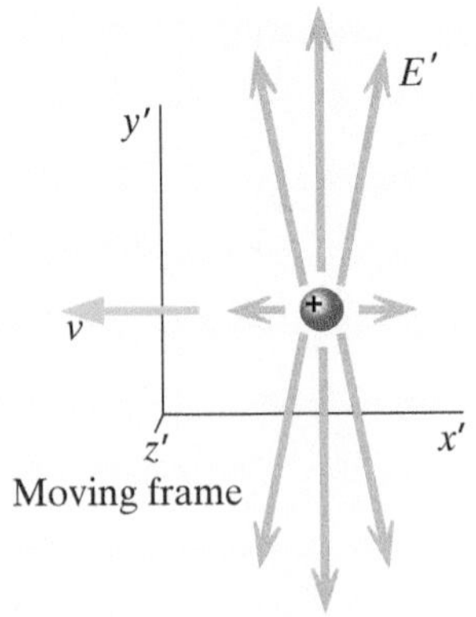

Figure 20.73 The electric field of the particle in the moving frame, where the particle is moving rapidly.

Moving through a Region of Uniform Magnetic Field

The previous examples involved a situation in which there was no magnetic field in the rest frame of the particle but there were both electric and magnetic fields in a frame where the particle was moving. Next we'll consider a region of uniform magnetic field in the lab frame, but no electric field, and see what fields are present in the moving reference frame. In the lab frame the magnetic field is $\langle 0, 0, -B \rangle$, so $B_z = -B$ (Figure 20.74).

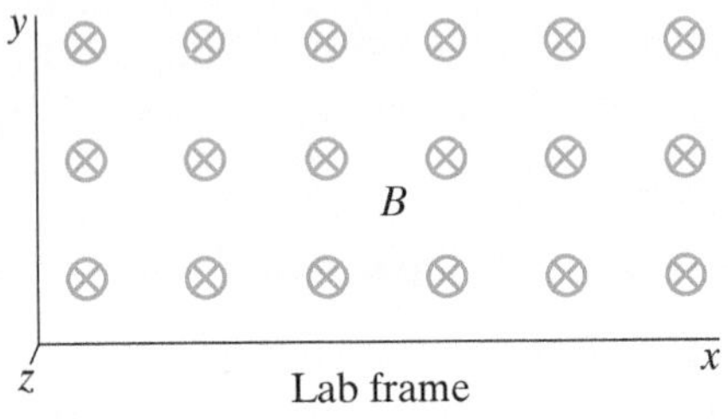

Figure 20.74 A region in the lab frame of uniform magnetic field in the $-z$ direction.

QUESTION Suppose that you move with speed v in the $+x$ direction through this region. Using the field transforms given above, what electric field would you observe?

The transforms show that in the moving reference frame, you will observe an upward electric field of magnitude $E = vB$ ($+y$ direction; see Figure 20.75), though there was no electric field in the lab frame:

$$E'_y = \frac{0 - vB_z}{\sqrt{1 - v^2/c^2}} = \frac{0 - v(-B)}{\sqrt{1 - v^2/c^2}} \approx vB \quad \text{if } v \ll c$$

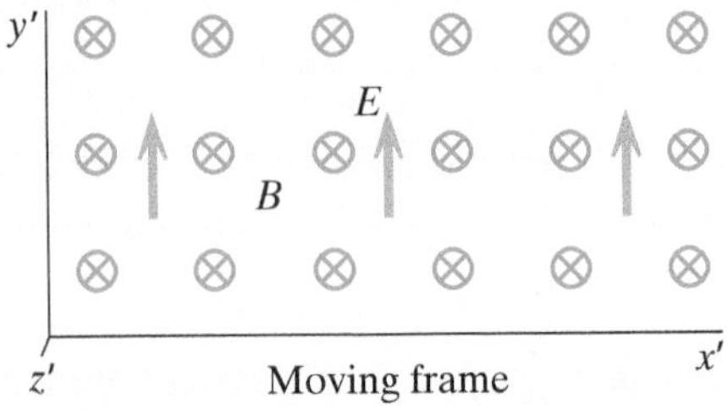

Figure 20.75 In the moving frame there is an electric field in the $+y$ direction of magnitude $E = vB$.

Suppose that we are in this moving frame, and we observe a metal bar that is at rest in this frame. The metal bar polarizes because there is an upward electric field of magnitude $E = vB$ everywhere in this region (Figure 20.76). The bar polarizes until there is sufficient charge buildup to produce a downward electric field of magnitude $E_{pol} = vB$, at which point the net electric field inside the metal is zero, and there is equilibrium.

In the lab frame, the metal bar moves at constant speed v in the $+x$ direction through a region of uniform magnetic field and zero electric field (Figure 20.77). We discussed this situation earlier in the chapter in connection with motional emf. In the lab frame the moving metal bar polarizes due to the magnetic force qvB on the moving mobile charges. The polarization charges produce a downward electric field E_{pol} ($-y$ direction). At equilibrium it must be that $qE_{pol} = qvB$, so $E_{pol} = vB$.

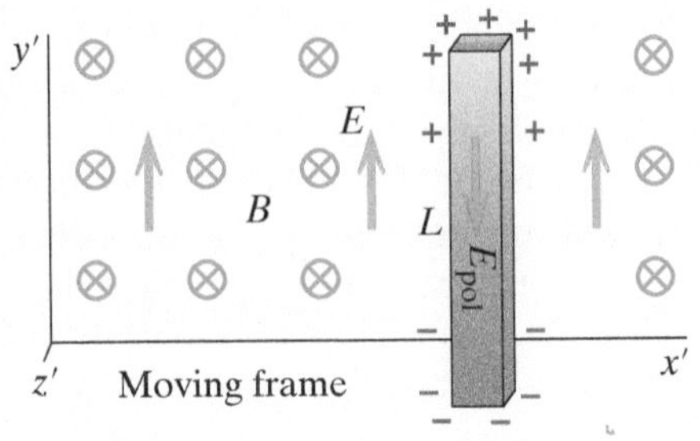

Figure 20.76 A metal bar at rest in the moving frame is polarized by the upward applied electric field. The polarization charges produce a downward electric field $E_{pol} = vB$.

The amount of charge buildup is the same in either reference frame, but the explanation that observers in the two frames give for the cause of the charge buildup is different. In the moving reference frame, where the bar is at rest, there is an upward electric field of magnitude $E = vB$ that polarizes the bar until this electric field becomes balanced by the downward electric field $E_{pol} = vB$ of the polarization charges. In the lab frame, where the bar is

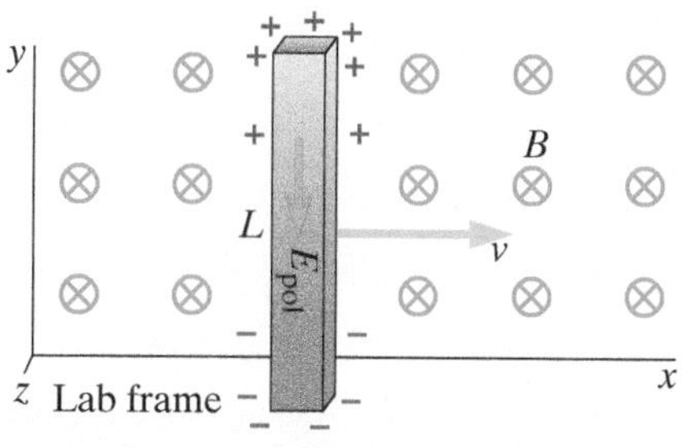

Figure 20.77 In the lab frame the moving metal bar is polarized by the magnetic force on the moving charges. The polarization charges produce a downward electric field $E_{pol} = vB$.

in motion, there is a magnetic force that polarizes the bar until the magnetic force becomes balanced by the electric force of the polarization charges; the downward electric field has magnitude $E_{pol} = vB$.

THE PRINCIPLE OF RELATIVITY

There may be different mechanisms for different observers in different reference frames, but all observers can correctly predict what will happen in their own frames, using the same relativistically correct physical laws.

Moving through a Velocity Selector

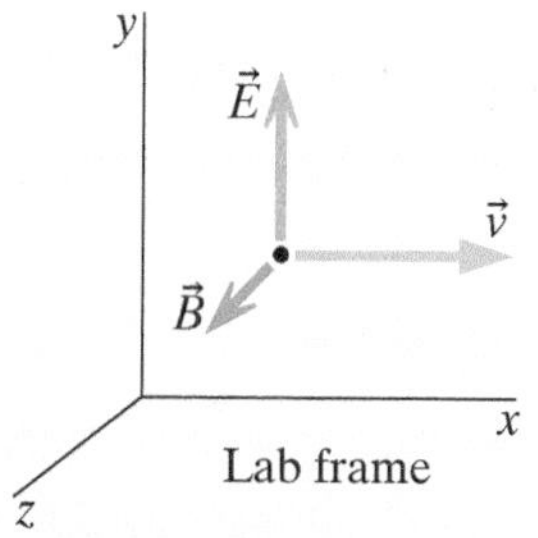

Figure 20.78 A velocity selector. If $E = vB$, the charged particle passes through with no deflection.

As another example, consider the velocity selector discussed earlier and shown in Figure 20.78. In the lab frame a charged particle traveling with speed v moves in the x direction through a region where there is an electric field of magnitude E in the $+y$ direction and a magnetic field of magnitude B in the $+z$ direction. If $E = vB$, the charged particle passes through with no deflection, because the electric force of magnitude qE is equal and opposite to the magnetic force of magnitude qvB.

QUESTION Suppose you move with speed v in the $+x$ direction through this region, without the charged particle present. Using the field transforms given above, what electric and magnetic fields would you observe?

The only fields you would observe in the moving frame are these:

$$E'_y = \frac{(E_y - vB_z)}{\sqrt{1-v^2/c^2}} = \frac{E - vB}{\sqrt{1-v^2/c^2}} = \frac{vB - vB}{\sqrt{1-v^2/c^2}} = 0$$

$$B'_z = \frac{\left(B_z - \frac{v}{c^2}E_y\right)}{\sqrt{1-v^2/c^2}} = \frac{B - \frac{v}{c^2}E}{\sqrt{1-v^2/c^2}} = \frac{B - \frac{v}{c^2}(vB)}{\sqrt{1-v^2/c^2}} = \frac{(1-v^2/c^2)}{\sqrt{1-v^2/c^2}}B$$

QUESTION Given that you observe these fields, what will happen to a charged particle initially at rest in your reference frame?

A charge at rest is unaffected by the magnetic field, and there is no electric field, so there is no force on the charged particle. This makes sense, because a particle at rest in your reference frame moves with constant velocity in the lab frame, straight through the velocity selector.

Note again that observers in the two reference frames differ in their analyses of the situation, but both correctly predict what happens to the charged particle. You predict that a charged particle at rest will remain at rest, because there is no electric field, and the net electric and magnetic force is zero. An observer in the lab frame calculates an electric force and a magnetic force, determines that they are equal and opposite to each other, and predicts that the particle will have a uniform velocity since the net force on the particle is zero.

An Alternative Form of the Transforms

You may have noticed terms that look like a cross product in the transforms for the y and z components of the fields—the components perpendicular to the motion. We can write the transforms for the fields parallel to the motion and perpendicular to the motion in the following way:

$$E'_{\parallel} = E_{\parallel} \qquad B'_{\parallel} = B_{\parallel}$$

$$E'_{\perp} = \frac{(\vec{E}+\vec{v}\times\vec{B})_{\perp}}{\sqrt{1-v^2/c^2}} \qquad B'_{\perp} = \frac{\left(\vec{B}-\dfrac{\vec{v}\times\vec{E}}{c^2}\right)_{\perp}}{\sqrt{1-v^2/c^2}}$$

Checkpoint 17 Try using these versions of the transforms to reanalyze one or more of the phenomena treated earlier: fields of a moving charge, the metal bar moving through a magnetic field, or the velocity selector.

If you would like to know more about relativity, there are many fine books on the subject, or you might like to take a course on relativity. Here are two excellent introductory books:

R. Resnick, *Introduction to Special Relativity* (Wiley, 1968)
E. Taylor and J. Wheeler, *Spacetime Physics* (Freeman, 1992)

SUMMARY

Net Electric and Magnetic Force

$$\vec{F} = q\vec{E} + q\vec{v}\times\vec{B}$$
(called the "Lorentz force")

Force on a Short Length of Current-Carrying Wire

$$\Delta\vec{F}_{\text{magnetic}} = I\Delta\vec{l}\times\vec{B}$$

Circular Motion in a Uniform Magnetic Field

$$\omega = \frac{|q||\vec{B}|}{\gamma m} \quad \text{at any speed}$$

$$\omega \approx \frac{|q||\vec{B}|}{m} \quad \text{if } v \ll c$$

Momentum $|\vec{p}| = |q||\vec{B}|R$, valid even for relativistic speeds.

The *Hall effect:* A sideways electric field in a current-carrying wire in a magnetic field due to the magnetic force on moving particles. The Hall effect allows us to determine the sign and number density of the mobile charges in a particular material.

Motional emf is the integral of the magnetic force that acts on mobile charges in a wire moving through a magnetic field.

A current-carrying coil or magnetic dipole moment experiences a torque in a magnetic field, and twists to align with the applied magnetic field.

Magnetic torque on a magnetic dipole is

$$\vec{\tau} = \vec{\mu}\times\vec{B}$$

Potential energy associated with a magnetic dipole in an applied magnetic field is $U_m = -\vec{\mu}\bullet\vec{B}$, and the force on a magnetic dipole is

$$F_x = -\frac{dU_m}{dx} = \frac{d(\vec{\mu}\bullet\vec{B})}{dx}$$

Electric field required to ionize air:

$$E_{\text{critical}} = 3\times 10^6\ \text{V/m}$$

When the applied electric field is sufficiently large to accelerate free electrons to the energy required to ionize a nitrogen or oxygen molecule, a chain reaction can occur.

The pattern of electric and magnetic fields is different in stationary and moving reference frames. A phenomenon such as the polarization of a metal bar may have different explanations in different reference frames. The following transforms relate the fields $\vec{E}'$ and $\vec{B}'$ in a reference frame moving at speed v to the fields $\vec{E}$ and $\vec{B}$ in a stationary reference frame.

$$E'_x = E_x \qquad E'_y = \frac{(E_y - vB_z)}{\sqrt{1-v^2/c^2}} \qquad E'_z = \frac{(E_z + vB_y)}{\sqrt{1-v^2/c^2}}$$

$$B'_x = B_x \qquad B'_y = \frac{\left(B_y + \dfrac{v}{c^2}E_z\right)}{\sqrt{1-v^2/c^2}} \qquad B'_z = \frac{\left(B_z - \dfrac{v}{c^2}E_y\right)}{\sqrt{1-v^2/c^2}}$$

EXPERIMENTS

EXP1 Observing magnetic force

Observe for yourself the force that a magnetic field exerts on a current-carrying wire. Short-circuit one or more batteries and observe how a current-carrying wire deflects near your bar magnet (make and break the circuit to see the direction of the deflection). Use your 2-m-long wire and let a portion of it hang as freely as possible over the edge of the desk with a weight on the top part, to keep the lower part of the wire from moving when you connect and disconnect the circuit (Figure 20.79). When you connect the wire you can see the wire jerk sideways slightly. The force is small, so observe closely.

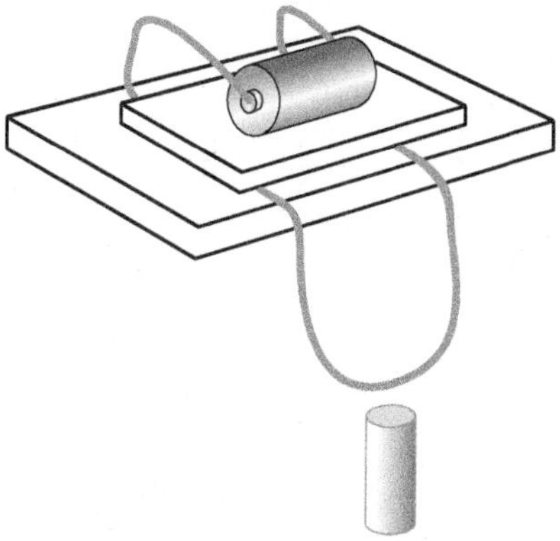

Figure 20.79

Try different directions of current and magnetic field (using the N and S markings on your magnet obtained previously, or by checking with your compass), and take turns with a partner in predicting the direction of the deflection. Make sketches of the various situations and use the equation to explain what you observe.

EXP2 Bar magnet and coil

Wrap your long wire into a coil, then hang the coil by its connecting wires and bring your bar magnet near the hanging coil.

(a) Under what conditions does the bar magnet repel the coil? Under what conditions does the bar magnet attract the coil?

(b) Sketch a situation in which the coil twists under the influence of the bar magnet. Explain briefly in terms of what forces act on what parts of the coil.

EXP3 Series and parallel batteries

(a) If you assume that all of your batteries have the same internal resistance, what would you predict would happen to the magnetic force if you had more batteries in series? In parallel? An ordinary flashlight battery has a resistance of about a quarter or a half an ohm; the resistance of your wire is much smaller than that.

(b) Using various battery combinations (including combining batteries with your partner), observe the deflection of a wire or a coil by your bar magnet. What do you observe, and how do your observations compare with your predictions? (*Hint on technique:* Compare the effects of different numbers of batteries in parallel by connecting some of them in and out of the circuit while observing the deflection of the wire.)

EXP4 Magnet picks up magnet

Prediction: Determine the magnetic dipole moments μ_1 and μ_2 of two magnets, and weigh the magnets. Predict the numerical value of the distance x at which the magnetic attraction between these two magnets just equals the weight Mg of one magnet.

Measurement: Place a magnet upright on the desk and try to lift it with the other magnet. Measure the distance x between their centers at the point where the lower magnet is lifted, and compare with your theoretical prediction. There is a ruler on the inside back cover of this textbook. In using the center-to-center distance we are averaging over distance, for a force that varies like $1/x^4$. This works out better than one might expect.

QUESTIONS

Q1 Suppose that a proton has a component of velocity parallel to the magnetic field as well as perpendicular to it (Figure 20.80). What is the effect of the magnetic field on this parallel component of the velocity? What will the trajectory of the proton look like?

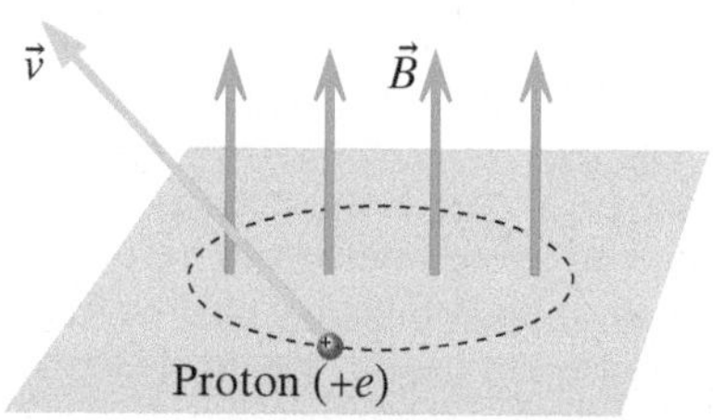

Figure 20.80

Q2 An electron is moving with a speed v in the plane of the page (Figure 20.81), and there is a uniform magnetic field B into the page throughout this region; the magnetic field is produced by some large coils that are not shown. Draw the trajectory of the electron, and explain qualitatively.

Figure 20.81

Q3 A proton and an electron are moving at a particular instant as shown in Figure 20.82.

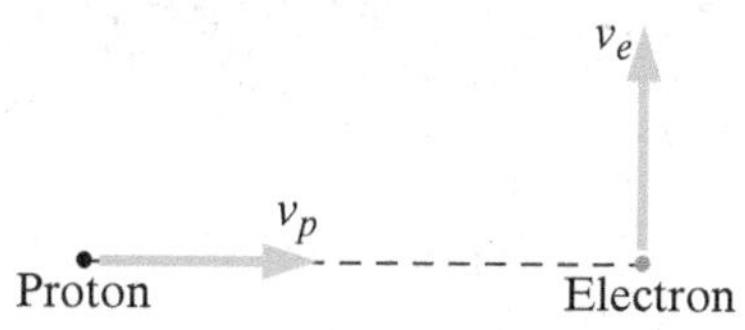

Figure 20.82

(a) On a similar diagram, show the directions of the electric field and magnetic field that the proton produces at the location of the electron. **(b)** Show the directions of the electric force and the magnetic force that the proton exerts on the electron. **(c)** Similarly, show the directions of the electric field and magnetic field that the electron produces at the location of the proton. **(d)** Show the directions of the electric force and the magnetic force that the electron exerts on the proton. **(e)** Consider carefully your results from parts (b) and (d). Are the magnetic forces on electron and proton in opposite directions? Does reciprocity hold for magnetic forces? **(f)** Will the total momentum of the two particles remain constant? Is this a violation of conservation of momentum for an isolated system?

There is momentum associated with the electric and magnetic fields of these particles, as we will see in a later chapter. Although the momentum of the proton plus the momentum of the electron does change, the sum of the particle momenta plus the field momentum does not change.

Q4 A copper wire with square cross section carries a conventional current I to the left (as in Figure 20.83). There is a magnetic field B perpendicular to the wire. Describe the direction of $E_{\perp}$, the transverse electric field inside the wire due to the Hall effect, and explain briefly.

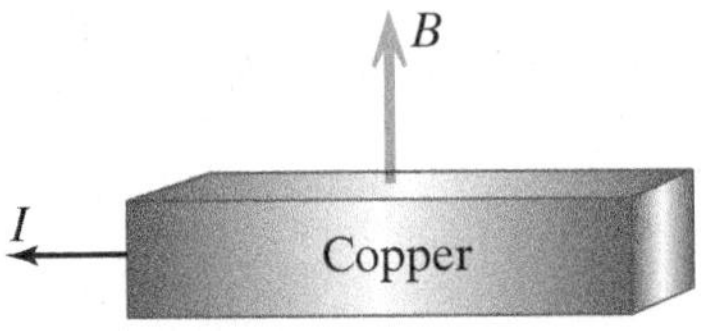

Figure 20.83

Q5 You hold your magnet perpendicular to a bar of copper that is connected to a battery (Figure 20.84). Describe the directions of the components of the electric field at the center of the copper bar, both parallel to the bar and perpendicular to the bar. Explain carefully.

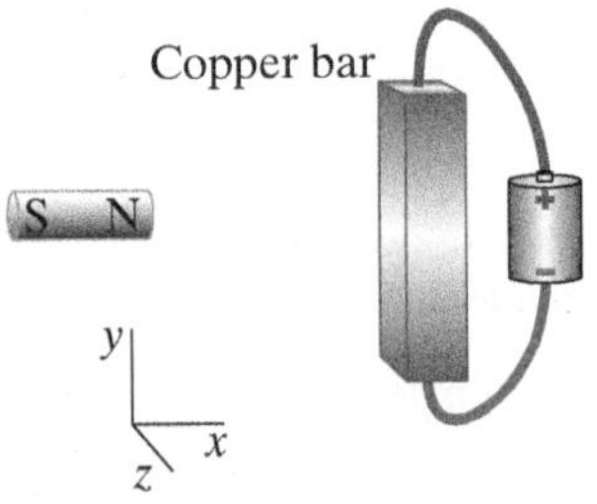

Figure 20.84

Q6 The Hall effect can be used to determine the sign of the mobile charges in a particular conducting material. A bar of a new kind of conducting material is connected to a battery as shown in Figure 20.85. In this diagram, the x axis runs to the right, the y axis runs up, and the z axis runs out of the page, toward you. A voltmeter is connected across the bar as shown, with the leads placed directly opposite each other along a vertical line. In order to answer the following questions, you should draw a careful diagram of the situation, including all relevant charges, electric fields, magnetic fields, and velocities.

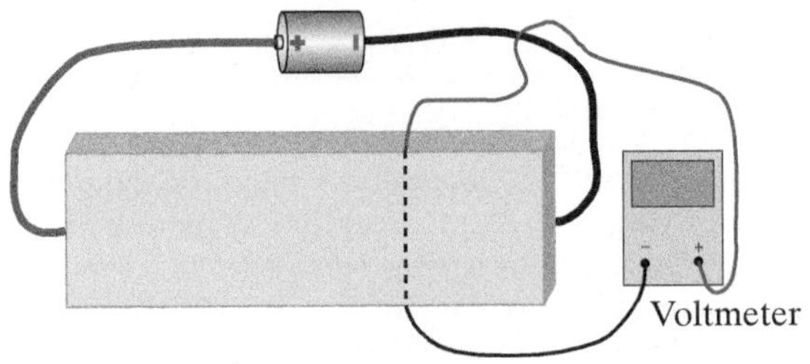

Figure 20.85

Initially, there is no magnetic field in the region of the bar. **(a)** Inside the bar, what is the direction of the electric field $\vec{E}$ due to the charges on the batteries and surface of the wires and the bar? This is the electric field that drives the current in the bar. **(b)** If the mobile charges in the bar are positive, in what direction do they move when current runs? **(c)** If the mobile charges in the bar are negative, in what direction do they move when current runs? **(d)** In this situation (zero magnetic field), what is the sign of the reading on the voltmeter?

Next, large coils (not shown) are moved near the bar, and current runs through the coils, making a magnetic field in the $-z$ direction (into the page). **(e)** If the mobile charges in the bar are negative, what is the direction of the magnetic force on the mobile charges? **(f)** If the mobile charges in the bar are negative, which of the following things will happen? (1) Positive charge will accumulate on the top of the bar. (2) The bar will not become polarized. (3) Negative charge will accumulate on the left end of the bar. (4) Negative charge will accumulate on the top of the bar. **(g)** If the mobile charges in the bar are positive, what is the direction of the magnetic force on the mobile charges? **(h)** If the mobile charges in the bar are positive, which of these things will happen? (1) Positive charge will accumulate on the top of the bar. (2) The bar will not become polarized. (3) Positive charge will accumulate on the right end of the bar. (4) Negative charge will accumulate on the top of the bar.

You look at the voltmeter and find that the reading on the meter is -5×10^{-4} volts. **(i)** What can you conclude from this observation? (Remember that a voltmeter gives a positive reading if the positive lead is attached to the higher potential location.) (1) There is not enough information to figure out the sign of the mobile charges. (2) The mobile charges are negative. (3) The mobile charges are positive.

Q7 A metal wire with mobile electrons and square cross section carries a conventional current I in the $-y$ direction. There is a magnetic field B in the $+x$ direction. Draw a diagram illustrating the situation, to help you answer the following questions. **(a)** What is the direction of E_{parallel}, the electric field that causes the current to flow? **(b)** What is the direction of the drift velocity of the mobile electrons? **(c)** What is the direction of the magnetic force on the moving electrons? **(d)** What is the direction of $E_{\text{transverse}}$, the electric field due to the Hall effect, inside the wire? **(e)** If the mobile charges had been positive (holes) instead of negative, what would have been the direction of the magnetic force on the moving positive charges? **(f)** If the mobile charges had been positive instead of negative, what would have been the direction of the transverse electric field?

Q8 You have a battery with known emf = K, a sensitive voltmeter, and a slab of high-resistance metal in which the mobile charges are electrons with mobility u; there are n electrons per cubic meter. The metal slab has length L, height h, and width w. You also have plenty of very low-resistance wires you can use to connect the components in Figure 20.86 together. There is a uniform magnetic field pointing up throughout the region where the slab is located; this magnetic field is produced by a large magnet that is not shown. The goal is to figure out the magnitude B of this magnetic field, using the available equipment.

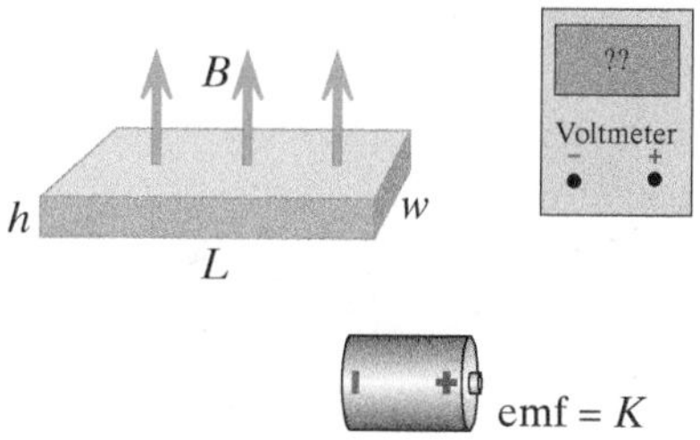

Figure 20.86

(a) Draw wires on the diagram to show how you would connect the equipment. **(b)** Explain in detail, including appropriate diagrams, how you will determine B, and give an algebraic expression for B involving the reading ΔV on the voltmeter and the known quantities (K, u, n, L, h, w).

Q9 In which direction will conventional current flow through the resistor in Figure 20.87? What will be the direction of the magnetic force on the moving bar?

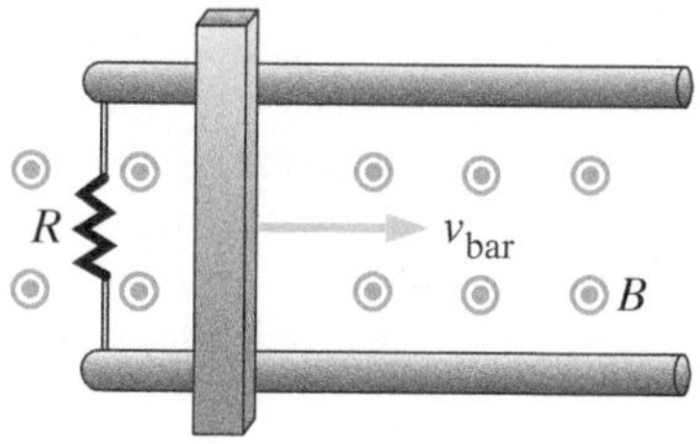

Figure 20.87

Q10 Two metal wires of length w are attached to two metal wires of length h to form a rectangle of dimensions w by h. The rectangle rotates counterclockwise on a frictionless axle (due to external forces) with constant angular velocity ω in a uniform upward magnetic field B (Figure 20.88).

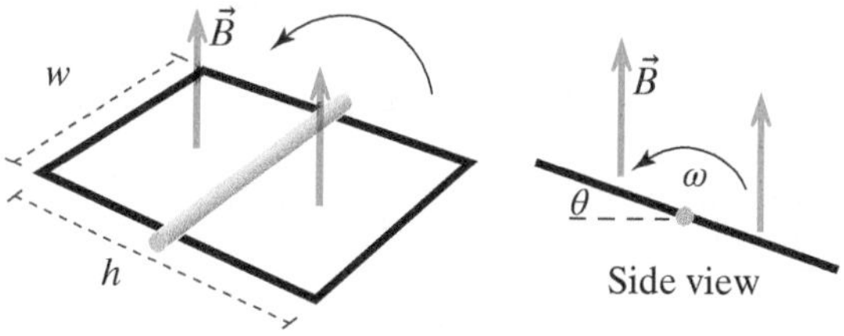

Figure 20.88

(a) At the moment when the rectangle is at an angle θ from the horizontal, what is the magnitude of the emf around the circuit and the direction of the current I? **(b)** Why are external forces required to rotate the loop, even though the rectangle rotates at a constant angular speed on a frictionless axle? Show the direction of the external forces on a diagram.

Q11 See Figure 20.89. If you travel at speed v along with the moving bar, in your reference frame the charged particles in the bar are not moving. Explain why, in your moving reference frame, the bar still polarizes.

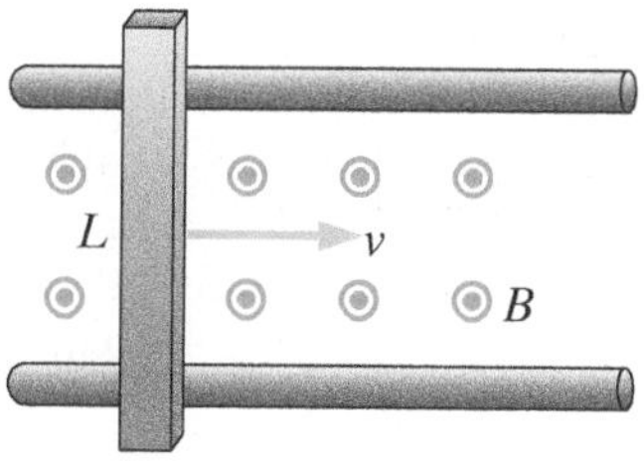

Figure 20.89

Q12 A current-carrying rectangular loop of wire is mounted on a pivot, held at an angle to a uniform magnetic field. If the loop is released, will it turn clockwise or counterclockwise? Explain briefly, and use Figure 20.90 in your explanation.

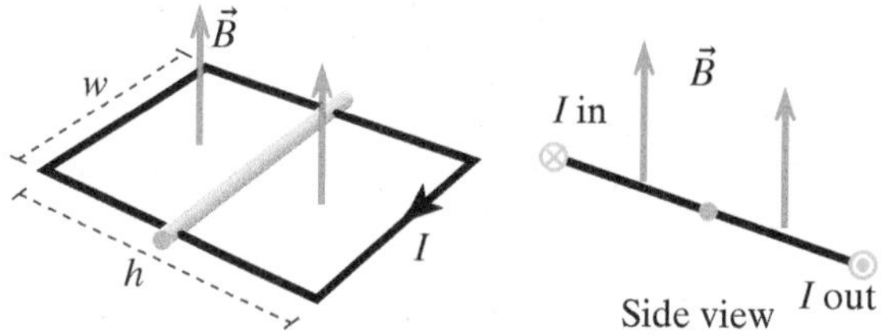

Figure 20.90

Q13 A bar magnet is mounted on a pivot and released in the presence of a magnetic field $\vec{B}$ made by coils that are not shown in Figure 20.91. Does the magnet initially turn clockwise or counterclockwise? Explain qualitatively in detail why it turns, and why it turns in the predicted direction.

Figure 20.91

Q14 Figure 20.92 shows a region of nonuniform magnetic field. The field is stronger to the right. What is the initial direction of the magnetic force on a magnetic dipole in each of the four orientations shown in the diagram? (*Hint:* A magnetic dipole if free to move goes to a region of lower potential energy. You can also check with what you know about the interaction of two magnets.)

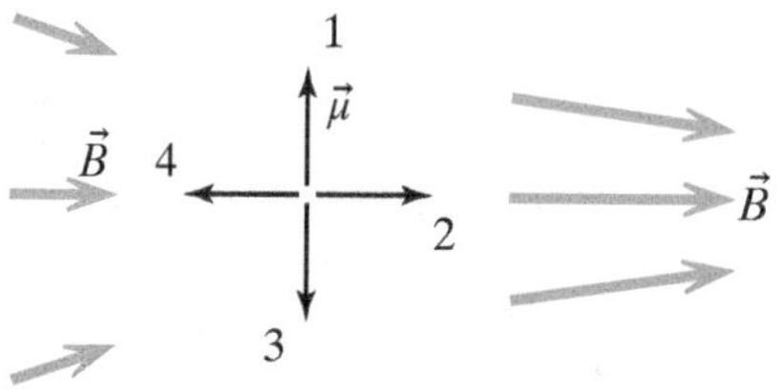

Figure 20.92

Q15 Put together what you know about ferromagnetic materials and about the forces on magnetic moments to explain qualitatively why a magnet can pick up a paper clip.

Q16 A student said, "During a spark, protons are pulled out of the positive ball and jump across to the negative ball." Give at least two reasons why this analysis cannot be correct.

Q17 "When you connect a metal wire between two oppositely charged metal blocks, electrons on the negatively charged block jump to the positively charged block." Explain briefly what is wrong with this statement.

Q18 What would happen if an electron with energy less than the ionization energy of a nitrogen molecule collided with a nitrogen molecule?

Q19 At high altitudes the air is less dense. If the density of air were half that at sea level, how would d, the mean free path of an electron in air, change? (1) d would be the same. (2) d would be twice as long. (3) d would be half as long. (4) d would be 1/4 as long. (5) d would be 4 times as long.

Q20 If the radius of an air molecule were twice as large, how would d, the mean free path of an electron in air, change? (1) d would be the same. (2) d would be twice as long. (3) d would be half as long. (4) d would be 1/4 as long. (5) d would be 4 times as long.

Q21 If you observed that under certain conditions an electric field six times as great as usual were required to ionize air, what might you conclude about the mean free path under these conditions? Give your answer as a ratio of the mean free path under these conditions to the mean free path at STP.

Q22 Metal sphere 1 is charged negatively, and metal sphere 2 is not charged in Figure 20.93.

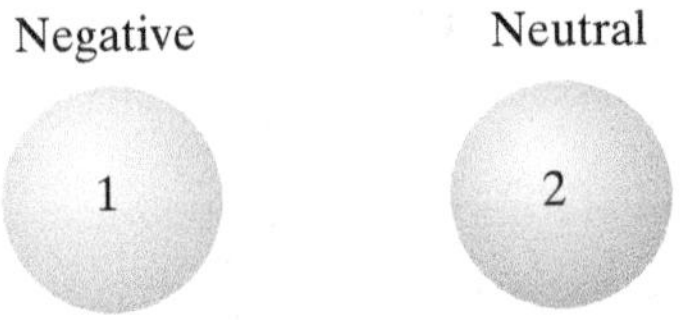

Figure 20.93

(a) Using the conventions for diagrams described and used in this textbook, show the charge distribution for both spheres. **(b)** The spheres are moved closer to each other (using insulating supports so as not to change their charge). When they are a certain short distance apart, a spark is seen for a brief instant. Explain qualitatively what determines this distance for the spark to be produced. (Why doesn't the spark occur when the spheres are farther apart? What's special about this distance?) **(c)** After the spark stops, show the charge distribution for both spheres, and explain briefly but completely how this new charge distribution came about.

Q23 In very dry weather, if you shuffle across the carpet wearing rubber-soled shoes and then bring your finger near a metal object such as a doorknob, you will probably get a shock and see a spark. How this can occur is puzzling, since rubber is an insulator, so charge can't move through the soles of your shoes. Explain this process in detail, with appropriate diagrams. Make sure that you answer the following questions: **(a)** Carry out an experiment to determine the sign of the charge on the sole of your shoe after rubbing it on carpet, wool, or other cloth. Explain the test you did. (If you are unable to obtain results, choose a sign to use in the rest of the discussion.) **(b)** Draw a diagram that includes both the shoe and the rest of your body. Include all relevant charges and fields. **(c)** Suppose that a spark occurred when your finger was 1 cm from the doorknob. Draw two diagrams showing your hand, the doorknob, all relevant charges, and all contributions to the electric field at relevant locations: when your finger is 1.5 cm from the doorknob (no spark yet), and when your finger is 1 cm from the doorknob (spark starts). On the basis of these diagrams, explain why the spark starts only when your finger is close enough (1 cm) to the doorknob and not farther away. **(d)** At what location do you think the spark starts, and why (why not somewhere else)? **(e)** Why does the spark stop?

PROBLEMS

Section 20.1

•P24 An electron is moving in the y direction at a speed $v = 2 \times 10^7$ m/s at a point where there is a magnetic field $B = 3.5$ T in the z direction (Figure 20.94). What are the magnitude and direction of the magnetic force on the moving electron? Draw the force vector on a diagram of the situation.

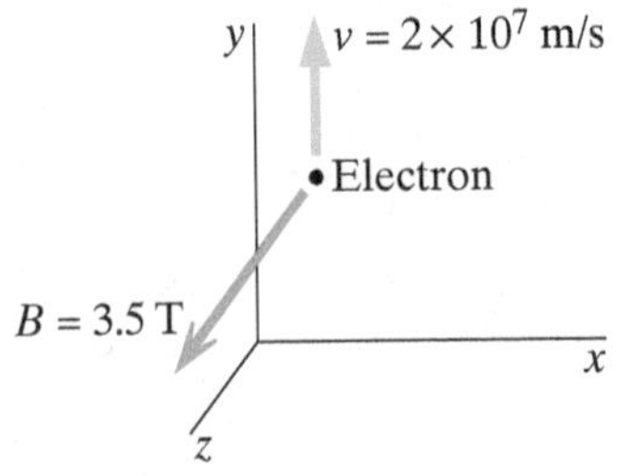

Figure 20.94

•P25 At a particular instant a proton is traveling in the $-x$ direction, with speed 5×10^5 m/s. At the location of the proton there is a magnetic field of magnitude 0.24 T in the $-y$ direction, due to current running in a nearby coil. What are the direction and magnitude of the magnetic force on the proton?

•P26 At a particular instant an electron is traveling in the $-x$ direction, with speed 4×10^5 m/s. At the location of the electron there is a magnetic field of magnitude 0.27 T in the $+z$ direction, due to a large bar magnet. What are the direction and magnitude of the magnetic force on the electron?

•P27 At a particular instant an electron is traveling in the $+z$ direction, with speed 8×10^5 m/s. At the location of the electron there is a magnetic field of magnitude 0.32 T in the $-z$ direction, due to a large bar magnet. What are the direction and magnitude of the magnetic force on the electron?

•P28 To become familiar with the order of magnitude of magnetic effects, consider the situation in a television cathode ray tube (not a flat-panel TV). In order to bend the electron trajectory so that the electron hits the top of the screen rather than going straight through to the center of the screen, you need a radius of curvature in the magnetic field of about 20 cm (Figure 20.95). If the electrons are accelerated through a 15,000 V potential difference, they have a speed of 0.7×10^8 m/s. Calculate the magnitude of the magnetic field required to make the electrons hit the top of the screen. Is the magnetic field into or out of the page?

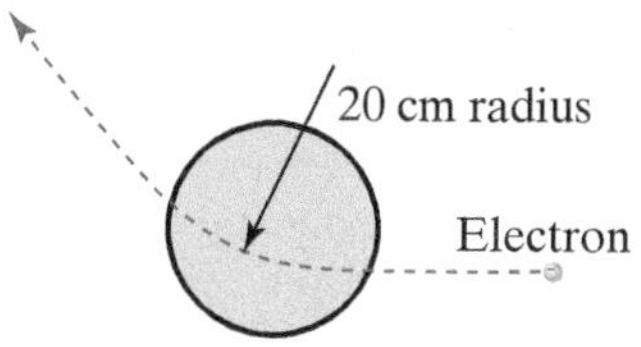

Figure 20.95

•**P29** An alpha particle (consisting of two protons and two neutrons) is moving at constant speed in a circle, perpendicular to a uniform magnetic field applied by some current-carrying coils, making one clockwise revolution every 80 ns (Figure 20.96). If the speed is small compared to the speed of light, what is the numerical magnitude B of the magnetic field made by the coils? What is the direction of this magnetic field?

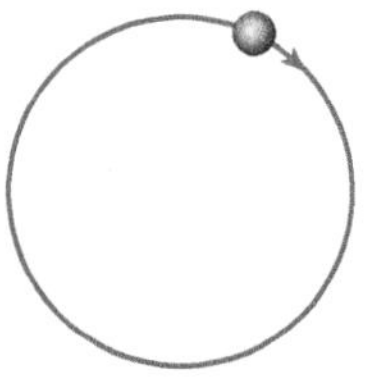

Figure 20.96

••**P30** An electron and a proton are both in motion near each other as shown in Figure 20.97; at this instant they are a distance r apart. The proton is moving with speed v_p at an angle θ to the horizontal, and the electron is moving straight up with speed v_e.

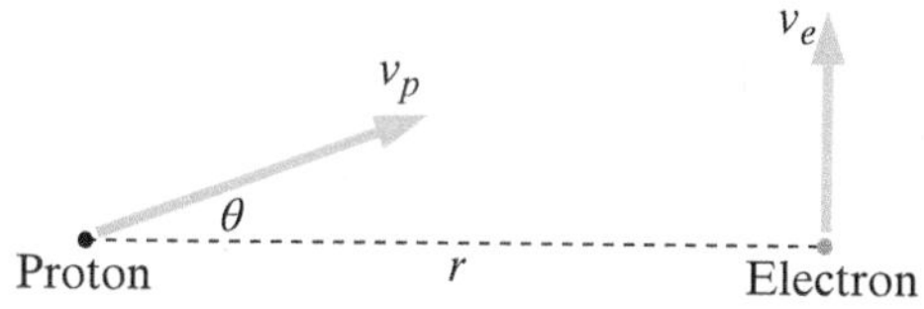

Figure 20.97

(a) Calculate the x and y components of the force that the proton exerts on the electron. First calculate the electric field and magnetic field that the proton produces at the location of the electron, then calculate the forces that these fields exert on the electron. **(b)** Calculate the x and y components of the force that the electron exerts on the proton. First calculate the electric field and magnetic field that the electron produces at the location of the proton, then calculate the forces that these fields exert on the proton. (In a completely correct quantitative calculation we would have to use the relativistically correct fields.) **(c)** Consider carefully your results. Are the magnetic forces on electron and proton equal and opposite? Does reciprocity hold for magnetic forces? **(d)** Will the total momentum of the two particles remain constant? Is this a violation of conservation of momentum for an isolated system?

See the comments at the end of Question Q3.

••**P31** A mass spectrometer is a tool used to determine accurately the mass of individual ionized atoms or molecules, or to separate atoms or molecules that have similar but slightly different masses. For example, you can deduce the age of a small sample of cloth from an ancient tomb by using a mass spectrometer to determine the relative abundances of carbon-14 (whose nucleus contains 6 protons and 8 neutrons) and carbon-12 (the most common isotope, whose nucleus contains 6 protons and 6 neutrons).

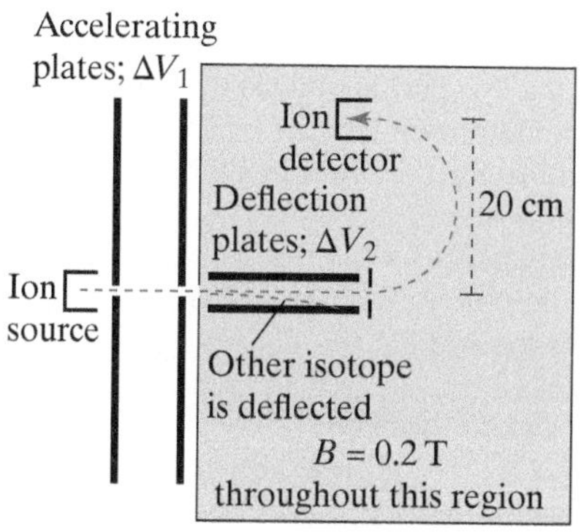

Figure 20.98

A particular kind of mass spectrometer is shown in Figure 20.98. Carbon from the sample is ionized in the ion source at the left. The resulting singly ionized $^{12}C^+$ and $^{14}C^+$ ions have negligibly small initial velocities (and can be considered to be at rest). They are accelerated through the potential difference ΔV_1. They then enter a region where the magnetic field has a fixed magnitude $B = 0.2$ T. The ions pass through electric deflection plates that are 1 cm apart and have a potential difference ΔV_2 that is adjusted so that the electric deflection and the magnetic deflection cancel each other for a particular isotope: one isotope goes straight through, and the other isotope is deflected and misses the entrance to the next section of the spectrometer. The distance from the entrance to the fixed ion detector is a distance of 20 cm.

There are controls that let you vary the accelerating potential ΔV_1 and the deflection potential ΔV_2 in order that only $^{12}C^+$ or $^{14}C^+$ ions go all the way through the system and reach the detector. You count each kind of ion for fixed times and thus determine the relative abundances. The various deflections ensure that you count only the desired type of ion for a particular setting of the two voltages.

Carry out the following calculations, and give brief explanations of your work: **(a)** Determine which accelerating plate is positive (left or right), which deflection plate is positive (upper or lower), and the direction of the magnetic field. **(b)** Determine the appropriate numerical values of ΔV_1 and ΔV_2 for ^{12}C. Carry out your intermediate calculations algebraically, so that you can use the algebraic results in part (c). **(c)** Determine the appropriate numerical values of ΔV_1 and ΔV_2 for ^{14}C.

Background: In organic material the ratio of ^{14}C to ^{12}C depends on how old the material is, which is the basis for carbon-14 dating. ^{14}C is continually produced in the upper atmosphere by nuclear reactions caused by cosmic rays (high-energy charged particles from outer space, mainly protons), and ^{14}C is radioactive with a half-life of 5700 y. When a cotton plant is growing, some of the CO_2 it extracts from the air to build tissue contains ^{14}C that has diffused down from the upper atmosphere. After the cotton has been harvested, however, there is no further intake of ^{14}C from the air, and the cosmic rays that create ^{14}C in the upper atmosphere can't penetrate the atmosphere and reach the cloth. Thus the amount of ^{14}C in cotton cloth continually decreases with time, while the amount of nonradioactive ^{12}C remains constant.

••**P32** The design and operation of a cyclotron (Figure 20.99) was discussed in Section 20.1. **(a)** Show that the period of the motion, the time between one kick to the right and the next kick in the same direction, does not depend on the current speed of the proton at speeds small compared to the speed of light. As a result, we can place across the dees a simple sinusoidal potential difference having this period and achieve continual acceleration out to the maximum radius of the cyclotron. **(b)** One of Ernest Lawrence's first cyclotrons, built in 1932, had a diameter of only about 30 cm and was placed in a magnetic field of about 1 T. What was the frequency (= 1/period, in hertz = cycles per second) of the sinusoidal potential difference placed across the dees to accelerate the protons?

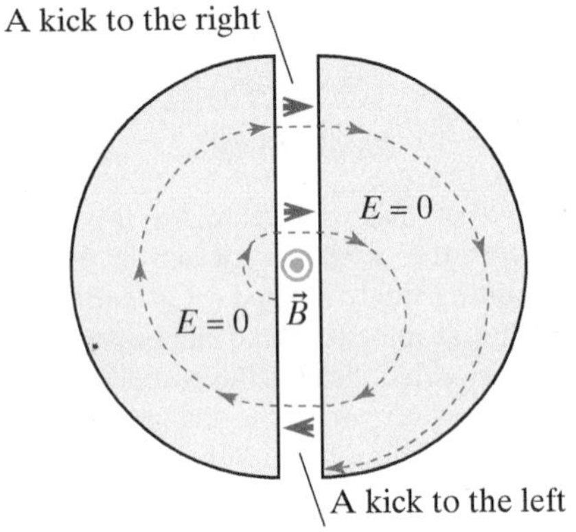

Figure 20.99

(c) Show that the equivalent accelerating potential of this little cyclotron was about a million volts! That is, the kinetic energy gain from the center to the outermost radius was $\Delta K = e\Delta V_{\text{eq}}$, with $\Delta V_{\text{eq}} = 1 \times 10^6$ V. (*Hint:* Calculate the final speed at the outermost radius.) **(d)** If the sinusoidal potential difference applied to the dees had an amplitude of 500 V (that is, it varied between +500 and −500 V), show that it took about 65 μs for a proton to move from the center to the outer radius.

••P33 The thin circular coil in Figure 20.100 has radius $r = 15$ cm and contains $N = 3$ turns of Nichrome wire. The coil has a resistance of 6 Ω. A small compass is placed at the center of the coil. With the battery disconnected, the compass needle points to the right, in the plane of the coil. The apparatus is located near the equator, where the Earth's magnetic field B_{Earth} is horizontal, with a magnitude of about 4×10^{-5} T.

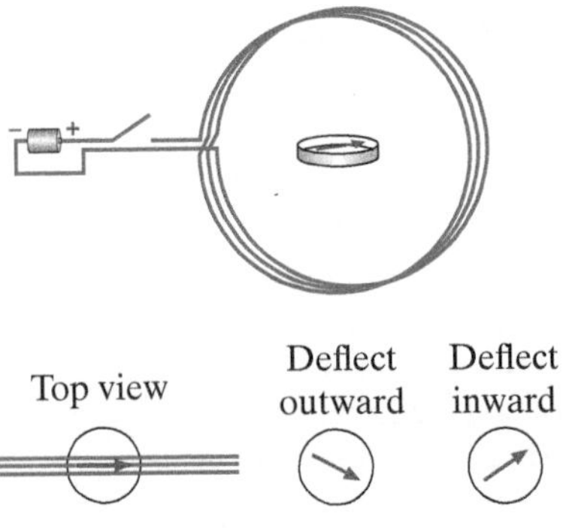

Figure 20.100

(a) When the 1.5 V battery is connected, predict the deflection of the compass needle. If you have to make any approximations, state what they are. Is the deflection outward or inward as seen from above? What is the magnitude of the deflection? **(b)** The compass is removed. The current in the coil continues to run. An electron is at the center of the coil and is moving with speed $v = 5 \times 10^6$ m/s into the page (perpendicular to the plane of the coil). In addition to the magnetic fields in the region due to the Earth and the coil, there is an electric field at the center of the coil (due to charges not shown in Figure 20.100), and this electric field points upward and has a magnitude $E = 250$ V/m. What are the magnitude and direction of the force on the electron at this moment? (You can neglect the gravitational force, which easily can be shown to be negligible.)

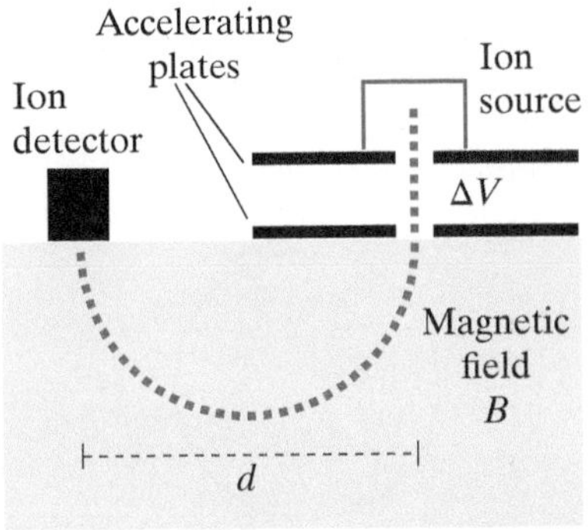

Figure 20.101

••P34 In the simple mass spectrometer shown in Figure 20.101, positive ions are generated in the ion source. They are released, traveling at very low speed, into the region between two accelerating plates between which there is a potential difference ΔV. In the shaded region there is a uniform magnetic field $\vec{B}$; outside this region there is negligible magnetic field. The semicircle traces the path of one singly charged positive ion of mass M, which travels through the accelerating plates into the magnetic field region, and hits the ion detector as shown. Determine the appropriate magnitude and direction of the magnetic field $\vec{B}$, in terms of the known quantities shown in Figure 20.101. Explain all steps in your reasoning.

••P35 A long solenoid with diameter 4 cm is in a vacuum, and a lithium nucleus (4 neutrons and 3 protons) is in a clockwise circular orbit inside the solenoid (Figure 20.102). It takes 50 ns (50×10^{-9} s) for the lithium nucleus to complete one orbit.

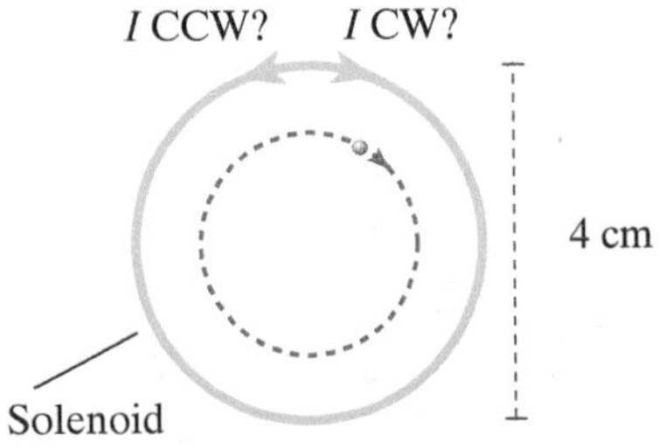

Figure 20.102

(a) Does the current in the solenoid run clockwise or counterclockwise? Explain, including physics diagrams. **(b)** What is the magnitude of the magnetic field made by the solenoid?

••P36 A long straight wire suspended in the air carries a conventional current of 8.2 A in the $-x$ direction as shown in Figure 20.103 (the wire runs along the x axis). At a particular instant an electron at location $\langle 0, -0.003, 0 \rangle$ m has velocity $\langle -1.5 \times 10^5, -1.8 \times 10^5, 0 \rangle$ m/s.

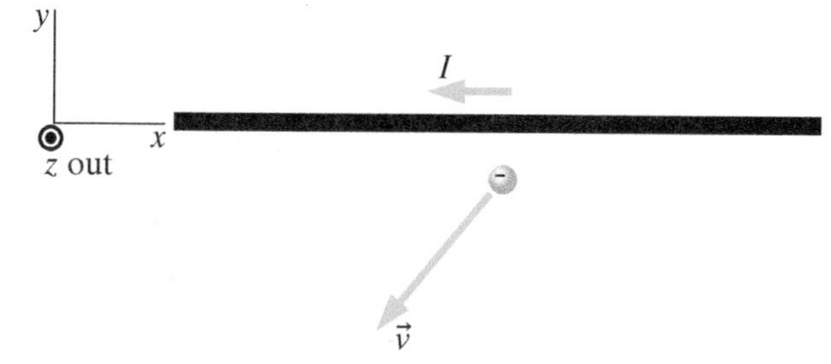

Figure 20.103

(a) What is the (vector) magnetic field due to the wire at the location of the electron? **(b)** What is the (vector) magnetic force on the electron due to the current in the wire? **(c)** If the moving particle were a proton instead of an electron, what would be the (vector) magnetic force on the proton?

Section 20.2

•**P37** A wire is oriented along the x axis. It is connected to two batteries, and a conventional current of 1.8 A runs through the wire in the $+x$ direction. Along 0.25 m of the length of the wire there is a magnetic field of 0.54 T in the $+y$ direction, due to a large magnet nearby. At other locations in the circuit, the magnetic field due to external sources is negligible. What are the direction and magnitude of the magnetic force on the wire?

•**P38** A 1.5 V battery is connected to a resistive wire whose resistance is 5 Ω. The wire is laid out in the form of a rectangle 50 cm by 3 cm (Figure 20.104). What are the approximate magnitude and direction of the magnetic force that the upper part of the wire exerts on the lower part of the wire in the diagram? Explain briefly.

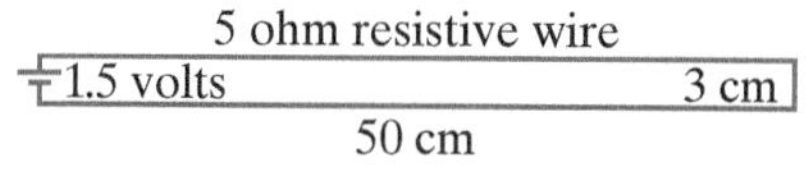

Figure 20.104

•**P39** A current-carrying wire is oriented along the y axis. It passes through a region 0.6 m long in which there is a magnetic field of 4.5 T in the $+z$ direction. The wire experiences a force of 14.9 N in the $-x$ direction. **(a)** What is the magnitude of the conventional current in the wire? **(b)** What is the direction of the conventional current in the wire?

•**P40** If your flashlight battery has an internal resistance of about 1/4 Ω, and your bar magnet produces a magnetic field of about 0.1 T near one end of the magnet, what is the approximate magnitude of the magnetic force on 5 cm of a wire that short-circuits the battery (Figure 20.105)? Indicate the direction of this force on a diagram. Check your direction experimentally.

Figure 20.105

••**P41** A Nichrome wire has the shape of two quarter-circles of radius a and b, connected by straight sections as shown in Figure 20.106. The wire carries conventional current I in the direction shown.

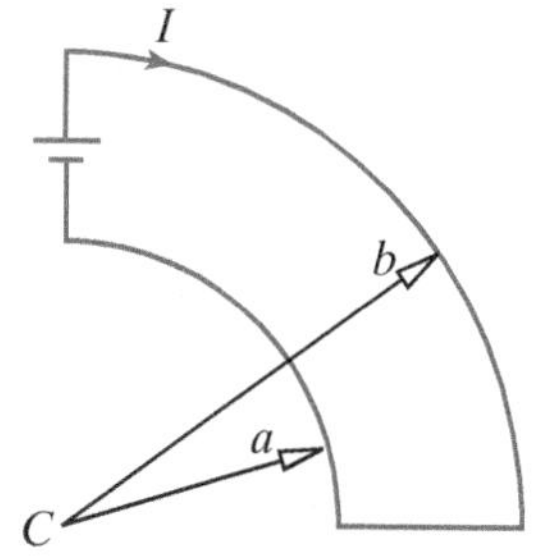

Figure 20.106

(a) Calculate the direction and magnitude of the magnetic field at point C, the center of the quarter-circles. Briefly explain your work, including a diagram. **(b)** At a particular instant, an electron is at point C, traveling to the right with speed v. Calculate the direction and magnitude of the magnetic force on the electron. Briefly explain your work, including a diagram.

••**P42** A long wire carries a current I_1 upward, and a rectangular loop of height h and width w carries a current I_2 clockwise (Figure 20.107). The loop is a distance d away from the long wire. The long wire and the rectangular loop are in the same plane. Find the magnitude and direction of the net magnetic force exerted by the long wire on the rectangular loop. Briefly explain your work, including a diagram.

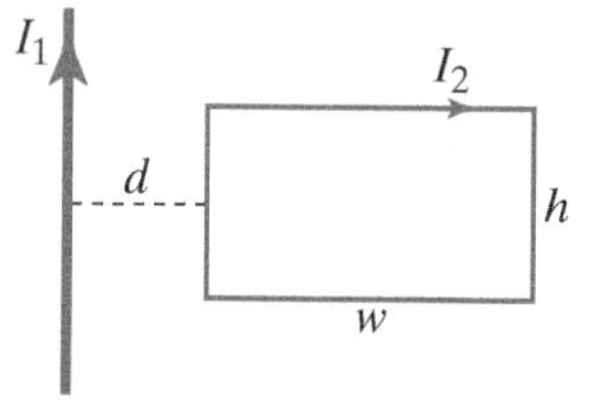

Figure 20.107

••**P43** A metal rod of length $L = 12$ cm and mass $m = 70$ g has metal loops at both ends, which go around two metal poles (Figure 20.108). The rod is in good electrical contact with the poles but can slide freely up and down. The metal poles are connected by wires to a battery, and a current of $I = 5$ A flows through the rod. A magnet supplies a large uniform magnetic field B in the region of the rod, large enough that you can neglect the magnetic fields due to the 5 A current. The magnetic field is oriented so as to have the maximum effect on the rod. The rod sits at rest a distance $d = 4$ cm above the table. What are the magnitude and direction of the magnetic field in the region of the rod? Briefly explain your work, including a diagram.

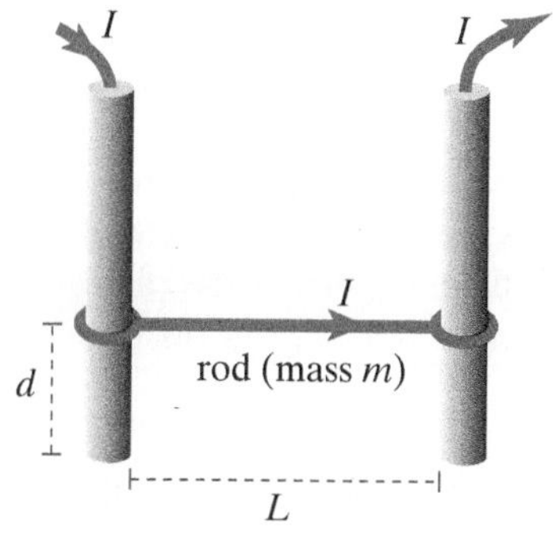

Figure 20.108

Section 20.3

•**P44** An object of charge q is moving in the x direction with speed v. Throughout the region there is a uniform electric field in the y direction of magnitude E (Figure 20.109). Determine a direction and magnitude B for a uniform magnetic field such that the net electric and magnetic force on the moving charge is zero. Draw $\vec{B}$ on a diagram of the situation.

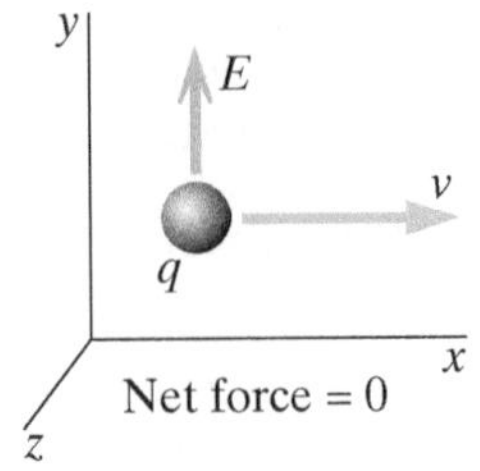

Figure 20.109

•**P45** In Figure 20.110 a particle with a charge of +9 nC travels to the left in a straight line at constant speed through a region where the electric field is 3800 V/m in the $-y$ direction and the magnetic field is 0.4 T in the $+z$ direction. What is the speed of the particle?

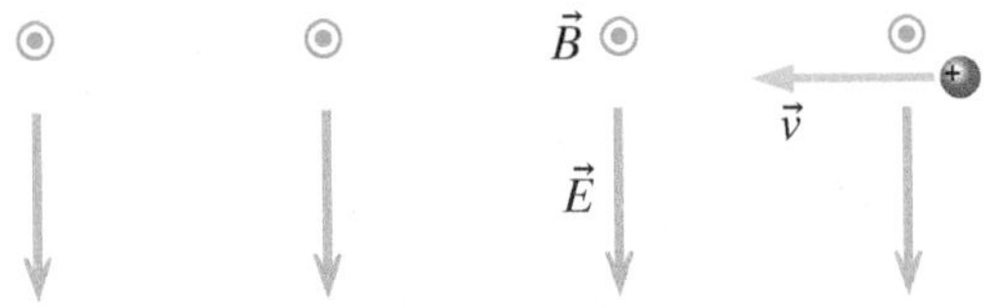

Figure 20.110

•**P46** An electron travels with velocity $\langle 5\times10^5, 0, 0\rangle$ m/s. It enters a region in which there is a uniform magnetic field of $\langle 0, 0.9, 0\rangle$ T. **(a)** What is the (vector) magnetic force on the electron? **(b)** Despite the magnetic force, the electron continues to travel in a straight line at constant speed. You conclude that there must be another force acting on the electron. Since you know there is also an electric field in this region, you decide that the other force must be an electric force. What is this (vector) electric force? **(c)** What is the (vector) electric field in this region that is responsible for the electric force?

•**P47** A proton traveling with speed 3105 m/s in the $+x$ direction passes through a region in which there is a uniform magnetic field of magnitude 0.6 T in the $+y$ direction. You want to keep the proton traveling in a straight line at constant speed. To do this, you can turn on an apparatus that can create a uniform electric field throughout the region. What (vector) electric field should you apply?

•**P48** A proton moves at constant velocity in the $+y$ direction, through a region in which there is an electric field and a magnetic field. The electric field is in the $+x$ direction and has magnitude 200 V/m. The magnetic field is in the $-z$ direction and has magnitude 0.95 T. **(a)** What is the magnitude of the net force on the proton? **(b)** What is the speed of the proton?

••**P49** Two long straight wires carrying 8 A of conventional current are connected by a three-quarter-circular arc of radius 0.035 m (Figure 20.111). (The rest of the circuit is far enough away that it contributes little magnetic field in this region.) Electric fields are also present in this region, due to charges not shown on the drawing. An electron is moving to the right with a speed of 5.7×10^5 m/s as it passes through the center C of the arc, and at that instant the net electric and magnetic force on the electron is zero.

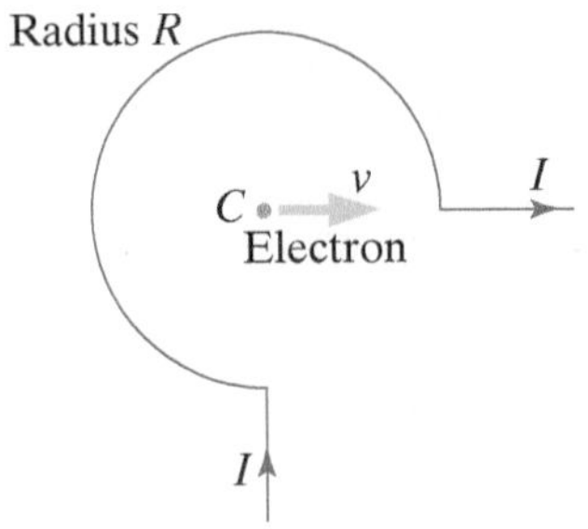

Figure 20.111

(a) What is the direction of the magnetic field at the center C due to the current? **(b)** What is the direction of the electric field at the center C? **(c)** What is the magnitude of the magnetic field at the center C due to the current? **(d)** What is the magnitude of the electric field at the center C?

••**P50** An electron travels with velocity $\langle 0, 0, -2.9\times10^6\rangle$ m/s in a region between a pair of very large charged plates, separated by a distance of 4.5 mm. In this region there is also a uniform magnetic field. The velocity of the electron remains constant throughout this region. The magnetic field, due to large coils outside the region, is measured and is found to be $\langle 0, -0.27, 0\rangle$ T.

On paper, draw a diagram showing the electron's velocity, the direction of the magnetic field in the region, the direction of the magnetic force on the electron, the direction of the electric force on the electron, and the direction of the electric field in the region. Use your diagram to answer the following questions. **(a)** What is the direction of the magnetic force on the electron? **(b)** What is the direction of the electric force on the electron? **(c)** What is the direction of the electric field in the region? **(d)** What is the magnitude of the magnetic force on the electron? **(e)** What is the absolute value of the potential difference between the plates?

••**P51** A cathode ray tube (CRT), as shown in Figures 20.112 and 20.113, is an evacuated glass tube. A current runs through a filament at the right end of the tube.

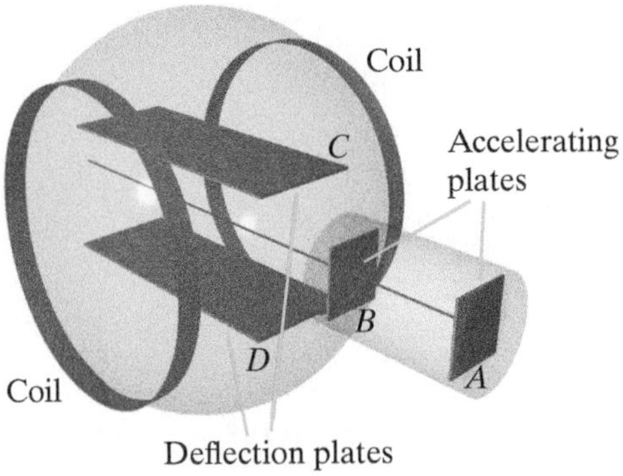

Figure 20.112

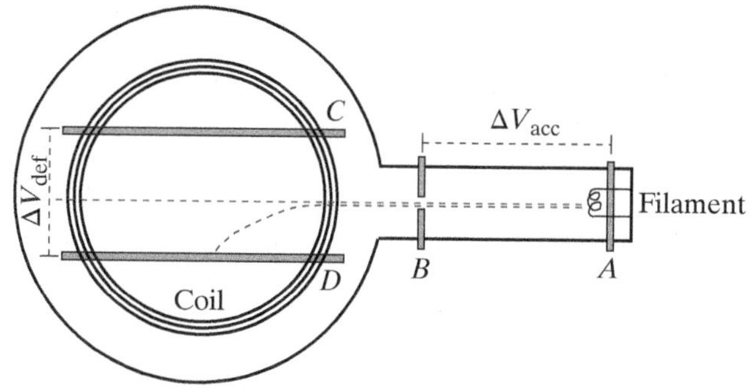

Figure 20.113

When the metal filament gets very hot, electrons occasionally escape from it. These electrons can be accelerated away from the filament by applying a potential difference ΔV_{acc} across the metal plates labeled A and B in the diagram.

The electrons pass through a hole in plate B and enter the glass sphere. There they pass between the two horizontal metal "deflection" plates labeled C and D. A potential difference ΔV_{def} can be applied across these plates, to deflect the beam of electrons.

In front of and in back of the glass sphere are two coils, through which current can be run to produce a magnetic field in the region between the deflection plates. The coils are oriented so that they both produce magnetic fields into the page in this region.

(a) Which of the accelerating plates, A or B, has a positive charge? **(b)** With no current in the coils, if the potential difference across the deflection plates, V_{def}, is zero, the electrons in the beam travel in a straight line, as indicated in Figure 20.113 by the dashed blue path. However, if V_{def} is not zero, the electron beam is deflected downward, following the path indicated on the diagram by the dashed red line. If the beam follows the dashed red line, what is the direction of the electric field between the deflection plates? **(c)** If a current runs through the coils, there will be a magnetic field in the region between the deflection plates. If the magnetic field made by the coils points into the page in the region between the plates, what is the direction of the magnetic force on an electron in the beam? (You can neglect the effect of the Earth's magnetic field, which is small.) **(d)** The accelerating potential difference ΔV_{acc} is measured to be 3.1 kV. What is the speed of an electron after it passes through the hole in plate B? Each of the two coils has 320 turns. The average radius of the coil is 6 cm. The distance from the center of one coil to the electron beam is 3 cm. If a current of 0.5 A runs through the coils, what is the magnitude of the magnetic field at a location on the axis of the coils, midway between the coils? (The electron beam passes through this location.) When deciding whether to use an exact or an approximate equation here, consider the relative magnitudes of the distances involved. **(e)** In a particular experiment, the accelerating potential difference ΔV_{acc} is set to 3.1 kV. The distance between the deflection plates is 8 mm. A current of 0.5 A runs through the coils, and the potential difference ΔV_{def} is adjusted until the electron beam again follows the straight line path indicated by the dashed blue line. In this situation (electron beam traveling in a straight line), which of the following statements are true? (1) The magnetic force on an electron in the beam is in the direction of its motion. (2) The electric and magnetic forces on an electron in the beam are in opposite directions. (3) The magnitude of the electric field in the region is equal to the magnitude of the magnetic field in the region. (4) The angle between the electric field and the magnetic field in the region is 180°. (5) The net force on an electron in the beam is zero. (6) The electric force on an electron in the beam is equal in magnitude to the magnetic force on the electron. **(f)** What is the value of ΔV_{def} required to make the electron beam travel in a straight line?

••P52 In Figure 20.114 two straight wires carrying conventional current I are connected by a three-quarter-circular arc of radius R_1 and a one-quarter-circular arc of radius R_2. Electric fields are also present in this region, due to charges not shown on the drawing.

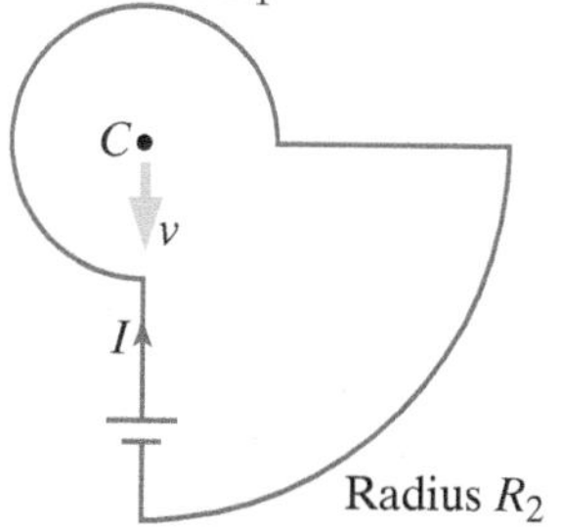

Figure 20.114

An electron is moving down with speed v as it passes through the center C of the arc, and at that instant the net electric and magnetic force on the electron is zero. (The Earth's magnetic field is negligible here.) **(a)** What is the direction of the magnetic field at the center C due to the current I? Explain briefly. **(b)** What is the direction of the electric field at the center C? Explain your reasoning clearly. **(c)** What is the magnitude of the magnetic field at the center C due to the current I? **(d)** What is the magnitude of the electric field at the center C? Explain briefly.

••P53 In Figure 20.115 two long straight wires carrying a large conventional current I are connected by one-and-a-quarter turns of wire of radius R. An electron is moving to the right with speed v at the instant that it passes through the center of the arc. You apply an electric field $\vec{E}$ at the center of the arc in such a way that the net force on the electron at this instant is zero. (You can neglect the gravitational force on the electron, which is easily shown to be negligible, and the magnetic field of the coil is much larger than the magnetic field of the Earth.)

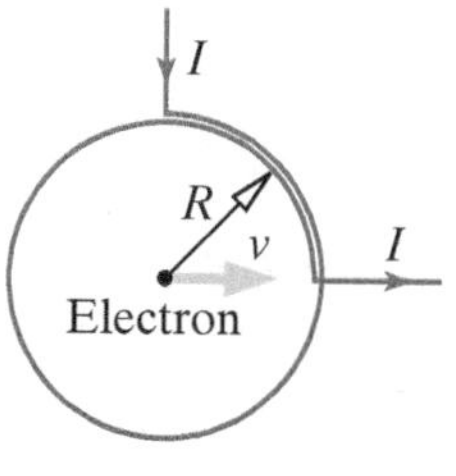

Figure 20.115

Determine the direction and magnitude of the electric field $\vec{E}$. Be sure to explain your work fully; draw and label any vectors you use.

••P54 In Figure 20.116 a battery with known emf = K is connected to two large parallel metal plates. Each plate has a length L and width W, and the plates are a very short distance s apart. The plates are surrounded by a vertical thin circular coil of radius R containing N turns through which runs a steady conventional current I. The center of the coil is at the center of the gap between the plates. At a certain instant, a proton (charge

$+e$, mass M) travels through the center of the coil to the right with speed v, and the net force on the proton at this instant is zero (neglecting the very weak gravitational force). What are the magnitude and direction of conventional current in the coil? Explain clearly.

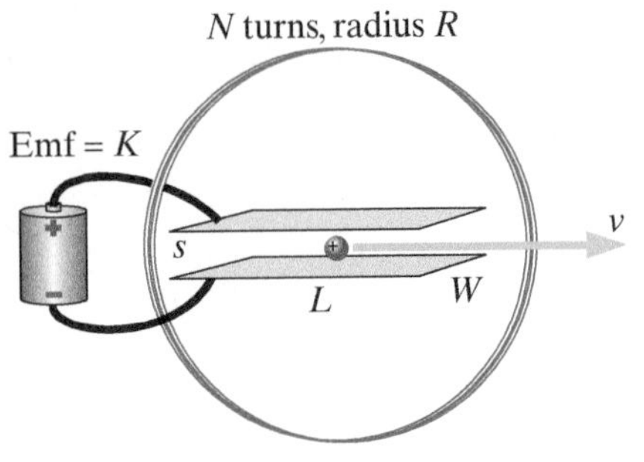

Figure 20.116

Section 20.4

••P55 A slab made of unknown material is connected to a power supply as shown in Figure 20.117. There is a uniform magnetic field of 0.7 T pointing upward throughout this region (perpendicular to the horizontal slab). Two voltmeters are connected to the slab and read steady voltages as shown. The connections across the slab are carefully placed directly across from each other. Assume that there is only one kind of mobile charges in this material, but we don't know whether they are positive or negative.

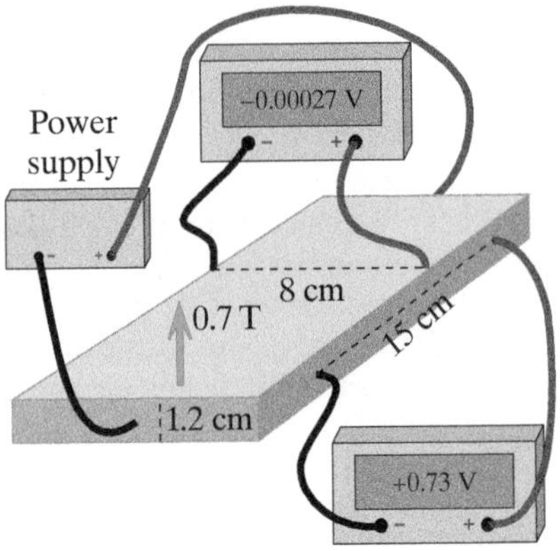

Figure 20.117

(a) Determine the (previously unknown) sign of the mobile charges, and state which way these charges move inside the slab. Explain carefully, using diagrams to support your explanation. **(b)** What is the drift speed $\bar{v}$ of the mobile charges? **(c)** What is the mobility u of the mobile charges? **(d)** The current running through the slab was measured to be 0.3 A. If each mobile charge is singly charged ($|q| = e$), how many mobile charges are there in 1 m^3 of this material? **(e)** What is the resistance in ohms of a 15 cm length of this slab?

••P56 An experiment was carried out to determine the electrical properties of a new conducting material. A bar was made out of the material, 18 cm long with a rectangular cross section 3 cm high and 0.8 cm deep. The bar was part of a circuit and carried a steady current (Figure 20.118 shows only part of the circuit). A uniform magnetic field of 1.5 T was applied perpendicular to the bar, coming out of the page (using some coils that are not shown).

Two voltmeters were connected along and across the bar and read steady voltages as shown (mV = millivolt = 1×10^{-3} V). The connections across the bar were carefully placed directly across from each other to eliminate false readings corresponding to the much larger voltage along the bar. Assume that there is only one kind of mobile charge in this material.

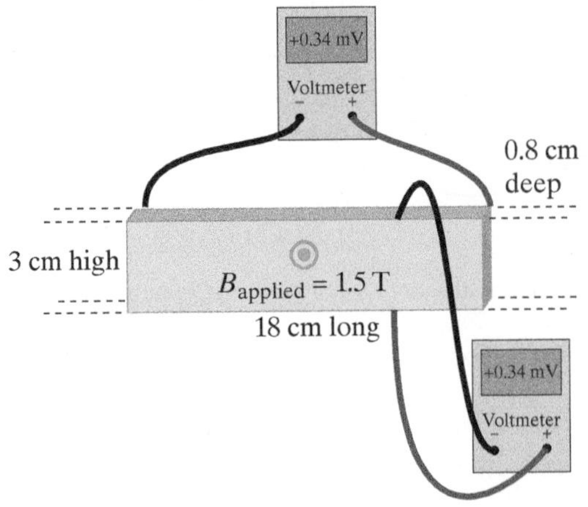

Figure 20.118

(a) What is the sign of the mobile charges, and which way do they move? Explain carefully, using diagrams to support your explanation. **(b)** What is the drift speed v of the mobile charges? Explain your reasoning. **(c)** What is the mobility u of the mobile charges? **(d)** The current running through the bar was measured to be 0.6 A. If each mobile charge is singly charged ($|q| = e$), how many mobile charges are there in 1 m^3 of this material? **(e)** What is the resistance in ohms of this length of bar?

••P57 In Figure 20.119 a bar of length L, width w, and thickness d is positioned a distance h underneath a long straight wire that carries a large but unknown steady current I_{wire} (w is small compared to h, and h is small compared to the length of the long straight wire). The material that the bar is made of is known to have n positive mobile charge carriers ("holes") per cubic meter, each of charge e, and these are the only mobile charge carriers in the bar. An ammeter measures the small current I passing through the bar.

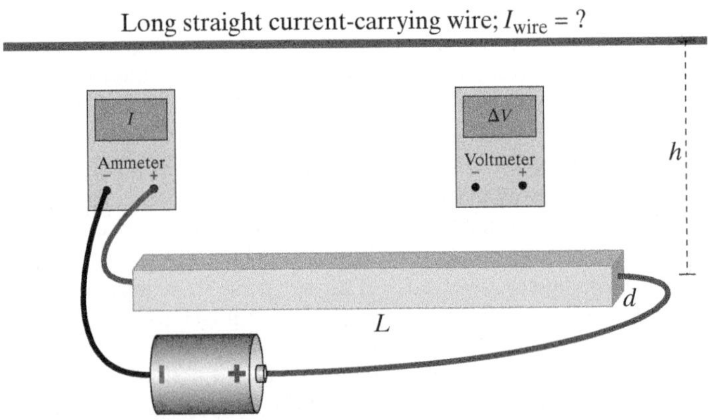

Figure 20.119

(a) Draw connecting wires from the voltmeter to the bar, showing clearly how the ends of the connecting wires must be positioned on the bar in order to permit determining the amount of current in the long straight wire I_{wire}. If the voltmeter reading ΔV is positive, does the current in the long straight wire run to the left or to the right? Explain briefly.

(b) If the voltmeter reading is ΔV, what is the current in the long straight wire, I_{wire}? Express your answer for the large current I_{wire} only in terms of the known quantities: h, L, w, d, n, $I_{ammeter}$, ΔV, and known physical constants such as e.

••P58 In Figure 20.120 the center of a large bar magnet is 20 cm from a thin plate of high-resistance material 12 cm long, 2.5 cm high, and 0.1 cm thick that is connected to a 240 V power supply whose internal resistance is negligible. The bar magnet is perpendicular to the plate.

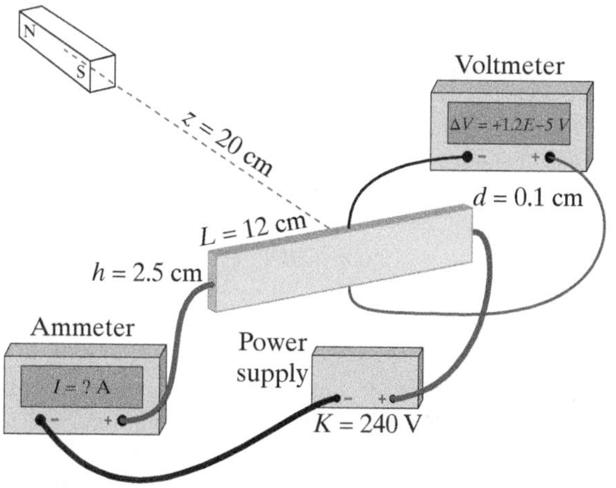

Figure 20.120

The mobility of charge carriers in the thin plate is 3.5×10^{-4} (m/s)/(V/m), and the number density of the singly charged charge carriers is 4×10^{23} m^{-3}. A voltmeter is connected precisely vertically across the thin plate and reads $+1.2 \times 10^{-5}$ V. A low-resistance ammeter is in series with the rest of the circuit. **(a)** Are the charge carriers electrons or holes? Explain, including a physics diagram. **(b)** What is the magnetic dipole moment of the bar magnet? Include units. **(c)** What does the ammeter read, including sign?

••P59 In Figure 20.121 a bar 11 cm long with a rectangular cross section 3 cm high and 2 cm deep is connected to a 1.2 V battery and an ammeter. The resistance of the copper connecting wires and the ammeter, and the internal resistance of the battery, are all negligible compared to the resistance of the bar.

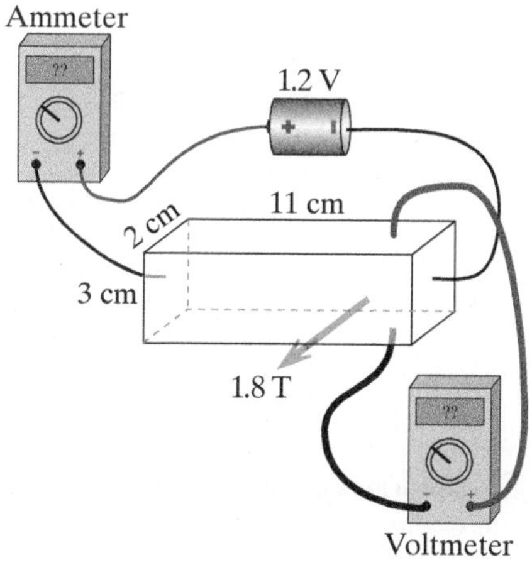

Figure 20.121

Using large coils not shown on the diagram, a uniform magnetic field of 1.8 T was applied perpendicular to the bar (out of the page, as shown). A voltmeter was connected across the bar, with the connections across the bar carefully placed directly across from each other.

The mobile charges in the bar have charge $+e$, their density is 7×10^{23}/m^3, and their mobility is 3×10^{-5} (m/s)/(V/m).

Predict the readings of the voltmeter and ammeter, including signs. Explain carefully, using diagrams to support your explanation. Remember that a voltmeter reads positive if the + terminal is connected to higher potential, and that an ammeter reads positive if conventional current enters the + terminal.

Section 20.5

•P60 A neutral iron bar is dragged to the left at speed v through a region with a magnetic field B that points out of the page (Figure 20.122). Which diagram (1–5) best shows the state of the bar?

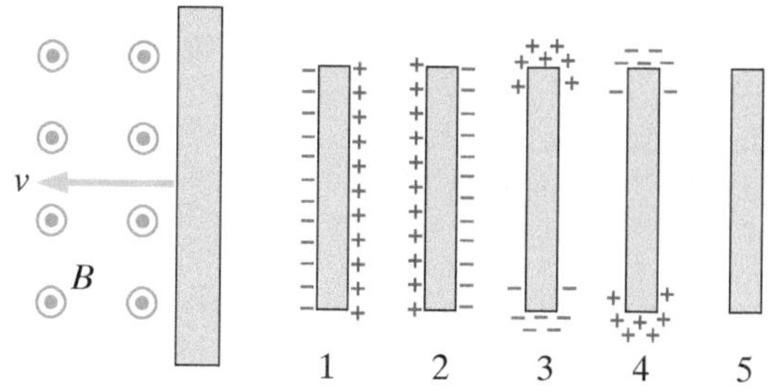

Figure 20.122

•P61 A neutral copper bar oriented horizontally moves upward through a region where there is a magnetic field out of the page. Which diagram (1–5) in Figure 20.123 correctly shows the distribution of charge on the bar?

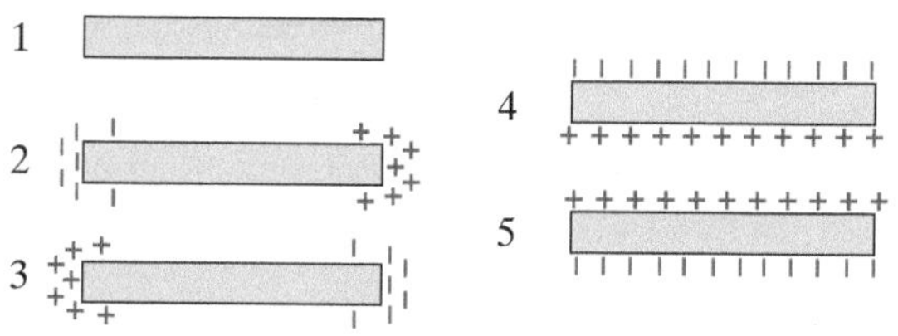

Figure 20.123

•P62 A metal rod of length L slides horizontally at constant speed v on frictionless insulating rails through a region of uniform upward magnetic field of magnitude B (Figure 20.124).

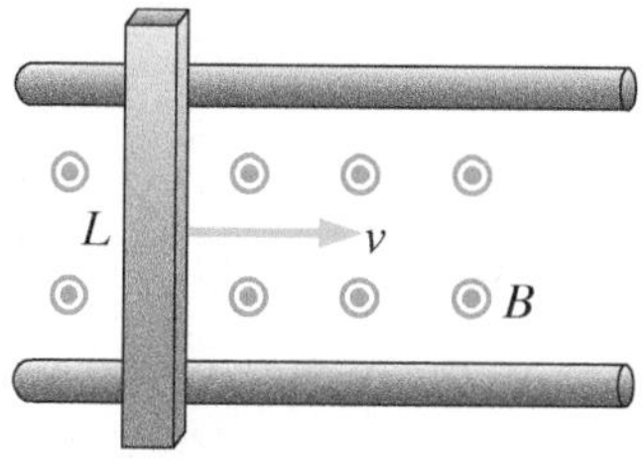

Figure 20.124

On a diagram, show the polarization of the rod and the direction of the Coulomb electric field inside the rod. Explain briefly. What is the magnitude E_C of the Coulomb electric field inside the rod? What is the potential difference across the rod? What is the emf across the rod? What are the magnitude and direction of the force you have to apply to keep the rod moving at a constant speed v?

•**P63** A neutral metal rod of length 0.45 m slides horizontally to the left at a constant speed of 7 m/s on frictionless conducting rails through a region of uniform magnetic field of magnitude 0.4 T, directed out of the page as shown in Figure 20.125.

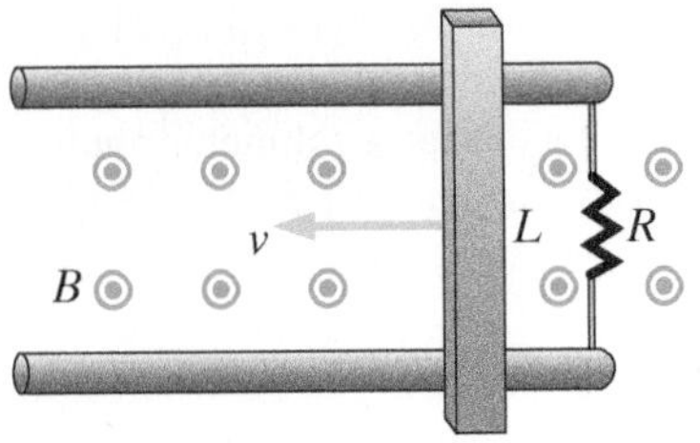

Figure 20.125

Before answering the following questions, draw a diagram showing the polarization of the rod, and the direction of the Coulomb electric field inside the rod.

(a) Which of the following statements is true? (1) The top of the moving rod is positive. (2) The top of the moving rod is negative. (3) The right side of the moving rod is positive. (4) The right side of the moving rod is negative. (5) The moving rod is not polarized. **(b)** After the initial transient, what is the magnitude of the net force on a mobile electron inside the rod? **(c)** What is the magnitude of the electric force on a mobile electron inside the rod? **(d)** What is the magnitude of the magnetic force on a mobile electron inside the rod? **(e)** What is the magnitude of the potential difference across the rod? **(f)** In what direction must you exert a force to keep the rod moving at constant speed?

••**P64** A metal bar of mass M and length L slides with negligible friction but with good electrical contact down between two vertical metal posts (Figure 20.126). The bar falls at a constant speed v. The falling bar and the vertical metal posts have negligible electrical resistance, but the bottom rod is a resistor with resistance R. Throughout the entire region there is a uniform magnetic field of magnitude B coming straight out of the page. Explain every step clearly and in detail.

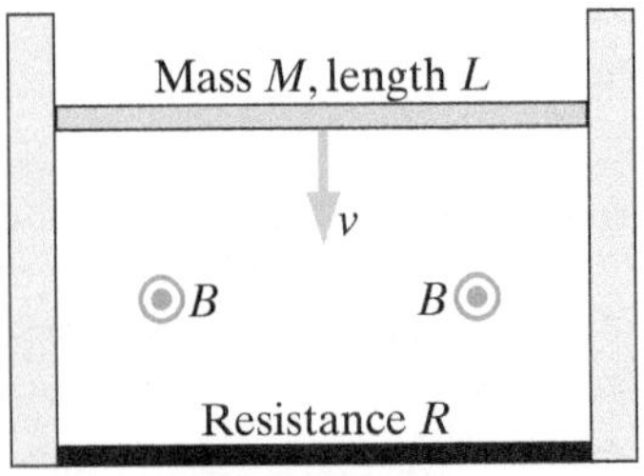

Figure 20.126

(a) Calculate the amount of current I running through the resistor. **(b)** On a diagram, clearly show the surface-charge distribution all the way around the circuit and the direction of the conventional current I. Explain. **(c)** Calculate the constant speed v of the falling bar. **(d)** Show that the rate of change of the gravitational energy of the Universe is equal to the rate of energy dissipated in the resistor.

••**P65** In Figure 20.127 a rectangular loop of wire with width w, height h, and resistance R is dragged at constant speed v to the right through a region where there is a uniform magnetic field of magnitude B out of the page. The magnetic field is negligibly small outside that region.

For each stage of the motion, determine the current I in the loop and the force $\vec{F}$ required to maintain the constant speed. There are five stages. **(a)** Before entering the magnetic field region. **(b)** While the loop is part way into the magnetic field region (this is the stage that is shown). **(c)** While the loop is entirely inside the magnetic field region. **(d)** While the loop is part way out of the magnetic field region. **(e)** After leaving the magnetic field region.

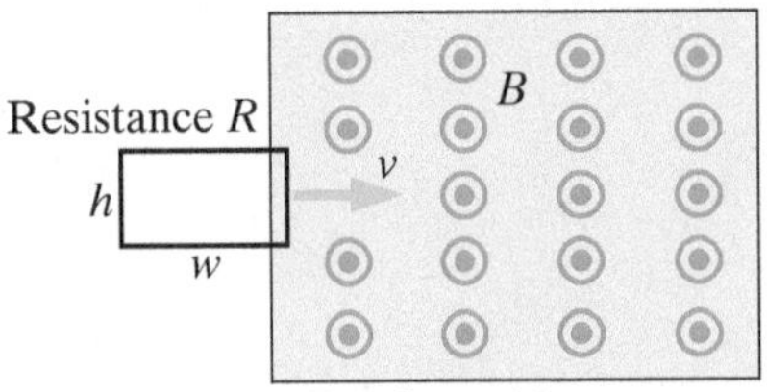

Figure 20.127

Explain briefly in each case. In addition to determining the magnitude of $\vec{F}$ and I for each stage, draw a diagram for each of the five stages and show the direction of $\vec{F}$ and I on the diagram. Also show the approximate surface-charge distribution on the loop.

••**P66** In Figure 20.128 on the left is a region of uniform magnetic field B_1 into the page, and adjacent on the right is a region of uniform magnetic field B_2 also into the page. The magnetic field B_2 is smaller than B_1 ($B_2 < B_1$). You pull a rectangular loop of wire of length w, height h, and resistance R from the first region into the second region, on a frictionless surface. While you do this you apply a constant force F to the right, and you notice that the loop doesn't accelerate but moves with a constant speed.

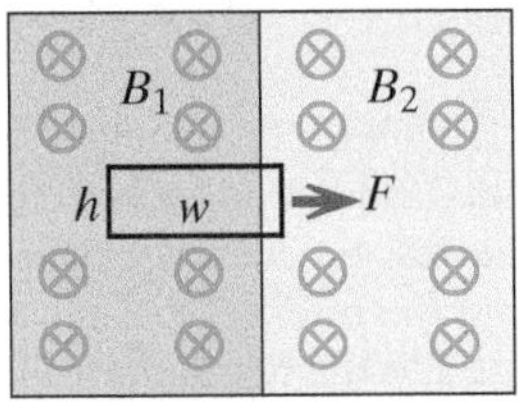

Figure 20.128

Calculate this constant speed v in terms of the known quantities B_1, B_2, w, h, R, and F, and explain your calculation carefully. Also show the approximate surface-charge distribution on the loop.

••**P67** A neutral metal rod of length 60 cm falls toward the Earth. The rod is horizontal and oriented east west. **(1)** Which end of the rod, east or west, has excess electrons? Explain, using physics diagrams. **(2)** At a moment when the rod's speed is 4 m/s, approximately how many excess electrons are at the negative end of the rod?

••**P68** A "unipolar" generator consists of a copper disk of radius R rotating in a uniform, steady magnetic field B perpendicular to the disk, out of the page (Figure 20.129). (The magnetic field is produced by large coils carrying constant current, not shown in the diagram.) Sliding contacts are made at the center (on the axle) and at the rim of the disk, and the wires are connected to a voltmeter. If the outer rim travels counterclockwise at a speed v, what does the voltmeter read (magnitude and sign)?

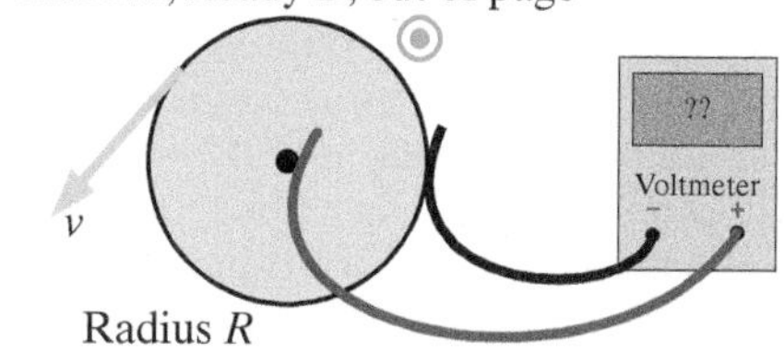

Figure 20.129

••**P69** In Figure 20.130 a rectangular loop of wire $L = 15$ cm long by $h = 3$ cm high, with a resistance of $R = 0.3\ \Omega$, moves with constant speed $v = 8$ m/s to the right, emerging from a rectangular region where there is a uniform magnetic field into a region where the magnetic field is negligibly small. In the region of magnetic field, the magnetic field points into the page, and its magnitude is $B = 1.2$ T.

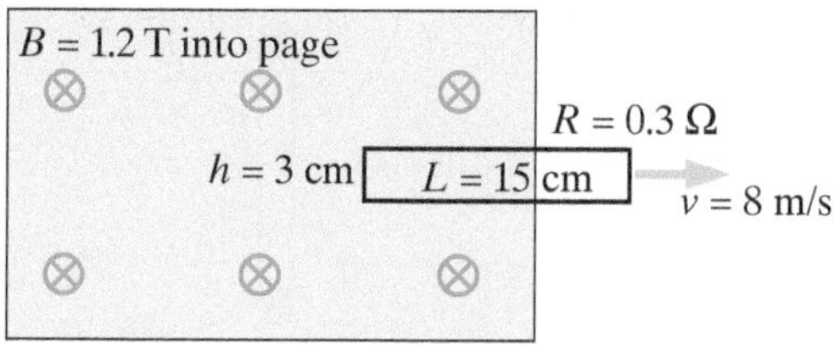

Figure 20.130

(a) On a diagram, show the approximate charge distribution on the moving loop. **(b)** What are the direction and magnitude of conventional current in the loop? Explain briefly. **(c)** Which of the following are true? (1) The magnetic force only stretches the loop; the net magnetic force on the loop is zero. (2) Because a current flows in the loop, there is a magnetic force on the loop. (3) The magnetic force on the loop is in the same direction as the velocity of the loop. **(d)** What are the direction and magnitude of the force that you have to apply in order to make the loop move at its constant speed?

Section 20.6

•**P70** A proton moves with speed v in the $+x$ direction. In the lab (stationary) reference frame, what electric and magnetic fields would you observe a distance r above the proton (in the $+y$ direction)? In a reference frame moving with the proton, what electric and magnetic fields would you observe at the same location? Be quantitative.

Section 20.7

•**P71** A bar magnet whose magnetic dipole moment is $\langle 4, 0, 1.5\rangle$ $\text{A}\cdot\text{m}^2$ is suspended from a thread in a region where external coils apply a magnetic field of $\langle 0.8, 0, 0\rangle$ T. What is the vector torque that acts on the bar magnet?

•••**P72** A bar magnet whose mass is 0.02 kg and whose magnetic dipole moment is $\langle 6, 0, 0\rangle$ $\text{A}\cdot\text{m}^2$ is suspended on a low-friction pivot in a region where external coils apply a magnetic field of $\langle 1.5, 0, 0\rangle$ T. You rotate the bar magnet slightly in the horizontal plane and release it. **(a)** What is the angular frequency of the oscillating magnet? **(b)** What would be the angular frequency if the applied magnetic field were $\langle 3, 0, 0\rangle$ T?

Section 20.8

•**P73** A bar magnet whose magnetic dipole moment is 15 $\text{A}\cdot\text{m}^2$ is aligned with an applied magnetic field of 4 T. How much work must you do to rotate the bar magnet 180° to point in the direction opposite to the magnetic field?

•**P74** The center of a bar magnet whose magnetic dipole moment is $\langle 8, 0, 0\rangle$ $\text{A}\cdot\text{m}^2$ is located at the origin. A second bar magnet whose magnetic dipole moment is $\langle 3, 0, 0\rangle$ $\text{A}\cdot\text{m}^2$ is located at $x = 0.12$ m. What is the vector force on the second magnet due to the first magnet?

Section 20.9

•**P75** A thin rectangular coil 8 cm by 5 cm has 50 turns of copper wire. It is made to rotate with angular frequency 100 rad/s in a magnetic field of 1.2 T. **(a)** What is the maximum emf produced in the coil? **(b)** What is the maximum power delivered to a 50 ohm resistor?

Section 20.10

•**P76** In order for a spark to occur, it is necessary to ionize the air, which is not usually a conductor. One possible model for the process by which air becomes ionized is this: if a sufficiently strong electric field were applied to an atom, it would be possible to pull an outer electron out of the atom, leaving a positively charged ion and a free electron. We will estimate the strength of the electric field required to pull an electron out of an atom. **(a)** Consider the interaction of a single outer electron in a nitrogen atom with the "atomic core" (all the other charged particles in the atom—the protons in the nucleus and all the other electrons). What is the net charge of the atomic core? **(b)** If the radius of the atom is approximately 1×10^{-10} m, what is the magnitude of the electric field due to the atomic core at the location of the outer electron? **(c)** What is the magnitude of the electric field you would have to apply in order to pull the outer electron out of the atom? **(d)** It is observed experimentally that an applied electric field of 3×10^6 V/m is sufficient to cause a spark in air. What is the ratio of the electric field you calculated to the observed field needed to start a spark? **(e)** What should we conclude about this model? (1) Using more significant figures would not improve the agreement. (2) We need to think of a different physical explanation of how air gets ionized. (3) If our calculation used more accurate values the numbers would probably agree. (4) Since the discrepancy is so large, this model must be wrong.

•**P77** A different model for how air becomes ionized in a spark focuses on energy. Suppose that a very energetic particle collided with a nitrogen or oxygen atom and knocked out an electron, ionizing the molecule. Estimate the kinetic energy required for this process. **(a)** Consider an outer electron in a nitrogen atom, interacting with the atomic core (nucleus plus all other electrons). What is the net charge of the atomic core? **(b)** Assuming the radius of the nitrogen atom is about 1×10^{-10} m, what is the electric potential at the location of this outer electron? **(c)** Ionizing the atom means moving the outer electron a very large distance away from the atomic core. What is the electric potential a very large distance from the atomic core? **(d)** What is the potential difference between these two locations? **(e)** What is the change in electric potential energy of the system of the electron plus the atomic core in this process? **(f)** How much kinetic energy would an incoming particle need in order to ionize the atom in a collision between the particle and the atom?

•**P78** We will consider the possibility that a free electron acted on by an electric field could gain enough energy to ionize an air molecule in a collision. **(a)** Consider an electron that starts from rest in a region where there is an electric field (due to some charged objects nearby) whose magnitude is nearly constant. If the electron travels a distance d and the magnitude of the electric field is E, what is the potential difference through which the electron travels? (Pay attention to signs: Is the electron traveling with the electric field or opposite to the electric field?) **(b)** What is the change in potential energy of the system in this process? **(c)** What is the change in the kinetic energy of the electron in this process? **(d)** We found the mean free path of an electron in air to be about 5×10^{-7} m, and in the previous question you calculated the energy required to knock an electron out of an atom. What is the magnitude of the electric field that would be required in order for an electron to gain sufficient kinetic energy to ionize a nitrogen molecule? **(e)** The electric field required to cause a spark in air is observed to be about 3×10^6 V/m at STP. What is the ratio of the magnitude of the field you calculated in the previous part to the observed value at STP? **(f)** What is it reasonable to conclude about this model of how air becomes ionized? (1) Since we used accurate numbers, this is a huge discrepancy, and the model is wrong. (2) Considering the approximations we made, this is pretty good agreement, and the model may be correct.

•P79 Explain briefly why there is a limit to how much charge can be placed on a metal sphere in the classroom. If the radius of the sphere is 15 cm, what is the maximum amount of charge you can place on the sphere? (Remember that a uniform sphere of charge makes an electric field outside the sphere as though all the charge were concentrated at the center of the sphere.)

••P80 A Van de Graaff generator pulls electrons out of the Earth and transports them on a conveyor belt onto a nearly spherical metal shell (Figure 20.131). The diameter of this generator's metal shell is 24 cm.

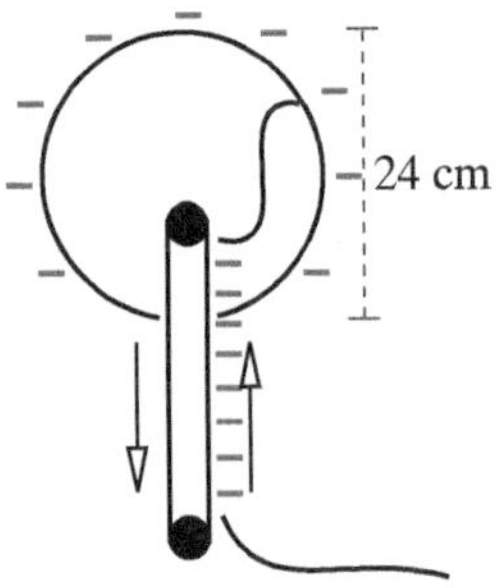

Figure 20.131

(a) Suppose that the conveyor belt in the Van de Graaff generator is running so fast that the generator succeeds in building up and maintaining just enough charge on the metal shell to cause the air to steadily glow bluish near the surface of the shell. Under these conditions, how much net charge $|Q|$ is on the metal shell? Calculate a numerical value for $|Q|$ and explain briefly. (Remember that the electric field outside a uniformly charged sphere is like that of a point charge located at the center of the sphere.) **(b)** Under these conditions (with the air steadily glowing), the Van de Graaff generator continually delivers additional electrons to the metal shell and yet the net charge Q on the shell does not change. Explain briefly but in detail why the charge on the shell does not change.

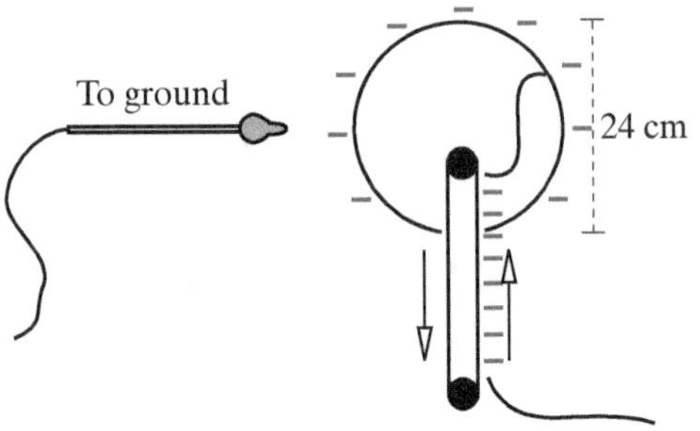

Figure 20.132

(c) Now assume that the Van de Graaff generator is run more slowly, and the buildup of charge on the metal shell is limited by the inability of the motor to force any more electrons onto the negatively charged shell. The air no longer glows. We have a pear-shaped piece of metal that is attached by a metal wire to the earth (grounded). When we bring the grounded piece of metal to a location near the Van de Graaff metal shell, as shown in Figure 20.132, we observe a big spark. Explain why a spark occurs now but doesn't occur without the additional piece of metal nearby. **(d)** This spark lasts only a very short time. Why? **(e)** If we keep holding the grounded piece of metal 5 cm from the metal shell, we observe that there are big, brief sparks every 2 s. If we reduce the distance to 2 cm, the sparks that we observe occur more frequently, and they are less intense. Explain briefly.

COMPUTATIONAL PROBLEMS

••P81 Starting with the skeleton program given below, write a program to predict the motion of a proton in a constant magnetic field. The skeleton program draws a "floor" and displays a uniform magnetic field, which initially has the value $\langle 0, 0.2, 0\rangle$ T. Before beginning it may be helpful to review a previous program in which you predicted the motion of an object under the influence of one or more forces. Recall that you must:

Repeat:
- Calculate the net force $\vec{F}_{\text{net}}$ acting on the system.
- Update the momentum of the system: $\vec{p}_f = \vec{p}_i + \vec{F}_{\text{net}}\,\Delta t$.
- Update the position: $\vec{r}_f = \vec{r}_i + \vec{v}_{\text{avg}}\,\Delta t$.

Use $\vec{p}_f/m$ to approximate the $\vec{v}_{\text{avg}}$ in each step.

Once your program is working, do the following: **(a)** Using an initial velocity of $\langle 2 \times 10^6, 0, 0\rangle$ m/s, change the `while` statement to `while t < 3.34e-7:` and run the program. The proton should just make one complete orbit; if it does not, review your code. **(b)** What is the approximate radius of the proton's path? **(c)** What happens to the radius of the path if you double the initial speed? **(d)** How does doubling the speed change the time to complete one orbit? **(e)** What happens to the radius of the path if you halve the initial speed of 2×10^6 m/s? **(f)** How does halving the speed change the time to complete one orbit? **(g)** Restore the proton's initial velocity to its original value, and double the magnetic field. What changes? **(h)** Restore the magnetic field to its original value. Change the proton's initial velocity so that it is parallel to the magnetic field. Describe the path followed by the proton, and explain. **(i)** Change the initial velocity back to its original value, then add a significant $+y$ component to the velocity. Increase the time allowed in the loop sufficiently to allow the proton to make five complete orbits. Describe the proton's path, and explain. **(j)** Change the proton to an antiproton, whose mass is the same as that of the proton, but whose charge is $-e$. What changes?

```
from visual import *
scene.width = 800
scene.height = 800
## CONSTANTS ##
mzofp = 1e-7  ## mu-zero-over-four-pi
qe = 1.6e-19
mproton = 1.7e-27
B0 = vector(0,0.2,0)
bscale = 1
#### THIS CODE DRAWS A GRID ##
#### AND DISPLAYS MAGNETIC FIELD ##
xmax = 0.4
dx = 0.1
yg = -0.1
x = -xmax
while x < xmax+dx:
    curve(pos=[(x,yg,-xmax),(x,yg,xmax)],
          color=(.7,.7,.7))
    x = x + dx
z = -xmax
while z < xmax+dx:
    curve(pos=[(-xmax,yg,z),(xmax,yg,z)],
          color=(.7,.7,.7))
    z = z + dx
x = -xmax
dx = 0.2
while x < xmax+dx:
    z = -xmax
    while z < xmax+dx:
        arrow(pos=(x,yg,z),
              axis=B0*bscale,
              color=(0,.8,.8))
        z = z + dx
    x = x + dx
#### OBJECTS AND INITIAL CONDITIONS ##
particle = sphere(pos=vector(0,0.15,0.3),
                  radius=1e-2,
                  color=color.yellow,
                  make_trail=True)
## make trail easier to see (thicker) ##
particle.trail_object.radius = particle.
radius/3
vparticle = vector(-2e6,0,0)
p = mproton*vparticle
qparticle = qe
deltat = 5e-11
t = 0
############################################
while True:
    rate(500)
    ## YOUR CODE GOES HERE ##
```

•••**P82** Write a computer program to compute and display the motion of a proton inside a cyclotron (see Section 20.1). The first cyclotrons were small desktop devices; we will model a small device in order to understand the basic physical phenomena. Consider a cyclotron whose radius is 5 cm, with a gap of 0.5 cm between the dees (Figure 20.133).

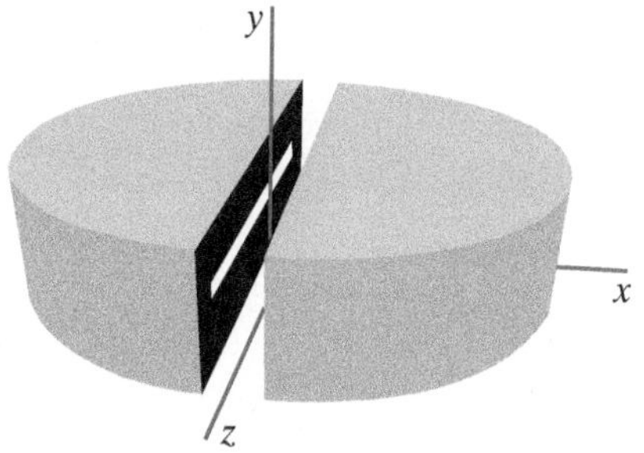

Figure 20.133

External magnets create a uniform magnetic field of 1.0 T in the $+y$ direction throughout the device. A proton starts essentially from rest at the center of the cyclotron. The maximum potential difference between the dees is 5000 V. Note that the electric field inside the dees is 0 at all times. Figure 20.133 shows that the metal dees form a nearly closed conducting enclosure.
(a) What is the angular frequency ω of the sinusoidally oscillating potential difference you will need to apply? **(b)** Compute and display the trajectory of a proton in the cyclotron; display the trail of the proton. You may assume that the proton does not reach a speed sufficiently close to the speed of light to require a relativistically correct calculation, but do check this assumption. To get your program going, you can use a large value of dt (such as 1E-10 s), but you will need to decrease this value later. **(c)** Plot the proton's kinetic energy K (in eV) vs. time. Explain the shape of the graph you observe. **(d)** What happens if you use a different angular frequency for the potential difference? For example, try an angular frequency of 0.6ω. **(e)** If the proton gains a significant amount of kinetic energy while inside the dees, your dt is too large. Adjust your dt appropriately. What value of dt gives reasonable results? **(f)** How much time does the proton take to reach the outer edge of the cyclotron? **(g)** What is K (in eV) when the proton reaches the outer edge of the cyclotron? **(h)** Given your answer to (g), how many orbits must the proton have made before reaching the outer edge of the cyclotron? **(i)** What is the final speed of the proton? Were you justified in assuming that $v \ll c$?

ANSWERS TO CHECKPOINTS

1 $F_{magnetic} \approx 1\times10^{-8}$ N; $F_{electric} \approx 9\times10^{-5}$ N, much larger.
2 Clockwise; $q\vec{v}\times\vec{B} = -e\vec{v}\times\vec{B}$ points inward and turns the electron's momentum.
3 $F = 2\times10^{-3}$ N per meter of wire.
4 (a) See Figure 20.134 below. **(b)** See Figure 20.135 below.
5 $\vec{E} = \langle 0, 0, 2.7\times10^{8}\rangle$ V/m. Note that a particle traveling that fast would presumably be in an evacuated chamber, so an electric field of this magnitude would not cause a spark.
6 $\Delta V_{Hall} = 1.5\times10^{-5}$ V; $\Delta V_{wire} = 3.3\times10^{-3}$ V $\gg \Delta V_{Hall}$
7 0.01 V ($\ll$ 1.5 V)
8 emf $= v_{bar}BL$; I counterclockwise; $I = v_{bar}BL/R$; $F = v_{bar}B^2L^2/R$ to the right
9 $F_{magnetic}/F_{electric} = 2.5\times10^{-11}$
10 Atomic magnetic dipole moments point north; swings toward north; magnetic dipoles twist to align with the magnetic field.
11 $2\mu B$
12 Net force would reverse and be repulsive; net force would be zero.
13 Emf is maximum when $\theta = 90°$ (velocity at right angle to magnetic field), minimum when $\theta = 0$ (velocity in the direction of magnetic field, so cross product is zero); 302 V
14 The mean free path would double; the mean free path would be one-quarter as long.
15 The mean free path must be one-third as long; 3.6×10^{6} V/m.
16 Below the charge in Figure 20.72, $E_y < 0$ and $E_z = 0$, so

$$B'_y = \frac{\left(B_y + \frac{v}{c^2}E_z\right)}{\sqrt{1-v^2/c^2}} = 0$$

$$B'_z = \frac{\left(B_z - \frac{v}{c^2}E_y\right)}{\sqrt{1-v^2/c^2}} = \frac{-\frac{v}{c^2}E_y}{\sqrt{1-v^2/c^2}} > 0$$

Therefore $B'_y = 0$ and $B'_z > 0$, as expected.

In front of the charge in Figure 20.72, $E_y = 0$ and $E_z > 0$, so

$$B'_y = \frac{\left(B_y + \frac{v}{c^2}E_z\right)}{\sqrt{1-v^2/c^2}} = \frac{\frac{v}{c^2}E_z}{\sqrt{1-v^2/c^2}} > 0$$

$$B'_z = \frac{\left(B_z - \frac{v}{c^2}E_y\right)}{\sqrt{1-v^2/c^2}} = 0$$

Behind the charge in Figure 20.72, $E_y = 0$ and $E_z < 0$, so

$$B'_y = \frac{\left(B_y + \frac{v}{c^2}E_z\right)}{\sqrt{1-v^2/c^2}} = \frac{\frac{v}{c^2}E_z}{\sqrt{1-v^2/c^2}} < 0$$

$$B'_z = \frac{\left(B_z - \frac{v}{c^2}E_y\right)}{\sqrt{1-v^2/c^2}} = 0$$

Therefore $B'_y < 0$ and $B'_z = 0$, as expected.

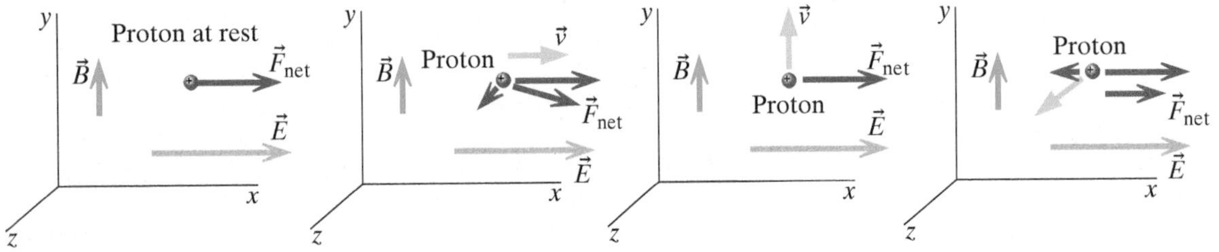

Figure 20.134 Checkpoint 4 (a).

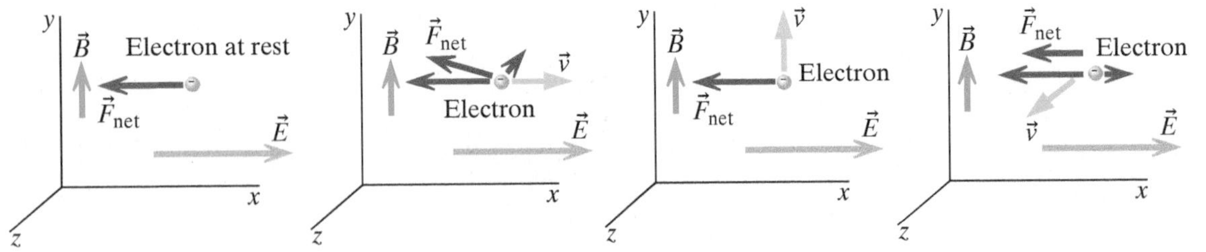

Figure 20.135 Checkpoint 4 (b).

CHAPTER

21

Patterns of Field in Space

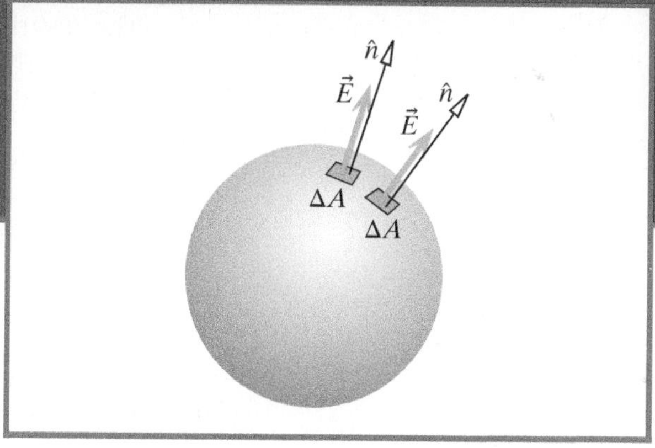

OBJECTIVES

After studying this chapter you should be able to

- Mathematically relate the net electric field over a closed surface to the net charge enclosed.
- Mathematically relate the net magnetic field around a closed loop to the net current enclosed.

Electric charges produce electric and magnetic fields, and we know how to calculate or estimate the magnitude and direction of the field due to a particular distribution of charges (including moving charges). Sometimes, however, it is useful to reason in the other direction: from an observed three-dimensional pattern of electric or magnetic field it may be possible to infer what charges are responsible for this pattern. Gauss's law relates patterns of electric field over a closed surface to the amount of charge inside that closed surface. Ampere's law relates patterns of magnetic field around a closed path to the amount of current passing through that closed path.

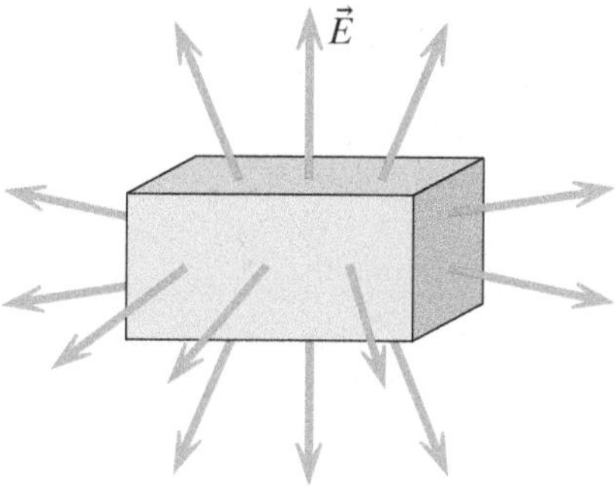

Figure 21.1 What's in this box?

21.1 PATTERNS OF ELECTRIC FIELD: GAUSS'S LAW

You are already familiar with some spatial patterns of electric field. Figure 21.1 shows an imaginary but opaque box that contains some charge. Suppose that you can't look inside the box, but you can measure the electric field all over the surface of the box.

> QUESTION Can you conclude that there is some charge in the box? Is it positive or negative?

The fact that the electric field points outward all over the box implies that there must be some positive charge inside the box.

Figure 21.2 shows another box (for clarity, the electric field on the surface is displayed with the head rather than the tail of the vector at the observation location). Again, suppose that you can't look inside the box, but you can measure the electric field all over the surface of the box.

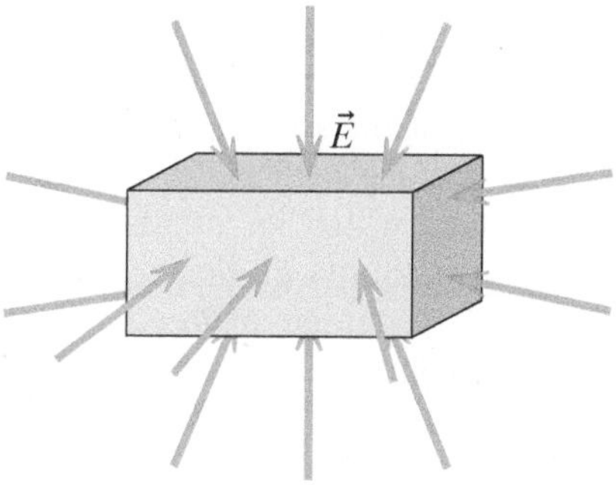

Figure 21.2 What's in this box?

> QUESTION Can you conclude that there is some charge in the box? Is it positive or negative?

Since the electric field points into the box, all over its surface, there must be some negative charge inside the box.

Suppose that you find a box on whose surface the electric field has a pattern like that in Figure 21.3.

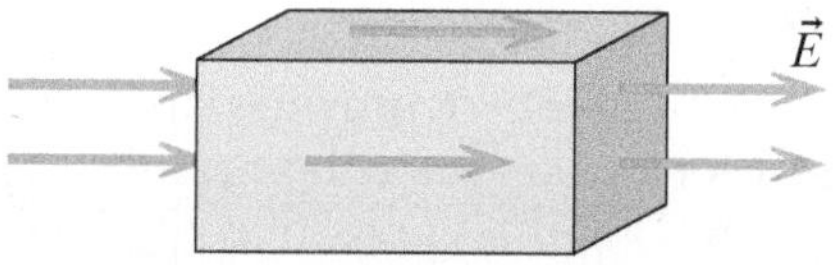

Figure 21.3 What's in this box?

> QUESTION Can you conclude that there is some charge in the box? Is it positive or negative? Can you think of a charge distribution that could produce this pattern of electric field?

The electric field points out of the box on part of the surface, and into the box on other parts of the surface. A possible charge distribution that would produce such a pattern of electric field is a very large positively charged plate to the left of the box, or a very large negatively charged plate to the right of the box, or plates of both kinds, in which case the box lies in the gap between the plates of a capacitor. (Another possibility would be that the box is a mathematical surface drawn inside a circuit wire, with a uniform electric field in the direction of the wire.)

It appears that if we find this kind of distribution of electric field over the surface of the box, we can conclude that there is no net charge inside the box. Although at first glance it might appear that there could be an electric dipole inside the box, a dipole would produce the observed pattern of electric field on the ends of the box ($\vec{E}$ to the right) but not on the sides of the box ($\vec{E}$ to the left), so this is not a possibility.

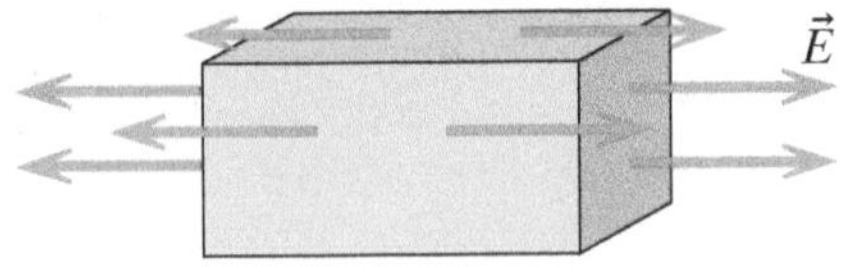

Figure 21.4 What's in this box?

Suppose that you find a pattern of electric field on a box as shown in Figure 21.4.

QUESTION Can you think of a distribution of charge that could produce such a pattern of electric field on the surface of the box? (*Hint:* There may be some charge outside the box in addition to the charge inside the box.)

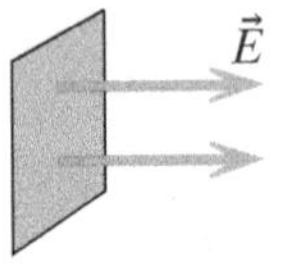

Figure 21.5 What can you conclude?

A charge distribution that could produce this pattern of field over the surface of the box is a very large vertical positively charged plate that cuts through the middle of the box, making electric field to the left and right of the plate. In this case there is positive charge inside the box (and outside, too).

Closed and Open Surfaces

On the other hand, if your measurements of the electric field don't surround a 3D region, you may not be able to conclude much. Figure 21.5 shows a pattern of electric field on just one face of a box.

QUESTION Give three different charge configurations that could be responsible for the field pattern observed in Figure 21.5.

Possibilities include a very large positively charge plate to the left of the surface, a very large negatively charged plate to the right of the surface, or both. We see an important difference between an "open" surface such as this one and a "closed" surface such as a box or a spherical surface, which surrounds a region in space. Qualitatively, it seems clear that measurements of electric field all over a closed surface can tell us quite a bit about the charge inside that closed surface. However, measurements of electric field on an open surface can't even tell us whether nearby charges are positive or negative.

Gauss's law is

- a quantitative relationship between
- measurements of electric field on a closed surface and
- the amount and sign of the charge inside that closed surface.

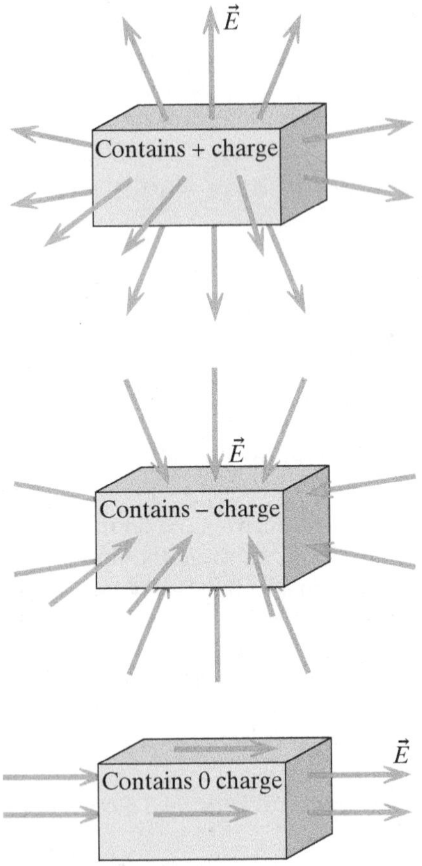

Figure 21.6 Compare patterns of electric field on these three boxes.

We will study Gauss's law for the additional insights it provides about the relationship between charge and field, and for certain important results that can be obtained using Gauss's law that are almost impossible to obtain using Coulomb's law alone, even though Gauss's law and Coulomb's law are essentially equivalent to each other (if the charges aren't moving).

Moreover, the concept of "flux" that arises in the context of Gauss's law plays an important role in our study of the effects of a time-varying magnetic field, Faraday's law, in the next chapter.

21.2 DEFINITION OF "ELECTRIC FLUX"

It is clear that there is some connection between the pattern of electric field on a closed surface and the amount of charge inside that surface, but how can we turn this into a quantitative relationship? We need to define a quantitative measure of the amount and direction of electric field over an entire surface. This measure is called "electric flux." We will develop a list of properties that a quantitative definition of electric flux should have.

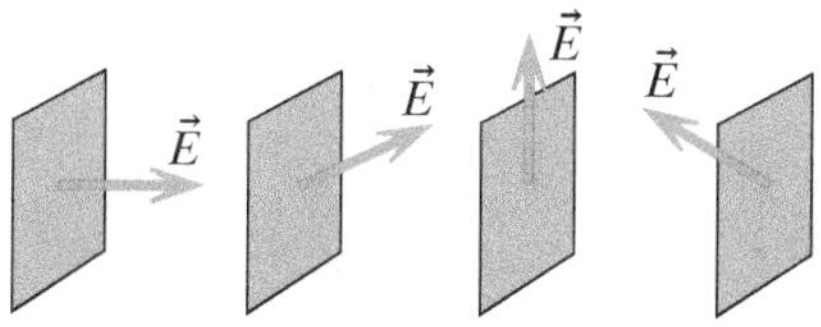

Figure 21.7 The electric field at different angles to a surface.

Property 1: Direction of Electric Field

Compare the electric field patterns on the surfaces of the three boxes shown in Figure 21.6, which contain positive, negative, and zero charge.

Evidently one property of "electric flux" is that it should be positive where the electric field points out of the box, negative where the electric field points into the box, and zero where the electric field does not pierce the surface. In that way, the total flux for the third box in Figure 21.6 would consist of +1 unit on the right face, −1 unit on the left face, and 0 on the other faces, for a total of zero, which corresponds to the zero charge that is inside the box.

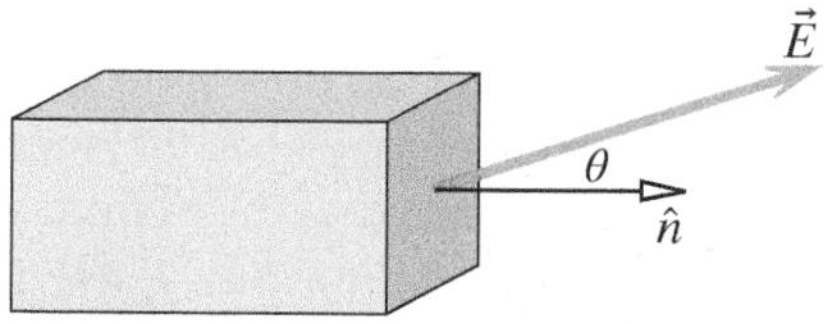

Figure 21.8 A unit vector pointing outward.

We want the sign of electric flux to be related to the angle the electric field makes with the surface. This angle factor varies between +1 (outward and perpendicular to the surface) and −1 (inward and perpendicular to the surface), and is zero if the electric field is parallel to the surface. These values of +1, −1, and 0 are the extremes, and we need some way to express varying degrees of "outward-goingness" in the range from −1 to +1.

Figure 21.7 shows a series of open surfaces where this angle factor varies smoothly from +1 down to 0, and on to negative values. This suggests parameterizing the angle factor as the cosine of an angle between the electric field vector and the "outward-going normal," a unit vector $\hat{n}$ perpendicular to the surface and pointing outward from the closed surface (Figure 21.8).

> QUESTION Show that the cosine of this angle θ has the appropriate properties we're looking for: +1 if $\vec{E}$ points straight outward, −1 if $\vec{E}$ points straight inward, 0 if $\vec{E}$ is parallel to the surface, and an intermediate value between +1 and −1 in other cases.

If $\vec{E}$ and $\hat{n}$ are in the same direction, $\theta = 0°$ and $\cos\theta = +1$. If $\vec{E} \perp \hat{n}$, $\theta = 90°$ and $\cos\theta = 0$. If $\vec{E}$ is opposite to $\hat{n}$, $\theta = 180°$ and $\cos\theta = -1$.

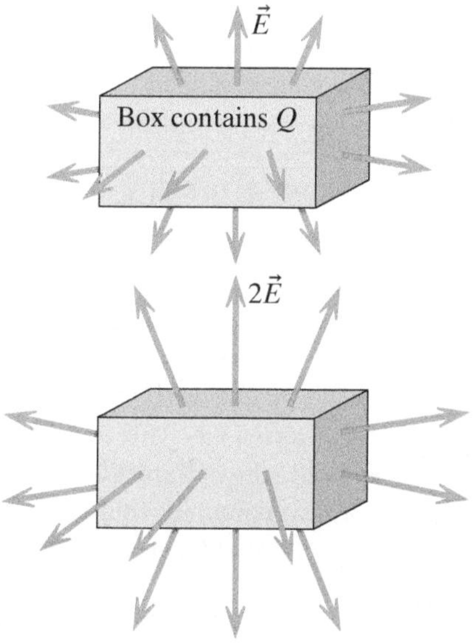

Figure 21.9 The magnitude of the field on the surface of the lower box is twice as large as the field on the upper box.

Property 2: Magnitude of Electric Field

How should the electric flux depend on the magnitude of the electric field? Consider the electric field patterns on the surfaces of the two boxes shown in Figure 21.9.

> QUESTION The upper box contains a charge Q. If the electric field vectors are twice as long on the surface of the lower box, how much charge is inside the lower box?

Doubling an amount of charge doubles the electric field made by that charge. Evidently the lower box contains an amount of charge $2Q$, so the definition of electric flux should have the property that twice the field gives twice the flux. Therefore electric flux ought to be proportional to $|\vec{E}|$ as well as being proportional to $\cos\theta$. Apparently the definition of electric flux should contain the product $E\cos\theta$.

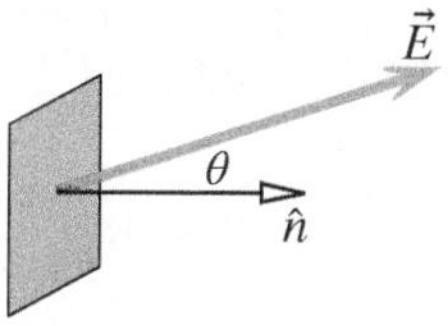

Figure 21.10 $E\cos\theta = \vec{E}\bullet\hat{n} = E|\hat{n}|\cos\theta$

We can write $E\cos\theta$ as $\vec{E}\bullet\hat{n} = E|\hat{n}|\cos\theta$, because the dot product involves the cosine of the angle between the two vectors, and the magnitude of a (dimensionless) unit vector $|\hat{n}|$ is 1 (Figure 21.10).

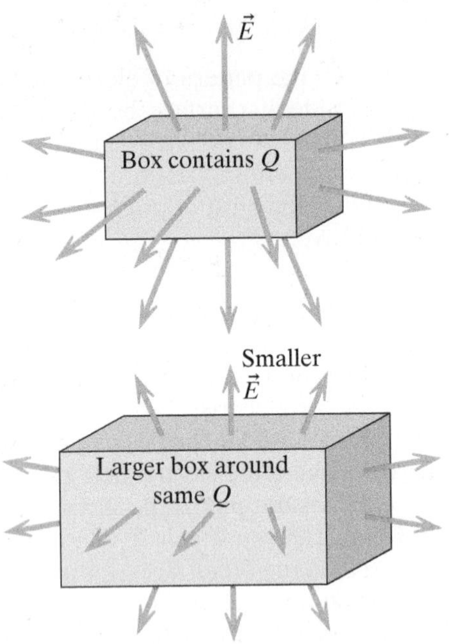

Figure 21.11 A larger box containing the same charge has smaller electric field on its surface.

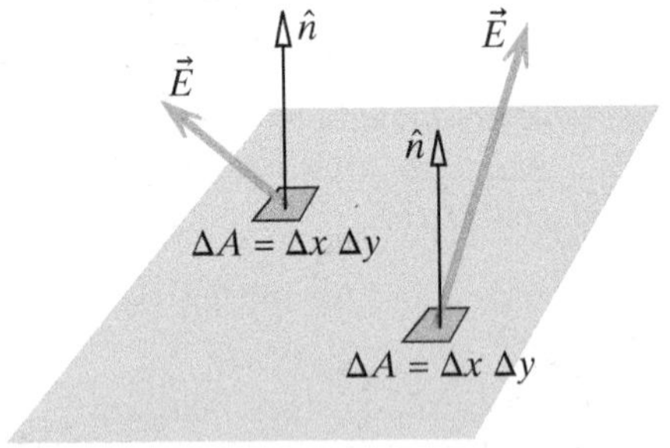

Figure 21.12 Electric flux on a surface is calculated by adding up the contributions $\vec{E} \bullet \hat{n}\,\Delta A$ on each little piece of area $\Delta A = \Delta x\,\Delta y$.

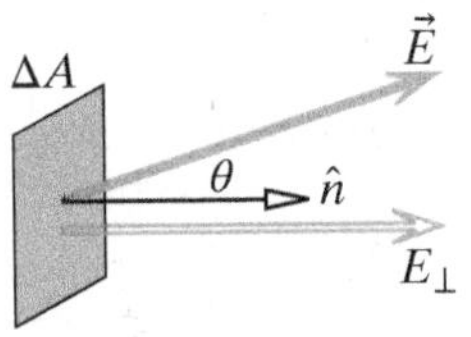

Figure 21.13 $\vec{E} \bullet \hat{n}\,\Delta A = E_\perp\,\Delta A$

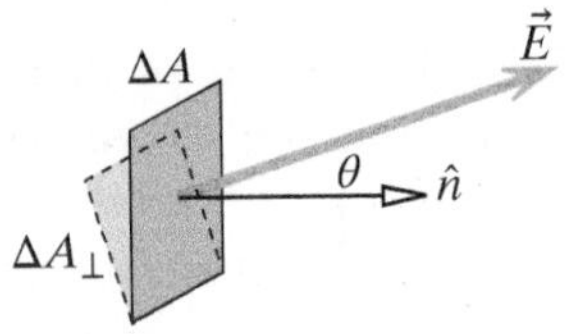

Figure 21.14 $\vec{E} \bullet \hat{n}\,\Delta A = E\,\Delta A_\perp$

QUESTION In Figure 21.10, show that the dot product $\vec{E} \bullet \hat{n}$ is $+E$ if the electric field is parallel to $\hat{n}$, $-E$ if the electric field is opposite to $\hat{n}$, and 0 if $\vec{E}$ is parallel to the surface.

Outward: $\vec{E} \bullet \hat{n} = \langle +E,0,0\rangle \bullet \langle 1,0,0\rangle = +E+0+0 = +E$

Inward: $\vec{E} \bullet \hat{n} = \langle -E,0,0\rangle \bullet \langle 1,0,0\rangle = -E+0+0 = -E$

Parallel: $\vec{E} \bullet \hat{n} = \langle 0,E,0\rangle \bullet \langle 1,0,0\rangle = 0+0+0 = 0$

Property 3: Surface Area

Finally, consider the two boxes shown in Figure 21.11. The small box contains a charge Q, and there are no charges outside the small box. If we enclose the same charge Q in a larger box, the electric field vectors on the surface of the larger box are smaller, because the surfaces are farther from the enclosed charge.

We need to define the electric flux in such a way that the flux is the same for the large electric fields on the surface of the small box or for the small electric fields on the surface of the large box. This suggests that the definition of electric flux ought to involve the area A of a surface, not just $\vec{E} \bullet \hat{n}$, in order to compensate for how big we draw the enclosing surface. Maybe electric flux should be defined as $\vec{E} \bullet \hat{n} A$. In fact, if you double all the distances, the surface area increases by a factor of four, while the magnitude of the electric field decreases by a factor of four. Thus including A in the definition ought to compensate exactly for changing the size of the box.

The magnitude and direction of $\vec{E}$ may vary from location to location on a large surface, so we should define electric flux on a small area ΔA over which there is little variation in the electric field.

Taking all these factors into account, perhaps a useful definition of electric flux over a small area $\Delta A = \Delta x\,\Delta y$ would be $\vec{E} \bullet \hat{n}\,\Delta A$, and we would add up all such contributions over an extended surface to get the electric flux over that entire surface (Figure 21.12), called Φ_{el} (Φ is uppercase Greek phi).

DEFINITION OF ELECTRIC FLUX ON A SURFACE

$$\Phi_{el} = \sum_{\text{surface}} \vec{E} \bullet \hat{n}\,\Delta A$$

Perpendicular Field or Perpendicular Area

We define electric flux as the sum of $\vec{E} \bullet \hat{n}\,\Delta A$ over a surface. This definition takes into account the direction of the electric field, the magnitude of the electric field, and the surface area. One of the ways to calculate the flux is to think of $E\cos\theta$ as the component of electric field perpendicular to the surface, $E_\perp = E\cos\theta$, so that the flux on a small area is $E_\perp \Delta A$ (Figure 21.13).

Alternatively, we can calculate the flux on a small area as the product of the electric field and the projection of the surface area ΔA perpendicular to the direction of the electric field (Figure 21.14). If we call this perpendicular projection of the area $\Delta A_\perp$, the flux is $E\Delta A_\perp$.

Thus we have three equivalent ways of writing the flux: $\vec{E} \bullet \hat{n}\,\Delta A$, $E_\perp\,\Delta A$, or $E\,\Delta A_\perp$. In many books you will find yet another notation for electric flux: $\vec{E} \bullet \Delta\vec{A}$. In this notation $\Delta\vec{A} = \hat{n}\Delta A$. $\Delta\vec{A}$ is considered to be a vector pointing in the direction of the outward-going normal $\hat{n}$, with a magnitude ΔA.

Checkpoint 1 In Figure 21.15 the magnitude of the electric field is 1000 V/m, and the field is at an angle of 30° to the outward-going normal. What is the flux on the small rectangle whose dimensions are 1 mm by 2 mm?

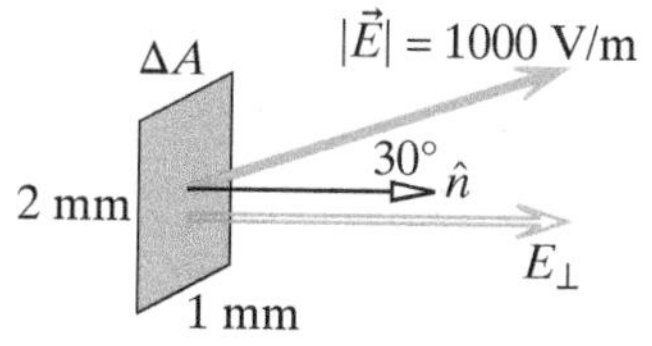

Figure 21.15 Find the electric flux on the rectangle.

Adding Up the Flux

The electric flux on a small area ΔA is $\vec{E} \bullet \hat{n} \Delta A$. To calculate the electric flux over an extended surface we need to divide the surface into little areas, calculate the flux on each little area, and then add up all these contributions to the flux. Symbolically, we have

$$\text{electric flux on a surface } = \sum \vec{E} \bullet \hat{n} \Delta A$$

If we divide the surface into an infinite number of infinitesimal areas, this summation becomes an integral. This is similar to what we did in Chapter 15 to calculate the electric field due to charges distributed over a ring or a disk:

$$\text{electric flux on a surface } = \Phi_{\text{el}} = \int \vec{E} \bullet \hat{n}\, dA$$

This "surface integral" can also be written in the following equivalent forms:

$$\Phi_{\text{el}} = \int E_{\perp} dA = \int E\, dA_{\perp} = \int \vec{E} \bullet d\vec{A}$$

We have seen that it is the electric flux on closed surfaces, such as a box, that tells us something about what charge is inside that closed surface. Whenever we need to add up the electric flux all over a closed surface, we can remind ourselves of this with the following notation:

$$\text{electric flux on a closed surface } = \oint \vec{E} \bullet \hat{n}\, dA = \sum_{\text{closed surface}} \vec{E} \bullet \hat{n} \Delta A$$

A circle superimposed on an integral sign is a standard notation for "completeness," in this case, that the flux should be integrated over an entire closed surface—that is, a surface that completely encloses a volume in space.

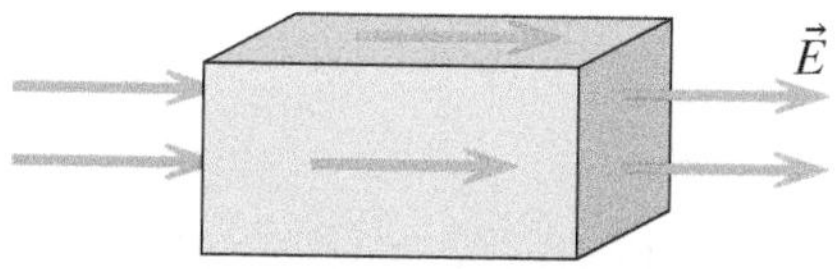

Figure 21.16 The electric field is uniform on each face of this closed surface.

> **QUESTION** What is the net flux Φ_{el} on the closed surface shown in Figure 21.16?

The net flux on this surface is zero. Because the electric field on each face is uniform, it is sufficient in this case to take each ΔA to be the entire area of a face. The flux on the right end is $+EA_{\text{end}}$, the flux on the left end is $-EA_{\text{end}}$, and the flux on all other sides is 0 because on those sides $\vec{E} \perp \hat{n}$.

21.3 GAUSS'S LAW

We can now state Gauss's law, which we will then prove:

GAUSS'S LAW

$$\sum_{\text{closed surface}} \vec{E} \bullet \hat{n} \Delta A = \frac{\sum q_{\text{inside}}}{\varepsilon_0}$$

or

$$\oint \vec{E} \bullet \hat{n}\, dA = \frac{\sum q_{\text{inside}}}{\varepsilon_0}$$

When we add up the electric flux on the entire area of a closed surface, this quantitative measure of the pattern of electric field does turn out to be proportional to the amount of charge inside that closed surface, as we had guessed from the patterns we saw at the beginning of this chapter.

Proving Gauss's Law

What do we need to do in order to prove Gauss's law?

- Show that the proportionality constant is indeed $1/\varepsilon_0$.
- Show that the size and shape of the surface don't matter.
- Show that it is true for any number of point charges inside the surface.
- Show that charges outside the surface contribute zero net flux.

The following sections give a formal proof of the correctness of Gauss's law, by addressing each of these issues. The proof is a geometry-based proof rather than an algebra-based proof. By now you should be convinced that there is indeed some kind of relationship between the charge inside a closed surface and the total electric flux on that closed surface. The formal proof nails this down, quantitatively.

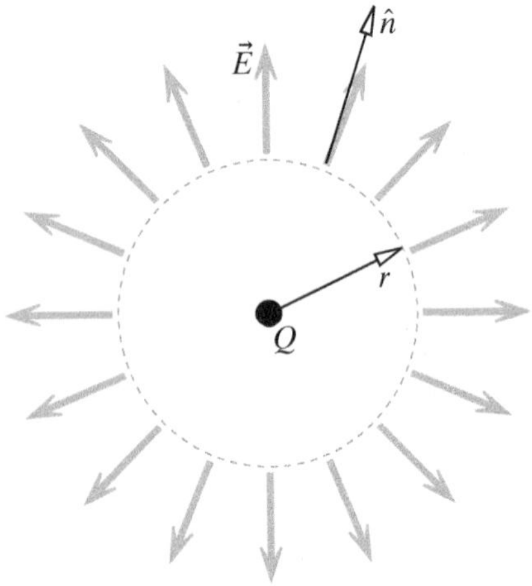

Figure 21.17 A point charge Q surrounded by a mathematical spherical surface of our choosing, of radius r.

Determining the Proportionality Constant

We have postulated that the net electric flux on a closed surface is proportional to the charge inside:

$$\sum_{\text{closed surface}} \vec{E} \bullet \hat{n} \Delta A = k \sum q_{\text{inside}}$$

and that the proportionality constant k is equal to $1/\varepsilon_0$.

To show this, we will calculate the net electric flux for a surface for which we can make the calculation easily: a closed spherical surface surrounding a positive point charge $+Q$ at the center of the sphere (Figure 21.17). It is extremely important to understand that this spherical surface is of our own choosing and is purely mathematical in nature: it has nothing to do with any actual spherical shell made of some material.

Note that because it is hard to draw and interpret three-dimensional diagrams, we will usually draw two-dimensional diagrams that are cross-sectional slices through the 3D surface. You must, however, keep in mind that the spherical surface we have drawn around the charge is a closed 3D surface that encloses the charge Q.

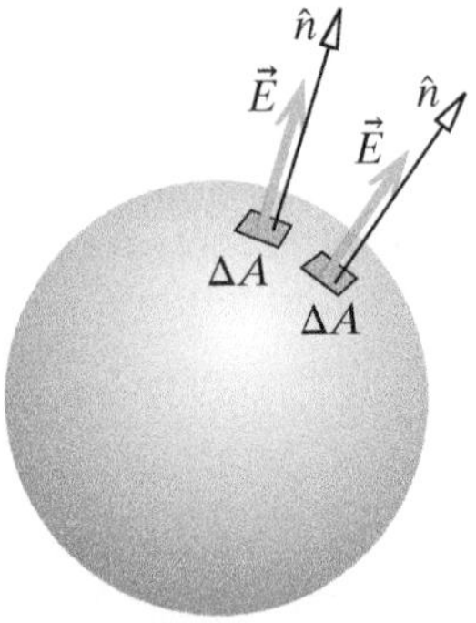

Figure 21.18 We need to calculate the flux on every little piece ΔA of the surface.

QUESTION We need to divide the spherical surface into small pieces ΔA and evaluate $\vec{E} \bullet \hat{n} \Delta A$ for each piece. Because the closed spherical surface is centered on the point charge, what can you say about the angle θ in $\vec{E} \bullet \hat{n} = E \cos\theta$, at all locations on the surface? (See Figure 21.18.)

Since both the electric field $\vec{E}$ and the outward-going normal $\hat{n}$ are perpendicular to our chosen spherical surface at all locations on the surface, the angle between the two vectors is zero, and $\cos\theta = +1$.

QUESTION We need to add up $E \cos\theta \Delta A$ over the entire surface. What can you say about the magnitude of E at all locations on the surface, if the radius of the surface is r?

Because we chose our mathematical surface to be centered on the charge Q, the magnitude of the electric field at all locations on the sphere is simply

$$E = \frac{1}{4\pi\varepsilon_0} \frac{Q}{r^2}$$

QUESTION The total surface area of a sphere is $4\pi r^2$. If you add up all the pieces $E \cos\theta \Delta A$ over the entire surface, what do you get, in terms of the charge Q at the center of the sphere?

$E\cos\theta\,\Delta A$ is simply equal to $E\Delta A$ (since $\cos\theta = +1$), and E is constant, so the net flux over the entire surface is given by

$$\sum_{\text{closed surface}} \vec{E} \bullet \hat{n}\,\Delta A = \left(\frac{1}{4\pi\varepsilon_0}\frac{Q}{r^2}\right)(+1)(4\pi r^2) = \frac{Q}{\varepsilon_0}$$

We expect the total sum of electric flux on our chosen closed surface to be proportional to the charge inside the closed surface:

$$\sum_{\text{closed surface}} \vec{E} \bullet \hat{n}\,\Delta A = kq_{\text{inside}}$$

QUESTION What is the proportionality factor k?

Evidently for the spherical surface we get $k = 1/\varepsilon_0$, since the charge inside is simply equal to Q. This looks promising. We've got a quantitative measure of the electric field pattern on a closed surface (the total electric flux), and at least in this one case of a spherical surface centered on a point charge the total electric flux is equal to $(1/\varepsilon_0)$ times the charge inside the surface.

QUESTION What if the charge at the center of our chosen sphere were negative $(-Q)$? Would the relationship still hold? Check that it does, by considering what $\cos\theta$ is when the inside charge is negative.

In this case the flux would be negative (due to the cosine factor being -1), and proportional to the negative charge inside $(-Q)$. This is additional evidence that the $\cos\theta$ term is an appropriate part of the definition of flux.

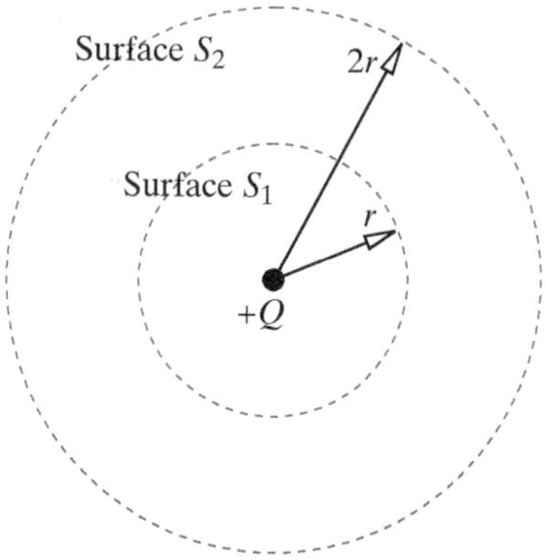

Figure 21.19 Consider the electric flux on two different spherical surfaces, chosen by us (and shown in cross section).

The Size of the Surface Doesn't Matter

Note that since r does not appear in the final result, apparently this relationship doesn't depend on the radius r we chose for the spherical surface. Why is this? Suppose that we consider another spherical surface with twice the radius. We draw cross sections of the two spherical surfaces S_1 and S_2 (Figure 21.19).

QUESTION How does E_2 on surface S_2 differ from E_1 on surface S_1? How does the total surface area differ (the surface area of a sphere is $4\pi r^2$)? How does the total electric flux differ? Is $\sum_{\text{closed surface}} \vec{E} \bullet \hat{n}\,\Delta A = q_{\text{inside}}/\varepsilon_0$ valid for both surfaces?

The magnitude of the electric field drops by a factor of 4, since it is proportional to $1/r^2$. The surface area increases by the same factor of 4, since the surface area is proportional to r^2. Therefore the electric flux on the outer surface is exactly the same as on the inner surface, and the same proportionality factor $k = 1/\varepsilon_0$ holds. This example shows why we needed to include the area in the definition of electric flux. Note that this result depends critically on the fact that the electric field of a point charge is proportional to $1/r^2$. If it were proportional to $1/r^{2.001}$, say, the electric flux would be smaller on the outer surface than on the inner surface, and the concept of electric flux wouldn't be very useful.

We have shown that

$$\sum_{\text{closed surface}} \vec{E} \bullet \hat{n}\,\Delta A = \frac{q_{\text{inside}}}{\varepsilon_0}$$

is true no matter what the sign of q_{inside} is, or how large we draw the enclosing spherical surface.

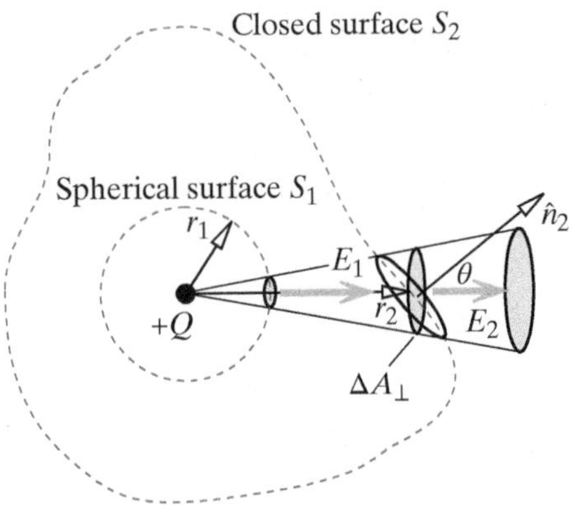

Figure 21.20 Compare the flux on an arbitrary closed surface with the flux on a small spherical surface.

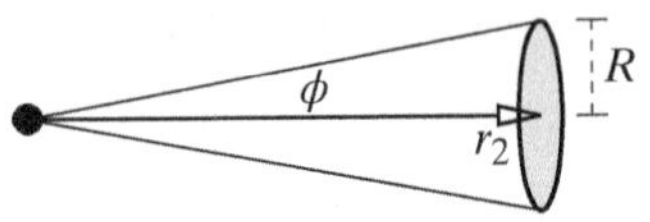

Figure 21.21 The perpendicular area $\Delta A_\perp = \pi R^2$, which is proportional to r_2^2.

The Shape of the Surface Doesn't Matter

Next consider an arbitrary closed surface surrounding the point charge, with a small spherical surface of radius r_1 for comparison (Figure 21.20). Remember that only cross sections of the surfaces are shown; the two dashed curves really represent closed surfaces that enclose the charge Q. We draw a cone that includes a small area of both surfaces.

The quantity $\Delta A_\perp = \Delta A \cos\theta$ in Figure 21.20 is the projection of the actual area ΔA onto a spherical surface drawn through that location, centered on the point charge (Figure 21.21). The radius R of this circular projected area is proportional to the distance r_2 from the charge Q; $R = r_2 \tan\phi$. Therefore the area $\Delta A_\perp$ is proportional to r_2^2.

QUESTION Explain why the electric flux on the portion of the outer surface enclosed by the cone in Figure 21.20 is equal to the electric flux on the corresponding portion of the inner spherical surface. That is, why is $E_2 \Delta A_{2\perp}$ equal to $E_1 \Delta A_{1\perp}$?

We have $E_2/E_1 = r_1^2/r_2^2$, since the magnitude of the electric field is proportional to $1/r^2$. We also have $\Delta A_{2\perp}/\Delta A_{1\perp} = r_2^2/r_1^2$, since the circular projected area is proportional to r^2. Therefore $(E_2 \Delta A_{2\perp})/(E_1 \Delta A_{1\perp})$ is equal to 1.

QUESTION If in every direction we find that the electric flux on any portion of the outer surface is the same as the electric flux on the projection of that area onto the small sphere, how does the total flux on the outer surface compare to the total flux on the inner sphere?

Evidently the total electric flux on the arbitrarily shaped outer surface is exactly equal to the total electric flux on the inner sphere. This is really encouraging!

We've shown that

$$\sum_{\text{closed surface}} \vec{E} \bullet \hat{n} \Delta A = \frac{q_{\text{inside}}}{\varepsilon_0}$$

is valid not only for an enclosing spherical surface but also that it holds true for any enclosing surface whatsoever. Again, the key reason why this works is the $1/r^2$ property of Coulomb's law: the electric field on a larger enclosing surface is exactly enough smaller to make the product of field and area come out to be exactly the same. If electric field didn't fall off exactly like $1/r^2$, electric flux wouldn't be a useful concept, because the total flux on differently shaped enclosing surfaces would differ.

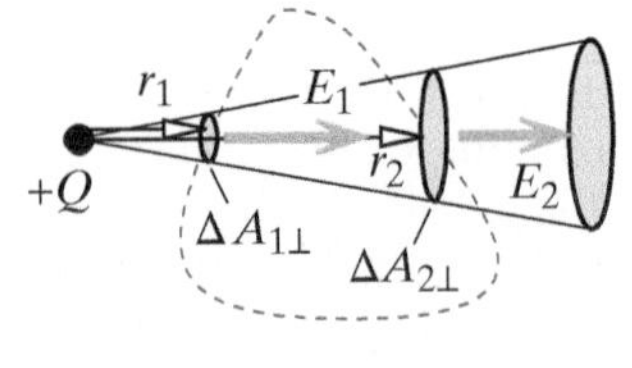

Figure 21.22 A point charge outside the surface (shown dashed) contributes zero net flux.

Charges Outside the Surface Contribute Zero Net Flux

For the equation to be valid in all cases, it should be true that the total flux is zero if there is no charge inside the closed surface, even if there are other charges *outside* the closed surface. Consider the situation in Figure 21.22, where a point charge $+Q$ is outside a closed surface (shown in cross section).

QUESTION Draw a cone from the charge, cutting through the near and far sides of the closed surface. Again note that $\Delta A_\perp = \Delta A \cos\theta$ is the projection of the actual area ΔA at the two locations where the cone pierces the surface. Show that the magnitude of the electric flux on the inner surface and on the outer surface is the same but that they differ in sign.

We can use the same argument we made before about the flux on two surfaces enclosed within a cone. However, in this case the flux on the first surface is negative (electric field pointing into the volume), whereas on the second surface the flux is positive (electric field pointing out of the volume). Therefore the fluxes add up to zero inside the cone.

> QUESTION Since this is true for any pair of inner and outer sections of the closed surface, what is the total electric flux over the entire surface due to the point charge outside the closed surface?

This cancellation holds true for any cone we draw, so the net flux added up over the entire surface is zero.

Point charges outside a closed surface certainly do produce electric field and electric flux on the surface, but when we add up all the contributions we find the total electric flux due to these outside charges is zero:

$$\sum_{\text{closed surface}} \vec{E} \bullet \hat{n} \Delta A = \frac{q_{\text{inside}}}{\varepsilon_0}$$

Superposition and Gauss's Law

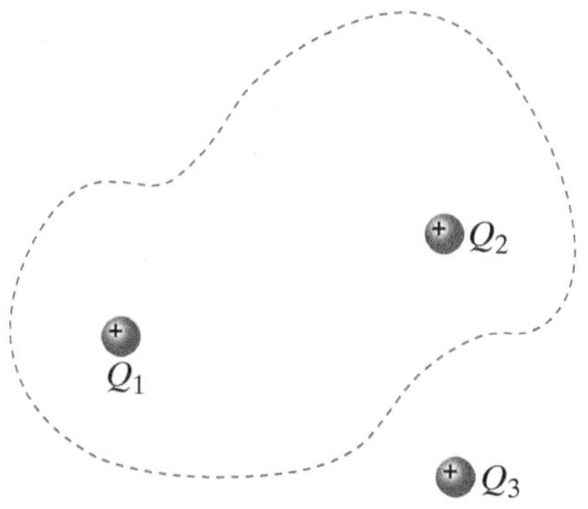

Figure 21.23 The charge outside the surface contributes zero to the net electric flux.

Consider three point charges, two inside and one outside of a mathematical closed surface (Figure 21.23). For each individual charge we have

$$\sum_{\text{closed surface}} \vec{E}_1 \bullet \hat{n} \Delta A = Q_1/\varepsilon_0$$

$$\sum_{\text{closed surface}} \vec{E}_2 \bullet \hat{n} \Delta A = Q_2/\varepsilon_0$$

$$\sum_{\text{closed surface}} \vec{E}_3 \bullet \hat{n} \Delta A = 0 \quad (\text{since } Q_3 \text{ is outside the surface})$$

If we add these up, we have

$$\sum_{\text{closed surface}} (\vec{E}_1 + \vec{E}_2 + \vec{E}_3) \bullet \hat{n} \Delta A = \frac{Q_1 + Q_2}{\varepsilon_0}$$

However, $\vec{E}_1 + \vec{E}_2 + \vec{E}_3$ is the net electric field due to all the charges, both inside and outside the surface, and it is this net field that we would measure on the surface. Generalizing to any number of point charges inside and outside a closed surface, we have proved what is called Gauss's law:

GAUSS'S LAW

$$\sum_{\text{closed surface}} \vec{E} \bullet \hat{n} \Delta A = \frac{\sum q_{\text{inside}}}{\varepsilon_0}$$

or

$$\oint \vec{E} \bullet \hat{n} \Delta A = \frac{\sum q_{\text{inside}}}{\varepsilon_0}$$

This is Gauss's law, the quantitative form of what we saw qualitatively: the pattern of electric field on a closed surface tells us something about the charges that are inside that closed surface. Now we see that by expressing the pattern of electric field on a surface quantitatively in terms of electric flux, we know exactly how much charge is inside our chosen surface, even though the electric field we measure may be partly due to charges outside the surface.

Gauss's Law and Coulomb's Law

Since we derived Gauss's law from Coulomb's law, it might seem that we should refer to "Gauss's theorem." However, it is possible in a similar way to derive

Coulomb's law from Gauss's law, so the two laws seem completely equivalent, despite their very different appearance.

However, Gauss's law and Coulomb's law are not equivalent if the charges are moving, and the two laws differ drastically in their predictions if the charges are moving at speeds that are comparable to the speed of light. The issue is retardation. The electric field at a location in space due to a fast-moving charge moving in a general way cannot be calculated from Coulomb's law but involves a complicated calculation that takes into account the past history of the charge's motion: the electric field at the observation location *now* depends on the location of the source charge *then*.

Unlike Coulomb's law, it turns out that Gauss's law is correct in all cases. Suppose that a large number of observers with synchronized watches are deployed all over a closed surface. By prior agreement, at 9:35 AM they all measure the electric flux on their own small piece of the surface and then report their measurements.

The sum of all these electric flux measurements is equal to the sum of all the charges (divided by ε_0) that were inside the surface at 9:35 AM, even if some of the charges were moving at speeds comparable to the speed of light. Thus Gauss's law is consistent with the special theory of relativity, but Coulomb's law isn't. (There is a practical limitation to this measurement: since the observers can't transmit their reports faster to us than the speed of light, by the time we add up all the flux contributions and thereby determine how much charge was inside the surface at 9:35 AM, the time is already quite a bit later!)

EXAMPLE **A Cylindrical Surface**

The electric field is measured over a cylindrical Gaussian surface of radius 5 cm and height 15 cm, as shown in Figure 21.24. Everywhere on the curved part of the surface, the electric field $\vec{E}_1$ is found to be constant in magnitude (550 N/C), and to point radially outward. On the flat end caps, the field $\vec{E}_2$ varies in magnitude, but also points radially outward. **(a)** Find the amount of charge enclosed by the surface. **(b)** If the source of the field is a long uniformly charged thin rod, part of which is inside the surface, what is the linear charge density (charge per unit length, Q/L) on the long rod?

Solution **(a)** On the curved portion of the surface, $\vec{E}_1 \| \hat{n}$ everywhere, so $\vec{E}_1 \cdot \hat{n} = \left|\vec{E}_1\right|$.

$$\Phi_{\text{side}} = \left|\vec{E}_1\right|(2\pi rh) = (550\,\text{N/C})(2\pi)(0.05\text{ m})(0.15\text{ m}) = 25.9\text{ V}\cdot\text{m}$$

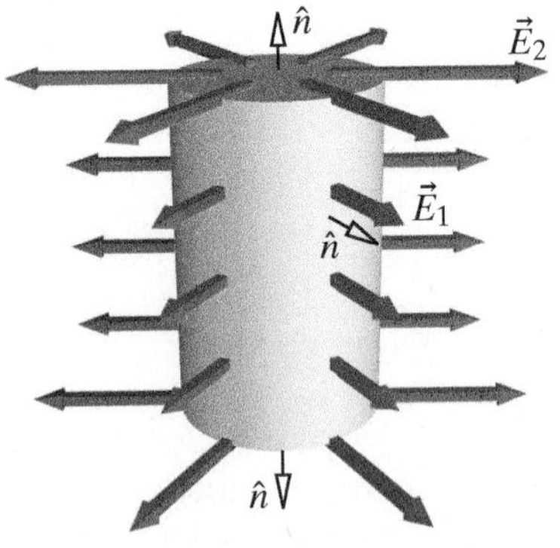

Figure 21.24 The electric field on a cylindrical Gaussian surface.

Everywhere on the flat top of the surface, $\vec{E}_2 \perp \hat{n}$, so $\vec{E}_2 \cdot \hat{n} = 0$ and $\Phi_{\text{top}} = 0$. Everywhere on the flat bottom of the surface, $\vec{E}_2 \perp \hat{n}$, so $\vec{E}_2 \cdot \hat{n} = 0$ and $\Phi_{\text{bottom}} = 0$.

$$\oint \vec{E}\cdot\hat{n}\,dA = (25.9 + 0 + 0)\text{ V}\cdot\text{m} = 25.9\text{ V}\cdot\text{m}$$

Applying Gauss's law:

$$\oint \vec{E}\cdot\hat{n}\,dA = \frac{q_{\text{inside}}}{\varepsilon_0}$$

$$q_{\text{inside}} = (25.9\text{ V}\cdot\text{m})\left(8.85\times10^{-12}\frac{\text{C}^2}{\text{N}\cdot\text{m}^2}\right)$$

$$= 2.29\times10^{-10}\text{ C}$$

(b) If the Gaussian cylinder encloses a section of a long uniformly charged rod, the charge density on the rod is

$$\frac{Q}{L} = \frac{2.29\times10^{-10}\text{ C}}{0.15\text{ m}} = 1.53\times10^{-9}\frac{\text{C}}{\text{m}}$$

21.4 REASONING FROM GAUSS'S LAW

We've used the total electric flux on a closed surface to find out how much charge is inside that closed surface. This means that we are using knowledge of the electric field to find out something about charges. We can in some cases use Gauss's law in the other direction, to derive the magnitude of the electric field due to a charge distribution. We'll study several such examples. To apply Gauss's law you

- choose a (mathematical) closed surface that passes through locations where you know $\vec{E}$ or want to find $\vec{E}$, and
- that encloses a region in which you know what charges there are or want to find out what charges there are.

The Electric Field of a Large Plate

Suppose we don't know or don't remember the equation for the electric field near the center of a large uniformly charged plate. All we know is that near the center of the plate the electric field must be perpendicular to the plate, as shown in Figure 21.25.

EXAMPLE **The Electric Field of a Large Charged Plate**

Use Gauss's law to determine the equation for the electric field near the center of a very large uniformly charged plate (Figure 21.25). The area of one surface of the plate is A, and the total amount of charge on the plate is Q.

Solution We choose a box-shaped Gaussian surface, centered on the plate, as shown in Figure 21.25. Let the area of the end face of the Gaussian surface be A_{box}. Assume that we don't know the magnitude of either E_{right} or E_{left}, but by symmetry we can assume that they have the same magnitude $E_{\text{right}} = E_{\text{left}} = E$.

First, we find the total flux in terms of the unknown field E. If we draw the unit normal vector $\hat{n}$ on each face of the box, we see that there is nonzero electric flux only on the ends of the box, so the total electric flux over the entire box is the flux on the two ends, $\Phi_{\text{net}} = \oint \vec{E}\cdot\hat{n}\,dA = 2E(\cos 0)A_{\text{box}}$.

Second, we find an expression for the amount of charge inside the box. The fraction of the total area of the plate that is inside the box is (A_{box}/A), so the amount of charge inside the box is $Q(A_{\text{box}}/A)$.

Now we can apply Gauss's Law:

$$2EA_{\text{box}} = \frac{(Q/A)A_{\text{box}}}{\varepsilon_0}$$

$$E = \frac{(Q/A)}{2\varepsilon_0}$$

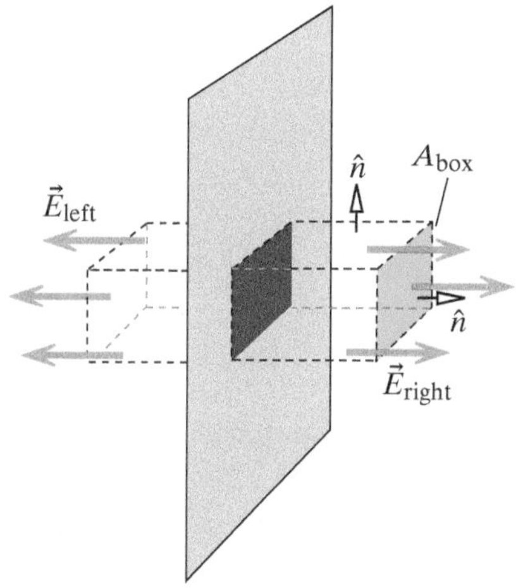

Figure 21.25 A portion of a large uniformly charged plate, with a Gaussian surface in the form of a box whose end area (shaded green) is A_{box}. The portion of the plate inside the box is shown darker than the rest of the plate, for clarity.

Contrast this simple calculation of the electric field near a large plate with the lengthy calculation for a disk that you did in Chapter 15!

Here you see the power of Gauss's law. In situations where we have some partial knowledge about the electric field (here it was the direction and symmetry of the field), it may be possible to determine the electric field due to a complex distribution of charges without carrying out a complicated integral.

Note that the area A_{box} of the end face of our chosen mathematical Gaussian surface drops out of the calculation. This must happen, because the physical result for the electric field cannot depend on the size of some mathematical surface that we happen to choose for our own convenience.

The Electric Field of a Uniform Spherical Shell

As a second example of using partial knowledge about the electric field pattern to deduce the magnitude of the electric field, consider a uniformly charged spherical shell. Recall from Chapter 15 that integrating Coulomb's law to get the electric field of this charge distribution is quite difficult.

EXAMPLE **The Electric Field of a Uniformly Charged Spherical Shell**

Use Gauss's law to find equations for the electric field due to a uniformly charged hollow sphere of charge Q, at locations both outside and inside the sphere.

Solution To find the field outside the shell, we draw a spherical Gaussian surface of radius r outside the sphere, centered on the sphere (Figure 21.26). The magnitude of the field on this surface is unknown, but because of the symmetry of the sphere and the Gaussian surface we reason that the field must be radial and uniform in magnitude. If the electric field weren't radial, by symmetry the pattern would have to look something like the one shown in Figure 21.27, but in that case the round-trip integral of electric field, $V_A - V_A$, wouldn't be zero; we can therefore rule out this field pattern.

Everywhere on the spherical Gaussian surface $\vec{E} \| \hat{n}$, so

$$\oint \vec{E} \cdot \hat{n}\, dA = EA = E(4\pi r^2)$$

All of the charge Q on the sphere is inside the surface. By Gauss's law,

$$\oint \vec{E} \cdot \hat{n}\, dA = \frac{q_{\text{inside}}}{\varepsilon_0}$$

$$E(4\pi r^2) = \frac{Q}{\varepsilon_0}$$

$$E = \frac{1}{4\pi\varepsilon_0}\frac{Q}{r^2}$$

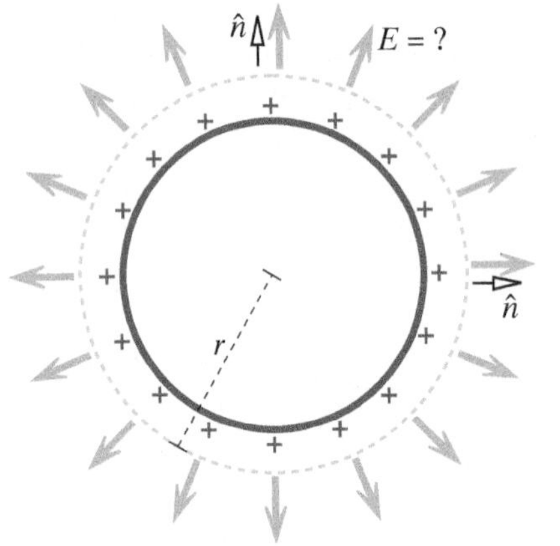

Figure 21.26 A spherical Gaussian surface (green) is drawn outside a uniformly charged spherical shell.

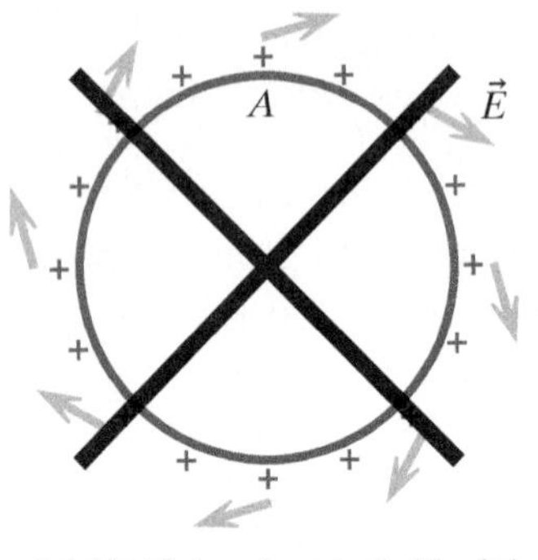

Figure 21.27 If the electric field of the charged sphere were not radial, it would have to look like this, which is impossible.

Since the radius r could have any value, this result is valid for any location outside the sphere.

To find the field inside the shell, we draw a spherical Gaussian surface of radius r inside the charged sphere, as shown in Figure 21.28. Again, we conclude that $\vec{E}$ must be radial, because of the symmetry of the situation. The electric flux on the inner surface is now $-E(\cos 0)(4\pi r^2)$ (because the $\vec{E}$ we have drawn is opposite to $\hat{n}$), but the charge inside the inner surface is zero. By Gauss's law,

$$\oint \vec{E} \cdot \hat{n}\, dA = \frac{q_{\text{inside}}}{\varepsilon_0}$$

$$E(4\pi r^2) = \frac{0}{\varepsilon_0}$$

$$E = 0$$

We see that E must be zero at any location inside the sphere.

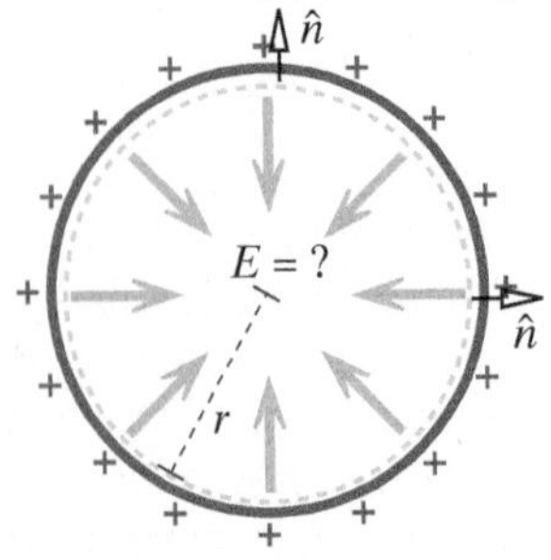

Figure 21.28 A spherical Gaussian surface (green) is drawn inside a uniformly charged spherical shell.

Note that we placed the Gaussian surface at the locations at which we wanted to determine the magnitude of the electric field—this is a key part of this kind of reasoning. In this example, Gauss's law allowed us to determine the magnitude of the electric field inside and outside of a uniformly charged spherical shell with very simple calculations. In contrast, determining the electric field of a uniformly charged spherical shell by integrating Coulomb's law over the charge distribution involves rather difficult trigonometry and integral calculus (see Chapter 15). The calculation using potential in Chapter 16 was also much more difficult than using Gauss's law.

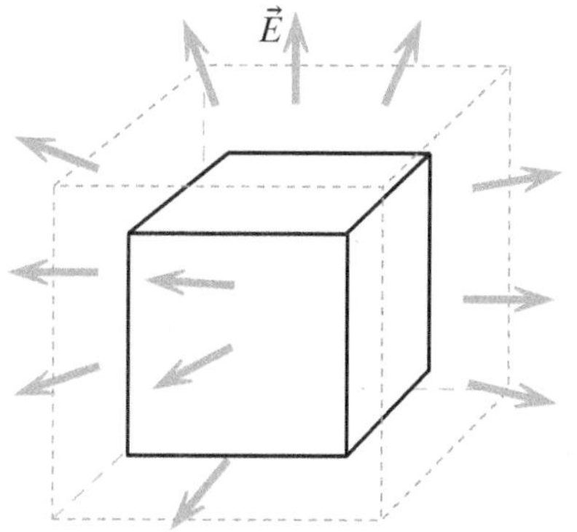

Figure 21.29 Trying to use Gauss's law to determine the electric field of a charged cube.

The Electric Field of a Uniform Cube

Next let's try to use Gauss's law to determine the electric field near a cube that is uniformly charged positive. This cube must be made of an insulator, because on a metal cube the charge would not be distributed uniformly. It seems plausible to choose a cubical box as the Gaussian surface surrounding the charged cube (Figure 21.29).

However, the electric field on a face of the cubical Gaussian surface is not uniform in direction or magnitude, as it was in the case of the large plate or the sphere.

QUESTION Is Gauss's law valid for this cubical Gaussian surface?

Gauss's law is always valid, no matter what the situation. It is definitely the case that the total electric flux over the entire Gaussian surface is numerically equal to the charge on the cube, divided by ε_0.

QUESTION Explain why we can't use Gauss's law to determine the electric field near a cube.

Although Gauss's law is valid, it doesn't help. The magnitude of the electric field varies all over the surface, and so does the angle between the electric field and the outward-going normal. There's no symmetry to simplify the integral, and neither E nor $\cos\theta$ can be factored out of the flux calculation. Therefore we can't carry out the flux calculation in terms of a single unknown quantity E.

Proving Some Important Properties of Metals

We have repeatedly stated that there cannot be any excess charge in the interior of a metal: any excess charge must be on a surface. Using Gauss's law, we can now prove this for the case of a metal in equilibrium. We will use "proof by contradiction," by assuming that there is some excess charge inside the metal and using Gauss's law to demonstrate a contradiction.

Consider a piece of metal that has been polarized by external charges and is in equilibrium. We assume (incorrectly) that there is some excess charge in the interior, as shown, and we draw a Gaussian surface (closed mathematical surface) that encloses that excess charge (Figure 21.30). The Gaussian surface is drawn in such a way that all elements of the surface are inside the metal. (Remember that this surface is just a mathematical construct, so it's fine to put it inside a solid object.)

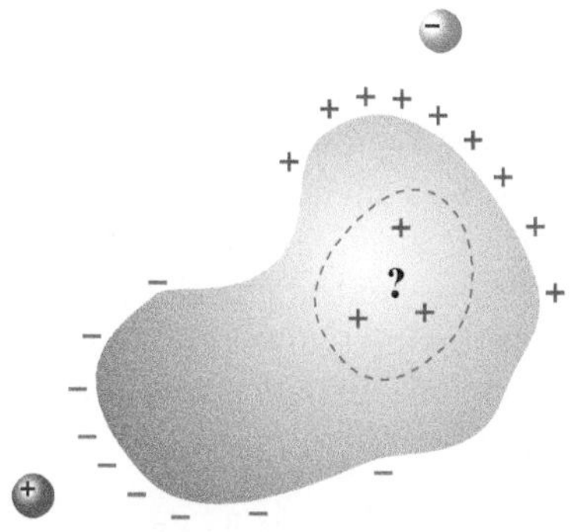

Figure 21.30 Can there be excess charge in the interior of a metal?

QUESTION Since the metal is in equilibrium, what is the magnitude of the electric field everywhere on the Gaussian surface? What is the total electric flux on that surface? Using Gauss's law, what can you conclude about the charge inside the Gaussian surface?

Everywhere inside a metal in equilibrium, the electric field is zero, because if it were nonzero, electrons would flow and increase the polarization of the metal, until eventually in equilibrium the net field in the interior would become zero. Since the field is zero on our mathematical surface, which is inside the metal, the total electric flux on that surface must be zero. Gauss's law tells us that the charge inside the surface must be zero.

Thus we reach a contradiction, and our initial assumption must be wrong: in equilibrium there cannot be any excess charge in the interior of a metal—any excess charge therefore must be on the surface. Note that in this analysis we knew something about the electric field ($E = 0$), and from that we could deduce something about the charges.

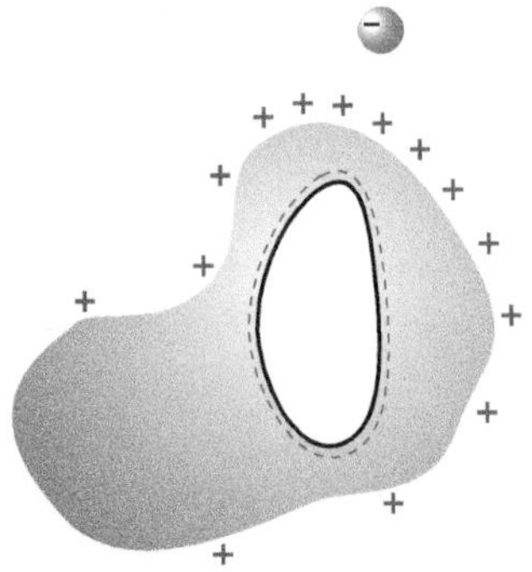

Figure 21.31 A hole inside a charged, polarized piece of metal.

QUESTION Suppose that there is a hole inside the metal as shown in Figure 21.31. Can there be charge on the inner surface? Using a Gaussian surface drawn in the metal but close to the boundary of the hole (Figure 21.31), what can you conclude about the net charge on the surface of the hole?

Again, all over a Gaussian surface inside a metal in equilibrium the electric field is zero, so the total electric flux is zero, and charge enclosed inside the Gaussian surface must be zero.

Is the electric field inside the hole zero or nonzero? It might conceivably be nonzero if there were + and − charges on this interior surface, as long as they add up to zero. However, we can show that E must be zero inside the hole. Consider the potential difference calculated along two different paths: A to C to B, and A to D to B (Figure 21.32).

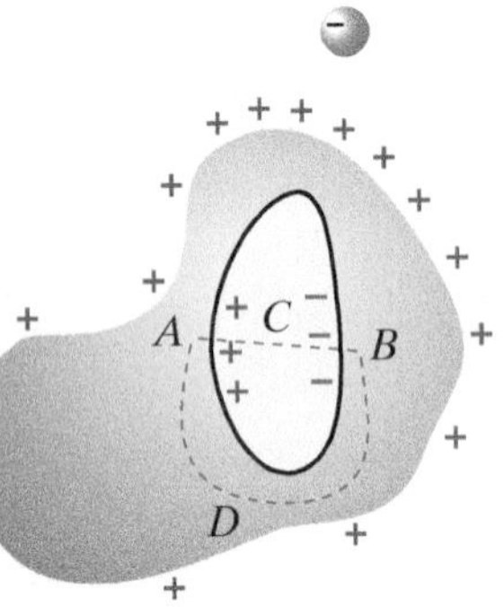

Figure 21.32 Could there be a nonzero electric field inside the hole, as long as the net charge on the surface of the hole is equal to zero?

QUESTION Along the path A to D to B, through the metal, what is the potential difference ΔV? Since potential difference is independent of path, what must be the potential difference along the path A to C to B?

All locations along the path A to D to B pass through the metal, where the electric field is zero, so the potential difference (path integral of the electric field) must be zero. There would be a potential difference along the path A to C to B if there were an electric field inside the hole made by charges on the surface of the hole, but the potential difference $V_B - V_A$ must be independent of path. Therefore we have reached a contradiction, and our assumption that there are charges on the surface of the hole must be false.

We have proved that in equilibrium

- there are no charges at all on the surface of an empty hole in a metal;
- the electric field is zero everywhere inside an empty hole in a metal.

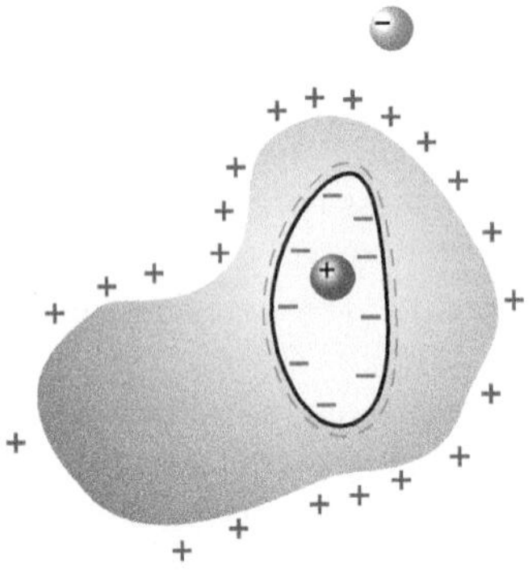

Figure 21.33 Now the hole is not empty but contains a +5 nC charge.

Having gone through formal reasoning to show that an empty hole in a metal in equilibrium contains no charges or electric field, it may be helpful to think about this important result in a less formal way. If the metal were solid, the electric field in equilibrium would be zero throughout, and there would be no excess charge anywhere in the interior, as we concluded earlier. Imagine somehow removing lots of neutral, unpolarized atoms from the interior, as though they could be eaten away with acid. These neutral, unpolarized atoms contributed nothing to the electric field, so removing them will not affect the distribution of charge or field. Hence we conclude that the hole that we've made inside the metal will contain neither field nor charge.

QUESTION On the other hand, suppose that there is a charge of +5 nC floating inside the hole (Figure 21.33). Using Gauss's law with the indicated Gaussian surface, what can you conclude about the charge on the inner surface of the hole?

Since the flux on the chosen surface must be zero (because the electric field is zero all over this surface in the metal), the net charge inside the surface must be zero. Therefore there must be −5 nC of charge distributed on the inner surface of the hole. Thus in this case, where we have deliberately placed extra charge in the hole, there are charges and fields inside the hole.

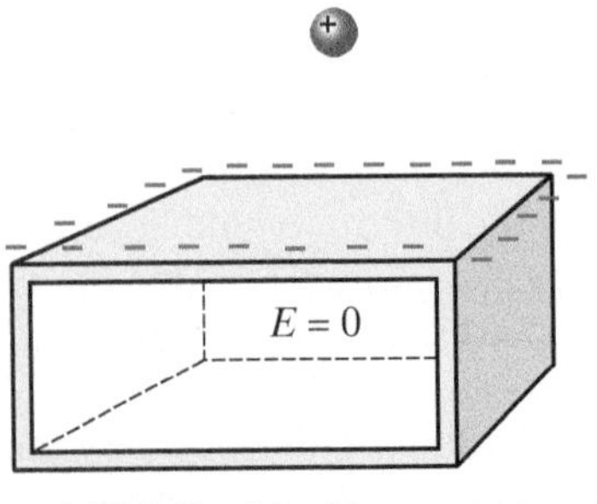

Figure 21.34 $E = 0$ inside an empty metal box, even though the box has a net charge and is polarized by an external charge.

"Screening"

We used Gauss's law (and potential theory) to show that in equilibrium the electric field is zero not only in the interior of a metal but even in an empty hole inside the metal. This has important practical consequences. A metal box "screens" its contents from the effects of charges outside the box (Figure 21.34).

For many purposes the box doesn't even have to have solid metal sides. A box with some holes in the sides, or a box with sides made of wire mesh or screen, can be quite effective in reducing the field inside the box.

QUESTION How does this "screening" really work? Are the electric fields produced by the outside charges blocked from getting inside the box? What about the principle that electric fields pass right through matter? Can you explain the seeming paradox?

Electric field cannot be blocked; according to the superposition principle, the field contributed by the outside charges is definitely present in the box. However, the electric field contributed by the surface charges on the polarized metal is equal and opposite to the field due to the outside charges, making a net field that is zero.

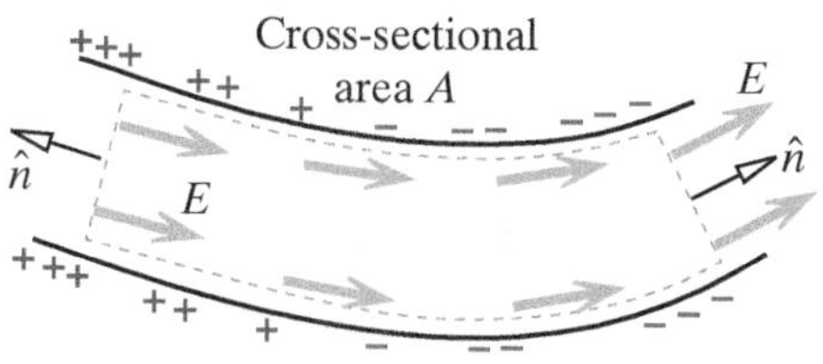

Figure 21.35 A Gaussian surface just inside a current-carrying wire in a circuit.

Gauss's Law Applied to Circuits

All of the examples so far have dealt with equilibrium. However, Gauss's law expresses a completely general relationship between charges and electric field, so it applies to electric circuits, too. Figure 21.35 shows a portion of a current-carrying wire of constant cross section A and uniform composition.

QUESTION In the steady state, excess charges arrange themselves on the surface of such a wire so that the electric field E has the same magnitude at different locations along the wire, and is everywhere parallel to the wire. In Figure 21.35 the dashed tube-shaped Gaussian surface nearly fills the wire. Use Gauss's law to prove that there is no excess charge in the interior of the wire.

Along the sides of the tube the electric field is parallel to the surface and contributes no flux. The right end of the tube contributes an amount of flux EA, but the left end contributes $-EA$, so the total electric flux on the Gaussian surface is zero, which means that there must not be any net charge inside the Gaussian surface; all excess charge is on the outer surface of the wire.

EXAMPLE **Charge at an Interface**

Figure 21.36 shows a portion of a circuit in which a steady-state electron current i is running as shown. A wire made of metal 1 joins a wire made of metal 2; both wires have uniform circular cross section of radius r. The electron mobility u_2 in metal 2 is much smaller than the electron mobility u_1 in metal 1. There are n_1 mobile electrons per unit volume in metal 1 and n_2 in metal 2; $n_1 > n_2$. Determine the amount of charge on the interface between metal 1 and metal 2, including sign. Explain carefully. Explicitly state all assumptions, and show how all quantities were evaluated.

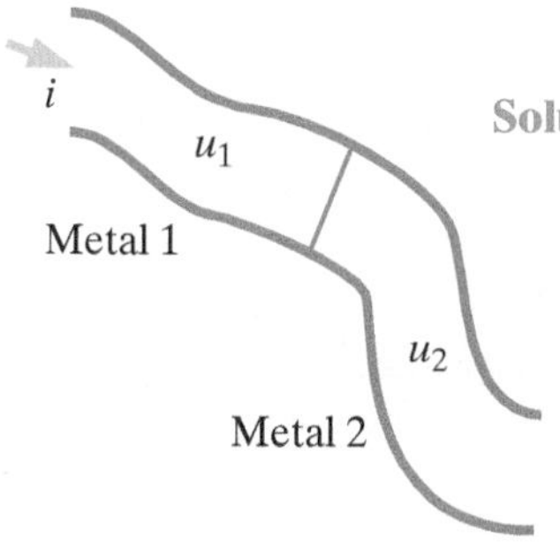

Figure 21.36 Current runs through a part of a circuit containing two different kinds of wire.

Solution Start from fundamental principles:
By the current node rule, $i_1 = i_2$, so $n_1 A u_1 E_1 = n_2 A u_2 E_2$ and therefore

$$E_2 = \frac{n_1 u_1}{n_2 u_2} E_1$$

This makes sense, because E_2 must be greater than E_1. In order to create a bigger electric field in metal 2, it is likely that there is some excess charge at the interface.

Since we don't want to include charge on the outer surface of the wire, we choose a Gaussian surface inside the wire. We consider a tube-shaped Gaussian surface just inside the surface of the wire, shown by the dashed line in

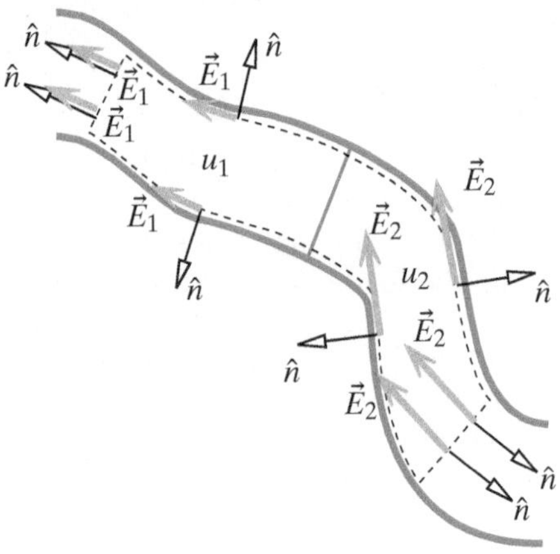

Figure 21.37 A Gaussian surface just inside the wire, with electric field and normal vector shown at several locations.

Figure 21.37. Any charge inside the Gaussian surface must be on the interface between the wires, since we just showed that in a uniform section of wire there could be no charge inside the wire.

By Gauss's law,

$$\oint \vec{E} \bullet d\vec{A} = \frac{\sum q_{\text{inside}}}{\varepsilon_0}$$

Assumptions: The electric field inside the wire is everywhere parallel to the wire, and is uniform in magnitude inside a wire of uniform composition. (We proved these assertions in Chapter 18.)

Given the direction of the electron current, the electric fields in the wire must be as shown in Figure 21.37. Evaluating the flux on different parts of the surface:

Outer surface of wire: flux $= 0$ because $\vec{E} \perp \hat{n}$
Top end (metal 1): flux $= E_1(\pi r^2)\cos 0^\circ = +E_1(\pi r^2)$
Bottom end (metal 2): flux $= E_2(\pi r^2)\cos 180^\circ = -E_2(\pi r^2)$

$$\oint \vec{E} \bullet d\vec{A} = 0 + E_1(\pi r^2) - E_2(\pi r^2)$$

Therefore the amount of charge on the interface is this, since $E = i/(nAu)$:

$$\sum q_{\text{inside}} = \varepsilon_0(E_1(\pi r^2) - E_2(\pi r^2))$$
$$\sum q_{\text{inside}} = \varepsilon_0 i\left(\frac{1}{n_1 u_1} - \frac{1}{n_2 u_2}\right)$$

This quantity is negative, which makes sense, because the charge on the interface should be negative in order to increase E_2 and decrease E_1.

21.5 GAUSS'S LAW FOR MAGNETISM

There is a dramatic difference between electric and magnetic dipoles. The individual positive and negative electric charges ("monopoles") making up an electric dipole can be separated from each other, and these point charges make outward-going or inward-going electric fields (which vary like $1/r^2$). One might expect a magnetic dipole such as a bar magnet to be made of positive and negative magnetic monopoles, but although scientists have searched carefully, no one has ever found an individual magnetic monopole.

Such a magnetic monopole would presumably make an outward-going or inward-going magnetic field (which would vary like $1/r^2$), but such a pattern of magnetic field has never been observed (Figure 21.38). The nonobservation of a monopole field pattern can be expressed as a "Gauss's law" for magnetism—that the net flux of the magnetic field $\vec{B}$ over a closed surface is equal to the (nonexistent) "magnetic charge" inside:

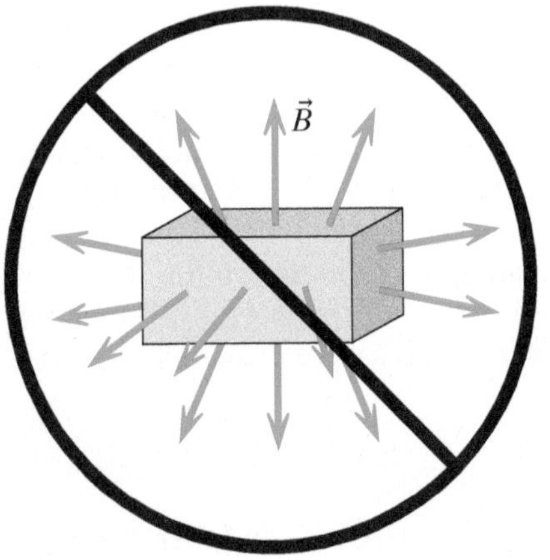

Figure 21.38 This pattern of magnetic field has never been observed.

GAUSS'S LAW FOR MAGNETISM

$$\oint \vec{B} \bullet \hat{n}\, dA = 0$$

As we saw in Chapter 17, a magnet can be thought of as consisting of atomic current loops (or electrons that act like current loops). If you cut a magnet in two, you don't get two magnetic monopoles—you just get two magnets!

21.6 PATTERNS OF MAGNETIC FIELD: AMPERE'S LAW

There is a relationship called Ampere's law that is an alternative version of the Biot–Savart law, just as Gauss's law is an alternative version of Coulomb's law. Ampere's law refers to a closed path rather than a closed surface.

We will study Ampere's law for the additional insights it provides about the relationship between current and magnetic field, and for certain important results that are most easily obtained using Ampere's law. Moreover, Ampere's law is relativistically correct, as is Gauss's law. Even more important, in Chapter 23 we will study Maxwell's extension of Ampere's law that makes a connection between time-varying electric and magnetic fields, which leads to understanding the nature of light.

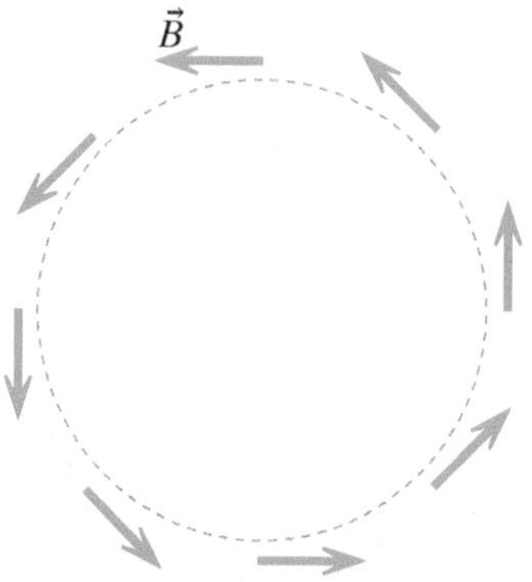

Figure 21.39 Is there current passing through this region? Does conventional current run into or out of the page?

Patterns of Magnetic Field

Moving charges produce magnetic fields, and we know how to calculate or estimate the magnitude and direction of the magnetic field due to a particular distribution of moving charges or currents. Sometimes, however, it is useful to reason in the other direction: from an observed pattern of magnetic field it may be possible to infer what currents are responsible for this pattern.

Suppose that you can't look inside the circular region shown in Figure 21.39, but you can measure the magnetic field all along the boundary of the region.

> QUESTION Can you conclude that there is some current inside the region? Does conventional current go into the page or out of the page?

A likely possibility is that a current-carrying wire is located at the center of the circle, with conventional current headed out of the page. This would produce the observed pattern of magnetic field.

Again, suppose that you can't look inside the circular region shown in Figure 21.40, but you can measure the magnetic field all along the boundary of the region.

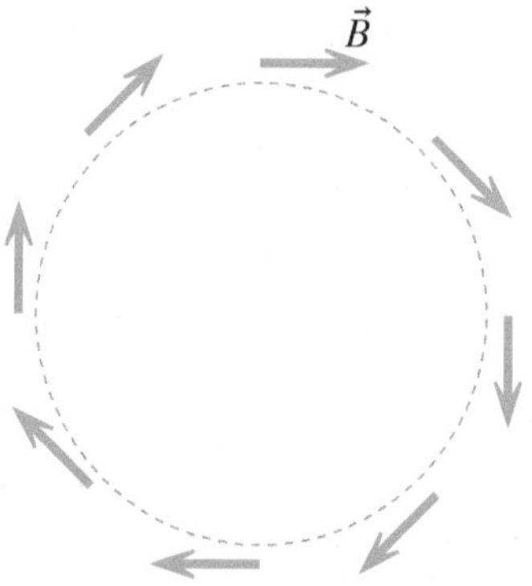

Figure 21.40 Is there current passing through this region? Does conventional current run into or out of the page?

> QUESTION Can you conclude that there is some current inside the region? Does conventional current go into the page or out of the page?

A wire carrying conventional current into the page at the center of the circle would produce the observed pattern of magnetic field.

Ampere's law is a quantitative relationship between measurements of magnetic field along a closed path and the amount and direction of the currents passing through that boundary. Note the close parallel with Gauss's law, which relates measurements of electric field on a closed surface to the amount of charge inside that closed surface.

Quantifying the Magnetic Field Pattern

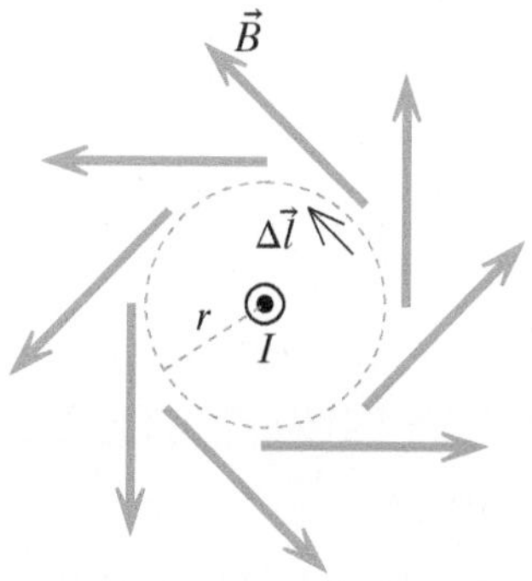

Figure 21.41 The magnetic field a distance r from a very long straight wire.

It is clear that there is some connection between the pattern of magnetic field along a closed path and the amount of current passing through the region bounded by that path, but how can we turn this into a quantitative relationship? We need to define a quantitative measure of the amount and direction of magnetic field along an entire closed path. Let's consider the magnetic field of a very long current-carrying wire a distance r from the wire (Figure 21.41), which we calculated using the Biot–Savart law in Chapter 17:

$$B_{\text{wire}} \approx \frac{\mu_0}{4\pi}\frac{2I}{r} \quad \text{(a distance } r \text{ from a very long wire)}$$

Given the curly character of the magnetic field around a wire, it is probably useful to integrate $\int \vec{B} \bullet d\vec{l}$ along a round-trip circular path around a wire and see what we get:

$$\oint \vec{B} \bullet d\vec{l} = \left(\frac{\mu_0}{4\pi}\frac{2I}{r}\right) \oint dl$$

because $\vec{B} \| d\vec{l}$ all along the path, and $r =$ constant,

$$\oint \vec{B} \bullet d\vec{l} = \left(\frac{\mu_0}{4\pi}\frac{2I}{r}\right)(2\pi r) = \mu_0 I$$

This is Ampere's law: the path integral of the magnetic field is equal to μ_0 times the current passing through the region enclosed by the path. (This is similar to Gauss's law, in which the surface integral of the electric field is equal to $1/\varepsilon_0$ times the charge inside the closed surface.)

In order to obtain a general proof of this quantitative relationship between current and magnetic field, we need to show that we can follow any path that surrounds the current, and that currents outside the path contribute zero to the path integral.

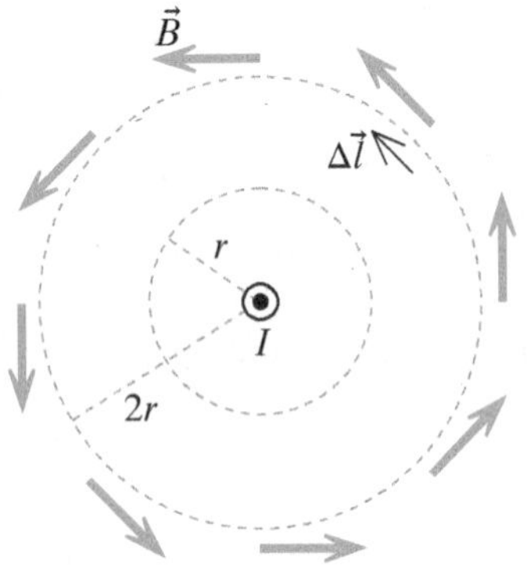

Figure 21.42 The magnetic field a distance $2r$ from a very long straight wire.

A Circular Path with Twice the Radius

It is easy to see that our result is independent of the radius r of the path we choose. For example, suppose that we choose a path along a larger circle of radius $2r$ (Figure 21.42).

QUESTION Why is the path integral the same along this longer path?

We know that the magnetic field near a very long wire is proportional to $1/r$, where r is the perpendicular distance to the wire. Therefore on a circle of twice the radius (and twice the circumference) the magnetic field is half as big, making the path integral be the same as before. The path integral of the magnetic field is independent of how big the circle is. This is similar to Gauss's law, where if you double the radius of a Gaussian sphere surrounding a charge, the surface area of the sphere goes up by a factor of four, but the electric field drops by a factor of four.

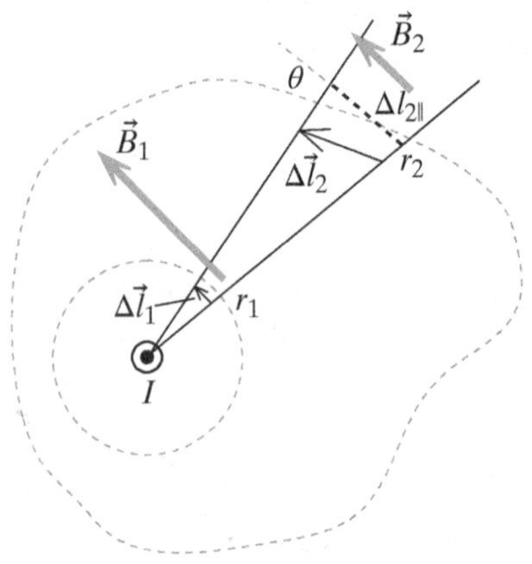

Figure 21.43 Comparing the path integral of magnetic field on a noncircular path to the path integral on a circular path.

A Noncircular Path

Suppose that we evaluate the path integral along a noncircular path. Focus on the contribution $\vec{B}_1 \bullet \Delta\vec{l}_1$ along a short arc of the circle and the contribution $\vec{B}_2 \bullet \Delta\vec{l}_2$ along the corresponding section of the noncircular path (Figure 21.43).

In Figure 21.43 the parallel projection $\Delta l_{2\|}$ of $\Delta\vec{l}_2$ along the direction of $\vec{B}_2$ is equal to $\Delta l_2 \cos\theta$, and by similar triangles it is also true that

$$\frac{\Delta l_{2\|}}{r_2} = \frac{\Delta l_1}{r_1}$$

Therefore

$$\vec{B}_2 \bullet \Delta\vec{l}_2 = B_2 \Delta l_2 \cos\theta = B_2 \Delta l_{2\|} = B_2(r_2/r_1)\Delta l_1$$

However, since the magnetic field near a long straight wire is inversely proportional to the distance from the wire, we have $B_2/B_1 = r_1/r_2$, so

$$\vec{B}_2 \bullet \Delta\vec{l}_2 = B_1 \Delta l_1 = \vec{B}_1 \bullet \Delta\vec{l}_1$$

Since this is true for any arc, it must be that along the noncircular path $\oint \vec{B} \bullet d\vec{l}$ has the same value it has along the circular path, which is $\mu_0 I$. We have almost enough to prove Ampere's law. All that remains is to show that any current-carrying wires outside the closed path don't contribute.

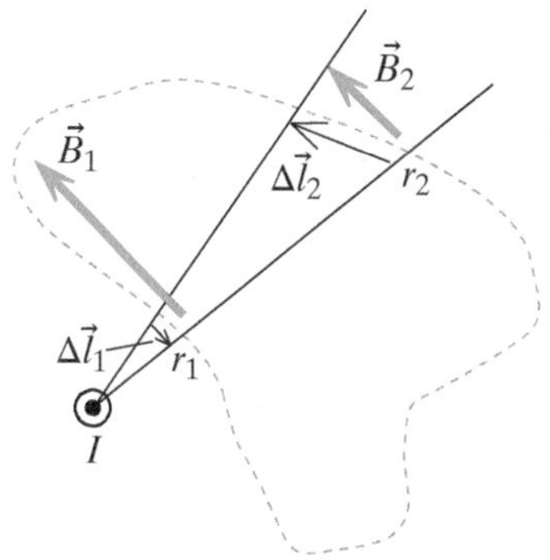

Figure 21.44 A current outside the path makes a zero net contribution to the path integral of the magnetic field.

Effect of Currents Outside the Path

Consider Figure 21.44, in which there is a current outside the path.

> QUESTION Briefly explain why $\vec{B}_2 \bullet \Delta\vec{l}_2 = -\vec{B}_1 \bullet \Delta\vec{l}_1$ as we go around counterclockwise, and why $\oint \vec{B} \bullet d\vec{l} = 0$ in this case.

If $\Delta\vec{l}_1$ and $\Delta\vec{l}_2$ are short enough that they approximate arc segments on the two circles, then the length $\Delta l_{2\perp} = \Delta l_2 \cos\theta$ goes up by r but B goes down by $1/r$, so the magnitude of the product $B\Delta l_\perp$ is the same at both locations. As we go around the path counterclockwise, $\vec{B}_1 \bullet \Delta\vec{l}_1$ is negative but $\vec{B}_2 \bullet \Delta\vec{l}_2$ is positive, so $\vec{B}_1 \bullet \Delta\vec{l}_1 + \vec{B}_2 \bullet \Delta\vec{l}_2 = 0$. Adding up all such pairs, we find that a current that is outside the chosen closed path adds zero to the path integral $\oint \vec{B} \bullet d\vec{l}$.

Let's consider a specific example of three current-carrying wires (Figure 21.45). We draw a mathematical boundary that encloses two of the wires, and we walk counterclockwise along this path. For each individual current we have

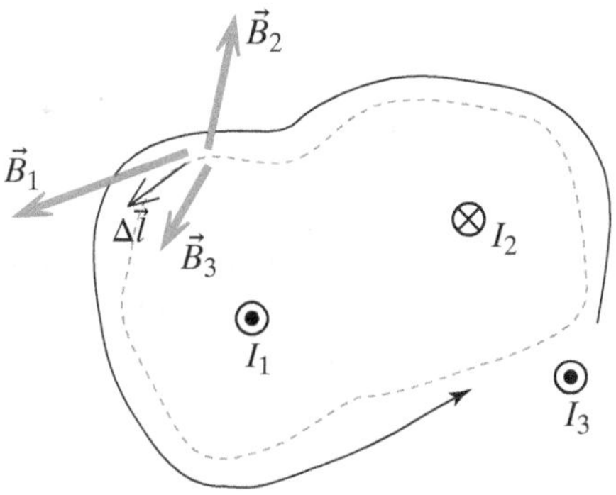

Figure 21.45 Add up all the $\vec{B} \bullet \Delta\vec{l}$ contributions along this counterclockwise path.

$$\oint \vec{B}_1 \bullet d\vec{l} = \mu_0 I_1 \quad \text{(since } I_1 \text{ heads out of the page)}$$

$$\oint \vec{B}_2 \bullet d\vec{l} = -\mu_0 I_2 \quad \text{(since } I_2 \text{ heads into the page)}$$

$$\oint \vec{B}_3 \bullet d\vec{l} = 0 \quad \text{(since } I_3 \text{ is outside the boundary)}$$

If we add these up, we have

$$\oint (\vec{B}_1 + \vec{B}_2 + \vec{B}_3) \bullet d\vec{l} = \mu_0 (I_1 - I_2)$$

However, $(\vec{B}_1 + \vec{B}_2 + \vec{B}_3)$ is the net magnetic field $\vec{B}$ due to *all* the currents, both inside and outside the boundary, and it is this net field that we would measure along the boundary. Note that we count a current positive or negative depending on whether it makes a magnetic field in the direction of our counterclockwise walk along the path, or opposite to our walk, and this is equivalent to specifying the sign of the current depending on whether the current comes out of the page or goes into the page.

Putting it all together, we see that the path integral of magnetic field is equal to μ_0 times the sum of the enclosed currents (plus or minus), and currents outside the path contribute zero to the sum. This relationship is called Ampere's law:

AMPERE'S LAW

$$\oint \vec{B} \bullet d\vec{l} = \mu_0 \sum I_{\text{inside path}}$$

All the currents in the universe contribute by superposition to the magnetic field, yet Ampere's law refers only to those currents inside the chosen path because the outside currents contribute zero to the path integral. There is a strong parallel with Gauss's law.

Despite its very different form, Ampere's law is essentially equivalent to the Biot–Savart law from which it was derived. However, in some cases it is much easier to use Ampere's law than the Biot–Savart law. Moreover, Ampere's law is relativistically correct, whereas the Biot–Savart law is not, since it doesn't include retardation. This is just like the relationship between Gauss's law and Coulomb's law.

Note that Ampere's law involves the component of $\vec{B}$ parallel to a path, whereas Gauss's law involves the component of $\vec{E}$ perpendicular to a surface.

Checkpoint 2 Along the path shown in Figure 21.46 the magnetic field is measured and is found to be uniform in magnitude and always tangent to the circular path. If the radius of the path is 3 cm and B along the path is 1.3×10^{-6} T, what are the magnitude and direction of the current enclosed by the path?

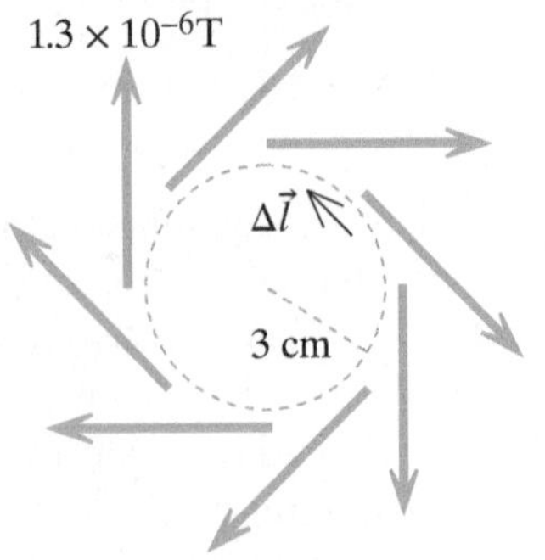

Figure 21.46 A magnetic field of 1.3×10^{-6} T is measured everywhere on a circular path of radius 3 cm.

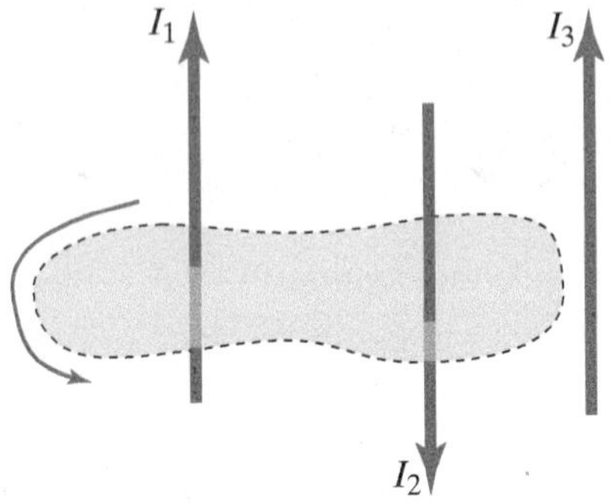

Figure 21.47 Curl the fingers of your right hand in the direction of the counterclockwise path. Piercings in the direction of your thumb (I_1) count positive; piercings in the opposite direction count negative (I_2). Don't count I_3, because it doesn't pierce the soap film.

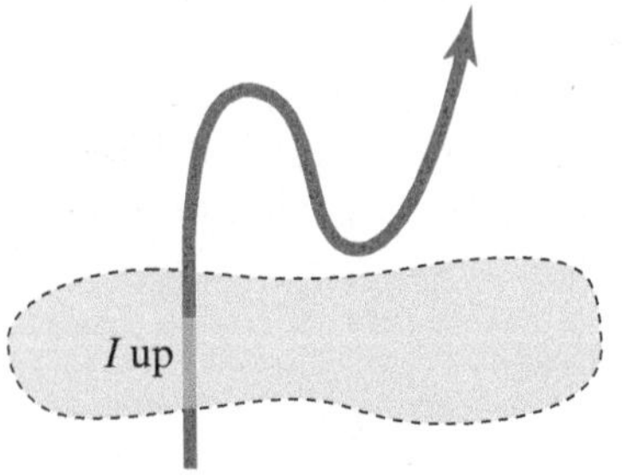

Figure 21.48 Just one piercing of the (flat) soap film.

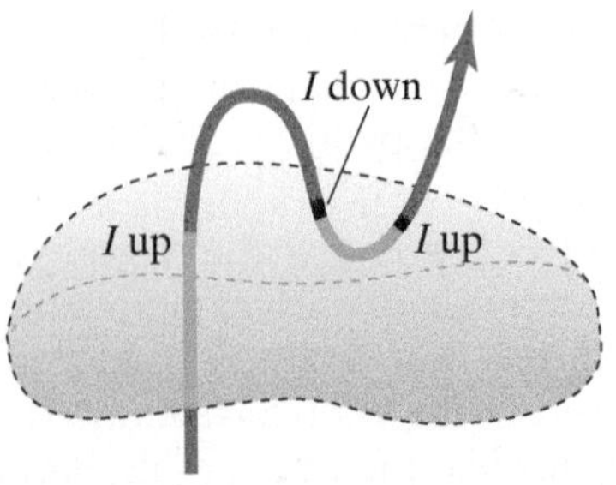

Figure 21.49 Here the dome-shaped soap film is pierced three times, two up and one down.

Which Currents Are "Inside" the Path?

Before we apply Ampere's law to some important situations, we should pause to be more precise about how to decide what currents are "inside" a closed boundary. In the simple case of a single wire it is easy to say, but in more complicated cases this can be a bit tricky. We offer a procedure that is easy to follow and that moreover leads in a later chapter to an important extension of Ampere's law, made by Maxwell.

Just as every application of Gauss's law must start with choosing a closed surface to which to apply the law, so every application of Ampere's law must start with choosing a closed path that is the boundary of the region of interest.

After choosing a closed path, imagine stretching a soap film over your chosen path. You calculate the net current through the boundary by counting the current-carrying wires that pierce your soap film. You walk counterclockwise around the path, adding up $\oint \vec{B} \bullet d\vec{l}$ as you go. You count as positive any currents that pierce the soap film coming up through the film. To be more precise, let the fingers of your right hand curl counterclockwise around the path, and count a current as positive if it goes in the direction of your thumb. Any currents that pierce the film going in the opposite direction are counted as negative.

In Figure 21.47 we have $\oint \vec{B} \bullet d\vec{l} = \mu_0 \sum I_{\text{inside path}} = \mu_0(I_1 - I_2)$, since I_1 pierces the imaginary soap film in the upward direction, I_2 in the downward direction, and I_3 doesn't pierce the film.

Though we will not carry out the proof (which is based on dividing up the soap film into little areas and applying Ampere's law to the boundary of each area, then adding up the results), it can be shown that Ampere's law is still valid even if the closed boundary path does not lie in a plane, or even if the soap film that is anchored on the boundary is deformed into a nonflat shape. Certainly in the case of the straight current-carrying wires shown above, deforming the soap film upward or downward would make no difference in counting the piercings.

For a more interesting case, consider a wire with an S-shaped bend in it, with a closed boundary drawn around the wire as shown in Figure 21.48. How much current is "inside" the boundary? If we count the current by counting the piercings of the soap film, we can show that we get the same (correct) answer for any and all deformations of the soap film, as long as the film stays anchored at its edges to the boundary path.

In Figure 21.48, there is just one piercing of the imaginary soap film, and $\sum I = +I$. In Figure 21.49, there are three piercings of the imaginary soap film, but when we add up all the currents we again get the result that $\sum I = (+I) + (-I) + (+I) = +I$. Therefore, deforming the imaginary soap film does not affect the accounting for the amount of current enclosed by the boundary of that film (our chosen closed path).

Applications of Ampere's Law

We'll apply Ampere's law to some situations in which there are lots of currents distributed over a region. Here is a summary of the key steps in applying Ampere's law:

1. Choose a (mathematical) closed path as a boundary.
2. Stretch an imaginary soap film over the boundary.

3. Walk around the boundary counterclockwise, integrating $\oint \vec{B} \bullet d\vec{l}$.
4. Add up the positive and negative currents that pierce the soap film; this is $\sum I_{\text{inside path}}$. Count as positive those currents that pierce the imaginary soap film coming out of the film (that is, in the direction of your right thumb with the fingers of your right hand curling around in the direction of your walk); currents that pierce the film going into the film are counted as negative.
5. Apply Ampere's law, equating the two sums:

$$\oint \vec{B} \bullet d\vec{l} = \mu_0 \sum I_{\text{inside path}}$$

We'll apply this procedure to find the magnetic field due to a long thick wire, a solenoid, and a toroid.

A Long Thick Wire

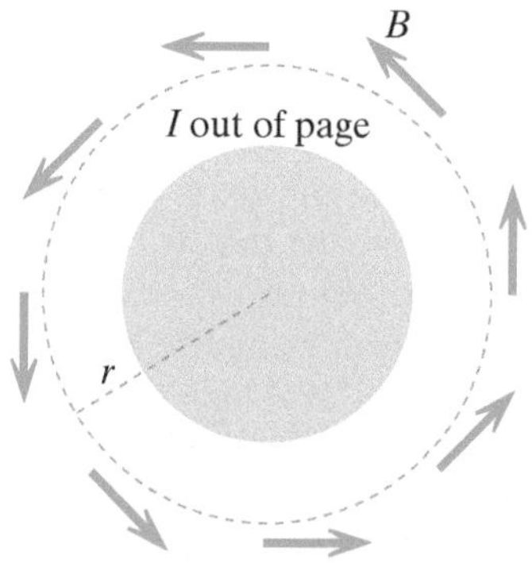

Figure 21.50 Using Ampere's law to find the magnetic field of a thick wire.

Consider the problem of determining the magnetic field near a long *thick* wire (Figure 21.50). If we use the Biot–Savart law, we face the difficult problem of setting up and evaluating an integral not only along the length of the wire but also across the thickness of the wire, with a varying distance to those current elements. However, the result is very easy to get using Ampere's law.

Draw a circular path of radius r around the thick circular wire, assuming by symmetry that the magnetic field B is tangent to the path and constant in magnitude. Stretch an imaginary soap film over this circular path.

> QUESTION Using Ampere's law, calculate the magnitude of the magnetic field at a distance r from the center of a long thick wire.

The path integral of the magnetic field is simply $B(2\pi r)$, since B is by symmetry constant in magnitude and is always parallel (tangential) to the circular path. The amount of current piercing the imaginary soap film is $\mu_0 I$, so we have this:

$$B(2\pi r) = \mu_0 I$$

$$B = \frac{\mu_0 I}{2\pi r} = \frac{\mu_0}{4\pi}\frac{2I}{r} \quad \text{(thick wire)}$$

This simple result (equivalent to the magnetic field of a thin wire) would have been very difficult to obtain using the Biot–Savart law directly. Even though we derived Ampere's law from the Biot–Savart law, there are symmetrical situations for which it is much easier to use Ampere's law.

A Solenoid

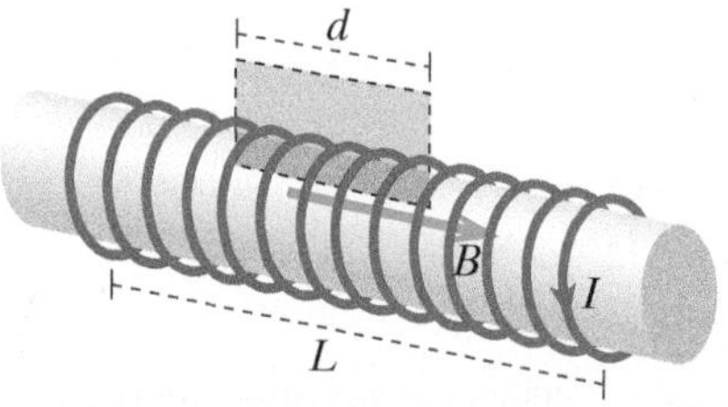

Figure 21.51 A solenoid, with a rectangular path chosen in such a way that the surface bounded by the path is pierced by some of the current-carrying wires.

A long coil of wire with a small diameter is called a "solenoid." A simple electromagnet made by winding a wire on a nail is an iron-filled solenoid. We can calculate the magnetic field along the axis of a solenoid consisting of lots of circular loops. It turns out that the magnetic field is nearly the same in magnitude and direction throughout the interior of the solenoid, as long as you don't get too near the ends. One of the uses of solenoids is to make a uniform magnetic field. Solenoids also play an important role in some kinds of electronic circuits.

Figure 21.51 shows a portion of a very long solenoid. We assume that if we are far from the ends, the magnetic field inside the solenoid is fairly uniform and the magnetic field outside is very small due to near cancellation of the contributions of the near and far sides of the loops. (See the analysis using the Biot–Savart law, at the end of Chapter 17.)

We choose a rectangular path as shown in Figure 21.51, and we stretch an imaginary soap film over the rectangular frame. If there are N loops of wire along the entire length L of the solenoid, there are $(N/L)d$ wires that pierce the soap film, coming out of the page.

> QUESTION Use Ampere's law to show that the magnetic field inside the solenoid is $B \approx \mu_0 NI/L$, which agrees with the much more difficult analysis at the end of Chapter 17 based on the Biot–Savart law.

The path integral is just Bd, because along the two ends of the rectangular path the magnetic field is perpendicular to the path, and along the outside portion of the path the magnetic field is assumed to be very small. The current piercing the soap film is the number of piercings times I, or $[(N/L)d]I$. Ampere's law says that

$$Bd = \mu_0[(N/L)d]I$$

from which we easily conclude this:

$$B = \frac{\mu_0 NI}{L} \quad \text{(solenoid)}$$

The length d does not appear in the result, because the magnetic field cannot depend on the length of the mathematical path that we chose.

No matter where we choose to place that part of the path that is inside the solenoid, we would get this same result. That means that we expect the magnetic field to be quite uniform inside the solenoid, as long as we are far from the ends. This is confirmed by a numerical computation using the Biot–Savart law, as shown in Figure 21.52 (see the computational problem at the end of Chapter 17). The computation includes locations outside the solenoid, but the arrows representing the field there are too small to be seen. Clearly, our assumption is justified that the field is small outside the solenoid, far from the ends.

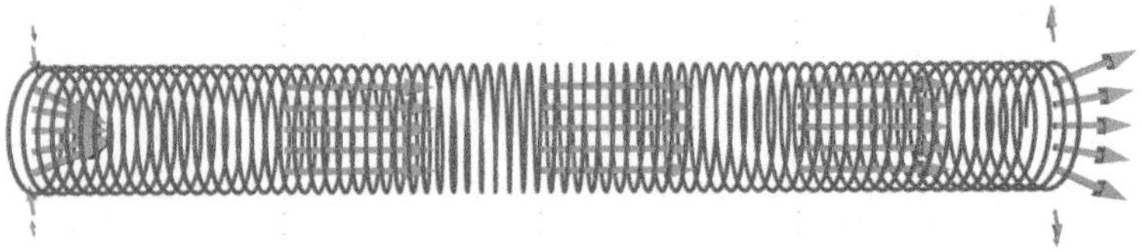

Figure 21.52

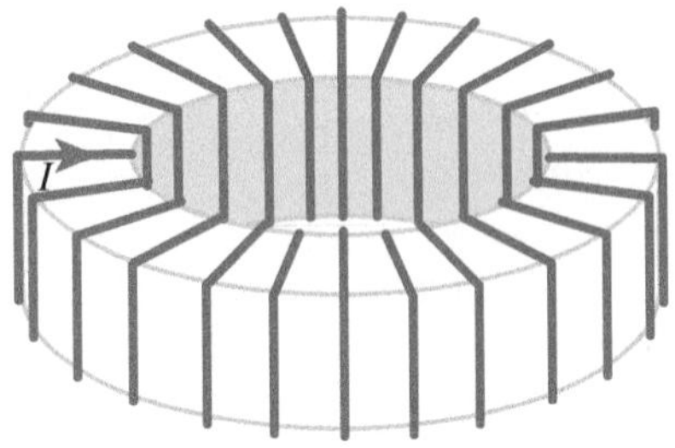

Figure 21.53 A toroid, which is essentially a solenoid bent around to make the ends meet.

A Toroid

Figure 21.53 shows a toroid, a "doughnut" wrapped with wire that is essentially a solenoid bent around to make the ends meet. By symmetry, the magnetic field on a circular path inside the toroid is tangent to the path and constant in magnitude, as shown in a top view in Figure 21.54.

We draw a closed path in the form of a circle of radius r inside the windings. If there are N loops of wire wound on the solenoid, there are N wires that pierce a soap film stretched over that closed path.

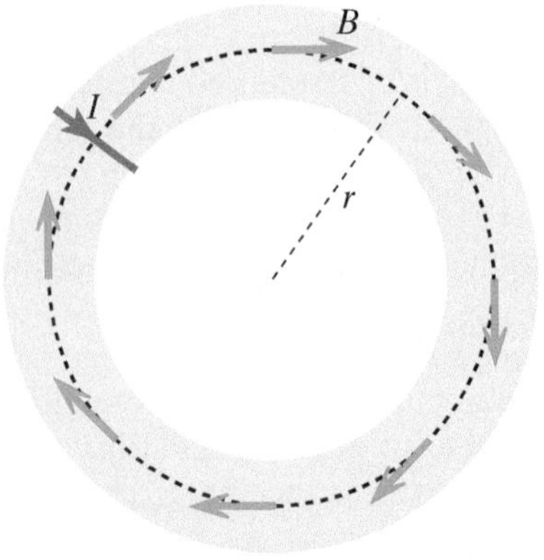

Figure 21.54 Top view of the toroid.

> QUESTION Use Ampere's law to show that the magnetic field inside the toroid, a distance r from the center, is $B = (\mu_0 NI)/(2\pi r)$.

The path integral of the magnetic field is simply $B(2\pi r)$, since $\vec{B}$ is constant in magnitude and always tangential (parallel to $\Delta\vec{l}$). The current piercing the soap film is NI, so Ampere's law is this:

$$B(2\pi r) = \mu_0 NI$$

From this we find the magnetic field:

$$B = \frac{\mu_0 NI}{2\pi r} \quad \text{(toroid)}$$

Note that B is proportional to $1/r$, and the magnetic field varies across the cross section of the toroid. We could not have easily used the Biot–Savart law to obtain this result, because the integration over all the current elements would have been extremely difficult. Yet Ampere's law gives us the result right away, with little calculation.

21.7 MAXWELL'S EQUATIONS

In this chapter we have developed three new equations: Gauss's law for electricity, Gauss's law for magnetism, and Ampere's law for magnetism. In a way, we already have an "Ampere's law for electricity." The round-trip path integral of the electric field produced by stationary or slowly moving point charges is zero:

$$\oint \vec{E} \bullet d\vec{l} = 0 \quad \text{since round-trip potential difference is zero}$$

The four equations taken together are the basis for the four equations known collectively as "Maxwell's equations," though two of the equations are incomplete. What we have just called "Ampere's law for electricity" when completed is called "Faraday's law" and is the subject of Chapter 22. What we have called "Ampere's law" when completed is called the "Ampere–Maxwell law" and is discussed in Chapter 23. In summary, here are the four equations:

MAXWELL'S EQUATIONS (INCOMPLETE)

$\oint \vec{E} \bullet \hat{n}\, dA = \frac{\sum q_{\text{inside}}}{\varepsilon_0}$	Gauss's law for electricity
$\oint \vec{B} \bullet \hat{n}\, dA = 0$	Gauss's law for magnetism
$\oint \vec{E} \bullet d\vec{l} = 0$	Incomplete version of Faraday's law
$\oint \vec{B} \bullet d\vec{l} = \mu_0 \sum I_{\text{inside path}}$	Ampere's law (incomplete version of Ampere–Maxwell law)

Note the structure of these laws. Two of the equations involve integrals over a surface (for electric field or magnetic field), and two involve integrals along a path (for electric field or magnetic field). The incompleteness in the last two laws comes from effects we have not yet studied. The complete forms of these laws include terms involving the time derivatives of electric and magnetic fields.

21.8 SEMICONDUCTOR DEVICES

On the student web site for this textbook, www.wiley.com/college/chabay, there is available at no charge an optional Supplement S2, "Semiconductor Devices." You now have the tools and concepts you need to be able to learn about these elements of modern electronic devices.

21.9 *THE DIFFERENTIAL FORM OF GAUSS'S LAW

This section deals with an advanced topic, which is not required as background for later chapters. It includes the important issue of the relativistic correctness of Gauss's law.

We have pointed out several times that unless charges are motionless or moving very slowly compared to the speed of light, there is a flaw in an attempt to calculate electric field at some location by using Coulomb's law. The problem is relativistic retardation. The electric field depends on where fast-moving particles were at some time in the past, not where they are now. Therefore, Coulomb's law is not consistent with the theory of special relativity. Gauss's law, however, *is* consistent with relativity. In this section we will see why.

One way we could avoid problems with retardation would be to come up with some physical law that would relate a property of the electric field at some location and time (x,y,z,t) to source charges at the *same* location and time (x,y,z,t). If we could write our law in this form, relating source and fields at the same location and time, retardation wouldn't be a problem, and our law could be consistent with special relativity, unlike Coulomb's law.

In brief, we're looking for

- a property of the electric field $\vec{E}$ at some (x,y,z,t)
- that is related to source charges at the *same* (x,y,z,t).

There is a difficulty in this scheme with point charges, since the electric field is infinite at the location of a point charge. After we discuss the general idea, we'll explain how point charges are dealt with.

Divergence

We will now introduce an appropriate property of the electric field, called the "divergence," which has the characteristics we want: we can relate the divergence to the charges at the same location and time.

Consider a very small Gaussian surface around a location in space, enclosing a very small volume ΔV. (Here V is volume, not potential!) In this region there is a charge density ρ measured in coulombs per cubic meter (ρ is lowercase Greek rho). The charge inside the tiny volume is $\rho \Delta V$ (multiply coulombs per cubic meter times volume in cubic meters). From Gauss's law, we know that the net flux on the surface of the tiny volume is:

$$\oint \vec{E} \bullet \hat{n}\, dA = \frac{\rho \Delta V}{\varepsilon_0}$$

Divide by the tiny volume ΔV:

$$\frac{\oint \vec{E} \bullet \hat{n}\, dA}{\Delta V} = \frac{\rho}{\varepsilon_0}$$

On the left side of this equation we have the net flux per unit volume. In the limit as the volume becomes arbitrarily small, $\Delta V \to 0$, this ratio is called the "divergence" (div) of the electric field at this location:

$$\text{div}(\vec{E}) \equiv \lim_{\Delta V \to 0} \frac{\oint \vec{E} \bullet \hat{n}\, dA}{\Delta V}$$

(Note that this is a scalar, not a vector.) This is the property we are looking for, because we now have an equation in a form that doesn't have retardation problems:

$$\text{div}(\vec{E}) = \frac{\rho}{\varepsilon_0}$$

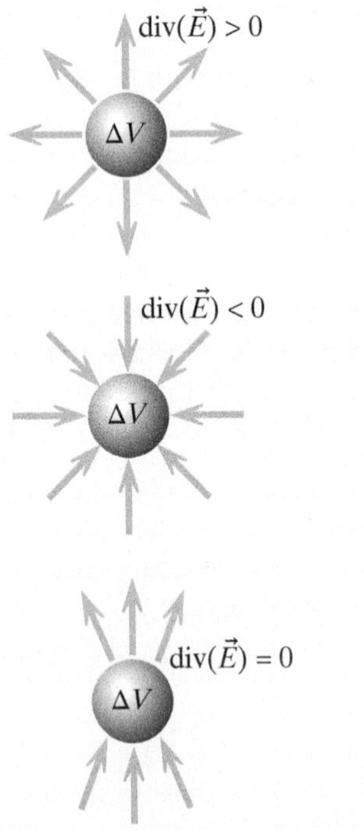

Figure 21.55 The divergence of the electric field at a location enclosed by a tiny volume $\Delta V \to 0$.

Why is this property of the electric field called "divergence"? Look again at the pictures at the beginning of this chapter and at Figure 21.55. If there is positive charge inside a (vanishingly small) volume, the electric field directions "diverge," the net electric flux is positive, and $\text{div}(\vec{E})$ is positive. If there is negative charge inside the tiny volume, the electric field directions converge ("antidiverge"), the net electric flux is negative, and $\text{div}(\vec{E})$ is negative. If there is no charge inside the volume, the electric field looks as though it simply flows through the volume at this location with neither net divergence nor net convergence, the net electric flux is zero, and div($\vec{E}$) is zero. In this introduction to the topic, we write the divergence as div($\vec{E}$), an older notation that reminds us of the definition. Later we will introduce a different notation, $\vec{\nabla} \bullet \vec{E}$, that is normally used for this property.

Divergence Is Relativistically Correct

The divergence (net electric flux per unit volume, in the limit as the volume goes to zero) is a measure of an important property of the electric field at some location and time, and this is proportional to the charge density at that *same* location and time. This is called a "local" relationship between charge and field, because the charge and the divergence property are evaluated at the same location and time, rather than at different locations and times. This avoids the problem of relativistic retardation and forms the basis for what is called a "local field theory" that is consistent with special relativity.

In the beginning of this chapter we developed what is called the "integral form" of Gauss's law, involving an integral of electric flux over a (possibly large) closed surface. What we have just developed is the "differential form" of Gauss's law:

DIFFERENTIAL FORM OF GAUSS'S LAW

$$\text{div}(\vec{E}) = \frac{\rho}{\varepsilon_0}$$

where the divergence is defined as

$$\text{div}(\vec{E}) \equiv \lim_{\Delta V \to 0} \frac{\oint \vec{E} \bullet \hat{n}\, dA}{\Delta V}$$

which is a scalar quantity.

You can show that the units of the divergence of the electric field are the same as the units of ρ/ε_0.

Checkpoint 3 A lead nucleus is spherical with a radius of about 7×10^{-15} m. The nucleus contains 82 protons (and typically 126 neutrons). Because of their motions the protons can be considered on average to be uniformly distributed throughout the nucleus. Based on the net flux at the surface of the nucleus, calculate the divergence of the electric field as electric flux per unit volume. Repeat the calculation at a radius of 3.5×10^{-15} m. (You can use Gauss's law to determine the magnitude of the electric field at this radius.) Also calculate the quantity ρ/ε_0 inside the nucleus.

Why the Integral Form of Gauss's Law Is Relativistically Correct

In the beginning of this chapter we derived the integral form of Gauss's law by starting from Coulomb's law for stationary charges. The differential form was in turn derived from the integral form, yet the differential form of Gauss's law has the important property of being consistent with the theory of relativity, so it is actually more general than the Coulomb's law that we started from. Now we start from the differential form of Gauss's law, which is relativistically correct, and show that the integral form is also relativistically correct.

There is a purely mathematical theorem, the "divergence theorem," that shows that the volume integral of the divergence of a vector field,

$$\int \text{div}(\vec{E})\, dV$$

throughout a large volume at a particular time is equal to a familiar surface integral of the perpendicular component of that vector field,

$$\oint \vec{E} \bullet \hat{n}\, dA$$

on the closed bounding surface of the volume. In other words, the mathematical "divergence theorem" gives this result:

$$\int \text{div}(\vec{E})\, dV = \oint \vec{E} \bullet \hat{n}\, dA$$

However, $\text{div}(\vec{E}) = \rho/\varepsilon_0$, so

$$\int \text{div}(\vec{E})\, dV = \int \frac{\rho}{\varepsilon_0} dV = \frac{Q_{\text{inside}}}{\varepsilon_0}$$

and the volume integral of the charge density ρ is the charge contained inside the bounding surface at this moment in time. You can think of this as simultaneously calculating the divergence of every tiny volume ΔV inside the surface to get the amount of charge in that tiny volume, then adding up all these charges to get the total charge inside the surface.

Combining these results we obtain the integral form of Gauss's law, starting from the differential form, and using only a purely mathematical theorem in the derivation:

$$\oint \vec{E} \bullet \hat{n}\, dA = \frac{Q_{\text{inside}}}{\varepsilon_0}$$

At a fixed moment in time we have integrated the differential form of Gauss's law, which is relativistically valid, and we have recovered the integral form of Gauss's law. This is why even the integral form of Gauss's law, unlike Coulomb's law, is consistent with the special theory of relativity (see the earlier discussion in Section 21.3).

Dealing with Point Charges

When calculating the divergence by taking the limit as the volume goes to zero, $\Delta V \to 0$, if ΔV shrinks down onto the exact location of a point charge such as an electron, the limiting process doesn't work—the flux per unit volume becomes infinite. (At a location arbitrarily close to the point charge, there is no difficulty, and the charge density at that location is zero.)

The mathematical difficulty with point charges is handled formally through the use of the "delta function." The delta function is defined to be zero everywhere except at one point, where it is infinite, and its integral is exactly 1. In other words, the delta function is an infinite spike with zero width, constructed in such a way that the product of the infinite height and the zero width is finite (Figure 21.56). This peculiar entity is not a true function but is an example of a class of mathematical objects called "generalized functions" or "distributions."

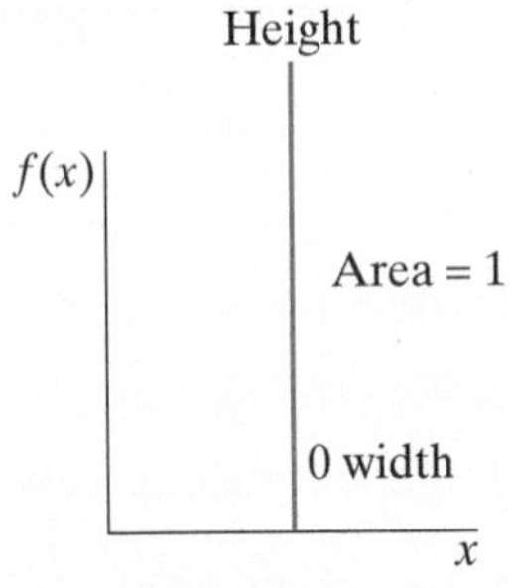

Figure 21.56 The one-dimensional delta function.

An equivalent way of dealing with the mathematical problem is to approximate the point charge as a sphere of charge with a very small radius and a very large charge density. In that case the divergence limit is well defined throughout the tiny sphere of charge, as it was in your calculations of the divergence inside a lead nucleus (Checkpoint 3).

In the practical case of the interior of ordinary atomic matter, we can stop shrinking the volume at a size that, though tiny, is still large enough to include lots of atoms. In this case we are averaging the divergence and the charge density over a tiny volume containing many protons and electrons. This is similar to what we did at the end of Chapter 16, where we found an average electric field inside an insulator, averaged over many atoms.

Another Way to Calculate the Divergence

It is possible to calculate the divergence directly from the definition, as a limit as $\Delta V \to 0$. In this section we present another way to calculate the divergence. Consider the tiny box-shaped surface shown in Figure 21.57. In this tiny region the electric field is in the x direction, but varying in magnitude. Let's use the definition of divergence to calculate this property at a location inside the surface, in the limit that the volume goes to zero:

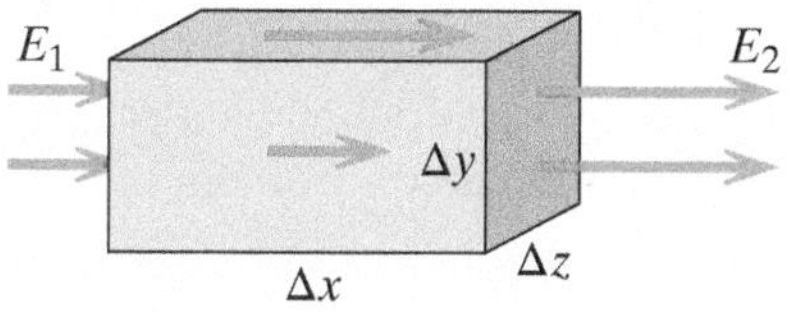

Figure 21.57 Calculating the divergence of the electric field on the closed surface of a tiny volume.

$$\text{div}(\vec{E}) \equiv \lim_{\Delta V \to 0} \frac{\oint \vec{E} \bullet \hat{n}\, dA}{\Delta V} = \lim_{\Delta V \to 0} \frac{(E_2 - E_1)\Delta y \Delta z}{\Delta x \Delta y \Delta z}$$

$$\text{div}(\vec{E}) = \lim_{\Delta x \to 0} \frac{(E_2 - E_1)}{\Delta x}$$

However, this limit is the definition of the derivative of the electric field with respect to x:

$$\text{div}(\vec{E}) = \frac{dE}{dx}$$

More precisely, the divergence in this case is equal to the "partial derivative" of E_x with respect to x, where the special partial derivative notation indicates that we hold y and z constant while taking the derivative:

$$\text{div}(\vec{E}) = \frac{\partial E_x}{\partial x}$$

More generally, there may be contributions to the divergence involving E_y or E_z, which we would calculate in the same way. Adding together all these contributions to the divergence, we find the following:

DIFFERENTIAL FORM OF GAUSS'S LAW

$$\text{div}(\vec{E}) = \frac{\partial E_x}{\partial x} + \frac{\partial E_y}{\partial y} + \frac{\partial E_z}{\partial z} = \frac{\rho}{\varepsilon_0}$$

Electric Potential and Gauss's Law

Remember that electric field can be expressed as the negative gradient of the electric potential (just as force can be expressed as the negative gradient of potential energy). Writing V for electric potential (not volume!), we have this:

$$E_x = -\frac{\partial V}{\partial x}, \quad E_y = -\frac{\partial V}{\partial y}, \quad E_z = -\frac{\partial V}{\partial z}$$

Putting these gradient forms into the differential form of Gauss's law gives the following important equation:

$$\frac{\partial^2 V}{\partial x^2} + \frac{\partial^2 V}{\partial y^2} + \frac{\partial^2 V}{\partial z^2} = -\frac{\rho}{\varepsilon_0}$$

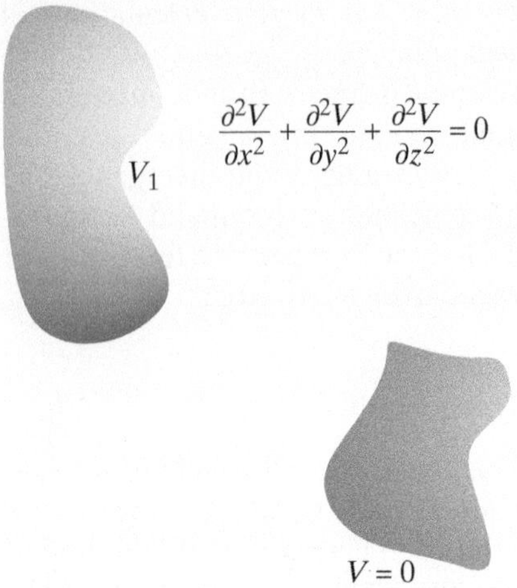

Figure 21.58 A boundary value problem. One piece of metal is maintained at a fixed potential V_1 relative to a potential defined to be $V=0$ for the other piece. In the empty space Gauss's law gives a differential equation to be solved for $V(x,y,z)$.

Here we have a differential equation relating the electric potential at some location to the charge density at that same location. A frequent use of this equation is in determining the electric potential in empty space (where the charge density ρ is zero) near some charged pieces of metal that are maintained at known potentials (a "boundary value" problem, Figure 21.58). If we knew the charge distribution on these pieces of metal, we could calculate the electric potential and electric field anywhere, through superposition. Often, however, we know only the fixed potentials of the pieces of metal, not the charge distribution. In such cases we can use the following differential equation, called "Laplace's equation":

$$\frac{\partial^2 V}{\partial x^2}+\frac{\partial^2 V}{\partial y^2}+\frac{\partial^2 V}{\partial z^2}=0 \quad \text{(in empty space)}$$

Throughout the empty space between the pieces of metal, the electric potential must be a solution of this differential equation. There exist both analytical and numerical methods for solving Laplace's equation. At locations very near the pieces of metal, the solution of Laplace's equation in the empty space must equal the known potentials. This is an example of what is called a "boundary value problem," which is an important class of problem in science and engineering.

Checkpoint 4 In a certain region of space, the electric potential depends on position in the following way, where V_0 and a are constants: $V=V_0+a(x^2+y^2+z^2)$. What is the charge density in this region? In a different region of space, the electric potential depends on position in the following way, where V_0 and b are constants: $V=V_0+b(3x+4y-2z)$. What is the charge density in this region?

Standard Notation

To help introduce the topic, we have written the divergence as $\text{div}(\vec{E})$. The standard notation normally used for divergence is this:

$$\vec{\nabla}\bullet\vec{E}=\frac{\partial E_x}{\partial x}+\frac{\partial E_y}{\partial y}+\frac{\partial E_z}{\partial z}=\frac{\rho}{\varepsilon_0}$$

The $\vec{\nabla}$ (or "del") operator is expressed in the following form:

$$\vec{\nabla}=\left\langle\frac{\partial}{\partial x},\frac{\partial}{\partial y},\frac{\partial}{\partial z}\right\rangle$$

An operator operates on some quantity to its right. In this case the operation involves partial differentiation. The usual rules of calculating the dot product give the following for the quantity $\vec{\nabla}\bullet\vec{E}$:

$$\left\langle\frac{\partial}{\partial x},\frac{\partial}{\partial y},\frac{\partial}{\partial z}\right\rangle\bullet\langle E_x,E_y,E_z\rangle=\frac{\partial E_x}{\partial x}+\frac{\partial E_y}{\partial y}+\frac{\partial E_z}{\partial z}$$

Also, note that the del operator can be used to express the concept of gradient:

$$\vec{E}=-\left(\frac{\partial V}{\partial x}\hat{\imath}+\frac{\partial V}{\partial y}\hat{\jmath}+\frac{\partial V}{\partial z}\hat{k}\right)=-\vec{\nabla}V$$

We can rewrite the differential form of Gauss's law like this:

$$\vec{\nabla}\bullet(\vec{\nabla}V)=\frac{\partial^2 V}{\partial x^2}+\frac{\partial^2 V}{\partial y^2}+\frac{\partial^2 V}{\partial z^2}=-\frac{\rho}{\varepsilon_0}$$

This combination, the divergence of a gradient, is called the "Laplacian" and is written like this:

$$\nabla^2 V = \frac{\partial^2 V}{\partial x^2} + \frac{\partial^2 V}{\partial y^2} + \frac{\partial^2 V}{\partial z^2} = -\frac{\rho}{\varepsilon_0}$$

In regions where there are no charges, we have Laplace's equation:

$$\nabla^2 V = \frac{\partial^2 V}{\partial x^2} + \frac{\partial^2 V}{\partial y^2} + \frac{\partial^2 V}{\partial z^2} = 0 \quad \text{(in empty space)}$$

21.10 *THE DIFFERENTIAL FORM OF AMPERE'S LAW

We have pointed out several times that unless charges are moving very slowly compared to the speed of light, there is a flaw in trying to calculate magnetic field at some location by using the Biot–Savart law. The problem is relativistic retardation. The magnetic field depends on where fast-moving particles were at some time in the past, not where they are now, so the Biot–Savart law is not consistent with the theory of special relativity. Ampere's law, however, is consistent with relativity. In this section we will see why.

The argument is very similar to the one presented in connection with Gauss's law. We look for

- a property of the magnetic field $\vec{B}$ at some (x,y,z,t)
- that is related to moving source charges at the *same* (x,y,z,t).

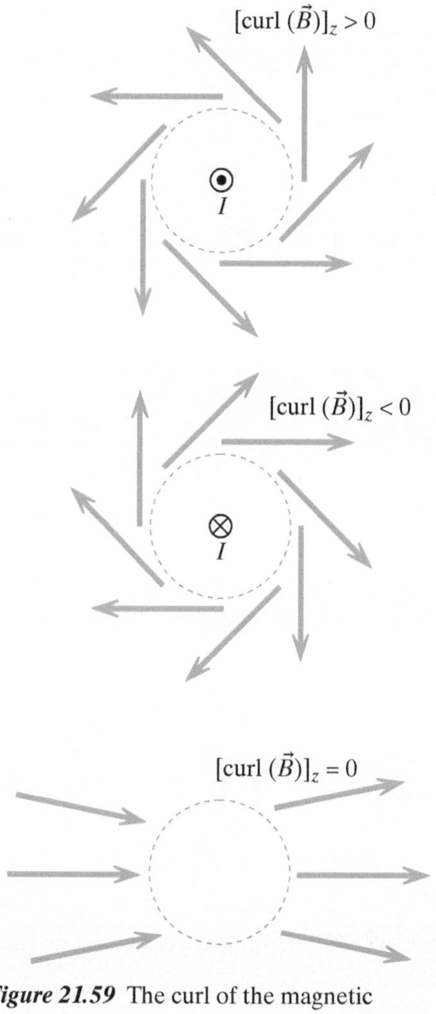

Figure 21.59 The curl of the magnetic field at a location surrounded by a tiny $\Delta A \to 0$.

Curl

We introduce an appropriate property of the magnetic field, called the "curl." Consider a very small Amperian path around a location in space, enclosing a very small area ΔA. In this region there is a current density $\vec{J}$ measured in amperes per square meter. The direction of the vector $\vec{J}$ is the direction of the conventional current at that location. The current passing through the tiny area is $\vec{J} \bullet \hat{n}\,\Delta A$ (multiply amperes per square meter times area in square meters), where $\hat{n}$ is a "normal" unit vector perpendicular to the area. From Ampere's law, we know that the path integral of magnetic field around the perimeter of the tiny area is this:

$$\oint \vec{B} \bullet d\vec{l} = \mu_0 \vec{J} \bullet \hat{n}\,\Delta A$$

Divide by the tiny area ΔA:

$$\frac{\oint \vec{B} \bullet d\vec{l}}{\Delta A} = \mu_0 \vec{J} \bullet \hat{n}$$

On the left side of this equation we have the path integral of magnetic field per unit area. In the limit as the area becomes arbitrarily small, $\Delta A \to 0$, this ratio is called the component in the $\hat{n}$ direction of the "curl" of the magnetic field at this location:

$$[\text{curl}(\vec{B})] \bullet \hat{n} \equiv \lim_{\Delta A \to 0} \frac{\oint \vec{B} \bullet d\vec{l}}{\Delta A}$$

This is the property we are looking for. We obtain an equation in a form that doesn't have retardation problems:

$$\text{curl}(\vec{B}) = \mu_0 \vec{J} \quad \text{(note that this is a \textit{vector} equation)}$$

Why is this property of the magnetic field called the curl of the magnetic field? Look at Figure 21.59. If there is a current crossing through a (vanishingly small) area, the magnetic field curls around the current.

Curl Is Relativistically Correct

The curl (net path integral of magnetic field per unit area, in the limit as the area goes to zero) is a measure of an important property of the magnetic field at some location and time, and this is proportional to the current density at that *same* location and time. This is called a "local" relationship between current and field, because the current and the curl property are evaluated at the same location and time, rather than at different locations and times. This avoids the problem of relativistic retardation and forms the basis for what is called a "local field theory" that is consistent with special relativity.

Earlier in this chapter we developed what is called the "integral form" of Ampere's law, involving a path integral of magnetic field over a (possibly long) closed path. What we have just developed is the "differential form" of Ampere's law:

DIFFERENTIAL FORM OF AMPERE'S LAW

$$\text{curl}(\vec{B}) = \mu_0 \vec{J}$$

where the component of the curl in the $\hat{n}$ direction, perpendicular to a tiny area ΔA, is defined as

$$[\text{curl}(\vec{B})] \bullet \hat{n} \equiv \lim_{\Delta A \to 0} \frac{\oint \vec{B} \bullet d\vec{l}}{\Delta A}$$

Using the del operator, Ampere's law is written like this:

$$\vec{\nabla} \times \vec{B} = \mu_0 \vec{J}$$

The differential form of Ampere's law is correct in linking current density to magnetic field at the same location and time. However, we will see later that Ampere's law is incomplete.

Checkpoint 5 Based on its definition as path integral of magnetic field per unit area, what are the units of the curl of the magnetic field? Show that these units are the same as the units of $\mu_0 J$.

SUMMARY

Fundamental Principles

Gauss's law

$$\oint \vec{E} \bullet \hat{n}\, dA = \frac{\sum q_{\text{inside}}}{\varepsilon_0}$$

Gauss's law for magnetism

$$\oint \vec{B} \bullet \hat{n}\, dA = 0$$

Ampere's law

$$\oint \vec{B} \bullet d\vec{l} = \mu_0 \sum I_{\text{inside path}}$$

Alternative Notations for Gauss's Law

Finite sum of electric flux

$$\sum_{\text{closed surface}} \vec{E} \bullet \hat{n}\, \Delta A = \frac{\sum q_{\text{inside}}}{\varepsilon_0}$$

$E_\perp = E$ perpendicular to surface:

$$\oint E_\perp dA = \frac{\sum q_{\text{inside}}}{\varepsilon_0}$$

$A_\perp = A$ perpendicular to electric field:

$$\oint E dA_\perp = \frac{\sum q_{\text{inside}}}{\varepsilon_0}$$

$d\vec{A} = \hat{n}\,dA$ for representing area and normal:

$$\oint \vec{E} \bullet d\vec{A} = \frac{\sum q_{\text{inside}}}{\varepsilon_0}$$

The circle superimposed on the integral sign means "integrate over the entire closed surface."

Electric flux on a surface (not necessarily closed)

$$\Phi_{\text{el}} = \int \vec{E} \bullet \hat{n}\,dA$$

Results

We used Gauss's law to find the electric field of a plate and a sphere. We also proved some important properties about metals, especially that excess charge is on the surface and that there is no field in an empty hole in a metal in equilibrium. We used Ampere's law to find the magnetic field of a thick wire, a solenoid, and a toroid.

Advanced *Differential Form of Gauss's Law

$$\vec{\nabla} \bullet \vec{E} = \frac{\partial E_x}{\partial x} + \frac{\partial E_y}{\partial y} + \frac{\partial E_z}{\partial z} = \frac{\rho}{\varepsilon_0}, \text{ or}$$

$$\nabla^2 V = \frac{\partial^2 V}{\partial x^2} + \frac{\partial^2 V}{\partial y^2} + \frac{\partial^2 V}{\partial z^2} = -\frac{\rho}{\varepsilon_0}$$

Advanced *Differential Form of Ampere's Law

$$\vec{\nabla} \times \vec{B} = \mu_0 \vec{J}$$

QUESTIONS

Q1 The electric field on a closed surface is due to all the charges in the Universe, including the charges outside the closed surface. Explain why the total flux nevertheless is proportional only to the charges that are inside the surface, with no apparent influence of the charges outside.

Q2 In Chapter 15 we calculated the electric field at a location on the axis of a uniformly charged ring. Without doing all those calculations, explain why we can't use Gauss's law to determine the electric field at that location.

Q3 Explain why you are safe and unaffected inside a car that is struck by lightning. Also explain why it might not be safe to step out of the car just after the lightning strike, with one foot in the car and one on the ground. (Note that it is current that kills; a rather small current passing through the region of the heart can be fatal.)

Q4 Based on its definition as electric flux per unit volume, what are the units of the divergence of the electric field?

Q5 In the region shown in Figure 21.60, the magnetic field is vertical and was measured to have the values shown on the surface of a cylinder. Why should you suspect something is wrong with these measurements?

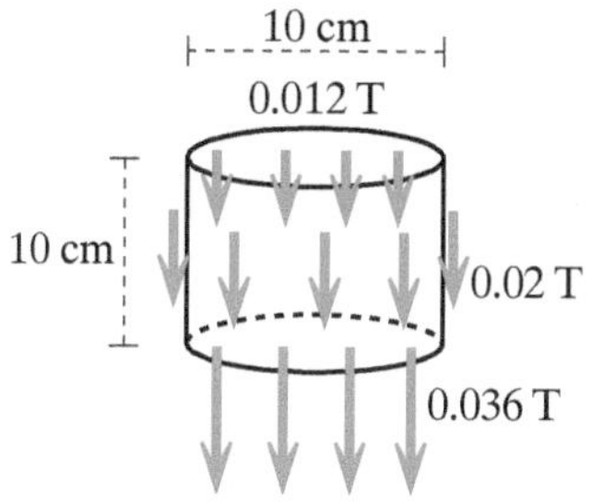

Figure 21.60

PROBLEMS

Section 21.2

•P6 Figure 21.61 shows a disk-shaped region of radius 2 cm on which there is a uniform electric field of magnitude 300 V/m at an angle of 30° to the plane of the disk. Assume that $\hat{n}$ points upward, in the $+y$ direction. Calculate the electric flux on the disk, and include the correct units.

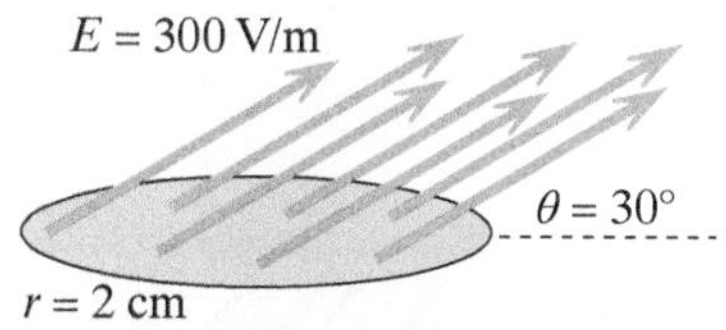

Figure 21.61

Section 21.3

•P7 Figure 21.62 shows a box on whose surfaces the electric field is measured to be horizontal and to the right. On the left face (3 cm by 2 cm) the magnitude of the electric field is 400 V/m, and on the right face the magnitude of the electric field is 1000 V/m. On the other faces only the direction is known (horizontal). Calculate the electric flux on every face of the box, the total flux, and the total amount of charge that is inside the box.

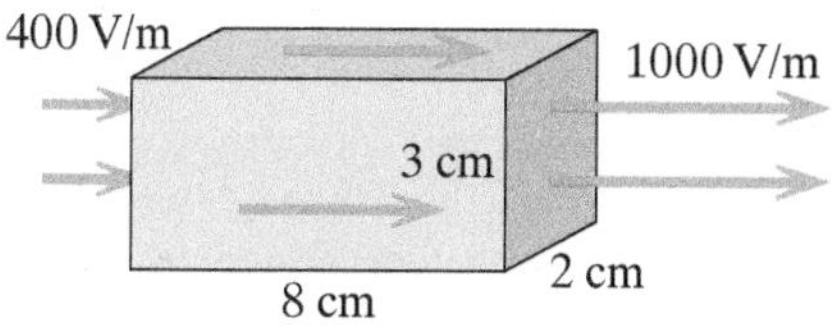

Figure 21.62

•P8 A Gaussian cylinder is drawn around some hidden charges, as shown in Figure 21.63. Given this pattern of electric field on the surface of the cylinder, how much net charge is inside? Invent a possible charge distribution inside the cylinder that could give this pattern of electric field.

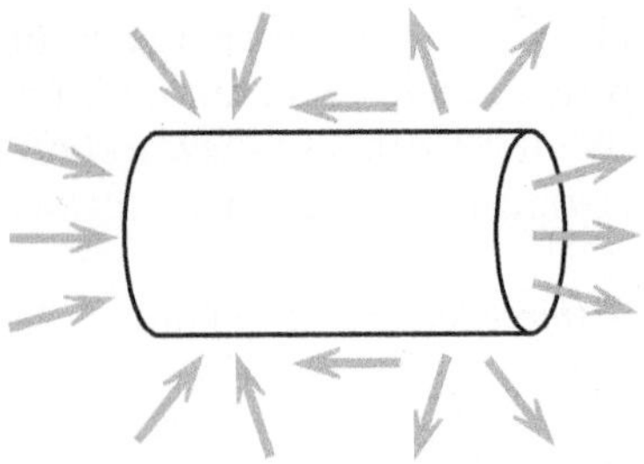

Figure 21.63

••**P9** The electric field has been measured to be vertically upward everywhere on the surface of a box 20 cm long, 4 cm high, and 3 cm deep, as shown in Figure 21.64. All over the bottom of the box $E_1 = 1500$ V/m, all over the sides $E_2 = 1000$ V/m, and all over the top $E_3 = 600$ V/m. What can you conclude about the contents of the box? Include a numerical result.

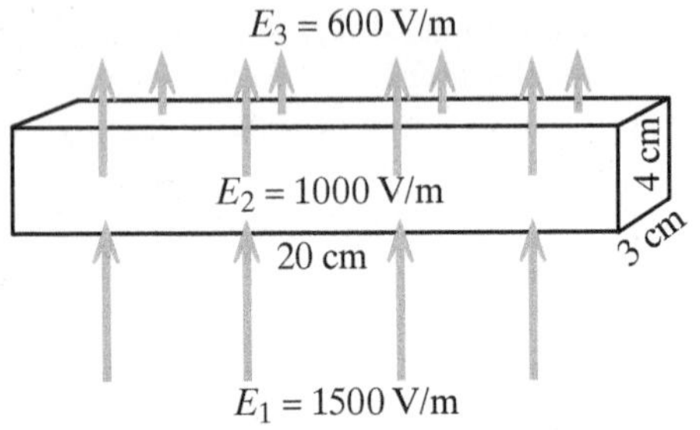

Figure 21.64

••**P10** The electric field is horizontal and has the values indicated on the surface of a cylinder as shown in Figure 21.65. What can you deduce from this pattern of electric field? Include a numerical result.

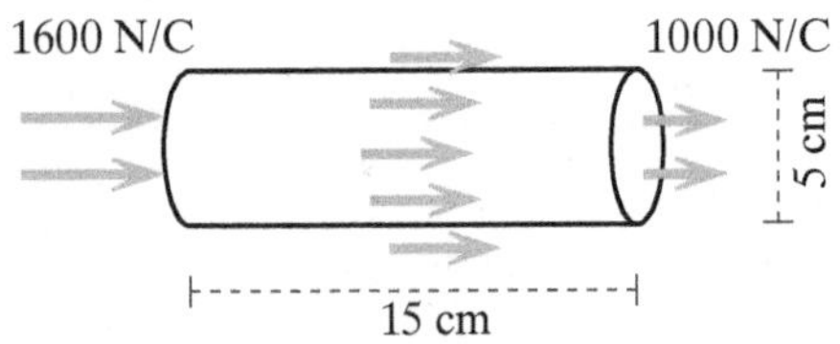

Figure 21.65

••**P11** Consider a cube with one corner at the origin and with sides of length 10 cm positioned along the xyz axes. There is an electric field $\vec{E} = \langle 50, 200y, 0 \rangle$ N/m throughout the region that has a constant x component and a y component that increases linearly with y. **(a)** Draw a diagram showing $\hat{n}$ and $\vec{E}$ for every side of the cube. **(b)** Determine how much charge is inside the cube.

••**P12** The electric field has been measured to be horizontal and to the right everywhere on the closed box as shown in Figure 21.66. All over the left side of the box $E_1 = 100$ V/m, and all over the right (slanting) side of the box $E_2 = 300$ V/m. On the top the average field is $E_3 = 150$ V/m, on the front and back the average field is $E_4 = 175$ V/m, and on the bottom the average field is $E_5 = 220$ V/m. How much charge is inside the box? Explain briefly.

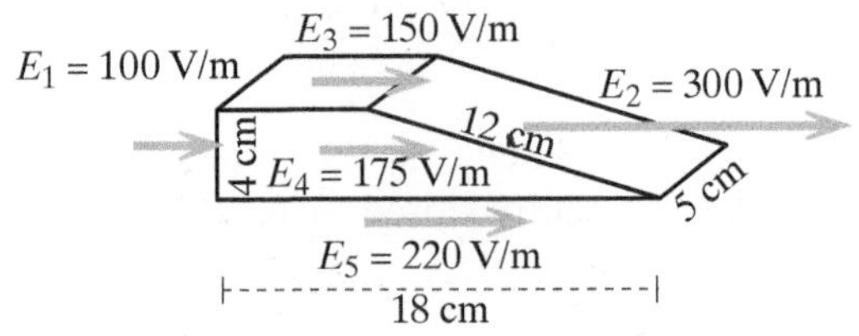

Figure 21.66

••**P13** The electric field is measured at locations on every surface of a rectangular box oriented with its long side on the x axis, as shown in Figure 21.67. The width w of the box is 6 cm, the height h of the box is 4 cm, and the depth d of the box is 2 cm. On the bottom of the box the electric field $\vec{E}_1$ is found to be $\langle 50, 40, 0 \rangle$ N/C. On every other surface of the box (front, back, top, left end, right end) the electric field $\vec{E}_2$ is found to be $\langle 120, 96, 0 \rangle$ N/C. (For clarity, the electric field on the front, back, and left faces has not been shown in the diagram.) **(a)** For every face of the box, specify $\hat{n}$. **(b)** Calculate the electric flux on every face of the box. **(c)** What is the net charge inside the box?

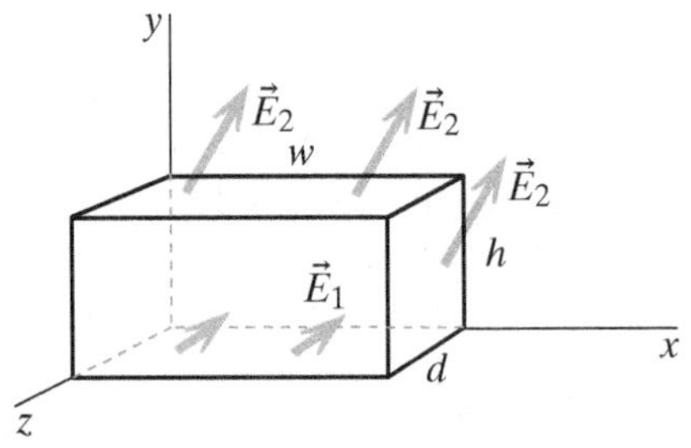

Figure 21.67

••**P14** The electric field is measured all over a cubical surface, and the pattern of field detected is shown in Figure 21.68. On the right side of the cube and the bottom of the cube, the electric field has the value $\langle 370, -370, 0 \rangle$ N/C. On the top of the cube and the left side of the cube, the electric field is zero. On half of the front and back faces, the electric field has the value $\langle 370, -370, 0 \rangle$ N/C; on the other half of the front and back faces, the electric field is zero. One edge of the cube is 55 cm long. **(a)** What is the net electric flux on this cubical surface? **(b)** What is the net charge inside this cubical surface? **(c)** Can you think of a charge distribution that could generate this pattern of electric field?

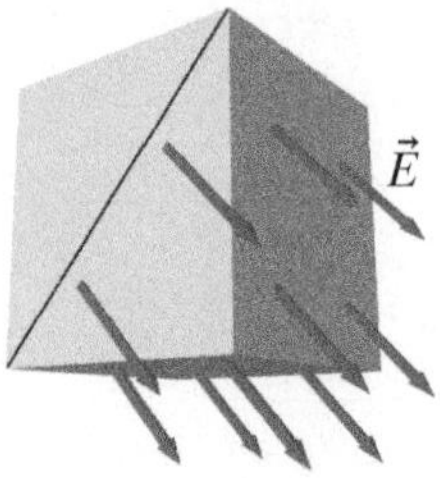

Figure 21.68

••**P15** The center of a thin paper cube in outer space is located at the origin. Each edge is 10 cm long. The only other objects in the neighborhood are some small charged particles whose charges and positions at this instant are the following: $+9$ nC at $\langle 4,2,-3\rangle$ cm, -6 nC at $\langle 1,-3,2\rangle$ cm, $+7$ nC at $\langle 15,0,4\rangle$ cm, $+4$ nC at $\langle -1,3,2\rangle$ cm, and -3 nC at $\langle -2,12,-3\rangle$ cm. The electric flux on five of the six faces of the cube totals 564 V·m. What is the flux on the other face?

Section 21.4

••P16 A negative point charge $-Q$ is at the center of a hollow insulating spherical shell, which has an inner radius R_1 and an outer radius R_2. There is a total charge of $+3Q$ spread uniformly throughout the volume of the insulating shell, not just on its surface. Determine the electric field for **(a)** $r < R_1$, **(b)** $R_1 < r < R_2$, and **(c)** $R_2 < r$.

••P17 You may have seen a coaxial cable connected to a television set. As shown in Figure 21.69, a coaxial cable consists of a central copper wire of radius r_1 surrounded by a hollow copper tube (typically made of braided copper wire) of inner radius r_2 and outer radius r_3. Normally the space between the central wire and the outer tube is filled with an insulator, but in this problem assume for simplicity that this space is filled with air. Assume that no current runs in the cable.

Suppose that a coaxial cable is straight and has a very long length L, and that the central wire carries a charge $+Q$ uniformly distributed along the wire (so that the charge per unit length is $+Q/L$ everywhere along the wire). Also suppose that the outer tube carries a charge $-Q$ uniformly distributed along its length L. The cylindrical symmetry of the situation indicates that the electric field must point radially outward or radially inward. The electric field cannot have any component parallel to the cable. In this problem, draw mathematical Gaussian cylinders of length d (with d much less than the cable length L) and appropriate radius r, centered on the central wire. **(a)** Use a mathematical Gaussian cylinder located inside the central wire ($r < r_1$) and another Gaussian cylinder with a radius in the interior of the outer tube ($r_2 < r < r_3$) to determine the exact amount and location of the charge on the inner and outer conductors. (*Hint:* What do you know about the electric field in the interior of the two conductors? What do you know about the flux on the ends of your Gaussian cylinders?) **(b)** Use a mathematical Gaussian cylinder whose radius is in the air gap ($r_1 < r < r_2$) to determine the electric field in the gap as a function of r. (Don't forget to consider the flux on the ends of your Gaussian cylinder.) **(c)** Use a mathematical Gaussian cylinder whose radius is outside the cable ($r > r_3$) to determine the electric field outside the cable. (Don't forget to consider the flux on the ends of your Gaussian cylinder.)

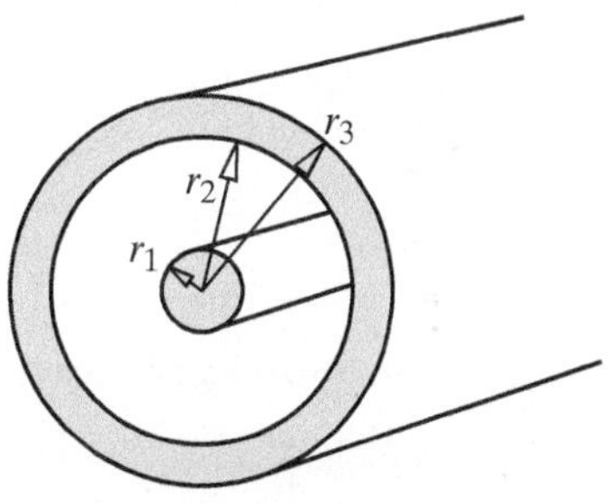

Figure 21.69

••P18 A straight circular plastic cylinder of length L and radius R (where $R \ll L$) is irradiated with a beam of protons so that there is a total excess charge Q distributed uniformly throughout the cylinder. Find the electric field inside the cylinder, a distance r from the center of the cylinder far from the ends, where $r < R$.

••P19 Figure 21.70 shows a close-up of the central region of a capacitor made of two large metal plates of area A, very close together and charged equally and oppositely. There are $+Q$ and $-Q$ on the inner surfaces of the plates and small amounts of charge $+q$ and $-q$ on the outer surfaces.

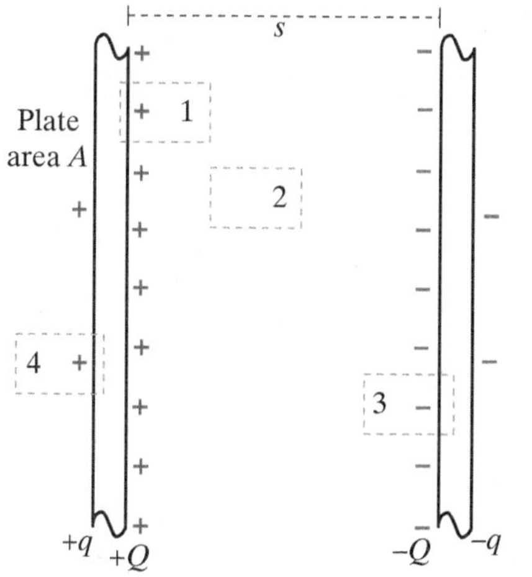

Figure 21.70

(a) Knowing Q, determine E: Consider a Gaussian surface in the shape of a shoe box, with one end of area A_{box} in the interior of the left plate and the other end in the air gap (surface 1 in the diagram). Using only the fact that the electric field is expected to be horizontal everywhere in this region, use Gauss's law to determine the magnitude of the field in the air gap. Check that your result agrees with our earlier calculations in Chapter 15. (Be sure to consider the flux on all faces of your Gaussian box.) **(b)** Proof by contradiction: Consider Gaussian box 2, with both ends in the gap. Use Gauss's law to prove that the electric field is constant in magnitude throughout the gap. **(c)** Knowing Q, determine E: Consider Gaussian box 3, with its left end in the gap and its right end in the interior of the right plate. Show that Gauss's law applied to this Gaussian box gives the correct magnitude of the electric field in the gap. **(d)** Knowing E, determine q/A: Finally, consider Gaussian box 4, with its left end outside the capacitor and its right end in the interior of the left plate. In Chapter 15 we determined that the fringe field is approximately given by

$$E_{\text{fringe}} \approx \frac{Q/A}{2\varepsilon_0} \frac{s}{R}$$

where s is the gap width and R is the radius of the circular plates. Use this information and Gaussian box 4 to determine the approximate amount of charge q on the outer surface of the plate.

•••P20 In Chapter 20 we showed that the nonuniform direction of the magnetic field of a magnet is responsible for the force $F = I(2\pi R)B\sin\theta$ exerted by the magnet on a current loop. However, we didn't know the angle θ, which is a measure of the nonuniformity of the direction of the magnetic field (Figure 21.71). Gauss's law for magnetism can be used to determine this angle θ. **(a)** Apply Gauss's law for magnetism to a thin disk of radius R and thickness Δx located where the current loop will be placed. Use Gauss's law to determine the component $B_3 \sin\theta$ of the magnetic field that is perpendicular to the axis of the magnet at a radius R from the axis. Assume that R is small enough that the x component of magnetic field is approximately uniform everywhere on the left face of the disk and also on the right face of the disk (but with a smaller value). The magnet has a magnetic dipole moment μ.

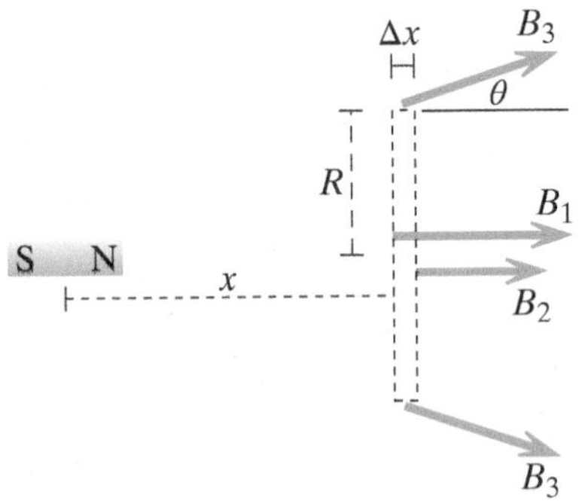

Figure 21.71

(b) Now imagine placing the current loop at this location, and show that $F = I(2\pi R)B_3 \sin\theta$ can be rewritten as $F = \mu_{loop}|dB/dx|$, which is the result we obtained from potential-energy arguments in Chapter 20.

•••**P21** A solid chunk of metal carries a net positive charge and is also polarized by other charges that are not on the metal, as shown in Figure 21.72. The electric field very near the surface of a metal in equilibrium is perpendicular to the surface: any component of the electric field parallel to the surface would drive currents in the metal. (Along a rectangular path, part inside and part outside the metal, the round-trip potential difference is zero, so the parallel component of electric field is the same just inside and just outside the surface.) Let S be the "local" surface charge density—the surface charge per unit area (C/m^2) at a particular location on the surface. **(a)** Calculate the magnitude E of the electric field in the air very near a location on the surface of the metal, in terms of the surface charge density S at that location. Use a cylindrical Gaussian surface as shown, with one face just inside the metal and one face just outside. Don't skip steps in your analysis!

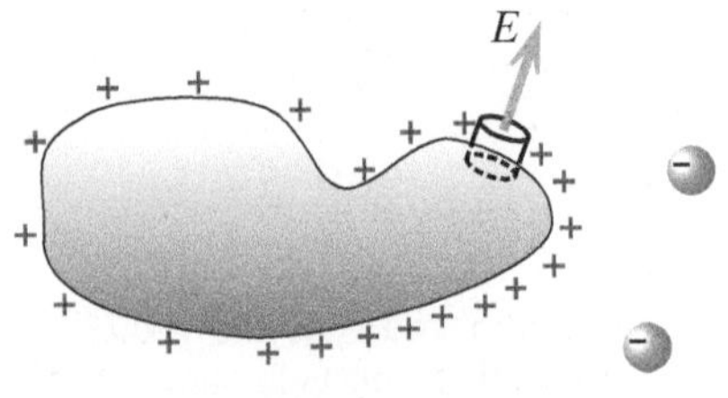

Figure 21.72

(b) Show that nearby charges contribute half of E, and distant charges contribute the other half.

Section 21.6

•**P22** The magnetic field has been measured to be horizontal everywhere along a rectangular path 20 cm long and 4 cm high, as shown in Figure 21.73. Along the bottom the average magnetic field $B_1 = 1.5 \times 10^{-4}$ T, along the sides the average magnetic field $B_2 = 1.0 \times 10^{-4}$ T, and along the top the average magnetic field $B_3 = 0.6 \times 10^{-4}$ T. What can you conclude about the electric currents in the area that is surrounded by the rectangular path?

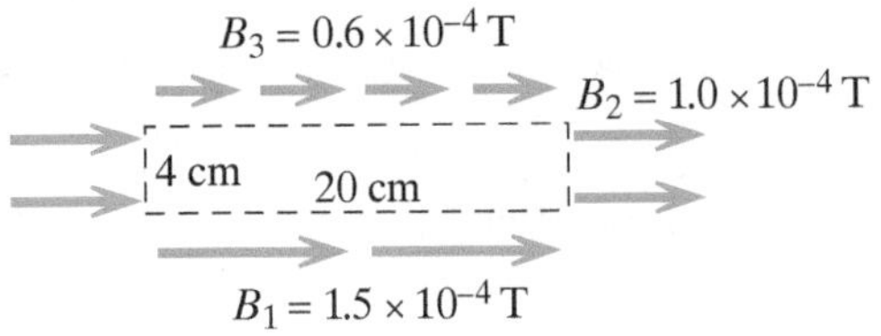

Figure 21.73

••**P23** Figure 21.74 shows a measured pattern of magnetic field in space. How much current I passes through the shaded area? In what direction?

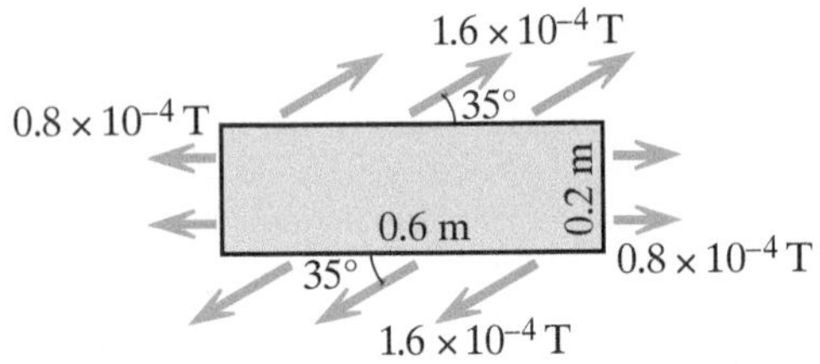

Figure 21.74

••**P24** A D-shaped frame is made out of plastic of small square cross section and tightly wrapped uniformly with N turns of wire as shown in Figure 21.75, so that the magnetic field has essentially the same magnitude throughout the plastic. (R is much bigger than w.)

With a current I flowing, what is the magnetic field inside the plastic? Show the direction of the magnetic field in the plastic at several locations.

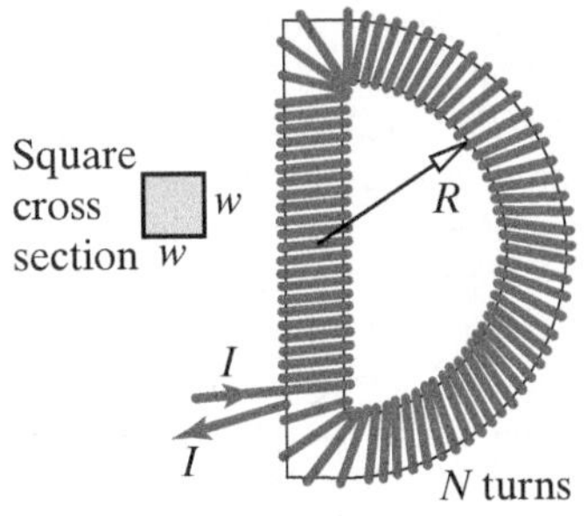

Figure 21.75

••**P25** Figure 21.76 shows a large number N of closely packed wires, each carrying a current I out of the page. The width of this sheet of wires is L.

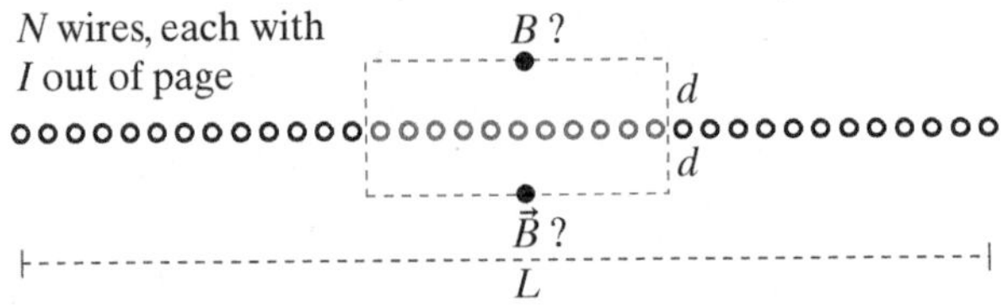

Figure 21.76

(a) Show (and explain) the direction of the magnetic field at the two indicated locations, a distance d from the middle wire, where

d is much less than $L(d \ll L)$. **(b)** Using the rectangular path shown on the diagram, calculate the magnitude of the magnetic field at the indicated locations, in terms of the given physical quantities N, I, and L. **(c)** Using your result from part (b), find $\vec{B}$ in the region between two parallel current sheets with equal currents running in opposite directions. **(d)** What is $\vec{B}$ outside both sheets?

••P26 A long plastic rod with rectangular cross section $h \times w$ is wound with a coil of length d consisting of N closely packed turns of wire with negligible resistance. A current I runs through the coil.

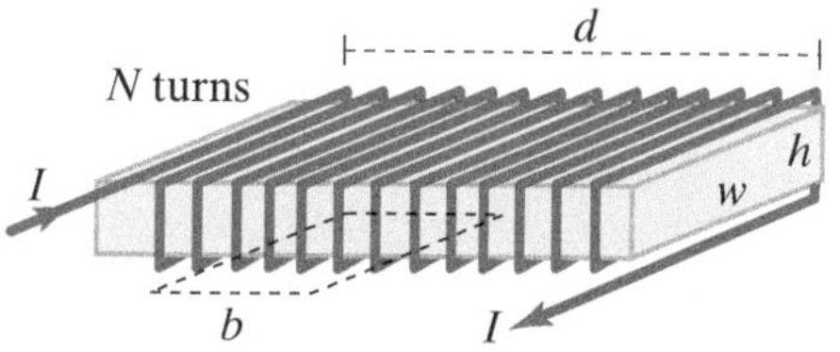

Figure 21.77

Using the rectangular path of length b shown on the diagram in Figure 21.77, prove that the magnetic field inside the coil is $B = \mu_0 NI/d$, and describe the direction of $\vec{B}$ inside the coil.

••P27 A coaxial cable consists of an inner metal wire of radius r_1 and an outer metal cylinder of radius r_2 (Figure 21.78). There is constant current I_1 to the right in the wire and constant current $I_2 < I_1$ to the left in the cylinder. Consider a circular Amperian path of radius R centered on the wire and perpendicular to the wire. Determine the magnitude and direction of the magnetic field at a location R above the wire, far from the ends of the coaxial cable. Explain in detail, including all assumptions, directions, calculations, and so on. Don't skip steps.

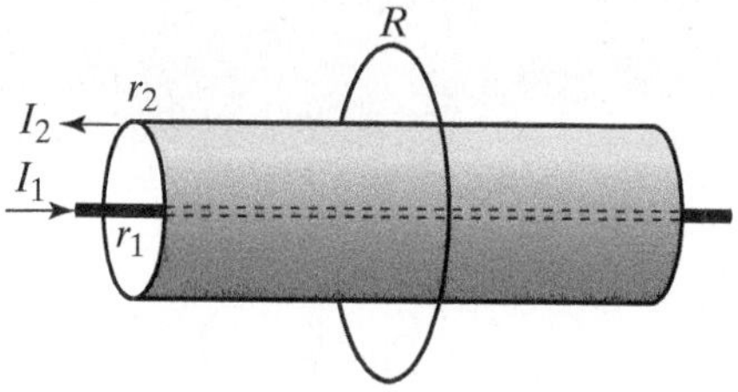

Figure 21.78

••P28 A thick wire of radius R carries a constant current I (Figure 21.79). Find the magnetic field inside the wire, a distance r from the center of the wire and far from the ends of the wire, where $r < R$ and R is much less than the length of the wire.

Figure 21.79

Section 21.9

•P29 A lead nucleus is spherical with a radius of about 7×10^{-15} m. The nucleus contains 82 protons (and typically 126 neutrons). Because of their motions the protons can be considered on average to be uniformly distributed throughout the nucleus. Based on the net flux at the surface of the nucleus, calculate the divergence of the electric field as electric flux per unit volume.

COMPUTATIONAL PROBLEM

•••P30 A point charge of 2×10^{-9} C is located at $\langle -0.03, 0, 0\rangle$ m. Write a program to calculate the net electric flux on a cubical Gaussian surface centered at $\langle 0.03, 0, 0\rangle$ m. The length of a side of the cube is 0.01 m. **(a)** Begin by considering each face of the cube as a single area element ΔA, and using the value of $\vec{E}$ at the center of each face. How does the net flux compare to $\sum q_{\text{inside}}/\varepsilon_0$? **(b)** Refine the calculation by dividing each face of the cube into four smaller squares. How does your result change? **(c)** Continue to subdivide each square into smaller squares until your result no longer changes. Now how does the net flux compare to $\sum q_{\text{inside}}/\varepsilon_0$? **(d)** Move the charged particle to location $\langle 0.035, 0, 0\rangle$ m, inside the cube, and recalculate the total flux. How does the net flux compare to $\sum q_{\text{inside}}/\varepsilon_0$?

ANSWERS TO CHECKPOINTS

1 1.73×10^{-3} V·m
2 0.195 A, into the page
3 1.03×10^{36} N/C/m; 1.03×10^{36} N/C/m; 1.03×10^{36} N/C/m
4 $-6\varepsilon_0 a$; 0
5 T/m

CHAPTER

22

Faraday's Law

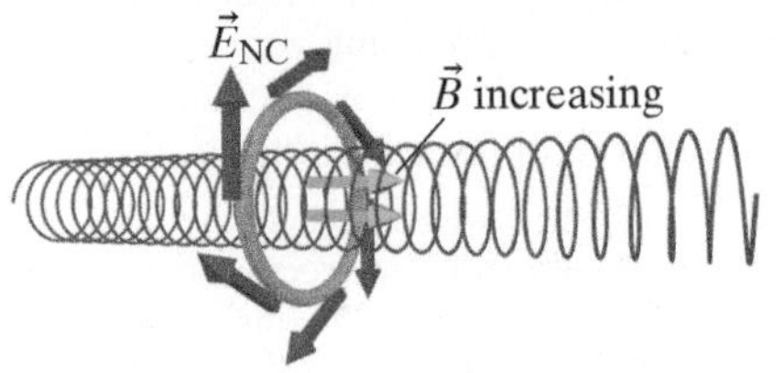

OBJECTIVES

After studying this chapter you should be able to

- Calculate the magnitude and direction of the electric field associated with a time-varying magnetic field.

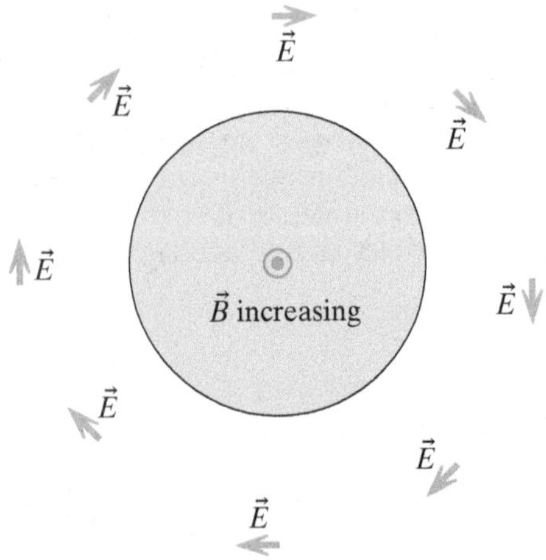

Figure 22.1 In a region where there is a time-varying magnetic field, there is also a curly electric field.

In previous chapters we have considered electric and magnetic fields as separate entities. Stationary charges make electric fields, and electric fields affect other charges. Moving charges make magnetic fields as well as electric fields, and magnetic fields affect other moving charges. In this chapter, however, we will see that there is a connection between electric and magnetic fields that we have not previously suspected. We find experimentally that if there is a time-varying magnetic field in a region of space, there is also a curly electric field throughout that region and surrounding regions (Figure 22.1). In Chapter 16, by considering the round-trip integral of electric field, we proved that a curly pattern of electric field cannot be made by a collection of stationary charges, so this is a particularly interesting phenomenon. No matter how an electric field comes into being, it has the same effect on a charge q (that is, $\vec{F} = q\vec{E}$).

The historical term "magnetic induction" is often used to describe this phenomenon, and one says that the time-varying magnetic field "induces" the curly electric field. This is somewhat misleading. It is more correct to say that anywhere we observe a time-varying magnetic field, we also observe a curly electric field. Faraday's law relates these observations quantitatively.

22.1 CURLY ELECTRIC FIELDS

In Chapter 17 using the Biot–Savart law and in Chapter 21 using Ampere's law we found that with a current I_1 in a solenoid, the magnetic field B_1 inside a tightly wound solenoid (not too near the ends) is $B_1 = \mu_0 N I_1/d$, where N is the number of turns of wire and d is the length of the solenoid. The magnetic field outside the solenoid (not too near the ends) is very small and for a very long solenoid can be taken to be zero.

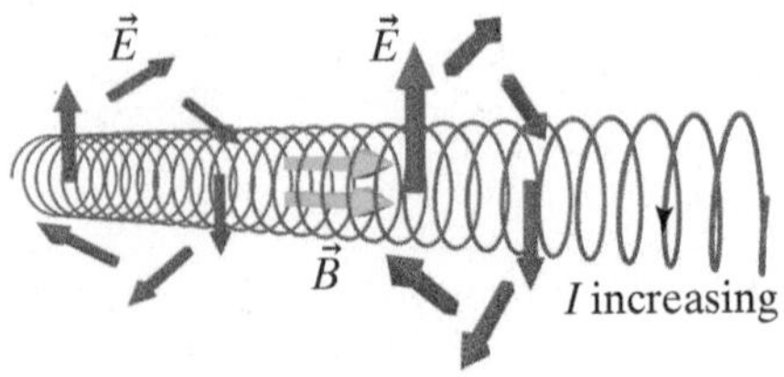

Figure 22.2 With a time-varying current in the solenoid there is a curly electric field.

If the current is constant, then B_1 is constant in time. In that case a charge that is in motion somewhere outside the solenoid will not experience an electric or magnetic force, because the electric and magnetic fields outside the solenoid are essentially zero. However, something remarkable happens if we vary the current, so that the magnetic field B_1 inside the solenoid varies with time (Figure 22.2). There is still almost no magnetic field outside the solenoid, but we observe a curly electric field both inside and outside of the solenoid. This unusual electric field curls around the axis of the solenoid (end view; Figure 22.3). The electric field is proportional to dB_1/dt, the rate of change of the magnetic field.

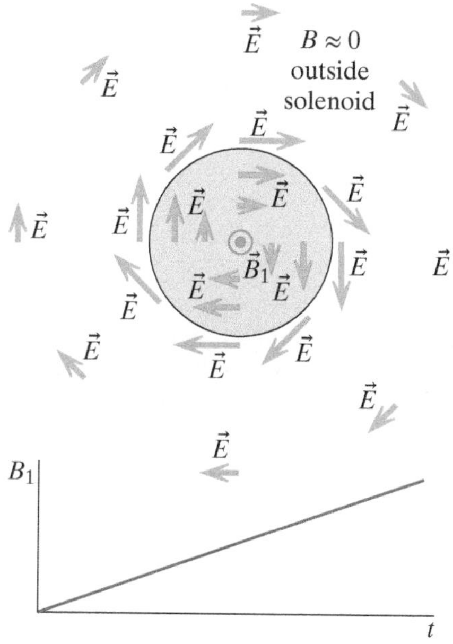

Figure 22.3 There is a curly electric field in the presence of a time-varying magnetic field (top). In this case B_1 is increasing with time (bottom).

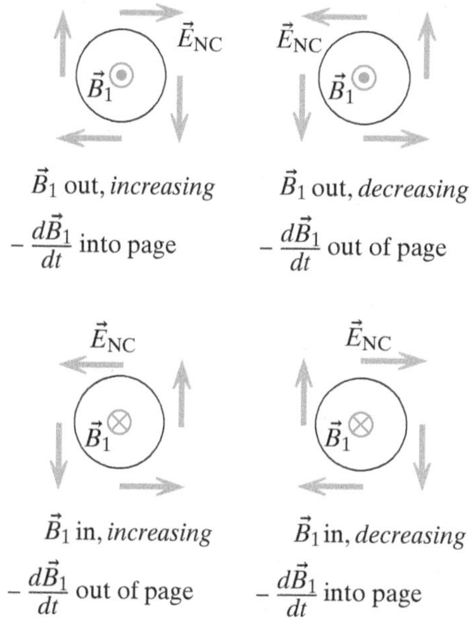

Figure 22.4 Four cases: magnetic field out or in, increasing or decreasing.

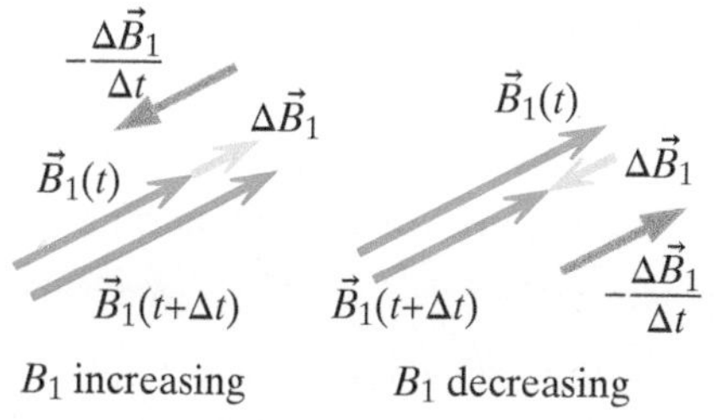

Figure 22.5 Find the change in the magnetic field as a basis for determining the direction of $-d\vec{B}_1/dt$.

Inside the solenoid the electric field is proportional to r, the distance from the axis (smaller near the axis). Outside the solenoid the curly electric field is proportional to $1/r$; the electric field gets smaller as you go farther away from the solenoid.

The curly electric field has the usual effect on charges: a charge q experiences a force $\vec{F} = q\vec{E}$. For example, a proton placed above the solenoid in Figure 22.3 will be initially pushed to the right. Because this curly electric field is associated with the time-varying current and magnetic field rather than with stationary charges, we call such an electric field a "non-Coulomb" field, $\vec{E}_{NC}$.

QUESTION Explain how you know that this pattern of curly electric field cannot be produced by an arrangement of stationary charges.

Traveling in a complete loop around the solenoid, $\oint \vec{E}_{NC} \bullet d\vec{l} \neq 0$. Since the round-trip integral of the electric field due to stationary charges is always zero, this pattern of electric field cannot be produced by stationary charges.

NON-COULOMB ELECTRIC FIELD

- A *Coulomb* electric field is produced by charges according to Coulomb's law:

$$\vec{E} = \frac{1}{4\pi\varepsilon_0}\frac{q}{r^2}\hat{r}$$

- A *non-Coulomb* electric field $\vec{E}_{NC}$ is associated with time-varying magnetic fields $d\vec{B}/dt$. Outside of a long solenoid inside of which the magnetic field is B_1, the non-Coulomb electric field is proportional to dB_1/dt and decreases with distance like $1/r$.

Figure 22.4 shows what is observed experimentally in four different cases, where the magnetic field points out of or into the page, and increases or decreases with time. From these results you can see that it is not the direction of $\vec{B}_1$ that determines the direction of $\vec{E}_{NC}$, but rather the direction of the rate of change of $\vec{B}_1$. The following right-hand rule summarizes the experimental observations:

DIRECTION OF THE CURLY ELECTRIC FIELD

With the thumb of your right hand pointing in the direction of $-d\vec{B}_1/dt$, your fingers curl around in the direction of $\vec{E}_{NC}$.

Finding the Direction of $-d\vec{B}/dt$

In order to use this right-hand rule, you need to be able to determine the direction of the vector quantity $-d\vec{B}_1/dt$. It helps to think about this quantity in the form $-\Delta\vec{B}_1/\Delta t$, where Δt is a small, finite time interval. From this form you can see that the direction of $-d\vec{B}_1/dt$ is the same as the direction of $-\Delta\vec{B}_1$, the negative of the change in direction of the magnetic field during a short time.

A good way to find the direction of $-d\vec{B}_1/dt$ is to draw the magnetic field $\vec{B}_1(t)$ at a time t, and the magnetic field $\vec{B}_1(t+\Delta t)$ at a slightly later time $t+\Delta t$, and observe the change $\Delta\vec{B}_1$ (Figure 22.5). Then $-\Delta\vec{B}_1$ is the direction of $-d\vec{B}_1/dt$.

Checkpoint 1 A magnetic field near the floor points up and is increasing. Looking down at the floor, does the non-Coulomb electric field curl clockwise or counterclockwise? A magnetic field near the ceiling points down and is decreasing. Looking up at the ceiling, does the non-Coulomb electric field curl clockwise or counterclockwise?

Driving Current with a Non-Coulomb Electric Field

Suppose that we place a circular metal ring of radius r_2 around a solenoid (Figure 22.6), with the magnetic field in the solenoid increasing with time (Figure 22.7). The non-Coulomb electric field inside the metal will drive conventional current clockwise around the ring ($-d\vec{B}_1/dt$ points into the page; point your right thumb into the page and see how your fingers curl clockwise). The technological importance of this effect is that it can make current run in a wire just as though a battery were present.

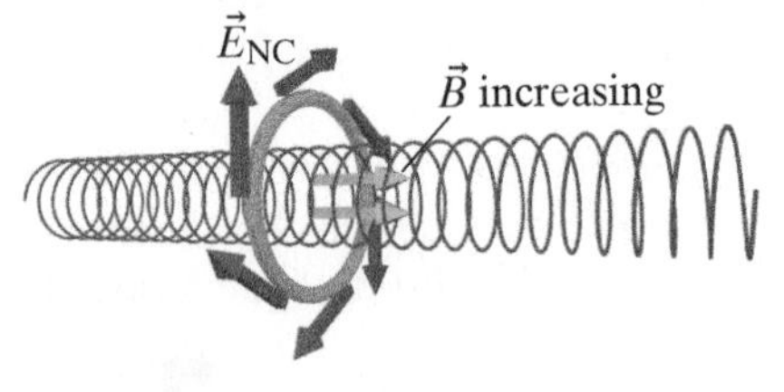

Figure 22.6 A metal ring is placed around the solenoid.

The current I_2 in the ring is proportional to the electric field E_{NC} inside the metal, as in an ordinary circuit. However, in ordinary circuits there are charges on the surface of the wires that, together with charges on the battery, produce the electric field inside the metal that drives the current.

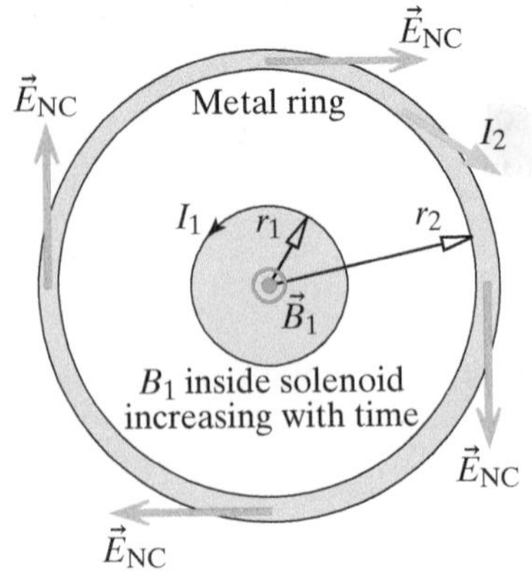

Figure 22.7 End view: The non-Coulomb electric field drives a current I_2 in the ring.

> **QUESTION** Think about a possible pattern of surface charge on this ring. Why is it impossible to draw a plausible gradient of surface charge on the ring in this situation?

The symmetry of the ring makes it impossible to have a surface charge gradient along the ring. We reason by contradiction: If you pick one point and draw positive surface charge there, then gradually decrease the amount of positive charge and increase the amount of negative surface charge, you find that when you get back to the starting point there is a huge change in surface charge (from – to +), which would produce a huge E, in the wrong direction (Figure 22.8). However, there is nothing special about the point you picked, so it can't have a different electric field from all other points on the ring, and there cannot be a varying surface charge around the ring. (You may have realized, however, that there must be a tiny pileup of electrons on the outside of the ring, which then provides the radially inward force that turns the electron current; the pileup is extremely small because the radial acceleration $\bar{v}^2/r$ is very small due to the low drift speed $\bar{v}$ of the mobile electrons.)

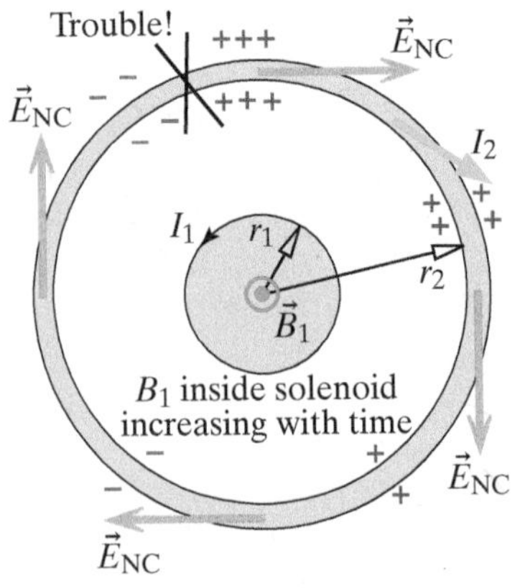

Figure 22.8 This pattern of surface charge is impossible, because it would imply a huge E at the marked location, and in the wrong direction!

The emf is the (non-Coulomb) energy input per unit charge. The (non-Coulomb) force per unit charge is the (non-Coulomb) field E_{NC}.

> **QUESTION** Therefore, what is the emf in terms of E_{NC} for this ring, if the ring has a radius r_2?

$$\text{emf} = \oint \vec{E}_{NC} \bullet d\vec{l} = E_{NC}(2\pi r_2)$$

since E_{NC} is constant and parallel to the path. The current in the ring of radius r_2 is $I_2 = \text{emf}/R$, where R is the resistance of the ring. It is as though we had inserted a battery into the ring.

> **QUESTION** If the metal ring had a radius r_2 twice as large, what would be the emf, since $\text{emf} = \oint \vec{E}_{NC} \bullet d\vec{l}$? Remember that the experimental observations show that the non-Coulomb electric field outside the solenoid is proportional to $1/r$.

Double the radius implies half the electric field, so the product $E_{NC}(2\pi r_2)$ stays the same. Apparently the emf in a ring encircling the solenoid is the same for any radius. (In fact, we get the same emf around any circuit surrounding the solenoid, not just a circular ring.)

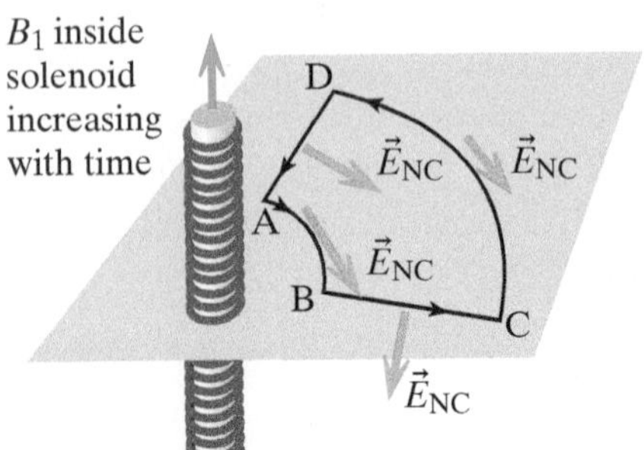

Figure 22.9 A path that does not encircle the solenoid.

> **QUESTION** Consider a round-trip path that does not encircle the solenoid (Figure 22.9). The electric field is shown all along the path A-B-C-D-A. Use the fact that $E_{NC} = \text{emf}/(2\pi r)$ to explain why we have the result $\text{emf} = \oint \vec{E}_{NC} \bullet d\vec{l} = 0$ around this path that does not encircle the solenoid.

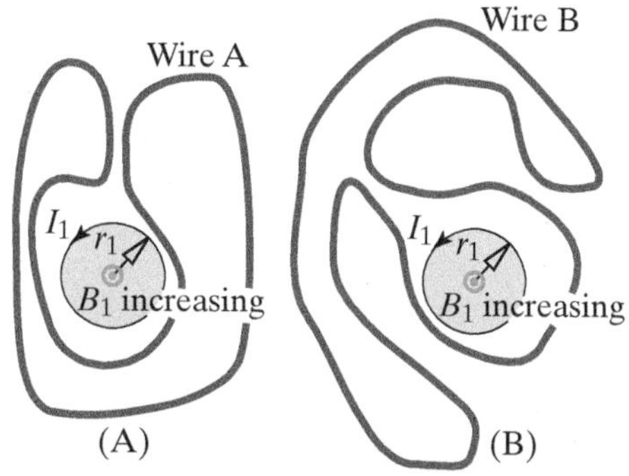

Figure 22.10 Will current run in these purple wires?

As you make a round trip around the path A-B-C-D-A, the contribution to $\oint \vec{E}_{NC} \bullet d\vec{l}$ is positive along A-B, zero along B-C (the parallel component of electric field is zero), negative along C-D, and zero along D-A. Moreover, the magnitude of the contribution along C-D is equal to that along A-B, because the longer path is compensated by the smaller electric field (which is proportional to $1/r$). Therefore we have a zero emf along this path. The non-Coulomb electric field would polarize a wire that followed this path but would not drive current around the loop.

We see that in order to drive current in a wire the wire must encircle a region where the magnetic field is changing. If the wire doesn't encircle the region of changing magnetic field, there is no emf.

Checkpoint 2 **(a)** In Figure 22.10, will current run in wire A? **(b)** In wire B? On a circular path of radius 10 cm in air around a solenoid with increasing magnetic field, the emf is 30 V. **(c)** What is the magnitude of the non-Coulomb electric field on this path? **(d)** A wire with resistance 4 Ω is placed along the path. What is the current in the wire?

22.2 FARADAY'S LAW

So far we have a right-hand rule for determining the direction of the curly electric field, but we don't have a way to calculate the magnitude of the non-Coulomb electric field. Faraday's law states a quantitative relationship between the rate of change of the magnetic field and the magnitude of the non-Coulomb electric field. To establish this quantitative relationship, in principle we could vary the magnetic field and measure E_{NC} by observing the effect of the curly electric field on an individual charged particle, but it can be difficult to track the path of an individual charged particle. Alternatively, we can construct a circuit following a path where we expect a non-Coulomb electric field and measure the current in the circuit as a function of dB/dt. We will describe circuit experiments that lead to Faraday's law.

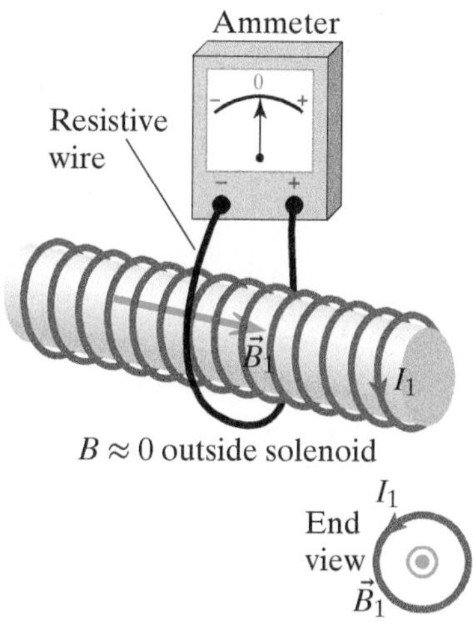

Figure 22.11 An ammeter measures current in a loop surrounding the solenoid. Initially I_1 is constant, so B_1 is constant, and no current runs through the ammeter. (The power supply and connections to the solenoid are not shown.)

Observing Current Caused by Non-Coulomb Electric Fields

In Figure 22.11, if we vary the magnetic field B_1 by varying I_1 and use the ammeter to measure the current in the loop around the solenoid, we can infer something about the resulting non-Coulomb electric field E_{NC} from its integral around that loop, which is the emf. Unfortunately you can't observe the phenomenon with simple equipment such as a compass; you need a sensitive ammeter. Perhaps your instructor will demonstrate the effects or arrange for you to experiment with appropriate equipment.

Initially the solenoid current I_1 is constant, so B_1 is constant, and no current runs through the resistive wire and the ammeter. Next we vary the current I_1 in the long solenoid, thus causing the magnetic field B_1 to vary, and we observe the current I_2 in the outer wire. In the solenoid we first increase the current rapidly, then hold the current constant, and then slowly decrease the current, at half the rate we used at first (Figure 22.12). If we know the resistance R of the circuit containing the ammeter we can determine the emf from the ammeter reading, since emf $- RI_2 = 0$; the wire acts as though a battery were inserted.

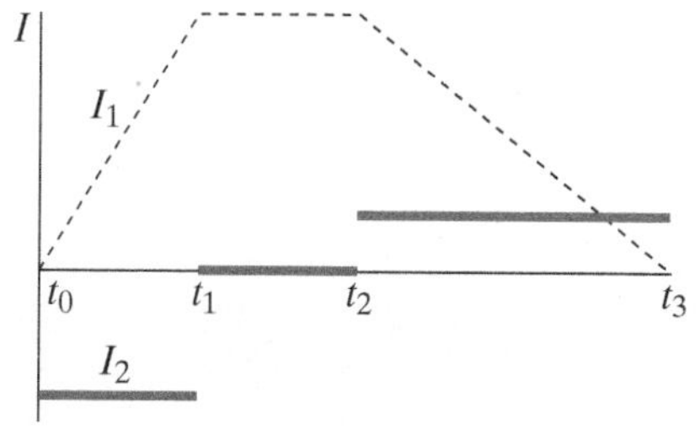

Figure 22.12 Vary the solenoid current I_1 and observe the current I_2 that runs in the outer wire, through the ammeter.

1. While the solenoid current I_1 is increasing from t_0 to t_1 (Figure 22.12), B_1 is increasing, and the current I_2 runs clockwise in the circuit, out of the "+" terminal of the ammeter. The ammeter is observed to read a negative current. Remember that conventional current flowing into the positive terminal of an ammeter gives a positive reading.

If $dB/dt \neq 0$, there is a curly E_{NC}.

2. While the solenoid current I_1 is held constant from t_1 to t_2 (Figure 22.12), the magnetic field in the solenoid isn't changing, and there is no emf in the circuit. The ammeter reading is zero.

    No dB/dt, no curly E_{NC}, even if there is a large (constant) B.

3. While the solenoid current I_1 is decreasing from t_2 to t_3 (at half the initial rate; see Figure 22.12), B_1 is decreasing, and the current I_2 runs counterclockwise in the circuit, into the + terminal of the ammeter. The ammeter is observed to read a positive current. The current I_2 is observed to be half what it was during the first interval.

    E_{NC} (and associated emf) outside the solenoid are proportional to dB/dt inside the solenoid.

4. There is one more crucial experiment. If we use a solenoid with twice the cross-sectional area but the same magnetic field, we find that the current I_2 is twice as big (which means that the emf is twice as big).

    E_{NC} (and associated emf) outside the solenoid are proportional to the cross-sectional area of the solenoid.

Magnetic Flux

Putting together these experimental observations and being quantitative about the emf, we find experimentally that the magnitude of the emf is numerically equal to this:

$$|\text{emf}| = \left|\frac{d}{dt}(B_1 \pi r_1^2)\right|$$

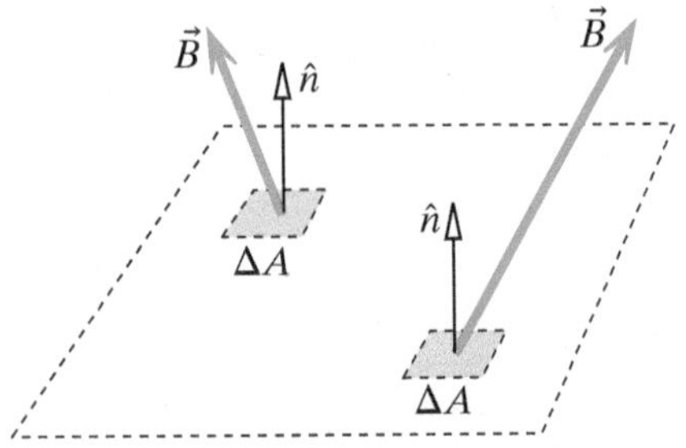

Figure 22.13 The magnetic flux on an area is the sum of the magnetic flux on each small subarea.

The quantity $(B_1 \pi r_1^2)$ is called the "magnetic flux" Φ_{mag} on the area encircled by the circuit (Φ is the uppercase Greek letter phi). Magnetic flux is calculated in the same way as the electric flux introduced in Chapter 21 on Gauss's law.

The magnetic flux on a small area ΔA is $\vec{B} \bullet \hat{n} \Delta A = B_\perp \Delta A$, where $\hat{n}$ is a dimensionless unit vector perpendicular to the area ΔA ($\hat{n}$ is called the "normal" to the surface); $B_\perp$ is the perpendicular component of magnetic field. We add up all such contributions over an extended surface to get the magnetic flux over that surface (Figure 22.13):

DEFINITION OF MAGNETIC FLUX

$$\Phi_{mag} = \int \vec{B} \bullet \hat{n} dA = \int B_\perp dA$$

Note that magnetic flux can be positive, negative, or zero, depending on the orientation of the magnetic field $\vec{B}$ relative to the normal $\hat{n}$.

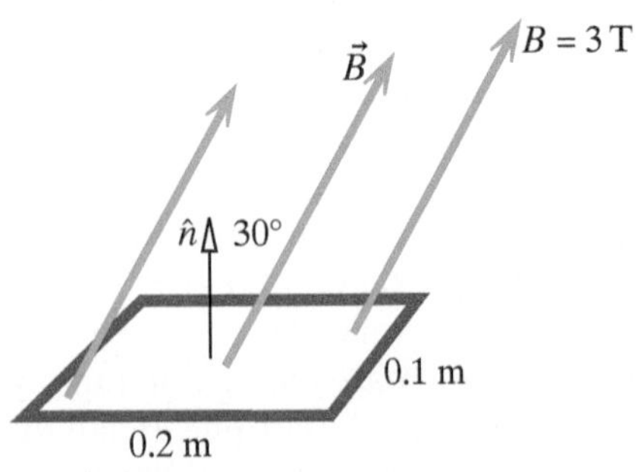

Figure 22.14 Calculate the magnetic flux on the area enclosed by this loop of wire.

Checkpoint 3 A uniform magnetic field of 3 T points 30° away from the perpendicular to the plane of a rectangular loop of wire 0.1 m by 0.2 m (Figure 22.14). What is the magnetic flux on this loop?

Faraday's Law—Quantitative

The experimental fact that in the presence of a time-varying magnetic field there is an emf whose magnitude is equal to the rate of change of magnetic flux is called "Faraday's law" (discovered by the British scientist Michael Faraday in 1831). Faraday's law summarizes a great variety of experimental data, not just the data for the experiments we have discussed. Faraday's law is a major physical law concerning the relationship between electric and magnetic fields. Unlike the motional emf we studied in Chapter 20, Faraday's law cannot be derived from any of the other fundamental principles we have studied so far.

FARADAY'S LAW

$$\text{emf} = -\frac{d\Phi_{\text{mag}}}{dt}$$

$$\text{where emf} = \oint \vec{E}_{\text{NC}} \bullet d\vec{l} \text{ and } \Phi_{\text{mag}} \equiv \int \vec{B} \bullet \hat{n}dA.$$

In words: The emf along a round-trip path is equal to the rate of change of the magnetic flux on the area encircled by the path.

Direction: With the thumb of your right hand pointing in the direction of $-d\vec{B}/dt$, your fingers curl around in the direction of $\vec{E}_{\text{NC}}$.

The meaning of the minus sign: If the thumb of your right hand points in the direction of $-d\vec{B}/dt$ (that is, the opposite of the direction in which the magnetic field is increasing), your fingers curl around in the direction along which the path integral of electric field is positive:

$$\oint \vec{E}_{\text{NC}} \bullet d\vec{l} = -\frac{d}{dt}\left[\int \vec{B} \bullet \hat{n}dA\right] \quad \text{(sign given by right-hand rule)}$$

For most of our work it is simplest to calculate the magnitude of the effect, ignoring signs and directions, and then give the appropriate sign or direction based on the right-hand rule. We will include the minus sign in Faraday's law, as a reminder of what we need to do to get directions.

What area exactly do we use when calculating the magnetic flux in Faraday's law? Imagine a soap film stretched over the closed path around which we are calculating $\oint \vec{E}_{\text{NC}} \bullet d\vec{l}$. We calculate Φ_{mag} on the area covered by that soap film.

Putting these pieces together, we have the following form:

FORMAL VERSION OF FARADAY'S LAW

$$\oint \vec{E}_{\text{NC}} \bullet d\vec{l} = -\frac{d}{dt}\int \vec{B} \bullet \hat{n}dA \quad \text{(sign given by right-hand rule)}$$

Faraday's law summarizes the experiments we've described so far:

- A bigger emf is associated with a faster rate of change of magnetic flux (if the magnetic flux is constant there is no emf).
- If there is a bigger area with the same perpendicular magnetic field there is a bigger emf.

Checkpoint 4 A wire of resistance 10 Ω and length 2.5 m is bent into a circle that is concentric with and encircles a solenoid in which the magnetic flux changes from 5 $\text{T} \cdot \text{m}^2$ to 3 $\text{T} \cdot \text{m}^2$ in 0.1 s. What is the emf in the wire? What is the non-Coulomb electric field in the wire? What is the current in the wire?

The Coulomb Electric Field Can Be Included in Faraday's Law

We will usually write emf $= \oint \vec{E}_{\text{NC}} \bullet d\vec{l}$ as a reminder that it is the curly non-Coulomb electric field that has a nonzero round-trip path integral, which we call the emf around that path. However, we could also calculate the emf in terms of the net electric field $\vec{E} = \vec{E}_{\text{C}} + \vec{E}_{\text{NC}}$, where $\vec{E}_{\text{C}}$ is the Coulomb electric field due to charges:

$$\text{emf} = \oint \vec{E} \bullet d\vec{l}$$

This is true because we have

$$\text{emf} = \oint \vec{E} \bullet d\vec{l} = \oint (\vec{E}_C + \vec{E}_{NC}) \bullet d\vec{l} = 0 + \oint \vec{E}_{NC} \bullet d\vec{l}$$

since $\oint \vec{E}_C \bullet d\vec{l} = 0$ (the round-trip integral of a Coulomb electric field is zero).

We can apply Faraday's law to analyze quantitatively the situation of a circuit with an ammeter surrounding a solenoid (Figure 22.15; a cross section of the solenoid is shown). There is magnetic field $\vec{B}_1$ only over the circle of radius r_1 (the solenoid), and it points in the same direction as $\hat{n}$, the unit vector perpendicular to the surface.

EXAMPLE

A Circuit Surrounding a Solenoid

In Figure 22.15 the magnetic field in the solenoid increases from 0.1 T to 0.7 T in 0.2 s, and the area of the solenoid is 3 cm^2. **(a)** Calculate the magnetic flux Φ_{mag} on the area enclosed by the solenoid, the flux on the other portions of the area encircled by the circuit, and the total flux on the area encircled by the circuit. **(b)** What is the emf around the circuit? **(c)** If the resistance of the wire plus ammeter is 0.5 Ω, what current will the ammeter display?

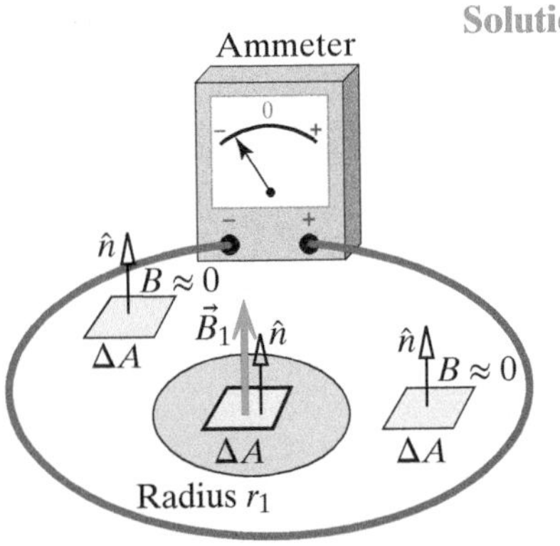

Figure 22.15 A cross section through the solenoid; calculate the flux enclosed by the wire.

Solution

(a) On the area enclosed by the solenoid, $\Phi_{mag} = B_1(\pi r_1^2)$, because $B_\perp$ is constant and equal to B_1 throughout the cross-section of the solenoid. On every small area outside the solenoid, the magnetic field is nearly zero, so $\Phi_{mag} = 0$. Therefore the total flux through the outer wire is $\Phi_{mag} = B_1(\pi r_1^2)$.

What counts is the magnetic flux encircled by the circuit, not the magnetic flux on a closed surface such as a box or sphere. Unlike electric flux, the total magnetic flux on a closed surface is always zero, because "magnetic monopoles" seem not to exist.

(b) Experimentally we find that the emf around the circuit is numerically equal to $d\Phi_{mag}/dt$, as predicted by Faraday's law. Faraday's law predicts the following average emf:

$$\text{emf} = \frac{\Delta\Phi_{mag}}{\Delta t} = \frac{(0.7\text{ T})(3\times10^{-4}\text{ m}^2) - (0.1\text{ T})(3\times10^{-4}\text{ m}^2)}{(0.2\text{ s})} = 9\times10^{-4}\text{ V}$$

(c) The circuit acts as though a battery were inserted, emf − $RI = 0$, so

$$I = \frac{\text{emf}}{R} = \frac{(9\times10^{-4}\text{ V})}{(0.5\ \Omega)} = 1.8\times10^{-3}\text{ A}$$

One more experiment: let's reconnect the wire and ammeter so that they don't encircle the solenoid (Figure 22.16).

QUESTION As we increase the current (and magnetic field) in the solenoid, what would you predict from Faraday's law about the ammeter reading? Why?

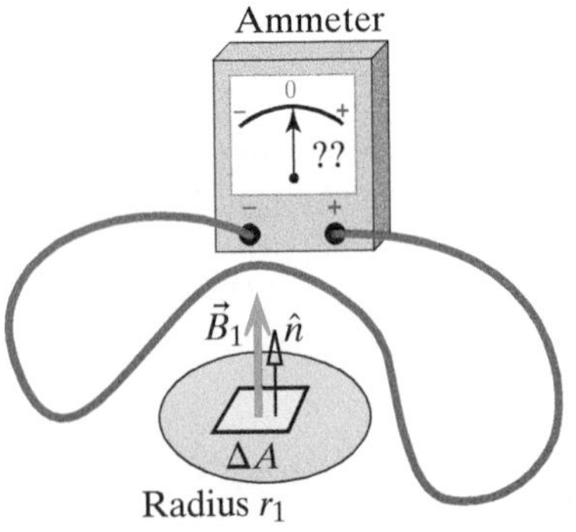

Figure 22.16 What happens if the circuit does not encircle the solenoid?

Within the area bounded by the wire outside the solenoid, there is practically no magnetic field, so no magnetic flux Φ_{mag}, and no rate of change of magnetic flux $d\Phi_{mag}/dt$. When we try the experiment, we do indeed find that the ammeter shows little or no current.

Voltmeter Readings

Since a voltmeter acts like an ammeter with a large resistance in series, a voltmeter may give a puzzling reading in the presence of time-varying magnetic flux. Consider the two voltmeters shown in Figure 22.17. You normally expect that when voltmeter leads are connected to each other, the voltmeter must read zero volts.

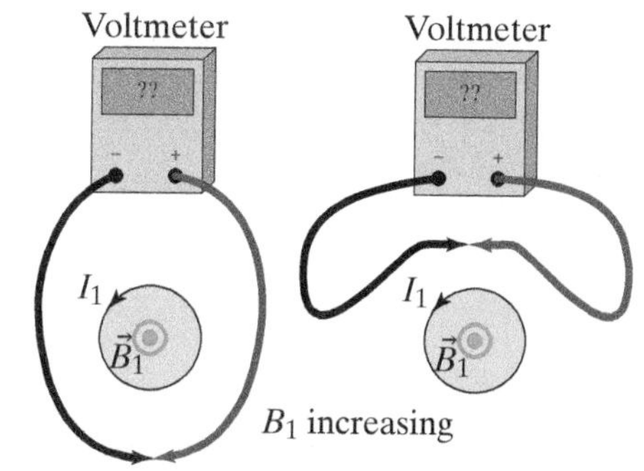

Figure 22.17 What do these voltmeters read?

QUESTION Do these voltmeters read zero?

The voltmeter leads on the left of Figure 22.17 encircle a region of changing magnetic field, so the voltmeter will read an emf equal to $d\Phi_{\text{mag}}/dt$. The leads of the other voltmeter don't encircle a region of changing magnetic field, so the voltmeter reads zero.

Because of the effect of time-varying magnetic fields, you have to be a bit careful in interpreting a voltmeter reading when there are varying magnetic fields around. For example, when there are sinusoidally alternating currents (AC) there are time-varying magnetic fields due to those time-varying currents. If the leads of an AC voltmeter happen to surround some AC magnetic flux, this will affect the voltmeter reading.

The Emf for a Coil with Multiple Turns

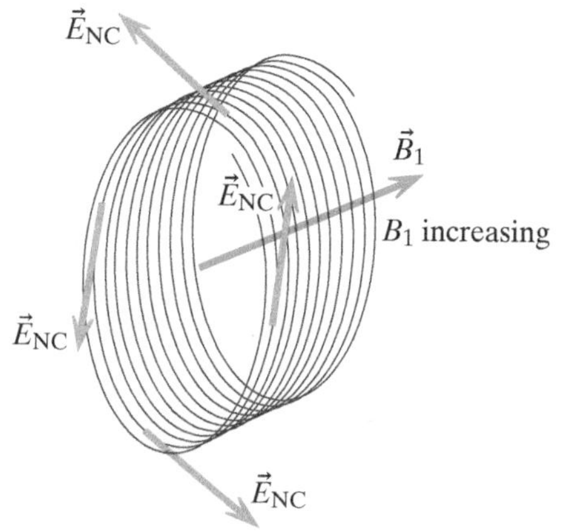

Figure 22.18 The emf of a thin coil is approximately N times the emf of one loop.

In many devices, instead of just one loop of wire surrounding a time-varying flux there is a coil with many turns, which increases the effect: the emf in a coil containing N turns is approximately N times the emf for one loop of the coil. This point needs some discussion.

In the space around the increasing flux in the coil, there is a curly pattern of non-Coulomb electric field. If the loops are tightly wound so that they are very close together, E_{NC} is about the same in each loop (Figure 22.18). The emf from one end of the coil to the other is the integral of the non-Coulomb field:

$$\text{emf} = \int_{N\text{ turns}} \vec{E}_{NC} \bullet d\vec{l} = N(E_{NC}L_{\text{one turn}}) = N(\text{emf}_{\text{one turn}})$$

Because of this, Faraday's law for a coil is written in the following form:

FARADAY'S LAW FOR A COIL

$$\text{emf} = -N\frac{d\Phi_{\text{mag}}}{dt} \quad \text{(sign given by right-hand rule)}$$

The emf in a coil of N turns is equal to N times the rate of change of the magnetic flux on one loop of the coil.

Faraday's Law and Moving Coils or Magnets

Figure 22.19 Move coil 1 toward coil 2, and there is a time-varying magnetic field inside coil 2.

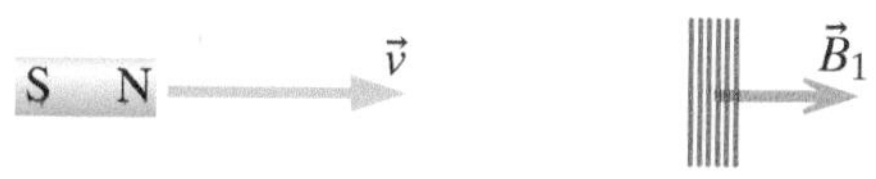

Figure 22.20 Moving a magnet toward coil 2 creates a time-varying magnetic field inside the coil.

Figure 22.21 Rotating a bar magnet (or coil 1) produces a time-varying magnetic field inside coil 2.

A time-varying magnetic field is associated with a curly electric field. One way to create a time-varying magnetic field is by varying the current in a stationary coil, but changes in the position of currents can also produce a time-varying magnetic field and the associated curly electric field. With a steady current in one coil in Figure 22.19, you can move that coil closer to a second coil. This increases the magnetic field (and magnetic flux) inside the second coil, and while the magnetic field is increasing there is an emf that can drive a current in the second coil.

You will also find an emf if you move a bar magnet toward or away from coil 2, since this creates a time-varying magnetic field (and magnetic flux) inside the coil (Figure 22.20).

You can also get an emf by rotating the first coil (or a bar magnet; Figure 22.21), since this changes the flux through the second coil as long as you are rotating and therefore changing the magnetic field (and magnetic flux) in the region of space surrounded by the second coil.

These various experiments give additional confirmation that the emf in one or more loops of wire is associated with a time-varying magnetic field, and numerically equal to the rate of change of the magnetic flux enclosed by the loops:

$$\text{emf} = -N\frac{d\Phi_{\text{mag}}}{dt} = -N\frac{d}{dt}\left(\sum \vec{B}_1 \bullet \hat{n}\,\Delta A_2\right)$$

An Interesting Complication

The current I_1 in a coil of wire makes a magnetic field B_1. If I_1 varies with time, there is a time-varying magnetic field dB_1/dt in a second coil and an emf and current I_2 in the second coil. This current I_2 *also* makes a magnetic field B_2, an effect that we have ignored until now. If I_2 is constant (due to a constant dB_1/dt), the additional magnetic field B_2 is constant and so does not contribute any additional non-Coulomb electric field.

However, if $d^2B_1/dt^2 \neq 0$ (there is a time-varying rate of change), then I_2 isn't constant, and there is a time-varying magnetic field dB_2/dt and a non-Coulomb electric field and emf of its own. In many cases this effect is small compared to the main effect and can be ignored. If the effect is sizable, you can see that a full analysis could be rather difficult, because you have a changing I_1 making a changing I_2 that contributes to the change, which ... Whew!

We will mostly avoid trying to analyze this complicated kind of situation in this introductory textbook, though we will see an example in the discussion of superconductors later in this chapter. A related issue is self-inductance, which we will also study later in this chapter.

EXAMPLE

Two Coils

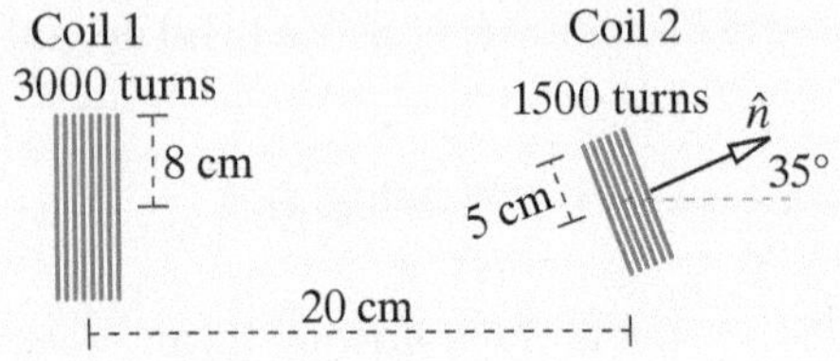

Figure 22.22 Two coils.

Two coils of wire are near each other, 20 cm apart, as shown in Figure 22.22. Coil 1 has a radius 8 cm and contains 3000 loops of wire. It is connected to a power supply whose output voltage can be changed, so that the current I_1 in coil 1 can be varied. Coil 2 contains 1500 loops of wire and has radius 5 cm, and is rotated 35° from the axis of coil 1, as shown in Figure 22.22. The current in coil 1 changes from 0 to 3 A in 4 ms. Calculate the emf in coil 2.

Solution Start from fundamental principles. For a given current I_1 in the first coil, we can find B_1, the magnetic field that this current makes at the location of coil 2. We know (dI_1/dt), so we can find $(d\Phi_{\text{mag}}/dt)$ in coil 2, which gives us the emf in coil 2.

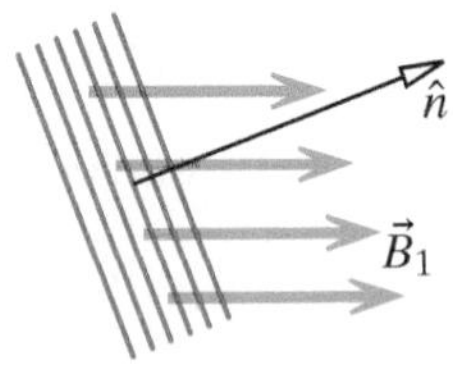

Figure 22.23 Assume the magnetic field made by coil 1 is uniform through the interior of coil 2.

Approximations: Assume that the coils are far enough apart that the magnetic field of the first coil is approximately uniform in direction and magnitude throughout the interior of the second coil (Figure 22.23), and can be approximated as a magnetic dipole field. Also assume that the length of the coils is small compared to the distance between them (essentially treating all loops as if they were located at the center of the coil).

Work out everything symbolically first; plug in numbers at the end.

Magnetic field B_1 at the location of coil 2:

$$B_1 \approx N_1 \frac{\mu_0}{4\pi} \frac{2I_1(\pi r_1^2)}{z^3}$$

Magnetic flux through one loop of coil 2:

$$\Phi_{\text{mag}} \approx B_1(\pi r_2^2)\cos\theta$$

$$= N_1 \frac{\mu_0}{4\pi} \frac{2I_1(\pi r_1^2)}{z^3}(\pi r_2^2)\cos\theta$$

I_1 is the only quantity that is changing, so the emf around one loop of coil 2 is:

$$\text{emf} = \left|\frac{d\Phi_{\text{mag}}}{dt}\right| \approx N_1 \frac{\mu_0}{4\pi} \frac{2(\pi r_1^2)(\pi r_2^2)}{z^3} \cos\theta \frac{dI_1}{dt}$$

Total emf for all of coil 2:

$$\text{emf} = N_2\left|\frac{d\Phi_{\text{mag}}}{dt}\right| \approx N_1N_2\frac{\mu_0}{4\pi}\frac{2(\pi r_1^2)(\pi r_2^2)}{z^3}\cos\theta\frac{dI_1}{dt}$$

$$\frac{(1500)(3000)(1\times10^{-7}\,\text{T}\cdot\text{m/A})(2)(0.08\text{ m})^2(\pi^2)(0.05\text{ m})^2\cos 35°(3\text{ A})}{(0.20\text{ m})^3(4\times10^{-3}\text{ s})}$$

$$= 10.9\text{ V}$$

EXAMPLE

Wire Loop and a Solenoid

A length of wire whose total resistance is R is made into a loop with two quarter-circle arcs of radius r_1 and radius r_2 and two straight radial sections as shown in Figure 22.24. A very long solenoid is positioned as shown, going into and out of the page. The solenoid consists of N turns of wire wound tightly onto a cylinder of radius r and length d, and it carries a counterclockwise current I_s that is increasing at a rate dI_s/dt.

What are the magnitude and direction of the magnetic field at C, at the center of the two arcs? The "fringe" magnetic field of the solenoid (that is, the magnetic field produced by the solenoid outside of the solenoid) is negligible at point C. Show all relevant quantities on a diagram.

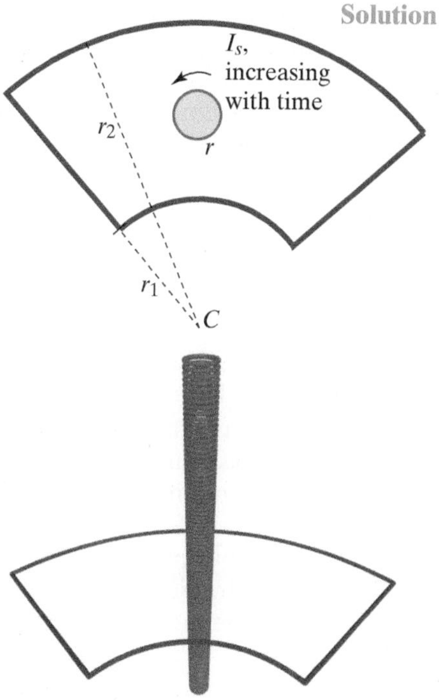

Figure 22.24 Upper: A long solenoid passes through a loop of wire. Lower: A 3D view.

Solution As usual, start from fundamentals. Initially the only sources of field are the moving charges in the solenoid (which make B_s in the solenoid) and dB_s/dt in the solenoid (with an associated curly non-Coulomb electric field around the solenoid). The curly electric field constitutes an emf that drives current I_{loop} in the surrounding wire. The current-carrying wires produce a magnetic field at location C.

Minor point: Because the wire is not circular, pileup leads to tiny amounts of surface charge on the wire. The net electric field in the wire (the superposition of the non-Coulomb and Coulomb fields) follows the wire and is uniform in magnitude throughout the wire.

(a) Inside the solenoid, $B_s = \frac{\mu_0 N I_s}{d}$ out of the page. See Figure 22.25.

(b) Flux through *outer* loop =

$$B_s A_s\cos(0°) + (0)A_{\text{rest of loop}} = \frac{\mu_0 N I_s}{d}\pi r^2$$

(c) $-d\vec{B}/dt$ into page, so $\vec{E}_{\text{NC}}$ and current I_{loop} are clockwise as shown.

$$|\text{emf}| = \frac{d}{dt}\left(\frac{\mu_0 N I_s}{d}\pi r^2\right) = \frac{\mu_0 N}{d}\pi r^2\frac{dI_s}{dt}$$

$$\text{so } I_{\text{loop}} = \frac{\text{emf}}{R} = \frac{\mu_0 N}{d}\frac{\pi r^2}{R}\frac{dI_s}{dt}.$$

(d) $B_{\text{at C}}$ due to straight segments, inner wire, and outer wire:

$$B_{\text{straight segments}} = 0 \text{ because } d\vec{l}\times\hat{r} = 0 \quad \text{(see diagram)}$$

$$B_{\text{inner wire}} = \left|\int_{\text{quarter loop}}\frac{\mu_0}{4\pi}\frac{I_{\text{loop}}\,d\vec{l}\times\hat{r}}{r_1^2}\right| = \frac{\mu_0}{4\pi}\frac{I_{\text{loop}}}{r_1^2}\int_{\text{quarter loop}} dl\sin(90°)$$

$$B_{\text{inner wire}} = \frac{\mu_0}{4\pi}\frac{I_{\text{loop}}}{r_1^2}\left(\frac{2\pi r_1}{4}\right) = \frac{\mu_0}{4\pi}\frac{\pi I_{\text{loop}}}{2r_1}, \text{ and } B_{\text{outer wire}} = \frac{\mu_0}{4\pi}\frac{\pi I_{\text{loop}}}{2r_2}$$

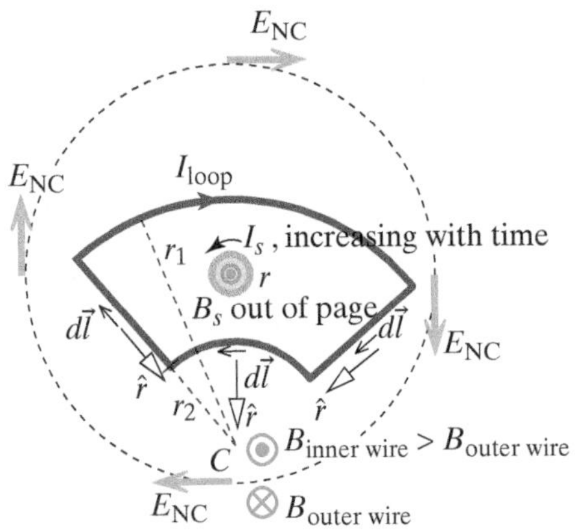

Figure 22.25 Fields and currents.

$B_{\text{inner wire}}$ is out of the page, and $B_{\text{outer wire}}$ is into the page at location C. Since B is proportional to $1/r$, $B_{\text{inner wire}}$ is larger than $B_{\text{outer wire}}$. The net field is out of the page, with this magnitude:

$$B_{\text{net at C}} = B_{\text{inner wire}} - B_{\text{outer wire}} = \frac{\mu_0}{4\pi}\frac{\pi I_{\text{loop}}}{2}\left(\frac{1}{r_1} - \frac{1}{r_2}\right)$$

$$B_{\text{net at C}} = \frac{\mu_0}{4\pi}\left(\frac{\pi}{2}\right)\left(\frac{\mu_0 N}{d}\frac{\pi r^2}{R}\frac{dI_s}{dt}\right)\left(\frac{1}{r_1} - \frac{1}{r_2}\right)$$

QUESTION Suppose that in this example the solenoid had been filled with an iron core. How would this have affected the emf in the wire?

Recall from Chapter 17 that iron has a multiplier effect. When you apply a magnetic field to iron, magnetic domains that are nearly aligned with the applied magnetic field tend to grow at the expense of domains that are not aligned, and within a domain the individual magnetic dipoles can rotate to be more aligned to the applied magnetic field. The new magnetic field due to the aligned magnetic dipoles in the iron can be much larger than the applied magnetic field, and as long as not all domains are fully aligned, this larger magnetic field will be approximately proportional to the applied field.

To be concrete, suppose that the magnetic field of a solenoid without an iron core varies from 0 to 0.1 T in 0.1 s, so that $dB/dt = 0.1/0.1 = 1$ T/s. Suppose that with an iron core present the magnetic field is 10 times greater and varies from 0 to 1 T in 0.1 s, so that $dB/dt = 1/0.1 = 10$ T/s. With 10 times the rate of change of magnetic field, the emf generated around the solenoid will be 10 times greater. This is the reason that you will often see coils of wire wrapped around iron cores in various kinds of electrical devices.

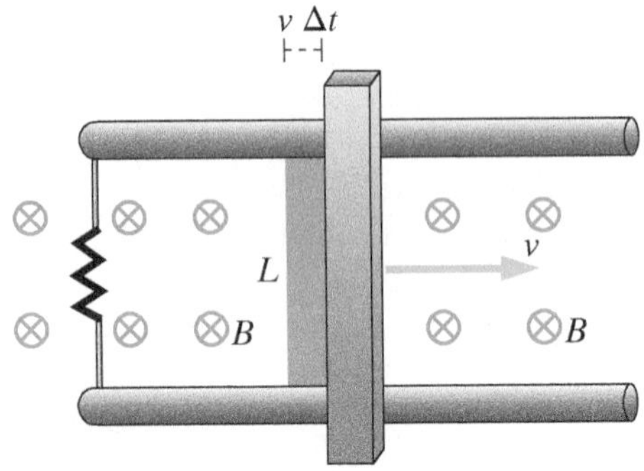

Figure 22.26 There is increased flux through the circuit as the bar moves.

22.3 FARADAY'S LAW AND MOTIONAL EMF

In Chapter 20 we studied motional emf: when a wire moves through a magnetic field, magnetic forces drive current along the wire. We calculated the motional emf in a moving bar of length L as $(q|\vec{v} \times \vec{B}|)L/q$, where $\vec{v}$ is the velocity of the bar. If $\vec{v}$ is perpendicular to the magnetic field, the motional emf is simply vBL. There is another way to calculate motional emf that is often mathematically easier, in terms of magnetic flux.

As shown in Figure 22.26, in a short time Δt the bar moved a distance $\Delta x = v\Delta t$, and the area surrounded by the current-carrying pieces of the circuit increased by an amount $\Delta A = L\Delta x = Lv\Delta t$. There is an increased magnetic flux through the circuit of amount

$$\Delta\Phi_{\text{mag}} = B_\perp \Delta A = B(Lv\Delta t)$$

Dividing by Δt we find this:

$$\frac{\Delta\Phi_{\text{mag}}}{\Delta t} = BLv$$

However, BLv is equal to the emf vBL associated with the magnetic force, and in the limit of small Δt we have the following:

MOTIONAL EMF = RATE OF CHANGE OF MAGNETIC FLUX

$$|\text{emf}| = \left|\frac{d\Phi_{\text{mag}}}{dt}\right| \quad \text{(sign given by direction of magnetic force)}$$

Although we have proved this result only for a particular special case, it can be shown that this result applies in general to calculating motional emf—that is, an emf associated with the motion of a piece of a circuit in a magnetic field.

As an example, consider a generator (Figure 22.27). When the normal to the loop is at an angle $\theta = \omega t$ to the magnetic field, the flux through the loop is $Bhw\cos(\omega t)$, so emf $= |d\Phi/dt| = \omega Bhw\sin(\omega t)$, which is the same result we obtained in Chapter 20 when we calculated the emf directly in terms of magnetic forces acting on electrons in the moving wires (motional emf).

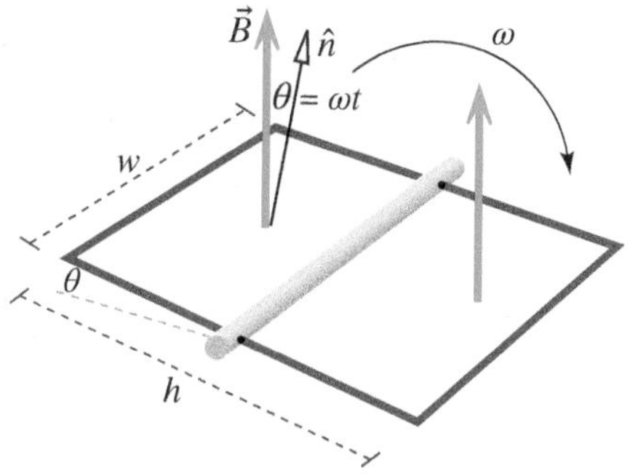

Figure 22.27 We can compute the emf of a generator from the rate of change of flux.

Checkpoint 5 A uniform, non-time-varying magnetic field of 3 T points 30° away from the perpendicular to the plane of a rectangular loop of wire 0.1 m by 0.2 m (Figure 22.28). The loop as a whole is moved in such a way that it maintains its shape and its orientation in the uniform magnetic field. What is the emf around the loop during this move? In 0.1 s the loop in Figure 22.28 is stretched to be 0.12 m by 0.22 m while keeping the center of the loop in one place. What is the average emf around the loop during this time?

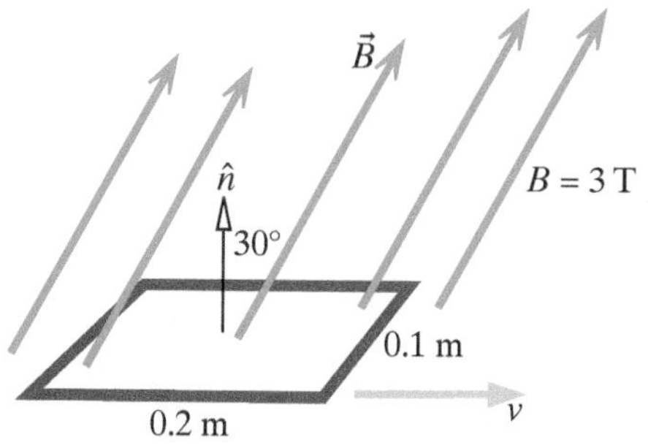

Figure 22.28 Move this rectangular circuit around in a uniform, constant magnetic field.

The Two Pieces of the Flux Derivative

A time-varying magnetic field is associated with a curly (non-Coulomb) electric field. We can quantitatively predict the behavior of circuits enclosing a time-varying magnetic field by calculating the rate of change of magnetic flux on the surface bounded by the circuit.

However, in the case of the sliding bar we just saw that there is another way that magnetic flux can change. If the magnetic field is constant but the shape or orientation of the loop changes so as to enclose more or less flux, the flux changes with time even though the magnetic field doesn't, and this is the phenomenon of motional emf associated with magnetic forces.

Any change in magnetic flux, whether due to a change in magnetic field or a change in shape or orientation, produces an emf equal to the rate of change of flux. Both effects can be present simultaneously. In Figure 22.29, suppose that as you drag a metal bar along rails the magnitude of the uniform magnetic field is also increasing with time. The flux is $B_\perp A$, and the rate of change of the flux is due in part to the change in the magnetic field, and in part to the change in the area enclosed by the changing path:

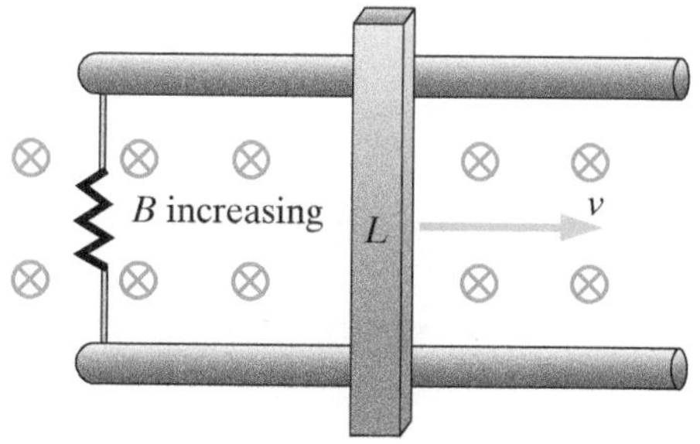

Figure 22.29 Because B is increasing there is a curly electric field $\vec{E}_{NC}$, and because the bar is moving there is also motional emf.

$$\begin{aligned}\frac{d\Phi_{mag}}{dt} &= \frac{d}{dt}(B_\perp A)\\ &= \frac{dB_\perp}{dt}A + B_\perp\frac{dA}{dt}\end{aligned}$$

The first term, $(dB_\perp/dt)A$, is nonzero if $B_\perp$ varies with time (Faraday's law). The second term, $B_\perp dA/dt$, is nonzero if there is a change in the area enclosing magnetic flux (motional emf).

Evaluating these two terms for the case of a bar moving along rails, with the magnetic field changing in time as well as there being a change in the area enclosing magnetic flux, we have this:

$$\begin{aligned}\frac{d\Phi_{mag}}{dt} &= \frac{dB_\perp}{dt}A + B_\perp\frac{dA}{dt}\\ &= \frac{dB_\perp}{dt}A + B_\perp\frac{Ldx}{dt}\\ &= \frac{dB_\perp}{dt}A + B_\perp Lv\end{aligned}$$

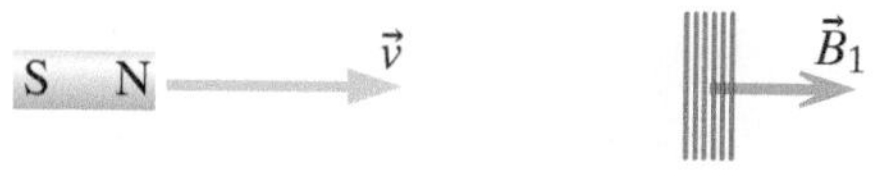

Figure 22.30 Move a magnet toward a stationary coil; there is a changing magnetic field in the coil, and a curly electric field $\vec{E}_{NC}$.

In some but not all cases the two effects can be related through a change of reference frame. If you move a magnet toward a stationary coil (Figure 22.30), the magnetic field in the coil is increasing with time, and there is a curly non-Coulomb electric field that drives current in the coil.

If on the other hand you move a coil toward a stationary magnet (Figure 22.31), there is no fixed location in space where the magnetic field is changing with time, but the coil is moving in a magnetic field, so there is a magnetic force on the mobile charges in the wire, and motional emf. (It is true that the magnetic field at the location of the moving coil is changing with time, but this is not a fixed location; there is no location where $dB/dt \neq 0$.)

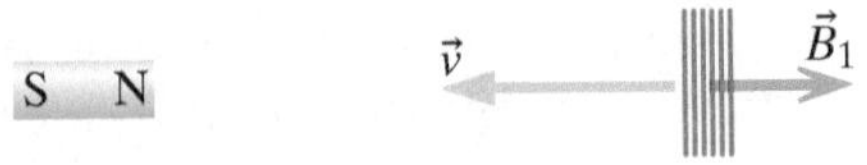

Figure 22.31 Move a coil toward a stationary magnet; there is no changing magnetic field at any fixed location in space, but there is motional emf in the coil.

In both situations you can calculate the same emf from the same rate of change of flux, and you observe the same current to run in the coil, though the mechanism looks different in the two cases. You can shift between one situation and the other simply by changing your reference frame (move with the magnet, or move with the coil). The principle of relativity implies that you should predict the same emf and the same current in either frame of reference, and you do.

It is not always possible to relate the two kinds of effects simply by changing reference frame. For example, consider a bar sliding along rails. In the reference frame of the bar, the bar is at rest but the resistor is moving. In neither reference frame is there a time-varying magnetic field.

(A full analysis in the reference frame of the bar is complicated. If the bar is at rest in this frame, why is it polarized? We saw in Chapter 20 that the answer is given by special relativity: when you transform to the frame of the bar in the presence of a uniform magnetic field, a transverse electric field of magnitude vB appears, and in this reference frame it is this new electric field that polarizes the bar.)

Summary: Non-Coulomb Fields and Forces

If the magnetic field changes ($d\vec{B}/dt$ nonzero), there is a non-Coulomb electric field $\vec{E}_{NC}$ that curls around the region of changing flux, and emf $= \oint \vec{E}_{NC} \bullet d\vec{l}$ is nonzero. In purely motional emf, however, where the path changes but the magnetic field doesn't change, the non-Coulomb forces are magnetic forces, not electric.

The non-Coulomb electric field $\vec{E}_{NC}$ affects stationary as well as moving charges, but the (non-Coulomb) magnetic forces only affect moving charges, such as electrons in the moving metal bar. It is somewhat odd that the magnitude of the emf in both of these situations is $d\Phi_{mag}/dt$, but that's how it works.

We have identified three different kinds of electric and magnetic forces on the electrons inside a wire:

1. Forces from Coulomb electric fields due to surface charges: superposition of contributions of the form $\vec{E} = \dfrac{1}{4\pi\varepsilon_0}\dfrac{q}{r^2}\hat{r}$
2. Forces from non-Coulomb electric fields associated with time-varying magnetic fields, $\dfrac{d\vec{B}}{dt}$
3. Magnetic forces $q\vec{v} \times \vec{B}$ if the wire is moving (yielding a motional $|\text{emf}| = \left|\dfrac{d\Phi_{mag}}{dt}\right|$ due to change of path)

All of these effects can be present simultaneously. For example, if the magnetic field is changing while you pull a bar along rails, there is a Coulomb electric field due to the charges that build up on the surfaces of the circuit elements, there is a non-Coulomb electric field due to the changing magnetic field, and there are non-Coulomb magnetic forces on electrons inside the moving bar. The round-trip integral of the Coulomb electric field is zero. The round-trip integral of the non-Coulomb electric field and the non-Coulomb magnetic force per unit charge together gives the emf, and the current I is given by $I = \text{emf}/R$, where $|\text{emf}| = |d\Phi_{mag}/dt|$.

22.4 MAXWELL'S EQUATIONS

In Chapter 21 we summarized the four Maxwell's equations that describe the patterning of electric and magnetic fields. We can update Maxwell's equations now that we know that the round-trip integral of electric field need not be zero (Faraday's law):

MAXWELL'S EQUATIONS

$\oint \vec{E} \bullet \hat{n} dA = \frac{1}{\varepsilon_0} \sum Q_{\text{inside the surface}}$	Gauss's law
$\oint \vec{B} \bullet \hat{n} dA = 0$	Gauss's law for magnetism
$\oint \vec{E} \bullet d\vec{l} = -\frac{d}{dt} \int \vec{B} \bullet \hat{n} dA$	Faraday's law
$\oint \vec{B} \bullet d\vec{l} = \mu_0 \sum I_{\text{inside path}}$	Ampere's law (incomplete)

In Chapter 23 we will see that Ampere's law needs to be modified, at which point Maxwell's equations will be complete.

Maxwell's equations are a summary of a large number of diverse experiments concerning the possible patterns of and relationships between $\vec{E}$ and $\vec{B}$ in three-dimensional space. No violation of these relations has ever been observed.

You might wonder, "Where is Coulomb's law in all this? Where is the Biot–Savart law?" The equation for the electric field of a stationary point charge can be deduced from Gauss's law. Draw a spherical Gaussian surface around the charge and assume by spherical symmetry that on that surface $\vec{E}$ is everywhere radial and everywhere has the same magnitude. Then the electric flux over the whole spherical surface is $E(4\pi r^2)$, which Gauss's law says should be equal to q/ε_0. Solving for the field, we find the familiar result:

$$\vec{E} = \frac{1}{4\pi\varepsilon_0} \frac{q}{r^2} \hat{r}$$

Moreover, Faraday's law assures us that $\oint \vec{E} \bullet d\vec{l} = 0$ (when there is no time-varying magnetic field), and this is consistent with the outward-going or inward-going pattern of electric field around a stationary point charge, with no curly electric field.

Similarly, starting from Ampere's law it is possible to derive the Biot–Savart law (the magnetic field made by moving charges) but the mathematics is much more difficult and beyond the scope of this introductory course. The curly nature of the magnetic field deduced from Ampere's law is consistent with Gauss's law for magnetism, which says that magnetic field cannot have an outward-going or inward-going pattern but must be purely curly. (Another way of saying this is that no one has ever found a magnetic monopole, whose magnetic field would be outward-going or inward-going, and Gauss's law for magnetism embodies this experimental null result.)

Maxwell's equations also can be used to predict the electric and magnetic fields of accelerated charges or time-varying currents. In this chapter we have seen that there is a curly electric field in the presence of a time-varying current, and this fact of nature is embedded in Faraday's law, which relates the path integral of electric field to the rate of change of magnetic field. In Chapter 23 we will see additional consequences of the fields of accelerated charges.

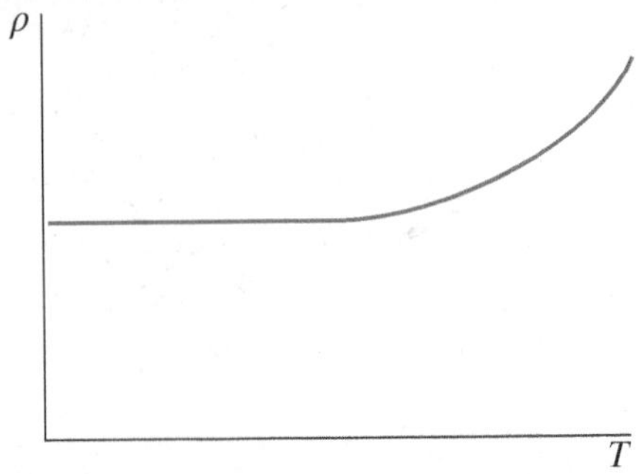

Figure 22.32 Resistivity vs. temperature for an ordinary metal.

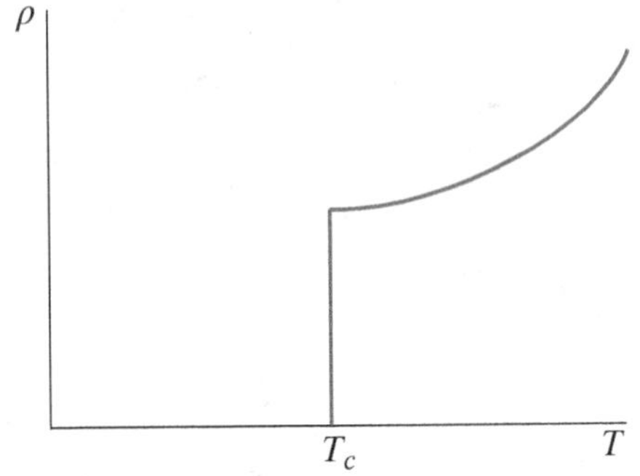

Figure 22.33 Resistivity vs. temperature for a superconductor.

Figure 22.34 A current in a superconducting ring persists for years.

22.5 SUPERCONDUCTORS

As you lower the temperature, the resistivity of an ordinary metal decreases but remains nonzero even at absolute zero (Figure 22.32), whereas some materials lose all resistance to electric current. Such materials are called "superconductors." It isn't that the resistance merely gets very small. Below a critical temperature T_C that is different for each superconducting material, the resistance vanishes completely, as indicated in Figure 22.33. To put it another way, the mobility becomes infinite. A nonzero electric field in a superconductor would produce an infinite current! As a result, the electric field inside a superconductor must always be zero, yet a nonzero current can run anyway.

In a ring made of lead that is kept at a temperature below 7.2 K (that is, only 7.2 kelvins above absolute zero), one can observe a current that persists undiminished for years (Figure 22.34)!

> QUESTION How can you check that the current in the lead ring hasn't changed since last month, without inserting an ammeter into the ring or touching the ring in any way?

Since the current loop produces a magnetic field, you could use a compass or other more sensitive device to measure the magnetic field produced by the current.

> QUESTION Does this persistent current violate the fundamental principle of conservation of energy? Why or why not?

No energy is being dissipated in the ring. Since its resistance is zero, the power used up is zero (power = RI^2). Energy conservation is not violated.

> QUESTION Do the persistent atomic currents in a bar magnet violate the principle of conservation of energy? Why or why not?

The "currents" in a bar magnet are due to the spin and orbital motion of electrons in the atoms of the magnet. The energy of these electrons remains constant—no energy is used up, so no energy input is required. Energy conservation is not violated.

A fruitful way to think about a superconductor is to say that the persistent current is similar to the persistent atomic currents in a magnet, but the diameter of a persistent current loop in a superconductor can be huge compared to an atom. In both the magnet and the superconductor the persistent currents can be explained only by means of quantum mechanics. In both cases the explanation hinges on the fact that these systems have discrete energy levels. Since the energy levels are separated by a gap, there can be no small energy changes, and hence there is no mechanism for energy dissipation.

High-Temperature Superconductors

The lack of resistance in a superconductor offers obvious advantages for power transmission and in electromagnets. However, for a long time after 1911, when the Dutch physicist Kamerlingh Onnes discovered superconductivity in mercury below 4.2 K, all superconductors required very low temperatures. Recently materials have been discovered that are superconducting at temperatures above the boiling point of liquid nitrogen (77 K, −196 °C). Because liquid nitrogen is rather inexpensive to produce, this opens up new opportunities for the use of superconductors. There is even hope that some material might be found that would be a superconductor at room temperature.

Magnetic Flux through a Superconducting Ring

The complete lack of electric resistance has some odd consequences for the effect of magnetic fields on a superconductor. Suppose you try to change

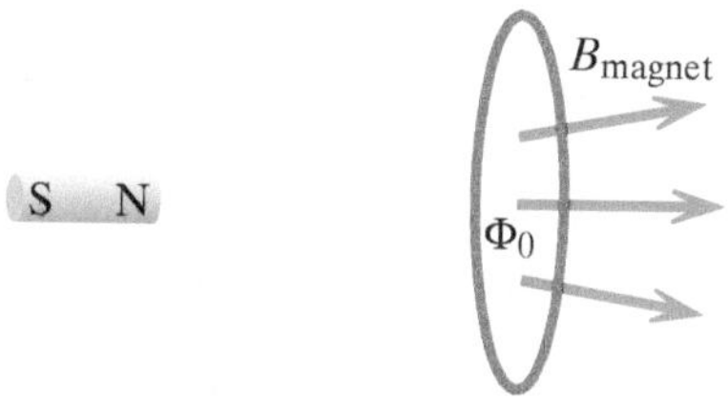

Figure 22.35 Start with a ring with initial flux Φ_0, then cool down the ring to make it become superconducting.

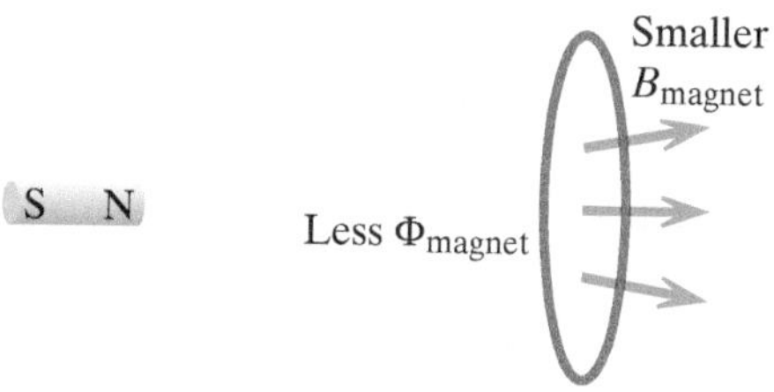

Figure 22.36 At low temperature, move the magnet away from the ring, reducing the flux in the ring contributed by the magnet.

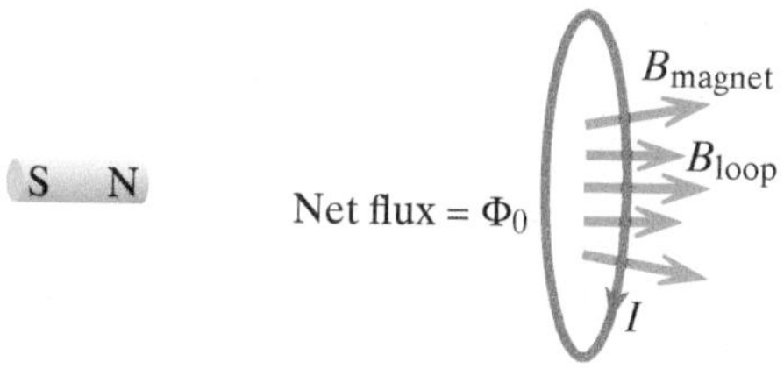

Figure 22.37 Current runs in the superconducting ring in such a way that the net change in the flux in the ring is zero.

the flux through a superconducting metal ring. This changing flux would be associated with a curly electric field (whose round-trip integral is an emf equal to $d\Phi_{mag}/dt$). This curly electric field would drive an infinite current in the ring, because there is no resistance! Because an infinite current is impossible, we conclude that it is not possible to change the flux through a superconducting ring, and this is what is observed.

What does happen when you try to change the flux through the ring by an amount $\Delta\Phi_{mag}$? Since the flux cannot change, what happens is that a (noninfinite) current runs in the ring in such a way as to produce an amount of flux $-\Delta\Phi_{mag}$, so that the net flux change is zero! We'll consider a specific experimental situation to illustrate the phenomenon.

Start with a ring and a magnet at room temperature, so that there is a certain amount of flux Φ_0 through the ring (Figure 22.35), then cool the ring below the critical temperature at which it becomes superconducting (obviously the ring has to be made of a material that does become superconducting at some low temperature).

Next move the magnet away from the ring, thus decreasing the flux made by the magnet inside the ring (Figure 22.36); there is a change in the flux due to the magnet, of some amount $-\Delta\Phi$.

In a nonsuperconducting material, there would be an emf in the ring, with a current emf $= RI$. However, since the resistance R of the superconducting ring is zero, the emf must be zero (0 times I is 0). Assuming that Faraday's law applies to superconductors, it must be that just enough current runs in the superconductor, in the appropriate direction, to increase the flux by an amount $+\Delta\Phi$. The net change in the flux is zero, and the flux remains equal to the original flux Φ_0.

Experimentally, that's exactly what happens (Figure 22.37). Enough current runs in the ring to make exactly the right amount of magnetic flux to make up for the decrease in the flux due to moving the magnet farther away, so that the flux through the ring remains constant. Even though a current runs, the electric field inside the superconductor is zero.

Checkpoint 6 When the magnet is moved very far away, how much flux is inside the ring compared to the original flux Φ_0? How much of this flux is due to the current in the ring?

The Meissner Effect

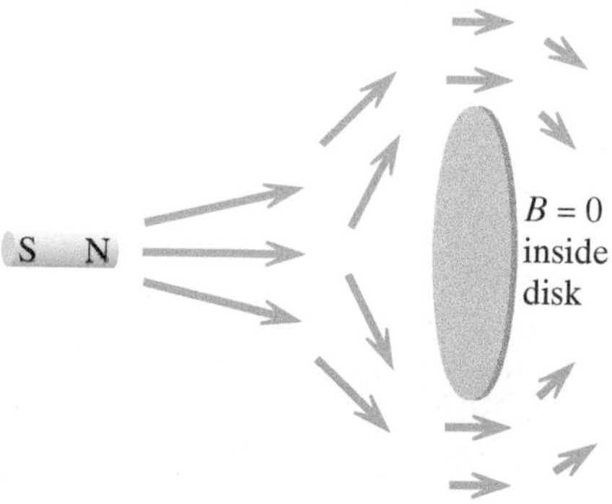

Figure 22.38 The Meissner effect: Currents run in the solid superconducting disk in such a way that the net magnetic field is zero inside the disk.

For many years it was assumed that a similar effect would happen with a solid disk instead of a hollow ring. It was expected that as you cooled the disk below its critical temperature the magnetic flux through the disk would stay the same, and that when you moved the magnet away there would be a current in the disk that would maintain the original flux. In 1933 physicists were astonished to discover that something else happened, something quite dramatic. As soon as the disk was cooled down below the critical temperature for the onset of superconductivity, the magnetic field throughout the interior of the solid disk suddenly went to zero!

This peculiar phenomenon is called the Meissner effect, and it applies to a particular class of materials called "Type I" superconductors. The configuration of net magnetic field (due to the magnet plus the currents inside the superconductor) looks something like Figure 22.38, with no net magnetic field inside the superconductor. Persistent currents are created in the superconductor, which make a magnetic field that exactly cancels the magnetic field due to the magnet everywhere inside the superconductor. Note that the currents in Figure 22.38 are oriented so that the disk and magnet repel each other.

The Meissner effect was totally unexpected, and it cannot be explained simply in terms of the lack of electric resistance in a superconductor (plus Faraday's law) as we did for the hollow ring. The Meissner effect is a special quantum-mechanical property of superconductors, and its explanation along with the explanation of other properties of superconductors was a triumph of the Bardeen–Cooper–Schrieffer (BCS) theory published in 1957.

Given that both the electric field and the magnetic field must be zero inside the superconductor, by using Gauss's law and Ampere's law it can be shown that the superconducting currents can run only on the surface of the superconductor, not in the interior. In particular, current in the interior would produce magnetic field throughout the material.

The electric field in an ordinary conductor in equilibrium goes to zero due to polarization. However, the fact that electric and magnetic fields must be zero in a superconductor even when not in equilibrium is quite a different, quantum-mechanical effect.

Magnetic Levitation

The Meissner effect is the basis for a dramatic kind of magnetic levitation. A magnet can hover above a superconducting disk, because no matter how the magnet is oriented, currents run in the superconducting disk in such a way as to repel the magnet. One design for maglev trains uses this superconducting effect.

This is in contrast to the situation with ordinary magnets. If you try to balance a magnet above another magnet, with the slightest misalignment the upper magnet flips over and is attracted downward rather than repelled upward.

The Levitron® toy achieves levitation with ordinary magnets by giving the upper magnet a high spin, so that the magnetic torque makes the spin direction precess rather than flip over. When a superconductor supports the upper magnet, gyroscopic stabilization is not needed.

22.6 INDUCTANCE

Changing the current in a coil can "induce" an emf in a second coil. A related effect happens even with a single coil: an attempt to change the current in a coil leads to an additional emf in the coil itself, because the coil surrounds a region of time-varying magnetic field (dB/dt), produced by itself. We can show that this self-induced emf acts in a direction to oppose the change in the current. As a result, there is a kind of sluggishness in any coil of wire: it is hard to change the current, either to increase it or to decrease it. It is not difficult to calculate this self-induction effect quantitatively for a solenoid.

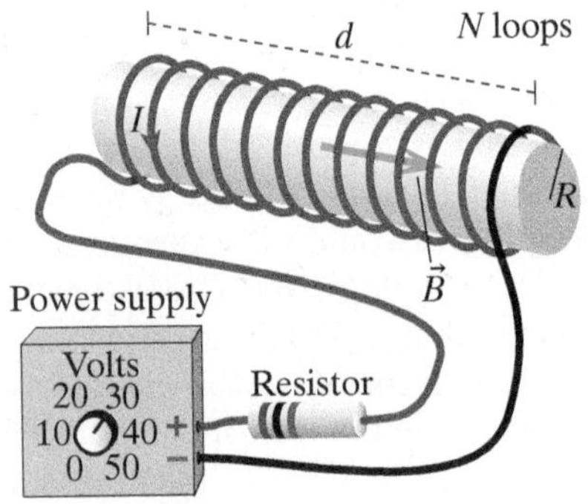

Figure 22.39 A circuit contains a solenoid.

A very long plastic cylinder with radius R and length $d \gg R$ is wound with N closely packed loops of wire with negligible resistance. The solenoid is placed in series with a power supply and a resistor, and current I runs through the solenoid in the direction shown in Figure 22.39. This current makes a magnetic field B inside the solenoid that points to the right.

If the current is steady, the electric field inside the metal of the solenoid coil is nearly zero (very small resistance of this wire). Now suppose that you try to increase the current by turning up the voltage of the power supply. If the current increases with time, there will be an increasing magnetic flux enclosed by each loop of the solenoid. In Chapters 17 and 21 we found that the magnetic field inside a solenoid of length d is $B = \mu_0 NI/d$.

QUESTION How much emf is generated in one loop of the coil if the current changes at a known rate dI/dt?

We have this:

$$\text{emf} = \left|\frac{d\Phi_{\text{mag}}}{dt}\right| = \frac{d}{dt}\left[\frac{\mu_0 NI}{d}\pi R^2\right] = \frac{\mu_0 N}{d}\pi R^2 \frac{dI}{dt} \quad \text{(one loop)}$$

QUESTION There are N loops, and each loop encloses approximately the same amount of time-varying magnetic flux. Therefore, what is the net emf along the full length of the solenoid?

Evidently we add up all the individual one-loop emfs in series to obtain the emf from one end of the coil to the other:

$$\text{emf} = N\left[\frac{\mu_0 N}{d}\pi R^2 \frac{dI}{dt}\right] = \frac{\mu_0 N^2}{d}\pi R^2 \frac{dI}{dt} \quad \text{(entire solenoid)}$$

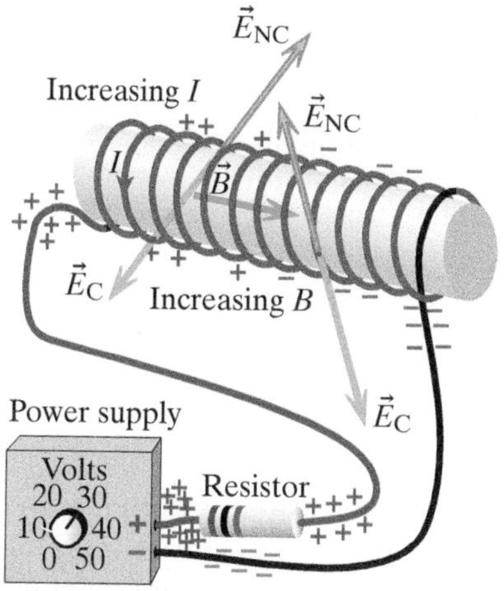

Figure 22.40 The non-Coulomb electric field opposes the increase in the current, and polarizes the solenoid.

QUESTION If we increase the current I, what is the direction of the associated curly electric field, clockwise or counterclockwise as seen from the right end of the solenoid? Does this curly electric field act to assist or oppose the change in I?

If the current is increasing, the non-Coulomb electric field $\vec{E}_{NC}$ curls clockwise around the solenoid, opposing the increase in the current and polarizing the solenoid (Figure 22.40). The new surface charges produce a Coulomb electric field $\vec{E}_C$ that follows the wire and points opposite to the non-Coulomb field. If the resistance of the solenoid wire is very small, the magnitude of the two electric fields is nearly equal ($E_C \approx E_{NC}$).

Note the similarity to a battery: a non-Coulomb emf_{ind} maintains a charge separation, and there is a potential drop ΔV_{sol} along the solenoid that is numerically equal to emf_{ind}. If the wire resistance r_{sol} is significant, we have $\Delta V_{\text{sol}} = \text{emf}_{\text{ind}} - r_{\text{sol}} I$, just like a battery with internal resistance. You can think of this self-induced emf as making the solenoid act like a battery that has been put in the circuit "backwards," opposing the change in the current (Figure 22.41).

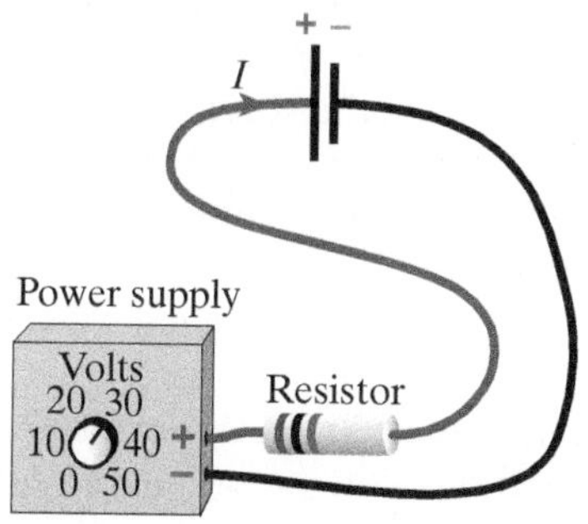

Figure 22.41 The coil acts temporarily like a battery inserted to oppose the change.

It is standard practice to lump the many constants together and write that the self-induced emf is proportional to the rate of change of the current, with proportionality constant L:

$$|\text{emf}_{\text{ind}}| = L\left|\frac{dI}{dt}\right|$$

The proportionality constant L is called the "inductance" or "self-inductance" of the device, which is called an "inductor."

QUESTION Use the results above to express the inductance L in terms of the properties of a solenoid.

Evidently the (self-)inductance of a solenoid is this:

$$L = \frac{\mu_0 N^2}{d}\pi R^2$$

The unit of inductance (V·s/A) is called the "henry" (H) in honor of the 19th-century American physicist Joseph Henry, who simultaneously with Michael Faraday in 1831 discovered the effects of time-varying magnetic fields.

QUESTION What if you decrease the current through the solenoid? What happens to the emf? Does it assist or oppose the change?

If you go through the previous analysis but with a decreasing current, you find that the emf goes in the other direction, tending to drive current in the original current direction and therefore opposing the decrease. The solenoid introduces some sluggishness into the circuit: it is more difficult to increase the current or to decrease the current. The sign of the effect should not be surprising, since if the self-induced emf assisted rather than opposed an increase in the current, the current would grow without limit!

Checkpoint 7 To get an idea of the order of magnitude of inductance, calculate the self-inductance in henries for a solenoid with 1000 loops of wire wound on a rod 10 cm long with radius 1 cm. If the solenoid were filled with iron so that the actual magnetic field were 10 times larger for the same current in the solenoid, what would be the inductance?

Transformers

Figure 22.42 A transformer with a 100-turn primary coil and a 20-turn secondary coil.

Electric energy is moved long distances with potential differences of hundreds of thousands of volts, but home appliances need around a hundred volts. Conversion from high voltage to low voltage can be accomplished with transformers in which a primary coil with many turns makes an emf in a secondary coil with few turns. Suppose that you make a solenoid by wrapping $N_1 = 100$ turns around a hollow cylinder $d = 0.1$ m long for the primary coil, and you wrap $N_2 = 20$ turns around the outside for the secondary coil (Figure 22.42). Connect the 100-turn primary coil to an AC power supply whose voltage is $\Delta V_{\text{max}} \cos(\omega t)$. What emf will develop along the secondary coil?

The magnetic field made by the primary coil is $B = \mu_0 N_1 I/d$, and if the cross-sectional area of the solenoid is A, the emf in one turn of the secondary coil is AdB/dt. The total emf in the secondary coil is $N_2 = 20$ times the emf in one turn, so the potential difference across the secondary coil is $N_2 A(\mu_0 N_1/d)dI/dt$.

What is the potential difference across the primary coil? We just saw that it is LdI/dt, where $L = A\,\mu_0 N_1^2/d$ (since $A = \pi R^2$), so the potential difference across the primary coil is $A\,(\mu_0 N_1^2/d)dI/dt$.

Comparing $\text{emf}_2 = N_2 A\,(\mu_0 N_1/d)dI/dt$ with $\text{emf}_1 = A\,(\mu_0 N_1^2/d)dI/dt$, we see that $\text{emf}_2 = (N_2/N_1)\text{emf}_1$, or 20/100 = 0.2 times the emf in the primary coil. The ratio of the number of turns determines the change in voltage. If ΔV_{max} is 500 V across the primary coil, it will be only 100 V across the secondary coil.

Because energy is conserved and power is $I\Delta V$, the smaller voltage in the secondary coil is accompanied by a larger current. This two-coil device, a "transformer," has reduced the voltage by a factor of 0.2 and increased the current by a factor of 5, assuming that the process is fully efficient.

We've just described briefly a "step-down" transformer, which reduces the voltage (and increases the current). There are also "step-up" transformers, in which the primary has few turns and the secondary many; the effect is to increase the voltage and decrease the current.

In actual practice the inside of the coils is normally iron rather than air, which has the effect of making the voltage across the secondary coil be less affected by the type of circuit elements connected to the secondary coil.

Checkpoint 8 A transformer has 500 turns in its primary coil and 20,000 turns in its secondary coil. An AC voltage with maximum voltage of 100 V is applied to the primary coil. What is the maximum voltage across the secondary coil?

Energy Density

In Chapter 16 we showed that there is an energy density associated with electric field, by expressing the energy in a capacitor in terms of the electric field in the capacitor:

$$\frac{\text{Electric energy}}{\text{Volume}} = \frac{1}{2}\varepsilon_0 E^2$$

Now we can show that there is energy density associated with magnetic field, by expressing the energy in an inductor in terms of the magnetic field in the inductor. We obtain an important result about magnetic energy that is quite general, despite the fact that the derivation is for the specific case of an inductor.

It is easy to show that the magnetic energy stored in an inductor is $\frac{1}{2}LI^2$. The power going into an inductor is $I\Delta V$ as for any device, and

$$P = I\Delta V = I(\text{emf}) = I\left(L\frac{dI}{dt}\right)$$

Integrating over time, we have

$$\text{Energy input} = \int (I\Delta V)dt = L\int_{t_i}^{t_f} I\frac{dI}{dt}dt = L\int_{I_i}^{I_f} IdI$$

$$= L\left[\frac{1}{2}I^2\right]_{I_i}^{I_f} = \Delta\left(\frac{1}{2}LI^2\right)$$

Since $B = \frac{\mu_0 NI}{d}$ and $L = \frac{\mu_0 N^2}{d}\pi R^2$, we have

$$\text{Magnetic energy} = \frac{1}{2}\left(\frac{\mu_0 N^2}{d}\pi R^2\right)\left(\frac{Bd}{\mu_0 N}\right)^2$$

$$= \frac{1}{2}\frac{1}{\mu_0}(\pi R^2 d)B^2$$

The magnetic field of the solenoid is large in the volume $(\pi R^2 d)$. Therefore we have the following result for the electric and magnetic energy densities:

ELECTRIC AND MAGNETIC ENERGY DENSITIES

$$\frac{\text{Energy}}{\text{Volume}} = \frac{1}{2}\varepsilon_0 E^2 + \frac{1}{2}\frac{1}{\mu_0}B^2$$

measured in J/m^3.

Although we have calculated the magnetic energy density for the specific case of a long solenoid, this is a general result. The interpretation is that where there are electric or magnetic fields in space, there is an associated energy density, proportional to the square of the field (E^2 and/or B^2).

22.7 *INDUCTOR CIRCUITS

Current in an RL Circuit

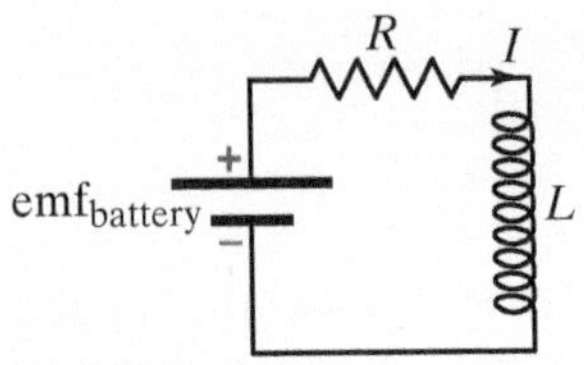

Figure 22.43 An RL circuit containing a resistor and an inductor.

A circuit containing a resistor R and an inductor L is called an RL circuit. Figure 22.43 shows a series RL circuit some time after a switch was closed. The energy conservation loop rule for this series circuit is this:

$$\Delta V_{\text{battery}} + \Delta V_{\text{resistor}} + \Delta V_{\text{inductor}} = 0$$
$$\text{emf}_{\text{battery}} - RI - L\frac{dI}{dt} = 0$$

The term $-LdI/dt$ in the loop equation has a minus sign because, as we have seen, the emf of the inductor opposes the attempt to increase the current when we close the switch. The properties of this differential equation are similar to those of the differential equation of a resistor-capacitor (RC) circuit analyzed in Chapter 19.

Let the time when we closed the switch be $t = 0$. The current was of course zero just before closing the switch, and it is also zero just after closing the switch, since the sluggishness of the inductor does not permit an instantaneous change from $I = 0$ to $I =$ nonzero. In fact, such an instantaneous change would mean that dI/dt would be infinite, which would require an infinite battery voltage to overcome the infinite self-induced emf in the inductor.

QUESTION Prove that

$$I = \frac{\text{emf}_{\text{battery}}}{R}\left[1 - e^{-(R/L)t}\right]$$

by substituting I and dI/dt into the energy conservation equation. What is the final current in the circuit (that is, after a very long time)? Does that make sense?

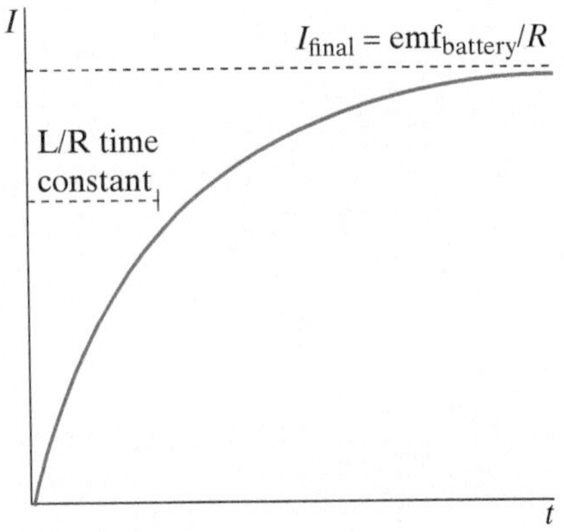

Figure 22.44 A graph of current vs. time in the RL circuit. The switch was closed at $t = 0$.

For large t, the exponential goes to zero, and we have $I = \text{emf}_{\text{battery}}/R$. This makes sense, because after a long time there is a steady-state current, and the flux does not change any more. With no change in the flux, no emf is generated in the inductor, and the current has the value that it would have if the inductor were not in the circuit. The current builds up slowly to this value. Figure 22.44 shows a graph of the current vs. time. One measure of the time it takes for the current to build up is the "time constant" L/R; at this time the exponential factor has dropped to $1/e$:

$$e^{-(R/L)t} = e^{-(R/L)(L/R)} = e^{-1} = \frac{1}{e} = 0.37$$

You see that having an inductor in a circuit makes the circuit somewhat sluggish. It takes a while for the current to reach the value that it would have reached almost immediately if the circuit had consisted just of a battery and a resistor. (Note, however, that any circuit has some inductance and some sluggishness, because changing the current from zero to nonzero involves a time-varying magnetic flux.)

QUESTION It is interesting to see what happens when you try to open the switch after the steady current has been attained. If the current were to drop from I to zero in a short time Δt, explain why the emf along the solenoid would be very large, and could be much larger than the battery emf.

The emf depends on dB/dt, which is proportional to dI/dt. If the time interval is very short, then dI/dt is very large, and the emf is very large. As a result of this large emf, you may see a spark jump across the opening switch. This phenomenon makes it dangerous to open a switch in an inductive circuit if there are explosive gases around.

Current in an LC Circuit

We have studied many examples of equilibrium and steady-state currents. We have also seen examples of a slow approach to equilibrium (an RC circuit) or to a steady-state current (the RL circuit we just examined).

There is another possibility: a circuit might oscillate, with charge sloshing back and forth forever—the system never settles down to a final equilibrium or a steady state. Of the systems we've analyzed or experimented with, none of them oscillated continuously, because there was always some resistance in the system that would have damped out any such oscillatory tendencies. It is possible that during the first few nanoseconds when surface charge is rearranging itself in a circuit, there may be some sloshing of charge back and forth, but this oscillation dies out due to dissipation in the resistive wires.

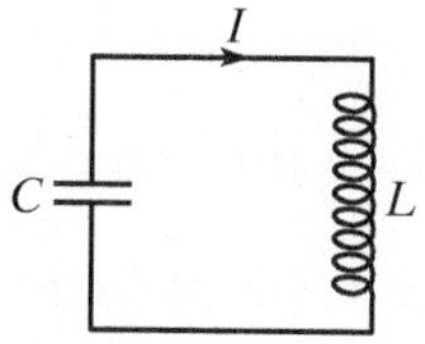

Figure 22.45 An LC circuit consists of an inductor and a capacitor.

A circuit containing an inductor L and a capacitor C is called an LC circuit (Figure 22.45). Such a circuit can oscillate if the resistance is small. In this circuit, the connecting wires are low-resistance thick copper wires. Suppose that the capacitor is initially charged, and then you close a switch, connecting the capacitor to the inductor.

At first it is difficult for charge on the capacitor to flow through the inductor, because the inductor opposes attempts to change the current (and the initial current was zero). However, the inductor can't completely prevent the current from changing, so little by little there is more and more current, which of course drains the capacitor of its charge.

We'll see that just at the moment when the capacitor runs out of charge, there is a current in the inductor that can't change to zero instantaneously, due to the sluggishness of the inductor. So the system doesn't come to equilibrium but overshoots the equilibrium condition and pours charge into the capacitor. When the capacitor gets fully charged (opposite in sign to the original condition), the capacitor starts discharging back through the inductor.

This oscillatory cycle repeats forever if there is no resistance in the circuit. If there is some resistance, the oscillations eventually die out, but the system may go through many cycles before equilibrium is reached.

The energy conservation loop rule for this circuit is

$$\Delta V_{\text{capacitor}} + \Delta V_{\text{inductor}} = \frac{1}{C}Q - L\frac{dI}{dt} = 0$$

where Q is the charge on the upper plate of the capacitor, and I is the conventional current leaving the upper plate and going through the inductor.

QUESTION Can you explain why $I = -dQ/dt$?

dQ/dt is the amount of charge flowing off of the capacitor plate per second. This is the same thing as the current. Because charge is leaving the plate, dQ is negative, so $I = -dQ/dt$. Since $I = -dQ/dt$, we can rewrite the energy conservation equation as

$$\frac{1}{C}Q + L\frac{d^2Q}{dt^2} = 0$$

QUESTION Show that

$$Q = Q_i \cos\left(\frac{1}{\sqrt{LC}}t\right)$$

is a possible solution of the rewritten energy conservation equation, by substituting Q and its second derivative into the equation.

Moreover, this solution fits the initial conditions at $t = 0$, just after closing the switch, if Q_i is the initial charge on the upper plate, since this expression reduces to $Q = Q_i \cos(0) = Q_i$. (For other initial conditions, the correct solution is a sine, or a combination of sine and cosine.)

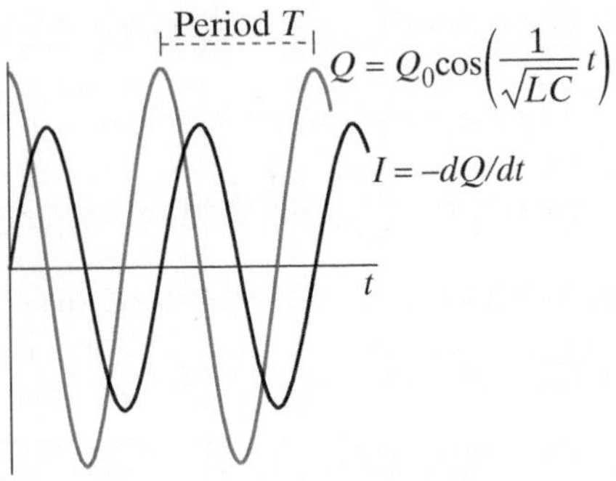

Figure 22.46 Capacitor charge and inductor current in an LC circuit with no resistance.

QUESTION Now that you know Q on the upper plate of the capacitor as a function of time, calculate the current I through the inductor as a function of time.

The current is given by the following:

$$I = -\frac{dQ}{dt} = \frac{Q_i}{\sqrt{LC}} \sin\left(\frac{1}{\sqrt{LC}}t\right)$$

Now that we have expressions for Q and I as a function of time, we can graph both quantities (Figure 22.46). This is very special: charge oscillates back and forth in the circuit forever (if there is negligible resistance). There is no equilibrium, and there is no steady state. No battery or other source of energy input is needed, because there is no dissipation. The charge just keeps going back and forth on its own. If there is some resistance, the oscillations slowly die out, as shown in Figure 22.47 (the loop equation must be modified to contain the term $-RI$).

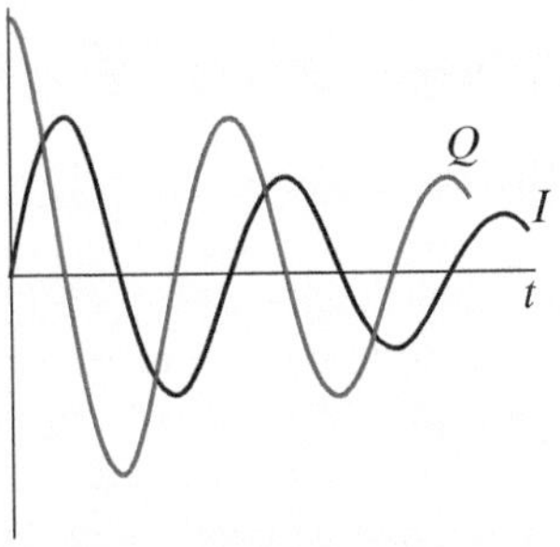

Figure 22.47 An LC circuit with resistance.

Note that the current I does not become large instantaneously (due to the sluggishness of the inductor). Also note that the current, which is the rate of depletion of the capacitor charge, reaches a maximum just when Q goes to zero. This maximum current starts charging the bottom plate of the capacitor positive, which makes the current decrease. When the current becomes zero, the system is in a state very similar to its original state, but inverted (the bottom plate is positive). Now the system runs the other way and eventually gets back after one complete cycle to a state in which the top plate is again positive, and the current is momentarily zero. Then the process repeats.

After one complete cycle, $(1/\sqrt{LC})t = 2\pi$, so the period of the oscillation is $T = 2\pi\sqrt{LC}$. The frequency $f = 1/T = 1/(2\pi\sqrt{LC})$.

Checkpoint 9 What is the oscillation frequency of an LC circuit whose capacitor has a capacitance of 1 μF and whose inductor has an inductance of 1 mH? (Both of these are fairly typical values for capacitors and inductors in electronic circuits. See Checkpoint 7 for a numerical example of inductance.)

Energy in an LC Circuit

Another way to talk about an LC circuit is in terms of stored energy. The original energy was the electric energy stored in the capacitor, equal to $Q^2/(2C)$; see Chapter 19. At the instant during the oscillation when the charge on the capacitor is momentarily zero, there is no energy stored in the capacitor; all the energy at that moment is stored in the inductor, in the form of magnetic energy. As the system oscillates, energy is passed back and forth between the capacitor and the inductor.

Earlier in this chapter we showed that the magnetic energy stored in an inductor is $\frac{1}{2}LI^2$. At $t = 0$, the current is zero, and all the energy in the system is the electric energy of the capacitor. At the instant when $Q = 0$ on the capacitor, the current is a maximum, and all the energy in the system is the magnetic energy of the inductor.

QUESTION If Q_i is the initial charge on the capacitor, use an energy argument to find the maximum current I_{max} in the circuit.

Initially, all of the stored energy is electric:

$$U = U_{electric} + U_{magnetic} = \frac{1}{2}\frac{Q_i^2}{C}$$

When $I = I_{max}$, $Q = 0$, so all of the stored energy is magnetic:

$$U = U_{electric} + U_{magnetic} = \frac{1}{2}LI_{max}^2$$

Since the total energy in the circuit does not change, because the circuit has no resistance and no energy is dissipated as heat, the following must be true:

$$\frac{1}{2}LI_{max}^2 = \frac{1}{2}\frac{Q_i^2}{C}, \quad \text{and} \quad I_{max} = \frac{Q_i}{\sqrt{LC}}$$

QUESTION Show that this result is consistent with $Q = Q_i \cos(t/\sqrt{LC})$ by calculating $I = -dQ/dt$ and seeing what the maximum I is.

Starting from the charge as a function of time, we have this:

$$I = -\frac{dQ}{dt} = -\frac{Q_i}{\sqrt{LC}}\sin\left(\frac{t}{\sqrt{LC}}\right) \quad \text{so that} \quad I_{max} = \frac{Q_i}{\sqrt{LC}}$$

This is the same result we obtained by using an energy argument.

We have been discussing "free" oscillations, in which the oscillations (perpetual or dying out) proceed with no inputs from the outside. If you try to force the oscillations by applying an AC voltage from the outside (with an AC power supply), you find that it is difficult to get much current to run unless you nearly match the free oscillation frequency $f = 1/(2\pi\sqrt{LC})$. This is an example of what is called a "resonance" phenomenon (see the end of Chapter 7), in which a system (the LC circuit in this case) responds significantly only when the forcing of the system is done at a frequency close to the free-oscillation frequency.

AC Circuits

For completeness, it should be mentioned that an important use of inductors is in AC circuits (sinusoidally alternating current). You have just seen that LC circuits are characterized by sinusoidally alternating currents, as are generators with loops rotating in a magnetic field. Whenever there is an AC current passing through an inductor, there is an AC emf generated in the inductor due to the time-varying flux. If the current is a sine function, the emf is a cosine function, and vice versa.

There is an elaborate mathematical formalism for dealing with AC circuits, including accounting for the fact that the current and voltage in an AC circuit need not reach their maxima at the same time (a "phase shift," which was 90° in the LC circuit we just analyzed). We will not go into these complications here, but it is worth noting that an important element in understanding such circuits is to recognize the role of the self-induced emf in an inductor, and the fact that in an inductor a sinusoidal current is associated with a cosinusoidal voltage.

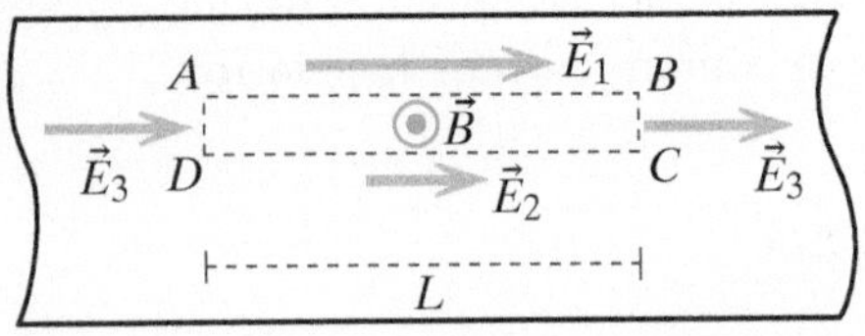

Figure 22.48 An AC current produces a time-varying magnetic field in a wire, and $E_1 > E_2$.

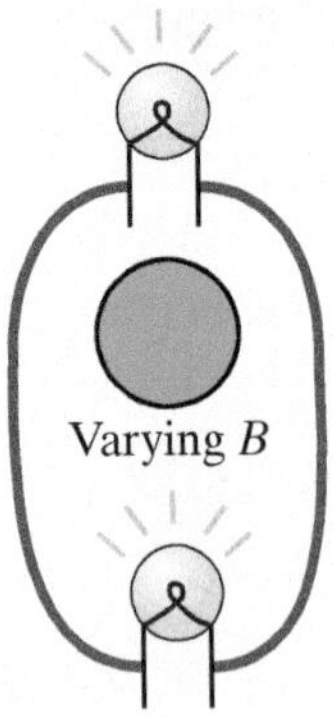

Figure 22.49 Two light bulbs connected around a long solenoid with varying B.

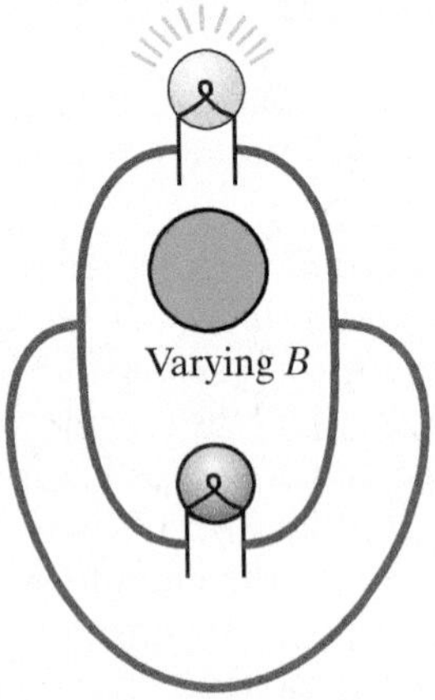

Figure 22.50 Add a thick copper wire to the two-bulb circuit.

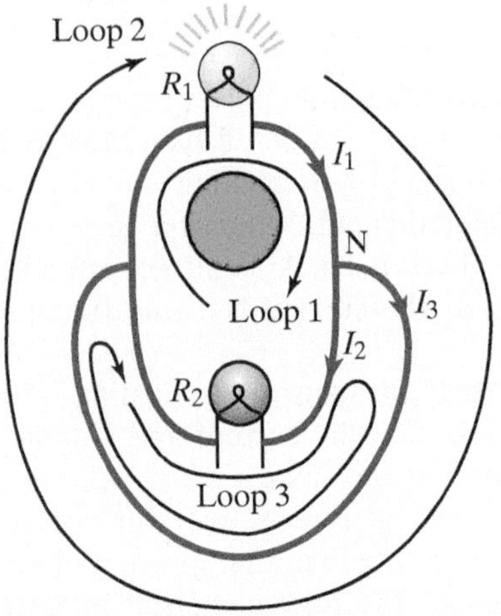

Figure 22.51 Write loop and node equations for this circuit.

Skin Effect

In Chapter 18 we showed that in a DC circuit the electric field was uniform across the cross section of a wire, because the round-trip path potential difference ΔV_{ABCDA} in Figure 22.48 had to be zero, so that $\vec{E}_1$ had to be equal to $\vec{E}_2$. This is not true in an AC circuit, because the magnetic field made by the AC current varies sinusoidally in time. The magnetic field inside the loop $ABCDA$ is out of the page. When the current is increasing, the magnetic field is increasing out of the page, and there will be a clockwise curly non-Coulomb electric field. Going around the path $ABCDA$, the path integral $\int \vec{E} \bullet d\vec{l} = E_1 L - E_2 L = d\Phi/dt$, which is positive. Therefore $E_1 > E_2$. The bigger $d\Phi/dt$, the larger the difference between E_1 and E_2.

The higher the AC frequency of the current, the bigger is $d\Phi/dt$, so that at very high frequencies E_2 is very small compared to E_1, and there is significant electric field only very near the surface of the wire. Since $J = I/A = \sigma E$, there is almost no current density anywhere in the interior of the wire: all of the current runs very near the surface of the wire. This high-frequency phenomenon is called "skin effect."

22.8 *SOME PECULIAR CIRCUITS

This section offers examples of surprising aspects of circuits when there is a time-varying magnetic field.

Two Bulbs Near a Solenoid

We can show you a very simple circuit that behaves in a rather surprising way. Consider two light bulbs connected in series around a solenoid with a varying magnetic field (Figure 22.49). The solenoid carries an alternating current (AC) in order to provide a time-varying magnetic flux through the circuit, so that there is an emf to light the bulbs. If the flux varies as $\Phi_{mag} = \Phi_0 \sin(\omega t)$, in the bulb circuit we have a varying emf $=$ emf$_0 \cos(\omega t)$, since the emf is proportional to the rate of change of the magnetic flux. If each bulb has a resistance R, there is an alternating current through the bulbs of this amount:

$$I = \frac{\text{emf}}{2R} = \frac{\text{emf}_0}{2R}\cos(\omega t)$$

Notice that a uniformly increasing flux ($\Phi_{mag} = \Phi_0 t$) would yield a constant emf, but a sinusoidally varying flux yields a cosinusoidally varying emf. It is very convenient to use AC currents to study the effects of time-varying magnetic fields, because both the initiating current and the induced current vary as sines and cosines, and AC voltmeters and ammeters can be used to measure both parts of the circuit.

Now we alter the circuit by connecting a thick copper wire across the circuit (Figure 22.50). You may find it surprising that the top bulb gets much brighter and the bottom bulb no longer glows. Let's try to understand how this happens.

As usual, we need to write charge-conservation node equations and energy conservation loop equations, so we need to label the nodes and loops as in Figure 22.51 (any two of these three loops would be enough, but we'd like to show how to handle all of them).

Here is the key to analyzing this peculiar circuit. The round trip around loop 1 or loop 2 takes in an emf (just as though there were a battery in the loop). However, the round trip around loop 3 does not enclose any time-varying flux and so has no emf. Therefore the loop and node rules are the following, where R_1 and R_2 are the resistances of the hot top bulb and cold bottom bulb, and the resistance of the added copper wire is essentially zero:

$$\text{loop 1: emf} - R_1 I_1 - R_2 I_2 = 0 \qquad \text{node N: } I_1 = I_2 + I_3$$
$$\text{loop 2: emf} - R_1 I_1 = 0$$
$$\text{loop 3: } R_2 I_2 = 0 \text{ (no flux enclosed)}$$

From the loop 3 equation we see that $I_2 = 0$, which is why the bottom bulb doesn't glow.

With $I_2 = 0$, the node N rule says that $I_1 = I_3$: all the current in the top bulb goes through the copper wire (and none through the bottom bulb). Both the loop 1 and loop 2 equations reduce to $R_1 I_1 = \text{emf}$, so the current through the top bulb is $I_1 = \text{emf}/R_1 = (\text{emf}_0/R_1)\cos(\omega t)$, which is nearly twice what it was before we added the extra copper wire ($R_1 > R$ due to higher temperature).

Checkpoint 10 Figure 22.52 shows four different ways to connect the copper wire. Based on the analysis we have just carried out, involving identifying whether or not there is a battery-like emf in a loop, what is the brightness of both bulbs in circuits 1, 2, 3, and 4?

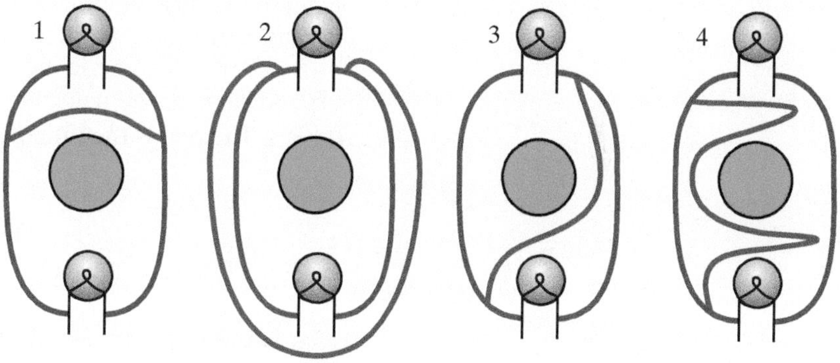

Figure 22.52

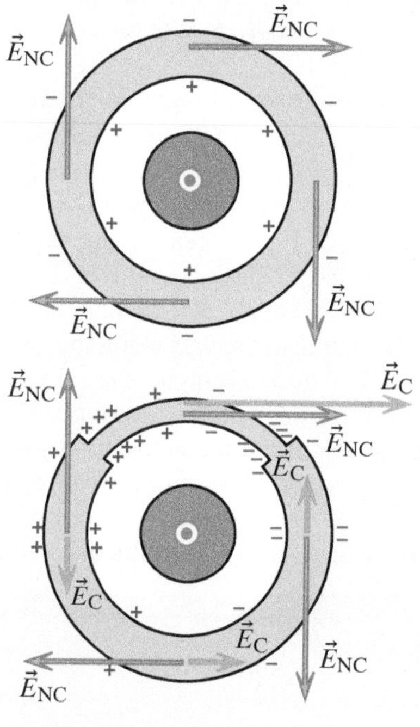

Figure 22.53 A uniform metal ring, and a metal ring with a thin section.

Coulomb and Non-Coulomb Electric Fields in a Nonuniform Ring

In a circular ring encircling a solenoid with a time-varying magnetic field, there is current driven by a non-Coulomb electric field. There is no gradient of surface charge along the ring, although there is a small transverse polarization required to turn the electrons in a circle (top of Figure 22.53).

If there is a thin section of the ring, however, charge will pile up at the ends of the thin section, and there will be surface charges all along the ring (bottom of Figure 22.53). These surface charges will produce a Coulomb electric field $\vec{E}_C$, which adds vectorially to the non-Coulomb electric field $\vec{E}_{NC}$ that is associated with the changing magnetic flux. The net electric field $\vec{E}_{NC} + \vec{E}_C$ is what drives the electrons around the ring.

Inside the thin section the net electric field is larger than it was for the uniform ring, and in the thick sections the net electric field is smaller than it was for the uniform ring.

QUESTION If both rings are made out of the same material, how do you know that the current must be smaller in the ring with the thin section than it was in the uniform ring?

In the thick section of the nonuniform ring, the only thing that changed is that the net electric field is smaller, so the current in the thick section must be smaller. (Of course this same smaller current runs through the thin section of this ring.)

Note that the Coulomb electric field $\vec{E}_C$ is large and goes clockwise in the thin section, but $\vec{E}_C$ is small and goes counterclockwise in the thick section. Since $\vec{E}_C$ is due to point charges, the round-trip integral of $\vec{E}_C$ is zero.

On the other hand, the non-Coulomb electric field $\vec{E}_{NC}$ due to the changing magnetic flux goes clockwise all around the ring, and its round-trip integral is emf $= d\Phi_{mag}/dt$, which is not zero.

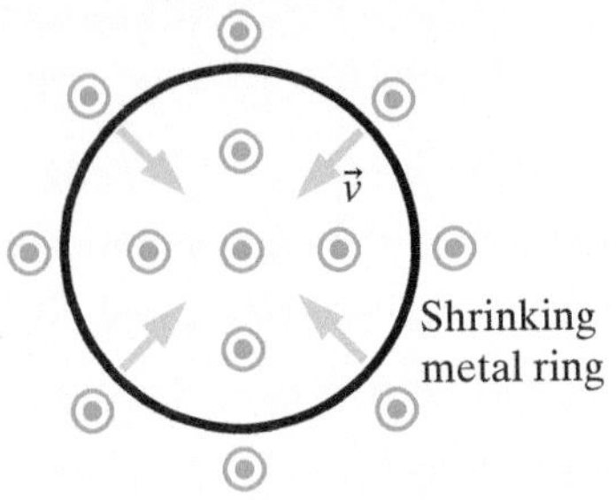

Figure 22.54 A springy metal ring that had been stretched and is now shrinking.

Two Competing Effects in a Shrinking Ring

Suppose that we stretch a springy metal ring of radius R and place it in a region of uniform magnetic field B pointing out of the page, which is increasing at a rate of dB/dt (Figure 22.54). When we let go of the ring, it contracts at a rate dR/dt (which is a negative number). What is the emf around the metal ring? Assume that the ring has high resistance and therefore carries only a small current, so we can neglect any magnetic field contributed by current in the ring.

In terms of flux, there are two competing effects here. The increasing magnetic field B tends to increase the flux, but the decreasing area tends to decrease the flux. Whether or not a current will flow depends on the relative magnitudes of these two contributions to the net flux.

Since the magnetic field is the same throughout the ring and is perpendicular to the page, the flux through the ring at any instant is $\Phi_{mag} = B(\pi R^2)$. The emf is the rate of change of the flux:

$$\text{emf} = \frac{d\Phi_{mag}}{dt} = \frac{d}{dt}[B(\pi R^2)] = \frac{dB}{dt}(\pi R^2) + B\frac{d(\pi R^2)}{dt}$$
$$= \frac{dB}{dt}(\pi R^2) + B(2\pi R)\frac{dR}{dt}$$

What is the physical significance of these two contributions to the emf? In the presence of a changing magnetic field, there is a clockwise non-Coulomb electric field in the wire. The first term, involving dB/dt, is associated with this non-Coulomb electric field in the wire.

Because the wire is moving inward, there is a magnetic force on the electrons in the wire. In this case, $\vec{v} \times \vec{B}$ (the magnetic force per unit charge) is counterclockwise in the ring. The second term, involving the negative quantity dR/dt, is associated with the magnetic force on the moving pieces of the ring (motional emf). Note that $|dR/dt|$ is the speed of pieces of ring toward the center, and $2\pi r$ is the circumference, so the contribution to the emf is similar to emf $= BLv$ for a rod of length L sliding on rails.

Note that if $(dB/dt)(\pi R^2) = B(2\pi R)|(dR/dt)|$, the emf would be zero even though we had both a changing magnetic field and a changing area. In this case the magnetic force on an electron inside the ring is equal and opposite to the non-Coulomb electric force, due to the non-Coulomb electric field.

There is no surface charge gradient around this symmetrical ring, and no Coulomb electric field around the ring. The forces on electrons in the metal ring are due to the non-Coulomb electric field and to the non-Coulomb magnetic force on the moving pieces of the ring.

22.9 *THE DIFFERENTIAL FORM OF FARADAY'S LAW

As with Gauss's law and Ampere's law, there is a differential form of Faraday's law, valid at a particular location and time. Using properties of vector calculus one can show that the integral form of Faraday's law is equivalent to the following, where the derivative is a "partial derivative" with respect to time (holding position constant):

DIFFERENTIAL FORM OF FARADAY'S LAW

$$\text{curl}(\vec{E}) = -\frac{\partial \vec{B}}{\partial t} \quad \text{or} \quad \vec{\nabla} \times \vec{E} = -\frac{\partial \vec{B}}{\partial t}$$

22.10 *LENZ'S RULE

With the thumb of your right hand pointing in the direction of $-d\vec{B}_1/dt$, your fingers curl around in the direction of $\vec{E}_{NC}$. Another rule, called "Lenz's rule," can also be used to get the direction of the non-Coulomb electric field. We describe Lenz's rule briefly, although it is unnecessary to learn and use Lenz's rule in this introductory textbook on electricity and magnetism, for reasons that we will give below.

Imagine that you place a wire around the changing flux, so that the non-Coulomb electric field drives current in the wire:

LENZ'S RULE

The non-Coulomb electric field would drive current in a direction to make a magnetic field that attempts to keep the flux constant.

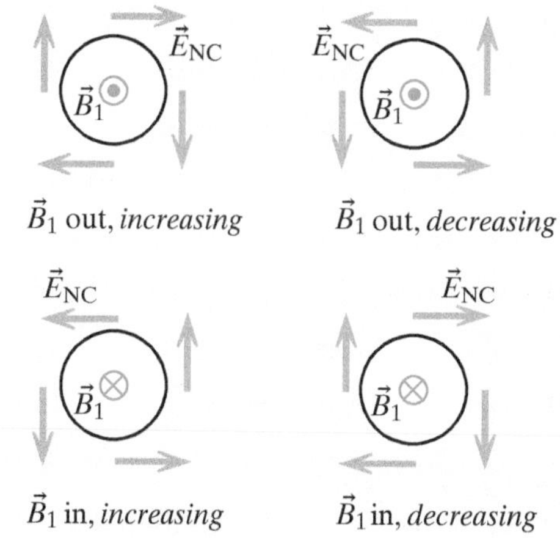

Figure 22.55 Using Lenz's rule to determine the direction of the non-Coulomb electric field.

To see how this works, consider the first example in Figure 22.55, where $\vec{B}_1$ points out of the page and is increasing, and the non-Coulomb electric field curls clockwise around the solenoid. If a wire encircles the solenoid, conventional current runs clockwise in the wire, and this current produces an additional magnetic field in the region, pointing into the page.

This additional magnetic field is in the opposite direction to the change in $\vec{B}_1$, so we say that the additional magnetic field attempts to keep the magnetic flux constant despite the increase in B_1. The induced current does not succeed at keeping the net flux constant (unless the wire is made of superconducting material). In particular, if B_1 is increasing at a constant rate, the current I_2 in the ring and the additional magnetic field are constant by Faraday's law, so the net flux does increase.

If on the other hand $\vec{B}_1$ points out of the page and is decreasing (the second example in Figure 22.55), there is a counterclockwise non-Coulomb electric field that would drive conventional current counterclockwise in a wire, which would make a magnetic field out of the page, in the direction to attempt to keep the magnetic flux constant despite the decrease in B_1. (Again, the net flux will nevertheless decrease, unless the ring is superconducting.)

> QUESTION Go through this analysis for the third and fourth examples shown in Figure 22.55 and convince yourself that Lenz's rule does correctly summarize the experimental data for the direction of the non-Coulomb electric field.

An interesting property of Lenz's rule is that it also correctly gives the direction of current in the case of motional emf, or in situations where both a time-varying magnetic field and motion contribute to a change in the flux enclosed by a part of a circuit.

We have not emphasized Lenz's rule for two reasons. We have emphasized time-varying magnetic field and curly non-Coulomb electric field (rather than just time-varying flux and emf), to give a stronger sense of mechanism and to distinguish the effects of time-varying magnetic fields from the effects of the motion of a wire in a magnetic field.

Also, it is very easy to make serious conceptual mistakes using Lenz's rule when first studying the effect of time-varying magnetic fields, due to a natural tendency to try to use it for predicting quantitatively the magnitude of the induced current flow, not just its direction. In particular, except for superconductors, it is not true that "enough current runs to cancel the change in the magnetic field."

SUMMARY

Faraday's Law

The new phenomenon introduced in this chapter is that a time-varying magnetic field is accompanied by a non-Coulomb electric field that curls around the region of varying field. Unlike the Coulomb electric field produced by point charges, the non-Coulomb electric field has a nonzero round-trip path integral if the path encircles a time-varying flux.

Faraday's Law

$$\text{emf} = -\frac{d\Phi_{mag}}{dt}$$

where $\text{emf} = \oint \vec{E}_{NC} \bullet d\vec{l}$ and $\Phi_{mag} \equiv \int \vec{B} \bullet \hat{n} dA$.

In words: The emf along a round-trip path is equal to the rate of change of the magnetic flux on the area encircled by the path.

Direction: With the thumb of your right hand pointing in the direction of $-d\vec{B}/dt$, your fingers curl around in the direction of $\vec{E}_{NC}$.

Formal version of Faraday's law

$$\oint \vec{E}_{NC} \bullet d\vec{l} = -\frac{d}{dt}\int \vec{B} \bullet \hat{n} dA$$

(with the sign given by the right-hand rule).

Faraday's law also covers the case where the flux changes due to changes in the path ("motional emf"), and the non-Coulomb force in this case is a magnetic force.

In general, there can be any combination of three effects acting on electrons in a circuit:

1. Coulomb electric field due to surface charges
2. Non-Coulomb electric field associated with time-varying magnetic field
3. Magnetic forces on moving portions of the circuit (motional emf)

Superconductors

Superconductors have zero resistance, and a consequence of this is that a superconducting ring maintains a constant flux in the area enclosed by the ring. The magnetic field inside a superconductor is always zero (the Meissner effect), and this important property cannot be explained merely in terms of zero resistance.

Inductance

When you vary the current in a coil, the varying flux enclosed by the coil is associated with an emf in the *same* coil, which makes it difficult to change the current. This sluggishness plays a role in RL and LC circuits, and it is quantified by the "inductance" L of the coil. The emf is proportional to the rate of change of the magnetic flux, which is proportional to the rate of change of the current:

$$|\text{emf}_{ind}| = L\left|\frac{dI}{dt}\right|$$

We calculated the inductance in one particular case, that of a long solenoid:

$$L_{solenoid} = \frac{\mu_0 N^2}{d}\pi R^2$$

Electric and Magnetic Energy Densities

$$\frac{\text{Energy}}{\text{Volume}} = \frac{1}{2}\varepsilon_0 E^2 + \frac{1}{2}\frac{1}{\mu_0}B^2$$

measured in J/m^3.

RL (Resistor–Inductor) Circuits

The current in an RL series circuit varies with time:

$$I = \frac{\text{emf}_{battery}}{R}\left[1 - e^{-(R/L)t}\right]$$

LC (Inductor–Capacitor) Circuits

An LC circuit oscillates sinusoidally with a frequency $f = 1/(2\pi\sqrt{LC})$:

$$Q = Q_i \cos\left(\frac{1}{\sqrt{LC}}t\right)$$

If there is no resistance, an LC circuit never reaches a final equilibrium or a final steady state.

****The Differential Form of Faraday's Law***

$$\text{curl}(\vec{E}) = -\frac{\partial \vec{B}}{\partial t} \quad \text{or} \quad \vec{\nabla} \times \vec{E} = -\frac{\partial \vec{B}}{\partial t}$$

QUESTIONS

Q1 In Figure 22.56 a bar magnet with its S pole on top moves upward toward the origin, $\langle 0,0,0 \rangle$.

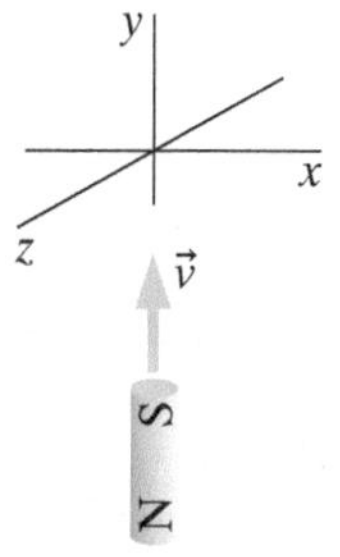

Figure 22.56

(a) At the origin, what is the direction of $\vec{B}$ at this instant, time t_1? **(b)** At the origin, what is the direction of $\vec{B}$ at a slightly later time t_2, when the bar magnet is still underneath the origin but closer to it? **(c)** At the origin, is the magnitude of the magnetic field $|\vec{B}|$ larger at time t_1 or time t_2? **(d)** At the origin, what is the direction of $\Delta\vec{B}$, the vector change in the magnetic field from time t_1 to time t_2? **(e)** At the origin, what is the direction of $-\Delta\vec{B}/\Delta t$? **(f)** Looking down at the origin from above (along the $+y$ axis), is the pattern of curly electric field clockwise or counterclockwise? **(g)** What is the direction of $\vec{E}_{NC}$ at the location $\langle 1,0,0 \rangle$?

Q2 Two coils of wire are near each other, positioned on a common axis (Figure 22.57). Coil 1 is connected to a power supply whose output voltage can be adjusted by turning a knob, so that the current I_1 in coil 1 can be varied, and I_1 is measured by ammeter 1.

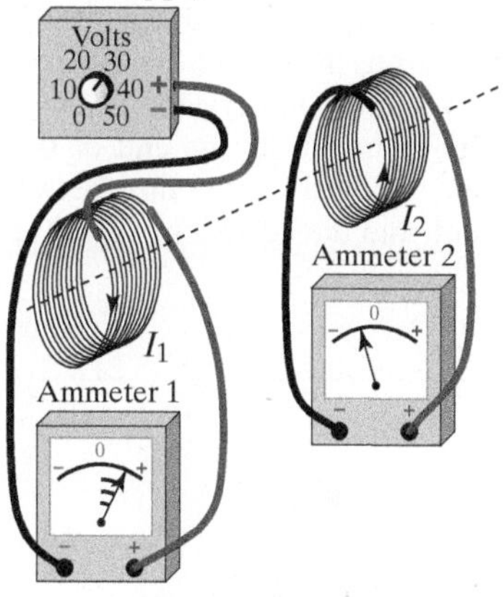

Figure 22.57

Current I_2 in coil 2 is measured by ammeter 2. The ammeters have needles that deflect positive or negative depending on the direction of current passing through the ammeter, and ammeters read positive if conventional current flows into the + terminal. Figure 22.58 is a graph of I_1 vs. time. Draw a graph of I_2 vs. time over the same time interval. Explain your reasoning.

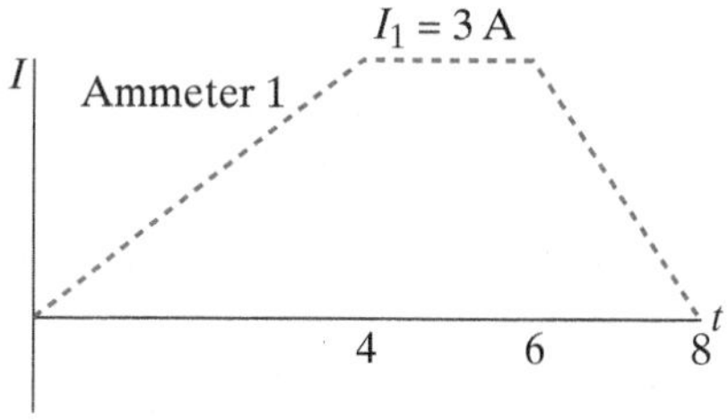

Figure 22.58

Q3 Two metal rings lie side-by-side on a table (Figure 22.59). Current in the left ring runs clockwise and is increasing with time, so a current runs in the right ring. Does this current run clockwise or counterclockwise? Explain, using a diagram. (*Hint:* Think carefully about the direction of magnetic field in the right ring produced by the left ring, taking into consideration what sections of the left ring are closest.)

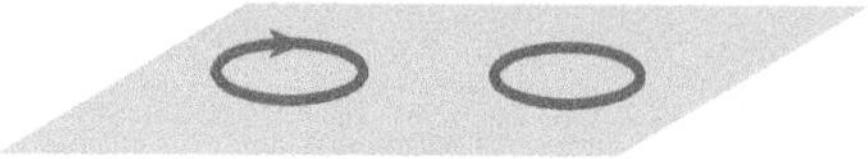

Figure 22.59

Q4 A very long straight wire (essentially infinite in length) carries a current of 6 ampere (Figure 22.60). The wire passes through the center of a circular metal ring of radius 4 cm and resistance 2 Ω that is perpendicular to the wire. If the current in the wire increases at a rate of 0.25 A/s, what is the current in the ring? Explain carefully.

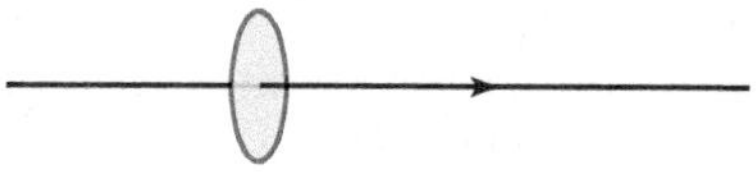

Figure 22.60

Q5 A bar magnet is dropped through a vertical copper tube and is observed to fall very slowly, despite the fact that mechanical friction between the magnet and the tube is negligible (Figure 22.61). Explain carefully, including adequate diagrams.

Figure 22.61

Q6 Suppose that you move a bar magnet toward the coil in Figure 22.62, with the S end of the bar magnet closest to the coil. **(a)** Will the ammeter read positive or negative? **(b)** Now move the bar magnet away from the coil, with the S end still closest to the coil. Will the ammeter read + or −?

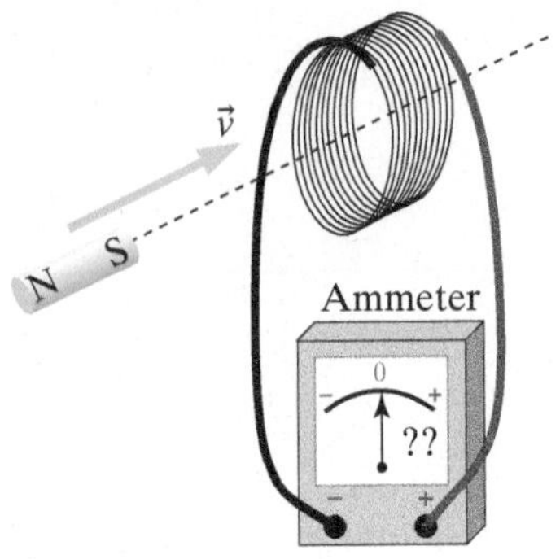

Figure 22.62

Q7 A bar magnet is held vertically above a horizontal metal ring, with the south pole of the magnet at the top (Figure 22.63). If the magnet is lifted straight up, will current run clockwise or counterclockwise in the ring, as seen from above?

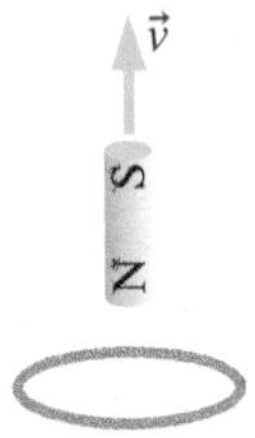

Figure 22.63

Q8 One of the methods physicists have used to search for magnetic monopoles is to monitor the current produced in a loop of wire. Draw graphs of current in the loop vs. time for an electrically uncharged magnetic monopole passing through the loop, and for an electrically uncharged magnetic dipole (such as a neutron) passing through the loop with its north end head-first. Don't worry too much about the details of the exact moment when the particle goes through the plane of the loop; concentrate on the times just before and just after this event. Explain the differences in the two graphs.

(A signal corresponding to a magnetic monopole was seen once in such an experiment, but no one has been able to reproduce this result, and most physicists seem to believe that the supposed event was due to extraneous noise in the system or other malfunction of the apparatus.)

Q9 A conventional current I runs through a coil in the direction shown in Figure 22.64. A single loop of copper wire is near the coil. The loop and the coil are stationary.

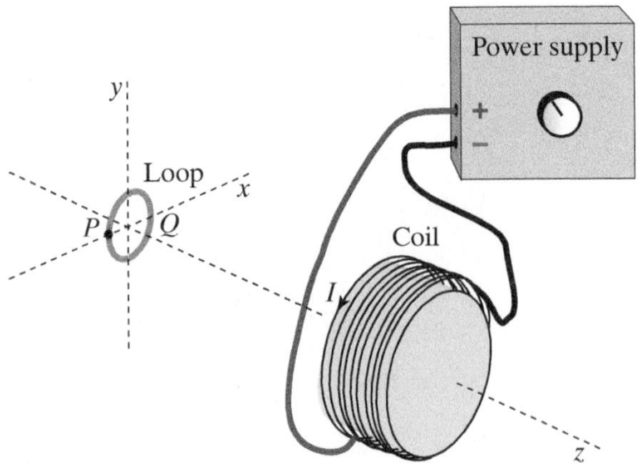

Figure 22.64

(a) In this initial state (constant current in coil), what is the direction of the magnetic field at the center of the copper loop, due to the current in the coil? **(b)** In this initial state, what is the direction of the electric field at location P inside the copper loop? **(c)** What is the direction of the electric field at location Q inside the copper loop?

Next the power supply is adjusted so the current in the coil *decreases* with time. **(d)** Now, what is the direction of $d\vec{B}/dt$ at the center of the copper loop? **(e)** What is the direction of $-d\vec{B}/dt$ at the center of the copper loop? **(f)** What is the direction of the electric field at location P in the copper wire? At location Q? **(g)** At this moment, is the magnitude of the magnetic flux inside the copper loop increasing, decreasing, or staying constant?

Q10 Would the inductance of a solenoid be larger or smaller if the solenoid is filled with iron? Explain briefly.

Q11 **(a)** When an RL circuit has been connected to a 1.5 V battery for a very *short* time, what is the potential difference from one end of the resistor to the other? **(b)** When an RL circuit has been connected to a 1.5 V battery for a very *long* time, what is the potential difference from one end of the resistor to the other? **(c)** Explain briefly the difference between equilibrium, steady-state current, and the behavior of an LC circuit.

PROBLEMS

Section 22.2

•P12 In Figure 22.65 the solenoid radius is 2 cm and the ring radius is 10 cm. $B = 0.5$ T inside the solenoid, and $B \approx 0$ outside the solenoid. What is the magnetic flux through the outer ring?

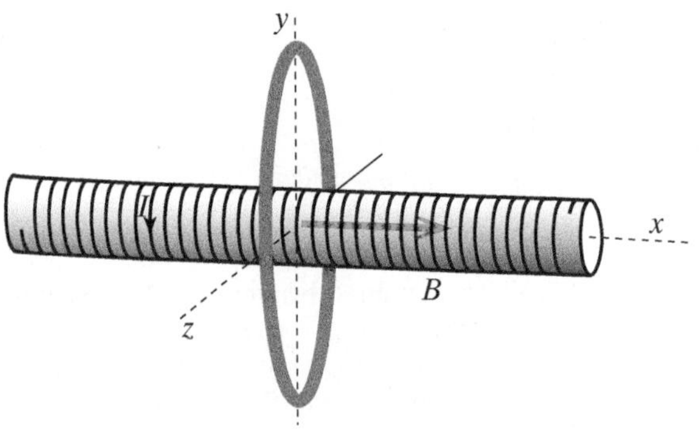

Figure 22.65

•**P13** In Figure 22.66 the north pole of a bar magnet points toward a thin circular coil of wire containing 40 turns. The magnet is moved away from the coil, so that the flux through one turn inside the coil decreases by $0.3\ \text{T}\cdot\text{m}^2$ in 0.2 s. What is the average emf in the coil during this time interval? Viewed from the right side (opposite the bar magnet), does the induced current run clockwise or counterclockwise? Explain briefly.

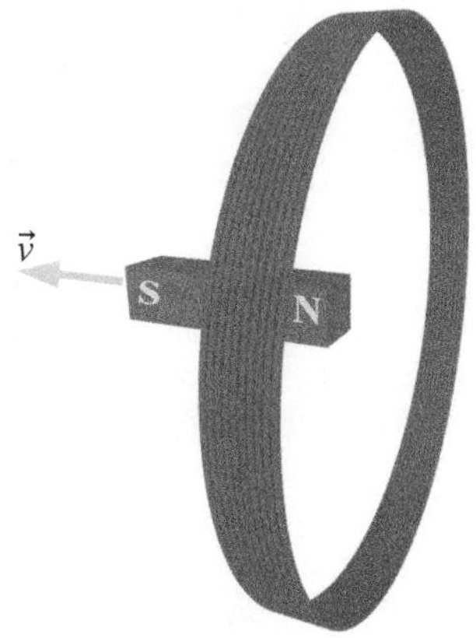

Figure 22.66

•**P14** The magnetic field in a solenoid is $B = \mu_0 NI/d$. A circular wire of radius 10 cm is concentric with a solenoid of radius 2 cm and length $d = 1$ m, containing 10,000 turns. The current increases at a rate of 50 A/s. What is the non-Coulomb electric field in the wire?

•**P15** In Figure 22.67 there is a circular region of radius r_1 in which there is a uniform magnetic field B pointing out of the paper (the magnetic field is essentially zero outside this region). The magnetic field is *decreasing* at this moment at a rate $|dB/dt|$. A wire of radius r_2 and resistance R lies entirely *outside* the magnetic-field region (where there is no magnetic field). In which direction does conventional current I flow in the ring? What is the magnitude of the current I? Show the pattern of (non-Coulomb) electric field in the ring. What is the magnitude E_{NC} of the non-Coulomb electric field?

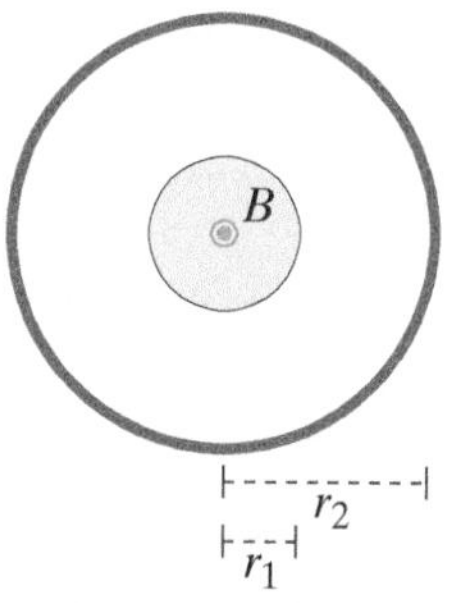

Figure 22.67

•**P16** A thick copper wire connected to a voltmeter surrounds a region of time-varying magnetic flux, and the voltmeter reads 10 V. If instead of a single wire we use a coil of thick copper wire containing 20 turns, what does the voltmeter read? A thin Nichrome wire connected to an ammeter surrounds a region of time-varying magnetic flux, and the ammeter reads 10 A. If instead of a single wire we use a coil of thin Nichrome wire containing 20 turns, what does the ammeter read?

••**P17** In Figure 22.64 a coil of wire with 300 turns is connected to a power supply, and a current runs in the coil. A single loop of copper wire is located near the coil, with its axis on the same line as the axis of the coil. The loop and the coil are stationary. The radius of the coil is 9 cm, and the radius of the loop is 4 cm. The center of the loop is 22 cm from the center of the coil. At time t_1 the conventional current I in the coil is 5 A, and it is not changing. **(a)** What is the absolute value of the magnetic flux through the loop at time t_1? **(b)** What approximations or assumptions did you make in calculating your answer to part (a)? **(c)** What is the direction of the electric field at location P in the copper wire of the loop at time t_1? (Remember that at this time the current in the coil is constant.)

At a later time t_2 the current in the coil begins to decrease. **(d)** Now what is the direction of the curly electric field in the loop?

At time t_2 the current in the coil is decreasing at a rate $dI/dt = -0.3$ A/s. **(e)** At this time, what is the absolute value of the rate of change of the magnetic flux through the loop? **(f)** At this time, what is the absolute value of the emf in the loop? **(g)** What is the magnitude of the electric field at location P, which is inside the wire? **(h)** Now the wire loop is removed. Everything else remains as it was at time t_2; the current is still changing at the same rate. What is the magnitude of the electric field at location P?

••**P18** In Figure 22.68 the magnetic field is uniform and out of the page inside a circle of radius R, and is essentially zero outside the circular region. The magnitude of the magnetic field is changing with time; as a function of time the magnitude of the magnetic field is $(B_0 + bt^3)$, where $B_0 = 1.5$ T, $b = 1.4\ \text{T/s}^3$, $r_1 = 3.6$ cm, $r_2 = 51$ cm, $t = 1.3$ s, and $R = 17$ cm.

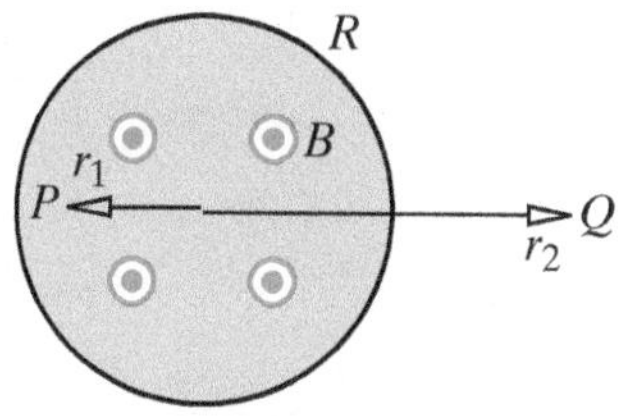

Figure 22.68

(a) What is the direction of the electric field at location P, a distance r_1 to the left of the center ($r_1 < R$)? **(b)** What is the magnitude of the electric field at location P? (*Hint:* Remember that "emf" is the integral of the non-Coulomb electric field around a closed path.) **(c)** What is the direction of the electric field at location Q, a distance r_2 to the right of the center ($r_2 > R$)? **(d)** What is the magnitude of the induced electric field at location Q?

••**P19** In Figure 22.69 a thin circular coil of radius r_1 with N_1 turns carries a current $I_1 = a + bt + ct^2$, where t is the time and a, b, and c are positive constants. A second thin circular coil of radius r_2 with N_2 turns is located a long distance x along the axis of the first coil. The second coil is connected to an oscilloscope, which has very high resistance.

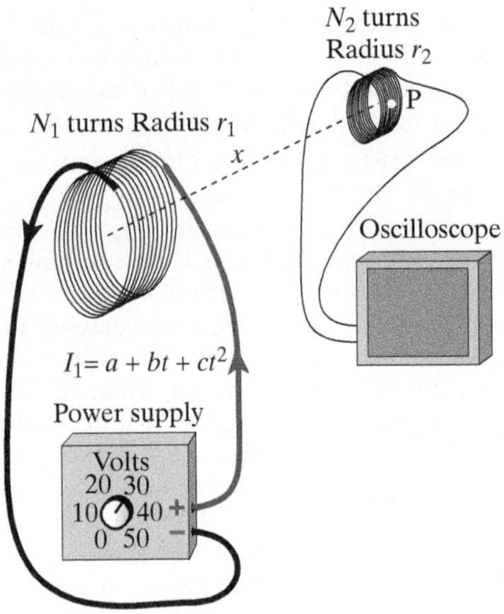

Figure 22.69

(a) As a function of time t, calculate the magnitude of the voltage that is displayed on the oscilloscope. Explain your work carefully, but you do not have to worry about signs. **(b)** At point P on the drawing (on the right side of the second coil), draw a vector representing the non-Coulomb electric field. **(c)** Calculate the magnitude of this non-Coulomb electric field.

••P20 In Figure 22.70 a long straight wire carrying current I is moving with speed v toward a small circular coil of radius r containing N turns, which is attached to a voltmeter as shown. The long wire is in the plane of the coil. (Only a small portion of the wire is shown in the diagram.)

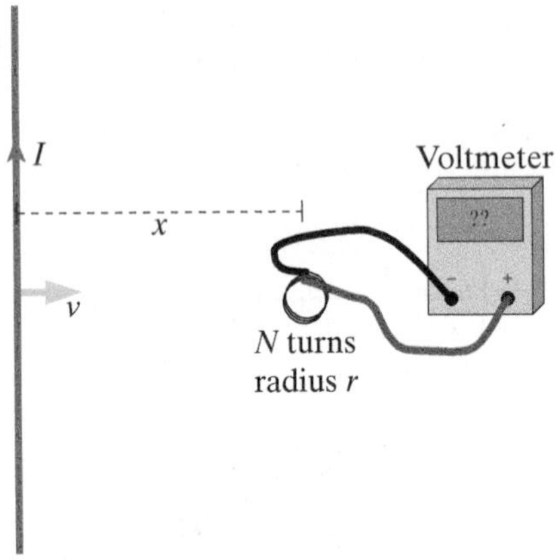

Figure 22.70

The radius of the coil is 0.02 m, and the coil has 11 turns. At a particular instant $I = 3$ A, $v = 3.2$ m/s, and the distance from the wire to the center of the coil is $x = 0.13$ m. **(a)** What is the magnitude of the rate of change of the magnetic field inside the coil? You will need to calculate this symbolically before you can get a number. Write an expression for the magnetic field due to the wire at the location of the coil. Use the approximate equation, since the wire is very long. Remember the chain rule, and remember that $v = dx/dt$. **(b)** What is the magnitude of the voltmeter reading? Remember that this includes all 11 turns of the coil. **(c)** What is the direction of $d\vec{B}/dt$? **(d)** What is the direction of the curly electric field in the coil?

••P21 A coil with 3000 turns and radius 5 cm is connected to an oscilloscope (Figure 22.71). You move a bar magnet toward the coil along the coil's axis, moving from 40 cm away to 30 cm away in 0.2 s. The bar magnet has a magnetic moment of 0.8 $A \cdot m^2$.

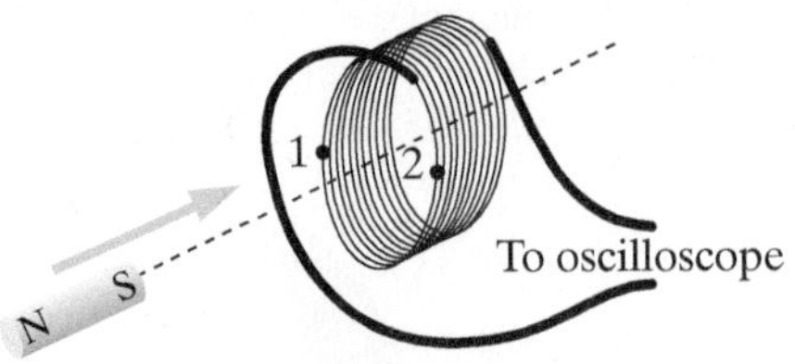

Figure 22.71 .

(a) On the diagram, draw the non-Coulomb electric field vectors at locations 1 and 2. Briefly explain your choices graphically. **(b)** What is the approximate magnitude of the signal observed on the oscilloscope? **(c)** What approximations or simplifying assumptions did you make?

••P22 In Figure 22.72 a toroid has a rectangular cross section with an inner radius $r_1 = 9$ cm, an outer radius $r_2 = 12$ cm, and a height $h = 5$ cm, and it is wrapped around by many densely packed turns of current-carrying wire (not shown in the diagram). The direction of the magnetic field inside the windings is shown on the diagram. There is essentially no magnetic field outside the windings. A wire is connected to a sensitive ammeter as shown. The resistance of the wire and ammeter is $R = 1.4$ ohms.

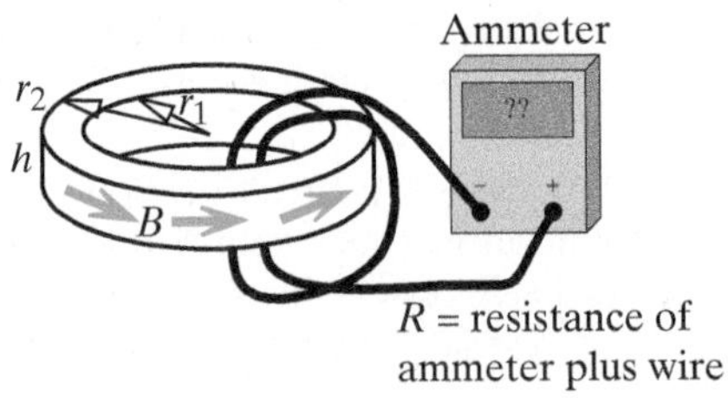

Figure 22.72

The current in the windings of the toroid is varied so that the magnetic field inside the windings, averaged over the cross section, varies with time as shown in Figure 22.73:

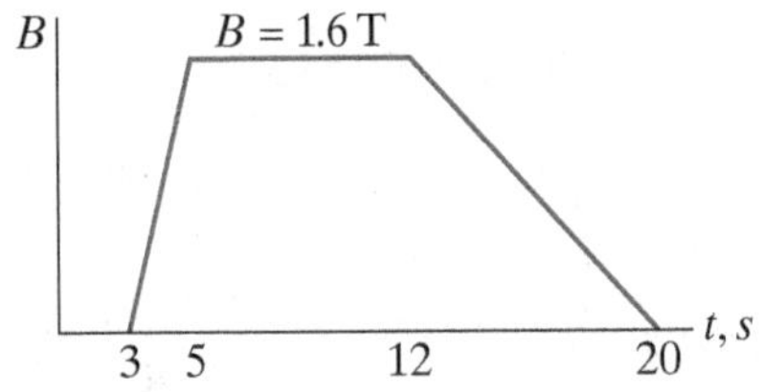

Figure 22.73

Make a careful graph of the ammeter reading, including sign, as a function of time. Label your graph, and explain the numerical aspects of the graph, including signs.

••P23 In Figure 22.74 a thin coil of radius $r_1 = 4$ cm, containing $N_1 = 5000$ turns, is connected through a resistor $R = 100\ \Omega$ to an AC power supply running at a frequency $f = 2500$ H, so that the current through the resistor (and coil) is $I_1 \sin(2\pi \times 2500t)$. The voltage across the resistor triggers an oscilloscope that also displays this voltage, which is 10 V peak-to-peak and therefore has an amplitude of 5 V (and therefore the amplitude of the current is $I_1 = 0.05$ A).

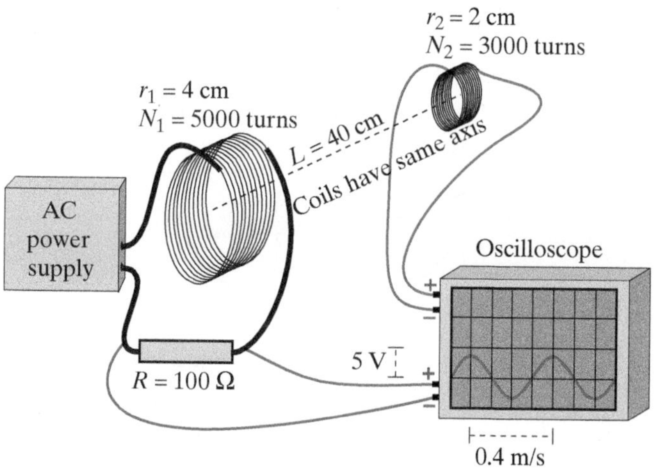

Figure 22.74

A second thin coil of radius $r_2 = 2$ cm, containing $N_2 = 3000$ turns, is a distance $L = 40$ cm from the first coil. The axes of the two coils are along the same line. The second coil is connected to the upper input of the oscilloscope, so that the voltage across the second coil can be displayed along with the voltage across the resistor. **(a)** Assume you have adjusted the VOLTS/DIVISION knob on the upper input so that you can easily see the signal from the second coil. Sketch the second coil voltage along with the resistor voltage on the oscilloscope display, paying careful attention to the shape and positioning of the two voltage curves with respect to each other. Explain briefly. **(b)** Calculate the amplitude (maximum voltage) of the second coil voltage. If you must make simplifying assumptions, state clearly what they are.

••P24 A thin rectangular coil lies flat on a low-friction table (Figure 22.75). A very long straight wire also lies flat on the table, a distance z from the coil. The wire carries a conventional current I to the right as shown, and this current is decreasing: $I = a - bt$, where t is the time in seconds, and a and b are positive constants. The coil has length L and width w, where $w \ll z$. It has N turns of wire with total resistance R.

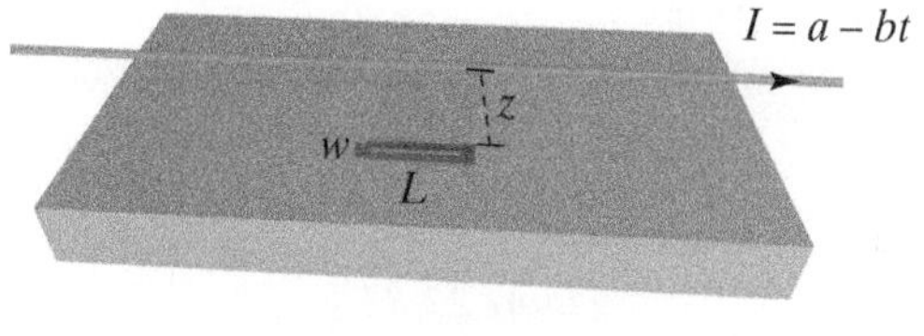

Figure 22.75

What are the initial magnitude and direction of the nonzero net force that is acting on the coil? You can neglect friction. Explain in detail. If you must make simplifying assumptions, state clearly what they are, but bear in mind that the net force is not zero.

••P25 There are a lot of numbers in this problem. Just about the only way to get it right is to work out each step symbolically first, and then plug numbers into the final symbolic result.

Two coils of wire are aligned with their axes along the z axis, as shown in Figure 22.76. Coil 1 is connected to a power supply and conventional current flows clockwise through coil 1, as seen from the location of coil 2. Coil 2 is connected to a voltmeter. The distance between the centers of the coils is 0.14 m.

Coil 1 has $N_1 = 570$ turns of wire, and its radius is $R_1 = 0.07$ m. The current through coil 1 is changing with time. At $t = 0$ s, the current through coil 1 is $I_0 = 18$ A. At $t = 0.4$ s, the current through coil 1 is $I_{0.4} = 6$ A.

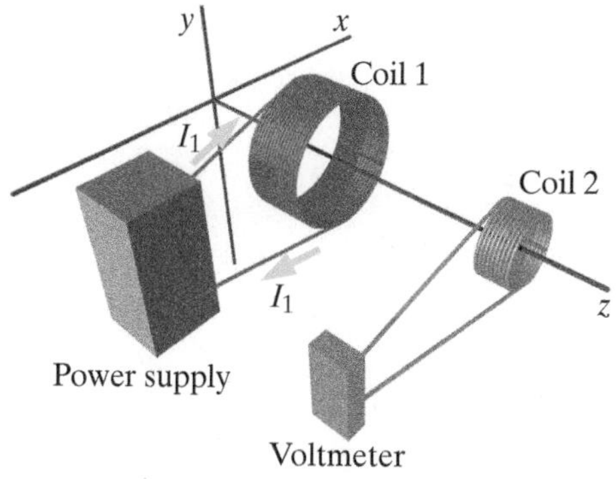

Figure 22.76

Coil 2 has $N_2 = 275$ turns of wire, and its radius is $R_2 = 0.03$ m. **(a)** Inside coil 2, what is the direction of $d\vec{B}/dt$ during this interval? **(b)** What is the direction of the electric field inside the wire of coil 2, at a location on the top of coil 2? **(c)** At time $t = 0$, what is the magnetic flux through one turn of coil 2? Remember that all turns of coil 1 contribute to the magnetic field. Note that the coils are not very far apart (compared to their radii). **(d)** What approximations did you make in calculating the flux through one turn of coil 2? **(e)** At $t = 0.4$ s, what is the magnetic flux through one turn of coil 2? **(f)** What is the emf in one turn of coil 2 during this time interval? **(g)** The voltmeter is connected across all turns of coil 2. What is the reading on the voltmeter during this time interval? **(h)** During this interval, what is the magnitude of the non-Coulomb electric field inside the wire of coil 2? Remember that the emf measured by the voltmeter involves the entire length of the wire making up coil 2. **(i)** At $t = 0.5$ s, the current in coil 1 becomes constant, at 5 A. Which of the following statements are true? (1) The voltmeter now reads 0 V. (2) The electric field inside the wire of coil 2 now points in the opposite direction. (3) The voltmeter reading is about the same as it was at $t = 0.4$ s. (4) The electric field inside the wire of coil 2 is now 0 V/m.

••P26 In Figure 22.77 a very small circular metal ring of radius $r = 0.5$ cm and resistance $x = 5\ \Omega$ is at the center of a large concentric circular metal ring of radius $R = 50$ cm. The two rings lie in the same plane. At $l = 3$ s, the large ring carries a *clockwise* current of 3 A. At $l = 3.2$ s, the large ring carries a *counterclockwise* current of 5 A.

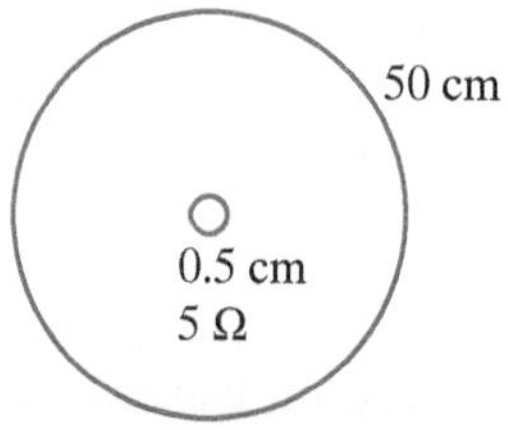

Figure 22.77

(a) What is the magnitude of the emf in the small ring during this time? **(b)** What are the magnitude and direction of the average current in the small ring? **(c)** What is the average electric field in the small ring during this time? Give both the magnitude and direction of the electric field. Draw a diagram, showing all relevant quantities.

••P27 A very long, tightly wound solenoid has a circular cross section of radius 3 cm (only a portion of the very long solenoid is shown in Figure 22.78). The magnetic field outside the solenoid is negligible. Throughout the inside of the solenoid the magnetic field B is uniform, to the left as shown, but varying with time t: $B = (0.07 + 0.03t^2)$ T. Surrounding the circular solenoid is a loop of 4 turns of wire in the shape of a rectangle 10 cm by 15 cm. The total resistance of the 4-turn loop is 0.1 ohm.

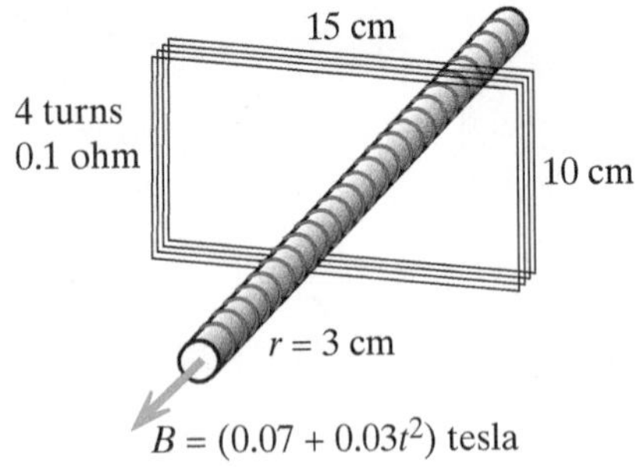

Figure 22.78

(a) At $t = 2$ s, what is the direction of the current in the 4-turn loop? Explain briefly. **(b)** At $t = 2$ s, what is the magnitude of the current in the 4-turn loop? Explain briefly.

••P28 Tall towers support power lines 50 m above the ground and 20 m apart that run from a hydroelectric plant to a large city, carrying 60 H alternating current with amplitude 1×10^4 A (Figure 22.79). That is, the current in both of the power lines is $I = (1 \times 10^4 \text{A}) \sin(2\pi \cdot 60 \text{ H} \cdot t)$.

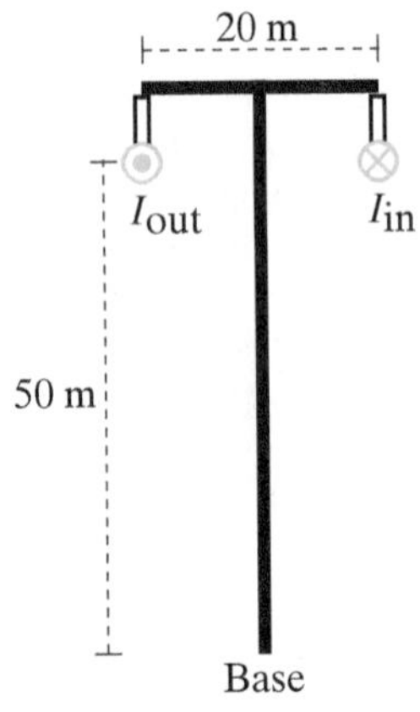

Figure 22.79

(a) Calculate the amplitude (largest magnitude) and direction of the magnetic field produced by the two power lines at the base of the tower, when a current of 1×10^4 A in the left power line is headed out of the page, and a current of 1×10^4 A in the right power line is headed into the page. **(b)** This magnetic field is not large compared to the Earth's magnetic field, but it varies in time and so might have different biological effects than the Earth's steady field. For a person lying on the ground at the base of the tower, approximately what is the maximum emf produced around the perimeter of the body, which is about 2 m long by half a meter wide?

••P29 In Figure 22.80 a very long solenoid of length d and radius r_1 is tightly wound uniformly with N turns of wire. A variable power supply forces a current to run in the solenoid of amount $I_1 = p - kt$, where p and k are positive constants (so that the current is initially equal to p but decreases with time). A circular metal ring of radius r_2 and resistance R is centered on the solenoid and located near the middle of the solenoid, very far from the ends of the solenoid (so that the solenoid contributes essentially no magnetic field outside the solenoid).

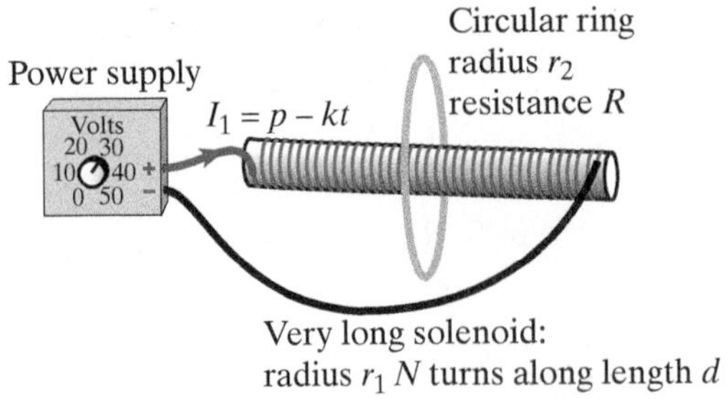

Figure 22.80

(a) What are the magnitude and the direction of the magnetic field at the center of the ring, due to the solenoid and the ring, as a function of the time t? Explain your work; part of the credit will be given for the clarity of your explanation, including clarity of appropriate diagrams. **(b)** Qualitatively, how would these results have been affected if an iron rod had been inserted into the solenoid? Very briefly, explain why.

••P30 You throw a bar magnet downward with its south end pointing down (Figure 22.81). The bar magnet has a magnetic dipole moment of 1.2 $\text{A} \cdot \text{m}^2$. Lying on the table is a nearly flat circular coil of 1000 turns of wire, with radius 5 cm. The coil is connected to an oscilloscope, which has a very large resistance.

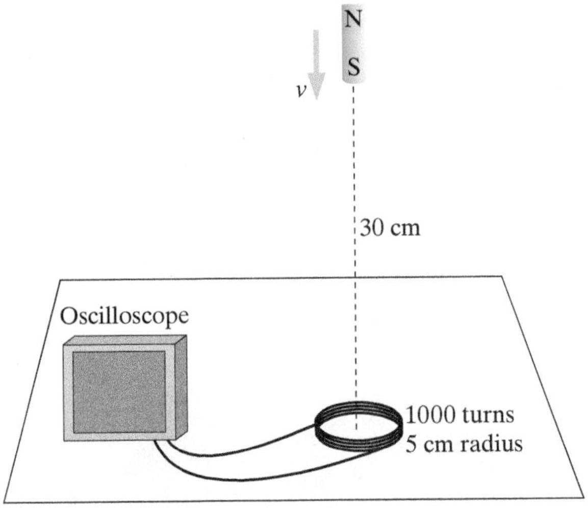

Figure 22.81

(a) On a diagram, show the pattern of non-Coulomb electric field in the coil. Explain briefly. **(b)** At the instant when the magnet is 30 cm above the table, the oscilloscope indicates a voltage of magnitude 2 mV. What is the speed v of the magnet at that instant?

••P31 Suppose that you have an electron moving with speed comparable to the speed of light in a circular orbit of radius r in a large region of uniform magnetic field B. **(a)** What must be the relativistic momentum p of the electron? **(b)** Now the uniform magnetic field begins to increase with time: $B = B_0 + bt$, where B_0 and b are positive constants. In one orbit, how much does the energy of the electron increase, assuming that in one orbit the radius doesn't change very much? (This effect was exploited in the "betatron," an electron accelerator invented in the 1940s.) *Note:* It turns out that the electron's energy increases by less than the amount you calculated, for reasons that will become clear in Chapter 23.

••P32 A thin coil is located at the origin; its radius is 3 cm and its axis lies on the x axis. It has 30 low-resistance turns and is connected to a 75 Ω resistor. A second thin coil is located at $\langle 15,0,0\rangle$ cm and is traveling toward the origin with a speed of 8 m/s; it has 50 low-resistance turns, its axis lies on the x axis, its radius is 4 cm, and it has a current of 12 A, powered by a battery. What is the magnitude of the current in the first coil?

••P33 A bar magnet whose dipole moment is $\langle 0,0,6\rangle$ A·m^2 has a constant velocity of $\langle 0,0,5\rangle$ m/s. When the center of the magnet is at location $\langle 1,5,3\rangle$ m, what is the (vector) electric field at location $\langle 1.04,5,2\rangle$ m?

••P34 A very long wire carries a current I_1 *upward* as shown in Figure 22.82, and this current is *decreasing with time*. Nearby is a rectangular loop of wire lying in the plane of the long wire, and containing a resistor R; the resistance of the rest of the loop is negligible compared to R. The loop has a width w and height h, and is located a distance d to the right of the long wire.

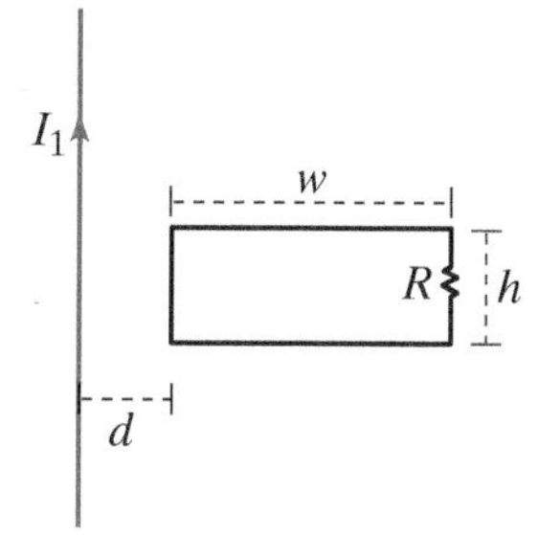

Figure 22.82

(a) Does the current I_2 in the loop run clockwise or counterclockwise? Explain, using a diagram. **(b)** Show that the magnitude of the current I_2 in the loop is this:

$$I_2 = \frac{\mu_0}{4\pi}\frac{(2h)\ln[(d+w)/d]}{R}\left|\frac{dI_1}{dt}\right|$$

(*Hint:* Divide the area into narrow vertical strips along which you know the magnetic field.)

••P35 In Figure 22.83 a single circular loop of wire with radius R carries a large clockwise constant current $I_{\text{loop}} = I_0$, which constrains a proton of mass M and charge e to travel in a small circle of radius r at constant speed around the center of the loop, in the plane of the loop. The orbit radius r is much smaller than the loop radius R: $r \ll R$.

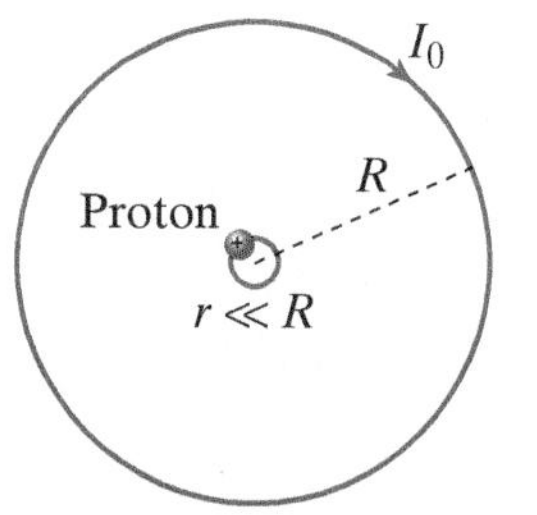

Figure 22.83

(a) Draw a diagram of the proton orbit, indicating the direction that the proton travels, clockwise or counterclockwise. Explain briefly. **(b)** What is the speed v of the proton, in terms of the known quantities I_0, R, r, M, and e? Explain your work, including any approximations you must make. **(c)** The current was constant for a while, but at a certain time $t = t_0$ it began to decrease slowly, so that after t_0 the current was $I_{\text{loop}} = I_0 - k(t - t_0)$. On your diagram of the proton orbit, draw electric field vectors at four locations (one in each quadrant) and explain briefly. **(d)** When the current decreases, does the proton speed up or slow down?

Section 22.3

••P36 A copper bar of length h and electric resistance R slides with negligible friction on metal rails that have negligible electric resistance (Figure 22.84). The rails are connected on the right with a wire of negligible electric resistance, and a magnetic compass is placed under this wire (the diagram is a top view). The compass needle deflects to the right of north, as shown on the diagram.

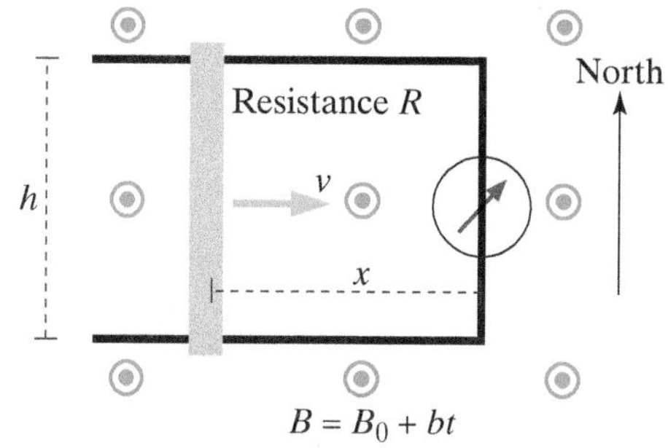

Figure 22.84

Throughout this region there is a uniform magnetic field B pointing out of the page, produced by large coils that are not shown. This magnetic field is increasing with time, and the magnitude is $B = B_0 + bt$, where B_0 and b are constants, and t is the time in seconds. You slide the copper bar to the right and at time $t = 0$ you release the bar when it is a distance x from the right end of the apparatus. At that instant the bar is moving to the right with a speed v. **(a)** Calculate the magnitude of the initial current I in this circuit. **(b)** Calculate the magnitude of the net force on the bar just after you release it. **(c)** Will the bar speed up, slow down, or slide at a constant speed? Explain briefly.

Section 22.5

••P37 A superconducting ring of radius R has a permanent conventional current I in the direction shown in Figure 22.85 and is oriented with its axis along the x axis. The ring is moving to the right along the x axis with uniform speed v. Calculate the electric field (magnitude and direction) at location $\langle x,y,0\rangle$ relative to the center of the ring. If you must make any simplifying assumptions or approximations, state them explicitly.

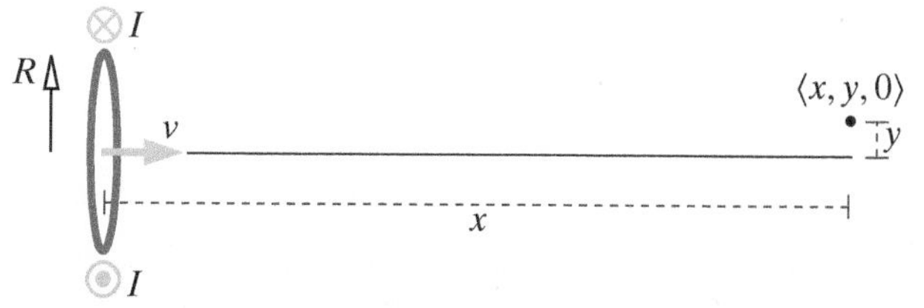

Figure 22.85

Section 22.6

•P38 A thin coil has 12 rectangular turns of wire. When a current of 3 A runs through the coil, there is a total flux of 1×10^{-3} T·m^2

enclosed by one turn of the coil (note that $\Phi = kI$, and you can calculate the proportionality constant k). Determine the inductance in henries.

•P39 A transformer has a primary coil with 100 turns and a secondary coil of 350 turns. The AC voltage across the primary coil has a maximum of 120 V and the AC current through the primary coil has a maximum of 4 A. What are the maximum values of the voltage and current for the secondary coil?

••P40 A 1000-turn solenoid has a radius of 1.4 cm and a length of 25 cm. The current in the solenoid is 8 A. **(a)** What is the inductance of this solenoid? **(b)** Inside the solenoid, what is the magnetic energy density (J/m^3) far from the ends of the solenoid? **(c)** What is the total magnetic energy, in the approximation that there is little magnetic field outside the solenoid and the magnetic field is nearly uniform inside the solenoid? **(d)** Show that the result in part (c) is equal to $\frac{1}{2}LI^2$.

••P41 It is now possible to buy capacitors that have a capacitance of one farad. **(a)** Design a solenoid so that when it is connected to a charged 1 F capacitor, the circuit will oscillate with a period of one second. Give all the relevant parameters of the solenoid (length, etc.) so that someone could build the solenoid from your design specifications. Assume that there is wire available with low enough resistance that the resistance of the solenoid is negligible, although this may be difficult to achieve in practice unless the wire is superconducting. **(b)** If the 1 F capacitor is initially charged to 3 V, what is the maximum current that will run through the inductor?

••P42 A cylindrical solenoid 40 cm long with a radius of 5 mm has 300 tightly wound turns of wire uniformly distributed along its length (Figure 22.86). Around the middle of the solenoid is a 2-turn rectangular loop 3 cm by 2 cm made of resistive wire having a resistance of 150 Ω.

One microsecond after connecting the loose wire to the battery to form a series circuit with the battery and a 20 Ω resistor, what is the magnitude of the current in the rectangular loop and its direction (clockwise or counterclockwise in the diagram)?

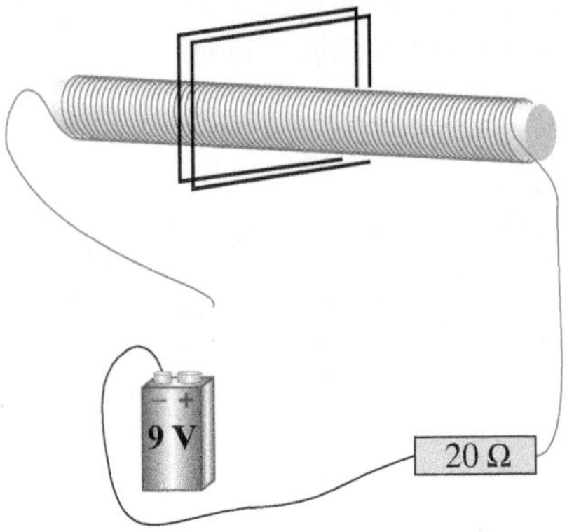

Figure 22.86

ANSWERS TO CHECKPOINTS

1 Clockwise; counterclockwise
2 (a) No current in wire A. **(b)** Current will run in wire B; **(c)** 47.7 V/m; **(d)** 7.5 A
3 0.052 T·m^2
4 20 V; 8 V/m; 2 A
5 0 (there is no flux change); 0.166 V
6 The flux must still be equal to Φ_0. Since the magnet is too far away to contribute, all the flux must be due to the current in the ring.
7 4×10^{-3} H; 4×10^{-2} H
8 4000 V
9 5033 Hz
10 In the first three cases, for the bulb that is bright a loop can be drawn that consists of just that bulb plus a wire that encircles the varying magnetic flux. For the bulb that is dark a loop can be drawn that consists of just that bulb plus a wire that does not encircle the varying magnetic flux. (1) Lower bulb bright, upper dark; (2) upper bulb bright, lower dark; (3) upper bulb bright, lower dark. For the fourth case, both bulbs are in a loop that encircles the varying magnetic flux, so they both glow. The current is smaller than in the other circuits, because the emf drives two bulbs in series.

CHAPTER

23

Electromagnetic Radiation

OBJECTIVES

After studying this chapter you should be able to

- Calculate the 3D radiative electric and magnetic fields produced by an accelerated charge.
- Mathematically relate the period, wavelength, and speed of a sinusoidal electromagnetic wave, and relate energy flux to amplitude.
- Explain physical phenomena involving the re-radiation of electromagnetic radiation.
- Mathematically relate index of refraction, frequency, speed of propagation, and angle of deflection of light in a medium.

We are approaching the climax of our study of electricity and magnetism: the classical model of electromagnetic radiation, or light. In Volume 1 we studied aspects of the quantum mechanical model of light, which involves particles called photons that have energy and momentum but zero rest mass. However, there are important properties of light, and aspects of the interaction of light and matter, that we cannot explain with our simple particle model. These phenomena, which have to do with the wavelike nature of light, can be explained by the classical model of electromagnetic radiation, which we will explore in this chapter. Aspects of this classical model are important to a fuller understanding of the quantum mechanical model of light, as well.

23.1 MAXWELL'S EQUATIONS

The four Maxwell equations provide a complete description of possible spatial patterns of electric and magnetic field in space. So far we have complete versions of three of these equations: Gauss's law, Gauss's law for magnetism, and Faraday's law. However, we have an incomplete version of the fourth equation, Ampere's law. Before we can go further, we need to complete this equation. Once we have all of Maxwell's equations, we will be able to show that some surprising configurations of electric and magnetic fields in space are not only possible but extremely important.

The Ampere–Maxwell Law

As we saw in Chapter 22, Faraday's law expresses a connection between a time-varying magnetic field in a region of space and a curly (non-Coulomb) electric field throughout space. You may have wondered whether a time-varying electric field in a region might be associated with a magnetic field throughout space. In fact, this does turn out to be the case. We will see that Ampere's law needs to be modified to include this relation.

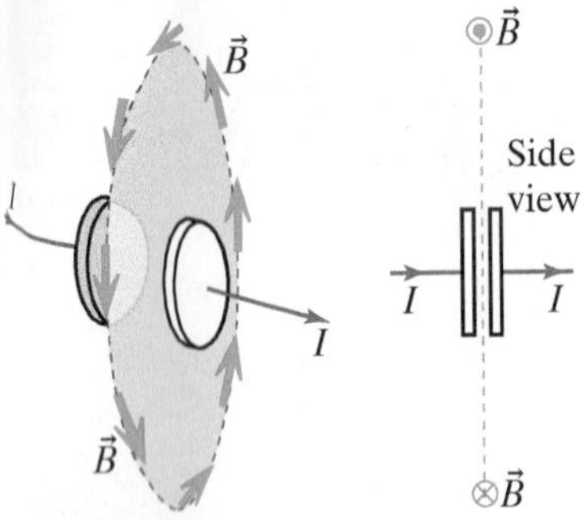

Figure 23.1 The soap film is stretched flat over the circular Amperian path.

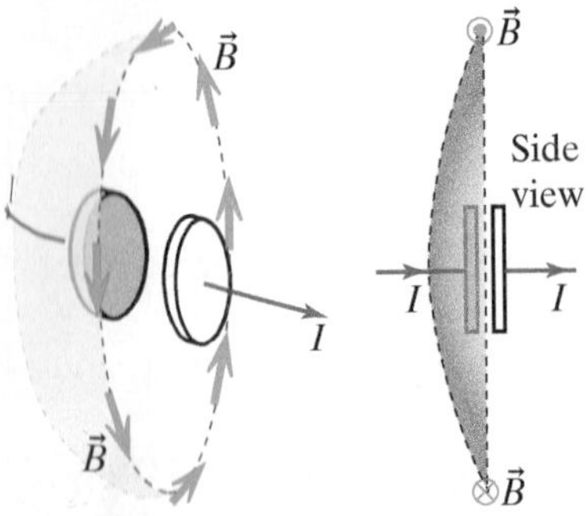

Figure 23.2 The soap film bulges out and is pierced by the current on the left.

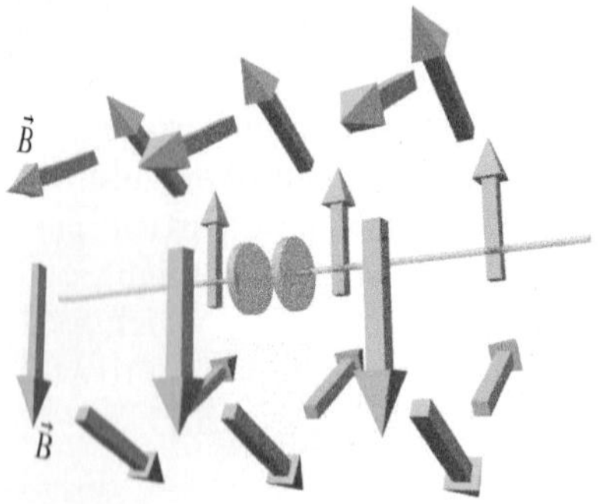

Figure 23.3 Probable magnetic field near the capacitor.

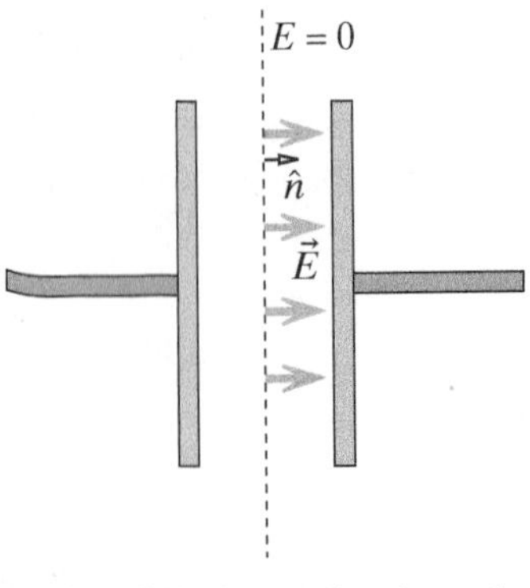

Figure 23.4 Calculating flux through a surface between the plates of the capacitor (side view).

If we try to apply Ampere's law to a circuit containing long straight wires carrying current to charge a capacitor, we will find what appears to be a contradiction. We choose a circular "Amperian" path around a charging capacitor, and stretch an imaginary soap film over the chosen path. Now consider two different soap film positions.

In Figure 23.1 the path does not enclose any current because the soap film is flat and passes between the capacitor plates. We must therefore conclude that $\oint \vec{B} \bullet d\vec{l}$ is zero along this circular path.

However, in Figure 23.2 the path does enclose the current I because the soap film bulges out and is pierced by the long straight wire on the left. In this case we conclude that $\oint \vec{B} \bullet d\vec{l}$ is equal to $\mu_0 I$. This contradicts our previous result.

> QUESTION Which result seems more reasonable? Would you expect the magnetic field along the chosen path to be zero or nonzero?

Presumably the magnetic field along the path is really nonzero, because the current in the wire contributes to the magnetic field at all locations. We would expect the magnetic field around the capacitor to be nearly the same as the magnetic field around a nearby section of the wire, as indicated in Figure 23.3.

In Chapter 17 we found by applying the Biot–Savart law applied to a long wire that the magnetic field near the wire was

$$B = \frac{\mu_0}{4\pi}\frac{2I}{r} \quad \text{(a distance } r \text{ from the center of a long wire)}$$

This should be approximately the magnetic field along the circular path in Figure 23.1, because there is only a very short piece of wire missing from the long straight wire. Therefore around the Amperian path in Figure 23.1 we should have

$$\oint \vec{B} \bullet d\vec{l} \approx \frac{\mu_0}{4\pi}\frac{2I}{r} 2\pi r = \mu_0 I$$

This dilemma was resolved by James Clerk Maxwell, a Scottish physicist. Knowing of Faraday's discovery that a time-varying magnetic field is accompanied by an electric field, he guessed that a time-varying electric field might be accompanied by a magnetic field. Between the plates of the capacitor there is a time-varying electric field, because the electric field between the plates is growing as the capacitor becomes more and more charged. Maxwell guessed that the time-varying electric field could account for the magnetic field around the capacitor.

Let's explore this idea quantitatively. By analogy to Faraday's law, we will guess that the time rate of change of the electric flux in a region is related to the integral of the magnetic field along a path surrounding the region.

First we need an expression for the electric flux through the surface shown in Figure 23.4. As we saw in Chapter 21,

$$\Phi_{\text{electric}} = \int \vec{E} \cdot \hat{n} dA$$

As shown in Figure 23.4, inside the capacitor the electric field is parallel to $\hat{n}$ and has magnitude $(Q/A)/\varepsilon_0$. Outside the capacitor the electric field is very small if the plates are large and close together. Therefore

$$\Phi_{\text{electric}} = \left(\frac{Q}{A\varepsilon_0}\right) A \cos 0° = \frac{Q}{\varepsilon_0}$$

QUESTION What is the time derivative of the electric flux in this situation?

In a time Δt the amount of charge ΔQ that flows onto the capacitor is $\Delta Q = I\Delta t$. Thus, the rate of change of Q, the amount of charge on the capacitor, is

$$\frac{dQ}{dt} = I$$

Therefore, the time rate of change of electric flux is

$$\frac{d\Phi_{\text{electric}}}{dt} = \frac{d}{dt}\left(\frac{Q}{\varepsilon_0}\right) = \frac{I}{\varepsilon_0}$$

The quantity $\varepsilon_0 \frac{d\Phi_{\text{electric}}}{dt}$ has units of amperes, so we will combine it with I in Ampere's law. We will call this more general form of Ampere's law the "Ampere–Maxwell law":

THE AMPERE–MAXWELL LAW

$$\oint \vec{B} \bullet d\vec{l} = \mu_0 \left[\sum I_{\text{inside path}} + \varepsilon_0 \frac{d\Phi_{\text{electric}}}{dt}\right]$$

$$\Phi_{\text{electric}} = \int \vec{E} \cdot \hat{n} dA$$

If our imaginary soap film is outside the capacitor, there is just the term involving I but no term involving changing electric flux. If our imaginary soap film goes between the capacitor plates, there is just the term involving changing electric flux but no term involving I. The Ampere–Maxwell law handles both cases.

It also handles the case of a bulge in the soap film near the center of the capacitor plate (Figure 23.5). In this case the film doesn't enclose all of the electric flux. The soap film is pierced not only by the incoming current I but also by a lesser amount of radial current I_2 flowing outward from the center of the capacitor plate toward the edge of the plate, so that the net current piercing the soap film is $(I - I_2)$. In this case the Ampere–Maxwell equation contains both a current term and a flux term, adding up to $\mu_0 I$.

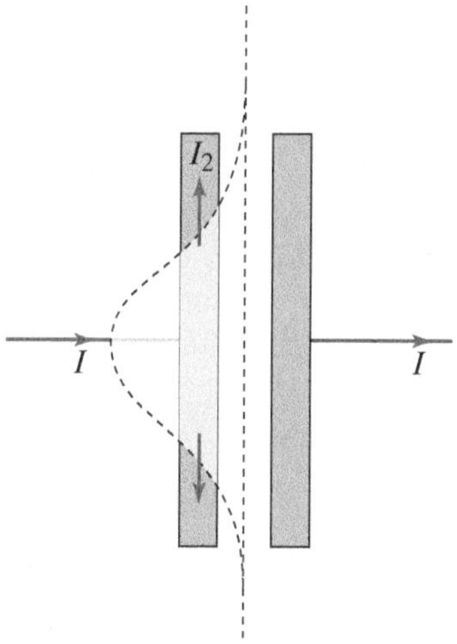

Figure 23.5 The soap film has a bulge near the center of the plate.

We interpret the Ampere–Maxwell law as saying that a time-varying electric field is always associated with a magnetic field. It is quite difficult to verify Maxwell's ingenious guess by a direct experiment, because in order to get a changing electric field you have to run a current to deposit a growing amount of charge, and this unavoidable current may be the source of the magnetic field you observe. Unless you are very clever in your geometrical arrangement, the real currents may mask the effects of the time-varying electric field.

There has in fact been direct experimental verification of the electric flux term, but the most compelling confirmation of Maxwell's inspired guess is indirect, in the explanation he was able to provide for the nature of light and radio waves, as we will soon see.

Maxwell's Equations (Integral Form)

We can now list the four equations that summarize all electromagnetic effects and are collectively called "Maxwell's equations" (although Maxwell himself did not write them in this modern form). Note that the first three are familiar equations from previous chapters:

MAXWELL'S EQUATIONS

$\oint \vec{E} \bullet \hat{n} dA = \frac{1}{\varepsilon_0} \sum Q_{\text{inside the surface}}$	Gauss's law
$\oint \vec{B} \bullet \hat{n} dA = 0$	Gauss's law for magnetism
$\oint \vec{E} \bullet d\vec{l} = -\frac{d}{dt} \int \vec{B} \bullet \hat{n} dA$	Faraday's law
$\oint \vec{B} \bullet d\vec{l} = \mu_0 \left[\sum I_{\text{inside path}} + \varepsilon_0 \frac{d}{dt} \int \vec{E} \bullet \hat{n} dA \right]$	Ampere–Maxwell law

These are the "integral" versions of Maxwell's equations. The "differential" versions of Maxwell's equations are given at the end of this chapter.

In addition to Maxwell's equations, to complete our list of fundamental relations of electricity and magnetism we need to add the equation for the net electric and magnetic force, the Lorentz force, which defines what we mean by electric and magnetic fields in terms of their effects on charges:

THE NET ELECTRIC AND MAGNETIC FORCE

$$\vec{F} = q\vec{E} + q\vec{v} \times \vec{B} \qquad (d\vec{F} = I d\vec{l} \times \vec{B} \text{ for currents})$$

23.2 FIELDS TRAVELING THROUGH SPACE

A time-varying magnetic field is associated with an electric field, and a time-varying electric field is associated with a magnetic field. Might it be possible for time-varying electric and magnetic fields in empty space to exist without any charges or currents nearby?

If we can show that this is possible (that is, if this would not be inconsistent with one or more of Maxwell's equations), we will still need to answer the question of how such a process might get started. Eventually we will show that such a configuration of fields can be produced by an accelerated charge, but first we will concentrate on establishing that such a configuration of electric and magnetic fields is indeed possible.

Solving Maxwell's Equations

In principle, a straightforward way to show how this could work would be to solve the partial differential equations represented by Maxwell's equations in their differential form, but the required mathematics is beyond the scope of this introductory textbook. However, we can do what we did in finding a solution to the differential equation for an RC circuit in Chapter 19. In that case we guessed that an exponential function would be a solution for the current as a function of time. We then showed that this function was a solution of the differential equation, by taking derivatives of the guessed function and plugging them into the differential equation. This is one of the most common schemes for solving a differential equation, and it can be used to solve integral equations as well.

Solution Plan

We will take the following approach to the problem:

- Propose a particular configuration of electric and magnetic fields in space and time.
- Show that this is consistent with all four Maxwell equations, and is therefore a possible solution.
- Show that an accelerated charge produces such fields.
- Identify the effects that such fields would have on matter.
- Analyze a variety of phenomena involving such fields.

A Simple Configuration of Traveling Fields

We could consider various configurations of time-varying electric and magnetic fields, in the absence of any charges, and we would find that most of them do not satisfy Maxwell's equations and hence are not possible. However, if we add the assumption that the region of nonzero field is "traveling" through space at a velocity $\vec{v}$, we can find a solution that works. This idea—that the region where the fields are nonzero must move through space—is the key to finding time-varying electric and magnetic fields that satisfy all of Maxwell's equations.

You may recognize this phenomenon—a disturbance moving through space—as a wave. To someone familiar with differential equations, the differential form of Maxwell's equations suggests a wave as a possible solution. Many waveforms are possible, including the familiar sinusoidal wave. To simplify our analysis, we will first consider the simplest possible wave: a single propagating pulse, as shown in Figure 23.6.

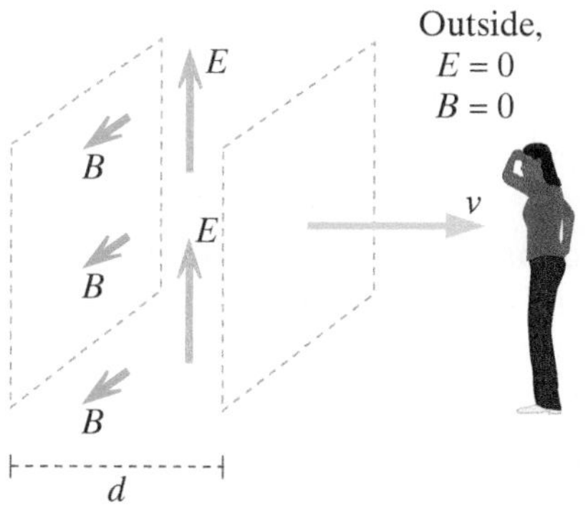

Figure 23.6 A slab of space contains electric and magnetic fields, with no fields outside the slab. The slab moves to the right with speed v.

A Moving Slab (or Pulse)

Consider a rectangular "slab" of space, of thickness d and very large height and depth, filled with a uniform electric field $\vec{E}$ pointing up and a uniform magnetic field $\vec{B}$ pointing out of the page (Figure 23.6). The electric and magnetic fields are zero outside the slab. We imagine the field-filled slab moving to the right with an unknown speed v, by which we mean that at a later time the slab-shaped region of space that contains fields is somewhere to the right of the present position (and at that time the fields are zero in the original position).

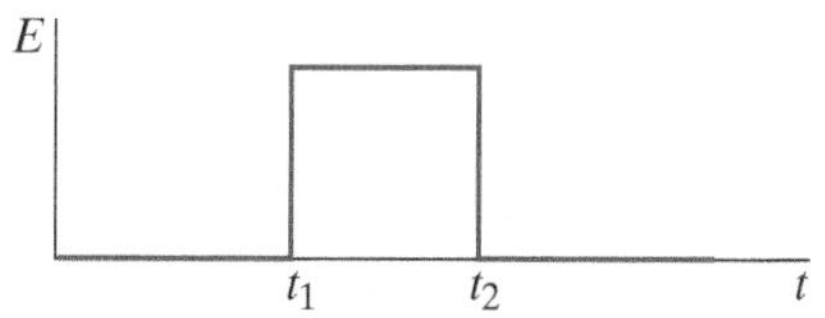

Figure 23.7 Magnitude of electric field observed at a particular location in space, as a function of time, as the slab passes. Note that at times t_1 and t_2, the electric field at this location is changing ($dE/dt \neq 0$).

Time-Varying Fields

If you stand at a location to the right of this region, as shown in Figure 23.6, for a while you don't experience any electric or magnetic fields. Then, as shown in Figure 23.7, the fields at your location change. For a short time you can detect an upward-pointing electric field, and a magnetic field coming out of the page. After the slab passes by, you again experience no electric or magnetic fields. You might describe your experience by saying that a short "pulse" of fields passed by. As the edges of the pulse pass a location, there are time-varying fields at that location. (In a real pulse, the magnitude of the field would not change instantaneously, so dE/dt and dB/dt at the edges of the pulse would not be infinite.)

Propagation

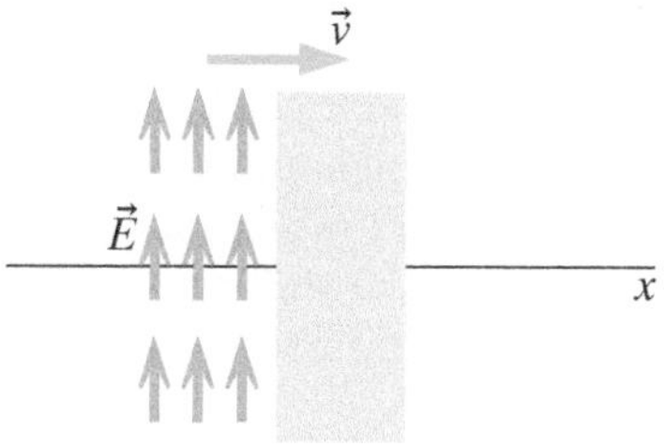

Figure 23.8 At a given instant the electric field is nonzero in the region shown. In the next instant, the region of nonzero field will propagate to the shaded area. For clarity, magnetic field is not shown here.

We say that in our example the fields "propagate" to the right, or that the "direction of propagation" is to the right. "Propagation" means the spreading of something (in this case, electric and magnetic fields) into a new region. Strictly speaking, the fields don't move. Rather, it's just that at one instant in time there are fields in a particular slab-shaped region of space, and at a later time there are fields in a different slab-shaped region of space, to the right of the original region (Figure 23.8). At a particular instant in time, the fields vary with location. At a particular location in space, the fields vary with time (Figure 23.7). Mathematically, we can say that the region of space in which $\vec{E}$ and $\vec{B}$ are nonzero is propagating to the right with velocity $\vec{v}$.

We will show that this configuration of electric and magnetic fields is consistent with all four of Maxwell's equations. Of course, we have chosen to consider these particular directions for $\vec{E}$, $\vec{B}$, and $\vec{v}$ because we know that they will work. However, if we try other sets of relative directions, say with $\vec{E}$ and $\vec{B}$ both pointing down and $\vec{v}$ up, we find that these directions are not consistent with all four of Maxwell's equations. Only the arrangement we treat here turns out to satisfy Maxwell's equations, with $\vec{E}$, $\vec{B}$, and $\vec{v}$ perpendicular to each other as shown in Figure 23.6. (Of course, $\vec{v}$ can be in any direction; it is the relative orientation of $\vec{E}$, $\vec{B}$, and $\vec{v}$ that is important.)

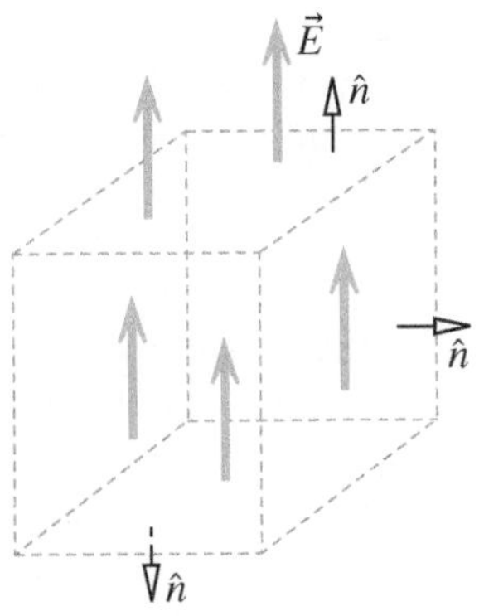

Figure 23.9 Applying Gauss's law to a box-shaped region within the slab, we see that the net electric flux on the box is 0. $\vec{B}$ is not shown.

Gauss's Law

First we'll consider Gauss's law,

$$\oint \vec{E} \bullet \hat{n} dA = \frac{1}{\varepsilon_0} \sum Q_{\text{inside the surface}}$$

We need to show that our chosen configuration of electric fields is permissible in a region of space where there are no charged particles. We choose an imaginary closed box, fixed in space, through which the field-filled slab is moving (Figure 23.9). The positive flux on the top of the box is exactly canceled by an equal negative flux on the bottom. On the sides $\vec{E}$ is perpendicular to $\hat{n}$, so the flux on the sides is zero. Since the net flux is zero, there must be no charge in the box, as we expected.

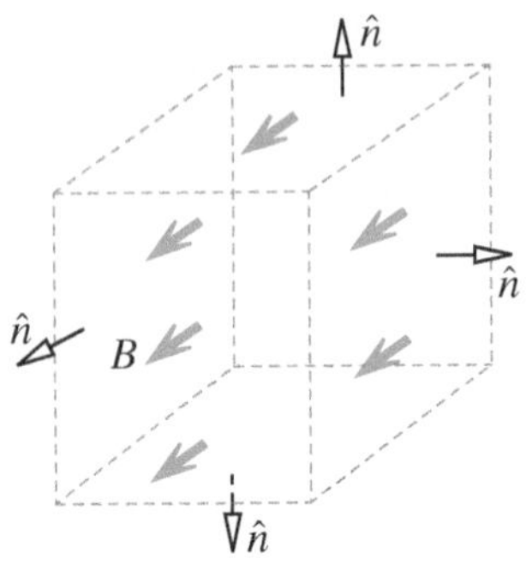

Figure 23.10 Applying Gauss's law for magnetism to a box-shaped surface within the slab, we find that the net magnetic flux on the box is 0. $\vec{E}$ is not shown.

Gauss's Law for Magnetism

Next we consider Gauss's law for magnetism,

$$\oint \vec{B} \bullet \hat{n} dA = 0$$

Using the same box-shaped surface (Figure 23.10), we find that the positive magnetic flux on the front of the box is canceled by an equal amount of negative flux on the back of the box. The other sides contribute no flux, because $\vec{B}$ is perpendicular to $\hat{n}$. Thus, this configuration of magnetic field can exist only in a region where there are no magnetic monopoles, which is fine, because as far as we know there are no magnetic monopoles anywhere!

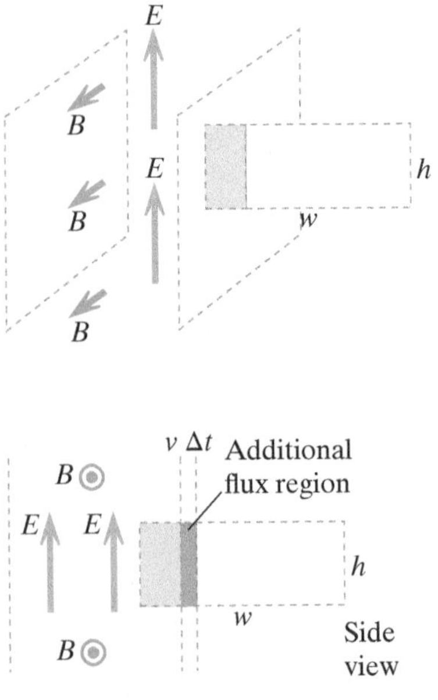

Figure 23.11 Applying Faraday's law to a rectangular loop partly inside the slab. The lower diagram shows a side view.

Faraday's Law

Faraday's law relates a changing magnetic flux to a non-Coulomb electric field,

$$\oint \vec{E} \bullet d\vec{l} = -\frac{d}{dt} \int \vec{B} \bullet \hat{n} dA$$

Because the region of space in which $\vec{E}$ and $\vec{B}$ are nonzero is "moving," the magnetic field at a particular location in space changes with time, and this time-varying magnetic field is associated with a curly electric field. To consider Faraday's law, we draw a vertical rectangular path, h high and w wide, partly inside the slab and partly outside as shown in Figure 23.11, which includes a side view. This rectangular path is fixed in space, whereas the slab "moves" to the right. Let's apply Faraday's law to the rectangular path. First we will consider the right-hand side of the equation.

> QUESTION If the slab is moving to the right at speed v, calculate the rate of change of magnetic flux enclosed by the path in terms of B, by calculating how much additional flux there is in a time Δt, then dividing by Δt. Note in the side view in Figure 23.11 the additional region of magnetic flux that appears in an amount of time Δt.

In a time Δt the area of magnetic field inside our chosen path increases by an amount $(v\Delta t)h$, and since the magnetic field B is uniform within the slab, the magnetic flux increases by an amount $B(v\Delta t)h\cos 0 = Bvh\Delta t$. Therefore the rate of change of magnetic flux inside our chosen path is this:

$$\frac{d\Phi_{\text{magnetic}}}{dt} = Bvh$$

By Faraday's law, this rate of change of magnetic flux is equal to the emf around the path, which is a path integral of the electric field (the left-hand side of the equation).

> **QUESTION** Keeping in mind the fact that part of the path is outside the slab, what is the path integral of the electric field in terms of the electric field E?

The only portion of the path along which there is a parallel component of nonzero electric field is along the piece of length h, so we have this:

$$|\text{emf}| = \left|\oint \vec{E} \bullet d\vec{l}\right| = Eh\cos 0 = Eh$$

> **QUESTION** Therefore Faraday's law establishes an algebraic relation between E and B. State this relation.

The emf is equal to the rate of change of magnetic flux, $Eh = Bvh$, so we have the following relationship between B and E in the slab:

$$E = vB$$

> **QUESTION** Explain how the direction of $\vec{E}$ is consistent with Faraday's law.

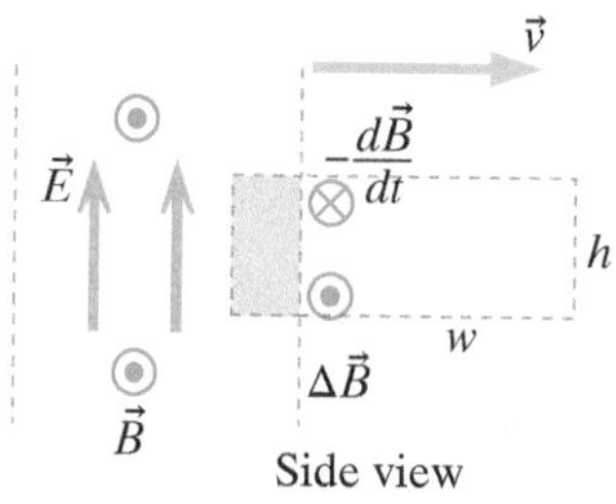

Figure 23.12 The direction of the electric field is consistent with Faraday's law.

In the region ahead of the slab, the magnetic field changes from being zero to pointing out of the page, so $\Delta\vec{B}$ points out of the page and $-d\vec{B}/dt$ points into the page (Figure 23.12). With the right thumb pointing in the direction of $-d\vec{B}/dt$, the fingers of the right hand curl clockwise, which is the sense of the electric field around the path.

We have shown that our chosen configuration of moving fields is consistent with Faraday's law if $E = vB$.

The Ampere–Maxwell Law

The Ampere–Maxwell law relates a changing electric flux to a magnetic field:

$$\oint \vec{B} \bullet d\vec{l} = \mu_0 \left[\sum I_{\text{inside path}} + \varepsilon_0 \frac{d\Phi_{\text{electric}}}{dt}\right]$$

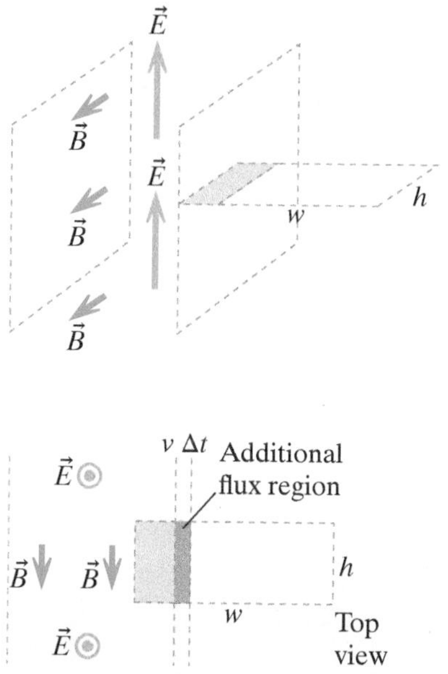

Figure 23.13 Applying the Ampere–Maxwell law to the slab. The lower diagram shows a view from above.

In this situation there are no currents. Because the field-containing slab is "moving," the electric field at a location in space changes with time, and by the Ampere–Maxwell law this changing electric field must be associated with a curly magnetic field. To consider the Ampere–Maxwell law, we draw a rectangular path, fixed in space, that is horizontal (so this rectangle is perpendicular to the electric field). Figure 23.13 includes a view from above, looking down on the upward-pointing electric field. First we will consider the right-hand side of the equation:

> **QUESTION** Let's apply the Ampere–Maxwell law to this path. If the slab is moving to the right at speed v, calculate the rate of change of electric flux enclosed by the path in terms of E, just as you calculated the rate of change of magnetic flux in the preceding section. Note in the view from above in Figure 23.13 the additional region of electric flux that appears in an amount of time Δt.

In a time Δt the area of electric field inside our chosen path increases by an amount $(v\Delta t)h$, and since the electric field E is uniform within the slab, the electric flux increases by an amount $E(v\Delta t)h\cos 0 = Evh\Delta t$. Therefore the rate of change of electric flux inside our chosen path is this:

$$\frac{d\Phi_{\text{electric}}}{dt} = Evh$$

QUESTION Keeping in mind the fact that part of the path is outside the slab, what is the path integral of the magnetic field (the left-hand side of the equation) in terms of B?

The only portion of the path along which there is a parallel component of nonzero magnetic field is along the piece of length h, so we have this:

$$\oint \vec{B} \bullet d\vec{l} = Bh\cos 0 = Bh$$

QUESTION Therefore application of the Ampere–Maxwell law establishes another algebraic relation between E and B. State this relation.

There is no current I, so the path integral of the magnetic field is equal to $\mu_0\varepsilon_0$ times the rate of change of electric flux, $Bh = \mu_0\varepsilon_0 Evh$, so we have the following relationship between B and E in the slab:

$$B = \mu_0\varepsilon_0 vE$$

Speed of Propagation

Since we have shown that our chosen configuration of fields is possible if the region of nonzero field is traveling with speed v, we are very interested in finding out what v must be. We have two different algebraic relations involving E and B, both of them involving the unknown speed v with which the pulse of electric and magnetic fields is traveling through space:

$$E = vB$$
$$B = \mu_0\varepsilon_0 vE$$

QUESTION Solve for v and evaluate the result numerically to see with what speed the fields must propagate in order that this pattern of fields be consistent with all four of Maxwell's equations.

By combining $E = vB$ and $B = \mu_0\varepsilon_0 vE$, we find that

$$v^2 = \frac{1}{(\mu_0\varepsilon_0)}$$

Evaluating this numerically we obtain this:

$$v = \sqrt{\frac{1}{\mu_0\varepsilon_0}} = \sqrt{\left(\frac{1}{4\pi\varepsilon_0}\right)\left(\frac{4\pi}{\mu_0}\right)}$$
$$v = \sqrt{(9\times 10^9\ \text{N}\cdot\text{m}^2/\text{C}^2)(1\times 10^7\ \text{A/T}\cdot\text{m})} = 3\times 10^8\ \text{m/s}$$

An electromagnetic wave propagates at the speed of light! This is a spectacular result: the pulse is predicted to advance at a speed that is the same as the known speed of light. At the time that Maxwell went through a similar line of reasoning in the mid-1800s, no one had any idea as to the real nature of light, but the speed of light was known to be 3×10^8 m/s. Maxwell went through a theoretical analysis leading to the possibility of "moving" fields, and he predicted the speed of propagation of such fields by using values for μ_0 and ε_0 that were determined from simple experiments on electricity and magnetism.

When the predicted speed turned out to be the known speed of light, Maxwell concluded that light must be a combination of time-varying electric and magnetic fields that can propagate through otherwise empty space, far from any charges or currents. This theoretical analysis was brilliantly confirmed in the period 1885–1889 by Heinrich Hertz, who accelerated charges in a spark and observed the first radio waves, which also propagate at the speed of light.

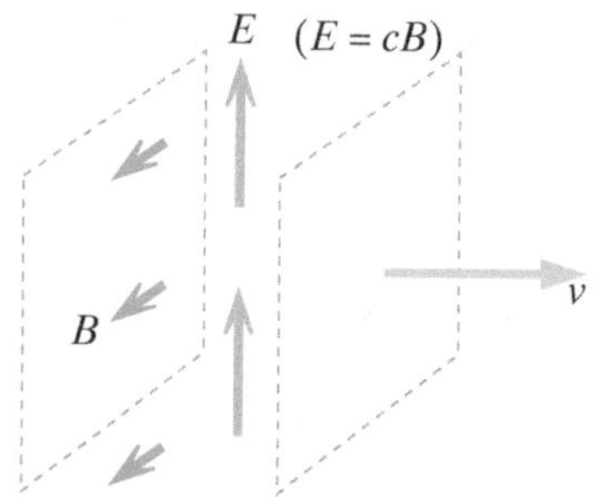

Figure 23.14 The direction of propagation of the electromagnetic pulse.

It is useful to note that the direction of propagation $\vec{v}$ of the pulse is the same as the direction of the cross product $\vec{E} \times \vec{B}$ (Figure 23.14). We can summarize the relation between E and B in a pulse:

RELATION OF E TO B IN AN ELECTROMAGNETIC WAVE

$$E = cB$$

where c is the speed of light. The direction of $\vec{v}$ is the direction of $\vec{E} \times \vec{B}$.

The fact that $E = cB$ is easy to remember if you note that electric force is $q\vec{E}$ while magnetic force is $q\vec{v} \times \vec{B}$, so cB has the same units as E.

Checkpoint 1 A pulse of radiation propagates with velocity $\vec{v} = \langle 0, c, 0 \rangle$. The electric field in the pulse is $\vec{E} = \langle 0, 0, 1 \times 10^6 \rangle$ N/C. What is the magnetic field (vector) in the pulse?

Electromagnetic Radiation

The pulse of propagating electric and magnetic fields is called "electromagnetic radiation"—"electromagnetic" because it involves both electric fields and magnetic fields, and "radiation" because it "radiates" outward from a source. Visible light—electromagnetic radiation that can be detected by the human eye—makes up only part of the spectrum of electromagnetic radiation. When we say "light," we often mean not only visible light, but also any kind of electromagnetic radiation: microwaves, radio waves, infrared, ultraviolet, x-rays, or gamma-rays. Later in this chapter we will discuss the electromagnetic spectrum in more detail.

In any case, there is only one arrangement of propagating electric and magnetic fields that satisfies Maxwell's equations in empty space: $\vec{E}$ and $\vec{B}$ at right angles to each other, with $E = cB$, and the region containing nonzero fields propagating in the direction of $\vec{E} \times \vec{B}$ at the speed of light. Now that we have successfully shown that these propagating fields are consistent with Maxwell's equations, our next task is to investigate how such fields are produced.

Other Waveforms

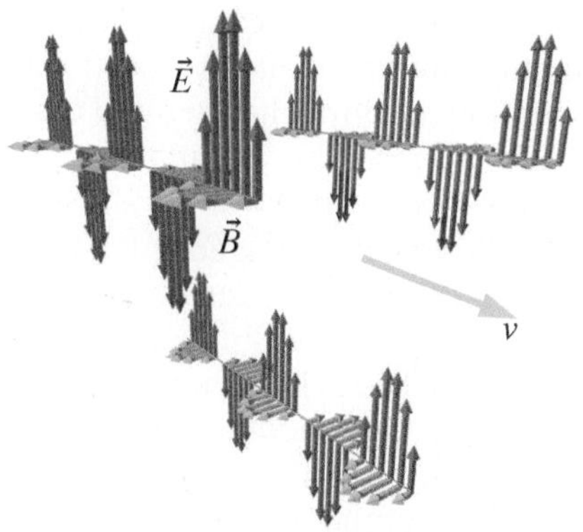

Figure 23.15 A sinusoidal electromagnetic wave.

The simple pulse of radiation that we've been considering is not the only possible kind of electromagnetic wave. In fact, any waveform that can be made up from successive short pulses of different magnitudes would also satisfy Maxwell's equations. For example, a square wave or a sinusoidal wave (Figure 23.15) would also be possible. We will see later that sinusoidal electromagnetic waves are particularly common.

23.3 ACCELERATED CHARGES PRODUCE RADIATION

We have seen that a pulse of electromagnetic radiation propagating with the speed of light is consistent with Maxwell's equations. The next question we need to answer is how such a configuration of propagating fields could be produced. It turns out that the way to produce electromagnetic radiation is to accelerate a charge.

Consider a positive charge q sitting at rest (Figure 23.16). The direction of the electric field produced by this charge is radially outward in all directions, which we indicate with "electric field lines" showing the direction (but not the magnitude) of the electric field at each point.

We kick the charge downward, so that in a very short time it picks up a little bit of speed and then coasts downward at that slow speed (there is no friction). Some time after the kick, we take a snapshot of the direction of electric field throughout this region, and the snapshot has the surprising and peculiar appearance shown in Figure 23.17. We will discuss what is observed by two people stationed at location 1 and location 2. Observer 1 sees an electric field that points away from the present location of the slowly moving charge.

Figure 23.16 Directions of electric field surrounding a stationary positive charge.

QUESTION Explain why at this instant in time observer 2 sees an electric field that points away from the original position of the charge rather than the present location of the charge.

The key issue is retardation. Information about the change in the electric field has not yet reached observer 2. Since changes in the electric field propagate at the speed of light, it takes a while for the new field to appear at observer 2's position.

Between observer 1 and observer 2 there is a kink in the electric field direction. At some later time this kink will be at location 2 (Figure 23.18).

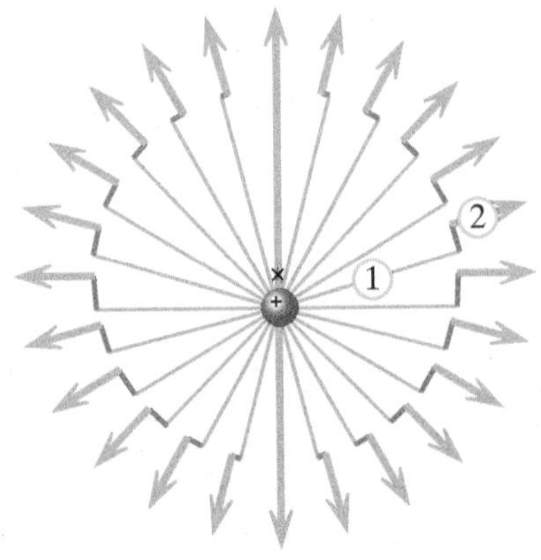

Figure 23.17 Directions of electric field shortly after the charge was kicked. The × marks the original location of the charge.

QUESTION If the distance from the original location of the charge to observer 2 is r, how long after we kicked the charge does observer 2 detect a sideways (transverse) electric field?

Since the transverse electric field propagates at the speed of light, we have $r = ct$ and $t = r/c$. If observer 2 is very far away from the original position of the charge, the radius of curvature r of the propagating radiation is very large, and observer 2 experiences a brief pulse of transverse electric field that looks very much like the transverse electric field associated with the flat moving slab we discussed earlier. An accelerated charge produces electromagnetic radiation. Of course, these flat diagrams show only a cross-sectional slice through a 3D pattern of radiation, shown in Figure 23.19.

It is clear from our application of Faraday's law and the Ampere–Maxwell law to the moving slab that a pulse of radiation must contain both electric and magnetic fields, so that the pulse can move forever through otherwise empty space, far from any charges. The transverse electric field is accompanied by a transverse magnetic field perpendicular to the electric field, with $\vec{E} \times \vec{B}$ in the direction of propagation, which means that the radiative magnetic field must point out of the page on the right of Figures 23.18 and 23.19, and into the page on the left of these figures.

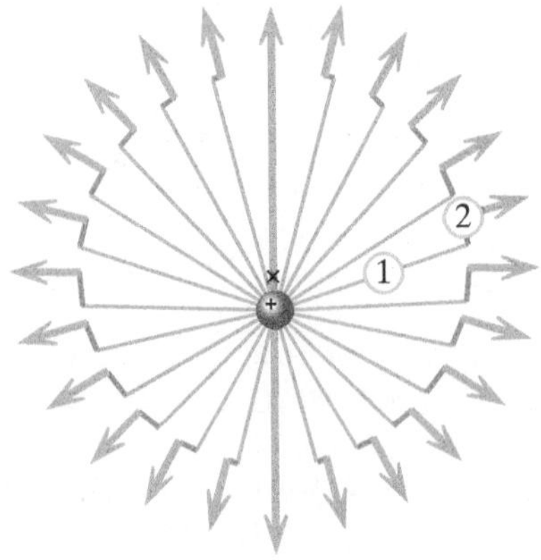

Figure 23.18 Information about the change has reached observer 2.

Outside the expanding sphere of radiation there is no magnetic field because the charge was initially at rest. Inside the expanding sphere there is a charge moving downward with constant velocity, and it makes an ordinary magnetic field (by the Biot–Savart law) that is out of the page on the right and into the page on the left. In this case the Biot–Savart magnetic field is in the same direction as the radiative magnetic field, but this need not be the case. If the charge's acceleration is opposite to its velocity (slowing down rather than speeding up), the two kinds of magnetic fields point in opposite directions.

QUESTION A charge is initially at rest. Then you move it to another position, leaving it again at rest. Is there any radiation?

Yes. You have to accelerate the charge to get it moving, then decelerate it later to bring it again to rest, so there will be at least two pulses of radiation.

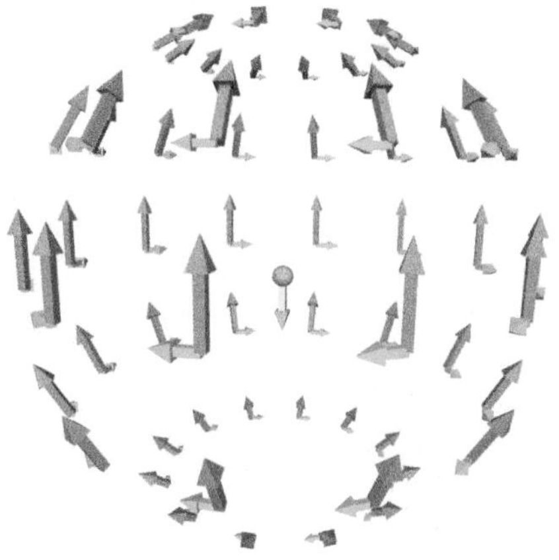

Figure 23.19 Radiation expands in a spherical pattern from a charge (center) that has been briefly accelerated downward. The orange arrows represent electric field, and the cyan arrows represent the accompanying magnetic field.

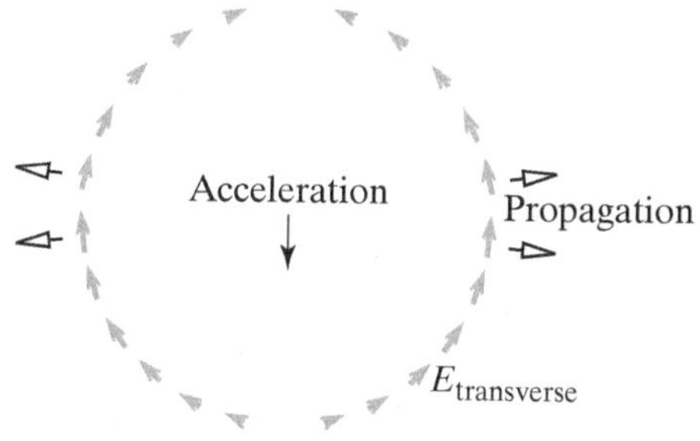

Figure 23.20 A snapshot at one instant of the transverse electric field, which propagates away from the accelerated charge. Magnetic field is not shown.

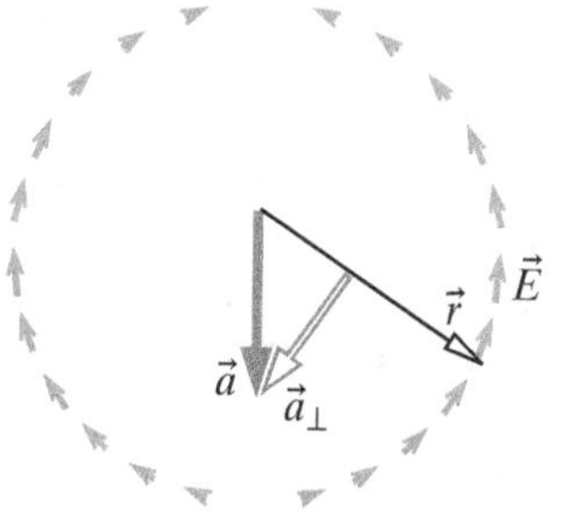

Figure 23.21 The transverse electric field is proportional to the transverse component of acceleration.

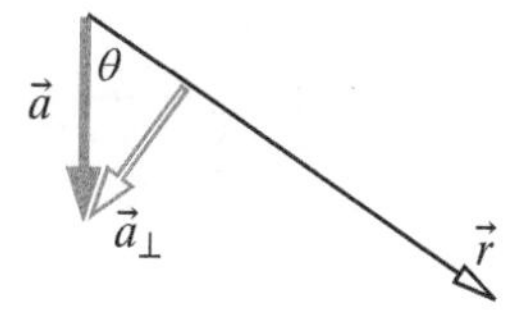

Figure 23.22 The magnitude of $\vec{a}_\perp$ is $a\sin\theta$.

Direction of the Transverse Electric Field

If we draw only the transverse electric field at some time after the kick, we see a pattern like that shown in Figure 23.20. Bear in mind that Figure 23.20 is a two-dimensional slice through the three-dimensional phenomenon that is shown in Figure 23.19, with the pulse headed out of the page and into the page as well as moving in the plane of the diagram.

> QUESTION In what directions is the electromagnetic radiation most intense? In what directions is there no electromagnetic radiation?

It is most intense to the sides; there is zero radiation in the direction of the acceleration (and the opposite direction).

Magnitude of the Transverse Electric Field

The qualitative argument we have just made gives the direction of the radiative fields. This argument can be made quantitative by an application of Gauss's law, to derive expressions for the magnitude and direction of the radiative electric and magnetic fields. This derivation, due originally to Purcell, is given in Section 23.11. The results of that derivation give the following description of the radiative fields.

The transverse electric field is associated with the kink in the electric field direction that is associated with accelerated motion of the source charge, and from the diagrams of this process it is plausible that the magnitude of the radiative field is proportional to $-q\vec{a}_\perp$, where $\vec{a}_\perp$ is the "projected" acceleration of the charge (projected onto a plane perpendicular to the line of sight, as in Figure 23.21). Note that the magnitude of $\vec{a}_\perp$ is just $a\sin\theta$ (Figure 23.22).

The direction of propagation of the radiation is outward from the source charge, in the direction of the unit vector $\hat{r}$, drawn as usual from the source charge toward the observation location. This is also the direction of $\vec{E}\times\vec{B}$.

The radiative (transverse) electric field is proportional to the charge q and the projected acceleration $\vec{a}_\perp$, and it turns out to be inversely proportional to the distance r. The full relationship is the following:

PRODUCING A RADIATIVE ELECTRIC FIELD

$$\vec{E}_{\text{radiative}} = \frac{1}{4\pi\varepsilon_0}\frac{-q\vec{a}_\perp}{c^2 r}$$

There are two important points to notice in this equation:

- The direction of the transverse electric field is opposite to $q\vec{a}_\perp$.
- The field falls off with distance at the rate of $1/r$.

This is very different from the $1/r^2$ behavior of the ordinary electric field of a stationary charge or of a charge moving at a constant velocity. It is this much slower fall-off with distance that makes it possible for electromagnetic radiation to affect matter that is very far from the accelerated charges that produced the radiation. We see extremely distant stars because of this slow fall-off with distance.

In earlier chapters we showed that there is energy density associated with electric and magnetic fields, proportional to the squares of the field magnitudes. Similarly, later in this chapter we will show that the radiative power flow in watts per square meter is proportional to the square of the electric field, so the power flow per square meter falls off like $1/r^2$. Energy outflow over larger and larger spherical surfaces is constant, because the surface area $4\pi r^2$ grows proportional to r^2.

Note that if the accelerated object is an electron or other negative charge, all the radiative electric field vectors in Figure 23.21 will point in the opposite direction, because the electric field is proportional to the charge q.

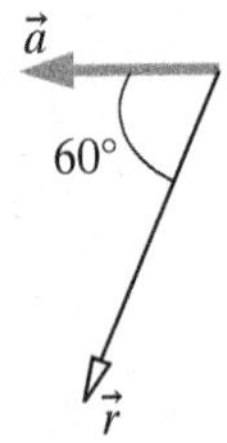

Figure 23.23 An electron is accelerated in the direction shown (Checkpoint 2).

Checkpoint 2 An electron is briefly accelerated in the direction shown in Figure 23.23. The resultant electromagnetic radiation is observed at a location indicated by the vector $\vec{r}$ in the diagram. On a copy of the diagram, draw two vectors: $\vec{a}_\perp$, and $\vec{E}$ at the observation location.

Stability of Atoms

Classically (before quantum mechanics), any accelerated charge must emit electromagnetic radiation. Since electromagnetic radiation can accelerate other charges, there must be a flow of energy associated with the radiation. This implies that an accelerated charge loses some mechanical energy simply by virtue of emitting radiation.

Before quantum mechanics, this posed a puzzle for the stability of atoms. If we think of an atom as a miniature solar system, with electrons orbiting the nucleus, the electrons are undergoing accelerated motion (radial acceleration $= v^2/r$) and ought to lose energy by radiation and spiral into the nucleus. When one calculates how long an atom should last, it is a tiny fraction of a second! This paradox also applies to the persistent atomic currents in a magnet or the persistent currents in a superconducting ring.

This paradox was resolved by quantum mechanics. The basic idea is that the stable energies of a system are discrete, and in some cases (atoms, magnets, superconductors) there is no lower-energy state to which the system can go, and so there is no way to radiate energy and in the process drop to a lower-energy state. On the other hand, an atom that is in a state with energy above the "ground state" can and does drop to a lower-energy state and emits energy in the form of light.

EXAMPLE **Electron Enters Region of Electric Field**

An electron is traveling at nonrelativistic speed along the z axis in the $+z$ direction in a region of nearly zero fields (Figure 23.24). At $t = 0$ it reaches the origin $\langle 0,0,0\rangle$ and enters a region where the electric field is $\langle 0,0,2\times 10^{16}\ \text{V/m}\rangle$. You observe fields at location $\langle 5,3,0\rangle$ m. In the following, explain your work fully. **(a)** When do you first observe radiative fields? Explain. **(b)** On the diagram, show the direction of the initial radiative electric field at your location. Explain. **(c)** On the diagram, show the direction of the initial radiative magnetic field at your location. Explain. **(d)** What is the magnitude of the initial radiative electric field at your location? **(e)** What is the magnitude of the initial radiative magnetic field at your location?

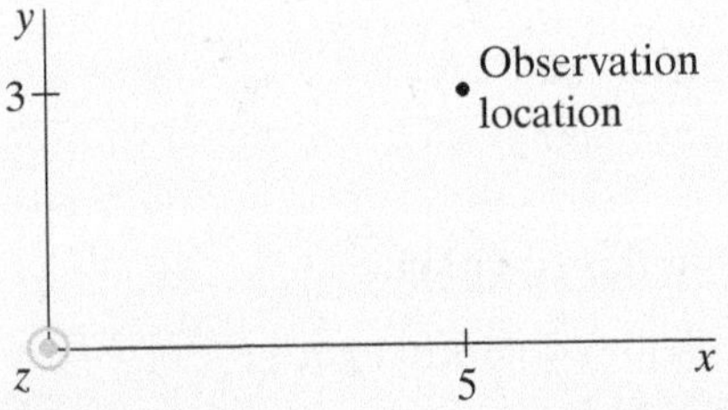

Figure 23.24 An electron enters a region of electric field.

Solution **(a)** It takes $\dfrac{\sqrt{(5\ \text{m})^2+(3\ \text{m})^2}}{3\times 10^8\ \text{m/s}} = 1.9\times 10^{-8}$ s for light to propagate from the origin to the observation location.

(b) See Figure 23.25. The electron undergoes an acceleration (deceleration) in the $-z$ direction, so $\vec{a}_\perp$ is in the $-z$ direction, and $-q\vec{a}_\perp = -(-e)\vec{a}_\perp$ is in the $-z$ direction. This produces a radiative electric field in the $-z$ direction, as shown.

(c) The direction of propagation is given by the direction of $\vec{E}\times\vec{B}$, which requires that $\vec{B}$ point in the direction shown.

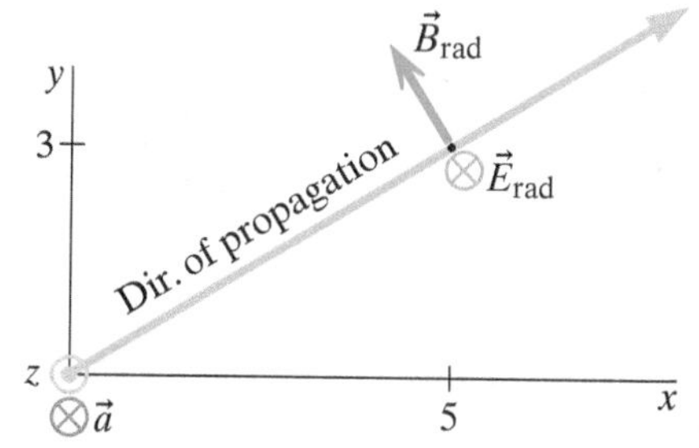

Figure 23.25 Directions of radiative fields.

It is important to note that the direction of the magnetic field predicted by the Biot–Savart law is in the opposite direction, because it depends on the velocity of the electron (in the $+z$ direction), not the acceleration (in the $-z$ direction). At a far distance the radiative magnetic field ($1/r$) is much larger than the Biot–Savart magnetic field ($1/r^2$).

(d) The speed is nonrelativistic, so $a_\perp = a = eE/m$, and

$$E_{rad} = \frac{1}{4\pi\varepsilon_0}\frac{e(eE/m)}{c^2 r}$$

$$E_{rad} = \left(9\times 10^9 \frac{\text{N}\cdot\text{m}^2}{\text{C}^2}\right)\frac{(1.6\times10^{-19}\ \text{C})^2(2\times10^6\ \text{V/m})}{(3\times10^8\ \text{m/s})^2(\sqrt{(5\ \text{m})^2+(3\ \text{m})^2})(9\times10^{-31}\ \text{kg})}$$

$$E_{rad} = 9.8\times10^{-10}\ \text{V/m}$$

(e) $$B_{rad} = \frac{E_{rad}}{c} = \frac{9.8\times10^{-10}\ \text{V/m}}{3\times10^8\ \text{m/s}} = 3.2\times10^{-18}\ \text{T}$$

Summary: Production of Electric and Magnetic Fields

Here is a summary of the most basic properties of how electric and magnetic fields are produced:

FIELDS MADE BY CHARGES

- A charge at rest makes a $1/r^2$ electric field but no magnetic field.
- A charge moving with constant velocity makes a $1/r^2$ electric field and a $1/r^2$ magnetic field.
- An accelerated charge in addition makes electromagnetic radiation, with a $1/r$ electric field and an accompanying $1/r$ magnetic field.

Remember that so far, no one has found an isolated "magnetic monopole," but if there were such objects, they would presumably produce fields as follows:

FIELDS MADE BY (HYPOTHETICAL) MAGNETIC MONOPOLES

- A magnetic monopole at rest would make a $1/r^2$ magnetic field, but no electric field.
- A magnetic monopole moving with constant velocity would make a $1/r^2$ magnetic field and a $1/r^2$ electric field.
- An accelerated magnetic monopole would in addition make electromagnetic radiation, with a $1/r$ magnetic field and an accompanying $1/r$ electric field.

23.4 SINUSOIDAL ELECTROMAGNETIC RADIATION

If a charge is given just one kick to get it moving, during the brief acceleration the charge emits a single brief pulse of electromagnetic radiation. However, the charge emits continuous radiation if it is moved sinusoidally, with the position of the charge described by

$$y = y_{max}\sin(\omega t)$$

where ω is the "angular frequency" in radians per second. Because the electric interaction between the electron cloud in an atom and the nucleus is spring-like (in the sense that the force is proportional to displacement), electrons in atoms can be usefully modeled as charges on springs; see Problem P62 in Chapter 15.

Ordinary matter emits excess energy in the form of sinusoidal electromagnetic radiation.

QUESTION If the charge moves sinusoidally, its y position is described by $y = y_{max} \sin(\omega t)$. Why will the electromagnetic radiation emitted also be sinusoidal?

$$v_y = \frac{dy}{dt} = \omega y_{max} \cos(\omega t)$$
$$a_y = \frac{dv_y}{dt} = -\omega^2 y_{max} \sin(\omega t)$$

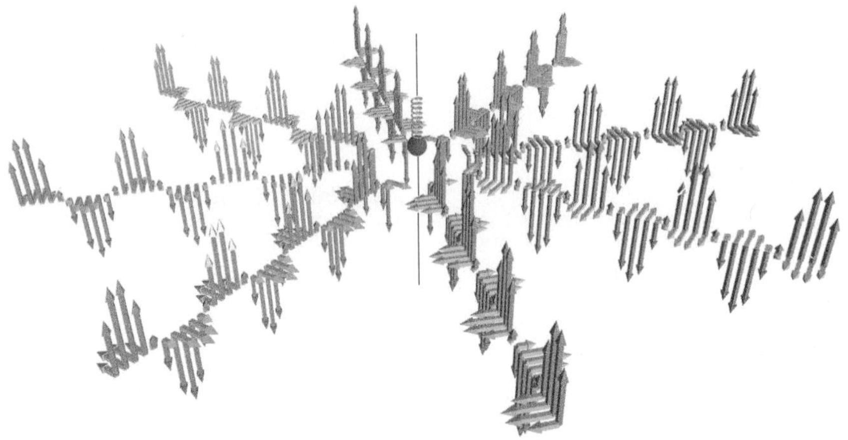

Figure 23.26 A charge moving sinusoidally in the $\pm y$ direction emits sinusoidal electromagnetic radiation, which propagates outward from the source. For clarity, this radiation is shown only in the xz plane, and the $1/r$ falloff in magnitude is not shown. As usual electric fields are represented by orange arrows, magnetic fields by cyan arrows.

The y component of the acceleration of the charge is also sinusoidal. The sinusoidal motion involves a sinusoidal acceleration which produces sinusoidal electromagnetic radiation. This radiation propagates away from the accelerated charge in all directions except $\pm y$ (Figure 23.26). For waves propagating in the xz plane, we have:

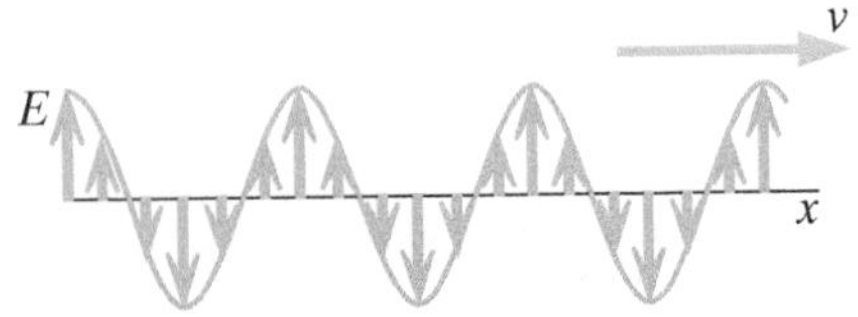

Figure 23.27 Radiation from a sinusoidally moving charge, at a single instant in time. Only the electric field is shown in this snapshot. Often only the outline of the wave is shown in diagrams (solid line).

$$E_y = \frac{1}{4\pi\epsilon_0}\frac{-qa_y}{c^2 r}$$
$$E_y = \frac{1}{4\pi\epsilon_0}\frac{q\omega^2 y_{max}}{c^2 r}\sin(\omega t)$$

A snapshot of such a wave at a single instant in time is shown in Figure 23.27. Often only the outline of the wave is shown in diagrams (Figure 23.28).

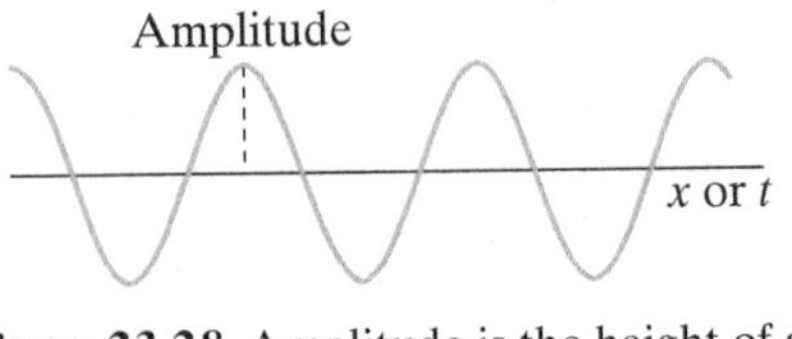

Figure 23.28 Amplitude is the height of a wave crest.

Amplitude

The *amplitude* of a sinusoidal wave is the height of a peak in the wave, measured from the zero line, as shown in Figure 23.28. In the case of an electromagnetic wave, the amplitude is the maximum magnitude of the electric field.

Period

Imagine that you are standing at a particular location, observing a sinusoidal wave passing by. The *period* of the wave is the time between crests at your

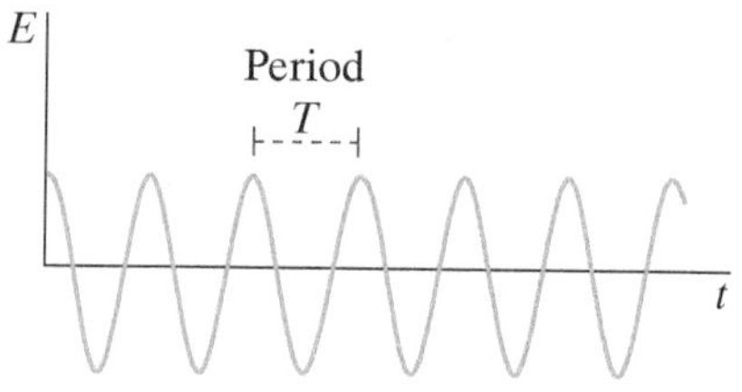

Figure 23.29 A plot of E *vs.* t at a single location in space. The period of a wave is the time between arrival of successive crests at a single location.

location. You could imagine measuring the period of a wave using a stopwatch: starting the timer when a crest (maximum amplitude) reaches your location, and stopping the timer when the next crest reaches your location. Figure 23.29 shows a plot of E vs. t at a specific location; the period is indicated on the graph.

Frequency

The frequency f of the sinusoidal electromagnetic wave is the number of peaks per second that pass a given location. It is the inverse of the period:

$$f = \frac{1}{T}$$

The units of frequency are s^{-1}, or hertz (abbreviated Hz). Frequency f is related to angular frequency ω in radians per second by

$$f = \frac{\omega}{2\pi}$$

Wavelength

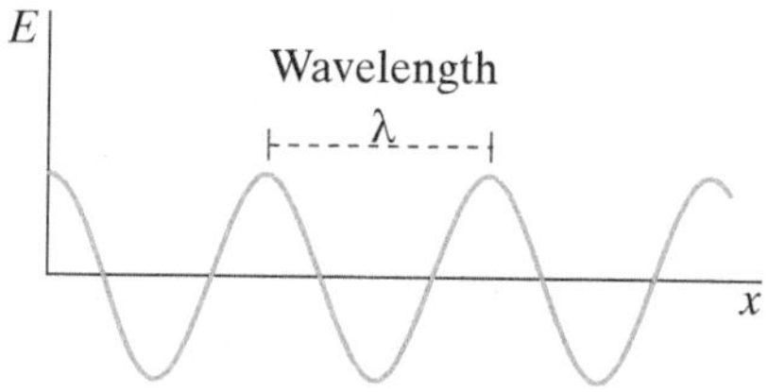

Figure 23.30 A plot of E vs. x at a single instant in time. The wavelength λ is the distance between two maxima.

Sinusoidal electromagnetic radiation is often described in terms of wavelength λ instead of frequency f, and $c = \lambda f$, because the wave advances a distance $\lambda = cT$ in one period $T = 1/f$, as illustrated in Figure 23.30. Electromagnetic radiation of different wavelength ranges has been given different names, such as *x-rays* or *microwaves*. Radiation with wavelengths of about 400–700 nm ($1 \text{ nm} = 1 \times 10^{-9}$ m) is special because it can be detected by molecules in the human eye; we refer to this as *visible light*.

It should also be evident that electromagnetic radiation need not be sinusoidal. The moving slab that we treated earlier represented a single pulse and was not at all like a sinusoid. However, sinusoidal radiation is of great practical importance because so many different kinds of systems emit sinusoidal radiation, from atoms to radio stations.

Speed

How might we measure the speed of propagation of an electromagnetic wave? One can think of two different approaches:

(a) Follow a wave crest: If you watch one particular wave crest, you will see that it travels a distance λ (one wavelength) in a time T (one period). Therefore the *speed* of the crest is

$$v = \frac{\lambda}{T}$$

(b) Time the arrival of a radiative electric field: One could imagine a different way of measuring the speed of an electromagnetic wave. Suppose that you and a friend synchronize your clocks, then travel to locations that are a distance d apart. Your friend aims a laser at your location, and precisely at time t_1, turns on the laser. You record the time t_2 at which you first detect the radiative electric field in the laser light, and knowing the distance between the locations and the elapsed time $\Delta t = t_2 - t_2$, you calculate the speed at which the laser light traveled toward you:

$$v = \frac{d}{\Delta t}$$

In a vacuum, these two ways of measuring the speed of a sinusoidal electromagnetic wave will give the same answer: 3×10^8 m/s. However, this will not necessarily be the case if part or all of the space through which the light wave travels is filled with a medium such as water, glass, or even air. In this case, method 2 (measuring the time required for information about a *change* in the electromagnetic field to travel a given distance) will still give 3×10^8 m/s. However, method 1 (timing the interval between crests in a steady state electromagnetic wave inside the medium) will give a different answer, which will almost always be less than 3×10^8 m/s. In Section 23.8 we consider the reason for this discrepancy.

Combining the equations defining speed, wavelength, and frequency, we get these relations:

$$v = \frac{\lambda}{T} \text{ and } v = \lambda f$$

Checkpoint 3 The wavelength of green light from a laser pointer is 532 nm. What is the frequency of this light?

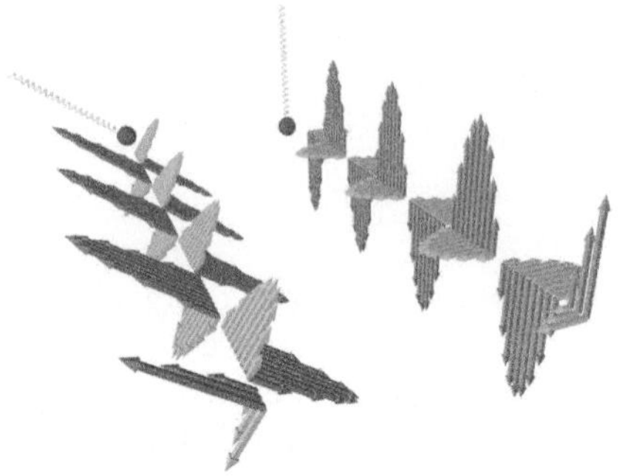

Figure 23.31 Radiation emitted by a collection of charges oscillating in different directions is not polarized; the electric field is not aligned in a single plane.

Polarized Radiation

The term *polarized* refers to a beam of electromagnetic radiation in which the electric field is aligned only along one axis. For example, in the sinusoidal radiation illustrated in Figure 23.27, the electric field is either in the $+y$ direction or in the $-y$ direction. Likewise, if you imagine standing at a location in the xz plane and observing the radiation emitted by the oscillating charge shown in Figure 23.26, you would detect an electric field in either the $+y$ or $-y$ direction.

How can radiation be unpolarized? A collection of oscillating source charges, in which the charges oscillate along different directions, will produce unpolarized radiation (Figure 23.31). Direct sunlight, or light from an ordinary light bulb, is unpolarized. It is possible to construct devices called polarizers, which allow only radiation with an electric field in a particular direction to pass through. Polarizers are discussed in Section 23.6.

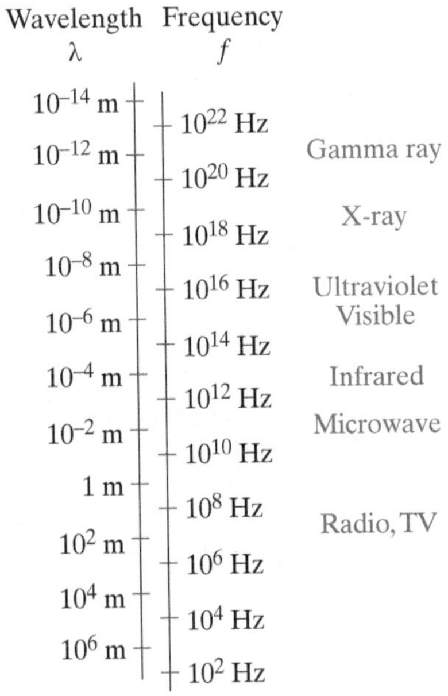

Figure 23.32 The electromagnetic spectrum: wavelength and frequency (logarithmic scale).

Transmitters

How can we produce electromagnetic radiation of a desired frequency? Radio transmitters have an LC circuit connected to the transmitting antenna, and either the inductance L or the capacitance C can be adjusted to choose the desired frequency (see Chapter 22). This circuit accelerates electrons in the antenna, and the accelerated electrons radiate.

Visible light is often produced by making some material so hot that the thermal oscillation of the atoms reaches frequencies in the visible region. This is the situation in the filament of a flashlight bulb. In this case there is a broad mixture of different frequencies that we perceive as white light.

Energy transitions between two quantized atomic levels produce light of a single frequency, as in a laser.

The Electromagnetic Spectrum

Figure 23.32 is a summary of the "electromagnetic spectrum," consisting of the ranges of wavelengths and frequencies that are characteristic of various mechanisms for producing and detecting electromagnetic radiation. The names given to radiation of different wavelengths are historical and were assigned before it was clear that all these entities were fundamentally the same. We still find these names useful for referring to electromagnetic radiation in different frequency or wavelength ranges.

23.5 ENERGY AND MOMENTUM IN RADIATION

In Chapter 8 we saw that according to the particle model of light, photons have both energy and momentum, which they can transfer to matter. In the classical model that we are now studying, it is also the case that electromagnetic radiation carries both momentum and energy, and can impart both energy and momentum to matter.

Energy in Electromagnetic Radiation

How would a pulse of electromagnetic radiation interact with ordinary matter? Since electromagnetic radiation is composed of electric and magnetic fields, we can analyze its effects straightforwardly, in terms of the effects of electric and magnetic fields on charged particles. We can see the main effects by examining what would happen to a single charge exposed to the pulse of electromagnetic radiation. We'll hang a small ball with a positive charge $+q$ from a spring, so that the ball can move up and down in the direction of the electric field, and we'll watch the ball to see what happens when the pulse of radiation goes by (Figure 23.33).

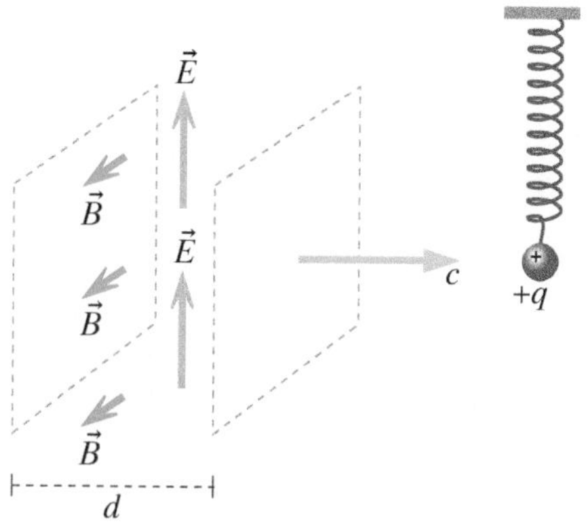

Figure 23.33 A hanging charged ball is affected by electromagnetic radiation.

A Charged Ball on a Spring

Of course nothing happens until the pulse reaches the ball, at which time there is a short pulse of electric and magnetic fields that act on the ball. If the slab containing the fields has a thickness d, the pulse lasts for a short time d/c (c is the speed of light).

For example, if $d = 30$ cm (1 ft), the pulse duration is only one nanosecond. During that brief time, the charged ball experiences an upward electric force $F = qE$, where E is the magnitude of the electric field in the electromagnetic pulse. This brief impulse $F\Delta t$ gives the ball some upward momentum:

$$\Delta p = p - 0 = F\Delta t = (qE)\left(\frac{d}{c}\right)$$

The interaction is over so quickly that the ball hardly has time to move a significant distance upward during the short duration of the pulse, but because the ball acquires some upward momentum, it will now oscillate up and down on the spring for a while after the pulse has passed by (until mechanical friction damps out the oscillations).

Energy Transfer

Since the kinetic energy of the ball has to come from somewhere, it must be the case that there is energy carried by electromagnetic radiation, and that when the ball is accelerated some of this energy is transferred from the radiation to the charged ball. Since the momentum of the ball is proportional to the magnitude of the electric field in the pulse,

$$\Delta K = K - 0 = \frac{p^2}{2m} = \left(qE\frac{d}{c}\right)^2 \left(\frac{1}{2m}\right)$$

assuming the speed of the ball remains low compared to the speed of light during this brief interaction. This implies that the amount of energy in a pulse of electromagnetic radiation is proportional to the square of the electric field:

$$\text{Radiative energy} \propto E^2$$

Conservation of Energy

Since the ball has gained energy, the radiation should have lost an equal amount of energy. Apparently the electric field E should be smaller after the pulse has accelerated the ball, and later in the chapter we will see how this can come about.

Energy Density

We can be quantitative about the amount of energy carried by electromagnetic radiation. Earlier we found that there is an energy density (energy per unit volume) associated with electric and magnetic fields:

ENERGY DENSITY IN ELECTRIC AND MAGNETIC FIELDS

$$\frac{\text{Energy}}{\text{Volume}} = \frac{1}{2}\varepsilon_0 E^2 + \frac{1}{2}\frac{1}{\mu_0}B^2, \quad \text{measured in joules/m}^3$$

For the particular case of electromagnetic radiation, $E = cB$, so

ENERGY DENSITY IN ELECTROMAGNETIC RADIATION

$$\frac{\text{Energy}}{\text{Volume}} = \frac{1}{2}\varepsilon_0 E^2 + \frac{1}{2}\frac{1}{\mu_0}\left(\frac{E}{c}\right)^2 = \frac{1}{2}\varepsilon_0 E^2\left(1 + \frac{1}{\mu_0\varepsilon_0 c^2}\right) = \varepsilon_0 E^2$$

since $\mu_0\varepsilon_0 = 1/c^2$; the electric and magnetic energy densities are equal.

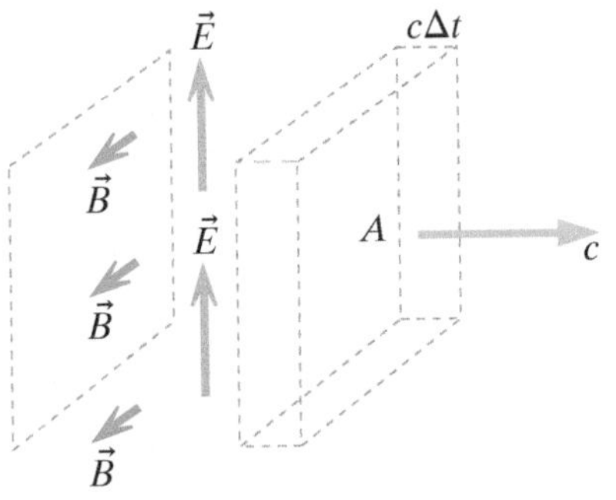

Figure 23.34 The advancing slab of radiation carries an energy flux in W/m^2.

The Poynting Vector

Energy density is related to energy flux in joules per second per square meter, or watts per square meter (in this case, flux denotes a flow of energy through a bounded surface). Consider a cross-sectional area A of an advancing slab of electromagnetic radiation (Figure 23.34). In a time Δt, a volume $A(c\Delta t)$ of electric and magnetic fields passes through the area A. The amount of energy passing through area A in the time Δt is $\varepsilon_0 E^2(Ac\Delta t)$, and therefore the energy flux is $\varepsilon_0 E^2 c$ in W/m^2. Since $E = cB$, and $\mu_0\varepsilon_0 = 1/c^2$, we can write this:

$$\text{energy flux} = \varepsilon_0 EBc^2 = \frac{1}{\mu_0}EB$$

Since the direction of the energy flow is given by $\vec{E}\times\vec{B}$, and since $\vec{E}$ and $\vec{B}$ are perpendicular to each other, the magnitude $|\vec{E}\times\vec{B}| = EB$. We define the Poynting vector $\vec{S}$, whose magnitude is the rate of energy flux in watts per square meter and whose direction is the direction of propagation of the electromagnetic radiation:

ENERGY FLUX (THE POYNTING VECTOR)

$$\vec{S} = \frac{1}{\mu_0}\vec{E}\times\vec{B}, \quad \text{in W/m}^2$$

Checkpoint 4 In the vicinity of Earth's orbit around the Sun, the energy intensity of sunlight is about 1400 W/m^2. What is the approximate magnitude of the electric field in sunlight? (What you calculate is actually the root-mean-square or rms magnitude of the electric field, because in sunlight the magnitude of the electric field at a fixed location varies sinusoidally, and the intensity is proportional to E^2.)

Momentum in Electromagnetic Radiation

The classical model of electromagnetic radiation can also account for the transfer of momentum from a pulse of radiation to an object.

In the previous section we saw that the electric field in a pulse of radiation exerted a force on a charged object—starting it moving in a direction perpendicular to the direction of propagation of the pulse. While the main effect of electromagnetic radiation is this sideways ("transverse") kick of the electric field, the magnetic field also has an interesting effect, although it is very small and hard to observe. Consider again the charged ball hanging from the spring (Figure 23.35). As soon as the electric field imparts some upward momentum Δp to the ball, the magnetic field will exert a force on the moving ball, which has a speed v upward.

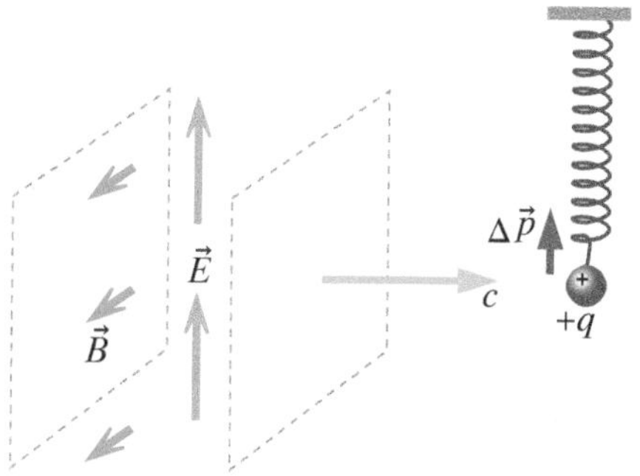

Figure 23.35 What is the direction of the magnetic force on the positively charged ball?

QUESTION What is the direction of the magnetic force on the ball, which carries a positive charge $+q$?

The direction of $+q\vec{v}\times\vec{B}$ is to the right.

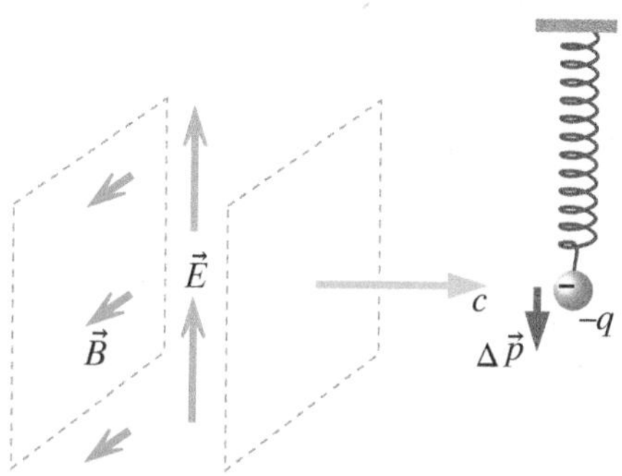

Figure 23.36 What is the direction of the magnetic force on the negatively charged ball?

QUESTION Now consider what happens if the ball carries a negative charge $-q$ (Figure 23.36). What is the direction of the magnetic force on the negatively charged ball?

The direction of $-q\vec{v}\times\vec{B}$ is again to the right, because the charge is negative (q is a positive value) and the velocity of the charge is downward. Thus the direction of the magnetic force is the same whether the ball is positively charged or negatively charged. This effect that electromagnetic radiation has on matter is called "radiation pressure." It even acts on neutral matter, because it affects the protons and electrons in the material in the same way. It is difficult to observe because it is a small effect.

We can estimate the magnitude. A charge q acquires a speed v due to the passage of the electromagnetic pulse. The average speed is $v/2$, and the average magnetic force is $|q\vec{v}\times\vec{B}| = q(v/2)B\sin(90°) = q(v/2)(E/c)$ since $B = E/c$ in the slab of radiation. Thus the ratio of the average magnetic force to the electric force is this:

$$\frac{\overline{F}_m}{F_e} = \frac{q(v/2)(E/c)}{qE} = \frac{1}{2}\frac{v}{c}$$

Normally the speed v acquired due to the passage of an electromagnetic pulse is much less than the speed of light, so the effect of the magnetic field is very small compared to the effect of the electric field.

Electromagnetic Radiation Carries Momentum

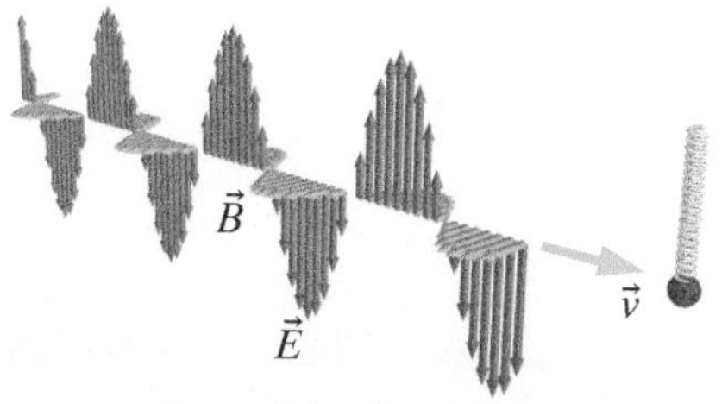

Figure 23.37 Sinusoidal radiation incident upon a charged particle.

A charged particle exposed to sinusoidal electromagnetic radiation (Figure 23.37) would gain no net momentum in a direction perpendicular to the direction of propagation of the radiation, since it would alternately be accelerated in opposite directions by the alternating electric field, and the average electric force would be zero. However, the average magnetic force would not be zero, so the particle's momentum in the direction of $\vec{v}$ would increase.

Recall that relativistic momentum p and energy E are related to each other for a particle of mass m by the equation $E^2 = (pc)^2 + (mc^2)^2$ (here E is energy, not electric field!). In the quantum view of light as photons, light can be considered to have particle-like properties with zero mass, in which case $E = pc$. This relation between energy E and momentum p is also valid in Maxwell's classical theory of light, and the energy and momentum carried by electromagnetic radiation are both proportional to the square of the electric field. The momentum flux is $1/c$ times the energy flux given by the Poynting vector:

MOMENTUM FLUX

$$\frac{\vec{S}}{c} = \frac{1}{\mu_0 c}\vec{E} \times \vec{B}, \quad \text{in N/m}^2$$

The units of the Poynting vector are W/m^2, and dividing by c means that the momentum flux is in W · s/m^3, which is J/m^3, or N · m/m^3, which is N/m^2. Hence the momentum flux has dimensions of pressure.

Conservation of Momentum

If electromagnetic radiation imparts momentum to a particle, the radiation should lose an equal amount of momentum. Again, we see that the electric and magnetic fields should be smaller in magnitude after an interaction with matter. In a moment we will see how this can come about through re-radiation.

Radiation Pressure

When electromagnetic radiation strikes neutral matter, both the protons and the electrons experience a force in the direction of propagation of the radiation (radiation pressure). Despite the fact that radiation pressure is a small effect, there have been serious proposals to build spaceships with huge solar sails, many kilometers in diameter, that could move around the solar system thanks to the radiation pressure of sunlight on the sails. The effect is doubled if the sails are highly reflective, and the plans call for sails made of very thin aluminized plastic.

Checkpoint 5 Since force is dp/dt, the force due to radiation pressure reflected off of a solar sail can be calculated as 2 times the radiative momentum striking the sail per second. In the vicinity of Earth's orbit around the Sun, the energy intensity of sunlight is about 1400 W/m^2. What is the approximate magnitude of the pressure (force per square meter) on the sail? For comparison, atmospheric pressure is about 1×10^5 N/m^2.

Re-radiation or "Scattering"

Consider once again the charged ball hanging from the spring, exposed to electromagnetic radiation and therefore oscillating at the radiation frequency (Figure 23.37).

Wait a minute! Won't the ball radiate? After all, the oscillating ball is a sinusoidally moving (and sinusoidally accelerating) charged object, so shouldn't it emit electromagnetic radiation? Yes!

The oscillating ball will re-radiate in (nearly) all directions (see Figure 23.19). The re-radiation by the charged ball redistributes the energy in new directions. This re-radiation is often called "scattering" because part of the original energy moving to the right has been "scattered" into other directions. Figure 23.38 illustrates the effect of the superposition of incident radiation and re-radiation by the charged ball.

QUESTION Explain why scattering (re-radiation) is never observed in the direction of or directly opposite to the direction of the original transverse electric field (that is, up or down in Figure 23.38).

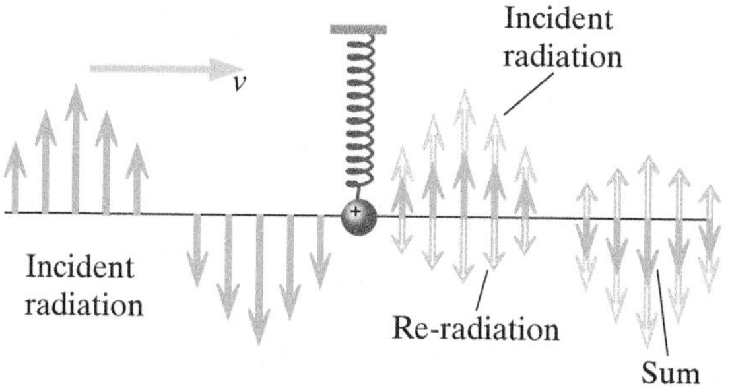

Figure 23.38 Re-radiation by the accelerated charged ball. The accelerated charge re-radiates in almost all directions; for clarity only re-radiation in the $+x$ direction has been shown in this diagram. The outgoing radiation is the superposition of the incident radiation and the re-radiated fields. E outgoing is smaller than E incoming because some energy has been scattered in other directions.

Straight above or below the accelerated charge, the projected acceleration $\vec{a}_\perp$ is zero, so there is no re-radiation in those directions.

We saw earlier that incoming radiation loses energy by giving kinetic energy to a charged ball, proportional to E^2. The re-radiation by the accelerated charge provides a mechanism for decreasing the energy of the original radiation. To the right of the charged ball, the outgoing electromagnetic radiation is the superposition of the original fields plus the fields radiated by the accelerated ball. If an incoming electric field points upward, it accelerates a positively charged ball upward, and to the right of the ball the re-radiated electric field points downward.

As a result, the net outgoing electric field to the right of the ball in Figure 23.38 is smaller than the original incoming electric field. The energy in the electromagnetic radiation, which is proportional to E^2, is smaller after giving kinetic energy to the ball.

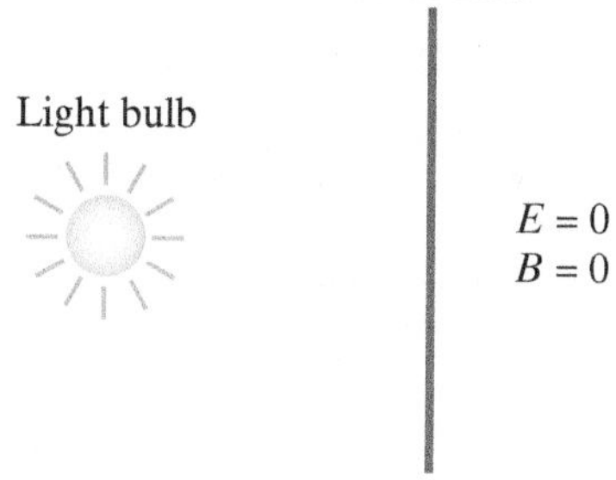

Figure 23.39 Why isn't there light behind the cardboard?

Checkpoint 6 Why doesn't light go through a piece of cardboard (Figure 23.39)? According to the superposition principle, the electric and magnetic fields radiated by the accelerated charges in the light bulb are definitely present to the right of the cardboard, so why isn't there any light there? What must be true of the light re-radiated by charges in the cardboard?

23.6 EFFECTS OF RADIATION ON MATTER

Because radiation pressure is such a small effect, the most important aspect of the interaction between electromagnetic radiation and matter is the effect that the electric field has on the charged particles inside atoms and molecules. For the rest of the chapter, we will concentrate on the electric field in electromagnetic radiation, and neglect the magnetic field. Of course the magnetic field must be present in electromagnetic radiation, but usually only the electric field need be considered when we try to understand the main effects of electromagnetic radiation on matter, in terms of our model of electromagnetic radiation.

In the following sections we will consider a wide range of important and interesting phenomena in terms of what we have learned about how to produce electromagnetic radiation consisting of propagating electric and magnetic fields, and what we know about the effects that electric and magnetic fields have on the charged particles in matter.

The Effect of Radiation on a Neutral Atom

The main effect that a pulse of radiation has on matter is that it gives a brief electric kick to charged objects. This kick is sideways or transverse to the direction of propagation of the fields. What would a pulse of electromagnetic radiation do to a neutral atom or molecule? A simple model of an atom or molecule is adequate to get the main idea. Consider a positive core (nucleus plus tightly bound inner electrons) and a single, loosely bound outer electron (Figure 23.40).

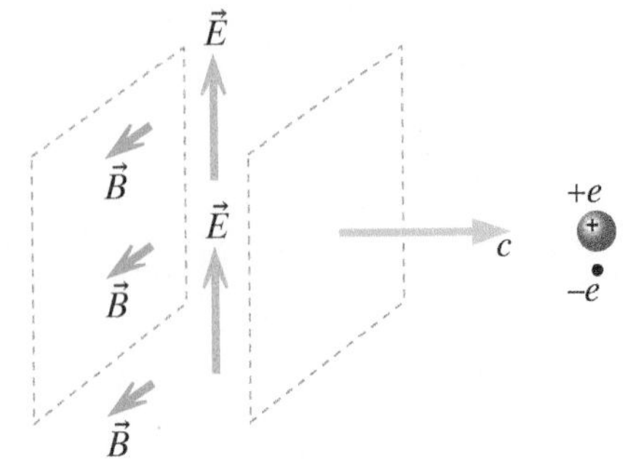

Figure 23.40 A neutral atom is affected by electromagnetic radiation.

QUESTION What happens to this atom when the pulse of electromagnetic radiation goes by? Which part of the atom picks up more speed: the electron or the positive core?

The electric field polarizes the atom. During this process the electron moves farther than the much more massive ion (nucleus plus inner electrons); the magnitude of the force eE is the same on both objects. For many purposes the interaction of electromagnetic radiation with an atom can be analyzed by modeling an outer electron as being connected to the rest of the atom by a spring; see Problem P62 in Chapter 15.

According to our model of electromagnetic radiation, this is the basic mechanism for the interaction of light and matter: the electric field of the light exerts forces that move charged particles inside atoms and molecules. Although electromagnetic radiation is rather unfamiliar compared to topics we studied earlier, there is nothing unfamiliar about the effect of the associated electric field on charged particles, which is simply $\vec{F} = q\vec{E}$.

Polarizers

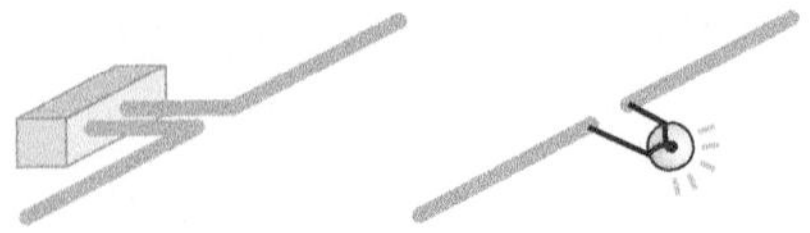

Figure 23.41 A high-frequency radio transmitter and receiver.

Figure 23.41 illustrates a demonstration of electromagnetic radiation that is sometimes presented in physics lectures. A radio transmitter consists of a high-frequency AC voltage supply (frequency around 300 MHz) connected to a transmitting antenna in the form of two metal rods, and a radio receiver is made of two metal rods connected by a flashlight bulb. When positioned as shown, the bulb lights although it is not connected to the circuit!

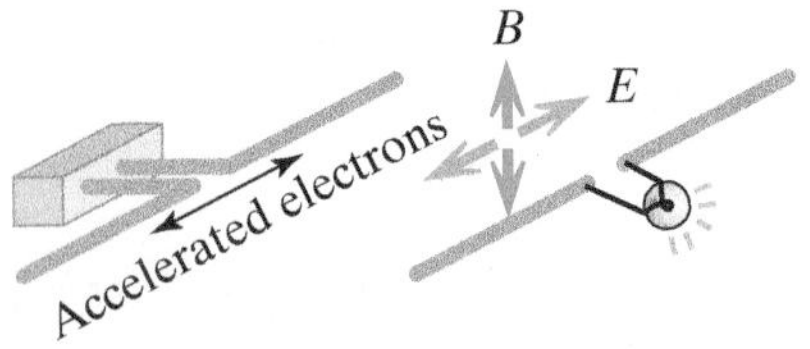

Figure 23.42 The radiative electric field drives current through the bulb.

The high-frequency AC voltage drives the mobile electrons back and forth in the transmitting antenna, and the sinusoidal motion of these electrons implies a sinusoidal acceleration, which produces electromagnetic radiation (Figure 23.42). For the radiation that propagates to the right, the oscillating electric field is horizontal, and it can accelerate the mobile electrons in the receiving antenna, driving them back and forth through the flashlight bulb, and lighting the bulb.

Figure 23.43 When the distance is greater, the bulb is dimmer.

If you move the receiver and transmitter farther apart (Figure 23.43), the bulb gets dimmer because of the $1/r$ fall-off of the magnitude of the radiative electric field.

If you turn the receiving antenna vertical, the bulb does not light (Figure 23.44).

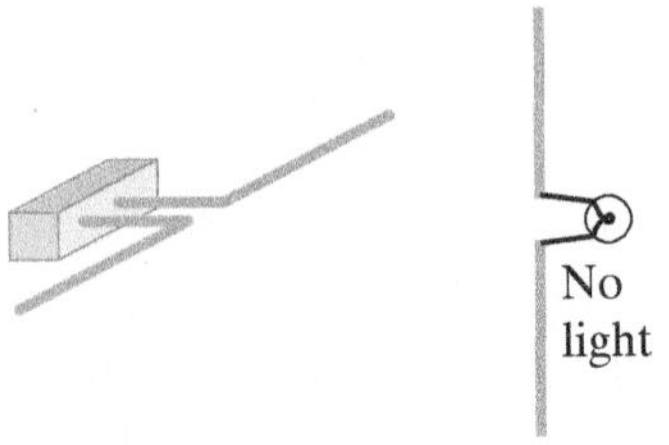

Figure 23.44 With the receiving antenna vertical, the bulb doesn't light.

QUESTION Explain why the bulb does not light when the receiving antenna is vertical.

The emitted radiation is polarized along the axis of the transmitting antenna, but the radiated electric field is perpendicular to the receiving antenna. The electrons in the antenna are accelerated back and forth across the width of the wire (not its length), so the current does not go through the bulb.

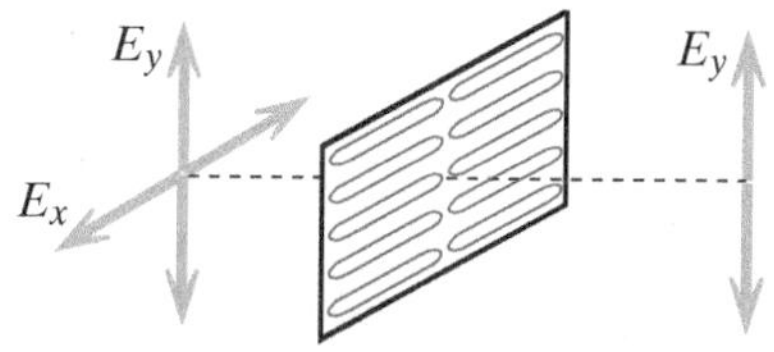

Figure 23.45 Making polarized light from unpolarized light. The incident beam contains unpolarized sinusoidal radiation consisting of waves with electric fields in all directions perpendicular to the direction of propagation.

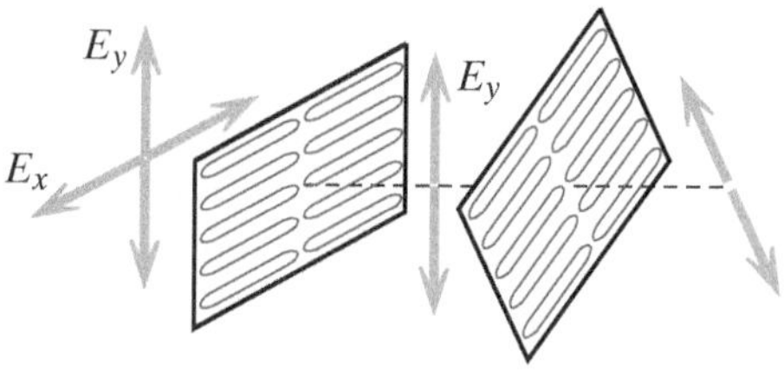

Figure 23.46 A second polarizer rotates the polarization direction.

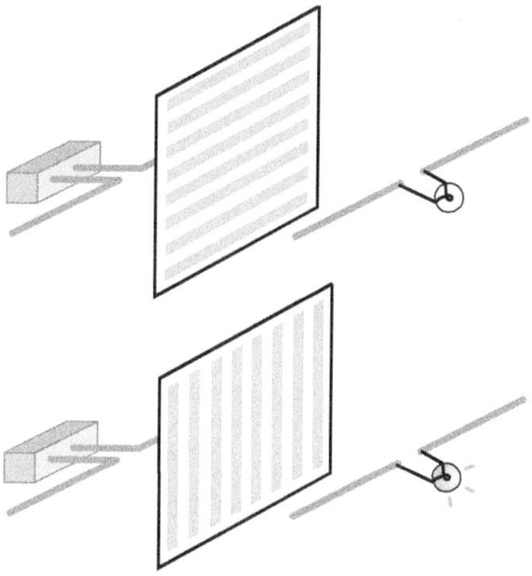

Figure 23.47 A polarizer for radio waves.

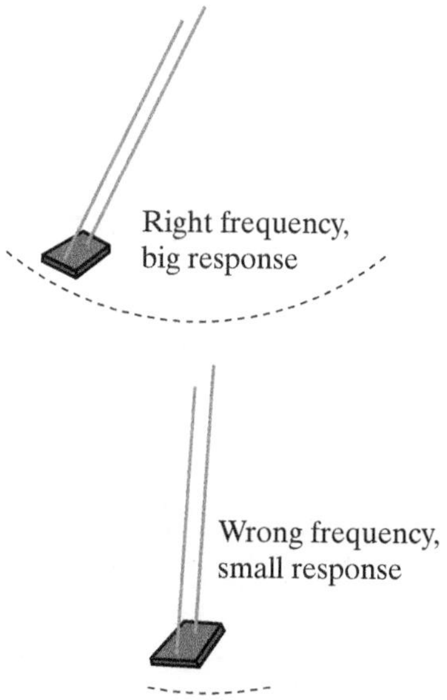

Figure 23.48 Resonance behavior in pushing a swing.

You can produce polarized visible light by passing unpolarized light through a sheet of a special plastic whose long molecules are aligned with each other, such as the material in Polaroid® sunglasses (Figure 23.45). Light that is polarized in the same direction as these long molecules is able to drive electrons along the length of the molecule and lose energy to this relatively large-scale motion. Light that is polarized perpendicular to these long molecules can't move the electrons very far and consequently loses much less energy in the plastic. For light that is polarized at some other angle only the component of electric field that is perpendicular to the long molecules manages to show up after the polarizer. The net effect is that polarized light is produced.

If you place a second polarizer after the first one, its orientation determines whether all, some, or none of the polarized light is seen after the polarizer, and if any component of the light is seen after the second polarizer it is polarized in alignment with the second polarizer (Figure 23.46).

> QUESTION Polaroid® sunglasses are useful at the beach because unpolarized sunlight that glances off the water happens to become polarized in the process, with the electric field horizontal. In order to block this glare but let other (unpolarized) aspects of the scene come through, should the long molecules in the plastic lenses be horizontal or vertical?

Since we want to diminish the horizontally polarized radiation (glare), the molecules should also be horizontal. Radiation with a horizontal electric field will be absorbed and re-radiated, with only a fraction seen in the original direction. Vertically polarized light will be seen unaffected.

You can observe these effects with radio waves, too. For a polarizer, use a piece of cardboard that has a series of parallel metal foils taped to the cardboard. When the foils are aligned with the antennas, the bulb does not glow; but when the foils are aligned perpendicular to the antennas, the bulb does glow (Figure 23.47).

> **Checkpoint 7** Briefly explain the observations in the previous paragraph (Figure 23.47). (*Hint:* Think about the re-radiation that may occur from the wires.)

Effects that Depend on Frequency

Let us again suspend a charged ball from a spring and watch what happens as this sinusoidally varying electromagnetic radiation goes by (Figure 23.37). In the vertical direction the electric force on the ball varies like $qE_{\text{max}} \sin(\omega t)$, and this forces the ball to oscillate up and down at the same angular frequency ω. How large an oscillation we get depends on how close ω is to the angular frequency at which the spring–ball system would freely oscillate in the absence of the radiation. If the radiation frequency nearly matches the free-oscillation frequency, the ball will oscillate up and down with a large motion. If on the other hand the radiation frequency is very different from the free-oscillation frequency, the ball will move up and down very little, even if the radiative electric field is quite large.

Resonance

This phenomenon is called "resonance." It is easy to observe resonance when you push someone on a swing (Figure 23.48). If you give little pushes at exactly the same frequency as the swing's free-oscillation frequency (for example, every time the swing comes all the way back), even these little pushes add up to produce a very large motion, and very small pushes are sufficient to maintain this large motion. However, if you push at a frequency different from the free-oscillation frequency, the swing is forced to oscillate at your frequency but will hardly move at all, because your inputs don't reinforce each other at the right times.

There is an extremely important consequence of the fact that systems respond strongly to sinusoidal radiation only if the frequency of the radiation closely matches the frequency at which these systems would oscillate freely: Different frequencies of sinusoidal electromagnetic radiation affect matter differently. For example:

- Sinusoidal electromagnetic radiation with a very high frequency, $f = \omega/2\pi$ of about 1×10^{15} Hz, has a big effect on the organic molecules in your retina, and you detect "visible light." However, a radio does not respond to visible light.
- Sinusoidal electromagnetic radiation with a frequency of about 1×10^6 hertz (1000 kHz) has a big effect on the mobile electrons in the metal of a radio antenna, and the radio receiver detects "radio waves." You can tune a radio to be sensitive to electromagnetic radiation at 1020 kHz and be insensitive to all other radio frequencies, so that you hear just one station at a time. However, your eye does not detect radio waves.
- The very high frequency of the sinusoidal electromagnetic radiation known as x-rays has very little effect on the atoms of your body and so mostly passes right through with no interaction; there is just enough interaction that shadows of the denser parts of your body appear on film sensitive to x-rays.

For many purposes involving the classical interaction of light and matter, a neutral atom can be modeled as though an electron in the atom acts like a mass on a spring, whose free-oscillation frequency determines the resonance response to electromagnetic radiation. See Problem P62 in Chapter 15.

Color Vision

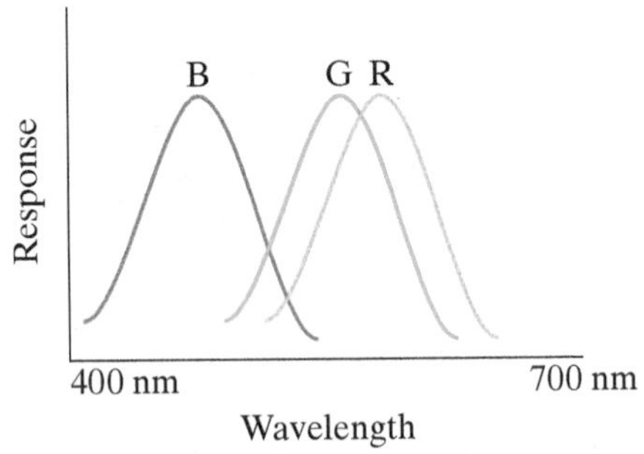

Figure 23.49 Approximate response of light-sensitive molecules in the human eye to visible light as a function of wavelength.

In your retina there are three kinds of organic molecules, which we might call R, G, and B. Type R responds strongly to electromagnetic radiation near 560 nm (0.56×10^{-6} m, or 5.4×10^{14} Hz); type G responds strongly to electromagnetic radiation near 530 nm (0.53×10^{-6} m, or 5.7×10^{14} Hz); and type B responds strongly to electromagnetic radiation near 420 nm (0.42×10^{-6} m, or 7.1×10^{14} Hz). Each of these molecules responds less or not at all to the wavelengths favored by the other molecules (Figure 23.49). Your brain interprets signals from R molecules as indicating red light; signals from G molecules indicate green light; and signals from B molecules indicate blue light.

Combinations of these responses are interpreted as all of the other colors that you can perceive, many of which are not in the rainbow. For example, a mixture of electromagnetic radiation at 700 nm (red) and at 400 nm (blue) is interpreted by the brain as magenta. In a rainbow, red and blue are separated from each other, so magenta doesn't occur. Equal stimulation of the R, G, and B molecules is interpreted by the brain as white light, another color that is not in the rainbow.

There is some overlap in wavelengths in the sensitivity of the R, G, and B receptors. For example, light containing the single wavelength of 588 nm (the yellow light of burning sodium) stimulates both the R and the G receptors, so you also perceive yellow with a suitable mixture of two wavelengths, one red and one green. Color television and computer screens use closely spaced red and green dots to fool the brain into seeing a yellow dot.

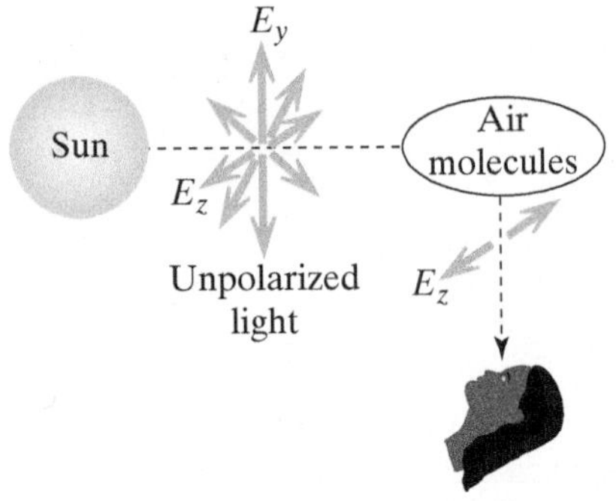

Figure 23.50 Looking up at the sky—why is there any light? (Unpolarized light is indicated schematically here.)

Why the Sky Is Blue (and Polarized)

Why is the light we see coming from the sky blue (and polarized)? Suppose that you look straight up at the blue sky, in the early morning or late afternoon, so that the Sun is near the horizon (Figure 23.50).

> QUESTION First, why do you see any light at all coming from the sky overhead? After all, if you stand on the Moon and look up, the sky is black (except for stars).

The reason you see any light at all coming down from straight overhead on the Earth is that electrons in the air molecules all through the atmosphere above your head are accelerated by the sunlight and re-radiate. The Moon has no atmosphere to scatter the sunlight. Interestingly, it is also critical that the air molecules are moving around randomly, so that in any cubic centimeter of air the number of molecules continually fluctuates. It can be shown that without this fluctuation the re-radiations from the fixed array of molecules would cancel each other out. This is discussed in Supplement S3, in the section "Further Applications of the Wave Model."

Second, the sky light coming from overhead is polarized, despite the fact that light from the Sun is unpolarized due to being emitted by many atoms at random orientations. This polarization is easy to observe for yourself. While wearing Polaroid® sunglasses on a cloudless day, look straight overhead in early morning or late afternoon and turn your head to vary the orientation of the polarizer. You will see a marked change in the intensity of the sky light.

> QUESTION Second, explain why the sky light is polarized, and state what the direction of polarization is.

The electric field of light from the Sun is perpendicular to the direction of propagation. The light from the Sun is not polarized, so the electric field can point anywhere in the yz plane (Figure 23.50). However, only charges in air molecules that are accelerated in the $\pm z$ direction will re-radiate in the $-y$ direction, downward toward you, so the sky light is polarized with the electric field in the $\pm z$ direction.

> QUESTION Third, why is the sky light blue?

Light from the Sun is a mixture of many different frequencies of sinusoidal electromagnetic radiation. The spectrum of sunlight runs from red to blue, and the mixture is perceived by you as nearly white (but somewhat yellowish). So why does the light from the sky look blue instead of yellow? After all, the electrons are forced to oscillate at the same frequency as the light that hits them.

The electrons in the atom are shaken by the oscillating electric field. If the position of the electron can be represented by $x = A\cos(\omega t)$, then the acceleration is $a = -\omega^2 A\cos(\omega t)$. The magnitude of the re-radiated electric field is proportional to the acceleration, and the power (energy/second) of the re-radiation is proportional to the square of the acceleration. We expect the re-radiation that comes to your eye to be proportional to ω^4. This implies that re-radiation of the high-frequency blue component of sunlight will be much greater than the re-radiation of the low-frequency red component of sunlight. The ratio of blue to red frequencies is nearly a factor of 2, so the ratio of re-radiated intensities is about a factor of $2^4 = 16$.

Why Distant Mountains Appear Bluish

On a very clear day, with little dust or haze in the air, distant mountains nevertheless look bluish. An indication that there is little haze is the fact that distant mountain ridges appear with crisp and sharp lines, not blurred as they would look in haze or fog. The explanation is essentially the same as the explanation of why the sky is blue! When you look up at the sky and see a brilliant blue color, you are looking through the equivalent of only a few kilometers of sea-level air (the atmosphere extends out to many tens of kilometers, but the air has exponentially decreasing density as you go higher).

When you look at a distant mountain through many kilometers of air, sunlight coming from the side scatters off that column of air into your eyes, adding a bluish light to the image of the mountain. You need a dark mountain as a background in order to be able to notice this relatively low-intensity bluish light.

23.7 LIGHT PROPAGATION THROUGH A MEDIUM

The propagation of electromagnetic radiation through matter is more complicated than the propagation of radiation through vacuum, because the electric and magnetic fields in the incident radiation affect charged particles in the medium. The affected particles may be accelerated and re-radiate. The net field is the superposition (sum) of the original field and the re-radiated field. In the following sections we will explore the optical consequences of this superposition process.

Wavefronts

In the sinusoidal electromagnetic radiation emitted by an oscillating charge, a *wavefront* is considered to be a set of peaks (amplitude maxima) all emitted at the same time. In Figure 23.51, the circles mark successive wavefronts in the radiation emitted by a sinusoidally oscillating charge. To simplify diagrams, often only the circles representing the wavefront are shown, as in Figure 23.52.

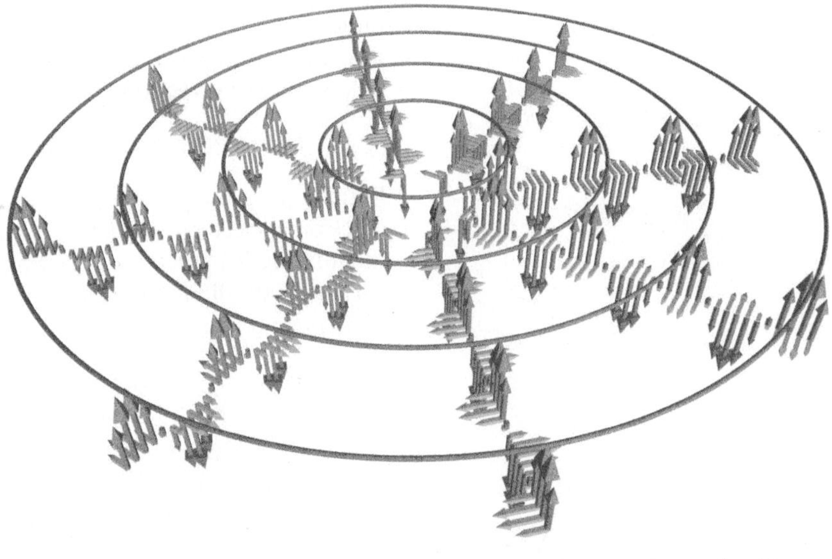

Figure 23.51 The circles mark successive wavefronts in the radiation emitted by a charge at the center (not shown) oscillating along the y axis. For clarity, only radiation emitted in the xz plane is shown (radiation is in fact emitted in all directions except the $\pm y$ direction).

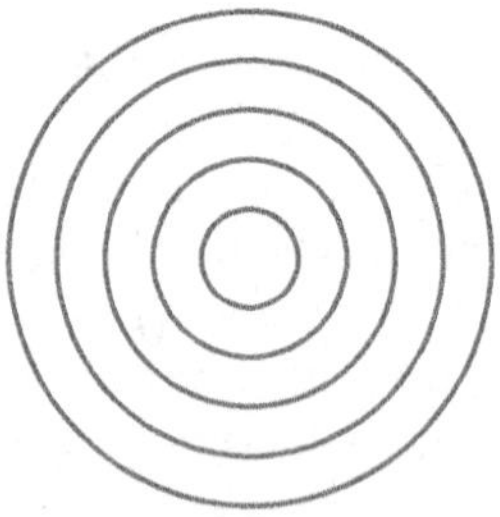

Figure 23.52 A schematic depiction of the wavefronts spreading out from an oscillating source charge.

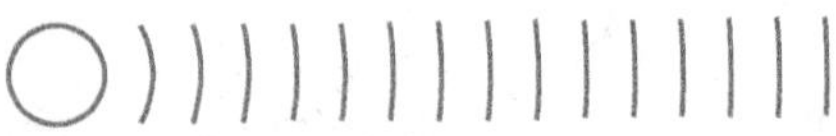

Figure 23.53 Far from the source, segments of wavefronts appear nearly planar, and parallel to each other.

Plane Waves

The wavefronts emanating from an oscillating charge are nearly spherical, as indicated in Figure 23.52. However, as the wave propagates outward, the radius of curvature increases so that to an observer far from the source (where *far* means a distance large compared to the wavelength of the radiation), successive wavefronts appear nearly planar, and parallel to each other, as indicated in Figure 23.53. In this situation, we describe the wave as a *plane wave*. The plane referred to is the plane defined by the $\vec{E}$ and $\vec{B}$ vectors, as indicated in

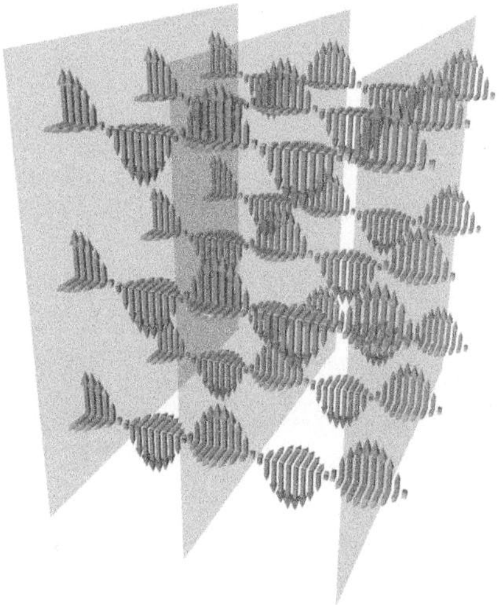

Figure 23.54 A plane wave. The gray planes indicate the location of parallel (or nearly parallel) wavefronts.

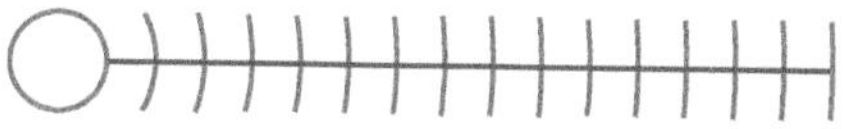

Figure 23.55 A line called a *ray* (the red line) is often used to describe a section of a wavefront propagating in a given direction. Rays are drawn perpendicular to wavefronts.

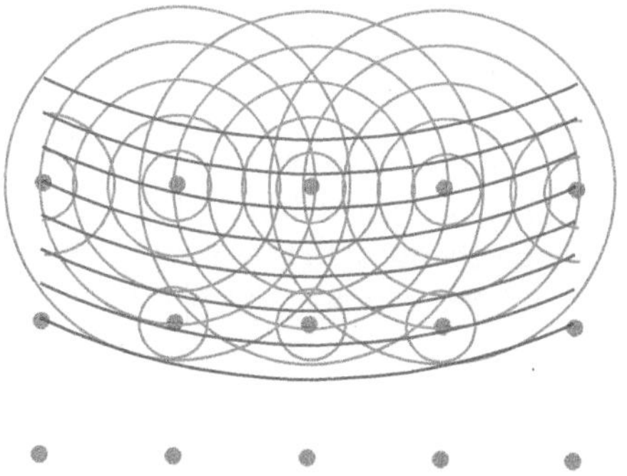

Figure 23.56 Gray dots represent atoms in a solid medium. The blue curves represent wavefronts in the incident radiation. The light red circles represent wavefronts in the re-radiation from charges in the atoms that were accelerated by the incident radiation. The net field at any location is the sum of all fields at that location.

Figure 23.54; this plane is perpendicular to the direction of propagation of the wave.

The radiation traveling in a particular direction of propagation is often described as a *ray* and represented by a single line in the direction of propagation, perpendicular to wavefronts (Figure 23.55). Although there are not really discrete rays—the radiation emitted by an oscillating charge is spatially continuous—imagining particular rays can be useful in understanding the geometric behavior of light as it travels through objects such as lenses or prisms.

Re-radiation and Superposition

When electromagnetic radiation travels through a region filled with a medium such as glass, air, water, metal, plastic, or anything other than a vacuum, it may interact with charged particles in the medium. As discussed previously, the electric field in the incoming radiation accelerates charged particles in the medium, causing these accelerated charges to emit electromagnetic radiation as well. (The magnitude of the electric force on a proton is the same as that on an electron, but the acceleration of a low-mass electron is much larger, so it is the electrons that contribute almost all of the re-radiation.) Therefore, at any location inside a medium, the net radiative electric and magnetic fields are the superposition of the incident (initial) sinusoidally oscillating fields and the re-radiated oscillating fields. As indicated schematically in Figure 23.56, this sum can be quite complex, especially in a dense medium such as a solid material, which is composed of many charged particles. The behavior of light in a material depends very much on the structure of the material at the atomic level.

One possible result of superposition is that the incident wave and re-radiated waves may superimpose in just such a way that troughs in one wave occur at the same locations as crests in the other wave. In this case, the net amplitude is zero, and no electromagnetic radiation emerges from the other side of the medium. This is called *destructive interference*. In this case the material is *opaque*. Opacity depends on the wavelength of the incident radiation. For example, cardboard can be opaque to visible light, but not to microwave radiation—a useful property if you wish to use a microwave oven to reheat leftover food in a cardboard box.

In a material that is transparent to electromagnetic radiation of a particular wavelength, incident radiation and re-radiation add *constructively*, so the amplitude of the waves inside the material is not zero. However, this addition process can result in changes in the wavelength of the net waves and in the speed at which a particular wave crest travels through the medium.

Index of Refraction

Most commonly, in ordinary solid materials, the incident wave and the re-radiated waves combine to create a resultant net wave whose crests travel through the medium at a speed slower than c. Because the frequency f of the re-radiation is the same as that of the incident radiation, the wavelength inside the medium is less than the wavelength in vacuum: $v = \lambda/T = \lambda f$, so $\lambda = v/f$, and smaller v means shorter λ (because f is unchanged).

The *index of refraction* of a medium is defined as the ratio of c, the speed of a wavefront in a vacuum to v, the speed of a wavefront in the medium, both measured by the time for one crest to travel a given distance:

$$n = \frac{c}{v}$$

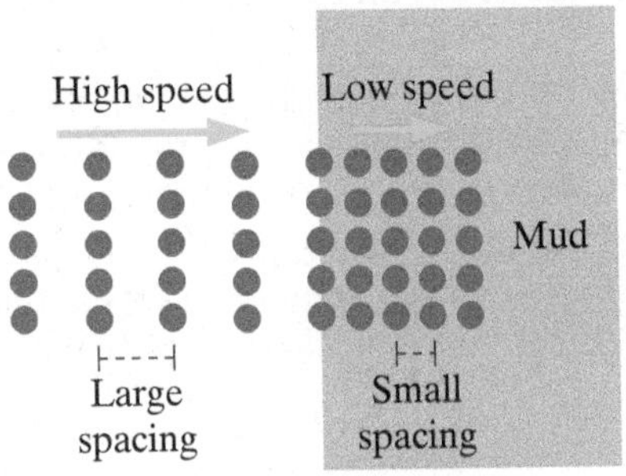

Figure 23.57 When members of a marching band encounter mud and slow down, the ranks get closer together.

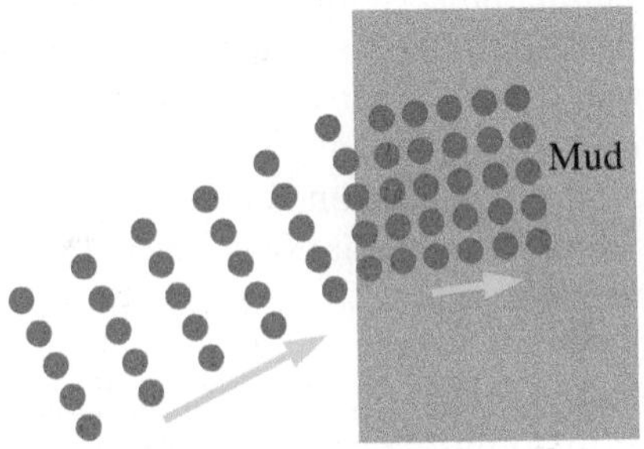

Figure 23.58 When the marching band hits the mud at an angle, the direction of march changes.

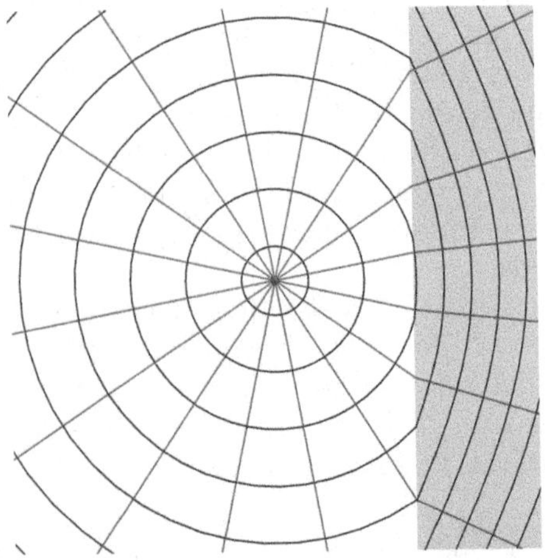

Figure 23.59 Light spreads out in air from a sinusoidally oscillating point source and encounters a slab of glass. The red rays indicate the directions of propagation, and the curving arcs are the wavefronts, perpendicular to the rays.

Here are indices of refraction for some typical materials, for a wavelength of 589 nm (yellow sodium light):

Material	n
Vacuum	1.0000
Air	1.00029
Water	1.33
Fused quartz	1.46
Plexiglas	1.51
Zircon	1.92

Checkpoint 8 Light propagates through gallium phosphide at a speed of 8.57×10^7 m/s. What is the index of refraction of gallium phosphide?

23.8 REFRACTION: BENDING OF LIGHT

The fact that the speed of a wavefront is different in different materials leads to bending of a ray when a wave moves from one medium into another.

A Marching Band Changes Direction

There is a useful analogy that helps explain what happens to wavefronts and rays of light at the surface between the two materials. Think of a marching band that is marching quickly across grass. When it comes to a muddy part of the field it has to slow down. In Figure 23.57 all members of each marching rank reach the mud at the same time, and the pileup on reaching the mud makes the ranks (wavefronts) closer together in the mud than on the grass. This is analogous to the decreased wavelength when light passes from air into glass.

In Figure 23.58 the band approaches the mud at an angle. When the first band members in a rank encounter the mud and slow down, there are other marchers in their rank who are still on the grass and still moving quickly. If the marchers are required to maintain the same spacing within their rank as before, the ranks (wavefronts) in the mud end up being at a different angle than the ranks still on grass. The direction of march changes; the outgoing "ray" is in a different direction than the incoming "ray." We'll look at this phenomenon more quantitatively in the case of light.

Bending of Light

In Figure 23.59 light from a sinusoidally oscillating point source in air encounters the surface of a slab of glass (the gray area). The wavefronts, indicated by the curving arcs, get closer together in the glass, as a result of the superposition of the fields of the point source and the additional fields reradiated by accelerated charges in the glass. The wavelength λ of the radiation is shorter inside the glass than it was in the air. Because the frequency and period of the oscillation don't change, the speed of a wavefront in the glass is slower than it was in air (speed $= \lambda/T$). The red rays represent the directions of propagation, which are perpendicular to the wavefronts. The rays bend inward at the surface between the air and the glass. The bending of rays of light at an interface between two materials is called "refraction."

Reflection

In addition to the refraction (bending of light), there is also some reflection of the light in situations like that shown in Figure 23.59. The accelerated electrons in the glass radiate in all directions and, in addition to making the wavelength shorter in the glass, they produce light in backward directions. It looks as though this "reflected" light bounces off the surface, but in fact this new light is emitted by all of the accelerated charges throughout the glass. The reflected light may be quite weak; you have probably noticed that reflections off a sheet of glass can be quite dim.

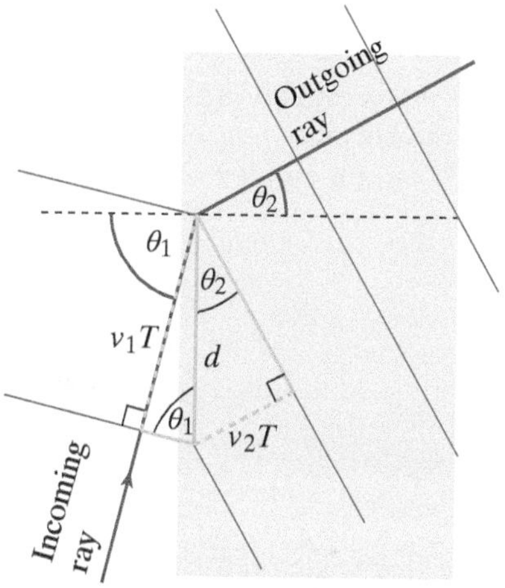

Figure 23.60 Snell's law. Red lines represent rays; cyan lines represent wavefronts. The dashed black line indicates the normal to the surface. Dashed orange lines indicate the distance traveled by a wavefront in time T. The gray medium (glass) has a higher index of refraction.

Snell's Law

We can derive a relationship called "Snell's law" between the bending of the light rays at an interface and the speed of wavefronts in the two media. Zoom in close enough that the wavefronts can be approximated by parallel straight lines (Figure 23.60). It is customary to express the directions of the incoming and outgoing rays as angles between the rays and "normals" to the surface (lines that are perpendicular to the surface).

In Figure 23.60 the incoming ray is at an angle θ_1 to the normal and the outgoing ray is at an angle θ_2 to the normal. The same angles occur in the orange triangles, since if the sides of an angle are perpendicular to the sides of another angle, these two angles are either equal to each other or their sum is 180° (in either case their sines will be equal).

The wavefront speed in air is v_1 and the wavefront speed in the glass is v_2. In a time T (the period), a wavefront in air travels a distance v_1T and a wavefront in glass travels a distance v_2T, indicated by dashed orange lines in the diagram. The two orange triangles share a side of length d.

In the triangle on the left (in air):

$$\sin\theta_1 = \frac{v_1T}{d}$$

In the triangle on the right (in glass):

$$\sin\theta_2 = \frac{v_2T}{d}$$

Since $\frac{T}{d}$ is the same for both triangles,

$$\frac{\sin\theta_1}{v_1} = \frac{\sin\theta_2}{v_2}$$

It is customary to express this result in terms of the index of refraction $n = c/v$, which can be written as $v = c/n$:

$$\frac{\sin\theta_1}{c/n_1} = \frac{\sin\theta_2}{c/n_2}$$

$$\frac{n_1\sin\theta_1}{c} = \frac{n_2\sin\theta_2}{c}$$

Canceling the common factor c, we obtain Snell's law:

SNELL'S LAW

$$n_1\sin\theta_1 = n_2\sin\theta_2$$

where n_1 and n_2 are the indices of refraction of the two media, and θ_1 and θ_2 are the angles the incident and transmitted rays, respectively, make with the normal to the surface.

For concreteness we've talked about air and glass, but this result is valid for any two materials whose indices of refraction are n_1 and n_2.

EXAMPLE **Ray Travels from Water into Glass**

In Figure 23.61 a beam of light travels through water and hits a glass surface. The angle θ_1 between the incident beam and the normal to the glass surface is 23°. The index of refraction of water is 1.33, and the index of refraction of this type of glass is 1.65. What is the angle θ_2?

Solution From Snell's law:

$$n_1 \sin\theta_1 = n_2 \sin\theta_2$$

$$\sin\theta_2 = \frac{n_1 \sin\theta_1}{n_2}$$

$$\theta_2 = \arcsin\frac{1.33 \sin 23°}{1.65} = 18.4°$$

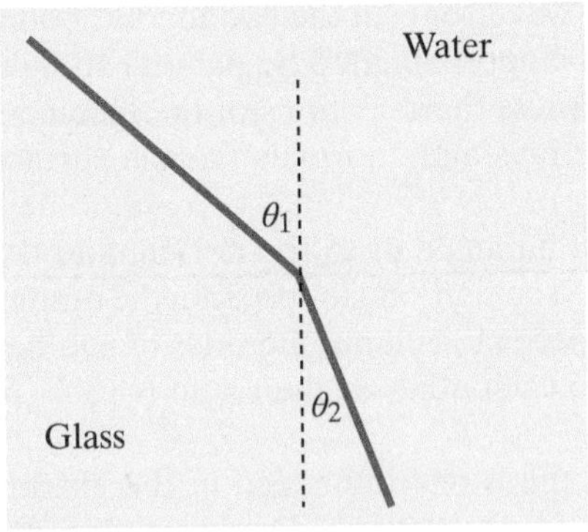

Figure 23.61 A beam of light travels from water into glass.

Total Internal Reflection

We've been considering light that passes from air to glass, with a decrease in the wavefront speed associated with an increase in the index of refraction. Something unusual happens in the other direction, going from a higher index of refraction to a lower index of refraction. For example, suppose that light goes from water (with an index of refraction of 1.33) into air, whose index of refraction is very close to 1.0. Consider a ray in the water at an angle of 60° to the normal, as shown in Figure 23.62. Let's use Snell's law to predict the angle the ray will make to the normal in the air:

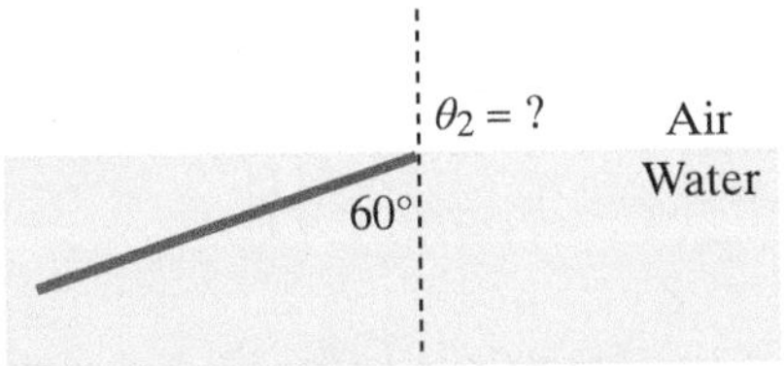

Figure 23.62 A beam of light travels through water to the water–air interface, at an angle of 60° to the normal.

$$1.33 \sin 60° = 1.0 \sin\theta_2$$

$$\sin\theta_2 = 1.33 \sin 60° = 1.15$$

QUESTION Try taking the arcsine of 1.15 on your calculator.

What happens? You get an error! The sine function varies between −1 and +1, and there is no angle whose sine is 1.15. Snell's law tells us that there is no possible angle for this ray to make in the air. What happens if a ray of light at an angle of 60° to the vertical tries to emerge from a pool of water into the air above? It doesn't emerge but reflects back down into the water, as indicated in Figure 23.63. This phenomenon is called "total internal reflection."

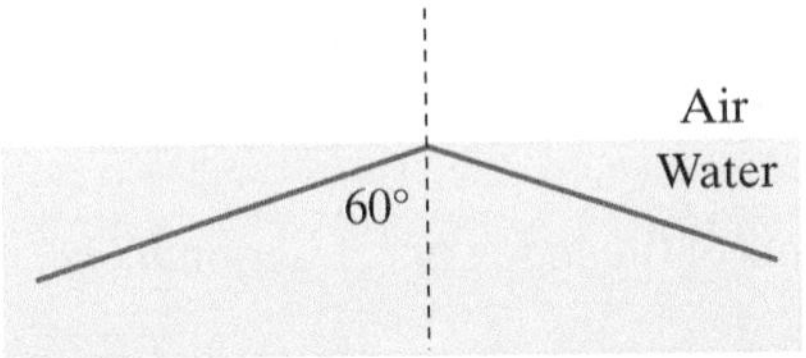

Figure 23.63 Total internal reflection.

Figure 23.64 shows rays of light emitted from a source that is underwater, emerging into the air above. The largest possible angle from the vertical (the normal to the surface) for a ray in the air is 90°.

QUESTION What is the angle to the vertical for a ray in the water that gives a ray in the air at 90° to the vertical? This is a horizontal ray that just skims the surface. (*Hint:* Use Snell's law.)

Snell's law says that $1.33 \sin\theta_1 = 1.0 \sin 90°$, so $\sin\theta_1 = 1.0/1.33 = 0.75$, and $\theta_1 = 49°$. Any underwater rays at more than 49° to the vertical cannot escape into the air but are "totally internally reflected" back down into the water, as you can see in Figure 23.64. If you look up at the surface while swimming underwater, at a large angle to the vertical you see bright reflections off the surface due to this effect.

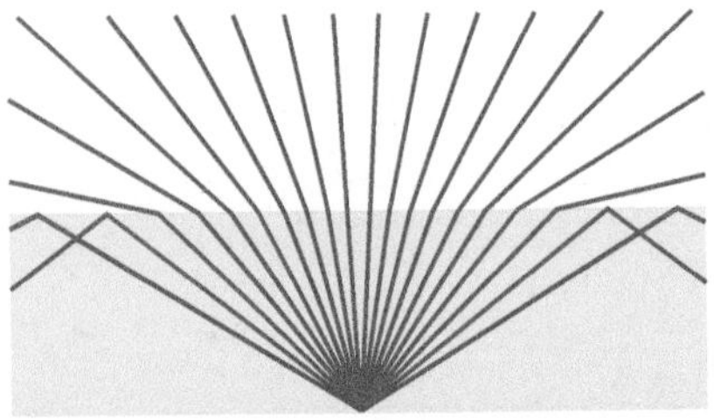

Figure 23.64 Rays of light coming from a source underwater bend outward when passing into the air, where the index of refraction is smaller. Underwater rays whose refracted angle would be greater than 90° cannot escape the water and are reflected downward ("total internal reflection").

Fiber Optics

An important application of the phenomenon of total internal reflection is the use of "fiber optics" for long-distance high-speed internet connections. It is possible to transmit much more data over long distances by means

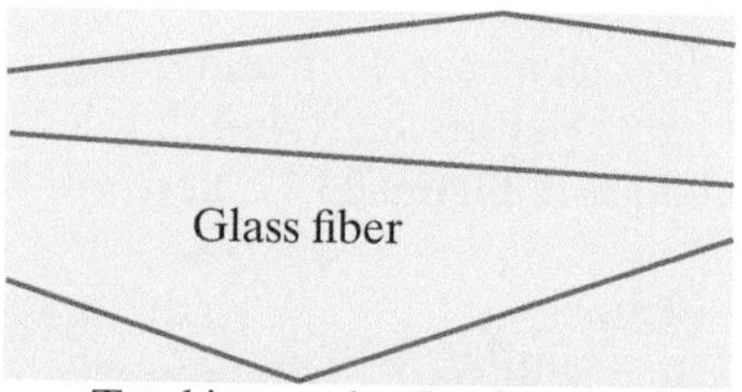

Figure 23.65 In glass fibers used for long-distance high-speed Internet communication, total internal reflection keeps rays from escaping.

of light propagating in glass fibers than by means of electrical signals propagating in copper wires, but there are two major design issues. The glass must be extremely pure in order to be extraordinarily transparent, so that signal strength is maintained over long distances. The second issue is that the glass fibers must be designed in such a way that light rays traveling through the glass undergo total internal reflection at the surface of the fiber rather than escaping. This is achieved by using glass with an appropriate index of refraction. In some designs there is a variation in the index of refraction as one approaches the surface. Figure 23.65 shows the basic idea: rays that are not quite along the axis are kept in the glass fiber by total internal reflection.

Checkpoint 9 A certain glass has an index of refraction of 1.6 and is in air. What is the largest angle from the vertical for which a ray originating in the glass can emerge into the air? (For larger angles there will be total internal reflection.)

23.9 LENSES

Prisms: Refraction at Two Surfaces

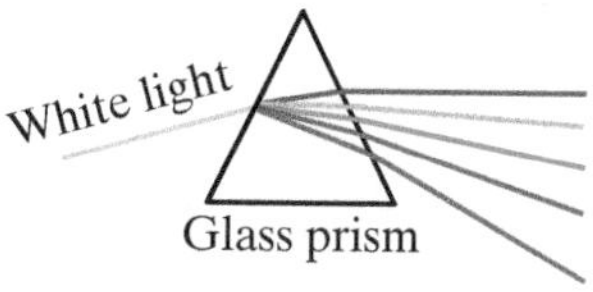

Figure 23.66 Because the index of refraction of some substances depends strongly on wavelength, light of different wavelengths is bent through significantly different angles when passing through a prism.

The index of refraction of most materials is not exactly the same for light of different wavelengths. This variation is due to differences in how light of different wavelengths interacts with the charged particles that compose the material. As a result of the wavelength dependence of index of refraction, a glass prism can be used to show that white light in fact contains a mixture of all colors of light, as shown schematically in Figure 23.66, where a few of the colors in white light are shown. Since the index of refraction depends on the wavelength, light of different wavelengths is bent by different amounts as it passes from air into glass, and again when it passes from glass back into the air.

Two Prisms Are Like a Lens

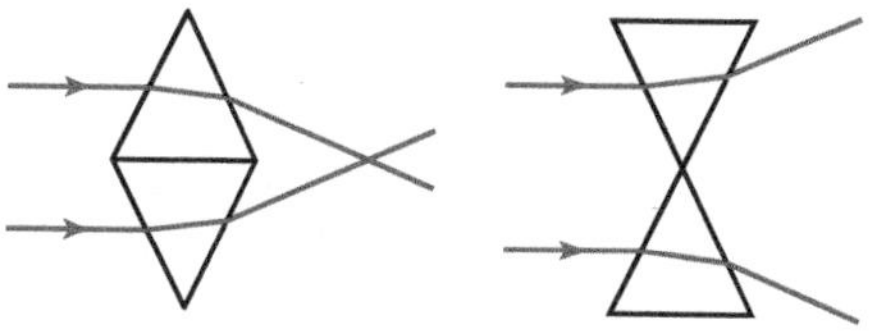

Figure 23.67 Two prisms can be used to make light rays converge or diverge.

If two prisms are placed base to base as shown at the left of Figure 23.67, light rays incident on the glass are bent toward the center line: they converge. If instead the two prisms are placed tip to tip as shown at the right of Figure 23.67, incoming rays are bent away from the center line: they diverge. These stacks of prisms, whose function is to bend incoming light rays, are like very simple lenses.

Lenses: Converging and Diverging

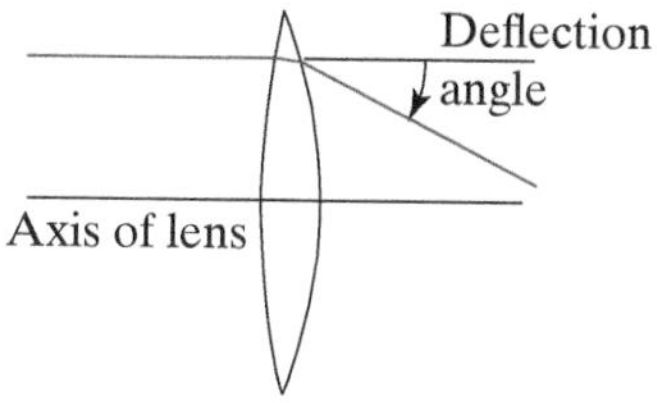

Figure 23.68 A converging lens deflects the incoming ray toward the axis of the lens.

Figure 23.68 shows the cross section of a lens made of transparent glass or plastic, whose index of refraction is greater than that of the air. The surfaces are spherical. A line through the center of the lens, perpendicular to the lens, is called the axis of the lens (or the "optic axis" of the lens). An incoming ray is refracted at the first spherical surface, and then again at the second spherical surface. The net effect is that the ray is deflected from its original direction toward the axis of the lens. We say this is a "converging" lens, because incoming rays "converge" toward the axis. Converging lenses are thicker in the middle than at the edge.

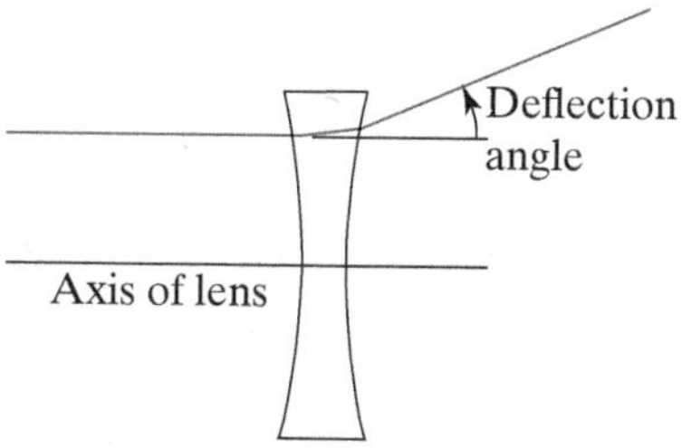

Figure 23.69 A diverging lens deflects the incoming ray away from the axis of the lens.

Figure 23.69 shows a "diverging" lens. Incoming rays are deflected away from the axis of the lens. Diverging lenses are thinner in the middle than at the edge. Figure 23.70 shows two more examples of lenses, one converging and one diverging.

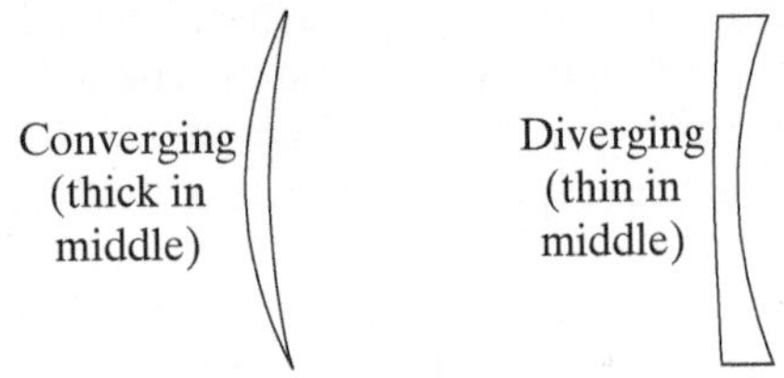

Figure 23.70 Additional examples of converging and diverging lenses.

Brief Review of Triangles

Sines, cosines, and tangents of small angles play a central role in understanding the details of how thin lenses bend rays of light. Before analyzing thin lenses we offer a very brief review. For the right triangle in Figure 23.71, here are the key relationships:

$$\sin\theta = \frac{y}{r} \qquad y = r\sin\theta$$

$$\cos\theta = \frac{x}{r} \qquad x = r\cos\theta$$

$$\tan\theta = \frac{y}{x} = \frac{\sin\theta}{\cos\theta} \qquad \theta + \phi = 90°$$

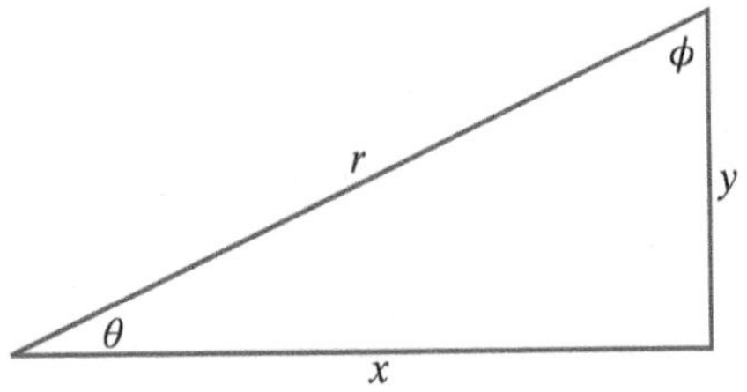

Figure 23.71 A right triangle.

The Small-Angle Approximation

Measuring angles in terms of radians plays an important role, especially when dealing with small angles. If an arc of a circle of radius R has a length s, the angle θ measured in radians is defined to be $\theta = s/R$. A full circle has an arc length of the circumference, $s = 2\pi R$, and the angle in radians for a full circle is $s/R = 2\pi$, or 6.28 rad in a circle. To convert radians to degrees, multiply the number of radians by $(360°)/(2\pi\text{ rad})$. For example, the number of degrees in one radian is

$$(1\text{ rad})\frac{(360°)}{(2\pi\text{ rad})} = \frac{360°}{2\pi} = 57.3°$$

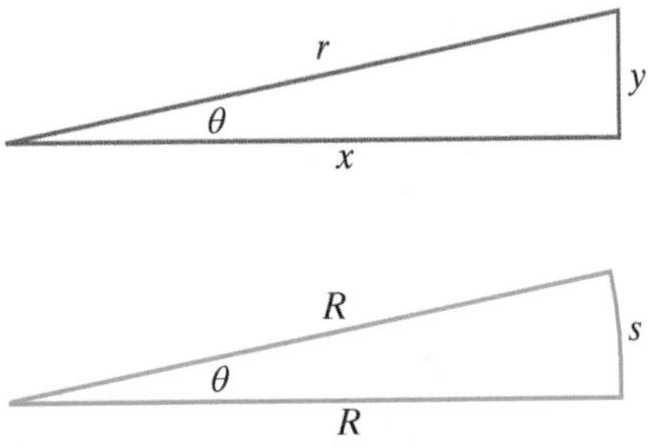

Figure 23.72 A right triangle (red) with a small angle θ, and a sector of a circle (green) with the same small angle θ.

In Figure 23.72 are two figures that look very similar: a right triangle (red) with a small angle θ, and a sector of a circle (green) with the same small angle θ. Evidently $s \approx y$ and $R \approx r$, so the sine, y/r, or tangent, y/x, of a small angle is nearly equal to the angle measured in radians, (s/R):

$$\tan\theta \approx \theta \quad \text{measured in radians}$$

$$\sin\theta \approx \theta \quad \text{measured in radians}$$

$$\cos\theta \approx 1 \quad \text{and} \quad x \approx R \text{ and } x \approx r$$

It is easy to verify the validity of these approximations using your calculator. For example:

> QUESTION 5° is $5(2\pi)/360° = 0.08727$ rad. Calculate $\tan 5°$ and $\sin 5°$.

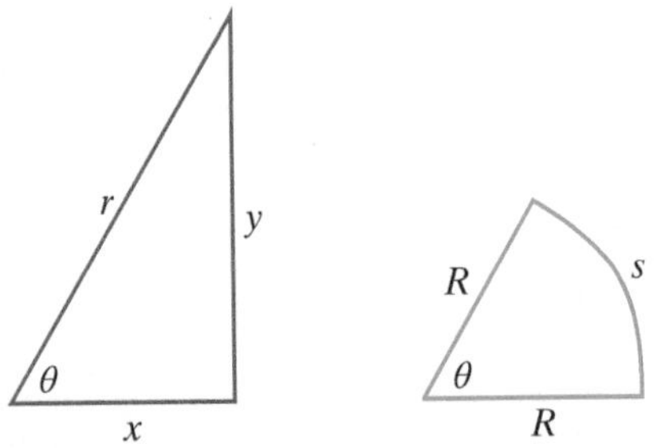

Figure 23.73 A right triangle (red) with a large angle θ, and a sector of a circle (green) with the same large angle θ.

You should find that $\tan 5°$ is 0.08749 and $\sin 5°$ is 0.08716; both of these numbers are approximately equal to 0.08727 rad. This approximation is called the "small-angle" approximation. As you can see, it is a very good approximation for small angles.

In contrast, in the large-angle triangle and sector shown in Figure 23.73, s and y are quite different in length, so $\theta = s/R$ is quite different from $\tan\theta = y/x$. Also, $\cos\theta$ is not approximately equal to 1, so x is not approximately equal to r, nor is $\sin\theta \approx \theta$.

Thin Lenses

If the y position of a light ray does not change very much as the ray travels through the lens, we call the lens a *thin lens*. Figure 23.74 is the output of a computer program that traced two rays through a thin converging lens, using Snell's law at both spherical surfaces. Notice that:

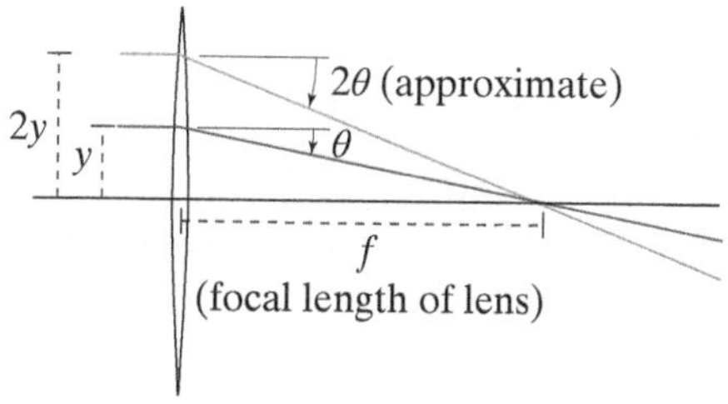

Figure 23.74 A ray that is twice as far from the axis is deflected twice as much (small-angle approximation).

- To a good approximation, the ray that is twice as far from the axis is deflected twice as much: Deflection is proportional to distance from the axis.
- Both rays converge to nearly the same point on the axis, a distance f from the center of the lens.

These results apply to all rays parallel to the axis that pass through thin lenses, not just the two rays computed in Figure 23.74.

The distance f is called the "focal length of the lens" and is a property of the radii of curvature of the two spherical surfaces and of the index of refraction of the material of the lens.

Deflection Angles for Thin Lenses

In Figure 23.74, consider the right triangle whose horizontal base is f and height is y. The angle opposite y is θ, so we have

$$\tan\theta = \frac{y}{f}$$

Now consider a second right triangle with the same base but with height $2y$; the angle opposite $2y$ is approximately 2θ:

$$\tan 2\theta = \frac{2y}{f}$$

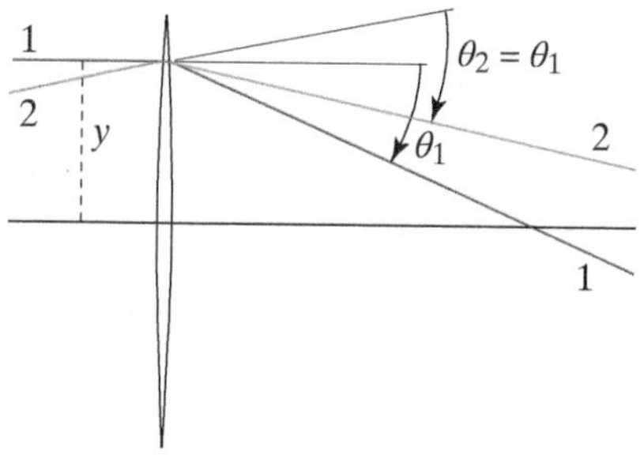

Figure 23.75 In the small-angle thin-lens approximation, all rays that hit the lens at a distance y from the axis are deflected through the same angle $\theta_1 y/f$.

Measuring angles in radians, if θ and 2θ are small:

$$\theta \approx \frac{y}{f}$$

$$2\theta \approx \frac{2y}{f}$$

We see that in the small-angle approximation, the deflection angle is simply $\theta \approx y/f$, where y is the distance from the axis where the ray hits the lens. Moreover, it doesn't matter what the angle of the incoming ray is. In Figure 23.75 two rays approach the same point on the lens but arrive along different angles to the axis. The computer ray tracing using Snell's law shows that the deflection angles are the same for these two rays. The approach angle doesn't affect the deflection angle, which is determined solely by the distance from the axis where the ray hits the lens.

THIN-LENS DEFLECTION ANGLES

> For small angles, the deflection angle for any ray passing through a thin lens is y/f, where y is the distance from the axis where the ray hits the lens, and f is the focal length of the lens.

To keep track of signs and directions, one says that a diverging lens has a negative focal length. Then the deflection angle $\theta \approx y/f$ is negative, and a negative deflection angle is interpreted as diverging away from the axis.

The Thin-Lens Equation

In the preceding discussion we noted that:

- A thin lens changes the angle of a ray by an amount proportional to how far from the axis the ray hits the lens.
- The deflection angle is y/f, where f is the focal length of the lens and y is the distance from the axis where the ray hits the lens.

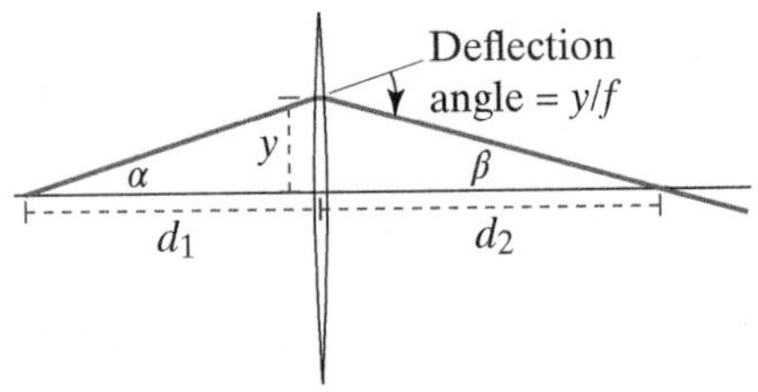

Figure 23.76 Proving the thin-lens equation.

From this basic property we can derive a "lens equation" that will be particularly useful in analyzing images, to be discussed in the next numbered section. In Figure 23.76 we consider a ray that starts from (or goes through) a point on the axis, a distance d_1 to the left of the (thin) lens and is deflected toward the axis, crossing the axis a distance d_2 from the lens. For clarity we have exaggerated the vertical dimension so that some angles look rather large, but in the small-angle approximation we have

$$\alpha \approx \tan\alpha = \frac{y}{d_1}$$

$$\beta \approx \tan\beta = \frac{y}{d_2}$$

The angles of the large triangle whose base is $d_1 + d_2$ add up to 180°, from which we deduce that the deflection angle y/f must be equal to $\alpha + \beta$. Using the results we just obtained for α and β, we have this:

$$\frac{y}{f} = \frac{y}{d_1} + \frac{y}{d_2}$$

Dividing through by y, we obtain the thin-lens equation:

THE THIN-LENS EQUATION

$$\frac{1}{f} = \frac{1}{d_1} + \frac{1}{d_2}$$

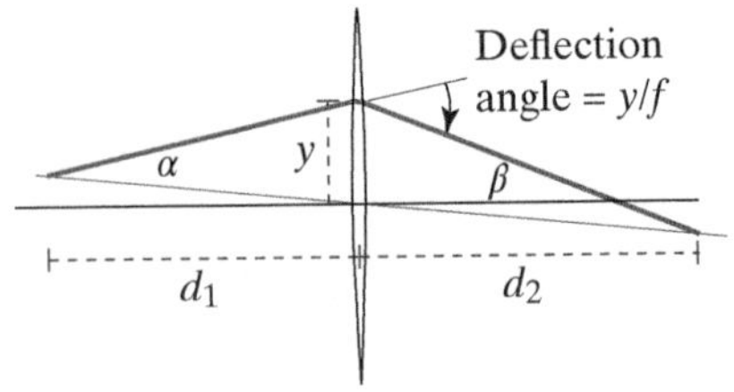

Figure 23.77 The thin-lens equation is also valid for rays that don't start on the axis.

When Rays Don't Start from the Axis

If we rotate the rays in Figure 23.76 through a small angle, as in Figure 23.77, the lens equation is still valid (in the small-angle approximation). The deflection angle is still the same, because that depends only on y, the distance from the axis where the ray hits the lens. Also, because the cosine of a small angle is nearly equal to 1, the distances along the axis of the lens hardly change when the source is a little off the axis. The result is that we can use the lens equation to find locations along the axis even for rays that don't start or end on the axis itself, as long as the angles are small.

Mathematical Proof: Two Approximations

We omit a mathematical proof that the deflection angle is in fact approximately equal to y/f, because it is algebraically and geometrically somewhat complicated. There are two approximations that must be made. One is the small-angle approximation. Another is the thin-lens approximation: the lens must be thin compared to its focal length. For clarity, in some of the figures the lenses are shown fatter than a "thin" lens would be.

Checkpoint 10 A small light bulb is 0.3 m to the left of a converging lens whose focal length is 0.2 m. How far to the right of the lens will rays from the bulb cross the axis?

23.10 IMAGE FORMATION

In this section we will discuss images that are made by lenses. An image is an *apparent* source of light. It is not a real light source, because there is no actual object at the location of an image.

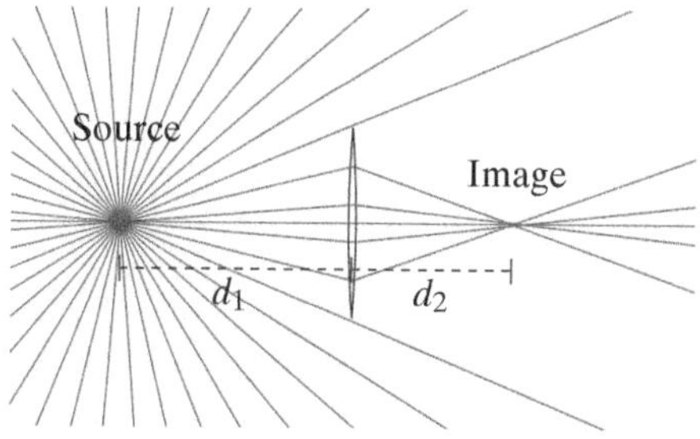

Figure 23.78 Forming a real image.

Real Images

In Figure 23.78 we see an example of the formation of what is called a "real image" of a point source (which could be a small light bulb). On the left, a small point-like lamp emits light in all directions. Those rays that hit the lens are deflected toward the axis; the other rays simply keep going straight. It is because the deflection angle y/f of a ray is proportional to how far from the axis the ray hits the lens (y) that those rays that hit the lens all pass (nearly) through the same point on the axis, a distance d_2 to the right of the lens.

The thin-lens equation

$$\frac{1}{f} = \frac{1}{d_1} + \frac{1}{d_2}$$

implies that all rays that start at a distance d_1 to the left of the lens will cross the axis at the same distance d_2 to the right of the lens. Of course these rays keep going, and they seem to be coming from this new point. In many ways this point acts just like a real source. If you stand farther to the right and look toward the lens, you will see what looks like a lamp at the new location.

We call this an "image," by which we mean an apparent source of light—apparent because there is no actual object at the location of the image. If you place a sheet of film at the location of the image, or an array of digital light sensors such as you find in a modern digital camera, you'll detect a spot of light at the location of the real image, because rays of light actually hit that spot. This kind of image is called a "real image" in contrast to a "virtual image," which we'll discuss soon.

A real image acts in some ways just like a real source, but there are differences. The real source emits light in all directions, but the image on the right emits light headed only to the right, in a cone whose angular size is determined by the diameter of the lens.

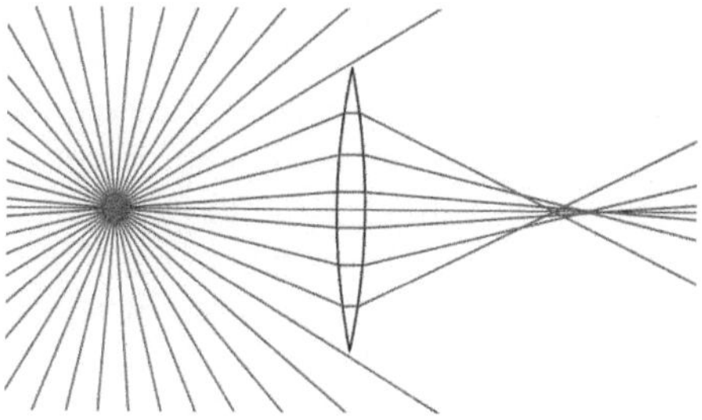

Figure 23.79 If the assumptions of a thin-lens and small angles are not valid, the predictions of the thin-lens model are inadequate. Here we see that not all rays converge to quite the same point. This is "spherical aberration."

Aberrations: Limitations of the Simple Thin-Lens Model

If we take a closer look, we can see limitations of the thin-lens model, due to the approximations of a thin lens and small angles. Figure 23.79 is again a computer calculation using Snell's law at the two spherical lens surfaces, with a somewhat thick lens and somewhat large angles. The rays do come to nearly the same point, but not quite. We say that this lens has significant "aberration" in its image formation. Note that it is the largest angles that deviate the most, reflecting the role of the small-angle approximation in our simplified model. It is a challenge to designers of optical systems to reduce aberrations to a minimum. The aberration associated with spherical lens surfaces is called "spherical aberration," and this particular kind of aberration can be minimized by making lens surfaces that are more complicated than spherical ones.

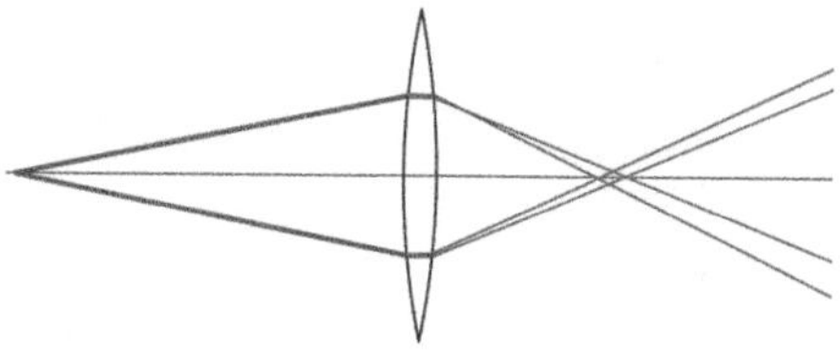

Figure 23.80 Because the index of refraction is typically larger for blue light than red light, a lens bends blue light more than red light, giving rise to "chromatic aberration."

Another kind of aberration is due to the fact that the index of refraction of a lens varies with the wavelength of the light, so that the focal length f of the lens is different for red light and blue light. The index of refraction in glass for blue light is larger than the index for red light, so blue light is bent more than red light. If a source emits a mixture of red and blue light the blue image will be closer to the lens than the red image, as shown in Figure 23.80. A white-light source, which emits a wide range of colors between red and blue, will form images that lie between the red and blue images.

This aberration associated with color is called "chromatic aberration." It can be minimized by combining two lenses made of different kinds of glass (with different indices of refraction), with different focal lengths, which partially compensate for each other's aberrations. Such a lens is called "achromatic" (Greek for "without color"), though the correction isn't perfect. One of the advantages of curved mirrors used in large telescopes is that angles of reflection depend only on the shape of the mirror, not on the color of the light, so reflecting telescopes are truly achromatic.

Virtual Images

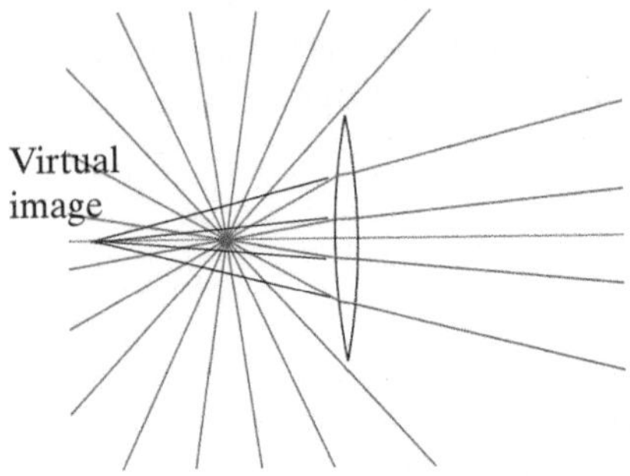

Figure 23.81 Forming a virtual image. To an observer on the right side of the lens, light appears to be coming from the location of the virtual image.

In Figure 23.81 we've traced rays through a weaker lens (it has a smaller index of refraction, though still larger than that of air) so that there isn't enough bending of the rays to bring them to a point on the axis to the right of the lens. If we follow these rays back they *seem* to come from a point to the left of the actual source, but they don't really come from there. If you go to that place, you won't find any light headed to the right from there. Contrast this virtual image with the real image in Figure 23.78. There are real rays of light emanating from a real image, and if you go to that place, you'll find those rays of light. If, however, you put detectors at the location of the virtual image in Figure 23.81 you won't detect a spot of light, because there are no actual rays of light converging to that point or emanating from it. In fact, in the situation shown, where the lamp emits light in all directions, your entire detector will be illuminated rather than one point.

This is not to say that virtual images are somehow not important; they are. In many optical instruments from magnifying glasses to microscopes, it is often a virtual image that we observe with our eyes. However, if you follow the rays all the way backward, there is no actual source of light at the point from which the rays seem to emanate.

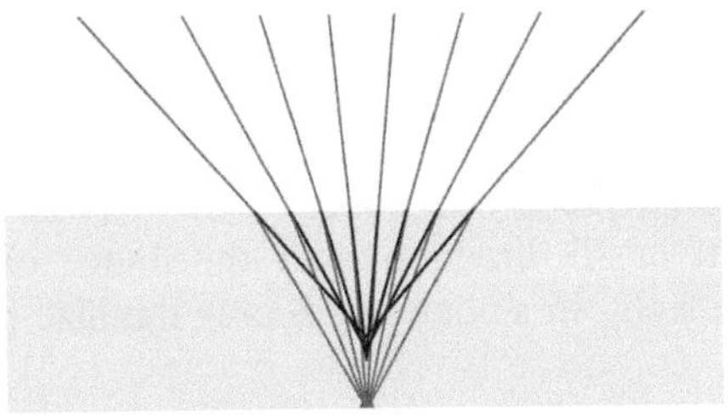

Figure 23.82 Rays emerging from the water seem to be coming from a location above the actual source. This location is a virtual image.

In Figure 23.82 you can see another example of a virtual image. The rays of light emanating from an object underwater seem to be coming from a location above the actual source of the light. There is a virtual image of the object at this higher location. Your eyes are fooled by the fact that the light rays you see seem to be radiating from the image point rather than from the actual point. This effect can be seen by looking at a coin at the bottom of a pan full of water; the coin seems to be higher than expected.

If you look carefully at Figure 23.82 you can see that the imaging isn't perfect: the rays seem to come from a small region rather than from a point. This is another example of aberrations. Typically, the wider the range of ray angles in an optical system, the worse the aberrations and the more difficult it is to compensate for the aberrations in a more complicated design. In Figure 23.83 you can see that if we include an even wider range of rays the virtual image is even less point-like. Also, you can see that if you look at a coin in a pan of water from off to the side, the bundle of rays that pass through your eye seem to come from a higher location than when you look straight down at the coin. This effect is easily observable.

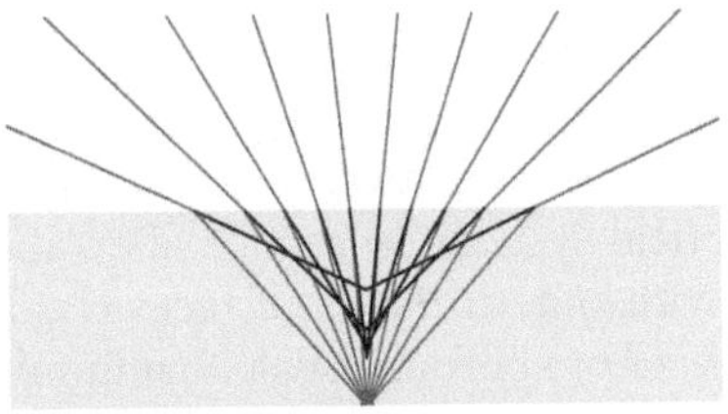

Figure 23.83 The wider the range of angles of the rays, the less point-like the virtual image.

Images of Extended Objects

So far we've been analyzing the images of single point sources. Next we'll consider the image of an extended object, a short stick that has a red lamp at one end and a blue lamp at the other. Figure 23.84 traces rays from the two ends of the stick.

You see two real images corresponding to the two lamps, one red and one blue. Note that there is an inversion: the upper (blue) source forms the lower image. If you place an array of digital light detectors, like those found in a camera, where the real images are you will get two spots, with the blue spot below the red spot. The inversion is typical of real images. If you have a converging lens available, hold it away from your eye and look through it at a distant object. As long as the distance from the lens to your eye is longer than the focal length of the lens, you'll see an inverted image.

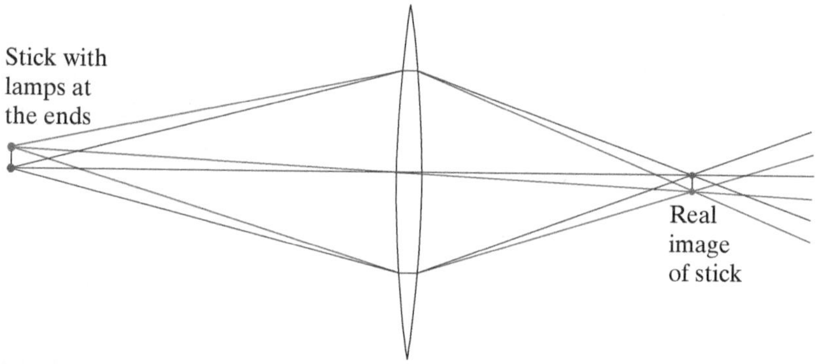

Figure 23.84 Two point sources of light at the end of a stick, a red lamp and a blue lamp, produce two real images (only a few rays are shown for clarity). Note the inversion: the upper, blue source forms the lower image and the lower, red source forms the upper image.

Focal Point

Let us apply the thin-lens equation $1/f = 1/d_1 + 1/d_2$ to the situation of a source far from the lens.

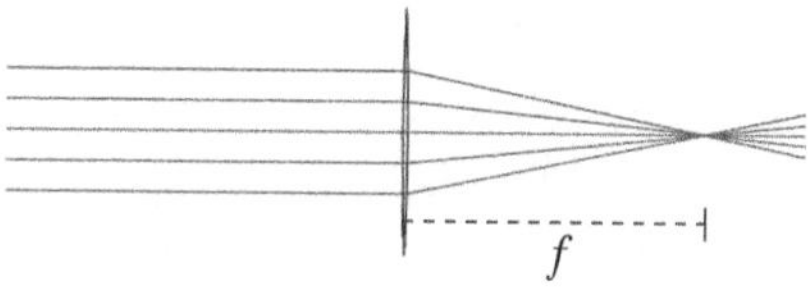

Figure 23.85 Rays from far away are nearly parallel and pass through the focal point.

> QUESTION If a source is very far away, rays hit the lens nearly parallel to the axis. In this case (source is very far away), approximately how big is the term $1/d_1$?

If d_1 is very large, $1/d_1$ is nearly zero, and the lens equation reduces to $1/f = 1/d_2$, so $d_2 = f$. Rays parallel to the axis, or nearly parallel when coming from a very distant source, are deflected in such a way as to pass through a point a distance f beyond a converging lens; this point is called a focal point of the lens (Figure 23.85).

DEFINITION OF FOCAL POINT

> The two points a distance f along the axis from the lens are called the "focal points" of a lens. They are the points on the axis through which initially parallel rays pass after being deflected by the lens.

Ray Tracing: Determining Image Locations Graphically

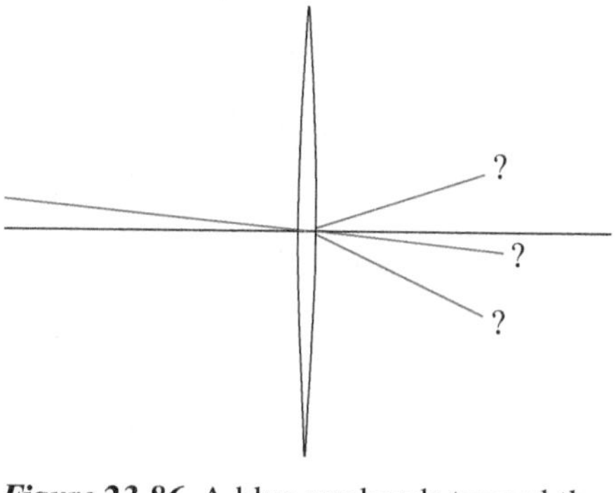

Figure 23.86 A blue ray heads toward the center of a lens. How will it be deflected?

Since the red lamp in Figure 23.84 lies on the axis, we know that its image will also lie on the axis—but what about the blue image?

We'll show how to determine the location of the blue image using graphical analysis. If we trace the paths of two or more different rays as they leave the source and pass through the lens, these deflected rays will intersect at the location of the image. Of the many rays we might draw in a diagram, there are two rays that are particularly easy to draw.

(a) First, draw a ray from the blue lamp toward the center of the lens (Figure 23.86).

> QUESTION How much will this ray be deflected in passing through the lens? How do you know?

Since the deflection angle for a ray that hits the lens at a perpendicular distance y from the axis is y/f, and since $y = 0$ for the ray you drew, the deflection will be zero. In the small-angle, thin-lens approximation, a ray that hits the center of a lens goes straight through without deflection, no matter what the angle of the ray is.

(b) Second, draw a ray that starts out parallel to the axis.

The path of this ray is also easy to determine. A ray that starts out parallel to the axis is deflected by a converging lens in such a way as to pass through a point on the axis to the right of the lens that is a distance f from the lens. This is shown in Figure 23.87, the key point being that the deflection of the parallel ray is y/f (small-angle approximation). The triangle whose base is f and height is y has an angle whose tangent (and, approximately, the angle in radians) is y/f, equal to the deflection angle.

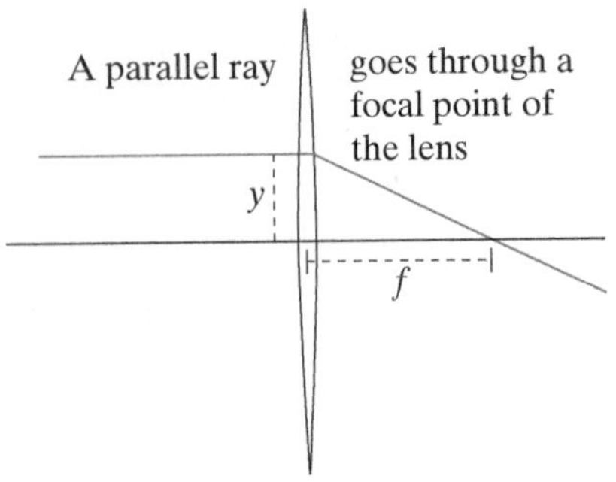

Figure 23.87 A parallel ray is deflected to cross the axis at a distance f from the lens. This is one of the two "focal points" of the lens.

The two rays that we drew intersect at the image location (Figure 23.88). We drew only two of the rays that leave the blue lamp. All other rays from the lamp that go through the lens will also go through the location of the image of the lamp, in the thin-lens, small-angle approximation. To the extent that these approximations are not quite valid, the many rays will pass nearly but not exactly through the same point, and the image of the lamp won't be a nice clean, compact point; it will be somewhat smeared out.

Determining Image Locations Algebraically

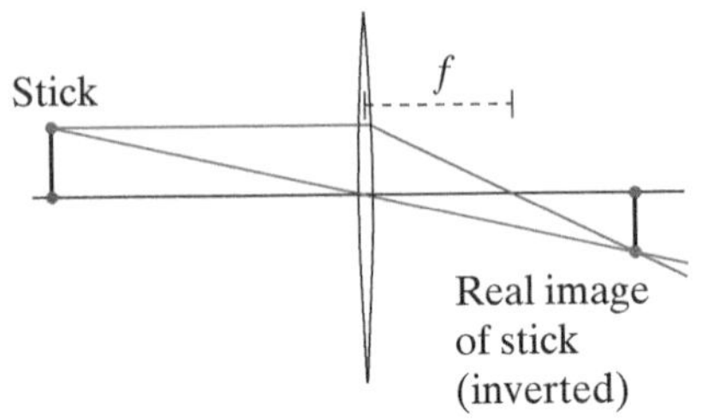

Figure 23.88 Draw the two "easy" rays from the blue lamp through the lens. The intersection is the location of the real image of the blue lamp.

How can we calculate the locations of the red and blue real images quantitatively? Use the lens equation, $1/f = 1/d_1 + 1/d_2$, to find the distance along the axis of these images. The lens equation is based on small-angle and thin-lens approximations, where all angles are assumed to be small compared to 1 rad. For that reason, the height of the blue lamp above the axis of the lens doesn't affect our use of the lens equation; we simply use the distances along the axis for d_1 and d_2. If we have a lens with known focal length f, and the red and blue lamps are a distance d_1 from the lens, we can calculate how far along the axis the images are from the lens: $1/d_2 = 1/f - 1/d_1$.

You know that all the rays from the blue lamp that hit the lens get deflected in such a way as to pass through the (real) image. One of those rays on its way to the image passes through the center of the lens, without deflection. You also know the distance along the axis d_2 of the blue image. This is enough information to determine the location of the blue image.

EXAMPLE **Using Algebra to Locate an Image**

Suppose that a converging lens is located at the origin $\langle 0,0,0 \rangle$ and has a focal length of 12 cm, with its axis pointing along the x axis. There is a stick 12.5 cm to the left of the lens. At one end of the stick is a red lamp, located at $\langle -12.5,0,0 \rangle$ cm, and at the other end is a blue lamp 1 cm above the red lamp, at $\langle -12.5,1,0 \rangle$ cm (Figure 23.89). Find the locations of the images of the red and blue lamps; the complete image of the stick will lie between the images of the two lamps that are at the ends of the stick.

Solution For the x coordinate of the image, use the lens equation:

$$\frac{1}{12} = \frac{1}{12.5} + \frac{1}{d_2}$$
$$d_2 = 300\,\text{cm}$$

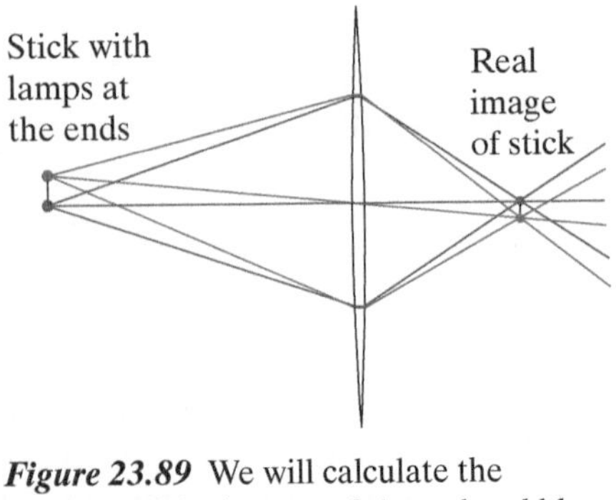

Figure 23.89 We will calculate the location of the images of the red and blue lamps, at the ends of the stick.

For the y coordinate of the image, we already know that the image of the red lamp lies on the x axis. The image location of the red lamp is $\langle 300,0,0 \rangle$ cm.

For the blue lamp, the ray passing through the center of the lens is undeflected. Therefore the slope of the ray is the same to the left and right of the lens:

$$\frac{-1}{12.5} = \frac{y_2}{300}$$
$$y_2 = -24\,\text{cm}$$

The blue lamp was only 1 cm above the axis and 12.5 cm along the axis from the lens, but its real image is 24 cm below the axis, 300 cm along the axis from the lens. As a vector, the image location of the blue lamp is $\langle 300, -24, 0\rangle$ cm. The complete image of the stick lies between $\langle 300, 0, 0\rangle$ cm and $\langle 300, -24, 0\rangle$ cm. The image is 24 times as big as the actual object. This factor is called the "magnification."

We didn't need to convert to meters in doing the analysis, since all the quantities were given in centimeters.

When the distance of a lamp from a lens is only slightly larger than the focal length f of the lens, as it is in this case (12.5 cm is only slightly larger than 12 cm), the image is very far from the lens (300 cm or 3 m in this case). Evidently Figure 23.89 is not to scale!

Computer and Movie Projectors

The effect we've just seen, that a source near the lens can make a real image far from the lens that is greatly enlarged, is the basis for computer and movie projection. A small glowing or illuminated object is placed a distance slightly greater than f from a converging lens, and an image of the object is displayed on a screen that in this case must be placed 300 cm from the lens. If the screen is not accurately placed 300 cm from the lens, the red and blue rays won't converge to a point on the screen, and there will be big blobs of red light and blue light on the misplaced screen. To put it in more practical terms, if the screen is placed 300 cm away but the stick and the lens aren't accurately positioned 12.5 cm apart, the image will be out of focus, with points turning into blobs. Projectors have a lens held in a screw mount to make it easy to focus the image on the screen by changing the object-to-lens distance simply by turning the lens in the screw mount.

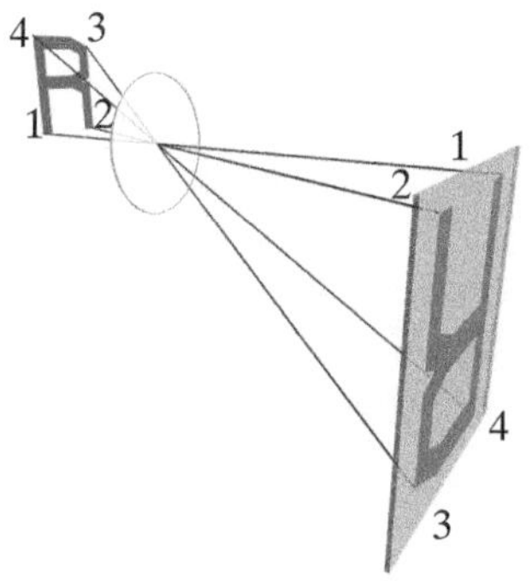

Figure 23.90 A lens makes a real image of an illuminated R on a sheet of ground glass. Rays from four numbered corners of the R are drawn through the center of the lens to show why the image is inverted.

Each illuminated point in the object is represented by a dot of light on the screen, making a complete picture of the object. The image is inverted, so the object must be mounted upside down in order to make a screen display that is right side up. In Figure 23.90 a lens makes a real image of an illuminated R on a sheet of ground glass. Rays are drawn from four numbered corners of the R through the center of the lens to the image. Study the Figure carefully to understand why the image is inverted.

> **Checkpoint 11** An object 5 cm tall is 20 cm to the left of a converging lens, which forms a real image of the object on a screen 500 cm to the right of the lens. **(a)** How tall is the image on the screen? **(b)** What is the magnification? **(c)** What is the focal length of this lens?

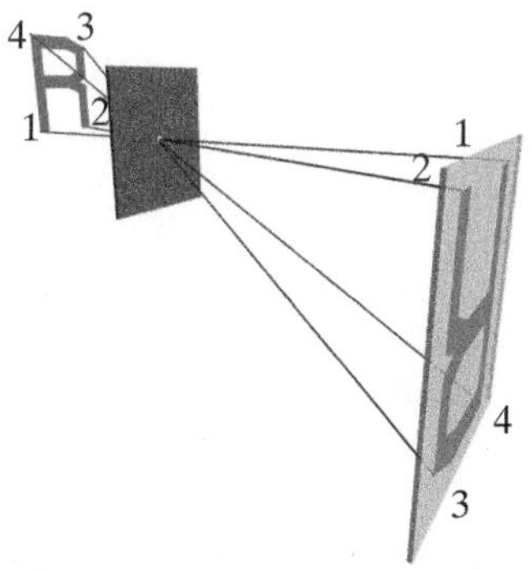

Figure 23.91 A pinhole makes a real image. The image would actually be very dim.

A Lens Gathers Light

In addition to forming images by bending rays, a major function of a lens is to gather light. In Figure 23.90, imagine replacing the lens by a sheet of cardboard with a pinhole at its center.

> QUESTION Would there still be an (inverted) image of an R on the ground glass?

Yes, because the rays traced in Figure 23.90 would still go through the pinhole to the same places on the glass as before. However, the image would be very dim, because very little light from any source point could pass through the pinhole. Figure 23.91 shows an image made by a pinhole. For clarity the inverted R is shown much brighter than it would really be.

In contrast, with the lens in place, a large bundle of rays from one source point is focused to one corresponding point on the ground glass, so the image is much brighter. Remember that in Figure 23.90 we show only those single rays from the four source points that go through the center of the lens; a large bundle of rays from each source point passes through the full area of the lens, but for clarity these rays are not shown in the figure.

The larger the lens, the brighter the image, though there is a practical limit, because the larger the lens, the larger the angles of the rays, and at some point the aberrations for a spherical lens would distort the image too much because the small-angle approximation would no longer be valid. (Moreover, the larger the lens, the more expensive it is to manufacture.)

There is another difference between a pinhole and a lens. With a pinhole, there will be a (dim) image on the ground glass no matter how far from the pinhole the glass is positioned. In contrast, a lens will focus a bundle of rays from one source point to one corresponding point on the glass only if the glass is placed at the correct distance from the lens, which we have calculated using the thin-lens equation.

The light-gathering property of a lens is indicated by the "*f*-number," which is the ratio of the focal length to the diameter of the lens, f/D; a larger diameter means a smaller *f*-number. The amount of light collected is proportional to the area of the lens and so is proportional to the diameter squared and to the inverse square of the *f*-number. Compare *f*-numbers of 2.8 and 4.0: a lens with an *f*-number of 2.8 will collect $(4.0/2.8)^2 = 2$ times as much light as a lens with an *f*-number of 4.0.

Cameras often have an adjustable circular aperture, called a "diaphragm," for reducing the effective size of the lens when the scene is too bright and one needs to reduce the brightness of the image. One sets the diaphragm by specifying the effective *f*-number. Reducing the effective size of the lens by making the opening in the diaphragm smaller (increasing the *f*-number) also reduces aberrations by reducing the maximum angles of rays that are accepted, thereby improving image quality.

Parallel Beams of Light

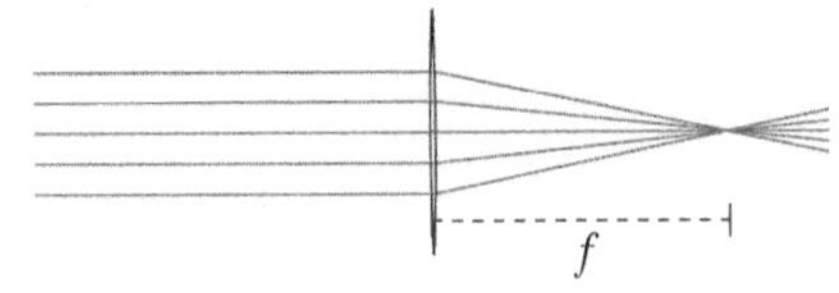

Figure 23.92 Parallel rays converge at the focal point, a distance *f* to the right of the lens.

An important special case is the image made by an object that is very far away from the lens. In the lens equation $1/f = 1/d_1 + 1/d_2$, if d_1 is very large, $1/d_1$ is very small compared to the other terms, so approximately we have $1/f = 1/d_2$, or $d_2 = f$; all the rays from the distant object pass through a point a distance f beyond the lens. The incoming rays are nearly parallel. Figure 23.92 shows this situation. A ray that hits the lens at a distance y from the axis is deflected through an angle y/f, and since the incoming ray is parallel to the axis, it will cross the axis a distance f beyond the lens.

A practical application of this is the concentration of solar energy. The Sun is so far away that rays from the center of the Sun are nearly parallel as they hit a lens, and they get concentrated at the focal point, a distance f beyond the lens. Rays from one edge of the Sun are also nearly parallel but at a slightly different angle (Figure 23.93). The radius of the Sun is 7×10^8 m, and its distance from the Earth is 1.5×10^{11} m, so the angular width corresponding to the radius of the Sun is $7 \times 10^8/1.5 \times 10^{11} = 4.7 \times 10^{-3}$ radians (about one-quarter of a degree).

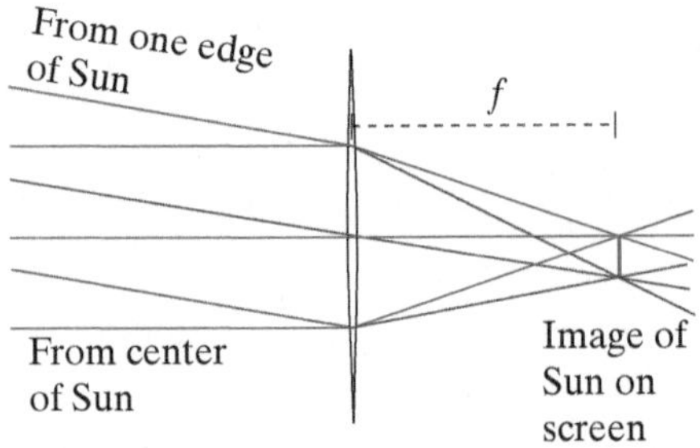

Figure 23.93 Nearly parallel rays arriving from the center of the Sun, and another set of nearly parallel rays arriving from the edge of the Sun. A real image of the Sun is formed on a screen.

Suppose that the focal length of the lens is $f = 20$ cm. Place a screen 20 cm behind the lens, so that the incoming rays will focus onto the screen. A ray that goes from the edge of the Sun through the center of the lens will be displaced $(4.7 \times 10^{-3})(20) = 9.4 \times 10^{-2}$ cm from the axis of the lens when it hits the screen. That is, the (real) image of the Sun on the screen will have a radius of 9.4×10^{-2} cm. If the lens has a radius of 5 cm, the effect is to concentrate solar energy from an area of $\pi r^2 = \pi(5)^2$ cm^2 into an area of $\pi y^2 = \pi(9.4 \times 10^{-2})^2$ cm^2. The energy concentration factor is $(5/(9.4 \times 10^{-2}))^2 \approx 3000$! The spot on the screen gets very hot.

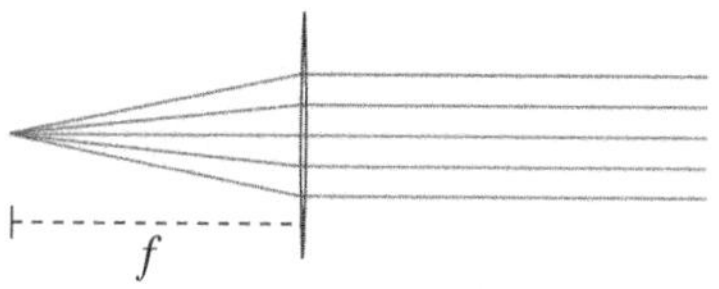

Figure 23.94 Rays emitted from a focal point of a converging lens produce a parallel beam.

A converging lens can also be used to make a parallel beam of light from a point source. In Figure 23.94 we see that if we place a point source at a focal point, a distance f from a converging lens, all rays that hit the lens will emerge parallel to the axis, because the deflection angle is y/f and the angle of the incoming ray is also y/f.

It's interesting to see what the lens equation predicts for this case, where d_1 is equal to the focal length f: $1/f = 1/f + 1/d_2$, so $1/d_2$ is zero or approximately zero.

QUESTION If $1/d_2$ is nearly zero, what does this say about d_2?

For $1/d_2$ to be very small, d_2 must be very large. If we place the source close to f from the lens, the place where the rays cross after passing through the lens will be at a very great distance. In the limit, those rays will cross only at infinity. That is, these rays are parallel.

Analyzing Virtual Images

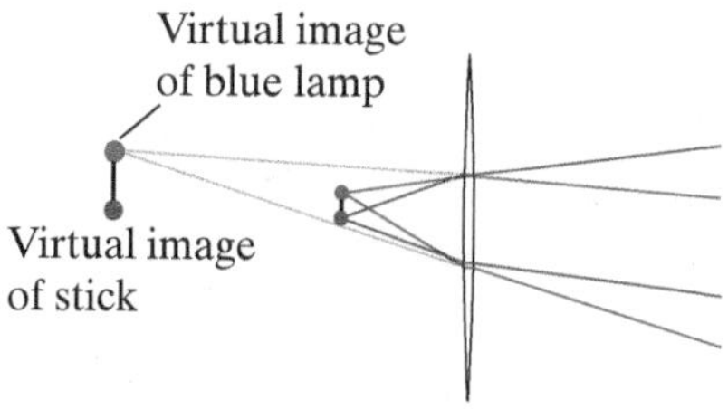

Figure 23.95 A virtual image of a stick, which is larger than the actual stick and right side up.

We've discussed how to predict where a real image will be formed. The prediction of the location of a virtual image is a little trickier, because no rays actually go through the virtual image. Rather, the lens deflects rays in such a way that those rays *seem* to come from an image, as shown in Figure 23.95. The object is again a short stick with a red lamp at one end and a blue lamp at the other. The virtual image of the stick is larger than the actual stick, and right side up.

First we'll analyze this situation graphically. As before, there are two rays whose deflections are easy to determine. A ray from the blue lamp that hits the center of the lens will pass straight through, with zero deflection (Figure 23.96). A ray from the blue lamp that starts out parallel to the axis is deflected by the converging lens in such a way as to pass through a point on the axis to the right of the lens that is a distance f from the lens. Draw these two particularly "easy" rays, the one through the center of the lens and the one that starts parallel to the axis and crosses the axis a distance f to the right of the lens. Follow these rays back to the point where they intersect; this is the location of the virtual image of the blue lamp. All other rays (that don't miss the lens) will also seem to come from the location of the image, in the thin-lens, small-angle approximation. The virtual image is upright, not inverted.

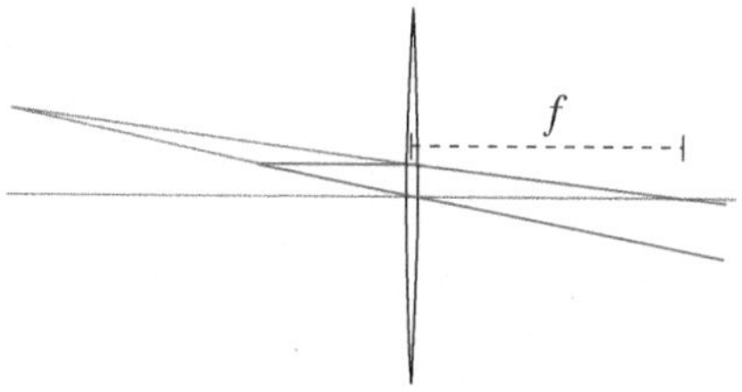

Figure 23.96 "Easy" rays emitted by the blue lamp: one that goes through the center of the lens, undeflected, and one that goes parallel to the axis then through the focal point.

Next, let's calculate quantitatively the location of the virtual image of the stick:

The converging lens is located at the origin $\langle 0,0,0\rangle$ and has a focal length of 12 cm, with its axis pointing along the x axis. The red lamp is located at $\langle -7,0,0\rangle$ cm and the blue lamp is 1 cm above the red lamp, at $\langle -7,1,0\rangle$ cm. Find the locations of the images of the red and blue lamps; the complete image of the stick will lie between the images of the two lamps that are at the ends of the stick.

1. Lens equation: $1/12 = 1/7 + 1/d_2$, which yields $d_2 = -16.8$ cm.

QUESTION Can you guess the physical significance of this distance turning out to be negative?

It means that the location along the axis is to the *left* of the lens, on the same side as the source. To the right of the lens, rays seem to be diverging from a location to the left of the actual source of light. The location of the virtual image of the red lamp is $\langle -16.8,0,0\rangle$ cm.

2. Ray through center of lens: On the lamp side of the lens the slope of this ray coming from the blue lamp is $-1/7$, and on the right side of the lens the slope is the same (no deflection). This is one of the rays that seem to be coming from the virtual source, so we follow it back to a location whose distance

along the axis is 16.8 cm to the left of the lens. The distance from the axis is 16.8 times the slope, which is (16.8)(1/7) = 2.4 cm above the axis. The vector position of the virtual image of the blue lamp is $\langle -16.8, 2.4, 0\rangle$ cm.

A Magnifying Glass

This particular arrangement we've just discussed is an example of a magnifying glass. If you look at a stick 1 cm long from a distance of 30 cm (a typical viewing distance), the angular width (the angle "subtended" by the stick) is 1/30 rad, or about 0.03 rad. Place the lens 7 cm in front of the stick, and the angular width is now 1/7 = 0.14 rad. That is, the difference in angle of the central ray coming from one end of the stick and the central ray coming from the other end of the stick is 1/7, or 0.14 rad. Without the lens, you would have had to be 0.14/0.03 = 4.3 times closer for the stick to appear to have that angular width in your field of view, and you might not be able to focus on the object if your eye were that close to it. We say that the magnification of the object is 4.3. The virtual image of the stick appears longer.

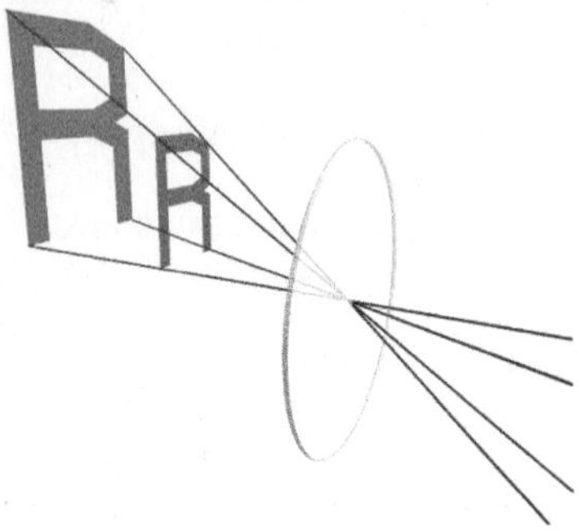

Figure 23.97 A converging lens makes an enlarged, right-side-up virtual image of an illuminated R. Rays from four corners of the R are drawn through the center of the lens to show why the image is right side up and larger.

If you have a converging lens available, hold it away from your eye and look through it at a nearby object. As long as the distance from the lens to the object is less than the focal length of the lens, you'll see a virtual image, which is right side up, as is typical of virtual images. Figure 23.97 shows the situation in 3D. Rays from the corners of the illuminated R that go through the center of the lens seem to be coming from the corners of the large right side up virtual image of the R.

Diverging Lenses

Another way to make a virtual image is with a diverging lens. In fact, because a diverging lens cannot make rays converge toward the axis, this kind of lens cannot make real images. In contrast, we've seen that a converging lens can produce either real or virtual images, depending on where the source is located.

In analyzing the effects of a diverging lens, two of the rays from a source are again particularly easy to draw. A ray that hits the center of the lens passes through without deflection. A ray that starts out parallel to the axis and hits the lens at a distance y from the axis is deflected *outward, away* from the axis, through a deflection angle $y/|f|$. To avoid having to use absolute-value notation, it is standard practice to say that a diverging lens has a negative focal length. That is, f is a negative number. Note that the virtual image of the stick is smaller than the stick and is right side up; if you have a diverging lens you can see these two effects simply by looking at an object through the lens.

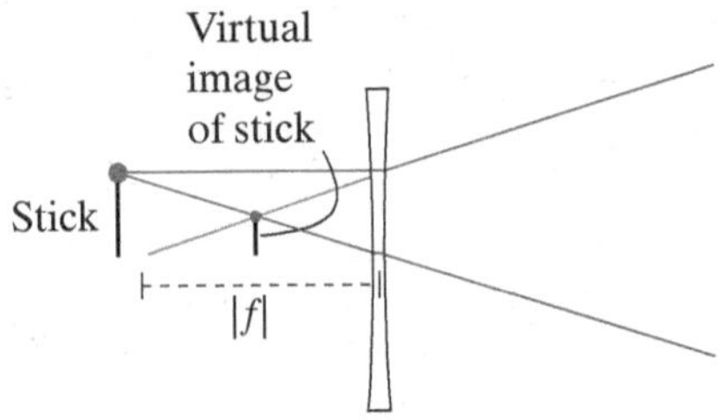

Figure 23.98 "Easy" rays emitted by the blue lamp: one that goes through the center of the lens, undeflected, and one that goes parallel to the axis, then is deflected outward through an angle that looks like it is coming from the focal point on the left.

In Figure 23.98 we see that for an initially parallel ray we draw the outgoing ray as though it came from a location a distance $|f|$ to the left of the diverging lens, whereas with a converging lens a parallel ray goes through the axis at a distance f to the right of the lens. Equivalently, we can say that for both kinds of lens an initially parallel ray crosses (or seems to cross) the axis at a location f to the right of the lens, where f has a negative value for a diverging lens and hence is to the left of the lens.

This sign convention carries over to the lens equation, which we continue to write as $1/f = 1/d_1 + 1/d_2$, but where f is a negative number for a diverging lens. We already saw that d_1 or d_2 could also be negative numbers.

Sign Conventions

Let's summarize the sign conventions that are implicit in the thin-lens equation $1/f = 1/d_1 + 1/d_2$:

- f is positive for a converging lens, negative for a diverging lens.

- We measure d_1 from the source to the lens and we measure d_2 from the lens to the image.
- The distances d_1 and d_2 are considered positive if they are in the same direction that the light is traveling and negative otherwise. Both d_1 and d_2 are positive for the simple case of a distant source and converging lens.

With these sign conventions the thin-lens equation applies to all combinations of thin lenses (and spherical mirrors) and all combinations of source or image locations. Implicit in these conventions is that if the source is to the left of the lens and d_2 is positive, the image is to the right of the lens and is real and inverted, whereas if d_2 is negative, the image seems to be to the left of the lens and is virtual and right side up.

What does it mean for d_1 to be negative? Suppose that lens 1 would make a real image 20 cm to the right of lens 2 if lens 2 were not there. The effect of lens 2 can be calculated by setting d_1 to -20 cm: the "source" can be considered to be a real image that would be to the right of lens 2.

EXAMPLE **Sign Conventions: A Converging Lens**

A point source of light is located to the left of a converging lens of focal length $+10$ cm. **(a)** If the source is 12 cm to the left of the lens, where is the image? **(b)** Is the image real or virtual? **(c)** The source is moved to be 8 cm to the left of the lens; now where is the image? **(d)** Is this image real or virtual?

Solution **(a)** The thin-lens equation is $1/10 = 1/12 + 1/d_2$, where we have used centimeters for f and d_1, so our answer will be in centimeters. Solving, $d_2 = 60$ cm.

(b) Because d_2 is positive, rays have converged to a point 60 cm to the right of the lens, where there is a real image from which light rays originate. If a screen is placed 60 cm to the right of the lens an inverted image will be seen on the screen.

(c) $1/10 = 1/8 + 1/d_2$, so $d_2 = -40$ cm.

(d) Because d_2 is negative, the image is to the left of the lens and is a virtual image from which light rays *seem* to originate, but in fact no rays come from there. If a screen is placed 40 cm to the left of the lens, no image will be seen on the screen.

In part (a) the source is to the left of the left focal point (12 cm to the left of the lens, 2 cm to the left of the focal point) and a real image is formed. In part (b) the source is to the right of the left focal point (8 cm to the left of the lens, 2 cm to the right of the focal point) and there is a virtual image. This is a general result that emerges from the thin-lens equation: sources farther than f from a converging lens form real images. Sources closer to a converging lens than f form virtual images.

EXAMPLE **Sign Conventions: A Diverging Lens**

A point source of light is located to the left of a diverging lens of focal length -10 cm. **(a)** If the source is 12 cm to the left of the lens, where is the image? **(b)** Is the image real or virtual? **(c)** The source is moved to be 8 cm to the left of the lens; now where is the image? **(d)** Is this image real or virtual?

Solution **(a)** The thin-lens equation is $1/(-10) = 1/12 + 1/d_2$. Solving, $d_2 = -5.45$ cm.

(b) Because d_2 is negative, the image is to the left of the lens and is a virtual image from which light rays *seem* to originate, but in fact no rays come from there. If a screen is placed −5.45 cm to the left of the lens, no image will be seen on the screen. Looking through the lens from the right, you would see a (virtual) image that seems to be 5.45 cm behind the lens.

(c) $1/(-10) = 1/8 + 1/d_2$, so $d_2 = -4.44$ cm.

(d) Because d_2 is negative, there is again a virtual image to the left of the lens, at a location not very different from what we found in part (a).

A single diverging lens always forms a virtual image and cannot form a real image, no matter where the source is placed.

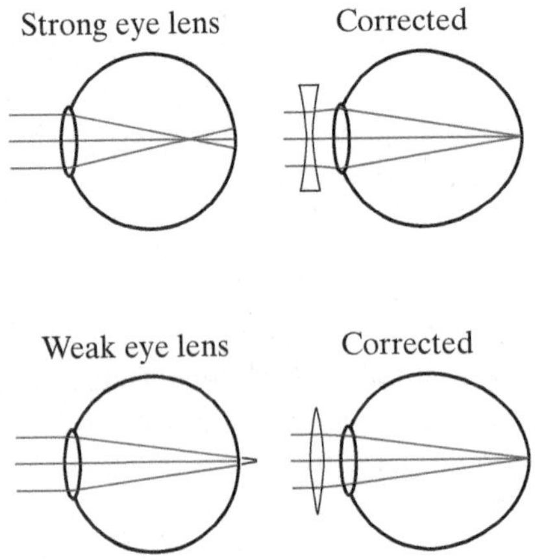

Figure 23.99 An eye lens that is too strong or too weak fails to focus rays onto the retina at the back of the eye. A diverging or converging lens can correct this.

Eyeglasses and Contact Lenses

An important use of diverging lenses is to correct defective vision (Figure 23.99). If the lens in the eye is too strong and bends rays too much, rays coming from a point source will form a (real) image in front of the retina, and the retina will detect a blob of light rather than a point. A diverging lens bends rays away from the axis and can compensate for too much bending toward the axis by the eye lens. The effect is to move the real image from in front of the retina onto the retina.

In contrast, if the eye lens is too weak and doesn't bend rays enough, rays coming from a point source would form a (real) image behind the retina, and the retina will detect a blob of light rather than a point. A converging lens, which bends rays toward the axis, can add to the bending by the eye lens and move the image from behind the retina onto the retina.

Eyeglass and contact lens properties are given in terms of "diopters," which is the inverse of the focal length in meters. A lens with +2.5 diopters has a focal length $f = 1/2.5 = 0.4$ m, or 40 cm. A lens of −2.5 diopters is a diverging lens with focal length −40 cm.

Multiple Lenses

Many optical instruments including binoculars, telescopes, and microscopes contain two or more lenses. In a telescope, the front lens produces a real image of distant objects, which you examine with another lens (the eyepiece) that acts as a magnifying glass. The front lens is made much larger than the lens in your eye, so it captures much more light than your own eye can, making the image much brighter than it would otherwise be.

The design and analysis of multilens systems depend on the fact that light coming from a real or virtual image acts like the light coming from an actual object (except for being restricted in the range of angles of the rays), so in analyzing what a second lens will do you can pretend that there is an actual light-emitting object at the location of the real or virtual image made by the first lens.

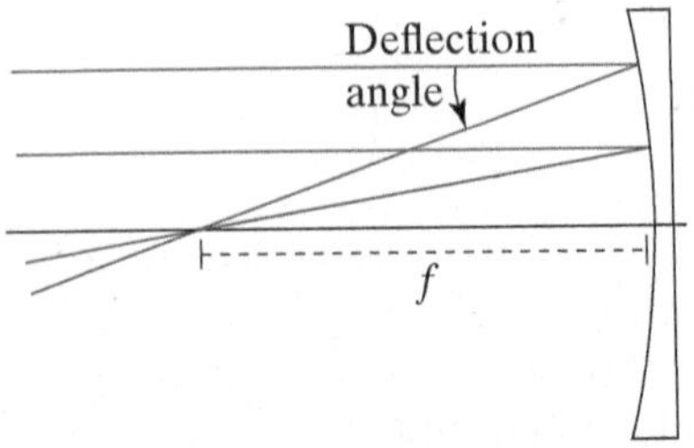

Figure 23.100 A spherical mirror acts in a way that is similar to a spherical lens.

Mirrors

The effect of a spherical mirror is essentially the same as that of a lens with spherical surfaces (Figure 23.100). The thin-lens equation still applies, with the same sign conventions. For example, with a distant source to the left of the mirror in Figure 23.100, what is the sign of d_2? We measure d_2 from the lens (or mirror in this case) to the image, and it is positive because that is the direction in which the light is propagating.

For rays at large angles, spherical mirrors exhibit the same kind of spherical aberration that is characteristic of spherical lenses. There is, however, a very important advantage of a mirror: all colors are reflected in the same way, so there is no chromatic aberration, whereas the variation of the index of refraction in a lens causes the refraction of a ray to differ for different colors.

A practical advantage for large astronomical telescopes is that a mirror can be very thin and weigh much less than a lens of the same diameter and focal length. The largest telescope lens ever made is only 1 m in diameter, but there are now telescope mirrors over 10 m in diameter. Moreover, a lens can be supported only around its edge, but a large mirror can be supported all over its back surface.

Many telescope mirrors are made with a paraboloid surface rather than a spherical surface, because incoming rays that are parallel to the axis are all reflected through the focal point no matter how far from the optical axis they are (or how large the deflection angle is). This is approximately true for a spherical lens (or mirror) only in the small-angle approximation, which is equivalent to the rays being near the optical axis.

23.11 *THE FIELD OF AN ACCELERATED CHARGE

One can use Gauss's law to determine the equation for the electric field of an accelerated charge. We will construct a Gaussian surface near to but not enclosing a charge that is briefly accelerated. Since no charge is enclosed by the surface, the net flux on the surface is zero. We can find the flux on the left and right portions of the surface, and knowing this we can draw conclusions about the magnitude and direction of the field on the side portion of the surface, which includes the radiative electric field.

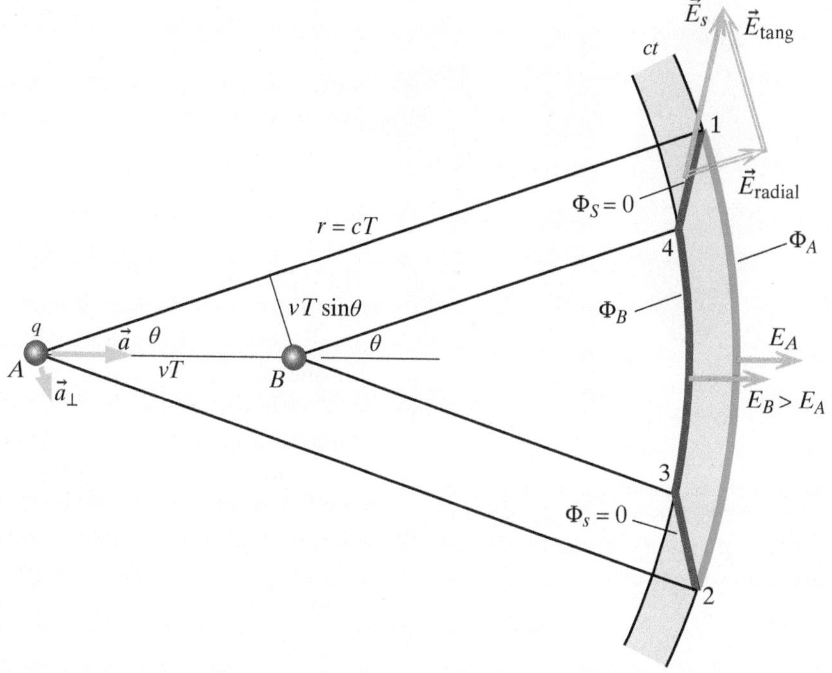

Figure 23.101 A positively charged particle was at rest at location A when it experienced a brief acceleration to the right and then coasted at constant speed. The diagram is not to scale.

In Figure 23.101 a positively charged particle was at rest at location A when it experienced a brief acceleration a to the right, lasting a very short time t. It then coasted at a constant low speed $v = at$, passing location B at time T, having gone a distance vT. The diagram is a slice through a 3D distribution of electric field. For clarity, the diagram is not drawn to scale: the distance vT is actually much shorter than the distance cT since $v \ll c$, and the acceleration time t is much smaller than the time T so the distance ct is much smaller than cT.

Observers at a distance greater than $r = cT$ still see the electric field originally associated with the stationary charge at location A, but other observers see the electric field of a slowly moving charge, which at time T is at location B. The cyan area is an expanding thin spherical shell whose thickness ct corresponds to the brief time t of the acceleration.

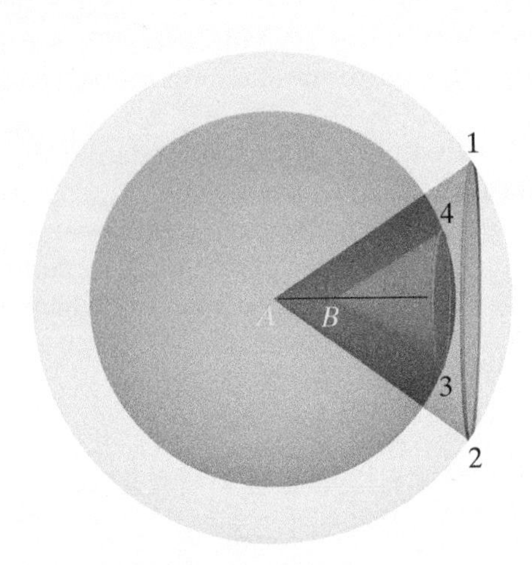

Figure 23.102 A 3D view of the geometry used to find the radiative field of an accelerated charge.

Figure 23.102 shows a 3D view of the situation. The cone with its apex at location A intersects the outer surface of the spherical shell at the locations 1 and 2 that are shown in Figure 23.101, and the cone with its apex at location B intersects the inner surface of the spherical shell at locations 3 and 4 that are shown in Figure 23.101.

We choose a Gaussian surface marked with red, green, and blue lines on Figure 23.101 (again, the diagram is a slice through the 3D closed Gaussian surface). This Gaussian surface does not contain any charge, so the total electric flux over this surface is zero. Three views of this Gaussian closed surface are shown in Figure 23.103. Locations 1, 2, 3, and 4 are shown. Also shown are the surfaces on which we will determine the fluxes Φ_A, Φ_B, and Φ_s.

The field E_A on the outside of the spherical shell is what it was when the charge was still at location A, but the field E_B on the inside of the expanding spherical shell is larger, because E_B is the field of the charge at location B, which is closer than location A.

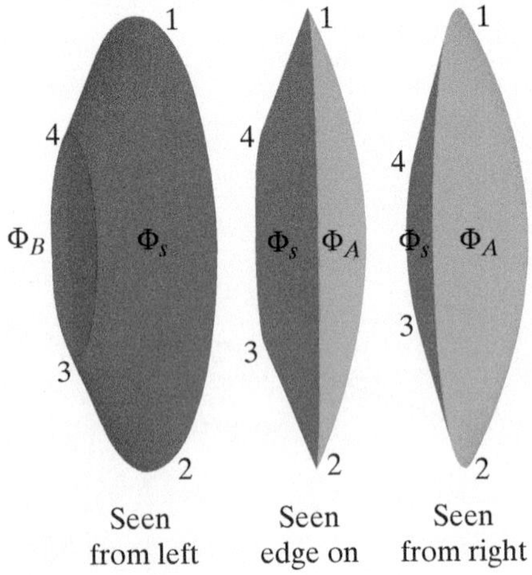

Figure 23.103 3D views of the Gaussian surface we have chosen. The blue portion on the left is concave, the green portion on the right is convex, and the slanted side portion is red.

Suppose that the size of the angle θ is such that the right (green) portion of our Gaussian surface represents one-tenth of the total area of a sphere around location A. Then the flux Φ_A on the right portion of our Gaussian surface is $+0.1q/\varepsilon_0$, since the total flux on a complete sphere would be $+q/\varepsilon_0$. We have the same angle θ at location B, so the flux on the left (blue) side of our Gaussian surface is $-0.1q/\varepsilon_0$, and $|\Phi_A| = |\Phi_B|$. Because B is closer than A, $E_B > E_A$, but the right side of our Gaussian surface has more area than the left side. These two effects compensate each other to make the magnitude of the flux on the left side equal to the magnitude of the flux on the right side.

Since $\Phi_A + \Phi_B = 0$, it must be that the flux Φ_s on the other, ring-shaped (red) portion of the Gaussian surface is also zero, and given the symmetry of the situation it must be that the electric field $\vec{E}_s$ on that surface is everywhere parallel to the surface. Note in Figure 23.101 that $\vec{E}_s$ has a tangential component $\vec{E}_{\text{tang}}$ that is parallel to $-q\vec{a}_\perp$, which is what we expect for a radiative electric field. The triangle in Figure 23.101 for which the upper red line is the hypotenuse is shown enlarged in Figure 23.104, and this triangle is similar to the triangle formed by $\vec{E}_s$ and its components. From $\tan\alpha$ in Figure 23.104 we obtain a relationship between the tangential and radial components of $\vec{E}_s = \vec{E}_{\text{tang}} + \vec{E}_{\text{radial}}$ that we'll need in a moment:

$$\tan\alpha = \frac{E_{\text{tang}}}{E_{\text{radial}}} = \frac{vT\sin\theta}{ct}$$

$$E_{\text{tang}} = E_{\text{radial}}\frac{vT\sin\theta}{ct}$$

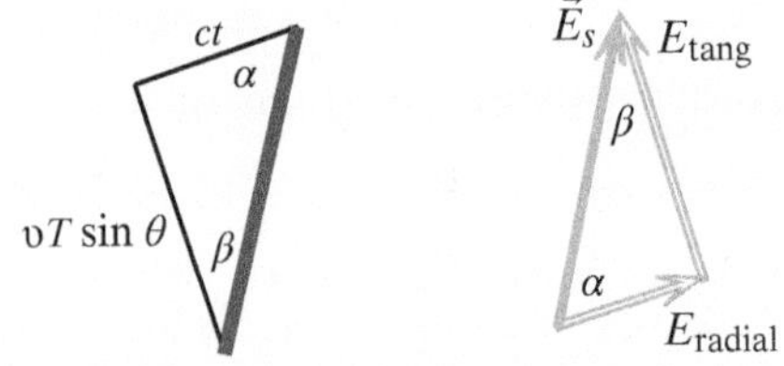

Figure 23.104 Two similar triangles. The triangle on the left is enlarged from the upper slanted surface in Figure 23.101.

Next, we can determine E_{radial} and from that determine E_{tang}. In Figure 23.105 is a close-up of the expanding spherical shell. We choose another Gaussian surface, this time one that is just inside and just outside the inner surface of the shell, with very small areas and therefore negligible flux on the top and bottom in the diagram. Because this Gaussian surface includes no charge, the magnitude of the flux $E_{\text{radial}}\Delta A$ on the left and right sides of this Gaussian

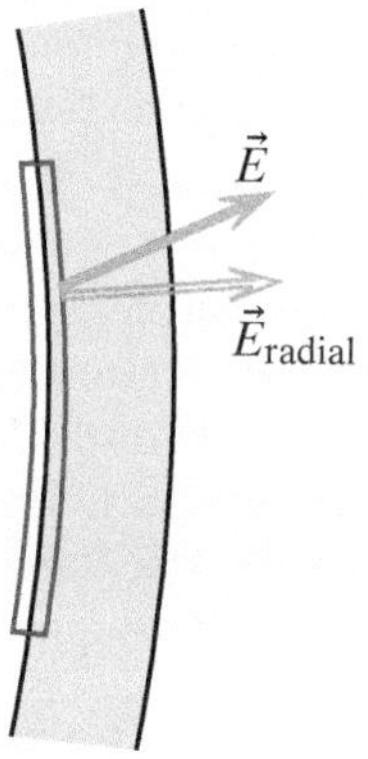

Figure 23.105 A close-up look at a small section of the expanding spherical shell.

surface must be equal, which proves that the radial component of the electric field doesn't change as we cross from just inside the inner surface of the shell to just outside the inner surface of the shell. In particular, the radial component of $\vec{E}_s$ must be the same as the electric field of a charge at location B (and we use the approximation that since $vT \ll cT$, the distance from B to the spherical shell is approximately r):

$$E_{\text{radial}} = \frac{1}{4\pi\varepsilon_0}\frac{q}{r^2}$$

Substitute this result into our earlier equation and simplify, noting that $r = cT$ (so $T = r/c$) and v/t is the acceleration a:

$$E_{\text{tang}} = \frac{1}{4\pi\varepsilon_0}\frac{q}{r^2}\frac{vT\sin\theta}{ct} = \frac{1}{4\pi\varepsilon_0}\frac{qa\sin\theta}{c^2 r}$$

Note that $a\sin\theta$ is the perpendicular component of the vector acceleration $\vec{a}$. Evidently $\vec{E}_{\text{tang}}$ is the transverse radiative field, and $\vec{E}_{\text{radial}}$ is the ordinary Coulomb electric field. Because the Coulomb electric field falls off as $1/r^2$, at large distances only the $1/r$ radiative field is significant. The radiative electric field is accompanied by a magnetic field whose magnitude is $B = E_{\text{rad}}/c$.

Taking into account the direction of $\vec{E}_{\text{tang}}$ as well as its magnitude, we have proved that the transverse radiative electric field is

$$\vec{E}_{\text{radiative}} = \frac{1}{4\pi\varepsilon_0}\frac{-q\vec{a}_\perp}{c^2 r}$$

This derivation of the radiative electric field is based on that given in Appendix B of E. M. Purcell, *Electricity and Magnetism*, Berkeley Physics Course Vol. 2, 2nd edition, 1985, McGraw-Hill.

23.12 *DIFFERENTIAL FORM OF MAXWELL'S EQUATIONS

Here are the differential forms of Maxwell's equations, where $\vec{J}$ is the current density in amperes per square meter:

MAXWELL'S EQUATIONS—DIFFERENTIAL FORMS

$\text{div}(\vec{E}) = \vec{\nabla} \bullet \vec{E} = \frac{\rho}{\varepsilon_0}$	Gauss's law
$\text{div}(\vec{B}) = \vec{\nabla} \bullet \vec{B} = 0$	Gauss's law for magnetism
$\text{curl}(\vec{E}) = \vec{\nabla} \times \vec{E} = -\frac{\partial \vec{B}}{\partial t}$	Faraday's law
$\text{curl}(\vec{B}) = \vec{\nabla} \times \vec{B} = \mu_0\left(\vec{J} + \varepsilon_0\frac{\partial \vec{E}}{\partial t}\right)$	Ampere–Maxwell law

SUMMARY

Maxwell's Equations

$$\oint \vec{E} \bullet \hat{n} dA = \frac{1}{\varepsilon_0} \sum Q_{\text{inside surface}} \quad \text{Gauss's law}$$

$$\oint \vec{B} \bullet \hat{n} dA = 0 \quad \text{Gauss's law for magnetism}$$

$$\oint \vec{E} \bullet d\vec{l} = -\frac{d}{dt} \int \vec{B} \bullet \hat{n} dA \quad \text{Faraday's law}$$

Ampere–Maxwell law:

$$\oint \vec{B} \bullet d\vec{l} = \mu_0 \left[\sum I_{\text{inside path}} + \varepsilon_0 \frac{d}{dt} \int \vec{E} \bullet \hat{n} dA \right]$$

Magnetic and electric flux terms

$$\Phi_{\text{mag}} = \int \vec{B} \bullet \hat{n} dA \qquad \Phi_{\text{el}} = \int \vec{E} \bullet \hat{n} dA$$

The force due to electric and magnetic interactions

$$\vec{F} = q\vec{E} + q\vec{v} \times \vec{B} \qquad (d\vec{F} = Id\vec{l} \times \vec{B} \text{ for currents})$$

Producing a radiative electric field (Figure 23.106)

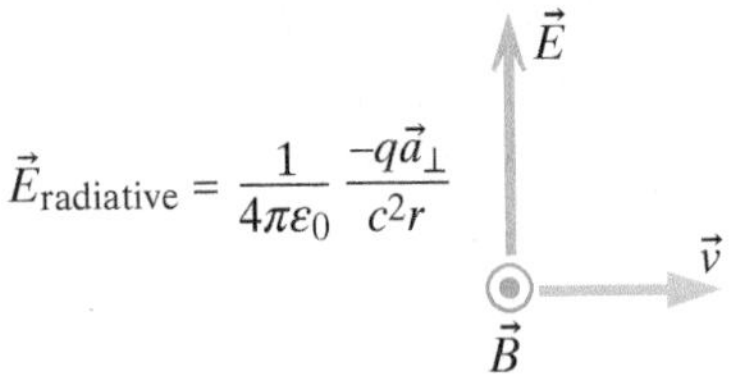

Figure 23.106

The direction of $\vec{v}$ is the direction of $\vec{E} \times \vec{B}$, $E = cB$ in electromagnetic radiation.

$$v = \frac{1}{\sqrt{\mu_0 \varepsilon_0}} = c = 3 \times 10^8 \text{ m/s}$$

Effects of electromagnetic radiation on matter

The main effect that electromagnetic radiation has on matter is due to $\vec{F} = q\vec{E}_{\text{radiative}}$, but the magnetic field is responsible for radiation pressure, a very small effect.

If $\vec{E}_{\text{radiative}}$ makes a charge accelerate, that charge radiates in turn ("re-radiation" or "scattering").

Different systems respond differently to sinusoidal electromagnetic radiation of a particular frequency $f (c = f\lambda)$. Systems that freely oscillate sinusoidally with a frequency f respond very strongly to sinusoidal electromagnetic radiation of the same frequency f ("resonance" phenomenon).

Energy flux of electromagnetic radiation

$$\vec{S} = \frac{1}{\mu_0} \vec{E} \times \vec{B}, \text{ in W/m}^2 \quad \text{(the "Poynting vector")}$$

Momentum flux of electromagnetic radiation

$$\frac{\vec{S}}{c} = \frac{1}{\mu_0 c} \vec{E} \times \vec{B}, \quad \text{in N/m}^2$$

The sky is blue and polarized due to re-radiation of sunlight, with blue light re-radiated more strongly than red light.

Geometric Optics

Snell's law and the index of refraction $n = c/v$:

$n_1 \sin\theta_1 = n_2 \sin\theta_2$

Converging and diverging lenses

The thin-lens equation: $\frac{1}{f} = \frac{1}{d_1} + \frac{1}{d_2}$

Real and virtual images

QUESTIONS

Q1 Match the physical situation with the Maxwell's equation that would be useful in analyzing the situation:

(a) Gauss's law	1. Find the magnetic field inside a current-carrying coaxial cable.
(b) Gauss's law for magnetism	2. Predict what an AC voltmeter reads whose leads encircle a coil.
(c) Faraday's law	3. Find the amount of charge inside a box.
(d) Ampere–Maxwell law	4. Relate the magnetic flux on one face of a cube to the magnetic flux on the other faces.

Q2 At a time t_0 an electric field is detected in the region shown on the left diagram in Figure 23.107. The electric field is zero at all other locations. At a later time t_1 an electric field is detected in the region shown on the right diagram; the electric field is zero elsewhere. You conclude that an electromagnetic wave is passing through the region. **(a)** What is the direction of propagation of the electromagnetic wave? **(b)** If you were to measure the magnetic field in the region where the electric field is nonzero, what would be the direction of the magnetic field?

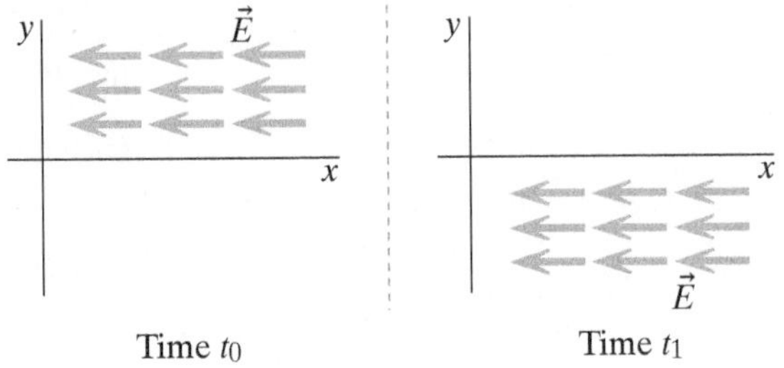

Figure 23.107

Q3 A pulse of electromagnetic radiation is detected at a particular location. In the pulse, the electric field is in the $+y$ direction and the magnetic field is in the $-z$ direction. What is the direction of propagation of this pulse of radiation?

Q4 Show that a moving region of space in which $\vec{E} = \langle 0, E, 0 \rangle$, $\vec{B} = \langle 0, -B, 0 \rangle$, and $\vec{v} = \langle -c, 0, 0 \rangle$ does not satisfy Maxwell's equations, and therefore is not a possible configuration of traveling fields.

Q5 A proton is accelerated in the direction shown in Figure 23.108. **(a)** Which arrow (a–j) best indicates the direction of propagation of radiation reaching observation location P? **(b)** Which arrow (a–j) best indicates the direction of $\vec{a}_\perp$ at observation location P? **(c)** Which arrow (a–j) best indicates the direction of the radiative electric field observed at location P? **(d)** Which arrow (a–j) best indicates the direction of the radiative electric field observed at location S?

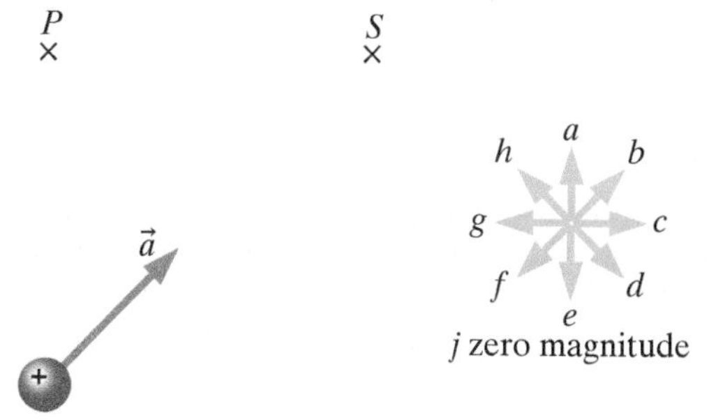

Figure 23.108

Q6 A proton is accelerated in the direction shown in Figure 23.109. At which location(s) (A–D) will electromagnetic radiation be detected a short time later?

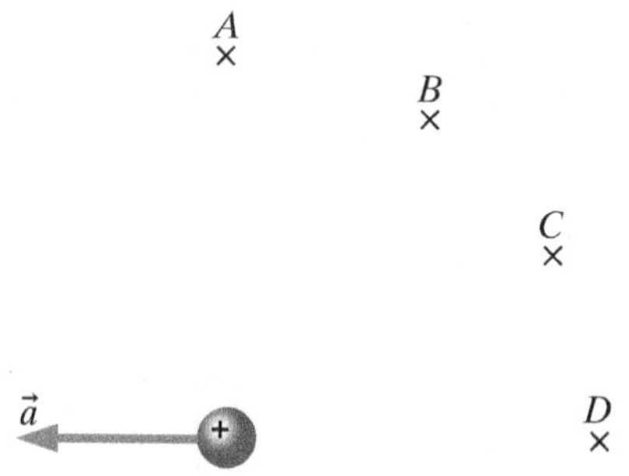

Figure 23.109

Q7 An electron is briefly accelerated in the direction shown in Figure 23.110. What will be the direction (a–j) of the radiative electric field observed at location S a short time later?

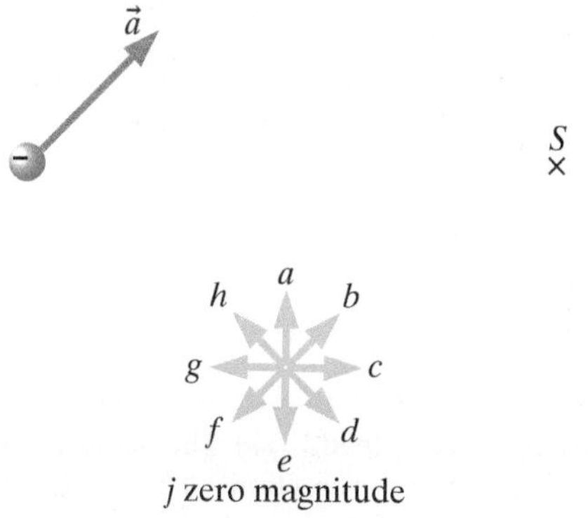

Figure 23.110

Q8 When informally discussing the effect of electromagnetic radiation on matter, physicists often talk about the effects that electromagnetic radiation has on the electrons in the matter. Why would they omit commenting on the effects on the nucleus, which is also charged?

Q9 Explain why electromagnetic radiation moving to the right applies radiation pressure to the right on a neutral grain of dust in the air, even though the grain has zero net charge.

Q10 Figure 23.111 shows several different orientations (A, B, and C) of the receiving antenna discussed earlier. In each case, predict the brightness of the bulb and explain briefly.

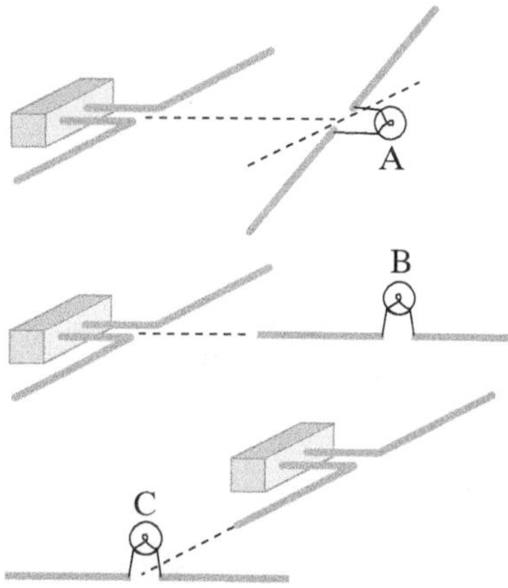

Figure 23.111

Q11 You may have noticed that television antennas on housetops have metal rods mounted horizontally (Figure 23.112), at least in the United States (we're not talking about satellite dishes). What does this imply about the construction of the transmitting antennas used by television stations? Why?

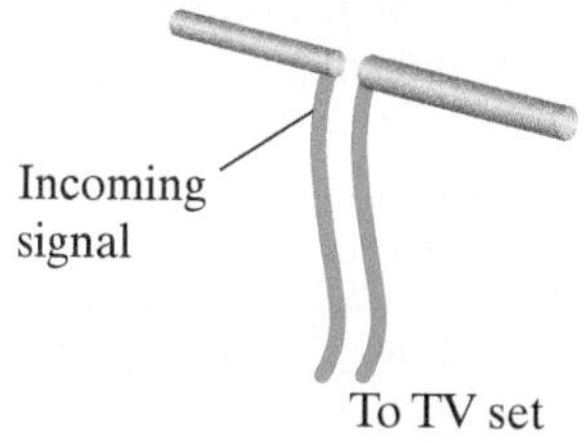

Figure 23.112

Q12 Television antennas often have one or more horizontal metal bars mounted behind the receiving antenna and insulated from it, as shown in Figure 23.113. Explain qualitatively why adding a second metal bar can make the television signal either stronger or weaker, depending on the distance between the front receiving antenna and the second bar.

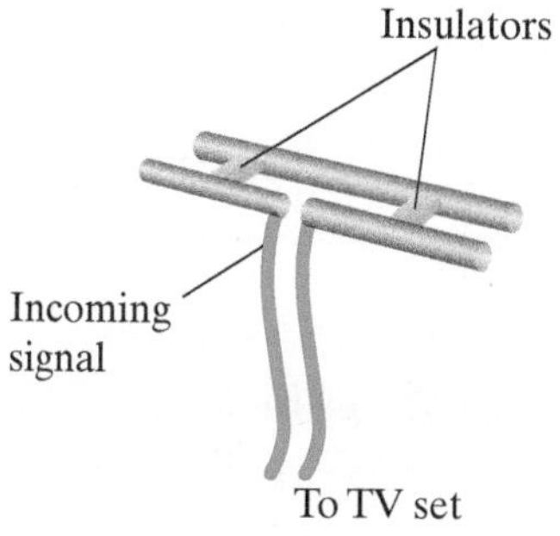

Figure 23.113

Q13 At dawn, with the Sun just rising in the east, you face the Sun and bend your head back to look straight up, and you examine the blue sky light with a Polaroid® filter. Why is the light polarized? What is the direction of the electric field, east–west or north–south? Explain carefully.

Q14 On a cloudless day, when you look away from the Sun at the rest of the sky, the sky is bright and you cannot see the stars. On the Moon, however, the sky is dark and you *can* see the stars even when the Sun is visible. Explain briefly.

Q15 If a beam of light from a medium with a higher index of refraction emerges into a medium with a lower index of refraction, what happens? (1) The emerging beam bends toward the normal (smaller angle). (2) The emerging beam bends away from the normal (larger angle). (3) The emerging beam does not bend at all.

PROBLEMS

Section 23.1

••P16 N closely spaced turns of wire are wound in the direction indicated on a hollow plastic ring of radius R, with circular cross section (Figure 23.114).

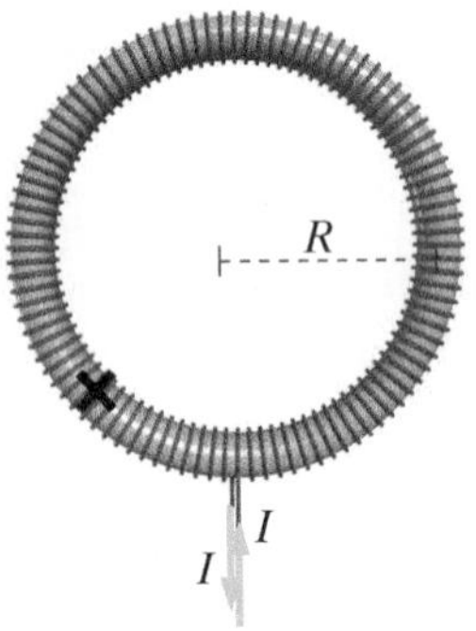

Figure 23.114

(a) If the current in the wire in Figure 23.114 is I, determine the magnetic field B at the location indicated by the $\times$, at the center of the cross section of the ring, and indicate the direction by drawing a vector at that location. Give a complete derivation, not just a final result. Use a circular path. **(b)** Throughout this region there is a uniform electric field E into the paper. This electric field begins to increase at a rate dE/dt, and there continues to be a current I in the wire. Now what is the magnitude of the magnetic field at the indicated location?

••P17 A rectangle 4 cm by 8 cm is drawn in empty space, in a region where there is no matter present, and the magnetic field is measured at all points along the rectangle as shown (Figure 23.115). What can you conclude about the region enclosed by the rectangle? Be as quantitative as you can.

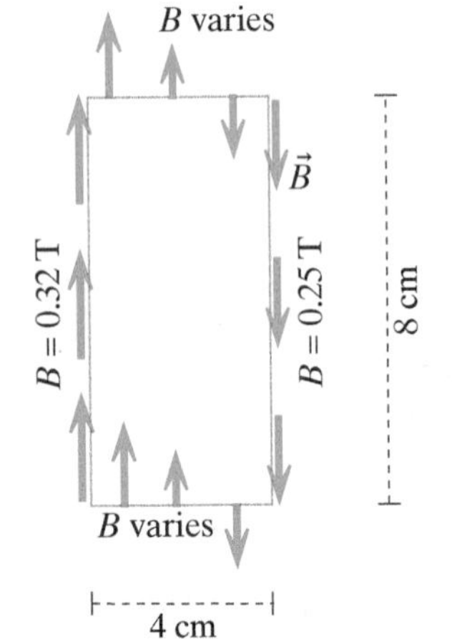

Figure 23.115

Section 23.2

•P18 If the magnetic field in a particular pulse has a magnitude of 1×10^{-5} T (comparable to the Earth's magnetic field), what is the magnitude of the associated electric field?

•P19 In Figure 23.116 electromagnetic radiation is moving to the right, and at this time and place the electric field is horizontal and points out of the page. The magnitude of the electric field is $E = 3000$ N/C. What are the magnitude and direction of the associated magnetic field at this time and place?

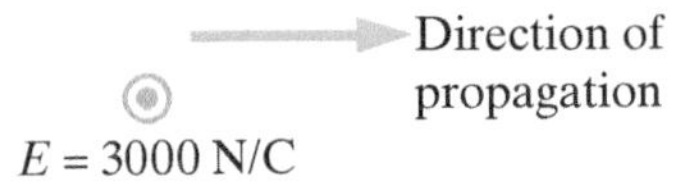

Figure 23.116

•P20 A pulse of electromagnetic radiation is propagating in the $+y$ direction. You have two devices that can detect electric and magnetic fields. You place detector 1 at location $\langle 0, -3, 0\rangle$ m and detector 2 at location $\langle 0, 3, 0\rangle$ m. **(a)** At time $t = 0$, detector 1 detects an electric field in the $-x$ direction. At that instant, what is the direction of the magnetic field at the location of detector 1? **(b)** At what time will detector 2 detect electric and magnetic fields?

•P21 A pulse of radiation propagates with velocity $\langle 0, 0, -c\rangle$. The electric field in the pulse is $\langle 7.2 \times 10^6, 0, 0\rangle$ N/C. What is the magnetic field in the pulse?

Section 23.3

•P22 An electric field of 1×10^6 N/C acts on an electron, resulting in an acceleration of 1.8×10^{17} m/s^2 for a short time. What is the magnitude of the radiative electric field observed at a location a distance of 2 cm away along a line perpendicular to the direction of the acceleration?

•P23 A proton is briefly accelerated in the direction of the arrow labeled $\vec{a}$ in Figure 23.117. The vector $\vec{r}$ indicates the location of an observation location relative to the initial position of the proton. The vector $\vec{a}_\perp$ for this observation location is indicated on the diagram. **(a)** If θ is 40°, and the magnitude of the acceleration is 1.8×10^{17} m/s^2, what is the magnitude of $\vec{a}_\perp$? **(b)** What is the magnitude of the radiative electric field at the indicated observation location, if the magnitude of $\vec{r}$ is 0.011 m? **(c)** At the observation location, what is the direction of the radiative electric field? (1) The same as the direction of $\vec{a}$, (2) Opposite to the direction of $\vec{a}$, (3) The same as the direction of $\vec{a}_\perp$, (4) Opposite to the direction of $\vec{a}_\perp$, (5) The same as the direction of $\vec{r}$, (6) Opposite to the direction of $\vec{r}$, (7) Into the page, (8) Out of the page. **(d)** At the observation location, what is the direction of the ordinary Coulomb electric field due to the proton? (Same choices as above.) **(e)** At the observation location, what is the direction of the radiative magnetic field? (Same choices as above.)

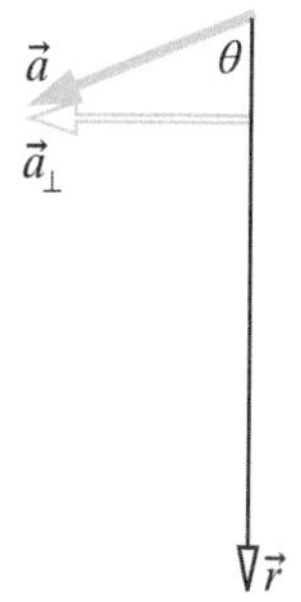

Figure 23.117

•P24 A proton is accelerated in the direction shown by the arrow labeled $\vec{a}$ in Figure 23.118. Which of the arrows labeled A–F correctly shows the direction of the vector $\vec{a}_\perp$ at the observation location indicated by the vector $\vec{r}$?

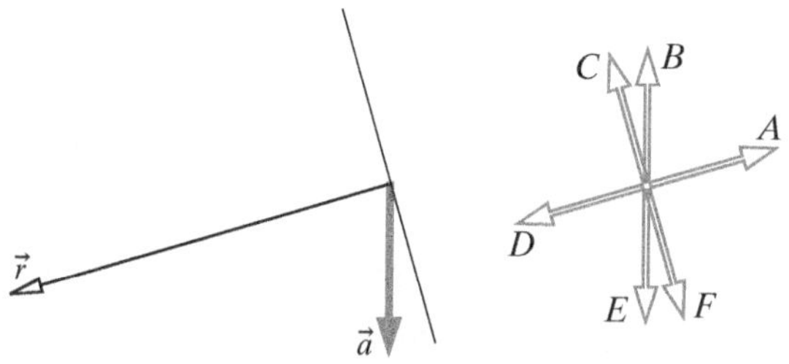

Figure 23.118

•P25 A proton located at the origin is accelerated in the $-y$ direction for a brief time. **(a)** How much time passes before a detector located at $\langle 0.15, 0, 0\rangle$ m detects a radiative electric field? **(b)** What is the direction of the radiative electric field observed at this location? **(c)** If the accelerated particle had been an electron instead of a proton, what would have been the direction of the radiative electric field at this location?

•P26 At time $t = 0$ an electron at the origin is subjected to a force that briefly accelerates it in the $+z$ direction, with an acceleration of 8×10^{17} m/s^2. Before trying to answer the following questions, draw a clear diagram. Location D is $\langle -26, 0, 0\rangle$ m, and location H is at $\langle 0, 16, 0\rangle$ m. **(a)** At what time is a radiative electric field first detected at location D? **(b)** What is the direction of propagation of electromagnetic radiation that is detected at location D? **(c)** What is the direction of the radiative electric field observed at location D? **(d)** What is the magnitude of the radiative electric field observed at location D? **(e)** What is the direction of the radiative magnetic field observed at location D? **(f)** What is the magnitude of the radiative magnetic field observed at location D? **(g)** At what time is a radiative electric field first detected at location H? **(h)** What is the direction of propagation of electromagnetic radiation that is detected at location H? **(i)** What is the direction of the radiative electric field observed at location H? **(j)** What is the magnitude of the radiative electric field observed at location H? **(k)** What is the direction of the radiative magnetic field observed at location H? **(l)** What is the magnitude of the radiative magnetic field observed at location H?

•P27 If the electric field inside a capacitor exceeds about 3×10^6 V/m, the few free electrons in the air are accelerated enough to trigger an avalanche and make a spark. In the spark shown in Figure 23.119, electrons are accelerated upward and positive ions are accelerated downward.

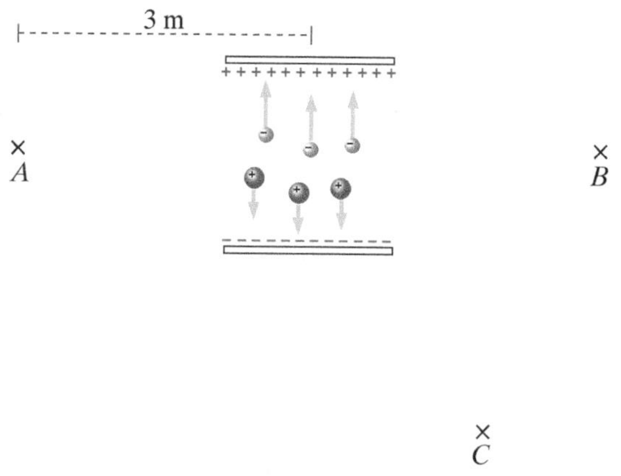

Figure 23.119

(a) Qualitatively, explain the directions and relative magnitudes of the radiative electric fields at location A, to the left of the capacitor, due to the indicated motions of the electrons and of the ions (actually, in a spark there are also accelerations in the opposite direction when the electrons and ions collide with air molecules). **(b)** Repeat the analysis for locations B (to the right) and C (to the right and down a bit). **(c)** If you are at location A, 3 m to the left of the capacitor, how long after the initiation of the spark could you first detect a magnetic field? What is the direction of the radiative magnetic field? (Before the spark occurs there is no magnetic field anywhere in this region.) **(d)** In addition to producing electromagnetic radiation, the moving charges in the spark produce electric fields according to Coulomb's law and magnetic fields according to the Biot–Savart law. Explain why these fields are much smaller than the radiative fields, if you are far away from the capacitor.

••P28 A positive charge coasts upward at a constant velocity for a long time. Then at $t = 0$ (Figure 23.120) a force acts downward on it for 1 ns (1×10^{-9} s). After this force stops acting, the charge coasts upward at a smaller constant speed for 1 ns; then a force acts upward for 1 ns and it resumes its original speed. The new position reached at $t = 3$ ns is much less than a millimeter from the original position.

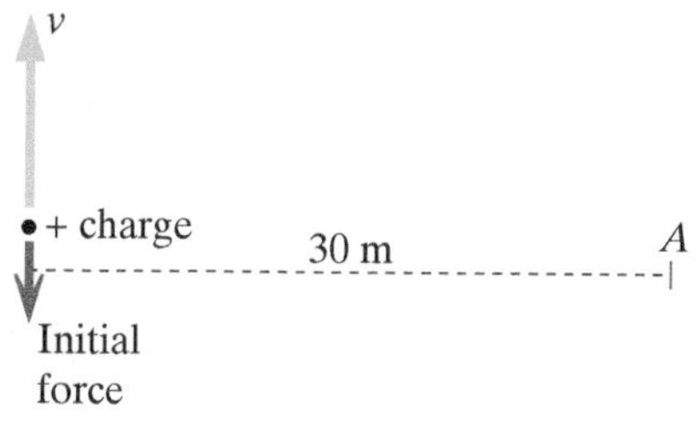

Figure 23.120

You stand at location A, 30 m to the right of the charge (Figure 23.120), with instruments for measuring electric and magnetic fields. What will you observe due to the motion of the positive charge, at what times? You do not need to calculate the magnitudes of the electric and magnetic fields, but you do need to specify their directions, and the times when these fields are observed.

••P29 An electron is initially at rest. At time $t_1 = 0$ it is accelerated upward with an acceleration of $1 \times 10^{18}\ \text{m/s}^2$ for a very short time (this large acceleration is possible because the electron has a very small mass). We make observations at location A, 15 m from the electron (Figure 23.121).

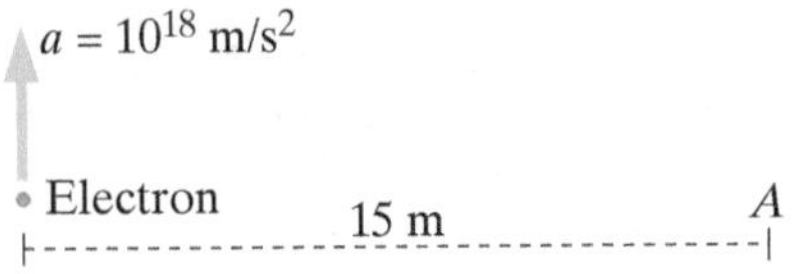

Figure 23.121

(a) At time $t_2 = 1$ ns (1×10^{-9} s), what are the magnitude and direction of the electric field at location A due to the electron? **(b)** At what time t_3 does the electric field at location A change? **(c)** What is the direction of the radiative electric field at location A at time t_3? **(d)** What is the magnitude of this electric field? How does it compare to the magnitude of the field at location A at time t_2? **(e)** Just after time t_3, what is the direction of the magnetic force on a positive charge that was initially at rest at location A? Explain with a diagram.

••P30 At time $t = 0$, electrons in a vertical copper wire (Figure 23.122) are accelerated downward for a very short time Δt (by a power supply that is not shown). A proton, initially at rest, is located a very large distance L from the wire.

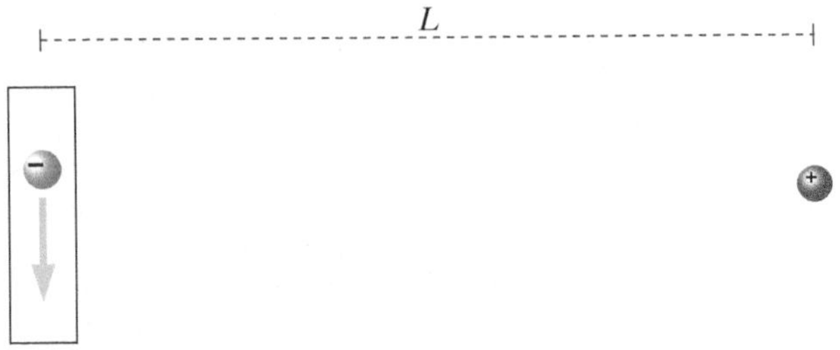

Figure 23.122

Explain in detail what happens to the proton (neglect gravity), and at what time.

••P31 A slab (pulse) of electromagnetic radiation that is 15 cm thick is propagating downward (in the $-y$ direction) toward a short horizontal copper wire located at the origin and oriented parallel to the z axis, as shown in Figure 23.123. **(a)** The direction of the electric field inside the slab is out of the page (in the $+z$ direction). On a diagram, show and describe clearly the direction of the magnetic field inside the slab.

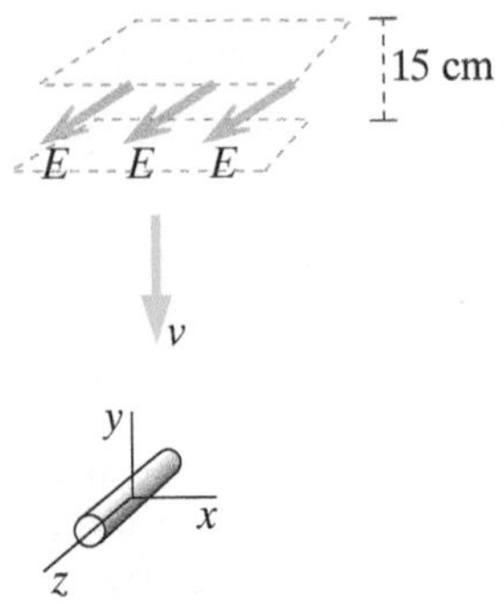

Figure 23.123

(b) You stand on the x axis at location $\langle 12,0,0 \rangle$ m, at right angles to the direction of propagation of the pulse. Your friend stands on the z axis at location $\langle 0,0,12 \rangle$ m. The pulse passes the copper wire at time $t_1 = 0$, and at a later time t_2 you observe new nonzero electric and magnetic fields at your location, but your friend does not. Explain. What is t_2? (Give a numerical answer.) How long a time do these new fields last for you? **(c)** On a diagram, show and describe clearly the directions of these new nonzero fields ($\vec{E}$ and $\vec{B}$) at your location. Explain briefly but carefully.

Section 23.4

•P32 A particular AM radio station broadcasts at a frequency of 990 kHz. **(a)** What is the wavelength of this electromagnetic radiation? **(b)** How much time is required for the radiation to propagate from the broadcasting antenna to a radio 3 km away?

•P33 What is the wavelength of ultraviolet light whose frequency is 5×10^{15} Hz?

•P34 The wavelength of red light is about 695 nm. What are the frequency and period of this radiation?

•P35 Calculate the wave length for several examples of sinusoidal electromagnetic radiation:

radio, 1000 kHz, $\lambda = ?$

television, 100 MHz, $\lambda = ?$

red light, 4.3×10^{14} Hz, $\lambda = ?$

blue light, 7.5×10^{14} Hz, $\lambda = ?$

(Note for comparison that an atomic diameter is about 1×10^{-10} m.)

Section 23.5

•P36 A small laser used as a pointer produces a beam of red light 5 mm in diameter and has a power output of 5 mW. What is the magnitude of the electric field in the laser beam?

••P37 A 100 W light bulb is placed in a fixture with a reflector that makes a spot of radius 20 cm. Calculate approximately the amplitude of the radiative electric field in the spot.

Section 23.6

•P38 A sinusoidal beam of electromagnetic radiation that is only 2 cm high propagates in the $+x$ direction, with the electric field in the $\pm z$ direction. The beam strikes a metal rod that is centered on the origin and aligned along the z axis, as shown in Figure 23.124. You place detectors sensitive to electromagnetic radiation at location $\langle 0,-17,0 \rangle$ m (detector A) and location $\langle 0,0,17 \rangle$ m (detector B). Answer the following questions for detector A and for detector B: Will the detector detect electromagnetic radiation? If so: **(a)** What is the direction of propagation of the electromagnetic radiation detected? **(b)** What is the direction of the electric field in the radiation? **(c)** What is the direction of the magnetic field in the radiation?

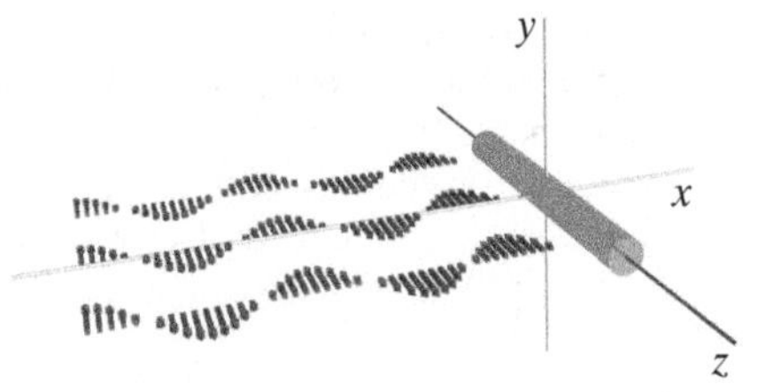

Figure 23.124

Section 23.8

•**P39** The index of refraction of water is 1.33. At what speed would a wave crest in a beam of light travel through this medium?

•**P40** If a diver who is underwater shines a flashlight upward toward the surface, at an angle of 29° from the normal, at what angle does the light emerge from the water? The indices of refraction are: water, 1.33; air, 1.00029.

•**P41** To get total internal reflection at the interface of water (refractive index 1.33) and a plastic whose refractive index is 1.46: **(a)** Which material must the light start in? (1) It doesn't matter, (2) Water, (3) Plastic, **(b)** What is the critical angle?

•**P42** For a particular liquid–air interface the critical angle (for total internal reflection) is 59.8°. The index of refraction of air is 1.00029. What is the index of refraction of the liquid?

••**P43** A spotlight mounted beside a swimming pool sends a beam of light at an angle into the pool, as shown in Figure 23.125. The light hits the water 2.4 m from the edge of the pool. The pool is 2 m deep and the light is mounted 1.2 m above the surface of the water. (The diagram is not drawn to scale.) The indices of refraction are: water, 1.33; air, 1.00029.

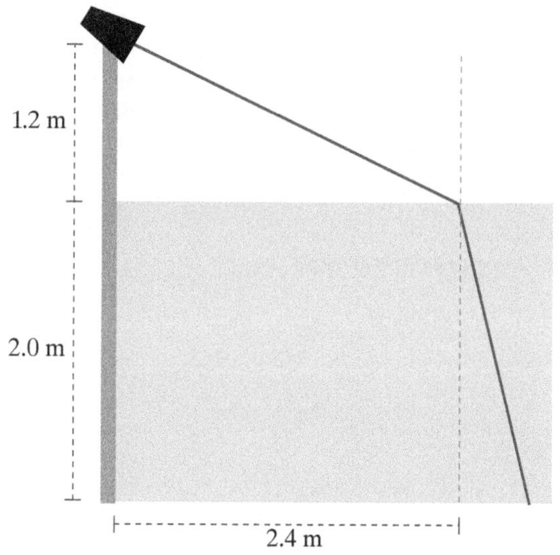

Figure 23.125

(a) What is the angle of the beam from the normal in the air? **(b)** What is the angle of the beam from the normal in the water? **(c)** How far from the edge of the pool is the spot of light on the bottom of the pool?

Section 23.10

•**P44** A point source of green light is placed on the axis of a thin converging lens, 15 cm to the left of the lens. The lens has a focal length of 10 cm. Where is the location of the image of the source? Is it a real or a virtual image? If you placed a sheet of paper at the location of the image, what would you see on the paper?

•**P45** A point source of green light is placed on the axis of a thin converging lens, 7 cm to the left of the lens. The lens has a focal length of 10 cm. Where is the location of the image of the source? Is it a real or a virtual image? If you placed a sheet of paper at the location of the image, what would you see on the paper?

•**P46** A thin diverging lens of focal length 25 cm is placed 18 cm to the right of a point source of blue light on the axis of the lens. Where is the image of the source? Is it a real or a virtual image? If you placed a sheet of paper at the location of the image, what would you see on the paper?

•**P47** A thin converging lens of focal length 20 cm is located at the origin and its axis lies on the x axis. A point source of red light is placed at location $\langle -25, 1, 0\rangle$ cm. Where is the location of the image of the source? Is it a real or a virtual image? It helps to draw a diagram and trace the "easy" rays.

•**P48** A thin converging lens of focal length 20 cm is located at the origin and its axis lies on the x axis. A point source of red light is placed at location $\langle -14, 1, 0\rangle$ cm. Where is the location of the image of the source? Is it a real or a virtual image? It helps to draw a diagram and trace the "easy" rays.

•**P49** You have a thin converging lens whose focal length is 0.3 m and a sheet of white paper on which to display a real image of a small tree that is 2 m tall and 40 m away. How far in back of the lens should you place the paper in order to get a sharp image of the tree? How tall is the image of the tree? Is it inverted or right side up?

••**P50** You have a small but bright computer display that is only 4 cm wide by 3 cm tall. You will place a lens near the display to project an image of the display about 2.5 m tall onto a large screen that is 6 m from the lens of the projector. **(a)** Should you use a converging lens or a diverging lens? **(b)** Approximately, what should the focal length of the lens be? **(c)** Suppose that you choose a lens that has the focal length that you calculated in part (b). What exactly should be the distance from the computer display to the lens? **(d)** In order that the screen display be right side up, should the computer display be inverted or right side up? **(e)** On another occasion the screen is moved close, so that it is only 4 m from the lens. To focus the image, you have to readjust the distance between the computer display and the lens. What is the new exact distance from the computer display to the lens? **(f)** How tall is the image on the screen, now that the screen is only 4 m from the lens?

COMPUTATIONAL PROBLEMS

•••**P51** **(a)** Write a computer program that displays a sinusoidal electromagnetic wave propagating through space. Make the wavelength 600 m (corresponding to a frequency of about 500 kHz, which is in the AM radio frequency band) and the amplitude of the electric field 1.0×10^4 V/m. (This is roughly the amplitude that would be measured a few meters from a 50,000 W radio transmitter. On average, the amplitude of the radiative field from the Sun at Earth's orbital radius is about 700 V/m.)

Animate the wave as a function of time, making it propagate in the positive x direction, with the electric field polarized in the y direction. Display both the electric and magnetic field vectors, and make sure they have the correct directions relative to each other. Display at least three full wavelengths, with enough observation locations per wavelength that you can clearly see the sinusoidal character of the wave. Think carefully about scaling. The length of the arrow objects must be scaled such that both the arrows and the wavelength can be seen. Also, to see smooth wave-like motion, the time step must be a small fraction of the period of the wave.

(b) Place a positron initially at rest in the presence of the electromagnetic wave from part (a). (Using a positron instead of an electron makes it easier to think about signs and directions.) Modify your program to model the motion of the positron due to its interaction with the electromagnetic wave. Leave a trail. This is a relativistic situation since the instantaneous speed of the particle can get quite high. To accurately model the positron's motion, you'll need to use the relativistic relationship between momentum and velocity:

$$\vec{p} = \frac{m\vec{v}}{\sqrt{1-(v^2/c^2)}} \quad \text{and therefore} \quad \vec{v} = \frac{\vec{p}}{m}\left(\frac{1}{\sqrt{1+(p/(mc))^2}}\right)$$

ANSWERS TO CHECKPOINTS

1 $\vec{B} = \langle 3.3\times 10^{-3}, 0, 0\rangle$ T

2 An electron accelerated to the left produces a radiative electric field as shown in Figure 23.126.

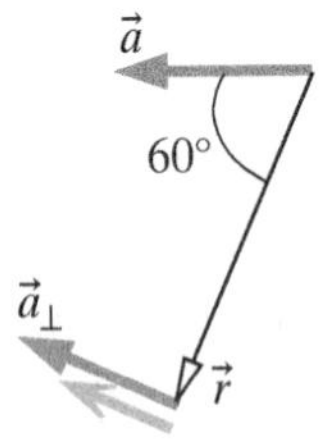

Figure 23.126

3 5.6×10^{14} Hz

4 726 V/m

5 9.3×10^{-6} N/m^2

6 Since electric fields are not blocked by matter, the radiated fields must be present on the other side of the cardboard. Yet there is clearly no light there, so the net field must be zero there. There must be an additional source of electric and magnetic field somewhere, producing a field that adds to the radiative field to produce a zero net field. The electrons in the atoms of the cardboard must be accelerated by the electric field of the incoming radiation in such a way that the re-radiated field is equal and opposite to the incident field. The net field on the other side of the cardboard is the superposition of the two fields, and is zero.

7 Top: Horizontally polarized radiation accelerates electrons back and forth along the horizontal metal foils. The electrons re-radiate, with much of the radiation going off in other directions. The net field reaching the antenna does accelerate electrons along the antenna, but does not produce enough current to light the bulb. Bottom: Horizontally polarized radiation across the narrow width of the foils is not very effective in causing a large response in the foils. There is minimal re-radiation, so most of the original radiation accelerates electrons in the receiver along its length, thus lighting the bulb.

8 $n = c/v = 3.5$

9 38.7°

10 0.6 m

11 **(a)** 125 cm; **(b)** 25; **(c)** 19.23 cm

Answers to Odd-Numbered Problems

Chapter 1

P13 2.15×10^7 m/s **P15** **(a)** $\vec{a} = \langle -4, -3, 0\rangle$ **(b)** $\vec{b} = \langle -4, -3, 0\rangle$ **(c)** True **(d)** $\vec{c} = \langle 4, 3, 0\rangle$ **(e)** True **(f)** False **P17** $\langle 0.04, -3.4, 60.0\rangle$ **P19** $\langle 0.58, 0.58, 0.58\rangle$; $\langle 0.58, 0.58, 0.58\rangle$ **P21** 458.26 m/s^2 $\langle 0.872872, 0.436436, -0.218218\rangle$ **P23** **(a)** $\langle 3 \times 10^{-10}, -3 \times 10^{-10}, 8 \times 10^{-10}\rangle$ m **(b)** 9.1×10^{-10} m **(c)** $\langle 0.33, -0.33, 0.88\rangle$ **P25** **(a)** $\langle -5.5, -20, 0\rangle$ m **(b)** 20.74 m **P27** **(a)** $\langle -10 \times 10^{10}, -17 \times 10^{10}, 0\rangle$ m **(b)** $\langle 10 \times 10^{10}, 17 \times 10^{10}, 0\rangle$ m **P29** $\langle x_p - x_e, y_p - y_e, z_p - z_e\rangle$; $\langle x_e - x_p, y_e - y_p, z_e - z_p\rangle$ **P31** **(a)** $\langle -2.01 \times 10^5, 5.2 \times 10^4, -1 \times 10^3\rangle$ m/s **(b)** 2.08×10^5 m/s **P33** **(a)** $\langle 974, 0, 684\rangle$ m **(b)** $\langle 1.50, 0, 1.05\rangle$ m/s **P35** 3.0 s; $\langle 5.33 \times 10^2, -7.33 \times 10^2, 5.67 \times 10^2\rangle$ m **P37** **(a)** $\langle 4.4, -6.4, 0\rangle$ m/s **(b)** $\langle 4, -7.7, 0\rangle$ m/s **(c)** The time interval from $t = 6.3$ s to 6.8 s **(d)** $\langle 0.132, -0.192, 0\rangle$ m **P39** **(a)** $\langle 0, 9 \times 10^5, -4 \times 10^5\rangle$ m/s **(b)** $\langle 0.02, 6.34, -2.86\rangle$ m **P41** 147 m **P43** **(a)** $\langle 22.3, 26.1, 0\rangle$ m/s **(b)** $\langle 44.6, 52.2, 0\rangle$ m **(c)** $\Delta t = 1.0$ s was too big. **P45** $6.2\text{ kg}\cdot\text{m/s}$ **P47** $2.24 \times 10^5\text{ kg}\cdot\text{m/s}$ **P49** $6.7\text{ kg}\cdot\text{m/s}$ **P51** **(a)** $\langle -2mv_x, 0, 0\rangle$ **(b)** 0 **P53** $\langle 0, 4.1, 0\rangle\text{ kg}\cdot\text{m/s}$ **P55** 0; $500\text{ kg}\cdot\text{m/s}$ **P57** $\langle 20, 50, -10\rangle$ m/s **P59** $\langle 0, 0, -11\rangle$ m **P61** $7.35 \times 10^{-19}\text{ kg}\cdot\text{m/s}$ **P63** $1.93 \times 10^{-20}\text{ kg}\cdot\text{m/s}$ **P65** 1250 **P67** $0.65c$

Chapter 2

P9 $\langle -9510, 0, 0\rangle$ N **P11** $\langle 180, -180, 700\rangle$ N **P13** **(a)** $\langle 0.011, -0.005, -0.003\rangle\text{ kg}\cdot\text{m/s}$ **(b)** $\langle 0.003, -0.049, -0.003\rangle\text{ kg}\cdot\text{m/s}$ **(c)** $\langle -0.004, -0.022, 0\rangle$ N **P15** $\langle 91, 106, 80\rangle$ m/s **P17** **(a)** 12 m/s **(b)** 0.06 s **(c)** 1.0×10^6 N **(d)** 41 **(e)** $\vec{v}_{\text{avg}} = (\vec{v}_i + \vec{v}_f)/2$ was valid **P19** Depends on what approximations you make. **P21** $\langle 0.15, 2.102, 0\rangle$ m, $\langle 3, 2.04, 0\rangle$ m/s; $\langle 0.3, 2.106, 0\rangle$ m, $\langle 3, 0.08, 0\rangle$ m/s; $\langle 0.45, 2.012, 0\rangle$ m, $\langle 3, -1.88, 0\rangle$ m/s **P23** $\langle -3.83, 0, 2.79\rangle$ m **P25** **(a)** 4 m/s **(b)** 12 m **P27** **(a)** 90 mi **(b)** 37.5 mi/h **(c)** 45 mi/h **P29** **(a)** none **(b)** 6 **(c)** 3 **(d)** 2 **(e)** none **(f)** none **(g)** 5 **(h)** 1 **(i)** 4 **P31** **(a)** 6 **(b)** 7 **(c)** 5 **P33** assuming the flower pot fell from rest, it started from 5.5 m above the top of the window **P35** **(a)** $\langle -11, 11.1, -6\rangle$ m/s **(b)** arithmetic **(c)** $\langle -11, 13.6, -6\rangle$ m/s **(d)** $\langle 3.5, 6.8, -9\rangle$ m **(e)** 0 **(f)** $0 = 16\text{ m/s} + (-9.8\text{ N/kg})\Delta t$ **(g)** 1.63 s **(h)** 13 m **P37** **(a)** $\Delta t = \sqrt{2h/g}$ **(b)** $v_{fy} = -g\sqrt{2h/g}$ **P39** **(a)** 1 **(b)** 9/8 **(c)** 3/4 **P41** 23 cm **P43** $\langle 0, 0.2807, 0\rangle$ m **P45** At $t = 0.04$ s, $s = 0.0157$ m and $L = 0.0843$ m. At $t = 0.08$ s, $s = 0.0466$ m and $L = 0.0554$ m **P47** **(a)** $\Delta t \approx 6.15 \times 10^{-10}$ s **(b)** 18 cm

Chapter 3

P11 3/16 **P13** **(a)** $\langle 2.8 \times 10^8, 0, -2.8 \times 10^8\rangle$ m **(b)** 4.0×10^8 m **(c)** $\langle 0.7, 0, -0.7\rangle$ **(d)** $\langle -4.9 \times 10^{28}, 0, 4.9 \times 10^{28}\rangle$ N **P15** 1.12×10^{24} N **P17** **(a)** 2×10^{22} N **(b)** 2×10^{22} N **P19** **(a)** $\langle -7 \times 10^{11}, 5 \times 10^{11}, 0\rangle$ m **(b)** 8.6×10^{11} m **(c)** $\langle -0.81, 0.58, 0\rangle$ **(d)** 1.8×10^{21} N **(e)** 1.8×10^{21} N **(f)** $\langle -1.5 \times 10^{21}, 1.0 \times 10^{21}, 0\rangle$ N **(g)** $\langle 1.5 \times 10^{21}, -1.0 \times 10^{21}, 0\rangle$ N **P21** about 1×10^{-6} N and about 10 N **P23** 3.2×10^4 m **P25** mg **P27** **(a)** $\langle 1.39 \times 10^{-4}, -2.32 \times 10^{-4}, -1.86 \times 10^{-4}\rangle$ N **(b)** $\langle 8.4 \times 10^{-4}, 1.40 \times 10^{-3}, 1.12 \times 10^{-3}\rangle\text{ kg}\cdot\text{m/s}$ **P29** **(a)** $\langle 3.04 \times 10^3, 1.496 \times 10^4, 0\rangle$ m/s **(b)** $\langle 3.004\times10^{12}, 4.01\times10^{12}, 0\rangle$ m **(c)** Force and velocity are not constant, so a large time interval will give inaccurate results. **P31** **(b)** about $5\text{ kg}\cdot\text{m/s}$ **(c)** about 500 m **P33** To the right **P35** c **P37** **(a)** $\langle -0.9, 0.6, 0\rangle$ m **(b)** 1.08 m **(c)** $\langle -0.833, 0.556, 0\rangle$ **(d)** 2.29×10^{-16} N **(e)** $\langle 1.91 \times 10^{-16}, 1.27 \times 10^{-16}, 0\rangle$ N **(f)** 6.17×10^{-8} N **(g)** $\langle -5.14 \times 10^{-8}, 3.43 \times 10^{-8}, 0\rangle$ N **(h)** 2.69×10^8 **(i)** 2.69×10^8 **P39** 2.3×10^{-8} N, 5.4×10^{-51} N; will repel **P41** **(a)** $\langle 2.2, -0.4, 2.6\rangle\text{ kg}\cdot\text{m/s}$ **(b)** $\langle 0, -7.84, 0\rangle$ N **(c)** $\langle 2.2, -1.2, 2.6\rangle\text{ kg}\cdot\text{m/s}$ **P43** 4.6×10^6 m, so inside the Earth **P45** **(a)** $\langle 1.875, 0.5, 0\rangle\text{ kg}\cdot\text{m/s}$ **(b)** $\langle 5.36, 1.43, 0\rangle$ m/s **P47** $\langle -2, 14, 7\rangle\text{ kg}\cdot\text{m/s}$ **P49** **(a)** $\langle 0, 0.09, -0.18\rangle\text{ kg}\cdot\text{m/s}$ **(b)** $\langle 0, 0.09, -0.18\rangle\text{ kg}\cdot\text{m/s}$ **(c)** $\langle 0, 1.5, -3\rangle$ m/s **P51** **(a)** $m\,|\vec{v}|$ **(b)** $\left|\vec{F}\right|$ **(c)** $m\,|\vec{v}|$ **(d)** $|\vec{v}_{\text{mosquito}}| \approx (M/m)\,|\vec{v}_{\text{car}}|$ **P53** **(a)** $\langle 25, 1, 0\rangle\text{ kg}\cdot\text{m/s}$ **(b)** approximately zero **(c)** $\langle 25, 1, 0\rangle\text{ kg}\cdot\text{m/s}$ **(d)** $\langle 7, -4, 0\rangle\text{ kg}\cdot\text{m/s}$ **P55** $\langle 3138, -1750, 4200\rangle$ m/s **P57** $\langle 13.4, -1.4, 0\rangle$ m/s **P59** $v_{fx} = [m(v_{1x} - v_{2x}) + Mv]/M$; $v_{fy} = [m(v_{1y} - v_{2y})]/M$ **P61** $-(m/M)\langle v\cos\theta, v\sin\theta, 0\rangle$ **P63** about 1×10^{-13} m/s

Chapter 4

P21 $1.7 \times 10^4\text{ kg/m}^3$ **P23** **(a)** 1.06×10^{-25} kg/atom **(b)** 2.02×10^8 atoms **(c)** 0.870 kg **P25** 1.35×10^4 N/m **P27** 450 N/m **P29** 1950 N/m **P31** 27 N/m **P33** 46 N/m **P35** **(a)** $5 \times 10^{-5}\text{ m}^2$ **(b)** $7.2 \times 10^{-5}\text{ m}^2$ **(c)** $3.6 \times 10^{-5}\text{ m}^2$ **P37** 0.21 mm **P39** $2.0\times10^{11}\text{ N/m}^2$ **P41** **(a)** $-F/(m_1 + m_2 + m_3)$ **(b)** $-m_3F/(m_1 + m_2 + m_3)$; $-(m_2 + m_3)F/(m_1 + m_2 + m_3)$ **(c)** $\langle -m_3F/(m_1 + m_2 + m_3), 0, 0\rangle$ **P43** 3 N **P45** $\langle 10.2, 0, 0\rangle$ m; $\langle 4.23, 0, 0\rangle$ m/s **P47** depends on your data **P49** $3Mg$ **P51** -0.105 m **P53** 1.6 s **P55** 0.4 s **P57** **(a)** 0.94 m/s **(b)** 15 m/s^2 **P61** **(a)** $dp/dt = -mg\sin(s/L)$ **(b)** $d^2s/dt^2 + (g/L)s = 0$ **(c)** $2\pi\sqrt{(L/g)}$ **P63** 2710 m/s **P65** about 1×10^{19} molecules **P67** 20.4 m; 19.8 m

Chapter 5

P7 (a) $\vec{0}$ **(b)** 3 **(c)** −392 N **(d)** 392 N **(e)** 497.4 N **(f)** 306.2 N **(g)** −306.2 N **P9 (a)** $\vec{0}$ **(b)** $\vec{0}$ **(c)** rope, floor **(d)** 195.3 N **(e)** −195.3 N **(f)** Earth, floor, rope **(g)** 163.9 N **(h)** 294 N **(i)** 130 N **P11 (a)** 6098 N, 3980 N, 7840 N **(b)** 0.049, 0.032, 0.063 **P13 (a)** 1040 N, 1420 N **(b)** 528 N, 726 N **P15 (a)** 2.95 m/s^2 **(b)** 39.3 N **P17** 2.24 m/s, tangential **P19 (a)** $(1.8 \times 10^{23}\text{ N})\hat{p}$ **(b)** 3.13×10^{29} kg · m/s **P21** $\langle -4.97 \times 1022, 1.88 \times 10^{22}, 0\rangle$ N, $\langle 6.47 \times 10^{22}, 1.71 \times 10^{23}, 0\rangle$ N **P23 (a)** 3.67 m/s **(b)** tangent to path **(c)** inward, toward center **P25 (a)** d **(b)** b **(c)** 8.74×10^{-15} N **P27** $\vec{0}$, 0.349 m/s, 14.0 kg · m/s, 0.977 N **P29** 1.76 m/s, 66.3 N **P31** 8.78 N **P33 (a)** a **(b)** $\vec{0}$ **(c)** 492 N **(d)** $+x$ **(e)** $\langle 0, 492, 0\rangle$ N **(f)** $\langle 296, 369, 0\rangle$ N **P35** $\sqrt{Rg}$ **P37 (a)** $\vec{0}$ **(b)** 8.71 N **(c)** 871 N **(d)** $\langle 0, 1130, 0\rangle$ N **(e)** 1.4×10^4 N/m **P39** 6.65×10^{12} kg **P41** 1.46 m/s **P43 (a)** 13.7 m/s **(b)** direction changes **(c)** spring exerts a force **(d)** 300 N **(e)** 32.8 kg **P45** $\sqrt{4\pi^2 mR/(k_s(R-L))}$ **P47** 3.0×10^4 m/s **P49 (a)** $\langle 0, 200, 0\rangle$ kg · m/s^2 **(b)** $\langle 0, -549, 0\rangle$ N **(c)** $\langle 0, 749, 0\rangle$ N **(d)** $\langle 0, -200, 0\rangle$ kg · m/s^2 **(e)** $\langle 0, -549, 0\rangle$ N **(f)** $\langle 0, 349, 0\rangle$ N **(g)** heavier **(h)** lighter **P51** 2.69 m/s; 1.21 s **P53 (a)** 4.24×10^7 m **(b)** 0,48 s **(c)** 85 min = 1.4 hr **(d)** 7.9×10^3 m/s **(e)** 3250 s = 0.9 hr **P55 (a)** 1.60×10^6 m/s$(0.005c)$ **(b)** yes **(c)** 1.11×10^{37} kg **(d)** 5.6 million solar masses

Chapter 6

P9 6 **P11** $0.99999985c$; 1.5×10^{-10} J **P13 (a)** 5.76×10^{-10} J **(b)** no **(c)** 1.48×10^{-10} J **(d)** C **P15** 81.6 J **P17** 0 **P19** 104 J **P21** 585 J **P23** $\langle 0, -4.9, 0\rangle$ N; 4.9 J **P25** −8.6 J **P27** 2.5×10^{-19} J **P29** 189 J **P31** 2.58×10^{-9} J; 4.59×10^{-10} J; 2.12×10^{-9} J; 7.28×10^{-9} J; 4.59×10^{-10} J; 6.82×10^{-9} J **P33** 10.2 m/s **P35** 4.5 m/s **P37** 9.81 m/s **P39** 1×10^{-13} N **P41** a, b, e, and f; 0.99 **P43** 0.5 MeV **P45 (a)** 3.499767×10^{-8} J **(b)** 3.499708×10^{-8} J **(c)** decreased **(d)** 3.72 MeV **P47 (b)** 1.194×10^{-12} J **(c)** 2.2×10^{-14} J **(d)** $\gamma = 1.002$ **P49** 19.8 m/s **P51** −2200 J **P53** 6.3 m/s **P55 (a)** 9.8 J **(b)** 6.3 m/s **P57** 102 m **P59 (a)** 2970 m/s **(b)** 2835 m/s **P61** 25.5 m/s **P63** 4.26×10^3 m/s **P65** $0.99c$ **P67** $mgR/2$, half **P69 (a)** $mgL(1-\cos\theta)$ **(c)** $-mg\sin\theta$ **(d)** $\sqrt{4gL}$ **P71 (b)** $\sqrt{1-m^2/M^2}c$, high speed **(c)** $0.99c$ **(d)** 8.4×10^{-18} m **P73 (a)** $\langle 0, 13, 0\rangle$ m**(b)** $\langle 0, -784, 0\rangle$ N **(c)** -1.02×10^4 J **(d)** 0 **(e)** -1.02×10^4 J **(f)** 0 **(g)** -1.02×10^4 J **(h)** 0 **(i)** -1.02×10^4 J **(j)** -1.02×10^4 J **(k)** all the same **P75 (a)** 1.26×10^7 m/s **(b)** 1.57×10^{-14} m **(c)** 6.38×10^{-15} m; gap is only 3×10^{-15} m **(d)** 1.9×10^{13} J

Chapter 7

P17 0.5 m/s **P19 (a)** Mg/k_s; $-\frac{1}{2}(Mg)^2/k_s$ **(b)** $-2Mg/k_s$ **(c)** $\sqrt{k_s s^2/M - 2g(L-s)}$ **P21** 2.37 m **P23** 0.0746 m **P25 (a)** −0.2 eV **(b)** 0.2 eV **(c)** $6a/r^7$ **P27** 80.5°C **P29** −516 J |31| **(a)** 58.9°C **(b)** energy transfer between system and surroundings is negligible; heat capacities independent of temperature **(c)** 1.81×10^4 J **P33** 490 J; 2040 kg/s **P35** 3 W **P37 (a)** 9 J **(b)** 9 J **(c)** 3 m/s **(d)** 0.2 s **(e)** 0 **(f)** −9 J **(g)** 15 m/s **(h)** 18 m/s **(i)** 225 J **(j)** 325 J **(k)** 99 J **(l)** 3.3 m **(m)** 99 J **(n)** 3 m **(o)** 90 J **(p)** −9 J **P39** 31.9 m; too large **P41 (a)** 353 N **(b)** 696 N **P43 (a)** same as hand **(b)** small **(c)** large **(d)** reduced

Chapter 8

P9 13.6 eV **P11** 6190 atoms **P13** 1.4 eV **P15 (a)** excited states of 1.91 eV and 2.48 eV above the ground state; excited states of 0.57 eV and 2.48 eV above the ground state **(b)** Bombard with electrons whose kinetic energy is greater than 0.57 eV and less than 1.91 eV and see whether there is a dark line. **P17 (a)** 0.8, 1.9, 1.1, 3.8, 3.0, and 1.9 eV (note two transitions of 1.9 eV) **(b)** 1.9, 3.0, and 3.8 eV **P19** 11, 5, 2, 9, 3, and 6 eV; 6, 9, and 11 eV **P21 (a)** 3, 2.5, 1.9, 0.5, 1.1, and 0.6 eV **(b)** 1.9, 2.5, and 3.0 eV **P23 (a)** one possible scheme: ground state, and 0.3, 0.8, and 2.8 eV above the ground state **(b)** no **(c)** 0.3, 0.8, and 2.8 eV **P25 (a)** 0.023 eV **(b)** 4.1×10^{-38} kg **(c)** 0.023, 0.046, and 0.069 eV **P27 (a)** 0.015 eV **(b)** 1.8 eV **(c)** 120 **(d)** 0.03 and 0.045 eV

Chapter 9

P9 4.6×10^6 m from center of Earth **P11** $\langle L/4, L/4, 0\rangle$ from lower left; $\langle -3L/4, L/4, 0\rangle$ from lower right **P13** 4.7×10^4 J **P15 (a)** $\langle 25, 97, 0\rangle$ kg · m/s **(b)** $\langle 3.125, 12.125, 0\rangle$ m/s **(c)** 809 J **(d)** 627.125 J **(e)** 181.875 J **(f)** 68.203 J and 113.672 J, which add to 181.875 J **P17 (a)** center of mass moves with constant velocity **(b)** the centers of the two orbits coincide at the center of mass; M_1 must be greater than M_2 **(c)** 7.5×10^{29} kg and 5×10^{29} kg **P19** 14.3 J **P21** 395 J **P23** 10.5 rad/s **P25 (a)** 2.99×10^4 m/s **(b)** 2.686×10^{33} J **(c)** 7.27×10^{-5} rad/s **(d)** 2.60×10^{29} J **(e)** 2.69×10^{33} J **P27 (a)** 0.007 J **(b)** 0.90 m/s **(c)** 0.207 J **(d)** 313 rad/s **P29 (a)** 6.75 m/s **(b)** 225 J **(c)** 65 J **P31** typical results are 1100 N, −350 J, 0.2 s **P33 (a)** $\sqrt{gh + v_i^2}$ **(b)** $\left((M+m)gh + (M+\frac{1}{2}m)v_i^2)/(M+\frac{1}{2}m)\right)^{\frac{1}{2}}$ **P35 (a)** $\sqrt{Fb/M}$ **(b)** $F(d-b)$ **P37 (a)** $\left(2Fw/M + v_i^2\right)^{\frac{1}{2}}$ **(b)** $\left((2Fd)/(3mr^2) + \omega_i^2\right)^{\frac{1}{2}}$ **P39 (a)** 24.2 J **(b)** 25.0 J **P41** $\sqrt{2Mg(y-y_0)/I + \omega_i^2}$ **P43 (a)** $\sqrt{2Fd/(M+4m)}$ **(b)** $\left(Fw/(2mb^2 + \frac{1}{4}MR^2)\right)^{\frac{1}{2}}$ **P45 (a)** $\sqrt{2Fd/(M+2m)}$ **(b)** $Fs - \frac{1}{4}\left(MR^2 + mL^2\right)\omega^2$

Chapter 10

P15 $v_{1,f} = v_{1,i}(m_1 - m_2)/(m_1 + m_2)$; $v_{2,f} = 2m_1 v_{1,i}/(m_1 + m_2)$ **P17 (a)** $\langle 312, -398, 185\rangle$ m/s **(b)** 4.3×10^6 J **(c)** 1 **P19 (a)** $\langle 2500, -3500, -1500\rangle$ kg · m/s **(b)** 8.02×10^7 J **(c)** 0 **(d)** 6.82×10^7 J **(e)** 1.20×10^7 J **(f)** 4.8×10^7 J **(g)** 0 **P21 (a)** car + truck; Momentum Principle **(b)** 0 **(c)** 2.73×10^6 J **(d)** inelastic **P23 (a)** $\langle -1.40 \times 10^{-19}, 0, 0\rangle$ kg · m/s **(b)** $\langle 2.86 \times 10^{-19}, 0, 0\rangle$ kg · m/s **(c)** 9.17 MeV **(d)** 0.78 MeV **(e)** 2.28×10^{-14} m **P25** $|\vec{p}_n| = |\vec{p}_{\pi^-}|$; $E_n + E_{\pi^-} = 1196$ MeV; $E_n^2 - (p_n c)^2 = (939\text{ MeV})^2$; $E_{\pi^-} - (p_{\pi^-}c)^2 = (140\text{ MeV})^2$ **P27 (a)** $0.5c$ **(b)** 1.87 m **P29 (a)** 3 m/s **(b)** 4.8×10^{-8} eV; negligible compared to 10.2 eV **P31** 259 MeV; 8×10^7 m/s **P33** (answers given for x components) **(a)** $mv_1/(m+M)$ **(b)** $Mv_1/(m+M)$; $-mv_1/(m+M)$ **(c)** The momenta have equal magnitudes before the collision, and after the collision they also have to have equal magnitudes in order for the total momentum to remain zero. If a magnitude changes, that would mean the energy changes, so the Energy Principle would be violated. **(d)** $-(M-m)v/(M+m)$; $2mv/(M+m)$

Chapter 11

P13 $22\,\hat{k}$ **P15** $\langle 0,0,144\rangle$ kg·m²/s **P17** **(a)** out of page **(b)** $d_1p_1 \sin\alpha$ **(c)** out of page **(d)** d_2p_2 **P19** 7.1×10^{33} kg·m²/s **P21** 34.1 J **P23** **(a)** 0.18 kg·m² **(b)** 14.2 J **(c)** 2.26 kg·m²/s **P25** **(a)** 0.0765 kg·m² **(b)** 0.91 kg·m²/s **(c)** 6.28 J **P27** **(a)** $\langle 29.4,0,0\rangle$ kg·m/s **(b)** $\langle 49,0,0\rangle$ m/s **(c)** $\langle 0,0,-25.9\rangle$ kg·m²/s **(d)** $\langle 0,0,4.95\rangle$ kg·m²/s **(e)** $\langle 0,0,-30.9\rangle$ kg·m²/s **(f)** $\langle 32.9,0,0\rangle$ kg·m/s **P29** **(a)** 1.44 kg·m²/s **(b)** 9.72 kg·m²/s, into page **(c)** 11.16 kg·m²/s, into page **P31** $\langle 4,3.8,0\rangle$ kg·m²/s **P33** $d/dt\langle 0,0,xmv_y\rangle = \langle 0,0,-xmg\rangle$ **P35** **(a)** -0.323 kg·m²/s **(b)** -0.323 kg·m²/s **(c)** -0.138 kg·m²/s **(d)** 1 **P37** **(a)** 1, 3, 4, and 5 **(b)** 0.189 m/s **P39** **(a)** 2.01 s **(b)** 1.64 s **P41** **(a)** 7.42 N **(b)** 0.177 m **P43** **(a)** about 0.5 s **(b)** about 100 J **P45** **(a)** about 0.2 s **(b)** about 800 J **P47** **(a)** $2m_1v_1 + (-m_2v_2\cos\theta_2) = m_1v_3\cos\theta_3 + m_2v_4\cos\theta_4$ **(b)** $-m_2v_2\sin\theta_2 = 2m_1v_3\sin\theta_3 - m_2v_4\sin\theta_4$ **(c)** $2m_1(L/2)^2\omega_1 + (-bm_2v_2\cos\theta_2) = 2m_1(L/2)^2\omega_2 + bm_2v_4\cos\theta_4$ **P49** **(a)** $\langle 2582,-9.6,0\rangle$ m/s, 7.375 rad/s out of page **(b)** 1.55×10^7 J **P51** **(a)** $\langle 8.68,1.3,0\rangle$ m/s **(b)** 166 rad/s **(c)** 803 J **P53** **(a)** 4.37×10^4 m/s **(b)** increases 1 hour **P55** **(a)** $\langle 0,0,-0.0256\rangle$ kg·m²/s **(b)** $L_z = -0.186$ kg·m²/s **(c)** $\omega_z = -145$ rad/s **P57** **(b)** 147 N·m **(c)** 1 and 3 **P59** 3.67 rad/s **P61** **(b)** $\frac{1}{2}g\sin\theta$; $\frac{2}{3}g\sin\theta$; $\frac{5}{7}g\sin\theta$ **(c)** same time; sphere **(d)** $\frac{1}{2}\left(M + I_{CM}/R^2\right)v_{CM}^2 = Mg\sin(\theta)\Delta x$; $\frac{1}{2}Mv_{CM}^2 = \left(Mg\sin\theta - I_{CM}/R^2\,(dv_{CM}/dt)\right)\Delta x$; time derivatives agree with force and torque analyses **P63** 12 rad/s **P65** **(a)** 20.9 rad/s **(b)** 104.7 rad **(c)** 6000° **P67** **(a)** 4 rad/s² **(b)** 30 rad/s **(c)** 90 rad **(d)** 5160° **P69** **(a)** 62.8 rad/s **(b)** 377 rad **(c)** 69.8 rad/s **(d)** 332 rad **(e)** 99.5 m **P71** 3.55 rad, 203° **P73** $0.543L$ **P75** 3.15×10^{-34} J·s **P77** -54.4 eV, -13.6 eV, 40.8 eV, all 4 times the values for hydrogen **P79** **(a)** 2.38×10^{-54} kg·m² **(b)** 4.7×10^{-15} m **(c)** 7×10^{-15} m; nucleus may not be spherical **P83** **(a)** clockwise **(b)** 32 s **P85** **(a)** clockwise as seen from above **(b)** slower precession **(c)** same

Chapter 12

P17 252; 1 **P19** 18000 **P21** 8.04×10^{-22} J **P23** 5.4×10^{88} year, or about 5×10^{78} times the age of the Universe **P25** **(a)** 5.65×10^{-23} J/K **(b)** 6.62×10^{-23} J/K**(c)** 1.23×10^{-23} J/K **P27** **(a)** 7.86 J/K **(b)** -7.74 J/K **(c)** 0.12 J/K **(d)** 0 **P29** $E = \frac{1}{4}a^2T^2$ **P31** $C = \frac{1}{2}a^2T/N_{atoms}$ (J/K/atom) **P33** **(a)** 215.55 K **(b)** 273.02 K **(c)** 3.48×10^{-23} J/K/atom **P35** **(a)** 3.61006×10^{-21} J **(b)** temperatures 118.7 K and 121.0 K **(c)** 2.62×10^{-23} J/K/atom **P37** about 250 K **P39** 25.1°C **P41** **(a)** $v_i + F\Delta t/(M+m)$ **(b)** $\omega_i + 2(F+f)\Delta t/(MR)$ **(c)** $F(d-x) - \frac{1}{4}MR^2(\omega_f^2 - \omega_i^2)$ **(d)** 500 K **P43** 4.04×10^{-21} J, 1/40 eV **P45** 8300 m **P47** 1350 m/s; 510 m/s **P49** similar to Figure 12.54, but with different break points **P51** about 2 cm **P53** about 600 m/s; rms speed is larger; roughly one-tenth **P55** **(a)** about 0.004 **(b)** 1×10^{17} **(c)** 120 km **P57** **(a)** helium **(b)** 8 K **P59** 108 K

Chapter 13

P17 $\langle 0,-4.48\times10^{-17},0\rangle$ N **P19** $\langle 3.2\times10^{-15}, 3.2\times10^{-15},0\rangle$ N **P21** **(a)** negative **(b)** down and to the left **(c)** $\langle 0.707,0.707,0\rangle$ **(d)** $\langle -1.4\times10^{-5},-1.4\times10^{-5},0\rangle$ N **(e)** $\langle -0.707,-0.707,0\rangle$ **P23** $\langle 0,0,-9.13\times10^4\rangle$ N/C **P25** 9.38×10^3 N/C **P27** **(a)** $\langle -0.6,-0.7,-0.2\rangle$ m **(b)** $\langle 0.5,-0.1,-0.5\rangle$ m **(c)** $\langle 1.1,0.6,-0.3\rangle$ m **(d)** 1.29 m **(e)** $\langle 0.854,0.466,-0.233\rangle$ **(f)** 48.8 N/C **(g)** $\langle 41.7,22.7,-11.4\rangle$ N/C **P29** $\langle -2550,-2550,0\rangle$ N/C **P31** 1.13×10^4 N/C **P33** $\langle -225,0,0\rangle$ N/C **P35** **(a)** $\langle 0.8,0.7,-0.8\rangle$ m **(b)** $\langle 0.5,1,-0.5\rangle$ m **(c)** $\langle -0.3,0.3,0.3\rangle$ m **(d)** 0.520 m **(e)** $\langle -0.577,0.577,0.577\rangle$ **(f)** -5.33×10^{-9} N/C **(g)** $\langle 3.08\times10^{-9},-3.08\times10^{-9},-3.08\times10^{-9}\rangle$ N/C **P37** $\langle 3,0,0\rangle$ m **P39** **(a)** $\langle 0,-5.92\times10^{-7},0\rangle$ m **(b)** $\langle 0,5.92\times10^{-7},0\rangle$ m **P41** $\langle -960,-960,0\rangle$ N/C **P43** $\langle 0,3\times10^{-6},0\rangle$ m **P45** **(a)** $+x$ **(b)** 7.31×10^5 N/C **(c)** -7.69×10^{-8} m to the left **P47** **(a)** $\langle -1.15\times10^7, 8.64\times10^6,0\rangle$ N/C **(b)** $\langle 0,-3\times10^7,0\rangle$ N/C **(c)** $\langle -1.15\times10^7,-2.13\times10^7,0\rangle$ N/C **(d)** $\langle 23.0, 42.6,0\rangle$ N/C **(e)** $\langle 0,4\times10^7,0\rangle$ N/C **(f)** $\langle -8.64\times10^6, -6.48\times10^6,0\rangle$ N/C **(g)** $\langle 1.125\times10^7,0,0\rangle$ N/C **(h)** $\langle 2.61\times10^6,3.35\times10^7,0\rangle$ N/C **(i)** $\langle -7.83\times10^{-3}, -0.101,0\rangle$ N/C **P49** **(a)** $\langle 1.92\times10^5,0,0\rangle$ N/C **(b)** $\langle -1.27\times10^5,0,0\rangle$ N/C **(c)** $\langle -3.07\times10^{-14},0,0\rangle$ N **P51** **(a)** $\vec{0}$ **(b)** 3.38×10^4 N/C **(c)** 2.22×10^3 N/C **P53** 1.07×10^4 N/C **P55** $\langle 0,-4.43\times10^{-15},0\rangle$ N **P57** **(a)** 64.8 N/C **(b)** 32.4 N/C **P59** $\langle 0,105,0\rangle$ N/C **P61** **(a)** right end **(b)** 8.3×10^{-8} C

Chapter 14

P27 4×10^{-10} C **P29** 2 and 3 **P31** 1 **P33** 1/243 **P35** **(a)** 2 **(b)** g **(c)** c **(d)** g **P37** **(a)** 894 m/s² **(b)** 1/32 **P41** c **P43** 1.25×10^{-4} m/s **P45** 3 **P47** b **P49** j **P51** **(a)** 1 **(b)** 3 **(c)** 2 **P53** **(a)** B **(b)** 3 **P55** 4.08×10^{-4} m/s **P57** **(a)** 675 N/C **(b)** 9.1×10^{-3} N/C **(c)** increase **(d)** no **P59** **(a)** $(1/(4\pi\epsilon_0)\langle 2p/(L+R+b)^3 - Q/(R+b)^2, -p/(L+R+b)^3, 0\rangle$ **(b)** $L \gg s$ **(c)** $\vec{E}_4$ points to the right and downward; $\vec{E}_{ball}$ points to the left and upward **(d)** positive surface charge on the lower right, negative surface charge on the upper left **(e)** $\vec{0}$ **P61** 1, 2, and 4 **P63** **(a)** 2.5 nC **(b)** 3

Chapter 15

P21 **(a)** $-Q/(2A)$ **(b)** $-Qdx/(2A)$ **(c)** $\langle -x,y,0\rangle$ **(d)** $\sqrt{x^2+y^2}$ **(e)** x **P23** **(a)** 0.024 m **(b)** -4×10^{-10} C **(c)** Q/N **(d)** L/dL **(e)** $(Q/L)dL$ **P25** **(a)** $-(1/(4\pi\epsilon_0))2Q\,|\vec{p}|\,/(Lx^2)$ **(b)** 2×10^4 m/s² **P27** **(a)** each piece is an arc of radius R and angle $\Delta\theta$ **(b)** $(1/(4\pi\epsilon_0))(|Q|/(\alpha R^2))\Delta\theta\,\langle\cos\theta,\sin\theta,0\rangle$ **(c)** $(1/(4\pi\epsilon_0))(|Q|/(\alpha R^2))\,(1-\cos\alpha)$ **(d)** units correct; field points toward rod midpoint (symmetry); for small α, field approaches that of a point charge **P29** **(a)** small-radius rings of charge; contribution of each ring is to the right **(b)** $(1/(4\pi\epsilon_0))(Q/L)\Delta x/\,(L+d-x)^2\,\langle 1,0,0\rangle$ **(c)** $(1/(4\pi\epsilon_0))Q/(d(L+d))$ **(d)** for $d \gg L$, field approaches that of a point charge **P31** **(a)** 3.048×10^4 N/C toward the negative ring **(b)** 2.743×10^{-4} N toward the positive ring **P33** 0 **P35** **(a)** 2.031×10^6 N/C **(b)** 2.012×10^6 N/C **(c)** differs by less than 1% **P37** **(a)** 2.795×10^6 N/C **(b)** 2.760×10^6 N/C **(c)** differs by only 1.3% **P39** **(a)** negative on left face, positive on right face **(b)** 0 **(c)** 2.66×10^{-9} C **P41** 1.84×10^{-5} C **P43** 1.01×10^{-4} C **P45** **(a)** curves downward between plates, then moves in a straight line down and to the right **(b)** 1.8×10^{16} m/s² **(c)** 3.18×10^{-9} C; negative **P47** $Q(s/R)/(2 - s/R - 2t/R)$ **P49** **(a)** metal ball polarizes with right side more negative than left side; molecules in plastic polarize with positive ends pointing toward metal ball **(b)** 7780 N/C to the left, ignoring the field produced by the polarized molecules in the plastic **(c)** 0 **(d)** 7780 N/C to the right **P51** 4.59×10^4 N/C inward; 3.51×10^4 N/C outward **P53 (a)** $\langle -5 \times 10^{-5}, 0, 0\rangle$ N/C **(b)** 0 **(c)** left side more positive than right side **P55 (a)** dipoles are radial with negative ends pointing toward center **(b)** no polarization **(c)** same **(d)** negative on inner surface, positive on outer surface **(e)** $-Q$ on inner surface, $+Q$ on outer surface **(f)** same **P57 (a)** 249 N/C; 0 **(b)** 364 N/C; −182 N/C **(c)** -5.82×10^{-17} N; 2.91×10^{-17} N **P59 (a)** $(1/(4\pi\epsilon_0))qs/L^3$ to the right **(b)** 0; right side of ball more positive than left side; dipole field points to the right and field of charges on ball points to left **P61** $(1/(4\pi\epsilon_0))^2(4\alpha Q^2/L)/|\vec{d}|^4$

Chapter 16

P21 3.11×10^6 m/s **P23 (a)** 1.64×10^{-25} J **(b)** 3.01×10^{-20} J **P25** -7.52×10^{-23} J **P27** 3.6×10^4 V/m **P29 (a)** $\langle -1, 0, 0\rangle$ m **(b)** 750 V **(c)** 1.2×10^{-16} J = 750 eV **(d)** -1.2×10^{-16} J = −750 eV **P31 (a)** $\langle 0.7, 0, 0\rangle$ m **(b)** 595 V **(c)** 9.52×10^{-17} J = 595 eV **(d)** -9.52×10^{-17} J = −595 eV **P33** 7860 V/m **P35** 9×10^3 V **P37** 1500 V **P39** B; 150 V/m to the left **P41 (a)** to the right **(b)** for example, $R = 0.1$ m, $Q_A = 2.78 \times 10^{-8}$ C and $Q_B = -2.78 \times 10^{-8}$ C **P43** −1200 V **P45** 1.92×10^{-19} J **P47 (a)** −205 V **(b)** 205 V **(c)** -3.28×10^{-17} J **(d)** 4.64×10^{-17} J **P49 (a)** $(q^2/(4\pi\epsilon_0))(1/((x-s/2)^2+y^2)^{1/2} - 1/((x+s/2)^2 + y^2)^{1/2})$ **(b)** $(1/(4\pi\epsilon_0))px/r^3$ **(c)** $(1/(4\pi\epsilon_0))2p/x^3$ **(d)** E is in x direction **P51** $(Q_2b - Q_1a)/(A\epsilon_0)$ **P53** $(1/(4\pi\epsilon_0))2(Q/L)b/r$ **P55** $(1/(4\pi\epsilon_0))Qe(1/R - 1/(R+h))$ **P57** **(a)** $(Q_1/(4\pi\epsilon_0))(1/L - 1/(L+R_2))$ **(b)** $(Q_2/(4\pi\epsilon_0))(1/(R_2 + d) - 1/R_2) + (Q_1/(4\pi\epsilon_0))(1/(L+R_2) - 1/(L+R_2+d)$ **(c)** decrease **P59 (a)** 37 V **(b)** 3214 V **P61** $(2Q/(4\pi\epsilon_0))(1/(R+d+L) - 1/(R+d))$ **P63** $(1/(4\pi\epsilon_0))(Q/R)\ln(d/(d-h)) + (Q/(2\epsilon_0))(h/A)$ **P65** 8 V **P67 (a)** 0 **(b)** 0 **P69** $(1/(4\pi\epsilon_0))(q_1/r_{1,A} - q_2/r_{2,A})$ **P71** −450 V **P73** 608 V/m **P75** $-(1/(4\pi\epsilon_0))Q/(a^2+h^2)^{1/2}$ **P77** $(11/8)(1/(4\pi\epsilon_0))(Q/R)$ **P79 (a)** $Q_2 = -Q$; $Q_3 = 5Q$ **(b)** $(1/(4\pi\epsilon_0))5Q/r_3$; $(1/(4\pi\epsilon_0))5Q/r_3$; $(1/(4\pi\epsilon_0))5Q/r_3$; $(1/(4\pi\epsilon_0))(5Q/r_3 - Q/r_2 + Q/r_1)$; $(1/(4\pi\epsilon_0))(5Q/r_3 - Q/r_2 + Q/r_1)$ **P81 (a)** no polarization **(b)** $-(1/(4\pi\epsilon_0))Q/R$ **P83** 250 V; 100 V; 250 V; 600 V **P85** $(1/(4\pi\epsilon_0))Qz/(z^2+R^2)^{3/2}$ **P87 (a)** 0 **(b)** $-(1/(4\pi\epsilon_0))qs(1/a^2 - 1/b^2)$ **(c)** $(1/(4\pi\epsilon_0))eqs(1/a^2 - 1/b^2)$

Chapter 17

P15 $\langle 0, 0, 7.5\rangle$; $\langle 0, 0, -7.5\rangle$ **P17** $\langle 15, 15, 6\rangle$ **P19** 0 at 2 and 5; $\langle 0, 0, -1.28 \times 10^{-17}\rangle$ T at 1 and 6; $\langle 0, 0, 1.28 \times 10^{-17}\rangle$ T at 3 and 4 **P21** 0 at 2 and 5; $\langle 0, 0, -5.28 \times 10^{-18}\rangle$ T at 3 and 4; $\langle 0, 0, 5.28 \times 10^{-18}\rangle$ T at 1 and 6 **P23 (a)** electric $\langle -1, 0, 0\rangle$; magnetic $\langle 0, 0, 1\rangle$ **(b)** electric 5.76×10^9 N/C; magnetic 0.66 T **P25 (a)** 7.5×10^{19} electrons/s **(b)** 3.15×10^{-5} m/s **P27** 8.98×10^{28} m^{-3} **P29 (a)** 3.5×10^{-6} T **(b)** north **P31** 5.14×10^{-6} T **P33 (a)** $+x$ **(b)** 0.1625 m **(c)** 0.1625 m **(d)** $\langle 0.1625, 0, 0\rangle$ m **(e)** $\langle -0.08125, 0, 0\rangle$ m **(f)** $\langle 0.16225, 0.178, 0\rangle$ m **(g)** $\langle 0.674, 0.739, 0\rangle$ **(h)** $\langle 0, 0, 0.120\rangle$ m **(i)** $\langle 0, 0, 2.95 \times 10^{-6}\rangle$ T **P35 (a)** conventional current down and to the left; magnetic field down and to the right **(b)** 0.088 A **P37 (a)** 0 **(b)** $\langle 0, 0, hdx\rangle$ **(c)** $+x$ **P39 (a)** 0.0917 A **(b)** $\langle 0, 0, -1.63 \times 10^{-6}\rangle$ T **(c)** assumed $r \ll L$; neglected curved part of wire **P41 (a)** 1.11×10^{-7} T **(b)** 1.11×10^{-5} T **P43** 1.5 A **P45** $(\mu_0/(4\pi))I\pi(1/R_1 + 1/R_2)$; direction $\langle 0, 0, -1\rangle$ **P49 (a)** into the page **(b)** $(\mu_0/(4\pi))I(\pi/R + 2/h)$ **P51** $(\mu_0/(4\pi))I\pi(1/R + 3/r)$; $\langle 0, 0, -1\rangle$ **P53** 4.3 A·m^2 **P55** 0.5 A·m^2 **P57 (a)** S end **(b)** 1 and 5 **(c)** 12

Chapter 18

P21 1, 3, 6, 8, and 9 **P23** $i_A = i_B + i_C$, $i_B + i_C = i_D$, $i_A = i_D$; i_D **P25 (a)** 11 A **(b)** yes **(c)** 5 A **(d)** no **P27** 0.016 V/m **P29** 2, 5, 8, and 9 **P31** 6.24×10^{19} electrons/s **P33** 1; 6×10^{18} electrons/sl 0.26 V/m **P35 (a)** downward **(b)** upward **(c)** 19.8 V/m; to the left **P37** 5 V/m; same **P39** 3.33 V/m; same **P41 (a)** 3.5 V **(b)** C **P43 (a)** 2 **(b)** $1.3\text{ V} + V_E - V_G + 1.3\text{ V} + V_B - V_D = 0$ **(c)** 5 V/m **(d)** 6.73×10^{18} s^{-1} **(e)** 5 V/m **(f)** (c) **P45 (a)** 1, 2, and 4 **(b)** $1.3\text{ V} + 2(V_A - V_C) + V_C - V_E = 0$ **(c)** $i_B = i_D = i_F$ **(d)** 21.9 V/m; 0.925 V/m **P47 (a)** 1.2×10^{18} s^{-1} **(b)** neglect connection wires; filaments same lengths; ideal battery (no internal resistance) **(c)** high positive density near + terminal of battery; small gradient along wires; large gradient across bulbs (largest along thin-filament bulb) **P49 (a)** 3×10^{17} s^{-1} **(b)** 4.5×10^{17} s^{-1} **(c)** do the experiment **(d)** 6×10^{17} s^{-1} **P51 (a)** 2.27 V/m **(b)** 2.27 V/m **(c)** 3 **P53** 18.7°

Chapter 19

P29 d **P31** c **P33** d **P35** 1, 3, and 4 **P37 (a)** 1, 2, 3, and 5 **(b)** 1, 2, 3, and 5 **(c)** c **(d)** a **P39** 1.85×10^{-8} F **P41** 8×10^{-10} C **P43** battery: $-2E$, 0; bulb: 0, 0; capacitor: $+E$, $-E$; surroundings: $+E$, $+E$ **P45** 4.06×10^7 $(\Omega\cdot\text{m})^{-1}$ **P47** 40 Ω **P49 (a)** 0.83 Ω, 3.6 A **(b)** 4.2×10^{-4} Ω, 15.0015 V **P51** 240 Ω; 280 Ω **P53** bulb resistance increases with temperature **P55** 3 C; 1.9×10^{19} electrons **P57** 5×10^{-3} V/m **P59 (a)** field is small in thick wires, large in thin wires; field goes counterclockwise **(b)** roughly, a small gradient along the thick wires, large along the thin wires **(c)** $n(1/4)\pi d_2^2 uK/\left((d_2/d_1)^2(2L_1 + L_3) + 2L_2\right)$ **(d)** $n(1/4)\pi d_2^2 uK/(2L_2)$ **(e)** $K/4$ **P61 (a)** E decreases with x **(b)** mobility increases with x (lower temperature) **(c)** $I/(ne(wh)(u_0 + kx))$ **(d)** negative **(e)** $(I/(ne(wh)k))\ln((u_0 + kd)/u_0)$ **(f)** $(1/(ne(wh)k))\ln((u_0 + kd)/u_0)$ **P63 (a)** $3.0\text{ V} - I_4(15\,\Omega) - I_3(30\,\Omega + 20\,\Omega) = 0$; $3.0\text{ V} - I_4(15\,\Omega) - I_1(20\,\Omega) = 0$; $I_4 = I_3 + I_1$ **(b)** compass 1 deflects 3° toward NE; compass 2 deflects 4.2° toward NW **(c)** 126 V/m **(d)** $E_1 > E_2$ **(e)** 4.6 mm **(f)** −2.4 V **(g)** 4.8×10^{-19} J **(h)** 0.153 W **P67** 4 A **P69 (a)** 34.3 Ω **(b)** 0 **(c)** 0.26 A **P71 (a)** 15 A **(b)** 0 **P73** 10 V **P75 (a)** $C\times$emf **(b)** field: polarized molecules reduce fringe field, so current runs again; potential: gap voltage reduced, so current runs again **(c)** (emf/R)$(1 - 1/K)$ **(d)** $KC\times$emf **P77 (a)** $E = 3600$ V/m; too small to make a spark **(b)** 3.83×10^{-7} C **(c)** 6.12 V **(d)** 3.02 V **P79 (a)** $ABCEFGA$: $20 - 10I_1 - 15I_4 - 12I_6 - 20I_1 = 0$; $DECD$: $5 + 15I_4 - 20I_2 = 0$; $DFED$: $-30I_3 + 12I_6 - 5 = 0$; C: $I_1 = I_2 + I_4$; D: $I_2 = I_5 + I_3$; F: $I_3 + I_6 = I_1$ **(b)** do the check **(c)** 33.2 V **(d)** 1.63 W **(e)** 1730 V/m

Chapter 20

P25 $\langle 0, 0, 1.92 \times 10^{-14}\rangle$ N **P27** $\vec{0}$ **P29** 1.64 T **P31 (a)** left accelerating plate is positive; top deflection plate is positive; B into page **(b)** 1600 V; 320 V **(c)** 1371 V; 274 V **P33 (a)** 8.9° **(b)** 4×10^{-17} N **P35 (a)** CCW **(b)** 9.2 T **P37** $\langle 0, 0, 0.243\rangle$ N

P39 **(a)** 5.52 A **(b)** $-y$ **P41** **(a)** $\frac{1}{4}(\mu_0/(4\pi))2\pi I(1/a - 1/b)$; $\langle 0, 0, 1\rangle$ **(b)** $\frac{1}{4}(\mu_0/(4\pi))2\pi Iev(1/a - 1/b)$; $\langle 0, 1, 0\rangle$ **P43** 1.14 T into the page **P45** 9500 m/s **P47** $\langle 0, 0, -1863\rangle$ V/m **P49** **(a)** $-z$ **(b)** $-y$ **(c)** 1.08×10^{-4} T **(d)** 61.6 V/m **P51** **(a)** B **(b)** upward **(c)** upward **(d)** 3.30×10^7 m/s; 2.40×10^{-3} T **(e)** 2, 5, 6 **(f)** 650 V **P53** $(5/4)(\mu_0/(4\pi))2\pi IV/R, -y$ **P55** **(a)** positive **(b)** 4.8×10^{-3} m/s **(c)** 9.8×10^{-4} (m/s)/(V/m) **(d)** 4×10^{23} m^{-3} **(e)** 2.4 Ω **P57** **(a)** If you connect the + lead to the bottom of the bar (with the − lead straight above it) and the voltmeter reads positive, conventional current in the long wire goes to the right. **(b)** $nedh\Delta V/((\mu_0/(4\pi))2I_{\text{ammeter}})$ **P59** 1.77×10^{-5} V; 0.022 A **P61** 2 **P63** **(a)** 1 **(b)** 0 **(c)** slightly less than 4.48×10^{-19} N **(d)** 4.48×10^{-19} N **(e)** 1.26 V **(f)** to the left **P65** **(a)** 0; 0 **(b)** vBh/R, vB^2h^2/R **(c)** 0, 0 **(d)** vBh/R, vB^2h^2/R **(e)** 0, 0 **P67** west end has excess electrons; about 2500 electrons **P69** **(a)** negative below, positive above **(b)** 0.96A, CCW **(c)** 2 **(d)** 0.035 N to the right **P71** $\langle 0, 1.2, 0\rangle$ N **P73** 120 J **P75** **(a)** 24 V **(b)** 11.5 W **P77** **(a)** $+1.6 \times 10^{-19}$ C **(b)** 14.4 V **(c)** 0 **(d)** −14.4 V **(e)** 14.4 eV **(f)** 14.4 eV **P79** 7.5×10^{-6} C

Chapter 21

P7 left -0.24 V·m; right 0.60 V·m; all other faces 0; total 0.36 V·m; 3.19×10^{-12} C **P9** contains -5×10^{-11} C **P11** contains 1.8×10^{-12} C **P13** total flux 0.0672 V·m; contains 5.95×10^{-13} C **P15** -227 V·m **P17** **(a)** $+Q$ on surface of inner wire; $-Q$ on inner surface of outer tube **(b)** $(Q/L)/(2\pi\epsilon_0 r)$ **(c)** 0 **P19** **(a)** $(Q/A)/\epsilon_0$ **(b)** assume E varies and show this is inconsistent with Gauss's law **(c)** $(Q/A)/\epsilon_0$ **(d)** $Qs/(2R)$ **P21** **(a)** S/ϵ_0 **(b)** the field of a large sheet would be $(1/2)S/\epsilon_0$ **P23** 125 A, out of page **P25** **(a)** left above, right below **(b)** $\mu_0 NI/(2L)$ **(c)** $\mu_0 NI/L$ **(d)** 0 **P27** $(\mu_0/(4\pi))(I_1 - I_2)/R$, out of page **P29** 1×10^{36} N/C/m

Chapter 22

P13 1.5 V; CCW **P15** CCW; $|dB/dt|\pi r_1^2/R$; tangent, CCW; $|dB/dt|r_1^2/(2r_2)$ **P17** **(a)** 2.6×10^{-6} T·m^2 **(b)** treat coil as magnetic dipole **(c)** there is no electric field **(d)** $-y$ **(e)** 2.16×10^{-7} V **(f)** 2.16×10^{-7} V **(g)** 8.59×10^{-7} V/m **(h)** 8.59×10^{-7} V/m **P19** **(a)** $(\mu_0/(4\pi))(2N_1N_2\pi^2r_1^2r_2^2/x^3)(b + 2ct)$ **(b)** downward **(c)** $(\mu_0/(4\pi))(N_1N_2\pi r_1^2 r_2/x^3)(b + 2ct)$ **P21** **(a)** upward at 1, downward at 2 **(b)** 0.404 mV **(c)** B uniform across coil; used average dB/dt **P23** **(a)** sine wave in coil 1, −cosine wave in coil 2 **(b)** 0.233 V; assume $z >> r_1$ and coils are thin **P25** **(a)** $+z$ **(b)** $+x$ **(c)** 2.33×10^{-5} T·m^2 **(d)** coils are thin; B uniform across coil 2 **(e)** 7.76×10^{-6} T·m^2 **(f)** 3.88×10^{-5} V **(g)** 10.7 mV **(h)** 2.06×10^{-4} V/m **(i)** 1 and 4 **P27** **(a)** CCW **(b)** 13.6 mA **P29** **(a)** $(\pi r_1^2/R)\,|(\mu_0 N(-k))/d|$; to the right **(b)** much larger current in ring **P31** **(a)** eBR **(b)** $\pi r^2 be$ **P33** 1.13×10^{-6} V/m **P35** **(a)** CCW **(b)** $(e/m)(\mu_0/(4\pi))2I_0Ar/R^3$ **(c)** tangent to orbit, CW **(d)** slows down **P37** $(\mu_0/(4\pi))3\pi R^2 Iyv/x^4$ **P39** 420 V, 1.14 A **P41** **(a)** radius = 0.01 m, length = 0.04 m, 1600 turns **(b)** 18.8 A

Chapter 23

P17 $dE/dt = 1.3 \times 10^{18}$ V/m/s inside the loop **P19** 1×10^{-5} T in $-y$ direction **P21** 0.024 T in $-y$ direction **P23** **(a)** 1.16×10^{17} m/s^2 **(b)** 1.69×10^{-7} V/m **(c)** 4 **(d)** 5 **(e)** 8 **P25** **(a)** 5×10^{-10} s **(b)** $+y$ **(c)** $-y$ **P27** **(a)** electrons have low mass so experience larger acceleration than the ions; electric field at A is in $+y$ direction **(b)** electric field at B is in $+y$ direction; electric field at C is up and to the right **(c)** 1×10^{-8} s; into page **(d)** proportional to $1/r^2$ instead of $1/r$ **P29** **(a)** 6.4×10^{-12} V/m to the left **(b)** 50 ns **(c)** upward **(d)** 1.07×10^{-9} V/m; much larger **(e)** to the right **P31** **(a)** to the left **(b)** 4×10^{-8} s; 5×10^{-10} s **(c)** E out of page, B up **P33** 60 nM **P35** 300 m; 3 m; 698 nm; 400 nm **P37** 550 V/m **P39** 2.66×10^8 m/s **P41** **(a)** 3 **(b)** 65.6° **P43** **(a)** 63.4° **(b)** 42.3° **(c)** 4.2 m **P45** 0.23 m to the left of the lens; virtual; no image **P47** $\langle 100, -4, 0\rangle$ cm; real **P49** 0.302 m; 1.5 cm; inverted

Supplement S1

P3 2.4×10^{22} molecules/s **P5** **(b)** down to 1 percent in 4.6 min **P7** about 2 hours **P9** **(a)** 1.0×10^5 N/m^2 **(b)** 1300 m/s **(c)** 4×10^{20} atoms **(d)** helium leaves faster **(e)** decrease **P11** 3.17 **P13** **(a)** 2910 J **(b)** −2910 J **(c)** 0 J **P15** **(a)** 214 K **(b)** 313.6 J **(c)** 2784 N **(d)** 2652 N **(e)** 2514 N **P17** 25 s; increase **P19** 0.05 J **P21** **(a)** $NkT_H \ln(V_2/V_1)$ **(b)** $(T_H/T_L)^{3/2} = V_3/V_2$ **(c)** $NkT_L \ln(V_4/V_3)$ **(d)** $(T_H/T_L)^{3/2} = V_3/V_2 = V_4/V_1$; putting the pieces together, one finds that Q_H/T_H does indeed equal Q_L/T_L **P23** 5.9 mm

Supplement S2

P1 5×10^{-8} m

Supplement S3

P13 7.5×10^{14} Hz; 1.3×10^{-15} s **P15** 1.33 m **P17** **(a)** reradiation; electric field in and out of page, magnetic field in the xy plane, perpendicular to the electric field **(b)** 23.6°; 11.5° **P19** 23.6°; 53.1° **P21** 1.2×10^{-10} m; 1.6×10^{-10} m **P23** 212 nm **P25** **(a)** 28 cm **(b)** wider **(c)** narrower **P27** 9.4 N **P29** 0.02 m; 0.6 m; 7.5 ms; 1.67 m^{-1}; 838 rad/s; 133 Hz **P31** $0.04 \cos((2\pi/0.2)(x + 45t))$ **P33** 8.2×10^{-14} m; extremely small compared to 1×10^{-10} m **P35** **(a)** 89.6 m/s **(b)** 128 Hz, 70 cm; 192 Hz, 46.7 cm; 256 Hz, 35 cm; 320 Hz, 28 cm **P37** 365 nm **P39** 774 V/m; 2.8×10^{20} photons/s

Index

The text of Supplements 1–3 can be found at www.wiley.com/college/chabay

T